Fundamentals of Physics

Physics I w/ Calculus.

Authors

David Halliday • Jearl Walker • Robert Resnick

ISBN 9781119315414

Printed in the United States of America 10 9 8 7 6 5 4 3 2 1

List of Titles

Fundamentals of Physics Extended, 10th edition
by David Halliday, Robert Resnick, and Jearl Walker
Copyright © 2014, ISBN: 978-1-118-23072-5

Table of Contents

EXTENDED

FUNDAMENTALS OF PHYSICS

TENTH EDITION

EXTENDED

Halliday & Resnick

FUNDAMENTALS OF PHYSICS

TENTH EDITION

JEARL WALKER
CLEVELAND STATE UNIVERSITY

WILEY

EXECUTIVE EDITOR Stuart Johnson
SENIOR PRODUCT DESIGNER Geraldine Osnato
CONTENT EDITOR Alyson Rentrop
ASSOCIATE MARKETING DIRECTOR Christine Kushner
TEXT and COVER DESIGNER Madelyn Lesure
PAGE MAKE-UP Lee Goldstein
PHOTO EDITOR Jennifer Atkins
COPYEDITOR Helen Walden
PROOFREADER Lilian Brady
SENIOR PRODUCTION EDITOR Elizabeth Swain

COVER IMAGE © 2007 CERN

This book was set in 10/12 Times Ten by cMPreparé, CSR Francesca Monaco, and was printed and bound by Quad Graphics.The cover was printed by Quad Graphics.

This book is printed on acid free paper.

Library of Congress Cataloging-in-Publication Data

Walker, Jearl
Fundamentals of physics / Jearl Walker, David Halliday, Robert Resnick—10th edition.
volumes cm
Includes index.
ISBN 978-1-118-23072-5 (Extended edition)
Binder-ready version ISBN 978-1-118-23061-9 (Extended edition)
1. Physics—Textbooks. I. Resnick, Robert. II. Halliday, David. III. Title.
QC21.3.H35 2014
530—dc23

2012035307

Printed in the United States of America
10 9 8 7 6 5 4 3 2 1

PREFACE

WHY I WROTE THIS BOOK

Fun with a big challenge. That is how I have regarded physics since the day when Sharon, one of the students in a class I taught as a graduate student, suddenly demanded of me, "What has any of this got to do with my life?" Of course I immediately responded, "Sharon, this has everything to do with your life—this is physics."

She asked me for an example. I thought and thought but could not come up with a single one. That night I began writing the book *The Flying Circus of Physics* (John Wiley & Sons Inc., 1975) for Sharon but also for me because I realized her complaint was mine. I had spent six years slugging my way through many dozens of physics textbooks that were carefully written with the best of pedagogical plans, but there was something missing. Physics is the most interesting subject in the world because it is about how the world works, and yet the textbooks had been thoroughly wrung of any connection with the real world. The fun was missing.

I have packed a lot of real-world physics into *Fundamentals of Physics*, connecting it with the new edition of *The Flying Circus of Physics*. Much of the material comes from the introductory physics classes I teach, where I can judge from the faces and blunt comments what material and presentations work and what do not. The notes I make on my successes and failures there help form the basis of this book. My message here is the same as I had with every student I've met since Sharon so long ago: "Yes, you *can* reason from basic physics concepts all the way to valid conclusions about the real world, and that understanding of the real world is where the fun is."

I have many goals in writing this book but the overriding one is to provide instructors with tools by which they can teach students how to effectively read scientific material, identify fundamental concepts, reason through scientific questions, and solve quantitative problems. This process is not easy for either students or instructors. Indeed, the course associated with this book may be one of the most challenging of all the courses taken by a student. However, it can also be one of the most rewarding because it reveals the world's fundamental clockwork from which all scientific and engineering applications spring.

Many users of the ninth edition (both instructors and students) sent in comments and suggestions to improve the book. These improvements are now incorporated into the narrative and problems throughout the book. The publisher John Wiley & Sons and I regard the book as an ongoing project and encourage more input from users. You can send suggestions, corrections, and positive or negative comments to John Wiley & Sons or Jearl Walker (mail address: Physics Department, Cleveland State University, Cleveland, OH 44115 USA; or the blog site at www.flyingcircusofphysics.com). We may not be able to respond to all suggestions, but we keep and study each of them.

WHAT'S NEW?

Modules and Learning Objectives "What was I supposed to learn from this section?" Students have asked me this question for decades, from the weakest student to the strongest. The problem is that even a thoughtful student may not feel confident that the important points were captured while reading a section. I felt the same way back when I was using the first edition of Halliday and Resnick while taking first-year physics.

To ease the problem in this edition, I restructured the chapters into concept modules based on a primary theme and begin each module with a list of the module's learning objectives. The list is an explicit statement of the skills and learning points that should be gathered in reading the module. Each list is following by a brief summary of the key ideas that should also be gathered. For example, check out the first module in Chapter 16, where a student faces a truck load of concepts and terms. Rather than depending on the student's ability to gather and sort those ideas, I now provide an explicit checklist that functions somewhat like the checklist a pilot works through before taxiing out to the runway for takeoff.

Links Between Homework Problems and Learning Objectives In *WileyPLUS*, every question and problem at the end of the chapter is linked to a learning objective, to answer the (usually unspoken) questions, "Why am I working this problem? What am I supposed to learn from it?" By being explicit about a problem's purpose, I believe that a student might better transfer the learning objective to other problems with a different wording but the same key idea. Such transference would help defeat the common trouble that a student learns to work a particular problem but cannot then apply its key idea to a problem in a different setting.

Rewritten Chapters My students have continued to be challenged by several key chapters and by spots in several other chapters and so, in this edition, I rewrote a lot of the material. For example, I redesigned the chapters on Gauss' law and electric potential, which have proved to be tough-going for my students. The presentations are now smoother and more direct to the key points. In the quantum chapters, I expanded the coverage of the Schrödinger equation, including reflection of matter waves from a step potential. At the request of several instructors, I decoupled the discussion of the Bohr atom from the Schrödinger solution for the hydrogen atom so that the historical account of Bohr's work can be bypassed. Also, there is now a module on Planck's blackbody radiation.

New Sample Problems and Homework Questions and Problems Sixteen new sample problems have been added to the chapters, written so as to spotlight some of the difficult areas for my students. Also, about 250 problems and 50 questions have been added to the homework sections of the chapters. Some of these problems come from earlier editions of the book, as requested by several instructors.

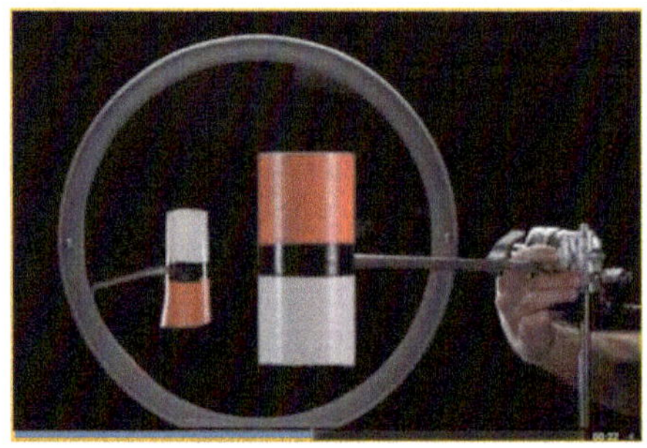

Video Illustrations In the eVersion of the text available in *WileyPLUS*, David Maiullo of Rutgers University has created video versions of approximately 30 of the photographs and figures from the text. Much of physics is the study of things that move and video can often provide a better representation than a static photo or figure.

Online Aid *WileyPLUS* is not just an online grading program. Rather, it is a dynamic learning center stocked with many different learning aids, including just-in-time problem-solving tutorials, embedded reading quizzes to encourage reading, animated figures, hundreds of sample problems, loads of simulations and demonstrations, and over 1500 videos ranging from math reviews to mini-lectures to examples. More of these learning aids are added every semester. For this 10th edition of HRW, some of the photos involving motion have been converted into videos so that the motion can be slowed and analyzed.

These thousands of learning aids are available 24/7 and can be repeated as many times as desired. Thus, if a student gets stuck on a homework problem at, say, 2:00 AM (which appears to be a popular time for doing physics homework), friendly and helpful resources are available at the click of a mouse.

LEARNING TOOLS

When I learned first-year physics in the first edition of Halliday and Resnick, I caught on by repeatedly rereading a chapter. These days we better understand that students have a wide range of learning styles. So, I have produced a wide range of learning tools, both in this new edition and online in *WileyPLUS*:

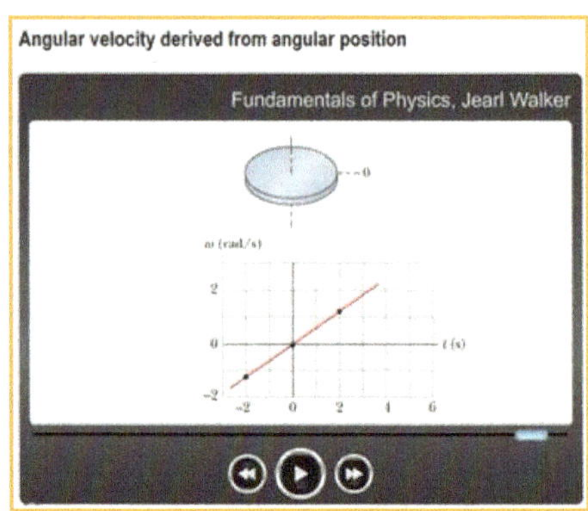

Animations of one of the key figures in each chapter. Here in the book, those figures are flagged with the swirling icon. In the online chapter in *WileyPLUS*, a mouse click begins the animation. I have chosen the figures that are rich in information so that a student can see the physics in action and played out over a minute or two instead of just being flat on a printed page. Not only does this give life to the physics, but the animation can be repeated as many times as a student wants.

Videos I have made well over 1500 instructional videos, with more coming each semester. Students can watch me draw or type on the screen as they hear me talk about a solution, tutorial, sample problem, or review, very much as they would experience were they sitting next to me in my office while I worked out something on a notepad. An instructor's lectures and tutoring will always be the most valuable learning tools, but my videos are available 24 hours a day, 7 days a week, and can be repeated indefinitely.

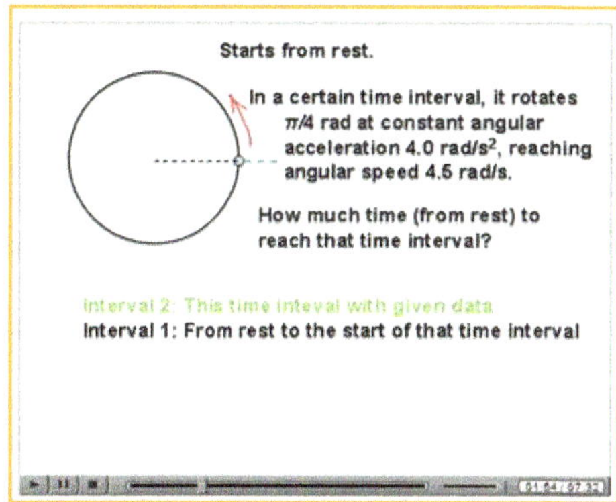

- **Video tutorials on subjects in the chapters.** I chose the subjects that challenge the students the most, the ones that my students scratch their heads about.
- **Video reviews of high school math**, such as basic algebraic manipulations, trig functions, and simultaneous equations.
- **Video introductions to math**, such as vector multiplication, that will be new to the students.
- **Video presentations of every Sample Problem** in the textbook chapters . My intent is to work out the physics, starting with the Key Ideas instead of just grabbing a formula. However, I also want to demonstrate how to read a sample problem, that is, how to read technical material to learn problem-solving procedures that can be transferred to other types of problems.
- **Video solutions to 20% of the end-of chapter problems.** The availability and timing of these solutions are controlled by the instructor. For example, they might be available after a homework deadline or a quiz. Each solution is not simply a plug-and-chug recipe. Rather I build a solution from the Key Ideas to the first step of reasoning and to a final solution. The student learns not just how to solve a particular problem but how to tackle any problem, even those that require *physics courage*.
- **Video examples of how to read data from graphs** (more than simply reading off a number with no comprehension of the physics).

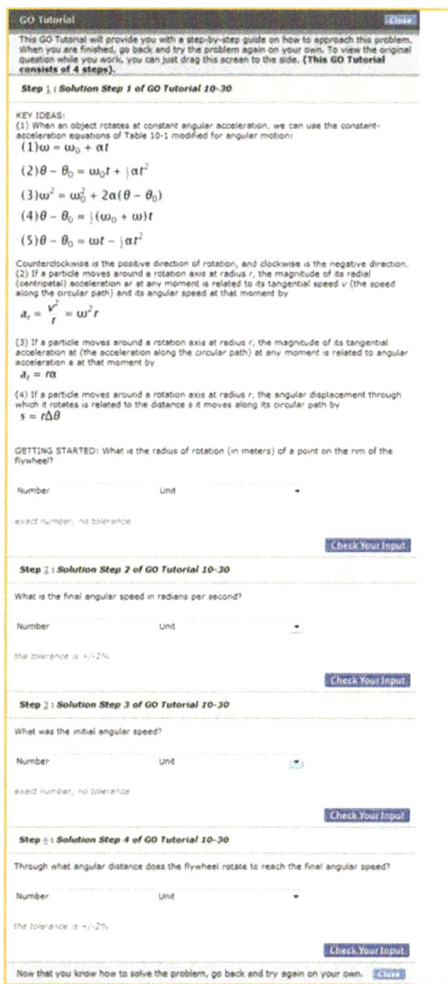

Problem-Solving Help I have written a large number of resources for *WileyPLUS* designed to help build the students' problem-solving skills.

- **Every sample problem in the textbook** is available online in both reading and video formats.
- **Hundreds of additional sample problems.** These are available as standalone resources but (at the discretion of the instructor) they are also linked out of the homework problems. So, if a homework problem deals with, say, forces on a block on a ramp, a link to a related sample problem is provided. However, the sample problem is not just a replica of the homework problem and thus does not provide a solution that can be merely duplicated without comprehension.
- **GO Tutorials** for 15% of the end-of-chapter homework problems. In multiple steps, I lead a student through a homework problem, starting with the Key Ideas and giving hints when wrong answers are submitted. However, I purposely leave the last step (for the final answer) to the student so that they are responsible at the end. Some online tutorial systems trap a student when wrong answers are given, which can generate a lot of frustration. My GO Tutorials are not traps, because at any step along the way, a student can return to the main problem.
- **Hints on every end-of-chapter homework problem** are available (at the discretion of the instructor). I wrote these as true hints about the main ideas and the general procedure for a solution, not as recipes that provide an answer without any comprehension.

Evaluation Materials

• **Reading questions are available within each online section.** I wrote these so that they do not require analysis or any deep understanding; rather they simply test whether a student has read the section. When a student opens up a section, a randomly chosen reading question (from a bank of questions) appears at the end. The instructor can decide whether the question is part of the grading for that section or whether it is just for the benefit of the student.

• **Checkpoints are available within most sections.** I wrote these so that they require analysis and decisions about the physics in the section. *Answers to all checkpoints are in the back of the book.*

Checkpoint 1

Here are three pairs of initial and final positions, respectively, along an x axis. Which pairs give a negative displacement: (a) -3 m, $+5$ m; (b) -3 m, -7 m; (c) 7 m, -3 m?

• **All end-of-chapter homework Problems** in the book (and many more problems) are available in *WileyPLUS*. The instructor can construct a homework assignment and control how it is graded when the answers are submitted online. For example, the instructor controls the deadline for submission and how many attempts a student is allowed on an answer. The instructor also controls which, if any, learning aids are available with each homework problem. Such links can include hints, sample problems, in-chapter reading materials, video tutorials, video math reviews, and even video solutions (which can be made available to the students after, say, a homework deadline).

• **Symbolic notation problems** that require algebraic answers are available in every chapter.

• **All end-of-chapter homework Questions** in the book are available for assignment in *WileyPLUS*. These Questions (in a multiple choice format) are designed to evaluate the students' conceptual understanding.

Icons for Additional Help When worked-out solutions are provided either in print or electronically for certain of the odd-numbered problems, the statements for those problems include an icon to alert both student and instructor as to where the solutions are located. There are also icons indicating which problems have GO Tutorial, an Interactive LearningWare, or a link to the *The Flying Circus of Physics*. An icon guide is provided here and at the beginning of each set of problems.

GO Tutoring problem available (at instructor's discretion) in *WileyPLUS* and WebAssign
SSM Worked-out solution available in Student Solutions Manual
• – ••• Number of dots indicates level of problem difficulty
WWW Worked-out solution is at http://www.wiley.com/college/halliday
ILW Interactive solution is at http://www.wiley.com/college/halliday
Additional information available in *The Flying Circus of Physics* and at flyingcircusofphysics.com

VERSIONS OF THE TEXT

To accommodate the individual needs of instructors and students, the ninth edition of *Fundamentals of Physics* is available in a number of different versions.

The **Regular Edition** consists of Chapters 1 through 37 (ISBN 9781118230718).

The **Extended Edition** contains seven additional chapters on quantum physics and cosmology, Chapters 1–44 (ISBN 9781118230725).

Volume 1 — Chapters 1–20 (Mechanics and Thermodynamics), hardcover, ISBN 9781118233764

Volume 2 — Chapters 21–44 (E&M, Optics, and Quantum Physics), hardcover, ISBN 9781118230732

INSTRUCTOR SUPPLEMENTS

Instructor's Solutions Manual by Sen-Ben Liao, Lawrence Livermore National Laboratory. This manual provides worked-out solutions for all problems found at the end of each chapter. It is available in both MSWord and PDF.

Instructor Companion Site http://www.wiley.com/college/halliday

- **Instructor's Manual** This resource contains lecture notes outlining the most important topics of each chapter; demonstration experiments; laboratory and computer projects; film and video sources; answers to all Questions, Exercises, Problems, and Checkpoints; and a correlation guide to the Questions, Exercises, and Problems in the previous edition. It also contains a complete list of all problems for which solutions are available to students (SSM, WWW, and ILW).
- **Lecture PowerPoint Slides** These PowerPoint slides serve as a helpful starter pack for instructors, outlining key concepts and incorporating figures and equations from the text.
- **Classroom Response Systems ("Clicker") Questions** by David Marx, Illinois State University. There are two sets of questions available: Reading Quiz questions and Interactive Lecture questions. The Reading Quiz questions are intended to be relatively straightforward for any student who reads the assigned material. The Interactive Lecture questions are intended for use in an interactive lecture setting.
- **Wiley Physics Simulations** by Andrew Duffy, Boston University and John Gastineau, Vernier Software. This is a collection of 50 interactive simulations (Java applets) that can be used for classroom demonstrations.
- **Wiley Physics Demonstrations** by David Maiullo, Rutgers University. This is a collection of digital videos of 80 standard physics demonstrations. They can be shown in class or accessed from *WileyPLUS*. There is an accompanying Instructor's Guide that includes "clicker" questions.
- **Test Bank** For the 10th edition, the Test Bank has been completely over-hauled by Suzanne Willis, Northern Illinois University. The Test Bank includes more than 2200 multiple-choice questions. These items are also available in the Computerized Test Bank which provides full editing features to help you customize tests (available in both IBM and Macintosh versions).
- **All text illustrations** suitable for both classroom projection and printing.

Online Homework and Quizzing. In addition to *WileyPLUS*, *Fundamentals of Physics*, tenth edition, also supports WebAssignPLUS and LON-CAPA, which are other programs that give instructors the ability to deliver and grade homework and quizzes online. WebAssign PLUS also offers students an online version of the text.

STUDENT SUPPLEMENTS

Student Companion Site. The web site http://www.wiley.com/college/halliday was developed specifically for *Fundamentals of Physics*, tenth edition, and is designed to further assist students in the study of physics. It includes solutions to selected end-of-chapter problems (which are identified with a www icon in the text); simulation exercises; tips on how to make best use of a programmable calculator; and the Interactive LearningWare tutorials that are described below.

Student Study Guide (ISBN 9781118230787) by Thomas Barrett of Ohio State University. The Student Study Guide consists of an overview of the chapter's important concepts, problem solving techniques and detailed examples.

Student Solutions Manual (ISBN 9781118230664) by Sen-Ben Liao, Lawrence Livermore National Laboratory. This manual provides students with complete worked-out solutions to 15 percent of the problems found at the end of each chapter within the text. The Student Solutions Manual for the 10th edition is written using an innovative approach called TEAL which stands for Think, Express, Analyze, and Learn. This learning strategy was originally developed at the Massachusetts Institute of Technology and has proven to be an effective learning tool for students. These problems with TEAL solutions are indicated with an SSM icon in the text.

Interactive Learningware. This software guides students through solutions to 200 of the end-of-chapter problems. These problems are indicated with an ILW icon in the text. The solutions process is developed interactively, with appropriate feedback and access to error-specific help for the most common mistakes.

Introductory Physics with Calculus as a Second Language: (ISBN 9780471739104) *Mastering Problem Solving* by Thomas Barrett of Ohio State University. This brief paperback teaches the student how to approach problems more efficiently and effectively. The student will learn how to recognize common patterns in physics problems, break problems down into manageable steps, and apply appropriate techniques. The book takes the student step by step through the solutions to numerous examples.

ACKNOWLEDGMENTS

A great many people have contributed to this book. Sen-Ben Liao of Lawrence Livermore National Laboratory, James Whitenton of Southern Polytechnic State University, and Jerry Shi, of Pasadena City College, performed the Herculean task of working out solutions for every one of the homework problems in the book. At John Wiley publishers, the book received support from Stuart Johnson, Geraldine Osnato and Aly Rentrop, the editors who oversaw the entire project from start to finish. We thank Elizabeth Swain, the production editor, for pulling all the pieces together during the complex production process. We also thank Maddy Lesure for her design of the text and the cover; Lee Goldstein for her page make-up; Helen Walden for her copyediting; and Lilian Brady for her proofreading. Jennifer Atkins was inspired in the search for unusual and interesting photographs. Both the publisher John Wiley & Sons, Inc. and Jearl Walker would like to thank the following for comments and ideas about the recent editions:

Jonathan Abramson, *Portland State University*; Omar Adawi, *Parkland College*; Edward Adelson, *The Ohio State University*; Steven R. Baker, *Naval Postgraduate School*; George Caplan, *Wellesley College*; Richard Kass, *The Ohio State University*; M. R. Khoshbin-e-Khoshnazar, *Research Institution for Curriculum Development & Educational Innovations (Tehran)*; Craig Kletzing, *University of Iowa*, Stuart Loucks, *American River College*; Laurence Lurio, *Northern Illinois University*; Ponn Maheswaranathan, *Winthrop University;* Joe McCullough, *Cabrillo College*; Carl E. Mungan, *U. S. Naval Academy*, Don N. Page, *University of Alberta*; Elie Riachi, *Fort Scott Community College*; Andrew G. Rinzler, *University of Florida*; Dubravka Rupnik, *Louisiana State University*; Robert Schabinger, *Rutgers University*; Ruth Schwartz, *Milwaukee School of Engineering*; Carol Strong, *University of Alabama at Huntsville*, Nora Thornber, *Raritan Valley Community College*; Frank Wang, *LaGuardia Community College*; Graham W. Wilson, *University of Kansas*; Roland Winkler, *Northern Illinois University*; William Zacharias, *Cleveland State University*; Ulrich Zurcher, *Cleveland State University*.

Finally, our external reviewers have been outstanding and we acknowledge here our debt to each member of that team.

Maris A. Abolins, *Michigan State University*
Edward Adelson, *Ohio State University*
Nural Akchurin, *Texas Tech*
Yildirim Aktas, *University of North Carolina-Charlotte*
Barbara Andereck, *Ohio Wesleyan University*
Tetyana Antimirova, *Ryerson University*
Mark Arnett, *Kirkwood Community College*
Arun Bansil, *Northeastern University*
Richard Barber, *Santa Clara University*
Neil Basecu, *Westchester Community College*
Anand Batra, *Howard University*
Kenneth Bolland, *The Ohio State University*
Richard Bone, *Florida International University*
Michael E. Browne, *University of Idaho*
Timothy J. Burns, *Leeward Community College*
Joseph Buschi, *Manhattan College*
Philip A. Casabella, *Rensselaer Polytechnic Institute*
Randall Caton, *Christopher Newport College*
Roger Clapp, *University of South Florida*
W. R. Conkie, *Queen's University*
Renate Crawford, *University of Massachusetts-Dartmouth*
Mike Crivello, *San Diego State University*
Robert N. Davie, Jr., *St. Petersburg Junior College*
Cheryl K. Dellai, *Glendale Community College*
Eric R. Dietz, *California State University at Chico*
N. John DiNardo, *Drexel University*
Eugene Dunnam, *University of Florida*
Robert Endorf, *University of Cincinnati*
F. Paul Esposito, *University of Cincinnati*
Jerry Finkelstein, *San Jose State University*
Robert H. Good, *California State University-Hayward*
Michael Gorman, *University of Houston*
Benjamin Grinstein, *University of California, San Diego*
John B. Gruber, *San Jose State University*
Ann Hanks, *American River College*
Randy Harris, *University of California-Davis*
Samuel Harris, *Purdue University*
Harold B. Hart, *Western Illinois University*
Rebecca Hartzler, *Seattle Central Community College*
John Hubisz, *North Carolina StateUniversity*
Joey Huston, *Michigan State University*
David Ingram, *Ohio University*
Shawn Jackson, *University of Tulsa*
Hector Jimenez, *University of Puerto Rico*
Sudhakar B. Joshi, *York University*
Leonard M. Kahn, *University of Rhode Island*
Sudipa Kirtley, *Rose-Hulman Institute*
Leonard Kleinman, *University of Texas at Austin*
Craig Kletzing, *University of Iowa*
Peter F. Koehler, *University of Pittsburgh*

Arthur Z. Kovacs, *Rochester Institute of Technology*
Kenneth Krane, *Oregon State University*
Hadley Lawler, *Vanderbilt University*
Priscilla Laws, *Dickinson College*
Edbertho Leal, *Polytechnic University of Puerto Rico*
Vern Lindberg, *Rochester Institute of Technology*
Peter Loly, *University of Manitoba*
James MacLaren, *Tulane University*
Andreas Mandelis, *University of Toronto*
Robert R. Marchini, *Memphis State University*
Andrea Markelz, *University at Buffalo, SUNY*
Paul Marquard, *Caspar College*
David Marx, *Illinois State University*
Dan Mazilu, *Washington and LeeUniversity*
James H. McGuire, *Tulane University*
David M. McKinstry, *Eastern Washington University*
Jordon Morelli, *Queen's University*
Eugene Mosca, *United States Naval Academy*
Eric R. Murray, *Georgia Institute of Technology, School of Physics*
James Napolitano, *Rensselaer Polytechnic Institute*
Blaine Norum, *University of Virginia*
Michael O'Shea, *Kansas State University*
Patrick Papin, *San Diego State University*
Kiumars Parvin, *San Jose State University*
Robert Pelcovits, *Brown University*
Oren P. Quist, *South Dakota State University*
Joe Redish, *University of Maryland*
Timothy M. Ritter, *University of North Carolina at Pembroke*
Dan Styer, *Oberlin College*
Frank Wang, *LaGuardia Community College*
Robert Webb, *Texas A&M University*
Suzanne Willis, *Northern Illinois University*
Shannon Willoughby, *Montana State University*

MATHEMATICAL FORMULAS*

Quadratic Formula

If $ax^2 + bx + c = 0$, then $x = \dfrac{-b \pm \sqrt{b^2 - 4ac}}{2a}$

Binomial Theorem

$$(1 + x)^n = 1 + \frac{nx}{1!} + \frac{n(n-1)x^2}{2!} + \cdots \qquad (x^2 < 1)$$

Products of Vectors

Let θ be the smaller of the two angles between $\vec{a}$ and $\vec{b}$. Then

$$\vec{a} \cdot \vec{b} = \vec{b} \cdot \vec{a} = a_x b_x + a_y b_y + a_z b_z = ab \cos \theta$$

$$\vec{a} \times \vec{b} = -\vec{b} \times \vec{a} = \begin{vmatrix} \hat{\mathrm{i}} & \hat{\mathrm{j}} & \hat{\mathrm{k}} \\ a_x & a_y & a_z \\ b_x & b_y & b_z \end{vmatrix}$$

$$= \hat{\mathrm{i}} \begin{vmatrix} a_y & a_z \\ b_y & b_z \end{vmatrix} - \hat{\mathrm{j}} \begin{vmatrix} a_x & a_z \\ b_x & b_z \end{vmatrix} + \hat{\mathrm{k}} \begin{vmatrix} a_x & a_y \\ b_x & b_y \end{vmatrix}$$

$$= (a_y b_z - b_y a_z)\hat{\mathrm{i}} + (a_z b_x - b_z a_x)\hat{\mathrm{j}} + (a_x b_y - b_x a_y)\hat{\mathrm{k}}$$

$$|\vec{a} \times \vec{b}| = ab \sin \theta$$

Trigonometric Identities

$$\sin \alpha \pm \sin \beta = 2 \sin \tfrac{1}{2}(\alpha \pm \beta) \cos \tfrac{1}{2}(\alpha \mp \beta)$$

$$\cos \alpha + \cos \beta = 2 \cos \tfrac{1}{2}(\alpha + \beta) \cos \tfrac{1}{2}(\alpha - \beta)$$

*See Appendix E for a more complete list.

Derivatives and Integrals

$$\frac{d}{dx} \sin x = \cos x \qquad \int \sin x\, dx = -\cos x$$

$$\frac{d}{dx} \cos x = -\sin x \qquad \int \cos x\, dx = \sin x$$

$$\frac{d}{dx} e^x = e^x \qquad \int e^x\, dx = e^x$$

$$\int \frac{dx}{\sqrt{x^2 + a^2}} = \ln(x + \sqrt{x^2 + a^2})$$

$$\int \frac{x\, dx}{(x^2 + a^2)^{3/2}} = -\frac{1}{(x^2 + a^2)^{1/2}}$$

$$\int \frac{dx}{(x^2 + a^2)^{3/2}} = \frac{x}{a^2(x^2 + a^2)^{1/2}}$$

Cramer's Rule

Two simultaneous equations in unknowns x and y,

$$a_1 x + b_1 y = c_1 \quad \text{and} \quad a_2 x + b_2 y = c_2,$$

have the solutions

$$x = \frac{\begin{vmatrix} c_1 & b_1 \\ c_2 & b_2 \end{vmatrix}}{\begin{vmatrix} a_1 & b_1 \\ a_2 & b_2 \end{vmatrix}} = \frac{c_1 b_2 - c_2 b_1}{a_1 b_2 - a_2 b_1}$$

and

$$y = \frac{\begin{vmatrix} a_1 & c_1 \\ a_2 & c_2 \end{vmatrix}}{\begin{vmatrix} a_1 & b_1 \\ a_2 & b_2 \end{vmatrix}} = \frac{a_1 c_2 - a_2 c_1}{a_1 b_2 - a_2 b_1}.$$

SI PREFIXES*

Factor	Prefix	Symbol	Factor	Prefix	Symbol
10^{24}	yotta	Y	10^{-1}	deci	d
10^{21}	zetta	Z	10^{-2}	centi	c
10^{18}	exa	E	10^{-3}	milli	m
10^{15}	peta	P	10^{-6}	micro	μ
10^{12}	tera	T	10^{-9}	nano	n
10^{9}	giga	G	10^{-12}	pico	p
10^{6}	mega	M	10^{-15}	femto	f
10^{3}	kilo	k	10^{-18}	atto	a
10^{2}	hecto	h	10^{-21}	zepto	z
10^{1}	deka	da	10^{-24}	yocto	y

*In all cases, the first syllable is accented, as in ná-no-mé-ter.

CHAPTER 1

Measurement

1-1 MEASURING THINGS, INCLUDING LENGTHS

Learning Objectives

After reading this module, you should be able to . . .

1.01 Identify the base quantities in the SI system.

1.02 Name the most frequently used prefixes for SI units.

1.03 Change units (here for length, area, and volume) by using chain-link conversions.

1.04 Explain that the meter is defined in terms of the speed of light in vacuum.

Key Ideas

- Physics is based on measurement of physical quantities. Certain physical quantities have been chosen as base quantities (such as length, time, and mass); each has been defined in terms of a standard and given a unit of measure (such as meter, second, and kilogram). Other physical quantities are defined in terms of the base quantities and their standards and units.
- The unit system emphasized in this book is the International System of Units (SI). The three physical quantities displayed in Table 1-1 are used in the early chapters. Standards, which must be both accessible and invariable, have been established for these base quantities by international agreement. These standards are used in all physical measurement, for both the base quantities and the quantities derived from them. Scientific notation and the prefixes of Table 1-2 are used to simplify measurement notation.
- Conversion of units may be performed by using chain-link conversions in which the original data are multiplied successively by conversion factors written as unity and the units are manipulated like algebraic quantities until only the desired units remain.
- The meter is defined as the distance traveled by light during a precisely specified time interval.

What Is Physics?

Science and engineering are based on measurements and comparisons. Thus, we need rules about how things are measured and compared, and we need experiments to establish the units for those measurements and comparisons. One purpose of physics (and engineering) is to design and conduct those experiments.

For example, physicists strive to develop clocks of extreme accuracy so that any time or time interval can be precisely determined and compared. You may wonder whether such accuracy is actually needed or worth the effort. Here is one example of the worth: Without clocks of extreme accuracy, the Global Positioning System (GPS) that is now vital to worldwide navigation would be useless.

Measuring Things

We discover physics by learning how to measure the quantities involved in physics. Among these quantities are length, time, mass, temperature, pressure, and electric current.

We measure each physical quantity in its own units, by comparison with a **standard**. The **unit** is a unique name we assign to measures of that quantity—for example, meter (m) for the quantity length. The standard corresponds to exactly 1.0 unit of the quantity. As you will see, the standard for length, which corresponds to exactly 1.0 m, is the distance traveled by light in a vacuum during a certain fraction of a second. We can define a unit and its standard in any way we care to. However, the important thing is to do so in such a way that scientists around the world will agree that our definitions are both sensible and practical.

Once we have set up a standard—say, for length—we must work out procedures by which any length whatever, be it the radius of a hydrogen atom, the wheelbase of a skateboard, or the distance to a star, can be expressed in terms of the standard. Rulers, which approximate our length standard, give us one such procedure for measuring length. However, many of our comparisons must be indirect. You cannot use a ruler, for example, to measure the radius of an atom or the distance to a star.

Base Quantities. There are so many physical quantities that it is a problem to organize them. Fortunately, they are not all independent; for example, speed is the ratio of a length to a time. Thus, what we do is pick out—by international agreement—a small number of physical quantities, such as length and time, and assign standards to them alone. We then define all other physical quantities in terms of these *base quantities* and their standards (called *base standards*). Speed, for example, is defined in terms of the base quantities length and time and their base standards.

Base standards must be both accessible and invariable. If we define the length standard as the distance between one's nose and the index finger on an outstretched arm, we certainly have an accessible standard—but it will, of course, vary from person to person. The demand for precision in science and engineering pushes us to aim first for invariability. We then exert great effort to make duplicates of the base standards that are accessible to those who need them.

Table 1-1 Units for Three SI Base Quantities

Quantity	Unit Name	Unit Symbol
Length	meter	m
Time	second	s
Mass	kilogram	kg

The International System of Units

In 1971, the 14th General Conference on Weights and Measures picked seven quantities as base quantities, thereby forming the basis of the International System of Units, abbreviated SI from its French name and popularly known as the *metric system*. Table 1-1 shows the units for the three base quantities—length, mass, and time—that we use in the early chapters of this book. These units were defined to be on a "human scale."

Many SI *derived units* are defined in terms of these base units. For example, the SI unit for power, called the **watt** (W), is defined in terms of the base units for mass, length, and time. Thus, as you will see in Chapter 7,

$$1 \text{ watt} = 1 \text{ W} = 1 \text{ kg} \cdot \text{m}^2/\text{s}^3, \tag{1-1}$$

where the last collection of unit symbols is read as kilogram-meter squared per second cubed.

To express the very large and very small quantities we often run into in physics, we use *scientific notation*, which employs powers of 10. In this notation,

$$3\,560\,000\,000 \text{ m} = 3.56 \times 10^9 \text{ m} \tag{1-2}$$

and

$$0.000\,000\,492 \text{ s} = 4.92 \times 10^{-7} \text{ s}. \tag{1-3}$$

Scientific notation on computers sometimes takes on an even briefer look, as in 3.56 E9 and 4.92 E–7, where E stands for "exponent of ten." It is briefer still on some calculators, where E is replaced with an empty space.

As a further convenience when dealing with very large or very small measurements, we use the prefixes listed in Table 1-2. As you can see, each prefix represents a certain power of 10, to be used as a multiplication factor. Attaching a prefix to an SI unit has the effect of multiplying by the associated factor. Thus, we can express a particular electric power as

$$1.27 \times 10^9 \text{ watts} = 1.27 \text{ gigawatts} = 1.27 \text{ GW} \tag{1-4}$$

Table 1-2 Prefixes for SI Units

Factor	Prefix[a]	Symbol
10^{24}	yotta-	Y
10^{21}	zetta-	Z
10^{18}	exa-	E
10^{15}	peta-	P
10^{12}	tera-	T
$\mathbf{10^9}$	**giga-**	**G**
$\mathbf{10^6}$	**mega-**	**M**
$\mathbf{10^3}$	**kilo-**	**k**
10^2	hecto-	h
10^1	deka-	da
10^{-1}	deci-	d
$\mathbf{10^{-2}}$	**centi-**	**c**
$\mathbf{10^{-3}}$	**milli-**	**m**
$\mathbf{10^{-6}}$	**micro-**	$\boldsymbol{\mu}$
$\mathbf{10^{-9}}$	**nano-**	**n**
$\mathbf{10^{-12}}$	**pico-**	**p**
10^{-15}	femto-	f
10^{-18}	atto-	a
10^{-21}	zepto-	z
10^{-24}	yocto-	y

[a]The most frequently used prefixes are shown in bold type.

or a particular time interval as

$$2.35 \times 10^{-9}\ \text{s} = 2.35\ \text{nanoseconds} = 2.35\ \text{ns}. \tag{1-5}$$

Some prefixes, as used in milliliter, centimeter, kilogram, and megabyte, are probably familiar to you.

Changing Units

We often need to change the units in which a physical quantity is expressed. We do so by a method called *chain-link conversion*. In this method, we multiply the original measurement by a **conversion factor** (a ratio of units that is equal to unity). For example, because 1 min and 60 s are identical time intervals, we have

$$\frac{1\ \text{min}}{60\ \text{s}} = 1 \quad \text{and} \quad \frac{60\ \text{s}}{1\ \text{min}} = 1.$$

Thus, the ratios (1 min)/(60 s) and (60 s)/(1 min) can be used as conversion factors. This is *not* the same as writing $\frac{1}{60} = 1$ or $60 = 1$; each *number* and its *unit* must be treated together.

Because multiplying any quantity by unity leaves the quantity unchanged, we can introduce conversion factors wherever we find them useful. In chain-link conversion, we use the factors to cancel unwanted units. For example, to convert 2 min to seconds, we have

$$2\ \text{min} = (2\ \text{min})(1) = (2\ \text{min})\left(\frac{60\ \text{s}}{1\ \text{min}}\right) = 120\ \text{s}. \tag{1-6}$$

If you introduce a conversion factor in such a way that unwanted units do *not* cancel, invert the factor and try again. In conversions, the units obey the same algebraic rules as variables and numbers.

Appendix D gives conversion factors between SI and other systems of units, including non-SI units still used in the United States. However, the conversion factors are written in the style of "1 min = 60 s" rather than as a ratio. So, you need to decide on the numerator and denominator in any needed ratio.

Length

In 1792, the newborn Republic of France established a new system of weights and measures. Its cornerstone was the meter, defined to be one ten-millionth of the distance from the north pole to the equator. Later, for practical reasons, this Earth standard was abandoned and the meter came to be defined as the distance between two fine lines engraved near the ends of a platinum–iridium bar, the **standard meter bar**, which was kept at the International Bureau of Weights and Measures near Paris. Accurate copies of the bar were sent to standardizing laboratories throughout the world. These **secondary standards** were used to produce other, still more accessible standards, so that ultimately every measuring device derived its authority from the standard meter bar through a complicated chain of comparisons.

Eventually, a standard more precise than the distance between two fine scratches on a metal bar was required. In 1960, a new standard for the meter, based on the wavelength of light, was adopted. Specifically, the standard for the meter was redefined to be 1 650 763.73 wavelengths of a particular orange-red light emitted by atoms of krypton-86 (a particular isotope, or type, of krypton) in a gas discharge tube that can be set up anywhere in the world. This awkward number of wavelengths was chosen so that the new standard would be close to the old meter-bar standard.

By 1983, however, the demand for higher precision had reached such a point that even the krypton-86 standard could not meet it, and in that year a bold step was taken. The meter was redefined as the distance traveled by light in a specified time interval. In the words of the 17th General Conference on Weights and Measures:

> The meter is the length of the path traveled by light in a vacuum during a time interval of 1/299 792 458 of a second.

This time interval was chosen so that the speed of light c is exactly

$$c = 299\,792\,458 \text{ m/s}.$$

Measurements of the speed of light had become extremely precise, so it made sense to adopt the speed of light as a defined quantity and to use it to redefine the meter.

Table 1-3 shows a wide range of lengths, from that of the universe (top line) to those of some very small objects.

Table 1-3 Some Approximate Lengths

Measurement	Length in Meters
Distance to the first galaxies formed	2×10^{26}
Distance to the Andromeda galaxy	2×10^{22}
Distance to the nearby star Proxima Centauri	4×10^{16}
Distance to Pluto	6×10^{12}
Radius of Earth	6×10^{6}
Height of Mt. Everest	9×10^{3}
Thickness of this page	1×10^{-4}
Length of a typical virus	1×10^{-8}
Radius of a hydrogen atom	5×10^{-11}
Radius of a proton	1×10^{-15}

Significant Figures and Decimal Places

Suppose that you work out a problem in which each value consists of two digits. Those digits are called **significant figures** and they set the number of digits that you can use in reporting your final answer. With data given in two significant figures, your final answer should have only two significant figures. However, depending on the mode setting of your calculator, many more digits might be displayed. Those extra digits are meaningless.

In this book, final results of calculations are often rounded to match the least number of significant figures in the given data. (However, sometimes an extra significant figure is kept.) When the leftmost of the digits to be discarded is 5 or more, the last remaining digit is rounded up; otherwise it is retained as is. For example, 11.3516 is rounded to three significant figures as 11.4 and 11.3279 is rounded to three significant figures as 11.3. (The answers to sample problems in this book are usually presented with the symbol = instead of ≈ even if rounding is involved.)

When a number such as 3.15 or 3.15×10^3 is provided in a problem, the number of significant figures is apparent, but how about the number 3000? Is it known to only one significant figure (3×10^3)? Or is it known to as many as four significant figures (3.000×10^3)? In this book, we assume that all the zeros in such given numbers as 3000 are significant, but you had better not make that assumption elsewhere.

Don't confuse *significant figures* with *decimal places*. Consider the lengths 35.6 mm, 3.56 m, and 0.00356 m. They all have three significant figures but they have one, two, and five decimal places, respectively.

Sample Problem 1.01 Estimating order of magnitude, ball of string

The world's largest ball of string is about 2 m in radius. To the nearest order of magnitude, what is the total length L of the string in the ball?

KEY IDEA

We could, of course, take the ball apart and measure the total length L, but that would take great effort and make the ball's builder most unhappy. Instead, because we want only the nearest order of magnitude, we can estimate any quantities required in the calculation.

Calculations: Let us assume the ball is spherical with radius $R = 2$ m. The string in the ball is not closely packed (there are uncountable gaps between adjacent sections of string). To allow for these gaps, let us somewhat overestimate the cross-sectional area of the string by assuming the cross section is square, with an edge length $d = 4$ mm. Then, with a cross-sectional area of d^2 and a length L, the string occupies a total volume of

$$V = \text{(cross-sectional area)(length)} = d^2L.$$

This is approximately equal to the volume of the ball, given by $\frac{4}{3}\pi R^3$, which is about $4R^3$ because π is about 3. Thus, we have the following

$$d^2L = 4R^3,$$

or
$$L = \frac{4R^3}{d^2} = \frac{4(2\text{ m})^3}{(4 \times 10^{-3}\text{ m})^2}$$
$$= 2 \times 10^6\text{ m} \approx 10^6\text{ m} = 10^3\text{ km}. \quad \text{(Answer)}$$

(Note that you do not need a calculator for such a simplified calculation.) To the nearest order of magnitude, the ball contains about 1000 km of string!

WILEY PLUS Additional examples, video, and practice available at *WileyPLUS*

1-2 TIME

Learning Objectives

After reading this module, you should be able to . . .

1.05 Change units for time by using chain-link conversions.

1.06 Use various measures of time, such as for motion or as determined on different clocks.

Key Idea

- The second is defined in terms of the oscillations of light emitted by an atomic (cesium-133) source. Accurate time signals are sent worldwide by radio signals keyed to atomic clocks in standardizing laboratories.

Time

Time has two aspects. For civil and some scientific purposes, we want to know the time of day so that we can order events in sequence. In much scientific work, we want to know how long an event lasts. Thus, any time standard must be able to answer two questions: "*When* did it happen?" and "What is its *duration*?" Table 1-4 shows some time intervals.

Any phenomenon that repeats itself is a possible time standard. Earth's rotation, which determines the length of the day, has been used in this way for centuries; Fig. 1-1 shows one novel example of a watch based on that rotation. A quartz clock, in which a quartz ring is made to vibrate continuously, can be calibrated against Earth's rotation via astronomical observations and used to measure time intervals in the laboratory. However, the calibration cannot be carried out with the accuracy called for by modern scientific and engineering technology.

Steven Pitkin

Figure 1-1 When the metric system was proposed in 1792, the hour was redefined to provide a 10-hour day. The idea did not catch on. The maker of this 10-hour watch wisely provided a small dial that kept conventional 12-hour time. Do the two dials indicate the same time?

Table 1-4 Some Approximate Time Intervals

Measurement	Time Interval in Seconds	Measurement	Time Interval in Seconds
Lifetime of the proton (predicted)	3×10^{40}	Time between human heartbeats	8×10^{-1}
Age of the universe	5×10^{17}	Lifetime of the muon	2×10^{-6}
Age of the pyramid of Cheops	1×10^{11}	Shortest lab light pulse	1×10^{-16}
Human life expectancy	2×10^{9}	Lifetime of the most unstable particle	1×10^{-23}
Length of a day	9×10^{4}	The Planck time[a]	1×10^{-43}

[a]This is the earliest time after the big bang at which the laws of physics as we know them can be applied.

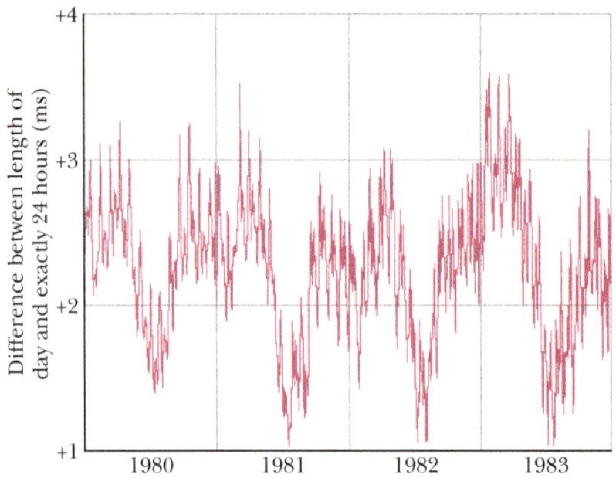

Figure 1-2 Variations in the length of the day over a 4-year period. Note that the entire vertical scale amounts to only 3 ms (= 0.003 s).

To meet the need for a better time standard, atomic clocks have been developed. An atomic clock at the National Institute of Standards and Technology (NIST) in Boulder, Colorado, is the standard for Coordinated Universal Time (UTC) in the United States. Its time signals are available by shortwave radio (stations WWV and WWVH) and by telephone (303-499-7111). Time signals (and related information) are also available from the United States Naval Observatory at website http://tycho.usno.navy.mil/time.html. (To set a clock extremely accurately at your particular location, you would have to account for the travel time required for these signals to reach you.)

Figure 1-2 shows variations in the length of one day on Earth over a 4-year period, as determined by comparison with a cesium (atomic) clock. Because the variation displayed by Fig. 1-2 is seasonal and repetitious, we suspect the rotating Earth when there is a difference between Earth and atom as timekeepers. The variation is due to tidal effects caused by the Moon and to large-scale winds.

The 13th General Conference on Weights and Measures in 1967 adopted a standard second based on the cesium clock:

> One second is the time taken by 9 192 631 770 oscillations of the light (of a specified wavelength) emitted by a cesium-133 atom.

Atomic clocks are so consistent that, in principle, two cesium clocks would have to run for 6000 years before their readings would differ by more than 1 s. Even such accuracy pales in comparison with that of clocks currently being developed; their precision may be 1 part in 10^{18}—that is, 1 s in 1×10^{18} s (which is about 3×10^{10} y).

1-3 MASS

Learning Objectives

After reading this module, you should be able to . . .

1.07 Change units for mass by using chain-link conversions.

1.08 Relate density to mass and volume when the mass is uniformly distributed.

Key Ideas

- The kilogram is defined in terms of a platinum–iridium standard mass kept near Paris. For measurements on an atomic scale, the atomic mass unit, defined in terms of the atom carbon-12, is usually used.
- The density ρ of a material is the mass per unit volume:

$$\rho = \frac{m}{V}.$$

Mass

The Standard Kilogram

The SI standard of mass is a cylinder of platinum and iridium (Fig. 1-3) that is kept at the International Bureau of Weights and Measures near Paris and assigned, by

Courtesy Bureau International des Poids et Mesures. Reproduced with permission of the BIPM.

Figure 1-3 The international 1 kg standard of mass, a platinum–iridium cylinder 3.9 cm in height and in diameter.

international agreement, a mass of 1 kilogram. Accurate copies have been sent to standardizing laboratories in other countries, and the masses of other bodies can be determined by balancing them against a copy. Table 1-5 shows some masses expressed in kilograms, ranging over about 83 orders of magnitude.

The U.S. copy of the standard kilogram is housed in a vault at NIST. It is removed, no more than once a year, for the purpose of checking duplicate copies that are used elsewhere. Since 1889, it has been taken to France twice for recomparison with the primary standard.

Table 1-5 Some Approximate Masses

Object	Mass in Kilograms
Known universe	1×10^{53}
Our galaxy	2×10^{41}
Sun	2×10^{30}
Moon	7×10^{22}
Asteroid Eros	5×10^{15}
Small mountain	1×10^{12}
Ocean liner	7×10^{7}
Elephant	5×10^{3}
Grape	3×10^{-3}
Speck of dust	7×10^{-10}
Penicillin molecule	5×10^{-17}
Uranium atom	4×10^{-25}
Proton	2×10^{-27}
Electron	9×10^{-31}

A Second Mass Standard

The masses of atoms can be compared with one another more precisely than they can be compared with the standard kilogram. For this reason, we have a second mass standard. It is the carbon-12 atom, which, by international agreement, has been assigned a mass of 12 **atomic mass units** (u). The relation between the two units is

$$1 \text{ u} = 1.660\,538\,86 \times 10^{-27} \text{ kg}, \tag{1-7}$$

with an uncertainty of ± 10 in the last two decimal places. Scientists can, with reasonable precision, experimentally determine the masses of other atoms relative to the mass of carbon-12. What we presently lack is a reliable means of extending that precision to more common units of mass, such as a kilogram.

Density

As we shall discuss further in Chapter 14, **density** ρ (lowercase Greek letter rho) is the mass per unit volume:

$$\rho = \frac{m}{V}. \tag{1-8}$$

Densities are typically listed in kilograms per cubic meter or grams per cubic centimeter. The density of water (1.00 gram per cubic centimeter) is often used as a comparison. Fresh snow has about 10% of that density; platinum has a density that is about 21 times that of water.

Sample Problem 1.02 Density and liquefaction

A heavy object can sink into the ground during an earthquake if the shaking causes the ground to undergo *liquefaction*, in which the soil grains experience little friction as they slide over one another. The ground is then effectively quicksand. The possibility of liquefaction in sandy ground can be predicted in terms of the *void ratio e* for a sample of the ground:

$$e = \frac{V_{\text{voids}}}{V_{\text{grains}}}. \tag{1-9}$$

Here, V_{grains} is the total volume of the sand grains in the sample and V_{voids} is the total volume between the grains (in the *voids*). If e exceeds a critical value of 0.80, liquefaction can occur during an earthquake. What is the corresponding sand density ρ_{sand}? Solid silicon dioxide (the primary component of sand) has a density of $\rho_{SiO_2} = 2.600 \times 10^3$ kg/m³.

KEY IDEA

The density of the sand ρ_{sand} in a sample is the mass per unit volume—that is, the ratio of the total mass m_{sand} of the sand grains to the total volume V_{total} of the sample:

$$\rho_{\text{sand}} = \frac{m_{\text{sand}}}{V_{\text{total}}}. \tag{1-10}$$

Calculations: The total volume V_{total} of a sample is

$$V_{\text{total}} = V_{\text{grains}} + V_{\text{voids}}.$$

Substituting for V_{voids} from Eq. 1-9 and solving for V_{grains} lead to

$$V_{\text{grains}} = \frac{V_{\text{total}}}{1 + e}. \tag{1-11}$$

From Eq. 1-8, the total mass m_{sand} of the sand grains is the product of the density of silicon dioxide and the total volume of the sand grains:

$$m_{sand} = \rho_{SiO_2} V_{grains}. \tag{1-12}$$

Substituting this expression into Eq. 1-10 and then substituting for V_{grains} from Eq. 1-11 lead to

$$\rho_{sand} = \frac{\rho_{SiO_2}}{V_{total}} \frac{V_{total}}{1+e} = \frac{\rho_{SiO_2}}{1+e}. \tag{1-13}$$

Substituting $\rho_{SiO_2} = 2.600 \times 10^3$ kg/m³ and the critical value of $e = 0.80$, we find that liquefaction occurs when the sand density is less than

$$\rho_{sand} = \frac{2.600 \times 10^3 \text{ kg/m}^3}{1.80} = 1.4 \times 10^3 \text{ kg/m}^3. \quad \text{(Answer)}$$

A building can sink several meters in such liquefaction.

WILEY PLUS Additional examples, video, and practice available at *WileyPLUS*

Review & Summary

Measurement in Physics Physics is based on measurement of physical quantities. Certain physical quantities have been chosen as **base quantities** (such as length, time, and mass); each has been defined in terms of a **standard** and given a **unit** of measure (such as meter, second, and kilogram). Other physical quantities are defined in terms of the base quantities and their standards and units.

SI Units The unit system emphasized in this book is the International System of Units (SI). The three physical quantities displayed in Table 1-1 are used in the early chapters. Standards, which must be both accessible and invariable, have been established for these base quantities by international agreement. These standards are used in all physical measurement, for both the base quantities and the quantities derived from them. Scientific notation and the prefixes of Table 1-2 are used to simplify measurement notation.

Changing Units Conversion of units may be performed by using *chain-link conversions* in which the original data are multiplied successively by conversion factors written as unity and the units are manipulated like algebraic quantities until only the desired units remain.

Length The meter is defined as the distance traveled by light during a precisely specified time interval.

Time The second is defined in terms of the oscillations of light emitted by an atomic (cesium-133) source. Accurate time signals are sent worldwide by radio signals keyed to atomic clocks in standardizing laboratories.

Mass The kilogram is defined in terms of a platinum–iridium standard mass kept near Paris. For measurements on an atomic scale, the atomic mass unit, defined in terms of the atom carbon-12, is usually used.

Density The density ρ of a material is the mass per unit volume:

$$\rho = \frac{m}{V}. \tag{1-8}$$

Problems

GO Tutoring problem available (at instructor's discretion) in *WileyPLUS* and WebAssign
SSM Worked-out solution available in Student Solutions Manual
• – ••• Number of dots indicates level of problem difficulty
WWW Worked-out solution is at http://www.wiley.com/college/halliday
ILW Interactive solution is at http://www.wiley.com/college/halliday
Additional information available in *The Flying Circus of Physics* and at flyingcircusofphysics.com

Module 1-1 Measuring Things, Including Lengths

•1 SSM Earth is approximately a sphere of radius 6.37×10^6 m. What are (a) its circumference in kilometers, (b) its surface area in square kilometers, and (c) its volume in cubic kilometers?

•2 A *gry* is an old English measure for length, defined as 1/10 of a line, where *line* is another old English measure for length, defined as 1/12 inch. A common measure for length in the publishing business is a *point*, defined as 1/72 inch. What is an area of 0.50 gry² in points squared (points²)?

•3 The micrometer (1 μm) is often called the *micron*. (a) How many microns make up 1.0 km? (b) What fraction of a centimeter equals 1.0 μm? (c) How many microns are in 1.0 yd?

•4 Spacing in this book was generally done in units of points and picas: 12 points = 1 pica, and 6 picas = 1 inch. If a figure was misplaced in the page proofs by 0.80 cm, what was the misplacement in (a) picas and (b) points?

•5 SSM WWW Horses are to race over a certain English meadow for a distance of 4.0 furlongs. What is the race distance in (a) rods and (b) chains? (1 furlong = 201.168 m, 1 rod = 5.0292 m, and 1 chain = 20.117 m.)

••6 You can easily convert common units and measures electronically, but you still should be able to use a conversion table, such as those in Appendix D. Table 1-6 is part of a conversion table for a system of volume measures once common in Spain; a volume of 1 fanega is equivalent to 55.501 dm^3 (cubic decimeters). To complete the table, what numbers (to three significant figures) should be entered in (a) the cahiz column, (b) the fanega column, (c) the cuartilla column, and (d) the almude column, starting with the top blank? Express 7.00 almudes in (e) medios, (f) cahizes, and (g) cubic centimeters (cm^3).

Table 1-6 Problem 6

	cahiz	fanega	cuartilla	almude	medio
1 cahiz =	1	12	48	144	288
1 fanega =		1	4	12	24
1 cuartilla =			1	3	6
1 almude =				1	2
1 medio =					1

••7 ILW Hydraulic engineers in the United States often use, as a unit of volume of water, the *acre-foot,* defined as the volume of water that will cover 1 acre of land to a depth of 1 ft. A severe thunderstorm dumped 2.0 in. of rain in 30 min on a town of area 26 km^2. What volume of water, in acre-feet, fell on the town?

••8 GO Harvard Bridge, which connects MIT with its fraternities across the Charles River, has a length of 364.4 Smoots plus one ear. The unit of one Smoot is based on the length of Oliver Reed Smoot, Jr., class of 1962, who was carried or dragged length by length across the bridge so that other pledge members of the Lambda Chi Alpha fraternity could mark off (with paint) 1-Smoot lengths along the bridge. The marks have been repainted biannually by fraternity pledges since the initial measurement, usually during times of traffic congestion so that the police cannot easily interfere. (Presumably, the police were originally upset because the Smoot is not an SI base unit, but these days they seem to have accepted the unit.) Figure 1-4 shows three parallel paths, measured in Smoots (S), Willies (W), and Zeldas (Z). What is the length of 50.0 Smoots in (a) Willies and (b) Zeldas?

Figure 1-4 Problem 8.

••9 Antarctica is roughly semicircular, with a radius of 2000 km (Fig. 1-5). The average thickness of its ice cover is 3000 m. How many cubic centimeters of ice does Antarctica contain? (Ignore the curvature of Earth.)

Figure 1-5 Problem 9.

Module 1-2 Time

•10 Until 1883, every city and town in the United States kept its own local time. Today, travelers reset their watches only when the time change equals 1.0 h. How far, on the average, must you travel in degrees of longitude between the time-zone boundaries at which your watch must be reset by 1.0 h? (*Hint:* Earth rotates 360° in about 24 h.)

•11 For about 10 years after the French Revolution, the French government attempted to base measures of time on multiples of ten: One week consisted of 10 days, one day consisted of 10 hours, one hour consisted of 100 minutes, and one minute consisted of 100 seconds. What are the ratios of (a) the French decimal week to the standard week and (b) the French decimal second to the standard second?

•12 The fastest growing plant on record is a *Hesperoyucca whipplei* that grew 3.7 m in 14 days. What was its growth rate in micrometers per second?

•13 GO Three digital clocks *A*, *B*, and *C* run at different rates and do not have simultaneous readings of zero. Figure 1-6 shows simultaneous readings on pairs of the clocks for four occasions. (At the earliest occasion, for example, *B* reads 25.0 s and *C* reads 92.0 s.) If two events are 600 s apart on clock *A*, how far apart are they on (a) clock *B* and (b) clock *C*? (c) When clock *A* reads 400 s, what does clock *B* read? (d) When clock *C* reads 15.0 s, what does clock *B* read? (Assume negative readings for prezero times.)

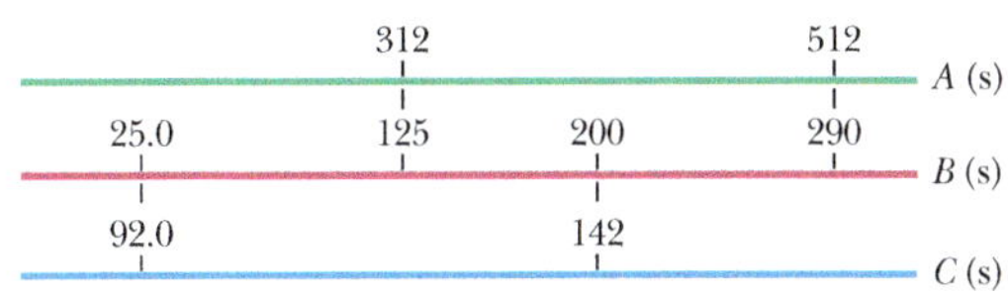

Figure 1-6 Problem 13.

•14 A lecture period (50 min) is close to 1 microcentury. (a) How long is a microcentury in minutes? (b) Using

$$\text{percentage difference} = \left(\frac{\text{actual} - \text{approximation}}{\text{actual}}\right) 100,$$

find the percentage difference from the approximation.

•15 A fortnight is a charming English measure of time equal to 2.0 weeks (the word is a contraction of "fourteen nights"). That is a nice amount of time in pleasant company but perhaps a painful string of microseconds in unpleasant company. How many microseconds are in a fortnight?

•16 Time standards are now based on atomic clocks. A promising second standard is based on *pulsars,* which are rotating neutron stars (highly compact stars consisting only of neutrons). Some rotate at a rate that is highly stable, sending out a radio beacon that sweeps briefly across Earth once with each rotation, like a lighthouse beacon. Pulsar PSR 1937 + 21 is an example; it rotates once every 1.557 806 448 872 75 ± 3 ms, where the trailing ±3 indicates the uncertainty in the last decimal place (it does *not* mean ±3 ms). (a) How many rotations does PSR 1937 + 21 make in 7.00 days? (b) How much time does the pulsar take to rotate exactly one million times and (c) what is the associated uncertainty?

•17 SSM Five clocks are being tested in a laboratory. Exactly at noon, as determined by the WWV time signal, on successive days of a week the clocks read as in the following table. Rank the five clocks according to their relative value as good timekeepers, best to worst. Justify your choice.

Clock	Sun.	Mon.	Tues.	Wed.	Thurs.	Fri.	Sat.
A	12:36:40	12:36:56	12:37:12	12:37:27	12:37:44	12:37:59	12:38:14
B	11:59:59	12:00:02	11:59:57	12:00:07	12:00:02	11:59:56	12:00:03
C	15:50:45	15:51:43	15:52:41	15:53:39	15:54:37	15:55:35	15:56:33
D	12:03:59	12:02:52	12:01:45	12:00:38	11:59:31	11:58:24	11:57:17
E	12:03:59	12:02:49	12:01:54	12:01:52	12:01:32	12:01:22	12:01:12

••18 Because Earth's rotation is gradually slowing, the length of each day increases: The day at the end of 1.0 century is 1.0 ms longer than the day at the start of the century. In 20 centuries, what is the total of the daily increases in time?

•••19 Suppose that, while lying on a beach near the equator watching the Sun set over a calm ocean, you start a stopwatch just as the top of the Sun disappears. You then stand, elevating your eyes by a height $H = 1.70$ m, and stop the watch when the top of the Sun again disappears. If the elapsed time is $t = 11.1$ s, what is the radius r of Earth?

Module 1-3 Mass

•20 GO The record for the largest glass bottle was set in 1992 by a team in Millville, New Jersey—they blew a bottle with a volume of 193 U.S. fluid gallons. (a) How much short of 1.0 million cubic centimeters is that? (b) If the bottle were filled with water at the leisurely rate of 1.8 g/min, how long would the filling take? Water has a density of 1000 kg/m^3.

•21 Earth has a mass of 5.98×10^{24} kg. The average mass of the atoms that make up Earth is 40 u. How many atoms are there in Earth?

•22 Gold, which has a density of 19.32 g/cm^3, is the most ductile metal and can be pressed into a thin leaf or drawn out into a long fiber. (a) If a sample of gold, with a mass of 27.63 g, is pressed into a leaf of 1.000 μm thickness, what is the area of the leaf? (b) If, instead, the gold is drawn out into a cylindrical fiber of radius 2.500 μm, what is the length of the fiber?

•23 SSM (a) Assuming that water has a density of exactly 1 g/cm^3, find the mass of one cubic meter of water in kilograms. (b) Suppose that it takes 10.0 h to drain a container of 5700 m^3 of water. What is the "mass flow rate," in kilograms per second, of water from the container?

••24 GO Grains of fine California beach sand are approximately spheres with an average radius of 50 μm and are made of silicon dioxide, which has a density of 2600 kg/m^3. What mass of sand grains would have a total surface area (the total area of all the individual spheres) equal to the surface area of a cube 1.00 m on an edge?

••25 During heavy rain, a section of a mountainside measuring 2.5 km horizontally, 0.80 km up along the slope, and 2.0 m deep slips into a valley in a mud slide. Assume that the mud ends up uniformly distributed over a surface area of the valley measuring 0.40 km $\times$ 0.40 km and that mud has a density of 1900 kg/m^3. What is the mass of the mud sitting above a 4.0 m^2 area of the valley floor?

••26 One cubic centimeter of a typical cumulus cloud contains 50 to 500 water drops, which have a typical radius of 10 μm. For that range, give the lower value and the higher value, respectively, for the following. (a) How many cubic meters of water are in a cylindrical cumulus cloud of height 3.0 km and radius 1.0 km? (b) How many 1-liter pop bottles would that water fill? (c) Water has a density of 1000 kg/m^3. How much mass does the water in the cloud have?

••27 Iron has a density of 7.87 g/cm^3, and the mass of an iron atom is 9.27×10^{-26} kg. If the atoms are spherical and tightly packed, (a) what is the volume of an iron atom and (b) what is the distance between the centers of adjacent atoms?

••28 A mole of atoms is 6.02×10^{23} atoms. To the nearest order of magnitude, how many moles of atoms are in a large domestic cat? The masses of a hydrogen atom, an oxygen atom, and a carbon atom are 1.0 u, 16 u, and 12 u, respectively. (*Hint:* Cats are sometimes known to kill a mole.)

••29 On a spending spree in Malaysia, you buy an ox with a weight of 28.9 piculs in the local unit of weights: 1 picul = 100 gins, 1 gin = 16 tahils, 1 tahil = 10 chees, and 1 chee = 10 hoons. The weight of 1 hoon corresponds to a mass of 0.3779 g. When you arrange to ship the ox home to your astonished family, how much mass in kilograms must you declare on the shipping manifest? (*Hint:* Set up multiple chain-link conversions.)

••30 GO Water is poured into a container that has a small leak. The mass m of the water is given as a function of time t by $m = 5.00t^{0.8} - 3.00t + 20.00$, with $t \geq 0$, m in grams, and t in seconds. (a) At what time is the water mass greatest, and (b) what is that greatest mass? In kilograms per minute, what is the rate of mass change at (c) $t = 2.00$ s and (d) $t = 5.00$ s?

•••31 A vertical container with base area measuring 14.0 cm by 17.0 cm is being filled with identical pieces of candy, each with a volume of 50.0 mm^3 and a mass of 0.0200 g. Assume that the volume of the empty spaces between the candies is negligible. If the height of the candies in the container increases at the rate of 0.250 cm/s, at what rate (kilograms per minute) does the mass of the candies in the container increase?

Additional Problems

32 In the United States, a doll house has the scale of 1:12 of a real house (that is, each length of the doll house is $\frac{1}{12}$ that of the real house) and a miniature house (a doll house to fit within a doll house) has the scale of 1:144 of a real house. Suppose a real house (Fig. 1-7) has a front length of 20 m, a depth of 12 m, a height of 6.0 m, and a standard sloped roof (vertical triangular faces on the ends) of height 3.0 m. In cubic meters, what are the volumes of the corresponding (a) doll house and (b) miniature house?

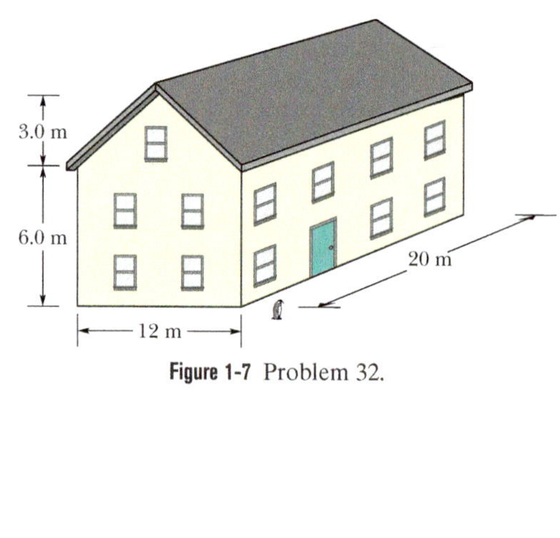

Figure 1-7 Problem 32.

33 SSM A ton is a measure of volume frequently used in shipping, but that use requires some care because there are at least three types of tons: A *displacement ton* is equal to 7 barrels bulk, a *freight ton* is equal to 8 barrels bulk, and a *register ton* is equal to 20 barrels bulk. A *barrel bulk* is another measure of volume: 1 barrel bulk = 0.1415 m^3. Suppose you spot a shipping order for "73 tons" of M&M candies, and you are certain that the client who sent the order intended "ton" to refer to volume (instead of weight or mass, as discussed in Chapter 5). If the client actually meant displacement tons, how many extra U.S. bushels of the candies will you erroneously ship if you interpret the order as (a) 73 freight tons and (b) 73 register tons? (1 m^3 = 28.378 U.S. bushels.)

34 Two types of *barrel* units were in use in the 1920s in the United States. The apple barrel had a legally set volume of 7056 cubic inches; the cranberry barrel, 5826 cubic inches. If a merchant sells 20 cranberry barrels of goods to a customer who thinks he is receiving apple barrels, what is the discrepancy in the shipment volume in liters?

35 An old English children's rhyme states, "Little Miss Muffet sat on a tuffet, eating her curds and whey, when along came a spider who sat down beside her. . . ." The spider sat down not because of the curds and whey but because Miss Muffet had a stash of 11 tuffets of dried flies. The volume measure of a tuffet is given by 1 tuffet = 2 pecks = 0.50 Imperial bushel, where 1 Imperial bushel = 36.3687 liters (L). What was Miss Muffet's stash in (a) pecks, (b) Imperial bushels, and (c) liters?

36 Table 1-7 shows some old measures of liquid volume. To complete the table, what numbers (to three significant figures) should be entered in (a) the wey column, (b) the chaldron column, (c) the bag column, (d) the pottle column, and (e) the gill column, starting from the top down? (f) The volume of 1 bag is equal to 0.1091 m^3. If an old story has a witch cooking up some vile liquid in a cauldron of volume 1.5 chaldrons, what is the volume in cubic meters?

Table 1-7 Problem 36

	wey	chaldron	bag	pottle	gill
1 wey =	1	10/9	40/3	640	120 240
1 chaldron =					
1 bag =					
1 pottle =					
1 gill =					

37 A typical sugar cube has an edge length of 1 cm. If you had a cubical box that contained a mole of sugar cubes, what would its edge length be? (One mole = 6.02×10^{23} units.)

38 An old manuscript reveals that a landowner in the time of King Arthur held 3.00 acres of plowed land plus a livestock area of 25.0 perches by 4.00 perches. What was the total area in (a) the old unit of roods and (b) the more modern unit of square meters? Here, 1 acre is an area of 40 perches by 4 perches, 1 rood is an area of 40 perches by 1 perch, and 1 perch is the length 16.5 ft.

39 SSM A tourist purchases a car in England and ships it home to the United States. The car sticker advertised that the car's fuel consumption was at the rate of 40 miles per gallon on the open road. The tourist does not realize that the U.K. gallon differs from the U.S. gallon:

$$1 \text{ U.K. gallon} = 4.546\,090\,0 \text{ liters}$$
$$1 \text{ U.S. gallon} = 3.785\,411\,8 \text{ liters.}$$

For a trip of 750 miles (in the United States), how many gallons of fuel does (a) the mistaken tourist believe she needs and (b) the car actually require?

40 Using conversions and data in the chapter, determine the number of hydrogen atoms required to obtain 1.0 kg of hydrogen. A hydrogen atom has a mass of 1.0 u.

41 SSM A *cord* is a volume of cut wood equal to a stack 8 ft long, 4 ft wide, and 4 ft high. How many cords are in 1.0 m^3?

42 One molecule of water (H_2O) contains two atoms of hydrogen and one atom of oxygen. A hydrogen atom has a mass of 1.0 u and an atom of oxygen has a mass of 16 u, approximately. (a) What is the mass in kilograms of one molecule of water? (b) How many molecules of water are in the world's oceans, which have an estimated total mass of 1.4×10^{21} kg?

43 A person on a diet might lose 2.3 kg per week. Express the mass loss rate in milligrams per second, as if the dieter could sense the second-by-second loss.

44 What mass of water fell on the town in Problem 7? Water has a density of 1.0×10^3 kg/m^3.

45 (a) A unit of time sometimes used in microscopic physics is the *shake*. One shake equals 10^{-8} s. Are there more shakes in a second than there are seconds in a year? (b) Humans have existed for about 10^6 years, whereas the universe is about 10^{10} years old. If the age of the universe is defined as 1 "universe day," where a universe day consists of "universe seconds" as a normal day consists of normal seconds, how many universe seconds have humans existed?

46 A unit of area often used in measuring land areas is the *hectare*, defined as 10^4 m^2. An open-pit coal mine consumes 75 hectares of land, down to a depth of 26 m, each year. What volume of earth, in cubic kilometers, is removed in this time?

47 SSM An astronomical unit (AU) is the average distance between Earth and the Sun, approximately 1.50×10^8 km. The speed of light is about 3.0×10^8 m/s. Express the speed of light in astronomical units per minute.

48 The common Eastern mole, a mammal, typically has a mass of 75 g, which corresponds to about 7.5 moles of atoms. (A mole of atoms is 6.02×10^{23} atoms.) In atomic mass units (u), what is the average mass of the atoms in the common Eastern mole?

49 A traditional unit of length in Japan is the ken (1 ken = 1.97 m). What are the ratios of (a) square kens to square meters and (b) cubic kens to cubic meters? What is the volume of a cylindrical water tank of height 5.50 kens and radius 3.00 kens in (c) cubic kens and (d) cubic meters?

50 You receive orders to sail due east for 24.5 mi to put your salvage ship directly over a sunken pirate ship. However, when your divers probe the ocean floor at that location and find no evidence of a ship, you radio back to your source of information, only to discover that the sailing distance was supposed to be 24.5 *nautical miles*, not regular miles. Use the Length table in Appendix D to calculate how far horizontally you are from the pirate ship in kilometers.

51 The cubit is an ancient unit of length based on the distance between the elbow and the tip of the middle finger of the measurer. Assume that the distance ranged from 43 to 53 cm, and suppose that ancient drawings indicate that a cylindrical pillar was to have a length of 9 cubits and a diameter of 2 cubits. For the stated range, what are the lower value and the upper value, respectively, for (a) the cylinder's length in meters, (b) the cylinder's length in millimeters, and (c) the cylinder's volume in cubic meters?

52 As a contrast between the old and the modern and between the large and the small, consider the following: In old rural England 1 hide (between 100 and 120 acres) was the area of land needed to sustain one family with a single plough for one year. (An area of 1 acre is equal to 4047 m^2.) Also, 1 wapentake was the area of land needed by 100 such families. In quantum physics, the cross-sectional area of a nucleus (defined in terms of the chance of a particle hitting and being absorbed by it) is measured in units of barns, where 1 barn is 1×10^{-28} m^2. (In nuclear physics jargon, if a nucleus is "large," then shooting a particle at it is like shooting a bullet at a barn door, which can hardly be missed.) What is the ratio of 25 wapentakes to 11 barns?

53 SSM An *astronomical unit* (AU) is equal to the average distance from Earth to the Sun, about 92.9×10^6 mi. A *parsec* (pc) is the distance at which a length of 1 AU would subtend an angle of exactly 1 second of arc (Fig. 1-8). A *light-year* (ly) is the distance that light, traveling through a vacuum with a speed of 186 000 mi/s, would cover in 1.0 year. Express the Earth–Sun distance in (a) parsecs and (b) light-years.

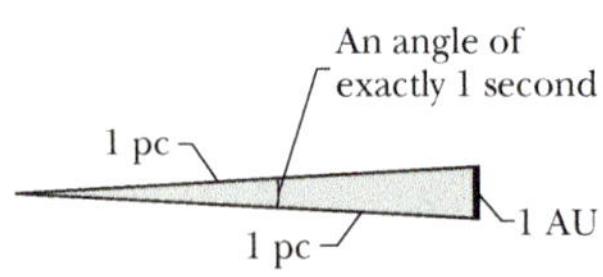

Figure 1-8 Problem 53.

54 The description for a certain brand of house paint claims a coverage of 460 ft^2/gal. (a) Express this quantity in square meters per liter. (b) Express this quantity in an SI unit (see Appendices A and D). (c) What is the inverse of the original quantity, and (d) what is its physical significance?

55 Strangely, the wine for a large wedding reception is to be served in a stunning cut-glass receptacle with the interior dimensions of 40 cm × 40 cm × 30 cm (height). The receptacle is to be initially filled to the top. The wine can be purchased in bottles of the sizes given in the following table. Purchasing a larger bottle instead of multiple smaller bottles decreases the overall cost of the wine. To minimize the cost, (a) which bottle sizes should be purchased and how many of each should be purchased and, once the receptacle is filled, how much wine is left over in terms of (b) standard bottles and (c) liters?

1 standard bottle
1 magnum = 2 standard bottles
1 jeroboam = 4 standard bottles
1 rehoboam = 6 standard bottles
1 methuselah = 8 standard bottles
1 salmanazar = 12 standard bottles
1 balthazar = 16 standard bottles = 11.356 L
1 nebuchadnezzar = 20 standard bottles

56 The *corn–hog ratio* is a financial term used in the pig market and presumably is related to the cost of feeding a pig until it is large enough for market. It is defined as the ratio of the market price of a pig with a mass of 3.108 slugs to the market price of a U.S. bushel of corn. (The word "slug" is derived from an old German word that means "to hit"; we have the same meaning for "slug" as a verb in modern English.) A U.S. bushel is equal to 35.238 L. If the corn–hog ratio is listed as 5.7 on the market exchange, what is it in the metric units of

$$\frac{\text{price of 1 kilogram of pig}}{\text{price of 1 liter of corn}}?$$

(*Hint:* See the Mass table in Appendix D.)

57 You are to fix dinners for 400 people at a convention of Mexican food fans. Your recipe calls for 2 jalapeño peppers per serving (one serving per person). However, you have only habanero peppers on hand. The spiciness of peppers is measured in terms of the *scoville heat unit* (SHU). On average, one jalapeño pepper has a spiciness of 4000 SHU and one habanero pepper has a spiciness of 300 000 SHU. To get the desired spiciness, how many habanero peppers should you substitute for the jalapeño peppers in the recipe for the 400 dinners?

58 A standard interior staircase has steps each with a rise (height) of 19 cm and a run (horizontal depth) of 23 cm. Research suggests that the stairs would be safer for descent if the run were, instead, 28 cm. For a particular staircase of total height 4.57 m, how much farther into the room would the staircase extend if this change in run were made?

59 In purchasing food for a political rally, you erroneously order shucked medium-size Pacific oysters (which come 8 to 12 per U.S. pint) instead of shucked medium-size Atlantic oysters (which come 26 to 38 per U.S. pint). The filled oyster container shipped to you has the interior measure of 1.0 m × 12 cm × 20 cm, and a U.S. pint is equivalent to 0.4732 liter. By how many oysters is the order short of your anticipated count?

60 An old English cookbook carries this recipe for cream of nettle soup: "Boil stock of the following amount: 1 breakfastcup plus 1 teacup plus 6 tablespoons plus 1 dessertspoon. Using gloves, separate nettle tops until you have 0.5 quart; add the tops to the boiling stock. Add 1 tablespoon of cooked rice and 1 saltspoon of salt. Simmer for 15 min." The following table gives some of the conversions among old (premetric) British measures and among common (still premetric) U.S. measures. (These measures just scream for metrication.) For liquid measures, 1 British teaspoon = 1 U.S. teaspoon. For dry measures, 1 British teaspoon = 2 U.S. teaspoons and 1 British quart = 1 U.S. quart. In U.S. measures, how much (a) stock, (b) nettle tops, (c) rice, and (d) salt are required in the recipe?

Old British Measures	U.S. Measures
teaspoon = 2 saltspoons	tablespoon = 3 teaspoons
dessertspoon = 2 teaspoons	half cup = 8 tablespoons
tablespoon = 2 dessertspoons	cup = 2 half cups
teacup = 8 tablespoons	
breakfastcup = 2 teacups	

CHAPTER 2

Motion Along a Straight Line

2-1 POSITION, DISPLACEMENT, AND AVERAGE VELOCITY

Learning Objectives

After reading this module, you should be able to ...

2.01 Identify that if all parts of an object move in the same direction and at the same rate, we can treat the object as if it were a (point-like) particle. (This chapter is about the motion of such objects.)

2.02 Identify that the position of a particle is its location as read on a scaled axis, such as an x axis.

2.03 Apply the relationship between a particle's displacement and its initial and final positions.

2.04 Apply the relationship between a particle's average velocity, its displacement, and the time interval for that displacement.

2.05 Apply the relationship between a particle's average speed, the total distance it moves, and the time interval for the motion.

2.06 Given a graph of a particle's position versus time, determine the average velocity between any two particular times.

Key Ideas

- The position x of a particle on an x axis locates the particle with respect to the origin, or zero point, of the axis.

- The position is either positive or negative, according to which side of the origin the particle is on, or zero if the particle is at the origin. The positive direction on an axis is the direction of increasing positive numbers; the opposite direction is the negative direction on the axis.

- The displacement Δx of a particle is the change in its position:

$$\Delta x = x_2 - x_1.$$

- Displacement is a vector quantity. It is positive if the particle has moved in the positive direction of the x axis and negative if the particle has moved in the negative direction.

- When a particle has moved from position x_1 to position x_2 during a time interval $\Delta t = t_2 - t_1$, its average velocity during that interval is

$$v_{avg} = \frac{\Delta x}{\Delta t} = \frac{x_2 - x_1}{t_2 - t_1}.$$

- The algebraic sign of v_{avg} indicates the direction of motion (v_{avg} is a vector quantity). Average velocity does not depend on the actual distance a particle moves, but instead depends on its original and final positions.

- On a graph of x versus t, the average velocity for a time interval Δt is the slope of the straight line connecting the points on the curve that represent the two ends of the interval.

- The average speed s_{avg} of a particle during a time interval Δt depends on the total distance the particle moves in that time interval:

$$s_{avg} = \frac{\text{total distance}}{\Delta t}.$$

What Is Physics?

One purpose of physics is to study the motion of objects—how fast they move, for example, and how far they move in a given amount of time. NASCAR engineers are fanatical about this aspect of physics as they determine the performance of their cars before and during a race. Geologists use this physics to measure tectonic-plate motion as they attempt to predict earthquakes. Medical researchers need this physics to map the blood flow through a patient when diagnosing a partially closed artery, and motorists use it to determine how they might slow sufficiently when their radar detector sounds a warning. There are countless other examples. In this chapter, we study the basic physics of motion where the object (race car, tectonic plate, blood cell, or any other object) moves along a single axis. Such motion is called *one-dimensional motion*.

Motion

The world, and everything in it, moves. Even seemingly stationary things, such as a roadway, move with Earth's rotation, Earth's orbit around the Sun, the Sun's orbit around the center of the Milky Way galaxy, and that galaxy's migration relative to other galaxies. The classification and comparison of motions (called **kinematics**) is often challenging. What exactly do you measure, and how do you compare?

Before we attempt an answer, we shall examine some general properties of motion that is restricted in three ways.

1. The motion is along a straight line only. The line may be vertical, horizontal, or slanted, but it must be straight.
2. Forces (pushes and pulls) cause motion but will not be discussed until Chapter 5. In this chapter we discuss only the motion itself and changes in the motion. Does the moving object speed up, slow down, stop, or reverse direction? If the motion does change, how is time involved in the change?
3. The moving object is either a **particle** (by which we mean a point-like object such as an electron) or an object that moves like a particle (such that every portion moves in the same direction and at the same rate). A stiff pig slipping down a straight playground slide might be considered to be moving like a particle; however, a tumbling tumbleweed would not.

Position and Displacement

To locate an object means to find its position relative to some reference point, often the **origin** (or zero point) of an axis such as the x axis in Fig. 2-1. The **positive direction** of the axis is in the direction of increasing numbers (coordinates), which is to the right in Fig. 2-1. The opposite is the **negative direction.**

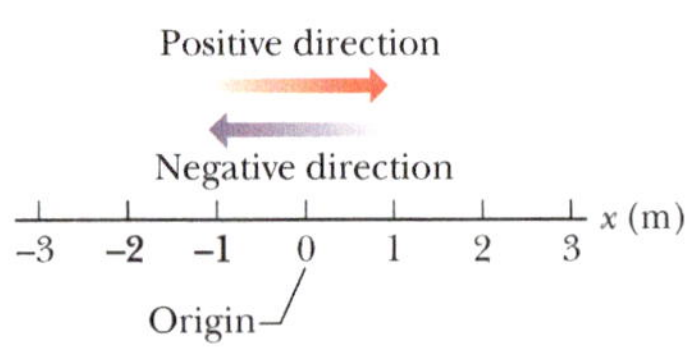

Figure 2-1 Position is determined on an axis that is marked in units of length (here meters) and that extends indefinitely in opposite directions. The axis name, here x, is always on the positive side of the origin.

For example, a particle might be located at $x = 5$ m, which means it is 5 m in the positive direction from the origin. If it were at $x = -5$ m, it would be just as far from the origin but in the opposite direction. On the axis, a coordinate of -5 m is less than a coordinate of -1 m, and both coordinates are less than a coordinate of $+5$ m. A plus sign for a coordinate need not be shown, but a minus sign must always be shown.

A change from position x_1 to position x_2 is called a **displacement** Δx, where

$$\Delta x = x_2 - x_1. \qquad (2\text{-}1)$$

(The symbol Δ, the Greek uppercase delta, represents a change in a quantity, and it means the final value of that quantity minus the initial value.) When numbers are inserted for the position values x_1 and x_2 in Eq. 2-1, a displacement in the positive direction (to the right in Fig. 2-1) always comes out positive, and a displacement in the opposite direction (left in the figure) always comes out negative. For example, if the particle moves from $x_1 = 5$ m to $x_2 = 12$ m, then the displacement is $\Delta x = (12 \text{ m}) - (5 \text{ m}) = +7$ m. The positive result indicates that the motion is in the positive direction. If, instead, the particle moves from $x_1 = 5$ m to $x_2 = 1$ m, then $\Delta x = (1 \text{ m}) - (5 \text{ m}) = -4$ m. The negative result indicates that the motion is in the negative direction.

The actual number of meters covered for a trip is irrelevant; displacement involves only the original and final positions. For example, if the particle moves from $x = 5$ m out to $x = 200$ m and then back to $x = 5$ m, the displacement from start to finish is $\Delta x = (5 \text{ m}) - (5 \text{ m}) = 0$.

Signs. A plus sign for a displacement need not be shown, but a minus sign must always be shown. If we ignore the sign (and thus the direction) of a displacement, we are left with the **magnitude** (or absolute value) of the displacement. For example, a displacement of $\Delta x = -4$ m has a magnitude of 4 m.

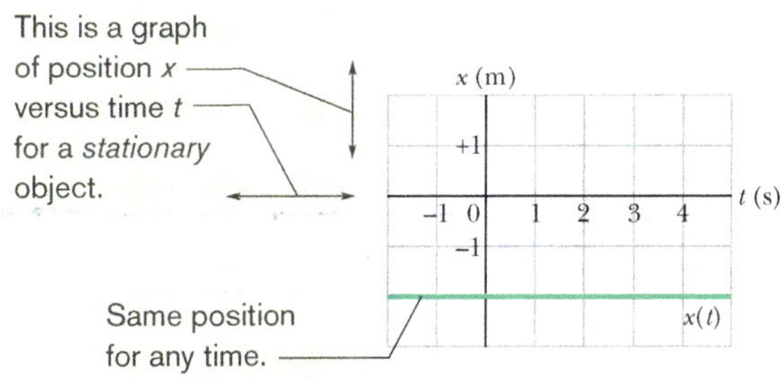

Figure 2-2 The graph of $x(t)$ for an armadillo that is stationary at $x = -2$ m. The value of x is -2 m for all times t.

Displacement is an example of a **vector quantity**, which is a quantity that has both a direction and a magnitude. We explore vectors more fully in Chapter 3, but here all we need is the idea that displacement has two features: (1) Its *magnitude* is the distance (such as the number of meters) between the original and final positions. (2) Its *direction*, from an original position to a final position, can be represented by a plus sign or a minus sign if the motion is along a single axis.

Here is the first of many checkpoints where you can check your understanding with a bit of reasoning. The answers are in the back of the book.

Checkpoint 1

Here are three pairs of initial and final positions, respectively, along an x axis. Which pairs give a negative displacement: (a) -3 m, $+5$ m; (b) -3 m, -7 m; (c) 7 m, -3 m?

Average Velocity and Average Speed

A compact way to describe position is with a graph of position x plotted as a function of time t—a graph of $x(t)$. (The notation $x(t)$ represents a function x of t, not the product x times t.) As a simple example, Fig. 2-2 shows the position function $x(t)$ for a stationary armadillo (which we treat as a particle) over a 7 s time interval. The animal's position stays at $x = -2$ m.

Figure 2-3 is more interesting, because it involves motion. The armadillo is apparently first noticed at $t = 0$ when it is at the position $x = -5$ m. It moves

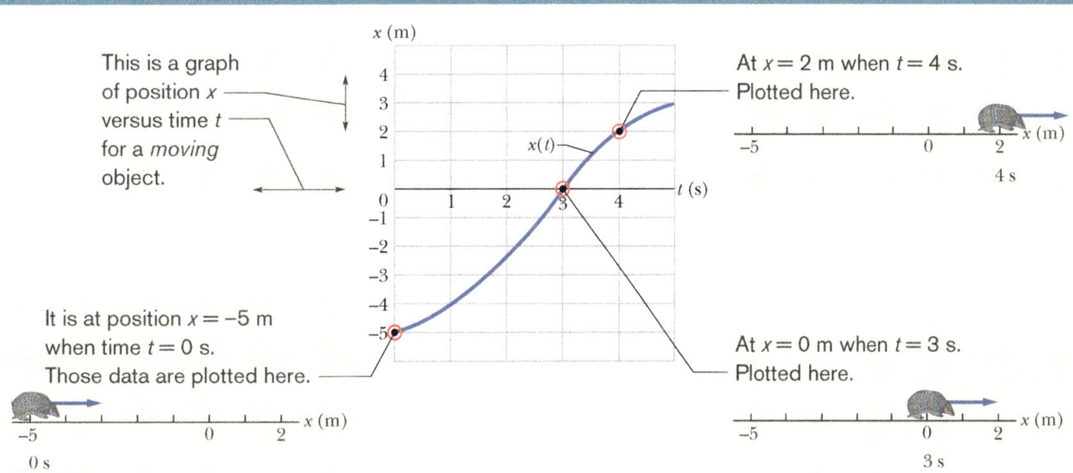

Figure 2-3 The graph of $x(t)$ for a moving armadillo. The path associated with the graph is also shown, at three times.

toward $x = 0$, passes through that point at $t = 3$ s, and then moves on to increasingly larger positive values of x. Figure 2-3 also depicts the straight-line motion of the armadillo (at three times) and is something like what you would see. The graph in Fig. 2-3 is more abstract, but it reveals how fast the armadillo moves.

Actually, several quantities are associated with the phrase "how fast." One of them is the **average velocity** v_{avg}, which is the ratio of the displacement Δx that occurs during a particular time interval Δt to that interval:

$$v_{avg} = \frac{\Delta x}{\Delta t} = \frac{x_2 - x_1}{t_2 - t_1}. \quad (2\text{-}2)$$

The notation means that the position is x_1 at time t_1 and then x_2 at time t_2. A common unit for v_{avg} is the meter per second (m/s). You may see other units in the problems, but they are always in the form of length/time.

Graphs. On a graph of x versus t, v_{avg} is the **slope** of the straight line that connects two particular points on the $x(t)$ curve: one is the point that corresponds to x_2 and t_2, and the other is the point that corresponds to x_1 and t_1. Like displacement, v_{avg} has both magnitude and direction (it is another vector quantity). Its magnitude is the magnitude of the line's slope. A positive v_{avg} (and slope) tells us that the line slants upward to the right; a negative v_{avg} (and slope) tells us that the line slants downward to the right. The average velocity v_{avg} always has the same sign as the displacement Δx because Δt in Eq. 2-2 is always positive.

Figure 2-4 shows how to find v_{avg} in Fig. 2-3 for the time interval $t = 1$ s to $t = 4$ s. We draw the straight line that connects the point on the position curve at the beginning of the interval and the point on the curve at the end of the interval. Then we find the slope $\Delta x/\Delta t$ of the straight line. For the given time interval, the average velocity is

$$v_{avg} = \frac{6\text{ m}}{3\text{ s}} = 2\text{ m/s}.$$

Average speed s_{avg} is a different way of describing "how fast" a particle moves. Whereas the average velocity involves the particle's displacement Δx, the average speed involves the total distance covered (for example, the number of meters moved), independent of direction; that is,

$$s_{avg} = \frac{\text{total distance}}{\Delta t}. \quad (2\text{-}3)$$

Because average speed does *not* include direction, it lacks any algebraic sign. Sometimes s_{avg} is the same (except for the absence of a sign) as v_{avg}. However, the two can be quite different.

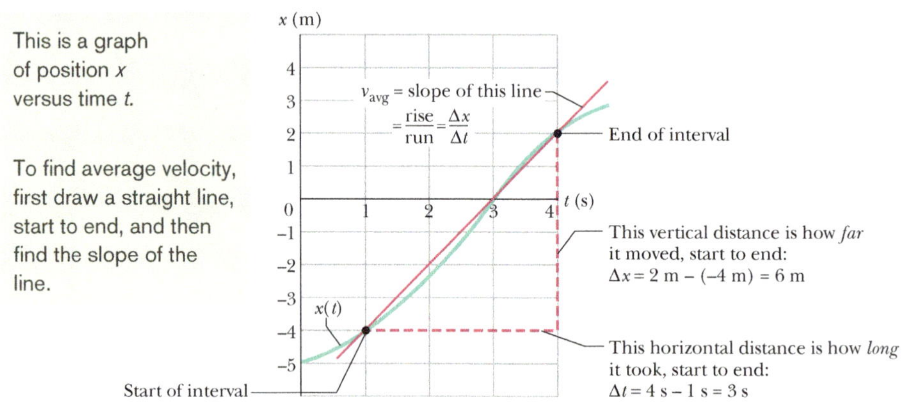

Figure 2-4 Calculation of the average velocity between $t = 1$ s and $t = 4$ s as the slope of the line that connects the points on the $x(t)$ curve representing those times. The swirling icon indicates that a figure is available in *WileyPLUS* as an animation with voiceover.

Sample Problem 2.01 Average velocity, beat-up pickup truck

You drive a beat-up pickup truck along a straight road for 8.4 km at 70 km/h, at which point the truck runs out of gasoline and stops. Over the next 30 min, you walk another 2.0 km farther along the road to a gasoline station.

(a) What is your overall displacement from the beginning of your drive to your arrival at the station?

KEY IDEA

Assume, for convenience, that you move in the positive direction of an x axis, from a first position of $x_1 = 0$ to a second position of x_2 at the station. That second position must be at $x_2 = 8.4\text{ km} + 2.0\text{ km} = 10.4\text{ km}$. Then your displacement Δx along the x axis is the second position minus the first position.

Calculation: From Eq. 2-1, we have

$$\Delta x = x_2 - x_1 = 10.4\text{ km} - 0 = 10.4\text{ km}. \qquad \text{(Answer)}$$

Thus, your overall displacement is 10.4 km in the positive direction of the x axis.

(b) What is the time interval Δt from the beginning of your drive to your arrival at the station?

KEY IDEA

We already know the walking time interval Δt_{wlk} (= 0.50 h), but we lack the driving time interval Δt_{dr}. However, we know that for the drive the displacement Δx_{dr} is 8.4 km and the average velocity $v_{\text{avg,dr}}$ is 70 km/h. Thus, this average velocity is the ratio of the displacement for the drive to the time interval for the drive.

Calculations: We first write

$$v_{\text{avg,dr}} = \frac{\Delta x_{\text{dr}}}{\Delta t_{\text{dr}}}.$$

Rearranging and substituting data then give us

$$\Delta t_{\text{dr}} = \frac{\Delta x_{\text{dr}}}{v_{\text{avg,dr}}} = \frac{8.4\text{ km}}{70\text{ km/h}} = 0.12\text{ h}.$$

So,

$$\begin{aligned}\Delta t &= \Delta t_{\text{dr}} + \Delta t_{\text{wlk}} \\ &= 0.12\text{ h} + 0.50\text{ h} = 0.62\text{ h}. \qquad \text{(Answer)}\end{aligned}$$

(c) What is your average velocity v_{avg} from the beginning of your drive to your arrival at the station? Find it both numerically and graphically.

KEY IDEA

From Eq. 2-2 we know that v_{avg} *for the entire trip* is the ratio of the displacement of 10.4 km *for the entire trip* to the time interval of 0.62 h *for the entire trip.*

Calculation: Here we find

$$\begin{aligned}v_{\text{avg}} &= \frac{\Delta x}{\Delta t} = \frac{10.4\text{ km}}{0.62\text{ h}} \\ &= 16.8\text{ km/h} \approx 17\text{ km/h}. \qquad \text{(Answer)}\end{aligned}$$

To find v_{avg} graphically, first we graph the function $x(t)$ as shown in Fig. 2-5, where the beginning and arrival points on the graph are the origin and the point labeled as "Station." Your average velocity is the slope of the straight line connecting those points; that is, v_{avg} is the ratio of the *rise* ($\Delta x = 10.4$ km) to the *run* ($\Delta t = 0.62$ h), which gives us $v_{\text{avg}} = 16.8$ km/h.

(d) Suppose that to pump the gasoline, pay for it, and walk back to the truck takes you another 45 min. What is your average speed from the beginning of your drive to your return to the truck with the gasoline?

KEY IDEA

Your average speed is the ratio of the total distance you move to the total time interval you take to make that move.

Calculation: The total distance is 8.4 km + 2.0 km + 2.0 km = 12.4 km. The total time interval is 0.12 h + 0.50 h + 0.75 h = 1.37 h. Thus, Eq. 2-3 gives us

$$s_{\text{avg}} = \frac{12.4\text{ km}}{1.37\text{ h}} = 9.1\text{ km/h}. \qquad \text{(Answer)}$$

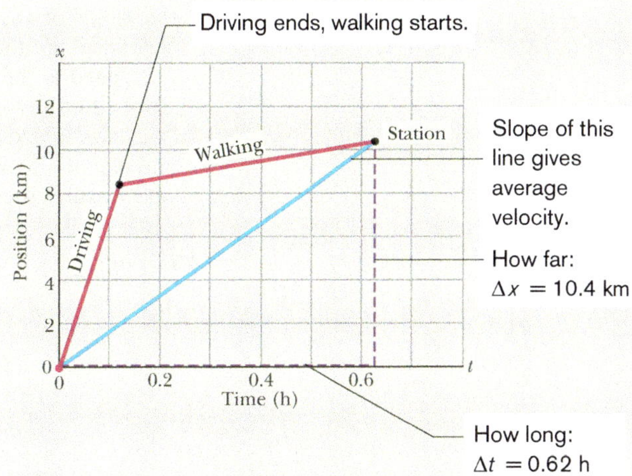

Figure 2-5 The lines marked "Driving" and "Walking" are the position–time plots for the driving and walking stages. (The plot for the walking stage assumes a constant rate of walking.) The slope of the straight line joining the origin and the point labeled "Station" is the average velocity for the trip, from the beginning to the station.

WILEY PLUS Additional examples, video, and practice available at *WileyPLUS*

2-2 INSTANTANEOUS VELOCITY AND SPEED

Learning Objectives

After reading this module, you should be able to . . .

2.07 Given a particle's position as a function of time, calculate the instantaneous velocity for any particular time.

2.08 Given a graph of a particle's position versus time, determine the instantaneous velocity for any particular time.

2.09 Identify speed as the magnitude of the instantaneous velocity.

Key Ideas

- The instantaneous velocity (or simply velocity) v of a moving particle is

$$v = \lim_{\Delta t \to 0} \frac{\Delta x}{\Delta t} = \frac{dx}{dt},$$

where $\Delta x = x_2 - x_1$ and $\Delta t = t_2 - t_1$.

- The instantaneous velocity (at a particular time) may be found as the slope (at that particular time) of the graph of x versus t.
- Speed is the magnitude of instantaneous velocity.

Instantaneous Velocity and Speed

You have now seen two ways to describe how fast something moves: average velocity and average speed, both of which are measured over a time interval Δt. However, the phrase "how fast" more commonly refers to how fast a particle is moving at a given instant—its **instantaneous velocity** (or simply **velocity**) v.

The velocity at any instant is obtained from the average velocity by shrinking the time interval Δt closer and closer to 0. As Δt dwindles, the average velocity approaches a limiting value, which is the velocity at that instant:

$$v = \lim_{\Delta t \to 0} \frac{\Delta x}{\Delta t} = \frac{dx}{dt}. \tag{2-4}$$

Note that v is the rate at which position x is changing with time at a given instant; that is, v is the derivative of x with respect to t. Also note that v at any instant is the slope of the position–time curve at the point representing that instant. Velocity is another vector quantity and thus has an associated direction.

Speed is the magnitude of velocity; that is, speed is velocity that has been stripped of any indication of direction, either in words or via an algebraic sign. (*Caution:* Speed and average speed can be quite different.) A velocity of $+5$ m/s and one of -5 m/s both have an associated speed of 5 m/s. The speedometer in a car measures speed, not velocity (it cannot determine the direction).

Checkpoint 2

The following equations give the position $x(t)$ of a particle in four situations (in each equation, x is in meters, t is in seconds, and $t > 0$): (1) $x = 3t - 2$; (2) $x = -4t^2 - 2$; (3) $x = 2/t^2$; and (4) $x = -2$. (a) In which situation is the velocity v of the particle constant? (b) In which is v in the negative x direction?

Sample Problem 2.02 Velocity and slope of *x* versus *t*, elevator cab

Figure 2-6*a* is an $x(t)$ plot for an elevator cab that is initially stationary, then moves upward (which we take to be the positive direction of x), and then stops. Plot $v(t)$.

KEY IDEA

We can find the velocity at any time from the slope of the $x(t)$ curve at that time.

Calculations: The slope of $x(t)$, and so also the velocity, is zero in the intervals from 0 to 1 s and from 9 s on, so then the cab is stationary. During the interval bc, the slope is constant and nonzero, so then the cab moves with constant velocity. We calculate the slope of $x(t)$ then as

$$\frac{\Delta x}{\Delta t} = v = \frac{24\text{ m} - 4.0\text{ m}}{8.0\text{ s} - 3.0\text{ s}} = +4.0\text{ m/s}. \tag{2-5}$$

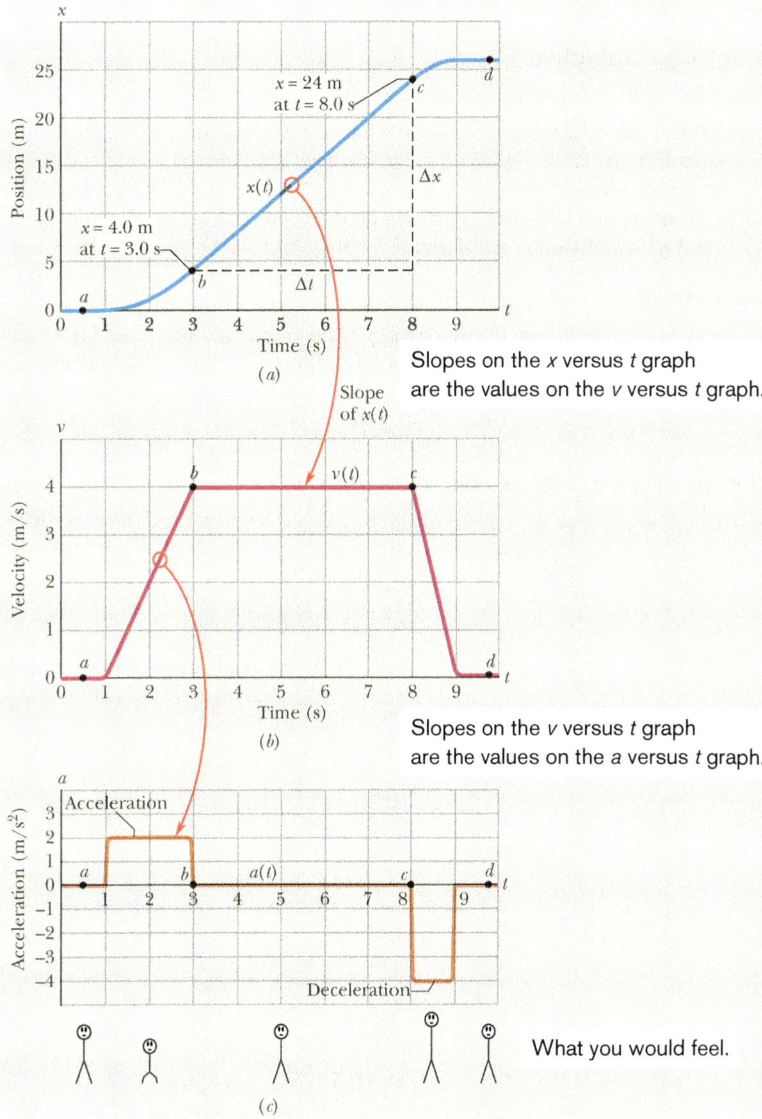

Figure 2-6 (*a*) The $x(t)$ curve for an elevator cab that moves upward along an x axis. (*b*) The $v(t)$ curve for the cab. Note that it is the derivative of the $x(t)$ curve ($v = dx/dt$). (*c*) The $a(t)$ curve for the cab. It is the derivative of the $v(t)$ curve ($a = dv/dt$). The stick figures along the bottom suggest how a passenger's body might feel during the accelerations.

The plus sign indicates that the cab is moving in the positive x direction. These intervals (where $v = 0$ and $v = 4$ m/s) are plotted in Fig. 2-6*b*. In addition, as the cab initially begins to move and then later slows to a stop, v varies as indicated in the intervals 1 s to 3 s and 8 s to 9 s. Thus, Fig. 2-6*b* is the required plot. (Figure 2-6*c* is considered in Module 2-3.)

Given a $v(t)$ graph such as Fig. 2-6*b*, we could "work backward" to produce the shape of the associated $x(t)$ graph (Fig. 2-6*a*). However, we would not know the actual values for x at various times, because the $v(t)$ graph indicates only *changes* in x. To find such a change in x during any interval, we must, in the language of calculus, calculate the area "under the curve" on the $v(t)$ graph for that interval. For example, during the interval 3 s to 8 s in which the cab has a velocity of 4.0 m/s, the change in x is

$$\Delta x = (4.0\text{ m/s})(8.0\text{ s} - 3.0\text{ s}) = +20\text{ m}. \qquad (2\text{-}6)$$

(This area is positive because the $v(t)$ curve is above the t axis.) Figure 2-6*a* shows that x does indeed increase by 20 m in that interval. However, Fig. 2-6*b* does not tell us the *values* of x at the beginning and end of the interval. For that, we need additional information, such as the value of x at some instant.

WILEY PLUS Additional examples, video, and practice available at *WileyPLUS*

2-3 ACCELERATION

Learning Objectives

After reading this module, you should be able to . . .

2.10 Apply the relationship between a particle's average acceleration, its change in velocity, and the time interval for that change.

2.11 Given a particle's velocity as a function of time, calculate the instantaneous acceleration for any particular time.

2.12 Given a graph of a particle's velocity versus time, determine the instantaneous acceleration for any particular time and the average acceleration between any two particular times.

Key Ideas

- Average acceleration is the ratio of a change in velocity Δv to the time interval Δt in which the change occurs:

$$a_{avg} = \frac{\Delta v}{\Delta t}.$$

The algebraic sign indicates the direction of a_{avg}.

- Instantaneous acceleration (or simply acceleration) a is the first time derivative of velocity $v(t)$ and the second time derivative of position $x(t)$:

$$a = \frac{dv}{dt} = \frac{d^2x}{dt^2}.$$

- On a graph of v versus t, the acceleration a at any time t is the slope of the curve at the point that represents t.

Acceleration

When a particle's velocity changes, the particle is said to undergo **acceleration** (or to accelerate). For motion along an axis, the **average acceleration** a_{avg} over a time interval Δt is

$$a_{avg} = \frac{v_2 - v_1}{t_2 - t_1} = \frac{\Delta v}{\Delta t}, \tag{2-7}$$

where the particle has velocity v_1 at time t_1 and then velocity v_2 at time t_2. The **instantaneous acceleration** (or simply **acceleration**) is

$$a = \frac{dv}{dt}. \tag{2-8}$$

In words, the acceleration of a particle at any instant is the rate at which its velocity is changing at that instant. Graphically, the acceleration at any point is the slope of the curve of $v(t)$ at that point. We can combine Eq. 2-8 with Eq. 2-4 to write

$$a = \frac{dv}{dt} = \frac{d}{dt}\left(\frac{dx}{dt}\right) = \frac{d^2x}{dt^2}. \tag{2-9}$$

In words, the acceleration of a particle at any instant is the second derivative of its position $x(t)$ with respect to time.

A common unit of acceleration is the meter per second per second: m/(s · s) or m/s^2. Other units are in the form of length/(time · time) or length/time^2. Acceleration has both magnitude and direction (it is yet another vector quantity). Its algebraic sign represents its direction on an axis just as for displacement and velocity; that is, acceleration with a positive value is in the positive direction of an axis, and acceleration with a negative value is in the negative direction.

Figure 2-6 gives plots of the position, velocity, and acceleration of an elevator moving up a shaft. Compare the $a(t)$ curve with the $v(t)$ curve—each point on the $a(t)$ curve shows the derivative (slope) of the $v(t)$ curve at the corresponding time. When v is constant (at either 0 or 4 m/s), the derivative is zero and so also is the acceleration. When the cab first begins to move, the $v(t)$

curve has a positive derivative (the slope is positive), which means that $a(t)$ is positive. When the cab slows to a stop, the derivative and slope of the $v(t)$ curve are negative; that is, $a(t)$ is negative.

Next compare the slopes of the $v(t)$ curve during the two acceleration periods. The slope associated with the cab's slowing down (commonly called "deceleration") is steeper because the cab stops in half the time it took to get up to speed. The steeper slope means that the magnitude of the deceleration is larger than that of the acceleration, as indicated in Fig. 2-6*c*.

Sensations. The sensations you would feel while riding in the cab of Fig. 2-6 are indicated by the sketched figures at the bottom. When the cab first accelerates, you feel as though you are pressed downward; when later the cab is braked to a stop, you seem to be stretched upward. In between, you feel nothing special. In other words, your body reacts to accelerations (it is an accelerometer) but not to velocities (it is not a speedometer). When you are in a car traveling at 90 km/h or an airplane traveling at 900 km/h, you have no bodily awareness of the motion. However, if the car or plane quickly changes velocity, you may become keenly aware of the change, perhaps even frightened by it. Part of the thrill of an amusement park ride is due to the quick changes of velocity that you undergo (you pay for the accelerations, not for the speed). A more extreme example is shown in the photographs of Fig. 2-7, which were taken while a rocket sled was rapidly accelerated along a track and then rapidly braked to a stop.

g Units. Large accelerations are sometimes expressed in terms of g units, with

$$1g = 9.8 \text{ m/s}^2 \quad (g \text{ unit}). \tag{2-10}$$

(As we shall discuss in Module 2-5, g is the magnitude of the acceleration of a falling object near Earth's surface.) On a roller coaster, you may experience brief accelerations up to $3g$, which is $(3)(9.8 \text{ m/s}^2)$, or about 29 m/s^2, more than enough to justify the cost of the ride.

Signs. In common language, the sign of an acceleration has a nonscientific meaning: positive acceleration means that the speed of an object is increasing, and negative acceleration means that the speed is decreasing (the object is decelerating). In this book, however, the sign of an acceleration indicates a direction, not

Figure 2-7 Colonel J. P. Stapp in a rocket sled as it is brought up to high speed (acceleration out of the page) and then very rapidly braked (acceleration into the page).

Courtesy U.S. Air Force

whether an object's speed is increasing or decreasing. For example, if a car with an initial velocity $v = -25$ m/s is braked to a stop in 5.0 s, then $a_{avg} = +5.0$ m/s^2. The acceleration is *positive*, but the car's speed has decreased. The reason is the difference in signs: the direction of the acceleration is opposite that of the velocity.

Here then is the proper way to interpret the signs:

> If the signs of the velocity and acceleration of a particle are the same, the speed of the particle increases. If the signs are opposite, the speed decreases.

Checkpoint 3

A wombat moves along an x axis. What is the sign of its acceleration if it is moving (a) in the positive direction with increasing speed, (b) in the positive direction with decreasing speed, (c) in the negative direction with increasing speed, and (d) in the negative direction with decreasing speed?

Sample Problem 2.03 Acceleration and dv/dt

A particle's position on the x axis of Fig. 2-1 is given by

$$x = 4 - 27t + t^3,$$

with x in meters and t in seconds.

(a) Because position x depends on time t, the particle must be moving. Find the particle's velocity function $v(t)$ and acceleration function $a(t)$.

KEY IDEAS

(1) To get the velocity function $v(t)$, we differentiate the position function $x(t)$ with respect to time. (2) To get the acceleration function $a(t)$, we differentiate the velocity function $v(t)$ with respect to time.

Calculations: Differentiating the position function, we find

$$v = -27 + 3t^2, \qquad \text{(Answer)}$$

with v in meters per second. Differentiating the velocity function then gives us

$$a = +6t, \qquad \text{(Answer)}$$

with a in meters per second squared.

(b) Is there ever a time when $v = 0$?

Calculation: Setting $v(t) = 0$ yields

$$0 = -27 + 3t^2,$$

which has the solution

$$t = \pm 3 \text{ s}. \qquad \text{(Answer)}$$

Thus, the velocity is zero both 3 s before and 3 s after the clock reads 0.

(c) Describe the particle's motion for $t \geq 0$.

Reasoning: We need to examine the expressions for $x(t)$, $v(t)$, and $a(t)$.

At $t = 0$, the particle is at $x(0) = +4$ m and is moving with a velocity of $v(0) = -27$ m/s—that is, in the negative direction of the x axis. Its acceleration is $a(0) = 0$ because just then the particle's velocity is not changing (Fig. 2-8*a*).

For $0 < t < 3$ s, the particle still has a negative velocity, so it continues to move in the negative direction. However, its acceleration is no longer 0 but is increasing and positive. Because the signs of the velocity and the acceleration are opposite, the particle must be slowing (Fig. 2-8*b*).

Indeed, we already know that it stops momentarily at $t = 3$ s. Just then the particle is as far to the left of the origin in Fig. 2-1 as it will ever get. Substituting $t = 3$ s into the expression for $x(t)$, we find that the particle's position just then is $x = -50$ m (Fig. 2-8*c*). Its acceleration is still positive.

For $t > 3$ s, the particle moves to the right on the axis. Its acceleration remains positive and grows progressively larger in magnitude. The velocity is now positive, and it too grows progressively larger in magnitude (Fig. 2-8*d*).

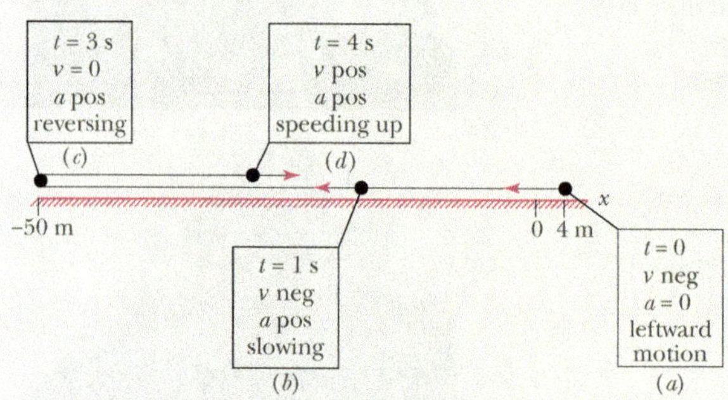

Figure 2-8 Four stages of the particle's motion.

WileyPLUS Additional examples, video, and practice available at *WileyPLUS*

2-4 CONSTANT ACCELERATION

Learning Objectives

After reading this module, you should be able to . . .

2.13 For constant acceleration, apply the relationships between position, displacement, velocity, acceleration, and elapsed time (Table 2-1).

2.14 Calculate a particle's change in velocity by integrating its acceleration function with respect to time.

2.15 Calculate a particle's change in position by integrating its velocity function with respect to time.

Key Ideas

- The following five equations describe the motion of a particle with constant acceleration:

$$v = v_0 + at, \qquad x - x_0 = v_0 t + \frac{1}{2}at^2,$$

$$v^2 = v_0^2 + 2a(x - x_0), \qquad x - x_0 = \frac{1}{2}(v_0 + v)t, \qquad x - x_0 = vt - \frac{1}{2}at^2.$$

These are *not* valid when the acceleration is not constant.

Constant Acceleration: A Special Case

In many types of motion, the acceleration is either constant or approximately so. For example, you might accelerate a car at an approximately constant rate when a traffic light turns from red to green. Then graphs of your position, velocity, and acceleration would resemble those in Fig. 2-9. (Note that $a(t)$ in Fig. 2-9*c* is constant, which requires that $v(t)$ in Fig. 2-9*b* have a constant slope.) Later when you brake the car to a stop, the acceleration (or deceleration in common language) might also be approximately constant.

Such cases are so common that a special set of equations has been derived for dealing with them. One approach to the derivation of these equations is given in this section. A second approach is given in the next section. Throughout both sections and later when you work on the homework problems, keep in mind that *these equations are valid only for constant acceleration (or situations in which you can approximate the acceleration as being constant).*

First Basic Equation. When the acceleration is constant, the average acceleration and instantaneous acceleration are equal and we can write Eq. 2-7, with some changes in notation, as

$$a = a_{avg} = \frac{v - v_0}{t - 0}.$$

Here v_0 is the velocity at time $t = 0$ and v is the velocity at any later time t. We can recast this equation as

$$v = v_0 + at. \tag{2-11}$$

As a check, note that this equation reduces to $v = v_0$ for $t = 0$, as it must. As a further check, take the derivative of Eq. 2-11. Doing so yields $dv/dt = a$, which is the definition of a. Figure 2-9*b* shows a plot of Eq. 2-11, the $v(t)$ function; the function is linear and thus the plot is a straight line.

Second Basic Equation. In a similar manner, we can rewrite Eq. 2-2 (with a few changes in notation) as

$$v_{avg} = \frac{x - x_0}{t - 0}$$

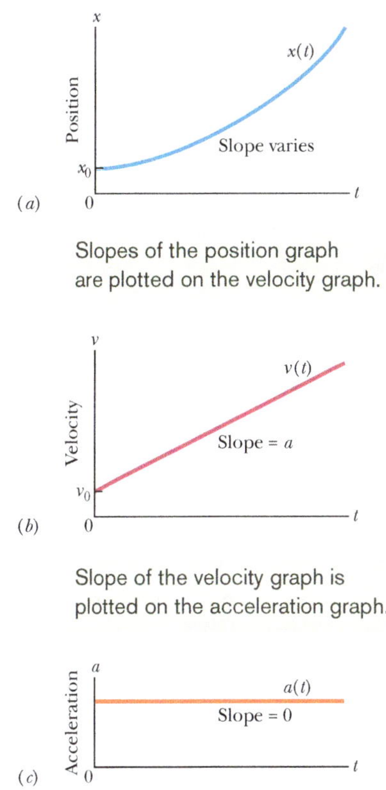

Figure 2-9 (*a*) The position $x(t)$ of a particle moving with constant acceleration. (*b*) Its velocity $v(t)$, given at each point by the slope of the curve of $x(t)$. (*c*) Its (constant) acceleration, equal to the (constant) slope of the curve of $v(t)$.

and then as

$$x = x_0 + v_{avg}t, \tag{2-12}$$

in which x_0 is the position of the particle at $t = 0$ and v_{avg} is the average velocity between $t = 0$ and a later time t.

For the linear velocity function in Eq. 2-11, the *average* velocity over any time interval (say, from $t = 0$ to a later time t) is the average of the velocity at the beginning of the interval ($= v_0$) and the velocity at the end of the interval ($= v$). For the interval from $t = 0$ to the later time t then, the average velocity is

$$v_{avg} = \tfrac{1}{2}(v_0 + v). \tag{2-13}$$

Substituting the right side of Eq. 2-11 for v yields, after a little rearrangement,

$$v_{avg} = v_0 + \tfrac{1}{2}at. \tag{2-14}$$

Finally, substituting Eq. 2-14 into Eq. 2-12 yields

$$x - x_0 = v_0t + \tfrac{1}{2}at^2. \tag{2-15}$$

As a check, note that putting $t = 0$ yields $x = x_0$, as it must. As a further check, taking the derivative of Eq. 2-15 yields Eq. 2-11, again as it must. Figure 2-9*a* shows a plot of Eq. 2-15; the function is quadratic and thus the plot is curved.

Three Other Equations. Equations 2-11 and 2-15 are the *basic equations for constant acceleration;* they can be used to solve any constant acceleration problem in this book. However, we can derive other equations that might prove useful in certain specific situations. First, note that as many as five quantities can possibly be involved in any problem about constant acceleration—namely, $x - x_0$, v, t, a, and v_0. Usually, one of these quantities is *not* involved in the problem, *either as a given or as an unknown.* We are then presented with three of the remaining quantities and asked to find the fourth.

Equations 2-11 and 2-15 each contain four of these quantities, but not the same four. In Eq. 2-11, the "missing ingredient" is the displacement $x - x_0$. In Eq. 2-15, it is the velocity v. These two equations can also be combined in three ways to yield three additional equations, each of which involves a different "missing variable." First, we can eliminate t to obtain

$$v^2 = v_0^2 + 2a(x - x_0). \tag{2-16}$$

This equation is useful if we do not know t and are not required to find it. Second, we can eliminate the acceleration a between Eqs. 2-11 and 2-15 to produce an equation in which a does not appear:

$$x - x_0 = \tfrac{1}{2}(v_0 + v)t. \tag{2-17}$$

Finally, we can eliminate v_0, obtaining

$$x - x_0 = vt - \tfrac{1}{2}at^2. \tag{2-18}$$

Note the subtle difference between this equation and Eq. 2-15. One involves the initial velocity v_0; the other involves the velocity v at time t.

Table 2-1 lists the basic constant acceleration equations (Eqs. 2-11 and 2-15) as well as the specialized equations that we have derived. To solve a simple constant acceleration problem, you can usually use an equation from this list (*if* you have the list with you). Choose an equation for which the only unknown variable is the variable requested in the problem. A simpler plan is to remember only Eqs. 2-11 and 2-15, and then solve them as simultaneous equations whenever needed.

Table 2-1 Equations for Motion with Constant Acceleration[a]

Equation Number	Equation	Missing Quantity
2-11	$v = v_0 + at$	$x - x_0$
2-15	$x - x_0 = v_0t + \frac{1}{2}at^2$	v
2-16	$v^2 = v_0^2 + 2a(x - x_0)$	t
2-17	$x - x_0 = \frac{1}{2}(v_0 + v)t$	a
2-18	$x - x_0 = vt - \frac{1}{2}at^2$	v_0

[a]Make sure that the acceleration is indeed constant before using the equations in this table.

Checkpoint 4

The following equations give the position $x(t)$ of a particle in four situations: (1) $x = 3t - 4$; (2) $x = -5t^3 + 4t^2 + 6$; (3) $x = 2/t^2 - 4/t$; (4) $x = 5t^2 - 3$. To which of these situations do the equations of Table 2-1 apply?

Sample Problem 2.04 Drag race of car and motorcycle

A popular web video shows a jet airplane, a car, and a motorcycle racing from rest along a runway (Fig. 2-10). Initially the motorcycle takes the lead, but then the jet takes the lead, and finally the car blows past the motorcycle. Here let's focus on the car and motorcycle and assign some reasonable values to the motion. The motorcycle first takes the lead because its (constant) acceleration $a_m = 8.40\ \text{m/s}^2$ is greater than the car's (constant) acceleration $a_c = 5.60\ \text{m/s}^2$, but it soon loses to the car because it reaches its greatest speed $v_m = 58.8$ m/s before the car reaches its greatest speed $v_c = 106$ m/s. How long does the car take to reach the motorcycle?

KEY IDEAS

We can apply the equations of constant acceleration to both vehicles, but for the motorcycle we must consider the motion in two stages: (1) First it travels through distance x_{m1} with zero initial velocity and acceleration $a_m = 8.40\ \text{m/s}^2$, reaching speed $v_m = 58.8$ m/s. (2) Then it travels through distance x_{m2} with constant velocity $v_m = 58.8$ m/s and zero acceleration (that, too, is a constant acceleration). (Note that we symbolized the distances even though we do not know their values. Symbolizing unknown quantities is often helpful in solving physics problems, but introducing such unknowns sometimes takes *physics courage*.)

Calculations: So that we can draw figures and do calculations, let's assume that the vehicles race along the positive direction of an x axis, starting from $x = 0$ at time $t = 0$. (We can choose any initial numbers because we are looking for the elapsed time, not a particular time in, say, the afternoon, but let's stick with these easy numbers.) We want the car to pass the motorcycle, but what does that mean mathematically?

It means that at some time t, the side-by-side vehicles are at the same coordinate: x_c for the car and the sum $x_{m1} + x_{m2}$ for the motorcycle. We can write this statement mathematically as

$$x_c = x_{m1} + x_{m2}. \tag{2-19}$$

(Writing this first step is the hardest part of the problem. That is true of most physics problems. How do you go from the problem statement (in words) to a mathematical expression? One purpose of this book is for you to build up that ability of writing the first step — it takes lots of practice just as in learning, say, tae-kwon-do.)

Now let's fill out both sides of Eq. 2-19, left side first. To reach the passing point at x_c, the car accelerates from rest. From Eq. 2-15 ($x - x_0 = v_0 t + \frac{1}{2}at^2$), with x_0 and $v_0 = 0$, we have

$$x_c = \tfrac{1}{2}a_c t^2. \tag{2-20}$$

To write an expression for x_{m1} for the motorcycle, we first find the time t_m it takes to reach its maximum speed v_m, using Eq. 2-11 ($v = v_0 + at$). Substituting $v_0 = 0$, $v = v_m = 58.8$ m/s, and $a = a_m = 8.40\ \text{m/s}^2$, that time is

$$t_m = \frac{v_m}{a_m} \tag{2-21}$$

$$= \frac{58.8\ \text{m/s}}{8.40\ \text{m/s}^2} = 7.00\ \text{s}.$$

To get the distance x_{m1} traveled by the motorcycle during the first stage, we again use Eq. 2-15 with $x_0 = 0$ and $v_0 = 0$, but we also substitute from Eq. 2-21 for the time. We find

$$x_{m1} = \tfrac{1}{2}a_m t_m^2 = \tfrac{1}{2}a_m\left(\frac{v_m}{a_m}\right)^2 = \frac{1}{2}\frac{v_m^2}{a_m}. \tag{2-22}$$

For the remaining time of $t - t_m$, the motorcycle travels at its maximum speed with zero acceleration. To get the distance, we use Eq. 2-15 for this second stage of the motion, but now the initial velocity is $v_0 = v_m$ (the speed at the end of the first stage) and the acceleration is $a = 0$. So, the distance traveled during the second stage is

$$x_{m2} = v_m(t - t_m) = v_m(t - 7.00\ \text{s}). \tag{2-23}$$

Figure 2-10 A jet airplane, a car, and a motorcycle just after accelerating from rest.

To finish the calculation, we substitute Eqs. 2-20, 2-22, and 2-23 into Eq. 2-19, obtaining

$$\tfrac{1}{2}a_c t^2 = \frac{1}{2}\frac{v_m^2}{a_m} + v_m(t - 7.00\text{ s}). \tag{2-24}$$

This is a quadratic equation. Substituting in the given data, we solve the equation (by using the usual quadratic-equation formula or a polynomial solver on a calculator), finding $t = 4.44$ s and $t = 16.6$ s.

But what do we do with two answers? Does the car pass the motorcycle twice? No, of course not, as we can see in the video. So, one of the answers is mathematically correct but not physically meaningful. Because we know that the car passes the motorcycle *after* the motorcycle reaches its maximum speed at $t = 7.00$ s, we discard the solution with $t < 7.00$ s as being the unphysical answer and conclude that the passing occurs at

$$t = 16.6\text{ s}. \quad \text{(Answer)}$$

Figure 2-11 is a graph of the position versus time for the two vehicles, with the passing point marked. Notice that at $t = 7.00$ s the plot for the motorcycle switches from being curved (because the speed had been increasing) to being straight (because the speed is thereafter constant).

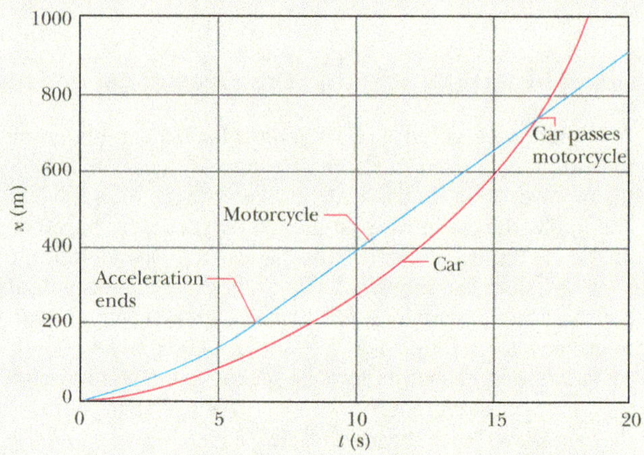

Figure 2-11 Graph of position versus time for car and motorcycle.

WILEY PLUS Additional examples, video, and practice available at *WileyPLUS*

Another Look at Constant Acceleration*

The first two equations in Table 2-1 are the basic equations from which the others are derived. Those two can be obtained by integration of the acceleration with the condition that a is constant. To find Eq. 2-11, we rewrite the definition of acceleration (Eq. 2-8) as

$$dv = a\,dt.$$

We next write the *indefinite integral* (or *antiderivative*) of both sides:

$$\int dv = \int a\,dt.$$

Since acceleration a is a constant, it can be taken outside the integration. We obtain

$$\int dv = a\int dt$$

or

$$v = at + C. \tag{2-25}$$

To evaluate the constant of integration C, we let $t = 0$, at which time $v = v_0$. Substituting these values into Eq. 2-25 (which must hold for all values of t, including $t = 0$) yields

$$v_0 = (a)(0) + C = C.$$

Substituting this into Eq. 2-25 gives us Eq. 2-11.

To derive Eq. 2-15, we rewrite the definition of velocity (Eq. 2-4) as

$$dx = v\,dt$$

and then take the indefinite integral of both sides to obtain

$$\int dx = \int v\,dt.$$

*This section is intended for students who have had integral calculus.

Next, we substitute for v with Eq. 2-11:

$$\int dx = \int (v_0 + at)\, dt.$$

Since v_0 is a constant, as is the acceleration a, this can be rewritten as

$$\int dx = v_0 \int dt + a \int t\, dt.$$

Integration now yields

$$x = v_0 t + \tfrac{1}{2} a t^2 + C', \tag{2-26}$$

where C' is another constant of integration. At time $t = 0$, we have $x = x_0$. Substituting these values in Eq. 2-26 yields $x_0 = C'$. Replacing C' with x_0 in Eq. 2-26 gives us Eq. 2-15.

2-5 FREE-FALL ACCELERATION

Learning Objectives

After reading this module, you should be able to . . .

2.16 Identify that if a particle is in free flight (whether upward or downward) and if we can neglect the effects of air on its motion, the particle has a constant downward acceleration with a magnitude g that we take to be 9.8 m/s^2.

2.17 Apply the constant-acceleration equations (Table 2-1) to free-fall motion.

Key Ideas

- An important example of straight-line motion with constant acceleration is that of an object rising or falling freely near Earth's surface. The constant acceleration equations describe this motion, but we make two changes in notation: (1) we refer the motion to the vertical y axis with $+y$ vertically up; (2) we replace a with $-g$, where g is the magnitude of the free-fall acceleration. Near Earth's surface,

$$g = 9.8 \text{ m/s}^2 = 32 \text{ ft/s}^2.$$

Free-Fall Acceleration

If you tossed an object either up or down and could somehow eliminate the effects of air on its flight, you would find that the object accelerates downward at a certain constant rate. That rate is called the **free-fall acceleration,** and its magnitude is represented by g. The acceleration is independent of the object's characteristics, such as mass, density, or shape; it is the same for all objects.

Two examples of free-fall acceleration are shown in Fig. 2-12, which is a series of stroboscopic photos of a feather and an apple. As these objects fall, they accelerate downward—both at the same rate g. Thus, their speeds increase at the same rate, and they fall together.

The value of g varies slightly with latitude and with elevation. At sea level in Earth's midlatitudes the value is 9.8 m/s^2 (or 32 ft/s^2), which is what you should use as an exact number for the problems in this book unless otherwise noted.

The equations of motion in Table 2-1 for constant acceleration also apply to free fall near Earth's surface; that is, they apply to an object in vertical flight, either up or down, when the effects of the air can be neglected. However, note that for free fall: (1) The directions of motion are now along a vertical y axis instead of the x axis, with the positive direction of y upward. (This is important for later chapters when combined horizontal and vertical motions are examined.) (2) The free-fall acceleration is negative—that is, downward on the y axis, toward Earth's center—and so it has the value $-g$ in the equations.

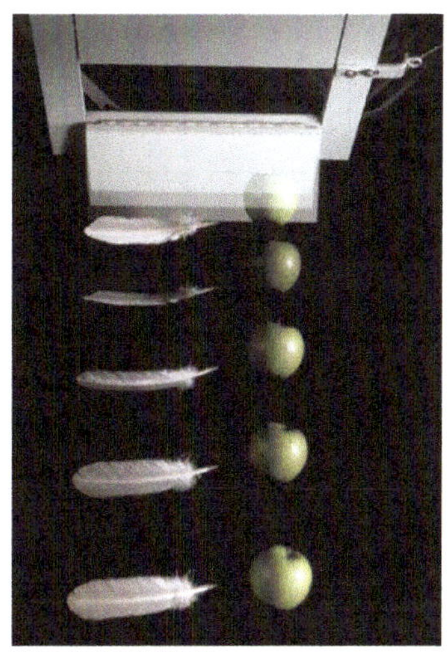

Figure 2-12 A feather and an apple free fall in vacuum at the same magnitude of acceleration g. The acceleration increases the distance between successive images. In the absence of air, the feather and apple fall together.

The free-fall acceleration near Earth's surface is $a = -g = -9.8\ \text{m/s}^2$, and the *magnitude* of the acceleration is $g = 9.8\ \text{m/s}^2$. Do not substitute $-9.8\ \text{m/s}^2$ for g.

Suppose you toss a tomato directly upward with an initial (positive) velocity v_0 and then catch it when it returns to the release level. During its *free-fall flight* (from just after its release to just before it is caught), the equations of Table 2-1 apply to its motion. The acceleration is always $a = -g = -9.8\ \text{m/s}^2$, negative and thus downward. The velocity, however, changes, as indicated by Eqs. 2-11 and 2-16: during the ascent, the magnitude of the positive velocity decreases, until it momentarily becomes zero. Because the tomato has then stopped, it is at its maximum height. During the descent, the magnitude of the (now negative) velocity increases.

Checkpoint 5

(a) If you toss a ball straight up, what is the sign of the ball's displacement for the ascent, from the release point to the highest point? (b) What is it for the descent, from the highest point back to the release point? (c) What is the ball's acceleration at its highest point?

Sample Problem 2.05 Time for full up-down flight, baseball toss

In Fig. 2-13, a pitcher tosses a baseball up along a y axis, with an initial speed of 12 m/s.

(a) How long does the ball take to reach its maximum height?

KEY IDEAS

(1) Once the ball leaves the pitcher and before it returns to his hand, its acceleration is the free-fall acceleration $a = -g$. Because this is constant, Table 2-1 applies to the motion. (2) The velocity v at the maximum height must be 0.

Calculation: Knowing v, a, and the initial velocity $v_0 = 12$ m/s, and seeking t, we solve Eq. 2-11, which contains those four variables. This yields

$$t = \frac{v - v_0}{a} = \frac{0 - 12\ \text{m/s}}{-9.8\ \text{m/s}^2} = 1.2\ \text{s}. \qquad \text{(Answer)}$$

(b) What is the ball's maximum height above its release point?

Calculation: We can take the ball's release point to be $y_0 = 0$. We can then write Eq. 2-16 in y notation, set $y - y_0 = y$ and $v = 0$ (at the maximum height), and solve for y. We get

$$y = \frac{v^2 - v_0^2}{2a} = \frac{0 - (12\ \text{m/s})^2}{2(-9.8\ \text{m/s}^2)} = 7.3\ \text{m}. \qquad \text{(Answer)}$$

(c) How long does the ball take to reach a point 5.0 m above its release point?

Calculations: We know v_0, $a = -g$, and displacement $y - y_0 = 5.0$ m, and we want t, so we choose Eq. 2-15. Rewriting it for y and setting $y_0 = 0$ give us

$$y = v_0 t - \tfrac{1}{2}gt^2,$$

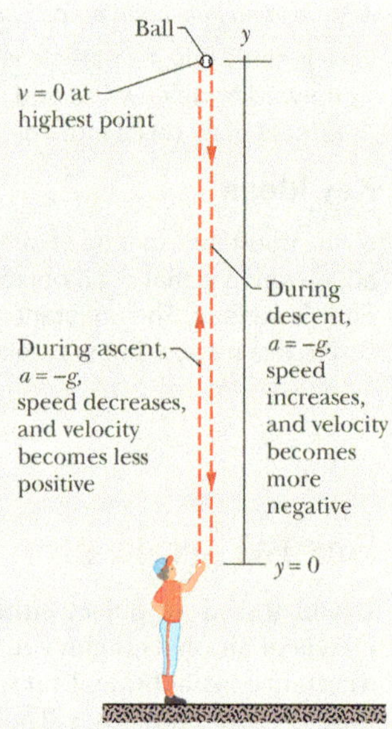

Figure 2-13 A pitcher tosses a baseball straight up into the air. The equations of free fall apply for rising as well as for falling objects, provided any effects from the air can be neglected.

or

$$5.0\ \text{m} = (12\ \text{m/s})t - (\tfrac{1}{2})(9.8\ \text{m/s}^2)t^2.$$

If we temporarily omit the units (having noted that they are consistent), we can rewrite this as

$$4.9t^2 - 12t + 5.0 = 0.$$

Solving this quadratic equation for t yields

$$t = 0.53\ \text{s} \quad \text{and} \quad t = 1.9\ \text{s}. \qquad \text{(Answer)}$$

There are two such times! This is not really surprising because the ball passes twice through $y = 5.0$ m, once on the way up and once on the way down.

Additional examples, video, and practice available at *WileyPLUS*

2-6 GRAPHICAL INTEGRATION IN MOTION ANALYSIS

Learning Objectives

After reading this module, you should be able to . . .

2.18 Determine a particle's change in velocity by graphical integration on a graph of acceleration versus time.

2.19 Determine a particle's change in position by graphical integration on a graph of velocity versus time.

Key Ideas

- On a graph of acceleration a versus time t, the change in the velocity is given by

$$v_1 - v_0 = \int_{t_0}^{t_1} a\,dt.$$

The integral amounts to finding an area on the graph:

$$\int_{t_0}^{t_1} a\,dt = \begin{pmatrix}\text{area between acceleration curve}\\ \text{and time axis, from } t_0 \text{ to } t_1\end{pmatrix}.$$

- On a graph of velocity v versus time t, the change in the position is given by

$$x_1 - x_0 = \int_{t_0}^{t_1} v\,dt,$$

where the integral can be taken from the graph as

$$\int_{t_0}^{t_1} v\,dt = \begin{pmatrix}\text{area between velocity curve}\\ \text{and time axis, from } t_0 \text{ to } t_1\end{pmatrix}.$$

Graphical Integration in Motion Analysis

Integrating Acceleration. When we have a graph of an object's acceleration a versus time t, we can integrate on the graph to find the velocity at any given time. Because a is defined as $a = dv/dt$, the Fundamental Theorem of Calculus tells us that

$$v_1 - v_0 = \int_{t_0}^{t_1} a\,dt. \tag{2-27}$$

The right side of the equation is a definite integral (it gives a numerical result rather than a function), v_0 is the velocity at time t_0, and v_1 is the velocity at later time t_1. The definite integral can be evaluated from an $a(t)$ graph, such as in Fig. 2-14*a*. In particular,

$$\int_{t_0}^{t_1} a\,dt = \begin{pmatrix}\text{area between acceleration curve}\\ \text{and time axis, from } t_0 \text{ to } t_1\end{pmatrix}. \tag{2-28}$$

If a unit of acceleration is 1 m/s^2 and a unit of time is 1 s, then the corresponding unit of area on the graph is

$$(1\ \text{m/s}^2)(1\ \text{s}) = 1\ \text{m/s},$$

which is (properly) a unit of velocity. When the acceleration curve is above the time axis, the area is positive; when the curve is below the time axis, the area is negative.

Integrating Velocity. Similarly, because velocity v is defined in terms of the position x as $v = dx/dt$, then

$$x_1 - x_0 = \int_{t_0}^{t_1} v\,dt, \tag{2-29}$$

where x_0 is the position at time t_0 and x_1 is the position at time t_1. The definite integral on the right side of Eq. 2-29 can be evaluated from a $v(t)$ graph, like that shown in Fig. 2-14*b*. In particular,

$$\int_{t_0}^{t_1} v\,dt = \begin{pmatrix}\text{area between velocity curve}\\ \text{and time axis, from } t_0 \text{ to } t_1\end{pmatrix}. \tag{2-30}$$

If the unit of velocity is 1 m/s and the unit of time is 1 s, then the corresponding unit of area on the graph is

$$(1\ \text{m/s})(1\ \text{s}) = 1\ \text{m},$$

which is (properly) a unit of position and displacement. Whether this area is positive or negative is determined as described for the $a(t)$ curve of Fig. 2-14*a*.

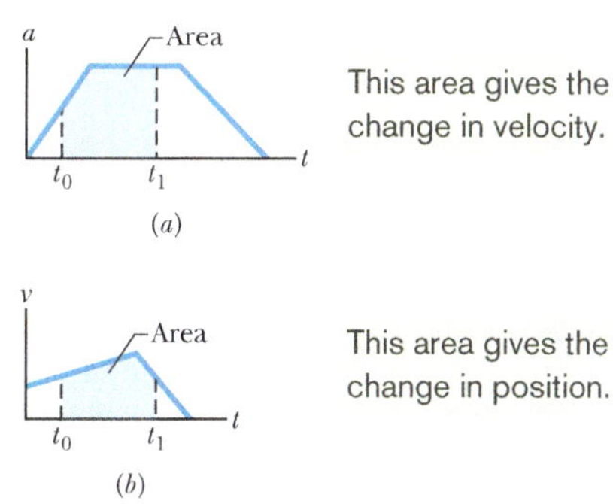

Figure 2-14 The area between a plotted curve and the horizontal time axis, from time t_0 to time t_1, is indicated for (*a*) a graph of acceleration a versus t and (*b*) a graph of velocity v versus t.

Sample Problem 2.06 Graphical integration a versus t, whiplash injury

"Whiplash injury" commonly occurs in a rear-end collision where a front car is hit from behind by a second car. In the 1970s, researchers concluded that the injury was due to the occupant's head being whipped back over the top of the seat as the car was slammed forward. As a result of this finding, head restraints were built into cars, yet neck injuries in rear-end collisions continued to occur.

In a recent test to study neck injury in rear-end collisions, a volunteer was strapped to a seat that was then moved abruptly to simulate a collision by a rear car moving at 10.5 km/h. Figure 2-15*a* gives the accelerations of the volunteer's torso and head during the collision, which began at time $t = 0$. The torso acceleration was delayed by 40 ms because during that time interval the seat back had to compress against the volunteer. The head acceleration was delayed by an additional 70 ms. What was the torso speed when the head began to accelerate?

KEY IDEA

We can calculate the torso speed at any time by finding an area on the torso $a(t)$ graph.

Calculations: We know that the initial torso speed is $v_0 = 0$ at time $t_0 = 0$, at the start of the "collision." We want the torso speed v_1 at time $t_1 = 110$ ms, which is when the head begins to accelerate.

Combining Eqs. 2-27 and 2-28, we can write

$$v_1 - v_0 = \left(\begin{array}{c}\text{area between acceleration curve}\\ \text{and time axis, from } t_0 \text{ to } t_1\end{array}\right). \quad (2\text{-}31)$$

For convenience, let us separate the area into three regions (Fig. 2-15*b*). From 0 to 40 ms, region *A* has no area:

$$\text{area}_A = 0.$$

From 40 ms to 100 ms, region *B* has the shape of a triangle, with area

$$\text{area}_B = \tfrac{1}{2}(0.060 \text{ s})(50 \text{ m/s}^2) = 1.5 \text{ m/s}.$$

From 100 ms to 110 ms, region *C* has the shape of a rectangle, with area

$$\text{area}_C = (0.010 \text{ s})(50 \text{ m/s}^2) = 0.50 \text{ m/s}.$$

Substituting these values and $v_0 = 0$ into Eq. 2-31 gives us

$$v_1 - 0 = 0 + 1.5 \text{ m/s} + 0.50 \text{ m/s},$$

or

$$v_1 = 2.0 \text{ m/s} = 7.2 \text{ km/h}. \quad \text{(Answer)}$$

Comments: When the head is just starting to move forward, the torso already has a speed of 7.2 km/h. Researchers argue that it is this difference in speeds during the early stage of a rear-end collision that injures the neck. The backward whipping of the head happens later and could, especially if there is no head restraint, increase the injury.

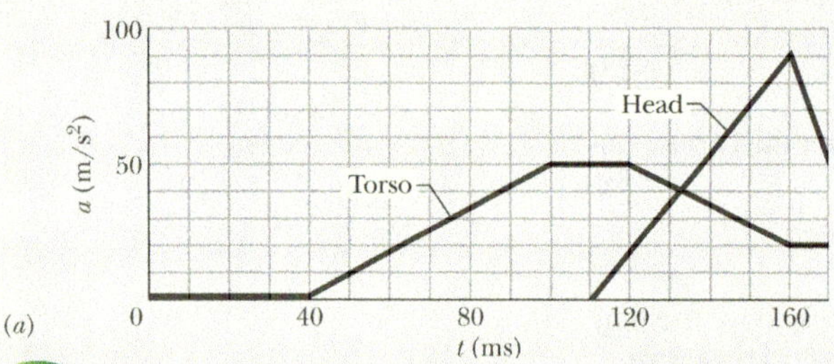

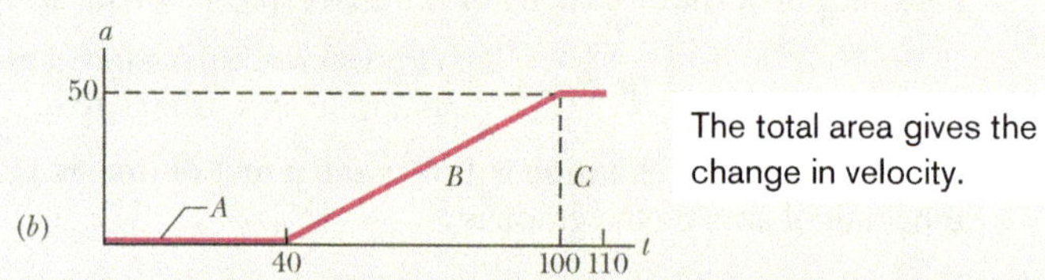

Figure 2-15 (*a*) The $a(t)$ curve of the torso and head of a volunteer in a simulation of a rear-end collision. (*b*) Breaking up the region between the plotted curve and the time axis to calculate the area.

WILEY PLUS Additional examples, video, and practice available at *WileyPLUS*

Review & Summary

Position The *position* x of a particle on an x axis locates the particle with respect to the **origin,** or zero point, of the axis. The position is either positive or negative, according to which side of the origin the particle is on, or zero if the particle is at the origin. The **positive direction** on an axis is the direction of increasing positive numbers; the opposite direction is the **negative direction** on the axis.

Displacement The *displacement* Δx of a particle is the change in its position:

$$\Delta x = x_2 - x_1. \quad (2\text{-}1)$$

Displacement is a vector quantity. It is positive if the particle has moved in the positive direction of the x axis and negative if the particle has moved in the negative direction.

Average Velocity When a particle has moved from position x_1 to position x_2 during a time interval $\Delta t = t_2 - t_1$, its *average velocity* during that interval is

$$v_{\text{avg}} = \frac{\Delta x}{\Delta t} = \frac{x_2 - x_1}{t_2 - t_1}. \quad (2\text{-}2)$$

The algebraic sign of v_{avg} indicates the direction of motion (v_{avg} is a vector quantity). Average velocity does not depend on the actual distance a particle moves, but instead depends on its original and final positions.

On a graph of x versus t, the average velocity for a time interval Δt is the slope of the straight line connecting the points on the curve that represent the two ends of the interval.

Average Speed The *average speed* s_{avg} of a particle during a time interval Δt depends on the total distance the particle moves in that time interval:

$$s_{avg} = \frac{\text{total distance}}{\Delta t}. \quad (2\text{-}3)$$

Instantaneous Velocity The *instantaneous velocity* (or simply **velocity**) v of a moving particle is

$$v = \lim_{\Delta t \to 0} \frac{\Delta x}{\Delta t} = \frac{dx}{dt}, \quad (2\text{-}4)$$

where Δx and Δt are defined by Eq. 2-2. The instantaneous velocity (at a particular time) may be found as the slope (at that particular time) of the graph of x versus t. **Speed** is the magnitude of instantaneous velocity.

Average Acceleration *Average acceleration* is the ratio of a change in velocity Δv to the time interval Δt in which the change occurs:

$$a_{avg} = \frac{\Delta v}{\Delta t}. \quad (2\text{-}7)$$

The algebraic sign indicates the direction of a_{avg}.

Instantaneous Acceleration *Instantaneous acceleration* (or simply **acceleration**) a is the first time derivative of velocity $v(t)$ and the second time derivative of position $x(t)$:

$$a = \frac{dv}{dt} = \frac{d^2x}{dt^2}. \quad (2\text{-}8, 2\text{-}9)$$

On a graph of v versus t, the acceleration a at any time t is the slope of the curve at the point that represents t.

Constant Acceleration The five equations in Table 2-1 describe the motion of a particle with constant acceleration:

$$v = v_0 + at, \quad (2\text{-}11)$$

$$x - x_0 = v_0t + \tfrac{1}{2}at^2, \quad (2\text{-}15)$$

$$v^2 = v_0^2 + 2a(x - x_0), \quad (2\text{-}16)$$

$$x - x_0 = \tfrac{1}{2}(v_0 + v)t, \quad (2\text{-}17)$$

$$x - x_0 = vt - \tfrac{1}{2}at^2. \quad (2\text{-}18)$$

These are *not* valid when the acceleration is not constant.

Free-Fall Acceleration An important example of straight-line motion with constant acceleration is that of an object rising or falling freely near Earth's surface. The constant acceleration equations describe this motion, but we make two changes in notation: (1) we refer the motion to the vertical y axis with $+y$ vertically *up*; (2) we replace a with $-g$, where g is the magnitude of the free-fall acceleration. Near Earth's surface, $g = 9.8\ \text{m/s}^2$ ($= 32\ \text{ft/s}^2$).

Questions

1 Figure 2-16 gives the velocity of a particle moving on an x axis. What are (a) the initial and (b) the final directions of travel? (c) Does the particle stop momentarily? (d) Is the acceleration positive or negative? (e) Is it constant or varying?

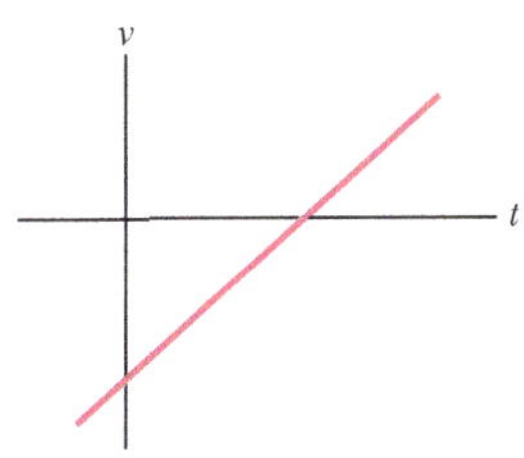

Figure 2-16 Question 1.

2 Figure 2-17 gives the acceleration $a(t)$ of a Chihuahua as it chases a German shepherd along an axis. In which of the time periods indicated does the Chihuahua move at constant speed?

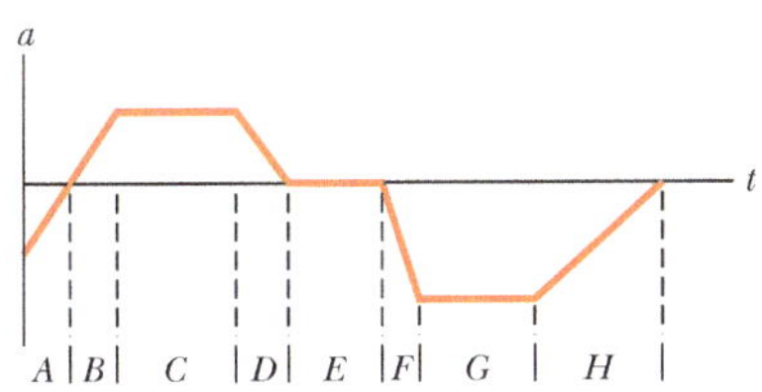

Figure 2-17 Question 2.

3 Figure 2-18 shows four paths along which objects move from a starting point to a final point, all in the same time interval. The paths pass over a grid of equally spaced straight lines. Rank the paths according to (a) the average velocity of the objects and (b) the average speed of the objects, greatest first.

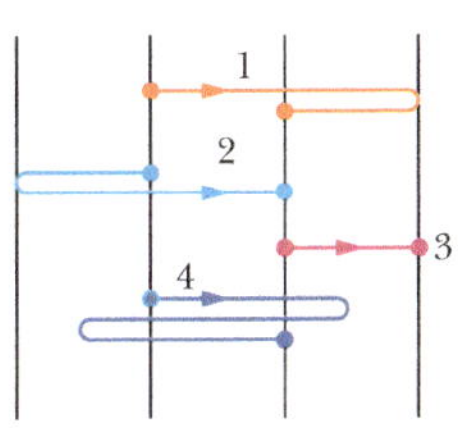

Figure 2-18 Question 3.

4 Figure 2-19 is a graph of a particle's position along an x axis versus time. (a) At time $t = 0$, what is the sign of the particle's position? Is the particle's velocity positive, negative, or 0 at (b) $t = 1$ s, (c) $t = 2$ s, and (d) $t = 3$ s? (e) How many times does the particle go through the point $x = 0$?

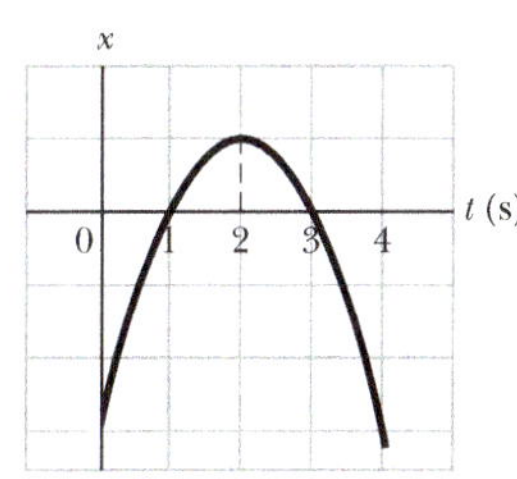

Figure 2-19 Question 4.

5 Figure 2-20 gives the velocity of a particle moving along an axis. Point 1 is at the highest point on the curve; point 4 is at the lowest point; and points 2 and 6 are at the same height. What is the direction of travel at (a) time $t = 0$ and (b) point 4? (c) At which of the six numbered points does the particle reverse its direction of travel? (d) Rank the six points according to the magnitude of the acceleration, greatest first.

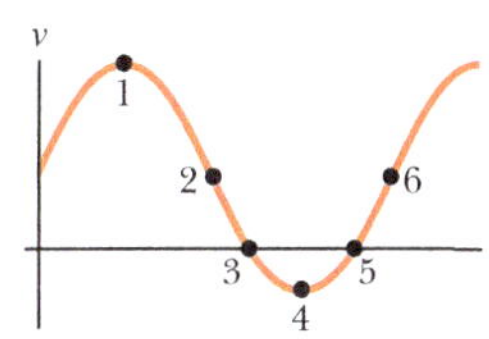

Figure 2-20 Question 5.

6 At $t = 0$, a particle moving along an x axis is at position $x_0 = -20$ m. The signs of the particle's initial velocity v_0 (at time t_0) and constant acceleration a are, respectively, for four situations: (1) $+, +$; (2) $+, -$; (3) $-, +$; (4) $-, -$. In which situations will the particle (a) stop momentarily, (b) pass through the origin, and (c) never pass through the origin?

7 Hanging over the railing of a bridge, you drop an egg (no initial velocity) as you throw a second egg downward. Which curves in Fig. 2-21

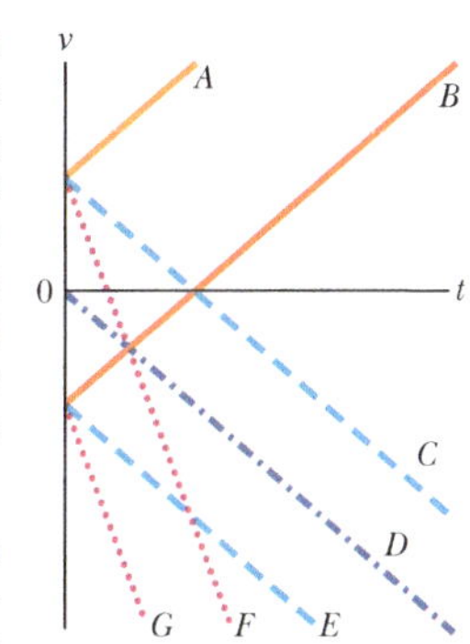

Figure 2-21 Question 7.

give the velocity $v(t)$ for (a) the dropped egg and (b) the thrown egg? (Curves A and B are parallel; so are C, D, and E; so are F and G.)

8 The following equations give the velocity $v(t)$ of a particle in four situations: (a) $v = 3$; (b) $v = 4t^2 + 2t - 6$; (c) $v = 3t - 4$; (d) $v = 5t^2 - 3$. To which of these situations do the equations of Table 2-1 apply?

9 In Fig. 2-22, a cream tangerine is thrown directly upward past three evenly spaced windows of equal heights. Rank the windows according to (a) the average speed of the cream tangerine while passing them, (b) the time the cream tangerine takes to pass them, (c) the magnitude of the acceleration of the cream tangerine while passing them, and (d) the change Δv in the speed of the cream tangerine during the passage, greatest first.

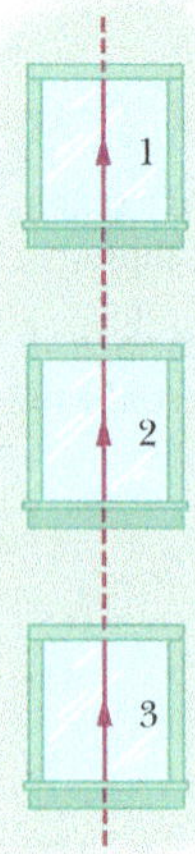

Figure 2-22 Question 9.

10 Suppose that a passenger intent on lunch during his first ride in a hot-air balloon accidently drops an apple over the side during the balloon's liftoff. At the moment of the apple's release, the balloon is accelerating upward with a magnitude of 4.0 m/s^2 and has an upward velocity of magnitude 2 m/s. What are the (a) magnitude and (b) direction of the acceleration of the apple just after it is released? (c) Just then, is the apple moving upward or downward, or is it stationary? (d) What is the magnitude of its velocity just then? (e) In the next few moments, does the speed of the apple increase, decrease, or remain constant?

11 Figure 2-23 shows that a particle moving along an x axis undergoes three periods of acceleration. Without written computation, rank the acceleration periods according to the increases they produce in the particle's velocity, greatest first.

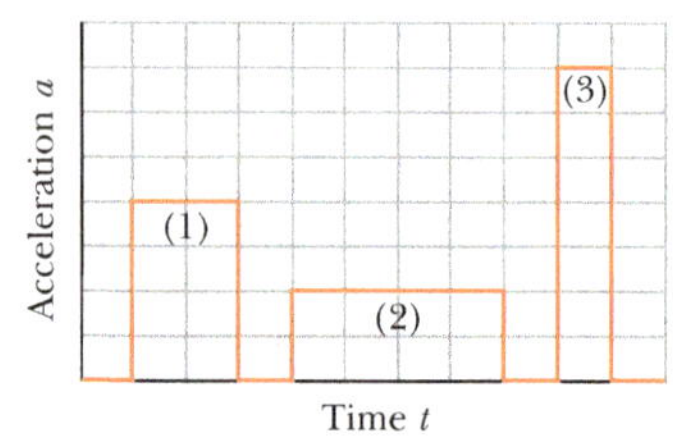

Figure 2-23 Question 11.

Problems

GO Tutoring problem available (at instructor's discretion) in *WileyPLUS* and WebAssign
SSM Worked-out solution available in Student Solutions Manual
• – ••• Number of dots indicates level of problem difficulty
WWW Worked-out solution is at
ILW Interactive solution is at
http://www.wiley.com/college/halliday
Additional information available in *The Flying Circus of Physics* and at flyingcircusofphysics.com

Module 2-1 Position, Displacement, and Average Velocity

•1 While driving a car at 90 km/h, how far do you move while your eyes shut for 0.50 s during a hard sneeze?

•2 Compute your average velocity in the following two cases: (a) You walk 73.2 m at a speed of 1.22 m/s and then run 73.2 m at a speed of 3.05 m/s along a straight track. (b) You walk for 1.00 min at a speed of 1.22 m/s and then run for 1.00 min at 3.05 m/s along a straight track. (c) Graph x versus t for both cases and indicate how the average velocity is found on the graph.

•3 SSM WWW An automobile travels on a straight road for 40 km at 30 km/h. It then continues in the same direction for another 40 km at 60 km/h. (a) What is the average velocity of the car during the full 80 km trip? (Assume that it moves in the positive x direction.) (b) What is the average speed? (c) Graph x versus t and indicate how the average velocity is found on the graph.

•4 A car moves uphill at 40 km/h and then back downhill at 60 km/h. What is the average speed for the round trip?

•5 SSM The position of an object moving along an x axis is given by $x = 3t - 4t^2 + t^3$, where x is in meters and t in seconds. Find the position of the object at the following values of t: (a) 1 s, (b) 2 s, (c) 3 s, and (d) 4 s. (e) What is the object's displacement between $t = 0$ and $t = 4$ s? (f) What is its average velocity for the time interval from $t = 2$ s to $t = 4$ s? (g) Graph x versus t for $0 \leq t \leq 4$ s and indicate how the answer for (f) can be found on the graph.

•6 The 1992 world speed record for a bicycle (human-powered vehicle) was set by Chris Huber. His time through the measured 200 m stretch was a sizzling 6.509 s, at which he commented, "Cogito ergo zoom!" (I think, therefore I go fast!). In 2001, Sam Whittingham beat Huber's record by 19.0 km/h. What was Whittingham's time through the 200 m?

••7 Two trains, each having a speed of 30 km/h, are headed at each other on the same straight track. A bird that can fly 60 km/h flies off the front of one train when they are 60 km apart and heads directly for the other train. On reaching the other train, the (crazy) bird flies directly back to the first train, and so forth. What is the total distance the bird travels before the trains collide?

••8 GO *Panic escape.* Figure 2-24 shows a general situation in which a stream of people attempt to escape through an exit door that turns out to be locked. The people move toward the door at speed $v_s = 3.50$ m/s, are each $d = 0.25$ m in depth, and are separated by $L = 1.75$ m. The arrangement in Fig. 2-24 occurs at time $t = 0$. (a) At what average rate does the layer of people at the door increase? (b) At what time does the layer's depth reach 5.0 m? (The answers reveal how quickly such a situation becomes dangerous.)

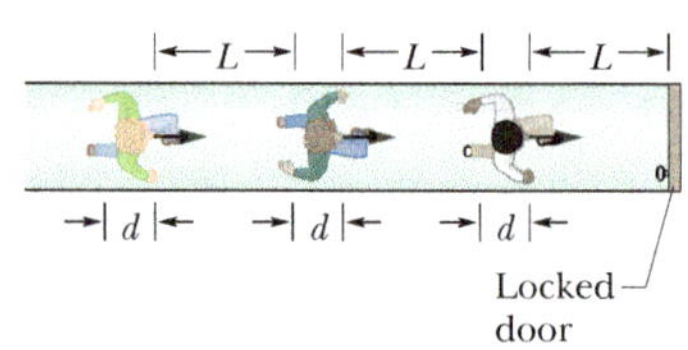

Figure 2-24 Problem 8.

••9 ILW In 1 km races, runner 1 on track 1 (with time 2 min, 27.95 s) appears to be faster than runner 2 on track 2 (2 min, 28.15 s). However, length L_2 of track 2 might be slightly greater than length L_1 of track 1. How large can $L_2 - L_1$ be for us still to conclude that runner 1 is faster?

••10 To set a speed record in a measured (straight-line) distance d, a race car must be driven first in one direction (in time t_1) and then in the opposite direction (in time t_2). (a) To eliminate the effects of the wind and obtain the car's speed v_c in a windless situation, should we find the average of d/t_1 and d/t_2 (method 1) or should we divide d by the average of t_1 and t_2? (b) What is the fractional difference in the two methods when a steady wind blows along the car's route and the ratio of the wind speed v_w to the car's speed v_c is 0.0240?

••11 You are to drive 300 km to an interview. The interview is at 11:15 A.M. You plan to drive at 100 km/h, so you leave at 8:00 A.M. to allow some extra time. You drive at that speed for the first 100 km, but then construction work forces you to slow to 40 km/h for 40 km. What would be the least speed needed for the rest of the trip to arrive in time for the interview?

•••12 *Traffic shock wave.* An abrupt slowdown in concentrated traffic can travel as a pulse, termed a *shock wave,* along the line of cars, either downstream (in the traffic direction) or upstream, or it can be stationary. Figure 2-25 shows a uniformly spaced line of cars moving at speed $v = 25.0$ m/s toward a uniformly spaced line of slow cars moving at speed $v_s = 5.00$ m/s. Assume that each faster car adds length $L = 12.0$ m (car length plus buffer zone) to the line of slow cars when it joins the line, and assume it slows abruptly at the last instant. (a) For what separation distance d between the faster cars does the shock wave remain stationary? If the separation is twice that amount, what are the (b) speed and (c) direction (upstream or downstream) of the shock wave?

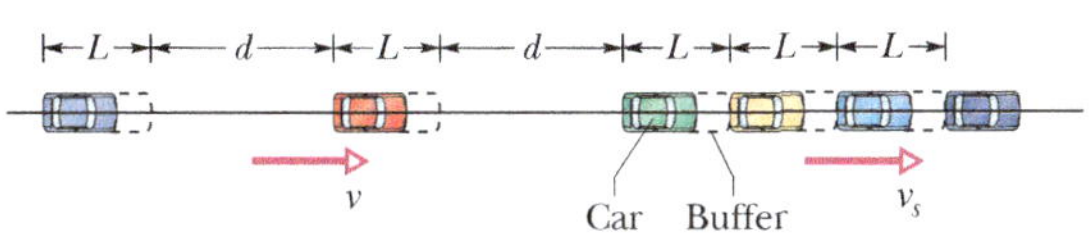

Figure 2-25 Problem 12.

•••13 ILW You drive on Interstate 10 from San Antonio to Houston, half the *time* at 55 km/h and the other half at 90 km/h. On the way back you travel half the *distance* at 55 km/h and the other half at 90 km/h. What is your average speed (a) from San Antonio to Houston, (b) from Houston back to San Antonio, and (c) for the entire trip? (d) What is your average velocity for the entire trip? (e) Sketch x versus t for (a), assuming the motion is all in the positive x direction. Indicate how the average velocity can be found on the sketch.

Module 2-2 Instantaneous Velocity and Speed

•14 An electron moving along the x axis has a position given by $x = 16te^{-t}$ m, where t is in seconds. How far is the electron from the origin when it momentarily stops?

•15 (a) If a particle's position is given by $x = 4 - 12t + 3t^2$ (where t is in seconds and x is in meters), what is its velocity at $t = 1$ s? (b) Is it moving in the positive or negative direction of x just then? (c) What is its speed just then? (d) Is the speed increasing or decreasing just then? (Try answering the next two questions without further calculation.) (e) Is there ever an instant when the velocity is zero? If so, give the time t; if not, answer no. (f) Is there a time after $t = 3$ s when the particle is moving in the negative direction of x? If so, give the time t; if not, answer no.

•16 The position function $x(t)$ of a particle moving along an x axis is $x = 4.0 - 6.0t^2$, with x in meters and t in seconds. (a) At what time and (b) where does the particle (momentarily) stop? At what (c) negative time and (d) positive time does the particle pass through the origin? (e) Graph x versus t for the range -5 s to $+5$ s. (f) To shift the curve rightward on the graph, should we include the term $+20t$ or the term $-20t$ in $x(t)$? (g) Does that inclusion increase or decrease the value of x at which the particle momentarily stops?

••17 The position of a particle moving along the x axis is given in centimeters by $x = 9.75 + 1.50t^3$, where t is in seconds. Calculate (a) the average velocity during the time interval $t = 2.00$ s to $t = 3.00$ s; (b) the instantaneous velocity at $t = 2.00$ s; (c) the instantaneous velocity at $t = 3.00$ s; (d) the instantaneous velocity at $t = 2.50$ s; and (e) the instantaneous velocity when the particle is midway between its positions at $t = 2.00$ s and $t = 3.00$ s. (f) Graph x versus t and indicate your answers graphically.

Module 2-3 Acceleration

•18 The position of a particle moving along an x axis is given by $x = 12t^2 - 2t^3$, where x is in meters and t is in seconds. Determine (a) the position, (b) the velocity, and (c) the acceleration of the particle at $t = 3.0$ s. (d) What is the maximum positive coordinate reached by the particle and (e) at what time is it reached? (f) What is the maximum positive velocity reached by the particle and (g) at what time is it reached? (h) What is the acceleration of the particle at the instant the particle is not moving (other than at $t = 0$)? (i) Determine the average velocity of the particle between $t = 0$ and $t = 3$ s.

•19 SSM At a certain time a particle had a speed of 18 m/s in the positive x direction, and 2.4 s later its speed was 30 m/s in the opposite direction. What is the average acceleration of the particle during this 2.4 s interval?

•20 (a) If the position of a particle is given by $x = 20t - 5t^3$, where x is in meters and t is in seconds, when, if ever, is the particle's velocity zero? (b) When is its acceleration a zero? (c) For what time range (positive or negative) is a negative? (d) Positive? (e) Graph $x(t)$, $v(t)$, and $a(t)$.

••21 From $t = 0$ to $t = 5.00$ min, a man stands still, and from $t = 5.00$ min to $t = 10.0$ min, he walks briskly in a straight line at a constant speed of 2.20 m/s. What are (a) his average velocity v_{avg} and (b) his average acceleration a_{avg} in the time interval 2.00 min to 8.00 min? What are (c) v_{avg} and (d) a_{avg} in the time interval 3.00 min to 9.00 min? (e) Sketch x versus t and v versus t, and indicate how the answers to (a) through (d) can be obtained from the graphs.

••22 The position of a particle moving along the x axis depends on the time according to the equation $x = ct^2 - bt^3$, where x is in meters and t in seconds. What are the units of (a) constant c and (b) constant b? Let their numerical values be 3.0 and 2.0, respectively. (c) At what time does the particle reach its maximum positive x position? From $t = 0.0$ s to $t = 4.0$ s, (d) what distance does the particle move and (e) what is its displacement? Find its velocity at times (f) 1.0 s, (g) 2.0 s, (h) 3.0 s, and (i) 4.0 s. Find its acceleration at times (j) 1.0 s, (k) 2.0 s, (l) 3.0 s, and (m) 4.0 s.

Module 2-4 Constant Acceleration

•23 SSM An electron with an initial velocity $v_0 = 1.50 \times 10^5$ m/s enters a region of length $L = 1.00$ cm where it is electrically accelerated (Fig. 2-26). It emerges with $v = 5.70 \times 10^6$ m/s. What is its acceleration, assumed constant?

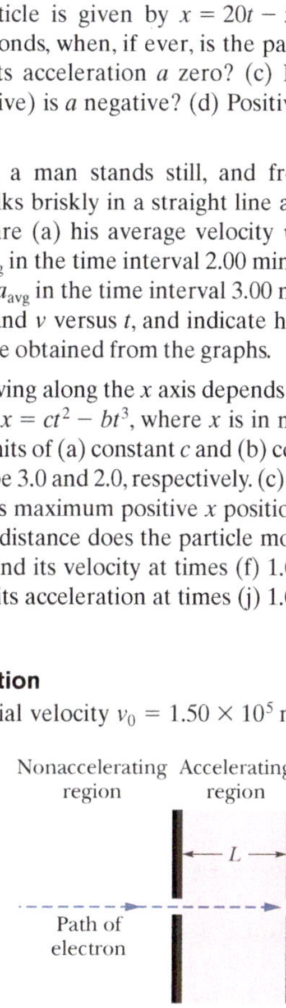

Figure 2-26 Problem 23.

•24 *Catapulting mushrooms.* Certain mushrooms launch their spores by a catapult mechanism. As water condenses from the air onto a spore that is attached to the mushroom, a drop grows on one side of the spore and a film grows on the other side. The spore is bent over by the drop's weight, but when the film reaches the drop, the drop's water suddenly spreads into the film and the spore springs upward so rapidly that it is slung off into the air. Typically, the spore reaches a speed of 1.6 m/s in a 5.0 μm launch; its speed is then reduced to zero in 1.0 mm by the air. Using those data and assuming constant accelerations, find the acceleration in terms of g during (a) the launch and (b) the speed reduction.

•25 An electric vehicle starts from rest and accelerates at a rate of 2.0 m/s^2 in a straight line until it reaches a speed of 20 m/s. The vehicle then slows at a constant rate of 1.0 m/s^2 until it stops. (a) How much time elapses from start to stop? (b) How far does the vehicle travel from start to stop?

•26 A muon (an elementary particle) enters a region with a speed of 5.00×10^6 m/s and then is slowed at the rate of 1.25×10^{14} m/s^2. (a) How far does the muon take to stop? (b) Graph x versus t and v versus t for the muon.

•27 An electron has a constant acceleration of +3.2 m/s^2. At a certain instant its velocity is +9.6 m/s. What is its velocity (a) 2.5 s earlier and (b) 2.5 s later?

•28 On a dry road, a car with good tires may be able to brake with a constant deceleration of 4.92 m/s^2. (a) How long does such a car, initially traveling at 24.6 m/s, take to stop? (b) How far does it travel in this time? (c) Graph x versus t and v versus t for the deceleration.

•29 ILW A certain elevator cab has a total run of 190 m and a maximum speed of 305 m/min, and it accelerates from rest and then back to rest at 1.22 m/s^2. (a) How far does the cab move while accelerating to full speed from rest? (b) How long does it take to make the nonstop 190 m run, starting and ending at rest?

•30 The brakes on your car can slow you at a rate of 5.2 m/s^2. (a) If you are going 137 km/h and suddenly see a state trooper, what is the minimum time in which you can get your car under the 90 km/h speed limit? (The answer reveals the futility of braking to keep your high speed from being detected with a radar or laser gun.) (b) Graph x versus t and v versus t for such a slowing.

•31 SSM Suppose a rocket ship in deep space moves with constant acceleration equal to 9.8 m/s^2, which gives the illusion of normal gravity during the flight. (a) If it starts from rest, how long will it take to acquire a speed one-tenth that of light, which travels at 3.0×10^8 m/s? (b) How far will it travel in so doing?

•32 A world's land speed record was set by Colonel John P. Stapp when in March 1954 he rode a rocket-propelled sled that moved along a track at 1020 km/h. He and the sled were brought to a stop in 1.4 s. (See Fig. 2-7.) In terms of g, what acceleration did he experience while stopping?

•33 SSM ILW A car traveling 56.0 km/h is 24.0 m from a barrier when the driver slams on the brakes. The car hits the barrier 2.00 s later. (a) What is the magnitude of the car's constant acceleration before impact? (b) How fast is the car traveling at impact?

••34 GO In Fig. 2-27, a red car and a green car, identical except for the color, move toward each other in adjacent lanes and parallel to an x axis. At time $t = 0$, the red car is at $x_r = 0$ and the green car is at $x_g = 220$ m. If the red car has a constant velocity of 20 km/h, the cars pass each other at $x = 44.5$ m, and if it has a constant velocity of 40 km/h, they pass each other at $x = 76.6$ m. What are (a) the initial velocity and (b) the constant acceleration of the green car?

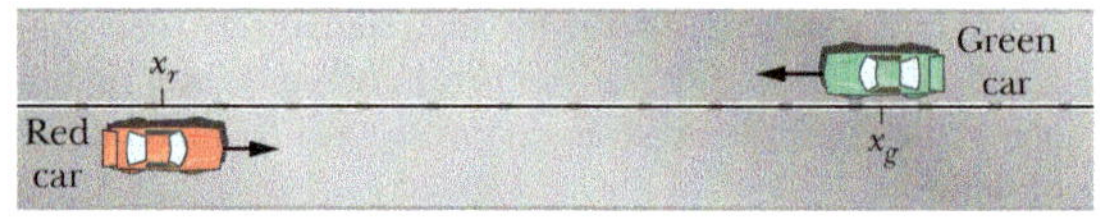

Figure 2-27 Problems 34 and 35.

••35 Figure 2-27 shows a red car and a green car that move toward each other. Figure 2-28 is a graph of their motion, showing the positions $x_{g0} = 270$ m and $x_{r0} = -35.0$ m at time $t = 0$. The green car has a constant speed of 20.0 m/s and the red car begins from rest. What is the acceleration magnitude of the red car?

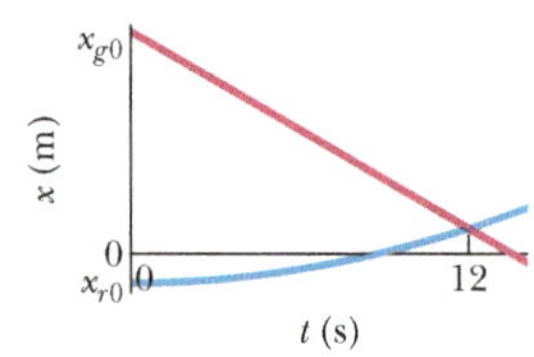

Figure 2-28 Problem 35.

••36 A car moves along an x axis through a distance of 900 m, starting at rest (at $x = 0$) and ending at rest (at $x = 900$ m). Through the first $\frac{1}{4}$ of that distance, its acceleration is +2.25 m/s^2. Through the rest of that distance, its acceleration is −0.750 m/s^2. What are (a) its travel time through the 900 m and (b) its maximum speed? (c) Graph position x, velocity v, and acceleration a versus time t for the trip.

••37 Figure 2-29 depicts the motion of a particle moving along an x axis with a constant acceleration. The figure's vertical scaling is set by $x_s = 6.0$ m. What are the (a) magnitude and (b) direction of the particle's acceleration?

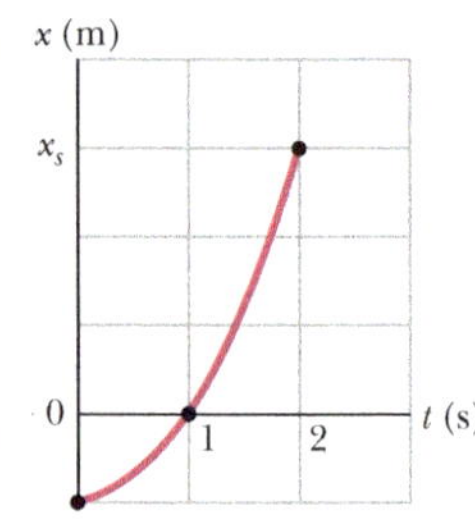

Figure 2-29 Problem 37.

••38 (a) If the maximum acceleration that is tolerable for passengers in a subway train is 1.34 m/s^2 and subway stations are located 806 m apart, what is the maximum speed a subway train can attain between stations? (b) What is the travel time between stations? (c) If a subway train stops for 20 s at each station, what is the maximum average speed of the train, from one start-up to the next? (d) Graph x, v, and a versus t for the interval from one start-up to the next.

••39 Cars A and B move in the same direction in adjacent lanes. The position x of car A is given in Fig. 2-30, from time $t = 0$ to $t = 7.0$ s. The figure's vertical scaling is set by $x_s = 32.0$ m. At $t = 0$, car B is at $x = 0$, with a velocity of 12 m/s and a negative constant acceleration a_B. (a) What must a_B be such that the cars are (momentarily) side by side (momentarily at the same value of x) at $t = 4.0$ s? (b) For that value of a_B, how many times are the cars side by side? (c) Sketch the position x of car B versus time t on Fig. 2-30. How many times will the cars be side by side if the magnitude of acceleration a_B is (d) more than and (e) less than the answer to part (a)?

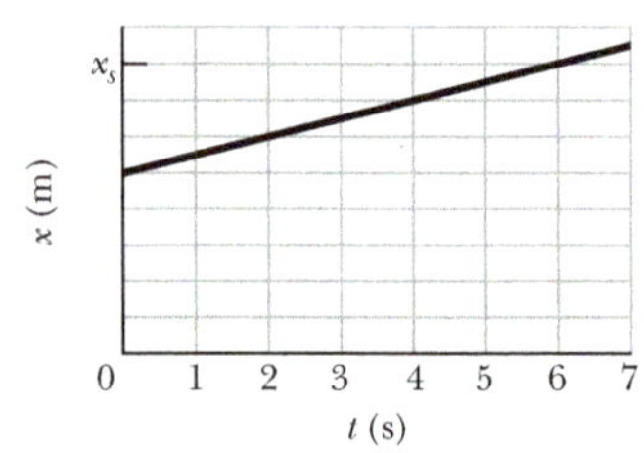

Figure 2-30 Problem 39.

••40 You are driving toward a traffic signal when it turns yellow. Your speed is the legal speed limit of $v_0 = 55$ km/h; your best deceleration rate has the magnitude $a = 5.18$ m/s^2. Your best reaction time to begin braking is $T = 0.75$ s. To avoid having the front of your car enter the intersection after the light turns red, should you brake to a stop or continue to move at 55 km/h if the distance to the intersection and the duration of the yellow light are (a) 40 m and 2.8 s, and (b) 32 m and 1.8 s? Give an answer of brake, continue, either (if either strategy works), or neither (if neither strategy works and the yellow duration is inappropriate).

••41 GO As two trains move along a track, their conductors suddenly notice that they are headed toward each other. Figure 2-31 gives their velocities v as functions of time t as the conductors slow the trains. The figure's vertical scaling is set by v_s = 40.0 m/s. The slowing processes begin when the trains are 200 m apart. What is their separation when both trains have stopped?

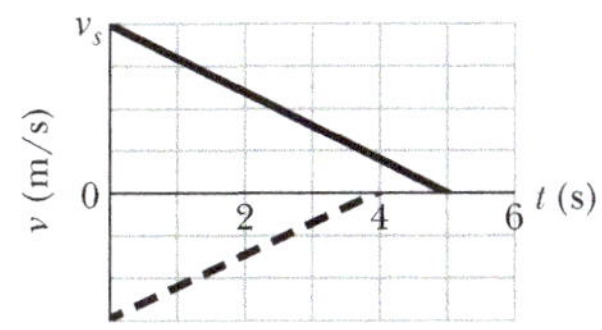

Figure 2-31 Problem 41.

•••42 GO You are arguing over a cell phone while trailing an unmarked police car by 25 m; both your car and the police car are traveling at 110 km/h. Your argument diverts your attention from the police car for 2.0 s (long enough for you to look at the phone and yell, "I won't do that!"). At the beginning of that 2.0 s, the police officer begins braking suddenly at 5.0 m/s^2. (a) What is the separation between the two cars when your attention finally returns? Suppose that you take another 0.40 s to realize your danger and begin braking. (b) If you too brake at 5.0 m/s^2, what is your speed when you hit the police car?

•••43 GO When a high-speed passenger train traveling at 161 km/h rounds a bend, the engineer is shocked to see that a locomotive has improperly entered onto the track from a siding and is a distance D = 676 m ahead (Fig. 2-32). The locomotive is moving at 29.0 km/h. The engineer of the high-speed train immediately applies the brakes. (a) What must be the magnitude of the resulting constant deceleration if a collision is to be just avoided? (b) Assume that the engineer is at x = 0 when, at t = 0, he first spots the locomotive. Sketch $x(t)$ curves for the locomotive and high-speed train for the cases in which a collision is just avoided and is not quite avoided.

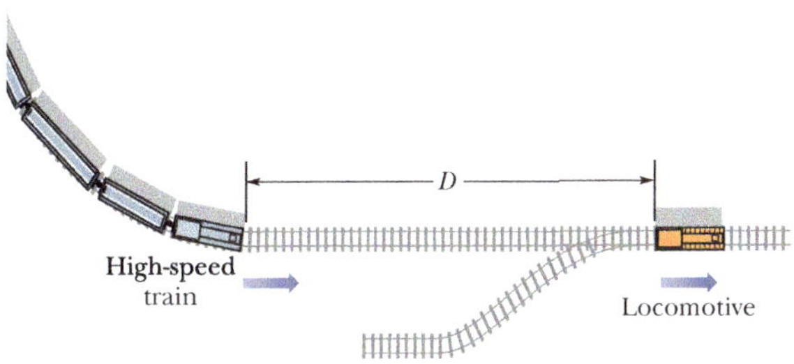

Figure 2-32 Problem 43.

Module 2-5 Free-Fall Acceleration

•44 When startled, an armadillo will leap upward. Suppose it rises 0.544 m in the first 0.200 s. (a) What is its initial speed as it leaves the ground? (b) What is its speed at the height of 0.544 m? (c) How much higher does it go?

•45 SSM WWW (a) With what speed must a ball be thrown vertically from ground level to rise to a maximum height of 50 m? (b) How long will it be in the air? (c) Sketch graphs of y, v, and a versus t for the ball. On the first two graphs, indicate the time at which 50 m is reached.

•46 Raindrops fall 1700 m from a cloud to the ground. (a) If they were not slowed by air resistance, how fast would the drops be moving when they struck the ground? (b) Would it be safe to walk outside during a rainstorm?

•47 SSM At a construction site a pipe wrench struck the ground with a speed of 24 m/s. (a) From what height was it inadvertently dropped? (b) How long was it falling? (c) Sketch graphs of y, v, and a versus t for the wrench.

•48 A hoodlum throws a stone vertically downward with an initial speed of 12.0 m/s from the roof of a building, 30.0 m above the ground. (a) How long does it take the stone to reach the ground? (b) What is the speed of the stone at impact?

•49 SSM A hot-air balloon is ascending at the rate of 12 m/s and is 80 m above the ground when a package is dropped over the side. (a) How long does the package take to reach the ground? (b) With what speed does it hit the ground?

••50 At time t = 0, apple 1 is dropped from a bridge onto a roadway beneath the bridge; somewhat later, apple 2 is thrown down from the same height. Figure 2-33 gives the vertical positions y of the apples versus t during the falling, until both apples have hit the roadway. The scaling is set by t_s = 2.0 s. With approximately what speed is apple 2 thrown down?

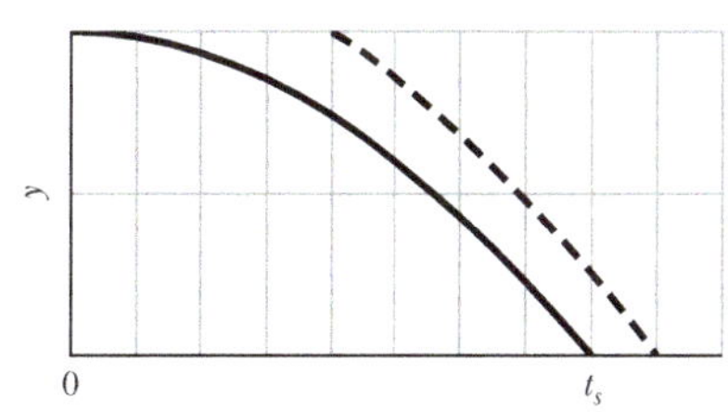

Figure 2-33 Problem 50.

••51 As a runaway scientific balloon ascends at 19.6 m/s, one of its instrument packages breaks free of a harness and free-falls. Figure 2-34 gives the vertical velocity of the package versus time, from before it breaks free to when it reaches the ground. (a) What maximum height above the break-free point does it rise? (b) How high is the break-free point above the ground?

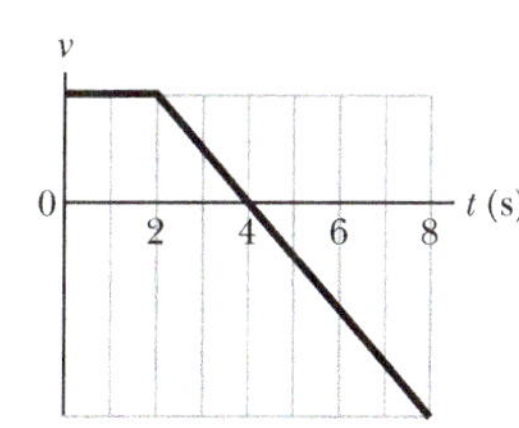

Figure 2-34 Problem 51.

••52 GO A bolt is dropped from a bridge under construction, falling 90 m to the valley below the bridge. (a) In how much time does it pass through the last 20% of its fall? What is its speed (b) when it begins that last 20% of its fall and (c) when it reaches the valley beneath the bridge?

••53 SSM ILW A key falls from a bridge that is 45 m above the water. It falls directly into a model boat, moving with constant velocity, that is 12 m from the point of impact when the key is released. What is the speed of the boat?

••54 GO A stone is dropped into a river from a bridge 43.9 m above the water. Another stone is thrown vertically down 1.00 s after the first is dropped. The stones strike the water at the same time. (a) What is the initial speed of the second stone? (b) Plot velocity versus time on a graph for each stone, taking zero time as the instant the first stone is released.

••55 SSM A ball of moist clay falls 15.0 m to the ground. It is in contact with the ground for 20.0 ms before stopping. (a) What is the magnitude of the average acceleration of the ball during the time it is in contact with the ground? (Treat the ball as a particle.) (b) Is the average acceleration up or down?

••56 GO Figure 2-35 shows the speed v versus height y of a ball tossed directly upward, along a y axis. Distance d is 0.40 m. The speed at height y_A is v_A. The speed at height y_B is $\frac{1}{3}v_A$. What is speed v_A?

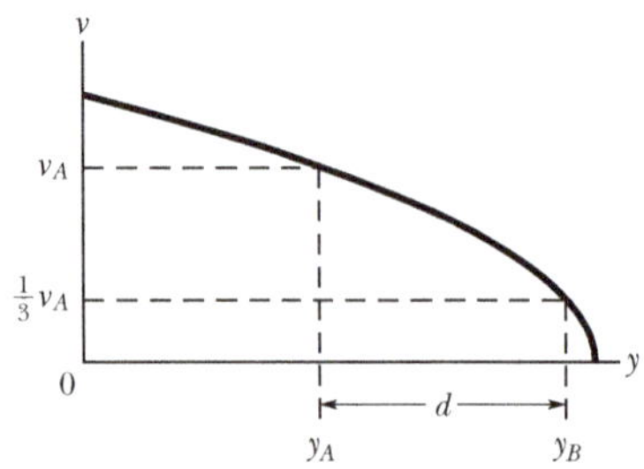

Figure 2-35 Problem 56.

••57 To test the quality of a tennis ball, you drop it onto the floor from a height of 4.00 m. It rebounds to a height of 2.00 m. If the ball is in contact with the floor for 12.0 ms, (a) what is the magnitude of its average acceleration during that contact and (b) is the average acceleration up or down?

••58 An object falls a distance h from rest. If it travels $0.50h$ in the last 1.00 s, find (a) the time and (b) the height of its fall. (c) Explain the physically unacceptable solution of the quadratic equation in t that you obtain.

••59 Water drips from the nozzle of a shower onto the floor 200 cm below. The drops fall at regular (equal) intervals of time, the first drop striking the floor at the instant the fourth drop begins to fall. When the first drop strikes the floor, how far below the nozzle are the (a) second and (b) third drops?

••60 GO A rock is thrown vertically upward from ground level at time $t = 0$. At $t = 1.5$ s it passes the top of a tall tower, and 1.0 s later it reaches its maximum height. What is the height of the tower?

•••61 GO A steel ball is dropped from a building's roof and passes a window, taking 0.125 s to fall from the top to the bottom of the window, a distance of 1.20 m. It then falls to a sidewalk and bounces back past the window, moving from bottom to top in 0.125 s. Assume that the upward flight is an exact reverse of the fall. The time the ball spends below the bottom of the window is 2.00 s. How tall is the building?

•••62 A basketball player grabbing a rebound jumps 76.0 cm vertically. How much total time (ascent and descent) does the player spend (a) in the top 15.0 cm of this jump and (b) in the bottom 15.0 cm? (The player seems to hang in the air at the top.)

•••63 GO A drowsy cat spots a flowerpot that sails first up and then down past an open window. The pot is in view for a total of 0.50 s, and the top-to-bottom height of the window is 2.00 m. How high above the window top does the flowerpot go?

•••64 A ball is shot vertically upward from the surface of another planet. A plot of y versus t for the ball is shown in Fig. 2-36, where y is the height of the ball above its starting point and $t = 0$ at the instant the ball is shot. The figure's vertical scaling is set by $y_s = 30.0$ m. What are the magnitudes of (a) the free-fall acceleration on the planet and (b) the initial velocity of the ball?

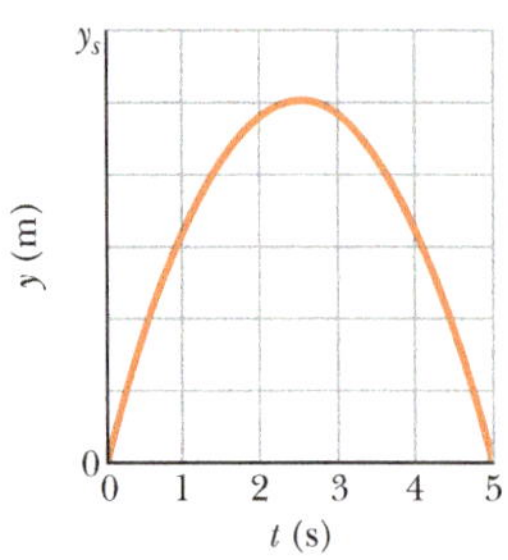

Figure 2-36 Problem 64.

Module 2-6 Graphical Integration in Motion Analysis

•65 Figure 2-15a gives the acceleration of a volunteer's head and torso during a rear-end collision. At maximum head acceleration, what is the speed of (a) the head and (b) the torso?

••66 In a forward punch in karate, the fist begins at rest at the waist and is brought rapidly forward until the arm is fully extended. The speed $v(t)$ of the fist is given in Fig. 2-37 for someone skilled in karate. The vertical scaling is set by $v_s = 8.0$ m/s. How far has the fist moved at (a) time $t = 50$ ms and (b) when the speed of the fist is maximum?

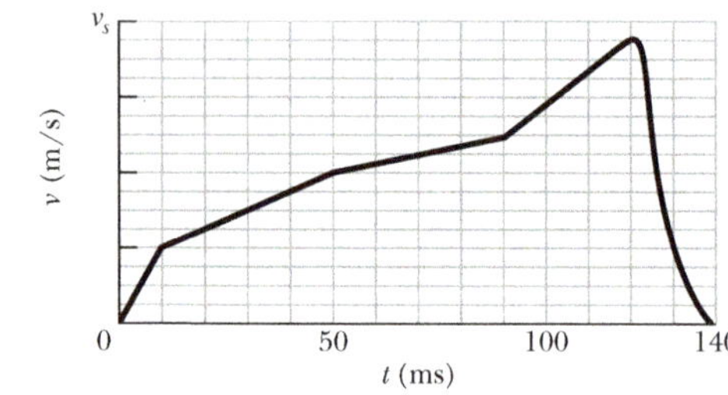

Figure 2-37 Problem 66.

••67 When a soccer ball is kicked toward a player and the player deflects the ball by "heading" it, the acceleration of the head during the collision can be significant. Figure 2-38 gives the measured acceleration $a(t)$ of a soccer player's head for a bare head and a helmeted head, starting from rest. The scaling on the vertical axis is set by $a_s = 200$ m/s². At time $t = 7.0$ ms, what is the difference in the speed acquired by the bare head and the speed acquired by the helmeted head?

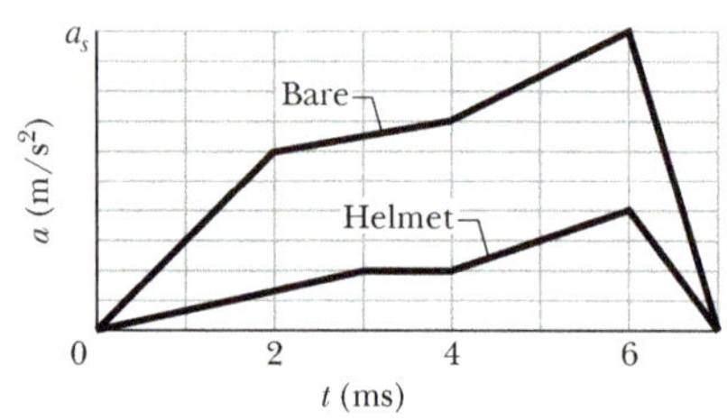

Figure 2-38 Problem 67.

••68 A salamander of the genus *Hydromantes* captures prey by launching its tongue as a projectile: The skeletal part of the tongue is shot forward, unfolding the rest of the tongue, until the outer portion lands on the prey, sticking to it. Figure 2-39 shows the acceleration magnitude a versus time t for the acceleration phase of the launch in a typical situation. The indicated accelerations are $a_2 = 400$ m/s² and $a_1 = 100$ m/s². What is the outward speed of the tongue at the end of the acceleration phase?

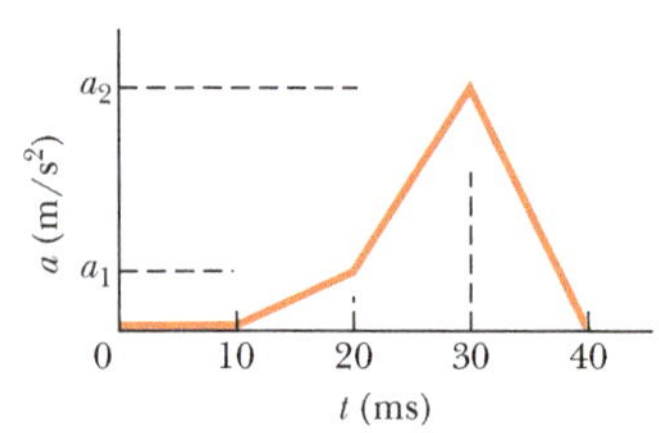

Figure 2-39 Problem 68.

••69 ILW How far does the runner whose velocity–time graph is shown in Fig. 2-40 travel in 16 s? The figure's vertical scaling is set by $v_s = 8.0$ m/s.

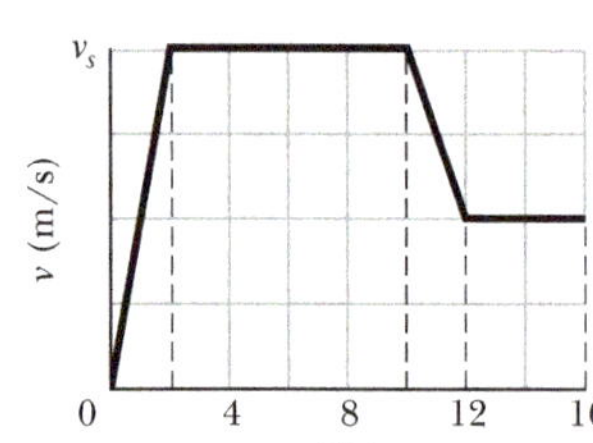

Figure 2-40 Problem 69.

•••70 Two particles move along an x axis. The position of particle 1 is given by $x = 6.00t^2 + 3.00t + 2.00$ (in meters and seconds); the acceleration of particle 2 is given by $a = -8.00t$ (in meters per second squared and seconds) and, at $t = 0$, its velocity is 20 m/s. When the velocities of the particles match, what is their velocity?

Additional Problems

71 In an arcade video game, a spot is programmed to move across the screen according to $x = 9.00t - 0.750t^3$, where x is distance in centimeters measured from the left edge of the screen and t is time in seconds. When the spot reaches a screen edge, at either $x = 0$ or $x = 15.0$ cm, t is reset to 0 and the spot starts moving again according to $x(t)$. (a) At what time after starting is the spot instantaneously at rest? (b) At what value of x does this occur? (c) What is the spot's acceleration (including sign) when this occurs? (d) Is it moving right or left just prior to coming to rest? (e) Just after? (f) At what time $t > 0$ does it first reach an edge of the screen?

72 A rock is shot vertically upward from the edge of the top of a tall building. The rock reaches its maximum height above the top of the building 1.60 s after being shot. Then, after barely missing the edge of the building as it falls downward, the rock strikes the ground 6.00 s after it is launched. In SI units: (a) with what upward velocity is the rock shot, (b) what maximum height above the top of the building is reached by the rock, and (c) how tall is the building?

73 GO At the instant the traffic light turns green, an automobile starts with a constant acceleration a of 2.2 m/s^2. At the same instant a truck, traveling with a constant speed of 9.5 m/s, overtakes and passes the automobile. (a) How far beyond the traffic signal will the automobile overtake the truck? (b) How fast will the automobile be traveling at that instant?

74 A pilot flies horizontally at 1300 km/h, at height $h = 35$ m above initially level ground. However, at time $t = 0$, the pilot begins to fly over ground sloping upward at angle $\theta = 4.3°$ (Fig. 2-41). If the pilot does not change the airplane's heading, at what time t does the plane strike the ground?

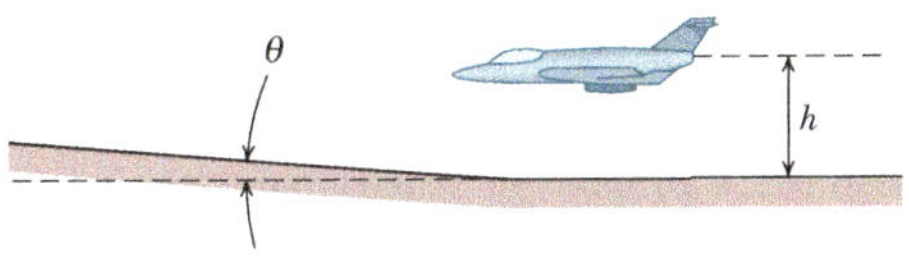

Figure 2-41 Problem 74.

75 GO To stop a car, first you require a certain reaction time to begin braking; then the car slows at a constant rate. Suppose that the total distance moved by your car during these two phases is 56.7 m when its initial speed is 80.5 km/h, and 24.4 m when its initial speed is 48.3 km/h. What are (a) your reaction time and (b) the magnitude of the acceleration?

76 GO Figure 2-42 shows part of a street where traffic flow is to be controlled to allow a *platoon* of cars to move smoothly along the street. Suppose that the platoon leaders have just reached intersection 2, where the green appeared when they were distance d from the intersection. They continue to travel at a certain speed v_p (the speed limit) to reach intersection 3, where the green appears when they are distance d from it. The intersections are separated by distances D_{23} and D_{12}. (a) What should be the time delay of the onset of green at intersection 3 relative to that at intersection 2 to keep the platoon moving smoothly?

Suppose, instead, that the platoon had been stopped by a red light at intersection 1. When the green comes on there, the leaders require a certain time t_r to respond to the change and an additional time to accelerate at some rate a to the cruising speed v_p. (b) If the green at intersection 2 is to appear when the leaders are distance d from that intersection, how long after the light at intersection 1 turns green should the light at intersection 2 turn green?

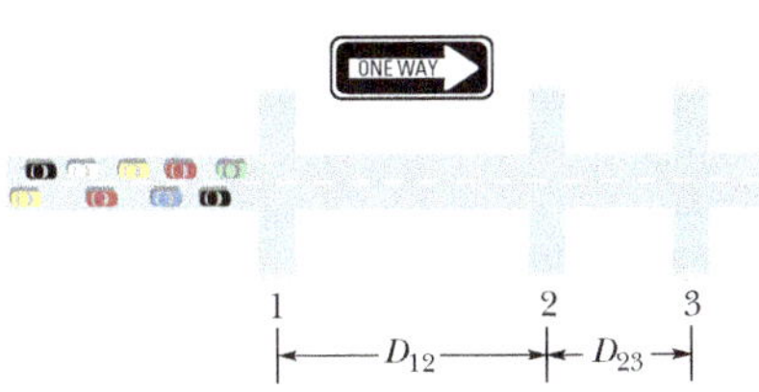

Figure 2-42 Problem 76.

77 SSM A hot rod can accelerate from 0 to 60 km/h in 5.4 s. (a) What is its average acceleration, in m/s^2, during this time? (b) How far will it travel during the 5.4 s, assuming its acceleration is constant? (c) From rest, how much time would it require to go a distance of 0.25 km if its acceleration could be maintained at the value in (a)?

78 GO A red train traveling at 72 km/h and a green train traveling at 144 km/h are headed toward each other along a straight, level track. When they are 950 m apart, each engineer sees the other's train and applies the brakes. The brakes slow each train at the rate of 1.0 m/s^2. Is there a collision? If so, answer yes and give the speed of the red train and the speed of the green train at impact, respectively. If not, answer no and give the separation between the trains when they stop.

79 GO At time $t = 0$, a rock climber accidentally allows a piton to fall freely from a high point on the rock wall to the valley below him. Then, after a short delay, his climbing partner, who is 10 m higher on the wall, throws a piton downward. The positions y of the pitons versus t during the falling are given in Fig. 2-43. With what speed is the second piton thrown?

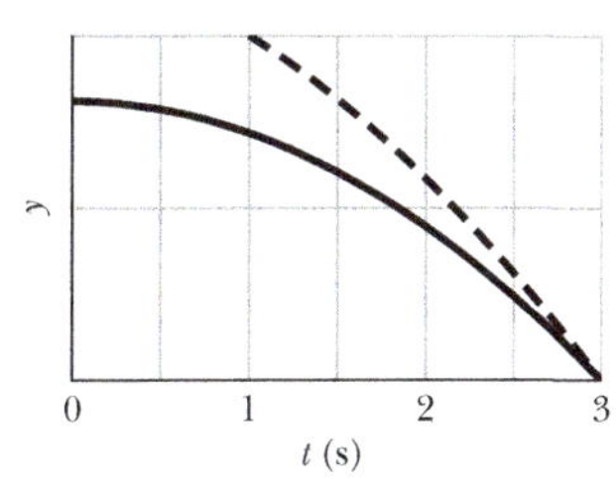

Figure 2-43 Problem 79.

80 A train started from rest and moved with constant acceleration. At one time it was traveling 30 m/s, and 160 m farther on it was traveling 50 m/s. Calculate (a) the acceleration, (b) the time required to travel the 160 m mentioned, (c) the time required to attain the speed of 30 m/s, and (d) the distance moved from rest to the time the train had a speed of 30 m/s. (e) Graph x versus t and v versus t for the train, from rest.

81 SSM A particle's acceleration along an x axis is $a = 5.0t$, with t in seconds and a in meters per second squared. At $t = 2.0$ s, its velocity is +17 m/s. What is its velocity at $t = 4.0$ s?

82 Figure 2-44 gives the acceleration a versus time t for a particle moving along an x axis. The a-axis scale is set by $a_s = 12.0$ m/s^2. At $t = -2.0$ s, the particle's velocity is 7.0 m/s. What is its velocity at $t = 6.0$ s?

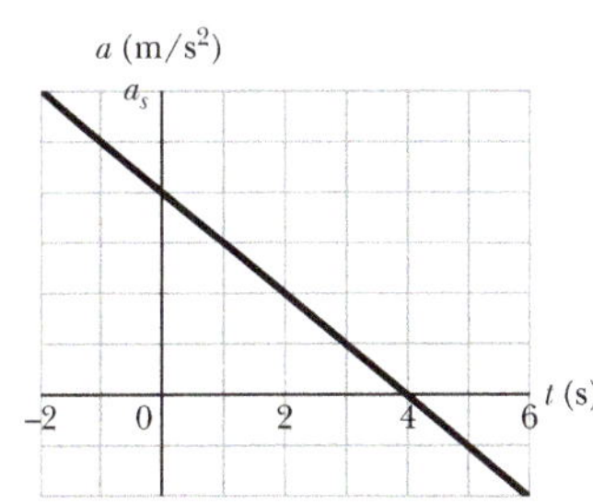

Figure 2-44 Problem 82.

83 Figure 2-45 shows a simple device for measuring your reaction time. It consists of a cardboard strip marked with a scale and two large dots. A friend holds the strip *vertically,* with thumb and forefinger at the dot on the right in Fig. 2-45. You then position your thumb and forefinger at the other dot (on the left in Fig. 2-45), being careful not to touch the strip. Your friend releases the strip, and you try to pinch it as soon as possible after you see it begin to fall. The mark at the place where you pinch the strip gives your reaction time. (a) How far from the lower dot should you place the 50.0 ms mark? How much higher should you place the marks for (b) 100, (c) 150, (d) 200, and (e) 250 ms? (For example, should the 100 ms marker be 2 times as far from the dot as the 50 ms marker? If so, give an answer of 2 times. Can you find any pattern in the answers?)

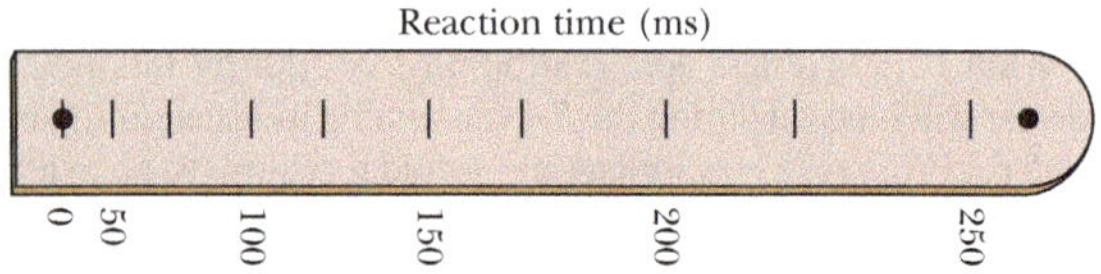

Figure 2-45 Problem 83.

84 A rocket-driven sled running on a straight, level track is used to investigate the effects of large accelerations on humans. One such sled can attain a speed of 1600 km/h in 1.8 s, starting from rest. Find (a) the acceleration (assumed constant) in terms of g and (b) the distance traveled.

85 A mining cart is pulled up a hill at 20 km/h and then pulled back down the hill at 35 km/h through its original level. (The time required for the cart's reversal at the top of its climb is negligible.) What is the average speed of the cart for its round trip, from its original level back to its original level?

86 A motorcyclist who is moving along an x axis directed toward the east has an acceleration given by $a = (6.1 - 1.2t)$ m/s^2 for $0 \leq t \leq 6.0$ s. At $t = 0$, the velocity and position of the cyclist are 2.7 m/s and 7.3 m. (a) What is the maximum speed achieved by the cyclist? (b) What total distance does the cyclist travel between $t = 0$ and 6.0 s?

87 SSM When the legal speed limit for the New York Thruway was increased from 55 mi/h to 65 mi/h, how much time was saved by a motorist who drove the 700 km between the Buffalo entrance and the New York City exit at the legal speed limit?

88 A car moving with constant acceleration covered the distance between two points 60.0 m apart in 6.00 s. Its speed as it passed the second point was 15.0 m/s. (a) What was the speed at the first point? (b) What was the magnitude of the acceleration? (c) At what prior distance from the first point was the car at rest? (d) Graph x versus t and v versus t for the car, from rest ($t = 0$).

89 SSM A certain juggler usually tosses balls vertically to a height H. To what height must they be tossed if they are to spend twice as much time in the air?

90 A particle starts from the origin at $t = 0$ and moves along the positive x axis. A graph of the velocity of the particle as a function of the time is shown in Fig. 2-46; the v-axis scale is set by $v_s = 4.0$ m/s. (a) What is the coordinate of the particle at $t = 5.0$ s? (b) What is the velocity of the particle at $t = 5.0$ s? (c) What is the acceleration of the particle at $t = 5.0$ s? (d) What is the average velocity of the particle between $t = 1.0$ s and $t = 5.0$ s? (e) What is the average acceleration of the particle between $t = 1.0$ s and $t = 5.0$ s?

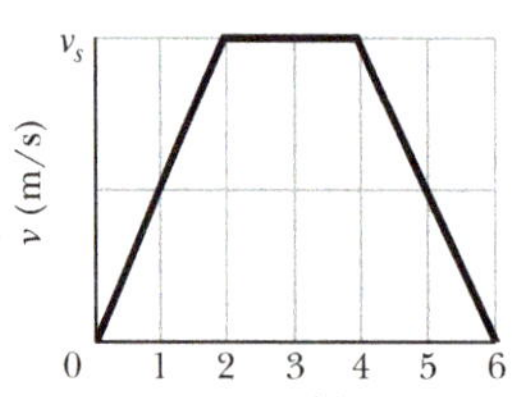

Figure 2-46 Problem 90.

91 A rock is dropped from a 100-m-high cliff. How long does it take to fall (a) the first 50 m and (b) the second 50 m?

92 Two subway stops are separated by 1100 m. If a subway train accelerates at $+1.2$ m/s^2 from rest through the first half of the distance and decelerates at -1.2 m/s^2 through the second half, what are (a) its travel time and (b) its maximum speed? (c) Graph x, v, and a versus t for the trip.

93 A stone is thrown vertically upward. On its way up it passes point A with speed v, and point B, 3.00 m higher than A, with speed $\frac{1}{2}v$. Calculate (a) the speed v and (b) the maximum height reached by the stone above point B.

94 A rock is dropped (from rest) from the top of a 60-m-tall building. How far above the ground is the rock 1.2 s before it reaches the ground?

95 SSM An iceboat has a constant velocity toward the east when a sudden gust of wind causes the iceboat to have a constant acceleration toward the east for a period of 3.0 s. A plot of x versus t is shown in Fig. 2-47, where $t = 0$ is taken to be the instant the wind starts to blow and the positive x axis is toward the east. (a) What is the acceleration of the iceboat during the 3.0 s interval? (b) What is the velocity of the iceboat at the end of the 3.0 s interval? (c) If the acceleration remains constant for an additional 3.0 s, how far does the iceboat travel during this second 3.0 s interval?

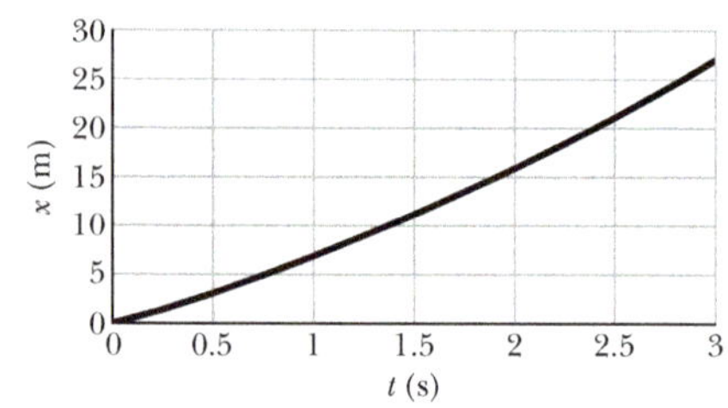

Figure 2-47 Problem 95.

96 A lead ball is dropped in a lake from a diving board 5.20 m above the water. It hits the water with a certain velocity and then sinks to the bottom with this same constant velocity. It reaches the bottom 4.80 s after it is dropped. (a) How deep is the lake? What are the (b) magnitude and (c) direction (up or down) of the average velocity of the ball for the entire fall? Suppose that all the water is drained from the lake. The ball is now thrown from the diving board so that it again reaches the bottom in 4.80 s. What are the (d) magnitude and (e) direction of the initial velocity of the ball?

97 The single cable supporting an unoccupied construction elevator breaks when the elevator is at rest at the top of a 120-m-high building. (a) With what speed does the elevator strike the ground? (b) How long is it falling? (c) What is its speed when it passes the halfway point on the way down? (d) How long has it been falling when it passes the halfway point?

98 Two diamonds begin a free fall from rest from the same height, 1.0 s apart. How long after the first diamond begins to fall will the two diamonds be 10 m apart?

99 A ball is thrown vertically downward from the top of a 36.6-m-tall building. The ball passes the top of a window that is 12.2 m above the ground 2.00 s after being thrown. What is the speed of the ball as it passes the top of the window?

100 A parachutist bails out and freely falls 50 m. Then the parachute opens, and thereafter she decelerates at 2.0 m/s^2. She reaches the ground with a speed of 3.0 m/s. (a) How long is the parachutist in the air? (b) At what height does the fall begin?

101 A ball is thrown *down* vertically with an initial *speed* of v_0 from a height of h. (a) What is its speed just before it strikes the ground? (b) How long does the ball take to reach the ground? What would be the answers to (c) part a and (d) part b if the ball were thrown *upward* from the same height and with the same initial speed? Before solving any equations, decide whether the answers to (c) and (d) should be greater than, less than, or the same as in (a) and (b).

102 The sport with the fastest moving ball is jai alai, where measured speeds have reached 303 km/h. If a professional jai alai player faces a ball at that speed and involuntarily blinks, he blacks out the scene for 100 ms. How far does the ball move during the blackout?

103 If a baseball pitcher throws a fastball at a horizontal speed of 160 km/h, how long does the ball take to reach home plate 18.4 m away?

104 A proton moves along the x axis according to the equation $x = 50t + 10t^2$, where x is in meters and t is in seconds. Calculate (a) the average velocity of the proton during the first 3.0 s of its motion, (b) the instantaneous velocity of the proton at $t = 3.0$ s, and (c) the instantaneous acceleration of the proton at $t = 3.0$ s. (d) Graph x versus t and indicate how the answer to (a) can be obtained from the plot. (e) Indicate the answer to (b) on the graph. (f) Plot v versus t and indicate on it the answer to (c).

105 A motorcycle is moving at 30 m/s when the rider applies the brakes, giving the motorcycle a constant deceleration. During the 3.0 s interval immediately after braking begins, the speed decreases to 15 m/s. What distance does the motorcycle travel from the instant braking begins until the motorcycle stops?

106 A shuffleboard disk is accelerated at a constant rate from rest to a speed of 6.0 m/s over a 1.8 m distance by a player using a cue. At this point the disk loses contact with the cue and slows at a constant rate of 2.5 m/s^2 until it stops. (a) How much time elapses from when the disk begins to accelerate until it stops? (b) What total distance does the disk travel?

107 The head of a rattlesnake can accelerate at 50 m/s^2 in striking a victim. If a car could do as well, how long would it take to reach a speed of 100 km/h from rest?

108 A jumbo jet must reach a speed of 360 km/h on the runway for takeoff. What is the lowest constant acceleration needed for takeoff from a 1.80 km runway?

109 An automobile driver increases the speed at a constant rate from 25 km/h to 55 km/h in 0.50 min. A bicycle rider speeds up at a constant rate from rest to 30 km/h in 0.50 min. What are the magnitudes of (a) the driver's acceleration and (b) the rider's acceleration?

110 On average, an eye blink lasts about 100 ms. How far does a MiG-25 "Foxbat" fighter travel during a pilot's blink if the plane's average velocity is 3400 km/h?

111 A certain sprinter has a top speed of 11.0 m/s. If the sprinter starts from rest and accelerates at a constant rate, he is able to reach his top speed in a distance of 12.0 m. He is then able to maintain this top speed for the remainder of a 100 m race. (a) What is his time for the 100 m race? (b) In order to improve his time, the sprinter tries to decrease the distance required for him to reach his top speed. What must this distance be if he is to achieve a time of 10.0 s for the race?

112 The speed of a bullet is measured to be 640 m/s as the bullet emerges from a barrel of length 1.20 m. Assuming constant acceleration, find the time that the bullet spends in the barrel after it is fired.

113 The Zero Gravity Research Facility at the NASA Glenn Research Center includes a 145 m drop tower. This is an evacuated vertical tower through which, among other possibilities, a 1-m-diameter sphere containing an experimental package can be dropped. (a) How long is the sphere in free fall? (b) What is its speed just as it reaches a catching device at the bottom of the tower? (c) When caught, the sphere experiences an average deceleration of $25g$ as its speed is reduced to zero. Through what distance does it travel during the deceleration?

114 A car can be braked to a stop from the autobahn-like speed of 200 km/h in 170 m. Assuming the acceleration is constant, find its magnitude in (a) SI units and (b) in terms of g. (c) How much time T_b is required for the braking? Your *reaction time* T_r is the time you require to perceive an emergency, move your foot to the brake, and begin the braking. If $T_r = 400$ ms, then (d) what is T_b in terms of T_r, and (e) is most of the full time required to stop spent in reacting or braking? Dark sunglasses delay the visual signals sent from the eyes to the visual cortex in the brain, increasing T_r. (f) In the extreme case in which T_r is increased by 100 ms, how much farther does the car travel during your reaction time?

115 In 1889, at Jubbulpore, India, a tug-of-war was finally won after 2 h 41 min, with the winning team displacing the center of the rope 3.7 m. In centimeters per minute, what was the magnitude of the average velocity of that center point during the contest?

116 Most important in an investigation of an airplane crash by the U.S. National Transportation Safety Board is the data stored on the airplane's flight-data recorder, commonly called the "black box" in spite of its orange coloring and reflective tape. The recorder is engineered to withstand a crash with an average deceleration of magnitude $3400g$ during a time interval of 6.50 ms. In such a crash, if the recorder and airplane have zero speed at the end of that time interval, what is their speed at the beginning of the interval?

117 From January 26, 1977, to September 18, 1983, George Meegan of Great Britain walked from Ushuaia, at the southern tip of South America, to Prudhoe Bay in Alaska, covering 30 600 km. In meters per second, what was the magnitude of his average velocity during that time period?

118 The wings on a stonefly do not flap, and thus the insect cannot fly. However, when the insect is on a water surface, it can sail across the surface by lifting its wings into a breeze. Suppose that you time stoneflies as they move at constant speed along a straight path of a certain length. On average, the trips each take 7.1 s with the wings set as sails and 25.0 s with the wings tucked in. (a) What is the ratio of the sailing speed v_s to the nonsailing speed v_{ns}? (b) In terms of v_s, what is the difference in the times the insects take to travel the first 2.0 m along the path with and without sailing?

119 The position of a particle as it moves along a y axis is given by

$$y = (2.0\text{ cm})\sin(\pi t/4),$$

with t in seconds and y in centimeters. (a) What is the average velocity of the particle between $t = 0$ and $t = 2.0$ s? (b) What is the instantaneous velocity of the particle at $t = 0$, 1.0, and 2.0 s? (c) What is the average acceleration of the particle between $t = 0$ and $t = 2.0$ s? (d) What is the instantaneous acceleration of the particle at $t = 0$, 1.0, and 2.0 s?

C H A P T E R 3

Vectors

3-1 VECTORS AND THEIR COMPONENTS

Learning Objectives

After reading this module, you should be able to . . .

3.01 Add vectors by drawing them in head-to-tail arrangements, applying the commutative and associative laws.

3.02 Subtract a vector from a second one.

3.03 Calculate the components of a vector on a given coordinate system, showing them in a drawing.

3.04 Given the components of a vector, draw the vector and determine its magnitude and orientation.

3.05 Convert angle measures between degrees and radians.

Key Ideas

- Scalars, such as temperature, have magnitude only. They are specified by a number with a unit (10°C) and obey the rules of arithmetic and ordinary algebra. Vectors, such as displacement, have both magnitude and direction (5 m, north) and obey the rules of vector algebra.
- Two vectors $\vec{a}$ and $\vec{b}$ may be added geometrically by drawing them to a common scale and placing them head to tail. The vector connecting the tail of the first to the head of the second is the vector sum $\vec{s}$. To subtract $\vec{b}$ from $\vec{a}$, reverse the direction of $\vec{b}$ to get $-\vec{b}$; then add $-\vec{b}$ to $\vec{a}$. Vector addition is commutative and obeys the associative law.
- The (scalar) components a_x and a_y of any two-dimensional vector $\vec{a}$ along the coordinate axes are found by dropping perpendicular lines from the ends of $\vec{a}$ onto the coordinate axes. The components are given by

$$a_x = a \cos \theta \quad \text{and} \quad a_y = a \sin \theta,$$

where θ is the angle between the positive direction of the x axis and the direction of $\vec{a}$. The algebraic sign of a component indicates its direction along the associated axis. Given its components, we can find the magnitude and orientation of the vector $\vec{a}$ with

$$a = \sqrt{a_x^2 + a_y^2} \quad \text{and} \quad \tan \theta = \frac{a_y}{a_x}.$$

What Is Physics?

Physics deals with a great many quantities that have both size and direction, and it needs a special mathematical language—the language of vectors—to describe those quantities. This language is also used in engineering, the other sciences, and even in common speech. If you have ever given directions such as "Go five blocks down this street and then hang a left," you have used the language of vectors. In fact, navigation of any sort is based on vectors, but physics and engineering also need vectors in special ways to explain phenomena involving rotation and magnetic forces, which we get to in later chapters. In this chapter, we focus on the basic language of vectors.

Vectors and Scalars

A particle moving along a straight line can move in only two directions. We can take its motion to be positive in one of these directions and negative in the other. For a particle moving in three dimensions, however, a plus sign or minus sign is no longer enough to indicate a direction. Instead, we must use a *vector.*

A **vector** has magnitude as well as direction, and vectors follow certain (vector) rules of combination, which we examine in this chapter. A **vector quantity** is a quantity that has both a magnitude and a direction and thus can be represented with a vector. Some physical quantities that are vector quantities are displacement, velocity, and acceleration. You will see many more throughout this book, so learning the rules of vector combination now will help you greatly in later chapters.

Not all physical quantities involve a direction. Temperature, pressure, energy, mass, and time, for example, do not "point" in the spatial sense. We call such quantities **scalars,** and we deal with them by the rules of ordinary algebra. A single value, with a sign (as in a temperature of $-40°F$), specifies a scalar.

The simplest vector quantity is displacement, or change of position. A vector that represents a displacement is called, reasonably, a **displacement vector.** (Similarly, we have velocity vectors and acceleration vectors.) If a particle changes its position by moving from A to B in Fig. 3-1a, we say that it undergoes a displacement from A to B, which we represent with an arrow pointing from A to B. The arrow specifies the vector graphically. To distinguish vector symbols from other kinds of arrows in this book, we use the outline of a triangle as the arrowhead.

In Fig. 3-1a, the arrows from A to B, from A' to B', and from A'' to B'' have the same magnitude and direction. Thus, they specify identical displacement vectors and represent the same *change of position* for the particle. A vector can be shifted without changing its value *if* its length and direction are not changed.

The displacement vector tells us nothing about the actual path that the particle takes. In Fig. 3-1b, for example, all three paths connecting points A and B correspond to the same displacement vector, that of Fig. 3-1a. Displacement vectors represent only the overall effect of the motion, not the motion itself.

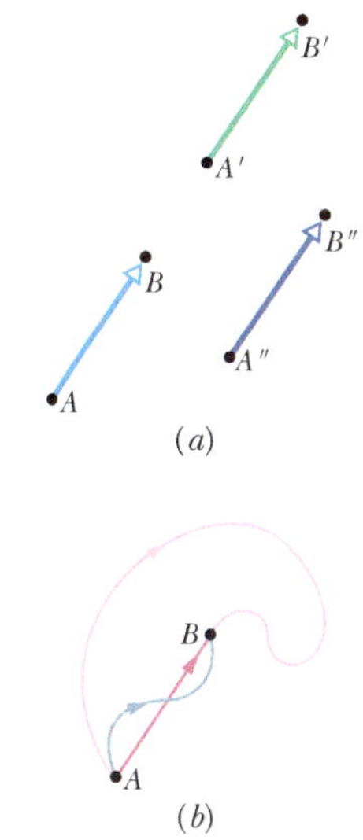

Figure 3-1 (a) All three arrows have the same magnitude and direction and thus represent the same displacement. (b) All three paths connecting the two points correspond to the same displacement vector.

Adding Vectors Geometrically

Suppose that, as in the vector diagram of Fig. 3-2a, a particle moves from A to B and then later from B to C. We can represent its overall displacement (no matter what its actual path) with two successive displacement vectors, AB and BC. The *net* displacement of these two displacements is a single displacement from A to C. We call AC the **vector sum** (or **resultant**) of the vectors AB and BC. This sum is not the usual algebraic sum.

In Fig. 3-2b, we redraw the vectors of Fig. 3-2a and relabel them in the way that we shall use from now on, namely, with an arrow over an italic symbol, as in $\vec{a}$. If we want to indicate only the magnitude of the vector (a quantity that lacks a sign or direction), we shall use the italic symbol, as in a, b, and s. (You can use just a handwritten symbol.) A symbol with an overhead arrow always implies both properties of a vector, magnitude and direction.

We can represent the relation among the three vectors in Fig. 3-2b with the *vector equation*

$$\vec{s} = \vec{a} + \vec{b}, \tag{3-1}$$

which says that the vector $\vec{s}$ is the vector sum of vectors $\vec{a}$ and $\vec{b}$. The symbol + in Eq. 3-1 and the words "sum" and "add" have different meanings for vectors than they do in the usual algebra because they involve both magnitude *and* direction.

Figure 3-2 suggests a procedure for adding two-dimensional vectors $\vec{a}$ and $\vec{b}$ geometrically. (1) On paper, sketch vector $\vec{a}$ to some convenient scale and at the proper angle. (2) Sketch vector $\vec{b}$ to the same scale, with its tail at the head of vector $\vec{a}$, again at the proper angle. (3) The vector sum $\vec{s}$ is the vector that extends from the tail of $\vec{a}$ to the head of $\vec{b}$.

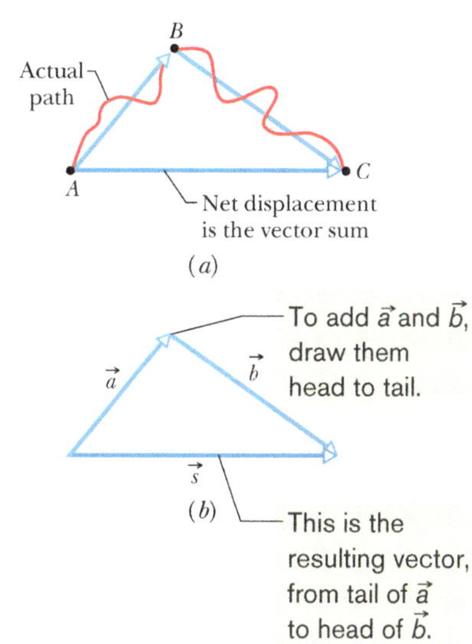

Figure 3-2 (a) AC is the vector sum of the vectors AB and BC. (b) The same vectors relabeled.

Properties. Vector addition, defined in this way, has two important properties. First, the order of addition does not matter. Adding $\vec{a}$ to $\vec{b}$ gives the same

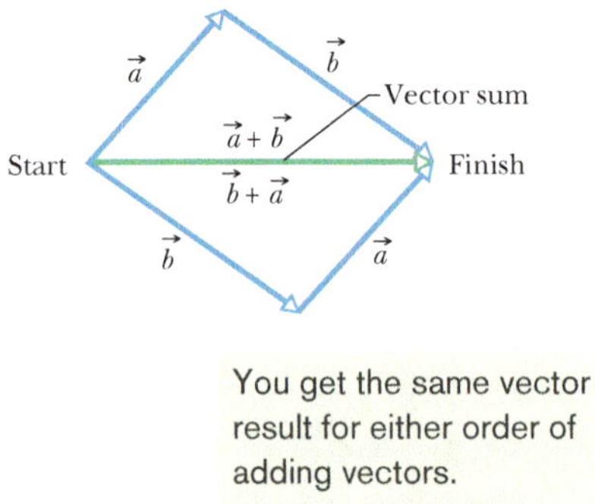

Figure 3-3 The two vectors $\vec{a}$ and $\vec{b}$ can be added in either order; see Eq. 3-2.

result as adding $\vec{b}$ to $\vec{a}$ (Fig. 3-3); that is,

$$\vec{a}+\vec{b}=\vec{b}+\vec{a} \quad \text{(commutative law).} \tag{3-2}$$

Second, when there are more than two vectors, we can group them in any order as we add them. Thus, if we want to add vectors $\vec{a}$, $\vec{b}$, and $\vec{c}$, we can add $\vec{a}$ and $\vec{b}$ first and then add their vector sum to $\vec{c}$. We can also add $\vec{b}$ and $\vec{c}$ first and then add *that* sum to $\vec{a}$. We get the same result either way, as shown in Fig. 3-4. That is,

$$(\vec{a}+\vec{b})+\vec{c}=\vec{a}+(\vec{b}+\vec{c}) \quad \text{(associative law).} \tag{3-3}$$

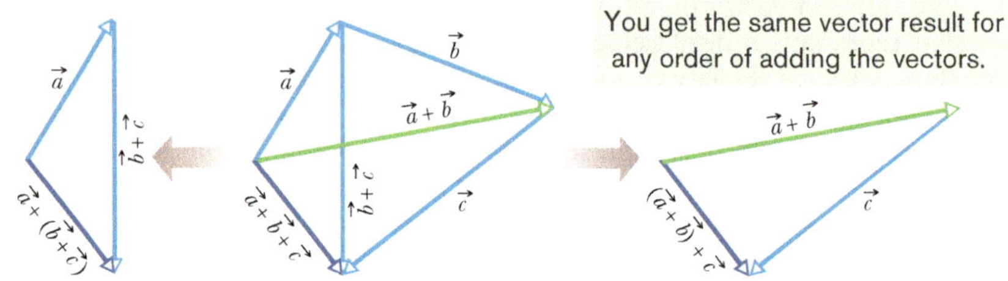

Figure 3-4 The three vectors $\vec{a}$, $\vec{b}$, and $\vec{c}$ can be grouped in any way as they are added; see Eq. 3-3.

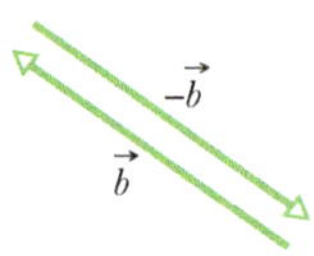

Figure 3-5 The vectors $\vec{b}$ and $-\vec{b}$ have the same magnitude and opposite directions.

The vector $-\vec{b}$ is a vector with the same magnitude as $\vec{b}$ but the opposite direction (see Fig. 3-5). Adding the two vectors in Fig. 3-5 would yield

$$\vec{b}+(-\vec{b})=0.$$

Thus, adding $-\vec{b}$ has the effect of subtracting $\vec{b}$. We use this property to define the difference between two vectors: let $\vec{d}=\vec{a}-\vec{b}$. Then

$$\vec{d}=\vec{a}-\vec{b}=\vec{a}+(-\vec{b}) \quad \text{(vector subtraction);} \tag{3-4}$$

that is, we find the difference vector $\vec{d}$ by adding the vector $-\vec{b}$ to the vector $\vec{a}$. Figure 3-6 shows how this is done geometrically.

As in the usual algebra, we can move a term that includes a vector symbol from one side of a vector equation to the other, but we must change its sign. For example, if we are given Eq. 3-4 and need to solve for $\vec{a}$, we can rearrange the equation as

$$\vec{d}+\vec{b}=\vec{a} \quad \text{or} \quad \vec{a}=\vec{d}+\vec{b}.$$

Remember that, although we have used displacement vectors here, the rules for addition and subtraction hold for vectors of all kinds, whether they represent velocities, accelerations, or any other vector quantity. However, we can add only vectors of the same kind. For example, we can add two displacements, or two velocities, but adding a displacement and a velocity makes no sense. In the arithmetic of scalars, that would be like trying to add 21 s and 12 m.

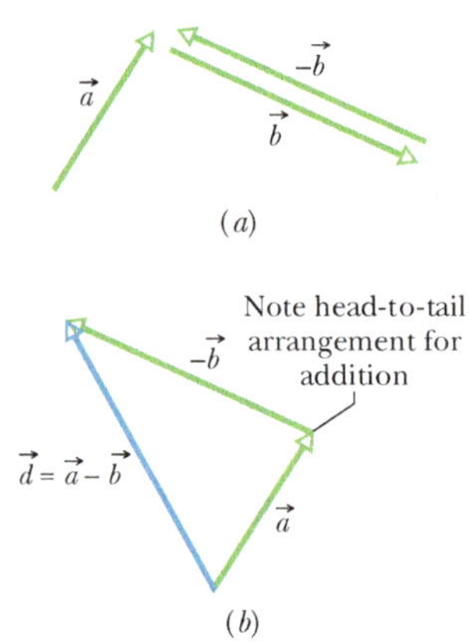

Figure 3-6 (*a*) Vectors $\vec{a}$, $\vec{b}$, and $-\vec{b}$. (*b*) To subtract vector $\vec{b}$ from vector $\vec{a}$, add vector $-\vec{b}$ to vector $\vec{a}$.

Checkpoint 1

The magnitudes of displacements $\vec{a}$ and $\vec{b}$ are 3 m and 4 m, respectively, and $\vec{c}=\vec{a}+\vec{b}$. Considering various orientations of $\vec{a}$ and $\vec{b}$, what are (a) the maximum possible magnitude for $\vec{c}$ and (b) the minimum possible magnitude?

Components of Vectors

Adding vectors geometrically can be tedious. A neater and easier technique involves algebra but requires that the vectors be placed on a rectangular coordinate system. The x and y axes are usually drawn in the plane of the page, as shown in Fig. 3-7a. The z axis comes directly out of the page at the origin; we ignore it for now and deal only with two-dimensional vectors.

A **component** of a vector is the projection of the vector on an axis. In Fig. 3-7a, for example, a_x is the component of vector $\vec{a}$ on (or along) the x axis and a_y is the component along the y axis. To find the projection of a vector along an axis, we draw perpendicular lines from the two ends of the vector to the axis, as shown. The projection of a vector on an x axis is its *x component*, and similarly the projection on the y axis is the *y component*. The process of finding the components of a vector is called **resolving the vector.**

A component of a vector has the same direction (along an axis) as the vector. In Fig. 3-7, a_x and a_y are both positive because $\vec{a}$ extends in the positive direction of both axes. (Note the small arrowheads on the components, to indicate their direction.) If we were to reverse vector $\vec{a}$, then both components would be negative and their arrowheads would point toward negative x and y. Resolving vector $\vec{b}$ in Fig. 3-8 yields a positive component b_x and a negative component b_y.

In general, a vector has three components, although for the case of Fig. 3-7a the component along the z axis is zero. As Figs. 3-7a and b show, if you shift a vector without changing its direction, its components do not change.

Finding the Components. We can find the components of $\vec{a}$ in Fig. 3-7a geometrically from the right triangle there:

$$a_x = a\cos\theta \quad \text{and} \quad a_y = a\sin\theta, \tag{3-5}$$

where θ is the angle that the vector $\vec{a}$ makes with the positive direction of the x axis, and a is the magnitude of $\vec{a}$. Figure 3-7c shows that $\vec{a}$ and its x and y components form a right triangle. It also shows how we can reconstruct a vector from its components: we arrange those components *head to tail*. Then we complete a right triangle with the vector forming the hypotenuse, from the tail of one component to the head of the other component.

Once a vector has been resolved into its components along a set of axes, the components themselves can be used in place of the vector. For example, $\vec{a}$ in Fig. 3-7a is given (completely determined) by a and θ. It can also be given by its components a_x and a_y. Both pairs of values contain the same information. If we know a vector in *component notation* (a_x and a_y) and want it in *magnitude-angle notation* (a and θ), we can use the equations

$$a = \sqrt{a_x^2 + a_y^2} \quad \text{and} \quad \tan\theta = \frac{a_y}{a_x} \tag{3-6}$$

to transform it.

In the more general three-dimensional case, we need a magnitude and two angles (say, a, θ, and ϕ) or three components (a_x, a_y, and a_z) to specify a vector.

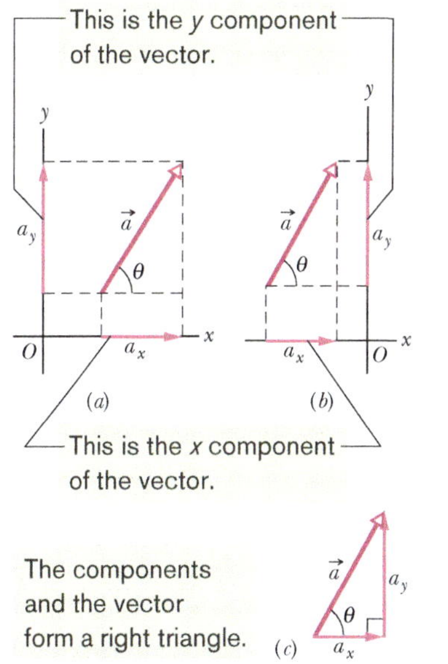

Figure 3-7 (a) The components a_x and a_y of vector $\vec{a}$. (b) The components are unchanged if the vector is shifted, as long as the magnitude and orientation are maintained. (c) The components form the legs of a right triangle whose hypotenuse is the magnitude of the vector.

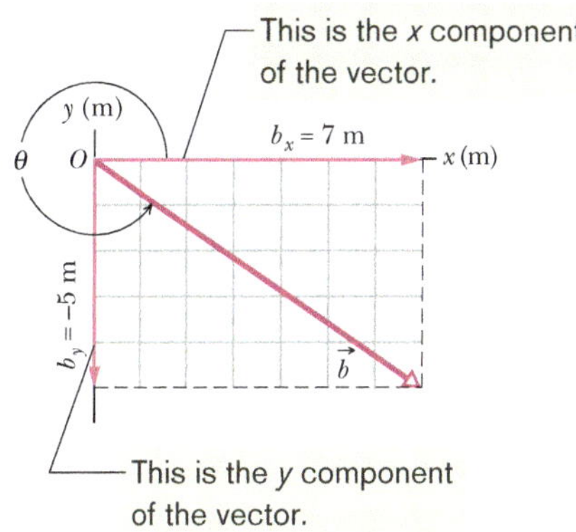

Figure 3-8 The component of $\vec{b}$ on the x axis is positive, and that on the y axis is negative.

Checkpoint 2

In the figure, which of the indicated methods for combining the x and y components of vector $\vec{a}$ are proper to determine that vector?

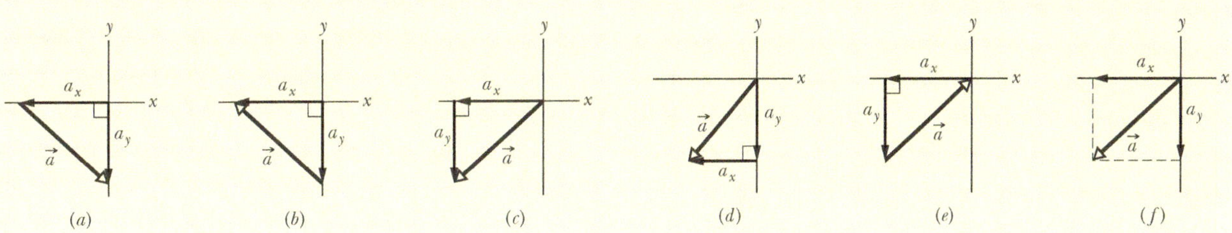

Sample Problem 3.01 Adding vectors in a drawing, orienteering

In an orienteering class, you have the goal of moving as far (straight-line distance) from base camp as possible by making three straight-line moves. You may use the following displacements in any order: (a) $\vec{a}$, 2.0 km due east (directly toward the east); (b) $\vec{b}$, 2.0 km 30° north of east (at an angle of 30° toward the north from due east); (c) $\vec{c}$, 1.0 km due west. Alternatively, you may substitute either $-\vec{b}$ for $\vec{b}$ or $-\vec{c}$ for $\vec{c}$. What is the greatest distance you can be from base camp at the end of the third displacement? (We are not concerned about the direction.)

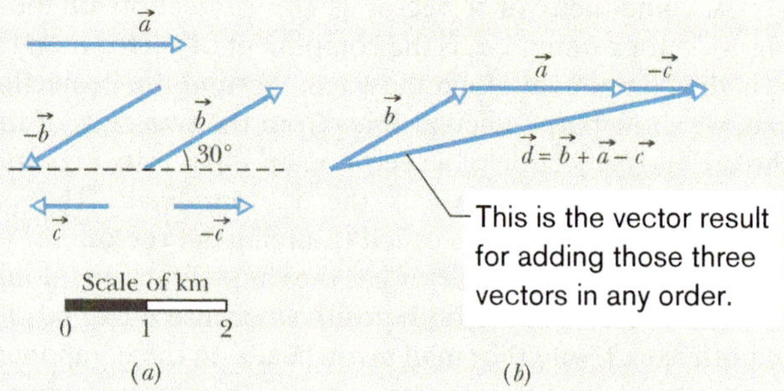

Figure 3-9 (*a*) Displacement vectors; three are to be used. (*b*) Your distance from base camp is greatest if you undergo displacements $\vec{a}$, $\vec{b}$, and $-\vec{c}$, in any order.

Reasoning: Using a convenient scale, we draw vectors $\vec{a}$, $\vec{b}$, $\vec{c}$, $-\vec{b}$, and $-\vec{c}$ as in Fig. 3-9*a*. We then mentally slide the vectors over the page, connecting three of them at a time in head-to-tail arrangements to find their vector sum $\vec{d}$. The tail of the first vector represents base camp. The head of the third vector represents the point at which you stop. The vector sum $\vec{d}$ extends from the tail of the first vector to the head of the third vector. Its magnitude d is your distance from base camp. Our goal here is to maximize that base-camp distance.

We find that distance d is greatest for a head-to-tail arrangement of vectors $\vec{a}$, $\vec{b}$, and $-\vec{c}$. They can be in any order, because their vector sum is the same for any order. (Recall from Eq. 3-2 that vectors commute.) The order shown in Fig. 3-9*b* is for the vector sum

$$\vec{d} = \vec{b} + \vec{a} + (-\vec{c}).$$

Using the scale given in Fig. 3-9*a*, we measure the length d of this vector sum, finding

$$d = 4.8 \text{ km}. \qquad \text{(Answer)}$$

Sample Problem 3.02 Finding components, airplane flight

A small airplane leaves an airport on an overcast day and is later sighted 215 km away, in a direction making an angle of 22° east of due north. This means that the direction is not due north (directly toward the north) but is rotated 22° toward the east from due north. How far east and north is the airplane from the airport when sighted?

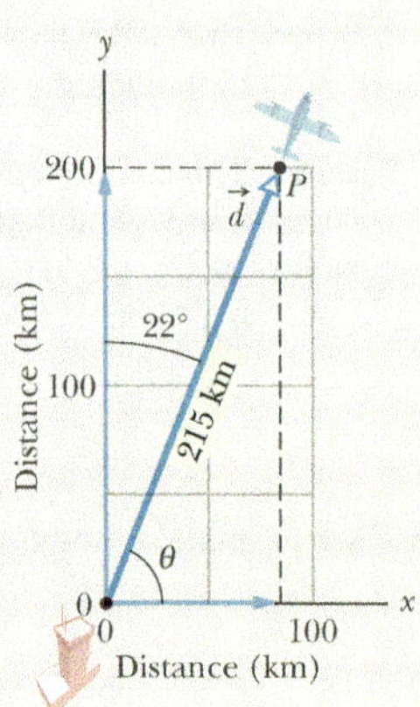

Figure 3-10 A plane takes off from an airport at the origin and is later sighted at *P*.

KEY IDEA

We are given the magnitude (215 km) and the angle (22° east of due north) of a vector and need to find the components of the vector.

Calculations: We draw an xy coordinate system with the positive direction of x due east and that of y due north (Fig. 3-10). For convenience, the origin is placed at the airport. (We don't have to do this. We could shift and misalign the coordinate system but, given a choice, why make the problem more difficult?) The airplane's displacement $\vec{d}$ points from the origin to where the airplane is sighted.

To find the components of $\vec{d}$, we use Eq. 3-5 with θ = 68° (= 90° − 22°):

$$d_x = d\cos\theta = (215 \text{ km})(\cos 68°)$$
$$= 81 \text{ km} \qquad \text{(Answer)}$$
$$d_y = d\sin\theta = (215 \text{ km})(\sin 68°)$$
$$= 199 \text{ km} \approx 2.0 \times 10^2 \text{ km}. \qquad \text{(Answer)}$$

Thus, the airplane is 81 km east and 2.0×10^2 km north of the airport.

WILEY PLUS Additional examples, video, and practice available at *WileyPLUS*

Problem-Solving Tactics Angles, trig functions, and inverse trig functions

Tactic 1: Angles—Degrees and Radians Angles that are measured relative to the positive direction of the x axis are positive if they are measured in the counterclockwise direction and negative if measured clockwise. For example, 210° and −150° are the same angle.

Angles may be measured in degrees or radians (rad). To relate the two measures, recall that a full circle is 360° and 2π rad. To convert, say, 40° to radians, write

$$40° \frac{2\pi \text{ rad}}{360°} = 0.70 \text{ rad}.$$

Tactic 2: Trig Functions You need to know the definitions of the common trigonometric functions—sine, cosine, and tangent—because they are part of the language of science and engineering. They are given in Fig. 3-11 in a form that does not depend on how the triangle is labeled.

You should also be able to sketch how the trig functions vary with angle, as in Fig. 3-12, in order to be able to judge whether a calculator result is reasonable. Even knowing the signs of the functions in the various quadrants can be of help.

Tactic 3: Inverse Trig Functions When the inverse trig functions $\sin^{-1}$, $\cos^{-1}$, and $\tan^{-1}$ are taken on a calculator, you must consider the reasonableness of the answer you get, because there is usually another possible answer that the calculator does not give. The range of operation for a calculator in taking each inverse trig function is indicated in Fig. 3-12. As an example, $\sin^{-1} 0.5$ has associated angles of 30° (which is displayed by the calculator, since 30° falls within its range of operation) and 150°. To see both values, draw a horizontal line through 0.5 in Fig. 3-12a and note where it cuts the sine curve. How do you distinguish a correct answer? It is the one that seems more reasonable for the given situation.

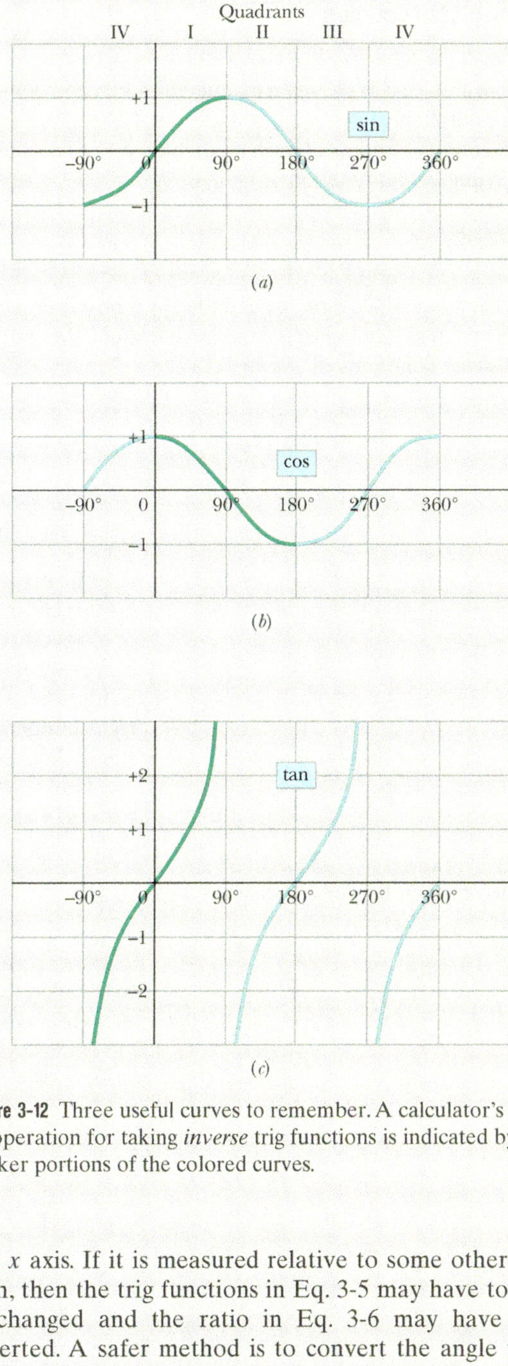

Figure 3-12 Three useful curves to remember. A calculator's range of operation for taking *inverse* trig functions is indicated by the darker portions of the colored curves.

Tactic 4: Measuring Vector Angles The equations for $\cos\theta$ and $\sin\theta$ in Eq. 3-5 and for $\tan\theta$ in Eq. 3-6 are valid only if the angle is measured from the positive direction of the x axis. If it is measured relative to some other direction, then the trig functions in Eq. 3-5 may have to be interchanged and the ratio in Eq. 3-6 may have to be inverted. A safer method is to convert the angle to one measured from the positive direction of the x axis. In *WileyPLUS*, the system expects you to report an angle of direction like this (and positive if counterclockwise and negative if clockwise).

$$\sin\theta = \frac{\text{leg opposite } \theta}{\text{hypotenuse}}$$

$$\cos\theta = \frac{\text{leg adjacent to } \theta}{\text{hypotenuse}}$$

$$\tan\theta = \frac{\text{leg opposite } \theta}{\text{leg adjacent to } \theta}$$

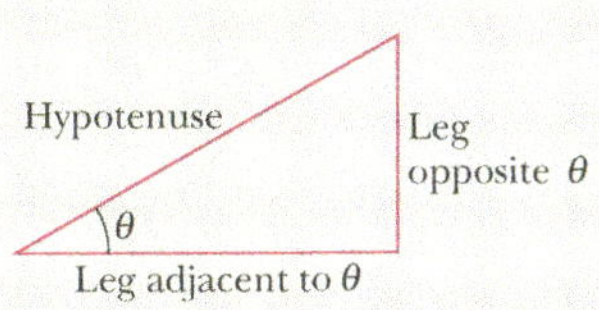

Figure 3-11 A triangle used to define the trigonometric functions. See also Appendix E.

3-2 UNIT VECTORS, ADDING VECTORS BY COMPONENTS

Learning Objectives

After reading this module, you should be able to . . .

3.06 Convert a vector between magnitude-angle and unit-vector notations.

3.07 Add and subtract vectors in magnitude-angle notation and in unit-vector notation.

3.08 Identify that, for a given vector, rotating the coordinate system about the origin can change the vector's components but not the vector itself.

Key Ideas

- Unit vectors $\hat{\mathrm{i}}$, $\hat{\mathrm{j}}$, and $\hat{\mathrm{k}}$ have magnitudes of unity and are directed in the positive directions of the x, y, and z axes, respectively, in a right-handed coordinate system. We can write a vector $\vec{a}$ in terms of unit vectors as

$$\vec{a} = a_x\hat{\mathrm{i}} + a_y\hat{\mathrm{j}} + a_z\hat{\mathrm{k}},$$

in which $a_x\hat{\mathrm{i}}$, $a_y\hat{\mathrm{j}}$, and $a_z\hat{\mathrm{k}}$ are the vector components of $\vec{a}$ and a_x, a_y, and a_z are its scalar components.

- To add vectors in component form, we use the rules

$$r_x = a_x + b_x \quad r_y = a_y + b_y \quad r_z = a_z + b_z.$$

Here $\vec{a}$ and $\vec{b}$ are the vectors to be added, and $\vec{r}$ is the vector sum. Note that we add components axis by axis.

Unit Vectors

A **unit vector** is a vector that has a magnitude of exactly 1 and points in a particular direction. It lacks both dimension and unit. Its sole purpose is to point—that is, to specify a direction. The unit vectors in the positive directions of the x, y, and z axes are labeled $\hat{\mathrm{i}}$, $\hat{\mathrm{j}}$, and $\hat{\mathrm{k}}$, where the hat ^ is used instead of an overhead arrow as for other vectors (Fig. 3-13). The arrangement of axes in Fig. 3-13 is said to be a **right-handed coordinate system.** The system remains right-handed if it is rotated rigidly. We use such coordinate systems exclusively in this book.

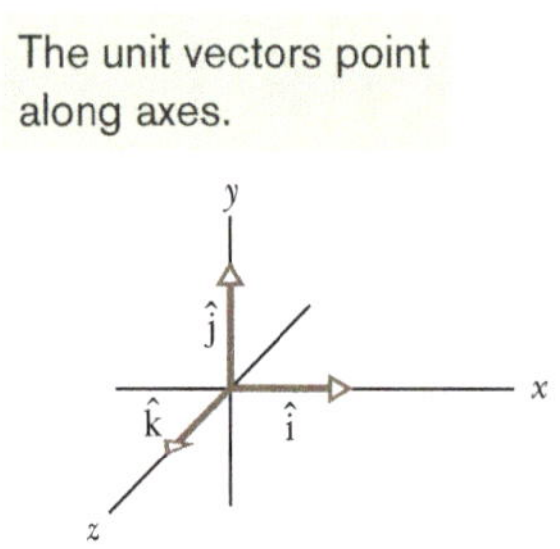

Figure 3.13 Unit vectors $\hat{\mathrm{i}}$, $\hat{\mathrm{j}}$, and $\hat{\mathrm{k}}$ define the directions of a right-handed coordinate system.

Unit vectors are very useful for expressing other vectors; for example, we can express $\vec{a}$ and $\vec{b}$ of Figs. 3-7 and 3-8 as

$$\vec{a} = a_x\hat{\mathrm{i}} + a_y\hat{\mathrm{j}} \tag{3-7}$$

and

$$\vec{b} = b_x\hat{\mathrm{i}} + b_y\hat{\mathrm{j}}. \tag{3-8}$$

These two equations are illustrated in Fig. 3-14. The quantities $a_x\hat{\mathrm{i}}$ and $a_y\hat{\mathrm{j}}$ are vectors, called the **vector components** of $\vec{a}$. The quantities a_x and a_y are scalars, called the **scalar components** of $\vec{a}$ (or, as before, simply its **components**).

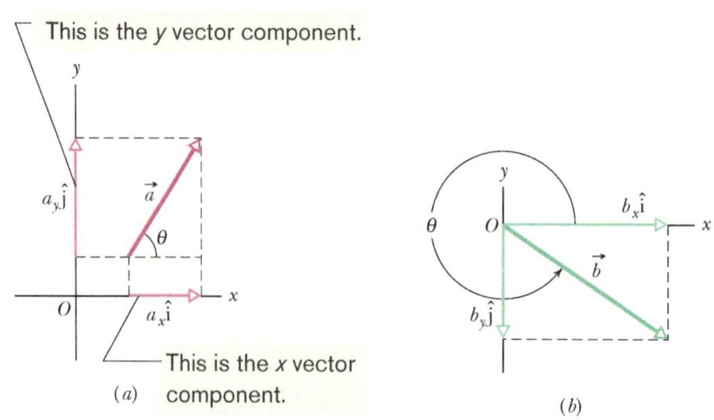

Figure 3-14 (*a*) The vector components of vector $\vec{a}$. (*b*) The vector components of vector $\vec{b}$.

Adding Vectors by Components

We can add vectors geometrically on a sketch or directly on a vector-capable calculator. A third way is to combine their components axis by axis.

To start, consider the statement

$$\vec{r} = \vec{a} + \vec{b}, \tag{3-9}$$

which says that the vector $\vec{r}$ is the same as the vector $(\vec{a} + \vec{b})$. Thus, each component of $\vec{r}$ must be the same as the corresponding component of $(\vec{a} + \vec{b})$:

$$r_x = a_x + b_x \tag{3-10}$$

$$r_y = a_y + b_y \tag{3-11}$$

$$r_z = a_z + b_z. \tag{3-12}$$

In other words, two vectors must be equal if their corresponding components are equal. Equations 3-9 to 3-12 tell us that to add vectors $\vec{a}$ and $\vec{b}$, we must (1) resolve the vectors into their scalar components; (2) combine these scalar components, axis by axis, to get the components of the sum $\vec{r}$; and (3) combine the components of $\vec{r}$ to get $\vec{r}$ itself. We have a choice in step 3. We can express $\vec{r}$ in unit-vector notation or in magnitude-angle notation.

This procedure for adding vectors by components also applies to vector subtractions. Recall that a subtraction such as $\vec{d} = \vec{a} - \vec{b}$ can be rewritten as an addition $\vec{d} = \vec{a} + (-\vec{b})$. To subtract, we add $\vec{a}$ and $-\vec{b}$ by components, to get

$$d_x = a_x - b_x, \quad d_y = a_y - b_y, \quad \text{and} \quad d_z = a_z - b_z,$$

where

$$\vec{d} = d_x\hat{\text{i}} + d_y\hat{\text{j}} + d_z\hat{\text{k}}. \tag{3-13}$$

Checkpoint 3

(a) In the figure here, what are the signs of the x components of $\vec{d}_1$ and $\vec{d}_2$? (b) What are the signs of the y components of $\vec{d}_1$ and $\vec{d}_2$? (c) What are the signs of the x and y components of $\vec{d}_1 + \vec{d}_2$?

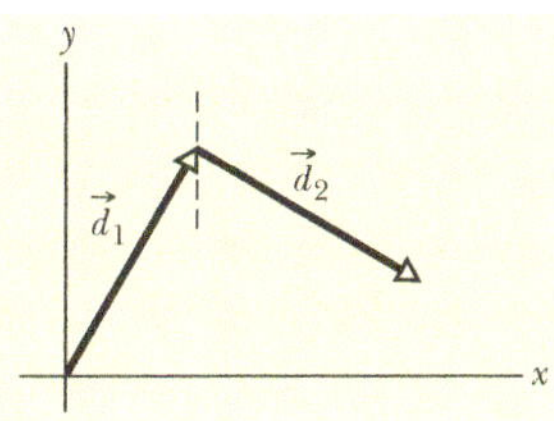

Vectors and the Laws of Physics

So far, in every figure that includes a coordinate system, the x and y axes are parallel to the edges of the book page. Thus, when a vector $\vec{a}$ is included, its components a_x and a_y are also parallel to the edges (as in Fig. 3-15*a*). The only reason for that orientation of the axes is that it looks "proper"; there is no deeper reason. We could, instead, rotate the axes (but not the vector $\vec{a}$) through an angle ϕ as in Fig. 3-15*b*, in which case the components would have new values, call them a'_x and a'_y. Since there are an infinite number of choices of ϕ, there are an infinite number of different pairs of components for $\vec{a}$.

Which then is the "right" pair of components? The answer is that they are all equally valid because each pair (with its axes) just gives us a different way of describing the same vector $\vec{a}$; all produce the same magnitude and direction for the vector. In Fig. 3-15 we have

$$a = \sqrt{a_x^2 + a_y^2} = \sqrt{a_x'^2 + a_y'^2} \tag{3-14}$$

and

$$\theta = \theta' + \phi. \tag{3-15}$$

The point is that we have great freedom in choosing a coordinate system, because the relations among vectors do not depend on the location of the origin or on the orientation of the axes. This is also true of the relations of physics; they are all independent of the choice of coordinate system. Add to that the simplicity and richness of the language of vectors and you can see why the laws of physics are almost always presented in that language: one equation, like Eq. 3-9, can represent three (or even more) relations, like Eqs. 3-10, 3-11, and 3-12.

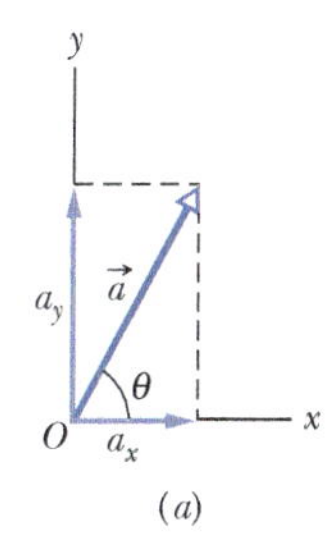

Rotating the axes changes the components but not the vector.

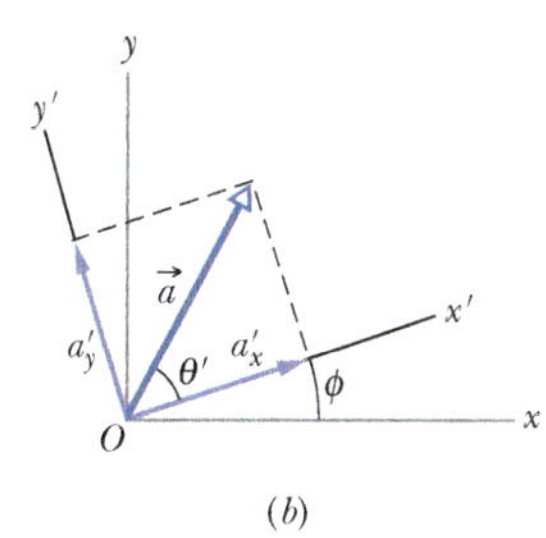

Figure 3-15 (*a*) The vector $\vec{a}$ and its components. (*b*) The same vector, with the axes of the coordinate system rotated through an angle ϕ.

Sample Problem 3.03 Searching through a hedge maze

A hedge maze is a maze formed by tall rows of hedge. After entering, you search for the center point and then for the exit. Figure 3-16*a* shows the entrance to such a maze and the first two choices we make at the junctions we encounter in moving from point *i* to point *c*. We undergo three displacements as indicated in the overhead view of Fig. 3-16*b*:

$$d_1 = 6.00 \text{ m} \qquad \theta_1 = 40°$$
$$d_2 = 8.00 \text{ m} \qquad \theta_2 = 30°$$
$$d_3 = 5.00 \text{ m} \qquad \theta_3 = 0°,$$

where the last segment is parallel to the superimposed *x* axis. When we reach point *c*, what are the magnitude and angle of our net displacement $\vec{d}_{net}$ from point *i*?

KEY IDEAS

(1) To find the net displacement $\vec{d}_{net}$, we need to sum the three individual displacement vectors:

$$\vec{d}_{net} = \vec{d}_1 + \vec{d}_2 + \vec{d}_3.$$

(2) To do this, we first evaluate this sum for the *x* components alone,

$$d_{net,x} = d_{1x} + d_{2x} + d_{3x}, \qquad (3\text{-}16)$$

and then the *y* components alone,

$$d_{net,y} = d_{1y} + d_{2y} + d_{3y}. \qquad (3\text{-}17)$$

(3) Finally, we construct $\vec{d}_{net}$ from its *x* and *y* components.

Calculations: To evaluate Eqs. 3-16 and 3-17, we find the *x* and *y* components of each displacement. As an example, the components for the first displacement are shown in Fig. 3-16*c*. We draw similar diagrams for the other two displacements and then we apply the *x* part of Eq. 3-5 to each displacement, using angles relative to the positive direction of the *x* axis:

$$d_{1x} = (6.00 \text{ m}) \cos 40° = 4.60 \text{ m}$$
$$d_{2x} = (8.00 \text{ m}) \cos(-60°) = 4.00 \text{ m}$$
$$d_{3x} = (5.00 \text{ m}) \cos 0° = 5.00 \text{ m}.$$

Equation 3-16 then gives us

$$d_{net,x} = +4.60 \text{ m} + 4.00 \text{ m} + 5.00 \text{ m} = 13.60 \text{ m}.$$

Similarly, to evaluate Eq. 3-17, we apply the *y* part of Eq. 3-5 to each displacement:

$$d_{1y} = (6.00 \text{ m}) \sin 40° = 3.86 \text{ m}$$
$$d_{2y} = (8.00 \text{ m}) \sin(-60°) = -6.93 \text{ m}$$
$$d_{3y} = (5.00 \text{ m}) \sin 0° = 0 \text{ m}.$$

Equation 3-17 then gives us

$$d_{net,y} = +3.86 \text{ m} - 6.93 \text{ m} + 0 \text{ m} = -3.07 \text{ m}.$$

Next we use these components of $\vec{d}_{net}$ to construct the vector as shown in Fig. 3-16*d*: the components are in a head-to-tail arrangement and form the legs of a right triangle, and

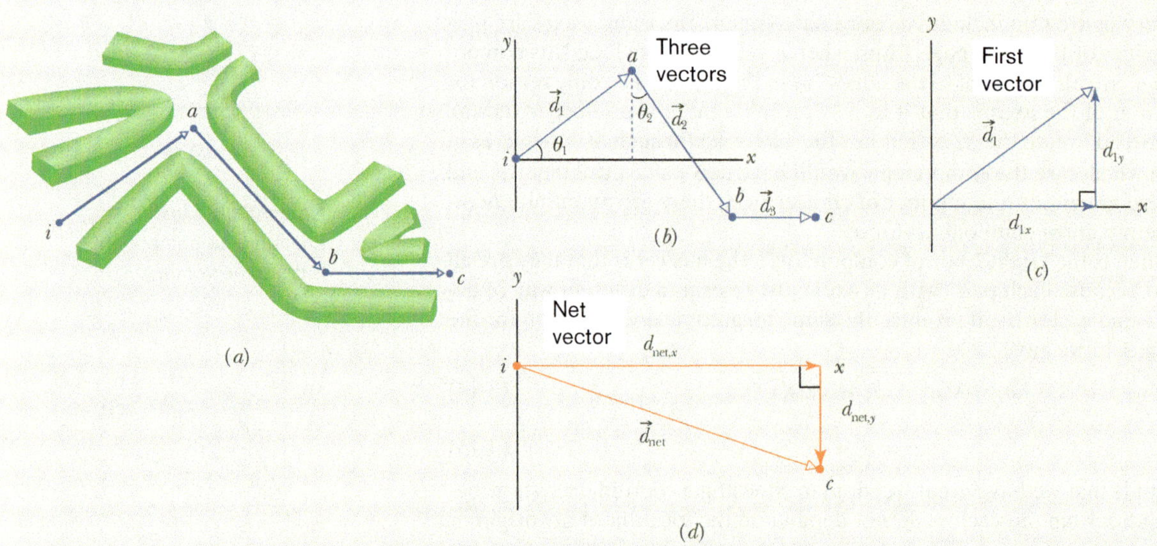

Figure 3-16 (*a*) Three displacements through a hedge maze. (*b*) The displacement vectors. (*c*) The first displacement vector and its components. (*d*) The net displacement vector and its components.

the vector forms the hypotenuse. We find the magnitude and angle of $\vec{d}_{net}$ with Eq. 3-6. The magnitude is

$$d_{net} = \sqrt{d_{net,x}^2 + d_{net,y}^2} \tag{3-18}$$

$$= \sqrt{(13.60 \text{ m})^2 + (-3.07 \text{ m})^2} = 13.9 \text{ m}. \quad \text{(Answer)}$$

To find the angle (measured from the positive direction of x), we take an inverse tangent:

$$\theta = \tan^{-1}\left(\frac{d_{net,y}}{d_{net,x}}\right) \tag{3-19}$$

$$= \tan^{-1}\left(\frac{-3.07 \text{ m}}{13.60 \text{ m}}\right) = -12.7°. \quad \text{(Answer)}$$

The angle is negative because it is measured clockwise from positive x. We must always be alert when we take an inverse tangent on a calculator. The answer it displays is mathematically correct but it may not be the correct answer for the physical situation. In those cases, we have to add 180° to the displayed answer, to reverse the vector. To check, we always need to draw the vector and its components as we did in Fig. 3-16*d*. In our physical situation, the figure shows us that $\theta = -12.7°$ is a reasonable answer, whereas $-12.7° + 180° = 167°$ is clearly not.

We can see all this on the graph of tangent versus angle in Fig. 3-12*c*. In our maze problem, the argument of the inverse tangent is $-3.07/13.60$, or -0.226. On the graph draw a horizontal line through that value on the vertical axis. The line cuts through the darker plotted branch at $-12.7°$ and also through the lighter branch at 167°. The first cut is what a calculator displays.

Sample Problem 3.04 Adding vectors, unit-vector components

Figure 3-17*a* shows the following three vectors:

$$\vec{a} = (4.2 \text{ m})\hat{i} - (1.5 \text{ m})\hat{j},$$

$$\vec{b} = (-1.6 \text{ m})\hat{i} + (2.9 \text{ m})\hat{j},$$

and

$$\vec{c} = (-3.7 \text{ m})\hat{j}.$$

What is their vector sum $\vec{r}$ which is also shown?

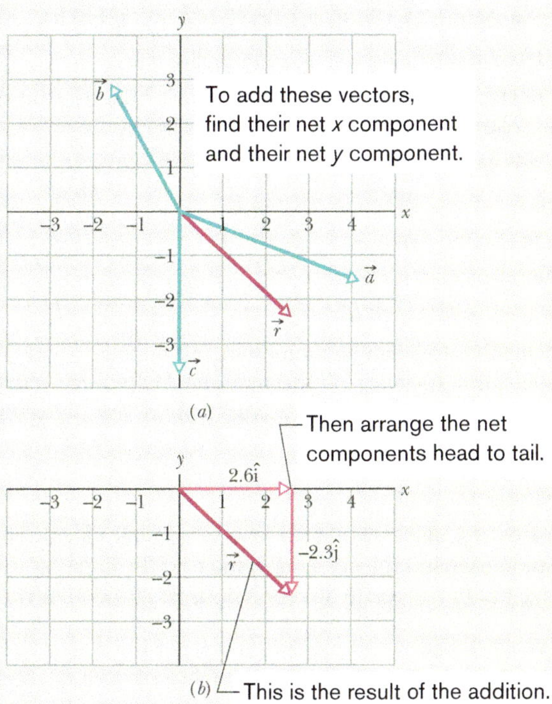

Figure 3-17 Vector $\vec{r}$ is the vector sum of the other three vectors.

KEY IDEA

We can add the three vectors by components, axis by axis, and then combine the components to write the vector sum $\vec{r}$.

Calculations: For the x axis, we add the x components of $\vec{a}$, $\vec{b}$, and $\vec{c}$, to get the x component of the vector sum $\vec{r}$:

$$r_x = a_x + b_x + c_x$$

$$= 4.2 \text{ m} - 1.6 \text{ m} + 0 = 2.6 \text{ m}.$$

Similarly, for the y axis,

$$r_y = a_y + b_y + c_y$$

$$= -1.5 \text{ m} + 2.9 \text{ m} - 3.7 \text{ m} = -2.3 \text{ m}.$$

We then combine these components of $\vec{r}$ to write the vector in unit-vector notation:

$$\vec{r} = (2.6 \text{ m})\hat{i} - (2.3 \text{ m})\hat{j}, \quad \text{(Answer)}$$

where $(2.6 \text{ m})\hat{i}$ is the vector component of $\vec{r}$ along the x axis and $-(2.3 \text{ m})\hat{j}$ is that along the y axis. Figure 3-17*b* shows one way to arrange these vector components to form $\vec{r}$. (Can you sketch the other way?)

We can also answer the question by giving the magnitude and an angle for $\vec{r}$. From Eq. 3-6, the magnitude is

$$r = \sqrt{(2.6 \text{ m})^2 + (-2.3 \text{ m})^2} \approx 3.5 \text{ m} \quad \text{(Answer)}$$

and the angle (measured from the $+x$ direction) is

$$\theta = \tan^{-1}\left(\frac{-2.3 \text{ m}}{2.6 \text{ m}}\right) = -41°, \quad \text{(Answer)}$$

where the minus sign means clockwise.

WileyPLUS Additional examples, video, and practice available at *WileyPLUS*

3-3 MULTIPLYING VECTORS

Learning Objectives

After reading this module, you should be able to . . .

3.09 Multiply vectors by scalars.

3.10 Identify that multiplying a vector by a scalar gives a vector, taking the dot (or scalar) product of two vectors gives a scalar, and taking the cross (or vector) product gives a new vector that is perpendicular to the original two.

3.11 Find the dot product of two vectors in magnitude-angle notation and in unit-vector notation.

3.12 Find the angle between two vectors by taking their dot product in both magnitude-angle notation and unit-vector notation.

3.13 Given two vectors, use a dot product to find how much of one vector lies along the other vector.

3.14 Find the cross product of two vectors in magnitude-angle and unit-vector notations.

3.15 Use the right-hand rule to find the direction of the vector that results from a cross product.

3.16 In nested products, where one product is buried inside another, follow the normal algebraic procedure by starting with the innermost product and working outward.

Key Ideas

- The product of a scalar s and a vector $\vec{v}$ is a new vector whose magnitude is sv and whose direction is the same as that of $\vec{v}$ if s is positive, and opposite that of $\vec{v}$ if s is negative. To divide $\vec{v}$ by s, multiply $\vec{v}$ by $1/s$.
- The scalar (or dot) product of two vectors $\vec{a}$ and $\vec{b}$ is written $\vec{a}\cdot\vec{b}$ and is the *scalar* quantity given by

$$\vec{a}\cdot\vec{b} = ab\cos\phi,$$

in which ϕ is the angle between the directions of $\vec{a}$ and $\vec{b}$. A scalar product is the product of the magnitude of one vector and the scalar component of the second vector along the direction of the first vector. In unit-vector notation,

$$\vec{a}\cdot\vec{b} = (a_x\hat{\text{i}} + a_y\hat{\text{j}} + a_z\hat{\text{k}})\cdot(b_x\hat{\text{i}} + b_y\hat{\text{j}} + b_z\hat{\text{k}}),$$

which may be expanded according to the distributive law. Note that $\vec{a}\cdot\vec{b} = \vec{b}\cdot\vec{a}$.

- The vector (or cross) product of two vectors $\vec{a}$ and $\vec{b}$ is written $\vec{a}\times\vec{b}$ and is a *vector* $\vec{c}$ whose magnitude c is given by

$$c = ab\sin\phi,$$

in which ϕ is the smaller of the angles between the directions of $\vec{a}$ and $\vec{b}$. The direction of $\vec{c}$ is perpendicular to the plane defined by $\vec{a}$ and $\vec{b}$ and is given by a right-hand rule, as shown in Fig. 3-19. Note that $\vec{a}\times\vec{b} = -(\vec{b}\times\vec{a})$. In unit-vector notation,

$$\vec{a}\times\vec{b} = (a_x\hat{\text{i}} + a_y\hat{\text{j}} + a_z\hat{\text{k}})\times(b_x\hat{\text{i}} + b_y\hat{\text{j}} + b_z\hat{\text{k}}),$$

which we may expand with the distributive law.

- In nested products, where one product is buried inside another, follow the normal algebraic procedure by starting with the innermost product and working outward.

Multiplying Vectors*

There are three ways in which vectors can be multiplied, but none is exactly like the usual algebraic multiplication. As you read this material, keep in mind that a vector-capable calculator will help you multiply vectors only if you understand the basic rules of that multiplication.

Multiplying a Vector by a Scalar

If we multiply a vector $\vec{a}$ by a scalar s, we get a new vector. Its magnitude is the product of the magnitude of $\vec{a}$ and the absolute value of s. Its direction is the direction of $\vec{a}$ if s is positive but the opposite direction if s is negative. To divide $\vec{a}$ by s, we multiply $\vec{a}$ by $1/s$.

Multiplying a Vector by a Vector

There are two ways to multiply a vector by a vector: one way produces a scalar (called the *scalar product*), and the other produces a new vector (called the *vector product*). (Students commonly confuse the two ways.)

*This material will not be employed until later (Chapter 7 for scalar products and Chapter 11 for vector products), and so your instructor may wish to postpone it.

The Scalar Product

The **scalar product** of the vectors $\vec{a}$ and $\vec{b}$ in Fig. 3-18*a* is written as $\vec{a}\cdot\vec{b}$ and defined to be

$$\vec{a}\cdot\vec{b} = ab\cos\phi, \tag{3-20}$$

where a is the magnitude of $\vec{a}$, b is the magnitude of $\vec{b}$, and ϕ is the angle between $\vec{a}$ and $\vec{b}$ (or, more properly, between the directions of $\vec{a}$ and $\vec{b}$). There are actually two such angles: ϕ and $360° - \phi$. Either can be used in Eq. 3-20, because their cosines are the same.

Note that there are only scalars on the right side of Eq. 3-20 (including the value of $\cos\phi$). Thus $\vec{a}\cdot\vec{b}$ on the left side represents a *scalar* quantity. Because of the notation, $\vec{a}\cdot\vec{b}$ is also known as the **dot product** and is spoken as "a dot b."

A dot product can be regarded as the product of two quantities: (1) the magnitude of one of the vectors and (2) the scalar component of the second vector along the direction of the first vector. For example, in Fig. 3-18*b*, $\vec{a}$ has a scalar component $a\cos\phi$ along the direction of $\vec{b}$; note that a perpendicular dropped from the head of $\vec{a}$ onto $\vec{b}$ determines that component. Similarly, $\vec{b}$ has a scalar component $b\cos\phi$ along the direction of $\vec{a}$.

If the angle ϕ between two vectors is 0°, the component of one vector along the other is maximum, and so also is the dot product of the vectors. If, instead, ϕ is 90°, the component of one vector along the other is zero, and so is the dot product.

Equation 3-20 can be rewritten as follows to emphasize the components:

$$\vec{a}\cdot\vec{b} = (a\cos\phi)(b) = (a)(b\cos\phi). \tag{3-21}$$

The commutative law applies to a scalar product, so we can write

$$\vec{a}\cdot\vec{b} = \vec{b}\cdot\vec{a}.$$

When two vectors are in unit-vector notation, we write their dot product as

$$\vec{a}\cdot\vec{b} = (a_x\hat{i} + a_y\hat{j} + a_z\hat{k})\cdot(b_x\hat{i} + b_y\hat{j} + b_z\hat{k}), \tag{3-22}$$

which we can expand according to the distributive law: Each vector component of the first vector is to be dotted with each vector component of the second vector. By doing so, we can show that

$$\vec{a}\cdot\vec{b} = a_xb_x + a_yb_y + a_zb_z. \tag{3-23}$$

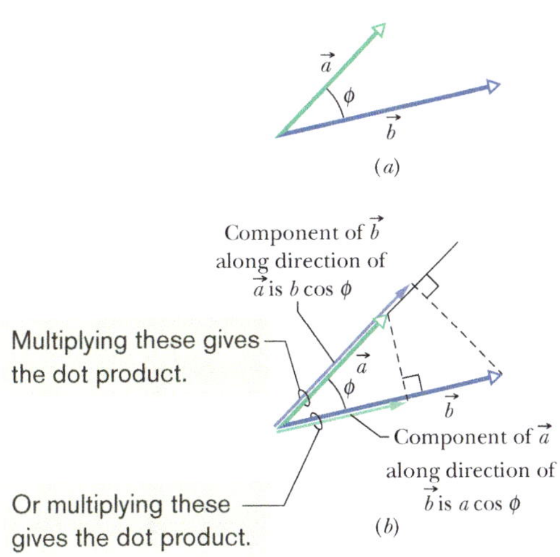

Figure 3-18 (*a*) Two vectors $\vec{a}$ and $\vec{b}$, with an angle ϕ between them. (*b*) Each vector has a component along the direction of the other vector.

Checkpoint 4

Vectors $\vec{C}$ and $\vec{D}$ have magnitudes of 3 units and 4 units, respectively. What is the angle between the directions of $\vec{C}$ and $\vec{D}$ if $\vec{C}\cdot\vec{D}$ equals (a) zero, (b) 12 units, and (c) −12 units?

The Vector Product

The **vector product** of $\vec{a}$ and $\vec{b}$, written $\vec{a}\times\vec{b}$, produces a third vector $\vec{c}$ whose magnitude is

$$c = ab\sin\phi, \tag{3-24}$$

where ϕ is the *smaller* of the two angles between $\vec{a}$ and $\vec{b}$. (You must use the smaller of the two angles between the vectors because $\sin\phi$ and $\sin(360° - \phi)$ differ in algebraic sign.) Because of the notation, $\vec{a}\times\vec{b}$ is also known as the **cross product,** and in speech it is "a cross b."

If $\vec{a}$ and $\vec{b}$ are parallel or antiparallel, $\vec{a}\times\vec{b} = 0$. The magnitude of $\vec{a}\times\vec{b}$, which can be written as $|\vec{a}\times\vec{b}|$, is maximum when $\vec{a}$ and $\vec{b}$ are perpendicular to each other.

The direction of $\vec{c}$ is perpendicular to the plane that contains $\vec{a}$ and $\vec{b}$. Figure 3-19*a* shows how to determine the direction of $\vec{c} = \vec{a}\times\vec{b}$ with what is known as a **right-hand rule.** Place the vectors $\vec{a}$ and $\vec{b}$ tail to tail without altering their orientations, and imagine a line that is perpendicular to their plane where they meet. Pretend to place your *right* hand around that line in such a way that your fingers would sweep $\vec{a}$ into $\vec{b}$ through the smaller angle between them. Your outstretched thumb points in the direction of $\vec{c}$.

The order of the vector multiplication is important. In Fig. 3-19*b*, we are determining the direction of $\vec{c}\,' = \vec{b}\times\vec{a}$, so the fingers are placed to sweep $\vec{b}$ into $\vec{a}$ through the smaller angle. The thumb ends up in the opposite direction from previously, and so it must be that $\vec{c}\,' = -\vec{c}$; that is,

$$\vec{b}\times\vec{a} = -(\vec{a}\times\vec{b}). \tag{3-25}$$

In other words, the commutative law does not apply to a vector product.

In unit-vector notation, we write

$$\vec{a}\times\vec{b} = (a_x\hat{\mathrm{i}} + a_y\hat{\mathrm{j}} + a_z\hat{\mathrm{k}})\times(b_x\hat{\mathrm{i}} + b_y\hat{\mathrm{j}} + b_z\hat{\mathrm{k}}), \tag{3-26}$$

which can be expanded according to the distributive law; that is, each component of the first vector is to be crossed with each component of the second vector. The cross products of unit vectors are given in Appendix E (see "Products of Vectors"). For example, in the expansion of Eq. 3-26, we have

$$a_x\hat{\mathrm{i}}\times b_x\hat{\mathrm{i}} = a_x b_x(\hat{\mathrm{i}}\times\hat{\mathrm{i}}) = 0,$$

because the two unit vectors $\hat{\mathrm{i}}$ and $\hat{\mathrm{i}}$ are parallel and thus have a zero cross product. Similarly, we have

$$a_x\hat{\mathrm{i}}\times b_y\hat{\mathrm{j}} = a_x b_y(\hat{\mathrm{i}}\times\hat{\mathrm{j}}) = a_x b_y\hat{\mathrm{k}}.$$

In the last step we used Eq. 3-24 to evaluate the magnitude of $\hat{\mathrm{i}}\times\hat{\mathrm{j}}$ as unity. (These vectors $\hat{\mathrm{i}}$ and $\hat{\mathrm{j}}$ each have a magnitude of unity, and the angle between them is 90°.) Also, we used the right-hand rule to get the direction of $\hat{\mathrm{i}}\times\hat{\mathrm{j}}$ as being in the positive direction of the z axis (thus in the direction of $\hat{\mathrm{k}}$).

Continuing to expand Eq. 3-26, you can show that

$$\vec{a} \times \vec{b} = (a_y b_z - b_y a_z)\hat{\text{i}} + (a_z b_x - b_z a_x)\hat{\text{j}} + (a_x b_y - b_x a_y)\hat{\text{k}}. \qquad (3\text{-}27)$$

A determinant (Appendix E) or a vector-capable calculator can also be used.

To check whether any xyz coordinate system is a right-handed coordinate system, use the right-hand rule for the cross product $\hat{\text{i}} \times \hat{\text{j}} = \hat{\text{k}}$ with that system. If your fingers sweep $\hat{\text{i}}$ (positive direction of x) into $\hat{\text{j}}$ (positive direction of y) with the outstretched thumb pointing in the positive direction of z (not the negative direction), then the system is right-handed.

Checkpoint 5

Vectors $\vec{C}$ and $\vec{D}$ have magnitudes of 3 units and 4 units, respectively. What is the angle between the directions of $\vec{C}$ and $\vec{D}$ if the magnitude of the vector product $\vec{C} \times \vec{D}$ is (a) zero and (b) 12 units?

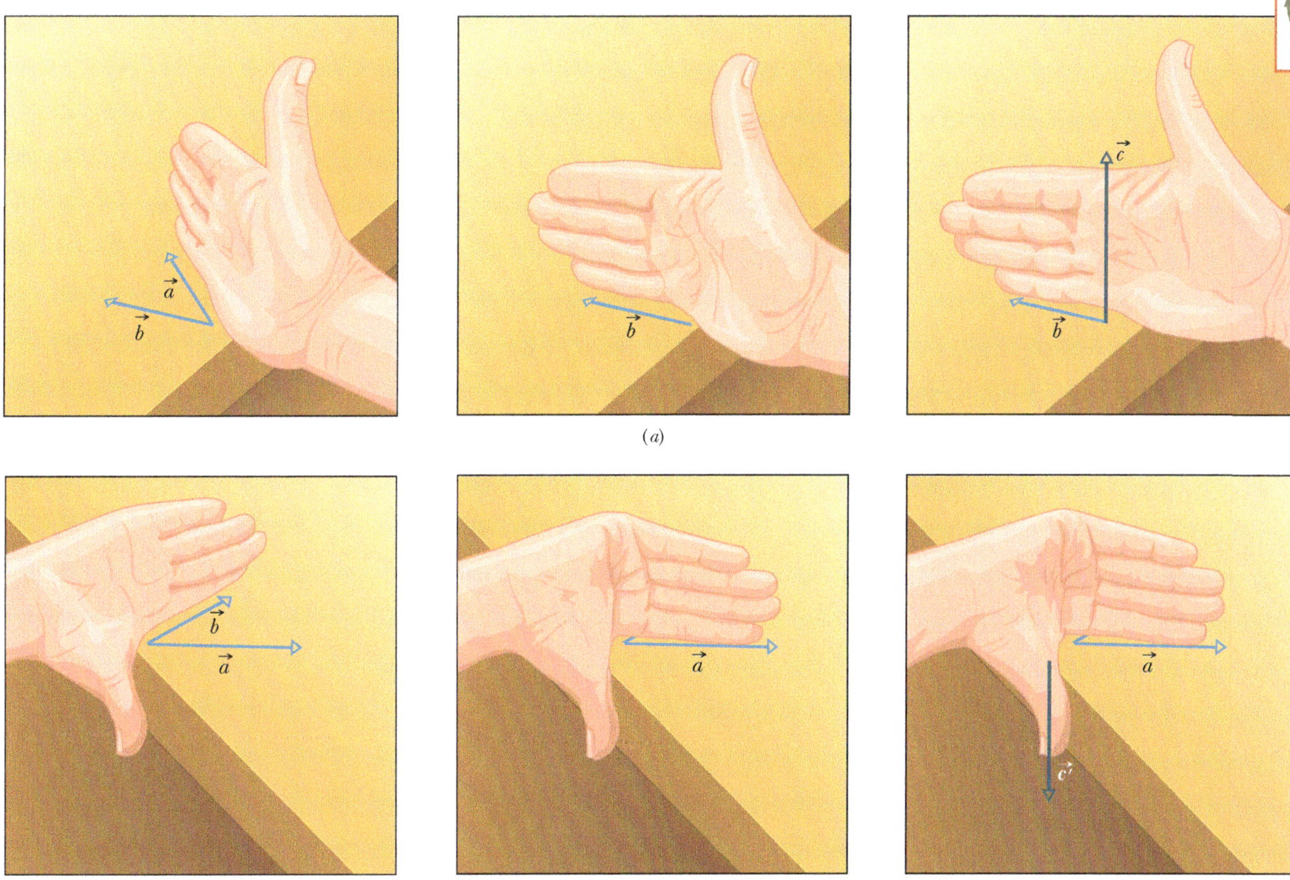

Figure 3-19 Illustration of the right-hand rule for vector products. (*a*) Sweep vector $\vec{a}$ into vector $\vec{b}$ with the fingers of your right hand. Your outstretched thumb shows the direction of vector $\vec{c} = \vec{a} \times \vec{b}$. (*b*) Showing that $\vec{b} \times \vec{a}$ is the reverse of $\vec{a} \times \vec{b}$.

Sample Problem 3.05 Angle between two vectors using dot products

What is the angle ϕ between $\vec{a} = 3.0\hat{i} - 4.0\hat{j}$ and $\vec{b} = -2.0\hat{i} + 3.0\hat{k}$? (*Caution:* Although many of the following steps can be bypassed with a vector-capable calculator, you will learn more about scalar products if, at least here, you use these steps.)

KEY IDEA

The angle between the directions of two vectors is included in the definition of their scalar product (Eq. 3-20):

$$\vec{a}\cdot\vec{b} = ab\cos\phi. \tag{3-28}$$

Calculations: In Eq. 3-28, a is the magnitude of $\vec{a}$, or

$$a = \sqrt{3.0^2 + (-4.0)^2} = 5.00, \tag{3-29}$$

and b is the magnitude of $\vec{b}$, or

$$b = \sqrt{(-2.0)^2 + 3.0^2} = 3.61. \tag{3-30}$$

We can separately evaluate the left side of Eq. 3-28 by writing the vectors in unit-vector notation and using the distributive law:

$$\begin{aligned}\vec{a}\cdot\vec{b} &= (3.0\hat{i} - 4.0\hat{j})\cdot(-2.0\hat{i} + 3.0\hat{k})\\ &= (3.0\hat{i})\cdot(-2.0\hat{i}) + (3.0\hat{i})\cdot(3.0\hat{k})\\ &\quad + (-4.0\hat{j})\cdot(-2.0\hat{i}) + (-4.0\hat{j})\cdot(3.0\hat{k}).\end{aligned}$$

We next apply Eq. 3-20 to each term in this last expression. The angle between the unit vectors in the first term ($\hat{i}$ and $\hat{i}$) is 0°, and in the other terms it is 90°. We then have

$$\begin{aligned}\vec{a}\cdot\vec{b} &= -(6.0)(1) + (9.0)(0) + (8.0)(0) - (12)(0)\\ &= -6.0.\end{aligned}$$

Substituting this result and the results of Eqs. 3-29 and 3-30 into Eq. 3-28 yields

$$-6.0 = (5.00)(3.61)\cos\phi,$$

so $$\phi = \cos^{-1}\frac{-6.0}{(5.00)(3.61)} = 109° \approx 110°. \quad \text{(Answer)}$$

Sample Problem 3.06 Cross product, right-hand rule

In Fig. 3-20, vector $\vec{a}$ lies in the xy plane, has a magnitude of 18 units, and points in a direction 250° from the positive direction of the x axis. Also, vector $\vec{b}$ has a magnitude of 12 units and points in the positive direction of the z axis. What is the vector product $\vec{c} = \vec{a} \times \vec{b}$?

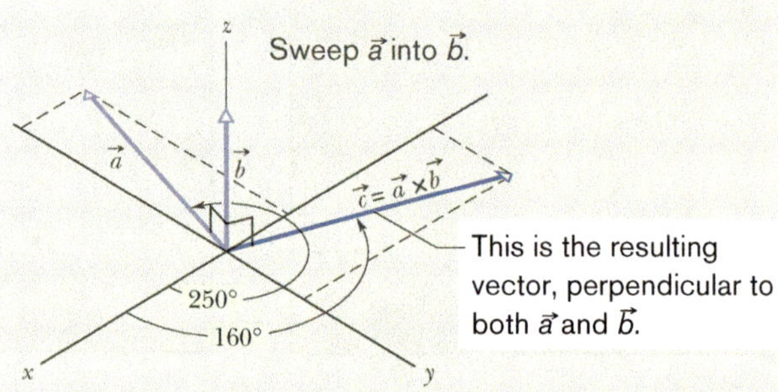

Figure 3-20 Vector $\vec{c}$ (in the xy plane) is the vector (or cross) product of vectors $\vec{a}$ and $\vec{b}$.

KEY IDEA

When we have two vectors in magnitude-angle notation, we find the magnitude of their cross product with Eq. 3-24 and the direction of their cross product with the right-hand rule of Fig. 3-19.

Calculations: For the magnitude we write

$$c = ab\sin\phi = (18)(12)(\sin 90°) = 216. \quad \text{(Answer)}$$

To determine the direction in Fig. 3-20, imagine placing the fingers of your right hand around a line perpendicular to the plane of $\vec{a}$ and $\vec{b}$ (the line on which $\vec{c}$ is shown) such that your fingers sweep $\vec{a}$ into $\vec{b}$. Your outstretched thumb then gives the direction of $\vec{c}$. Thus, as shown in the figure, $\vec{c}$ lies in the xy plane. Because its direction is perpendicular to the direction of $\vec{a}$ (a cross product always gives a perpendicular vector), it is at an angle of

$$250° - 90° = 160° \quad \text{(Answer)}$$

from the positive direction of the x axis.

Sample Problem 3.07 Cross product, unit-vector notation

If $\vec{a} = 3\hat{i} - 4\hat{j}$ and $\vec{b} = -2\hat{i} + 3\hat{k}$, what is $\vec{c} = \vec{a} \times \vec{b}$?

KEY IDEA

When two vectors are in unit-vector notation, we can find their cross product by using the distributive law.

Calculations: Here we write

$$\begin{aligned}\vec{c} &= (3\hat{i} - 4\hat{j}) \times (-2\hat{i} + 3\hat{k})\\ &= 3\hat{i} \times (-2\hat{i}) + 3\hat{i} \times 3\hat{k} + (-4\hat{j}) \times (-2\hat{i})\\ &\quad + (-4\hat{j}) \times 3\hat{k}.\end{aligned}$$

We next evaluate each term with Eq. 3-24, finding the direction with the right-hand rule. For the first term here, the angle ϕ between the two vectors being crossed is 0. For the other terms, ϕ is 90°. We find

$$\vec{c} = -6(0) + 9(-\hat{j}) + 8(-\hat{k}) - 12\hat{i}$$
$$= -12\hat{i} - 9\hat{j} - 8\hat{k}. \qquad \text{(Answer)}$$

This vector $\vec{c}$ is perpendicular to both $\vec{a}$ and $\vec{b}$, a fact you can check by showing that $\vec{c}\cdot\vec{a} = 0$ and $\vec{c}\cdot\vec{b} = 0$; that is, there is no component of $\vec{c}$ along the direction of either $\vec{a}$ or $\vec{b}$.

In general: A cross product gives a perpendicular vector, two perpendicular vectors have a zero dot product, and two vectors along the same axis have a zero cross product.

WILEY PLUS Additional examples, video, and practice available at *WileyPLUS*

Review & Summary

Scalars and Vectors *Scalars,* such as temperature, have magnitude only. They are specified by a number with a unit (10°C) and obey the rules of arithmetic and ordinary algebra. *Vectors,* such as displacement, have both magnitude and direction (5 m, north) and obey the rules of vector algebra.

Adding Vectors Geometrically Two vectors $\vec{a}$ and $\vec{b}$ may be added geometrically by drawing them to a common scale and placing them head to tail. The vector connecting the tail of the first to the head of the second is the vector sum $\vec{s}$. To subtract $\vec{b}$ from $\vec{a}$, reverse the direction of $\vec{b}$ to get $-\vec{b}$; then add $-\vec{b}$ to $\vec{a}$. Vector addition is commutative

$$\vec{a} + \vec{b} = \vec{b} + \vec{a} \qquad (3\text{-}2)$$

and obeys the associative law

$$(\vec{a} + \vec{b}) + \vec{c} = \vec{a} + (\vec{b} + \vec{c}). \qquad (3\text{-}3)$$

Components of a Vector The (scalar) *components* a_x and a_y of any two-dimensional vector $\vec{a}$ along the coordinate axes are found by dropping perpendicular lines from the ends of $\vec{a}$ onto the coordinate axes. The components are given by

$$a_x = a\cos\theta \quad \text{and} \quad a_y = a\sin\theta, \qquad (3\text{-}5)$$

where θ is the angle between the positive direction of the x axis and the direction of $\vec{a}$. The algebraic sign of a component indicates its direction along the associated axis. Given its components, we can find the magnitude and orientation (direction) of the vector $\vec{a}$ by using

$$a = \sqrt{a_x^2 + a_y^2} \quad \text{and} \quad \tan\theta = \frac{a_y}{a_x} \qquad (3\text{-}6)$$

Unit-Vector Notation *Unit vectors* $\hat{i}, \hat{j}$, and $\hat{k}$ have magnitudes of unity and are directed in the positive directions of the x, y, and z axes, respectively, in a right-handed coordinate system (as defined by the vector products of the unit vectors). We can write a vector $\vec{a}$ in terms of unit vectors as

$$\vec{a} = a_x\hat{i} + a_y\hat{j} + a_z\hat{k}, \qquad (3\text{-}7)$$

in which $a_x\hat{i}, a_y\hat{j}$, and $a_z\hat{k}$ are the **vector components** of $\vec{a}$ and a_x, a_y, and a_z are its **scalar components.**

Adding Vectors in Component Form To add vectors in component form, we use the rules

$$r_x = a_x + b_x \qquad r_y = a_y + b_y \qquad r_z = a_z + b_z. \qquad (3\text{-}10 \text{ to } 3\text{-}12)$$

Here $\vec{a}$ and $\vec{b}$ are the vectors to be added, and $\vec{r}$ is the vector sum. Note that we add components axis by axis. We can then express the sum in unit-vector notation or magnitude-angle notation.

Product of a Scalar and a Vector The product of a scalar s and a vector $\vec{v}$ is a new vector whose magnitude is sv and whose direction is the same as that of $\vec{v}$ if s is positive, and opposite that of $\vec{v}$ if s is negative. (The negative sign reverses the vector.) To divide $\vec{v}$ by s, multiply $\vec{v}$ by $1/s$.

The Scalar Product The **scalar** (or **dot**) **product** of two vectors $\vec{a}$ and $\vec{b}$ is written $\vec{a}\cdot\vec{b}$ and is the *scalar* quantity given by

$$\vec{a}\cdot\vec{b} = ab\cos\phi, \qquad (3\text{-}20)$$

in which ϕ is the angle between the directions of $\vec{a}$ and $\vec{b}$. A scalar product is the product of the magnitude of one vector and the scalar component of the second vector along the direction of the first vector. Note that $\vec{a}\cdot\vec{b} = \vec{b}\cdot\vec{a}$, which means that the scalar product obeys the commutative law.

In unit-vector notation,

$$\vec{a}\cdot\vec{b} = (a_x\hat{i} + a_y\hat{j} + a_z\hat{k})\cdot(b_x\hat{i} + b_y\hat{j} + b_z\hat{k}), \qquad (3\text{-}22)$$

which may be expanded according to the distributive law.

The Vector Product The **vector** (or **cross**) **product** of two vectors $\vec{a}$ and $\vec{b}$ is written $\vec{a}\times\vec{b}$ and is a *vector* $\vec{c}$ whose magnitude c is given by

$$c = ab\sin\phi, \qquad (3\text{-}24)$$

in which ϕ is the smaller of the angles between the directions of $\vec{a}$ and $\vec{b}$. The direction of $\vec{c}$ is perpendicular to the plane defined by $\vec{a}$ and $\vec{b}$ and is given by a right-hand rule, as shown in Fig. 3-19. Note that $\vec{a}\times\vec{b} = -(\vec{b}\times\vec{a})$, which means that the vector product does not obey the commutative law.

In unit-vector notation,

$$\vec{a}\times\vec{b} = (a_x\hat{i} + a_y\hat{j} + a_z\hat{k})\times(b_x\hat{i} + b_y\hat{j} + b_z\hat{k}), \qquad (3\text{-}26)$$

which we may expand with the distributive law.

Questions

1 Can the sum of the magnitudes of two vectors ever be equal to the magnitude of the sum of the same two vectors? If no, why not? If yes, when?

2 The two vectors shown in Fig. 3-21 lie in an xy plane. What are the signs of the x and y components, respectively, of (a) $\vec{d}_1 + \vec{d}_2$, (b) $\vec{d}_1 - \vec{d}_2$, and (c) $\vec{d}_2 - \vec{d}_1$?

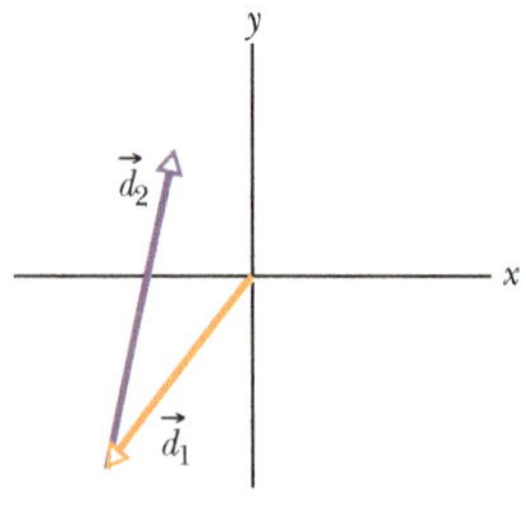

Figure 3-21 Question 2.

3 Being part of the "Gators," the University of Florida golfing team must play on a putting green with an alligator pit. Figure 3-22 shows an overhead view of one putting challenge of the team; an xy coordinate system is superimposed. Team members must putt from the origin to the hole, which is at xy coordinates (8 m, 12 m), but they can putt the golf ball using only one or more of the following displacements, one or more times:

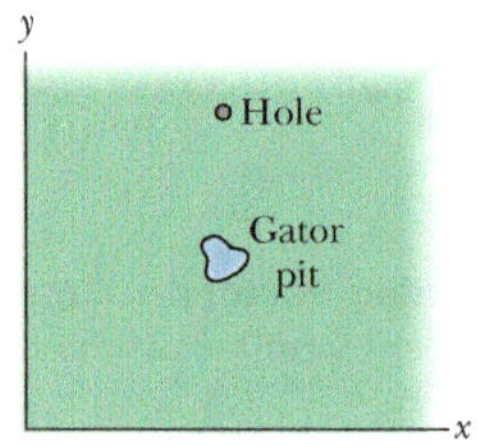

Figure 3-22 Question 3.

$$\vec{d}_1 = (8\text{ m})\hat{i} + (6\text{ m})\hat{j}, \qquad \vec{d}_2 = (6\text{ m})\hat{j}, \qquad \vec{d}_3 = (8\text{ m})\hat{i}.$$

The pit is at coordinates (8 m, 6 m). If a team member putts the ball into or through the pit, the member is automatically transferred to Florida State University, the arch rival. What sequence of displacements should a team member use to avoid the pit and the school transfer?

4 Equation 3-2 shows that the addition of two vectors $\vec{a}$ and $\vec{b}$ is commutative. Does that mean subtraction is commutative, so that $\vec{a} - \vec{b} = \vec{b} - \vec{a}$?

5 Which of the arrangements of axes in Fig. 3-23 can be labeled "right-handed coordinate system"? As usual, each axis label indicates the positive side of the axis.

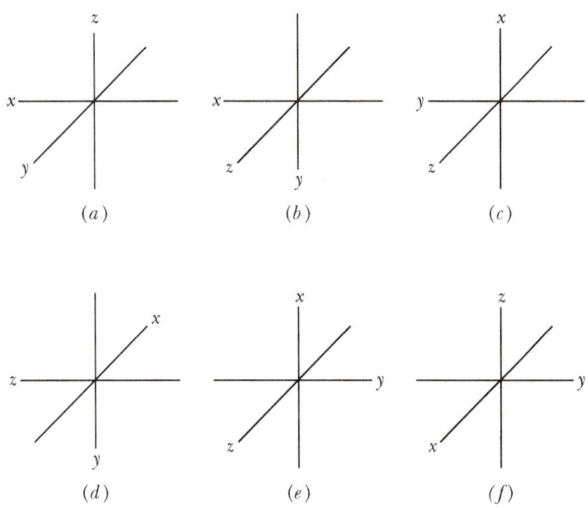

Figure 3-23 Question 5.

6 Describe two vectors $\vec{a}$ and $\vec{b}$ such that

(a) $\vec{a} + \vec{b} = \vec{c}$ and $a + b = c$;

(b) $\vec{a} + \vec{b} = \vec{a} - \vec{b}$;

(c) $\vec{a} + \vec{b} = \vec{c}$ and $a^2 + b^2 = c^2$.

7 If $\vec{d} = \vec{a} + \vec{b} + (-\vec{c})$, does (a) $\vec{a} + (-\vec{d}) = \vec{c} + (-\vec{b})$, (b) $\vec{a} = (-\vec{b}) + \vec{d} + \vec{c}$, and (c) $\vec{c} + (-\vec{d}) = \vec{a} + \vec{b}$?

8 If $\vec{a} \cdot \vec{b} = \vec{a} \cdot \vec{c}$, must $\vec{b}$ equal $\vec{c}$?

9 If $\vec{F} = q(\vec{v} \times \vec{B})$ and $\vec{v}$ is perpendicular to $\vec{B}$, then what is the direction of $\vec{B}$ in the three situations shown in Fig. 3-24 when constant q is (a) positive and (b) negative?

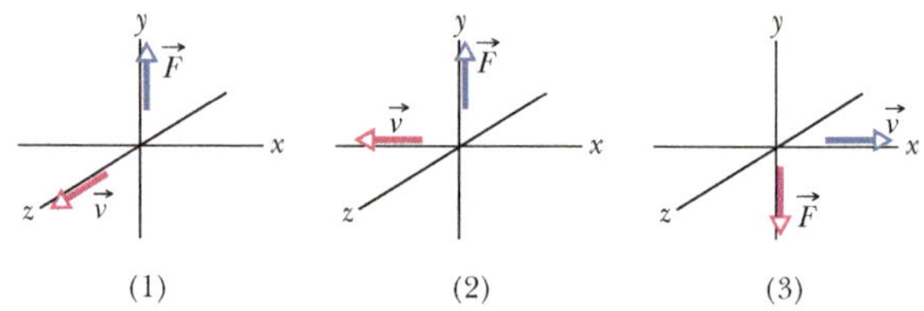

Figure 3-24 Question 9.

10 Figure 3-25 shows vector $\vec{A}$ and four other vectors that have the same magnitude but differ in orientation. (a) Which of those other four vectors have the same dot product with $\vec{A}$? (b) Which have a negative dot product with $\vec{A}$?

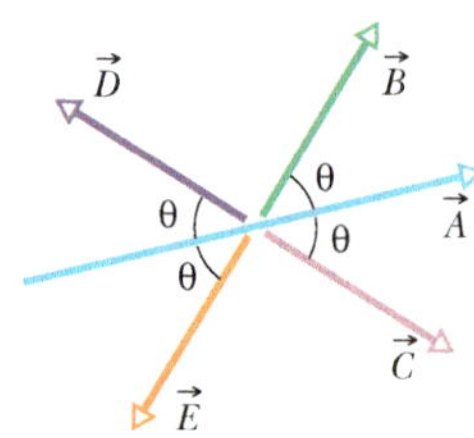

Figure 3-25 Question 10.

11 In a game held within a three-dimensional maze, you must move your game piece from *start*, at xyz coordinates (0, 0, 0), to *finish*, at coordinates (−2 cm, 4 cm, −4 cm). The game piece can undergo only the displacements (in centimeters) given below. If, along the way, the game piece lands at coordinates (−5 cm, −1 cm, −1 cm) or (5 cm, 2 cm, −1 cm), you lose the game. Which displacements and in what sequence will get your game piece to *finish*?

$$\vec{p} = -7\hat{i} + 2\hat{j} - 3\hat{k} \qquad \vec{r} = 2\hat{i} - 3\hat{j} + 2\hat{k}$$
$$\vec{q} = 2\hat{i} - \hat{j} + 4\hat{k} \qquad \vec{s} = 3\hat{i} + 5\hat{j} - 3\hat{k}.$$

12 The x and y components of four vectors $\vec{a}$, $\vec{b}$, $\vec{c}$, and $\vec{d}$ are given below. For which vectors will your calculator give you the correct angle θ when you use it to find θ with Eq. 3-6? Answer first by examining Fig. 3-12, and then check your answers with your calculator.

$$a_x = 3 \qquad a_y = 3 \qquad c_x = -3 \qquad c_y = -3$$
$$b_x = -3 \qquad b_y = 3 \qquad d_x = 3 \qquad d_y = -3.$$

13 Which of the following are correct (meaningful) vector expressions? What is wrong with any incorrect expression?

(a) $\vec{A} \cdot (\vec{B} \cdot \vec{C})$ (f) $\vec{A} + (\vec{B} \times \vec{C})$

(b) $\vec{A} \times (\vec{B} \cdot \vec{C})$ (g) $5 + \vec{A}$

(c) $\vec{A} \cdot (\vec{B} \times \vec{C})$ (h) $5 + (\vec{B} \cdot \vec{C})$

(d) $\vec{A} \times (\vec{B} \times \vec{C})$ (i) $5 + (\vec{B} \times \vec{C})$

(e) $\vec{A} + (\vec{B} \cdot \vec{C})$ (j) $(\vec{A} \cdot \vec{B}) + (\vec{B} \times \vec{C})$

Problems

GO Tutoring problem available (at instructor's discretion) in *WileyPLUS* and WebAssign

SSM Worked-out solution available in Student Solutions Manual

• – ••• Number of dots indicates level of problem difficulty

WWW Worked-out solution is at

ILW Interactive solution is at

http://www.wiley.com/college/halliday

Additional information available in *The Flying Circus of Physics* and at flyingcircusofphysics.com

Module 3-1 Vectors and Their Components

•1 SSM What are (a) the x component and (b) the y component of a vector $\vec{a}$ in the xy plane if its direction is 250° counterclockwise from the positive direction of the x axis and its magnitude is 7.3 m?

•2 A displacement vector $\vec{r}$ in the xy plane is 15 m long and directed at angle $\theta = 30°$ in Fig. 3-26. Determine (a) the x component and (b) the y component of the vector.

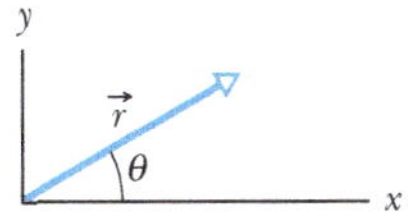

Figure 3-26 Problem 2.

•3 SSM The x component of vector $\vec{A}$ is −25.0 m and the y component is +40.0 m. (a) What is the magnitude of $\vec{A}$? (b) What is the angle between the direction of $\vec{A}$ and the positive direction of x?

•4 Express the following angles in radians: (a) 20.0°, (b) 50.0°, (c) 100°. Convert the following angles to degrees: (d) 0.330 rad, (e) 2.10 rad, (f) 7.70 rad.

•5 A ship sets out to sail to a point 120 km due north. An unexpected storm blows the ship to a point 100 km due east of its starting point. (a) How far and (b) in what direction must it now sail to reach its original destination?

•6 In Fig. 3-27, a heavy piece of machinery is raised by sliding it a distance $d = 12.5$ m along a plank oriented at angle $\theta = 20.0°$ to the horizontal. How far is it moved (a) vertically and (b) horizontally?

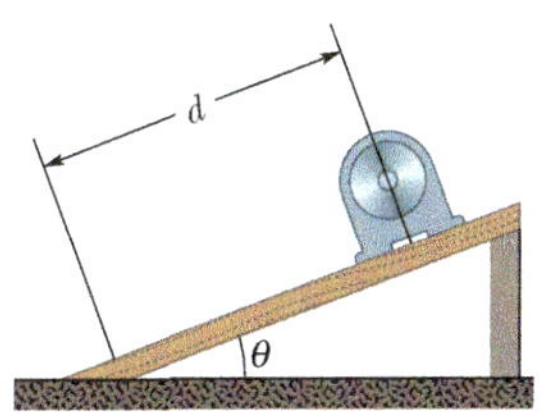

Figure 3-27 Problem 6.

•7 Consider two displacements, one of magnitude 3 m and another of magnitude 4 m. Show how the displacement vectors may be combined to get a resultant displacement of magnitude (a) 7 m, (b) 1 m, and (c) 5 m.

Module 3-2 Unit Vectors, Adding Vectors by Components

•8 A person walks in the following pattern: 3.1 km north, then 2.4 km west, and finally 5.2 km south. (a) Sketch the vector diagram that represents this motion. (b) How far and (c) in what direction would a bird fly in a straight line from the same starting point to the same final point?

•9 Two vectors are given by

$$\vec{a} = (4.0\text{ m})\hat{i} - (3.0\text{ m})\hat{j} + (1.0\text{ m})\hat{k}$$

and

$$\vec{b} = (-1.0\text{ m})\hat{i} + (1.0\text{ m})\hat{j} + (4.0\text{ m})\hat{k}.$$

In unit-vector notation, find (a) $\vec{a} + \vec{b}$, (b) $\vec{a} - \vec{b}$, and (c) a third vector $\vec{c}$ such that $\vec{a} - \vec{b} + \vec{c} = 0$.

•10 Find the (a) x, (b) y, and (c) z components of the sum $\vec{r}$ of the displacements $\vec{c}$ and $\vec{d}$ whose components in meters are $c_x = 7.4, c_y = -3.8, c_z = -6.1; d_x = 4.4, d_y = -2.0, d_z = 3.3$.

•11 SSM (a) In unit-vector notation, what is the sum $\vec{a} + \vec{b}$ if $\vec{a} = (4.0\text{ m})\hat{i} + (3.0\text{ m})\hat{j}$ and $\vec{b} = (-13.0\text{ m})\hat{i} + (7.0\text{ m})\hat{j}$? What are the (b) magnitude and (c) direction of $\vec{a} + \vec{b}$?

•12 A car is driven east for a distance of 50 km, then north for 30 km, and then in a direction 30° east of north for 25 km. Sketch the vector diagram and determine (a) the magnitude and (b) the angle of the car's total displacement from its starting point.

•13 A person desires to reach a point that is 3.40 km from her present location and in a direction that is 35.0° north of east. However, she must travel along streets that are oriented either north–south or east–west. What is the minimum distance she could travel to reach her destination?

•14 You are to make four straight-line moves over a flat desert floor, starting at the origin of an xy coordinate system and ending at the xy coordinates (−140 m, 30 m). The x component and y component of your moves are the following, respectively, in meters: (20 and 60), then (b_x and −70), then (−20 and c_y), then (−60 and −70). What are (a) component b_x and (b) component c_y? What are (c) the magnitude and (d) the angle (relative to the positive direction of the x axis) of the overall displacement?

•15 SSM ILW WWW The two vectors $\vec{a}$ and $\vec{b}$ in Fig. 3-28 have equal magnitudes of 10.0 m and the angles are $\theta_1 = 30°$ and $\theta_2 = 105°$. Find the (a) x and (b) y components of their vector sum $\vec{r}$, (c) the magnitude of $\vec{r}$, and (d) the angle $\vec{r}$ makes with the positive direction of the x axis.

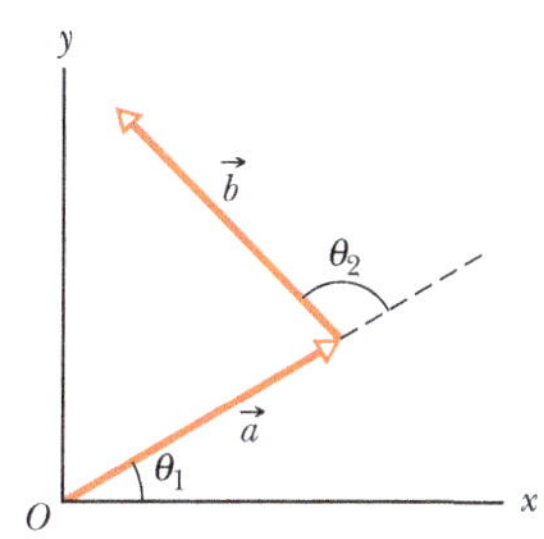

Figure 3-28 Problem 15.

•16 For the displacement vectors $\vec{a} = (3.0\text{ m})\hat{i} + (4.0\text{ m})\hat{j}$ and $\vec{b} = (5.0\text{ m})\hat{i} + (-2.0\text{ m})\hat{j}$, give $\vec{a} + \vec{b}$ in (a) unit-vector notation, and as (b) a magnitude and (c) an angle (relative to $\hat{i}$). Now give $\vec{b} - \vec{a}$ in (d) unit-vector notation, and as (e) a magnitude and (f) an angle.

•17 GO ILW Three vectors $\vec{a}$, $\vec{b}$, and $\vec{c}$ each have a magnitude of 50 m and lie in an xy plane. Their directions relative to the positive direction of the x axis are 30°, 195°, and 315°, respectively. What are (a) the magnitude and (b) the angle of the vector $\vec{a} + \vec{b} + \vec{c}$, and (c) the magnitude and (d) the angle of $\vec{a} - \vec{b} + \vec{c}$? What are the (e) magnitude and (f) angle of a fourth vector $\vec{d}$ such that $(\vec{a} + \vec{b}) - (\vec{c} + \vec{d}) = 0$?

•18 In the sum $\vec{A} + \vec{B} = \vec{C}$, vector $\vec{A}$ has a magnitude of 12.0 m and is angled 40.0° counterclockwise from the $+x$ direction, and vector $\vec{C}$ has a magnitude of 15.0 m and is angled 20.0° counterclockwise from the $-x$ direction. What are (a) the magnitude and (b) the angle (relative to $+x$) of $\vec{B}$?

•19 In a game of lawn chess, where pieces are moved between the centers of squares that are each 1.00 m on edge, a knight is moved in the following way: (1) two squares forward, one square rightward; (2) two squares leftward, one square forward; (3) two squares forward, one square leftward. What are (a) the magnitude and (b) the angle (relative to "forward") of the knight's overall displacement for the series of three moves?

••20 An explorer is caught in a whiteout (in which the snowfall is so thick that the ground cannot be distinguished from the sky) while returning to base camp. He was supposed to travel due north for 5.6 km, but when the snow clears, he discovers that he actually traveled 7.8 km at 50° north of due east. (a) How far and (b) in what direction must he now travel to reach base camp?

••21 GO An ant, crazed by the Sun on a hot Texas afternoon, darts over an xy plane scratched in the dirt. The x and y components of four consecutive darts are the following, all in centimeters: (30.0, 40.0), (b_x, −70.0), (−20.0, c_y), (−80.0, −70.0). The overall displacement of the four darts has the xy components (−140, −20.0). What are (a) b_x and (b) c_y? What are the (c) magnitude and (d) angle (relative to the positive direction of the x axis) of the overall displacement?

••22 (a) What is the sum of the following four vectors in unit-vector notation? For that sum, what are (b) the magnitude, (c) the angle in degrees, and (d) the angle in radians?

$\vec{E}$: 6.00 m at +0.900 rad $\vec{F}$: 5.00 m at −75.0°

$\vec{G}$: 4.00 m at +1.20 rad $\vec{H}$: 6.00 m at −210°

••23 If $\vec{B}$ is added to $\vec{C} = 3.0\hat{i} + 4.0\hat{j}$, the result is a vector in the positive direction of the y axis, with a magnitude equal to that of $\vec{C}$. What is the magnitude of $\vec{B}$?

••24 GO Vector $\vec{A}$, which is directed along an x axis, is to be added to vector $\vec{B}$, which has a magnitude of 7.0 m. The sum is a third vector that is directed along the y axis, with a magnitude that is 3.0 times that of $\vec{A}$. What is that magnitude of $\vec{A}$?

••25 GO Oasis B is 25 km due east of oasis A. Starting from oasis A, a camel walks 24 km in a direction 15° south of east and then walks 8.0 km due north. How far is the camel then from oasis B?

••26 What is the sum of the following four vectors in (a) unit-vector notation, and as (b) a magnitude and (c) an angle?

$\vec{A} = (2.00\text{ m})\hat{i} + (3.00\text{ m})\hat{j}$ $\vec{B}$: 4.00 m, at +65.0°

$\vec{C} = (-4.00\text{ m})\hat{i} + (-6.00\text{ m})\hat{j}$ $\vec{D}$: 5.00 m, at −235°

••27 GO If $\vec{d}_1 + \vec{d}_2 = 5\vec{d}_3$, $\vec{d}_1 - \vec{d}_2 = 3\vec{d}_3$, and $\vec{d}_3 = 2\hat{i} + 4\hat{j}$, then what are, in unit-vector notation, (a) $\vec{d}_1$ and (b) $\vec{d}_2$?

••28 Two beetles run across flat sand, starting at the same point. Beetle 1 runs 0.50 m due east, then 0.80 m at 30° north of due east. Beetle 2 also makes two runs; the first is 1.6 m at 40° east of due north. What must be (a) the magnitude and (b) the direction of its second run if it is to end up at the new location of beetle 1?

••29 GO Typical backyard ants often create a network of chemical trails for guidance. Extending outward from the nest, a trail branches (*bifurcates*) repeatedly, with 60° between the branches. If a roaming ant chances upon a trail, it can tell the way to the nest at any branch point: If it is moving away from the nest, it has two choices of path requiring a small turn in its travel direction, either 30° leftward or 30° rightward. If it is moving toward the nest, it has only one such choice. Figure 3-29 shows a typical ant trail, with lettered straight sections of 2.0 cm length and symmetric bifurcation of 60°. Path v is parallel to the y axis. What are the (a) magnitude and (b) angle (relative to the positive direction of the superimposed x axis) of an ant's displacement from the nest (find it in the figure) if the ant enters the trail at point A? What are the (c) magnitude and (d) angle if it enters at point B?

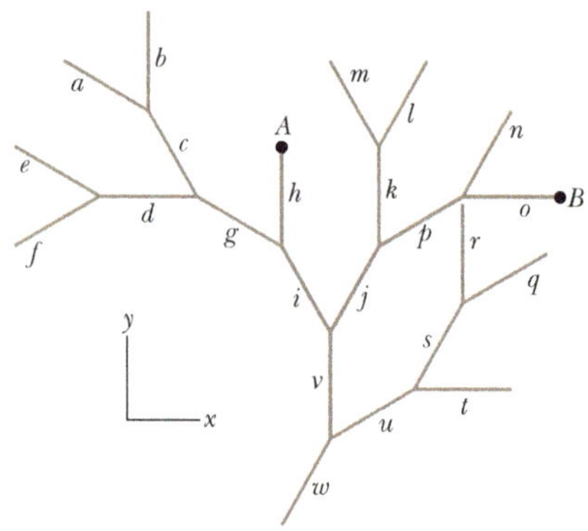

Figure 3-29 Problem 29.

••30 GO Here are two vectors:

$$\vec{a} = (4.0\text{ m})\hat{i} - (3.0\text{ m})\hat{j} \quad \text{and} \quad \vec{b} = (6.0\text{ m})\hat{i} + (8.0\text{ m})\hat{j}.$$

What are (a) the magnitude and (b) the angle (relative to $\hat{i}$) of $\vec{a}$? What are (c) the magnitude and (d) the angle of $\vec{b}$? What are (e) the magnitude and (f) the angle of $\vec{a} + \vec{b}$; (g) the magnitude and (h) the angle of $\vec{b} - \vec{a}$; and (i) the magnitude and (j) the angle of $\vec{a} - \vec{b}$? (k) What is the angle between the directions of $\vec{b} - \vec{a}$ and $\vec{a} - \vec{b}$?

••31 In Fig. 3-30, a vector $\vec{a}$ with a magnitude of 17.0 m is directed at angle $\theta = 56.0°$ counterclockwise from the $+x$ axis. What are the components (a) a_x and (b) a_y of the vector? A second coordinate system is inclined by angle $\theta' = 18.0°$ with respect to the first. What are the components (c) a'_x and (d) a'_y in this primed coordinate system?

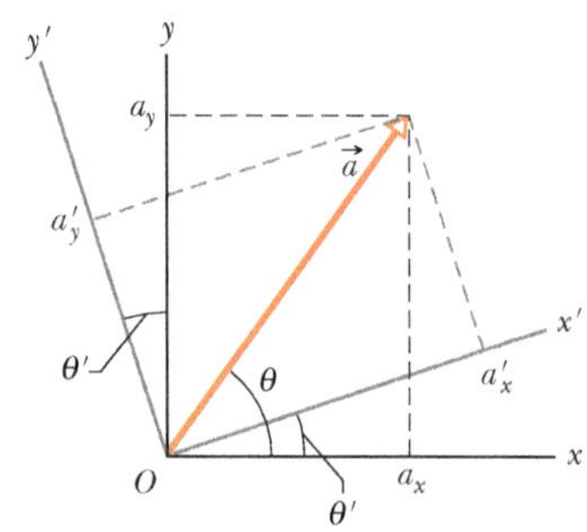

Figure 3-30 Problem 31.

•••32 In Fig. 3-31, a cube of edge length a sits with one corner at the origin of an xyz coordinate system. A *body diagonal* is a line that extends from one corner to another through the center. In unit-vector notation, what is the body diagonal that extends from the corner at (a) coordinates (0, 0, 0), (b) coordinates (a, 0, 0), (c) coordinates (0, a, 0), and (d) coordinates (a, a, 0)? (e) Determine the

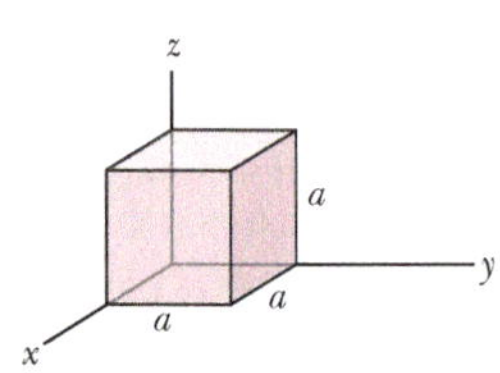

Figure 3-31 Problem 32.

angles that the body diagonals make with the adjacent edges. (f) Determine the length of the body diagonals in terms of a.

Module 3-3 Multiplying Vectors

•33 For the vectors in Fig. 3-32, with $a = 4$, $b = 3$, and $c = 5$, what are (a) the magnitude and (b) the direction of $\vec{a} \times \vec{b}$, (c) the magnitude and (d) the direction of $\vec{a} \times \vec{c}$, and (e) the magnitude and (f) the direction of $\vec{b} \times \vec{c}$? (The z axis is not shown.)

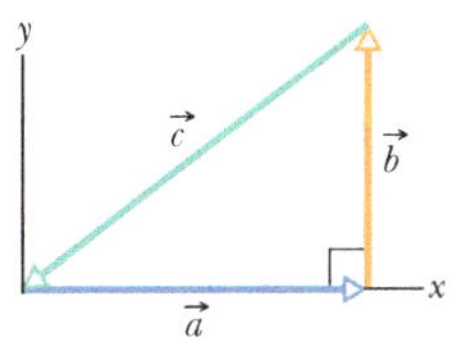

Figure 3-32 Problems 33 and 54.

•34 Two vectors are presented as $\vec{a} = 3.0\hat{i} + 5.0\hat{j}$ and $\vec{b} = 2.0\hat{i} + 4.0\hat{j}$. Find (a) $\vec{a} \times \vec{b}$, (b) $\vec{a} \cdot \vec{b}$, (c) $(\vec{a} + \vec{b}) \cdot \vec{b}$, and (d) the component of $\vec{a}$ along the direction of $\vec{b}$. (*Hint:* For (d), consider Eq. 3-20 and Fig. 3-18.)

•35 Two vectors, $\vec{r}$ and $\vec{s}$, lie in the xy plane. Their magnitudes are 4.50 and 7.30 units, respectively, and their directions are 320° and 85.0°, respectively, as measured counterclockwise from the positive x axis. What are the values of (a) $\vec{r} \cdot \vec{s}$ and (b) $\vec{r} \times \vec{s}$?

•36 If $\vec{d}_1 = 3\hat{i} - 2\hat{j} + 4\hat{k}$ and $\vec{d}_2 = -5\hat{i} + 2\hat{j} - \hat{k}$, then what is $(\vec{d}_1 + \vec{d}_2) \cdot (\vec{d}_1 \times 4\vec{d}_2)$?

•37 Three vectors are given by $\vec{a} = 3.0\hat{i} + 3.0\hat{j} - 2.0\hat{k}$, $\vec{b} = -1.0\hat{i} - 4.0\hat{j} + 2.0\hat{k}$, and $\vec{c} = 2.0\hat{i} + 2.0\hat{j} + 1.0\hat{k}$. Find (a) $\vec{a} \cdot (\vec{b} \times \vec{c})$, (b) $\vec{a} \cdot (\vec{b} + \vec{c})$, and (c) $\vec{a} \times (\vec{b} + \vec{c})$.

••38 GO For the following three vectors, what is $3\vec{C} \cdot (2\vec{A} \times \vec{B})$?

$$\vec{A} = 2.00\hat{i} + 3.00\hat{j} - 4.00\hat{k}$$

$$\vec{B} = -3.00\hat{i} + 4.00\hat{j} + 2.00\hat{k} \qquad \vec{C} = 7.00\hat{i} - 8.00\hat{j}$$

••39 Vector $\vec{A}$ has a magnitude of 6.00 units, vector $\vec{B}$ has a magnitude of 7.00 units, and $\vec{A} \cdot \vec{B}$ has a value of 14.0. What is the angle between the directions of $\vec{A}$ and $\vec{B}$?

••40 GO Displacement $\vec{d}_1$ is in the yz plane 63.0° from the positive direction of the y axis, has a positive z component, and has a magnitude of 4.50 m. Displacement $\vec{d}_2$ is in the xz plane 30.0° from the positive direction of the x axis, has a positive z component, and has magnitude 1.40 m. What are (a) $\vec{d}_1 \cdot \vec{d}_2$, (b) $\vec{d}_1 \times \vec{d}_2$, and (c) the angle between $\vec{d}_1$ and $\vec{d}_2$?

••41 SSM ILW WWW Use the definition of scalar product, $\vec{a} \cdot \vec{b} = ab \cos \theta$, and the fact that $\vec{a} \cdot \vec{b} = a_x b_x + a_y b_y + a_z b_z$ to calculate the angle between the two vectors given by $\vec{a} = 3.0\hat{i} + 3.0\hat{j} + 3.0\hat{k}$ and $\vec{b} = 2.0\hat{i} + 1.0\hat{j} + 3.0\hat{k}$.

••42 In a meeting of mimes, mime 1 goes through a displacement $\vec{d}_1 = (4.0\text{ m})\hat{i} + (5.0\text{ m})\hat{j}$ and mime 2 goes through a displacement $\vec{d}_2 = (-3.0\text{ m})\hat{i} + (4.0\text{ m})\hat{j}$. What are (a) $\vec{d}_1 \times \vec{d}_2$, (b) $\vec{d}_1 \cdot \vec{d}_2$, (c) $(\vec{d}_1 + \vec{d}_2) \cdot \vec{d}_2$, and (d) the component of $\vec{d}_1$ along the direction of $\vec{d}_2$? (*Hint:* For (d), see Eq. 3-20 and Fig. 3-18.)

••43 SSM ILW The three vectors in Fig. 3-33 have magnitudes $a = 3.00$ m, $b = 4.00$ m, and $c = 10.0$ m and angle $\theta = 30.0°$. What are (a) the x component and (b) the y component of $\vec{a}$; (c) the x component and (d) the y component of $\vec{b}$; and (e) the x component and (f) the y component of $\vec{c}$? If $\vec{c} = p\vec{a} + q\vec{b}$, what are the values of (g) p and (h) q?

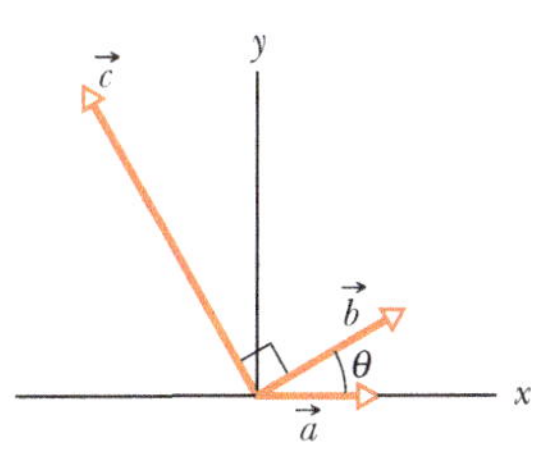

Figure 3-33 Problem 43.

••44 GO In the product $\vec{F} = q\vec{v} \times \vec{B}$, take $q = 2$,

$$\vec{v} = 2.0\hat{i} + 4.0\hat{j} + 6.0\hat{k} \quad \text{and} \quad \vec{F} = 4.0\hat{i} - 20\hat{j} + 12\hat{k}.$$

What then is $\vec{B}$ in unit-vector notation if $B_x = B_y$?

Additional Problems

45 Vectors $\vec{A}$ and $\vec{B}$ lie in an xy plane. $\vec{A}$ has magnitude 8.00 and angle 130°; $\vec{B}$ has components $B_x = -7.72$ and $B_y = -9.20$. (a) What is $5\vec{A} \cdot \vec{B}$? What is $4\vec{A} \times 3\vec{B}$ in (b) unit-vector notation and (c) magnitude-angle notation with spherical coordinates (see Fig. 3-34)? (d) What is the angle between the directions of $\vec{A}$ and $4\vec{A} \times 3\vec{B}$? (*Hint:* Think a bit before you resort to a calculation.) What is $\vec{A} + 3.00\hat{k}$ in (e) unit-vector notation and (f) magnitude-angle notation with spherical coordinates?

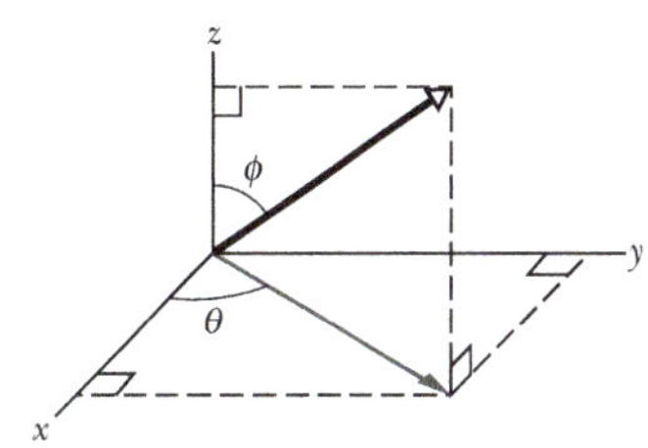

Figure 3-34 Problem 45.

46 GO Vector $\vec{a}$ has a magnitude of 5.0 m and is directed east. Vector $\vec{b}$ has a magnitude of 4.0 m and is directed 35° west of due north. What are (a) the magnitude and (b) the direction of $\vec{a} + \vec{b}$? What are (c) the magnitude and (d) the direction of $\vec{b} - \vec{a}$? (e) Draw a vector diagram for each combination.

47 Vectors $\vec{A}$ and $\vec{B}$ lie in an xy plane. $\vec{A}$ has magnitude 8.00 and angle 130°; $\vec{B}$ has components $B_x = -7.72$ and $B_y = -9.20$. What are the angles between the negative direction of the y axis and (a) the direction of $\vec{A}$, (b) the direction of the product $\vec{A} \times \vec{B}$, and (c) the direction of $\vec{A} \times (\vec{B} + 3.00\hat{k})$?

48 GO Two vectors $\vec{a}$ and $\vec{b}$ have the components, in meters, $a_x = 3.2$, $a_y = 1.6$, $b_x = 0.50$, $b_y = 4.5$. (a) Find the angle between the directions of $\vec{a}$ and $\vec{b}$. There are two vectors in the xy plane that are perpendicular to $\vec{a}$ and have a magnitude of 5.0 m. One, vector $\vec{c}$, has a positive x component and the other, vector $\vec{d}$, a negative x component. What are (b) the x component and (c) the y component of vector $\vec{c}$, and (d) the x component and (e) the y component of vector $\vec{d}$?

49 SSM A sailboat sets out from the U.S. side of Lake Erie for a point on the Canadian side, 90.0 km due north. The sailor, however, ends up 50.0 km due east of the starting point. (a) How far and (b) in what direction must the sailor now sail to reach the original destination?

50 Vector $\vec{d}_1$ is in the negative direction of a y axis, and vector $\vec{d}_2$ is in the positive direction of an x axis. What are the directions of (a) $\vec{d}_2/4$ and (b) $\vec{d}_1/(-4)$? What are the magnitudes of products (c) $\vec{d}_1 \cdot \vec{d}_2$ and (d) $\vec{d}_1 \cdot (\vec{d}_2/4)$? What is the direction of the vector resulting from (e) $\vec{d}_1 \times \vec{d}_2$ and (f) $\vec{d}_2 \times \vec{d}_1$? What is the magnitude of the vector product in (g) part (e) and (h) part (f)? What are the (i) magnitude and (j) direction of $\vec{d}_1 \times (\vec{d}_2/4)$?

51 Rock *faults* are ruptures along which opposite faces of rock have slid past each other. In Fig. 3-35, points A and B coincided before the rock in the foreground slid down to the right. The net displacement $\vec{AB}$ is along the plane of the fault. The horizontal component of $\vec{AB}$ is the *strike-slip AC*. The component of $\vec{AB}$ that is directed down the plane of the fault is the *dip-slip AD*. (a) What is the magnitude of the net displacement $\vec{AB}$ if the strike-slip is 22.0 m and the dip-slip is 17.0 m? (b) If the plane of the fault is inclined at angle $\phi = 52.0°$ to the horizontal, what is the vertical component of $\vec{AB}$?

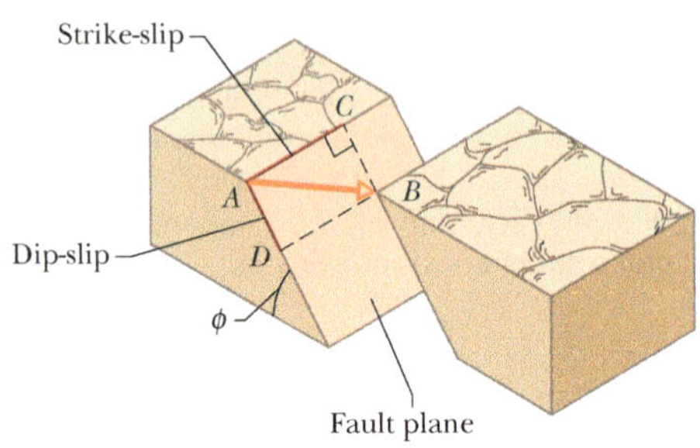

Figure 3-35 Problem 51.

52 Here are three displacements, each measured in meters: $\vec{d}_1 = 4.0\hat{i} + 5.0\hat{j} - 6.0\hat{k}$, $\vec{d}_2 = -1.0\hat{i} + 2.0\hat{j} + 3.0\hat{k}$, and $\vec{d}_3 = 4.0\hat{i} + 3.0\hat{j} + 2.0\hat{k}$. (a) What is $\vec{r} = \vec{d}_1 - \vec{d}_2 + \vec{d}_3$? (b) What is the angle between $\vec{r}$ and the positive z axis? (c) What is the component of $\vec{d}_1$ along the direction of $\vec{d}_2$? (d) What is the component of $\vec{d}_1$ that is perpendicular to the direction of $\vec{d}_2$ and in the plane of $\vec{d}_1$ and $\vec{d}_2$? (*Hint:* For (c), consider Eq. 3-20 and Fig. 3-18; for (d), consider Eq. 3-24.)

53 SSM A vector $\vec{a}$ of magnitude 10 units and another vector $\vec{b}$ of magnitude 6.0 units differ in directions by 60°. Find (a) the scalar product of the two vectors and (b) the magnitude of the vector product $\vec{a} \times \vec{b}$.

54 For the vectors in Fig. 3-32, with $a = 4$, $b = 3$, and $c = 5$, calculate (a) $\vec{a} \cdot \vec{b}$, (b) $\vec{a} \cdot \vec{c}$, and (c) $\vec{b} \cdot \vec{c}$.

55 A particle undergoes three successive displacements in a plane, as follows: $\vec{d}_1$, 4.00 m southwest; then $\vec{d}_2$, 5.00 m east; and finally $\vec{d}_3$, 6.00 m in a direction 60.0° north of east. Choose a coordinate system with the y axis pointing north and the x axis pointing east. What are (a) the x component and (b) the y component of $\vec{d}_1$? What are (c) the x component and (d) the y component of $\vec{d}_2$? What are (e) the x component and (f) the y component of $\vec{d}_3$? Next, consider the *net* displacement of the particle for the three successive displacements. What are (g) the x component, (h) the y component, (i) the magnitude, and (j) the direction of the net displacement? If the particle is to return directly to the starting point, (k) how far and (l) in what direction should it move?

56 Find the sum of the following four vectors in (a) unit-vector notation, and as (b) a magnitude and (c) an angle relative to $+x$.

$\vec{P}$: 10.0 m, at 25.0° counterclockwise from $+x$
$\vec{Q}$: 12.0 m, at 10.0° counterclockwise from $+y$
$\vec{R}$: 8.00 m, at 20.0° clockwise from $-y$
$\vec{S}$: 9.00 m, at 40.0° counterclockwise from $-y$

57 SSM If $\vec{B}$ is added to $\vec{A}$, the result is $6.0\hat{i} + 1.0\hat{j}$. If $\vec{B}$ is subtracted from $\vec{A}$, the result is $-4.0\hat{i} + 7.0\hat{j}$. What is the magnitude of $\vec{A}$?

58 A vector $\vec{d}$ has a magnitude of 2.5 m and points north. What are (a) the magnitude and (b) the direction of $4.0\vec{d}$? What are (c) the magnitude and (d) the direction of $-3.0\vec{d}$?

59 $\vec{A}$ has the magnitude 12.0 m and is angled 60.0° counterclockwise from the positive direction of the x axis of an xy coordinate system. Also, $\vec{B} = (12.0 \text{ m})\hat{i} + (8.00 \text{ m})\hat{j}$ on that same coordinate system. We now rotate the system counterclockwise about the origin by 20.0° to form an $x'y'$ system. On this new system, what are (a) $\vec{A}$ and (b) $\vec{B}$, both in unit-vector notation?

60 If $\vec{a} - \vec{b} = 2\vec{c}$, $\vec{a} + \vec{b} = 4\vec{c}$, and $\vec{c} = 3\hat{i} + 4\hat{j}$, then what are (a) $\vec{a}$ and (b) $\vec{b}$?

61 (a) In unit-vector notation, what is $\vec{r} = \vec{a} - \vec{b} + \vec{c}$ if $\vec{a} = 5.0\hat{i} + 4.0\hat{j} - 6.0\hat{k}$, $\vec{b} = -2.0\hat{i} + 2.0\hat{j} + 3.0\hat{k}$, and $\vec{c} = 4.0\hat{i} + 3.0\hat{j} + 2.0\hat{k}$? (b) Calculate the angle between $\vec{r}$ and the positive z axis. (c) What is the component of $\vec{a}$ along the direction of $\vec{b}$? (d) What is the component of $\vec{a}$ perpendicular to the direction of $\vec{b}$ but in the plane of $\vec{a}$ and $\vec{b}$? (*Hint:* For (c), see Eq. 3-20 and Fig. 3-18; for (d), see Eq. 3-24.)

62 A golfer takes three putts to get the ball into the hole. The first putt displaces the ball 3.66 m north, the second 1.83 m southeast, and the third 0.91 m southwest. What are (a) the magnitude and (b) the direction of the displacement needed to get the ball into the hole on the first putt?

63 Here are three vectors in meters:

$$\vec{d}_1 = -3.0\hat{i} + 3.0\hat{j} + 2.0\hat{k}$$
$$\vec{d}_2 = -2.0\hat{i} - 4.0\hat{j} + 2.0\hat{k}$$
$$\vec{d}_3 = 2.0\hat{i} + 3.0\hat{j} + 1.0\hat{k}.$$

What results from (a) $\vec{d}_1 \cdot (\vec{d}_2 + \vec{d}_3)$, (b) $\vec{d}_1 \cdot (\vec{d}_2 \times \vec{d}_3)$, and (c) $\vec{d}_1 \times (\vec{d}_2 + \vec{d}_3)$?

64 SSM WWW A room has dimensions 3.00 m (height) × 3.70 m × 4.30 m. A fly starting at one corner flies around, ending up at the diagonally opposite corner. (a) What is the magnitude of its displacement? (b) Could the length of its path be less than this magnitude? (c) Greater? (d) Equal? (e) Choose a suitable coordinate system and express the components of the displacement vector in that system in unit-vector notation. (f) If the fly walks, what is the length of the shortest path? (*Hint:* This can be answered without calculus. The room is like a box. Unfold its walls to flatten them into a plane.)

65 A protester carries his sign of protest, starting from the origin of an xyz coordinate system, with the xy plane horizontal. He moves 40 m in the negative direction of the x axis, then 20 m along a perpendicular path to his left, and then 25 m up a water tower. (a) In unit-vector notation, what is the displacement of the sign from start to end? (b) The sign then falls to the foot of the tower. What is the magnitude of the displacement of the sign from start to this new end?

66 Consider $\vec{a}$ in the positive direction of x, $\vec{b}$ in the positive direction of y, and a scalar d. What is the direction of $\vec{b}/d$ if d is (a) positive and (b) negative? What is the magnitude of (c) $\vec{a} \cdot \vec{b}$ and (d) $\vec{a} \cdot \vec{b}/d$? What is the direction of the vector resulting from (e) $\vec{a} \times \vec{b}$ and (f) $\vec{b} \times \vec{a}$? (g) What is the magnitude of the vector product in (e)? (h) What is the magnitude of the vector product in (f)? What are (i) the magnitude and (j) the direction of $\vec{a} \times \vec{b}/d$ if d is positive?

67 Let $\hat{i}$ be directed to the east, $\hat{j}$ be directed to the north, and $\hat{k}$ be directed upward. What are the values of products (a) $\hat{i} \cdot \hat{k}$, (b) $(-\hat{k}) \cdot (-\hat{j})$, and (c) $\hat{j} \cdot (-\hat{j})$? What are the directions (such as east or down) of products (d) $\hat{k} \times \hat{j}$, (e) $(-\hat{i}) \times (-\hat{j})$, and (f) $(-\hat{k}) \times (-\hat{j})$?

68 A bank in downtown Boston is robbed (see the map in Fig. 3-36). To elude police, the robbers escape by helicopter, making three successive flights described by the following displacements: 32 km, 45° south of east; 53 km, 26° north of west; 26 km, 18° east of south. At the end of the third flight they are captured. In what town are they apprehended?

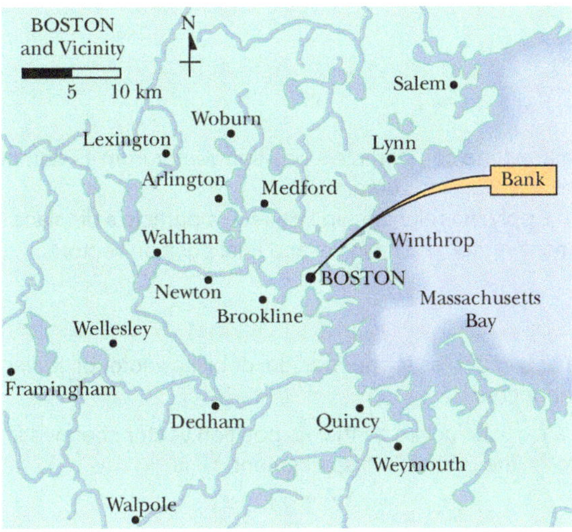

Figure 3-36 Problem 68.

69 A wheel with a radius of 45.0 cm rolls without slipping along a horizontal floor (Fig. 3-37). At time t_1, the dot P painted on the rim of the wheel is at the point of contact between the wheel and the floor. At a later time t_2, the wheel has rolled through one-half of a revolution. What are (a) the magnitude and (b) the angle (relative to the floor) of the displacement of P?

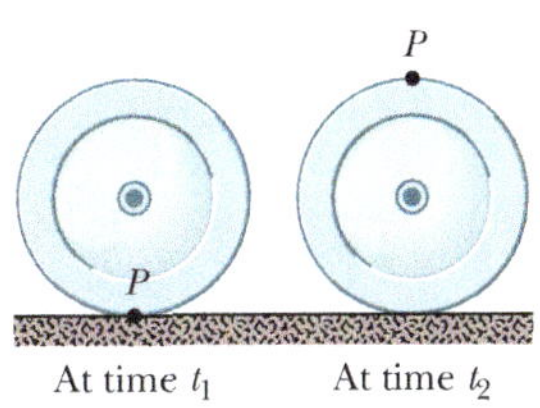

Figure 3-37 Problem 69.

70 A woman walks 250 m in the direction 30° east of north, then 175 m directly east. Find (a) the magnitude and (b) the angle of her final displacement from the starting point. (c) Find the distance she walks. (d) Which is greater, that distance or the magnitude of her displacement?

71 A vector $\vec{d}$ has a magnitude 3.0 m and is directed south. What are (a) the magnitude and (b) the direction of the vector $5.0\vec{d}$? What are (c) the magnitude and (d) the direction of the vector $-2.0\vec{d}$?

72 A fire ant, searching for hot sauce in a picnic area, goes through three displacements along level ground: $\vec{d}_1$ for 0.40 m southwest (that is, at 45° from directly south and from directly west), $\vec{d}_2$ for 0.50 m due east, $\vec{d}_3$ for 0.60 m at 60° north of east. Let the positive x direction be east and the positive y direction be north. What are (a) the x component and (b) the y component of $\vec{d}_1$? Next, what are (c) the x component and (d) the y component of $\vec{d}_2$? Also, what are (e) the x component and (f) the y component of $\vec{d}_3$?

What are (g) the x component, (h) the y component, (i) the magnitude, and (j) the direction of the ant's net displacement? If the ant is to return directly to the starting point, (k) how far and (l) in what direction should it move?

73 Two vectors are given by $\vec{a} = 3.0\hat{i} + 5.0\hat{j}$ and $\vec{b} = 2.0\hat{i} + 4.0\hat{j}$. Find (a) $\vec{a} \times \vec{b}$, (b) $\vec{a} \cdot \vec{b}$, (c) $(\vec{a} + \vec{b}) \cdot \vec{b}$, and (d) the component of $\vec{a}$ along the direction of $\vec{b}$.

74 Vector $\vec{a}$ lies in the yz plane 63.0° from the positive direction of the y axis, has a positive z component, and has magnitude 3.20 units. Vector $\vec{b}$ lies in the xz plane 48.0° from the positive direction of the x axis, has a positive z component, and has magnitude 1.40 units. Find (a) $\vec{a} \cdot \vec{b}$, (b) $\vec{a} \times \vec{b}$, and (c) the angle between $\vec{a}$ and $\vec{b}$.

75 Find (a) "north cross west," (b) "down dot south," (c) "east cross up," (d) "west dot west," and (e) "south cross south." Let each "vector" have unit magnitude.

76 A vector $\vec{B}$, with a magnitude of 8.0 m, is added to a vector $\vec{A}$, which lies along an x axis. The sum of these two vectors is a third vector that lies along the y axis and has a magnitude that is twice the magnitude of $\vec{A}$. What is the magnitude of $\vec{A}$?

77 A man goes for a walk, starting from the origin of an xyz coordinate system, with the xy plane horizontal and the x axis eastward. Carrying a bad penny, he walks 1300 m east, 2200 m north, and then drops the penny from a cliff 410 m high. (a) In unit-vector notation, what is the displacement of the penny from start to its landing point? (b) When the man returns to the origin, what is the magnitude of his displacement for the return trip?

78 What is the magnitude of $\vec{a} \times (\vec{b} \times \vec{a})$ if $a = 3.90$, $b = 2.70$, and the angle between the two vectors is 63.0°?

79 In Fig. 3-38, the magnitude of $\vec{a}$ is 4.3, the magnitude of $\vec{b}$ is 5.4, and $\phi = 46°$. Find the area of the triangle contained between the two vectors and the thin diagonal line.

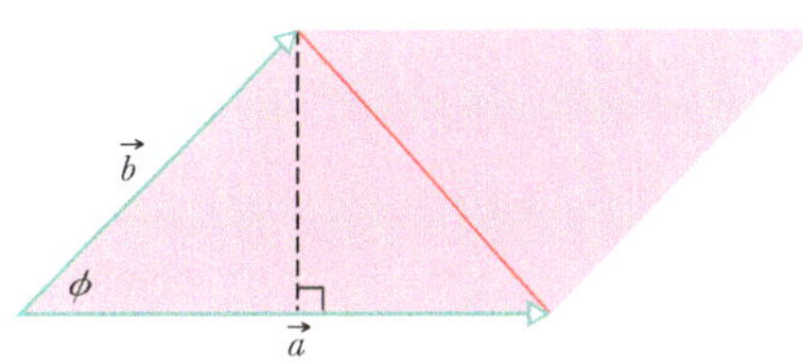

Figure 3-38 Problem 79.

C H A P T E R 4

Motion in Two and Three Dimensions

4-1 POSITION AND DISPLACEMENT

Learning Objectives

After reading this module, you should be able to . . .

4.01 Draw two-dimensional and three-dimensional position vectors for a particle, indicating the components along the axes of a coordinate system.

4.02 On a coordinate system, determine the direction and magnitude of a particle's position vector from its components, and vice versa.

4.03 Apply the relationship between a particle's displacement vector and its initial and final position vectors.

Key Ideas

- The location of a particle relative to the origin of a coordinate system is given by a position vector $\vec{r}$, which in unit-vector notation is

$$\vec{r} = x\hat{\mathrm{i}} + y\hat{\mathrm{j}} + z\hat{\mathrm{k}}.$$

Here $x\hat{\mathrm{i}}$, $y\hat{\mathrm{j}}$, and $z\hat{\mathrm{k}}$ are the vector components of position vector $\vec{r}$, and x, y, and z are its scalar components (as well as the coordinates of the particle).

- A position vector is described either by a magnitude and one or two angles for orientation, or by its vector or scalar components.

- If a particle moves so that its position vector changes from $\vec{r}_1$ to $\vec{r}_2$, the particle's displacement $\Delta\vec{r}$ is

$$\Delta\vec{r} = \vec{r}_2 - \vec{r}_1.$$

The displacement can also be written as

$$\Delta\vec{r} = (x_2 - x_1)\hat{\mathrm{i}} + (y_2 - y_1)\hat{\mathrm{j}} + (z_2 - z_1)\hat{\mathrm{k}}$$
$$= \Delta x\hat{\mathrm{i}} + \Delta y\hat{\mathrm{j}} + \Delta z\hat{\mathrm{k}}.$$

What Is Physics?

In this chapter we continue looking at the aspect of physics that analyzes motion, but now the motion can be in two or three dimensions. For example, medical researchers and aeronautical engineers might concentrate on the physics of the two- and three-dimensional turns taken by fighter pilots in dogfights because a modern high-performance jet can take a tight turn so quickly that the pilot immediately loses consciousness. A sports engineer might focus on the physics of basketball. For example, in a *free throw* (where a player gets an uncontested shot at the basket from about 4.3 m), a player might employ the *overhand push shot,* in which the ball is pushed away from about shoulder height and then released. Or the player might use an *underhand loop shot,* in which the ball is brought upward from about the belt-line level and released. The first technique is the overwhelming choice among professional players, but the legendary Rick Barry set the record for free-throw shooting with the underhand technique.

Motion in three dimensions is not easy to understand. For example, you are probably good at driving a car along a freeway (one-dimensional motion) but would probably have a difficult time in landing an airplane on a runway (three-dimensional motion) without a lot of training.

In our study of two- and three-dimensional motion, we start with position and displacement.

Position and Displacement

One general way of locating a particle (or particle-like object) is with a **position vector** $\vec{r}$, which is a vector that extends from a reference point (usually the origin) to the particle. In the unit-vector notation of Module 3-2, $\vec{r}$ can be written

$$\vec{r} = x\hat{\mathrm{i}} + y\hat{\mathrm{j}} + z\hat{\mathrm{k}}, \tag{4-1}$$

where $x\hat{\mathrm{i}}$, $y\hat{\mathrm{j}}$, and $z\hat{\mathrm{k}}$ are the vector components of $\vec{r}$ and the coefficients x, y, and z are its scalar components.

The coefficients x, y, and z give the particle's location along the coordinate axes and relative to the origin; that is, the particle has the rectangular coordinates (x, y, z). For instance, Fig. 4-1 shows a particle with position vector

$$\vec{r} = (-3\text{ m})\hat{\mathrm{i}} + (2\text{ m})\hat{\mathrm{j}} + (5\text{ m})\hat{\mathrm{k}}$$

and rectangular coordinates $(-3\text{ m}, 2\text{ m}, 5\text{ m})$. Along the x axis the particle is 3 m from the origin, in the $-\hat{\mathrm{i}}$ direction. Along the y axis it is 2 m from the origin, in the $+\hat{\mathrm{j}}$ direction. Along the z axis it is 5 m from the origin, in the $+\hat{\mathrm{k}}$ direction.

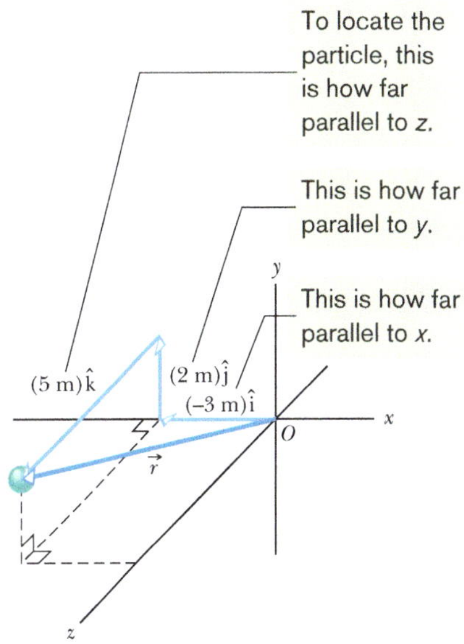

Figure 4-1 The position vector $\vec{r}$ for a particle is the vector sum of its vector components.

As a particle moves, its position vector changes in such a way that the vector always extends to the particle from the reference point (the origin). If the position vector changes—say, from $\vec{r}_1$ to $\vec{r}_2$ during a certain time interval—then the particle's **displacement** $\Delta\vec{r}$ during that time interval is

$$\Delta\vec{r} = \vec{r}_2 - \vec{r}_1. \tag{4-2}$$

Using the unit-vector notation of Eq. 4-1, we can rewrite this displacement as

$$\Delta\vec{r} = (x_2\hat{\mathrm{i}} + y_2\hat{\mathrm{j}} + z_2\hat{\mathrm{k}}) - (x_1\hat{\mathrm{i}} + y_1\hat{\mathrm{j}} + z_1\hat{\mathrm{k}})$$

or as

$$\Delta\vec{r} = (x_2 - x_1)\hat{\mathrm{i}} + (y_2 - y_1)\hat{\mathrm{j}} + (z_2 - z_1)\hat{\mathrm{k}}, \tag{4-3}$$

where coordinates (x_1, y_1, z_1) correspond to position vector $\vec{r}_1$ and coordinates (x_2, y_2, z_2) correspond to position vector $\vec{r}_2$. We can also rewrite the displacement by substituting Δx for $(x_2 - x_1)$, Δy for $(y_2 - y_1)$, and Δz for $(z_2 - z_1)$:

$$\Delta\vec{r} = \Delta x\hat{\mathrm{i}} + \Delta y\hat{\mathrm{j}} + \Delta z\hat{\mathrm{k}}. \tag{4-4}$$

Sample Problem 4.01 Two-dimensional position vector, rabbit run

A rabbit runs across a parking lot on which a set of coordinate axes has, strangely enough, been drawn. The coordinates (meters) of the rabbit's position as functions of time t (seconds) are given by

$$x = -0.31t^2 + 7.2t + 28 \tag{4-5}$$

and

$$y = 0.22t^2 - 9.1t + 30. \tag{4-6}$$

(a) At $t = 15$ s, what is the rabbit's position vector $\vec{r}$ in unit-vector notation and in magnitude-angle notation?

KEY IDEA

The x and y coordinates of the rabbit's position, as given by Eqs. 4-5 and 4-6, are the scalar components of the rabbit's position vector $\vec{r}$. Let's evaluate those coordinates at the given time, and then we can use Eq. 3-6 to evaluate the magnitude and orientation of the position vector.

Calculations: We can write

$$\vec{r}(t) = x(t)\hat{\mathrm{i}} + y(t)\hat{\mathrm{j}}. \tag{4-7}$$

(We write $\vec{r}(t)$ rather than $\vec{r}$ because the components are functions of t, and thus $\vec{r}$ is also.)

At $t = 15$ s, the scalar components are

$$x = (-0.31)(15)^2 + (7.2)(15) + 28 = 66\text{ m}$$

and

$$y = (0.22)(15)^2 - (9.1)(15) + 30 = -57\text{ m},$$

so

$$\vec{r} = (66\text{ m})\hat{\mathrm{i}} - (57\text{ m})\hat{\mathrm{j}}, \qquad \text{(Answer)}$$

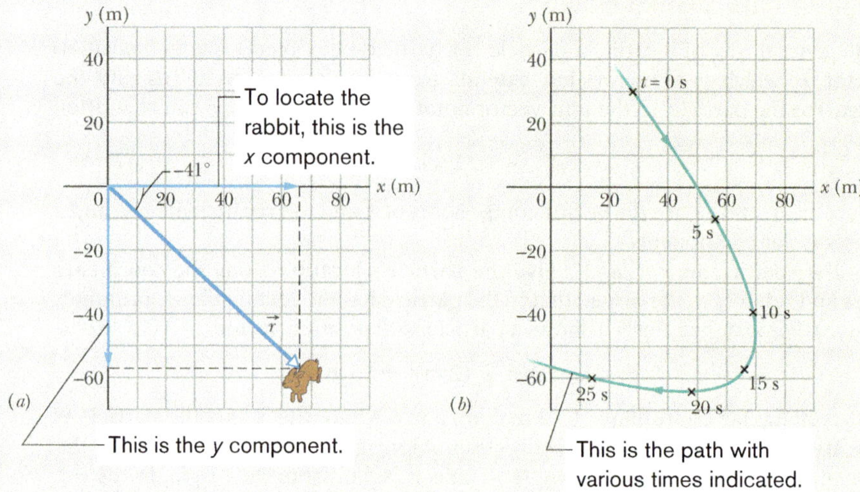

Figure 4-2 (*a*) A rabbit's position vector $\vec{r}$ at time $t = 15$ s. The scalar components of $\vec{r}$ are shown along the axes. (*b*) The rabbit's path and its position at six values of *t*.

which is drawn in Fig. 4-2*a*. To get the magnitude and angle of $\vec{r}$, notice that the components form the legs of a right triangle and *r* is the hypotenuse. So, we use Eq. 3-6:

$$r = \sqrt{x^2 + y^2} = \sqrt{(66\text{ m})^2 + (-57\text{ m})^2} = 87\text{ m}, \quad \text{(Answer)}$$

and $$\theta = \tan^{-1}\frac{y}{x} = \tan^{-1}\left(\frac{-57\text{ m}}{66\text{ m}}\right) = -41°. \quad \text{(Answer)}$$

Check: Although $\theta = 139°$ has the same tangent as $-41°$, the components of position vector $\vec{r}$ indicate that the desired angle is $139° - 180° = -41°$.

(b) Graph the rabbit's path for $t = 0$ to $t = 25$ s.

Graphing: We have located the rabbit at one instant, but to see its path we need a graph. So we repeat part (a) for several values of *t* and then plot the results. Figure 4-2*b* shows the plots for six values of *t* and the path connecting them.

4-2 AVERAGE VELOCITY AND INSTANTANEOUS VELOCITY

Learning Objectives

After reading this module, you should be able to . . .

4.04 Identify that velocity is a vector quantity and thus has both magnitude and direction and also has components.

4.05 Draw two-dimensional and three-dimensional velocity vectors for a particle, indicating the components along the axes of the coordinate system.

4.06 In magnitude-angle and unit-vector notations, relate a particle's initial and final position vectors, the time interval between those positions, and the particle's average velocity vector.

4.07 Given a particle's position vector as a function of time, determine its (instantaneous) velocity vector.

Key Ideas

- If a particle undergoes a displacement $\Delta\vec{r}$ in time interval Δt, its average velocity $\vec{v}_{avg}$ for that time interval is

$$\vec{v}_{avg} = \frac{\Delta\vec{r}}{\Delta t}.$$

- As Δt is shrunk to 0, $\vec{v}_{avg}$ reaches a limit called either the velocity or the instantaneous velocity $\vec{v}$:

$$\vec{v} = \frac{d\vec{r}}{dt},$$

which can be rewritten in unit-vector notation as

$$\vec{v} = v_x\hat{\mathrm{i}} + v_y\hat{\mathrm{j}} + v_z\hat{\mathrm{k}},$$

where $v_x = dx/dt$, $v_y = dy/dt$, and $v_z = dz/dt$.

- The instantaneous velocity $\vec{v}$ of a particle is always directed along the tangent to the particle's path at the particle's position.

Average Velocity and Instantaneous Velocity

If a particle moves from one point to another, we might need to know how fast it moves. Just as in Chapter 2, we can define two quantities that deal with "how fast": *average velocity* and *instantaneous velocity*. However, here we must consider these quantities as vectors and use vector notation.

If a particle moves through a displacement $\Delta\vec{r}$ in a time interval Δt, then its **average velocity** $\vec{v}_{avg}$ is

$$\text{average velocity} = \frac{\text{displacement}}{\text{time interval}},$$

or

$$\vec{v}_{avg} = \frac{\Delta\vec{r}}{\Delta t}. \tag{4-8}$$

This tells us that the direction of $\vec{v}_{avg}$ (the vector on the left side of Eq. 4-8) must be the same as that of the displacement $\Delta\vec{r}$ (the vector on the right side). Using Eq. 4-4, we can write Eq. 4-8 in vector components as

$$\vec{v}_{avg} = \frac{\Delta x\hat{i} + \Delta y\hat{j} + \Delta z\hat{k}}{\Delta t} = \frac{\Delta x}{\Delta t}\hat{i} + \frac{\Delta y}{\Delta t}\hat{j} + \frac{\Delta z}{\Delta t}\hat{k}. \tag{4-9}$$

For example, if a particle moves through displacement $(12\text{ m})\hat{i} + (3.0\text{ m})\hat{k}$ in 2.0 s, then its average velocity during that move is

$$\vec{v}_{avg} = \frac{\Delta\vec{r}}{\Delta t} = \frac{(12\text{ m})\hat{i} + (3.0\text{ m})\hat{k}}{2.0\text{ s}} = (6.0\text{ m/s})\hat{i} + (1.5\text{ m/s})\hat{k}.$$

That is, the average velocity (a vector quantity) has a component of 6.0 m/s along the x axis and a component of 1.5 m/s along the z axis.

When we speak of the **velocity** of a particle, we usually mean the particle's **instantaneous velocity** $\vec{v}$ at some instant. This $\vec{v}$ is the value that $\vec{v}_{avg}$ approaches in the limit as we shrink the time interval Δt to 0 about that instant. Using the language of calculus, we may write $\vec{v}$ as the derivative

$$\vec{v} = \frac{d\vec{r}}{dt}. \tag{4-10}$$

Figure 4-3 shows the path of a particle that is restricted to the xy plane. As the particle travels to the right along the curve, its position vector sweeps to the right. During time interval Δt, the position vector changes from $\vec{r}_1$ to $\vec{r}_2$ and the particle's displacement is $\Delta\vec{r}$.

To find the instantaneous velocity of the particle at, say, instant t_1 (when the particle is at position 1), we shrink interval Δt to 0 about t_1. Three things happen as we do so. (1) Position vector $\vec{r}_2$ in Fig. 4-3 moves toward $\vec{r}_1$ so that $\Delta\vec{r}$ shrinks

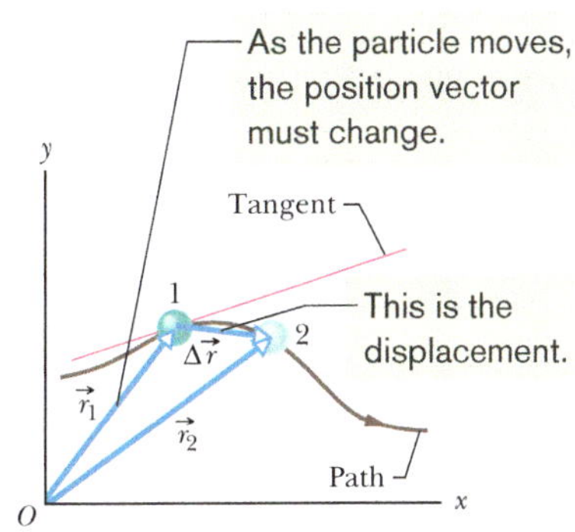

Figure 4-3 The displacement $\Delta\vec{r}$ of a particle during a time interval Δt, from position 1 with position vector $\vec{r}_1$ at time t_1 to position 2 with position vector $\vec{r}_2$ at time t_2. The tangent to the particle's path at position 1 is shown.

toward zero. (2) The direction of $\Delta\vec{r}/\Delta t$ (and thus of $\vec{v}_{avg}$) approaches the direction of the line tangent to the particle's path at position 1. (3) The average velocity $\vec{v}_{avg}$ approaches the instantaneous velocity $\vec{v}$ at t_1.

In the limit as $\Delta t \to 0$, we have $\vec{v}_{avg} \to \vec{v}$ and, most important here, $\vec{v}_{avg}$ takes on the direction of the tangent line. Thus, $\vec{v}$ has that direction as well:

The direction of the instantaneous velocity $\vec{v}$ of a particle is always tangent to the particle's path at the particle's position.

The result is the same in three dimensions: $\vec{v}$ is always tangent to the particle's path.

To write Eq. 4-10 in unit-vector form, we substitute for $\vec{r}$ from Eq. 4-1:

$$\vec{v} = \frac{d}{dt}(x\hat{i} + y\hat{j} + z\hat{k}) = \frac{dx}{dt}\hat{i} + \frac{dy}{dt}\hat{j} + \frac{dz}{dt}\hat{k}.$$

This equation can be simplified somewhat by writing it as

$$\vec{v} = v_x\hat{i} + v_y\hat{j} + v_z\hat{k}, \tag{4-11}$$

where the scalar components of $\vec{v}$ are

$$v_x = \frac{dx}{dt}, \quad v_y = \frac{dy}{dt}, \quad \text{and} \quad v_z = \frac{dz}{dt}. \tag{4-12}$$

For example, dx/dt is the scalar component of $\vec{v}$ along the x axis. Thus, we can find the scalar components of $\vec{v}$ by differentiating the scalar components of $\vec{r}$.

Figure 4-4 shows a velocity vector $\vec{v}$ and its scalar x and y components. Note that $\vec{v}$ is tangent to the particle's path at the particle's position. *Caution:* When a position vector is drawn, as in Figs. 4-1 through 4-3, it is an arrow that extends from one point (a "here") to another point (a "there"). However, when a velocity vector is drawn, as in Fig. 4-4, it does *not* extend from one point to another. Rather, it shows the instantaneous direction of travel of a particle at the tail, and its length (representing the velocity magnitude) can be drawn to any scale.

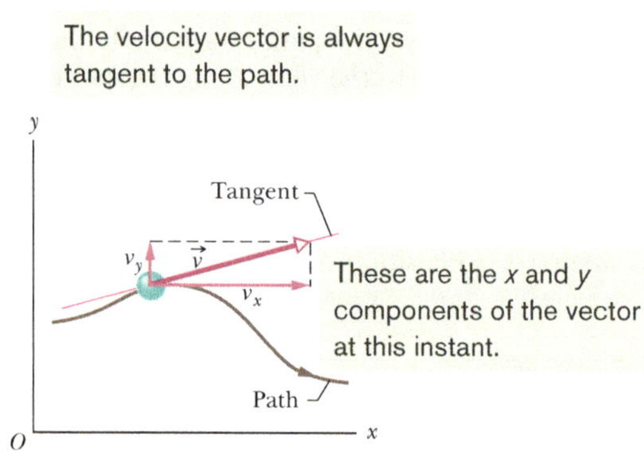

Figure 4-4 The velocity $\vec{v}$ of a particle, along with the scalar components of $\vec{v}$.

Checkpoint 1

The figure shows a circular path taken by a particle. If the instantaneous velocity of the particle is $\vec{v} = (2\text{ m/s})\hat{i} - (2\text{ m/s})\hat{j}$, through which quadrant is the particle moving at that instant if it is traveling (a) clockwise and (b) counterclockwise around the circle? For both cases, draw $\vec{v}$ on the figure.

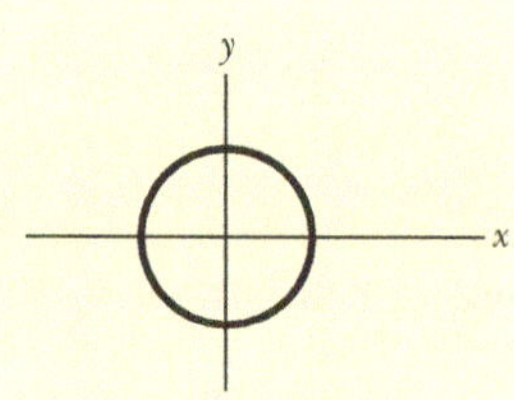

Sample Problem 4.02 Two-dimensional velocity, rabbit run

For the rabbit in the preceding sample problem, find the velocity $\vec{v}$ at time $t = 15$ s.

KEY IDEA

We can find $\vec{v}$ by taking derivatives of the components of the rabbit's position vector.

Calculations: Applying the v_x part of Eq. 4-12 to Eq. 4-5, we find the x component of $\vec{v}$ to be

$$v_x = \frac{dx}{dt} = \frac{d}{dt}(-0.31t^2 + 7.2t + 28)$$
$$= -0.62t + 7.2. \tag{4-13}$$

At $t = 15$ s, this gives $v_x = -2.1$ m/s. Similarly, applying the v_y part of Eq. 4-12 to Eq. 4-6, we find

$$v_y = \frac{dy}{dt} = \frac{d}{dt}(0.22t^2 - 9.1t + 30)$$
$$= 0.44t - 9.1. \tag{4-14}$$

At $t = 15$ s, this gives $v_y = -2.5$ m/s. Equation 4-11 then yields

$$\vec{v} = (-2.1 \text{ m/s})\hat{\text{i}} + (-2.5 \text{ m/s})\hat{\text{j}}, \qquad \text{(Answer)}$$

which is shown in Fig. 4-5, tangent to the rabbit's path and in the direction the rabbit is running at $t = 15$ s.

To get the magnitude and angle of $\vec{v}$, either we use a vector-capable calculator or we follow Eq. 3-6 to write

$$v = \sqrt{v_x^2 + v_y^2} = \sqrt{(-2.1 \text{ m/s})^2 + (-2.5 \text{ m/s})^2}$$
$$= 3.3 \text{ m/s} \qquad \text{(Answer)}$$

and

$$\theta = \tan^{-1}\frac{v_y}{v_x} = \tan^{-1}\left(\frac{-2.5 \text{ m/s}}{-2.1 \text{ m/s}}\right)$$
$$= \tan^{-1} 1.19 = -130°. \qquad \text{(Answer)}$$

Check: Is the angle $-130°$ or $-130° + 180° = 50°$?

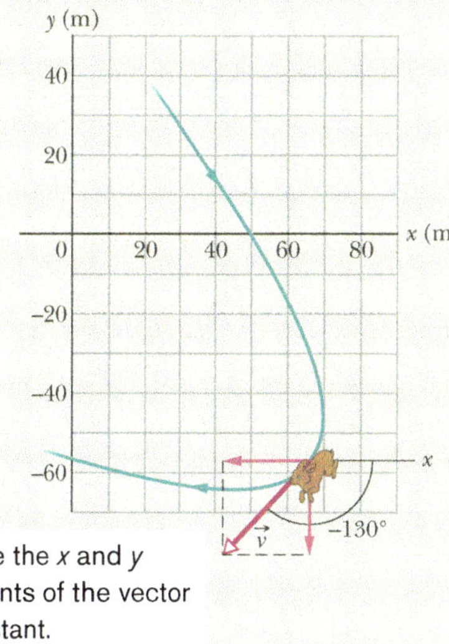

Figure 4-5 The rabbit's velocity $\vec{v}$ at $t = 15$ s.

WILEY PLUS Additional examples, video, and practice available at *WileyPLUS*

4-3 AVERAGE ACCELERATION AND INSTANTANEOUS ACCELERATION

Learning Objectives

After reading this module, you should be able to . . .

4.08 Identify that acceleration is a vector quantity and thus has both magnitude and direction and also has components.

4.09 Draw two-dimensional and three-dimensional acceleration vectors for a particle, indicating the components.

4.10 Given the initial and final velocity vectors of a particle and the time interval between those velocities, determine the average acceleration vector in magnitude-angle and unit-vector notations.

4.11 Given a particle's velocity vector as a function of time, determine its (instantaneous) acceleration vector.

4.12 For each dimension of motion, apply the constant-acceleration equations (Chapter 2) to relate acceleration, velocity, position, and time.

Key Ideas

- If a particle's velocity changes from $\vec{v}_1$ to $\vec{v}_2$ in time interval Δt, its average acceleration during Δt is

$$\vec{a}_{\text{avg}} = \frac{\vec{v}_2 - \vec{v}_1}{\Delta t} = \frac{\Delta\vec{v}}{\Delta t}.$$

- As Δt is shrunk to 0, $\vec{a}_{\text{avg}}$ reaches a limiting value called either the acceleration or the instantaneous acceleration $\vec{a}$:

$$\vec{a} = \frac{d\vec{v}}{dt}.$$

- In unit-vector notation,

$$\vec{a} = a_x\hat{\text{i}} + a_y\hat{\text{j}} + a_z\hat{\text{k}},$$

where $a_x = dv_x/dt$, $a_y = dv_y/dt$, and $a_z = dv_z/dt$.

Average Acceleration and Instantaneous Acceleration

When a particle's velocity changes from $\vec{v}_1$ to $\vec{v}_2$ in a time interval Δt, its **average acceleration** $\vec{a}_{avg}$ during Δt is

$$\text{average acceleration} = \frac{\text{change in velocity}}{\text{time interval}},$$

or

$$\vec{a}_{avg} = \frac{\vec{v}_2 - \vec{v}_1}{\Delta t} = \frac{\Delta \vec{v}}{\Delta t}. \tag{4-15}$$

If we shrink Δt to zero about some instant, then in the limit $\vec{a}_{avg}$ approaches the **instantaneous acceleration** (or **acceleration**) $\vec{a}$ at that instant; that is,

$$\vec{a} = \frac{d\vec{v}}{dt}. \tag{4-16}$$

If the velocity changes in *either* magnitude *or* direction (or both), the particle must have an acceleration.

We can write Eq. 4-16 in unit-vector form by substituting Eq. 4-11 for $\vec{v}$ to obtain

$$\begin{aligned}\vec{a} &= \frac{d}{dt}(v_x\hat{\imath} + v_y\hat{\jmath} + v_z\hat{k}) \\ &= \frac{dv_x}{dt}\hat{\imath} + \frac{dv_y}{dt}\hat{\jmath} + \frac{dv_z}{dt}\hat{k}.\end{aligned}$$

We can rewrite this as

$$\vec{a} = a_x\hat{\imath} + a_y\hat{\jmath} + a_z\hat{k}, \tag{4-17}$$

where the scalar components of $\vec{a}$ are

$$a_x = \frac{dv_x}{dt}, \quad a_y = \frac{dv_y}{dt}, \quad \text{and} \quad a_z = \frac{dv_z}{dt}. \tag{4-18}$$

To find the scalar components of $\vec{a}$, we differentiate the scalar components of $\vec{v}$.

Figure 4-6 shows an acceleration vector $\vec{a}$ and its scalar components for a particle moving in two dimensions. *Caution:* When an acceleration vector is drawn, as in Fig. 4-6, it does *not* extend from one position to another. Rather, it shows the direction of acceleration for a particle located at its tail, and its length (representing the acceleration magnitude) can be drawn to any scale.

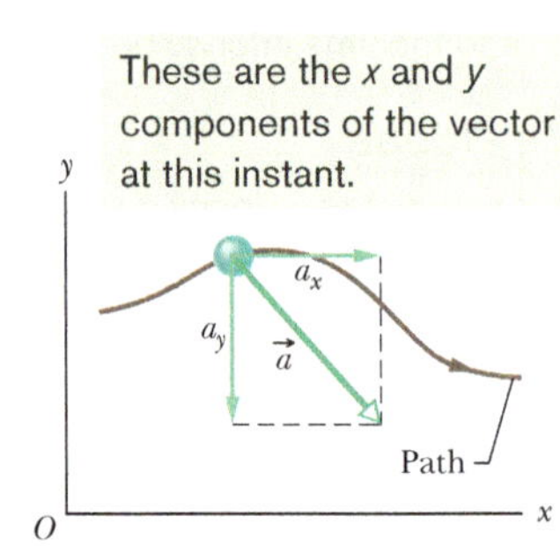

Figure 4-6 The acceleration $\vec{a}$ of a particle and the scalar components of $\vec{a}$.

Checkpoint 2

Here are four descriptions of the position (in meters) of a puck as it moves in an xy plane:

(1) $x = -3t^2 + 4t - 2$ and $y = 6t^2 - 4t$ (3) $\vec{r} = 2t^2\hat{i} - (4t + 3)\hat{j}$

(2) $x = -3t^3 - 4t$ and $y = -5t^2 + 6$ (4) $\vec{r} = (4t^3 - 2t)\hat{i} + 3\hat{j}$

Are the x and y acceleration components constant? Is acceleration $\vec{a}$ constant?

Sample Problem 4.03 Two-dimensional acceleration, rabbit run

For the rabbit in the preceding two sample problems, find the acceleration $\vec{a}$ at time $t = 15$ s.

KEY IDEA

We can find $\vec{a}$ by taking derivatives of the rabbit's velocity components.

Calculations: Applying the a_x part of Eq. 4-18 to Eq. 4-13, we find the x component of $\vec{a}$ to be

$$a_x = \frac{dv_x}{dt} = \frac{d}{dt}(-0.62t + 7.2) = -0.62 \text{ m/s}^2.$$

Similarly, applying the a_y part of Eq. 4-18 to Eq. 4-14 yields the y component as

$$a_y = \frac{dv_y}{dt} = \frac{d}{dt}(0.44t - 9.1) = 0.44 \text{ m/s}^2.$$

We see that the acceleration does not vary with time (it is a constant) because the time variable t does not appear in the expression for either acceleration component. Equation 4-17 then yields

$$\vec{a} = (-0.62 \text{ m/s}^2)\hat{i} + (0.44 \text{ m/s}^2)\hat{j}, \quad \text{(Answer)}$$

which is superimposed on the rabbit's path in Fig. 4-7.

To get the magnitude and angle of $\vec{a}$, either we use a vector-capable calculator or we follow Eq. 3-6. For the magnitude we have

$$a = \sqrt{a_x^2 + a_y^2} = \sqrt{(-0.62 \text{ m/s}^2)^2 + (0.44 \text{ m/s}^2)^2}$$
$$= 0.76 \text{ m/s}^2. \quad \text{(Answer)}$$

For the angle we have

$$\theta = \tan^{-1}\frac{a_y}{a_x} = \tan^{-1}\left(\frac{0.44 \text{ m/s}^2}{-0.62 \text{ m/s}^2}\right) = -35°.$$

However, this angle, which is the one displayed on a calculator, indicates that $\vec{a}$ is directed to the right and downward in Fig. 4-7. Yet, we know from the components that $\vec{a}$ must be directed to the left and upward. To find the other angle that has the same tangent as $-35°$ but is not displayed on a calculator, we add 180°:

$$-35° + 180° = 145°. \quad \text{(Answer)}$$

This *is* consistent with the components of $\vec{a}$ because it gives a vector that is to the left and upward. Note that $\vec{a}$ has the same magnitude and direction throughout the rabbit's run because the acceleration is constant. That means that we could draw the very same vector at any other point along the rabbit's path (just shift the vector to put its tail at some other point on the path without changing the length or orientation).

This has been the second sample problem in which we needed to take the derivative of a vector that is written in unit-vector notation. One common error is to neglect the unit vectors themselves, with a result of only a set of numbers and symbols. Keep in mind that a derivative of a vector is always another vector.

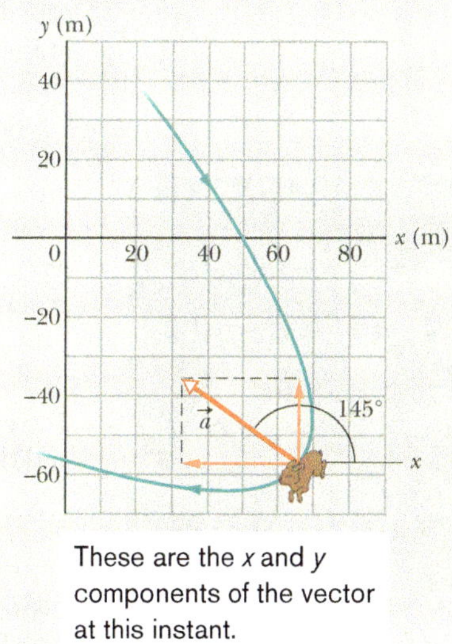

Figure 4-7 The acceleration $\vec{a}$ of the rabbit at $t = 15$ s. The rabbit happens to have this same acceleration at all points on its path.

WILEY PLUS Additional examples, video, and practice available at *WileyPLUS*

4-4 PROJECTILE MOTION

Learning Objectives

After reading this module, you should be able to . . .

4.13 On a sketch of the path taken in projectile motion, explain the magnitudes and directions of the velocity and acceleration components during the flight.

4.14 Given the launch velocity in either magnitude-angle or unit-vector notation, calculate the particle's position, displacement, and velocity at a given instant during the flight.

4.15 Given data for an instant during the flight, calculate the launch velocity.

Key Ideas

- In projectile motion, a particle is launched into the air with a speed v_0 and at an angle θ_0 (as measured from a horizontal x axis). During flight, its horizontal acceleration is zero and its vertical acceleration is $-g$ (downward on a vertical y axis).
- The equations of motion for the particle (while in flight) can be written as

$$x - x_0 = (v_0 \cos \theta_0)t,$$
$$y - y_0 = (v_0 \sin \theta_0)t - \tfrac{1}{2}gt^2,$$
$$v_y = v_0 \sin \theta_0 - gt,$$
$$v_y^2 = (v_0 \sin \theta_0)^2 - 2g(y - y_0).$$

- The trajectory (path) of a particle in projectile motion is parabolic and is given by

$$y = (\tan \theta_0)x - \frac{gx^2}{2(v_0 \cos \theta_0)^2},$$

if x_0 and y_0 are zero.
- The particle's horizontal range R, which is the horizontal distance from the launch point to the point at which the particle returns to the launch height, is

$$R = \frac{v_0^2}{g} \sin 2\theta_0.$$

Projectile Motion

We next consider a special case of two-dimensional motion: A particle moves in a vertical plane with some initial velocity $\vec{v}_0$ but its acceleration is always the free-fall acceleration $\vec{g}$, which is downward. Such a particle is called a **projectile** (meaning that it is projected or launched), and its motion is called **projectile motion.** A projectile might be a tennis ball (Fig. 4-8) or baseball in flight, but it is not a duck in flight. Many sports involve the study of the projectile motion of a ball. For example, the racquetball player who discovered the Z-shot in the 1970s easily won his games because of the ball's perplexing flight to the rear of the court.

Our goal here is to analyze projectile motion using the tools for two-dimensional motion described in Modules 4-1 through 4-3 and making the assumption that air has no effect on the projectile. Figure 4-9, which we shall analyze soon, shows the path followed by a projectile when the air has no effect. The projectile is launched with an initial velocity $\vec{v}_0$ that can be written as

$$\vec{v}_0 = v_{0x}\hat{\mathrm{i}} + v_{0y}\hat{\mathrm{j}}. \tag{4-19}$$

The components v_{0x} and v_{0y} can then be found if we know the angle θ_0 between $\vec{v}_0$ and the positive x direction:

$$v_{0x} = v_0 \cos \theta_0 \quad \text{and} \quad v_{0y} = v_0 \sin \theta_0. \tag{4-20}$$

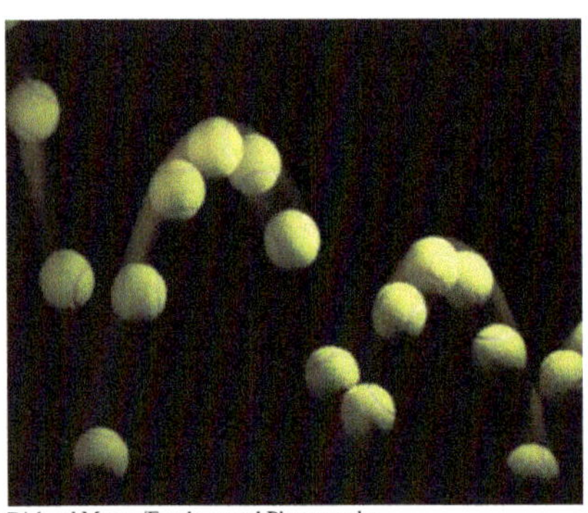

Richard Megna/Fundamental Photographs

Figure 4-8 A stroboscopic photograph of a yellow tennis ball bouncing off a hard surface. Between impacts, the ball has projectile motion.

During its two-dimensional motion, the projectile's position vector $\vec{r}$ and velocity vector $\vec{v}$ change continuously, but its acceleration vector $\vec{a}$ is constant and *always* directed vertically downward. The projectile has *no* horizontal acceleration.

Projectile motion, like that in Figs. 4-8 and 4-9, looks complicated, but we have the following simplifying feature (known from experiment):

> In projectile motion, the horizontal motion and the vertical motion are independent of each other; that is, neither motion affects the other.

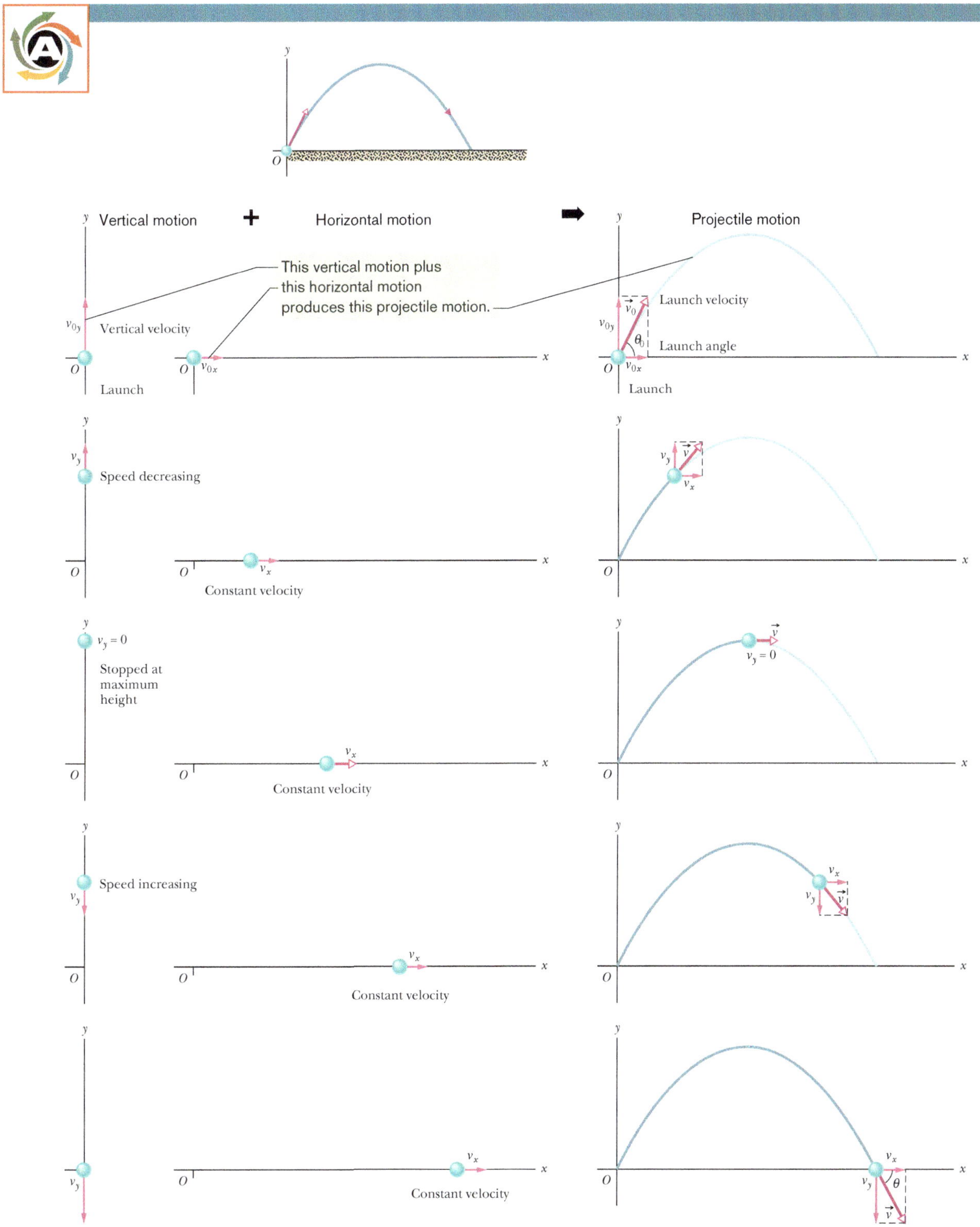

Figure 4-9 The *projectile motion* of an object launched into the air at the origin of a coordinate system and with launch velocity $\vec{v}_0$ at angle θ_0. The motion is a combination of vertical motion (constant acceleration) and horizontal motion (constant velocity), as shown by the velocity components.

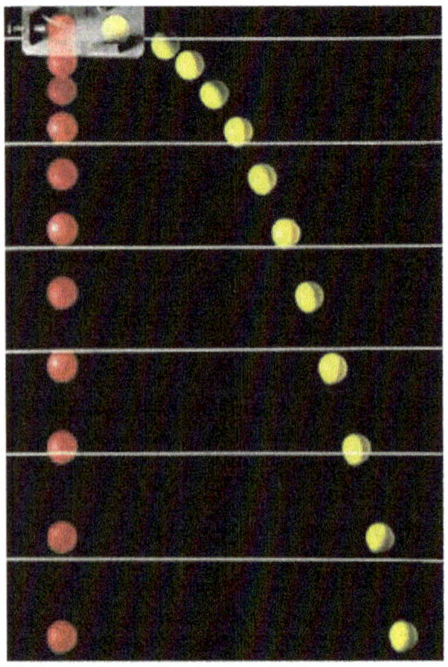

Richard Megna/Fundamental Photographs

Figure 4-10 One ball is released from rest at the same instant that another ball is shot horizontally to the right. Their vertical motions are identical.

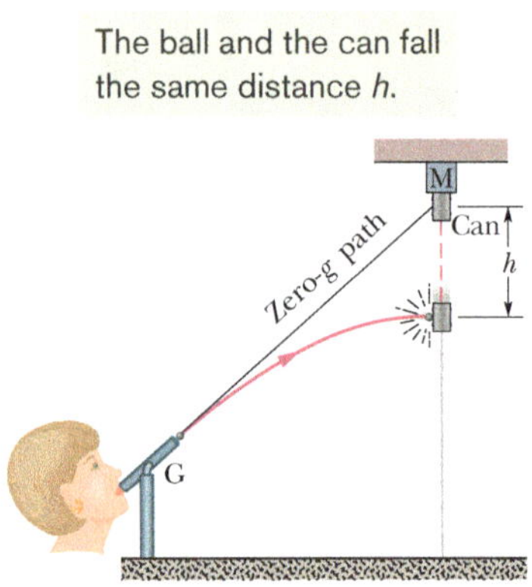

Figure 4-11 The projectile ball always hits the falling can. Each falls a distance h from where it would be were there no free-fall acceleration.

This feature allows us to break up a problem involving two-dimensional motion into two separate and easier one-dimensional problems, one for the horizontal motion (with *zero acceleration*) and one for the vertical motion (with *constant downward acceleration*). Here are two experiments that show that the horizontal motion and the vertical motion are independent.

Two Golf Balls

Figure 4-10 is a stroboscopic photograph of two golf balls, one simply released and the other shot horizontally by a spring. The golf balls have the same vertical motion, both falling through the same vertical distance in the same interval of time. *The fact that one ball is moving horizontally while it is falling has no effect on its vertical motion;* that is, the horizontal and vertical motions are independent of each other.

A Great Student Rouser

In Fig. 4-11, a blowgun G using a ball as a projectile is aimed directly at a can suspended from a magnet M. Just as the ball leaves the blowgun, the can is released. If g (the magnitude of the free-fall acceleration) were zero, the ball would follow the straight-line path shown in Fig. 4-11 and the can would float in place after the magnet released it. The ball would certainly hit the can. However, g is *not* zero, but the ball *still* hits the can! As Fig. 4-11 shows, during the time of flight of the ball, both ball and can fall the same distance h from their zero-g locations. The harder the demonstrator blows, the greater is the ball's initial speed, the shorter the flight time, and the smaller the value of h.

Checkpoint 3

At a certain instant, a fly ball has velocity $\vec{v} = 25\hat{i} - 4.9\hat{j}$ (the x axis is horizontal, the y axis is upward, and $\vec{v}$ is in meters per second). Has the ball passed its highest point?

The Horizontal Motion

Now we are ready to analyze projectile motion, horizontally and vertically. We start with the horizontal motion. Because there is *no acceleration* in the horizontal direction, the horizontal component v_x of the projectile's velocity remains unchanged from its initial value v_{0x} throughout the motion, as demonstrated in Fig. 4-12. At any time t, the projectile's horizontal displacement $x - x_0$ from an initial position x_0 is given by Eq. 2-15 with $a = 0$, which we write as

$$x - x_0 = v_{0x}t.$$

Because $v_{0x} = v_0 \cos \theta_0$, this becomes

$$x - x_0 = (v_0 \cos \theta_0)t. \qquad (4\text{-}21)$$

The Vertical Motion

The vertical motion is the motion we discussed in Module 2-5 for a particle in free fall. Most important is that the acceleration is constant. Thus, the equations of Table 2-1 apply, provided we substitute $-g$ for a and switch to y notation. Then, for example, Eq. 2-15 becomes

$$\begin{aligned} y - y_0 &= v_{0y}t - \tfrac{1}{2}gt^2 \\ &= (v_0 \sin \theta_0)t - \tfrac{1}{2}gt^2, \end{aligned} \qquad (4\text{-}22)$$

where the initial vertical velocity component v_{0y} is replaced with the equivalent $v_0 \sin \theta_0$. Similarly, Eqs. 2-11 and 2-16 become

$$v_y = v_0 \sin \theta_0 - gt \qquad (4\text{-}23)$$

and

$$v_y^2 = (v_0 \sin \theta_0)^2 - 2g(y - y_0). \qquad (4\text{-}24)$$

As is illustrated in Fig. 4-9 and Eq. 4-23, the vertical velocity component behaves just as for a ball thrown vertically upward. It is directed upward initially, and its magnitude steadily decreases to zero, *which marks the maximum height of the path.* The vertical velocity component then reverses direction, and its magnitude becomes larger with time.

The Equation of the Path

We can find the equation of the projectile's path (its **trajectory**) by eliminating time t between Eqs. 4-21 and 4-22. Solving Eq. 4-21 for t and substituting into Eq. 4-22, we obtain, after a little rearrangement,

$$y = (\tan \theta_0)x - \frac{gx^2}{2(v_0 \cos \theta_0)^2} \quad \text{(trajectory)}. \tag{4-25}$$

This is the equation of the path shown in Fig. 4-9. In deriving it, for simplicity we let $x_0 = 0$ and $y_0 = 0$ in Eqs. 4-21 and 4-22, respectively. Because g, θ_0, and v_0 are constants, Eq. 4-25 is of the form $y = ax + bx^2$, in which a and b are constants. This is the equation of a parabola, so the path is *parabolic*.

The Horizontal Range

The *horizontal range* R of the projectile is the *horizontal* distance the projectile has traveled when it returns to its initial height (the height at which it is launched). To find range R, let us put $x - x_0 = R$ in Eq. 4-21 and $y - y_0 = 0$ in Eq. 4-22, obtaining

$$R = (v_0 \cos \theta_0)t$$

and

$$0 = (v_0 \sin \theta_0)t - \tfrac{1}{2}gt^2.$$

Eliminating t between these two equations yields

$$R = \frac{2v_0^2}{g} \sin \theta_0 \cos \theta_0.$$

Using the identity $\sin 2\theta_0 = 2 \sin \theta_0 \cos \theta_0$ (see Appendix E), we obtain

$$R = \frac{v_0^2}{g} \sin 2\theta_0. \tag{4-26}$$

This equation does *not* give the horizontal distance traveled by a projectile when the final height is not the launch height. Note that R in Eq. 4-26 has its maximum value when $\sin 2\theta_0 = 1$, which corresponds to $2\theta_0 = 90°$ or $\theta_0 = 45°$.

The horizontal range R is maximum for a launch angle of 45°.

However, when the launch and landing heights differ, as in many sports, a launch angle of 45° does not yield the maximum horizontal distance.

The Effects of the Air

We have assumed that the air through which the projectile moves has no effect on its motion. However, in many situations, the disagreement between our calculations and the actual motion of the projectile can be large because the air resists (opposes) the motion. Figure 4-13, for example, shows two paths for a fly ball that leaves the bat at an angle of 60° with the horizontal and an initial speed of 44.7 m/s. Path I (the baseball player's fly ball) is a calculated path that approximates normal conditions of play, in air. Path II (the physics professor's fly ball) is the path the ball would follow in a vacuum.

Jamie Budge

Figure 4-12 The vertical component of this skateboarder's velocity is changing but not the horizontal component, which matches the skateboard's velocity. As a result, the skateboard stays underneath him, allowing him to land on it.

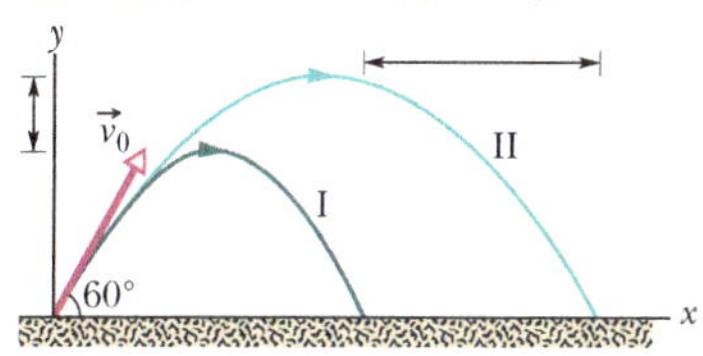

Figure 4-13 (I) The path of a fly ball calculated by taking air resistance into account. (II) The path the ball would follow in a vacuum, calculated by the methods of this chapter. See Table 4-1 for corresponding data. (Based on "The Trajectory of a Fly Ball," by Peter J. Brancazio, *The Physics Teacher,* January 1985.)

Table 4-1 Two Fly Balls[a]

	Path I (Air)	Path II (Vacuum)
Range	98.5 m	177 m
Maximum height	53.0 m	76.8 m
Time of flight	6.6 s	7.9 s

[a]See Fig. 4-13. The launch angle is 60° and the launch speed is 44.7 m/s.

Checkpoint 4

A fly ball is hit to the outfield. During its flight (ignore the effects of the air), what happens to its (a) horizontal and (b) vertical components of velocity? What are the (c) horizontal and (d) vertical components of its acceleration during ascent, during descent, and at the topmost point of its flight?

Sample Problem 4.04 Projectile dropped from airplane

In Fig. 4-14, a rescue plane flies at 198 km/h (= 55.0 m/s) and constant height $h = 500$ m toward a point directly over a victim, where a rescue capsule is to land.

(a) What should be the angle ϕ of the pilot's line of sight to the victim when the capsule release is made?

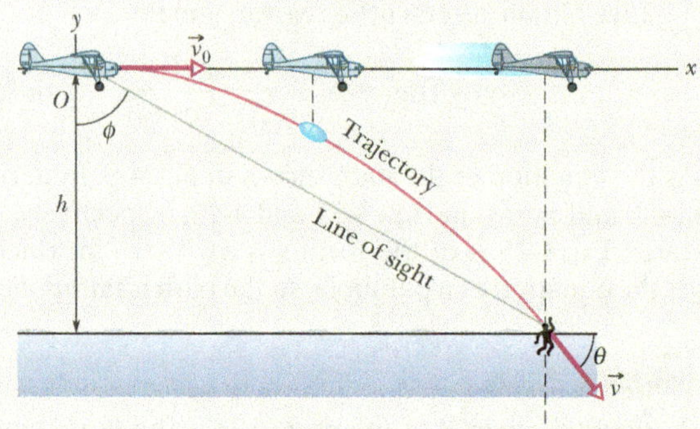

Figure 4-14 A plane drops a rescue capsule while moving at constant velocity in level flight. While falling, the capsule remains under the plane.

KEY IDEAS

Once released, the capsule is a projectile, so its horizontal and vertical motions can be considered separately (we need not consider the actual curved path of the capsule).

Calculations: In Fig. 4-14, we see that ϕ is given by

$$\phi = \tan^{-1}\frac{x}{h}, \tag{4-27}$$

where x is the horizontal coordinate of the victim (and of the capsule when it hits the water) and $h = 500$ m. We should be able to find x with Eq. 4-21:

$$x - x_0 = (v_0 \cos \theta_0)t. \tag{4-28}$$

Here we know that $x_0 = 0$ because the origin is placed at the point of release. Because the capsule is *released* and not shot from the plane, its initial velocity $\vec{v}_0$ is equal to the plane's velocity. Thus, we know also that the initial velocity has magnitude $v_0 = 55.0$ m/s and angle $\theta_0 = 0°$ (measured relative to the positive direction of the x axis). However, we do not know the time t the capsule takes to move from the plane to the victim.

To find t, we next consider the *vertical* motion and specifically Eq. 4-22:

$$y - y_0 = (v_0 \sin \theta_0)t - \tfrac{1}{2}gt^2. \tag{4-29}$$

Here the vertical displacement $y - y_0$ of the capsule is -500 m (the negative value indicates that the capsule moves *downward*). So,

$$-500 \text{ m} = (55.0 \text{ m/s})(\sin 0°)t - \tfrac{1}{2}(9.8 \text{ m/s}^2)t^2. \tag{4-30}$$

Solving for t, we find $t = 10.1$ s. Using that value in Eq. 4-28 yields

$$x - 0 = (55.0 \text{ m/s})(\cos 0°)(10.1 \text{ s}), \tag{4-31}$$

or

$$x = 555.5 \text{ m}.$$

Then Eq. 4-27 gives us

$$\phi = \tan^{-1}\frac{555.5 \text{ m}}{500 \text{ m}} = 48.0°. \quad \text{(Answer)}$$

(b) As the capsule reaches the water, what is its velocity $\vec{v}$?

KEY IDEAS

(1) The horizontal and vertical components of the capsule's velocity are independent. (2) Component v_x does not change from its initial value $v_{0x} = v_0 \cos \theta_0$ because there is no horizontal acceleration. (3) Component v_y changes from its initial value $v_{0y} = v_0 \sin \theta_0$ because there is a vertical acceleration.

Calculations: When the capsule reaches the water,

$$v_x = v_0 \cos \theta_0 = (55.0 \text{ m/s})(\cos 0°) = 55.0 \text{ m/s}.$$

Using Eq. 4-23 and the capsule's time of fall $t = 10.1$ s, we also find that when the capsule reaches the water,

$$\begin{aligned} v_y &= v_0 \sin \theta_0 - gt \\ &= (55.0 \text{ m/s})(\sin 0°) - (9.8 \text{ m/s}^2)(10.1 \text{ s}) \\ &= -99.0 \text{ m/s}. \end{aligned}$$

Thus, at the water

$$\vec{v} = (55.0 \text{ m/s})\hat{i} - (99.0 \text{ m/s})\hat{j}. \quad \text{(Answer)}$$

From Eq. 3-6, the magnitude and the angle of $\vec{v}$ are

$$v = 113 \text{ m/s} \quad \text{and} \quad \theta = -60.9°. \quad \text{(Answer)}$$

WILEY PLUS Additional examples, video, and practice available at *WileyPLUS*

Sample Problem 4.05 Launched into the air from a water slide

One of the most dramatic videos on the web (but entirely fictitious) supposedly shows a man sliding along a long water slide and then being launched into the air to land in a water pool. Let's attach some reasonable numbers to such a flight to calculate the velocity with which the man would have hit the water. Figure 4-15*a* indicates the launch and landing sites and includes a superimposed coordinate system with its origin conveniently located at the launch site. From the video we take the horizontal flight distance as $D = 20.0$ m, the flight time as $t = 2.50$ s, and the launch angle as $\theta_0 = 40.0°$. Find the magnitude of the velocity at launch and at landing.

KEY IDEAS

(1) For projectile motion, we can apply the equations for constant acceleration along the horizontal and vertical axes *separately*. (2) Throughout the flight, the vertical acceleration is $a_y = -g = -9.8$ m/s² and the horizontal acceleration is $a_x = 0$.

Calculations: In most projectile problems, the initial challenge is to figure out where to start. There is nothing wrong with trying out various equations, to see if we can somehow get to the velocities. But here is a clue. Because we are going to apply the constant-acceleration equations separately to the x and y motions, we should find the horizontal and vertical components of the velocities at launch and at landing. For each site, we can then combine the velocity components to get the velocity.

Because we know the horizontal displacement $D = 20.0$ m, let's start with the horizontal motion. Since $a_x = 0$, we know that the horizontal velocity component v_x is constant during the flight and thus is always equal to the horizontal component v_{0x} at launch. We can relate that component, the displacement $x - x_0$, and the flight time $t = 2.50$ s with Eq. 2-15:

$$x - x_0 = v_{0x}t + \tfrac{1}{2}a_x t^2. \tag{4-32}$$

Substituting $a_x = 0$, this becomes Eq. 4-21. With $x - x_0 = D$, we then write

$$20 \text{ m} = v_{0x}(2.50 \text{ s}) + \tfrac{1}{2}(0)(2.50 \text{ s})^2$$
$$v_{0x} = 8.00 \text{ m/s}.$$

That is a component of the launch velocity, but we need the magnitude of the full vector, as shown in Fig. 4-15*b*, where the components form the legs of a right triangle and the full vector forms the hypotenuse. We can then apply a trig definition to find the magnitude of the full velocity at launch:

$$\cos\theta_0 = \frac{v_{0x}}{v_0},$$

and so

$$v_0 = \frac{v_{0x}}{\cos\theta_0} = \frac{8.00 \text{ m/s}}{\cos 40°}$$
$$= 10.44 \text{ m/s} \approx 10.4 \text{ m/s}. \quad \text{(Answer)}$$

Now let's go after the magnitude v of the landing velocity. We already know the horizontal component, which does not change from its initial value of 8.00 m/s. To find the vertical component v_y and because we know the elapsed time $t = 2.50$ s and the vertical acceleration $a_y = -9.8$ m/s², let's rewrite Eq. 2-11 as

$$v_y = v_{0y} + a_y t$$

and then (from Fig. 4-15*b*) as

$$v_y = v_0 \sin\theta_0 + a_y t. \tag{4-33}$$

Substituting $a_y = -g$, this becomes Eq. 4-23. We can then write

$$v_y = (10.44 \text{ m/s}) \sin(40.0°) - (9.8 \text{ m/s}^2)(2.50 \text{ s})$$
$$= -17.78 \text{ m/s}.$$

Now that we know both components of the landing velocity, we use Eq. 3-6 to find the velocity magnitude:

$$v = \sqrt{v_x^2 + v_y^2}$$
$$= \sqrt{(8.00 \text{ m/s})^2 + (-17.78 \text{ m/s})^2}$$
$$= 19.49 \text{ m/s} \approx 19.5 \text{ m/s}. \quad \text{(Answer)}$$

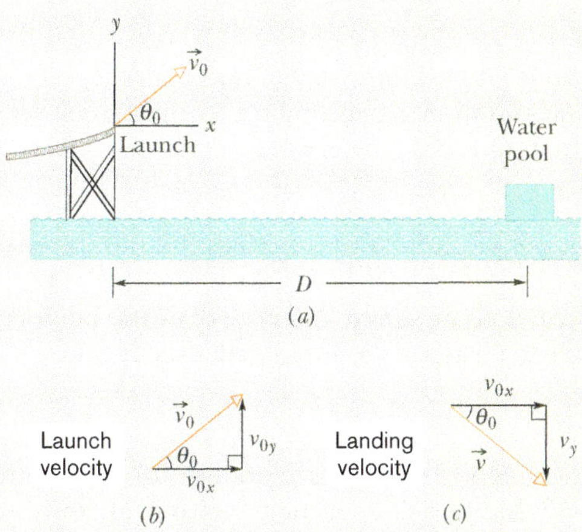

Figure 4-15 (*a*) Launch from a water slide, to land in a water pool. The velocity at (*b*) launch and (*c*) landing.

4-5 UNIFORM CIRCULAR MOTION

Learning Objectives

After reading this module, you should be able to . . .

4.16 Sketch the path taken in uniform circular motion and explain the velocity and acceleration vectors (magnitude and direction) during the motion.

4.17 Apply the relationships between the radius of the circular path, the period, the particle's speed, and the particle's acceleration magnitude.

Key Ideas

- If a particle travels along a circle or circular arc of radius r at constant speed v, it is said to be in uniform circular motion and has an acceleration $\vec{a}$ of constant magnitude

$$a = \frac{v^2}{r}.$$

The direction of $\vec{a}$ is toward the center of the circle or circular arc, and $\vec{a}$ is said to be centripetal. The time for the particle to complete a circle is

$$T = \frac{2\pi r}{v}.$$

T is called the period of revolution, or simply the period, of the motion.

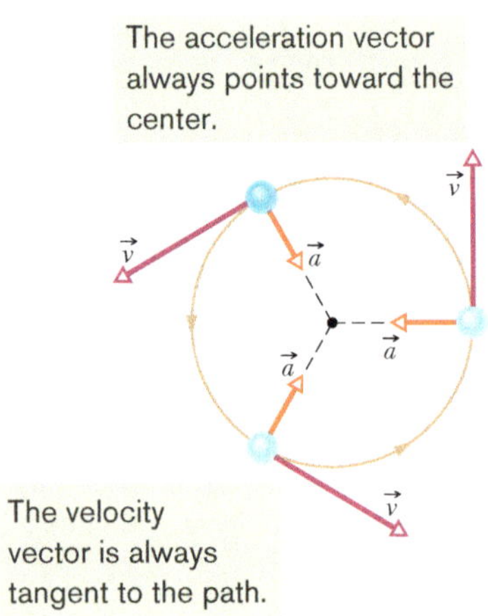

Figure 4-16 Velocity and acceleration vectors for uniform circular motion.

Uniform Circular Motion

A particle is in **uniform circular motion** if it travels around a circle or a circular arc at constant (*uniform*) speed. Although the speed does not vary, *the particle is accelerating* because the velocity changes in direction.

Figure 4-16 shows the relationship between the velocity and acceleration vectors at various stages during uniform circular motion. Both vectors have constant magnitude, but their directions change continuously. The velocity is always directed tangent to the circle in the direction of motion. The acceleration is always directed *radially inward.* Because of this, the acceleration associated with uniform circular motion is called a **centripetal** (meaning "center seeking") **acceleration.** As we prove next, the magnitude of this acceleration $\vec{a}$ is

$$a = \frac{v^2}{r} \quad \text{(centripetal acceleration)}, \tag{4-34}$$

where r is the radius of the circle and v is the speed of the particle.

In addition, during this acceleration at constant speed, the particle travels the circumference of the circle (a distance of $2\pi r$) in time

$$T = \frac{2\pi r}{v} \quad \text{(period)}. \tag{4-35}$$

T is called the *period of revolution,* or simply the *period,* of the motion. It is, in general, the time for a particle to go around a closed path exactly once.

Proof of Eq. 4-34

To find the magnitude and direction of the acceleration for uniform circular motion, we consider Fig. 4-17. In Fig. 4-17*a*, particle p moves at constant speed v around a circle of radius r. At the instant shown, p has coordinates x_p and y_p.

Recall from Module 4-2 that the velocity $\vec{v}$ of a moving particle is always tangent to the particle's path at the particle's position. In Fig. 4-17*a*, that means $\vec{v}$ is perpendicular to a radius r drawn to the particle's position. Then the angle θ that $\vec{v}$ makes with a vertical at p equals the angle θ that radius r makes with the x axis.

The scalar components of $\vec{v}$ are shown in Fig. 4-17*b*. With them, we can write the velocity $\vec{v}$ as

$$\vec{v} = v_x\hat{\mathrm{i}} + v_y\hat{\mathrm{j}} = (-v \sin\theta)\hat{\mathrm{i}} + (v\cos\theta)\hat{\mathrm{j}}. \tag{4-36}$$

Now, using the right triangle in Fig. 4-17*a*, we can replace $\sin\theta$ with y_p/r and $\cos\theta$ with x_p/r to write

$$\vec{v} = \left(-\frac{vy_p}{r}\right)\hat{\mathrm{i}} + \left(\frac{vx_p}{r}\right)\hat{\mathrm{j}}. \tag{4-37}$$

To find the acceleration $\vec{a}$ of particle p, we must take the time derivative of this equation. Noting that speed v and radius r do not change with time, we obtain

$$\vec{a} = \frac{d\vec{v}}{dt} = \left(-\frac{v}{r}\frac{dy_p}{dt}\right)\hat{\mathrm{i}} + \left(\frac{v}{r}\frac{dx_p}{dt}\right)\hat{\mathrm{j}}. \tag{4-38}$$

Now note that the rate dy_p/dt at which y_p changes is equal to the velocity component v_y. Similarly, $dx_p/dt = v_x$, and, again from Fig. 4-17*b*, we see that $v_x = -v\sin\theta$ and $v_y = v\cos\theta$. Making these substitutions in Eq. 4-38, we find

$$\vec{a} = \left(-\frac{v^2}{r}\cos\theta\right)\hat{\mathrm{i}} + \left(-\frac{v^2}{r}\sin\theta\right)\hat{\mathrm{j}}. \tag{4-39}$$

This vector and its components are shown in Fig. 4-17*c*. Following Eq. 3-6, we find

$$a = \sqrt{a_x^2 + a_y^2} = \frac{v^2}{r}\sqrt{(\cos\theta)^2 + (\sin\theta)^2} = \frac{v^2}{r}\sqrt{1} = \frac{v^2}{r},$$

as we wanted to prove. To orient $\vec{a}$, we find the angle ϕ shown in Fig. 4-17*c*:

$$\tan\phi = \frac{a_y}{a_x} = \frac{-(v^2/r)\sin\theta}{-(v^2/r)\cos\theta} = \tan\theta.$$

Thus, $\phi = \theta$, which means that $\vec{a}$ is directed along the radius r of Fig. 4-17*a*, toward the circle's center, as we wanted to prove.

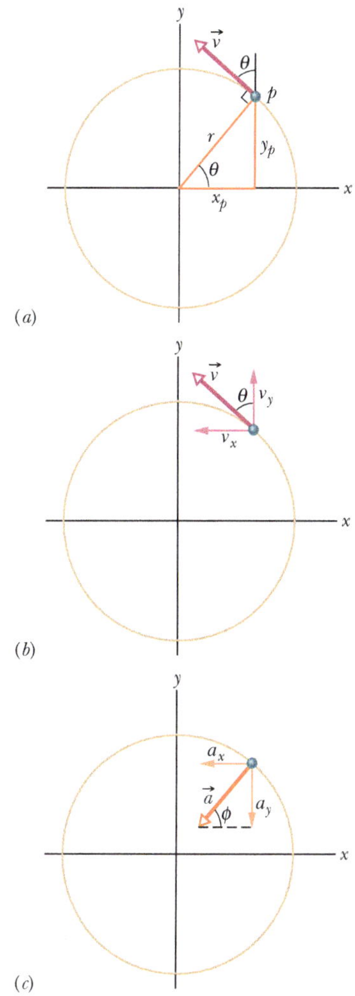

Figure 4-17 Particle p moves in counterclockwise uniform circular motion. (*a*) Its position and velocity $\vec{v}$ at a certain instant. (*b*) Velocity $\vec{v}$. (*c*) Acceleration $\vec{a}$.

Checkpoint 5

An object moves at constant speed along a circular path in a horizontal xy plane, with the center at the origin. When the object is at $x = -2$ m, its velocity is $-(4 \text{ m/s})\hat{\mathrm{j}}$. Give the object's (a) velocity and (b) acceleration at $y = 2$ m.

Sample Problem 4.06 Top gun pilots in turns

"Top gun" pilots have long worried about taking a turn too tightly. As a pilot's body undergoes centripetal acceleration, with the head toward the center of curvature, the blood pressure in the brain decreases, leading to loss of brain function.

There are several warning signs. When the centripetal acceleration is 2*g* or 3*g*, the pilot feels heavy. At about 4*g*, the pilot's vision switches to black and white and narrows to "tunnel vision." If that acceleration is sustained or increased, vision ceases and, soon after, the pilot is unconscious—a condition known as *g*-LOC for "*g*-induced loss of consciousness."

What is the magnitude of the acceleration, in *g* units, of a pilot whose aircraft enters a horizontal circular turn with a velocity of $\vec{v}_i = (400\hat{\mathrm{i}} + 500\hat{\mathrm{j}})$ m/s and 24.0 s later leaves the turn with a velocity of $\vec{v}_f = (-400\hat{\mathrm{i}} - 500\hat{\mathrm{j}})$ m/s?

KEY IDEAS

We assume the turn is made with uniform circular motion. Then the pilot's acceleration is centripetal and has magnitude a given by Eq. 4-34 ($a = v^2/R$), where R is the circle's radius. Also, the time required to complete a full circle is the period given by Eq. 4-35 ($T = 2\pi R/v$).

Calculations: Because we do not know radius R, let's solve Eq. 4-35 for R and substitute into Eq. 4-34. We find

$$a = \frac{2\pi v}{T}.$$

To get the constant speed v, let's substitute the components of the initial velocity into Eq. 3-6:

$$v = \sqrt{(400 \text{ m/s})^2 + (500 \text{ m/s})^2} = 640.31 \text{ m/s}.$$

To find the period T of the motion, first note that the final velocity is the reverse of the initial velocity. This means the aircraft leaves on the opposite side of the circle from the initial point and must have completed half a circle in the given 24.0 s. Thus a full circle would have taken $T = 48.0$ s. Substituting these values into our equation for a, we find

$$a = \frac{2\pi(640.31 \text{ m/s})}{48.0 \text{ s}} = 83.81 \text{ m/s}^2 \approx 8.6g. \quad \text{(Answer)}$$

WILEY PLUS Additional examples, video, and practice available at *WileyPLUS*

4-6 RELATIVE MOTION IN ONE DIMENSION

Learning Objective

After reading this module, you should be able to . . .

4.18 Apply the relationship between a particle's position, velocity, and acceleration as measured from two reference frames that move relative to each other at constant velocity and along a single axis.

Key Idea

- When two frames of reference A and B are moving relative to each other at constant velocity, the velocity of a particle P as measured by an observer in frame A usually differs from that measured from frame B. The two measured velocities are related by

$$\vec{v}_{PA} = \vec{v}_{PB} + \vec{v}_{BA},$$

where $\vec{v}_{BA}$ is the velocity of B with respect to A. Both observers measure the same acceleration for the particle:

$$\vec{a}_{PA} = \vec{a}_{PB}.$$

Relative Motion in One Dimension

Suppose you see a duck flying north at 30 km/h. To another duck flying alongside, the first duck seems to be stationary. In other words, the velocity of a particle depends on the **reference frame** of whoever is observing or measuring the velocity. For our purposes, a reference frame is the physical object to which we attach our coordinate system. In everyday life, that object is the ground. For example, the speed listed on a speeding ticket is always measured relative to the ground. The speed relative to the police officer would be different if the officer were moving while making the speed measurement.

Suppose that Alex (at the origin of frame A in Fig. 4-18) is parked by the side of a highway, watching car P (the "particle") speed past. Barbara (at the origin of frame B) is driving along the highway at constant speed and is also watching car P. Suppose that they both measure the position of the car at a given moment. From Fig. 4-18 we see that

$$x_{PA} = x_{PB} + x_{BA}. \tag{4-40}$$

Frame B moves past frame A while both observe P.

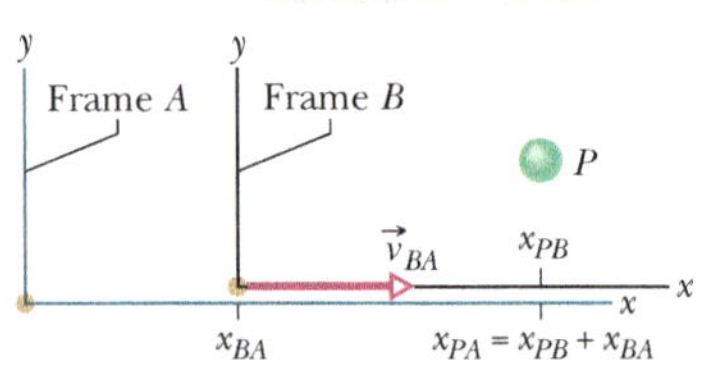

Figure 4-18 Alex (frame A) and Barbara (frame B) watch car P, as both B and P move at different velocities along the common x axis of the two frames. At the instant shown, x_{BA} is the coordinate of B in the A frame. Also, P is at coordinate x_{PB} in the B frame and coordinate $x_{PA} = x_{PB} + x_{BA}$ in the A frame.

The equation is read: "The coordinate x_{PA} of P as measured by A *is equal to* the coordinate x_{PB} of P as measured by B *plus* the coordinate x_{BA} of B as measured by A." Note how this reading is supported by the sequence of the subscripts.

Taking the time derivative of Eq. 4-40, we obtain

$$\frac{d}{dt}(x_{PA}) = \frac{d}{dt}(x_{PB}) + \frac{d}{dt}(x_{BA}).$$

Thus, the velocity components are related by

$$v_{PA} = v_{PB} + v_{BA}. \tag{4-41}$$

This equation is read: "The velocity v_{PA} of P as measured by A *is equal to* the velocity v_{PB} of P as measured by B *plus* the velocity v_{BA} of B as measured by A." The term v_{BA} is the velocity of frame B relative to frame A.

Here we consider only frames that move at constant velocity relative to each other. In our example, this means that Barbara (frame B) drives always at constant velocity v_{BA} relative to Alex (frame A). Car P (the moving particle), however, can change speed and direction (that is, it can accelerate).

To relate an acceleration of P as measured by Barbara and by Alex, we take the time derivative of Eq. 4-41:

$$\frac{d}{dt}(v_{PA}) = \frac{d}{dt}(v_{PB}) + \frac{d}{dt}(v_{BA}).$$

Because v_{BA} is constant, the last term is zero and we have

$$a_{PA} = a_{PB}. \quad (4\text{-}42)$$

In other words,

Observers on different frames of reference that move at constant velocity relative to each other will measure the same acceleration for a moving particle.

Sample Problem 4.07 Relative motion, one dimensional, Alex and Barbara

In Fig. 4-18, suppose that Barbara's velocity relative to Alex is a constant $v_{BA} = 52$ km/h and car P is moving in the negative direction of the x axis.

(a) If Alex measures a constant $v_{PA} = -78$ km/h for car P, what velocity v_{PB} will Barbara measure?

KEY IDEAS

We can attach a frame of reference A to Alex and a frame of reference B to Barbara. Because the frames move at constant velocity relative to each other along one axis, we can use Eq. 4-41 ($v_{PA} = v_{PB} + v_{BA}$) to relate v_{PB} to v_{PA} and v_{BA}.

Calculation: We find

$$-78 \text{ km/h} = v_{PB} + 52 \text{ km/h}.$$

Thus, $$v_{PB} = -130 \text{ km/h}. \quad \text{(Answer)}$$

Comment: If car P were connected to Barbara's car by a cord wound on a spool, the cord would be unwinding at a speed of 130 km/h as the two cars separated.

(b) If car P brakes to a stop relative to Alex (and thus relative to the ground) in time $t = 10$ s at constant acceleration, what is its acceleration a_{PA} relative to Alex?

KEY IDEAS

To calculate the acceleration of car P *relative to Alex,* we must use the car's velocities *relative to Alex.* Because the acceleration is constant, we can use Eq. 2-11 ($v = v_0 + at$) to relate the acceleration to the initial and final velocities of P.

Calculation: The initial velocity of P relative to Alex is $v_{PA} = -78$ km/h and the final velocity is 0. Thus, the acceleration relative to Alex is

$$a_{PA} = \frac{v - v_0}{t} = \frac{0 - (-78 \text{ km/h})}{10 \text{ s}} \frac{1 \text{ m/s}}{3.6 \text{ km/h}}$$
$$= 2.2 \text{ m/s}^2. \quad \text{(Answer)}$$

(c) What is the acceleration a_{PB} of car P relative to Barbara during the braking?

KEY IDEA

To calculate the acceleration of car P *relative to Barbara,* we must use the car's velocities *relative to Barbara.*

Calculation: We know the initial velocity of P relative to Barbara from part (a) ($v_{PB} = -130$ km/h). The final velocity of P relative to Barbara is -52 km/h (because this is the velocity of the stopped car relative to the moving Barbara). Thus,

$$a_{PB} = \frac{v - v_0}{t} = \frac{-52 \text{ km/h} - (-130 \text{ km/h})}{10 \text{ s}} \frac{1 \text{ m/s}}{3.6 \text{ km/h}}$$
$$= 2.2 \text{ m/s}^2. \quad \text{(Answer)}$$

Comment: We should have foreseen this result: Because Alex and Barbara have a constant relative velocity, they must measure the same acceleration for the car.

WileyPLUS Additional examples, video, and practice available at *WileyPLUS*

4-7 RELATIVE MOTION IN TWO DIMENSIONS

Learning Objective

After reading this module, you should be able to . . .

4.19 Apply the relationship between a particle's position, velocity, and acceleration as measured from two reference frames that move relative to each other at constant velocity and in two dimensions.

Key Idea

- When two frames of reference A and B are moving relative to each other at constant velocity, the velocity of a particle P as measured by an observer in frame A usually differs from that measured from frame B. The two measured velocities are related by

$$\vec{v}_{PA} = \vec{v}_{PB} + \vec{v}_{BA},$$

where $\vec{v}_{BA}$ is the velocity of B with respect to A. Both observers measure the same acceleration for the particle:

$$\vec{a}_{PA} = \vec{a}_{PB}.$$

Relative Motion in Two Dimensions

Our two observers are again watching a moving particle P from the origins of reference frames A and B, while B moves at a constant velocity $\vec{v}_{BA}$ relative to A. (The corresponding axes of these two frames remain parallel.) Figure 4-19 shows a certain instant during the motion. At that instant, the position vector of the origin of B relative to the origin of A is $\vec{r}_{BA}$. Also, the position vectors of particle P are $\vec{r}_{PA}$ relative to the origin of A and $\vec{r}_{PB}$ relative to the origin of B. From the arrangement of heads and tails of those three position vectors, we can relate the vectors with

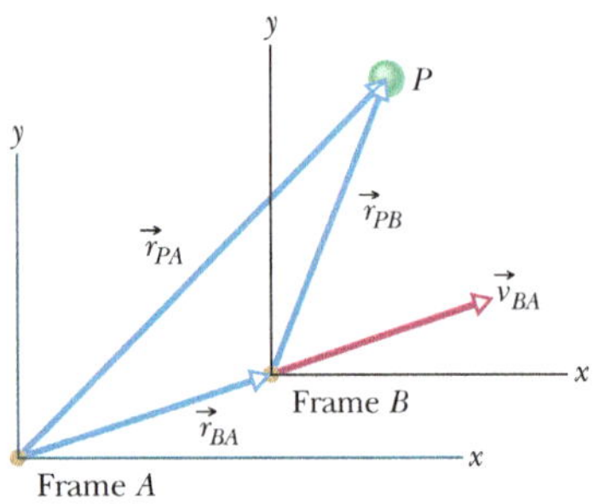

Figure 4-19 Frame B has the constant two-dimensional velocity $\vec{v}_{BA}$ relative to frame A. The position vector of B relative to A is $\vec{r}_{BA}$. The position vectors of particle P are $\vec{r}_{PA}$ relative to A and $\vec{r}_{PB}$ relative to B.

$$\vec{r}_{PA} = \vec{r}_{PB} + \vec{r}_{BA}. \tag{4-43}$$

By taking the time derivative of this equation, we can relate the velocities $\vec{v}_{PA}$ and $\vec{v}_{PB}$ of particle P relative to our observers:

$$\vec{v}_{PA} = \vec{v}_{PB} + \vec{v}_{BA}. \tag{4-44}$$

By taking the time derivative of this relation, we can relate the accelerations $\vec{a}_{PA}$ and $\vec{a}_{PB}$ of the particle P relative to our observers. However, note that because $\vec{v}_{BA}$ is constant, its time derivative is zero. Thus, we get

$$\vec{a}_{PA} = \vec{a}_{PB}. \tag{4-45}$$

As for one-dimensional motion, we have the following rule: Observers on different frames of reference that move at constant velocity relative to each other will measure the *same* acceleration for a moving particle.

Sample Problem 4.08 Relative motion, two dimensional, airplanes

In Fig. 4-20*a*, a plane moves due east while the pilot points the plane somewhat south of east, toward a steady wind that blows to the northeast. The plane has velocity $\vec{v}_{PW}$ relative to the wind, with an airspeed (speed relative to the wind) of 215 km/h, directed at angle θ south of east. The wind has velocity $\vec{v}_{WG}$ relative to the ground with speed 65.0 km/h, directed 20.0° east of north. What is the magnitude of the velocity $\vec{v}_{PG}$ of the plane relative to the ground, and what is θ?

KEY IDEAS

The situation is like the one in Fig. 4-19. Here the moving particle P is the plane, frame A is attached to the ground (call it G), and frame B is "attached" to the wind (call it W). We need a vector diagram like Fig. 4-19 but with three velocity vectors.

Calculations: First we construct a sentence that relates the three vectors shown in Fig. 4-20*b*:

$$\underset{(PG)}{\begin{array}{c}\text{velocity of plane}\\ \text{relative to ground}\end{array}} = \underset{(PW)}{\begin{array}{c}\text{velocity of plane}\\ \text{relative to wind}\end{array}} + \underset{(WG)}{\begin{array}{c}\text{velocity of wind}\\ \text{relative to ground.}\end{array}}$$

This relation is written in vector notation as

$$\vec{v}_{PG} = \vec{v}_{PW} + \vec{v}_{WG}. \tag{4-46}$$

We need to resolve the vectors into components on the coordinate system of Fig. 4-20*b* and then solve Eq. 4-46 axis by axis. For the *y* components, we find

$$v_{PG,y} = v_{PW,y} + v_{WG,y}$$

or $\quad 0 = -(215 \text{ km/h}) \sin \theta + (65.0 \text{ km/h})(\cos 20.0°).$

Solving for θ gives us

$$\theta = \sin^{-1} \frac{(65.0 \text{ km/h})(\cos 20.0°)}{215 \text{ km/h}} = 16.5°. \qquad \text{(Answer)}$$

Similarly, for the *x* components we find

$$v_{PG,x} = v_{PW,x} + v_{WG,x}.$$

Here, because $\vec{v}_{PG}$ is parallel to the *x* axis, the component $v_{PG,x}$ is equal to the magnitude v_{PG}. Substituting this notation and the value $\theta = 16.5°$, we find

$$v_{PG} = (215 \text{ km/h})(\cos 16.5°) + (65.0 \text{ km/h})(\sin 20.0°)$$
$$= 228 \text{ km/h}. \qquad \text{(Answer)}$$

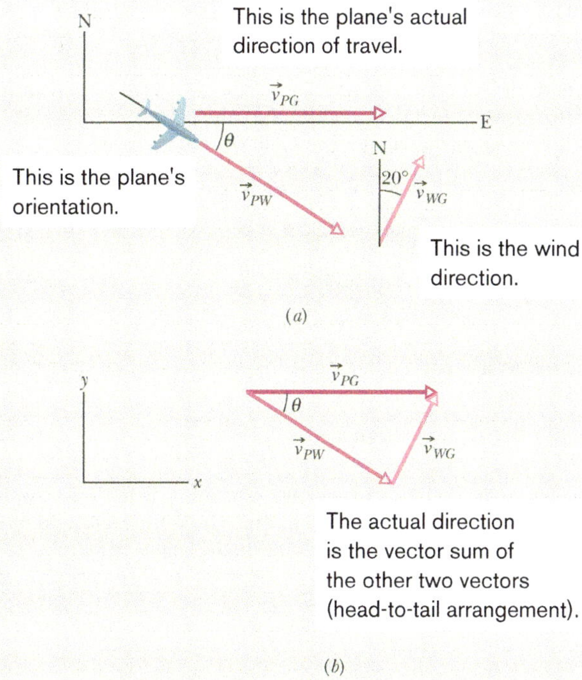

Figure 4-20 A plane flying in a wind.

Review & Summary

Position Vector The location of a particle relative to the origin of a coordinate system is given by a *position vector* $\vec{r}$, which in unit-vector notation is

$$\vec{r} = x\hat{i} + y\hat{j} + z\hat{k}. \tag{4-1}$$

Here $x\hat{i}$, $y\hat{j}$, and $z\hat{k}$ are the vector components of position vector $\vec{r}$, and x, y, and z are its scalar components (as well as the coordinates of the particle). A position vector is described either by a magnitude and one or two angles for orientation, or by its vector or scalar components.

Displacement If a particle moves so that its position vector changes from $\vec{r}_1$ to $\vec{r}_2$, the particle's *displacement* $\Delta\vec{r}$ is

$$\Delta\vec{r} = \vec{r}_2 - \vec{r}_1. \tag{4-2}$$

The displacement can also be written as

$$\Delta\vec{r} = (x_2 - x_1)\hat{i} + (y_2 - y_1)\hat{j} + (z_2 - z_1)\hat{k} \tag{4-3}$$
$$= \Delta x\hat{i} + \Delta y\hat{j} + \Delta z\hat{k}. \tag{4-4}$$

Average Velocity and Instantaneous Velocity If a particle undergoes a displacement $\Delta\vec{r}$ in time interval Δt, its *average velocity* $\vec{v}_{avg}$ for that time interval is

$$\vec{v}_{avg} = \frac{\Delta\vec{r}}{\Delta t}. \tag{4-8}$$

As Δt in Eq. 4-8 is shrunk to 0, $\vec{v}_{avg}$ reaches a limit called either the *velocity* or the *instantaneous velocity* $\vec{v}$:

$$\vec{v} = \frac{d\vec{r}}{dt}, \tag{4-10}$$

which can be rewritten in unit-vector notation as

$$\vec{v} = v_x\hat{i} + v_y\hat{j} + v_z\hat{k}, \tag{4-11}$$

where $v_x = dx/dt$, $v_y = dy/dt$, and $v_z = dz/dt$. The instantaneous velocity $\vec{v}$ of a particle is always directed along the tangent to the particle's path at the particle's position.

Average Acceleration and Instantaneous Acceleration If a particle's velocity changes from $\vec{v}_1$ to $\vec{v}_2$ in time interval Δt, its *average acceleration* during Δt is

$$\vec{a}_{avg} = \frac{\vec{v}_2 - \vec{v}_1}{\Delta t} = \frac{\Delta\vec{v}}{\Delta t}. \tag{4-15}$$

As Δt in Eq. 4-15 is shrunk to 0, $\vec{a}_{avg}$ reaches a limiting value called either the *acceleration* or the *instantaneous acceleration* $\vec{a}$:

$$\vec{a} = \frac{d\vec{v}}{dt}. \tag{4-16}$$

In unit-vector notation,

$$\vec{a} = a_x\hat{i} + a_y\hat{j} + a_z\hat{k}, \tag{4-17}$$

where $a_x = dv_x/dt$, $a_y = dv_y/dt$, and $a_z = dv_z/dt$.

Projectile Motion *Projectile motion* is the motion of a particle that is launched with an initial velocity $\vec{v}_0$. During its flight, the particle's horizontal acceleration is zero and its vertical acceleration is the free-fall acceleration $-g$. (Upward is taken to be a positive direction.) If $\vec{v}_0$ is expressed as a magnitude (the speed v_0) and an angle θ_0 (measured from the horizontal), the particle's equations of motion along the horizontal x axis and vertical y axis are

$$x - x_0 = (v_0 \cos \theta_0)t, \tag{4-21}$$

$$y - y_0 = (v_0 \sin \theta_0)t - \tfrac{1}{2}gt^2, \tag{4-22}$$

$$v_y = v_0 \sin \theta_0 - gt, \tag{4-23}$$

$$v_y^2 = (v_0 \sin \theta_0)^2 - 2g(y - y_0). \tag{4-24}$$

The **trajectory** (path) of a particle in projectile motion is parabolic and is given by

$$y = (\tan \theta_0)x - \frac{gx^2}{2(v_0 \cos \theta_0)^2}, \tag{4-25}$$

if x_0 and y_0 of Eqs. 4-21 to 4-24 are zero. The particle's **horizontal range** R, which is the horizontal distance from the launch point to the point at which the particle returns to the launch height, is

$$R = \frac{v_0^2}{g} \sin 2\theta_0. \tag{4-26}$$

Uniform Circular Motion If a particle travels along a circle or circular arc of radius r at constant speed v, it is said to be in *uniform circular motion* and has an acceleration $\vec{a}$ of constant magnitude

$$a = \frac{v^2}{r}. \tag{4-34}$$

The direction of $\vec{a}$ is toward the center of the circle or circular arc, and $\vec{a}$ is said to be *centripetal*. The time for the particle to complete a circle is

$$T = \frac{2\pi r}{v}. \tag{4-35}$$

T is called the *period of revolution*, or simply the *period*, of the motion.

Relative Motion When two frames of reference A and B are moving relative to each other at constant velocity, the velocity of a particle P as measured by an observer in frame A usually differs from that measured from frame B. The two measured velocities are related by

$$\vec{v}_{PA} = \vec{v}_{PB} + \vec{v}_{BA}, \tag{4-44}$$

where $\vec{v}_{BA}$ is the velocity of B with respect to A. Both observers measure the same acceleration for the particle:

$$\vec{a}_{PA} = \vec{a}_{PB}. \tag{4-45}$$

Questions

1 Figure 4-21 shows the path taken by a skunk foraging for trash food, from initial point i. The skunk took the same time T to go from each labeled point to the next along its path. Rank points a, b, and c according to the magnitude of the average velocity of the skunk to reach them from initial point i, greatest first.

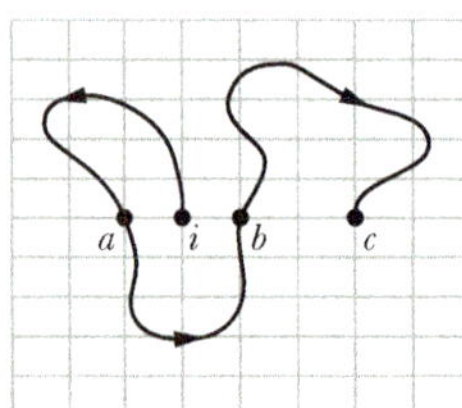

Figure 4-21 Question 1.

2 Figure 4-22 shows the initial position i and the final position f of a particle. What are the (a) initial position vector $\vec{r}_i$ and (b) final position vector $\vec{r}_f$, both in unit-vector notation? (c) What is the x component of displacement $\Delta\vec{r}$?

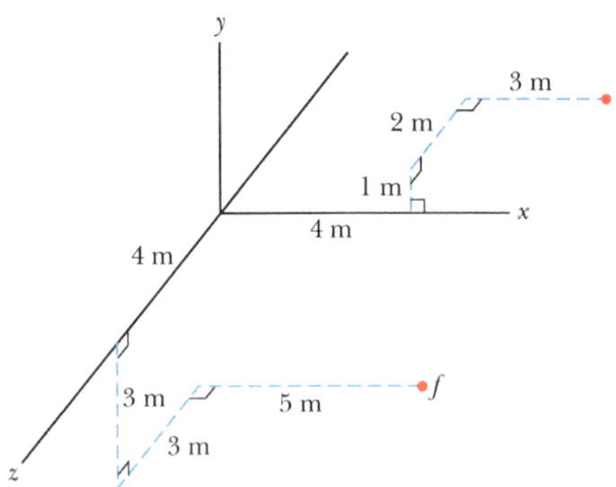

Figure 4-22 Question 2.

3 When Paris was shelled from 100 km away with the WWI long-range artillery piece "Big Bertha," the shells were fired at an angle greater than 45° to give them a greater range, possibly even twice as long as at 45°. Does that result mean that the air density at high altitudes increases with altitude or decreases?

4 You are to launch a rocket, from just above the ground, with one of the following initial velocity vectors: (1) $\vec{v}_0 = 20\hat{i} + 70\hat{j}$, (2) $\vec{v}_0 = -20\hat{i} + 70\hat{j}$, (3) $\vec{v}_0 = 20\hat{i} - 70\hat{j}$, (4) $\vec{v}_0 = -20\hat{i} - 70\hat{j}$. In your coordinate system, x runs along level ground and y increases upward. (a) Rank the vectors according to the launch speed of the projectile, greatest first. (b) Rank the vectors according to the time of flight of the projectile, greatest first.

5 Figure 4-23 shows three situations in which identical projectiles are launched (at the same level) at identical initial speeds and angles. The projectiles do not land on the same terrain, however. Rank the situations according to the final speeds of the projectiles just before they land, greatest first.

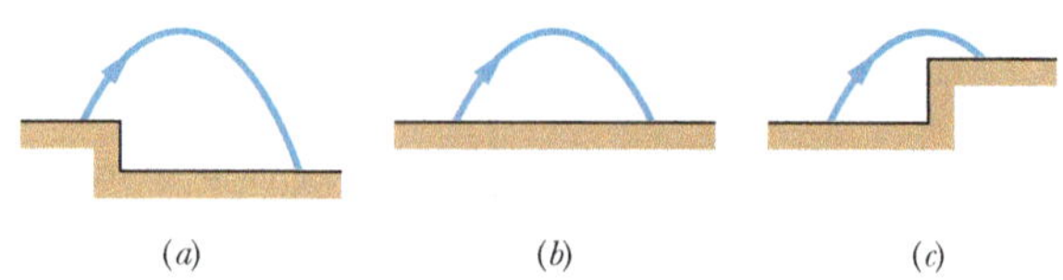

Figure 4-23 Question 5.

6 The only good use of a fruitcake is in catapult practice. Curve 1 in Fig. 4-24 gives the height y of a catapulted fruitcake versus the angle θ between its velocity vector and its acceleration vector during flight. (a) Which of the lettered points on that curve corresponds to the landing of the fruitcake on the ground? (b) Curve 2 is a similar plot for the same

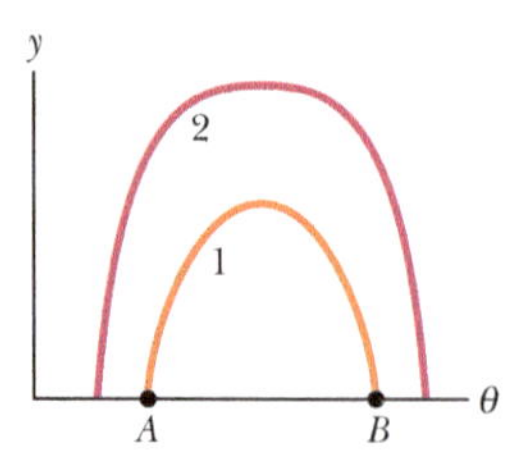

Figure 4-24 Question 6.

launch speed but for a different launch angle. Does the fruitcake now land farther away or closer to the launch point?

7 An airplane flying horizontally at a constant speed of 350 km/h over level ground releases a bundle of food supplies. Ignore the effect of the air on the bundle. What are the bundle's initial (a) vertical and (b) horizontal components of velocity? (c) What is its horizontal component of velocity just before hitting the ground? (d) If the airplane's speed were, instead, 450 km/h, would the time of fall be longer, shorter, or the same?

8 In Fig. 4-25, a cream tangerine is thrown up past windows 1, 2, and 3, which are identical in size and regularly spaced vertically. Rank those three windows according to (a) the time the cream tangerine takes to pass them and (b) the average speed of the cream tangerine during the passage, greatest first.

The cream tangerine then moves down past windows 4, 5, and 6, which are identical in size and irregularly spaced horizontally. Rank those three windows according to (c) the time the cream tangerine takes to pass them and (d) the average speed of the cream tangerine during the passage, greatest first.

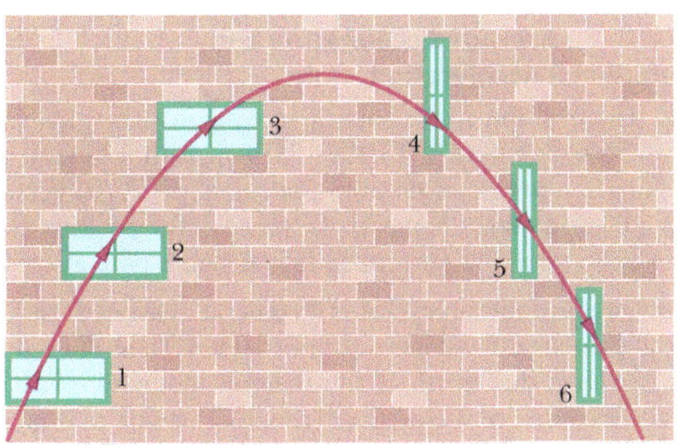

Figure 4-25 Question 8.

9 Figure 4-26 shows three paths for a football kicked from ground level. Ignoring the effects of air, rank the paths according to (a) time of flight, (b) initial vertical velocity component, (c) initial horizontal velocity component, and (d) initial speed, greatest first.

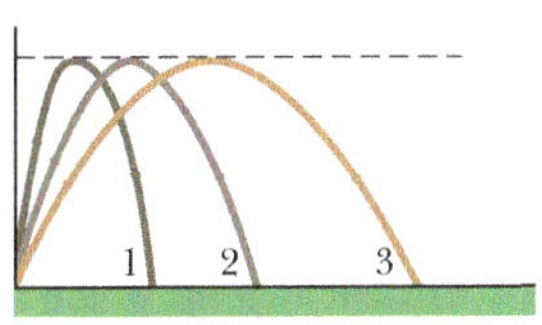

Figure 4-26 Question 9.

10 A ball is shot from ground level over level ground at a certain initial speed. Figure 4-27 gives the range R of the ball versus its launch angle θ_0. Rank the three lettered points on the plot according to (a) the total flight time of the ball and (b) the ball's speed at maximum height, greatest first.

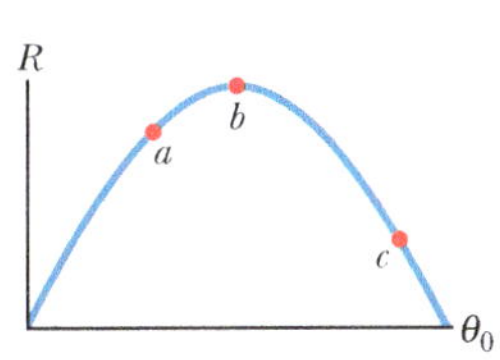

Figure 4-27 Question 10.

11 Figure 4-28 shows four tracks (either half- or quarter-circles) that can be taken by a train, which moves at a constant speed. Rank the tracks according to the magnitude of a train's acceleration on the curved portion, greatest first.

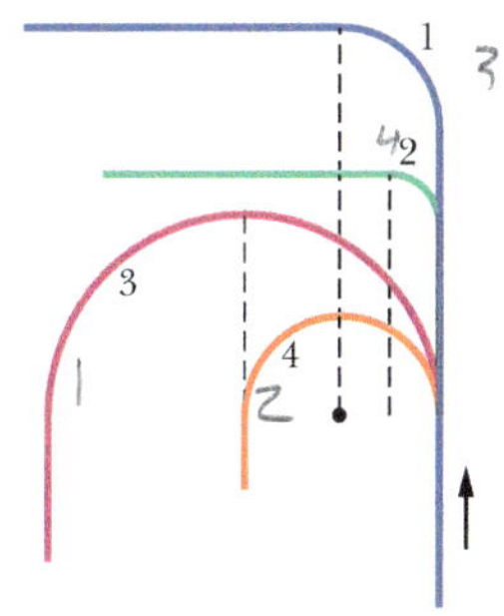

Figure 4-28 Question 11.

12 In Fig. 4-29, particle P is in uniform circular motion, centered on the origin of an xy coordinate system. (a) At what values of θ is the vertical component r_y of the position vector greatest in magnitude? (b) At what values of θ is the vertical component v_y of the particle's velocity greatest in magnitude? (c) At what values of θ is the vertical component a_y of the particle's acceleration greatest in magnitude?

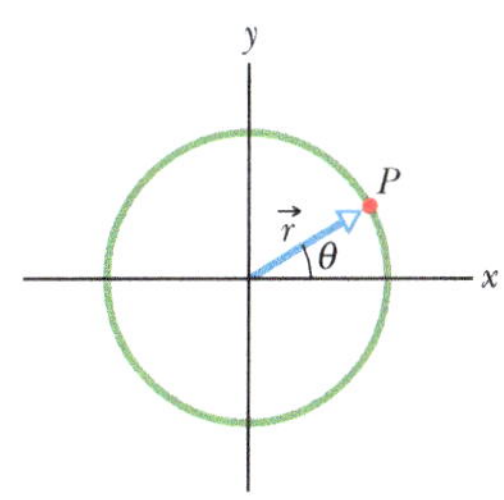

Figure 4-29 Question 12.

13 (a) Is it possible to be accelerating while traveling at constant speed? Is it possible to round a curve with (b) zero acceleration and (c) a constant magnitude of acceleration?

14 While riding in a moving car, you toss an egg directly upward. Does the egg tend to land behind you, in front of you, or back in your hands if the car is (a) traveling at a constant speed, (b) increasing in speed, and (c) decreasing in speed?

15 A snowball is thrown from ground level (by someone in a hole) with initial speed v_0 at an angle of 45° relative to the (level) ground, on which the snowball later lands. If the launch angle is increased, do (a) the range and (b) the flight time increase, decrease, or stay the same?

16 You are driving directly behind a pickup truck, going at the same speed as the truck. A crate falls from the bed of the truck to the road. (a) Will your car hit the crate before the crate hits the road if you neither brake nor swerve? (b) During the fall, is the horizontal speed of the crate more than, less than, or the same as that of the truck?

17 At what point in the path of a projectile is the speed a minimum?

18 In shot put, the shot is put (thrown) from above the athlete's shoulder level. Is the launch angle that produces the greatest range 45°, less than 45°, or greater than 45°?

Problems

GO Tutoring problem available (at instructor's discretion) in *WileyPLUS* and WebAssign

SSM Worked-out solution available in Student Solutions Manual

• – ••• Number of dots indicates level of problem difficulty

Additional information available in *The Flying Circus of Physics* and at flyingcircusofphysics.com

WWW Worked-out solution is at

ILW Interactive solution is at

http://www.wiley.com/college/halliday

Module 4-1 Position and Displacement

•1 The position vector for an electron is $\vec r = (5.0\text{ m})\hat i - (3.0\text{ m})\hat j + (2.0\text{ m})\hat k$. (a) Find the magnitude of $\vec r$. (b) Sketch the vector on a right-handed coordinate system.

•2 A watermelon seed has the following coordinates: $x = -5.0$ m, $y = 8.0$ m, and $z = 0$ m. Find its position vector (a) in unit-vector notation and as (b) a magnitude and (c) an angle relative to the positive direction of the x axis. (d) Sketch the vector on a right-handed coordinate system. If the seed is moved to the xyz coordinates (3.00 m, 0 m, 0 m), what is its displacement (e) in unit-vector notation and as (f) a magnitude and (g) an angle relative to the positive x direction?

•3 A positron undergoes a displacement $\Delta\vec r = 2.0\hat i - 3.0\hat j + 6.0\hat k$, ending with the position vector $\vec r = 3.0\hat j - 4.0\hat k$, in meters. What was the positron's initial position vector?

••4 The minute hand of a wall clock measures 10 cm from its tip to the axis about which it rotates. The magnitude and angle of the displacement vector of the tip are to be determined for three time intervals. What are the (a) magnitude and (b) angle from a quarter after the hour to half past, the (c) magnitude and (d) angle for the next half hour, and the (e) magnitude and (f) angle for the hour after that?

Module 4-2 Average Velocity and Instantaneous Velocity

•5 SSM A train at a constant 60.0 km/h moves east for 40.0 min, then in a direction 50.0° east of due north for 20.0 min, and then west for 50.0 min. What are the (a) magnitude and (b) angle of its average velocity during this trip?

•6 An electron's position is given by $\vec r = 3.00t\hat i - 4.00t^2\hat j + 2.00\hat k$, with t in seconds and $\vec r$ in meters. (a) In unit-vector notation, what is the electron's velocity $\vec v(t)$? At $t =$ 2.00 s, what is $\vec v$ (b) in unit-vector notation and as (c) a magnitude and (d) an angle relative to the positive direction of the x axis?

•7 An ion's position vector is initially $\vec r = 5.0\hat i - 6.0\hat j + 2.0\hat k$, and 10 s later it is $\vec r = -2.0\hat i + 8.0\hat j - 2.0\hat k$, all in meters. In unit-vector notation, what is its $\vec v_{avg}$ during the 10 s?

••8 A plane flies 483 km east from city A to city B in 45.0 min and then 966 km south from city B to city C in 1.50 h. For the total trip, what are the (a) magnitude and (b) direction of the plane's displacement, the (c) magnitude and (d) direction of its average velocity, and (e) its average speed?

••9 Figure 4-30 gives the path of a squirrel moving about on level ground, from point A (at time $t = 0$), to points B (at $t = 5.00$ min), C (at $t = 10.0$ min), and finally D (at $t = 15.0$ min). Consider the average velocities of the squirrel from point A to each of the other three points. Of them, what are the (a) magnitude and (b) angle of the one with the least magnitude and the (c) magnitude and (d) angle of the one with the greatest magnitude?

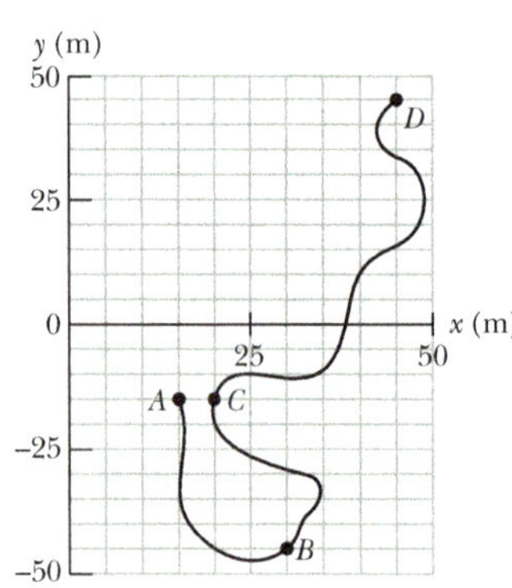

Figure 4-30 Problem 9.

•••10 The position vector $\vec r = 5.00t\hat i + (et + ft^2)\hat j$ locates a particle as a function of time t. Vector $\vec r$ is in meters, t is in seconds, and factors e and f are constants. Figure 4-31 gives the angle θ of the particle's direction of travel as a function of t (θ is measured from the positive x direction). What are (a) e and (b) f, including units?

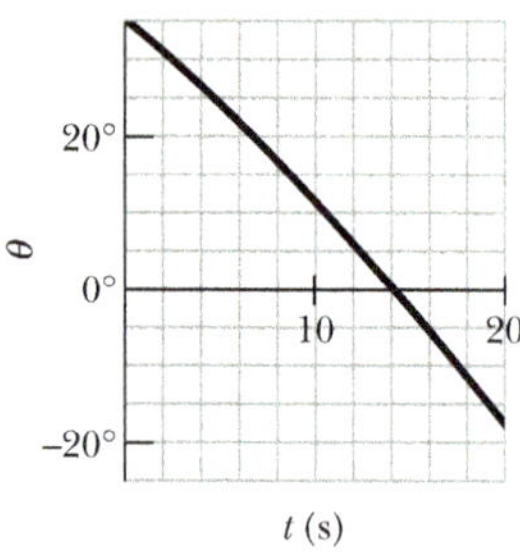

Figure 4-31 Problem 10.

Module 4-3 Average Acceleration and Instantaneous Acceleration

•11 GO The position $\vec r$ of a particle moving in an xy plane is given by $\vec r = (2.00t^3 - 5.00t)\hat i + (6.00 - 7.00t^4)\hat j$, with $\vec r$ in meters and t in seconds. In unit-vector notation, calculate (a) $\vec r$, (b) $\vec v$, and (c) $\vec a$ for $t = 2.00$ s. (d) What is the angle between the positive direction of the x axis and a line tangent to the particle's path at $t = 2.00$ s?

•12 At one instant a bicyclist is 40.0 m due east of a park's flagpole, going due south with a speed of 10.0 m/s. Then 30.0 s later, the cyclist is 40.0 m due north of the flagpole, going due east with a speed of 10.0 m/s. For the cyclist in this 30.0 s interval, what are the (a) magnitude and (b) direction of the displacement, the (c) magnitude and (d) direction of the average velocity, and the (e) magnitude and (f) direction of the average acceleration?

•13 SSM A particle moves so that its position (in meters) as a function of time (in seconds) is $\vec r = \hat i + 4t^2\hat j + t\hat k$. Write expressions for (a) its velocity and (b) its acceleration as functions of time.

•14 A proton initially has $\vec v = 4.0\hat i - 2.0\hat j + 3.0\hat k$ and then 4.0 s later has $\vec v = -2.0\hat i - 2.0\hat j + 5.0\hat k$ (in meters per second). For that 4.0 s, what are (a) the proton's average acceleration $\vec a_{avg}$ in unit-vector notation, (b) the magnitude of $\vec a_{avg}$, and (c) the angle between $\vec a_{avg}$ and the positive direction of the x axis?

••15 SSM ILW A particle leaves the origin with an initial velocity $\vec v = (3.00\hat i)$ m/s and a constant acceleration $\vec a = (-1.00\hat i - 0.500\hat j)$ m/s^2. When it reaches its maximum x coordinate, what are its (a) velocity and (b) position vector?

••16 GO The velocity $\vec v$ of a particle moving in the xy plane is given by $\vec v = (6.0t - 4.0t^2)\hat i + 8.0\hat j$, with $\vec v$ in meters per second and t (> 0) in seconds. (a) What is the acceleration when $t = 3.0$ s? (b) When (if ever) is the acceleration zero? (c) When (if ever) is the velocity zero? (d) When (if ever) does the speed equal 10 m/s?

••17 A cart is propelled over an xy plane with acceleration components $a_x = 4.0$ m/s^2 and $a_y = -2.0$ m/s^2. Its initial velocity has components $v_{0x} = 8.0$ m/s and $v_{0y} = 12$ m/s. In unit-vector notation, what is the velocity of the cart when it reaches its greatest y coordinate?

••18 A moderate wind accelerates a pebble over a horizontal xy plane with a constant acceleration $\vec a = (5.00\text{ m/s}^2)\hat i + (7.00\text{ m/s}^2)\hat j$.

At time $t = 0$, the velocity is $(4.00\text{ m/s})\hat{i}$. What are the (a) magnitude and (b) angle of its velocity when it has been displaced by 12.0 m parallel to the x axis?

•••19 The acceleration of a particle moving only on a horizontal xy plane is given by $\vec{a} = 3t\hat{i} + 4t\hat{j}$, where $\vec{a}$ is in meters per second-squared and t is in seconds. At $t = 0$, the position vector $\vec{r} = (20.0\text{ m})\hat{i} + (40.0\text{ m})\hat{j}$ locates the particle, which then has the velocity vector $\vec{v} = (5.00\text{ m/s})\hat{i} + (2.00\text{ m/s})\hat{j}$. At $t = 4.00$ s, what are (a) its position vector in unit-vector notation and (b) the angle between its direction of travel and the positive direction of the x axis?

•••20 GO In Fig. 4-32, particle A moves along the line $y = 30$ m with a constant velocity $\vec{v}$ of magnitude 3.0 m/s and parallel to the x axis. At the instant particle A passes the y axis, particle B leaves the origin with a zero initial speed and a constant acceleration $\vec{a}$ of magnitude 0.40 m/s^2. What angle θ between $\vec{a}$ and the positive direction of the y axis would result in a collision?

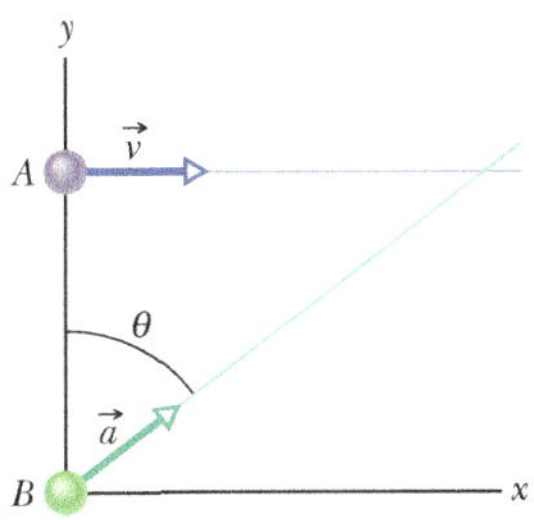

Figure 4-32 Problem 20.

Module 4-4 Projectile Motion

•21 A dart is thrown horizontally with an initial speed of 10 m/s toward point P, the bull's-eye on a dart board. It hits at point Q on the rim, vertically below P, 0.19 s later. (a) What is the distance PQ? (b) How far away from the dart board is the dart released?

•22 A small ball rolls horizontally off the edge of a tabletop that is 1.20 m high. It strikes the floor at a point 1.52 m horizontally from the table edge. (a) How long is the ball in the air? (b) What is its speed at the instant it leaves the table?

•23 A projectile is fired horizontally from a gun that is 45.0 m above flat ground, emerging from the gun with a speed of 250 m/s. (a) How long does the projectile remain in the air? (b) At what horizontal distance from the firing point does it strike the ground? (c) What is the magnitude of the vertical component of its velocity as it strikes the ground?

•24 In the 1991 World Track and Field Championships in Tokyo, Mike Powell jumped 8.95 m, breaking by a full 5 cm the 23-year long-jump record set by Bob Beamon. Assume that Powell's speed on takeoff was 9.5 m/s (about equal to that of a sprinter) and that $g = 9.80$ m/s^2 in Tokyo. How much less was Powell's range than the maximum possible range for a particle launched at the same speed?

•25 The current world-record motorcycle jump is 77.0 m, set by Jason Renie. Assume that he left the take-off ramp at 12.0° to the horizontal and that the take-off and landing heights are the same. Neglecting air drag, determine his take-off speed.

•26 A stone is catapulted at time $t = 0$, with an initial velocity of magnitude 20.0 m/s and at an angle of 40.0° above the horizontal. What are the magnitudes of the (a) horizontal and (b) vertical components of its displacement from the catapult site at $t = 1.10$ s? Repeat for the (c) horizontal and (d) vertical components at $t = 1.80$ s, and for the (e) horizontal and (f) vertical components at $t = 5.00$ s.

••27 ILW A certain airplane has a speed of 290.0 km/h and is diving at an angle of $\theta = 30.0°$ below the horizontal when the pilot releases a radar decoy (Fig. 4-33). The horizontal distance between the release point and the point where the decoy strikes the ground is $d = 700$ m. (a) How long is the decoy in the air? (b) How high was the release point?

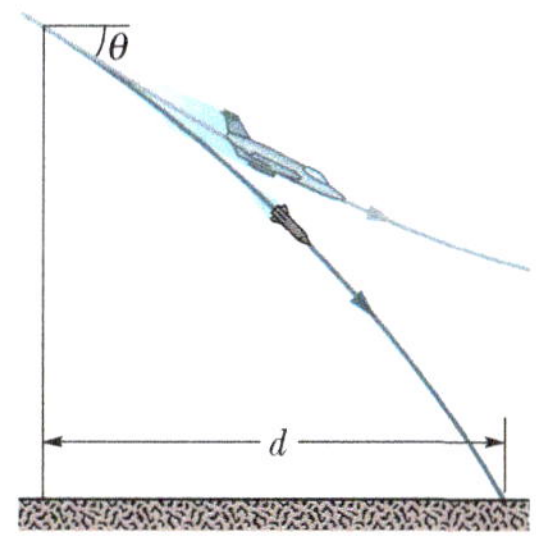

Figure 4-33 Problem 27.

••28 GO In Fig. 4-34, a stone is projected at a cliff of height h with an initial speed of 42.0 m/s directed at angle $\theta_0 = 60.0°$ above the horizontal. The stone strikes at A, 5.50 s after launching. Find (a) the height h of the cliff, (b) the speed of the stone just before impact at A, and (c) the maximum height H reached above the ground.

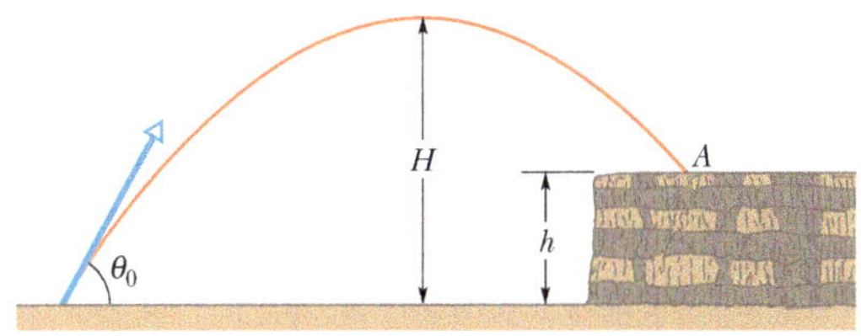

Figure 4-34 Problem 28.

••29 A projectile's launch speed is five times its speed at maximum height. Find launch angle θ_0.

••30 GO A soccer ball is kicked from the ground with an initial speed of 19.5 m/s at an upward angle of 45°. A player 55 m away in the direction of the kick starts running to meet the ball at that instant. What must be his average speed if he is to meet the ball just before it hits the ground?

••31 In a jump spike, a volleyball player slams the ball from overhead and toward the opposite floor. Controlling the angle of the spike is difficult. Suppose a ball is spiked from a height of 2.30 m with an initial speed of 20.0 m/s at a downward angle of 18.00°. How much farther on the opposite floor would it have landed if the downward angle were, instead, 8.00°?

••32 GO You throw a ball toward a wall at speed 25.0 m/s and at angle $\theta_0 = 40.0°$ above the horizontal (Fig. 4-35). The wall is distance $d = 22.0$ m from the release point of the ball. (a) How far above the release point does the ball hit the wall? What are the (b) horizontal and (c) vertical components of its velocity as it hits the wall? (d) When it hits, has it passed the highest point on its trajectory?

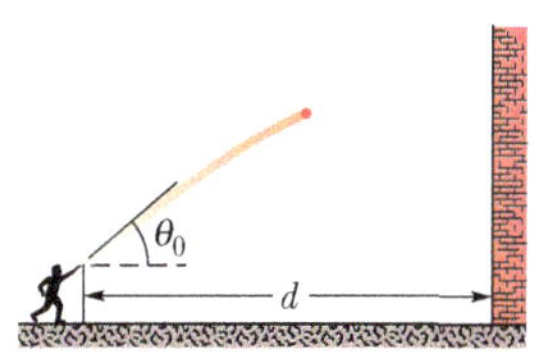

Figure 4-35 Problem 32.

••33 SSM A plane, diving with constant speed at an angle of 53.0° with the vertical, releases a projectile at an altitude of 730 m. The projectile hits the ground 5.00 s after release. (a) What is the speed of the plane? (b) How far does the projectile travel horizontally during its flight? What are the (c) horizontal and (d) vertical components of its velocity just before striking the ground?

••34 A trebuchet was a hurling machine built to attack the walls of a castle under siege. A large stone could be hurled against a wall to break apart the wall. The machine was not placed near the wall because then arrows could reach it from the castle wall. Instead, it was positioned so that the stone hit the wall during the second half of its flight. Suppose a stone is launched with a speed of $v_0 = 28.0$ m/s and at an angle of $\theta_0 = 40.0°$. What is the speed of the stone if it hits the wall (a) just as it reaches the top of its parabolic path and (b) when it has descended to half that height? (c) As a percentage, how much faster is it moving in part (b) than in part (a)?

••35 SSM A rifle that shoots bullets at 460 m/s is to be aimed at a target 45.7 m away. If the center of the target is level with the rifle, how high above the target must the rifle barrel be pointed so that the bullet hits dead center?

••36 GO During a tennis match, a player serves the ball at 23.6 m/s, with the center of the ball leaving the racquet horizontally 2.37 m above the court surface. The net is 12 m away and 0.90 m high. When the ball reaches the net, (a) does the ball clear it and (b) what is the distance between the center of the ball and the top of the net? Suppose that, instead, the ball is served as before but now it leaves the racquet at 5.00° below the horizontal. When the ball reaches the net, (c) does the ball clear it and (d) what now is the distance between the center of the ball and the top of the net?

••37 SSM WWW A lowly high diver pushes off horizontally with a speed of 2.00 m/s from the platform edge 10.0 m above the surface of the water. (a) At what horizontal distance from the edge is the diver 0.800 s after pushing off? (b) At what vertical distance above the surface of the water is the diver just then? (c) At what horizontal distance from the edge does the diver strike the water?

••38 A golf ball is struck at ground level. The speed of the golf ball as a function of the time is shown in Fig. 4-36, where $t = 0$ at the instant the ball is struck. The scaling on the vertical axis is set by $v_a = 19$ m/s and $v_b = 31$ m/s. (a) How far does the golf ball travel horizontally before returning to ground level? (b) What is the maximum height above ground level attained by the ball?

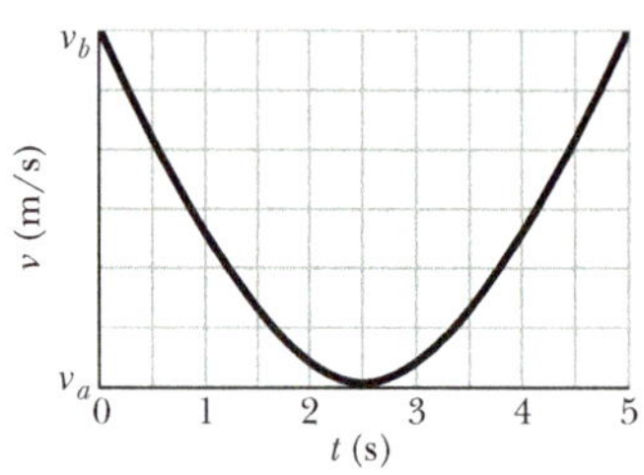

Figure 4-36 Problem 38.

••39 In Fig. 4-37, a ball is thrown leftward from the left edge of the roof, at height h above the ground. The ball hits the ground 1.50 s later, at distance $d = 25.0$ m from the building and at angle $\theta = 60.0°$ with the horizontal. (a) Find h. (*Hint:* One way is to reverse the motion, as if on video.) What are the (b) magnitude and (c) angle relative to the horizontal of the velocity at which the ball is thrown? (d) Is the angle above or below the horizontal?

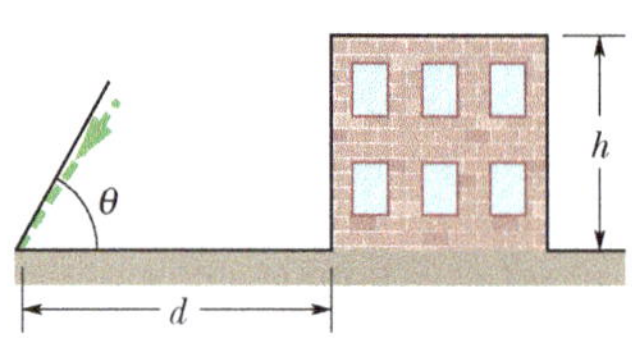

Figure 4-37 Problem 39.

••40 Suppose that a shot putter can put a shot at the world-class speed $v_0 = 15.00$ m/s and at a height of 2.160 m. What horizontal distance would the shot travel if the launch angle θ_0 is (a) 45.00° and (b) 42.00°? The answers indicate that the angle of 45°, which maximizes the range of projectile motion, does not maximize the horizontal distance when the launch and landing are at different heights.

••41 GO Upon spotting an insect on a twig overhanging water, an archer fish squirts water drops at the insect to knock it into the water (Fig. 4-38). Although the insect is located along a straight-line path at angle ϕ and distance d, a drop must be launched at a different angle θ_0 if its parabolic path is to intersect the insect. If $\phi = 36.0°$ and $d = 0.900$ m, what launch angle θ_0 is required for the drop to be at the top of the parabolic path when it reaches the insect?

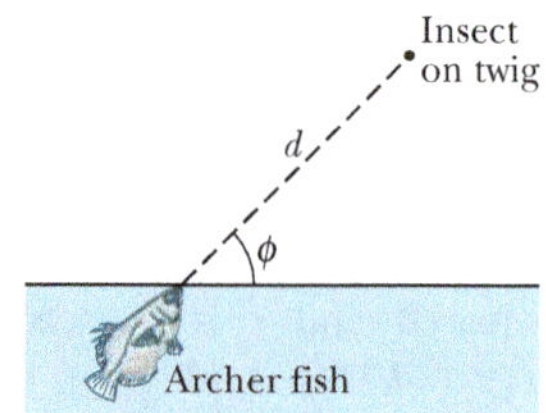

Figure 4-38 Problem 41.

••42 In 1939 or 1940, Emanuel Zacchini took his human-cannonball act to an extreme: After being shot from a cannon, he soared over three Ferris wheels and into a net (Fig. 4-39). Assume that he is launched with a speed of 26.5 m/s and at an angle of 53.0°. (a) Treating him as a particle, calculate his clearance over the first wheel. (b) If he reached maximum height over the middle wheel, by how much did he clear it? (c) How far from the cannon should the net's center have been positioned (neglect air drag)?

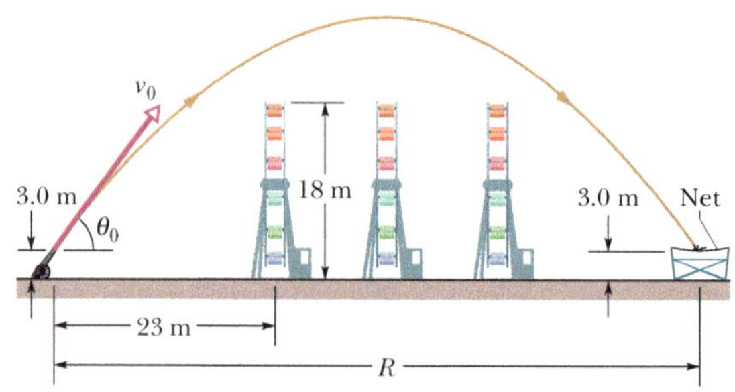

Figure 4-39 Problem 42.

••43 ILW A ball is shot from the ground into the air. At a height of 9.1 m, its velocity is $\vec{v} = (7.6\hat{i} + 6.1\hat{j})$ m/s, with $\hat{i}$ horizontal and $\hat{j}$ upward. (a) To what maximum height does the ball rise? (b) What total horizontal distance does the ball travel? What are the (c) magnitude and (d) angle (below the horizontal) of the ball's velocity just before it hits the ground?

••44 A baseball leaves a pitcher's hand horizontally at a speed of 161 km/h. The distance to the batter is 18.3 m. (a) How long does the ball take to travel the first half of that distance? (b) The second half? (c) How far does the ball fall freely during the first half? (d) During the second half? (e) Why aren't the quantities in (c) and (d) equal?

••45 In Fig. 4-40, a ball is launched with a velocity of magnitude 10.0 m/s, at an angle of 50.0° to the horizontal. The launch point is at the base of a ramp of horizontal length $d_1 = 6.00$ m and height $d_2 = 3.60$ m. A plateau is located at the top of the ramp. (a) Does the ball land on the ramp or the plateau? When it lands, what are the (b) magnitude and (c) angle of its displacement from the launch point?

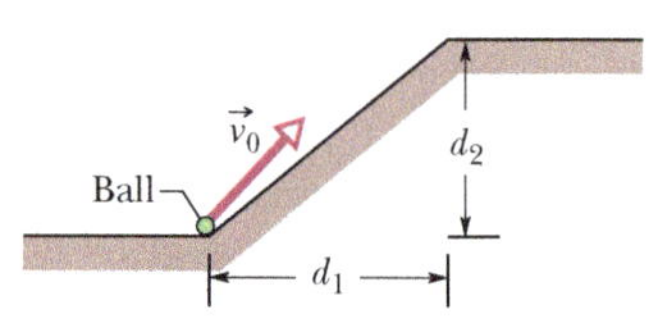

Figure 4-40 Problem 45.

••46 GO In basketball, *hang* is an illusion in which a player seems to weaken the gravitational acceleration while in midair. The illusion depends much on a skilled player's ability to rapidly shift the ball between hands during the flight, but it might also be supported by the longer horizontal distance the player travels in the upper part of the jump than in the lower part. If a player jumps with an initial speed of $v_0 = 7.00$ m/s at an angle of $\theta_0 = 35.0°$, what percent of the jump's range does the player spend in the upper half of the jump (between maximum height and half maximum height)?

••47 SSM WWW A batter hits a pitched ball when the center of the ball is 1.22 m above the ground. The ball leaves the bat at an angle of 45° with the ground. With that launch, the ball should have a horizontal range (returning to the *launch* level) of 107 m. (a) Does the ball clear a 7.32-m-high fence that is 97.5 m horizontally from the launch point? (b) At the fence, what is the distance between the fence top and the ball center?

••48 GO In Fig. 4-41, a ball is thrown up onto a roof, landing 4.00 s later at height $h = 20.0$ m above the release level. The ball's path just before landing is angled at $\theta = 60.0°$ with the roof. (a) Find the horizontal distance d it travels. (See the hint to Problem 39.) What are the (b) magnitude and (c) angle (relative to the horizontal) of the ball's initial velocity?

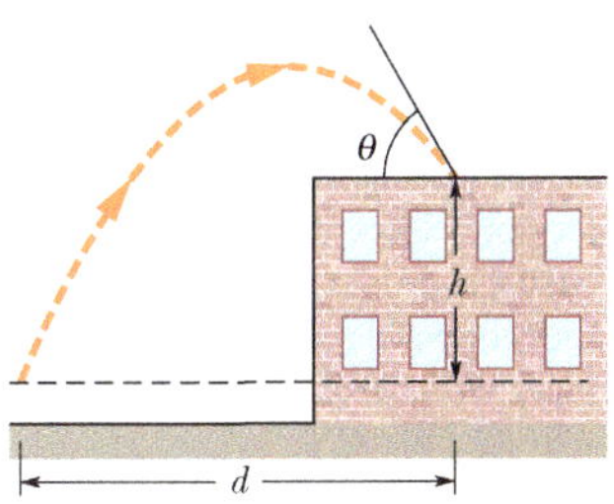

Figure 4-41 Problem 48.

•••49 SSM A football kicker can give the ball an initial speed of 25 m/s. What are the (a) least and (b) greatest elevation angles at which he can kick the ball to score a field goal from a point 50 m in front of goalposts whose horizontal bar is 3.44 m above the ground?

•••50 GO Two seconds after being projected from ground level, a projectile is displaced 40 m horizontally and 53 m vertically above its launch point. What are the (a) horizontal and (b) vertical components of the initial velocity of the projectile? (c) At the instant the projectile achieves its maximum height above ground level, how far is it displaced horizontally from the launch point?

•••51 A skilled skier knows to jump upward before reaching a downward slope. Consider a jump in which the launch speed is $v_0 = 10$ m/s, the launch angle is $\theta_0 = 11.3°$, the initial course is approximately flat, and the steeper track has a slope of 9.0°. Figure 4-42*a* shows a *prejump* that allows the skier to land on the top portion of the steeper track. Figure 4-42*b* shows a jump at the edge of the steeper track. In Fig. 4-42*a*, the skier lands at approximately the launch level. (a) In the landing, what is the angle ϕ between the skier's path and the slope? In Fig. 4-42*b*, (b) how far below the launch level does the skier land and (c) what is ϕ? (The greater fall and greater ϕ can result in loss of control in the landing.)

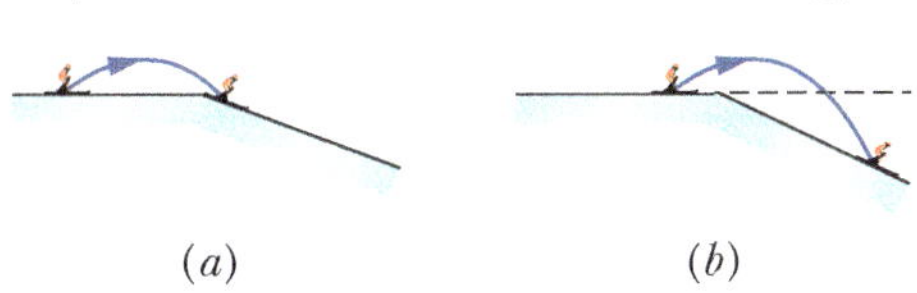

Figure 4-42 Problem 51.

•••52 A ball is to be shot from level ground toward a wall at distance x (Fig. 4-43*a*). Figure 4-43*b* shows the y component v_y of the ball's velocity just as it would reach the wall, as a function of that distance x. The scaling is set by $v_{ys} = 5.0$ m/s and $x_s = 20$ m. What is the launch angle?

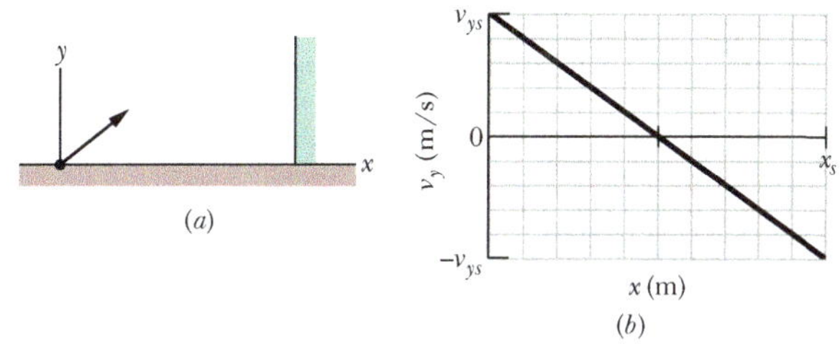

Figure 4-43 Problem 52.

•••53 GO In Fig. 4-44, a baseball is hit at a height $h = 1.00$ m and then caught at the same height. It travels alongside a wall, moving up past the top of the wall 1.00 s after it is hit and then down past the top of the wall 4.00 s later, at distance $D = 50.0$ m farther along the wall. (a) What horizontal distance is traveled by the ball from hit to catch? What are the (b) magnitude and (c) angle (relative to the horizontal) of the ball's velocity just after being hit? (d) How high is the wall?

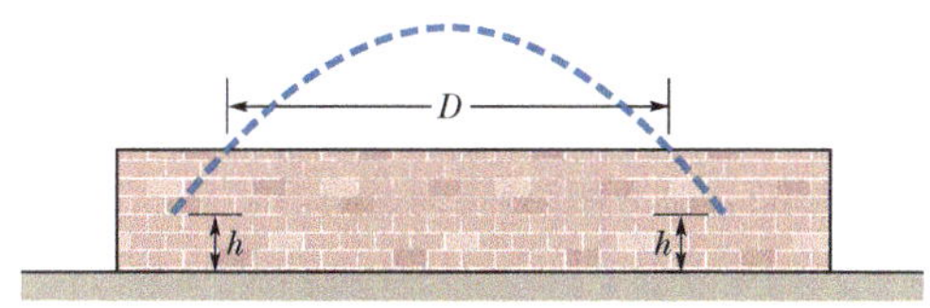

Figure 4-44 Problem 53.

•••54 GO A ball is to be shot from level ground with a certain speed. Figure 4-45 shows the range R it will have versus the launch angle θ_0. The value of θ_0 determines the flight time; let t_{max} represent the maximum flight time. What is the least speed the ball will have during its flight if θ_0 is chosen such that the flight time is $0.500t_{max}$?

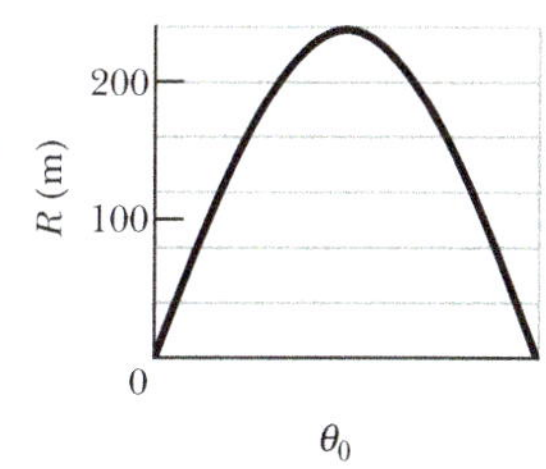

Figure 4-45 Problem 54.

•••55 SSM A ball rolls horizontally off the top of a stairway with a speed of 1.52 m/s. The steps are 20.3 cm high and 20.3 cm wide. Which step does the ball hit first?

Module 4-5 Uniform Circular Motion

•56 An Earth satellite moves in a circular orbit 640 km (uniform circular motion) above Earth's surface with a period of 98.0 min. What are (a) the speed and (b) the magnitude of the centripetal acceleration of the satellite?

•57 A carnival merry-go-round rotates about a vertical axis at a constant rate. A man standing on the edge has a constant speed of 3.66 m/s and a centripetal acceleration $\vec{a}$ of magnitude 1.83 m/s^2. Position vector $\vec{r}$ locates him relative to the rotation axis. (a) What is the magnitude of $\vec{r}$? What is the direction of $\vec{r}$ when $\vec{a}$ is directed (b) due east and (c) due south?

•58 A rotating fan completes 1200 revolutions every minute. Consider the tip of a blade, at a radius of 0.15 m. (a) Through what distance does the tip move in one revolution? What are (b) the tip's speed and (c) the magnitude of its acceleration? (d) What is the period of the motion?

•59 ILW A woman rides a carnival Ferris wheel at radius 15 m, completing five turns about its horizontal axis every minute. What are (a) the period of the motion, the (b) magnitude and (c) direction of her centripetal acceleration at the highest point, and the (d) magnitude and (e) direction of her centripetal acceleration at the lowest point?

•60 A centripetal-acceleration addict rides in uniform circular motion with radius $r = 3.00$ m. At one instant his acceleration is $\vec{a} = (6.00 \text{ m/s}^2)\hat{i} + (-4.00 \text{ m/s}^2)\hat{j}$. At that instant, what are the values of (a) $\vec{v} \cdot \vec{a}$ and (b) $\vec{r} \times \vec{a}$?

•61 When a large star becomes a *supernova,* its core may be compressed so tightly that it becomes a *neutron star,* with a radius of about 20 km (about the size of the San Francisco area). If a neutron star rotates once every second, (a) what is the speed of a particle on the star's equator and (b) what is the magnitude of the particle's centripetal acceleration? (c) If the neutron star rotates faster, do the answers to (a) and (b) increase, decrease, or remain the same?

•62 What is the magnitude of the acceleration of a sprinter running at 10 m/s when rounding a turn of radius 25 m?

••63 GO At $t_1 = 2.00$ s, the acceleration of a particle in counterclockwise circular motion is $(6.00 \text{ m/s}^2)\hat{i} + (4.00 \text{ m/s}^2)\hat{j}$. It moves at constant speed. At time $t_2 = 5.00$ s, the particle's acceleration is $(4.00 \text{ m/s}^2)\hat{i} + (-6.00 \text{ m/s}^2)\hat{j}$. What is the radius of the path taken by the particle if $t_2 - t_1$ is less than one period?

••64 GO A particle moves horizontally in uniform circular motion, over a horizontal xy plane. At one instant, it moves through the point at coordinates (4.00 m, 4.00 m) with a velocity of $-5.00\hat{i}$ m/s and an acceleration of $+12.5\hat{j}$ m/s^2. What are the (a) x and (b) y coordinates of the center of the circular path?

••65 A purse at radius 2.00 m and a wallet at radius 3.00 m travel in uniform circular motion on the floor of a merry-go-round as the ride turns. They are on the same radial line. At one instant, the acceleration of the purse is $(2.00 \text{ m/s}^2)\hat{i} + (4.00 \text{ m/s}^2)\hat{j}$. At that instant and in unit-vector notation, what is the acceleration of the wallet?

••66 A particle moves along a circular path over a horizontal xy coordinate system, at constant speed. At time $t_1 = 4.00$ s, it is at point (5.00 m, 6.00 m) with velocity $(3.00 \text{ m/s})\hat{j}$ and acceleration in the positive x direction. At time $t_2 = 10.0$ s, it has velocity $(-3.00 \text{ m/s})\hat{i}$ and acceleration in the positive y direction. What are the (a) x and (b) y coordinates of the center of the circular path if $t_2 - t_1$ is less than one period?

•••67 SSM WWW A boy whirls a stone in a horizontal circle of radius 1.5 m and at height 2.0 m above level ground. The string breaks, and the stone flies off horizontally and strikes the ground after traveling a horizontal distance of 10 m. What is the magnitude of the centripetal acceleration of the stone during the circular motion?

•••68 GO A cat rides a merry-go-round turning with uniform circular motion. At time $t_1 = 2.00$ s, the cat's velocity is $\vec{v}_1 = (3.00 \text{ m/s})\hat{i} + (4.00 \text{ m/s})\hat{j}$, measured on a horizontal xy coordinate system. At $t_2 = 5.00$ s, the cat's velocity is $\vec{v}_2 = (-3.00 \text{ m/s})\hat{i} + (-4.00 \text{ m/s})\hat{j}$. What are (a) the magnitude of the cat's centripetal acceleration and (b) the cat's average acceleration during the time interval $t_2 - t_1$, which is less than one period?

Module 4-6 Relative Motion in One Dimension

•69 A cameraman on a pickup truck is traveling westward at 20 km/h while he records a cheetah that is moving westward 30 km/h faster than the truck. Suddenly, the cheetah stops, turns, and then runs at 45 km/h eastward, as measured by a suddenly nervous crew member who stands alongside the cheetah's path. The change in the animal's velocity takes 2.0 s. What are the (a) magnitude and (b) direction of the animal's acceleration according to the cameraman and the (c) magnitude and (d) direction according to the nervous crew member?

•70 A boat is traveling upstream in the positive direction of an x axis at 14 km/h with respect to the water of a river. The water is flowing at 9.0 km/h with respect to the ground. What are the (a) magnitude and (b) direction of the boat's velocity with respect to the ground? A child on the boat walks from front to rear at 6.0 km/h with respect to the boat. What are the (c) magnitude and (d) direction of the child's velocity with respect to the ground?

••71 A suspicious-looking man runs as fast as he can along a moving sidewalk from one end to the other, taking 2.50 s. Then security agents appear, and the man runs as fast as he can back along the sidewalk to his starting point, taking 10.0 s. What is the ratio of the man's running speed to the sidewalk's speed?

Module 4-7 Relative Motion in Two Dimensions

•72 A rugby player runs with the ball directly toward his opponent's goal, along the positive direction of an x axis. He can legally pass the ball to a teammate as long as the ball's velocity relative to the field does not have a positive x component. Suppose the player runs at speed 4.0 m/s relative to the field while he passes the ball with velocity $\vec{v}_{BP}$ relative to himself. If $\vec{v}_{BP}$ has magnitude 6.0 m/s, what is the smallest angle it can have for the pass to be legal?

••73 Two highways intersect as shown in Fig. 4-46. At the instant shown, a police car P is distance $d_P = 800$ m from the intersection and moving at speed $v_P = 80$ km/h. Motorist M is distance $d_M = 600$ m from the intersection and moving at speed $v_M = 60$ km/h.

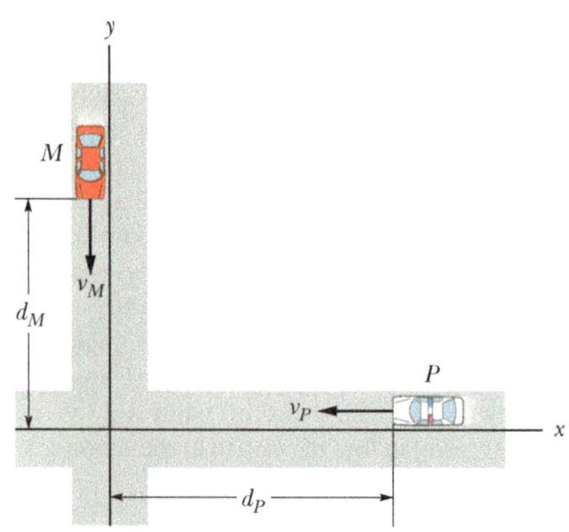

Figure 4-46 Problem 73.

(a) In unit-vector notation, what is the velocity of the motorist with respect to the police car? (b) For the instant shown in Fig. 4-46, what is the angle between the velocity found in (a) and the line of sight between the two cars? (c) If the cars maintain their velocities, do the answers to (a) and (b) change as the cars move nearer the intersection?

••74 After flying for 15 min in a wind blowing 42 km/h at an angle of 20° south of east, an airplane pilot is over a town that is 55 km due north of the starting point. What is the speed of the airplane relative to the air?

••75 SSM A train travels due south at 30 m/s (relative to the ground) in a rain that is blown toward the south by the wind. The path of each raindrop makes an angle of 70° with the vertical, as measured by an observer stationary on the ground. An observer on the train, however, sees the drops fall perfectly vertically. Determine the speed of the raindrops relative to the ground.

••76 A light plane attains an airspeed of 500 km/h. The pilot sets out for a destination 800 km due north but discovers that the plane must be headed 20.0° east of due north to fly there directly. The plane arrives in 2.00 h. What were the (a) magnitude and (b) direction of the wind velocity?

••77 SSM Snow is falling vertically at a constant speed of 8.0 m/s. At what angle from the vertical do the snowflakes appear to be falling as viewed by the driver of a car traveling on a straight, level road with a speed of 50 km/h?

••78 In the overhead view of Fig. 4-47, Jeeps P and B race along straight lines, across flat terrain, and past stationary border guard A. Relative to the guard, B travels at a constant speed of 20.0 m/s, at the angle $\theta_2 = 30.0°$. Relative to the guard, P has accelerated from rest at a constant rate of 0.400 m/s^2 at the angle $\theta_1 = 60.0°$. At a certain time during the acceleration, P has a speed of 40.0 m/s. At that time, what are the (a) magnitude and (b) direction of the velocity of P relative to B and the (c) magnitude and (d) direction of the acceleration of P relative to B?

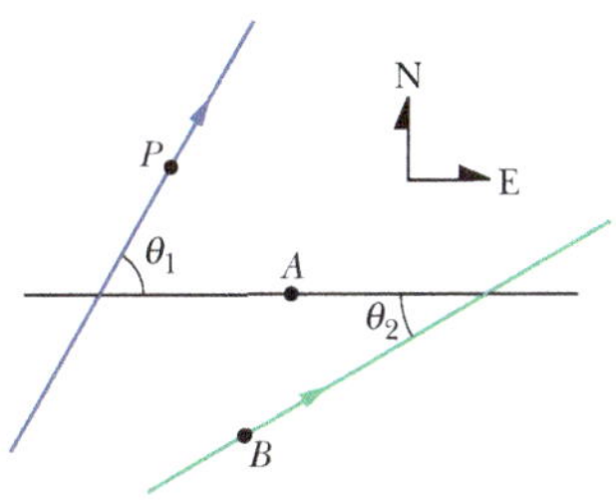

Figure 4-47 Problem 78.

••79 SSM ILW Two ships, A and B, leave port at the same time. Ship A travels northwest at 24 knots, and ship B travels at 28 knots in a direction 40° west of south. (1 knot = 1 nautical mile per hour; see Appendix D.) What are the (a) magnitude and (b) direction of the velocity of ship A relative to B? (c) After what time will the ships be 160 nautical miles apart? (d) What will be the bearing of B (the direction of B's position) relative to A at that time?

••80 GO A 200-m-wide river flows due east at a uniform speed of 2.0 m/s. A boat with a speed of 8.0 m/s relative to the water leaves the south bank pointed in a direction 30° west of north. What are the (a) magnitude and (b) direction of the boat's velocity relative to the ground? (c) How long does the boat take to cross the river?

•••81 GO Ship A is located 4.0 km north and 2.5 km east of ship B. Ship A has a velocity of 22 km/h toward the south, and ship B has a velocity of 40 km/h in a direction 37° north of east. (a) What is the velocity of A relative to B in unit-vector notation with $\hat{i}$ toward the east? (b) Write an expression (in terms of $\hat{i}$ and $\hat{j}$) for the position of A relative to B as a function of t, where $t = 0$ when the ships are in the positions described above. (c) At what time is the separation between the ships least? (d) What is that least separation?

•••82 GO A 200-m-wide river has a uniform flow speed of 1.1 m/s through a jungle and toward the east. An explorer wishes to leave a small clearing on the south bank and cross the river in a powerboat that moves at a constant speed of 4.0 m/s with respect to the water. There is a clearing on the north bank 82 m upstream from a point directly opposite the clearing on the south bank. (a) In what direction must the boat be pointed in order to travel in a straight line and land in the clearing on the north bank? (b) How long will the boat take to cross the river and land in the clearing?

Additional Problems

83 A woman who can row a boat at 6.4 km/h in still water faces a long, straight river with a width of 6.4 km and a current of 3.2 km/h. Let $\hat{i}$ point directly across the river and $\hat{j}$ point directly downstream. If she rows in a straight line to a point directly opposite her starting position, (a) at what angle to $\hat{i}$ must she point the boat and (b) how long will she take? (c) How long will she take if, instead, she rows 3.2 km *down* the river and then back to her starting point? (d) How long if she rows 3.2 km *up* the river and then back to her starting point? (e) At what angle to $\hat{i}$ should she point the boat if she wants to cross the river in the shortest possible time? (f) How long is that shortest time?

84 In Fig. 4-48*a*, a sled moves in the negative x direction at constant speed v_s while a ball of ice is shot from the sled with a velocity $\vec{v}_0 = v_{0x}\hat{i} + v_{0y}\hat{j}$ relative to the sled. When the ball lands, its horizontal displacement Δx_{bg} relative to the ground (from its launch position to its landing position) is measured. Figure 4-48*b* gives Δx_{bg} as a function of v_s. Assume the ball lands at approximately its launch height. What are the values of (a) v_{0x} and (b) v_{0y}? The ball's displacement Δx_{bs} relative to the sled can also be measured. Assume that the sled's velocity is not changed when the ball is shot. What is Δx_{bs} when v_s is (c) 5.0 m/s and (d) 15 m/s?

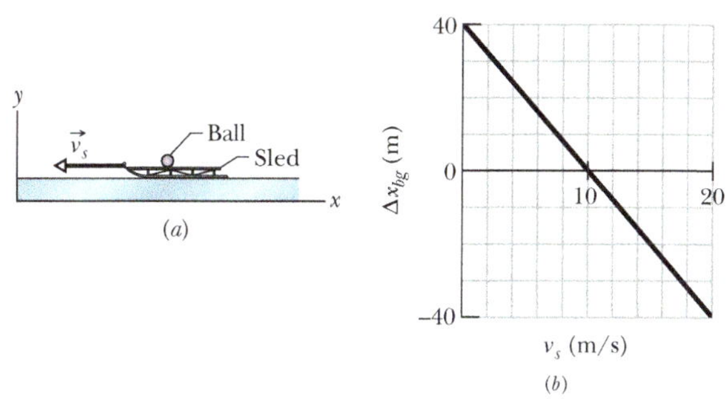

Figure 4-48 Problem 84.

85 You are kidnapped by political-science majors (who are upset because you told them political science is not a real science). Although blindfolded, you can tell the speed of their car (by the whine of the engine), the time of travel (by mentally counting off seconds), and the direction of travel (by turns along the rectangular street system). From these clues, you know that you are taken along the following course: 50 km/h for 2.0 min, turn 90° to the right, 20 km/h for 4.0 min, turn 90° to the right, 20 km/h for 60 s, turn 90° to the left, 50 km/h for 60 s, turn 90° to the right, 20 km/h for 2.0 min, turn 90° to the left, 50 km/h for 30 s. At that point, (a) how far are you from your starting point, and (b) in what direction relative to your initial direction of travel are you?

86 A radar station detects an airplane approaching directly from the east. At first observation, the airplane is at distance $d_1 = 360$ m from the station and at angle $\theta_1 = 40°$ above the horizon (Fig. 4-49). The airplane is tracked through an angular change $\Delta\theta = 123°$ in the vertical east–west plane; its distance is then $d_2 = 790$ m. Find the (a) magnitude and (b) direction of the airplane's displacement during this period.

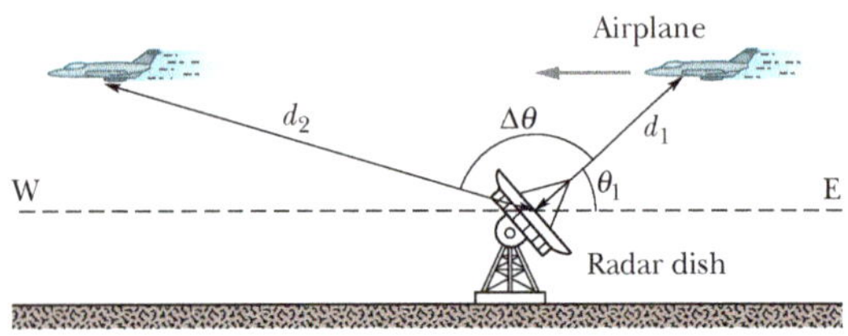

Figure 4-49 Problem 86.

87 SSM A baseball is hit at ground level. The ball reaches its maximum height above ground level 3.0 s after being hit. Then 2.5 s after reaching its maximum height, the ball barely clears a fence that is 97.5 m from where it was hit. Assume the ground is level. (a) What maximum height above ground level is reached by the ball? (b) How high is the fence? (c) How far beyond the fence does the ball strike the ground?

88 Long flights at midlatitudes in the Northern Hemisphere encounter the jet stream, an eastward airflow that can affect a plane's speed relative to Earth's surface. If a pilot maintains a certain speed relative to the air (the plane's *airspeed*), the speed relative to the surface (the plane's *ground speed*) is more when the flight is in the direction of the jet stream and less when the flight is opposite the jet stream. Suppose a round-trip flight is scheduled between two cities separated by 4000 km, with the outgoing flight in the direction of the jet stream and the return flight opposite it. The airline computer advises an airspeed of 1000 km/h, for which the difference in flight times for the outgoing and return flights is 70.0 min. What jet-stream speed is the computer using?

89 SSM A particle starts from the origin at $t = 0$ with a velocity of $8.0\hat{j}$ m/s and moves in the xy plane with constant acceleration $(4.0\hat{i} + 2.0\hat{j})$ m/s^2. When the particle's x coordinate is 29 m, what are its (a) y coordinate and (b) speed?

90 At what initial speed must the basketball player in Fig. 4-50 throw the ball, at angle $\theta_0 = 55°$ above the horizontal, to make the foul shot? The horizontal distances are $d_1 = 1.0$ ft and $d_2 = 14$ ft, and the heights are $h_1 = 7.0$ ft and $h_2 = 10$ ft.

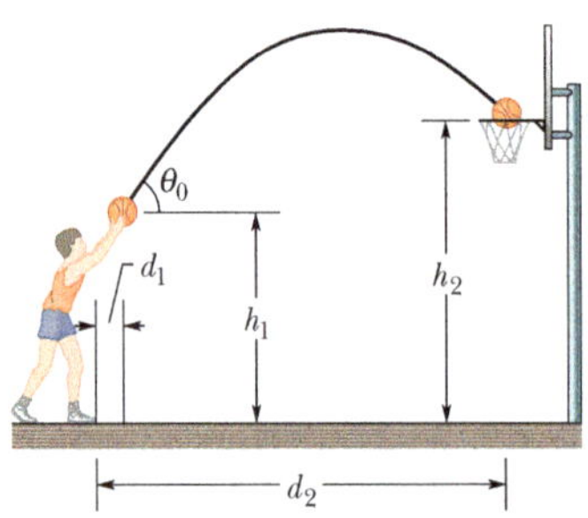

Figure 4-50 Problem 90.

91 During volcanic eruptions, chunks of solid rock can be blasted out of the volcano; these projectiles are called *volcanic bombs.* Figure 4-51 shows a cross section of Mt. Fuji, in Japan. (a) At what initial speed would a bomb have to be ejected, at angle $\theta_0 = 35°$ to the horizontal, from the vent at A in order to fall at the foot of the volcano at B, at vertical distance $h = 3.30$ km and horizontal distance $d = 9.40$ km? Ignore, for the moment, the effects of air on the bomb's travel. (b) What would be the time of flight? (c) Would the effect of the air increase or decrease your answer in (a)?

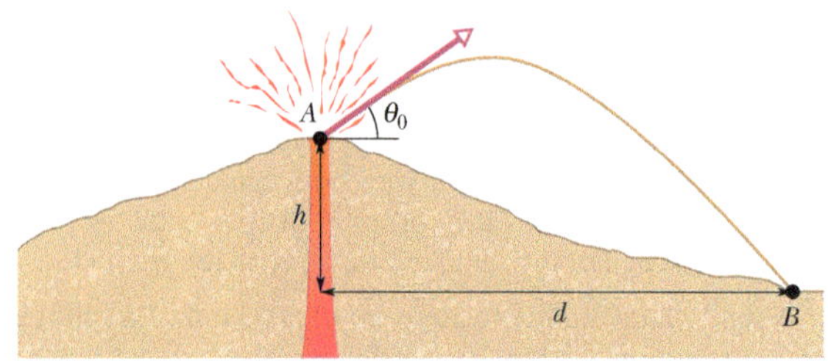

Figure 4-51 Problem 91.

92 An astronaut is rotated in a horizontal centrifuge at a radius of 5.0 m. (a) What is the astronaut's speed if the centripetal acceleration has a magnitude of 7.0g? (b) How many revolutions per minute are required to produce this acceleration? (c) What is the period of the motion?

93 SSM Oasis A is 90 km due west of oasis B. A desert camel leaves A and takes 50 h to walk 75 km at 37° north of due east. Next it takes 35 h to walk 65 km due south. Then it rests for 5.0 h. What are the (a) magnitude and (b) direction of the camel's displacement relative to A at the resting point? From the time the camel leaves A until the end of the rest period, what are the (c) magnitude and (d) direction of its average velocity and (e) its average speed? The camel's last drink was at A; it must be at B no more than 120 h later for its next drink. If it is to reach B just in time, what must be the (f) magnitude and (g) direction of its average velocity after the rest period?

94 *Curtain of death.* A large metallic asteroid strikes Earth and quickly digs a crater into the rocky material below ground level by launching rocks upward and outward. The following table gives five pairs of launch speeds and angles (from the horizontal) for such rocks, based on a model of crater formation. (Other rocks, with intermediate speeds and angles, are also launched.) Suppose that you are at $x = 20$ km when the asteroid strikes the ground at time $t = 0$ and position $x = 0$ (Fig. 4-52). (a) At $t = 20$ s, what are the x and y coordinates of the rocks headed in your direction from launches A through E? (b) Plot these coordinates and then sketch a curve through the points to include rocks with intermediate launch speeds and angles. The curve should indicate what you would see as you look up into the approaching rocks.

Launch	Speed (m/s)	Angle (degrees)
A	520	14.0
B	630	16.0
C	750	18.0
D	870	20.0
E	1000	22.0

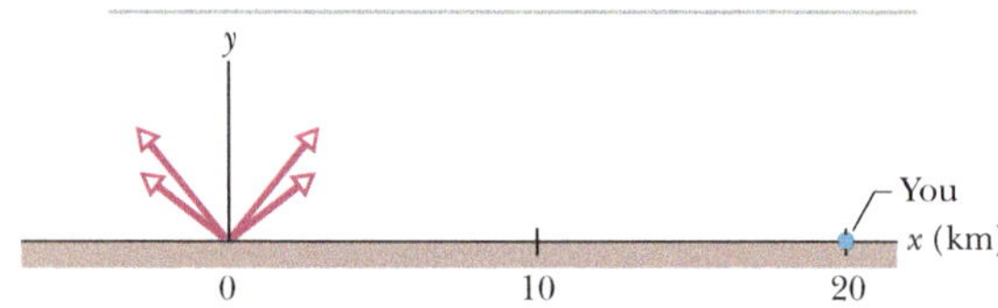

Figure 4-52 Problem 94.

95 Figure 4-53 shows the straight path of a particle across an xy coordinate system as the particle is accelerated from rest during time interval Δt_1. The acceleration is constant. The xy coordinates for point A are (4.00 m, 6.00 m); those for point B are (12.0 m, 18.0 m). (a) What is the ratio a_y/a_x of the acceleration components? (b) What are the coordinates of the particle if the motion is continued for another interval equal to Δt_1?

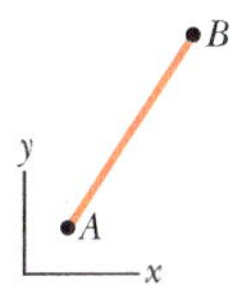

Figure 4-53 Problem 95.

96 For women's volleyball the top of the net is 2.24 m above the floor and the court measures 9.0 m by 9.0 m on each side of the net. Using a jump serve, a player strikes the ball at a point that is 3.0 m above the floor and a horizontal distance of 8.0 m from the net. If the initial velocity of the ball is horizontal, (a) what minimum magnitude must it have if the ball is to clear the net and (b) what maximum magnitude can it have if the ball is to strike the floor inside the back line on the other side of the net?

97 SSM A rifle is aimed horizontally at a target 30 m away. The bullet hits the target 1.9 cm below the aiming point. What are (a) the bullet's time of flight and (b) its speed as it emerges from the rifle?

98 A particle is in uniform circular motion about the origin of an xy coordinate system, moving clockwise with a period of 7.00 s. At one instant, its position vector (measured from the origin) is $\vec{r} = (2.00\text{ m})\hat{i} - (3.00\text{ m})\hat{j}$. At that instant, what is its velocity in unit-vector notation?

99 In Fig. 4-54, a lump of wet putty moves in uniform circular motion as it rides at a radius of 20.0 cm on the rim of a wheel rotating counterclockwise with a period of 5.00 ms. The lump then happens to fly off the rim at the 5 o'clock position (as if on a clock face). It leaves the rim at a height of $h = 1.20$ m from the floor and at a distance $d = 2.50$ m from a wall. At what height on the wall does the lump hit?

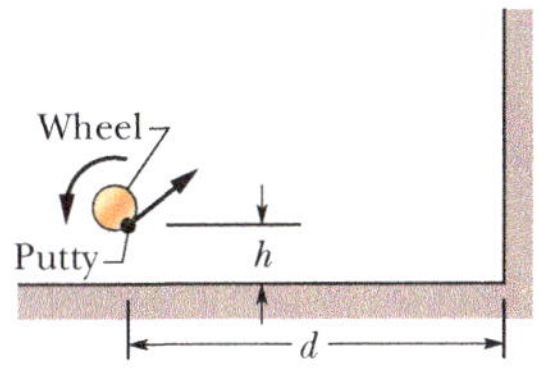

Figure 4-54 Problem 99.

100 An iceboat sails across the surface of a frozen lake with constant acceleration produced by the wind. At a certain instant the boat's velocity is $(6.30\hat{i} - 8.42\hat{j})$ m/s. Three seconds later, because of a wind shift, the boat is instantaneously at rest. What is its average acceleration for this 3.00 s interval?

101 In Fig. 4-55, a ball is shot directly upward from the ground with an initial speed of $v_0 = 7.00$ m/s. Simultaneously, a construction elevator cab begins to move upward from the ground with a constant speed of $v_c = 3.00$ m/s. What maximum height does the ball reach relative to (a) the ground and (b) the cab floor? At what rate does the speed of the ball change relative to (c) the ground and (d) the cab floor?

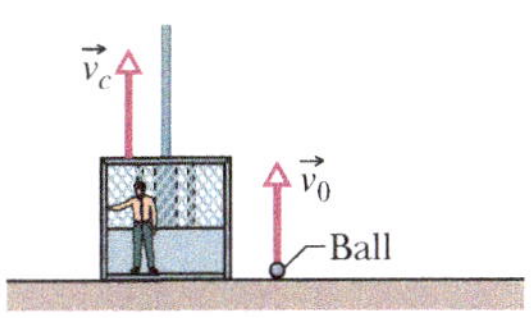

Figure 4-55 Problem 101.

102 A magnetic field forces an electron to move in a circle with radial acceleration 3.0×10^{14} m/s^2. (a) What is the speed of the electron if the radius of its circular path is 15 cm? (b) What is the period of the motion?

103 In 3.50 h, a balloon drifts 21.5 km north, 9.70 km east, and 2.88 km upward from its release point on the ground. Find (a) the magnitude of its average velocity and (b) the angle its average velocity makes with the horizontal.

104 A ball is thrown horizontally from a height of 20 m and hits the ground with a speed that is three times its initial speed. What is the initial speed?

105 A projectile is launched with an initial speed of 30 m/s at an angle of 60° above the horizontal. What are the (a) magnitude and (b) angle of its velocity 2.0 s after launch, and (c) is the angle above or below the horizontal? What are the (d) magnitude and (e) angle of its velocity 5.0 s after launch, and (f) is the angle above or below the horizontal?

106 The position vector for a proton is initially $\vec{r} = 5.0\hat{i} - 6.0\hat{j} + 2.0\hat{k}$ and then later is $\vec{r} = -2.0\hat{i} + 6.0\hat{j} + 2.0\hat{k}$, all in meters. (a) What is the proton's displacement vector, and (b) to what plane is that vector parallel?

107 A particle P travels with constant speed on a circle of radius $r = 3.00$ m (Fig. 4-56) and completes one revolution in 20.0 s. The particle passes through O at time $t = 0$. State the following vectors in magnitude-angle notation (angle relative to the positive direction of x). With respect to O, find the particle's position vector at the times t of (a) 5.00 s, (b) 7.50 s, and (c) 10.0 s. (d) For the 5.00 s interval from the end of the fifth second to the end of the tenth second, find the particle's displacement. For that interval, find (e) its average velocity and its velocity at the (f) beginning and (g) end. Next, find the acceleration at the (h) beginning and (i) end of that interval.

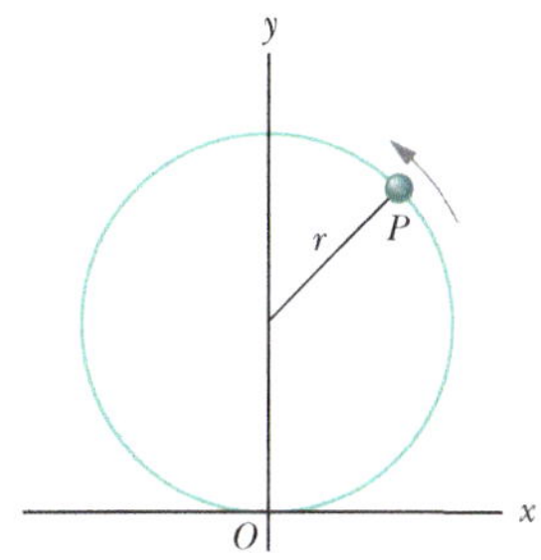

Figure 4-56 Problem 107.

108 The fast French train known as the TGV (Train à Grande Vitesse) has a scheduled average speed of 216 km/h. (a) If the train goes around a curve at that speed and the magnitude of the acceleration experienced by the passengers is to be limited to 0.050g, what is the smallest radius of curvature for the track that can be tolerated? (b) At what speed must the train go around a curve with a 1.00 km radius to be at the acceleration limit?

109 (a) If an electron is projected horizontally with a speed of 3.0×10^6 m/s, how far will it fall in traversing 1.0 m of horizontal distance? (b) Does the answer increase or decrease if the initial speed is increased?

110 A person walks up a stalled 15-m-long escalator in 90 s. When standing on the same escalator, now moving, the person is carried up in 60 s. How much time would it take that person to walk up the moving escalator? Does the answer depend on the length of the escalator?

111 (a) What is the magnitude of the centripetal acceleration of an object on Earth's equator due to the rotation of Earth? (b) What would Earth's rotation period have to be for objects on the equator to have a centripetal acceleration of magnitude 9.8 m/s^2?

112 The range of a projectile depends not only on v_0 and θ_0 but also on the value g of the free-fall acceleration, which varies from place to place. In 1936, Jesse Owens established a world's running broad jump record of 8.09 m at the Olympic Games at Berlin (where $g = 9.8128$ m/s^2). Assuming the same values of v_0 and θ_0, by how much would his record have differed if he had competed instead in 1956 at Melbourne (where $g = 9.7999$ m/s^2)?

113 Figure 4-57 shows the path taken by a drunk skunk over level ground, from initial point i to final point f. The angles are $\theta_1 = 30.0°$, $\theta_2 = 50.0°$, and $\theta_3 = 80.0°$, and the distances are $d_1 = 5.00$ m, $d_2 = 8.00$ m, and $d_3 = 12.0$ m. What are the (a) magnitude and (b) angle of the skunk's displacement from i to f?

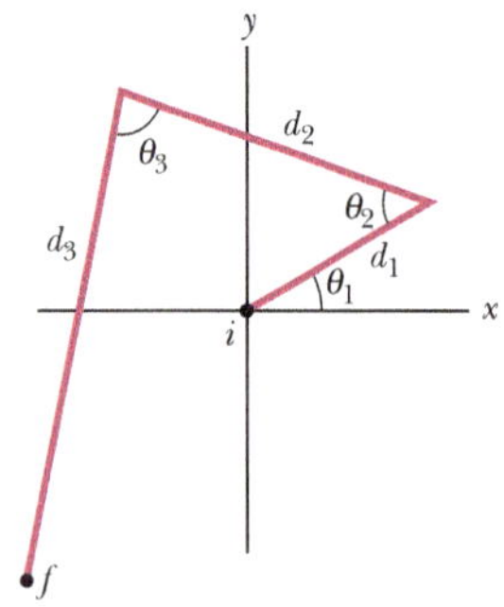

Figure 4-57 Problem 113.

114 The position vector $\vec{r}$ of a particle moving in the xy plane is $\vec{r} = 2t\hat{i} + 2\sin[(\pi/4 \text{ rad/s})t]\hat{j}$, with $\vec{r}$ in meters and t in seconds. (a) Calculate the x and y components of the particle's position at $t = 0$, 1.0, 2.0, 3.0, and 4.0 s and sketch the particle's path in the xy plane for the interval $0 \le t \le 4.0$ s. (b) Calculate the components of the particle's velocity at $t = 1.0$, 2.0, and 3.0 s. Show that the velocity is tangent to the path of the particle and in the direction the particle is moving at each time by drawing the velocity vectors on the plot of the particle's path in part (a). (c) Calculate the components of the particle's acceleration at $t = 1.0$, 2.0, and 3.0 s.

115 An electron having an initial horizontal velocity of magnitude 1.00×10^9 cm/s travels into the region between two horizontal metal plates that are electrically charged. In that region, the electron travels a horizontal distance of 2.00 cm and has a constant downward acceleration of magnitude 1.00×10^{17} cm/s^2 due to the charged plates. Find (a) the time the electron takes to travel the 2.00 cm, (b) the vertical distance it travels during that time, and the magnitudes of its (c) horizontal and (d) vertical velocity components as it emerges from the region.

116 An elevator without a ceiling is ascending with a constant speed of 10 m/s. A boy on the elevator shoots a ball directly upward, from a height of 2.0 m above the elevator floor, just as the elevator floor is 28 m above the ground. The initial speed of the ball with respect to the elevator is 20 m/s. (a) What maximum height above the ground does the ball reach? (b) How long does the ball take to return to the elevator floor?

117 A football player punts the football so that it will have a "hang time" (time of flight) of 4.5 s and land 46 m away. If the ball leaves the player's foot 150 cm above the ground, what must be the (a) magnitude and (b) angle (relative to the horizontal) of the ball's initial velocity?

118 An airport terminal has a moving sidewalk to speed passengers through a long corridor. Larry does not use the moving sidewalk; he takes 150 s to walk through the corridor. Curly, who simply stands on the moving sidewalk, covers the same distance in 70 s. Moe boards the sidewalk and walks along it. How long does Moe take to move through the corridor? Assume that Larry and Moe walk at the same speed.

119 A wooden boxcar is moving along a straight railroad track at speed v_1. A sniper fires a bullet (initial speed v_2) at it from a high-powered rifle. The bullet passes through both lengthwise walls of the car, its entrance and exit holes being exactly opposite each other as viewed from within the car. From what direction, relative to the track, is the bullet fired? Assume that the bullet is not deflected upon entering the car, but that its speed decreases by 20%. Take $v_1 = 85$ km/h and $v_2 = 650$ m/s. (Why don't you need to know the width of the boxcar?)

120 A sprinter running on a circular track has a velocity of constant magnitude 9.20 m/s and a centripetal acceleration of magnitude 3.80 m/s^2. What are (a) the track radius and (b) the period of the circular motion?

121 Suppose that a space probe can withstand the stresses of a $20g$ acceleration. (a) What is the minimum turning radius of such a craft moving at a speed of one-tenth the speed of light? (b) How long would it take to complete a 90° turn at this speed?

122 GO You are to throw a ball with a speed of 12.0 m/s at a target that is height $h = 5.00$ m above the level at which you release the ball (Fig. 4-58). You want the ball's velocity to be horizontal at the instant it reaches the target. (a) At what angle θ above the horizontal must you throw the ball? (b) What is the horizontal distance from the release point to the target? (c) What is the speed of the ball just as it reaches the target?

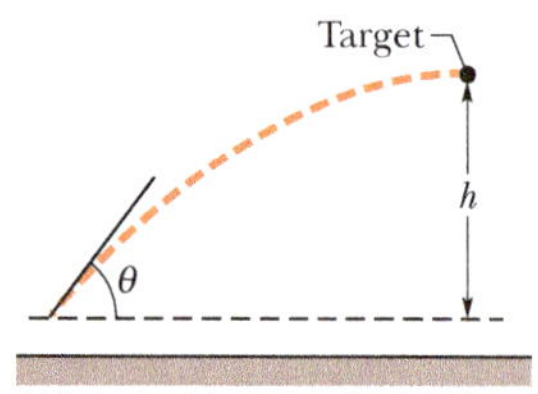

Figure 4-58 Problem 122.

123 A projectile is fired with an initial speed $v_0 = 30.0$ m/s from level ground at a target that is on the ground, at distance $R = 20.0$ m, as shown in Fig. 4-59. What are the (a) least and (b) greatest launch angles that will allow the projectile to hit the target?

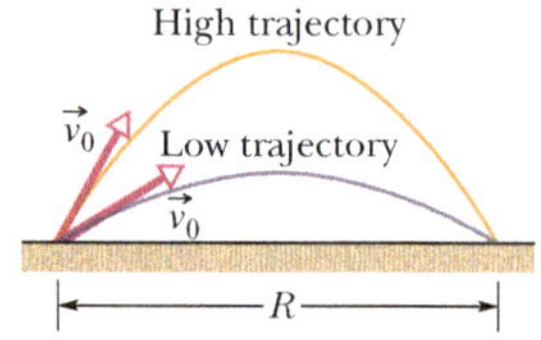

Figure 4-59 Problem 123.

124 *A graphing surprise.* At time $t = 0$, a burrito is launched from level ground, with an initial speed of 16.0 m/s and launch angle θ_0. Imagine a position vector $\vec{r}$ continuously directed from the launching point to the burrito during the flight. Graph the magnitude r of the position vector for (a) $\theta_0 = 40.0°$ and (b) $\theta_0 = 80.0°$. For $\theta_0 = 40.0°$, (c) when does r reach its maximum value, (d) what is that value, and how far (e) horizontally and (f) vertically is the burrito from the launch point? For $\theta_0 = 80.0°$, (g) when does r reach its maximum value, (h) what is that value, and how far (i) horizontally and (j) vertically is the burrito from the launch point?

125 A cannon located at sea level fires a ball with initial speed 82 m/s and initial angle 45°. The ball lands in the water after traveling a horizontal distance 686 m. How much greater would the horizontal distance have been had the cannon been 30 m higher?

126 The magnitude of the velocity of a projectile when it is at its maximum height above ground level is 10.0 m/s. (a) What is the magnitude of the velocity of the projectile 1.00 s before it achieves its maximum height? (b) What is the magnitude of the velocity of the projectile 1.00 s after it achieves its maximum height? If we take $x = 0$ and $y = 0$ to be at the point of maximum height and positive x to be in the direction of the velocity there, what are the (c) x coordinate and (d) y coordinate of the projectile 1.00 s before it reaches its maximum height and the (e) x coordinate and (f) y coordinate 1.0 s after it reaches its maximum height?

127 A frightened rabbit moving at 6.00 m/s due east runs onto a large area of level ice of negligible friction. As the rabbit slides across the ice, the force of the wind causes it to have a constant acceleration of 1.40 m/s^2, due north. Choose a coordinate system with the origin at the rabbit's initial position on the ice and the positive x axis directed toward the east. In unit-vector notation, what are the rabbit's (a) velocity and (b) position when it has slid for 3.00 s?

128 The pilot of an aircraft flies due east relative to the ground in a wind blowing 20.0 km/h toward the south. If the speed of the aircraft in the absence of wind is 70.0 km/h, what is the speed of the aircraft relative to the ground?

129 The pitcher in a slow-pitch softball game releases the ball at a point 3.0 ft above ground level. A stroboscopic plot of the position of the ball is shown in Fig. 4-60, where the readings are 0.25 s apart and the ball is released at $t = 0$. (a) What is the initial speed of the ball? (b) What is the speed of the ball at the instant it reaches its maximum height above ground level? (c) What is that maximum height?

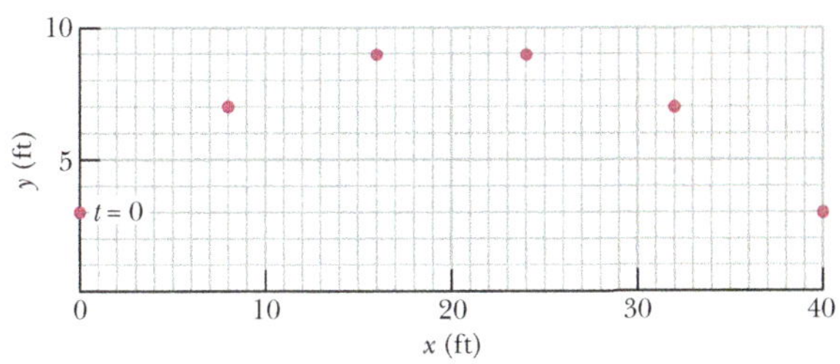

Figure 4-60 Problem 129.

130 Some state trooper departments use aircraft to enforce highway speed limits. Suppose that one of the airplanes has a speed of 135 mi/h in still air. It is flying straight north so that it is at all times directly above a north–south highway. A ground observer tells the pilot by radio that a 70.0 mi/h wind is blowing but neglects to give the wind direction. The pilot observes that in spite of the wind the plane can travel 135 mi along the highway in 1.00 h. In other words, the ground speed is the same as if there were no wind. (a) From what direction is the wind blowing? (b) What is the heading of the plane; that is, in what direction does it point?

131 A golfer tees off from the top of a rise, giving the golf ball an initial velocity of 43.0 m/s at an angle of 30.0° above the horizontal. The ball strikes the fairway a horizontal distance of 180 m from the tee. Assume the fairway is level. (a) How high is the rise above the fairway? (b) What is the speed of the ball as it strikes the fairway?

132 A track meet is held on a planet in a distant solar system. A shot-putter releases a shot at a point 2.0 m above ground level. A stroboscopic plot of the position of the shot is shown in Fig. 4-61,

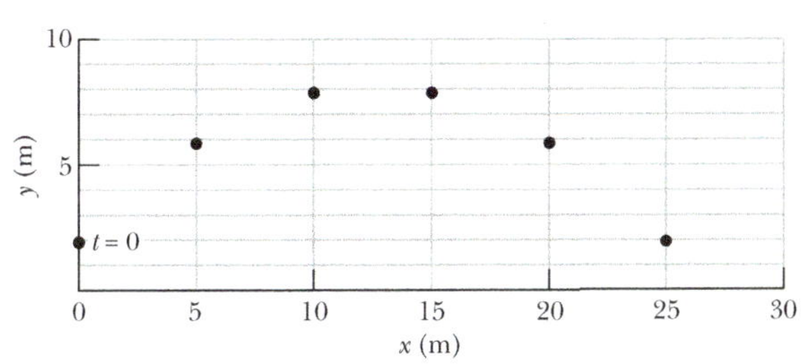

Figure 4-61 Problem 132.

where the readings are 0.50 s apart and the shot is released at time $t = 0$. (a) What is the initial velocity of the shot in unit-vector notation? (b) What is the magnitude of the free-fall acceleration on the planet? (c) How long after it is released does the shot reach the ground? (d) If an identical throw of the shot is made on the surface of Earth, how long after it is released does it reach the ground?

133 A helicopter is flying in a straight line over a level field at a constant speed of 6.20 m/s and at a constant altitude of 9.50 m. A package is ejected horizontally from the helicopter with an initial velocity of 12.0 m/s relative to the helicopter and in a direction opposite the helicopter's motion. (a) Find the initial speed of the package relative to the ground. (b) What is the horizontal distance between the helicopter and the package at the instant the package strikes the ground? (c) What angle does the velocity vector of the package make with the ground at the instant before impact, as seen from the ground?

134 A car travels around a flat circle on the ground, at a constant speed of 12.0 m/s. At a certain instant the car has an acceleration of 3.00 m/s^2 toward the east. What are its distance and direction from the center of the circle at that instant if it is traveling (a) clockwise around the circle and (b) counterclockwise around the circle?

135 You throw a ball from a cliff with an initial velocity of 15.0 m/s at an angle of 20.0° below the horizontal. Find (a) its horizontal displacement and (b) its vertical displacement 2.30 s later.

136 A baseball is hit at Fenway Park in Boston at a point 0.762 m above home plate with an initial velocity of 33.53 m/s directed 55.0° above the horizontal. The ball is observed to clear the 11.28-m-high wall in left field (known as the "green monster") 5.00 s after it is hit, at a point just inside the left-field foul-line pole. Find (a) the horizontal distance down the left-field foul line from home plate to the wall; (b) the vertical distance by which the ball clears the wall; (c) the horizontal and vertical displacements of the ball with respect to home plate 0.500 s before it clears the wall.

137 A transcontinental flight of 4350 km is scheduled to take 50 min longer westward than eastward. The airspeed of the airplane is 966 km/h, and the jet stream it will fly through is presumed to move due east. What is the assumed speed of the jet stream?

138 A woman can row a boat at 6.40 km/h in still water. (a) If she is crossing a river where the current is 3.20 km/h, in what direction must her boat be headed if she wants to reach a point directly opposite her starting point? (b) If the river is 6.40 km wide, how long will she take to cross the river? (c) Suppose that instead of crossing the river she rows 3.20 km *down* the river and then back to her starting point. How long will she take? (d) How long will she take to row 3.20 km *up* the river and then back to her starting point? (e) In what direction should she head the boat if she wants to cross in the shortest possible time, and what is that time?

CHAPTER 5

Force and Motion—I

5-1 NEWTON'S FIRST AND SECOND LAWS

Learning Objectives

After reading this module, you should be able to . . .

5.01 Identify that a force is a vector quantity and thus has both magnitude and direction and also components.

5.02 Given two or more forces acting on the same particle, add the forces *as vectors* to get the net force.

5.03 Identify Newton's first and second laws of motion.

5.04 Identify inertial reference frames.

5.05 Sketch a free-body diagram for an object, showing the object as a particle and drawing the forces acting on it as vectors with their tails anchored on the particle.

5.06 Apply the relationship (Newton's second law) between the net force on an object, the mass of the object, and the acceleration produced by the net force.

5.07 Identify that only *external* forces on an object can cause the object to accelerate.

Key Ideas

- The velocity of an object can change (the object can accelerate) when the object is acted on by one or more forces (pushes or pulls) from other objects. Newtonian mechanics relates accelerations and forces.

- Forces are vector quantities. Their magnitudes are defined in terms of the acceleration they would give the standard kilogram. A force that accelerates that standard body by exactly 1 m/s^2 is defined to have a magnitude of 1 N. The direction of a force is the direction of the acceleration it causes. Forces are combined according to the rules of vector algebra. The net force on a body is the vector sum of all the forces acting on the body.

- If there is no net force on a body, the body remains at rest if it is initially at rest or moves in a straight line at constant speed if it is in motion.

- Reference frames in which Newtonian mechanics holds are called inertial reference frames or inertial frames. Reference frames in which Newtonian mechanics does not hold are called noninertial reference frames or noninertial frames.

- The mass of a body is the characteristic of that body that relates the body's acceleration to the net force causing the acceleration. Masses are scalar quantities.

- The net force $\vec{F}_{\text{net}}$ on a body with mass m is related to the body's acceleration $\vec{a}$ by

$$\vec{F}_{\text{net}} = m\vec{a},$$

which may be written in the component versions

$$F_{\text{net},x} = ma_x \quad F_{\text{net},y} = ma_y \quad \text{and} \quad F_{\text{net},z} = ma_z.$$

The second law indicates that in SI units

$$1\text{ N} = 1\text{ kg}\cdot\text{m/s}^2.$$

- A free-body diagram is a stripped-down diagram in which only *one* body is considered. That body is represented by either a sketch or a dot. The external forces on the body are drawn, and a coordinate system is superimposed, oriented so as to simplify the solution.

What Is Physics?

We have seen that part of physics is a study of motion, including accelerations, which are changes in velocities. Physics is also a study of what can *cause* an object to accelerate. That cause is a **force,** which is, loosely speaking, a push or pull on the object. The force is said to *act* on the object to change its velocity. For example, when a dragster accelerates, a force from the track acts on the rear tires to cause the dragster's acceleration. When a defensive guard knocks down a quarterback, a force from the guard acts on the quarterback to cause the quarterback's backward acceleration. When a car slams into a telephone pole, a force on the car from the pole causes the car to stop. Science, engineering, legal, and medical journals are filled with articles about forces on objects, including people.

A Heads Up. Many students find this chapter to be more challenging than the preceding ones. One reason is that we need to use vectors in setting up equations—we cannot just sum some scalars. So, we need the vector rules from Chapter 3. Another reason is that we shall see a lot of different arrangements: objects will move along floors, ceilings, walls, and ramps. They will move upward on ropes looped around pulleys or by sitting in ascending or descending elevators. Sometimes, objects will even be tied together.

However, in spite of the variety of arrangements, we need only a single key idea (Newton's second law) to solve most of the homework problems. The purpose of this chapter is for us to explore how we can apply that single key idea to any given arrangement. The application will take experience—we need to solve lots of problems, not just read words. So, let's go through some of the words and then get to the sample problems.

Newtonian Mechanics

The relation between a force and the acceleration it causes was first understood by Isaac Newton (1642–1727) and is the subject of this chapter. The study of that relation, as Newton presented it, is called *Newtonian mechanics.* We shall focus on its three primary laws of motion.

Newtonian mechanics does not apply to all situations. If the speeds of the interacting bodies are very large—an appreciable fraction of the speed of light—we must replace Newtonian mechanics with Einstein's special theory of relativity, which holds at any speed, including those near the speed of light. If the interacting bodies are on the scale of atomic structure (for example, they might be electrons in an atom), we must replace Newtonian mechanics with quantum mechanics. Physicists now view Newtonian mechanics as a special case of these two more comprehensive theories. Still, it is a very important special case because it applies to the motion of objects ranging in size from the very small (almost on the scale of atomic structure) to astronomical (galaxies and clusters of galaxies).

Newton's First Law

Before Newton formulated his mechanics, it was thought that some influence, a "force," was needed to keep a body moving at constant velocity. Similarly, a body was thought to be in its "natural state" when it was at rest. For a body to move with constant velocity, it seemingly had to be propelled in some way, by a push or a pull. Otherwise, it would "naturally" stop moving.

These ideas were reasonable. If you send a puck sliding across a wooden floor, it does indeed slow and then stop. If you want to make it move across the floor with constant velocity, you have to continuously pull or push it.

Send a puck sliding over the ice of a skating rink, however, and it goes a lot farther. You can imagine longer and more slippery surfaces, over which the puck would slide farther and farther. In the limit you can think of a long, extremely slippery surface (said to be a **frictionless surface**), over which the puck would hardly slow. (We can in fact come close to this situation by sending a puck sliding over a horizontal air table, across which it moves on a film of air.)

From these observations, we can conclude that a body will keep moving with constant velocity if no force acts on it. That leads us to the first of Newton's three laws of motion:

> **Newton's First Law:** If no force acts on a body, the body's velocity cannot change; that is, the body cannot accelerate.

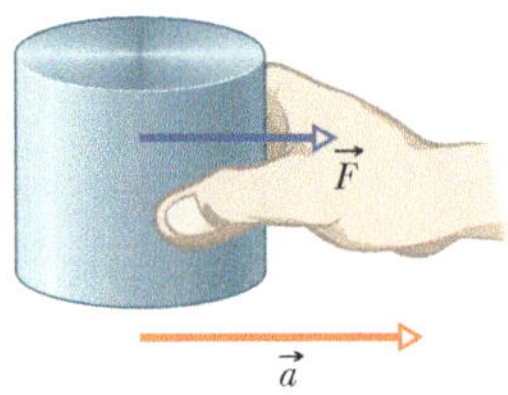

Figure 5-1 A force $\vec{F}$ on the standard kilogram gives that body an acceleration $\vec{a}$.

In other words, if the body is at rest, it stays at rest. If it is moving, it continues to move with the same velocity (same magnitude *and* same direction).

Force

Before we begin working problems with forces, we need to discuss several features of forces, such as the force unit, the vector nature of forces, the combining of forces, and the circumstances in which we can measure forces (without being fooled by a fictitious force).

Unit. We can define the unit of force in terms of the acceleration a force would give to the standard kilogram (Fig. 1-3), which has a mass defined to be exactly 1 kg. Suppose we put that body on a horizontal, frictionless surface and pull horizontally (Fig. 5-1) such that the body has an acceleration of 1 m/s^2. Then we can define our applied force as having a magnitude of 1 newton (abbreviated N). If we then pulled with a force magnitude of 2 N, we would find that the acceleration is 2 m/s^2. Thus, the acceleration is proportional to the force. If the standard body of 1 kg has an acceleration of magnitude a (in meters per second per second), then the force (in newtons) producing the acceleration has a magnitude equal to a. We now have a workable definition of the force unit.

Vectors. Force is a vector quantity and thus has not only magnitude but also direction. So, if two or more forces act on a body, we find the **net force** (or **resultant force**) by adding them as vectors, following the rules of Chapter 3. A single force that has the same magnitude and direction as the calculated net force would then have the same effect as all the individual forces. This fact, called the **principle of superposition for forces,** makes everyday forces reasonable and predictable. The world would indeed be strange and unpredictable if, say, you and a friend each pulled on the standard body with a force of 1 N and somehow the net pull was 14 N and the resulting acceleration was 14 m/s^2.

In this book, forces are most often represented with a vector symbol such as $\vec{F}$, and a net force is represented with the vector symbol $\vec{F}_{net}$. As with other vectors, a force or a net force can have components along coordinate axes. When forces act only along a single axis, they are single-component forces. Then we can drop the overhead arrows on the force symbols and just use signs to indicate the directions of the forces along that axis.

The First Law. Instead of our previous wording, the more proper statement of Newton's First Law is in terms of a *net* force:

Newton's First Law: If no *net* force acts on a body ($\vec{F}_{net} = 0$), the body's velocity cannot change; that is, the body cannot accelerate.

There may be multiple forces acting on a body, but if their net force is zero, the body cannot accelerate. So, if we happen to know that a body's velocity is constant, we can immediately say that the net force on it is zero.

Inertial Reference Frames

Newton's first law is not true in all reference frames, but we can always find reference frames in which it (as well as the rest of Newtonian mechanics) is true. Such special frames are referred to as **inertial reference frames,** or simply **inertial frames.**

An inertial reference frame is one in which Newton's laws hold.

For example, we can assume that the ground is an inertial frame provided we can neglect Earth's astronomical motions (such as its rotation).

That assumption works well if, say, a puck is sent sliding along a *short* strip of frictionless ice—we would find that the puck's motion obeys Newton's laws. However, suppose the puck is sent sliding along a *long* ice strip extending from the north pole (Fig. 5-2*a*). If we view the puck from a stationary frame in space, the puck moves south along a simple straight line because Earth's rotation around the north pole merely slides the ice beneath the puck. However, if we view the puck from a point on the ground so that we rotate with Earth, the puck's path is not a simple straight line. Because the eastward speed of the ground beneath the puck is greater the farther south the puck slides, from our ground-based view the puck appears to be deflected westward (Fig. 5-2*b*). However, this apparent deflection is caused not by a force as required by Newton's laws but by the fact that we see the puck from a rotating frame. In this situation, the ground is a **noninertial frame,** and trying to explain the deflection in terms of a force would lead us to a fictitious force. A more common example of inventing such a nonexistent force can occur in a car that is rapidly increasing in speed. You might claim that a force to the rear shoves you hard into the seat back.

In this book we usually assume that the ground is an inertial frame and that measured forces and accelerations are from this frame. If measurements are made in, say, a vehicle that is accelerating relative to the ground, then the measurements are being made in a noninertial frame and the results can be surprising.

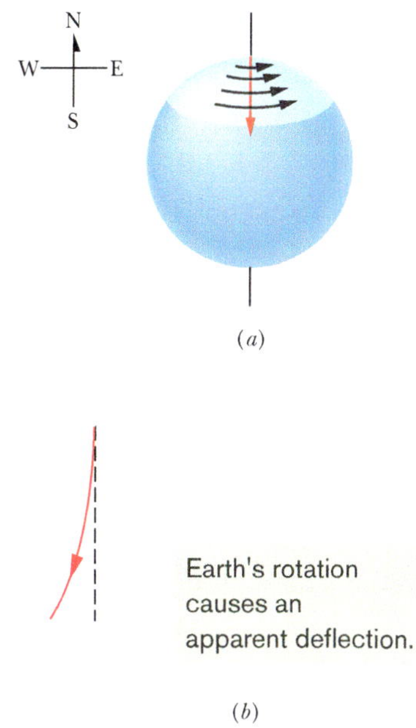

Figure 5-2 (*a*) The path of a puck sliding from the north pole as seen from a stationary point in space. Earth rotates to the east. (*b*) The path of the puck as seen from the ground.

Checkpoint 1

Which of the figure's six arrangements correctly show the vector addition of forces $\vec{F}_1$ and $\vec{F}_2$ to yield the third vector, which is meant to represent their net force $\vec{F}_{net}$?

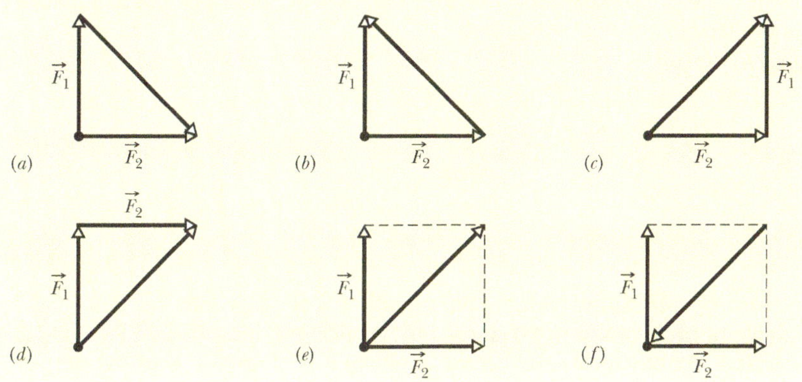

Mass

From everyday experience you already know that applying a given force to bodies (say, a baseball and a bowling ball) results in different accelerations. The common explanation is correct: The object with the larger mass is accelerated less. But we can be more precise. The acceleration is actually inversely related to the mass (rather than, say, the square of the mass).

Let's justify that inverse relationship. Suppose, as previously, we push on the standard body (defined to have a mass of exactly 1 kg) with a force of magnitude 1 N. The body accelerates with a magnitude of 1 m/s^2. Next we push on body X with the same force and find that it accelerates at 0.25 m/s^2. Let's make the (correct) assumption that with the same force,

$$\frac{m_X}{m_0} = \frac{a_0}{a_X},$$

and thus

$$m_X = m_0 \frac{a_0}{a_X} = (1.0\ \text{kg}) \frac{1.0\ \text{m/s}^2}{0.25\ \text{m/s}^2} = 4.0\ \text{kg}.$$

Defining the mass of X in this way is useful only if the procedure is consistent. Suppose we apply an 8.0 N force first to the standard body (getting an acceleration of 8.0 m/s^2) and then to body X (getting an acceleration of 2.0 m/s^2). We would then calculate the mass of X as

$$m_X = m_0 \frac{a_0}{a_X} = (1.0\ \text{kg}) \frac{8.0\ \text{m/s}^2}{2.0\ \text{m/s}^2} = 4.0\ \text{kg},$$

which means that our procedure is consistent and thus usable.

The results also suggest that mass is an intrinsic characteristic of a body—it automatically comes with the existence of the body. Also, it is a scalar quantity. However, the nagging question remains: What, exactly, is mass?

Since the word *mass* is used in everyday English, we should have some intuitive understanding of it, maybe something that we can physically sense. Is it a body's size, weight, or density? The answer is no, although those characteristics are sometimes confused with mass. We can say only that *the mass of a body is the characteristic that relates a force on the body to the resulting acceleration.* Mass has no more familiar definition; you can have a physical sensation of mass only when you try to accelerate a body, as in the kicking of a baseball or a bowling ball.

Newton's Second Law

All the definitions, experiments, and observations we have discussed so far can be summarized in one neat statement:

> **Newton's Second Law:** The net force on a body is equal to the product of the body's mass and its acceleration.

In equation form,

$$\vec{F}_{\text{net}} = m\vec{a} \quad \text{(Newton's second law).} \tag{5-1}$$

Identify the Body. This simple equation is the key idea for nearly all the homework problems in this chapter, but we must use it cautiously. First, we must be certain about which body we are applying it to. Then $\vec{F}_{\text{net}}$ must be the vector sum of *all* the forces that act on *that* body. Only forces that act on *that* body are to be included in the vector sum, not forces acting on other bodies that might be involved in the given situation. For example, if you are in a rugby scrum, the net force on *you* is the vector sum of all the pushes and pulls on *your* body. It does not include any push or pull on another player from you or from anyone else. Every time you work a force problem, your first step is to clearly state the body to which you are applying Newton's law.

Separate Axes. Like other vector equations, Eq. 5-1 is equivalent to three component equations, one for each axis of an xyz coordinate system:

$$F_{\text{net},x} = ma_x, \quad F_{\text{net},y} = ma_y, \quad \text{and} \quad F_{\text{net},z} = ma_z. \tag{5-2}$$

Each of these equations relates the net force component along an axis to the acceleration along that same axis. For example, the first equation tells us that the sum of all the force components along the x axis causes the x component a_x of the body's acceleration, but causes no acceleration in the y and z directions. Turned around, the acceleration component a_x is caused only by the sum of the force components along the x axis and is *completely* unrelated to force components along another axis. In general,

> The acceleration component along a given axis is caused *only by* the sum of the force components along that *same* axis, and not by force components along any other axis.

Forces in Equilibrium. Equation 5-1 tells us that if the net force on a body is zero, the body's acceleration $\vec{a} = 0$. If the body is at rest, it stays at rest; if it is moving, it continues to move at constant velocity. In such cases, any forces on the body *balance* one another, and both the forces and the body are said to be in *equilibrium*. Commonly, the forces are also said to *cancel* one another, but the term "cancel" is tricky. It does *not* mean that the forces cease to exist (canceling forces is not like canceling dinner reservations). The forces still act on the body but cannot change the velocity.

Units. For SI units, Eq. 5-1 tells us that

$$1 \text{ N} = (1 \text{ kg})(1 \text{ m/s}^2) = 1 \text{ kg} \cdot \text{m/s}^2. \tag{5-3}$$

Some force units in other systems of units are given in Table 5-1 and Appendix D.

Diagrams. To solve problems with Newton's second law, we often draw a **free-body diagram** in which the only body shown is the one for which we are summing forces. A sketch of the body itself is preferred by some teachers but, to save space in these chapters, we shall usually represent the body with a dot. Also, each force on the body is drawn as a vector arrow with its tail anchored on the body. A coordinate system is usually included, and the acceleration of the body is sometimes shown with a vector arrow (labeled as an acceleration). This whole procedure is designed to focus our attention on the body of interest.

Table 5-1 Units in Newton's Second Law (Eqs. 5-1 and 5-2)

System	Force	Mass	Acceleration
SI	newton (N)	kilogram (kg)	m/s^2
CGS[a]	dyne	gram (g)	cm/s^2
British[b]	pound (lb)	slug	ft/s^2

[a]1 dyne = 1 g·cm/s^2.
[b]1 lb = 1 slug·ft/s^2.

External Forces Only. A **system** consists of one or more bodies, and any force on the bodies inside the system from bodies outside the system is called an **external force.** If the bodies making up a system are rigidly connected to one another, we can treat the system as one composite body, and the net force $\vec{F}_{net}$ on it is the vector sum of all external forces. (We do not include **internal forces**—that is, forces between two bodies inside the system. Internal forces cannot accelerate the system.) For example, a connected railroad engine and car form a system. If, say, a tow line pulls on the front of the engine, the force due to the tow line acts on the whole engine–car system. Just as for a single body, we can relate the net external force on a system to its acceleration with Newton's second law, $\vec{F}_{net} = m\vec{a}$, where m is the total mass of the system.

Checkpoint 2

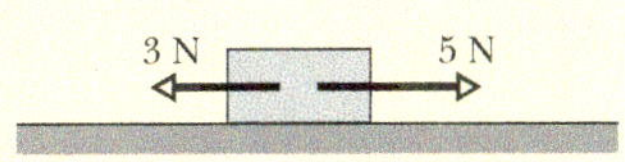

The figure here shows two horizontal forces acting on a block on a frictionless floor. If a third horizontal force $\vec{F}_3$ also acts on the block, what are the magnitude and direction of $\vec{F}_3$ when the block is (a) stationary and (b) moving to the left with a constant speed of 5 m/s?

Sample Problem 5.01 One- and two-dimensional forces, puck

Here are examples of how to use Newton's second law for a puck when one or two forces act on it. Parts A, B, and C of Fig. 5-3 show three situations in which one or two forces act on a puck that moves over frictionless ice along an x axis, in one-dimensional motion. The puck's mass is $m = 0.20$ kg. Forces $\vec{F}_1$ and $\vec{F}_2$ are directed along the axis and have magnitudes $F_1 = 4.0$ N and $F_2 = 2.0$ N. Force $\vec{F}_3$ is directed at angle $\theta = 30°$ and has magnitude $F_3 = 1.0$ N. In each situation, what is the acceleration of the puck?

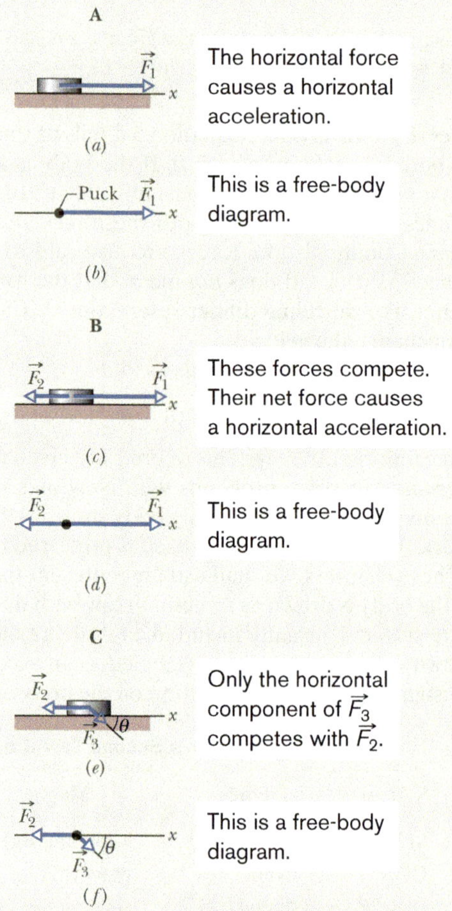

Figure 5-3 In three situations, forces act on a puck that moves along an x axis. Free-body diagrams are also shown.

KEY IDEA

In each situation we can relate the acceleration $\vec{a}$ to the net force $\vec{F}_{net}$ acting on the puck with Newton's second law, $\vec{F}_{net} = m\vec{a}$. However, because the motion is along only the x axis, we can simplify each situation by writing the second law for x components only:

$$F_{net,x} = ma_x. \tag{5-4}$$

The free-body diagrams for the three situations are also given in Fig. 5-3, with the puck represented by a dot.

Situation A: For Fig. 5-3b, where only one horizontal force acts, Eq. 5-4 gives us

$$F_1 = ma_x,$$

which, with given data, yields

$$a_x = \frac{F_1}{m} = \frac{4.0 \text{ N}}{0.20 \text{ kg}} = 20 \text{ m/s}^2. \qquad \text{(Answer)}$$

The positive answer indicates that the acceleration is in the positive direction of the x axis.

Situation B: In Fig. 5-3d, two horizontal forces act on the puck, $\vec{F}_1$ in the positive direction of x and $\vec{F}_2$ in the negative direction. Now Eq. 5-4 gives us

$$F_1 - F_2 = ma_x,$$

which, with given data, yields

$$a_x = \frac{F_1 - F_2}{m} = \frac{4.0 \text{ N} - 2.0 \text{ N}}{0.20 \text{ kg}} = 10 \text{ m/s}^2. \qquad \text{(Answer)}$$

Thus, the net force accelerates the puck in the positive direction of the x axis.

Situation C: In Fig. 5-3f, force $\vec{F}_3$ is not directed along the direction of the puck's acceleration; only x component $F_{3,x}$ is. (Force $\vec{F}_3$ is two-dimensional but the motion is only one-dimensional.) Thus, we write Eq. 5-4 as

$$F_{3,x} - F_2 = ma_x. \tag{5-5}$$

From the figure, we see that $F_{3,x} = F_3 \cos\theta$. Solving for the acceleration and substituting for $F_{3,x}$ yield

$$a_x = \frac{F_{3,x} - F_2}{m} = \frac{F_3 \cos\theta - F_2}{m}$$
$$= \frac{(1.0 \text{ N})(\cos 30°) - 2.0 \text{ N}}{0.20 \text{ kg}} = -5.7 \text{ m/s}^2. \qquad \text{(Answer)}$$

Thus, the net force accelerates the puck in the negative direction of the x axis.

WILEY PLUS Additional examples, video, and practice available at *WileyPLUS*

Sample Problem 5.02 Two-dimensional forces, cookie tin

Here we find a missing force by using the acceleration. In the overhead view of Fig. 5-4*a*, a 2.0 kg cookie tin is accelerated at 3.0 m/s^2 in the direction shown by $\vec{a}$, over a frictionless horizontal surface. The acceleration is caused by three horizontal forces, only two of which are shown: $\vec{F}_1$ of magnitude 10 N and $\vec{F}_2$ of magnitude 20 N. What is the third force $\vec{F}_3$ in unit-vector notation and in magnitude-angle notation?

KEY IDEA

The net force $\vec{F}_{net}$ on the tin is the sum of the three forces and is related to the acceleration $\vec{a}$ via Newton's second law ($\vec{F}_{net} = m\vec{a}$). Thus,

$$\vec{F}_1 + \vec{F}_2 + \vec{F}_3 = m\vec{a}, \tag{5-6}$$

which gives us

$$\vec{F}_3 = m\vec{a} - \vec{F}_1 - \vec{F}_2. \tag{5-7}$$

Calculations: Because this is a two-dimensional problem, we *cannot* find $\vec{F}_3$ merely by substituting the magnitudes for the vector quantities on the right side of Eq. 5-7. Instead, we must vectorially add $m\vec{a}$, $-\vec{F}_1$ (the reverse of $\vec{F}_1$), and $-\vec{F}_2$ (the reverse of $\vec{F}_2$), as shown in Fig. 5-4*b*. This addition can be done directly on a vector-capable calculator because we know both magnitude and angle for all three vectors. However, here we shall evaluate the right side of Eq. 5-7 in terms of components, first along the *x* axis and then along the *y* axis. *Caution:* Use only one axis at a time.

x components: Along the *x* axis we have

$$F_{3,x} = ma_x - F_{1,x} - F_{2,x}$$
$$= m(a\cos 50°) - F_1\cos(-150°) - F_2\cos 90°.$$

Then, substituting known data, we find

$$F_{3,x} = (2.0\text{ kg})(3.0\text{ m/s}^2)\cos 50° - (10\text{ N})\cos(-150°) - (20\text{ N})\cos 90°$$
$$= 12.5\text{ N}.$$

y components: Similarly, along the *y* axis we find

$$F_{3,y} = ma_y - F_{1,y} - F_{2,y}$$
$$= m(a\sin 50°) - F_1\sin(-150°) - F_2\sin 90°$$
$$= (2.0\text{ kg})(3.0\text{ m/s}^2)\sin 50° - (10\text{ N})\sin(-150°) - (20\text{ N})\sin 90°$$
$$= -10.4\text{ N}.$$

Vector: In unit-vector notation, we can write

$$\vec{F}_3 = F_{3,x}\hat{i} + F_{3,y}\hat{j} = (12.5\text{ N})\hat{i} - (10.4\text{ N})\hat{j}$$
$$\approx (13\text{ N})\hat{i} - (10\text{ N})\hat{j}. \quad \text{(Answer)}$$

We can now use a vector-capable calculator to get the magnitude and the angle of $\vec{F}_3$. We can also use Eq. 3-6 to obtain the magnitude and the angle (from the positive direction of the *x* axis) as

$$F_3 = \sqrt{F_{3,x}^2 + F_{3,y}^2} = 16\text{ N}$$

and

$$\theta = \tan^{-1}\frac{F_{3,y}}{F_{3,x}} = -40°. \quad \text{(Answer)}$$

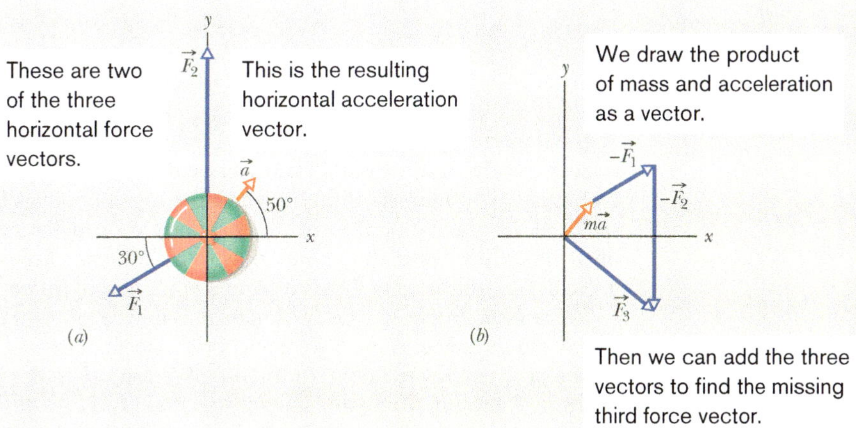

Figure 5-4 (*a*) An overhead view of two of three horizontal forces that act on a cookie tin, resulting in acceleration $\vec{a}$. $\vec{F}_3$ is not shown. (*b*) An arrangement of vectors $m\vec{a}$, $-\vec{F}_1$, and $-\vec{F}_2$ to find force $\vec{F}_3$.

5-2 SOME PARTICULAR FORCES

Learning Objectives

After reading this module, you should be able to . . .

5.08 Determine the magnitude and direction of the gravitational force acting on a body with a given mass, at a location with a given free-fall acceleration.

5.09 Identify that the weight of a body is the magnitude of the net force required to prevent the body from falling freely, as measured from the reference frame of the ground.

5.10 Identify that a scale gives an object's weight when the measurement is done in an inertial frame but not in an accelerating frame, where it gives an apparent weight.

5.11 Determine the magnitude and direction of the normal force on an object when the object is pressed or pulled onto a surface.

5.12 Identify that the force parallel to the surface is a frictional force that appears when the object slides or attempts to slide along the surface.

5.13 Identify that a tension force is said to pull at both ends of a cord (or a cord-like object) when the cord is taut.

Key Ideas

- A gravitational force $\vec{F}_g$ on a body is a pull by another body. In most situations in this book, the other body is Earth or some other astronomical body. For Earth, the force is directed down toward the ground, which is assumed to be an inertial frame. With that assumption, the magnitude of $\vec{F}_g$ is

$$F_g = mg,$$

where m is the body's mass and g is the magnitude of the free-fall acceleration.

- The weight W of a body is the magnitude of the upward force needed to balance the gravitational force on the body. A body's weight is related to the body's mass by

$$W = mg.$$

- A normal force $\vec{F}_N$ is the force on a body from a surface against which the body presses. The normal force is always perpendicular to the surface.

- A frictional force $\vec{f}$ is the force on a body when the body slides or attempts to slide along a surface. The force is always parallel to the surface and directed so as to oppose the sliding. On a frictionless surface, the frictional force is negligible.

- When a cord is under tension, each end of the cord pulls on a body. The pull is directed along the cord, away from the point of attachment to the body. For a massless cord (a cord with negligible mass), the pulls at both ends of the cord have the same magnitude T, even if the cord runs around a massless, frictionless pulley (a pulley with negligible mass and negligible friction on its axle to oppose its rotation).

Some Particular Forces

The Gravitational Force

A **gravitational force** $\vec{F}_g$ on a body is a certain type of pull that is directed toward a second body. In these early chapters, we do not discuss the nature of this force and usually consider situations in which the second body is Earth. Thus, when we speak of *the* gravitational force $\vec{F}_g$ on a body, we usually mean a force that pulls on it directly toward the center of Earth—that is, directly down toward the ground. We shall assume that the ground is an inertial frame.

Free Fall. Suppose a body of mass m is in free fall with the free-fall acceleration of magnitude g. Then, if we neglect the effects of the air, the only force acting on the body is the gravitational force $\vec{F}_g$. We can relate this downward force and downward acceleration with Newton's second law ($\vec{F} = m\vec{a}$). We place a vertical y axis along the body's path, with the positive direction upward. For this axis, Newton's second law can be written in the form $F_{net,y} = ma_y$, which, in our situation, becomes

$$-F_g = m(-g)$$

or

$$F_g = mg. \tag{5-8}$$

In words, the magnitude of the gravitational force is equal to the product mg.

At Rest. This same gravitational force, with the same magnitude, still acts on the body even when the body is not in free fall but is, say, at rest on a pool table or moving across the table. (For the gravitational force to disappear, Earth would have to disappear.)

We can write Newton's second law for the gravitational force in these vector forms:

$$\vec{F}_g = -F_g\hat{j} = -mg\hat{j} = m\vec{g}, \tag{5-9}$$

where $\hat{j}$ is the unit vector that points upward along a y axis, directly away from the ground, and $\vec{g}$ is the free-fall acceleration (written as a vector), directed downward.

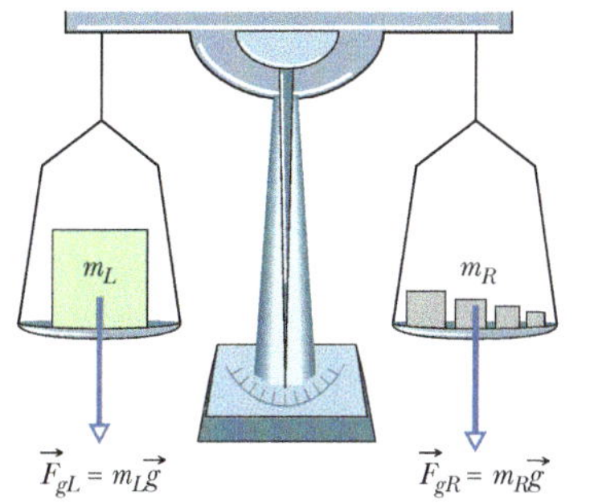

Figure 5-5 An equal-arm balance. When the device is in balance, the gravitational force $\vec{F}_{gL}$ on the body being weighed (on the left pan) and the total gravitational force $\vec{F}_{gR}$ on the reference bodies (on the right pan) are equal. Thus, the mass m_L of the body being weighed is equal to the total mass m_R of the reference bodies.

Weight

The **weight** W of a body is the magnitude of the net force required to prevent the body from falling freely, as measured by someone on the ground. For example, to keep a ball at rest in your hand while you stand on the ground, you must provide an upward force to balance the gravitational force on the ball from Earth. Suppose the magnitude of the gravitational force is 2.0 N. Then the magnitude of your upward force must be 2.0 N, and thus the weight W of the ball is 2.0 N. We also say that the ball *weighs* 2.0 N and speak about the ball *weighing* 2.0 N.

A ball with a weight of 3.0 N would require a greater force from you—namely, a 3.0 N force—to keep it at rest. The reason is that the gravitational force you must balance has a greater magnitude—namely, 3.0 N. We say that this second ball is *heavier* than the first ball.

Now let us generalize the situation. Consider a body that has an acceleration $\vec{a}$ of zero relative to the ground, which we again assume to be an inertial frame. Two forces act on the body: a downward gravitational force $\vec{F}_g$ and a balancing upward force of magnitude W. We can write Newton's second law for a vertical y axis, with the positive direction upward, as

$$F_{\text{net},y} = ma_y.$$

In our situation, this becomes

$$W - F_g = m(0) \tag{5-10}$$

or

$$W = F_g \quad \text{(weight, with ground as inertial frame).} \tag{5-11}$$

This equation tells us (assuming the ground is an inertial frame) that

The weight W of a body is equal to the magnitude F_g of the gravitational force on the body.

Substituting mg for F_g from Eq. 5-8, we find

$$W = mg \quad \text{(weight)}, \tag{5-12}$$

which relates a body's weight to its mass.

Weighing. To *weigh* a body means to measure its weight. One way to do this is to place the body on one of the pans of an equal-arm balance (Fig. 5-5) and then place reference bodies (whose masses are known) on the other pan until we strike a balance (so that the gravitational forces on the two sides match). The masses on the pans then match, and we know the mass of the body. If we know the value of g for the location of the balance, we can also find the weight of the body with Eq. 5-12.

We can also weigh a body with a spring scale (Fig. 5-6). The body stretches a spring, moving a pointer along a scale that has been calibrated and marked in

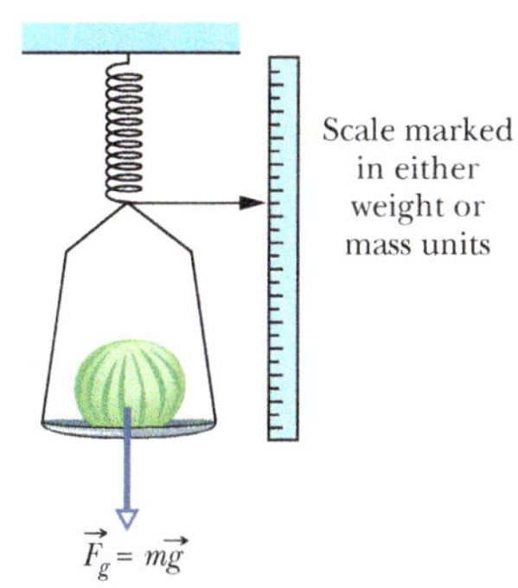

Figure 5-6 A spring scale. The reading is proportional to the *weight* of the object on the pan, and the scale gives that weight if marked in weight units. If, instead, it is marked in mass units, the reading is the object's weight only if the value of g at the location where the scale is being used is the same as the value of g at the location where the scale was calibrated.

either mass or weight units. (Most bathroom scales in the United States work this way and are marked in the force unit pounds.) If the scale is marked in mass units, it is accurate only where the value of g is the same as where the scale was calibrated.

The weight of a body must be measured when the body is not accelerating vertically relative to the ground. For example, you can measure your weight on a scale in your bathroom or on a fast train. However, if you repeat the measurement with the scale in an accelerating elevator, the reading differs from your weight because of the acceleration. Such a measurement is called an *apparent weight.*

Caution: A body's weight is not its mass. Weight is the magnitude of a force and is related to mass by Eq. 5-12. If you move a body to a point where the value of g is different, the body's mass (an intrinsic property) is not different but the weight is. For example, the weight of a bowling ball having a mass of 7.2 kg is 71 N on Earth but only 12 N on the Moon. The mass is the same on Earth and Moon, but the free-fall acceleration on the Moon is only 1.6 m/s^2.

The Normal Force

If you stand on a mattress, Earth pulls you downward, but you remain stationary. The reason is that the mattress, because it deforms downward due to you, pushes up on you. Similarly, if you stand on a floor, it deforms (it is compressed, bent, or buckled ever so slightly) and pushes up on you. Even a seemingly rigid concrete floor does this (if it is not sitting directly on the ground, enough people on the floor could break it).

The push on you from the mattress or floor is a **normal force** $\vec{F}_N$. The name comes from the mathematical term *normal,* meaning perpendicular: The force on you from, say, the floor is perpendicular to the floor.

When a body presses against a surface, the surface (even a seemingly rigid one) deforms and pushes on the body with a normal force $\vec{F}_N$ that is perpendicular to the surface.

Figure 5-7a shows an example. A block of mass m presses down on a table, deforming it somewhat because of the gravitational force $\vec{F}_g$ on the block. The table pushes up on the block with normal force $\vec{F}_N$. The free-body diagram for the block is given in Fig. 5-7b. Forces $\vec{F}_g$ and $\vec{F}_N$ are the only two forces on the block and they are both vertical. Thus, for the block we can write Newton's second law for a positive-upward y axis ($F_{\text{net},y} = ma_y$) as

$$F_N - F_g = ma_y.$$

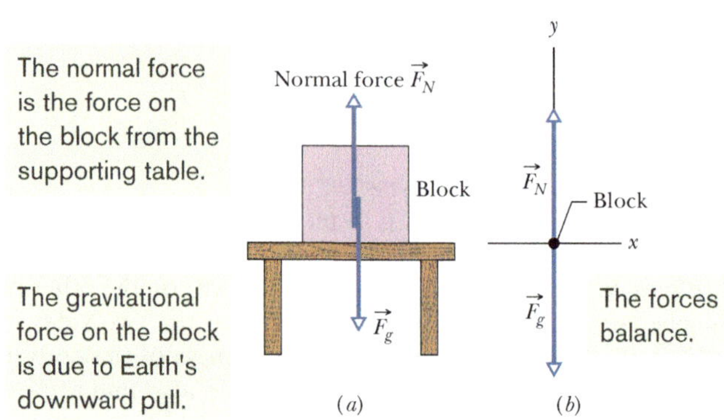

Figure 5-7 (a) A block resting on a table experiences a normal force $\vec{F}_N$ perpendicular to the tabletop. (b) The free-body diagram for the block.

From Eq. 5-8, we substitute mg for F_g, finding

$$F_N - mg = ma_y.$$

Then the magnitude of the normal force is

$$F_N = mg + ma_y = m(g + a_y) \tag{5-13}$$

for any vertical acceleration a_y of the table and block (they might be in an accelerating elevator). (*Caution:* We have already included the sign for g but a_y can be positive or negative here.) If the table and block are not accelerating relative to the ground, then $a_y = 0$ and Eq. 5-13 yields

$$F_N = mg. \tag{5-14}$$

Checkpoint 3

In Fig. 5-7, is the magnitude of the normal force $\vec{F}_N$ greater than, less than, or equal to mg if the block and table are in an elevator moving upward (a) at constant speed and (b) at increasing speed?

Friction

If we either slide or attempt to slide a body over a surface, the motion is resisted by a bonding between the body and the surface. (We discuss this bonding more in the next chapter.) The resistance is considered to be a single force $\vec{f}$, called either the **frictional force** or simply **friction.** This force is directed along the surface, opposite the direction of the intended motion (Fig. 5-8). Sometimes, to simplify a situation, friction is assumed to be negligible (the surface, or even the body, is said to be *frictionless*).

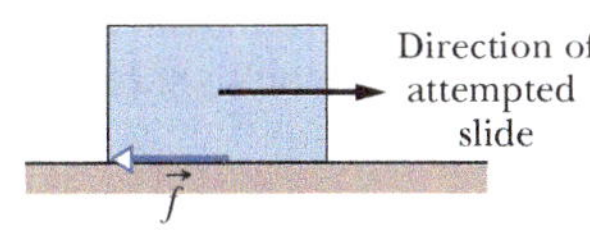

Figure 5-8 A frictional force $\vec{f}$ opposes the attempted slide of a body over a surface.

Tension

When a cord (or a rope, cable, or other such object) is attached to a body and pulled taut, the cord pulls on the body with a force $\vec{T}$ directed away from the body and along the cord (Fig. 5-9*a*). The force is often called a *tension force* because the cord is said to be in a state of *tension* (or to be *under tension*), which means that it is being pulled taut. The *tension in the cord* is the magnitude T of the force on the body. For example, if the force on the body from the cord has magnitude $T = 50$ N, the tension in the cord is 50 N.

A cord is often said to be *massless* (meaning its mass is negligible compared to the body's mass) and *unstretchable.* The cord then exists only as a connection between two bodies. It pulls on both bodies with the same force magnitude T,

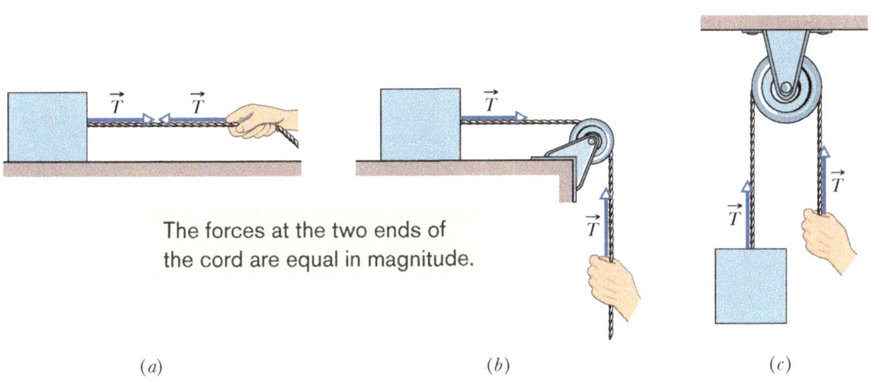

Figure 5-9 (*a*) The cord, pulled taut, is under tension. If its mass is negligible, the cord pulls on the body and the hand with force $\vec{T}$, even if the cord runs around a massless, frictionless pulley as in (*b*) and (*c*).

even if the bodies and the cord are accelerating and even if the cord runs around a *massless, frictionless pulley* (Figs. 5-9*b* and *c*). Such a pulley has negligible mass compared to the bodies and negligible friction on its axle opposing its rotation. If the cord wraps halfway around a pulley, as in Fig. 5-9*c*, the net force on the pulley from the cord has the magnitude $2T$.

Checkpoint 4

The suspended body in Fig. 5-9*c* weighs 75 N. Is T equal to, greater than, or less than 75 N when the body is moving upward (a) at constant speed, (b) at increasing speed, and (c) at decreasing speed?

5-3 APPLYING NEWTON'S LAWS

Learning Objectives

After reading this module, you should be able to . . .

5.14 Identify Newton's third law of motion and third-law force pairs.

5.15 For an object that moves vertically or on a horizontal or inclined plane, apply Newton's second law to a free-body diagram of the object.

5.16 For an arrangement where a system of several objects moves rigidly together, draw a free-body diagram and apply Newton's second law for the individual objects and also for the system taken as a composite object.

Key Ideas

- The net force $\vec{F}_{net}$ on a body with mass m is related to the body's acceleration $\vec{a}$ by

$$\vec{F}_{net} = m\vec{a},$$

which may be written in the component versions

$$F_{net,x} = ma_x \quad F_{net,y} = ma_y \quad \text{and} \quad F_{net,z} = ma_z.$$

- If a force $\vec{F}_{BC}$ acts on body B due to body C, then there is a force $\vec{F}_{CB}$ on body C due to body B:

$$\vec{F}_{BC} = -\vec{F}_{CB}.$$

The forces are equal in magnitude but opposite in directions.

Newton's Third Law

Two bodies are said to *interact* when they push or pull on each other—that is, when a force acts on each body due to the other body. For example, suppose you position a book B so it leans against a crate C (Fig. 5-10*a*). Then the book and crate interact: There is a horizontal force $\vec{F}_{BC}$ on the book from the crate (or due to the crate) and a horizontal force $\vec{F}_{CB}$ on the crate from the book (or due to the book). This pair of forces is shown in Fig. 5-10*b*. Newton's third law states that

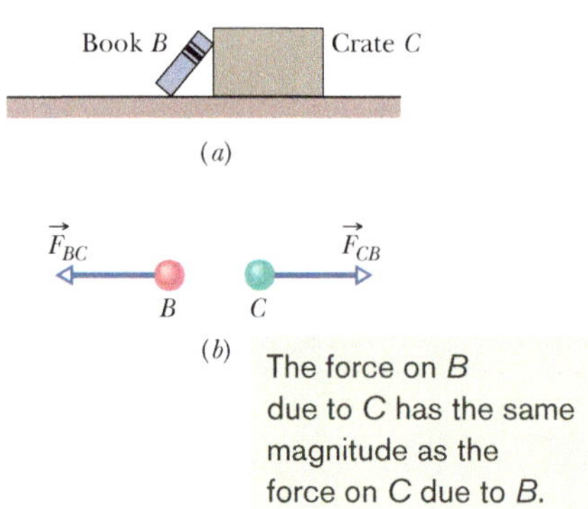

Figure 5-10 (*a*) Book B leans against crate C. (*b*) Forces $\vec{F}_{BC}$ (the force on the book from the crate) and $\vec{F}_{CB}$ (the force on the crate from the book) have the same magnitude and are opposite in direction.

Newton's Third Law: When two bodies interact, the forces on the bodies from each other are always equal in magnitude and opposite in direction.

For the book and crate, we can write this law as the scalar relation

$$F_{BC} = F_{CB} \quad \text{(equal magnitudes)}$$

or as the vector relation

$$\vec{F}_{BC} = -\vec{F}_{CB} \quad \text{(equal magnitudes and opposite directions)}, \tag{5-15}$$

where the minus sign means that these two forces are in opposite directions. We can call the forces between two interacting bodies a **third-law force pair.** When

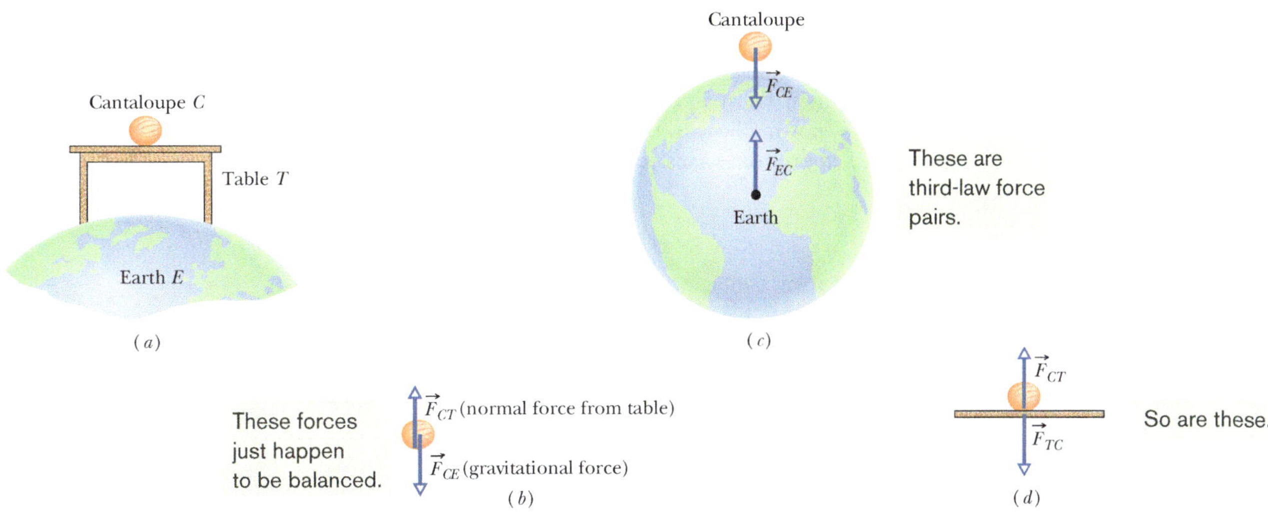

Figure 5-11 (*a*) A cantaloupe lies on a table that stands on Earth. (*b*) The forces *on the cantaloupe* are $\vec{F}_{CT}$ and $\vec{F}_{CE}$. (*c*) The third-law force pair for the cantaloupe–Earth interaction. (*d*) The third-law force pair for the cantaloupe–table interaction.

any two bodies interact in any situation, a third-law force pair is present. The book and crate in Fig. 5-10*a* are stationary, but the third law would still hold if they were moving and even if they were accelerating.

As another example, let us find the third-law force pairs involving the cantaloupe in Fig. 5-11*a*, which lies on a table that stands on Earth. The cantaloupe interacts with the table and with Earth (this time, there are three bodies whose interactions we must sort out).

Let's first focus on the forces acting on the cantaloupe (Fig. 5-11*b*). Force $\vec{F}_{CT}$ is the normal force on the cantaloupe from the table, and force $\vec{F}_{CE}$ is the gravitational force on the cantaloupe due to Earth. Are they a third-law force pair? No, because they are forces on a single body, the cantaloupe, and not on two interacting bodies.

To find a third-law pair, we must focus not on the cantaloupe but on the interaction between the cantaloupe and one other body. In the cantaloupe–Earth interaction (Fig. 5-11*c*), Earth pulls on the cantaloupe with a gravitational force $\vec{F}_{CE}$ and the cantaloupe pulls on Earth with a gravitational force $\vec{F}_{EC}$. Are these forces a third-law force pair? Yes, because they are forces on two interacting bodies, the force on each due to the other. Thus, by Newton's third law,

$$\vec{F}_{CE} = -\vec{F}_{EC} \qquad \text{(cantaloupe–Earth interaction).}$$

Next, in the cantaloupe–table interaction, the force on the cantaloupe from the table is $\vec{F}_{CT}$ and, conversely, the force on the table from the cantaloupe is $\vec{F}_{TC}$ (Fig. 5-11*d*). These forces are also a third-law force pair, and so

$$\vec{F}_{CT} = -\vec{F}_{TC} \qquad \text{(cantaloupe–table interaction).}$$

Checkpoint 5

Suppose that the cantaloupe and table of Fig. 5-11 are in an elevator cab that begins to accelerate upward. (a) Do the magnitudes of $\vec{F}_{TC}$ and $\vec{F}_{CT}$ increase, decrease, or stay the same? (b) Are those two forces still equal in magnitude and opposite in direction? (c) Do the magnitudes of $\vec{F}_{CE}$ and $\vec{F}_{EC}$ increase, decrease, or stay the same? (d) Are those two forces still equal in magnitude and opposite in direction?

Applying Newton's Laws

The rest of this chapter consists of sample problems. You should pore over them, learning their procedures for attacking a problem. Especially important is knowing how to translate a sketch of a situation into a free-body diagram with appropriate axes, so that Newton's laws can be applied.

Sample Problem 5.03 Block on table, block hanging

Figure 5-12 shows a block S (the *sliding block*) with mass $M = 3.3$ kg. The block is free to move along a horizontal frictionless surface and connected, by a cord that wraps over a frictionless pulley, to a second block H (the *hanging block*), with mass $m = 2.1$ kg. The cord and pulley have negligible masses compared to the blocks (they are "massless"). The hanging block H falls as the sliding block S accelerates to the right. Find (a) the acceleration of block S, (b) the acceleration of block H, and (c) the tension in the cord.

Q *What is this problem all about?*

You are given two bodies—sliding block and hanging block—but must also consider *Earth*, which pulls on both bodies. (Without Earth, nothing would happen here.) A total of five forces act on the blocks, as shown in Fig. 5-13:

1. The cord pulls to the right on sliding block S with a force of magnitude T.
2. The cord pulls upward on hanging block H with a force of the same magnitude T. This upward force keeps block H from falling freely.
3. Earth pulls down on block S with the gravitational force $\vec{F}_{gS}$, which has a magnitude equal to Mg.
4. Earth pulls down on block H with the gravitational force $\vec{F}_{gH}$, which has a magnitude equal to mg.
5. The table pushes up on block S with a normal force $\vec{F}_N$.

There is another thing you should note. We assume that the cord does not stretch, so that if block H falls 1 mm in a certain time, block S moves 1 mm to the right in that same time. This means that the blocks move together and their accelerations have the same magnitude a.

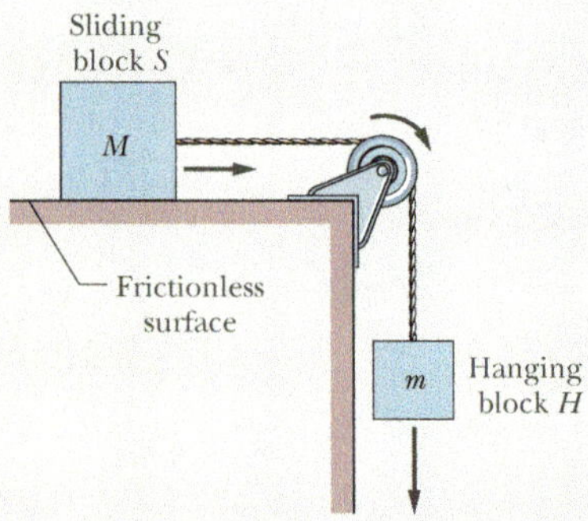

Figure 5-12 A block S of mass M is connected to a block H of mass m by a cord that wraps over a pulley.

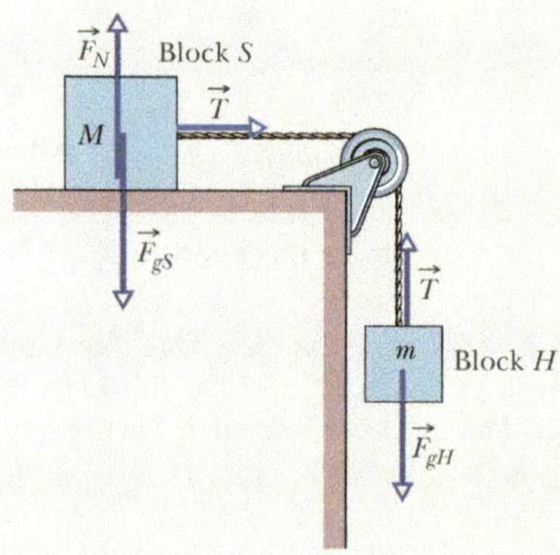

Figure 5-13 The forces acting on the two blocks of Fig. 5-12.

Q *How do I classify this problem? Should it suggest a particular law of physics to me?*

Yes. Forces, masses, and accelerations are involved, and they should suggest Newton's second law of motion, $\vec{F}_{net} = m\vec{a}$. That is our starting key idea.

Q *If I apply Newton's second law to this problem, to which body should I apply it?*

We focus on two bodies, the sliding block and the hanging block. Although they are *extended objects* (they are not points), we can still treat each block as a particle because every part of it moves in exactly the same way. A second key idea is to apply Newton's second law separately to each block.

Q *What about the pulley?*

We cannot represent the pulley as a particle because different parts of it move in different ways. When we discuss rotation, we shall deal with pulleys in detail. Meanwhile, we eliminate the pulley from consideration by assuming its mass to be negligible compared with the masses of the two blocks. Its only function is to change the cord's orientation.

Q *OK. Now how do I apply $\vec{F}_{net} = m\vec{a}$ to the sliding block?*

Represent block S as a particle of mass M and draw *all* the forces that act *on* it, as in Fig. 5-14*a*. This is the block's free-body diagram. Next, draw a set of axes. It makes sense

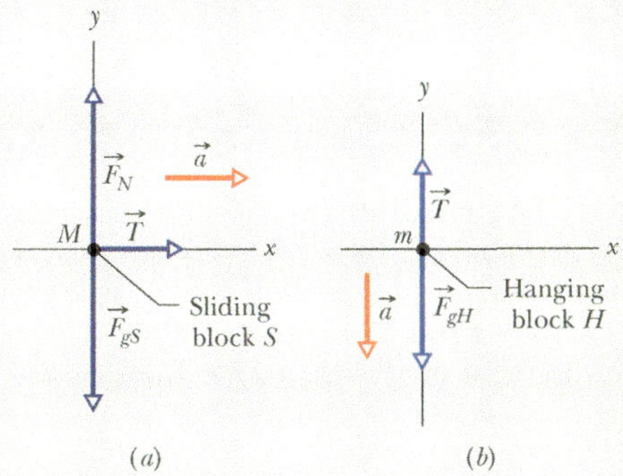

Figure 5-14 (*a*) A free-body diagram for block *S* of Fig. 5-12. (*b*) A free-body diagram for block *H* of Fig. 5-12.

to draw the x axis parallel to the table, in the direction in which the block moves.

Q *Thanks, but you still haven't told me how to apply $\vec{F}_{net} = m\vec{a}$ to the sliding block. All you've done is explain how to draw a free-body diagram.*

You are right, and here's the third key idea: The expression $\vec{F}_{net} = M\vec{a}$ is a vector equation, so we can write it as three component equations:

$$F_{net,x} = Ma_x \quad F_{net,y} = Ma_y \quad F_{net,z} = Ma_z \tag{5-16}$$

in which $F_{net,x}$, $F_{net,y}$, and $F_{net,z}$ are the components of the net force along the three axes. Now we apply each component equation to its corresponding direction. Because block S does not accelerate vertically, $F_{net,y} = Ma_y$ becomes

$$F_N - F_{gS} = 0 \quad \text{or} \quad F_N = F_{gS}. \tag{5-17}$$

Thus in the y direction, the magnitude of the normal force is equal to the magnitude of the gravitational force.

No force acts in the z direction, which is perpendicular to the page.

In the x direction, there is only one force component, which is T. Thus, $F_{net,x} = Ma_x$ becomes

$$T = Ma. \tag{5-18}$$

This equation contains two unknowns, T and a; so we cannot yet solve it. Recall, however, that we have not said anything about the hanging block.

Q *I agree. How do I apply $\vec{F}_{net} = m\vec{a}$ to the hanging block?*

We apply it just as we did for block S: Draw a free-body diagram for block H, as in Fig. 5-14b. Then apply $\vec{F}_{net} = m\vec{a}$ in component form. This time, because the acceleration is along the y axis, we use the y part of Eq. 5-16 ($F_{net,y} = ma_y$) to write

$$T - F_{gH} = ma_y. \tag{5-19}$$

We can now substitute mg for F_{gH} and $-a$ for a_y (negative because block H accelerates in the negative direction of the y axis). We find

$$T - mg = -ma. \tag{5-20}$$

Now note that Eqs. 5-18 and 5-20 are simultaneous equations with the same two unknowns, T and a. Subtracting these equations eliminates T. Then solving for a yields

$$a = \frac{m}{M + m} g. \tag{5-21}$$

Substituting this result into Eq. 5-18 yields

$$T = \frac{Mm}{M + m} g. \tag{5-22}$$

Putting in the numbers gives, for these two quantities,

$$a = \frac{m}{M + m} g = \frac{2.1 \text{ kg}}{3.3 \text{ kg} + 2.1 \text{ kg}} (9.8 \text{ m/s}^2)$$
$$= 3.8 \text{ m/s}^2 \qquad \text{(Answer)}$$

and
$$T = \frac{Mm}{M + m} g = \frac{(3.3 \text{ kg})(2.1 \text{ kg})}{3.3 \text{ kg} + 2.1 \text{ kg}} (9.8 \text{ m/s}^2)$$
$$= 13 \text{ N}. \qquad \text{(Answer)}$$

Q *The problem is now solved, right?*

That's a fair question, but the problem is not really finished until we have examined the results to see whether they make sense. (If you made these calculations on the job, wouldn't you want to see whether they made sense before you turned them in?)

Look first at Eq. 5-21. Note that it is dimensionally correct and that the acceleration a will always be less than g (because of the cord, the hanging block is not in free fall).

Look now at Eq. 5-22, which we can rewrite in the form

$$T = \frac{M}{M + m} mg. \tag{5-23}$$

In this form, it is easier to see that this equation is also dimensionally correct, because both T and mg have dimensions of forces. Equation 5-23 also lets us see that the tension in the cord is always less than mg, and thus is always less than the gravitational force on the hanging block. That is a comforting thought because, if T were *greater* than mg, the hanging block would accelerate upward.

We can also check the results by studying special cases, in which we can guess what the answers must be. A simple example is to put $g = 0$, as if the experiment were carried out in interstellar space. We know that in that case, the blocks would not move from rest, there would be no forces on the ends of the cord, and so there would be no tension in the cord. Do the formulas predict this? Yes, they do. If you put $g = 0$ in Eqs. 5-21 and 5-22, you find $a = 0$ and $T = 0$. Two more special cases you might try are $M = 0$ and $m \rightarrow \infty$.

Sample Problem 5.04 Cord accelerates box up a ramp

Many students consider problems involving ramps (inclined planes) to be especially hard. The difficulty is probably visual because we work with (a) a tilted coordinate system and (b) the components of the gravitational force, not the full force. Here is a typical example with all the tilting and angles explained. (In *WileyPLUS*, the figure is available as an animation with voiceover.) In spite of the tilt, the key idea is to apply Newton's second law to the axis along which the motion occurs.

In Fig. 5-15*a*, a cord pulls a box of sea biscuits up along a frictionless plane inclined at angle $\theta = 30.0°$. The box has mass $m = 5.00$ kg, and the force from the cord has magnitude $T = 25.0$ N. What is the box's acceleration a along the inclined plane?

KEY IDEA

The acceleration along the plane is set by the force components along the plane (not by force components perpendicular to the plane), as expressed by Newton's second law (Eq. 5-1).

Calculations: We need to write Newton's second law for motion along an axis. Because the box moves along the inclined plane, placing an x axis along the plane seems reasonable (Fig. 5-15*b*). (There is nothing wrong with using our usual coordinate system, but the expressions for components would be a lot messier because of the misalignment of the x axis with the motion.)

After choosing a coordinate system, we draw a free-body diagram with a dot representing the box (Fig. 5-15*b*). Then we draw all the vectors for the forces acting on the box, with the tails of the vectors anchored on the dot. (Drawing the vectors willy-nilly on the diagram can easily lead to errors, especially on exams, so always anchor the tails.)

Force $\vec{T}$ from the cord is up the plane and has magnitude $T = 25.0$ N. The gravitational force $\vec{F}_g$ is downward (of

Figure 5-15 (*a*) A box is pulled up a plane by a cord. (*b*) The three forces acting on the box: the cord's force $\vec{T}$, the gravitational force $\vec{F}_g$, and the normal force $\vec{F}_N$. (*c*)–(*i*) Finding the force components along the plane and perpendicular to it. **In *WileyPLUS*, this figure is available as an animation with voiceover.**

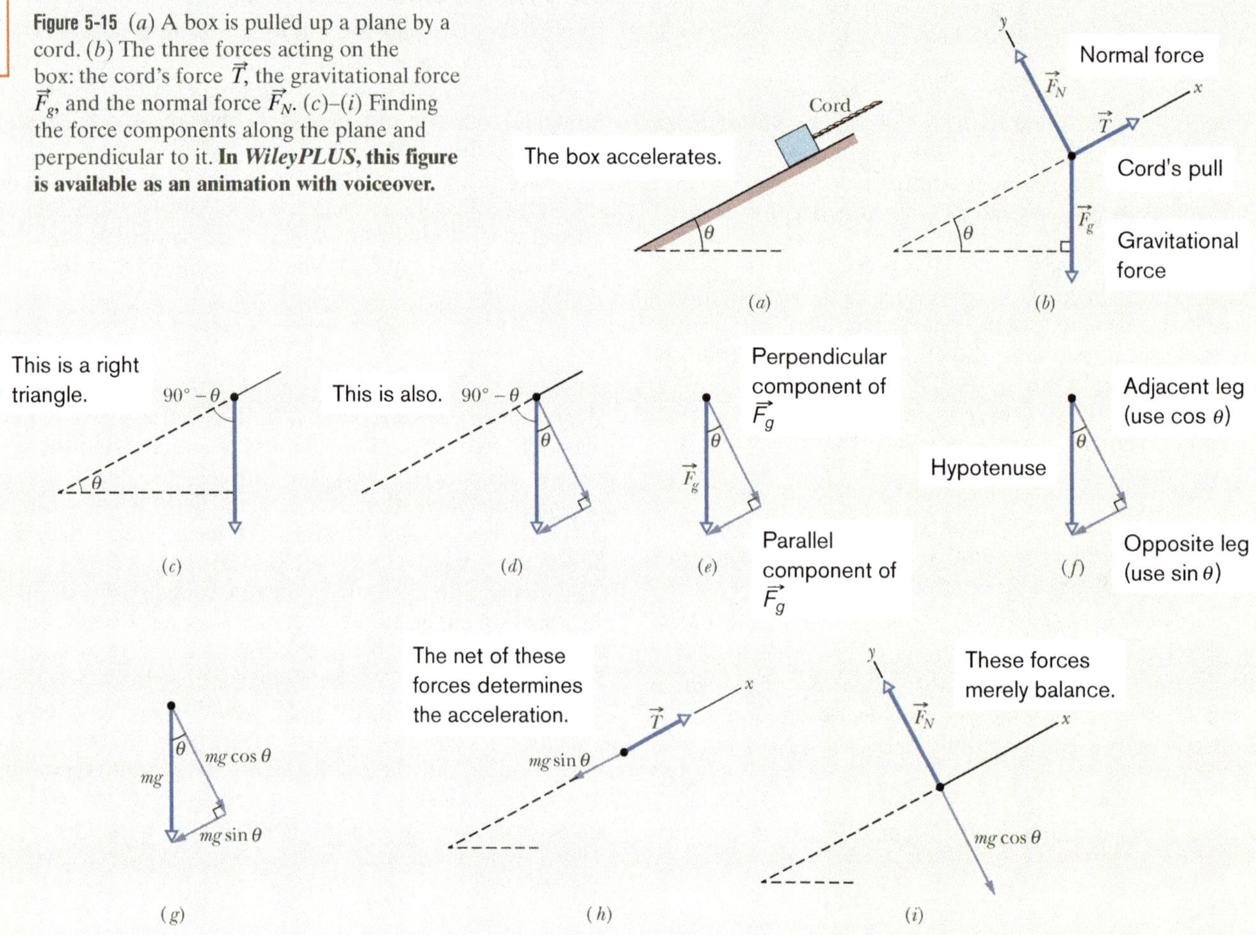

course) and has magnitude $mg = (5.00\text{ kg})(9.80\text{ m/s}^2) = 49.0\text{ N}$. That direction means that only a component of the force is along the plane, and only that component (not the full force) affects the box's acceleration along the plane. Thus, before we can write Newton's second law for motion along the x axis, we need to find an expression for that important component.

Figures 5-15c to h indicate the steps that lead to the expression. We start with the given angle of the plane and work our way to a triangle of the force components (they are the legs of the triangle and the full force is the hypotenuse). Figure 5-15c shows that the angle between the ramp and $\vec{F}_g$ is $90° - \theta$. (Do you see a right triangle there?) Next, Figs. 5-15d to f show $\vec{F}_g$ and its components: One component is parallel to the plane (that is the one we want) and the other is perpendicular to the plane.

Because the perpendicular component *is* perpendicular, the angle between it and $\vec{F}_g$ must be θ (Fig. 5-15d). The component we want is the far leg of the component right triangle. The magnitude of the hypotenuse is mg (the magnitude of the gravitational force). Thus, the component we want has magnitude $mg \sin \theta$ (Fig. 5-15g).

We have one more force to consider, the normal force $\vec{F}_N$ shown in Fig. 5-15b. However, it is perpendicular to the plane and thus cannot affect the motion along the plane. (It has no component along the plane to accelerate the box.)

We are now ready to write Newton's second law for motion along the tilted x axis:

$$F_{\text{net},x} = ma_x.$$

The component a_x is the only component of the acceleration (the box is not leaping up from the plane, which would be strange, or descending into the plane, which would be even stranger). So, let's simply write a for the acceleration along the plane. Because $\vec{T}$ is in the positive x direction and the component $mg \sin \theta$ is in the negative x direction, we next write

$$T - mg \sin \theta = ma. \quad (5\text{-}24)$$

Substituting data and solving for a, we find

$$a = 0.100\text{ m/s}^2. \quad \text{(Answer)}$$

The result is positive, indicating that the box accelerates up the inclined plane, in the positive direction of the tilted x axis. If we decreased the magnitude of $\vec{T}$ enough to make $a = 0$, the box would move up the plane at constant speed. And if we decreased the magnitude of $\vec{T}$ even more, the acceleration would be negative in spite of the cord's pull.

Sample Problem 5.05 Reading a force graph

Here is an example of where you must dig information out of a graph, not just read off a number. In Fig. 5-16a, two forces are applied to a 4.00 kg block on a frictionless floor, but only force $\vec{F}_1$ is indicated. That force has a fixed magnitude but can be applied at an adjustable angle θ to the positive direction of the x axis. Force $\vec{F}_2$ is horizontal and fixed in both magnitude and angle. Figure 5-16b gives the horizontal acceleration a_x of the block for any given value of θ from 0° to 90°. What is the value of a_x for $\theta = 180°$?

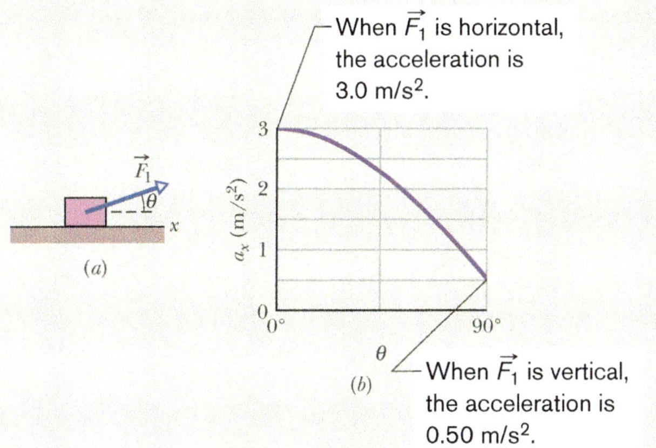

Figure 5-16 (a) One of the two forces applied to a block is shown. Its angle θ can be varied. (b) The block's acceleration component a_x versus θ.

KEY IDEAS

(1) The horizontal acceleration a_x depends on the net horizontal force $F_{\text{net},x}$, as given by Newton's second law. (2) The net horizontal force is the sum of the horizontal components of forces $\vec{F}_1$ and $\vec{F}_2$.

Calculations: The x component of $\vec{F}_2$ is F_2 because the vector is horizontal. The x component of $\vec{F}_1$ is $F_1 \cos \theta$. Using these expressions and a mass m of 4.00 kg, we can write Newton's second law ($\vec{F}_{\text{net}} = m\vec{a}$) for motion along the x axis as

$$F_1 \cos \theta + F_2 = 4.00a_x. \quad (5\text{-}25)$$

From this equation we see that when angle $\theta = 90°$, $F_1 \cos \theta$ is zero and $F_2 = 4.00a_x$. From the graph we see that the corresponding acceleration is 0.50 m/s^2. Thus, $F_2 = 2.00$ N and $\vec{F}_2$ must be in the positive direction of the x axis.

From Eq. 5-25, we find that when $\theta = 0°$,

$$F_1 \cos 0° + 2.00 = 4.00a_x. \quad (5\text{-}26)$$

From the graph we see that the corresponding acceleration is 3.0 m/s^2. From Eq. 5-26, we then find that $F_1 = 10$ N.

Substituting $F_1 = 10$ N, $F_2 = 2.00$ N, and $\theta = 180°$ into Eq. 5-25 leads to

$$a_x = -2.00\text{ m/s}^2. \quad \text{(Answer)}$$

Sample Problem 5.06 Forces within an elevator cab

Although people would surely avoid getting into the elevator with you, suppose that you weigh yourself while on an elevator that is moving. Would you weigh more than, less than, or the same as when the scale is on a stationary floor?

In Fig. 5-17*a*, a passenger of mass $m = 72.2$ kg stands on a platform scale in an elevator cab. We are concerned with the scale readings when the cab is stationary and when it is moving up or down.

(a) Find a general solution for the scale reading, whatever the vertical motion of the cab.

KEY IDEAS

(1) The reading is equal to the magnitude of the normal force $\vec{F}_N$ on the passenger from the scale. The only other force acting on the passenger is the gravitational force $\vec{F}_g$, as shown in the free-body diagram of Fig. 5-17*b*. (2) We can relate the forces on the passenger to his acceleration $\vec{a}$ by using Newton's second law ($\vec{F}_{net} = m\vec{a}$). However, recall that we can use this law only in an inertial frame. If the cab accelerates, then it is *not* an inertial frame. So we choose the ground to be our inertial frame and make any measure of the passenger's acceleration relative to it.

Calculations: Because the two forces on the passenger and his acceleration are all directed vertically, along the y axis in Fig. 5-17*b*, we can use Newton's second law written for y components ($F_{net,y} = ma_y$) to get

$$F_N - F_g = ma$$

or

$$F_N = F_g + ma. \qquad (5\text{-}27)$$

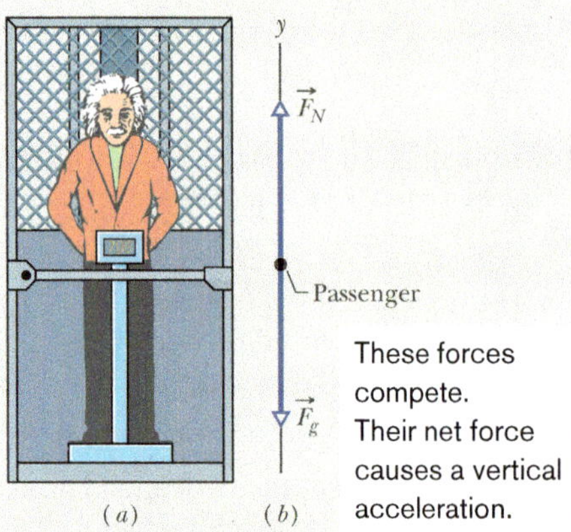

Figure 5-17 (*a*) A passenger stands on a platform scale that indicates either his weight or his apparent weight. (*b*) The free-body diagram for the passenger, showing the normal force $\vec{F}_N$ on him from the scale and the gravitational force $\vec{F}_g$.

This tells us that the scale reading, which is equal to normal force magnitude F_N, depends on the vertical acceleration. Substituting mg for F_g gives us

$$F_N = m(g + a) \quad \text{(Answer)} \qquad (5\text{-}28)$$

for any choice of acceleration a. If the acceleration is upward, a is positive; if it is downward, a is negative.

(b) What does the scale read if the cab is stationary or moving upward at a constant 0.50 m/s?

KEY IDEA

For any constant velocity (zero or otherwise), the acceleration a of the passenger is zero.

Calculation: Substituting this and other known values into Eq. 5-28, we find

$$F_N = (72.2 \text{ kg})(9.8 \text{ m/s}^2 + 0) = 708 \text{ N}. \quad \text{(Answer)}$$

This is the weight of the passenger and is equal to the magnitude F_g of the gravitational force on him.

(c) What does the scale read if the cab accelerates upward at 3.20 m/s² and downward at 3.20 m/s²?

Calculations: For $a = 3.20$ m/s², Eq. 5-28 gives

$$F_N = (72.2 \text{ kg})(9.8 \text{ m/s}^2 + 3.20 \text{ m/s}^2) = 939 \text{ N}, \quad \text{(Answer)}$$

and for $a = -3.20$ m/s², it gives

$$F_N = (72.2 \text{ kg})(9.8 \text{ m/s}^2 - 3.20 \text{ m/s}^2) = 477 \text{ N}. \quad \text{(Answer)}$$

For an upward acceleration (either the cab's upward speed is increasing or its downward speed is decreasing), the scale reading is greater than the passenger's weight. That reading is a measurement of an apparent weight, because it is made in a noninertial frame. For a downward acceleration (either decreasing upward speed or increasing downward speed), the scale reading is less than the passenger's weight.

(d) During the upward acceleration in part (c), what is the magnitude F_{net} of the net force on the passenger, and what is the magnitude $a_{p,cab}$ of his acceleration as measured in the frame of the cab? Does $\vec{F}_{net} = m\vec{a}_{p,cab}$?

Calculation: The magnitude F_g of the gravitational force on the passenger does not depend on the motion of the passenger or the cab; so, from part (b), F_g is 708 N. From part (c), the magnitude F_N of the normal force on the passenger during the upward acceleration is the 939 N reading on the scale. Thus, the net force on the passenger is

$$F_{net} = F_N - F_g = 939\text{ N} - 708\text{ N} = 231\text{ N}, \qquad \text{(Answer)}$$

during the upward acceleration. However, his acceleration $a_{p,cab}$ relative to the frame of the cab is zero. Thus, in the noninertial frame of the accelerating cab, F_{net} is not equal to $ma_{p,cab}$, and Newton's second law does not hold.

Sample Problem 5.07 Acceleration of block pushing on block

Some homework problems involve objects that move together, because they are either shoved together or tied together. Here is an example in which you apply Newton's second law to the composite of two blocks and then to the individual blocks.

In Fig. 5-18*a*, a constant horizontal force $\vec{F}_{app}$ of magnitude 20 N is applied to block *A* of mass $m_A = 4.0$ kg, which pushes against block *B* of mass $m_B = 6.0$ kg. The blocks slide over a frictionless surface, along an *x* axis.

(a) What is the acceleration of the blocks?

Serious Error: Because force $\vec{F}_{app}$ is applied directly to block *A*, we use Newton's second law to relate that force to the acceleration $\vec{a}$ of block *A*. Because the motion is along the *x* axis, we use that law for *x* components ($F_{net,x} = ma_x$), writing it as

$$F_{app} = m_A a.$$

However, this is seriously wrong because $\vec{F}_{app}$ is not the only horizontal force acting on block *A*. There is also the force $\vec{F}_{AB}$ from block *B* (Fig. 5-18*b*).

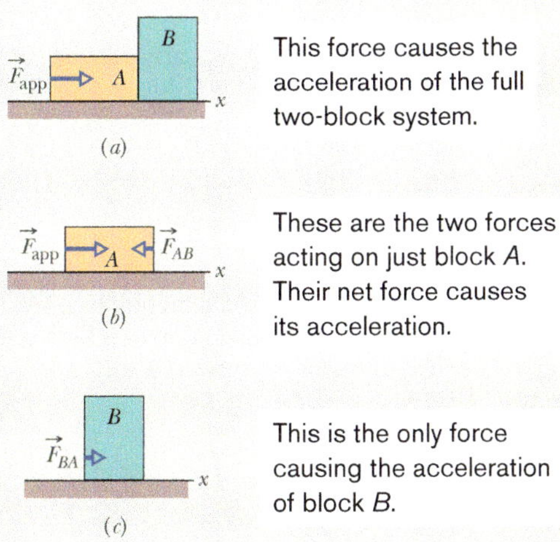

Figure 5-18 (*a*) A constant horizontal force $\vec{F}_{app}$ is applied to block *A*, which pushes against block *B*. (*b*) Two horizontal forces act on block *A*. (*c*) Only one horizontal force acts on block *B*.

Dead-End Solution: Let us now include force $\vec{F}_{AB}$ by writing, again for the *x* axis,

$$F_{app} - F_{AB} = m_A a.$$

(We use the minus sign to include the direction of $\vec{F}_{AB}$.) Because F_{AB} is a second unknown, we cannot solve this equation for *a*.

Successful Solution: Because of the direction in which force $\vec{F}_{app}$ is applied, the two blocks form a rigidly connected system. We can relate the net force *on the system* to the acceleration *of the system* with Newton's second law. Here, once again for the *x* axis, we can write that law as

$$F_{app} = (m_A + m_B)a,$$

where now we properly apply $\vec{F}_{app}$ to the system with total mass $m_A + m_B$. Solving for *a* and substituting known values, we find

$$a = \frac{F_{app}}{m_A + m_B} = \frac{20\text{ N}}{4.0\text{ kg} + 6.0\text{ kg}} = 2.0\text{ m/s}^2. \qquad \text{(Answer)}$$

Thus, the acceleration of the system and of each block is in the positive direction of the *x* axis and has the magnitude 2.0 m/s^2.

(b) What is the (horizontal) force $\vec{F}_{BA}$ on block *B* from block *A* (Fig. 5-18*c*)?

KEY IDEA

We can relate the net force on block *B* to the block's acceleration with Newton's second law.

Calculation: Here we can write that law, still for components along the *x* axis, as

$$F_{BA} = m_B a,$$

which, with known values, gives

$$F_{BA} = (6.0\text{ kg})(2.0\text{ m/s}^2) = 12\text{ N}. \qquad \text{(Answer)}$$

Thus, force $\vec{F}_{BA}$ is in the positive direction of the *x* axis and has a magnitude of 12 N.

WILEY PLUS Additional examples, video, and practice available at *WileyPLUS*

Review & Summary

Newtonian Mechanics The velocity of an object can change (the object can accelerate) when the object is acted on by one or more **forces** (pushes or pulls) from other objects. *Newtonian mechanics* relates accelerations and forces.

Force Forces are vector quantities. Their magnitudes are defined in terms of the acceleration they would give the standard kilogram. A force that accelerates that standard body by exactly 1 m/s^2 is defined to have a magnitude of 1 N. The direction of a force is the direction of the acceleration it causes. Forces are combined according to the rules of vector algebra. The **net force** on a body is the vector sum of all the forces acting on the body.

Newton's First Law If there is no net force on a body, the body remains at rest if it is initially at rest or moves in a straight line at constant speed if it is in motion.

Inertial Reference Frames Reference frames in which Newtonian mechanics holds are called *inertial reference frames* or *inertial frames.* Reference frames in which Newtonian mechanics does not hold are called *noninertial reference frames* or *noninertial frames.*

Mass The **mass** of a body is the characteristic of that body that relates the body's acceleration to the net force causing the acceleration. Masses are scalar quantities.

Newton's Second Law The net force $\vec{F}_{\text{net}}$ on a body with mass m is related to the body's acceleration $\vec{a}$ by

$$\vec{F}_{\text{net}} = m\vec{a}, \tag{5-1}$$

which may be written in the component versions

$$F_{\text{net},x} = ma_x \quad F_{\text{net},y} = ma_y \quad \text{and} \quad F_{\text{net},z} = ma_z. \tag{5-2}$$

The second law indicates that in SI units

$$1\text{ N} = 1\text{ kg}\cdot\text{m/s}^2. \tag{5-3}$$

A **free-body diagram** is a stripped-down diagram in which only *one* body is considered. That body is represented by either a sketch or a dot. The external forces on the body are drawn, and a coordinate system is superimposed, oriented so as to simplify the solution.

Some Particular Forces A **gravitational force** $\vec{F}_g$ on a body is a pull by another body. In most situations in this book, the other body is Earth or some other astronomical body. For Earth, the force is directed down toward the ground, which is assumed to be an inertial frame. With that assumption, the magnitude of $\vec{F}_g$ is

$$F_g = mg, \tag{5-8}$$

where m is the body's mass and g is the magnitude of the free-fall acceleration.

The **weight** W of a body is the magnitude of the upward force needed to balance the gravitational force on the body. A body's weight is related to the body's mass by

$$W = mg. \tag{5-12}$$

A **normal force** $\vec{F}_N$ is the force on a body from a surface against which the body presses. The normal force is always perpendicular to the surface.

A **frictional force** $\vec{f}$ is the force on a body when the body slides or attempts to slide along a surface. The force is always parallel to the surface and directed so as to oppose the sliding. On a *frictionless surface,* the frictional force is negligible.

When a cord is under **tension,** each end of the cord pulls on a body. The pull is directed along the cord, away from the point of attachment to the body. For a *massless* cord (a cord with negligible mass), the pulls at both ends of the cord have the same magnitude T, even if the cord runs around a *massless, frictionless pulley* (a pulley with negligible mass and negligible friction on its axle to oppose its rotation).

Newton's Third Law If a force $\vec{F}_{BC}$ acts on body B due to body C, then there is a force $\vec{F}_{CB}$ on body C due to body B:

$$\vec{F}_{BC} = -\vec{F}_{CB}.$$

Questions

1 Figure 5-19 gives the free-body diagram for four situations in which an object is pulled by several forces across a frictionless floor, as seen from overhead. In which situations does the acceleration $\vec{a}$ of the object have (a) an x component and (b) a y component? (c) In each situation, give the direction of $\vec{a}$ by naming either a quadrant or a direction along an axis. (Don't reach for the calculator because this can be answered with a few mental calculations.)

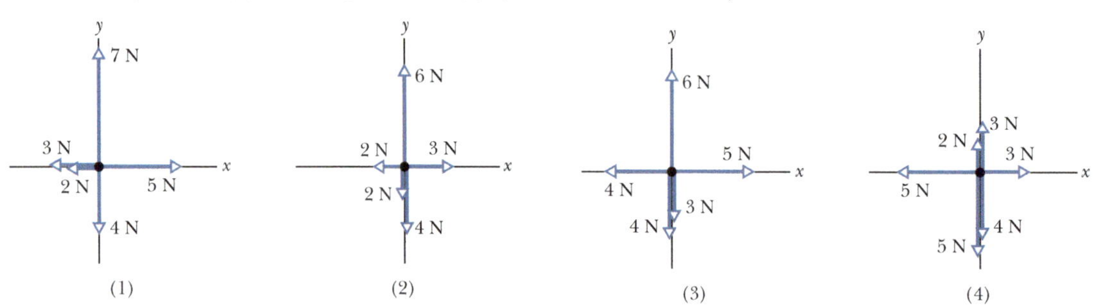

Figure 5-19 Question 1.

2 Two horizontal forces,

$$\vec{F}_1 = (3\text{ N})\hat{\text{i}} - (4\text{ N})\hat{\text{j}} \quad \text{and} \quad \vec{F}_2 = -(1\text{ N})\hat{\text{i}} - (2\text{ N})\hat{\text{j}}$$

pull a banana split across a frictionless lunch counter. Without using a calculator, determine which of the vectors in the free-body diagram of Fig. 5-20 best represent (a) $\vec{F}_1$ and (b) $\vec{F}_2$. What is the net-force component along (c) the x axis and (d) the y axis? Into which quadrants do (e) the net-force vector and (f) the split's acceleration vector point?

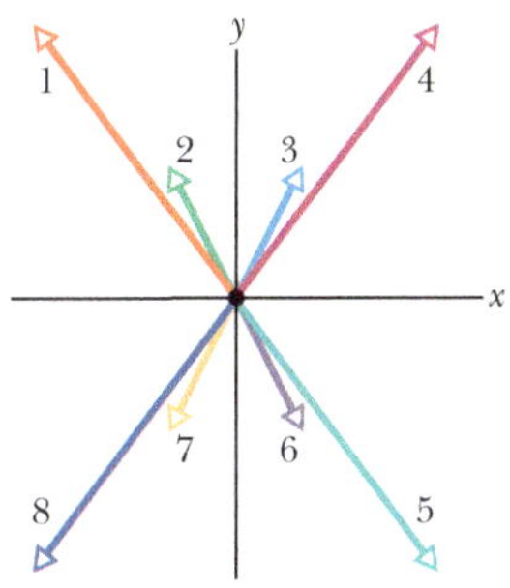

Figure 5-20 Question 2.

3 In Fig. 5-21, forces $\vec{F}_1$ and $\vec{F}_2$ are applied to a lunchbox as it slides at constant velocity over a frictionless floor. We are to decrease angle θ without changing the magnitude of $\vec{F}_1$. For constant velocity, should we increase, decrease, or maintain the magnitude of $\vec{F}_2$?

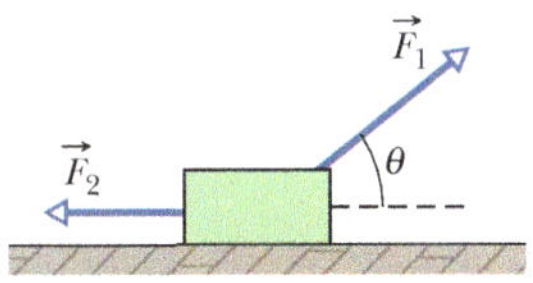

Figure 5-21 Question 3.

4 At time $t = 0$, constant $\vec{F}$ begins to act on a rock moving through deep space in the $+x$ direction. (a) For time $t > 0$, which are possible functions $x(t)$ for the rock's position: (1) $x = 4t - 3$, (2) $x = -4t^2 + 6t - 3$, (3) $x = 4t^2 + 6t - 3$? (b) For which function is $\vec{F}$ directed opposite the rock's initial direction of motion?

5 Figure 5-22 shows overhead views of four situations in which forces act on a block that lies on a frictionless floor. If the force magnitudes are chosen properly, in which situations is it possible that the block is (a) stationary and (b) moving with a constant velocity?

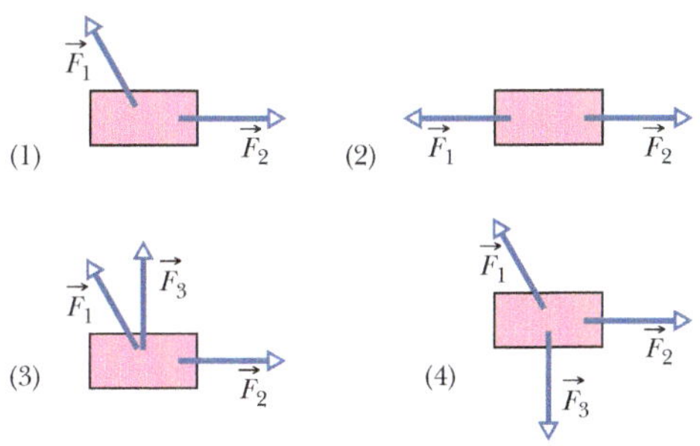

Figure 5-22 Question 5.

6 Figure 5-23 shows the same breadbox in four situations where horizontal forces are applied. Rank the situations according to the magnitude of the box's acceleration, greatest first.

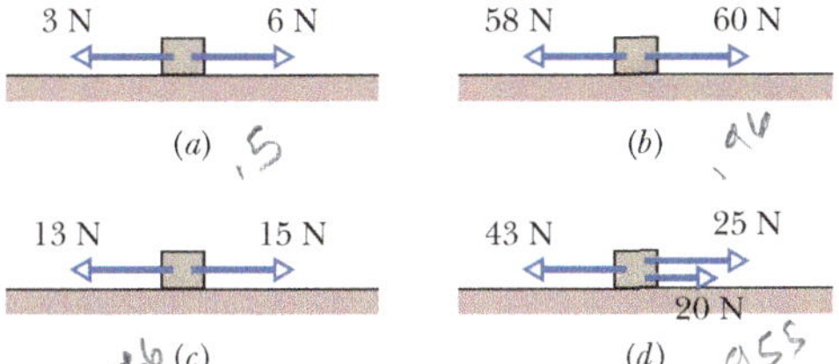

Figure 5-23 Question 6.

7 July 17, 1981, Kansas City: The newly opened Hyatt Regency is packed with people listening and dancing to a band playing favorites from the 1940s. Many of the people are crowded onto the walkways that hang like bridges across the wide atrium. Suddenly two of the walkways collapse, falling onto the merrymakers on the main floor.

The walkways were suspended one above another on vertical rods and held in place by nuts threaded onto the rods. In the original design, only two long rods were to be used, each extending through all three walkways (Fig. 5-24*a*). If each walkway and the merrymakers on it have a combined mass of M, what is the total mass supported by the threads and two nuts on (a) the lowest walkway and (b) the highest walkway?

Apparently someone responsible for the actual construction realized that threading nuts on a rod is impossible except at the ends, so the design was changed: Instead, six rods were used, each connecting two walkways (Fig. 5-24*b*). What now is the total mass supported by the threads and two nuts on (c) the lowest walkway, (d) the upper side of the highest walkway, and (e) the lower side of the highest walkway? It was this design that failed on that tragic night—a simple engineering error.

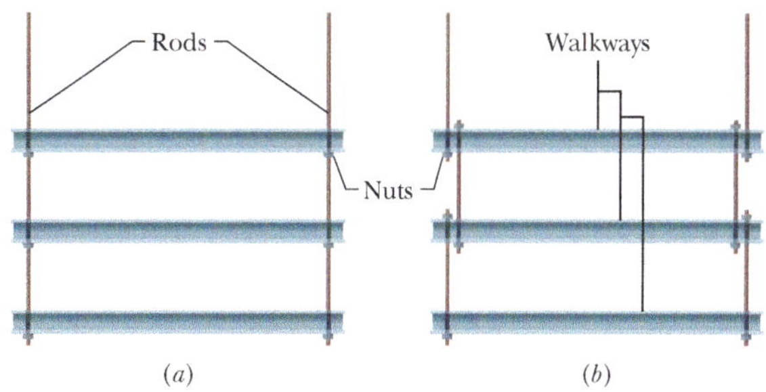

Figure 5-24 Question 7.

8 Figure 5-25 gives three graphs of velocity component $v_x(t)$ and three graphs of velocity component $v_y(t)$. The graphs are not to scale. Which $v_x(t)$ graph and which $v_y(t)$ graph best correspond to each of the four situations in Question 1 and Fig. 5-19?

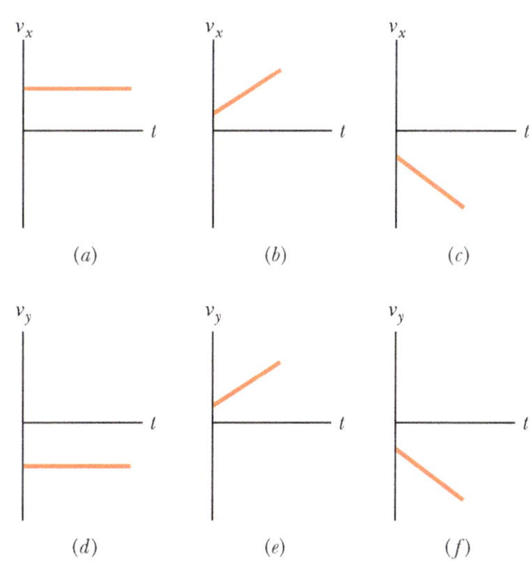

Figure 5-25 Question 8.

9 Figure 5-26 shows a train of four blocks being pulled across a frictionless floor by force $\vec{F}$. What total mass is accelerated to the right by (a) force $\vec{F}$, (b) cord 3, and (c) cord 1? (d) Rank the blocks according to their accelerations, greatest first. (e) Rank the cords according to their tension, greatest first.

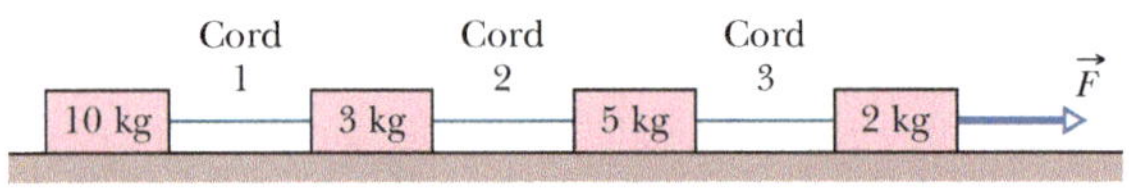

Figure 5-26 Question 9.

10 Figure 5-27 shows three blocks being pushed across a frictionless floor by horizontal force $\vec{F}$. What total mass is accelerated to the right by (a) force $\vec{F}$, (b) force $\vec{F}_{21}$ on block 2 from block 1, and (c) force $\vec{F}_{32}$ on block 3 from block 2? (d) Rank the blocks according to their acceleration magnitudes, greatest first. (e) Rank forces $\vec{F}$, $\vec{F}_{21}$, and $\vec{F}_{32}$ according to magnitude, greatest first.

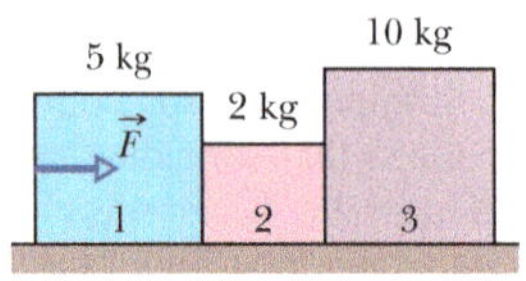

Figure 5-27 Question 10.

11 A vertical force $\vec{F}$ is applied to a block of mass m that lies on a floor. What happens to the magnitude of the normal force $\vec{F}_N$ on the block from the floor as magnitude F is increased from zero if force $\vec{F}$ is (a) downward and (b) upward?

12 Figure 5-28 shows four choices for the direction of a force of magnitude F to be applied to a block on an inclined plane. The directions are either horizontal or vertical. (For choice b, the force is not enough to lift the block off the plane.) Rank the choices according to the magnitude of the normal force acting on the block from the plane, greatest first.

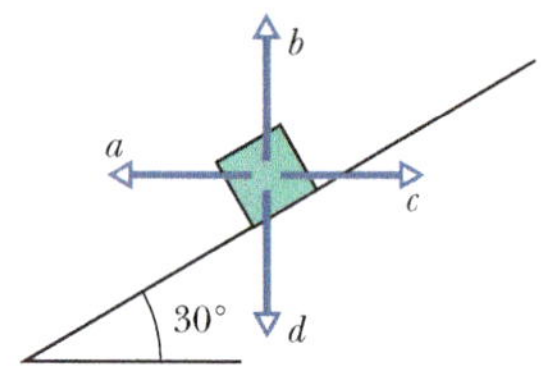

Figure 5-28 Question 12.

Problems

GO Tutoring problem available (at instructor's discretion) in *WileyPLUS* and WebAssign

SSM Worked-out solution available in Student Solutions Manual

WWW Worked-out solution is at

• – ••• Number of dots indicates level of problem difficulty

ILW Interactive solution is at

http://www.wiley.com/college/halliday

Additional information available in *The Flying Circus of Physics* and at flyingcircusofphysics.com

Module 5-1 Newton's First and Second Laws

•1 Only two horizontal forces act on a 3.0 kg body that can move over a frictionless floor. One force is 9.0 N, acting due east, and the other is 8.0 N, acting 62° north of west. What is the magnitude of the body's acceleration?

•2 Two horizontal forces act on a 2.0 kg chopping block that can slide over a frictionless kitchen counter, which lies in an xy plane. One force is $\vec{F}_1 = (3.0\text{ N})\hat{i} + (4.0\text{ N})\hat{j}$. Find the acceleration of the chopping block in unit-vector notation when the other force is (a) $\vec{F}_2 = (-3.0\text{ N})\hat{i} + (-4.0\text{ N})\hat{j}$, (b) $\vec{F}_2 = (-3.0\text{ N})\hat{i} + (4.0\text{ N})\hat{j}$, and (c) $\vec{F}_2 = (3.0\text{ N})\hat{i} + (-4.0\text{ N})\hat{j}$.

•3 If the 1 kg standard body has an acceleration of 2.00 m/s^2 at 20.0° to the positive direction of an x axis, what are (a) the x component and (b) the y component of the net force acting on the body, and (c) what is the net force in unit-vector notation?

••4 While two forces act on it, a particle is to move at the constant velocity $\vec{v} = (3\text{ m/s})\hat{i} - (4\text{ m/s})\hat{j}$. One of the forces is $\vec{F}_1 = (2\text{ N})\hat{i} + (-6\text{ N})\hat{j}$. What is the other force?

••5 GO Three astronauts, propelled by jet backpacks, push and guide a 120 kg asteroid toward a processing dock, exerting the forces shown in Fig. 5-29, with $F_1 = 32$ N, $F_2 = 55$ N, $F_3 = 41$ N, $\theta_1 = 30°$, and $\theta_3 = 60°$. What is the asteroid's acceleration (a) in unit-vector notation and as (b) a magnitude and (c) a direction relative to the positive direction of the x axis?

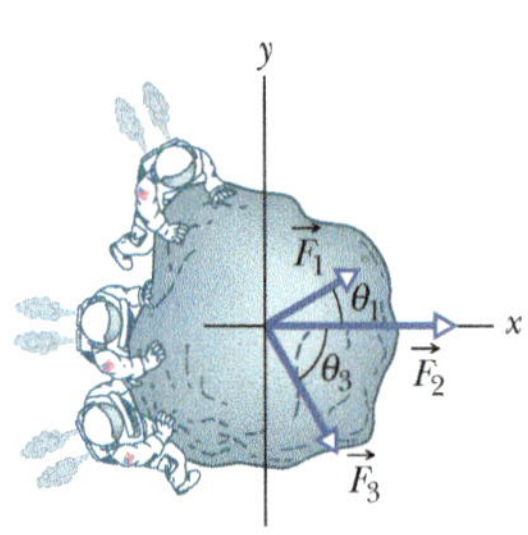

Figure 5-29 Problem 5.

••6 In a two-dimensional tug-of-war, Alex, Betty, and Charles pull horizontally on an automobile tire at the angles shown in the overhead view of Fig. 5-30. The tire remains stationary in spite of the three pulls. Alex pulls with force $\vec{F}_A$ of magnitude 220 N, and Charles pulls with force $\vec{F}_C$ of magnitude 170 N. Note that the direction of $\vec{F}_C$ is not given. What is the magnitude of Betty's force $\vec{F}_B$?

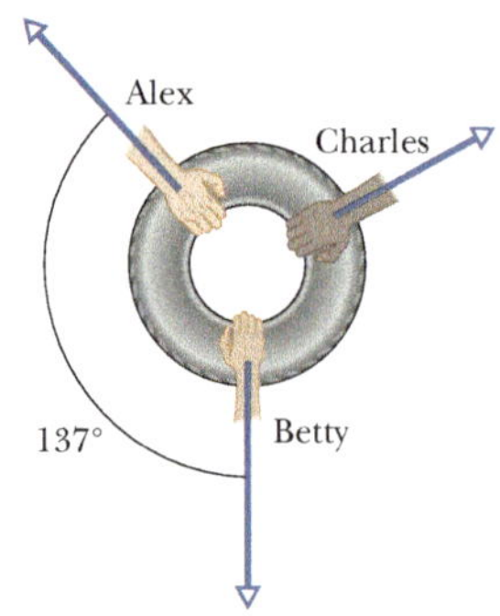

Figure 5-30 Problem 6.

••7 SSM There are two forces on the 2.00 kg box in the overhead view of Fig. 5-31, but only one is shown. For $F_1 = 20.0$ N, $a = 12.0$ m/s^2, and $\theta = 30.0°$, find the second force (a) in unit-vector notation and as (b) a magnitude and (c) an angle relative to the positive direction of the x axis.

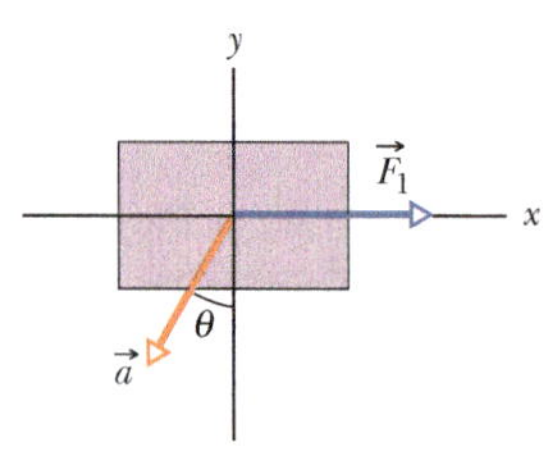

Figure 5-31 Problem 7.

••8 A 2.00 kg object is subjected to three forces that give it an acceleration $\vec{a} = -(8.00\text{ m/s}^2)\hat{i} + (6.00\text{ m/s}^2)\hat{j}$. If two of the three forces are $\vec{F}_1 = (30.0\text{ N})\hat{i} + (16.0\text{ N})\hat{j}$ and $\vec{F}_2 = -(12.0\text{ N})\hat{i} + (8.00\text{ N})\hat{j}$, find the third force.

••9 A 0.340 kg particle moves in an xy plane according to $x(t) = -15.00 + 2.00t - 4.00t^3$ and $y(t) = 25.00 + 7.00t - 9.00t^2$, with x and y in meters and t in seconds. At $t = 0.700$ s, what are

(a) the magnitude and (b) the angle (relative to the positive direction of the x axis) of the net force on the particle, and (c) what is the angle of the particle's direction of travel?

••10 A 0.150 kg particle moves along an x axis according to $x(t) = -13.00 + 2.00t + 4.00t^2 - 3.00t^3$, with x in meters and t in seconds. In unit-vector notation, what is the net force acting on the particle at $t = 3.40$ s?

••11 A 2.0 kg particle moves along an x axis, being propelled by a variable force directed along that axis. Its position is given by $x = 3.0\text{ m} + (4.0\text{ m/s})t + ct^2 - (2.0\text{ m/s}^3)t^3$, with x in meters and t in seconds. The factor c is a constant. At $t = 3.0$ s, the force on the particle has a magnitude of 36 N and is in the negative direction of the axis. What is c?

•••12 Two horizontal forces $\vec{F}_1$ and $\vec{F}_2$ act on a 4.0 kg disk that slides over frictionless ice, on which an xy coordinate system is laid out. Force $\vec{F}_1$ is in the positive direction of the x axis and has a magnitude of 7.0 N. Force $\vec{F}_2$ has a magnitude of 9.0 N. Figure 5-32 gives the x component v_x of the velocity of the disk as a function of time t during the sliding. What is the angle between the constant directions of forces $\vec{F}_1$ and $\vec{F}_2$?

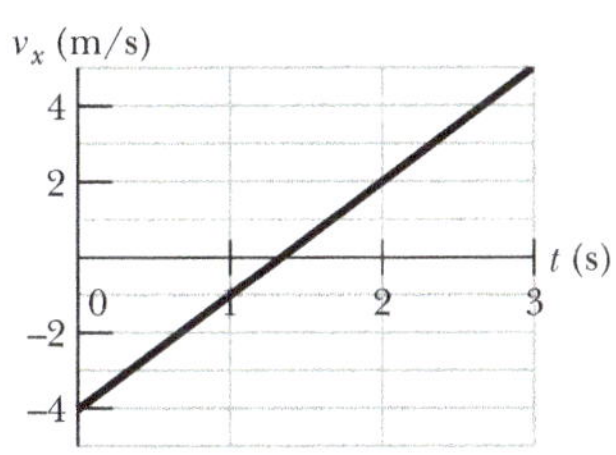

Figure 5-32 Problem 12.

Module 5-2 Some Particular Forces

•13 Figure 5-33 shows an arrangement in which four disks are suspended by cords. The longer, top cord loops over a frictionless pulley and pulls with a force of magnitude 98 N on the wall to which it is attached. The tensions in the three shorter cords are $T_1 = 58.8$ N, $T_2 = 49.0$ N, and $T_3 = 9.8$ N. What are the masses of (a) disk A, (b) disk B, (c) disk C, and (d) disk D?

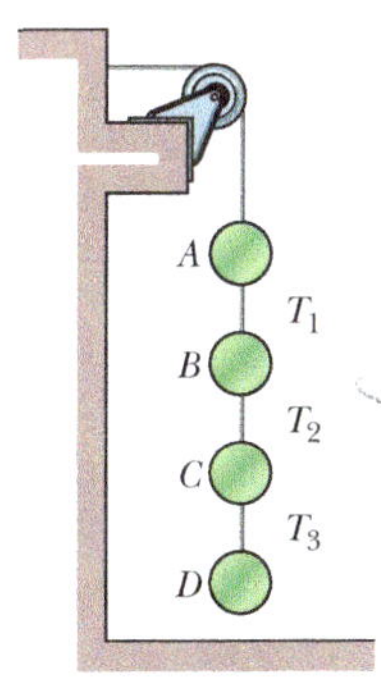

Figure 5-33 Problem 13.

•14 A block with a weight of 3.0 N is at rest on a horizontal surface. A 1.0 N upward force is applied to the block by means of an attached vertical string. What are the (a) magnitude and (b) direction of the force of the block on the horizontal surface?

•15 SSM (a) An 11.0 kg salami is supported by a cord that runs to a spring scale, which is supported by a cord hung from the ceiling (Fig. 5-34a). What is the reading on the scale, which is marked in SI weight units? (This is a way to measure weight by a deli owner.) (b) In Fig. 5-34b the salami is supported by a cord that runs around a pulley and to a scale. The opposite end of the scale is attached by a cord to a wall. What is the reading on the scale? (This is the way by a physics major.) (c) In Fig. 5-34c the wall has been replaced with a second 11.0 kg salami, and the assembly is stationary. What is the reading on the scale? (This is the way by a deli owner who was once a physics major.)

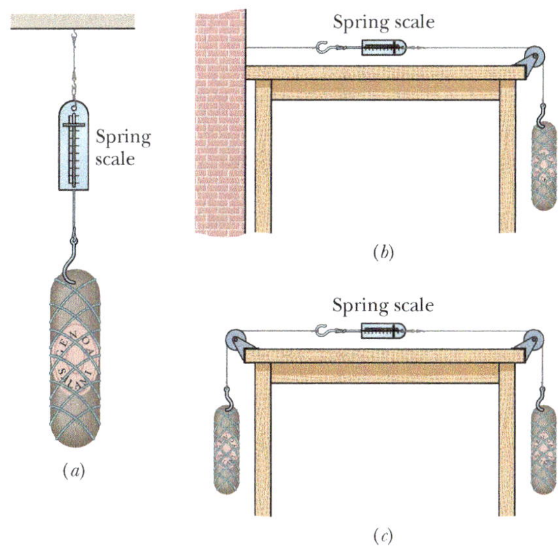

Figure 5-34 Problem 15.

••16 Some insects can walk below a thin rod (such as a twig) by hanging from it. Suppose that such an insect has mass m and hangs from a horizontal rod as shown in Fig. 5-35, with angle $\theta = 40°$. Its six legs are all under the same tension, and the leg sections nearest the body are horizontal. (a) What is the ratio of the tension in each tibia (forepart of a leg) to the insect's weight? (b) If the insect straightens out its legs somewhat, does the tension in each tibia increase, decrease, or stay the same?

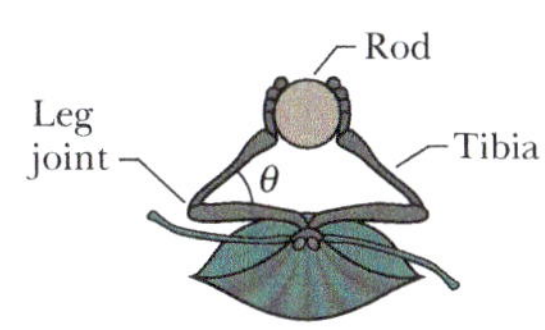

Figure 5-35 Problem 16.

Module 5-3 Applying Newton's Laws

•17 SSM WWW In Fig. 5-36, let the mass of the block be 8.5 kg and the angle θ be 30°. Find (a) the tension in the cord and (b) the normal force acting on the block. (c) If the cord is cut, find the magnitude of the resulting acceleration of the block.

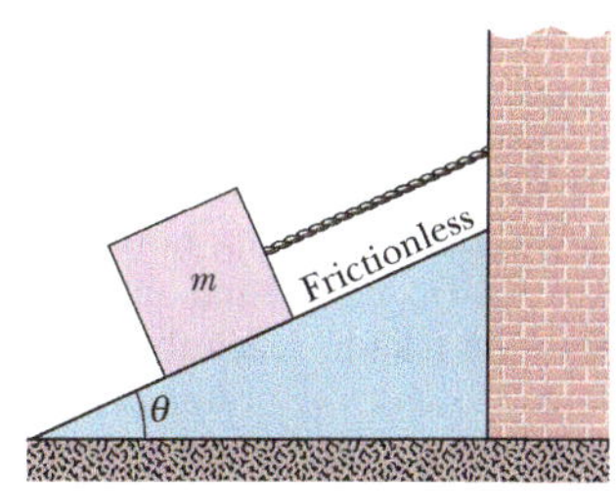

Figure 5-36 Problem 17.

•18 In April 1974, John Massis of Belgium managed to move two passenger railroad cars. He did so by clamping his teeth down on a bit that was attached to the cars with a rope and then leaning backward while pressing his feet against the railway ties. The cars together weighed 700 kN (about 80 tons). Assume that he pulled with a constant force that was 2.5 times his body weight, at an upward angle θ of 30° from the horizontal. His mass was 80 kg, and he moved the cars by 1.0 m. Neglecting any retarding force from the wheel rotation, find the speed of the cars at the end of the pull.

•**19** SSM A 500 kg rocket sled can be accelerated at a constant rate from rest to 1600 km/h in 1.8 s. What is the magnitude of the required net force?

•**20** A car traveling at 53 km/h hits a bridge abutment. A passenger in the car moves forward a distance of 65 cm (with respect to the road) while being brought to rest by an inflated air bag. What magnitude of force (assumed constant) acts on the passenger's upper torso, which has a mass of 41 kg?

•**21** A constant horizontal force $\vec{F}_a$ pushes a 2.00 kg FedEx package across a frictionless floor on which an xy coordinate system has been drawn. Figure 5-37 gives the package's x and y velocity components versus time t. What are the (a) magnitude and (b) direction of $\vec{F}_a$?

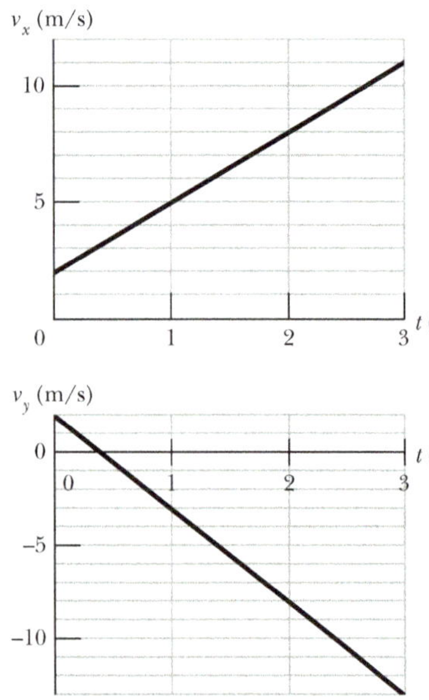

Figure 5-37 Problem 21.

•**22** A customer sits in an amusement park ride in which the compartment is to be pulled downward in the negative direction of a y axis with an acceleration magnitude of 1.24g, with $g = 9.80$ m/s^2. A 0.567 g coin rests on the customer's knee. Once the motion begins and in unit-vector notation, what is the coin's acceleration relative to (a) the ground and (b) the customer? (c) How long does the coin take to reach the compartment ceiling, 2.20 m above the knee? In unit-vector notation, what are (d) the actual force on the coin and (e) the apparent force according to the customer's measure of the coin's acceleration?

•**23** Tarzan, who weighs 820 N, swings from a cliff at the end of a 20.0 m vine that hangs from a high tree limb and initially makes an angle of 22.0° with the vertical. Assume that an x axis extends horizontally away from the cliff edge and a y axis extends upward. Immediately after Tarzan steps off the cliff, the tension in the vine is 760 N. Just then, what are (a) the force on him from the vine in unit-vector notation and the net force on him (b) in unit-vector notation and as (c) a magnitude and (d) an angle relative to the positive direction of the x axis? What are the (e) magnitude and (f) angle of Tarzan's acceleration just then?

•**24** There are two horizontal forces on the 2.0 kg box in the overhead view of Fig. 5-38 but only one (of magnitude $F_1 = 20$ N) is shown. The box moves along the x axis. For each of the following values for the acceleration a_x of the box, find the second force in unit-vector notation: (a) 10 m/s^2, (b) 20 m/s^2, (c) 0, (d) −10 m/s^2, and (e) −20 m/s^2.

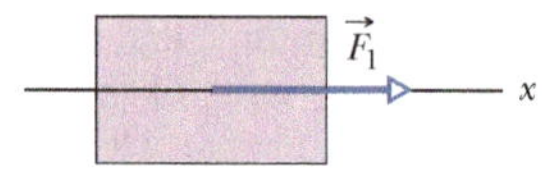

Figure 5-38 Problem 24.

•**25** *Sunjamming.* A "sun yacht" is a spacecraft with a large sail that is pushed by sunlight. Although such a push is tiny in everyday circumstances, it can be large enough to send the spacecraft outward from the Sun on a cost-free but slow trip. Suppose that the spacecraft has a mass of 900 kg and receives a push of 20 N. (a) What is the magnitude of the resulting acceleration? If the craft starts from rest, (b) how far will it travel in 1 day and (c) how fast will it then be moving?

•**26** The tension at which a fishing line snaps is commonly called the line's "strength." What minimum strength is needed for a line that is to stop a salmon of weight 85 N in 11 cm if the fish is initially drifting at 2.8 m/s? Assume a constant deceleration.

•**27** SSM An electron with a speed of 1.2×10^7 m/s moves horizontally into a region where a constant vertical force of 4.5×10^{-16} N acts on it. The mass of the electron is 9.11×10^{-31} kg. Determine the vertical distance the electron is deflected during the time it has moved 30 mm horizontally.

•**28** A car that weighs 1.30×10^4 N is initially moving at 40 km/h when the brakes are applied and the car is brought to a stop in 15 m. Assuming the force that stops the car is constant, find (a) the magnitude of that force and (b) the time required for the change in speed. If the initial speed is doubled, and the car experiences the same force during the braking, by what factors are (c) the stopping distance and (d) the stopping time multiplied? (There could be a lesson here about the danger of driving at high speeds.)

•**29** A firefighter who weighs 712 N slides down a vertical pole with an acceleration of 3.00 m/s^2, directed downward. What are the (a) magnitude and (b) direction (up or down) of the vertical force on the firefighter from the pole and the (c) magnitude and (d) direction of the vertical force on the pole from the firefighter?

•**30** The high-speed winds around a tornado can drive projectiles into trees, building walls, and even metal traffic signs. In a laboratory simulation, a standard wood toothpick was shot by pneumatic gun into an oak branch. The toothpick's mass was 0.13 g, its speed before entering the branch was 220 m/s, and its penetration depth was 15 mm. If its speed was decreased at a uniform rate, what was the magnitude of the force of the branch on the toothpick?

••**31** SSM WWW A block is projected up a frictionless inclined plane with initial speed $v_0 = 3.50$ m/s. The angle of incline is $\theta = 32.0°$. (a) How far up the plane does the block go? (b) How long does it take to get there? (c) What is its speed when it gets back to the bottom?

••**32** Figure 5-39 shows an overhead view of a 0.0250 kg lemon half and

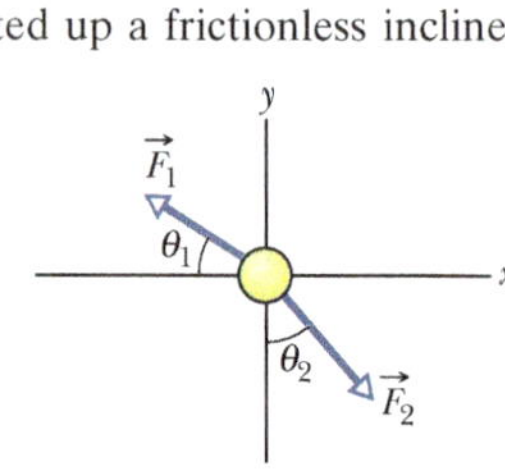

Figure 5-39 Problem 32.

two of the three horizontal forces that act on it as it is on a frictionless table. Force $\vec{F}_1$ has a magnitude of 6.00 N and is at $\theta_1 = 30.0°$. Force $\vec{F}_2$ has a magnitude of 7.00 N and is at $\theta_2 = 30.0°$. In unit-vector notation, what is the third force if the lemon half (a) is stationary, (b) has the constant velocity $\vec{v} = (13.0\hat{i} - 14.0\hat{j})$ m/s, and (c) has the varying velocity $\vec{v} = (13.0t\hat{i} - 14.0t\hat{j})$ m/s², where t is time?

••33 An elevator cab and its load have a combined mass of 1600 kg. Find the tension in the supporting cable when the cab, originally moving downward at 12 m/s, is brought to rest with constant acceleration in a distance of 42 m.

••34 GO In Fig. 5-40, a crate of mass $m = 100$ kg is pushed at constant speed up a frictionless ramp ($\theta = 30.0°$) by a horizontal force $\vec{F}$. What are the magnitudes of (a) $\vec{F}$ and (b) the force on the crate from the ramp?

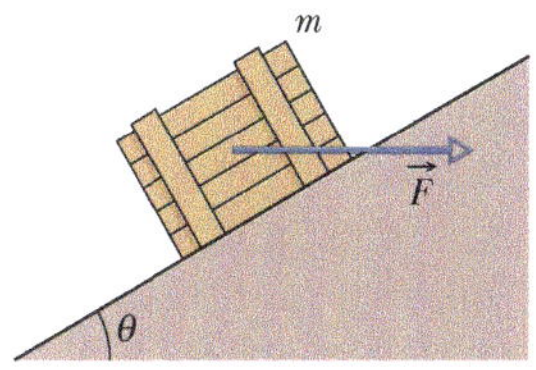

Figure 5-40 Problem 34.

••35 The velocity of a 3.00 kg particle is given by $\vec{v} = (8.00t\hat{i} + 3.00t^2\hat{j})$ m/s, with time t in seconds. At the instant the net force on the particle has a magnitude of 35.0 N, what are the direction (relative to the positive direction of the x axis) of (a) the net force and (b) the particle's direction of travel?

••36 Holding on to a towrope moving parallel to a frictionless ski slope, a 50 kg skier is pulled up the slope, which is at an angle of 8.0° with the horizontal. What is the magnitude F_{rope} of the force on the skier from the rope when (a) the magnitude v of the skier's velocity is constant at 2.0 m/s and (b) $v = 2.0$ m/s as v increases at a rate of 0.10 m/s²?

••37 A 40 kg girl and an 8.4 kg sled are on the frictionless ice of a frozen lake, 15 m apart but connected by a rope of negligible mass. The girl exerts a horizontal 5.2 N force on the rope. What are the acceleration magnitudes of (a) the sled and (b) the girl? (c) How far from the girl's initial position do they meet?

••38 A 40 kg skier skis directly down a frictionless slope angled at 10° to the horizontal. Assume the skier moves in the negative direction of an x axis along the slope. A wind force with component F_x acts on the skier. What is F_x if the magnitude of the skier's velocity is (a) constant, (b) increasing at a rate of 1.0 m/s², and (c) increasing at a rate of 2.0 m/s²?

••39 ILW A sphere of mass 3.0×10^{-4} kg is suspended from a cord. A steady horizontal breeze pushes the sphere so that the cord makes a constant angle of 37° with the vertical. Find (a) the push magnitude and (b) the tension in the cord.

••40 GO A dated box of dates, of mass 5.00 kg, is sent sliding up a frictionless ramp at an angle of θ to the horizontal. Figure 5-41 gives, as a function of time t, the component v_x of the box's velocity along an x axis that extends directly up the ramp. What is the magnitude of the normal force on the box from the ramp?

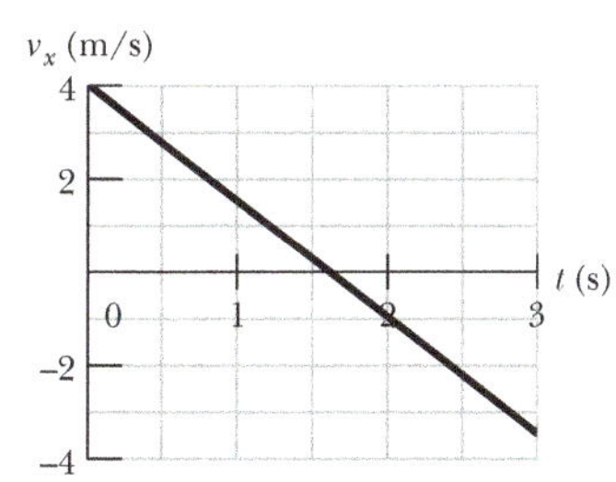

Figure 5-41 Problem 40.

••41 Using a rope that will snap if the tension in it exceeds 387 N, you need to lower a bundle of old roofing material weighing 449 N from a point 6.1 m above the ground. Obviously if you hang the bundle on the rope, it will snap. So, you allow the bundle to accelerate downward. (a) What magnitude of the bundle's acceleration will put the rope on the verge of snapping? (b) At that acceleration, with what speed would the bundle hit the ground?

••42 GO In earlier days, horses pulled barges down canals in the manner shown in Fig. 5-42. Suppose the horse pulls on the rope with a force of 7900 N at an angle of $\theta = 18°$ to the direction of motion of the barge, which is headed straight along the positive direction of an x axis. The mass of the barge is 9500 kg, and the magnitude of its acceleration is 0.12 m/s². What are the (a) magnitude and (b) direction (relative to positive x) of the force on the barge from the water?

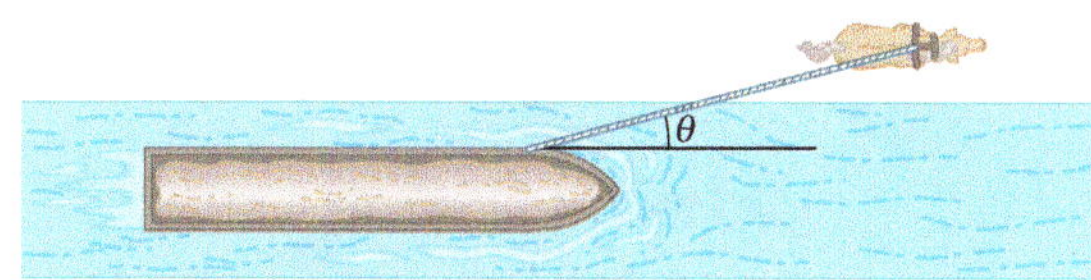

Figure 5-42 Problem 42.

••43 SSM In Fig. 5-43, a chain consisting of five links, each of mass 0.100 kg, is lifted vertically with constant acceleration of magnitude $a = 2.50$ m/s². Find the magnitudes of (a) the force on link 1 from link 2, (b) the force on link 2 from link 3, (c) the force on link 3 from link 4, and (d) the force on link 4 from link 5. Then find the magnitudes of (e) the force $\vec{F}$ on the top link from the person lifting the chain and (f) the *net* force accelerating each link.

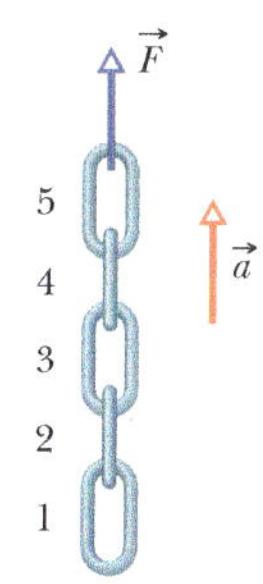

Figure 5-43 Problem 43.

••44 A lamp hangs vertically from a cord in a descending elevator that decelerates at 2.4 m/s². (a) If the tension in the cord is 89 N, what is the lamp's mass? (b) What is the cord's tension when the elevator ascends with an upward acceleration of 2.4 m/s²?

••45 An elevator cab that weighs 27.8 kN moves upward. What is the tension in the cable if the cab's speed is (a) increasing at a rate of 1.22 m/s² and (b) decreasing at a rate of 1.22 m/s²?

••46 An elevator cab is pulled upward by a cable. The cab and its single occupant have a combined mass of 2000 kg. When that occupant drops a coin, its acceleration relative to the cab is 8.00 m/s² downward. What is the tension in the cable?

••47 GO The Zacchini family was renowned for their human-cannonball act in which a family member was shot from a cannon using either elastic bands or compressed air. In one version of the act, Emanuel Zacchini was shot over three Ferris wheels to land in a net at the same height as the open end of the cannon and at a range of 69 m. He was propelled inside the barrel for 5.2 m and launched at an angle of 53°. If his mass was 85 kg and he underwent constant acceleration inside the barrel, what was the magnitude of the force propelling him? (*Hint:* Treat the launch as though it were along a ramp at 53°. Neglect air drag.)

••48 GO In Fig. 5-44, elevator cabs A and B are connected by a short cable and can be pulled upward or lowered by the cable above cab A. Cab A has mass 1700 kg; cab B has mass 1300 kg. A 12.0 kg box of catnip lies on the floor of cab A. The tension in the cable connecting the cabs is 1.91×10^4 N. What is the magnitude of the normal force on the box from the floor?

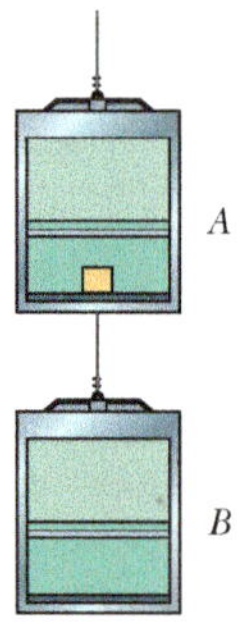

Figure 5-44 Problem 48.

••49 In Fig. 5-45, a block of mass $m = 5.00$ kg is pulled along a horizontal frictionless floor by a cord that exerts a force of magnitude $F = 12.0$ N at an angle $\theta = 25.0°$. (a) What is the magnitude of the block's acceleration? (b) The force magnitude F is slowly increased. What is its value just before the block is lifted (completely) off the floor? (c) What is the magnitude of the block's acceleration just before it is lifted (completely) off the floor?

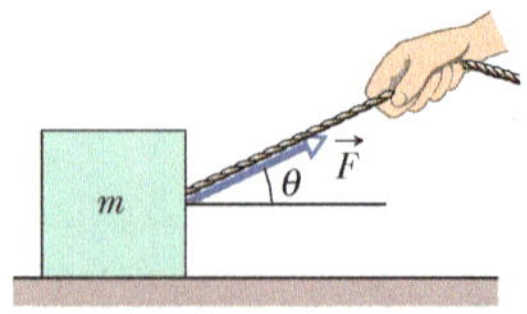

Figure 5-45 Problems 49 and 60.

••50 GO In Fig. 5-46, three ballot boxes are connected by cords, one of which wraps over a pulley having negligible friction on its axle and negligible mass. The three masses are $m_A = 30.0$ kg, $m_B = 40.0$ kg, and $m_C = 10.0$ kg. When the assembly is released from rest, (a) what is the tension in the cord connecting B and C, and (b) how far does A move in the first 0.250 s (assuming it does not reach the pulley)?

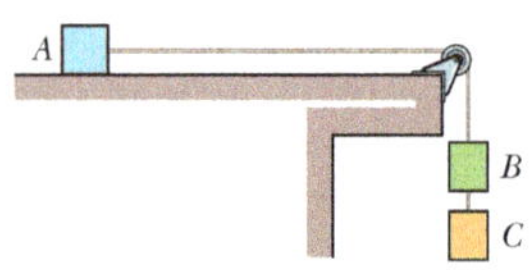

Figure 5-46 Problem 50.

••51 GO Figure 5-47 shows two blocks connected by a cord (of negligible mass) that passes over a frictionless pulley (also of negligible mass). The arrangement is known as *Atwood's machine*. One block has mass $m_1 = 1.30$ kg; the other has mass $m_2 = 2.80$ kg. What are (a) the magnitude of the blocks' acceleration and (b) the tension in the cord?

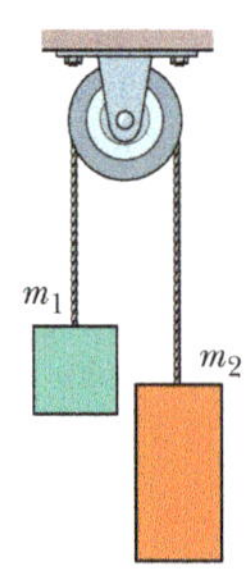

Figure 5-47 Problems 51 and 65.

••52 An 85 kg man lowers himself to the ground from a height of 10.0 m by holding onto a rope that runs over a frictionless pulley to a 65 kg sandbag. With what speed does the man hit the ground if he started from rest?

••53 In Fig. 5-48, three connected blocks are pulled to the right on a horizontal frictionless table by a force of magnitude $T_3 = 65.0$ N. If $m_1 = 12.0$ kg, $m_2 = 24.0$ kg, and $m_3 = 31.0$ kg, calculate (a) the magnitude of the system's acceleration, (b) the tension T_1, and (c) the tension T_2.

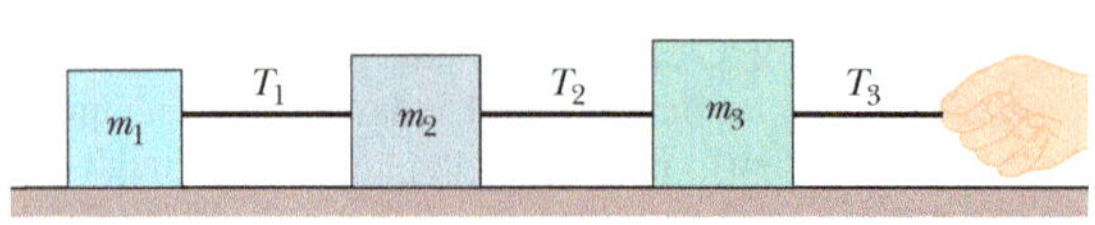

Figure 5-48 Problem 53.

••54 GO Figure 5-49 shows four penguins that are being playfully pulled along very slippery (frictionless) ice by a curator. The masses of three penguins and the tension in two of the cords are $m_1 = 12$ kg, $m_3 = 15$ kg, $m_4 = 20$ kg, $T_2 = 111$ N, and $T_4 = 222$ N. Find the penguin mass m_2 that is not given.

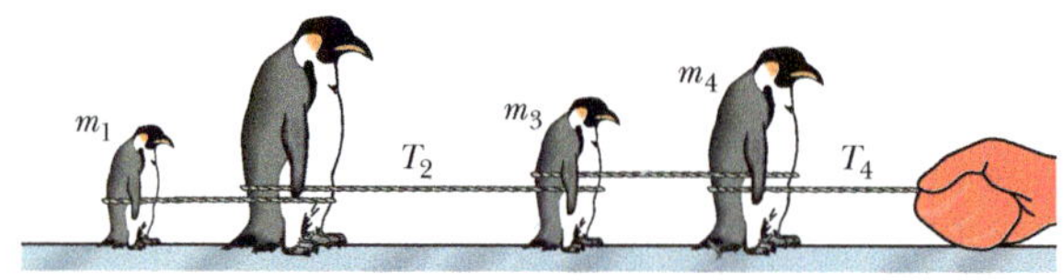

Figure 5-49 Problem 54.

••55 SSM ILW WWW Two blocks are in contact on a frictionless table. A horizontal force is applied to the larger block, as shown in Fig. 5-50. (a) If $m_1 = 2.3$ kg, $m_2 = 1.2$ kg, and $F = 3.2$ N, find the magnitude of the force between the two blocks. (b) Show that if a force of the same magnitude F is applied to the smaller block but in the opposite direction, the magnitude of the force between the blocks is 2.1 N, which is not the same value calculated in (a). (c) Explain the difference.

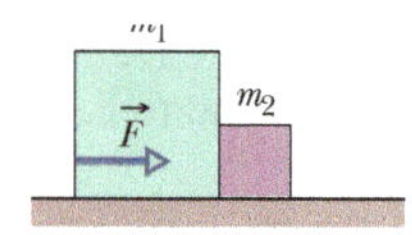

Figure 5-50 Problem 55.

••56 GO In Fig. 5-51a, a constant horizontal force $\vec{F}_a$ is applied to block A, which pushes against block B with a 20.0 N force directed horizontally to the right. In Fig. 5-51b, the same force $\vec{F}_a$ is applied to block B; now block A pushes on block B with a 10.0 N force directed horizontally to the left. The blocks have a combined mass of 12.0 kg. What are the magnitudes of (a) their acceleration in Fig. 5-51a and (b) force $\vec{F}_a$?

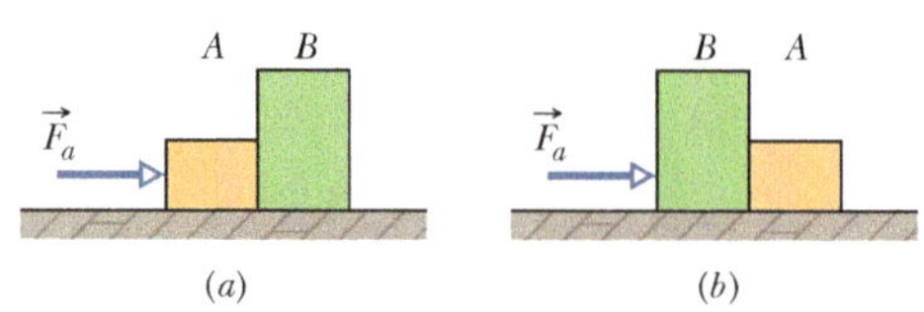

Figure 5-51 Problem 56.

••57 ILW A block of mass $m_1 = 3.70$ kg on a frictionless plane inclined at angle $\theta = 30.0°$ is connected by a cord over a massless, frictionless pulley to a second block of mass $m_2 = 2.30$ kg (Fig. 5-52). What are (a) the magnitude of the acceleration of each block, (b) the direction of the acceleration of the hanging block, and (c) the tension in the cord?

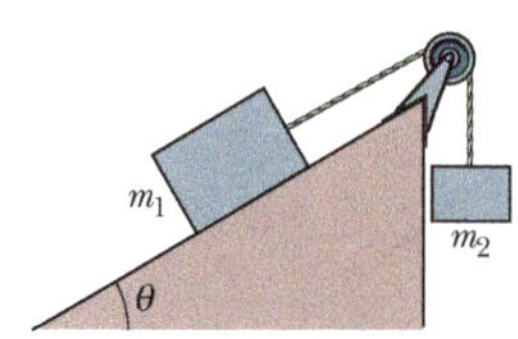

Figure 5-52 Problem 57.

••58 Figure 5-53 shows a man sitting in a bosun's chair that dangles from a massless rope, which runs over a massless, frictionless pulley and back down to the man's hand. The combined mass of man and chair is 95.0 kg. With what force magnitude must the man pull on the rope if he is to rise (a) with a constant velocity and

(b) with an upward acceleration of 1.30 m/s²? (*Hint:* A free-body diagram can really help.) If the rope on the right extends to the ground and is pulled by a co-worker, with what force magnitude must the co-worker pull for the man to rise (c) with a constant velocity and (d) with an upward acceleration of 1.30 m/s²? What is the magnitude of the force on the ceiling from the pulley system in (e) part a, (f) part b, (g) part c, and (h) part d?

Figure 5-53 Problem 58.

••59 SSM A 10 kg monkey climbs up a massless rope that runs over a frictionless tree limb and back down to a 15 kg package on the ground (Fig. 5-54). (a) What is the magnitude of the least acceleration the monkey must have if it is to lift the package off the ground? If, after the package has been lifted, the monkey stops its climb and holds onto the rope, what are the (b) magnitude and (c) direction of the monkey's acceleration and (d) the tension in the rope?

Figure 5-54 Problem 59.

••60 Figure 5-45 shows a 5.00 kg block being pulled along a frictionless floor by a cord that applies a force of constant magnitude 20.0 N but with an angle $\theta(t)$ that varies with time. When angle $\theta = 25.0°$, at what rate is the acceleration of the block changing if (a) $\theta(t) = (2.00 \times 10^{-2}\text{ deg/s})t$ and (b) $\theta(t) = -(2.00 \times 10^{-2}\text{ deg/s})t$? (*Hint:* The angle should be in radians.)

••61 SSM ILW A hot-air balloon of mass M is descending vertically with downward acceleration of magnitude a. How much mass (ballast) must be thrown out to give the balloon an upward acceleration of magnitude a? Assume that the upward force from the air (the lift) does not change because of the decrease in mass.

•••62 In shot putting, many athletes elect to launch the shot at an angle that is smaller than the theoretical one (about 42°) at which the distance of a projected ball at the same speed and height is greatest. One reason has to do with the speed the athlete can give the shot during the acceleration phase of the throw. Assume that a 7.260 kg shot is accelerated along a straight path of length 1.650 m by a constant applied force of magnitude 380.0 N, starting with an initial speed of 2.500 m/s (due to the athlete's preliminary motion). What is the shot's speed at the end of the acceleration phase if the angle between the path and the horizontal is (a) 30.00° and (b) 42.00°? (*Hint:* Treat the motion as though it were along a ramp at the given angle.) (c) By what percent is the launch speed decreased if the athlete increases the angle from 30.00° to 42.00°?

•••63 GO Figure 5-55 gives, as a function of time t, the force component F_x that acts on a 3.00 kg ice block that can move only along the x axis. At $t = 0$, the block is moving in the positive direction of the axis, with a speed of 3.0 m/s. What are its (a) speed and (b) direction of travel at $t = 11$ s?

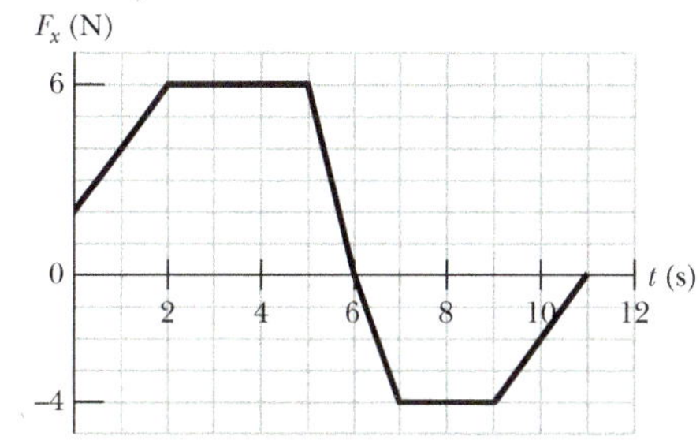

Figure 5-55 Problem 63.

•••64 GO Figure 5-56 shows a box of mass $m_2 = 1.0$ kg on a frictionless plane inclined at angle $\theta = 30°$. It is connected by a cord of negligible mass to a box of mass $m_1 = 3.0$ kg on a horizontal frictionless surface. The pulley is frictionless and massless. (a) If the magnitude of horizontal force $\vec{F}$ is 2.3 N, what is the tension in the connecting cord? (b) What is the largest value the magnitude of $\vec{F}$ may have without the cord becoming slack?

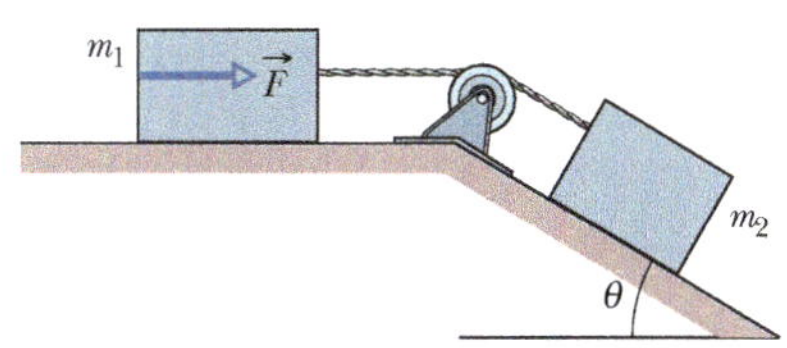

Figure 5-56 Problem 64.

•••65 GO Figure 5-47 shows *Atwood's machine,* in which two containers are connected by a cord (of negligible mass) passing over a frictionless pulley (also of negligible mass). At time $t = 0$, container 1 has mass 1.30 kg and container 2 has mass 2.80 kg, but container 1 is losing mass (through a leak) at the constant rate of 0.200 kg/s. At what rate is the acceleration magnitude of the containers changing at (a) $t = 0$ and (b) $t = 3.00$ s? (c) When does the acceleration reach its maximum value?

•••66 GO Figure 5-57 shows a section of a cable-car system. The maximum permissible mass of each car with occupants is 2800 kg. The cars, riding on a support cable, are pulled by a second cable attached to the support tower on each car. Assume that the cables

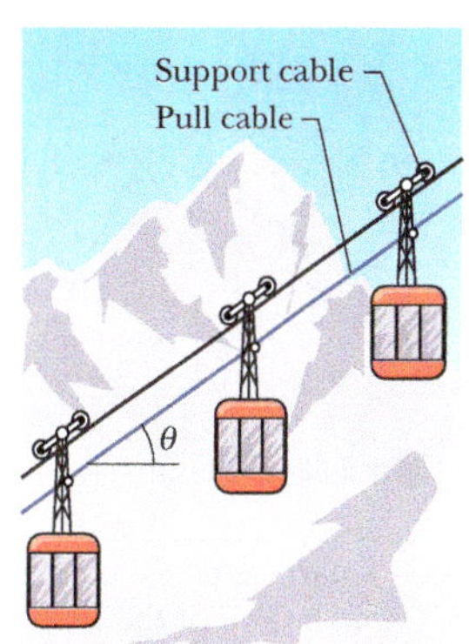

Figure 5-57 Problem 66.

are taut and inclined at angle $\theta = 35°$. What is the difference in tension between adjacent sections of pull cable if the cars are at the maximum permissible mass and are being accelerated up the incline at 0.81 m/s^2?

•••67 Figure 5-58 shows three blocks attached by cords that loop over frictionless pulleys. Block B lies on a frictionless table; the masses are m_A = 6.00 kg, m_B = 8.00 kg, and m_C = 10.0 kg. When the blocks are released, what is the tension in the cord at the right?

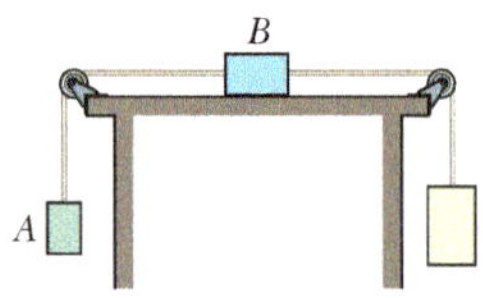

Figure 5-58 Problem 67.

•••68 A shot putter launches a 7.260 kg shot by pushing it along a straight line of length 1.650 m and at an angle of 34.10° from the horizontal, accelerating the shot to the launch speed from its initial speed of 2.500 m/s (which is due to the athlete's preliminary motion). The shot leaves the hand at a height of 2.110 m and at an angle of 34.10°, and it lands at a horizontal distance of 15.90 m. What is the magnitude of the athlete's average force on the shot during the acceleration phase? (*Hint:* Treat the motion during the acceleration phase as though it were along a ramp at the given angle.)

Additional Problems

69 In Fig. 5-59, 4.0 kg block A and 6.0 kg block B are connected by a string of negligible mass. Force $\vec{F}_A = (12\text{ N})\hat{i}$ acts on block A; force $\vec{F}_B = (24\text{ N})\hat{i}$ acts on block B. What is the tension in the string?

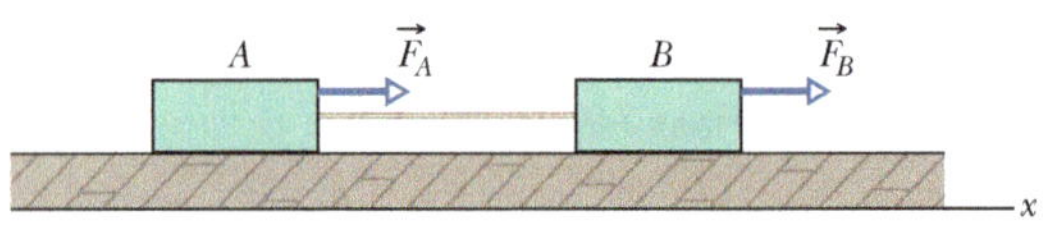

Figure 5-59 Problem 69.

70 An 80 kg man drops to a concrete patio from a window 0.50 m above the patio. He neglects to bend his knees on landing, taking 2.0 cm to stop. (a) What is his average acceleration from when his feet first touch the patio to when he stops? (b) What is the magnitude of the average stopping force exerted on him by the patio?

71 SSM Figure 5-60 shows a box of dirty money (mass m_1 = 3.0 kg) on a frictionless plane inclined at angle $\theta_1 = 30°$. The box is connected via a cord of negligible mass to a box of laundered money (mass m_2 = 2.0 kg) on a frictionless plane inclined at angle $\theta_2 = 60°$. The pulley is frictionless and has negligible mass. What is the tension in the cord?

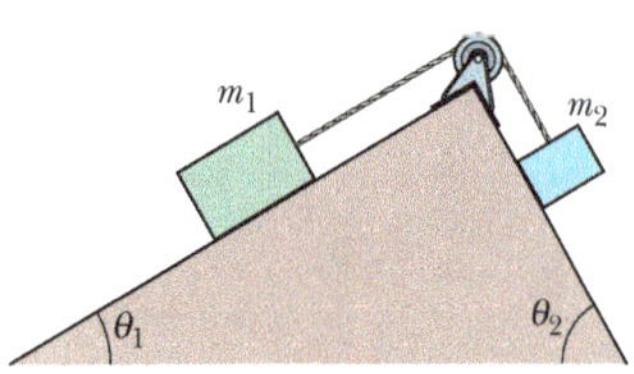

Figure 5-60 Problem 71.

72 Three forces act on a particle that moves with unchanging velocity $\vec{v} = (2\text{ m/s})\hat{i} - (7\text{ m/s})\hat{j}$. Two of the forces are $\vec{F}_1 = (2\text{ N})\hat{i} + (3\text{ N})\hat{j} + (-2\text{ N})\hat{k}$ and $\vec{F}_2 = (-5\text{ N})\hat{i} + (8\text{ N})\hat{j} + (-2\text{ N})\hat{k}$. What is the third force?

73 SSM In Fig. 5-61, a tin of antioxidants (m_1 = 1.0 kg) on a frictionless inclined surface is connected to a tin of corned beef (m_2 = 2.0 kg). The pulley is massless and frictionless. An upward force of magnitude F = 6.0 N acts on the corned beef tin, which has a downward acceleration of 5.5 m/s^2. What are (a) the tension in the connecting cord and (b) angle β?

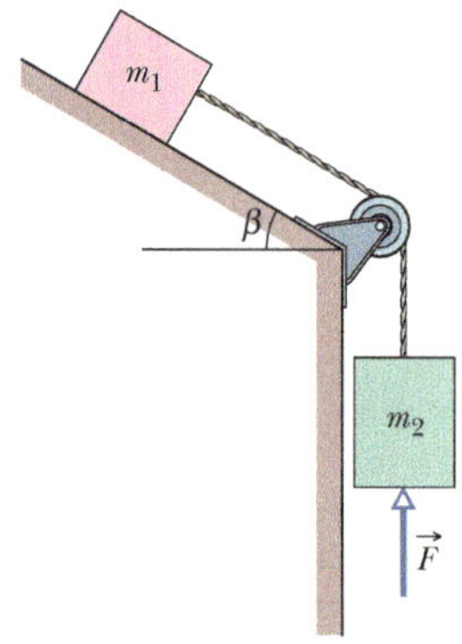

Figure 5-61 Problem 73.

74 The only two forces acting on a body have magnitudes of 20 N and 35 N and directions that differ by 80°. The resulting acceleration has a magnitude of 20 m/s^2. What is the mass of the body?

75 Figure 5-62 is an overhead view of a 12 kg tire that is to be pulled by three horizontal ropes. One rope's force (F_1 = 50 N) is indicated. The forces from the other ropes are to be oriented such that the tire's acceleration magnitude a is least. What is that least a if (a) F_2 = 30 N, F_3 = 20 N; (b) F_2 = 30 N, F_3 = 10 N; and (c) $F_2 = F_3$ = 30 N?

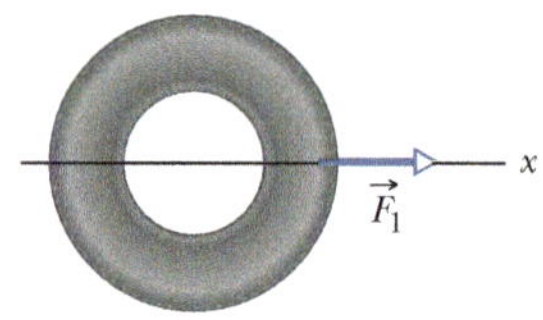

Figure 5-62 Problem 75.

76 A block of mass M is pulled along a horizontal frictionless surface by a rope of mass m, as shown in Fig. 5-63. A horizontal force $\vec{F}$ acts on one end of the rope. (a) Show that the rope *must* sag, even if only by an imperceptible amount. Then, assuming that the sag is negligible, find (b) the acceleration of rope and block, (c) the force on the block from the rope, and (d) the tension in the rope at its midpoint.

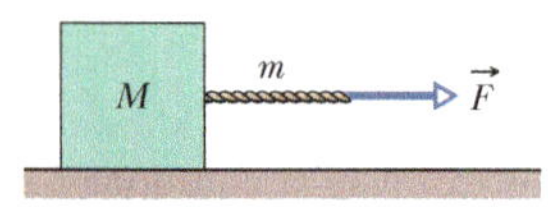

Figure 5-63 Problem 76.

77 SSM A worker drags a crate across a factory floor by pulling on a rope tied to the crate. The worker exerts a force of magnitude F = 450 N on the rope, which is inclined at an upward angle $\theta = 38°$ to the horizontal, and the floor exerts a horizontal force of magnitude f = 125 N that opposes the motion. Calculate the magnitude of the acceleration of the crate if (a) its mass is 310 kg and (b) its weight is 310 N.

78 In Fig. 5-64, a force $\vec{F}$ of magnitude 12 N is applied to a FedEx box of mass m_2 = 1.0 kg. The force is directed up a plane tilted by θ = 37°. The box is connected by a cord to a UPS box of mass m_1 = 3.0 kg on the floor. The floor, plane, and pulley are frictionless, and the masses of the pulley and cord are negligible. What is the tension in the cord?

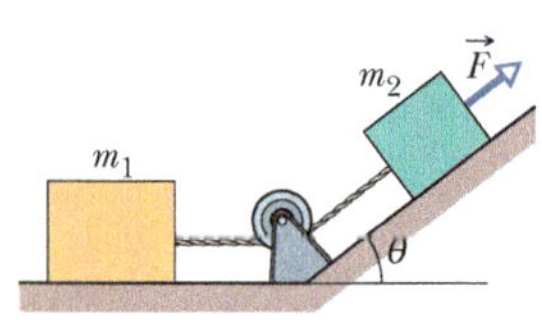

Figure 5-64 Problem 78.

79 A certain particle has a weight of 22 N at a point where g = 9.8 m/s^2. What are its (a) weight and (b) mass at a point where g = 4.9 m/s^2? What are its (c) weight and (d) mass if it is moved to a point in space where g = 0?

80 An 80 kg person is parachuting and experiencing a downward acceleration of 2.5 m/s^2. The mass of the parachute is 5.0 kg. (a)

What is the upward force on the open parachute from the air? (b) What is the downward force on the parachute from the person?

81 A spaceship lifts off vertically from the Moon, where $g = 1.6\ m/s^2$. If the ship has an upward acceleration of $1.0\ m/s^2$ as it lifts off, what is the magnitude of the force exerted by the ship on its pilot, who weighs 735 N on Earth?

82 In the overhead view of Fig. 5-65, five forces pull on a box of mass $m = 4.0$ kg. The force magnitudes are $F_1 = 11$ N, $F_2 = 17$ N, $F_3 = 3.0$ N, $F_4 = 14$ N, and $F_5 = 5.0$ N, and angle θ_4 is 30°. Find the box's acceleration (a) in unit-vector notation and as (b) a magnitude and (c) an angle relative to the positive direction of the x axis.

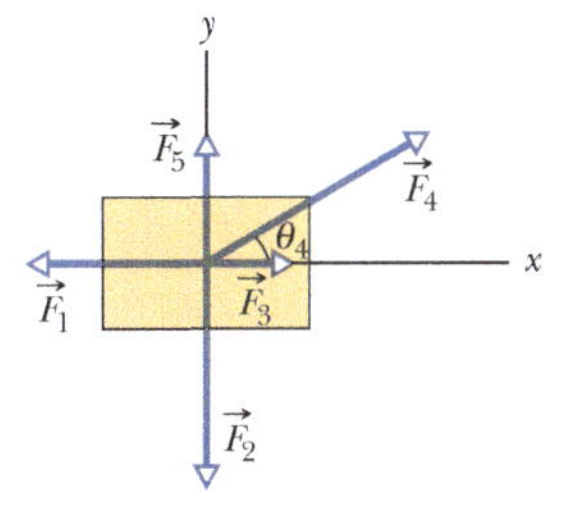

Figure 5-65 Problem 82.

83 SSM A certain force gives an object of mass m_1 an acceleration of $12.0\ m/s^2$ and an object of mass m_2 an acceleration of $3.30\ m/s^2$. What acceleration would the force give to an object of mass (a) $m_2 - m_1$ and (b) $m_2 + m_1$?

84 You pull a short refrigerator with a constant force $\vec{F}$ across a greased (frictionless) floor, either with $\vec{F}$ horizontal (case 1) or with $\vec{F}$ tilted upward at an angle θ (case 2). (a) What is the ratio of the refrigerator's speed in case 2 to its speed in case 1 if you pull for a certain time t? (b) What is this ratio if you pull for a certain distance d?

85 A 52 kg circus performer is to slide down a rope that will break if the tension exceeds 425 N. (a) What happens if the performer hangs stationary on the rope? (b) At what magnitude of acceleration does the performer just avoid breaking the rope?

86 Compute the weight of a 75 kg space ranger (a) on Earth, (b) on Mars, where $g = 3.7\ m/s^2$, and (c) in interplanetary space, where $g = 0$. (d) What is the ranger's mass at each location?

87 An object is hung from a spring balance attached to the ceiling of an elevator cab. The balance reads 65 N when the cab is standing still. What is the reading when the cab is moving upward (a) with a constant speed of 7.6 m/s and (b) with a speed of 7.6 m/s while decelerating at a rate of $2.4\ m/s^2$?

88 Imagine a landing craft approaching the surface of Callisto, one of Jupiter's moons. If the engine provides an upward force (thrust) of 3260 N, the craft descends at constant speed; if the engine provides only 2200 N, the craft accelerates downward at $0.39\ m/s^2$. (a) What is the weight of the landing craft in the vicinity of Callisto's surface? (b) What is the mass of the craft? (c) What is the magnitude of the free-fall acceleration near the surface of Callisto?

89 A 1400 kg jet engine is fastened to the fuselage of a passenger jet by just three bolts (this is the usual practice). Assume that each bolt supports one-third of the load. (a) Calculate the force on each bolt as the plane waits in line for clearance to take off. (b) During flight, the plane encounters turbulence, which suddenly imparts an upward vertical acceleration of $2.6\ m/s^2$ to the plane. Calculate the force on each bolt now.

90 An interstellar ship has a mass of 1.20×10^6 kg and is initially at rest relative to a star system. (a) What constant acceleration is needed to bring the ship up to a speed of $0.10c$ (where c is the speed of light, 3.0×10^8 m/s) relative to the star system in 3.0 days? (b) What is that acceleration in g units? (c) What force is required for the acceleration? (d) If the engines are shut down when $0.10c$ is reached (the speed then remains constant), how long does the ship take (start to finish) to journey 5.0 light-months, the distance that light travels in 5.0 months?

91 SSM A motorcycle and 60.0 kg rider accelerate at $3.0\ m/s^2$ up a ramp inclined 10° above the horizontal. What are the magnitudes of (a) the net force on the rider and (b) the force on the rider from the motorcycle?

92 Compute the initial upward acceleration of a rocket of mass 1.3×10^4 kg if the initial upward force produced by its engine (the thrust) is 2.6×10^5 N. Do not neglect the gravitational force on the rocket.

93 SSM Figure 5-66a shows a mobile hanging from a ceiling; it consists of two metal pieces ($m_1 = 3.5$ kg and $m_2 = 4.5$ kg) that are strung together by cords of negligible mass. What is the tension in (a) the bottom cord and (b) the top cord? Figure 5-66b shows a mobile consisting of three metal pieces. Two of the masses are $m_3 = 4.8$ kg and $m_5 = 5.5$ kg. The tension in the top cord is 199 N. What is the tension in (c) the lowest cord and (d) the middle cord?

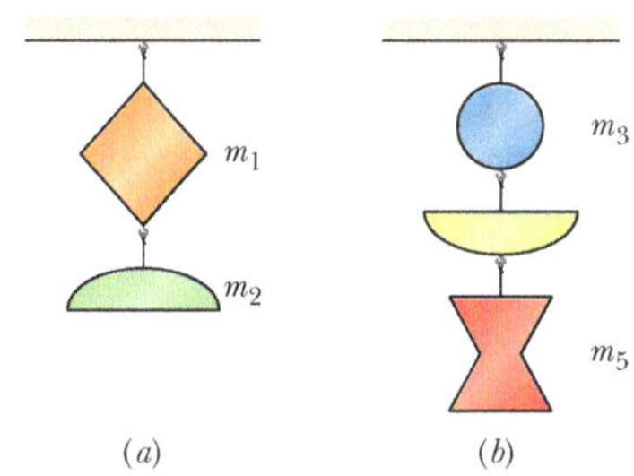

Figure 5-66 Problem 93.

94 For sport, a 12 kg armadillo runs onto a large pond of level, frictionless ice. The armadillo's initial velocity is 5.0 m/s along the positive direction of an x axis. Take its initial position on the ice as being the origin. It slips over the ice while being pushed by a wind with a force of 17 N in the positive direction of the y axis. In unit-vector notation, what are the animal's (a) velocity and (b) position vector when it has slid for 3.0 s?

95 Suppose that in Fig. 5-12, the masses of the blocks are 2.0 kg and 4.0 kg. (a) Which mass should the hanging block have if the magnitude of the acceleration is to be as large as possible? What then are (b) the magnitude of the acceleration and (c) the tension in the cord?

96 A nucleus that captures a stray neutron must bring the neutron to a stop within the diameter of the nucleus by means of the *strong force*. That force, which "glues" the nucleus together, is approximately zero outside the nucleus. Suppose that a stray neutron with an initial speed of 1.4×10^7 m/s is just barely captured by a nucleus with diameter $d = 1.0 \times 10^{-14}$ m. Assuming the strong force on the neutron is constant, find the magnitude of that force. The neutron's mass is 1.67×10^{-27} kg.

97 If the 1 kg standard body is accelerated by only $\vec{F}_1 = (3.0\ N)\hat{i} + (4.0\ N)\hat{j}$ and $\vec{F}_2 = (-2.0\ N)\hat{i} + (-6.0\ N)\hat{j}$, then what is $\vec{F}_{net}$ (a) in unit-vector notation and as (b) a magnitude and (c) an angle relative to the positive x direction? What are the (d) magnitude and (e) angle of $\vec{a}$?

CHAPTER 6

Force and Motion—II

6-1 FRICTION

Learning Objectives

After reading this module, you should be able to . . .

6.01 Distinguish between friction in a static situation and a kinetic situation.

6.02 Determine direction and magnitude of a frictional force.

6.03 For objects on horizontal, vertical, or inclined planes in situations involving friction, draw free-body diagrams and apply Newton's second law.

Key Ideas

- When a force $\vec{F}$ tends to slide a body along a surface, a frictional force from the surface acts on the body. The frictional force is parallel to the surface and directed so as to oppose the sliding. It is due to bonding between the body and the surface.

  If the body does not slide, the frictional force is a static frictional force $\vec{f}_s$. If there is sliding, the frictional force is a kinetic frictional force $\vec{f}_k$.

- If a body does not move, the static frictional force $\vec{f}_s$ and the component of $\vec{F}$ parallel to the surface are equal in magnitude, and $\vec{f}_s$ is directed opposite that component. If the component increases, f_s also increases.

- The magnitude of $\vec{f}_s$ has a maximum value $\vec{f}_{s,\max}$ given by

$$f_{s,\max} = \mu_s F_N,$$

where μ_s is the coefficient of static friction and F_N is the magnitude of the normal force. If the component of $\vec{F}$ parallel to the surface exceeds $f_{s,\max}$, the body slides on the surface.

- If the body begins to slide on the surface, the magnitude of the frictional force rapidly decreases to a constant value f_k given by

$$f_k = \mu_k F_N,$$

where μ_k is the coefficient of kinetic friction.

What Is Physics?

In this chapter we focus on the physics of three common types of force: frictional force, drag force, and centripetal force. An engineer preparing a car for the Indianapolis 500 must consider all three types. Frictional forces acting on the tires are crucial to the car's acceleration out of the pit and out of a curve (if the car hits an oil slick, the friction is lost and so is the car). Drag forces acting on the car from the passing air must be minimized or else the car will consume too much fuel and have to pit too early (even one 14 s pit stop can cost a driver the race). Centripetal forces are crucial in the turns (if there is insufficient centripetal force, the car slides into the wall). We start our discussion with frictional forces.

Friction

Frictional forces are unavoidable in our daily lives. If we were not able to counteract them, they would stop every moving object and bring to a halt every rotating shaft. About 20% of the gasoline used in an automobile is needed to counteract friction in the engine and in the drive train. On the other hand, if friction were totally absent, we could not get an automobile to go anywhere, and we could not walk or ride a bicycle. We could not hold a pencil, and, if we could, it would not write. Nails and screws would be useless, woven cloth would fall apart, and knots would untie.

Three Experiments. Here we deal with the frictional forces that exist between dry solid surfaces, either stationary relative to each other or moving across each other at slow speeds. Consider three simple thought experiments:

1. Send a book sliding across a long horizontal counter. As expected, the book slows and then stops. This means the book must have an acceleration parallel to the counter surface, in the direction opposite the book's velocity. From Newton's second law, then, a force must act on the book parallel to the counter surface, in the direction opposite its velocity. That force is a frictional force.
2. Push horizontally on the book to make it travel at constant velocity along the counter. Can the force from you be the only horizontal force on the book? No, because then the book would accelerate. From Newton's second law, there must be a second force, directed opposite your force but with the same magnitude, so that the two forces balance. That second force is a frictional force, directed parallel to the counter.
3. Push horizontally on a heavy crate. The crate does not move. From Newton's second law, a second force must also be acting on the crate to counteract your force. Moreover, this second force must be directed opposite your force and have the same magnitude as your force, so that the two forces balance. That second force is a frictional force. Push even harder. The crate still does not move. Apparently the frictional force can change in magnitude so that the two forces still balance. Now push with all your strength. The crate begins to slide. Evidently, there is a maximum magnitude of the frictional force. When you exceed that maximum magnitude, the crate slides.

Two Types of Friction. Figure 6-1 shows a similar situation. In Fig. 6-1*a*, a block rests on a tabletop, with the gravitational force $\vec{F}_g$ balanced by a normal force $\vec{F}_N$. In Fig. 6-1*b*, you exert a force $\vec{F}$ on the block, attempting to pull it to the left. In response, a frictional force $\vec{f}_s$ is directed to the right, exactly balancing your force. The force $\vec{f}_s$ is called the **static frictional force.** The block does not move.

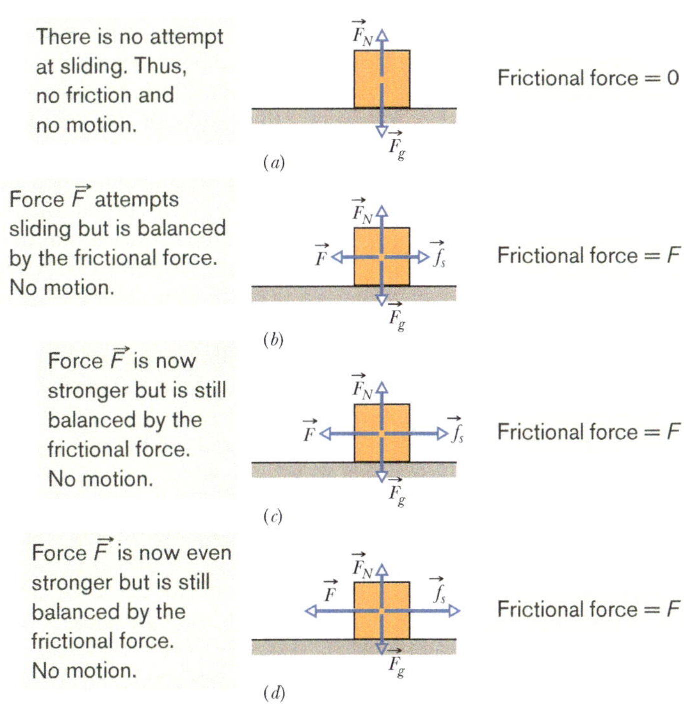

Figure 6-1 (*a*) The forces on a stationary block. (*b*–*d*) An external force $\vec{F}$, applied to the block, is balanced by a static frictional force $\vec{f}_s$. As F is increased, f_s also increases, until f_s reaches a certain maximum value. (*Figure continues*)

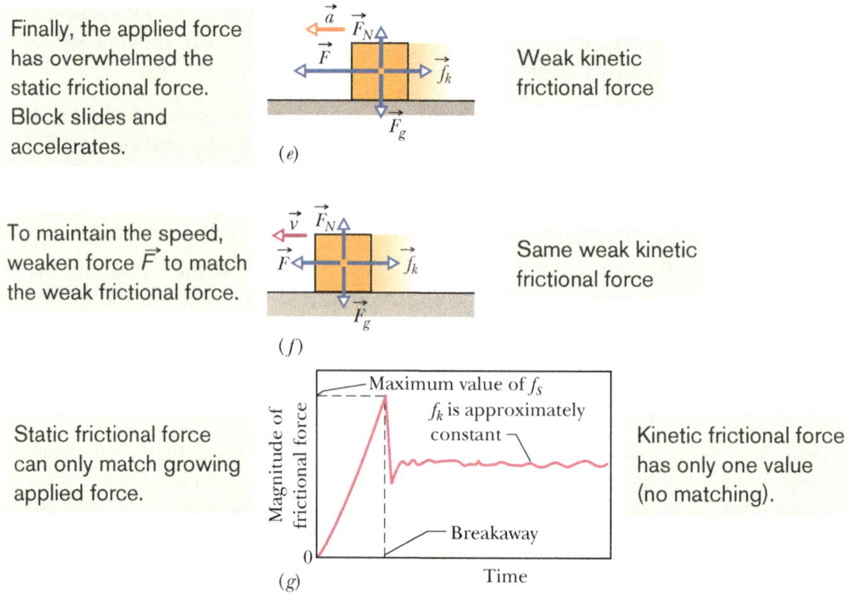

Figure 6-1 (*Continued*) (*e*) Once f_s reaches its maximum value, the block "breaks away," accelerating suddenly in the direction of $\vec{F}$. (*f*) If the block is now to move with constant velocity, F must be reduced from the maximum value it had just before the block broke away. (*g*) Some experimental results for the sequence (*a*) through (*f*). **In *WileyPLUS*, this figure is available as an animation with voiceover.**

Figures 6-1*c* and 6-1*d* show that as you increase the magnitude of your applied force, the magnitude of the static frictional force $\vec{f}_s$ also increases and the block remains at rest. When the applied force reaches a certain magnitude, however, the block "breaks away" from its intimate contact with the tabletop and accelerates leftward (Fig. 6-1*e*). The frictional force that then opposes the motion is called the **kinetic frictional force** $\vec{f}_k$.

Usually, the magnitude of the kinetic frictional force, which acts when there is motion, is less than the maximum magnitude of the static frictional force, which acts when there is no motion. Thus, if you wish the block to move across the surface with a constant speed, you must usually decrease the magnitude of the applied force once the block begins to move, as in Fig. 6-1*f*. As an example, Fig. 6-1*g* shows the results of an experiment in which the force on a block was slowly increased until breakaway occurred. Note the reduced force needed to keep the block moving at constant speed after breakaway.

Microscopic View. A frictional force is, in essence, the vector sum of many forces acting between the surface atoms of one body and those of another body. If two highly polished and carefully cleaned metal surfaces are brought together in a very good vacuum (to keep them clean), they cannot be made to slide over each other. Because the surfaces are so smooth, many atoms of one surface contact many atoms of the other surface, and the surfaces *cold-weld* together instantly, forming a single piece of metal. If a machinist's specially polished gage blocks are brought together in air, there is less atom-to-atom contact, but the blocks stick firmly to each other and can be separated only by means of a wrenching motion. Usually, however, this much atom-to-atom contact is not possible. Even a highly polished metal surface is far from being flat on the atomic scale. Moreover, the surfaces of everyday objects have layers of oxides and other contaminants that reduce cold-welding.

When two ordinary surfaces are placed together, only the high points touch each other. (It is like having the Alps of Switzerland turned over and placed down on the Alps of Austria.) The actual *microscopic* area of contact is much less than the apparent *macroscopic* contact area, perhaps by a factor of 10^4. Nonetheless, many contact points do cold-weld together. These welds produce static friction when an applied force attempts to slide the surfaces relative to each other.

If the applied force is great enough to pull one surface across the other, there is first a tearing of welds (at breakaway) and then a continuous re-forming and tearing of welds as movement occurs and chance contacts are made (Fig. 6-2). The kinetic frictional force $\vec{f}_k$ that opposes the motion is the vector sum of the forces at those many chance contacts.

If the two surfaces are pressed together harder, many more points cold-weld. Now getting the surfaces to slide relative to each other requires a greater applied force: The static frictional force $\vec{f}_s$ has a greater maximum value. Once the surfaces are sliding, there are many more points of momentary cold-welding, so the kinetic frictional force $\vec{f}_k$ also has a greater magnitude.

Often, the sliding motion of one surface over another is "jerky" because the two surfaces alternately stick together and then slip. Such repetitive *stick-and-slip* can produce squeaking or squealing, as when tires skid on dry pavement, fingernails scratch along a chalkboard, or a rusty hinge is opened. It can also produce beautiful and captivating sounds, as in music when a bow is drawn properly across a violin string.

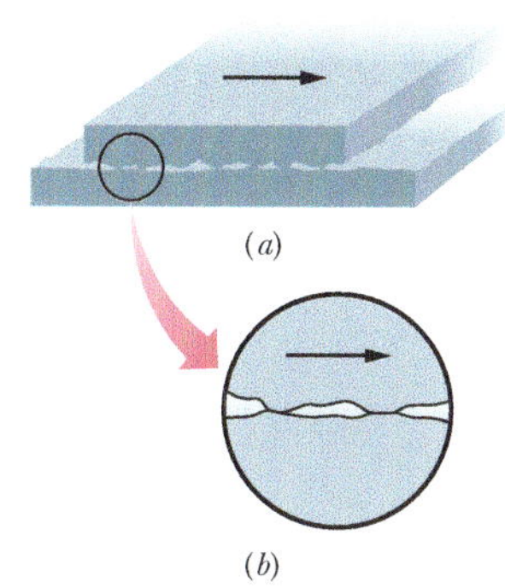

Figure 6-2 The mechanism of sliding friction. (*a*) The upper surface is sliding to the right over the lower surface in this enlarged view. (*b*) A detail, showing two spots where cold-welding has occurred. Force is required to break the welds and maintain the motion.

Properties of Friction

Experiment shows that when a dry and unlubricated body presses against a surface in the same condition and a force $\vec{F}$ attempts to slide the body along the surface, the resulting frictional force has three properties:

Property 1. If the body does not move, then the static frictional force $\vec{f}_s$ and the component of $\vec{F}$ that is parallel to the surface balance each other. They are equal in magnitude, and $\vec{f}_s$ is directed opposite that component of $\vec{F}$.

Property 2. The magnitude of $\vec{f}_s$ has a maximum value $f_{s,\max}$ that is given by

$$f_{s,\max} = \mu_s F_N, \tag{6-1}$$

where μ_s is the **coefficient of static friction** and F_N is the magnitude of the normal force on the body from the surface. If the magnitude of the component of $\vec{F}$ that is parallel to the surface exceeds $f_{s,\max}$, then the body begins to slide along the surface.

Property 3. If the body begins to slide along the surface, the magnitude of the frictional force rapidly decreases to a value f_k given by

$$f_k = \mu_k F_N, \tag{6-2}$$

where μ_k is the **coefficient of kinetic friction.** Thereafter, during the sliding, a kinetic frictional force $\vec{f}_k$ with magnitude given by Eq. 6-2 opposes the motion.

The magnitude F_N of the normal force appears in properties 2 and 3 as a measure of how firmly the body presses against the surface. If the body presses harder, then, by Newton's third law, F_N is greater. Properties 1 and 2 are worded in terms of a single applied force $\vec{F}$, but they also hold for the net force of several applied forces acting on the body. Equations 6-1 and 6-2 are *not* vector equations; the direction of $\vec{f}_s$ or $\vec{f}_k$ is always parallel to the surface and opposed to the attempted sliding, and the normal force $\vec{F}_N$ is perpendicular to the surface.

The coefficients μ_s and μ_k are dimensionless and must be determined experimentally. Their values depend on certain properties of both the body and the surface; hence, they are usually referred to with the preposition "between," as in "the value of μ_s *between* an egg and a Teflon-coated skillet is 0.04, but that *between* rock-climbing shoes and rock is as much as 1.2." We assume that the value of μ_k does not depend on the speed at which the body slides along the surface.

Checkpoint 1

A block lies on a floor. (a) What is the magnitude of the frictional force on it from the floor? (b) If a horizontal force of 5 N is now applied to the block, but the block does not move, what is the magnitude of the frictional force on it? (c) If the maximum value $f_{s,\max}$ of the static frictional force on the block is 10 N, will the block move if the magnitude of the horizontally applied force is 8 N? (d) If it is 12 N? (e) What is the magnitude of the frictional force in part (c)?

Sample Problem 6.01 Angled force applied to an initially stationary block

This sample problem involves a tilted applied force, which requires that we work with components to find a frictional force. The main challenge is to sort out all the components. Figure 6-3*a* shows a force of magnitude $F =$ 12.0 N applied to an 8.00 kg block at a downward angle of $\theta = 30.0°$. The coefficient of static friction between block and floor is $\mu_s = 0.700$; the coefficient of kinetic friction is $\mu_k = 0.400$. Does the block begin to slide or does it remain stationary? What is the magnitude of the frictional force on the block?

KEY IDEAS

(1) When the object is stationary on a surface, the static frictional force balances the force component that is attempting to slide the object along the surface. (2) The maximum possible magnitude of that force is given by Eq. 6-1 ($f_{s,\max} = \mu_s F_N$). (3) If the component of the applied force along the surface exceeds this limit on the static friction, the block begins to slide. (4) If the object slides, the kinetic frictional force is given by Eq. 6-2 ($f_k = \mu_k F_N$).

Calculations: To see if the block slides (and thus to calculate the magnitude of the frictional force), we must compare the applied force component F_x with the maximum magnitude $f_{s,\max}$ that the static friction can have. From the triangle of components and full force shown in Fig. 6-3*b*, we see that

$$\begin{aligned} F_x &= F\cos\theta \\ &= (12.0\text{ N})\cos 30° = 10.39\text{ N}. \end{aligned} \tag{6-3}$$

From Eq. 6-1, we know that $f_{s,\max} = \mu_s F_N$, but we need the magnitude F_N of the normal force to evaluate $f_{s,\max}$. Because the normal force is vertical, we need to write Newton's second law ($F_{\text{net},y} = ma_y$) for the vertical force components acting on the block, as displayed in Fig. 6-3*c*. The gravitational force with magnitude mg acts downward. The applied force has a downward component $F_y = F\sin\theta$. And the vertical acceleration a_y is just zero. Thus, we can write Newton's second law as

$$F_N - mg - F\sin\theta = m(0), \tag{6-4}$$

which gives us

$$F_N = mg + F\sin\theta. \tag{6-5}$$

Now we can evaluate $f_{s,\max} = \mu_s F_N$:

$$\begin{aligned} f_{s,\max} &= \mu_s(mg + F\sin\theta) \\ &= (0.700)((8.00\text{ kg})(9.8\text{ m/s}^2) + (12.0\text{ N})(\sin 30°)) \\ &= 59.08\text{ N}. \end{aligned} \tag{6-6}$$

Because the magnitude F_x (= 10.39 N) of the force component attempting to slide the block is less than $f_{s,\max}$ (= 59.08 N), the block remains stationary. That means that the magnitude f_s of the frictional force *matches* F_x. From Fig. 6-3*d*, we can write Newton's second law for *x* components as

$$F_x - f_s = m(0), \tag{6-7}$$

and thus $\quad f_s = F_x = 10.39\text{ N} \approx 10.4\text{ N}.$ (Answer)

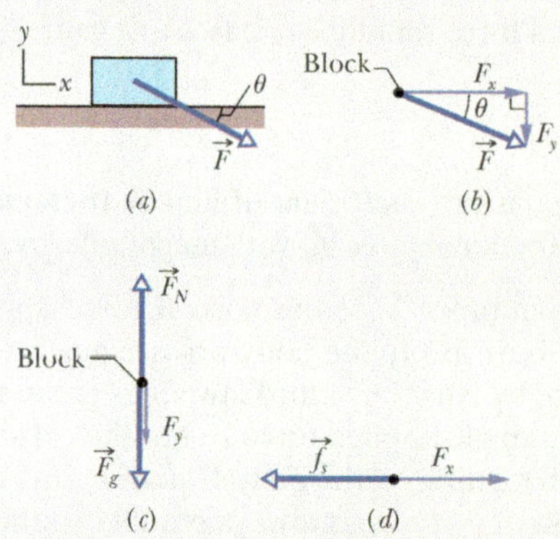

Figure 6-3 (*a*) A force is applied to an initially stationary block. (*b*) The components of the applied force. (*c*) The vertical force components. (*d*) The horizontal force components.

WileyPLUS Additional examples, video, and practice available at *WileyPLUS*

Sample Problem 6.02 Sliding to a stop on icy roads, horizontal and inclined

Some of the funniest videos on the web involve motorists sliding uncontrollably on icy roads. Here let's compare the typical stopping distances for a car sliding to a stop from an initial speed of 10.0 m/s on a dry horizontal road, an icy horizontal road, and (everyone's favorite) an icy hill.

(a) How far does the car take to slide to a stop on a horizontal road (Fig. 6-4*a*) if the coefficient of kinetic friction is $\mu_k = 0.60$, which is typical of regular tires on dry pavement? Let's neglect any effect of the air on the car, assume that the wheels lock up and the tires slide, and extend an x axis in the car's direction of motion.

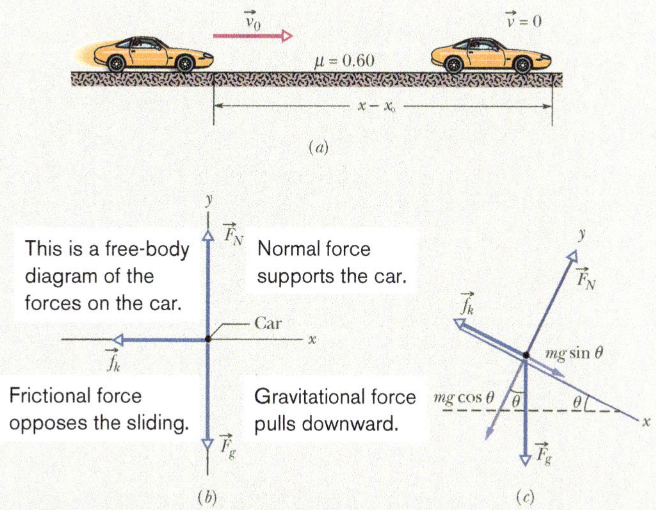

Figure 6-4 (*a*) A car sliding to the right and finally stopping. A free-body diagram for the car on (*b*) the same horizontal road and (*c*) a hill.

KEY IDEAS

(1) The car accelerates (its speed decreases) because a horizontal frictional force acts against the motion, in the negative direction of the x axis. (2) The frictional force is a kinetic frictional force with a magnitude given by Eq. 6-2 ($f_k = \mu_k F_N$), in which F_N is the magnitude of the normal force on the car from the road. (3) We can relate the frictional force to the resulting acceleration by writing Newton's second law ($F_{net,x} = ma_x$) for motion along the road.

Calculations: Figure 6-4*b* shows the free-body diagram for the car. The normal force is upward, the gravitational force is downward, and the frictional force is horizontal. Because the frictional force is the only force with an x component, Newton's second law written for motion along the x axis becomes

$$-f_k = ma_x. \tag{6-8}$$

Substituting $f_k = \mu_k F_N$ gives us

$$-\mu_k F_N = ma_x. \tag{6-9}$$

From Fig. 6-4*b* we see that the upward normal force balances the downward gravitational force, so in Eq. 6-9 let's replace magnitude F_N with magnitude mg. Then we can cancel m (the stopping distance is thus independent of the car's mass—the car can be heavy or light, it does not matter). Solving for a_x we find

$$a_x = -\mu_k g. \tag{6-10}$$

Because this acceleration is constant, we can use the constant-acceleration equations of Table 2-1. The easiest choice for finding the sliding distance $x - x_0$ is Eq. 2-16 ($v^2 = v_0^2 + 2a(x - x_0)$), which gives us

$$x - x_0 = \frac{v^2 - v_0^2}{2a_x}. \tag{6-11}$$

Substituting from Eq. 6-10, we then have

$$x - x_0 = \frac{v^2 - v_0^2}{-2\mu_k g}. \tag{6-12}$$

Inserting the initial speed $v_0 = 10.0$ m/s, the final speed $v = 0$, and the coefficient of kinetic friction $\mu_k = 0.60$, we find that the car's stopping distance is

$$x - x_0 = 8.50 \text{ m} \approx 8.5 \text{ m}. \qquad \text{(Answer)}$$

(b) What is the stopping distance if the road is covered with ice with $\mu_k = 0.10$?

Calculation: Our solution is perfectly fine through Eq. 6-12 but now we substitute this new μ_k, finding

$$x - x_0 = 51 \text{ m}. \qquad \text{(Answer)}$$

Thus, a much longer clear path would be needed to avoid the car hitting something along the way.

(c) Now let's have the car sliding down an icy hill with an inclination of $\theta = 5.00°$ (a mild incline, nothing like the hills of San Francisco). The free-body diagram shown in Fig. 6-4*c* is like the ramp in Sample Problem 5.04 except, to be consistent with Fig. 6-4*b*, the positive direction of the x axis is *down* the ramp. What now is the stopping distance?

Calculations: Switching from Fig. 6-4*b* to *c* involves two major changes. (1) Now a component of the gravitational force is along the tilted x axis, pulling the car down the hill. From Sample Problem 5.04 and Fig. 5-15, that down-the-hill component is $mg \sin\theta$, which is in the positive direction of the x axis in Fig. 6-4*c*. (2) The normal force (still perpendicular to the road) now balances only a component of the gravitational force, not the full force. From Sample Problem 5.04 (see Fig. 5-15i), we write that balance as

$$F_N = mg\cos\theta.$$

In spite of these changes, we still want to write Newton's second law ($F_{\text{net},x} = ma_x$) for the motion along the (now tilted) x axis. We have

$$-f_k + mg\sin\theta = ma_x,$$
$$-\mu_k F_N + mg\sin\theta = ma_x,$$

and $$-\mu_k mg\cos\theta + mg\sin\theta = ma_x.$$

Solving for the acceleration and substituting the given data now give us

$$\begin{aligned} a_x &= -\mu_k g\cos\theta + g\sin\theta \\ &= -(0.10)(9.8\ \text{m/s}^2)\cos 5.00° + (9.8\ \text{m/s}^2)\sin 5.00° \\ &= -0.122\ \text{m/s}^2. \end{aligned} \tag{6-13}$$

Substituting this result into Eq. 6-11 gives us the stopping distance down the hill:

$$x - x_0 = 409\ \text{m} \approx 400\ \text{m}, \quad \text{(Answer)}$$

which is about $\frac{1}{4}$ mi! Such icy hills separate people who can do this calculation (and thus know to stay home) from people who cannot (and thus end up in web videos).

WILEY PLUS Additional examples, video, and practice available at *WileyPLUS*

6-2 THE DRAG FORCE AND TERMINAL SPEED

Learning Objectives

After reading this module, you should be able to . . .

6.04 Apply the relationship between the drag force on an object moving through air and the speed of the object.

6.05 Determine the terminal speed of an object falling through air.

Key Ideas

- When there is relative motion between air (or some other fluid) and a body, the body experiences a drag force $\vec{D}$ that opposes the relative motion and points in the direction in which the fluid flows relative to the body. The magnitude of $\vec{D}$ is related to the relative speed v by an experimentally determined drag coefficient C according to

$$D = \tfrac{1}{2}C\rho Av^2,$$

where ρ is the fluid density (mass per unit volume) and A is the effective cross-sectional area of the body (the area of a cross section taken perpendicular to the relative velocity $\vec{v}$).

- When a blunt object has fallen far enough through air, the magnitudes of the drag force $\vec{D}$ and the gravitational force $\vec{F}_g$ on the body become equal. The body then falls at a constant terminal speed v_t given by

$$v_t = \sqrt{\frac{2F_g}{C\rho A}}.$$

The Drag Force and Terminal Speed

A **fluid** is anything that can flow—generally either a gas or a liquid. When there is a relative velocity between a fluid and a body (either because the body moves through the fluid or because the fluid moves past the body), the body experiences a **drag force** $\vec{D}$ that opposes the relative motion and points in the direction in which the fluid flows relative to the body.

Here we examine only cases in which air is the fluid, the body is blunt (like a baseball) rather than slender (like a javelin), and the relative motion is fast enough so that the air becomes turbulent (breaks up into swirls) behind the body. In such cases, the magnitude of the drag force $\vec{D}$ is related to the relative speed v by an experimentally determined **drag coefficient** C according to

$$D = \tfrac{1}{2}C\rho Av^2, \tag{6-14}$$

Table 6-1 Some Terminal Speeds in Air

Object	Terminal Speed (m/s)	95% Distance[a] (m)
Shot (from shot put)	145	2500
Sky diver (typical)	60	430
Baseball	42	210
Tennis ball	31	115
Basketball	20	47
Ping-Pong ball	9	10
Raindrop (radius = 1.5 mm)	7	6
Parachutist (typical)	5	3

[a]This is the distance through which the body must fall from rest to reach 95% of its terminal speed.
Based on Peter J. Brancazio, *Sport Science,* 1984, Simon & Schuster, New York.

Karl-Josef Hildenbrand/dpa/Landov LLC

Figure 6-5 This skier crouches in an "egg position" so as to minimize her effective cross-sectional area and thus minimize the air drag acting on her.

where ρ is the air density (mass per volume) and A is the **effective cross-sectional area** of the body (the area of a cross section taken perpendicular to the velocity $\vec{v}$). The drag coefficient C (typical values range from 0.4 to 1.0) is not truly a constant for a given body because if v varies significantly, the value of C can vary as well. Here, we ignore such complications.

Downhill speed skiers know well that drag depends on A and v^2. To reach high speeds a skier must reduce D as much as possible by, for example, riding the skis in the "egg position" (Fig. 6-5) to minimize A.

Falling. When a blunt body falls from rest through air, the drag force $\vec{D}$ is directed upward; its magnitude gradually increases from zero as the speed of the body increases. This upward force $\vec{D}$ opposes the downward gravitational force $\vec{F}_g$ on the body. We can relate these forces to the body's acceleration by writing Newton's second law for a vertical y axis ($F_{\text{net},y} = ma_y$) as

$$D - F_g = ma, \tag{6-15}$$

where m is the mass of the body. As suggested in Fig. 6-6, if the body falls long enough, D eventually equals F_g. From Eq. 6-15, this means that $a = 0$, and so the body's speed no longer increases. The body then falls at a constant speed, called the **terminal speed** v_t.

To find v_t, we set $a = 0$ in Eq. 6-15 and substitute for D from Eq. 6-14, obtaining

$$\tfrac{1}{2}C\rho A v_t^2 - F_g = 0,$$

which gives

$$v_t = \sqrt{\frac{2F_g}{C\rho A}}. \tag{6-16}$$

Table 6-1 gives values of v_t for some common objects.

According to calculations* based on Eq. 6-14, a cat must fall about six floors to reach terminal speed. Until it does so, $F_g > D$ and the cat accelerates downward because of the net downward force. Recall from Chapter 2 that your body is an accelerometer, not a speedometer. Because the cat also senses the acceleration, it is frightened and keeps its feet underneath its body, its head tucked in, and its spine bent upward, making A small, v_t large, and injury likely.

However, if the cat does reach v_t during a longer fall, the acceleration vanishes and the cat relaxes somewhat, stretching its legs and neck horizontally outward and

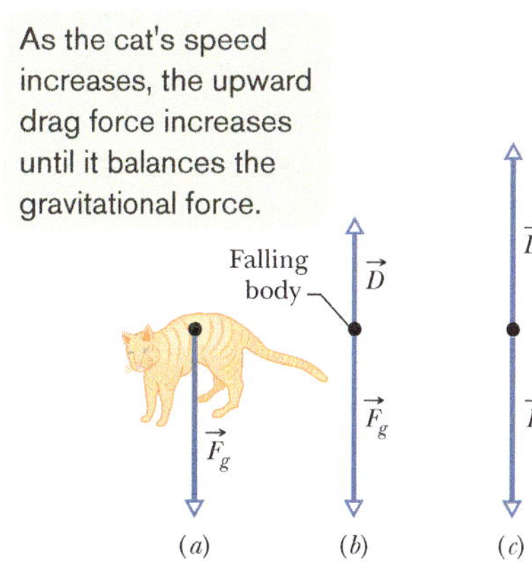

Figure 6-6 The forces that act on a body falling through air: (*a*) the body when it has just begun to fall and (*b*) the free-body diagram a little later, after a drag force has developed. (*c*) The drag force has increased until it balances the gravitational force on the body. The body now falls at its constant terminal speed.

*W. O. Whitney and C. J. Mehlhaff, "High-Rise Syndrome in Cats." *The Journal of the American Veterinary Medical Association,* 1987.

Steve Fitchett/Taxi/Getty Images

Figure 6-7 Sky divers in a horizontal "spread eagle" maximize air drag.

straightening its spine (it then resembles a flying squirrel). These actions increase area A and thus also, by Eq. 6-14, the drag D. The cat begins to slow because now $D > F_g$ (the net force is upward), until a new, smaller v_t is reached. The decrease in v_t reduces the possibility of serious injury on landing. Just before the end of the fall, when it sees it is nearing the ground, the cat pulls its legs back beneath its body to prepare for the landing.

Humans often fall from great heights for the fun of skydiving. However, in April 1987, during a jump, sky diver Gregory Robertson noticed that fellow sky diver Debbie Williams had been knocked unconscious in a collision with a third sky diver and was unable to open her parachute. Robertson, who was well above Williams at the time and who had not yet opened his parachute for the 4 km plunge, reoriented his body head-down so as to minimize A and maximize his downward speed. Reaching an estimated v_t of 320 km/h, he caught up with Williams and then went into a horizontal "spread eagle" (as in Fig. 6-7) to increase D so that he could grab her. He opened her parachute and then, after releasing her, his own, a scant 10 s before impact. Williams received extensive internal injuries due to her lack of control on landing but survived.

Sample Problem 6.03 Terminal speed of falling raindrop

A raindrop with radius $R = 1.5$ mm falls from a cloud that is at height $h = 1200$ m above the ground. The drag coefficient C for the drop is 0.60. Assume that the drop is spherical throughout its fall. The density of water ρ_w is 1000 kg/m^3, and the density of air ρ_a is 1.2 kg/m^3.

(a) As Table 6-1 indicates, the raindrop reaches terminal speed after falling just a few meters. What is the terminal speed?

KEY IDEA

The drop reaches a terminal speed v_t when the gravitational force on it is balanced by the air drag force on it, so its acceleration is zero. We could then apply Newton's second law and the drag force equation to find v_t, but Eq. 6-16 does all that for us.

Calculations: To use Eq. 6-16, we need the drop's effective cross-sectional area A and the magnitude F_g of the gravitational force. Because the drop is spherical, A is the area of a circle (πR^2) that has the same radius as the sphere. To find F_g, we use three facts: (1) $F_g = mg$, where m is the drop's mass; (2) the (spherical) drop's volume is $V = \frac{4}{3}\pi R^3$; and (3) the density of the water in the drop is the mass per volume, or $\rho_w = m/V$. Thus, we find

$$F_g = V\rho_w g = \tfrac{4}{3}\pi R^3 \rho_w g.$$

We next substitute this, the expression for A, and the given data into Eq. 6-16. Being careful to distinguish between the air density ρ_a and the water density ρ_w, we obtain

$$v_t = \sqrt{\frac{2F_g}{C\rho_a A}} = \sqrt{\frac{8\pi R^3 \rho_w g}{3C\rho_a \pi R^2}} = \sqrt{\frac{8R\rho_w g}{3C\rho_a}}$$

$$= \sqrt{\frac{(8)(1.5 \times 10^{-3}\text{ m})(1000\text{ kg/m}^3)(9.8\text{ m/s}^2)}{(3)(0.60)(1.2\text{ kg/m}^3)}}$$

$$= 7.4\text{ m/s} \approx 27\text{ km/h.} \qquad \text{(Answer)}$$

Note that the height of the cloud does not enter into the calculation.

(b) What would be the drop's speed just before impact if there were no drag force?

KEY IDEA

With no drag force to reduce the drop's speed during the fall, the drop would fall with the constant free-fall acceleration g, so the constant-acceleration equations of Table 2-1 apply.

Calculation: Because we know the acceleration is g, the initial velocity v_0 is 0, and the displacement $x - x_0$ is $-h$, we use Eq. 2-16 to find v:

$$v = \sqrt{2gh} = \sqrt{(2)(9.8\text{ m/s}^2)(1200\text{ m})}$$

$$= 153\text{ m/s} \approx 550\text{ km/h.} \qquad \text{(Answer)}$$

Had he known this, Shakespeare would scarcely have written, "it droppeth as the gentle rain from heaven, upon the place beneath." In fact, the speed is close to that of a bullet from a large-caliber handgun!

WILEY PLUS Additional examples, video, and practice available at *WileyPLUS*

6-3 UNIFORM CIRCULAR MOTION

Learning Objectives

After reading this module, you should be able to. . .

6.06 Sketch the path taken in uniform circular motion and explain the velocity, acceleration, and force vectors (magnitudes and directions) during the motion.

6.07 Identify that unless there is a radially inward net force (a centripetal force), an object cannot move in circular motion.

6.08 For a particle in uniform circular motion, apply the relationship between the radius of the path, the particle's speed and mass, and the net force acting on the particle.

Key Ideas

- If a particle moves in a circle or a circular arc of radius R at constant speed v, the particle is said to be in uniform circular motion. It then has a centripetal acceleration $\vec{a}$ with magnitude given by

$$a = \frac{v^2}{R}.$$

- This acceleration is due to a net centripetal force on the particle, with magnitude given by

$$F = \frac{mv^2}{R},$$

where m is the particle's mass. The vector quantities $\vec{a}$ and $\vec{F}$ are directed toward the center of curvature of the particle's path.

Uniform Circular Motion

From Module 4-5, recall that when a body moves in a circle (or a circular arc) at constant speed v, it is said to be in uniform circular motion. Also recall that the body has a centripetal acceleration (directed toward the center of the circle) of constant magnitude given by

$$a = \frac{v^2}{R} \quad \text{(centripetal acceleration)}, \qquad (6\text{-}17)$$

where R is the radius of the circle. Here are two examples:

1. *Rounding a curve in a car.* You are sitting in the center of the rear seat of a car moving at a constant high speed along a flat road. When the driver suddenly turns left, rounding a corner in a circular arc, you slide across the seat toward the right and then jam against the car wall for the rest of the turn. What is going on?

   While the car moves in the circular arc, it is in uniform circular motion; that is, it has an acceleration that is directed toward the center of the circle. By Newton's second law, a force must cause this acceleration. Moreover, the force must also be directed toward the center of the circle. Thus, it is a **centripetal force,** where the adjective indicates the direction. In this example, the centripetal force is a frictional force on the tires from the road; it makes the turn possible.

   If you are to move in uniform circular motion along with the car, there must also be a centripetal force on you. However, apparently the frictional force on you from the seat was not great enough to make you go in a circle with the car. Thus, the seat slid beneath you, until the right wall of the car jammed into you. Then its push on you provided the needed centripetal force on you, and you joined the car's uniform circular motion.

2. *Orbiting Earth.* This time you are a passenger in the space shuttle *Atlantis.* As it and you orbit Earth, you float through your cabin. What is going on?

   Both you and the shuttle are in uniform circular motion and have accelerations directed toward the center of the circle. Again by Newton's second law, centripetal forces must cause these accelerations. This time the centripetal forces are gravitational pulls (the pull on you and the pull on the shuttle) exerted by Earth and directed radially inward, toward the center of Earth.

In both car and shuttle you are in uniform circular motion, acted on by a centripetal force—yet your sensations in the two situations are quite different. In the car, jammed up against the wall, you are aware of being compressed by the wall. In the orbiting shuttle, however, you are floating around with no sensation of any force acting on you. Why this difference?

The difference is due to the nature of the two centripetal forces. In the car, the centripetal force is the push on the part of your body touching the car wall. You can sense the compression on that part of your body. In the shuttle, the centripetal force is Earth's gravitational pull on every atom of your body. Thus, there is no compression (or pull) on any one part of your body and no sensation of a force acting on you. (The sensation is said to be one of "weightlessness," but that description is tricky. The pull on you by Earth has certainly not disappeared and, in fact, is only a little less than it would be with you on the ground.)

Another example of a centripetal force is shown in Fig. 6-8. There a hockey puck moves around in a circle at constant speed v while tied to a string looped around a central peg. This time the centripetal force is the radially inward pull on the puck from the string. Without that force, the puck would slide off in a straight line instead of moving in a circle.

Note again that a centripetal force is not a new kind of force. The name merely indicates the direction of the force. It can, in fact, be a frictional force, a gravitational force, the force from a car wall or a string, or any other force. For any situation:

> A centripetal force accelerates a body by changing the direction of the body's velocity without changing the body's speed.

From Newton's second law and Eq. 6-17 ($a = v^2/R$), we can write the magnitude F of a centripetal force (or a net centripetal force) as

$$F = m\frac{v^2}{R} \quad \text{(magnitude of centripetal force).} \tag{6-18}$$

Because the speed v here is constant, the magnitudes of the acceleration and the force are also constant.

However, the directions of the centripetal acceleration and force are not constant; they vary continuously so as to always point toward the center of the circle. For this reason, the force and acceleration vectors are sometimes drawn along a radial axis r that moves with the body and always extends from the center of the circle to the body, as in Fig. 6-8. The positive direction of the axis is radially outward, but the acceleration and force vectors point radially inward.

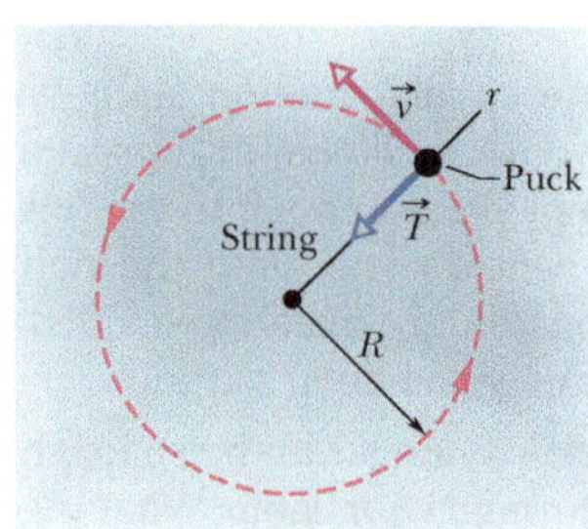

The puck moves in uniform circular motion only because of a toward-the-center force.

Figure 6-8 An overhead view of a hockey puck moving with constant speed v in a circular path of radius R on a horizontal frictionless surface. The centripetal force on the puck is $\vec{T}$, the pull from the string, directed inward along the radial axis r extending through the puck.

Checkpoint 2

As every amusement park fan knows, a Ferris wheel is a ride consisting of seats mounted on a tall ring that rotates around a horizontal axis. When you ride in a Ferris wheel at constant speed, what are the directions of your acceleration $\vec{a}$ and the normal force $\vec{F}_N$ on you (from the always upright seat) as you pass through (a) the highest point and (b) the lowest point of the ride? (c) How does the magnitude of the acceleration at the highest point compare with that at the lowest point? (d) How do the magnitudes of the normal force compare at those two points?

Sample Problem 6.04 Vertical circular loop, Diavolo

Largely because of riding in cars, you are used to horizontal circular motion. Vertical circular motion would be a novelty. In this sample problem, such motion seems to defy the gravitational force.

In a 1901 circus performance, Allo "Dare Devil" Diavolo introduced the stunt of riding a bicycle in a loop-the-loop (Fig. 6-9a). Assuming that the loop is a circle with radius $R = 2.7$ m, what is the least speed v that Diavolo and his bicycle could have at the top of the loop to remain in contact with it there?

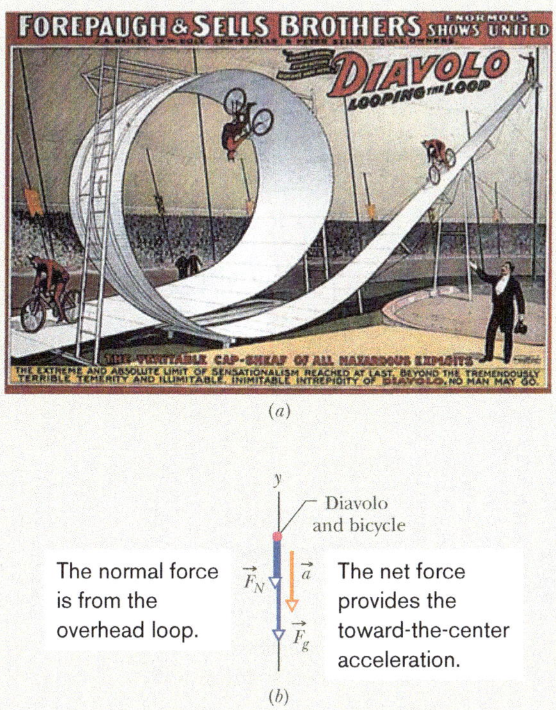

Figure 6-9 (a) Contemporary advertisement for Diavolo and (b) free-body diagram for the performer at the top of the loop.

KEY IDEA

We can assume that Diavolo and his bicycle travel through the top of the loop as a single particle in uniform circular motion. Thus, at the top, the acceleration $\vec{a}$ of this particle must have the magnitude $a = v^2/R$ given by Eq. 6-17 and be directed downward, toward the center of the circular loop.

Calculations: The forces on the particle when it is at the top of the loop are shown in the free-body diagram of Fig 6-9b. The gravitational force $\vec{F}_g$ is downward along a y axis; so is the normal force $\vec{F}_N$ on the particle from the loop (the loop can push down, not pull up); so also is the centripetal acceleration of the particle. Thus, Newton's second law for y components ($F_{net,y} = ma_y$) gives us

$$-F_N - F_g = m(-a)$$

and

$$-F_N - mg = m\left(-\frac{v^2}{R}\right). \tag{6-19}$$

If the particle has the *least speed* v needed to remain in contact, then it is on the *verge of losing contact* with the loop (falling away from the loop), which means that $F_N = 0$ at the top of the loop (the particle and loop touch but without any normal force). Substituting 0 for F_N in Eq. 6-19, solving for v, and then substituting known values give us

$$v = \sqrt{gR} = \sqrt{(9.8 \text{ m/s}^2)(2.7 \text{ m})}$$
$$= 5.1 \text{ m/s.} \quad \text{(Answer)}$$

Comments: Diavolo made certain that his speed at the top of the loop was greater than 5.1 m/s so that he did not lose contact with the loop and fall away from it. Note that this speed requirement is independent of the mass of Diavolo and his bicycle. Had he feasted on, say, pierogies before his performance, he still would have had to exceed only 5.1 m/s to maintain contact as he passed through the top of the loop.

WileyPLUS Additional examples, video, and practice available at *WileyPLUS*

Sample Problem 6.05 Car in flat circular turn

Upside-down racing: A modern race car is designed so that the passing air pushes down on it, allowing the car to travel much faster through a flat turn in a Grand Prix without friction failing. This downward push is called *negative lift.* Can a race car have so much negative lift that it could be driven upside down on a long ceiling, as done fictionally by a sedan in the first *Men in Black* movie?

Figure 6-10*a* represents a Grand Prix race car of mass $m = 600$ kg as it travels on a flat track in a circular arc of radius $R = 100$ m. Because of the shape of the car and the wings on it, the passing air exerts a negative lift $\vec{F}_L$ downward on the car. The coefficient of static friction between the tires and the track is 0.75. (Assume that the forces on the four tires are identical.)

(a) If the car is on the verge of sliding out of the turn when its speed is 28.6 m/s, what is the magnitude of the negative lift $\vec{F}_L$ acting downward on the car?

KEY IDEAS

1. A centripetal force must act on the car because the car is moving around a circular arc; that force must be directed toward the center of curvature of the arc (here, that is horizontally).
2. The only horizontal force acting on the car is a frictional force on the tires from the road. So the required centripetal force is a frictional force.
3. Because the car is not sliding, the frictional force must be a *static* frictional force $\vec{f}_s$ (Fig. 6-10*a*).
4. Because the car is on the verge of sliding, the magnitude f_s is equal to the maximum value $f_{s,\max} = \mu_s F_N$, where F_N is the magnitude of the normal force $\vec{F}_N$ acting on the car from the track.

Radial calculations: The frictional force $\vec{f}_s$ is shown in the free-body diagram of Fig. 6-10*b*. It is in the negative direction of a radial axis r that always extends from the center of curvature through the car as the car moves. The force produces a centripetal acceleration of magnitude v^2/R. We can relate the force and acceleration by writing Newton's second law for components along the r axis ($F_{\text{net},r} = ma_r$) as

$$-f_s = m\left(-\frac{v^2}{R}\right). \tag{6-20}$$

Substituting $f_{s,\max} = \mu_s F_N$ for f_s leads us to

$$\mu_s F_N = m\left(\frac{v^2}{R}\right). \tag{6-21}$$

Vertical calculations: Next, let's consider the vertical forces on the car. The normal force $\vec{F}_N$ is directed up, in the positive direction of the y axis in Fig. 6-10*b*. The gravitational force $\vec{F}_g = m\vec{g}$ and the negative lift $\vec{F}_L$ are directed down. The acceleration of the car along the y axis is zero. Thus we can write Newton's second law for components along the y axis ($F_{\text{net},y} = ma_y$) as

$$F_N - mg - F_L = 0,$$

or

$$F_N = mg + F_L. \tag{6-22}$$

Combining results: Now we can combine our results along the two axes by substituting Eq. 6-22 for F_N in Eq. 6-21. Doing so and then solving for F_L lead to

$$F_L = m\left(\frac{v^2}{\mu_s R} - g\right)$$

$$= (600\text{ kg})\left(\frac{(28.6\text{ m/s})^2}{(0.75)(100\text{ m})} - 9.8\text{ m/s}^2\right)$$

$$= 663.7\text{ N} \approx 660\text{ N}. \quad \text{(Answer)}$$

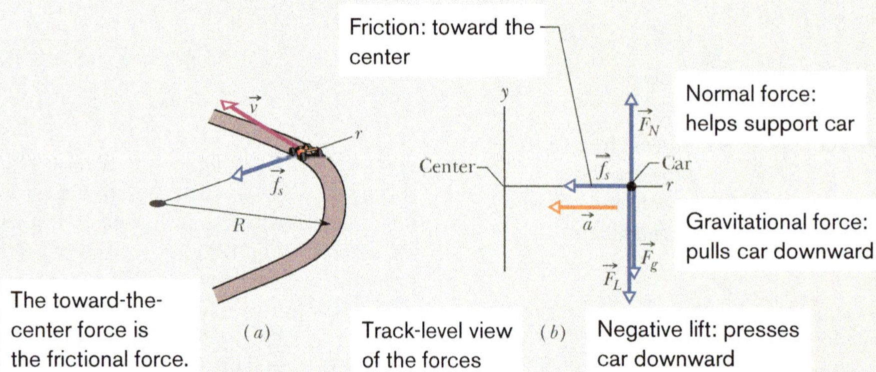

Figure 6-10 (*a*) A race car moves around a flat curved track at constant speed v. The frictional force $\vec{f}_s$ provides the necessary centripetal force along a radial axis r. (*b*) A free-body diagram (not to scale) for the car, in the vertical plane containing r.

(b) The magnitude F_L of the negative lift on a car depends on the square of the car's speed v^2, just as the drag force does (Eq. 6-14). Thus, the negative lift on the car here is greater when the car travels faster, as it does on a straight section of track. What is the magnitude of the negative lift for a speed of 90 m/s?

KEY IDEA

F_L is proportional to v^2.

Calculations: Thus we can write a ratio of the negative lift $F_{L,90}$ at $v = 90$ m/s to our result for the negative lift F_L at $v = 28.6$ m/s as

$$\frac{F_{L,90}}{F_L} = \frac{(90\text{ m/s})^2}{(28.6\text{ m/s})^2}.$$

Substituting our known negative lift of $F_L = 663.7$ N and solving for $F_{L,90}$ give us

$$F_{L,90} = 6572\text{ N} \approx 6600\text{ N}. \qquad \text{(Answer)}$$

Upside-down racing: The gravitational force is, of course, the force to beat if there is a chance of racing upside down:

$$F_g = mg = (600\text{ kg})(9.8\text{ m/s}^2)$$
$$= 5880\text{ N}.$$

With the car upside down, the negative lift is an *upward* force of 6600 N, which exceeds the downward 5880 N. Thus, the car could run on a long ceiling *provided* that it moves at about 90 m/s (= 324 km/h = 201 mi/h). However, moving that fast while right side up on a horizontal track is dangerous enough, so you are not likely to see upside-down racing except in the movies.

Sample Problem 6.06 Car in banked circular turn

This problem is quite challenging in setting up but takes only a few lines of algebra to solve. We deal with not only uniformly circular motion but also a ramp. However, we will not need a tilted coordinate system as with other ramps. Instead we can take a freeze-frame of the motion and work with simply horizontal and vertical axes. As always in this chapter, the starting point will be to apply Newton's second law, but that will require us to identify the force component that is responsible for the uniform circular motion.

Curved portions of highways are always banked (tilted) to prevent cars from sliding off the highway. When a highway is dry, the frictional force between the tires and the road surface may be enough to prevent sliding. When the highway is wet, however, the frictional force may be negligible, and banking is then essential. Figure 6-11*a* represents a car of mass *m* as it moves at a constant speed *v* of 20 m/s around a banked circular track of radius $R = 190$ m. (It is a normal car, rather than a race car, which means that any vertical force from the passing air is negligible.) If the frictional force from the track is negligible, what bank angle θ prevents sliding?

KEY IDEAS

Here the track is banked so as to tilt the normal force $\vec{F}_N$ on the car toward the center of the circle (Fig. 6-11*b*). Thus, $\vec{F}_N$ now has a centripetal component of magnitude F_{Nr}, directed inward along a radial axis *r*. We want to find the value of the bank angle θ such that this centripetal component keeps the car on the circular track without need of friction.

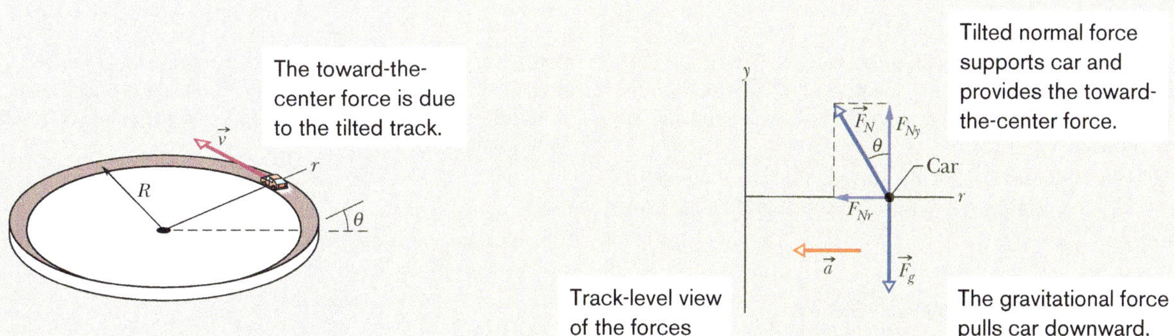

Figure 6-11 (*a*) A car moves around a curved banked road at constant speed *v*. The bank angle is exaggerated for clarity. (*b*) A free-body diagram for the car, assuming that friction between tires and road is zero and that the car lacks negative lift. The radially inward component F_{Nr} of the normal force (along radial axis *r*) provides the necessary centripetal force and radial acceleration.

Radial calculation: As Fig. 6-11*b* shows (and as you should verify), the angle that force $\vec{F}_N$ makes with the vertical is equal to the bank angle θ of the track. Thus, the radial component F_{Nr} is equal to $F_N \sin \theta$. We can now write Newton's second law for components along the r axis ($F_{net,r} = ma_r$) as

$$-F_N \sin \theta = m\left(-\frac{v^2}{R}\right). \tag{6-23}$$

We cannot solve this equation for the value of θ because it also contains the unknowns F_N and m.

Vertical calculations: We next consider the forces and acceleration along the y axis in Fig. 6-11*b*. The vertical component of the normal force is $F_{Ny} = F_N \cos \theta$, the gravitational force $\vec{F}_g$ on the car has the magnitude mg, and the acceleration of the car along the y axis is zero. Thus we can write Newton's second law for components along the y axis ($F_{net,y} = ma_y$) as

$$F_N \cos \theta - mg = m(0),$$

from which

$$F_N \cos \theta = mg. \tag{6-24}$$

Combining results: Equation 6-24 also contains the unknowns F_N and m, but note that dividing Eq. 6-23 by Eq. 6-24 neatly eliminates both those unknowns. Doing so, replacing $(\sin \theta)/(\cos \theta)$ with $\tan \theta$, and solving for θ then yield

$$\theta = \tan^{-1}\frac{v^2}{gR}$$

$$= \tan^{-1}\frac{(20 \text{ m/s})^2}{(9.8 \text{ m/s}^2)(190 \text{ m})} = 12°. \quad \text{(Answer)}$$

WILEY PLUS Additional examples, video, and practice available at *WileyPLUS*

Review & Summary

Friction When a force $\vec{F}$ tends to slide a body along a surface, a **frictional force** from the surface acts on the body. The frictional force is parallel to the surface and directed so as to oppose the sliding. It is due to bonding between the atoms on the body and the atoms on the surface, an effect called cold-welding.

If the body does not slide, the frictional force is a **static frictional force** $\vec{f}_s$. If there is sliding, the frictional force is a **kinetic frictional force** $\vec{f}_k$.

1. If a body does not move, the static frictional force $\vec{f}_s$ and the component of $\vec{F}$ parallel to the surface are equal in magnitude, and $\vec{f}_s$ is directed opposite that component. If the component increases, f_s also increases.
2. The magnitude of $\vec{f}_s$ has a maximum value $f_{s,max}$ given by
$$f_{s,max} = \mu_s F_N, \tag{6-1}$$
where μ_s is the **coefficient of static friction** and F_N is the magnitude of the normal force. If the component of $\vec{F}$ parallel to the surface exceeds $f_{s,max}$, the static friction is overwhelmed and the body slides on the surface.
3. If the body begins to slide on the surface, the magnitude of the frictional force rapidly decreases to a constant value f_k given by
$$f_k = \mu_k F_N, \tag{6-2}$$
where μ_k is the **coefficient of kinetic friction.**

Drag Force When there is relative motion between air (or some other fluid) and a body, the body experiences a **drag force** $\vec{D}$ that opposes the relative motion and points in the direction in which the fluid flows relative to the body. The magnitude of $\vec{D}$ is related to the relative speed v by an experimentally determined **drag coefficient** C according to

$$D = \tfrac{1}{2}C\rho A v^2, \tag{6-14}$$

where ρ is the fluid density (mass per unit volume) and A is the **effective cross-sectional area** of the body (the area of a cross section taken perpendicular to the relative velocity $\vec{v}$).

Terminal Speed When a blunt object has fallen far enough through air, the magnitudes of the drag force $\vec{D}$ and the gravitational force $\vec{F}_g$ on the body become equal. The body then falls at a constant **terminal speed** v_t given by

$$v_t = \sqrt{\frac{2F_g}{C\rho A}}. \tag{6-16}$$

Uniform Circular Motion If a particle moves in a circle or a circular arc of radius R at constant speed v, the particle is said to be in **uniform circular motion.** It then has a **centripetal acceleration** $\vec{a}$ with magnitude given by

$$a = \frac{v^2}{R}. \tag{6-17}$$

This acceleration is due to a net **centripetal force** on the particle, with magnitude given by

$$F = \frac{mv^2}{R}, \tag{6-18}$$

where m is the particle's mass. The vector quantities $\vec{a}$ and $\vec{F}$ are directed toward the center of curvature of the particle's path. A particle can move in circular motion only if a net centripetal force acts on it.

Questions

1 In Fig. 6-12, if the box is stationary and the angle θ between the horizontal and force $\vec{F}$ is increased somewhat, do the following quantities increase, decrease, or remain the same: (a) F_x; (b) f_s; (c) F_N; (d) $f_{s,\max}$? (e) If, instead, the box is sliding and θ is increased, does the magnitude of the frictional force on the box increase, decrease, or remain the same?

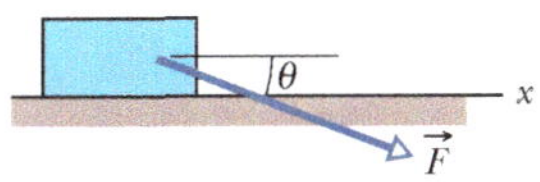

Figure 6-12 Question 1.

2 Repeat Question 1 for force $\vec{F}$ angled upward instead of downward as drawn.

3 In Fig. 6-13, horizontal force $\vec{F}_1$ of magnitude 10 N is applied to a box on a floor, but the box does not slide. Then, as the magnitude of vertical force $\vec{F}_2$ is increased from zero, do the following quantities increase, decrease, or stay the same: (a) the magnitude of the frictional force $\vec{f}_s$ on the box; (b) the magnitude of the normal force $\vec{F}_N$ on the box from the floor; (c) the maximum value $f_{s,\max}$ of the magnitude of the static frictional force on the box? (d) Does the box eventually slide?

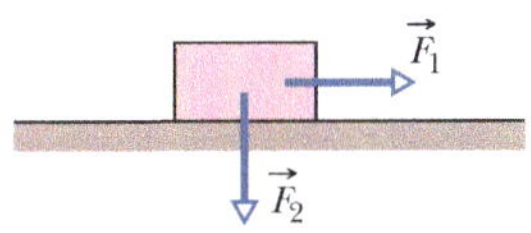

Figure 6-13 Question 3.

4 In three experiments, three different horizontal forces are applied to the same block lying on the same countertop. The force magnitudes are $F_1 = 12$ N, $F_2 = 8$ N, and $F_3 = 4$ N. In each experiment, the block remains stationary in spite of the applied force. Rank the forces according to (a) the magnitude f_s of the static frictional force on the block from the countertop and (b) the maximum value $f_{s,\max}$ of that force, greatest first.

5 If you press an apple crate against a wall so hard that the crate cannot slide down the wall, what is the direction of (a) the static frictional force $\vec{f}_s$ on the crate from the wall and (b) the normal force $\vec{F}_N$ on the crate from the wall? If you increase your push, what happens to (c) f_s, (d) F_N, and (e) $f_{s,\max}$?

6 In Fig. 6-14, a block of mass m is held stationary on a ramp by the frictional force on it from the ramp. A force $\vec{F}$, directed up the ramp, is then applied to the block and gradually increased in magnitude from zero. During the increase, what happens to the direction and magnitude of the frictional force on the block?

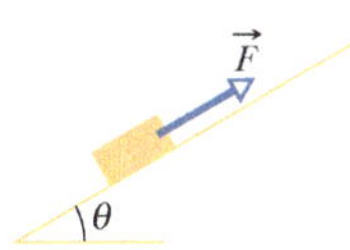

Figure 6-14 Question 6.

7 Reconsider Question 6 but with the force $\vec{F}$ now directed down the ramp. As the magnitude of $\vec{F}$ is increased from zero, what happens to the direction and magnitude of the frictional force on the block?

8 In Fig. 6-15, a horizontal force of 100 N is to be applied to a 10 kg slab that is initially stationary on a frictionless floor, to accelerate the slab. A 10 kg block lies on top of the slab; the coefficient of friction μ between the block and the slab is not known, and the block might slip. In fact, the contact between the block and the slab might even be frictionless. (a) Considering that possibility, what is the possible range of values for the magnitude of the slab's acceleration a_{slab}? (*Hint:* You don't need written calculations; just consider extreme values for μ.) (b) What is the possible range for the magnitude a_{block} of the block's acceleration?

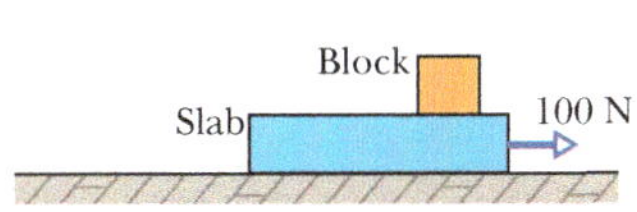

Figure 6-15 Question 8.

9 Figure 6-16 shows the overhead view of the path of an amusement-park ride that travels at constant speed through five circular arcs of radii R_0, $2R_0$, and $3R_0$. Rank the arcs according to the magnitude of the centripetal force on a rider traveling in the arcs, greatest first.

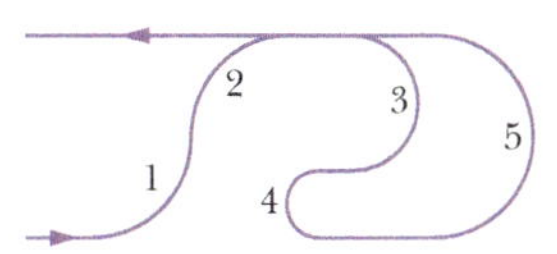

Figure 6-16 Question 9.

10 In 1987, as a Halloween stunt, two sky divers passed a pumpkin back and forth between them while they were in free fall just west of Chicago. The stunt was great fun until the last sky diver with the pumpkin opened his parachute. The pumpkin broke free from his grip, plummeted about 0.5 km, ripped through the roof of a house, slammed into the kitchen floor, and splattered all over the newly remodeled kitchen. From the sky diver's viewpoint and from the pumpkin's viewpoint, why did the sky diver lose control of the pumpkin?

11 A person riding a Ferris wheel moves through positions at (1) the top, (2) the bottom, and (3) midheight. If the wheel rotates at a constant rate, rank these three positions according to (a) the magnitude of the person's centripetal acceleration, (b) the magnitude of the net centripetal force on the person, and (c) the magnitude of the normal force on the person, greatest first.

12 During a routine flight in 1956, test pilot Tom Attridge put his jet fighter into a 20° dive for a test of the aircraft's 20 mm machine cannons. While traveling faster than sound at 4000 m altitude, he shot a burst of rounds. Then, after allowing the cannons to cool, he shot another burst at 2000 m; his speed was then 344 m/s, the speed of the rounds relative to him was 730 m/s, and he was still in a dive.

Almost immediately the canopy around him was shredded and his right air intake was damaged. With little flying capability left, the jet crashed into a wooded area, but Attridge managed to escape the resulting explosion. Explain what apparently happened just after the second burst of cannon rounds. (Attridge has been the only pilot who has managed to shoot himself down.)

13 A box is on a ramp that is at angle θ to the horizontal. As θ is increased from zero, and before the box slips, do the following increase, decrease, or remain the same: (a) the component of the gravitational force on the box, along the ramp, (b) the magnitude of the static frictional force on the box from the ramp, (c) the component of the gravitational force on the box, perpendicular to the ramp, (d) the magnitude of the normal force on the box from the ramp, and (e) the maximum value $f_{s,\max}$ of the static frictional force?

Problems

GO Tutoring problem available (at instructor's discretion) in *WileyPLUS* and WebAssign
SSM Worked-out solution available in Student Solutions Manual
WWW Worked-out solution is at http://www.wiley.com/college/halliday
• – ••• Number of dots indicates level of problem difficulty
ILW Interactive solution is at http://www.wiley.com/college/halliday
Additional information available in *The Flying Circus of Physics* and at flyingcircusofphysics.com

Module 6-1 Friction

•1 The floor of a railroad flatcar is loaded with loose crates having a coefficient of static friction of 0.25 with the floor. If the train is initially moving at a speed of 48 km/h, in how short a distance can the train be stopped at constant acceleration without causing the crates to slide over the floor?

•2 In a pickup game of dorm shuffleboard, students crazed by final exams use a broom to propel a calculus book along the dorm hallway. If the 3.5 kg book is pushed from rest through a distance of 0.90 m by the horizontal 25 N force from the broom and then has a speed of 1.60 m/s, what is the coefficient of kinetic friction between the book and floor?

•3 SSM WWW A bedroom bureau with a mass of 45 kg, including drawers and clothing, rests on the floor. (a) If the coefficient of static friction between the bureau and the floor is 0.45, what is the magnitude of the minimum horizontal force that a person must apply to start the bureau moving? (b) If the drawers and clothing, with 17 kg mass, are removed before the bureau is pushed, what is the new minimum magnitude?

•4 A slide-loving pig slides down a certain 35° slide in twice the time it would take to slide down a frictionless 35° slide. What is the coefficient of kinetic friction between the pig and the slide?

•5 GO A 2.5 kg block is initially at rest on a horizontal surface. A horizontal force $\vec{F}$ of magnitude 6.0 N and a vertical force $\vec{P}$ are then applied to the block (Fig. 6-17). The coefficients of friction for the block and surface are $\mu_s = 0.40$ and $\mu_k = 0.25$. Determine the magnitude of the frictional force acting on the block if the magnitude of $\vec{P}$ is (a) 8.0 N, (b) 10 N, and (c) 12 N.

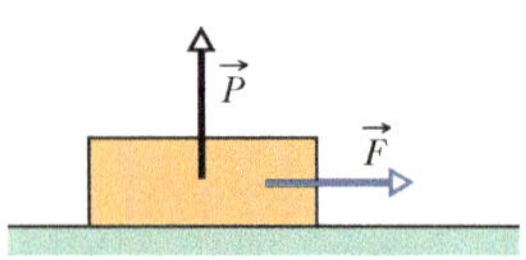

Figure 6-17 Problem 5.

•6 A baseball player with mass $m = 79$ kg, sliding into second base, is retarded by a frictional force of magnitude 470 N. What is the coefficient of kinetic friction μ_k between the player and the ground?

•7 SSM ILW A person pushes horizontally with a force of 220 N on a 55 kg crate to move it across a level floor. The coefficient of kinetic friction between the crate and the floor is 0.35. What is the magnitude of (a) the frictional force and (b) the acceleration of the crate?

•8 *The mysterious sliding stones.* Along the remote Racetrack Playa in Death Valley, California, stones sometimes gouge out prominent trails in the desert floor, as if the stones had been migrating (Fig. 6-18). For years curiosity mounted about why the stones moved. One explanation was that strong winds during occasional rainstorms would drag the rough stones over ground softened by rain. When the desert dried out, the trails behind the stones were hard-baked in place. According to measurements, the coefficient of kinetic friction between the stones and the wet playa ground is about 0.80. What horizontal force must act on a 20 kg stone (a typical mass) to maintain the stone's motion once a gust has started it moving? (Story continues with Problem 37.)

Jerry Schad/Photo Researchers, Inc.

Figure 6-18 Problem 8. What moved the stone?

•9 GO A 3.5 kg block is pushed along a horizontal floor by a force $\vec{F}$ of magnitude 15 N at an angle $\theta = 40°$ with the horizontal (Fig. 6-19). The coefficient of kinetic friction between the block and the floor is 0.25. Calculate the magnitudes of (a) the frictional force on the block from the floor and (b) the block's acceleration.

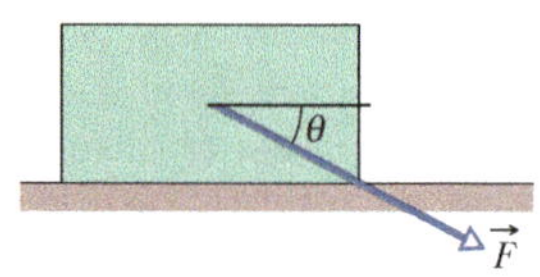

Figure 6-19 Problems 9 and 32.

•10 Figure 6-20 shows an initially stationary block of mass m on a floor. A force of magnitude $0.500mg$ is then applied at upward angle $\theta = 20°$. What is the magnitude of the acceleration of the block across the floor if the friction coefficients are (a) $\mu_s = 0.600$ and $\mu_k = 0.500$ and (b) $\mu_s = 0.400$ and $\mu_k = 0.300$?

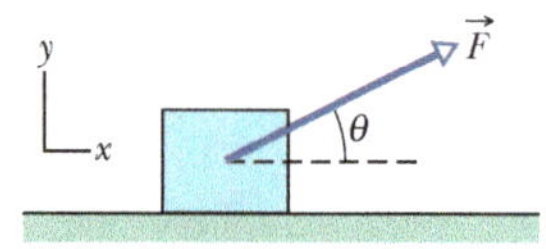

Figure 6-20 Problem 10.

•11 SSM A 68 kg crate is dragged across a floor by pulling on a rope attached to the crate and inclined 15° above the horizontal. (a) If the coefficient of static friction is 0.50, what minimum force magnitude is required from the rope to start the crate moving? (b) If $\mu_k = 0.35$, what is the magnitude of the initial acceleration of the crate?

•12 In about 1915, Henry Sincosky of Philadelphia suspended himself from a rafter by gripping the rafter with the thumb of each hand on one side and the fingers on the opposite side (Fig. 6-21). Sincosky's mass was 79 kg. If the coefficient of static friction between hand and rafter was 0.70, what was the least magnitude of the normal force on the rafter from each thumb or opposite fingers? (After suspending himself, Sincosky chinned himself on the rafter and then moved hand-over-hand along the rafter. If you do not think Sincosky's grip was remarkable, try to repeat his stunt.)

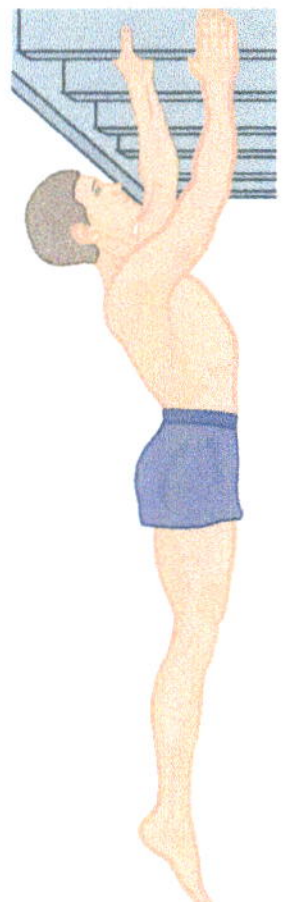
Figure 6-21 Problem 12.

•**13** A worker pushes horizontally on a 35 kg crate with a force of magnitude 110 N. The coefficient of static friction between the crate and the floor is 0.37. (a) What is the value of $f_{s,\max}$ under the circumstances? (b) Does the crate move? (c) What is the frictional force on the crate from the floor? (d) Suppose, next, that a second worker pulls directly upward on the crate to help out. What is the least vertical pull that will allow the first worker's 110 N push to move the crate? (e) If, instead, the second worker pulls horizontally to help out, what is the least pull that will get the crate moving?

•**14** Figure 6-22 shows the cross section of a road cut into the side of a mountain. The solid line AA' represents a weak bedding plane along which sliding is possible. Block B directly above the highway is separated from uphill rock by a large crack (called a *joint*), so that only friction between the block and the bedding plane prevents sliding. The mass of the block is 1.8×10^7 kg, the *dip angle* θ of the bedding plane is 24°, and the coefficient of static friction between block and plane is 0.63. (a) Show that the block will not slide under these circumstances. (b) Next, water seeps into the joint and expands upon freezing, exerting on the block a force $\vec{F}$ parallel to AA'. What minimum value of force magnitude F will trigger a slide down the plane?

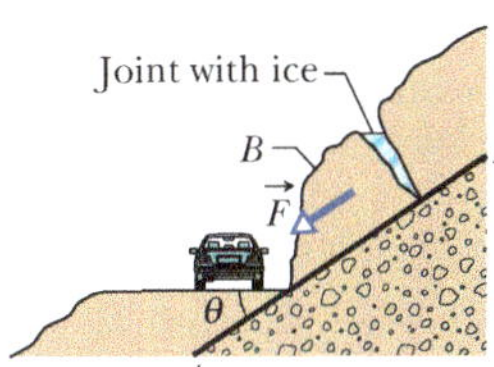

Figure 6-22 Problem 14.

•**15** The coefficient of static friction between Teflon and scrambled eggs is about 0.04. What is the smallest angle from the horizontal that will cause the eggs to slide across the bottom of a Teflon-coated skillet?

••**16** A loaded penguin sled weighing 80 N rests on a plane inclined at angle $\theta = 20°$ to the horizontal (Fig. 6-23). Between the sled and the plane, the coefficient of static friction is 0.25, and the coefficient of kinetic friction is 0.15. (a) What is the least magnitude of the force $\vec{F}$, parallel to the plane, that will prevent the sled from slipping down the plane? (b) What is the minimum magnitude F that will start the sled moving up the plane? (c) What value of F is required to move the sled up the plane at constant velocity?

Figure 6-23 Problems 16 and 22.

••**17** In Fig. 6-24, a force $\vec{P}$ acts on a block weighing 45 N. The block is initially at rest on a plane inclined at angle $\theta = 15°$ to the horizontal. The positive direction of the x axis is up the plane. Between block and plane, the coefficient of static friction is $\mu_s = 0.50$ and the coefficient of kinetic friction is $\mu_k = 0.34$. In unit-vector notation, what is the frictional force on the block from the plane when $\vec{P}$ is (a) $(-5.0\text{ N})\hat{i}$, (b) $(-8.0\text{ N})\hat{i}$, and (c) $(-15\text{ N})\hat{i}$?

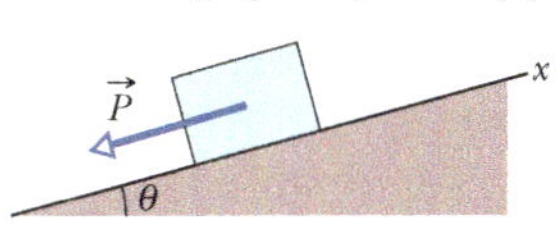

Figure 6-24 Problem 17.

••**18** GO You testify as an *expert witness* in a case involving an accident in which car A slid into the rear of car B, which was stopped at a red light along a road headed down a hill (Fig. 6-25). You find that the slope of the hill is $\theta = 12.0°$, that the cars were separated by distance $d = 24.0$ m when the driver of car A put the car into a slide (it lacked any automatic anti-brake-lock system), and that the speed of car A at the onset of braking was $v_0 = 18.0$ m/s. With what speed did car A hit car B if the coefficient of kinetic friction was (a) 0.60 (dry road surface) and (b) 0.10 (road surface covered with wet leaves)?

Figure 6-25 Problem 18.

••**19** A 12 N horizontal force $\vec{F}$ pushes a block weighing 5.0 N against a vertical wall (Fig. 6-26). The coefficient of static friction between the wall and the block is 0.60, and the coefficient of kinetic friction is 0.40. Assume that the block is not moving initially. (a) Will the block move? (b) In unit-vector notation, what is the force on the block from the wall?

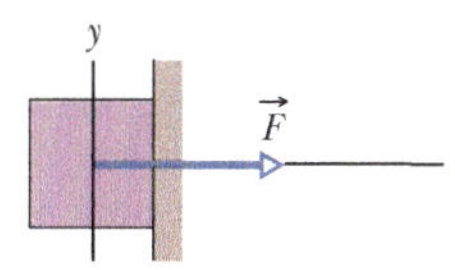

Figure 6-26 Problem 19.

••**20** GO In Fig. 6-27, a box of Cheerios (mass $m_C = 1.0$ kg) and a box of Wheaties (mass $m_W = 3.0$ kg) are accelerated across a horizontal surface by a horizontal force $\vec{F}$ applied to the Cheerios box. The magnitude of the frictional force on the Cheerios box is 2.0 N, and the magnitude of the frictional force on the Wheaties box is 4.0 N. If the magnitude of $\vec{F}$ is 12 N, what is the magnitude of the force on the Wheaties box from the Cheerios box?

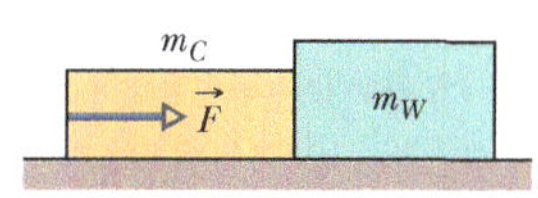

Figure 6-27 Problem 20.

••**21** An initially stationary box of sand is to be pulled across a floor by means of a cable in which the tension should not exceed 1100 N. The coefficient of static friction between the box and the floor is 0.35. (a) What should be the angle between the cable and the horizontal in order to pull the greatest possible amount of sand, and (b) what is the weight of the sand and box in that situation?

••**22** GO In Fig. 6-23, a sled is held on an inclined plane by a cord pulling directly up the plane. The sled is to be on the verge of moving up the plane. In Fig. 6-28, the magnitude F required of the cord's force on the sled is plotted versus a range of values for the coefficient of static friction μ_s between sled and plane: $F_1 = 2.0$ N, $F_2 = 5.0$ N, and $\mu_2 = 0.50$. At what angle θ is the plane inclined?

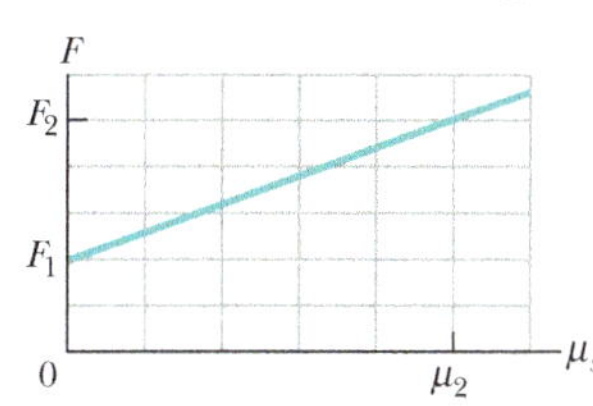

Figure 6-28 Problem 22.

••23 When the three blocks in Fig. 6-29 are released from rest, they accelerate with a magnitude of 0.500 m/s^2. Block 1 has mass M, block 2 has $2M$, and block 3 has $2M$. What is the coefficient of kinetic friction between block 2 and the table?

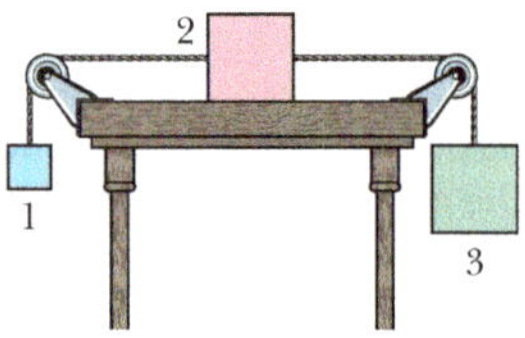

Figure 6-29 Problem 23.

••24 A 4.10 kg block is pushed along a floor by a constant applied force that is horizontal and has a magnitude of 40.0 N. Figure 6-30 gives the block's speed v versus time t as the block moves along an x axis on the floor. The scale of the figure's vertical axis is set by v_s = 5.0 m/s. What is the coefficient of kinetic friction between the block and the floor?

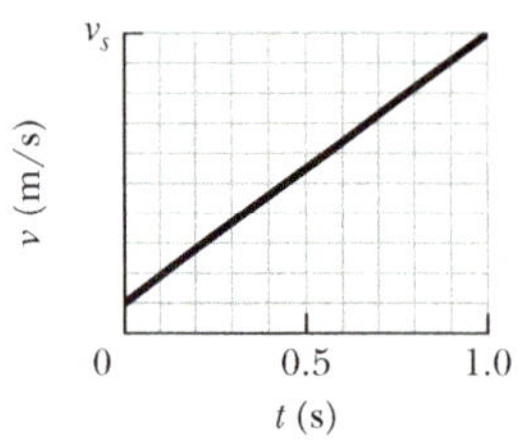

Figure 6-30 Problem 24.

••25 SSM WWW Block B in Fig. 6-31 weighs 711 N. The coefficient of static friction between block and table is 0.25; angle θ is 30°; assume that the cord between B and the knot is horizontal. Find the maximum weight of block A for which the system will be stationary.

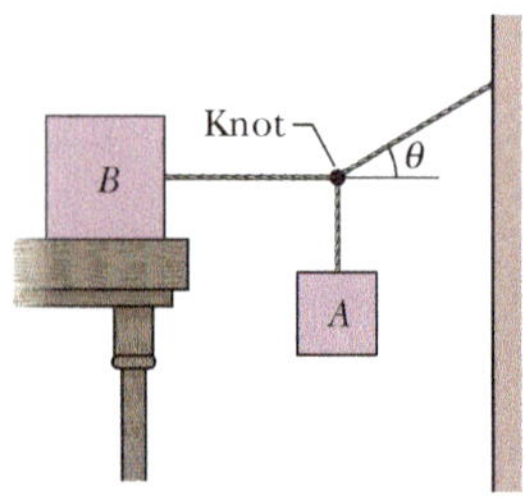

Figure 6-31 Problem 25.

••26 GO Figure 6-32 shows three crates being pushed over a concrete floor by a horizontal force $\vec{F}$ of magnitude 440 N. The masses of the crates are m_1 = 30.0 kg, m_2 = 10.0 kg, and m_3 = 20.0 kg. The coefficient of kinetic friction between the floor and each of the crates is 0.700. (a) What is the magnitude F_{32} of the force on crate 3 from crate 2? (b) If the crates then slide onto a polished floor, where the coefficient of kinetic friction is less than 0.700, is magnitude F_{32} more than, less than, or the same as it was when the coefficient was 0.700?

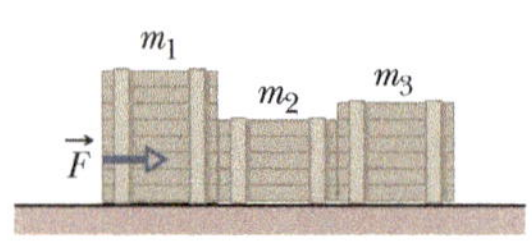

Figure 6-32 Problem 26.

••27 GO Body A in Fig. 6-33 weighs 102 N, and body B weighs 32 N. The coefficients of friction between A and the incline are μ_s = 0.56 and μ_k = 0.25. Angle θ is 40°. Let the positive direction of an x axis be up the incline. In unit-vector notation, what is the acceleration of A if A is initially (a) at rest, (b) moving up the incline, and (c) moving down the incline?

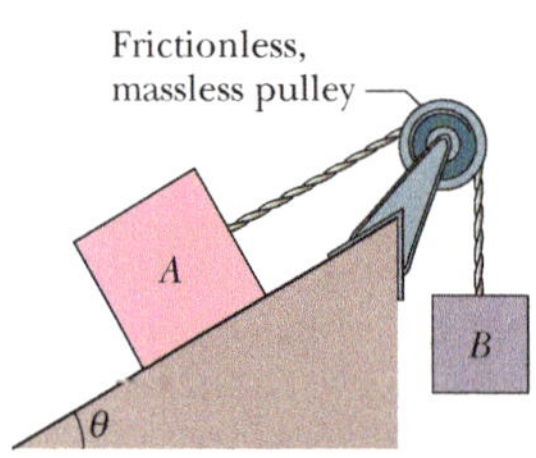

Figure 6-33 Problems 27 and 28.

••28 In Fig. 6-33, two blocks are connected over a pulley. The mass of block A is 10 kg, and the coefficient of kinetic friction between A and the incline is 0.20. Angle θ of the incline is 30°. Block A slides down the incline at constant speed. What is the mass of block B? Assume the connecting rope has negligible mass. (The pulley's function is only to redirect the rope.)

••29 GO In Fig. 6-34, blocks A and B have weights of 44 N and 22 N, respectively. (a) Determine the minimum weight of block C to keep A from sliding if μ_s between A and the table is 0.20. (b) Block C suddenly is lifted off A. What is the acceleration of block A if μ_k between A and the table is 0.15?

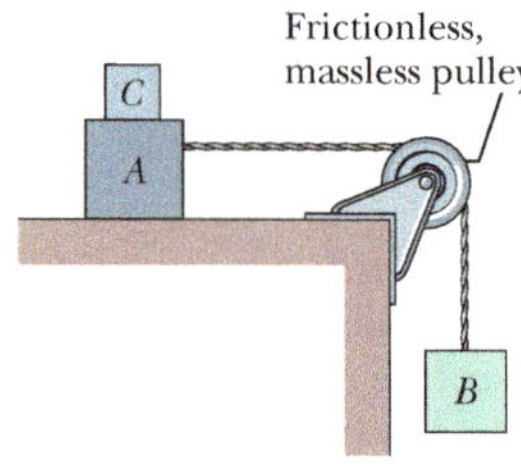

Figure 6-34 Problem 29.

••30 A toy chest and its contents have a combined weight of 180 N. The coefficient of static friction between toy chest and floor is 0.42. The child in Fig. 6-35 attempts to move the chest across the floor by pulling on an attached rope. (a) If θ is 42°, what is the magnitude of the force $\vec{F}$ that the child must exert on the rope to put the chest on the verge of moving? (b) Write an expression for the magnitude F required to put the chest on the verge of moving as a function of the angle θ. Determine (c) the value of θ for which F is a minimum and (d) that minimum magnitude.

Figure 6-35 Problem 30.

••31 SSM Two blocks, of weights 3.6 N and 7.2 N, are connected by a massless string and slide down a 30° inclined plane. The coefficient of kinetic friction between the lighter block and the plane is 0.10, and the coefficient between the heavier block and the plane is 0.20. Assuming that the lighter block leads, find (a) the magnitude of the acceleration of the blocks and (b) the tension in the taut string.

••32 GO A block is pushed across a floor by a constant force that is applied at downward angle θ (Fig. 6-19). Figure 6-36 gives the acceleration magnitude a versus a range of values for the coefficient of kinetic friction μ_k between block and floor: a_1 = 3.0 m/s^2, μ_{k2} = 0.20, and μ_{k3} = 0.40. What is the value of θ?

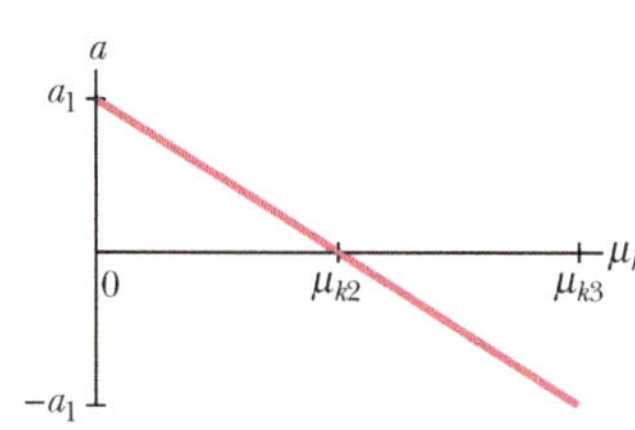

Figure 6-36 Problem 32.

•••33 SSM A 1000 kg boat is traveling at 90 km/h when its engine is shut off. The magnitude of the frictional force $\vec{f}_k$ between boat and water is proportional to the speed v of the boat: $f_k = 70v$, where v is in meters per second and f_k is in newtons. Find the time required for the boat to slow to 45 km/h.

•••34 GO In Fig. 6-37, a slab of mass $m_1 = 40$ kg rests on a frictionless floor, and a block of mass $m_2 = 10$ kg rests on top of the slab. Between block and slab, the coefficient of static friction is 0.60, and the coefficient of kinetic friction is 0.40. A horizontal force $\vec{F}$ of magnitude 100 N begins to pull directly on the block, as shown. In unit-vector notation, what are the resulting accelerations of (a) the block and (b) the slab?

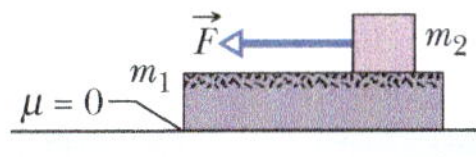

Figure 6-37 Problem 34.

•••35 ILW The two blocks ($m = 16$ kg and $M = 88$ kg) in Fig. 6-38 are not attached to each other. The coefficient of static friction between the blocks is $\mu_s = 0.38$, but the surface beneath the larger block is frictionless. What is the minimum magnitude of the horizontal force $\vec{F}$ required to keep the smaller block from slipping down the larger block?

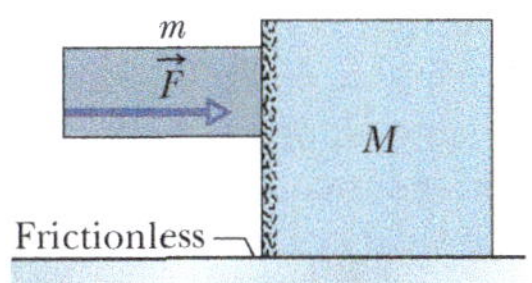

Figure 6-38 Problem 35.

Module 6-2 The Drag Force and Terminal Speed

•36 The terminal speed of a sky diver is 160 km/h in the spread-eagle position and 310 km/h in the nosedive position. Assuming that the diver's drag coefficient C does not change from one position to the other, find the ratio of the effective cross-sectional area A in the slower position to that in the faster position.

••37 *Continuation of Problem 8.* Now assume that Eq. 6-14 gives the magnitude of the air drag force on the typical 20 kg stone, which presents to the wind a vertical cross-sectional area of 0.040 m^2 and has a drag coefficient C of 0.80. Take the air density to be 1.21 kg/m^3, and the coefficient of kinetic friction to be 0.80. (a) In kilometers per hour, what wind speed V along the ground is needed to maintain the stone's motion once it has started moving? Because winds along the ground are retarded by the ground, the wind speeds reported for storms are often measured at a height of 10 m. Assume wind speeds are 2.00 times those along the ground. (b) For your answer to (a), what wind speed would be reported for the storm? (c) Is that value reasonable for a high-speed wind in a storm? (Story continues with Problem 65.)

••38 Assume Eq. 6-14 gives the drag force on a pilot plus ejection seat just after they are ejected from a plane traveling horizontally at 1300 km/h. Assume also that the mass of the seat is equal to the mass of the pilot and that the drag coefficient is that of a sky diver. Making a reasonable guess of the pilot's mass and using the appropriate v_t value from Table 6-1, estimate the magnitudes of (a) the drag force on the *pilot + seat* and (b) their horizontal deceleration (in terms of g), both just after ejection. (The result of (a) should indicate an engineering requirement: The seat must include a protective barrier to deflect the initial wind blast away from the pilot's head.)

••39 Calculate the ratio of the drag force on a jet flying at 1000 km/h at an altitude of 10 km to the drag force on a prop-driven transport flying at half that speed and altitude. The density of air is 0.38 kg/m^3 at 10 km and 0.67 kg/m^3 at 5.0 km. Assume that the airplanes have the same effective cross-sectional area and drag coefficient C.

••40 In downhill speed skiing a skier is retarded by both the air drag force on the body and the kinetic frictional force on the skis. (a) Suppose the slope angle is $\theta = 40.0°$, the snow is dry snow with a coefficient of kinetic friction $\mu_k = 0.0400$, the mass of the skier and equipment is $m = 85.0$ kg, the cross-sectional area of the (tucked) skier is $A = 1.30$ m^2, the drag coefficient is $C = 0.150$, and the air density is 1.20 kg/m^3. (a) What is the terminal speed? (b) If a skier can vary C by a slight amount dC by adjusting, say, the hand positions, what is the corresponding variation in the terminal speed?

Module 6-3 Uniform Circular Motion

•41 A cat dozes on a stationary merry-go-round in an amusement park, at a radius of 5.4 m from the center of the ride. Then the operator turns on the ride and brings it up to its proper turning rate of one complete rotation every 6.0 s. What is the least coefficient of static friction between the cat and the merry-go-round that will allow the cat to stay in place, without sliding (or the cat clinging with its claws)?

•42 Suppose the coefficient of static friction between the road and the tires on a car is 0.60 and the car has no negative lift. What speed will put the car on the verge of sliding as it rounds a level curve of 30.5 m radius?

•43 ILW What is the smallest radius of an unbanked (flat) track around which a bicyclist can travel if her speed is 29 km/h and the μ_s between tires and track is 0.32?

•44 During an Olympic bobsled run, the Jamaican team makes a turn of radius 7.6 m at a speed of 96.6 km/h. What is their acceleration in terms of g?

••45 SSM ILW A student of weight 667 N rides a steadily rotating Ferris wheel (the student sits upright). At the highest point, the magnitude of the normal force $\vec{F}_N$ on the student from the seat is 556 N. (a) Does the student feel "light" or "heavy" there? (b) What is the magnitude of $\vec{F}_N$ at the lowest point? If the wheel's speed is doubled, what is the magnitude F_N at the (c) highest and (d) lowest point?

••46 A police officer in hot pursuit drives her car through a circular turn of radius 300 m with a constant speed of 80.0 km/h. Her mass is 55.0 kg. What are (a) the magnitude and (b) the angle (relative to vertical) of the *net* force of the officer on the car seat? (*Hint:* Consider both horizontal and vertical forces.)

••47 A circular-motion addict of mass 80 kg rides a Ferris wheel around in a vertical circle of radius 10 m at a constant speed of 6.1 m/s. (a) What is the period of the motion? What is the magnitude of the normal force on the addict from the seat when both go through (b) the highest point of the circular path and (c) the lowest point?

••48 A roller-coaster car at an amusement park has a mass of 1200 kg when fully loaded with passengers. As the car passes over the top of a circular hill of radius 18 m, assume that its speed is not changing. At the top of the hill, what are the (a) magnitude F_N and (b) direction (up or down) of the normal force on the car from the track if the car's speed is $v = 11$ m/s? What are (c) F_N and (d) the direction if $v = 14$ m/s?

••49 GO In Fig. 6-39, a car is driven at constant speed over a circular hill and then into a circular valley with the same radius. At the top of the hill, the normal force on the driver from the car seat is 0. The driver's mass is 70.0 kg. What is the magnitude of the normal force on the driver from the seat when the car passes through the bottom of the valley?

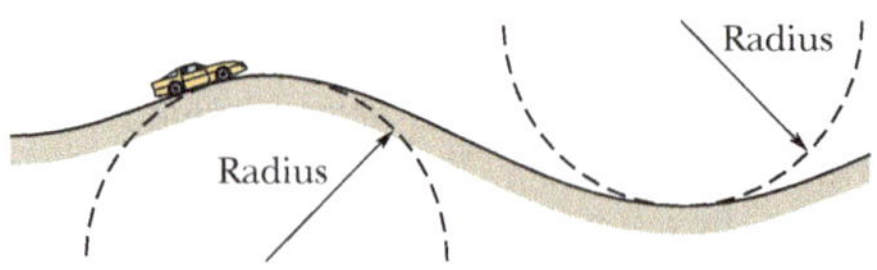

Figure 6-39 Problem 49.

••50 An 85.0 kg passenger is made to move along a circular path of radius $r = 3.50$ m in uniform circular motion. (a) Figure 6-40*a* is a plot of the required magnitude F of the net centripetal force for a range of possible values of the passenger's speed v. What is the plot's slope at $v = 8.30$ m/s? (b) Figure 6-40*b* is a plot of F for a range of possible values of T, the period of the motion. What is the plot's slope at $T = 2.50$ s?

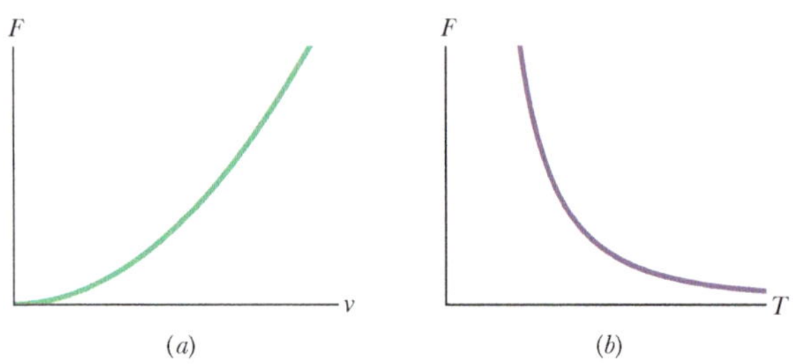

Figure 6-40 Problem 50.

••51 SSM WWW An airplane is flying in a horizontal circle at a speed of 480 km/h (Fig. 6-41). If its wings are tilted at angle $\theta = 40°$ to the horizontal, what is the radius of the circle in which the plane is flying? Assume that the required force is provided entirely by an "aerodynamic lift" that is perpendicular to the wing surface.

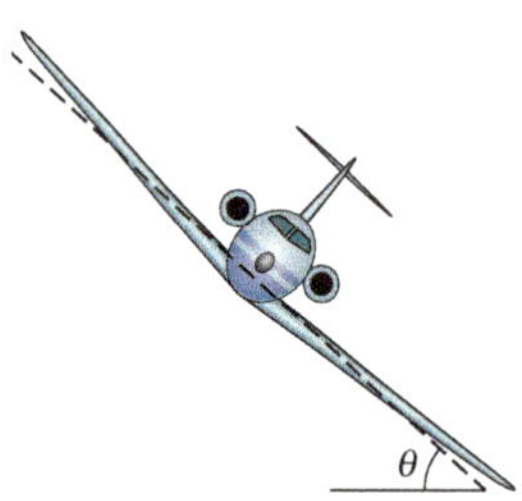

Figure 6-41 Problem 51.

••52 An amusement park ride consists of a car moving in a vertical circle on the end of a rigid boom of negligible mass. The combined weight of the car and riders is 5.0 kN, and the circle's radius is 10 m. At the top of the circle, what are the (a) magnitude F_B and (b) direction (up or down) of the force on the car from the boom if the car's speed is $v = 5.0$ m/s? What are (c) F_B and (d) the direction if $v = 12$ m/s?

••53 An old streetcar rounds a flat corner of radius 9.1 m, at 16 km/h. What angle with the vertical will be made by the loosely hanging hand straps?

••54 In designing circular rides for amusement parks, mechanical engineers must consider how small variations in certain parameters can alter the net force on a passenger. Consider a passenger of mass m riding around a horizontal circle of radius r at speed v. What is the variation dF in the net force magnitude for (a) a variation dr in the radius with v held constant, (b) a variation dv in the speed with r held constant, and (c) a variation dT in the period with r held constant?

••55 A bolt is threaded onto one end of a thin horizontal rod, and the rod is then rotated horizontally about its other end. An engineer monitors the motion by flashing a strobe lamp onto the rod and bolt, adjusting the strobe rate until the bolt appears to be in the same eight places during each full rotation of the rod (Fig. 6-42). The strobe rate is 2000 flashes per second; the bolt has mass 30 g and is at radius 3.5 cm. What is the magnitude of the force on the bolt from the rod?

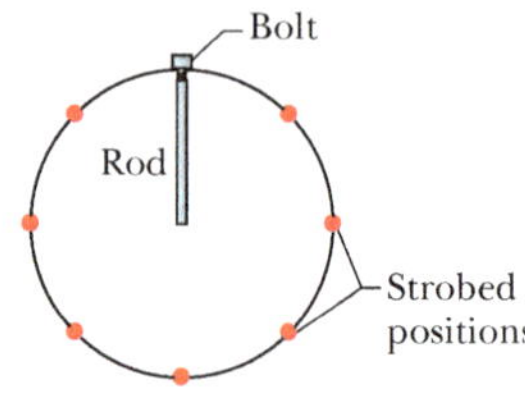

Figure 6-42 Problem 55.

••56 GO A banked circular highway curve is designed for traffic moving at 60 km/h. The radius of the curve is 200 m. Traffic is moving along the highway at 40 km/h on a rainy day. What is the minimum coefficient of friction between tires and road that will allow cars to take the turn without sliding off the road? (Assume the cars do not have negative lift.)

••57 GO A puck of mass $m = 1.50$ kg slides in a circle of radius $r = 20.0$ cm on a frictionless table while attached to a hanging cylinder of mass $M = 2.50$ kg by means of a cord that extends through a hole in the table (Fig. 6-43). What speed keeps the cylinder at rest?

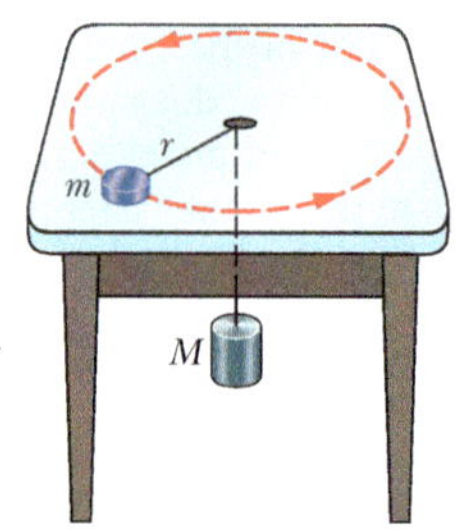

Figure 6-43 Problem 57.

••58 *Brake or turn?* Figure 6-44 depicts an overhead view of a car's path as the car travels toward a wall. Assume that the driver begins to brake the car when the distance to the wall is $d = 107$ m, and take the car's mass as $m = 1400$ kg, its initial speed as $v_0 = 35$ m/s, and the coefficient of static friction as $\mu_s = 0.50$. Assume that the car's weight is distributed evenly on the four wheels, even during braking. (a) What magnitude of static friction is needed (between tires and road) to stop the car just as it reaches the wall? (b) What is the maximum possible static friction $f_{s,max}$? (c) If the coefficient of kinetic friction between the (sliding) tires and the road is $\mu_k = 0.40$, at what speed will the car hit the wall? To avoid the crash, a driver could elect to turn the car so that it just barely misses the wall, as shown in the figure. (d) What magnitude of frictional force would be required to keep the car in a circular path of radius d and at the given speed v_0, so that the car moves in a quarter circle and then parallel to the wall? (e) Is the required force less than $f_{s,max}$ so that a circular path is possible?

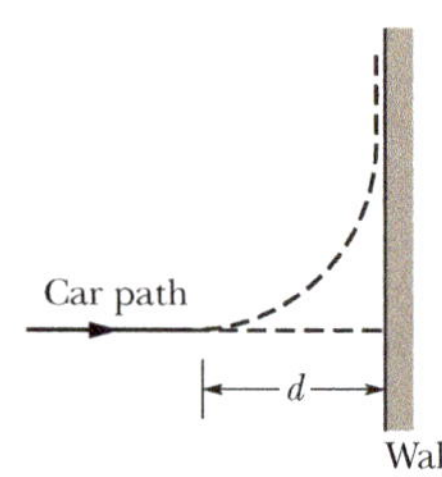

Figure 6-44 Problem 58.

•••59 SSM ILW In Fig. 6-45, a 1.34 kg ball is connected by means of two massless strings, each of length $L = 1.70$ m, to a vertical, rotating rod. The strings are tied to the rod with separation $d = 1.70$ m and are taut. The tension in the upper string is 35 N. What are the (a) tension in the lower string, (b) magnitude of the net force $\vec{F}_{net}$ on the ball, and (c) speed of the ball? (d) What is the direction of $\vec{F}_{net}$?

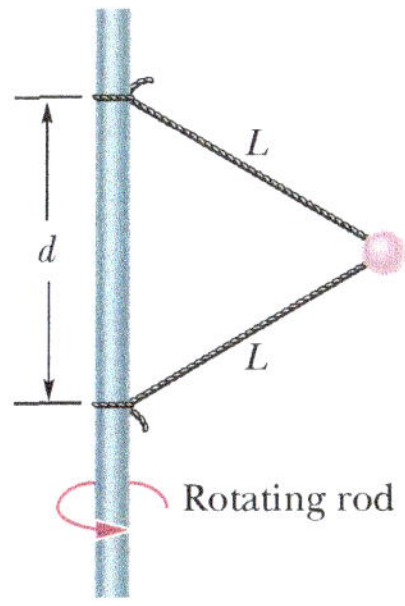

Figure 6-45 Problem 59.

Additional Problems

60 GO In Fig. 6-46, a box of ant aunts (total mass $m_1 = 1.65$ kg) and a box of ant uncles (total mass $m_2 = 3.30$ kg) slide down an inclined plane while attached by a massless rod parallel to the plane. The angle of incline is $\theta = 30.0°$. The coefficient of kinetic friction between the aunt box and the incline is $\mu_1 = 0.226$; that between the uncle box and the incline is $\mu_2 = 0.113$. Compute (a) the tension in the rod and (b) the magnitude of the common acceleration of the two boxes. (c) How would the answers to (a) and (b) change if the uncles trailed the aunts?

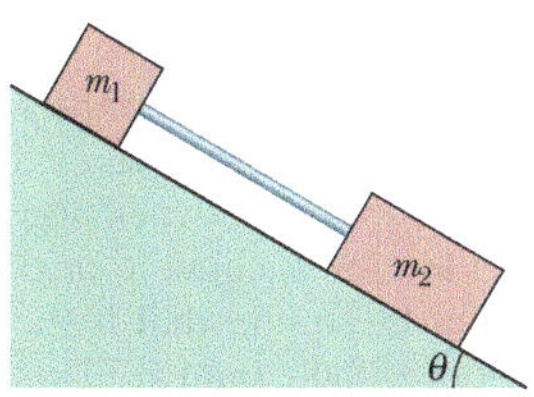

Figure 6-46 Problem 60.

61 SSM A block of mass $m_t = 4.0$ kg is put on top of a block of mass $m_b = 5.0$ kg. To cause the top block to slip on the bottom one while the bottom one is held fixed, a horizontal force of at least 12 N must be applied to the top block. The assembly of blocks is now placed on a horizontal, frictionless table (Fig. 6-47). Find the magnitudes of (a) the maximum horizontal force $\vec{F}$ that can be applied to the lower block so that the blocks will move together and (b) the resulting acceleration of the blocks.

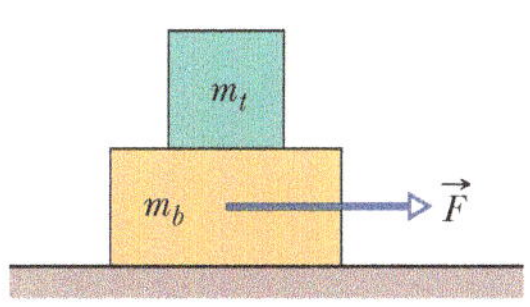

Figure 6-47 Problem 61.

62 A 5.00 kg stone is rubbed across the horizontal ceiling of a cave passageway (Fig. 6-48). If the coefficient of kinetic friction is 0.65 and the force applied to the stone is angled at $\theta = 70.0°$, what must the magnitude of the force be for the stone to move at constant velocity?

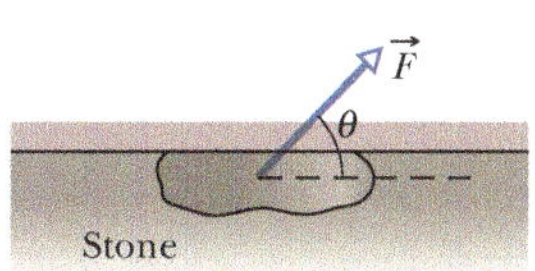

Figure 6-48 Problem 62.

63 In Fig. 6-49, a 49 kg rock climber is climbing a "chimney." The coefficient of static friction between her shoes and the rock is 1.2; between her back and the rock is 0.80. She has reduced her push against the rock until her back and her shoes are on the verge of slipping. (a) Draw a free-body diagram of her. (b) What is the magnitude of her push against the rock? (c) What fraction of her weight is supported by the frictional force on her shoes?

Figure 6-49 Problem 63.

64 A high-speed railway car goes around a flat, horizontal circle of radius 470 m at a constant speed. The magnitudes of the horizontal and vertical components of the force of the car on a 51.0 kg passenger are 210 N and 500 N, respectively. (a) What is the magnitude of the net force (of *all* the forces) on the passenger? (b) What is the speed of the car?

65 *Continuation of Problems 8 and 37.* Another explanation is that the stones move only when the water dumped on the playa during a storm freezes into a large, thin sheet of ice. The stones are trapped in place in the ice. Then, as air flows across the ice during a wind, the air-drag forces on the ice and stones move them both, with the stones gouging out the trails. The magnitude of the air-drag force on this horizontal "ice sail" is given by $D_{ice} = 4C_{ice}\rho A_{ice}v^2$, where C_{ice} is the drag coefficient (2.0×10^{-3}), ρ is the air density (1.21 kg/m^3), A_{ice} is the horizontal area of the ice, and v is the wind speed along the ice.

Assume the following: The ice sheet measures 400 m by 500 m by 4.0 mm and has a coefficient of kinetic friction of 0.10 with the ground and a density of 917 kg/m^3. Also assume that 100 stones identical to the one in Problem 8 are trapped in the ice. To maintain the motion of the sheet, what are the required wind speeds (a) near the sheet and (b) at a height of 10 m? (c) Are these reasonable values for high-speed winds in a storm?

66 GO In Fig. 6-50, block 1 of mass $m_1 = 2.0$ kg and block 2 of mass $m_2 = 3.0$ kg are connected by a string of negligible mass and are initially held in place. Block 2 is on a frictionless surface tilted at $\theta = 30°$. The coefficient of kinetic friction between block 1 and the horizontal surface is 0.25. The pulley has negligible mass and friction. Once they are released, the blocks move. What then is the tension in the string?

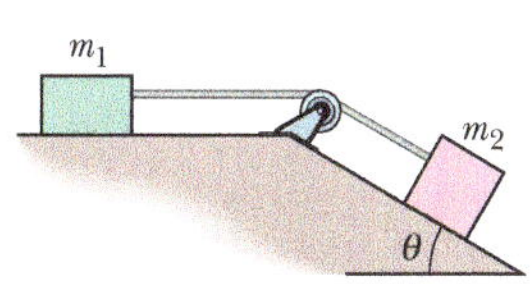

Figure 6-50 Problem 66.

67 In Fig. 6-51, a crate slides down an inclined right-angled trough. The coefficient of kinetic friction between the crate and the trough is μ_k. What is the acceleration of the crate in terms of μ_k, θ, and g?

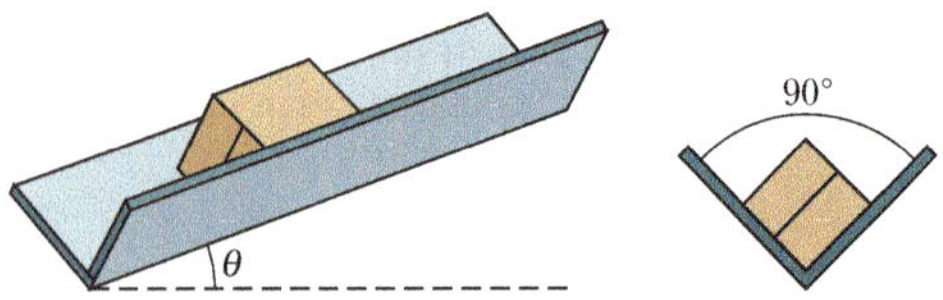

Figure 6-51 Problem 67.

68 *Engineering a highway curve.* If a car goes through a curve too fast, the car tends to slide out of the curve. For a banked curve with friction, a frictional force acts on a fast car to oppose the tendency to slide out of the curve; the force is directed down the bank (in the direction water would drain). Consider a circular curve of radius $R = 200$ m and bank angle θ, where the coefficient of static friction between tires and pavement is μ_s. A car (without negative lift) is driven around the curve as shown in Fig. 6-11. (a) Find an expression for the car speed v_{max} that puts the car on the verge of sliding out. (b) On the same graph, plot v_{max} versus angle θ for the range 0° to 50°, first for $\mu_s = 0.60$ (dry pavement) and then for $\mu_s = 0.050$ (wet or icy pavement). In kilometers per hour, evaluate v_{max} for a bank angle of $\theta = 10°$ and for (c) $\mu_s = 0.60$ and (d) $\mu_s = 0.050$. (Now you can see why accidents occur in highway curves when icy conditions are not obvious to drivers, who tend to drive at normal speeds.)

69 A student, crazed by final exams, uses a force $\vec{P}$ of magnitude 80 N and angle $\theta = 70°$ to push a 5.0 kg block across the ceiling of his room (Fig. 6-52). If the coefficient of kinetic friction between the block and the ceiling is 0.40, what is the magnitude of the block's acceleration?

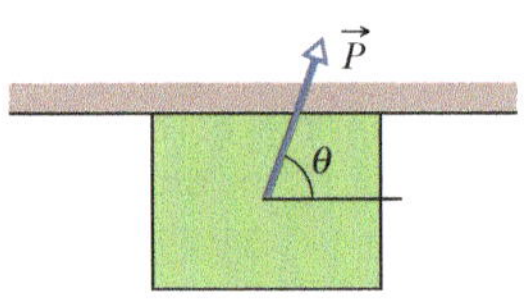

Figure 6-52 Problem 69.

70 GO Figure 6-53 shows a *conical pendulum,* in which the bob (the small object at the lower end of the cord) moves in a horizontal circle at constant speed. (The cord sweeps out a cone as the bob rotates.) The bob has a mass of 0.040 kg, the string has length $L = 0.90$ m and negligible mass, and the bob follows a circular path of circumference 0.94 m. What are (a) the tension in the string and (b) the period of the motion?

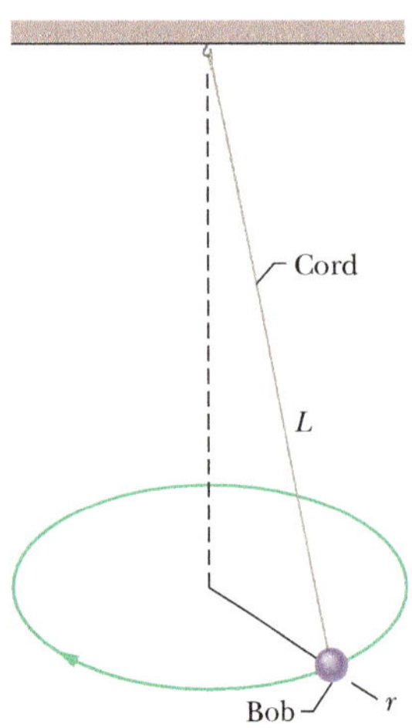

Figure 6-53 Problem 70.

71 An 8.00 kg block of steel is at rest on a horizontal table. The coefficient of static friction between the block and the table is 0.450. A force is to be applied to the block. To three significant figures, what is the magnitude of that applied force if it puts the block on the verge of sliding when the force is directed (a) horizontally, (b) upward at 60.0° from the horizontal, and (c) downward at 60.0° from the horizontal?

72 A box of canned goods slides down a ramp from street level into the basement of a grocery store with acceleration 0.75 m/s² directed down the ramp. The ramp makes an angle of 40° with the horizontal. What is the coefficient of kinetic friction between the box and the ramp?

73 In Fig. 6-54, the coefficient of kinetic friction between the block and inclined plane is 0.20, and angle θ is 60°. What are the (a) magnitude a and (b) direction (up or down the plane) of the block's acceleration if the block is sliding down the plane? What are (c) a and (d) the direction if the block is sent sliding up the plane?

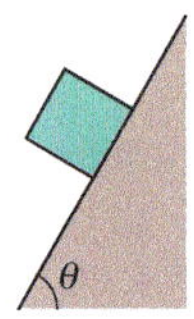

Figure 6-54 Problem 73.

74 A 110 g hockey puck sent sliding over ice is stopped in 15 m by the frictional force on it from the ice. (a) If its initial speed is 6.0 m/s, what is the magnitude of the frictional force? (b) What is the coefficient of friction between the puck and the ice?

75 A locomotive accelerates a 25-car train along a level track. Every car has a mass of 5.0×10^4 kg and is subject to a frictional force $f = 250v$, where the speed v is in meters per second and the force f is in newtons. At the instant when the speed of the train is 30 km/h, the magnitude of its acceleration is 0.20 m/s². (a) What is the tension in the coupling between the first car and the locomotive? (b) If this tension is equal to the maximum force the locomotive can exert on the train, what is the steepest grade up which the locomotive can pull the train at 30 km/h?

76 A house is built on the top of a hill with a nearby slope at angle $\theta = 45°$ (Fig. 6-55). An engineering study indicates that the slope angle should be reduced because the top layers of soil along the slope might slip past the lower layers. If the coefficient of static friction between two such layers is 0.5, what is the least angle ϕ through which the present slope should be reduced to prevent slippage?

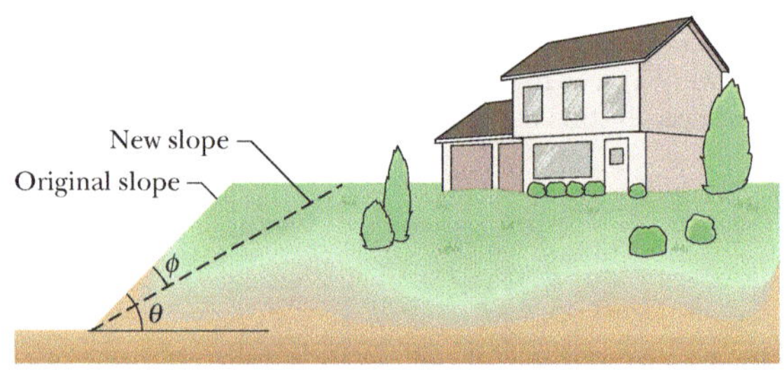

Figure 6-55 Problem 76.

77 What is the terminal speed of a 6.00 kg spherical ball that has a radius of 3.00 cm and a drag coefficient of 1.60? The density of the air through which the ball falls is 1.20 kg/m³.

78 A student wants to determine the coefficients of static friction and kinetic friction between a box and a plank. She places the box on the plank and gradually raises one end of the plank. When the angle of inclination with the horizontal reaches 30°, the box starts to slip, and it then slides 2.5 m down the plank in 4.0 s at constant acceleration. What are (a) the coefficient of static friction and (b) the coefficient of kinetic friction between the box and the plank?

79 SSM Block A in Fig. 6-56 has mass $m_A = 4.0$ kg, and block B has mass $m_B = 2.0$ kg. The coefficient of kinetic friction between block B and the horizontal plane is $\mu_k = 0.50$. The inclined plane is frictionless and at angle $\theta = 30°$. The pulley serves only to change the direction of the cord connecting the blocks. The cord has negligible mass. Find (a) the tension in the cord and (b) the magnitude of the acceleration of the blocks.

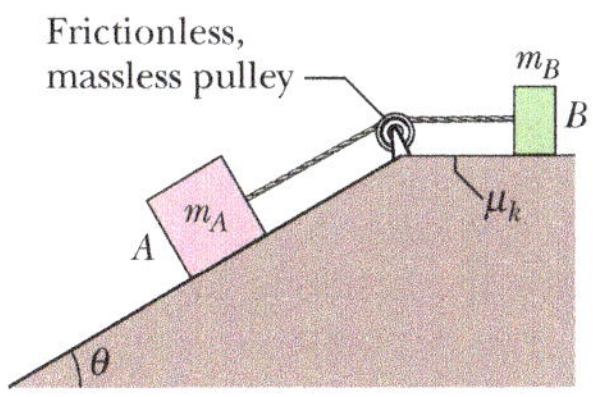

Figure 6-56 Problem 79.

80 Calculate the magnitude of the drag force on a missile 53 cm in diameter cruising at 250 m/s at low altitude, where the density of air is 1.2 kg/m^3. Assume $C = 0.75$.

81 SSM A bicyclist travels in a circle of radius 25.0 m at a constant speed of 9.00 m/s. The bicycle–rider mass is 85.0 kg. Calculate the magnitudes of (a) the force of friction on the bicycle from the road and (b) the *net* force on the bicycle from the road.

82 In Fig. 6-57, a stuntman drives a car (without negative lift) over the top of a hill, the cross section of which can be approximated by a circle of radius $R = 250$ m. What is the greatest speed at which he can drive without the car leaving the road at the top of the hill?

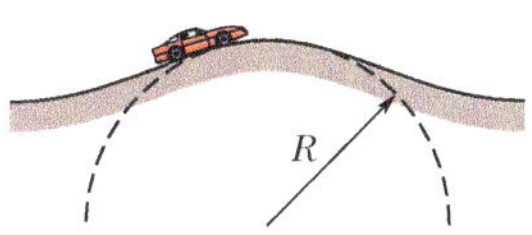

Figure 6-57 Problem 82.

83 You must push a crate across a floor to a docking bay. The crate weighs 165 N. The coefficient of static friction between crate and floor is 0.510, and the coefficient of kinetic friction is 0.32. Your force on the crate is directed horizontally. (a) What magnitude of your push puts the crate on the verge of sliding? (b) With what magnitude must you then push to keep the crate moving at a constant velocity? (c) If, instead, you then push with the same magnitude as the answer to (a), what is the magnitude of the crate's acceleration?

84 In Fig. 6-58, force $\vec{F}$ is applied to a crate of mass m on a floor where the coefficient of static friction between crate and floor is μ_s. Angle θ is initially 0° but is gradually increased so that the force vector rotates clockwise in the figure. During the rotation, the magnitude F of the force is continuously adjusted so that the crate is always on the verge of sliding. For $\mu_s = 0.70$, (a) plot the ratio F/mg versus θ and (b) determine the angle θ_{inf} at which the ratio approaches an infinite value. (c) Does lubricating the floor increase or decrease θ_{inf}, or is the value unchanged? (d) What is θ_{inf} for $\mu_s = 0.60$?

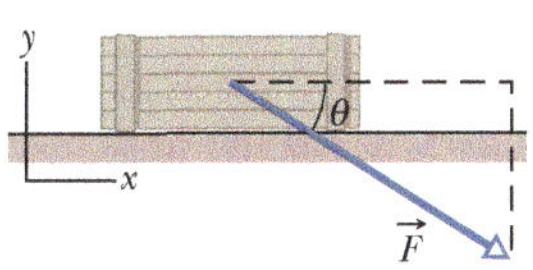

Figure 6-58 Problem 84.

85 In the early afternoon, a car is parked on a street that runs down a steep hill, at an angle of 35.0° relative to the horizontal. Just then the coefficient of static friction between the tires and the street surface is 0.725. Later, after nightfall, a sleet storm hits the area, and the coefficient decreases due to both the ice and a chemical change in the road surface because of the temperature decrease. By what percentage must the coefficient decrease if the car is to be in danger of sliding down the street?

86 A sling-thrower puts a stone (0.250 kg) in the sling's pouch (0.010 kg) and then begins to make the stone and pouch move in a vertical circle of radius 0.650 m. The cord between the pouch and the person's hand has negligible mass and will break when the tension in the cord is 33.0 N or more. Suppose the sling-thrower could gradually increase the speed of the stone. (a) Will the breaking occur at the lowest point of the circle or at the highest point? (b) At what speed of the stone will that breaking occur?

87 SSM A car weighing 10.7 kN and traveling at 13.4 m/s without negative lift attempts to round an unbanked curve with a radius of 61.0 m. (a) What magnitude of the frictional force on the tires is required to keep the car on its circular path? (b) If the coefficient of static friction between the tires and the road is 0.350, is the attempt at taking the curve successful?

88 In Fig. 6-59, block 1 of mass $m_1 = 2.0$ kg and block 2 of mass $m_2 = 1.0$ kg are connected by a string of negligible mass. Block 2 is pushed by force $\vec{F}$ of magnitude 20 N and angle $\theta = 35°$. The coefficient of kinetic friction between each block and the horizontal surface is 0.20. What is the tension in the string?

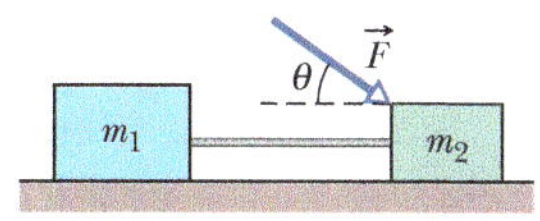

Figure 6-59 Problem 88.

89 SSM A filing cabinet weighing 556 N rests on the floor. The coefficient of static friction between it and the floor is 0.68, and the coefficient of kinetic friction is 0.56. In four different attempts to move it, it is pushed with horizontal forces of magnitudes (a) 222 N, (b) 334 N, (c) 445 N, and (d) 556 N. For each attempt, calculate the magnitude of the frictional force on it from the floor. (The cabinet is initially at rest.) (e) In which of the attempts does the cabinet move?

90 In Fig. 6-60, a block weighing 22 N is held at rest against a vertical wall by a horizontal force $\vec{F}$ of magnitude 60 N. The coefficient of static friction between the wall and the block is 0.55, and the coefficient of kinetic friction between them is 0.38. In six experiments, a second force $\vec{P}$ is applied to the block and directed parallel to the wall with these magnitudes and directions: (a) 34 N, up, (b) 12 N, up, (c) 48 N, up, (d) 62 N, up, (e) 10 N, down, and (f) 18 N, down. In each experiment, what is the magnitude of the frictional force on the block? In which does the block move (g) up the wall and (h) down the wall? (i) In which is the frictional force directed down the wall?

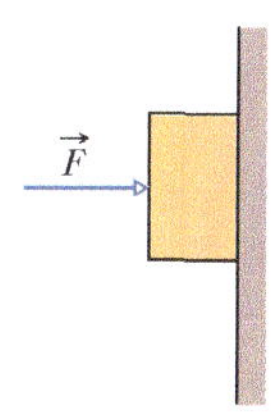

Figure 6-60 Problem 90.

91 SSM A block slides with constant velocity down an inclined plane that has slope angle θ. The block is then projected up the same plane with an initial speed v_0. (a) How far up the plane will it move before coming to rest? (b) After the block comes to rest, will it slide down the plane again? Give an argument to back your answer.

92 A circular curve of highway is designed for traffic moving at 60 km/h. Assume the traffic consists of cars without negative lift. (a) If the radius of the curve is 150 m, what is the correct angle of banking of the road? (b) If the curve were not banked, what would be the minimum coefficient of friction between tires and road that would keep traffic from skidding out of the turn when traveling at 60 km/h?

93 A 1.5 kg box is initially at rest on a horizontal surface when at $t = 0$ a horizontal force $\vec{F} = (1.8t)\hat{\text{i}}$ N (with t in seconds) is applied to the box. The acceleration of the box as a function of time t is given by $\vec{a} = 0$ for $0 \leq t \leq 2.8$ s and $\vec{a} = (1.2t - 2.4)\hat{\text{i}}$ m/s^2 for $t >$ 2.8 s. (a) What is the coefficient of static friction between the box and the surface? (b) What is the coefficient of kinetic friction between the box and the surface?

94 A child weighing 140 N sits at rest at the top of a playground slide that makes an angle of 25° with the horizontal. The child keeps from sliding by holding onto the sides of the slide. After letting go of the sides, the child has a constant acceleration of 0.86 m/s^2 (down the slide, of course). (a) What is the coefficient of kinetic friction between the child and the slide? (b) What maximum and minimum values for the coefficient of static friction between the child and the slide are consistent with the information given here?

95 In Fig. 6-61 a fastidious worker pushes directly along the handle of a mop with a force $\vec{F}$. The handle is at an angle θ with the vertical, and μ_s and μ_k are the coefficients of static and kinetic friction between the head of the mop and the floor. Ignore the mass of the handle and assume that all the mop's mass m is in its head. (a) If the mop head moves along the floor with a constant velocity, then what is F? (b) Show that if θ is less than a certain value θ_0, then $\vec{F}$ (still directed along the handle) is unable to move the mop head. Find θ_0.

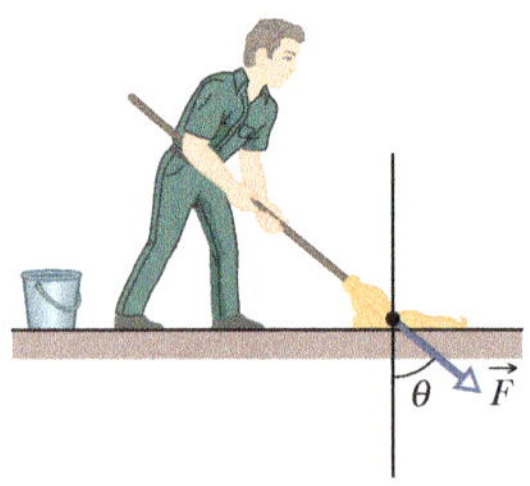

Figure 6-61 Problem 95.

96 A child places a picnic basket on the outer rim of a merry-go-round that has a radius of 4.6 m and revolves once every 30 s. (a) What is the speed of a point on that rim? (b) What is the lowest value of the coefficient of static friction between basket and merry-go-round that allows the basket to stay on the ride?

97 SSM A warehouse worker exerts a constant horizontal force of magnitude 85 N on a 40 kg box that is initially at rest on the horizontal floor of the warehouse. When the box has moved a distance of 1.4 m, its speed is 1.0 m/s. What is the coefficient of kinetic friction between the box and the floor?

98 In Fig. 6-62, a 5.0 kg block is sent sliding up a plane inclined at $\theta = 37°$ while a horizontal force $\vec{F}$ of magnitude 50 N acts on it. The coefficient of kinetic friction between block and plane is 0.30. What are the (a) magnitude and (b) direction (up or down the plane) of the block's acceleration? The block's initial speed is 4.0 m/s. (c) How far up the plane does the block go? (d) When it reaches its highest point, does it remain at rest or slide back down the plane?

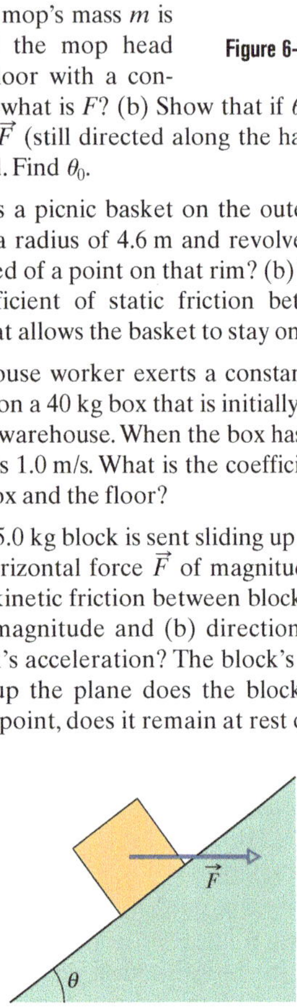

Figure 6-62 Problem 98.

99 An 11 kg block of steel is at rest on a horizontal table. The coefficient of static friction between block and table is 0.52. (a) What is the magnitude of the horizontal force that will put the block on the verge of moving? (b) What is the magnitude of a force acting upward 60° from the horizontal that will put the block on the verge of moving? (c) If the force acts downward at 60° from the horizontal, how large can its magnitude be without causing the block to move?

100 A ski that is placed on snow will stick to the snow. However, when the ski is moved along the snow, the rubbing warms and partially melts the snow, reducing the coefficient of kinetic friction and promoting sliding. Waxing the ski makes it water repellent and reduces friction with the resulting layer of water. A magazine reports that a new type of plastic ski is especially water repellent and that, on a gentle 200 m slope in the Alps, a skier reduced his top-to-bottom time from 61 s with standard skis to 42 s with the new skis. Determine the magnitude of his average acceleration with (a) the standard skis and (b) the new skis. Assuming a 3.0° slope, compute the coefficient of kinetic friction for (c) the standard skis and (d) the new skis.

101 Playing near a road construction site, a child falls over a barrier and down onto a dirt slope that is angled downward at 35° to the horizontal. As the child slides *down* the slope, he has an acceleration that has a magnitude of 0.50 m/s^2 and that is directed *up* the slope. What is the coefficient of kinetic friction between the child and the slope?

102 A 100 N force, directed at an angle θ above a horizontal floor, is applied to a 25.0 kg chair sitting on the floor. If $\theta = 0°$, what are (a) the horizontal component F_h of the applied force and (b) the magnitude F_N of the normal force of the floor on the chair? If $\theta = 30.0°$, what are (c) F_h and (d) F_N? If $\theta = 60.0°$, what are (e) F_h and (f) F_N? Now assume that the coefficient of static friction between chair and floor is 0.420. Does the chair slide or remain at rest if θ is (g) 0°, (h) 30.0°, and (i) 60.0°?

103 A certain string can withstand a maximum tension of 40 N without breaking. A child ties a 0.37 kg stone to one end and, holding the other end, whirls the stone in a vertical circle of radius 0.91 m, slowly increasing the speed until the string breaks. (a) Where is the stone on its path when the string breaks? (b) What is the speed of the stone as the string breaks?

104 A four-person bobsled (total mass = 630 kg) comes down a straightaway at the start of a bobsled run. The straightaway is 80.0 m long and is inclined at a constant angle of 10.2° with the horizontal. Assume that the combined effects of friction and air drag produce on the bobsled a constant force of 62.0 N that acts parallel to the incline and up the incline. Answer the following questions to three significant digits. (a) If the speed of the bobsled at the start of the run is 6.20 m/s, how long does the bobsled take to come down the straightaway? (b) Suppose the crew is able to reduce the effects of friction and air drag to 42.0 N. For the same initial velocity, how long does the bobsled now take to come down the straightaway?

105 As a 40 N block slides down a plane that is inclined at 25° to the horizontal, its acceleration is 0.80 m/s^2, directed up the plane. What is the coefficient of kinetic friction between the block and the plane?

C H A P T E R 7

Kinetic Energy and Work

7-1 KINETIC ENERGY

Learning Objectives

After reading this module, you should be able to . . .

7.01 Apply the relationship between a particle's kinetic energy, mass, and speed.

7.02 Identify that kinetic energy is a scalar quantity.

Key Idea

- The kinetic energy K associated with the motion of a particle of mass m and speed v, where v is well below the speed of light, is

$$K = \tfrac{1}{2}mv^2 \quad \text{(kinetic energy)}.$$

What Is Physics?

One of the fundamental goals of physics is to investigate something that everyone talks about: energy. The topic is obviously important. Indeed, our civilization is based on acquiring and effectively using energy.

For example, everyone knows that any type of motion requires energy: Flying across the Pacific Ocean requires it. Lifting material to the top floor of an office building or to an orbiting space station requires it. Throwing a fastball requires it. We spend a tremendous amount of money to acquire and use energy. Wars have been started because of energy resources. Wars have been ended because of a sudden, overpowering use of energy by one side. Everyone knows many examples of energy and its use, but what does the term *energy* really mean?

What Is Energy?

The term *energy* is so broad that a clear definition is difficult to write. Technically, energy is a scalar quantity associated with the state (or condition) of one or more objects. However, this definition is too vague to be of help to us now.

A looser definition might at least get us started. Energy is a number that we associate with a system of one or more objects. If a force changes one of the objects by, say, making it move, then the energy number changes. After countless experiments, scientists and engineers realized that if the scheme by which we assign energy numbers is planned carefully, the numbers can be used to predict the outcomes of experiments and, even more important, to build machines, such as flying machines. This success is based on a wonderful property of our universe: Energy can be transformed from one type to another and transferred from one object to another, but the total amount is always the same (energy is *conserved*). No exception to this *principle of energy conservation* has ever been found.

Money. Think of the many types of energy as being numbers representing money in many types of bank accounts. Rules have been made about what such money numbers mean and how they can be changed. You can transfer money numbers from one account to another or from one system to another, perhaps electronically with nothing material actually moving. However, the total amount (the total of all the money numbers) can always be accounted for: It is always conserved. In this chapter we focus on only one type of energy (*kinetic energy*) and on only one way in which energy can be transferred (*work*).

Kinetic Energy

Kinetic energy K is energy associated with the *state of motion* of an object. The faster the object moves, the greater is its kinetic energy. When the object is stationary, its kinetic energy is zero.

For an object of mass m whose speed v is well below the speed of light,

$$K = \tfrac{1}{2}mv^2 \quad \text{(kinetic energy).} \tag{7-1}$$

For example, a 3.0 kg duck flying past us at 2.0 m/s has a kinetic energy of 6.0 kg·m²/s²; that is, we associate that number with the duck's motion.

The SI unit of kinetic energy (and all types of energy) is the **joule** (J), named for James Prescott Joule, an English scientist of the 1800s and defined as

$$1 \text{ joule} = 1 \text{ J} = 1 \text{ kg}\cdot\text{m}^2/\text{s}^2. \tag{7-2}$$

Thus, the flying duck has a kinetic energy of 6.0 J.

Sample Problem 7.01 Kinetic energy, train crash

In 1896 in Waco, Texas, William Crush parked two locomotives at opposite ends of a 6.4-km-long track, fired them up, tied their throttles open, and then allowed them to crash head-on at full speed (Fig. 7-1) in front of 30,000 spectators. Hundreds of people were hurt by flying debris; several were killed. Assuming each locomotive weighed 1.2×10^6 N and its acceleration was a constant 0.26 m/s², what was the total kinetic energy of the two locomotives just before the collision?

KEY IDEAS

(1) We need to find the kinetic energy of each locomotive with Eq. 7-1, but that means we need each locomotive's speed just before the collision and its mass. (2) Because we can assume each locomotive had constant acceleration, we can use the equations in Table 2-1 to find its speed v just before the collision.

Calculations: We choose Eq. 2-16 because we know values for all the variables except v:

$$v^2 = v_0^2 + 2a(x - x_0).$$

With $v_0 = 0$ and $x - x_0 = 3.2 \times 10^3$ m (half the initial separation), this yields

$$v^2 = 0 + 2(0.26 \text{ m/s}^2)(3.2 \times 10^3 \text{ m}),$$

or

$$v = 40.8 \text{ m/s} = 147 \text{ km/h}.$$

We can find the mass of each locomotive by dividing its given weight by g:

$$m = \frac{1.2 \times 10^6 \text{ N}}{9.8 \text{ m/s}^2} = 1.22 \times 10^5 \text{ kg}.$$

Now, using Eq. 7-1, we find the total kinetic energy of the two locomotives just before the collision as

$$K = 2(\tfrac{1}{2}mv^2) = (1.22 \times 10^5 \text{ kg})(40.8 \text{ m/s})^2$$
$$= 2.0 \times 10^8 \text{ J.} \qquad \text{(Answer)}$$

This collision was like an exploding bomb.

Courtesy Library of Congress

Figure 7-1 The aftermath of an 1896 crash of two locomotives.

Additional examples, video, and practice available at *WileyPLUS*

7-2 WORK AND KINETIC ENERGY

Learning Objectives

After reading this module, you should be able to . . .

7.03 Apply the relationship between a force (magnitude and direction) and the work done on a particle by the force when the particle undergoes a displacement.

7.04 Calculate work by taking a dot product of the force vector and the displacement vector, in either magnitude-angle or unit-vector notation.

7.05 If multiple forces act on a particle, calculate the net work done by them.

7.06 Apply the work–kinetic energy theorem to relate the work done by a force (or the net work done by multiple forces) and the resulting change in kinetic energy.

Key Ideas

- Work W is energy transferred to or from an object via a force acting on the object. Energy transferred to the object is positive work, and from the object, negative work.
- The work done on a particle by a constant force $\vec{F}$ during displacement $\vec{d}$ is

$$W = Fd\cos\phi = \vec{F}\cdot\vec{d} \quad \text{(work, constant force)},$$

in which ϕ is the constant angle between the directions of $\vec{F}$ and $\vec{d}$.

- Only the component of $\vec{F}$ that is along the displacement $\vec{d}$ can do work on the object.
- When two or more forces act on an object, their net work is the sum of the individual works done by the forces, which is also equal to the work that would be done on the object by the net force $\vec{F}_{\text{net}}$ of those forces.
- For a particle, a change ΔK in the kinetic energy equals the net work W done on the particle:

$$\Delta K = K_f - K_i = W \quad \text{(work–kinetic energy theorem)},$$

in which K_i is the initial kinetic energy of the particle and K_f is the kinetic energy after the work is done. The equation rearranged gives us

$$K_f = K_i + W.$$

Work

If you accelerate an object to a greater speed by applying a force to the object, you increase the kinetic energy $K\,(=\frac{1}{2}mv^2)$ of the object. Similarly, if you decelerate the object to a lesser speed by applying a force, you decrease the kinetic energy of the object. We account for these changes in kinetic energy by saying that your force has transferred energy *to* the object from yourself or *from* the object to yourself. In such a transfer of energy via a force, **work** W is said to be *done on the object by the force.* More formally, we define work as follows:

> Work W is energy transferred to or from an object by means of a force acting on the object. Energy transferred to the object is positive work, and energy transferred from the object is negative work.

"Work," then, is transferred energy; "doing work" is the act of transferring the energy. Work has the same units as energy and is a scalar quantity.

The term *transfer* can be misleading. It does not mean that anything material flows into or out of the object; that is, the transfer is not like a flow of water. Rather, it is like the electronic transfer of money between two bank accounts: The number in one account goes up while the number in the other account goes down, with nothing material passing between the two accounts.

Note that we are not concerned here with the common meaning of the word "work," which implies that *any* physical or mental labor is work. For example, if you push hard against a wall, you tire because of the continuously repeated muscle contractions that are required, and you are, in the common sense, working. However, such effort does not cause an energy transfer to or from the wall and thus is not work done on the wall as defined here.

To avoid confusion in this chapter, we shall use the symbol W only for work and shall represent a weight with its equivalent mg.

Work and Kinetic Energy

Finding an Expression for Work

Let us find an expression for work by considering a bead that can slide along a frictionless wire that is stretched along a horizontal x axis (Fig. 7-2). A constant force $\vec{F}$, directed at an angle ϕ to the wire, accelerates the bead along the wire. We can relate the force and the acceleration with Newton's second law, written for components along the x axis:

$$F_x = ma_x, \tag{7-3}$$

where m is the bead's mass. As the bead moves through a displacement $\vec{d}$, the force changes the bead's velocity from an initial value $\vec{v}_0$ to some other value $\vec{v}$. Because the force is constant, we know that the acceleration is also constant. Thus, we can use Eq. 2-16 to write, for components along the x axis,

$$v^2 = v_0^2 + 2a_x d. \tag{7-4}$$

Solving this equation for a_x, substituting into Eq. 7-3, and rearranging then give us

$$\tfrac{1}{2}mv^2 - \tfrac{1}{2}mv_0^2 = F_x d. \tag{7-5}$$

The first term is the kinetic energy K_f of the bead at the end of the displacement d, and the second term is the kinetic energy K_i of the bead at the start. Thus, the left side of Eq. 7-5 tells us the kinetic energy has been changed by the force, and the right side tells us the change is equal to $F_x d$. Therefore, the work W done on the bead by the force (the energy transfer due to the force) is

$$W = F_x d. \tag{7-6}$$

If we know values for F_x and d, we can use this equation to calculate the work W.

To calculate the work a force does on an object as the object moves through some displacement, we use only the force component along the object's displacement. The force component perpendicular to the displacement does zero work.

From Fig. 7-2, we see that we can write F_x as $F \cos \phi$, where ϕ is the angle between the directions of the displacement $\vec{d}$ and the force $\vec{F}$. Thus,

$$W = Fd \cos \phi \quad \text{(work done by a constant force).} \tag{7-7}$$

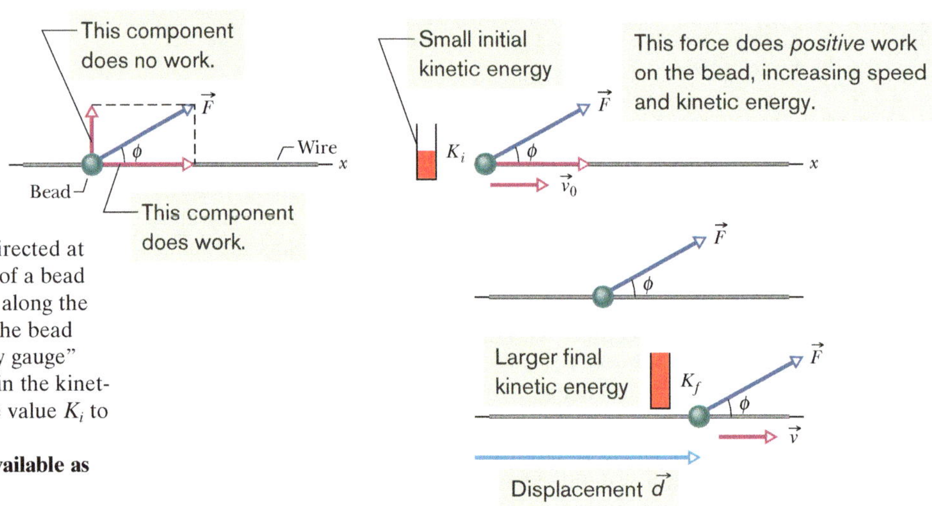

Figure 7-2 A constant force $\vec{F}$ directed at angle ϕ to the displacement $\vec{d}$ of a bead on a wire accelerates the bead along the wire, changing the velocity of the bead from $\vec{v}_0$ to $\vec{v}$. A "kinetic energy gauge" indicates the resulting change in the kinetic energy of the bead, from the value K_i to the value K_f.
In *WileyPLUS*, this figure is available as an animation with voiceover.

We can use the definition of the scaler (dot) product (Eq. 3-20) to write

$$W = \vec{F} \cdot \vec{d} \quad \text{(work done by a constant force),} \qquad (7\text{-}8)$$

where F is the magnitude of $\vec{F}$. (You may wish to review the discussion of scaler products in Module 3-3.) Equation 7-8 is especially useful for calculating the work when $\vec{F}$ and $\vec{d}$ are given in unit-vector notation.

Cautions. There are two restrictions to using Eqs. 7-6 through 7-8 to calculate work done on an object by a force. First, the force must be a *constant force;* that is, it must not change in magnitude or direction as the object moves. (Later, we shall discuss what to do with a *variable force* that changes in magnitude.) Second, the object must be *particle-like.* This means that the object must be *rigid;* all parts of it must move together, in the same direction. In this chapter we consider only particle-like objects, such as the bed and its occupant being pushed in Fig. 7-3.

Signs for Work. The work done on an object by a force can be either positive work or negative work. For example, if angle ϕ in Eq. 7-7 is less than 90°, then cos ϕ is positive and thus so is the work. However, if ϕ is greater than 90° (up to 180°), then cos ϕ is negative and thus so is the work. (Can you see that the work is zero when $\phi = 90°$?) These results lead to a simple rule. To find the sign of the work done by a force, consider the force vector component that is parallel to the displacement:

> A force does positive work when it has a vector component in the same direction as the displacement, and it does negative work when it has a vector component in the opposite direction. It does zero work when it has no such vector component.

Units for Work. Work has the SI unit of the joule, the same as kinetic energy. However, from Eqs. 7-6 and 7-7 we can see that an equivalent unit is the newton-meter (N·m). The corresponding unit in the British system is the foot-pound (ft·lb). Extending Eq. 7-2, we have

$$1\ \text{J} = 1\ \text{kg}\cdot\text{m}^2/\text{s}^2 = 1\ \text{N}\cdot\text{m} = 0.738\ \text{ft}\cdot\text{lb}. \qquad (7\text{-}9)$$

Net Work. When two or more forces act on an object, the **net work** done on the object is the sum of the works done by the individual forces. We can calculate the net work in two ways. (1) We can find the work done by each force and then sum those works. (2) Alternatively, we can first find the net force $\vec{F}_{net}$ of those forces. Then we can use Eq. 7-7, substituting the magnitude F_{net} for F and also the angle between the directions of $\vec{F}_{net}$ and $\vec{d}$ for ϕ. Similarly, we can use Eq. 7-8 with $\vec{F}_{net}$ substituted for $\vec{F}$.

Work–Kinetic Energy Theorem

Equation 7-5 relates the change in kinetic energy of the bead (from an initial $K_i = \frac{1}{2}mv_0^2$ to a later $K_f = \frac{1}{2}mv^2$) to the work W (= $F_x d$) done on the bead. For such particle-like objects, we can generalize that equation. Let ΔK be the change in the kinetic energy of the object, and let W be the net work done on it. Then

$$\Delta K = K_f - K_i = W, \qquad (7\text{-}10)$$

which says that

$$\begin{pmatrix}\text{change in the kinetic}\\ \text{energy of a particle}\end{pmatrix} = \begin{pmatrix}\text{net work done on}\\ \text{the particle}\end{pmatrix}.$$

We can also write

$$K_f = K_i + W, \qquad (7\text{-}11)$$

which says that

$$\begin{pmatrix}\text{kinetic energy after}\\ \text{the net work is done}\end{pmatrix} = \begin{pmatrix}\text{kinetic energy}\\ \text{before the net work}\end{pmatrix} + \begin{pmatrix}\text{the net}\\ \text{work done}\end{pmatrix}.$$

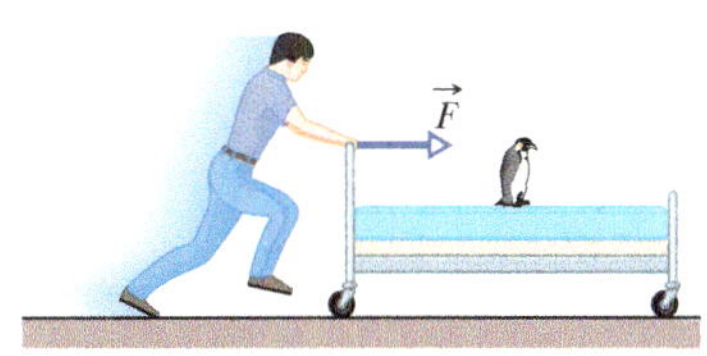

Figure 7-3 A contestant in a bed race. We can approximate the bed and its occupant as being a particle for the purpose of calculating the work done on them by the force applied by the contestant.

These statements are known traditionally as the **work–kinetic energy theorem** for particles. They hold for both positive and negative work: If the net work done on a particle is positive, then the particle's kinetic energy increases by the amount of the work. If the net work done is negative, then the particle's kinetic energy decreases by the amount of the work.

For example, if the kinetic energy of a particle is initially 5 J and there is a net transfer of 2 J to the particle (positive net work), the final kinetic energy is 7 J. If, instead, there is a net transfer of 2 J from the particle (negative net work), the final kinetic energy is 3 J.

Checkpoint 1

A particle moves along an x axis. Does the kinetic energy of the particle increase, decrease, or remain the same if the particle's velocity changes (a) from -3 m/s to -2 m/s and (b) from -2 m/s to 2 m/s? (c) In each situation, is the work done on the particle positive, negative, or zero?

Sample Problem 7.02 Work done by two constant forces, industrial spies

Figure 7-4a shows two industrial spies sliding an initially stationary 225 kg floor safe a displacement $\vec{d}$ of magnitude 8.50 m. The push $\vec{F}_1$ of spy 001 is 12.0 N at an angle of 30.0° downward from the horizontal; the pull $\vec{F}_2$ of spy 002 is 10.0 N at 40.0° above the horizontal. The magnitudes and directions of these forces do not change as the safe moves, and the floor and safe make frictionless contact.

(a) What is the net work done on the safe by forces $\vec{F}_1$ and $\vec{F}_2$ during the displacement $\vec{d}$?

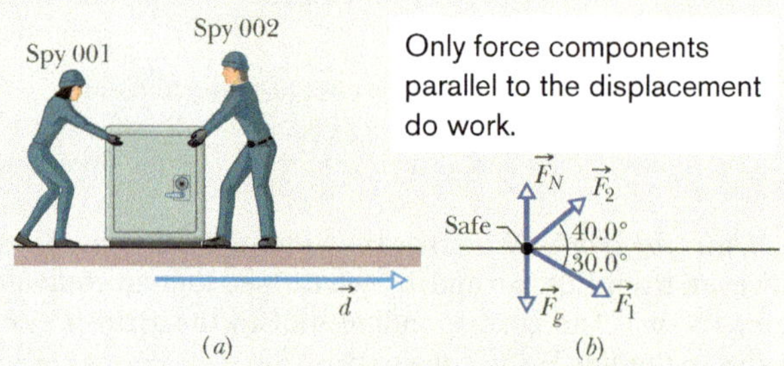

Figure 7-4 (a) Two spies move a floor safe through a displacement $\vec{d}$. (b) A free-body diagram for the safe.

KEY IDEAS

(1) The net work W done on the safe by the two forces is the sum of the works they do individually. (2) Because we can treat the safe as a particle and the forces are constant in both magnitude and direction, we can use either Eq. 7-7 ($W = Fd\cos\phi$) or Eq. 7-8 ($W = \vec{F}\cdot\vec{d}$) to calculate those works. Let's choose Eq. 7-7.

Calculations: From Eq. 7-7 and the free-body diagram for the safe in Fig. 7-4b, the work done by $\vec{F}_1$ is

$$W_1 = F_1 d\cos\phi_1 = (12.0\text{ N})(8.50\text{ m})(\cos 30.0°)$$
$$= 88.33\text{ J},$$

and the work done by $\vec{F}_2$ is

$$W_2 = F_2 d\cos\phi_2 = (10.0\text{ N})(8.50\text{ m})(\cos 40.0°)$$
$$= 65.11\text{ J}.$$

Thus, the net work W is

$$W = W_1 + W_2 = 88.33\text{ J} + 65.11\text{ J}$$
$$= 153.4\text{ J} \approx 153\text{ J}. \qquad \text{(Answer)}$$

During the 8.50 m displacement, therefore, the spies transfer 153 J of energy to the kinetic energy of the safe.

(b) During the displacement, what is the work W_g done on the safe by the gravitational force $\vec{F}_g$ and what is the work W_N done on the safe by the normal force $\vec{F}_N$ from the floor?

KEY IDEA

Because these forces are constant in both magnitude and direction, we can find the work they do with Eq. 7-7.

Calculations: Thus, with mg as the magnitude of the gravitational force, we write

$$W_g = mgd\cos 90° = mgd(0) = 0 \qquad \text{(Answer)}$$

and

$$W_N = F_N d\cos 90° = F_N d(0) = 0. \qquad \text{(Answer)}$$

We should have known this result. Because these forces are perpendicular to the displacement of the safe, they do zero work on the safe and do not transfer any energy to or from it.

(c) The safe is initially stationary. What is its speed v_f at the end of the 8.50 m displacement?

KEY IDEA

The speed of the safe changes because its kinetic energy is changed when energy is transferred to it by $\vec{F}_1$ and $\vec{F}_2$.

Calculations: We relate the speed to the work done by combining Eqs. 7-10 (the work–kinetic energy theorem) and 7-1 (the definition of kinetic energy):

$$W = K_f - K_i = \tfrac{1}{2}mv_f^2 - \tfrac{1}{2}mv_i^2.$$

The initial speed v_i is zero, and we now know that the work done is 153.4 J. Solving for v_f and then substituting known data, we find that

$$v_f = \sqrt{\frac{2W}{m}} = \sqrt{\frac{2(153.4\text{ J})}{225\text{ kg}}} = 1.17\text{ m/s}. \qquad \text{(Answer)}$$

Sample Problem 7.03 Work done by a constant force in unit-vector notation

During a storm, a crate of crepe is sliding across a slick, oily parking lot through a displacement $\vec{d} = (-3.0\text{ m})\hat{i}$ while a steady wind pushes against the crate with a force $\vec{F} = (2.0\text{ N})\hat{i} + (-6.0\text{ N})\hat{j}$. The situation and coordinate axes are shown in Fig. 7-5.

(a) How much work does this force do on the crate during the displacement?

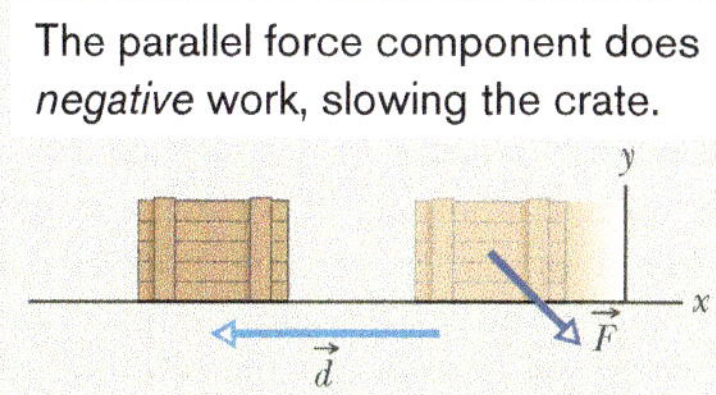

Figure 7-5 Force $\vec{F}$ slows a crate during displacement $\vec{d}$.

KEY IDEA

Because we can treat the crate as a particle and because the wind force is constant ("steady") in both magnitude and direction during the displacement, we can use either Eq. 7-7 ($W = Fd\cos\phi$) or Eq. 7-8 ($W = \vec{F}\cdot\vec{d}$) to calculate the work. Since we know $\vec{F}$ and $\vec{d}$ in unit-vector notation, we choose Eq. 7-8.

Calculations: We write

$$W = \vec{F}\cdot\vec{d} = [(2.0\text{ N})\hat{i} + (-6.0\text{ N})\hat{j}]\cdot[(-3.0\text{ m})\hat{i}].$$

Of the possible unit-vector dot products, only $\hat{i}\cdot\hat{i}$, $\hat{j}\cdot\hat{j}$, and $\hat{k}\cdot\hat{k}$ are nonzero (see Appendix E). Here we obtain

$$W = (2.0\text{ N})(-3.0\text{ m})\hat{i}\cdot\hat{i} + (-6.0\text{ N})(-3.0\text{ m})\hat{j}\cdot\hat{i}$$
$$= (-6.0\text{ J})(1) + 0 = -6.0\text{ J}. \qquad \text{(Answer)}$$

Thus, the force does a negative 6.0 J of work on the crate, transferring 6.0 J of energy from the kinetic energy of the crate.

(b) If the crate has a kinetic energy of 10 J at the beginning of displacement $\vec{d}$, what is its kinetic energy at the end of $\vec{d}$?

KEY IDEA

Because the force does negative work on the crate, it reduces the crate's kinetic energy.

Calculation: Using the work–kinetic energy theorem in the form of Eq. 7-11, we have

$$K_f = K_i + W = 10\text{ J} + (-6.0\text{ J}) = 4.0\text{ J}. \qquad \text{(Answer)}$$

Less kinetic energy means that the crate has been slowed.

WILEY PLUS Additional examples, video, and practice available at *WileyPLUS*

7-3 WORK DONE BY THE GRAVITATIONAL FORCE

Learning Objectives

After reading this module, you should be able to . . .

7.07 Calculate the work done by the gravitational force when an object is lifted or lowered.

7.08 Apply the work–kinetic energy theorem to situations where an object is lifted or lowered.

Key Ideas

- The work W_g done by the gravitational force $\vec{F}_g$ on a particle-like object of mass m as the object moves through a displacement $\vec{d}$ is given by

$$W_g = mgd\cos\phi,$$

in which ϕ is the angle between $\vec{F}_g$ and $\vec{d}$.

- The work W_a done by an applied force as a particle-like object is either lifted or lowered is related to the work W_g done by the gravitational force and the change ΔK in the object's kinetic energy by

$$\Delta K = K_f - K_i = W_a + W_g.$$

If $K_f = K_i$, then the equation reduces to

$$W_a = -W_g,$$

which tells us that the applied force transfers as much energy to the object as the gravitational force transfers from it.

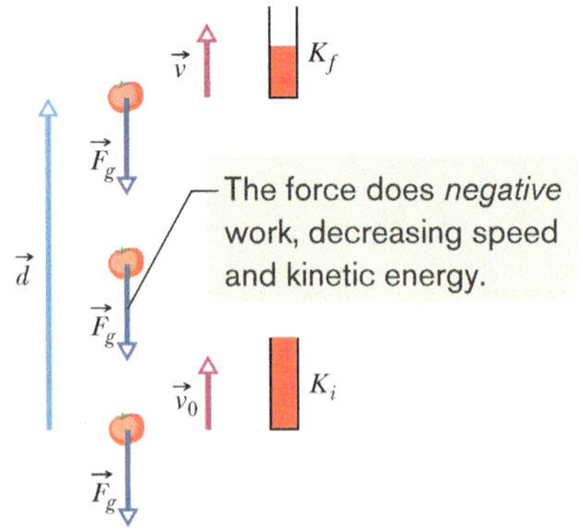

Figure 7-6 Because the gravitational force $\vec{F}_g$ acts on it, a particle-like tomato of mass m thrown upward slows from velocity $\vec{v}_0$ to velocity $\vec{v}$ during displacement $\vec{d}$. A kinetic energy gauge indicates the resulting change in the kinetic energy of the tomato, from $K_i (= \frac{1}{2}mv_0^2)$ to $K_f (= \frac{1}{2}mv^2)$.

Work Done by the Gravitational Force

We next examine the work done on an object by the gravitational force acting on it. Figure 7-6 shows a particle-like tomato of mass m that is thrown upward with initial speed v_0 and thus with initial kinetic energy $K_i = \frac{1}{2}mv_0^2$. As the tomato rises, it is slowed by a gravitational force $\vec{F}_g$; that is, the tomato's kinetic energy decreases because $\vec{F}_g$ does work on the tomato as it rises. Because we can treat the tomato as a particle, we can use Eq. 7-7 ($W = Fd \cos \phi$) to express the work done during a displacement $\vec{d}$. For the force magnitude F, we use mg as the magnitude of $\vec{F}_g$. Thus, the work W_g done by the gravitational force $\vec{F}_g$ is

$$W_g = mgd \cos \phi \quad \text{(work done by gravitational force).} \tag{7-12}$$

For a rising object, force $\vec{F}_g$ is directed opposite the displacement $\vec{d}$, as indicated in Fig. 7-6. Thus, $\phi = 180°$ and

$$W_g = mgd \cos 180° = mgd(-1) = -mgd. \tag{7-13}$$

The minus sign tells us that during the object's rise, the gravitational force acting on the object transfers energy in the amount mgd from the kinetic energy of the object. This is consistent with the slowing of the object as it rises.

After the object has reached its maximum height and is falling back down, the angle ϕ between force $\vec{F}_g$ and displacement $\vec{d}$ is zero. Thus,

$$W_g = mgd \cos 0° = mgd(+1) = +mgd. \tag{7-14}$$

The plus sign tells us that the gravitational force now transfers energy in the amount mgd to the kinetic energy of the falling object (it speeds up, of course).

Work Done in Lifting and Lowering an Object

Now suppose we lift a particle-like object by applying a vertical force $\vec{F}$ to it. During the upward displacement, our applied force does positive work W_a on the object while the gravitational force does negative work W_g on it. Our applied force tends to transfer energy to the object while the gravitational force tends to transfer energy from it. By Eq. 7-10, the change ΔK in the kinetic energy of the object due to these two energy transfers is

$$\Delta K = K_f - K_i = W_a + W_g, \tag{7-15}$$

in which K_f is the kinetic energy at the end of the displacement and K_i is that at the start of the displacement. This equation also applies if we lower the object, but then the gravitational force tends to transfer energy *to* the object while our force tends to transfer energy *from* it.

If an object is stationary before and after a lift (as when you lift a book from the floor to a shelf), then K_f and K_i are both zero, and Eq. 7-15 reduces to

$$W_a + W_g = 0$$

or

$$W_a = -W_g. \tag{7-16}$$

Note that we get the same result if K_f and K_i are not zero but are still equal. Either way, the result means that the work done by the applied force is the negative of the work done by the gravitational force; that is, the applied force transfers the same amount of energy to the object as the gravitational force transfers from the object. Using Eq. 7-12, we can rewrite Eq. 7-16 as

$$W_a = -mgd \cos \phi \quad \text{(work done in lifting and lowering; } K_f = K_i\text{),} \tag{7-17}$$

with ϕ being the angle between $\vec{F}_g$ and $\vec{d}$. If the displacement is vertically upward (Fig. 7-7*a*), then $\phi = 180°$ and the work done by the applied force equals mgd.

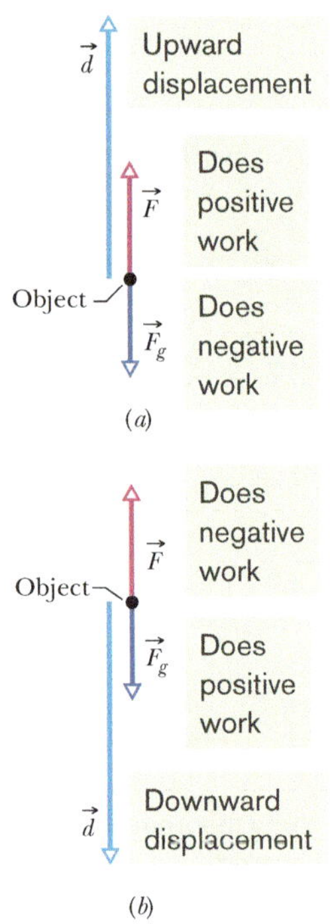

Figure 7-7 (*a*) An applied force $\vec{F}$ lifts an object. The object's displacement $\vec{d}$ makes an angle $\phi = 180°$ with the gravitational force $\vec{F}_g$ on the object. The applied force does positive work on the object. (*b*) An applied force $\vec{F}$ lowers an object. The displacement $\vec{d}$ of the object makes an angle $\phi = 0°$ with the gravitational force $\vec{F}_g$. The applied force does negative work on the object.

If the displacement is vertically downward (Fig. 7-7*b*), then $\phi = 0°$ and the work done by the applied force equals $-mgd$.

Equations 7-16 and 7-17 apply to any situation in which an object is lifted or lowered, with the object stationary before and after the lift. They are independent of the magnitude of the force used. For example, if you lift a mug from the floor to over your head, your force on the mug varies considerably during the lift. Still, because the mug is stationary before and after the lift, the work your force does on the mug is given by Eqs. 7-16 and 7-17, where, in Eq. 7-17, mg is the weight of the mug and d is the distance you lift it.

Sample Problem 7.04 Work in pulling a sleigh up a snowy slope

In this problem an object is pulled along a ramp but the object starts and ends at rest and thus has no overall change in its kinetic energy (that is important). Figure 7-8*a* shows the situation. A rope pulls a 200 kg sleigh (which you may know) up a slope at incline angle $\theta = 30°$, through distance $d = 20$ m. The sleigh and its contents have a total mass of 200 kg. The snowy slope is so slippery that we take it to be frictionless. How much work is done by each force acting on the sleigh?

KEY IDEAS

(1) During the motion, the forces are constant in magnitude and direction and thus we can calculate the work done by each with Eq. 7-7 ($W = Fd \cos \phi$) in which ϕ is the angle between the force and the displacement. We reach the same result with Eq. 7-8 ($W = \vec{F} \cdot \vec{d}$) in which we take a dot product of the force vector and displacement vector. (2) We can relate the net work done by the forces to the change in kinetic energy (or lack of a change, as here) with the work–kinetic energy theorem of Eq. 7-10 ($\Delta K = W$).

Calculations: The first thing to do with most physics problems involving forces is to draw a free-body diagram to organize our thoughts. For the sleigh, Fig. 7-8*b* is our free-body diagram, showing the gravitational force $\vec{F}_g$, the force $\vec{T}$ from the rope, and the normal force $\vec{F}_N$ from the slope.

Work W_N by the normal force. Let's start with this easy calculation. The normal force is perpendicular to the slope and thus also to the sleigh's displacement. Thus the normal force does not affect the sleigh's motion and does zero work. To be more formal, we can apply Eq. 7-7 to write

$$W_N = F_N d \cos 90° = 0. \qquad \text{(Answer)}$$

Work W_g by the gravitational force. We can find the work done by the gravitational force in either of two ways (you pick the more appealing way). From an earlier discussion about ramps (Sample Problem 5.04 and Fig. 5-15), we know that the component of the gravitational force along the slope has magnitude $mg \sin \theta$ and is directed down the slope. Thus the magnitude is

$$F_{gx} = mg \sin \theta = (200 \text{ kg})(9.8 \text{ m/s}^2) \sin 30° = 980 \text{ N}.$$

The angle ϕ between the displacement and this force component is 180°. So we can apply Eq. 7-7 to write

$$W_g = F_{gx} d \cos 180° = (980 \text{ N})(20 \text{ m})(-1) = -1.96 \times 10^4 \text{ J}. \qquad \text{(Answer)}$$

The negative result means that the gravitational force removes energy from the sleigh.

The second (equivalent) way to get this result is to use the full gravitational force $\vec{F}_g$ instead of a component. The angle between $\vec{F}_g$ and $\vec{d}$ is 120° (add the incline angle 30° to 90°). So, Eq. 7-7 gives us

$$W_g = F_g d \cos 120° = mgd \cos 120° = (200 \text{ kg})(9.8 \text{ m/s}^2)(20 \text{ m}) \cos 120° = -1.96 \times 10^4 \text{ J}. \qquad \text{(Answer)}$$

Work W_T by the rope's force. We have two ways of calculating this work. The quickest way is to use the work–kinetic energy theorem of Eq. 7-10 ($\Delta K = W$), where the net work W done by the forces is $W_N + W_g + W_T$ and the change ΔK in the kinetic energy is just zero (because the initial and final kinetic energies are the same—namely, zero). So, Eq. 7-10 gives us

$$0 = W_N + W_g + W_T = 0 - 1.96 \times 10^4 \text{ J} + W_T$$

and

$$W_T = 1.96 \times 10^4 \text{ J}. \qquad \text{(Answer)}$$

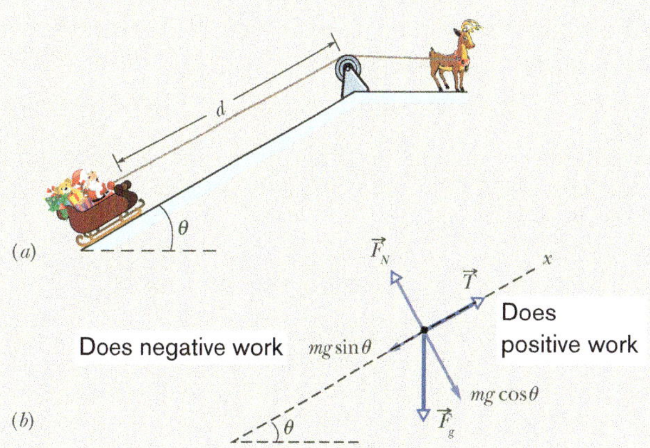

Figure 7-8 (*a*) A sleigh is pulled up a snowy slope. (*b*) The free-body diagram for the sleigh.

Instead of doing this, we can apply Newton's second law for motion along the x axis to find the magnitude F_T of the rope's force. Assuming that the acceleration along the slope is zero (except for the brief starting and stopping), we can write

$$F_{\text{net},x} = ma_x,$$
$$F_T - mg \sin 30° = m(0),$$

to find

$$F_T = mg \sin 30°.$$

This is the magnitude. Because the force and the displacement are both up the slope, the angle between those two vectors is zero. So, we can now write Eq. 7-7 to find the work done by the rope's force:

$$W_T = F_T d \cos 0° = (mg \sin 30°)d \cos 0°$$
$$= (200 \text{ kg})(9.8 \text{ m/s}^2)(\sin 30°)(20 \text{ m}) \cos 0°$$
$$= 1.96 \times 10^4 \text{ J}. \quad \text{(Answer)}$$

Sample Problem 7.05 Work done on an accelerating elevator cab

An elevator cab of mass $m = 500$ kg is descending with speed $v_i = 4.0$ m/s when its supporting cable begins to slip, allowing it to fall with constant acceleration $\vec{a} = \vec{g}/5$ (Fig. 7-9a).

(a) During the fall through a distance $d = 12$ m, what is the work W_g done on the cab by the gravitational force $\vec{F}_g$?

KEY IDEA

We can treat the cab as a particle and thus use Eq. 7-12 ($W_g = mgd \cos \phi$) to find the work W_g.

Calculation: From Fig. 7-9b, we see that the angle between the directions of $\vec{F}_g$ and the cab's displacement $\vec{d}$ is 0°. So,

$$W_g = mgd \cos 0° = (500 \text{ kg})(9.8 \text{ m/s}^2)(12 \text{ m})(1)$$
$$= 5.88 \times 10^4 \text{ J} \approx 59 \text{ kJ}. \quad \text{(Answer)}$$

(b) During the 12 m fall, what is the work W_T done on the cab by the upward pull $\vec{T}$ of the elevator cable?

KEY IDEA

We can calculate work W_T with Eq. 7-7 ($W = Fd \cos \phi$) by first writing $F_{\text{net},y} = ma_y$ for the components in Fig. 7-9b.

Calculations: We get

$$T - F_g = ma. \quad (7\text{-}18)$$

Solving for T, substituting mg for F_g, and then substituting the result in Eq. 7-7, we obtain

$$W_T = Td \cos \phi = m(a + g)d \cos \phi. \quad (7\text{-}19)$$

Next, substituting $-g/5$ for the (downward) acceleration a and then 180° for the angle ϕ between the directions of forces $\vec{T}$ and $m\vec{g}$, we find

$$W_T = m\left(-\frac{g}{5} + g\right)d \cos \phi = \frac{4}{5} mgd \cos \phi$$
$$= \frac{4}{5}(500 \text{ kg})(9.8 \text{ m/s}^2)(12 \text{ m}) \cos 180°$$
$$= -4.70 \times 10^4 \text{ J} \approx -47 \text{ kJ}. \quad \text{(Answer)}$$

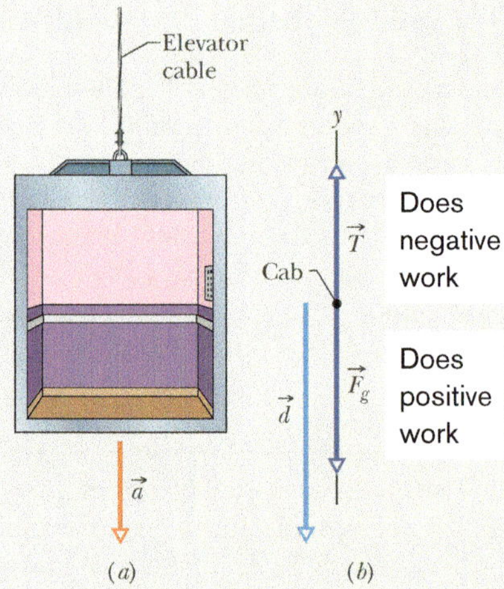

Figure 7-9 An elevator cab, descending with speed v_i, suddenly begins to accelerate downward. (a) It moves through a displacement $\vec{d}$ with constant acceleration $\vec{a} = \vec{g}/5$. (b) A free-body diagram for the cab, displacement included.

Caution: Note that W_T is not simply the negative of W_g because the cab accelerates during the fall. Thus, Eq. 7-16 (which assumes that the initial and final kinetic energies are equal) does not apply here.

(c) What is the net work W done on the cab during the fall?

Calculation: The net work is the sum of the works done by the forces acting on the cab:

$$W = W_g + W_T = 5.88 \times 10^4 \text{ J} - 4.70 \times 10^4 \text{ J}$$
$$= 1.18 \times 10^4 \text{ J} \approx 12 \text{ kJ}. \quad \text{(Answer)}$$

(d) What is the cab's kinetic energy at the end of the 12 m fall?

KEY IDEA

The kinetic energy changes *because* of the net work done on the cab, according to Eq. 7-11 ($K_f = K_i + W$).

Calculation: From Eq. 7-1, we write the initial kinetic energy as $K_i = \frac{1}{2}mv_i^2$. We then write Eq. 7-11 as

$$K_f = K_i + W = \tfrac{1}{2}mv_i^2 + W$$
$$= \tfrac{1}{2}(500 \text{ kg})(4.0 \text{ m/s})^2 + 1.18 \times 10^4 \text{ J}$$
$$= 1.58 \times 10^4 \text{ J} \approx 16 \text{ kJ}. \quad \text{(Answer)}$$

WILEY PLUS Additional examples, video, and practice available at *WileyPLUS*

7-4 WORK DONE BY A SPRING FORCE

Learning Objectives

After reading this module, you should be able to . . .

7.09 Apply the relationship (Hooke's law) between the force on an object due to a spring, the stretch or compression of the spring, and the spring constant of the spring.

7.10 Identify that a spring force is a variable force.

7.11 Calculate the work done on an object by a spring force by integrating the force from the initial position to the final position of the object or by using the known generic result of that integration.

7.12 Calculate work by graphically integrating on a graph of force versus position of the object.

7.13 Apply the work–kinetic energy theorem to situations in which an object is moved by a spring force.

Key Ideas

- The force $\vec{F}_s$ from a spring is

$$\vec{F}_s = -k\vec{d} \quad \text{(Hooke's law)},$$

where $\vec{d}$ is the displacement of the spring's free end from its position when the spring is in its relaxed state (neither compressed nor extended), and k is the spring constant (a measure of the spring's stiffness). If an x axis lies along the spring, with the origin at the location of the spring's free end when the spring is in its relaxed state, we can write

$$F_x = -kx \quad \text{(Hooke's law)}.$$

- A spring force is thus a variable force: It varies with the displacement of the spring's free end.
- If an object is attached to the spring's free end, the work W_s done on the object by the spring force when the object is moved from an initial position x_i to a final position x_f is

$$W_s = \tfrac{1}{2}kx_i^2 - \tfrac{1}{2}kx_f^2.$$

If $x_i = 0$ and $x_f = x$, then the equation becomes

$$W_s = -\tfrac{1}{2}kx^2.$$

Work Done by a Spring Force

We next want to examine the work done on a particle-like object by a particular type of *variable force*—namely, a **spring force**, the force from a spring. Many forces in nature have the same mathematical form as the spring force. Thus, by examining this one force, you can gain an understanding of many others.

The Spring Force

Figure 7-10*a* shows a spring in its **relaxed state**—that is, neither compressed nor extended. One end is fixed, and a particle-like object—a block, say—is attached to the other, free end. If we stretch the spring by pulling the block to the right as in Fig. 7-10*b*, the spring pulls on the block toward the left. (Because a spring force acts to restore the relaxed state, it is sometimes said to be a *restoring force.*) If we compress the spring by pushing the block to the left as in Fig. 7-10*c*, the spring now pushes on the block toward the right.

To a good approximation for many springs, the force $\vec{F}_s$ from a spring is proportional to the displacement $\vec{d}$ of the free end from its position when the spring is in the relaxed state. The *spring force* is given by

$$\vec{F}_s = -k\vec{d} \quad \text{(Hooke's law)}, \tag{7-20}$$

which is known as **Hooke's law** after Robert Hooke, an English scientist of the late 1600s. The minus sign in Eq. 7-20 indicates that the direction of the spring force is always opposite the direction of the displacement of the spring's free end. The constant k is called the **spring constant** (or **force constant**) and is a measure of the stiffness of the spring. The larger k is, the stiffer the spring; that is, the larger k is, the stronger the spring's pull or push for a given displacement. The SI unit for k is the newton per meter.

In Fig. 7-10 an x axis has been placed parallel to the length of the spring, with the origin ($x = 0$) at the position of the free end when the spring is in its relaxed

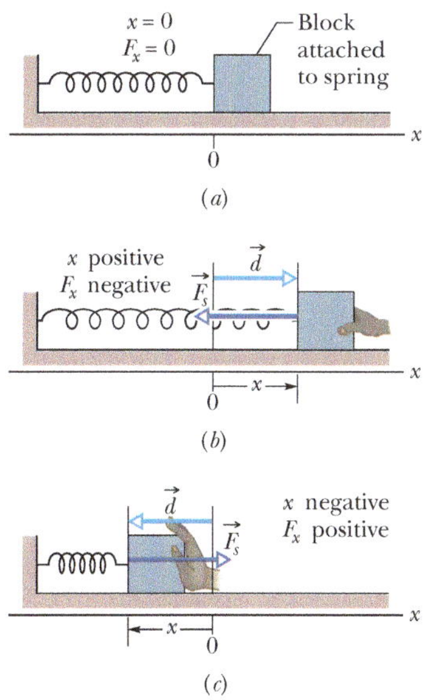

Figure 7-10 (*a*) A spring in its relaxed state. The origin of an x axis has been placed at the end of the spring that is attached to a block. (*b*) The block is displaced by $\vec{d}$, and the spring is stretched by a positive amount x. Note the restoring force $\vec{F}_s$ exerted by the spring. (*c*) The spring is compressed by a negative amount x. Again, note the restoring force.

state. For this common arrangement, we can write Eq. 7-20 as

$$F_x = -kx \quad \text{(Hooke's law)}, \tag{7-21}$$

where we have changed the subscript. If x is positive (the spring is stretched toward the right on the x axis), then F_x is negative (it is a pull toward the left). If x is negative (the spring is compressed toward the left), then F_x is positive (it is a push toward the right). Note that a spring force is a *variable force* because it is a function of x, the position of the free end. Thus F_x can be symbolized as $F(x)$. Also note that Hooke's law is a *linear* relationship between F_x and x.

The Work Done by a Spring Force

To find the work done by the spring force as the block in Fig. 7-10*a* moves, let us make two simplifying assumptions about the spring. (1) It is *massless;* that is, its mass is negligible relative to the block's mass. (2) It is an *ideal spring;* that is, it obeys Hooke's law exactly. Let us also assume that the contact between the block and the floor is frictionless and that the block is particle-like.

We give the block a rightward jerk to get it moving and then leave it alone. As the block moves rightward, the spring force F_x does work on the block, decreasing the kinetic energy and slowing the block. However, we *cannot* find this work by using Eq. 7-7 ($W = Fd \cos \phi$) because there is no one value of F to plug into that equation—the value of F increases as the block stretches the spring.

There is a neat way around this problem. (1) We break up the block's displacement into tiny segments that are so small that we can neglect the variation in F in each segment. (2) Then in each segment, the force has (approximately) a single value and thus we *can* use Eq. 7-7 to find the work in that segment. (3) Then we add up the work results for all the segments to get the total work. Well, that is our intent, but we don't really want to spend the next several days adding up a great many results and, besides, they would be only approximations. Instead, let's make the segments *infinitesimal* so that the error in each work result goes to zero. And then let's add up all the results by integration instead of by hand. Through the ease of calculus, we can do all this in minutes instead of days.

Let the block's initial position be x_i and its later position be x_f. Then divide the distance between those two positions into many segments, each of tiny length Δx. Label these segments, starting from x_i, as segments 1, 2, and so on. As the block moves through a segment, the spring force hardly varies because the segment is so short that x hardly varies. Thus, we can approximate the force magnitude as being constant within the segment. Label these magnitudes as F_{x1} in segment 1, F_{x2} in segment 2, and so on.

With the force now constant in each segment, we *can* find the work done within each segment by using Eq. 7-7. Here $\phi = 180°$, and so $\cos \phi = -1$. Then the work done is $-F_{x1}\,\Delta x$ in segment 1, $-F_{x2}\,\Delta x$ in segment 2, and so on. The net work W_s done by the spring, from x_i to x_f, is the sum of all these works:

$$W_s - \sum -F_{xj}\,\Delta x, \tag{7-22}$$

where j labels the segments. In the limit as Δx goes to zero, Eq. 7-22 becomes

$$W_s = \int_{x_i}^{x_f} -F_x\,dx. \tag{7-23}$$

From Eq. 7-21, the force magnitude F_x is kx. Thus, substitution leads to

$$\begin{aligned} W_s &= \int_{x_i}^{x_f} -kx\,dx = -k\int_{x_i}^{x_f} x\,dx \\ &= (-\tfrac{1}{2}k)[x^2]_{x_i}^{x_f} = (-\tfrac{1}{2}k)(x_f^2 - x_i^2). \end{aligned} \tag{7-24}$$

Multiplied out, this yields

$$W_s = \tfrac{1}{2}kx_i^2 - \tfrac{1}{2}kx_f^2 \quad \text{(work by a spring force).} \tag{7-25}$$

This work W_s done by the spring force can have a positive or negative value, depending on whether the *net* transfer of energy is to or from the block as the block moves from x_i to x_f. *Caution:* The final position x_f appears in the *second* term on the right side of Eq. 7-25. Therefore, Eq. 7-25 tells us:

Work W_s is positive if the block ends up closer to the relaxed position ($x = 0$) than it was initially. It is negative if the block ends up farther away from $x = 0$. It is zero if the block ends up at the same distance from $x = 0$.

If $x_i = 0$ and if we call the final position x, then Eq. 7-25 becomes

$$W_s = -\tfrac{1}{2}kx^2 \quad \text{(work by a spring force).} \tag{7-26}$$

The Work Done by an Applied Force

Now suppose that we displace the block along the x axis while continuing to apply a force $\vec{F}_a$ to it. During the displacement, our applied force does work W_a on the block while the spring force does work W_s. By Eq. 7-10, the change ΔK in the kinetic energy of the block due to these two energy transfers is

$$\Delta K = K_f - K_i = W_a + W_s, \tag{7-27}$$

in which K_f is the kinetic energy at the end of the displacement and K_i is that at the start of the displacement. If the block is stationary before and after the displacement, then K_f and K_i are both zero and Eq. 7-27 reduces to

$$W_a = -W_s. \tag{7-28}$$

If a block that is attached to a spring is stationary before and after a displacement, then the work done on it by the applied force displacing it is the negative of the work done on it by the spring force.

Caution: If the block is not stationary before and after the displacement, then this statement is *not* true.

Checkpoint 2

For three situations, the initial and final positions, respectively, along the x axis for the block in Fig. 7-10 are (a) -3 cm, 2 cm; (b) 2 cm, 3 cm; and (c) -2 cm, 2 cm. In each situation, is the work done by the spring force on the block positive, negative, or zero?

Sample Problem 7.06 Work done by a spring to change kinetic energy

When a spring does work on an object, we *cannot* find the work by simply multiplying the spring force by the object's displacement. The reason is that there is no one value for the force—it changes. However, we can split the displacement up into an infinite number of tiny parts and then approximate the force in each as being constant. Integration sums the work done in all those parts. Here we use the generic result of the integration.

In Fig. 7-11, a cumin canister of mass $m = 0.40$ kg slides across a horizontal frictionless counter with speed $v = 0.50$ m/s.

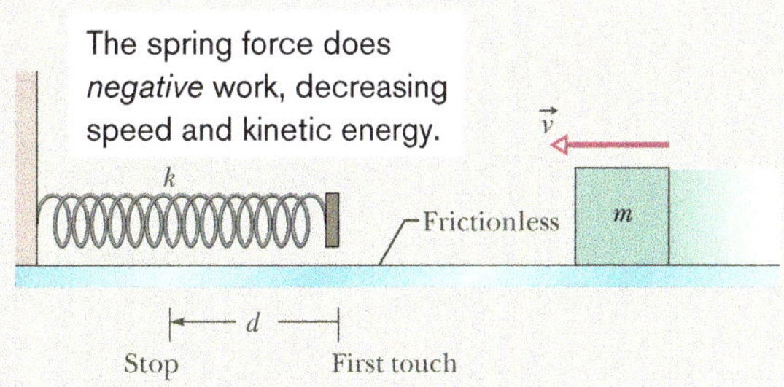

Figure 7-11 A canister moves toward a spring.

It then runs into and compresses a spring of spring constant $k = 750$ N/m. When the canister is momentarily stopped by the spring, by what distance d is the spring compressed?

KEY IDEAS

1. The work W_s done on the canister by the spring force is related to the requested distance d by Eq. 7-26 ($W_s = -\frac{1}{2}kx^2$), with d replacing x.
2. The work W_s is also related to the kinetic energy of the canister by Eq. 7-10 ($K_f - K_i = W$).
3. The canister's kinetic energy has an initial value of $K = \frac{1}{2}mv^2$ and a value of zero when the canister is momentarily at rest.

Calculations: Putting the first two of these ideas together, we write the work–kinetic energy theorem for the canister as

$$K_f - K_i = -\tfrac{1}{2}kd^2.$$

Substituting according to the third key idea gives us this expression:

$$0 - \tfrac{1}{2}mv^2 = -\tfrac{1}{2}kd^2.$$

Simplifying, solving for d, and substituting known data then give us

$$d = v\sqrt{\frac{m}{k}} = (0.50\text{ m/s})\sqrt{\frac{0.40\text{ kg}}{750\text{ N/m}}}$$
$$= 1.2 \times 10^{-2}\text{ m} = 1.2\text{ cm}. \quad \text{(Answer)}$$

7-5 WORK DONE BY A GENERAL VARIABLE FORCE

Learning Objectives

After reading this module, you should be able to . . .

7.14 Given a variable force as a function of position, calculate the work done by it on an object by integrating the function from the initial to the final position of the object, in one or more dimensions.

7.15 Given a graph of force versus position, calculate the work done by graphically integrating from the initial position to the final position of the object.

7.16 Convert a graph of acceleration versus position to a graph of force versus position.

7.17 Apply the work–kinetic energy theorem to situations where an object is moved by a variable force.

Key Ideas

- When the force $\vec{F}$ on a particle-like object depends on the position of the object, the work done by $\vec{F}$ on the object while the object moves from an initial position r_i with coordinates (x_i, y_i, z_i) to a final position r_f with coordinates (x_f, y_f, z_f) must be found by integrating the force. If we assume that component F_x may depend on x but not on y or z, component F_y may depend on y but not on x or z, and component F_z may depend on z but not on x or y, then the work is

$$W = \int_{x_i}^{x_f} F_x\,dx + \int_{y_i}^{y_f} F_y\,dy + \int_{z_i}^{z_f} F_z\,dz.$$

- If $\vec{F}$ has only an x component, then this reduces to

$$W = \int_{x_i}^{x_f} F(x)\,dx.$$

Work Done by a General Variable Force

One-Dimensional Analysis

Let us return to the situation of Fig. 7-2 but now consider the force to be in the positive direction of the x axis and the force magnitude to vary with position x. Thus, as the bead (particle) moves, the magnitude $F(x)$ of the force doing work on it changes. Only the magnitude of this variable force changes, not its direction, and the magnitude at any position does not change with time.

Figure 7-12*a* shows a plot of such a *one-dimensional variable force.* We want an expression for the work done on the particle by this force as the particle moves from an initial point x_i to a final point x_f. However, we *cannot* use Eq. 7-7 ($W = Fd\cos\phi$) because it applies only for a constant force $\vec{F}$. Here, again, we shall use calculus. We divide the area under the curve of Fig. 7-12*a* into a number of narrow strips of width Δx (Fig. 7-12*b*). We choose Δx small enough to permit us to take the force $F(x)$ as being reasonably constant over that interval. We let $F_{j,\text{avg}}$ be the average value of $F(x)$ within the jth interval. Then in Fig. 7-12*b*, $F_{j,\text{avg}}$ is the height of the jth strip.

With $F_{j,\text{avg}}$ considered constant, the increment (small amount) of work ΔW_j done by the force in the jth interval is now approximately given by Eq. 7-7 and is

$$\Delta W_j = F_{j,\text{avg}}\,\Delta x. \tag{7-29}$$

In Fig. 7-12*b*, ΔW_j is then equal to the area of the jth rectangular, shaded strip.

To approximate the total work W done by the force as the particle moves from x_i to x_f, we add the areas of all the strips between x_i and x_f in Fig. 7-12*b*:

$$W = \sum \Delta W_j = \sum F_{j,\text{avg}}\,\Delta x. \tag{7-30}$$

Equation 7-30 is an approximation because the broken "skyline" formed by the tops of the rectangular strips in Fig. 7-12*b* only approximates the actual curve of $F(x)$.

We can make the approximation better by reducing the strip width Δx and using more strips (Fig. 7-12*c*). In the limit, we let the strip width approach zero; the number of strips then becomes infinitely large and we have, as an exact result,

$$W = \lim_{\Delta x \to 0} \sum F_{j,\text{avg}}\,\Delta x. \tag{7-31}$$

This limit is exactly what we mean by the integral of the function $F(x)$ between the limits x_i and x_f. Thus, Eq. 7-31 becomes

$$W = \int_{x_i}^{x_f} F(x)\,dx \quad \text{(work: variable force).} \tag{7-32}$$

If we know the function $F(x)$, we can substitute it into Eq. 7-32, introduce the proper limits of integration, carry out the integration, and thus find the work. (Appendix E contains a list of common integrals.) Geometrically, the work is equal to the area between the $F(x)$ curve and the x axis, between the limits x_i and x_f (shaded in Fig. 7-12*d*).

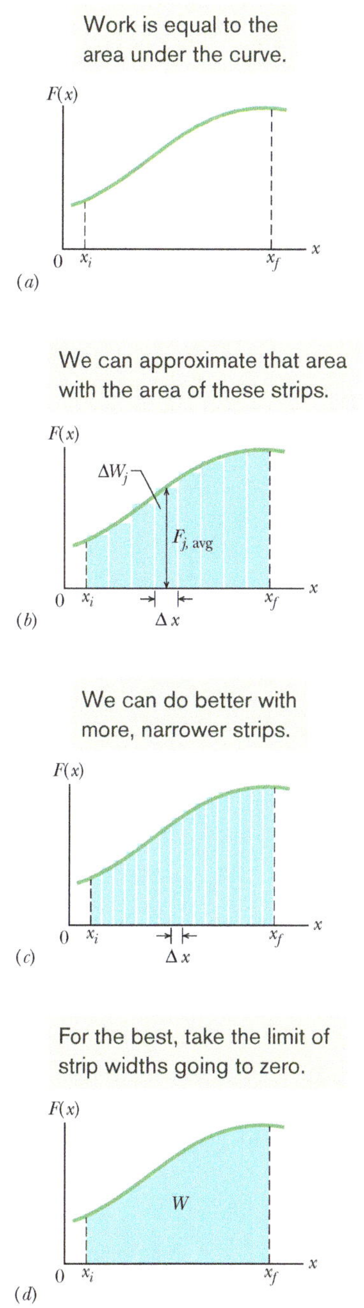

Figure 7-12 (*a*) A one-dimensional force $\vec{F}(x)$ plotted against the displacement x of a particle on which it acts. The particle moves from x_i to x_f. (*b*) Same as (*a*) but with the area under the curve divided into narrow strips. (*c*) Same as (*b*) but with the area divided into narrower strips. (*d*) The limiting case. The work done by the force is given by Eq. 7-32 and is represented by the shaded area between the curve and the x axis and between x_i and x_f.

Three-Dimensional Analysis

Consider now a particle that is acted on by a three-dimensional force

$$\vec{F} = F_x\hat{\mathrm{i}} + F_y\hat{\mathrm{j}} + F_z\hat{\mathrm{k}}, \tag{7-33}$$

in which the components F_x, F_y, and F_z can depend on the position of the particle; that is, they can be functions of that position. However, we make three simplifications: F_x may depend on x but not on y or z, F_y may depend on y but not on x or z, and F_z may depend on z but not on x or y. Now let the particle move through an incremental displacement

$$d\vec{r} = dx\hat{\mathrm{i}} + dy\hat{\mathrm{j}} + dz\hat{\mathrm{k}}. \tag{7-34}$$

The increment of work dW done on the particle by $\vec{F}$ during the displacement $d\vec{r}$ is, by Eq. 7-8,

$$dW = \vec{F}\cdot d\vec{r} = F_x\,dx + F_y\,dy + F_z\,dz. \tag{7-35}$$

The work W done by $\vec{F}$ while the particle moves from an initial position r_i having coordinates (x_i, y_i, z_i) to a final position r_f having coordinates (x_f, y_f, z_f) is then

$$W = \int_{r_i}^{r_f} dW = \int_{x_i}^{x_f} F_x\,dx + \int_{y_i}^{y_f} F_y\,dy + \int_{z_i}^{z_f} F_z\,dz. \qquad (7\text{-}36)$$

If $\vec{F}$ has only an x component, then the y and z terms in Eq. 7-36 are zero and the equation reduces to Eq. 7-32.

Work–Kinetic Energy Theorem with a Variable Force

Equation 7-32 gives the work done by a variable force on a particle in a one-dimensional situation. Let us now make certain that the work is equal to the change in kinetic energy, as the work–kinetic energy theorem states.

Consider a particle of mass m, moving along an x axis and acted on by a net force $F(x)$ that is directed along that axis. The work done on the particle by this force as the particle moves from position x_i to position x_f is given by Eq. 7-32 as

$$W = \int_{x_i}^{x_f} F(x)\,dx = \int_{x_i}^{x_f} ma\,dx, \qquad (7\text{-}37)$$

in which we use Newton's second law to replace $F(x)$ with ma. We can write the quantity $ma\,dx$ in Eq. 7-37 as

$$ma\,dx = m\frac{dv}{dt}\,dx. \qquad (7\text{-}38)$$

From the chain rule of calculus, we have

$$\frac{dv}{dt} = \frac{dv}{dx}\frac{dx}{dt} = \frac{dv}{dx}v, \qquad (7\text{-}39)$$

and Eq. 7-38 becomes

$$ma\,dx = m\frac{dv}{dx}v\,dx = mv\,dv. \qquad (7\text{-}40)$$

Substituting Eq. 7-40 into Eq. 7-37 yields

$$W = \int_{v_i}^{v_f} mv\,dv = m\int_{v_i}^{v_f} v\,dv$$

$$= \tfrac{1}{2}mv_f^2 - \tfrac{1}{2}mv_i^2. \qquad (7\text{-}41)$$

Note that when we change the variable from x to v we are required to express the limits on the integral in terms of the new variable. Note also that because the mass m is a constant, we are able to move it outside the integral.

Recognizing the terms on the right side of Eq. 7-41 as kinetic energies allows us to write this equation as

$$W = K_f - K_i = \Delta K,$$

which is the work–kinetic energy theorem.

Sample Problem 7.07 Work calculated by graphical integration

In Fig. 7-13b, an 8.0 kg block slides along a frictionless floor as a force acts on it, starting at $x_1 = 0$ and ending at $x_3 = 6.5$ m. As the block moves, the magnitude and direction of the force varies according to the graph shown in Fig. 7-13a. For example, from $x = 0$ to $x = 1$ m, the force is positive (in the positive direction of the x axis) and increases in magnitude from 0 to 40 N. And from $x = 4$ m to $x = 5$ m, the force is negative and increases in magnitude from 0 to 20 N.

(Note that this latter value is displayed as −20 N.) The block's kinetic energy at x_1 is K_1 = 280 J. What is the block's speed at $x_1 = 0$, $x_2 = 4.0$ m, and $x_3 = 6.5$ m?

KEY IDEAS

(1) At any point, we can relate the speed of the block to its kinetic energy with Eq. 7-1 ($K = \frac{1}{2}mv^2$). (2) We can relate the kinetic energy K_f at a later point to the initial kinetic K_i and the work W done on the block by using the work–kinetic energy theorem of Eq. 7-10 ($K_f - K_i = W$). (3) We can calculate the work W done by a variable force $F(x)$ by integrating the force versus position x. Equation 7-32 tells us that

$$W = \int_{x_i}^{x_f} F(x)\,dx.$$

We don't have a function $F(x)$ to carry out the integration, but we do have a graph of $F(x)$ where we can integrate by finding the area between the plotted line and the x axis. Where the plot is above the axis, the work (which is equal to the area) is positive. Where it is below the axis, the work is negative.

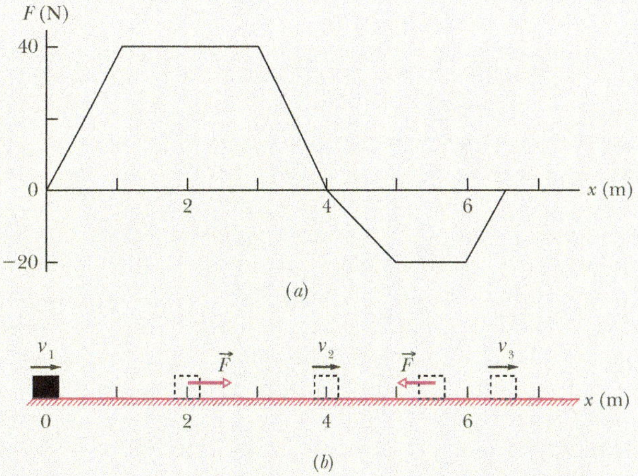

Figure 7-13 (*a*) A graph indicating the magnitude and direction of a variable force that acts on a block as it moves along an *x* axis on a floor. (*b*) The location of the block at several times.

Calculations: The requested speed at $x = 0$ is easy because we already know the kinetic energy. So, we just plug the kinetic energy into the formula for kinetic energy:

$$K_1 = \tfrac{1}{2}mv_1^2,$$

$$280\ \text{J} = \tfrac{1}{2}(8.0\ \text{kg})v_1^2,$$

and then

$$v_1 = 8.37\ \text{m/s} \approx 8.4\ \text{m/s}. \qquad \text{(Answer)}$$

As the block moves from $x = 0$ to $x = 4.0$ m, the plot in Figure 7-13*a* is above the x axis, which means that positive work is being done on the block. We split the area under the plot into a triangle at the left, a rectangle in the center, and a triangle at the right. Their total area is

$$\tfrac{1}{2}(40\ \text{N})(1\ \text{m}) + (40\ \text{N})(2\ \text{m}) + \tfrac{1}{2}(40\ \text{N})(1\ \text{m}) = 120\ \text{N}\cdot\text{m}$$
$$= 120\ \text{J}.$$

This means that between $x = 0$ and $x = 4.0$ m, the force does 120 J of work on the block, increasing the kinetic energy and speed of the block. So, when the block reaches $x = 4.0$ m, the work–kinetic energy theorem tells us that the kinetic energy is

$$K_2 = K_1 + W$$
$$= 280\ \text{J} + 120\ \text{J} = 400\ \text{J}.$$

Again using the definition of kinetic energy, we find

$$K_2 = \tfrac{1}{2}mv_2^2,$$

$$400\ \text{J} = \tfrac{1}{2}(8.0\ \text{kg})v_2^2,$$

and then

$$v_2 = 10\ \text{m/s}. \qquad \text{(Answer)}$$

This is the block's greatest speed because from $x = 4.0$ m to $x = 6.5$ m the force is negative, meaning that it opposes the block's motion, doing negative work on the block and thus decreasing the kinetic energy and speed. In that range, the area between the plot and the x axis is

$$\tfrac{1}{2}(20\ \text{N})(1\ \text{m}) + (20\ \text{N})(1\ \text{m}) + \tfrac{1}{2}(20\ \text{N})(0.5\ \text{m}) = 35\ \text{N}\cdot\text{m}$$
$$= 35\ \text{J}.$$

This means that the work done by the force in that range is −35 J. At $x = 4.0$ m, the block's $K = 400$ J. At $x = 6.5$ m, the work–kinetic energy theorem tells us that its kinetic energy is

$$K_3 = K_2 + W$$
$$= 400\ \text{J} - 35\ \text{J} = 365\ \text{J}.$$

Again using the definition of kinetic energy, we find

$$K_3 = \tfrac{1}{2}mv_3^2,$$

$$365\ \text{J} = \tfrac{1}{2}(8.0\ \text{kg})v_3^2,$$

and then

$$v_3 = 9.55\ \text{m/s} \approx 9.6\ \text{m/s}. \qquad \text{(Answer)}$$

The block is still moving in the positive direction of the x axis, a bit faster than initially.

WILEY PLUS Additional examples, video, and practice available at *WileyPLUS*

Sample Problem 7.08 Work, two-dimensional integration

When the force on an object depends on the position of the object, we *cannot* find the work done by it on the object by simply multiplying the force by the displacement. The reason is that there is no one value for the force—it changes. So, we must find the work in tiny little displacements and then add up all the work results. We effectively say, "Yes, the force varies over any given tiny little displacement, but the variation is so small we can approximate the force as being constant during the displacement." Sure, it is not precise, but if we make the displacements infinitesimal, then our error becomes infinitesimal and the result becomes precise. But, to add an infinite number of work contributions by hand would take us forever, longer than a semester. So, we add them up via an integration, which allows us to do all this in minutes (much less than a semester).

Force $\vec{F} = (3x^2\text{ N})\hat{i} + (4\text{ N})\hat{j}$, with x in meters, acts on a particle, changing only the kinetic energy of the particle. How much work is done on the particle as it moves from coordinates (2 m, 3 m) to (3 m, 0 m)? Does the speed of the particle increase, decrease, or remain the same?

KEY IDEA

The force is a variable force because its x component depends on the value of x. Thus, we cannot use Eqs. 7-7 and 7-8 to find the work done. Instead, we must use Eq. 7-36 to integrate the force.

Calculation: We set up two integrals, one along each axis:

$$W = \int_2^3 3x^2\,dx + \int_3^0 4\,dy = 3\int_2^3 x^2\,dx + 4\int_3^0 dy$$

$$= 3[\tfrac{1}{3}x^3]_2^3 + 4[y]_3^0 = [3^3 - 2^3] + 4[0 - 3]$$

$$= 7.0\text{ J}. \qquad \text{(Answer)}$$

The positive result means that energy is transferred to the particle by force $\vec{F}$. Thus, the kinetic energy of the particle increases and, because $K = \frac{1}{2}mv^2$, its speed must also increase. If the work had come out negative, the kinetic energy and speed would have decreased.

WileyPLUS Additional examples, video, and practice available at *WileyPLUS*

7-6 POWER

Learning Objectives

After reading this module, you should be able to . . .

7.18 Apply the relationship between average power, the work done by a force, and the time interval in which that work is done.

7.19 Given the work as a function of time, find the instantaneous power.

7.20 Determine the instantaneous power by taking a dot product of the force vector and an object's velocity vector, in magnitude-angle and unit-vector notations.

Key Ideas

- The power due to a force is the *rate* at which that force does work on an object.
- If the force does work W during a time interval Δt, the average power due to the force over that time interval is

$$P_{\text{avg}} = \frac{W}{\Delta t}.$$

- Instantaneous power is the instantaneous rate of doing work:

$$P = \frac{dW}{dt}.$$

- For a force $\vec{F}$ at an angle ϕ to the direction of travel of the instantaneous velocity $\vec{v}$, the instantaneous power is

$$P = Fv\cos\phi = \vec{F}\cdot\vec{v}.$$

Power

The time rate at which work is done by a force is said to be the **power** due to the force. If a force does an amount of work W in an amount of time Δt, the **average power** due to the force during that time interval is

$$P_{\text{avg}} = \frac{W}{\Delta t} \quad \text{(average power)}. \tag{7-42}$$

The **instantaneous power** P is the instantaneous time rate of doing work, which we can write as

$$P = \frac{dW}{dt} \quad \text{(instantaneous power).} \tag{7-43}$$

Suppose we know the work $W(t)$ done by a force as a function of time. Then to get the instantaneous power P at, say, time $t = 3.0$ s during the work, we would first take the time derivative of $W(t)$ and then evaluate the result for $t = 3.0$ s.

The SI unit of power is the joule per second. This unit is used so often that it has a special name, the **watt** (W), after James Watt, who greatly improved the rate at which steam engines could do work. In the British system, the unit of power is the foot-pound per second. Often the horsepower is used. These are related by

$$1 \text{ watt} = 1 \text{ W} = 1 \text{ J/s} = 0.738 \text{ ft}\cdot\text{lb/s} \tag{7-44}$$

and

$$1 \text{ horsepower} = 1 \text{ hp} = 550 \text{ ft}\cdot\text{lb/s} = 746 \text{ W}. \tag{7-45}$$

Inspection of Eq. 7-42 shows that work can be expressed as power multiplied by time, as in the common unit kilowatt-hour. Thus,

$$1 \text{ kilowatt-hour} = 1 \text{ kW}\cdot\text{h} = (10^3 \text{ W})(3600 \text{ s})$$
$$= 3.60 \times 10^6 \text{ J} = 3.60 \text{ MJ}. \tag{7-46}$$

Perhaps because they appear on our utility bills, the watt and the kilowatt-hour have become identified as electrical units. They can be used equally well as units for other examples of power and energy. Thus, if you pick up a book from the floor and put it on a tabletop, you are free to report the work that you have done as, say, 4×10^{-6} kW·h (or more conveniently as 4 mW·h).

We can also express the rate at which a force does work on a particle (or particle-like object) in terms of that force and the particle's velocity. For a particle that is moving along a straight line (say, an x axis) and is acted on by a constant force $\vec{F}$ directed at some angle ϕ to that line, Eq. 7-43 becomes

$$P = \frac{dW}{dt} = \frac{F\cos\phi\, dx}{dt} = F\cos\phi\left(\frac{dx}{dt}\right),$$

or

$$P = Fv\cos\phi. \tag{7-47}$$

Reorganizing the right side of Eq. 7-47 as the dot product $\vec{F}\cdot\vec{v}$, we may also write the equation as

$$P = \vec{F}\cdot\vec{v} \quad \text{(instantaneous power).} \tag{7-48}$$

For example, the truck in Fig. 7-14 exerts a force $\vec{F}$ on the trailing load, which has velocity $\vec{v}$ at some instant. The instantaneous power due to $\vec{F}$ is the rate at which $\vec{F}$ does work on the load at that instant and is given by Eqs. 7-47 and 7-48. Saying that this power is "the power of the truck" is often acceptable, but keep in mind what is meant: Power is the rate at which the applied *force* does work.

Figure 7-14 The power due to the truck's applied force on the trailing load is the rate at which that force does work on the load.

Checkpoint 3

A block moves with uniform circular motion because a cord tied to the block is anchored at the center of a circle. Is the power due to the force on the block from the cord positive, negative, or zero?

Sample Problem 7.09 Power, force, and velocity

Here we calculate an instantaneous work—that is, the rate at which work is being done at any given instant rather than averaged over a time interval. Figure 7-15 shows constant forces $\vec{F}_1$ and $\vec{F}_2$ acting on a box as the box slides rightward across a frictionless floor. Force $\vec{F}_1$ is horizontal, with magnitude 2.0 N; force $\vec{F}_2$ is angled upward by 60° to the floor and has magnitude 4.0 N. The speed v of the box at a certain instant is 3.0 m/s. What is the power due to each force acting on the box at that instant, and what is the net power? Is the net power changing at that instant?

KEY IDEA

We want an instantaneous power, not an average power over a time period. Also, we know the box's velocity (rather than the work done on it).

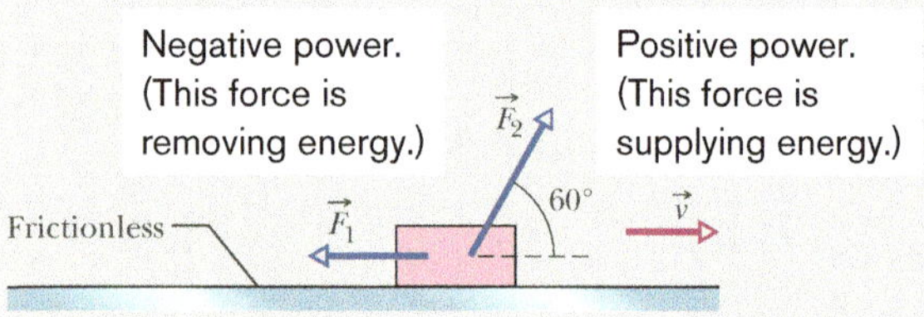

Figure 7-15 Two forces $\vec{F}_1$ and $\vec{F}_2$ act on a box that slides rightward across a frictionless floor. The velocity of the box is $\vec{v}$.

Calculation: We use Eq. 7-47 for each force. For force $\vec{F}_1$, at angle $\phi_1 = 180°$ to velocity $\vec{v}$, we have

$$P_1 = F_1 v \cos\phi_1 = (2.0\text{ N})(3.0\text{ m/s})\cos 180° = -6.0\text{ W}. \quad \text{(Answer)}$$

This negative result tells us that force $\vec{F}_1$ is transferring energy *from* the box at the rate of 6.0 J/s.

For force $\vec{F}_2$, at angle $\phi_2 = 60°$ to velocity $\vec{v}$, we have

$$P_2 = F_2 v \cos\phi_2 = (4.0\text{ N})(3.0\text{ m/s})\cos 60° = 6.0\text{ W}. \quad \text{(Answer)}$$

This positive result tells us that force $\vec{F}_2$ is transferring energy *to* the box at the rate of 6.0 J/s.

The net power is the sum of the individual powers (complete with their algebraic signs):

$$P_{net} = P_1 + P_2 = -6.0\text{ W} + 6.0\text{ W} = 0, \quad \text{(Answer)}$$

which tells us that the net rate of transfer of energy to or from the box is zero. Thus, the kinetic energy ($K = \frac{1}{2}mv^2$) of the box is not changing, and so the speed of the box will remain at 3.0 m/s. With neither the forces $\vec{F}_1$ and $\vec{F}_2$ nor the velocity $\vec{v}$ changing, we see from Eq. 7-48 that P_1 and P_2 are constant and thus so is P_{net}.

WILEY PLUS Additional examples, video, and practice available at *WileyPLUS*

Review & Summary

Kinetic Energy The **kinetic energy** K associated with the motion of a particle of mass m and speed v, where v is well below the speed of light, is

$$K = \tfrac{1}{2}mv^2 \quad \text{(kinetic energy)}. \qquad (7\text{-}1)$$

Work **Work** W is energy transferred to or from an object via a force acting on the object. Energy transferred to the object is positive work, and from the object, negative work.

Work Done by a Constant Force The work done on a particle by a constant force $\vec{F}$ during displacement $\vec{d}$ is

$$W = Fd\cos\phi = \vec{F}\cdot\vec{d} \quad \text{(work, constant force)}, \qquad (7\text{-}7, 7\text{-}8)$$

in which ϕ is the constant angle between the directions of $\vec{F}$ and $\vec{d}$. Only the component of $\vec{F}$ that is along the displacement $\vec{d}$ can do work on the object. When two or more forces act on an object, their **net work** is the sum of the individual works done by the forces, which is also equal to the work that would be done on the object by the net force $\vec{F}_{net}$ of those forces.

Work and Kinetic Energy For a particle, a change ΔK in the kinetic energy equals the net work W done on the particle:

$$\Delta K = K_f - K_i = W \quad \text{(work–kinetic energy theorem)}, \qquad (7\text{-}10)$$

in which K_i is the initial kinetic energy of the particle and K_f is the kinetic energy after the work is done. Equation 7-10 rearranged gives us

$$K_f = K_i + W. \qquad (7\text{-}11)$$

Work Done by the Gravitational Force The work W_g done by the gravitational force $\vec{F}_g$ on a particle-like object of mass m as the object moves through a displacement $\vec{d}$ is given by

$$W_g = mgd\cos\phi, \qquad (7\text{-}12)$$

in which ϕ is the angle between $\vec{F}_g$ and $\vec{d}$.

Work Done in Lifting and Lowering an Object The work W_a done by an applied force as a particle-like object is either lifted or lowered is related to the work W_g done by the gravitational force and the change ΔK in the object's kinetic energy by

$$\Delta K = K_f - K_i = W_a + W_g. \qquad (7\text{-}15)$$

If $K_f = K_i$, then Eq. 7-15 reduces to

$$W_a = -W_g, \qquad (7\text{-}16)$$

which tells us that the applied force transfers as much energy to the object as the gravitational force transfers from it.

Spring Force The force $\vec{F}_s$ from a spring is

$$\vec{F}_s = -k\vec{d} \quad \text{(Hooke's law)}, \tag{7-20}$$

where $\vec{d}$ is the displacement of the spring's free end from its position when the spring is in its **relaxed state** (neither compressed nor extended), and k is the **spring constant** (a measure of the spring's stiffness). If an x axis lies along the spring, with the origin at the location of the spring's free end when the spring is in its relaxed state, Eq. 7-20 can be written as

$$F_x = -kx \quad \text{(Hooke's law)}. \tag{7-21}$$

A spring force is thus a variable force: It varies with the displacement of the spring's free end.

Work Done by a Spring Force If an object is attached to the spring's free end, the work W_s done on the object by the spring force when the object is moved from an initial position x_i to a final position x_f is

$$W_s = \tfrac{1}{2}kx_i^2 - \tfrac{1}{2}kx_f^2. \tag{7-25}$$

If $x_i = 0$ and $x_f = x$, then Eq. 7-25 becomes

$$W_s = -\tfrac{1}{2}kx^2. \tag{7-26}$$

Work Done by a Variable Force When the force $\vec{F}$ on a particle-like object depends on the position of the object, the work done by $\vec{F}$ on the object while the object moves from an initial position r_i with coordinates (x_i, y_i, z_i) to a final position r_f with coordinates (x_f, y_f, z_f) must be found by integrating the force. If we assume that component F_x may depend on x but not on y or z, component F_y may depend on y but not on x or z, and component F_z may depend on z but not on x or y, then the work is

$$W = \int_{x_i}^{x_f} F_x\,dx + \int_{y_i}^{y_f} F_y\,dy + \int_{z_i}^{z_f} F_z\,dz. \tag{7-36}$$

If $\vec{F}$ has only an x component, then Eq. 7-36 reduces to

$$W = \int_{x_i}^{x_f} F(x)\,dx. \tag{7-32}$$

Power The **power** due to a force is the *rate* at which that force does work on an object. If the force does work W during a time interval Δt, the *average power* due to the force over that time interval is

$$P_{\text{avg}} = \frac{W}{\Delta t}. \tag{7-42}$$

Instantaneous power is the instantaneous rate of doing work:

$$P = \frac{dW}{dt}. \tag{7-43}$$

For a force $\vec{F}$ at an angle ϕ to the direction of travel of the instantaneous velocity $\vec{v}$, the instantaneous power is

$$P = Fv\cos\phi = \vec{F}\cdot\vec{v}. \tag{7-47, 7-48}$$

Questions

1 Rank the following velocities according to the kinetic energy a particle will have with each velocity, greatest first: (a) $\vec{v} = 4\hat{i} + 3\hat{j}$, (b) $\vec{v} = -4\hat{i} + 3\hat{j}$, (c) $\vec{v} = -3\hat{i} + 4\hat{j}$, (d) $\vec{v} = 3\hat{i} - 4\hat{j}$, (e) $\vec{v} = 5\hat{i}$, and (f) $v = 5$ m/s at 30° to the horizontal.

2 Figure 7-16*a* shows two horizontal forces that act on a block that is sliding to the right across a frictionless floor. Figure 7-16*b* shows three plots of the block's kinetic energy K versus time t. Which of the plots best corresponds to the following three situations: (a) $F_1 = F_2$, (b) $F_1 > F_2$, (c) $F_1 < F_2$?

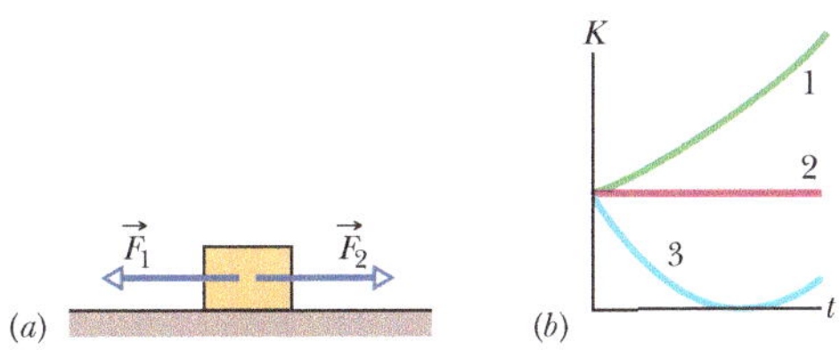

Figure 7-16 Question 2.

3 Is positive or negative work done by a constant force $\vec{F}$ on a particle during a straight-line displacement $\vec{d}$ if (a) the angle between $\vec{F}$ and $\vec{d}$ is 30°; (b) the angle is 100°; (c) $\vec{F} = 2\hat{i} - 3\hat{j}$ and $\vec{d} = -4\hat{i}$?

4 In three situations, a briefly applied horizontal force changes the velocity of a hockey puck that slides over frictionless ice. The overhead views of Fig. 7-17 indicate, for each situation, the puck's initial speed v_i, its final speed v_f, and the directions of the corresponding velocity vectors. Rank the situations according to the work done on the puck by the applied force, most positive first and most negative last.

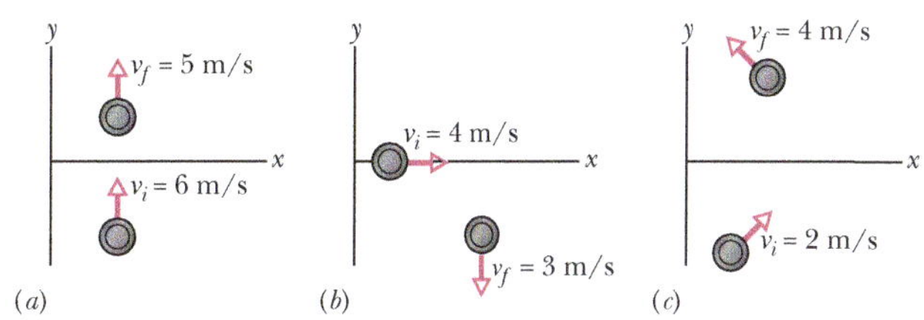

Figure 7-17 Question 4.

5 The graphs in Fig. 7-18 give the x component F_x of a force acting on a particle moving along an x axis. Rank them according to the work done by the force on the particle from $x = 0$ to $x = x_1$, from most positive work first to most negative work last.

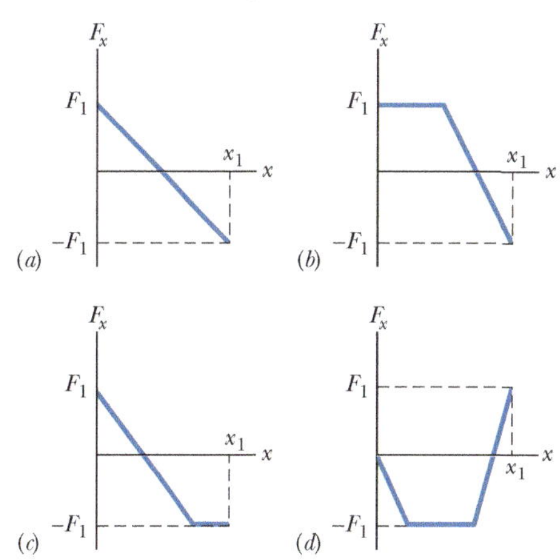

Figure 7-18 Question 5.

6 Figure 7-19 gives the x component F_x of a force that can act on a particle. If the particle begins at rest at $x = 0$, what is its coordinate when it has (a) its greatest kinetic energy, (b) its greatest speed, and (c) zero speed? (d) What is the particle's direction of travel after it reaches $x = 6$ m?

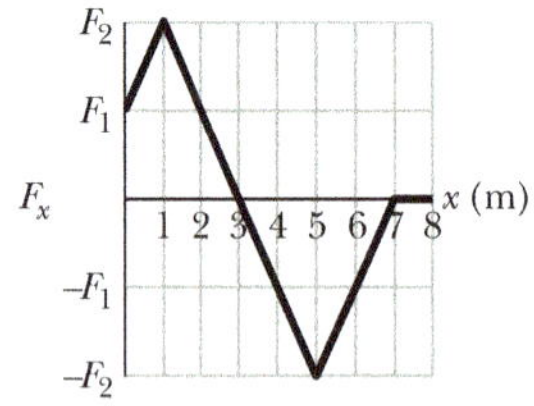

Figure 7-19 Question 6.

7 In Fig. 7-20, a greased pig has a choice of three frictionless slides along which to slide to the ground. Rank the slides according to how much work the gravitational force does on the pig during the descent, greatest first.

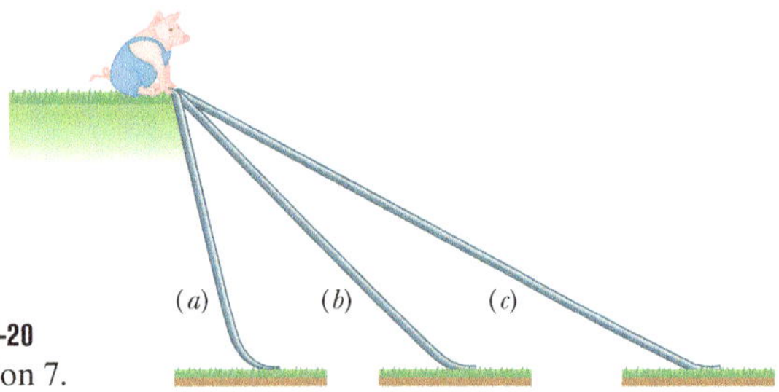

Figure 7-20 Question 7.

8 Figure 7-21a shows four situations in which a horizontal force acts on the same block, which is initially at rest. The force magnitudes are $F_2 = F_4 = 2F_1 = 2F_3$. The horizontal component v_x of the block's velocity is shown in Fig. 7-21b for the four situations. (a) Which plot in Fig. 7-21b best corresponds to which force in Fig. 7-21a? (b) Which plot in Fig. 7-21c (for kinetic energy K versus time t) best corresponds to which plot in Fig. 7-21b?

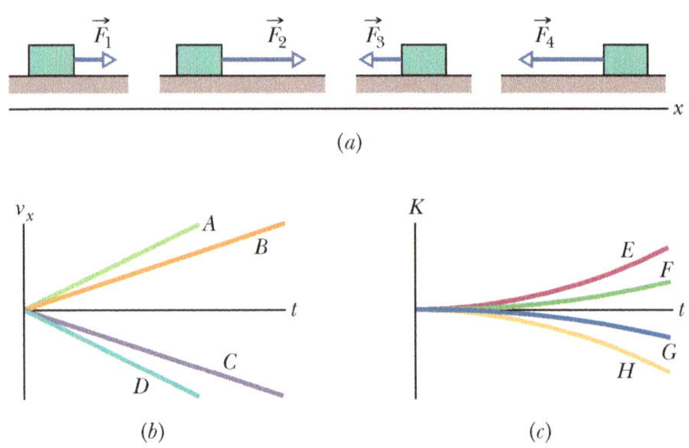

Figure 7-21 Question 8.

9 Spring A is stiffer than spring B ($k_A > k_B$). The spring force of which spring does more work if the springs are compressed (a) the same distance and (b) by the same applied force?

10 A glob of slime is launched or dropped from the edge of a cliff. Which of the graphs in Fig. 7-22 could possibly show how the kinetic energy of the glob changes during its flight?

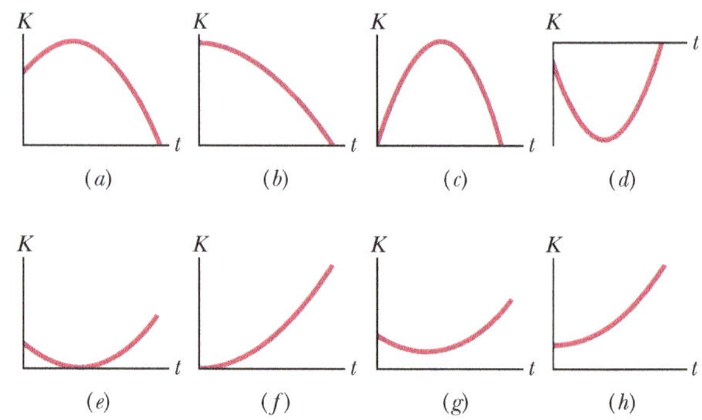

Figure 7-22 Question 10.

11 In three situations, a single force acts on a moving particle. Here are the velocities (at that instant) and the forces: (1) $\vec{v} = (-4\hat{i})$ m/s, $\vec{F} = (6\hat{i} - 20\hat{j})$ N; (2) $\vec{v} = (2\hat{i} - 3\hat{j})$ m/s, $\vec{F} = (-2\hat{j} + 7\hat{k})$ N; (3) $\vec{v} = (-3\hat{i} + \hat{j})$ m/s, $\vec{F} = (2\hat{i} + 6\hat{j})$ N. Rank the situations according to the rate at which energy is being transferred, greatest transfer to the particle ranked first, greatest transfer from the particle ranked last.

12 Figure 7-23 shows three arrangements of a block attached to identical springs that are in their relaxed state when the block is centered as shown. Rank the arrangements according to the magnitude of the net force on the block, largest first, when the block is displaced by distance d (a) to the right and (b) to the left. Rank the arrangements according to the work done on the block by the spring forces, greatest first, when the block is displaced by d (c) to the right and (d) to the left.

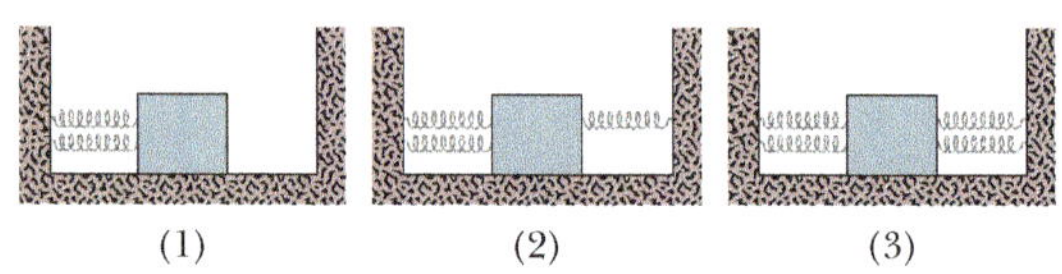

Figure 7-23 Question 12.

Problems

GO Tutoring problem available (at instructor's discretion) in *WileyPLUS* and WebAssign
SSM Worked-out solution available in Student Solutions Manual
• – ••• Number of dots indicates level of problem difficulty
Additional information available in *The Flying Circus of Physics* and at flyingcircusofphysics.com
WWW Worked-out solution is at
ILW Interactive solution is at
http://www.wiley.com/college/halliday

Module 7-1 Kinetic Energy

•1 SSM A proton (mass $m = 1.67 \times 10^{-27}$ kg) is being accelerated along a straight line at 3.6×10^{15} m/s^2 in a machine. If the proton has an initial speed of 2.4×10^{7} m/s and travels 3.5 cm, what then is (a) its speed and (b) the increase in its kinetic energy?

•2 If a Saturn V rocket with an Apollo spacecraft attached had a combined mass of 2.9×10^{5} kg and reached a speed of 11.2 km/s, how much kinetic energy would it then have?

•3 On August 10, 1972, a large meteorite skipped across the atmosphere above the western United States and western Canada, much like a stone skipped across water. The accompanying fireball was so bright that it could be seen in the daytime sky and was brighter than the usual meteorite trail. The meteorite's mass was about 4×10^6 kg; its speed was about 15 km/s. Had it entered the atmosphere vertically, it would have hit Earth's surface with about the same speed. (a) Calculate the meteorite's loss of kinetic energy (in joules) that would have been associated with the vertical impact. (b) Express the energy as a multiple of the explosive energy of 1 megaton of TNT, which is 4.2×10^{15} J. (c) The energy associated with the atomic bomb explosion over Hiroshima was equivalent to 13 kilotons of TNT. To how many Hiroshima bombs would the meteorite impact have been equivalent?

•4 An explosion at ground level leaves a crater with a diameter that is proportional to the energy of the explosion raised to the $\frac{1}{3}$ power; an explosion of 1 megaton of TNT leaves a crater with a 1 km diameter. Below Lake Huron in Michigan there appears to be an ancient impact crater with a 50 km diameter. What was the kinetic energy associated with that impact, in terms of (a) megatons of TNT (1 megaton yields 4.2×10^{15} J) and (b) Hiroshima bomb equivalents (13 kilotons of TNT each)? (Ancient meteorite or comet impacts may have significantly altered the climate, killing off the dinosaurs and other life-forms.)

••5 A father racing his son has half the kinetic energy of the son, who has half the mass of the father. The father speeds up by 1.0 m/s and then has the same kinetic energy as the son. What are the original speeds of (a) the father and (b) the son?

••6 A bead with mass 1.8×10^{-2} kg is moving along a wire in the positive direction of an x axis. Beginning at time $t = 0$, when the bead passes through $x = 0$ with speed 12 m/s, a constant force acts on the bead. Figure 7-24 indicates the bead's position at these four times: $t_0 = 0$, $t_1 = 1.0$ s, $t_2 = 2.0$ s, and $t_3 = 3.0$ s. The bead momentarily stops at $t = 3.0$ s. What is the kinetic energy of the bead at $t = 10$ s?

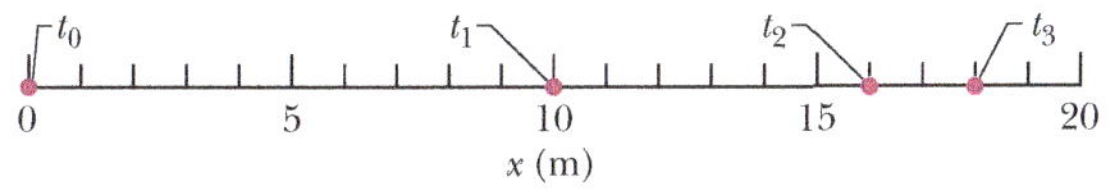

Figure 7-24 Problem 6.

Module 7-2 Work and Kinetic Energy

•7 A 3.0 kg body is at rest on a frictionless horizontal air track when a constant horizontal force $\vec{F}$ acting in the positive direction of an x axis along the track is applied to the body. A stroboscopic graph of the position of the body as it slides to the right is shown in Fig. 7-25. The force $\vec{F}$ is applied to the body at $t = 0$, and the graph records the position of the body at 0.50 s intervals. How much work is done on the body by the applied force $\vec{F}$ between $t = 0$ and $t = 2.0$ s?

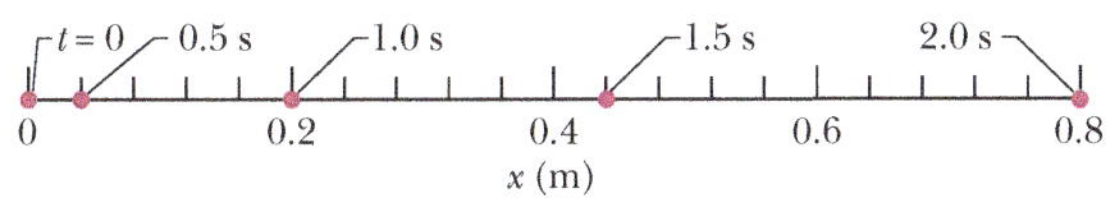

Figure 7-25 Problem 7.

•8 A ice block floating in a river is pushed through a displacement $\vec{d} = (15\text{ m})\hat{i} - (12\text{ m})\hat{j}$ along a straight embankment by rushing water, which exerts a force $\vec{F} = (210\text{ N})\hat{i} - (150\text{ N})\hat{j}$ on the block. How much work does the force do on the block during the displacement?

•9 The only force acting on a 2.0 kg canister that is moving in an xy plane has a magnitude of 5.0 N. The canister initially has a velocity of 4.0 m/s in the positive x direction and some time later has a velocity of 6.0 m/s in the positive y direction. How much work is done on the canister by the 5.0 N force during this time?

•10 A coin slides over a frictionless plane and across an xy coordinate system from the origin to a point with xy coordinates (3.0 m, 4.0 m) while a constant force acts on it. The force has magnitude 2.0 N and is directed at a counterclockwise angle of 100° from the positive direction of the x axis. How much work is done by the force on the coin during the displacement?

••11 A 12.0 N force with a fixed orientation does work on a particle as the particle moves through the three-dimensional displacement $\vec{d} = (2.00\hat{i} - 4.00\hat{j} + 3.00\hat{k})$ m. What is the angle between the force and the displacement if the change in the particle's kinetic energy is (a) +30.0 J and (b) −30.0 J?

••12 A can of bolts and nuts is pushed 2.00 m along an x axis by a broom along the greasy (frictionless) floor of a car repair shop in a version of shuffleboard. Figure 7-26 gives the work W done on the can by the constant horizontal force from the broom, versus the can's position x. The scale of the figure's vertical axis is set by $W_s = 6.0$ J. (a) What is the magnitude of that force? (b) If the can had an initial kinetic energy of 3.00 J, moving in the positive direction of the x axis, what is its kinetic energy at the end of the 2.00 m?

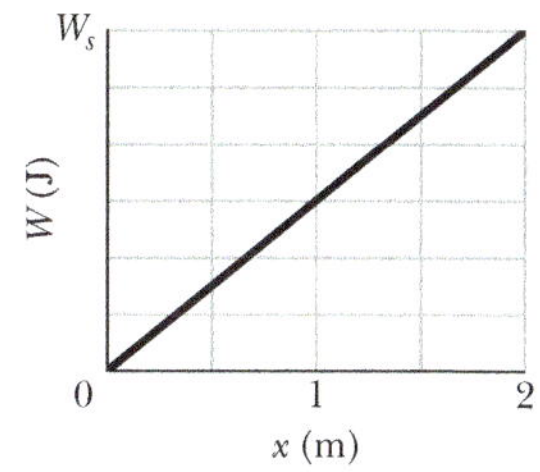

Figure 7-26 Problem 12.

••13 A luge and its rider, with a total mass of 85 kg, emerge from a downhill track onto a horizontal straight track with an initial speed of 37 m/s. If a force slows them to a stop at a constant rate of 2.0 m/s², (a) what magnitude F is required for the force, (b) what distance d do they travel while slowing, and (c) what work W is done on them by the force? What are (d) F, (e) d, and (f) W if they, instead, slow at 4.0 m/s²?

••14 Figure 7-27 shows an overhead view of three horizontal forces acting on a cargo canister that was initially stationary but now moves across a frictionless floor. The force magnitudes are $F_1 = 3.00$ N, $F_2 = 4.00$ N, and $F_3 = 10.0$ N, and the indicated angles are $\theta_2 = 50.0°$ and $\theta_3 = 35.0°$. What is the net work done on the canister by the three forces during the first 4.00 m of displacement?

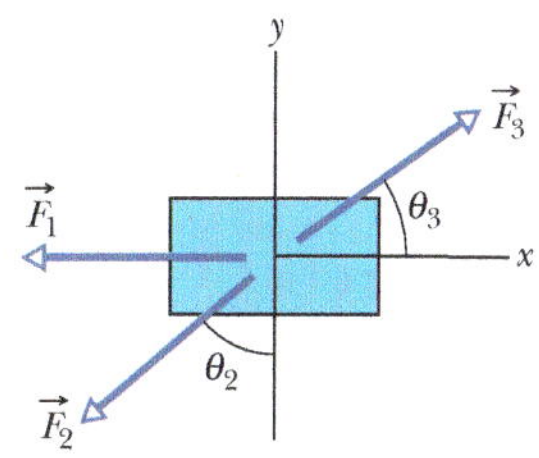

Figure 7-27 Problem 14.

••15 Figure 7-28 shows three forces applied to a trunk that moves leftward by 3.00 m over a frictionless floor. The force magnitudes are $F_1 = 5.00$ N, $F_2 = 9.00$ N, and $F_3 = 3.00$ N, and the indicated angle is $\theta = 60.0°$. During the displacement, (a) what is the net work done on the trunk by the three forces and (b) does the kinetic energy of the trunk increase or decrease?

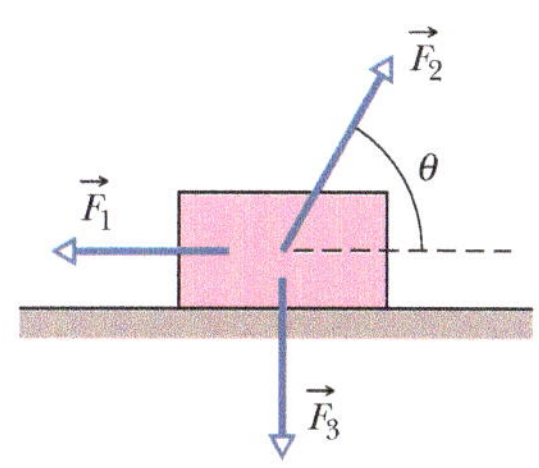

Figure 7-28 Problem 15.

••16 An 8.0 kg object is moving in the positive direction of an x axis. When it passes through $x = 0$, a constant force directed along the axis begins to act on it. Figure 7-29 gives its kinetic energy K versus position x as it moves from $x = 0$ to $x = 5.0$ m; $K_0 = 30.0$ J. The force continues to act. What is v when the object moves back through $x = -3.0$ m?

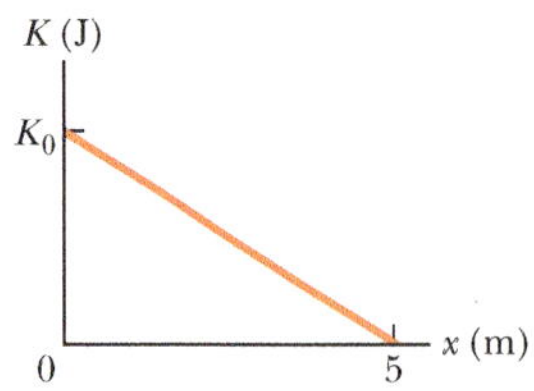

Figure 7-29 Problem 16.

Module 7-3 Work Done by the Gravitational Force

•17 SSM WWW A helicopter lifts a 72 kg astronaut 15 m vertically from the ocean by means of a cable. The acceleration of the astronaut is $g/10$. How much work is done on the astronaut by (a) the force from the helicopter and (b) the gravitational force on her? Just before she reaches the helicopter, what are her (c) kinetic energy and (d) speed?

•18 (a) In 1975 the roof of Montreal's Velodrome, with a weight of 360 kN, was lifted by 10 cm so that it could be centered. How much work was done on the roof by the forces making the lift? (b) In 1960 a Tampa, Florida, mother reportedly raised one end of a car that had fallen onto her son when a jack failed. If her panic lift effectively raised 4000 N (about $\frac{1}{4}$ of the car's weight) by 5.0 cm, how much work did her force do on the car?

••19 GO In Fig. 7-30, a block of ice slides down a frictionless ramp at angle $\theta = 50°$ while an ice worker pulls on the block (via a rope) with a force $\vec{F}_r$ that has a magnitude of 50 N and is directed up the ramp. As the block slides through distance $d = 0.50$ m along the ramp, its kinetic energy increases by 80 J. How much greater would its kinetic energy have been if the rope had not been attached to the block?

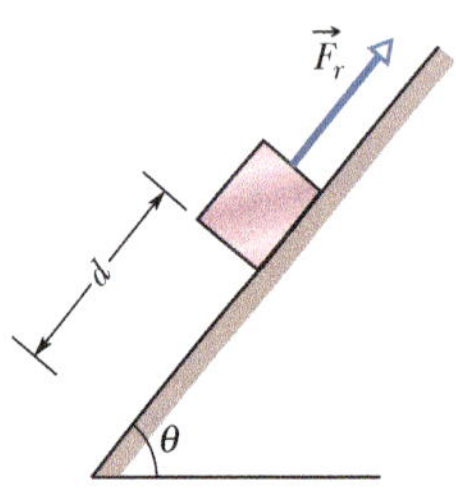

Figure 7-30 Problem 19.

••20 A block is sent up a frictionless ramp along which an x axis extends upward. Figure 7-31 gives the kinetic energy of the block as a function of position x; the scale of the figure's vertical axis is set by $K_s = 40.0$ J. If the block's initial speed is 4.00 m/s, what is the normal force on the block?

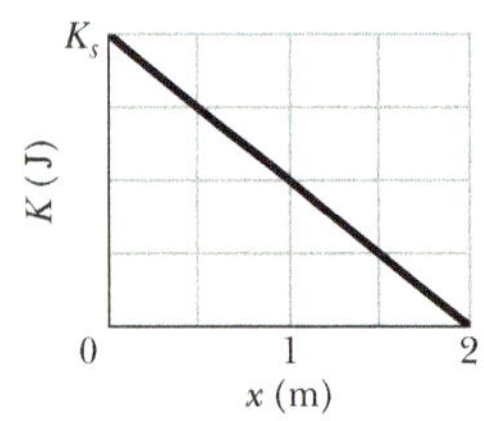

Figure 7-31 Problem 20.

••21 SSM A cord is used to vertically lower an initially stationary block of mass M at a constant downward acceleration of $g/4$. When the block has fallen a distance d, find (a) the work done by the cord's force on the block, (b) the work done by the gravitational force on the block, (c) the kinetic energy of the block, and (d) the speed of the block.

••22 A cave rescue team lifts an injured spelunker directly upward and out of a sinkhole by means of a motor-driven cable. The lift is performed in three stages, each requiring a vertical distance of 10.0 m: (a) the initially stationary spelunker is accelerated to a speed of 5.00 m/s; (b) he is then lifted at the constant speed of 5.00 m/s; (c) finally he is decelerated to zero speed. How much work is done on the 80.0 kg rescuee by the force lifting him during each stage?

••23 In Fig. 7-32, a constant force $\vec{F}_a$ of magnitude 82.0 N is applied to a 3.00 kg shoe box at angle $\phi = 53.0°$, causing the box to move up a frictionless ramp at constant speed. How much work is done on the box by $\vec{F}_a$ when the box has moved through vertical distance $h = 0.150$ m?

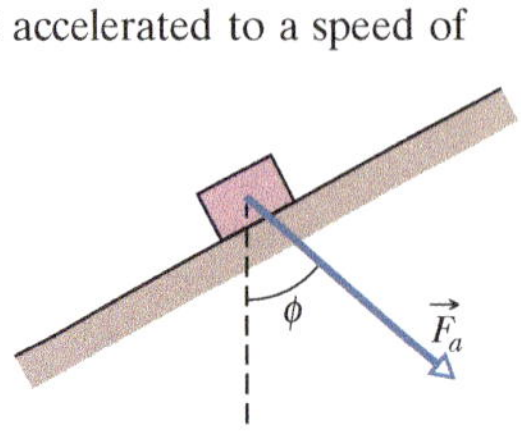

Figure 7-32 Problem 23.

••24 GO In Fig. 7-33, a horizontal force $\vec{F}_a$ of magnitude 20.0 N is applied to a 3.00 kg psychology book as the book slides a distance $d = 0.500$ m up a frictionless ramp at angle $\theta = 30.0°$. (a) During the displacement, what is the net work done on the book by $\vec{F}_a$, the gravitational force on the book, and the normal force on the book? (b) If the book has zero kinetic energy at the start of the displacement, what is its speed at the end of the displacement?

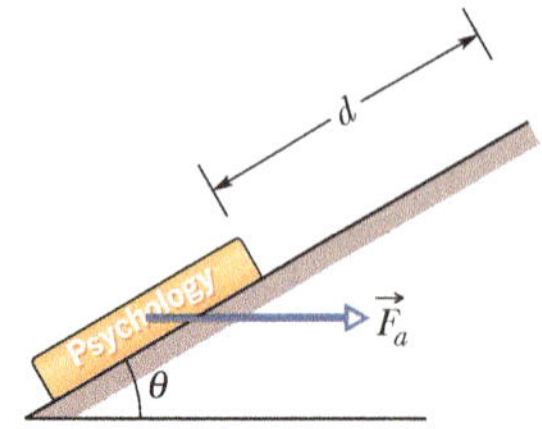

Figure 7-33 Problem 24.

•••25 GO In Fig. 7-34, a 0.250 kg block of cheese lies on the floor of a 900 kg elevator cab that is being pulled upward by a cable through distance $d_1 = 2.40$ m and then through distance $d_2 = 10.5$ m. (a) Through d_1, if the normal force on the block from the floor has constant magnitude $F_N = 3.00$ N, how much work is done on the cab by the force from the cable? (b) Through d_2, if the work done on the cab by the (constant) force from the cable is 92.61 kJ, what is the magnitude of F_N?

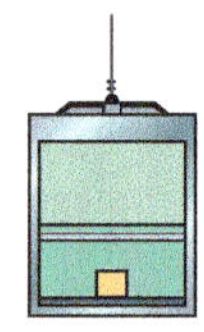
Figure 7-34 Problem 25.

Module 7-4 Work Done by a Spring Force

•26 In Fig. 7-10, we must apply a force of magnitude 80 N to hold the block stationary at $x = -2.0$ cm. From that position, we then slowly move the block so that our force does +4.0 J of work on the spring–block system; the block is then again stationary. What is the block's position? (*Hint:* There are two answers.)

•27 A spring and block are in the arrangement of Fig. 7-10. When the block is pulled out to $x = +4.0$ cm, we must apply a force of magnitude 360 N to hold it there. We pull the block to $x = 11$ cm and then release it. How much work does the spring do on the block as the block moves from $x_i = +5.0$ cm to (a) $x = +3.0$ cm, (b) $x = -3.0$ cm, (c) $x = -5.0$ cm, and (d) $x = -9.0$ cm?

•28 During spring semester at MIT, residents of the parallel buildings of the East Campus dorms battle one another with large catapults that are made with surgical hose mounted on a window frame. A balloon filled with dyed water is placed in a pouch attached to the hose, which is then stretched through the width of the room. Assume that the stretching of the hose obeys Hooke's law with a spring constant of 100 N/m. If the hose is stretched by 5.00 m and then released, how much work does the force from the hose do on the balloon in the pouch by the time the hose reaches its relaxed length?

••29 In the arrangement of Fig. 7-10, we gradually pull the block from $x = 0$ to $x = +3.0$ cm, where it is stationary. Figure 7-35 gives

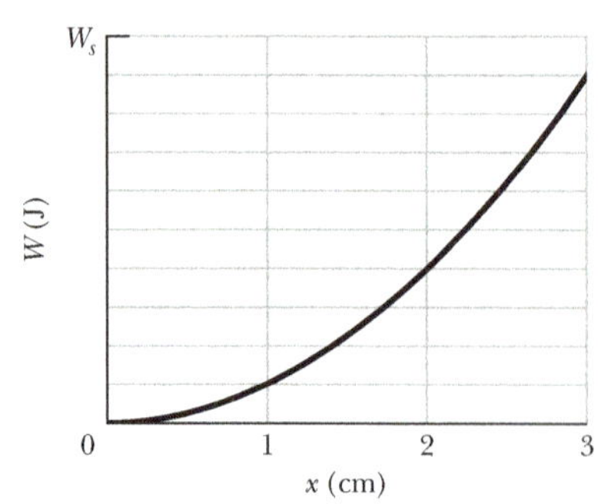

Figure 7-35 Problem 29.

the work that our force does on the block. The scale of the figure's vertical axis is set by $W_s = 1.0$ J. We then pull the block out to $x = +5.0$ cm and release it from rest. How much work does the spring do on the block when the block moves from $x_i = +5.0$ cm to (a) $x = +4.0$ cm, (b) $x = -2.0$ cm, and (c) $x = -5.0$ cm?

••30 In Fig. 7-10*a*, a block of mass m lies on a horizontal frictionless surface and is attached to one end of a horizontal spring (spring constant k) whose other end is fixed. The block is initially at rest at the position where the spring is unstretched ($x = 0$) when a constant horizontal force $\vec{F}$ in the positive direction of the x axis is applied to it. A plot of the resulting kinetic energy of the block versus its position x is shown in Fig. 7-36. The scale of the figure's vertical axis is set by $K_s = 4.0$ J. (a) What is the magnitude of $\vec{F}$? (b) What is the value of k?

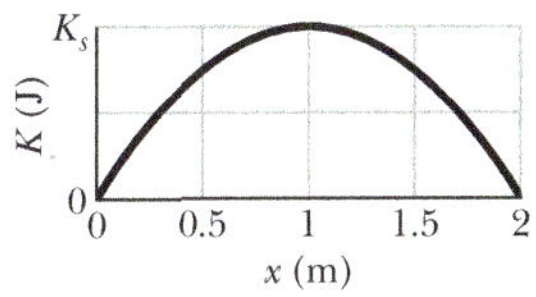

Figure 7-36 Problem 30.

••31 SSM WWW The only force acting on a 2.0 kg body as it moves along a positive x axis has an x component $F_x = -6x$ N, with x in meters. The velocity at $x = 3.0$ m is 8.0 m/s. (a) What is the velocity of the body at $x = 4.0$ m? (b) At what positive value of x will the body have a velocity of 5.0 m/s?

••32 Figure 7-37 gives spring force F_x versus position x for the spring–block arrangement of Fig. 7-10. The scale is set by $F_s = 160.0$ N. We release the block at $x = 12$ cm. How much work does the spring do on the block when the block moves from $x_i = +8.0$ cm to (a) $x = +5.0$ cm, (b) $x = -5.0$ cm, (c) $x = -8.0$ cm, and (d) $x = -10.0$ cm?

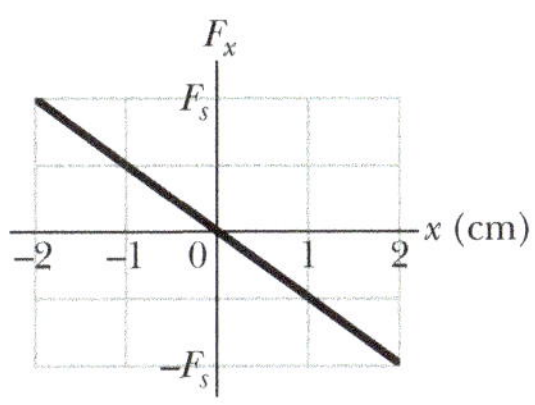

Figure 7-37 Problem 32.

•••33 GO The block in Fig. 7-10*a* lies on a horizontal frictionless surface, and the spring constant is 50 N/m. Initially, the spring is at its relaxed length and the block is stationary at position $x = 0$. Then an applied force with a constant magnitude of 3.0 N pulls the block in the positive direction of the x axis, stretching the spring until the block stops. When that stopping point is reached, what are (a) the position of the block, (b) the work that has been done on the block by the applied force, and (c) the work that has been done on the block by the spring force? During the block's displacement, what are (d) the block's position when its kinetic energy is maximum and (e) the value of that maximum kinetic energy?

Module 7-5 Work Done by a General Variable Force

•34 ILW A 10 kg brick moves along an x axis. Its acceleration as a function of its position is shown in Fig. 7-38. The scale of the figure's vertical axis is set by $a_s = 20.0$ m/s^2. What is the net work performed on the brick by the force causing the acceleration as the brick moves from $x = 0$ to $x = 8.0$ m?

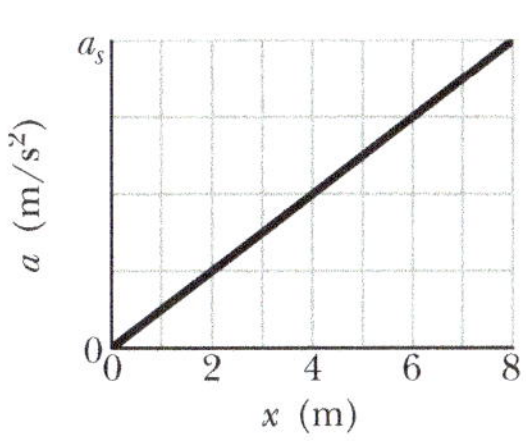

Figure 7-38 Problem 34.

•35 SSM WWW The force on a particle is directed along an x axis and given by $F = F_0(x/x_0 - 1)$. Find the work done by the force in moving the particle from $x = 0$ to $x = 2x_0$ by (a) plotting $F(x)$ and measuring the work from the graph and (b) integrating $F(x)$.

•36 GO A 5.0 kg block moves in a straight line on a horizontal frictionless surface under the influence of a force that varies with position as shown in Fig. 7-39. The scale of the figure's vertical axis is set by $F_s = 10.0$ N. How much work is done by the force as the block moves from the origin to $x = 8.0$ m?

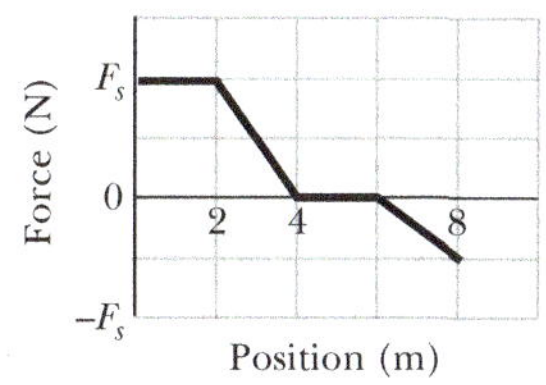

Figure 7-39 Problem 36.

••37 GO Figure 7-40 gives the acceleration of a 2.00 kg particle as an applied force $\vec{F}_a$ moves it from rest along an x axis from $x = 0$ to $x = 9.0$ m. The scale of the figure's vertical axis is set by $a_s = 6.0$ m/s^2. How much work has the force done on the particle when the particle reaches (a) $x = 4.0$ m, (b) $x = 7.0$ m, and (c) $x = 9.0$ m? What is the particle's speed and direction of travel when it reaches (d) $x = 4.0$ m, (e) $x = 7.0$ m, and (f) $x = 9.0$ m?

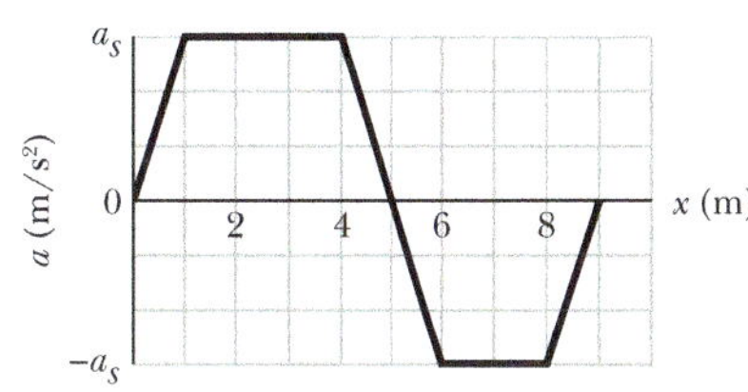

Figure 7-40 Problem 37.

••38 A 1.5 kg block is initially at rest on a horizontal frictionless surface when a horizontal force along an x axis is applied to the block. The force is given by $\vec{F}(x) = (2.5 - x^2)\hat{i}$ N, where x is in meters and the initial position of the block is $x = 0$. (a) What is the kinetic energy of the block as it passes through $x = 2.0$ m? (b) What is the maximum kinetic energy of the block between $x = 0$ and $x = 2.0$ m?

••39 GO A force $\vec{F} = (cx - 3.00x^2)\hat{i}$ acts on a particle as the particle moves along an x axis, with $\vec{F}$ in newtons, x in meters, and c a constant. At $x = 0$, the particle's kinetic energy is 20.0 J; at $x = 3.00$ m, it is 11.0 J. Find c.

••40 A can of sardines is made to move along an x axis from $x = 0.25$ m to $x = 1.25$ m by a force with a magnitude given by $F = \exp(-4x^2)$, with x in meters and F in newtons. (Here exp is the exponential function.) How much work is done on the can by the force?

••41 A single force acts on a 3.0 kg particle-like object whose position is given by $x = 3.0t - 4.0t^2 + 1.0t^3$, with x in meters and t in seconds. Find the work done by the force from $t = 0$ to $t = 4.0$ s.

•••42 GO Figure 7-41 shows a cord attached to a cart that can slide along a frictionless horizontal rail aligned along an x axis. The left

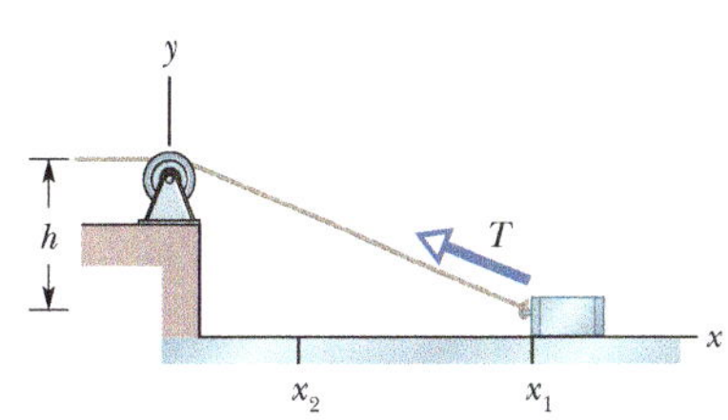

Figure 7-41 Problem 42.

end of the cord is pulled over a pulley, of negligible mass and friction and at cord height $h = 1.20$ m, so the cart slides from $x_1 = 3.00$ m to $x_2 = 1.00$ m. During the move, the tension in the cord is a constant 25.0 N. What is the change in the kinetic energy of the cart during the move?

Module 7-6 Power

•43 SSM A force of 5.0 N acts on a 15 kg body initially at rest. Compute the work done by the force in (a) the first, (b) the second, and (c) the third seconds and (d) the instantaneous power due to the force at the end of the third second.

•44 A skier is pulled by a towrope up a frictionless ski slope that makes an angle of 12° with the horizontal. The rope moves parallel to the slope with a constant speed of 1.0 m/s. The force of the rope does 900 J of work on the skier as the skier moves a distance of 8.0 m up the incline. (a) If the rope moved with a constant speed of 2.0 m/s, how much work would the force of the rope do on the skier as the skier moved a distance of 8.0 m up the incline? At what rate is the force of the rope doing work on the skier when the rope moves with a speed of (b) 1.0 m/s and (c) 2.0 m/s?

•45 SSM ILW A 100 kg block is pulled at a constant speed of 5.0 m/s across a horizontal floor by an applied force of 122 N directed 37° above the horizontal. What is the rate at which the force does work on the block?

•46 The loaded cab of an elevator has a mass of 3.0×10^3 kg and moves 210 m up the shaft in 23 s at constant speed. At what average rate does the force from the cable do work on the cab?

••47 A machine carries a 4.0 kg package from an initial position of $\vec{d}_i = (0.50\text{ m})\hat{i} + (0.75\text{ m})\hat{j} + (0.20\text{ m})\hat{k}$ at $t = 0$ to a final position of $\vec{d}_f = (7.50\text{ m})\hat{i} + (12.0\text{ m})\hat{j} + (7.20\text{ m})\hat{k}$ at $t = 12$ s. The constant force applied by the machine on the package is $\vec{F} = (2.00\text{ N})\hat{i} + (4.00\text{ N})\hat{j} + (6.00\text{ N})\hat{k}$. For that displacement, find (a) the work done on the package by the machine's force and (b) the average power of the machine's force on the package.

••48 A 0.30 kg ladle sliding on a horizontal frictionless surface is attached to one end of a horizontal spring ($k = 500$ N/m) whose other end is fixed. The ladle has a kinetic energy of 10 J as it passes through its equilibrium position (the point at which the spring force is zero). (a) At what rate is the spring doing work on the ladle as the ladle passes through its equilibrium position? (b) At what rate is the spring doing work on the ladle when the spring is compressed 0.10 m and the ladle is moving away from the equilibrium position?

••49 SSM A fully loaded, slow-moving freight elevator has a cab with a total mass of 1200 kg, which is required to travel upward 54 m in 3.0 min, starting and ending at rest. The elevator's counterweight has a mass of only 950 kg, and so the elevator motor must help. What average power is required of the force the motor exerts on the cab via the cable?

••50 (a) At a certain instant, a particle-like object is acted on by a force $\vec{F} = (4.0\text{ N})\hat{i} - (2.0\text{ N})\hat{j} + (9.0\text{ N})\hat{k}$ while the object's velocity is $\vec{v} = -(2.0\text{ m/s})\hat{i} + (4.0\text{ m/s})\hat{k}$. What is the instantaneous rate at which the force does work on the object? (b) At some other time, the velocity consists of only a y component. If the force is unchanged and the instantaneous power is -12 W, what is the velocity of the object?

••51 A force $\vec{F} = (3.00\text{ N})\hat{i} + (7.00\text{ N})\hat{j} + (7.00\text{ N})\hat{k}$ acts on a 2.00 kg mobile object that moves from an initial position of $\vec{d}_i = (3.00\text{ m})\hat{i} - (2.00\text{ m})\hat{j} + (5.00\text{ m})\hat{k}$ to a final position of $\vec{d}_f = -(5.00\text{ m})\hat{i} + (4.00\text{ m})\hat{j} + (7.00\text{ m})\hat{k}$ in 4.00 s. Find (a) the work done on the object by the force in the 4.00 s interval, (b) the average power due to the force during that interval, and (c) the angle between vectors $\vec{d}_i$ and $\vec{d}_f$.

•••52 A funny car accelerates from rest through a measured track distance in time T with the engine operating at a constant power P. If the track crew can increase the engine power by a differential amount dP, what is the change in the time required for the run?

Additional Problems

53 Figure 7-42 shows a cold package of hot dogs sliding rightward across a frictionless floor through a distance $d = 20.0$ cm while three forces act on the package. Two of them are horizontal and have the magnitudes $F_1 = 5.00$ N and $F_2 = 1.00$ N; the third is angled down by $\theta = 60.0°$ and has the magnitude $F_3 = 4.00$ N. (a) For the 20.0 cm displacement, what is the *net* work done on the package by the three applied forces, the gravitational force on the package, and the normal force on the package? (b) If the package has a mass of 2.0 kg and an initial kinetic energy of 0, what is its speed at the end of the displacement?

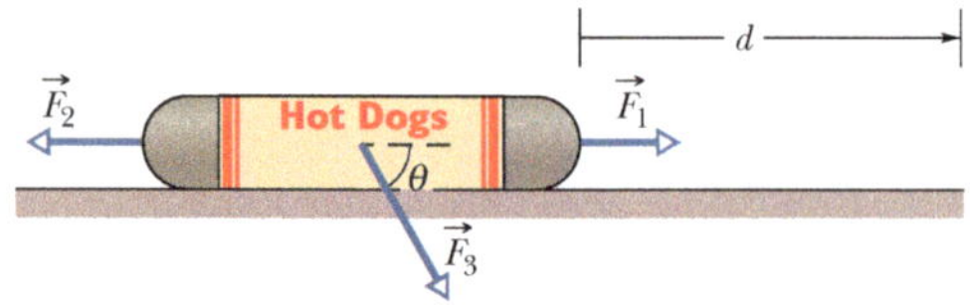

Figure 7-42 Problem 53.

54 GO The only force acting on a 2.0 kg body as the body moves along an x axis varies as shown in Fig. 7-43. The scale of the figure's vertical axis is set by $F_s = 4.0$ N. The velocity of the body at $x = 0$ is 4.0 m/s. (a) What is the kinetic energy of the body at $x = 3.0$ m? (b) At what value of x will the body have a kinetic energy of 8.0 J? (c) What is the maximum kinetic energy of the body between $x = 0$ and $x = 5.0$ m?

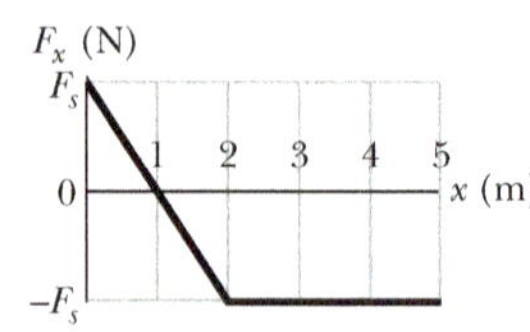

Figure 7-43 Problem 54.

55 SSM A horse pulls a cart with a force of 40 lb at an angle of 30° above the horizontal and moves along at a speed of 6.0 mi/h. (a) How much work does the force do in 10 min? (b) What is the average power (in horsepower) of the force?

56 An initially stationary 2.0 kg object accelerates horizontally and uniformly to a speed of 10 m/s in 3.0 s. (a) In that 3.0 s interval, how much work is done on the object by the force accelerating it? What is the instantaneous power due to that force (b) at the end of the interval and (c) at the end of the first half of the interval?

57 A 230 kg crate hangs from the end of a rope of length $L = 12.0$ m. You push horizontally on the crate with a varying force $\vec{F}$ to move it distance $d = 4.00$ m to the side (Fig. 7-44). (a) What is the magnitude of $\vec{F}$ when the crate is in this final position? During the crate's displacement, what are (b) the total

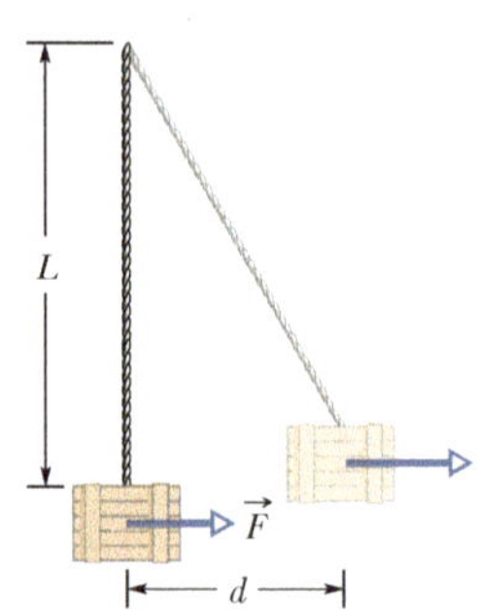

Figure 7-44 Problem 57.

work done on it, (c) the work done by the gravitational force on the crate, and (d) the work done by the pull on the crate from the rope? (e) Knowing that the crate is motionless before and after its displacement, use the answers to (b), (c), and (d) to find the work your force $\vec{F}$ does on the crate. (f) Why is the work of your force not equal to the product of the horizontal displacement and the answer to (a)?

58 To pull a 50 kg crate across a horizontal frictionless floor, a worker applies a force of 210 N, directed 20° above the horizontal. As the crate moves 3.0 m, what work is done on the crate by (a) the worker's force, (b) the gravitational force, and (c) the normal force? (d) What is the total work?

59 A force $\vec{F}_a$ is applied to a bead as the bead is moved along a straight wire through displacement +5.0 cm. The magnitude of $\vec{F}_a$ is set at a certain value, but the angle ϕ between $\vec{F}_a$ and the bead's displacement can be chosen. Figure 7-45 gives the work W done by $\vec{F}_a$ on the bead for a range of ϕ values; $W_0 = 25$ J. How much work is done by $\vec{F}_a$ if ϕ is (a) 64° and (b) 147°?

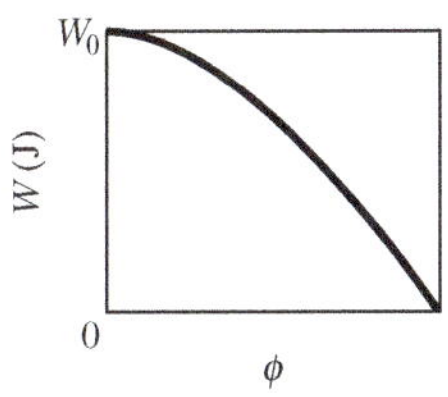

Figure 7-45 Problem 59.

60 A frightened child is restrained by her mother as the child slides down a frictionless playground slide. If the force on the child from the mother is 100 N up the slide, the child's kinetic energy increases by 30 J as she moves down the slide a distance of 1.8 m. (a) How much work is done on the child by the gravitational force during the 1.8 m descent? (b) If the child is not restrained by her mother, how much will the child's kinetic energy increase as she comes down the slide that same distance of 1.8 m?

61 How much work is done by a force $\vec{F} = (2x\text{ N})\hat{\text{i}} + (3\text{ N})\hat{\text{j}}$, with x in meters, that moves a particle from a position $\vec{r}_i = (2\text{ m})\hat{\text{i}} + (3\text{ m})\hat{\text{j}}$ to a position $\vec{r}_f = -(4\text{ m})\hat{\text{i}} - (3\text{ m})\hat{\text{j}}$?

62 A 250 g block is dropped onto a relaxed vertical spring that has a spring constant of $k = 2.5$ N/cm (Fig. 7-46). The block becomes attached to the spring and compresses the spring 12 cm before momentarily stopping. While the spring is being compressed, what work is done on the block by (a) the gravitational force on it and (b) the spring force? (c) What is the speed of the block just before it hits the spring? (Assume that friction is negligible.) (d) If the speed at impact is doubled, what is the maximum compression of the spring?

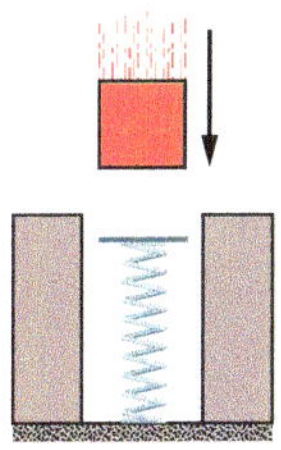

Figure 7-46 Problem 62.

63 SSM To push a 25.0 kg crate up a frictionless incline, angled at 25.0° to the horizontal, a worker exerts a force of 209 N parallel to the incline. As the crate slides 1.50 m, how much work is done on the crate by (a) the worker's applied force, (b) the gravitational force on the crate, and (c) the normal force exerted by the incline on the crate? (d) What is the total work done on the crate?

64 Boxes are transported from one location to another in a warehouse by means of a conveyor belt that moves with a constant speed of 0.50 m/s. At a certain location the conveyor belt moves for 2.0 m up an incline that makes an angle of 10° with the horizontal, then for 2.0 m horizontally, and finally for 2.0 m down an incline that makes an angle of 10° with the horizontal. Assume that a 2.0 kg box rides on the belt without slipping. At what rate is the force of the conveyor belt doing work on the box as the box moves (a) up the 10° incline, (b) horizontally, and (c) down the 10° incline?

65 In Fig. 7-47, a cord runs around two massless, frictionless pulleys. A canister with mass $m = 20$ kg hangs from one pulley, and you exert a force $\vec{F}$ on the free end of the cord. (a) What must be the magnitude of $\vec{F}$ if you are to lift the canister at a constant speed? (b) To lift the canister by 2.0 cm, how far must you pull the free end of the cord? During that lift, what is the work done on the canister by (c) your force (via the cord) and (d) the gravitational force? (*Hint:* When a cord loops around a pulley as shown, it pulls on the pulley with a net force that is twice the tension in the cord.)

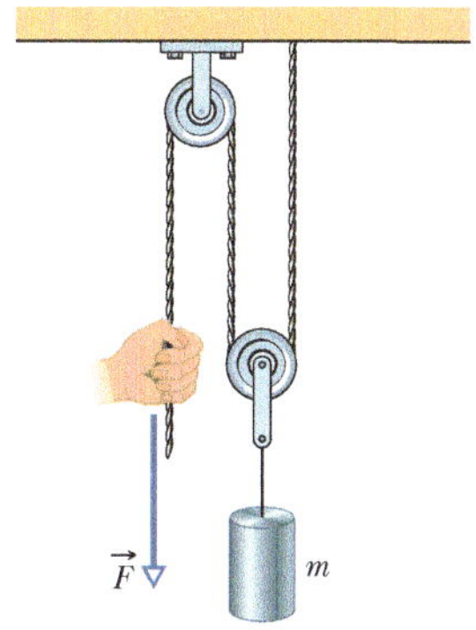

Figure 7-47 Problem 65.

66 If a car of mass 1200 kg is moving along a highway at 120 km/h, what is the car's kinetic energy as determined by someone standing alongside the highway?

67 SSM A spring with a pointer attached is hanging next to a scale marked in millimeters. Three different packages are hung from the spring, in turn, as shown in Fig. 7-48. (a) Which mark on the scale will the pointer indicate when no package is hung from the spring? (b) What is the weight W of the third package?

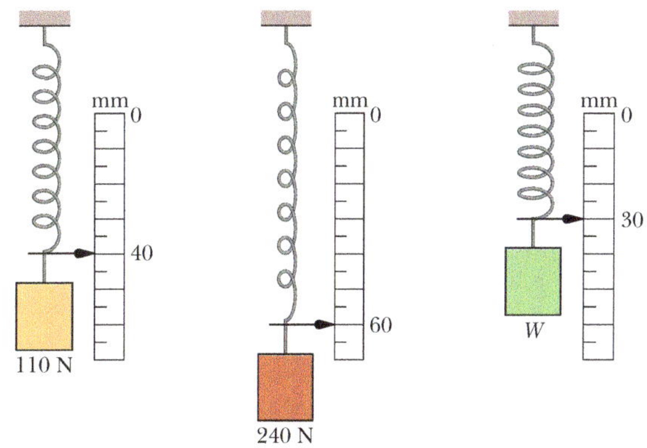

Figure 7-48 Problem 67.

68 An iceboat is at rest on a frictionless frozen lake when a sudden wind exerts a constant force of 200 N, toward the east, on the boat. Due to the angle of the sail, the wind causes the boat to slide in a straight line for a distance of 8.0 m in a direction 20° north of east. What is the kinetic energy of the iceboat at the end of that 8.0 m?

69 If a ski lift raises 100 passengers averaging 660 N in weight to a height of 150 m in 60.0 s, at constant speed, what average power is required of the force making the lift?

70 A force $\vec{F} = (4.0\text{ N})\hat{\text{i}} + c\hat{\text{j}}$ acts on a particle as the particle goes through displacement $\vec{d} = (3.0\text{ m})\hat{\text{i}} - (2.0\text{ m})\hat{\text{j}}$. (Other forces also act on the particle.) What is c if the work done on the particle by force $\vec{F}$ is (a) 0, (b) 17 J, and (c) −18 J?

71 A constant force of magnitude 10 N makes an angle of 150° (measured counterclockwise) with the positive x direction as it acts on a 2.0 kg object moving in an xy plane. How much work is done on the object by the force as the object moves from the origin to the point having position vector $(2.0\text{ m})\hat{\text{i}} - (4.0\text{ m})\hat{\text{j}}$?

72 In Fig. 7-49*a*, a 2.0 N force is applied to a 4.0 kg block at a downward angle θ as the block moves rightward through 1.0 m across a frictionless floor. Find an expression for the speed v_f of the block at the end of that distance if the block's initial velocity is (a) 0 and (b) 1.0 m/s to the right. (c) The situation in Fig. 7-49*b* is similar in that the block is initially moving at 1.0 m/s to the right, but now the 2.0 N force is directed downward to the left. Find an expression for the speed v_f of the block at the end of the 1.0 m distance. (d) Graph all three expressions for v_f versus downward angle θ for $\theta = 0°$ to $\theta = 90°$. Interpret the graphs.

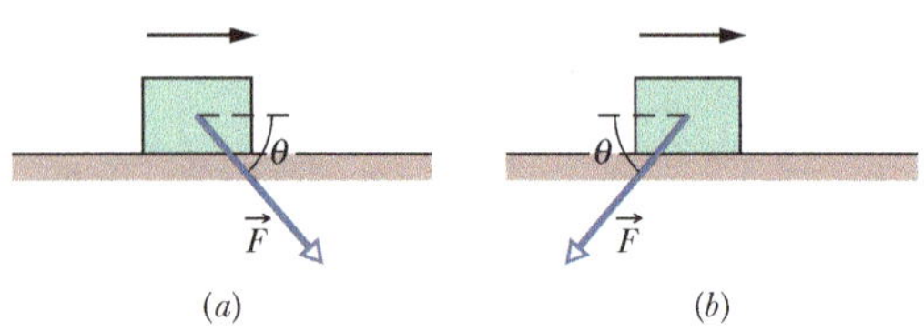

Figure 7-49 Problem 72.

73 A force $\vec{F}$ in the positive direction of an x axis acts on an object moving along the axis. If the magnitude of the force is $F = 10e^{-x/2.0}$ N, with x in meters, find the work done by $\vec{F}$ as the object moves from $x = 0$ to $x = 2.0$ m by (a) plotting $F(x)$ and estimating the area under the curve and (b) integrating to find the work analytically.

74 A particle moves along a straight path through displacement $\vec{d} = (8\text{ m})\hat{i} + c\hat{j}$ while force $\vec{F} = (2\text{ N})\hat{i} - (4\text{ N})\hat{j}$ acts on it. (Other forces also act on the particle.) What is the value of c if the work done by $\vec{F}$ on the particle is (a) zero, (b) positive, and (c) negative?

75 SSM What is the power of the force required to move a 4500 kg elevator cab with a load of 1800 kg upward at constant speed 3.80 m/s?

76 A 45 kg block of ice slides down a frictionless incline 1.5 m long and 0.91 m high. A worker pushes up against the ice, parallel to the incline, so that the block slides down at constant speed. (a) Find the magnitude of the worker's force. How much work is done on the block by (b) the worker's force, (c) the gravitational force on the block, (d) the normal force on the block from the surface of the incline, and (e) the net force on the block?

77 As a particle moves along an x axis, a force in the positive direction of the axis acts on it. Figure 7-50 shows the magnitude F of the force versus position x of the particle. The curve is given by $F = a/x^2$, with $a = 9.0\text{ N}\cdot\text{m}^2$. Find the work done on the particle by the force as the particle moves from $x = 1.0$ m to $x = 3.0$ m by (a) estimating the work from the graph and (b) integrating the force function.

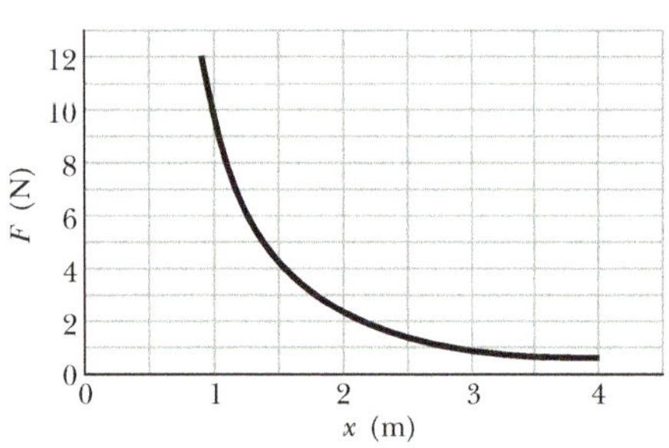

Figure 7-50 Problem 77.

78 A CD case slides along a floor in the positive direction of an x axis while an applied force $\vec{F}_a$ acts on the case. The force is directed along the x axis and has the x component $F_{ax} = 9x - 3x^2$, with x in meters and F_{ax} in newtons. The case starts at rest at the position $x = 0$, and it moves until it is again at rest. (a) Plot the work $\vec{F}_a$ does on the case as a function of x. (b) At what position is the work maximum, and (c) what is that maximum value? (d) At what position has the work decreased to zero? (e) At what position is the case again at rest?

79 SSM A 2.0 kg lunchbox is sent sliding over a frictionless surface, in the positive direction of an x axis along the surface. Beginning at time $t = 0$, a steady wind pushes on the lunchbox in the negative direction of the x axis. Figure 7-51 shows the position x of the lunchbox as a function of time t as the wind pushes on the lunchbox. From the graph, estimate the kinetic energy of the lunchbox at (a) $t = 1.0$ s and (b) $t = 5.0$ s. (c) How much work does the force from the wind do on the lunchbox from $t = 1.0$ s to $t = 5.0$ s?

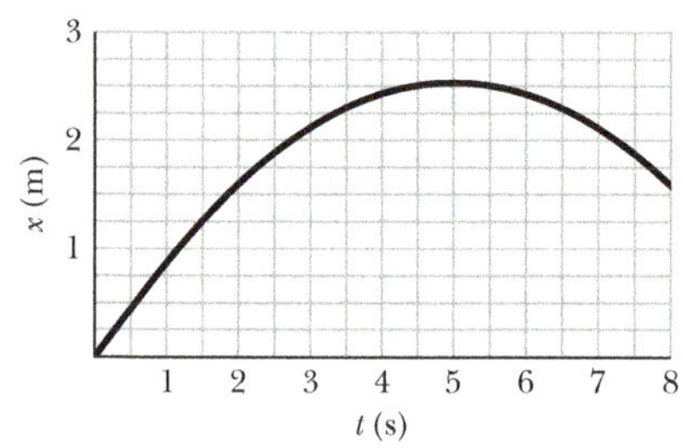

Figure 7-51 Problem 79.

80 *Numerical integration.* A breadbox is made to move along an x axis from $x = 0.15$ m to $x = 1.20$ m by a force with a magnitude given by $F = \exp(-2x^2)$, with x in meters and F in newtons. (Here exp is the exponential function.) How much work is done on the breadbox by the force?

81 In the block–spring arrangement of Fig. 7-10, the block's mass is 4.00 kg and the spring constant is 500 N/m. The block is released from position $x_i = 0.300$ m. What are (a) the block's speed at $x = 0$, (b) the work done by the spring when the block reaches $x = 0$, (c) the instantaneous power due to the spring at the release point x_i, (d) the instantaneous power at $x = 0$, and (e) the block's position when the power is maximum?

82 A 4.00 kg block is pulled up a frictionless inclined plane by a 50.0 N force that is parallel to the plane, starting from rest. The normal force on the block from the plane has magnitude 13.41 N. What is the block's speed when its displacement up the ramp is 3.00 m?

83 A spring with a spring constant of 18.0 N/cm has a cage attached to its free end. (a) How much work does the spring force do on the cage when the spring is stretched from its relaxed length by 7.60 mm? (b) How much additional work is done by the spring force when the spring is stretched by an additional 7.60 mm?

84 A force $\vec{F} = (2.00\hat{i} + 9.00\hat{j} + 5.30\hat{k})$ N acts on a 2.90 kg object that moves in time interval 2.10 s from an initial position $\vec{r}_1 = (2.70\hat{i} - 2.90\hat{j} + 5.50\hat{k})$ m to a final position $\vec{r}_2 = (-4.10\hat{i} + 3.30\hat{j} + 5.40\hat{k})$ m. Find (a) the work done on the object by the force in that time interval, (b) the average power due to the force during that time interval, and (c) the angle between vectors $\vec{r}_1$ and $\vec{r}_2$.

85 At $t = 0$, force $\vec{F} = (-5.00\hat{i} + 5.00\hat{j} + 4.00\hat{k})$ N begins to act on a 2.00 kg particle with an initial speed of 4.00 m/s. What is the particle's speed when its displacement from the initial point is $\vec{d} = (2.00\hat{i} + 2.00\hat{j} + 7.00\hat{k})$ m?

CHAPTER 8

Potential Energy and Conservation of Energy

8-1 POTENTIAL ENERGY

Learning Objectives

After reading this module, you should be able to . . .

8.01 Distinguish a conservative force from a nonconservative force.

8.02 For a particle moving between two points, identify that the work done by a conservative force does not depend on which path the particle takes.

8.03 Calculate the gravitational potential energy of a particle (or, more properly, a particle–Earth system).

8.04 Calculate the elastic potential energy of a block–spring system.

Key Ideas

- A force is a conservative force if the net work it does on a particle moving around any closed path, from an initial point and then back to that point, is zero. Equivalently, a force is conservative if the net work it does on a particle moving between two points does not depend on the path taken by the particle. The gravitational force and the spring force are conservative forces; the kinetic frictional force is a nonconservative force.

- Potential energy is energy that is associated with the configuration of a system in which a conservative force acts. When the conservative force does work W on a particle within the system, the change ΔU in the potential energy of the system is

$$\Delta U = -W.$$

If the particle moves from point x_i to point x_f, the change in the potential energy of the system is

$$\Delta U = -\int_{x_i}^{x_f} F(x)\, dx.$$

- The potential energy associated with a system consisting of Earth and a nearby particle is gravitational potential energy. If the particle moves from height y_i to height y_f, the change in the gravitational potential energy of the particle–Earth system is

$$\Delta U = mg(y_f - y_i) = mg\,\Delta y.$$

- If the reference point of the particle is set as $y_i = 0$ and the corresponding gravitational potential energy of the system is set as $U_i = 0$, then the gravitational potential energy U when the particle is at any height y is

$$U(y) = mgy.$$

- Elastic potential energy is the energy associated with the state of compression or extension of an elastic object. For a spring that exerts a spring force $F = -kx$ when its free end has displacement x, the elastic potential energy is

$$U(x) = \tfrac{1}{2}kx^2.$$

- The reference configuration has the spring at its relaxed length, at which $x = 0$ and $U = 0$.

What Is Physics?

One job of physics is to identify the different types of energy in the world, especially those that are of common importance. One general type of energy is **potential energy** U. Technically, potential energy is energy that can be associated with the configuration (arrangement) of a system of objects that exert forces on one another.

Rough Guides/Greg Roden/Getty Images, Inc.

Figure 8-1 The kinetic energy of a bungee-cord jumper increases during the free fall, and then the cord begins to stretch, slowing the jumper.

This is a pretty formal definition of something that is actually familiar to you. An example might help better than the definition: A bungee-cord jumper plunges from a staging platform (Fig. 8-1). The system of objects consists of Earth and the jumper. The force between the objects is the gravitational force. The configuration of the system changes (the separation between the jumper and Earth decreases—that is, of course, the thrill of the jump). We can account for the jumper's motion and increase in kinetic energy by defining a **gravitational potential energy** U. This is the energy associated with the state of separation between two objects that attract each other by the gravitational force, here the jumper and Earth.

When the jumper begins to stretch the bungee cord near the end of the plunge, the system of objects consists of the cord and the jumper. The force between the objects is an elastic (spring-like) force. The configuration of the system changes (the cord stretches). We can account for the jumper's decrease in kinetic energy and the cord's increase in length by defining an **elastic potential energy** U. This is the energy associated with the state of compression or extension of an elastic object, here the bungee cord.

Physics determines how the potential energy of a system can be calculated so that energy might be stored or put to use. For example, before any particular bungee-cord jumper takes the plunge, someone (probably a mechanical engineer) must determine the correct cord to be used by calculating the gravitational and elastic potential energies that can be expected. Then the jump is only thrilling and not fatal.

Work and Potential Energy

In Chapter 7 we discussed the relation between work and a change in kinetic energy. Here we discuss the relation between work and a change in potential energy.

Let us throw a tomato upward (Fig. 8-2). We already know that as the tomato rises, the work W_g done on the tomato by the gravitational force is negative because the force transfers energy *from* the kinetic energy of the tomato. We can now finish the story by saying that this energy is transferred by the gravitational force *to* the gravitational potential energy of the tomato–Earth system.

The tomato slows, stops, and then begins to fall back down because of the gravitational force. During the fall, the transfer is reversed: The work W_g done on the tomato by the gravitational force is now positive—that force transfers energy *from* the gravitational potential energy of the tomato–Earth system *to* the kinetic energy of the tomato.

For either rise or fall, the change ΔU in gravitational potential energy is defined as being equal to the negative of the work done on the tomato by the gravitational force. Using the general symbol W for work, we write this as

$$\Delta U = -W. \tag{8-1}$$

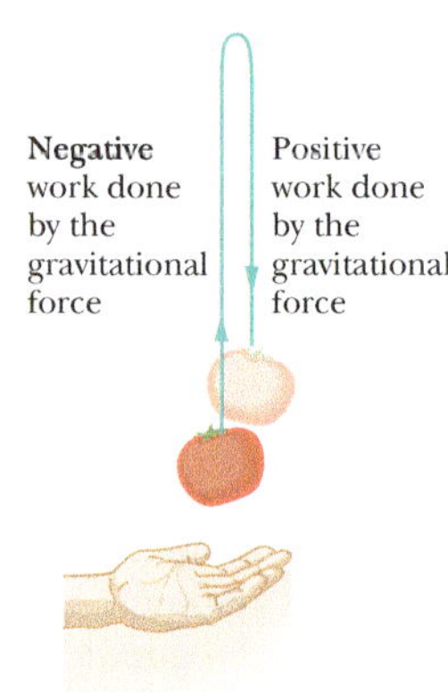

Figure 8-2 A tomato is thrown upward. As it rises, the gravitational force does negative work on it, decreasing its kinetic energy. As the tomato descends, the gravitational force does positive work on it, increasing its kinetic energy.

This equation also applies to a block–spring system, as in Fig. 8-3. If we abruptly shove the block to send it moving rightward, the spring force acts leftward and thus does negative work on the block, transferring energy from the kinetic energy of the block to the elastic potential energy of the spring–block system. The block slows and eventually stops, and then begins to move leftward because the spring force is still leftward. The transfer of energy is then reversed—it is from potential energy of the spring–block system to kinetic energy of the block.

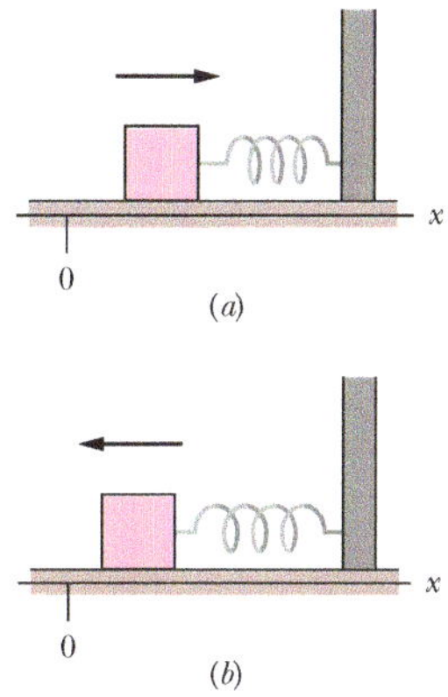

Figure 8-3 A block, attached to a spring and initially at rest at $x = 0$, is set in motion toward the right. (*a*) As the block moves rightward (as indicated by the arrow), the spring force does negative work on it. (*b*) Then, as the block moves back toward $x = 0$, the spring force does positive work on it.

Conservative and Nonconservative Forces

Let us list the key elements of the two situations we just discussed:

1. The *system* consists of two or more objects.
2. A *force* acts between a particle-like object (tomato or block) in the system and the rest of the system.
3. When the system configuration changes, the force does *work* (call it W_1) on the particle-like object, transferring energy between the kinetic energy K of the object and some other type of energy of the system.
4. When the configuration change is reversed, the force reverses the energy transfer, doing work W_2 in the process.

In a situation in which $W_1 = -W_2$ is always true, the other type of energy is a potential energy and the force is said to be a **conservative force.** As you might suspect, the gravitational force and the spring force are both conservative (since otherwise we could not have spoken of gravitational potential energy and elastic potential energy, as we did previously).

A force that is not conservative is called a **nonconservative force.** The kinetic frictional force and drag force are nonconservative. For an example, let us send a block sliding across a floor that is not frictionless. During the sliding, a kinetic frictional force from the floor slows the block by transferring energy from its kinetic energy to a type of energy called *thermal energy* (which has to do with the random motions of atoms and molecules). We know from experiment that this energy transfer cannot be reversed (thermal energy cannot be transferred back to kinetic energy of the block by the kinetic frictional force). Thus, although we have a system (made up of the block and the floor), a force that acts between parts of the system, and a transfer of energy by the force, the force is not conservative. Therefore, thermal energy is not a potential energy.

When only conservative forces act on a particle-like object, we can greatly simplify otherwise difficult problems involving motion of the object. Let's next develop a test for identifying conservative forces, which will provide one means for simplifying such problems.

Path Independence of Conservative Forces

The primary test for determining whether a force is conservative or nonconservative is this: Let the force act on a particle that moves along any *closed path,* beginning at some initial position and eventually returning to that position (so that the particle makes a *round trip* beginning and ending at the initial position). The force is conservative only if the total energy it transfers to and from the particle during the round trip along this and any other closed path is zero. In other words:

> The net work done by a conservative force on a particle moving around any closed path is zero.

We know from experiment that the gravitational force passes this *closed-path test.* An example is the tossed tomato of Fig. 8-2. The tomato leaves the launch point with speed v_0 and kinetic energy $\frac{1}{2}mv_0^2$. The gravitational force acting

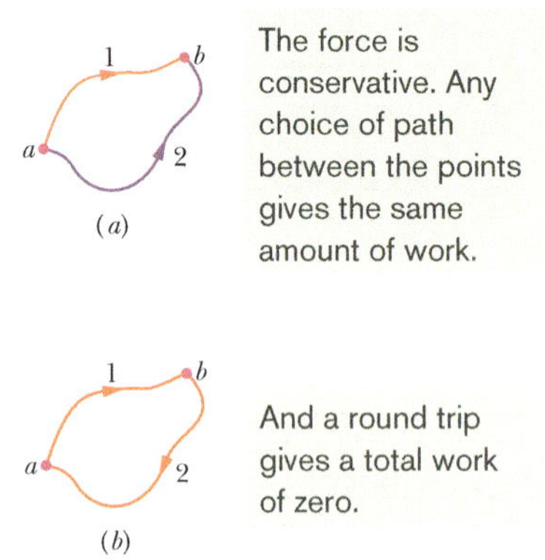

Figure 8-4 (*a*) As a conservative force acts on it, a particle can move from point *a* to point *b* along either path 1 or path 2. (*b*) The particle moves in a round trip, from point *a* to point *b* along path 1 and then back to point *a* along path 2.

on the tomato slows it, stops it, and then causes it to fall back down. When the tomato returns to the launch point, it again has speed v_0 and kinetic energy $\frac{1}{2}mv_0^2$. Thus, the gravitational force transfers as much energy *from* the tomato during the ascent as it transfers *to* the tomato during the descent back to the launch point. The net work done on the tomato by the gravitational force during the round trip is zero.

An important result of the closed-path test is that:

> The work done by a conservative force on a particle moving between two points does not depend on the path taken by the particle.

For example, suppose that a particle moves from point *a* to point *b* in Fig. 8-4*a* along either path 1 or path 2. If only a conservative force acts on the particle, then the work done on the particle is the same along the two paths. In symbols, we can write this result as

$$W_{ab,1} = W_{ab,2}, \tag{8-2}$$

where the subscript *ab* indicates the initial and final points, respectively, and the subscripts 1 and 2 indicate the path.

This result is powerful because it allows us to simplify difficult problems when only a conservative force is involved. Suppose you need to calculate the work done by a conservative force along a given path between two points, and the calculation is difficult or even impossible without additional information. You can find the work by substituting some other path between those two points for which the calculation is easier and possible.

Proof of Equation 8-2

Figure 8-4*b* shows an arbitrary round trip for a particle that is acted upon by a single force. The particle moves from an initial point *a* to point *b* along path 1 and then back to point *a* along path 2. The force does work on the particle as the particle moves along each path. Without worrying about where positive work is done and where negative work is done, let us just represent the work done from *a* to *b* along path 1 as $W_{ab,1}$ and the work done from *b* back to *a* along path 2 as $W_{ba,2}$. If the force is conservative, then the net work done during the round trip must be zero:

$$W_{ab,1} + W_{ba,2} = 0,$$

and thus

$$W_{ab,1} = -W_{ba,2}. \tag{8-3}$$

In words, the work done along the outward path must be the negative of the work done along the path back.

Let us now consider the work $W_{ab,2}$ done on the particle by the force when the particle moves from *a* to *b* along path 2, as indicated in Fig. 8-4*a*. If the force is conservative, that work is the negative of $W_{ba,2}$:

$$W_{ab,2} = -W_{ba,2}. \tag{8-4}$$

Substituting $W_{ab,2}$ for $-W_{ba,2}$ in Eq. 8-3, we obtain

$$W_{ab,1} = W_{ab,2},$$

which is what we set out to prove.

Checkpoint 1

The figure shows three paths connecting points *a* and *b*. A single force $\vec{F}$ does the indicated work on a particle moving along each path in the indicated direction. On the basis of this information, is force $\vec{F}$ conservative?

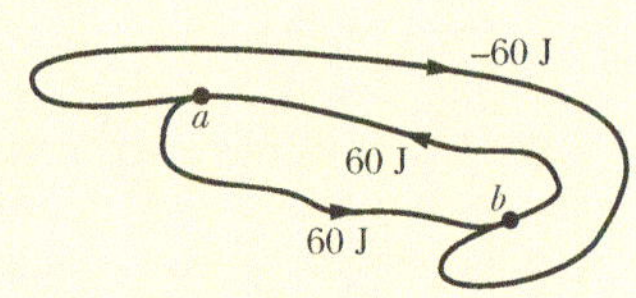

Sample Problem 8.01 Equivalent paths for calculating work, slippery cheese

The main lesson of this sample problem is this: It is perfectly all right to choose an easy path instead of a hard path. Figure 8-5*a* shows a 2.0 kg block of slippery cheese that slides along a frictionless track from point *a* to point *b*. The cheese travels through a total distance of 2.0 m along the track, and a net vertical distance of 0.80 m. How much work is done on the cheese by the gravitational force during the slide?

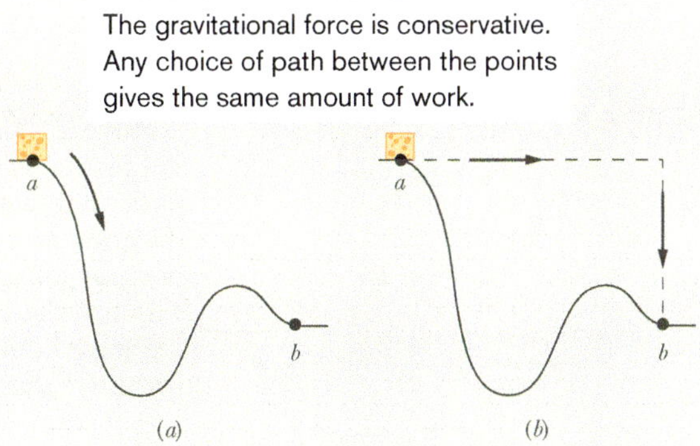

Figure 8-5 (*a*) A block of cheese slides along a frictionless track from point *a* to point *b*. (*b*) Finding the work done on the cheese by the gravitational force is easier along the dashed path than along the actual path taken by the cheese; the result is the same for both paths.

KEY IDEAS

(1) We *cannot* calculate the work by using Eq. 7-12 ($W_g = mgd\cos\phi$). The reason is that the angle ϕ between the directions of the gravitational force $\vec{F}_g$ and the displacement $\vec{d}$ varies along the track in an unknown way. (Even if we did know the shape of the track and could calculate ϕ along it, the calculation could be very difficult.) (2) Because $\vec{F}_g$ is a conservative force, we can find the work by choosing some other path between *a* and *b*—one that makes the calculation easy.

Calculations: Let us choose the dashed path in Fig. 8-5*b*; it consists of two straight segments. Along the horizontal segment, the angle ϕ is a constant 90°. Even though we do not know the displacement along that horizontal segment, Eq. 7-12 tells us that the work W_h done there is

$$W_h = mgd\cos 90° = 0.$$

Along the vertical segment, the displacement *d* is 0.80 m and, with $\vec{F}_g$ and $\vec{d}$ both downward, the angle ϕ is a constant 0°. Thus, Eq. 7-12 gives us, for the work W_v done along the vertical part of the dashed path,

$$W_v = mgd\cos 0°$$
$$= (2.0\text{ kg})(9.8\text{ m/s}^2)(0.80\text{ m})(1) = 15.7\text{ J}.$$

The total work done on the cheese by $\vec{F}_g$ as the cheese moves from point *a* to point *b* along the dashed path is then

$$W = W_h + W_v = 0 + 15.7\text{ J} \approx 16\text{ J}. \qquad \text{(Answer)}$$

This is also the work done as the cheese slides along the track from *a* to *b*.

Determining Potential Energy Values

Here we find equations that give the value of the two types of potential energy discussed in this chapter: gravitational potential energy and elastic potential energy. However, first we must find a general relation between a conservative force and the associated potential energy.

Consider a particle-like object that is part of a system in which a conservative force $\vec{F}$ acts. When that force does work *W* on the object, the change ΔU in the potential energy associated with the system is the negative of the work done. We wrote this fact as Eq. 8-1 ($\Delta U = -W$). For the most general case, in which the force may vary with position, we may write the work *W* as in Eq. 7-32:

$$W = \int_{x_i}^{x_f} F(x)\,dx. \qquad (8\text{-}5)$$

This equation gives the work done by the force when the object moves from point x_i to point x_f, changing the configuration of the system. (Because the force is conservative, the work is the same for all paths between those two points.)

Substituting Eq. 8-5 into Eq. 8-1, we find that the change in potential energy due to the change in configuration is, in general notation,

$$\Delta U = -\int_{x_i}^{x_f} F(x)\,dx. \tag{8-6}$$

Gravitational Potential Energy

We first consider a particle with mass m moving vertically along a y axis (the positive direction is upward). As the particle moves from point y_i to point y_f, the gravitational force $\vec{F}_g$ does work on it. To find the corresponding change in the gravitational potential energy of the particle–Earth system, we use Eq. 8-6 with two changes: (1) We integrate along the y axis instead of the x axis, because the gravitational force acts vertically. (2) We substitute $-mg$ for the force symbol F, because $\vec{F}_g$ has the magnitude mg and is directed down the y axis. We then have

$$\Delta U = -\int_{y_i}^{y_f} (-mg)\,dy = mg\int_{y_i}^{y_f} dy = mg\Big[y\Big]_{y_i}^{y_f},$$

which yields

$$\Delta U = mg(y_f - y_i) = mg\,\Delta y. \tag{8-7}$$

Only *changes* ΔU in gravitational potential energy (or any other type of potential energy) are physically meaningful. However, to simplify a calculation or a discussion, we sometimes would like to say that a certain gravitational potential value U is associated with a certain particle–Earth system when the particle is at a certain height y. To do so, we rewrite Eq. 8-7 as

$$U - U_i = mg(y - y_i). \tag{8-8}$$

Then we take U_i to be the gravitational potential energy of the system when it is in a **reference configuration** in which the particle is at a **reference point** y_i. Usually we take $U_i = 0$ and $y_i = 0$. Doing this changes Eq. 8-8 to

$$U(y) = mgy \quad \text{(gravitational potential energy).} \tag{8-9}$$

This equation tells us:

The gravitational potential energy associated with a particle–Earth system depends only on the vertical position y (or height) of the particle relative to the reference position $y = 0$, not on the horizontal position.

Elastic Potential Energy

We next consider the block–spring system shown in Fig. 8-3, with the block moving on the end of a spring of spring constant k. As the block moves from point x_i to point x_f, the spring force $F_x = -kx$ does work on the block. To find the corresponding change in the elastic potential energy of the block–spring system, we substitute $-kx$ for $F(x)$ in Eq. 8-6. We then have

$$\Delta U = -\int_{x_i}^{x_f} (-kx)\,dx = k\int_{x_i}^{x_f} x\,dx = \tfrac{1}{2}k\Big[x^2\Big]_{x_i}^{x_f},$$

or

$$\Delta U = \tfrac{1}{2}kx_f^2 - \tfrac{1}{2}kx_i^2. \tag{8-10}$$

To associate a potential energy value U with the block at position x, we choose the reference configuration to be when the spring is at its relaxed length and the block is at $x_i = 0$. Then the elastic potential energy U_i is 0, and Eq. 8-10 becomes

$$U - 0 = \tfrac{1}{2}kx^2 - 0,$$

which gives us

$$U(x) = \tfrac{1}{2}kx^2 \quad \text{(elastic potential energy).} \tag{8-11}$$

Checkpoint 2

A particle is to move along an x axis from $x = 0$ to x_1 while a conservative force, directed along the x axis, acts on the particle. The figure shows three situations in which the x component of that force varies with x. The force has the same maximum magnitude F_1 in all three situations. Rank the situations according to the change in the associated potential energy during the particle's motion, most positive first.

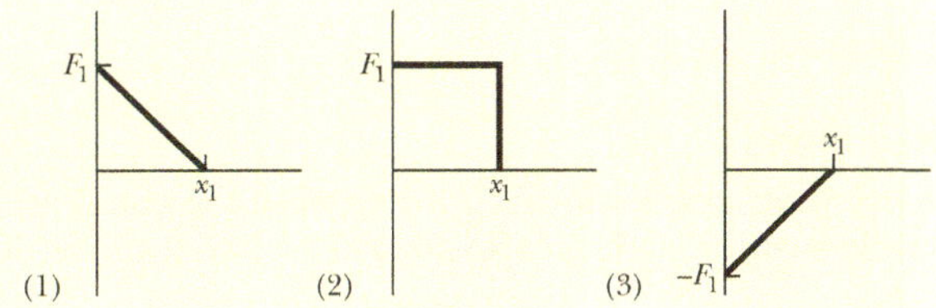

Sample Problem 8.02 Choosing reference level for gravitational potential energy, sloth

Here is an example with this lesson plan: Generally you can choose any level to be the reference level, but once chosen, be consistent. A 2.0 kg sloth hangs 5.0 m above the ground (Fig. 8-6).

(a) What is the gravitational potential energy U of the sloth–Earth system if we take the reference point $y = 0$ to be (1) at the ground, (2) at a balcony floor that is 3.0 m above the ground, (3) at the limb, and (4) 1.0 m above the limb? Take the gravitational potential energy to be zero at $y = 0$.

Figure 8-6 Four choices of reference point $y = 0$. Each y axis is marked in units of meters. The choice affects the value of the potential energy U of the sloth–Earth system. However, it does not affect the change ΔU in potential energy of the system if the sloth moves by, say, falling.

KEY IDEA

Once we have chosen the reference point for $y = 0$, we can calculate the gravitational potential energy U of the system *relative to that reference point* with Eq. 8-9.

Calculations: For choice (1) the sloth is at $y = 5.0$ m, and

$$U = mgy = (2.0 \text{ kg})(9.8 \text{ m/s}^2)(5.0 \text{ m})$$
$$= 98 \text{ J.} \qquad \text{(Answer)}$$

For the other choices, the values of U are

(2) $U = mgy = mg(2.0 \text{ m}) = 39 \text{ J},$

(3) $U = mgy = mg(0) = 0 \text{ J},$

(4) $U = mgy = mg(-1.0 \text{ m})$
$= -19.6 \text{ J} \approx -20 \text{ J.}$ (Answer)

(b) The sloth drops to the ground. For each choice of reference point, what is the change ΔU in the potential energy of the sloth–Earth system due to the fall?

KEY IDEA

The *change* in potential energy does not depend on the choice of the reference point for $y = 0$; instead, it depends on the change in height Δy.

Calculation: For all four situations, we have the same $\Delta y = -5.0$ m. Thus, for (1) to (4), Eq. 8-7 tells us that

$$\Delta U = mg\,\Delta y = (2.0 \text{ kg})(9.8 \text{ m/s}^2)(-5.0 \text{ m})$$
$$= -98 \text{ J.} \qquad \text{(Answer)}$$

Additional examples, video, and practice available at *WileyPLUS*

8-2 CONSERVATION OF MECHANICAL ENERGY

Learning Objectives

After reading this module, you should be able to . . .

8.05 After first clearly defining which objects form a system, identify that the mechanical energy of the system is the sum of the kinetic energies and potential energies of those objects.

8.06 For an isolated system in which only conservative forces act, apply the conservation of mechanical energy to relate the initial potential and kinetic energies to the potential and kinetic energies at a later instant.

Key Ideas

- The mechanical energy E_{mec} of a system is the sum of its kinetic energy K and potential energy U:

$$E_{mec} = K + U.$$

- An isolated system is one in which no external force causes energy changes. If only conservative forces do work within an isolated system, then the mechanical energy E_{mec} of the system cannot change. This principle of conservation of mechanical energy is written as

$$K_2 + U_2 = K_1 + U_1,$$

in which the subscripts refer to different instants during an energy transfer process. This conservation principle can also be written as

$$\Delta E_{mec} = \Delta K + \Delta U = 0.$$

Conservation of Mechanical Energy

The **mechanical energy** E_{mec} of a system is the sum of its potential energy U and the kinetic energy K of the objects within it:

$$E_{mec} = K + U \quad \text{(mechanical energy).} \tag{8-12}$$

In this module, we examine what happens to this mechanical energy when only conservative forces cause energy transfers within the system—that is, when frictional and drag forces do not act on the objects in the system. Also, we shall assume that the system is *isolated* from its environment; that is, no *external force* from an object outside the system causes energy changes inside the system.

When a conservative force does work W on an object within the system, that force transfers energy between kinetic energy K of the object and potential energy U of the system. From Eq. 7-10, the change ΔK in kinetic energy is

$$\Delta K = W \tag{8-13}$$

and from Eq. 8-1, the change ΔU in potential energy is

$$\Delta U = -W. \tag{8-14}$$

Combining Eqs. 8-13 and 8-14, we find that

$$\Delta K = -\Delta U. \tag{8-15}$$

In words, one of these energies increases exactly as much as the other decreases.

We can rewrite Eq. 8-15 as

$$K_2 - K_1 = -(U_2 - U_1), \tag{8-16}$$

where the subscripts refer to two different instants and thus to two different arrangements of the objects in the system. Rearranging Eq. 8-16 yields

$$K_2 + U_2 = K_1 + U_1 \quad \text{(conservation of mechanical energy).} \tag{8-17}$$

In words, this equation says:

$$\begin{pmatrix}\text{the sum of } K \text{ and } U \text{ for} \\ \text{any state of a system}\end{pmatrix} = \begin{pmatrix}\text{the sum of } K \text{ and } U \text{ for} \\ \text{any other state of the system}\end{pmatrix},$$

©AP/Wide World Photos

In olden days, a person would be tossed via a blanket to be able to see farther over the flat terrain. Nowadays, it is done just for fun. During the ascent of the person in the photograph, energy is transferred from kinetic energy to gravitational potential energy. The maximum height is reached when that transfer is complete. Then the transfer is reversed during the fall.

when the system is isolated and only conservative forces act on the objects in the system. In other words:

> In an isolated system where only conservative forces cause energy changes, the kinetic energy and potential energy can change, but their sum, the mechanical energy E_{mec} of the system, cannot change.

This result is called the **principle of conservation of mechanical energy.** (Now you can see where *conservative* forces got their name.) With the aid of Eq. 8-15, we can write this principle in one more form, as

$$\Delta E_{mec} = \Delta K + \Delta U = 0. \tag{8-18}$$

The principle of conservation of mechanical energy allows us to solve problems that would be quite difficult to solve using only Newton's laws:

> When the mechanical energy of a system is conserved, we can relate the sum of kinetic energy and potential energy at one instant to that at another instant *without considering the intermediate motion* and *without finding the work done by the forces involved.*

Figure 8-7 shows an example in which the principle of conservation of mechanical energy can be applied: As a pendulum swings, the energy of the

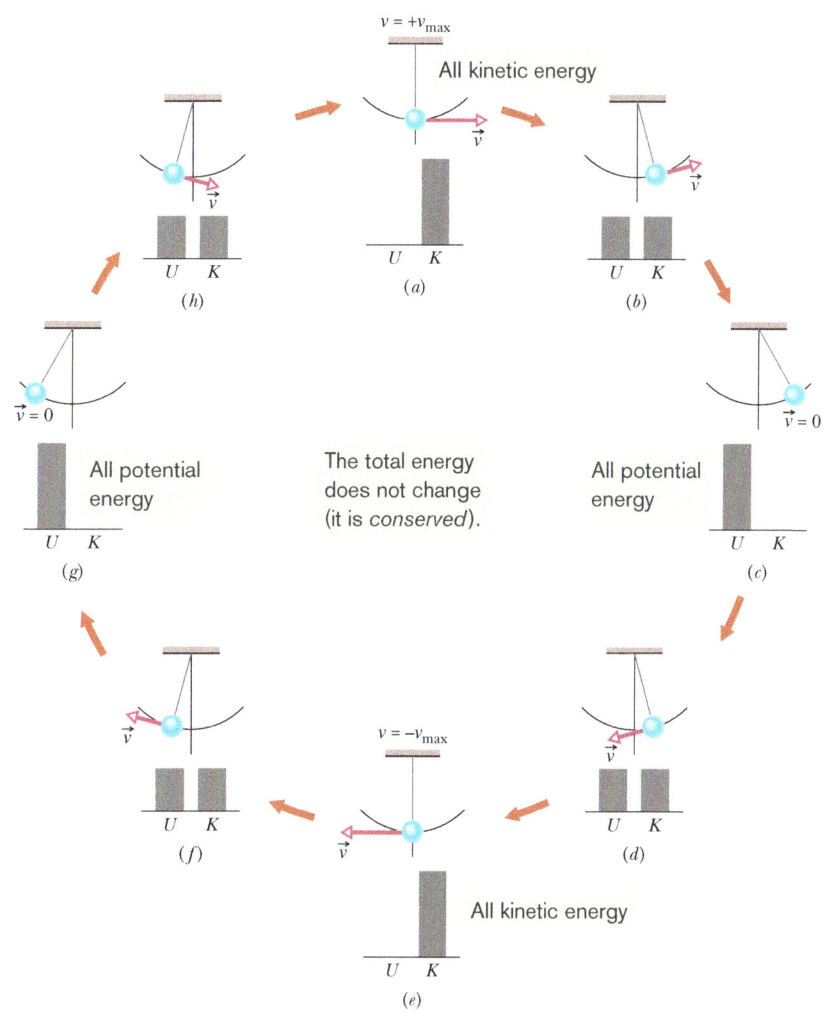

Figure 8-7 A pendulum, with its mass concentrated in a bob at the lower end, swings back and forth. One full cycle of the motion is shown. During the cycle the values of the potential and kinetic energies of the pendulum–Earth system vary as the bob rises and falls, but the mechanical energy E_{mec} of the system remains constant. The energy E_{mec} can be described as continuously shifting between the kinetic and potential forms. In stages (*a*) and (*e*), all the energy is kinetic energy. The bob then has its greatest speed and is at its lowest point. In stages (*c*) and (*g*), all the energy is potential energy. The bob then has zero speed and is at its highest point. In stages (*b*), (*d*), (*f*), and (*h*), half the energy is kinetic energy and half is potential energy. If the swinging involved a frictional force at the point where the pendulum is attached to the ceiling, or a drag force due to the air, then E_{mec} would not be conserved, and eventually the pendulum would stop.

pendulum–Earth system is transferred back and forth between kinetic energy K and gravitational potential energy U, with the sum $K + U$ being constant. If we know the gravitational potential energy when the pendulum bob is at its highest point (Fig. 8-7*c*), Eq. 8-17 gives us the kinetic energy of the bob at the lowest point (Fig. 8-7*e*).

For example, let us choose the lowest point as the reference point, with the gravitational potential energy $U_2 = 0$. Suppose then that the potential energy at the highest point is $U_1 = 20$ J relative to the reference point. Because the bob momentarily stops at its highest point, the kinetic energy there is $K_1 = 0$. Putting these values into Eq. 8-17 gives us the kinetic energy K_2 at the lowest point:

$$K_2 + 0 = 0 + 20\text{ J} \qquad \text{or} \qquad K_2 = 20\text{ J}.$$

Note that we get this result without considering the motion between the highest and lowest points (such as in Fig. 8-7*d*) and without finding the work done by any forces involved in the motion.

Checkpoint 3

The figure shows four situations—one in which an initially stationary block is dropped and three in which the block is allowed to slide down frictionless ramps. (a) Rank the situations according to the kinetic energy of the block at point B, greatest first. (b) Rank them according to the speed of the block at point B, greatest first.

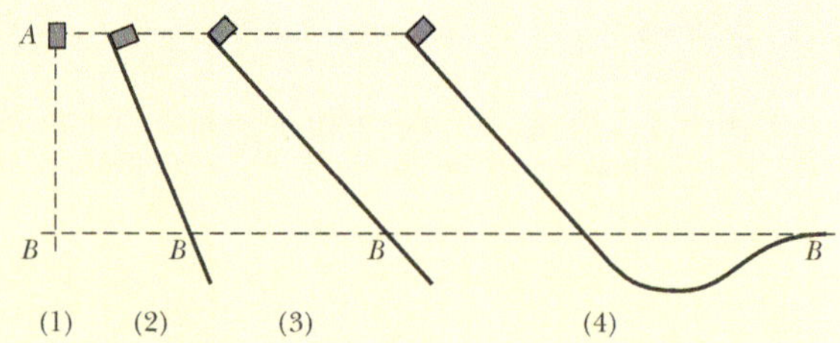

Sample Problem 8.03 Conservation of mechanical energy, water slide

The huge advantage of using the conservation of energy instead of Newton's laws of motion is that we can jump from the initial state to the final state without considering all the intermediate motion. Here is an example. In Fig. 8-8, a child of mass m is released from rest at the top of a water slide, at height $h = 8.5$ m above the bottom of the slide. Assuming that the slide is frictionless because of the water on it, find the child's speed at the bottom of the slide.

KEY IDEAS

(1) We cannot find her speed at the bottom by using her acceleration along the slide as we might have in earlier chapters because we do not know the slope (angle) of the slide. However, because that speed is related to her kinetic energy, perhaps we can use the principle of conservation of mechanical energy to get the speed. Then we would not need to know the slope. (2) Mechanical energy is conserved in a system *if* the system is isolated and *if* only conservative forces cause energy transfers within it. Let's check.

Forces: Two forces act on the child. The *gravitational force,* a conservative force, does work on her. The *normal force* on her from the slide does no work because its direction at any point during the descent is always perpendicular to the direction in which the child moves.

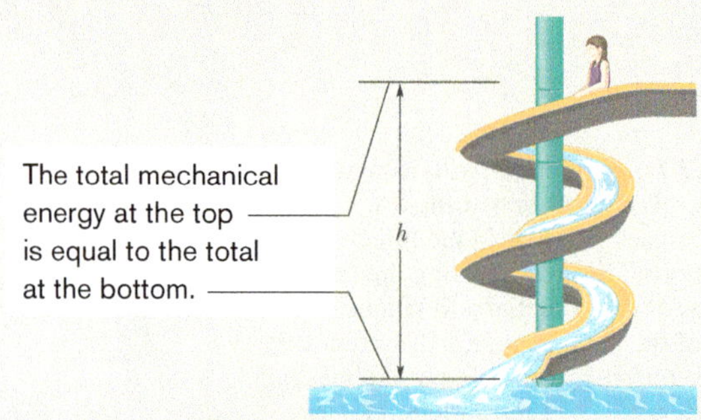

Figure 8-8 A child slides down a water slide as she descends a height h.

System: Because the only force doing work on the child is the gravitational force, we choose the child–Earth system as our system, which we can take to be isolated.

Thus, we have only a conservative force doing work in an isolated system, so we *can* use the principle of conservation of mechanical energy.

Calculations: Let the mechanical energy be $E_{mec,t}$ when the child is at the top of the slide and $E_{mec,b}$ when she is at the bottom. Then the conservation principle tells us

$$E_{mec,b} = E_{mec,t}. \qquad (8\text{-}19)$$

To show both kinds of mechanical energy, we have

$$K_b + U_b = K_t + U_t, \quad (8\text{-}20)$$

or $$\tfrac{1}{2}mv_b^2 + mgy_b = \tfrac{1}{2}mv_t^2 + mgy_t.$$

Dividing by m and rearranging yield

$$v_b^2 = v_t^2 + 2g(y_t - y_b).$$

Putting $v_t = 0$ and $y_t - y_b = h$ leads to

$$v_b = \sqrt{2gh} = \sqrt{(2)(9.8 \text{ m/s}^2)(8.5 \text{ m})}$$
$$= 13 \text{ m/s}. \quad \text{(Answer)}$$

This is the same speed that the child would reach if she fell 8.5 m vertically. On an actual slide, some frictional forces would act and the child would not be moving quite so fast.

Comments: Although this problem is hard to solve directly with Newton's laws, using conservation of mechanical energy makes the solution much easier. However, if we were asked to find the time taken for the child to reach the bottom of the slide, energy methods would be of no use; we would need to know the shape of the slide, and we would have a difficult problem.

8-3 READING A POTENTIAL ENERGY CURVE

Learning Objectives

After reading this module, you should be able to . . .

8.07 Given a particle's potential energy as a function of its position x, determine the force on the particle.

8.08 Given a graph of potential energy versus x, determine the force on a particle.

8.09 On a graph of potential energy versus x, superimpose a line for a particle's mechanical energy and determine the particle's kinetic energy for any given value of x.

8.10 If a particle moves along an x axis, use a potential-energy graph for that axis and the conservation of mechanical energy to relate the energy values at one position to those at another position.

8.11 On a potential-energy graph, identify any turning points and any regions where the particle is not allowed because of energy requirements.

8.12 Explain neutral equilibrium, stable equilibrium, and unstable equilibrium.

Key Ideas

- If we know the potential energy function $U(x)$ for a system in which a one-dimensional force $F(x)$ acts on a particle, we can find the force as

$$F(x) = -\frac{dU(x)}{dx}.$$

- If $U(x)$ is given on a graph, then at any value of x, the force $F(x)$ is the negative of the slope of the curve there and the kinetic energy of the particle is given by

$$K(x) = E_{\text{mec}} - U(x),$$

where E_{mec} is the mechanical energy of the system.

- A turning point is a point x at which the particle reverses its motion (there, $K = 0$).
- The particle is in equilibrium at points where the slope of the $U(x)$ curve is zero (there, $F(x) = 0$).

Reading a Potential Energy Curve

Once again we consider a particle that is part of a system in which a conservative force acts. This time suppose that the particle is constrained to move along an x axis while the conservative force does work on it. We want to plot the potential energy $U(x)$ that is associated with that force and the work that it does, and then we want to consider how we can relate the plot back to the force and to the kinetic energy of the particle. However, before we discuss such plots, we need one more relationship between the force and the potential energy.

Finding the Force Analytically

Equation 8-6 tells us how to find the change ΔU in potential energy between two points in a one-dimensional situation if we know the force $F(x)$. Now we want to go the other way; that is, we know the potential energy function $U(x)$ and want to find the force.

For one-dimensional motion, the work W done by a force that acts on a particle as the particle moves through a distance Δx is $F(x)\ \Delta x$. We can then write Eq. 8-1 as

$$\Delta U(x) = -W = -F(x)\,\Delta x. \tag{8-21}$$

Solving for $F(x)$ and passing to the differential limit yield

$$F(x) = -\frac{dU(x)}{dx} \quad \text{(one-dimensional motion)}, \tag{8-22}$$

which is the relation we sought.

We can check this result by putting $U(x) = \frac{1}{2}kx^2$, which is the elastic potential energy function for a spring force. Equation 8-22 then yields, as expected, $F(x) = -kx$, which is Hooke's law. Similarly, we can substitute $U(x) = mgx$, which is the gravitational potential energy function for a particle–Earth system, with a particle of mass m at height x above Earth's surface. Equation 8-22 then yields $F = -mg$, which is the gravitational force on the particle.

The Potential Energy Curve

Figure 8-9*a* is a plot of a potential energy function $U(x)$ for a system in which a particle is in one-dimensional motion while a conservative force $F(x)$ does work on it. We can easily find $F(x)$ by (graphically) taking the slope of the $U(x)$ curve at various points. (Equation 8-22 tells us that $F(x)$ is the negative of the slope of the $U(x)$ curve.) Figure 8-9*b* is a plot of $F(x)$ found in this way.

Turning Points

In the absence of a nonconservative force, the mechanical energy E of a system has a constant value given by

$$U(x) + K(x) = E_{\text{mec}}. \tag{8-23}$$

Here $K(x)$ is the *kinetic energy function* of a particle in the system (this $K(x)$ gives the kinetic energy as a function of the particle's location x). We may rewrite Eq. 8-23 as

$$K(x) = E_{\text{mec}} - U(x). \tag{8-24}$$

Suppose that E_{mec} (which has a constant value, remember) happens to be 5.0 J. It would be represented in Fig. 8-9*c* by a horizontal line that runs through the value 5.0 J on the energy axis. (It is, in fact, shown there.)

Equation 8-24 and Fig. 8-9*d* tell us how to determine the kinetic energy K for any location x of the particle: On the $U(x)$ curve, find U for that location x and then subtract U from E_{mec}. In Fig. 8-9*e* for example, if the particle is at any point to the right of x_5, then $K = 1.0$ J. The value of K is greatest (5.0 J) when the particle is at x_2 and least (0 J) when the particle is at x_1.

Since K can never be negative (because v^2 is always positive), the particle can never move to the left of x_1, where $E_{\text{mec}} - U$ is negative. Instead, as the particle moves toward x_1 from x_2, K decreases (the particle slows) until $K = 0$ at x_1 (the particle stops there).

Note that when the particle reaches x_1, the force on the particle, given by Eq. 8-22, is positive (because the slope dU/dx is negative). This means that the particle does not remain at x_1 but instead begins to move to the right, opposite its earlier motion. Hence x_1 is a **turning point,** a place where $K = 0$ (because $U = E$) and the particle changes direction. There is no turning point (where $K = 0$) on the right side of the graph. When the particle heads to the right, it will continue indefinitely.

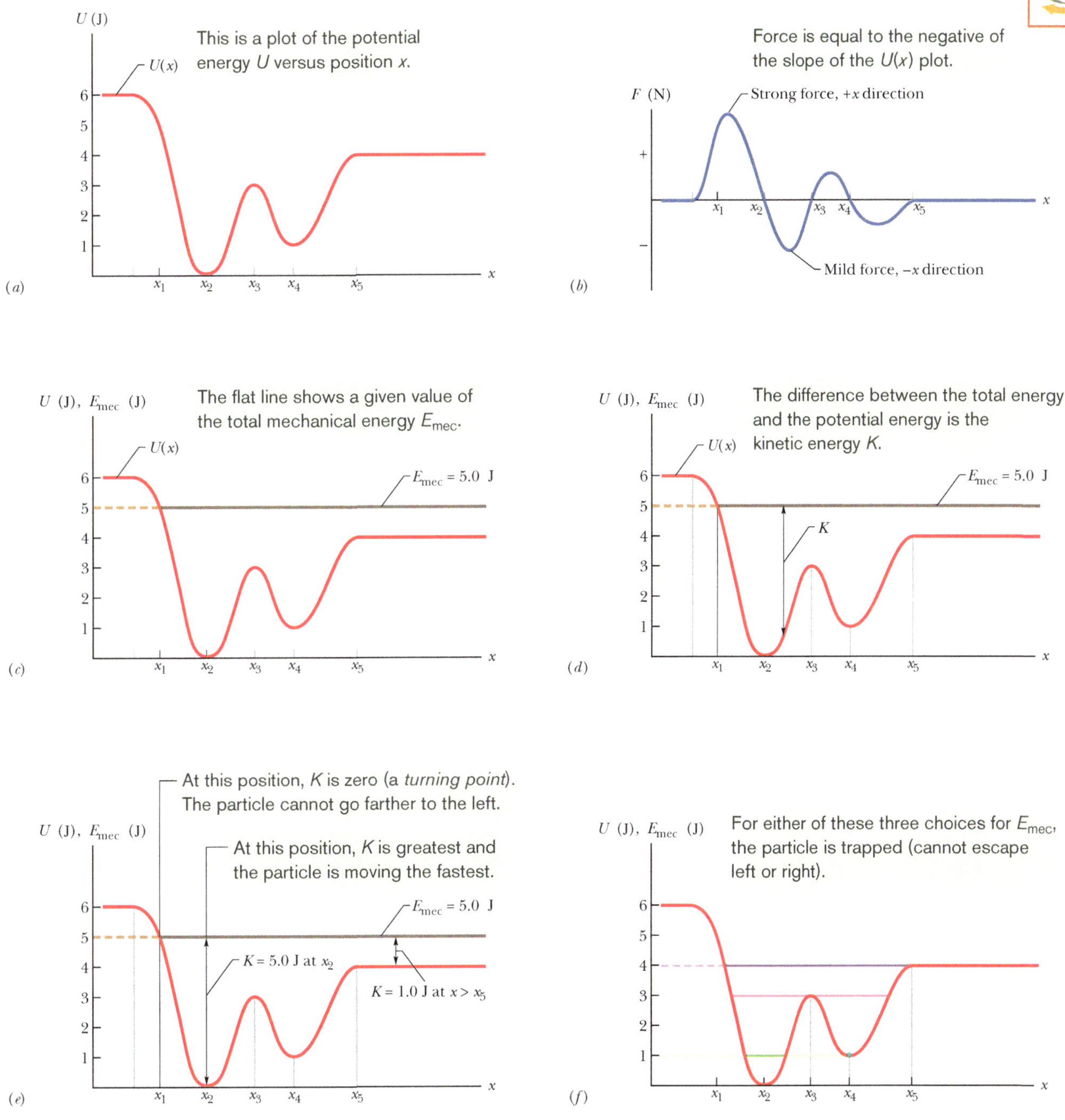

Figure 8-9 (*a*) A plot of $U(x)$, the potential energy function of a system containing a particle confined to move along an x axis. There is no friction, so mechanical energy is conserved. (*b*) A plot of the force $F(x)$ acting on the particle, derived from the potential energy plot by taking its slope at various points. (*c*)–(*e*) How to determine the kinetic energy. (*f*) The $U(x)$ plot of (*a*) with three possible values of E_{mec} shown. **In *WileyPLUS*, this figure is available as an animation with voiceover.**

Equilibrium Points

Figure 8-9*f* shows three different values for E_{mec} superposed on the plot of the potential energy function $U(x)$ of Fig. 8-9*a*. Let us see how they change the situation. If $E_{mec} = 4.0$ J (purple line), the turning point shifts from x_1 to a point between x_1 and x_2. Also, at any point to the right of x_5, the system's mechanical energy is equal to its potential energy; thus, the particle has no kinetic energy and (by Eq. 8-22) no force acts on it, and so it must be stationary. A particle at such a position is said to be in **neutral equilibrium.** (A marble placed on a horizontal tabletop is in that state.)

If $E_{mec} = 3.0$ J (pink line), there are two turning points: One is between x_1 and x_2, and the other is between x_4 and x_5. In addition, x_3 is a point at which $K = 0$. If the particle is located exactly there, the force on it is also zero, and the particle remains stationary. However, if it is displaced even slightly in either direction, a nonzero force pushes it farther in the same direction, and the particle continues to move. A particle at such a position is said to be in **unstable equilibrium.** (A marble balanced on top of a bowling ball is an example.)

Next consider the particle's behavior if $E_{mec} = 1.0$ J (green line). If we place it at x_4, it is stuck there. It cannot move left or right on its own because to do so would require a negative kinetic energy. If we push it slightly left or right, a restoring force appears that moves it back to x_4. A particle at such a position is said to be in **stable equilibrium.** (A marble placed at the bottom of a hemispherical bowl is an example.) If we place the particle in the cup-like *potential well* centered at x_2, it is between two turning points. It can still move somewhat, but only partway to x_1 or x_3.

Checkpoint 4

The figure gives the potential energy function $U(x)$ for a system in which a particle is in one-dimensional motion. (a) Rank regions AB, BC, and CD according to the magnitude of the force on the particle, greatest first. (b) What is the direction of the force when the particle is in region AB?

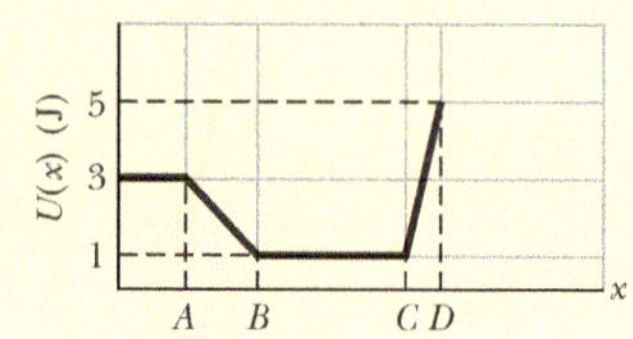

Sample Problem 8.04 Reading a potential energy graph

A 2.00 kg particle moves along an x axis in one-dimensional motion while a conservative force along that axis acts on it. The potential energy $U(x)$ associated with the force is plotted in Fig. 8-10*a*. That is, if the particle were placed at any position between $x = 0$ and $x = 7.00$ m, it would have the plotted value of U. At $x = 6.5$ m, the particle has velocity $\vec{v}_0 = (-4.00 \text{ m/s})\hat{\text{i}}$.

(a) From Fig. 8-10*a*, determine the particle's speed at $x_1 = 4.5$ m.

KEY IDEAS

(1) The particle's kinetic energy is given by Eq. 7-1 ($K = \frac{1}{2}mv^2$). (2) Because only a conservative force acts on the particle, the mechanical energy $E_{mec}(= K + U)$ is conserved as the particle moves. (3) Therefore, on a plot of $U(x)$ such as Fig. 8-10*a*, the kinetic energy is equal to the difference between E_{mec} and U.

Calculations: At $x = 6.5$ m, the particle has kinetic energy

$$K_0 = \tfrac{1}{2}mv_0^2 = \tfrac{1}{2}(2.00 \text{ kg})(4.00 \text{ m/s})^2$$
$$= 16.0 \text{ J}.$$

Because the potential energy there is $U = 0$, the mechanical energy is

$$E_{mec} = K_0 + U_0 = 16.0 \text{ J} + 0 = 16.0 \text{ J}.$$

This value for E_{mec} is plotted as a horizontal line in Fig. 8-10*a*. From that figure we see that at $x = 4.5$ m, the potential energy is $U_1 = 7.0$ J. The kinetic energy K_1 is the difference between E_{mec} and U_1:

$$K_1 = E_{mec} - U_1 = 16.0 \text{ J} - 7.0 \text{ J} = 9.0 \text{ J}.$$

Because $K_1 = \frac{1}{2}mv_1^2$, we find

$$v_1 = 3.0 \text{ m/s}. \quad \text{(Answer)}$$

(b) Where is the particle's turning point located?

KEY IDEA

The turning point is where the force momentarily stops and then reverses the particle's motion. That is, it is where the particle momentarily has $v = 0$ and thus $K = 0$.

Calculations: Because K is the difference between E_{mec} and U, we want the point in Fig. 8-10*a* where the plot of U rises to meet the horizontal line of E_{mec}, as shown in Fig. 8-10*b*. Because the plot of U is a straight line in Fig. 8-10*b*, we can draw nested right triangles as shown and then write the proportionality of distances

$$\frac{16 - 7.0}{d} = \frac{20 - 7.0}{4.0 - 1.0},$$

which gives us $d = 2.08$ m. Thus, the turning point is at

$$x = 4.0 \text{ m} - d = 1.9 \text{ m}. \qquad \text{(Answer)}$$

(c) Evaluate the force acting on the particle when it is in the region $1.9 \text{ m} < x < 4.0 \text{ m}$.

KEY IDEA

The force is given by Eq. 8-22 ($F(x) = -dU(x)/dx$): The force is equal to the negative of the slope on a graph of $U(x)$.

Calculations: For the graph of Fig. 8-10*b*, we see that for the range $1.0 \text{ m} < x < 4.0 \text{ m}$ the force is

$$F = -\frac{20 \text{ J} - 7.0 \text{ J}}{1.0 \text{ m} - 4.0 \text{ m}} = 4.3 \text{ N}. \qquad \text{(Answer)}$$

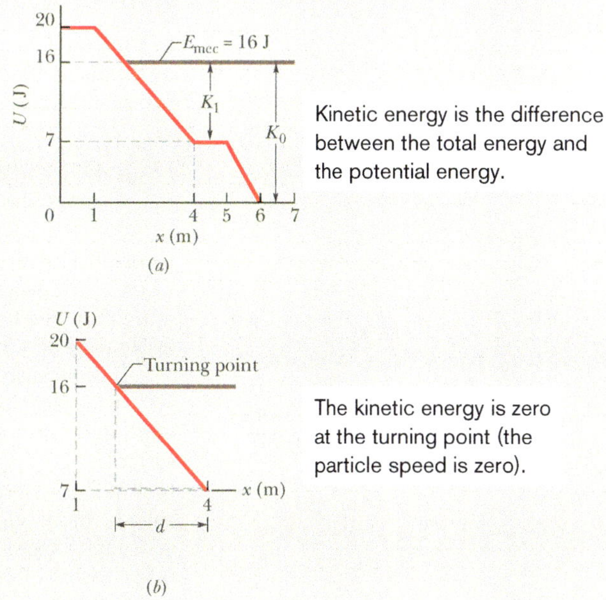

Figure 8-10 (*a*) A plot of potential energy U versus position x. (*b*) A section of the plot used to find where the particle turns around.

Thus, the force has magnitude 4.3 N and is in the positive direction of the x axis. This result is consistent with the fact that the initially leftward-moving particle is stopped by the force and then sent rightward.

WILEY PLUS Additional examples, video, and practice available at *WileyPLUS*

8-4 WORK DONE ON A SYSTEM BY AN EXTERNAL FORCE

Learning Objectives

After reading this module, you should be able to . . .

8.13 When work is done on a system by an external force with no friction involved, determine the changes in kinetic energy and potential energy.

8.14 When work is done on a system by an external force with friction involved, relate that work to the changes in kinetic energy, potential energy, and thermal energy.

Key Ideas

- Work W is energy transferred to or from a system by means of an external force acting on the system.
- When more than one force acts on a system, their net work is the transferred energy.
- When friction is not involved, the work done on the system and the change ΔE_{mec} in the mechanical energy of the system are equal:

$$W = \Delta E_{mec} = \Delta K + \Delta U.$$

- When a kinetic frictional force acts within the system, then the thermal energy E_{th} of the system changes. (This energy is associated with the random motion of atoms and molecules in the system.) The work done on the system is then

$$W = \Delta E_{mec} + \Delta E_{th}.$$

- The change ΔE_{th} is related to the magnitude f_k of the frictional force and the magnitude d of the displacement caused by the external force by

$$\Delta E_{th} = f_k d.$$

Work Done on a System by an External Force

In Chapter 7, we defined work as being energy transferred to or from an object by means of a force acting on the object. We can now extend that definition to an external force acting on a system of objects.

> Work is energy transferred to or from a system by means of an external force acting on that system.

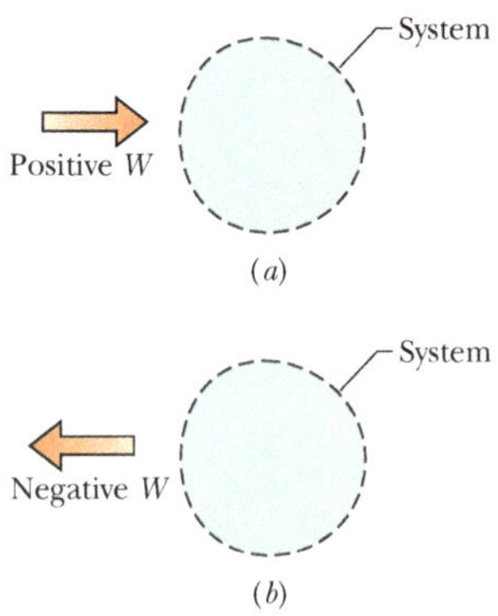

Figure 8-11 (*a*) Positive work W done on an arbitrary system means a transfer of energy to the system. (*b*) Negative work W means a transfer of energy from the system.

Figure 8-11*a* represents positive work (a transfer of energy *to* a system), and Fig. 8-11*b* represents negative work (a transfer of energy *from* a system). When more than one force acts on a system, their *net work* is the energy transferred to or from the system.

These transfers are like transfers of money to and from a bank account. If a system consists of a single particle or particle-like object, as in Chapter 7, the work done on the system by a force can change only the kinetic energy of the system. The energy statement for such transfers is the work–kinetic energy theorem of Eq. 7-10 ($\Delta K = W$); that is, a single particle has only one energy account, called kinetic energy. External forces can transfer energy into or out of that account. If a system is more complicated, however, an external force can change other forms of energy (such as potential energy); that is, a more complicated system can have multiple energy accounts.

Let us find energy statements for such systems by examining two basic situations, one that does not involve friction and one that does.

No Friction Involved

To compete in a bowling-ball-hurling contest, you first squat and cup your hands under the ball on the floor. Then you rapidly straighten up while also pulling your hands up sharply, launching the ball upward at about face level. During your upward motion, your applied force on the ball obviously does work; that is, it is an external force that transfers energy, but to what system?

To answer, we check to see which energies change. There is a change ΔK in the ball's kinetic energy and, because the ball and Earth become more separated, there is a change ΔU in the gravitational potential energy of the ball–Earth system. To include both changes, we need to consider the ball–Earth system. Then your force is an external force doing work on that system, and the work is

$$W = \Delta K + \Delta U, \tag{8-25}$$

or

$$W = \Delta E_{\text{mec}} \quad \text{(work done on system, no friction involved)}, \tag{8-26}$$

where ΔE_{mec} is the change in the mechanical energy of the system. These two equations, which are represented in Fig. 8-12, are equivalent energy statements for work done on a system by an external force when friction is not involved.

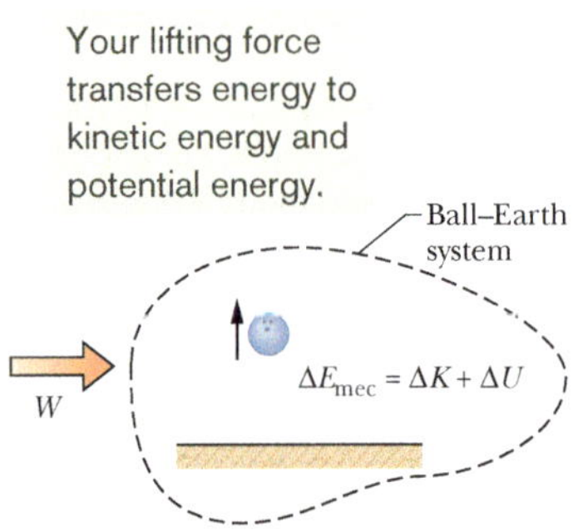

Figure 8-12 Positive work W is done on a system of a bowling ball and Earth, causing a change ΔE_{mec} in the mechanical energy of the system, a change ΔK in the ball's kinetic energy, and a change ΔU in the system's gravitational potential energy.

Friction Involved

We next consider the example in Fig. 8-13*a*. A constant horizontal force $\vec{F}$ pulls a block along an x axis and through a displacement of magnitude d, increasing the block's velocity from $\vec{v}_0$ to $\vec{v}$. During the motion, a constant kinetic frictional force $\vec{f}_k$ from the floor acts on the block. Let us first choose the block as our system and apply Newton's second law to it. We can write that law for components along the x axis ($F_{\text{net},x} = ma_x$) as

$$F - f_k = ma. \tag{8-27}$$

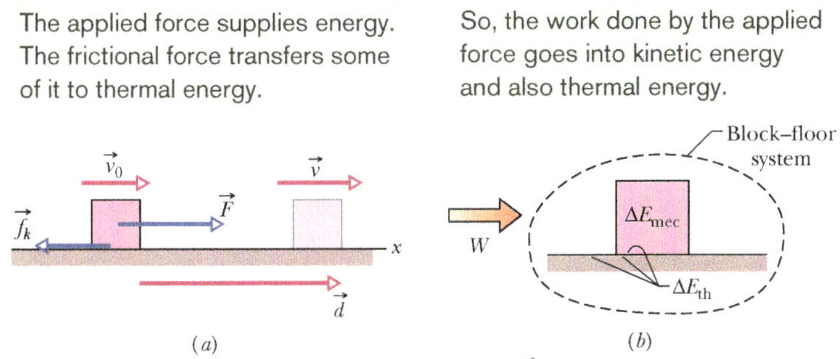

Figure 8-13 (*a*) A block is pulled across a floor by force $\vec{F}$ while a kinetic frictional force $\vec{f}_k$ opposes the motion. The block has velocity $\vec{v}_0$ at the start of a displacement $\vec{d}$ and velocity $\vec{v}$ at the end of the displacement. (*b*) Positive work W is done on the block–floor system by force $\vec{F}$, resulting in a change ΔE_{mec} in the block's mechanical energy and a change ΔE_{th} in the thermal energy of the block and floor.

Because the forces are constant, the acceleration $\vec{a}$ is also constant. Thus, we can use Eq. 2-16 to write

$$v^2 = v_0^2 + 2ad.$$

Solving this equation for a, substituting the result into Eq. 8-27, and rearranging then give us

$$Fd = \tfrac{1}{2}mv^2 - \tfrac{1}{2}mv_0^2 + f_k d \tag{8-28}$$

or, because $\frac{1}{2}mv^2 - \frac{1}{2}mv_0^2 = \Delta K$ for the block,

$$Fd = \Delta K + f_k d. \tag{8-29}$$

In a more general situation (say, one in which the block is moving up a ramp), there can be a change in potential energy. To include such a possible change, we generalize Eq. 8-29 by writing

$$Fd = \Delta E_{mec} + f_k d. \tag{8-30}$$

By experiment we find that the block and the portion of the floor along which it slides become warmer as the block slides. As we shall discuss in Chapter 18, the temperature of an object is related to the object's thermal energy E_{th} (the energy associated with the random motion of the atoms and molecules in the object). Here, the thermal energy of the block and floor increases because (1) there is friction between them and (2) there is sliding. Recall that friction is due to the cold-welding between two surfaces. As the block slides over the floor, the sliding causes repeated tearing and re-forming of the welds between the block and the floor, which makes the block and floor warmer. Thus, the sliding increases their thermal energy E_{th}.

Through experiment, we find that the increase ΔE_{th} in thermal energy is equal to the product of the magnitudes f_k and d:

$$\Delta E_{th} = f_k d \quad \text{(increase in thermal energy by sliding).} \tag{8-31}$$

Thus, we can rewrite Eq. 8-30 as

$$Fd = \Delta E_{mec} + \Delta E_{th}. \tag{8-32}$$

Fd is the work W done by the external force $\vec{F}$ (the energy transferred by the force), but on which system is the work done (where are the energy transfers made)? To answer, we check to see which energies change. The block's mechanical energy changes, and the thermal energies of the block and floor also change. Therefore, the work done by force $\vec{F}$ is done on the block–floor system. That work is

$$W = \Delta E_{mec} + \Delta E_{th} \quad \text{(work done on system, friction involved).} \tag{8-33}$$

This equation, which is represented in Fig. 8-13*b*, is the energy statement for the work done on a system by an external force when friction is involved.

Checkpoint 5

In three trials, a block is pushed by a horizontal applied force across a floor that is not frictionless, as in Fig. 8-13*a*. The magnitudes F of the applied force and the results of the pushing on the block's speed are given in the table. In all three trials, the block is pushed through the same distance d. Rank the three trials according to the change in the thermal energy of the block and floor that occurs in that distance d, greatest first.

Trial	F	Result on Block's Speed
a	5.0 N	decreases
b	7.0 N	remains constant
c	8.0 N	increases

Sample Problem 8.05 Work, friction, change in thermal energy, cabbage heads

A food shipper pushes a wood crate of cabbage heads (total mass $m = 14$ kg) across a concrete floor with a constant horizontal force $\vec{F}$ of magnitude 40 N. In a straight-line displacement of magnitude $d = 0.50$ m, the speed of the crate decreases from $v_0 = 0.60$ m/s to $v = 0.20$ m/s.

(a) How much work is done by force $\vec{F}$, and on what system does it do the work?

KEY IDEA

Because the applied force $\vec{F}$ is constant, we can calculate the work it does by using Eq. 7-7 ($W = Fd\cos\phi$).

Calculation: Substituting given data, including the fact that force $\vec{F}$ and displacement $\vec{d}$ are in the same direction, we find

$$W = Fd\cos\phi = (40\text{ N})(0.50\text{ m})\cos 0°$$
$$= 20\text{ J}. \quad \text{(Answer)}$$

Reasoning: To determine the system on which the work is done, let's check which energies change. Because the crate's speed changes, there is certainly a change ΔK in the crate's kinetic energy. Is there friction between the floor and the crate, and thus a change in thermal energy? Note that $\vec{F}$ and the crate's velocity have the same direction. Thus, if there is no friction, then $\vec{F}$ should be accelerating the crate to a *greater* speed. However, the crate is *slowing*, so there must be friction and a change ΔE_{th} in thermal energy of the crate and the floor. Therefore, the system on which the work is done is the crate–floor system, because both energy changes occur in that system.

(b) What is the increase ΔE_{th} in the thermal energy of the crate and floor?

KEY IDEA

We can relate ΔE_{th} to the work W done by $\vec{F}$ with the energy statement of Eq. 8-33 for a system that involves friction:

$$W = \Delta E_{mec} + \Delta E_{th}. \tag{8-34}$$

Calculations: We know the value of W from (a). The change ΔE_{mec} in the crate's mechanical energy is just the change in its kinetic energy because no potential energy changes occur, so we have

$$\Delta E_{mec} = \Delta K = \tfrac{1}{2}mv^2 - \tfrac{1}{2}mv_0^2.$$

Substituting this into Eq. 8-34 and solving for ΔE_{th}, we find

$$\Delta E_{th} = W - (\tfrac{1}{2}mv^2 - \tfrac{1}{2}mv_0^2) = W - \tfrac{1}{2}m(v^2 - v_0^2)$$
$$= 20\text{ J} - \tfrac{1}{2}(14\text{ kg})[(0.20\text{ m/s})^2 - (0.60\text{ m/s})^2]$$
$$= 22.2\text{ J} \approx 22\text{ J}. \quad \text{(Answer)}$$

Without further experiments, we cannot say how much of this thermal energy ends up in the crate and how much in the floor. We simply know the total amount.

WILEY PLUS Additional examples, video, and practice available at *WileyPLUS*

8-5 CONSERVATION OF ENERGY

Learning Objectives

After reading this module, you should be able to . . .

8.15 For an isolated system (no net external force), apply the conservation of energy to relate the initial total energy (energies of all kinds) to the total energy at a later instant.

8.16 For a nonisolated system, relate the work done on the system by a net external force to the changes in the various types of energies within the system.

8.17 Apply the relationship between average power, the associated energy transfer, and the time interval in which that transfer is made.

8.18 Given an energy transfer as a function of time (either as an equation or a graph), determine the instantaneous power (the transfer at any given instant).

Key Ideas

- The total energy E of a system (the sum of its mechanical energy and its internal energies, including thermal energy) can change only by amounts of energy that are transferred to or from the system. This experimental fact is known as the law of conservation of energy.
- If work W is done on the system, then

$$W = \Delta E = \Delta E_{mec} + \Delta E_{th} + \Delta E_{int}.$$

If the system is isolated ($W = 0$), this gives

$$\Delta E_{mec} + \Delta E_{th} + \Delta E_{int} = 0$$

and $$E_{mec,2} = E_{mec,1} - \Delta E_{th} - \Delta E_{int},$$

where the subscripts 1 and 2 refer to two different instants.

- The power due to a force is the *rate* at which that force transfers energy. If an amount of energy ΔE is transferred by a force in an amount of time Δt, the average power of the force is

$$P_{avg} = \frac{\Delta E}{\Delta t}.$$

- The instantaneous power due to a force is

$$P = \frac{dE}{dt}.$$

On a graph of energy E versus time t, the power is the slope of the plot at any given time.

Conservation of Energy

We now have discussed several situations in which energy is transferred to or from objects and systems, much like money is transferred between accounts. In each situation we assume that the energy that was involved could always be accounted for; that is, energy could not magically appear or disappear. In more formal language, we assumed (correctly) that energy obeys a law called the **law of conservation of energy,** which is concerned with the **total energy** E of a system. That total is the sum of the system's mechanical energy, thermal energy, and any type of *internal energy* in addition to thermal energy. (We have not yet discussed other types of internal energy.) The law states that

> The total energy E of a system can change only by amounts of energy that are transferred to or from the system.

The only type of energy transfer that we have considered is work W done on a system by an external force. Thus, for us at this point, this law states that

$$W = \Delta E = \Delta E_{mec} + \Delta E_{th} + \Delta E_{int}, \tag{8-35}$$

where ΔE_{mec} is any change in the mechanical energy of the system, ΔE_{th} is any change in the thermal energy of the system, and ΔE_{int} is any change in any other type of internal energy of the system. Included in ΔE_{mec} are changes ΔK in kinetic energy and changes ΔU in potential energy (elastic, gravitational, or any other type we might find).

This law of conservation of energy is *not* something we have derived from basic physics principles. Rather, it is a law based on countless experiments.

Tyler Stableford/The Image Bank/Getty Images

Figure 8-14 To descend, the rock climber must transfer energy from the gravitational potential energy of a system consisting of him, his gear, and Earth. He has wrapped the rope around metal rings so that the rope rubs against the rings. This allows most of the transferred energy to go to the thermal energy of the rope and rings rather than to his kinetic energy.

Scientists and engineers have never found an exception to it. Energy simply cannot magically appear or disappear.

Isolated System

If a system is isolated from its environment, there can be no energy transfers to or from it. For that case, the law of conservation of energy states:

> The total energy E of an isolated system cannot change.

Many energy transfers may be going on *within* an isolated system—between, say, kinetic energy and a potential energy or between kinetic energy and thermal energy. However, the total of all the types of energy in the system cannot change. Here again, energy cannot magically appear or disappear.

We can use the rock climber in Fig. 8-14 as an example, approximating him, his gear, and Earth as an isolated system. As he rappels down the rock face, changing the configuration of the system, he needs to control the transfer of energy from the gravitational potential energy of the system. (That energy cannot just disappear.) Some of it is transferred to his kinetic energy. However, he obviously does not want very much transferred to that type or he will be moving too quickly, so he has wrapped the rope around metal rings to produce friction between the rope and the rings as he moves down. The sliding of the rings on the rope then transfers the gravitational potential energy of the system to thermal energy of the rings and rope in a way that he can control. The total energy of the climber–gear–Earth system (the total of its gravitational potential energy, kinetic energy, and thermal energy) does not change during his descent.

For an isolated system, the law of conservation of energy can be written in two ways. First, by setting $W = 0$ in Eq. 8-35, we get

$$\Delta E_{\text{mec}} + \Delta E_{\text{th}} + \Delta E_{\text{int}} = 0 \quad \text{(isolated system)}. \tag{8-36}$$

We can also let $\Delta E_{\text{mec}} = E_{\text{mec},2} - E_{\text{mec},1}$, where the subscripts 1 and 2 refer to two different instants—say, before and after a certain process has occurred. Then Eq. 8-36 becomes

$$E_{\text{mec},2} = E_{\text{mec},1} - \Delta E_{\text{th}} - \Delta E_{\text{int}}. \tag{8-37}$$

Equation 8-37 tells us:

> In an isolated system, we can relate the total energy at one instant to the total energy at another instant *without considering the energies at intermediate times.*

This fact can be a very powerful tool in solving problems about isolated systems when you need to relate energies of a system before and after a certain process occurs in the system.

In Module 8-2, we discussed a special situation for isolated systems—namely, the situation in which nonconservative forces (such as a kinetic frictional force) do not act within them. In that special situation, ΔE_{th} and ΔE_{int} are both zero, and so Eq. 8-37 reduces to Eq. 8-18. In other words, the mechanical energy of an isolated system is conserved when nonconservative forces do not act in it.

External Forces and Internal Energy Transfers

An external force can change the kinetic energy or potential energy of an object without doing work on the object—that is, without transferring energy to the object. Instead, the force is responsible for transfers of energy from one type to another inside the object.

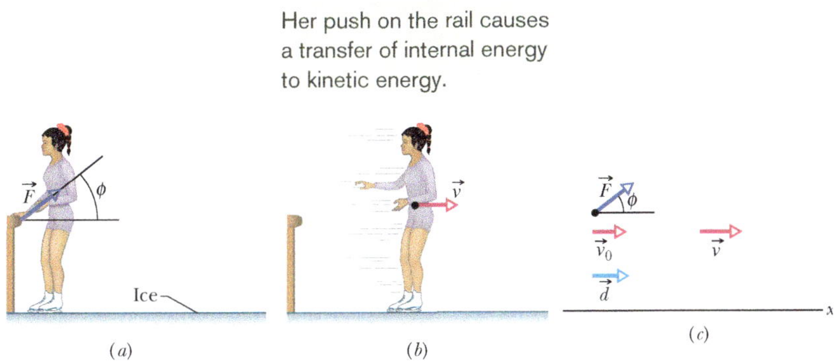

Figure 8-15 (a) As a skater pushes herself away from a railing, the force on her from the railing is $\vec{F}$. (b) After the skater leaves the railing, she has velocity $\vec{v}$. (c) External force $\vec{F}$ acts on the skater, at angle ϕ with a horizontal x axis. When the skater goes through displacement $\vec{d}$, her velocity is changed from $\vec{v}_0$ (= 0) to $\vec{v}$ by the horizontal component of $\vec{F}$.

Figure 8-15 shows an example. An initially stationary ice-skater pushes away from a railing and then slides over the ice (Figs. 8-15a and b). Her kinetic energy increases because of an external force $\vec{F}$ on her from the rail. However, that force does not transfer energy from the rail to her. Thus, the force does no work on her. Rather, her kinetic energy increases as a result of internal transfers from the biochemical energy in her muscles.

Figure 8-16 shows another example. An engine increases the speed of a car with four-wheel drive (all four wheels are made to turn by the engine). During the acceleration, the engine causes the tires to push backward on the road surface. This push produces frictional forces $\vec{f}$ that act on each tire in the forward direction. The net external force $\vec{F}$ from the road, which is the sum of these frictional forces, accelerates the car, increasing its kinetic energy. However, $\vec{F}$ does not transfer energy from the road to the car and so does no work on the car. Rather, the car's kinetic energy increases as a result of internal transfers from the energy stored in the fuel.

In situations like these two, we can sometimes relate the external force $\vec{F}$ on an object to the change in the object's mechanical energy if we can simplify the situation. Consider the ice-skater example. During her push through distance d in Fig. 8-15c, we can simplify by assuming that the acceleration is constant, her speed changing from $v_0 = 0$ to v. (That is, we assume $\vec{F}$ has constant magnitude F and angle ϕ.) After the push, we can simplify the skater as being a particle and neglect the fact that the exertions of her muscles have increased the thermal energy in her muscles and changed other physiological features. Then we can apply Eq. 7-5 ($\frac{1}{2}mv^2 - \frac{1}{2}mv_0^2 = F_x d$) to write

$$K - K_0 = (F\cos\phi)d,$$

or

$$\Delta K = Fd\cos\phi. \tag{8-38}$$

If the situation also involves a change in the elevation of an object, we can include the change ΔU in gravitational potential energy by writing

$$\Delta U + \Delta K = Fd\cos\phi. \tag{8-39}$$

The force on the right side of this equation does no work on the object but is still responsible for the changes in energy shown on the left side.

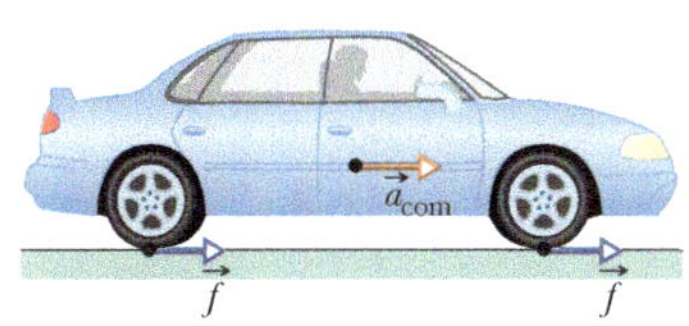

Figure 8-16 A vehicle accelerates to the right using four-wheel drive. The road exerts four frictional forces (two of them shown) on the bottom surfaces of the tires. Taken together, these four forces make up the net external force $\vec{F}$ acting on the car.

Power

Now that you have seen how energy can be transferred from one type to another, we can expand the definition of power given in Module 7-6. There power is defined as the rate at which work is done by a force. In a more general sense, power P is the rate at which energy is transferred by a force from one type to another. If an amount of energy ΔE is transferred in an amount of time Δt, the **average power** due to the force is

$$P_{avg} = \frac{\Delta E}{\Delta t}. \tag{8-40}$$

Similarly, the **instantaneous power** due to the force is

$$P = \frac{dE}{dt}. \tag{8-41}$$

Sample Problem 8.06 Lots of energies at an amusement park water slide

Figure 8-17 shows a water-slide ride in which a glider is shot by a spring along a water-drenched (frictionless) track that takes the glider from a horizontal section down to ground level. As the glider then moves along ground-level track, it is gradually brought to rest by friction. The total mass of the glider and its rider is m = 200 kg, the initial compression of the spring is d = 5.00 m, the spring constant is $k = 3.20 \times 10^3$ N/m, the initial height is h = 35.0 m, and the coefficient of kinetic friction along the ground-level track is $\mu_k = 0.800$. Through what distance L does the glider slide along the ground-level track until it stops?

KEY IDEAS

Before we touch a calculator and start plugging numbers into equations, we need to examine all the forces and then determine what our system should be. Only then can we decide what equation to write. Do we have an isolated system (our equation would be for the conservation of energy) or a system on which an external force does work (our equation would relate that work to the system's change in energy)?

Forces: The normal force on the glider from the track does no work on the glider because the direction of this force is always perpendicular to the direction of the glider's displacement. The gravitational force does work on the glider, and because the force is conservative we can associate a potential energy with it. As the spring pushes on the glider to get it moving, a spring force does work on it, transferring energy from the elastic potential energy of the compressed spring to kinetic energy of the glider. The spring force also pushes against a rigid wall. Because there is friction between the glider and the ground-level track, the sliding of the glider along that track section increases their thermal energies.

System: Let's take the system to contain all the interacting bodies: glider, track, spring, Earth, and wall. Then, because all the force interactions are *within* the system, the system is *isolated* and thus its total energy cannot change. So, the equation we should use is not that of some external force doing work on the system. Rather, it is a conservation of energy. We write this in the form of Eq. 8-37:

$$E_{mec,2} = E_{mec,1} - \Delta E_{th}. \tag{8-42}$$

This is like a money equation: The final money is equal to the initial money *minus* the amount stolen away by a thief. Here, the final mechanical energy is equal to the initial mechanical energy *minus* the amount stolen away by friction. None has magically appeared or disappeared.

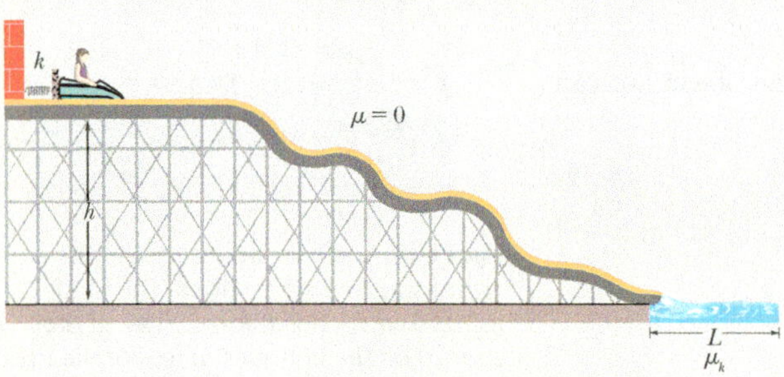

Figure 8-17 A spring-loaded amusement park water slide.

Calculations: Now that we have an equation, let's find distance L. Let subscript 1 correspond to the initial state of the glider (when it is still on the compressed spring) and subscript 2 correspond to the final state of the glider (when it has come to rest on the ground-level track). For both states, the mechanical energy of the system is the sum of any potential energy and any kinetic energy.

We have two types of potential energy: the elastic potential energy ($U_e = \frac{1}{2}kx^2$) associated with the compressed spring and the gravitational potential energy ($U_g = mgy$) associated with the glider's elevation. For the latter, let's take ground level as the reference level. That means that the glider is initially at height $y = h$ and finally at height $y = 0$.

In the initial state, with the glider stationary and elevated and the spring compressed, the energy is

$$E_{mec,1} = K_1 + U_{e1} + U_{g1}$$
$$= 0 + \tfrac{1}{2}kd^2 + mgh. \tag{8-43}$$

In the final state, with the spring now in its relaxed state and the glider again stationary but no longer elevated, the final mechanical energy of the system is

$$E_{mec,2} = K_2 + U_{e2} + U_{g2}$$
$$= 0 + 0 + 0. \quad (8\text{-}44)$$

Let's next go after the change ΔE_{th} of the thermal energy of the glider and ground-level track. From Eq. 8-31, we can substitute for ΔE_{th} with $f_k L$ (the product of the frictional force magnitude and the distance of rubbing). From Eq. 6-2, we know that $f_k = \mu_k F_N$, where F_N is the normal force. Because the glider moves horizontally through the region with friction, the magnitude of F_N is equal to mg (the upward force matches the downward force). So, the friction's theft from the mechanical energy amounts to

$$\Delta E_{th} = \mu_k mgL. \quad (8\text{-}45)$$

(By the way, without further experiments, we *cannot* say how much of this thermal energy ends up in the glider and how much in the track. We simply know the total amount.)

Substituting Eqs. 8-43 through 8-45 into Eq. 8-42, we find

$$0 = \tfrac{1}{2}kd^2 + mgh - \mu_k mgL, \quad (8\text{-}46)$$

and

$$L = \frac{kd^2}{2\mu_k mg} + \frac{h}{\mu_k}$$
$$= \frac{(3.20 \times 10^3\ \text{N/m})(5.00\ \text{m})^2}{2(0.800)(200\ \text{kg})(9.8\ \text{m/s}^2)} + \frac{35\ \text{m}}{0.800}$$
$$= 69.3\ \text{m}. \quad \text{(Answer)}$$

Finally, note how algebraically simple our solution is. By carefully defining a system and realizing that we have an isolated system, we get to use the law of the conservation of energy. That means we can relate the initial and final states of the system with no consideration of the intermediate states. In particular, we did not need to consider the glider as it slides over the uneven track. If we had, instead, applied Newton's second law to the motion, we would have had to know the details of the track and would have faced a far more difficult calculation.

WILEY PLUS Additional examples, video, and practice available at *WileyPLUS*

Review & Summary

Conservative Forces A force is a **conservative force** if the net work it does on a particle moving around any closed path, from an initial point and then back to that point, is zero. Equivalently, a force is conservative if the net work it does on a particle moving between two points does not depend on the path taken by the particle. The gravitational force and the spring force are conservative forces; the kinetic frictional force is a **nonconservative force.**

Potential Energy A **potential energy** is energy that is associated with the configuration of a system in which a conservative force acts. When the conservative force does work W on a particle within the system, the change ΔU in the potential energy of the system is

$$\Delta U = -W. \quad (8\text{-}1)$$

If the particle moves from point x_i to point x_f, the change in the potential energy of the system is

$$\Delta U = -\int_{x_i}^{x_f} F(x)\,dx. \quad (8\text{-}6)$$

Gravitational Potential Energy The potential energy associated with a system consisting of Earth and a nearby particle is **gravitational potential energy.** If the particle moves from height y_i to height y_f, the change in the gravitational potential energy of the particle–Earth system is

$$\Delta U = mg(y_f - y_i) = mg\,\Delta y. \quad (8\text{-}7)$$

If the **reference point** of the particle is set as $y_i = 0$ and the corresponding gravitational potential energy of the system is set as $U_i = 0$, then the gravitational potential energy U when the particle is at any height y is

$$U(y) = mgy. \quad (8\text{-}9)$$

Elastic Potential Energy **Elastic potential energy** is the energy associated with the state of compression or extension of an elastic object. For a spring that exerts a spring force $F = -kx$ when its free end has displacement x, the elastic potential energy is

$$U(x) = \tfrac{1}{2}kx^2. \quad (8\text{-}11)$$

The **reference configuration** has the spring at its relaxed length, at which $x = 0$ and $U = 0$.

Mechanical Energy The **mechanical energy** E_{mec} of a system is the sum of its kinetic energy K and potential energy U:

$$E_{mec} = K + U. \quad (8\text{-}12)$$

An *isolated system* is one in which no *external force* causes energy changes. If only conservative forces do work within an isolated system, then the mechanical energy E_{mec} of the system cannot change. This **principle of conservation of mechanical energy** is written as

$$K_2 + U_2 = K_1 + U_1, \quad (8\text{-}17)$$

in which the subscripts refer to different instants during an energy transfer process. This conservation principle can also be written as

$$\Delta E_{mec} = \Delta K + \Delta U = 0. \quad (8\text{-}18)$$

Potential Energy Curves If we know the potential energy function $U(x)$ for a system in which a one-dimensional force $F(x)$

acts on a particle, we can find the force as

$$F(x) = -\frac{dU(x)}{dx}. \quad (8\text{-}22)$$

If $U(x)$ is given on a graph, then at any value of x, the force $F(x)$ is the negative of the slope of the curve there and the kinetic energy of the particle is given by

$$K(x) = E_{mec} - U(x), \quad (8\text{-}24)$$

where E_{mec} is the mechanical energy of the system. A **turning point** is a point x at which the particle reverses its motion (there, $K = 0$). The particle is in **equilibrium** at points where the slope of the $U(x)$ curve is zero (there, $F(x) = 0$).

Work Done on a System by an External Force Work W is energy transferred to or from a system by means of an external force acting on the system. When more than one force acts on a system, their *net work* is the transferred energy. When friction is not involved, the work done on the system and the change ΔE_{mec} in the mechanical energy of the system are equal:

$$W = \Delta E_{mec} = \Delta K + \Delta U. \quad (8\text{-}26, 8\text{-}25)$$

When a kinetic frictional force acts within the system, then the thermal energy E_{th} of the system changes. (This energy is associated with the random motion of atoms and molecules in the system.) The work done on the system is then

$$W = \Delta E_{mec} + \Delta E_{th}. \quad (8\text{-}33)$$

The change ΔE_{th} is related to the magnitude f_k of the frictional force and the magnitude d of the displacement caused by the external force by

$$\Delta E_{th} = f_k d. \quad (8\text{-}31)$$

Conservation of Energy The **total energy** E of a system (the sum of its mechanical energy and its internal energies, including thermal energy) can change only by amounts of energy that are transferred to or from the system. This experimental fact is known as the **law of conservation of energy.** If work W is done on the system, then

$$W = \Delta E = \Delta E_{mec} + \Delta E_{th} + \Delta E_{int}. \quad (8\text{-}35)$$

If the system is isolated ($W = 0$), this gives

$$\Delta E_{mec} + \Delta E_{th} + \Delta E_{int} = 0 \quad (8\text{-}36)$$

and

$$E_{mec,2} = E_{mec,1} - \Delta E_{th} - \Delta E_{int}, \quad (8\text{-}37)$$

where the subscripts 1 and 2 refer to two different instants.

Power The **power** due to a force is the *rate* at which that force transfers energy. If an amount of energy ΔE is transferred by a force in an amount of time Δt, the **average power** of the force is

$$P_{avg} = \frac{\Delta E}{\Delta t}. \quad (8\text{-}40)$$

The **instantaneous power** due to a force is

$$P = \frac{dE}{dt}. \quad (8\text{-}41)$$

Questions

1 In Fig. 8-18, a horizontally moving block can take three frictionless routes, differing only in elevation, to reach the dashed finish line. Rank the routes according to (a) the speed of the block at the finish line and (b) the travel time of the block to the finish line, greatest first.

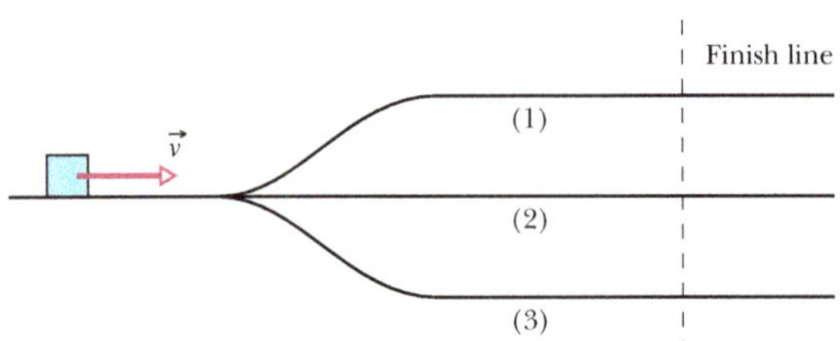

Figure 8-18 Question 1.

2 Figure 8-19 gives the potential energy function of a particle. (a) Rank regions AB, BC, CD, and DE according to the magnitude of the force on the particle, greatest first. What value must the mechanical energy E_{mec} of the particle not exceed if the particle is to be (b) trapped in the potential well at the left, (c) trapped in the potential well at the right, and (d) able to move between the two potential wells but not to the right of point H? For the situation of (d), in which of regions BC, DE, and FG will the particle have (e) the greatest kinetic energy and (f) the least speed?

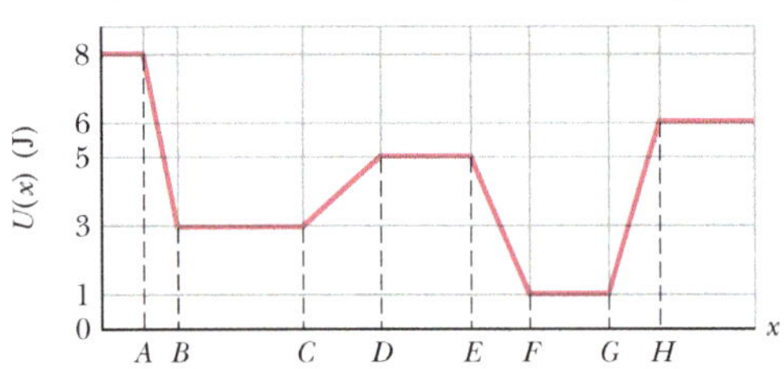

Figure 8-19 Question 2.

3 Figure 8-20 shows one direct path and four indirect paths from point i to point f. Along the direct path and three of the indirect paths, only a conservative force F_c acts on a certain object. Along the fourth indirect path, both F_c and a nonconservative force F_{nc} act on the object. The change ΔE_{mec} in the object's mechanical energy (in joules) in going from i to f is indicated along each straight-line segment of the indirect paths. What is ΔE_{mec} (a) from i to f along the direct path and (b) due to F_{nc} along the one path where it acts?

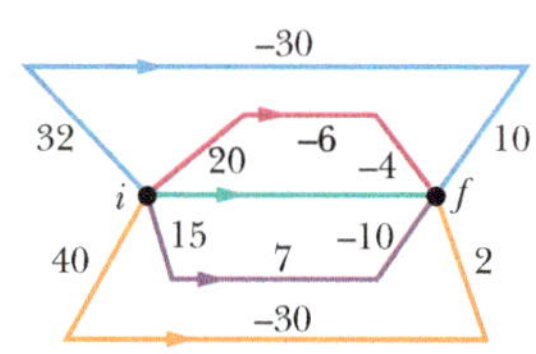

Figure 8-20 Question 3.

4 In Fig. 8-21, a small, initially stationary block is released on a frictionless ramp at a height of 3.0 m. Hill heights along the ramp are as shown in the figure. The hills have identical circular tops, and the block does not fly off any hill. (a) Which hill is the first the block cannot cross? (b) What does the block do after failing to cross that hill? Of the hills that the block can cross, on which hilltop is (c) the centripetal acceleration of the block greatest and (d) the normal force on the block least?

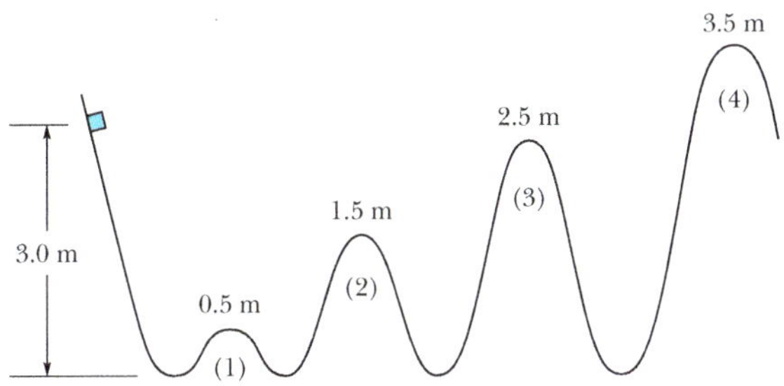

Figure 8-21 Question 4.

5 In Fig. 8-22, a block slides from A to C along a frictionless ramp, and then it passes through horizontal region CD, where a frictional force acts on it. Is the block's kinetic energy increasing, decreasing, or constant in (a) region AB, (b) region BC, and (c) region CD? (d) Is the block's mechanical energy increasing, decreasing, or constant in those regions?

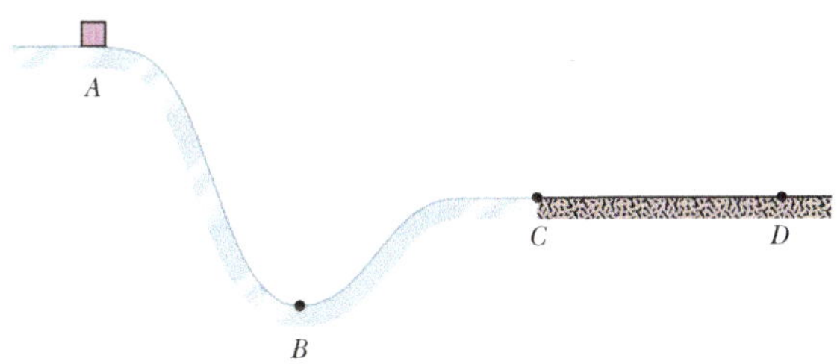

Figure 8-22 Question 5.

6 In Fig. 8-23a, you pull upward on a rope that is attached to a cylinder on a vertical rod. Because the cylinder fits tightly on the rod, the cylinder slides along the rod with considerable friction. Your force does work $W = +100$ J on the cylinder–rod–Earth system (Fig. 8-23b). An "energy statement" for the system is shown in Fig. 8-23c: the kinetic energy K increases by 50 J, and the gravitational potential energy U_g increases by 20 J. The only other change in energy within the system is for the thermal energy E_{th}. What is the change ΔE_{th}?

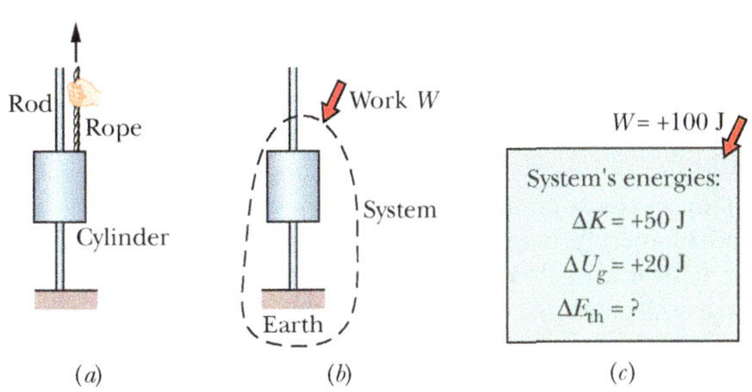

Figure 8-23 Question 6.

7 The arrangement shown in Fig. 8-24 is similar to that in Question 6. Here you pull downward on the rope that is attached to the cylinder, which fits tightly on the rod. Also, as the cylinder descends, it pulls on a block via a second rope, and the block slides over a lab table. Again consider the cylinder–rod–Earth system, similar to that shown in Fig. 8-23b. Your work on the system is 200 J. The system does work of 60 J on the block. Within the system, the kinetic energy increases by 130 J and the gravitational potential energy decreases by 20 J. (a) Draw an "energy statement" for the system, as in Fig. 8-23c. (b) What is the change in the thermal energy within the system?

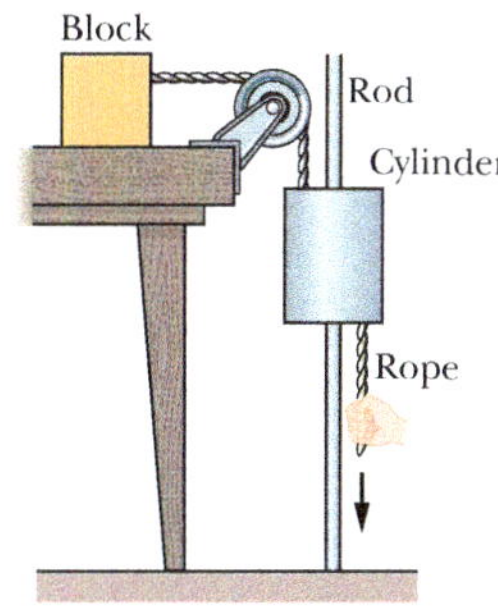

Figure 8-24 Question 7.

8 In Fig. 8-25, a block slides along a track that descends through distance h. The track is frictionless except for the lower section. There the block slides to a stop in a certain distance D because of friction. (a) If we decrease h, will the block now slide to a stop in a distance that is greater than, less than, or equal to D? (b) If, instead, we increase the mass of the block, will the stopping distance now be greater than, less than, or equal to D?

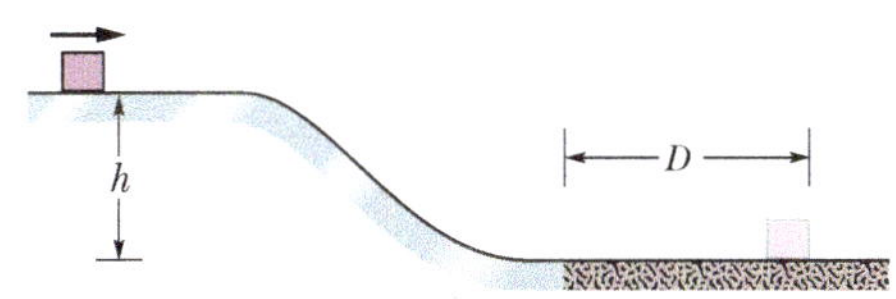

Figure 8-25 Question 8.

9 Figure 8-26 shows three situations involving a plane that is not frictionless and a block sliding along the plane. The block begins with the same speed in all three situations and slides until the kinetic frictional force has stopped it. Rank the situations according to the increase in thermal energy due to the sliding, greatest first.

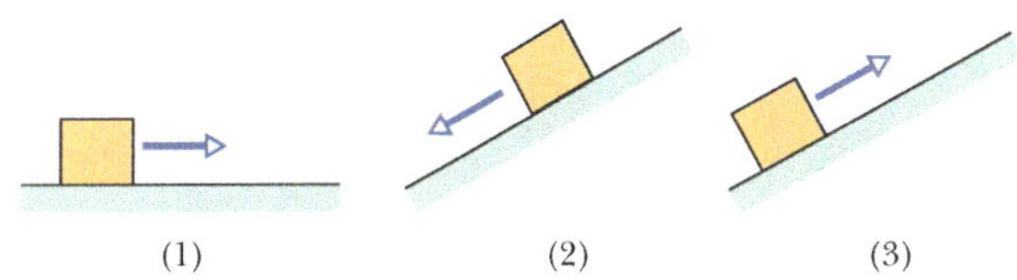

Figure 8-26 Question 9.

10 Figure 8-27 shows three plums that are launched from the same level with the same speed. One moves straight upward, one is launched at a small angle to the vertical, and one is launched along a frictionless incline. Rank the plums according to their speed when they reach the level of the dashed line, greatest first.

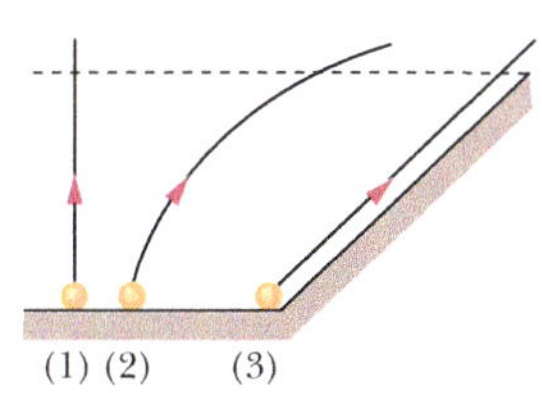

Figure 8-27 Question 10.

11 When a particle moves from f to i and from j to i along the paths shown in Fig. 8-28, and in the indicated directions, a conservative force $\vec{F}$ does the indicated amounts of work on it. How much work is done on the particle by $\vec{F}$ when the particle moves directly from f to j?

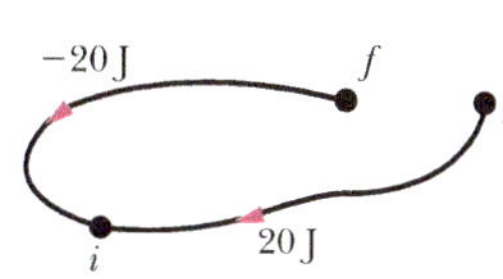

Figure 8-28 Question 11.

Problems

GO Tutoring problem available (at instructor's discretion) in *WileyPLUS* and WebAssign

SSM Worked-out solution available in Student Solutions Manual

• – ••• Number of dots indicates level of problem difficulty

Additional information available in *The Flying Circus of Physics* and at flyingcircusofphysics.com

WWW Worked-out solution is at

ILW Interactive solution is at

http://www.wiley.com/college/halliday

Module 8-1 Potential Energy

•1 SSM What is the spring constant of a spring that stores 25 J of elastic potential energy when compressed by 7.5 cm?

•2 In Fig. 8-29, a single frictionless roller-coaster car of mass $m = 825$ kg tops the first hill with speed $v_0 = 17.0$ m/s at height $h = 42.0$ m. How much work does the gravitational force do on the car from that point to (a) point A, (b) point B, and (c) point C? If the gravitational potential energy of the car–Earth system is taken to be zero at C, what is its value when the car is at (d) B and (e) A? (f) If mass m were doubled, would the change in the gravitational potential energy of the system between points A and B increase, decrease, or remain the same?

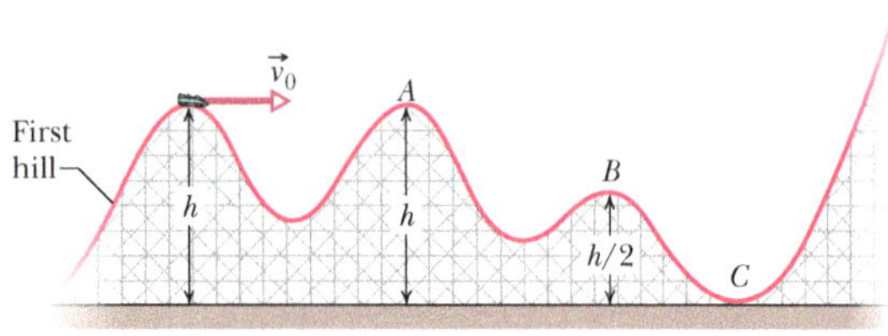

Figure 8-29 Problems 2 and 9.

•3 You drop a 2.00 kg book to a friend who stands on the ground at distance $D = 10.0$ m below. If your friend's outstretched hands are at distance $d = 1.50$ m above the ground (Fig. 8-30), (a) how much work W_g does the gravitational force do on the book as it drops to her hands? (b) What is the change ΔU in the gravitational potential energy of the book–Earth system during the drop? If the gravitational potential energy U of that system is taken to be zero at ground level, what is U (c) when the book is released and (d) when it reaches her hands? Now take U to be 100 J at ground level and again find (e) W_g, (f) ΔU, (g) U at the release point, and (h) U at her hands.

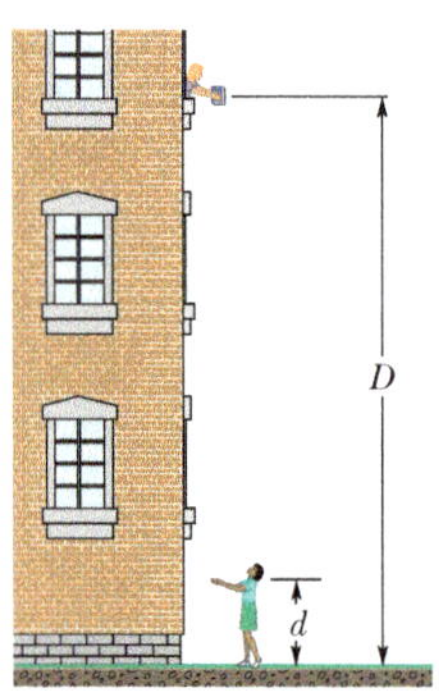

Figure 8-30 Problems 3 and 10.

•4 Figure 8-31 shows a ball with mass $m = 0.341$ kg attached to the end of a thin rod with length $L = 0.452$ m and negligible mass. The other end of the rod is pivoted so that the ball can move in a vertical circle. The rod is held horizontally as shown and then given enough of a downward push to cause the ball to swing down and around and just reach the vertically up position, with zero speed there. How much work is done on the ball by the gravitational force from the initial point to (a) the lowest point, (b) the highest point, and (c) the point on the right level with the initial point? If the gravitational potential energy of the ball–Earth system is taken to be zero at the initial point, what is it when the ball reaches (d) the lowest point, (e) the highest point, and (f) the point on the right level with the initial point? (g) Suppose the rod were pushed harder so that the ball passed through the highest point with a nonzero speed. Would ΔU_g from the lowest point to the highest point then be greater than, less than, or the same as it was when the ball stopped at the highest point?

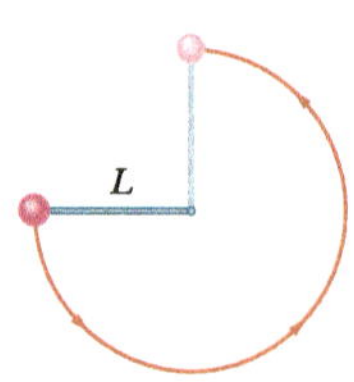

Figure 8-31 Problems 4 and 14.

•5 SSM In Fig. 8-32, a 2.00 g ice flake is released from the edge of a hemispherical bowl whose radius r is 22.0 cm. The flake–bowl contact is frictionless. (a) How much work is done on the flake by the gravitational force during the flake's descent to the bottom of the bowl? (b) What is the change in the potential energy of the flake–Earth system during that descent? (c) If that potential energy is taken to be zero at the bottom of the bowl, what is its value when the flake is released? (d) If, instead, the potential energy is taken to be zero at the release point, what is its value when the flake reaches the bottom of the bowl? (e) If the mass of the flake were doubled, would the magnitudes of the answers to (a) through (d) increase, decrease, or remain the same?

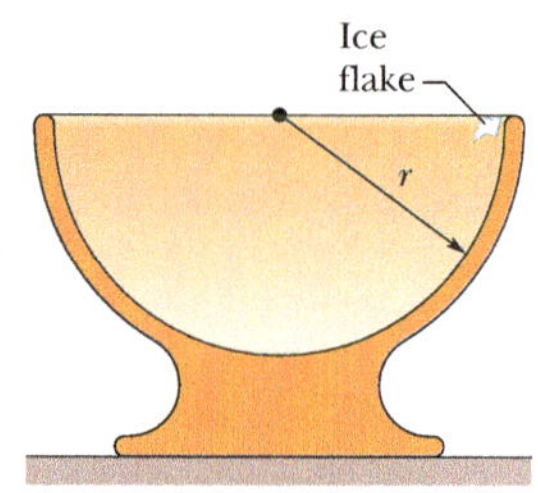

Figure 8-32 Problems 5 and 11.

••6 In Fig. 8-33, a small block of mass $m = 0.032$ kg can slide along the frictionless loop-the-loop, with loop radius $R = 12$ cm. The block is released from rest at point P, at height $h = 5.0R$ above the bottom of the loop. How much work does the gravitational force do on the block as the block travels from point P to (a) point Q and (b) the top of the loop? If the gravitational potential energy of the block–Earth system is taken to be zero at the bottom of the loop, what is that potential energy when the block is (c) at point P, (d) at point Q, and (e) at the top of the loop? (f) If, instead of merely being released, the block is given some initial speed downward along the track, do the answers to (a) through (e) increase, decrease, or remain the same?

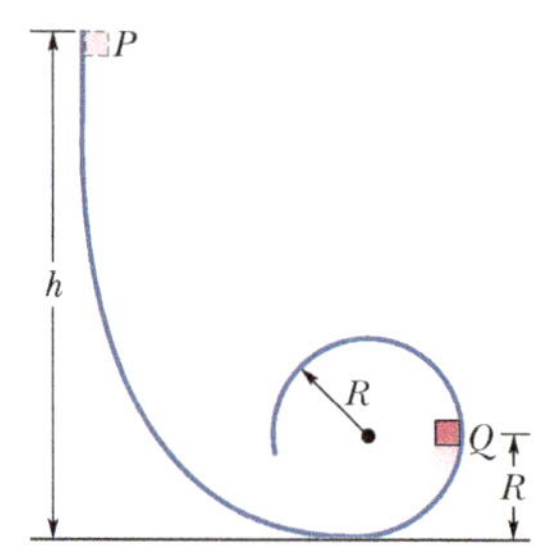

Figure 8-33 Problems 6 and 17.

••7 Figure 8-34 shows a thin rod, of length $L = 2.00$ m and negligible mass, that can pivot about one end to rotate in a vertical circle. A ball of mass $m = 5.00$ kg is attached to the other end. The rod is pulled aside to angle $\theta_0 = 30.0°$ and released with initial velocity $\vec{v}_0 = 0$. As the ball descends to its lowest point, (a) how much work does the gravitational force do on it and (b) what is the change in the gravitational potential energy of the ball–Earth system? (c) If the gravitational potential energy is taken to be zero at the lowest point, what is its value just as the ball is released? (d) Do the magnitudes of the answers to (a) through (c) increase, decrease, or remain the same if angle θ_0 is increased?

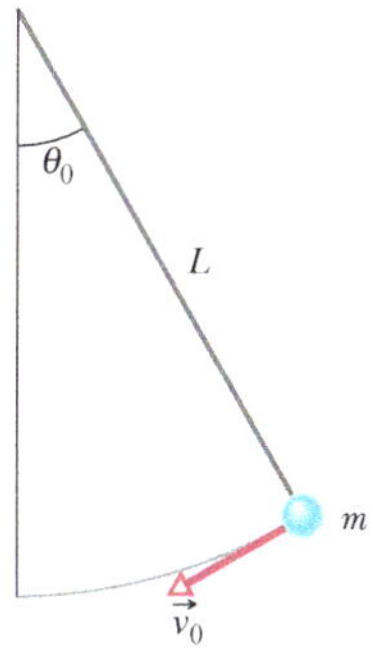

Figure 8-34 Problems 7, 18, and 21.

••8 A 1.50 kg snowball is fired from a cliff 12.5 m high. The snowball's initial velocity is 14.0 m/s, directed 41.0° above the horizontal. (a) How much work is done on the snowball by the gravitational force during its flight to the flat ground below the cliff? (b) What is the change in the gravitational potential energy of the snowball–Earth system during the flight? (c) If that gravitational potential energy is taken to be zero at the height of the cliff, what is its value when the snowball reaches the ground?

Module 8-2 Conservation of Mechanical Energy

•9 GO In Problem 2, what is the speed of the car at (a) point A, (b) point B, and (c) point C? (d) How high will the car go on the last hill, which is too high for it to cross? (e) If we substitute a second car with twice the mass, what then are the answers to (a) through (d)?

•10 (a) In Problem 3, what is the speed of the book when it reaches the hands? (b) If we substituted a second book with twice the mass, what would its speed be? (c) If, instead, the book were thrown down, would the answer to (a) increase, decrease, or remain the same?

•11 SSM WWW (a) In Problem 5, what is the speed of the flake when it reaches the bottom of the bowl? (b) If we substituted a second flake with twice the mass, what would its speed be? (c) If, instead, we gave the flake an initial downward speed along the bowl, would the answer to (a) increase, decrease, or remain the same?

•12 (a) In Problem 8, using energy techniques rather than the techniques of Chapter 4, find the speed of the snowball as it reaches the ground below the cliff. What is that speed (b) if the launch angle is changed to 41.0° *below* the horizontal and (c) if the mass is changed to 2.50 kg?

•13 SSM A 5.0 g marble is fired vertically upward using a spring gun. The spring must be compressed 8.0 cm if the marble is to just reach a target 20 m above the marble's position on the compressed spring. (a) What is the change ΔU_g in the gravitational potential energy of the marble–Earth system during the 20 m ascent? (b) What is the change ΔU_s in the elastic potential energy of the spring during its launch of the marble? (c) What is the spring constant of the spring?

•14 (a) In Problem 4, what initial speed must be given the ball so that it reaches the vertically upward position with zero speed? What then is its speed at (b) the lowest point and (c) the point on the right at which the ball is level with the initial point? (d) If the ball's mass were doubled, would the answers to (a) through (c) increase, decrease, or remain the same?

•15 SSM In Fig. 8-35, a runaway truck with failed brakes is moving downgrade at 130 km/h just before the driver steers the truck up a frictionless emergency escape ramp with an inclination of $\theta = 15°$. The truck's mass is 1.2×10^4 kg. (a) What minimum length L must the ramp have if the truck is to stop (momentarily) along it? (Assume the truck is a particle, and justify that assumption.) Does the minimum length L increase, decrease, or remain the same if (b) the truck's mass is decreased and (c) its speed is decreased?

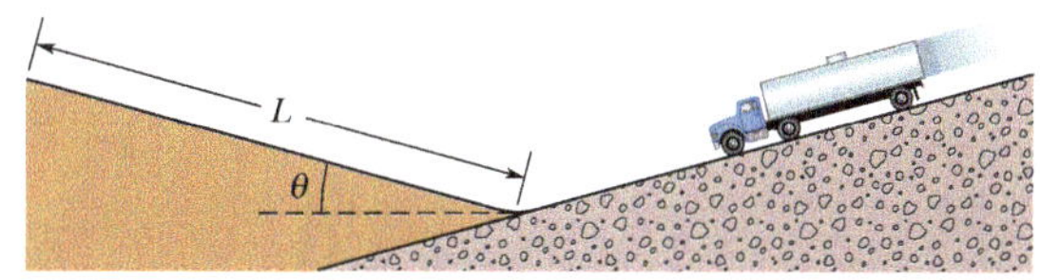

Figure 8-35 Problem 15.

••16 A 700 g block is released from rest at height h_0 above a vertical spring with spring constant $k = 400$ N/m and negligible mass. The block sticks to the spring and momentarily stops after compressing the spring 19.0 cm. How much work is done (a) by the block on the spring and (b) by the spring on the block? (c) What is the value of h_0? (d) If the block were released from height $2.00h_0$ above the spring, what would be the maximum compression of the spring?

••17 In Problem 6, what are the magnitudes of (a) the horizontal component and (b) the vertical component of the *net* force acting on the block at point Q? (c) At what height h should the block be released from rest so that it is on the verge of losing contact with the track at the top of the loop? (*On the verge of losing contact* means that the normal force on the block from the track has just then become zero.) (d) Graph the magnitude of the normal force on the block at the top of the loop versus initial height h, for the range $h = 0$ to $h = 6R$.

••18 (a) In Problem 7, what is the speed of the ball at the lowest point? (b) Does the speed increase, decrease, or remain the same if the mass is increased?

••19 GO Figure 8-36 shows an 8.00 kg stone at rest on a spring. The spring is compressed 10.0 cm by the stone. (a) What is the spring constant? (b) The stone is pushed down an additional 30.0 cm and released. What is the elastic potential energy of the compressed spring just before that release? (c) What is the change in the gravitational potential energy of the stone–Earth system when the stone moves from the release point to its maximum height? (d) What is that maximum height, measured from the release point?

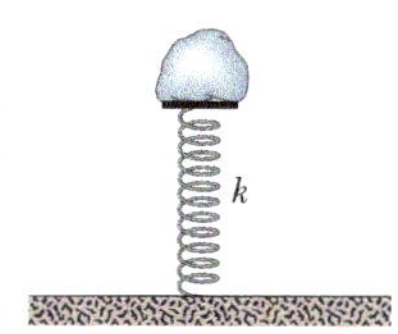

Figure 8-36 Problem 19.

••20 GO A pendulum consists of a 2.0 kg stone swinging on a 4.0 m string of negligible mass. The stone has a speed of 8.0 m/s when it passes its lowest point. (a) What is the speed when the string is at 60° to the vertical? (b) What is the greatest angle with the vertical that the string will reach during the stone's motion? (c) If the potential energy of the pendulum–Earth system is taken to be zero at the stone's lowest point, what is the total mechanical energy of the system?

••21 Figure 8-34 shows a pendulum of length $L = 1.25$ m. Its bob (which effectively has all the mass) has speed v_0 when the cord makes an angle $\theta_0 = 40.0°$ with the vertical. (a) What is the speed of the bob when it is in its lowest position if $v_0 = 8.00$ m/s? What is the least value that v_0 can have if the pendulum is to swing down and then up (b) to a horizontal position, and (c) to a vertical position with the cord remaining straight? (d) Do the answers to (b) and (c) increase, decrease, or remain the same if θ_0 is increased by a few degrees?

••22 A 60 kg skier starts from rest at height $H = 20$ m above the end of a ski-jump ramp (Fig. 8-37) and leaves the ramp at angle $\theta = 28°$. Neglect the effects of air resistance and assume the ramp is frictionless. (a) What is the maximum height h of his jump above the end of the ramp? (b) If he increased his weight by putting on a backpack, would h then be greater, less, or the same?

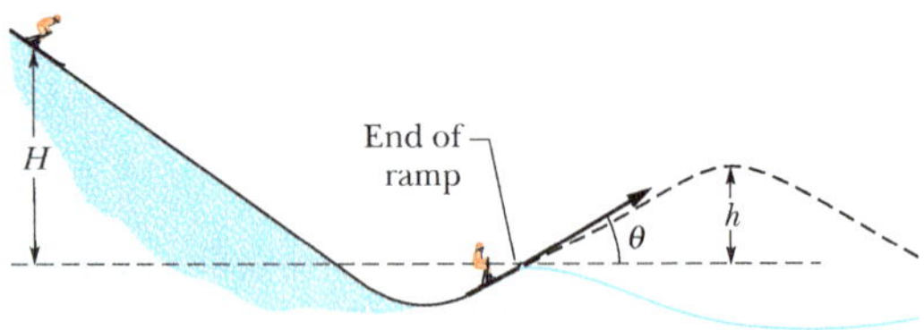

Figure 8-37 Problem 22.

••23 ILW The string in Fig. 8-38 is $L = 120$ cm long, has a ball attached to one end, and is fixed at its other end. The distance d from the fixed end to a fixed peg at point P is 75.0 cm. When the initially stationary ball is released with the string horizontal as shown, it will swing along the dashed arc. What is its speed when it reaches (a) its lowest point and (b) its highest point after the string catches on the peg?

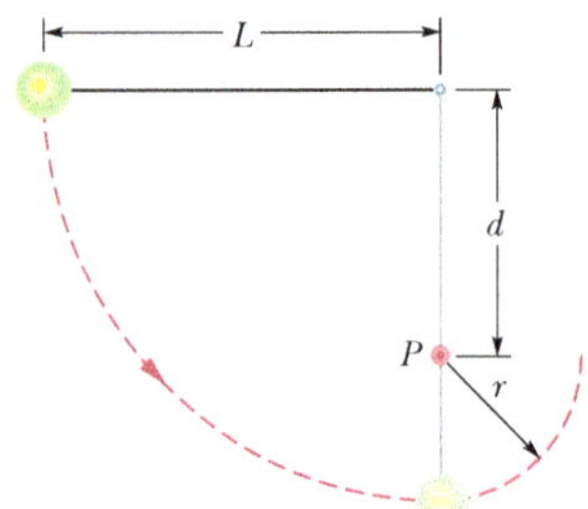

Figure 8-38 Problems 23 and 70.

••24 A block of mass $m = 2.0$ kg is dropped from height $h = 40$ cm onto a spring of spring constant $k = 1960$ N/m (Fig. 8-39). Find the maximum distance the spring is compressed.

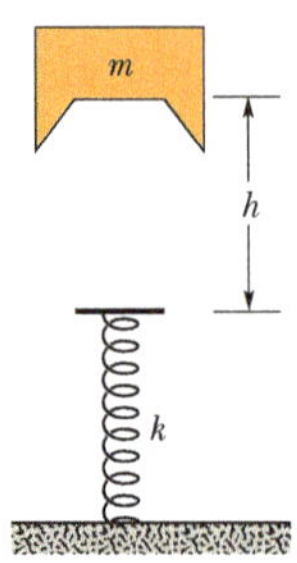

Figure 8-39 Problem 24.

••25 At $t = 0$ a 1.0 kg ball is thrown from a tall tower with $\vec{v} = (18 \text{ m/s})\hat{i} + (24 \text{ m/s})\hat{j}$. What is ΔU of the ball–Earth system between $t = 0$ and $t = 6.0$ s (still free fall)?

••26 A conservative force $\vec{F} = (6.0x - 12)\hat{i}$ N, where x is in meters, acts on a particle moving along an x axis. The potential energy U associated with this force is assigned a value of 27 J at $x = 0$. (a) Write an expression for U as a function of x, with U in joules and x in meters. (b) What is the maximum positive potential energy? At what (c) negative value and (d) positive value of x is the potential energy equal to zero?

••27 Tarzan, who weighs 688 N, swings from a cliff at the end of a vine 18 m long (Fig. 8-40). From the top of the cliff to the bottom of the swing, he descends by 3.2 m. The vine will break if the force on it exceeds 950 N. (a) Does the vine break? (b) If no, what is the greatest force on it during the swing? If yes, at what angle with the vertical does it break?

Figure 8-40 Problem 27.

••28 Figure 8-41a applies to the spring in a cork gun (Fig. 8-41b); it shows the spring force as a function of the stretch or compression of the spring. The spring is compressed by 5.5 cm and used to propel a 3.8 g cork from the gun. (a) What is the speed of the cork if it is released as the spring passes through its relaxed position? (b) Suppose, instead, that the cork sticks to the spring and stretches it 1.5 cm before separation occurs. What now is the speed of the cork at the time of release?

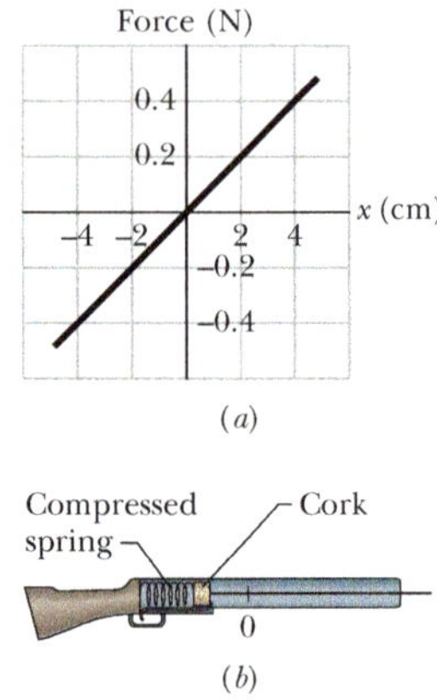

Figure 8-41 Problem 28.

••29 SSM WWW In Fig. 8-42, a block of mass $m = 12$ kg is released from rest on a frictionless incline of angle $\theta = 30°$. Below the block is a spring that can be compressed 2.0 cm by a force of 270 N. The block momentarily stops when it compresses the spring by 5.5 cm. (a) How far does the block move down the incline from its rest position to this stopping point? (b) What is the speed of the block just as it touches the spring?

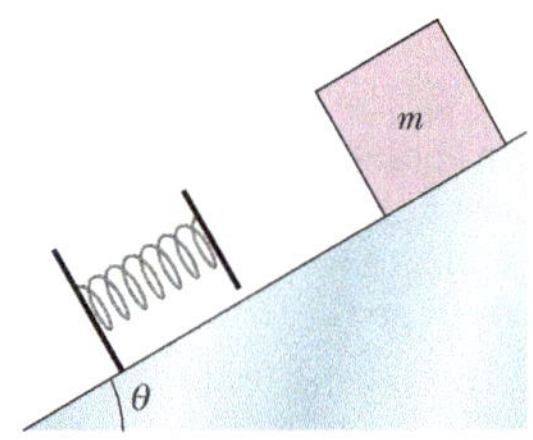

Figure 8-42 Problems 29 and 35.

••30 GO A 2.0 kg breadbox on a frictionless incline of angle $\theta = 40°$ is connected, by a cord that runs over a pulley, to a light spring of spring constant $k = 120$ N/m, as shown in Fig. 8-43. The box is released from rest when the spring is unstretched. Assume that the pulley is massless and frictionless. (a) What is the speed of the box when it has moved 10 cm down the incline? (b) How far down the incline from its point of release does the box slide before momentarily stopping, and what are the (c) magnitude and (d) direction (up or down the incline) of the box's acceleration at the instant the box momentarily stops?

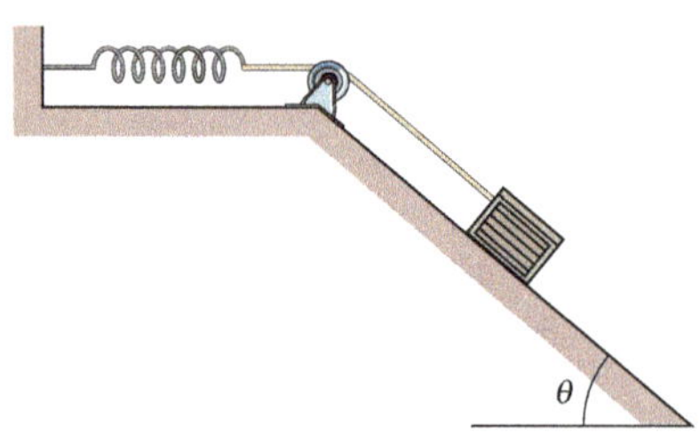

Figure 8-43 Problem 30.

••31 ILW A block with mass $m = 2.00$ kg is placed against a spring on a frictionless incline with angle $\theta = 30.0°$ (Fig. 8-44). (The block is not attached to the spring.) The spring, with spring constant $k = 19.6$ N/cm, is compressed 20.0 cm and then released. (a) What is the elastic potential energy of the compressed spring? (b) What is the change in the gravitational potential energy of the block–Earth system as the block moves from the release point to its highest point on the incline? (c) How far along the incline is the highest point from the release point?

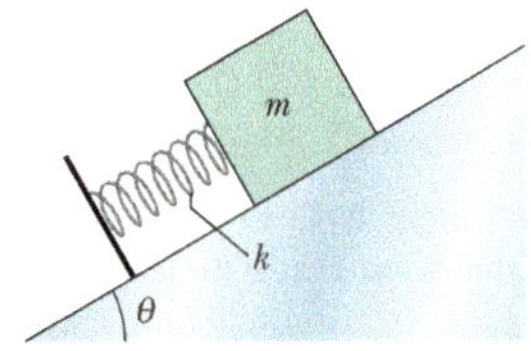

Figure 8-44 Problem 31.

••32 In Fig. 8-45, a chain is held on a frictionless table with one-fourth of its length hanging over the edge. If the chain has length $L = 28$ cm and mass $m = 0.012$ kg, how much work is required to pull the hanging part back onto the table?

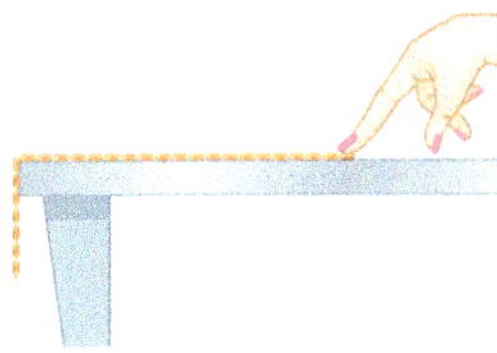

Figure 8-45 Problem 32.

•••33 In Fig. 8-46, a spring with $k = 170$ N/m is at the top of a frictionless incline of angle $\theta = 37.0°$. The lower end of the incline is distance $D = 1.00$ m from the end of the spring, which is at its relaxed length. A 2.00 kg canister is pushed against the spring until the spring is compressed 0.200 m and released from rest. (a) What is the speed of the canister at the instant the spring returns to its relaxed length (which is when the canister loses contact with the spring)? (b) What is the speed of the canister when it reaches the lower end of the incline?

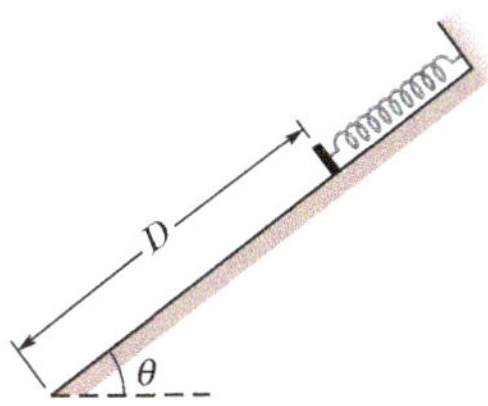

Figure 8-46 Problem 33.

•••34 A boy is initially seated on the top of a hemispherical ice mound of radius $R = 13.8$ m. He begins to slide down the ice, with a negligible initial speed (Fig. 8-47). Approximate the ice as being frictionless. At what height does the boy lose contact with the ice?

Figure 8-47 Problem 34.

•••35 In Fig. 8-42, a block of mass $m = 3.20$ kg slides from rest a distance d down a frictionless incline at angle $\theta = 30.0°$ where it runs into a spring of spring constant 431 N/m. When the block momentarily stops, it has compressed the spring by 21.0 cm. What are (a) distance d and (b) the distance between the point of the first block–spring contact and the point where the block's speed is greatest?

•••36 Two children are playing a game in which they try to hit a small box on the floor with a marble fired from a spring-loaded gun that is mounted on a table. The target box is horizontal distance $D = 2.20$ m from the edge of the table; see Fig. 8-48. Bobby compresses the spring 1.10 cm, but the center of the marble falls 27.0 cm short of the center of the box. How far should Rhoda compress the spring to score a direct hit? Assume that neither the spring nor the ball encounters friction in the gun.

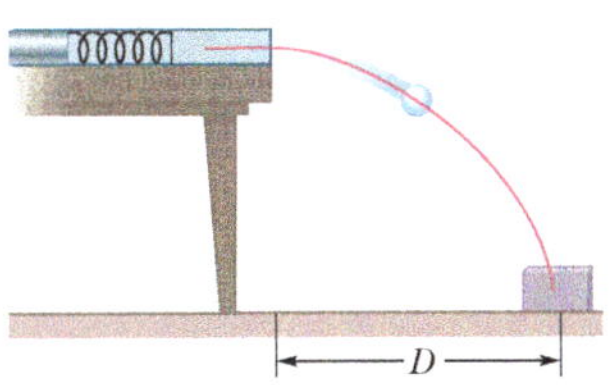

Figure 8-48 Problem 36.

•••37 A uniform cord of length 25 cm and mass 15 g is initially stuck to a ceiling. Later, it hangs vertically from the ceiling with only one end still stuck. What is the change in the gravitational potential energy of the cord with this change in orientation? (*Hint:* Consider a differential slice of the cord and then use integral calculus.)

Module 8-3 Reading a Potential Energy Curve

••38 Figure 8-49 shows a plot of potential energy U versus position x of a 0.200 kg particle that can travel only along an x axis under the influence of a conservative force. The graph has these values: $U_A = 9.00$ J, $U_C = 20.00$ J, and $U_D = 24.00$ J. The particle is released at the point where U forms a "potential hill" of "height" $U_B = 12.00$ J, with kinetic energy 4.00 J. What is the speed of the particle at (a) $x = 3.5$ m and (b) $x = 6.5$ m? What is the position of the turning point on (c) the right side and (d) the left side?

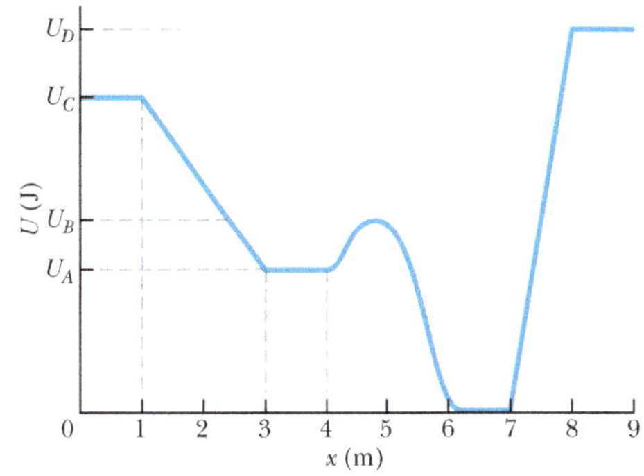

Figure 8-49 Problem 38.

••39 Figure 8-50 shows a plot of potential energy U versus position x of a 0.90 kg particle that can travel only along an x axis. (Nonconservative forces are not involved.) Three values are $U_A = 15.0$ J, $U_B = 35.0$ J, and $U_C = 45.0$ J. The particle is released at $x = 4.5$ m with an initial speed of 7.0 m/s, headed in the negative x direction. (a) If the particle can reach $x = 1.0$ m, what is its speed there, and if it cannot, what is its turning point? What are the (b) magnitude and (c) direction of the force on the particle as it begins to move to the left of $x = 4.0$ m? Suppose, instead, the particle is headed in the positive x direction when it is released at $x = 4.5$ m at speed 7.0 m/s. (d) If the particle can reach $x = 7.0$ m, what is its speed there, and if it cannot, what is its turning point? What are the (e) magnitude and (f) direction of the force on the particle as it begins to move to the right of $x = 5.0$ m?

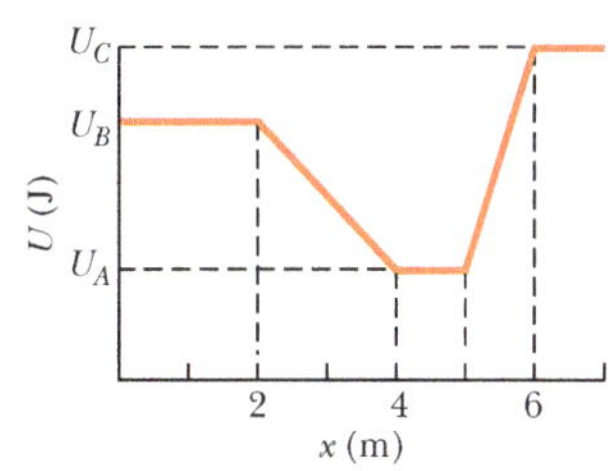

Figure 8-50 Problem 39.

••40 The potential energy of a diatomic molecule (a two-atom system like H_2 or O_2) is given by

$$U = \frac{A}{r^{12}} - \frac{B}{r^6},$$

where r is the separation of the two atoms of the molecule and A and B are positive constants. This potential energy is associated with the force that binds the two atoms together. (a) Find the *equilibrium separation*—that is, the distance between the atoms at which the force on each atom is zero. Is the force repulsive (the atoms are pushed apart) or attractive (they are pulled together) if their separation is (b) smaller and (c) larger than the equilibrium separation?

•••41 A single conservative force $F(x)$ acts on a 1.0 kg particle that moves along an x axis. The potential energy $U(x)$ associated with $F(x)$ is given by

$$U(x) = -4x\,e^{-x/4}\text{ J},$$

where x is in meters. At $x = 5.0$ m the particle has a kinetic energy of 2.0 J. (a) What is the mechanical energy of the system? (b) Make a plot of $U(x)$ as a function of x for $0 \le x \le 10$ m, and on the same graph draw the line that represents the mechanical energy of the system. Use part (b) to determine (c) the least value of x the particle can reach and (d) the greatest value of x the particle can reach. Use part (b) to determine (e) the maximum kinetic energy of the particle and (f) the value of x at which it occurs. (g) Determine an expression in newtons and meters for $F(x)$ as a function of x. (h) For what (finite) value of x does $F(x) = 0$?

Module 8-4 Work Done on a System by an External Force

•42 A worker pushed a 27 kg block 9.2 m along a level floor at constant speed with a force directed 32° below the horizontal. If the coefficient of kinetic friction between block and floor was 0.20, what were (a) the work done by the worker's force and (b) the increase in thermal energy of the block–floor system?

•43 A collie drags its bed box across a floor by applying a horizontal force of 8.0 N. The kinetic frictional force acting on the box has magnitude 5.0 N. As the box is dragged through 0.70 m along the way, what are (a) the work done by the collie's applied force and (b) the increase in thermal energy of the bed and floor?

••44 A horizontal force of magnitude 35.0 N pushes a block of mass 4.00 kg across a floor where the coefficient of kinetic friction is 0.600. (a) How much work is done by that applied force on the block–floor system when the block slides through a displacement of 3.00 m across the floor? (b) During that displacement, the thermal energy of the block increases by 40.0 J. What is the increase in thermal energy of the floor? (c) What is the increase in the kinetic energy of the block?

••45 SSM A rope is used to pull a 3.57 kg block at constant speed 4.06 m along a horizontal floor. The force on the block from the rope is 7.68 N and directed 15.0° above the horizontal. What are (a) the work done by the rope's force, (b) the increase in thermal energy of the block–floor system, and (c) the coefficient of kinetic friction between the block and floor?

Module 8-5 Conservation of Energy

•46 An outfielder throws a baseball with an initial speed of 81.8 mi/h. Just before an infielder catches the ball at the same level, the ball's speed is 110 ft/s. In foot-pounds, by how much is the mechanical energy of the ball–Earth system reduced because of air drag? (The weight of a baseball is 9.0 oz.)

•47 A 75 g Frisbee is thrown from a point 1.1 m above the ground with a speed of 12 m/s. When it has reached a height of 2.1 m, its speed is 10.5 m/s. What was the reduction in E_{mec} of the Frisbee–Earth system because of air drag?

•48 In Fig. 8-51, a block slides down an incline. As it moves from point A to point B, which are 5.0 m apart, force $\vec{F}$ acts on the block, with magnitude 2.0 N and directed down the incline. The magnitude of the frictional force acting on the block is 10 N. If the kinetic energy of the block increases by 35 J between A and B, how much work is done on the block by the gravitational force as the block moves from A to B?

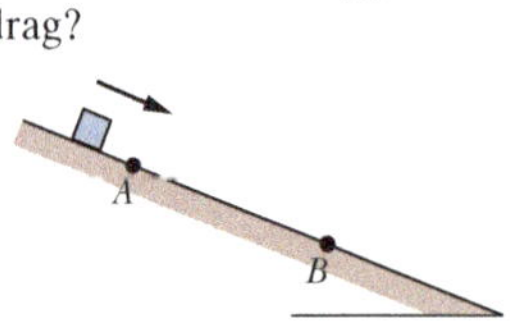

Figure 8-51 Problems 48 and 71.

•49 SSM ILW A 25 kg bear slides, from rest, 12 m down a lodgepole pine tree, moving with a speed of 5.6 m/s just before hitting the ground. (a) What change occurs in the gravitational potential energy of the bear–Earth system during the slide? (b) What is the kinetic energy of the bear just before hitting the ground? (c) What is the average frictional force that acts on the sliding bear?

•50 A 60 kg skier leaves the end of a ski-jump ramp with a velocity of 24 m/s directed 25° above the horizontal. Suppose that as a result of air drag the skier returns to the ground with a speed of 22 m/s, landing 14 m vertically below the end of the ramp. From the launch to the return to the ground, by how much is the mechanical energy of the skier–Earth system reduced because of air drag?

•51 During a rockslide, a 520 kg rock slides from rest down a hillside that is 500 m long and 300 m high. The coefficient of kinetic friction between the rock and the hill surface is 0.25. (a) If the gravitational potential energy U of the rock–Earth system is zero at the bottom of the hill, what is the value of U just before the slide? (b) How much energy is transferred to thermal energy during the slide? (c) What is the kinetic energy of the rock as it reaches the bottom of the hill? (d) What is its speed then?

••52 A large fake cookie sliding on a horizontal surface is attached to one end of a horizontal spring with spring constant $k = 400$ N/m; the other end of the spring is fixed in place. The cookie has a kinetic energy of 20.0 J as it passes through the spring's equilibrium position. As the cookie slides, a frictional force of magnitude 10.0 N acts on it. (a) How far will the cookie slide from the equilibrium position before coming momentarily to rest? (b) What will be the kinetic energy of the cookie as it slides back through the equilibrium position?

••53 In Fig. 8-52, a 3.5 kg block is accelerated from rest by a compressed spring of spring constant 640 N/m. The block leaves the spring at the spring's relaxed length and then travels over a horizontal floor with a coefficient of kinetic friction $\mu_k = 0.25$. The frictional force stops the block in distance $D = 7.8$ m. What are (a) the increase in the thermal energy of the block–floor system, (b) the maximum kinetic energy of the block, and (c) the original compression distance of the spring?

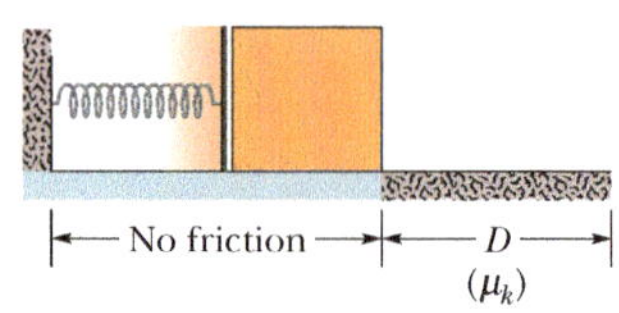

Figure 8-52 Problem 53.

••54 A child whose weight is 267 N slides down a 6.1 m playground slide that makes an angle of 20° with the horizontal. The coefficient of kinetic friction between slide and child is 0.10. (a) How much energy is transferred to thermal energy? (b) If she starts at the top with a speed of 0.457 m/s, what is her speed at the bottom?

••55 ILW In Fig. 8-53, a block of mass $m = 2.5$ kg slides head on into a spring of spring constant $k = 320$ N/m. When the block stops, it has compressed the spring by 7.5 cm. The coefficient of kinetic friction between block and floor is 0.25. While the block is in contact with the spring and being brought to rest, what are (a) the work done by the spring force and (b) the increase in thermal energy of the block–floor system? (c) What is the block's speed just as it reaches the spring?

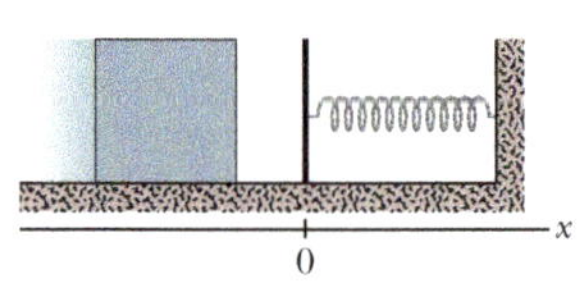

Figure 8-53 Problem 55.

••56 You push a 2.0 kg block against a horizontal spring, compressing the spring by 15 cm. Then you release the block, and the spring sends it sliding across a tabletop. It stops 75 cm from where you released it. The spring constant is 200 N/m. What is the block–table coefficient of kinetic friction?

••57 GO In Fig. 8-54, a block slides along a track from one level to a higher level after passing through an intermediate valley. The track is frictionless until the block reaches the higher level. There a frictional force stops the block in a distance d. The block's initial speed v_0 is 6.0 m/s, the height difference h is 1.1 m, and μ_k is 0.60. Find d.

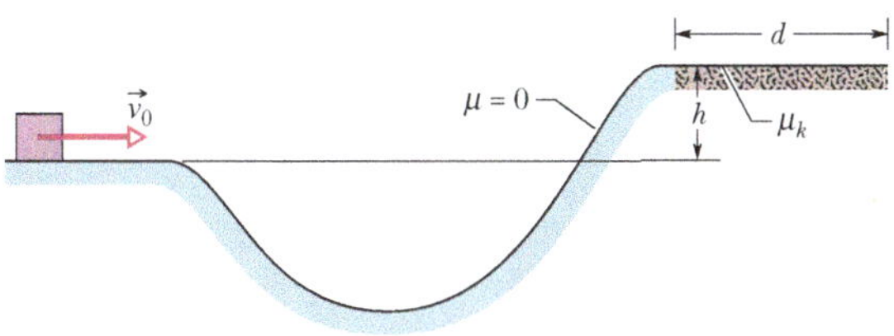

Figure 8-54 Problem 57.

••58 A cookie jar is moving up a 40° incline. At a point 55 cm from the bottom of the incline (measured along the incline), the jar has a speed of 1.4 m/s. The coefficient of kinetic friction between jar and incline is 0.15. (a) How much farther up the incline will the jar move? (b) How fast will it be going when it has slid back to the bottom of the incline? (c) Do the answers to (a) and (b) increase, decrease, or remain the same if we decrease the coefficient of kinetic friction (but do not change the given speed or location)?

••59 A stone with a weight of 5.29 N is launched vertically from ground level with an initial speed of 20.0 m/s, and the air drag on it is 0.265 N throughout the flight. What are (a) the maximum height reached by the stone and (b) its speed just before it hits the ground?

••60 A 4.0 kg bundle starts up a 30° incline with 128 J of kinetic energy. How far will it slide up the incline if the coefficient of kinetic friction between bundle and incline is 0.30?

••61 When a click beetle is upside down on its back, it jumps upward by suddenly arching its back, transferring energy stored in a muscle to mechanical energy. This launching mechanism produces an audible click, giving the beetle its name. Videotape of a certain click-beetle jump shows that a beetle of mass $m = 4.0 \times 10^{-6}$ kg moved directly upward by 0.77 mm during the launch and then to a maximum height of $h = 0.30$ m. During the launch, what are the average magnitudes of (a) the external force on the beetle's back from the floor and (b) the acceleration of the beetle in terms of g?

•••62 GO In Fig. 8-55, a block slides along a path that is without friction until the block reaches the section of length $L = 0.75$ m, which begins at height $h = 2.0$ m on a ramp of angle $\theta = 30°$. In that section, the coefficient of kinetic friction is 0.40. The block passes through point A with a speed of 8.0 m/s. If the block can reach point B (where the friction ends), what is its speed there, and if it cannot, what is its greatest height above A?

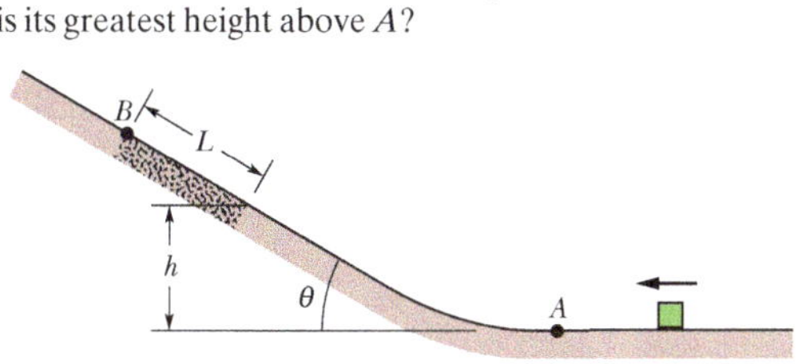

Figure 8-55 Problem 62.

•••63 The cable of the 1800 kg elevator cab in Fig. 8-56 snaps when the cab is at rest at the first floor, where the cab bottom is a distance $d = 3.7$ m above a spring of spring constant $k = 0.15$ MN/m. A safety device clamps the cab against guide rails so that a constant frictional force of 4.4 kN opposes the cab's motion. (a) Find the speed of the cab just before it hits the spring. (b) Find the maximum distance x that the spring is compressed (the frictional force still acts during this compression). (c) Find the distance that the cab will bounce back up the shaft. (d) Using conservation of energy, find the approximate total distance that the cab will move before coming to rest. (Assume that the frictional force on the cab is negligible when the cab is stationary.)

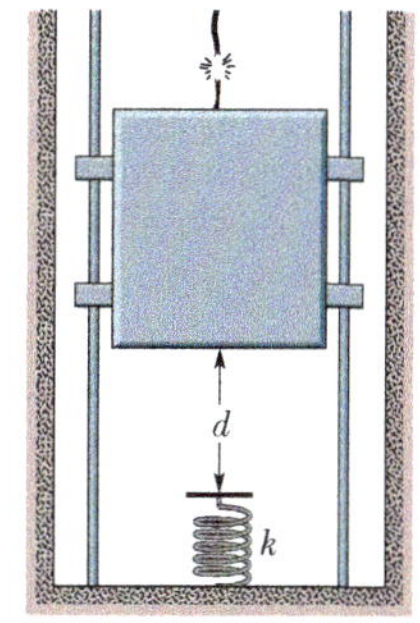

Figure 8-56 Problem 63.

•••64 GO In Fig. 8-57, a block is released from rest at height $d = 40$ cm and slides down a frictionless ramp and onto a first plateau, which has length d and where the coefficient of kinetic friction is 0.50. If the block is still moving, it then slides down a second frictionless ramp through height $d/2$ and onto a lower plateau, which has length $d/2$ and where the coefficient of kinetic friction is again 0.50. If the block is still moving, it then slides up a frictionless ramp until it (momentarily) stops. Where does the block stop? If its final stop is on a plateau, state which one and give the distance L from the left edge of that plateau. If the block reaches the ramp, give the height H above the lower plateau where it momentarily stops.

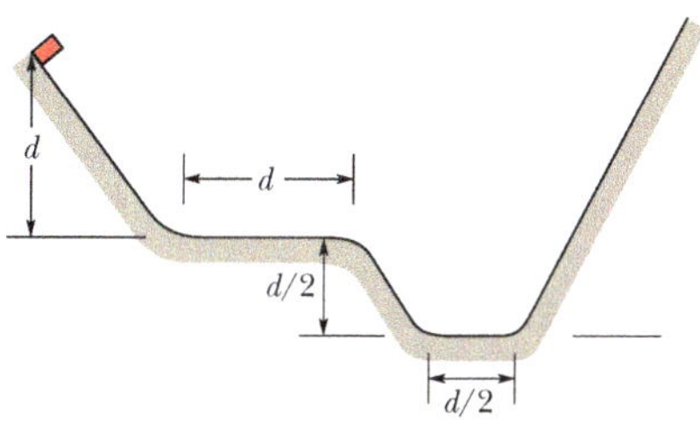

Figure 8-57 Problem 64.

•••65 GO A particle can slide along a track with elevated ends and a flat central part, as shown in Fig. 8-58. The flat part has length $L = 40$ cm. The curved portions of the track are frictionless, but for the flat part the coefficient of kinetic friction is $\mu_k = 0.20$. The particle is released from rest at point A, which is at height $h = L/2$. How far from the left edge of the flat part does the particle finally stop?

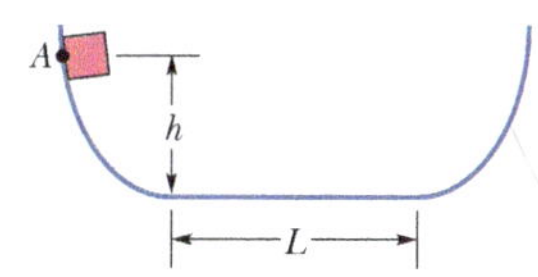

Figure 8-58 Problem 65.

Additional Problems

66 A 3.2 kg sloth hangs 3.0 m above the ground. (a) What is the gravitational potential energy of the sloth–Earth system if we take the reference point $y = 0$ to be at the ground? If the sloth drops to the ground and air drag on it is assumed to be negligible, what are the (b) kinetic energy and (c) speed of the sloth just before it reaches the ground?

67 SSM A spring ($k = 200$ N/m) is fixed at the top of a frictionless plane inclined at angle $\theta = 40°$ (Fig. 8-59). A 1.0 kg block is projected up the plane, from an initial position that is distance $d = 0.60$ m from the end of the relaxed spring, with an initial kinetic energy of 16 J. (a) What is the kinetic energy of the block at the instant it has compressed the spring 0.20 m? (b) With what kinetic energy must the block be projected up the plane if it is to stop momentarily when it has compressed the spring by 0.40 m?

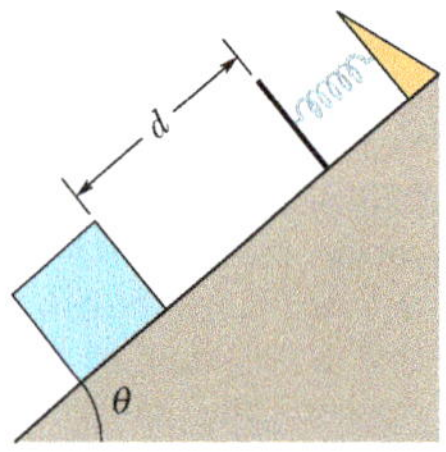

Figure 8-59 Problem 67.

68 From the edge of a cliff, a 0.55 kg projectile is launched with an initial kinetic energy of 1550 J. The projectile's maximum upward displacement from the launch point is +140 m. What are the (a) horizontal and (b) vertical components of its launch velocity? (c) At the instant the vertical component of its velocity is 65 m/s, what is its vertical displacement from the launch point?

69 SSM In Fig. 8-60, the pulley has negligible mass, and both it and the inclined plane are frictionless. Block A has a mass of 1.0 kg, block B has a mass of 2.0 kg, and angle θ is 30°. If the blocks are released from rest with the connecting cord taut, what is their total kinetic energy when block B has fallen 25 cm?

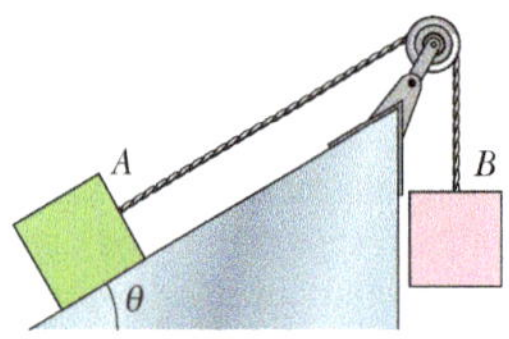

Figure 8-60 Problem 69.

70 GO In Fig. 8-38, the string is $L = 120$ cm long, has a ball attached to one end, and is fixed at its other end. A fixed peg is at point P. Released from rest, the ball swings down until the string catches on the peg; then the ball swings up, around the peg. If the ball is to swing completely around the peg, what value must distance d exceed? (*Hint:* The ball must still be moving at the top of its swing. Do you see why?)

71 SSM In Fig. 8-51, a block is sent sliding down a frictionless ramp. Its speeds at points A and B are 2.00 m/s and 2.60 m/s, respectively. Next, it is again sent sliding down the ramp, but this time its speed at point A is 4.00 m/s. What then is its speed at point B?

72 Two snowy peaks are at heights $H = 850$ m and $h = 750$ m above the valley between them. A ski run extends between the peaks, with a total length of 3.2 km and an average slope of $\theta = 30°$ (Fig. 8-61). (a) A skier starts from rest at the top of the higher peak. At what speed will he arrive at the top of the lower peak if he coasts without using ski poles? Ignore friction. (b) Approximately what coefficient of kinetic friction between snow and skis would make him stop just at the top of the lower peak?

Figure 8-61 Problem 72.

73 SSM The temperature of a plastic cube is monitored while the cube is pushed 3.0 m across a floor at constant speed by a horizontal force of 15 N. The thermal energy of the cube increases by 20 J. What is the increase in the thermal energy of the floor along which the cube slides?

74 A skier weighing 600 N goes over a frictionless circular hill of radius $R = 20$ m (Fig. 8-62). Assume that the effects of air resistance on the skier are negligible. As she comes up the hill, her speed is 8.0 m/s at point B, at angle $\theta = 20°$. (a) What is her speed at the hilltop (point A) if she coasts without using her poles? (b) What minimum speed can she have at B and still coast to the hilltop? (c) Do the answers to these two questions increase, decrease, or remain the same if the skier weighs 700 N instead of 600 N?

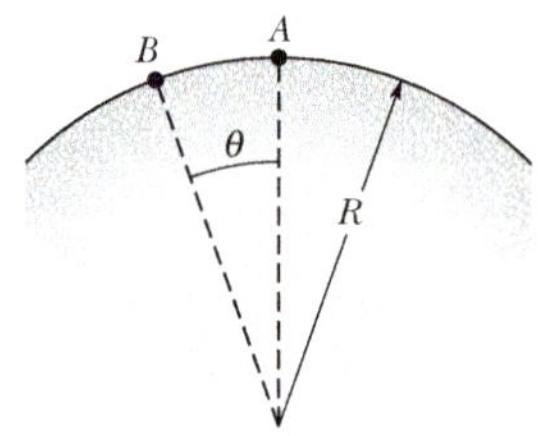

Figure 8-62 Problem 74.

75 SSM To form a pendulum, a 0.092 kg ball is attached to one end of a rod of length 0.62 m and negligible mass, and the other end of the rod is mounted on a pivot. The rod is rotated until it is straight up, and then it is released from rest so that it swings down around the pivot. When the ball reaches its lowest point, what are (a) its speed and (b) the tension in the rod? Next, the rod is rotated until it is horizontal, and then it is again released from rest. (c) At what angle from the vertical does the tension in the rod equal the weight of the ball? (d) If the mass of the ball is increased, does the answer to (c) increase, decrease, or remain the same?

76 We move a particle along an x axis, first outward from $x = 1.0$ m to $x = 4.0$ m and then back to $x = 1.0$ m, while an external force acts on it. That force is directed along the x axis, and its x component can have different values for the outward trip and for the return trip. Here are the values (in newtons) for four situations, where x is in meters:

Outward	Inward
(a) +3.0	−3.0
(b) +5.0	+5.0
(c) $+2.0x$	$-2.0x$
(d) $+3.0x^2$	$+3.0x^2$

Find the net work done on the particle by the external force *for the round trip* for each of the four situations. (e) For which, if any, is the external force conservative?

77 SSM A conservative force $F(x)$ acts on a 2.0 kg particle that moves along an x axis. The potential energy $U(x)$ associated with $F(x)$ is graphed in Fig. 8-63. When the particle is at $x = 2.0$ m, its velocity is −1.5 m/s. What are the (a) magnitude and (b) direction of $F(x)$ at this position? Between what positions on the (c) left and (d) right does the particle move? (e) What is the particle's speed at $x = 7.0$ m?

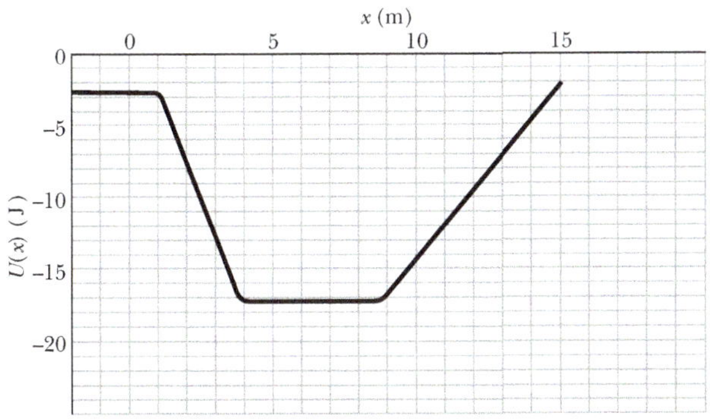

Figure 8-63 Problem 77.

78 At a certain factory, 300 kg crates are dropped vertically from a packing machine onto a conveyor belt moving at 1.20 m/s (Fig. 8-64). (A motor maintains the belt's constant speed.) The coefficient of kinetic friction between the belt and each crate is 0.400. After a short time, slipping between the belt and the crate ceases, and the crate then moves along with the belt. For the period of time during which the crate is being brought to rest relative to the belt, calculate, for a coordinate system at rest in the factory, (a) the kinetic energy supplied to the crate, (b) the magnitude of the kinetic frictional force acting on the crate, and (c) the energy supplied by the motor. (d) Explain why answers (a) and (c) differ.

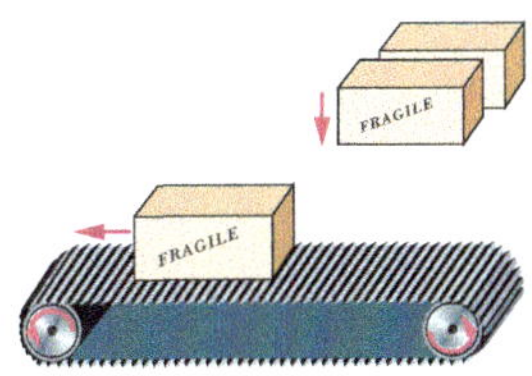

Figure 8-64 Problem 78.

79 SSM A 1500 kg car begins sliding down a 5.0° inclined road with a speed of 30 km/h. The engine is turned off, and the only forces acting on the car are a net frictional force from the road and the gravitational force. After the car has traveled 50 m along the road, its speed is 40 km/h. (a) How much is the mechanical energy of the car reduced because of the net frictional force? (b) What is the magnitude of that net frictional force?

80 GO In Fig. 8-65, a 1400 kg block of granite is pulled up an incline at a constant speed of 1.34 m/s by a cable and winch. The indicated distances are $d_1 = 40$ m and $d_2 = 30$ m. The coefficient of kinetic friction between the block and the incline is 0.40. What is the power due to the force applied to the block by the cable?

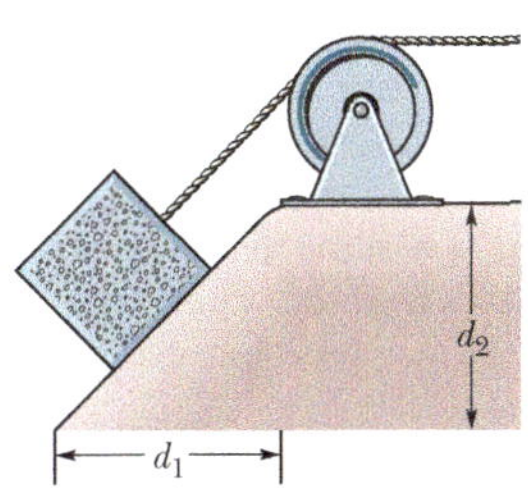

Figure 8-65 Problem 80.

81 A particle can move along only an x axis, where conservative forces act on it (Fig. 8-66 and the following table). The particle is released at $x = 5.00$ m with a kinetic energy of $K = 14.0$ J and a potential energy of $U = 0$. If its motion is in the negative direction of the x axis, what are its (a) K and (b) U at $x = 2.00$ m and its (c) K and (d) U at $x = 0$? If its motion is in the positive direction of the x axis, what are its (e) K and (f) U at $x = 11.0$ m, its (g) K and (h) U at $x = 12.0$ m, and its (i) K and (j) U at $x = 13.0$ m? (k) Plot $U(x)$ versus x for the range $x = 0$ to $x = 13.0$ m.

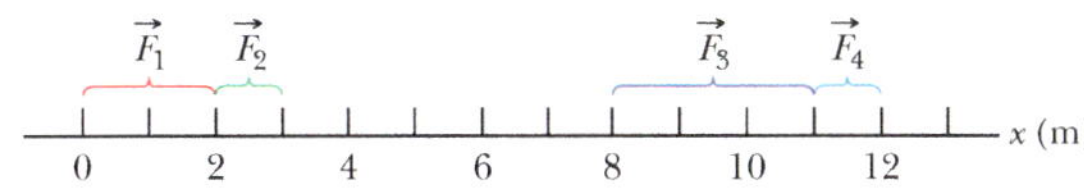

Figure 8-66 Problems 81 and 82.

Next, the particle is released from rest at $x = 0$. What are (l) its kinetic energy at $x = 5.0$ m and (m) the maximum positive position x_{max} it reaches? (n) What does the particle do after it reaches x_{max}?

Range	Force
0 to 2.00 m	$\vec{F}_1 = +(3.00\text{ N})\hat{\mathrm{i}}$
2.00 m to 3.00 m	$\vec{F}_2 = +(5.00\text{ N})\hat{\mathrm{i}}$
3.00 m to 8.00 m	$F = 0$
8.00 m to 11.0 m	$\vec{F}_3 = -(4.00\text{ N})\hat{\mathrm{i}}$
11.0 m to 12.0 m	$\vec{F}_4 = -(1.00\text{ N})\hat{\mathrm{i}}$
12.0 m to 15.0 m	$F = 0$

82 For the arrangement of forces in Problem 81, a 2.00 kg particle is released at $x = 5.00$ m with an initial velocity of 3.45 m/s in the negative direction of the x axis. (a) If the particle can reach $x = 0$ m, what is its speed there, and if it cannot, what is its turning point? Suppose, instead, the particle is headed in the positive x direction when it is released at $x = 5.00$ m at speed 3.45 m/s. (b) If the particle can reach $x = 13.0$ m, what is its speed there, and if it cannot, what is its turning point?

83 SSM A 15 kg block is accelerated at 2.0 m/s^2 along a horizontal frictionless surface, with the speed increasing from 10 m/s to 30 m/s. What are (a) the change in the block's mechanical energy and (b) the average rate at which energy is transferred to the block? What is the instantaneous rate of that transfer when the block's speed is (c) 10 m/s and (d) 30 m/s?

84 A certain spring is found *not* to conform to Hooke's law. The force (in newtons) it exerts when stretched a distance x (in meters) is found to have magnitude $52.8x + 38.4x^2$ in the direction opposing the stretch. (a) Compute the work required to stretch the spring from $x = 0.500$ m to $x = 1.00$ m. (b) With one end of the spring fixed, a particle of mass 2.17 kg is attached to the other end of the spring when it is stretched by an amount $x = 1.00$ m. If the particle is then released from rest, what is its speed at the instant the stretch in the spring is $x = 0.500$ m? (c) Is the force exerted by the spring conservative or nonconservative? Explain.

85 SSM Each second, 1200 m^3 of water passes over a waterfall 100 m high. Three-fourths of the kinetic energy gained by the water in falling is transferred to electrical energy by a hydroelectric generator. At what rate does the generator produce electrical energy? (The mass of 1 m^3 of water is 1000 kg.)

86 GO In Fig. 8-67, a small block is sent through point A with a speed of 7.0 m/s. Its path is without friction until it reaches the section of length $L = 12$ m, where the coefficient of kinetic friction is 0.70. The indicated heights are $h_1 = 6.0$ m and $h_2 = 2.0$ m. What are the speeds of the block at (a) point B and (b) point C? (c) Does the block reach point D? If so, what is its speed there; if not, how far through the section of friction does it travel?

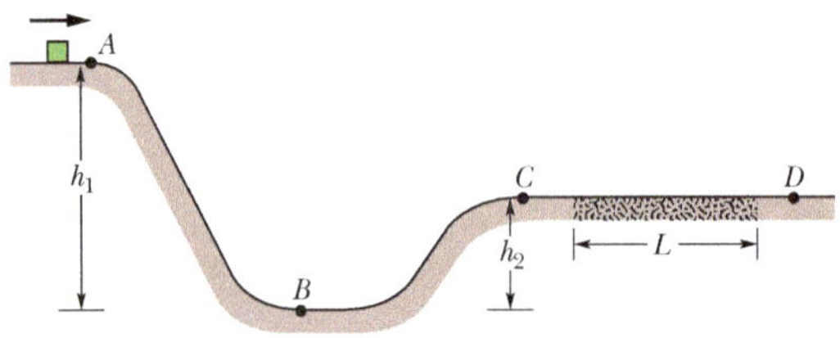

Figure 8-67 Problem 86.

87 SSM A massless rigid rod of length L has a ball of mass m attached to one end (Fig. 8-68). The other end is pivoted in such a way that the ball will move in a vertical circle. First, assume that there is no friction at the pivot. The system is launched downward from the horizontal position A with initial speed v_0. The ball just barely reaches point D and then stops. (a) Derive an expression for v_0 in terms of L, m, and g. (b) What is the tension in the rod when the ball passes through B? (c) A little grit is placed on the pivot to increase the friction there. Then the ball just barely reaches C when launched from A with the same speed as before. What is the decrease in the mechanical energy during this motion? (d) What is the decrease in the mechanical energy by the time the ball finally comes to rest at B after several oscillations?

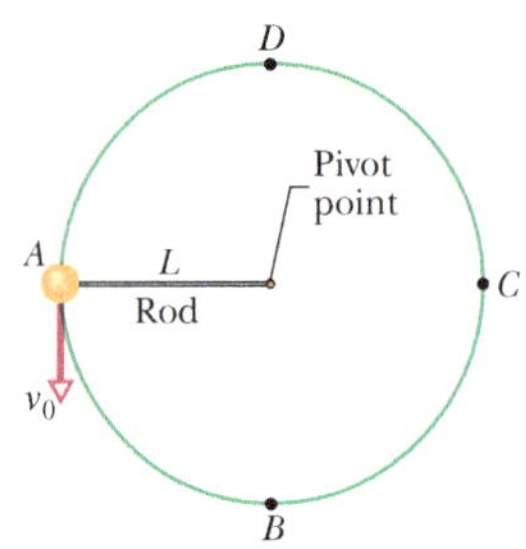

Figure 8-68 Problem 87.

88 A 1.50 kg water balloon is shot straight up with an initial speed of 3.00 m/s. (a) What is the kinetic energy of the balloon just as it is launched? (b) How much work does the gravitational force do on the balloon during the balloon's full ascent? (c) What is the change in the gravitational potential energy of the balloon–Earth system during the full ascent? (d) If the gravitational potential energy is taken to be zero at the launch point, what is its value when the balloon reaches its maximum height? (e) If, instead, the gravitational potential energy is taken to be zero at the maximum height, what is its value at the launch point? (f) What is the maximum height?

89 A 2.50 kg beverage can is thrown directly downward from a height of 4.00 m, with an initial speed of 3.00 m/s. The air drag on the can is negligible. What is the kinetic energy of the can (a) as it reaches the ground at the end of its fall and (b) when it is halfway to the ground? What are (c) the kinetic energy of the can and (d) the gravitational potential energy of the can–Earth system 0.200 s before the can reaches the ground? For the latter, take the reference point $y = 0$ to be at the ground.

90 A constant horizontal force moves a 50 kg trunk 6.0 m up a 30° incline at constant speed. The coefficient of kinetic friction is 0.20. What are (a) the work done by the applied force and (b) the increase in the thermal energy of the trunk and incline?

91 GO Two blocks, of masses $M = 2.0$ kg and $2M$, are connected to a spring of spring constant $k = 200$ N/m that has one end fixed, as shown in Fig. 8-69. The horizontal surface and the pulley are frictionless, and the pulley has negligible mass. The blocks are released from rest with the spring relaxed. (a) What is the combined kinetic energy of the two blocks when the hanging block has fallen 0.090 m? (b) What is the kinetic energy of the hanging block when it has fallen that 0.090 m? (c) What maximum distance does the hanging block fall before momentarily stopping?

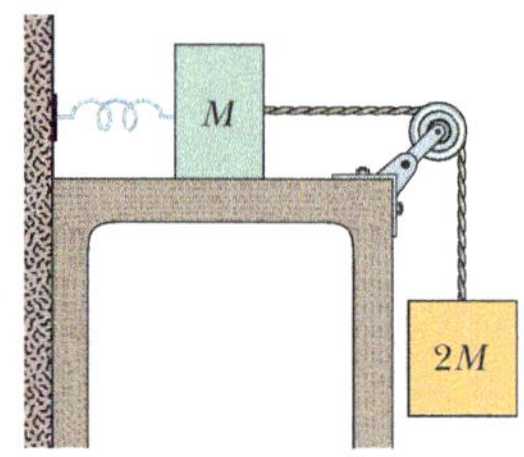

Figure 8-69 Problem 91.

92 A volcanic ash flow is moving across horizontal ground when it encounters a 10° upslope. The front of the flow then travels 920 m up the slope before stopping. Assume that the gases entrapped in the flow lift the flow and thus make the frictional force from the ground negligible; assume also that the mechanical energy of the front of the flow is conserved. What was the initial speed of the front of the flow?

93 A playground slide is in the form of an arc of a circle that has a radius of 12 m. The maximum height of the slide is $h = 4.0$ m, and the ground is tangent to the circle (Fig. 8-70). A 25 kg child starts from rest at the top of the slide and has a speed of 6.2 m/s at the bottom. (a) What is the length of the slide? (b) What average frictional force acts on the child over this distance? If, instead of the ground, a vertical line through the *top of the slide* is tangent to the circle, what are (c) the length of the slide and (d) the average frictional force on the child?

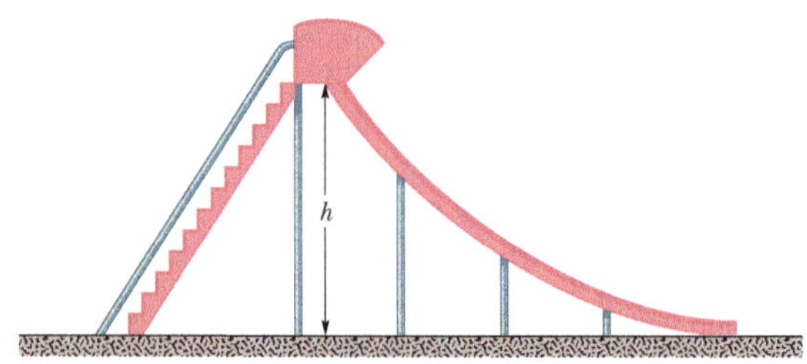

Figure 8-70 Problem 93.

94 The luxury liner *Queen Elizabeth 2* has a diesel-electric power plant with a maximum power of 92 MW at a cruising speed of 32.5 knots. What forward force is exerted on the ship at this speed? (1 knot = 1.852 km/h.)

95 A factory worker accidentally releases a 180 kg crate that was being held at rest at the top of a ramp that is 3.7 m long and inclined at 39° to the horizontal. The coefficient of kinetic friction between the crate and the ramp, and between the crate and the horizontal factory floor, is 0.28. (a) How fast is the crate moving as it reaches the bottom of the ramp? (b) How far will it subsequently slide across the floor? (Assume that the crate's kinetic energy does not change as it moves from the ramp onto the floor.) (c) Do the answers to (a) and (b) increase, decrease, or remain the same if we halve the mass of the crate?

96 If a 70 kg baseball player steals home by sliding into the plate with an initial speed of 10 m/s just as he hits the ground, (a) what is the decrease in the player's kinetic energy and (b) what is the increase in the thermal energy of his body and the ground along which he slides?

97 A 0.50 kg banana is thrown directly upward with an initial speed of 4.00 m/s and reaches a maximum height of 0.80 m. What change does air drag cause in the mechanical energy of the banana–Earth system during the ascent?

98 A metal tool is sharpened by being held against the rim of a wheel on a grinding machine by a force of 180 N. The frictional forces between the rim and the tool grind off small pieces of the tool. The wheel has a radius of 20.0 cm and rotates at 2.50 rev/s. The coefficient of kinetic friction between the wheel and the tool is 0.320. At what rate is energy being transferred from the motor driving the wheel to the thermal energy of the wheel and tool and to the kinetic energy of the material thrown from the tool?

99 A swimmer moves through the water at an average speed of 0.22 m/s. The average drag force is 110 N. What average power is required of the swimmer?

100 An automobile with passengers has weight 16 400 N and is moving at 113 km/h when the driver brakes, sliding to a stop. The frictional force on the wheels from the road has a magnitude of 8230 N. Find the stopping distance.

101 A 0.63 kg ball thrown directly upward with an initial speed of 14 m/s reaches a maximum height of 8.1 m. What is the change in the mechanical energy of the ball–Earth system during the ascent of the ball to that maximum height?

102 The summit of Mount Everest is 8850 m above sea level. (a) How much energy would a 90 kg climber expend against the gravitational force on him in climbing to the summit from sea level? (b) How many candy bars, at 1.25 MJ per bar, would supply an energy equivalent to this? Your answer should suggest that work done against the gravitational force is a very small part of the energy expended in climbing a mountain.

103 A sprinter who weighs 670 N runs the first 7.0 m of a race in 1.6 s, starting from rest and accelerating uniformly. What are the sprinter's (a) speed and (b) kinetic energy at the end of the 1.6 s? (c) What average power does the sprinter generate during the 1.6 s interval?

104 A 20 kg object is acted on by a conservative force given by $F = -3.0x - 5.0x^2$, with F in newtons and x in meters. Take the potential energy associated with the force to be zero when the object is at $x = 0$. (a) What is the potential energy of the system associated with the force when the object is at $x = 2.0$ m? (b) If the object has a velocity of 4.0 m/s in the negative direction of the x axis when it is at $x = 5.0$ m, what is its speed when it passes through the origin? (c) What are the answers to (a) and (b) if the potential energy of the system is taken to be -8.0 J when the object is at $x = 0$?

105 A machine pulls a 40 kg trunk 2.0 m up a 40° ramp at constant velocity, with the machine's force on the trunk directed parallel to the ramp. The coefficient of kinetic friction between the trunk and the ramp is 0.40. What are (a) the work done on the trunk by the machine's force and (b) the increase in thermal energy of the trunk and the ramp?

106 The spring in the muzzle of a child's spring gun has a spring constant of 700 N/m. To shoot a ball from the gun, first the spring is compressed and then the ball is placed on it. The gun's trigger then releases the spring, which pushes the ball through the muzzle. The ball leaves the spring just as it leaves the outer end of the muzzle. When the gun is inclined upward by 30° to the horizontal, a 57 g ball is shot to a maximum height of 1.83 m above the gun's muzzle. Assume air drag on the ball is negligible. (a) At what speed does the spring launch the ball? (b) Assuming that friction on the ball within the gun can be neglected, find the spring's initial compression distance.

107 The only force acting on a particle is conservative force $\vec{F}$. If the particle is at point A, the potential energy of the system associated with $\vec{F}$ and the particle is 40 J. If the particle moves from point A to point B, the work done on the particle by $\vec{F}$ is +25 J. What is the potential energy of the system with the particle at B?

108 In 1981, Daniel Goodwin climbed 443 m up the *exterior* of the Sears Building in Chicago using suction cups and metal clips. (a) Approximate his mass and then compute how much energy he had to transfer from biomechanical (internal) energy to the gravitational potential energy of the Earth–Goodwin system to lift himself to that height. (b) How much energy would he have had to transfer if he had, instead, taken the stairs inside the building (to the same height)?

109 A 60.0 kg circus performer slides 4.00 m down a pole to the circus floor, starting from rest. What is the kinetic energy of the performer as she reaches the floor if the frictional force on her from the pole (a) is negligible (she will be hurt) and (b) has a magnitude of 500 N?

110 A 5.0 kg block is projected at 5.0 m/s up a plane that is inclined at 30° with the horizontal. How far up along the plane does the block go (a) if the plane is frictionless and (b) if the coefficient of kinetic friction between the block and the plane is 0.40? (c) In the latter case, what is the increase in thermal energy of block and plane during the block's ascent? (d) If the block then slides back down against the frictional force, what is the block's speed when it reaches the original projection point?

111 A 9.40 kg projectile is fired vertically upward. Air drag decreases the mechanical energy of the projectile–Earth system by 68.0 kJ during the projectile's ascent. How much higher would the projectile have gone were air drag negligible?

112 A 70.0 kg man jumping from a window lands in an elevated fire rescue net 11.0 m below the window. He momentarily stops when he has stretched the net by 1.50 m. Assuming that mechanical energy is conserved during this process and that the net functions like an ideal spring, find the elastic potential energy of the net when it is stretched by 1.50 m.

113 A 30 g bullet moving a horizontal velocity of 500 m/s comes to a stop 12 cm within a solid wall. (a) What is the change in the bullet's mechanical energy? (b) What is the magnitude of the average force from the wall stopping it?

114 A 1500 kg car starts from rest on a horizontal road and gains a speed of 72 km/h in 30 s. (a) What is its kinetic energy at the end of the 30 s? (b) What is the average power required of the car during the 30 s interval? (c) What is the instantaneous power at the end of the 30 s interval, assuming that the acceleration is constant?

115 A 1.50 kg snowball is shot upward at an angle of 34.0° to the horizontal with an initial speed of 20.0 m/s. (a) What is its initial kinetic energy? (b) By how much does the gravitational potential

energy of the snowball–Earth system change as the snowball moves from the launch point to the point of maximum height? (c) What is that maximum height?

116 A 68 kg sky diver falls at a constant terminal speed of 59 m/s. (a) At what rate is the gravitational potential energy of the Earth–sky diver system being reduced? (b) At what rate is the system's mechanical energy being reduced?

117 A 20 kg block on a horizontal surface is attached to a horizontal spring of spring constant $k = 4.0$ kN/m. The block is pulled to the right so that the spring is stretched 10 cm beyond its relaxed length, and the block is then released from rest. The frictional force between the sliding block and the surface has a magnitude of 80 N. (a) What is the kinetic energy of the block when it has moved 2.0 cm from its point of release? (b) What is the kinetic energy of the block when it first slides back through the point at which the spring is relaxed? (c) What is the maximum kinetic energy attained by the block as it slides from its point of release to the point at which the spring is relaxed?

118 Resistance to the motion of an automobile consists of road friction, which is almost independent of speed, and air drag, which is proportional to speed-squared. For a certain car with a weight of 12 000 N, the total resistant force F is given by $F = 300 + 1.8v^2$, with F in newtons and v in meters per second. Calculate the power (in horsepower) required to accelerate the car at 0.92 m/s^2 when the speed is 80 km/h.

119 SSM A 50 g ball is thrown from a window with an initial velocity of 8.0 m/s at an angle of 30° above the horizontal. Using energy methods, determine (a) the kinetic energy of the ball at the top of its flight and (b) its speed when it is 3.0 m below the window. Does the answer to (b) depend on either (c) the mass of the ball or (d) the initial angle?

120 A spring with a spring constant of 3200 N/m is initially stretched until the elastic potential energy of the spring is 1.44 J. ($U = 0$ for the relaxed spring.) What is ΔU if the initial stretch is changed to (a) a stretch of 2.0 cm, (b) a compression of 2.0 cm, and (c) a compression of 4.0 cm?

121 A locomotive with a power capability of 1.5 MW can accelerate a train from a speed of 10 m/s to 25 m/s in 6.0 min. (a) Calculate the mass of the train. Find (b) the speed of the train and (c) the force accelerating the train as functions of time (in seconds) during the 6.0 min interval. (d) Find the distance moved by the train during the interval.

122 SSM A 0.42 kg shuffleboard disk is initially at rest when a player uses a cue to increase its speed to 4.2 m/s at constant acceleration. The acceleration takes place over a 2.0 m distance, at the end of which the cue loses contact with the disk. Then the disk slides an additional 12 m before stopping. Assume that the shuffleboard court is level and that the force of friction on the disk is constant. What is the increase in the thermal energy of the disk–court system (a) for that additional 12 m and (b) for the entire 14 m distance? (c) How much work is done on the disk by the cue?

123 A river descends 15 m through rapids. The speed of the water is 3.2 m/s upon entering the rapids and 13 m/s upon leaving. What percentage of the gravitational potential energy of the water–Earth system is transferred to kinetic energy during the descent? (*Hint:* Consider the descent of, say, 10 kg of water.)

124 The magnitude of the gravitational force between a particle of mass m_1 and one of mass m_2 is given by

$$F(x) = G\frac{m_1 m_2}{x^2},$$

where G is a constant and x is the distance between the particles. (a) What is the corresponding potential energy function $U(x)$? Assume that $U(x) \to 0$ as $x \to \infty$ and that x is positive. (b) How much work is required to increase the separation of the particles from $x = x_1$ to $x = x_1 + d$?

125 Approximately 5.5×10^6 kg of water falls 50 m over Niagara Falls each second. (a) What is the decrease in the gravitational potential energy of the water–Earth system each second? (b) If all this energy could be converted to electrical energy (it cannot be), at what rate would electrical energy be supplied? (The mass of 1 m^3 of water is 1000 kg.) (c) If the electrical energy were sold at 1 cent/kW·h, what would be the yearly income?

126 To make a pendulum, a 300 g ball is attached to one end of a string that has a length of 1.4 m and negligible mass. (The other end of the string is fixed.) The ball is pulled to one side until the string makes an angle of 30.0° with the vertical; then (with the string taut) the ball is released from rest. Find (a) the speed of the ball when the string makes an angle of 20.0° with the vertical and (b) the maximum speed of the ball. (c) What is the angle between the string and the vertical when the speed of the ball is one-third its maximum value?

127 In a circus act, a 60 kg clown is shot from a cannon with an initial velocity of 16 m/s at some unknown angle above the horizontal. A short time later the clown lands in a net that is 3.9 m vertically above the clown's initial position. Disregard air drag. What is the kinetic energy of the clown as he lands in the net?

128 A 70 kg firefighter slides, from rest, 4.3 m down a vertical pole. (a) If the firefighter holds onto the pole lightly, so that the frictional force of the pole on her is negligible, what is her speed just before reaching the ground floor? (b) If the firefighter grasps the pole more firmly as she slides, so that the average frictional force of the pole on her is 500 N upward, what is her speed just before reaching the ground floor?

129 The surface of the continental United States has an area of about 8×10^6 km^2 and an average elevation of about 500 m (above sea level). The average yearly rainfall is 75 cm. The fraction of this rainwater that returns to the atmosphere by evaporation is $\frac{2}{3}$; the rest eventually flows into the ocean. If the decrease in gravitational potential energy of the water–Earth system associated with that flow could be fully converted to electrical energy, what would be the average power? (The mass of 1 m^3 of water is 1000 kg.)

130 A spring with spring constant $k = 200$ N/m is suspended vertically with its upper end fixed to the ceiling and its lower end at position $y = 0$. A block of weight 20 N is attached to the lower end, held still for a moment, and then released. What are (a) the kinetic energy K, (b) the change (from the initial value) in the gravitational potential energy ΔU_g, and (c) the change in the elastic potential energy ΔU_e of the spring–block system when the block is at $y = -5.0$ cm? What are (d) K, (e) ΔU_g, and (f) ΔU_e when $y = -10$ cm, (g) K, (h) ΔU_g, and (i) ΔU_e when $y = -15$ cm, and (j) K, (k) ΔU_g, and (l) ΔU_e when $y = -20$ cm?

131 Fasten one end of a vertical spring to a ceiling, attach a cabbage to the other end, and then slowly lower the cabbage until the upward force on it from the spring balances the gravitational force on it. Show that the loss of gravitational potential energy of the cabbage–Earth system equals twice the gain in the spring's potential energy.

132 The maximum force you can exert on an object with one of your back teeth is about 750 N. Suppose that as you gradually bite on a clump of licorice, the licorice resists compression by one of your teeth by acting like a spring for which $k = 2.5 \times 10^5$ N/m. Find (a) the distance the licorice is compressed by your tooth and (b) the work the tooth does on the licorice during the compression. (c) Plot the magnitude of your force versus the compression distance. (d) If there is a potential energy associated with this compression, plot it versus compression distance.

In the 1990s the pelvis of a particular *Triceratops* dinosaur was found to have deep bite marks. The shape of the marks suggested that they were made by a *Tyrannosaurus rex* dinosaur. To test the idea, researchers made a replica of a *T. rex* tooth from bronze and aluminum and then used a hydraulic press to gradually drive the replica into cow bone to the depth seen in the *Triceratops* bone. A graph of the force required versus depth of penetration is given in Fig. 8-71 for one trial; the required force increased with depth because, as the nearly conical tooth penetrated the bone, more of the tooth came in contact with the bone. (e) How much work was done by the hydraulic press—and thus presumably by the *T. rex*—in such a penetration? (f) Is there a potential energy associated with this penetration? (The large biting force and energy expenditure attributed to the *T. rex* by this research suggest that the animal was a predator and not a scavenger.)

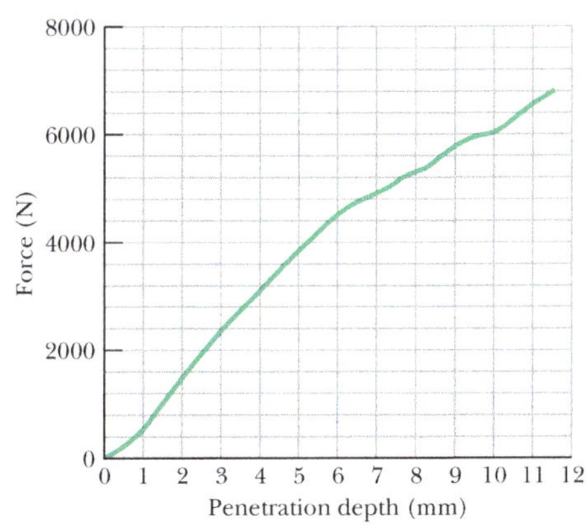

Figure 8-71 Problem 132.

133 Conservative force $F(x)$ acts on a particle that moves along an x axis. Figure 8-72 shows how the potential energy $U(x)$ associated with force $F(x)$ varies with the position of the particle, (a) Plot $F(x)$ for the range $0 < x < 6$ m. (b) The mechanical energy E of the system is 4.0 J. Plot the kinetic energy $K(x)$ of the particle directly on Fig. 8-72.

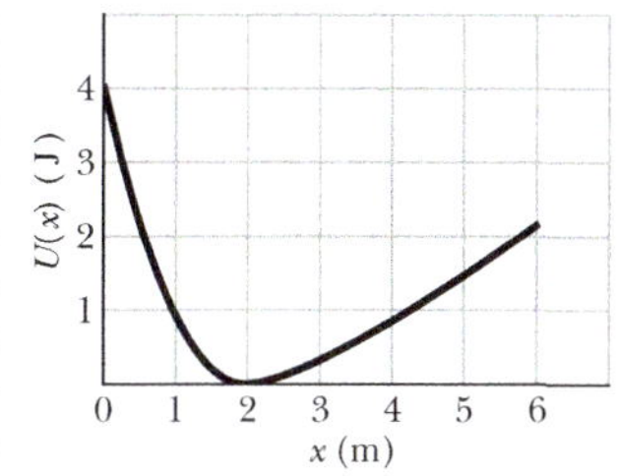

Figure 8-72 Problem 133.

134 Figure 8-73*a* shows a molecule consisting of two atoms of masses m and M (with $m \ll M$) and separation r. Figure 8-73*b* shows the potential energy $U(r)$ of the molecule as a function of r. Describe the motion of the atoms (a) if the total mechanical energy E of the two-atom system is greater than zero (as is E_1), and (b) if E is less than zero (as is E_2). For $E_1 = 1 \times 10^{-19}$ J and $r = 0.3$ nm, find (c) the potential energy of the system, (d) the total kinetic energy of the atoms, and (e) the force (magnitude and direction) acting on each atom. For what values of r is the force (f) repulsive, (g) attractive, and (h) zero?

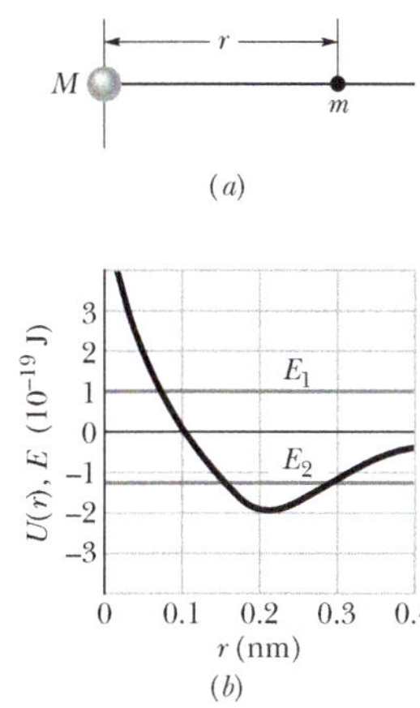

Figure 8-73 Problem 134.

135 Repeat Problem 83, but now with the block accelerated up a frictionless plane inclined at 5.0° to the horizontal.

136 A spring with spring constant $k = 620$ N/m is placed in a vertical orientation with its lower end supported by a horizontal surface. The upper end is depressed 25 cm, and a block with a weight of 50 N is placed (unattached) on the depressed spring. The system is then released from rest. Assume that the gravitational potential energy U_g of the block is zero at the release point ($y = 0$) and calculate the kinetic energy K of the block for y equal to (a) 0, (b) 0.050 m, (c) 0.10 m, (d) 0.15 m, and (e) 0.20 m. Also, (f) how far above its point of release does the block rise?

C H A P T E R 9

Center of Mass and Linear Momentum

9-1 CENTER OF MASS

Learning Objectives

After reading this module, you should be able to . . .

9.01 Given the positions of several particles along an axis or a plane, determine the location of their center of mass.

9.02 Locate the center of mass of an extended, symmetric object by using the symmetry.

9.03 For a two-dimensional or three-dimensional extended object with a uniform distribution of mass, determine the center of mass by (a) mentally dividing the object into simple geometric figures, each of which can be replaced by a particle at its center and (b) finding the center of mass of those particles.

Key Idea

- The center of mass of a system of n particles is defined to be the point whose coordinates are given by

$$x_{\text{com}} = \frac{1}{M}\sum_{i=1}^{n} m_i x_i, \quad y_{\text{com}} = \frac{1}{M}\sum_{i=1}^{n} m_i y_i, \quad z_{\text{com}} = \frac{1}{M}\sum_{i=1}^{n} m_i z_i,$$

or

$$\vec{r}_{\text{com}} = \frac{1}{M}\sum_{i=1}^{n} m_i \vec{r}_i,$$

where M is the total mass of the system.

What Is Physics?

Every mechanical engineer who is hired as a courtroom expert witness to reconstruct a traffic accident uses physics. Every dance trainer who coaches a ballerina on how to leap uses physics. Indeed, analyzing complicated motion of any sort requires simplification via an understanding of physics. In this chapter we discuss how the complicated motion of a system of objects, such as a car or a ballerina, can be simplified if we determine a special point of the system—the *center of mass* of that system.

Here is a quick example. If you toss a ball into the air without much spin on the ball (Fig. 9-1*a*), its motion is simple—it follows a parabolic path, as we discussed in Chapter 4, and the ball can be treated as a particle. If, instead, you flip a baseball bat into the air (Fig. 9-1*b*), its motion is more complicated. Because every part of the bat moves differently, along paths of many different shapes, you cannot represent the bat as a particle. Instead, it is a system of particles each of which follows its own path through the air. However, the bat has one special point—the center of mass—that *does* move in a simple parabolic path. The other parts of the bat move around the center of mass. (To locate the center of mass, balance the bat on an outstretched finger; the point is above your finger, on the bat's central axis.)

You cannot make a career of flipping baseball bats into the air, but you can make a career of advising long-jumpers or dancers on how to leap properly into the air while either moving their arms and legs or rotating their torso. Your starting point would be to determine the person's center of mass because of its simple motion.

The Center of Mass

We define the **center of mass** (com) of a system of particles (such as a person) in order to predict the possible motion of the system.

> The center of mass of a system of particles is the point that moves as though (1) all of the system's mass were concentrated there and (2) all external forces were applied there.

Here we discuss how to determine where the center of mass of a system of particles is located. We start with a system of only a few particles, and then we consider a system of a great many particles (a solid body, such as a baseball bat). Later in the chapter, we discuss how the center of mass of a system moves when external forces act on the system.

Systems of Particles

Two Particles. Figure 9-2*a* shows two particles of masses m_1 and m_2 separated by distance d. We have arbitrarily chosen the origin of an x axis to coincide with the particle of mass m_1. We *define* the position of the center of mass (com) of this two-particle system to be

$$x_{com} = \frac{m_2}{m_1 + m_2} d. \quad (9\text{-}1)$$

Suppose, as an example, that $m_2 = 0$. Then there is only one particle, of mass m_1, and the center of mass must lie at the position of that particle; Eq. 9-1 dutifully reduces to $x_{com} = 0$. If $m_1 = 0$, there is again only one particle (of mass m_2), and we have, as we expect, $x_{com} = d$. If $m_1 = m_2$, the center of mass should be halfway between the two particles; Eq. 9-1 reduces to $x_{com} = \frac{1}{2}d$, again as we expect. Finally, Eq. 9-1 tells us that if neither m_1 nor m_2 is zero, x_{com} can have only values that lie between zero and d; that is, the center of mass must lie somewhere between the two particles.

We are not required to place the origin of the coordinate system on one of the particles. Figure 9-2*b* shows a more generalized situation, in which the coordinate system has been shifted leftward. The position of the center of mass is now defined

as
$$x_{com} = \frac{m_1x_1 + m_2x_2}{m_1 + m_2}. \quad (9\text{-}2)$$

Note that if we put $x_1 = 0$, then x_2 becomes d and Eq. 9-2 reduces to Eq. 9-1, as it must. Note also that in spite of the shift of the coordinate system, the center

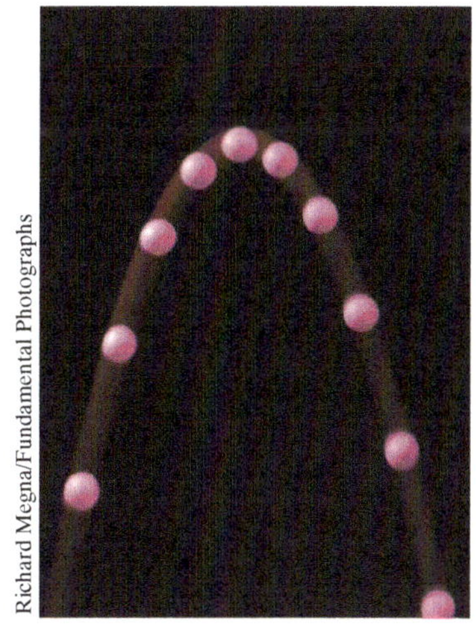

(*a*)

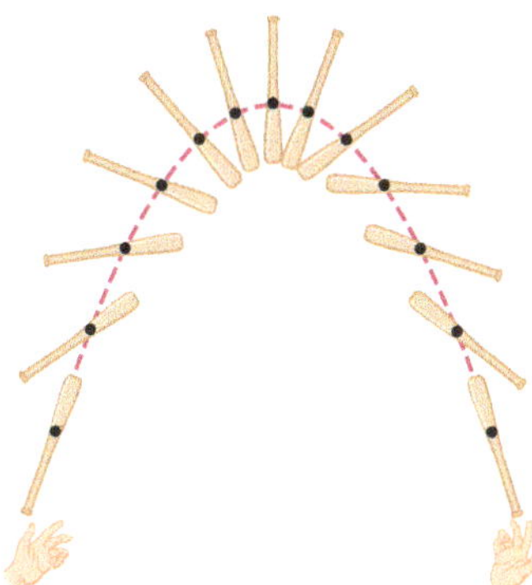

(*b*)

Figure 9-1 (*a*) A ball tossed into the air follows a parabolic path. (*b*) The center of mass (black dot) of a baseball bat flipped into the air follows a parabolic path, but all other points of the bat follow more complicated curved paths.

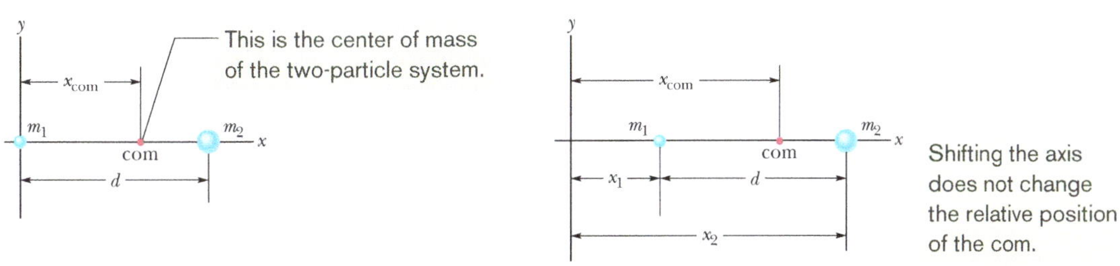

Figure 9-2 (*a*) Two particles of masses m_1 and m_2 are separated by distance d. The dot labeled com shows the position of the center of mass, calculated from Eq. 9-1. (*b*) The same as (*a*) except that the origin is located farther from the particles. The position of the center of mass is calculated from Eq. 9-2. The location of the center of mass with respect to the particles is the same in both cases.

of mass is still the same distance from each particle. The com is a property of the physical particles, not the coordinate system we happen to use.

We can rewrite Eq. 9-2 as

$$x_{\text{com}} = \frac{m_1 x_1 + m_2 x_2}{M}, \tag{9-3}$$

in which M is the total mass of the system. (Here, $M = m_1 + m_2$.)

Many Particles. We can extend this equation to a more general situation in which n particles are strung out along the x axis. Then the total mass is $M = m_1 + m_2 + \cdots + m_n$, and the location of the center of mass is

$$\begin{aligned} x_{\text{com}} &= \frac{m_1 x_1 + m_2 x_2 + m_3 x_3 + \cdots + m_n x_n}{M} \\ &= \frac{1}{M} \sum_{i=1}^{n} m_i x_i. \end{aligned} \tag{9-4}$$

The subscript i is an index that takes on all integer values from 1 to n.

Three Dimensions. If the particles are distributed in three dimensions, the center of mass must be identified by three coordinates. By extension of Eq. 9-4, they are

$$x_{\text{com}} = \frac{1}{M} \sum_{i=1}^{n} m_i x_i, \qquad y_{\text{com}} = \frac{1}{M} \sum_{i=1}^{n} m_i y_i, \qquad z_{\text{com}} = \frac{1}{M} \sum_{i=1}^{n} m_i z_i. \tag{9-5}$$

We can also define the center of mass with the language of vectors. First recall that the position of a particle at coordinates x_i, y_i, and z_i is given by a position vector (it points from the origin to the particle):

$$\vec{r}_i = x_i \hat{\mathrm{i}} + y_i \hat{\mathrm{j}} + z_i \hat{\mathrm{k}}. \tag{9-6}$$

Here the index identifies the particle, and $\hat{\mathrm{i}}$, $\hat{\mathrm{j}}$, and $\hat{\mathrm{k}}$ are unit vectors pointing, respectively, in the positive direction of the x, y, and z axes. Similarly, the position of the center of mass of a system of particles is given by a position vector:

$$\vec{r}_{\text{com}} = x_{\text{com}} \hat{\mathrm{i}} + y_{\text{com}} \hat{\mathrm{j}} + z_{\text{com}} \hat{\mathrm{k}}. \tag{9-7}$$

If you are a fan of concise notation, the three scalar equations of Eq. 9-5 can now be replaced by a single vector equation,

$$\vec{r}_{\text{com}} = \frac{1}{M} \sum_{i=1}^{n} m_i \vec{r}_i, \tag{9-8}$$

where again M is the total mass of the system. You can check that this equation is correct by substituting Eqs. 9-6 and 9-7 into it, and then separating out the x, y, and z components. The scalar relations of Eq. 9-5 result.

Solid Bodies

An ordinary object, such as a baseball bat, contains so many particles (atoms) that we can best treat it as a continuous distribution of matter. The "particles" then become differential mass elements dm, the sums of Eq. 9-5 become integrals, and the coordinates of the center of mass are defined as

$$x_{\text{com}} = \frac{1}{M} \int x\, dm, \qquad y_{\text{com}} = \frac{1}{M} \int y\, dm, \qquad z_{\text{com}} = \frac{1}{M} \int z\, dm, \tag{9-9}$$

where M is now the mass of the object. The integrals effectively allow us to use Eq. 9-5 for a huge number of particles, an effort that otherwise would take many years.

Evaluating these integrals for most common objects (such as a television set or a moose) would be difficult, so here we consider only *uniform* objects. Such objects have uniform *density*, or mass per unit volume; that is, the density ρ (Greek letter rho) is the same for any given element of an object as for the whole object. From Eq. 1-8, we can write

$$\rho = \frac{dm}{dV} = \frac{M}{V}, \tag{9-10}$$

where dV is the volume occupied by a mass element dm, and V is the total volume of the object. Substituting $dm = (M/V)\,dV$ from Eq. 9-10 into Eq. 9-9 gives

$$x_{com} = \frac{1}{V}\int x\,dV, \qquad y_{com} = \frac{1}{V}\int y\,dV, \qquad z_{com} = \frac{1}{V}\int z\,dV. \tag{9-11}$$

Symmetry as a Shortcut. You can bypass one or more of these integrals if an object has a point, a line, or a plane of symmetry. The center of mass of such an object then lies at that point, on that line, or in that plane. For example, the center of mass of a uniform sphere (which has a point of symmetry) is at the center of the sphere (which is the point of symmetry). The center of mass of a uniform cone (whose axis is a line of symmetry) lies on the axis of the cone. The center of mass of a banana (which has a plane of symmetry that splits it into two equal parts) lies somewhere in the plane of symmetry.

The center of mass of an object need not lie within the object. There is no dough at the com of a doughnut, and no iron at the com of a horseshoe.

Sample Problem 9.01 com of three particles

Three particles of masses $m_1 = 1.2$ kg, $m_2 = 2.5$ kg, and $m_3 = 3.4$ kg form an equilateral triangle of edge length $a = 140$ cm. Where is the center of mass of this system?

KEY IDEA

We are dealing with particles instead of an extended solid body, so we can use Eq. 9-5 to locate their center of mass. The particles are in the plane of the equilateral triangle, so we need only the first two equations.

Calculations: We can simplify the calculations by choosing the x and y axes so that one of the particles is located at the origin and the x axis coincides with one of the triangle's sides (Fig. 9-3). The three particles then have the following coordinates:

Particle	Mass (kg)	x (cm)	y (cm)
1	1.2	0	0
2	2.5	140	0
3	3.4	70	120

The total mass M of the system is 7.1 kg.

From Eq. 9-5, the coordinates of the center of mass are

$$x_{com} = \frac{1}{M}\sum_{i=1}^{3} m_i x_i = \frac{m_1x_1 + m_2x_2 + m_3x_3}{M}$$

$$= \frac{(1.2\text{ kg})(0) + (2.5\text{ kg})(140\text{ cm}) + (3.4\text{ kg})(70\text{ cm})}{7.1\text{ kg}}$$

$$= 83\text{ cm} \qquad \text{(Answer)}$$

and $$y_{com} = \frac{1}{M}\sum_{i=1}^{3} m_i y_i = \frac{m_1y_1 + m_2y_2 + m_3y_3}{M}$$

$$= \frac{(1.2\text{ kg})(0) + (2.5\text{ kg})(0) + (3.4\text{ kg})(120\text{ cm})}{7.1\text{ kg}}$$

$$= 58\text{ cm}. \qquad \text{(Answer)}$$

In Fig. 9-3, the center of mass is located by the position vector $\vec{r}_{com}$, which has components x_{com} and y_{com}. If we had chosen some other orientation of the coordinate system, these coordinates would be different but the location of the com relative to the particles would be the same.

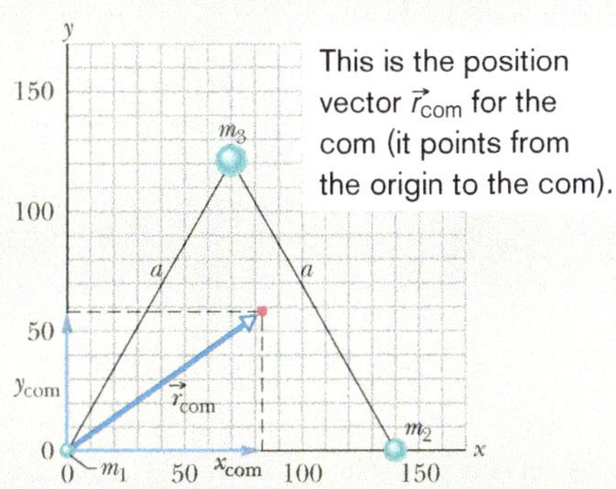

Figure 9-3 Three particles form an equilateral triangle of edge length a. The center of mass is located by the position vector $\vec{r}_{com}$.

Sample Problem 9.02 com of plate with missing piece

This sample problem has lots of words to read, but they will allow you to calculate a com using easy algebra instead of challenging integral calculus. Figure 9-4*a* shows a uniform metal plate *P* of radius $2R$ from which a disk of radius *R* has been stamped out (removed) in an assembly line. The disk is shown in Fig. 9-4*b*. Using the *xy* coordinate system shown, locate the center of mass com_P of the remaining plate.

KEY IDEAS

(1) Let us roughly locate the center of plate *P* by using symmetry. We note that the plate is symmetric about the *x* axis (we get the portion below that axis by rotating the upper portion about the axis). Thus, com_P must be on the *x* axis. The plate (with the disk removed) is not symmetric about the *y* axis. However, because there is somewhat more mass on the right of the *y* axis, com_P must be somewhat to the right of that axis. Thus, the location of com_P should be roughly as indicated in Fig. 9-4*a*.

(2) Plate *P* is an extended solid body, so in principle we can use Eqs. 9-11 to find the actual coordinates of the center of mass of plate *P*. Here we want the *xy* coordinates of the center of mass because the plate is thin and uniform. If it had any appreciable thickness, we would just say that the center of mass is midway across the thickness. Still, using Eqs. 9-11 would be challenging because we would need a function for the shape of the plate with its hole, and then we would need to integrate the function in two dimensions.

(3) Here is a much easier way: In working with centers of mass, we can assume that the mass of a uniform object (as we have here) is concentrated in a particle at the object's center of mass. Thus we can treat the object as a particle and avoid any two-dimensional integration.

Calculations: First, put the stamped-out disk (call it disk *S*) back into place (Fig. 9-4*c*) to form the original composite plate (call it plate *C*). Because of its circular symmetry, the center of mass com_S for disk *S* is at the center of *S*, at $x = -R$ (as shown). Similarly, the center of mass com_C for composite plate *C* is at the center of *C*, at the origin (as shown). We then have the following:

Plate	Center of Mass	Location of com	Mass
P	com_P	$x_P = ?$	m_P
S	com_S	$x_S = -R$	m_S
C	com_C	$x_C = 0$	$m_C = m_S + m_P$

Assume that mass m_S of disk *S* is concentrated in a particle at $x_S = -R$, and mass m_P is concentrated in a particle at x_P (Fig. 9-4*d*). Next we use Eq. 9-2 to find the center of mass x_{S+P} of the two-particle system:

$$x_{S+P} = \frac{m_S x_S + m_P x_P}{m_S + m_P}. \tag{9-12}$$

Next note that the combination of disk *S* and plate *P* is composite plate *C*. Thus, the position x_{S+P} of com_{S+P} must coincide with the position x_C of com_C, which is at the origin; so $x_{S+P} = x_C = 0$. Substituting this into Eq. 9-12, we get

$$x_P = -x_S \frac{m_S}{m_P}. \tag{9-13}$$

We can relate these masses to the face areas of *S* and *P* by noting that

$$\begin{aligned} \text{mass} &= \text{density} \times \text{volume} \\ &= \text{density} \times \text{thickness} \times \text{area}. \end{aligned}$$

Then $$\frac{m_S}{m_P} = \frac{\text{density}_S}{\text{density}_P} \times \frac{\text{thickness}_S}{\text{thickness}_P} \times \frac{\text{area}_S}{\text{area}_P}.$$

Because the plate is uniform, the densities and thicknesses are equal; we are left with

$$\begin{aligned} \frac{m_S}{m_P} &= \frac{\text{area}_S}{\text{area}_P} = \frac{\text{area}_S}{\text{area}_C - \text{area}_S} \\ &= \frac{\pi R^2}{\pi(2R)^2 - \pi R^2} = \frac{1}{3}. \end{aligned}$$

Substituting this and $x_S = -R$ into Eq. 9-13, we have

$$x_P = \tfrac{1}{3}R. \qquad \text{(Answer)}$$

WileyPLUS Additional examples, video, and practice available at *WileyPLUS*

Checkpoint 1

The figure shows a uniform square plate from which four identical squares at the corners will be removed. (a) Where is the center of mass of the plate originally? Where is it after the removal of (b) square 1; (c) squares 1 and 2; (d) squares 1 and 3; (e) squares 1, 2, and 3; (f) all four squares? Answer in terms of quadrants, axes, or points (without calculation, of course).

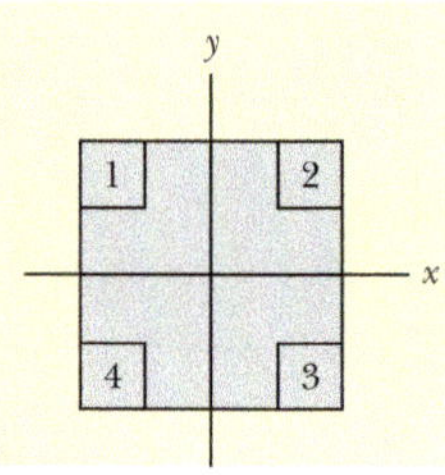

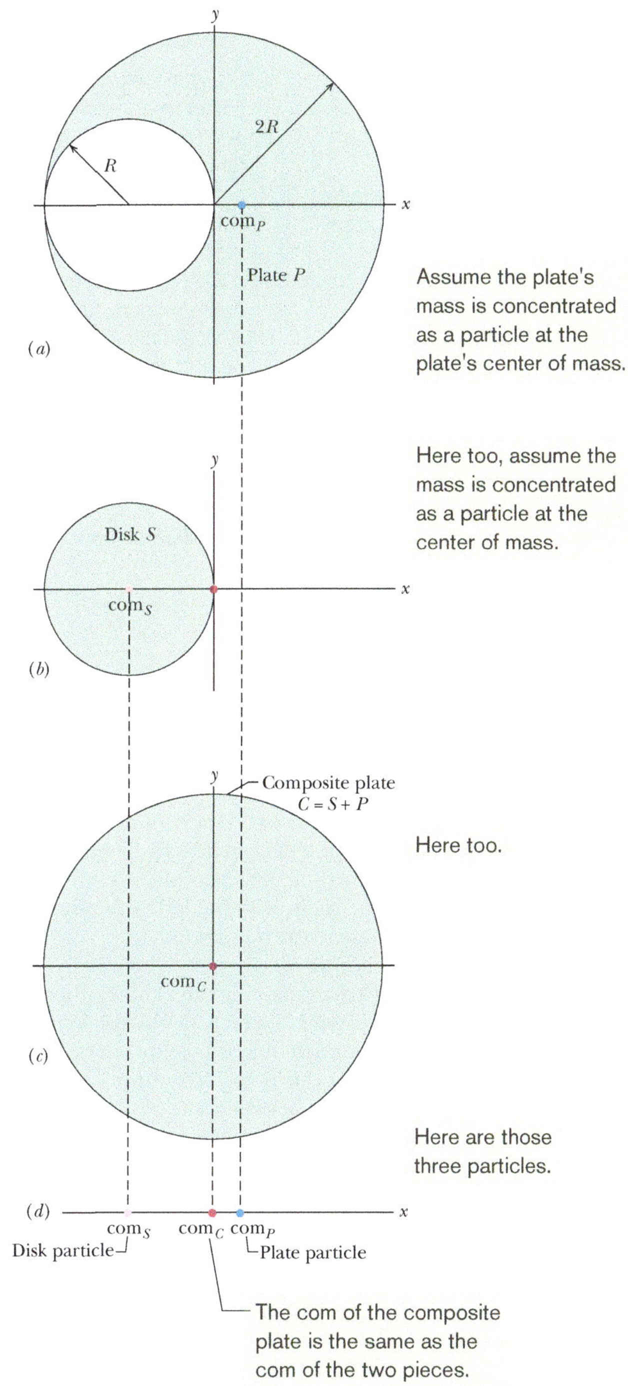

Figure 9-4 (*a*) Plate P is a metal plate of radius $2R$, with a circular hole of radius R. The center of mass of P is at point com_P. (*b*) Disk S. (*c*) Disk S has been put back into place to form a composite plate C. The center of mass com_S of disk S and the center of mass com_C of plate C are shown. (*d*) The center of mass com_{S+P} of the combination of S and P coincides with com_C, which is at $x = 0$.

9-2 NEWTON'S SECOND LAW FOR A SYSTEM OF PARTICLES

Learning Objectives

After reading this module, you should be able to . . .

9.04 Apply Newton's second law to a system of particles by relating the net force (of the forces acting on the particles) to the acceleration of the system's center of mass.

9.05 Apply the constant-acceleration equations to the motion of the individual particles in a system and to the motion of the system's center of mass.

9.06 Given the mass and velocity of the particles in a system, calculate the velocity of the system's center of mass.

9.07 Given the mass and acceleration of the particles in a system, calculate the acceleration of the system's center of mass.

9.08 Given the position of a system's center of mass as a function of time, determine the velocity of the center of mass.

9.09 Given the velocity of a system's center of mass as a function of time, determine the acceleration of the center of mass.

9.10 Calculate the change in the velocity of a com by integrating the com's acceleration function with respect to time.

9.11 Calculate a com's displacement by integrating the com's velocity function with respect to time.

9.12 When the particles in a two-particle system move without the system's com moving, relate the displacements of the particles and the velocities of the particles.

Key Idea

- The motion of the center of mass of any system of particles is governed by Newton's second law for a system of particles, which is

$$\vec{F}_{\text{net}} = M\vec{a}_{\text{com}}.$$

Here $\vec{F}_{\text{net}}$ is the net force of all the *external* forces acting on the system, M is the total mass of the system, and $\vec{a}_{\text{com}}$ is the acceleration of the system's center of mass.

Newton's Second Law for a System of Particles

Now that we know how to locate the center of mass of a system of particles, we discuss how external forces can move a center of mass. Let us start with a simple system of two billiard balls.

If you roll a cue ball at a second billiard ball that is at rest, you expect that the two-ball system will continue to have some forward motion after impact. You would be surprised, for example, if both balls came back toward you or if both moved to the right or to the left. You already have an intuitive sense that *something* continues to move forward.

What continues to move forward, its steady motion completely unaffected by the collision, is the center of mass of the two-ball system. If you focus on this point—which is always halfway between these bodies because they have identical masses—you can easily convince yourself by trial at a billiard table that this is so. No matter whether the collision is glancing, head-on, or somewhere in between, the center of mass continues to move forward, as if the collision had never occurred. Let us look into this center-of-mass motion in more detail.

Motion of a System's com. To do so, we replace the pair of billiard balls with a system of n particles of (possibly) different masses. We are interested not in the individual motions of these particles but *only* in the motion of the center of mass of the system. Although the center of mass is just a point, it moves like a particle whose mass is equal to the total mass of the system; we can assign a position, a velocity, and an acceleration to it. We state (and shall prove next) that the vector equation that governs the motion of the center of mass of such a system of particles is

$$\vec{F}_{\text{net}} = M\vec{a}_{\text{com}} \quad \text{(system of particles).} \tag{9-14}$$

This equation is Newton's second law for the motion of the center of mass of a system of particles. Note that its form is the same as the form of the equation

($\vec{F}_{net} = m\vec{a}$) for the motion of a single particle. However, the three quantities that appear in Eq. 9-14 must be evaluated with some care:

1. $\vec{F}_{net}$ is the net force of *all external forces* that act on the system. Forces on one part of the system from another part of the system (*internal forces*) are not included in Eq. 9-14.
2. M is the *total mass* of the system. We assume that no mass enters or leaves the system as it moves, so that M remains constant. The system is said to be **closed.**
3. $\vec{a}_{com}$ is the acceleration of the *center of mass* of the system. Equation 9-14 gives no information about the acceleration of any other point of the system.

Equation 9-14 is equivalent to three equations involving the components of $\vec{F}_{net}$ and $\vec{a}_{com}$ along the three coordinate axes. These equations are

$$F_{net,x} = Ma_{com,x} \qquad F_{net,y} = Ma_{com,y} \qquad F_{net,z} = Ma_{com,z}. \tag{9-15}$$

Billiard Balls. Now we can go back and examine the behavior of the billiard balls. Once the cue ball has begun to roll, no net external force acts on the (two-ball) system. Thus, because $\vec{F}_{net} = 0$, Eq. 9-14 tells us that $\vec{a}_{com} = 0$ also. Because acceleration is the rate of change of velocity, we conclude that the velocity of the center of mass of the system of two balls does not change. When the two balls collide, the forces that come into play are *internal* forces, on one ball from the other. Such forces do not contribute to the net force $\vec{F}_{net}$, which remains zero. Thus, the center of mass of the system, which was moving forward before the collision, must continue to move forward after the collision, with the same speed and in the same direction.

Solid Body. Equation 9-14 applies not only to a system of particles but also to a solid body, such as the bat of Fig. 9-1*b*. In that case, M in Eq. 9-14 is the mass of the bat and $\vec{F}_{net}$ is the gravitational force on the bat. Equation 9-14 then tells us that $\vec{a}_{com} = \vec{g}$. In other words, the center of mass of the bat moves as if the bat were a single particle of mass M, with force $\vec{F}_g$ acting on it.

Exploding Bodies. Figure 9-5 shows another interesting case. Suppose that at a fireworks display, a rocket is launched on a parabolic path. At a certain point, it explodes into fragments. If the explosion had not occurred, the rocket would have continued along the trajectory shown in the figure. The forces of the explosion are *internal* to the system (at first the system is just the rocket, and later it is its fragments); that is, they are forces on parts of the system from other parts. If we ignore air drag, the net *external* force $\vec{F}_{net}$ acting on the system is the gravitational force on the system, regardless of whether the rocket explodes. Thus, from Eq. 9-14, the acceleration $\vec{a}_{com}$ of the center of mass of the fragments (while they are in flight) remains equal to $\vec{g}$. This means that the center of mass of the fragments follows the same parabolic trajectory that the rocket would have followed had it not exploded.

Ballet Leap. When a ballet dancer leaps across the stage in a grand jeté, she raises her arms and stretches her legs out horizontally as soon as her feet leave the

Figure 9-5 A fireworks rocket explodes in flight. In the absence of air drag, the center of mass of the fragments would continue to follow the original parabolic path, until fragments began to hit the ground.

Figure 9-6 A grand jeté. (Based on *The Physics of Dance,* by Kenneth Laws, Schirmer Books, 1984.)

stage (Fig. 9-6). These actions shift her center of mass upward through her body. Although the shifting center of mass faithfully follows a parabolic path across the stage, its movement relative to the body decreases the height that is attained by her head and torso, relative to that of a normal jump. The result is that the head and torso follow a nearly horizontal path, giving an illusion that the dancer is floating.

Proof of Equation 9-14

Now let us prove this important equation. From Eq. 9-8 we have, for a system of n particles,

$$M\vec{r}_{\text{com}} = m_1\vec{r}_1 + m_2\vec{r}_2 + m_3\vec{r}_3 + \cdots + m_n\vec{r}_n, \tag{9-16}$$

in which M is the system's total mass and $\vec{r}_{\text{com}}$ is the vector locating the position of the system's center of mass.

Differentiating Eq. 9-16 with respect to time gives

$$M\vec{v}_{\text{com}} = m_1\vec{v}_1 + m_2\vec{v}_2 + m_3\vec{v}_3 + \cdots + m_n\vec{v}_n. \tag{9-17}$$

Here $\vec{v}_i$ $(= d\vec{r}_i/dt)$ is the velocity of the ith particle, and $\vec{v}_{\text{com}}$ $(= d\vec{r}_{\text{com}}/dt)$ is the velocity of the center of mass.

Differentiating Eq. 9-17 with respect to time leads to

$$M\vec{a}_{\text{com}} = m_1\vec{a}_1 + m_2\vec{a}_2 + m_3\vec{a}_3 + \cdots + m_n\vec{a}_n. \tag{9-18}$$

Here $\vec{a}_i$ $(= d\vec{v}_i/dt)$ is the acceleration of the ith particle, and $\vec{a}_{\text{com}}$ $(= d\vec{v}_{\text{com}}/dt)$ is the acceleration of the center of mass. Although the center of mass is just a geometrical point, it has a position, a velocity, and an acceleration, as if it were a particle.

From Newton's second law, $m_i\vec{a}_i$ is equal to the resultant force $\vec{F}_i$ that acts on the ith particle. Thus, we can rewrite Eq. 9-18 as

$$M\vec{a}_{\text{com}} = \vec{F}_1 + \vec{F}_2 + \vec{F}_3 + \cdots + \vec{F}_n. \tag{9-19}$$

Among the forces that contribute to the right side of Eq. 9-19 will be forces that the particles of the system exert on each other (internal forces) and forces exerted on the particles from outside the system (external forces). By Newton's third law, the internal forces form third-law force pairs and cancel out in the sum that appears on the right side of Eq. 9-19. What remains is the vector sum of all the *external* forces that act on the system. Equation 9-19 then reduces to Eq. 9-14, the relation that we set out to prove.

Checkpoint 2

Two skaters on frictionless ice hold opposite ends of a pole of negligible mass. An axis runs along it, with the origin at the center of mass of the two-skater system. One skater, Fred, weighs twice as much as the other skater, Ethel. Where do the skaters meet if (a) Fred pulls hand over hand along the pole so as to draw himself to Ethel, (b) Ethel pulls hand over hand to draw herself to Fred, and (c) both skaters pull hand over hand?

Sample Problem 9.03 Motion of the com of three particles

If the particles in a system all move together, the com moves with them—no trouble there. But what happens when they move in different directions with different accelerations? Here is an example.

The three particles in Fig. 9-7*a* are initially at rest. Each experiences an *external* force due to bodies outside the three-particle system. The directions are indicated, and the magnitudes are $F_1 = 6.0$ N, $F_2 = 12$ N, and $F_3 = 14$ N. What is the acceleration of the center of mass of the system, and in what direction does it move?

KEY IDEAS

The position of the center of mass is marked by a dot in the figure. We can treat the center of mass as if it were a real particle, with a mass equal to the system's total mass $M = 16$ kg. We can also treat the three external forces as if they act at the center of mass (Fig. 9-7*b*).

Calculations: We can now apply Newton's second law ($\vec{F}_{\text{net}} = m\vec{a}$) to the center of mass, writing

$$\vec{F}_{\text{net}} = M\vec{a}_{\text{com}} \tag{9-20}$$

or

$$\vec{F}_1 + \vec{F}_2 + \vec{F}_3 = M\vec{a}_{\text{com}}$$

so

$$\vec{a}_{\text{com}} = \frac{\vec{F}_1 + \vec{F}_2 + \vec{F}_3}{M}. \tag{9-21}$$

Equation 9-20 tells us that the acceleration $\vec{a}_{\text{com}}$ of the center of mass is in the same direction as the net external force $\vec{F}_{\text{net}}$ on the system (Fig. 9-7*b*). Because the particles are initially at rest, the center of mass must also be at rest. As the center of mass then begins to accelerate, it must move off in the common direction of $\vec{a}_{\text{com}}$ and $\vec{F}_{\text{net}}$.

We can evaluate the right side of Eq. 9-21 directly on a vector-capable calculator, or we can rewrite Eq. 9-21 in component form, find the components of $\vec{a}_{\text{com}}$, and then find $\vec{a}_{\text{com}}$. Along the x axis, we have

$$a_{\text{com},x} = \frac{F_{1x} + F_{2x} + F_{3x}}{M}$$
$$= \frac{-6.0\text{ N} + (12\text{ N})\cos 45° + 14\text{ N}}{16\text{ kg}} = 1.03\text{ m/s}^2.$$

Along the y axis, we have

$$a_{\text{com},y} = \frac{F_{1y} + F_{2y} + F_{3y}}{M}$$
$$= \frac{0 + (12\text{ N})\sin 45° + 0}{16\text{ kg}} = 0.530\text{ m/s}^2.$$

From these components, we find that $\vec{a}_{\text{com}}$ has the magnitude

$$a_{\text{com}} = \sqrt{(a_{\text{com},x})^2 + (a_{\text{com},y})^2}$$
$$= 1.16\text{ m/s}^2 \approx 1.2\text{ m/s}^2 \qquad \text{(Answer)}$$

and the angle (from the positive direction of the x axis)

$$\theta = \tan^{-1}\frac{a_{\text{com},y}}{a_{\text{com},x}} = 27°. \qquad \text{(Answer)}$$

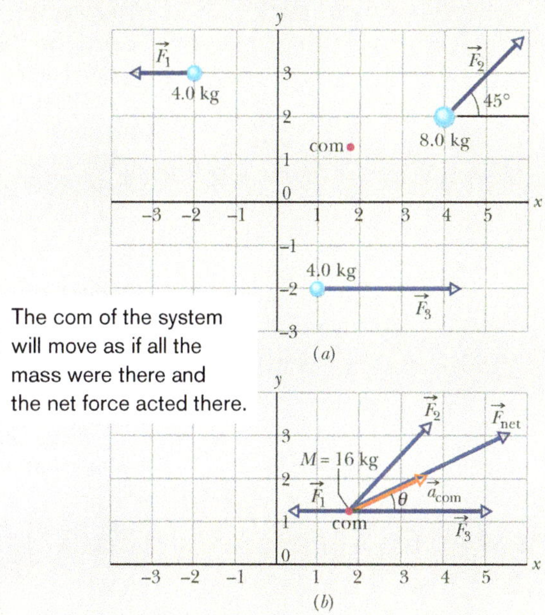

Figure 9-7 (*a*) Three particles, initially at rest in the positions shown, are acted on by the external forces shown. The center of mass (com) of the system is marked. (*b*) The forces are now transferred to the center of mass of the system, which behaves like a particle with a mass M equal to the total mass of the system. The net external force $\vec{F}_{\text{net}}$ and the acceleration $\vec{a}_{\text{com}}$ of the center of mass are shown.

WileyPLUS Additional examples, video, and practice available at *WileyPLUS*

9-3 LINEAR MOMENTUM

Learning Objectives

After reading this module, you should be able to . . .

9.13 Identify that momentum is a vector quantity and thus has both magnitude and direction and also components.

9.14 Calculate the (linear) momentum of a particle as the product of the particle's mass and velocity.

9.15 Calculate the change in momentum (magnitude and direction) when a particle changes its speed and direction of travel.

9.16 Apply the relationship between a particle's momentum and the (net) force acting on the particle.

9.17 Calculate the momentum of a system of particles as the product of the system's total mass and its center-of-mass velocity.

9.18 Apply the relationship between a system's center-of-mass momentum and the net force acting on the system.

Key Ideas

- For a single particle, we define a quantity $\vec{p}$ called its linear momentum as

$$\vec{p} = m\vec{v},$$

which is a vector quantity that has the same direction as the particle's velocity. We can write Newton's second law in terms of this momentum:

$$\vec{F}_{\text{net}} = \frac{d\vec{p}}{dt}.$$

- For a system of particles these relations become

$$\vec{P} = M\vec{v}_{\text{com}} \quad \text{and} \quad \vec{F}_{\text{net}} = \frac{d\vec{P}}{dt}.$$

Linear Momentum

Here we discuss only a single particle instead of a system of particles, in order to define two important quantities. Then we shall extend those definitions to systems of many particles.

The first definition concerns a familiar word—*momentum*—that has several meanings in everyday language but only a single precise meaning in physics and engineering. The **linear momentum** of a particle is a vector quantity $\vec{p}$ that is defined as

$$\vec{p} = m\vec{v} \quad \text{(linear momentum of a particle)}, \tag{9-22}$$

in which m is the mass of the particle and $\vec{v}$ is its velocity. (The adjective *linear* is often dropped, but it serves to distinguish $\vec{p}$ from *angular* momentum, which is introduced in Chapter 11 and which is associated with rotation.) Since m is always a positive scalar quantity, Eq. 9-22 tells us that $\vec{p}$ and $\vec{v}$ have the same direction. From Eq. 9-22, the SI unit for momentum is the kilogram-meter per second (kg · m/s).

Force and Momentum. Newton expressed his second law of motion in terms of momentum:

> The time rate of change of the momentum of a particle is equal to the net force acting on the particle and is in the direction of that force.

In equation form this becomes

$$\vec{F}_{\text{net}} = \frac{d\vec{p}}{dt}. \tag{9-23}$$

In words, Eq. 9-23 says that the net external force $\vec{F}_{\text{net}}$ on a particle changes the particle's linear momentum $\vec{p}$. Conversely, the linear momentum can be changed only by a net external force. If there is no net external force, $\vec{p}$ *cannot* change. As we shall see in Module 9-5, this last fact can be an extremely powerful tool in solving problems.

Manipulating Eq. 9-23 by substituting for $\vec{p}$ from Eq. 9-22 gives, for constant mass m,

$$\vec{F}_{\text{net}} = \frac{d\vec{p}}{dt} = \frac{d}{dt}(m\vec{v}) = m\frac{d\vec{v}}{dt} = m\vec{a}.$$

Thus, the relations $\vec{F}_{\text{net}} = d\vec{p}/dt$ and $\vec{F}_{\text{net}} = m\vec{a}$ are equivalent expressions of Newton's second law of motion for a particle.

Checkpoint 3

The figure gives the magnitude p of the linear momentum versus time t for a particle moving along an axis. A force directed along the axis acts on the particle. (a) Rank the four regions indicated according to the magnitude of the force, greatest first. (b) In which region is the particle slowing?

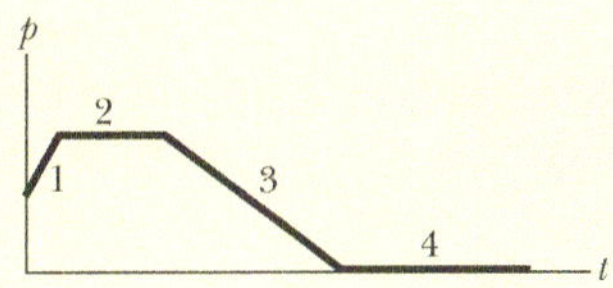

The Linear Momentum of a System of Particles

Let's extend the definition of linear momentum to a system of particles. Consider a system of n particles, each with its own mass, velocity, and linear momentum. The particles may interact with each other, and external forces may act on them. The system as a whole has a total linear momentum $\vec{P}$, which is defined to be the vector sum of the individual particles' linear momenta. Thus,

$$\begin{aligned}\vec{P} &= \vec{p}_1 + \vec{p}_2 + \vec{p}_3 + \cdots + \vec{p}_n \\ &= m_1\vec{v}_1 + m_2\vec{v}_2 + m_3\vec{v}_3 + \cdots + m_n\vec{v}_n.\end{aligned} \tag{9-24}$$

If we compare this equation with Eq. 9-17, we see that

$$\vec{P} = M\vec{v}_{\text{com}} \quad \text{(linear momentum, system of particles)}, \tag{9-25}$$

which is another way to define the linear momentum of a system of particles:

> The linear momentum of a system of particles is equal to the product of the total mass M of the system and the velocity of the center of mass.

Force and Momentum. If we take the time derivative of Eq. 9-25 (the velocity can change but not the mass), we find

$$\frac{d\vec{P}}{dt} = M\frac{d\vec{v}_{\text{com}}}{dt} = M\vec{a}_{\text{com}}. \tag{9-26}$$

Comparing Eqs. 9-14 and 9-26 allows us to write Newton's second law for a system of particles in the equivalent form

$$\vec{F}_{\text{net}} = \frac{d\vec{P}}{dt} \quad \text{(system of particles)}, \tag{9-27}$$

where $\vec{F}_{\text{net}}$ is the net external force acting on the system. This equation is the generalization of the single-particle equation $\vec{F}_{\text{net}} = d\vec{p}/dt$ to a system of many particles. In words, the equation says that the net external force $\vec{F}_{\text{net}}$ on a system of particles changes the linear momentum $\vec{P}$ of the system. Conversely, the linear momentum can be changed only by a net external force. If there is no net external force, $\vec{P}$ *cannot* change. Again, this fact gives us an extremely powerful tool for solving problems.

9-4 COLLISION AND IMPULSE

Learning Objectives

After reading this module, you should be able to . . .

9.19 Identify that impulse is a vector quantity and thus has both magnitude and direction and also components.

9.20 Apply the relationship between impulse and momentum change.

9.21 Apply the relationship between impulse, average force, and the time interval taken by the impulse.

9.22 Apply the constant-acceleration equations to relate impulse to average force.

9.23 Given force as a function of time, calculate the impulse (and thus also the momentum change) by integrating the function.

9.24 Given a graph of force versus time, calculate the impulse (and thus also the momentum change) by graphical integration.

9.25 In a continuous series of collisions by projectiles, calculate the average force on the target by relating it to the rate at which mass collides and to the velocity change experienced by each projectile.

Key Ideas

- Applying Newton's second law in momentum form to a particle-like body involved in a collision leads to the impulse–linear momentum theorem:

$$\vec{p}_f - \vec{p}_i = \Delta\vec{p} = \vec{J},$$

where $\vec{p}_f - \vec{p}_i = \Delta\vec{p}$ is the change in the body's linear momentum, and $\vec{J}$ is the impulse due to the force $\vec{F}(t)$ exerted on the body by the other body in the collision:

$$\vec{J} = \int_{t_i}^{t_f} \vec{F}(t)\, dt.$$

- If F_{avg} is the average magnitude of $\vec{F}(t)$ during the collision and Δt is the duration of the collision, then for one-dimensional motion

$$J = F_{avg}\, \Delta t.$$

- When a steady stream of bodies, each with mass m and speed v, collides with a body whose position is fixed, the average force on the fixed body is

$$F_{avg} = -\frac{n}{\Delta t}\Delta p = -\frac{n}{\Delta t} m\, \Delta v,$$

where $n/\Delta t$ is the rate at which the bodies collide with the fixed body, and Δv is the change in velocity of each colliding body. This average force can also be written as

$$F_{avg} = -\frac{\Delta m}{\Delta t}\Delta v,$$

where $\Delta m/\Delta t$ is the rate at which mass collides with the fixed body. The change in velocity is $\Delta v = -v$ if the bodies stop upon impact and $\Delta v = -2v$ if they bounce directly backward with no change in their speed.

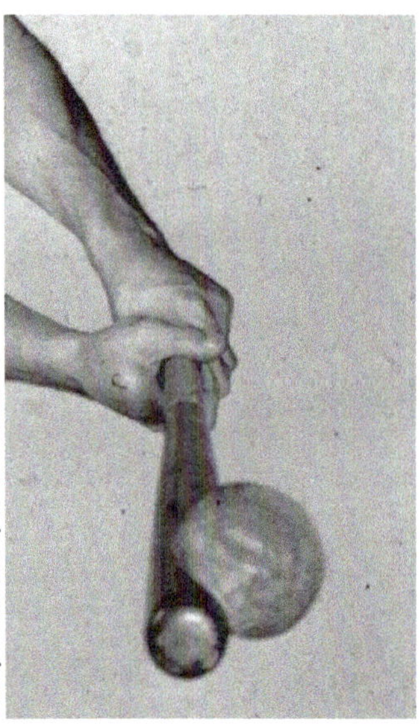

Photo by Harold E. Edgerton. © The Harold and Esther Edgerton Family Trust, courtesy of Palm Press, Inc.

The collision of a ball with a bat collapses part of the ball.

Collision and Impulse

The momentum $\vec{p}$ of any particle-like body cannot change unless a net external force changes it. For example, we could push on the body to change its momentum. More dramatically, we could arrange for the body to collide with a baseball bat. In such a *collision* (or *crash*), the external force on the body is brief, has large magnitude, and suddenly changes the body's momentum. Collisions occur commonly in our world, but before we get to them, we need to consider a simple collision in which a moving particle-like body (a *projectile*) collides with some other body (a *target*).

Single Collision

Let the projectile be a ball and the target be a bat. The collision is brief, and the ball experiences a force that is great enough to slow, stop, or even reverse its motion. Figure 9-8 depicts the collision at one instant. The ball experiences a force $\vec{F}(t)$ that varies during the collision and changes the linear momentum $\vec{p}$ of the ball. That change is related to the force by Newton's second law written in the form $\vec{F} = d\vec{p}/dt$. By rearranging this second-law expression, we see that, in time interval dt, the change in the ball's momentum is

$$d\vec{p} = \vec{F}(t)\, dt. \tag{9-28}$$

We can find the net change in the ball's momentum due to the collision if we integrate both sides of Eq. 9-28 from a time t_i just before the collision to a time t_f just after the collision:

$$\int_{t_i}^{t_f} d\vec{p} = \int_{t_i}^{t_f} \vec{F}(t)\, dt. \tag{9-29}$$

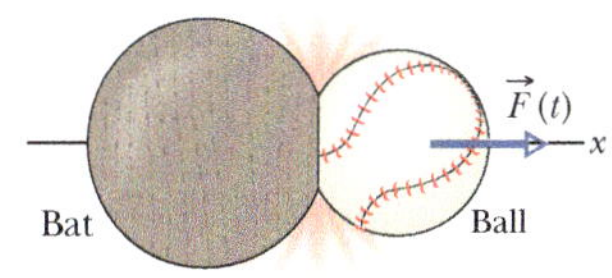

Figure 9-8 Force $\vec{F}(t)$ acts on a ball as the ball and a bat collide.

The left side of this equation gives us the change in momentum: $\vec{p}_f - \vec{p}_i = \Delta\vec{p}$. The right side, which is a measure of both the magnitude and the duration of the collision force, is called the **impulse** $\vec{J}$ of the collision:

$$\vec{J} = \int_{t_i}^{t_f} \vec{F}(t)\, dt \quad \text{(impulse defined).} \tag{9-30}$$

Thus, the change in an object's momentum is equal to the impulse on the object:

$$\Delta\vec{p} = \vec{J} \quad \text{(linear momentum–impulse theorem).} \tag{9-31}$$

This expression can also be written in the vector form

$$\vec{p}_f - \vec{p}_i = \vec{J} \tag{9-32}$$

and in such component forms as

$$\Delta p_x = J_x \tag{9-33}$$

and

$$p_{fx} - p_{ix} = \int_{t_i}^{t_f} F_x\, dt. \tag{9-34}$$

Integrating the Force. If we have a function for $\vec{F}(t)$, we can evaluate $\vec{J}$ (and thus the change in momentum) by integrating the function. If we have a plot of $\vec{F}$ versus time t, we can evaluate $\vec{J}$ by finding the area between the curve and the t axis, such as in Fig. 9-9*a*. In many situations we do not know how the force varies with time but we do know the average magnitude F_{avg} of the force and the duration Δt $(= t_f - t_i)$ of the collision. Then we can write the magnitude of the impulse as

$$J = F_{avg}\,\Delta t. \tag{9-35}$$

The average force is plotted versus time as in Fig. 9-9*b*. The area under that curve is equal to the area under the curve for the actual force $F(t)$ in Fig. 9-9*a* because both areas are equal to impulse magnitude J.

Instead of the ball, we could have focused on the bat in Fig. 9-8. At any instant, Newton's third law tells us that the force on the bat has the same magnitude but the opposite direction as the force on the ball. From Eq. 9-30, this means that the impulse on the bat has the same magnitude but the opposite direction as the impulse on the ball.

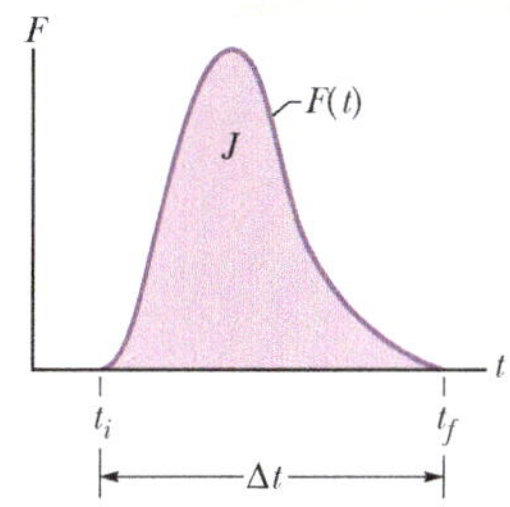

(*a*)

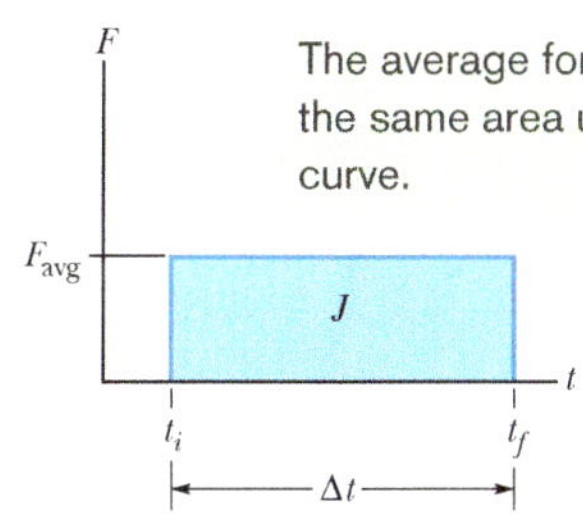

(*b*)

Figure 9-9 (*a*) The curve shows the magnitude of the time-varying force $F(t)$ that acts on the ball in the collision of Fig. 9-8. The area under the curve is equal to the magnitude of the impulse $\vec{J}$ on the ball in the collision. (*b*) The height of the rectangle represents the average force F_{avg} acting on the ball over the time interval Δt. The area within the rectangle is equal to the area under the curve in (*a*) and thus is also equal to the magnitude of the impulse $\vec{J}$ in the collision.

Checkpoint 4

A paratrooper whose chute fails to open lands in snow; he is hurt slightly. Had he landed on bare ground, the stopping time would have been 10 times shorter and the collision lethal. Does the presence of the snow increase, decrease, or leave unchanged the values of (a) the paratrooper's change in momentum, (b) the impulse stopping the paratrooper, and (c) the force stopping the paratrooper?

Series of Collisions

Now let's consider the force on a body when it undergoes a series of identical, repeated collisions. For example, as a prank, we might adjust one of those machines that fire tennis balls to fire them at a rapid rate directly at a wall. Each collision would produce a force on the wall, but that is not the force we are seeking. We

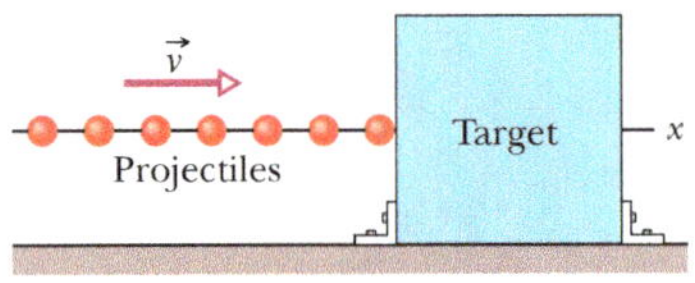

Figure 9-10 A steady stream of projectiles, with identical linear momenta, collides with a target, which is fixed in place. The average force F_{avg} on the target is to the right and has a magnitude that depends on the rate at which the projectiles collide with the target or, equivalently, the rate at which mass collides with the target.

want the average force F_{avg} on the wall during the bombardment—that is, the average force during a large number of collisions.

In Fig. 9-10, a steady stream of projectile bodies, with identical mass m and linear momenta $m\vec{v}$, moves along an x axis and collides with a target body that is fixed in place. Let n be the number of projectiles that collide in a time interval Δt. Because the motion is along only the x axis, we can use the components of the momenta along that axis. Thus, each projectile has initial momentum mv and undergoes a change Δp in linear momentum because of the collision. The total change in linear momentum for n projectiles during interval Δt is $n\,\Delta p$. The resulting impulse $\vec{J}$ on the target during Δt is along the x axis and has the same magnitude of $n\,\Delta p$ but is in the opposite direction. We can write this relation in component form as

$$J = -n\,\Delta p, \qquad (9\text{-}36)$$

where the minus sign indicates that J and Δp have opposite directions.

Average Force. By rearranging Eq. 9-35 and substituting Eq. 9-36, we find the average force F_{avg} acting on the target during the collisions:

$$F_{avg} = \frac{J}{\Delta t} = -\frac{n}{\Delta t}\Delta p = -\frac{n}{\Delta t} m\,\Delta v. \qquad (9\text{-}37)$$

This equation gives us F_{avg} in terms of $n/\Delta t$, the rate at which the projectiles collide with the target, and Δv, the change in the velocity of those projectiles.

Velocity Change. If the projectiles stop upon impact, then in Eq. 9-37 we can substitute, for Δv,

$$\Delta v = v_f - v_i = 0 - v = -v, \qquad (9\text{-}38)$$

where v_i $(= v)$ and v_f $(= 0)$ are the velocities before and after the collision, respectively. If, instead, the projectiles bounce (rebound) directly backward from the target with no change in speed, then $v_f = -v$ and we can substitute

$$\Delta v = v_f - v_i = -v - v = -2v. \qquad (9\text{-}39)$$

In time interval Δt, an amount of mass $\Delta m = nm$ collides with the target. With this result, we can rewrite Eq. 9-37 as

$$F_{avg} = -\frac{\Delta m}{\Delta t}\Delta v. \qquad (9\text{-}40)$$

This equation gives the average force F_{avg} in terms of $\Delta m/\Delta t$, the rate at which mass collides with the target. Here again we can substitute for Δv from Eq. 9-38 or 9-39 depending on what the projectiles do.

Checkpoint 5

The figure shows an overhead view of a ball bouncing from a vertical wall without any change in its speed. Consider the change $\Delta\vec{p}$ in the ball's linear momentum. (a) Is Δp_x positive, negative, or zero? (b) Is Δp_y positive, negative, or zero? (c) What is the direction of $\Delta\vec{p}$?

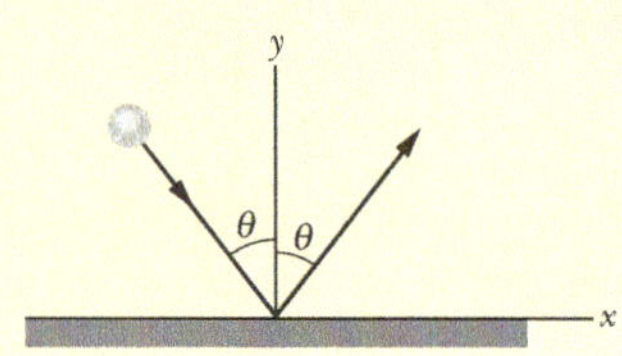

Sample Problem 9.04 Two-dimensional impulse, race car–wall collision

Race car–wall collision. Figure 9-11*a* is an overhead view of the path taken by a race car driver as his car collides with the racetrack wall. Just before the collision, he is traveling at speed $v_i = 70$ m/s along a straight line at 30° from the wall. Just after the collision, he is traveling at speed $v_f = 50$ m/s along a straight line at 10° from the wall. His mass m is 80 kg.

(a) What is the impulse $\vec{J}$ on the driver due to the collision?

KEY IDEAS

We can treat the driver as a particle-like body and thus apply the physics of this module. However, we cannot calculate $\vec{J}$ directly from Eq. 9-30 because we do not know anything about the force $\vec{F}(t)$ on the driver during the collision. That is, we do not have a function of $\vec{F}(t)$ or a plot for it and thus cannot integrate to find $\vec{J}$. However, we *can* find $\vec{J}$ from the change in the driver's linear momentum $\vec{p}$ via Eq. 9-32 ($\vec{J} = \vec{p}_f - \vec{p}_i$).

Calculations: Figure 9-11*b* shows the driver's momentum $\vec{p}_i$ before the collision (at angle 30° from the positive x direction) and his momentum $\vec{p}_f$ after the collision (at angle −10°). From Eqs. 9-32 and 9-22 ($\vec{p} = m\vec{v}$), we can write

$$\vec{J} = \vec{p}_f - \vec{p}_i = m\vec{v}_f - m\vec{v}_i = m(\vec{v}_f - \vec{v}_i). \qquad (9\text{-}41)$$

We could evaluate the right side of this equation directly on a vector-capable calculator because we know m is 80 kg, $\vec{v}_f$ is 50 m/s at −10°, and $\vec{v}_i$ is 70 m/s at 30°. Instead, here we evaluate Eq. 9-41 in component form.

x component: Along the x axis we have

$$\begin{aligned} J_x &= m(v_{fx} - v_{ix}) \\ &= (80\text{ kg})[(50\text{ m/s})\cos(-10°) - (70\text{ m/s})\cos 30°] \\ &= -910\text{ kg}\cdot\text{m/s}. \end{aligned}$$

y component: Along the y axis,

$$\begin{aligned} J_y &= m(v_{fy} - v_{iy}) \\ &= (80\text{ kg})[(50\text{ m/s})\sin(-10°) - (70\text{ m/s})\sin 30°] \\ &= -3495\text{ kg}\cdot\text{m/s} \approx -3500\text{ kg}\cdot\text{m/s}. \end{aligned}$$

Impulse: The impulse is then

$$\vec{J} = (-910\hat{\mathrm{i}} - 3500\hat{\mathrm{j}})\text{ kg}\cdot\text{m/s}, \qquad \text{(Answer)}$$

which means the impulse magnitude is

$$J = \sqrt{J_x^2 + J_y^2} = 3616\text{ kg}\cdot\text{m/s} \approx 3600\text{ kg}\cdot\text{m/s}.$$

The angle of $\vec{J}$ is given by

$$\theta = \tan^{-1}\frac{J_y}{J_x}, \qquad \text{(Answer)}$$

which a calculator evaluates as 75.4°. Recall that the physically correct result of an inverse tangent might be the displayed answer plus 180°. We can tell which is correct here by drawing the components of $\vec{J}$ (Fig. 9-11*c*). We find that θ is actually 75.4° + 180° = 255.4°, which we can write as

$$\theta = -105°. \qquad \text{(Answer)}$$

(b) The collision lasts for 14 ms. What is the magnitude of the average force on the driver during the collision?

KEY IDEA

From Eq. 9-35 ($J = F_{avg}\,\Delta t$), the magnitude F_{avg} of the average force is the ratio of the impulse magnitude J to the duration Δt of the collision.

Calculations: We have

$$\begin{aligned} F_{avg} &= \frac{J}{\Delta t} = \frac{3616\text{ kg}\cdot\text{m/s}}{0.014\text{ s}} \\ &= 2.583 \times 10^5\text{ N} \approx 2.6 \times 10^5\text{ N}. \end{aligned} \qquad \text{(Answer)}$$

Using $F = ma$ with $m = 80$ kg, you can show that the magnitude of the driver's average acceleration during the collision is about $3.22 \times 10^3\text{ m/s}^2 = 329g$, which is fatal.

Surviving: Mechanical engineers attempt to reduce the chances of a fatality by designing and building racetrack walls with more "give," so that a collision lasts longer. For example, if the collision here lasted 10 times longer and the other data remained the same, the magnitudes of the average force and average acceleration would be 10 times less and probably survivable.

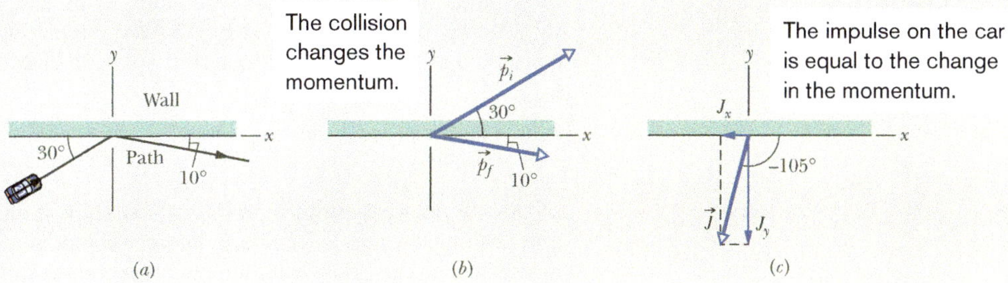

Figure 9-11 (*a*) Overhead view of the path taken by a race car and its driver as the car slams into the racetrack wall. (*b*) The initial momentum $\vec{p}_i$ and final momentum $\vec{p}_f$ of the driver. (*c*) The impulse $\vec{J}$ on the driver during the collision.

WileyPLUS Additional examples, video, and practice available at *WileyPLUS*

9-5 CONSERVATION OF LINEAR MOMENTUM

Learning Objectives

After reading this module, you should be able to . . .

9.26 For an isolated system of particles, apply the conservation of linear momenta to relate the initial momenta of the particles to their momenta at a later instant.

9.27 Identify that the conservation of linear momentum can be done along an individual axis by using components along that axis, *provided* that there is no net external force component along that axis.

Key Ideas

- If a system is closed and isolated so that no net *external* force acts on it, then the linear momentum $\vec{P}$ must be constant even if there are internal changes:

$$\vec{P} = \text{constant} \quad \text{(closed, isolated system).}$$

- This conservation of linear momentum can also be written in terms of the system's initial momentum and its momentum at some later instant:

$$\vec{P}_i = \vec{P}_f \quad \text{(closed, isolated system),}$$

Conservation of Linear Momentum

Suppose that the net external force $\vec{F}_{\text{net}}$ (and thus the net impulse $\vec{J}$) acting on a system of particles is zero (the system is isolated) and that no particles leave or enter the system (the system is closed). Putting $\vec{F}_{\text{net}} = 0$ in Eq. 9-27 then yields $d\vec{P}/dt = 0$, which means that

$$\vec{P} = \text{constant} \quad \text{(closed, isolated system).} \tag{9-42}$$

In words,

> If no net external force acts on a system of particles, the total linear momentum $\vec{P}$ of the system cannot change.

This result is called the **law of conservation of linear momentum** and is an extremely powerful tool in solving problems. In the homework we usually write the law as

$$\vec{P}_i = \vec{P}_f \quad \text{(closed, isolated system).} \tag{9-43}$$

In words, this equation says that, for a closed, isolated system,

$$\begin{pmatrix}\text{total linear momentum} \\ \text{at some initial time } t_i\end{pmatrix} = \begin{pmatrix}\text{total linear momentum} \\ \text{at some later time } t_f\end{pmatrix}.$$

Caution: Momentum should not be confused with energy. In the sample problems of this module, momentum is conserved but energy is definitely not.

Equations 9-42 and 9-43 are vector equations and, as such, each is equivalent to three equations corresponding to the conservation of linear momentum in three mutually perpendicular directions as in, say, an xyz coordinate system. Depending on the forces acting on a system, linear momentum might be conserved in one or two directions but not in all directions. However,

> If the component of the net *external* force on a closed system is zero along an axis, then the component of the linear momentum of the system along that axis cannot change.

In a homework problem, how can you know if linear momentum can be conserved along, say, an x axis? Check the force components along that axis. If the net of any such components is zero, then the conservation applies. As an example, suppose that you toss a grapefruit across a room. During its flight, the only external force acting on the grapefruit (which we take as the system) is the gravitational force $\vec{F}_g$, which is directed vertically downward. Thus, the vertical component of the linear momentum of the grapefruit changes, but since no horizontal external force acts on the grapefruit, the horizontal component of the linear momentum cannot change.

Note that we focus on the external forces acting on a closed system. Although internal forces can change the linear momentum of portions of the system, they cannot change the total linear momentum of the entire system. For example, there are plenty of forces acting between the organs of your body, but they do not propel you across the room (thankfully).

The sample problems in this module involve explosions that are either one-dimensional (meaning that the motions before and after the explosion are along a single axis) or two-dimensional (meaning that they are in a plane containing two axes). In the following modules we consider collisions.

Checkpoint 6

An initially stationary device lying on a frictionless floor explodes into two pieces, which then slide across the floor, one of them in the positive x direction. (a) What is the sum of the momenta of the two pieces after the explosion? (b) Can the second piece move at an angle to the x axis? (c) What is the direction of the momentum of the second piece?

Sample Problem 9.05 One-dimensional explosion, relative velocity, space hauler

One-dimensional explosion: Figure 9-12*a* shows a space hauler and cargo module, of total mass M, traveling along an x axis in deep space. They have an initial velocity $\vec{v}_i$ of magnitude 2100 km/h relative to the Sun. With a small explosion, the hauler ejects the cargo module, of mass $0.20M$ (Fig. 9-12*b*). The hauler then travels 500 km/h faster than the module along the x axis; that is, the relative speed v_{rel} between the hauler and the module is 500 km/h. What then is the velocity $\vec{v}_{HS}$ of the hauler relative to the Sun?

KEY IDEA

Because the hauler–module system is closed and isolated, its total linear momentum is conserved; that is,

$$\vec{P}_i = \vec{P}_f, \tag{9-44}$$

where the subscripts i and f refer to values before and after the ejection, respectively. (We need to be careful here: Although the momentum of the *system* does not change, the momenta of the hauler and module certainly do.)

Calculations: Because the motion is along a single axis, we can write momenta and velocities in terms of their x components, using a sign to indicate direction. Before the ejection, we have

$$P_i = Mv_i. \tag{9-45}$$

Let v_{MS} be the velocity of the ejected module relative to the Sun. The total linear momentum of the system after the ejection is then

$$P_f = (0.20M)v_{MS} + (0.80M)v_{HS}, \tag{9-46}$$

where the first term on the right is the linear momentum of the module and the second term is that of the hauler.

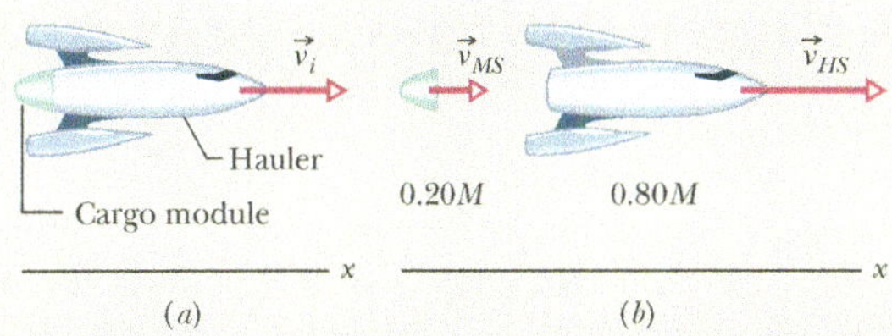

Figure 9-12 (*a*) A space hauler, with a cargo module, moving at initial velocity $\vec{v}_i$. (*b*) The hauler has ejected the cargo module. Now the velocities relative to the Sun are $\vec{v}_{MS}$ for the module and $\vec{v}_{HS}$ for the hauler.

We can relate the v_{MS} to the known velocities with

$$\begin{pmatrix}\text{velocity of}\\ \text{hauler relative}\\ \text{to Sun}\end{pmatrix} = \begin{pmatrix}\text{velocity of}\\ \text{hauler relative}\\ \text{to module}\end{pmatrix} + \begin{pmatrix}\text{velocity of}\\ \text{module relative}\\ \text{to Sun}\end{pmatrix}.$$

In symbols, this gives us

$$v_{HS} = v_{rel} + v_{MS} \tag{9-47}$$

or

$$v_{MS} = v_{HS} - v_{rel}.$$

Substituting this expression for v_{MS} into Eq. 9-46, and then substituting Eqs. 9-45 and 9-46 into Eq. 9-44, we find

$$Mv_i = 0.20M(v_{HS} - v_{rel}) + 0.80Mv_{HS},$$

which gives us

$$v_{HS} = v_i + 0.20v_{rel},$$

or

$$v_{HS} = 2100 \text{ km/h} + (0.20)(500 \text{ km/h})$$
$$= 2200 \text{ km/h}. \quad \text{(Answer)}$$

WILEY PLUS Additional examples, video, and practice available at *WileyPLUS*

Sample Problem 9.06 Two-dimensional explosion, momentum, coconut

Two-dimensional explosion: A firecracker placed inside a coconut of mass M, initially at rest on a frictionless floor, blows the coconut into three pieces that slide across the floor. An overhead view is shown in Fig. 9-13a. Piece C, with mass $0.30M$, has final speed $v_{fC} = 5.0$ m/s.

(a) What is the speed of piece B, with mass $0.20M$?

KEY IDEA

First we need to see whether linear momentum is conserved. We note that (1) the coconut and its pieces form a closed system, (2) the explosion forces are internal to that system, and (3) no net external force acts on the system. Therefore, the linear momentum of the system is conserved. (We need to be careful here: Although the momentum of the system does not change, the momenta of the pieces certainly do.)

Calculations: To get started, we superimpose an xy coordinate system as shown in Fig. 9-13b, with the negative direction of the x axis coinciding with the direction of $\vec{v}_{fA}$. The x axis is at 80° with the direction of $\vec{v}_{fC}$ and 50° with the direction of $\vec{v}_{fB}$.

Linear momentum is conserved separately along each axis. Let's use the y axis and write

$$P_{iy} = P_{fy}, \qquad (9\text{-}48)$$

where subscript i refers to the initial value (before the explosion), and subscript y refers to the y component of $\vec{P}_i$ or $\vec{P}_f$.

The component P_{iy} of the initial linear momentum is zero, because the coconut is initially at rest. To get an expression for P_{fy}, we find the y component of the final linear momentum of each piece, using the y-component version of Eq. 9-22 ($p_y = mv_y$):

$$p_{fA,y} = 0,$$
$$p_{fB,y} = -0.20Mv_{fB,y} = -0.20Mv_{fB}\sin 50°,$$
$$p_{fC,y} = 0.30Mv_{fC,y} = 0.30Mv_{fC}\sin 80°.$$

(Note that $p_{fA,y} = 0$ because of our nice choice of axes.) Equation 9-48 can now be written as

$$P_{iy} = P_{fy} = p_{fA,y} + p_{fB,y} + p_{fC,y}.$$

Then, with $v_{fC} = 5.0$ m/s, we have

$$0 = 0 - 0.20Mv_{fB}\sin 50° + (0.30M)(5.0\text{ m/s})\sin 80°,$$

from which we find

$$v_{fB} = 9.64\text{ m/s} \approx 9.6\text{ m/s}. \qquad \text{(Answer)}$$

(b) What is the speed of piece A?

Calculations: Linear momentum is also conserved along the x axis because there is no net external force acting on the coconut and pieces along that axis. Thus we have

$$P_{ix} = P_{fx}, \qquad (9\text{-}49)$$

where $P_{ix} = 0$ because the coconut is initially at rest. To get P_{fx}, we find the x components of the final momenta, using the fact that piece A must have a mass of $0.50M$ ($= M - 0.20M - 0.30M$):

$$p_{fA,x} = -0.50Mv_{fA},$$
$$p_{fB,x} = 0.20Mv_{fB,x} = 0.20Mv_{fB}\cos 50°,$$
$$p_{fC,x} = 0.30Mv_{fC,x} = 0.30Mv_{fC}\cos 80°.$$

Equation 9-49 for the conservation of momentum along the x axis can now be written as

$$P_{ix} = P_{fx} = p_{fA,x} + p_{fB,x} + p_{fC,x}.$$

Then, with $v_{fC} = 5.0$ m/s and $v_{fB} = 9.64$ m/s, we have

$$0 = -0.50Mv_{fA} + 0.20M(9.64\text{ m/s})\cos 50° + 0.30M(5.0\text{ m/s})\cos 80°,$$

from which we find

$$v_{fA} = 3.0\text{ m/s}. \qquad \text{(Answer)}$$

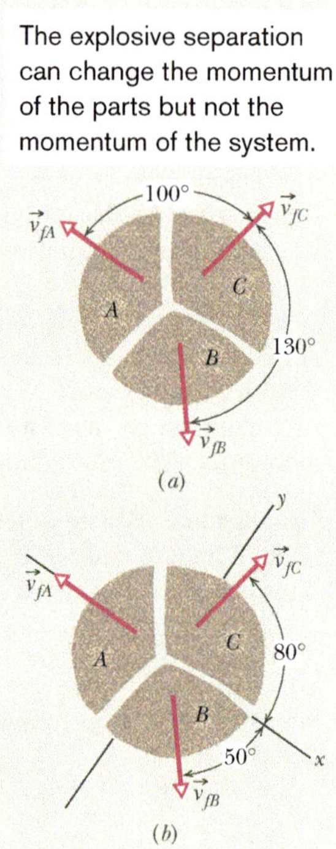

Figure 9-13 Three pieces of an exploded coconut move off in three directions along a frictionless floor. (a) An overhead view of the event. (b) The same with a two-dimensional axis system imposed.

Additional examples, video, and practice available at *WileyPLUS*

9-6 MOMENTUM AND KINETIC ENERGY IN COLLISIONS

Learning Objectives

After reading this module, you should be able to . . .

9.28 Distinguish between elastic collisions, inelastic collisions, and completely inelastic collisions.

9.29 Identify a one-dimensional collision as one where the objects move along a single axis, both before and after the collision.

9.30 Apply the conservation of momentum for an isolated one-dimensional collision to relate the initial momenta of the objects to their momenta after the collision.

9.31 Identify that in an isolated system, the momentum and velocity of the center of mass are not changed even if the objects collide.

Key Ideas

- In an inelastic collision of two bodies, the kinetic energy of the two-body system is not conserved. If the system is closed and isolated, the total linear momentum of the system *must* be conserved, which we can write in vector form as

$$\vec{p}_{1i} + \vec{p}_{2i} = \vec{p}_{1f} + \vec{p}_{2f},$$

where subscripts i and f refer to values just before and just after the collision, respectively.

- If the motion of the bodies is along a single axis, the collision is one-dimensional and we can write the equation in terms of velocity components along that axis:

$$m_1 v_{1i} + m_2 v_{2i} = m_1 v_{1f} + m_2 v_{2f}.$$

- If the bodies stick together, the collision is a completely inelastic collision and the bodies have the same final velocity V (because they *are* stuck together).

- The center of mass of a closed, isolated system of two colliding bodies is not affected by a collision. In particular, the velocity $\vec{v}_{com}$ of the center of mass cannot be changed by the collision.

Momentum and Kinetic Energy in Collisions

In Module 9-4, we considered the collision of two particle-like bodies but focused on only one of the bodies at a time. For the next several modules we switch our focus to the system itself, with the assumption that the system is closed and isolated. In Module 9-5, we discussed a rule about such a system: The total linear momentum $\vec{P}$ of the system cannot change because there is no net external force to change it. This is a very powerful rule because it can allow us to determine the results of a collision *without* knowing the details of the collision (such as how much damage is done).

We shall also be interested in the total kinetic energy of a system of two colliding bodies. If that total happens to be unchanged by the collision, then the kinetic energy of the system is *conserved* (it is the same before and after the collision). Such a collision is called an **elastic collision.** In everyday collisions of common bodies, such as two cars or a ball and a bat, some energy is always transferred from kinetic energy to other forms of energy, such as thermal energy or energy of sound. Thus, the kinetic energy of the system is *not* conserved. Such a collision is called an **inelastic collision.**

However, in some situations, we can *approximate* a collision of common bodies as elastic. Suppose that you drop a Superball onto a hard floor. If the collision between the ball and floor (or Earth) were elastic, the ball would lose no kinetic energy because of the collision and would rebound to its original height. However, the actual rebound height is somewhat short, showing that at least some kinetic energy is lost in the collision and thus that the collision is somewhat inelastic. Still, we might choose to neglect that small loss of kinetic energy to approximate the collision as elastic.

The inelastic collision of two bodies always involves a loss in the kinetic energy of the system. The greatest loss occurs if the bodies stick together, in which case the collision is called a **completely inelastic collision.** The collision of a baseball and a bat is inelastic. However, the collision of a wet putty ball and a bat is completely inelastic because the putty sticks to the bat.

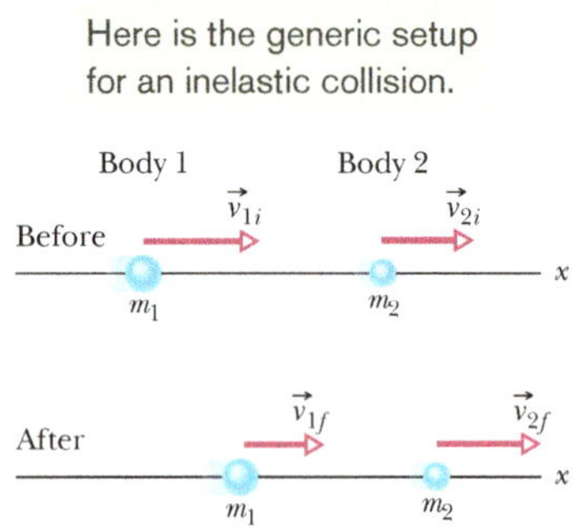

Figure 9-14 Bodies 1 and 2 move along an x axis, before and after they have an inelastic collision.

Inelastic Collisions in One Dimension

One-Dimensional Inelastic Collision

Figure 9-14 shows two bodies just before and just after they have a one-dimensional collision. The velocities before the collision (subscript i) and after the collision (subscript f) are indicated. The two bodies form our system, which is closed and isolated. We can write the law of conservation of linear momentum for this two-body system as

$$\begin{pmatrix}\text{total momentum } \vec{P}_i \\ \text{before the collision}\end{pmatrix} = \begin{pmatrix}\text{total momentum } \vec{P}_f \\ \text{after the collision}\end{pmatrix},$$

which we can symbolize as

$$\vec{p}_{1i} + \vec{p}_{2i} = \vec{p}_{1f} + \vec{p}_{2f} \quad \text{(conservation of linear momentum).} \tag{9-50}$$

Because the motion is one-dimensional, we can drop the overhead arrows for vectors and use only components along the axis, indicating direction with a sign. Thus, from $p = mv$, we can rewrite Eq. 9-50 as

$$m_1v_{1i} + m_2v_{2i} = m_1v_{1f} + m_2v_{2f}. \tag{9-51}$$

If we know values for, say, the masses, the initial velocities, and one of the final velocities, we can find the other final velocity with Eq. 9-51.

One-Dimensional Completely Inelastic Collision

Figure 9-15 shows two bodies before and after they have a completely inelastic collision (meaning they stick together). The body with mass m_2 happens to be initially at rest ($v_{2i} = 0$). We can refer to that body as the *target* and to the incoming body as the *projectile*. After the collision, the stuck-together bodies move with velocity V. For this situation, we can rewrite Eq. 9-51 as

$$m_1v_{1i} = (m_1 + m_2)V \tag{9-52}$$

or

$$V = \frac{m_1}{m_1 + m_2}v_{1i}. \tag{9-53}$$

If we know values for, say, the masses and the initial velocity v_{1i} of the projectile, we can find the final velocity V with Eq. 9-53. Note that V must be less than v_{1i} because the mass ratio $m_1/(m_1 + m_2)$ must be less than unity.

Velocity of the Center of Mass

In a closed, isolated system, the velocity $\vec{v}_{com}$ of the center of mass of the system cannot be changed by a collision because, with the system isolated, there is no net external force to change it. To get an expression for $\vec{v}_{com}$, let us return to the

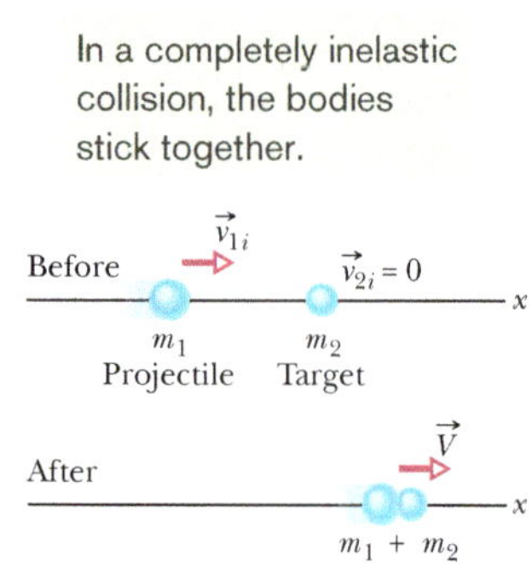

Figure 9-15 A completely inelastic collision between two bodies. Before the collision, the body with mass m_2 is at rest and the body with mass m_1 moves directly toward it. After the collision, the stuck-together bodies move with the same velocity $\vec{V}$.

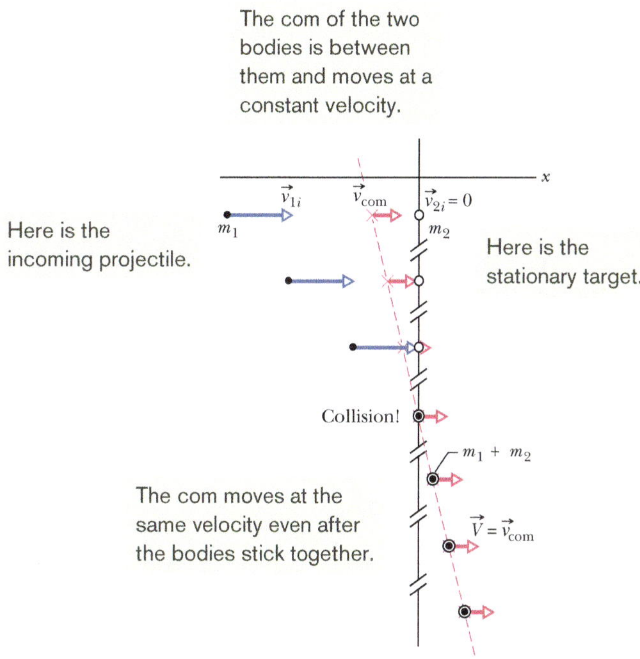

Figure 9-16 Some freeze-frames of the two-body system in Fig. 9-15, which undergoes a completely inelastic collision. The system's center of mass is shown in each freeze-frame. The velocity $\vec{v}_{com}$ of the center of mass is unaffected by the collision. Because the bodies stick together after the collision, their common velocity $\vec{V}$ must be equal to $\vec{v}_{com}$.

two-body system and one-dimensional collision of Fig. 9-14. From Eq. 9-25 ($\vec{P} = M\vec{v}_{com}$), we can relate $\vec{v}_{com}$ to the total linear momentum $\vec{P}$ of that two-body system by writing

$$\vec{P} = M\vec{v}_{com} = (m_1 + m_2)\vec{v}_{com}. \tag{9-54}$$

The total linear momentum $\vec{P}$ is conserved during the collision; so it is given by either side of Eq. 9-50. Let us use the left side to write

$$\vec{P} = \vec{p}_{1i} + \vec{p}_{2i}. \tag{9-55}$$

Substituting this expression for $\vec{P}$ in Eq. 9-54 and solving for $\vec{v}_{com}$ give us

$$\vec{v}_{com} = \frac{\vec{P}}{m_1 + m_2} = \frac{\vec{p}_{1i} + \vec{p}_{2i}}{m_1 + m_2}. \tag{9-56}$$

The right side of this equation is a constant, and $\vec{v}_{com}$ has that same constant value before and after the collision.

For example, Fig. 9-16 shows, in a series of freeze-frames, the motion of the center of mass for the completely inelastic collision of Fig. 9-15. Body 2 is the target, and its initial linear momentum in Eq. 9-56 is $\vec{p}_{2i} = m_2\vec{v}_{2i} = 0$. Body 1 is the projectile, and its initial linear momentum in Eq. 9-56 is $\vec{p}_{1i} = m_1\vec{v}_{1i}$. Note that as the series of freeze-frames progresses to and then beyond the collision, the center of mass moves at a constant velocity to the right. After the collision, the common final speed V of the bodies is equal to $\vec{v}_{com}$ because then the center of mass travels with the stuck-together bodies.

Checkpoint 7

Body 1 and body 2 are in a completely inelastic one-dimensional collision. What is their final momentum if their initial momenta are, respectively, (a) 10 kg · m/s and 0; (b) 10 kg · m/s and 4 kg · m/s; (c) 10 kg · m/s and −4 kg · m/s?

Sample Problem 9.07 Conservation of momentum, ballistic pendulum

Here is an example of a common technique in physics. We have a demonstration that cannot be worked out as a whole (we don't have a workable equation for it). So, we break it up into steps that can be worked separately (we have equations for them).

The *ballistic pendulum* was used to measure the speeds of bullets before electronic timing devices were developed. The version shown in Fig. 9-17 consists of a large block of wood of mass $M = 5.4$ kg, hanging from two long cords. A bullet of mass $m = 9.5$ g is fired into the block, coming quickly to rest. The *block + bullet* then swing upward, their center of mass rising a vertical distance $h = 6.3$ cm before the pendulum comes momentarily to rest at the end of its arc. What is the speed of the bullet just prior to the collision?

KEY IDEAS

We can see that the bullet's speed v must determine the rise height h. However, we cannot use the conservation of mechanical energy to relate these two quantities because surely energy is transferred from mechanical energy to other forms (such as thermal energy and energy to break apart the wood) as the bullet penetrates the block. Nevertheless, we can split this complicated motion into two steps that we can separately analyze: (1) the bullet–block collision and (2) the bullet–block rise, during which mechanical energy *is* conserved.

Reasoning step 1: Because the collision within the bullet–block system is so brief, we can make two important assumptions: (1) During the collision, the gravitational force on the block and the force on the block from the cords are still balanced. Thus, during the collision, the net external impulse on the bullet–block system is zero. Therefore, the system is isolated and its total linear momentum is conserved:

$$\begin{pmatrix}\text{total momentum}\\ \text{before the collision}\end{pmatrix} = \begin{pmatrix}\text{total momentum}\\ \text{after the collision}\end{pmatrix}. \quad (9\text{-}57)$$

(2) The collision is one-dimensional in the sense that the direction of the bullet and block *just after the collision* is in the bullet's original direction of motion.

Because the collision is one-dimensional, the block is initially at rest, and the bullet sticks in the block, we use Eq. 9-53 to express the conservation of linear momentum. Replacing the symbols there with the corresponding symbols here, we have

$$V = \frac{m}{m + M}v. \quad (9\text{-}58)$$

Reasoning step 2: As the bullet and block now swing up together, the mechanical energy of the bullet–block–Earth system is conserved:

$$\begin{pmatrix}\text{mechanical energy}\\ \text{at bottom}\end{pmatrix} = \begin{pmatrix}\text{mechanical energy}\\ \text{at top}\end{pmatrix}. \quad (9\text{-}59)$$

(This mechanical energy is not changed by the force of the cords on the block, because that force is always directed perpendicular to the block's direction of travel.) Let's take the block's initial level as our reference level of zero gravitational potential energy. Then conservation of mechanical energy means that the system's kinetic energy at the start of the swing must equal its gravitational potential energy at the highest point of the swing. Because the speed of the bullet and block at the start of the swing is the speed V immediately after the collision, we may write this conservation as

$$\tfrac{1}{2}(m + M)V^2 = (m + M)gh. \quad (9\text{-}60)$$

Combining steps: Substituting for V from Eq. 9-58 leads to

$$v = \frac{m + M}{m}\sqrt{2gh} \quad (9\text{-}61)$$

$$= \left(\frac{0.0095 \text{ kg} + 5.4 \text{ kg}}{0.0095 \text{ kg}}\right)\sqrt{(2)(9.8 \text{ m/s}^2)(0.063 \text{ m})}$$

$$= 630 \text{ m/s}. \quad \text{(Answer)}$$

The ballistic pendulum is a kind of "transformer," exchanging the high speed of a light object (the bullet) for the low—and thus more easily measurable—speed of a massive object (the block).

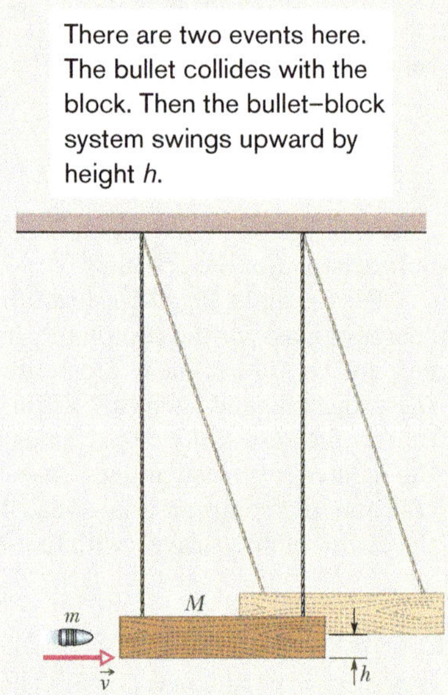

Figure 9-17 A ballistic pendulum, used to measure the speeds of bullets.

Additional examples, video, and practice available at *WileyPLUS*

9-7 ELASTIC COLLISIONS IN ONE DIMENSION

Learning Objectives

After reading this module, you should be able to . . .

9.32 For isolated elastic collisions in one dimension, apply the conservation laws for both the total energy and the net momentum of the colliding bodies to relate the initial values to the values after the collision.

9.33 For a projectile hitting a stationary target, identify the resulting motion for the three general cases: equal masses, target more massive than projectile, projectile more massive than target.

Key Idea

- An elastic collision is a special type of collision in which the kinetic energy of a system of colliding bodies is conserved. If the system is closed and isolated, its linear momentum is also conserved. For a one-dimensional collision in which body 2 is a target and body 1 is an incoming projectile, conservation of kinetic energy and linear momentum yield the following expressions for the velocities immediately after the collision:

$$v_{1f} = \frac{m_1 - m_2}{m_1 + m_2} v_{1i}$$

and

$$v_{2f} = \frac{2m_1}{m_1 + m_2} v_{1i}.$$

Elastic Collisions in One Dimension

As we discussed in Module 9-6, everyday collisions are inelastic but we can approximate some of them as being elastic; that is, we can approximate that the total kinetic energy of the colliding bodies is conserved and is not transferred to other forms of energy:

$$\begin{pmatrix}\text{total kinetic energy}\\ \text{before the collision}\end{pmatrix} = \begin{pmatrix}\text{total kinetic energy}\\ \text{after the collision}\end{pmatrix}. \tag{9-62}$$

This means:

> In an elastic collision, the kinetic energy of each colliding body may change, but the total kinetic energy of the system does not change.

For example, the collision of a cue ball with an object ball in a game of pool can be approximated as being an elastic collision. If the collision is head-on (the cue ball heads directly toward the object ball), the kinetic energy of the cue ball can be transferred almost entirely to the object ball. (Still, the collision transfers some of the energy to the sound you hear.)

Stationary Target

Figure 9-18 shows two bodies before and after they have a one-dimensional collision, like a head-on collision between pool balls. A projectile body of mass m_1 and initial velocity v_{1i} moves toward a target body of mass m_2 that is initially at rest ($v_{2i} = 0$). Let's assume that this two-body system is closed and isolated. Then the net linear momentum of the system is conserved, and from Eq. 9-51 we can write that conservation as

$$m_1 v_{1i} = m_1 v_{1f} + m_2 v_{2f} \quad \text{(linear momentum).} \tag{9-63}$$

If the collision is also elastic, then the total kinetic energy is conserved and we can write that conservation as

$$\tfrac{1}{2} m_1 v_{1i}^2 = \tfrac{1}{2} m_1 v_{1f}^2 + \tfrac{1}{2} m_2 v_{2f}^2 \quad \text{(kinetic energy).} \tag{9-64}$$

In each of these equations, the subscript i identifies the initial velocities and the subscript f the final velocities of the bodies. If we know the masses of the bodies and if we also know v_{1i}, the initial velocity of body 1, the only unknown quantities are v_{1f} and v_{2f}, the final velocities of the two bodies. With two equations at our disposal, we should be able to find these two unknowns.

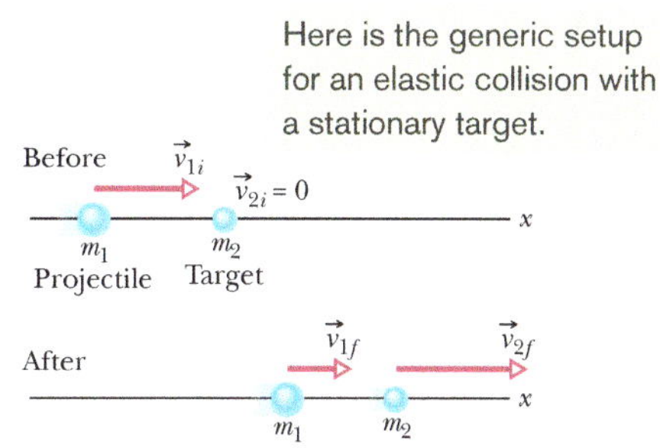

Figure 9-18 Body 1 moves along an x axis before having an elastic collision with body 2, which is initially at rest. Both bodies move along that axis after the collision.

To do so, we rewrite Eq. 9-63 as

$$m_1(v_{1i} - v_{1f}) = m_2 v_{2f} \tag{9-65}$$

and Eq. 9-64 as*

$$m_1(v_{1i} - v_{1f})(v_{1i} + v_{1f}) = m_2 v_{2f}^2. \tag{9-66}$$

After dividing Eq. 9-66 by Eq. 9-65 and doing some more algebra, we obtain

$$v_{1f} = \frac{m_1 - m_2}{m_1 + m_2} v_{1i} \tag{9-67}$$

and

$$v_{2f} = \frac{2m_1}{m_1 + m_2} v_{1i}. \tag{9-68}$$

Note that v_{2f} is always positive (the initially stationary target body with mass m_2 always moves forward). From Eq. 9-67 we see that v_{1f} may be of either sign (the projectile body with mass m_1 moves forward if $m_1 > m_2$ but rebounds if $m_1 < m_2$).

Let us look at a few special situations.

1. ***Equal masses*** If $m_1 = m_2$, Eqs. 9-67 and 9-68 reduce to

$$v_{1f} = 0 \quad \text{and} \quad v_{2f} = v_{1i},$$

which we might call a pool player's result. It predicts that after a head-on collision of bodies with equal masses, body 1 (initially moving) stops dead in its tracks and body 2 (initially at rest) takes off with the initial speed of body 1. In head-on collisions, bodies of equal mass simply exchange velocities. This is true even if body 2 is not initially at rest.

2. ***A massive target*** In Fig. 9-18, a massive target means that $m_2 \gg m_1$. For example, we might fire a golf ball at a stationary cannonball. Equations 9-67 and 9-68 then reduce to

$$v_{1f} \approx -v_{1i} \quad \text{and} \quad v_{2f} \approx \left(\frac{2m_1}{m_2}\right) v_{1i}. \tag{9-69}$$

This tells us that body 1 (the golf ball) simply bounces back along its incoming path, its speed essentially unchanged. Initially stationary body 2 (the cannonball) moves forward at a low speed, because the quantity in parentheses in Eq. 9-69 is much less than unity. All this is what we should expect.

3. ***A massive projectile*** This is the opposite case; that is, $m_1 \gg m_2$. This time, we fire a cannonball at a stationary golf ball. Equations 9-67 and 9-68 reduce to

$$v_{1f} \approx v_{1i} \quad \text{and} \quad v_{2f} \approx 2v_{1i}. \tag{9-70}$$

Equation 9-70 tells us that body 1 (the cannonball) simply keeps on going, scarcely slowed by the collision. Body 2 (the golf ball) charges ahead at twice the speed of the cannonball. Why twice the speed? Recall the collision described by Eq. 9-69, in which the velocity of the incident light body (the golf ball) changed from $+v$ to $-v$, a velocity *change* of $2v$. The same *change* in velocity (but now from zero to $2v$) occurs in this example also.

Moving Target

Now that we have examined the elastic collision of a projectile and a stationary target, let us examine the situation in which both bodies are moving before they undergo an elastic collision.

For the situation of Fig. 9-19, the conservation of linear momentum is written as

$$m_1 v_{1i} + m_2 v_{2i} = m_1 v_{1f} + m_2 v_{2f}, \tag{9-71}$$

*In this step, we use the identity $a^2 - b^2 = (a - b)(a + b)$. It reduces the amount of algebra needed to solve the simultaneous equations Eqs. 9-65 and 9-66.

and the conservation of kinetic energy is written as

$$\tfrac{1}{2}m_1v_{1i}^2 + \tfrac{1}{2}m_2v_{2i}^2 = \tfrac{1}{2}m_1v_{1f}^2 + \tfrac{1}{2}m_2v_{2f}^2. \tag{9-72}$$

To solve these simultaneous equations for v_{1f} and v_{2f}, we first rewrite Eq. 9-71 as

$$m_1(v_{1i} - v_{1f}) = -m_2(v_{2i} - v_{2f}), \tag{9-73}$$

and Eq. 9-72 as

$$m_1(v_{1i} - v_{1f})(v_{1i} + v_{1f}) = -m_2(v_{2i} - v_{2f})(v_{2i} + v_{2f}). \tag{9-74}$$

After dividing Eq. 9-74 by Eq. 9-73 and doing some more algebra, we obtain

$$v_{1f} = \frac{m_1 - m_2}{m_1 + m_2}v_{1i} + \frac{2m_2}{m_1 + m_2}v_{2i} \tag{9-75}$$

and

$$v_{2f} = \frac{2m_1}{m_1 + m_2}v_{1i} + \frac{m_2 - m_1}{m_1 + m_2}v_{2i}. \tag{9-76}$$

Note that the assignment of subscripts 1 and 2 to the bodies is arbitrary. If we exchange those subscripts in Fig. 9-19 and in Eqs. 9-75 and 9-76, we end up with the same set of equations. Note also that if we set $v_{2i} = 0$, body 2 becomes a stationary target as in Fig. 9-18, and Eqs. 9-75 and 9-76 reduce to Eqs. 9-67 and 9-68, respectively.

Here is the generic setup for an elastic collision with a moving target.

Figure 9-19 Two bodies headed for a one-dimensional elastic collision.

Checkpoint 8

What is the final linear momentum of the target in Fig. 9-18 if the initial linear momentum of the projectile is 6 kg · m/s and the final linear momentum of the projectile is (a) 2 kg · m/s and (b) −2 kg · m/s? (c) What is the final kinetic energy of the target if the initial and final kinetic energies of the projectile are, respectively, 5 J and 2 J?

Sample Problem 9.08 Chain reaction of elastic collisions

In Fig. 9-20*a*, block 1 approaches a line of two stationary blocks with a velocity of $v_{1i} = 10$ m/s. It collides with block 2, which then collides with block 3, which has mass $m_3 = 6.0$ kg. After the second collision, block 2 is again stationary and block 3 has velocity $v_{3f} = 5.0$ m/s (Fig. 9-20*b*). Assume that the collisions are elastic. What are the masses of blocks 1 and 2? What is the final velocity v_{1f} of block 1?

KEY IDEAS

Because we assume that the collisions are elastic, we are to conserve mechanical energy (thus energy losses to sound, heating, and oscillations of the blocks are negligible). Because no external horizontal force acts on the blocks, we are to conserve linear momentum along the *x* axis. For these two reasons, we can apply Eqs. 9-67 and 9-68 to each of the collisions.

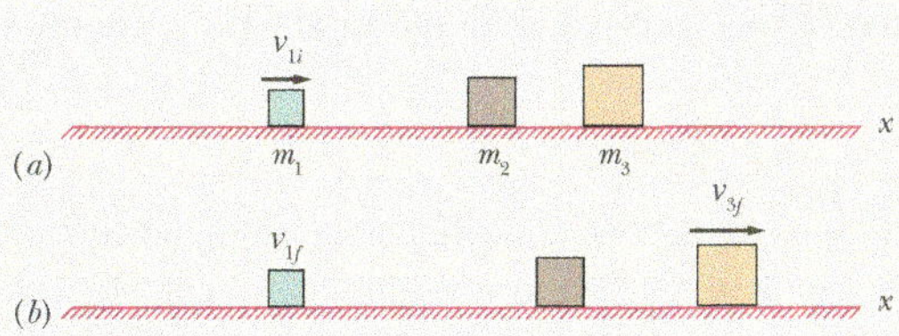

Figure 9-20 Block 1 collides with stationary block 2, which then collides with stationary block 3.

Calculations: If we start with the first collision, we have too many unknowns to make any progress: we do not know the masses or the final velocities of the blocks. So, let's start with the second collision in which block 2 stops because of its collision with block 3. Applying Eq. 9-67 to this collision, with changes in notation, we have

$$v_{2f} = \frac{m_2 - m_3}{m_2 + m_3}v_{2i},$$

where v_{2i} is the velocity of block 2 just before the collision and v_{2f} is the velocity just afterward. Substituting $v_{2f} = 0$ (block 2 stops) and then $m_3 = 6.0$ kg gives us

$$m_2 = m_3 = 6.00 \text{ kg}. \qquad \text{(Answer)}$$

With similar notation changes, we can rewrite Eq. 9-68 for the second collision as

$$v_{3f} = \frac{2m_2}{m_2 + m_3}v_{2i},$$

where v_{3f} is the final velocity of block 3. Substituting $m_2 = m_3$ and the given $v_{3f} = 5.0$ m/s, we find

$$v_{2i} = v_{3f} = 5.0 \text{ m/s}.$$

Next, let's reconsider the first collision, but we have to be careful with the notation for block 2: its velocity v_{2f} just after the first collision is the same as its velocity v_{2i} (= 5.0 m/s) just before the second collision. Applying Eq. 9-68 to the first collision and using the given v_{1i} = 10 m/s, we have

$$v_{2f} = \frac{2m_1}{m_1 + m_2} v_{1i},$$

$$5.0 \text{ m/s} = \frac{2m_1}{m_1 + m_2} (10 \text{ m/s}),$$

which leads to

$$m_1 = \tfrac{1}{3}m_2 = \tfrac{1}{3}(6.0 \text{ kg}) = 2.0 \text{ kg}. \qquad \text{(Answer)}$$

Finally, applying Eq. 9-67 to the first collision with this result and the given v_{1i}, we write

$$v_{1f} = \frac{m_1 - m_2}{m_1 + m_2} v_{1i},$$

$$= \frac{\frac{1}{3}m_2 - m_2}{\frac{1}{3}m_2 + m_2}(10 \text{ m/s}) = -5.0 \text{ m/s}. \qquad \text{(Answer)}$$

9-8 COLLISIONS IN TWO DIMENSIONS

Learning Objectives

After reading this module, you should be able to . . .

9.34 For an isolated system in which a two-dimensional collision occurs, apply the conservation of momentum along each axis of a coordinate system to relate the momentum components along an axis before the collision to the momentum components *along the same axis* after the collision.

9.35 For an isolated system in which a two-dimensional *elastic* collision occurs, (a) apply the conservation of momentum along each axis of a coordinate system to relate the momentum components along an axis before the collision to the momentum components *along the same axis* after the collision and (b) apply the conservation of total kinetic energy to relate the kinetic energies before and after the collision.

Key Idea

- If two bodies collide and their motion is not along a single axis (the collision is not head-on), the collision is two-dimensional. If the two-body system is closed and isolated, the law of conservation of momentum applies to the collision and can be written as

$$\vec{P}_{1i} + \vec{P}_{2i} = \vec{P}_{1f} + \vec{P}_{2f}.$$

In component form, the law gives two equations that describe the collision (one equation for each of the two dimensions). If the collision is also elastic (a special case), the conservation of kinetic energy during the collision gives a third equation:

$$K_{1i} + K_{2i} = K_{1f} + K_{2f}.$$

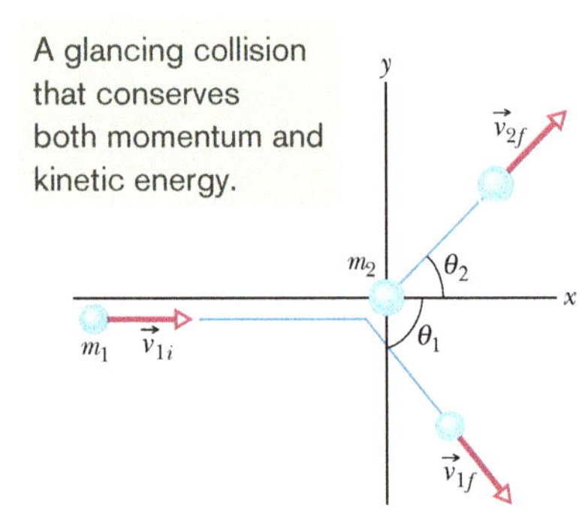

Figure 9-21 An elastic collision between two bodies in which the collision is not head-on. The body with mass m_2 (the target) is initially at rest.

Collisions in Two Dimensions

When two bodies collide, the impulse between them determines the directions in which they then travel. In particular, when the collision is not head-on, the bodies do not end up traveling along their initial axis. For such two-dimensional collisions in a closed, isolated system, the total linear momentum must still be conserved:

$$\vec{P}_{1i} + \vec{P}_{2i} = \vec{P}_{1f} + \vec{P}_{2f}. \qquad (9\text{-}77)$$

If the collision is also elastic (a special case), then the total kinetic energy is also conserved:

$$K_{1i} + K_{2i} = K_{1f} + K_{2f}. \qquad (9\text{-}78)$$

Equation 9-77 is often more useful for analyzing a two-dimensional collision if we write it in terms of components on an *xy* coordinate system. For example, Fig. 9-21 shows a *glancing collision* (it is not head-on) between a projectile body and a target body initially at rest. The impulses between the bodies have sent the bodies off at angles θ_1 and θ_2 to the *x* axis, along which the projectile initially traveled. In this situation we would rewrite Eq. 9-77 for components along the x axis as

$$m_1 v_{1i} = m_1 v_{1f} \cos\theta_1 + m_2 v_{2f} \cos\theta_2, \tag{9-79}$$

and along the y axis as

$$0 = -m_1 v_{1f} \sin\theta_1 + m_2 v_{2f} \sin\theta_2. \tag{9-80}$$

We can also write Eq. 9-78 (for the special case of an elastic collision) in terms of speeds:

$$\tfrac{1}{2}m_1 v_{1i}^2 = \tfrac{1}{2}m_1 v_{1f}^2 + \tfrac{1}{2}m_2 v_{2f}^2 \quad \text{(kinetic energy)}. \tag{9-81}$$

Equations 9-79 to 9-81 contain seven variables: two masses, m_1 and m_2; three speeds, v_{1i}, v_{1f}, and v_{2f}; and two angles, θ_1 and θ_2. If we know any four of these quantities, we can solve the three equations for the remaining three quantities.

Checkpoint 9

In Fig. 9-21, suppose that the projectile has an initial momentum of 6 kg·m/s, a final x component of momentum of 4 kg·m/s, and a final y component of momentum of −3 kg·m/s. For the target, what then are (a) the final x component of momentum and (b) the final y component of momentum?

9-9 SYSTEMS WITH VARYING MASS: A ROCKET

Learning Objectives

After reading this module, you should be able to . . .

9.36 Apply the first rocket equation to relate the rate at which the rocket loses mass, the speed of the exhaust products relative to the rocket, the mass of the rocket, and the acceleration of the rocket.

9.37 Apply the second rocket equation to relate the change in the rocket's speed to the relative speed of the exhaust products and the initial and final mass of the rocket.

9.38 For a moving system undergoing a change in mass at a given rate, relate that rate to the change in momentum.

Key Ideas

- In the absence of external forces a rocket accelerates at an instantaneous rate given by

$$Rv_{rel} = Ma \quad \text{(first rocket equation)},$$

in which M is the rocket's instantaneous mass (including unexpended fuel), R is the fuel consumption rate, and v_{rel} is the fuel's exhaust speed relative to the rocket. The term Rv_{rel} is the thrust of the rocket engine.

- For a rocket with constant R and v_{rel}, whose speed changes from v_i to v_f when its mass changes from M_i to M_f,

$$v_f - v_i = v_{rel} \ln \frac{M_i}{M_f} \quad \text{(second rocket equation)}.$$

Systems with Varying Mass: A Rocket

So far, we have assumed that the total mass of the system remains constant. Sometimes, as in a rocket, it does not. Most of the mass of a rocket on its launching pad is fuel, all of which will eventually be burned and ejected from the nozzle of the rocket engine. We handle the variation of the mass of the rocket as the rocket accelerates by applying Newton's second law, not to the rocket alone but to the rocket and its ejected combustion products taken together. The mass of *this* system does *not* change as the rocket accelerates.

Finding the Acceleration

Assume that we are at rest relative to an inertial reference frame, watching a rocket accelerate through deep space with no gravitational or atmospheric drag

The ejection of mass from the rocket's rear increases the rocket's speed.

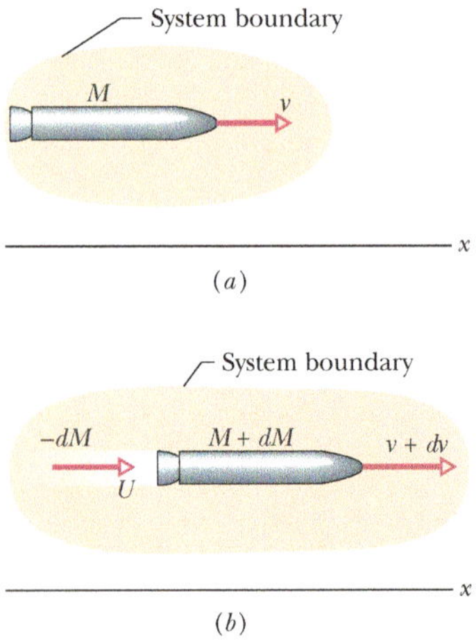

Figure 9-22 (*a*) An accelerating rocket of mass M at time t, as seen from an inertial reference frame. (*b*) The same but at time $t + dt$. The exhaust products released during interval dt are shown.

forces acting on it. For this one-dimensional motion, let M be the mass of the rocket and v its velocity at an arbitrary time t (see Fig. 9-22*a*).

Figure 9-22*b* shows how things stand a time interval dt later. The rocket now has velocity $v + dv$ and mass $M + dM$, where the change in mass dM is a *negative quantity.* The exhaust products released by the rocket during interval dt have mass $-dM$ and velocity U relative to our inertial reference frame.

Conserve Momentum. Our system consists of the rocket and the exhaust products released during interval dt. The system is closed and isolated, so the linear momentum of the system must be conserved during dt; that is,

$$P_i = P_f, \tag{9-82}$$

where the subscripts i and f indicate the values at the beginning and end of time interval dt. We can rewrite Eq. 9-82 as

$$Mv = -dM\, U + (M + dM)(v + dv), \tag{9-83}$$

where the first term on the right is the linear momentum of the exhaust products released during interval dt and the second term is the linear momentum of the rocket at the end of interval dt.

Use Relative Speed. We can simplify Eq. 9-83 by using the relative speed v_{rel} between the rocket and the exhaust products, which is related to the velocities relative to the frame with

$$\begin{pmatrix}\text{velocity of rocket}\\ \text{relative to frame}\end{pmatrix} = \begin{pmatrix}\text{velocity of rocket}\\ \text{relative to products}\end{pmatrix} + \begin{pmatrix}\text{velocity of products}\\ \text{relative to frame}\end{pmatrix}.$$

In symbols, this means

$$(v + dv) = v_{rel} + U,$$

or

$$U = v + dv - v_{rel}. \tag{9-84}$$

Substituting this result for U into Eq. 9-83 yields, with a little algebra,

$$-dM\, v_{rel} = M\, dv. \tag{9-85}$$

Dividing each side by dt gives us

$$-\frac{dM}{dt} v_{rel} = M \frac{dv}{dt}. \tag{9-86}$$

We replace dM/dt (the rate at which the rocket loses mass) by $-R$, where R is the (positive) mass rate of fuel consumption, and we recognize that dv/dt is the acceleration of the rocket. With these changes, Eq. 9-86 becomes

$$Rv_{rel} = Ma \quad \text{(first rocket equation)}. \tag{9-87}$$

Equation 9-87 holds for the values at any given instant.

Note the left side of Eq. 9-87 has the dimensions of force ($\text{kg/s} \cdot \text{m/s} = \text{kg} \cdot \text{m/s}^2 = \text{N}$) and depends only on design characteristics of the rocket engine—namely, the rate R at which it consumes fuel mass and the speed v_{rel} with which that mass is ejected relative to the rocket. We call this term Rv_{rel} the **thrust** of the rocket engine and represent it with T. Newton's second law emerges if we write Eq. 9-87 as $T = Ma$, in which a is the acceleration of the rocket at the time that its mass is M.

Finding the Velocity

How will the velocity of a rocket change as it consumes its fuel? From Eq. 9-85 we have

$$dv = -v_{rel} \frac{dM}{M}.$$

Integrating leads to

$$\int_{v_i}^{v_f} dv = -v_{rel} \int_{M_i}^{M_f} \frac{dM}{M},$$

in which M_i is the initial mass of the rocket and M_f its final mass. Evaluating the integrals then gives

$$v_f - v_i = v_{rel} \ln \frac{M_i}{M_f} \quad \text{(second rocket equation)} \tag{9-88}$$

for the increase in the speed of the rocket during the change in mass from M_i to M_f. (The symbol "ln" in Eq. 9-88 means the *natural logarithm.*) We see here the advantage of multistage rockets, in which M_f is reduced by discarding successive stages when their fuel is depleted. An ideal rocket would reach its destination with only its payload remaining.

Sample Problem 9.09 Rocket engine, thrust, acceleration

In all previous examples in this chapter, the mass of a system is constant (fixed as a certain number). Here is an example of a system (a rocket) that is losing mass. A rocket whose initial mass M_i is 850 kg consumes fuel at the rate $R = 2.3$ kg/s. The speed v_{rel} of the exhaust gases relative to the rocket engine is 2800 m/s. What thrust does the rocket engine provide?

KEY IDEA

Thrust T is equal to the product of the fuel consumption rate R and the relative speed v_{rel} at which exhaust gases are expelled, as given by Eq. 9-87.

Calculation: Here we find

$$T = Rv_{rel} = (2.3 \text{ kg/s})(2800 \text{ m/s}) = 6440 \text{ N} \approx 6400 \text{ N}. \quad \text{(Answer)}$$

(b) What is the initial acceleration of the rocket?

KEY IDEA

We can relate the thrust T of a rocket to the magnitude a of the resulting acceleration with $T = Ma$, where M is the rocket's mass. However, M decreases and a increases as fuel is consumed. Because we want the initial value of a here, we must use the intial value M_i of the mass.

Calculation: We find

$$a = \frac{T}{M_i} = \frac{6440 \text{ N}}{850 \text{ kg}} = 7.6 \text{ m/s}^2. \quad \text{(Answer)}$$

To be launched from Earth's surface, a rocket must have an initial acceleration greater than $g = 9.8$ m/s². That is, it must be greater than the gravitational acceleration at the surface. Put another way, the thrust T of the rocket engine must exceed the initial gravitational force on the rocket, which here has the magnitude M_ig, which gives us

$$(850 \text{ kg})(9.8 \text{ m/s}^2) = 8330 \text{ N}.$$

Because the acceleration or thrust requirement is not met (here $T = 6400$ N), our rocket could not be launched from Earth's surface by itself; it would require another, more powerful, rocket.

WILEY PLUS Additional examples, video, and practice available at *WileyPLUS*

Review & Summary

Center of Mass The **center of mass** of a system of n particles is defined to be the point whose coordinates are given by

$$x_{com} = \frac{1}{M}\sum_{i=1}^{n} m_i x_i, \quad y_{com} = \frac{1}{M}\sum_{i=1}^{n} m_i y_i, \quad z_{com} = \frac{1}{M}\sum_{i=1}^{n} m_i z_i, \tag{9-5}$$

or

$$\vec{r}_{com} = \frac{1}{M}\sum_{i=1}^{n} m_i \vec{r}_i, \tag{9-8}$$

where M is the total mass of the system.

Newton's Second Law for a System of Particles The motion of the center of mass of any system of particles is governed by **Newton's second law for a system of particles,** which is

$$\vec{F}_{net} = M\vec{a}_{com}. \tag{9-14}$$

Here $\vec{F}_{net}$ is the net force of all the *external* forces acting on the system, M is the total mass of the system, and $\vec{a}_{com}$ is the acceleration of the system's center of mass.

Linear Momentum and Newton's Second Law For a single particle, we define a quantity $\vec{p}$ called its **linear momentum** as

$$\vec{p} = m\vec{v}, \tag{9-22}$$

and can write Newton's second law in terms of this momentum:

$$\vec{F}_{\text{net}} = \frac{d\vec{p}}{dt}. \tag{9-23}$$

For a system of particles these relations become

$$\vec{P} = M\vec{v}_{\text{com}} \quad \text{and} \quad \vec{F}_{\text{net}} = \frac{d\vec{P}}{dt}. \qquad (9\text{-}25, 9\text{-}27)$$

Collision and Impulse Applying Newton's second law in momentum form to a particle-like body involved in a collision leads to the **impulse–linear momentum theorem:**

$$\vec{p}_f - \vec{p}_i = \Delta\vec{p} = \vec{J}, \qquad (9\text{-}31, 9\text{-}32)$$

where $\vec{p}_f - \vec{p}_i = \Delta\vec{p}$ is the change in the body's linear momentum, and $\vec{J}$ is the **impulse** due to the force $\vec{F}(t)$ exerted on the body by the other body in the collision:

$$\vec{J} = \int_{t_i}^{t_f} \vec{F}(t)\, dt. \tag{9-30}$$

If F_{avg} is the average magnitude of $\vec{F}(t)$ during the collision and Δt is the duration of the collision, then for one-dimensional motion

$$J = F_{\text{avg}}\, \Delta t. \tag{9-35}$$

When a steady stream of bodies, each with mass m and speed v, collides with a body whose position is fixed, the average force on the fixed body is

$$F_{\text{avg}} = -\frac{n}{\Delta t}\Delta p = -\frac{n}{\Delta t} m\, \Delta v, \tag{9-37}$$

where $n/\Delta t$ is the rate at which the bodies collide with the fixed body, and Δv is the change in velocity of each colliding body. This average force can also be written as

$$F_{\text{avg}} = -\frac{\Delta m}{\Delta t}\Delta v, \tag{9-40}$$

where $\Delta m/\Delta t$ is the rate at which mass collides with the fixed body. In Eqs. 9-37 and 9-40, $\Delta v = -v$ if the bodies stop upon impact and $\Delta v = -2v$ if they bounce directly backward with no change in their speed.

Conservation of Linear Momentum If a system is isolated so that no net *external* force acts on it, the linear momentum $\vec{P}$ of the system remains constant:

$$\vec{P} = \text{constant} \quad \text{(closed, isolated system)}. \tag{9-42}$$

This can also be written as

$$\vec{P}_i = \vec{P}_f \quad \text{(closed, isolated system)}, \tag{9-43}$$

where the subscripts refer to the values of $\vec{P}$ at some initial time and at a later time. Equations 9-42 and 9-43 are equivalent statements of the **law of conservation of linear momentum.**

Inelastic Collision in One Dimension In an *inelastic collision* of two bodies, the kinetic energy of the two-body system is not conserved (it is not a constant). If the system is closed and isolated, the total linear momentum of the system *must* be conserved (it *is* a constant), which we can write in vector form as

$$\vec{p}_{1i} + \vec{p}_{2i} = \vec{p}_{1f} + \vec{p}_{2f}, \tag{9-50}$$

where subscripts i and f refer to values just before and just after the collision, respectively.

If the motion of the bodies is along a single axis, the collision is one-dimensional and we can write Eq. 9-50 in terms of velocity components along that axis:

$$m_1v_{1i} + m_2v_{2i} = m_1v_{1f} + m_2v_{2f}. \tag{9-51}$$

If the bodies stick together, the collision is a *completely inelastic collision* and the bodies have the same final velocity V (because they *are* stuck together).

Motion of the Center of Mass The center of mass of a closed, isolated system of two colliding bodies is not affected by a collision. In particular, the velocity $\vec{v}_{\text{com}}$ of the center of mass cannot be changed by the collision.

Elastic Collisions in One Dimension An *elastic collision* is a special type of collision in which the kinetic energy of a system of colliding bodies is conserved. If the system is closed and isolated, its linear momentum is also conserved. For a one-dimensional collision in which body 2 is a target and body 1 is an incoming projectile, conservation of kinetic energy and linear momentum yield the following expressions for the velocities immediately after the collision:

$$v_{1f} = \frac{m_1 - m_2}{m_1 + m_2} v_{1i} \tag{9-67}$$

and

$$v_{2f} = \frac{2m_1}{m_1 + m_2} v_{1i}. \tag{9-68}$$

Collisions in Two Dimensions If two bodies collide and their motion is not along a single axis (the collision is not head-on), the collision is two-dimensional. If the two-body system is closed and isolated, the law of conservation of momentum applies to the collision and can be written as

$$\vec{P}_{1i} + \vec{P}_{2i} = \vec{P}_{1f} + \vec{P}_{2f}. \tag{9-77}$$

In component form, the law gives two equations that describe the collision (one equation for each of the two dimensions). If the collision is also elastic (a special case), the conservation of kinetic energy during the collision gives a third equation:

$$K_{1i} + K_{2i} = K_{1f} + K_{2f}. \tag{9-78}$$

Variable-Mass Systems In the absence of external forces a rocket accelerates at an instantaneous rate given by

$$Rv_{\text{rel}} = Ma \quad \text{(first rocket equation)}, \tag{9-87}$$

in which M is the rocket's instantaneous mass (including unexpended fuel), R is the fuel consumption rate, and v_{rel} is the fuel's exhaust speed relative to the rocket. The term Rv_{rel} is the **thrust** of the rocket engine. For a rocket with constant R and v_{rel}, whose speed changes from v_i to v_f when its mass changes from M_i to M_f,

$$v_f - v_i = v_{\text{rel}} \ln \frac{M_i}{M_f} \quad \text{(second rocket equation)}. \tag{9-88}$$

Questions

1 Figure 9-23 shows an overhead view of three particles on which external forces act. The magnitudes and directions of the forces on two of the particles are indicated. What are the magnitude and direction of the force acting on the third particle if the center of mass of the three-particle system is (a) stationary, (b) moving at a constant velocity rightward, and (c) accelerating rightward?

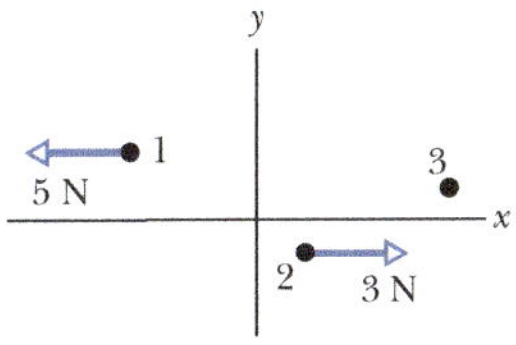

Figure 9-23 Question 1.

2 Figure 9-24 shows an overhead view of four particles of equal mass sliding over a frictionless surface at constant velocity. The directions of the velocities are indicated; their magnitudes are equal. Consider pairing the particles. Which pairs form a system with a center of mass that (a) is stationary, (b) is stationary and at the origin, and (c) passes through the origin?

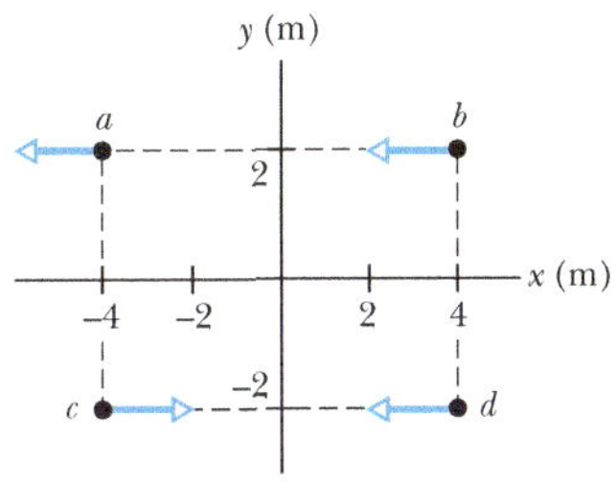

Figure 9-24 Question 2.

3 Consider a box that explodes into two pieces while moving with a constant positive velocity along an x axis. If one piece, with mass m_1, ends up with positive velocity $\vec{v}_1$, then the second piece, with mass m_2, could end up with (a) a positive velocity $\vec{v}_2$ (Fig. 9-25*a*), (b) a negative velocity $\vec{v}_2$ (Fig. 9-25*b*), or (c) zero velocity (Fig. 9-25*c*). Rank those three possible results for the second piece according to the corresponding magnitude of $\vec{v}_1$, greatest first.

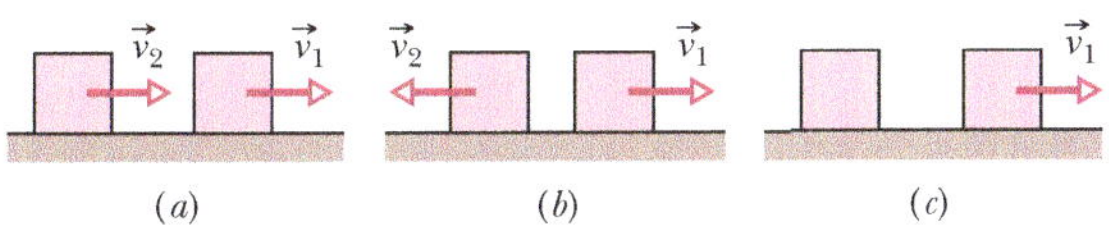

Figure 9-25 Question 3.

4 Figure 9-26 shows graphs of force magnitude versus time for a body involved in a collision. Rank the graphs according to the magnitude of the impulse on the body, greatest first.

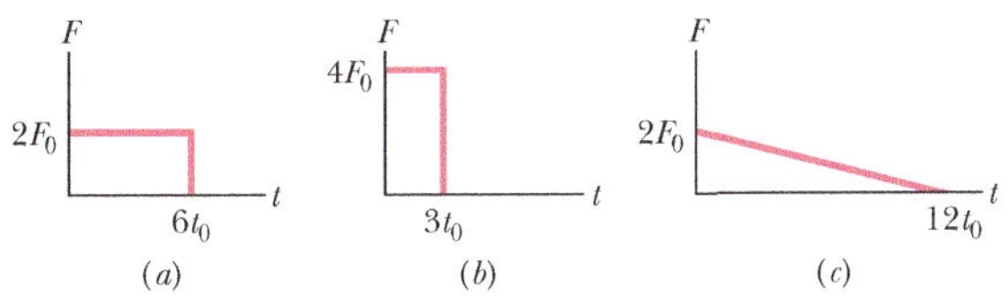

Figure 9-26 Question 4.

5 The free-body diagrams in Fig. 9-27 give, from overhead views, the horizontal forces acting on three boxes of chocolates as the boxes move over a frictionless confectioner's counter. For each box, is its linear momentum conserved along the x axis and the y axis?

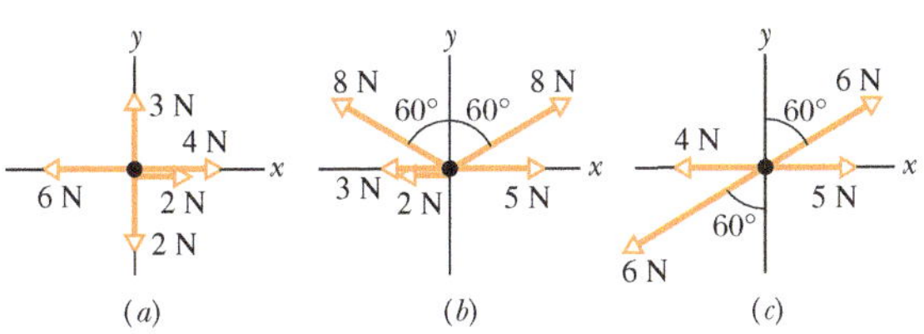

Figure 9-27 Question 5.

6 Figure 9-28 shows four groups of three or four identical particles that move parallel to either the x axis or the y axis, at identical speeds. Rank the groups according to center-of-mass speed, greatest first.

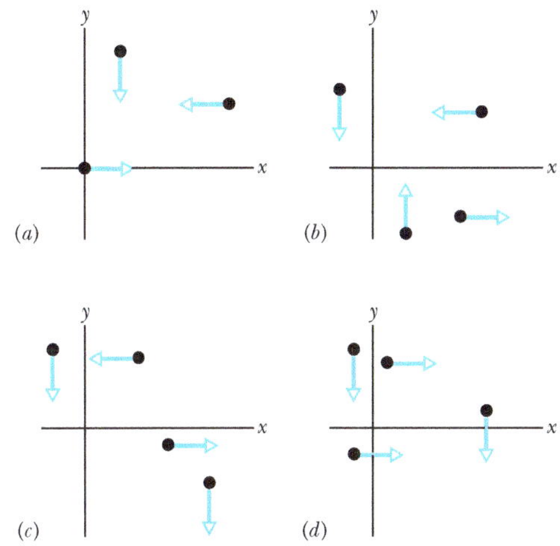

Figure 9-28 Question 6.

7 A block slides along a frictionless floor and into a stationary second block with the same mass. Figure 9-29 shows four choices for a graph of the kinetic energies K of the blocks. (a) Determine which represent physically impossible situations. Of the others, which best represents (b) an elastic collision and (c) an inelastic collision?

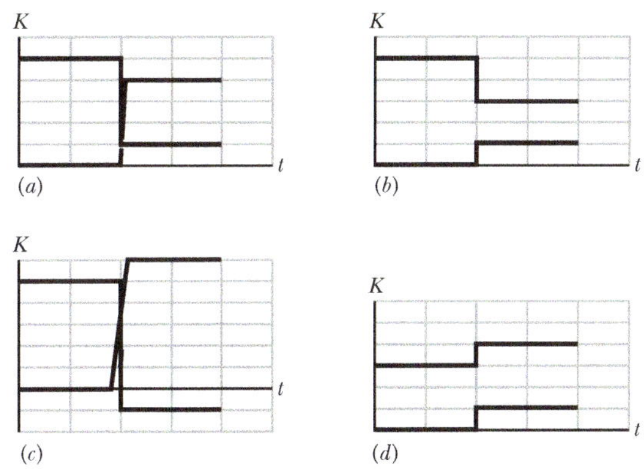

Figure 9-29 Question 7.

8 Figure 9-30 shows a snapshot of block 1 as it slides along an x axis on a frictionless floor, before it undergoes an elastic collision with stationary block 2. The figure also shows three possible positions of the center of mass (com) of the two-block system at the time of the snapshot. (Point B is halfway between the centers of the two blocks.) Is block 1 stationary, moving forward, or moving backward after the collision if the com is located in the snapshot at (a) A, (b) B, and (c) C?

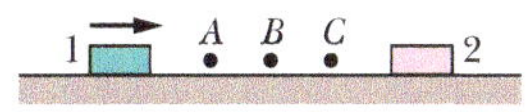

Figure 9-30 Question 8.

9 Two bodies have undergone an elastic one-dimensional collision along an x axis. Figure 9-31 is a graph of position versus time for those bodies and for their center of mass. (a) Were both bodies initially moving, or was one initially stationary? Which line segment corresponds to the motion of the center of mass (b) before the collision and (c) after the collision? (d) Is the mass of the body that was moving faster before the collision greater than, less than, or equal to that of the other body?

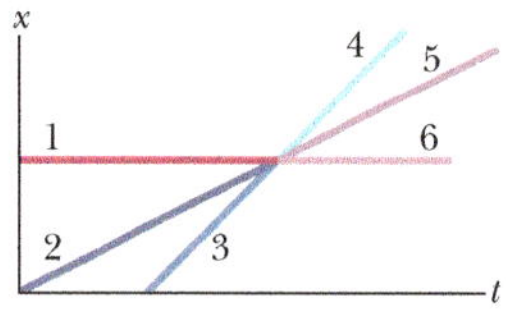

Figure 9-31 Question 9.

10 Figure 9-32: A block on a horizontal floor is initially either stationary, sliding in the positive direction of an x axis, or sliding in the negative direction of that axis. Then the block explodes into two pieces that slide along the x axis. Assume the block and the two pieces form a closed, isolated system. Six choices for a graph of the momenta of the block and the pieces are given, all versus time t. Determine which choices represent physically impossible situations and explain why.

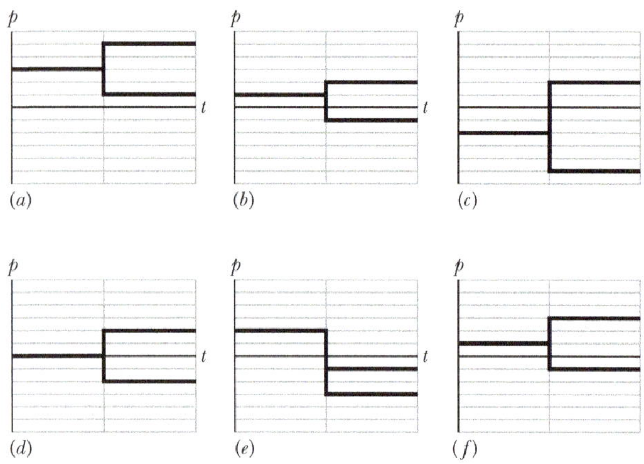

Figure 9-32 Question 10.

11 Block 1 with mass m_1 slides along an x axis across a frictionless floor and then undergoes an elastic collision with a stationary block 2 with mass m_2. Figure 9-33 shows a plot of position x versus time t of block 1 until the collision occurs at position x_c and time t_c. In which of the lettered regions on the graph will the plot be continued (after the collision) if (a) $m_1 < m_2$ and (b) $m_1 > m_2$? (c) Along which of the numbered dashed lines will the plot be continued if $m_1 = m_2$?

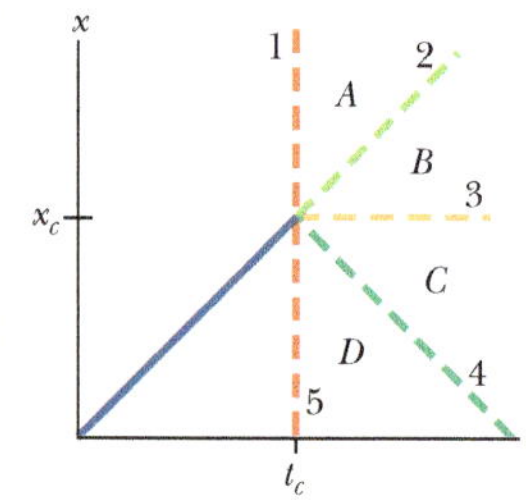

Figure 9-33 Question 11.

12 Figure 9-34 shows four graphs of position versus time for two bodies and their center of mass. The two bodies form a closed, isolated system and undergo a completely inelastic, one-dimensional collision on an x axis. In graph 1, are (a) the two bodies and (b) the center of mass moving in the positive or negative direction of the x axis? (c) Which of the graphs correspond to a physically impossible situation? Explain.

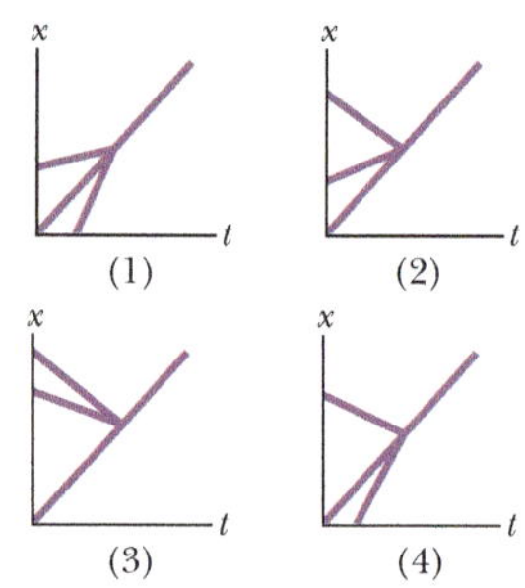

Figure 9-34 Question 12.

Problems

GO Tutoring problem available (at instructor's discretion) in *WileyPLUS* and WebAssign

SSM Worked-out solution available in Student Solutions Manual

• – ••• Number of dots indicates level of problem difficulty

Additional information available in *The Flying Circus of Physics* and at flyingcircusofphysics.com

WWW Worked-out solution is at

ILW Interactive solution is at

http://www.wiley.com/college/halliday

Module 9-1 Center of Mass

•1 A 2.00 kg particle has the xy coordinates $(-1.20\text{ m}, 0.500\text{ m})$, and a 4.00 kg particle has the xy coordinates $(0.600\text{ m}, -0.750\text{ m})$. Both lie on a horizontal plane. At what (a) x and (b) y coordinates must you place a 3.00 kg particle such that the center of mass of the three-particle system has the coordinates $(-0.500\text{ m}, -0.700\text{ m})$?

•2 Figure 9-35 shows a three-particle system, with masses $m_1 = 3.0$ kg, $m_2 = 4.0$ kg, and $m_3 = 8.0$ kg. The scales on the axes are set by $x_s = 2.0$ m and $y_s = 2.0$ m. What are (a) the x coordinate and (b) the y coordinate of the system's center of mass? (c) If m_3 is gradually increased, does the center of mass of the system shift toward or away from that particle, or does it remain stationary?

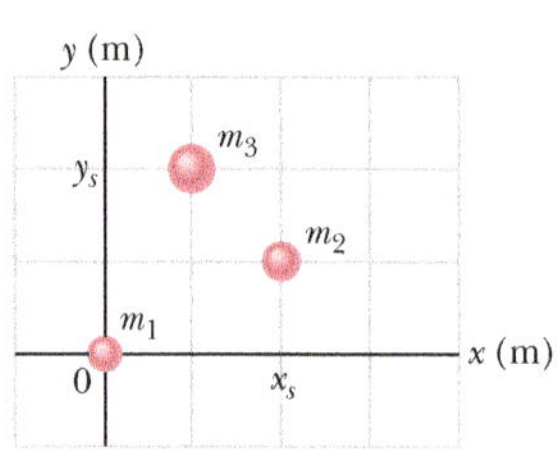

Figure 9-35 Problem 2.

••3 Figure 9-36 shows a slab with dimensions $d_1 = 11.0$ cm, $d_2 = 2.80$ cm, and $d_3 = 13.0$ cm. Half the slab consists of aluminum (density $= 2.70\text{ g/cm}^3$) and half consists of iron (density $= 7.85\text{ g/cm}^3$). What are (a) the x coordinate, (b) the y coordinate, and (c) the z coordinate of the slab's center of mass?

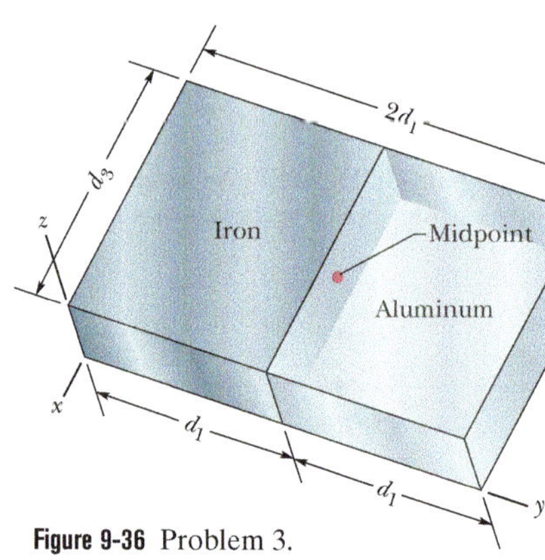

Figure 9-36 Problem 3.

••4 In Fig. 9-37, three uniform thin rods, each of length $L = 22$ cm, form an inverted U. The vertical rods each have a mass of 14 g; the horizontal rod has a mass of 42 g. What are (a) the x coordinate and (b) the y coordinate of the system's center of mass?

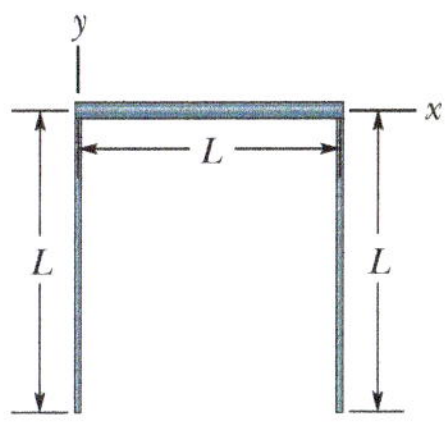

Figure 9-37 Problem 4.

••5 GO What are (a) the x coordinate and (b) the y coordinate of the center of mass for the uniform plate shown in Fig. 9-38 if $L = 5.0$ cm?

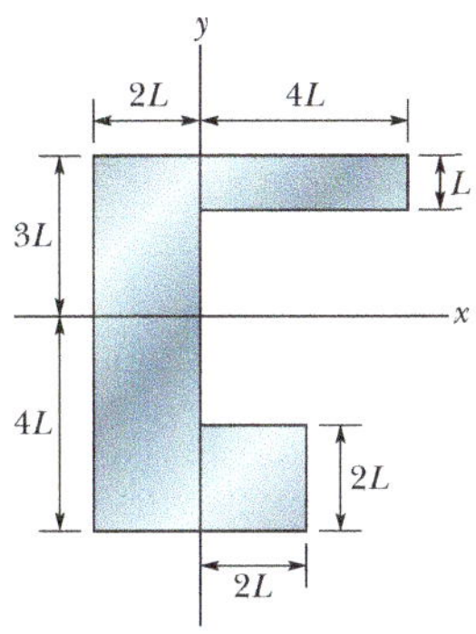

Figure 9-38 Problem 5.

••6 Figure 9-39 shows a cubical box that has been constructed from uniform metal plate of negligible thickness. The box is open at the top and has edge length $L =$ 40 cm. Find (a) the x coordinate, (b) the y coordinate, and (c) the z coordinate of the center of mass of the box.

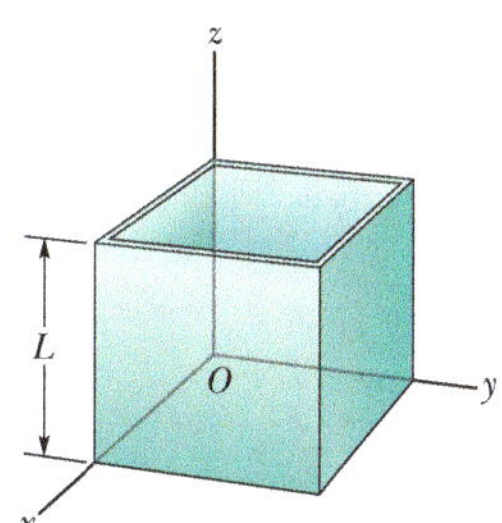

Figure 9-39 Problem 6.

•••7 ILW In the ammonia (NH_3) molecule of Fig. 9-40, three hydrogen (H) atoms form an equilateral triangle, with the center of the triangle at distance $d = 9.40 \times 10^{-11}$ m from each hydrogen atom. The nitrogen (N) atom is at the apex of a pyramid, with the three hydrogen atoms forming the base. The nitrogen-to-hydrogen atomic mass ratio is 13.9, and the nitrogen-to-hydrogen distance is $L = 10.14 \times 10^{-11}$ m. What are the (a) x and (b) y coordinates of the molecule's center of mass?

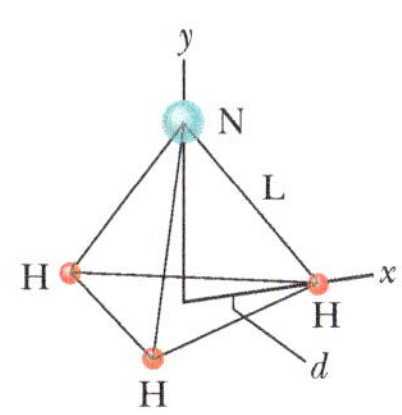

Figure 9-40 Problem 7.

•••8 GO A uniform soda can of mass 0.140 kg is 12.0 cm tall and filled with 0.354 kg of soda (Fig. 9-41). Then small holes are drilled in the top and bottom (with negligible loss of metal) to drain the soda. What is the height h of the com of the can and contents (a) initially and (b) after the can loses all the soda? (c) What happens to h as the soda drains out? (d) If x is the height of the remaining soda at any given instant, find x when the com reaches its lowest point.

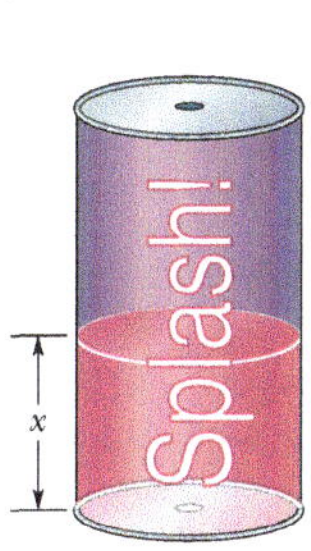

Figure 9-41 Problem 8.

Module 9-2 Newton's Second Law for a System of Particles

•9 ILW A stone is dropped at $t = 0$. A second stone, with twice the mass of the first, is dropped from the same point at $t = 100$ ms. (a) How far below the release point is the center of mass of the two stones at $t = 300$ ms? (Neither stone has yet reached the ground.) (b) How fast is the center of mass of the two-stone system moving at that time?

•10 GO A 1000 kg automobile is at rest at a traffic signal. At the instant the light turns green, the automobile starts to move with a constant acceleration of 4.0 m/s^2. At the same instant a 2000 kg truck, traveling at a constant speed of 8.0 m/s, overtakes and passes the automobile. (a) How far is the com of the automobile–truck system from the traffic light at $t = 3.0$ s? (b) What is the speed of the com then?

•11 A big olive ($m = 0.50$ kg) lies at the origin of an xy coordinate system, and a big Brazil nut ($M = 1.5$ kg) lies at the point (1.0, 2.0) m. At $t = 0$, a force $\vec{F}_o = (2.0\hat{i} + 3.0\hat{j})$ N begins to act on the olive, and a force $\vec{F}_n = (-3.0\hat{i} - 2.0\hat{j})$ N begins to act on the nut. In unit-vector notation, what is the displacement of the center of mass of the olive–nut system at $t = 4.0$ s, with respect to its position at $t = 0$?

•12 Two skaters, one with mass 65 kg and the other with mass 40 kg, stand on an ice rink holding a pole of length 10 m and negligible mass. Starting from the ends of the pole, the skaters pull themselves along the pole until they meet. How far does the 40 kg skater move?

••13 SSM A shell is shot with an initial velocity $\vec{v}_0$ of 20 m/s, at an angle of $\theta_0 = 60°$ with the horizontal. At the top of the trajectory, the shell explodes into two fragments of equal mass (Fig. 9-42). One fragment, whose speed immediately after the explosion is zero, falls vertically. How far from the gun does the other fragment land, assuming that the terrain is level and that air drag is negligible?

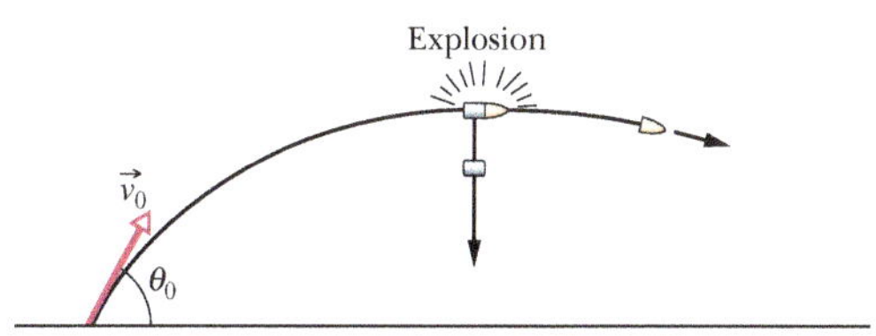

Figure 9-42 Problem 13.

••14 In Figure 9-43, two particles are launched from the origin of the coordinate system at time $t = 0$. Particle 1 of mass $m_1 = 5.00$ g is shot directly along the x axis on a frictionless floor, with constant speed 10.0 m/s. Particle 2 of mass $m_2 = 3.00$ g is shot with a velocity of magnitude 20.0 m/s, at an upward angle such that it always stays directly above particle 1. (a) What is the maximum height H_{max} reached by the com of the two-particle system? In unit-vector notation, what are the (b) velocity and (c) acceleration of the com when the com reaches H_{max}?

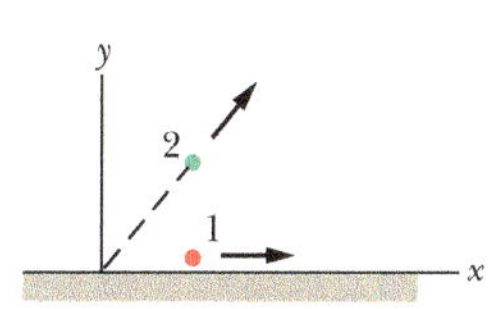

Figure 9-43 Problem 14.

••15 Figure 9-44 shows an arrangement with an air track, in which a cart is connected by a cord to a hanging block. The cart has mass $m_1 = 0.600$ kg, and its center is initially at xy coordinates $(-0.500$ m, 0 m); the block has mass $m_2 = 0.400$ kg, and its center is initially at xy coordinates $(0, -0.100$ m). The mass of the cord and pulley are negligible. The cart is released from rest, and both cart and block move until the cart hits the pulley. The friction between the cart and the air track and between the pulley and its axle is negligible. (a) In unit-vector notation, what is the acceleration of the center of mass of the cart–block system? (b) What is the velocity of the com as a function of time t? (c) Sketch the path taken by the com. (d) If the path is curved, determine whether it bulges upward to the right or downward to the left, and if it is straight, find the angle between it and the x axis.

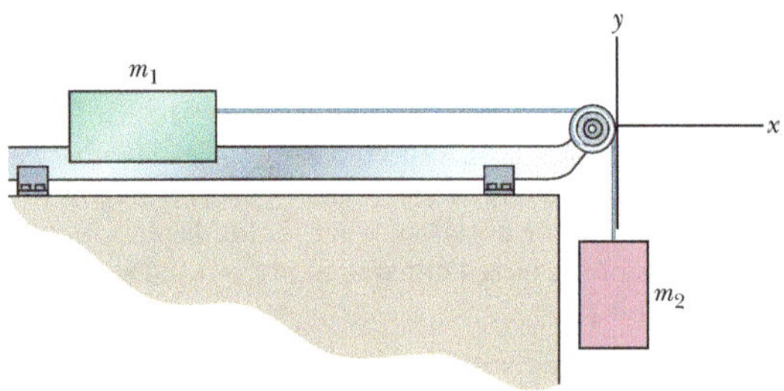

Figure 9-44 Problem 15.

•••16 GO Ricardo, of mass 80 kg, and Carmelita, who is lighter, are enjoying Lake Merced at dusk in a 30 kg canoe. When the canoe is at rest in the placid water, they exchange seats, which are 3.0 m apart and symmetrically located with respect to the canoe's center. If the canoe moves 40 cm horizontally relative to a pier post, what is Carmelita's mass?

•••17 GO In Fig. 9-45a, a 4.5 kg dog stands on an 18 kg flatboat at distance $D = 6.1$ m from the shore. It walks 2.4 m along the boat toward shore and then stops. Assuming no friction between the boat and the water, find how far the dog is then from the shore. (*Hint:* See Fig. 9-45b.)

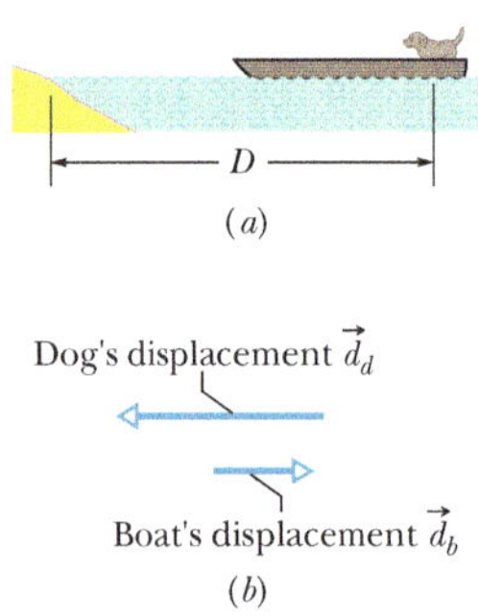

Figure 9-45 Problem 17.

Module 9-3 Linear Momentum

•18 A 0.70 kg ball moving horizontally at 5.0 m/s strikes a vertical wall and rebounds with speed 2.0 m/s. What is the magnitude of the change in its linear momentum?

•19 ILW A 2100 kg truck traveling north at 41 km/h turns east and accelerates to 51 km/h. (a) What is the change in the truck's kinetic energy? What are the (b) magnitude and (c) direction of the change in its momentum?

••20 GO At time $t = 0$, a ball is struck at ground level and sent over level ground. The momentum p versus t during the flight is given by Fig. 9-46 (with $p_0 = 6.0$ kg·m/s and $p_1 = 4.0$ kg·m/s). At what initial angle is the ball launched? (*Hint:* Find a solution that does not require you to read the time of the low point of the plot.)

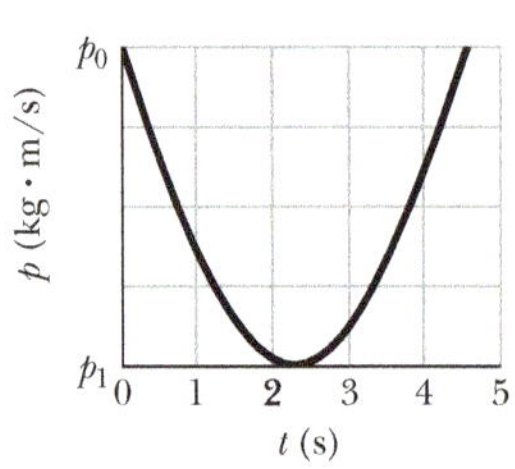

Figure 9-46 Problem 20.

••21 A 0.30 kg softball has a velocity of 15 m/s at an angle of 35° below the horizontal just before making contact with the bat. What is the magnitude of the change in momentum of the ball while in contact with the bat if the ball leaves with a velocity of (a) 20 m/s, vertically downward, and (b) 20 m/s, horizontally back toward the pitcher?

••22 Figure 9-47 gives an overhead view of the path taken by a 0.165 kg cue ball as it bounces from a rail of a pool table. The ball's initial speed is 2.00 m/s, and the angle θ_1 is 30.0°. The bounce reverses the y component of the ball's velocity but does not alter the x component. What are (a) angle θ_2 and (b) the change in the ball's linear momentum in unit-vector notation? (The fact that the ball rolls is irrelevant to the problem.)

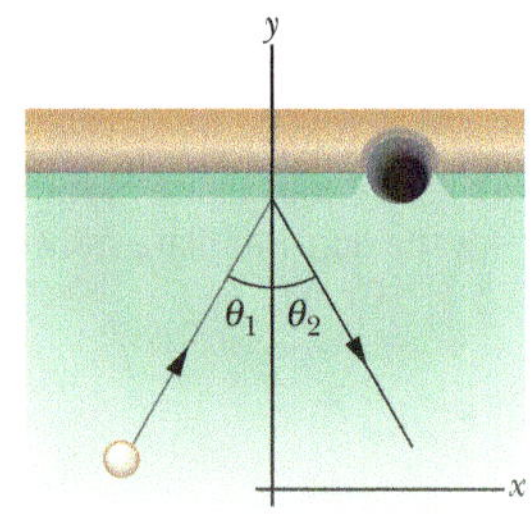

Figure 9-47 Problem 22.

Module 9-4 Collision and Impulse

•23 Until his seventies, Henri LaMothe (Fig. 9-48) excited audiences by belly-flopping from a height of 12 m into 30 cm of water. Assuming that he stops just as he reaches the bottom of the water and estimating his mass, find the magnitude of the impulse on him from the water.

George Long/Getty Images, Inc.

Figure 9-48 Problem 23. Belly-flopping into 30 cm of water.

•24 In February 1955, a paratrooper fell 370 m from an airplane without being able to open his chute but happened to land in snow, suffering only minor injuries. Assume that his speed at impact was 56 m/s (terminal speed), that his mass (including gear) was 85 kg, and that the magnitude of the force on him from the snow was at the survivable limit of 1.2×10^5 N. What are (a) the minimum depth of snow that would have stopped him safely and (b) the magnitude of the impulse on him from the snow?

•25 A 1.2 kg ball drops vertically onto a floor, hitting with a speed of 25 m/s. It rebounds with an initial speed of 10 m/s. (a) What impulse acts on the ball during the contact? (b) If the ball is in contact with the floor for 0.020 s, what is the magnitude of the average force on the floor from the ball?

•26 In a common but dangerous prank, a chair is pulled away as a person is moving downward to sit on it, causing the victim to land hard on the floor. Suppose the victim falls by 0.50 m, the mass that moves downward is 70 kg, and the collision on the floor lasts 0.082 s. What are the magnitudes of the (a) impulse and (b) average force acting on the victim from the floor during the collision?

•27 SSM A force in the negative direction of an x axis is applied for 27 ms to a 0.40 kg ball initially moving at 14 m/s in the positive direction of the axis. The force varies in magnitude, and the impulse has magnitude 32.4 N · s. What are the ball's (a) speed and (b) direction of travel just after the force is applied? What are (c) the average magnitude of the force and (d) the direction of the impulse on the ball?

•28 In tae-kwon-do, a hand is slammed down onto a target at a speed of 13 m/s and comes to a stop during the 5.0 ms collision. Assume that during the impact the hand is independent of the arm and has a mass of 0.70 kg. What are the magnitudes of the (a) impulse and (b) average force on the hand from the target?

•29 Suppose a gangster sprays Superman's chest with 3 g bullets at the rate of 100 bullets/min, and the speed of each bullet is 500 m/s. Suppose too that the bullets rebound straight back with no change in speed. What is the magnitude of the average force on Superman's chest?

••30 *Two average forces.* A steady stream of 0.250 kg snowballs is shot perpendicularly into a wall at a speed of 4.00 m/s. Each ball sticks to the wall. Figure 9-49 gives the magnitude F of the force on the wall as a function of time t for two of the snowball impacts. Impacts occur with a repetition time interval $\Delta t_r = 50.0$ ms, last a duration time interval $\Delta t_d = 10$ ms, and produce isosceles triangles on the graph, with each impact reaching a force maximum $F_{max} = 200$ N. During each impact, what are the magnitudes of (a) the impulse and (b) the average force on the wall? (c) During a time interval of many impacts, what is the magnitude of the average force on the wall?

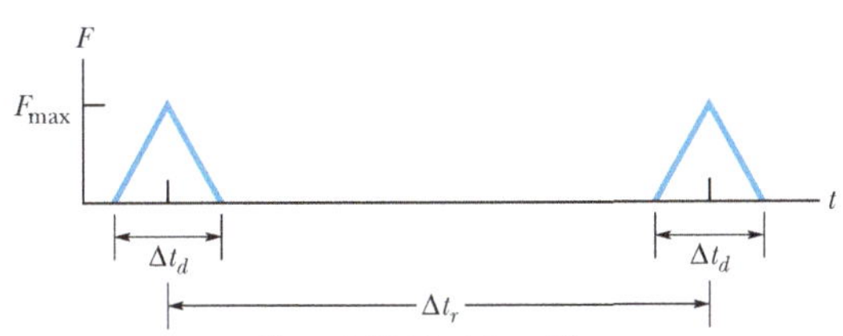

Figure 9-49 Problem 30.

••31 *Jumping up before the elevator hits.* After the cable snaps and the safety system fails, an elevator cab free-falls from a height of 36 m. During the collision at the bottom of the elevator shaft, a 90 kg passenger is stopped in 5.0 ms. (Assume that neither the passenger nor the cab rebounds.) What are the magnitudes of the (a) impulse and (b) average force on the passenger during the collision? If the passenger were to jump upward with a speed of 7.0 m/s relative to the cab floor just before the cab hits the bottom of the shaft, what are the magnitudes of the (c) impulse and (d) average force (assuming the same stopping time)?

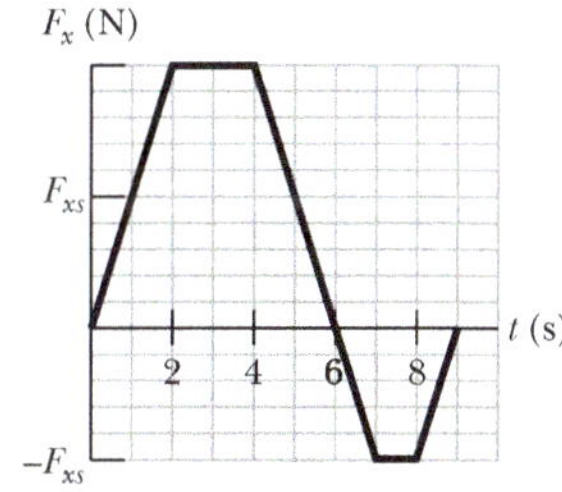

Figure 9-50 Problem 32.

••32 A 5.0 kg toy car can move along an x axis; Fig. 9-50 gives F_x of the force acting on the car, which begins at rest at time $t = 0$. The scale on the F_x axis is set by $F_{xs} = 5.0$ N. In unit-vector notation, what is $\vec{p}$ at (a) $t = 4.0$ s and (b) $t = 7.0$ s, and (c) what is $\vec{v}$ at $t = 9.0$ s?

••33 GO Figure 9-51 shows a 0.300 kg baseball just before and just after it collides with a bat. Just before, the ball has velocity $\vec{v}_1$ of magnitude 12.0 m/s and angle $\theta_1 = 35.0°$. Just after, it is traveling directly upward with velocity $\vec{v}_2$ of magnitude 10.0 m/s. The duration of the collision is 2.00 ms. What are the (a) magnitude and (b) direction (relative to the positive direction of the x axis) of the impulse on the ball from the bat? What are the (c) magnitude and (d) direction of the average force on the ball from the bat?

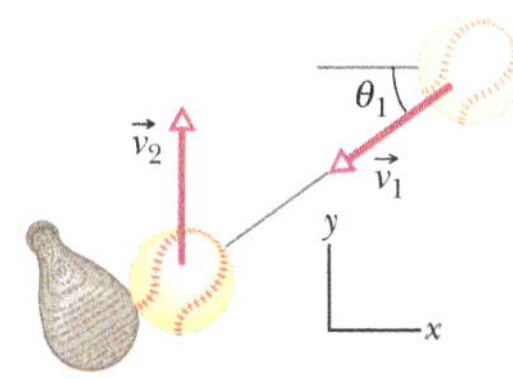

Figure 9-51 Problem 33.

••34 Basilisk lizards can run across the top of a water surface (Fig. 9-52). With each step, a lizard first slaps its foot against the water and then pushes it down into the water rapidly enough to form an air cavity around the top of the foot. To avoid having to pull the foot back up against water drag in order to complete the step, the lizard withdraws the foot before water can flow into the air cavity. If the lizard is not to sink, the average upward impulse on the lizard during this full action of slap, downward push, and withdrawal must match the downward impulse due to the gravitational force. Suppose the mass of a basilisk lizard is 90.0 g, the mass of each foot is 3.00 g, the speed of a foot as it slaps the water is 1.50 m/s, and the time for a single step is 0.600 s. (a) What is the magnitude of the impulse on the lizard during the slap? (Assume this impulse is directly upward.) (b) During the 0.600 s duration of a step, what is the downward impulse on the lizard due to the gravitational force? (c) Which action, the slap or the push, provides the primary support for the lizard, or are they approximately equal in their support?

Stephen Dalton/Photo Researchers, Inc.

Figure 9-52 Problem 34. Lizard running across water.

••35 GO Figure 9-53 shows an approximate plot of force magnitude F versus time t during the collision of a 58 g Superball with a wall. The initial velocity of the ball is 34 m/s perpendicular to the wall; the ball rebounds directly back with approximately the same speed, also perpendicular to the wall. What is F_{max}, the maximum magnitude of the force on the ball from the wall during the collision?

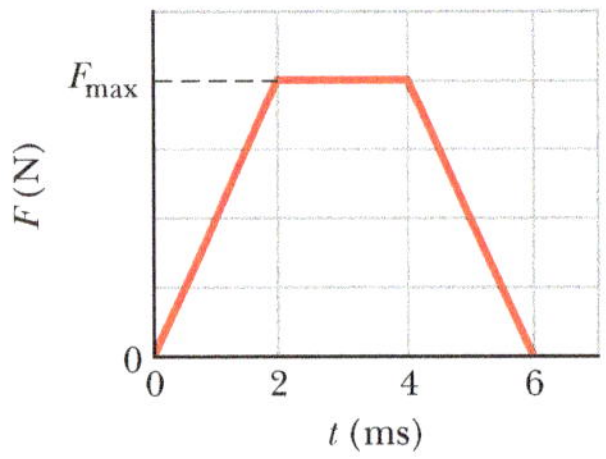

Figure 9-53 Problem 35.

••36 A 0.25 kg puck is initially stationary on an ice surface with negligible friction. At time $t = 0$, a horizontal force begins to move the puck. The force is given by $\vec{F} = (12.0 - 3.00t^2)\hat{i}$, with $\vec{F}$ in newtons and t in seconds, and it acts until its magnitude is zero. (a) What is the magnitude of the impulse on the puck from the force between $t = 0.500$ s and $t = 1.25$ s? (b) What is the change in momentum of the puck between $t = 0$ and the instant at which $F = 0$?

••37 SSM A soccer player kicks a soccer ball of mass 0.45 kg that is initially at rest. The foot of the player is in contact with the ball for 3.0×10^{-3} s, and the force of the kick is given by

$$F(t) = [(6.0 \times 10^6)t - (2.0 \times 10^9)t^2]\ \text{N}$$

for $0 \leq t \leq 3.0 \times 10^{-3}$ s, where t is in seconds. Find the magnitudes of (a) the impulse on the ball due to the kick, (b) the average force on the ball from the player's foot during the period of contact, (c) the maximum force on the ball from the player's foot during the period of contact, and (d) the ball's velocity immediately after it loses contact with the player's foot.

••38 In the overhead view of Fig. 9-54, a 300 g ball with a speed v of 6.0 m/s strikes a wall at an angle θ of 30° and then rebounds with the same speed and angle. It is in contact with the wall for 10 ms. In unit-vector notation, what are (a) the impulse on the ball from the wall and (b) the average force on the wall from the ball?

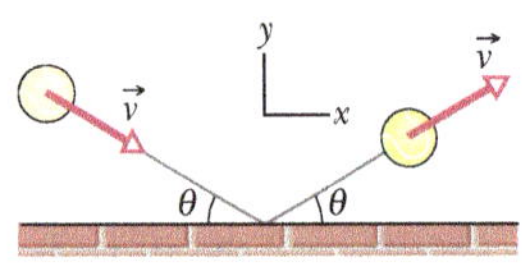

Figure 9-54 Problem 38.

Module 9-5 Conservation of Linear Momentum

•39 SSM A 91 kg man lying on a surface of negligible friction shoves a 68 g stone away from himself, giving it a speed of 4.0 m/s. What speed does the man acquire as a result?

•40 A space vehicle is traveling at 4300 km/h relative to Earth when the exhausted rocket motor (mass $4m$) is disengaged and sent backward with a speed of 82 km/h relative to the command module (mass m). What is the speed of the command module relative to Earth just after the separation?

••41 Figure 9-55 shows a two-ended "rocket" that is initially stationary on a frictionless floor, with its center at the origin of an x axis. The rocket consists of a central block C (of mass $M = 6.00$ kg) and blocks L and R (each of mass $m = 2.00$ kg) on the left and right sides. Small explosions can shoot either of the side blocks away from block C and along the x axis. Here is the sequence: (1) At time $t = 0$, block L is shot to the left with a speed of 3.00 m/s *relative* to the velocity that the explosion gives the rest of the rocket. (2) Next, at time $t = 0.80$ s, block R is shot to the right with a speed of 3.00 m/s *relative* to the velocity that block C then has. At $t = 2.80$ s, what are (a) the velocity of block C and (b) the position of its center?

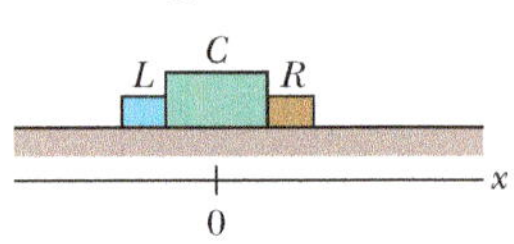

Figure 9-55 Problem 41.

••42 An object, with mass m and speed v relative to an observer, explodes into two pieces, one three times as massive as the other; the explosion takes place in deep space. The less massive piece stops relative to the observer. How much kinetic energy is added to the system during the explosion, as measured in the observer's reference frame?

••43 In the Olympiad of 708 B.C., some athletes competing in the standing long jump used handheld weights called *halteres* to lengthen their jumps (Fig. 9-56). The weights were swung up in front just before liftoff and then swung down and thrown backward during the flight. Suppose a modern 78 kg long jumper similarly uses two 5.50 kg halteres, throwing them horizontally to the rear at his maximum height such that their horizontal velocity is zero relative to the ground. Let his liftoff velocity be $\vec{v} = (9.5\hat{i} + 4.0\hat{j})$ m/s with or without the halteres, and assume that he lands at the liftoff level. What distance would the use of the halteres add to his range?

Réunion des Musées Nationaux/
Art Resource

Figure 9-56 Problem 43.

••44 GO In Fig. 9-57, a stationary block explodes into two pieces L and R that slide across a frictionless floor and then into regions with friction, where they stop. Piece L, with a mass of 2.0 kg, encounters a coefficient of kinetic friction $\mu_L = 0.40$ and slides to a stop in distance $d_L = 0.15$ m. Piece R encounters a coefficient of kinetic friction $\mu_R = 0.50$ and slides to a stop in distance $d_R = 0.25$ m. What was the mass of the block?

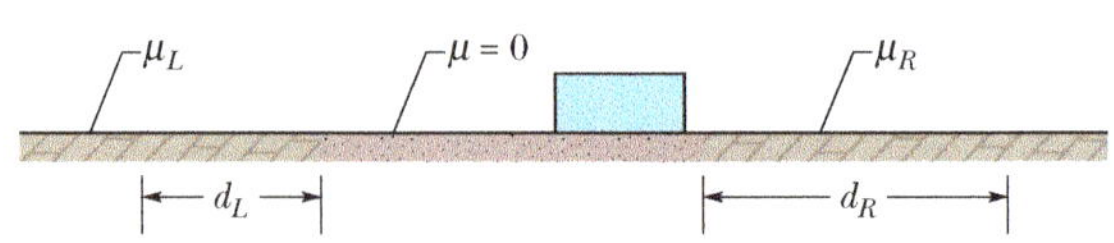

Figure 9-57 Problem 44.

••45 SSM WWW A 20.0 kg body is moving through space in the positive direction of an x axis with a speed of 200 m/s when, due to an internal explosion, it breaks into three parts. One part, with a mass of 10.0 kg, moves away from the point of explosion with a speed of 100 m/s in the positive y direction. A second part, with a mass of 4.00 kg, moves in the negative x direction with a speed of 500 m/s. (a) In unit-vector notation, what is the velocity of the third part? (b) How much energy is released in the explosion? Ignore effects due to the gravitational force.

••46 A 4.0 kg mess kit sliding on a frictionless surface explodes into two 2.0 kg parts: 3.0 m/s, due north, and 5.0 m/s, 30° north of east. What is the original speed of the mess kit?

••47 A vessel at rest at the origin of an xy coordinate system explodes into three pieces. Just after the explosion, one piece, of mass m, moves with velocity $(-30 \text{ m/s})\hat{i}$ and a second piece, also of mass m, moves with velocity $(-30 \text{ m/s})\hat{j}$. The third piece has mass $3m$. Just after the explosion, what are the (a) magnitude and (b) direction of the velocity of the third piece?

•••48 GO Particle A and particle B are held together with a compressed spring between them. When they are released, the spring pushes them apart, and they then fly off in opposite directions, free of the spring. The mass of A is 2.00 times the mass of B, and the energy stored in the spring was 60 J. Assume that the spring has negligible mass and that all its stored energy is transferred to the particles. Once that transfer is complete, what are the kinetic energies of (a) particle A and (b) particle B?

Module 9-6 Momentum and Kinetic Energy in Collisions

•49 A bullet of mass 10 g strikes a ballistic pendulum of mass 2.0 kg. The center of mass of the pendulum rises a vertical distance of 12 cm. Assuming that the bullet remains embedded in the pendulum, calculate the bullet's initial speed.

•50 A 5.20 g bullet moving at 672 m/s strikes a 700 g wooden block at rest on a frictionless surface. The bullet emerges, traveling in the same direction with its speed reduced to 428 m/s. (a) What is the resulting speed of the block? (b) What is the speed of the bullet–block center of mass?

••51 GO In Fig. 9-58a, a 3.50 g bullet is fired horizontally at two blocks at rest on a frictionless table. The bullet passes through block 1 (mass 1.20 kg) and embeds itself in block 2 (mass 1.80 kg). The blocks end up with speeds $v_1 = 0.630$ m/s and $v_2 = 1.40$ m/s (Fig. 9-58b). Neglecting the material removed from block 1 by the bullet, find the speed of the bullet as it (a) leaves and (b) enters block 1.

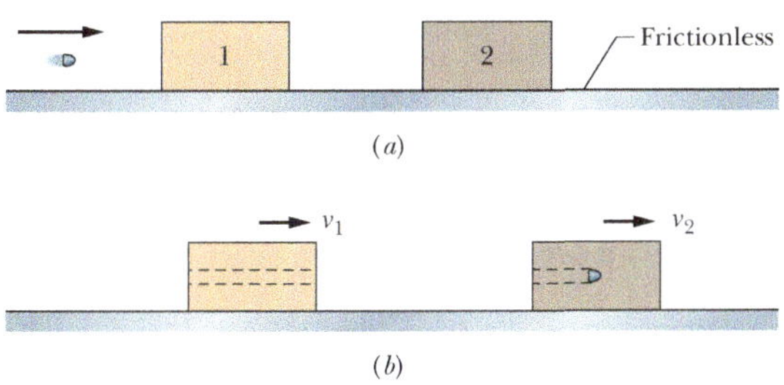

Figure 9-58 Problem 51.

••52 GO In Fig. 9-59, a 10 g bullet moving directly upward at 1000 m/s strikes and passes through the center of mass of a 5.0 kg block initially at rest. The bullet emerges from the block moving directly upward at 400 m/s. To what maximum height does the block then rise above its initial position?

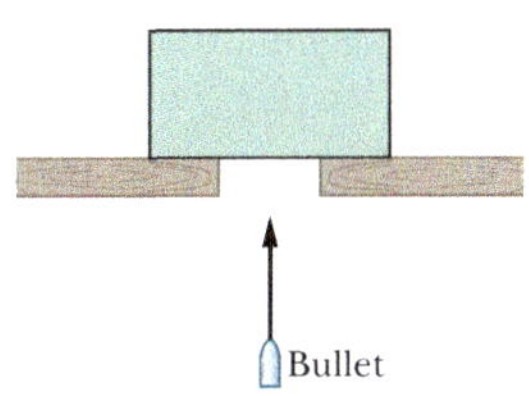

Figure 9-59 Problem 52.

••53 In Anchorage, collisions of a vehicle with a moose are so common that they are referred to with the abbreviation MVC. Suppose a 1000 kg car slides into a stationary 500 kg moose on a very slippery road, with the moose being thrown through the windshield (a common MVC result). (a) What percent of the original kinetic energy is lost in the collision to other forms of energy? A similar danger occurs in Saudi Arabia because of camel–vehicle collisions (CVC). (b) What percent of the original kinetic energy is lost if the car hits a 300 kg camel? (c) Generally, does the percent loss increase or decrease if the animal mass decreases?

••54 A completely inelastic collision occurs between two balls of wet putty that move directly toward each other along a vertical axis. Just before the collision, one ball, of mass 3.0 kg, is moving upward at 20 m/s and the other ball, of mass 2.0 kg, is moving downward at 12 m/s. How high do the combined two balls of putty rise above the collision point? (Neglect air drag.)

••55 ILW A 5.0 kg block with a speed of 3.0 m/s collides with a 10 kg block that has a speed of 2.0 m/s in the same direction. After the collision, the 10 kg block travels in the original direction with a speed of 2.5 m/s. (a) What is the velocity of the 5.0 kg block immediately after the collision? (b) By how much does the total kinetic energy of the system of two blocks change because of the collision? (c) Suppose, instead, that the 10 kg block ends up with a speed of 4.0 m/s. What then is the change in the total kinetic energy? (d) Account for the result you obtained in (c).

••56 In the "before" part of Fig. 9-60, car A (mass 1100 kg) is stopped at a traffic light when it is rear-ended by car B (mass 1400 kg). Both cars then slide with locked wheels until the frictional force from the slick road (with a low μ_k of 0.13) stops them, at distances $d_A = 8.2$ m and $d_B = 6.1$ m. What are the speeds of (a) car A and (b) car B at the start of the sliding, just after the collision? (c) Assuming that linear momentum is conserved during the collision, find the speed of car B just before the collision. (d) Explain why this assumption may be invalid.

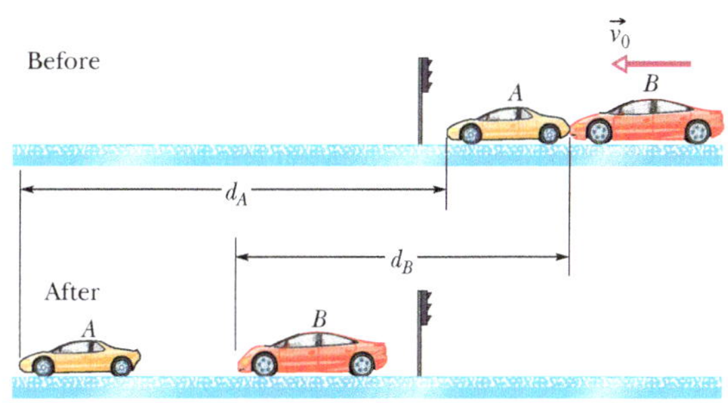

Figure 9-60 Problem 56.

••57 GO In Fig. 9-61, a ball of mass $m = 60$ g is shot with speed $v_i = 22$ m/s into the barrel of a spring gun of mass $M = 240$ g initially at rest on a frictionless surface. The ball sticks in the barrel at the point of maximum compression of the spring. Assume that the increase in thermal energy due to friction between the ball and the barrel is negligible. (a) What is the speed of the spring gun after the ball stops in the barrel? (b) What fraction of the initial kinetic energy of the ball is stored in the spring?

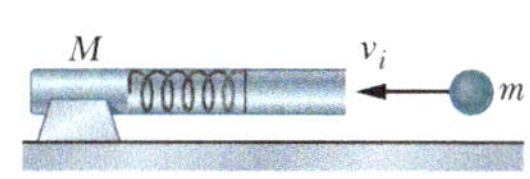

Figure 9-61 Problem 57.

•••58 In Fig. 9-62, block 2 (mass 1.0 kg) is at rest on a frictionless surface and touching the end of an unstretched spring of spring constant 200 N/m. The other end of the spring is fixed to a wall. Block 1 (mass 2.0 kg), traveling at speed $v_1 = 4.0$ m/s, collides with block 2, and the two blocks stick together. When the blocks momentarily stop, by what distance is the spring compressed?

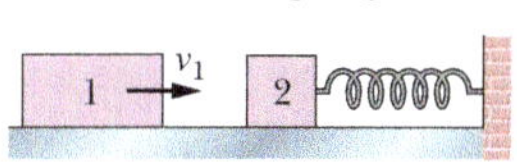

Figure 9-62 Problem 58.

•••59 ILW In Fig. 9-63, block 1 (mass 2.0 kg) is moving rightward at 10 m/s and block 2 (mass 5.0 kg) is moving rightward at 3.0 m/s. The surface is frictionless, and a spring with a spring constant of 1120 N/m is fixed to block 2. When the blocks collide, the compression of the spring is maximum at the instant the blocks have the same velocity. Find the maximum compression.

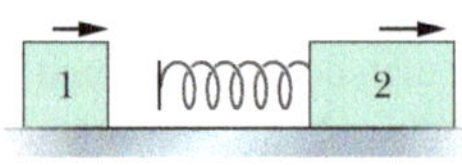

Figure 9-63 Problem 59.

Module 9-7 Elastic Collisions in One Dimension

•60 In Fig. 9-64, block A (mass 1.6 kg) slides into block B (mass 2.4 kg), along a frictionless surface. The directions of three velocities before (i) and after (f) the collision are indicated; the corresponding speeds are v_{Ai} = 5.5 m/s, v_{Bi} = 2.5 m/s, and v_{Bf} = 4.9 m/s. What are the (a) speed and (b) direction (left or right) of velocity $\vec{v}_{Af}$? (c) Is the collision elastic?

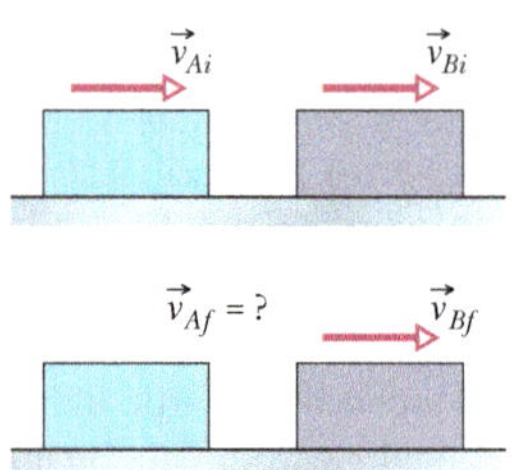

Figure 9-64 Problem 60.

•61 SSM A cart with mass 340 g moving on a frictionless linear air track at an initial speed of 1.2 m/s undergoes an elastic collision with an initially stationary cart of unknown mass. After the collision, the first cart continues in its original direction at 0.66 m/s. (a) What is the mass of the second cart? (b) What is its speed after impact? (c) What is the speed of the two-cart center of mass?

•62 Two titanium spheres approach each other head-on with the same speed and collide elastically. After the collision, one of the spheres, whose mass is 300 g, remains at rest. (a) What is the mass of the other sphere? (b) What is the speed of the two-sphere center of mass if the initial speed of each sphere is 2.00 m/s?

••63 Block 1 of mass m_1 slides along a frictionless floor and into a one-dimensional elastic collision with stationary block 2 of mass $m_2 = 3m_1$. Prior to the collision, the center of mass of the two-block system had a speed of 3.00 m/s. Afterward, what are the speeds of (a) the center of mass and (b) block 2?

••64 GO A steel ball of mass 0.500 kg is fastened to a cord that is 70.0 cm long and fixed at the far end. The ball is then released when the cord is horizontal (Fig. 9-65). At the bottom of its path, the ball strikes a 2.50 kg steel block initially at rest on a frictionless surface. The collision is elastic. Find (a) the speed of the ball and (b) the speed of the block, both just after the collision.

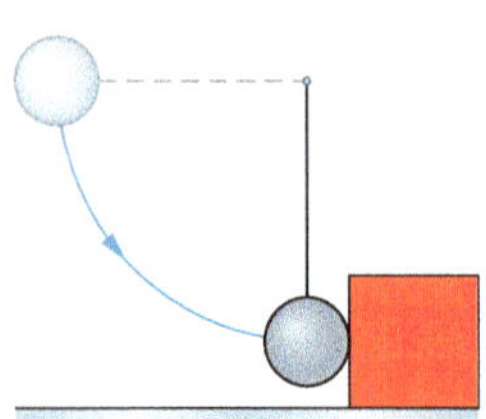

Figure 9-65 Problem 64.

••65 SSM A body of mass 2.0 kg makes an elastic collision with another body at rest and continues to move in the original direction but with one-fourth of its original speed. (a) What is the mass of the other body? (b) What is the speed of the two-body center of mass if the initial speed of the 2.0 kg body was 4.0 m/s?

••66 Block 1, with mass m_1 and speed 4.0 m/s, slides along an x axis on a frictionless floor and then undergoes a one-dimensional elastic collision with stationary block 2, with mass $m_2 = 0.40m_1$. The two blocks then slide into a region where the coefficient of kinetic friction is 0.50; there they stop. How far into that region do (a) block 1 and (b) block 2 slide?

••67 In Fig. 9-66, particle 1 of mass m_1 = 0.30 kg slides rightward along an x axis on a frictionless floor with a speed of 2.0 m/s. When it reaches x = 0, it undergoes a one-dimensional elastic collision with stationary particle 2 of mass m_2 = 0.40 kg. When particle 2 then reaches a wall at x_w = 70 cm, it bounces from the wall with no loss of speed. At what position on the x axis does particle 2 then collide with particle 1?

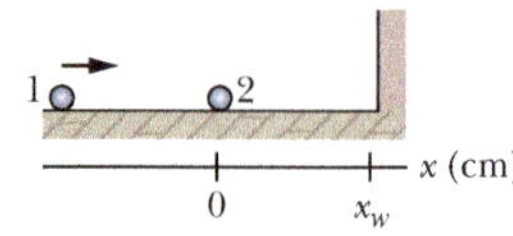

Figure 9-66 Problem 67.

••68 GO In Fig. 9-67, block 1 of mass m_1 slides from rest along a frictionless ramp from height h = 2.50 m and then collides with stationary block 2, which has mass $m_2 = 2.00m_1$. After the collision, block 2 slides into a region where the coefficient of kinetic friction μ_k is 0.500 and comes to a stop in distance d within that region. What is the value of distance d if the collision is (a) elastic and (b) completely inelastic?

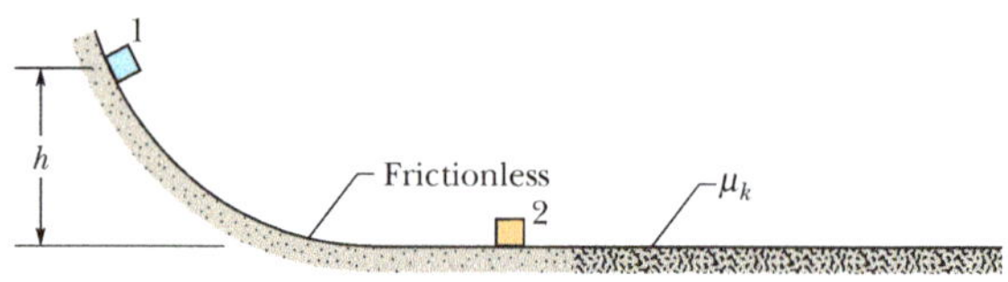

Figure 9-67 Problem 68.

•••69 GO A small ball of mass m is aligned above a larger ball of mass M = 0.63 kg (with a slight separation, as with the baseball and basketball of Fig. 9-68a), and the two are dropped simultaneously from a height of h = 1.8 m. (Assume the radius of each ball is negligible relative to h.) (a) If the larger ball rebounds elastically from the floor and then the small ball rebounds elastically from the larger ball, what value of m results in the larger ball stopping when it collides with the small ball? (b) What height does the small ball then reach (Fig. 9-68b)?

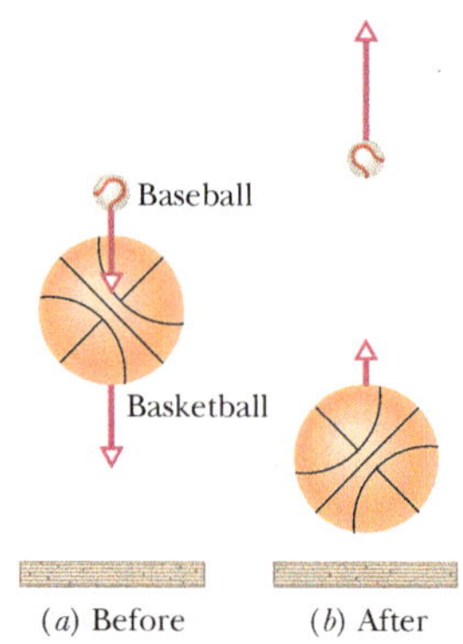

Figure 9-68 Problem 69.

•••70 GO In Fig. 9-69, puck 1 of mass m_1 = 0.20 kg is sent sliding across a frictionless lab bench, to undergo a one-dimensional elastic collision with stationary puck 2. Puck 2 then slides off the bench and lands a distance d from the base of the bench. Puck 1 rebounds from the collision and slides off the opposite edge of the bench, landing a distance 2d from the base of the bench. What is the mass of puck 2? (*Hint:* Be careful with signs.)

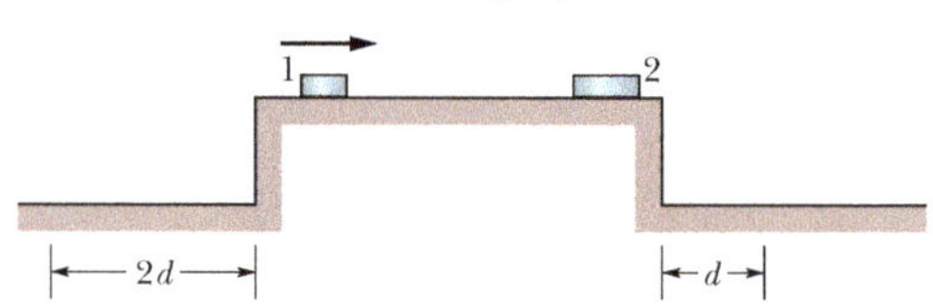

Figure 9-69 Problem 70.

Module 9-8 Collisions in Two Dimensions

••71 ILW In Fig. 9-21, projectile particle 1 is an alpha particle and target particle 2 is an oxygen nucleus. The alpha particle is scattered at angle $\theta_1 = 64.0°$ and the oxygen nucleus recoils with speed 1.20×10^5 m/s and at angle $\theta_2 = 51.0°$. In atomic mass units, the mass of the alpha particle is 4.00 u and the mass of the oxygen nucleus is 16.0 u. What are the (a) final and (b) initial speeds of the alpha particle?

••72 Ball B, moving in the positive direction of an x axis at speed v, collides with stationary ball A at the origin. A and B have different masses. After the collision, B moves in the negative direction of the y axis at speed $v/2$. (a) In what direction does A move? (b) Show that the speed of A cannot be determined from the given information.

••73 After a completely inelastic collision, two objects of the same mass and same initial speed move away together at half their initial speed. Find the angle between the initial velocities of the objects.

••74 Two 2.0 kg bodies, A and B, collide. The velocities before the collision are $\vec{v}_A = (15\hat{i} + 30\hat{j})$ m/s and $\vec{v}_B = (-10\hat{i} + 5.0\hat{j})$ m/s. After the collision, $\vec{v}\,'_A = (-5.0\hat{i} + 20\hat{j})$ m/s. What are (a) the final velocity of B and (b) the change in the total kinetic energy (including sign)?

••75 GO A projectile proton with a speed of 500 m/s collides elastically with a target proton initially at rest. The two protons then move along perpendicular paths, with the projectile path at 60° from the original direction. After the collision, what are the speeds of (a) the target proton and (b) the projectile proton?

Module 9-9 Systems with Varying Mass: A Rocket

•76 A 6090 kg space probe moving nose-first toward Jupiter at 105 m/s relative to the Sun fires its rocket engine, ejecting 80.0 kg of exhaust at a speed of 253 m/s relative to the space probe. What is the final velocity of the probe?

•77 SSM In Fig. 9-70, two long barges are moving in the same direction in still water, one with a speed of 10 km/h and the other with a speed of 20 km/h. While they are passing each other, coal is shoveled from the slower to the faster one at a rate of 1000 kg/min. How much additional force must be provided by the driving engines of (a) the faster barge and (b) the slower barge if neither is to change speed? Assume that the shoveling is always perfectly sideways and that the frictional forces between the barges and the water do not depend on the mass of the barges.

Figure 9-70 Problem 77.

•78 Consider a rocket that is in deep space and at rest relative to an inertial reference frame. The rocket's engine is to be fired for a certain interval. What must be the rocket's *mass ratio* (ratio of initial to final mass) over that interval if the rocket's original speed relative to the inertial frame is to be equal to (a) the exhaust speed (speed of the exhaust products relative to the rocket) and (b) 2.0 times the exhaust speed?

•79 SSM ILW A rocket that is in deep space and initially at rest relative to an inertial reference frame has a mass of 2.55×10^5 kg, of which 1.81×10^5 kg is fuel. The rocket engine is then fired for 250 s while fuel is consumed at the rate of 480 kg/s. The speed of the exhaust products relative to the rocket is 3.27 km/s. (a) What is the rocket's thrust? After the 250 s firing, what are (b) the mass and (c) the speed of the rocket?

Additional Problems

80 An object is tracked by a radar station and determined to have a position vector given by $\vec{r} = (3500 - 160t)\hat{i} + 2700\hat{j} + 300\hat{k}$, with $\vec{r}$ in meters and t in seconds. The radar station's x axis points east, its y axis north, and its z axis vertically up. If the object is a 250 kg meteorological missile, what are (a) its linear momentum, (b) its direction of motion, and (c) the net force on it?

81 The last stage of a rocket, which is traveling at a speed of 7600 m/s, consists of two parts that are clamped together: a rocket case with a mass of 290.0 kg and a payload capsule with a mass of 150.0 kg. When the clamp is released, a compressed spring causes the two parts to separate with a relative speed of 910.0 m/s. What are the speeds of (a) the rocket case and (b) the payload after they have separated? Assume that all velocities are along the same line. Find the total kinetic energy of the two parts (c) before and (d) after they separate. (e) Account for the difference.

82 *Pancake collapse of a tall building.* In the section of a tall building shown in Fig. 9-71a, the infrastructure of any given floor K must support the weight W of all higher floors. Normally the infrastructure is constructed with a safety factor s so that it can withstand an even greater downward force of sW. If, however, the support columns between K and L suddenly collapse and allow the higher floors to free-fall together onto floor K (Fig. 9-71b), the force in the collision can exceed sW and, after a brief pause, cause K to collapse onto floor J, which collapses on floor I, and so on until the ground is reached. Assume that the floors are separated by $d = 4.0$ m and have the same mass. Also assume that when the floors above K free-fall onto K, the collision lasts 1.5 ms. Under these simplified conditions, what value must the safety factor s exceed to prevent pancake collapse of the building?

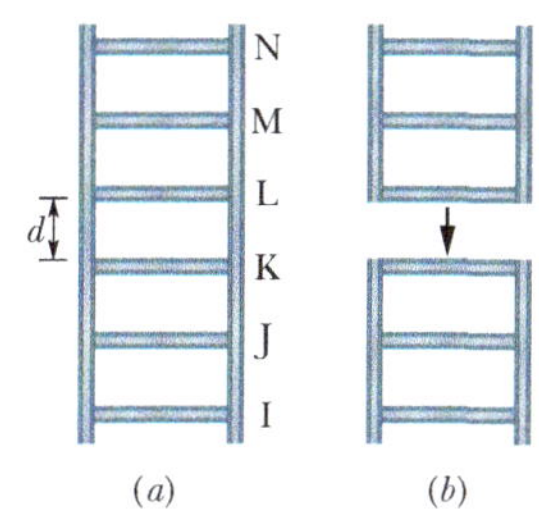

Figure 9-71 Problem 82.

83 *"Relative" is an important word.* In Fig. 9-72, block L of mass $m_L = 1.00$ kg and block R of mass $m_R = 0.500$ kg are held in place with a compressed spring between them. When the blocks are released, the spring sends them sliding across a frictionless floor. (The spring has negligible mass and falls to the floor after the blocks leave it.) (a) If the spring gives block L a release speed of 1.20 m/s *relative* to the floor, how far does block R travel in the next 0.800 s? (b) If, instead, the spring gives block L a release speed of 1.20 m/s *relative* to the velocity that the spring gives block R, how far does block R travel in the next 0.800 s?

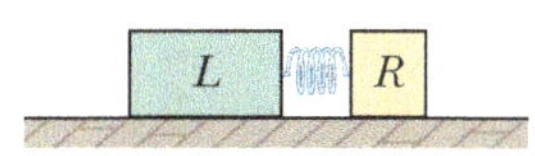

Figure 9-72 Problem 83.

84 Figure 9-73 shows an overhead view of two particles sliding at constant velocity over a frictionless surface. The particles have the same mass and the same initial speed $v = 4.00$ m/s, and they collide where their paths intersect. An x axis is arranged to bisect the angle between their incoming paths, such that $\theta = 40.0°$. The region to the right of the collision is divided into four lettered sections by the x axis and four numbered dashed lines. In what region or along what line do the particles travel if the collision is (a) completely inelastic, (b) elastic, and (c) inelastic? What are their final speeds if the collision is (d) completely inelastic and (e) elastic?

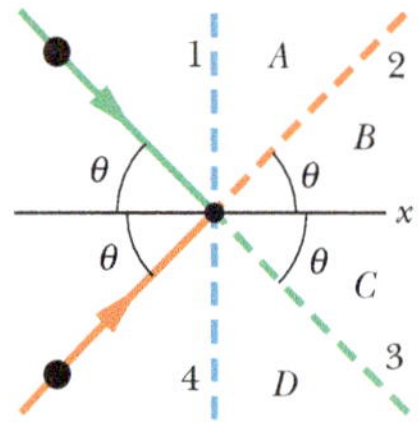

Figure 9-73 Problem 84.

85 *Speed deamplifier.* In Fig. 9-74, block 1 of mass m_1 slides along an x axis on a frictionless floor at speed 4.00 m/s. Then it undergoes a one-dimensional elastic collision with stationary block 2 of mass $m_2 = 2.00m_1$. Next, block 2 undergoes a one-dimensional elastic collision with stationary block 3 of mass $m_3 = 2.00m_2$. (a) What then is the speed of block 3? Are (b) the speed, (c) the kinetic energy, and (d) the momentum of block 3 greater than, less than, or the same as the initial values for block 1?

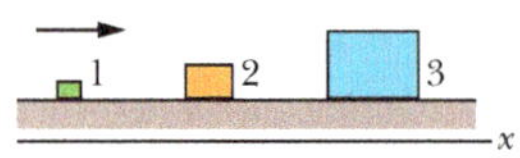

Figure 9-74 Problem 85.

86 *Speed amplifier.* In Fig. 9-75, block 1 of mass m_1 slides along an x axis on a frictionless floor with a speed of $v_{1i} = 4.00$ m/s. Then it undergoes a one-dimensional elastic collision with stationary block 2 of mass $m_2 = 0.500m_1$. Next, block 2 undergoes a one-dimensional elastic collision with stationary block 3 of mass $m_3 = 0.500m_2$. (a) What then is the speed of block 3? Are (b) the speed, (c) the kinetic energy, and (d) the momentum of block 3 greater than, less than, or the same as the initial values for block 1?

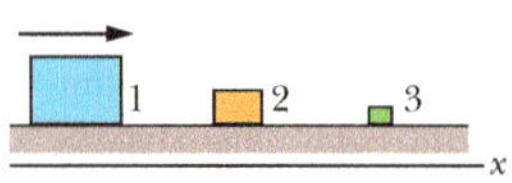

Figure 9-75 Problem 86.

87 A ball having a mass of 150 g strikes a wall with a speed of 5.2 m/s and rebounds with only 50% of its initial kinetic energy. (a) What is the speed of the ball immediately after rebounding? (b) What is the magnitude of the impulse on the wall from the ball? (c) If the ball is in contact with the wall for 7.6 ms, what is the magnitude of the average force on the ball from the wall during this time interval?

88 A spacecraft is separated into two parts by detonating the explosive bolts that hold them together. The masses of the parts are 1200 kg and 1800 kg; the magnitude of the impulse on each part from the bolts is 300 N·s. With what relative speed do the two parts separate because of the detonation?

89 SSM A 1400 kg car moving at 5.3 m/s is initially traveling north along the positive direction of a y axis. After completing a 90° right-hand turn in 4.6 s, the inattentive operator drives into a tree, which stops the car in 350 ms. In unit-vector notation, what is the impulse on the car (a) due to the turn and (b) due to the collision? What is the magnitude of the average force that acts on the car (c) during the turn and (d) during the collision? (e) What is the direction of the average force during the turn?

90 ILW A certain radioactive (parent) nucleus transforms to a different (daughter) nucleus by emitting an electron and a neutrino. The parent nucleus was at rest at the origin of an xy coordinate system. The electron moves away from the origin with linear momentum $(-1.2 \times 10^{-22}\ \text{kg}\cdot\text{m/s})\hat{\text{i}}$; the neutrino moves away from the origin with linear momentum $(-6.4 \times 10^{-23}\ \text{kg}\cdot\text{m/s})\hat{\text{j}}$. What are the (a) magnitude and (b) direction of the linear momentum of the daughter nucleus? (c) If the daughter nucleus has a mass of 5.8×10^{-26} kg, what is its kinetic energy?

91 A 75 kg man rides on a 39 kg cart moving at a velocity of 2.3 m/s. He jumps off with zero horizontal velocity relative to the ground. What is the resulting change in the cart's velocity, including sign?

92 Two blocks of masses 1.0 kg and 3.0 kg are connected by a spring and rest on a frictionless surface. They are given velocities toward each other such that the 1.0 kg block travels initially at 1.7 m/s toward the center of mass, which remains at rest. What is the initial speed of the other block?

93 SSM A railroad freight car of mass 3.18×10^4 kg collides with a stationary caboose car. They couple together, and 27.0% of the initial kinetic energy is transferred to thermal energy, sound, vibrations, and so on. Find the mass of the caboose.

94 An old Chrysler with mass 2400 kg is moving along a straight stretch of road at 80 km/h. It is followed by a Ford with mass 1600 kg moving at 60 km/h. How fast is the center of mass of the two cars moving?

95 SSM In the arrangement of Fig. 9-21, billiard ball 1 moving at a speed of 2.2 m/s undergoes a glancing collision with identical billiard ball 2 that is at rest. After the collision, ball 2 moves at speed 1.1 m/s, at an angle of $\theta_2 = 60°$. What are (a) the magnitude and (b) the direction of the velocity of ball 1 after the collision? (c) Do the given data suggest the collision is elastic or inelastic?

96 A rocket is moving away from the solar system at a speed of 6.0×10^3 m/s. It fires its engine, which ejects exhaust with a speed of 3.0×10^3 m/s relative to the rocket. The mass of the rocket at this time is 4.0×10^4 kg, and its acceleration is 2.0 m/s². (a) What is the thrust of the engine? (b) At what rate, in kilograms per second, is exhaust ejected during the firing?

97 The three balls in the overhead view of Fig. 9-76 are identical. Balls 2 and 3 touch each other and are aligned perpendicular to the path of ball 1. The velocity of ball 1 has magnitude $v_0 = 10$ m/s and is directed at the contact point of balls 1 and 2. After the collision, what are the (a) speed and (b) direction of the velocity of ball 2, the (c) speed and (d) direction of the velocity of ball 3, and the (e) speed and (f) direction of the velocity of ball 1? (*Hint:* With friction absent, each impulse is directed along the line connecting the centers of the colliding balls, normal to the colliding surfaces.)

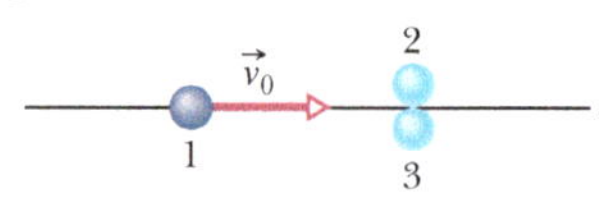

Figure 9-76 Problem 97.

98 A 0.15 kg ball hits a wall with a velocity of $(5.00\ \text{m/s})\hat{\text{i}} + (6.50\ \text{m/s})\hat{\text{j}} + (4.00\ \text{m/s})\hat{\text{k}}$. It rebounds from the wall with a velocity of $(2.00\ \text{m/s})\hat{\text{i}} + (3.50\ \text{m/s})\hat{\text{j}} + (-3.20\ \text{m/s})\hat{\text{k}}$. What are (a) the change in the ball's momentum, (b) the impulse on the ball, and (c) the impulse on the wall?

99 In Fig. 9-77, two identical containers of sugar are connected by a cord that passes over a frictionless pulley. The cord and pulley have negligible mass, each container and its sugar together have a mass of 500 g, the centers of the containers are separated by 50 mm, and the containers are held fixed at the same height. What is the horizontal distance between the center of container 1 and the center of mass of the two-container system (a) initially and

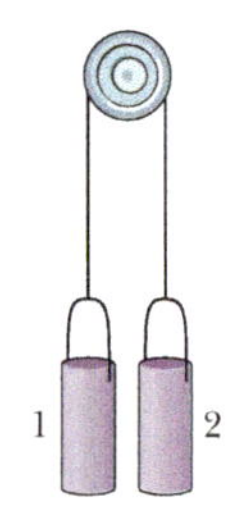

Figure 9-77 Problem 99.

(b) after 20 g of sugar is transferred from container 1 to container 2? After the transfer and after the containers are released, (c) in what direction and (d) at what acceleration magnitude does the center of mass move?

100 In a game of pool, the cue ball strikes another ball of the same mass and initially at rest. After the collision, the cue ball moves at 3.50 m/s along a line making an angle of 22.0° with the cue ball's original direction of motion, and the second ball has a speed of 2.00 m/s. Find (a) the angle between the direction of motion of the second ball and the original direction of motion of the cue ball and (b) the original speed of the cue ball. (c) Is kinetic energy (of the centers of mass, don't consider the rotation) conserved?

101 In Fig. 9-78, a 3.2 kg box of running shoes slides on a horizontal frictionless table and collides with a 2.0 kg box of ballet slippers initially at rest on the edge of the table, at height $h = 0.40$ m. The speed of the 3.2 kg box is 3.0 m/s just before the collision. If the two boxes stick together because of packing tape on their sides, what is their kinetic energy just before they strike the floor?

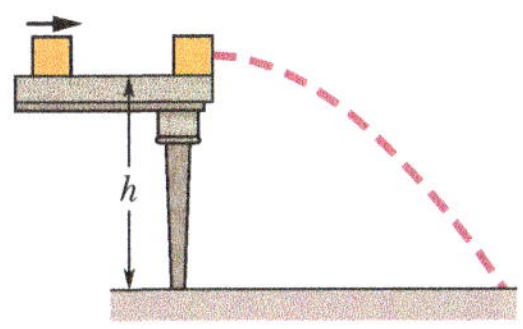

Figure 9-78 Problem 101.

102 In Fig. 9-79, an 80 kg man is on a ladder hanging from a balloon that has a total mass of 320 kg (including the basket passenger). The balloon is initially stationary relative to the ground. If the man on the ladder begins to climb at 2.5 m/s relative to the ladder, (a) in what direction and (b) at what speed does the balloon move? (c) If the man then stops climbing, what is the speed of the balloon?

Figure 9-79 Problem 102.

103 In Fig. 9-80, block 1 of mass $m_1 = 6.6$ kg is at rest on a long frictionless table that is up against a wall. Block 2 of mass m_2 is placed between block 1 and the wall and sent sliding to the left, toward block 1, with constant speed v_{2i}. Find the value of m_2 for which both blocks move with the same velocity after block 2 has collided once with block 1 and once with the wall. Assume all collisions are elastic (the collision with the wall does not change the speed of block 2).

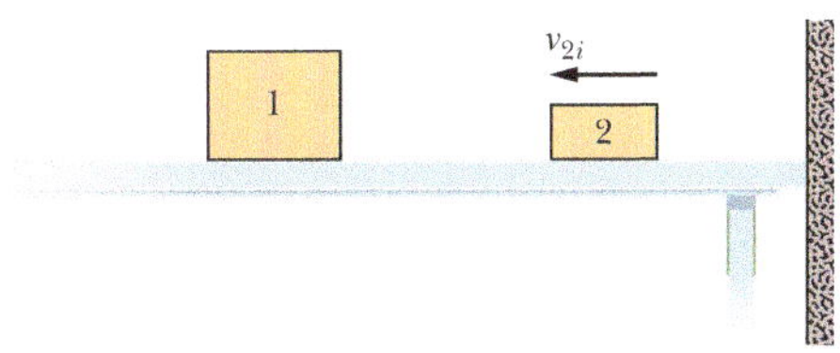

Figure 9-80 Problem 103.

104 The script for an action movie calls for a small race car (of mass 1500 kg and length 3.0 m) to accelerate along a flattop boat (of mass 4000 kg and length 14 m), from one end of the boat to the other, where the car will then jump the gap between the boat and a somewhat lower dock. You are the technical advisor for the movie. The boat will initially touch the dock, as in Fig. 9-81; the boat can slide through the water without significant resistance; both the car and the boat can be approximated as uniform in their mass distribution. Determine what the width of the gap will be just as the car is about to make the jump.

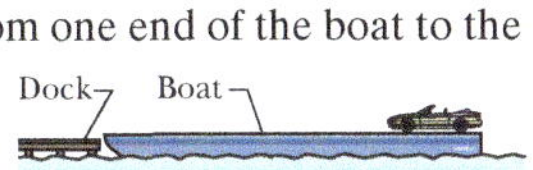

Figure 9-81 Problem 104.

105 SSM A 3.0 kg object moving at 8.0 m/s in the positive direction of an x axis has a one-dimensional elastic collision with an object of mass M, initially at rest. After the collision the object of mass M has a velocity of 6.0 m/s in the positive direction of the axis. What is mass M?

106 A 2140 kg railroad flatcar, which can move with negligible friction, is motionless next to a platform. A 242 kg sumo wrestler runs at 5.3 m/s along the platform (parallel to the track) and then jumps onto the flatcar. What is the speed of the flatcar if he then (a) stands on it, (b) runs at 5.3 m/s relative to it in his original direction, and (c) turns and runs at 5.3 m/s relative to the flatcar opposite his original direction?

107 SSM A 6100 kg rocket is set for vertical firing from the ground. If the exhaust speed is 1200 m/s, how much gas must be ejected each second if the thrust (a) is to equal the magnitude of the gravitational force on the rocket and (b) is to give the rocket an initial upward acceleration of 21 m/s^2?

108 A 500.0 kg module is attached to a 400.0 kg shuttle craft, which moves at 1000 m/s relative to the stationary main spaceship. Then a small explosion sends the module backward with speed 100.0 m/s relative to the new speed of the shuttle craft. As measured by someone on the main spaceship, by what fraction did the kinetic energy of the module and shuttle craft increase because of the explosion?

109 SSM (a) How far is the center of mass of the Earth–Moon system from the center of Earth? (Appendix C gives the masses of Earth and the Moon and the distance between the two.) (b) What percentage of Earth's radius is that distance?

110 A 140 g ball with speed 7.8 m/s strikes a wall perpendicularly and rebounds in the opposite direction with the same speed. The collision lasts 3.80 ms. What are the magnitudes of the (a) impulse and (b) average force on the wall from the ball during the elastic collision?

111 SSM A rocket sled with a mass of 2900 kg moves at 250 m/s on a set of rails. At a certain point, a scoop on the sled dips into a trough of water located between the tracks and scoops water into an empty tank on the sled. By applying the principle of conservation of linear momentum, determine the speed of the sled after 920 kg of water has been scooped up. Ignore any retarding force on the scoop.

112 SSM A pellet gun fires ten 2.0 g pellets per second with a speed of 500 m/s. The pellets are stopped by a rigid wall. What are (a) the magnitude of the momentum of each pellet, (b) the kinetic energy of each pellet, and (c) the magnitude of the average force on the wall from the stream of pellets? (d) If each pellet is in contact with the wall for 0.60 ms, what is the magnitude of the average force on the wall from each pellet during contact? (e) Why is this average force so different from the average force calculated in (c)?

113 A railroad car moves under a grain elevator at a constant speed of 3.20 m/s. Grain drops into the car at the rate of 540 kg/min. What is the magnitude of the force needed to keep the car moving at constant speed if friction is negligible?

114 Figure 9-82 shows a uniform square plate of edge length $6d = 6.0$ m from which a square piece of edge length $2d$ has been removed. What are (a) the x coordinate and (b) the y coordinate of the center of mass of the remaining piece?

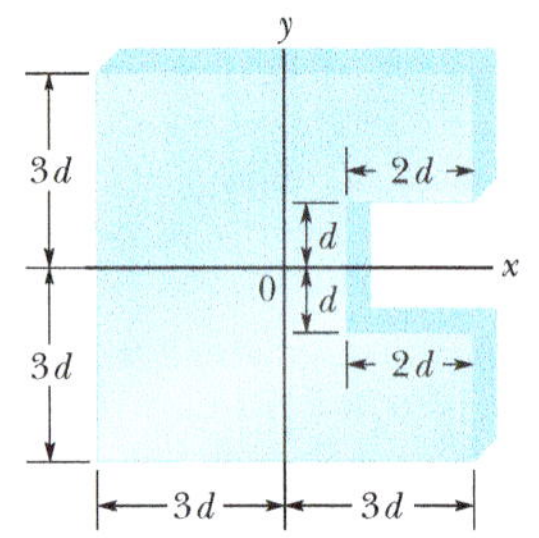

Figure 9-82 Problem 114.

115 SSM At time $t = 0$, force $\vec{F}_1 = (-4.00\hat{i} + 5.00\hat{j})$ N acts on an initially stationary particle of mass 2.00×10^{-3} kg and force $\vec{F}_2 = (2.00\hat{i} - 4.00\hat{j})$ N acts on an initially stationary particle of mass 4.00×10^{-3} kg. From time $t = 0$ to $t = 2.00$ ms, what are the (a) magnitude and (b) angle (relative to the positive direction of the x axis) of the displacement of the center of mass of the two-particle system? (c) What is the kinetic energy of the center of mass at $t = 2.00$ ms?

116 Two particles P and Q are released from rest 1.0 m apart. P has a mass of 0.10 kg, and Q a mass of 0.30 kg. P and Q attract each other with a constant force of 1.0×10^{-2} N. No external forces act on the system. (a) What is the speed of the center of mass of P and Q when the separation is 0.50 m? (b) At what distance from P's original position do the particles collide?

117 A collision occurs between a 2.00 kg particle traveling with velocity $\vec{v}_1 = (-4.00 \text{ m/s})\hat{i} + (-5.00 \text{ m/s})\hat{j}$ and a 4.00 kg particle traveling with velocity $\vec{v}_2 = (6.00 \text{ m/s})\hat{i} + (-2.00 \text{ m/s})\hat{j}$. The collision connects the two particles. What then is their velocity in (a) unit-vector notation and as a (b) magnitude and (c) angle?

118 In the two-sphere arrangement of Fig. 9-20, assume that sphere 1 has a mass of 50 g and an initial height of $h_1 = 9.0$ cm, and that sphere 2 has a mass of 85 g. After sphere 1 is released and collides elastically with sphere 2, what height is reached by (a) sphere 1 and (b) sphere 2? After the next (elastic) collision, what height is reached by (c) sphere 1 and (d) sphere 2? (*Hint:* Do not use rounded-off values.)

119 In Fig. 9-83, block 1 slides along an x axis on a frictionless floor with a speed of 0.75 m/s. When it reaches stationary block 2, the two blocks undergo an elastic collision. The following table gives the mass and length of the (uniform) blocks and also the locations of their centers at time $t = 0$. Where is the center of mass of the two-block system located (a) at $t = 0$, (b) when the two blocks first touch, and (c) at $t = 4.0$ s?

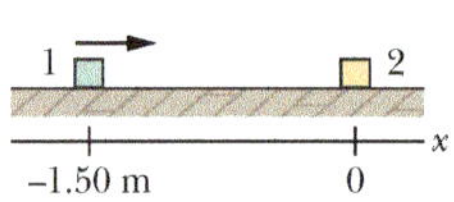

Figure 9-83 Problem 119.

Block	Mass (kg)	Length (cm)	Center at $t = 0$
1	0.25	5.0	$x = -1.50$ m
2	0.50	6.0	$x = 0$

120 A body is traveling at 2.0 m/s along the positive direction of an x axis; no net force acts on the body. An internal explosion separates the body into two parts, each of 4.0 kg, and increases the total kinetic energy by 16 J. The forward part continues to move in the original direction of motion. What are the speeds of (a) the rear part and (b) the forward part?

121 An electron undergoes a one-dimensional elastic collision with an initially stationary hydrogen atom. What percentage of the electron's initial kinetic energy is transferred to kinetic energy of the hydrogen atom? (The mass of the hydrogen atom is 1840 times the mass of the electron.)

122 A man (weighing 915 N) stands on a long railroad flatcar (weighing 2415 N) as it rolls at 18.2 m/s in the positive direction of an x axis, with negligible friction. Then the man runs along the flatcar in the negative x direction at 4.00 m/s relative to the flatcar. What is the resulting increase in the speed of the flatcar?

123 An unmanned space probe (of mass m and speed v relative to the Sun) approaches the planet Jupiter (of mass M and speed V_J relative to the Sun) as shown in Fig. 9-84. The spacecraft rounds the planet and departs in the opposite direction. What is its speed (in kilometers per second), relative to the Sun, after this slingshot encounter, which can be analyzed as a collision? Assume $v = 10.5$ km/s and $V_J = 13.0$ km/s (the orbital speed of Jupiter). The mass of Jupiter is very much greater than the mass of the spacecraft ($M \gg m$).

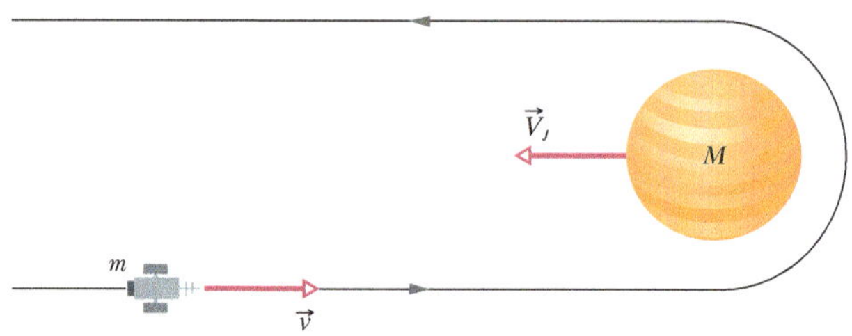

Figure 9-84 Problem 123.

124 A 0.550 kg ball falls directly down onto concrete, hitting it with a speed of 12.0 m/s and rebounding directly upward with a speed of 3.00 m/s. Extend a y axis upward. In unit-vector notation, what are (a) the change in the ball's momentum, (b) the impulse on the ball, and (c) the impulse on the concrete?

125 An atomic nucleus at rest at the origin of an xy coordinate system transforms into three particles. Particle 1, mass 16.7×10^{-27} kg, moves away from the origin at velocity $(6.00 \times 10^6 \text{ m/s})\hat{i}$; particle 2, mass 8.35×10^{-27} kg, moves away at velocity $(-8.00 \times 10^6 \text{ m/s})\hat{j}$. (a) In unit-vector notation, what is the linear momentum of the third particle, mass 11.7×10^{-27} kg? (b) How much kinetic energy appears in this transformation?

126 Particle 1 of mass 200 g and speed 3.00 m/s undergoes a one-dimensional collision with stationary particle 2 of mass 400 g. What is the magnitude of the impulse on particle 1 if the collision is (a) elastic and (b) completely inelastic?

127 During a lunar mission, it is necessary to increase the speed of a spacecraft by 2.2 m/s when it is moving at 400 m/s relative to the Moon. The speed of the exhaust products from the rocket engine is 1000 m/s relative to the spacecraft. What fraction of the initial mass of the spacecraft must be burned and ejected to accomplish the speed increase?

128 A cue stick strikes a stationary pool ball, with an average force of 32 N over a time of 14 ms. If the ball has mass 0.20 kg, what speed does it have just after impact?

CHAPTER 10

Rotation

10-1 ROTATIONAL VARIABLES

Learning Objectives

After reading this module, you should be able to . . .

10.01 Identify that if all parts of a body rotate around a fixed axis locked together, the body is a rigid body. (This chapter is about the motion of such bodies.)

10.02 Identify that the angular position of a rotating rigid body is the angle that an internal reference line makes with a fixed, external reference line.

10.03 Apply the relationship between angular displacement and the initial and final angular positions.

10.04 Apply the relationship between average angular velocity, angular displacement, and the time interval for that displacement.

10.05 Apply the relationship between average angular acceleration, change in angular velocity, and the time interval for that change.

10.06 Identify that counterclockwise motion is in the positive direction and clockwise motion is in the negative direction.

10.07 Given angular position as a *function of time*, calculate the instantaneous angular velocity at any particular time and the average angular velocity between any two particular times.

10.08 Given a *graph* of angular position versus time, determine the instantaneous angular velocity at a particular time and the average angular velocity between any two particular times.

10.09 Identify instantaneous angular speed as the magnitude of the instantaneous angular velocity.

10.10 Given angular velocity as a *function of time*, calculate the instantaneous angular acceleration at any particular time and the average angular acceleration between any two particular times.

10.11 Given a *graph* of angular velocity versus time, determine the instantaneous angular acceleration at any particular time and the average angular acceleration between any two particular times.

10.12 Calculate a body's change in angular velocity by integrating its angular acceleration function with respect to time.

10.13 Calculate a body's change in angular position by integrating its angular velocity function with respect to time.

Key Ideas

- To describe the rotation of a rigid body about a fixed axis, called the rotation axis, we assume a reference line is fixed in the body, perpendicular to that axis and rotating with the body. We measure the angular position θ of this line relative to a fixed direction. When θ is measured in radians,

$$\theta = \frac{s}{r} \quad \text{(radian measure)},$$

where s is the arc length of a circular path of radius r and angle θ.

- Radian measure is related to angle measure in revolutions and degrees by

$$1 \text{ rev} = 360° = 2\pi \text{ rad}.$$

- A body that rotates about a rotation axis, changing its angular position from θ_1 to θ_2, undergoes an angular displacement

$$\Delta\theta = \theta_2 - \theta_1,$$

where $\Delta\theta$ is positive for counterclockwise rotation and negative for clockwise rotation.

- If a body rotates through an angular displacement $\Delta\theta$ in a time interval Δt, its average angular velocity ω_{avg} is

$$\omega_{avg} = \frac{\Delta\theta}{\Delta t}.$$

The (instantaneous) angular velocity ω of the body is

$$\omega = \frac{d\theta}{dt}.$$

Both ω_{avg} and ω are vectors, with directions given by a right-hand rule. They are positive for counterclockwise rotation and negative for clockwise rotation. The magnitude of the body's angular velocity is the angular speed.

- If the angular velocity of a body changes from ω_1 to ω_2 in a time interval $\Delta t = t_2 - t_1$, the average angular acceleration α_{avg} of the body is

$$\alpha_{avg} = \frac{\omega_2 - \omega_1}{t_2 - t_1} = \frac{\Delta\omega}{\Delta t}.$$

The (instantaneous) angular acceleration α of the body is

$$\alpha = \frac{d\omega}{dt}.$$

Both α_{avg} and α are vectors.

What Is Physics?

As we have discussed, one focus of physics is motion. However, so far we have examined only the motion of **translation,** in which an object moves along a straight or curved line, as in Fig. 10-1*a*. We now turn to the motion of **rotation,** in which an object turns about an axis, as in Fig. 10-1*b*.

You see rotation in nearly every machine, you use it every time you open a beverage can with a pull tab, and you pay to experience it every time you go to an amusement park. Rotation is the key to many fun activities, such as hitting a long drive in golf (the ball needs to rotate in order for the air to keep it aloft longer) and throwing a curveball in baseball (the ball needs to rotate in order for the air to push it left or right). Rotation is also the key to more serious matters, such as metal failure in aging airplanes.

We begin our discussion of rotation by defining the variables for the motion, just as we did for translation in Chapter 2. As we shall see, the variables for rotation are analogous to those for one-dimensional motion and, as in Chapter 2, an important special situation is where the acceleration (here the rotational acceleration) is constant. We shall also see that Newton's second law can be written for rotational motion, but we must use a new quantity called *torque* instead of just force. Work and the work–kinetic energy theorem can also be applied to rotational motion, but we must use a new quantity called *rotational inertia* instead of just mass. In short, much of what we have discussed so far can be applied to rotational motion with, perhaps, a few changes.

Caution: In spite of this repetition of physics ideas, many students find this and the next chapter very challenging. Instructors have a variety of reasons as to why, but two reasons stand out: (1) There are a lot of symbols (with Greek

(*a*) (*b*)

Figure 10-1 Figure skater Sasha Cohen in motion of (*a*) pure translation in a fixed direction and (*b*) pure rotation about a vertical axis.

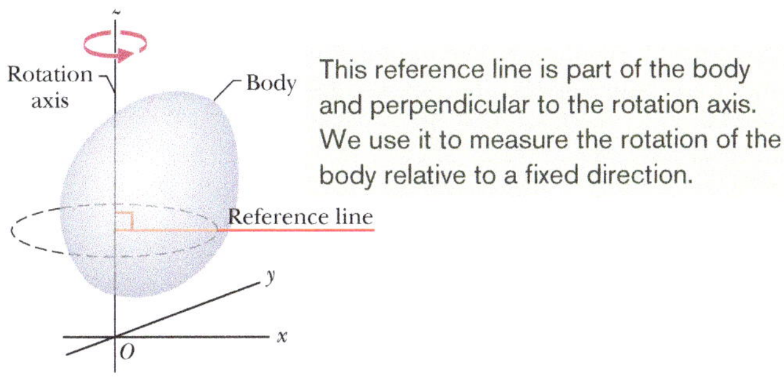

Figure 10-2 A rigid body of arbitrary shape in pure rotation about the z axis of a coordinate system. The position of the *reference line* with respect to the rigid body is arbitrary, but it is perpendicular to the rotation axis. It is fixed in the body and rotates with the body.

letters) to sort out. (2) Although you are very familiar with linear motion (you can get across the room and down the road just fine), you are probably very unfamiliar with rotation (and that is one reason why you are willing to pay so much for amusement park rides). If a homework problem looks like a foreign language to you, see if translating it into the one-dimensional linear motion of Chapter 2 helps. For example, if you are to find, say, an *angular* distance, temporarily delete the word *angular* and see if you can work the problem with the Chapter 2 notation and ideas.

Rotational Variables

We wish to examine the rotation of a rigid body about a fixed axis. A **rigid body** is a body that can rotate with all its parts locked together and without any change in its shape. A **fixed axis** means that the rotation occurs about an axis that does not move. Thus, we shall not examine an object like the Sun, because the parts of the Sun (a ball of gas) are not locked together. We also shall not examine an object like a bowling ball rolling along a lane, because the ball rotates about a moving axis (the ball's motion is a mixture of rotation and translation).

Figure 10-2 shows a rigid body of arbitrary shape in rotation about a fixed axis, called the **axis of rotation** or the **rotation axis.** In pure rotation (*angular motion*), every point of the body moves in a circle whose center lies on the axis of rotation, and every point moves through the same angle during a particular time interval. In pure translation (*linear motion*), every point of the body moves in a straight line, and every point moves through the same *linear distance* during a particular time interval.

We deal now—one at a time—with the angular equivalents of the linear quantities position, displacement, velocity, and acceleration.

Angular Position

Figure 10-2 shows a *reference line*, fixed in the body, perpendicular to the rotation axis and rotating with the body. The **angular position** of this line is the angle of the line relative to a fixed direction, which we take as the **zero angular position.** In Fig. 10-3, the angular position θ is measured relative to the positive direction of the x axis. From geometry, we know that θ is given by

$$\theta = \frac{s}{r} \quad \text{(radian measure).} \tag{10-1}$$

Here s is the length of a circular arc that extends from the x axis (the zero angular position) to the reference line, and r is the radius of the circle.

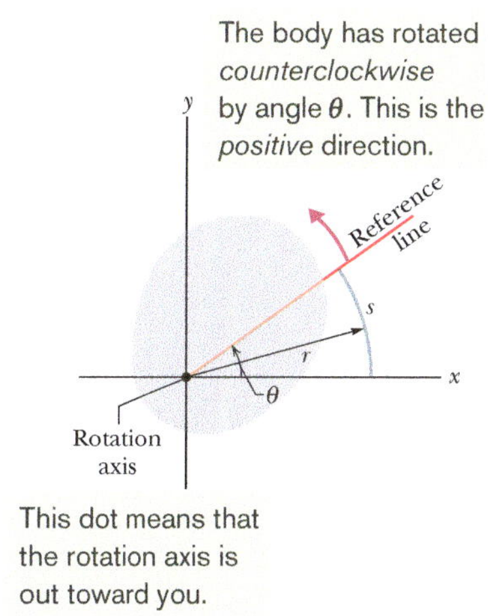

Figure 10-3 The rotating rigid body of Fig. 10-2 in cross section, viewed from above. The plane of the cross section is perpendicular to the rotation axis, which now extends out of the page, toward you. In this position of the body, the reference line makes an angle θ with the x axis.

An angle defined in this way is measured in **radians** (rad) rather than in revolutions (rev) or degrees. The radian, being the ratio of two lengths, is a pure number and thus has no dimension. Because the circumference of a circle of radius r is $2\pi r$, there are 2π radians in a complete circle:

$$1 \text{ rev} = 360° = \frac{2\pi r}{r} = 2\pi \text{ rad}, \tag{10-2}$$

and thus

$$1 \text{ rad} = 57.3° = 0.159 \text{ rev}. \tag{10-3}$$

We do *not* reset θ to zero with each complete rotation of the reference line about the rotation axis. If the reference line completes two revolutions from the zero angular position, then the angular position θ of the line is $\theta = 4\pi$ rad.

For pure translation along an x axis, we can know all there is to know about a moving body if we know $x(t)$, its position as a function of time. Similarly, for pure rotation, we can know all there is to know about a rotating body if we know $\theta(t)$, the angular position of the body's reference line as a function of time.

Angular Displacement

If the body of Fig. 10-3 rotates about the rotation axis as in Fig. 10-4, changing the angular position of the reference line from θ_1 to θ_2, the body undergoes an **angular displacement** $\Delta\theta$ given by

$$\Delta\theta = \theta_2 - \theta_1. \tag{10-4}$$

This definition of angular displacement holds not only for the rigid body as a whole but also for *every particle within that body.*

Clocks Are Negative. If a body is in translational motion along an x axis, its displacement Δx is either positive or negative, depending on whether the body is moving in the positive or negative direction of the axis. Similarly, the angular displacement $\Delta\theta$ of a rotating body is either positive or negative, according to the following rule:

An angular displacement in the counterclockwise direction is positive, and one in the clockwise direction is negative.

The phrase "*clocks are negative*" can help you remember this rule (they certainly are negative when their alarms sound off early in the morning).

Checkpoint 1

A disk can rotate about its central axis like a merry-go-round. Which of the following pairs of values for its initial and final angular positions, respectively, give a negative angular displacement: (a) 3 rad, 5 rad, (b) 3 rad, 7 rad, (c) 7 rad, 3 rad?

Angular Velocity

Suppose that our rotating body is at angular position θ_1 at time t_1 and at angular position θ_2 at time t_2 as in Fig. 10-4. We define the **average angular velocity** of the body in the time interval Δt from t_1 to t_2 to be

$$\omega_{\text{avg}} = \frac{\theta_2 - \theta_1}{t_2 - t_1} = \frac{\Delta\theta}{\Delta t}, \tag{10-5}$$

where $\Delta\theta$ is the angular displacement during Δt (ω is the lowercase omega).

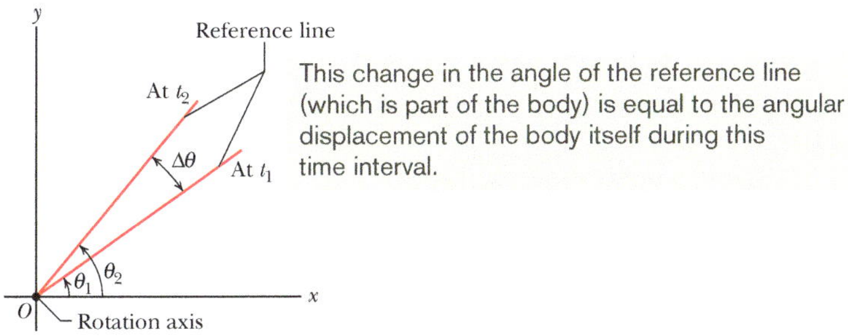

Figure 10-4 The reference line of the rigid body of Figs. 10-2 and 10-3 is at angular position θ_1 at time t_1 and at angular position θ_2 at a later time t_2. The quantity $\Delta\theta\ (= \theta_2 - \theta_1)$ is the angular displacement that occurs during the interval $\Delta t\ (= t_2 - t_1)$. The body itself is not shown.

The **(instantaneous) angular velocity** ω, with which we shall be most concerned, is the limit of the ratio in Eq. 10-5 as Δt approaches zero. Thus,

$$\omega = \lim_{\Delta t \to 0} \frac{\Delta\theta}{\Delta t} = \frac{d\theta}{dt}. \tag{10-6}$$

If we know $\theta(t)$, we can find the angular velocity ω by differentiation.

Equations 10-5 and 10-6 hold not only for the rotating rigid body as a whole but also for *every particle of that body* because the particles are all locked together. The unit of angular velocity is commonly the radian per second (rad/s) or the revolution per second (rev/s). Another measure of angular velocity was used during at least the first three decades of rock: Music was produced by vinyl (phonograph) records that were played on turntables at "$33\frac{1}{3}$ **rpm**" or "45 rpm," meaning at $33\frac{1}{3}$ **rev/min** or 45 rev/min.

If a particle moves in translation along an x axis, its linear velocity v is either positive or negative, depending on its direction along the axis. Similarly, the angular velocity ω of a rotating rigid body is either positive or negative, depending on whether the body is rotating counterclockwise (positive) or clockwise (negative). ("Clocks are negative" still works.) The magnitude of an angular velocity is called the **angular speed,** which is also represented with ω.

Angular Acceleration

If the angular velocity of a rotating body is not constant, then the body has an angular acceleration. Let ω_2 and ω_1 be its angular velocities at times t_2 and t_1, respectively. The **average angular acceleration** of the rotating body in the interval from t_1 to t_2 is defined as

$$\alpha_{\text{avg}} = \frac{\omega_2 - \omega_1}{t_2 - t_1} = \frac{\Delta\omega}{\Delta t}, \tag{10-7}$$

in which $\Delta\omega$ is the change in the angular velocity that occurs during the time interval Δt. The **(instantaneous) angular acceleration** α, with which we shall be most concerned, is the limit of this quantity as Δt approaches zero. Thus,

$$\alpha = \lim_{\Delta t \to 0} \frac{\Delta\omega}{\Delta t} = \frac{d\omega}{dt}. \tag{10-8}$$

As the name suggests, this is the angular acceleration of the body at a given instant. Equations 10-7 and 10-8 also hold for *every particle of that body.* The unit of angular acceleration is commonly the radian per second-squared (rad/s^2) or the revolution per second-squared (rev/s^2).

Sample Problem 10.01 Angular velocity derived from angular position

The disk in Fig. 10-5*a* is rotating about its central axis like a merry-go-round. The angular position $\theta(t)$ of a reference line on the disk is given by

$$\theta = -1.00 - 0.600t + 0.250t^2, \qquad (10\text{-}9)$$

with t in seconds, θ in radians, and the zero angular position as indicated in the figure. (If you like, you can translate all this into Chapter 2 notation by momentarily dropping the word "angular" from "angular position" and replacing the symbol θ with the symbol x. What you then have is an equation that gives the position as a function of time, for the one-dimensional motion of Chapter 2.)

(a) Graph the angular position of the disk versus time from $t = -3.0$ s to $t = 5.4$ s. Sketch the disk and its angular position reference line at $t = -2.0$ s, 0 s, and 4.0 s, and when the curve crosses the t axis.

KEY IDEA

The angular position of the disk is the angular position $\theta(t)$ of its reference line, which is given by Eq. 10-9 as a function of time t. So we graph Eq. 10-9; the result is shown in Fig. 10-5*b*.

Calculations: To sketch the disk and its reference line at a particular time, we need to determine θ for that time. To do so, we substitute the time into Eq. 10-9. For $t = -2.0$ s, we get

$$\theta = -1.00 - (0.600)(-2.0) + (0.250)(-2.0)^2$$

$$= 1.2 \text{ rad} = 1.2 \text{ rad}\,\frac{360°}{2\pi \text{ rad}} = 69°.$$

This means that at $t = -2.0$ s the reference line on the disk is rotated counterclockwise from the zero position by angle 1.2 rad = 69° (counterclockwise because θ is positive). Sketch 1 in Fig. 10-5*b* shows this position of the reference line.

Similarly, for $t = 0$, we find $\theta = -1.00$ rad $= -57°$, which means that the reference line is rotated clockwise from the zero angular position by 1.0 rad, or 57°, as shown in sketch 3. For $t = 4.0$ s, we find $\theta = 0.60$ rad $= 34°$ (sketch 5). Drawing sketches for when the curve crosses the t axis is easy, because then $\theta = 0$ and the reference line is momentarily aligned with the zero angular position (sketches 2 and 4).

(b) At what time t_{min} does $\theta(t)$ reach the minimum value shown in Fig. 10-5*b*? What is that minimum value?

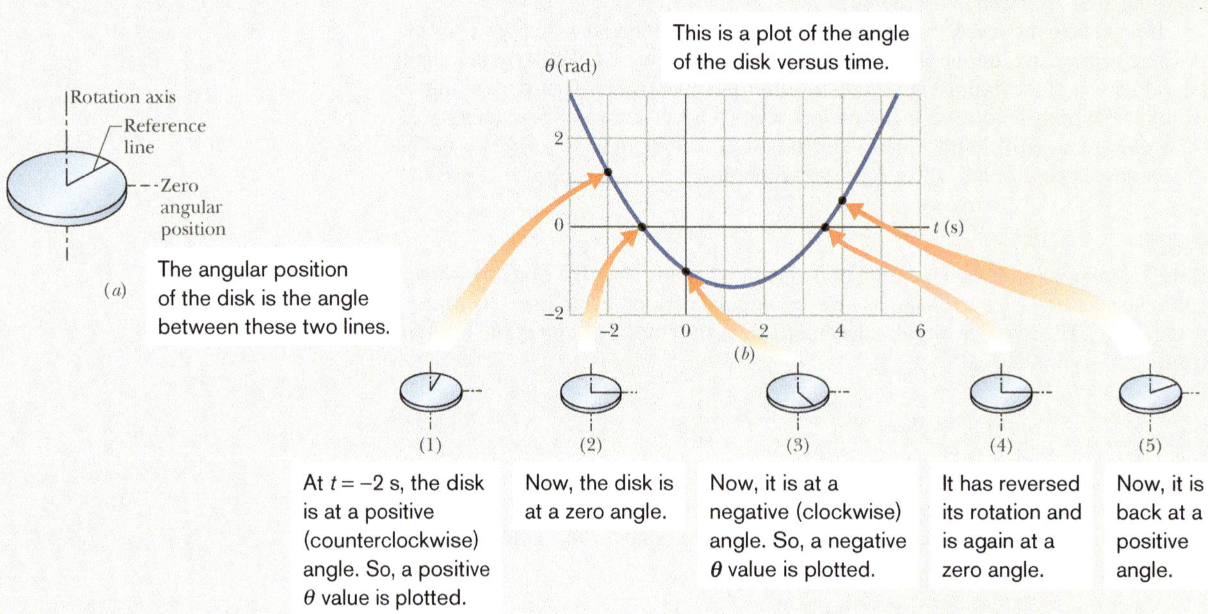

Figure 10-5 (*a*) A rotating disk. (*b*) A plot of the disk's angular position $\theta(t)$. Five sketches indicate the angular position of the reference line on the disk for five points on the curve. (*c*) A plot of the disk's angular velocity $\omega(t)$. Positive values of ω correspond to counterclockwise rotation, and negative values to clockwise rotation.

KEY IDEA

To find the extreme value (here the minimum) of a function, we take the first derivative of the function and set the result to zero.

Calculations: The first derivative of $\theta(t)$ is

$$\frac{d\theta}{dt} = -0.600 + 0.500t. \tag{10-10}$$

Setting this to zero and solving for t give us the time at which $\theta(t)$ is minimum:

$$t_{min} = 1.20 \text{ s}. \qquad \text{(Answer)}$$

To get the minimum value of θ, we next substitute t_{min} into Eq. 10-9, finding

$$\theta = -1.36 \text{ rad} \approx -77.9°. \qquad \text{(Answer)}$$

This *minimum* of $\theta(t)$ (the bottom of the curve in Fig. 10-5*b*) corresponds to the *maximum clockwise* rotation of the disk from the zero angular position, somewhat more than is shown in sketch 3.

(c) Graph the angular velocity ω of the disk versus time from $t = -3.0$ s to $t = 6.0$ s. Sketch the disk and indicate the direction of turning and the sign of ω at $t = -2.0$ s, 4.0 s, and t_{min}.

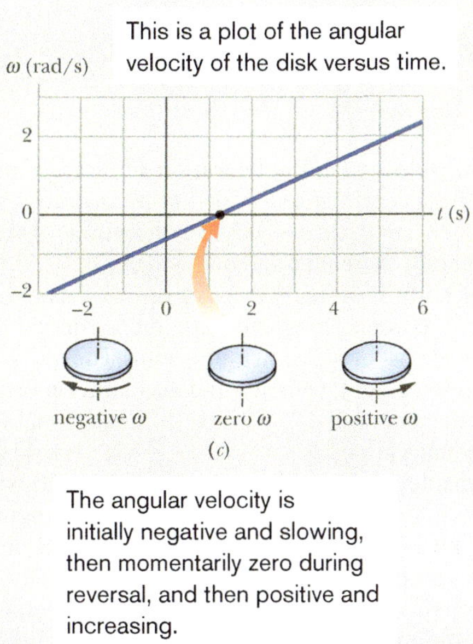

(*c*)

KEY IDEA

From Eq. 10-6, the angular velocity ω is equal to $d\theta/dt$ as given in Eq. 10-10. So, we have

$$\omega = -0.600 + 0.500t. \tag{10-11}$$

The graph of this function $\omega(t)$ is shown in Fig. 10-5*c*. Because the function is linear, the plot is a straight line. The slope is 0.500 rad/s^2 and the intercept with the vertical axis (not shown) is -0.600 rad/s.

Calculations: To sketch the disk at $t = -2.0$ s, we substitute that value into Eq. 10-11, obtaining

$$\omega = -1.6 \text{ rad/s}. \qquad \text{(Answer)}$$

The minus sign here tells us that at $t = -2.0$ s, the disk is turning clockwise (as indicated by the left-hand sketch in Fig. 10-5*c*).

Substituting $t = 4.0$ s into Eq. 10-11 gives us

$$\omega = 1.4 \text{ rad/s}. \qquad \text{(Answer)}$$

The implied plus sign tells us that now the disk is turning counterclockwise (the right-hand sketch in Fig. 10-5*c*).

For t_{min}, we already know that $d\theta/dt = 0$. So, we must also have $\omega = 0$. That is, the disk momentarily stops when the reference line reaches the minimum value of θ in Fig. 10-5*b*, as suggested by the center sketch in Fig. 10-5*c*. On the graph of ω versus t in Fig. 10-5*c*, this momentary stop is the zero point where the plot changes from the negative clockwise motion to the positive counterclockwise motion.

(d) Use the results in parts (a) through (c) to describe the motion of the disk from $t = -3.0$ s to $t = 6.0$ s.

Description: When we first observe the disk at $t = -3.0$ s, it has a positive angular position and is turning clockwise but slowing. It stops at angular position $\theta = -1.36$ rad and then begins to turn counterclockwise, with its angular position eventually becoming positive again.

Sample Problem 10.02 Angular velocity derived from angular acceleration

A child's top is spun with angular acceleration

$$\alpha = 5t^3 - 4t,$$

with t in seconds and α in radians per second-squared. At $t = 0$, the top has angular velocity 5 rad/s, and a reference line on it is at angular position $\theta = 2$ rad.

(a) Obtain an expression for the angular velocity $\omega(t)$ of the top. That is, find an expression that explicitly indicates how the angular velocity depends on time. (We can tell that there *is* such a dependence because the top is undergoing an angular acceleration, which means that its angular velocity *is* changing.)

KEY IDEA

By definition, $\alpha(t)$ is the derivative of $\omega(t)$ with respect to time. Thus, we can find $\omega(t)$ by integrating $\alpha(t)$ with respect to time.

Calculations: Equation 10-8 tells us

$$d\omega = \alpha\, dt,$$

so

$$\int d\omega = \int \alpha\, dt.$$

From this we find

$$\omega = \int (5t^3 - 4t)\, dt = \tfrac{5}{4}t^4 - \tfrac{4}{2}t^2 + C.$$

To evaluate the constant of integration C, we note that $\omega = 5$ rad/s at $t = 0$. Substituting these values in our expression for ω yields

$$5 \text{ rad/s} = 0 - 0 + C,$$

so $C = 5$ rad/s. Then

$$\omega = \tfrac{5}{4}t^4 - 2t^2 + 5. \qquad \text{(Answer)}$$

(b) Obtain an expression for the angular position $\theta(t)$ of the top.

KEY IDEA

By definition, $\omega(t)$ is the derivative of $\theta(t)$ with respect to time. Therefore, we can find $\theta(t)$ by integrating $\omega(t)$ with respect to time.

Calculations: Since Eq. 10-6 tells us that

$$d\theta = \omega\, dt,$$

we can write

$$\theta = \int \omega\, dt = \int (\tfrac{5}{4}t^4 - 2t^2 + 5)\, dt$$
$$= \tfrac{1}{4}t^5 - \tfrac{2}{3}t^3 + 5t + C'$$
$$= \tfrac{1}{4}t^5 - \tfrac{2}{3}t^3 + 5t + 2, \qquad \text{(Answer)}$$

where C' has been evaluated by noting that $\theta = 2$ rad at $t = 0$.

Are Angular Quantities Vectors?

We can describe the position, velocity, and acceleration of a single particle by means of vectors. If the particle is confined to a straight line, however, we do not really need vector notation. Such a particle has only two directions available to it, and we can indicate these directions with plus and minus signs.

In the same way, a rigid body rotating about a fixed axis can rotate only clockwise or counterclockwise as seen along the axis, and again we can select between the two directions by means of plus and minus signs. The question arises: "Can we treat the angular displacement, velocity, and acceleration of a rotating body as vectors?" The answer is a qualified "yes" (see the caution below, in connection with angular displacements).

Angular Velocities. Consider the angular velocity. Figure 10-6*a* shows a vinyl record rotating on a turntable. The record has a constant angular speed ω (= $33\frac{1}{3}$ rev/min) in the clockwise direction. We can represent its angular velocity as a vector $\vec{\omega}$ pointing along the axis of rotation, as in Fig. 10-6*b*. Here's how: We choose the length of this vector according to some convenient scale, for example, with 1 cm corresponding to 10 rev/min. Then we establish a direction for the vector $\vec{\omega}$ by using a **right-hand rule,** as Fig. 10-6*c* shows: Curl your right hand about the rotating record, your fingers pointing *in the direction of rotation.* Your extended thumb will then point in the direction of the angular velocity vector. If the record were to rotate in the opposite sense, the right-

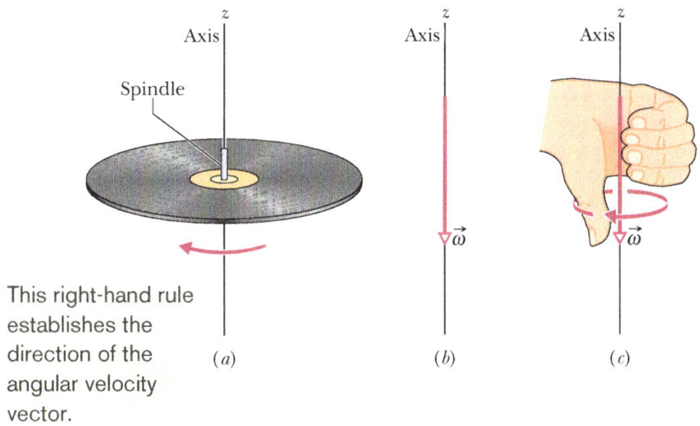

Figure 10-6 (*a*) A record rotating about a vertical axis that coincides with the axis of the spindle. (*b*) The angular velocity of the rotating record can be represented by the vector $\vec{\omega}$, lying along the axis and pointing down, as shown. (*c*) We establish the direction of the angular velocity vector as downward by using a right-hand rule. When the fingers of the right hand curl around the record and point the way it is moving, the extended thumb points in the direction of $\vec{\omega}$.

hand rule would tell you that the angular velocity vector then points in the opposite direction.

It is not easy to get used to representing angular quantities as vectors. We instinctively expect that something should be moving *along* the direction of a vector. That is not the case here. Instead, something (the rigid body) is rotating *around* the direction of the vector. In the world of pure rotation, a vector defines an axis of rotation, not a direction in which something moves. Nonetheless, the vector also defines the motion. Furthermore, it obeys all the rules for vector manipulation discussed in Chapter 3. The angular acceleration $\vec{\alpha}$ is another vector, and it too obeys those rules.

In this chapter we consider only rotations that are about a fixed axis. For such situations, we need not consider vectors—we can represent angular velocity with ω and angular acceleration with α, and we can indicate direction with an implied plus sign for counterclockwise or an explicit minus sign for clockwise.

Angular Displacements. Now for the caution: Angular *displacements* (unless they are very small) *cannot* be treated as vectors. Why not? We can certainly give them both magnitude and direction, as we did for the angular velocity vector in Fig. 10-6. However, to be represented as a vector, a quantity must *also* obey the rules of vector addition, one of which says that if you add two vectors, the order in which you add them does not matter. Angular displacements fail this test.

Figure 10-7 gives an example. An initially horizontal book is given two 90° angular displacements, first in the order of Fig. 10-7*a* and then in the order of Fig. 10-7*b*. Although the two angular displacements are identical, their order is not, and the book ends up with different orientations. Here's another example. Hold your right arm downward, palm toward your thigh. Keeping your wrist rigid, (1) lift the arm forward until it is horizontal, (2) move it horizontally until it points toward the right, and (3) then bring it down to your side. Your palm faces forward. If you start over, but reverse the steps, which way does your palm end up facing? From either example, we must conclude that the addition of two angular displacements depends on their order and they cannot be vectors.

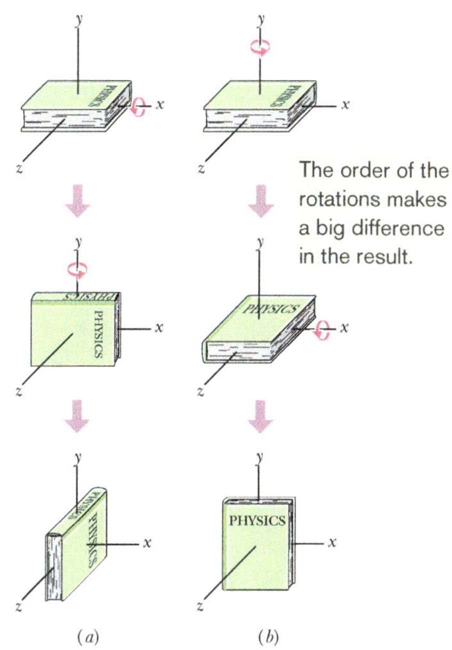

Figure 10-7 (*a*) From its initial position, at the top, the book is given two successive 90° rotations, first about the (horizontal) *x* axis and then about the (vertical) *y* axis. (*b*) The book is given the same rotations, but in the reverse order.

10-2 ROTATION WITH CONSTANT ANGULAR ACCELERATION

Learning Objective

After reading this module, you should be able to . . .

10.14 For constant angular acceleration, apply the relationships between angular position, angular displacement, angular velocity, angular acceleration, and elapsed time (Table 10-1).

Key Idea

- Constant angular acceleration (α = constant) is an important special case of rotational motion. The appropriate kinematic equations are

$$\omega = \omega_0 + \alpha t,$$
$$\theta - \theta_0 = \omega_0 t + \tfrac{1}{2}\alpha t^2,$$
$$\omega^2 = \omega_0^2 + 2\alpha(\theta - \theta_0),$$
$$\theta - \theta_0 = \tfrac{1}{2}(\omega_0 + \omega)t,$$
$$\theta - \theta_0 = \omega t - \tfrac{1}{2}\alpha t^2.$$

Rotation with Constant Angular Acceleration

In pure translation, motion with a *constant linear acceleration* (for example, that of a falling body) is an important special case. In Table 2-1, we displayed a series of equations that hold for such motion.

In pure rotation, the case of *constant angular acceleration* is also important, and a parallel set of equations holds for this case also. We shall not derive them here, but simply write them from the corresponding linear equations, substituting equivalent angular quantities for the linear ones. This is done in Table 10-1, which lists both sets of equations (Eqs. 2-11 and 2-15 to 2-18; 10-12 to 10-16).

Recall that Eqs. 2-11 and 2-15 are basic equations for constant linear acceleration—the other equations in the Linear list can be derived from them. Similarly, Eqs. 10-12 and 10-13 are the basic equations for constant angular acceleration, and the other equations in the Angular list can be derived from them. To solve a simple problem involving constant angular acceleration, you can usually use an equation from the Angular list (*if* you have the list). Choose an equation for which the only unknown variable will be the variable requested in the problem. A better plan is to remember only Eqs. 10-12 and 10-13, and then solve them as simultaneous equations whenever needed.

Checkpoint 2

In four situations, a rotating body has angular position $\theta(t)$ given by (a) $\theta = 3t - 4$, (b) $\theta = -5t^3 + 4t^2 + 6$, (c) $\theta = 2/t^2 - 4/t$, and (d) $\theta = 5t^2 - 3$. To which situations do the angular equations of Table 10-1 apply?

Table 10-1 Equations of Motion for Constant Linear Acceleration and for Constant Angular Acceleration

Equation Number	Linear Equation	Missing Variable		Angular Equation	Equation Number
(2-11)	$v = v_0 + at$	$x - x_0$	$\theta - \theta_0$	$\omega = \omega_0 + \alpha t$	(10-12)
(2-15)	$x - x_0 = v_0 t + \frac{1}{2}at^2$	v	ω	$\theta - \theta_0 = \omega_0 t + \frac{1}{2}\alpha t^2$	(10-13)
(2-16)	$v^2 = v_0^2 + 2a(x - x_0)$	t	t	$\omega^2 = \omega_0^2 + 2\alpha(\theta - \theta_0)$	(10-14)
(2-17)	$x - x_0 = \frac{1}{2}(v_0 + v)t$	a	α	$\theta - \theta_0 = \frac{1}{2}(\omega_0 + \omega)t$	(10-15)
(2-18)	$x - x_0 = vt - \frac{1}{2}at^2$	v_0	ω_0	$\theta - \theta_0 = \omega t - \frac{1}{2}\alpha t^2$	(10-16)

Sample Problem 10.03 Constant angular acceleration, grindstone

A grindstone (Fig. 10-8) rotates at constant angular acceleration $\alpha = 0.35\ \text{rad/s}^2$. At time $t = 0$, it has an angular velocity of $\omega_0 = -4.6$ rad/s and a reference line on it is horizontal, at the angular position $\theta_0 = 0$.

(a) At what time after $t = 0$ is the reference line at the angular position $\theta = 5.0$ rev?

KEY IDEA

The angular acceleration is constant, so we can use the rotation equations of Table 10-1. We choose Eq. 10-13,

$$\theta - \theta_0 = \omega_0 t + \tfrac{1}{2}\alpha t^2,$$

because the only unknown variable it contains is the desired time t.

Calculations: Substituting known values and setting $\theta_0 = 0$ and $\theta = 5.0\ \text{rev} = 10\pi\ \text{rad}$ give us

$$10\pi\ \text{rad} = (-4.6\ \text{rad/s})t + \tfrac{1}{2}(0.35\ \text{rad/s}^2)t^2.$$

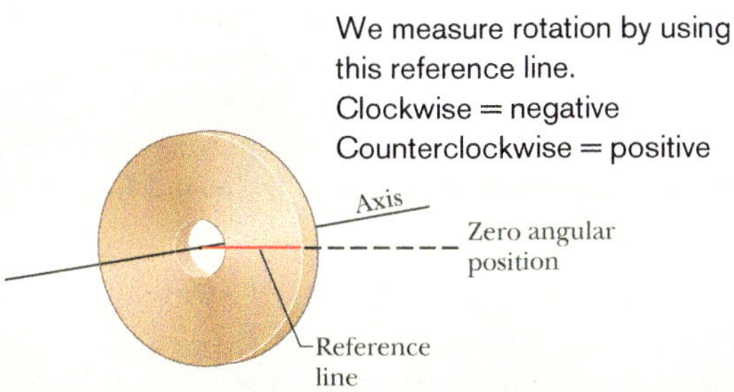

Figure 10-8 A grindstone. At $t = 0$ the reference line (which we imagine to be marked on the stone) is horizontal.

(We converted 5.0 rev to 10π rad to keep the units consistent.) Solving this quadratic equation for t, we find

$$t = 32\ \text{s}. \qquad \text{(Answer)}$$

Now notice something a bit strange. We first see the wheel when it is rotating in the negative direction and through the $\theta = 0$ orientation. Yet, we just found out that 32 s later it is at the positive orientation of $\theta = 5.0$ rev. What happened in that time interval so that it could be at a positive orientation?

(b) Describe the grindstone's rotation between $t = 0$ and $t = 32$ s.

Description: The wheel is initially rotating in the negative (clockwise) direction with angular velocity $\omega_0 = -4.6$ rad/s, but its angular acceleration α is positive. This initial opposition of the signs of angular velocity and angular acceleration means that the wheel slows in its rotation in the negative direction, stops, and then reverses to rotate in the positive direction. After the reference line comes back through its initial orientation of $\theta = 0$, the wheel turns an additional 5.0 rev by time $t = 32$ s.

(c) At what time t does the grindstone momentarily stop?

Calculation: We again go to the table of equations for constant angular acceleration, and again we need an equation that contains only the desired unknown variable t. However, now the equation must also contain the variable ω, so that we can set it to 0 and then solve for the corresponding time t. We choose Eq. 10-12, which yields

$$t = \frac{\omega - \omega_0}{\alpha} = \frac{0 - (-4.6\ \text{rad/s})}{0.35\ \text{rad/s}^2} = 13\ \text{s}. \qquad \text{(Answer)}$$

Sample Problem 10.04 Constant angular acceleration, riding a Rotor

While you are operating a Rotor (a large, vertical, rotating cylinder found in amusement parks), you spot a passenger in acute distress and decrease the angular velocity of the cylinder from 3.40 rad/s to 2.00 rad/s in 20.0 rev, at constant angular acceleration. (The passenger is obviously more of a "translation person" than a "rotation person.")

(a) What is the constant angular acceleration during this decrease in angular speed?

KEY IDEA

Because the cylinder's angular acceleration is constant, we can relate it to the angular velocity and angular displacement via the basic equations for constant angular acceleration (Eqs. 10-12 and 10-13).

Calculations: Let's first do a quick check to see if we can solve the basic equations. The initial angular velocity is $\omega_0 = 3.40$ rad/s, the angular displacement is $\theta - \theta_0 = 20.0$ rev, and the angular velocity at the end of that displacement is $\omega = 2.00$ rad/s. In addition to the angular acceleration α that we want, both basic equations also contain time t, which we do not necessarily want.

To eliminate the unknown t, we use Eq. 10-12 to write

$$t = \frac{\omega - \omega_0}{\alpha},$$

which we then substitute into Eq. 10-13 to write

$$\theta - \theta_0 = \omega_0\left(\frac{\omega - \omega_0}{\alpha}\right) + \tfrac{1}{2}\alpha\left(\frac{\omega - \omega_0}{\alpha}\right)^2.$$

Solving for α, substituting known data, and converting 20 rev to 125.7 rad, we find

$$\alpha = \frac{\omega^2 - \omega_0^2}{2(\theta - \theta_0)} = \frac{(2.00\ \text{rad/s})^2 - (3.40\ \text{rad/s})^2}{2(125.7\ \text{rad})}$$

$$= -0.0301\ \text{rad/s}^2. \qquad \text{(Answer)}$$

(b) How much time did the speed decrease take?

Calculation: Now that we know α, we can use Eq. 10-12 to solve for t:

$$t = \frac{\omega - \omega_0}{\alpha} = \frac{2.00 \text{ rad/s} - 3.40 \text{ rad/s}}{-0.0301 \text{ rad/s}^2} = 46.5 \text{ s.} \qquad \text{(Answer)}$$

10-3 RELATING THE LINEAR AND ANGULAR VARIABLES

Learning Objectives

After reading this module, you should be able to . . .

10.15 For a rigid body rotating about a fixed axis, relate the angular variables of the body (angular position, angular velocity, and angular acceleration) and the linear variables of a particle on the body (position, velocity, and acceleration) at any given radius.

10.16 Distinguish between tangential acceleration and radial acceleration, and draw a vector for each in a sketch of a particle on a body rotating about an axis, for both an increase in angular speed and a decrease.

Key Ideas

- A point in a rigid rotating body, at a perpendicular distance r from the rotation axis, moves in a circle with radius r. If the body rotates through an angle θ, the point moves along an arc with length s given by

$$s = \theta r \quad \text{(radian measure)},$$

where θ is in radians.

- The linear velocity $\vec{v}$ of the point is tangent to the circle; the point's linear speed v is given by

$$v = \omega r \quad \text{(radian measure)},$$

where ω is the angular speed (in radians per second) of the body, and thus also the point.

- The linear acceleration $\vec{a}$ of the point has both tangential and radial components. The tangential component is

$$a_t = \alpha r \quad \text{(radian measure)},$$

where α is the magnitude of the angular acceleration (in radians per second-squared) of the body. The radial component of $\vec{a}$ is

$$a_r = \frac{v^2}{r} = \omega^2 r \quad \text{(radian measure)}.$$

- If the point moves in uniform circular motion, the period T of the motion for the point and the body is

$$T = \frac{2\pi r}{v} = \frac{2\pi}{\omega} \quad \text{(radian measure)}.$$

Relating the Linear and Angular Variables

In Module 4-5, we discussed uniform circular motion, in which a particle travels at constant linear speed v along a circle and around an axis of rotation. When a rigid body, such as a merry-go-round, rotates around an axis, each particle in the body moves in its own circle around that axis. Since the body is rigid, all the particles make one revolution in the same amount of time; that is, they all have the same angular speed ω.

However, the farther a particle is from the axis, the greater the circumference of its circle is, and so the faster its linear speed v must be. You can notice this on a merry-go-round. You turn with the same angular speed ω regardless of your distance from the center, but your linear speed v increases noticeably if you move to the outside edge of the merry-go-round.

We often need to relate the linear variables s, v, and a for a particular point in a rotating body to the angular variables θ, ω, and α for that body. The two sets of variables are related by r, the *perpendicular distance* of the point from the rotation axis. This perpendicular distance is the distance between the point and the rotation axis, measured along a perpendicular to the axis. It is also the radius r of the circle traveled by the point around the axis of rotation.

The Position

If a reference line on a rigid body rotates through an angle θ, a point within the body at a position r from the rotation axis moves a distance s along a circular arc, where s is given by Eq. 10-1:

$$s = \theta r \quad \text{(radian measure)}. \tag{10-17}$$

This is the first of our linear–angular relations. *Caution:* The angle θ here must be measured in radians because Eq. 10-17 is itself the definition of angular measure in radians.

The Speed

Differentiating Eq. 10-17 with respect to time—with r held constant—leads to

$$\frac{ds}{dt} = \frac{d\theta}{dt} r.$$

However, ds/dt is the linear speed (the magnitude of the linear velocity) of the point in question, and $d\theta/dt$ is the angular speed ω of the rotating body. So

$$v = \omega r \quad \text{(radian measure)}. \tag{10-18}$$

Caution: The angular speed ω must be expressed in radian measure.

Equation 10-18 tells us that since all points within the rigid body have the same angular speed ω, points with greater radius r have greater linear speed v. Figure 10-9*a* reminds us that the linear velocity is always tangent to the circular path of the point in question.

If the angular speed ω of the rigid body is constant, then Eq. 10-18 tells us that the linear speed v of any point within it is also constant. Thus, each point within the body undergoes uniform circular motion. The period of revolution T for the motion of each point and for the rigid body itself is given by Eq. 4-35:

$$T = \frac{2\pi r}{v}. \tag{10-19}$$

This equation tells us that the time for one revolution is the distance $2\pi r$ traveled in one revolution divided by the speed at which that distance is traveled. Substituting for v from Eq. 10-18 and canceling r, we find also that

$$T = \frac{2\pi}{\omega} \quad \text{(radian measure)}. \tag{10-20}$$

This equivalent equation says that the time for one revolution is the angular distance 2π rad traveled in one revolution divided by the angular speed (or rate) at which that angle is traveled.

The Acceleration

Differentiating Eq. 10-18 with respect to time—again with r held constant—leads to

$$\frac{dv}{dt} = \frac{d\omega}{dt} r. \tag{10-21}$$

Here we run up against a complication. In Eq. 10-21, dv/dt represents only the part of the linear acceleration that is responsible for changes in the *magnitude* v of the linear velocity $\vec{v}$. Like $\vec{v}$, that part of the linear acceleration is tangent to the path of the point in question. We call it the *tangential component* a_t of the linear acceleration of the point, and we write

$$a_t = \alpha r \quad \text{(radian measure)}, \tag{10-22}$$

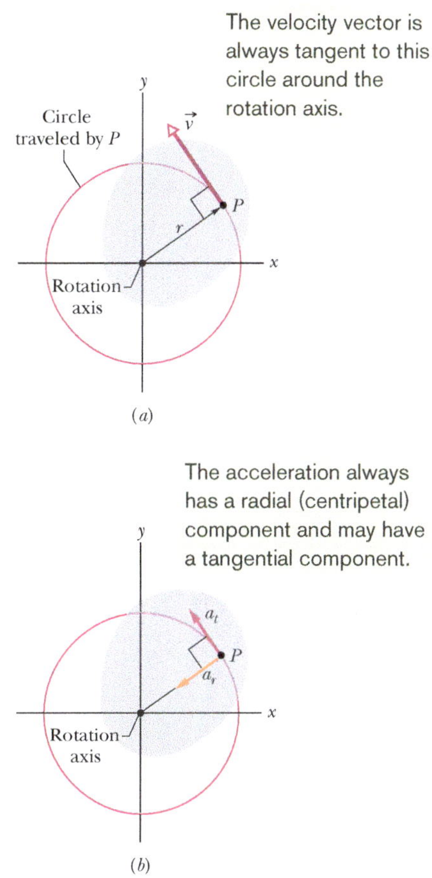

Figure 10-9 The rotating rigid body of Fig. 10-2, shown in cross section viewed from above. Every point of the body (such as P) moves in a circle around the rotation axis. (*a*) The linear velocity $\vec{v}$ of every point is tangent to the circle in which the point moves. (*b*) The linear acceleration $\vec{a}$ of the point has (in general) two components: tangential a_t and radial a_r.

where $\alpha = d\omega/dt$. *Caution:* The angular acceleration α in Eq. 10-22 must be expressed in radian measure.

In addition, as Eq. 4-34 tells us, a particle (or point) moving in a circular path has a *radial component* of linear acceleration, $a_r = v^2/r$ (directed radially inward), that is responsible for changes in the *direction* of the linear velocity $\vec{v}$. By substituting for v from Eq. 10-18, we can write this component as

$$a_r = \frac{v^2}{r} = \omega^2 r \quad \text{(radian measure).} \tag{10-23}$$

Thus, as Fig. 10-9*b* shows, the linear acceleration of a point on a rotating rigid body has, in general, two components. The radially inward component a_r (given by Eq. 10-23) is present whenever the angular velocity of the body is not zero. The tangential component a_t (given by Eq. 10-22) is present whenever the angular acceleration is not zero.

Checkpoint 3

A cockroach rides the rim of a rotating merry-go-round. If the angular speed of this system (*merry-go-round + cockroach*) is constant, does the cockroach have (a) radial acceleration and (b) tangential acceleration? If ω is decreasing, does the cockroach have (c) radial acceleration and (d) tangential acceleration?

Sample Problem 10.05 Designing The Giant Ring, a large-scale amusement park ride

We are given the job of designing a large horizontal ring that will rotate around a vertical axis and that will have a radius of $r = 33.1$ m (matching that of Beijing's The Great Observation Wheel, the largest Ferris wheel in the world). Passengers will enter through a door in the outer wall of the ring and then stand next to that wall (Fig. 10-10*a*). We decide that for the time interval $t = 0$ to $t = 2.30$ s, the angular position $\theta(t)$ of a reference line on the ring will be given by

$$\theta = ct^3, \tag{10-24}$$

with $c = 6.39 \times 10^{-2}$ rad/s^3. After $t = 2.30$ s, the angular speed will be held constant until the end of the ride. Once the ring begins to rotate, the floor of the ring will drop away from the riders but the riders will not fall—indeed, they feel as though they are pinned to the wall. For the time $t = 2.20$ s, let's determine a rider's angular speed ω, linear speed v, angular acceleration α, tangential acceleration a_t, radial acceleration a_r, and acceleration $\vec{a}$.

KEY IDEAS

(1) The angular speed ω is given by Eq. 10-6 ($\omega = d\theta/dt$). (2) The linear speed v (along the circular path) is related to the angular speed (around the rotation axis) by Eq. 10-18 ($v = \omega r$). (3) The angular acceleration α is given by Eq. 10-8 ($\alpha = d\omega/dt$). (4) The tangential acceleration a_t (along the circular path) is related to the angular acceleration (around the rotation axis) by Eq. 10-22 ($a_t = \alpha r$). (5) The radial acceleration a_r is given Eq. 10-23 ($a_r = \omega^2 r$). (6) The tangential and radial accelerations are the (perpendicular) components of the (full) acceleration $\vec{a}$.

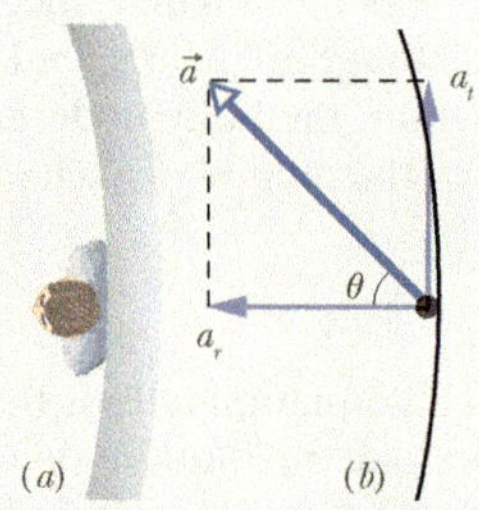

Figure 10-10 (*a*) Overhead view of a passenger ready to ride The Giant Ring. (*b*) The radial and tangential acceleration components of the (full) acceleration.

Calculations: Let's go through the steps. We first find the angular velocity by taking the time derivative of the given angular position function and then substituting the given time of $t = 2.20$ s:

$$\omega = \frac{d\theta}{dt} = \frac{d}{dt}(ct^3) = 3ct^2 \tag{10-25}$$

$$= 3(6.39 \times 10^{-2} \text{ rad/s}^3)(2.20 \text{ s})^2$$

$$= 0.928 \text{ rad/s.} \qquad \text{(Answer)}$$

From Eq. 10-18, the linear speed just then is

$$v = \omega r = 3ct^2 r \tag{10-26}$$

$$= 3(6.39 \times 10^{-2} \text{ rad/s}^3)(2.20 \text{ s})^2(33.1 \text{ m})$$

$$= 30.7 \text{ m/s.} \qquad \text{(Answer)}$$

Although this is fast (111 km/h or 68.7 mi/h), such speeds are common in amusement parks and not alarming because (as mentioned in Chapter 2) your body reacts to accelerations but not to velocities. (It is an accelerometer, not a speedometer.) From Eq. 10-26 we see that the linear speed is increasing as the square of the time (but this increase will cut off at $t = 2.30$ s).

Next, let's tackle the angular acceleration by taking the time derivative of Eq. 10-25:

$$\alpha = \frac{d\omega}{dt} = \frac{d}{dt}(3ct^2) = 6ct$$
$$= 6(6.39 \times 10^{-2}\ \text{rad/s}^3)(2.20\ \text{s}) = 0.843\ \text{rad/s}^2. \quad \text{(Answer)}$$

The tangential acceleration then follows from Eq. 10-22:

$$a_t = \alpha r = 6ctr \tag{10-27}$$
$$= 6(6.39 \times 10^{-2}\ \text{rad/s}^3)(2.20\ \text{s})(33.1\ \text{m})$$
$$= 27.91\ \text{m/s}^2 \approx 27.9\ \text{m/s}^2, \quad \text{(Answer)}$$

or $2.8g$ (which is reasonable and a bit exciting). Equation 10-27 tells us that the tangential acceleration is increasing with time (but it will cut off at $t = 2.30$ s). From Eq. 10-23, we write the radial acceleration as

$$a_r = \omega^2 r.$$

Substituting from Eq. 10-25 leads us to

$$a_r = (3ct^2)^2 r = 9c^2t^4r \tag{10-28}$$
$$= 9(6.39 \times 10^{-2}\ \text{rad/s}^3)^2(2.20\ \text{s})^4(33.1\ \text{m})$$
$$= 28.49\ \text{m/s}^2 \approx 28.5\ \text{m/s}^2, \quad \text{(Answer)}$$

or $2.9g$ (which is also reasonable and a bit exciting).

The radial and tangential accelerations are perpendicular to each other and form the components of the rider's acceleration $\vec{a}$ (Fig. 10-10*b*). The magnitude of $\vec{a}$ is given by

$$a = \sqrt{a_r^2 + a_t^2} \tag{10-29}$$
$$= \sqrt{(28.49\ \text{m/s}^2)^2 + (27.91\ \text{m/s}^2)^2}$$
$$\approx 39.9\ \text{m/s}^2, \quad \text{(Answer)}$$

or $4.1g$ (which is really exciting!). All these values are acceptable.

To find the orientation of $\vec{a}$, we can calculate the angle θ shown in Fig. 10-10*b*:

$$\tan\theta = \frac{a_t}{a_r}.$$

However, instead of substituting our numerical results, let's use the algebraic results from Eqs. 10-27 and 10-28:

$$\theta = \tan^{-1}\left(\frac{6ctr}{9c^2t^4r}\right) = \tan^{-1}\left(\frac{2}{3ct^3}\right). \tag{10-30}$$

The big advantage of solving for the angle algebraically is that we can then see that the angle (1) does not depend on the ring's radius and (2) decreases as t goes from 0 to 2.20 s. That is, the acceleration vector $\vec{a}$ swings toward being radially inward because the radial acceleration (which depends on t^4) quickly dominates over the tangential acceleration (which depends on only t). At our given time $t = 2.20$ s, we have

$$\theta = \tan^{-1}\frac{2}{3(6.39 \times 10^{-2}\ \text{rad/s}^3)(2.20\ \text{s})^3} = 44.4°. \quad \text{(Answer)}$$

WileyPLUS Additional examples, video, and practice available at *WileyPLUS*

10-4 KINETIC ENERGY OF ROTATION

Learning Objectives

After reading this module, you should be able to . . .

10.17 Find the rotational inertia of a particle about a point.

10.18 Find the total rotational inertia of many particles moving around the same fixed axis.

10.19 Calculate the rotational kinetic energy of a body in terms of its rotational inertia and its angular speed.

Key Idea

- The kinetic energy K of a rigid body rotating about a fixed axis is given by

$$K = \tfrac{1}{2}I\omega^2 \quad \text{(radian measure)},$$

in which I is the rotational inertia of the body, defined as

$$I = \sum m_i r_i^2$$

for a system of discrete particles.

Kinetic Energy of Rotation

The rapidly rotating blade of a table saw certainly has kinetic energy due to that rotation. How can we express the energy? We cannot apply the familiar formula $K = \frac{1}{2}mv^2$ to the saw as a whole because that would give us the kinetic energy only of the saw's center of mass, which is zero.

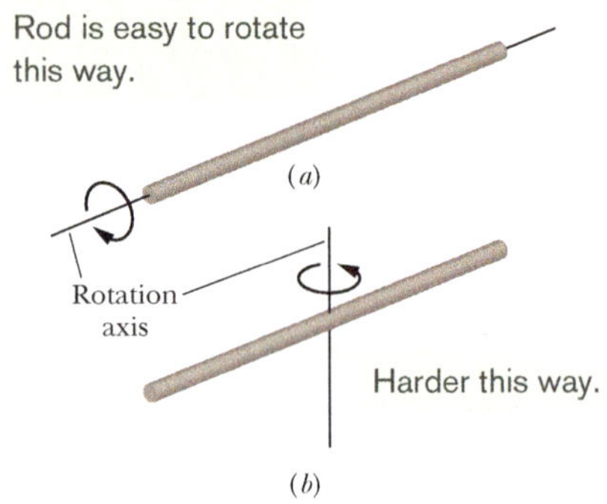

Figure 10-11 A long rod is much easier to rotate about (a) its central (longitudinal) axis than about (b) an axis through its center and perpendicular to its length. The reason for the difference is that the mass is distributed closer to the rotation axis in (a) than in (b).

Instead, we shall treat the table saw (and any other rotating rigid body) as a collection of particles with different speeds. We can then add up the kinetic energies of all the particles to find the kinetic energy of the body as a whole. In this way we obtain, for the kinetic energy of a rotating body,

$$K = \tfrac{1}{2}m_1v_1^2 + \tfrac{1}{2}m_2v_2^2 + \tfrac{1}{2}m_3v_3^2 + \cdots$$
$$= \sum \tfrac{1}{2}m_iv_i^2, \tag{10-31}$$

in which m_i is the mass of the ith particle and v_i is its speed. The sum is taken over all the particles in the body.

The problem with Eq. 10-31 is that v_i is not the same for all particles. We solve this problem by substituting for v from Eq. 10-18 ($v = \omega r$), so that we have

$$K = \sum \tfrac{1}{2}m_i(\omega r_i)^2 = \tfrac{1}{2}\left(\sum m_ir_i^2\right)\omega^2, \tag{10-32}$$

in which ω *is* the same for all particles.

The quantity in parentheses on the right side of Eq. 10-32 tells us how the mass of the rotating body is distributed about its axis of rotation. We call that quantity the **rotational inertia** (or **moment of inertia**) I of the body with respect to the axis of rotation. It is a constant for a particular rigid body and a particular rotation axis. (*Caution:* That axis must always be specified if the value of I is to be meaningful.)

We may now write

$$I = \sum m_ir_i^2 \quad \text{(rotational inertia)} \tag{10-33}$$

and substitute into Eq. 10-32, obtaining

$$K = \tfrac{1}{2}I\omega^2 \quad \text{(radian measure)} \tag{10-34}$$

as the expression we seek. Because we have used the relation $v = \omega r$ in deriving Eq. 10-34, ω must be expressed in radian measure. The SI unit for I is the kilogram–square meter ($\text{kg}\cdot\text{m}^2$).

The Plan. If we have a few particles and a specified rotation axis, we find mr^2 for each particle and then add the results as in Eq. 10-33 to get the total rotational inertia I. If we want the total rotational kinetic energy, we can then substitute that I into Eq. 10-34. That is the plan for a few particles, but suppose we have a huge number of particles such as in a rod. In the next module we shall see how to handle such *continuous bodies* and do the calculation in only a few minutes.

Equation 10-34, which gives the kinetic energy of a rigid body in pure rotation, is the angular equivalent of the formula $K = \tfrac{1}{2}Mv_{\text{com}}^2$, which gives the kinetic energy of a rigid body in pure translation. In both formulas there is a factor of $\tfrac{1}{2}$. Where mass M appears in one equation, I (which involves both mass and its distribution) appears in the other. Finally, each equation contains as a factor the square of a speed—translational or rotational as appropriate. The kinetic energies of translation and of rotation are not different kinds of energy. They are both kinetic energy, expressed in ways that are appropriate to the motion at hand.

We noted previously that the rotational inertia of a rotating body involves not only its mass but also how that mass is distributed. Here is an example that you can literally feel. Rotate a long, fairly heavy rod (a pole, a length of lumber, or something similar), first around its central (longitudinal) axis (Fig. 10-11a) and then around an axis perpendicular to the rod and through the center (Fig. 10-11b). Both rotations involve the very same mass, but the first rotation is much easier than the second. The reason is that the mass is distributed much closer to the rotation axis in the first rotation. As a result, the rotational inertia of the rod is much smaller in Fig. 10-11a than in Fig. 10-11b. In general, smaller rotational inertia means easier rotation.

Checkpoint 4

The figure shows three small spheres that rotate about a vertical axis. The perpendicular distance between the axis and the center of each sphere is given. Rank the three spheres according to their rotational inertia about that axis, greatest first.

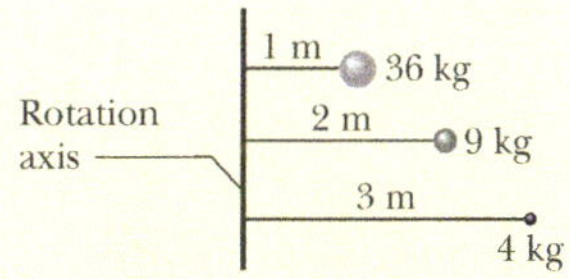

10-5 CALCULATING THE ROTATIONAL INERTIA

Learning Objectives

After reading this module, you should be able to . . .

10.20 Determine the rotational inertia of a body if it is given in Table 10-2.

10.21 Calculate the rotational inertia of a body by integration over the mass elements of the body.

10.22 Apply the parallel-axis theorem for a rotation axis that is displaced from a parallel axis through the center of mass of a body.

Key Ideas

- I is the rotational inertia of the body, defined as

$$I = \sum m_i r_i^2$$

for a system of discrete particles and defined as

$$I = \int r^2\, dm$$

for a body with continuously distributed mass. The r and r_i in these expressions represent the perpendicular distance from the axis of rotation to each mass element in the body, and the integration is carried out over the entire body so as to include every mass element.

- The parallel-axis theorem relates the rotational inertia I of a body about any axis to that of the same body about a parallel axis through the center of mass:

$$I = I_{com} + Mh^2.$$

Here h is the perpendicular distance between the two axes, and I_{com} is the rotational inertia of the body about the axis through the com. We can describe h as being the distance the actual rotation axis has been shifted from the rotation axis through the com.

Calculating the Rotational Inertia

If a rigid body consists of a few particles, we can calculate its rotational inertia about a given rotation axis with Eq. 10-33 ($I = \Sigma\, m_i r_i^2$); that is, we can find the product mr^2 for each particle and then sum the products. (Recall that r is the perpendicular distance a particle is from the given rotation axis.)

If a rigid body consists of a great many adjacent particles (it is *continuous*, like a Frisbee), using Eq. 10-33 would require a computer. Thus, instead, we replace the sum in Eq. 10-33 with an integral and define the rotational inertia of the body as

$$I = \int r^2\, dm \quad \text{(rotational inertia, continuous body).} \tag{10-35}$$

Table 10-2 gives the results of such integration for nine common body shapes and the indicated axes of rotation.

Parallel-Axis Theorem

Suppose we want to find the rotational inertia I of a body of mass M about a given axis. In principle, we can always find I with the integration of Eq. 10-35. However, there is a neat shortcut if we happen to already know the rotational inertia I_{com} of the body about a *parallel* axis that extends through the body's center of mass. Let h be the perpendicular distance between the given axis and the axis

Table 10-2 Some Rotational Inertias

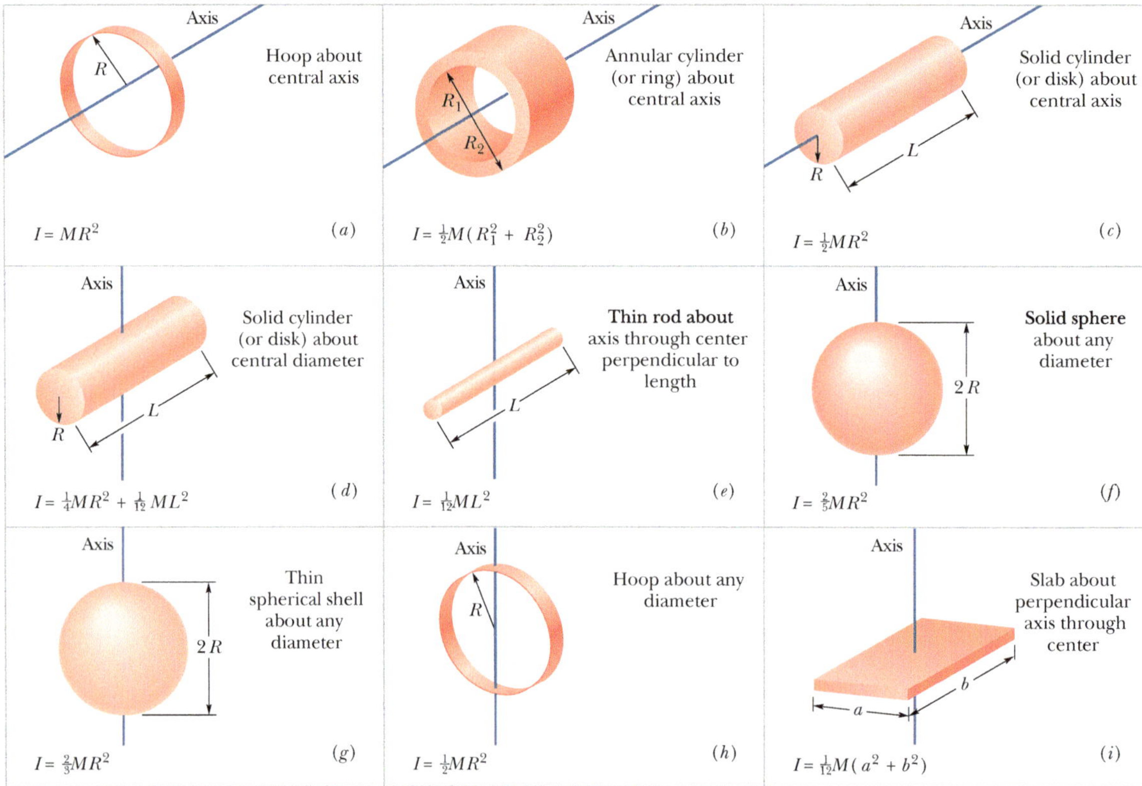

(a) Hoop about central axis $I = MR^2$	(b) Annular cylinder (or ring) about central axis $I = \frac{1}{2}M(R_1^2 + R_2^2)$	(c) Solid cylinder (or disk) about central axis $I = \frac{1}{2}MR^2$
(d) Solid cylinder (or disk) about central diameter $I = \frac{1}{4}MR^2 + \frac{1}{12}ML^2$	(e) Thin rod about axis through center perpendicular to length $I = \frac{1}{12}ML^2$	(f) Solid sphere about any diameter $I = \frac{2}{5}MR^2$
(g) Thin spherical shell about any diameter $I = \frac{2}{3}MR^2$	(h) Hoop about any diameter $I = \frac{1}{2}MR^2$	(i) Slab about perpendicular axis through center $I = \frac{1}{12}M(a^2 + b^2)$

We need to relate the rotational inertia around the axis at P to that around the axis at the com.

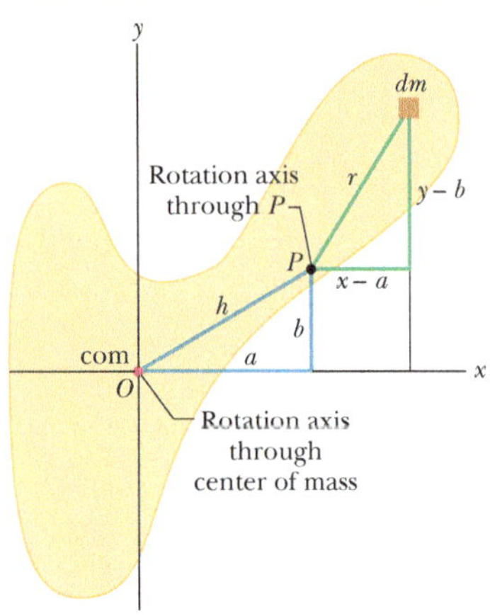

Figure 10-12 A rigid body in cross section, with its center of mass at O. The parallel-axis theorem (Eq. 10-36) relates the rotational inertia of the body about an axis through O to that about a parallel axis through a point such as P, a distance h from the body's center of mass.

through the center of mass (remember these two axes must be parallel). Then the rotational inertia I about the given axis is

$$I = I_{com} + Mh^2 \quad \text{(parallel-axis theorem)}. \tag{10-36}$$

Think of the distance h as being the distance we have shifted the rotation axis from being through the com. This equation is known as the **parallel-axis theorem.** We shall now prove it.

Proof of the Parallel-Axis Theorem

Let O be the center of mass of the arbitrarily shaped body shown in cross section in Fig. 10-12. Place the origin of the coordinates at O. Consider an axis through O perpendicular to the plane of the figure, and another axis through point P parallel to the first axis. Let the x and y coordinates of P be a and b.

Let dm be a mass element with the general coordinates x and y. The rotational inertia of the body about the axis through P is then, from Eq. 10-35,

$$I = \int r^2\, dm = \int [(x - a)^2 + (y - b)^2]\, dm,$$

which we can rearrange as

$$I = \int (x^2 + y^2)\, dm - 2a \int x\, dm - 2b \int y\, dm + \int (a^2 + b^2)\, dm. \tag{10-37}$$

From the definition of the center of mass (Eq. 9-9), the middle two integrals of Eq. 10-37 give the coordinates of the center of mass (multiplied by a constant) and thus must each be zero. Because $x^2 + y^2$ is equal to R^2, where R is the distance from O to dm, the first integral is simply I_{com}, the rotational inertia of the body about an axis through its center of mass. Inspection of Fig. 10-12 shows that the last term in Eq. 10-37 is Mh^2, where M is the body's total mass. Thus, Eq. 10-37 reduces to Eq. 10-36, which is the relation that we set out to prove.

Checkpoint 5

The figure shows a book-like object (one side is longer than the other) and four choices of rotation axes, all perpendicular to the face of the object. Rank the choices according to the rotational inertia of the object about the axis, greatest first.

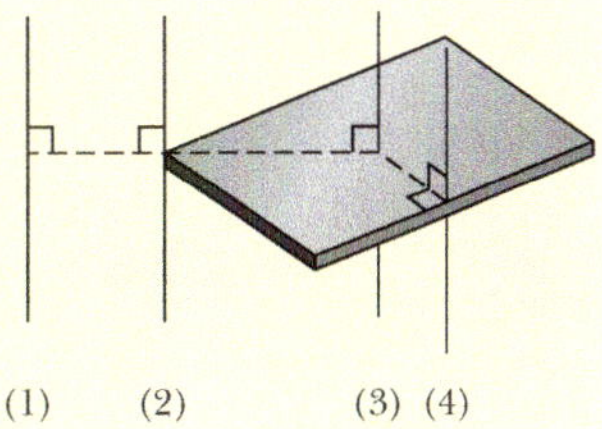

Sample Problem 10.06 Rotational inertia of a two-particle system

Figure 10-13a shows a rigid body consisting of two particles of mass m connected by a rod of length L and negligible mass.

(a) What is the rotational inertia I_{com} about an axis through the center of mass, perpendicular to the rod as shown?

KEY IDEA

Because we have only two particles with mass, we can find the body's rotational inertia I_{com} by using Eq. 10-33 rather than by integration. That is, we find the rotational inertia of each particle and then just add the results.

Calculations: For the two particles, each at perpendicular distance $\frac{1}{2}L$ from the rotation axis, we have

$$I = \sum m_i r_i^2 = (m)(\tfrac{1}{2}L)^2 + (m)(\tfrac{1}{2}L)^2$$
$$= \tfrac{1}{2}mL^2. \qquad \text{(Answer)}$$

(b) What is the rotational inertia I of the body about an axis through the left end of the rod and parallel to the first axis (Fig. 10-13b)?

KEY IDEAS

This situation is simple enough that we can find I using either of two techniques. The first is similar to the one used in part (a). The other, more powerful one is to apply the parallel-axis theorem.

First technique: We calculate I as in part (a), except here the perpendicular distance r_i is zero for the particle on the left and L for the particle on the right. Now Eq. 10-33 gives us

$$I = m(0)^2 + mL^2 = mL^2. \qquad \text{(Answer)}$$

Second technique: Because we already know I_{com} about an axis through the center of mass and because the axis here is parallel to that "com axis," we can apply the parallel-axis theorem (Eq. 10-36). We find

$$I = I_{com} + Mh^2 = \tfrac{1}{2}mL^2 + (2m)(\tfrac{1}{2}L)^2$$
$$= mL^2. \qquad \text{(Answer)}$$

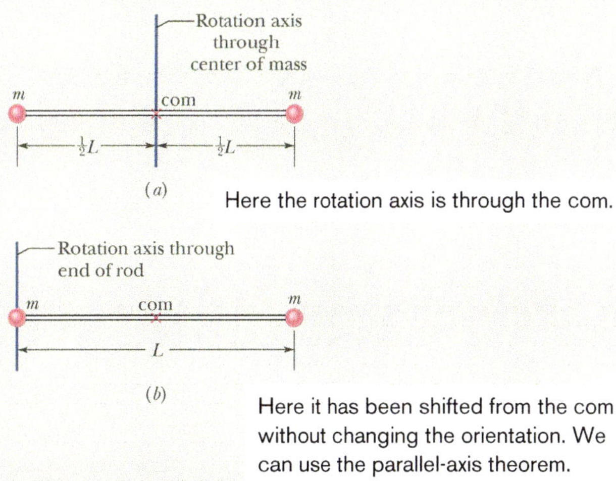

Figure 10-13 A rigid body consisting of two particles of mass m joined by a rod of negligible mass.

WileyPLUS Additional examples, video, and practice available at *WileyPLUS*

Sample Problem 10.07 Rotational inertia of a uniform rod, integration

Figure 10-14 shows a thin, uniform rod of mass M and length L, on an x axis with the origin at the rod's center.

(a) What is the rotational inertia of the rod about the perpendicular rotation axis through the center?

KEY IDEAS

(1) The rod consists of a huge number of particles at a great many different distances from the rotation axis. We certainly don't want to sum their rotational inertias individually. So, we first write a general expression for the rotational inertia of a mass element dm at distance r from the rotation axis: $r^2\, dm$. (2) Then we sum all such rotational inertias by integrating the expression (rather than adding them up one by one). From Eq. 10-35, we write

$$I = \int r^2\, dm. \tag{10-38}$$

(3) Because the rod is uniform and the rotation axis is at the center, we are actually calculating the rotational inertia I_{com} about the center of mass.

Calculations: We want to integrate with respect to coordinate x (not mass m as indicated in the integral), so we must relate the mass dm of an element of the rod to its length dx along the rod. (Such an element is shown in Fig. 10-14.) Because the rod is uniform, the ratio of mass to length is the same for all the elements and for the rod as a whole. Thus, we can write

$$\frac{\text{element's mass } dm}{\text{element's length } dx} = \frac{\text{rod's mass } M}{\text{rod's length } L}$$

or

$$dm = \frac{M}{L}\, dx.$$

We can now substitute this result for dm and x for r in Eq. 10-38. Then we integrate from end to end of the rod (from $x = -L/2$ to $x = L/2$) to include all the elements. We find

$$\begin{aligned} I &= \int_{x=-L/2}^{x=+L/2} x^2 \left(\frac{M}{L}\right) dx \\ &= \frac{M}{3L}\Big[x^3\Big]_{-L/2}^{+L/2} = \frac{M}{3L}\left[\left(\frac{L}{2}\right)^3 - \left(-\frac{L}{2}\right)^3\right] \\ &= \tfrac{1}{12} M L^2. \end{aligned} \qquad \text{(Answer)}$$

(b) What is the rod's rotational inertia I about a new rotation axis that is perpendicular to the rod and through the left end?

KEY IDEAS

We can find I by shifting the origin of the x axis to the left end of the rod and then integrating from $x = 0$ to $x = L$. However, here we shall use a more powerful (and easier) technique by applying the parallel-axis theorem (Eq. 10-36), in which we shift the rotation axis without changing its orientation.

Calculations: If we place the axis at the rod's end so that it is parallel to the axis through the center of mass, then we can use the parallel-axis theorem (Eq. 10-36). We know from part (a) that I_{com} is $\frac{1}{12}ML^2$. From Fig. 10-14, the perpendicular distance h between the new rotation axis and the center of mass is $\frac{1}{2}L$. Equation 10-36 then gives us

$$\begin{aligned} I &= I_{com} + Mh^2 = \tfrac{1}{12} ML^2 + (M)(\tfrac{1}{2}L)^2 \\ &= \tfrac{1}{3} ML^2. \end{aligned} \qquad \text{(Answer)}$$

Actually, this result holds for any axis through the left or right end that is perpendicular to the rod.

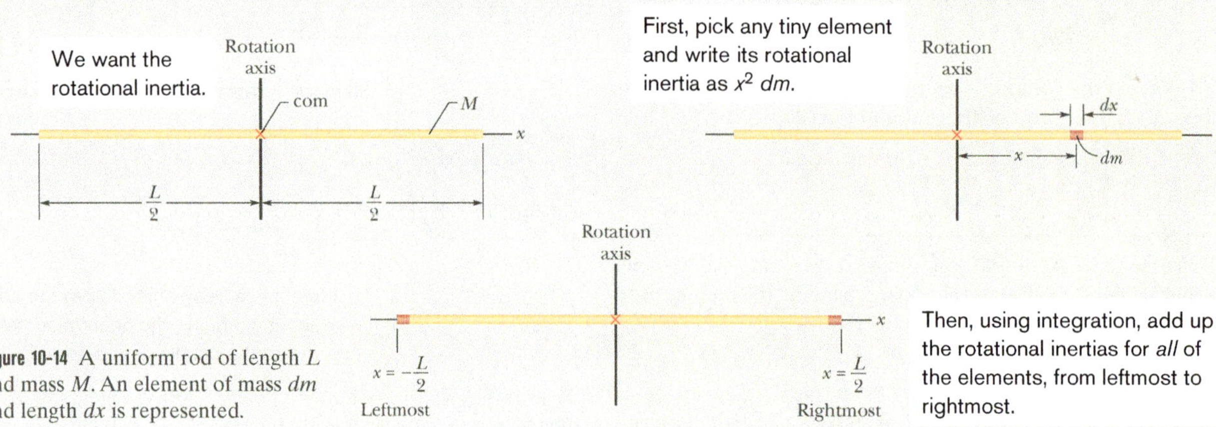

Figure 10-14 A uniform rod of length L and mass M. An element of mass dm and length dx is represented.

Sample Problem 10.08 Rotational kinetic energy, spin test explosion

Large machine components that undergo prolonged, high-speed rotation are first examined for the possibility of failure in a *spin test system.* In this system, a component is *spun up* (brought up to high speed) while inside a cylindrical arrangement of lead bricks and containment liner, all within a steel shell that is closed by a lid clamped into place. If the rotation causes the component to shatter, the soft lead bricks are supposed to catch the pieces for later analysis.

In 1985, Test Devices, Inc. (www.testdevices.com) was spin testing a sample of a solid steel rotor (a disk) of mass $M = 272$ kg and radius $R = 38.0$ cm. When the sample reached an angular speed ω of 14 000 rev/min, the test engineers heard a dull thump from the test system, which was located one floor down and one room over from them. Investigating, they found that lead bricks had been thrown out in the hallway leading to the test room, a door to the room had been hurled into the adjacent parking lot, one lead brick had shot from the test site through the wall of a neighbor's kitchen, the structural beams of the test building had been damaged, the concrete floor beneath the spin chamber had been shoved downward by about 0.5 cm, and the 900 kg lid had been blown upward through the ceiling and had then crashed back onto the test equipment (Fig. 10-15). The exploding pieces had not penetrated the room of the test engineers only by luck.

How much energy was released in the explosion of the rotor?

Figure 10-15 Some of the destruction caused by the explosion of a rapidly rotating steel disk.

Courtesy Test Devices, Inc.

KEY IDEA

The released energy was equal to the rotational kinetic energy K of the rotor just as it reached the angular speed of 14 000 rev/min.

Calculations: We can find K with Eq. 10-34 ($K = \frac{1}{2}I\omega^2$), but first we need an expression for the rotational inertia I. Because the rotor was a disk that rotated like a merry-go-round, I is given in Table 10-2*c* ($I = \frac{1}{2}MR^2$). Thus,

$$I = \tfrac{1}{2}MR^2 = \tfrac{1}{2}(272 \text{ kg})(0.38 \text{ m})^2 = 19.64 \text{ kg}\cdot\text{m}^2.$$

The angular speed of the rotor was

$$\omega = (14\,000 \text{ rev/min})(2\pi \text{ rad/rev})\left(\frac{1 \text{ min}}{60 \text{ s}}\right)$$
$$= 1.466 \times 10^3 \text{ rad/s}.$$

Then, with Eq. 10-34, we find the (huge) energy release:

$$K = \tfrac{1}{2}I\omega^2 = \tfrac{1}{2}(19.64 \text{ kg}\cdot\text{m}^2)(1.466 \times 10^3 \text{ rad/s})^2$$
$$= 2.1 \times 10^7 \text{ J}. \qquad \text{(Answer)}$$

WILEY PLUS Additional examples, video, and practice available at *WileyPLUS*

10-6 TORQUE

Learning Objectives

After reading this module, you should be able to . . .

10.23 Identify that a torque on a body involves a force and a position vector, which extends from a rotation axis to the point where the force is applied.

10.24 Calculate the torque by using (a) the angle between the position vector and the force vector, (b) the line of action and the moment arm of the force, and (c) the force component perpendicular to the position vector.

10.25 Identify that a rotation axis must always be specified to calculate a torque.

10.26 Identify that a torque is assigned a positive or negative sign depending on the direction it tends to make the body rotate about a specified rotation axis: "clocks are negative."

10.27 When more than one torque acts on a body about a rotation axis, calculate the net torque.

Key Ideas

- Torque is a turning or twisting action on a body about a rotation axis due to a force $\vec{F}$. If $\vec{F}$ is exerted at a point given by the position vector $\vec{r}$ relative to the axis, then the magnitude of the torque is

$$\tau = rF_t = r_\perp F = rF\sin\phi,$$

where F_t is the component of $\vec{F}$ perpendicular to $\vec{r}$ and ϕ is the angle between $\vec{r}$ and $\vec{F}$. The quantity $r_\perp$ is the perpendicular distance between the rotation axis and an extended line running through the $\vec{F}$ vector. This line is called the line of action of $\vec{F}$, and $r_\perp$ is called the moment arm of $\vec{F}$. Similarly, r is the moment arm of F_t.

- The SI unit of torque is the newton-meter (N·m). A torque τ is positive if it tends to rotate a body at rest counterclockwise and negative if it tends to rotate the body clockwise.

Torque

A doorknob is located as far as possible from the door's hinge line for a good reason. If you want to open a heavy door, you must certainly apply a force, but that is not enough. Where you apply that force and in what direction you push are also important. If you apply your force nearer to the hinge line than the knob, or at any angle other than 90° to the plane of the door, you must use a greater force than if you apply the force at the knob and perpendicular to the door's plane.

Figure 10-16*a* shows a cross section of a body that is free to rotate about an axis passing through O and perpendicular to the cross section. A force $\vec{F}$ is applied at point P, whose position relative to O is defined by a position vector $\vec{r}$. The directions of vectors $\vec{F}$ and $\vec{r}$ make an angle ϕ with each other. (For simplicity, we consider only forces that have no component parallel to the rotation axis; thus, $\vec{F}$ is in the plane of the page.)

To determine how $\vec{F}$ results in a rotation of the body around the rotation axis, we resolve $\vec{F}$ into two components (Fig. 10-16*b*). One component, called the *radial component* F_r, points along $\vec{r}$. This component does not cause rotation, because it acts along a line that extends through O. (If you pull on a door parallel to the plane of the door, you do not rotate the door.) The other component of $\vec{F}$, called the *tangential component* F_t, is perpendicular to $\vec{r}$ and has magnitude $F_t = F \sin \phi$. This component *does* cause rotation.

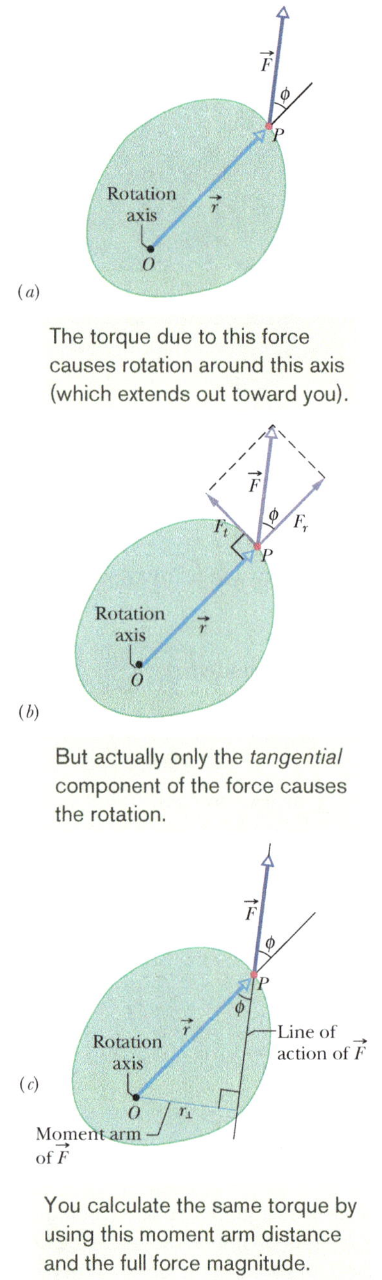

Figure 10-16 (*a*) A force $\vec{F}$ acts on a rigid body, with a rotation axis perpendicular to the page. The torque can be found with (*a*) angle ϕ, (*b*) tangential force component F_t, or (*c*) moment arm $r_\perp$.

Calculating Torques. The ability of $\vec{F}$ to rotate the body depends not only on the magnitude of its tangential component F_t, but also on just how far from O the force is applied. To include both these factors, we define a quantity called **torque** τ as the product of the two factors and write it as

$$\tau = (r)(F \sin \phi). \tag{10-39}$$

Two equivalent ways of computing the torque are

$$\tau = (r)(F \sin \phi) = rF_t \tag{10-40}$$

and

$$\tau = (r \sin \phi)(F) = r_\perp F, \tag{10-41}$$

where $r_\perp$ is the perpendicular distance between the rotation axis at O and an extended line running through the vector $\vec{F}$ (Fig. 10-16*c*). This extended line is called the **line of action** of $\vec{F}$, and $r_\perp$ is called the **moment arm** of $\vec{F}$. Figure 10-16*b* shows that we can describe r, the magnitude of $\vec{r}$, as being the moment arm of the force component F_t.

Torque, which comes from the Latin word meaning "to twist," may be loosely identified as the turning or twisting action of the force $\vec{F}$. When you apply a force to an object—such as a screwdriver or torque wrench—with the purpose of turning that object, you are applying a torque. The SI unit of torque is the newton-meter (N·m). *Caution:* The newton-meter is also the unit of work. Torque and work, however, are quite different quantities and must not be confused. Work is often expressed in joules (1 J = 1 N·m), but torque never is.

Clocks Are Negative. In Chapter 11 we shall use vector notation for torques, but here, with rotation around a single axis, we use only an algebraic sign. If a torque would cause counterclockwise rotation, it is positive. If it would cause clockwise rotation, it is negative. (The phrase "clocks are negative" from Module 10-1 still works.)

Torques obey the superposition principle that we discussed in Chapter 5 for forces: When several torques act on a body, the **net torque** (or **resultant torque**) is the sum of the individual torques. The symbol for net torque is τ_{net}.

Checkpoint 6

The figure shows an overhead view of a meter stick that can pivot about the dot at the position marked 20 (for 20 cm). All five forces on the stick are horizontal and have the same magnitude. Rank the forces according to the magnitude of the torque they produce, greatest first.

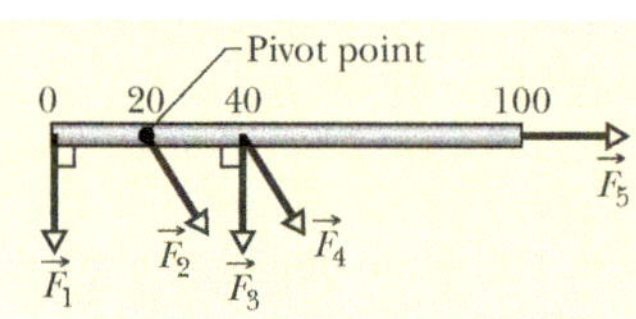

10-7 NEWTON'S SECOND LAW FOR ROTATION

Learning Objective

After reading this module, you should be able to . . .

10.28 Apply Newton's second law for rotation to relate the net torque on a body to the body's rotational inertia and rotational acceleration, all calculated relative to a specified rotation axis.

Key Idea

- The rotational analog of Newton's second law is

$$\tau_{net} = I\alpha,$$

where τ_{net} is the net torque acting on a particle or rigid body, I is the rotational inertia of the particle or body about the rotation axis, and α is the resulting angular acceleration about that axis.

Newton's Second Law for Rotation

A torque can cause rotation of a rigid body, as when you use a torque to rotate a door. Here we want to relate the net torque τ_{net} on a rigid body to the angular acceleration α that torque causes about a rotation axis. We do so by analogy with Newton's second law ($F_{net} = ma$) for the acceleration a of a body of mass m due to a net force F_{net} along a coordinate axis. We replace F_{net} with τ_{net}, m with I, and a with α in radian measure, writing

$$\tau_{net} = I\alpha \quad \text{(Newton's second law for rotation).} \tag{10-42}$$

Proof of Equation 10-42

We prove Eq. 10-42 by first considering the simple situation shown in Fig. 10-17. The rigid body there consists of a particle of mass m on one end of a massless rod of length r. The rod can move only by rotating about its other end, around a rotation axis (an axle) that is perpendicular to the plane of the page. Thus, the particle can move only in a circular path that has the rotation axis at its center.

A force $\vec{F}$ acts on the particle. However, because the particle can move only along the circular path, only the tangential component F_t of the force (the component that is tangent to the circular path) can accelerate the particle along the path. We can relate F_t to the particle's tangential acceleration a_t along the path with Newton's second law, writing

$$F_t = ma_t.$$

The torque acting on the particle is, from Eq. 10-40,

$$\tau = F_t r = ma_t r.$$

From Eq. 10-22 ($a_t = \alpha r$) we can write this as

$$\tau = m(\alpha r)r = (mr^2)\alpha. \tag{10-43}$$

The quantity in parentheses on the right is the rotational inertia of the particle about the rotation axis (see Eq. 10-33, but here we have only a single particle). Thus, using I for the rotational inertia, Eq. 10-43 reduces to

$$\tau = I\alpha \quad \text{(radian measure).} \tag{10-44}$$

If more than one force is applied to the particle, Eq. 10-44 becomes

$$\tau_{net} = I\alpha \quad \text{(radian measure),} \tag{10-45}$$

which we set out to prove. We can extend this equation to any rigid body rotating about a fixed axis, because any such body can always be analyzed as an assembly of single particles.

The torque due to the tangential component of the force causes an angular acceleration around the rotation axis.

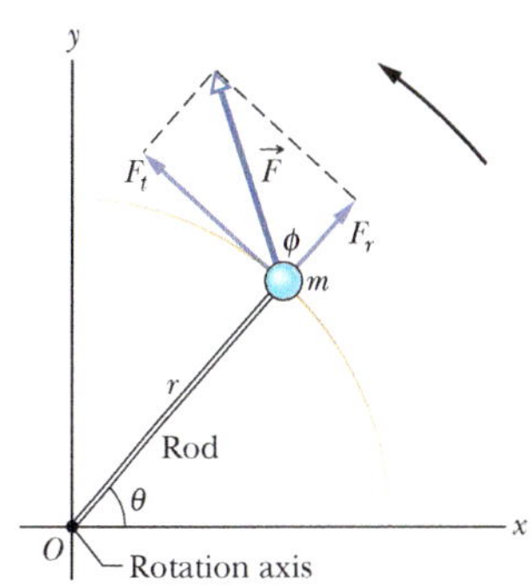

Figure 10-17 A simple rigid body, free to rotate about an axis through O, consists of a particle of mass m fastened to the end of a rod of length r and negligible mass. An applied force $\vec{F}$ causes the body to rotate.

Checkpoint 7

The figure shows an overhead view of a meter stick that can pivot about the point indicated, which is to the left of the stick's midpoint. Two horizontal forces, $\vec{F}_1$ and $\vec{F}_2$, are applied to the stick. Only $\vec{F}_1$ is shown. Force $\vec{F}_2$ is perpendicular to the stick and is applied at the right end. If the stick is not to turn, (a) what should be the direction of $\vec{F}_2$, and (b) should F_2 be greater than, less than, or equal to F_1?

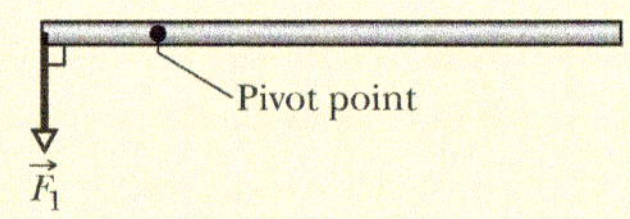

Sample Problem 10.09 Using Newton's second law for rotation in a basic judo hip throw

To throw an 80 kg opponent with a basic judo hip throw, you intend to pull his uniform with a force $\vec{F}$ and a moment arm $d_1 = 0.30$ m from a pivot point (rotation axis) on your right hip (Fig. 10-18). You wish to rotate him about the pivot point with an angular acceleration α of -6.0 rad/s²—that is, with an angular acceleration that is *clockwise* in the figure. Assume that his rotational inertia I relative to the pivot point is 15 kg·m².

(a) What must the magnitude of $\vec{F}$ be if, before you throw him, you bend your opponent forward to bring his center of mass to your hip (Fig. 10-18*a*)?

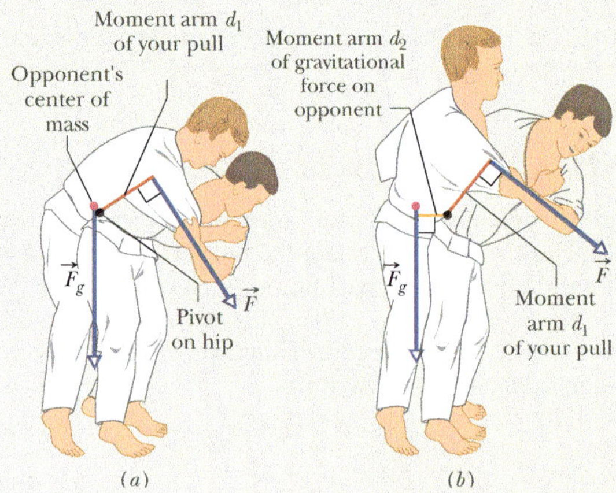

Figure 10-18 A judo hip throw (*a*) correctly executed and (*b*) incorrectly executed.

KEY IDEA

We can relate your pull $\vec{F}$ on your opponent to the given angular acceleration α via Newton's second law for rotation ($\tau_{\text{net}} = I\alpha$).

Calculations: As his feet leave the floor, we can assume that only three forces act on him: your pull $\vec{F}$, a force $\vec{N}$ on him from you at the pivot point (this force is not indicated in Fig. 10-18), and the gravitational force $\vec{F}_g$. To use $\tau_{\text{net}} = I\alpha$, we need the corresponding three torques, each about the pivot point.

From Eq. 10-41 ($\tau = r_\perp F$), the torque due to your pull $\vec{F}$ is equal to $-d_1F$, where d_1 is the moment arm $r_\perp$ and the sign indicates the clockwise rotation this torque tends to cause. The torque due to $\vec{N}$ is zero, because $\vec{N}$ acts at the pivot point and thus has moment arm $r_\perp = 0$.

To evaluate the torque due to $\vec{F}_g$, we can assume that $\vec{F}_g$ acts at your opponent's center of mass. With the center of mass at the pivot point, $\vec{F}_g$ has moment arm $r_\perp = 0$ and thus the torque due to $\vec{F}_g$ is zero. So, the only torque on your opponent is due to your pull $\vec{F}$, and we can write $\tau_{\text{net}} = I\alpha$ as

$$-d_1F = I\alpha.$$

We then find

$$F = \frac{-I\alpha}{d_1} = \frac{-(15\ \text{kg}\cdot\text{m}^2)(-6.0\ \text{rad/s}^2)}{0.30\ \text{m}}$$
$$= 300\ \text{N}. \qquad \text{(Answer)}$$

(b) What must the magnitude of $\vec{F}$ be if your opponent remains upright before you throw him, so that $\vec{F}_g$ has a moment arm $d_2 = 0.12$ m (Fig. 10-18*b*)?

KEY IDEA

Because the moment arm for $\vec{F}_g$ is no longer zero, the torque due to $\vec{F}_g$ is now equal to d_2mg and is positive because the torque attempts counterclockwise rotation.

Calculations: Now we write $\tau_{\text{net}} = I\alpha$ as

$$-d_1F + d_2mg = I\alpha,$$

which gives

$$F = -\frac{I\alpha}{d_1} + \frac{d_2mg}{d_1}.$$

From (a), we know that the first term on the right is equal to 300 N. Substituting this and the given data, we have

$$F = 300\ \text{N} + \frac{(0.12\ \text{m})(80\ \text{kg})(9.8\ \text{m/s}^2)}{0.30\ \text{m}}$$
$$= 613.6\ \text{N} \approx 610\ \text{N}. \qquad \text{(Answer)}$$

The results indicate that you will have to pull much harder if you do not initially bend your opponent to bring his center of mass to your hip. A good judo fighter knows this lesson from physics. Indeed, physics is the basis of most of the martial arts, figured out after countless hours of trial and error over the centuries.

WILEY PLUS Additional examples, video, and practice available at *WileyPLUS*

Sample Problem 10.10 Newton's second law, rotation, torque, disk

Figure 10-19a shows a uniform disk, with mass $M = 2.5$ kg and radius $R = 20$ cm, mounted on a fixed horizontal axle. A block with mass $m = 1.2$ kg hangs from a massless cord that is wrapped around the rim of the disk. Find the acceleration of the falling block, the angular acceleration of the disk, and the tension in the cord. The cord does not slip, and there is no friction at the axle.

KEY IDEAS

(1) Taking the block as a system, we can relate its acceleration a to the forces acting on it with Newton's second law ($\vec{F}_{\text{net}} = m\vec{a}$). (2) Taking the disk as a system, we can relate its angular acceleration α to the torque acting on it with Newton's second law for rotation ($\tau_{\text{net}} = I\alpha$). (3) To combine the motions of block and disk, we use the fact that the linear acceleration a of the block and the (tangential) linear acceleration a_t of the disk rim are equal. (To avoid confusion about signs, let's work with acceleration magnitudes and explicit algebraic signs.)

Forces on block: The forces are shown in the block's free-body diagram in Fig. 10-19b: The force from the cord is $\vec{T}$, and the gravitational force is $\vec{F}_g$, of magnitude mg. We can now write Newton's second law for components along a vertical y axis ($F_{\text{net},y} = ma_y$) as

$$T - mg = m(-a), \tag{10-46}$$

where a is the magnitude of the acceleration (down the y axis). However, we cannot solve this equation for a because it also contains the unknown T.

Torque on disk: Previously, when we got stuck on the y axis, we switched to the x axis. Here, we switch to the rotation of the disk and use Newton's second law in angular form. To calculate the torques and the rotational inertia I, we take the rotation axis to be perpendicular to the disk and through its center, at point O in Fig. 10-19c.

The torques are then given by Eq. 10-40 ($\tau = rF_t$). The gravitational force on the disk and the force on the disk from the axle both act at the center of the disk and thus at distance $r = 0$, so their torques are zero. The force $\vec{T}$ on the disk due to the cord acts at distance $r = R$ and is tangent to the rim of the disk. Therefore, its torque is $-RT$, negative because the torque rotates the disk clockwise from rest. Let α be the magnitude of the negative (clockwise) angular acceleration. From Table 10-2c, the rotational inertia I of the disk is $\frac{1}{2}MR^2$. Thus we can write the general equation $\tau_{\text{net}} = I\alpha$ as

$$-RT = \tfrac{1}{2}MR^2(-\alpha). \tag{10-47}$$

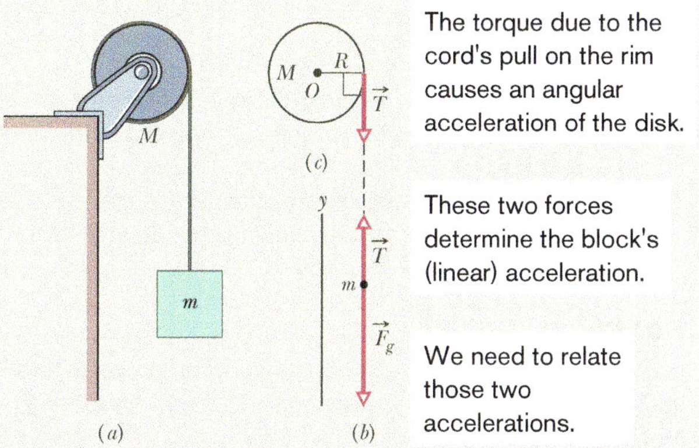

Figure 10-19 (a) The falling block causes the disk to rotate. (b) A free-body diagram for the block. (c) An incomplete free-body diagram for the disk.

This equation seems useless because it has two unknowns, α and T, neither of which is the desired a. However, mustering physics courage, we can make it useful with this fact: Because the cord does not slip, the magnitude a of the block's linear acceleration and the magnitude a_t of the (tangential) linear acceleration of the rim of the disk are equal. Then, by Eq. 10-22 ($a_t = \alpha r$) we see that here $\alpha = a/R$. Substituting this in Eq. 10-47 yields

$$T = \tfrac{1}{2}Ma. \tag{10-48}$$

Combining results: Combining Eqs. 10-46 and 10-48 leads to

$$a = g\frac{2m}{M + 2m} = (9.8 \text{ m/s}^2)\frac{(2)(1.2 \text{ kg})}{2.5 \text{ kg} + (2)(1.2 \text{ kg})}$$

$$= 4.8 \text{ m/s}^2. \qquad \text{(Answer)}$$

We then use Eq. 10-48 to find T:

$$T = \tfrac{1}{2}Ma = \tfrac{1}{2}(2.5 \text{ kg})(4.8 \text{ m/s}^2)$$

$$= 6.0 \text{ N}. \qquad \text{(Answer)}$$

As we should expect, acceleration a of the falling block is less than g, and tension T in the cord ($= 6.0$ N) is less than the gravitational force on the hanging block ($= mg = 11.8$ N). We see also that a and T depend on the mass of the disk but not on its radius.

As a check, we note that the formulas derived above predict $a = g$ and $T = 0$ for the case of a massless disk ($M = 0$). This is what we would expect; the block simply falls as a free body. From Eq. 10-22, the magnitude of the angular acceleration of the disk is

$$\alpha = \frac{a}{R} = \frac{4.8 \text{ m/s}^2}{0.20 \text{ m}} = 24 \text{ rad/s}^2. \qquad \text{(Answer)}$$

WILEY PLUS Additional examples, video, and practice available at *WileyPLUS*

10-8 WORK AND ROTATIONAL KINETIC ENERGY

Learning Objectives

After reading this module, you should be able to . . .

10.29 Calculate the work done by a torque acting on a rotating body by integrating the torque with respect to the angle of rotation.

10.30 Apply the work–kinetic energy theorem to relate the work done by a torque to the resulting change in the rotational kinetic energy of the body.

10.31 Calculate the work done by a *constant* torque by relating the work to the angle through which the body rotates.

10.32 Calculate the power of a torque by finding the rate at which work is done.

10.33 Calculate the power of a torque at any given instant by relating it to the torque and the angular velocity at that instant.

Key Ideas

- The equations used for calculating work and power in rotational motion correspond to equations used for translational motion and are

$$W = \int_{\theta_i}^{\theta_f} \tau \, d\theta$$

and

$$P = \frac{dW}{dt} = \tau\omega.$$

- When τ is constant, the integral reduces to

$$W = \tau(\theta_f - \theta_i).$$

- The form of the work–kinetic energy theorem used for rotating bodies is

$$\Delta K = K_f - K_i = \tfrac{1}{2}I\omega_f^2 - \tfrac{1}{2}I\omega_i^2 = W.$$

Work and Rotational Kinetic Energy

As we discussed in Chapter 7, when a force F causes a rigid body of mass m to accelerate along a coordinate axis, the force does work W on the body. Thus, the body's kinetic energy ($K = \frac{1}{2}mv^2$) can change. Suppose it is the only energy of the body that changes. Then we relate the change ΔK in kinetic energy to the work W with the work–kinetic energy theorem (Eq. 7-10), writing

$$\Delta K = K_f - K_i = \tfrac{1}{2}mv_f^2 - \tfrac{1}{2}mv_i^2 = W \quad \text{(work–kinetic energy theorem).} \tag{10-49}$$

For motion confined to an x axis, we can calculate the work with Eq. 7-32,

$$W = \int_{x_i}^{x_f} F \, dx \quad \text{(work, one-dimensional motion).} \tag{10-50}$$

This reduces to $W = Fd$ when F is constant and the body's displacement is d. The rate at which the work is done is the power, which we can find with Eqs. 7-43 and 7-48,

$$P = \frac{dW}{dt} = Fv \quad \text{(power, one-dimensional motion).} \tag{10-51}$$

Now let us consider a rotational situation that is similar. When a torque accelerates a rigid body in rotation about a fixed axis, the torque does work W on the body. Therefore, the body's rotational kinetic energy ($K = \frac{1}{2}I\omega^2$) can change. Suppose that it is the only energy of the body that changes. Then we can still relate the change ΔK in kinetic energy to the work W with the work–kinetic energy theorem, except now the kinetic energy is a rotational kinetic energy:

$$\Delta K = K_f - K_i = \tfrac{1}{2}I\omega_f^2 - \tfrac{1}{2}I\omega_i^2 = W \quad \text{(work–kinetic energy theorem).} \tag{10-52}$$

Here, I is the rotational inertia of the body about the fixed axis and ω_i and ω_f are the angular speeds of the body before and after the work is done.

Also, we can calculate the work with a rotational equivalent of Eq. 10-50,

$$W = \int_{\theta_i}^{\theta_f} \tau\, d\theta \quad \text{(work, rotation about fixed axis)}, \tag{10-53}$$

where τ is the torque doing the work W, and θ_i and θ_f are the body's angular positions before and after the work is done, respectively. When τ is constant, Eq. 10-53 reduces to

$$W = \tau(\theta_f - \theta_i) \quad \text{(work, constant torque)}. \tag{10-54}$$

The rate at which the work is done is the power, which we can find with the rotational equivalent of Eq. 10-51,

$$P = \frac{dW}{dt} = \tau\omega \quad \text{(power, rotation about fixed axis)}. \tag{10-55}$$

Table 10-3 summarizes the equations that apply to the rotation of a rigid body about a fixed axis and the corresponding equations for translational motion.

Proof of Eqs. 10-52 through 10-55

Let us again consider the situation of Fig. 10-17, in which force $\vec{F}$ rotates a rigid body consisting of a single particle of mass m fastened to the end of a massless rod. During the rotation, force $\vec{F}$ does work on the body. Let us assume that the only energy of the body that is changed by $\vec{F}$ is the kinetic energy. Then we can apply the work–kinetic energy theorem of Eq. 10-49:

$$\Delta K = K_f - K_i = W. \tag{10-56}$$

Using $K = \frac{1}{2}mv^2$ and Eq. 10-18 ($v = \omega r$), we can rewrite Eq. 10-56 as

$$\Delta K = \tfrac{1}{2}mr^2\omega_f^2 - \tfrac{1}{2}mr^2\omega_i^2 = W. \tag{10-57}$$

From Eq. 10-33, the rotational inertia for this one-particle body is $I = mr^2$. Substituting this into Eq. 10-57 yields

$$\Delta K = \tfrac{1}{2}I\omega_f^2 - \tfrac{1}{2}I\omega_i^2 = W,$$

which is Eq. 10-52. We derived it for a rigid body with one particle, but it holds for any rigid body rotated about a fixed axis.

We next relate the work W done on the body in Fig. 10-17 to the torque τ on the body due to force $\vec{F}$. When the particle moves a distance ds along its

Table 10-3 Some Corresponding Relations for Translational and Rotational Motion

Pure Translation (Fixed Direction)		Pure Rotation (Fixed Axis)	
Position	x	Angular position	θ
Velocity	$v = dx/dt$	Angular velocity	$\omega = d\theta/dt$
Acceleration	$a = dv/dt$	Angular acceleration	$\alpha = d\omega/dt$
Mass	m	Rotational inertia	I
Newton's second law	$F_{net} = ma$	Newton's second law	$\tau_{net} = I\alpha$
Work	$W = \int F\,dx$	Work	$W = \int \tau\, d\theta$
Kinetic energy	$K = \frac{1}{2}mv^2$	Kinetic energy	$K = \frac{1}{2}I\omega^2$
Power (constant force)	$P = Fv$	Power (constant torque)	$P = \tau\omega$
Work–kinetic energy theorem	$W = \Delta K$	Work–kinetic energy theorem	$W = \Delta K$

circular path, only the tangential component F_t of the force accelerates the particle along the path. Therefore, only F_t does work on the particle. We write that work dW as $F_t\,ds$. However, we can replace ds with $r\,d\theta$, where $d\theta$ is the angle through which the particle moves. Thus we have

$$dW = F_t r\,d\theta. \tag{10-58}$$

From Eq. 10-40, we see that the product $F_t r$ is equal to the torque τ, so we can rewrite Eq. 10-58 as

$$dW = \tau\,d\theta. \tag{10-59}$$

The work done during a finite angular displacement from θ_i to θ_f is then

$$W = \int_{\theta_i}^{\theta_f} \tau\,d\theta,$$

which is Eq. 10-53. It holds for any rigid body rotating about a fixed axis. Equation 10-54 comes directly from Eq. 10-53.

We can find the power P for rotational motion from Eq. 10-59:

$$P = \frac{dW}{dt} = \tau\frac{d\theta}{dt} = \tau\omega,$$

which is Eq. 10-55.

Sample Problem 10.11 Work, rotational kinetic energy, torque, disk

Let the disk in Fig. 10-19 start from rest at time $t = 0$ and also let the tension in the massless cord be 6.0 N and the angular acceleration of the disk be $-24\ \text{rad/s}^2$. What is its rotational kinetic energy K at $t = 2.5$ s?

KEY IDEA

We can find K with Eq. 10-34 ($K = \frac{1}{2}I\omega^2$). We already know that $I = \frac{1}{2}MR^2$, but we do not yet know ω at $t = 2.5$ s. However, because the angular acceleration α has the constant value of $-24\ \text{rad/s}^2$, we can apply the equations for constant angular acceleration in Table 10-1.

Calculations: Because we want ω and know α and ω_0 ($= 0$), we use Eq. 10-12:

$$\omega = \omega_0 + \alpha t = 0 + \alpha t = \alpha t.$$

Substituting $\omega = \alpha t$ and $I = \frac{1}{2}MR^2$ into Eq. 10-34, we find

$$\begin{aligned} K &= \tfrac{1}{2}I\omega^2 = \tfrac{1}{2}(\tfrac{1}{2}MR^2)(\alpha t)^2 = \tfrac{1}{4}M(R\alpha t)^2 \\ &= \tfrac{1}{4}(2.5\ \text{kg})[(0.20\ \text{m})(-24\ \text{rad/s}^2)(2.5\ \text{s})]^2 \\ &= 90\ \text{J}. \end{aligned} \quad \text{(Answer)}$$

KEY IDEA

We can also get this answer by finding the disk's kinetic energy from the work done on the disk.

Calculations: First, we relate the *change* in the kinetic energy of the disk to the net work W done on the disk, using the work–kinetic energy theorem of Eq. 10-52 ($K_f - K_i = W$). With K substituted for K_f and 0 for K_i, we get

$$K = K_i + W = 0 + W = W. \tag{10-60}$$

Next we want to find the work W. We can relate W to the torques acting on the disk with Eq. 10-53 or 10-54. The only torque causing angular acceleration and doing work is the torque due to force $\vec{T}$ on the disk from the cord, which is equal to $-TR$. Because α is constant, this torque also must be constant. Thus, we can use Eq. 10-54 to write

$$W = \tau(\theta_f - \theta_i) = -TR(\theta_f - \theta_i). \tag{10-61}$$

Because α is constant, we can use Eq. 10-13 to find $\theta_f - \theta_i$. With $\omega_i = 0$, we have

$$\theta_f - \theta_i = \omega_i t + \tfrac{1}{2}\alpha t^2 = 0 + \tfrac{1}{2}\alpha t^2 = \tfrac{1}{2}\alpha t^2.$$

Now we substitute this into Eq. 10-61 and then substitute the result into Eq. 10-60. Inserting the given values $T = 6.0$ N and $\alpha = -24\ \text{rad/s}^2$, we have

$$\begin{aligned} K = W &= -TR(\theta_f - \theta_i) = -TR(\tfrac{1}{2}\alpha t^2) = -\tfrac{1}{2}TR\alpha t^2 \\ &= -\tfrac{1}{2}(6.0\ \text{N})(0.20\ \text{m})(-24\ \text{rad/s}^2)(2.5\ \text{s})^2 \\ &= 90\ \text{J}. \end{aligned} \quad \text{(Answer)}$$

Review & Summary

Angular Position To describe the rotation of a rigid body about a fixed axis, called the **rotation axis,** we assume a **reference line** is fixed in the body, perpendicular to that axis and rotating with the body. We measure the **angular position** θ of this line relative to a fixed direction. When θ is measured in **radians,**

$$\theta = \frac{s}{r} \quad \text{(radian measure)}, \tag{10-1}$$

where s is the arc length of a circular path of radius r and angle θ. Radian measure is related to angle measure in revolutions and degrees by

$$1 \text{ rev} = 360° = 2\pi \text{ rad}. \tag{10-2}$$

Angular Displacement A body that rotates about a rotation axis, changing its angular position from θ_1 to θ_2, undergoes an **angular displacement**

$$\Delta\theta = \theta_2 - \theta_1, \tag{10-4}$$

where $\Delta\theta$ is positive for counterclockwise rotation and negative for clockwise rotation.

Angular Velocity and Speed If a body rotates through an angular displacement $\Delta\theta$ in a time interval Δt, its **average angular velocity** ω_{avg} is

$$\omega_{avg} = \frac{\Delta\theta}{\Delta t}. \tag{10-5}$$

The **(instantaneous) angular velocity** ω of the body is

$$\omega = \frac{d\theta}{dt}. \tag{10-6}$$

Both ω_{avg} and ω are vectors, with directions given by the **right-hand rule** of Fig. 10-6. They are positive for counterclockwise rotation and negative for clockwise rotation. The magnitude of the body's angular velocity is the **angular speed.**

Angular Acceleration If the angular velocity of a body changes from ω_1 to ω_2 in a time interval $\Delta t = t_2 - t_1$, the **average angular acceleration** α_{avg} of the body is

$$\alpha_{avg} = \frac{\omega_2 - \omega_1}{t_2 - t_1} = \frac{\Delta\omega}{\Delta t}. \tag{10-7}$$

The **(instantaneous) angular acceleration** α of the body is

$$\alpha = \frac{d\omega}{dt}. \tag{10-8}$$

Both α_{avg} and α are vectors.

The Kinematic Equations for Constant Angular Acceleration *Constant angular acceleration* (α = constant) is an important special case of rotational motion. The appropriate kinematic equations, given in Table 10-1, are

$$\omega = \omega_0 + \alpha t, \tag{10-12}$$

$$\theta - \theta_0 = \omega_0 t + \tfrac{1}{2}\alpha t^2, \tag{10-13}$$

$$\omega^2 = \omega_0^2 + 2\alpha(\theta - \theta_0), \tag{10-14}$$

$$\theta - \theta_0 = \tfrac{1}{2}(\omega_0 + \omega)t, \tag{10-15}$$

$$\theta - \theta_0 = \omega t - \tfrac{1}{2}\alpha t^2. \tag{10-16}$$

Linear and Angular Variables Related A point in a rigid rotating body, at a *perpendicular distance* r from the rotation axis, moves in a circle with radius r. If the body rotates through an angle θ, the point moves along an arc with length s given by

$$s = \theta r \quad \text{(radian measure)}, \tag{10-17}$$

where θ is in radians.

The linear velocity $\vec{v}$ of the point is tangent to the circle; the point's linear speed v is given by

$$v = \omega r \quad \text{(radian measure)}, \tag{10-18}$$

where ω is the angular speed (in radians per second) of the body.

The linear acceleration $\vec{a}$ of the point has both *tangential* and *radial* components. The tangential component is

$$a_t = \alpha r \quad \text{(radian measure)}, \tag{10-22}$$

where α is the magnitude of the angular acceleration (in radians per second-squared) of the body. The radial component of $\vec{a}$ is

$$a_r = \frac{v^2}{r} = \omega^2 r \quad \text{(radian measure)}. \tag{10-23}$$

If the point moves in uniform circular motion, the period T of the motion for the point and the body is

$$T = \frac{2\pi r}{v} = \frac{2\pi}{\omega} \quad \text{(radian measure)}. \tag{10-19, 10-20}$$

Rotational Kinetic Energy and Rotational Inertia The kinetic energy K of a rigid body rotating about a fixed axis is given by

$$K = \tfrac{1}{2}I\omega^2 \quad \text{(radian measure)}, \tag{10-34}$$

in which I is the **rotational inertia** of the body, defined as

$$I = \sum m_i r_i^2 \tag{10-33}$$

for a system of discrete particles and defined as

$$I = \int r^2\, dm \tag{10-35}$$

for a body with continuously distributed mass. The r and r_i in these expressions represent the perpendicular distance from the axis of rotation to each mass element in the body, and the integration is carried out over the entire body so as to include every mass element.

The Parallel-Axis Theorem The *parallel-axis theorem* relates the rotational inertia I of a body about any axis to that of the same body about a parallel axis through the center of mass:

$$I = I_{com} + Mh^2. \tag{10-36}$$

Here h is the perpendicular distance between the two axes, and I_{com} is the rotational inertia of the body about the axis through the com. We can describe h as being the distance the actual rotation axis has been shifted from the rotation axis through the com.

Torque *Torque* is a turning or twisting action on a body about a rotation axis due to a force $\vec{F}$. If $\vec{F}$ is exerted at a point given by the position vector $\vec{r}$ relative to the axis, then the magnitude of the torque is

$$\tau = rF_t = r_\perp F = rF \sin\phi, \tag{10-40, 10-41, 10-39}$$

where F_t is the component of $\vec{F}$ perpendicular to $\vec{r}$ and ϕ is the angle between $\vec{r}$ and $\vec{F}$. The quantity $r_\perp$ is the perpendicular distance between the rotation axis and an extended line running through the $\vec{F}$ vector. This line is called the **line of action** of $\vec{F}$, and $r_\perp$ is called the **moment arm** of $\vec{F}$. Similarly, r is the moment arm of F_t.

The SI unit of torque is the newton-meter (N·m). A torque τ is positive if it tends to rotate a body at rest counterclockwise and negative if it tends to rotate the body clockwise.

Newton's Second Law in Angular Form The rotational analog of Newton's second law is

$$\tau_{net} = I\alpha, \tag{10-45}$$

where τ_{net} is the net torque acting on a particle or rigid body, I is the rotational inertia of the particle or body about the rotation axis, and α is the resulting angular acceleration about that axis.

Work and Rotational Kinetic Energy The equations used for calculating work and power in rotational motion correspond to equations used for translational motion and are

$$W = \int_{\theta_i}^{\theta_f} \tau \, d\theta \tag{10-53}$$

and

$$P = \frac{dW}{dt} = \tau\omega. \tag{10-55}$$

When τ is constant, Eq. 10-53 reduces to

$$W = \tau(\theta_f - \theta_i). \tag{10-54}$$

The form of the work–kinetic energy theorem used for rotating bodies is

$$\Delta K = K_f - K_i = \tfrac{1}{2}I\omega_f^2 - \tfrac{1}{2}I\omega_i^2 = W. \tag{10-52}$$

Questions

1 Figure 10-20 is a graph of the angular velocity versus time for a disk rotating like a merry-go-round. For a point on the disk rim, rank the instants a, b, c, and d according to the magnitude of the (a) tangential and (b) radial acceleration, greatest first.

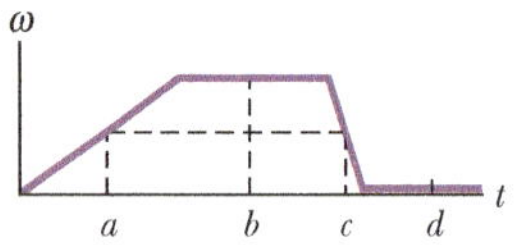

Figure 10-20 Question 1.

2 Figure 10-21 shows plots of angular position θ versus time t for three cases in which a disk is rotated like a merry-go-round. In each case, the rotation direction changes at a certain angular position θ_{change}. (a) For each case, determine whether θ_{change} is clockwise or counterclockwise from $\theta = 0$, or whether it is at $\theta = 0$. For each case, determine (b) whether ω is zero before, after, or at $t = 0$ and (c) whether α is positive, negative, or zero.

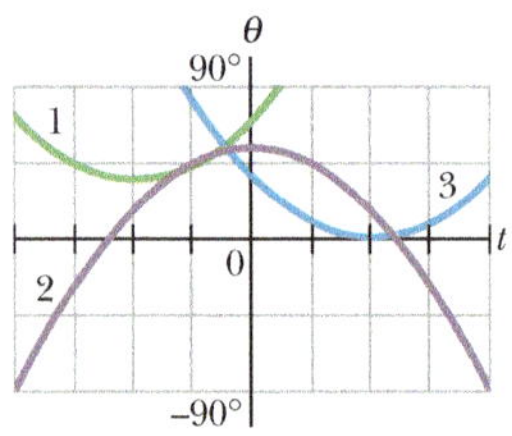

Figure 10-21 Question 2.

3 A force is applied to the rim of a disk that can rotate like a merry-go-round, so as to change its angular velocity. Its initial and final angular velocities, respectively, for four situations are: (a) −2 rad/s, 5 rad/s; (b) 2 rad/s, 5 rad/s; (c) −2 rad/s, −5 rad/s; and (d) 2 rad/s, −5 rad/s. Rank the situations according to the work done by the torque due to the force, greatest first.

4 Figure 10-22b is a graph of the angular position of the rotating disk of Fig. 10-22a. Is the angular velocity of the disk positive, negative, or zero at (a) $t = 1$ s, (b) $t = 2$ s, and (c) $t = 3$ s? (d) Is the angular acceleration positive or negative?

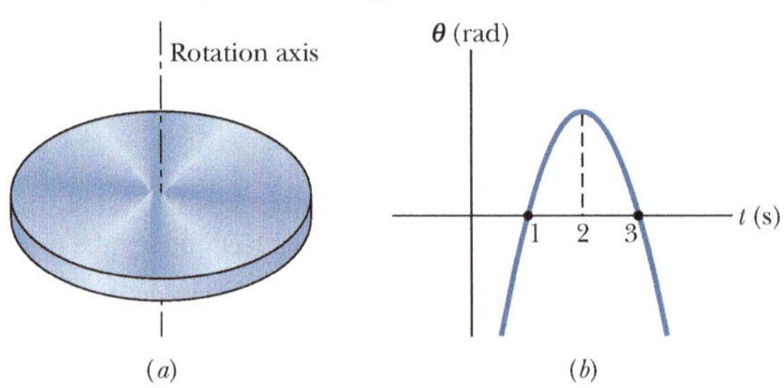

Figure 10-22 Question 4.

5 In Fig. 10-23, two forces $\vec{F}_1$ and $\vec{F}_2$ act on a disk that turns about its center like a merry-go-round. The forces maintain the indicated angles during the rotation, which is counterclockwise and at a constant rate. However, we are to decrease the angle θ of $\vec{F}_1$ without changing the magnitude of $\vec{F}_1$. (a) To keep the angular speed constant, should we increase, decrease, or maintain the magnitude of $\vec{F}_2$? Do forces (b) $\vec{F}_1$ and (c) $\vec{F}_2$ tend to rotate the disk clockwise or counterclockwise?

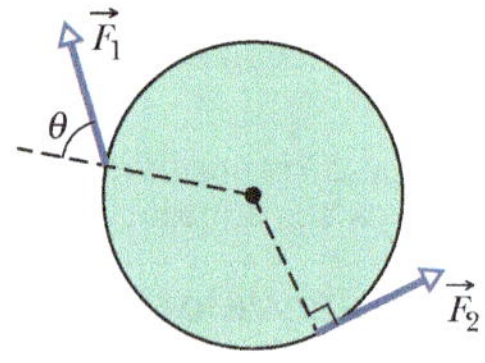

Figure 10-23 Question 5.

6 In the overhead view of Fig. 10-24, five forces of the same magnitude act on a strange merry-go-round; it is a square that can rotate about point P, at midlength along one of the edges. Rank the forces according to the magnitude of the torque they create about point P, greatest first.

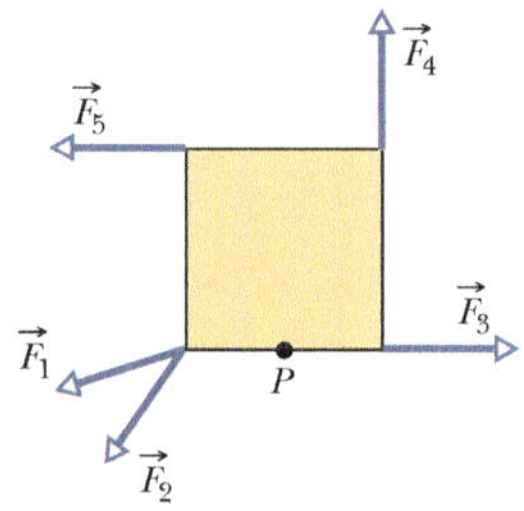

Figure 10-24 Question 6.

7 Figure 10-25a is an overhead view of a horizontal bar that can pivot; two horizontal forces act on the bar, but it is stationary. If the angle between the bar and $\vec{F}_2$ is now decreased from 90° and the bar is still not to turn, should F_2 be made larger, made smaller, or left the same?

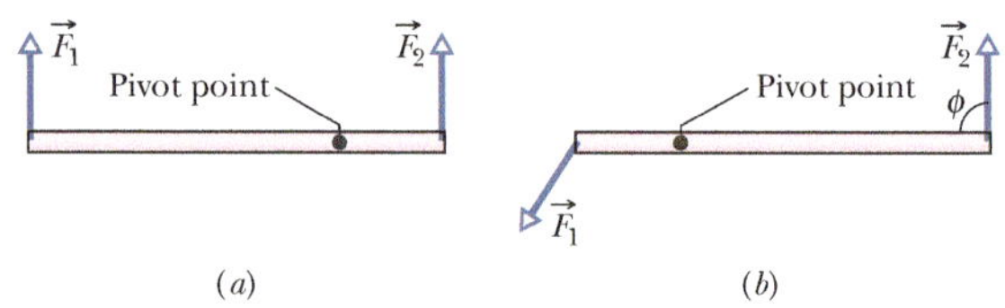

Figure 10-25 Questions 7 and 8.

8 Figure 10-25b shows an overhead view of a horizontal bar that is rotated about the pivot point by two horizontal forces, $\vec{F}_1$ and $\vec{F}_2$, with $\vec{F}_2$ at angle ϕ to the bar. Rank the following values of ϕ according to the magnitude of the angular acceleration of the bar, greatest first: 90°, 70°, and 110°.

9 Figure 10-26 shows a uniform metal plate that had been square before 25% of it was snipped off. Three lettered points are indicated. Rank them according to the rotational inertia of the plate around a perpendicular axis through them, greatest first.

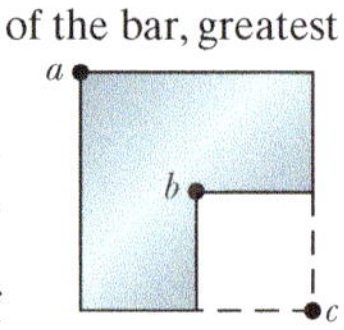

Figure 10-26 Question 9.

10 Figure 10-27 shows three flat disks (of the same radius) that can rotate about their centers like merry-go-rounds. Each disk consists of the same two materials, one denser than the other (density is mass per unit volume). In disks 1 and 3, the denser material forms the outer half of the disk area. In disk 2, it forms the inner half of the disk area. Forces with identical magnitudes are applied tangentially to the disk, either at the outer edge or at the interface of the two materials, as shown. Rank the disks according to (a) the torque about the disk center, (b) the rotational inertia about the disk center, and (c) the angular acceleration of the disk, greatest first.

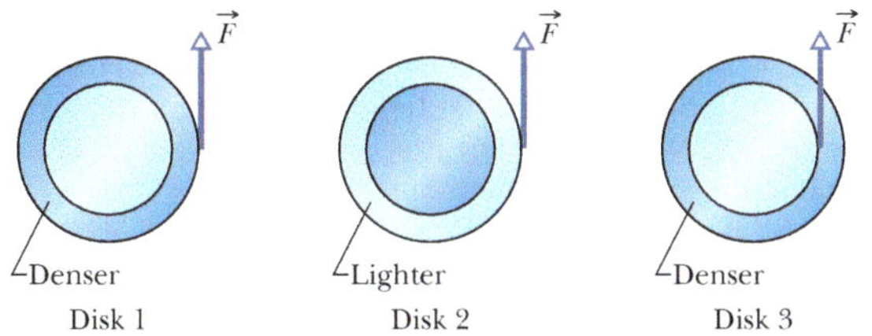

Figure 10-27 Question 10.

11 Figure 10-28a shows a meter stick, half wood and half steel, that is pivoted at the wood end at O. A force $\vec{F}$ is applied to the steel end at a. In Fig. 10-28b, the stick is reversed and pivoted at the steel end at O', and the same force is applied at the wood end at a'. Is the resulting angular acceleration of Fig. 10-28a greater than, less than, or the same as that of Fig. 10-28b?

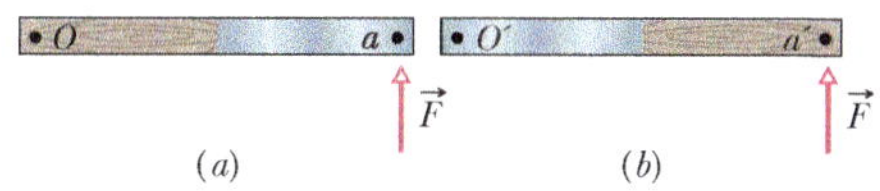

Figure 10-28 Question 11.

12 Figure 10-29 shows three disks, each with a uniform distribution of mass. The radii R and masses M are indicated. Each disk can rotate around its central axis (perpendicular to the disk face and through the center). Rank the disks according to their rotational inertias calculated about their central axes, greatest first.

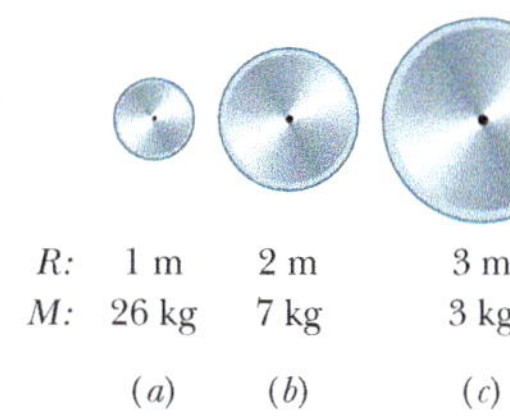

Figure 10-29 Question 12.

Problems

GO Tutoring problem available (at instructor's discretion) in *WileyPLUS* and WebAssign

SSM Worked-out solution available in Student Solutions Manual

• – ••• Number of dots indicates level of problem difficulty

Additional information available in *The Flying Circus of Physics* and at flyingcircusofphysics.com

WWW Worked-out solution is at http://www.wiley.com/college/halliday

ILW Interactive solution is at http://www.wiley.com/college/halliday

Module 10-1 Rotational Variables

•1 A good baseball pitcher can throw a baseball toward home plate at 85 mi/h with a spin of 1800 rev/min. How many revolutions does the baseball make on its way to home plate? For simplicity, assume that the 60 ft path is a straight line.

•2 What is the angular speed of (a) the second hand, (b) the minute hand, and (c) the hour hand of a smoothly running analog watch? Answer in radians per second.

••3 When a slice of buttered toast is accidentally pushed over the edge of a counter, it rotates as it falls. If the distance to the floor is 76 cm and for rotation less than 1 rev, what are the (a) smallest and (b) largest angular speeds that cause the toast to hit and then topple to be butter-side down?

••4 The angular position of a point on a rotating wheel is given by $\theta = 2.0 + 4.0t^2 + 2.0t^3$, where θ is in radians and t is in seconds. At $t = 0$, what are (a) the point's angular position and (b) its angular velocity? (c) What is its angular velocity at $t = 4.0$ s? (d) Calculate its angular acceleration at $t = 2.0$ s. (e) Is its angular acceleration constant?

••5 ILW A diver makes 2.5 revolutions on the way from a 10-m-high platform to the water. Assuming zero initial vertical velocity, find the average angular velocity during the dive.

••6 The angular position of a point on the rim of a rotating wheel is given by $\theta = 4.0t - 3.0t^2 + t^3$, where θ is in radians and t is in seconds. What are the angular velocities at (a) $t = 2.0$ s and (b) $t = 4.0$ s? (c) What is the average angular acceleration for the time interval that begins at $t = 2.0$ s and ends at $t = 4.0$ s? What are the instantaneous angular accelerations at (d) the beginning and (e) the end of this time interval?

•••7 The wheel in Fig. 10-30 has eight equally spaced spokes and a radius of 30 cm. It is mounted on a fixed axle and is spinning at 2.5 rev/s. You want to shoot a 20-cm-long arrow parallel to this axle and through the wheel without hitting any of the spokes. Assume that the arrow and the spokes are very thin. (a) What minimum speed must the arrow have? (b) Does it matter where between the axle and rim of the wheel you aim? If so, what is the best location?

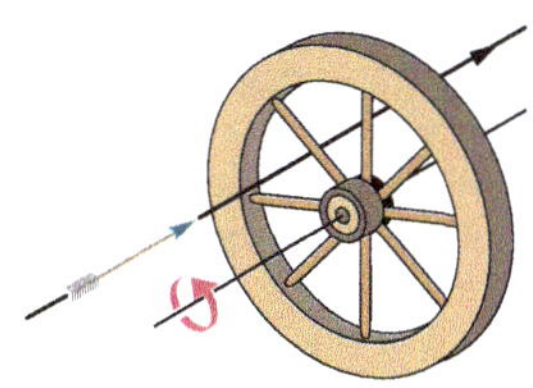

Figure 10-30 Problem 7.

•••8 The angular acceleration of a wheel is $\alpha = 6.0t^4 - 4.0t^2$, with α in radians per second-squared and t in seconds. At time $t = 0$, the wheel has an angular velocity of +2.0 rad/s and an angular position of +1.0 rad. Write expressions for (a) the angular velocity (rad/s) and (b) the angular position (rad) as functions of time (s).

Module 10-2 Rotation with Constant Angular Acceleration

•9 A drum rotates around its central axis at an angular velocity of 12.60 rad/s. If the drum then slows at a constant rate of 4.20 rad/s^2, (a) how much time does it take and (b) through what angle does it rotate in coming to rest?

•10 Starting from rest, a disk rotates about its central axis with constant angular acceleration. In 5.0 s, it rotates 25 rad. During that time, what are the magnitudes of (a) the angular acceleration and (b) the average angular velocity? (c) What is the instantaneous angular velocity of the disk at the end of the 5.0 s? (d) With the angular acceleration unchanged, through what additional angle will the disk turn during the next 5.0 s?

•11 A disk, initially rotating at 120 rad/s, is slowed down with a constant angular acceleration of magnitude 4.0 rad/s^2. (a) How much time does the disk take to stop? (b) Through what angle does the disk rotate during that time?

•12 The angular speed of an automobile engine is increased at a constant rate from 1200 rev/min to 3000 rev/min in 12 s. (a) What is its angular acceleration in revolutions per minute-squared? (b) How many revolutions does the engine make during this 12 s interval?

••13 ILW A flywheel turns through 40 rev as it slows from an angular speed of 1.5 rad/s to a stop. (a) Assuming a constant angular acceleration, find the time for it to come to rest. (b) What is its angular acceleration? (c) How much time is required for it to complete the first 20 of the 40 revolutions?

••14 GO A disk rotates about its central axis starting from rest and accelerates with constant angular acceleration. At one time it is rotating at 10 rev/s; 60 revolutions later, its angular speed is 15 rev/s. Calculate (a) the angular acceleration, (b) the time required to complete the 60 revolutions, (c) the time required to reach the 10 rev/s angular speed, and (d) the number of revolutions from rest until the time the disk reaches the 10 rev/s angular speed.

••15 SSM Starting from rest, a wheel has constant $\alpha = 3.0$ rad/s^2. During a certain 4.0 s interval, it turns through 120 rad. How much time did it take to reach that 4.0 s interval?

••16 A merry-go-round rotates from rest with an angular acceleration of 1.50 rad/s^2. How long does it take to rotate through (a) the first 2.00 rev and (b) the next 2.00 rev?

••17 At $t = 0$, a flywheel has an angular velocity of 4.7 rad/s, a constant angular acceleration of -0.25 rad/s^2, and a reference line at $\theta_0 = 0$. (a) Through what maximum angle θ_{max} will the reference line turn in the positive direction? What are the (b) first and (c) second times the reference line will be at $\theta = \frac{1}{2}\theta_{max}$? At what (d) negative time and (e) positive time will the reference line be at $\theta = 10.5$ rad? (f) Graph θ versus t, and indicate your answers.

•••18 A pulsar is a rapidly rotating neutron star that emits a radio beam the way a lighthouse emits a light beam. We receive a radio pulse for each rotation of the star. The period T of rotation is found by measuring the time between pulses. The pulsar in the Crab nebula has a period of rotation of $T = 0.033$ s that is increasing at the rate of 1.26×10^{-5} s/y. (a) What is the pulsar's angular acceleration α? (b) If α is constant, how many years from now will the pulsar stop rotating? (c) The pulsar originated in a supernova explosion seen in the year 1054. Assuming constant α, find the initial T.

Module 10-3 Relating the Linear and Angular Variables

•19 What are the magnitudes of (a) the angular velocity, (b) the radial acceleration, and (c) the tangential acceleration of a spaceship taking a circular turn of radius 3220 km at a speed of 29 000 km/h?

•20 An object rotates about a fixed axis, and the angular position of a reference line on the object is given by $\theta = 0.40e^{2t}$, where θ is in radians and t is in seconds. Consider a point on the object that is 4.0 cm from the axis of rotation. At $t = 0$, what are the magnitudes of the point's (a) tangential component of acceleration and (b) radial component of acceleration?

•21 Between 1911 and 1990, the top of the leaning bell tower at Pisa, Italy, moved toward the south at an average rate of 1.2 mm/y. The tower is 55 m tall. In radians per second, what is the average angular speed of the tower's top about its base?

•22 An astronaut is tested in a centrifuge with radius 10 m and rotating according to $\theta = 0.30t^2$. At $t = 5.0$ s, what are the magnitudes of the (a) angular velocity, (b) linear velocity, (c) tangential acceleration, and (d) radial acceleration?

•23 SSM WWW A flywheel with a diameter of 1.20 m is rotating at an angular speed of 200 rev/min. (a) What is the angular speed of the flywheel in radians per second? (b) What is the linear speed of a point on the rim of the flywheel? (c) What constant angular acceleration (in revolutions per minute-squared) will increase the wheel's angular speed to 1000 rev/min in 60.0 s? (d) How many revolutions does the wheel make during that 60.0 s?

•24 A vinyl record is played by rotating the record so that an approximately circular groove in the vinyl slides under a stylus. Bumps in the groove run into the stylus, causing it to oscillate. The equipment converts those oscillations to electrical signals and then to sound. Suppose that a record turns at the rate of $33\frac{1}{3}$ **rev/min**, the groove being played is at a radius of 10.0 cm, and the bumps in the groove are uniformly separated by 1.75 mm. At what rate (hits per second) do the bumps hit the stylus?

••25 SSM (a) What is the angular speed ω about the polar axis of a point on Earth's surface at latitude 40° N? (Earth rotates about that axis.) (b) What is the linear speed v of the point? What are (c) ω and (d) v for a point at the equator?

••26 The flywheel of a steam engine runs with a constant angular velocity of 150 rev/min. When steam is shut off, the friction of the bearings and of the air stops the wheel in 2.2 h. (a) What is the constant angular acceleration, in revolutions per minute-squared, of the wheel during the slowdown? (b) How many revolutions does the wheel make before stopping? (c) At the instant the flywheel is turning at 75 rev/min, what is the tangential component of the linear acceleration of a flywheel particle that is 50 cm from the axis of rotation? (d) What is the magnitude of the net linear acceleration of the particle in (c)?

••27 A seed is on a turntable rotating at $33\frac{1}{3}$ rev/min, 6.0 cm from the rotation axis. What are (a) the seed's acceleration and (b) the least coefficient of static friction to avoid slippage? (c) If the turntable had undergone constant angular acceleration from rest in 0.25 s, what is the least coefficient to avoid slippage?

••28 In Fig. 10-31, wheel A of radius $r_A = 10$ cm is coupled by belt B to wheel C of radius $r_C = 25$ cm. The angular speed of wheel A is increased from rest at a constant rate of 1.6 rad/s^2. Find the time needed for wheel C to reach an angular speed of 100 rev/min, assuming the belt does not slip. (*Hint:* If the belt does not slip, the linear speeds at the two rims must be equal.)

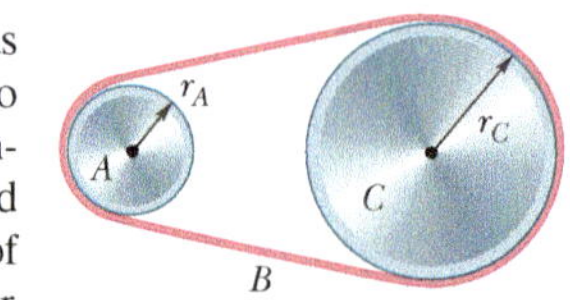

Figure 10-31 Problem 28.

••29 Figure 10-32 shows an early method of measuring the speed of light that makes use of a rotating slotted wheel. A beam of

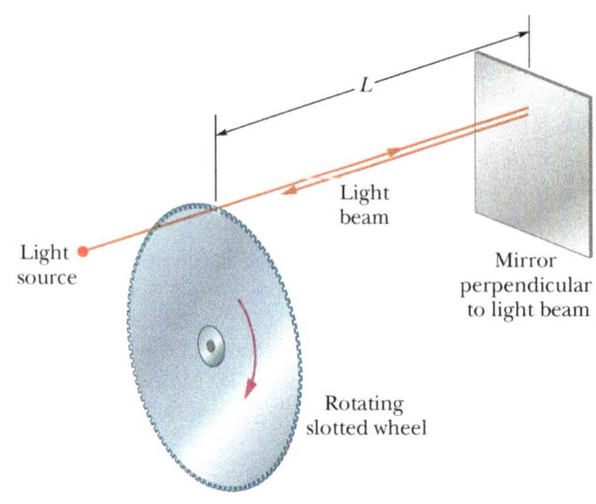

Figure 10-32 Problem 29.

light passes through one of the slots at the outside edge of the wheel, travels to a distant mirror, and returns to the wheel just in time to pass through the next slot in the wheel. One such slotted wheel has a radius of 5.0 cm and 500 slots around its edge. Measurements taken when the mirror is $L = 500$ m from the wheel indicate a speed of light of 3.0×10^5 km/s. (a) What is the (constant) angular speed of the wheel? (b) What is the linear speed of a point on the edge of the wheel?

••30 A gyroscope flywheel of radius 2.83 cm is accelerated from rest at 14.2 rad/s^2 until its angular speed is 2760 rev/min. (a) What is the tangential acceleration of a point on the rim of the flywheel during this spin-up process? (b) What is the radial acceleration of this point when the flywheel is spinning at full speed? (c) Through what distance does a point on the rim move during the spin-up?

••31 GO A disk, with a radius of 0.25 m, is to be rotated like a merry-go-round through 800 rad, starting from rest, gaining angular speed at the constant rate α_1 through the first 400 rad and then losing angular speed at the constant rate $-\alpha_1$ until it is again at rest. The magnitude of the centripetal acceleration of any portion of the disk is not to exceed 400 m/s^2. (a) What is the least time required for the rotation? (b) What is the corresponding value of α_1?

••32 A car starts from rest and moves around a circular track of radius 30.0 m. Its speed increases at the constant rate of 0.500 m/s^2. (a) What is the magnitude of its *net* linear acceleration 15.0 s later? (b) What angle does this net acceleration vector make with the car's velocity at this time?

Module 10-4 Kinetic Energy of Rotation

•33 SSM Calculate the rotational inertia of a wheel that has a kinetic energy of 24 400 J when rotating at 602 rev/min.

•34 Figure 10-33 gives angular speed versus time for a thin rod that rotates around one end. The scale on the ω axis is set by $\omega_s = 6.0$ rad/s. (a) What is the magnitude of the rod's angular acceleration? (b) At $t = 4.0$ s, the rod has a rotational kinetic energy of 1.60 J. What is its kinetic energy at $t = 0$?

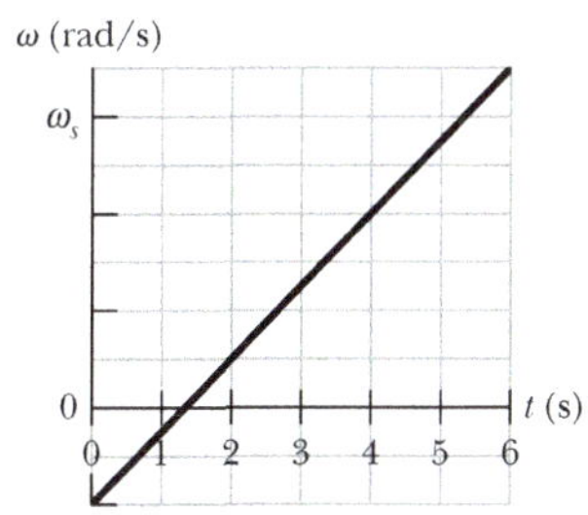

Figure 10-33 Problem 34.

Module 10-5 Calculating the Rotational Inertia

•35 SSM Two uniform solid cylinders, each rotating about its central (longitudinal) axis at 235 rad/s, have the same mass of 1.25 kg but differ in radius. What is the rotational kinetic energy of (a) the smaller cylinder, of radius 0.25 m, and (b) the larger cylinder, of radius 0.75 m?

•36 Figure 10-34*a* shows a disk that can rotate about an axis at a radial distance h from the center of the disk. Figure 10-34*b* gives the rotational inertia I of the disk about the axis as a function of that distance h, from the center out to the edge of the disk. The scale on the I axis is set by $I_A = 0.050$ kg·m^2 and $I_B = 0.150$ kg·m^2. What is the mass of the disk?

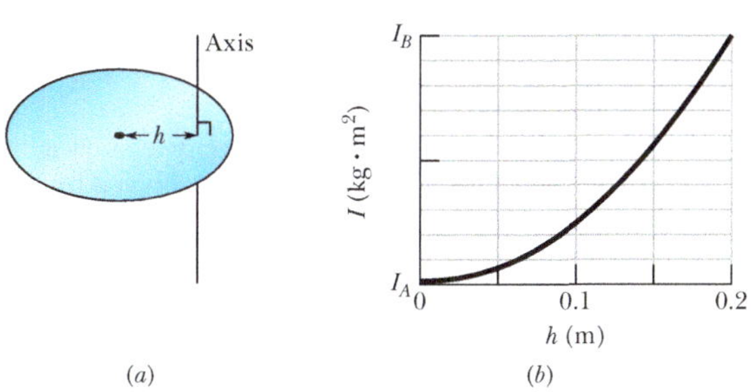

Figure 10-34 Problem 36.

•37 SSM Calculate the rotational inertia of a meter stick, with mass 0.56 kg, about an axis perpendicular to the stick and located at the 20 cm mark. (Treat the stick as a thin rod.)

•38 Figure 10-35 shows three 0.0100 kg particles that have been glued to a rod of length $L = 6.00$ cm and negligible mass. The assembly can rotate around a perpendicular axis through point O at the left end. If we remove one particle (that is, 33% of the mass), by what percentage does the rotational inertia of the assembly around the rotation axis decrease when that removed particle is (a) the innermost one and (b) the outermost one?

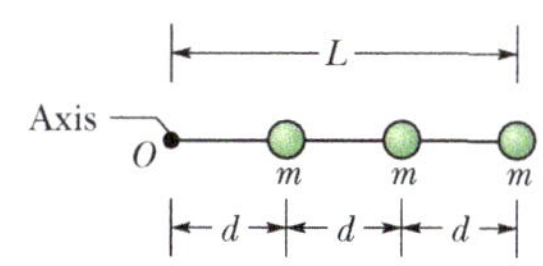

Figure 10-35 Problems 38 and 62.

••39 Trucks can be run on energy stored in a rotating flywheel, with an electric motor getting the flywheel up to its top speed of 200π rad/s. Suppose that one such flywheel is a solid, uniform cylinder with a mass of 500 kg and a radius of 1.0 m. (a) What is the kinetic energy of the flywheel after charging? (b) If the truck uses an average power of 8.0 kW, for how many minutes can it operate between chargings?

••40 Figure 10-36 shows an arrangement of 15 identical disks that have been glued together in a rod-like shape of length $L = 1.0000$ m and (total) mass $M = 100.0$ mg. The disks are uniform, and the disk arrangement can rotate about a perpendicular axis through its central disk at point O. (a) What is the rotational inertia of the arrangement about that axis? (b) If we approximated the arrangement as being a uniform rod of mass M and length L, what percentage error would we make in using the formula in Table 10-2*e* to calculate the rotational inertia?

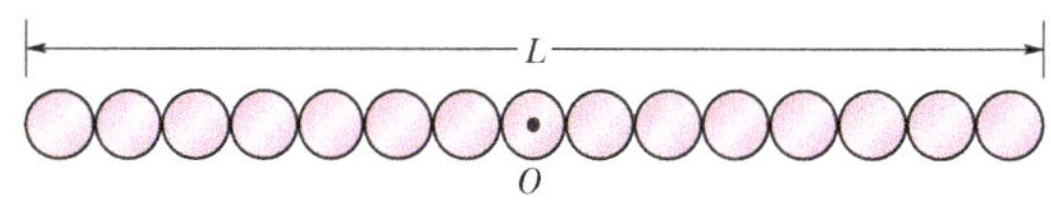

Figure 10-36 Problem 40.

••41 GO In Fig. 10-37, two particles, each with mass $m = 0.85$ kg, are fastened to each other, and to a rotation axis at O, by two thin rods, each with length $d = 5.6$ cm and mass $M = 1.2$ kg. The combination rotates around the rotation axis with the angular speed $\omega = 0.30$ rad/s. Measured about O, what are the combination's (a) rotational inertia and (b) kinetic energy?

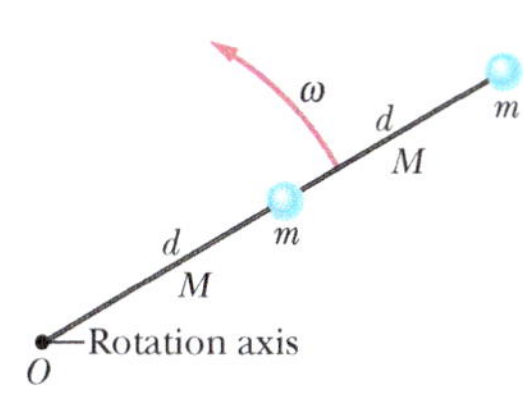

Figure 10-37 Problem 41.

••42 The masses and coordinates of four particles are as follows: 50 g, $x = 2.0$ cm, $y = 2.0$ cm; 25 g, $x = 0$, $y = 4.0$ cm; 25 g, $x = -3.0$ cm, $y = -3.0$ cm; 30 g, $x = -2.0$ cm, $y = 4.0$ cm. What are the rotational inertias of this collection about the (a) x, (b) y, and (c) z axes? (d) Suppose that we symbolize the answers to (a) and (b) as A and B, respectively. Then what is the answer to (c) in terms of A and B?

••43 SSM WWW The uniform solid block in Fig. 10-38 has mass 0.172 kg and edge lengths $a = 3.5$ cm, $b = 8.4$ cm, and $c = 1.4$ cm. Calculate its rotational inertia about an axis through one corner and perpendicular to the large faces.

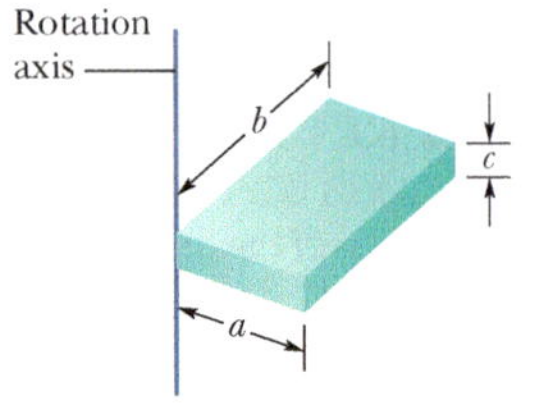

Figure 10-38 Problem 43.

••44 Four identical particles of mass 0.50 kg each are placed at the vertices of a 2.0 m × 2.0 m square and held there by four massless rods, which form the sides of the square. What is the rotational inertia of this rigid body about an axis that (a) passes through the midpoints of opposite sides and lies in the plane of the square, (b) passes through the midpoint of one of the sides and is perpendicular to the plane of the square, and (c) lies in the plane of the square and passes through two diagonally opposite particles?

Module 10-6 Torque

•45 SSM ILW The body in Fig. 10-39 is pivoted at O, and two forces act on it as shown. If $r_1 = 1.30$ m, $r_2 = 2.15$ m, $F_1 = 4.20$ N, $F_2 = 4.90$ N, $\theta_1 = 75.0°$, and $\theta_2 = 60.0°$, what is the net torque about the pivot?

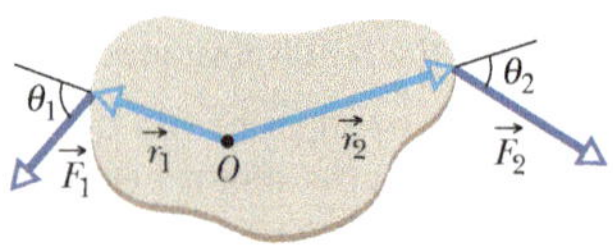

Figure 10-39 Problem 45.

•46 The body in Fig. 10-40 is pivoted at O. Three forces act on it: $F_A = 10$ N at point A, 8.0 m from O; $F_B = 16$ N at B, 4.0 m from O; and $F_C = 19$ N at C, 3.0 m from O. What is the net torque about O?

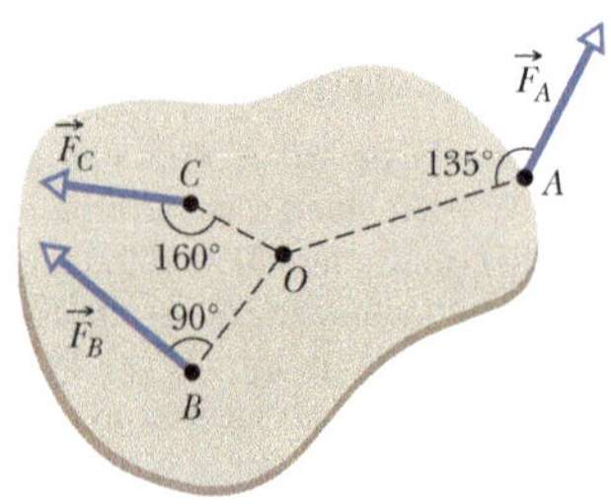

Figure 10-40 Problem 46.

•47 SSM A small ball of mass 0.75 kg is attached to one end of a 1.25-m-long massless rod, and the other end of the rod is hung from a pivot. When the resulting pendulum is 30° from the vertical, what is the magnitude of the gravitational torque calculated about the pivot?

•48 The length of a bicycle pedal arm is 0.152 m, and a downward force of 111 N is applied to the pedal by the rider. What is the magnitude of the torque about the pedal arm's pivot when the arm is at angle (a) 30°, (b) 90°, and (c) 180° with the vertical?

Module 10-7 Newton's Second Law for Rotation

•49 SSM ILW During the launch from a board, a diver's angular speed about her center of mass changes from zero to 6.20 rad/s in 220 ms. Her rotational inertia about her center of mass is 12.0 kg·m². During the launch, what are the magnitudes of (a) her average angular acceleration and (b) the average external torque on her from the board?

•50 If a 32.0 N·m torque on a wheel causes angular acceleration 25.0 rad/s², what is the wheel's rotational inertia?

••51 GO In Fig. 10-41, block 1 has mass $m_1 = 460$ g, block 2 has mass $m_2 = 500$ g, and the pulley, which is mounted on a horizontal axle with negligible friction, has radius $R = 5.00$ cm. When released from rest, block 2 falls 75.0 cm in 5.00 s without the cord slipping on the pulley. (a) What is the magnitude of the acceleration of the blocks? What are (b) tension T_2 and (c) tension T_1? (d) What is the magnitude of the pulley's angular acceleration? (e) What is its rotational inertia?

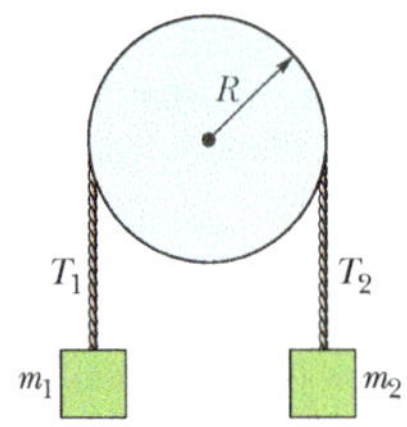

Figure 10-41 Problems 51 and 83.

••52 GO In Fig. 10-42, a cylinder having a mass of 2.0 kg can rotate about its central axis through point O. Forces are applied as shown: $F_1 = 6.0$ N, $F_2 = 4.0$ N, $F_3 = 2.0$ N, and $F_4 = 5.0$ N. Also, $r = 5.0$ cm and $R = 12$ cm. Find the (a) magnitude and (b) direction of the angular acceleration of the cylinder. (During the rotation, the forces maintain their same angles relative to the cylinder.)

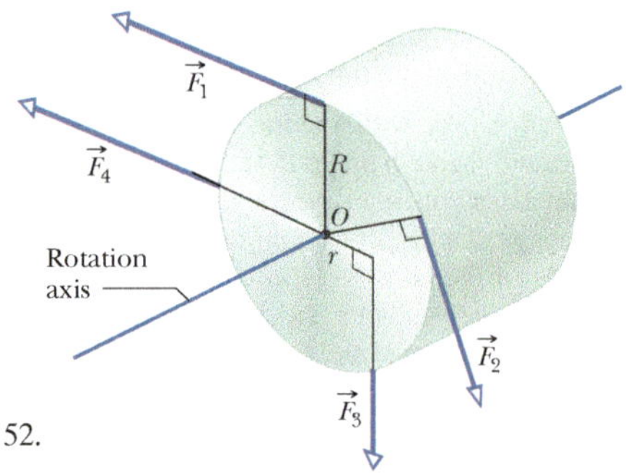

Figure 10-42 Problem 52.

••53 GO Figure 10-43 shows a uniform disk that can rotate around its center like a merry-go-round. The disk has a radius of 2.00 cm and a mass of 20.0 grams and is initially at rest. Starting at time $t = 0$, two forces are to be applied tangentially to the rim as indicated, so that at time $t = 1.25$ s the disk has an angular velocity of 250 rad/s counterclockwise. Force $\vec{F}_1$ has a magnitude of 0.100 N. What is magnitude F_2?

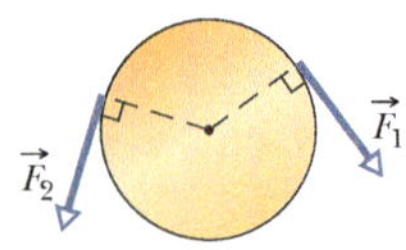

Figure 10-43 Problem 53.

••54 In a judo foot-sweep move, you sweep your opponent's left foot out from under him while pulling on his gi (uniform) toward that side. As a result, your opponent rotates around his right foot and onto the mat. Figure 10-44 shows a simplified diagram of your opponent as you face him, with his left foot swept out. The rotational axis is through point O. The gravitational force $\vec{F}_g$ on him effectively acts at his center of mass, which is a horizontal distance $d = 28$ cm from point O. His mass is 70 kg, and his rotational inertia about point O is 65 kg·m². What is the magnitude of his initial angular acceleration about point O if your pull $\vec{F}_a$ on his gi is (a) negligible and (b) horizontal with a magnitude of 300 N and applied at height $h = 1.4$ m?

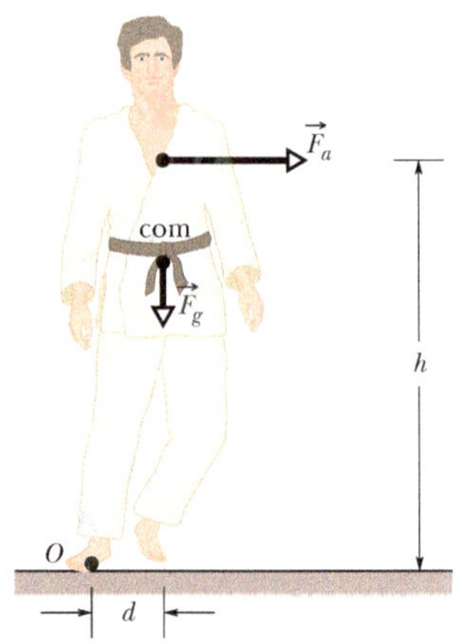

Figure 10-44 Problem 54.

••55 GO In Fig. 10-45a, an irregularly shaped plastic plate with uniform thickness and density (mass per unit volume) is to be rotated around an axle that is perpendicular to the plate face and through point O. The rotational inertia of the plate about that axle is measured with the following method. A circular disk of mass 0.500 kg and radius 2.00 cm is glued to the plate, with its center aligned with point O (Fig. 10-45b). A string is wrapped around the edge of the disk the way a string is wrapped around a top. Then the string is pulled for 5.00 s. As a result, the disk and plate are rotated by a constant force of 0.400 N that is applied by the string tangentially to the edge of the disk. The resulting angular speed is 114 rad/s. What is the rotational inertia of the plate about the axle?

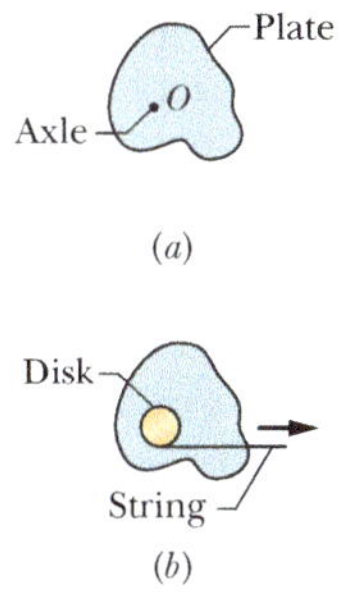

Figure 10-45 Problem 55.

••56 GO Figure 10-46 shows particles 1 and 2, each of mass m, fixed to the ends of a rigid massless rod of length $L_1 + L_2$, with $L_1 = 20$ cm and $L_2 = 80$ cm. The rod is held horizontally on the fulcrum and then released. What are the magnitudes of the initial accelerations of (a) particle 1 and (b) particle 2?

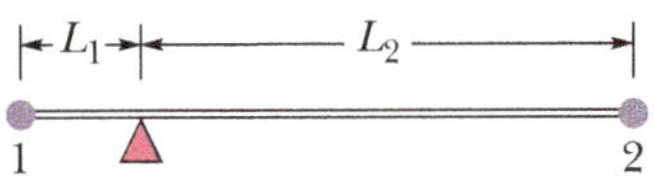

Figure 10-46 Problem 56.

•••57 GO A pulley, with a rotational inertia of 1.0×10^{-3} kg·m² about its axle and a radius of 10 cm, is acted on by a force applied tangentially at its rim. The force magnitude varies in time as $F = 0.50t + 0.30t^2$, with F in newtons and t in seconds. The pulley is initially at rest. At $t = 3.0$ s what are its (a) angular acceleration and (b) angular speed?

Module 10-8 Work and Rotational Kinetic Energy

•58 (a) If $R = 12$ cm, $M = 400$ g, and $m = 50$ g in Fig. 10-19, find the speed of the block after it has descended 50 cm starting from rest. Solve the problem using energy conservation principles. (b) Repeat (a) with $R = 5.0$ cm.

•59 An automobile crankshaft transfers energy from the engine to the axle at the rate of 100 hp (= 74.6 kW) when rotating at a speed of 1800 rev/min. What torque (in newton-meters) does the crankshaft deliver?

•60 A thin rod of length 0.75 m and mass 0.42 kg is suspended freely from one end. It is pulled to one side and then allowed to swing like a pendulum, passing through its lowest position with angular speed 4.0 rad/s. Neglecting friction and air resistance, find (a) the rod's kinetic energy at its lowest position and (b) how far above that position the center of mass rises.

•61 A 32.0 kg wheel, essentially a thin hoop with radius 1.20 m, is rotating at 280 rev/min. It must be brought to a stop in 15.0 s. (a) How much work must be done to stop it? (b) What is the required average power?

••62 In Fig. 10-35, three 0.0100 kg particles have been glued to a rod of length $L = 6.00$ cm and negligible mass and can rotate around a perpendicular axis through point O at one end. How much work is required to change the rotational rate (a) from 0 to 20.0 rad/s, (b) from 20.0 rad/s to 40.0 rad/s, and (c) from 40.0 rad/s to 60.0 rad/s? (d) What is the slope of a plot of the assembly's kinetic energy (in joules) versus the square of its rotation rate (in radians-squared per second-squared)?

••63 SSM ILW A meter stick is held vertically with one end on the floor and is then allowed to fall. Find the speed of the other end just before it hits the floor, assuming that the end on the floor does not slip. (*Hint:* Consider the stick to be a thin rod and use the conservation of energy principle.)

••64 A uniform cylinder of radius 10 cm and mass 20 kg is mounted so as to rotate freely about a horizontal axis that is parallel to and 5.0 cm from the central longitudinal axis of the cylinder. (a) What is the rotational inertia of the cylinder about the axis of rotation? (b) If the cylinder is released from rest with its central longitudinal axis at the same height as the axis about which the cylinder rotates, what is the angular speed of the cylinder as it passes through its lowest position?

•••65 GO A tall, cylindrical chimney falls over when its base is ruptured. Treat the chimney as a thin rod of length 55.0 m. At the instant it makes an angle of 35.0° with the vertical as it falls, what are (a) the radial acceleration of the top, and (b) the tangential acceleration of the top. (*Hint:* Use energy considerations, not a torque.) (c) At what angle θ is the tangential acceleration equal to g?

•••66 GO A uniform spherical shell of mass $M = 4.5$ kg and radius $R = 8.5$ cm can rotate about a vertical axis on frictionless bearings (Fig. 10-47). A massless cord passes around the equator of the shell, over a pulley of rotational inertia $I = 3.0 \times 10^{-3}$ kg·m² and radius $r = 5.0$ cm, and is attached to a small object of mass $m = 0.60$ kg. There is no friction on the pulley's axle; the cord does not slip on the pulley. What is the speed of the object when it has fallen 82 cm after being released from rest? Use energy considerations.

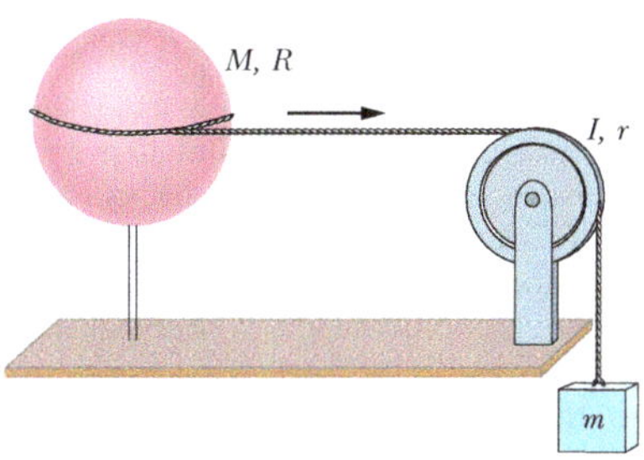

Figure 10-47 Problem 66.

•••67 GO Figure 10-48 shows a rigid assembly of a thin hoop (of mass m and radius $R = 0.150$ m) and a thin radial rod (of mass m and length $L = 2.00R$). The assembly is upright, but if we give it a slight nudge, it will rotate around a horizontal axis in the plane of the rod and hoop, through the lower end of the rod. Assuming that the energy given to the assembly in such a nudge is negligible, what would be the assembly's angular speed about the rotation axis when it passes through the upside-down (inverted) orientation?

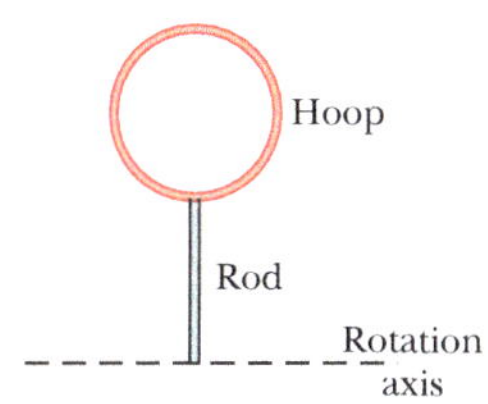

Figure 10-48 Problem 67.

Additional Problems

68 Two uniform solid spheres have the same mass of 1.65 kg, but one has a radius of 0.226 m and the other has a radius of 0.854 m. Each can rotate about an axis through its center. (a) What is the magnitude τ of the torque required to bring the smaller sphere from rest to an angular speed of 317 rad/s in 15.5 s? (b) What is the magnitude F of the force that must be applied tangentially at the sphere's equator to give that torque? What are the corresponding values of (c) τ and (d) F for the larger sphere?

69 In Fig. 10-49, a small disk of radius $r = 2.00$ cm has been glued to the edge of a larger disk of radius $R = 4.00$ cm so that

Figure 10-49 Problem 69.

the disks lie in the same plane. The disks can be rotated around a perpendicular axis through point O at the center of the larger disk. The disks both have a uniform density (mass per unit volume) of 1.40×10^3 kg/m^3 and a uniform thickness of 5.00 mm. What is the rotational inertia of the two-disk assembly about the rotation axis through O?

70 A wheel, starting from rest, rotates with a constant angular acceleration of 2.00 rad/s^2. During a certain 3.00 s interval, it turns through 90.0 rad. (a) What is the angular velocity of the wheel at the start of the 3.00 s interval? (b) How long has the wheel been turning before the start of the 3.00 s interval?

71 SSM In Fig. 10-50, two 6.20 kg blocks are connected by a massless string over a pulley of radius 2.40 cm and rotational inertia 7.40×10^{-4} kg·m^2. The string does not slip on the pulley; it is not known whether there is friction between the table and the sliding block; the pulley's axis is frictionless. When this system is released from rest, the pulley turns through 0.130 rad in 91.0 ms and the acceleration of the blocks is constant. What are (a) the magnitude of the pulley's angular acceleration, (b) the magnitude of either block's acceleration, (c) string tension T_1, and (d) string tension T_2?

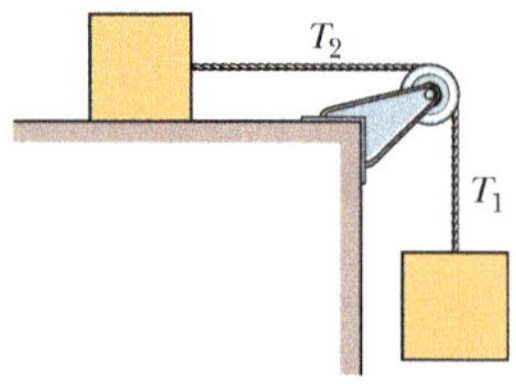

Figure 10-50 Problem 71.

72 Attached to each end of a thin steel rod of length 1.20 m and mass 6.40 kg is a small ball of mass 1.06 kg. The rod is constrained to rotate in a horizontal plane about a vertical axis through its midpoint. At a certain instant, it is rotating at 39.0 rev/s. Because of friction, it slows to a stop in 32.0 s. Assuming a constant retarding torque due to friction, compute (a) the angular acceleration, (b) the retarding torque, (c) the total energy transferred from mechanical energy to thermal energy by friction, and (d) the number of revolutions rotated during the 32.0 s. (e) Now suppose that the retarding torque is known not to be constant. If any of the quantities (a), (b), (c), and (d) can still be computed without additional information, give its value.

73 A uniform helicopter rotor blade is 7.80 m long, has a mass of 110 kg, and is attached to the rotor axle by a single bolt. (a) What is the magnitude of the force on the bolt from the axle when the rotor is turning at 320 rev/min? (*Hint:* For this calculation the blade can be considered to be a point mass at its center of mass. Why?) (b) Calculate the torque that must be applied to the rotor to bring it to full speed from rest in 6.70 s. Ignore air resistance. (The blade cannot be considered to be a point mass for this calculation. Why not? Assume the mass distribution of a uniform thin rod.) (c) How much work does the torque do on the blade in order for the blade to reach a speed of 320 rev/min?

74 *Racing disks.* Figure 10-51 shows two disks that can rotate about their centers like a merry-go-round. At time $t = 0$, the reference lines of the two disks have the same orientation. Disk A is already rotating, with a constant angular velocity of 9.5 rad/s. Disk B has been stationary but now begins to rotate at a constant angular acceleration of 2.2 rad/s^2. (a) At what time t will the reference lines of the two disks momentarily have the same angular displacement θ? (b) Will that time t be the first time since $t = 0$ that the reference lines are momentarily aligned?

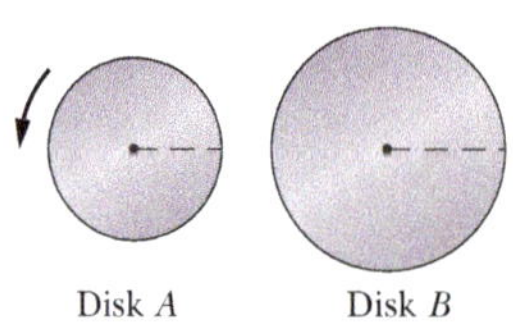

Figure 10-51 Problem 74.

75 A high-wire walker always attempts to keep his center of mass over the wire (or rope). He normally carries a long, heavy pole to help: If he leans, say, to his right (his com moves to the right) and is in danger of rotating around the wire, he moves the pole to his left (its com moves to the left) to slow the rotation and allow himself time to adjust his balance. Assume that the walker has a mass of 70.0 kg and a rotational inertia of **15.0 kg·m^2** about the wire. What is the magnitude of his angular acceleration about the wire if his com is 5.0 cm to the right of the wire and (a) he carries no pole and (b) the 14.0 kg pole he carries has its com 10 cm to the left of the wire?

76 Starting from rest at $t = 0$, a wheel undergoes a constant angular acceleration. When $t = 2.0$ s, the angular velocity of the wheel is 5.0 rad/s. The acceleration continues until $t = 20$ s, when it abruptly ceases. Through what angle does the wheel rotate in the interval $t = 0$ to $t = 40$ s?

77 SSM A record turntable rotating at $33\frac{1}{3}$ **rev/min** slows down and stops in 30 s after the motor is turned off. (a) Find its (constant) angular acceleration in revolutions per minute-squared. (b) How many revolutions does it make in this time?

78 GO A rigid body is made of three identical thin rods, each with length $L = 0.600$ m, fastened together in the form of a letter **H** (Fig. 10-52). The body is free to rotate about a horizontal axis that runs along the length of one of the legs of the **H**. The body is allowed to fall from rest from a position in which the plane of the **H** is horizontal. What is the angular speed of the body when the plane of the **H** is vertical?

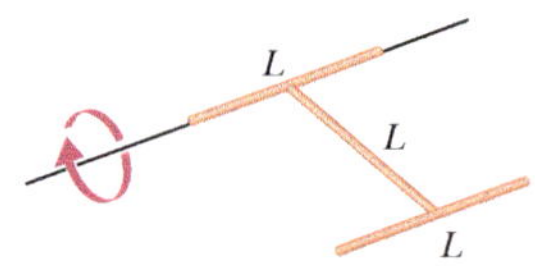

Figure 10-52 Problem 78.

79 SSM (a) Show that the rotational inertia of a solid cylinder of mass M and radius R about its central axis is equal to the rotational inertia of a thin hoop of mass M and radius $R/\sqrt{2}$ about its central axis. (b) Show that the rotational inertia I of any given body of mass M about any given axis is equal to the rotational inertia of an *equivalent hoop* about that axis, if the hoop has the same mass M and a radius k given by

$$k = \sqrt{\frac{I}{M}}.$$

The radius k of the equivalent hoop is called the *radius of gyration* of the given body.

80 A disk rotates at constant angular acceleration, from angular position $\theta_1 = 10.0$ rad to angular position $\theta_2 = 70.0$ rad in 6.00 s. Its angular velocity at θ_2 is 15.0 rad/s. (a) What was its angular velocity at θ_1? (b) What is the angular acceleration? (c) At what angular position was the disk initially at rest? (d) Graph θ versus time t and angular speed ω versus t for the disk, from the beginning of the motion (let $t = 0$ then).

81 GO The thin uniform rod in Fig. 10-53 has length 2.0 m and can pivot about a horizontal, frictionless pin through one end. It is released from rest at angle $\theta = 40°$ above the horizontal. Use the principle of conservation of energy to determine the angular speed of the rod as it passes through the horizontal position.

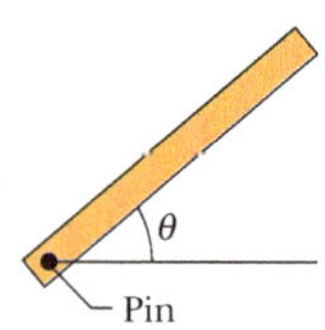

Figure 10-53 Problem 81.

82 George Washington Gale Ferris, Jr., a civil engineering graduate from Rensselaer Polytechnic Institute, built the original Ferris wheel for the 1893 World's Columbian Exposition in Chicago. The wheel, an astounding engineering construction at the time, carried 36 wooden cars, each holding up to 60 passengers, around a circle 76 m in diameter. The cars were loaded 6 at a time, and once all 36 cars were full, the wheel made a complete rotation at constant angular speed in about 2 min. Estimate the amount of work that was required of the machinery to rotate the passengers alone.

83 In Fig. 10-41, two blocks, of mass $m_1 = 400$ g and $m_2 = 600$ g, are connected by a massless cord that is wrapped around a uniform disk of mass $M = 500$ g and radius $R = 12.0$ cm. The disk can rotate without friction about a fixed horizontal axis through its center; the cord cannot slip on the disk. The system is released from rest. Find (a) the magnitude of the acceleration of the blocks, (b) the tension T_1 in the cord at the left, and (c) the tension T_2 in the cord at the right.

84 At 7:14 A.M. on June 30, 1908, a huge explosion occurred above remote central Siberia, at latitude 61° N and longitude 102° E; the fireball thus created was the brightest flash seen by anyone before nuclear weapons. The *Tunguska Event,* which according to one chance witness "covered an enormous part of the sky," was probably the explosion of a *stony asteroid* about 140 m wide. (a) Considering only Earth's rotation, determine how much later the asteroid would have had to arrive to put the explosion above Helsinki at longitude 25° E. This would have obliterated the city. (b) If the asteroid had, instead, been a *metallic asteroid,* it could have reached Earth's surface. How much later would such an asteroid have had to arrive to put the impact in the Atlantic Ocean at longitude 20° W? (The resulting tsunamis would have wiped out coastal civilization on both sides of the Atlantic.)

85 A golf ball is launched at an angle of 20° to the horizontal, with a speed of 60 m/s and a rotation rate of 90 rad/s. Neglecting air drag, determine the number of revolutions the ball makes by the time it reaches maximum height.

86 Figure 10-54 shows a flat construction of two circular rings that have a common center and are held together by three rods of negligible mass. The construction, which is initially at rest, can rotate around the common center (like a merry-go-round), where another rod of negligible mass lies. The mass, inner radius, and outer radius of the rings are given in the following table. A tangential force of magnitude 12.0 N is applied to the outer edge of the outer ring for 0.300 s. What is the change in the angular speed of the construction during the time interval?

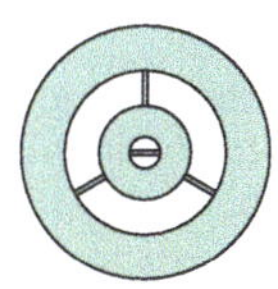

Figure 10-54 Problem 86.

Ring	Mass (kg)	Inner Radius (m)	Outer Radius (m)
1	0.120	0.0160	0.0450
2	0.240	0.0900	0.1400

87 In Fig. 10-55, a wheel of radius 0.20 m is mounted on a frictionless horizontal axle. A massless cord is wrapped around the wheel and attached to a 2.0 kg box that slides on a frictionless surface inclined at an angle $\theta = 20°$ with the horizontal. The box accelerates down the surface at 2.0 m/s². What is the rotational inertia of the wheel about the axle?

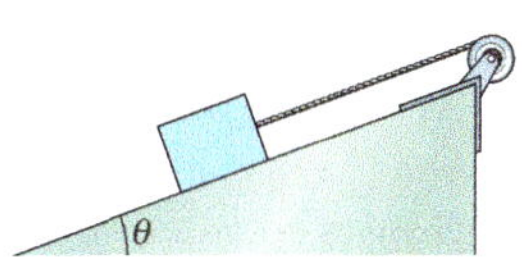

Figure 10-55 Problem 87.

88 A thin spherical shell has a radius of 1.90 m. An applied torque of 960 N·m gives the shell an angular acceleration of 6.20 rad/s² about an axis through the center of the shell. What are (a) the rotational inertia of the shell about that axis and (b) the mass of the shell?

89 A bicyclist of mass 70 kg puts all his mass on each downward-moving pedal as he pedals up a steep road. Take the diameter of the circle in which the pedals rotate to be 0.40 m, and determine the magnitude of the maximum torque he exerts about the rotation axis of the pedals.

90 The flywheel of an engine is rotating at 25.0 rad/s. When the engine is turned off, the flywheel slows at a constant rate and stops in 20.0 s. Calculate (a) the angular acceleration of the flywheel, (b) the angle through which the flywheel rotates in stopping, and (c) the number of revolutions made by the flywheel in stopping.

91 SSM In Fig. 10-19*a*, a wheel of radius 0.20 m is mounted on a frictionless horizontal axis. The rotational inertia of the wheel about the axis is 0.40 kg·m². A massless cord wrapped around the wheel's circumference is attached to a 6.0 kg box. The system is released from rest. When the box has a kinetic energy of 6.0 J, what are (a) the wheel's rotational kinetic energy and (b) the distance the box has fallen?

92 Our Sun is 2.3×10^4 ly (light-years) from the center of our Milky Way galaxy and is moving in a circle around that center at a speed of 250 km/s. (a) How long does it take the Sun to make one revolution about the galactic center? (b) How many revolutions has the Sun completed since it was formed about 4.5×10^9 years ago?

93 SSM A wheel of radius 0.20 m is mounted on a frictionless horizontal axis. The rotational inertia of the wheel about the axis is 0.050 kg·m². A massless cord wrapped around the wheel is attached to a 2.0 kg block that slides on a horizontal frictionless surface. If a horizontal force of magnitude $P = 3.0$ N is applied to the block as shown in Fig. 10-56, what is the magnitude of the angular acceleration of the wheel? Assume the cord does not slip on the wheel.

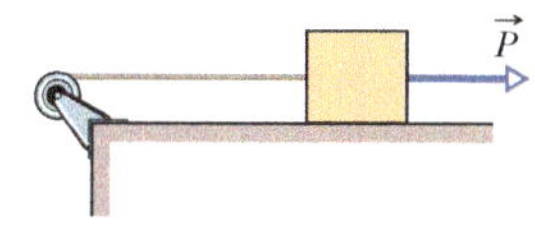

Figure 10-56 Problem 93.

94 If an airplane propeller rotates at 2000 rev/min while the airplane flies at a speed of 480 km/h relative to the ground, what is the linear speed of a point on the tip of the propeller, at radius 1.5 m, as seen by (a) the pilot and (b) an observer on the ground? The plane's velocity is parallel to the propeller's axis of rotation.

95 The rigid body shown in Fig. 10-57 consists of three particles connected by massless rods. It is to be rotated about an axis perpendicular to its plane through point P. If $M = 0.40$ kg, $a = 30$ cm, and $b = 50$ cm, how much work is required to take the body from rest to an angular speed of 5.0 rad/s?

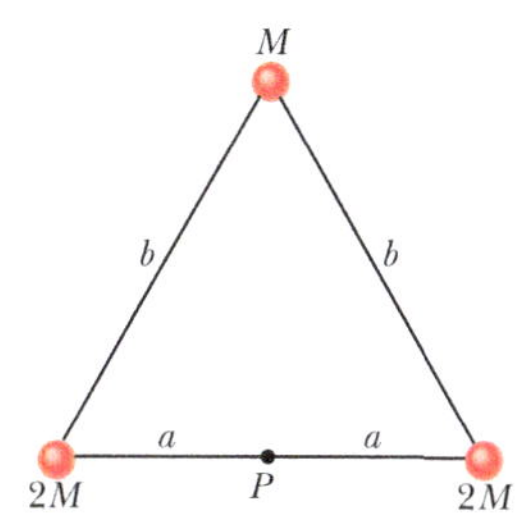

Figure 10-57 Problem 95.

96 *Beverage engineering.* The pull tab was a major advance in the engineering design of beverage containers. The tab pivots on a central bolt in the can's top. When you pull upward on one end of the tab, the other end presses downward on a portion of the can's top that has been scored. If you pull upward with a 10 N force, what force magnitude acts on the scored section? (You will need to examine a can with a pull tab.)

97 Figure 10-58 shows a propeller blade that rotates at 2000 rev/min about a perpendicular axis at point B. Point A is at the outer tip of the blade, at radial distance 1.50 m. (a) What is the difference in the magnitudes a of the centripetal acceleration of point A and of a point at radial distance 0.150 m? (b) Find the slope of a plot of a versus radial distance along the blade.

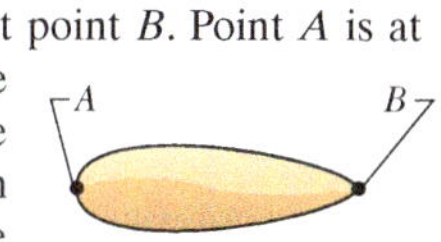

Figure 10-58 Problem 97.

98 A yo-yo-shaped device mounted on a horizontal frictionless axis is used to lift a 30 kg box as shown in Fig. 10-59. The outer radius R of the device is 0.50 m, and the radius r of the hub is 0.20 m. When a constant horizontal force $\vec{F}_{app}$ of magnitude 140 N is applied to a rope wrapped around the outside of the device, the box, which is suspended from a rope wrapped around the hub, has an upward acceleration of magnitude 0.80 m/s². What is the rotational inertia of the device about its axis of rotation?

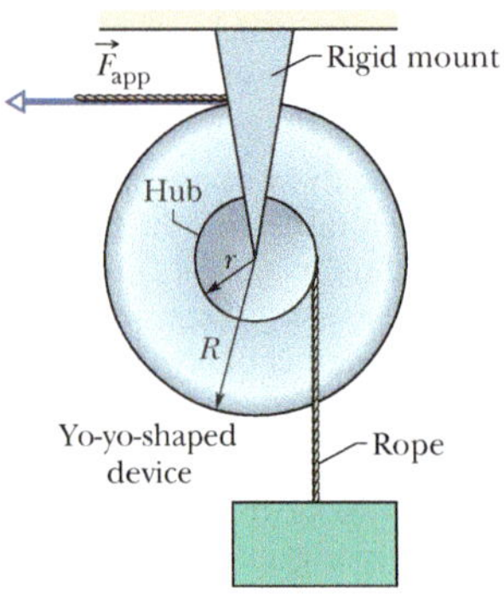

Figure 10-59 Problem 98.

99 A small ball with mass 1.30 kg is mounted on one end of a rod 0.780 m long and of negligible mass. The system rotates in a horizontal circle about the other end of the rod at 5010 rev/min. (a) Calculate the rotational inertia of the system about the axis of rotation. (b) There is an air drag of 2.30×10^{-2} N on the ball, directed opposite its motion. What torque must be applied to the system to keep it rotating at constant speed?

100 Two thin rods (each of mass 0.20 kg) are joined together to form a rigid body as shown in Fig. 10-60. One of the rods has length $L_1 = 0.40$ m, and the other has length $L_2 = 0.50$ m. What is the rotational inertia of this rigid body about (a) an axis that is perpendicular to the plane of the paper and passes through the center of the shorter rod and (b) an axis that is perpendicular to the plane of the paper and passes through the center of the longer rod?

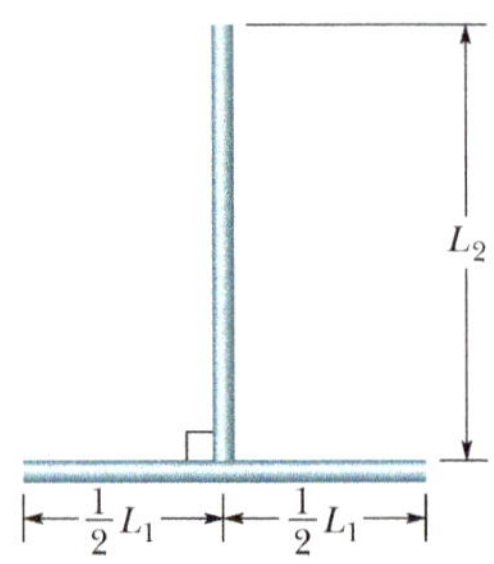

Figure 10-60 Problem 100.

101 In Fig. 10-61, four pulleys are connected by two belts. Pulley A (radius 15 cm) is the drive pulley, and it rotates at 10 rad/s. Pulley B (radius 10 cm) is connected by belt 1 to pulley A. Pulley B' (radius 5 cm) is concentric with pulley B and is rigidly attached to it. Pulley C (radius 25 cm) is connected by belt 2 to pulley B'. Calculate (a) the linear speed of a point on belt 1, (b) the angular speed of pulley B, (c) the angular speed of pulley B', (d) the linear speed of a point on belt 2, and (e) the angular speed of pulley C. (*Hint:* If the belt between two pulleys does not slip, the linear speeds at the rims of the two pulleys must be equal.)

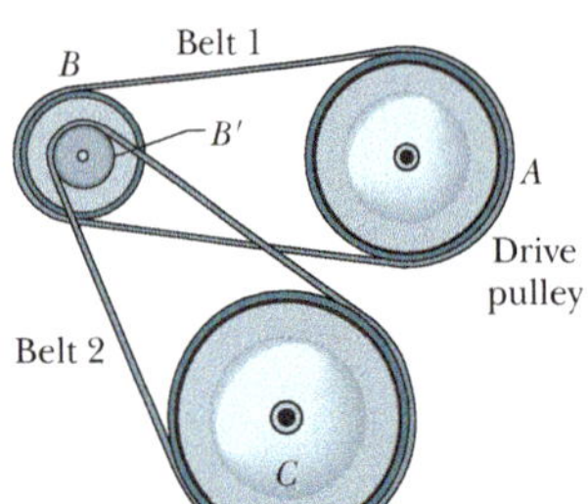

Figure 10-61 Problem 101.

102 The rigid object shown in Fig. 10-62 consists of three balls and three connecting rods, with $M = 1.6$ kg, $L = 0.60$ m, and $\theta = 30°$. The balls may be treated as particles, and the connecting rods have negligible mass. Determine the rotational kinetic energy of the object if it has an angular speed of 1.2 rad/s about (a) an axis that passes through point P and is perpendicular to the plane of the figure and (b) an axis that passes through point P, is perpendicular to the rod of length $2L$, and lies in the plane of the figure.

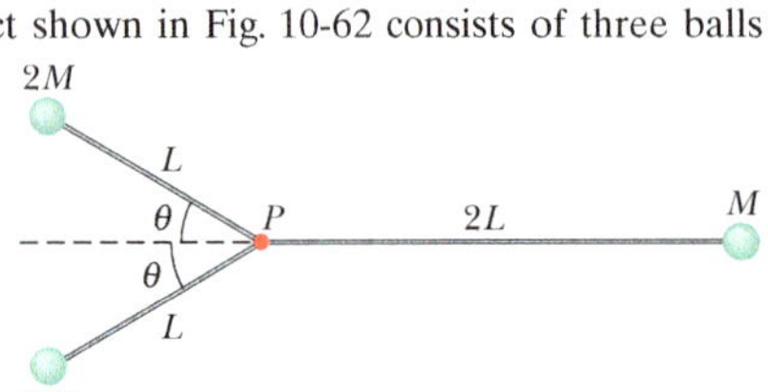

Figure 10-62 Problem 102.

103 In Fig. 10-63, a thin uniform rod (mass 3.0 kg, length 4.0 m) rotates freely about a horizontal axis A that is perpendicular to the rod and passes through a point at distance $d = 1.0$ m from the end of the rod. The kinetic energy of the rod as it passes through the vertical position is 20 J. (a) What is the rotational inertia of the rod about axis A? (b) What is the (linear) speed of the end B of the rod as the rod passes through the vertical position? (c) At what angle θ will the rod momentarily stop in its upward swing?

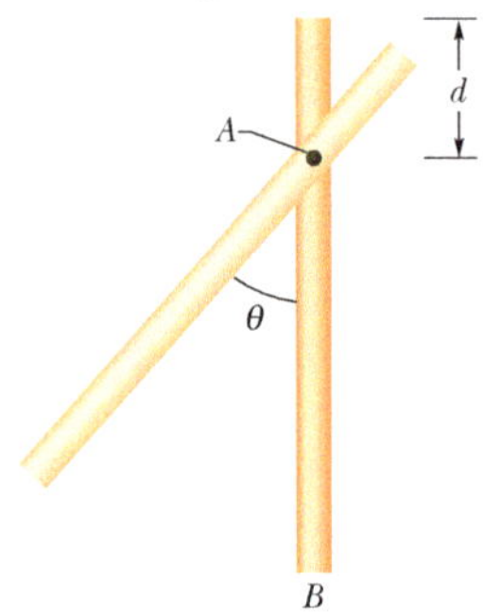

Figure 10-63 Problem 103.

104 Four particles, each of mass, 0.20 kg, are placed at the vertices of a square with sides of length 0.50 m. The particles are connected by rods of negligible mass. This rigid body can rotate in a vertical plane about a horizontal axis A that passes through one of the particles. The body is released from rest with rod AB horizontal (Fig. 10-64). (a) What is the rotational inertia of the body about axis A? (b) What is the angular speed of the body about axis A when rod AB swings through the vertical position?

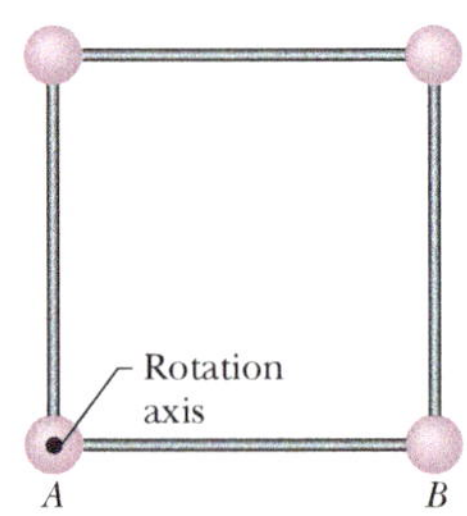

Figure 10-64 Problem 104.

105 Cheetahs running at top speed have been reported at an astounding 114 km/h (about 71 mi/h) by observers driving alongside the animals. Imagine trying to measure a cheetah's speed by keeping your vehicle abreast of the animal while also glancing at your speedometer, which is registering 114 km/h. You keep the vehicle a constant 8.0 m from the cheetah, but the noise of the vehicle causes the cheetah to continuously veer away from you along a circular path of radius 92 m. Thus, you travel along a circular path of radius 100 m. (a) What is the angular speed of you and the cheetah around the circular paths? (b) What is the linear speed of the cheetah along its path? (If you did not account for the circular motion, you would conclude erroneously that the cheetah's speed is 114 km/h, and that type of error was apparently made in the published reports.)

106 A point on the rim of a 0.75-m-diameter grinding wheel changes speed at a constant rate from 12 m/s to 25 m/s in 6.2 s. What is the average angular acceleration of the wheel?

107 A pulley wheel that is 8.0 cm in diameter has a 5.6-m-long cord wrapped around its periphery. Starting from rest, the wheel is given a constant angular acceleration of 1.5 rad/s². (a) Through what angle must the wheel turn for the cord to unwind completely? (b) How long will this take?

108 A vinyl record on a turntable rotates at $33\frac{1}{3}$ rev/min. (a) What is its angular speed in radians per second? What is the linear speed of a point on the record (b) 15 cm and (c) 7.4 cm from the turntable axis?

CHAPTER 11

Rolling, Torque, and Angular Momentum

11-1 ROLLING AS TRANSLATION AND ROTATION COMBINED

Learning Objectives

After reading this module, you should be able to . . .

11.01 Identify that smooth rolling can be considered as a combination of pure translation and pure rotation.

11.02 Apply the relationship between the center-of-mass speed and the angular speed of a body in smooth rolling.

Key Ideas

- For a wheel of radius R rolling smoothly,

$$v_{com} = \omega R,$$

where v_{com} is the linear speed of the wheel's center of mass and ω is the angular speed of the wheel about its center.

- The wheel may also be viewed as rotating instantaneously about the point P of the "road" that is in contact with the wheel. The angular speed of the wheel about this point is the same as the angular speed of the wheel about its center.

What Is Physics?

As we discussed in Chapter 10, physics includes the study of rotation. Arguably, the most important application of that physics is in the rolling motion of wheels and wheel-like objects. This applied physics has long been used. For example, when the prehistoric people of Easter Island moved their gigantic stone statues from the quarry and across the island, they dragged them over logs acting as rollers. Much later, when settlers moved westward across America in the 1800s, they rolled their possessions first by wagon and then later by train. Today, like it or not, the world is filled with cars, trucks, motorcycles, bicycles, and other rolling vehicles.

The physics and engineering of rolling have been around for so long that you might think no fresh ideas remain to be developed. However, skateboards and inline skates were invented and engineered fairly recently, to become huge financial successes. Street luge is now catching on, and the self-righting Segway (Fig. 11-1) may change the way people move around in large cities. Applying the physics of rolling can still lead to surprises and rewards. Our starting point in exploring that physics is to simplify rolling motion.

Justin Sullivan/Getty Images, Inc.

Figure 11-1 The self-righting Segway Human Transporter.

Rolling as Translation and Rotation Combined

Here we consider only objects that *roll smoothly* along a surface; that is, the objects roll without slipping or bouncing on the surface. Figure 11-2 shows how complicated smooth rolling motion can be: Although the center of the object moves in a straight line parallel to the surface, a point on the rim certainly does not. However, we can study this motion by treating it as a combination of translation of the center of mass and rotation of the rest of the object around that center.

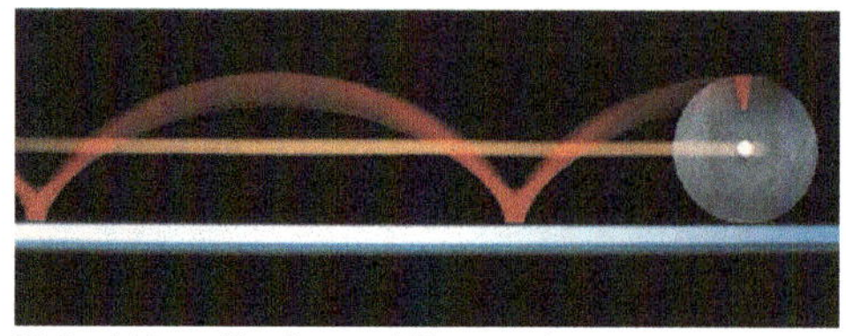

Richard Megna/Fundamental Photographs

Figure 11-2 A time-exposure photograph of a rolling disk. Small lights have been attached to the disk, one at its center and one at its edge. The latter traces out a curve called a *cycloid.*

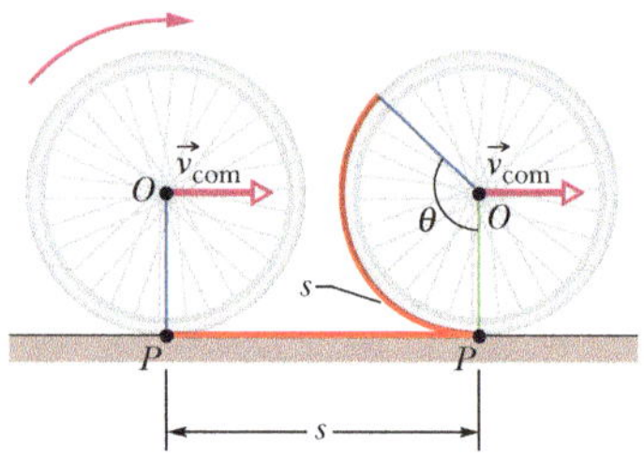

Figure 11-3 The center of mass O of a rolling wheel moves a distance s at velocity $\vec{v}_{com}$ while the wheel rotates through angle θ. The point P at which the wheel makes contact with the surface over which the wheel rolls also moves a distance s.

To see how we do this, pretend you are standing on a sidewalk watching the bicycle wheel of Fig. 11-3 as it rolls along a street. As shown, you see the center of mass O of the wheel move forward at constant speed v_{com}. The point P on the street where the wheel makes contact with the street surface also moves forward at speed v_{com}, so that P always remains directly below O.

During a time interval t, you see both O and P move forward by a distance s. The bicycle rider sees the wheel rotate through an angle θ about the center of the wheel, with the point of the wheel that was touching the street at the beginning of t moving through arc length s. Equation 10-17 relates the arc length s to the rotation angle θ:

$$s = \theta R, \tag{11-1}$$

where R is the radius of the wheel. The linear speed v_{com} of the center of the wheel (the center of mass of this uniform wheel) is ds/dt. The angular speed ω of the wheel about its center is $d\theta/dt$. Thus, differentiating Eq. 11-1 with respect to time (with R held constant) gives us

$$v_{com} = \omega R \quad \text{(smooth rolling motion).} \tag{11-2}$$

A Combination. Figure 11-4 shows that the rolling motion of a wheel is a combination of purely translational and purely rotational motions. Figure 11-4*a* shows the purely rotational motion (as if the rotation axis through the center were stationary): Every point on the wheel rotates about the center with angular speed ω. (This is the type of motion we considered in Chapter 10.) Every point on the outside edge of the wheel has linear speed v_{com} given by Eq. 11-2. Figure 11-4*b* shows the purely translational motion (as if the wheel did not rotate at all): Every point on the wheel moves to the right with speed v_{com}.

The combination of Figs. 11-4*a* and 11-4*b* yields the actual rolling motion of the wheel, Fig. 11-4*c*. Note that in this combination of motions, the portion of the wheel at the bottom (at point P) is stationary and the portion at the top

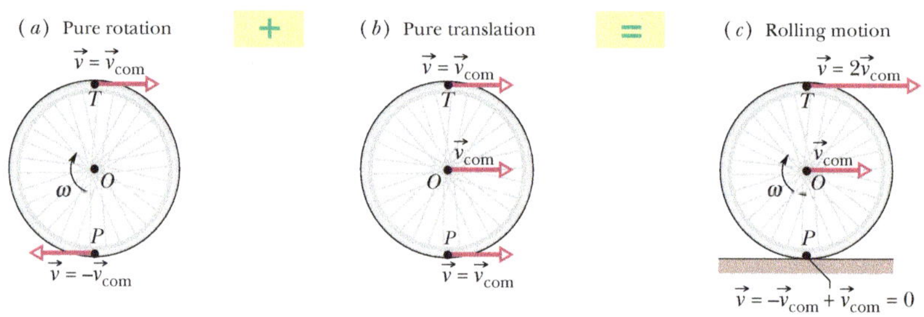

Figure 11-4 Rolling motion of a wheel as a combination of purely rotational motion and purely translational motion. (*a*) The purely rotational motion: All points on the wheel move with the same angular speed ω. Points on the outside edge of the wheel all move with the same linear speed $v = v_{com}$. The linear velocities $\vec{v}$ of two such points, at top (T) and bottom (P) of the wheel, are shown. (*b*) The purely translational motion: All points on the wheel move to the right with the same linear velocity $\vec{v}_{com}$. (*c*) The rolling motion of the wheel is the combination of (*a*) and (*b*).

Courtesy Alice Halliday

Figure 11-5 A photograph of a rolling bicycle wheel. The spokes near the wheel's top are more blurred than those near the bottom because the top ones are moving faster, as Fig. 11-4*c* shows.

(at point T) is moving at speed $2v_{\text{com}}$, faster than any other portion of the wheel. These results are demonstrated in Fig. 11-5, which is a time exposure of a rolling bicycle wheel. You can tell that the wheel is moving faster near its top than near its bottom because the spokes are more blurred at the top than at the bottom.

The motion of any round body rolling smoothly over a surface can be separated into purely rotational and purely translational motions, as in Figs. 11-4*a* and 11-4*b*.

Rolling as Pure Rotation

Figure 11-6 suggests another way to look at the rolling motion of a wheel—namely, as pure rotation about an axis that always extends through the point where the wheel contacts the street as the wheel moves. We consider the rolling motion to be pure rotation about an axis passing through point P in Fig. 11-4*c* and perpendicular to the plane of the figure. The vectors in Fig. 11-6 then represent the instantaneous velocities of points on the rolling wheel.

Question: What angular speed about this new axis will a stationary observer assign to a rolling bicycle wheel?

Answer: The same ω that the rider assigns to the wheel as she or he observes it in pure rotation about an axis through its center of mass.

To verify this answer, let us use it to calculate the linear speed of the top of the rolling wheel from the point of view of a stationary observer. If we call the wheel's radius R, the top is a distance $2R$ from the axis through P in Fig. 11-6, so the linear speed at the top should be (using Eq. 11-2)

$$v_{\text{top}} = (\omega)(2R) = 2(\omega R) = 2v_{\text{com}},$$

in exact agreement with Fig. 11-4*c*. You can similarly verify the linear speeds shown for the portions of the wheel at points O and P in Fig. 11-4*c*.

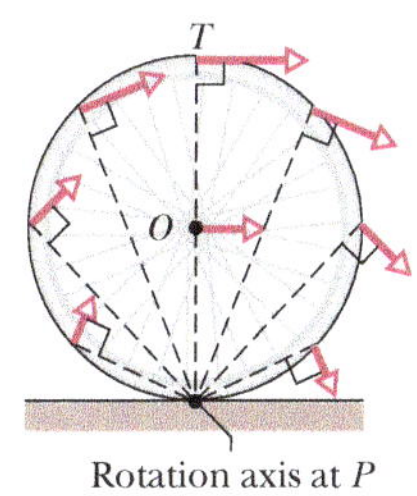

Figure 11-6 Rolling can be viewed as pure rotation, with angular speed ω, about an axis that always extends through P. The vectors show the instantaneous linear velocities of selected points on the rolling wheel. You can obtain the vectors by combining the translational and rotational motions as in Fig. 11-4.

Checkpoint 1

The rear wheel on a clown's bicycle has twice the radius of the front wheel. (a) When the bicycle is moving, is the linear speed at the very top of the rear wheel greater than, less than, or the same as that of the very top of the front wheel? (b) Is the angular speed of the rear wheel greater than, less than, or the same as that of the front wheel?

11-2 FORCES AND KINETIC ENERGY OF ROLLING

Learning Objectives

After reading this module, you should be able to . . .

11.03 Calculate the kinetic energy of a body in smooth rolling as the sum of the translational kinetic energy of the center of mass and the rotational kinetic energy around the center of mass.

11.04 Apply the relationship between the work done on a smoothly rolling object and the change in its kinetic energy.

11.05 For smooth rolling (and thus no sliding), conserve mechanical energy to relate initial energy values to the values at a later point.

11.06 Draw a free-body diagram of an accelerating body that is smoothly rolling on a horizontal surface or up or down a ramp.

11.07 Apply the relationship between the center-of-mass acceleration and the angular acceleration.

11.08 For smooth rolling of an object up or down a ramp, apply the relationship between the object's acceleration, its rotational inertia, and the angle of the ramp.

Key Ideas

- A smoothly rolling wheel has kinetic energy

$$K = \tfrac{1}{2}I_{com}\omega^2 + \tfrac{1}{2}Mv_{com}^2,$$

where I_{com} is the rotational inertia of the wheel about its center of mass and M is the mass of the wheel.

- If the wheel is being accelerated but is still rolling smoothly, the acceleration of the center of mass $\vec{a}_{com}$ is related to the angular acceleration α about the center with

$$a_{com} = \alpha R.$$

- If the wheel rolls smoothly down a ramp of angle θ, its acceleration along an x axis extending up the ramp is

$$a_{com,x} = -\frac{g\sin\theta}{1 + I_{com}/MR^2}.$$

The Kinetic Energy of Rolling

Let us now calculate the kinetic energy of the rolling wheel as measured by the stationary observer. If we view the rolling as pure rotation about an axis through P in Fig. 11-6, then from Eq. 10-34 we have

$$K = \tfrac{1}{2}I_P\omega^2, \tag{11-3}$$

in which ω is the angular speed of the wheel and I_P is the rotational inertia of the wheel about the axis through P. From the parallel-axis theorem of Eq. 10-36 ($I = I_{com} + Mh^2$), we have

$$I_P = I_{com} + MR^2, \tag{11-4}$$

in which M is the mass of the wheel, I_{com} is its rotational inertia about an axis through its center of mass, and R (the wheel's radius) is the perpendicular distance h. Substituting Eq. 11-4 into Eq. 11-3, we obtain

$$K = \tfrac{1}{2}I_{com}\omega^2 + \tfrac{1}{2}MR^2\omega^2,$$

and using the relation $v_{com} = \omega R$ (Eq. 11-2) yields

$$K = \tfrac{1}{2}I_{com}\omega^2 + \tfrac{1}{2}Mv_{com}^2. \tag{11-5}$$

We can interpret the term $\frac{1}{2}I_{com}\omega^2$ as the kinetic energy associated with the rotation of the wheel about an axis through its center of mass (Fig. 11-4a), and the term $\frac{1}{2}Mv_{com}^2$ as the kinetic energy associated with the translational motion of the wheel's center of mass (Fig. 11-4b). Thus, we have the following rule:

A rolling object has two types of kinetic energy: a rotational kinetic energy ($\frac{1}{2}I_{com}\omega^2$) due to its rotation about its center of mass and a translational kinetic energy ($\frac{1}{2}Mv_{com}^2$) due to translation of its center of mass.

The Forces of Rolling

Friction and Rolling

If a wheel rolls at constant speed, as in Fig. 11-3, it has no tendency to slide at the point of contact P, and thus no frictional force acts there. However, if a net force acts on the rolling wheel to speed it up or to slow it, then that net force causes acceleration $\vec{a}_{com}$ of the center of mass along the direction of travel. It also causes the wheel to rotate faster or slower, which means it causes an angular acceleration α. These accelerations tend to make the wheel slide at P. Thus, a frictional force must act on the wheel at P to oppose that tendency.

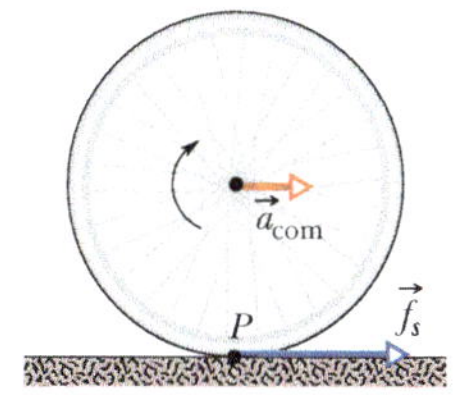

Figure 11-7 A wheel rolls horizontally without sliding while accelerating with linear acceleration $\vec{a}_{com}$, as on a bicycle at the start of a race. A static frictional force $\vec{f}_s$ acts on the wheel at P, opposing its tendency to slide.

If the wheel *does not* slide, the force is a *static* frictional force $\vec{f}_s$ and the motion is smooth rolling. We can then relate the magnitudes of the linear acceleration $\vec{a}_{com}$ and the angular acceleration α by differentiating Eq. 11-2 with respect to time (with R held constant). On the left side, dv_{com}/dt is a_{com}, and on the right side $d\omega/dt$ is α. So, for smooth rolling we have

$$a_{com} = \alpha R \quad \text{(smooth rolling motion).} \tag{11-6}$$

If the wheel *does* slide when the net force acts on it, the frictional force that acts at P in Fig. 11-3 is a *kinetic* frictional force $\vec{f}_k$. The motion then is not smooth rolling, and Eq. 11-6 does not apply to the motion. In this chapter we discuss only smooth rolling motion.

Figure 11-7 shows an example in which a wheel is being made to rotate faster while rolling to the right along a flat surface, as on a bicycle at the start of a race. The faster rotation tends to make the bottom of the wheel slide to the left at point P. A frictional force at P, directed to the right, opposes this tendency to slide. If the wheel does not slide, that frictional force is a static frictional force $\vec{f}_s$ (as shown), the motion is smooth rolling, and Eq. 11-6 applies to the motion. (Without friction, bicycle races would be stationary and very boring.)

If the wheel in Fig. 11-7 were made to rotate slower, as on a slowing bicycle, we would change the figure in two ways: The directions of the center-of-mass acceleration $\vec{a}_{com}$ and the frictional force $\vec{f}_s$ at point P would now be to the left.

Rolling Down a Ramp

Figure 11-8 shows a round uniform body of mass M and radius R rolling smoothly down a ramp at angle θ, along an x axis. We want to find an expression for the body's

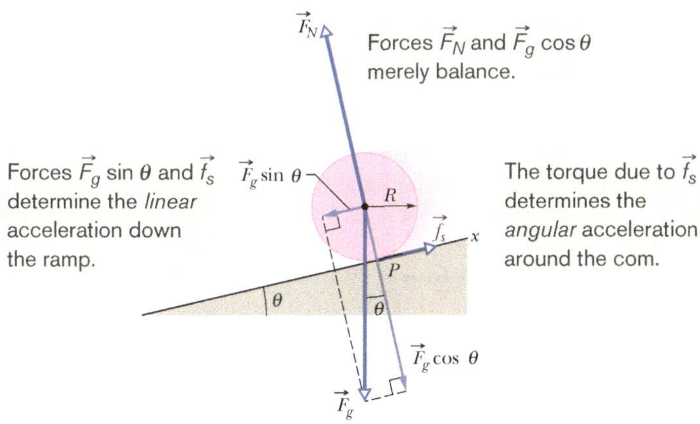

Figure 11-8 A round uniform body of radius R rolls down a ramp. The forces that act on it are the gravitational force $\vec{F}_g$, a normal force $\vec{F}_N$, and a frictional force $\vec{f}_s$ pointing up the ramp. (For clarity, vector $\vec{F}_N$ has been shifted in the direction it points until its tail is at the center of the body.)

acceleration $a_{com,x}$ down the ramp. We do this by using Newton's second law in both its linear version ($F_{net} = Ma$) and its angular version ($\tau_{net} = I\alpha$).

We start by drawing the forces on the body as shown in Fig. 11-8:

1. The gravitational force $\vec{F}_g$ on the body is directed downward. The tail of the vector is placed at the center of mass of the body. The component along the ramp is $F_g \sin\theta$, which is equal to $Mg \sin\theta$.
2. A normal force $\vec{F}_N$ is perpendicular to the ramp. It acts at the point of contact P, but in Fig. 11-8 the vector has been shifted along its direction until its tail is at the body's center of mass.
3. A static frictional force $\vec{f}_s$ acts at the point of contact P and is directed up the ramp. (Do you see why? If the body were to slide at P, it would slide *down* the ramp. Thus, the frictional force opposing the sliding must be *up* the ramp.)

We can write Newton's second law for components along the x axis in Fig. 11-8 ($F_{net,x} = ma_x$) as

$$f_s - Mg \sin\theta = Ma_{com,x}. \tag{11-7}$$

This equation contains two unknowns, f_s and $a_{com,x}$. (We should *not* assume that f_s is at its maximum value $f_{s,max}$. All we know is that the value of f_s is just right for the body to roll smoothly down the ramp, without sliding.)

We now wish to apply Newton's second law in angular form to the body's rotation about its center of mass. First, we shall use Eq. 10-41 ($\tau = r_\perp F$) to write the torques on the body about that point. The frictional force $\vec{f}_s$ has moment arm R and thus produces a torque Rf_s, which is positive because it tends to rotate the body counterclockwise in Fig. 11-8. Forces $\vec{F}_g$ and $\vec{F}_N$ have zero moment arms about the center of mass and thus produce zero torques. So we can write the angular form of Newton's second law ($\tau_{net} = I\alpha$) about an axis through the body's center of mass as

$$Rf_s = I_{com}\alpha. \tag{11-8}$$

This equation contains two unknowns, f_s and α.

Because the body is rolling smoothly, we can use Eq. 11-6 ($a_{com} = \alpha R$) to relate the unknowns $a_{com,x}$ and α. But we must be cautious because here $a_{com,x}$ is negative (in the negative direction of the x axis) and α is positive (counterclockwise). Thus we substitute $-a_{com,x}/R$ for α in Eq. 11-8. Then, solving for f_s, we obtain

$$f_s = -I_{com}\frac{a_{com,x}}{R^2}. \tag{11-9}$$

Substituting the right side of Eq. 11-9 for f_s in Eq. 11-7, we then find

$$a_{com,x} = -\frac{g \sin\theta}{1 + I_{com}/MR^2}. \tag{11-10}$$

We can use this equation to find the linear acceleration $a_{com,x}$ of any body rolling along an incline of angle θ with the horizontal.

Note that the pull by the gravitational force causes the body to come down the ramp, but it is the frictional force that causes the body to rotate and thus roll. If you eliminate the friction (by, say, making the ramp slick with ice or grease) or arrange for $Mg \sin\theta$ to exceed $f_{s,max}$, then you eliminate the smooth rolling and the body slides down the ramp.

Checkpoint 2

Disks A and B are identical and roll across a floor with equal speeds. Then disk A rolls up an incline, reaching a maximum height h, and disk B moves up an incline that is identical except that it is frictionless. Is the maximum height reached by disk B greater than, less than, or equal to h?

Sample Problem 11.01 Ball rolling down a ramp

A uniform ball, of mass $M = 6.00$ kg and radius R, rolls smoothly from rest down a ramp at angle $\theta = 30.0°$ (Fig. 11-8).

(a) The ball descends a vertical height $h = 1.20$ m to reach the bottom of the ramp. What is its speed at the bottom?

KEY IDEAS

The mechanical energy E of the ball–Earth system is conserved as the ball rolls down the ramp. The reason is that the only force doing work on the ball is the gravitational force, a conservative force. The normal force on the ball from the ramp does zero work because it is perpendicular to the ball's path. The frictional force on the ball from the ramp does not transfer any energy to thermal energy because the ball does not slide (it *rolls smoothly*).

Thus, we conserve mechanical energy ($E_f = E_i$):

$$K_f + U_f = K_i + U_i, \tag{11-11}$$

where subscripts f and i refer to the final values (at the bottom) and initial values (at rest), respectively. The gravitational potential energy is initially $U_i = Mgh$ (where M is the ball's mass) and finally $U_f = 0$. The kinetic energy is initially $K_i = 0$. For the final kinetic energy K_f, we need an additional idea: Because the ball rolls, the kinetic energy involves both translation *and* rotation, so we include them both by using the right side of Eq. 11-5.

Calculations: Substituting into Eq. 11-11 gives us

$$(\tfrac{1}{2}I_{com}\omega^2 + \tfrac{1}{2}Mv_{com}^2) + 0 = 0 + Mgh, \tag{11-12}$$

where I_{com} is the ball's rotational inertia about an axis through its center of mass, v_{com} is the requested speed at the bottom, and ω is the angular speed there.

Because the ball rolls smoothly, we can use Eq. 11-2 to substitute v_{com}/R for ω to reduce the unknowns in Eq. 11-12. Doing so, substituting $\frac{2}{5}MR^2$ for I_{com} (from Table 10-2*f*), and then solving for v_{com} give us

$$v_{com} = \sqrt{(\tfrac{10}{7})gh} = \sqrt{(\tfrac{10}{7})(9.8\text{ m/s}^2)(1.20\text{ m})}$$
$$= 4.10\text{ m/s}. \qquad \text{(Answer)}$$

Note that the answer does not depend on M or R.

(b) What are the magnitude and direction of the frictional force on the ball as it rolls down the ramp?

KEY IDEA

Because the ball rolls smoothly, Eq. 11-9 gives the frictional force on the ball.

Calculations: Before we can use Eq. 11-9, we need the ball's acceleration $a_{com,x}$ from Eq. 11-10:

$$a_{com,x} = -\frac{g\sin\theta}{1 + I_{com}/MR^2} = -\frac{g\sin\theta}{1 + \frac{2}{5}MR^2/MR^2}$$
$$= -\frac{(9.8\text{ m/s}^2)\sin 30.0°}{1 + \frac{2}{5}} = -3.50\text{ m/s}^2.$$

Note that we needed neither mass M nor radius R to find $a_{com,x}$. Thus, any size ball with any uniform mass would have this smoothly rolling acceleration down a 30.0° ramp.

We can now solve Eq. 11-9 as

$$f_s = -I_{com}\frac{a_{com,x}}{R^2} = -\tfrac{2}{5}MR^2\frac{a_{com,x}}{R^2} = -\tfrac{2}{5}Ma_{com,x}$$
$$= -\tfrac{2}{5}(6.00\text{ kg})(-3.50\text{ m/s}^2) = 8.40\text{ N}. \qquad \text{(Answer)}$$

Note that we needed mass M but not radius R. Thus, the frictional force on any 6.00 kg ball rolling smoothly down a 30.0° ramp would be 8.40 N regardless of the ball's radius but would be larger for a larger mass.

11-3 THE YO-YO

Learning Objectives

After reading this module, you should be able to . . .

11.09 Draw a free-body diagram of a yo-yo moving up or down its string.

11.10 Identify that a yo-yo is effectively an object that rolls smoothly up or down a ramp with an incline angle of 90°.

11.11 For a yo-yo moving up or down its string, apply the relationship between the yo-yo's acceleration and its rotational inertia.

11.12 Determine the tension in a yo-yo's string as the yo-yo moves up or down its string.

Key Idea

- A yo-yo, which travels vertically up or down a string, can be treated as a wheel rolling along an inclined plane at angle $\theta = 90°$.

The Yo-Yo

A yo-yo is a physics lab that you can fit in your pocket. If a yo-yo rolls down its string for a distance h, it loses potential energy in amount mgh but gains kinetic energy in both translational ($\frac{1}{2}Mv_{com}^2$) and rotational ($\frac{1}{2}I_{com}\omega^2$) forms. As it climbs back up, it loses kinetic energy and regains potential energy.

In a modern yo-yo, the string is not tied to the axle but is looped around it. When the yo-yo "hits" the bottom of its string, an upward force on the axle from the string stops the descent. The yo-yo then spins, axle inside loop, with only rotational kinetic energy. The yo-yo keeps spinning ("sleeping") until you "wake it" by jerking on the string, causing the string to catch on the axle and the yo-yo to climb back up. The rotational kinetic energy of the yo-yo at the bottom of its string (and thus the sleeping time) can be considerably increased by throwing the yo-yo downward so that it starts down the string with initial speeds v_{com} and ω instead of rolling down from rest.

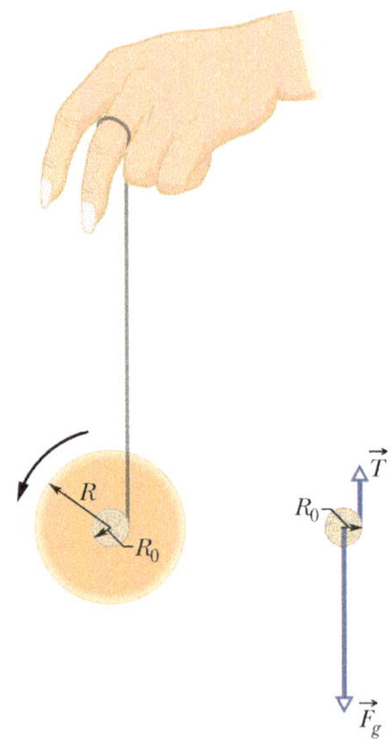

Figure 11-9 (*a*) A yo-yo, shown in cross section. The string, of assumed negligible thickness, is wound around an axle of radius R_0. (*b*) A free-body diagram for the falling yo-yo. Only the axle is shown.

To find an expression for the linear acceleration a_{com} of a yo-yo rolling down a string, we could use Newton's second law (in linear and angular forms) just as we did for the body rolling down a ramp in Fig. 11-8. The analysis is the same except for the following:

1. Instead of rolling down a ramp at angle θ with the horizontal, the yo-yo rolls down a string at angle $\theta = 90°$ with the horizontal.
2. Instead of rolling on its outer surface at radius R, the yo-yo rolls on an axle of radius R_0 (Fig. 11-9*a*).
3. Instead of being slowed by frictional force $\vec{f}_s$, the yo-yo is slowed by the force $\vec{T}$ on it from the string (Fig. 11-9*b*).

The analysis would again lead us to Eq. 11-10. Therefore, let us just change the notation in Eq. 11-10 and set $\theta = 90°$ to write the linear acceleration as

$$a_{com} = -\frac{g}{1 + I_{com}/MR_0^2}, \qquad (11\text{-}13)$$

where I_{com} is the yo-yo's rotational inertia about its center and M is its mass. A yo-yo has the same downward acceleration when it is climbing back up.

11-4 TORQUE REVISITED

Learning Objectives

After reading this module, you should be able to . . .

11.13 Identify that torque is a vector quantity.

11.14 Identify that the point about which a torque is calculated must always be specified.

11.15 Calculate the torque due to a force on a particle by taking the cross product of the particle's position vector and the force vector, in either unit-vector notation or magnitude-angle notation.

11.16 Use the right-hand rule for cross products to find the direction of a torque vector.

Key Ideas

- In three dimensions, torque $\vec{\tau}$ is a vector quantity defined relative to a fixed point (usually an origin); it is

$$\vec{\tau} = \vec{r} \times \vec{F},$$

where $\vec{F}$ is a force applied to a particle and $\vec{r}$ is a position vector locating the particle relative to the fixed point.

- The magnitude of $\vec{\tau}$ is given by

$$\tau = rF\sin\phi = rF_\perp = r_\perp F,$$

where ϕ is the angle between $\vec{F}$ and $\vec{r}$, $F_\perp$ is the component of $\vec{F}$ perpendicular to $\vec{r}$, and $r_\perp$ is the moment arm of $\vec{F}$.

- The direction of $\vec{\tau}$ is given by the right-hand rule for cross products.

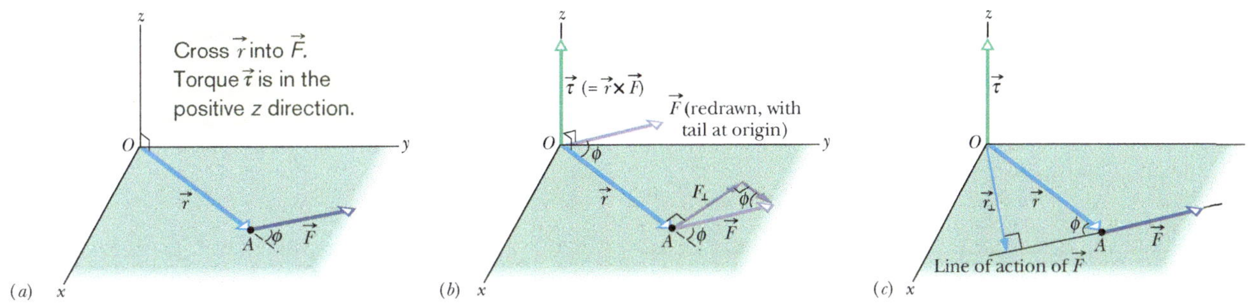

Figure 11-10 Defining torque. (*a*) A force $\vec{F}$, lying in an *xy* plane, acts on a particle at point *A*. (*b*) This force produces a torque $\vec{\tau}\ (=\vec{r}\times\vec{F})$ on the particle with respect to the origin *O*. By the right-hand rule for vector (cross) products, the torque vector points in the positive direction of *z*. Its magnitude is given by $rF_\perp$ in (*b*) and by $r_\perp F$ in (*c*).

Torque Revisited

In Chapter 10 we defined torque τ for a rigid body that can rotate around a fixed axis. We now expand the definition of torque to apply it to an individual particle that moves along any path relative to a fixed *point* (rather than a fixed axis). The path need no longer be a circle, and we must write the torque as a vector $\vec{\tau}$ that may have any direction. We can calculate the magnitude of the torque with a formula and determine its direction with the right-hand rule for cross products.

Figure 11-10*a* shows such a particle at point *A* in an *xy* plane. A single force $\vec{F}$ in that plane acts on the particle, and the particle's position relative to the origin *O* is given by position vector $\vec{r}$. The torque $\vec{\tau}$ acting on the particle relative to the fixed point *O* is a vector quantity defined as

$$\vec{\tau} = \vec{r}\times\vec{F} \quad \text{(torque defined).} \tag{11-14}$$

We can evaluate the vector (or cross) product in this definition of $\vec{\tau}$ by using the rules in Module 3-3. To find the direction of $\vec{\tau}$, we slide the vector $\vec{F}$ (without changing its direction) until its tail is at the origin *O*, so that the two vectors in the vector product are tail to tail as in Fig. 11-10*b*. We then use the right-hand rule in Fig. 3-19*a*, sweeping the fingers of the right hand from $\vec{r}$ (the first vector in the product) into $\vec{F}$ (the second vector). The outstretched right thumb then gives the direction of $\vec{\tau}$. In Fig. 11-10*b*, it is in the positive direction of the *z* axis.

To determine the magnitude of $\vec{\tau}$, we apply the general result of Eq. 3-27 ($c = ab\sin\phi$), finding

$$\tau = rF\sin\phi, \tag{11-15}$$

where ϕ is the smaller angle between the directions of $\vec{r}$ and $\vec{F}$ when the vectors are tail to tail. From Fig. 11-10*b*, we see that Eq. 11-15 can be rewritten as

$$\tau = rF_\perp, \tag{11-16}$$

where $F_\perp\ (=F\sin\phi)$ is the component of $\vec{F}$ perpendicular to $\vec{r}$. From Fig. 11-10*c*, we see that Eq. 11-15 can also be rewritten as

$$\tau = r_\perp F, \tag{11-17}$$

where $r_\perp\ (=r\sin\phi)$ is the moment arm of $\vec{F}$ (the perpendicular distance between *O* and the line of action of $\vec{F}$).

Checkpoint 3

The position vector $\vec{r}$ of a particle points along the positive direction of a *z* axis. If the torque on the particle is (a) zero, (b) in the negative direction of *x*, and (c) in the negative direction of *y*, in what direction is the force causing the torque?

Sample Problem 11.02 Torque on a particle due to a force

In Fig. 11-11*a*, three forces, each of magnitude 2.0 N, act on a particle. The particle is in the *xz* plane at point *A* given by position vector $\vec{r}$, where $r = 3.0$ m and $\theta = 30°$. What is the torque, about the origin *O*, due to each force?

KEY IDEA

Because the three force vectors do not lie in a plane, we must use cross products, with magnitudes given by Eq. 11-15 ($\tau = rF \sin \phi$) and directions given by the right-hand rule.

Calculations: Because we want the torques with respect to the origin *O*, the vector $\vec{r}$ required for each cross product is the given position vector. To determine the angle ϕ between $\vec{r}$ and each force, we shift the force vectors of Fig. 11-11*a*, each in turn, so that their tails are at the origin. Figures 11-11*b*, *c*, and *d*, which are direct views of the *xz* plane, show the shifted force vectors $\vec{F}_1$, $\vec{F}_2$, and $\vec{F}_3$, respectively. (Note how much easier the angles between the force vectors and the position vector are to see.) In Fig. 11-11*d*, the angle between the directions of $\vec{r}$ and $\vec{F}_3$ is 90° and the symbol ⊗ means $\vec{F}_3$ is directed into the page. (For out of the page, we would use ⊙.)

Now, applying Eq. 11-15, we find

$$\tau_1 = rF_1 \sin \phi_1 = (3.0 \text{ m})(2.0 \text{ N})(\sin 150°) = 3.0 \text{ N}\cdot\text{m},$$

$$\tau_2 = rF_2 \sin \phi_2 = (3.0 \text{ m})(2.0 \text{ N})(\sin 120°) = 5.2 \text{ N}\cdot\text{m},$$

and

$$\tau_3 = rF_3 \sin \phi_3 = (3.0 \text{ m})(2.0 \text{ N})(\sin 90°) = 6.0 \text{ N}\cdot\text{m}. \quad \text{(Answer)}$$

Next, we use the right-hand rule, placing the fingers of the right hand so as to rotate $\vec{r}$ into $\vec{F}$ through the *smaller* of the two angles between their directions. The thumb points in the direction of the torque. Thus $\vec{\tau}_1$ is directed into the page in Fig. 11-11*b*; $\vec{\tau}_2$ is directed out of the page in Fig. 11-11*c*; and $\vec{\tau}_3$ is directed as shown in Fig. 11-11*d*. All three torque vectors are shown in Fig. 11-11*e*.

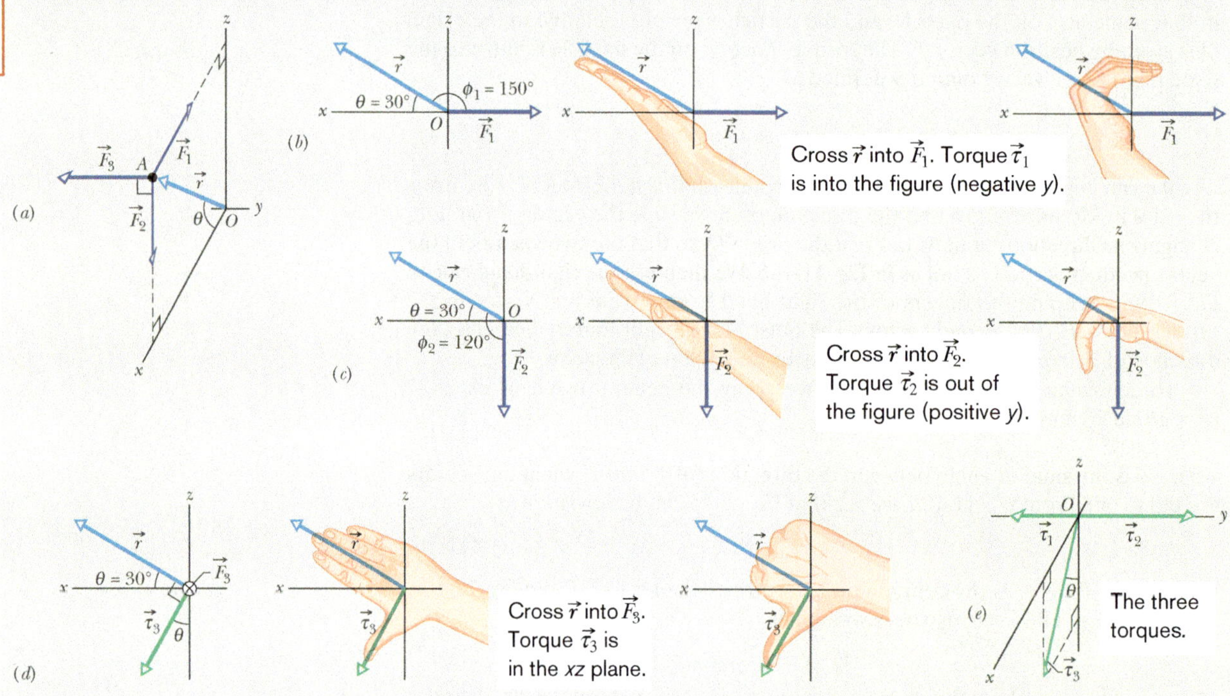

Figure 11-11 (*a*) A particle at point *A* is acted on by three forces, each parallel to a coordinate axis. The angle ϕ (used in finding torque) is shown (*b*) for $\vec{F}_1$ and (*c*) for $\vec{F}_2$. (*d*) Torque $\vec{\tau}_3$ is perpendicular to both $\vec{r}$ and $\vec{F}_3$ (force $\vec{F}_3$ is directed into the plane of the figure). (*e*) The torques.

Additional examples, video, and practice available at *WileyPLUS*

11-5 ANGULAR MOMENTUM

Learning Objectives

After reading this module, you should be able to . . .

11.17 Identify that angular momentum is a vector quantity.

11.18 Identify that the fixed point about which an angular momentum is calculated must always be specified.

11.19 Calculate the angular momentum of a particle by taking the cross product of the particle's position vector and its momentum vector, in either unit-vector notation or magnitude-angle notation.

11.20 Use the right-hand rule for cross products to find the direction of an angular momentum vector.

Key Ideas

- The angular momentum $\vec{\ell}$ of a particle with linear momentum $\vec{p}$, mass m, and linear velocity $\vec{v}$ is a vector quantity defined relative to a fixed point (usually an origin) as

$$\vec{\ell} = \vec{r} \times \vec{p} = m(\vec{r} \times \vec{v}).$$

- The magnitude of $\vec{\ell}$ is given by

$$\begin{aligned} \ell &= rmv \sin \phi \\ &= rp_\perp = rmv_\perp \\ &= r_\perp p = r_\perp mv, \end{aligned}$$

where ϕ is the angle between $\vec{r}$ and $\vec{p}$, $p_\perp$ and $v_\perp$ are the components of $\vec{p}$ and $\vec{v}$ perpendicular to $\vec{r}$, and $r_\perp$ is the perpendicular distance between the fixed point and the extension of $\vec{p}$.

- The direction of $\vec{\ell}$ is given by the right-hand rule: Position your right hand so that the fingers are in the direction of $\vec{r}$. Then rotate them around the palm to be in the direction of $\vec{p}$. Your outstretched thumb gives the direction of $\vec{\ell}$.

Angular Momentum

Recall that the concept of linear momentum $\vec{p}$ and the principle of conservation of linear momentum are extremely powerful tools. They allow us to predict the outcome of, say, a collision of two cars without knowing the details of the collision. Here we begin a discussion of the angular counterpart of $\vec{p}$, winding up in Module 11-8 with the angular counterpart of the conservation principle, which can lead to beautiful (almost magical) feats in ballet, fancy diving, ice skating, and many other activities.

Figure 11-12 shows a particle of mass m with linear momentum $\vec{p}\,(= m\vec{v})$ as it passes through point A in an xy plane. The **angular momentum** $\vec{\ell}$ of this particle with respect to the origin O is a vector quantity defined as

$$\vec{\ell} = \vec{r} \times \vec{p} = m(\vec{r} \times \vec{v}) \quad \text{(angular momentum defined)}, \qquad (11\text{-}18)$$

where $\vec{r}$ is the position vector of the particle with respect to O. As the particle moves relative to O in the direction of its momentum $\vec{p}\,(= m\vec{v})$, position vector $\vec{r}$ rotates around O. Note carefully that to have angular momentum about O, the particle does *not* itself have to rotate around O. Comparison of Eqs. 11-14 and 11-18 shows that angular momentum bears the same relation to linear momentum that torque does to force. The SI unit of angular momentum is the kilogram-meter-squared per second ($\text{kg} \cdot \text{m}^2/\text{s}$), equivalent to the joule-second ($\text{J} \cdot \text{s}$).

Direction. To find the direction of the angular momentum vector $\vec{\ell}$ in Fig. 11-12, we slide the vector $\vec{p}$ until its tail is at the origin O. Then we use the right-hand rule for vector products, sweeping the fingers from $\vec{r}$ into $\vec{p}$. The outstretched thumb then shows that the direction of $\vec{\ell}$ is in the positive direction of the z axis in Fig. 11-12. This positive direction is consistent with the counterclockwise rotation of position vector $\vec{r}$ about the z axis, as the particle moves. (A negative direction of $\vec{\ell}$ would be consistent with a clockwise rotation of $\vec{r}$ about the z axis.)

Magnitude. To find the magnitude of $\vec{\ell}$, we use the general result of Eq. 3-27 to write

$$\ell = rmv \sin \phi, \qquad (11\text{-}19)$$

where ϕ is the smaller angle between $\vec{r}$ and $\vec{p}$ when these two vectors are tail

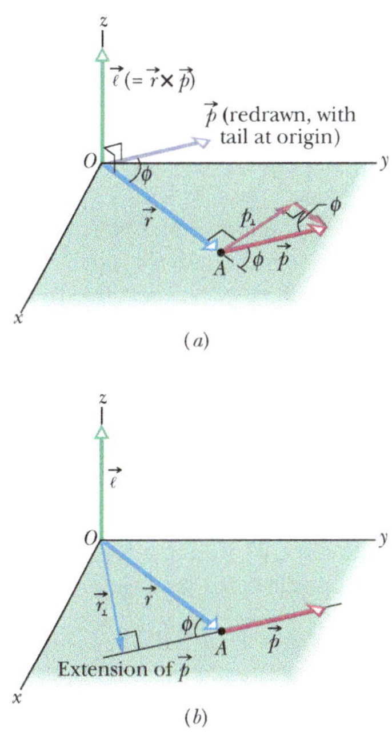

Figure 11-12 Defining angular momentum. A particle passing through point A has linear momentum $\vec{p}\,(= m\vec{v})$, with the vector $\vec{p}$ lying in an xy plane. The particle has angular momentum $\vec{\ell}\,(= \vec{r} \times \vec{p})$ with respect to the origin O. By the right-hand rule, the angular momentum vector points in the positive direction of z. (*a*) The magnitude of $\vec{\ell}$ is given by $\ell = rp_\perp = rmv_\perp$. (*b*) The magnitude of $\vec{\ell}$ is also given by $\ell = r_\perp p = r_\perp mv$.

to tail. From Fig. 11-12*a*, we see that Eq. 11-19 can be rewritten as

$$\ell = rp_\perp = rmv_\perp, \tag{11-20}$$

where $p_\perp$ is the component of $\vec{p}$ perpendicular to $\vec{r}$ and $v_\perp$ is the component of $\vec{v}$ perpendicular to $\vec{r}$. From Fig. 11-12*b*, we see that Eq. 11-19 can also be rewritten as

$$\ell = r_\perp p = r_\perp mv, \tag{11-21}$$

where $r_\perp$ is the perpendicular distance between O and the extension of $\vec{p}$.

Important. Note two features here: (1) angular momentum has meaning only with respect to a specified origin and (2) its direction is always perpendicular to the plane formed by the position and linear momentum vectors $\vec{r}$ and $\vec{p}$.

Checkpoint 4

In part *a* of the figure, particles 1 and 2 move around point O in circles with radii 2 m and 4 m. In part *b*, particles 3 and 4 travel along straight lines at perpendicular distances of 4 m and 2 m from point O. Particle 5 moves directly away from O. All five particles have the same mass and the same constant speed. (a) Rank the particles according to the magnitudes of their angular momentum about point O, greatest first. (b) Which particles have negative angular momentum about point O?

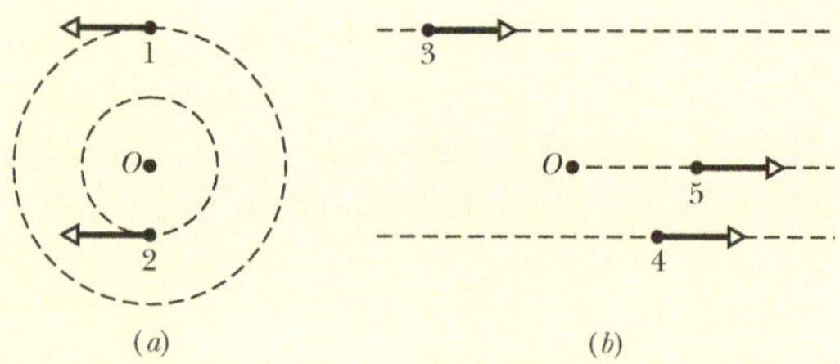

Sample Problem 11.03 Angular momentum of a two-particle system

Figure 11-13 shows an overhead view of two particles moving at constant momentum along horizontal paths. Particle 1, with momentum magnitude $p_1 = 5.0$ kg·m/s, has position vector $\vec{r}_1$ and will pass 2.0 m from point O. Particle 2, with momentum magnitude $p_2 = 2.0$ kg·m/s, has position vector $\vec{r}_2$ and will pass 4.0 m from point O. What are the magnitude and direction of the net angular momentum $\vec{L}$ about point O of the two-particle system?

KEY IDEA

To find $\vec{L}$, we can first find the individual angular momenta $\vec{\ell}_1$ and $\vec{\ell}_2$ and then add them. To evaluate their magnitudes, we can use any one of Eqs. 11-18 through 11-21. However, Eq. 11-21 is easiest, because we are given the perpendicular distances $r_{\perp 1}$ (= 2.0 m) and $r_{\perp 2}$ (= 4.0 m) and the momentum magnitudes p_1 and p_2.

Calculations: For particle 1, Eq. 11-21 yields

$$\ell_1 = r_{\perp 1}p_1 = (2.0\text{ m})(5.0\text{ kg}\cdot\text{m/s}) = 10\text{ kg}\cdot\text{m}^2/\text{s}.$$

To find the direction of vector $\vec{\ell}_1$, we use Eq. 11-18 and the right-hand rule for vector products. For $\vec{r}_1 \times \vec{p}_1$, the vector product is out of the page, perpendicular to the plane of Fig. 11-13. This is the positive direction, consistent with the counterclockwise rotation of the particle's position vector $\vec{r}_1$ around O as particle 1 moves. Thus, the angular momentum vector for particle 1 is

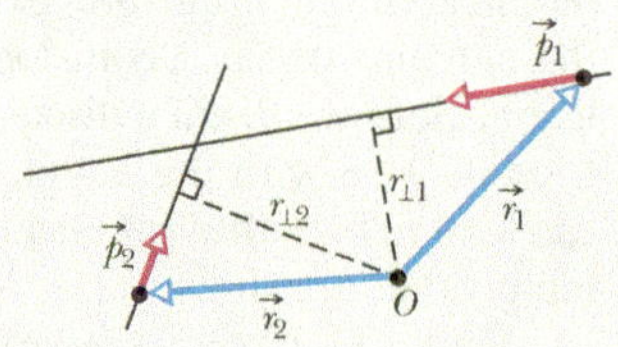

Figure 11-13 Two particles pass near point O.

$$\ell_1 = +10\text{ kg}\cdot\text{m}^2/\text{s}.$$

Similarly, the magnitude of $\vec{\ell}_2$ is

$$\ell_2 = r_{\perp 2}p_2 = (4.0\text{ m})(2.0\text{ kg}\cdot\text{m/s}) = 8.0\text{ kg}\cdot\text{m}^2/\text{s},$$

and the vector product $\vec{r}_2 \times \vec{p}_2$ is into the page, which is the negative direction, consistent with the clockwise rotation of $\vec{r}_2$ around O as particle 2 moves. Thus, the angular momentum vector for particle 2 is

$$\ell_2 = -8.0\text{ kg}\cdot\text{m}^2/\text{s}.$$

The net angular momentum for the two-particle system is

$$L = \ell_1 + \ell_2 = +10\text{ kg}\cdot\text{m}^2/\text{s} + (-8.0\text{ kg}\cdot\text{m}^2/\text{s}) = +2.0\text{ kg}\cdot\text{m}^2/\text{s}. \quad \text{(Answer)}$$

The plus sign means that the system's net angular momentum about point O is out of the page.

WILEY PLUS Additional examples, video, and practice available at *WileyPLUS*

11-6 NEWTON'S SECOND LAW IN ANGULAR FORM

Learning Objective

After reading this module, you should be able to . . .

11.21 Apply Newton's second law in angular form to relate the torque acting on a particle to the resulting rate of change of the particle's angular momentum, all relative to a specified point.

Key Idea

- Newton's second law for a particle can be written in angular form as

$$\vec{\tau}_{\text{net}} = \frac{d\vec{\ell}}{dt},$$

where $\vec{\tau}_{\text{net}}$ is the net torque acting on the particle and $\vec{\ell}$ is the angular momentum of the particle.

Newton's Second Law in Angular Form

Newton's second law written in the form

$$\vec{F}_{\text{net}} = \frac{d\vec{p}}{dt} \quad \text{(single particle)} \tag{11-22}$$

expresses the close relation between force and linear momentum for a single particle. We have seen enough of the parallelism between linear and angular quantities to be pretty sure that there is also a close relation between torque and angular momentum. Guided by Eq. 11-22, we can even guess that it must be

$$\vec{\tau}_{\text{net}} = \frac{d\vec{\ell}}{dt} \quad \text{(single particle).} \tag{11-23}$$

Equation 11-23 is indeed an angular form of Newton's second law for a single particle:

> The (vector) sum of all the torques acting on a particle is equal to the time rate of change of the angular momentum of that particle.

Equation 11-23 has no meaning unless the torques $\vec{\tau}$ and the angular momentum $\vec{\ell}$ are defined with respect to the same point, usually the origin of the coordinate system being used.

Proof of Equation 11-23

We start with Eq. 11-18, the definition of the angular momentum of a particle:

$$\vec{\ell} = m(\vec{r} \times \vec{v}),$$

where $\vec{r}$ is the position vector of the particle and $\vec{v}$ is the velocity of the particle. Differentiating* each side with respect to time t yields

$$\frac{d\vec{\ell}}{dt} = m\left(\vec{r} \times \frac{d\vec{v}}{dt} + \frac{d\vec{r}}{dt} \times \vec{v}\right). \tag{11-24}$$

However, $d\vec{v}/dt$ is the acceleration $\vec{a}$ of the particle, and $d\vec{r}/dt$ is its velocity $\vec{v}$. Thus, we can rewrite Eq. 11-24 as

$$\frac{d\vec{\ell}}{dt} = m(\vec{r} \times \vec{a} + \vec{v} \times \vec{v}).$$

*In differentiating a vector product, be sure not to change the order of the two quantities (here $\vec{r}$ and $\vec{v}$) that form that product. (See Eq. 3-25.)

Now $\vec{v} \times \vec{v} = 0$ (the vector product of any vector with itself is zero because the angle between the two vectors is necessarily zero). Thus, the last term of this expression is eliminated and we then have

$$\frac{d\vec{\ell}}{dt} = m(\vec{r} \times \vec{a}) = \vec{r} \times m\vec{a}.$$

We now use Newton's second law ($\vec{F}_{net} = m\vec{a}$) to replace $m\vec{a}$ with its equal, the vector sum of the forces that act on the particle, obtaining

$$\frac{d\vec{\ell}}{dt} = \vec{r} \times \vec{F}_{net} = \sum(\vec{r} \times \vec{F}). \tag{11-25}$$

Here the symbol Σ indicates that we must sum the vector products $\vec{r} \times \vec{F}$ for all the forces. However, from Eq. 11-14, we know that each one of those vector products is the torque associated with one of the forces. Therefore, Eq. 11-25 tells us that

$$\vec{\tau}_{net} = \frac{d\vec{\ell}}{dt}.$$

This is Eq. 11-23, the relation that we set out to prove.

Checkpoint 5

The figure shows the position vector $\vec{r}$ of a particle at a certain instant, and four choices for the direction of a force that is to accelerate the particle. All four choices lie in the *xy* plane. (a) Rank the choices according to the magnitude of the time rate of change ($d\vec{\ell}/dt$) they produce in the angular momentum of the particle about point *O*, greatest first. (b) Which choice results in a negative rate of change about *O*?

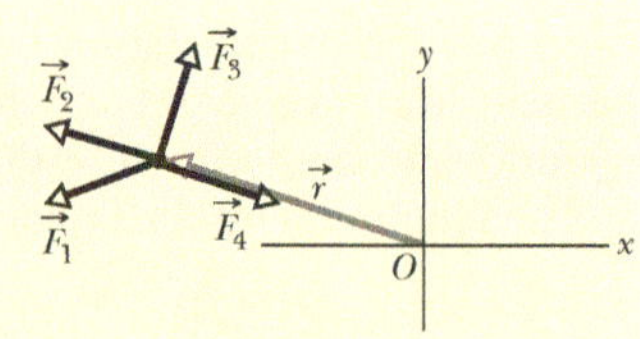

Sample Problem 11.04 Torque and the time derivative of angular momentum

Figure 11-14*a* shows a freeze-frame of a 0.500 kg particle moving along a straight line with a position vector given by

$$\vec{r} = (-2.00t^2 - t)\hat{i} + 5.00\hat{j},$$

with $\vec{r}$ in meters and *t* in seconds, starting at $t = 0$. The position vector points from the origin to the particle. In unit-vector notation, find expressions for the angular momentum $\vec{\ell}$ of the particle and the torque $\vec{\tau}$ acting on the particle, both with respect to (or about) the origin. Justify their algebraic signs in terms of the particle's motion.

KEY IDEAS

(1) The point about which an angular momentum of a particle is to be calculated must always be specified. Here it is the origin. (2) The angular momentum $\vec{\ell}$ of a particle is given by Eq. 11-18 ($\vec{\ell} = \vec{r} \times \vec{p} = m(\vec{r} \times \vec{v})$). (3) The sign associated with a particle's angular momentum is set by the sense of rotation of the particle's position vector (around the rotation axis) as the particle moves: clockwise is negative and counterclockwise is positive. (4) If the torque acting on a particle and the angular momentum of the particle are calculated around the *same* point, then the torque is related to angular momentum by Eq. 11-23 ($\vec{\tau} = d\vec{\ell}/dt$).

Calculations: In order to use Eq. 11-18 to find the angular momentum about the origin, we first must find an expression for the particle's velocity by taking a time derivative of its position vector. Following Eq. 4-10 ($\vec{v} = d\vec{r}/dt$), we write

$$\vec{v} = \frac{d}{dt}((-2.00t^2 - t)\hat{i} + 5.00\hat{j})$$
$$= (-4.00t - 1.00)\hat{i},$$

with $\vec{v}$ in meters per second.

Next, let's take the cross product of $\vec{r}$ and $\vec{v}$ using the template for cross products displayed in Eq. 3-27:

$$\vec{a} \times \vec{b} = (a_y b_z - b_y a_z)\hat{i} + (a_z b_x - b_z a_x)\hat{j} + (a_x b_y - b_x a_y)\hat{k}.$$

Here the generic $\vec{a}$ is $\vec{r}$ and the generic $\vec{b}$ is $\vec{v}$. However, because we really don't want to do more work than needed, let's first just think about our substitutions into

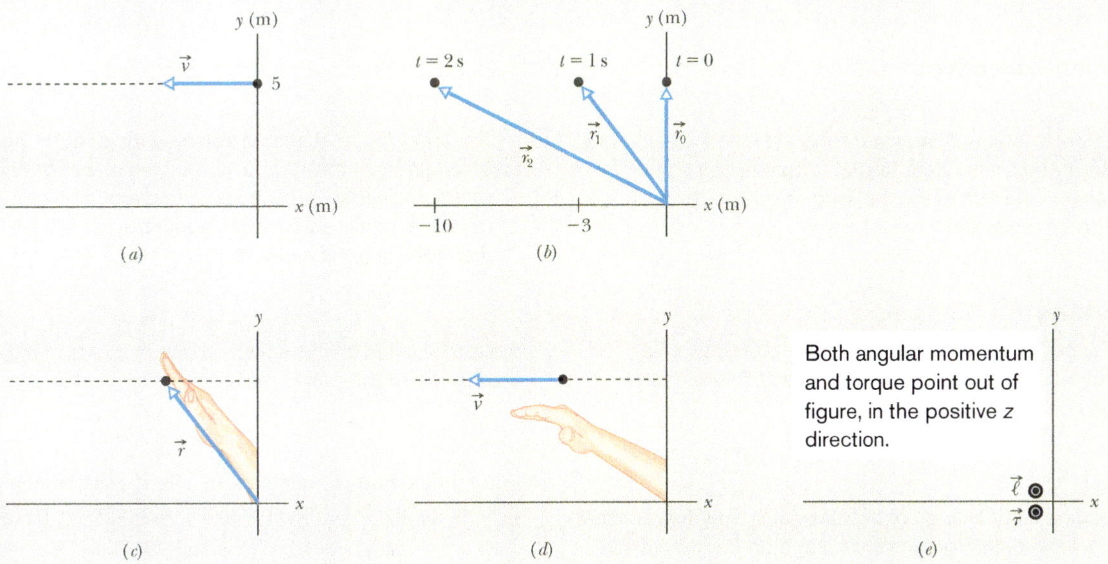

Figure 11-14 (*a*) A particle moving in a straight line, shown at time $t = 0$. (*b*) The position vector at $t = 0$, 1.00 s, and 2.00 s. (*c*) The first step in applying the right-hand rule for cross products. (*d*) The second step. (*e*) The angular momentum vector and the torque vector are along the z axis, which extends out of the plane of the figure.

the generic cross product. Because $\vec{r}$ lacks any z component and because $\vec{v}$ lacks any y or z component, the only nonzero term in the generic cross product is the very last one $(-b_x a_y)\hat{k}$. So, let's cut to the (mathematical) chase by writing

$$\vec{r} \times \vec{v} = -(-4.00t - 1.00)(5.00)\hat{k} = (20.0t + 5.00)\hat{k}\ \text{m}^2/\text{s}.$$

Note that, as always, the cross product produces a vector that is perpendicular to the original vectors.

To finish up Eq. 11-18, we multiply by the mass, finding

$$\vec{\ell} = (0.500\ \text{kg})[(20.0t + 5.00)\hat{k}\ \text{m}^2/\text{s}]$$
$$= (10.0t + 2.50)\hat{k}\ \text{kg}\cdot\text{m}^2/\text{s}. \qquad \text{(Answer)}$$

The torque about the origin then immediately follows from Eq. 11-23:

$$\vec{\tau} = \frac{d}{dt}(10.0t + 2.50)\hat{k}\ \text{kg}\cdot\text{m}^2/\text{s}$$
$$= 10.0\hat{k}\ \text{kg}\cdot\text{m}^2/\text{s}^2 = 10.0\hat{k}\ \text{N}\cdot\text{m}, \qquad \text{(Answer)}$$

which is in the positive direction of the z axis.

Our result for $\vec{\ell}$ tells us that the angular momentum is in the positive direction of the z axis. To make sense of that positive result in terms of the rotation of the position vector, let's evaluate that vector for several times:

$$t = 0, \qquad \vec{r}_0 = 5.00\hat{j}\ \text{m}$$
$$t = 1.00\ \text{s}, \qquad \vec{r}_1 = -3.00\hat{i} + 5.00\hat{j}\ \text{m}$$
$$t = 2.00\ \text{s}, \qquad \vec{r}_2 = -10.0\hat{i} + 5.00\hat{j}\ \text{m}$$

By drawing these results as in Fig. 11-14*b*, we see that $\vec{r}$ rotates counterclockwise in order to keep up with the particle. That is the positive direction of rotation. Thus, even though the particle is moving in a straight line, it is still moving counterclockwise around the origin and thus has a positive angular momentum.

We can also make sense of the direction of $\vec{\ell}$ by applying the right-hand rule for cross products (here $\vec{r} \times \vec{v}$, or, if you like, $m\vec{r} \times \vec{v}$, which gives the same direction). For any moment during the particle's motion, the fingers of the right hand are first extended in the direction of the first vector in the cross product ($\vec{r}$) as indicated in Fig. 11-14*c*. The orientation of the hand (on the page or viewing screen) is then adjusted so that the fingers can be comfortably rotated about the palm to be in the direction of the second vector in the cross product ($\vec{v}$) as indicated in Fig. 11-14*d*. The outstretched thumb then points in the direction of the result of the cross product. As indicated in Fig. 11-14*e*, the vector is in the positive direction of the z axis (which is directly out of the plane of the figure), consistent with our previous result. Figure 11-14*e* also indicates the direction of $\vec{\tau}$, which is also in the positive direction of the z axis because the angular momentum is in that direction and is increasing in magnitude.

11-7 ANGULAR MOMENTUM OF A RIGID BODY

Learning Objectives

After reading this module, you should be able to . . .

11.22 For a system of particles, apply Newton's second law in angular form to relate the net torque acting on the system to the rate of the resulting change in the system's angular momentum.

11.23 Apply the relationship between the angular momentum of a rigid body rotating around a fixed axis and the body's rotational inertia and angular speed around that axis.

11.24 If two rigid bodies rotate about the same axis, calculate their total angular momentum.

Key Ideas

- The angular momentum $\vec{L}$ of a system of particles is the vector sum of the angular momenta of the individual particles:

$$\vec{L} = \vec{\ell}_1 + \vec{\ell}_2 + \cdots + \vec{\ell}_n = \sum_{i=1}^{n} \vec{\ell}_i.$$

- The time rate of change of this angular momentum is equal to the net external torque on the system (the vector sum of the torques due to interactions of the particles of the system with particles external to the system):

$$\vec{\tau}_{\text{net}} = \frac{d\vec{L}}{dt} \quad \text{(system of particles).}$$

- For a rigid body rotating about a fixed axis, the component of its angular momentum parallel to the rotation axis is

$$L = I\omega \quad \text{(rigid body, fixed axis).}$$

The Angular Momentum of a System of Particles

Now we turn our attention to the angular momentum of a system of particles with respect to an origin. The total angular momentum $\vec{L}$ of the system is the (vector) sum of the angular momenta $\vec{\ell}$ of the individual particles (here with label i):

$$\vec{L} = \vec{\ell}_1 + \vec{\ell}_2 + \vec{\ell}_3 + \cdots + \vec{\ell}_n = \sum_{i=1}^{n} \vec{\ell}_i. \tag{11-26}$$

With time, the angular momenta of individual particles may change because of interactions between the particles or with the outside. We can find the resulting change in $\vec{L}$ by taking the time derivative of Eq. 11-26. Thus,

$$\frac{d\vec{L}}{dt} = \sum_{i=1}^{n} \frac{d\vec{\ell}_i}{dt}. \tag{11-27}$$

From Eq. 11-23, we see that $d\vec{\ell}_i/dt$ is equal to the net torque $\vec{\tau}_{\text{net},i}$ on the ith particle. We can rewrite Eq. 11-27 as

$$\frac{d\vec{L}}{dt} = \sum_{i=1}^{n} \vec{\tau}_{\text{net},i}. \tag{11-28}$$

That is, the rate of change of the system's angular momentum $\vec{L}$ is equal to the vector sum of the torques on its individual particles. Those torques include *internal torques* (due to forces between the particles) and *external torques* (due to forces on the particles from bodies external to the system). However, the forces between the particles always come in third-law force pairs so their torques sum to zero. Thus, the only torques that can change the total angular momentum $\vec{L}$ of the system are the external torques acting on the system.

Net External Torque. Let $\vec{\tau}_{\text{net}}$ represent the net external torque, the vector sum of all external torques on all particles in the system. Then we can write Eq. 11-28 as

$$\vec{\tau}_{\text{net}} = \frac{d\vec{L}}{dt} \quad \text{(system of particles),} \tag{11-29}$$

which is Newton's second law in angular form. It says:

> The net external torque $\vec{\tau}_{net}$ acting on a system of particles is equal to the time rate of change of the system's total angular momentum $\vec{L}$.

Equation 11-29 is analogous to $\vec{F}_{net} = d\vec{P}/dt$ (Eq. 9-27) but requires extra caution: Torques and the system's angular momentum must be measured relative to the same origin. If the center of mass of the system is not accelerating relative to an inertial frame, that origin can be any point. However, if it *is* accelerating, then it *must* be the origin. For example, consider a wheel as the system of particles. If it is rotating about an axis that is fixed relative to the ground, then the origin for applying Eq. 11-29 can be any point that is stationary relative to the ground. However, if it is rotating about an axis that is accelerating (such as when it rolls down a ramp), then the origin can be only at its center of mass.

The Angular Momentum of a Rigid Body Rotating About a Fixed Axis

We next evaluate the angular momentum of a system of particles that form a rigid body that rotates about a fixed axis. Figure 11-15*a* shows such a body. The fixed axis of rotation is a z axis, and the body rotates about it with constant angular speed ω. We wish to find the angular momentum of the body about that axis.

We can find the angular momentum by summing the z components of the angular momenta of the mass elements in the body. In Fig. 11-15*a*, a typical mass element, of mass Δm_i, moves around the z axis in a circular path. The position of the mass element is located relative to the origin O by position vector $\vec{r}_i$. The radius of the mass element's circular path is $r_{\perp i}$, the perpendicular distance between the element and the z axis.

The magnitude of the angular momentum $\vec{\ell}_i$ of this mass element, with respect to O, is given by Eq. 11-19:

$$\ell_i = (r_i)(p_i)(\sin 90°) = (r_i)(\Delta m_i\, v_i),$$

where p_i and v_i are the linear momentum and linear speed of the mass element, and 90° is the angle between $\vec{r}_i$ and $\vec{p}_i$. The angular momentum vector $\vec{\ell}_i$ for the mass element in Fig. 11-15*a* is shown in Fig. 11-15*b*; its direction must be perpendicular to those of $\vec{r}_i$ and $\vec{p}_i$.

The z Components. We are interested in the component of $\vec{\ell}_i$ that is parallel to the rotation axis, here the z axis. That z component is

$$\ell_{iz} = \ell_i \sin\theta = (r_i \sin\theta)(\Delta m_i\, v_i) = r_{\perp i}\, \Delta m_i\, v_i.$$

The z component of the angular momentum for the rotating rigid body as a whole is found by adding up the contributions of all the mass elements that make up the body. Thus, because $v = \omega r_\perp$, we may write

$$L_z = \sum_{i=1}^{n} \ell_{iz} = \sum_{i=1}^{n} \Delta m_i\, v_i r_{\perp i} = \sum_{i=1}^{n} \Delta m_i (\omega r_{\perp i}) r_{\perp i}$$
$$= \omega\left(\sum_{i=1}^{n} \Delta m_i\, r_{\perp i}^2\right). \quad (11\text{-}30)$$

We can remove ω from the summation here because it has the same value for all points of the rotating rigid body.

The quantity $\Sigma \Delta m_i\, r_{\perp i}^2$ in Eq. 11-30 is the rotational inertia I of the body about the fixed axis (see Eq. 10-33). Thus Eq. 11-30 reduces to

$$L = I\omega \quad \text{(rigid body, fixed axis)}. \quad (11\text{-}31)$$

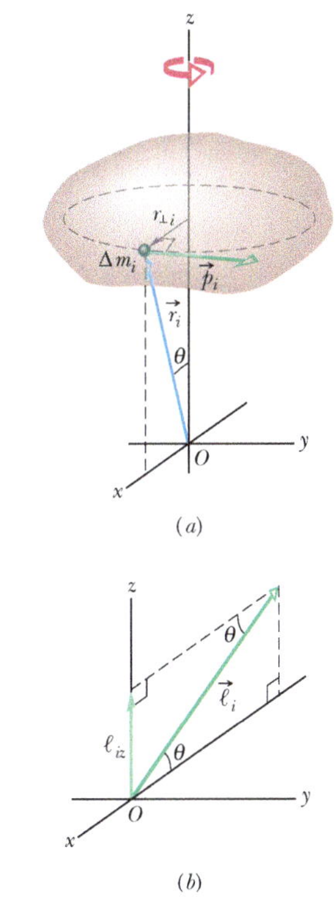

Figure 11-15 (*a*) A rigid body rotates about a z axis with angular speed ω. A mass element of mass Δm_i within the body moves about the z axis in a circle with radius $r_{\perp i}$. The mass element has linear momentum $\vec{p}_i$, and it is located relative to the origin O by position vector $\vec{r}_i$. Here the mass element is shown when $r_{\perp i}$ is parallel to the x axis. (*b*) The angular momentum $\vec{\ell}_i$, with respect to O, of the mass element in (*a*). The z component ℓ_{iz} is also shown.

Table 11-1 More Corresponding Variables and Relations for Translational and Rotational Motion[a]

Translational		Rotational	
Force	$\vec{F}$	Torque	$\vec{\tau}\ (= \vec{r} \times \vec{F})$
Linear momentum	$\vec{p}$	Angular momentum	$\vec{\ell}\ (= \vec{r} \times \vec{p})$
Linear momentum[b]	$\vec{P}\ (= \Sigma \vec{p}_i)$	Angular momentum[b]	$\vec{L}\ (= \Sigma \vec{\ell}_i)$
Linear momentum[b]	$\vec{P} = M\vec{v}_{com}$	Angular momentum[c]	$L = I\omega$
Newton's second law[b]	$\vec{F}_{net} = \dfrac{d\vec{P}}{dt}$	Newton's second law[b]	$\vec{\tau}_{net} = \dfrac{d\vec{L}}{dt}$
Conservation law[d]	$\vec{P}$ = a constant	Conservation law[d]	$\vec{L}$ = a constant

[a]See also Table 10-3.
[b]For systems of particles, including rigid bodies.
[c]For a rigid body about a fixed axis, with L being the component along that axis.
[d]For a closed, isolated system.

We have dropped the subscript z, but you must remember that the angular momentum defined by Eq. 11-31 is the angular momentum about the rotation axis. Also, I in that equation is the rotational inertia about that same axis.

Table 11-1, which supplements Table 10-3, extends our list of corresponding linear and angular relations.

Checkpoint 6

In the figure, a disk, a hoop, and a solid sphere are made to spin about fixed central axes (like a top) by means of strings wrapped around them, with the strings producing the same constant tangential force $\vec{F}$ on all three objects. The three objects have the same mass and radius, and they are initially stationary. Rank the objects according to (a) their angular momentum about their central axes and (b) their angular speed, greatest first, when the strings have been pulled for a certain time t.

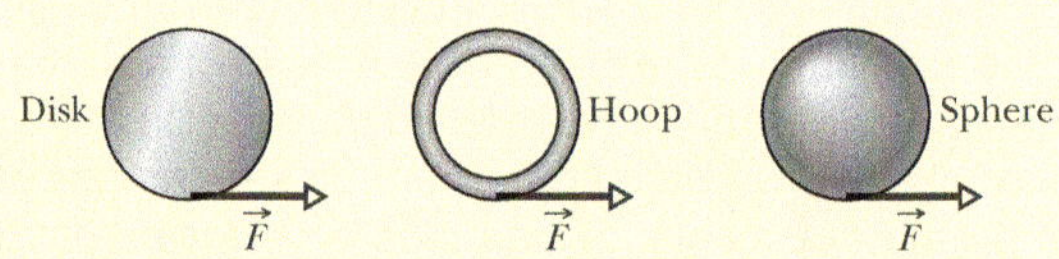

11-8 CONSERVATION OF ANGULAR MOMENTUM

Learning Objective

After reading this module, you should be able to . . .

11.25 When no external net torque acts on a system along a specified axis, apply the conservation of angular momentum to relate the initial angular momentum value along *that axis* to the value at a later instant.

Key Idea

- The angular momentum $\vec{L}$ of a system remains constant if the net external torque acting on the system is zero:

$$\vec{L} = \text{a constant} \quad \text{(isolated system)}$$

or

$$\vec{L}_i = \vec{L}_f \quad \text{(isolated system)}.$$

This is the law of conservation of angular momentum.

Conservation of Angular Momentum

So far we have discussed two powerful conservation laws, the conservation of energy and the conservation of linear momentum. Now we meet a third law of this type, involving the conservation of angular momentum. We start from

Eq. 11-29 ($\vec{\tau}_{net} = d\vec{L}/dt$), which is Newton's second law in angular form. If no net external torque acts on the system, this equation becomes $d\vec{L}/dt = 0$, or

$$\vec{L} = \text{a constant} \quad \text{(isolated system)}. \tag{11-32}$$

This result, called the **law of conservation of angular momentum,** can also be written as

$$\begin{pmatrix}\text{net angular momentum}\\ \text{at some initial time } t_i\end{pmatrix} = \begin{pmatrix}\text{net angular momentum}\\ \text{at some later time } t_f\end{pmatrix},$$

or

$$\vec{L}_i = \vec{L}_f \quad \text{(isolated system)}. \tag{11-33}$$

Equations 11-32 and 11-33 tell us:

If the net external torque acting on a system is zero, the angular momentum $\vec{L}$ of the system remains constant, no matter what changes take place within the system.

Equations 11-32 and 11-33 are vector equations; as such, they are equivalent to three component equations corresponding to the conservation of angular momentum in three mutually perpendicular directions. Depending on the torques acting on a system, the angular momentum of the system might be conserved in only one or two directions but not in all directions:

If the component of the net *external* torque on a system along a certain axis is zero, then the component of the angular momentum of the system along that axis cannot change, no matter what changes take place within the system.

This is a powerful statement: In this situation we are concerned with only the initial and final states of the system; we do not need to consider any intermediate state.

We can apply this law to the isolated body in Fig. 11-15, which rotates around the z axis. Suppose that the initially rigid body somehow redistributes its mass relative to that rotation axis, changing its rotational inertia about that axis. Equations 11-32 and 11-33 state that the angular momentum of the body cannot change. Substituting Eq. 11-31 (for the angular momentum along the rotational axis) into Eq. 11-33, we write this conservation law as

$$I_i\omega_i = I_f\omega_f. \tag{11-34}$$

Here the subscripts refer to the values of the rotational inertia I and angular speed ω before and after the redistribution of mass.

Like the other two conservation laws that we have discussed, Eqs. 11-32 and 11-33 hold beyond the limitations of Newtonian mechanics. They hold for particles whose speeds approach that of light (where the theory of special relativity reigns), and they remain true in the world of subatomic particles (where quantum physics reigns). No exceptions to the law of conservation of angular momentum have ever been found.

We now discuss four examples involving this law.

1. ***The spinning volunteer*** Figure 11-16 shows a student seated on a stool that can rotate freely about a vertical axis. The student, who has been set into rotation at a modest initial angular speed ω_i, holds two dumbbells in his outstretched hands. His angular momentum vector $\vec{L}$ lies along the vertical rotation axis, pointing upward.

   The instructor now asks the student to pull in his arms; this action reduces his rotational inertia from its initial value I_i to a smaller value I_f because he moves mass closer to the rotation axis. His rate of rotation increases markedly,

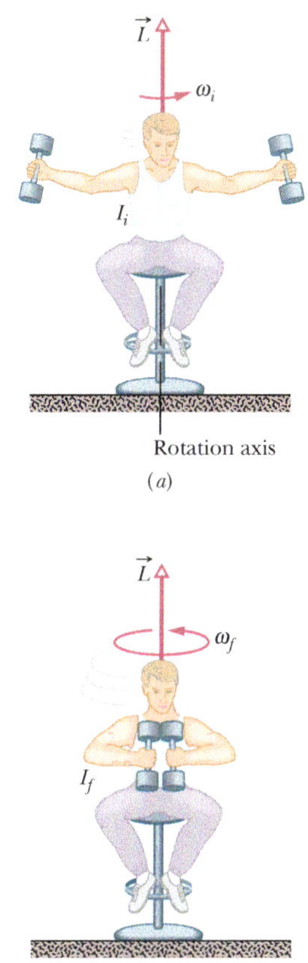

Figure 11-16 (*a*) The student has a relatively large rotational inertia about the rotation axis and a relatively small angular speed. (*b*) By decreasing his rotational inertia, the student automatically increases his angular speed. The angular momentum $\vec{L}$ of the rotating system remains unchanged.

from ω_i to ω_f. The student can then slow down by extending his arms once more, moving the dumbbells outward.

No net external torque acts on the system consisting of the student, stool, and dumbbells. Thus, the angular momentum of that system about the rotation axis must remain constant, no matter how the student maneuvers the dumbbells. In Fig. 11-16*a*, the student's angular speed ω_i is relatively low and his rotational inertia I_i is relatively high. According to Eq. 11-34, his angular speed in Fig. 11-16*b* must be greater to compensate for the decreased I_f.

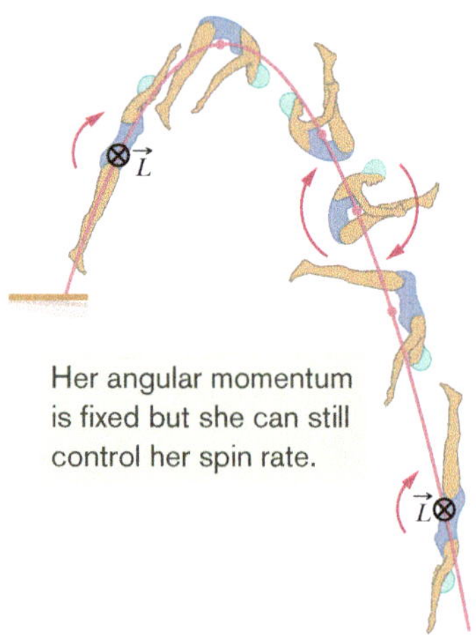

Figure 11-17 The diver's angular momentum $\vec{L}$ is constant throughout the dive, being represented by the tail ⊗ of an arrow that is perpendicular to the plane of the figure. Note also that her center of mass (see the dots) follows a parabolic path.

2. ***The springboard diver*** Figure 11-17 shows a diver doing a forward one-and-a-half-somersault dive. As you should expect, her center of mass follows a parabolic path. She leaves the springboard with a definite angular momentum $\vec{L}$ about an axis through her center of mass, represented by a vector pointing into the plane of Fig. 11-17, perpendicular to the page. When she is in the air, no net external torque acts on her about her center of mass, so her angular momentum about her center of mass cannot change. By pulling her arms and legs into the closed *tuck position,* she can considerably reduce her rotational inertia about the same axis and thus, according to Eq. 11-34, considerably increase her angular speed. Pulling out of the tuck position (into the *open layout position*) at the end of the dive increases her rotational inertia and thus slows her rotation rate so she can enter the water with little splash. Even in a more complicated dive involving both twisting and somersaulting, the angular momentum of the diver must be conserved, in both magnitude *and* direction, throughout the dive.

3. ***Long jump*** When an athlete takes off from the ground in a running long jump, the forces on the launching foot give the athlete an angular momentum with a forward rotation around a horizontal axis. Such rotation would not allow the jumper to land properly: In the landing, the legs should be together and extended forward at an angle so that the heels mark the sand at the greatest distance. Once airborne, the angular momentum cannot change (it is conserved) because no external torque acts to change it. However, the jumper can shift most of the angular momentum to the arms by rotating them in windmill fashion (Fig. 11-18). Then the body remains upright and in the proper orientation for landing.

Figure 11-18 Windmill motion of the arms during a long jump helps maintain body orientation for a proper landing.

4. ***Tour jeté*** In a tour jeté, a ballet performer leaps with a small twisting motion on the floor with one foot while holding the other leg perpendicular to the body (Fig. 11-19*a*). The angular speed is so small that it may not be perceptible

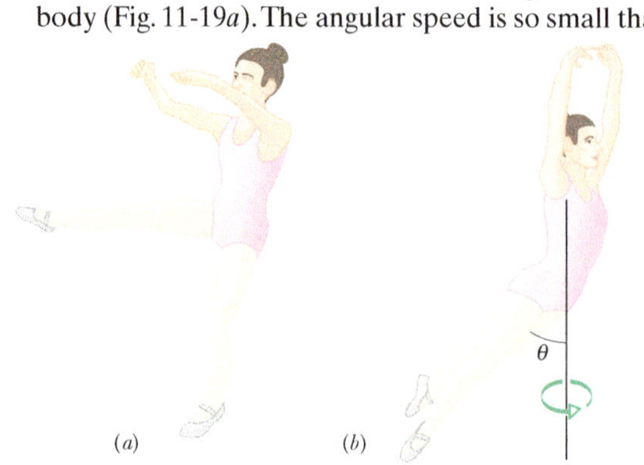

Figure 11-19 (*a*) Initial phase of a tour jeté: large rotational inertia and small angular speed. (*b*) Later phase: smaller rotational inertia and larger angular speed.

to the audience. As the performer ascends, the outstretched leg is brought down and the other leg is brought up, with both ending up at angle θ to the body (Fig. 11-19*b*). The motion is graceful, but it also serves to increase the rotation because bringing in the initially outstretched leg decreases the performer's rotational inertia. Since no external torque acts on the airborne performer, the angular momentum cannot change. Thus, with a decrease in rotational inertia, the angular speed must increase. When the jump is well executed, the performer seems to suddenly begin to spin and rotates 180° before the initial leg orientations are reversed in preparation for the landing. Once a leg is again outstretched, the rotation seems to vanish.

Checkpoint 7

A rhinoceros beetle rides the rim of a small disk that rotates like a merry-go-round. If the beetle crawls toward the center of the disk, do the following (each relative to the central axis) increase, decrease, or remain the same for the beetle–disk system: (a) rotational inertia, (b) angular momentum, and (c) angular speed?

Sample Problem 11.05 Conservation of angular momentum, rotating wheel demo

Figure 11-20*a* shows a student, again sitting on a stool that can rotate freely about a vertical axis. The student, initially at rest, is holding a bicycle wheel whose rim is loaded with lead and whose rotational inertia I_{wh} about its central axis is 1.2 kg · m^2. (The rim contains lead in order to make the value of I_{wh} substantial.)

The wheel is rotating at an angular speed ω_{wh} of 3.9 rev/s; as seen from overhead, the rotation is counterclockwise. The axis of the wheel is vertical, and the angular momentum $\vec{L}_{wh}$ of the wheel points vertically upward.

The student now inverts the wheel (Fig. 11-20*b*) so that, as seen from overhead, it is rotating clockwise. Its angular momentum is now $-\vec{L}_{wh}$. The inversion results in the student, the stool, and the wheel's center rotating together as a composite rigid body about the stool's rotation axis, with rotational inertia I_b = 6.8 kg · m^2. (The fact that the wheel is also rotating about its center does not affect the mass distribution of this composite body; thus, I_b has the same value whether or not the wheel rotates.) With what angular speed ω_b and in what direction does the composite body rotate after the inversion of the wheel?

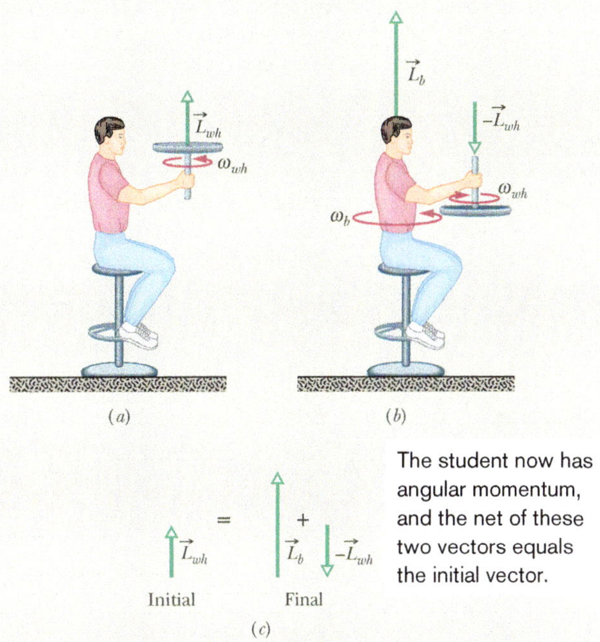

Figure 11-20 (*a*) A student holds a bicycle wheel rotating around a vertical axis. (*b*) The student inverts the wheel, setting himself into rotation. (*c*) The net angular momentum of the system must remain the same in spite of the inversion.

KEY IDEAS

1. The angular speed ω_b we seek is related to the final angular momentum $\vec{L}_b$ of the composite body about the stool's rotation axis by Eq. 11-31 ($L = I\omega$).
2. The initial angular speed ω_{wh} of the wheel is related to the angular momentum $\vec{L}_{wh}$ of the wheel's rotation about its center by the same equation.
3. The vector addition of $\vec{L}_b$ and $\vec{L}_{wh}$ gives the total angular momentum $\vec{L}_{tot}$ of the system of the student, stool, and wheel.
4. As the wheel is inverted, no net *external* torque acts on that system to change $\vec{L}_{tot}$ about any vertical axis. (Torques due to forces between the student and the wheel as the student inverts the wheel are *internal* to the system.) So, the system's total angular momentum is conserved about any vertical axis, including the rotation axis through the stool.

Calculations: The conservation of $\vec{L}_{tot}$ is represented with vectors in Fig. 11-20*c*. We can also write this conservation in terms of components along a vertical axis as

$$L_{b,f} + L_{wh,f} = L_{b,i} + L_{wh,i}, \tag{11-35}$$

where *i* and *f* refer to the initial state (before inversion of the wheel) and the final state (after inversion). Because inversion of the wheel inverted the angular momentum vector of the wheel's rotation, we substitute $-L_{wh,i}$ for $L_{wh,f}$. Then, if we set $L_{b,i} = 0$ (because the student, the stool, and the wheel's center were initially at rest), Eq. 11-35 yields

$$L_{b,f} = 2L_{wh,i}.$$

Using Eq. 11-31, we next substitute $I_b\omega_b$ for $L_{b,f}$ and $I_{wh}\omega_{wh}$ for $L_{wh,i}$ and solve for ω_b, finding

$$\omega_b = \frac{2I_{wh}}{I_b}\omega_{wh}$$

$$= \frac{(2)(1.2\ \text{kg}\cdot\text{m}^2)(3.9\ \text{rev/s})}{6.8\ \text{kg}\cdot\text{m}^2} = 1.4\ \text{rev/s}. \qquad \text{(Answer)}$$

This positive result tells us that the student rotates counterclockwise about the stool axis as seen from overhead. If the student wishes to stop rotating, he has only to invert the wheel once more.

Sample Problem 11.06 Conservation of angular momentum, cockroach on disk

In Fig. 11-21, a cockroach with mass m rides on a disk of mass $6.00m$ and radius R. The disk rotates like a merry-go-round around its central axis at angular speed $\omega_i = 1.50$ rad/s. The cockroach is initially at radius $r = 0.800R$, but then it crawls out to the rim of the disk. Treat the cockroach as a particle. What then is the angular speed?

KEY IDEAS

(1) The cockroach's crawl changes the mass distribution (and thus the rotational inertia) of the cockroach–disk system. (2) The angular momentum of the system does not change because there is no external torque to change it. (The forces and torques due to the cockroach's crawl are internal to the system.) (3) The magnitude of the angular momentum of a rigid body or a particle is given by Eq. 11-31 ($L = I\omega$).

Calculations: We want to find the final angular speed. Our key is to equate the final angular momentum L_f to the initial angular momentum L_i, because both involve angular speed. They also involve rotational inertia I. So, let's start by finding the rotational inertia of the system of cockroach and disk before and after the crawl.

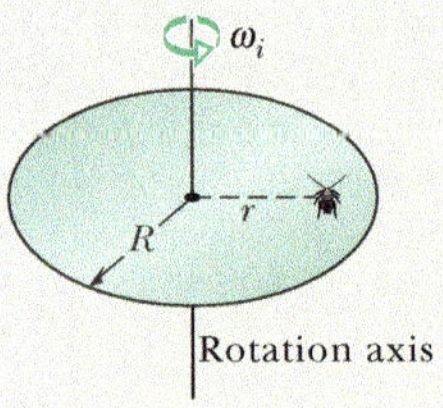

Figure 11-21 A cockroach rides at radius r on a disk rotating like a merry-go-round.

The rotational inertia of a disk rotating about its central axis is given by Table 10-2*c* as $\frac{1}{2}MR^2$. Substituting $6.00m$ for the mass M, our disk here has rotational inertia

$$I_d = 3.00mR^2. \tag{11-36}$$

(We don't have values for m and R, but we shall continue with physics courage.)

From Eq. 10-33, we know that the rotational inertia of the cockroach (a particle) is equal to mr^2. Substituting the cockroach's initial radius ($r = 0.800R$) and final radius ($r = R$), we find that its initial rotational inertia about the rotation axis is

$$I_{ci} = 0.64mR^2 \tag{11-37}$$

and its final rotational inertia about the rotation axis is

$$I_{cf} = mR^2. \tag{11-38}$$

So, the cockroach–disk system initially has the rotational inertia

$$I_i = I_d + I_{ci} = 3.64mR^2, \tag{11-39}$$

and finally has the rotational inertia

$$I_f = I_d + I_{cf} = 4.00mR^2. \tag{11-40}$$

Next, we use Eq. 11-31 ($L = I\omega$) to write the fact that the system's final angular momentum L_f is equal to the system's initial angular momentum L_i:

$$I_f\omega_f = I_i\omega_i$$

or

$$4.00mR^2\omega_f = 3.64mR^2(1.50\ \text{rad/s}).$$

After canceling the unknowns m and R, we come to

$$\omega_f = 1.37\ \text{rad/s}. \qquad \text{(Answer)}$$

Note that ω decreased because part of the mass moved outward, thus increasing that system's rotational inertia.

Additional examples, video, and practice available at *WileyPLUS*

11-9 PRECESSION OF A GYROSCOPE

Learning Objectives

After reading this module, you should be able to . . .

11.26 Identify that the gravitational force acting on a spinning gyroscope causes the spin angular momentum vector (and thus the gyroscope) to rotate about the vertical axis in a motion called precession.

11.27 Calculate the precession rate of a gyroscope.

11.28 Identify that a gyroscope's precession rate is independent of the gyroscope's mass.

Key Idea

- A spinning gyroscope can precess about a vertical axis through its support at the rate

$$\Omega = \frac{Mgr}{I\omega},$$

where M is the gyroscope's mass, r is the moment arm, I is the rotational inertia, and ω is the spin rate.

Precession of a Gyroscope

A simple gyroscope consists of a wheel fixed to a shaft and free to spin about the axis of the shaft. If one end of the shaft of a *nonspinning* gyroscope is placed on a support as in Fig. 11-22*a* and the gyroscope is released, the gyroscope falls by rotating downward about the tip of the support. Since the fall involves rotation, it is governed by Newton's second law in angular form, which is given by Eq. 11-29:

$$\vec{\tau} = \frac{d\vec{L}}{dt}. \qquad (11\text{-}41)$$

This equation tells us that the torque causing the downward rotation (the fall) changes the angular momentum $\vec{L}$ of the gyroscope from its initial value of zero. The torque $\vec{\tau}$ is due to the gravitational force $M\vec{g}$ acting at the gyroscope's center of mass, which we take to be at the center of the wheel. The moment arm relative to the support tip, located at O in Fig. 11-22*a*, is $\vec{r}$. The magnitude of $\vec{\tau}$ is

$$\tau = Mgr \sin 90° = Mgr \qquad (11\text{-}42)$$

(because the angle between $M\vec{g}$ and $\vec{r}$ is 90°), and its direction is as shown in Fig. 11-22*a*.

A rapidly spinning gyroscope behaves differently. Assume it is released with the shaft angled slightly upward. It first rotates slightly downward but then, while it is still spinning about its shaft, it begins to rotate horizontally about a vertical axis through support point O in a motion called **precession.**

Why Not Just Fall Over? Why does the spinning gyroscope stay aloft instead of falling over like the nonspinning gyroscope? The clue is that when the spinning gyroscope is released, the torque due to $M\vec{g}$ must change not an initial angular momentum of zero but rather some already existing nonzero angular momentum due to the spin.

To see how this nonzero initial angular momentum leads to precession, we first consider the angular momentum $\vec{L}$ of the gyroscope due to its spin. To simplify the situation, we assume the spin rate is so rapid that the angular momentum due to precession is negligible relative to $\vec{L}$. We also assume the shaft is horizontal when precession begins, as in Fig. 11-22*b*. The magnitude of $\vec{L}$ is given by Eq. 11-31:

$$L = I\omega, \qquad (11\text{-}43)$$

where I is the rotational moment of the gyroscope about its shaft and ω is the angular speed at which the wheel spins about the shaft. The vector $\vec{L}$ points along the shaft, as in Fig. 11-22*b*. Since $\vec{L}$ is parallel to $\vec{r}$, torque $\vec{\tau}$ must be perpendicular to $\vec{L}$.

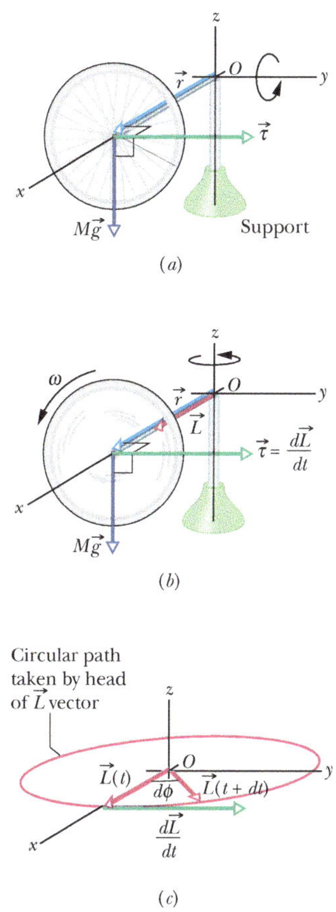

Figure 11-22 (*a*) A nonspinning gyroscope falls by rotating in an *xz* plane because of torque $\vec{\tau}$. (*b*) A rapidly spinning gyroscope, with angular momentum $\vec{L}$, precesses around the *z* axis. Its precessional motion is in the *xy* plane. (*c*) The change $d\vec{L}/dt$ in angular momentum leads to a rotation of $\vec{L}$ about O.

According to Eq. 11-41, torque $\vec{\tau}$ causes an incremental change $d\vec{L}$ in the angular momentum of the gyroscope in an incremental time interval dt; that is,

$$d\vec{L} = \vec{\tau}\,dt. \tag{11-44}$$

However, for a *rapidly spinning* gyroscope, the magnitude of $\vec{L}$ is fixed by Eq. 11-43. Thus the torque can change only the direction of $\vec{L}$, not its magnitude.

From Eq. 11-44 we see that the direction of $d\vec{L}$ is in the direction of $\vec{\tau}$, perpendicular to $\vec{L}$. The only way that $\vec{L}$ can be changed in the direction of $\vec{\tau}$ without the magnitude L being changed is for $\vec{L}$ to rotate around the z axis as shown in Fig. 11-22c. $\vec{L}$ maintains its magnitude, the head of the $\vec{L}$ vector follows a circular path, and $\vec{\tau}$ is always tangent to that path. Since $\vec{L}$ must always point along the shaft, the shaft must rotate about the z axis in the direction of $\vec{\tau}$. Thus we have precession. Because the spinning gyroscope must obey Newton's law in angular form in response to any change in its initial angular momentum, it must precess instead of merely toppling over.

Precession. We can find the **precession rate** Ω by first using Eqs. 11-44 and 11-42 to get the magnitude of $d\vec{L}$:

$$dL = \tau\,dt = Mgr\,dt. \tag{11-45}$$

As $\vec{L}$ changes by an incremental amount in an incremental time interval dt, the shaft and $\vec{L}$ precess around the z axis through incremental angle $d\phi$. (In Fig. 11-22c, angle $d\phi$ is exaggerated for clarity.) With the aid of Eqs. 11-43 and 11-45, we find that $d\phi$ is given by

$$d\phi = \frac{dL}{L} = \frac{Mgr\,dt}{I\omega}.$$

Dividing this expression by dt and setting the rate $\Omega = d\phi/dt$, we obtain

$$\Omega = \frac{Mgr}{I\omega} \quad \text{(precession rate).} \tag{11-46}$$

This result is valid under the assumption that the spin rate ω is rapid. Note that Ω decreases as ω is increased. Note also that there would be no precession if the gravitational force $M\vec{g}$ did not act on the gyroscope, but because I is a function of M, mass cancels from Eq. 11-46; thus Ω is independent of the mass.

Equation 11-46 also applies if the shaft of a spinning gyroscope is at an angle to the horizontal. It holds as well for a spinning top, which is essentially a spinning gyroscope at an angle to the horizontal.

Review & Summary

Rolling Bodies For a wheel of radius R rolling smoothly,

$$v_{com} = \omega R, \tag{11-2}$$

where v_{com} is the linear speed of the wheel's center of mass and ω is the angular speed of the wheel about its center. The wheel may also be viewed as rotating instantaneously about the point P of the "road" that is in contact with the wheel. The angular speed of the wheel about this point is the same as the angular speed of the wheel about its center. The rolling wheel has kinetic energy

$$K = \tfrac{1}{2}I_{com}\omega^2 + \tfrac{1}{2}Mv_{com}^2, \tag{11-5}$$

where I_{com} is the rotational inertia of the wheel about its center of mass and M is the mass of the wheel. If the wheel is being accelerated but is still rolling smoothly, the acceleration of the center of mass $\vec{a}_{com}$ is related to the angular acceleration α about the center with

$$a_{com} = \alpha R. \tag{11-6}$$

If the wheel rolls smoothly down a ramp of angle θ, its acceleration along an x axis extending up the ramp is

$$a_{com,x} = -\frac{g\sin\theta}{1 + I_{com}/MR^2}. \tag{11-10}$$

Torque as a Vector In three dimensions, *torque* $\vec{\tau}$ is a vector quantity defined relative to a fixed point (usually an origin); it is

$$\vec{\tau} = \vec{r} \times \vec{F}, \tag{11-14}$$

where $\vec{F}$ is a force applied to a particle and $\vec{r}$ is a position vector locating the particle relative to the fixed point. The magnitude of $\vec{\tau}$ is

$$\tau = rF\sin\phi = rF_\perp = r_\perp F, \tag{11-15, 11-16, 11-17}$$

where ϕ is the angle between $\vec{F}$ and $\vec{r}$, $F_\perp$ is the component of $\vec{F}$ perpendicular to $\vec{r}$, and $r_\perp$ is the moment arm of $\vec{F}$. The direction of $\vec{\tau}$ is given by the right-hand rule.

Angular Momentum of a Particle The *angular momentum* $\vec{\ell}$ of a particle with linear momentum $\vec{p}$, mass m, and linear velocity $\vec{v}$ is a vector quantity defined relative to a fixed point (usually an origin) as

$$\vec{\ell} = \vec{r} \times \vec{p} = m(\vec{r} \times \vec{v}). \quad (11\text{-}18)$$

The magnitude of $\vec{\ell}$ is given by

$$\ell = rmv \sin \phi \quad (11\text{-}19)$$
$$= rp_\perp = rmv_\perp \quad (11\text{-}20)$$
$$= r_\perp p = r_\perp mv, \quad (11\text{-}21)$$

where ϕ is the angle between $\vec{r}$ and $\vec{p}$, $p_\perp$ and $v_\perp$ are the components of $\vec{p}$ and $\vec{v}$ perpendicular to $\vec{r}$, and $r_\perp$ is the perpendicular distance between the fixed point and the extension of $\vec{p}$. The direction of $\vec{\ell}$ is given by the right-hand rule for cross products.

Newton's Second Law in Angular Form Newton's second law for a particle can be written in angular form as

$$\vec{\tau}_{net} = \frac{d\vec{\ell}}{dt}, \quad (11\text{-}23)$$

where $\vec{\tau}_{net}$ is the net torque acting on the particle and $\vec{\ell}$ is the angular momentum of the particle.

Angular Momentum of a System of Particles The angular momentum $\vec{L}$ of a system of particles is the vector sum of the angular momenta of the individual particles:

$$\vec{L} = \vec{\ell}_1 + \vec{\ell}_2 + \cdots + \vec{\ell}_n = \sum_{i=1}^{n} \vec{\ell}_i. \quad (11\text{-}26)$$

The time rate of change of this angular momentum is equal to the net external torque on the system (the vector sum of the torques due to interactions with particles external to the system):

$$\vec{\tau}_{net} = \frac{d\vec{L}}{dt} \quad \text{(system of particles)}. \quad (11\text{-}29)$$

Angular Momentum of a Rigid Body For a rigid body rotating about a fixed axis, the component of its angular momentum parallel to the rotation axis is

$$L = I\omega \quad \text{(rigid body, fixed axis)}. \quad (11\text{-}31)$$

Conservation of Angular Momentum The angular momentum $\vec{L}$ of a system remains constant if the net external torque acting on the system is zero:

$$\vec{L} = \text{a constant} \quad \text{(isolated system)} \quad (11\text{-}32)$$

or

$$\vec{L}_i = \vec{L}_f \quad \text{(isolated system)}. \quad (11\text{-}33)$$

This is the **law of conservation of angular momentum.**

Precession of a Gyroscope A spinning gyroscope can precess about a vertical axis through its support at the rate

$$\Omega = \frac{Mgr}{I\omega}, \quad (11\text{-}46)$$

where M is the gyroscope's mass, r is the moment arm, I is the rotational inertia, and ω is the spin rate.

Questions

1 Figure 11-23 shows three particles of the same mass and the same constant speed moving as indicated by the velocity vectors. Points a, b, c, and d form a square, with point e at the center. Rank the points according to the magnitude of the net angular momentum of the three-particle system when measured about the points, greatest first.

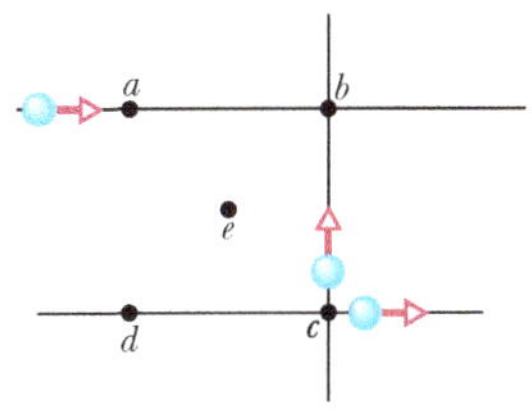

Figure 11-23 Question 1.

2 Figure 11-24 shows two particles A and B at xyz coordinates (1 m, 1 m, 0) and (1 m, 0, 1 m). Acting on each particle are three numbered forces, all of the same magnitude and each directed parallel to an axis. (a) Which of the forces produce a torque about the origin that is directed parallel to y? (b) Rank the forces according to the magnitudes of the torques they produce on the particles about the origin, greatest first.

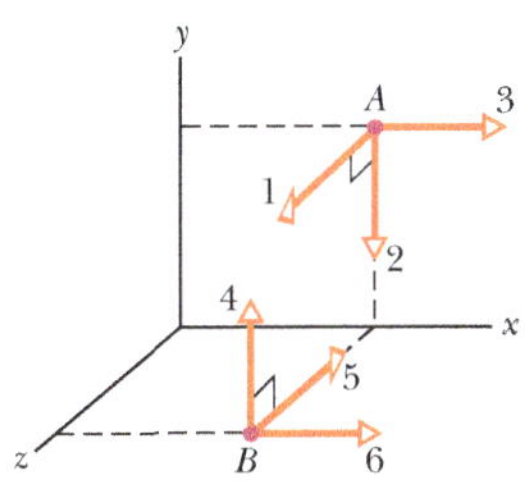

Figure 11-24 Question 2.

3 What happens to the initially stationary yo-yo in Fig. 11-25 if you pull it via its string with (a) force $\vec{F}_2$ (the line of action passes through the point of contact on the table, as indicated), (b) force $\vec{F}_1$ (the line of action passes above the point of contact), and (c) force $\vec{F}_3$ (the line of action passes to the right of the point of contact)?

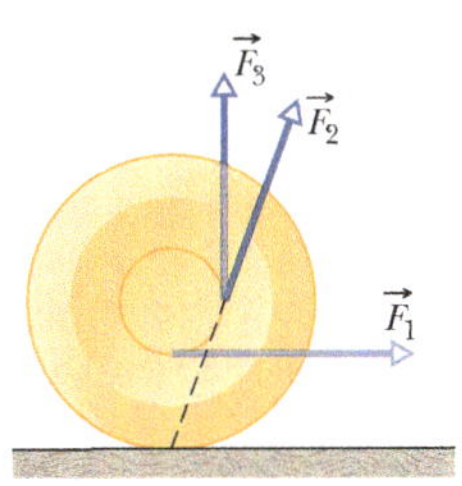

Figure 11-25 Question 3.

4 The position vector $\vec{r}$ of a particle relative to a certain point has a magnitude of 3 m, and the force $\vec{F}$ on the particle has a magnitude of 4 N. What is the angle between the directions of $\vec{r}$ and $\vec{F}$ if the magnitude of the associated torque equals (a) zero and (b) 12 N·m?

5 In Fig. 11-26, three forces of the same magnitude are applied to a particle at the origin ($\vec{F}_1$ acts directly into the plane of the figure). Rank the forces according to the magnitudes of the torques they create about (a) point P_1, (b) point P_2, and (c) point P_3, greatest first.

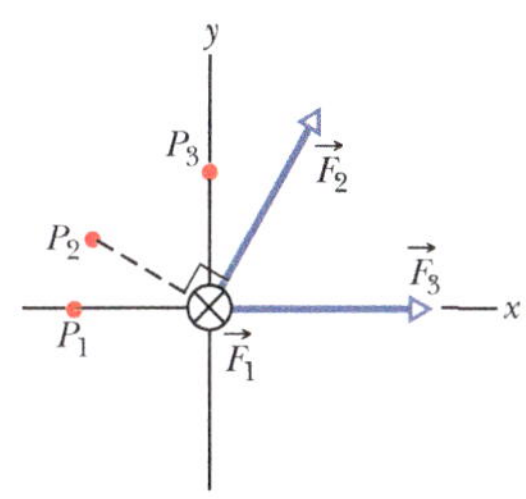

Figure 11-26 Question 5.

6 The angular momenta $\ell(t)$ of a particle in four situations are (1) $\ell = 3t + 4$; (2) $\ell = -6t^2$; (3) $\ell = 2$; (4) $\ell = 4/t$. In which situation is the net torque on the particle (a) zero, (b) positive and constant, (c) negative and increasing in magnitude ($t > 0$), and (d) negative and decreasing in magnitude ($t > 0$)?

7 A rhinoceros beetle rides the rim of a horizontal disk rotating counterclockwise like a merry-go-round. If the beetle then walks along the rim in the direction of the rotation, will the magnitudes of the following quantities (each measured about the rotation axis) increase, decrease, or remain the same (the disk is still rotating in the counterclockwise direction): (a) the angular momentum of the beetle–disk system, (b) the angular momentum and angular velocity of the beetle, and (c) the angular momentum and angular velocity of the disk? (d) What are your answers if the beetle walks in the direction opposite the rotation?

8 Figure 11-27 shows an overhead view of a rectangular slab that can spin like a merry-go-round about its center at O. Also shown are seven paths along which wads of bubble gum can be thrown (all with the same speed and mass) to stick onto the stationary slab. (a) Rank the paths according to the angular speed that the slab (and gum) will have after the gum sticks, greatest first. (b) For which paths will the angular momentum of the slab (and gum) about O be negative from the view of Fig. 11-27?

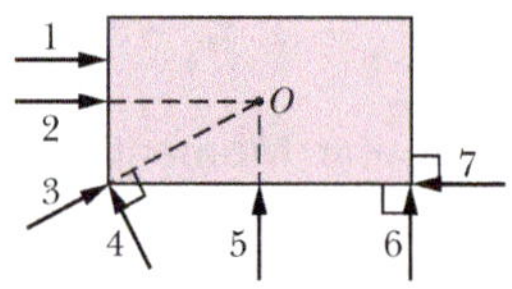

Figure 11-27 Question 8.

9 Figure 11-28 gives the angular momentum magnitude L of a wheel versus time t. Rank the four lettered time intervals according to the magnitude of the torque acting on the wheel, greatest first.

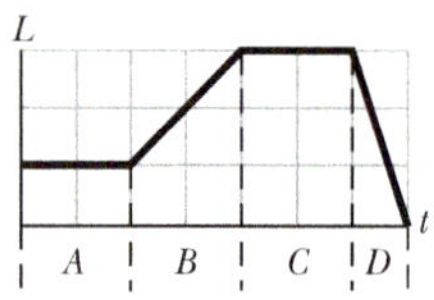

Figure 11-28 Question 9.

10 Figure 11-29 shows a particle moving at constant velocity $\vec{v}$ and five points with their xy coordinates. Rank the points according to the magnitude of the angular momentum of the particle measured about them, greatest first.

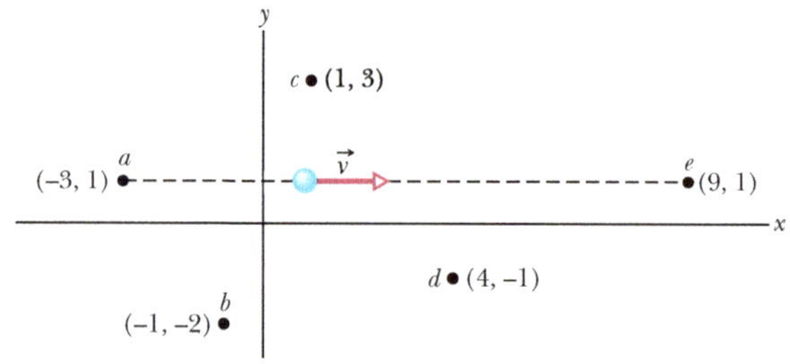

Figure 11-29 Question 10.

11 A cannonball and a marble roll smoothly from rest down an incline. Is the cannonball's (a) time to the bottom and (b) translational kinetic energy at the bottom more than, less than, or the same as the marble's?

12 A solid brass cylinder and a solid wood cylinder have the same radius and mass (the wood cylinder is longer). Released together from rest, they roll down an incline. (a) Which cylinder reaches the bottom first, or do they tie? (b) The wood cylinder is then shortened to match the length of the brass cylinder, and the brass cylinder is drilled out along its long (central) axis to match the mass of the wood cylinder. Which cylinder now wins the race, or do they tie?

Problems

GO Tutoring problem available (at instructor's discretion) in *WileyPLUS* and WebAssign
SSM Worked-out solution available in Student Solutions Manual
WWW Worked-out solution is at
• – ••• Number of dots indicates level of problem difficulty
ILW Interactive solution is at
http://www.wiley.com/college/halliday
Additional information available in *The Flying Circus of Physics* and at flyingcircusofphysics.com

Module 11-1 Rolling as Translation and Rotation Combined

•1 A car travels at 80 km/h on a level road in the positive direction of an x axis. Each tire has a diameter of 66 cm. Relative to a woman riding in the car and in unit-vector notation, what are the velocity $\vec{v}$ at the (a) center, (b) top, and (c) bottom of the tire and the magnitude a of the acceleration at the (d) center, (e) top, and (f) bottom of each tire? Relative to a hitchhiker sitting next to the road and in unit-vector notation, what are the velocity $\vec{v}$ at the (g) center, (h) top, and (i) bottom of the tire and the magnitude a of the acceleration at the (j) center, (k) top, and (l) bottom of each tire?

•2 An automobile traveling at 80.0 km/h has tires of 75.0 cm diameter. (a) What is the angular speed of the tires about their axles? (b) If the car is brought to a stop uniformly in 30.0 complete turns of the tires (without skidding), what is the magnitude of the angular acceleration of the wheels? (c) How far does the car move during the braking?

Module 11-2 Forces and Kinetic Energy of Rolling

•3 SSM A 140 kg hoop rolls along a horizontal floor so that the hoop's center of mass has a speed of 0.150 m/s. How much work must be done on the hoop to stop it?

•4 A uniform solid sphere rolls down an incline. (a) What must be the incline angle if the linear acceleration of the center of the sphere is to have a magnitude of $0.10g$? (b) If a frictionless block were to slide down the incline at that angle, would its acceleration magnitude be more than, less than, or equal to $0.10g$? Why?

•5 ILW A 1000 kg car has four 10 kg wheels. When the car is moving, what fraction of its total kinetic energy is due to rotation of the wheels about their axles? Assume that the wheels are uniform disks of the same mass and size. Why do you not need to know the radius of the wheels?

••6 Figure 11-30 gives the speed v versus time t for a 0.500 kg object of radius 6.00 cm that rolls smoothly down a 30° ramp. The scale on the velocity axis is set by $v_s = 4.0$ m/s. What is the rotational inertia of the object?

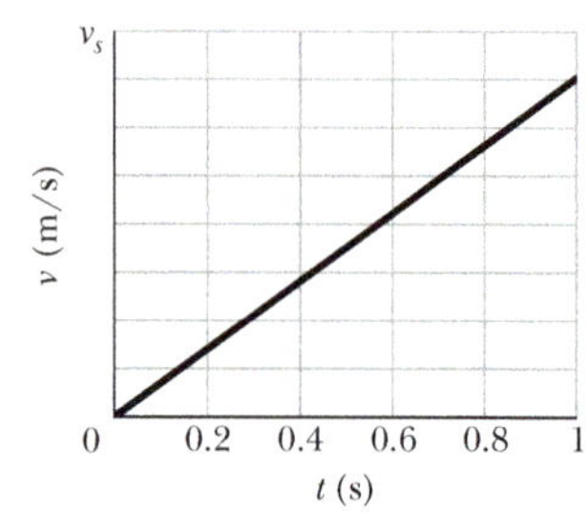

Figure 11-30 Problem 6.

••7 ILW In Fig. 11-31, a solid cylinder of radius 10 cm and mass 12 kg starts from rest and rolls without slipping a distance $L = 6.0$ m down a roof that is inclined at angle $\theta = 30°$. (a) What is the angular speed of the cylinder about its center as it leaves the roof? (b) The roof's edge is at height $H = 5.0$ m. How far horizontally from the roof's edge does the cylinder hit the level ground?

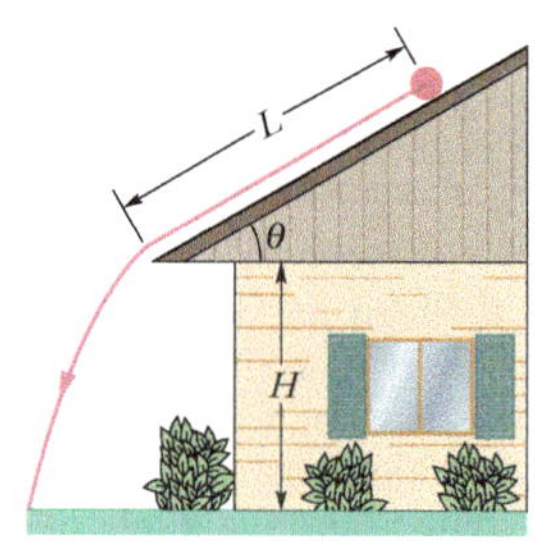

Figure 11-31 Problem 7.

••8 Figure 11-32 shows the potential energy $U(x)$ of a solid ball that can roll along an x axis. The scale on the U axis is set by $U_s = 100$ J. The ball is uniform, rolls smoothly, and has a mass of 0.400 kg. It is released at $x = 7.0$ m headed in the negative direction of the x axis with a mechanical energy of 75 J. (a) If the ball can reach $x = 0$ m, what is its speed there, and if it cannot, what is its turning point? Suppose, instead, it is headed in the positive direction of the x axis when it is released at $x = 7.0$ m with 75 J. (b) If the ball can reach $x = 13$ m, what is its speed there, and if it cannot, what is its turning point?

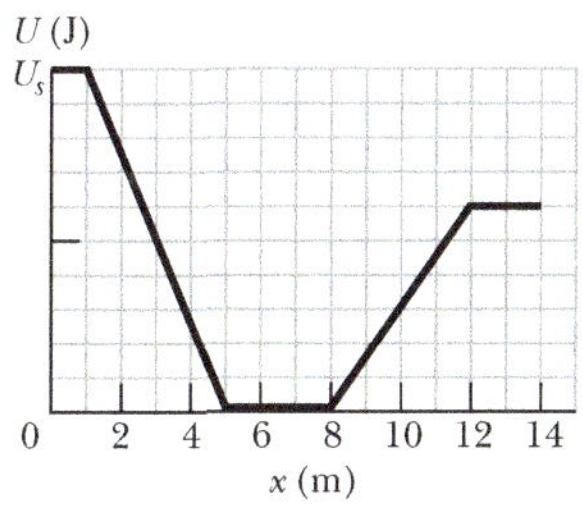

Figure 11-32 Problem 8.

••9 In Fig. 11-33, a solid ball rolls smoothly from rest (starting at height $H = 6.0$ m) until it leaves the horizontal section at the end of the track, at height $h = 2.0$ m. How far horizontally from point A does the ball hit the floor?

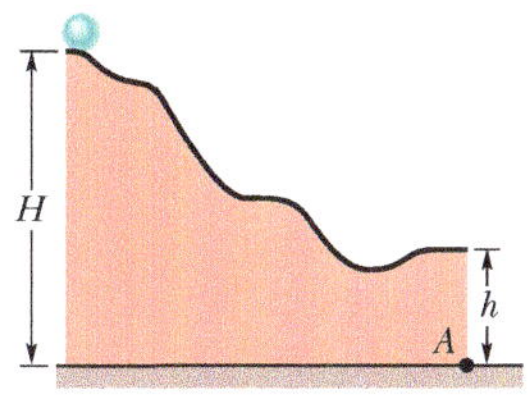

Figure 11-33 Problem 9.

••10 A hollow sphere of radius 0.15 m, with rotational inertia $I = 0.040\ \mathrm{kg \cdot m^2}$ about a line through its center of mass, rolls without slipping up a surface inclined at 30° to the horizontal. At a certain initial position, the sphere's total kinetic energy is 20 J. (a) How much of this initial kinetic energy is rotational? (b) What is the speed of the center of mass of the sphere at the initial position? When the sphere has moved 1.0 m up the incline from its initial position, what are (c) its total kinetic energy and (d) the speed of its center of mass?

••11 In Fig. 11-34, a constant horizontal force $\vec{F}_{app}$ of magnitude 10 N is applied to a wheel of mass 10 kg and radius 0.30 m. The wheel rolls smoothly on the horizontal surface, and the acceleration of its center of mass has magnitude 0.60 m/s². (a) In unit-vector notation, what is the frictional force on the wheel? (b) What is the rotational inertia of the wheel about the rotation axis through its center of mass?

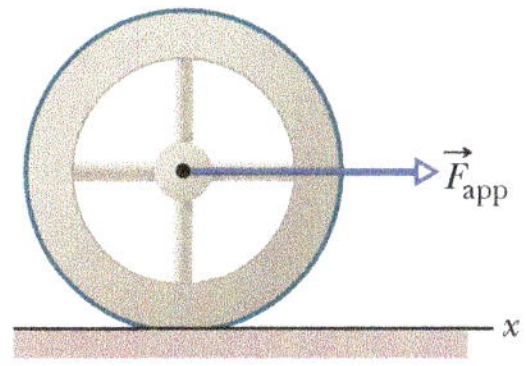

Figure 11-34 Problem 11.

••12 In Fig. 11-35, a solid brass ball of mass 0.280 g will roll smoothly along a loop-the-loop track when released from rest along the straight section. The circular loop has radius $R = 14.0$ cm, and the ball has radius $r \ll R$. (a) What is h if the ball is on the verge of leaving the track when it reaches the top of the loop? If the ball is released at height $h = 6.00R$, what are the (b) magnitude and (c) direction of the horizontal force component acting on the ball at point Q?

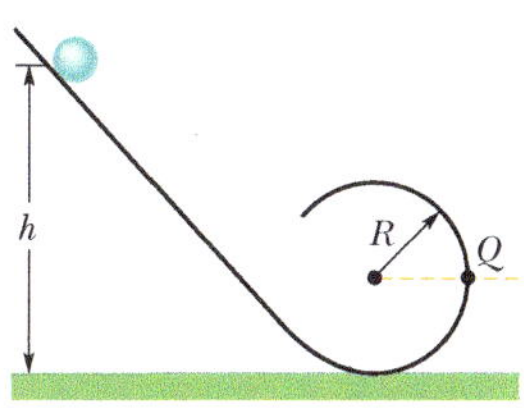

Figure 11-35 Problem 12.

•••13 *Nonuniform ball.* In Fig. 11-36, a ball of mass M and radius R rolls smoothly from rest down a ramp and onto a circular loop of radius 0.48 m. The initial height of the ball is $h = 0.36$ m. At the loop bottom, the magnitude of the normal force on the ball is $2.00Mg$. The ball consists of an outer spherical shell (of a certain uniform density) that is glued to a central sphere (of a different uniform density). The rotational inertia of the ball can be expressed in the general form $I = \beta MR^2$, but β is not 0.4 as it is for a ball of uniform density. Determine β.

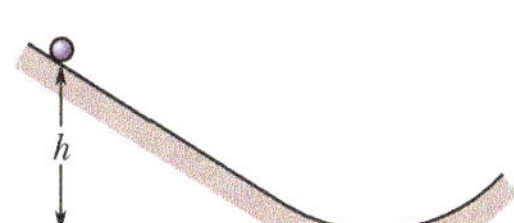

Figure 11-36 Problem 13.

•••14 In Fig. 11-37, a small, solid, uniform ball is to be shot from point P so that it rolls smoothly along a horizontal path, up along a ramp, and onto a plateau. Then it leaves the plateau horizontally to land on a game board, at a horizontal distance d from the right edge of the plateau. The vertical heights are $h_1 = 5.00$ cm and $h_2 = 1.60$ cm. With what speed must the ball be shot at point P for it to land at $d = 6.00$ cm?

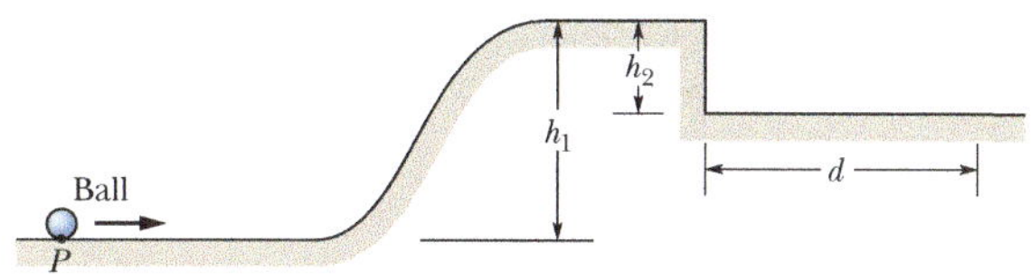

Figure 11-37 Problem 14.

•••15 A bowler throws a bowling ball of radius $R = 11$ cm along a lane. The ball (Fig. 11-38) slides on the lane with initial speed $v_{com,0} = 8.5$ m/s and initial angular speed $\omega_0 = 0$. The coefficient of kinetic friction between the ball and the lane is 0.21. The kinetic frictional force $\vec{f}_k$ acting on the ball causes a linear acceleration of the ball while producing a torque that causes an angular acceleration of the ball. When speed v_{com} has decreased enough and angular speed ω has increased enough, the ball stops sliding and then rolls smoothly. (a) What then is v_{com} in terms of ω? During the sliding, what are the ball's (b) linear acceleration and (c) angular acceleration? (d) How long does the ball slide? (e) How far does the ball slide? (f) What is the linear speed of the ball when smooth rolling begins?

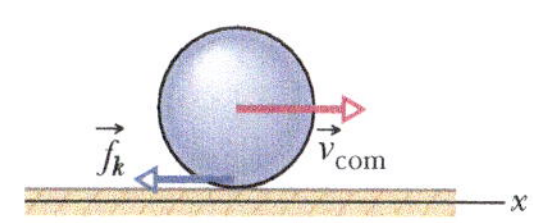

Figure 11-38 Problem 15.

•••16 *Nonuniform cylindrical object.* In Fig. 11-39, a cylindrical object of mass M and radius R rolls smoothly from rest down a ramp and onto a horizontal section. From there it rolls off the ramp and onto the floor, landing a horizontal distance $d = 0.506$ m from the end of the ramp. The initial height of the object is $H = 0.90$ m; the end of the ramp is at height $h = 0.10$ m. The object consists of an outer cylindrical shell (of a certain uniform density) that is glued to a central cylinder (of a different uniform density). The rotational inertia of the object can be expressed in the general form $I = \beta MR^2$, but β is not 0.5 as it is for a cylinder of uniform density. Determine β.

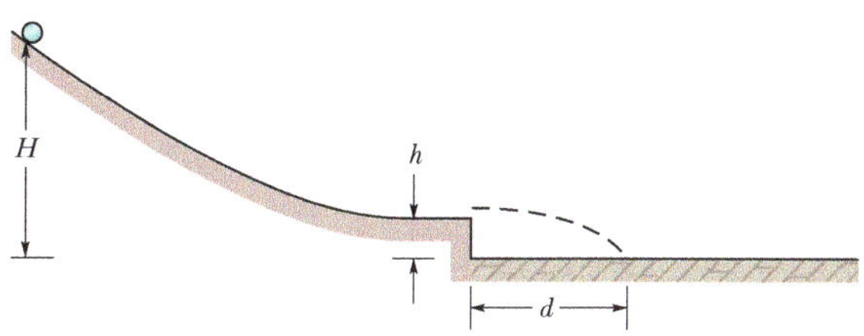

Figure 11-39 Problem 16.

Module 11-3 The Yo-Yo

•17 SSM A yo-yo has a rotational inertia of 950 g·cm^2 and a mass of 120 g. Its axle radius is 3.2 mm, and its string is 120 cm long. The yo-yo rolls from rest down to the end of the string. (a) What is the magnitude of its linear acceleration? (b) How long does it take to reach the end of the string? As it reaches the end of the string, what are its (c) linear speed, (d) translational kinetic energy, (e) rotational kinetic energy, and (f) angular speed?

•18 In 1980, over San Francisco Bay, a large yo-yo was released from a crane. The 116 kg yo-yo consisted of two uniform disks of radius 32 cm connected by an axle of radius 3.2 cm. What was the magnitude of the acceleration of the yo-yo during (a) its fall and (b) its rise? (c) What was the tension in the cord on which it rolled? (d) Was that tension near the cord's limit of 52 kN? Suppose you build a scaled-up version of the yo-yo (same shape and materials but larger). (e) Will the magnitude of your yo-yo's acceleration as it falls be greater than, less than, or the same as that of the San Francisco yo-yo? (f) How about the tension in the cord?

Module 11-4 Torque Revisited

•19 In unit-vector notation, what is the net torque about the origin on a flea located at coordinates (0, −4.0 m, 5.0 m) when forces $\vec{F}_1 = (3.0\text{ N})\hat{k}$ and $\vec{F}_2 = (-2.0\text{ N})\hat{j}$ act on the flea?

•20 A plum is located at coordinates (−2.0 m, 0, 4.0 m). In unit-vector notation, what is the torque about the origin on the plum if that torque is due to a force $\vec{F}$ whose only component is (a) $F_x = 6.0$ N, (b) $F_x = -6.0$ N, (c) $F_z = 6.0$ N, and (d) $F_z = -6.0$ N?

•21 In unit-vector notation, what is the torque about the origin on a particle located at coordinates (0, −4.0 m, 3.0 m) if that torque is due to (a) force $\vec{F}_1$ with components $F_{1x} = 2.0$ N, $F_{1y} = F_{1z} = 0$, and (b) force $\vec{F}_2$ with components $F_{2x} = 0$, $F_{2y} = 2.0$ N, $F_{2z} = 4.0$ N?

••22 A particle moves through an *xyz* coordinate system while a force acts on the particle. When the particle has the position vector $\vec{r} = (2.00\text{ m})\hat{i} - (3.00\text{ m})\hat{j} + (2.00\text{ m})\hat{k}$, the force is given by $\vec{F} = F_x\hat{i} + (7.00\text{ N})\hat{j} - (6.00\text{ N})\hat{k}$ and the corresponding torque about the origin is $\vec{\tau} = (4.00\text{ N}\cdot\text{m})\hat{i} + (2.00\text{ N}\cdot\text{m})\hat{j} - (1.00\text{ N}\cdot\text{m})\hat{k}$. Determine F_x.

••23 Force $\vec{F} = (2.0\text{ N})\hat{i} - (3.0\text{ N})\hat{k}$ acts on a pebble with position vector $\vec{r} = (0.50\text{ m})\hat{j} - (2.0\text{ m})\hat{k}$ relative to the origin. In unit-vector notation, what is the resulting torque on the pebble about (a) the origin and (b) the point (2.0 m, 0, −3.0 m)?

••24 In unit-vector notation, what is the torque about the origin on a jar of jalapeño peppers located at coordinates (3.0 m, −2.0 m, 4.0 m) due to (a) force $\vec{F}_1 = (3.0\text{ N})\hat{i} - (4.0\text{ N})\hat{j} + (5.0\text{ N})\hat{k}$, (b) force $\vec{F}_2 = (-3.0\text{ N})\hat{i} - (4.0\text{ N})\hat{j} - (5.0\text{ N})\hat{k}$, and (c) the vector sum of $\vec{F}_1$ and $\vec{F}_2$? (d) Repeat part (c) for the torque about the point with coordinates (3.0 m, 2.0 m, 4.0 m).

••25 SSM Force $\vec{F} = (-8.0\text{ N})\hat{i} + (6.0\text{ N})\hat{j}$ acts on a particle with position vector $\vec{r} = (3.0\text{ m})\hat{i} + (4.0\text{ m})\hat{j}$. What are (a) the torque on the particle about the origin, in unit-vector notation, and (b) the angle between the directions of $\vec{r}$ and $\vec{F}$?

Module 11-5 Angular Momentum

•26 At the instant of Fig. 11-40, a 2.0 kg particle *P* has a position vector $\vec{r}$ of magnitude 3.0 m and angle $\theta_1 = 45°$ and a velocity vector $\vec{v}$ of magnitude 4.0 m/s and angle $\theta_2 = 30°$. Force $\vec{F}$, of magnitude 2.0 N and angle $\theta_3 = 30°$, acts on *P*. All three vectors lie in the *xy* plane. About the origin, what are the (a) magnitude and (b) direction of the angular momentum of *P* and the (c) magnitude and (d) direction of the torque acting on *P*?

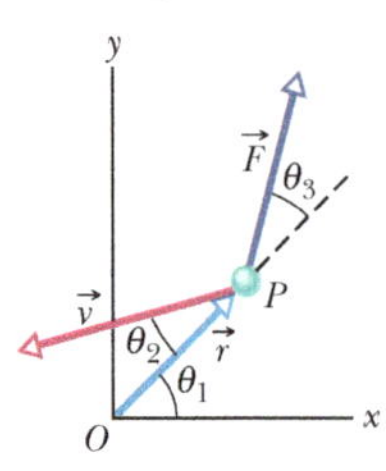

Figure 11-40 Problem 26.

•27 SSM At one instant, force $\vec{F} = 4.0\hat{j}$ N acts on a 0.25 kg object that has position vector $\vec{r} = (2.0\hat{i} - 2.0\hat{k})$ m and velocity vector $\vec{v} = (-5.0\hat{i} + 5.0\hat{k})$ m/s. About the origin and in unit-vector notation, what are (a) the object's angular momentum and (b) the torque acting on the object?

•28 A 2.0 kg particle-like object moves in a plane with velocity components $v_x = 30$ m/s and $v_y = 60$ m/s as it passes through the point with (*x*, *y*) coordinates of (3.0, −4.0) m. Just then, in unit-vector notation, what is its angular momentum relative to (a) the origin and (b) the point located at (−2.0, −2.0) m?

•29 ILW In the instant of Fig. 11-41, two particles move in an *xy* plane. Particle P_1 has mass 6.5 kg and speed $v_1 = 2.2$ m/s, and it is at distance $d_1 = 1.5$ m from point *O*. Particle P_2 has mass 3.1 kg and speed $v_2 = 3.6$ m/s, and it is at distance $d_2 = 2.8$ m from point *O*. What are the (a) magnitude and (b) direction of the net angular momentum of the two particles about *O*?

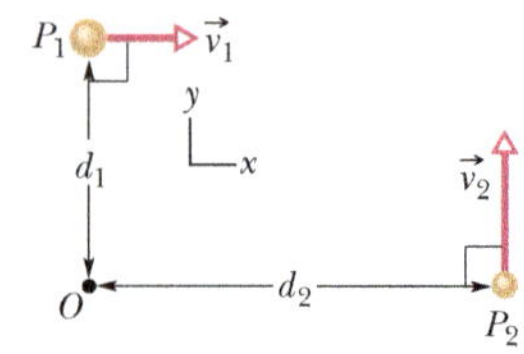

Figure 11-41 Problem 29.

••30 At the instant the displacement of a 2.00 kg object relative to the origin is $\vec{d} = (2.00\text{ m})\hat{i} + (4.00\text{ m})\hat{j} - (3.00\text{ m})\hat{k}$, its velocity is $\vec{v} = -(6.00\text{ m/s})\hat{i} + (3.00\text{ m/s})\hat{j} + (3.00\text{ m/s})\hat{k}$ and it is subject to a force $\vec{F} = (6.00\text{ N})\hat{i} - (8.00\text{ N})\hat{j} + (4.00\text{ N})\hat{k}$. Find (a) the acceleration of the object, (b) the angular momentum of the object about the origin, (c) the torque about the origin acting on the object, and (d) the angle between the velocity of the object and the force acting on the object.

••31 In Fig. 11-42, a 0.400 kg ball is shot directly upward at initial speed 40.0 m/s. What is its angular momentum about *P*, 2.00 m horizontally from the launch point, when the ball is (a) at maximum height and (b) halfway back to the ground? What is the torque on the ball about *P* due to the gravitational force when the ball is (c) at maximum height and (d) halfway back to the ground?

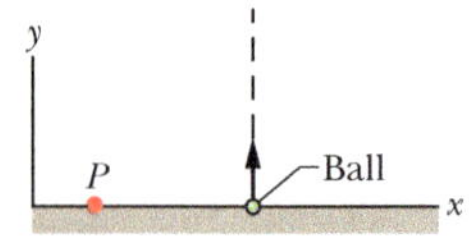

Figure 11-42 Problem 31.

Module 11-6 Newton's Second Law in Angular Form

•32 A particle is acted on by two torques about the origin: $\vec{\tau}_1$ has a magnitude of 2.0 N·m and is directed in the positive direction of the *x* axis, and $\vec{\tau}_2$ has a magnitude of 4.0 N·m and is directed in the negative direction of the *y* axis. In unit-vector notation, find $d\vec{\ell}/dt$, where $\vec{\ell}$ is the angular momentum of the particle about the origin.

•33 SSM WWW ILW At time $t = 0$, a 3.0 kg particle with velocity $\vec{v} = (5.0\text{ m/s})\hat{i} - (6.0\text{ m/s})\hat{j}$ is at $x = 3.0$ m, $y = 8.0$ m. It is pulled by a 7.0 N force in the negative *x* direction. About the origin, what are (a) the particle's angular momentum, (b) the torque acting on the particle, and (c) the rate at which the angular momentum is changing?

•34 A particle is to move in an *xy* plane, clockwise around the origin as seen from the positive side of the *z* axis. In unit-vector notation, what torque acts on the particle if the magnitude of its angular momentum about the origin is (a) 4.0 kg·m^2/s, (b) $4.0t^2$ kg·m^2/s, (c) $4.0\sqrt{t}$ kg·m^2/s, and (d) $4.0/t^2$ kg·m^2/s?

••35 At time t, the vector $\vec{r} = 4.0t^2\hat{\text{i}} - (2.0t + 6.0t^2)\hat{\text{j}}$ gives the position of a 3.0 kg particle relative to the origin of an xy coordinate system ($\vec{r}$ is in meters and t is in seconds). (a) Find an expression for the torque acting on the particle relative to the origin. (b) Is the magnitude of the particle's angular momentum relative to the origin increasing, decreasing, or unchanging?

Module 11-7 Angular Momentum of a Rigid Body

•36 Figure 11-43 shows three rotating, uniform disks that are coupled by belts. One belt runs around the rims of disks A and C. Another belt runs around a central hub on disk A and the rim of disk B. The belts move smoothly without slippage on the rims and hub. Disk A has radius R; its hub has radius $0.5000R$; disk B has radius $0.2500R$; and disk C has radius $2.000R$. Disks B and C have the same density (mass per unit volume) and thickness. What is the ratio of the magnitude of the angular momentum of disk C to that of disk B?

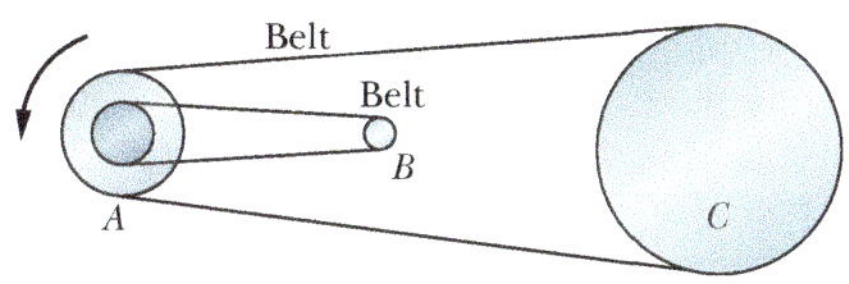

Figure 11-43 Problem 36.

•37 In Fig. 11-44, three particles of mass $m = 23$ g are fastened to three rods of length $d = 12$ cm and negligible mass. The rigid assembly rotates around point O at the angular speed $\omega = 0.85$ rad/s. About O, what are (a) the rotational inertia of the assembly, (b) the magnitude of the angular momentum of the middle particle, and (c) the magnitude of the angular momentum of the asssembly?

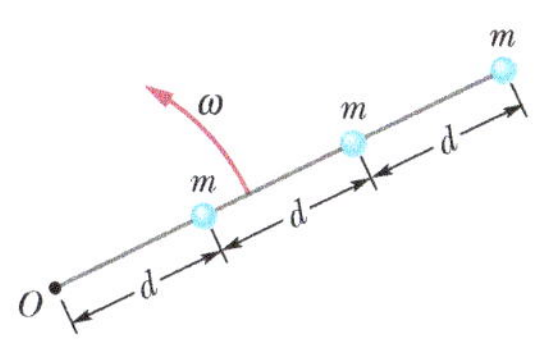

Figure 11-44 Problem 37.

•38 A sanding disk with rotational inertia 1.2×10^{-3} kg·m² is attached to an electric drill whose motor delivers a torque of magnitude 16 N·m about the central axis of the disk. About that axis and with the torque applied for 33 ms, what is the magnitude of the (a) angular momentum and (b) angular velocity of the disk?

•39 SSM The angular momentum of a flywheel having a rotational inertia of 0.140 kg·m² about its central axis decreases from 3.00 to 0.800 kg·m²/s in 1.50 s. (a) What is the magnitude of the average torque acting on the flywheel about its central axis during this period? (b) Assuming a constant angular acceleration, through what angle does the flywheel turn? (c) How much work is done on the wheel? (d) What is the average power of the flywheel?

••40 A disk with a rotational inertia of **7.00 kg·m²** rotates like a merry-go-round while undergoing a time-dependent torque given by $\tau = (5.00 + 2.00t)$ N·m. At time $t = 1.00$ s, its angular momentum is **5.00 kg·m²/s**. What is its angular momentum at $t = 3.00$ s?

••41 Figure 11-45 shows a rigid structure consisting of a circular hoop of radius R and mass m, and a square made of four thin bars, each of length R and mass m. The rigid structure rotates at a constant speed about a vertical axis, with a period of rotation of 2.5 s. Assuming $R = 0.50$ m and $m = 2.0$ kg, calculate (a) the structure's rotational inertia about the axis of rotation and (b) its angular momentum about that axis.

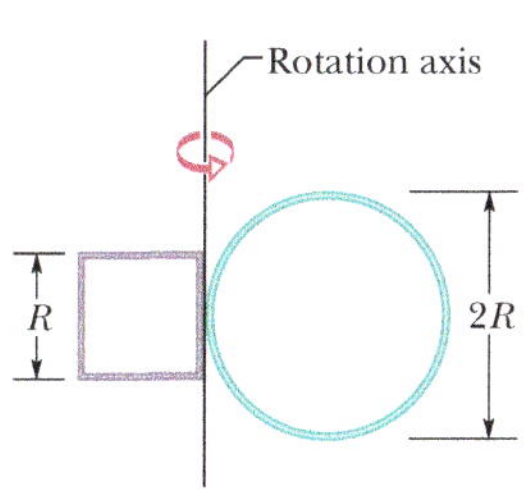

Figure 11-45 Problem 41.

••42 Figure 11-46 gives the torque τ that acts on an initially stationary disk that can rotate about its center like a merry-go-round. The scale on the τ axis is set by $\tau_s = 4.0$ N·m. What is the angular momentum of the disk about the rotation axis at times (a) $t = 7.0$ s and (b) $t = 20$ s?

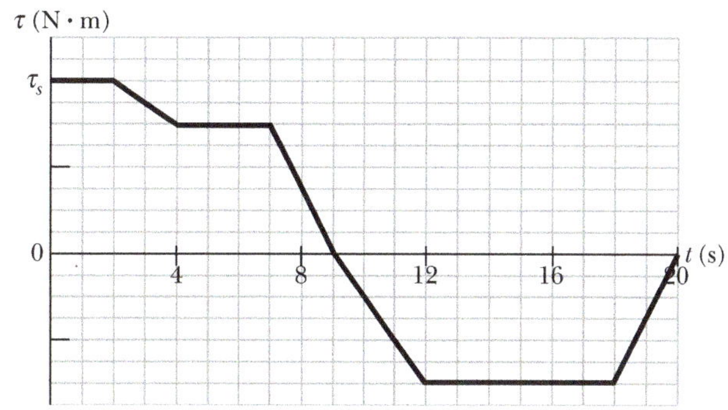

Figure 11-46 Problem 42.

Module 11-8 Conservation of Angular Momentum

•43 In Fig. 11-47, two skaters, each of mass 50 kg, approach each other along parallel paths separated by 3.0 m. They have opposite velocities of 1.4 m/s each. One skater carries one end of a long pole of negligible mass, and the other skater grabs the other end as she passes. The skaters then rotate around the center of the pole. Assume that the friction between skates and ice is negligible. What are (a) the radius of the circle, (b) the angular speed of the skaters, and (c) the kinetic energy of the two-skater system? Next, the skaters pull along the pole until they are separated by 1.0 m. What then are (d) their angular speed and (e) the kinetic energy of the system? (f) What provided the energy for the increased kinetic energy?

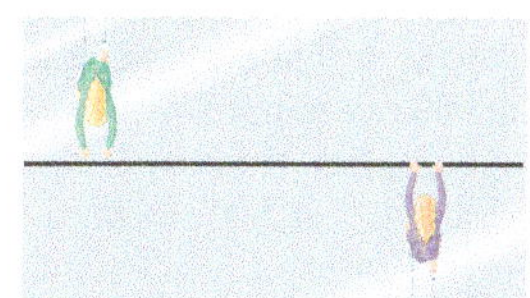

Figure 11-47 Problem 43.

•44 A Texas cockroach of mass 0.17 kg runs counterclockwise around the rim of a lazy Susan (a circular disk mounted on a vertical axle) that has radius 15 cm, rotational inertia 5.0×10^{-3} kg·m², and frictionless bearings. The cockroach's speed (relative to the ground) is 2.0 m/s, and the lazy Susan turns clockwise with angular speed $\omega_0 =$ 2.8 rad/s. The cockroach finds a bread crumb on the rim and, of course, stops. (a) What is the angular speed of the lazy Susan after the cockroach stops? (b) Is mechanical energy conserved as it stops?

•45 SSM WWW A man stands on a platform that is rotating (without friction) with an angular speed of 1.2 rev/s; his arms are outstretched and he holds a brick in each hand. The rotational inertia of the system consisting of the man, bricks, and platform about the central vertical axis of the platform is 6.0 kg·m². If by moving the bricks the man decreases the rotational inertia of the system to 2.0 kg·m², what are (a) the resulting angular speed of the platform and (b) the ratio of the new kinetic energy of the system to the original kinetic energy? (c) What source provided the added kinetic energy?

•46 The rotational inertia of a collapsing spinning star drops to $\frac{1}{3}$ its initial value. What is the ratio of the new rotational kinetic energy to the initial rotational kinetic energy?

•47 SSM A track is mounted on a large wheel that is free to turn with negligible friction about a vertical axis (Fig. 11-48). A toy train of mass m is placed on the track and, with the system initially at rest, the train's electrical power is turned on. The train reaches speed 0.15 m/s with respect to the track. What is the wheel's angular speed if its mass is $1.1m$ and its radius is 0.43 m? (Treat it as a hoop, and neglect the mass of the spokes and hub.)

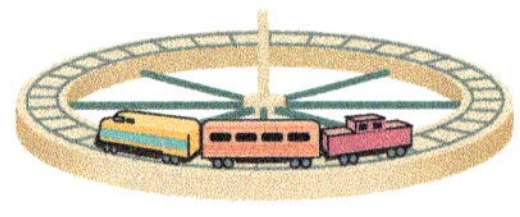

Figure 11-48 Problem 47.

•48 A Texas cockroach walks from the center of a circular disk (that rotates like a merry-go-round without external torques) out to the edge at radius R. The angular speed of the cockroach–disk system for the walk is given in Fig. 11-49 ($\omega_a = 5.0$ rad/s and $\omega_b = 6.0$ rad/s). After reaching R, what fraction of the rotational inertia of the disk does the cockroach have?

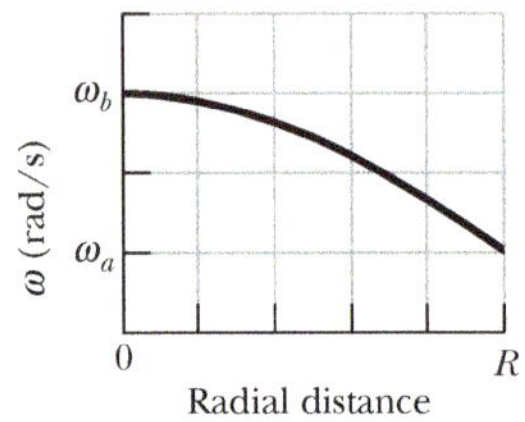

Figure 11-49 Problem 48.

•49 Two disks are mounted (like a merry-go-round) on low-friction bearings on the same axle and can be brought together so that they couple and rotate as one unit. The first disk, with rotational inertia 3.30 kg · m^2 about its central axis, is set spinning counterclockwise at 450 rev/min. The second disk, with rotational inertia 6.60 kg · m^2 about its central axis, is set spinning counterclockwise at 900 rev/min. They then couple together. (a) What is their angular speed after coupling? If instead the second disk is set spinning clockwise at 900 rev/min, what are their (b) angular speed and (c) direction of rotation after they couple together?

•50 The rotor of an electric motor has rotational inertia $I_m = 2.0 \times 10^{-3}$ kg · m^2 about its central axis. The motor is used to change the orientation of the space probe in which it is mounted. The motor axis is mounted along the central axis of the probe; the probe has rotational inertia $I_p = 12$ kg · m^2 about this axis. Calculate the number of revolutions of the rotor required to turn the probe through 30° about its central axis.

•51 SSM ILW A wheel is rotating freely at angular speed 800 rev/min on a shaft whose rotational inertia is negligible. A second wheel, initially at rest and with twice the rotational inertia of the first, is suddenly coupled to the same shaft. (a) What is the angular speed of the resultant combination of the shaft and two wheels? (b) What fraction of the original rotational kinetic energy is lost?

••52 GO A cockroach of mass m lies on the rim of a uniform disk of mass $4.00m$ that can rotate freely about its center like a merry-go-round. Initially the cockroach and disk rotate together with an angular velocity of 0.260 rad/s. Then the cockroach walks halfway to the center of the disk. (a) What then is the angular velocity of the cockroach–disk system? (b) What is the ratio K/K_0 of the new kinetic energy of the system to its initial kinetic energy? (c) What accounts for the change in the kinetic energy?

••53 GO In Fig. 11-50 (an overhead view), a uniform thin rod of length 0.500 m and mass 4.00 kg can rotate in a horizontal plane about a vertical axis through its center. The rod is at rest when a 3.00 g bullet traveling in the rotation plane is fired into one end of the rod. In the view from above, the bullet's path makes angle $\theta = 60.0°$ with the rod (Fig. 11-50). If the bullet lodges in the rod and the angular velocity of the rod is 10 rad/s immediately after the collision, what is the bullet's speed just before impact?

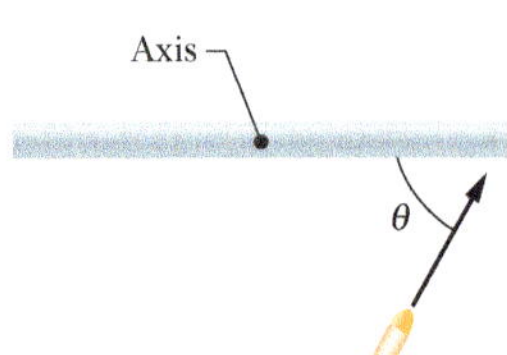

Figure 11-50 Problem 53.

••54 GO Figure 11-51 shows an overhead view of a ring that can rotate about its center like a merry-go-round. Its outer radius R_2 is 0.800 m, its inner radius R_1 is $R_2/2.00$, its mass M is 8.00 kg, and the mass of the crossbars at its center is negligible. It initially rotates at an angular speed of 8.00 rad/s with a cat of mass $m = M/4.00$ on its outer edge, at radius R_2. By how much does the cat increase the kinetic energy of the cat–ring system if the cat crawls to the inner edge, at radius R_1?

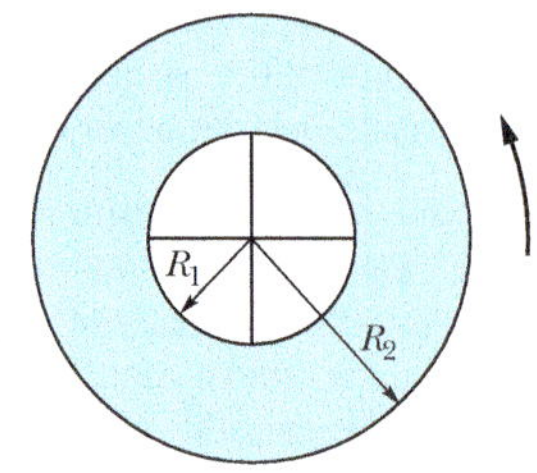

Figure 11-51 Problem 54.

••55 A horizontal vinyl record of mass 0.10 kg and radius 0.10 m rotates freely about a vertical axis through its center with an angular speed of 4.7 rad/s and a rotational inertia of 5.0×10^{-4} kg · m^2. Putty of mass 0.020 kg drops vertically onto the record from above and sticks to the edge of the record. What is the angular speed of the record immediately afterwards?

••56 In a long jump, an athlete leaves the ground with an initial angular momentum that tends to rotate her body forward, threatening to ruin her landing. To counter this tendency, she rotates her outstretched arms to "take up" the angular momentum (Fig. 11-18). In 0.700 s, one arm sweeps through 0.500 rev and the other arm sweeps through 1.000 rev. Treat each arm as a thin rod of mass 4.0 kg and length 0.60 m, rotating around one end. In the athlete's reference frame, what is the magnitude of the total angular momentum of the arms around the common rotation axis through the shoulders?

••57 A uniform disk of mass $10m$ and radius $3.0r$ can rotate freely about its fixed center like a merry-go-round. A smaller uniform disk of mass m and radius r lies on top of the larger disk, concentric with it. Initially the two disks rotate together with an angular velocity of 20 rad/s. Then a slight disturbance causes the smaller disk to slide outward across the larger disk, until the outer edge of the smaller disk catches on the outer edge of the larger disk. Afterward, the two disks again rotate together (without further sliding). (a) What then is their angular velocity about the center of the larger disk? (b) What is the ratio K/K_0 of the new kinetic energy of the two-disk system to the system's initial kinetic energy?

••58 A horizontal platform in the shape of a circular disk rotates on a frictionless bearing about a vertical axle through the center of the disk. The platform has a mass of 150 kg, a radius of 2.0 m, and a rotational inertia of 300 kg · m^2 about the axis of rotation. A 60 kg student walks slowly from the rim of the platform toward the center. If the angular speed of the system is 1.5 rad/s when the student starts at the rim, what is the angular speed when she is 0.50 m from the center?

••59 Figure 11-52 is an overhead view of a thin uniform rod of length 0.800 m and mass M rotating horizontally at angular speed 20.0 rad/s about an axis through its center. A particle of mass $M/3.00$ initially attached to one end is ejected from the rod and travels along a path that is perpendicular to the rod at the instant of ejection. If the particle's speed v_p is 6.00 m/s greater than the speed of the rod end just after ejection, what is the value of v_p?

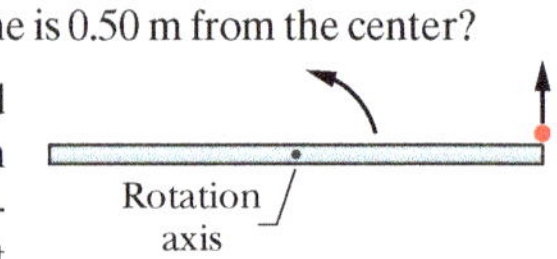

Figure 11-52 Problem 59.

••60 In Fig. 11-53, a 1.0 g bullet is fired into a 0.50 kg block attached to the end of a 0.60 m nonuniform rod of mass 0.50 kg. The block–rod–bullet system then rotates in the plane of the figure, about a fixed axis at A. The rotational inertia of the rod alone about that axis at A is 0.060 kg·m^2. Treat the block as a particle. (a) What then is the rotational inertia of the block–rod–bullet system about point A? (b) If the angular speed of the system about A just after impact is 4.5 rad/s, what is the bullet's speed just before impact?

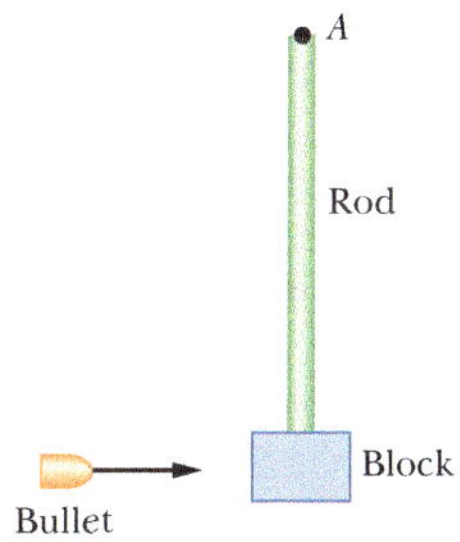

Figure 11-53 Problem 60.

••61 The uniform rod (length 0.60 m, mass 1.0 kg) in Fig. 11-54 rotates in the plane of the figure about an axis through one end, with a rotational inertia of 0.12 kg·m^2. As the rod swings through its lowest position, it collides with a 0.20 kg putty wad that sticks to the end of the rod. If the rod's angular speed just before collision is 2.4 rad/s, what is the angular speed of the rod–putty system immediately after collision?

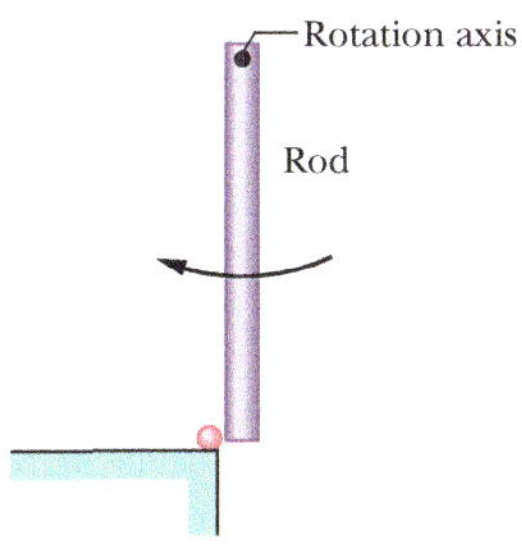

Figure 11-54 Problem 61.

•••62 During a jump to his partner, an aerialist is to make a quadruple somersault lasting a time t = 1.87 s. For the first and last quarter-revolution, he is in the extended orientation shown in Fig. 11-55, with rotational inertia I_1 = 19.9 kg·m^2 around his center of mass (the dot). During the rest of the flight he is in a tight tuck, with rotational inertia I_2 = 3.93 kg·m^2. What must be his angular speed ω_2 around his center of mass during the tuck?

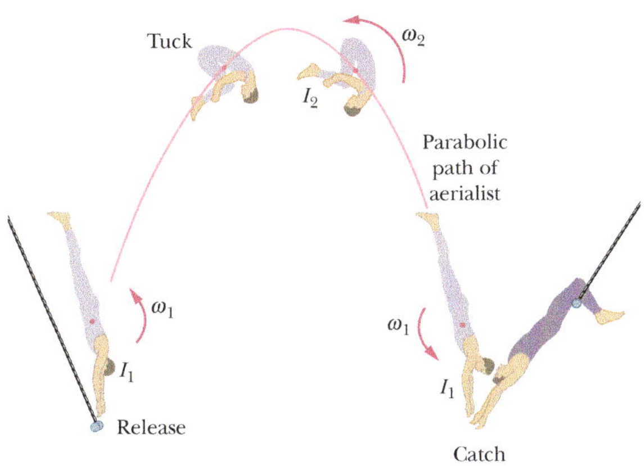

Figure 11-55 Problem 62.

•••63 In Fig. 11-56, a 30 kg child stands on the edge of a stationary merry-go-round of radius 2.0 m. The rotational inertia of the merry-go-round about its rotation axis is 150 kg·m^2. The child catches a ball of mass 1.0 kg thrown by a friend. Just before the ball is caught, it has a horizontal velocity $\vec{v}$ of magnitude 12 m/s, at angle ϕ = 37° with a line tangent to the outer edge of the merry-go-round, as shown. What is the angular speed of the merry-go-round just after the ball is caught?

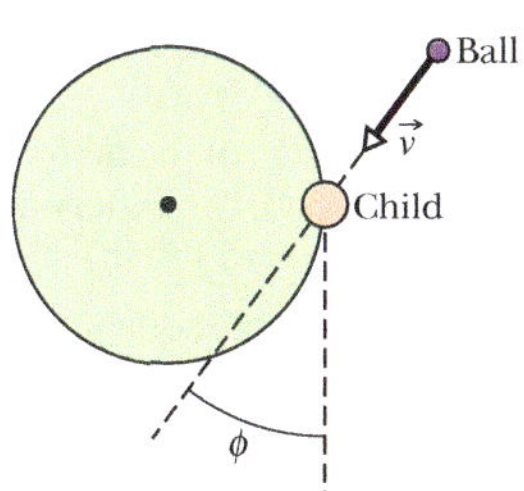

Figure 11-56 Problem 63.

•••64 A ballerina begins a tour jeté (Fig. 11-19a) with angular speed ω_i and a rotational inertia consisting of two parts: I_{leg} = 1.44 kg·m^2 for her leg extended outward at angle θ = 90.0° to her body and I_{trunk} = 0.660 kg·m^2 for the rest of her body (primarily her trunk). Near her maximum height she holds both legs at angle θ = 30.0° to her body and has angular speed ω_f (Fig. 11-19b). Assuming that I_{trunk} has not changed, what is the ratio ω_f/ω_i?

•••65 SSM WWW Two 2.00 kg balls are attached to the ends of a thin rod of length 50.0 cm and negligible mass. The rod is free to rotate in a vertical plane without friction about a horizontal axis through its center. With the rod initially horizontal (Fig. 11-57), a 50.0 g wad of wet putty drops onto one of the balls, hitting it with a speed of 3.00 m/s and then sticking to it. (a) What is the angular speed of the system just after the putty wad hits? (b) What is the ratio of the kinetic energy of the system after the collision to that of the putty wad just before? (c) Through what angle will the system rotate before it momentarily stops?

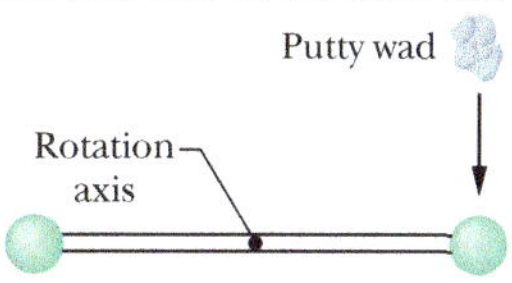

Figure 11-57 Problem 65.

•••66 In Fig. 11-58, a small 50 g block slides down a frictionless surface through height h = 20 cm and then sticks to a uniform rod of mass 100 g and length 40 cm. The rod pivots about point O through angle θ before momentarily stopping. Find θ.

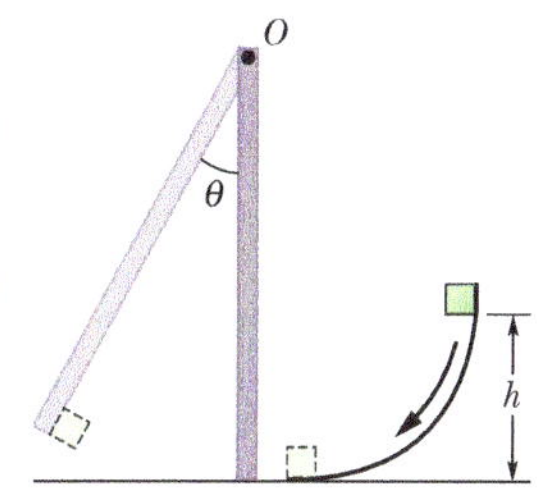

Figure 11-58 Problem 66.

•••67 Figure 11-59 is an overhead view of a thin uniform rod of length 0.600 m and mass M rotating horizontally at 80.0 rad/s counterclockwise about an axis through its center. A particle of mass M/3.00 and traveling horizontally at speed 40.0 m/s hits the rod and sticks. The particle's path is perpendicular to the rod at the instant of the hit, at a distance d from the rod's center. (a) At what value of d are rod and particle stationary after the hit? (b) In which direction do rod and particle rotate if d is greater than this value?

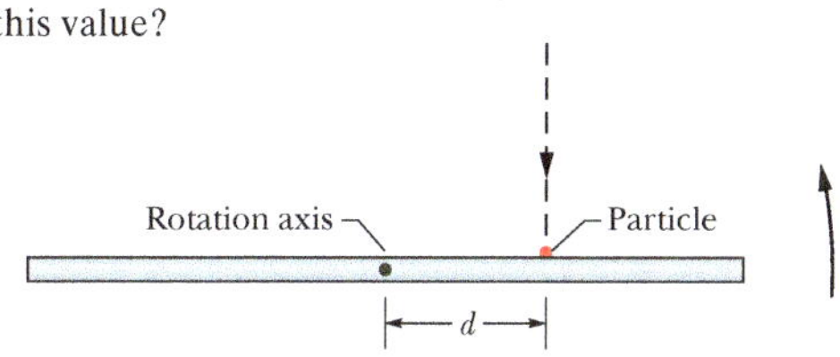

Figure 11-59 Problem 67.

Module 11-9 Precession of a Gyroscope

••68 A top spins at 30 rev/s about an axis that makes an angle of 30° with the vertical. The mass of the top is 0.50 kg, its rotational inertia about its central axis is 5.0 × 10^{-4} kg·m^2, and its center of mass is 4.0 cm from the pivot point. If the spin is clockwise from an overhead view, what are the (a) precession rate and (b) direction of the precession as viewed from overhead?

••69 A certain gyroscope consists of a uniform disk with a 50 cm radius mounted at the center of an axle that is 11 cm long and of negligible mass. The axle is horizontal and supported at one end. If the spin rate is 1000 rev/min, what is the precession rate?

Additional Problems

70 A uniform solid ball rolls smoothly along a floor, then up a ramp inclined at 15.0°. It momentarily stops when it has rolled 1.50 m along the ramp. What was its initial speed?

71 SSM In Fig. 11-60, a constant horizontal force $\vec{F}_{app}$ of magnitude 12 N is applied to a uniform solid cylinder by fishing line wrapped around the cylinder. The mass of the cylinder is 10 kg, its radius is 0.10 m, and the cylinder rolls smoothly on the horizontal surface. (a) What is the magnitude of the acceleration of the center of mass of the cylinder? (b) What is the magnitude of the angular acceleration of the cylinder about the center of mass? (c) In unit-vector notation, what is the frictional force acting on the cylinder?

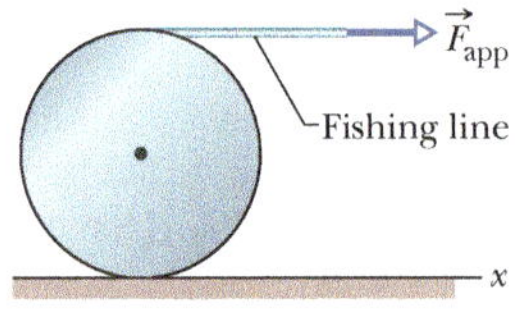

Figure 11-60 Problem 71.

72 A thin-walled pipe rolls along the floor. What is the ratio of its translational kinetic energy to its rotational kinetic energy about the central axis parallel to its length?

73 SSM A 3.0 kg toy car moves along an x axis with a velocity given by $\vec{v} = -2.0t^3\hat{i}$ m/s, with t in seconds. For $t > 0$, what are (a) the angular momentum $\vec{L}$ of the car and (b) the torque $\vec{\tau}$ on the car, both calculated about the origin? What are (c) $\vec{L}$ and (d) $\vec{\tau}$ about the point (2.0 m, 5.0 m, 0)? What are (e) $\vec{L}$ and (f) $\vec{\tau}$ about the point (2.0 m, −5.0 m, 0)?

74 A wheel rotates clockwise about its central axis with an angular momentum of 600 kg·m²/s. At time $t = 0$, a torque of magnitude 50 N·m is applied to the wheel to reverse the rotation. At what time t is the angular speed zero?

75 SSM In a playground, there is a small merry-go-round of radius 1.20 m and mass 180 kg. Its radius of gyration (see Problem 79 of Chapter 10) is 91.0 cm. A child of mass 44.0 kg runs at a speed of 3.00 m/s along a path that is tangent to the rim of the initially stationary merry-go-round and then jumps on. Neglect friction between the bearings and the shaft of the merry-go-round. Calculate (a) the rotational inertia of the merry-go-round about its axis of rotation, (b) the magnitude of the angular momentum of the running child about the axis of rotation of the merry-go-round, and (c) the angular speed of the merry-go-round and child after the child has jumped onto the merry-go-round.

76 A uniform block of granite in the shape of a book has face dimensions of 20 cm and 15 cm and a thickness of 1.2 cm. The density (mass per unit volume) of granite is 2.64 g/cm³. The block rotates around an axis that is perpendicular to its face and halfway between its center and a corner. Its angular momentum about that axis is 0.104 kg·m²/s. What is its rotational kinetic energy about that axis?

77 SSM Two particles, each of mass 2.90×10^{-4} kg and speed 5.46 m/s, travel in opposite directions along parallel lines separated by 4.20 cm. (a) What is the magnitude L of the angular momentum of the two-particle system around a point midway between the two lines? (b) Is the value different for a different location of the point? If the direction of either particle is reversed, what are the answers for (c) part (a) and (d) part (b)?

78 A wheel of radius 0.250 m, moving initially at 43.0 m/s, rolls to a stop in 225 m. Calculate the magnitudes of its (a) linear acceleration and (b) angular acceleration. (c) Its rotational inertia is 0.155 kg·m² about its central axis. Find the magnitude of the torque about the central axis due to friction on the wheel.

79 Wheels A and B in Fig. 11-61 are connected by a belt that does not slip. The radius of B is 3.00 times the radius of A. What would be the ratio of the rotational inertias I_A/I_B if the two wheels had (a) the same angular momentum about their central axes and (b) the same rotational kinetic energy?

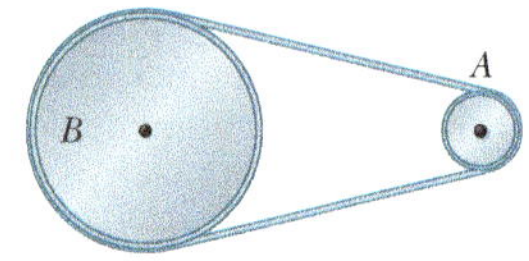

Figure 11-61 Problem 79.

80 A 2.50 kg particle that is moving horizontally over a floor with velocity $(-3.00 \text{ m/s})\hat{j}$ undergoes a completely inelastic collision with a 4.00 kg particle that is moving horizontally over the floor with velocity $(4.50 \text{ m/s})\hat{i}$. The collision occurs at xy coordinates (−0.500 m, −0.100 m). After the collision and in unit-vector notation, what is the angular momentum of the stuck-together particles with respect to the origin?

81 SSM A uniform wheel of mass 10.0 kg and radius 0.400 m is mounted rigidly on a massless axle through its center (Fig. 11-62). The radius of the axle is 0.200 m, and the rotational inertia of the wheel–axle combination about its central axis is 0.600 kg·m². The wheel is initially at rest at the top of a surface that is inclined at angle θ = 30.0° with the horizontal; the axle rests on the surface while the wheel extends into a groove in the surface without touching the surface. Once released, the axle rolls down along the surface smoothly and without slipping. When the wheel–axle combination has moved down the surface by 2.00 m, what are (a) its rotational kinetic energy and (b) its translational kinetic energy?

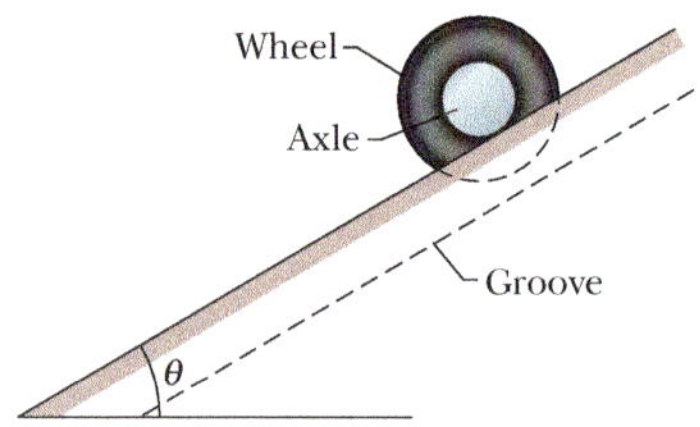

Figure 11-62 Problem 81.

82 A uniform rod rotates in a horizontal plane about a vertical axis through one end. The rod is 6.00 m long, weighs 10.0 N, and rotates at 240 rev/min. Calculate (a) its rotational inertia about the axis of rotation and (b) the magnitude of its angular momentum about that axis.

83 A solid sphere of weight 36.0 N rolls up an incline at an angle of 30.0°. At the bottom of the incline the center of mass of the sphere has a translational speed of 4.90 m/s. (a) What is the kinetic energy of the sphere at the bottom of the incline? (b) How far does the sphere travel up along the incline? (c) Does the answer to (b) depend on the sphere's mass?

84 Suppose that the yo-yo in Problem 17, instead of rolling from rest, is thrown so that its initial speed down the string is 1.3 m/s. (a) How long does the yo-yo take to reach the end of the string? As it reaches the end of the string, what are its (b) total kinetic energy, (c) linear speed, (d) translational kinetic energy, (e) angular speed, and (f) rotational kinetic energy?

85 A girl of mass M stands on the rim of a frictionless merry-go-round of radius R and rotational inertia I that is not moving. She throws a rock of mass m horizontally in a direction that is tangent to the outer edge of the merry-go-round. The speed of the rock, relative to the ground, is v. Afterward, what are (a) the angular speed of the merry-go-round and (b) the linear speed of the girl?

86 A body of radius R and mass m is rolling smoothly with speed v on a horizontal surface. It then rolls up a hill to a maximum height h. (a) If $h = 3v^2/4g$, what is the body's rotational inertia about the rotational axis through its center of mass? (b) What might the body be?

CHAPTER 12

Equilibrium and Elasticity

12-1 EQUILIBRIUM

Learning Objectives

After reading this module, you should be able to . . .

12.01 Distinguish between equilibrium and static equilibrium.

12.02 Specify the four conditions for static equilibrium.

12.03 Explain center of gravity and how it relates to center of mass.

12.04 For a given distribution of particles, calculate the coordinates of the center of gravity and the center of mass.

Key Ideas

- A rigid body at rest is said to be in static equilibrium. For such a body, the vector sum of the external forces acting on it is zero:

$$\vec{F}_{\text{net}} = 0 \quad \text{(balance of forces).}$$

If all the forces lie in the xy plane, this vector equation is equivalent to two component equations:

$$F_{\text{net},x} = 0 \quad \text{and} \quad F_{\text{net},y} = 0 \quad \text{(balance of forces).}$$

- Static equilibrium also implies that the vector sum of the external torques acting on the body about *any* point is zero, or

$$\vec{\tau}_{\text{net}} = 0 \quad \text{(balance of torques).}$$

If the forces lie in the xy plane, all torque vectors are parallel to the z axis, and the balance-of-torques equation is equivalent to the single component equation

$$\tau_{\text{net},z} = 0 \quad \text{(balance of torques).}$$

- The gravitational force acts individually on each element of a body. The net effect of all individual actions may be found by imagining an equivalent total gravitational force $\vec{F}_g$ acting at the center of gravity. If the gravitational acceleration $\vec{g}$ is the same for all the elements of the body, the center of gravity is at the center of mass.

What Is Physics?

Human constructions are supposed to be stable in spite of the forces that act on them. A building, for example, should be stable in spite of the gravitational force and wind forces on it, and a bridge should be stable in spite of the gravitational force pulling it downward and the repeated jolting it receives from cars and trucks.

One focus of physics is on what allows an object to be stable in spite of any forces acting on it. In this chapter we examine the two main aspects of stability: the *equilibrium* of the forces and torques acting on rigid objects and the *elasticity* of nonrigid objects, a property that governs how such objects can deform. When this physics is done correctly, it is the subject of countless articles in physics and engineering journals; when it is done incorrectly, it is the subject of countless articles in newspapers and legal journals.

Equilibrium

Consider these objects: (1) a book resting on a table, (2) a hockey puck sliding with constant velocity across a frictionless surface, (3) the rotating blades of a ceiling fan, and (4) the wheel of a bicycle that is traveling along a straight path at constant speed. For each of these four objects,

Kanwarjit Singh Boparai/Shutterstock

Figure 12-1 A balancing rock. Although its perch seems precarious, the rock is in static equilibrium.

1. The linear momentum $\vec{P}$ of its center of mass is constant.
2. Its angular momentum $\vec{L}$ about its center of mass, or about any other point, is also constant.

We say that such objects are in **equilibrium.** The two requirements for equilibrium are then

$$\vec{P} = \text{a constant} \quad \text{and} \quad \vec{L} = \text{a constant}. \tag{12-1}$$

Our concern in this chapter is with situations in which the constants in Eq. 12-1 are zero; that is, we are concerned largely with objects that are not moving in any way—either in translation or in rotation—in the reference frame from which we observe them. Such objects are in **static equilibrium.** Of the four objects mentioned near the beginning of this module, only one—the book resting on the table—is in static equilibrium.

The balancing rock of Fig. 12-1 is another example of an object that, for the present at least, is in static equilibrium. It shares this property with countless other structures, such as cathedrals, houses, filing cabinets, and taco stands, that remain stationary over time.

As we discussed in Module 8-3, if a body returns to a state of static equilibrium after having been displaced from that state by a force, the body is said to be in *stable* static equilibrium. A marble placed at the bottom of a hemispherical bowl is an example. However, if a small force can displace the body and end the equilibrium, the body is in *unstable* static equilibrium.

A Domino. For example, suppose we balance a domino with the domino's center of mass vertically above the supporting edge, as in Fig. 12-2*a*. The torque about the supporting edge due to the gravitational force $\vec{F}_g$ on the domino is zero because the line of action of $\vec{F}_g$ is through that edge. Thus, the domino is in equilibrium. Of course, even a slight force on it due to some chance disturbance ends the equilibrium. As the line of action of $\vec{F}_g$ moves to one side of the supporting edge (as in Fig. 12-2*b*), the torque due to $\vec{F}_g$ increases the rotation of the domino. Therefore, the domino in Fig. 12-2*a* is in unstable static equilibrium.

The domino in Fig. 12-2*c* is not quite as unstable. To topple this domino, a force would have to rotate it through and then beyond the balance position of Fig. 12-2*a*, in which the center of mass is above a supporting edge. A slight force will not topple this domino, but a vigorous flick of the finger against the domino certainly will. (If we arrange a chain of such upright dominos, a finger flick against the first can cause the whole chain to fall.)

A Block. The child's square block in Fig. 12-2*d* is even more stable because its center of mass would have to be moved even farther to get it to pass above a supporting edge. A flick of the finger may not topple the block. (This is why you

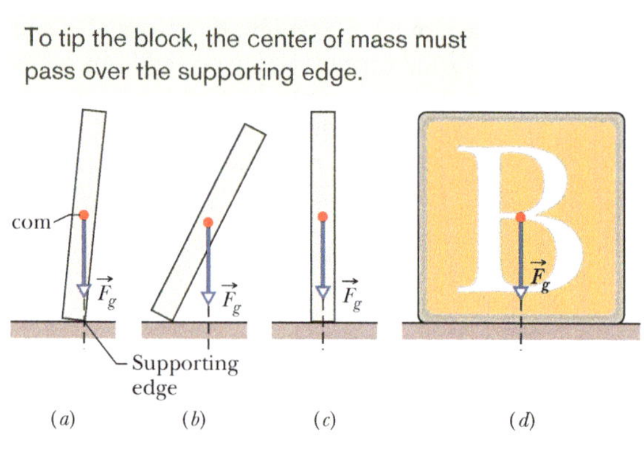

Figure 12-2 (*a*) A domino balanced on one edge, with its center of mass vertically above that edge. The gravitational force $\vec{F}_g$ on the domino is directed through the supporting edge. (*b*) If the domino is rotated even slightly from the balanced orientation, then $\vec{F}_g$ causes a torque that increases the rotation. (*c*) A domino upright on a narrow side is somewhat more stable than the domino in (*a*). (*d*) A square block is even more stable.

never see a chain of toppling square blocks.) The worker in Fig. 12-3 is like both the domino and the square block: Parallel to the beam, his stance is wide and he is stable; perpendicular to the beam, his stance is narrow and he is unstable (and at the mercy of a chance gust of wind).

The analysis of static equilibrium is very important in engineering practice. The design engineer must isolate and identify all the external forces and torques that may act on a structure and, by good design and wise choice of materials, ensure that the structure will remain stable under these loads. Such analysis is necessary to ensure, for example, that bridges do not collapse under their traffic and wind loads and that the landing gear of aircraft will function after the shock of rough landings.

Robert Brenner/PhotoEdit

Figure 12-3 A construction worker balanced on a steel beam is in static equilibrium but is more stable parallel to the beam than perpendicular to it.

The Requirements of Equilibrium

The translational motion of a body is governed by Newton's second law in its linear momentum form, given by Eq. 9-27 as

$$\vec{F}_{\text{net}} = \frac{d\vec{P}}{dt}. \tag{12-2}$$

If the body is in translational equilibrium—that is, if $\vec{P}$ is a constant—then $d\vec{P}/dt = 0$ and we must have

$$\vec{F}_{\text{net}} = 0 \quad \text{(balance of forces)}. \tag{12-3}$$

The rotational motion of a body is governed by Newton's second law in its angular momentum form, given by Eq. 11-29 as

$$\vec{\tau}_{\text{net}} = \frac{d\vec{L}}{dt}. \tag{12-4}$$

If the body is in rotational equilibrium—that is, if $\vec{L}$ is a constant—then $d\vec{L}/dt = 0$ and we must have

$$\vec{\tau}_{\text{net}} = 0 \quad \text{(balance of torques)}. \tag{12-5}$$

Thus, the two requirements for a body to be in equilibrium are as follows:

1. The vector sum of all the external forces that act on the body must be zero.
2. The vector sum of all external torques that act on the body, measured about *any* possible point, must also be zero.

These requirements obviously hold for *static* equilibrium. They also hold for the more general equilibrium in which $\vec{P}$ and $\vec{L}$ are constant but not zero.

Equations 12-3 and 12-5, as vector equations, are each equivalent to three independent component equations, one for each direction of the coordinate axes:

Balance of forces	Balance of torques	
$F_{\text{net},x} = 0$	$\tau_{\text{net},x} = 0$	
$F_{\text{net},y} = 0$	$\tau_{\text{net},y} = 0$	(12-6)
$F_{\text{net},z} = 0$	$\tau_{\text{net},z} = 0$	

The Main Equations. We shall simplify matters by considering only situations in which the forces that act on the body lie in the xy plane. This means that the only torques that can act on the body must tend to cause rotation around an axis parallel to the z axis. With this assumption, we eliminate one force equation and two torque equations from Eqs. 12-6, leaving

$$F_{\text{net},x} = 0 \quad \text{(balance of forces)}, \tag{12-7}$$

$$F_{\text{net},y} = 0 \quad \text{(balance of forces)}, \tag{12-8}$$

$$\tau_{\text{net},z} = 0 \quad \text{(balance of torques)}. \tag{12-9}$$

Here, $\tau_{\text{net},z}$ is the net torque that the external forces produce either about the z axis or about *any* axis parallel to it.

A hockey puck sliding at constant velocity over ice satisfies Eqs. 12-7, 12-8, and 12-9 and is thus in equilibrium *but not in static equilibrium.* For static equilibrium, the linear momentum $\vec{P}$ of the puck must be not only constant but also zero; the puck must be at rest on the ice. Thus, there is another requirement for static equilibrium:

3. The linear momentum $\vec{P}$ of the body must be zero.

Checkpoint 1

The figure gives six overhead views of a uniform rod on which two or more forces act perpendicularly to the rod. If the magnitudes of the forces are adjusted properly (but kept nonzero), in which situations can the rod be in static equilibrium?

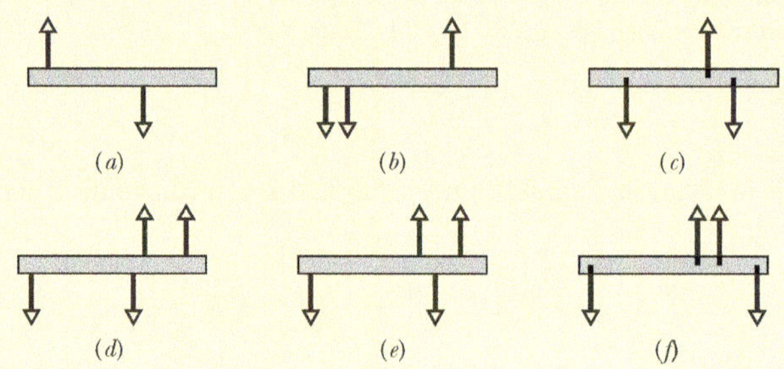

The Center of Gravity

The gravitational force on an extended body is the vector sum of the gravitational forces acting on the individual elements (the atoms) of the body. Instead of considering all those individual elements, we can say that

The gravitational force $\vec{F}_g$ on a body effectively acts at a single point, called the **center of gravity** (cog) of the body.

Here the word "effectively" means that if the gravitational forces on the individual elements were somehow turned off and the gravitational force $\vec{F}_g$ at the center of gravity were turned on, the net force and the net torque (about any point) acting on the body would not change.

Until now, we have assumed that the gravitational force $\vec{F}_g$ acts at the center of mass (com) of the body. This is equivalent to assuming that the center of gravity is at the center of mass. Recall that, for a body of mass M, the force $\vec{F}_g$ is equal to $M\vec{g}$, where $\vec{g}$ is the acceleration that the force would produce if the body were to fall freely. In the proof that follows, we show that

> If $\vec{g}$ is the same for all elements of a body, then the body's center of gravity (cog) is coincident with the body's center of mass (com).

This is approximately true for everyday objects because $\vec{g}$ varies only a little along Earth's surface and decreases in magnitude only slightly with altitude. Thus, for objects like a mouse or a moose, we have been justified in assuming that the gravitational force acts at the center of mass. After the following proof, we shall resume that assumption.

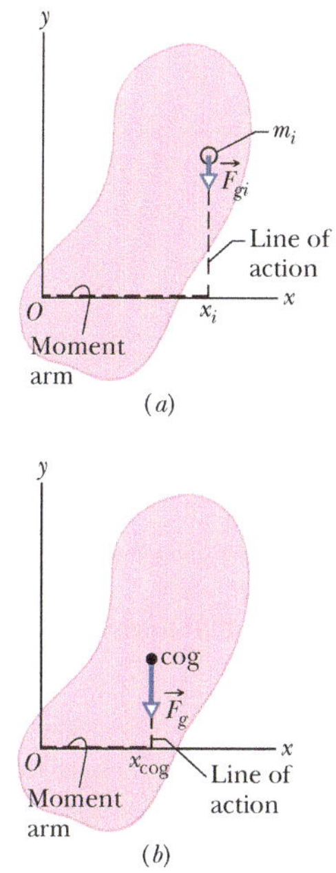

Figure 12-4 (a) An element of mass m_i in an extended body. The gravitational force $\vec{F}_{gi}$ on the element has moment arm x_i about the origin O of the coordinate system. (b) The gravitational force $\vec{F}_g$ on a body is said to act at the center of gravity (cog) of the body. Here $\vec{F}_g$ has moment arm x_{cog} about origin O.

Proof

First, we consider the individual elements of the body. Figure 12-4a shows an extended body, of mass M, and one of its elements, of mass m_i. A gravitational force $\vec{F}_{gi}$ acts on each such element and is equal to $m_i\vec{g}_i$. The subscript on $\vec{g}_i$ means $\vec{g}_i$ is the gravitational acceleration *at the location of the element i* (it can be different for other elements).

For the body in Fig. 12-4a, each force $\vec{F}_{gi}$ acting on an element produces a torque τ_i on the element about the origin O, with a moment arm x_i. Using Eq. 10-41 ($\tau = r_\perp F$) as a guide, we can write each torque τ_i as

$$\tau_i = x_i F_{gi}. \tag{12-10}$$

The net torque on all the elements of the body is then

$$\tau_{net} = \sum \tau_i = \sum x_i F_{gi}. \tag{12-11}$$

Next, we consider the body as a whole. Figure 12-4b shows the gravitational force $\vec{F}_g$ acting at the body's center of gravity. This force produces a torque τ on the body about O, with moment arm x_{cog}. Again using Eq. 10-41, we can write this torque as

$$\tau = x_{cog} F_g. \tag{12-12}$$

The gravitational force $\vec{F}_g$ on the body is equal to the sum of the gravitational forces $\vec{F}_{gi}$ on all its elements, so we can substitute ΣF_{gi} for F_g in Eq. 12-12 to write

$$\tau = x_{cog} \sum F_{gi}. \tag{12-13}$$

Now recall that the torque due to force $\vec{F}_g$ acting at the center of gravity is equal to the net torque due to all the forces $\vec{F}_{gi}$ acting on all the elements of the body. (That is how we defined the center of gravity.) Thus, τ in Eq. 12-13 is equal to τ_{net} in Eq. 12-11. Putting those two equations together, we can write

$$x_{cog} \sum F_{gi} = \sum x_i F_{gi}.$$

Substituting $m_i g_i$ for F_{gi} gives us

$$x_{cog} \sum m_i g_i = \sum x_i m_i g_i. \tag{12-14}$$

Now here is a key idea: If the accelerations g_i at all the locations of the elements are the same, we can cancel g_i from this equation to write

$$x_{cog} \sum m_i = \sum x_i m_i. \tag{12-15}$$

The sum Σm_i of the masses of all the elements is the mass M of the body. Therefore, we can rewrite Eq. 12-15 as

$$x_{cog} = \frac{1}{M} \sum x_i m_i. \tag{12-16}$$

The right side of this equation gives the coordinate x_{com} of the body's center of mass (Eq. 9-4). We now have what we sought to prove. If the acceleration of gravity is the same at all locations of the elements in a body, then the coordinates of the body's com and cog are identical:

$$x_{cog} = x_{com}. \tag{12-17}$$

12-2 SOME EXAMPLES OF STATIC EQUILIBRIUM

Learning Objectives

After reading this module, you should be able to . . .

12.05 Apply the force and torque conditions for static equilibrium.

12.06 Identify that a wise choice about the placement of the origin (about which to calculate torques) can simplify the calculations by eliminating one or more unknown forces from the torque equation.

Key Ideas

- A rigid body at rest is said to be in static equilibrium. For such a body, the vector sum of the external forces acting on it is zero:

$$\vec{F}_{net} = 0 \quad \text{(balance of forces).}$$

If all the forces lie in the xy plane, this vector equation is equivalent to two component equations:

$$F_{net,x} = 0 \quad \text{and} \quad F_{net,y} = 0 \quad \text{(balance of forces).}$$

- Static equilibrium also implies that the vector sum of the external torques acting on the body about *any* point is zero, or

$$\vec{\tau}_{net} = 0 \quad \text{(balance of torques).}$$

If the forces lie in the xy plane, all torque vectors are parallel to the z axis, and the balance-of-torques equation is equivalent to the single component equation

$$\tau_{net,z} = 0 \quad \text{(balance of torques).}$$

Some Examples of Static Equilibrium

Here we examine several sample problems involving static equilibrium. In each, we select a system of one or more objects to which we apply the equations of equilibrium (Eqs. 12-7, 12-8, and 12-9). The forces involved in the equilibrium are all in the xy plane, which means that the torques involved are parallel to the z axis. Thus, in applying Eq. 12-9, the balance of torques, we select an axis parallel to the z axis about which to calculate the torques. Although Eq. 12-9 is satisfied for *any* such choice of axis, you will see that certain choices simplify the application of Eq. 12-9 by eliminating one or more unknown force terms.

Checkpoint 2

The figure gives an overhead view of a uniform rod in static equilibrium. (a) Can you find the magnitudes of unknown forces $\vec{F}_1$ and $\vec{F}_2$ by balancing the forces? (b) If you wish to find the magnitude of force $\vec{F}_2$ by using a balance of torques equation, where should you place a rotation axis to eliminate $\vec{F}_1$ from the equation? (c) The magnitude of $\vec{F}_2$ turns out to be 65 N. What then is the magnitude of $\vec{F}_1$?

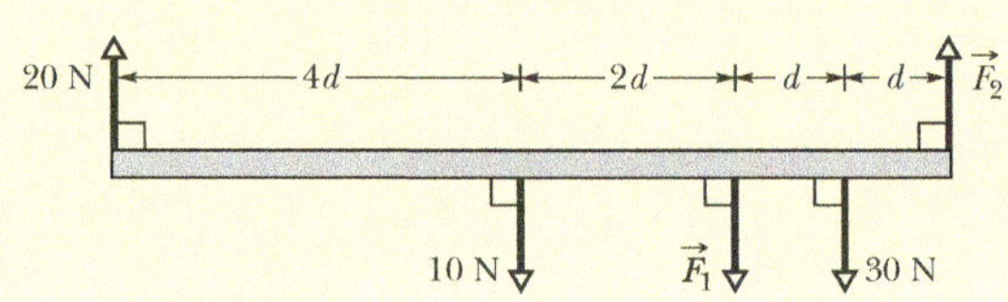

Sample Problem 12.01 Balancing a horizontal beam

In Fig. 12-5*a*, a uniform beam, of length L and mass $m = 1.8$ kg, is at rest on two scales. A uniform block, with mass $M = 2.7$ kg, is at rest on the beam, with its center a distance $L/4$ from the beam's left end. What do the scales read?

KEY IDEAS

The first steps in the solution of *any* problem about static equilibrium are these: Clearly define the system to be analyzed and then draw a free-body diagram of it, indicating all the forces on the system. Here, let us choose the system as the beam and block taken together. Then the forces on the system are shown in the free-body diagram of Fig. 12-5*b*. (Choosing the system takes experience, and often there can be more than one good choice.) Because the system is in static equilibrium, we can apply the balance of forces equations (Eqs. 12-7 and 12-8) and the balance of torques equation (Eq. 12-9) to it.

Calculations: The normal forces on the beam from the scales are $\vec{F}_l$ on the left and $\vec{F}_r$ on the right. The scale readings that we want are equal to the magnitudes of those forces. The gravitational force $\vec{F}_{g,beam}$ on the beam acts at the beam's center of mass and is equal to $m\vec{g}$. Similarly, the gravitational force $\vec{F}_{g,block}$ on the block acts at the block's center of mass and is equal to $M\vec{g}$. However, to simplify Fig. 12-5*b*, the block is represented by a dot within the boundary of the beam and vector $\vec{F}_{g,block}$ is drawn with its tail on that dot. (This shift of the vector $\vec{F}_{g,block}$ along its line of action does not alter the torque due to $\vec{F}_{g,block}$ about any axis perpendicular to the figure.)

The forces have no x components, so Eq. 12-7 ($F_{net,x} = 0$) provides no information. For the y components, Eq. 12-8 ($F_{net,y} = 0$) gives us

$$F_l + F_r - Mg - mg = 0. \tag{12-18}$$

This equation contains two unknowns, the forces F_l and F_r, so we also need to use Eq. 12-9, the balance of torques equation. We can apply it to *any* rotation axis perpendicular to the plane of Fig. 12-5. Let us choose a rotation axis through the left end of the beam. We shall also use our general rule for assigning signs to torques: If a torque would cause an initially stationary body to rotate clockwise about the rotation axis, the torque is negative. If the rotation would be counterclockwise, the torque is positive. Finally, we shall write the torques in the form $r_\perp F$, where the moment arm $r_\perp$ is 0 for $\vec{F}_l$, $L/4$ for $M\vec{g}$, $L/2$ for $m\vec{g}$, and L for $\vec{F}_r$.

We now can write the balancing equation ($\tau_{net,z} = 0$) as

$$(0)(F_l) - (L/4)(Mg) - (L/2)(mg) + (L)(F_r) = 0,$$

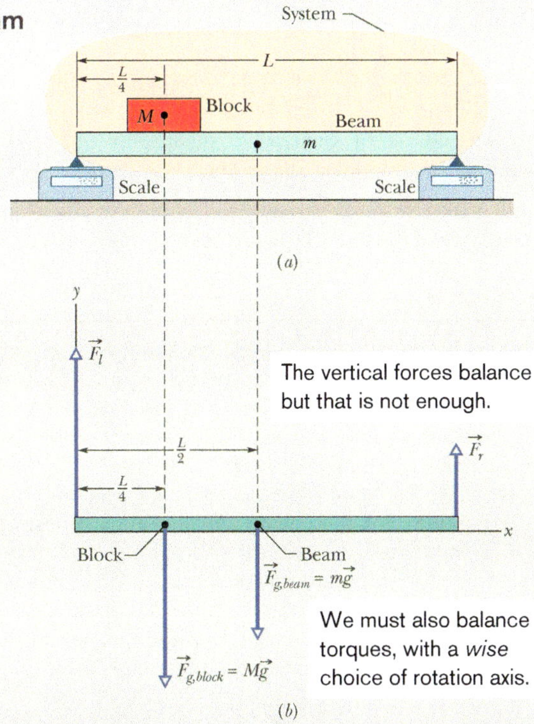

Figure 12-5 (*a*) A beam of mass m supports a block of mass M. (*b*) A free-body diagram, showing the forces that act on the system *beam + block*.

which gives us

$$\begin{aligned} F_r &= \tfrac{1}{4}Mg + \tfrac{1}{2}mg \\ &= \tfrac{1}{4}(2.7\text{ kg})(9.8\text{ m/s}^2) + \tfrac{1}{2}(1.8\text{ kg})(9.8\text{ m/s}^2) \\ &= 15.44\text{ N} \approx 15\text{ N}. \qquad \text{(Answer)} \end{aligned}$$

Now, solving Eq. 12-18 for F_l and substituting this result, we find

$$\begin{aligned} F_l &= (M + m)g - F_r \\ &= (2.7\text{ kg} + 1.8\text{ kg})(9.8\text{ m/s}^2) - 15.44\text{ N} \\ &= 28.66\text{ N} \approx 29\text{ N}. \qquad \text{(Answer)} \end{aligned}$$

Notice the strategy in the solution: When we wrote an equation for the balance of force components, we got stuck with two unknowns. If we had written an equation for the balance of torques around some *arbitrary* axis, we would have again gotten stuck with those two unknowns. However, because we chose the axis to pass through the point of application of one of the unknown forces, here $\vec{F}_l$, we did not get stuck. Our choice neatly eliminated that force from the torque equation, allowing us to solve for the other unknown force magnitude F_r. Then we returned to the equation for the balance of force components to find the remaining unknown force magnitude.

Sample Problem 12.02 Balancing a leaning boom

Figure 12-6*a* shows a safe (mass $M = 430$ kg) hanging by a rope (negligible mass) from a boom ($a = 1.9$ m and $b = 2.5$ m) that consists of a uniform hinged beam ($m = 85$ kg) and horizontal cable (negligible mass).

(a) What is the tension T_c in the cable? In other words, what is the magnitude of the force $\vec{T}_c$ on the beam from the cable?

KEY IDEAS

The system here is the beam alone, and the forces on it are shown in the free-body diagram of Fig. 12-6*b*. The force from the cable is $\vec{T}_c$. The gravitational force on the beam acts at the beam's center of mass (at the beam's center) and is represented by its equivalent $m\vec{g}$. The vertical component of the force on the beam from the hinge is $\vec{F}_v$, and the horizontal component of the force from the hinge is $\vec{F}_h$. The force from the rope supporting the safe is $\vec{T}_r$. Because beam, rope, and safe are stationary, the magnitude of $\vec{T}_r$ is equal to the weight of the safe: $T_r = Mg$. We place the origin O of an xy coordinate system at the hinge. Because the system is in static equilibrium, the balancing equations apply to it.

Calculations: Let us start with Eq. 12-9 ($\tau_{\text{net},z} = 0$). Note that we are asked for the magnitude of force $\vec{T}_c$ and not of forces $\vec{F}_h$ and $\vec{F}_v$ acting at the hinge, at point O. To eliminate $\vec{F}_h$ and $\vec{F}_v$ from the torque calculation, we should calculate torques about an axis that is perpendicular to the figure at point O. Then $\vec{F}_h$ and $\vec{F}_v$ will have moment arms of zero. The lines of action for $\vec{T}_c$, $\vec{T}_r$, and $m\vec{g}$ are dashed in Fig. 12-6*b*. The corresponding moment arms are a, b, and $b/2$.

Writing torques in the form of $r_\perp F$ and using our rule about signs for torques, the balancing equation $\tau_{\text{net},z} = 0$ becomes

$$(a)(T_c) - (b)(T_r) - (\tfrac{1}{2}b)(mg) = 0. \qquad (12\text{-}19)$$

Substituting Mg for T_r and solving for T_c, we find that

$$T_c = \frac{gb(M + \frac{1}{2}m)}{a}$$

$$= \frac{(9.8 \text{ m/s}^2)(2.5 \text{ m})(430 \text{ kg} + 85/2 \text{ kg})}{1.9 \text{ m}}$$

$$= 6093 \text{ N} \approx 6100 \text{ N}. \qquad \text{(Answer)}$$

(b) Find the magnitude F of the net force on the beam from the hinge.

KEY IDEA

Now we want the horizontal component F_h and vertical component F_v so that we can combine them to get the magnitude F of the net force. Because we know T_c, we apply the force balancing equations to the beam.

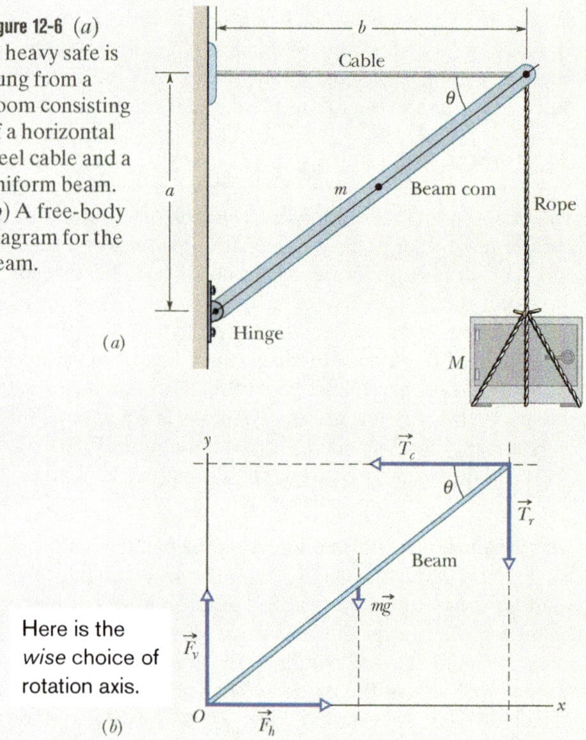

Figure 12-6 (*a*) A heavy safe is hung from a boom consisting of a horizontal steel cable and a uniform beam. (*b*) A free-body diagram for the beam.

Calculations: For the horizontal balance, we can rewrite $F_{\text{net},x} = 0$ as

$$F_h - T_c = 0, \qquad (12\text{-}20)$$

and so $\quad F_h = T_c = 6093$ N.

For the vertical balance, we write $F_{\text{net},y} = 0$ as

$$F_v - mg - T_r = 0.$$

Substituting Mg for T_r and solving for F_v, we find that

$$F_v = (m + M)g = (85 \text{ kg} + 430 \text{ kg})(9.8 \text{ m/s}^2)$$

$$= 5047 \text{ N}.$$

From the Pythagorean theorem, we now have

$$F = \sqrt{F_h^2 + F_v^2}$$

$$= \sqrt{(6093 \text{ N})^2 + (5047 \text{ N})^2} \approx 7900 \text{ N}. \qquad \text{(Answer)}$$

Note that F is substantially greater than either the combined weights of the safe and the beam, 5000 N, or the tension in the horizontal wire, 6100 N.

WILEY PLUS Additional examples, video, and practice available at *WileyPLUS*

Sample Problem 12.03 Balancing a leaning ladder

In Fig. 12-7*a*, a ladder of length $L = 12$ m and mass $m = 45$ kg leans against a slick wall (that is, there is no friction between the ladder and the wall). The ladder's upper end is at height $h = 9.3$ m above the pavement on which the lower end is supported (the pavement is not frictionless). The ladder's center of mass is $L/3$ from the lower end, along the length of the ladder. A firefighter of mass $M = 72$ kg climbs the ladder until her center of mass is $L/2$ from the lower end. What then are the magnitudes of the forces on the ladder from the wall and the pavement?

KEY IDEAS

First, we choose our system as being the firefighter and ladder, together, and then we draw the free-body diagram of Fig. 12-7*b* to show the forces acting on the system. Because the system is in static equilibrium, the balancing equations for both forces and torques (Eqs. 12-7 through 12-9) can be applied to it.

Calculations: In Fig. 12-7*b*, the firefighter is represented with a dot within the boundary of the ladder. The gravitational force on her is represented with its equivalent expression $M\vec{g}$, and that vector has been shifted along its line of action (the line extending through the force vector), so that its tail is on the dot. (The shift does not alter a torque due to $M\vec{g}$ about any axis perpendicular to the figure. Thus, the shift does not affect the torque balancing equation that we shall be using.)

The only force on the ladder from the wall is the horizontal force $\vec{F}_w$ (there cannot be a frictional force along a frictionless wall, so there is no vertical force on the ladder from the wall). The force $\vec{F}_p$ on the ladder from the pavement has two components: a horizontal component $\vec{F}_{px}$ that is a static frictional force and a vertical component $\vec{F}_{py}$ that is a normal force.

To apply the balancing equations, let's start with the torque balancing of Eq. 12-9 ($\tau_{\text{net},z} = 0$). To choose an axis about which to calculate the torques, note that we have unknown forces ($\vec{F}_w$ and $\vec{F}_p$) at the two ends of the ladder. To eliminate, say, $\vec{F}_p$ from the calculation, we place the axis at point O, perpendicular to the figure (Fig. 12-7*b*). We also place the origin of an xy coordinate system at O. We can find torques about O with any of Eqs. 10-39 through 10-41, but Eq. 10-41 ($\tau = r_\perp F$) is easiest to use here. *Making a wise choice about the placement of the origin can make our torque calculation much easier.*

To find the moment arm $r_\perp$ of the horizontal force $\vec{F}_w$ from the wall, we draw a line of action through that vector

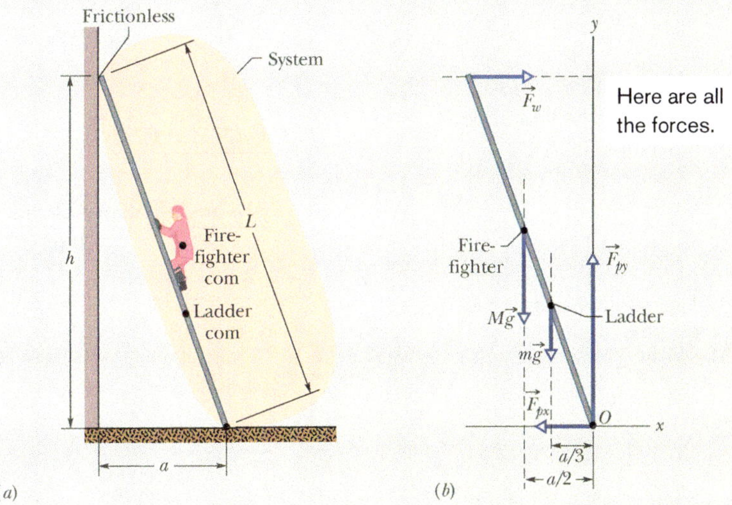

Figure 12-7 (*a*) A firefighter climbs halfway up a ladder that is leaning against a frictionless wall. The pavement beneath the ladder is not frictionless. (*b*) A free-body diagram, showing the forces that act on the *firefighter + ladder* system. The origin O of a coordinate system is placed at the point of application of the unknown force $\vec{F}_p$ (whose vector components $\vec{F}_{px}$ and $\vec{F}_{py}$ are shown). (*Figure 12-7 continues on following page.*)

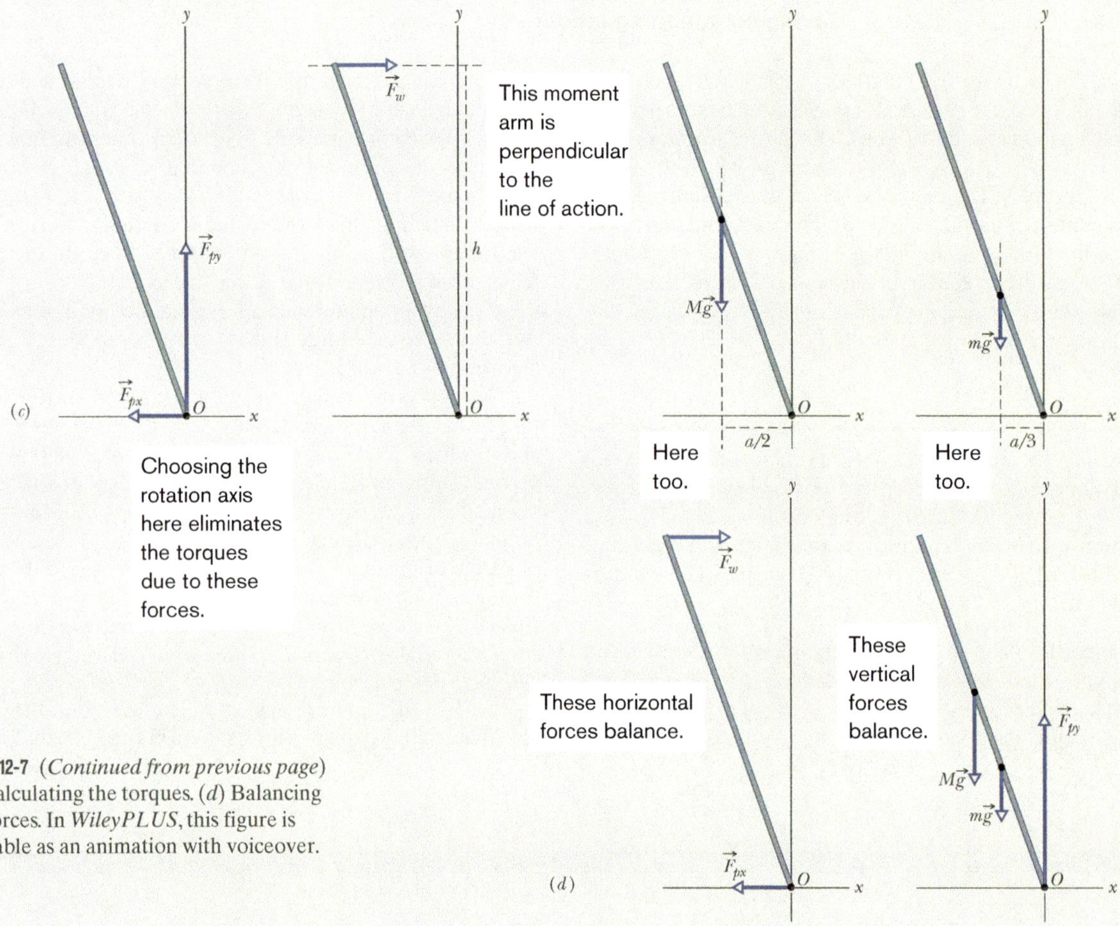

Figure 12-7 (*Continued from previous page*) (*c*) Calculating the torques. (*d*) Balancing the forces. In *WileyPLUS*, this figure is available as an animation with voiceover.

(it is the horizontal dashed line shown in Fig. 12-7*c*). Then $r_\perp$ is the perpendicular distance between O and the line of action. In Fig. 12-7*c*, $r_\perp$ extends along the y axis and is equal to the height h. We similarly draw lines of action for the gravitational force vectors $M\vec{g}$ and $m\vec{g}$ and see that their moment arms extend along the x axis. For the distance a shown in Fig. 12-7*a*, the moment arms are $a/2$ (the firefighter is halfway up the ladder) and $a/3$ (the ladder's center of mass is one-third of the way up the ladder), respectively. The moment arms for $\vec{F}_{px}$ and $\vec{F}_{py}$ are zero because the forces act at the origin.

Now, with torques written in the form $r_\perp F$, the balancing equation $\tau_{\text{net},z} = 0$ becomes

$$-(h)(F_w) + (a/2)(Mg) + (a/3)(mg) + (0)(F_{px}) + (0)(F_{py}) = 0. \quad (12\text{-}21)$$

(A positive torque corresponds to counterclockwise rotation and a negative torque corresponds to clockwise rotation.)

Using the Pythagorean theorem for the right triangle made by the ladder in Fig. 11-7*a*, we find that

$$a = \sqrt{L^2 - h^2} = 7.58 \text{ m}.$$

Then Eq. 12-21 gives us

$$F_w = \frac{ga(M/2 + m/3)}{h}$$

$$= \frac{(9.8 \text{ m/s}^2)(7.58 \text{ m})(72/2 \text{ kg} + 45/3 \text{ kg})}{9.3 \text{ m}}$$

$$= 407 \text{ N} \approx 410 \text{ N}. \quad \text{(Answer)}$$

Now we need to use the force balancing equations and Fig. 12-7*d*. The equation $F_{\text{net},x} = 0$ gives us

$$F_w - F_{px} = 0,$$

so

$$F_{px} = F_w = 410 \text{ N}. \quad \text{(Answer)}$$

The equation $F_{\text{net},y} = 0$ gives us

$$F_{py} - Mg - mg = 0,$$

so

$$F_{py} = (M + m)g = (72 \text{ kg} + 45 \text{ kg})(9.8 \text{ m/s}^2)$$

$$= 1146.6 \text{ N} \approx 1100 \text{ N}. \quad \text{(Answer)}$$

WILEY PLUS Additional examples, video, and practice available at *WileyPLUS*

Sample Problem 12.04 Balancing the leaning Tower of Pisa

Let's assume that the Tower of Pisa is a uniform hollow cylinder of radius $R = 9.8$ m and height $h = 60$ m. The center of mass is located at height $h/2$, along the cylinder's central axis. In Fig. 12-8*a*, the cylinder is upright. In Fig. 12-8*b*, it leans rightward (toward the tower's southern wall) by $\theta = 5.5°$, which shifts the com by a distance d. Let's assume that the ground exerts only two forces on the tower. A normal force $\vec{F}_{NL}$ acts on the left (northern) wall, and a normal force $\vec{F}_{NR}$ acts on the right (southern) wall. By what percent does the magnitude F_{NR} increase because of the leaning?

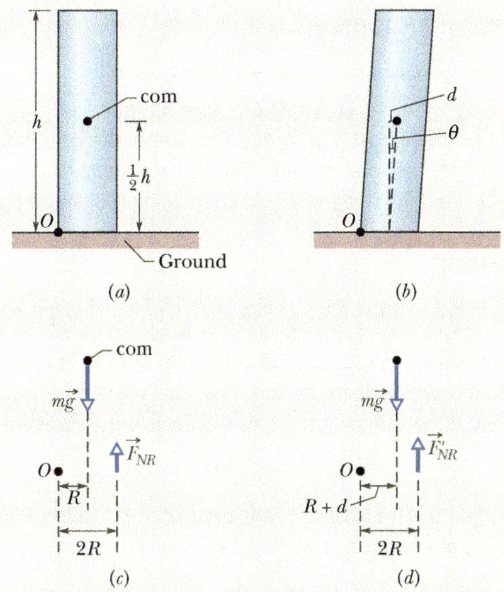

Figure 12-8 A cylinder modeling the Tower of Pisa: (*a*) upright and (*b*) leaning, with the center of mass shifted rightward. The forces and moment arms to find torques about a pivot at point O for the cylinder (*c*) upright and (*d*) leaning.

KEY IDEA

Because the tower is still standing, it is in equilibrium and thus the sum of torques calculated around any point must be zero.

Calculations: Because we want to calculate F_{NR} on the right side and do not know or want F_{NL} on the left side, we use a pivot point on the left side to calculate torques. The forces on the upright tower are represented in Fig. 12-8*c*. The gravitational force $m\vec{g}$, taken to act at the com, has a vertical line of action and a moment arm of R (the perpendicular distance from the pivot to the line of action). About the pivot, the torque associated with this force would tend to create clockwise rotation and thus is negative. The normal force $\vec{F}_{NR}$ on the southern wall also has a vertical line of action, and its moment arm is $2R$. About the pivot, the torque associated with this force would tend to create counterclockwise rotation and thus is positive. We now can write the torque-balancing equation ($\tau_{\text{net},z} = 0$) as

$$-(R)(mg) + (2R)(F_{NR}) = 0,$$

which yields

$$F_{NR} = \tfrac{1}{2}mg.$$

We should have been able to guess this result: With the center of mass located on the central axis (the cylinder's line of symmetry), the right side supports half the cylinder's weight.

In Fig. 12-8*b*, the com is shifted rightward by distance

$$d = \tfrac{1}{2}h \tan\theta.$$

The only change in the balance of torques equation is that the moment arm for the gravitational force is now $R + d$ and the normal force at the right has a new magnitude F'_{NR} (Fig. 12-8*d*). Thus, we write

$$-(R + d)(mg) + (2R)(F'_{NR}) = 0,$$

which gives us

$$F'_{NR} = \frac{(R + d)}{2R}mg.$$

Dividing this new result for the normal force at the right by the original result and then substituting for d, we obtain

$$\frac{F'_{NR}}{F_{NR}} = \frac{R + d}{R} = 1 + \frac{d}{R} = 1 + \frac{0.5h\tan\theta}{R}.$$

Substituting the values of $h = 60$ m, $R = 9.8$ m, and $\theta = 5.5°$ leads to

$$\frac{F'_{NR}}{F_{NR}} = 1.29.$$

Thus, our simple model predicts that, although the tilt is modest, the normal force on the tower's southern wall has increased by about 30%. One danger to the tower is that the force may cause the southern wall to buckle and explode outward. The cause of the leaning is the compressible soil beneath the tower, which worsened with each rainfall. Recently engineers have stabilized the tower and partially reversed the leaning by installing a drainage system.

12-3 ELASTICITY

Learning Objectives

After reading this module, you should be able to . . .

12.07 Explain what an indeterminate situation is.

12.08 For tension and compression, apply the equation that relates stress to strain and Young's modulus.

12.09 Distinguish between yield strength and ultimate strength.

12.10 For shearing, apply the equation that relates stress to strain and the shear modulus.

12.11 For hydraulic stress, apply the equation that relates fluid pressure to strain and the bulk modulus.

Key Ideas

- Three elastic moduli are used to describe the elastic behavior (deformations) of objects as they respond to forces that act on them. The strain (fractional change in length) is linearly related to the applied stress (force per unit area) by the proper modulus, according to the general stress–strain relation

$$\text{stress} = \text{modulus} \times \text{strain}.$$

- When an object is under tension or compression, the stress–strain relation is written as

$$\frac{F}{A} = E\frac{\Delta L}{L},$$

where $\Delta L/L$ is the tensile or compressive strain of the object, F is the magnitude of the applied force $\vec{F}$ causing the strain, A is the cross-sectional area over which $\vec{F}$ is applied (perpendicular to A), and E is the Young's modulus for the object. The stress is F/A.

- When an object is under a shearing stress, the stress–strain relation is written as

$$\frac{F}{A} = G\frac{\Delta x}{L},$$

where $\Delta x/L$ is the shearing strain of the object, Δx is the displacement of one end of the object in the direction of the applied force $\vec{F}$, and G is the shear modulus of the object. The stress is F/A.

- When an object undergoes hydraulic compression due to a stress exerted by a surrounding fluid, the stress–strain relation is written as

$$p = B\frac{\Delta V}{V},$$

where p is the pressure (hydraulic stress) on the object due to the fluid, $\Delta V/V$ (the strain) is the absolute value of the fractional change in the object's volume due to that pressure, and B is the bulk modulus of the object.

Indeterminate Structures

For the problems of this chapter, we have only three independent equations at our disposal, usually two balance of forces equations and one balance-of-torques equation about a given rotation axis. Thus, if a problem has more than three unknowns, we cannot solve it.

Consider an unsymmetrically loaded car. What are the forces—all different—on the four tires? Again, we cannot find them because we have only three independent equations. Similarly, we can solve an equilibrium problem for a table with three legs but not for one with four legs. Problems like these, in which there are more unknowns than equations, are called **indeterminate.**

Yet solutions to indeterminate problems exist in the real world. If you rest the tires of the car on four platform scales, each scale will register a definite reading, the sum of the readings being the weight of the car. What is eluding us in our efforts to find the individual forces by solving equations?

The problem is that we have assumed—without making a great point of it—that the bodies to which we apply the equations of static equilibrium are perfectly rigid. By this we mean that they do not deform when forces are applied to them. Strictly, there are no such bodies. The tires of the car, for example, deform easily under load until the car settles into a position of static equilibrium.

We have all had experience with a wobbly restaurant table, which we usually level by putting folded paper under one of the legs. If a big enough elephant sat on such a table, however, you may be sure that if the table did not collapse, it would deform just like the tires of a car. Its legs would all touch the floor, the forces acting upward on the table legs would all assume definite (and different) values as in Fig. 12-9, and the table would no longer wobble. Of course, we (and the elephant) would be thrown out onto the street but, in principle, how do we find the individual values of those forces acting on the legs in this or similar situations where there is deformation?

To solve such indeterminate equilibrium problems, we must supplement equilibrium equations with some knowledge of *elasticity,* the branch of physics and engineering that describes how real bodies deform when forces are applied to them.

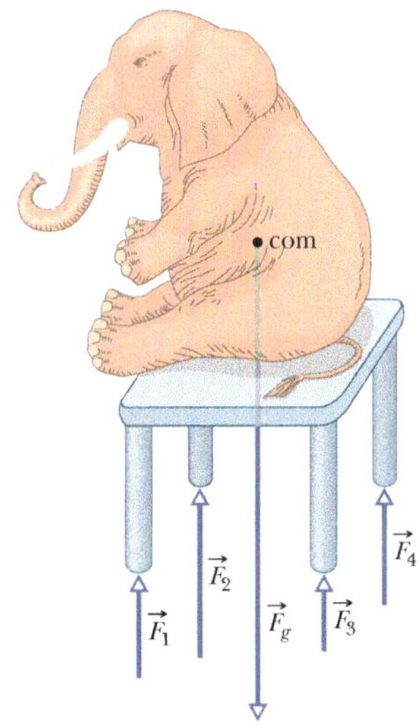

Figure 12-9 The table is an indeterminate structure. The four forces on the table legs differ from one another in magnitude and cannot be found from the laws of static equilibrium alone.

Checkpoint 3

A horizontal uniform bar of weight 10 N is to hang from a ceiling by two wires that exert upward forces $\vec{F}_1$ and $\vec{F}_2$ on the bar. The figure shows four arrangements for the wires. Which arrangements, if any, are indeterminate (so that we cannot solve for numerical values of $\vec{F}_1$ and $\vec{F}_2$)?

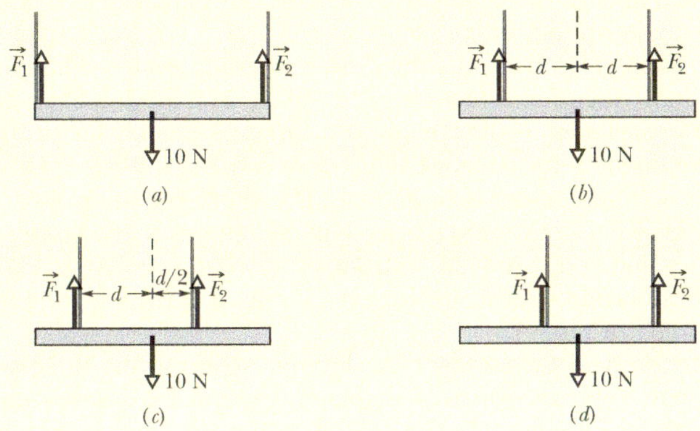

Elasticity

When a large number of atoms come together to form a metallic solid, such as an iron nail, they settle into equilibrium positions in a three-dimensional *lattice,* a repetitive arrangement in which each atom is a well-defined equilibrium distance from its nearest neighbors. The atoms are held together by interatomic forces that are modeled as tiny springs in Fig. 12-10. The lattice is remarkably rigid, which is another way of saying that the "interatomic springs" are extremely stiff. It is for this reason that we perceive many ordinary objects, such as metal ladders, tables, and spoons, as perfectly rigid. Of course, some ordinary objects, such as garden hoses or rubber gloves, do not strike us as rigid at all. The atoms that make up these objects *do not* form a rigid lattice like that of Fig. 12-10 but are aligned in long, flexible molecular chains, each chain being only loosely bound to its neighbors.

All real "rigid" bodies are to some extent **elastic,** which means that we can change their dimensions slightly by pulling, pushing, twisting, or compressing them. To get a feeling for the orders of magnitude involved, consider a vertical steel rod 1 m long and 1 cm in diameter attached to a factory ceiling. If you hang a subcompact car from the free end of such a rod, the rod will stretch but only by about 0.5 mm, or 0.05%. Furthermore, the rod will return to its original length when the car is removed.

If you hang two cars from the rod, the rod will be permanently stretched and will not recover its original length when you remove the load. If you hang three cars from the rod, the rod will break. Just before rupture, the elongation of the

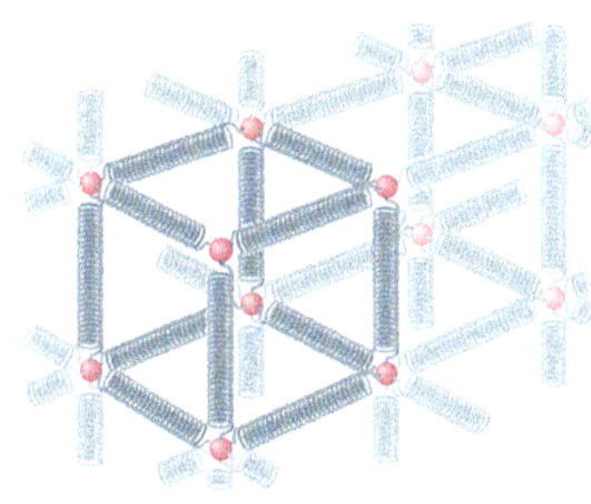

Figure 12-10 The atoms of a metallic solid are distributed on a repetitive three-dimensional lattice. The springs represent interatomic forces.

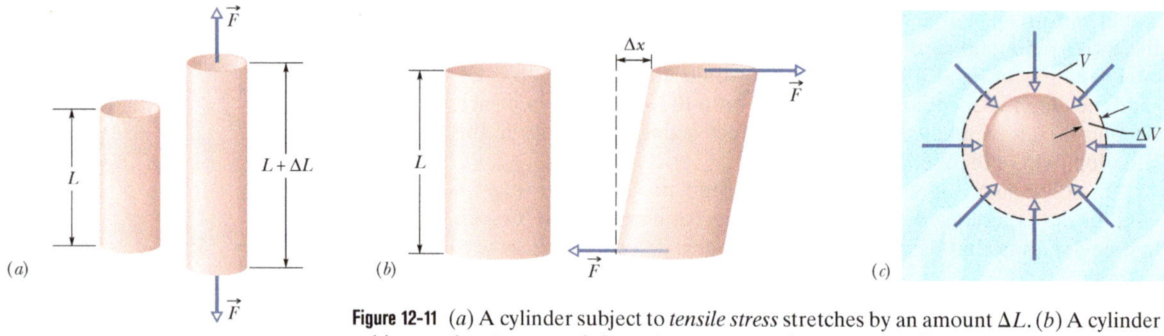

Figure 12-11 (*a*) A cylinder subject to *tensile stress* stretches by an amount ΔL. (*b*) A cylinder subject to *shearing stress* deforms by an amount Δx, somewhat like a pack of playing cards would. (*c*) A solid sphere subject to uniform *hydraulic stress* from a fluid shrinks in volume by an amount ΔV. All the deformations shown are greatly exaggerated.

rod will be less than 0.2%. Although deformations of this size seem small, they are important in engineering practice. (Whether a wing under load will stay on an airplane is obviously important.)

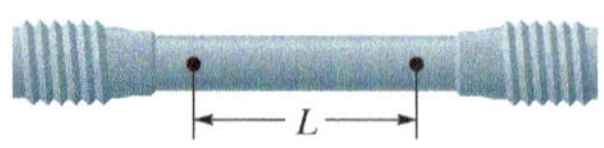

Figure 12-12 A test specimen used to determine a stress–strain curve such as that of Fig. 12-13. The change ΔL that occurs in a certain length L is measured in a tensile stress–strain test.

Three Ways. Figure 12-11 shows three ways in which a solid might change its dimensions when forces act on it. In Fig. 12-11*a*, a cylinder is stretched. In Fig. 12-11*b*, a cylinder is deformed by a force perpendicular to its long axis, much as we might deform a pack of cards or a book. In Fig. 12-11*c*, a solid object placed in a fluid under high pressure is compressed uniformly on all sides. What the three deformation types have in common is that a **stress,** or deforming force per unit area, produces a **strain,** or unit deformation. In Fig. 12-11, *tensile stress* (associated with stretching) is illustrated in (*a*), *shearing stress* in (*b*), and *hydraulic stress* in (*c*).

The stresses and the strains take different forms in the three situations of Fig. 12-11, but—over the range of engineering usefulness—stress and strain are proportional to each other. The constant of proportionality is called a **modulus of elasticity,** so that

$$\text{stress} = \text{modulus} \times \text{strain}. \tag{12-22}$$

In a standard test of tensile properties, the tensile stress on a test cylinder (like that in Fig. 12-12) is slowly increased from zero to the point at which the cylinder fractures, and the strain is carefully measured and plotted. The result is a graph of stress versus strain like that in Fig. 12-13. For a substantial range of applied stresses, the stress–strain relation is linear, and the specimen recovers its original dimensions when the stress is removed; it is here that Eq. 12-22 applies. If the stress is increased beyond the **yield strength** S_y of the specimen, the specimen becomes permanently deformed. If the stress continues to increase, the specimen eventually ruptures, at a stress called the **ultimate strength** S_u.

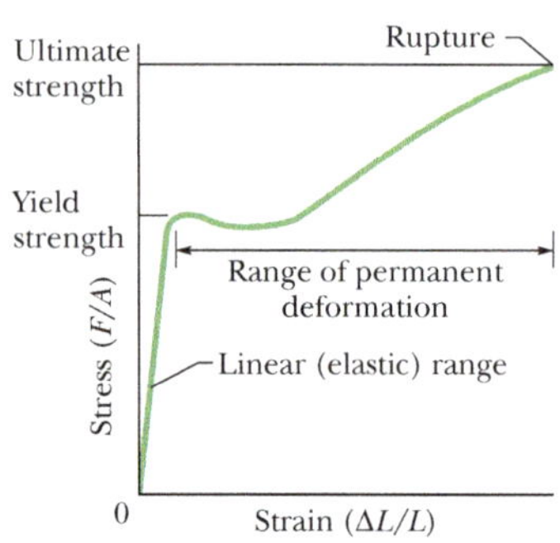

Figure 12-13 A stress–strain curve for a steel test specimen such as that of Fig. 12-12. The specimen deforms permanently when the stress is equal to the *yield strength* of the specimen's material. It ruptures when the stress is equal to the *ultimate strength* of the material.

Tension and Compression

For simple tension or compression, the stress on the object is defined as F/A, where F is the magnitude of the force applied perpendicularly to an area A on the object. The strain, or unit deformation, is then the dimensionless quantity $\Delta L/L$, the fractional (or sometimes percentage) change in a length of the specimen. If the specimen is a long rod and the stress does not exceed the yield strength, then not only the entire rod but also every section of it experiences the same strain when a given stress is applied. Because the strain is dimensionless, the modulus in Eq. 12-22 has the same dimensions as the stress—namely, force per unit area.

The modulus for tensile and compressive stresses is called the **Young's modulus** and is represented in engineering practice by the symbol E. Equation 12-22 becomes

$$\frac{F}{A} = E\frac{\Delta L}{L}. \tag{12-23}$$

The strain $\Delta L/L$ in a specimen can often be measured conveniently with a *strain gage* (Fig. 12-14), which can be attached directly to operating machinery with an adhesive. Its electrical properties are dependent on the strain it undergoes.

Although the Young's modulus for an object may be almost the same for tension and compression, the object's ultimate strength may well be different for the two types of stress. Concrete, for example, is very strong in compression but is so weak in tension that it is almost never used in that manner. Table 12-1 shows the Young's modulus and other elastic properties for some materials of engineering interest.

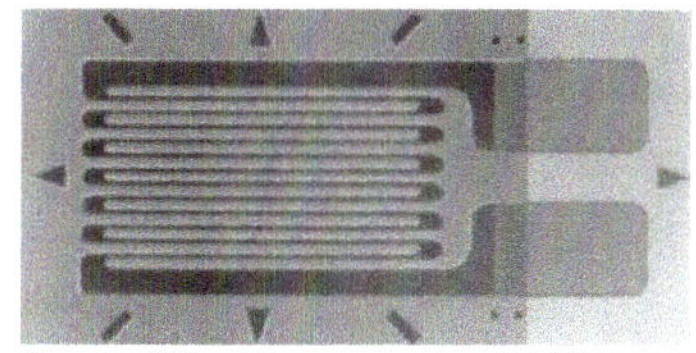

Courtesy Micro Measurements, a Division of Vishay Precision Group, Raleigh, NC

Figure 12-14 A strain gage of overall dimensions 9.8 mm by 4.6 mm. The gage is fastened with adhesive to the object whose strain is to be measured; it experiences the same strain as the object. The electrical resistance of the gage varies with the strain, permitting strains up to 3% to be measured.

Shearing

In the case of shearing, the stress is also a force per unit area, but the force vector lies in the plane of the area rather than perpendicular to it. The strain is the dimensionless ratio $\Delta x/L$, with the quantities defined as shown in Fig. 12-11*b*. The corresponding modulus, which is given the symbol G in engineering practice, is called the **shear modulus.** For shearing, Eq. 12-22 is written as

$$\frac{F}{A} = G\frac{\Delta x}{L}. \tag{12-24}$$

Shearing occurs in rotating shafts under load and in bone fractures due to bending.

Hydraulic Stress

In Fig. 12-11*c*, the stress is the fluid pressure p on the object, and, as you will see in Chapter 14, pressure is a force per unit area. The strain is $\Delta V/V$, where V is the original volume of the specimen and ΔV is the absolute value of the change in volume. The corresponding modulus, with symbol B, is called the **bulk modulus** of the material. The object is said to be under *hydraulic compression,* and the pressure can be called the *hydraulic stress.* For this situation, we write Eq. 12-22 as

$$p = B\frac{\Delta V}{V}. \tag{12-25}$$

The bulk modulus is 2.2×10^9 N/m^2 for water and 1.6×10^{11} N/m^2 for steel. The pressure at the bottom of the Pacific Ocean, at its average depth of about 4000 m, is 4.0×10^7 N/m^2. The fractional compression $\Delta V/V$ of a volume of water due to this pressure is 1.8%; that for a steel object is only about 0.025%. In general, solids—with their rigid atomic lattices—are less compressible than liquids, in which the atoms or molecules are less tightly coupled to their neighbors.

Table 12-1 Some Elastic Properties of Selected Materials of Engineering Interest

Material	Density ρ (kg/m^3)	Young's Modulus E (10^9 N/m^2)	Ultimate Strength S_u (10^6 N/m^2)	Yield Strength S_y (10^6 N/m^2)
Steel[a]	7860	200	400	250
Aluminum	2710	70	110	95
Glass	2190	65	50[b]	—
Concrete[c]	2320	30	40[b]	—
Wood[d]	525	13	50[b]	—
Bone	1900	9[b]	170[b]	—
Polystyrene	1050	3	48	—

[a]Structural steel (ASTM-A36).
[b]In compression.
[c]High strength
[d]Douglas fir.

Sample Problem 12.05 Stress and strain of elongated rod

One end of a steel rod of radius $R = 9.5$ mm and length $L = 81$ cm is held in a vise. A force of magnitude $F = 62$ kN is then applied perpendicularly to the end face (uniformly across the area) at the other end, pulling directly away from the vise. What are the stress on the rod and the elongation ΔL and strain of the rod?

KEY IDEAS

(1) Because the force is perpendicular to the end face and uniform, the stress is the ratio of the magnitude F of the force to the area A. The ratio is the left side of Eq. 12-23. (2) The elongation ΔL is related to the stress and Young's modulus E by Eq. 12-23 ($F/A = E\,\Delta L/L$). (3) Strain is the ratio of the elongation to the initial length L.

Calculations: To find the stress, we write

$$\text{stress} = \frac{F}{A} = \frac{F}{\pi R^2} = \frac{6.2 \times 10^4\ \text{N}}{(\pi)(9.5 \times 10^{-3}\ \text{m})^2}$$
$$= 2.2 \times 10^8\ \text{N/m}^2. \qquad \text{(Answer)}$$

The yield strength for structural steel is 2.5×10^8 N/m², so this rod is dangerously close to its yield strength.

We find the value of Young's modulus for steel in Table 12-1. Then from Eq. 12-23 we find the elongation:

$$\Delta L = \frac{(F/A)L}{E} = \frac{(2.2 \times 10^8\ \text{N/m}^2)(0.81\ \text{m})}{2.0 \times 10^{11}\ \text{N/m}^2}$$
$$= 8.9 \times 10^{-4}\ \text{m} = 0.89\ \text{mm}. \qquad \text{(Answer)}$$

For the strain, we have

$$\frac{\Delta L}{L} = \frac{8.9 \times 10^{-4}\ \text{m}}{0.81\ \text{m}}$$
$$= 1.1 \times 10^{-3} = 0.11\%. \qquad \text{(Answer)}$$

Sample Problem 12.06 Balancing a wobbly table

A table has three legs that are 1.00 m in length and a fourth leg that is longer by $d = 0.50$ mm, so that the table wobbles slightly. A steel cylinder with mass $M = 290$ kg is placed on the table (which has a mass much less than M) so that all four legs are compressed but unbuckled and the table is level but no longer wobbles. The legs are wooden cylinders with cross-sectional area $A = 1.0$ cm²; Young's modulus is $E = 1.3 \times 10^{10}$ N/m². What are the magnitudes of the forces on the legs from the floor?

KEY IDEAS

We take the table plus steel cylinder as our system. The situation is like that in Fig. 12-9, except now we have a steel cylinder on the table. If the tabletop remains level, the legs must be compressed in the following ways: Each of the short legs must be compressed by the same amount (call it ΔL_3) and thus by the same force of magnitude F_3. The single long leg must be compressed by a larger amount ΔL_4 and thus by a force with a larger magnitude F_4. In other words, for a level tabletop, we must have

$$\Delta L_4 = \Delta L_3 + d. \qquad (12\text{-}26)$$

From Eq. 12-23, we can relate a change in length to the force causing the change with $\Delta L = FL/AE$, where L is the original length of a leg. We can use this relation to replace ΔL_4 and ΔL_3 in Eq. 12-26. However, note that we can approximate the original length L as being the same for all four legs.

Calculations: Making those replacements and that approximation gives us

$$\frac{F_4 L}{AE} = \frac{F_3 L}{AE} + d. \qquad (12\text{-}27)$$

We cannot solve this equation because it has two unknowns, F_4 and F_3.

To get a second equation containing F_4 and F_3, we can use a vertical y axis and then write the balance of vertical forces ($F_{\text{net},y} = 0$) as

$$3F_3 + F_4 - Mg = 0, \qquad (12\text{-}28)$$

where Mg is equal to the magnitude of the gravitational force on the system. (*Three* legs have force $\vec{F}_3$ on them.) To solve the simultaneous equations 12-27 and 12-28 for, say, F_3, we first use Eq. 12-28 to find that $F_4 = Mg - 3F_3$. Substituting that into Eq. 12-27 then yields, after some algebra,

$$F_3 = \frac{Mg}{4} - \frac{dAE}{4L}$$
$$= \frac{(290\ \text{kg})(9.8\ \text{m/s}^2)}{4} - \frac{(5.0 \times 10^{-4}\ \text{m})(10^{-4}\ \text{m}^2)(1.3 \times 10^{10}\ \text{N/m}^2)}{(4)(1.00\ \text{m})}$$
$$= 548\ \text{N} \approx 5.5 \times 10^2\ \text{N}. \qquad \text{(Answer)}$$

From Eq. 12-28 we then find

$$F_4 = Mg - 3F_3 = (290\ \text{kg})(9.8\ \text{m/s}^2) - 3(548\ \text{N})$$
$$\approx 1.2\ \text{kN}. \qquad \text{(Answer)}$$

You can show that the three short legs are each compressed by 0.42 mm and the single long leg by 0.92 mm.

WILEY PLUS Additional examples, video, and practice available at *WileyPLUS*

Review & Summary

Static Equilibrium A rigid body at rest is said to be in **static equilibrium.** For such a body, the vector sum of the external forces acting on it is zero:

$$\vec{F}_{\text{net}} = 0 \quad \text{(balance of forces).} \tag{12-3}$$

If all the forces lie in the xy plane, this vector equation is equivalent to two component equations:

$$F_{\text{net},x} = 0 \quad \text{and} \quad F_{\text{net},y} = 0 \quad \text{(balance of forces).} \tag{12-7, 12-8}$$

Static equilibrium also implies that the vector sum of the external torques acting on the body about *any* point is zero, or

$$\vec{\tau}_{\text{net}} = 0 \quad \text{(balance of torques).} \tag{12-5}$$

If the forces lie in the xy plane, all torque vectors are parallel to the z axis, and Eq. 12-5 is equivalent to the single component equation

$$\tau_{\text{net},z} = 0 \quad \text{(balance of torques).} \tag{12-9}$$

Center of Gravity The gravitational force acts individually on each element of a body. The net effect of all individual actions may be found by imagining an equivalent total gravitational force $\vec{F}_g$ acting at the **center of gravity.** If the gravitational acceleration $\vec{g}$ is the same for all the elements of the body, the center of gravity is at the center of mass.

Elastic Moduli Three **elastic moduli** are used to describe the elastic behavior (deformations) of objects as they respond to forces that act on them. The **strain** (fractional change in length) is linearly related to the applied **stress** (force per unit area) by the proper modulus, according to the general relation

$$\text{stress} = \text{modulus} \times \text{strain.} \tag{12-22}$$

Tension and Compression When an object is under tension or compression, Eq. 12-22 is written as

$$\frac{F}{A} = E\frac{\Delta L}{L}, \tag{12-23}$$

where $\Delta L/L$ is the tensile or compressive strain of the object, F is the magnitude of the applied force $\vec{F}$ causing the strain, A is the cross-sectional area over which $\vec{F}$ is applied (perpendicular to A, as in Fig. 12-11*a*), and E is the **Young's modulus** for the object. The stress is F/A.

Shearing When an object is under a shearing stress, Eq. 12-22 is written as

$$\frac{F}{A} = G\frac{\Delta x}{L}, \tag{12-24}$$

where $\Delta x/L$ is the shearing strain of the object, Δx is the displacement of one end of the object in the direction of the applied force $\vec{F}$ (as in Fig. 12-11*b*), and G is the **shear modulus** of the object. The stress is F/A.

Hydraulic Stress When an object undergoes *hydraulic compression* due to a stress exerted by a surrounding fluid, Eq. 12-22 is written as

$$p = B\frac{\Delta V}{V}, \tag{12-25}$$

where p is the pressure (*hydraulic stress*) on the object due to the fluid, $\Delta V/V$ (the strain) is the absolute value of the fractional change in the object's volume due to that pressure, and B is the **bulk modulus** of the object.

Questions

1 Figure 12-15 shows three situations in which the same horizontal rod is supported by a hinge on a wall at one end and a cord at its other end. Without written calculation, rank the situations according to the magnitudes of (a) the force on the rod from the cord, (b) the vertical force on the rod from the hinge, and (c) the horizontal force on the rod from the hinge, greatest first.

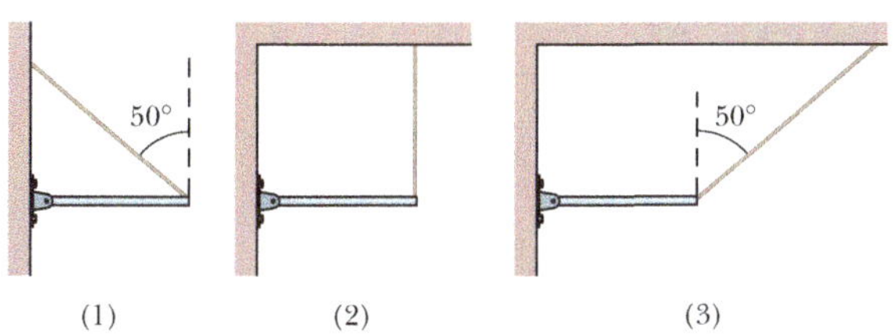

Figure 12-15 Question 1.

2 In Fig. 12-16, a rigid beam is attached to two posts that are fastened to a floor. A small but heavy safe is placed at the six positions indicated, in turn. Assume that the mass of the beam is negligible compared to that of the safe. (a) Rank the positions according to the force on post A due to the safe, greatest compression first, greatest tension last, and indicate where, if anywhere, the force is zero. (b) Rank them according to the force on post B.

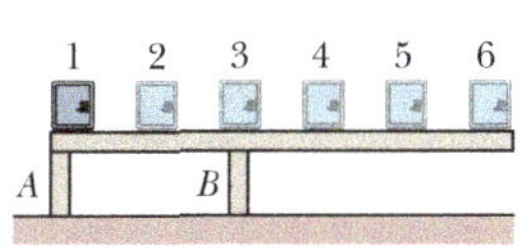

Figure 12-16 Question 2.

3 Figure 12-17 shows four overhead views of rotating uniform disks that are sliding across a frictionless floor. Three forces, of magnitude F, $2F$, or $3F$, act on each disk, either at the rim, at the center, or halfway between rim and center. The force vectors rotate along with the disks, and, in the "snapshots" of Fig. 12-17, point left or right. Which disks are in equilibrium?

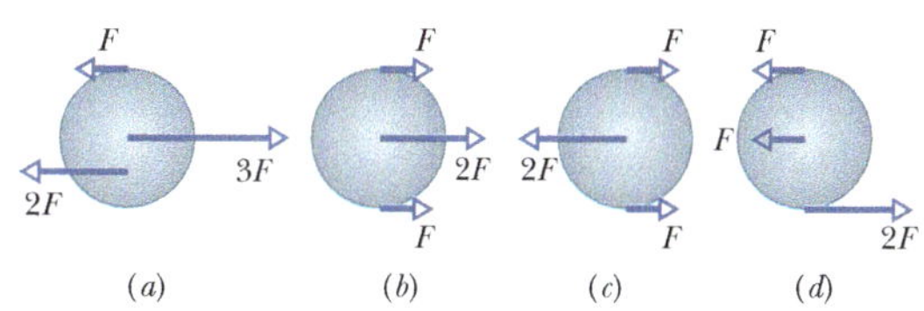

Figure 12-17 Question 3.

4 A ladder leans against a frictionless wall but is prevented from falling because of friction between it and the ground. Suppose you shift the base of the ladder toward the wall. Determine whether the following become larger, smaller, or stay the same (in magnitude): (a) the normal force on the ladder from the ground, (b) the force on the ladder from the wall, (c) the static frictional force on the ladder from the ground, and (d) the maximum value $f_{s,\max}$ of the static frictional force.

5 Figure 12-18 shows a mobile of toy penguins hanging from a ceiling. Each crossbar is horizontal, has negligible mass, and extends three times as far to the right of the wire supporting it as to the left. Penguin 1 has mass $m_1 = 48$ kg. What are the masses of (a) penguin 2, (b) penguin 3, and (c) penguin 4?

Figure 12-18 Question 5.

6 Figure 12-19 shows an overhead view of a uniform stick on which four forces act. Suppose we choose a rotation axis through point O, calculate the torques about that axis due to the forces, and find that these torques balance. Will the torques balance if, instead, the rotation axis is chosen to be at (a) point A (on the stick), (b) point B (on line with the stick), or (c) point C (off to one side of the stick)? (d) Suppose, instead, that we find that the torques about point O do not balance. Is there another point about which the torques will balance?

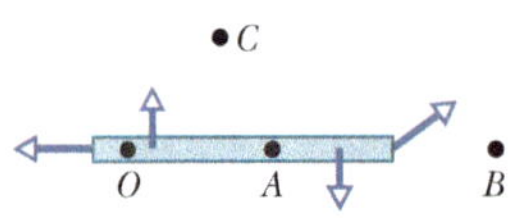

Figure 12-19 Question 6.

7 In Fig. 12-20, a stationary 5 kg rod AC is held against a wall by a rope and friction between rod and wall. The uniform rod is 1 m long, and angle $\theta = 30°$. (a) If you are to find the magnitude of the force $\vec{T}$ on the rod from the rope with a single equation, at what labeled point should a rotation axis be placed? With that choice of axis and counterclockwise torques positive, what is the sign of (b) the torque τ_w due to the rod's weight and (c) the torque τ_r due to the pull on the rod by the rope? (d) Is the magnitude of τ_r greater than, less than, or equal to the magnitude of τ_w?

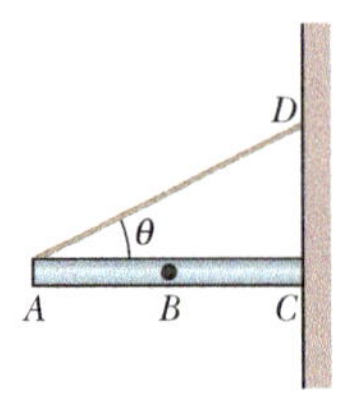

Figure 12-20 Question 7.

8 Three piñatas hang from the (stationary) assembly of massless pulleys and cords seen in Fig. 12-21. One long cord runs from the ceiling at the right to the lower pulley at the left, looping halfway around all the pulleys. Several shorter cords suspend pulleys from the ceiling or piñatas from the pulleys. The weights (in newtons) of two piñatas are given. (a) What is the weight of the third piñata? (*Hint:* A cord that loops halfway around a pulley pulls on the pulley with a net force that is twice the tension in the cord.) (b) What is the tension in the short cord labeled with T?

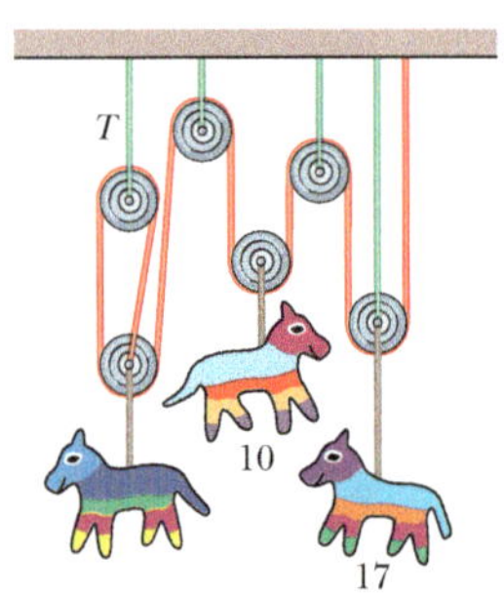

Figure 12-21 Question 8.

9 In Fig. 12-22, a vertical rod is hinged at its lower end and attached to a cable at its upper end. A horizontal force $\vec{F}_a$ is to be applied to the rod as shown. If the point at which the force is applied is moved up the rod, does the tension in the cable increase, decrease, or remain the same?

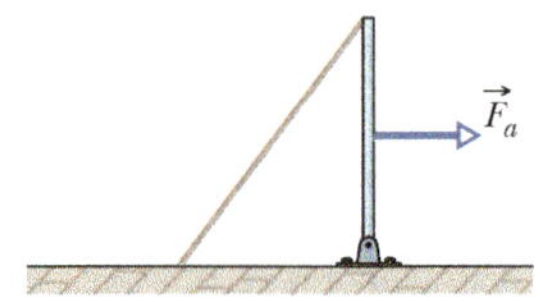

Figure 12-22 Question 9.

10 Figure 12-23 shows a horizontal block that is suspended by two wires, A and B, which are identical except for their original lengths. The center of mass of the block is closer to wire B than to wire A. (a) Measuring torques about the block's center of mass, state whether the magnitude of the torque due to wire A is greater than, less than, or equal to the magnitude of the torque due to wire B. (b) Which wire exerts more force on the block? (c) If the wires are now equal in length, which one was originally shorter (before the block was suspended)?

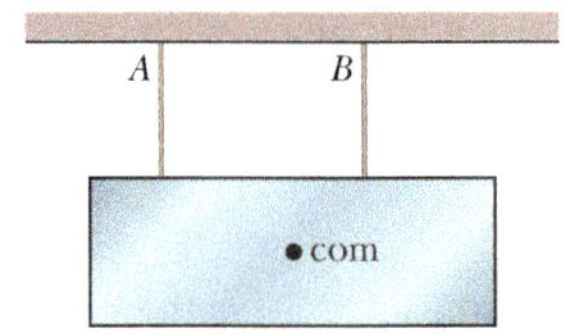

Figure 12-23 Question 10.

11 The table gives the initial lengths of three rods and the changes in their lengths when forces are applied to their ends to put them under strain. Rank the rods according to their strain, greatest first.

	Initial Length	Change in Length
Rod A	$2L_0$	ΔL_0
Rod B	$4L_0$	$2\Delta L_0$
Rod C	$10L_0$	$4\Delta L_0$

12 A physical therapist gone wild has constructed the (stationary) assembly of massless pulleys and cords seen in Fig. 12-24. One long cord wraps around all the pulleys, and shorter cords suspend pulleys from the ceiling or weights from the pulleys. Except for one, the weights (in newtons) are indicated. (a) What is that last weight? (*Hint:* When a cord loops halfway around a pulley as here, it pulls on the pulley with a net force that is twice the tension in the cord.) (b) What is the tension in the short cord labeled T?

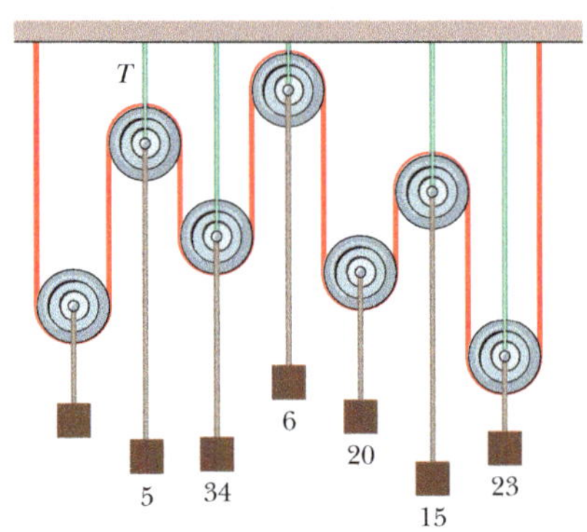

Figure 12-24 Question 12.

Problems

GO Tutoring problem available (at instructor's discretion) in *WileyPLUS* and WebAssign
SSM Worked-out solution available in Student Solutions Manual
WWW Worked-out solution is at http://www.wiley.com/college/halliday
• – ••• Number of dots indicates level of problem difficulty
ILW Interactive solution is at http://www.wiley.com/college/halliday
Additional information available in *The Flying Circus of Physics* and at flyingcircusofphysics.com

Module 12-1 Equilibrium

•1 Because g varies so little over the extent of most structures, any structure's center of gravity effectively coincides with its center of mass. Here is a fictitious example where g varies more significantly. Figure 12-25 shows an array of six particles, each with mass m, fixed to the edge of a rigid structure of negligible mass. The distance between adjacent particles along the edge is 2.00 m. The following table gives the value of g (m/s^2) at each particle's location. Using the coordinate system shown, find (a) the x coordinate x_{com} and (b) the y coordinate y_{com} of the center of mass of the six-particle system. Then find (c) the x coordinate x_{cog} and (d) the y coordinate y_{cog} of the center of gravity of the six-particle system.

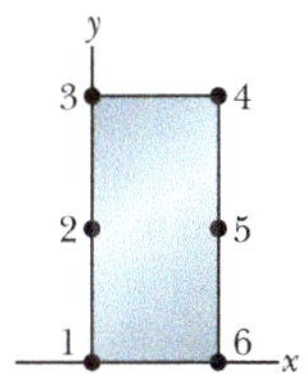

Figure 12-25 Problem 1.

Particle	g	Particle	g
1	8.00	4	7.40
2	7.80	5	7.60
3	7.60	6	7.80

Module 12-2 Some Examples of Static Equilibrium

•2 An automobile with a mass of 1360 kg has 3.05 m between the front and rear axles. Its center of gravity is located 1.78 m behind the front axle. With the automobile on level ground, determine the magnitude of the force from the ground on (a) each front wheel (assuming equal forces on the front wheels) and (b) each rear wheel (assuming equal forces on the rear wheels).

•3 SSM WWW In Fig. 12-26, a uniform sphere of mass $m = 0.85$ kg and radius $r = 4.2$ cm is held in place by a massless rope attached to a frictionless wall a distance $L = 8.0$ cm above the center of the sphere. Find (a) the tension in the rope and (b) the force on the sphere from the wall.

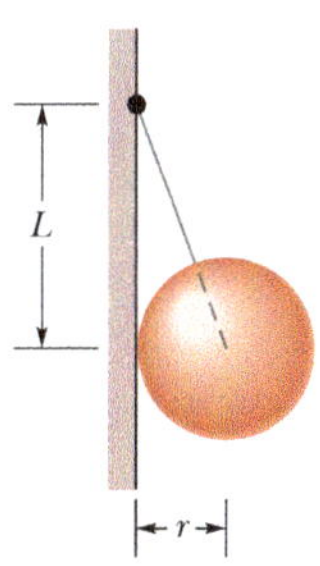

Figure 12-26 Problem 3.

•4 An archer's bow is drawn at its midpoint until the tension in the string is equal to the force exerted by the archer. What is the angle between the two halves of the string?

•5 ILW A rope of negligible mass is stretched horizontally between two supports that are 3.44 m apart. When an object of weight 3160 N is hung at the center of the rope, the rope is observed to sag by 35.0 cm. What is the tension in the rope?

•6 A scaffold of mass 60 kg and length 5.0 m is supported in a horizontal position by a vertical cable at each end. A window washer of mass 80 kg stands at a point 1.5 m from one end. What is the tension in (a) the nearer cable and (b) the farther cable?

•7 A 75 kg window cleaner uses a 10 kg ladder that is 5.0 m long. He places one end on the ground 2.5 m from a wall, rests the upper end against a cracked window, and climbs the ladder. He is 3.0 m up along the ladder when the window breaks. Neglect friction between the ladder and window and assume that the base of the ladder does not slip. When the window is on the verge of breaking, what are (a) the magnitude of the force on the window from the ladder, (b) the magnitude of the force on the ladder from the ground, and (c) the angle (relative to the horizontal) of that force on the ladder?

•8 A physics Brady Bunch, whose weights in newtons are indicated in Fig. 12-27, is balanced on a seesaw. What is the number of the person who causes the largest torque about the rotation axis at *fulcrum f* directed (a) out of the page and (b) into the page?

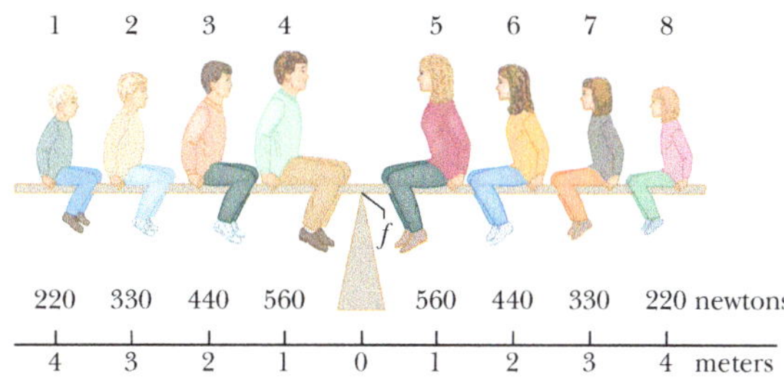

Figure 12-27 Problem 8.

•9 SSM A meter stick balances horizontally on a knife-edge at the 50.0 cm mark. With two 5.00 g coins stacked over the 12.0 cm mark, the stick is found to balance at the 45.5 cm mark. What is the mass of the meter stick?

•10 GO The system in Fig. 12-28 is in equilibrium, with the string in the center exactly horizontal. Block A weighs 40 N, block B weighs 50 N, and angle ϕ is 35°. Find (a) tension T_1, (b) tension T_2, (c) tension T_3, and (d) angle θ.

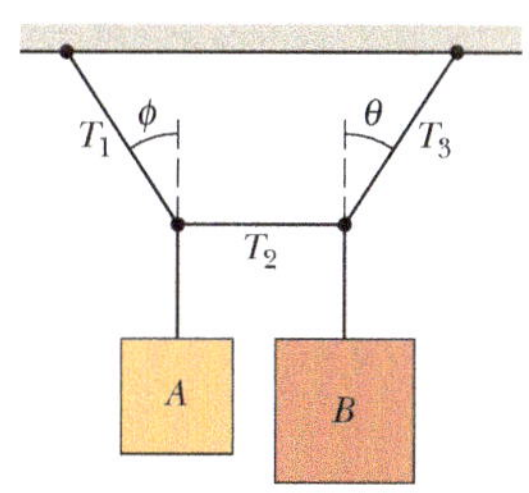

Figure 12-28 Problem 10.

•11 SSM Figure 12-29 shows a diver of weight 580 N standing at the end of a diving board with a length of $L = 4.5$ m and negligible mass. The board is fixed to two pedestals (supports) that are separated by distance $d = 1.5$ m. Of the forces acting on the board, what are the (a) magnitude and (b) direction (up or down) of the force from the left pedestal and the (c) magnitude and (d) direction (up or down) of the force from the right pedestal? (e) Which pedestal (left or right) is being stretched, and (f) which pedestal is being compressed?

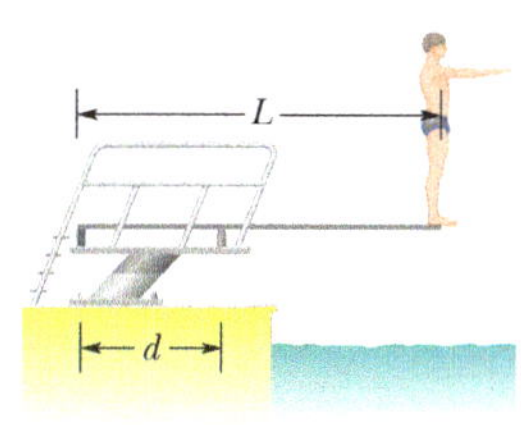

Figure 12-29 Problem 11.

•12 In Fig. 12-30, trying to get his car out of mud, a man ties one end of a rope around the front bumper and the other end tightly around a utility pole 18 m away. He then pushes sideways on the rope at its midpoint with a force of 550 N, displacing the center of the rope 0.30 m, but the car barely moves. What is the magnitude of the force on the car from the rope? (The rope stretches somewhat.)

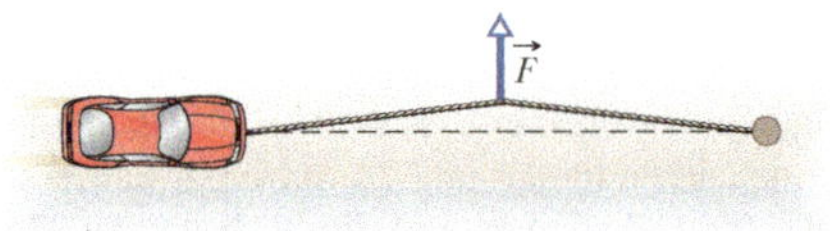

Figure 12-30 Problem 12.

•13 Figure 12-31 shows the anatomical structures in the lower leg and foot that are involved in standing on tiptoe, with the heel raised slightly off the floor so that the foot effectively contacts the floor only at point P. Assume distance $a = 5.0$ cm, distance $b = 15$ cm, and the person's weight $W = 900$ N. Of the forces acting on the foot, what are the (a) magnitude and (b) direction (up or down) of the force at point A from the calf muscle and the (c) magnitude and (d) direction (up or down) of the force at point B from the lower leg bones?

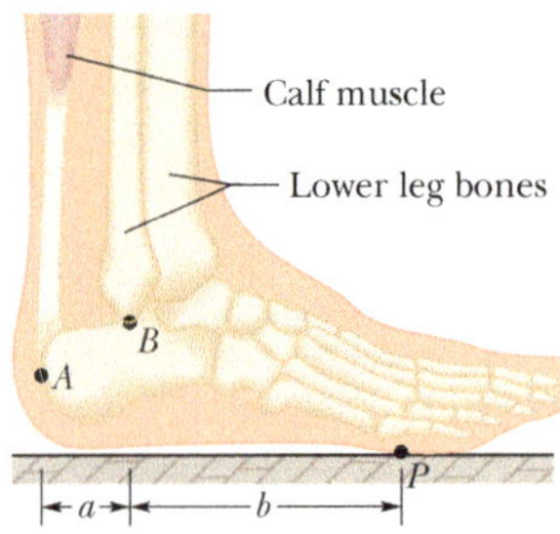

Figure 12-31 Problem 13.

•14 In Fig. 12-32, a horizontal scaffold, of length 2.00 m and uniform mass 50.0 kg, is suspended from a building by two cables. The scaffold has dozens of paint cans stacked on it at various points. The total mass of the paint cans is 75.0 kg. The tension in the cable at the right is 722 N. How far horizontally from *that* cable is the center of mass of the system of paint cans?

Figure 12-32 Problem 14.

•15 ILW Forces $\vec{F}_1$, $\vec{F}_2$, and $\vec{F}_3$ act on the structure of Fig. 12-33, shown in an overhead view. We wish to put the structure in equilibrium by applying a fourth force, at a point such as P. The fourth force has vector components $\vec{F}_h$ and $\vec{F}_v$. We are given that $a = 2.0$ m, $b = 3.0$ m, $c = 1.0$ m, $F_1 = 20$ N, $F_2 = 10$ N, and $F_3 = 5.0$ N. Find (a) F_h, (b) F_v, and (c) d.

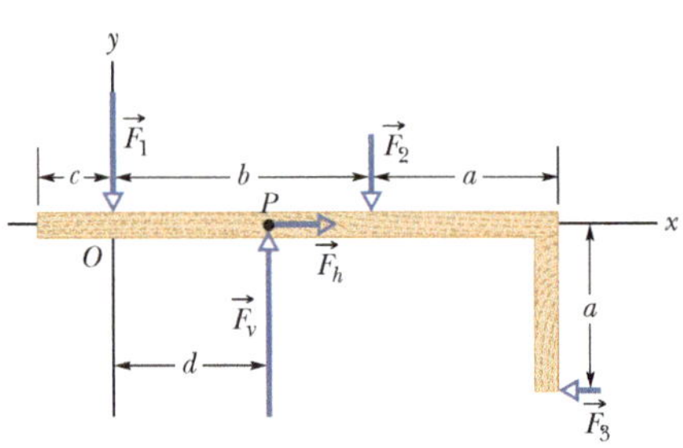

Figure 12-33 Problem 15.

•16 A uniform cubical crate is 0.750 m on each side and weighs 500 N. It rests on a floor with one edge against a very small, fixed obstruction. At what least height above the floor must a horizontal force of magnitude 350 N be applied to the crate to tip it?

•17 In Fig. 12-34, a uniform beam of weight 500 N and length 3.0 m is suspended horizontally. On the left it is hinged to a wall; on the right it is supported by a cable bolted to the wall at distance D above the beam. The least tension that will snap the cable is 1200 N. (a) What value of D corresponds to that tension? (b) To prevent the cable from snapping, should D be increased or decreased from that value?

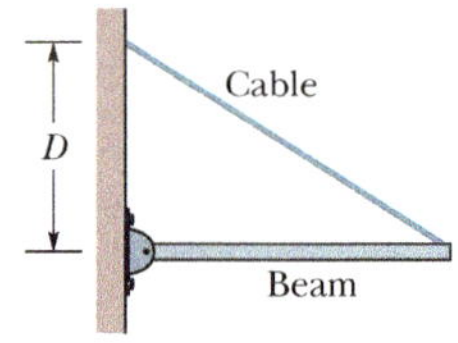

Figure 12-34 Problem 17.

•18 GO In Fig. 12-35, horizontal scaffold 2, with uniform mass $m_2 = 30.0$ kg and length $L_2 = 2.00$ m, hangs from horizontal scaffold 1, with uniform mass $m_1 = 50.0$ kg. A 20.0 kg box of nails lies on scaffold 2, centered at distance $d = 0.500$ m from the left end. What is the tension T in the cable indicated?

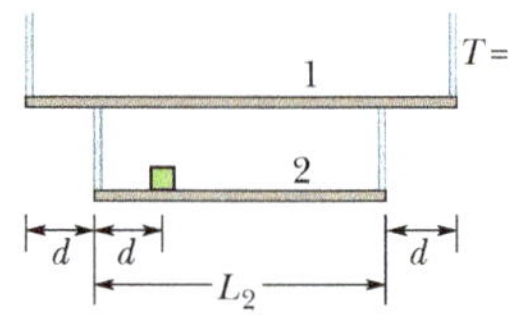

Figure 12-35 Problem 18.

•19 To crack a certain nut in a nutcracker, forces with magnitudes of at least 40 N must act on its shell from both sides. For the nutcracker of Fig. 12-36, with distances $L = 12$ cm and $d = 2.6$ cm, what are the force components $F_\perp$ (perpendicular to the handles) corresponding to that 40 N?

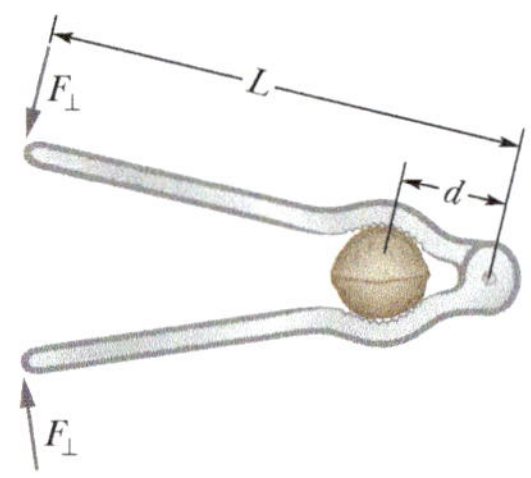

Figure 12-36 Problem 19.

•20 A bowler holds a bowling ball ($M = 7.2$ kg) in the palm of his hand (Fig. 12-37). His upper arm is vertical; his lower arm (1.8 kg) is horizontal. What is the magnitude of (a) the force of the biceps muscle on the lower arm and (b) the force between the bony structures at the elbow contact point?

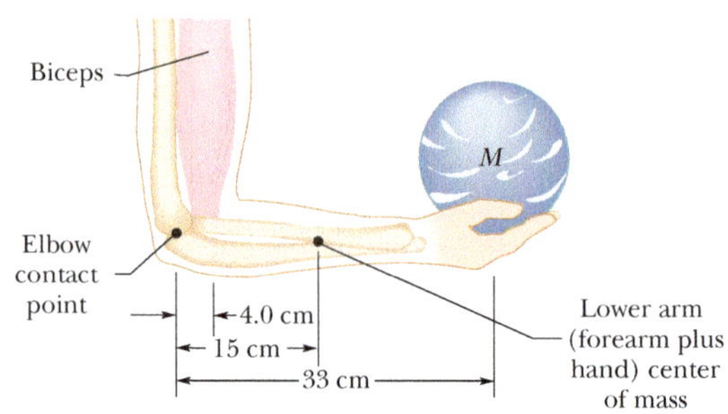

Figure 12-37 Problem 20.

••21 ILW The system in Fig. 12-38 is in equilibrium. A concrete block of mass 225 kg hangs from the end of the uniform strut of mass 45.0 kg. A cable runs from the ground, over the top of the strut, and down to the block, holding the block in place. For angles $\phi = 30.0°$ and $\theta = 45.0°$, find (a) the tension T in the cable and the (b) horizontal and (c) vertical components of the force on the strut from the hinge.

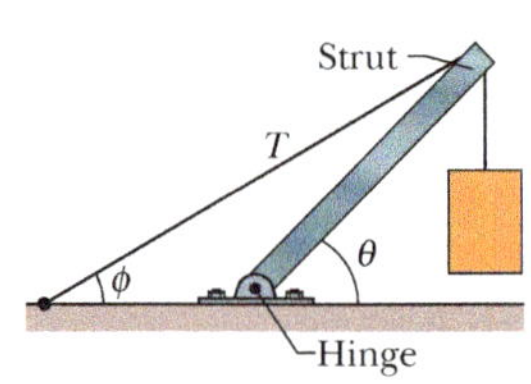

Figure 12-38 Problem 21.

••22 In Fig. 12-39, a 55 kg rock climber is in a lie-back climb along a fissure, with hands pulling on one side of the fissure and feet pressed against the opposite side. The fissure has width $w = 0.20$ m, and the center of mass of the climber is a horizontal distance $d = 0.40$ m from the fissure. The coefficient of static friction between hands and rock is $\mu_1 = 0.40$, and between boots and rock it is $\mu_2 = 1.2$. (a) What is the least horizontal pull by the hands and push by the feet that will keep the climber stable? (b) For the horizontal pull of (a), what must be the vertical distance h between hands and feet? If the climber encounters wet rock, so that μ_1 and μ_2 are reduced, what happens to (c) the answer to (a) and (d) the answer to (b)?

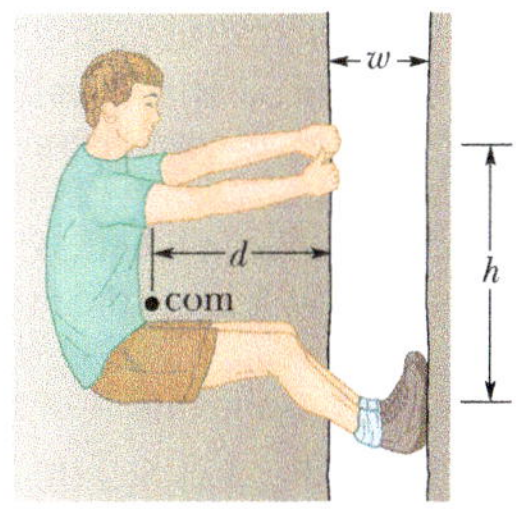

Figure 12-39 Problem 22.

••23 In Fig. 12-40, one end of a uniform beam of weight 222 N is hinged to a wall; the other end is supported by a wire that makes angles $\theta = 30.0°$ with both wall and beam. Find (a) the tension in the wire and the (b) horizontal and (c) vertical components of the force of the hinge on the beam.

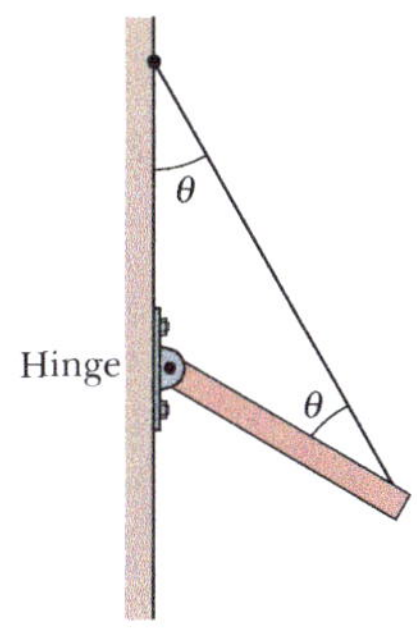

Figure 12-40 Problem 23.

••24 In Fig. 12-41, a climber with a weight of 533.8 N is held by a belay rope connected to her climbing harness and belay device; the force of the rope on her has a line of action through her center of mass. The indicated angles are $\theta = 40.0°$ and $\phi = 30.0°$. If her feet are on the verge of sliding on the vertical wall, what is the coefficient of static friction between her climbing shoes and the wall?

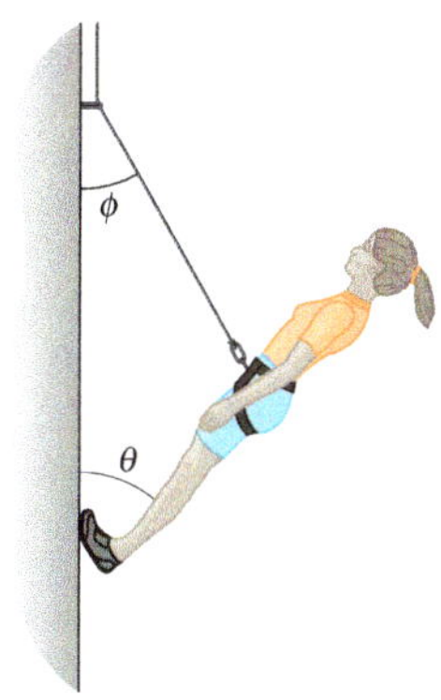

Figure 12-41 Problem 24.

••25 SSM WWW In Fig. 12-42, what magnitude of (constant) force $\vec{F}$ applied horizontally at the axle of the wheel is necessary to raise the wheel over a step obstacle of height $h = 3.00$ cm? The wheel's radius is $r = 6.00$ cm, and its mass is $m = 0.800$ kg.

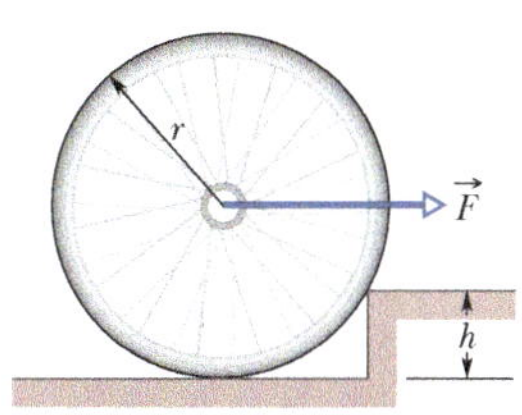

Figure 12-42 Problem 25.

••26 In Fig. 12-43, a climber leans out against a vertical ice wall that has negligible friction. Distance a is 0.914 m and distance L is 2.10 m. His center of mass is distance $d = 0.940$ m from the feet–ground contact point. If he is on the verge of sliding, what is the coefficient of static friction between feet and ground?

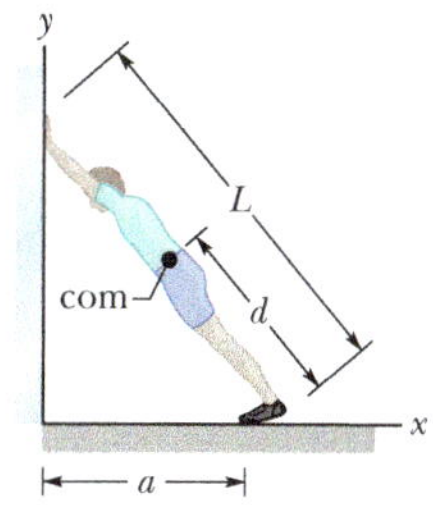

Figure 12-43 Problem 26.

••27 In Fig. 12-44, a 15 kg block is held in place via a pulley system. The person's upper arm is vertical; the forearm is at angle $\theta = 30°$ with the horizontal. Forearm and hand together have a mass of 2.0 kg, with a center of mass at distance $d_1 = 15$ cm from the contact point of the forearm bone and the upper-arm bone (humerus). The triceps muscle pulls vertically upward on the forearm at distance $d_2 = 2.5$ cm behind that contact point. Distance d_3 is 35 cm. What are the (a) magnitude and (b) direction (up or down) of the force on the forearm from the triceps muscle and the (c) magnitude and (d) direction (up or down) of the force on the forearm from the humerus?

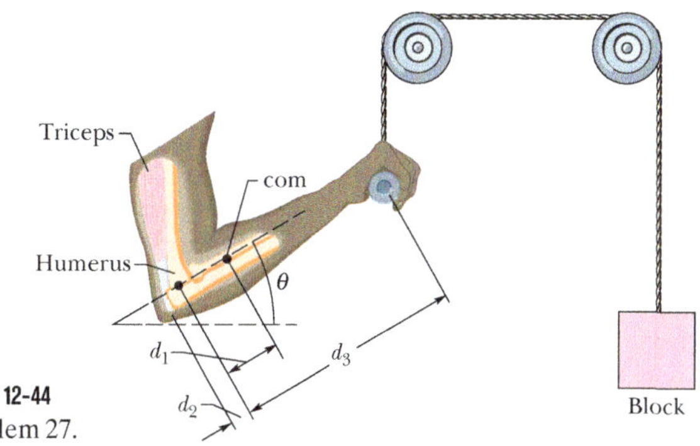

Figure 12-44 Problem 27.

••28 In Fig. 12-45, suppose the length L of the uniform bar is 3.00 m and its weight is 200 N. Also, let the block's weight $W = 300$ N and the angle $\theta = 30.0°$. The wire can withstand a maximum tension of 500 N. (a) What is the maximum possible distance x before the wire breaks? With the block placed at this maximum x, what are the (b) horizontal and (c) vertical components of the force on the bar from the hinge at A?

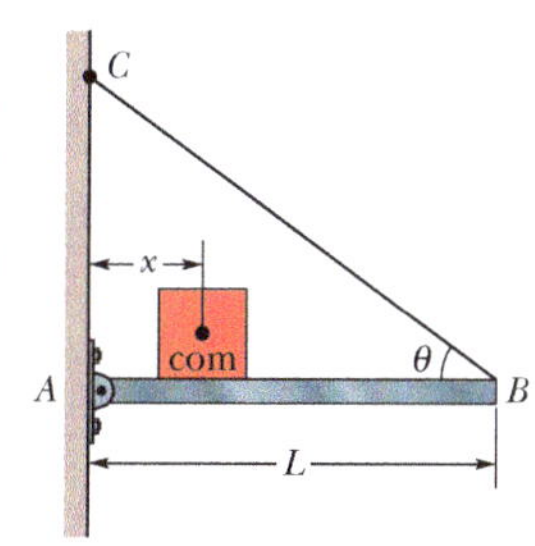

Figure 12-45 Problems 28 and 34.

••29 A door has a height of 2.1 m along a y axis that extends vertically upward and a width of 0.91 m along an x axis that extends outward from the hinged edge of the door. A hinge 0.30 m from the top and a hinge 0.30 m from the bottom each support half the door's mass, which is 27 kg. In unit-vector notation, what are the forces on the door at (a) the top hinge and (b) the bottom hinge?

••30 In Fig. 12-46, a 50.0 kg uniform square sign, of edge length $L = 2.00$ m, is hung from a horizontal rod of length $d_h = 3.00$ m and negligible mass. A cable is attached to the end of the rod

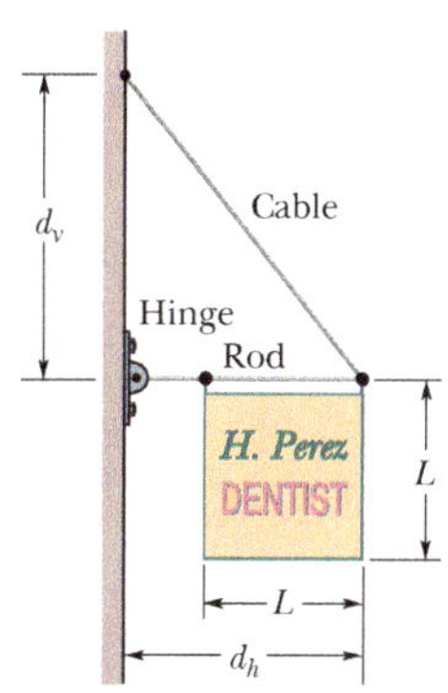

Figure 12-46 Problem 30.

and to a point on the wall at distance $d_v = 4.00$ m above the point where the rod is hinged to the wall. (a) What is the tension in the cable? What are the (b) magnitude and (c) direction (left or right) of the horizontal component of the force on the rod from the wall, and the (d) magnitude and (e) direction (up or down) of the vertical component of this force?

••31 GO In Fig. 12-47, a nonuniform bar is suspended at rest in a horizontal position by two massless cords. One cord makes the angle $\theta = 36.9°$ with the vertical; the other makes the angle $\phi = 53.1°$ with the vertical. If the length L of the bar is 6.10 m, compute the distance x from the left end of the bar to its center of mass.

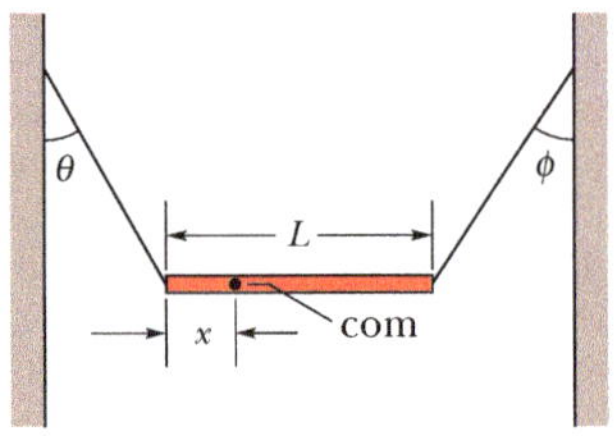

Figure 12-47 Problem 31.

••32 In Fig. 12-48, the driver of a car on a horizontal road makes an emergency stop by applying the brakes so that all four wheels lock and skid along the road. The coefficient of kinetic friction between tires and road is 0.40. The separation between the front and rear axles is $L = 4.2$ m, and the center of mass of the car is located at distance $d = 1.8$ m behind the front axle and distance $h = 0.75$ m above the road. The car weighs 11 kN. Find the magnitude of (a) the braking acceleration of the car, (b) the normal force on each rear wheel, (c) the normal force on each front wheel, (d) the braking force on each rear wheel, and (e) the braking force on each front wheel. (*Hint:* Although the car is not in translational equilibrium, it *is* in rotational equilibrium.)

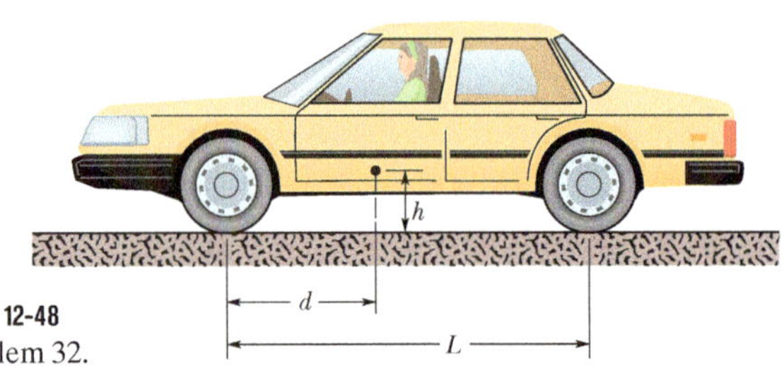

Figure 12-48 Problem 32.

••33 Figure 12-49*a* shows a vertical uniform beam of length L that is hinged at its lower end. A horizontal force $\vec{F}_a$ is applied to the beam at distance y from the lower end. The beam remains vertical because of a cable attached at the upper end, at angle θ with the horizontal. Figure 12-49*b* gives the tension T in the cable as a function of the position of the applied force given as a fraction y/L of the beam length. The scale of the T axis is set by $T_s = 600$ N. Figure 12-49*c* gives the magnitude F_h of the horizontal force on the beam from the hinge, also as a function of y/L. Evaluate (a) angle θ and (b) the magnitude of $\vec{F}_a$.

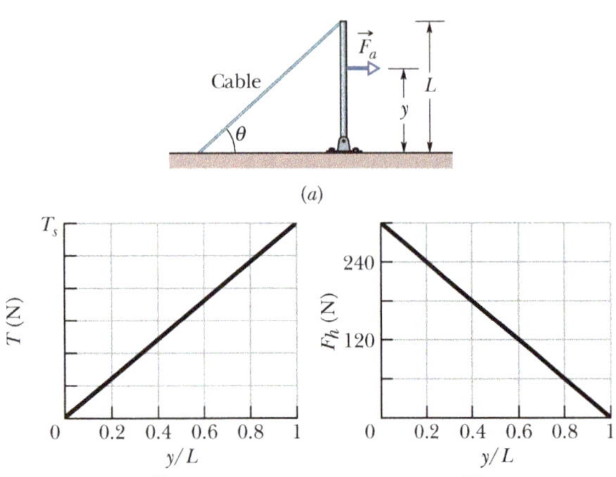

Figure 12-49 Problem 33.

••34 In Fig. 12-45, a thin horizontal bar AB of negligible weight and length L is hinged to a vertical wall at A and supported at B by a thin wire BC that makes an angle θ with the horizontal. A block of weight W can be moved anywhere along the bar; its position is defined by the distance x from the wall to its center of mass. As a function of x, find (a) the tension in the wire, and the (b) horizontal and (c) vertical components of the force on the bar from the hinge at A.

••35 SSM WWW A cubical box is filled with sand and weighs 890 N. We wish to "roll" the box by pushing horizontally on one of the upper edges. (a) What minimum force is required? (b) What minimum coefficient of static friction between box and floor is required? (c) If there is a more efficient way to roll the box, find the smallest possible force that would have to be applied directly to the box to roll it. (*Hint:* At the onset of tipping, where is the normal force located?)

••36 Figure 12-50 shows a 70 kg climber hanging by only the *crimp hold* of one hand on the edge of a shallow horizontal ledge in a rock wall. (The fingers are pressed down to gain purchase.) Her feet touch the rock wall at distance $H = 2.0$ m directly below her crimped fingers but do not provide any support. Her center of mass is distance $a = 0.20$ m from the wall. Assume that the force from the ledge supporting her fingers is equally shared by the four fingers. What are the values of the (a) horizontal component F_h and (b) vertical component F_v of the force on *each* fingertip?

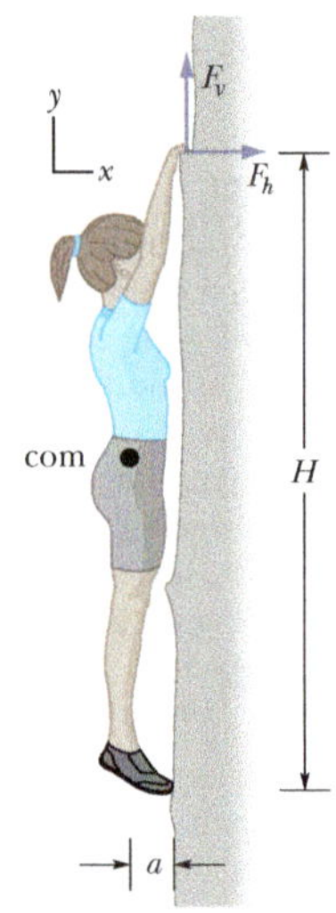

Figure 12-50 Problem 36.

••37 GO In Fig. 12-51, a uniform plank, with a length L of 6.10 m and a weight of 445 N, rests on the ground and against a frictionless roller at the top of a wall of height $h = 3.05$ m. The plank remains in equilibrium for any value of $\theta \geq 70°$ but slips if $\theta < 70°$. Find the coefficient of static friction between the plank and the ground.

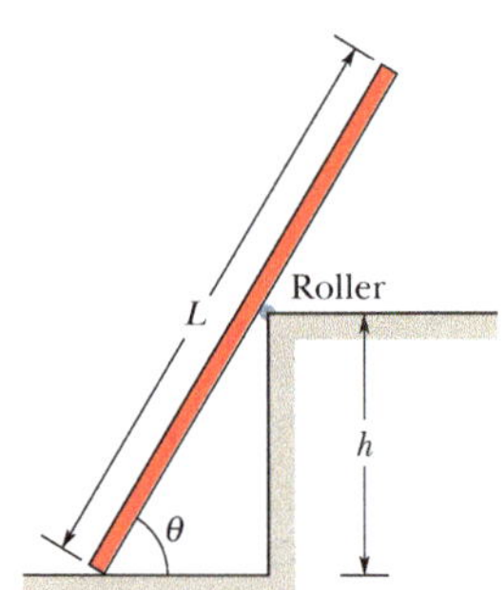

Figure 12-51 Problem 37.

••38 In Fig. 12-52, uniform beams A and B are attached to a wall with hinges and loosely bolted together (there is no torque of one on the other). Beam A has length $L_A = 2.40$ m and mass 54.0 kg; beam B has mass 68.0 kg. The two hinge points are separated by distance $d = 1.80$ m. In unit-vector notation, what is the force on (a) beam A due to its hinge, (b) beam A due to the bolt, (c) beam B due to its hinge, and (d) beam B due to the bolt?

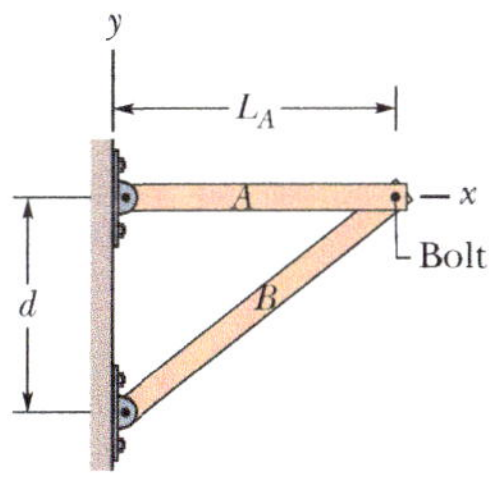

Figure 12-52 Problem 38.

•••39 For the stepladder shown in Fig. 12-53, sides AC and CE are each 2.44 m long and hinged at C. Bar BD is a tie-rod 0.762 m long, halfway up. A man weighing 854 N climbs 1.80 m along the ladder. Assuming that the floor is frictionless and neglecting the mass of the ladder, find (a) the tension in the tie-rod and the magnitudes of the forces on the ladder from the floor at (b) A and (c) E. (*Hint:* Isolate parts of the ladder in applying the equilibrium conditions.)

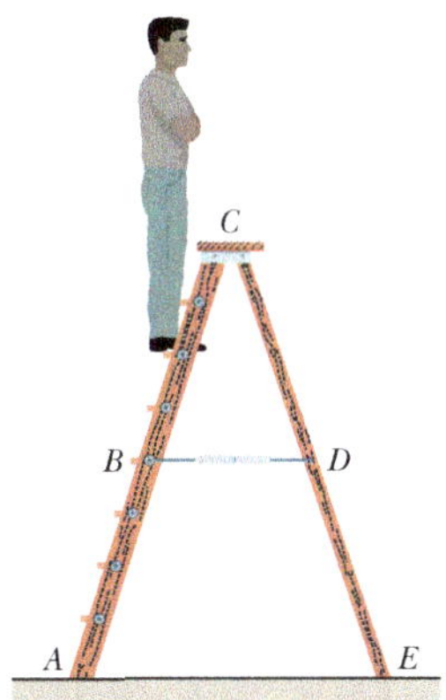

Figure 12-53 Problem 39.

•••40 Figure 12-54*a* shows a horizontal uniform beam of mass m_b and length L that is supported on the left by a hinge attached to a wall and on the right by a cable at angle θ with the horizontal. A package of mass m_p is positioned on the beam at a distance x from the left end. The total mass is $m_b + m_p = 61.22$ kg. Figure 12-54*b* gives the tension T in the cable as a function of the package's position given as a fraction x/L of the beam length. The scale of the T axis is set by $T_a = 500$ N and $T_b = 700$ N. Evaluate (a) angle θ, (b) mass m_b, and (c) mass m_p.

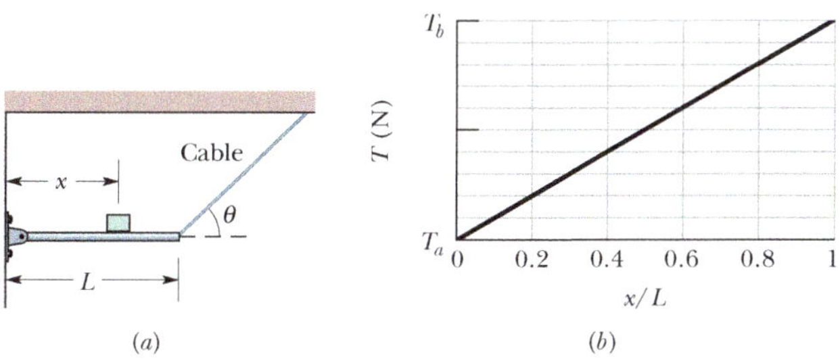

Figure 12-54 Problem 40.

•••41 A crate, in the form of a cube with edge lengths of 1.2 m, contains a piece of machinery; the center of mass of the crate and its contents is located 0.30 m above the crate's geometrical center. The crate rests on a ramp that makes an angle θ with the horizontal. As θ is increased from zero, an angle will be reached at which the crate will either tip over or start to slide down the ramp. If the coefficient of static friction μ_s between ramp and crate is 0.60, (a) does the crate tip or slide and (b) at what angle θ does this occur? If $\mu_s = 0.70$, (c) does the crate tip or slide and (d) at what angle θ does this occur? (*Hint:* At the onset of tipping, where is the normal force located?)

•••42 In Fig. 12-7 and the associated sample problem, let the coefficient of static friction μ_s between the ladder and the pavement be 0.53. How far (in percent) up the ladder must the firefighter go to put the ladder on the verge of sliding?

Module 12-3 Elasticity

•43 SSM ILW A horizontal aluminum rod 4.8 cm in diameter projects 5.3 cm from a wall. A 1200 kg object is suspended from the end of the rod. The shear modulus of aluminum is 3.0×10^{10} N/m^2. Neglecting the rod's mass, find (a) the shear stress on the rod and (b) the vertical deflection of the end of the rod.

•44 Figure 12-55 shows the stress–strain curve for a material. The scale of the stress axis is set by $s = 300$, in units of 10^6 N/m^2. What are (a) the Young's modulus and (b) the approximate yield strength for this material?

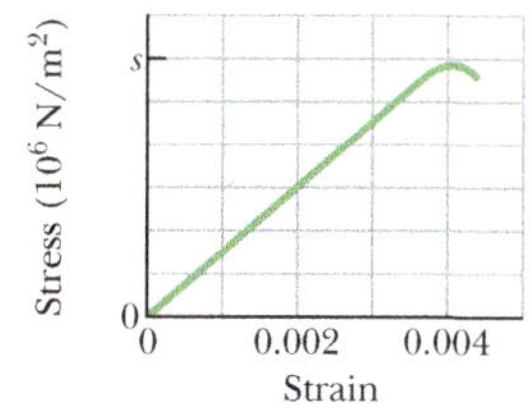

Figure 12-55 Problem 44.

••45 In Fig. 12-56, a lead brick rests horizontally on cylinders A and B. The areas of the top faces of the cylinders are related by $A_A = 2A_B$; the Young's moduli of the cylinders are related by $E_A = 2E_B$. The cylinders had identical lengths before the brick was placed on them. What fraction of the brick's mass is supported (a) by cylinder A and (b) by cylinder B? The horizontal distances between the center of mass of the brick and the centerlines of the cylinders are d_A for cylinder A and d_B for cylinder B. (c) What is the ratio d_A/d_B?

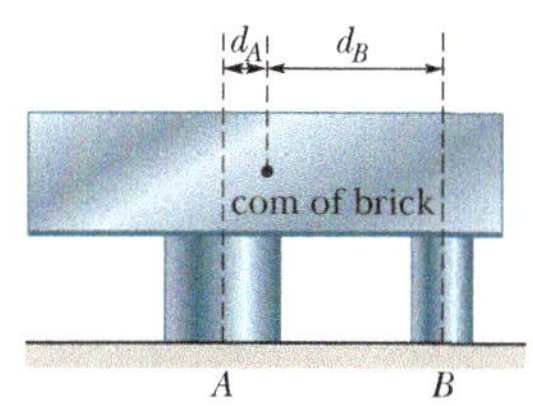

Figure 12-56 Problem 45.

••46 Figure 12-57 shows an approximate plot of stress versus strain for a spider-web thread, out to the point of breaking at a strain of 2.00. The vertical axis scale is set by values $a = 0.12$ GN/m^2, $b = 0.30$ GN/m^2, and $c = 0.80$ GN/m^2. Assume that the thread has an initial length of 0.80 cm, an initial cross-sectional area of 8.0×10^{-12} m^2, and (during stretching) a constant volume. The strain on the thread is the ratio of the change in the thread's length to that initial length, and the stress on the thread is the ratio of the collision force to that initial cross-sectional area. Assume that the work done on the thread by the collision force is given by the area under the curve on the graph. Assume also that when the single thread snares a flying insect, the insect's kinetic energy is transferred to the stretching of the thread. (a) How much kinetic energy would put the thread on the verge of breaking? What is the kinetic energy of (b) a fruit fly of mass 6.00 mg and speed 1.70 m/s and (c) a bumble bee of mass 0.388 g and speed 0.420 m/s? Would (d) the fruit fly and (e) the bumble bee break the thread?

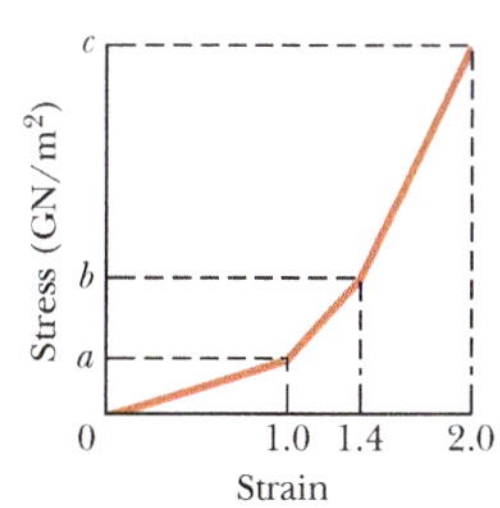

Figure 12-57 Problem 46.

••47 A tunnel of length $L = 150$ m, height $H = 7.2$ m, and width 5.8 m (with a flat roof) is to be constructed at distance $d = 60$ m beneath the ground. (See Fig. 12-58.) The tunnel roof is to be supported entirely by square steel columns, each with a cross-sectional area of 960 cm^2. The mass of 1.0 cm^3 of the ground material is 2.8 g. (a) What is the total weight of the ground material the columns must support? (b) How many columns are needed to keep the compressive stress on each column at one-half its ultimate strength?

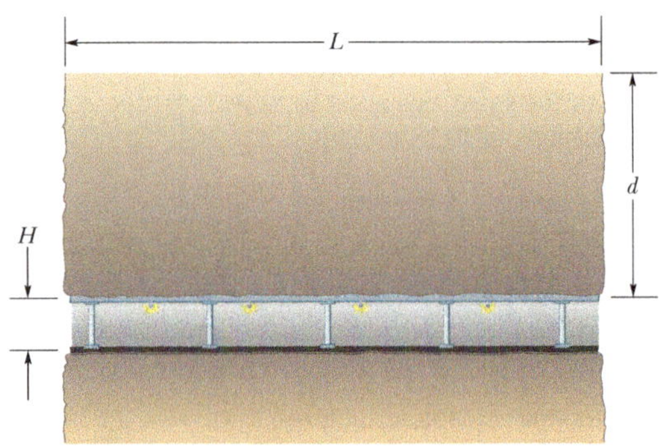

Figure 12-58 Problem 47.

••48 Figure 12-59 shows the stress versus strain plot for an aluminum wire that is stretched by a machine pulling in opposite directions at the two ends of the wire. The scale of the stress axis is set by $s = 7.0$, in units of 10^7 N/m^2. The wire has an initial length of 0.800 m and an initial cross-sectional area of 2.00×10^{-6} m^2. How much work does the force from the machine do on the wire to produce a strain of 1.00×10^{-3}?

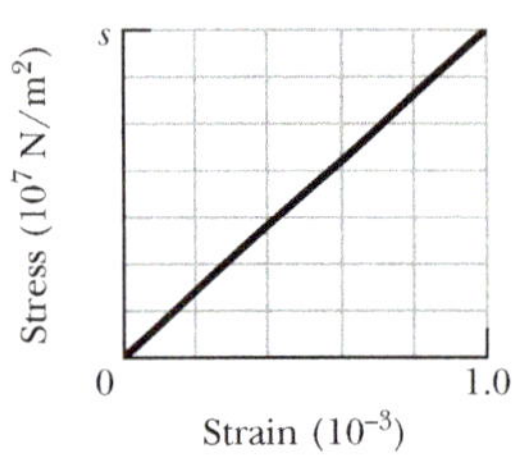

Figure 12-59 Problem 48.

••49 GO In Fig. 12-60, a 103 kg uniform log hangs by two steel wires, A and B, both of radius 1.20 mm. Initially, wire A was 2.50 m long and 2.00 mm shorter than wire B. The log is now horizontal. What are the magnitudes of the forces on it from (a) wire A and (b) wire B? (c) What is the ratio d_A/d_B?

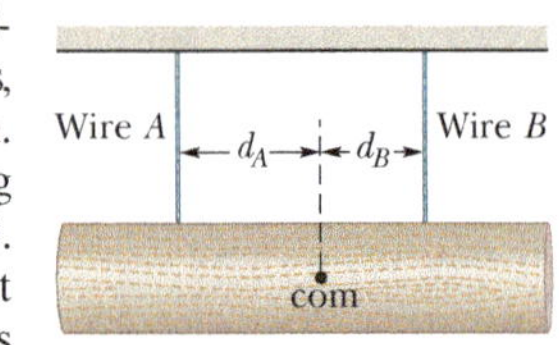

Figure 12-60 Problem 49.

•••50 GO Figure 12-61 represents an insect caught at the midpoint of a spider-web thread. The thread breaks under a stress of 8.20×10^8 N/m^2 and a strain of 2.00. Initially, it was horizontal and had a length of 2.00 cm and a cross-sectional area of 8.00×10^{-12} m^2. As the thread was stretched under the weight of the insect, its volume remained constant. If the weight of the insect puts the thread on the verge of breaking, what is the insect's mass? (A spider's web is built to break if a potentially harmful insect, such as a bumble bee, becomes snared in the web.)

Figure 12-61 Problem 50.

•••51 GO Figure 12-62 is an overhead view of a rigid rod that turns about a vertical axle until the identical rubber stoppers A and B are forced against rigid walls at distances $r_A = 7.0$ cm and $r_B = 4.0$ cm from the axle. Initially the stoppers touch the walls without being compressed. Then force $\vec{F}$ of magnitude 220 N is applied perpendicular to the rod at a distance $R = 5.0$ cm from the axle. Find the magnitude of the force compressing (a) stopper A and (b) stopper B.

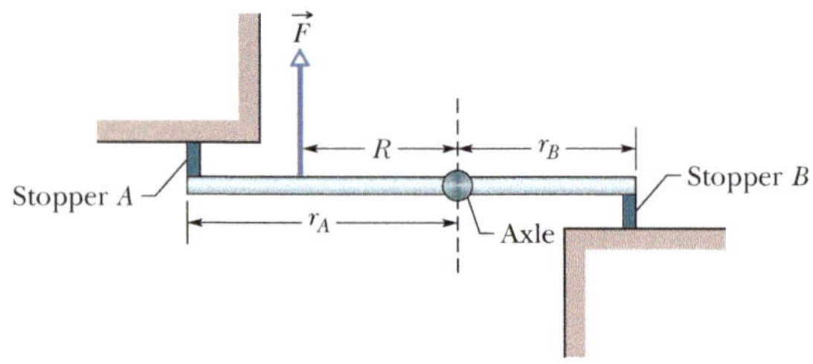

Figure 12-62 Problem 51.

Additional Problems

52 After a fall, a 95 kg rock climber finds himself dangling from the end of a rope that had been 15 m long and 9.6 mm in diameter but has stretched by 2.8 cm. For the rope, calculate (a) the strain, (b) the stress, and (c) the Young's modulus.

53 SSM In Fig. 12-63, a rectangular slab of slate rests on a bedrock surface inclined at angle $\theta = 26°$. The slab has length $L = 43$ m, thickness $T = 2.5$ m, and width $W = 12$ m, and 1.0 cm^3 of it has a mass of 3.2 g. The coefficient of static friction between slab and bedrock is 0.39. (a) Calculate the component of the gravitational force on the slab parallel to the bedrock surface. (b) Calculate the magnitude of the static frictional force on the slab. By comparing (a) and (b), you can see that the slab is in danger of sliding. This is prevented only by chance protrusions of bedrock. (c) To stabilize the slab, bolts are to be driven perpendicular to the bedrock surface (two bolts are shown). If each bolt has a cross-sectional area of 6.4 cm^2 and will snap under a shearing stress of 3.6×10^8 N/m^2, what is the minimum number of bolts needed? Assume that the bolts do not affect the normal force.

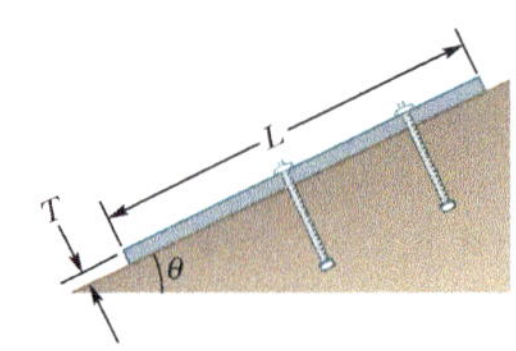

Figure 12-63 Problem 53.

54 A uniform ladder whose length is 5.0 m and whose weight is 400 N leans against a frictionless vertical wall. The coefficient of static friction between the level ground and the foot of the ladder is 0.46. What is the greatest distance the foot of the ladder can be placed from the base of the wall without the ladder immediately slipping?

55 SSM In Fig. 12-64, block A (mass 10 kg) is in equilibrium, but it would slip if block B (mass 5.0 kg) were any heavier. For angle $\theta = 30°$, what is the coefficient of static friction between block A and the surface below it?

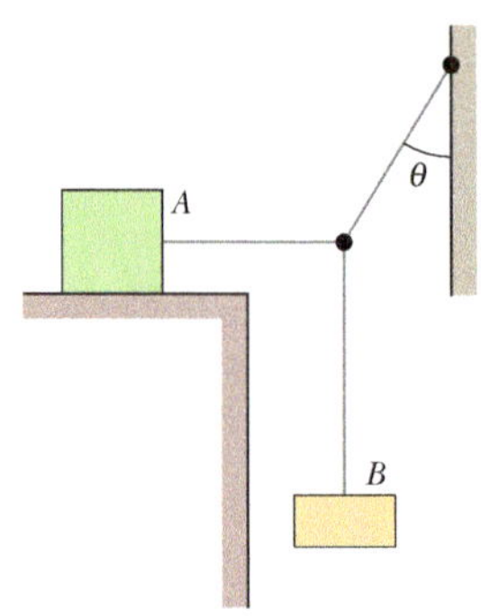

Figure 12-64 Problem 55.

56 Figure 12-65a shows a uniform ramp between two buildings that allows for motion between the buildings due to strong winds.

At its left end, it is hinged to the building wall; at its right end, it has a roller that can roll along the building wall. There is no vertical force on the roller from the building, only a horizontal force with magnitude F_h. The horizontal distance between the buildings is $D = 4.00$ m. The rise of the ramp is $h = 0.490$ m. A man walks across the ramp from the left. Figure 12-65*b* gives F_h as a function of the horizontal distance x of the man from the building at the left. The scale of the F_h axis is set by $a = 20$ kN and $b = 25$ kN. What are the masses of (a) the ramp and (b) the man?

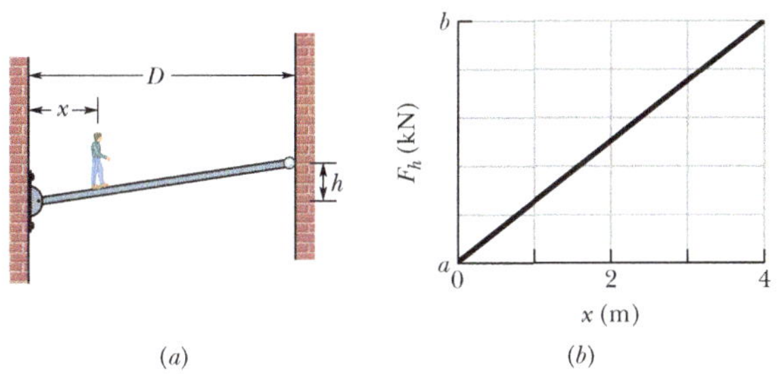

Figure 12-65 Problem 56.

57 GO In Fig. 12-66, a 10 kg sphere is supported on a frictionless plane inclined at angle $\theta = 45°$ from the horizontal. Angle ϕ is 25°. Calculate the tension in the cable.

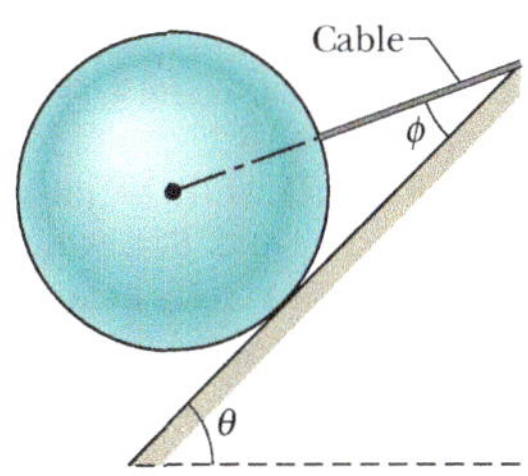

Figure 12-66 Problem 57.

58 In Fig. 12-67*a*, a uniform 40.0 kg beam is centered over two rollers. Vertical lines across the beam mark off equal lengths. Two of the lines are centered over the rollers; a 10.0 kg package of tamales is centered over roller *B*. What are the magnitudes of the forces on the beam from (a) roller *A* and (b) roller *B*? The beam is then rolled to the left until the right-hand end is centered over roller *B* (Fig. 12-67*b*). What now are the magnitudes of the forces on the beam from (c) roller *A* and (d) roller *B*? Next, the beam is rolled to the right. Assume that it has a length of 0.800 m. (e) What horizontal distance between the package and roller *B* puts the beam on the verge of losing contact with roller *A*?

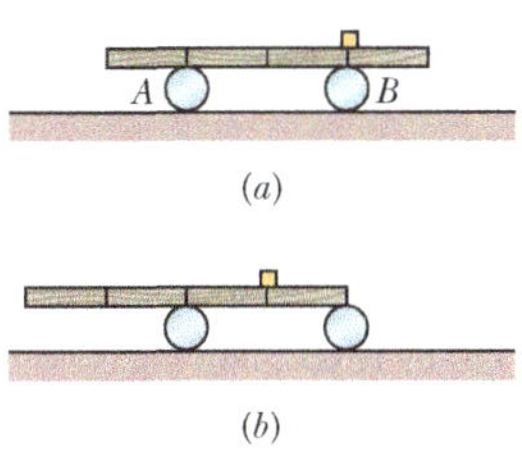

Figure 12-67 Problem 58.

59 SSM In Fig. 12-68, an 817 kg construction bucket is suspended by a cable *A* that is attached at *O* to two other cables *B* and *C*, making angles $\theta_1 = 51.0°$ and $\theta_2 = 66.0°$ with the horizontal. Find the tensions in (a) cable *A*, (b) cable *B*, and (c) cable *C*. (*Hint:* To avoid solving two equations in two unknowns, position the axes as shown in the figure.)

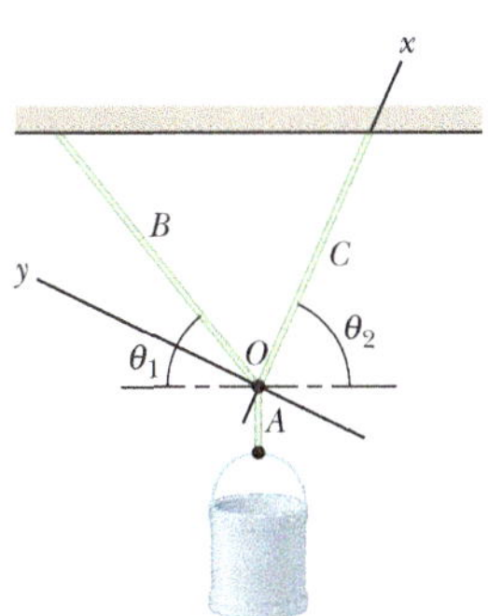

Figure 12-68 Problem 59.

60 In Fig. 12-69, a package of mass m hangs from a short cord that is tied to the wall via cord 1 and to the ceiling via cord 2. Cord 1 is at angle $\phi = 40°$ with the horizontal; cord 2 is at angle θ. (a) For what value of θ is the tension in cord 2 minimized? (b) In terms of mg, what is the minimum tension in cord 2?

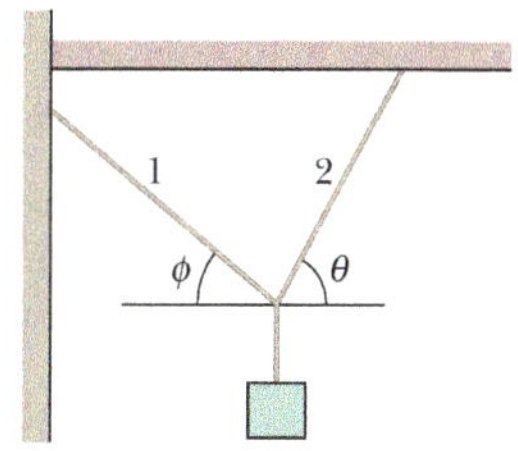

Figure 12-69 Problem 60.

61 ILW The force $\vec{F}$ in Fig. 12-70 keeps the 6.40 kg block and the pulleys in equilibrium. The pulleys have negligible mass and friction. Calculate the tension T in the upper cable. (*Hint:* When a cable wraps halfway around a pulley as here, the magnitude of its net force on the pulley is twice the tension in the cable.)

62 A mine elevator is supported by a single steel cable 2.5 cm in diameter. The total mass of the elevator cage and occupants is 670 kg. By how much does the cable stretch when the elevator hangs by (a) 12 m of cable and (b) 362 m of cable? (Neglect the mass of the cable.)

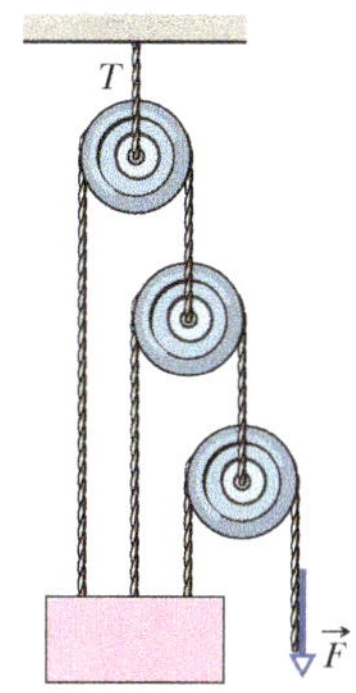

Figure 12-70 Problem 61.

63 Four bricks of length L, identical and uniform, are stacked on top of one another (Fig. 12-71) in such a way that part of each extends beyond the one beneath. Find, in terms of L, the maximum values of (a) a_1, (b) a_2, (c) a_3, (d) a_4, and (e) h, such that the stack is in equilibrium, on the verge of falling.

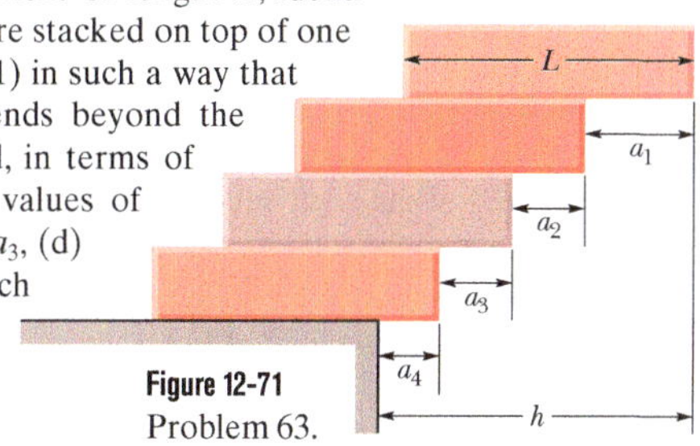

Figure 12-71 Problem 63.

64 In Fig. 12-72, two identical, uniform, and frictionless spheres, each of mass m, rest in a rigid rectangular container. A line connecting their centers is at 45° to the horizontal. Find the magnitudes of the forces on the spheres from (a) the bottom of the container, (b) the left side of the container, (c) the right side of the container, and (d) each other. (*Hint:* The force of one sphere on the other is directed along the center–center line.)

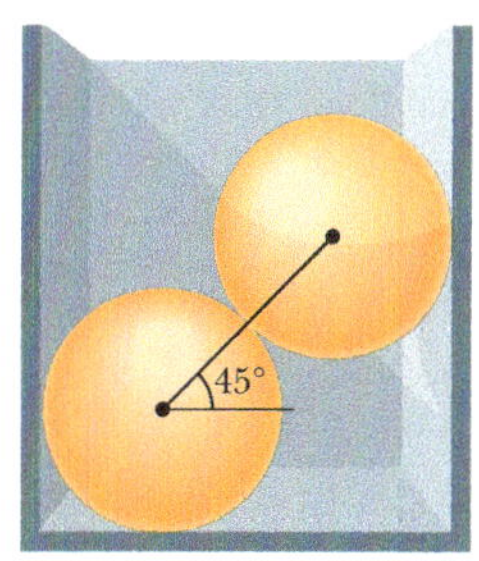

Figure 12-72 Problem 64.

65 In Fig. 12-73, a uniform beam with a weight of 60 N and a length of 3.2 m is hinged at its lower end, and a horizontal force $\vec{F}$ of magnitude 50 N acts at its upper end. The beam is held vertical by a cable that makes angle $\theta = 25°$ with the ground and is attached to the beam at height $h = 2.0$ m. What are (a) the tension in the cable and (b) the force on the beam from the hinge in unit-vector notation?

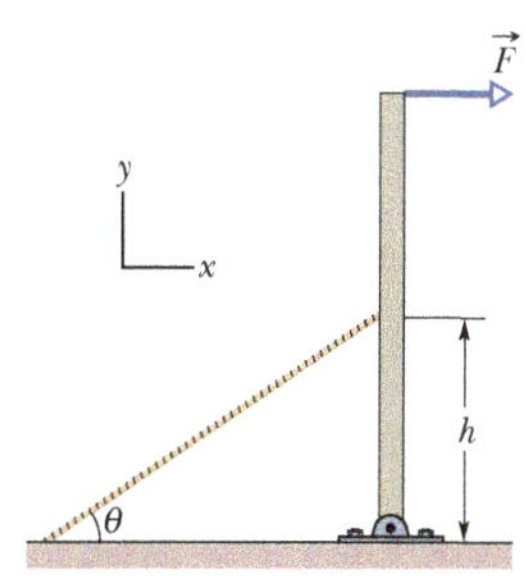

Figure 12-73 Problem 65.

66 A uniform beam is 5.0 m long and has a mass of 53 kg. In Fig. 12-74, the beam is supported in a horizontal position by a hinge and a cable, with angle $\theta = 60°$. In unit-vector notation, what is the force on the beam from the hinge?

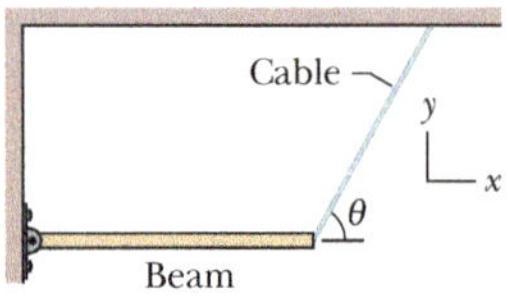

Figure 12-74 Problem 66.

67 A solid copper cube has an edge length of 85.5 cm. How much stress must be applied to the cube to reduce the edge length to 85.0 cm? The bulk modulus of copper is 1.4×10^{11} N/m^2.

68 A construction worker attempts to lift a uniform beam off the floor and raise it to a vertical position. The beam is 2.50 m long and weighs 500 N. At a certain instant the worker holds the beam momentarily at rest with one end at distance d = 1.50 m above the floor, as shown in Fig. 12-75, by exerting a force $\vec{P}$ on the beam, perpendicular to the beam. (a) What is the magnitude P? (b) What is the magnitude of the (net) force of the floor on the beam? (c) What is the minimum value the coefficient of static friction between beam and floor can have in order for the beam not to slip at this instant?

Figure 12-75 Problem 68.

69 SSM In Fig. 12-76, a uniform rod of mass m is hinged to a building at its lower end, while its upper end is held in place by a rope attached to the wall. If angle $\theta_1 = 60°$, what value must angle θ_2 have so that the tension in the rope is equal to $mg/2$?

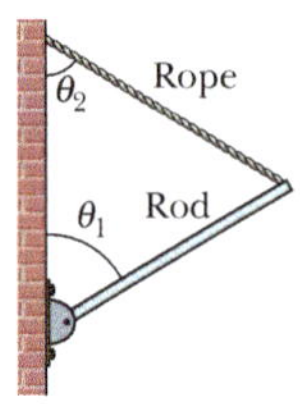

Figure 12-76 Problem 69.

70 A 73 kg man stands on a level bridge of length L. He is at distance $L/4$ from one end. The bridge is uniform and weighs 2.7 kN. What are the magnitudes of the vertical forces on the bridge from its supports at (a) the end farther from him and (b) the nearer end?

71 SSM A uniform cube of side length 8.0 cm rests on a horizontal floor. The coefficient of static friction between cube and floor is μ. A horizontal pull $\vec{P}$ is applied perpendicular to one of the vertical faces of the cube, at a distance 7.0 cm above the floor on the vertical midline of the cube face. The magnitude of $\vec{P}$ is gradually increased. During that increase, for what values of μ will the cube eventually (a) begin to slide and (b) begin to tip? (*Hint:* At the onset of tipping, where is the normal force located?)

72 The system in Fig. 12-77 is in equilibrium. The angles are $\theta_1 = 60°$ and $\theta_2 = 20°$, and the ball has mass M = 2.0 kg. What is the tension in (a) string ab and (b) string bc?

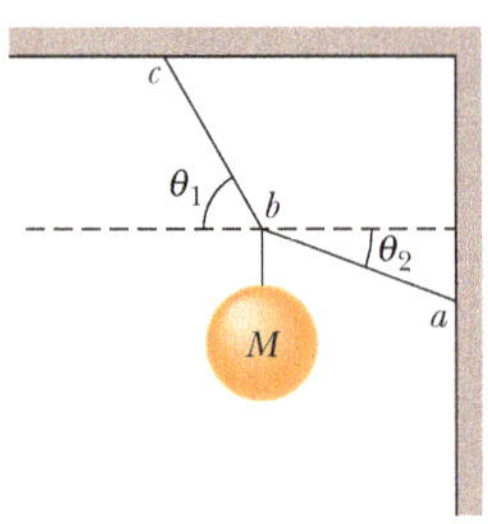

Figure 12-77 Problem 72.

73 SSM A uniform ladder is 10 m long and weighs 200 N. In Fig. 12-78, the ladder leans against a vertical, frictionless wall at height h = 8.0 m above the ground. A horizontal force $\vec{F}$ is applied to the ladder at distance d = 2.0 m from its base (measured along the ladder). (a) If force magnitude F = 50 N, what is the force of the ground on the ladder, in unit-vector notation? (b) If F = 150 N, what is the force of the ground on the ladder, also in unit-vector notation? (c) Suppose the coefficient of static friction between the ladder and the ground is 0.38; for what minimum value of the force magnitude F will the base of the ladder just barely start to move toward the wall?

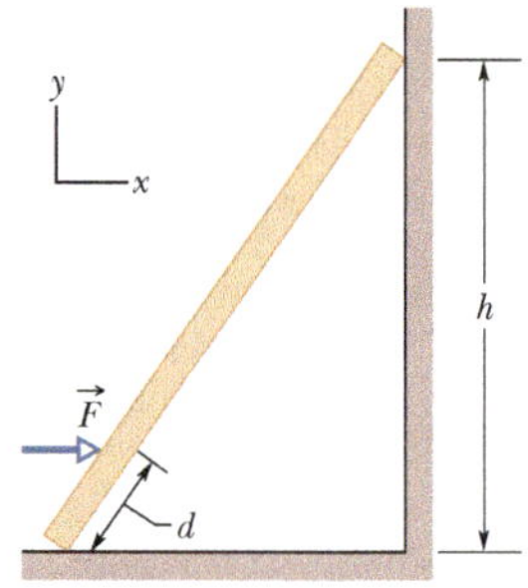

Figure 12-78 Problem 73.

74 A pan balance is made up of a rigid, massless rod with a hanging pan attached at each end. The rod is supported at and free to rotate about a point not at its center. It is balanced by unequal masses placed in the two pans. When an unknown mass m is placed in the left pan, it is balanced by a mass m_1 placed in the right pan; when the mass m is placed in the right pan, it is balanced by a mass m_2 in the left pan. Show that $m = \sqrt{m_1 m_2}$.

75 The rigid square frame in Fig. 12-79 consists of the four side bars AB, BC, CD, and DA plus two diagonal bars AC and BD, which pass each other freely at E. By means of the turnbuckle G, bar AB is put under tension, as if its ends were subject to horizontal, outward forces $\vec{T}$ of magnitude 535 N. (a) Which of the other bars are in tension? What are the magnitudes of (b) the forces causing the tension in those bars and (c) the forces causing compression in the other bars? (*Hint:* Symmetry considerations can lead to considerable simplification in this problem.)

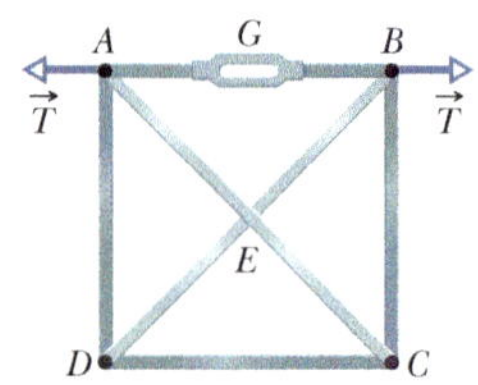

Figure 12-79 Problem 75.

76 A gymnast with mass 46.0 kg stands on the end of a uniform balance beam as shown in Fig. 12-80. The beam is 5.00 m long and has a mass of 250 kg (excluding the mass of the two supports). Each support is 0.540 m from its end of the beam. In unit-vector notation, what are the forces on the beam due to (a) support 1 and (b) support 2?

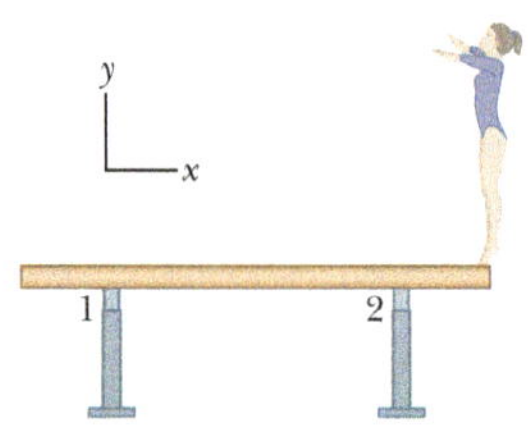

Figure 12-80 Problem 76.

77 Figure 12-81 shows a 300 kg cylinder that is horizontal. Three steel wires support the cylinder from a ceiling. Wires 1 and 3 are attached at the ends of the cylinder, and wire 2 is attached at the center. The wires each have a cross-sectional area of 2.00×10^{-6} m^2. Initially (before the cylinder was put in place) wires 1 and 3 were 2.0000 m long and wire 2 was 6.00 mm longer than that. Now (with the cylinder in place) all three wires have been stretched. What is the tension in (a) wire 1 and (b) wire 2?

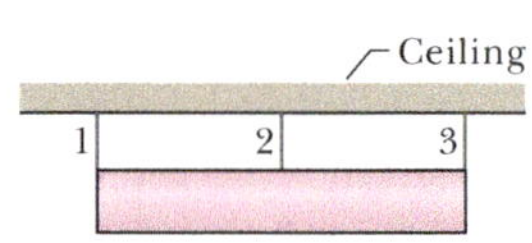

Figure 12-81 Problem 77.

78 In Fig. 12-82, a uniform beam of length 12.0 m is supported by a horizontal cable and a hinge at angle $\theta = 50.0°$. The tension in the cable is 400 N. In unit-vector notation, what are (a) the gravitational force on the beam and (b) the force on the beam from the hinge?

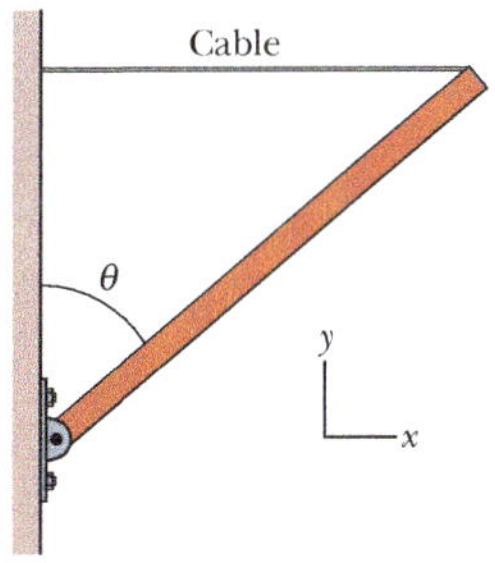

Figure 12-82 Problem 78.

79 Four bricks of length L, identical and uniform, are stacked on a table in two ways, as shown in Fig. 12-83 (compare with Problem 63). We seek to maximize the overhang distance h in both arrangements. Find the optimum distances a_1, a_2, b_1, and b_2, and calculate h for the two arrangements.

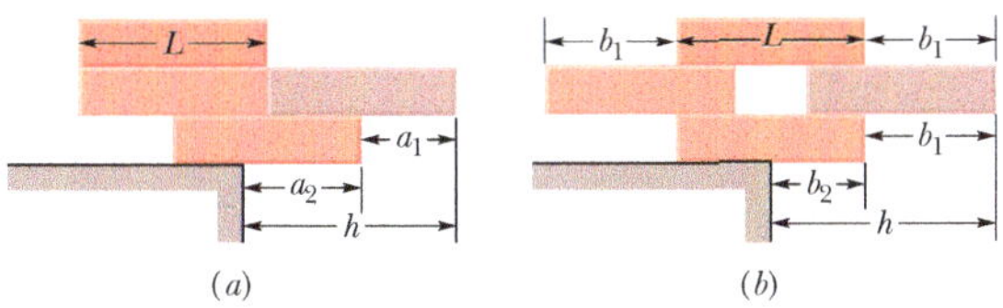

Figure 12-83 Problem 79.

80 A cylindrical aluminum rod, with an initial length of 0.8000 m and radius 1000.0 μm, is clamped in place at one end and then stretched by a machine pulling parallel to its length at its other end. Assuming that the rod's density (mass per unit volume) does not change, find the force magnitude that is required of the machine to decrease the radius to 999.9 μm. (The yield strength is not exceeded.)

81 A beam of length L is carried by three men, one man at one end and the other two supporting the beam between them on a crosspiece placed so that the load of the beam is equally divided among the three men. How far from the beam's free end is the crosspiece placed? (Neglect the mass of the crosspiece.)

82 If the (square) beam in Fig. 12-6*a* and the associated sample problem is of Douglas fir, what must be its thickness to keep the compressive stress on it to $\frac{1}{6}$ of its ultimate strength?

83 Figure 12-84 shows a stationary arrangement of two crayon boxes and three cords. Box A has a mass of 11.0 kg and is on a ramp at angle $\theta = 30.0°$; box B has a mass of 7.00 kg and hangs on a cord. The cord connected to box A is parallel to the ramp, which is frictionless. (a) What is the tension in the upper cord, and (b) what angle does that cord make with the horizontal?

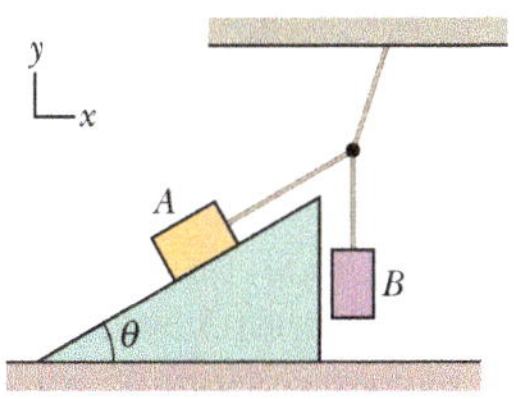

Figure 12-84 Problem 83.

84 A makeshift swing is constructed by making a loop in one end of a rope and tying the other end to a tree limb. A child is sitting in the loop with the rope hanging vertically when the child's father pulls on the child with a horizontal force and displaces the child to one side. Just before the child is released from rest, the rope makes an angle of 15° with the vertical and the tension in the rope is 280 N. (a) How much does the child weigh? (b) What is the magnitude of the (horizontal) force of the father on the child just before the child is released? (c) If the maximum horizontal force the father can exert on the child is 93 N, what is the maximum angle with the vertical the rope can make while the father is pulling horizontally?

85 Figure 12-85*a* shows details of a finger in the crimp hold of the climber in Fig. 12-50. A tendon that runs from muscles in the forearm is attached to the far bone in the finger. Along the way, the tendon runs through several guiding sheaths called *pulleys*. The A2 pulley is attached to the first finger bone; the A4 pulley is attached to the second finger bone. To pull the finger toward the palm, the forearm muscles pull the tendon through the pulleys, much like strings on a marionette can be pulled to move parts of the marionette. Figure 12-85*b* is a simplified diagram of the second finger bone, which has length d. The tendon's pull $\vec{F}_t$ on the bone acts at the point where the tendon enters the A4 pulley, at distance $d/3$ along the bone. If the force components on each of the four crimped fingers in Fig. 12-50 are $F_h = 13.4$ N and $F_v = 162.4$ N, what is the magnitude of $\vec{F}_t$? The result is probably tolerable, but if the climber hangs by only one or two fingers, the A2 and A4 pulleys can be ruptured, a common ailment among rock climbers.

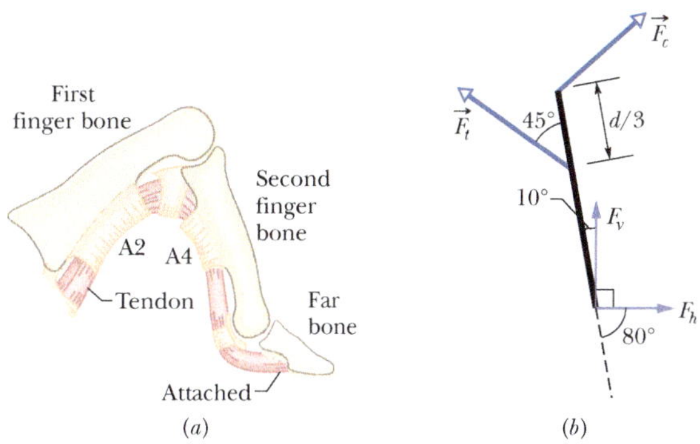

Figure 12-85 Problem 85.

86 A trap door in a ceiling is 0.91 m square, has a mass of 11 kg, and is hinged along one side, with a catch at the opposite side. If the center of gravity of the door is 10 cm toward the hinged side from the door's center, what are the magnitudes of the forces exerted by the door on (a) the catch and (b) the hinge?

87 A particle is acted on by forces given, in newtons, by $\vec{F}_1 = 8.40\hat{i} - 5.70\hat{j}$ and $\vec{F}_2 = 16.0\hat{i} + 4.10\hat{j}$. (a) What are the x component and (b) y component of the force $\vec{F}_3$ that balances the sum of these forces? (c) What angle does $\vec{F}_3$ have relative to the $+x$ axis?

88 The leaning Tower of Pisa is 59.1 m high and 7.44 m in diameter. The top of the tower is displaced 4.01 m from the vertical. Treat the tower as a uniform, circular cylinder. (a) What additional displacement, measured at the top, would bring the tower to the verge of toppling? (b) What angle would the tower then make with the vertical?

CHAPTER 13

Gravitation

13-1 NEWTON'S LAW OF GRAVITATION

Learning Objectives

After reading this module, you should be able to . . .

13.01 Apply Newton's law of gravitation to relate the gravitational force between two particles to their masses and their separation.

13.02 Identify that a uniform spherical shell of matter attracts a particle that is outside the shell as if all the shell's mass were concentrated as a particle at its center.

13.03 Draw a free-body diagram to indicate the gravitational force on a particle due to another particle or a uniform, spherical distribution of matter.

Key Ideas

- Any particle in the universe attracts any other particle with a gravitational force whose magnitude is

$$F = G\frac{m_1 m_2}{r^2} \quad \text{(Newton's law of gravitation)},$$

where m_1 and m_2 are the masses of the particles, r is their separation, and $G\ (= 6.67 \times 10^{-11}\ \text{N} \cdot \text{m}^2/\text{kg}^2)$ is the gravitational constant.

- The gravitational force between extended bodies is found by adding (integrating) the individual forces on individual particles within the bodies. However, if either of the bodies is a uniform spherical shell or a spherically symmetric solid, the net gravitational force it exerts on an *external* object may be computed as if all the mass of the shell or body were located at its center.

What Is Physics?

One of the long-standing goals of physics is to understand the gravitational force—the force that holds you to Earth, holds the Moon in orbit around Earth, and holds Earth in orbit around the Sun. It also reaches out through the whole of our Milky Way galaxy, holding together the billions and billions of stars in the Galaxy and the countless molecules and dust particles between stars. We are located somewhat near the edge of this disk-shaped collection of stars and other matter, 2.6×10^4 light-years (2.5×10^{20} m) from the galactic center, around which we slowly revolve.

The gravitational force also reaches across intergalactic space, holding together the Local Group of galaxies, which includes, in addition to the Milky Way, the Andromeda Galaxy (Fig. 13-1) at a distance of 2.3×10^6 light-years away from Earth, plus several closer dwarf galaxies, such as the Large Magellanic Cloud. The Local Group is part of the Local Supercluster of galaxies that is being drawn by the gravitational force toward an exceptionally massive region of space called the Great Attractor. This region appears to be about 3.0×10^8 light-years from Earth, on the opposite side of the Milky Way. And the gravitational force is even more far-reaching because it attempts to hold together the entire universe, which is expanding.

This force is also responsible for some of the most mysterious structures in the universe: *black holes.* When a star considerably larger than our Sun burns out, the gravitational force between all its particles can cause the star to collapse in on itself and thereby to form a black hole. The gravitational force at the surface of such a collapsed star is so strong that neither particles nor light can escape from the surface (thus the term "black hole"). Any star coming too near a black hole can be ripped apart by the strong gravitational force and pulled into the hole. Enough captures like this yields a *supermassive black hole.* Such mysterious monsters appear to be common in the universe. Indeed, such a monster lurks at the center of our Milky Way galaxy—the black hole there, called Sagittarius A*, has a mass of about 3.7×10^6 solar masses. The gravitational force near this black hole is so strong that it causes orbiting stars to whip around the black hole, completing an orbit in as little as 15.2 y.

Although the gravitational force is still not fully understood, the starting point in our understanding of it lies in the *law of gravitation* of Isaac Newton.

Courtesy NASA

Figure 13-1 The Andromeda Galaxy. Located 2.3×10^6 light-years from us, and faintly visible to the naked eye, it is very similar to our home galaxy, the Milky Way.

Newton's Law of Gravitation

Before we get to the equations, let's just think for a moment about something that we take for granted. We are held to the ground just about right, not so strongly that we have to crawl to get to school (though an occasional exam may leave you crawling home) and not so lightly that we bump our heads on the ceiling when we take a step. It is also just about right so that we are held to the ground but not to each other (that would be awkward in any classroom) or to the objects around us (the phrase "catching a bus" would then take on a new meaning). The attraction obviously depends on how much "stuff" there is in ourselves and other objects: Earth has lots of "stuff" and produces a big attraction but another person has less "stuff" and produces a smaller (even negligible) attraction. Moreover, this "stuff" always attracts other "stuff," never repelling it (or a hard sneeze could put us into orbit).

In the past people obviously knew that they were being pulled downward (especially if they tripped and fell over), but they figured that the downward force was unique to Earth and unrelated to the apparent movement of astronomical bodies across the sky. But in 1665, the 23-year-old Isaac Newton recognized that this force is responsible for holding the Moon in its orbit. Indeed he showed that every body in the universe attracts every other body. This tendency of bodies to move toward one another is called **gravitation**, and the "stuff" that is involved is the mass of each body. If the myth were true that a falling apple inspired Newton to his **law of gravitation**, then the attraction is between the mass of the apple and the mass of Earth. It is appreciable because the mass of Earth is so large, but even then it is only about 0.8 N. The attraction between two people standing near each other on a bus is (thankfully) much less (less than 1 μN) and imperceptible.

The gravitational attraction between extended objects such as two people can be difficult to calculate. Here we shall focus on Newton's force law between two *particles* (which have no size). Let the masses be m_1 and m_2 and r be their separation. Then the magnitude of the gravitational force acting on each due to the presence of the other is given by

$$F = G\frac{m_1 m_2}{r^2} \quad \text{(Newton's law of gravitation).} \tag{13-1}$$

G is the **gravitational constant:**

$$\begin{aligned} G &= 6.67 \times 10^{-11}\ \text{N}\cdot\text{m}^2/\text{kg}^2 \\ &= 6.67 \times 10^{-11}\ \text{m}^3/\text{kg}\cdot\text{s}^2. \end{aligned} \tag{13-2}$$

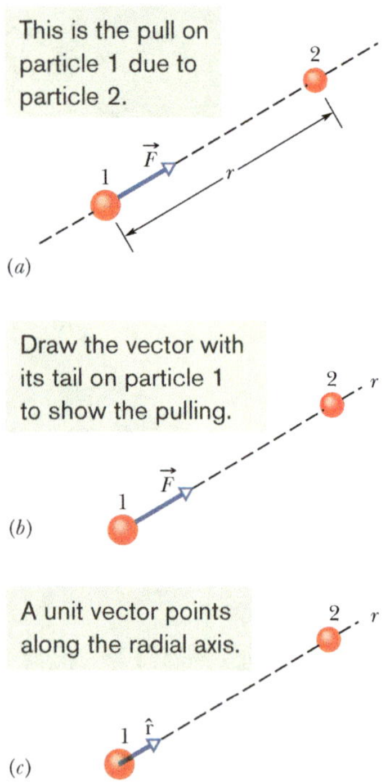

Figure 13-2 (*a*) The gravitational force $\vec{F}$ on particle 1 due to particle 2 is an attractive force because particle 1 is attracted to particle 2. (*b*) Force $\vec{F}$ is directed along a radial coordinate axis r extending from particle 1 through particle 2. (*c*) $\vec{F}$ is in the direction of a unit vector $\hat{r}$ along the r axis.

In Fig. 13-2*a*, $\vec{F}$ is the gravitational force acting on particle 1 (mass m_1) due to particle 2 (mass m_2). The force is directed toward particle 2 and is said to be an *attractive force* because particle 1 is attracted toward particle 2. The magnitude of the force is given by Eq. 13-1. We can describe $\vec{F}$ as being in the positive direction of an r axis extending radially from particle 1 through particle 2 (Fig. 13-2*b*). We can also describe $\vec{F}$ by using a radial unit vector $\hat{r}$ (a dimensionless vector of magnitude 1) that is directed away from particle 1 along the r axis (Fig. 13-2*c*). From Eq. 13-1, the force on particle 1 is then

$$\vec{F} = G\frac{m_1 m_2}{r^2}\hat{r}. \tag{13-3}$$

The gravitational force on particle 2 due to particle 1 has the same magnitude as the force on particle 1 but the opposite direction. These two forces form a third-law force pair, and we can speak of the gravitational force *between* the two particles as having a magnitude given by Eq. 13-1. This force between two particles is not altered by other objects, even if they are located between the particles. Put another way, no object can shield either particle from the gravitational force due to the other particle.

The strength of the gravitational force—that is, how strongly two particles with given masses at a given separation attract each other—depends on the value of the gravitational constant G. If G—by some miracle—were suddenly multiplied by a factor of 10, you would be crushed to the floor by Earth's attraction. If G were divided by this factor, Earth's attraction would be so weak that you could jump over a building.

Nonparticles. Although Newton's law of gravitation applies strictly to particles, we can also apply it to real objects as long as the sizes of the objects are small relative to the distance between them. The Moon and Earth are far enough apart so that, to a good approximation, we can treat them both as particles—but what about an apple and Earth? From the point of view of the apple, the broad and level Earth, stretching out to the horizon beneath the apple, certainly does not look like a particle.

Newton solved the apple–Earth problem with the *shell theorem:*

> A uniform spherical shell of matter attracts a particle that is outside the shell as if all the shell's mass were concentrated at its center.

Earth can be thought of as a nest of such shells, one within another and each shell attracting a particle outside Earth's surface as if the mass of that shell were at the center of the shell. Thus, from the apple's point of view, Earth *does* behave like a particle, one that is located at the center of Earth and has a mass equal to that of Earth.

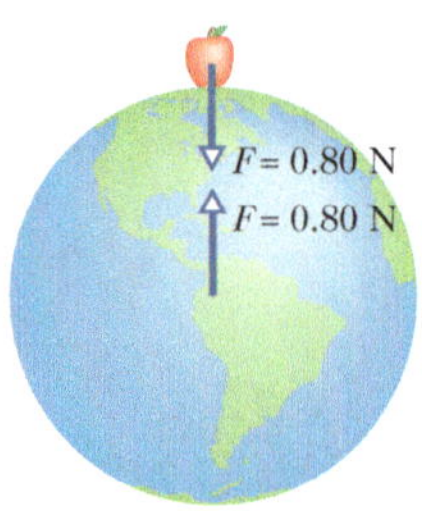

Figure 13-3 The apple pulls up on Earth just as hard as Earth pulls down on the apple.

Third-Law Force Pair. Suppose that, as in Fig. 13-3, Earth pulls down on an apple with a force of magnitude 0.80 N. The apple must then pull up on Earth with a force of magnitude 0.80 N, which we take to act at the center of Earth. In the language of Chapter 5, these forces form a force pair in Newton's third law. Although they are matched in magnitude, they produce different accelerations when the apple is released. The acceleration of the apple is about 9.8 m/s^2, the familiar acceleration of a falling body near Earth's surface. The acceleration of Earth, however, measured in a reference frame attached to the center of mass of the apple–Earth system, is only about 1×10^{-25} m/s^2.

Checkpoint 1

A particle is to be placed, in turn, outside four objects, each of mass m: (1) a large uniform solid sphere, (2) a large uniform spherical shell, (3) a small uniform solid sphere, and (4) a small uniform shell. In each situation, the distance between the particle and the center of the object is d. Rank the objects according to the magnitude of the gravitational force they exert on the particle, greatest first.

13-2 GRAVITATION AND THE PRINCIPLE OF SUPERPOSITION

Learning Objectives

After reading this module, you should be able to . . .

13.04 If more than one gravitational force acts on a particle, draw a free-body diagram showing those forces, with the tails of the force vectors anchored on the particle.

13.05 If more than one gravitational force acts on a particle, find the net force by adding the individual forces as vectors.

Key Ideas

- Gravitational forces obey the principle of superposition; that is, if n particles interact, the net force $\vec{F}_{1,\text{net}}$ on a particle labeled particle 1 is the sum of the forces on it from all the other particles taken one at a time:

$$\vec{F}_{1,\text{net}} = \sum_{i=2}^{n} \vec{F}_{1i},$$

in which the sum is a vector sum of the forces $\vec{F}_{1i}$ on particle 1 from particles 2, 3, . . . , n.

- The gravitational force $\vec{F}_1$ on a particle from an extended body is found by first dividing the body into units of differential mass dm, each of which produces a differential force $d\vec{F}$ on the particle, and then integrating over all those units to find the sum of those forces:

$$\vec{F}_1 = \int d\vec{F}.$$

Gravitation and the Principle of Superposition

Given a group of particles, we find the net (or resultant) gravitational force on any one of them from the others by using the **principle of superposition.** This is a general principle that says a net effect is the sum of the individual effects. Here, the principle means that we first compute the individual gravitational forces that act on our selected particle due to each of the other particles. We then find the net force by adding these forces vectorially, just as we have done when adding forces in earlier chapters.

Let's look at two important points in that last (probably quickly read) sentence. (1) Forces are vectors and can be in different directions, and thus we must *add them as vectors*, taking into account their directions. (If two people pull on you in the opposite direction, their net force on you is clearly different than if they pull in the same direction.) (2) We *add* the individual forces. Think how impossible the world would be if the net force depended on some multiplying factor that varied from force to force depending on the situation, or if the presence of one force somehow amplified the magnitude of another force. No, thankfully, the world requires only simple vector addition of the forces.

For n interacting particles, we can write the principle of superposition for the gravitational forces on particle 1 as

$$\vec{F}_{1,\text{net}} = \vec{F}_{12} + \vec{F}_{13} + \vec{F}_{14} + \vec{F}_{15} + \cdots + \vec{F}_{1n}. \tag{13-4}$$

Here $\vec{F}_{1,\text{net}}$ is the net force on particle 1 due to the other particles and, for example, $\vec{F}_{13}$ is the force on particle 1 from particle 3. We can express this equation more compactly as a vector sum:

$$\vec{F}_{1,\text{net}} = \sum_{i=2}^{n} \vec{F}_{1i}. \tag{13-5}$$

Real Objects. What about the gravitational force on a particle from a real (extended) object? This force is found by dividing the object into parts small enough to treat as particles and then using Eq. 13-5 to find the vector sum of the forces on the particle from all the parts. In the limiting case, we can divide the extended object into differential parts each of mass dm and each producing a differential force $d\vec{F}$

Sample Problem 13.01 Net gravitational force, 2D, three particles

Figure 13-4*a* shows an arrangement of three particles, particle 1 of mass $m_1 = 6.0$ kg and particles 2 and 3 of mass $m_2 = m_3 = 4.0$ kg, and distance $a = 2.0$ cm. What is the net gravitational force $\vec{F}_{1,\text{net}}$ on particle 1 due to the other particles?

KEY IDEAS

(1) Because we have particles, the magnitude of the gravitational force on particle 1 due to either of the other particles is given by Eq. 13-1 ($F = Gm_1m_2/r^2$). (2) The direction of either gravitational force on particle 1 is toward the particle responsible for it. (3) Because the forces are not along a single axis, we *cannot* simply add or subtract their magnitudes or their components to get the net force. Instead, we must add them as vectors.

Calculations: From Eq. 13-1, the magnitude of the force $\vec{F}_{12}$ on particle 1 from particle 2 is

$$F_{12} = \frac{Gm_1m_2}{a^2} \tag{13-7}$$

$$= \frac{(6.67 \times 10^{-11}\ \text{m}^3/\text{kg}\cdot\text{s}^2)(6.0\ \text{kg})(4.0\ \text{kg})}{(0.020\ \text{m})^2}$$

$$= 4.00 \times 10^{-6}\ \text{N}.$$

Similarly, the magnitude of force $\vec{F}_{13}$ on particle 1 from particle 3 is

$$F_{13} = \frac{Gm_1m_3}{(2a)^2} \tag{13-8}$$

$$= \frac{(6.67 \times 10^{-11}\ \text{m}^3/\text{kg}\cdot\text{s}^2)(6.0\ \text{kg})(4.0\ \text{kg})}{(0.040\ \text{m})^2}$$

$$= 1.00 \times 10^{-6}\ \text{N}.$$

Force $\vec{F}_{12}$ is directed in the positive direction of the y axis (Fig. 13-4*b*) and has only the y component F_{12}. Similarly, $\vec{F}_{13}$ is directed in the negative direction of the x axis and has only the x component $-F_{13}$ (Fig. 13-4*c*). (Note something important: We draw the force diagrams with the tail of a force vector anchored on the particle experiencing the force. Drawing them in other ways invites errors, especially on exams.)

To find the net force $\vec{F}_{1,\text{net}}$ on particle 1, we must add the two forces as vectors (Figs. 13-4*d* and *e*). We can do so on a vector-capable calculator. However, here we note that $-F_{13}$ and F_{12} are actually the x and y components of $\vec{F}_{1,\text{net}}$. Therefore, we can use Eq. 3-6 to find first the magnitude and then the direction of $\vec{F}_{1,\text{net}}$. The magnitude is

$$F_{1,\text{net}} = \sqrt{(F_{12})^2 + (-F_{13})^2}$$

$$= \sqrt{(4.00 \times 10^{-6}\ \text{N})^2 + (-1.00 \times 10^{-6}\ \text{N})^2}$$

$$= 4.1 \times 10^{-6}\ \text{N}. \qquad \text{(Answer)}$$

Relative to the positive direction of the x axis, Eq. 3-6 gives the direction of $\vec{F}_{1,\text{net}}$ as

$$\theta = \tan^{-1}\frac{F_{12}}{-F_{13}} = \tan^{-1}\frac{4.00 \times 10^{-6}\ \text{N}}{-1.00 \times 10^{-6}\ \text{N}} = -76°.$$

Is this a reasonable direction (Fig. 13-4*f*)? No, because the direction of $\vec{F}_{1,\text{net}}$ must be between the directions of $\vec{F}_{12}$ and $\vec{F}_{13}$. Recall from Chapter 3 that a calculator displays only one of the two possible answers to a $\tan^{-1}$ function. We find the other answer by adding 180°:

$$-76° + 180° = 104°, \qquad \text{(Answer)}$$

which *is* a reasonable direction for $\vec{F}_{1,\text{net}}$ (Fig. 13-4*g*).

Additional examples, video, and practice available at *WileyPLUS*

on the particle. In this limit, the sum of Eq. 13-5 becomes an integral and we have

$$\vec{F}_1 = \int d\vec{F}, \tag{13-6}$$

in which the integral is taken over the entire extended object and we drop the subscript "net." If the extended object is a uniform sphere or a spherical shell, we can avoid the integration of Eq. 13-6 by assuming that the object's mass is concentrated at the object's center and using Eq. 13-1.

Checkpoint 2

The figure shows four arrangements of three particles of equal masses. (a) Rank the arrangements according to the magnitude of the net gravitational force on the particle labeled m, greatest first. (b) In arrangement 2, is the direction of the net force closer to the line of length d or to the line of length D?

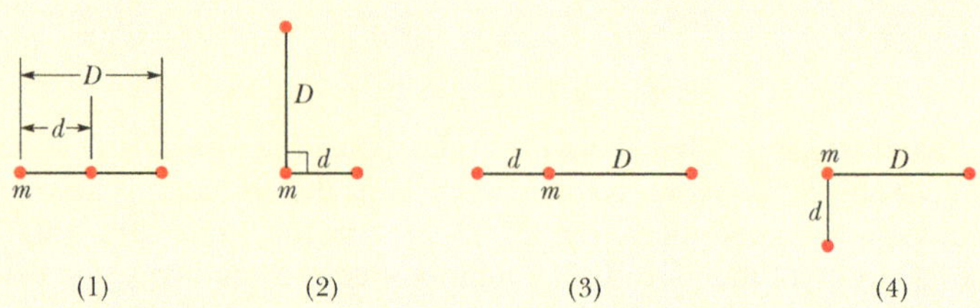

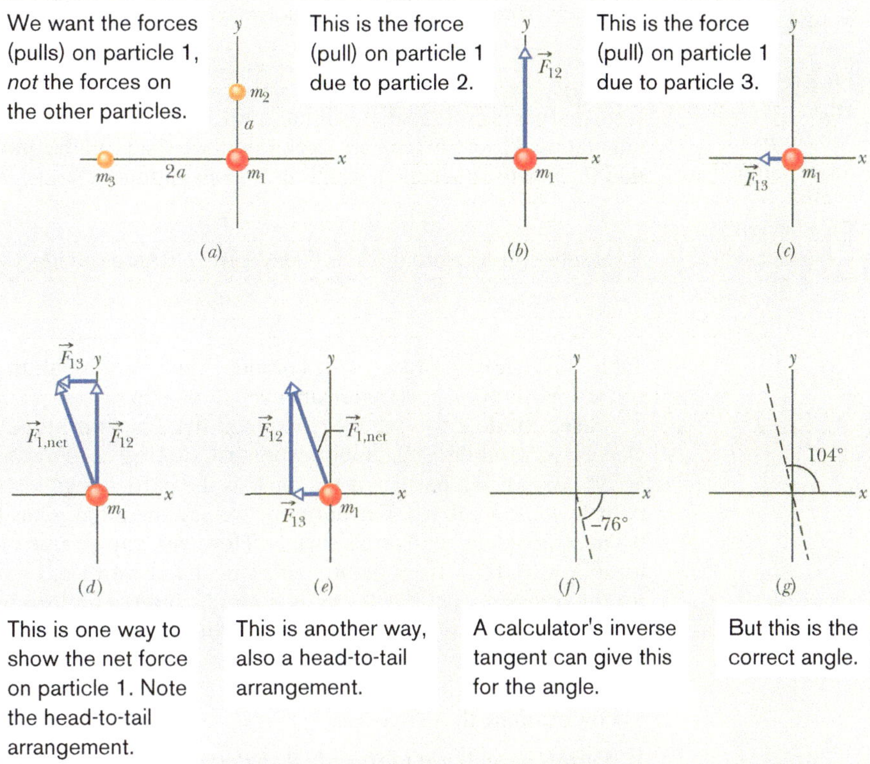

Figure 13-4 (*a*) An arrangement of three particles. The force on particle 1 due to (*b*) particle 2 and (*c*) particle 3. (*d*)–(*g*) Ways to combine the forces to get the net force magnitude and orientation. In *WileyPLUS*, this figure is available as an animation with voiceover.

13-3 GRAVITATION NEAR EARTH'S SURFACE

Learning Objectives

After reading this module, you should be able to . . .

13.06 Distinguish between the free-fall acceleration and the gravitational acceleration.

13.07 Calculate the gravitational acceleration near but outside a uniform, spherical astronomical body.

13.08 Distinguish between measured weight and the magnitude of the gravitational force.

Key Ideas

- The gravitational acceleration a_g of a particle (of mass m) is due solely to the gravitational force acting on it. When the particle is at distance r from the center of a uniform, spherical body of mass M, the magnitude F of the gravitational force on the particle is given by Eq. 13-1. Thus, by Newton's second law,

$$F = ma_g,$$

which gives

$$a_g = \frac{GM}{r^2}.$$

- Because Earth's mass is not distributed uniformly, because the planet is not perfectly spherical, and because it rotates, the actual free-fall acceleration $\vec{g}$ of a particle near Earth differs slightly from the gravitational acceleration $\vec{a}_g$, and the particle's weight (equal to mg) differs from the magnitude of the gravitational force on it.

Table 13-1 **Variation of a_g with Altitude**

Altitude (km)	a_g (m/s^2)	Altitude Example
0	9.83	Mean Earth surface
8.8	9.80	Mt. Everest
36.6	9.71	Highest crewed balloon
400	8.70	Space shuttle orbit
35 700	0.225	Communications satellite

Gravitation Near Earth's Surface

Let us assume that Earth is a uniform sphere of mass M. The magnitude of the gravitational force from Earth on a particle of mass m, located outside Earth a distance r from Earth's center, is then given by Eq. 13-1 as

$$F = G\frac{Mm}{r^2}. \quad (13\text{-}9)$$

If the particle is released, it will fall toward the center of Earth, as a result of the gravitational force $\vec{F}$, with an acceleration we shall call the **gravitational acceleration** $\vec{a}_g$. Newton's second law tells us that magnitudes F and a_g are related by

$$F = ma_g. \quad (13\text{-}10)$$

Now, substituting F from Eq. 13-9 into Eq. 13-10 and solving for a_g, we find

$$a_g = \frac{GM}{r^2}. \quad (13\text{-}11)$$

Table 13-1 shows values of a_g computed for various altitudes above Earth's surface. Notice that a_g is significant even at 400 km.

Since Module 5-1, we have assumed that Earth is an inertial frame by neglecting its rotation. This simplification has allowed us to assume that the free-fall acceleration g of a particle is the same as the particle's gravitational acceleration (which we now call a_g). Furthermore, we assumed that g has the constant value 9.8 m/s^2 any place on Earth's surface. However, any g value measured at a given location will differ from the a_g value calculated with Eq. 13-11 for that location for three reasons: (1) Earth's mass is not distributed uniformly, (2) Earth is not a perfect sphere, and (3) Earth rotates. Moreover, because g differs from a_g, the same three reasons mean that the measured weight mg of a particle differs from the magnitude of the gravitational force on the particle as given by Eq. 13-9. Let us now examine those reasons.

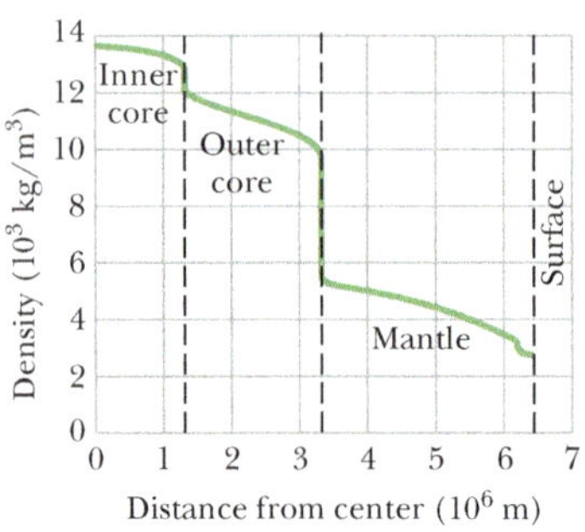

Figure 13-5 The density of Earth as a function of distance from the center. The limits of the solid inner core, the largely liquid outer core, and the solid mantle are shown, but the crust of Earth is too thin to show clearly on this plot.

1. ***Earth's mass is not uniformly distributed.*** The density (mass per unit volume) of Earth varies radially as shown in Fig. 13-5, and the density of the crust (outer section) varies from region to region over Earth's surface. Thus, g varies from region to region over the surface.
2. ***Earth is not a sphere.*** Earth is approximately an ellipsoid, flattened at the poles and bulging at the equator. Its equatorial radius (from its center point out to the equator) is greater than its polar radius (from its center point out to either north or south pole) by 21 km. Thus, a point at the poles is closer to the dense core of Earth than is a point on the equator. This is one reason the free-fall acceleration g increases if you were to measure it while moving at sea level from the equator toward the north or south pole. As you move, you are actually getting closer to the center of Earth and thus, by Newton's law of gravitation, g increases.
3. ***Earth is rotating.*** The rotation axis runs through the north and south poles of Earth. An object located on Earth's surface anywhere except at those poles must rotate in a circle about the rotation axis and thus must have a centripetal acceleration directed toward the center of the circle. This centripetal acceleration requires a centripetal net force that is also directed toward that center.

To see how Earth's rotation causes g to differ from a_g, let us analyze a simple situation in which a crate of mass m is on a scale at the equator. Figure 13-6*a* shows this situation as viewed from a point in space above the north pole.

Figure 13-6*b*, a free-body diagram for the crate, shows the two forces on the crate, both acting along a radial r axis that extends from Earth's center. The normal force $\vec{F}_N$ on the crate from the scale is directed outward, in the positive direction of the r axis. The gravitational force, represented with its equivalent $m\vec{a}_g$, is directed inward. Because it travels in a circle about the center of Earth

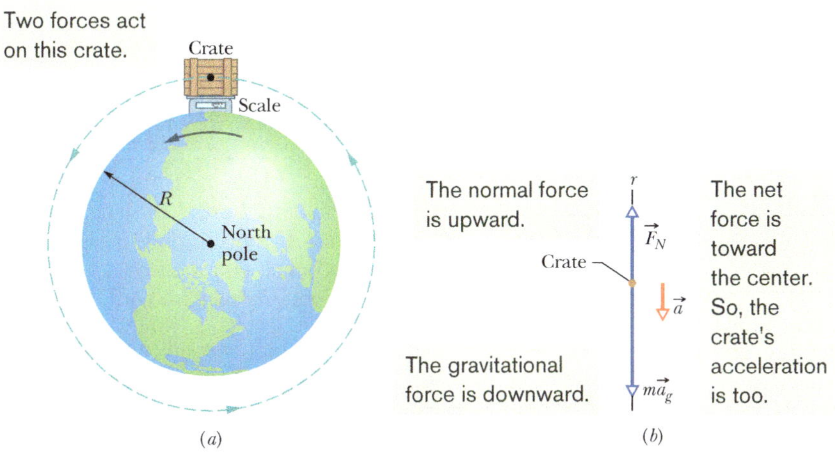

Figure 13-6 (*a*) A crate sitting on a scale at Earth's equator, as seen by an observer positioned on Earth's rotation axis at some point above the north pole. (*b*) A free-body diagram for the crate, with a radial r axis extending from Earth's center. The gravitational force on the crate is represented with its equivalent $m\vec{a}_g$. The normal force on the crate from the scale is $\vec{F}_N$. Because of Earth's rotation, the crate has a centripetal acceleration $\vec{a}$ that is directed toward Earth's center.

as Earth turns, the crate has a centripetal acceleration $\vec{a}$ directed toward Earth's center. From Eq. 10-23 ($a_r = \omega^2 r$), we know this acceleration is equal to $\omega^2 R$, where ω is Earth's angular speed and R is the circle's radius (approximately Earth's radius). Thus, we can write Newton's second law for forces along the r axis ($F_{\text{net},r} = ma_r$) as

$$F_N - ma_g = m(-\omega^2 R). \tag{13-12}$$

The magnitude F_N of the normal force is equal to the weight mg read on the scale. With mg substituted for F_N, Eq. 13-12 gives us

$$mg = ma_g - m(\omega^2 R), \tag{13-13}$$

which says

$$\begin{pmatrix}\text{measured}\\ \text{weight}\end{pmatrix} = \begin{pmatrix}\text{magnitude of}\\ \text{gravitational force}\end{pmatrix} - \begin{pmatrix}\text{mass times}\\ \text{centripetal acceleration}\end{pmatrix}.$$

Thus, the measured weight is less than the magnitude of the gravitational force on the crate, because of Earth's rotation.

Acceleration Difference. To find a corresponding expression for g and a_g, we cancel m from Eq. 13-13 to write

$$g = a_g - \omega^2 R, \tag{13-14}$$

which says

$$\begin{pmatrix}\text{free-fall}\\ \text{acceleration}\end{pmatrix} = \begin{pmatrix}\text{gravitational}\\ \text{acceleration}\end{pmatrix} - \begin{pmatrix}\text{centripetal}\\ \text{acceleration}\end{pmatrix}.$$

Thus, the measured free-fall acceleration is less than the gravitational acceleration because of Earth's rotation.

Equator. The difference between accelerations g and a_g is equal to $\omega^2 R$ and is greatest on the equator (for one reason, the radius of the circle traveled by the crate is greatest there). To find the difference, we can use Eq. 10-5 ($\omega = \Delta\theta/\Delta t$) and Earth's radius $R = 6.37 \times 10^6$ m. For one rotation of Earth, θ is 2π rad and the time period Δt is about 24 h. Using these values (and converting hours to seconds), we find that g is less than a_g by only about 0.034 m/s^2 (small compared to 9.8 m/s^2). Therefore, neglecting the difference in accelerations g and a_g is often justified. Similarly, neglecting the difference between weight and the magnitude of the gravitational force is also often justified.

Sample Problem 13.02 Difference in acceleration at head and feet

(a) An astronaut whose height h is 1.70 m floats "feet down" in an orbiting space shuttle at distance $r = 6.77 \times 10^6$ m away from the center of Earth. What is the difference between the gravitational acceleration at her feet and at her head?

KEY IDEAS

We can approximate Earth as a uniform sphere of mass M_E. Then, from Eq. 13-11, the gravitational acceleration at any distance r from the center of Earth is

$$a_g = \frac{GM_E}{r^2}. \tag{13-15}$$

We might simply apply this equation twice, first with $r = 6.77 \times 10^6$ m for the location of the feet and then with $r = 6.77 \times 10^6$ m + 1.70 m for the location of the head. However, a calculator may give us the same value for a_g twice, and thus a difference of zero, because h is so much smaller than r. Here's a more promising approach: Because we have a differential change dr in r between the astronaut's feet and head, we should differentiate Eq. 13-15 with respect to r.

Calculations: The differentiation gives us

$$da_g = -2\frac{GM_E}{r^3}\,dr, \tag{13-16}$$

where da_g is the differential change in the gravitational acceleration due to the differential change dr in r. For the astronaut, $dr = h$ and $r = 6.77 \times 10^6$ m. Substituting data into Eq. 13-16, we find

$$da_g = -2\frac{(6.67 \times 10^{-11}\ \text{m}^3/\text{kg}\cdot\text{s}^2)(5.98 \times 10^{24}\ \text{kg})}{(6.77 \times 10^6\ \text{m})^3}(1.70\ \text{m})$$
$$= -4.37 \times 10^{-6}\ \text{m/s}^2, \quad \text{(Answer)}$$

where the M_E value is taken from Appendix C. This result means that the gravitational acceleration of the astronaut's feet toward Earth is slightly greater than the gravitational acceleration of her head toward Earth. This difference in acceleration (often called a *tidal effect*) tends to stretch her body, but the difference is so small that she would never even sense the stretching, much less suffer pain from it.

(b) If the astronaut is now "feet down" at the same orbital radius $r = 6.77 \times 10^6$ m about a black hole of mass $M_h = 1.99 \times 10^{31}$ kg (10 times our Sun's mass), what is the difference between the gravitational acceleration at her feet and at her head? The black hole has a mathematical surface (*event horizon*) of radius $R_h = 2.95 \times 10^4$ m. Nothing, not even light, can escape from that surface or anywhere inside it. Note that the astronaut is well outside the surface (at $r = 229R_h$).

Calculations: We again have a differential change dr in r between the astronaut's feet and head, so we can again use Eq. 13-16. However, now we substitute $M_h = 1.99 \times 10^{31}$ kg for M_E. We find

$$da_g = -2\frac{(6.67 \times 10^{-11}\ \text{m}^3/\text{kg}\cdot\text{s}^2)(1.99 \times 10^{31}\ \text{kg})}{(6.77 \times 10^6\ \text{m})^3}(1.70\ \text{m})$$
$$= -14.5\ \text{m/s}^2. \quad \text{(Answer)}$$

This means that the gravitational acceleration of the astronaut's feet toward the black hole is noticeably larger than that of her head. The resulting tendency to stretch her body would be bearable but quite painful. If she drifted closer to the black hole, the stretching tendency would increase drastically.

WILEY PLUS Additional examples, video, and practice available at *WileyPLUS*

13-4 GRAVITATION INSIDE EARTH

Learning Objectives

After reading this module, you should be able to . . .

13.09 Identify that a uniform shell of matter exerts no net gravitational force on a particle located inside it.

13.10 Calculate the gravitational force that is exerted on a particle at a given radius inside a nonrotating uniform sphere of matter.

Key Ideas

- A uniform shell of matter exerts no *net* gravitational force on a particle located inside it.
- The gravitational force $\vec{F}$ on a particle inside a uniform solid sphere, at a distance r from the center, is due only to mass M_{ins} in an "inside sphere" with that radius r:

$$M_{\text{ins}} = \tfrac{4}{3}\pi r^3\rho = \frac{M}{R^3}r^3,$$

where ρ is the solid sphere's density, R is its radius, and M is its mass. We can assign this inside mass to be that of a particle at the center of the solid sphere and then apply Newton's law of gravitation for particles. We find that the magnitude of the force acting on mass m is

$$F = \frac{GmM}{R^3}r.$$

Gravitation Inside Earth

Newton's shell theorem can also be applied to a situation in which a particle is located *inside* a uniform shell, to show the following:

> A uniform shell of matter exerts no net gravitational force on a particle located inside it.

Caution: This statement does *not* mean that the gravitational forces on the particle from the various elements of the shell magically disappear. Rather, it means that the *sum* of the force vectors on the particle from all the elements is zero.

If Earth's mass were uniformly distributed, the gravitational force acting on a particle would be a maximum at Earth's surface and would decrease as the particle moved outward, away from the planet. If the particle were to move inward, perhaps down a deep mine shaft, the gravitational force would change for two reasons. (1) It would tend to increase because the particle would be moving closer to the center of Earth. (2) It would tend to decrease because the thickening shell of material lying outside the particle's radial position would not exert any net force on the particle.

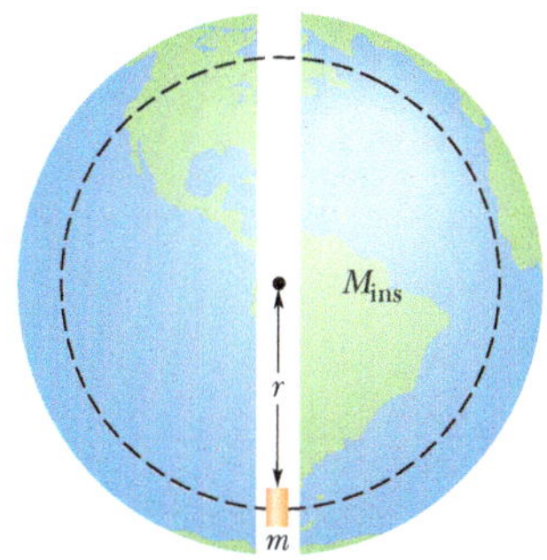

Figure 13-7 A capsule of mass m falls from rest through a tunnel that connects Earth's south and north poles. When the capsule is at distance r from Earth's center, the portion of Earth's mass that is contained in a sphere of that radius is M_{ins}.

To find an expression for the gravitational force inside a uniform Earth, let's use the plot in *Pole to Pole*, an early science fiction story by George Griffith. Three explorers attempt to travel by capsule through a naturally formed (and, of course, fictional) tunnel directly from the south pole to the north pole. Figure 13-7 shows the capsule (mass m) when it has fallen to a distance r from Earth's center. At that moment, the *net* gravitational force on the capsule is due to the mass M_{ins} inside the sphere with radius r (the mass enclosed by the dashed outline), not the mass in the outer spherical shell (outside the dashed outline). Moreover, we can assume that the inside mass M_{ins} is concentrated as a particle at Earth's center. Thus, we can write Eq. 13-1, for the magnitude of the gravitational force on the capsule, as

$$F = \frac{GmM_{ins}}{r^2}. \tag{13-17}$$

Because we assume a uniform density ρ, we can write this inside mass in terms of Earth's total mass M and its radius R:

$$\text{density} = \frac{\text{inside mass}}{\text{inside volume}} = \frac{\text{total mass}}{\text{total volume}},$$

$$\rho = \frac{M_{ins}}{\frac{4}{3}\pi r^3} = \frac{M}{\frac{4}{3}\pi R^3}.$$

Solving for M_{ins} we find

$$M_{ins} = \tfrac{4}{3}\pi r^3 \rho = \frac{M}{R^3} r^3. \tag{13-18}$$

Substituting the second expression for M_{ins} into Eq. 13-17 gives us the magnitude of the gravitational force on the capsule as a function of the capsule's distance r from Earth's center:

$$F = \frac{GmM}{R^3} r. \tag{13-19}$$

According to Griffith's story, as the capsule approaches Earth's center, the gravitational force on the explorers becomes alarmingly large and, exactly at the center, it suddenly but only momentarily disappears. From Eq. 13-19 we see that, in fact, the force magnitude decreases linearly as the capsule approaches the center, until it is zero at the center. At least Griffith got the zero-at-the-center detail correct.

Equation 13-19 can also be written in terms of the force vector $\vec{F}$ and the capsule's position vector $\vec{r}$ along a radial axis extending from Earth's center. Letting K represent the collection of constants in Eq. 13-19, we can rewrite the force in vector form as

$$\vec{F} = -K\vec{r}, \tag{13-20}$$

in which we have inserted a minus sign to indicate that $\vec{F}$ and $\vec{r}$ have opposite directions. Equation 13-20 has the form of Hooke's law (Eq. 7-20, $\vec{F} = -k\vec{d}$). Thus, under the idealized conditions of the story, the capsule would oscillate like a block on a spring, with the center of the oscillation at Earth's center. After the capsule had fallen from the south pole to Earth's center, it would travel from the center to the north pole (as Griffith said) and then back again, repeating the cycle forever.

For the real Earth, which certainly has a nonuniform distribution of mass (Fig. 13-5), the force on the capsule would initially *increase* as the capsule descends. The force would then reach a maximum at a certain depth, and only then would it begin to decrease as the capsule further descends.

13-5 GRAVITATIONAL POTENTIAL ENERGY

Learning Objectives

After reading this module, you should be able to . . .

13.11 Calculate the gravitational potential energy of a system of particles (or uniform spheres that can be treated as particles).

13.12 Identify that if a particle moves from an initial point to a final point while experiencing a gravitational force, the work done by that force (and thus the change in gravitational potential energy) is independent of which path is taken.

13.13 Using the gravitational force on a particle near an astronomical body (or some second body that is fixed in place), calculate the work done by the force when the body moves.

13.14 Apply the conservation of mechanical energy (including gravitational potential energy) to a particle moving relative to an astronomical body (or some second body that is fixed in place).

13.15 Explain the energy requirements for a particle to escape from an astronomical body (usually assumed to be a uniform sphere).

13.16 Calculate the escape speed of a particle in leaving an astronomical body.

Key Ideas

- The gravitational potential energy $U(r)$ of a system of two particles, with masses M and m and separated by a distance r, is the negative of the work that would be done by the gravitational force of either particle acting on the other if the separation between the particles were changed from infinite (very large) to r. This energy is

$$U = -\frac{GMm}{r} \quad \text{(gravitational potential energy).}$$

- If a system contains more than two particles, its total gravitational potential energy U is the sum of the terms representing the potential energies of all the pairs. As an example, for three particles, of masses m_1, m_2, and m_3,

$$U = -\left(\frac{Gm_1m_2}{r_{12}} + \frac{Gm_1m_3}{r_{13}} + \frac{Gm_2m_3}{r_{23}}\right).$$

- An object will escape the gravitational pull of an astronomical body of mass M and radius R (that is, it will reach an infinite distance) if the object's speed near the body's surface is at least equal to the escape speed, given by

$$v = \sqrt{\frac{2GM}{R}}.$$

Gravitational Potential Energy

In Module 8-1, we discussed the gravitational potential energy of a particle–Earth system. We were careful to keep the particle near Earth's surface, so that we could regard the gravitational force as constant. We then chose some reference configuration of the system as having a gravitational potential energy of zero. Often, in this configuration the particle was on Earth's surface. For particles not on Earth's surface, the gravitational potential energy decreased when the separation between the particle and Earth decreased.

Here, we broaden our view and consider the gravitational potential energy U of two particles, of masses m and M, separated by a distance r. We again choose a reference configuration with U equal to zero. However, to simplify the equations, the separation distance r in the reference configuration is now large enough to be approximated as *infinite.* As before, the gravitational potential energy decreases when the separation decreases. Since $U = 0$ for $r = \infty$, the potential energy is negative for any finite separation and becomes progressively more negative as the particles move closer together.

With these facts in mind and as we shall justify next, we take the gravitational potential energy of the two-particle system to be

$$U = -\frac{GMm}{r} \quad \text{(gravitational potential energy).} \tag{13-21}$$

Note that $U(r)$ approaches zero as r approaches infinity and that for any finite value of r, the value of $U(r)$ is negative.

Language. The potential energy given by Eq. 13-21 is a property of the system of two particles rather than of either particle alone. There is no way to divide this energy and say that so much belongs to one particle and so much to the other. However, if $M \gg m$, as is true for Earth (mass M) and a baseball (mass m), we often speak of "the potential energy of the baseball." We can get away with this because, when a baseball moves in the vicinity of Earth, changes in the potential energy of the baseball–Earth system appear almost entirely as changes in the kinetic energy of the baseball, since changes in the kinetic energy of Earth are too small to be measured. Similarly, in Module 13-7 we shall speak of "the potential energy of an artificial satellite" orbiting Earth, because the satellite's mass is so much smaller than Earth's mass. When we speak of the potential energy of bodies of comparable mass, however, we have to be careful to treat them as a system.

Multiple Particles. If our system contains more than two particles, we consider each pair of particles in turn, calculate the gravitational potential energy of that pair with Eq. 13-21 as if the other particles were not there, and then algebraically sum the results. Applying Eq. 13-21 to each of the three pairs of Fig. 13-8, for example, gives the potential energy of the system as

$$U = -\left(\frac{Gm_1m_2}{r_{12}} + \frac{Gm_1m_3}{r_{13}} + \frac{Gm_2m_3}{r_{23}}\right). \tag{13-22}$$

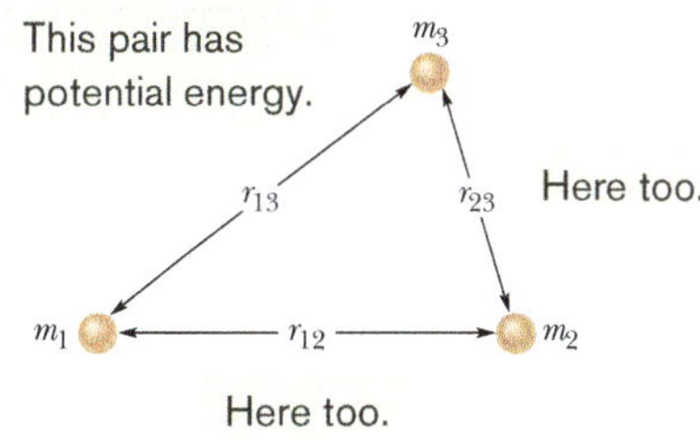

Figure 13-8 A system consisting of three particles. The gravitational potential energy *of the system* is the sum of the gravitational potential energies of all three pairs of particles.

Proof of Equation 13-21

Let us shoot a baseball directly away from Earth along the path in Fig. 13-9. We want to find an expression for the gravitational potential energy U of the ball at point P along its path, at radial distance R from Earth's center. To do so, we first find the work W done on the ball by the gravitational force as the ball travels from point P to a great (infinite) distance from Earth. Because the gravitational force $\vec{F}(r)$ is a variable force (its magnitude depends on r), we must use the techniques of Module 7-5 to find the work. In vector notation, we can write

$$W = \int_R^\infty \vec{F}(r) \cdot d\vec{r}. \tag{13-23}$$

The integral contains the scalar (or dot) product of the force $\vec{F}(r)$ and the differential displacement vector $d\vec{r}$ along the ball's path. We can expand that product as

$$\vec{F}(r) \cdot d\vec{r} = F(r)\, dr \cos\phi, \tag{13-24}$$

where ϕ is the angle between the directions of $\vec{F}(r)$ and $d\vec{r}$. When we substitute

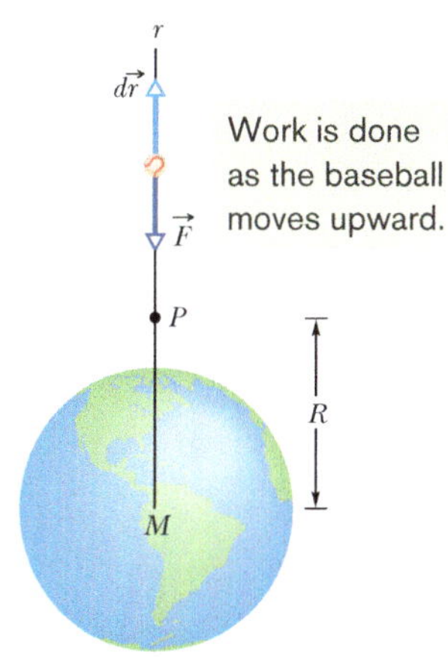

Figure 13-9 A baseball is shot directly away from Earth, through point P at radial distance R from Earth's center. The gravitational force $\vec{F}$ on the ball and a differential displacement vector $d\vec{r}$ are shown, both directed along a radial r axis.

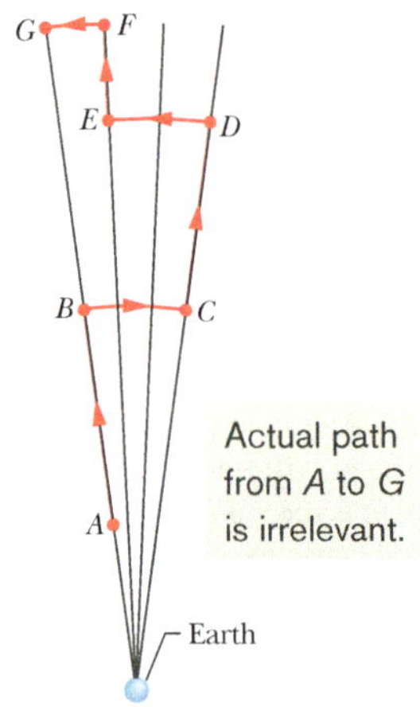

Figure 13-10 Near Earth, a baseball is moved from point A to point G along a path consisting of radial lengths and circular arcs.

180° for ϕ and Eq. 13-1 for $F(r)$, Eq. 13-24 becomes

$$\vec{F}(r) \cdot d\vec{r} = -\frac{GMm}{r^2}\,dr,$$

where M is Earth's mass and m is the mass of the ball.

Substituting this into Eq. 13-23 and integrating give us

$$W = -GMm \int_R^\infty \frac{1}{r^2}\,dr = \left[\frac{GMm}{r}\right]_R^\infty$$
$$= 0 - \frac{GMm}{R} = -\frac{GMm}{R}, \tag{13-25}$$

where W is the work required to move the ball from point P (at distance R) to infinity. Equation 8-1 ($\Delta U = -W$) tells us that we can also write that work in terms of potential energies as

$$U_\infty - U = -W.$$

Because the potential energy U_∞ at infinity is zero, U is the potential energy at P, and W is given by Eq. 13-25, this equation becomes

$$U = W = -\frac{GMm}{R}.$$

Switching R to r gives us Eq. 13-21, which we set out to prove.

Path Independence

In Fig. 13-10, we move a baseball from point A to point G along a path consisting of three radial lengths and three circular arcs (centered on Earth). We are interested in the total work W done by Earth's gravitational force $\vec{F}$ on the ball as it moves from A to G. The work done along each circular arc is zero, because the direction of $\vec{F}$ is perpendicular to the arc at every point. Thus, W is the sum of only the works done by $\vec{F}$ along the three radial lengths.

Now, suppose we mentally shrink the arcs to zero. We would then be moving the ball directly from A to G along a single radial length. Does that change W? No. Because no work was done along the arcs, eliminating them does not change the work. The path taken from A to G now is clearly different, but the work done by $\vec{F}$ is the same.

We discussed such a result in a general way in Module 8-1. Here is the point: The gravitational force is a conservative force. Thus, the work done by the gravitational force on a particle moving from an initial point i to a final point f is independent of the path taken between the points. From Eq. 8-1, the change ΔU in the gravitational potential energy from point i to point f is given by

$$\Delta U = U_f - U_i = -W. \tag{13-26}$$

Since the work W done by a conservative force is independent of the actual path taken, the change ΔU in gravitational potential energy is *also independent* of the path taken.

Potential Energy and Force

In the proof of Eq. 13-21, we derived the potential energy function $U(r)$ from the force function $\vec{F}(r)$. We should be able to go the other way—that is, to start from the potential energy function and derive the force function. Guided by Eq. 8-22 ($F(x) = -dU(x)/dx$), we can write

$$F = -\frac{dU}{dr} = -\frac{d}{dr}\left(-\frac{GMm}{r}\right)$$
$$= -\frac{GMm}{r^2}. \tag{13-27}$$

This is Newton's law of gravitation (Eq. 13-1). The minus sign indicates that the force on mass m points radially inward, toward mass M.

Escape Speed

If you fire a projectile upward, usually it will slow, stop momentarily, and return to Earth. There is, however, a certain minimum initial speed that will cause it to move upward forever, theoretically coming to rest only at infinity. This minimum initial speed is called the (Earth) **escape speed.**

Consider a projectile of mass m, leaving the surface of a planet (or some other astronomical body or system) with escape speed v. The projectile has a kinetic energy K given by $\frac{1}{2}mv^2$ and a potential energy U given by Eq. 13-21:

$$U = -\frac{GMm}{R},$$

in which M is the mass of the planet and R is its radius.

When the projectile reaches infinity, it stops and thus has no kinetic energy. It also has no potential energy because an infinite separation between two bodies is our zero-potential-energy configuration. Its total energy at infinity is therefore zero. From the principle of conservation of energy, its total energy at the planet's surface must also have been zero, and so

$$K + U = \tfrac{1}{2}mv^2 + \left(-\frac{GMm}{R}\right) = 0.$$

This yields

$$v = \sqrt{\frac{2GM}{R}}. \tag{13-28}$$

Note that v does not depend on the direction in which a projectile is fired from a planet. However, attaining that speed is easier if the projectile is fired in the direction the launch site is moving as the planet rotates about its axis. For example, rockets are launched eastward at Cape Canaveral to take advantage of the Cape's eastward speed of 1500 km/h due to Earth's rotation.

Equation 13-28 can be applied to find the escape speed of a projectile from any astronomical body, provided we substitute the mass of the body for M and the radius of the body for R. Table 13-2 shows some escape speeds.

Table 13-2 Some Escape Speeds

Body	Mass (kg)	Radius (m)	Escape Speed (km/s)
Ceres[a]	1.17×10^{21}	3.8×10^5	0.64
Earth's moon[a]	7.36×10^{22}	1.74×10^6	2.38
Earth	5.98×10^{24}	6.37×10^6	11.2
Jupiter	1.90×10^{27}	7.15×10^7	59.5
Sun	1.99×10^{30}	6.96×10^8	618
Sirius B[b]	2×10^{30}	1×10^7	5200
Neutron star[c]	2×10^{30}	1×10^4	2×10^5

[a]The most massive of the asteroids.
[b]A *white dwarf* (a star in a final stage of evolution) that is a companion of the bright star Sirius.
[c]The collapsed core of a star that remains after that star has exploded in a *supernova* event.

Checkpoint 3

You move a ball of mass m away from a sphere of mass M. (a) Does the gravitational potential energy of the system of ball and sphere increase or decrease? (b) Is positive work or negative work done by the gravitational force between the ball and the sphere?

Sample Problem 13.03 Asteroid falling from space, mechanical energy

An asteroid, headed directly toward Earth, has a speed of 12 km/s relative to the planet when the asteroid is 10 Earth radii from Earth's center. Neglecting the effects of Earth's atmosphere on the asteroid, find the asteroid's speed v_f when it reaches Earth's surface.

KEY IDEAS

Because we are to neglect the effects of the atmosphere on the asteroid, the mechanical energy of the asteroid–Earth system is conserved during the fall. Thus, the final mechanical energy (when the asteroid reaches Earth's surface) is equal to the initial mechanical energy. With kinetic energy K and gravitational potential energy U, we can write this as

$$K_f + U_f = K_i + U_i. \tag{13-29}$$

Also, if we assume the system is isolated, the system's linear momentum must be conserved during the fall. Therefore, the momentum change of the asteroid and that of Earth must be equal in magnitude and opposite in sign. However, because Earth's mass is so much greater than the asteroid's mass, the change in Earth's speed is negligible relative to the change in the asteroid's speed. So, the change in Earth's kinetic energy is also negligible. Thus, we can assume that the kinetic energies in Eq. 13-29 are those of the asteroid alone.

Calculations: Let m represent the asteroid's mass and M represent Earth's mass (5.98×10^{24} kg). The asteroid is initially at distance $10R_E$ and finally at distance R_E, where R_E is Earth's radius (6.37×10^6 m). Substituting Eq. 13-21 for U and $\frac{1}{2}mv^2$ for K, we rewrite Eq. 13-29 as

$$\tfrac{1}{2}mv_f^2 - \frac{GMm}{R_E} = \tfrac{1}{2}mv_i^2 - \frac{GMm}{10R_E}.$$

Rearranging and substituting known values, we find

$$\begin{aligned} v_f^2 &= v_i^2 + \frac{2GM}{R_E}\left(1 - \frac{1}{10}\right) \\ &= (12 \times 10^3 \text{ m/s})^2 \\ &\quad + \frac{2(6.67 \times 10^{-11} \text{ m}^3/\text{kg}\cdot\text{s}^2)(5.98 \times 10^{24} \text{ kg})}{6.37 \times 10^6 \text{ m}} 0.9 \\ &= 2.567 \times 10^8 \text{ m}^2/\text{s}^2, \end{aligned}$$

and $$v_f = 1.60 \times 10^4 \text{ m/s} = 16 \text{ km/s}. \quad \text{(Answer)}$$

At this speed, the asteroid would not have to be particularly large to do considerable damage at impact. If it were only 5 m across, the impact could release about as much energy as the nuclear explosion at Hiroshima. Alarmingly, about 500 million asteroids of this size are near Earth's orbit, and in 1994 one of them apparently penetrated Earth's atmosphere and exploded 20 km above the South Pacific (setting off nuclear-explosion warnings on six military satellites).

13-6 PLANETS AND SATELLITES: KEPLER'S LAWS

Learning Objectives

After reading this module, you should be able to . . .

13.17 Identify Kepler's three laws.

13.18 Identify which of Kepler's laws is equivalent to the law of conservation of angular momentum.

13.19 On a sketch of an elliptical orbit, identify the semimajor axis, the eccentricity, the perihelion, the aphelion, and the focal points.

13.20 For an elliptical orbit, apply the relationships between the semimajor axis, the eccentricity, the perihelion, and the aphelion.

13.21 For an orbiting natural or artificial satellite, apply Kepler's relationship between the orbital period and radius and the mass of the astronomical body being orbited.

Key Ideas

- The motion of satellites, both natural and artificial, is governed by Kepler's laws:

1. *The law of orbits.* All planets move in elliptical orbits with the Sun at one focus.
2. *The law of areas.* A line joining any planet to the Sun sweeps out equal areas in equal time intervals. (This statement is equivalent to conservation of angular momentum.)
3. *The law of periods.* The square of the period T of any planet is proportional to the cube of the semimajor axis a of its orbit. For circular orbits with radius r,

$$T^2 = \left(\frac{4\pi^2}{GM}\right)r^3 \quad \text{(law of periods),}$$

where M is the mass of the attracting body—the Sun in the case of the solar system. For elliptical planetary orbits, the semimajor axis a is substituted for r.

Planets and Satellites: Kepler's Laws

The motions of the planets, as they seemingly wander against the background of the stars, have been a puzzle since the dawn of history. The "loop-the-loop" motion of Mars, shown in Fig. 13-11, was particularly baffling. Johannes Kepler (1571–1630), after a lifetime of study, worked out the empirical laws that govern these motions. Tycho Brahe (1546–1601), the last of the great astronomers to make observations without the help of a telescope, compiled the extensive data from which Kepler was able to derive the three laws of planetary motion that now bear Kepler's name. Later, Newton (1642–1727) showed that his law of gravitation leads to Kepler's laws.

In this section we discuss each of Kepler's three laws. Although here we apply the laws to planets orbiting the Sun, they hold equally well for satellites, either natural or artificial, orbiting Earth or any other massive central body.

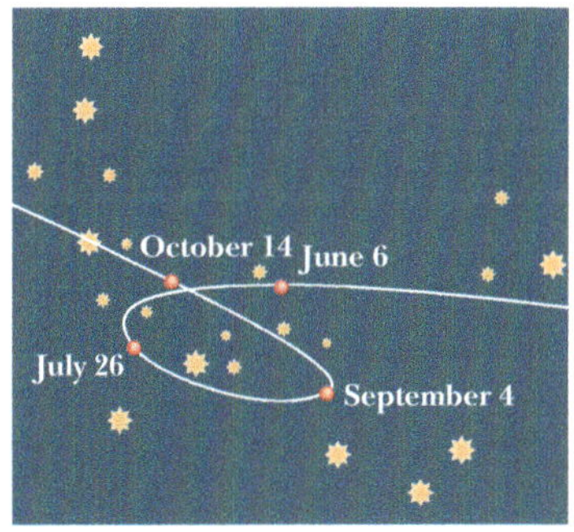

Figure 13-11 The path seen from Earth for the planet Mars as it moved against a background of the constellation Capricorn during 1971. The planet's position on four days is marked. Both Mars and Earth are moving in orbits around the Sun so that we see the position of Mars relative to us; this relative motion sometimes results in an apparent loop in the path of Mars.

1. **THE LAW OF ORBITS:** All planets move in elliptical orbits, with the Sun at one focus.

Figure 13-12 shows a planet of mass m moving in such an orbit around the Sun, whose mass is M. We assume that $M \gg m$, so that the center of mass of the planet–Sun system is approximately at the center of the Sun.

The orbit in Fig. 13-12 is described by giving its **semimajor axis** a and its **eccentricity** e, the latter defined so that ea is the distance from the center of the ellipse to either focus F or F'. *An eccentricity of zero corresponds to a circle,* in which the two foci merge to a single central point. The eccentricities of the planetary orbits are not large; so if the orbits are drawn to scale, they look circular. The eccentricity of the ellipse of Fig. 13-12, which has been exaggerated for clarity, is 0.74. The eccentricity of Earth's orbit is only 0.0167.

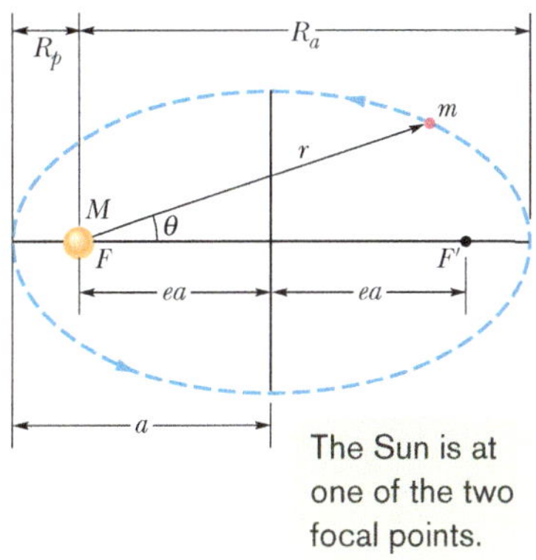

Figure 13-12 A planet of mass m moving in an elliptical orbit around the Sun. The Sun, of mass M, is at one focus F of the ellipse. The other focus is F', which is located in empty space. The semimajor axis a of the ellipse, the perihelion (nearest the Sun) distance R_p, and the aphelion (farthest from the Sun) distance R_a are also shown.

2. **THE LAW OF AREAS:** A line that connects a planet to the Sun sweeps out equal areas in the plane of the planet's orbit in equal time intervals; that is, the rate dA/dt at which it sweeps out area A is constant.

Qualitatively, this second law tells us that the planet will move most slowly when it is farthest from the Sun and most rapidly when it is nearest to the Sun. As it turns out, Kepler's second law is totally equivalent to the law of conservation of angular momentum. Let us prove it.

The area of the shaded wedge in Fig. 13-13a closely approximates the area swept out in time Δt by a line connecting the Sun and the planet, which are separated by distance r. The area ΔA of the wedge is approximately the area of

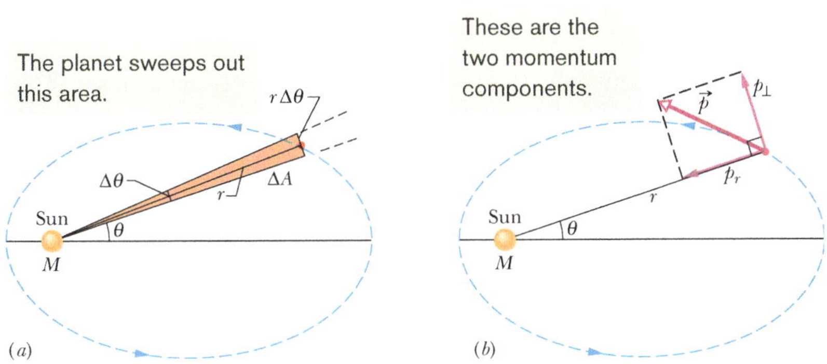

Figure 13-13 (a) In time Δt, the line r connecting the planet to the Sun moves through an angle $\Delta\theta$, sweeping out an area ΔA (shaded). (b) The linear momentum $\vec{p}$ of the planet and the components of $\vec{p}$.

a triangle with base $r\,\Delta\theta$ and height r. Since the area of a triangle is one-half of the base times the height, $\Delta A \approx \frac{1}{2}r^2\Delta\theta$. This expression for ΔA becomes more exact as Δt (hence $\Delta\theta$) approaches zero. The instantaneous rate at which area is being swept out is then

$$\frac{dA}{dt} = \frac{1}{2}r^2\frac{d\theta}{dt} = \frac{1}{2}r^2\omega, \tag{13-30}$$

in which ω is the angular speed of the line connecting Sun and planet, as the line rotates around the Sun.

Figure 13-13*b* shows the linear momentum $\vec{p}$ of the planet, along with the radial and perpendicular components of $\vec{p}$. From Eq. 11-20 ($L = rp_\perp$), the magnitude of the angular momentum $\vec{L}$ of the planet about the Sun is given by the product of r and $p_\perp$, the component of $\vec{p}$ perpendicular to r. Here, for a planet of mass m,

$$\begin{aligned} L &= rp_\perp = (r)(mv_\perp) = (r)(m\omega r) \\ &= mr^2\omega, \end{aligned} \tag{13-31}$$

where we have replaced $v_\perp$ with its equivalent ωr (Eq. 10-18). Eliminating $r^2\omega$ between Eqs. 13-30 and 13-31 leads to

$$\frac{dA}{dt} = \frac{L}{2m}. \tag{13-32}$$

If dA/dt is constant, as Kepler said it is, then Eq. 13-32 means that L must also be constant—angular momentum is conserved. Kepler's second law is indeed equivalent to the law of conservation of angular momentum.

3. THE LAW OF PERIODS: The square of the period of any planet is proportional to the cube of the semimajor axis of its orbit.

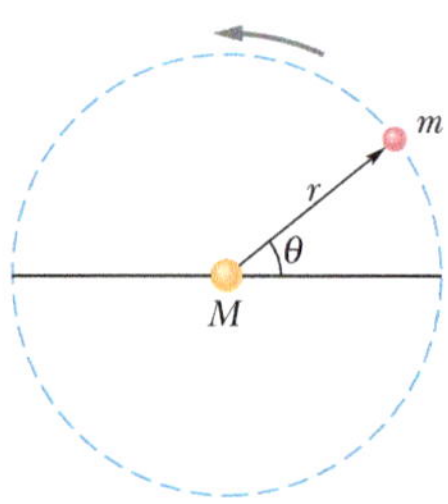

Figure 13-14 A planet of mass m moving around the Sun in a circular orbit of radius r.

To see this, consider the circular orbit of Fig. 13-14, with radius r (the radius of a circle is equivalent to the semimajor axis of an ellipse). Applying Newton's second law ($F = ma$) to the orbiting planet in Fig. 13-14 yields

$$\frac{GMm}{r^2} = (m)(\omega^2 r). \tag{13-33}$$

Here we have substituted from Eq. 13-1 for the force magnitude F and used Eq. 10-23 to substitute $\omega^2 r$ for the centripetal acceleration. If we now use Eq. 10-20 to replace ω with $2\pi/T$, where T is the period of the motion, we obtain Kepler's third law:

$$T^2 = \left(\frac{4\pi^2}{GM}\right)r^3 \quad \text{(law of periods).} \tag{13-34}$$

The quantity in parentheses is a constant that depends only on the mass M of the central body about which the planet orbits.

Equation 13-34 holds also for elliptical orbits, provided we replace r with a, the semimajor axis of the ellipse. This law predicts that the ratio T^2/a^3 has essentially the same value for every planetary orbit around a given massive body. Table 13-3 shows how well it holds for the orbits of the planets of the solar system.

Table 13-3 Kepler's Law of Periods for the Solar System

Planet	Semimajor Axis a (10^{10} m)	Period T (y)	T^2/a^3 (10^{-34} y^2/m^3)
Mercury	5.79	0.241	2.99
Venus	10.8	0.615	3.00
Earth	15.0	1.00	2.96
Mars	22.8	1.88	2.98
Jupiter	77.8	11.9	3.01
Saturn	143	29.5	2.98
Uranus	287	84.0	2.98
Neptune	450	165	2.99
Pluto	590	248	2.99

Checkpoint 4

Satellite 1 is in a certain circular orbit around a planet, while satellite 2 is in a larger circular orbit. Which satellite has (a) the longer period and (b) the greater speed?

Sample Problem 13.04 Kepler's law of periods, Comet Halley

Comet Halley orbits the Sun with a period of 76 years and, in 1986, had a distance of closest approach to the Sun, its *perihelion distance* R_p, of 8.9×10^{10} m. Table 13-3 shows that this is between the orbits of Mercury and Venus.

(a) What is the comet's farthest distance from the Sun, which is called its *aphelion distance* R_a?

KEY IDEAS

From Fig. 13-12, we see that $R_a + R_p = 2a$, where a is the semimajor axis of the orbit. Thus, we can find R_a if we first find a. We can relate a to the given period via the law of periods (Eq. 13-34) if we simply substitute the semimajor axis a for r.

Calculations: Making that substitution and then solving for a, we have

$$a = \left(\frac{GMT^2}{4\pi^2}\right)^{1/3}. \quad (13\text{-}35)$$

If we substitute the mass M of the Sun, 1.99×10^{30} kg, and the period T of the comet, 76 years or 2.4×10^9 s, into Eq. 13-35, we find that $a = 2.7 \times 10^{12}$ m. Now we have

$$\begin{aligned} R_a &= 2a - R_p \\ &= (2)(2.7 \times 10^{12}\text{ m}) - 8.9 \times 10^{10}\text{ m} \\ &= 5.3 \times 10^{12}\text{ m}. \end{aligned} \quad \text{(Answer)}$$

Table 13-3 shows that this is a little less than the semimajor axis of the orbit of Pluto. Thus, the comet does not get farther from the Sun than Pluto.

(b) What is the eccentricity e of the orbit of comet Halley?

KEY IDEA

We can relate e, a, and R_p via Fig. 13-12, in which we see that $ea = a - R_p$.

Calculation: We have

$$e = \frac{a - R_p}{a} = 1 - \frac{R_p}{a} \quad (13\text{-}36)$$

$$= 1 - \frac{8.9 \times 10^{10}\text{ m}}{2.7 \times 10^{12}\text{ m}} = 0.97. \quad \text{(Answer)}$$

This tells us that, with an eccentricity approaching unity, this orbit must be a long thin ellipse.

WileyPLUS Additional examples, video, and practice available at *WileyPLUS*

13-7 SATELLITES: ORBITS AND ENERGY

Learning Objectives

After reading this module, you should be able to . . .

13.22 For a satellite in a circular orbit around an astronomical body, calculate the gravitational potential energy, the kinetic energy, and the total energy.

13.23 For a satellite in an elliptical orbit, calculate the total energy.

Key Ideas

- When a planet or satellite with mass m moves in a circular orbit with radius r, its potential energy U and kinetic energy K are given by

$$U = -\frac{GMm}{r} \quad \text{and} \quad K = \frac{GMm}{2r}.$$

The mechanical energy $E = K + U$ is then

$$E = -\frac{GMm}{2r}.$$

For an elliptical orbit of semimajor axis a,

$$E = -\frac{GMm}{2a}.$$

Satellites: Orbits and Energy

As a satellite orbits Earth in an elliptical path, both its speed, which fixes its kinetic energy K, and its distance from the center of Earth, which fixes its gravitational potential energy U, fluctuate with fixed periods. However, the mechanical energy E of the satellite remains constant. (Since the satellite's mass is so much smaller than Earth's mass, we assign U and E for the Earth–satellite system to the satellite alone.)

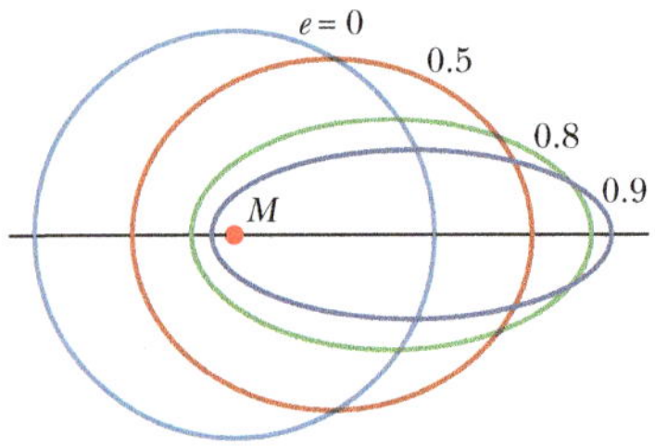

Figure 13-15 Four orbits with different eccentricities e about an object of mass M. All four orbits have the same semimajor axis a and thus correspond to the same total mechanical energy E.

The potential energy of the system is given by Eq. 13-21:

$$U = -\frac{GMm}{r}$$

(with $U = 0$ for infinite separation). Here r is the radius of the satellite's orbit, assumed for the time being to be circular, and M and m are the masses of Earth and the satellite, respectively.

To find the kinetic energy of a satellite in a circular orbit, we write Newton's second law ($F = ma$) as

$$\frac{GMm}{r^2} = m\frac{v^2}{r}, \tag{13-37}$$

where v^2/r is the centripetal acceleration of the satellite. Then, from Eq. 13-37, the kinetic energy is

$$K = \tfrac{1}{2}mv^2 = \frac{GMm}{2r}, \tag{13-38}$$

which shows us that for a satellite in a circular orbit,

$$K = -\frac{U}{2} \quad \text{(circular orbit)}. \tag{13-39}$$

The total mechanical energy of the orbiting satellite is

$$E = K + U = \frac{GMm}{2r} - \frac{GMm}{r}$$

or

$$E = -\frac{GMm}{2r} \quad \text{(circular orbit)}. \tag{13-40}$$

This tells us that for a satellite in a circular orbit, the total energy E is the negative of the kinetic energy K:

$$E = -K \quad \text{(circular orbit)}. \tag{13-41}$$

For a satellite in an elliptical orbit of semimajor axis a, we can substitute a for r in Eq. 13-40 to find the mechanical energy:

$$E = -\frac{GMm}{2a} \quad \text{(elliptical orbit)}. \tag{13-42}$$

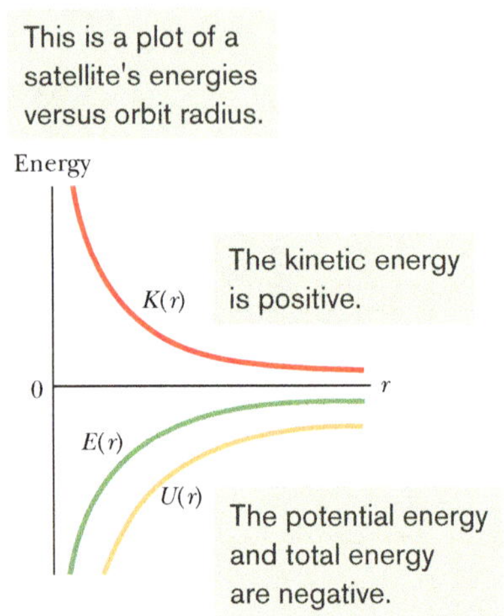

Figure 13-16 The variation of kinetic energy K, potential energy U, and total energy E with radius r for a satellite in a circular orbit. For any value of r, the values of U and E are negative, the value of K is positive, and $E = -K$. As $r \rightarrow \infty$, all three energy curves approach a value of zero.

Equation 13-42 tells us that the total energy of an orbiting satellite depends only on the semimajor axis of its orbit and not on its eccentricity e. For example, four orbits with the same semimajor axis are shown in Fig. 13-15; the same satellite would have the same total mechanical energy E in all four orbits. Figure 13-16 shows the variation of K, U, and E with r for a satellite moving in a circular orbit about a massive central body. Note that as r is increased, the kinetic energy (and thus also the orbital speed) decreases.

Checkpoint 5

In the figure here, a space shuttle is initially in a circular orbit of radius r about Earth. At point P, the pilot briefly fires a forward-pointing thruster to decrease the shuttle's kinetic energy K and mechanical energy E. (a) Which of the dashed elliptical orbits shown in the figure will the shuttle then take? (b) Is the orbital period T of the shuttle (the time to return to P) then greater than, less than, or the same as in the circular orbit?

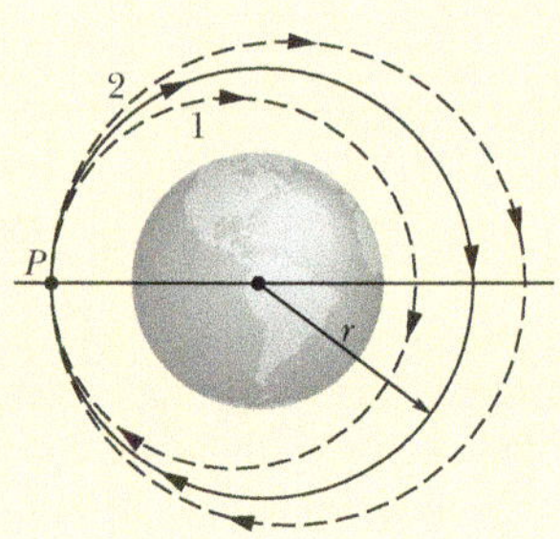

Sample Problem 13.05 Mechanical energy of orbiting bowling ball

A playful astronaut releases a bowling ball, of mass m = 7.20 kg, into circular orbit about Earth at an altitude h of 350 km.

(a) What is the mechanical energy E of the ball in its orbit?

KEY IDEA

We can get E from the orbital energy, given by Eq. 13-40 ($E = -GMm/2r$), if we first find the orbital radius r. (It is *not* simply the given altitude.)

Calculations: The orbital radius must be

$$r = R + h = 6370 \text{ km} + 350 \text{ km} = 6.72 \times 10^6 \text{ m},$$

in which R is the radius of Earth. Then, from Eq. 13-40 with Earth mass $M = 5.98 \times 10^{24}$ kg, the mechanical energy is

$$E = -\frac{GMm}{2r}$$
$$= -\frac{(6.67 \times 10^{-11} \text{ N}\cdot\text{m}^2/\text{kg}^2)(5.98 \times 10^{24} \text{ kg})(7.20 \text{ kg})}{(2)(6.72 \times 10^6 \text{ m})}$$
$$= -2.14 \times 10^8 \text{ J} = -214 \text{ MJ}. \quad \text{(Answer)}$$

(b) What is the mechanical energy E_0 of the ball on the launchpad at the Kennedy Space Center (before launch)? From there to the orbit, what is the change ΔE in the ball's mechanical energy?

KEY IDEA

On the launchpad, the ball is *not* in orbit and thus Eq. 13-40 does *not* apply. Instead, we must find $E_0 = K_0 + U_0$, where K_0 is the ball's kinetic energy and U_0 is the gravitational potential energy of the ball–Earth system.

Calculations: To find U_0, we use Eq. 13-21 to write

$$U_0 = -\frac{GMm}{R}$$
$$= -\frac{(6.67 \times 10^{-11} \text{ N}\cdot\text{m}^2/\text{kg}^2)(5.98 \times 10^{24} \text{ kg})(7.20 \text{ kg})}{6.37 \times 10^6 \text{ m}}$$
$$= -4.51 \times 10^8 \text{ J} = -451 \text{ MJ}.$$

The kinetic energy K_0 of the ball is due to the ball's motion with Earth's rotation. You can show that K_0 is less than 1 MJ, which is negligible relative to U_0. Thus, the mechanical energy of the ball on the launchpad is

$$E_0 = K_0 + U_0 \approx 0 - 451 \text{ MJ} = -451 \text{ MJ}. \quad \text{(Answer)}$$

The *increase* in the mechanical energy of the ball from launchpad to orbit is

$$\Delta E = E - E_0 = (-214 \text{ MJ}) - (-451 \text{ MJ})$$
$$= 237 \text{ MJ}. \quad \text{(Answer)}$$

This is worth a few dollars at your utility company. Obviously the high cost of placing objects into orbit is not due to their required mechanical energy.

Sample Problem 13.06 Transforming a circular orbit into an elliptical orbit

A spaceship of mass $m = 4.50 \times 10^3$ kg is in a circular Earth orbit of radius $r = 8.00 \times 10^6$ m and period $T_0 = 118.6$ min $= 7.119 \times 10^3$ s when a thruster is fired in the forward direction to decrease the speed to 96.0% of the original speed. What is the period T of the resulting elliptical orbit (Fig. 13-17)?

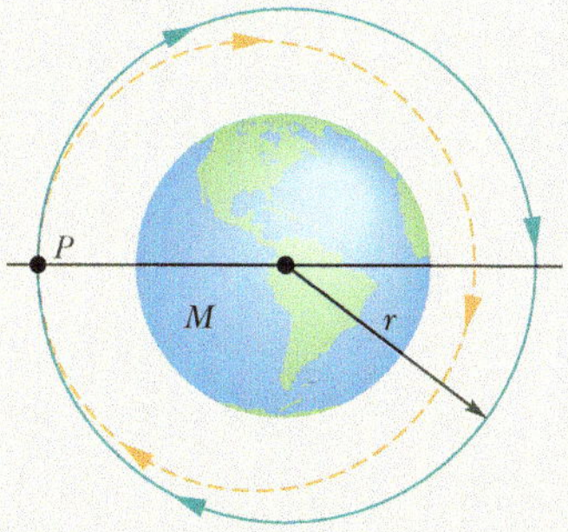

Figure 13-17 At point P a thruster is fired, changing a ship's orbit from circular to elliptical.

KEY IDEAS

(1) An elliptical orbit period is related to the semimajor axis a by Kepler's third law, written as Eq. 13-34 ($T^2 = 4\pi^2 r^3/GM$) but with a replacing r. (2) The semimajor axis a is related to the total mechanical energy E of the ship by Eq. 13-42 ($E = -GMm/2a$), in which Earth's mass is $M = 5.98 \times 10^{24}$ kg. (3) The potential energy of the ship at radius r from Earth's center is given by Eq. 13-21 ($U = -GMm/r$).

Calculations: Looking over the Key Ideas, we see that we need to calculate the total energy E to find the semimajor axis a, so that we can then determine the period of the elliptical orbit. Let's start with the kinetic energy, calculating it just after the thruster is fired. The speed v just then is 96% of the initial speed v_0, which was equal to the ratio of the circumference of the initial circular orbit to the initial period of the orbit. Thus, just after the thruster is fired, the kinetic energy is

$$K = \tfrac{1}{2}mv^2 = \tfrac{1}{2}m(0.96v_0)^2 = \tfrac{1}{2}m(0.96)^2\left(\frac{2\pi r}{T_0}\right)^2$$
$$= \tfrac{1}{2}(4.50 \times 10^3 \text{ kg})(0.96)^2\left(\frac{2\pi(8.00 \times 10^6 \text{ m})}{7.119 \times 10^3 \text{ s}}\right)^2$$
$$= 1.0338 \times 10^{11} \text{ J}.$$

Just after the thruster is fired, the ship is still at orbital radius r, and thus its gravitational potential energy is

$$U = -\frac{GMm}{r} = -\frac{(6.67 \times 10^{-11}\,\text{N}\cdot\text{m}^2/\text{kg}^2)(5.98 \times 10^{24}\,\text{kg})(4.50 \times 10^3\,\text{kg})}{8.00 \times 10^6\,\text{m}} = -2.2436 \times 10^{11}\,\text{J}.$$

We can now find the semimajor axis by rearranging Eq. 13-42, substituting a for r, and then substituting in our energy results:

$$a = -\frac{GMm}{2E} = -\frac{GMm}{2(K+U)} = -\frac{(6.67 \times 10^{-11}\,\text{N}\cdot\text{m}^2/\text{kg}^2)(5.98 \times 10^{24}\,\text{kg})(4.50 \times 10^3\,\text{kg})}{2(1.0338 \times 10^{11}\,\text{J} - 2.2436 \times 10^{11}\,\text{J})} = 7.418 \times 10^6\,\text{m}.$$

OK, one more step to go. We substitute a for r in Eq. 13-34 and then solve for the period T, substituting our result for a:

$$T = \left(\frac{4\pi^2 a^3}{GM}\right)^{1/2} = \left(\frac{4\pi^2(7.418 \times 10^6\,\text{m})^3}{(6.67 \times 10^{-11}\,\text{N}\cdot\text{m}^2/\text{kg}^2)(5.98 \times 10^{24}\,\text{kg})}\right)^{1/2} = 6.356 \times 10^3\,\text{s} = 106\,\text{min}. \quad \text{(Answer)}$$

This is the period of the elliptical orbit that the ship takes after the thruster is fired. It is less than the period T_0 for the circular orbit for two reasons. (1) The orbital path length is now less. (2) The elliptical path takes the ship closer to Earth everywhere except at the point of firing (Fig. 13-17). The resulting decrease in gravitational potential energy increases the kinetic energy and thus also the speed of the ship.

WILEY PLUS Additional examples, video, and practice available at *WileyPLUS*

13-8 EINSTEIN AND GRAVITATION

Learning Objectives

After reading this module, you should be able to . . .

13.24 Explain Einstein's principle of equivalence.

13.25 Identify Einstein's model for gravitation as being due to the curvature of spacetime.

Key Idea

- Einstein pointed out that gravitation and acceleration are equivalent. This principle of equivalence led him to a theory of gravitation (the general theory of relativity) that explains gravitational effects in terms of a curvature of space.

Einstein and Gravitation

Principle of Equivalence

Albert Einstein once said: "I was . . . in the patent office at Bern when all of a sudden a thought occurred to me: 'If a person falls freely, he will not feel his own weight.' I was startled. This simple thought made a deep impression on me. It impelled me toward a theory of gravitation."

Thus Einstein tells us how he began to form his **general theory of relativity.** The fundamental postulate of this theory about gravitation (the gravitating of objects toward each other) is called the **principle of equivalence,** which says that gravitation and acceleration are equivalent. If a physicist were locked up in a small box as in Fig. 13-18, he would not be able to tell whether the box was at

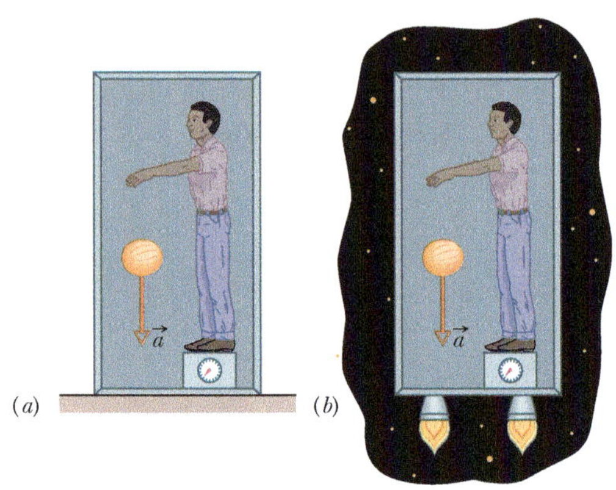

Figure 13-18 (a) A physicist in a box resting on Earth sees a cantaloupe falling with acceleration $a = 9.8\,\text{m/s}^2$. (b) If he and the box accelerate in deep space at $9.8\,\text{m/s}^2$, the cantaloupe has the same acceleration relative to him. It is not possible, by doing experiments within the box, for the physicist to tell which situation he is in. For example, the platform scale on which he stands reads the same weight in both situations.

rest on Earth (and subject only to Earth's gravitational force), as in Fig. 13-18*a*, or accelerating through interstellar space at 9.8 m/s^2 (and subject only to the force producing that acceleration), as in Fig. 13-18*b*. In both situations he would feel the same and would read the same value for his weight on a scale. Moreover, if he watched an object fall past him, the object would have the same acceleration relative to him in both situations.

Curvature of Space

We have thus far explained gravitation as due to a force between masses. Einstein showed that, instead, gravitation is due to a curvature of space that is caused by the masses. (As is discussed later in this book, space and time are entangled, so the curvature of which Einstein spoke is really a curvature of *spacetime,* the combined four dimensions of our universe.)

Picturing how space (such as vacuum) can have curvature is difficult. An analogy might help: Suppose that from orbit we watch a race in which two boats begin on Earth's equator with a separation of 20 km and head due south (Fig. 13-19*a*). To the sailors, the boats travel along flat, parallel paths. However, with time the boats draw together until, nearer the south pole, they touch. The sailors in the boats can interpret this drawing together in terms of a force acting on the boats. Looking on from space, however, we can see that the boats draw together simply because of the curvature of Earth's surface. We can see this because we are viewing the race from "outside" that surface.

Figure 13-19*b* shows a similar race: Two horizontally separated apples are dropped from the same height above Earth. Although the apples may appear to travel along parallel paths, they actually move toward each other because they both fall toward Earth's center. We can interpret the motion of the apples in terms of the gravitational force on the apples from Earth. We can also interpret the motion in terms of a curvature of the space near Earth, a curvature due to the presence of Earth's mass. This time we cannot see the curvature because we cannot get "outside" the curved space, as we got "outside" the curved Earth in the boat example. However, we can depict the curvature with a drawing like Fig. 13-19*c*; there the apples would move along a surface that curves toward Earth because of Earth's mass.

When light passes near Earth, the path of the light bends slightly because of the curvature of space there, an effect called *gravitational lensing.* When light passes a more massive structure, like a galaxy or a black hole having large mass, its path can be bent more. If such a massive structure is between us and a quasar (an extremely bright, extremely distant source of light), the light from the quasar

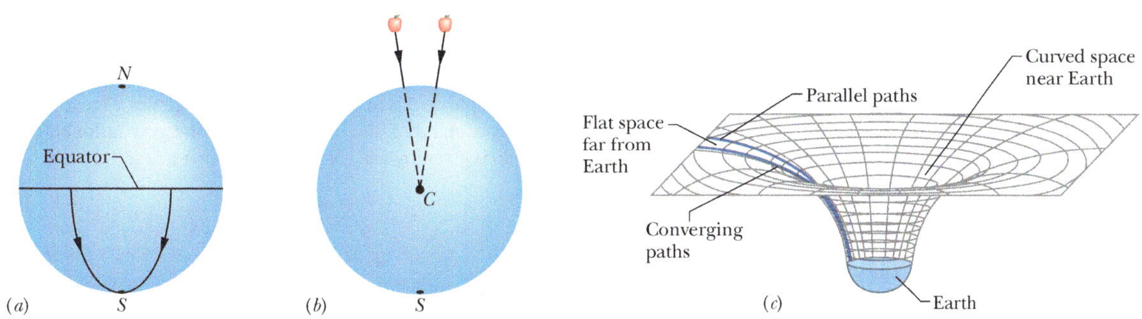

Figure 13-19 (*a*) Two objects moving along lines of longitude toward the south pole converge because of the curvature of Earth's surface. (*b*) Two objects falling freely near Earth move along lines that converge toward the center of Earth because of the curvature of space near Earth. (*c*) Far from Earth (and other masses), space is flat and parallel paths remain parallel. Close to Earth, the parallel paths begin to converge because space is curved by Earth's mass.

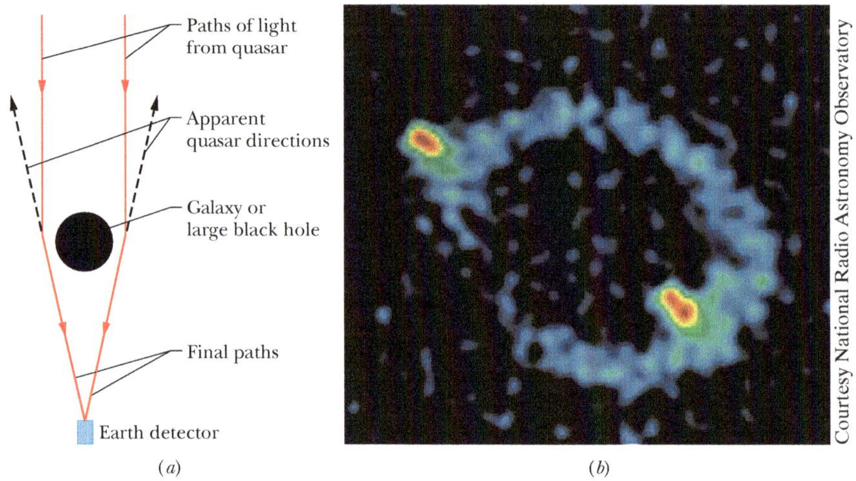

Figure 13-20 (*a*) Light from a distant quasar follows curved paths around a galaxy or a large black hole because the mass of the galaxy or black hole has curved the adjacent space. If the light is detected, it appears to have originated along the backward extensions of the final paths (dashed lines). (*b*) The Einstein ring known as MG1131+0456 on the computer screen of a telescope. The source of the light (actually, radio waves, which are a form of invisible light) is far behind the large, unseen galaxy that produces the ring; a portion of the source appears as the two bright spots seen along the ring.

can bend around the massive structure and toward us (Fig. 13-20*a*). Then, because the light seems to be coming to us from a number of slightly different directions in the sky, we see the same quasar in all those different directions. In some situations, the quasars we see blend together to form a giant luminous arc, which is called an *Einstein ring* (Fig. 13-20*b*).

Should we attribute gravitation to the curvature of spacetime due to the presence of masses or to a force between masses? Or should we attribute it to the actions of a type of fundamental particle called a *graviton,* as conjectured in some modern physics theories? Although our theories about gravitation have been enormously successful in describing everything from falling apples to planetary and stellar motions, we still do not fully understand it on either the cosmological scale or the quantum physics scale.

Review & Summary

The Law of Gravitation Any particle in the universe attracts any other particle with a **gravitational force** whose magnitude is

$$F = G\frac{m_1 m_2}{r^2} \quad \text{(Newton's law of gravitation)}, \tag{13-1}$$

where m_1 and m_2 are the masses of the particles, r is their separation, and G ($= 6.67 \times 10^{-11}\ \text{N}\cdot\text{m}^2/\text{kg}^2$) is the *gravitational constant.*

Gravitational Behavior of Uniform Spherical Shells The gravitational force between extended bodies is found by adding (integrating) the individual forces on individual particles within the bodies. However, if either of the bodies is a uniform spherical shell or a spherically symmetric solid, the net gravitational force it exerts on an *external* object may be computed as if all the mass of the shell or body were located at its center.

Superposition Gravitational forces obey the **principle of superposition;** that is, if n particles interact, the net force $\vec{F}_{1,\text{net}}$ on a particle labeled particle 1 is the sum of the forces on it from all the other particles taken one at a time:

$$\vec{F}_{1,\text{net}} = \sum_{i=2}^{n} \vec{F}_{1i}, \tag{13-5}$$

in which the sum is a vector sum of the forces $\vec{F}_{1i}$ on particle 1 from particles 2, 3, . . . , n. The gravitational force $\vec{F}_1$ on a particle from an extended body is found by dividing the body into units of differential mass dm, each of which produces a differential force $d\vec{F}$ on the particle, and then integrating to find the sum of those forces:

$$\vec{F}_1 = \int d\vec{F}. \tag{13-6}$$

Gravitational Acceleration The *gravitational acceleration* a_g of a particle (of mass m) is due solely to the gravitational force acting on it. When the particle is at distance r from the center of a uniform, spherical body of mass M, the magnitude F of the gravitational force on the particle is given by Eq. 13-1. Thus, by Newton's second law,

$$F = ma_g, \tag{13-10}$$

which gives

$$a_g = \frac{GM}{r^2}. \tag{13-11}$$

Free-Fall Acceleration and Weight Because Earth's mass is not distributed uniformly, because the planet is not perfectly spherical, and because it rotates, the actual free-fall acceleration $\vec{g}$ of a particle near Earth differs slightly from the gravitational acceleration $\vec{a}_g$, and the particle's weight (equal to mg) differs from the magnitude of the gravitational force on it as calculated by Newton's law of gravitation (Eq. 13-1).

Gravitation Within a Spherical Shell A uniform shell of matter exerts no net gravitational force on a particle located inside it. This means that if a particle is located inside a uniform solid sphere at distance r from its center, the gravitational force exerted on the particle is due only to the mass that lies inside a sphere of radius r (the *inside sphere*). The force magnitude is given by

$$F = \frac{GmM}{R^3} r, \tag{13-19}$$

where M is the sphere's mass and R is its radius.

Gravitational Potential Energy The gravitational potential energy $U(r)$ of a system of two particles, with masses M and m and separated by a distance r, is the negative of the work that would be done by the gravitational force of either particle acting on the other if the separation between the particles were changed from infinite (very large) to r. This energy is

$$U = -\frac{GMm}{r} \quad \text{(gravitational potential energy)}. \tag{13-21}$$

Potential Energy of a System If a system contains more than two particles, its total gravitational potential energy U is the sum of the terms representing the potential energies of all the pairs. As an example, for three particles, of masses m_1, m_2, and m_3,

$$U = -\left(\frac{Gm_1m_2}{r_{12}} + \frac{Gm_1m_3}{r_{13}} + \frac{Gm_2m_3}{r_{23}}\right). \tag{13-22}$$

Escape Speed An object will escape the gravitational pull of an astronomical body of mass M and radius R (that is, it will reach an infinite distance) if the object's speed near the body's surface is at least equal to the **escape speed,** given by

$$v = \sqrt{\frac{2GM}{R}}. \tag{13-28}$$

Kepler's Laws The motion of satellites, both natural and artificial, is governed by these laws:

1. ***The law of orbits.*** All planets move in elliptical orbits with the Sun at one focus.
2. ***The law of areas.*** A line joining any planet to the Sun sweeps out equal areas in equal time intervals. (This statement is equivalent to conservation of angular momentum.)
3. ***The law of periods.*** The square of the period T of any planet is proportional to the cube of the semimajor axis a of its orbit. For circular orbits with radius r,

$$T^2 = \left(\frac{4\pi^2}{GM}\right) r^3 \quad \text{(law of periods)}, \tag{13-34}$$

where M is the mass of the attracting body—the Sun in the case of the solar system. For elliptical planetary orbits, the semimajor axis a is substituted for r.

Energy in Planetary Motion When a planet or satellite with mass m moves in a circular orbit with radius r, its potential energy U and kinetic energy K are given by

$$U = -\frac{GMm}{r} \quad \text{and} \quad K = \frac{GMm}{2r}. \tag{13-21, 13-38}$$

The mechanical energy $E = K + U$ is then

$$E = -\frac{GMm}{2r}. \tag{13-40}$$

For an elliptical orbit of semimajor axis a,

$$E = -\frac{GMm}{2a}. \tag{13-42}$$

Einstein's View of Gravitation Einstein pointed out that gravitation and acceleration are equivalent. This **principle of equivalence** led him to a theory of gravitation (the **general theory of relativity**) that explains gravitational effects in terms of a curvature of space.

Questions

1 In Fig. 13-21, a central particle of mass M is surrounded by a square array of other particles, separated by either distance d or distance $d/2$ along the perimeter of the square. What are the magnitude and direction of the net gravitational force on the central particle due to the other particles?

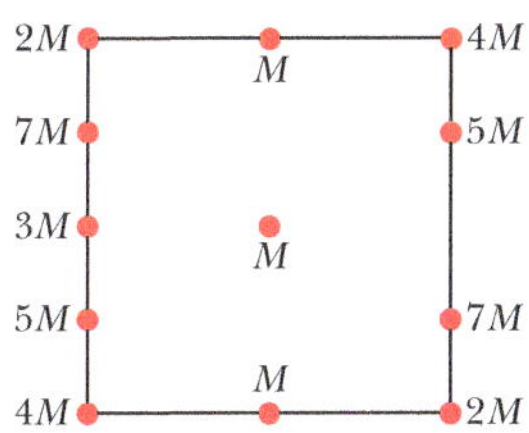

Figure 13-21 Question 1.

2 Figure 13-22 shows three arrangements of the same identical particles, with three of them placed on a circle of radius 0.20 m and the fourth one placed at the center of the circle. (a) Rank the arrangements according to the magnitude of the net gravitational force on the central particle due to the other three particles, greatest first. (b) Rank them according to the gravitational potential energy of the four-particle system, least negative first.

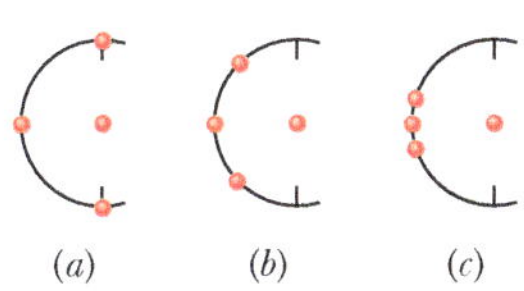

Figure 13-22 Question 2.

3 In Fig. 13-23, a central particle is surrounded by two circular rings of particles, at radii r and R, with $R > r$. All the particles have mass m. What are the magnitude and direction of the net gravitational force on the central particle due to the particles in the rings?

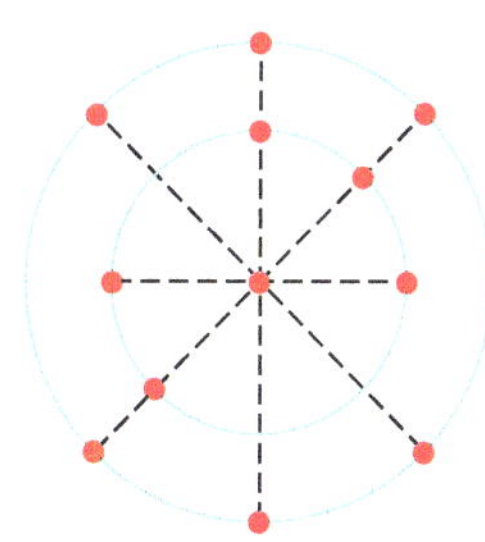

Figure 13-23 Question 3.

4 In Fig. 13-24, two particles, of masses m and $2m$, are fixed in place on an axis. (a) Where on the axis can a third particle of mass $3m$ be placed (other than at infinity) so that the net gravitational force on it from the first two particles is zero: to the left of the first two particles, to their right, between them but closer to the more massive particle, or between them but closer to the less massive particle? (b) Does the answer change if the third particle has, instead, a mass of $16m$? (c) Is there a point off the axis (other than infinity) at which the net force on the third particle would be zero?

Figure 13-24 Question 4.

5 Figure 13-25 shows three situations involving a point particle P with mass m and a spherical shell with a uniformly distributed mass M. The radii of the shells are given. Rank the situations according to the magnitude of the gravitational force on particle P due to the shell, greatest first.

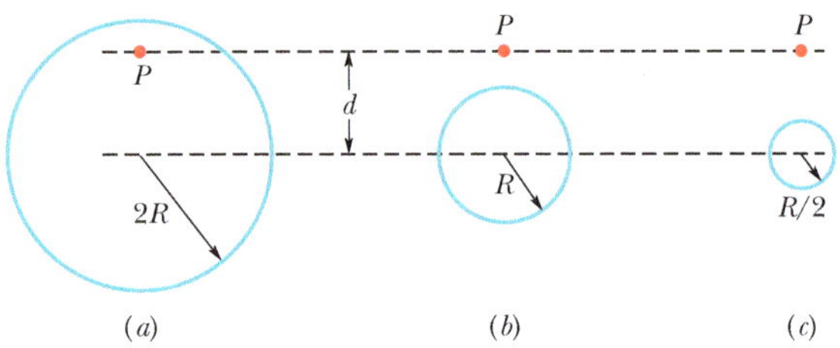

Figure 13-25 Question 5.

6 In Fig. 13-26, three particles are fixed in place. The mass of B is greater than the mass of C. Can a fourth particle (particle D) be placed somewhere so that the net gravitational force on particle A from particles B, C, and D is zero? If so, in which quadrant should it be placed and which axis should it be near?

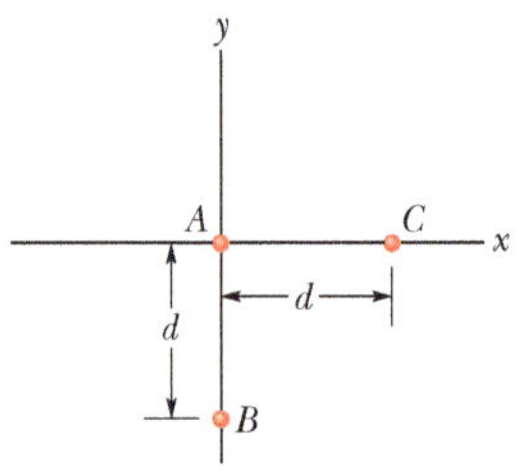

Figure 13-26 Question 6.

7 Rank the four systems of equal-mass particles shown in Checkpoint 2 according to the absolute value of the gravitational potential energy of the system, greatest first.

8 Figure 13-27 gives the gravitational acceleration a_g for four planets as a function of the radial distance r from the center of the planet, starting at the surface of the planet (at radius R_1, R_2, R_3, or R_4). Plots 1 and 2 coincide for $r \geq R_2$; plots 3 and 4 coincide for $r \geq R_4$. Rank the four planets according to (a) mass and (b) mass per unit volume, greatest first.

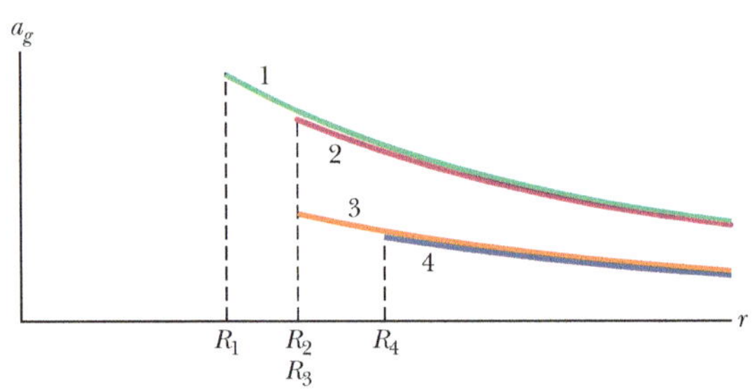

Figure 13-27 Question 8.

9 Figure 13-28 shows three particles initially fixed in place, with B and C identical and positioned symmetrically about the y axis, at distance d from A. (a) In what direction is the net gravitational force $\vec{F}_{net}$ on A? (b) If we move C directly away from the origin, does $\vec{F}_{net}$ change in direction? If so, how and what is the limit of the change?

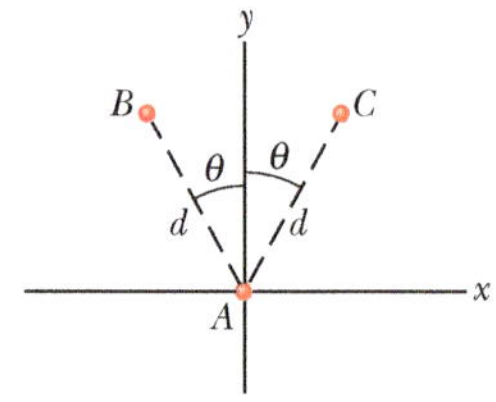

Figure 13-28 Question 9.

10 Figure 13-29 shows six paths by which a rocket orbiting a moon might move from point a to point b. Rank the paths according to (a) the corresponding change in the gravitational potential energy of the rocket–moon system and (b) the net work done on the rocket by the gravitational force from the moon, greatest first.

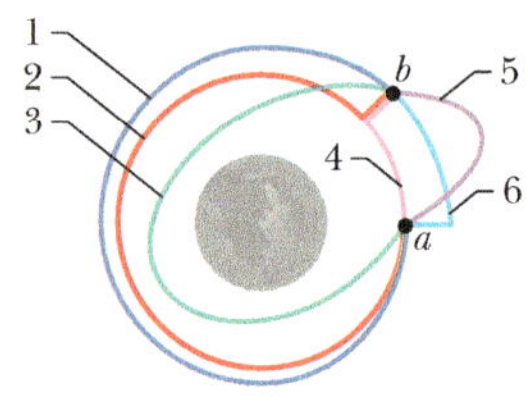

Figure 13-29 Question 10.

11 Figure 13-30 shows three uniform spherical planets that are identical in size and mass. The periods of rotation T for the planets are given, and six lettered points are indicated—three points are on the equators of the planets and three points are on the north poles. Rank the points according to the value of the free-fall acceleration g at them, greatest first.

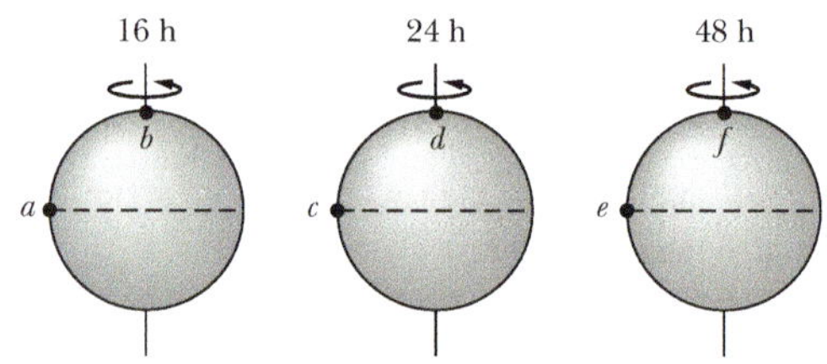

Figure 13-30 Question 11.

12 In Fig. 13-31, a particle of mass m (which is not shown) is to be moved from an infinite distance to one of the three possible locations a, b, and c. Two other particles, of masses m and $2m$, are already fixed in place on the axis, as shown. Rank the three possible locations according to the work done by the net gravitational force on the moving particle due to the fixed particles, greatest first.

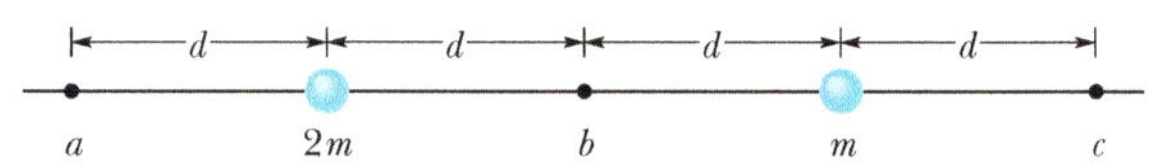

Figure 13-31 Question 12.

Problems

GO Tutoring problem available (at instructor's discretion) in *WileyPLUS* and WebAssign
SSM Worked-out solution available in Student Solutions Manual
• – ••• Number of dots indicates level of problem difficulty
WWW Worked-out solution is at
ILW Interactive solution is at
http://www.wiley.com/college/halliday
Additional information available in *The Flying Circus of Physics* and at flyingcircusofphysics.com

Module 13-1 Newton's Law of Gravitation

•**1** ILW A mass M is split into two parts, m and $M - m$, which are then separated by a certain distance. What ratio m/M maximizes the magnitude of the gravitational force between the parts?

•**2** *Moon effect.* Some people believe that the Moon controls their activities. If the Moon moves from being directly on the opposite side of Earth from you to being directly overhead, by what percent does (a) the Moon's gravitational pull on you increase and (b) your weight (as measured on a scale) decrease? Assume that the Earth–Moon (center-to-center) distance is 3.82×10^8 m and Earth's radius is 6.37×10^6 m.

•3 SSM What must the separation be between a 5.2 kg particle and a 2.4 kg particle for their gravitational attraction to have a magnitude of 2.3×10^{-12} N?

•4 The Sun and Earth each exert a gravitational force on the Moon. What is the ratio F_{Sun}/F_{Earth} of these two forces? (The average Sun–Moon distance is equal to the Sun–Earth distance.)

•5 *Miniature black holes.* Left over from the big-bang beginning of the universe, tiny black holes might still wander through the universe. If one with a mass of 1×10^{11} kg (and a radius of only 1×10^{-16} m) reached Earth, at what distance from your head would its gravitational pull on you match that of Earth's?

Module 13-2 Gravitation and the Principle of Superposition

•6 GO In Fig. 13-32, a square of edge length 20.0 cm is formed by four spheres of masses $m_1 = 5.00$ g, $m_2 = 3.00$ g, $m_3 = 1.00$ g, and $m_4 = 5.00$ g. In unit-vector notation, what is the net gravitational force from them on a central sphere with mass $m_5 = 2.50$ g?

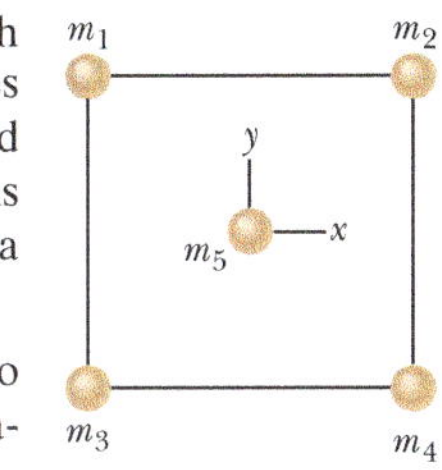

Figure 13-32 Problem 6.

•7 *One dimension.* In Fig. 13-33, two point particles are fixed on an x axis separated by distance d. Particle A has mass m_A and particle B has mass $3.00m_A$. A third particle C, of mass $75.0m_A$, is to be placed on the x axis and near particles A and B. In terms of distance d, at what x coordinate should C be placed so that the net gravitational force on particle A from particles B and C is zero?

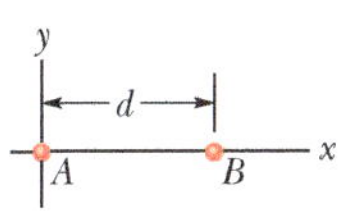

Figure 13-33 Problem 7.

•8 In Fig. 13-34, three 5.00 kg spheres are located at distances $d_1 = 0.300$ m and $d_2 = 0.400$ m. What are the (a) magnitude and (b) direction (relative to the positive direction of the x axis) of the net gravitational force on sphere B due to spheres A and C?

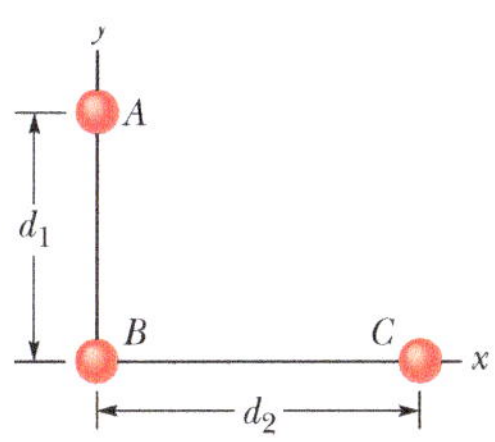

Figure 13-34 Problem 8.

•9 SSM WWW We want to position a space probe along a line that extends directly toward the Sun in order to monitor solar flares. How far from Earth's center is the point on the line where the Sun's gravitational pull on the probe balances Earth's pull?

••10 GO *Two dimensions.* In Fig. 13-35, three point particles are fixed in place in an xy plane. Particle A has mass m_A, particle B has mass $2.00m_A$, and particle C has mass $3.00m_A$. A fourth particle D, with mass $4.00m_A$, is to be placed near the other three particles. In terms of distance d, at what (a) x coordinate and (b) y coordinate should particle D be placed so that the net gravitational force on particle A from particles B, C, and D is zero?

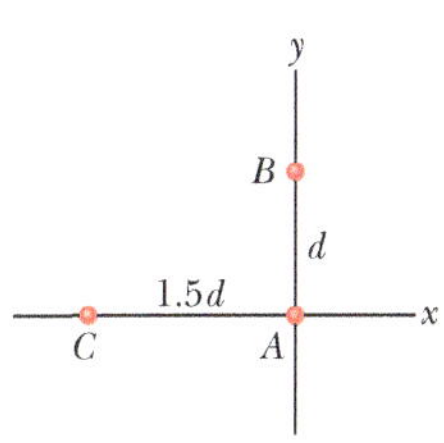

Figure 13-35 Problem 10.

••11 As seen in Fig. 13-36, two spheres of mass m and a third sphere of mass M form an equilateral triangle, and a fourth sphere of mass m_4 is at the center of the triangle. The net gravitational force on that central sphere from the three other spheres is zero. (a) What is M in terms of m? (b) If we double the value of m_4, what then is the magnitude of the net gravitational force on the central sphere?

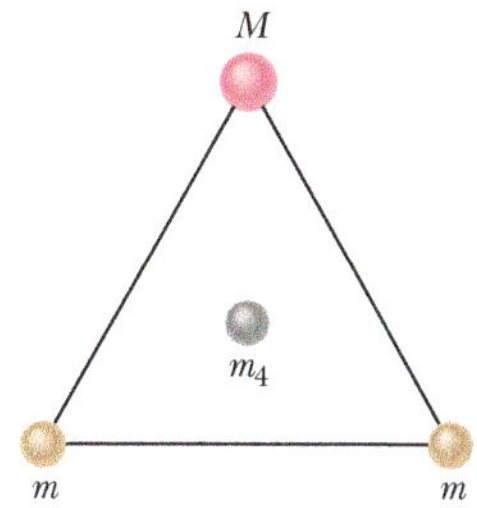

Figure 13-36 Problem 11.

••12 GO In Fig. 13-37*a*, particle A is fixed in place at $x = -0.20$ m on the x axis and particle B, with a mass of 1.0 kg, is fixed in place at the origin. Particle C (not shown) can be moved along the x axis, between particle B and $x = \infty$. Figure 13-37*b* shows the x component $F_{net,x}$ of the net gravitational force on particle B due to particles A and C, as a function of position x of particle C. The plot actually extends to the right, approaching an asymptote of -4.17×10^{-10} N as $x \to \infty$. What are the masses of (a) particle A and (b) particle C?

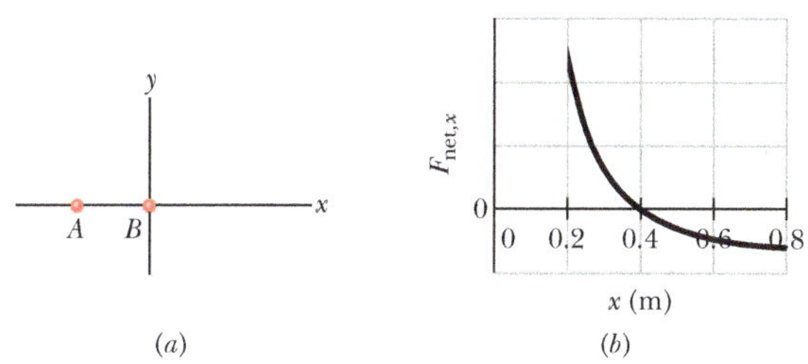

Figure 13-37 Problem 12.

••13 Figure 13-38 shows a spherical hollow inside a lead sphere of radius $R = 4.00$ cm; the surface of the hollow passes through the center of the sphere and "touches" the right side of the sphere. The mass of the sphere before hollowing was $M = 2.95$ kg. With what gravitational force does the hollowed-out lead sphere attract a small sphere of mass $m = 0.431$ kg that lies at a distance $d = 9.00$ cm from the center of the lead sphere, on the straight line connecting the centers of the spheres and of the hollow?

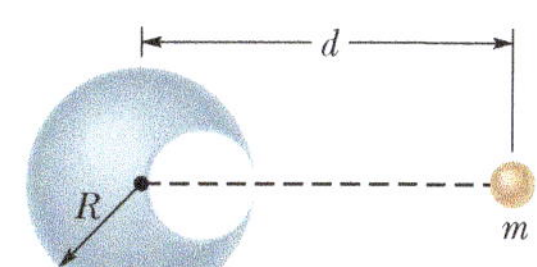

Figure 13-38 Problem 13.

••14 GO Three point particles are fixed in position in an xy plane. Two of them, particle A of mass 6.00 g and particle B of mass 12.0 g, are shown in Fig. 13-39, with a separation of $d_{AB} = 0.500$ m at angle $\theta = 30°$. Particle C, with mass 8.00 g, is not shown. The net gravitational force acting on particle A due to particles B and C is 2.77×10^{-14} N at an angle of $-163.8°$ from the positive direction of the x axis. What are (a) the x coordinate and (b) the y coordinate of particle C?

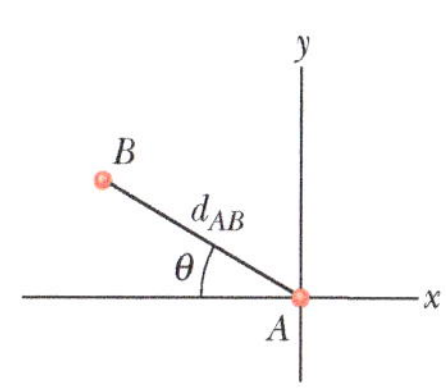

Figure 13-39 Problem 14.

•••15 GO *Three dimensions.* Three point particles are fixed in place in an xyz coordinate system. Particle A, at the origin, has mass m_A.

Particle B, at xyz coordinates $(2.00d, 1.00d, 2.00d)$, has mass $2.00m_A$, and particle C, at coordinates $(-1.00d, 2.00d, -3.00d)$, has mass $3.00m_A$. A fourth particle D, with mass $4.00m_A$, is to be placed near the other particles. In terms of distance d, at what (a) x, (b) y, and (c) z coordinate should D be placed so that the net gravitational force on A from B, C, and D is zero?

•••16 GO In Fig. 13-40, a particle of mass $m_1 = 0.67$ kg is a distance $d = 23$ cm from one end of a uniform rod with length $L = 3.0$ m and mass $M = 5.0$ kg. What is the magnitude of the gravitational force $\vec{F}$ on the particle from the rod?

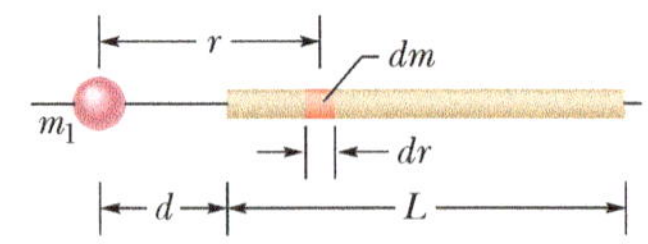

Figure 13-40 Problem 16.

Module 13-3 Gravitation Near Earth's Surface

•17 (a) What will an object weigh on the Moon's surface if it weighs 100 N on Earth's surface? (b) How many Earth radii must this same object be from the center of Earth if it is to weigh the same as it does on the Moon?

•18 *Mountain pull.* A large mountain can slightly affect the direction of "down" as determined by a plumb line. Assume that we can model a mountain as a sphere of radius $R = 2.00$ km and density (mass per unit volume) 2.6×10^3 kg/m^3. Assume also that we hang a 0.50 m plumb line at a distance of $3R$ from the sphere's center and such that the sphere pulls horizontally on the lower end. How far would the lower end move toward the sphere?

•19 SSM At what altitude above Earth's surface would the gravitational acceleration be 4.9 m/s^2?

•20 *Mile-high building.* In 1956, Frank Lloyd Wright proposed the construction of a mile-high building in Chicago. Suppose the building had been constructed. Ignoring Earth's rotation, find the change in your weight if you were to ride an elevator from the street level, where you weigh 600 N, to the top of the building.

••21 ILW Certain neutron stars (extremely dense stars) are believed to be rotating at about 1 rev/s. If such a star has a radius of 20 km, what must be its minimum mass so that material on its surface remains in place during the rapid rotation?

••22 The radius R_h and mass M_h of a black hole are related by $R_h = 2GM_h/c^2$, where c is the speed of light. Assume that the gravitational acceleration a_g of an object at a distance $r_o = 1.001R_h$ from the center of a black hole is given by Eq. 13-11 (it is, for large black holes). (a) In terms of M_h, find a_g at r_o. (b) Does a_g at r_o increase or decrease as M_h increases? (c) What is a_g at r_o for a very large black hole whose mass is 1.55×10^{12} times the solar mass of 1.99×10^{30} kg? (d) If an astronaut of height 1.70 m is at r_o with her feet down, what is the difference in gravitational acceleration between her head and feet? (e) Is the tendency to stretch the astronaut severe?

••23 One model for a certain planet has a core of radius R and mass M surrounded by an outer shell of inner radius R, outer radius $2R$, and mass $4M$. If $M = 4.1 \times 10^{24}$ kg and $R = 6.0 \times 10^6$ m, what is the gravitational acceleration of a particle at points (a) R and (b) $3R$ from the center of the planet?

Module 13-4 Gravitation Inside Earth

•24 Two concentric spherical shells with uniformly distributed masses M_1 and M_2 are situated as shown in Fig. 13-41. Find the magnitude of the net gravitational force on a particle of mass m, due to the shells, when the particle is located at radial distance (a) a, (b) b, and (c) c.

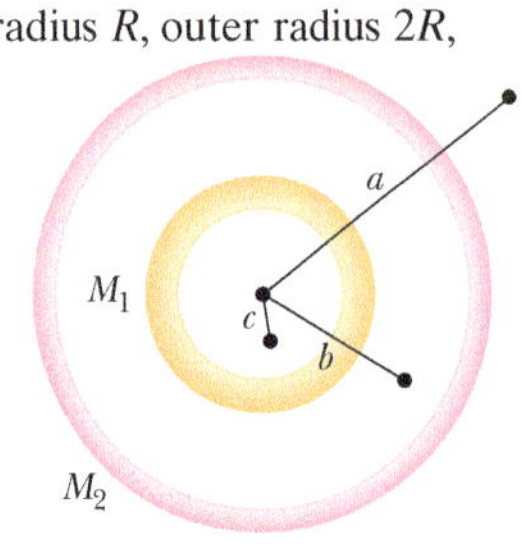

Figure 13-41 Problem 24.

••25 A solid sphere has a uniformly distributed mass of 1.0×10^4 kg and a radius of 1.0 m. What is the magnitude of the gravitational force due to the sphere on a particle of mass m when the particle is located at a distance of (a) 1.5 m and (b) 0.50 m from the center of the sphere? (c) Write a general expression for the magnitude of the gravitational force on the particle at a distance $r \leq 1.0$ m from the center of the sphere.

••26 A uniform solid sphere of radius R produces a gravitational acceleration of a_g on its surface. At what distance from the sphere's center are there points (a) inside and (b) outside the sphere where the gravitational acceleration is $a_g/3$?

••27 Figure 13-42 shows, not to scale, a cross section through the interior of Earth. Rather than being uniform throughout, Earth is divided into three zones: an outer *crust*, a *mantle*, and an inner *core*. The dimensions of these zones and the masses contained within them are shown on the figure. Earth has a total mass of 5.98×10^{24} kg and a radius of 6370 km. Ignore rotation and assume that Earth is spherical. (a) Calculate a_g at the surface. (b) Suppose that a bore hole (the *Mohole*) is driven to the crust–mantle interface at a depth of 25.0 km; what would be the value of a_g at the bottom of the hole? (c) Suppose that Earth were a uniform sphere with the same total mass and size. What would be the value of a_g at a depth of 25.0 km? (Precise measurements of a_g are sensitive probes of the interior structure of Earth, although results can be clouded by local variations in mass distribution.)

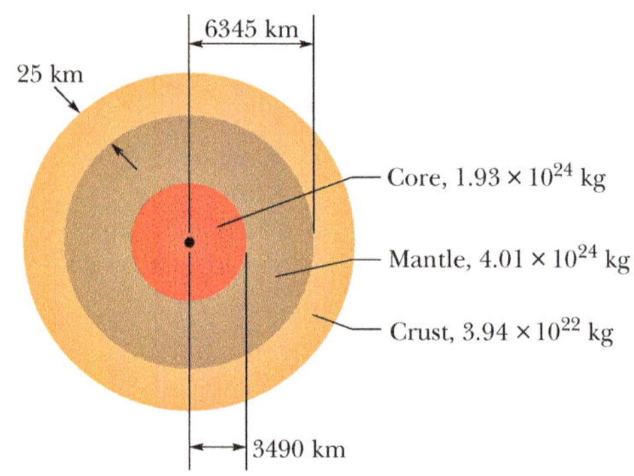

Figure 13-42 Problem 27.

••28 GO Assume a planet is a uniform sphere of radius R that (somehow) has a narrow radial tunnel through its center (Fig. 13-7). Also assume we can position an apple anywhere along the tunnel or outside the sphere. Let F_R be the magnitude of the gravitational force on the apple when it is located at the planet's surface. How far from the surface is there a point where the magnitude is $\frac{1}{2}F_R$ if we move the apple (a) away from the planet and (b) into the tunnel?

Module 13-5 Gravitational Potential Energy

•29 Figure 13-43 gives the potential energy function $U(r)$ of a projectile, plotted outward from

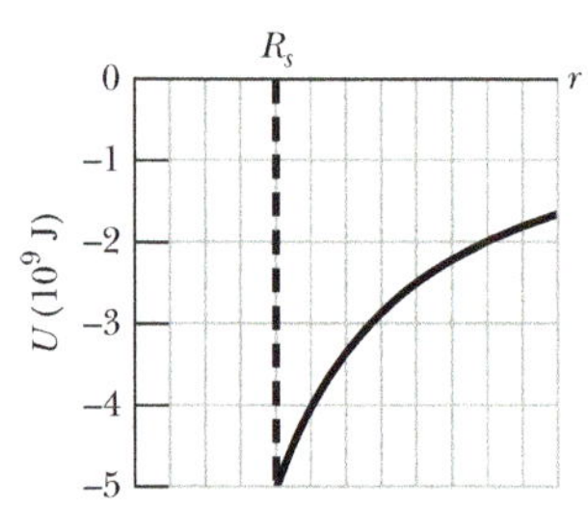

Figure 13-43 Problems 29 and 34.

the surface of a planet of radius R_s. What least kinetic energy is required of a projectile launched at the surface if the projectile is to "escape" the planet?

•30 In Problem 1, what ratio m/M gives the least gravitational potential energy for the system?

•31 SSM The mean diameters of Mars and Earth are 6.9×10^3 km and 1.3×10^4 km, respectively. The mass of Mars is 0.11 times Earth's mass. (a) What is the ratio of the mean density (mass per unit volume) of Mars to that of Earth? (b) What is the value of the gravitational acceleration on Mars? (c) What is the escape speed on Mars?

•32 (a) What is the gravitational potential energy of the two-particle system in Problem 3? If you triple the separation between the particles, how much work is done (b) by the gravitational force between the particles and (c) by you?

•33 What multiple of the energy needed to escape from Earth gives the energy needed to escape from (a) the Moon and (b) Jupiter?

•34 Figure 13-43 gives the potential energy function $U(r)$ of a projectile, plotted outward from the surface of a planet of radius R_s. If the projectile is launched radially outward from the surface with a mechanical energy of -2.0×10^9 J, what are (a) its kinetic energy at radius $r = 1.25R_s$ and (b) its *turning point* (see Module 8-3) in terms of R_s?

••35 GO Figure 13-44 shows four particles, each of mass 20.0 g, that form a square with an edge length of $d = 0.600$ m. If d is reduced to 0.200 m, what is the change in the gravitational potential energy of the four-particle system?

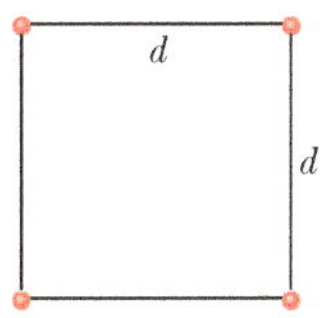

Figure 13-44 Problem 35.

••36 GO Zero, a hypothetical planet, has a mass of 5.0×10^{23} kg, a radius of 3.0×10^6 m, and no atmosphere. A 10 kg space probe is to be launched vertically from its surface. (a) If the probe is launched with an initial energy of 5.0×10^7 J, what will be its kinetic energy when it is 4.0×10^6 m from the center of Zero? (b) If the probe is to achieve a maximum distance of 8.0×10^6 m from the center of Zero, with what initial kinetic energy must it be launched from the surface of Zero?

••37 GO The three spheres in Fig. 13-45, with masses $m_A = 80$ g, $m_B = 10$ g, and $m_C = 20$ g, have their centers on a common line, with $L = 12$ cm and $d = 4.0$ cm. You move sphere B along the line until its center-to-center separation from C is $d = 4.0$ cm. How much work is done on sphere B (a) by you and (b) by the net gravitational force on B due to spheres A and C?

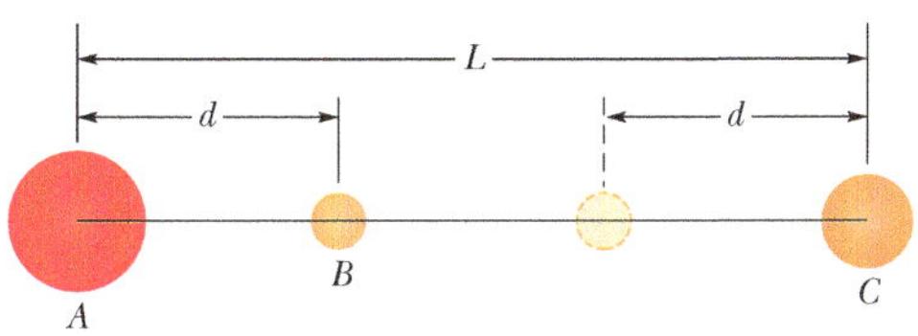

Figure 13-45 Problem 37.

••38 In deep space, sphere A of mass 20 kg is located at the origin of an x axis and sphere B of mass 10 kg is located on the axis at $x = 0.80$ m. Sphere B is released from rest while sphere A is held at the origin. (a) What is the gravitational potential energy of the two-sphere system just as B is released? (b) What is the kinetic energy of B when it has moved 0.20 m toward A?

••39 SSM (a) What is the escape speed on a spherical asteroid whose radius is 500 km and whose gravitational acceleration at the surface is 3.0 m/s^2? (b) How far from the surface will a particle go if it leaves the asteroid's surface with a radial speed of 1000 m/s? (c) With what speed will an object hit the asteroid if it is dropped from 1000 km above the surface?

••40 A projectile is shot directly away from Earth's surface. Neglect the rotation of Earth. What multiple of Earth's radius R_E gives the radial distance a projectile reaches if (a) its initial speed is 0.500 of the escape speed from Earth and (b) its initial kinetic energy is 0.500 of the kinetic energy required to escape Earth? (c) What is the least initial mechanical energy required at launch if the projectile is to escape Earth?

••41 SSM Two neutron stars are separated by a distance of 1.0×10^{10} m. They each have a mass of 1.0×10^{30} kg and a radius of 1.0×10^5 m. They are initially at rest with respect to each other. As measured from that rest frame, how fast are they moving when (a) their separation has decreased to one-half its initial value and (b) they are about to collide?

••42 GO Figure 13-46*a* shows a particle A that can be moved along a y axis from an infinite distance to the origin. That origin lies at the midpoint between particles B and C, which have identical masses, and the y axis is a perpendicular bisector between them. Distance D is 0.3057 m. Figure 13-46*b* shows the potential energy U of the three-particle system as a function of the position of particle A along the y axis. The curve actually extends rightward and approaches an asymptote of -2.7×10^{-11} J as $y \to \infty$. What are the masses of (a) particles B and C and (b) particle A?

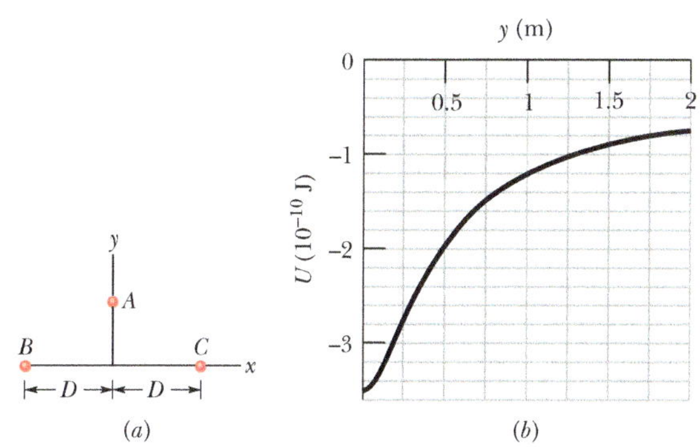

Figure 13-46 Problem 42.

Module 13-6 Planets and Satellites: Kepler's Laws

•43 (a) What linear speed must an Earth satellite have to be in a circular orbit at an altitude of 160 km above Earth's surface? (b) What is the period of revolution?

•44 A satellite is put in a circular orbit about Earth with a radius equal to one-half the radius of the Moon's orbit. What is its period of revolution in lunar months? (A lunar month is the period of revolution of the Moon.)

•45 The Martian satellite Phobos travels in an approximately circular orbit of radius 9.4×10^6 m with a period of 7 h 39 min. Calculate the mass of Mars from this information.

•46 The first known collision between space debris and a functioning satellite occurred in 1996: At an altitude of 700 km, a year-old French spy satellite was hit by a piece of an Ariane rocket. A stabilizing boom on the satellite was demolished, and the satellite was sent spinning out of control. Just before the collision and in kilometers per hour, what was the speed of the rocket piece relative to the satellite if both were in circular orbits and the collision was (a) head-on and (b) along perpendicular paths?

•47 SSM WWW The Sun, which is 2.2×10^{20} m from the center of the Milky Way galaxy, revolves around that center once every 2.5×10^8 years. Assuming each star in the Galaxy has a mass equal to the Sun's mass of 2.0×10^{30} kg, the stars are distributed uniformly in a sphere about the galactic center, and the Sun is at the edge of that sphere, estimate the number of stars in the Galaxy.

•48 The mean distance of Mars from the Sun is 1.52 times that of Earth from the Sun. From Kepler's law of periods, calculate the number of years required for Mars to make one revolution around the Sun; compare your answer with the value given in Appendix C.

•49 A comet that was seen in April 574 by Chinese astronomers on a day known by them as the Woo Woo day was spotted again in May 1994. Assume the time between observations is the period of the Woo Woo day comet and its eccentricity is 0.9932. What are (a) the semimajor axis of the comet's orbit and (b) its greatest distance from the Sun in terms of the mean orbital radius R_P of Pluto?

•50 An orbiting satellite stays over a certain spot on the equator of (rotating) Earth. What is the altitude of the orbit (called a *geosynchronous orbit*)?

•51 SSM A satellite, moving in an elliptical orbit, is 360 km above Earth's surface at its farthest point and 180 km above at its closest point. Calculate (a) the semimajor axis and (b) the eccentricity of the orbit.

•52 The Sun's center is at one focus of Earth's orbit. How far from this focus is the other focus, (a) in meters and (b) in terms of the solar radius, 6.96×10^8 m? The eccentricity is 0.0167, and the semimajor axis is 1.50×10^{11} m.

••53 A 20 kg satellite has a circular orbit with a period of 2.4 h and a radius of 8.0×10^6 m around a planet of unknown mass. If the magnitude of the gravitational acceleration on the surface of the planet is 8.0 m/s^2, what is the radius of the planet?

••54 GO *Hunting a black hole.* Observations of the light from a certain star indicate that it is part of a binary (two-star) system. This visible star has orbital speed $v = 270$ km/s, orbital period $T = 1.70$ days, and approximate mass $m_1 = 6M_s$, where M_s is the Sun's mass, 1.99×10^{30} kg. Assume that the visible star and its companion star, which is dark and unseen, are both in circular orbits (Fig. 13-47). What integer multiple of M_s gives the *approximate* mass m_2 of the dark star?

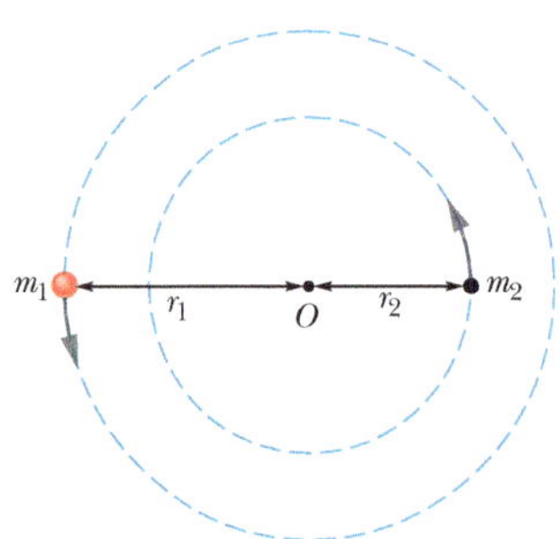

Figure 13-47 Problem 54.

••55 In 1610, Galileo used his telescope to discover four moons around Jupiter, with these mean orbital radii a and periods T:

Name	a (10^8 m)	T (days)
Io	4.22	1.77
Europa	6.71	3.55
Ganymede	10.7	7.16
Callisto	18.8	16.7

(a) Plot log a (y axis) against log T (x axis) and show that you get a straight line. (b) Measure the slope of the line and compare it with the value that you expect from Kepler's third law. (c) Find the mass of Jupiter from the intercept of this line with the y axis.

••56 In 1993 the spacecraft *Galileo* sent an image (Fig. 13-48) of asteroid 243 Ida and a tiny orbiting moon (now known as Dactyl), the first confirmed example of an asteroid–moon system. In the image, the moon, which is 1.5 km wide, is 100 km from the center of the asteroid, which is 55 km long. Assume the moon's orbit is circular with a period of 27 h. (a) What is the mass of the asteroid? (b) The volume of the asteroid, measured from the *Galileo* images, is 14 100 km^3. What is the density (mass per unit volume) of the asteroid?

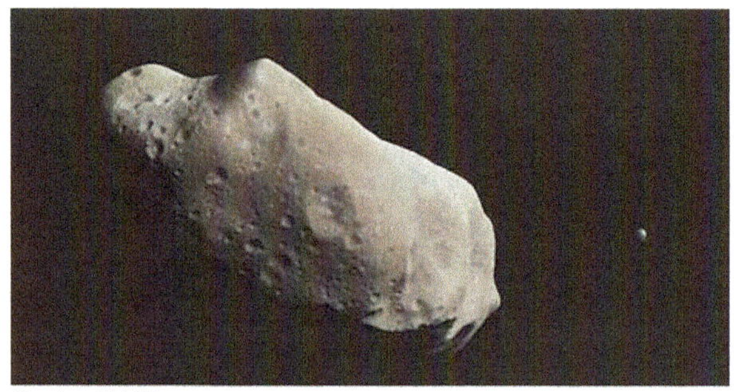

Courtesy NASA

Figure 13-48 Problem 56. A tiny moon (at right) orbits asteroid 243 Ida.

••57 ILW In a certain binary-star system, each star has the same mass as our Sun, and they revolve about their center of mass. The distance between them is the same as the distance between Earth and the Sun. What is their period of revolution in years?

•••58 GO The presence of an unseen planet orbiting a distant star can sometimes be inferred from the motion of the star as we see it. As the star and planet orbit the center of mass of the star–planet system, the star moves toward and away from us with what is called the *line of sight velocity,* a motion that can be detected. Figure 13-49 shows a graph of the line of sight velocity versus time for the star 14 Herculis. The star's mass is believed to be 0.90 of the mass of our Sun. Assume that only one planet orbits the star and that our view is along the plane of the orbit. Then approximate (a) the planet's mass in terms of Jupiter's mass m_J and (b) the planet's orbital radius in terms of Earth's orbital radius r_E.

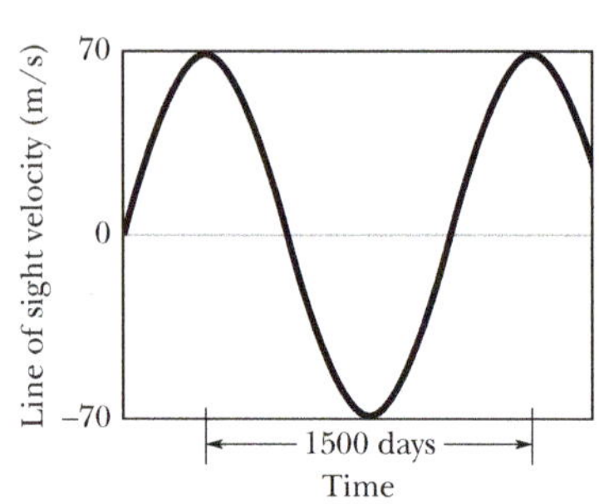

Figure 13-49 Problem 58.

•••59 Three identical stars of mass M form an equilateral triangle that rotates around the triangle's center as the stars move in a common circle about that center. The triangle has edge length L. What is the speed of the stars?

Module 13-7 Satellites: Orbits and Energy

•**60** In Fig. 13-50, two satellites, A and B, both of mass $m = 125$ kg, move in the same circular orbit of radius $r = 7.87 \times 10^6$ m around Earth but in opposite senses of rotation and therefore on a collision course. (a) Find the total mechanical energy $E_A + E_B$ of the *two satellites + Earth* system before the collision. (b) If the collision is completely inelastic so that the wreckage remains as one piece of tangled material (mass $= 2m$), find the total mechanical energy immediately after the collision. (c) Just after the collision, is the wreckage falling directly toward Earth's center or orbiting around Earth?

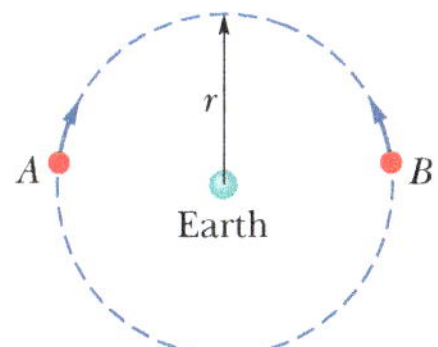

Figure 13-50 Problem 60.

•**61** (a) At what height above Earth's surface is the energy required to lift a satellite to that height equal to the kinetic energy required for the satellite to be in orbit at that height? (b) For greater heights, which is greater, the energy for lifting or the kinetic energy for orbiting?

•**62** Two Earth satellites, A and B, each of mass m, are to be launched into circular orbits about Earth's center. Satellite A is to orbit at an altitude of 6370 km. Satellite B is to orbit at an altitude of 19 110 km. The radius of Earth R_E is 6370 km. (a) What is the ratio of the potential energy of satellite B to that of satellite A, in orbit? (b) What is the ratio of the kinetic energy of satellite B to that of satellite A, in orbit? (c) Which satellite has the greater total energy if each has a mass of 14.6 kg? (d) By how much?

•**63** SSM WWW An asteroid, whose mass is 2.0×10^{-4} times the mass of Earth, revolves in a circular orbit around the Sun at a distance that is twice Earth's distance from the Sun. (a) Calculate the period of revolution of the asteroid in years. (b) What is the ratio of the kinetic energy of the asteroid to the kinetic energy of Earth?

•**64** A satellite orbits a planet of unknown mass in a circle of radius 2.0×10^7 m. The magnitude of the gravitational force on the satellite from the planet is $F = 80$ N. (a) What is the kinetic energy of the satellite in this orbit? (b) What would F be if the orbit radius were increased to 3.0×10^7 m?

••**65** A satellite is in a circular Earth orbit of radius r. The area A enclosed by the orbit depends on r^2 because $A = \pi r^2$. Determine how the following properties of the satellite depend on r: (a) period, (b) kinetic energy, (c) angular momentum, and (d) speed.

••**66** One way to attack a satellite in Earth orbit is to launch a swarm of pellets in the same orbit as the satellite but in the opposite direction. Suppose a satellite in a circular orbit 500 km above Earth's surface collides with a pellet having mass 4.0 g. (a) What is the kinetic energy of the pellet in the reference frame of the satellite just before the collision? (b) What is the ratio of this kinetic energy to the kinetic energy of a 4.0 g bullet from a modern army rifle with a muzzle speed of 950 m/s?

•••**67** What are (a) the speed and (b) the period of a 220 kg satellite in an approximately circular orbit 640 km above the surface of Earth? Suppose the satellite loses mechanical energy at the average rate of 1.4×10^5 J per orbital revolution. Adopting the reasonable approximation that the satellite's orbit becomes a "circle of slowly diminishing radius," determine the satellite's (c) altitude, (d) speed, and (e) period at the end of its 1500th revolution. (f) What is the magnitude of the average retarding force on the satellite? Is angular momentum around Earth's center conserved for (g) the satellite and (h) the satellite–Earth system (assuming that system is isolated)?

•••**68** GO Two small spaceships, each with mass $m = 2000$ kg, are in the circular Earth orbit of Fig. 13-51, at an altitude h of 400 km. Igor, the commander of one of the ships, arrives at any fixed point in the orbit 90 s ahead of Picard, the commander of the other ship. What are the (a) period T_0 and (b) speed v_0 of the ships? At point P in Fig. 13-51, Picard fires an instantaneous burst in the forward direction, *reducing* his ship's speed by 1.00%. After this burst, he follows the elliptical orbit shown dashed in the figure. What are the (c) kinetic energy and (d) potential energy of his ship immediately after the burst? In Picard's new elliptical orbit, what are (e) the total energy E, (f) the semimajor axis a, and (g) the orbital period T? (h) How much earlier than Igor will Picard return to P?

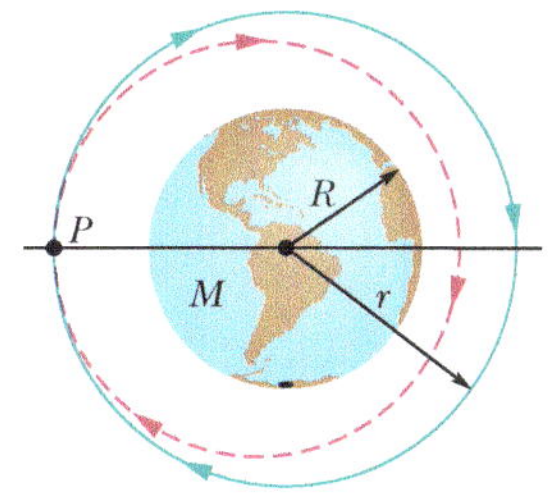

Figure 13-51 Problem 68.

Module 13-8 Einstein and Gravitation

•**69** In Fig. 13-18*b*, the scale on which the 60 kg physicist stands reads 220 N. How long will the cantaloupe take to reach the floor if the physicist drops it (from rest relative to himself) at a height of 2.1 m above the floor?

Additional Problems

70 GO The radius R_h of a black hole is the radius of a mathematical sphere, called the event horizon, that is centered on the black hole. Information from events inside the event horizon cannot reach the outside world. According to Einstein's general theory of relativity, $R_h = 2GM/c^2$, where M is the mass of the black hole and c is the speed of light.

Suppose that you wish to study a black hole near it, at a radial distance of $50R_h$. However, you do not want the difference in gravitational acceleration between your feet and your head to exceed 10 m/s² when you are feet down (or head down) toward the black hole. (a) As a multiple of our Sun's mass M_S, approximately what is the limit to the mass of the black hole you can tolerate at the given radial distance? (You need to estimate your height.) (b) Is the limit an upper limit (you can tolerate smaller masses) or a lower limit (you can tolerate larger masses)?

71 Several planets (Jupiter, Saturn, Uranus) are encircled by rings, perhaps composed of material that failed to form a satellite. In addition, many galaxies contain ring-like structures. Consider a homogeneous thin ring of mass M and outer radius R (Fig. 13-52). (a) What gravitational attraction does it exert on a particle of mass m located on the ring's central axis a distance x from the ring center? (b) Suppose the particle falls from rest as a result of the attraction of the ring of matter. What is the speed with which it passes through the center of the ring?

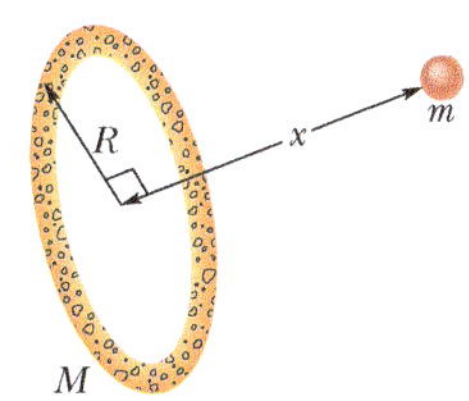

Figure 13-52 Problem 71.

72 A typical neutron star may have a mass equal to that of the Sun but a radius of only 10 km. (a) What is the gravitational acceleration at the surface of such a star? (b) How fast would an object be moving if it fell from rest through a distance of 1.0 m on such a star? (Assume the star does not rotate.)

73 Figure 13-53 is a graph of the kinetic energy K of an asteroid versus its distance r from Earth's center, as the asteroid falls directly in toward that center. (a) What is the (approximate) mass of the asteroid? (b) What is its speed at $r = 1.945 \times 10^7$ m?

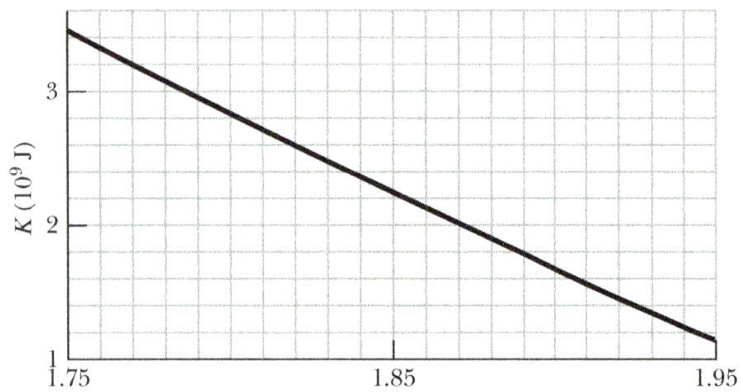

Figure 13-53 Problem 73.

74 The mysterious visitor that appears in the enchanting story *The Little Prince* was said to come from a planet that "was scarcely any larger than a house!" Assume that the mass per unit volume of the planet is about that of Earth and that the planet does not appreciably spin. Approximate (a) the free-fall acceleration on the planet's surface and (b) the escape speed from the planet.

75 ILW The masses and coordinates of three spheres are as follows: 20 kg, $x = 0.50$ m, $y = 1.0$ m; 40 kg, $x = -1.0$ m, $y = -1.0$ m; 60 kg, $x = 0$ m, $y = -0.50$ m. What is the magnitude of the gravitational force on a 20 kg sphere located at the origin due to these three spheres?

76 SSM A very early, simple satellite consisted of an inflated spherical aluminum balloon 30 m in diameter and of mass 20 kg. Suppose a meteor having a mass of 7.0 kg passes within 3.0 m of the surface of the satellite. What is the magnitude of the gravitational force on the meteor from the satellite at the closest approach?

77 GO Four uniform spheres, with masses $m_A = 40$ kg, $m_B = 35$ kg, $m_C = 200$ kg, and $m_D = 50$ kg, have (x, y) coordinates of (0, 50 cm), (0, 0), (−80 cm, 0), and (40 cm, 0), respectively. In unit-vector notation, what is the net gravitational force on sphere B due to the other spheres?

78 (a) In Problem 77, remove sphere A and calculate the gravitational potential energy of the remaining three-particle system. (b) If A is then put back in place, is the potential energy of the four-particle system more or less than that of the system in (a)? (c) In (a), is the work done by you to remove A positive or negative? (d) In (b), is the work done by you to replace A positive or negative?

79 SSM A certain triple-star system consists of two stars, each of mass m, revolving in the same circular orbit of radius r around a central star of mass M (Fig. 13-54). The two orbiting stars are always at opposite ends of a diameter of the orbit. Derive an expression for the period of revolution of the stars.

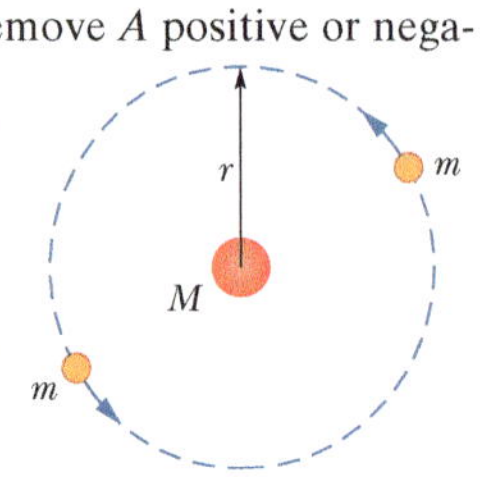

Figure 13-54 Problem 79.

80 The fastest possible rate of rotation of a planet is that for which the gravitational force on material at the equator just barely provides the centripetal force needed for the rotation. (Why?) (a) Show that the corresponding shortest period of rotation is

$$T = \sqrt{\frac{3\pi}{G\rho}},$$

where ρ is the uniform density (mass per unit volume) of the spherical planet. (b) Calculate the rotation period assuming a density of 3.0 g/cm^3, typical of many planets, satellites, and asteroids. No astronomical object has ever been found to be spinning with a period shorter than that determined by this analysis.

81 SSM In a double-star system, two stars of mass 3.0×10^{30} kg each rotate about the system's center of mass at radius 1.0×10^{11} m. (a) What is their common angular speed? (b) If a meteoroid passes through the system's center of mass perpendicular to their orbital plane, what minimum speed must it have at the center of mass if it is to escape to "infinity" from the two-star system?

82 A satellite is in elliptical orbit with a period of 8.00×10^4 s about a planet of mass 7.00×10^{24} kg. At aphelion, at radius 4.5×10^7 m, the satellite's angular speed is 7.158×10^{-5} rad/s. What is its angular speed at perihelion?

83 SSM In a shuttle craft of mass $m = 3000$ kg, Captain Janeway orbits a planet of mass $M = 9.50 \times 10^{25}$ kg, in a circular orbit of radius $r = 4.20 \times 10^7$ m. What are (a) the period of the orbit and (b) the speed of the shuttle craft? Janeway briefly fires a forward-pointing thruster, reducing her speed by 2.00%. Just then, what are (c) the speed, (d) the kinetic energy, (e) the gravitational potential energy, and (f) the mechanical energy of the shuttle craft? (g) What is the semimajor axis of the elliptical orbit now taken by the craft? (h) What is the difference between the period of the original circular orbit and that of the new elliptical orbit? (i) Which orbit has the smaller period?

84 Consider a pulsar, a collapsed star of extremely high density, with a mass M equal to that of the Sun (1.98×10^{30} kg), a radius R of only 12 km, and a rotational period T of 0.041 s. By what percentage does the free-fall acceleration g differ from the gravitational acceleration a_g at the equator of this spherical star?

85 ILW A projectile is fired vertically from Earth's surface with an initial speed of 10 km/s. Neglecting air drag, how far above the surface of Earth will it go?

86 An object lying on Earth's equator is accelerated (a) toward the center of Earth because Earth rotates, (b) toward the Sun because Earth revolves around the Sun in an almost circular orbit, and (c) toward the center of our galaxy because the Sun moves around the galactic center. For the latter, the period is 2.5×10^8 y and the radius is 2.2×10^{20} m. Calculate these three accelerations as multiples of $g = 9.8$ m/s^2.

87 (a) If the legendary apple of Newton could be released from rest at a height of 2 m from the surface of a neutron star with a mass 1.5 times that of our Sun and a radius of 20 km, what would be the apple's speed when it reached the surface of the star? (b) If the apple could rest on the surface of the star, what would be the approximate difference between the gravitational acceleration at the top and at the bottom of the apple? (Choose a reasonable size for an apple; the answer indicates that an apple would never survive near a neutron star.)

88 With what speed would mail pass through the center of Earth if falling in a tunnel through the center?

89 SSM The orbit of Earth around the Sun is *almost* circular: The closest and farthest distances are 1.47×10^8 km and 1.52×10^8 km respectively. Determine the corresponding variations in (a) total energy, (b) gravitational potential energy, (c) kinetic energy, and (d) orbital speed. (*Hint:* Use conservation of energy and conservation of angular momentum.)

90 A 50 kg satellite circles planet Cruton every 6.0 h. The magnitude of the gravitational force exerted on the satellite by Cruton is 80 N. (a) What is the radius of the orbit? (b) What is the kinetic energy of the satellite? (c) What is the mass of planet Cruton?

91 We watch two identical astronomical bodies A and B, each of mass m, fall toward each other from rest because of the gravitational force on each from the other. Their initial center-to-center separation is R_i. Assume that we are in an inertial reference frame that is stationary with respect to the center of mass of this two-body system. Use the principle of conservation of mechanical energy ($K_f + U_f = K_i + U_i$) to find the following when the center-to-center separation is $0.5R_i$: (a) the total kinetic energy of the system, (b) the kinetic energy of each body, (c) the speed of each body relative to us, and (d) the speed of body B relative to body A.

Next assume that we are in a reference frame attached to body A (we ride on the body). Now we see body B fall from rest toward us. From this reference frame, again use $K_f + U_f = K_i + U_i$ to find the following when the center-to-center separation is $0.5R_i$: (e) the kinetic energy of body B and (f) the speed of body B relative to body A. (g) Why are the answers to (d) and (f) different? Which answer is correct?

92 A 150.0 kg rocket moving radially outward from Earth has a speed of 3.70 km/s when its engine shuts off 200 km above Earth's surface. (a) Assuming negligible air drag acts on the rocket, find the rocket's kinetic energy when the rocket is 1000 km above Earth's surface. (b) What maximum height above the surface is reached by the rocket?

93 Planet Roton, with a mass of 7.0×10^{24} kg and a radius of 1600 km, gravitationally attracts a meteorite that is initially at rest relative to the planet, at a distance great enough to take as infinite. The meteorite falls toward the planet. Assuming the planet is airless, find the speed of the meteorite when it reaches the planet's surface.

94 Two 20 kg spheres are fixed in place on a y axis, one at $y = 0.40$ m and the other at $y = -0.40$ m. A 10 kg ball is then released from rest at a point on the x axis that is at a great distance (effectively infinite) from the spheres. If the only forces acting on the ball are the gravitational forces from the spheres, then when the ball reaches the (x, y) point (0.30 m, 0), what are (a) its kinetic energy and (b) the net force on it from the spheres, in unit-vector notation?

95 Sphere A with mass 80 kg is located at the origin of an xy coordinate system; sphere B with mass 60 kg is located at coordinates (0.25 m, 0); sphere C with mass 0.20 kg is located in the first quadrant 0.20 m from A and 0.15 m from B. In unit-vector notation, what is the gravitational force on C due to A and B?

96 In his 1865 science fiction novel *From the Earth to the Moon,* Jules Verne described how three astronauts are shot to the Moon by means of a huge gun. According to Verne, the aluminum capsule containing the astronauts is accelerated by ignition of nitrocellulose to a speed of 11 km/s along the gun barrel's length of 220 m. (a) In g units, what is the average acceleration of the capsule and astronauts in the gun barrel? (b) Is that acceleration tolerable or deadly to the astronauts?

A modern version of such gun-launched spacecraft (although without passengers) has been proposed. In this modern version, called the SHARP (Super High Altitude Research Project) gun, ignition of methane and air shoves a piston along the gun's tube, compressing hydrogen gas that then launches a rocket. During this launch, the rocket moves 3.5 km and reaches a speed of 7.0 km/s. Once launched, the rocket can be fired to gain additional speed. (c) In g units, what would be the average acceleration of the rocket within the launcher? (d) How much additional speed is needed (via the rocket engine) if the rocket is to orbit Earth at an altitude of 700 km?

97 An object of mass m is initially held in place at radial distance $r = 3R_E$ from the center of Earth, where R_E is the radius of Earth. Let M_E be the mass of Earth. A force is applied to the object to move it to a radial distance $r = 4R_E$, where it again is held in place. Calculate the work done by the applied force during the move by integrating the force magnitude.

98 To alleviate the traffic congestion between two cities such as Boston and Washington, D.C., engineers have proposed building a rail tunnel along a chord line connecting the cities (Fig. 13-55). A train, unpropelled by any engine and starting from rest, would fall through the first half of the tunnel and then move up the second half. Assuming Earth is a uniform sphere and ignoring air drag and friction, find the city-to-city travel time.

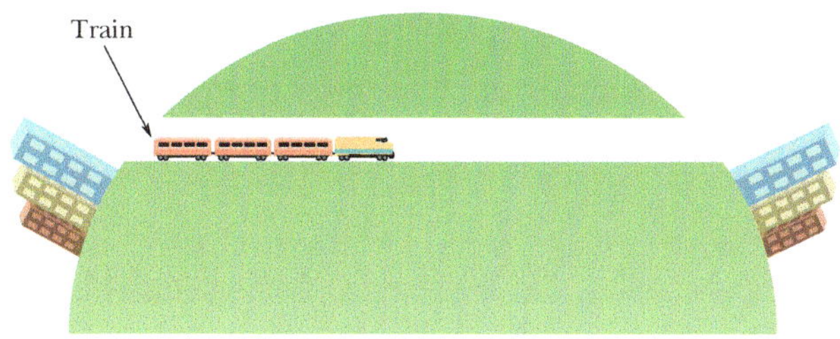

Figure 13-55 Problem 98.

99 A thin rod with mass $M = 5.00$ kg is bent in a semicircle of radius $R = 0.650$ m (Fig. 13-56). (a) What is its gravitational force (both magnitude and direction) on a particle with mass $m = 3.0 \times 10^{-3}$ kg at P, the center of curvature? (b) What would be the force on the particle if the rod were a complete circle?

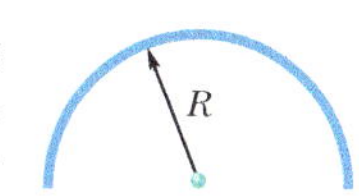

Figure 13-56 Problem 99.

100 In Fig. 13-57, identical blocks with identical masses $m = 2.00$ kg hang from strings of different lengths on a balance at Earth's surface. The strings have negligible mass and differ in length by $h = 5.00$ cm. Assume Earth is spherical with a uniform density $\rho = 5.50$ g/cm^3. What is the difference in the weight of the blocks due to one being closer to Earth than the other?

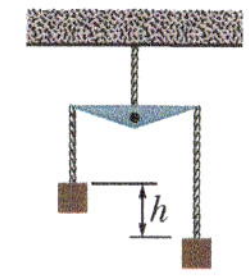

Figure 13-57 Problem 100.

101 A spaceship is on a straight-line path between Earth and the Moon. At what distance from Earth is the net gravitational force on the spaceship zero?

CHAPTER 14

Fluids

14-1 FLUIDS, DENSITY, AND PRESSURE

Learning Objectives

After reading this module, you should be able to . . .

14.01 Distinguish fluids from solids.

14.02 When mass is uniformly distributed, relate density to mass and volume.

14.03 Apply the relationship between hydrostatic pressure, force, and the surface area over which that force acts.

Key Ideas

- The density ρ of any material is defined as the material's mass per unit volume:

$$\rho = \frac{\Delta m}{\Delta V}.$$

Usually, where a material sample is much larger than atomic dimensions, we can write this as

$$\rho = \frac{m}{V}.$$

- A fluid is a substance that can flow; it conforms to the boundaries of its container because it cannot withstand shearing stress. It can, however, exert a force perpendicular to its surface. That force is described in terms of pressure p:

$$p = \frac{\Delta F}{\Delta A},$$

in which ΔF is the force acting on a surface element of area ΔA. If the force is uniform over a flat area, this can be written as

$$p = \frac{F}{A}.$$

- The force resulting from fluid pressure at a particular point in a fluid has the same magnitude in all directions.

What Is Physics?

The physics of fluids is the basis of hydraulic engineering, a branch of engineering that is applied in a great many fields. A nuclear engineer might study the fluid flow in the hydraulic system of an aging nuclear reactor, while a medical engineer might study the blood flow in the arteries of an aging patient. An environmental engineer might be concerned about the drainage from waste sites or the efficient irrigation of farmlands. A naval engineer might be concerned with the dangers faced by a deep-sea diver or with the possibility of a crew escaping from a downed submarine. An aeronautical engineer might design the hydraulic systems controlling the wing flaps that allow a jet airplane to land. Hydraulic engineering is also applied in many Broadway and Las Vegas shows, where huge sets are quickly put up and brought down by hydraulic systems.

Before we can study any such application of the physics of fluids, we must first answer the question "What is a fluid?"

What Is a Fluid?

A **fluid,** in contrast to a solid, is a substance that can flow. Fluids conform to the boundaries of any container in which we put them. They do so because a fluid cannot sustain a force that is tangential to its surface. (In the more formal language of Module 12-3, a fluid is a substance that flows because it cannot withstand a shearing stress. It can, however, exert a force in the direction perpendicular to its surface.) Some materials, such as pitch, take a long time to conform to the boundaries of a container, but they do so eventually; thus, we classify even those materials as fluids.

You may wonder why we lump liquids and gases together and call them fluids. After all (you may say), liquid water is as different from steam as it is from ice. Actually, it is not. Ice, like other crystalline solids, has its constituent atoms organized in a fairly rigid three-dimensional array called a crystalline lattice. In neither steam nor liquid water, however, is there any such orderly long-range arrangement.

Table 14-1 Some Densities

Material or Object	Density (kg/m^3)
Interstellar space	10^{-20}
Best laboratory vacuum	10^{-17}
Air: 20°C and 1 atm pressure	1.21
20°C and 50 atm	60.5
Styrofoam	1×10^2
Ice	0.917×10^3
Water: 20°C and 1 atm	0.998×10^3
20°C and 50 atm	1.000×10^3
Seawater: 20°C and 1 atm	1.024×10^3
Whole blood	1.060×10^3
Iron	7.9×10^3
Mercury (the metal, not the planet)	13.6×10^3
Earth: average	5.5×10^3
core	9.5×10^3
crust	2.8×10^3
Sun: average	1.4×10^3
core	1.6×10^5
White dwarf star (core)	10^{10}
Uranium nucleus	3×10^{17}
Neutron star (core)	10^{18}

Density and Pressure

When we discuss rigid bodies, we are concerned with particular lumps of matter, such as wooden blocks, baseballs, or metal rods. Physical quantities that we find useful, and in whose terms we express Newton's laws, are mass and force. We might speak, for example, of a 3.6 kg block acted on by a 25 N force.

With fluids, we are more interested in the extended substance and in properties that can vary from point to point in that substance. It is more useful to speak of **density** and **pressure** than of mass and force.

Density

To find the density ρ of a fluid at any point, we isolate a small volume element ΔV around that point and measure the mass Δm of the fluid contained within that element. The **density** is then

$$\rho = \frac{\Delta m}{\Delta V}. \tag{14-1}$$

In theory, the density at any point in a fluid is the limit of this ratio as the volume element ΔV at that point is made smaller and smaller. In practice, we assume that a fluid sample is large relative to atomic dimensions and thus is "smooth" (with uniform density), rather than "lumpy" with atoms. This assumption allows us to write the density in terms of the mass m and volume V of the sample:

$$\rho = \frac{m}{V} \quad \text{(uniform density)}. \tag{14-2}$$

Density is a scalar property; its SI unit is the kilogram per cubic meter. Table 14-1 shows the densities of some substances and the average densities of some objects. Note that the density of a gas (see Air in the table) varies considerably with pressure, but the density of a liquid (see Water) does not; that is, gases are readily *compressible* but liquids are not.

Pressure

Let a small pressure-sensing device be suspended inside a fluid-filled vessel, as in Fig. 14-1*a*. The sensor (Fig. 14-1*b*) consists of a piston of surface area ΔA riding in a close-fitting cylinder and resting against a spring. A readout arrangement allows us to record the amount by which the (calibrated) spring is compressed by the surrounding fluid, thus indicating the magnitude ΔF of the force that acts normal to the piston. We define the **pressure** on the piston as

$$p = \frac{\Delta F}{\Delta A}. \tag{14-3}$$

In theory, the pressure at any point in the fluid is the limit of this ratio as the surface area ΔA of the piston, centered on that point, is made smaller and smaller. However, if the force is uniform over a flat area A (it is evenly distributed over every point of

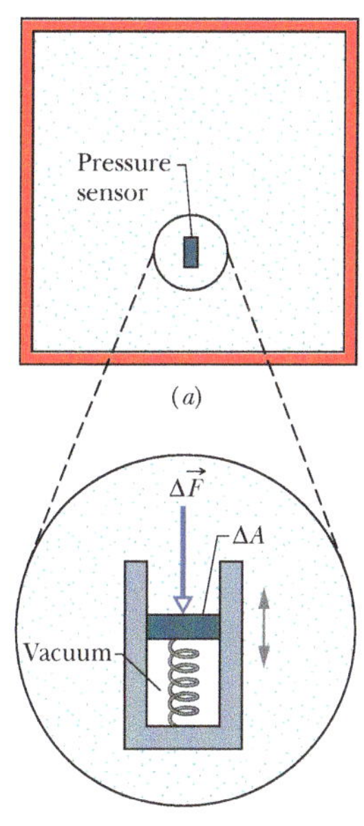

Figure 14-1 (*a*) A fluid-filled vessel containing a small pressure sensor, shown in (*b*). The pressure is measured by the relative position of the movable piston in the sensor.

Table 14-2 Some Pressures

	Pressure (Pa)
Center of the Sun	2×10^{16}
Center of Earth	4×10^{11}
Highest sustained laboratory pressure	1.5×10^{10}
Deepest ocean trench (bottom)	1.1×10^{8}
Spike heels on a dance floor	10^{6}
Automobile tire[a]	2×10^{5}
Atmosphere at sea level	1.0×10^{5}
Normal blood systolic pressure[a,b]	1.6×10^{4}
Best laboratory vacuum	10^{-12}

[a]Pressure in excess of atmospheric pressure.
[b]Equivalent to 120 torr on the physician's pressure gauge.

the area), we can write Eq. 14-3 as

$$p = \frac{F}{A} \quad \text{(pressure of uniform force on flat area)}, \tag{14-4}$$

where F is the magnitude of the normal force on area A.

We find by experiment that at a given point in a fluid at rest, the pressure p defined by Eq. 14-4 has the same value no matter how the pressure sensor is oriented. Pressure is a scalar, having no directional properties. It is true that the force acting on the piston of our pressure sensor is a vector quantity, but Eq. 14-4 involves only the *magnitude* of that force, a scalar quantity.

The SI unit of pressure is the newton per square meter, which is given a special name, the **pascal** (Pa). In metric countries, tire pressure gauges are calibrated in kilopascals. The pascal is related to some other common (non-SI) pressure units as follows:

$$1 \text{ atm} = 1.01 \times 10^5 \text{ Pa} = 760 \text{ torr} = 14.7 \text{ lb/in.}^2.$$

The *atmosphere* (atm) is, as the name suggests, the approximate average pressure of the atmosphere at sea level. The *torr* (named for Evangelista Torricelli, who invented the mercury barometer in 1674) was formerly called the *millimeter of mercury* (mm Hg). The pound per square inch is often abbreviated psi. Table 14-2 shows some pressures.

Sample Problem 14.01 Atmospheric pressure and force

A living room has floor dimensions of 3.5 m and 4.2 m and a height of 2.4 m.

(a) What does the air in the room weigh when the air pressure is 1.0 atm?

KEY IDEAS

(1) The air's weight is equal to mg, where m is its mass. (2) Mass m is related to the air density ρ and the air volume V by Eq. 14-2 ($\rho = m/V$).

Calculation: Putting the two ideas together and taking the density of air at 1.0 atm from Table 14-1, we find

$$\begin{aligned} mg &= (\rho V)g \\ &= (1.21 \text{ kg/m}^3)(3.5 \text{ m} \times 4.2 \text{ m} \times 2.4 \text{ m})(9.8 \text{ m/s}^2) \\ &= 418 \text{ N} \approx 420 \text{ N}. \qquad \text{(Answer)} \end{aligned}$$

This is the weight of about 110 cans of Pepsi.

(b) What is the magnitude of the atmosphere's downward force on the top of your head, which we take to have an area of 0.040 m^2?

KEY IDEA

When the fluid pressure p on a surface of area A is uniform, the fluid force on the surface can be obtained from Eq. 14-4 ($p = F/A$).

Calculation: Although air pressure varies daily, we can approximate that $p = 1.0$ atm. Then Eq. 14-4 gives

$$\begin{aligned} F = pA &= (1.0 \text{ atm})\left(\frac{1.01 \times 10^5 \text{ N/m}^2}{1.0 \text{ atm}}\right)(0.040 \text{ m}^2) \\ &= 4.0 \times 10^3 \text{ N}. \qquad \text{(Answer)} \end{aligned}$$

This large force is equal to the weight of the air column from the top of your head to the top of the atmosphere.

WILEY PLUS Additional examples, video, and practice available at *WileyPLUS*

14-2 FLUIDS AT REST

Learning Objectives

After reading this module, you should be able to . . .

14.04 Apply the relationship between the hydrostatic pressure, fluid density, and the height above or below a reference level.

14.05 Distinguish between total pressure (absolute pressure) and gauge pressure.

Key Ideas

- Pressure in a fluid at rest varies with vertical position y. For y measured positive upward,

$$p_2 = p_1 + \rho g(y_1 - y_2).$$

If h is the *depth* of a fluid sample *below* some reference level at which the pressure is p_0, this equation becomes

$$p = p_0 + \rho g h,$$

where p is the pressure in the sample.

- The pressure in a fluid is the same for all points at the same level.
- Gauge pressure is the difference between the actual pressure (or absolute pressure) at a point and the atmospheric pressure.

Fluids at Rest

Figure 14-2*a* shows a tank of water—or other liquid—open to the atmosphere. As every diver knows, the pressure *increases* with depth below the air–water interface. The diver's depth gauge, in fact, is a pressure sensor much like that of Fig. 14-1*b*. As every mountaineer knows, the pressure *decreases* with altitude as one ascends into the atmosphere. The pressures encountered by the diver and the mountaineer are usually called *hydrostatic pressures,* because they are due to fluids that are static (at rest). Here we want to find an expression for hydrostatic pressure as a function of depth or altitude.

Let us look first at the increase in pressure with depth below the water's surface. We set up a vertical y axis in the tank, with its origin at the air–water interface and the positive direction upward. We next consider a water sample con-

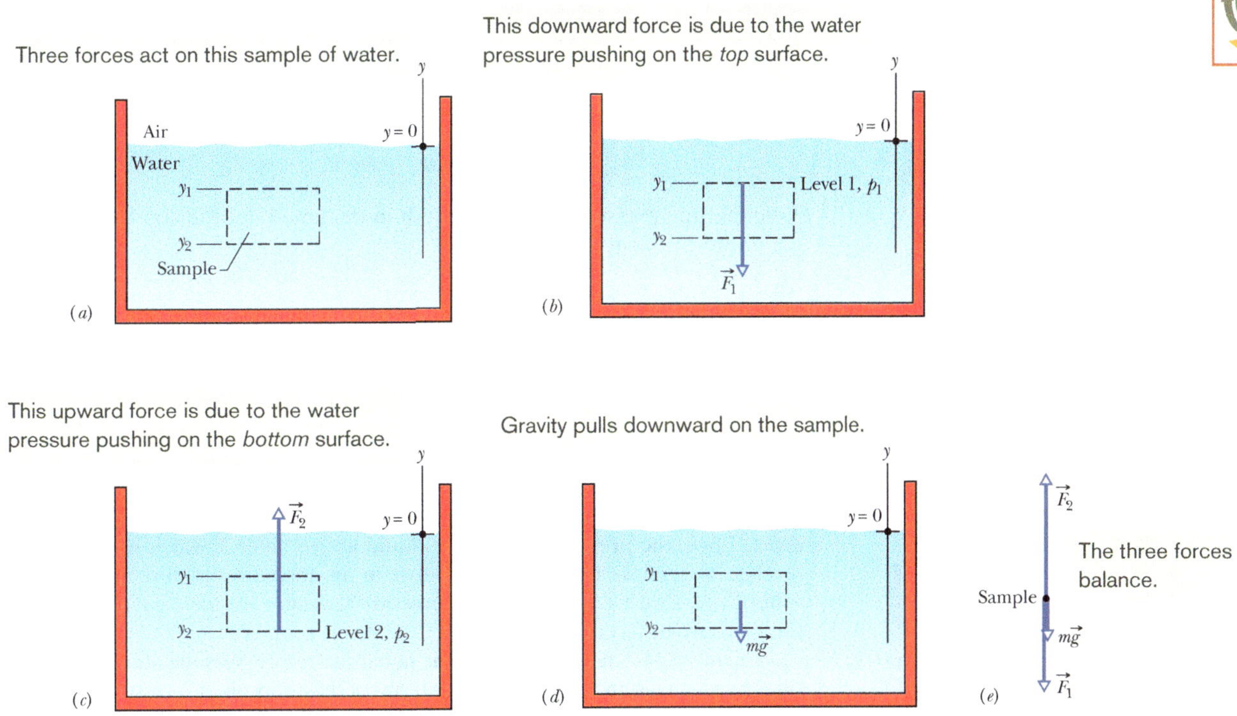

Figure 14-2 (*a*) A tank of water in which a sample of water is contained in an imaginary cylinder of horizontal base area A. (*b*)–(*d*) Force $\vec{F}_1$ acts at the top surface of the cylinder; force $\vec{F}_2$ acts at the bottom surface of the cylinder; the gravitational force on the water in the cylinder is represented by $m\vec{g}$. (*e*) A free-body diagram of the water sample. In *WileyPLUS*, this figure is available as an animation with voiceover.

tained in an imaginary right circular cylinder of horizontal base (or face) area A, such that y_1 and y_2 (both of which are *negative* numbers) are the depths below the surface of the upper and lower cylinder faces, respectively.

Figure 14-2*e* is a free-body diagram for the water in the cylinder. The water is in *static equilibrium;* that is, it is stationary and the forces on it balance. Three forces act on it vertically: Force $\vec{F}_1$ acts at the top surface of the cylinder and is due to the water above the cylinder (Fig. 14-2*b*). Force $\vec{F}_2$ acts at the bottom surface of the cylinder and is due to the water just below the cylinder (Fig. 14-2*c*). The gravitational force on the water is $m\vec{g}$, where m is the mass of the water in the cylinder (Fig. 14-2*d*). The balance of these forces is written as

$$F_2 = F_1 + mg. \tag{14-5}$$

To involve pressures, we use Eq. 14-4 to write

$$F_1 = p_1A \quad \text{and} \quad F_2 = p_2A. \tag{14-6}$$

The mass m of the water in the cylinder is, from Eq. 14-2, $m = \rho V$, where the cylinder's volume V is the product of its face area A and its height $y_1 - y_2$. Thus, m is equal to $\rho A(y_1 - y_2)$. Substituting this and Eq. 14-6 into Eq. 14-5, we find

$$p_2A = p_1A + \rho Ag(y_1 - y_2)$$

or

$$p_2 = p_1 + \rho g(y_1 - y_2). \tag{14-7}$$

This equation can be used to find pressure both in a liquid (as a function of depth) and in the atmosphere (as a function of altitude or height). For the former, suppose we seek the pressure p at a depth h below the liquid surface. Then we choose level 1 to be the surface, level 2 to be a distance h below it (as in Fig. 14-3), and p_0 to represent the atmospheric pressure on the surface. We then substitute

$$y_1 = 0, \quad p_1 = p_0 \quad \text{and} \quad y_2 = -h, \quad p_2 = p$$

into Eq. 14-7, which becomes

$$p = p_0 + \rho gh \quad \text{(pressure at depth } h\text{)}. \tag{14-8}$$

Note that the pressure at a given depth in the liquid depends on that depth but not on any horizontal dimension.

The pressure at a point in a fluid in static equilibrium depends on the depth of that point but not on any horizontal dimension of the fluid or its container.

Thus, Eq. 14-8 holds no matter what the shape of the container. If the bottom surface of the container is at depth h, then Eq. 14-8 gives the pressure p there.

In Eq. 14-8, p is said to be the total pressure, or **absolute pressure,** at level 2. To see why, note in Fig. 14-3 that the pressure p at level 2 consists of two contributions: (1) p_0, the pressure due to the atmosphere, which bears down on the liquid, and (2) ρgh, the pressure due to the liquid above level 2, which bears down on level 2. In general, the difference between an absolute pressure and an atmospheric pressure is called the **gauge pressure** (because we use a gauge to measure this pressure difference). For Fig. 14-3, the gauge pressure is ρgh.

Equation 14-7 also holds above the liquid surface: It gives the atmospheric pressure at a given distance above level 1 in terms of the atmospheric pressure p_1 at level 1 (*assuming* that the atmospheric density is uniform over that distance). For example, to find the atmospheric pressure at a distance d above level 1 in Fig. 14-3, we substitute

$$y_1 = 0, \quad p_1 = p_0 \quad \text{and} \quad y_2 = d, \quad p_2 = p.$$

Then with $\rho = \rho_{air}$, we obtain

$$p = p_0 - \rho_{air}gd.$$

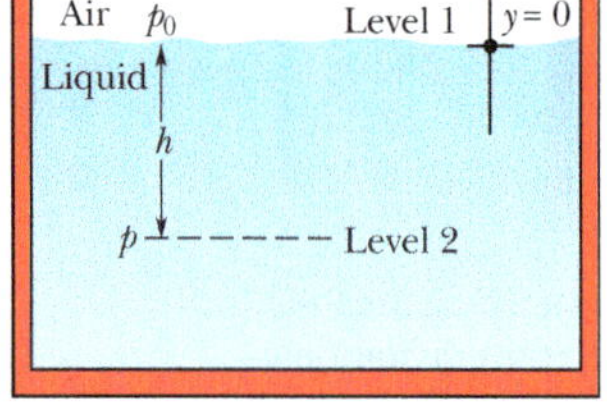

Figure 14-3 The pressure p increases with depth h below the liquid surface according to Eq. 14-8.

Checkpoint 1

The figure shows four containers of olive oil. Rank them according to the pressure at depth h, greatest first.

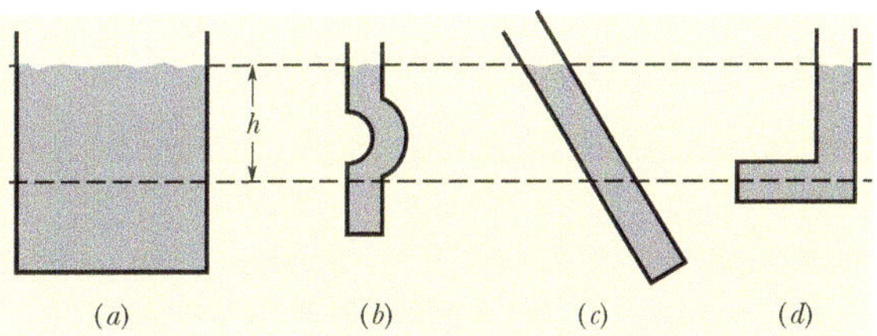

Sample Problem 14.02 Gauge pressure on a scuba diver

A novice scuba diver practicing in a swimming pool takes enough air from his tank to fully expand his lungs before abandoning the tank at depth L and swimming to the surface, failing to exhale during his ascent. At the surface, the difference Δp between the external pressure on him and the air pressure in his lungs is 9.3 kPa. From what depth does he start? What potentially lethal danger does he face?

KEY IDEA

The pressure at depth h in a liquid of density ρ is given by Eq. 14-8 ($p = p_0 + \rho gh$), where the gauge pressure ρgh is added to the atmospheric pressure p_0.

Calculations: Here, when the diver fills his lungs at depth L, the external pressure on him (and thus the air pressure within his lungs) is greater than normal and given by Eq. 14-8 as

$$p = p_0 + \rho gL,$$

where ρ is the water's density (998 kg/m^3, Table 14-1). As he ascends, the external pressure on him decreases, until it is atmospheric pressure p_0 at the surface. His blood pressure also decreases, until it is normal. However, because he does not exhale, the air pressure in his lungs remains at the value it had at depth L. At the surface, the pressure difference Δp is

$$\Delta p = p - p_0 = \rho gL,$$

so

$$L = \frac{\Delta p}{\rho g} = \frac{9300 \text{ Pa}}{(998 \text{ kg/m}^3)(9.8 \text{ m/s}^2)}$$

$$= 0.95 \text{ m.} \qquad \text{(Answer)}$$

This is not deep! Yet, the pressure difference of 9.3 kPa (about 9% of atmospheric pressure) is sufficient to rupture the diver's lungs and force air from them into the depressurized blood, which then carries the air to the heart, killing the diver. If the diver follows instructions and gradually exhales as he ascends, he allows the pressure in his lungs to equalize with the external pressure, and then there is no danger.

Sample Problem 14.03 Balancing of pressure in a U-tube

The U-tube in Fig. 14-4 contains two liquids in static equilibrium: Water of density ρ_w (= 998 kg/m^3) is in the right arm, and oil of unknown density ρ_x is in the left. Measurement gives l = 135 mm and d = 12.3 mm. What is the density of the oil?

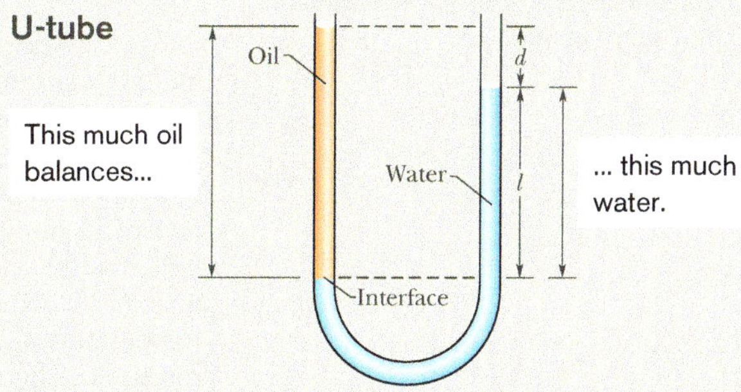

Figure 14-4 The oil in the left arm stands higher than the water.

KEY IDEAS

(1) The pressure p_{int} at the level of the oil–water interface in the left arm depends on the density ρ_x and height of the oil above the interface. (2) The water in the right arm *at the same level* must be at the same pressure p_{int}. The reason is that, because the water is in static equilibrium, pressures at points in the water at the same level must be the same.

Calculations: In the right arm, the interface is a distance l below the free surface of the *water,* and we have, from Eq. 14-8,

$$p_{int} = p_0 + \rho_w gl \quad \text{(right arm).}$$

In the left arm, the interface is a distance $l + d$ below the free surface of the *oil,* and we have, again from Eq. 14-8,

$$p_{int} = p_0 + \rho_x g(l + d) \quad \text{(left arm).}$$

Equating these two expressions and solving for the unknown density yield

$$\rho_x = \rho_w \frac{l}{l + d} = (998 \text{ kg/m}^3) \frac{135 \text{ mm}}{135 \text{ mm} + 12.3 \text{ mm}}$$

$$= 915 \text{ kg/m}^3. \qquad \text{(Answer)}$$

Note that the answer does not depend on the atmospheric pressure p_0 or the free-fall acceleration g.

WILEY PLUS Additional examples, video, and practice available at *WileyPLUS*

14-3 MEASURING PRESSURE

Learning Objectives

After reading this module, you should be able to . . .

14.06 Describe how a barometer can measure atmospheric pressure.

14.07 Describe how an open-tube manometer can measure the gauge pressure of a gas.

Key Ideas

- A mercury barometer can be used to measure atmospheric pressure.
- An open-tube manometer can be used to measure the gauge pressure of a confined gas.

Measuring Pressure

The Mercury Barometer

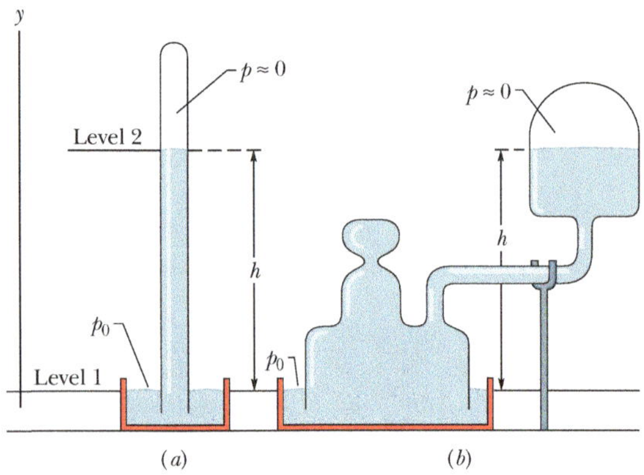

Figure 14-5 (*a*) A mercury barometer. (*b*) Another mercury barometer. The distance *h* is the same in both cases.

Figure 14-5*a* shows a very basic *mercury barometer,* a device used to measure the pressure of the atmosphere. The long glass tube is filled with mercury and inverted with its open end in a dish of mercury, as the figure shows. The space above the mercury column contains only mercury vapor, whose pressure is so small at ordinary temperatures that it can be neglected.

We can use Eq. 14-7 to find the atmospheric pressure p_0 in terms of the height *h* of the mercury column. We choose level 1 of Fig. 14-2 to be that of the air–mercury interface and level 2 to be that of the top of the mercury column, as labeled in Fig. 14-5*a*. We then substitute

$$y_1 = 0, \quad p_1 = p_0 \quad \text{and} \quad y_2 = h, \quad p_2 = 0$$

into Eq. 14-7, finding that

$$p_0 = \rho g h, \tag{14-9}$$

where ρ is the density of the mercury.

For a given pressure, the height *h* of the mercury column does not depend on the cross-sectional area of the vertical tube. The fanciful mercury barometer of Fig. 14-5*b* gives the same reading as that of Fig. 14-5*a*; all that counts is the vertical distance *h* between the mercury levels.

Equation 14-9 shows that, for a given pressure, the height of the column of mercury depends on the value of *g* at the location of the barometer and on the density of mercury, which varies with temperature. The height of the column (in millimeters) is numerically equal to the pressure (in torr) *only* if the barometer is at a place where *g* has its accepted standard value of 9.80665 m/s^2 *and* the temperature of the mercury is 0°C. If these conditions do not prevail (and they rarely do), small corrections must be made before the height of the mercury column can be transformed into a pressure.

The Open-Tube Manometer

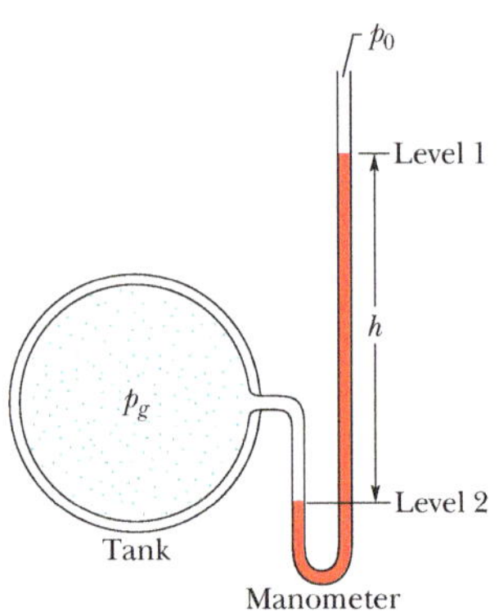

Figure 14-6 An open-tube manometer, connected to measure the gauge pressure of the gas in the tank on the left. The right arm of the **U**-tube is open to the atmosphere.

An *open-tube manometer* (Fig. 14-6) measures the gauge pressure p_g of a gas. It consists of a **U**-tube containing a liquid, with one end of the tube connected to the vessel whose gauge pressure we wish to measure and the other end open to the atmosphere. We can use Eq. 14-7 to find the gauge pressure in terms of the height *h* shown in Fig. 14-6. Let us choose levels 1 and 2 as shown in Fig. 14-6. With

$$y_1 = 0, \quad p_1 = p_0 \quad \text{and} \quad y_2 = -h, \quad p_2 = p$$

substituted into Eq. 14-7, we find that

$$p_g = p - p_0 = \rho g h, \tag{14-10}$$

where ρ is the liquid's density. The gauge pressure p_g is directly proportional to *h*.

The gauge pressure can be positive or negative, depending on whether $p > p_0$ or $p < p_0$. In inflated tires or the human circulatory system, the (absolute) pressure is greater than atmospheric pressure, so the gauge pressure is a positive quantity, sometimes called the *overpressure.* If you suck on a straw to pull fluid up the straw, the (absolute) pressure in your lungs is actually less than atmospheric pressure. The gauge pressure in your lungs is then a negative quantity.

14-4 PASCAL'S PRINCIPLE

Learning Objectives

After reading this module, you should be able to . . .

14.08 Identify Pascal's principle.

14.09 For a hydraulic lift, apply the relationship between the input area and displacement and the output area and displacement.

Key Idea

- Pascal's principle states that a change in the pressure applied to an enclosed fluid is transmitted undiminished to every portion of the fluid and to the walls of the containing vessel.

Pascal's Principle

When you squeeze one end of a tube to get toothpaste out the other end, you are watching **Pascal's principle** in action. This principle is also the basis for the Heimlich maneuver, in which a sharp pressure increase properly applied to the abdomen is transmitted to the throat, forcefully ejecting food lodged there. The principle was first stated clearly in 1652 by Blaise Pascal (for whom the unit of pressure is named):

> A change in the pressure applied to an enclosed incompressible fluid is transmitted undiminished to every portion of the fluid and to the walls of its container.

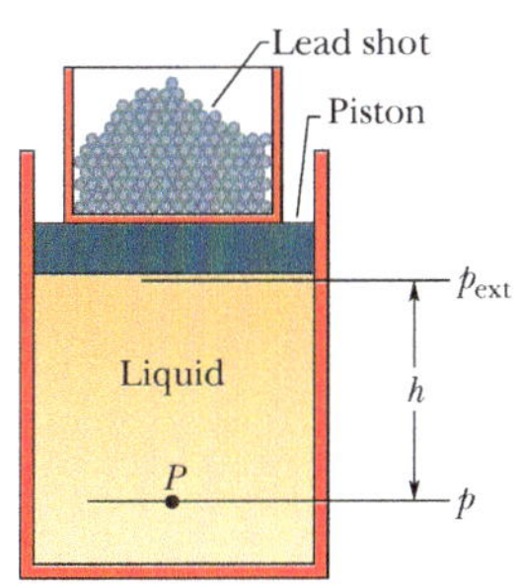

Figure 14-7 Lead shot (small balls of lead) loaded onto the piston create a pressure p_{ext} at the top of the enclosed (incompressible) liquid. If p_{ext} is increased, by adding more lead shot, the pressure increases by the same amount at all points within the liquid.

Demonstrating Pascal's Principle

Consider the case in which the incompressible fluid is a liquid contained in a tall cylinder, as in Fig. 14-7. The cylinder is fitted with a piston on which a container of lead shot rests. The atmosphere, container, and shot exert pressure p_{ext} on the piston and thus on the liquid. The pressure p at any point P in the liquid is then

$$p = p_{ext} + \rho g h. \tag{14-11}$$

Let us add a little more lead shot to the container to increase p_{ext} by an amount Δp_{ext}. The quantities ρ, g, and h in Eq. 14-11 are unchanged, so the pressure change at P is

$$\Delta p = \Delta p_{ext}. \tag{14-12}$$

This pressure change is independent of h, so it must hold for all points within the liquid, as Pascal's principle states.

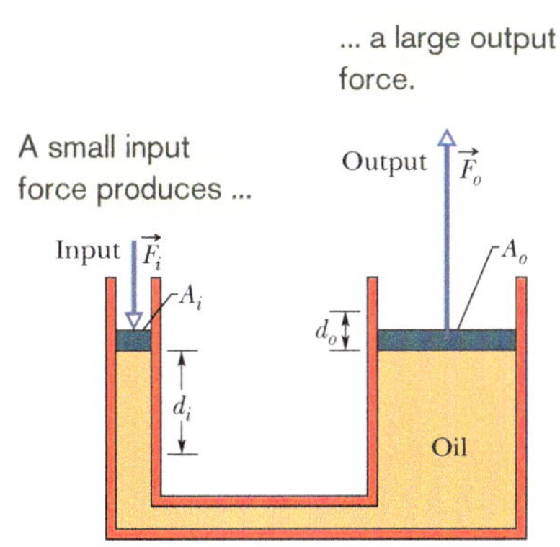

Figure 14-8 A hydraulic arrangement that can be used to magnify a force $\vec{F}_i$. The work done is, however, not magnified and is the same for both the input and output forces.

Pascal's Principle and the Hydraulic Lever

Figure 14-8 shows how Pascal's principle can be made the basis of a hydraulic lever. In operation, let an external force of magnitude F_i be directed downward on the left-hand (or input) piston, whose surface area is A_i. An incompressible liquid in the device then produces an upward force of magnitude F_o on the right-hand (or output) piston, whose surface area is A_o. To keep the system in equilibrium, there must be a downward force of magnitude F_o on the output piston from an external load (not shown). The force $\vec{F}_i$ applied on the left and the downward force $\vec{F}_o$ from the load on the right produce a change Δp in the pressure of the liquid that is given by

$$\Delta p = \frac{F_i}{A_i} = \frac{F_o}{A_o},$$

so

$$F_o = F_i \frac{A_o}{A_i}. \tag{14-13}$$

Equation 14-13 shows that the output force F_o on the load must be greater than the input force F_i if $A_o > A_i$, as is the case in Fig. 14-8.

If we move the input piston downward a distance d_i, the output piston moves upward a distance d_o, such that the same volume V of the incompressible liquid is displaced at both pistons. Then

$$V = A_i d_i = A_o d_o,$$

which we can write as

$$d_o = d_i \frac{A_i}{A_o}. \tag{14-14}$$

This shows that, if $A_o > A_i$ (as in Fig. 14-8), the output piston moves a smaller distance than the input piston moves.

From Eqs. 14-13 and 14-14 we can write the output work as

$$W = F_o d_o = \left(F_i \frac{A_o}{A_i}\right)\left(d_i \frac{A_i}{A_o}\right) = F_i d_i, \tag{14-15}$$

which shows that the work W done *on* the input piston by the applied force is equal to the work W done *by* the output piston in lifting the load placed on it.

The advantage of a hydraulic lever is this:

> With a hydraulic lever, a given force applied over a given distance can be transformed to a greater force applied over a smaller distance.

The product of force and distance remains unchanged so that the same work is done. However, there is often tremendous advantage in being able to exert the larger force. Most of us, for example, cannot lift an automobile directly but can with a hydraulic jack, even though we have to pump the handle farther than the automobile rises and in a series of small strokes.

14-5 ARCHIMEDES' PRINCIPLE

Learning Objectives

After reading this module, you should be able to . . .

14.10 Describe Archimedes' principle.

14.11 Apply the relationship between the buoyant force on a body and the mass of the fluid displaced by the body.

14.12 For a floating body, relate the buoyant force to the gravitational force.

14.13 For a floating body, relate the gravitational force to the mass of the fluid displaced by the body.

14.14 Distinguish between apparent weight and actual weight.

14.15 Calculate the apparent weight of a body that is fully or partially submerged.

Key Ideas

- Archimedes' principle states that when a body is fully or partially submerged in a fluid, the fluid pushes upward with a buoyant force with magnitude

$$F_b = m_f g,$$

where m_f is the mass of the fluid that has been pushed out of the way by the body.

- When a body floats in a fluid, the magnitude F_b of the (upward) buoyant force on the body is equal to the magnitude F_g of the (downward) gravitational force on the body.
- The apparent weight of a body on which a buoyant force acts is related to its actual weight by

$$\text{weight}_{\text{app}} = \text{weight} - F_b.$$

Archimedes' Principle

Figure 14-9 shows a student in a swimming pool, manipulating a very thin plastic sack (of negligible mass) that is filled with water. She finds that the sack and its contained water are in static equilibrium, tending neither to rise nor to sink. The downward gravitational force $\vec{F}_g$ on the contained water must be balanced by a net upward force from the water surrounding the sack.

This net upward force is a **buoyant force** $\vec{F}_b$. It exists because the pressure in the surrounding water increases with depth below the surface. Thus, the pressure near the bottom of the sack is greater than the pressure near the top, which means the forces on the sack due to this pressure are greater in magnitude near the bottom of the sack than near the top. Some of the forces are represented in Fig. 14-10*a*, where the space occupied by the sack has been left empty. Note that the force vectors drawn near the bottom of that space (with upward components) have longer lengths than those drawn near the top of the sack (with downward components). If we vectorially add all the forces on the sack from the water, the horizontal components cancel and the vertical components add to yield the upward buoyant force $\vec{F}_b$ on the sack. (Force $\vec{F}_b$ is shown to the right of the pool in Fig. 14-10*a*.)

Because the sack of water is in static equilibrium, the magnitude of $\vec{F}_b$ is equal to the magnitude $m_f g$ of the gravitational force $\vec{F}_g$ on the sack of water: $F_b = m_f g$. (Subscript f refers to *fluid*, here the water.) In words, the magnitude of the buoyant force is equal to the weight of the water in the sack.

In Fig. 14-10*b*, we have replaced the sack of water with a stone that exactly fills the hole in Fig. 14-10*a*. The stone is said to *displace* the water, meaning that the stone occupies space that would otherwise be occupied by water. We have changed nothing about the shape of the hole, so the forces at the hole's surface must be the same as when the water-filled sack was in place. Thus, the same upward buoyant force that acted on the water-filled sack now acts on the stone; that is, the magnitude F_b of the buoyant force is equal to $m_f g$, the weight of the water displaced by the stone.

Unlike the water-filled sack, the stone is not in static equilibrium. The downward gravitational force $\vec{F}_g$ on the stone is greater in magnitude than the upward buoyant force (Fig. 14-10*b*). The stone thus accelerates downward, sinking.

Let us next exactly fill the hole in Fig. 14-10*a* with a block of lightweight wood, as in Fig. 14-10*c*. Again, nothing has changed about the forces at the hole's surface, so the magnitude F_b of the buoyant force is still equal to $m_f g$, the weight

Figure 14-9 A thin-walled plastic sack of water is in static equilibrium in the pool. The gravitational force on the sack must be balanced by a net upward force on it from the surrounding water.

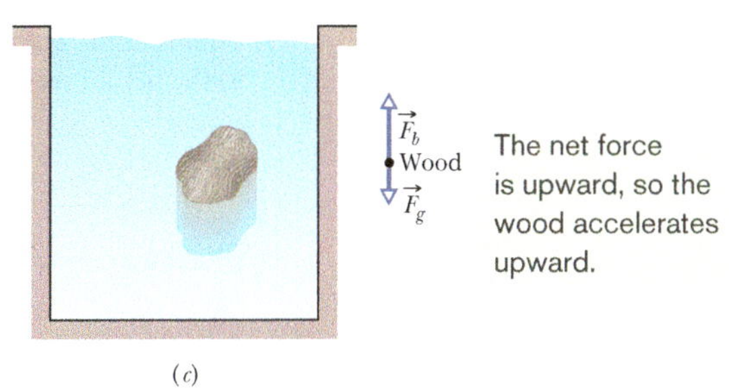

Figure 14-10 (*a*) The water surrounding the hole in the water produces a net upward buoyant force on whatever fills the hole. (*b*) For a stone of the same volume as the hole, the gravitational force exceeds the buoyant force in magnitude. (*c*) For a lump of wood of the same volume, the gravitational force is less than the buoyant force in magnitude.

of the displaced water. Like the stone, the block is not in static equilibrium. However, this time the gravitational force $\vec{F}_g$ is lesser in magnitude than the buoyant force (as shown to the right of the pool), and so the block accelerates upward, rising to the top surface of the water.

Our results with the sack, stone, and block apply to all fluids and are summarized in **Archimedes' principle:**

When a body is fully or partially submerged in a fluid, a buoyant force $\vec{F}_b$ from the surrounding fluid acts on the body. The force is directed upward and has a magnitude equal to the weight $m_f g$ of the fluid that has been displaced by the body.

The buoyant force on a body in a fluid has the magnitude

$$F_b = m_f g \quad \text{(buoyant force)}, \tag{14-16}$$

where m_f is the mass of the fluid that is displaced by the body.

Floating

When we release a block of lightweight wood just above the water in a pool, the block moves into the water because the gravitational force on it pulls it downward. As the block displaces more and more water, the magnitude F_b of the upward buoyant force acting on it increases. Eventually, F_b is large enough to equal the magnitude F_g of the downward gravitational force on the block, and the block comes to rest. The block is then in static equilibrium and is said to be *floating* in the water. In general,

When a body floats in a fluid, the magnitude F_b of the buoyant force on the body is equal to the magnitude F_g of the gravitational force on the body.

We can write this statement as

$$F_b = F_g \quad \text{(floating)}. \tag{14-17}$$

From Eq. 14-16, we know that $F_b = m_f g$. Thus,

When a body floats in a fluid, the magnitude F_g of the gravitational force on the body is equal to the weight $m_f g$ of the fluid that has been displaced by the body.

We can write this statement as

$$F_g = m_f g \quad \text{(floating)}. \tag{14-18}$$

In other words, a floating body displaces its own weight of fluid.

Apparent Weight in a Fluid

If we place a stone on a scale that is calibrated to measure weight, then the reading on the scale is the stone's weight. However, if we do this underwater, the upward buoyant force on the stone from the water decreases the reading. That reading is then an apparent weight. In general, an **apparent weight** is related to the actual weight of a body and the buoyant force on the body by

$$\begin{pmatrix}\text{apparent}\\ \text{weight}\end{pmatrix} = \begin{pmatrix}\text{actual}\\ \text{weight}\end{pmatrix} - \begin{pmatrix}\text{magnitude of}\\ \text{buoyant force}\end{pmatrix},$$

which we can write as

$$\text{weight}_{\text{app}} = \text{weight} - F_b \quad \text{(apparent weight)}. \tag{14-19}$$

If, in some test of strength, you had to lift a heavy stone, you could do it more easily with the stone underwater. Then your applied force would need to exceed only the stone's apparent weight, not its larger actual weight.

The magnitude of the buoyant force on a floating body is equal to the body's weight. Equation 14-19 thus tells us that a floating body has an apparent weight of zero—the body would produce a reading of zero on a scale. For example, when astronauts prepare to perform a complex task in space, they practice the task floating underwater, where their suits are adjusted to give them an apparent weight of zero.

Checkpoint 2

A penguin floats first in a fluid of density ρ_0, then in a fluid of density $0.95\rho_0$, and then in a fluid of density $1.1\rho_0$. (a) Rank the densities according to the magnitude of the buoyant force on the penguin, greatest first. (b) Rank the densities according to the amount of fluid displaced by the penguin, greatest first.

Sample Problem 14.04 Floating, buoyancy, and density

In Fig. 14-11, a block of density $\rho = 800\ \text{kg/m}^3$ floats face down in a fluid of density $\rho_f = 1200\ \text{kg/m}^3$. The block has height $H = 6.0$ cm.

(a) By what depth h is the block submerged?

KEY IDEAS

(1) Floating requires that the upward buoyant force on the block match the downward gravitational force on the block. (2) The buoyant force is equal to the weight $m_f g$ of the fluid displaced by the submerged portion of the block.

Calculations: From Eq. 14-16, we know that the buoyant force has the magnitude $F_b = m_f g$, where m_f is the mass of the fluid displaced by the block's submerged volume V_f. From Eq. 14-2 ($\rho = m/V$), we know that the mass of the displaced fluid is $m_f = \rho_f V_f$. We don't know V_f but if we symbolize the block's face length as L and its width as W, then from Fig. 14-11 we see that the submerged volume must be $V_f = LWh$. If we now combine our three expressions, we find that the upward buoyant force has magnitude

$$F_b = m_f g = \rho_f V_f g = \rho_f LWhg. \tag{14-20}$$

Similarly, we can write the magnitude F_g of the gravitational force on the block, first in terms of the block's mass m, then in terms of the block's density ρ and (full) volume V, and then in terms of the block's dimensions L, W, and H (the full height):

$$F_g = mg = \rho Vg = \rho_f LWHg. \tag{14-21}$$

The floating block is stationary. Thus, writing Newton's second law for components along a vertical y axis with the positive direction upward ($F_{\text{net},y} = ma_y$), we have

$$F_b - F_g = m(0),$$

Floating means that the buoyant force matches the gravitational force.

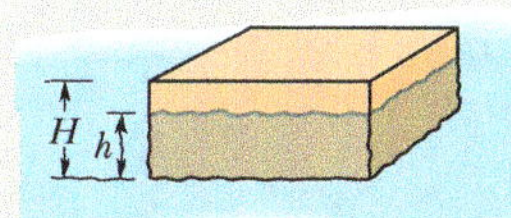

Figure 14-11 Block of height H floats in a fluid, to a depth of h.

or from Eqs. 14-20 and 14-21,

$$\rho_f LWhg - \rho LWHg = 0,$$

which gives us

$$h = \frac{\rho}{\rho_f} H = \frac{800\ \text{kg/m}^3}{1200\ \text{kg/m}^3}(6.0\ \text{cm})$$

$$= 4.0\ \text{cm}. \qquad \text{(Answer)}$$

(b) If the block is held fully submerged and then released, what is the magnitude of its acceleration?

Calculations: The gravitational force on the block is the same but now, with the block fully submerged, the volume of the displaced water is $V = LWH$. (The full height of the block is used.) This means that the value of F_b is now larger, and the block will no longer be stationary but will accelerate upward. Now Newton's second law yields

$$F_b - F_g = ma,$$

or

$$\rho_f LWHg - \rho LWHg = \rho LWHa,$$

where we inserted ρLWH for the mass m of the block. Solving for a leads to

$$a = \left(\frac{\rho_f}{\rho} - 1\right)g = \left(\frac{1200\ \text{kg/m}^3}{800\ \text{kg/m}^3} - 1\right)(9.8\ \text{m/s}^2)$$

$$= 4.9\ \text{m/s}^2. \qquad \text{(Answer)}$$

Additional examples, video, and practice available at *WileyPLUS*

14-6 THE EQUATION OF CONTINUITY

Learning Objectives

After reading this module, you should be able to . . .

14.16 Describe steady flow, incompressible flow, nonviscous flow, and irrotational flow.

14.17 Explain the term streamline.

14.18 Apply the equation of continuity to relate the cross-sectional area and flow speed at one point in a tube to those quantities at a different point.

14.19 Identify and calculate volume flow rate.

14.20 Identify and calculate mass flow rate.

Key Ideas

- An ideal fluid is incompressible and lacks viscosity, and its flow is steady and irrotational.
- A *streamline* is the path followed by an individual fluid particle.
- A *tube of flow* is a bundle of streamlines.
- The flow within any tube of flow obeys the equation of continuity:

$$R_V = Av = \text{a constant},$$

in which R_V is the volume flow rate, A is the cross-sectional area of the tube of flow at any point, and v is the speed of the fluid at that point.

- The mass flow rate R_m is

$$R_m = \rho R_V = \rho Av = \text{a constant}.$$

Will McIntyre/Photo Researchers, Inc.

Figure 14-12 At a certain point, the rising flow of smoke and heated gas changes from steady to turbulent.

Ideal Fluids in Motion

The motion of *real fluids* is very complicated and not yet fully understood. Instead, we shall discuss the motion of an **ideal fluid,** which is simpler to handle mathematically and yet provides useful results. Here are four assumptions that we make about our ideal fluid; they all are concerned with *flow:*

1. ***Steady flow*** In *steady* (or *laminar*) *flow,* the velocity of the moving fluid at any fixed point does not change with time. The gentle flow of water near the center of a quiet stream is steady; the flow in a chain of rapids is not. Figure 14-12 shows a transition from steady flow to *nonsteady* (or *nonlaminar* or *turbulent*) *flow* for a rising stream of smoke. The speed of the smoke particles increases as they rise and, at a certain critical speed, the flow changes from steady to nonsteady.
2. ***Incompressible flow*** We assume, as for fluids at rest, that our ideal fluid is incompressible; that is, its density has a constant, uniform value.
3. ***Nonviscous flow*** Roughly speaking, the viscosity of a fluid is a measure of how resistive the fluid is to flow. For example, thick honey is more resistive to flow than water, and so honey is said to be more viscous than water. Viscosity is the fluid analog of friction between solids; both are mechanisms by which the kinetic energy of moving objects can be transferred to thermal energy. In the absence of friction, a block could glide at constant speed along a horizontal surface. In the same way, an object moving through a nonviscous fluid would experience no *viscous drag force*—that is, no resistive force due to viscosity; it could move at constant speed through the fluid. The British scientist Lord Rayleigh noted that in an ideal fluid a ship's propeller would not work, but, on the other hand, in an ideal fluid a ship (once set into motion) would not need a propeller!
4. ***Irrotational flow*** Although it need not concern us further, we also assume that the flow is *irrotational.* To test for this property, let a tiny grain of dust move with the fluid. Although this test body may (or may not) move in a circular path, in irrotational flow the test body will not rotate about an axis through its own center of mass. For a loose analogy, the motion of a Ferris wheel is rotational; that of its passengers is irrotational.

We can make the flow of a fluid visible by adding a *tracer.* This might be a dye injected into many points across a liquid stream (Fig. 14-13) or smoke

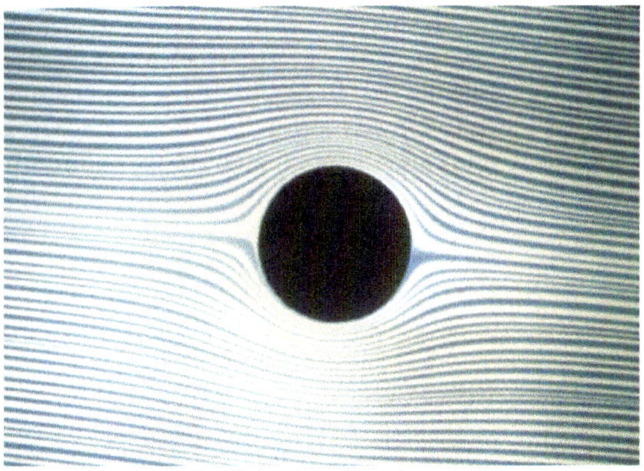

Courtesy D. H. Peregrine, University of Bristol

Figure 14-13 The steady flow of a fluid around a cylinder, as revealed by a dye tracer that was injected into the fluid upstream of the cylinder.

particles added to a gas flow (Fig. 14-12). Each bit of a tracer follows a *streamline,* which is the path that a tiny element of the fluid would take as the fluid flows. Recall from Chapter 4 that the velocity of a particle is always tangent to the path taken by the particle. Here the particle is the fluid element, and its velocity $\vec{v}$ is always tangent to a streamline (Fig. 14-14). For this reason, two streamlines can never intersect; if they did, then an element arriving at their intersection would have two different velocities simultaneously—an impossibility.

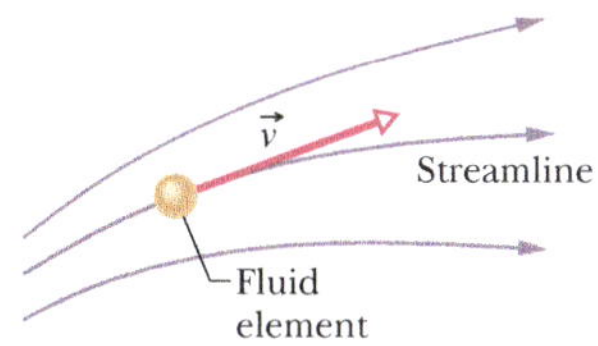

Figure 14-14 A fluid element traces out a streamline as it moves. The velocity vector of the element is tangent to the streamline at every point.

The Equation of Continuity

You may have noticed that you can increase the speed of the water emerging from a garden hose by partially closing the hose opening with your thumb. Apparently the speed v of the water depends on the cross-sectional area A through which the water flows.

Here we wish to derive an expression that relates v and A for the steady flow of an ideal fluid through a tube with varying cross section, like that in Fig. 14-15. The flow there is toward the right, and the tube segment shown (part of a longer tube) has length L. The fluid has speeds v_1 at the left end of the segment and v_2 at the right end. The tube has cross-sectional areas A_1 at the left end and A_2 at the right end. Suppose that in a time interval Δt a volume ΔV of fluid enters the tube segment at its left end (that volume is colored purple in Fig. 14-15). Then, because the fluid is incompressible, an identical volume ΔV must emerge from the right end of the segment (it is colored green in Fig. 14-15).

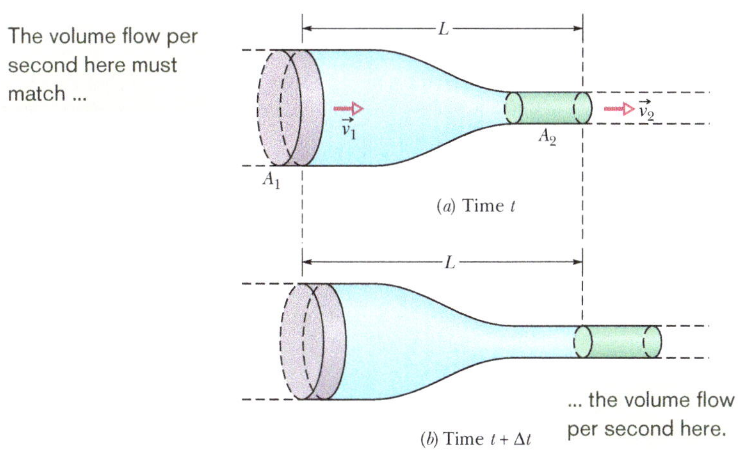

Figure 14-15 Fluid flows from left to right at a steady rate through a tube segment of length L. The fluid's speed is v_1 at the left side and v_2 at the right side. The tube's cross-sectional area is A_1 at the left side and A_2 at the right side. From time t in (a) to time $t + \Delta t$ in (b), the amount of fluid shown in purple enters at the left side and the equal amount of fluid shown in green emerges at the right side.

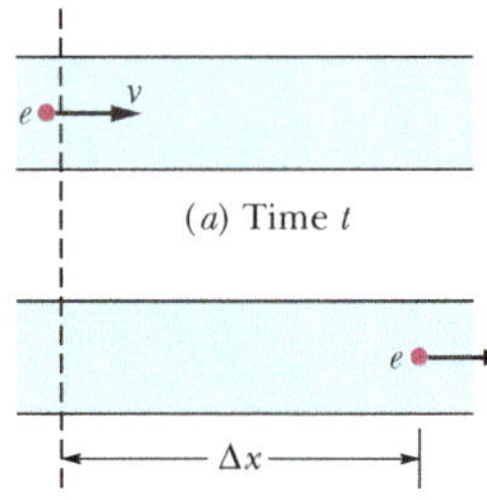

Figure 14-16 Fluid flows at a constant speed v through a tube. (a) At time t, fluid element e is about to pass the dashed line. (b) At time $t + \Delta t$, element e is a distance $\Delta x = v\,\Delta t$ from the dashed line.

We can use this common volume ΔV to relate the speeds and areas. To do so, we first consider Fig. 14-16, which shows a side view of a tube of *uniform* cross-sectional area A. In Fig. 14-16a, a fluid element e is about to pass through the dashed line drawn across the tube width. The element's speed is v, so during a time interval Δt, the element moves along the tube a distance $\Delta x = v\,\Delta t$. The volume ΔV of fluid that has passed through the dashed line in that time interval Δt is

$$\Delta V = A\,\Delta x = Av\,\Delta t. \tag{14-22}$$

Applying Eq. 14-22 to both the left and right ends of the tube segment in Fig. 14-15, we have

$$\Delta V = A_1v_1\,\Delta t = A_2v_2\,\Delta t$$

or

$$A_1v_1 = A_2v_2 \quad \text{(equation of continuity).} \tag{14-23}$$

This relation between speed and cross-sectional area is called the **equation of continuity** for the flow of an ideal fluid. It tells us that the flow speed increases when we decrease the cross-sectional area through which the fluid flows.

Equation 14-23 applies not only to an actual tube but also to any so-called *tube of flow,* or imaginary tube whose boundary consists of streamlines. Such a tube acts like a real tube because no fluid element can cross a streamline; thus, all the fluid within a tube of flow must remain within its boundary. Figure 14-17 shows a tube of flow in which the cross-sectional area increases from area A_1 to area A_2 along the flow direction. From Eq. 14-23 we know that, with the increase in area, the speed must decrease, as is indicated by the greater spacing between streamlines at the right in Fig. 14-17. Similarly, you can see that in Fig. 14-13 the speed of the flow is greatest just above and just below the cylinder.

We can rewrite Eq. 14-23 as

$$R_V = Av = \text{a constant} \quad \text{(volume flow rate, equation of continuity),} \tag{14-24}$$

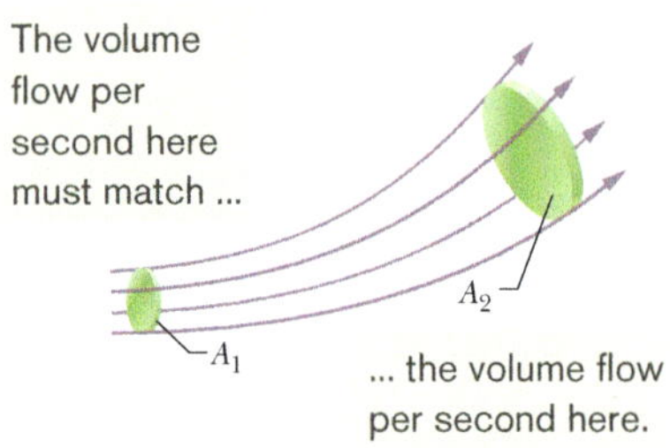

Figure 14-17 A tube of flow is defined by the streamlines that form the boundary of the tube. The volume flow rate must be the same for all cross sections of the tube of flow.

in which R_V is the **volume flow rate** of the fluid (volume past a given point per unit time). Its SI unit is the cubic meter per second (m^3/s). If the density ρ of the fluid is uniform, we can multiply Eq. 14-24 by that density to get the **mass flow rate** R_m (mass per unit time):

$$R_m = \rho R_V = \rho Av = \text{a constant} \quad \text{(mass flow rate).} \tag{14-25}$$

The SI unit of mass flow rate is the kilogram per second (kg/s). Equation 14-25 says that the mass that flows into the tube segment of Fig. 14-15 each second must be equal to the mass that flows out of that segment each second.

Checkpoint 3

The figure shows a pipe and gives the volume flow rate (in cm^3/s) and the direction of flow for all but one section. What are the volume flow rate and the direction of flow for that section?

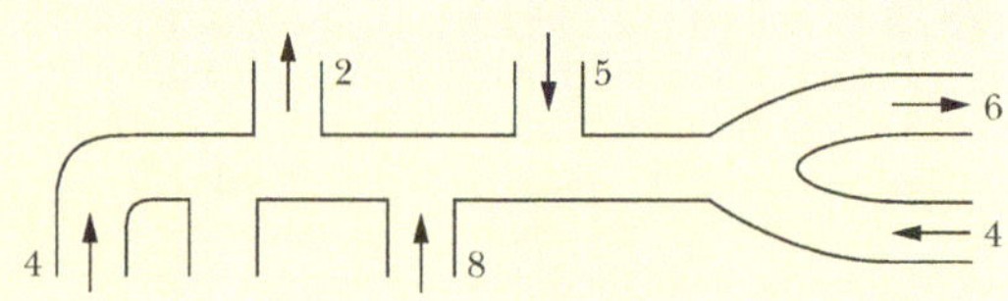

Sample Problem 14.05 A water stream narrows as it falls

Figure 14-18 shows how the stream of water emerging from a faucet "necks down" as it falls. This change in the horizontal cross-sectional area is characteristic of any laminar (non-turbulant) falling stream because the gravitational force increases the speed of the stream. Here the indicated cross-sectional areas are $A_0 = 1.2\ \text{cm}^2$ and $A = 0.35\ \text{cm}^2$. The two levels are separated by a vertical distance $h = 45$ mm. What is the volume flow rate from the tap?

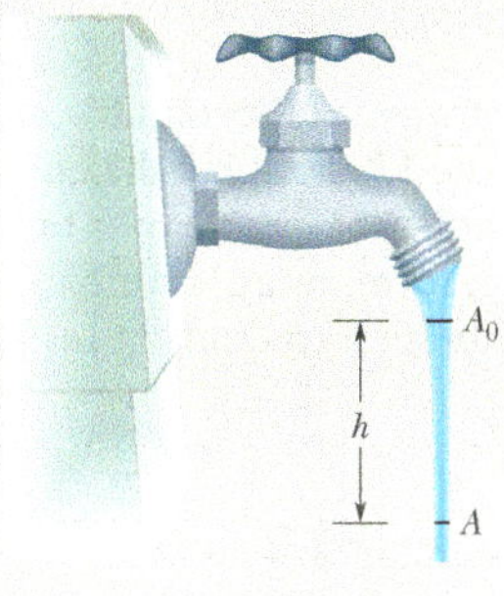

Figure 14-18 As water falls from a tap, its speed increases. Because the volume flow rate must be the same at all horizontal cross sections of the stream, the stream must "neck down" (narrow).

KEY IDEA

The volume flow rate through the higher cross section must be the same as that through the lower cross section.

Calculations: From Eq. 14-24, we have

$$A_0v_0 = Av, \tag{14-26}$$

where v_0 and v are the water speeds at the levels corresponding to A_0 and A. From Eq. 2-16 we can also write, because the water is falling freely with acceleration g,

$$v^2 = v_0^2 + 2gh. \tag{14-27}$$

Eliminating v between Eqs. 14-26 and 14-27 and solving for v_0, we obtain

$$v_0 = \sqrt{\frac{2ghA^2}{A_0^2 - A^2}}$$

$$= \sqrt{\frac{(2)(9.8\ \text{m/s}^2)(0.045\ \text{m})(0.35\ \text{cm}^2)^2}{(1.2\ \text{cm}^2)^2 - (0.35\ \text{cm}^2)^2}}$$

$$= 0.286\ \text{m/s} = 28.6\ \text{cm/s}.$$

From Eq. 14-24, the volume flow rate R_V is then

$$R_V = A_0v_0 = (1.2\ \text{cm}^2)(28.6\ \text{cm/s})$$

$$= 34\ \text{cm}^3/\text{s}. \qquad \text{(Answer)}$$

WILEY PLUS Additional examples, video, and practice available at *WileyPLUS*

14-7 BERNOULLI'S EQUATION

Learning Objectives

After reading this module, you should be able to . . .

14.21 Calculate the kinetic energy density in terms of a fluid's density and flow speed.
14.22 Identify the fluid pressure as being a type of energy density.
14.23 Calculate the gravitational potential energy density.
14.24 Apply Bernoulli's equation to relate the total energy density at one point on a streamline to the value at another point.
14.25 Identify that Bernoulli's equation is a statement of the conservation of energy.

Key Idea

- Applying the principle of conservation of mechanical energy to the flow of an ideal fluid leads to Bernoulli's equation:

$$p + \tfrac{1}{2}\rho v^2 + \rho gy = \text{a constant}$$

along any tube of flow.

Bernoulli's Equation

Figure 14-19 represents a tube through which an ideal fluid is flowing at a steady rate. In a time interval Δt, suppose that a volume of fluid ΔV, colored purple in Fig. 14-19, enters the tube at the left (or input) end and an identical volume,

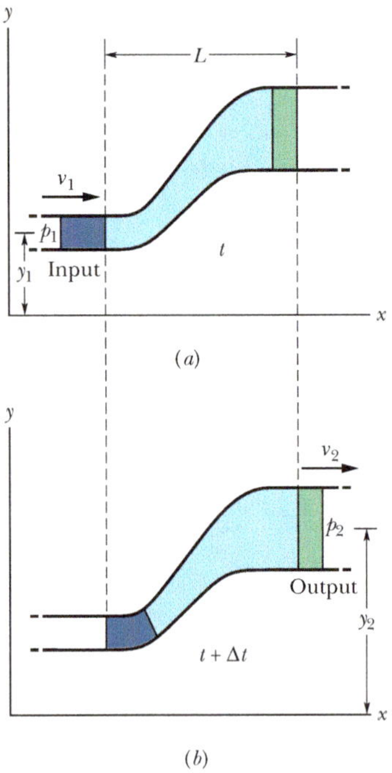

Figure 14-19 Fluid flows at a steady rate through a length L of a tube, from the input end at the left to the output end at the right. From time t in (a) to time $t + \Delta t$ in (b), the amount of fluid shown in purple enters the input end and the equal amount shown in green emerges from the output end.

colored green in Fig. 14-19, emerges at the right (or output) end. The emerging volume must be the same as the entering volume because the fluid is incompressible, with an assumed constant density ρ.

Let y_1, v_1, and p_1 be the elevation, speed, and pressure of the fluid entering at the left, and y_2, v_2, and p_2 be the corresponding quantities for the fluid emerging at the right. By applying the principle of conservation of energy to the fluid, we shall show that these quantities are related by

$$p_1 + \tfrac{1}{2}\rho v_1^2 + \rho g y_1 = p_2 + \tfrac{1}{2}\rho v_2^2 + \rho g y_2. \qquad (14\text{-}28)$$

In general, the term $\frac{1}{2}\rho v^2$ is called the fluid's **kinetic energy density** (kinetic energy per unit volume). We can also write Eq. 14-28 as

$$p + \tfrac{1}{2}\rho v^2 + \rho g y = \text{a constant} \quad \text{(Bernoulli's equation)}. \qquad (14\text{-}29)$$

Equations 14-28 and 14-29 are equivalent forms of **Bernoulli's equation,** after Daniel Bernoulli, who studied fluid flow in the 1700s.* Like the equation of continuity (Eq. 14-24), Bernoulli's equation is not a new principle but simply the reformulation of a familiar principle in a form more suitable to fluid mechanics. As a check, let us apply Bernoulli's equation to fluids at rest, by putting $v_1 = v_2 = 0$ in Eq. 14-28. The result is Eq. 14-7:

$$p_2 = p_1 + \rho g(y_1 - y_2).$$

A major prediction of Bernoulli's equation emerges if we take y to be a constant ($y = 0$, say) so that the fluid does not change elevation as it flows. Equation 14-28 then becomes

$$p_1 + \tfrac{1}{2}\rho v_1^2 = p_2 + \tfrac{1}{2}\rho v_2^2, \qquad (14\text{-}30)$$

which tells us that:

> If the speed of a fluid element increases as the element travels along a horizontal streamline, the pressure of the fluid must decrease, and conversely.

Put another way, where the streamlines are relatively close together (where the velocity is relatively great), the pressure is relatively low, and conversely.

The link between a change in speed and a change in pressure makes sense if you consider a fluid element that travels through a tube of various widths. Recall that the element's speed in the narrower regions is fast and its speed in the wider regions is slow. By Newton's second law, forces (or pressures) must cause the changes in speed (the accelerations). When the element nears a narrow region, the higher pressure behind it accelerates it so that it then has a greater speed in the narrow region. When it nears a wide region, the higher pressure ahead of it decelerates it so that it then has a lesser speed in the wide region.

Bernoulli's equation is strictly valid only to the extent that the fluid is ideal. If viscous forces are present, thermal energy will be involved, which here we neglect.

Proof of Bernoulli's Equation

Let us take as our system the entire volume of the (ideal) fluid shown in Fig. 14-19. We shall apply the principle of conservation of energy to this system as it moves from its initial state (Fig. 14-19a) to its final state (Fig. 14-19b). The fluid lying between the two vertical planes separated by a distance L in Fig. 14-19 does not change its properties during this process; we need be concerned only with changes that take place at the input and output ends.

*For irrotational flow (which we assume), the constant in Eq. 14-29 has the same value for all points within the tube of flow; the points do not have to lie along the same streamline. Similarly, the points 1 and 2 in Eq. 14-28 can lie anywhere within the tube of flow.

First, we apply energy conservation in the form of the work–kinetic energy theorem,

$$W = \Delta K, \tag{14-31}$$

which tells us that the change in the kinetic energy of our system must equal the net work done on the system. The change in kinetic energy results from the change in speed between the ends of the tube and is

$$\begin{aligned}\Delta K &= \tfrac{1}{2}\Delta m\, v_2^2 - \tfrac{1}{2}\Delta m\, v_1^2 \\ &= \tfrac{1}{2}\rho\,\Delta V(v_2^2 - v_1^2),\end{aligned} \tag{14-32}$$

in which Δm ($= \rho\,\Delta V$) is the mass of the fluid that enters at the input end and leaves at the output end during a small time interval Δt.

The work done on the system arises from two sources. The work W_g done by the gravitational force ($\Delta m\,\vec{g}$) on the fluid of mass Δm during the vertical lift of the mass from the input level to the output level is

$$\begin{aligned}W_g &= -\Delta m\, g(y_2 - y_1) \\ &= -\rho g\,\Delta V(y_2 - y_1).\end{aligned} \tag{14-33}$$

This work is negative because the upward displacement and the downward gravitational force have opposite directions.

Work must also be done *on* the system (at the input end) to push the entering fluid into the tube and *by* the system (at the output end) to push forward the fluid that is located ahead of the emerging fluid. In general, the work done by a force of magnitude F, acting on a fluid sample contained in a tube of area A to move the fluid through a distance Δx, is

$$F\,\Delta x = (pA)(\Delta x) = p(A\,\Delta x) = p\,\Delta V.$$

The work done on the system is then $p_1\,\Delta V$, and the work done by the system is $-p_2\,\Delta V$. Their sum W_p is

$$\begin{aligned}W_p &= -p_2\,\Delta V + p_1\,\Delta V \\ &= -(p_2 - p_1)\,\Delta V.\end{aligned} \tag{14-34}$$

The work–kinetic energy theorem of Eq. 14-31 now becomes

$$W = W_g + W_p = \Delta K.$$

Substituting from Eqs. 14-32, 14-33, and 14-34 yields

$$-\rho g\,\Delta V(y_2 - y_1) - \Delta V(p_2 - p_1) = \tfrac{1}{2}\rho\,\Delta V(v_2^2 - v_1^2).$$

This, after a slight rearrangement, matches Eq. 14-28, which we set out to prove.

Checkpoint 4

Water flows smoothly through the pipe shown in the figure, descending in the process. Rank the four numbered sections of pipe according to (a) the volume flow rate R_V through them, (b) the flow speed v through them, and (c) the water pressure p within them, greatest first.

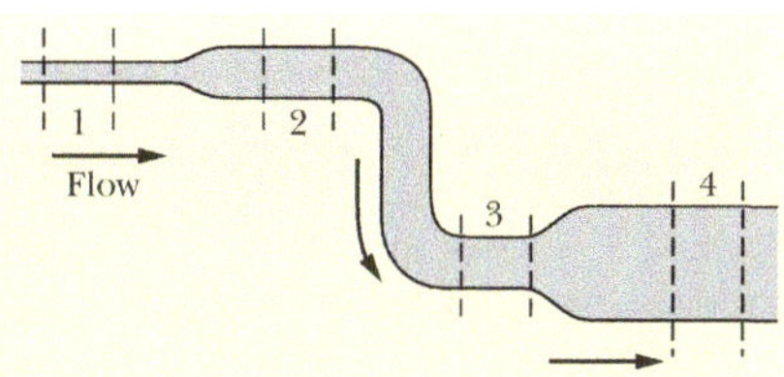

Sample Problem 14.06 Bernoulli principle of fluid through a narrowing pipe

Ethanol of density $\rho = 791\ \text{kg/m}^3$ flows smoothly through a horizontal pipe that tapers (as in Fig. 14-15) in cross-sectional area from $A_1 = 1.20 \times 10^{-3}\ \text{m}^2$ to $A_2 = A_1/2$. The pressure difference between the wide and narrow sections of pipe is 4120 Pa. What is the volume flow rate R_V of the ethanol?

KEY IDEAS

(1) Because the fluid flowing through the wide section of pipe must entirely pass through the narrow section, the volume flow rate R_V must be the same in the two sections. Thus, from Eq. 14-24,

$$R_V = v_1 A_1 = v_2 A_2. \quad (14\text{-}35)$$

However, with two unknown speeds, we cannot evaluate this equation for R_V. (2) Because the flow is smooth, we can apply Bernoulli's equation. From Eq. 14-28, we can write

$$p_1 + \tfrac{1}{2}\rho v_1^2 + \rho g y = p_2 + \tfrac{1}{2}\rho v_2^2 + \rho g y, \quad (14\text{-}36)$$

where subscripts 1 and 2 refer to the wide and narrow sections of pipe, respectively, and y is their common elevation. This equation hardly seems to help because it does not contain the desired R_V and it contains the unknown speeds v_1 and v_2.

Calculations: There is a neat way to make Eq. 14-36 work for us: First, we can use Eq. 14-35 and the fact that $A_2 = A_1/2$ to write

$$v_1 = \frac{R_V}{A_1} \quad \text{and} \quad v_2 = \frac{R_V}{A_2} = \frac{2R_V}{A_1}. \quad (14\text{-}37)$$

Then we can substitute these expressions into Eq. 14-36 to eliminate the unknown speeds and introduce the desired volume flow rate. Doing this and solving for R_V yield

$$R_V = A_1\sqrt{\frac{2(p_1 - p_2)}{3\rho}}. \quad (14\text{-}38)$$

We still have a decision to make: We know that the pressure difference between the two sections is 4120 Pa, but does that mean that $p_1 - p_2$ is 4120 Pa or −4120 Pa? We could guess the former is true, or otherwise the square root in Eq. 14-38 would give us an imaginary number. However, let's try some reasoning. From Eq. 14-35 we see that speed v_2 in the narrow section (small A_2) must be greater than speed v_1 in the wider section (larger A_1). Recall that if the speed of a fluid increases as the fluid travels along a horizontal path (as here), the pressure of the fluid must decrease. Thus, p_1 is greater than p_2, and $p_1 - p_2 = 4120$ Pa. Inserting this and known data into Eq. 14-38 gives

$$R_V = 1.20 \times 10^{-3}\,\text{m}^2\sqrt{\frac{(2)(4120\ \text{Pa})}{(3)(791\ \text{kg/m}^3)}}$$
$$= 2.24 \times 10^{-3}\ \text{m}^3/\text{s}. \quad \text{(Answer)}$$

Sample Problem 14.07 Bernoulli principle for a leaky water tank

In the old West, a desperado fires a bullet into an open water tank (Fig. 14-20), creating a hole a distance h below the water surface. What is the speed v of the water exiting the tank?

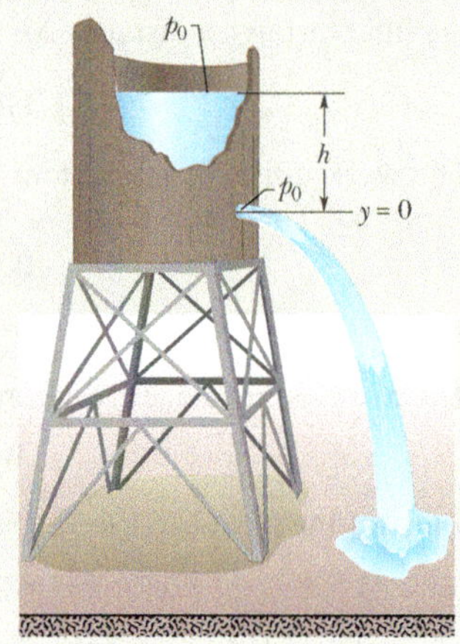

Figure 14-20 Water pours through a hole in a water tank, at a distance h below the water surface. The pressure at the water surface and at the hole is atmospheric pressure p_0.

KEY IDEAS

(1) This situation is essentially that of water moving (downward) with speed v_0 through a wide pipe (the tank) of cross-sectional area A and then moving (horizontally) with speed v through a narrow pipe (the hole) of cross-sectional area a. (2) Because the water flowing through the wide pipe must entirely pass through the narrow pipe, the volume flow rate R_V must be the same in the two "pipes." (3) We can also relate v to v_0 (and to h) through Bernoulli's equation (Eq. 14-28).

Calculations: From Eq. 14-24,

$$R_V = av = Av_0$$

and thus

$$v_0 = \frac{a}{A}v.$$

Because $a \ll A$, we see that $v_0 \ll v$. To apply Bernoulli's equation, we take the level of the hole as our reference level for measuring elevations (and thus gravitational potential energy). Noting that the pressure at the top of the tank and at the bullet hole is the atmospheric pressure p_0 (because both places are exposed to the atmosphere), we write Eq. 14-28 as

$$p_0 + \tfrac{1}{2}\rho v_0^2 + \rho g h = p_0 + \tfrac{1}{2}\rho v^2 + \rho g(0). \quad (14\text{-}39)$$

(Here the top of the tank is represented by the left side of the equation and the hole by the right side. The zero on the right indicates that the hole is at our reference level.) Before we solve Eq. 14-39 for v, we can use our result that $v_0 \ll v$ to simplify it: We assume that v_0^2, and thus the term $\tfrac{1}{2}\rho v_0^2$ in Eq. 14-39, is negligible relative to the other terms, and we drop it. Solving the remaining equation for v then yields

$$v = \sqrt{2gh}. \quad \text{(Answer)}$$

This is the same speed that an object would have when falling a height h from rest.

WILEY PLUS Additional examples, video, and practice available at *WileyPLUS*

Review & Summary

Density The **density** ρ of any material is defined as the material's mass per unit volume:

$$\rho = \frac{\Delta m}{\Delta V}. \qquad (14\text{-}1)$$

Usually, where a material sample is much larger than atomic dimensions, we can write Eq. 14-1 as

$$\rho = \frac{m}{V}. \qquad (14\text{-}2)$$

Fluid Pressure A **fluid** is a substance that can flow; it conforms to the boundaries of its container because it cannot withstand shearing stress. It can, however, exert a force perpendicular to its surface. That force is described in terms of **pressure** p:

$$p = \frac{\Delta F}{\Delta A}, \qquad (14\text{-}3)$$

in which ΔF is the force acting on a surface element of area ΔA. If the force is uniform over a flat area, Eq. 14-3 can be written as

$$p = \frac{F}{A}. \qquad (14\text{-}4)$$

The force resulting from fluid pressure at a particular point in a fluid has the same magnitude in all directions. **Gauge pressure** is the difference between the actual pressure (or *absolute pressure*) at a point and the atmospheric pressure.

Pressure Variation with Height and Depth Pressure in a fluid at rest varies with vertical position y. For y measured positive upward,

$$p_2 = p_1 + \rho g(y_1 - y_2). \qquad (14\text{-}7)$$

The pressure in a fluid is the same for all points at the same level. If h is the *depth* of a fluid sample below some reference level at which the pressure is p_0, then the pressure in the sample is

$$p = p_0 + \rho g h. \qquad (14\text{-}8)$$

Pascal's Principle A change in the pressure applied to an enclosed fluid is transmitted undiminished to every portion of the fluid and to the walls of the containing vessel.

Archimedes' Principle When a body is fully or partially submerged in a fluid, a buoyant force $\vec{F}_b$ from the surrounding fluid acts on the body. The force is directed upward and has a magnitude given by

$$F_b = m_f g, \qquad (14\text{-}16)$$

where m_f is the mass of the fluid that has been displaced by the body (that is, the fluid that has been pushed out of the way by the body).

When a body floats in a fluid, the magnitude F_b of the (upward) buoyant force on the body is equal to the magnitude F_g of the (downward) gravitational force on the body. The **apparent weight** of a body on which a buoyant force acts is related to its actual weight by

$$\text{weight}_{\text{app}} = \text{weight} - F_b. \qquad (14\text{-}19)$$

Flow of Ideal Fluids An **ideal fluid** is incompressible and lacks viscosity, and its flow is steady and irrotational. A *streamline* is the path followed by an individual fluid particle. A *tube of flow* is a bundle of streamlines. The flow within any tube of flow obeys the **equation of continuity:**

$$R_V = Av = \text{a constant}, \qquad (14\text{-}24)$$

in which R_V is the **volume flow rate,** A is the cross-sectional area of the tube of flow at any point, and v is the speed of the fluid at that point. The **mass flow rate** R_m is

$$R_m = \rho R_V = \rho A v = \text{a constant}. \qquad (14\text{-}25)$$

Bernoulli's Equation Applying the principle of conservation of mechanical energy to the flow of an ideal fluid leads to **Bernoulli's equation** along any tube of flow:

$$p + \tfrac{1}{2}\rho v^2 + \rho g y = \text{a constant}. \qquad (14\text{-}29)$$

Questions

1 We fully submerge an irregular 3 kg lump of material in a certain fluid. The fluid that would have been in the space now occupied by the lump has a mass of 2 kg. (a) When we release the lump, does it move upward, move downward, or remain in place? (b) If we next fully submerge the lump in a less dense fluid and again release it, what does it do?

2 Figure 14-21 shows four situations in which a red liquid and a gray liquid are in a **U**-tube. In one situation the liquids cannot be in static equilibrium. (a) Which situation is that? (b) For the other three situations, assume static equilibrium. For each of them, is the density of the red liquid greater than, less than, or equal to the density of the gray liquid?

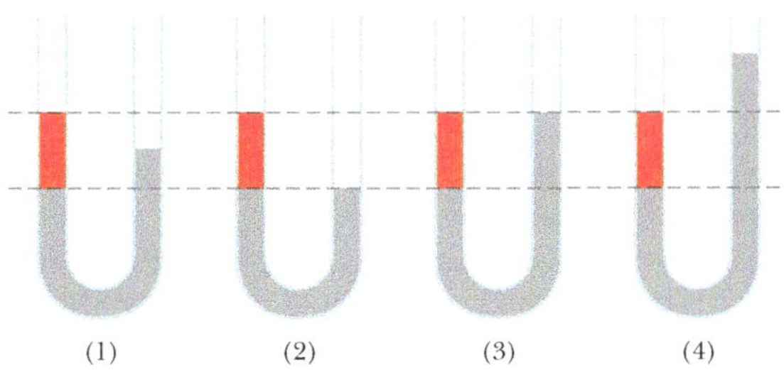

Figure 14-21 Question 2.

3 A boat with an anchor on board floats in a swimming pool that is somewhat wider than the boat. Does the pool water level move up, move down, or remain the same if the anchor is (a) dropped into the water or (b) thrown onto the surrounding ground? (c) Does the water level in the pool move upward, move downward, or remain the same if, instead, a cork is dropped from the boat into the water, where it floats?

4 Figure 14-22 shows a tank filled with water. Five horizontal floors and ceilings are indicated; all have the same area and are located at distances L, $2L$, or $3L$ below the top of the tank. Rank them according to the force on them due to the water, greatest first.

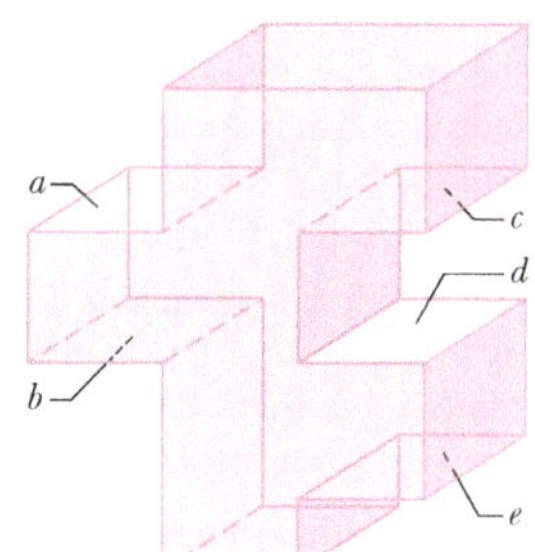

Figure 14-22 Question 4.

5 *The teapot effect.* Water poured slowly from a teapot spout can double back under the spout for a considerable distance (held there by atmospheric pressure) before detaching and falling. In Fig. 14-23, the four points are at the top or bottom of the water layers, inside or outside. Rank those four points according to the gauge pressure in the water there, most positive first.

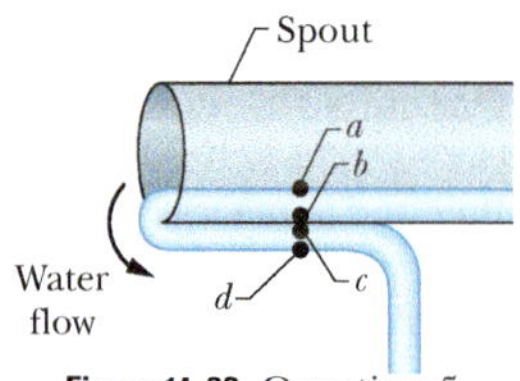

Figure 14-23 Question 5.

6 Figure 14-24 shows three identical open-top containers filled to the brim with water; toy ducks float in two of them. Rank the containers and contents according to their weight, greatest first.

Figure 14-24 Question 6.

7 Figure 14-25 shows four arrangements of pipes through which water flows smoothly toward the right. The radii of the pipe sections are indicated. In which arrangements is the net work done on a unit volume of water moving from the leftmost section to the rightmost section (a) zero, (b) positive, and (c) negative?

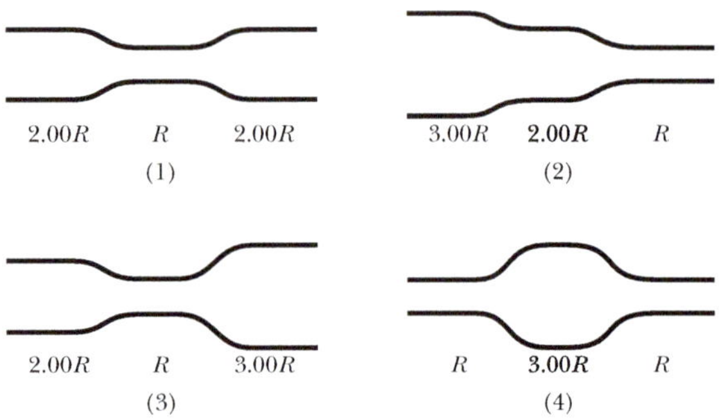

Figure 14-25 Question 7.

8 A rectangular block is pushed face-down into three liquids, in turn. The apparent weight W_{app} of the block versus depth h in the three liquids is plotted in Fig. 14-26. Rank the liquids according to their weight per unit volume, greatest first.

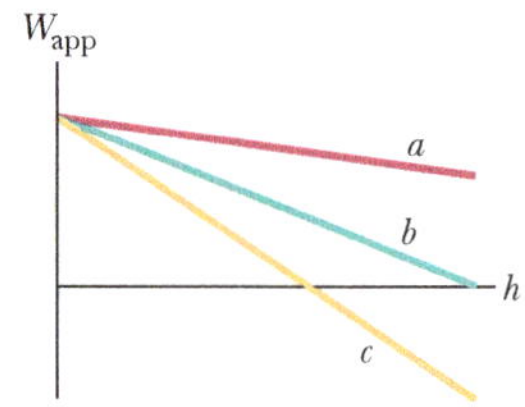

Figure 14-26 Question 8.

9 Water flows smoothly in a horizontal pipe. Figure 14-27 shows the kinetic energy K of a water element as it moves along an x axis that runs along the pipe. Rank the three lettered sections of the pipe according to the pipe radius, greatest first.

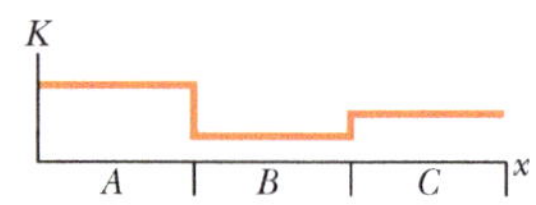

Figure 14-27 Question 9.

10 We have three containers with different liquids. The gauge pressure p_g versus depth h is plotted in Fig. 14-28 for the liquids. In each container, we will fully submerge a rigid plastic bead. Rank the plots according to the magnitude of the buoyant force on the bead, greatest first.

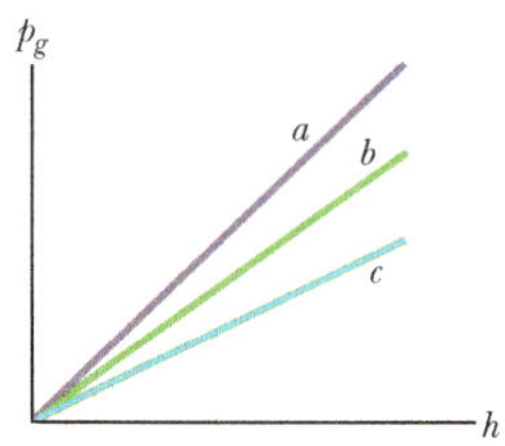

Figure 14-28 Question 10.

Problems

GO Tutoring problem available (at instructor's discretion) in *WileyPLUS* and WebAssign

SSM Worked-out solution available in Student Solutions Manual

WWW Worked-out solution is at

ILW Interactive solution is at

http://www.wiley.com/college/halliday

• – ••• Number of dots indicates level of problem difficulty

Additional information available in *The Flying Circus of Physics* and at flyingcircusofphysics.com

Module 14-1 Fluids, Density, and Pressure

•1 ILW A fish maintains its depth in fresh water by adjusting the air content of porous bone or air sacs to make its average density the same as that of the water. Suppose that with its air sacs collapsed, a fish has a density of 1.08 g/cm^3. To what fraction of its expanded body volume must the fish inflate the air sacs to reduce its density to that of water?

•2 A partially evacuated airtight container has a tight-fitting lid of surface area 77 m^2 and negligible mass. If the force required to remove the lid is 480 N and the atmospheric pressure is 1.0×10^5 Pa, what is the internal air pressure?

•3 SSM Find the pressure increase in the fluid in a syringe when a nurse applies a force of 42 N to the syringe's circular piston, which has a radius of 1.1 cm.

•4 Three liquids that will not mix are poured into a cylindrical container. The volumes and densities of the liquids are 0.50 L, 2.6 g/cm^3; 0.25 L, 1.0 g/cm^3; and 0.40 L, 0.80 g/cm^3. What is the force on the bottom of the container due to these liquids? One liter = 1 L = 1000 cm^3. (Ignore the contribution due to the atmosphere.)

•5 SSM An office window has dimensions 3.4 m by 2.1 m. As a result of the passage of a storm, the outside air pressure drops to 0.96 atm, but inside the pressure is held at 1.0 atm. What net force pushes out on the window?

•6 You inflate the front tires on your car to 28 psi. Later, you measure your blood pressure, obtaining a reading of 120/80, the readings being in mm Hg. In metric countries (which is to say, most of the world), these pressures are customarily reported in kilopascals (kPa). In kilopascals, what are (a) your tire pressure and (b) your blood pressure?

••7 In 1654 Otto von Guericke, inventor of the air pump, gave a demonstration before the noblemen of the Holy Roman Empire in which two teams of eight horses could not pull apart two evacuated brass hemispheres. (a) Assuming the hemispheres have (strong) thin walls, so that R in Fig. 14-29 may be considered both the inside and outside radius, show that the force $\vec{F}$ required to pull apart the hemispheres has magnitude $F = \pi R^2 \Delta p$, where Δp is the difference between the pressures outside and inside the sphere. (b) Taking R as 30 cm, the inside pressure as 0.10 atm, and the outside pressure as 1.00 atm, find the force magnitude the teams of horses would have had to exert to pull apart the hemispheres. (c) Explain why one team of horses could have proved the point just as well if the hemispheres were attached to a sturdy wall.

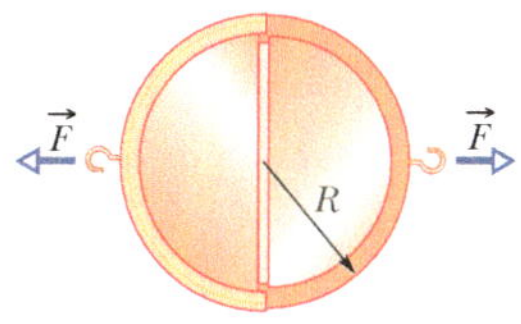

Figure 14-29 Problem 7.

Module 14-2 Fluids at Rest

•8 *The bends during flight.* Anyone who scuba dives is advised not to fly within the next 24 h because the air mixture for diving can introduce nitrogen to the bloodstream. Without allowing the nitrogen to come out of solution slowly, any sudden air-pressure reduction (such as during airplane ascent) can result in the nitrogen forming bubbles in the blood, creating the *bends*, which can be painful and even fatal. Military special operation forces are especially at risk. What is the change in pressure on such a special-op soldier who must scuba dive at a depth of 20 m in seawater one day and parachute at an altitude of 7.6 km the next day? Assume that the average air density within the altitude range is 0.87 kg/m^3.

•9 *Blood pressure in Argentinosaurus.* (a) If this long-necked, gigantic sauropod had a head height of 21 m and a heart height of 9.0 m, what (hydrostatic) gauge pressure in its blood was required at the heart such that the blood pressure at the brain was 80 torr (just enough to perfuse the brain with blood)? Assume the blood had a density of 1.06×10^3 kg/m^3. (b) What was the blood pressure (in torr or mm Hg) at the feet?

•10 The plastic tube in Fig. 14-30 has a cross-sectional area of 5.00 cm^2. The tube is filled with water until the short arm (of length d = 0.800 m) is full. Then the short arm is sealed and more water is gradually poured into the long arm. If the seal will pop off when the force on it exceeds 9.80 N, what total height of water in the long arm will put the seal on the verge of popping?

Figure 14-30 Problems 10 and 81.

•11 *Giraffe bending to drink.* In a giraffe with its head 2.0 m above its heart, and its heart 2.0 m above its feet, the (hydrostatic) gauge pressure in the blood at its heart is 250 torr. Assume that the giraffe stands upright and the blood density is 1.06×10^3 kg/m^3. In torr (or mm Hg), find the (gauge) blood pressure (a) at the brain (the pressure is enough to perfuse the brain with blood, to keep the giraffe from fainting) and (b) at the feet (the pressure must be countered by tight-fitting skin acting like a pressure stocking). (c) If the giraffe were to lower its head to drink from a pond without splaying its legs and moving slowly, what would be the increase in the blood pressure in the brain? (Such action would probably be lethal.)

•12 The maximum depth d_{max} that a diver can snorkel is set by the density of the water and the fact that human lungs can function against a maximum pressure difference (between inside and outside the chest cavity) of 0.050 atm. What is the difference in d_{max} for fresh water and the water of the Dead Sea (the saltiest natural water in the world, with a density of 1.5×10^3 kg/m^3)?

•13 At a depth of 10.9 km, the Challenger Deep in the Marianas Trench of the Pacific Ocean is the deepest site in any ocean. Yet, in 1960, Donald Walsh and Jacques Piccard reached the Challenger Deep in the bathyscaph *Trieste*. Assuming that seawater has a uniform density of 1024 kg/m^3, approximate the hydrostatic pressure (in atmospheres) that the *Trieste* had to withstand. (Even a slight defect in the *Trieste* structure would have been disastrous.)

•14 Calculate the hydrostatic difference in blood pressure between the brain and the foot in a person of height 1.83 m. The density of blood is 1.06×10^3 kg/m^3.

•15 What gauge pressure must a machine produce in order to suck mud of density 1800 kg/m^3 up a tube by a height of 1.5 m?

•16 *Snorkeling by humans and elephants.* When a person snorkels, the lungs are connected directly to the atmosphere through the snorkel tube and thus are at atmospheric pressure. In atmospheres, what is the difference Δp between this internal air pressure and the water pressure against the body if the length of the snorkel tube is (a) 20 cm (standard situation) and (b) 4.0 m (probably lethal situation)? In the latter, the pressure difference causes blood vessels on the walls of the lungs to rupture, releasing blood into the lungs. As depicted in Fig. 14-31, an elephant can safely snorkel through its trunk while swimming with its lungs 4.0 m below the water surface because the membrane around its lungs contains connective tissue that holds and protects the blood vessels, preventing rupturing.

Figure 14-31 Problem 16.

•17 SSM Crew members attempt to escape from a damaged submarine 100 m below the surface. What force must be applied to a pop-out hatch, which is 1.2 m by 0.60 m, to push it out at that depth? Assume that the density of the ocean water is 1024 kg/m^3 and the internal air pressure is at 1.00 atm.

•18 In Fig. 14-32, an open tube of length L = 1.8 m and cross-sectional area A = 4.6 cm^2 is fixed to the top of a cylindrical barrel of diameter D = 1.2 m and height H = 1.8 m. The barrel and tube are filled with water (to the top of the tube). Calculate the ratio of the hydrostatic force on the bottom of the barrel to the gravitational force on the water contained in the barrel. Why is that ratio not equal to 1.0? (You need not consider the atmospheric pressure.)

••19 A large aquarium of height 5.00 m is filled with fresh water to a depth of 2.00 m. One wall of the aquarium consists of thick plastic 8.00 m wide. By how much does the total force on that wall increase if the aquarium is next filled to a depth of 4.00 m?

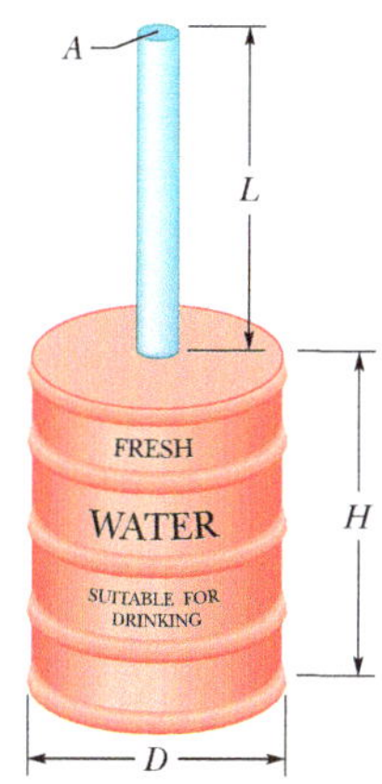

Figure 14-32 Problem 18.

••20 The **L**-shaped fish tank shown in Fig. 14-33 is filled with water and is open at the top. If $d = 5.0$ m, what is the (total) force exerted by the water (a) on face A and (b) on face B?

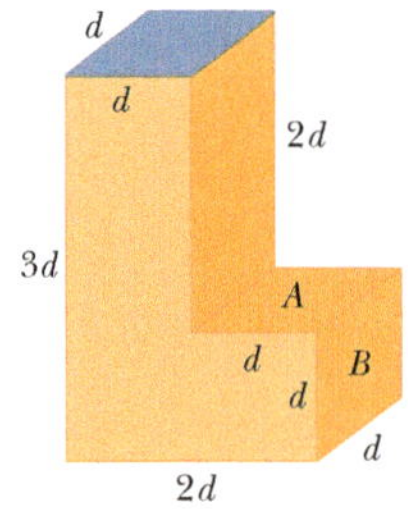

Figure 14-33 Problem 20.

••21 SSM Two identical cylindrical vessels with their bases at the same level each contain a liquid of density 1.30×10^3 kg/m^3. The area of each base is 4.00 cm^2, but in one vessel the liquid height is 0.854 m and in the other it is 1.560 m. Find the work done by the gravitational force in equalizing the levels when the two vessels are connected.

••22 *g-LOC in dogfights.* When a pilot takes a tight turn at high speed in a modern fighter airplane, the blood pressure at the brain level decreases, blood no longer perfuses the brain, and the blood in the brain drains. If the heart maintains the (hydrostatic) gauge pressure in the aorta at 120 torr (or mm Hg) when the pilot undergoes a horizontal centripetal acceleration of $4g$, what is the blood pressure (in torr) at the brain, 30 cm radially inward from the heart? The perfusion in the brain is small enough that the vision switches to black and white and narrows to "tunnel vision" and the pilot can undergo g-LOC ("g-induced loss of consciousness"). Blood density is 1.06×10^3 kg/m^3.

••23 GO In analyzing certain geological features, it is often appropriate to assume that the pressure at some horizontal *level of compensation,* deep inside Earth, is the same over a large region and is equal to the pressure due to the gravitational force on the overlying material. Thus, the pressure on the level of compensation is given by the fluid pressure formula. This model requires, for one thing, that mountains have *roots* of continental rock extending into the denser mantle (Fig. 14-34). Consider a mountain of height $H = 6.0$ km on a continent of thickness $T = 32$ km. The continental rock has a density of 2.9 g/cm^3, and beneath this rock the mantle has a density of 3.3 g/cm^3. Calculate the depth D of the root. (*Hint:* Set the pressure at points a and b equal; the depth y of the level of compensation will cancel out.)

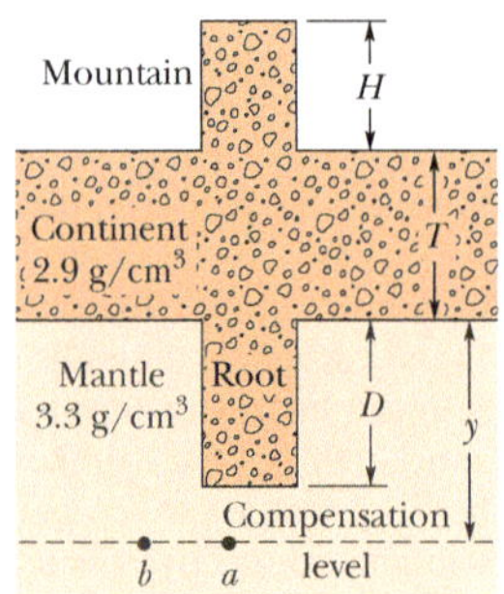

Figure 14-34 Problem 23.

•••24 GO In Fig. 14-35, water stands at depth $D = 35.0$ m behind the vertical upstream face of a dam of width $W = 314$ m. Find (a) the net horizontal force on the dam from the gauge pressure of the water and (b) the net torque due to that force about a horizontal line through O parallel to the (long) width of the dam. This torque tends to rotate the dam around that line, which would cause the dam to fail. (c) Find the moment arm of the torque.

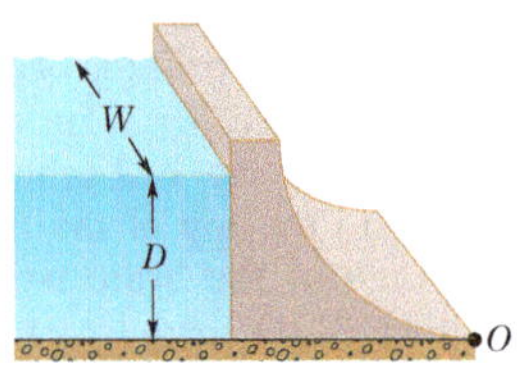

Figure 14-35 Problem 24.

Module 14-3 Measuring Pressure

•25 In one observation, the column in a mercury barometer (as is shown in Fig. 14-5*a*) has a measured height h of 740.35 mm. The temperature is −5.0°C, at which temperature the density of mercury ρ is 1.3608×10^4 kg/m^3. The free-fall acceleration g at the site of the barometer is 9.7835 m/s^2. What is the atmospheric pressure at that site in pascals and in torr (which is the common unit for barometer readings)?

•26 To suck lemonade of density 1000 kg/m^3 up a straw to a maximum height of 4.0 cm, what minimum gauge pressure (in atmospheres) must you produce in your lungs?

••27 SSM What would be the height of the atmosphere if the air density (a) were uniform and (b) decreased linearly to zero with height? Assume that at sea level the air pressure is 1.0 atm and the air density is 1.3 kg/m^3.

Module 14-4 Pascal's Principle

•28 A piston of cross-sectional area a is used in a hydraulic press to exert a small force of magnitude f on the enclosed liquid. A connecting pipe leads to a larger piston of cross-sectional area A (Fig. 14-36). (a) What force magnitude F will the larger piston sustain without moving? (b) If the piston diameters are 3.80 cm and 53.0 cm, what force magnitude on the small piston will balance a 20.0 kN force on the large piston?

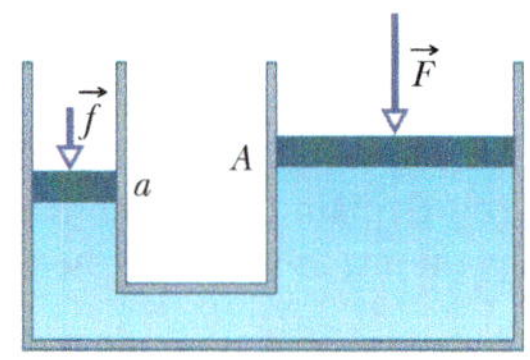

Figure 14-36 Problem 28.

••29 In Fig. 14-37, a spring of spring constant 3.00×10^4 N/m is between a rigid beam and the output piston of a hydraulic lever. An empty container with negligible mass sits on the input piston. The input piston has area A_i, and the output piston has area $18.0A_i$. Initially the spring is at its rest length. How many kilograms of sand must be (slowly) poured into the container to compress the spring by 5.00 cm?

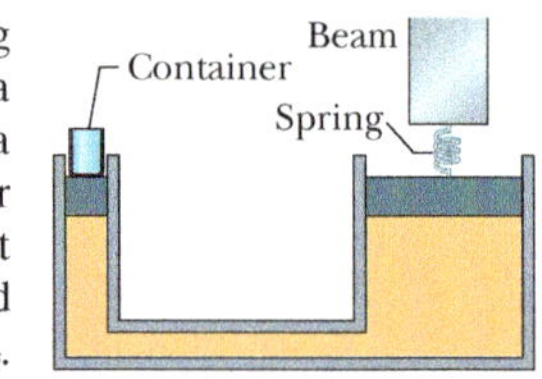

Figure 14-37 Problem 29.

Module 14-5 Archimedes' Principle

•30 A 5.00 kg object is released from rest while fully submerged in a liquid. The liquid displaced by the submerged object has a mass of 3.00 kg. How far and in what direction does the object move in 0.200 s, assuming that it moves freely and that the drag force on it from the liquid is negligible?

•31 SSM A block of wood floats in fresh water with two-thirds of its volume V submerged and in oil with $0.90V$ submerged. Find the density of (a) the wood and (b) the oil.

•32 In Fig. 14-38, a cube of edge length $L = 0.600$ m and mass 450 kg is suspended by a rope in an open tank of liquid of density 1030 kg/m^3. Find (a) the magnitude of the total downward force on the top of the cube from the liquid and the atmosphere, assuming atmospheric pressure is 1.00 atm, (b) the magnitude of the total upward force on the bottom of the cube, and (c) the tension in the rope. (d) Calculate the magnitude of the buoyant force on the cube using Archimedes' principle. What relation exists among all these quantities?

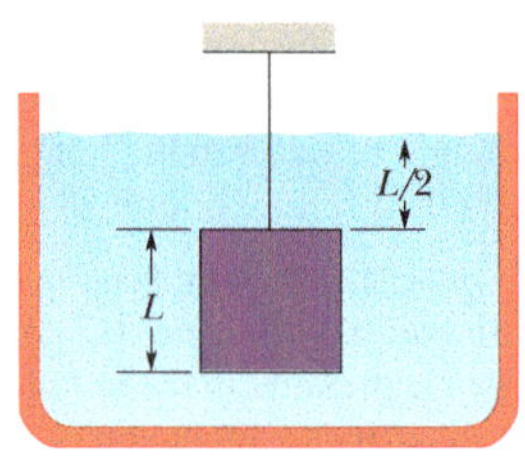

Figure 14-38 Problem 32.

•33 SSM An iron anchor of density 7870 kg/m^3 appears 200 N lighter in water than in air. (a) What is the volume of the anchor? (b) How much does it weigh in air?

•34 A boat floating in fresh water displaces water weighing

35.6 kN. (a) What is the weight of the water this boat displaces when floating in salt water of density 1.10×10^3 kg/m^3? (b) What is the difference between the volume of fresh water displaced and the volume of salt water displaced?

•35 Three children, each of weight 356 N, make a log raft by lashing together logs of diameter 0.30 m and length 1.80 m. How many logs will be needed to keep them afloat in fresh water? Take the density of the logs to be 800 kg/m^3.

••36 In Fig. 14-39*a*, a rectangular block is gradually pushed face-down into a liquid. The block has height d; on the bottom and top the face area is $A = 5.67$ cm^2. Figure 14-39*b* gives the apparent weight W_{app} of the block as a function of the depth h of its lower face. The scale on the vertical axis is set by $W_s = 0.20$ N. What is the density of the liquid?

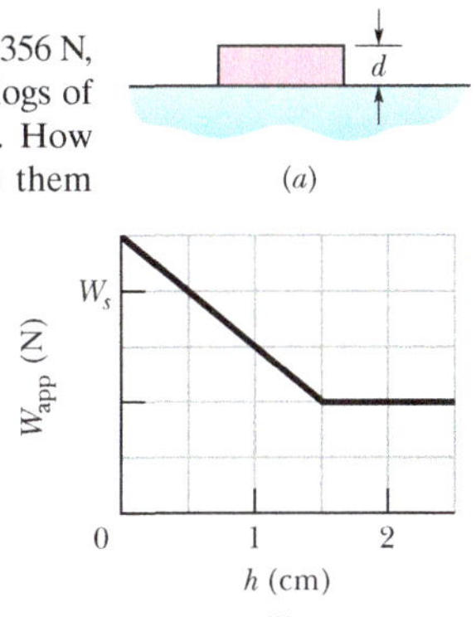

Figure 14-39 Problem 36.

••37 ILW A hollow spherical iron shell floats almost completely submerged in water. The outer diameter is 60.0 cm, and the density of iron is 7.87 g/cm^3. Find the inner diameter.

••38 A small solid ball is released from rest while fully submerged in a liquid and then its kinetic energy is measured when it has moved 4.0 cm in the liquid. Figure 14-40 gives the results after many liquids are used: The kinetic energy K is plotted versus the liquid density ρ_{liq}, and $K_s = 1.60$ J sets the scale on the vertical axis. What are (a) the density and (b) the volume of the ball?

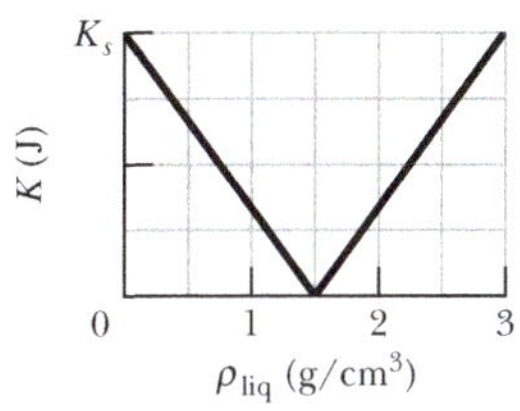

Figure 14-40 Problem 38.

••39 SSM WWW A hollow sphere of inner radius 8.0 cm and outer radius 9.0 cm floats half-submerged in a liquid of density 800 kg/m^3. (a) What is the mass of the sphere? (b) Calculate the density of the material of which the sphere is made.

••40 *Lurking alligators.* An alligator waits for prey by floating with only the top of its head exposed, so that the prey cannot easily see it. One way it can adjust the extent of sinking is by controlling the size of its lungs. Another way may be by swallowing stones (*gastrolithes*) that then reside in the stomach. Figure 14-41 shows a highly simplified model (a "rhombohedron gater") of mass 130 kg that roams with its head partially exposed. The top head surface has area 0.20 m^2. If the alligator were to swallow stones with a total mass of 1.0% of its body mass (a typical amount), how far would it sink?

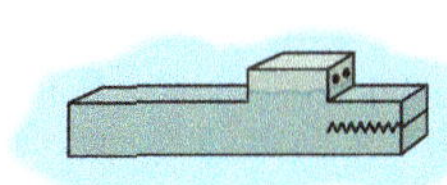

Figure 14-41 Problem 40.

••41 What fraction of the volume of an iceberg (density 917 kg/m^3) would be visible if the iceberg floats (a) in the ocean (salt water, density 1024 kg/m^3) and (b) in a river (fresh water, density 1000 kg/m^3)? (When salt water freezes to form ice, the salt is excluded. So, an iceberg could provide fresh water to a community.)

••42 A flotation device is in the shape of a right cylinder, with a height of 0.500 m and a face area of 4.00 m^2 on top and bottom, and its density is 0.400 times that of fresh water. It is initially held fully submerged in fresh water, with its top face at the water surface. Then it is allowed to ascend gradually until it begins to float. How much work does the buoyant force do on the device during the ascent?

••43 When researchers find a reasonably complete fossil of a dinosaur, they can determine the mass and weight of the living dinosaur with a scale model sculpted from plastic and based on the dimensions of the fossil bones. The scale of the model is 1/20; that is, lengths are 1/20 actual length, areas are $(1/20)^2$ actual areas, and volumes are $(1/20)^3$ actual volumes. First, the model is suspended from one arm of a balance and weights are added to the other arm until equilibrium is reached. Then the model is fully submerged in water and enough weights are removed from the second arm to reestablish equilibrium (Fig. 14-42). For a model of a particular *T. rex* fossil, 637.76 g had to be removed to reestablish equilibrium. What was the volume of (a) the model and (b) the actual *T. rex*? (c) If the density of *T. rex* was approximately the density of water, what was its mass?

Figure 14-42 Problem 43.

••44 A wood block (mass 3.67 kg, density 600 kg/m^3) is fitted with lead (density 1.14×10^4 kg/m^3) so that it floats in water with 0.900 of its volume submerged. Find the lead mass if the lead is fitted to the block's (a) top and (b) bottom.

••45 An iron casting containing a number of cavities weighs 6000 N in air and 4000 N in water. What is the total cavity volume in the casting? The density of solid iron is 7.87 g/cm^3.

••46 Suppose that you release a small ball from rest at a depth of 0.600 m below the surface in a pool of water. If the density of the ball is 0.300 that of water and if the drag force on the ball from the water is negligible, how high above the water surface will the ball shoot as it emerges from the water? (Neglect any transfer of energy to the splashing and waves produced by the emerging ball.)

••47 The volume of air space in the passenger compartment of an 1800 kg car is 5.00 m^3. The volume of the motor and front wheels is 0.750 m^3, and the volume of the rear wheels, gas tank, and trunk is 0.800 m^3; water cannot enter these two regions. The car rolls into a lake. (a) At first, no water enters the passenger compartment. How much of the car, in cubic meters, is below the water surface with the car floating (Fig. 14-43)? (b) As water slowly enters, the car sinks. How many cubic meters of water are in the car as it disappears below the water surface? (The car, with a heavy load in the trunk, remains horizontal.)

Figure 14-43 Problem 47.

•••48 Figure 14-44 shows an iron ball suspended by thread of negligible mass from an upright cylinder that floats partially submerged in water. The cylinder has a height of 6.00 cm, a face area of 12.0 cm^2 on the top and bottom, and a density of 0.30 g/cm^3, and 2.00 cm of its height is above the water surface. What is the radius of the iron ball?

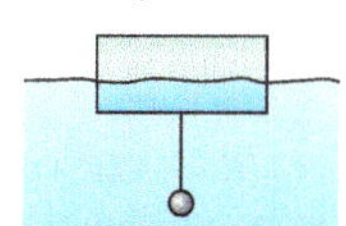

Figure 14-44 Problem 48.

Module 14-6 The Equation of Continuity

•49 *Canal effect.* Figure 14-45 shows an anchored barge that extends across a canal by distance $d = 30$ m and into the water by distance $b = 12$ m. The canal has a width $D = 55$ m, a water depth $H = 14$ m, and a uniform water-flow speed $v_i = 1.5$ m/s. Assume that the flow around the barge is uniform. As the water passes the bow, the water level undergoes a dramatic dip known as the canal effect. If the dip has depth $h = 0.80$ m, what is the water speed alongside the boat through the vertical cross sections at (a) point a and (b) point b? The erosion due to the speed increase is a common concern to hydraulic engineers.

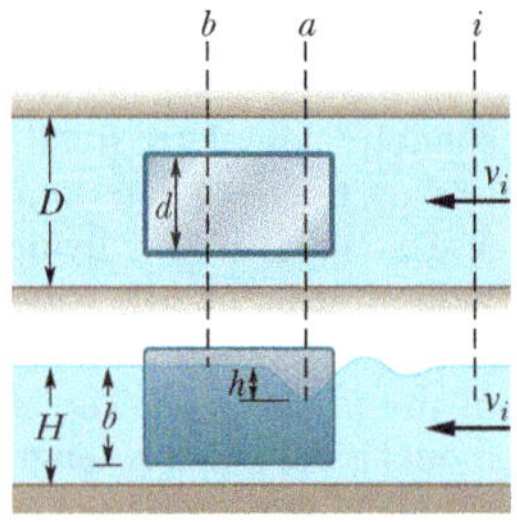

Figure 14-45 Problem 49.

•50 Figure 14-46 shows two sections of an old pipe system that runs through a hill, with distances $d_A = d_B = 30$ m and $D = 110$ m. On each side of the hill, the pipe radius is 2.00 cm. However, the radius of the pipe inside the hill is no longer known. To determine it, hydraulic engineers first establish that water flows through the left and right sections at 2.50 m/s. Then they release a dye in the water at point A and find that it takes 88.8 s to reach point B. What is the average radius of the pipe within the hill?

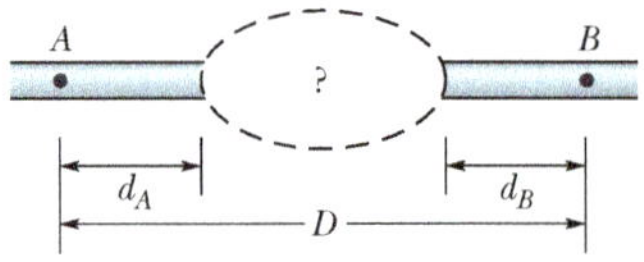

Figure 14-46 Problem 50.

•51 SSM A garden hose with an internal diameter of 1.9 cm is connected to a (stationary) lawn sprinkler that consists merely of a container with 24 holes, each 0.13 cm in diameter. If the water in the hose has a speed of 0.91 m/s, at what speed does it leave the sprinkler holes?

•52 Two streams merge to form a river. One stream has a width of 8.2 m, depth of 3.4 m, and current speed of 2.3 m/s. The other stream is 6.8 m wide and 3.2 m deep, and flows at 2.6 m/s. If the river has width 10.5 m and speed 2.9 m/s, what is its depth?

••53 SSM Water is pumped steadily out of a flooded basement at 5.0 m/s through a hose of radius 1.0 cm, passing through a window 3.0 m above the waterline. What is the pump's power?

••54 GO The water flowing through a 1.9 cm (inside diameter) pipe flows out through three 1.3 cm pipes. (a) If the flow rates in the three smaller pipes are 26, 19, and 11 L/min, what is the flow rate in the 1.9 cm pipe? (b) What is the ratio of the speed in the 1.9 cm pipe to that in the pipe carrying 26 L/min?

Module 14-7 Bernoulli's Equation

•55 How much work is done by pressure in forcing 1.4 m^3 of water through a pipe having an internal diameter of 13 mm if the difference in pressure at the two ends of the pipe is 1.0 atm?

•56 Suppose that two tanks, 1 and 2, each with a large opening at the top, contain different liquids. A small hole is made in the side of each tank at the same depth h below the liquid surface, but the hole in tank 1 has half the cross-sectional area of the hole in tank 2. (a) What is the ratio ρ_1/ρ_2 of the densities of the liquids if the mass flow rate is the same for the two holes? (b) What is the ratio R_{V1}/R_{V2} of the volume flow rates from the two tanks? (c) At one instant, the liquid in tank 1 is 12.0 cm above the hole. If the tanks are to have *equal* volume flow rates, what height above the hole must the liquid in tank 2 be just then?

•57 SSM A cylindrical tank with a large diameter is filled with water to a depth $D = 0.30$ m. A hole of cross-sectional area $A = 6.5$ cm^2 in the bottom of the tank allows water to drain out. (a) What is the drainage rate in cubic meters per second? (b) At what distance below the bottom of the tank is the cross-sectional area of the stream equal to one-half the area of the hole?

•58 The intake in Fig. 14-47 has cross-sectional area of 0.74 m^2 and water flow at 0.40 m/s. At the outlet, distance $D = 180$ m below the intake, the cross-sectional area is smaller than at the intake and the water flows out at 9.5 m/s into equipment. What is the pressure difference between inlet and outlet?

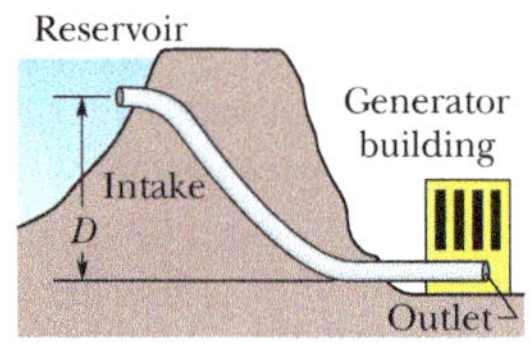

Figure 14-47 Problem 58.

•59 SSM Water is moving with a speed of 5.0 m/s through a pipe with a cross-sectional area of 4.0 cm^2. The water gradually descends 10 m as the pipe cross-sectional area increases to 8.0 cm^2. (a) What is the speed at the lower level? (b) If the pressure at the upper level is 1.5×10^5 Pa, what is the pressure at the lower level?

•60 Models of torpedoes are sometimes tested in a horizontal pipe of flowing water, much as a wind tunnel is used to test model airplanes. Consider a circular pipe of internal diameter 25.0 cm and a torpedo model aligned along the long axis of the pipe. The model has a 5.00 cm diameter and is to be tested with water flowing past it at 2.50 m/s. (a) With what speed must the water flow in the part of the pipe that is unconstricted by the model? (b) What will the pressure difference be between the constricted and unconstricted parts of the pipe?

•61 ILW A water pipe having a 2.5 cm inside diameter carries water into the basement of a house at a speed of 0.90 m/s and a pressure of 170 kPa. If the pipe tapers to 1.2 cm and rises to the second floor 7.6 m above the input point, what are the (a) speed and (b) water pressure at the second floor?

••62 A pitot tube (Fig. 14-48) is used to determine the airspeed of an airplane. It consists of an outer tube with a number of small holes B (four are shown) that allow air into the tube; that tube is connected to one arm of a **U**-tube. The other arm of the **U**-tube is connected to hole A at the front end of the device, which points in the direction the plane is headed. At A the air becomes stagnant so that $v_A = 0$. At B, however, the speed of the air presumably equals the airspeed v of the plane. (a) Use Bernoulli's equation to show that

$$v = \sqrt{\frac{2\rho g h}{\rho_{air}}},$$

where ρ is the density of the liquid in the **U**-tube and h is the difference in the liquid levels in that tube. (b) Suppose that the tube contains alcohol and the level difference h is 26.0 cm. What is the plane's speed relative to the air? The density of the air is 1.03 kg/m^3 and that of alcohol is 810 kg/m^3.

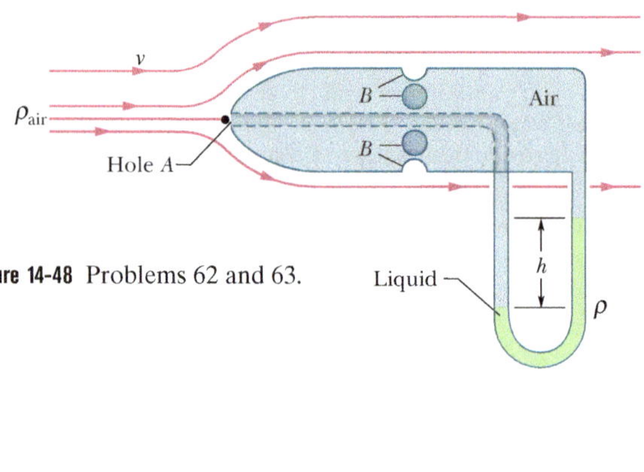

Figure 14-48 Problems 62 and 63.

••63 A pitot tube (see Problem 62) on a high-altitude aircraft measures a differential pressure of 180 Pa. What is the aircraft's airspeed if the density of the air is 0.031 kg/m^3?

••64 GO In Fig. 14-49, water flows through a horizontal pipe and then out into the atmosphere at a speed $v_1 = 15$ m/s. The diameters of the left and right sections of the pipe are 5.0 cm and 3.0 cm. (a) What volume of water flows into the atmosphere during a 10 min period? In the left section of the pipe, what are (b) the speed v_2 and (c) the gauge pressure?

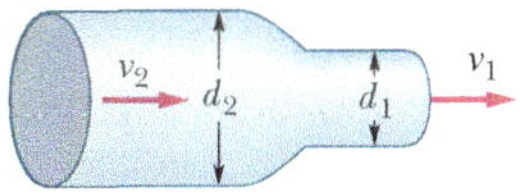

Figure 14-49 Problem 64.

••65 SSM WWW A *venturi meter* is used to measure the flow speed of a fluid in a pipe. The meter is connected between two sections of the pipe (Fig. 14-50); the cross-sectional area A of the entrance and exit of the meter matches the pipe's cross-sectional area. Between the entrance and exit, the fluid flows from the pipe with speed V and then through a narrow "throat" of cross-sectional area a with speed v. A manometer connects the wider portion of the meter to the narrower portion. The change in the fluid's speed is accompanied by a change Δp in the fluid's pressure, which causes a height difference h of the liquid in the two arms of the manometer. (Here Δp means pressure in the throat minus pressure in the pipe.) (a) By applying Bernoulli's equation and the equation of continuity to points 1 and 2 in Fig. 14-50, show that

$$V = \sqrt{\frac{2a^2\,\Delta p}{\rho(a^2 - A^2)}},$$

where ρ is the density of the fluid. (b) Suppose that the fluid is fresh water, that the cross-sectional areas are 64 cm^2 in the pipe and 32 cm^2 in the throat, and that the pressure is 55 kPa in the pipe and 41 kPa in the throat. What is the rate of water flow in cubic meters per second?

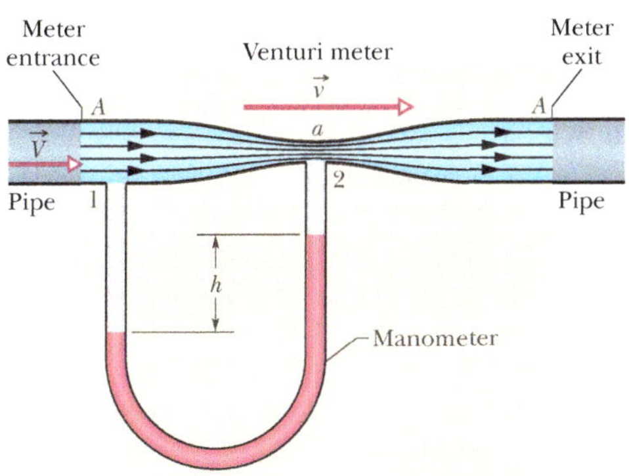

Figure 14-50 Problems 65 and 66.

••66 Consider the venturi tube of Problem 65 and Fig. 14-50 without the manometer. Let A equal $5a$. Suppose the pressure p_1 at A is 2.0 atm. Compute the values of (a) the speed V at A and (b) the speed v at a that make the pressure p_2 at a equal to zero. (c) Compute the corresponding volume flow rate if the diameter at A is 5.0 cm. The phenomenon that occurs at a when p_2 falls to nearly zero is known as cavitation. The water vaporizes into small bubbles.

••67 ILW In Fig. 14-51, the fresh water behind a reservoir dam has depth $D = 15$ m. A horizontal pipe 4.0 cm in diameter passes through the dam at depth $d = 6.0$ m. A plug secures the pipe opening. (a) Find the magnitude of the frictional force between plug and pipe wall. (b) The plug is removed. What water volume exits the pipe in 3.0 h?

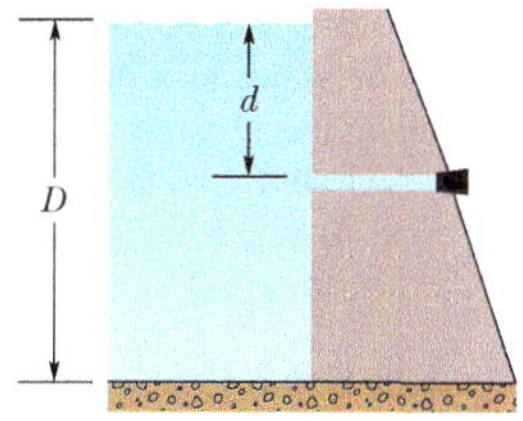

Figure 14-51 Problem 67.

••68 GO Fresh water flows horizontally from pipe section 1 of cross-sectional area A_1 into pipe section 2 of cross-sectional area A_2. Figure 14-52 gives a plot of the pressure difference $p_2 - p_1$ versus the inverse area squared A_1^{-2} that would be expected for a volume flow rate of a certain value if the water flow were laminar under all circumstances. The scale on the vertical axis is set by $\Delta p_s = 300$ kN/m^2. For the conditions of the figure, what are the values of (a) A_2 and (b) the volume flow rate?

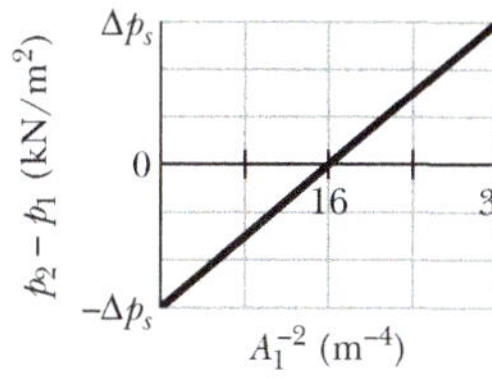

Figure 14-52 Problem 68.

••69 A liquid of density 900 kg/m^3 flows through a horizontal pipe that has a cross-sectional area of 1.90×10^{-2} m^2 in region A and a cross-sectional area of 9.50×10^{-2} m^2 in region B. The pressure difference between the two regions is 7.20×10^3 Pa. What are (a) the volume flow rate and (b) the mass flow rate?

••70 GO In Fig. 14-53, water flows steadily from the left pipe section (radius $r_1 = 2.00R$), through the middle section (radius R), and into the right section (radius $r_3 = 3.00R$). The speed of the water in the middle section is 0.500 m/s. What is the net work done on 0.400 m^3 of the water as it moves from the left section to the right section?

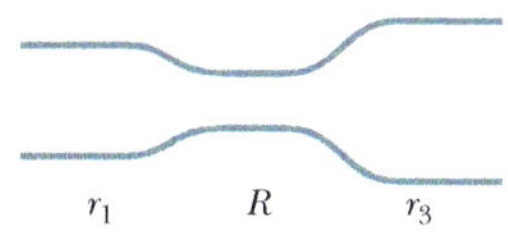

Figure 14-53 Problem 70.

••71 Figure 14-54 shows a stream of water flowing through a hole at depth $h = 10$ cm in a tank holding water to height $H = 40$ cm. (a) At what distance x does the stream strike the floor? (b) At what depth should a second hole be made to give the same value of x? (c) At what depth should a hole be made to maximize x?

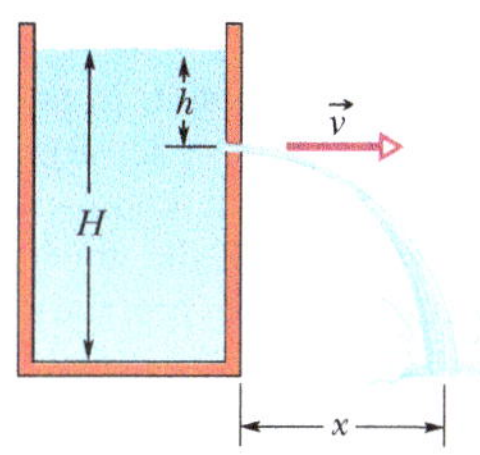

Figure 14-54 Problem 71.

•••72 GO A very simplified schematic of the rain drainage system for a home is shown in Fig. 14-55. Rain falling on the slanted roof runs off into gutters around the roof edge; it then drains through downspouts (only one is shown) into a main drainage pipe M below the basement, which carries the water to an even larger pipe below the street. In Fig. 14-55, a floor drain in the basement is also connected to drainage pipe M. Suppose the following apply:

(1) the downspouts have height $h_1 = 11$ m, (2) the floor drain has height $h_2 = 1.2$ m, (3) pipe M has radius 3.0 cm, (4) the house has side width $w = 30$ m and front length $L = 60$ m, (5) all

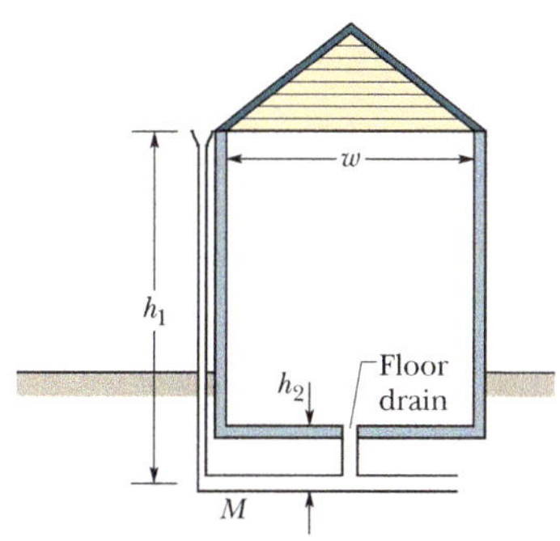

Figure 14-55 Problem 72.

the water striking the roof goes through pipe M, (6) the initial speed of the water in a downspout is negligible, and (7) the wind speed is negligible (the rain falls vertically).

At what rainfall rate, in centimeters per hour, will water from pipe M reach the height of the floor drain and threaten to flood the basement?

Additional Problems

73 About one-third of the body of a person floating in the Dead Sea will be above the waterline. Assuming that the human body density is 0.98 g/cm^3, find the density of the water in the Dead Sea. (Why is it so much greater than 1.0 g/cm^3?)

74 A simple open **U**-tube contains mercury. When 11.2 cm of water is poured into the right arm of the tube, how high above its initial level does the mercury rise in the left arm?

75 If a bubble in sparkling water accelerates upward at the rate of 0.225 m/s^2 and has a radius of 0.500 mm, what is its mass? Assume that the drag force on the bubble is negligible.

76 Suppose that your body has a uniform density of 0.95 times that of water. (a) If you float in a swimming pool, what fraction of your body's volume is above the water surface?

Quicksand is a fluid produced when water is forced up into sand, moving the sand grains away from one another so they are no longer locked together by friction. Pools of quicksand can form when water drains underground from hills into valleys where there are sand pockets. (b) If you float in a deep pool of quicksand that has a density 1.6 times that of water, what fraction of your body's volume is above the quicksand surface? (c) Are you unable to breathe?

77 A glass ball of radius 2.00 cm sits at the bottom of a container of milk that has a density of 1.03 g/cm^3. The normal force on the ball from the container's lower surface has magnitude 9.48×10^{-2} N. What is the mass of the ball?

78 Caught in an avalanche, a skier is fully submerged in flowing snow of density 96 kg/m^3. Assume that the average density of the skier, clothing, and skiing equipment is 1020 kg/m^3. What percentage of the gravitational force on the skier is offset by the buoyant force from the snow?

79 An object hangs from a spring balance. The balance registers 30 N in air, 20 N when this object is immersed in water, and 24 N when the object is immersed in another liquid of unknown density. What is the density of that other liquid?

80 In an experiment, a rectangular block with height h is allowed to float in four separate liquids. In the first liquid, which is water, it floats fully submerged. In liquids A, B, and C, it floats with heights $h/2$, $2h/3$, and $h/4$ above the liquid surface, respectively. What are the *relative densities* (the densities relative to that of water) of (a) A, (b) B, and (c) C?

81 SSM Figure 14-30 shows a modified **U**-tube: the right arm is shorter than the left arm. The open end of the right arm is height d = 10.0 cm above the laboratory bench. The radius throughout the tube is 1.50 cm. Water is gradually poured into the open end of the left arm until the water begins to flow out the open end of the right arm. Then a liquid of density 0.80 g/cm^3 is gradually added to the left arm until its height in that arm is 8.0 cm (it does not mix with the water). How much water flows out of the right arm?

82 What is the acceleration of a rising hot-air balloon if the ratio of the air density outside the balloon to that inside is 1.39? Neglect the mass of the balloon fabric and the basket.

83 Figure 14-56 shows a *siphon,* which is a device for removing liquid from a container. Tube ABC must initially be filled, but once this has been done, liquid will flow through the tube until the liquid surface in the container is level with the tube opening at A. The liquid has density 1000 kg/m^3 and negligible viscosity. The distances shown are h_1 = 25 cm, d = 12 cm, and h_2 = 40 cm. (a) With what speed does the liquid emerge from the tube at C? (b) If the atmospheric pressure is 1.0×10^5 Pa, what is the pressure in the liquid at the topmost point B? (c) Theoretically, what is the greatest possible height h_1 that a siphon can lift water?

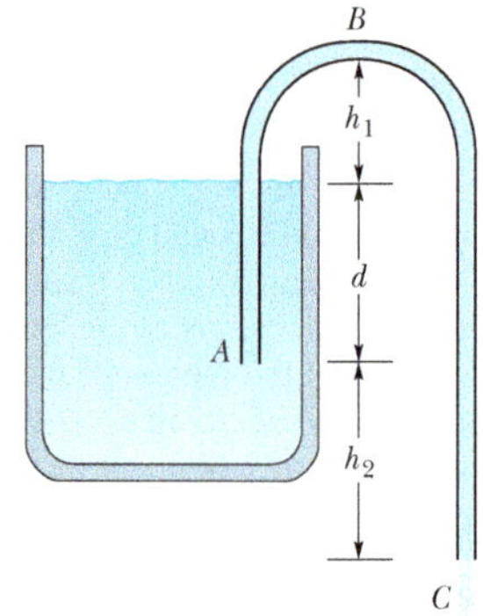

Figure 14-56 Problem 83.

84 When you cough, you expel air at high speed through the trachea and upper bronchi so that the air will remove excess mucus lining the pathway. You produce the high speed by this procedure: You breathe in a large amount of air, trap it by closing the glottis (the narrow opening in the larynx), increase the air pressure by contracting the lungs, partially collapse the trachea and upper bronchi to narrow the pathway, and then expel the air through the pathway by suddenly reopening the glottis. Assume that during the expulsion the volume flow rate is 7.0×10^{-3} m^3/s. What multiple of 343 m/s (the speed of sound v_s) is the airspeed through the trachea if the trachea diameter (a) remains its normal value of 14 mm and (b) contracts to 5.2 mm?

85 A tin can has a total volume of 1200 cm^3 and a mass of 130 g. How many grams of lead shot of density 11.4 g/cm^3 could it carry without sinking in water?

86 The tension in a string holding a solid block below the surface of a liquid (of density greater than the block) is T_0 when the container (Fig. 14-57) is at rest. When the container is given an upward acceleration of 0.250g, what multiple of T_0 gives the tension in the string?

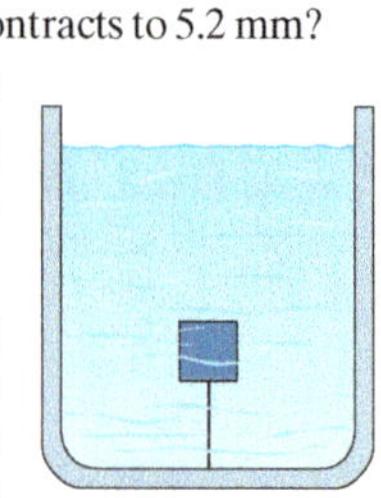

Figure 14-57 Problem 86.

87 What is the minimum area (in square meters) of the top surface of an ice slab 0.441 m thick floating on fresh water that will hold up a 938 kg automobile? Take the densities of ice and fresh water to be 917 kg/m^3 and 998 kg/m^3, respectively.

88 A 8.60 kg sphere of radius 6.22 cm is at a depth of 2.22 km in seawater that has an average density of 1025 kg/m^3. What are the (a) gauge pressure, (b) total pressure, and (c) corresponding total force compressing the sphere's surface? What are (d) the magnitude of the buoyant force on the sphere and (e) the magnitude of the sphere's acceleration if it is free to move? Take atmospheric pressure to be 1.01×10^5 Pa.

89 (a) For seawater of density 1.03 g/cm^3, find the weight of water on top of a submarine at a depth of 255 m if the horizontal cross-sectional hull area is 2200.0 m^2. (b) In atmospheres, what water pressure would a diver experience at this depth?

90 The sewage outlet of a house constructed on a slope is 6.59 m below street level. If the sewer is 2.16 m below street level, find the minimum pressure difference that must be created by the sewage pump to transfer waste of average density 1000.00 kg/m^3 from outlet to sewer.

C H A P T E R 1 5

Oscillations

15-1 SIMPLE HARMONIC MOTION

Learning Objectives

After reading this module, you should be able to . . .

15.01 Distinguish simple harmonic motion from other types of periodic motion.

15.02 For a simple harmonic oscillator, apply the relationship between position x and time t to calculate either if given a value for the other.

15.03 Relate period T, frequency f, and angular frequency ω.

15.04 Identify (displacement) amplitude x_m, phase constant (or phase angle) ϕ, and phase $\omega t + \phi$.

15.05 Sketch a graph of the oscillator's position x versus time t, identifying amplitude x_m and period T.

15.06 From a graph of position versus time, velocity versus time, or acceleration versus time, determine the amplitude of the plot and the value of the phase constant ϕ.

15.07 On a graph of position x versus time t describe the effects of changing period T, frequency f, amplitude x_m, or phase constant ϕ.

15.08 Identify the phase constant ϕ that corresponds to the starting time ($t = 0$) being set when a particle in SHM is at an extreme point or passing through the center point.

15.09 Given an oscillator's position $x(t)$ as a function of time, find its velocity $v(t)$ as a function of time, identify the velocity amplitude v_m in the result, and calculate the velocity at any given time.

15.10 Sketch a graph of an oscillator's velocity v versus time t, identifying the velocity amplitude v_m.

15.11 Apply the relationship between velocity amplitude v_m, angular frequency ω, and (displacement) amplitude x_m.

15.12 Given an oscillator's velocity $v(t)$ as a function of time, calculate its acceleration $a(t)$ as a function of time, identify the acceleration amplitude a_m in the result, and calculate the acceleration at any given time.

15.13 Sketch a graph of an oscillator's acceleration a versus time t, identifying the acceleration amplitude a_m.

15.14 Identify that for a simple harmonic oscillator the acceleration a at any instant is *always* given by the product of a negative constant and the displacement x just then.

15.15 For any given instant in an oscillation, apply the relationship between acceleration a, angular frequency ω, and displacement x.

15.16 Given data about the position x and velocity v at one instant, determine the phase $\omega t + \phi$ and phase constant ϕ.

15.17 For a spring–block oscillator, apply the relationships between spring constant k and mass m and either period T or angular frequency ω.

15.18 Apply Hooke's law to relate the force F on a simple harmonic oscillator at any instant to the displacement x of the oscillator at that instant.

Key Ideas

- The frequency f of periodic, or oscillatory, motion is the number of oscillations per second. In the SI system, it is measured in hertz: $1\ \text{Hz} = 1\ \text{s}^{-1}$.
- The period T is the time required for one complete oscillation, or cycle. It is related to the frequency by $T = 1/f$.
- In simple harmonic motion (SHM), the displacement $x(t)$ of a particle from its equilibrium position is described by the equation

$$x = x_m \cos(\omega t + \phi) \quad \text{(displacement)},$$

in which x_m is the amplitude of the displacement, $\omega t + \phi$ is the phase of the motion, and ϕ is the phase constant. The angular frequency ω is related to the period and frequency of the motion by $\omega = 2\pi/T = 2\pi f$.
- Differentiating $x(t)$ leads to equations for the particle's SHM velocity and acceleration as functions of time:

$$v = -\omega x_m \sin(\omega t + \phi) \quad \text{(velocity)}$$

and

$$a = -\omega^2 x_m \cos(\omega t + \phi) \quad \text{(acceleration)}.$$

In the velocity function, the positive quantity ωx_m is the velocity amplitude v_m. In the acceleration function, the positive quantity $\omega^2 x_m$ is the acceleration amplitude a_m.
- A particle with mass m that moves under the influence of a Hooke's law restoring force given by $F = -kx$ is a linear simple harmonic oscillator with

$$\omega = \sqrt{\frac{k}{m}} \quad \text{(angular frequency)}$$

and

$$T = 2\pi\sqrt{\frac{m}{k}} \quad \text{(period)}.$$

What Is Physics?

Our world is filled with oscillations in which objects move back and forth repeatedly. Many oscillations are merely amusing or annoying, but many others are dangerous or financially important. Here are a few examples: When a bat hits a baseball, the bat may oscillate enough to sting the batter's hands or even to break apart. When wind blows past a power line, the line may oscillate ("gallop" in electrical engineering terms) so severely that it rips apart, shutting off the power supply to a community. When an airplane is in flight, the turbulence of the air flowing past the wings makes them oscillate, eventually leading to metal fatigue and even failure. When a train travels around a curve, its wheels oscillate horizontally ("hunt" in mechanical engineering terms) as they are forced to turn in new directions (you can hear the oscillations).

When an earthquake occurs near a city, buildings may be set oscillating so severely that they are shaken apart. When an arrow is shot from a bow, the feathers at the end of the arrow manage to snake around the bow staff without hitting it because the arrow oscillates. When a coin drops into a metal collection plate, the coin oscillates with such a familiar ring that the coin's denomination can be determined from the sound. When a rodeo cowboy rides a bull, the cowboy oscillates wildly as the bull jumps and turns (at least the cowboy hopes to be oscillating).

The study and control of oscillations are two of the primary goals of both physics and engineering. In this chapter we discuss a basic type of oscillation called *simple harmonic motion.*

Heads Up. This material is quite challenging to most students. One reason is that there is a truckload of definitions and symbols to sort out, but the main reason is that we need to relate an object's oscillations (something that we can see or even experience) to the equations and graphs for the oscillations. Relating the real, visible motion to the abstraction of an equation or graph requires a lot of hard work.

Simple Harmonic Motion

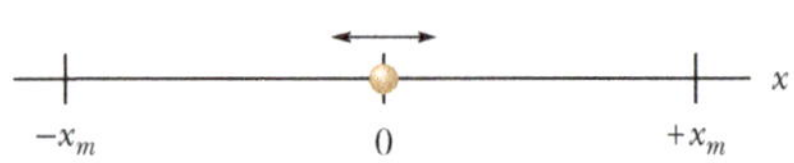

Figure 15-1 A particle repeatedly oscillates left and right along an x axis, between extreme points x_m and $-x_m$.

Figure 15-1 shows a particle that is oscillating about the origin of an x axis, repeatedly going left and right by identical amounts. The **frequency** f of the oscillation is the number of times per second that it completes a full oscillation (a *cycle*) and has the unit of hertz (abbreviated Hz), where

$$1 \text{ hertz} = 1 \text{ Hz} = 1 \text{ oscillation per second} = 1 \text{ s}^{-1}. \tag{15-1}$$

The time for one full cycle is the **period** T of the oscillation, which is

$$T = \frac{1}{f}. \tag{15-2}$$

Any motion that repeats at regular intervals is called periodic motion or harmonic motion. However, here we are interested in a particular type of periodic motion called **simple harmonic motion** (SHM). Such motion is a sinusoidal function of time t. That is, it can be written as a sine or a cosine of time t. Here we arbitrarily choose the cosine function and write the displacement (or position) of the particle in Fig. 15-1 as

$$x(t) = x_m \cos(\omega t + \phi) \quad \text{(displacement)}, \tag{15-3}$$

in which x_m, ω, and ϕ are quantities that we shall define.

Freeze-Frames. Let's take some freeze-frames of the motion and then arrange them one after another down the page (Fig. 15-2*a*). Our first freeze-frame is at $t = 0$ when the particle is at its rightmost position on the x axis. We label that coordinate as x_m (the subscript means *maximum*); it is the symbol in front of the cosine

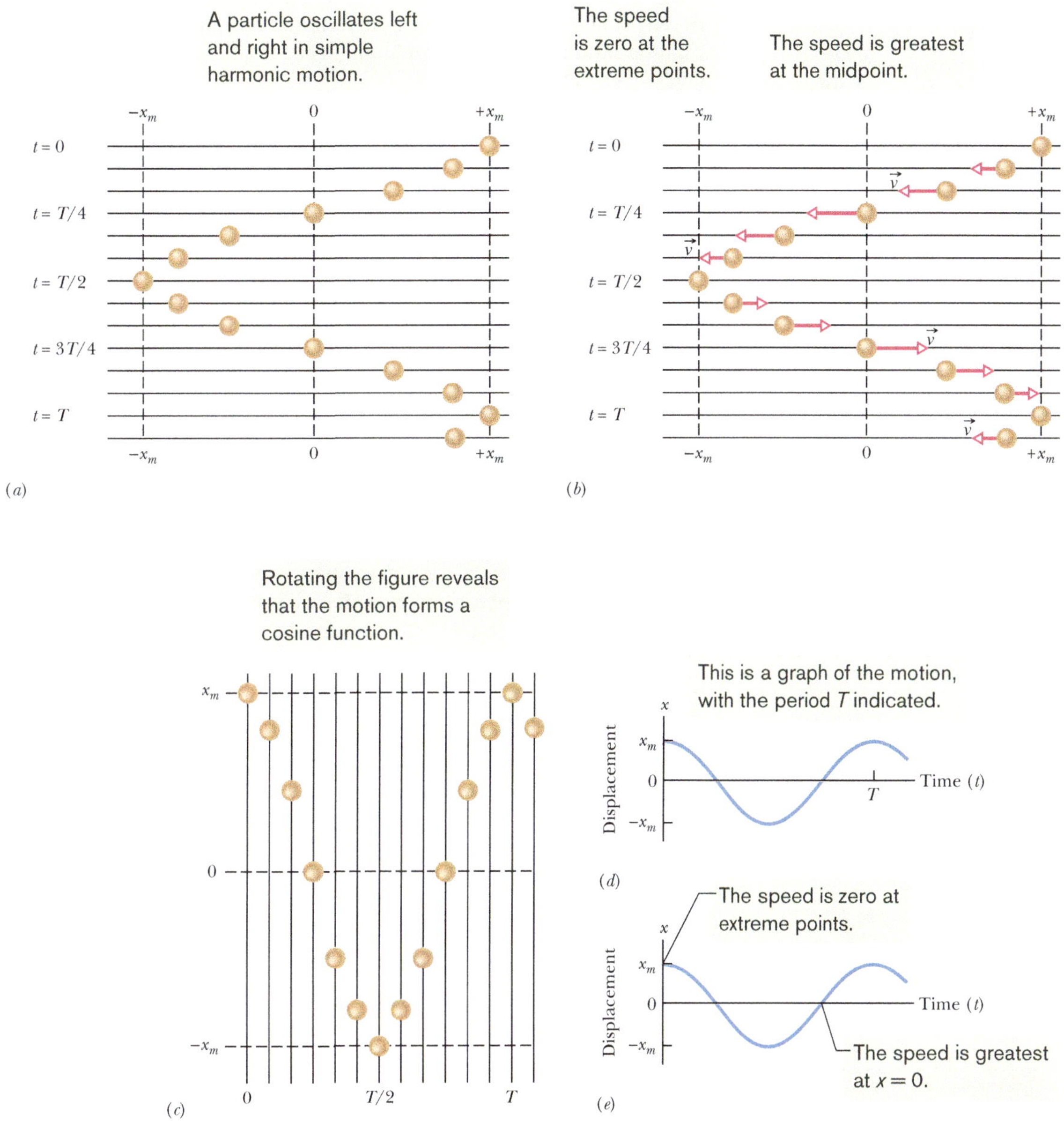

Figure 15-2 (a) A sequence of "freeze-frames" (taken at equal time intervals) showing the position of a particle as it oscillates back and forth about the origin of an x axis, between the limits $+x_m$ and $-x_m$. (b) The vector arrows are scaled to indicate the speed of the particle. The speed is maximum when the particle is at the origin and zero when it is at $\pm x_m$. If the time t is chosen to be zero when the particle is at $+x_m$, then the particle returns to $+x_m$ at $t = T$, where T is the period of the motion. The motion is then repeated. (c) Rotating the figure reveals the motion forms a cosine function of time, as shown in (d). (e) The speed (the slope) changes.

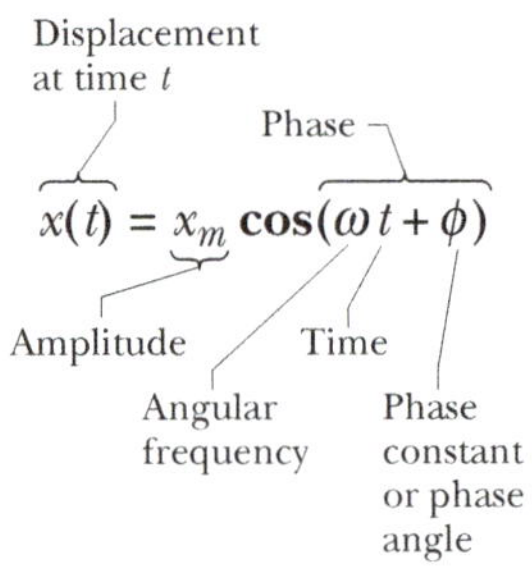

Figure 15-3 A handy guide to the quantities in Eq. 15-3 for simple harmonic motion.

function in Eq. 15-3. In the next freeze-frame, the particle is a bit to the left of x_m. It continues to move in the negative direction of x until it reaches the leftmost position, at coordinate $-x_m$. Thereafter, as time takes us down the page through more freeze-frames, the particle moves back to x_m and thereafter repeatedly oscillates between x_m and $-x_m$. In Eq. 15-3, the cosine function itself oscillates between +1 and −1. The value of x_m determines how far the particle moves in *its* oscillations and is called the **amplitude** of the oscillations (as labeled in the handy guide of Fig. 15-3).

Figure 15-2*b* indicates the velocity of the particle with respect to time, in the series of freeze-frames. We'll get to a function for the velocity soon, but for now just notice that the particle comes to a momentary stop at the extreme points and has its greatest speed (longest velocity vector) as it passes through the center point.

Mentally rotate Fig. 15-2*a* counterclockwise by 90°, so that the freeze-frames then progress rightward with time. We set time $t = 0$ when the particle is at x_m. The particle is back at x_m at time $t = T$ (the period of the oscillation), when it starts the next cycle of oscillation. If we filled in lots of the intermediate freeze-frames and drew a line through the particle positions, we would have the cosine curve shown in Fig. 15-2*d*. What we already noted about the speed is displayed in Fig. 15-2*e*. What we have in the whole of Fig. 15-2 is a transformation of what we can see (the reality of an oscillating particle) into the abstraction of a graph. (In *WileyPLUS* the transformation of Fig. 15-2 is available as an animation with voiceover.) Equation 15-3 is a concise way to capture the motion in the abstraction of an equation.

More Quantities. The handy guide of Fig. 15-3 defines more quantities about the motion. The argument of the cosine function is called the **phase** of the motion. As it varies with time, the value of the cosine function varies. The constant ϕ is called the **phase angle** or **phase constant.** It is in the argument only because we want to use Eq. 15-3 to describe the motion *regardless* of where the particle is in its oscillation when we happen to set the clock time to 0. In Fig. 15-2, we set $t = 0$ when the particle is at x_m. For that choice, Eq. 15-3 works just fine if we also set $\phi = 0$. However, if we set $t = 0$ when the particle happens to be at some other location, we need a different value of ϕ. A few values are indicated in Fig. 15-4. For example, suppose the particle is at its leftmost position when we happen to start the clock at $t = 0$. Then Eq. 15-3 describes the motion if $\phi = \pi$ rad. To check, substitute $t = 0$ and $\phi = \pi$ rad into Eq. 15-3. See, it gives $x = -x_m$ just then. Now check the other examples in Fig. 15-4.

The quantity ω in Eq. 15-3 is the **angular frequency** of the motion. To relate it to the frequency f and the period T, let's first note that the position $x(t)$ of the particle must (by definition) return to its initial value at the end of a period. That is, if $x(t)$ is the position at some chosen time t, then the particle must return to that same position at time $t + T$. Let's use Eq. 15-3 to express this condition, but let's also just set $\phi = 0$ to get it out of the way. Returning to the same position can then be written as

$$x_m \cos \omega t = x_m \cos \omega(t + T). \tag{15-4}$$

The cosine function first repeats itself when its argument (the *phase*, remember) has increased by 2π rad. So, Eq. 15-4 tells us that

$$\omega(t + T) = \omega t + 2\pi$$

or

$$\omega T = 2\pi.$$

Thus, from Eq. 15-2 the angular frequency is

$$\omega = \frac{2\pi}{T} = 2\pi f. \tag{15-5}$$

The SI unit of angular frequency is the radian per second.

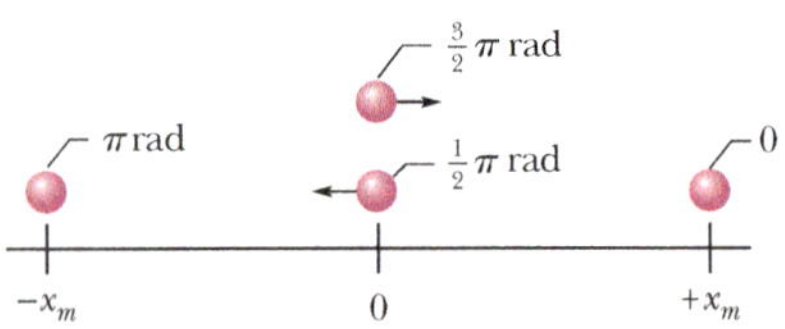

Figure 15-4 Values of ϕ corresponding to the position of the particle at time $t = 0$.

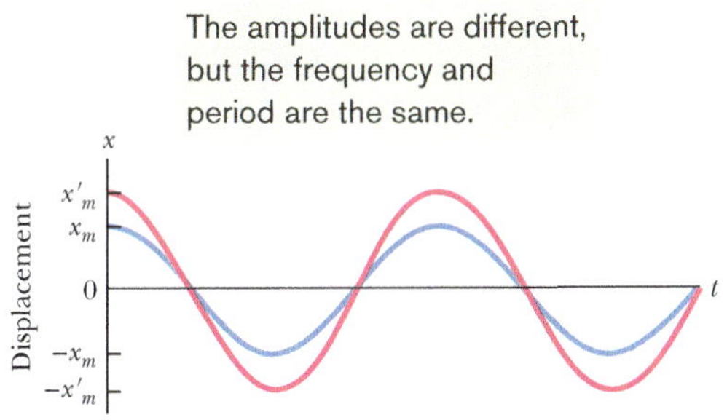

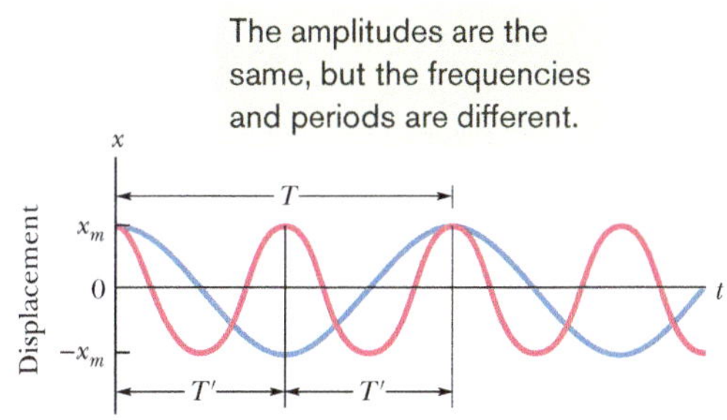

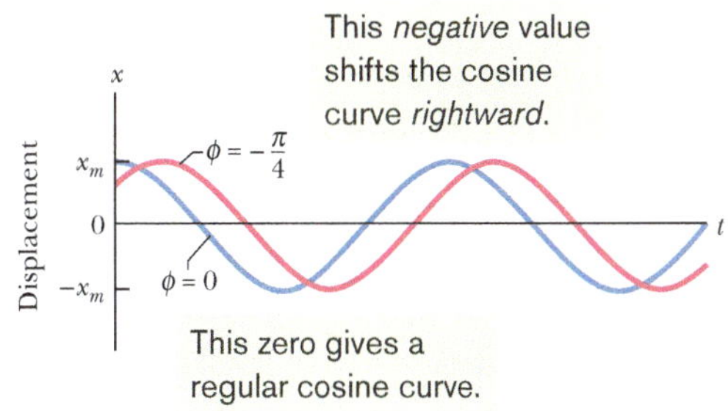

Figure 15-5 In all three cases, the blue curve is obtained from Eq. 15-3 with $\phi = 0$. (*a*) The red curve differs from the blue curve *only* in that the red-curve amplitude x'_m is greater (the red-curve extremes of displacement are higher and lower). (*b*) The red curve differs from the blue curve *only* in that the red-curve period is $T' = T/2$ (the red curve is compressed horizontally). (*c*) The red curve differs from the blue curve *only* in that for the red curve $\phi = -\pi/4$ rad rather than zero (the negative value of ϕ shifts the red curve to the right).

We've had a lot of quantities here, quantities that we could experimentally change to see the effects on the particle's SHM. Figure 15-5 gives some examples. The curves in Fig. 15-5*a* show the effect of changing the amplitude. Both curves have the same period. (See how the "peaks" line up?) And both are for $\phi = 0$. (See how the maxima of the curves both occur at $t = 0$?) In Fig. 15-5*b*, the two curves have the same amplitude x_m but one has twice the period as the other (and thus half the frequency as the other). Figure 15-5*c* is probably more difficult to understand. The curves have the same amplitude and same period but one is shifted relative to the other because of the different ϕ values. See how the one with $\phi = 0$ is just a regular cosine curve? The one with the negative ϕ is shifted rightward from it. That is a general result: negative ϕ values shift the regular cosine curve rightward and positive ϕ values shift it leftward. (Try this on a graphing calculator.)

Checkpoint 1

A particle undergoing simple harmonic oscillation of period T (like that in Fig. 15-2) is at $-x_m$ at time $t = 0$. Is it at $-x_m$, at $+x_m$, at 0, between $-x_m$ and 0, or between 0 and $+x_m$ when (a) $t = 2.00T$, (b) $t = 3.50T$, and (c) $t = 5.25T$?

The Velocity of SHM

We briefly discussed velocity as shown in Fig. 15-2*b*, finding that it varies in magnitude and direction as the particle moves between the extreme points (where the speed is momentarily zero) and through the central point (where the speed is maximum). To find the velocity $v(t)$ as a function of time, let's take a time derivative of the position function $x(t)$ in Eq. 15-3:

$$v(t) = \frac{dx(t)}{dt} = \frac{d}{dt}[x_m \cos(\omega t + \phi)]$$

or

$$v(t) = -\omega x_m \sin(\omega t + \phi) \quad \text{(velocity)}. \tag{15-6}$$

The velocity depends on time because the sine function varies with time, between the values of $+1$ and -1. The quantities in front of the sine function

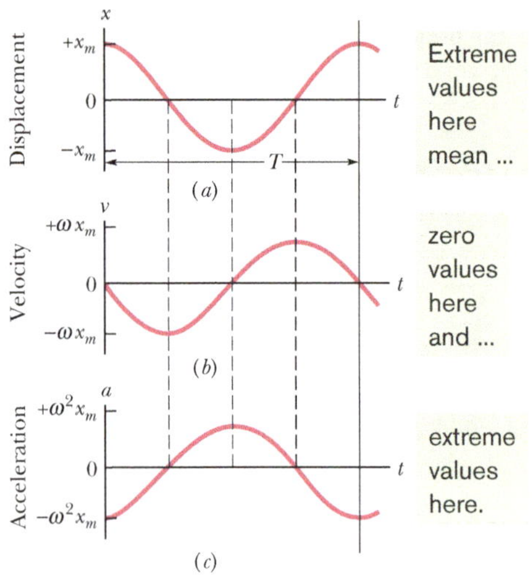

Figure 15-6 (*a*) The displacement $x(t)$ of a particle oscillating in SHM with phase angle ϕ equal to zero. The period T marks one complete oscillation. (*b*) The velocity $v(t)$ of the particle. (*c*) The acceleration $a(t)$ of the particle.

determine the extent of the variation in the velocity, between $+\omega x_m$ and $-\omega x_m$. We say that ωx_m is the **velocity amplitude** v_m of the velocity variation. When the particle is moving rightward through $x = 0$, its velocity is positive and the magnitude is at this greatest value. When it is moving leftward through $x = 0$, its velocity is negative and the magnitude is again at this greatest value. This variation with time (a negative sine function) is displayed in the graph of Fig. 15-6*b* for a phase constant of $\phi = 0$, which corresponds to the cosine function for the displacement versus time shown in Fig. 15-6*a*.

Recall that we use a cosine function for $x(t)$ regardless of the particle's position at $t = 0$. We simply choose an appropriate value of ϕ so that Eq. 15-3 gives us the correct position at $t = 0$. That decision about the cosine function leads us to a negative sine function for the velocity in Eq. 15-6, and the value of ϕ now gives the correct velocity at $t = 0$.

The Acceleration of SHM

Let's go one more step by differentiating the velocity function of Eq. 15-6 with respect to time to get the acceleration function of the particle in simple harmonic motion:

$$a(t) = \frac{dv(t)}{dt} = \frac{d}{dt}[-\omega x_m \sin(\omega t + \phi)]$$

or

$$a(t) = -\omega^2 x_m \cos(\omega t + \phi) \quad \text{(acceleration).} \tag{15-7}$$

We are back to a cosine function but with a minus sign out front. We know the drill by now. The acceleration varies because the cosine function varies with time, between $+1$ and -1. The variation in the magnitude of the acceleration is set by the **acceleration amplitude** a_m, which is the product $\omega^2 x_m$ that multiplies the cosine function.

Figure 15-6*c* displays Eq. 15-7 for a phase constant $\phi = 0$, consistent with Figs. 15-6*a* and 15-6*b*. Note that the acceleration magnitude is zero when the cosine is zero, which is when the particle is at $x = 0$. And the acceleration magnitude is maximum when the cosine magnitude is maximum, which is when the particle is at an extreme point, where it has been slowed to a stop so that its motion can be reversed. Indeed, comparing Eqs. 15-3 and 15-7 we see an extremely neat relationship:

$$a(t) = -\omega^2 x(t). \tag{15-8}$$

This is the hallmark of SHM: (1) The particle's acceleration is always opposite its displacement (hence the minus sign) and (2) the two quantities are always related by a constant (ω^2). If you ever see such a relationship in an oscillating situation (such as with, say, the current in an electrical circuit, or the rise and fall of water in a tidal bay), you can immediately say that the motion is SHM and immediately identify the angular frequency ω of the motion. In a nutshell:

> In SHM, the acceleration a is proportional to the displacement x but opposite in sign, and the two quantities are related by the square of the angular frequency ω.

Checkpoint 2

Which of the following relationships between a particle's acceleration a and its position x indicates simple harmonic oscillation: (a) $a = 3x^2$, (b) $a = 5x$, (c) $a = -4x$, (d) $a = -2/x$? For the SHM, what is the angular frequency (assume the unit of rad/s)?

The Force Law for Simple Harmonic Motion

Now that we have an expression for the acceleration in terms of the displacement in Eq. 15-8, we can apply Newton's second law to describe the force responsible for SHM:

$$F = ma = m(-\omega^2 x) = -(m\omega^2)x. \tag{15-9}$$

The minus sign means that the direction of the force on the particle is *opposite* the direction of the displacement of the particle. That is, in SHM the force is a *restoring force* in the sense that it fights against the displacement, attempting to restore the particle to the center point at $x = 0$. We've seen the general form of Eq. 15-9 back in Chapter 8 when we discussed a block on a spring as in Fig. 15-7. There we wrote Hooke's law,

$$F = -kx, \tag{15-10}$$

for the force acting on the block. Comparing Eqs. 15-9 and 15-10, we can now relate the spring constant k (a measure of the stiffness of the spring) to the mass of the block and the resulting angular frequency of the SHM:

$$k = m\omega^2. \tag{15-11}$$

Equation 15-10 is another way to write the hallmark equation for SHM.

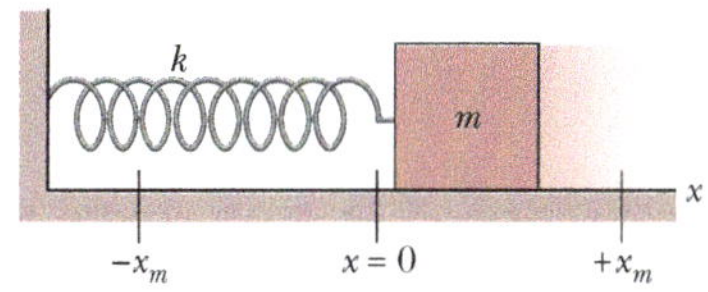

Figure 15-7 A linear simple harmonic oscillator. The surface is frictionless. Like the particle of Fig. 15-2, the block moves in simple harmonic motion once it has been either pulled or pushed away from the $x = 0$ position and released. Its displacement is then given by Eq. 15-3.

> Simple harmonic motion is the motion of a particle when the force acting on it is proportional to the particle's displacement but in the opposite direction.

The block–spring system of Fig. 15-7 is called a **linear simple harmonic oscillator** (linear oscillator, for short), where *linear* indicates that F is proportional to x to the *first* power (and not to some other power).

If you ever see a situation in which the force in an oscillation is always proportional to the displacement but in the opposite direction, you can immediately say that the oscillation is SHM. You can also immediately identify the associated spring constant k. If you know the oscillating mass, you can then determine the angular frequency of the motion by rewriting Eq. 15-11 as

$$\omega = \sqrt{\frac{k}{m}} \quad \text{(angular frequency).} \tag{15-12}$$

(This is usually more important than the value of k.) Further, you can determine the period of the motion by combining Eqs. 15-5 and 15-12 to write

$$T = 2\pi\sqrt{\frac{m}{k}} \quad \text{(period).} \tag{15-13}$$

Let's make a bit of physical sense of Eqs. 15-12 and 15-13. Can you see that a stiff spring (large k) tends to produce a large ω (rapid oscillations) and thus a small period T? Can you also see that a large mass m tends to result in a small ω (sluggish oscillations) and thus a large period T?

Every oscillating system, be it a diving board or a violin string, has some element of "springiness" and some element of "inertia" or mass. In Fig. 15-7, these elements are separated: The springiness is entirely in the spring, which we assume to be massless, and the inertia is entirely in the block, which we assume to be rigid. In a violin string, however, the two elements are both within the string.

Checkpoint 3

Which of the following relationships between the force F on a particle and the particle's position x gives SHM: (a) $F = -5x$, (b) $F = -400x^2$, (c) $F = 10x$, (d) $F = 3x^2$?

Sample Problem 15.01 Block–spring SHM, amplitude, acceleration, phase constant

A block whose mass m is 680 g is fastened to a spring whose spring constant k is 65 N/m. The block is pulled a distance $x = 11$ cm from its equilibrium position at $x = 0$ on a frictionless surface and released from rest at $t = 0$.

(a) What are the angular frequency, the frequency, and the period of the resulting motion?

KEY IDEA

The block–spring system forms a linear simple harmonic oscillator, with the block undergoing SHM.

Calculations: The angular frequency is given by Eq. 15-12:

$$\omega = \sqrt{\frac{k}{m}} = \sqrt{\frac{65\text{ N/m}}{0.68\text{ kg}}} = 9.78\text{ rad/s}$$
$$\approx 9.8\text{ rad/s}. \qquad \text{(Answer)}$$

The frequency follows from Eq. 15-5, which yields

$$f = \frac{\omega}{2\pi} = \frac{9.78\text{ rad/s}}{2\pi\text{ rad}} = 1.56\text{ Hz} \approx 1.6\text{ Hz}. \qquad \text{(Answer)}$$

The period follows from Eq. 15-2, which yields

$$T = \frac{1}{f} = \frac{1}{1.56\text{ Hz}} = 0.64\text{ s} = 640\text{ ms}. \qquad \text{(Answer)}$$

(b) What is the amplitude of the oscillation?

KEY IDEA

With no friction involved, the mechanical energy of the spring–block system is conserved.

Reasoning: The block is released from rest 11 cm from its equilibrium position, with zero kinetic energy and the elastic potential energy of the system at a maximum. Thus, the block will have zero kinetic energy whenever it is again 11 cm from its equilibrium position, which means it will never be farther than 11 cm from that position. Its maximum displacement is 11 cm:

$$x_m = 11\text{ cm}. \qquad \text{(Answer)}$$

(c) What is the maximum speed v_m of the oscillating block, and where is the block when it has this speed?

KEY IDEA

The maximum speed v_m is the velocity amplitude ωx_m in Eq. 15-6.

Calculation: Thus, we have

$$v_m = \omega x_m = (9.78\text{ rad/s})(0.11\text{ m})$$
$$= 1.1\text{ m/s}. \qquad \text{(Answer)}$$

This maximum speed occurs when the oscillating block is rushing through the origin; compare Figs. 15-6*a* and 15-6*b*, where you can see that the speed is a maximum whenever $x = 0$.

(d) What is the magnitude a_m of the maximum acceleration of the block?

KEY IDEA

The magnitude a_m of the maximum acceleration is the acceleration amplitude $\omega^2 x_m$ in Eq. 15-7.

Calculation: So, we have

$$a_m = \omega^2 x_m = (9.78\text{ rad/s})^2(0.11\text{ m})$$
$$= 11\text{ m/s}^2. \qquad \text{(Answer)}$$

This maximum acceleration occurs when the block is at the ends of its path, where the block has been slowed to a stop so that its motion can be reversed. At those extreme points, the force acting on the block has its maximum magnitude; compare Figs. 15-6*a* and 15-6*c*, where you can see that the magnitudes of the displacement and acceleration are maximum at the same times, when the speed is zero, as you can see in Fig. 15-6*b*.

(e) What is the phase constant ϕ for the motion?

Calculations: Equation 15-3 gives the displacement of the block as a function of time. We know that at time $t = 0$, the block is located at $x = x_m$. Substituting these *initial conditions*, as they are called, into Eq. 15-3 and canceling x_m give us

$$1 = \cos\phi. \qquad (15\text{-}14)$$

Taking the inverse cosine then yields

$$\phi = 0\text{ rad}. \qquad \text{(Answer)}$$

(Any angle that is an integer multiple of 2π rad also satisfies Eq. 15-14; we chose the smallest angle.)

(f) What is the displacement function $x(t)$ for the spring–block system?

Calculation: The function $x(t)$ is given in general form by Eq. 15-3. Substituting known quantities into that equation gives us

$$x(t) = x_m \cos(\omega t + \phi)$$
$$= (0.11\text{ m})\cos[(9.8\text{ rad/s})t + 0]$$
$$= 0.11\cos(9.8t), \qquad \text{(Answer)}$$

where x is in meters and t is in seconds.

WILEY PLUS Additional examples, video, and practice available at *WileyPLUS*

Sample Problem 15.02 Finding SHM phase constant from displacement and velocity

At $t = 0$, the displacement $x(0)$ of the block in a linear oscillator like that of Fig. 15-7 is -8.50 cm. (Read $x(0)$ as "x at time zero.") The block's velocity $v(0)$ then is -0.920 m/s, and its acceleration $a(0)$ is $+47.0$ m/s^2.

(a) What is the angular frequency ω of this system?

KEY IDEA

With the block in SHM, Eqs. 15-3, 15-6, and 15-7 give its displacement, velocity, and acceleration, respectively, and each contains ω.

Calculations: Let's substitute $t = 0$ into each to see whether we can solve any one of them for ω. We find

$$x(0) = x_m \cos \phi, \tag{15-15}$$

$$v(0) = -\omega x_m \sin \phi, \tag{15-16}$$

and

$$a(0) = -\omega^2 x_m \cos \phi. \tag{15-17}$$

In Eq. 15-15, ω has disappeared. In Eqs. 15-16 and 15-17, we know values for the left sides, but we do not know x_m and ϕ. However, if we divide Eq. 15-17 by Eq. 15-15, we neatly eliminate both x_m and ϕ and can then solve for ω as

$$\omega = \sqrt{-\frac{a(0)}{x(0)}} = \sqrt{-\frac{47.0 \text{ m/s}^2}{-0.0850 \text{ m}}}$$

$$= 23.5 \text{ rad/s.} \qquad \text{(Answer)}$$

(b) What are the phase constant ϕ and amplitude x_m?

Calculations: We know ω and want ϕ and x_m. If we divide Eq. 15-16 by Eq. 15-15, we eliminate one of those unknowns and reduce the other to a single trig function:

$$\frac{v(0)}{x(0)} = \frac{-\omega x_m \sin \phi}{x_m \cos \phi} = -\omega \tan \phi.$$

Solving for $\tan \phi$, we find

$$\tan \phi = -\frac{v(0)}{\omega x(0)} = -\frac{-0.920 \text{ m/s}}{(23.5 \text{ rad/s})(-0.0850 \text{ m})}$$

$$= -0.461.$$

This equation has two solutions:

$$\phi = -25° \quad \text{and} \quad \phi = 180° + (-25°) = 155°.$$

Normally only the first solution here is displayed by a calculator, but it may not be the physically possible solution. To choose the proper solution, we test them both by using them to compute values for the amplitude x_m. From Eq. 15-15, we find that if $\phi = -25°$, then

$$x_m = \frac{x(0)}{\cos \phi} = \frac{-0.0850 \text{ m}}{\cos(-25°)} = -0.094 \text{ m.}$$

We find similarly that if $\phi = 155°$, then $x_m = 0.094$ m. Because the amplitude of SHM must be a positive constant, the correct phase constant and amplitude here are

$$\phi = 155° \quad \text{and} \quad x_m = 0.094 \text{ m} = 9.4 \text{ cm.} \qquad \text{(Answer)}$$

WILEY PLUS Additional examples, video, and practice available at *WileyPLUS*

15-2 ENERGY IN SIMPLE HARMONIC MOTION

Learning Objectives

After reading this module, you should be able to . . .

15.19 For a spring–block oscillator, calculate the kinetic energy and elastic potential energy at any given time.

15.20 Apply the conservation of energy to relate the total energy of a spring–block oscillator at one instant to the total energy at another instant.

15.21 Sketch a graph of the kinetic energy, potential energy, and total energy of a spring–block oscillator, first as a function of time and then as a function of the oscillator's position.

15.22 For a spring–block oscillator, determine the block's position when the total energy is entirely kinetic energy and when it is entirely potential energy.

Key Ideas

- A particle in simple harmonic motion has, at any time, kinetic energy $K = \frac{1}{2}mv^2$ and potential energy $U = \frac{1}{2}kx^2$. If no friction is present, the mechanical energy $E = K + U$ remains constant even though K and U change.

Energy in Simple Harmonic Motion

Let's now examine the linear oscillator of Chapter 8, where we saw that the energy transfers back and forth between kinetic energy and potential energy, while the sum of the two—the mechanical energy E of the oscillator—remains constant. The

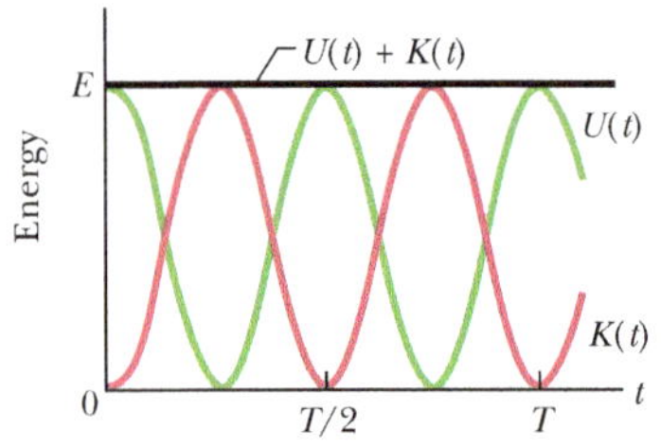

(a) As *time* changes, the energy shifts between the two types, but the total is constant.

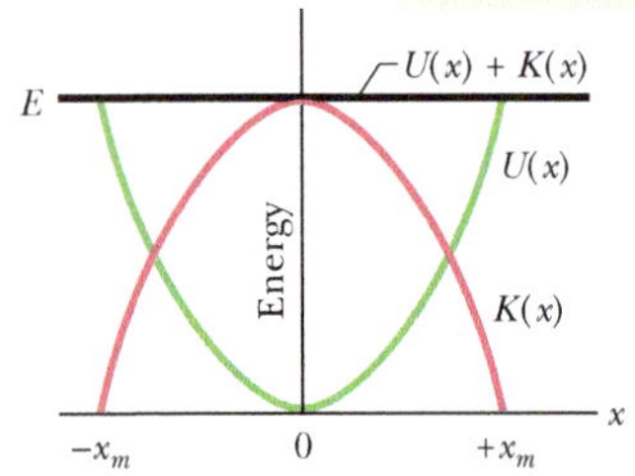

(b) As *position* changes, the energy shifts between the two types, but the total is constant.

Figure 15-8 (*a*) Potential energy $U(t)$, kinetic energy $K(t)$, and mechanical energy E as functions of time t for a linear harmonic oscillator. Note that all energies are positive and that the potential energy and the kinetic energy peak twice during every period. (*b*) Potential energy $U(x)$, kinetic energy $K(x)$, and mechanical energy E as functions of position x for a linear harmonic oscillator with amplitude x_m. For $x = 0$ the energy is all kinetic, and for $x = \pm x_m$ it is all potential.

potential energy of a linear oscillator like that of Fig. 15-7 is associated entirely with the spring. Its value depends on how much the spring is stretched or compressed—that is, on $x(t)$. We can use Eqs. 8-11 and 15-3 to find

$$U(t) = \tfrac{1}{2}kx^2 = \tfrac{1}{2}kx_m^2 \cos^2(\omega t + \phi). \tag{15-18}$$

Caution: A function written in the form $\cos^2 A$ (as here) means $(\cos A)^2$ and is *not* the same as one written $\cos A^2$, which means $\cos(A^2)$.

The kinetic energy of the system of Fig. 15-7 is associated entirely with the block. Its value depends on how fast the block is moving—that is, on $v(t)$. We can use Eq. 15-6 to find

$$K(t) = \tfrac{1}{2}mv^2 = \tfrac{1}{2}m\omega^2 x_m^2 \sin^2(\omega t + \phi). \tag{15-19}$$

If we use Eq. 15-12 to substitute k/m for ω^2, we can write Eq. 15-19 as

$$K(t) = \tfrac{1}{2}mv^2 = \tfrac{1}{2}kx_m^2 \sin^2(\omega t + \phi). \tag{15-20}$$

The mechanical energy follows from Eqs. 15-18 and 15-20 and is

$$\begin{aligned} E &= U + K \\ &= \tfrac{1}{2}kx_m^2 \cos^2(\omega t + \phi) + \tfrac{1}{2}kx_m^2 \sin^2(\omega t + \phi) \\ &= \tfrac{1}{2}kx_m^2 [\cos^2(\omega t + \phi) + \sin^2(\omega t + \phi)]. \end{aligned}$$

For any angle α,

$$\cos^2 \alpha + \sin^2 \alpha = 1.$$

Thus, the quantity in the square brackets above is unity and we have

$$E = U + K = \tfrac{1}{2}kx_m^2. \tag{15-21}$$

The mechanical energy of a linear oscillator is indeed constant and independent of time. The potential energy and kinetic energy of a linear oscillator are shown as functions of time t in Fig. 15-8*a* and as functions of displacement x in Fig. 15-8*b*. In any oscillating system, an element of springiness is needed to store the potential energy and an element of inertia is needed to store the kinetic energy.

Checkpoint 4

In Fig. 15-7, the block has a kinetic energy of 3 J and the spring has an elastic potential energy of 2 J when the block is at $x = +2.0$ cm. (a) What is the kinetic energy when the block is at $x = 0$? What is the elastic potential energy when the block is at (b) $x = -2.0$ cm and (c) $x = -x_m$?

Sample Problem 15.03 SHM potential energy, kinetic energy, mass dampers

Many tall buildings have *mass dampers*, which are anti-sway devices to prevent them from oscillating in a wind. The device might be a block oscillating at the end of a spring and on a lubricated track. If the building sways, say, eastward, the block also moves eastward but delayed enough so that when it finally moves, the building is then moving back westward. Thus, the motion of the oscillator is out of step with the motion of the building.

Suppose the block has mass $m = 2.72 \times 10^5$ kg and is designed to oscillate at frequency $f = 10.0$ Hz and with amplitude $x_m = 20.0$ cm.

(a) What is the total mechanical energy E of the spring–block system?

KEY IDEA

The mechanical energy E (the sum of the kinetic energy $K = \frac{1}{2}mv^2$ of the block and the potential energy $U = \frac{1}{2}kx^2$ of the spring) is constant throughout the motion of the oscillator. Thus, we can evaluate E at any point during the motion.

Calculations: Because we are given amplitude x_m of the oscillations, let's evaluate E when the block is at position $x = x_m$, where it has velocity $v = 0$. However, to evaluate U at that point, we first need to find the spring constant k. From Eq. 15-12 ($\omega = \sqrt{k/m}$) and Eq. 15-5 ($\omega = 2\pi f$), we find

$$k = m\omega^2 = m(2\pi f)^2$$
$$= (2.72 \times 10^5 \text{ kg})(2\pi)^2(10.0 \text{ Hz})^2$$
$$= 1.073 \times 10^9 \text{ N/m}.$$

We can now evaluate E as

$$E = K + U = \tfrac{1}{2}mv^2 + \tfrac{1}{2}kx^2$$
$$= 0 + \tfrac{1}{2}(1.073 \times 10^9 \text{ N/m})(0.20 \text{ m})^2$$
$$= 2.147 \times 10^7 \text{ J} \approx 2.1 \times 10^7 \text{ J}. \quad \text{(Answer)}$$

(b) What is the block's speed as it passes through the equilibrium point?

Calculations: We want the speed at $x = 0$, where the potential energy is $U = \frac{1}{2}kx^2 = 0$ and the mechanical energy is entirely kinetic energy. So, we can write

$$E = K + U = \tfrac{1}{2}mv^2 + \tfrac{1}{2}kx^2$$
$$2.147 \times 10^7 \text{ J} = \tfrac{1}{2}(2.72 \times 10^5 \text{ kg})v^2 + 0,$$

or

$$v = 12.6 \text{ m/s}. \quad \text{(Answer)}$$

Because E is entirely kinetic energy, this is the maximum speed v_m.

WILEY PLUS Additional examples, video, and practice available at *WileyPLUS*

15-3 AN ANGULAR SIMPLE HARMONIC OSCILLATOR

Learning Objectives

After reading this module, you should be able to . . .

15.23 Describe the motion of an angular simple harmonic oscillator.

15.24 For an angular simple harmonic oscillator, apply the relationship between the torque τ and the angular displacement θ (from equilibrium).

15.25 For an angular simple harmonic oscillator, apply the relationship between the period T (or frequency f), the rotational inertia I, and the torsion constant κ.

15.26 For an angular simple harmonic oscillator at any instant, apply the relationship between the angular acceleration α, the angular frequency ω, and the angular displacement θ.

Key Idea

- A torsion pendulum consists of an object suspended on a wire. When the wire is twisted and then released, the object oscillates in angular simple harmonic motion with a period given by

$$T = 2\pi\sqrt{\frac{I}{\kappa}},$$

where I is the rotational inertia of the object about the axis of rotation and κ is the torsion constant of the wire.

An Angular Simple Harmonic Oscillator

Figure 15-9 shows an angular version of a simple harmonic oscillator; the element of springiness or elasticity is associated with the twisting of a suspension wire rather than the extension and compression of a spring as we previously had. The device is called a **torsion pendulum,** with *torsion* referring to the twisting.

If we rotate the disk in Fig. 15-9 by some angular displacement θ from its rest position (where the reference line is at $\theta = 0$) and release it, it will oscillate about that position in **angular simple harmonic motion.** Rotating the disk through an angle θ in either direction introduces a restoring torque given by

$$\tau = -\kappa\theta. \quad (15\text{-}22)$$

Here κ (Greek *kappa*) is a constant, called the **torsion constant,** that depends on the length, diameter, and material of the suspension wire.

Comparison of Eq. 15-22 with Eq. 15-10 leads us to suspect that Eq. 15-22 is the angular form of Hooke's law, and that we can transform Eq. 15-13, which gives the period of linear SHM, into an equation for the period of angular SHM: We replace the spring constant k in Eq. 15-13 with its equivalent, the constant

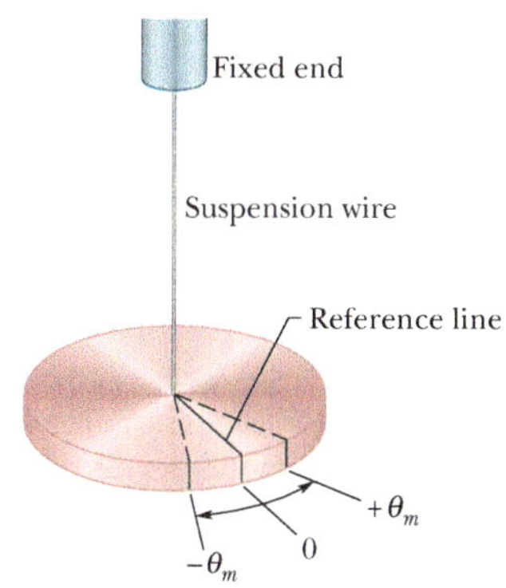

Figure 15-9 A torsion pendulum is an angular version of a linear simple harmonic oscillator. The disk oscillates in a horizontal plane; the reference line oscillates with angular amplitude θ_m. The twist in the suspension wire stores potential energy as a spring does and provides the restoring torque.

κ of Eq. 15-22, and we replace the mass m in Eq. 15-13 with *its* equivalent, the rotational inertia I of the oscillating disk. These replacements lead to

$$T = 2\pi\sqrt{\frac{I}{\kappa}} \quad \text{(torsion pendulum).} \tag{15-23}$$

Sample Problem 15.04 Angular simple harmonic oscillator, rotational inertia, period

Figure 15-10a shows a thin rod whose length L is 12.4 cm and whose mass m is 135 g, suspended at its midpoint from a long wire. Its period T_a of angular SHM is measured to be 2.53 s. An irregularly shaped object, which we call object X, is then hung from the same wire, as in Fig. 15-10b, and its period T_b is found to be 4.76 s. What is the rotational inertia of object X about its suspension axis?

KEY IDEA

The rotational inertia of either the rod or object X is related to the measured period by Eq. 15-23.

Calculations: In Table 10-2e, the rotational inertia of a thin rod about a perpendicular axis through its midpoint is given as $\frac{1}{12}mL^2$. Thus, we have, for the rod in Fig. 15-10a,

$$I_a = \tfrac{1}{12}mL^2 = (\tfrac{1}{12})(0.135 \text{ kg})(0.124 \text{ m})^2$$
$$= 1.73 \times 10^{-4} \text{ kg}\cdot\text{m}^2.$$

Now let us write Eq. 15-23 twice, once for the rod and once for object X:

$$T_a = 2\pi\sqrt{\frac{I_a}{\kappa}} \quad \text{and} \quad T_b = 2\pi\sqrt{\frac{I_b}{\kappa}}.$$

The constant κ, which is a property of the wire, is the same for both figures; only the periods and the rotational inertias differ.

Let us square each of these equations, divide the second by the first, and solve the resulting equation for I_b. The result is

$$I_b = I_a\frac{T_b^2}{T_a^2} = (1.73 \times 10^{-4} \text{ kg}\cdot\text{m}^2)\frac{(4.76 \text{ s})^2}{(2.53 \text{ s})^2}$$
$$= 6.12 \times 10^{-4} \text{ kg}\cdot\text{m}^2. \quad \text{(Answer)}$$

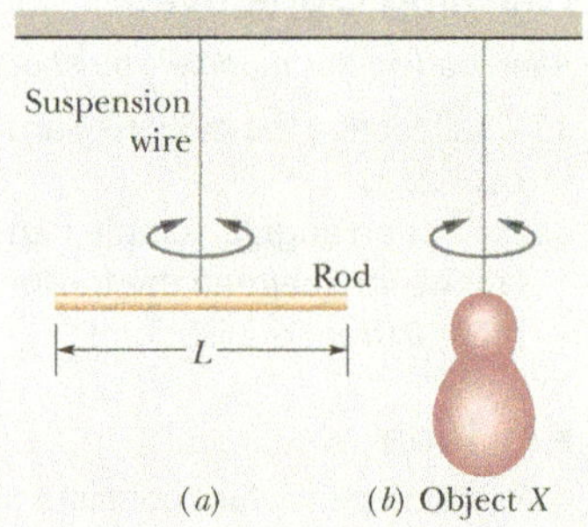

Figure 15-10 Two torsion pendulums, consisting of (a) a wire and a rod and (b) the same wire and an irregularly shaped object.

15-4 PENDULUMS, CIRCULAR MOTION

Learning Objectives

After reading this module, you should be able to . . .

15.27 Describe the motion of an oscillating simple pendulum.

15.28 Draw a free-body diagram of a pendulum bob with the pendulum at angle θ to the vertical.

15.29 For small-angle oscillations of a *simple pendulum*, relate the period T (or frequency f) to the pendulum's length L.

15.30 Distinguish between a simple pendulum and a physical pendulum.

15.31 For small-angle oscillations of a *physical pendulum*, relate the period T (or frequency f) to the distance h between the pivot and the center of mass.

15.32 For an angular oscillating system, determine the angular frequency ω from either an equation relating torque τ and angular displacement θ or an equation relating angular acceleration α and angular displacement θ.

15.33 Distinguish between a pendulum's angular frequency ω (having to do with the rate at which cycles are completed) and its $d\theta/dt$ (the rate at which its angle with the vertical changes).

15.34 Given data about the angular position θ and rate of change $d\theta/dt$ at one instant, determine the phase constant ϕ and amplitude θ_m.

15.35 Describe how the free-fall acceleration can be measured with a simple pendulum.

15.36 For a given physical pendulum, determine the location of the center of oscillation and identify the meaning of that phrase in terms of a simple pendulum.

15.37 Describe how simple harmonic motion is related to uniform circular motion.

Key Ideas

- A simple pendulum consists of a rod of negligible mass that pivots about its upper end, with a particle (the bob) attached at its lower end. If the rod swings through only small angles, its motion is approximately simple harmonic motion with a period given by

$$T = 2\pi\sqrt{\frac{I}{mgL}} \quad \text{(simple pendulum)},$$

where I is the particle's rotational inertia about the pivot, m is the particle's mass, and L is the rod's length.

- A physical pendulum has a more complicated distribution of mass. For small angles of swinging, its motion is simple harmonic motion with a period given by

$$T = 2\pi\sqrt{\frac{I}{mgh}} \quad \text{(physical pendulum)},$$

where I is the pendulum's rotational inertia about the pivot, m is the pendulum's mass, and h is the distance between the pivot and the pendulum's center of mass.

- Simple harmonic motion corresponds to the projection of uniform circular motion onto a diameter of the circle.

Pendulums

We turn now to a class of simple harmonic oscillators in which the springiness is associated with the gravitational force rather than with the elastic properties of a twisted wire or a compressed or stretched spring.

The Simple Pendulum

If an apple swings on a long thread, does it have simple harmonic motion? If so, what is the period T? To answer, we consider a **simple pendulum,** which consists of a particle of mass m (called the *bob* of the pendulum) suspended from one end of an unstretchable, massless string of length L that is fixed at the other end, as in Fig. 15-11*a*. The bob is free to swing back and forth in the plane of the page, to the left and right of a vertical line through the pendulum's pivot point.

The Restoring Torque. The forces acting on the bob are the force $\vec{T}$ from the string and the gravitational force $\vec{F}_g$, as shown in Fig. 15-11*b*, where the string makes an angle θ with the vertical. We resolve $\vec{F}_g$ into a radial component $F_g \cos\theta$ and a component $F_g \sin\theta$ that is tangent to the path taken by the bob. This tangential component produces a restoring torque about the pendulum's pivot point because the component always acts opposite the displacement of the bob so as to bring the bob back toward its central location. That location is called the *equilibrium position* ($\theta = 0$) because the pendulum would be at rest there were it not swinging.

From Eq. 10-41 ($\tau = r_\perp F$), we can write this restoring torque as

$$\tau = -L(F_g \sin\theta), \tag{15-24}$$

where the minus sign indicates that the torque acts to reduce θ and L is the moment arm of the force component $F_g \sin\theta$ about the pivot point. Substituting Eq. 15-24 into Eq. 10-44 ($\tau = I\alpha$) and then substituting mg as the magnitude of F_g, we obtain

$$-L(mg\sin\theta) = I\alpha, \tag{15-25}$$

where I is the pendulum's rotational inertia about the pivot point and α is its angular acceleration about that point.

We can simplify Eq. 15-25 if we assume the angle θ is small, for then we can approximate $\sin\theta$ with θ (expressed in radian measure). (As an example, if $\theta = 5.00° = 0.0873$ rad, then $\sin\theta = 0.0872$, a difference of only about 0.1%.) With that approximation and some rearranging, we then have

$$\alpha = -\frac{mgL}{I}\theta. \tag{15-26}$$

This equation is the angular equivalent of Eq. 15-8, the hallmark of SHM. It tells us that the angular acceleration α of the pendulum is proportional to the angular displacement θ but opposite in sign. Thus, as the pendulum bob moves to the right, as in Fig. 15-11*a*, its acceleration *to the left* increases until the bob stops and

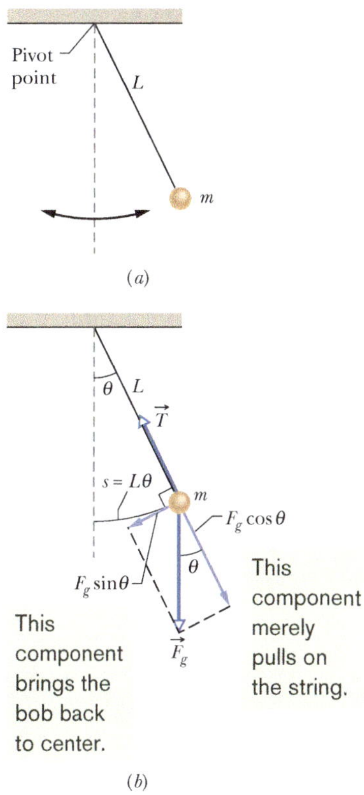

Figure 15-11 (*a*) A simple pendulum. (*b*) The forces acting on the bob are the gravitational force $\vec{F}_g$ and the force $\vec{T}$ from the string. The tangential component $F_g \sin\theta$ of the gravitational force is a restoring force that tends to bring the pendulum back to its central position.

begins moving to the left. Then, when it is to the left of the equilibrium position, its acceleration to the right tends to return it to the right, and so on, as it swings back and forth in SHM. More precisely, the motion of a *simple pendulum swinging through only small angles* is approximately SHM. We can state this restriction to small angles another way: The **angular amplitude** θ_m of the motion (the maximum angle of swing) must be small.

Angular Frequency. Here is a neat trick. Because Eq. 15-26 has the same form as Eq. 15-8 for SHM, we can immediately identify the pendulum's angular frequency as being the square root of the constants in front of the displacement:

$$\omega = \sqrt{\frac{mgL}{I}}.$$

In the homework problems you might see oscillating systems that do not seem to resemble pendulums. However, if you can relate the acceleration (linear or angular) to the displacement (linear or angular), you can then immediately identify the angular frequency as we have just done here.

Period. Next, if we substitute this expression for ω into Eq. 15-5 ($\omega = 2\pi/T$), we see that the period of the pendulum may be written as

$$T = 2\pi\sqrt{\frac{I}{mgL}}. \qquad (15\text{-}27)$$

All the mass of a simple pendulum is concentrated in the mass m of the particle-like bob, which is at radius L from the pivot point. Thus, we can use Eq. 10-33 ($I = mr^2$) to write $I = mL^2$ for the rotational inertia of the pendulum. Substituting this into Eq. 15-27 and simplifying then yield

$$T = 2\pi\sqrt{\frac{L}{g}} \quad \text{(simple pendulum, small amplitude).} \qquad (15\text{-}28)$$

We assume small-angle swinging in this chapter.

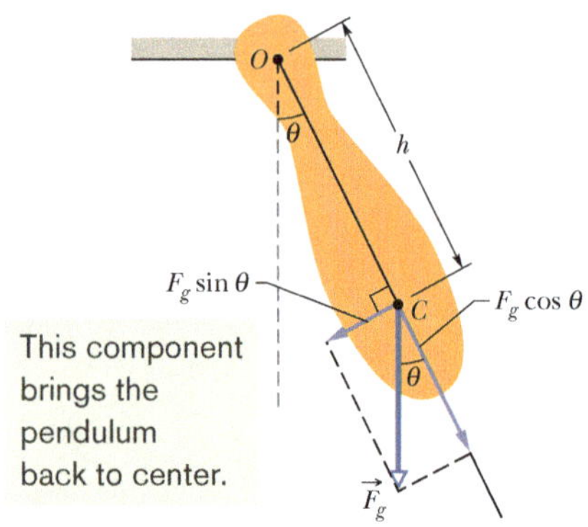

Figure 15-12 A physical pendulum. The restoring torque is $hF_g \sin\theta$. When $\theta = 0$, center of mass C hangs directly below pivot point O.

The Physical Pendulum

A real pendulum, usually called a **physical pendulum,** can have a complicated distribution of mass. Does it also undergo SHM? If so, what is its period?

Figure 15-12 shows an arbitrary physical pendulum displaced to one side by angle θ. The gravitational force $\vec{F}_g$ acts at its center of mass C, at a distance h from the pivot point O. Comparison of Figs. 15-12 and 15-11*b* reveals only one important difference between an arbitrary physical pendulum and a simple pendulum. For a physical pendulum the restoring component $F_g \sin\theta$ of the gravitational force has a moment arm of distance h about the pivot point, rather than of string length L. In all other respects, an analysis of the physical pendulum would duplicate our analysis of the simple pendulum up through Eq. 15-27. Again (for small θ_m), we would find that the motion is approximately SHM.

If we replace L with h in Eq. 15-27, we can write the period as

$$T = 2\pi\sqrt{\frac{I}{mgh}} \quad \text{(physical pendulum, small amplitude).} \qquad (15\text{-}29)$$

As with the simple pendulum, I is the rotational inertia of the pendulum about O. However, now I is not simply mL^2 (it depends on the shape of the physical pendulum), but it is still proportional to m.

A physical pendulum will not swing if it pivots at its center of mass. Formally, this corresponds to putting $h = 0$ in Eq. 15-29. That equation then predicts $T \to \infty$, which implies that such a pendulum will never complete one swing.

Corresponding to any physical pendulum that oscillates about a given pivot point O with period T is a simple pendulum of length L_0 with the same period T. We can find L_0 with Eq. 15-28. The point along the physical pendulum at distance L_0 from point O is called the *center of oscillation* of the physical pendulum for the given suspension point.

Measuring g

We can use a physical pendulum to measure the free-fall acceleration g at a particular location on Earth's surface. (Countless thousands of such measurements have been made during geophysical prospecting.)

To analyze a simple case, take the pendulum to be a uniform rod of length L, suspended from one end. For such a pendulum, h in Eq. 15-29, the distance between the pivot point and the center of mass, is $\frac{1}{2}L$. Table 10-2*e* tells us that the rotational inertia of this pendulum about a perpendicular axis through its center of mass is $\frac{1}{12}mL^2$. From the parallel-axis theorem of Eq. 10-36 ($I = I_{com} + Mh^2$), we then find that the rotational inertia about a perpendicular axis through one end of the rod is

$$I = I_{com} + mh^2 = \tfrac{1}{12}mL^2 + m(\tfrac{1}{2}L)^2 = \tfrac{1}{3}mL^2. \qquad (15\text{-}30)$$

If we put $h = \frac{1}{2}L$ and $I = \frac{1}{3}mL^2$ in Eq. 15-29 and solve for g, we find

$$g = \frac{8\pi^2 L}{3T^2}. \qquad (15\text{-}31)$$

Thus, by measuring L and the period T, we can find the value of g at the pendulum's location. (If precise measurements are to be made, a number of refinements are needed, such as swinging the pendulum in an evacuated chamber.)

Checkpoint 5

Three physical pendulums, of masses m_0, $2m_0$, and $3m_0$, have the same shape and size and are suspended at the same point. Rank the masses according to the periods of the pendulums, greatest first.

Sample Problem 15.05 Physical pendulum, period and length

In Fig. 15-13*a*, a meter stick swings about a pivot point at one end, at distance h from the stick's center of mass.

(a) What is the period of oscillation T?

KEY IDEA

The stick is not a simple pendulum because its mass is not concentrated in a bob at the end opposite the pivot point—so the stick is a physical pendulum.

Calculations: The period for a physical pendulum is given by Eq. 15-29, for which we need the rotational inertia I of the stick about the pivot point. We can treat the stick as a uniform rod of length L and mass m. Then Eq. 15-30 tells us that $I = \frac{1}{3}mL^2$, and the distance h in Eq. 15-29 is $\frac{1}{2}L$. Substituting these quantities into Eq. 15-29,

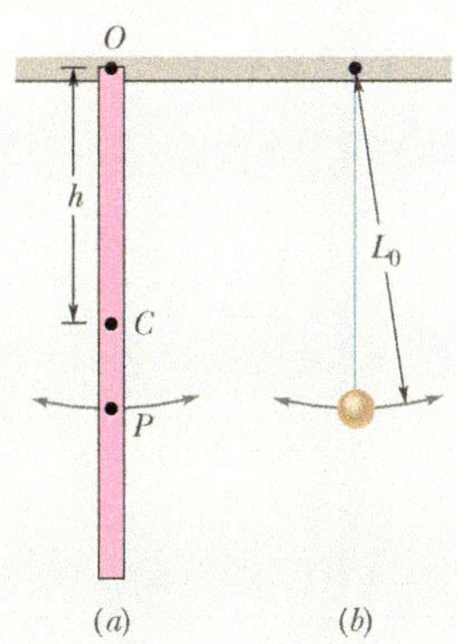

Figure 15-13 (*a*) A meter stick suspended from one end as a physical pendulum. (*b*) A simple pendulum whose length L_0 is chosen so that the periods of the two pendulums are equal. Point P on the pendulum of (*a*) marks the center of oscillation.

we find

$$T = 2\pi\sqrt{\frac{I}{mgh}} = 2\pi\sqrt{\frac{\frac{1}{3}mL^2}{mg(\frac{1}{2}L)}} \quad (15\text{-}32)$$

$$= 2\pi\sqrt{\frac{2L}{3g}} \quad (15\text{-}33)$$

$$= 2\pi\sqrt{\frac{(2)(1.00\text{ m})}{(3)(9.8\text{ m/s}^2)}} = 1.64\text{ s.} \quad \text{(Answer)}$$

Note the result is independent of the pendulum's mass m.

(b) What is the distance L_0 between the pivot point O of the stick and the center of oscillation of the stick?

Calculations: We want the length L_0 of the simple pendulum (drawn in Fig. 15-13b) that has the same period as the physical pendulum (the stick) of Fig. 15-13a. Setting Eqs. 15-28 and 15-33 equal yields

$$T = 2\pi\sqrt{\frac{L_0}{g}} = 2\pi\sqrt{\frac{2L}{3g}}. \quad (15\text{-}34)$$

You can see by inspection that

$$L_0 = \tfrac{2}{3}L \quad (15\text{-}35)$$

$$= (\tfrac{2}{3})(100\text{ cm}) = 66.7\text{ cm.} \quad \text{(Answer)}$$

In Fig. 15-13a, point P marks this distance from suspension point O. Thus, point P is the stick's center of oscillation for the given suspension point. Point P would be different for a different suspension choice.

WILEY PLUS Additional examples, video, and practice available at *WileyPLUS*

Simple Harmonic Motion and Uniform Circular Motion

In 1610, Galileo, using his newly constructed telescope, discovered the four principal moons of Jupiter. Over weeks of observation, each moon seemed to him to be moving back and forth relative to the planet in what today we would call simple harmonic motion; the disk of the planet was the midpoint of the motion. The record of Galileo's observations, written in his own hand, is actually still available. A. P. French of MIT used Galileo's data to work out the position of the moon Callisto relative to Jupiter (actually, the angular distance from Jupiter as seen from Earth) and found that the data approximates the curve shown in Fig. 15-14. The curve strongly suggests Eq. 15-3, the displacement function for simple harmonic motion. A period of about 16.8 days can be measured from the plot, but it is a period of what exactly? After all, a moon cannot possibly be oscillating back and forth like a block on the end of a spring, and so why would Eq. 15-3 have anything to do with it?

Actually, Callisto moves with essentially constant speed in an essentially circular orbit around Jupiter. Its true motion—far from being simple harmonic—is uniform circular motion along that orbit. What Galileo saw—and what you can see with a good pair of binoculars and a little patience—is the projection of this uniform circular motion on a line in the plane of the motion. We are led by Galileo's remarkable observations to the conclusion that simple harmonic

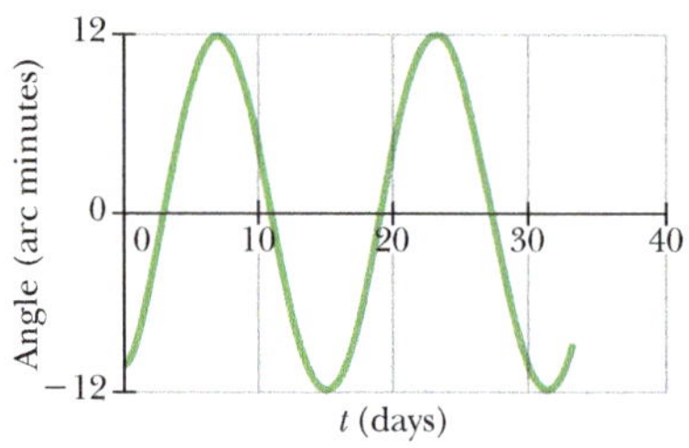

Figure 15-14 The angle between Jupiter and its moon Callisto as seen from Earth. Galileo's 1610 measurements approximate this curve, which suggests simple harmonic motion. At Jupiter's mean distance from Earth, 10 minutes of arc corresponds to about 2×10^6 km. (Based on A. P. French, *Newtonian Mechanics,* W. W. Norton & Company, New York, 1971, p. 288.)

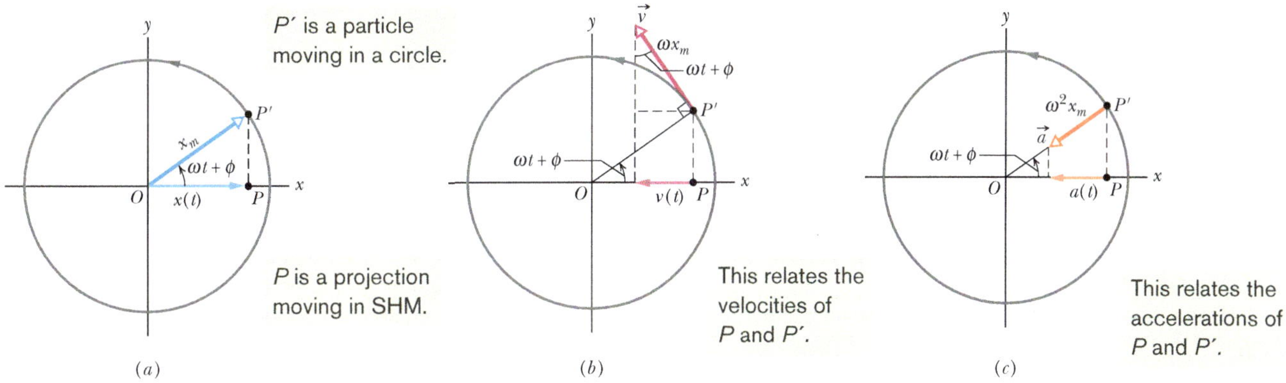

Figure 15-15 (*a*) A reference particle P' moving with uniform circular motion in a reference circle of radius x_m. Its projection P on the x axis executes simple harmonic motion. (*b*) The projection of the velocity $\vec{v}$ of the reference particle is the velocity of SHM. (*c*) The projection of the radial acceleration $\vec{a}$ of the reference particle is the acceleration of SHM.

motion is uniform circular motion viewed edge-on. In more formal language:

Simple harmonic motion is the projection of uniform circular motion on a diameter of the circle in which the circular motion occurs.

Figure 15-15*a* gives an example. It shows a *reference particle* P' moving in uniform circular motion with (constant) angular speed ω in a *reference circle.* The radius x_m of the circle is the magnitude of the particle's position vector. At any time t, the angular position of the particle is $\omega t + \phi$, where ϕ is its angular position at $t = 0$.

Position. The projection of particle P' onto the x axis is a point P, which we take to be a second particle. The projection of the position vector of particle P' onto the x axis gives the location $x(t)$ of P. (Can you see the x component in the triangle in Fig. 15-15*a*?) Thus, we find

$$x(t) = x_m \cos(\omega t + \phi), \tag{15-36}$$

which is precisely Eq. 15-3. Our conclusion is correct. If reference particle P' moves in uniform circular motion, its projection particle P moves in simple harmonic motion along a diameter of the circle.

Velocity. Figure 15-15*b* shows the velocity $\vec{v}$ of the reference particle. From Eq. 10-18 ($v = \omega r$), the magnitude of the velocity vector is ωx_m; its projection on the x axis is

$$v(t) = -\omega x_m \sin(\omega t + \phi), \tag{15-37}$$

which is exactly Eq. 15-6. The minus sign appears because the velocity component of P in Fig. 15-15*b* is directed to the left, in the negative direction of x. (The minus sign is consistent with the derivative of Eq. 15-36 with respect to time.)

Acceleration. Figure 15-15*c* shows the radial acceleration $\vec{a}$ of the reference particle. From Eq. 10-23 ($a_r = \omega^2 r$), the magnitude of the radial acceleration vector is $\omega^2 x_m$; its projection on the x axis is

$$a(t) = -\omega^2 x_m \cos(\omega t + \phi), \tag{15-38}$$

which is exactly Eq. 15-7. Thus, whether we look at the displacement, the velocity, or the acceleration, the projection of uniform circular motion is indeed simple harmonic motion.

15-5 DAMPED SIMPLE HARMONIC MOTION

Learning Objectives

After reading this module, you should be able to . . .

15.38 Describe the motion of a damped simple harmonic oscillator and sketch a graph of the oscillator's position as a function of time.

15.39 For any particular time, calculate the position of a damped simple harmonic oscillator.

15.40 Determine the amplitude of a damped simple harmonic oscillator at any given time.

15.41 Calculate the angular frequency of a damped simple harmonic oscillator in terms of the spring constant, the damping constant, and the mass, and approximate the angular frequency when the damping constant is small.

15.42 Apply the equation giving the (approximate) total energy of a damped simple harmonic oscillator as a function of time.

Key Ideas

- The mechanical energy E in a real oscillating system decreases during the oscillations because external forces, such as a drag force, inhibit the oscillations and transfer mechanical energy to thermal energy. The real oscillator and its motion are then said to be damped.
- If the damping force is given by $\vec{F}_d = -b\vec{v}$, where $\vec{v}$ is the velocity of the oscillator and b is a damping constant, then the displacement of the oscillator is given by

$$x(t) = x_m e^{-bt/2m} \cos(\omega' t + \phi),$$

where ω', the angular frequency of the damped oscillator, is given by

$$\omega' = \sqrt{\frac{k}{m} - \frac{b^2}{4m^2}}.$$

- If the damping constant is small ($b \ll \sqrt{km}$), then $\omega' \approx \omega$, where ω is the angular frequency of the undamped oscillator. For small b, the mechanical energy E of the oscillator is given by

$$E(t) \approx \tfrac{1}{2}kx_m^2 e^{-bt/m}.$$

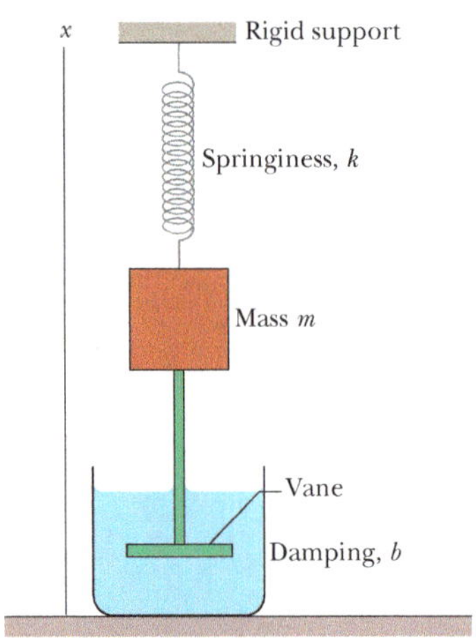

Figure 15-16 An idealized damped simple harmonic oscillator. A vane immersed in a liquid exerts a damping force on the block as the block oscillates parallel to the x axis.

Damped Simple Harmonic Motion

A pendulum will swing only briefly underwater, because the water exerts on the pendulum a drag force that quickly eliminates the motion. A pendulum swinging in air does better, but still the motion dies out eventually, because the air exerts a drag force on the pendulum (and friction acts at its support point), transferring energy from the pendulum's motion.

When the motion of an oscillator is reduced by an external force, the oscillator and its motion are said to be **damped.** An idealized example of a damped oscillator is shown in Fig. 15-16, where a block with mass m oscillates vertically on a spring with spring constant k. From the block, a rod extends to a vane (both assumed massless) that is submerged in a liquid. As the vane moves up and down, the liquid exerts an inhibiting drag force on it and thus on the entire oscillating system. With time, the mechanical energy of the block–spring system decreases, as energy is transferred to thermal energy of the liquid and vane.

Let us assume the liquid exerts a **damping force** $\vec{F}_d$ that is proportional to the velocity $\vec{v}$ of the vane and block (an assumption that is accurate if the vane moves slowly). Then, for force and velocity components along the x axis in Fig. 15-16, we have

$$F_d = -bv, \tag{15-39}$$

where b is a **damping constant** that depends on the characteristics of both the vane and the liquid and has the SI unit of kilogram per second. The minus sign indicates that $\vec{F}_d$ opposes the motion.

Damped Oscillations. The force on the block from the spring is $F_s = -kx$. Let us assume that the gravitational force on the block is negligible relative to F_d and F_s. Then we can write Newton's second law for components along the x axis ($F_{\text{net},x} = ma_x$) as

$$-bv - kx = ma. \tag{15-40}$$

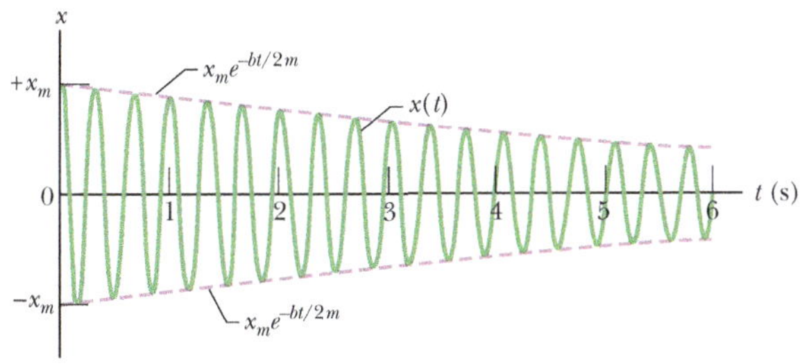

Figure 15-17 The displacement function $x(t)$ for the damped oscillator of Fig. 15-16. The amplitude, which is $x_m e^{-bt/2m}$, decreases exponentially with time.

Substituting dx/dt for v and d^2x/dt^2 for a and rearranging give us the differential equation

$$m\frac{d^2x}{dt^2} + b\frac{dx}{dt} + kx = 0. \tag{15-41}$$

The solution of this equation is

$$x(t) = x_m e^{-bt/2m}\cos(\omega' t + \phi), \tag{15-42}$$

where x_m is the amplitude and ω' is the angular frequency of the damped oscillator. This angular frequency is given by

$$\omega' = \sqrt{\frac{k}{m} - \frac{b^2}{4m^2}}. \tag{15-43}$$

If $b = 0$ (there is no damping), then Eq. 15-43 reduces to Eq. 15-12 ($\omega = \sqrt{k/m}$) for the angular frequency of an undamped oscillator, and Eq. 15-42 reduces to Eq. 15-3 for the displacement of an undamped oscillator. If the damping constant is small but not zero (so that $b \ll \sqrt{km}$), then $\omega' \approx \omega$.

Damped Energy. We can regard Eq. 15-42 as a cosine function whose amplitude, which is $x_m e^{-bt/2m}$, gradually decreases with time, as Fig. 15-17 suggests. For an undamped oscillator, the mechanical energy is constant and is given by Eq. 15-21 ($E = \frac{1}{2}kx_m^2$). If the oscillator is damped, the mechanical energy is not constant but decreases with time. If the damping is small, we can find $E(t)$ by replacing x_m in Eq. 15-21 with $x_m e^{-bt/2m}$, the amplitude of the damped oscillations. By doing so, we find that

$$E(t) \approx \tfrac{1}{2}kx_m^2 e^{-bt/m}, \tag{15-44}$$

which tells us that, like the amplitude, the mechanical energy decreases exponentially with time.

Checkpoint 6

Here are three sets of values for the spring constant, damping constant, and mass for the damped oscillator of Fig. 15-16. Rank the sets according to the time required for the mechanical energy to decrease to one-fourth of its initial value, greatest first.

Set 1	$2k_0$	b_0	m_0
Set 2	k_0	$6b_0$	$4m_0$
Set 3	$3k_0$	$3b_0$	m_0

Sample Problem 15.06 Damped harmonic oscillator, time to decay, energy

For the damped oscillator of Fig. 15-16, $m = 250$ g, $k = 85$ N/m, and $b = 70$ g/s.

(a) What is the period of the motion?

KEY IDEA

Because $b \ll \sqrt{km} = 4.6$ kg/s, the period is approximately that of the undamped oscillator.

Calculation: From Eq. 15-13, we then have

$$T = 2\pi\sqrt{\frac{m}{k}} = 2\pi\sqrt{\frac{0.25 \text{ kg}}{85 \text{ N/m}}} = 0.34 \text{ s.} \qquad \text{(Answer)}$$

(b) How long does it take for the amplitude of the damped oscillations to drop to half its initial value?

KEY IDEA

The amplitude at time t is displayed in Eq. 15-42 as $x_m e^{-bt/2m}$.

Calculations: The amplitude has the value x_m at $t = 0$. Thus, we must find the value of t for which

$$x_m e^{-bt/2m} = \tfrac{1}{2}x_m.$$

Canceling x_m and taking the natural logarithm of the equation that remains, we have $\ln\frac{1}{2}$ on the right side and

$$\ln(e^{-bt/2m}) = -bt/2m$$

on the left side. Thus,

$$t = \frac{-2m\ln\frac{1}{2}}{b} = \frac{-(2)(0.25 \text{ kg})(\ln\frac{1}{2})}{0.070 \text{ kg/s}}$$
$$= 5.0 \text{ s.} \qquad \text{(Answer)}$$

Because $T = 0.34$ s, this is about 15 periods of oscillation.

(c) How long does it take for the mechanical energy to drop to one-half its initial value?

KEY IDEA

From Eq. 15-44, the mechanical energy at time t is $\frac{1}{2}kx_m^2 e^{-bt/m}$.

Calculations: The mechanical energy has the value $\frac{1}{2}kx_m^2$ at $t = 0$. Thus, we must find the value of t for which

$$\tfrac{1}{2}kx_m^2 e^{-bt/m} = \tfrac{1}{2}(\tfrac{1}{2}kx_m^2).$$

If we divide both sides of this equation by $\frac{1}{2}kx_m^2$ and solve for t as we did above, we find

$$t = \frac{-m\ln\frac{1}{2}}{b} = \frac{-(0.25 \text{ kg})(\ln\frac{1}{2})}{0.070 \text{ kg/s}} = 2.5 \text{ s.} \qquad \text{(Answer)}$$

This is exactly half the time we calculated in (b), or about 7.5 periods of oscillation. Figure 15-17 was drawn to illustrate this sample problem.

WileyPLUS Additional examples, video, and practice available at *WileyPLUS*

15-6 FORCED OSCILLATIONS AND RESONANCE

Learning Objectives

After reading this module, you should be able to . . .

15.43 Distinguish between natural angular frequency ω and driving angular frequency ω_d.

15.44 For a forced oscillator, sketch a graph of the oscillation amplitude versus the ratio ω_d/ω of driving angular frequency to natural angular frequency, identify the approximate location of resonance, and indicate the effect of increasing the damping constant.

15.45 For a given natural angular frequency ω, identify the approximate driving angular frequency ω_d that gives resonance.

Key Ideas

- If an external driving force with angular frequency ω_d acts on an oscillating system with natural angular frequency ω, the system oscillates with angular frequency ω_d.
- The velocity amplitude v_m of the system is greatest when

$$\omega_d = \omega,$$

a condition called resonance. The amplitude x_m of the system is (approximately) greatest under the same condition.

Forced Oscillations and Resonance

A person swinging in a swing without anyone pushing it is an example of *free oscillation.* However, if someone pushes the swing periodically, the swing has *forced*, or *driven*, *oscillations*. *Two* angular frequencies are associated with a system undergoing driven oscillations: (1) the *natural* angular frequency ω of the system, which is the angular frequency at which it would oscillate if it were suddenly disturbed and then left to oscillate freely, and (2) the angular frequency ω_d of the external driving force causing the driven oscillations.

We can use Fig. 15-16 to represent an idealized forced simple harmonic oscillator if we allow the structure marked "rigid support" to move up and down at a variable angular frequency ω_d. Such a forced oscillator oscillates at the angular frequency ω_d of the driving force, and its displacement $x(t)$ is given by

$$x(t) = x_m \cos(\omega_d t + \phi), \qquad (15\text{-}45)$$

where x_m is the amplitude of the oscillations.

How large the displacement amplitude x_m is depends on a complicated function of ω_d and ω. The velocity amplitude v_m of the oscillations is easier to describe: it is greatest when

$$\omega_d = \omega \quad \text{(resonance)}, \qquad (15\text{-}46)$$

a condition called **resonance.** Equation 15-46 is also *approximately* the condition at which the displacement amplitude x_m of the oscillations is greatest. Thus, if you push a swing at its natural angular frequency, the displacement and velocity amplitudes will increase to large values, a fact that children learn quickly by trial and error. If you push at other angular frequencies, either higher or lower, the displacement and velocity amplitudes will be smaller.

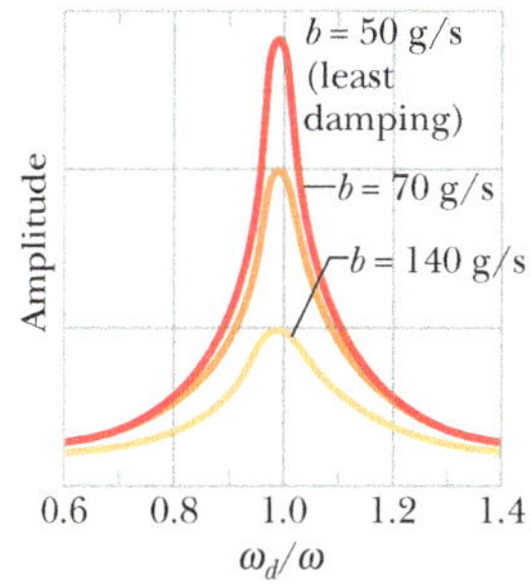

Figure 15-18 The displacement amplitude x_m of a forced oscillator varies as the angular frequency ω_d of the driving force is varied. The curves here correspond to three values of the damping constant b.

Figure 15-18 shows how the displacement amplitude of an oscillator depends on the angular frequency ω_d of the driving force, for three values of the damping coefficient b. Note that for all three the amplitude is approximately greatest when $\omega_d/\omega = 1$ (the resonance condition of Eq. 15-46). The curves of Fig. 15-18 show that less damping gives a taller and narrower *resonance peak.*

Examples. All mechanical structures have one or more natural angular frequencies, and if a structure is subjected to a strong external driving force that matches one of these angular frequencies, the resulting oscillations of the structure may rupture it. Thus, for example, aircraft designers must make sure that none of the natural angular frequencies at which a wing can oscillate matches the angular frequency of the engines in flight. A wing that flaps violently at certain engine speeds would obviously be dangerous.

Resonance appears to be one reason buildings in Mexico City collapsed in September 1985 when a major earthquake (8.1 on the Richter scale) occurred on the western coast of Mexico. The seismic waves from the earthquake should have been too weak to cause extensive damage when they reached Mexico City about 400 km away. However, Mexico City is largely built on an ancient lake bed, where the soil is still soft with water. Although the amplitude of the seismic waves was small in the firmer ground en route to Mexico City, their amplitude substantially increased in the loose soil of the city. Acceleration amplitudes of the waves were as much as $0.20g$, and the angular frequency was (surprisingly) concentrated around 3 rad/s. Not only was the ground severely oscillated, but many intermediate-height buildings had resonant angular frequencies of about 3 rad/s. Most of those buildings collapsed during the shaking (Fig. 15-19), while shorter buildings (with higher resonant angular frequencies) and taller buildings (with lower resonant angular frequencies) remained standing.

John T. Barr/Getty Images, Inc.

Figure 15-19 In 1985, buildings of intermediate height collapsed in Mexico City as a result of an earthquake far from the city. Taller and shorter buildings remained standing.

During a 1989 earthquake in the San Francisco–Oakland area, a similar resonant oscillation collapsed part of a freeway, dropping an upper deck onto a lower deck. That section of the freeway had been constructed on a loosely structured mudfill.

Review & Summary

Frequency The *frequency* f of periodic, or oscillatory, motion is the number of oscillations per second. In the SI system, it is measured in hertz:

$$1 \text{ hertz} = 1 \text{ Hz} = 1 \text{ oscillation per second} = 1 \text{ s}^{-1}. \tag{15-1}$$

Period The *period* T is the time required for one complete oscillation, or **cycle.** It is related to the frequency by

$$T = \frac{1}{f}. \tag{15-2}$$

Simple Harmonic Motion In *simple harmonic motion* (SHM), the displacement $x(t)$ of a particle from its equilibrium position is described by the equation

$$x = x_m \cos(\omega t + \phi) \quad \text{(displacement)}, \tag{15-3}$$

in which x_m is the **amplitude** of the displacement, $\omega t + \phi$ is the **phase** of the motion, and ϕ is the **phase constant.** The **angular frequency** ω is related to the period and frequency of the motion by

$$\omega = \frac{2\pi}{T} = 2\pi f \quad \text{(angular frequency)}. \tag{15-5}$$

Differentiating Eq. 15-3 leads to equations for the particle's SHM velocity and acceleration as functions of time:

$$v = -\omega x_m \sin(\omega t + \phi) \quad \text{(velocity)} \tag{15-6}$$

and

$$a = -\omega^2 x_m \cos(\omega t + \phi) \quad \text{(acceleration)}. \tag{15-7}$$

In Eq. 15-6, the positive quantity ωx_m is the **velocity amplitude** v_m of the motion. In Eq. 15-7, the positive quantity $\omega^2 x_m$ is the **acceleration amplitude** a_m of the motion.

The Linear Oscillator A particle with mass m that moves under the influence of a Hooke's law restoring force given by $F = -kx$ exhibits simple harmonic motion with

$$\omega = \sqrt{\frac{k}{m}} \quad \text{(angular frequency)} \tag{15-12}$$

and

$$T = 2\pi\sqrt{\frac{m}{k}} \quad \text{(period)}. \tag{15-13}$$

Such a system is called a **linear simple harmonic oscillator.**

Energy A particle in simple harmonic motion has, at any time, kinetic energy $K = \frac{1}{2}mv^2$ and potential energy $U = \frac{1}{2}kx^2$. If no friction is present, the mechanical energy $E = K + U$ remains constant even though K and U change.

Pendulums Examples of devices that undergo simple harmonic motion are the **torsion pendulum** of Fig. 15-9, the **simple pendulum** of Fig. 15-11, and the **physical pendulum** of Fig. 15-12. Their periods of oscillation for small oscillations are, respectively,

$$T = 2\pi\sqrt{I/\kappa} \quad \text{(torsion pendulum)}, \tag{15-23}$$

$$T = 2\pi\sqrt{L/g} \quad \text{(simple pendulum)}, \tag{15-28}$$

$$T = 2\pi\sqrt{I/mgh} \quad \text{(physical pendulum)}. \tag{15-29}$$

Simple Harmonic Motion and Uniform Circular Motion Simple harmonic motion is the projection of uniform circular motion onto the diameter of the circle in which the circular motion occurs. Figure 15-15 shows that all parameters of circular motion (position, velocity, and acceleration) project to the corresponding values for simple harmonic motion.

Damped Harmonic Motion The mechanical energy E in a real oscillating system decreases during the oscillations because external forces, such as a drag force, inhibit the oscillations and transfer mechanical energy to thermal energy. The real oscillator and its motion are then said to be **damped.** If the **damping force** is given by $\vec{F}_d = -b\vec{v}$, where $\vec{v}$ is the velocity of the oscillator and b is a **damping constant,** then the displacement of the oscillator is given by

$$x(t) = x_m e^{-bt/2m} \cos(\omega' t + \phi), \tag{15-42}$$

where ω', the angular frequency of the damped oscillator, is given by

$$\omega' = \sqrt{\frac{k}{m} - \frac{b^2}{4m^2}}. \tag{15-43}$$

If the damping constant is small ($b \ll \sqrt{km}$), then $\omega' \approx \omega$, where ω is the angular frequency of the undamped oscillator. For small b, the mechanical energy E of the oscillator is given by

$$E(t) \approx \tfrac{1}{2}kx_m^2 e^{-bt/m}. \tag{15-44}$$

Forced Oscillations and Resonance If an external driving force with angular frequency ω_d acts on an oscillating system with *natural* angular frequency ω, the system oscillates with angular frequency ω_d. The velocity amplitude v_m of the system is greatest when

$$\omega_d = \omega, \tag{15-46}$$

a condition called **resonance.** The amplitude x_m of the system is (approximately) greatest under the same condition.

Questions

1 Which of the following describe ϕ for the SHM of Fig. 15-20*a*:

(a) $-\pi < \phi < -\pi/2$,

(b) $\pi < \phi < 3\pi/2$,

(c) $-3\pi/2 < \phi < -\pi$?

2 The velocity $v(t)$ of a particle undergoing SHM is graphed in Fig. 15-20*b*. Is the particle momentarily stationary, headed toward $-x_m$, or headed toward $+x_m$ at (a) point A on the graph and (b) point B? Is the particle at $-x_m$, at $+x_m$, at 0, between $-x_m$ and 0, or between 0 and $+x_m$ when its velocity is represented by (c) point A and (d) point B? Is the speed of the particle increasing or decreasing at (e) point A and (f) point B?

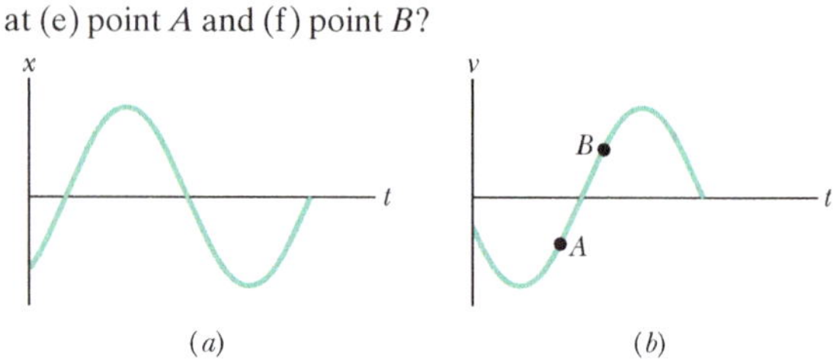

Figure 15-20 Questions 1 and 2.

3 The acceleration $a(t)$ of a particle undergoing SHM is graphed in Fig. 15-21. (a) Which of the labeled points corresponds to the particle at $-x_m$? (b) At point 4, is the velocity of the particle positive, negative, or zero? (c) At point 5, is the particle at $-x_m$, at $+x_m$, at 0, between $-x_m$ and 0, or between 0 and $+x_m$?

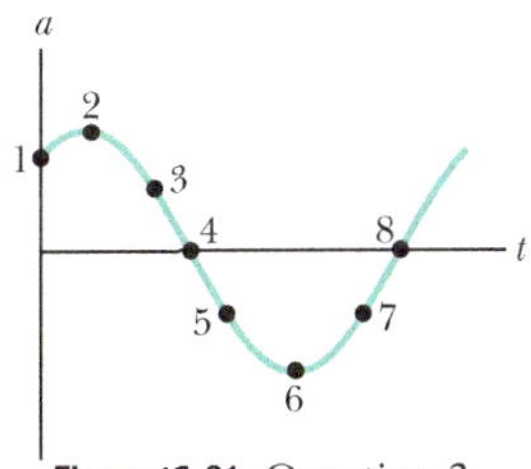

Figure 15-21 Question 3.

4 Which of the following relationships between the acceleration a and the displacement x of a particle involve SHM: (a) $a = 0.5x$, (b) $a = 400x^2$, (c) $a = -20x$, (d) $a = -3x^2$?

5 You are to complete Fig. 15-22a so that it is a plot of velocity v versus time t for the spring–block oscillator that is shown in Fig. 15-22b for $t = 0$. (a) In Fig. 15-22a, at which lettered point or in what region between the points should the (vertical) v axis intersect the t axis? (For example, should it intersect at point A, or maybe in the region between points A and B?) (b) If the block's velocity is given by $v = -v_m \sin(\omega t + \phi)$, what is the value of ϕ? Make it positive, and if you cannot specify the value (such as $+\pi/2$ rad), then give a range of values (such as between 0 and $\pi/2$ rad).

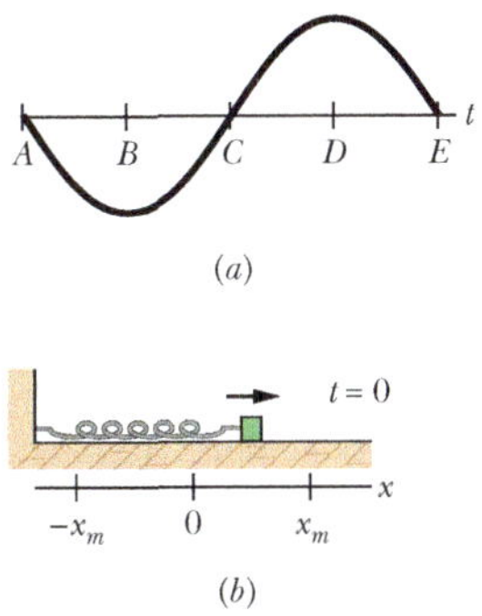

Figure 15-22 Question 5.

6 You are to complete Fig. 15-23a so that it is a plot of acceleration a versus time t for the spring–block oscillator that is shown in Fig. 15-23b for $t = 0$. (a) In Fig. 15-23a, at which lettered point or in what region between the points should the (vertical) a axis intersect the t axis? (For example, should it intersect at point A, or maybe in the region between points A and B?) (b) If the block's acceleration is given by $a = -a_m \cos(\omega t + \phi)$, what is the value of ϕ? Make it positive, and if you cannot specify the value (such as $+\pi/2$ rad), then give a range of values (such as between 0 and $\pi/2$).

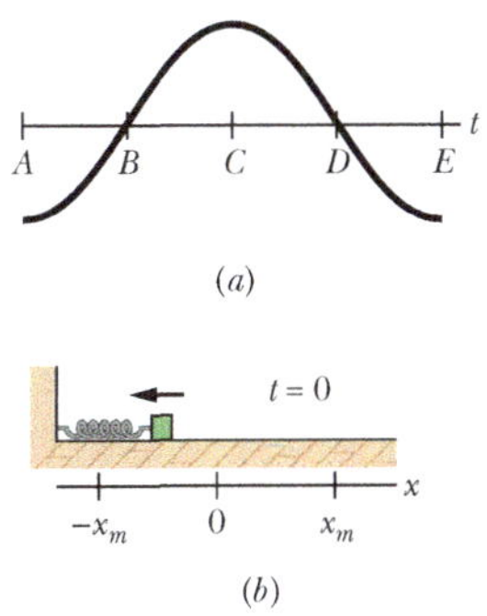

Figure 15-23 Question 6.

7 Figure 15-24 shows the $x(t)$ curves for three experiments involving a particular spring–box system oscillating in SHM. Rank the curves according to (a) the system's angular frequency, (b) the spring's potential energy at time $t = 0$, (c) the box's kinetic energy at $t = 0$, (d) the box's speed at $t = 0$, and (e) the box's maximum kinetic energy, greatest first.

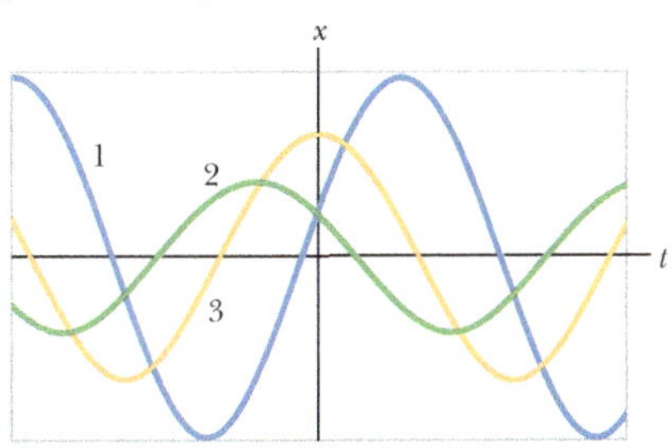

Figure 15-24 Question 7.

8 Figure 15-25 shows plots of the kinetic energy K versus position x for three harmonic oscillators that have the same mass. Rank the plots according to (a) the corresponding spring constant and (b) the corresponding period of the oscillator, greatest first.

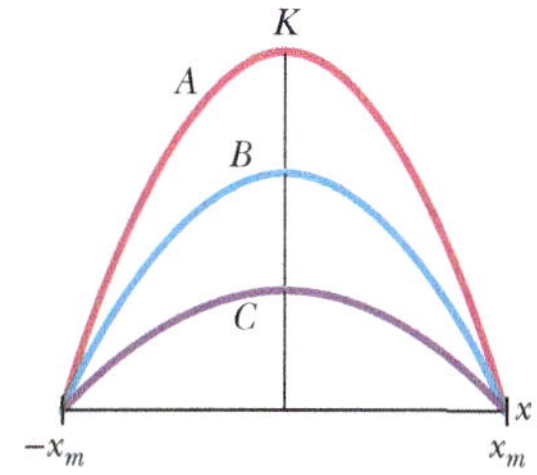

Figure 15-25 Question 8.

9 Figure 15-26 shows three physical pendulums consisting of identical uniform spheres of the same mass that are rigidly connected by identical rods of negligible mass. Each pendulum is vertical and can pivot about suspension point O. Rank the pendulums according to their period of oscillation, greatest first.

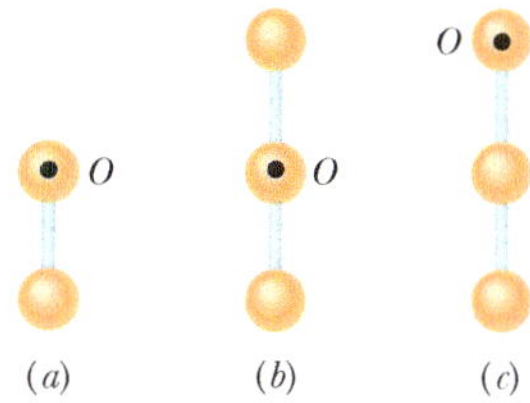

Figure 15-26 Question 9.

10 You are to build the oscillation transfer device shown in Fig. 15-27. It consists of two spring–block systems hanging from a flexible rod. When the spring of system 1 is stretched and then released, the resulting SHM of system 1 at frequency f_1 oscillates the rod. The rod then exerts a driving force on system 2, at the same frequency f_1. You can choose from four springs with spring constants k of 1600, 1500, 1400, and 1200 N/m, and four blocks with masses m of 800, 500, 400, and 200 kg. Mentally determine which spring should go with which block in each of the two systems to maximize the amplitude of oscillations in system 2.

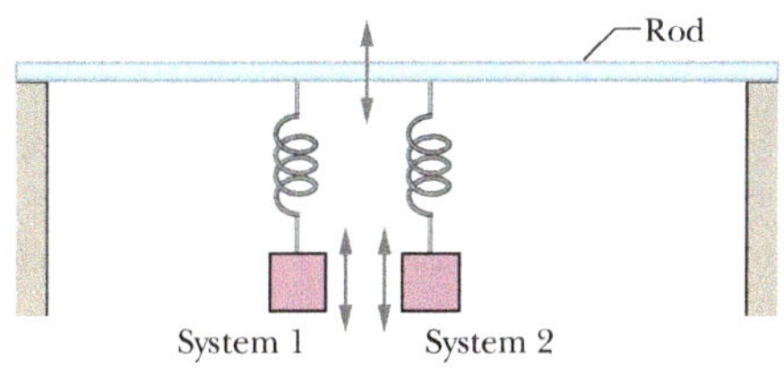

Figure 15-27 Question 10.

11 In Fig. 15-28, a spring–block system is put into SHM in two experiments. In the first, the block is pulled from the equilibrium position through a displacement d_1 and then released. In the second, it is pulled from the equilibrium position through a greater displacement d_2 and then released. Are the (a) amplitude, (b) period, (c) frequency, (d) maximum kinetic energy, and (e) maximum potential energy in the second experiment greater than, less than, or the same as those in the first experiment?

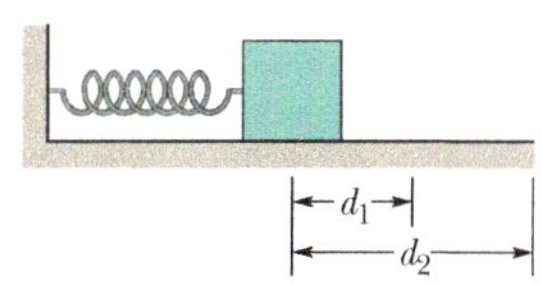

Figure 15-28 Question 11.

12 Figure 15-29 gives, for three situations, the displacements $x(t)$ of a pair of simple harmonic oscillators (A and B) that are identical except for phase. For each pair, what phase shift (in radians and in degrees) is needed to shift the curve for A to coincide with the curve for B? Of the many possible answers, choose the shift with the smallest absolute magnitude.

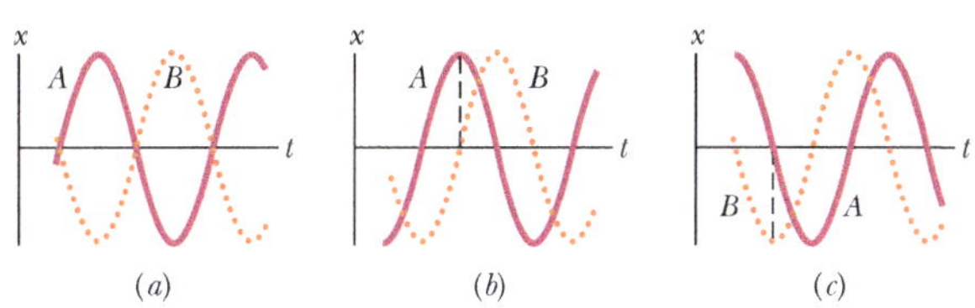

Figure 15-29 Question 12.

Problems

GO Tutoring problem available (at instructor's discretion) in *WileyPLUS* and WebAssign
SSM Worked-out solution available in Student Solutions Manual
• – ••• Number of dots indicates level of problem difficulty
WWW Worked-out solution is at
ILW Interactive solution is at
http://www.wiley.com/college/halliday
Additional information available in *The Flying Circus of Physics* and at flyingcircusofphysics.com

Module 15-1 Simple Harmonic Motion

•1 An object undergoing simple harmonic motion takes 0.25 s to travel from one point of zero velocity to the next such point. The distance between those points is 36 cm. Calculate the (a) period, (b) frequency, and (c) amplitude of the motion.

•2 A 0.12 kg body undergoes simple harmonic motion of amplitude 8.5 cm and period 0.20 s. (a) What is the magnitude of the maximum force acting on it? (b) If the oscillations are produced by a spring, what is the spring constant?

•3 What is the maximum acceleration of a platform that oscillates at amplitude 2.20 cm and frequency 6.60 Hz?

•4 An automobile can be considered to be mounted on four identical springs as far as vertical oscillations are concerned. The springs of a certain car are adjusted so that the oscillations have a frequency of 3.00 Hz. (a) What is the spring constant of each spring if the mass of the car is 1450 kg and the mass is evenly distributed over the springs? (b) What will be the oscillation frequency if five passengers, averaging 73.0 kg each, ride in the car with an even distribution of mass?

•5 SSM In an electric shaver, the blade moves back and forth over a distance of 2.0 mm in simple harmonic motion, with frequency 120 Hz. Find (a) the amplitude, (b) the maximum blade speed, and (c) the magnitude of the maximum blade acceleration.

•6 A particle with a mass of 1.00×10^{-20} kg is oscillating with simple harmonic motion with a period of 1.00×10^{-5} s and a maximum speed of 1.00×10^{3} m/s. Calculate (a) the angular frequency and (b) the maximum displacement of the particle.

•7 SSM A loudspeaker produces a musical sound by means of the oscillation of a diaphragm whose amplitude is limited to 1.00 μm. (a) At what frequency is the magnitude a of the diaphragm's acceleration equal to g? (b) For greater frequencies, is a greater than or less than g?

•8 What is the phase constant for the harmonic oscillator with the position function $x(t)$ given in Fig. 15-30 if the position function has the form $x = x_m \cos(\omega t + \phi)$? The vertical axis scale is set by $x_s = 6.0$ cm.

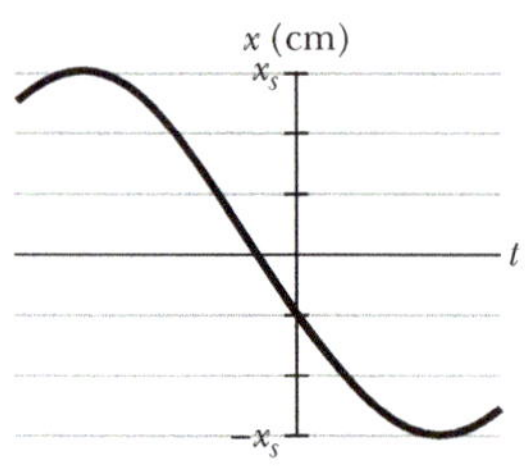

Figure 15-30 Problem 8.

•9 The position function $x = (6.0 \text{ m}) \cos[(3\pi \text{ rad/s})t + \pi/3 \text{ rad}]$ gives the simple harmonic motion of a body. At $t = 2.0$ s, what are the (a) displacement, (b) velocity, (c) acceleration, and (d) phase of the motion? Also, what are the (e) frequency and (f) period of the motion?

•10 An oscillating block–spring system takes 0.75 s to begin repeating its motion. Find (a) the period, (b) the frequency in hertz, and (c) the angular frequency in radians per second.

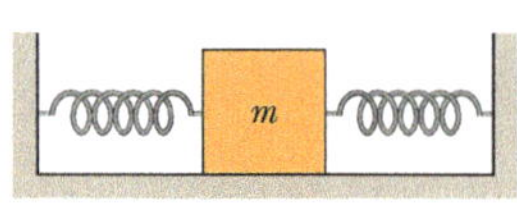

Figure 15-31 Problems 11 and 21.

•11 In Fig. 15-31, two identical springs of spring constant 7580 N/m are attached to a block of mass 0.245 kg. What is the frequency of oscillation on the frictionless floor?

•12 What is the phase constant for the harmonic oscillator with the velocity function $v(t)$ given in Fig. 15-32 if the position function $x(t)$ has the form $x = x_m \cos(\omega t + \phi)$? The vertical axis scale is set by $v_s = 4.0$ cm/s.

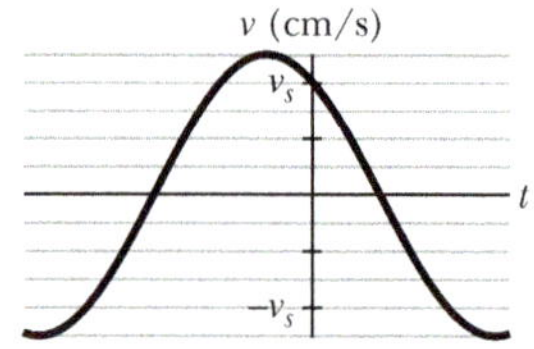

Figure 15-32 Problem 12.

•13 SSM An oscillator consists of a block of mass 0.500 kg connected to a spring. When set into oscillation with amplitude 35.0 cm, the oscillator repeats its motion every 0.500 s. Find the (a) period, (b) frequency, (c) angular frequency, (d) spring constant, (e) maximum speed, and (f) magnitude of the maximum force on the block from the spring.

••14 A simple harmonic oscillator consists of a block of mass 2.00 kg attached to a spring of spring constant 100 N/m. When $t = 1.00$ s, the position and velocity of the block are $x = 0.129$ m and $v = 3.415$ m/s. (a) What is the amplitude of the oscillations? What were the (b) position and (c) velocity of the block at $t = 0$ s?

••15 SSM Two particles oscillate in simple harmonic motion along a common straight-line segment of length A. Each particle has a period of 1.5 s, but they differ in phase by $\pi/6$ rad. (a) How far apart are they (in terms of A) 0.50 s after the lagging particle leaves one end of the path? (b) Are they then moving in the same direction, toward each other, or away from each other?

••16 Two particles execute simple harmonic motion of the same amplitude and frequency along close parallel lines. They pass each other moving in opposite directions each time their displacement is half their amplitude. What is their phase difference?

••17 ILW An oscillator consists of a block attached to a spring (k = 400 N/m). At some time t, the position (measured from the system's equilibrium location), velocity, and acceleration of the block are $x = 0.100$ m, $v = -13.6$ m/s, and $a = -123$ m/s^2. Calculate (a) the frequency of oscillation, (b) the mass of the block, and (c) the amplitude of the motion.

••18 GO At a certain harbor, the tides cause the ocean surface to rise and fall a distance d (from highest level to lowest level) in simple harmonic motion, with a period of 12.5 h. How long does it take for the water to fall a distance $0.250d$ from its highest level?

••19 A block rides on a piston (a squat cylindrical piece) that is moving vertically with simple harmonic motion. (a) If the SHM has period 1.0 s, at what amplitude of motion will the block and piston separate? (b) If the piston has an amplitude of 5.0 cm, what is the maximum frequency for which the block and piston will be in contact continuously?

••20 GO Figure 15-33a is a partial graph of the position function $x(t)$ for a simple harmonic oscillator with an angular frequency of

1.20 rad/s; Fig. 15-33b is a partial graph of the corresponding velocity function $v(t)$. The vertical axis scales are set by x_s = 5.0 cm and v_s = 5.0 cm/s. What is the phase constant of the SHM if the position function $x(t)$ is in the general form $x = x_m \cos(\omega t + \phi)$?

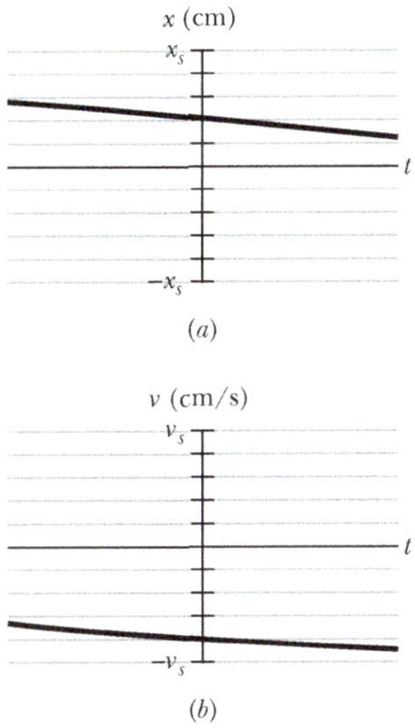

Figure 15-33 Problem 20.

••21 ILW In Fig. 15-31, two springs are attached to a block that can oscillate over a frictionless floor. If the left spring is removed, the block oscillates at a frequency of 30 Hz. If, instead, the spring on the right is removed, the block oscillates at a frequency of 45 Hz. At what frequency does the block oscillate with both springs attached?

••22 GO Figure 15-34 shows block 1 of mass 0.200 kg sliding to the right over a frictionless elevated surface at a speed of 8.00 m/s. The block undergoes an elastic collision with stationary block 2, which is attached to a spring of spring constant 1208.5 N/m. (Assume that the spring does not affect the collision.) After the collision, block 2 oscillates in SHM with a period of 0.140 s, and block 1 slides off the opposite end of the elevated surface, landing a distance d from the base of that surface after falling height h = 4.90 m. What is the value of d?

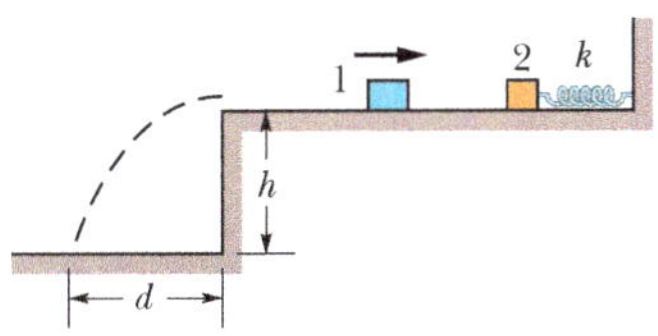

Figure 15-34 Problem 22.

••23 SSM WWW A block is on a horizontal surface (a shake table) that is moving back and forth horizontally with simple harmonic motion of frequency 2.0 Hz. The coefficient of static friction between block and surface is 0.50. How great can the amplitude of the SHM be if the block is not to slip along the surface?

•••24 In Fig. 15-35, two springs are joined and connected to a block of mass 0.245 kg that is set oscillating over a frictionless floor. The springs each have spring constant k = 6430 N/m. What is the frequency of the oscillations?

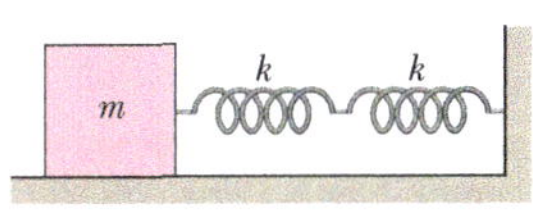

Figure 15-35 Problem 24.

•••25 GO In Fig. 15-36, a block weighing 14.0 N, which can slide without friction on an incline at angle θ = 40.0°, is connected to the top of the incline by a massless spring of unstretched length 0.450 m and spring constant 120 N/m. (a) How far from the top of the incline is the block's equilibrium point? (b) If the block is pulled slightly down the incline and released, what is the period of the resulting oscillations?

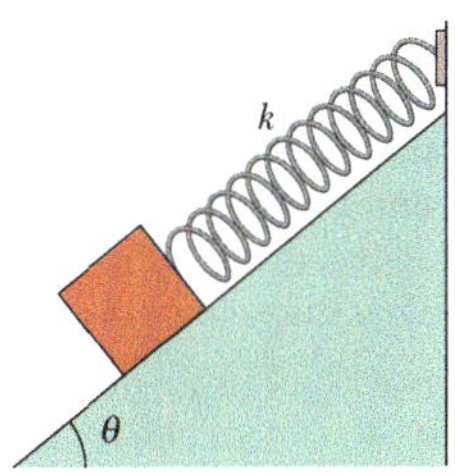

Figure 15-36 Problem 25.

•••26 GO In Fig. 15-37, two blocks (m = 1.8 kg and M = 10 kg) and a spring (k = 200 N/m) are arranged on a horizontal, frictionless surface. The coefficient of static friction between the two blocks is 0.40. What amplitude of simple harmonic motion of the spring–blocks system puts the smaller block on the verge of slipping over the larger block?

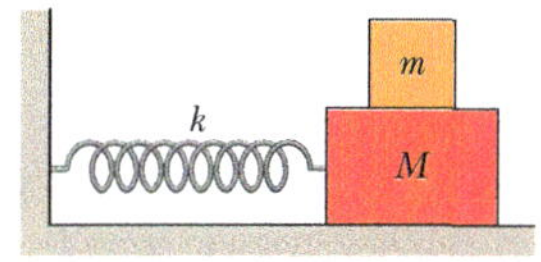

Figure 15-37 Problem 26.

Module 15-2 Energy in Simple Harmonic Motion

•27 SSM When the displacement in SHM is one-half the amplitude x_m, what fraction of the total energy is (a) kinetic energy and (b) potential energy? (c) At what displacement, in terms of the amplitude, is the energy of the system half kinetic energy and half potential energy?

•28 Figure 15-38 gives the one-dimensional potential energy well for a 2.0 kg particle (the function $U(x)$ has the form bx^2 and the vertical axis scale is set by U_s = 2.0 J). (a) If the particle passes through the equilibrium position with a velocity of 85 cm/s, will it be turned back before it reaches x = 15 cm? (b) If yes, at what position, and if no, what is the speed of the particle at x = 15 cm?

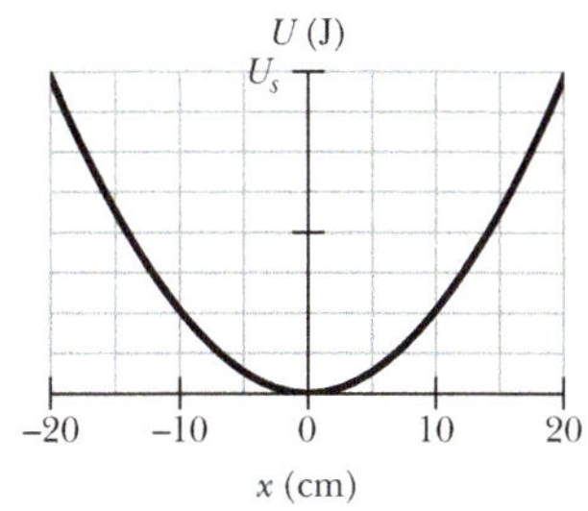

Figure 15-38 Problem 28.

•29 SSM Find the mechanical energy of a block–spring system with a spring constant of 1.3 N/cm and an amplitude of 2.4 cm.

•30 An oscillating block–spring system has a mechanical energy of 1.00 J, an amplitude of 10.0 cm, and a maximum speed of 1.20 m/s. Find (a) the spring constant, (b) the mass of the block, and (c) the frequency of oscillation.

•31 ILW A 5.00 kg object on a horizontal frictionless surface is attached to a spring with k = 1000 N/m. The object is displaced from equilibrium 50.0 cm horizontally and given an initial velocity of 10.0 m/s back toward the equilibrium position. What are (a) the motion's frequency, (b) the initial potential energy of the block–spring system, (c) the initial kinetic energy, and (d) the motion's amplitude?

•32 Figure 15-39 shows the kinetic energy K of a simple harmonic oscillator versus its position x. The vertical axis scale is set by K_s = 4.0 J. What is the spring constant?

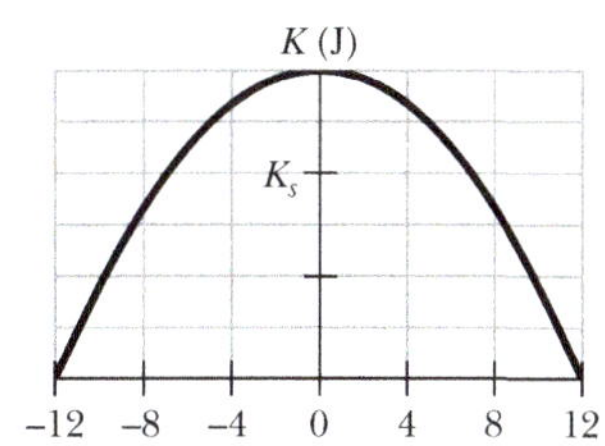

Figure 15-39 Problem 32.

••33 GO A block of mass M = 5.4 kg, at rest on a horizontal frictionless table, is attached to a rigid support by a spring of constant k = 6000 N/m. A bullet of mass m = 9.5 g and velocity $\vec{v}$ of magnitude 630 m/s strikes and is embedded in the block (Fig. 15-40). Assuming the compression of the spring is negligible until the bullet is embedded, determine (a) the speed of the block immediately after the collision and (b) the amplitude of the resulting simple harmonic motion.

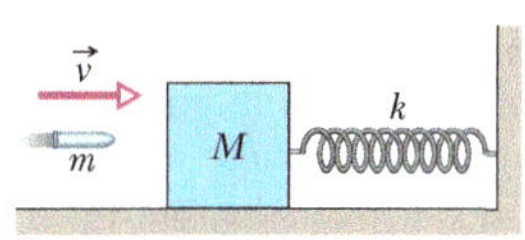

Figure 15-40 Problem 33.

••34 GO In Fig. 15-41, block 2 of mass 2.0 kg oscillates on the end of a spring in SHM with a period of 20 ms. The block's position is given by $x = (1.0 \text{ cm}) \cos(\omega t + \pi/2)$. Block 1 of mass 4.0 kg slides toward block 2 with a velocity of magnitude 6.0 m/s, directed along the spring's length. The two blocks undergo a completely inelastic collision at time $t = 5.0$ ms. (The duration of the collision is much less than the period of motion.) What is the amplitude of the SHM after the collision?

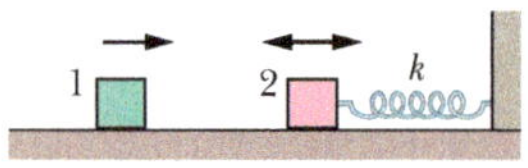

Figure 15-41 Problem 34.

••35 A 10 g particle undergoes SHM with an amplitude of 2.0 mm, a maximum acceleration of magnitude 8.0×10^3 m/s^2, and an unknown phase constant ϕ. What are (a) the period of the motion, (b) the maximum speed of the particle, and (c) the total mechanical energy of the oscillator? What is the magnitude of the force on the particle when the particle is at (d) its maximum displacement and (e) half its maximum displacement?

••36 If the phase angle for a block–spring system in SHM is $\pi/6$ rad and the block's position is given by $x = x_m \cos(\omega t + \phi)$, what is the ratio of the kinetic energy to the potential energy at time $t = 0$?

•••37 GO A massless spring hangs from the ceiling with a small object attached to its lower end. The object is initially held at rest in a position y_i such that the spring is at its rest length. The object is then released from y_i and oscillates up and down, with its lowest position being 10 cm below y_i. (a) What is the frequency of the oscillation? (b) What is the speed of the object when it is 8.0 cm below the initial position? (c) An object of mass 300 g is attached to the first object, after which the system oscillates with half the original frequency. What is the mass of the first object? (d) How far below y_i is the new equilibrium (rest) position with both objects attached to the spring?

Module 15-3 An Angular Simple Harmonic Oscillator

•38 A 95 kg solid sphere with a 15 cm radius is suspended by a vertical wire. A torque of 0.20 N·m is required to rotate the sphere through an angle of 0.85 rad and then maintain that orientation. What is the period of the oscillations that result when the sphere is then released?

••39 SSM WWW The balance wheel of an old-fashioned watch oscillates with angular amplitude π rad and period 0.500 s. Find (a) the maximum angular speed of the wheel, (b) the angular speed at displacement $\pi/2$ rad, and (c) the magnitude of the angular acceleration at displacement $\pi/4$ rad.

Module 15-4 Pendulums, Circular Motion

•40 ILW A physical pendulum consists of a meter stick that is pivoted at a small hole drilled through the stick a distance d from the 50 cm mark. The period of oscillation is 2.5 s. Find d.

•41 SSM In Fig. 15-42, the pendulum consists of a uniform disk with radius $r = 10.0$ cm and mass 500 g attached to a uniform rod with length $L = 500$ mm and mass 270 g. (a) Calculate the rotational inertia of the pendulum about the pivot point. (b) What is the distance between the pivot point and the center of mass of the pendulum? (c) Calculate the period of oscillation.

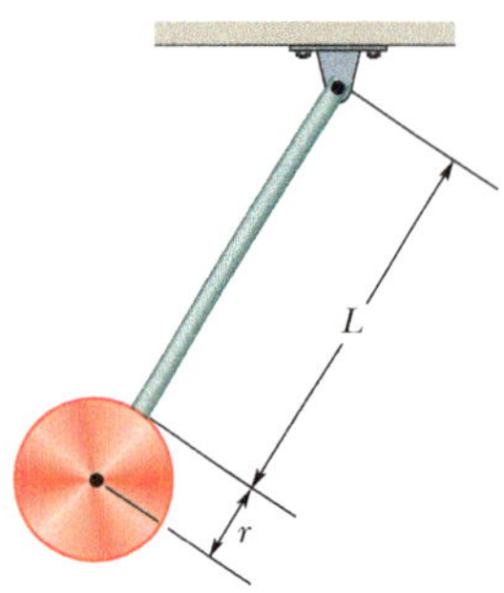

Figure 15-42 Problem 41.

•42 Suppose that a simple pendulum consists of a small 60.0 g bob at the end of a cord of negligible mass. If the angle θ between the cord and the vertical is given by

$$\theta = (0.0800 \text{ rad}) \cos[(4.43 \text{ rad/s})t + \phi],$$

what are (a) the pendulum's length and (b) its maximum kinetic energy?

•43 (a) If the physical pendulum of Fig. 15-13 and the associated sample problem is inverted and suspended at point P, what is its period of oscillation? (b) Is the period now greater than, less than, or equal to its previous value?

•44 A physical pendulum consists of two meter-long sticks joined together as shown in Fig. 15-43. What is the pendulum's period of oscillation about a pin inserted through point A at the center of the horizontal stick?

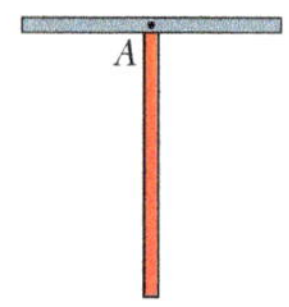

Figure 15-43 Problem 44.

•45 A performer seated on a trapeze is swinging back and forth with a period of 8.85 s. If she stands up, thus raising the center of mass of the *trapeze + performer* system by 35.0 cm, what will be the new period of the system? Treat *trapeze + performer* as a simple pendulum.

•46 A physical pendulum has a center of oscillation at distance $2L/3$ from its point of suspension. Show that the distance between the point of suspension and the center of oscillation for a physical pendulum of any form is I/mh, where I and h have the meanings in Eq. 15-29 and m is the mass of the pendulum.

•47 In Fig. 15-44, a physical pendulum consists of a uniform solid disk (of radius $R = 2.35$ cm) supported in a vertical plane by a pivot located a distance $d = 1.75$ cm from the center of the disk. The disk is displaced by a small angle and released. What is the period of the resulting simple harmonic motion?

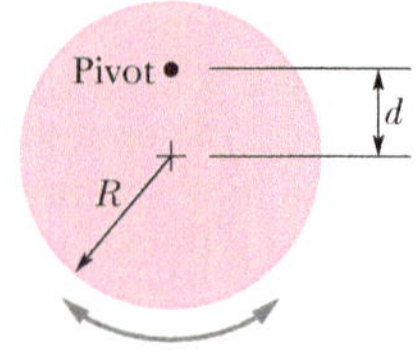

Figure 15-44 Problem 47.

••48 GO A rectangular block, with face lengths $a = 35$ cm and $b = 45$ cm, is to be suspended on a thin horizontal rod running through a narrow hole in the block. The block is then to be set swinging about the rod like a pendulum, through small angles so that it is in SHM. Figure 15-45 shows one possible position of the hole, at distance r from the block's center, along a line connecting the center with a corner. (a) Plot the period versus distance r along that line such that the minimum in the curve is apparent. (b) For what value of r does that minimum occur? There is a line of points around the block's center for which the period of swinging has the same minimum value. (c) What shape does that line make?

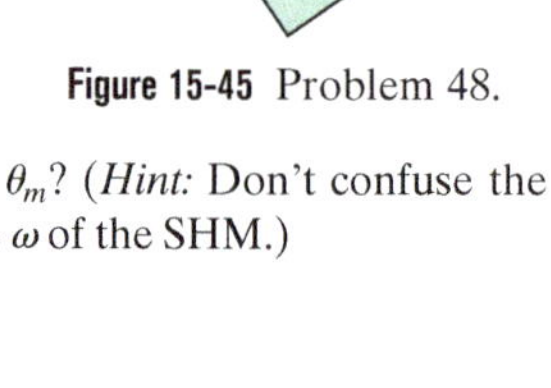

Figure 15-45 Problem 48.

••49 GO The angle of the pendulum of Fig. 15-11b is given by $\theta = \theta_m \cos[(4.44 \text{ rad/s})t + \phi]$. If at $t = 0$, $\theta = 0.040$ rad and $d\theta/dt = -0.200$ rad/s, what are (a) the phase constant ϕ and (b) the maximum angle θ_m? (*Hint:* Don't confuse the rate $d\theta/dt$ at which θ changes with the ω of the SHM.)

••**50** A thin uniform rod (mass = 0.50 kg) swings about an axis that passes through one end of the rod and is perpendicular to the plane of the swing. The rod swings with a period of 1.5 s and an angular amplitude of 10°. (a) What is the length of the rod? (b) What is the maximum kinetic energy of the rod as it swings?

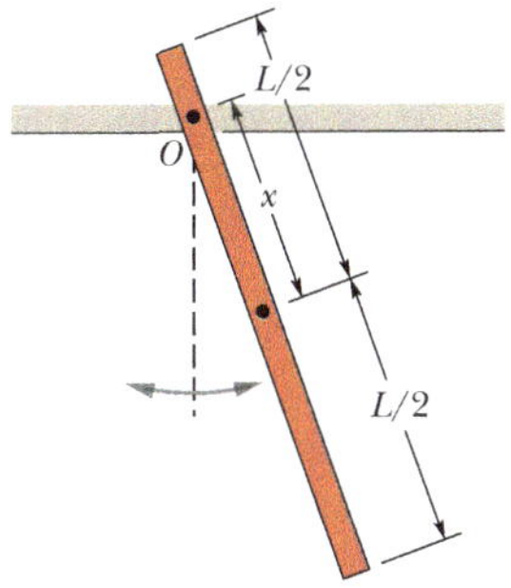

Figure 15-46 Problem 51.

••**51** GO In Fig. 15-46, a stick of length $L = 1.85$ m oscillates as a physical pendulum. (a) What value of distance x between the stick's center of mass and its pivot point O gives the least period? (b) What is that least period?

••**52** GO The 3.00 kg cube in Fig. 15-47 has edge lengths $d = 6.00$ cm and is mounted on an axle through its center. A spring ($k = 1200$ N/m) connects the cube's upper corner to a rigid wall. Initially the spring is at its rest length. If the cube is rotated 3° and released, what is the period of the resulting SHM?

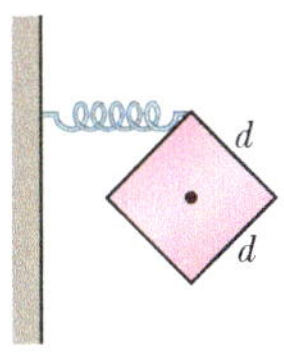

Figure 15-47 Problem 52.

••**53** SSM ILW In the overhead view of Fig. 15-48, a long uniform rod of mass 0.600 kg is free to rotate in a horizontal plane about a vertical axis through its center. A spring with force constant $k = 1850$ N/m is connected horizontally between one end of the rod and a fixed wall. When the rod is in equilibrium, it is parallel to the wall. What is the period of the small oscillations that result when the rod is rotated slightly and released?

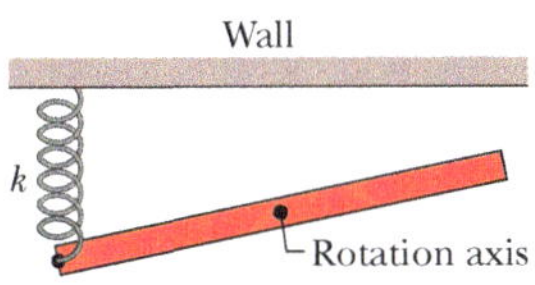

Figure 15-48 Problem 53.

••**54** GO In Fig. 15-49a, a metal plate is mounted on an axle through its center of mass. A spring with $k = 2000$ N/m connects a wall with a point on the rim a distance $r = 2.5$ cm from the center of mass. Initially the spring is at its rest length. If the plate is rotated by 7° and released, it rotates about the axle in SHM, with its angular position given by Fig. 15-49b. The horizontal axis scale is set by $t_s = 20$ ms. What is the rotational inertia of the plate about its center of mass?

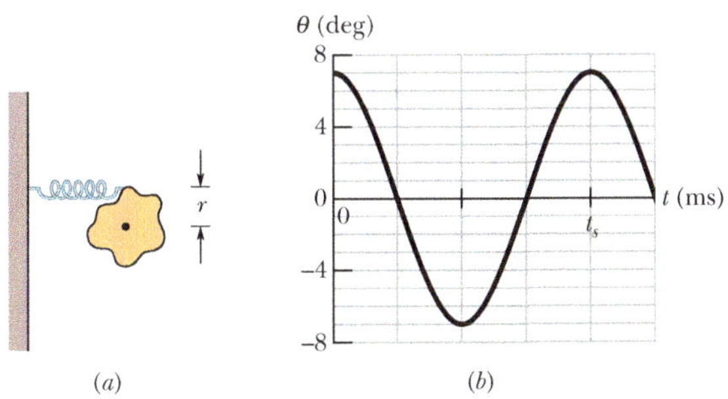

Figure 15-49 Problem 54.

•••**55** GO A pendulum is formed by pivoting a long thin rod about a point on the rod. In a series of experiments, the period is measured as a function of the distance x between the pivot point and the rod's center. (a) If the rod's length is $L = 2.20$ m and its mass is $m = 22.1$ g, what is the minimum period? (b) If x is chosen to minimize the period and then L is increased, does the period increase, decrease, or remain the same? (c) If, instead, m is increased without L increasing, does the period increase, decrease, or remain the same?

•••**56** GO In Fig. 15-50, a 2.50 kg disk of diameter $D = 42.0$ cm is supported by a rod of length $L = 76.0$ cm and negligible mass that is pivoted at its end. (a) With the massless torsion spring unconnected, what is the period of oscillation? (b) With the torsion spring connected, the rod is vertical at equilibrium. What is the torsion constant of the spring if the period of oscillation has been decreased by 0.500 s?

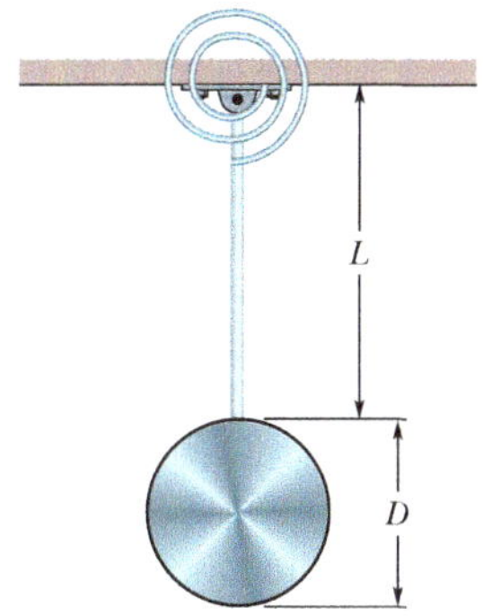

Figure 15-50 Problem 56.

Module 15-5 Damped Simple Harmonic Motion

•**57** The amplitude of a lightly damped oscillator decreases by 3.0% during each cycle. What percentage of the mechanical energy of the oscillator is lost in each cycle?

•**58** For the damped oscillator system shown in Fig. 15-16, with $m = 250$ g, $k = 85$ N/m, and $b = 70$ g/s, what is the ratio of the oscillation amplitude at the end of 20 cycles to the initial oscillation amplitude?

•**59** SSM WWW For the damped oscillator system shown in Fig. 15-16, the block has a mass of 1.50 kg and the spring constant is 8.00 N/m. The damping force is given by $-b(dx/dt)$, where $b = 230$ g/s. The block is pulled down 12.0 cm and released. (a) Calculate the time required for the amplitude of the resulting oscillations to fall to one-third of its initial value. (b) How many oscillations are made by the block in this time?

••**60** The suspension system of a 2000 kg automobile "sags" 10 cm when the chassis is placed on it. Also, the oscillation amplitude decreases by 50% each cycle. Estimate the values of (a) the spring constant k and (b) the damping constant b for the spring and shock absorber system of one wheel, assuming each wheel supports 500 kg.

Module 15-6 Forced Oscillations and Resonance

•**61** For Eq. 15-45, suppose the amplitude x_m is given by

$$x_m = \frac{F_m}{[m^2(\omega_d^2 - \omega^2)^2 + b^2\omega_d^2]^{1/2}},$$

where F_m is the (constant) amplitude of the external oscillating force exerted on the spring by the rigid support in Fig. 15-16. At resonance, what are the (a) amplitude and (b) velocity amplitude of the oscillating object?

•**62** Hanging from a horizontal beam are nine simple pendulums of the following lengths: (a) 0.10, (b) 0.30, (c) 0.40, (d) 0.80, (e) 1.2, (f) 2.8, (g) 3.5, (h) 5.0, and (i) 6.2 m. Suppose the beam undergoes horizontal oscillations with angular frequencies in the range from 2.00 rad/s to 4.00 rad/s. Which of the pendulums will be (strongly) set in motion?

••**63** A 1000 kg car carrying four 82 kg people travels over a "washboard" dirt road with corrugations 4.0 m apart. The car bounces with maximum amplitude when its speed is 16 km/h. When the car stops, and the people get out, by how much does the car body rise on its suspension?

Additional Problems

64 Although California is known for earthquakes, it has large regions dotted with precariously balanced rocks that would be easily toppled by even a mild earthquake. Apparently no major earthquakes have occurred in those regions. If an earthquake were to put such a rock into sinusoidal oscillation (parallel to the ground) with a frequency of 2.2 Hz, an oscillation amplitude of 1.0 cm would cause the rock to topple. What would be the magnitude of the maximum acceleration of the oscillation, in terms of g?

65 A loudspeaker diaphragm is oscillating in simple harmonic motion with a frequency of 440 Hz and a maximum displacement of 0.75 mm. What are the (a) angular frequency, (b) maximum speed, and (c) magnitude of the maximum acceleration?

66 A uniform spring with $k = 8600$ N/m is cut into pieces 1 and 2 of unstretched lengths $L_1 = 7.0$ cm and $L_2 = 10$ cm. What are (a) k_1 and (b) k_2? A block attached to the original spring as in Fig. 15-7 oscillates at 200 Hz. What is the oscillation frequency of the block attached to (c) piece 1 and (d) piece 2?

67 GO In Fig. 15-51, three 10 000 kg ore cars are held at rest on a mine railway using a cable that is parallel to the rails, which are inclined at angle $\theta = 30°$. The cable stretches 15 cm just before the coupling between the two lower cars breaks, detaching the lowest car. Assuming that the cable obeys Hooke's law, find the (a) frequency and (b) amplitude of the resulting oscillations of the remaining two cars.

Figure 15-51 Problem 67.

68 A 2.00 kg block hangs from a spring. A 300 g body hung below the block stretches the spring 2.00 cm farther. (a) What is the spring constant? (b) If the 300 g body is removed and the block is set into oscillation, find the period of the motion.

69 SSM In the engine of a locomotive, a cylindrical piece known as a piston oscillates in SHM in a cylinder head (cylindrical chamber) with an angular frequency of 180 rev/min. Its stroke (twice the amplitude) is 0.76 m. What is its maximum speed?

70 GO A wheel is free to rotate about its fixed axle. A spring is attached to one of its spokes a distance r from the axle, as shown in Fig. 15-52. (a) Assuming that the wheel is a hoop of mass m and radius R, what is the angular frequency ω of small oscillations of this system in terms of m, R, r, and the spring constant k? What is ω if (b) $r = R$ and (c) $r = 0$?

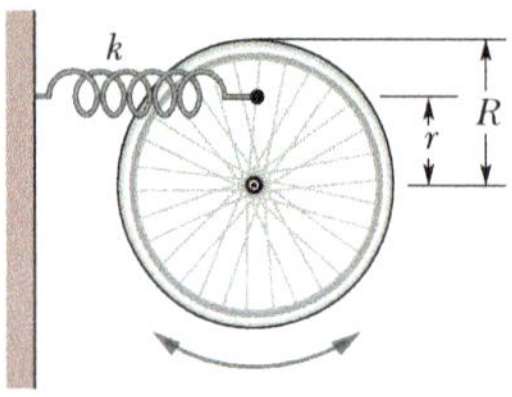

Figure 15-52 Problem 70.

71 A 50.0 g stone is attached to the bottom of a vertical spring and set vibrating. If the maximum speed of the stone is 15.0 cm/s and the period is 0.500 s, find the (a) spring constant of the spring, (b) amplitude of the motion, and (c) frequency of oscillation.

72 A uniform circular disk whose radius R is 12.6 cm is suspended as a physical pendulum from a point on its rim. (a) What is its period? (b) At what radial distance $r < R$ is there a pivot point that gives the same period?

73 SSM A vertical spring stretches 9.6 cm when a 1.3 kg block is hung from its end. (a) Calculate the spring constant. This block is then displaced an additional 5.0 cm downward and released from rest. Find the (b) period, (c) frequency, (d) amplitude, and (e) maximum speed of the resulting SHM.

74 A massless spring with spring constant 19 N/m hangs vertically. A body of mass 0.20 kg is attached to its free end and then released. Assume that the spring was unstretched before the body was released. Find (a) how far below the initial position the body descends, and the (b) frequency and (c) amplitude of the resulting SHM.

75 A 4.00 kg block is suspended from a spring with $k = 500$ N/m. A 50.0 g bullet is fired into the block from directly below with a speed of 150 m/s and becomes embedded in the block. (a) Find the amplitude of the resulting SHM. (b) What percentage of the original kinetic energy of the bullet is transferred to mechanical energy of the oscillator?

76 A 55.0 g block oscillates in SHM on the end of a spring with $k = 1500$ N/m according to $x = x_m \cos(\omega t + \phi)$. How long does the block take to move from position $+0.800x_m$ to (a) position $+0.600x_m$ and (b) position $-0.800x_m$?

77 Figure 15-53 gives the position of a 20 g block oscillating in SHM on the end of a spring. The horizontal axis scale is set by $t_s = 40.0$ ms. What are (a) the maximum kinetic energy of the block and (b) the number of times per second that maximum is reached? (*Hint:* Measuring a slope will probably not be very accurate. Find another approach.)

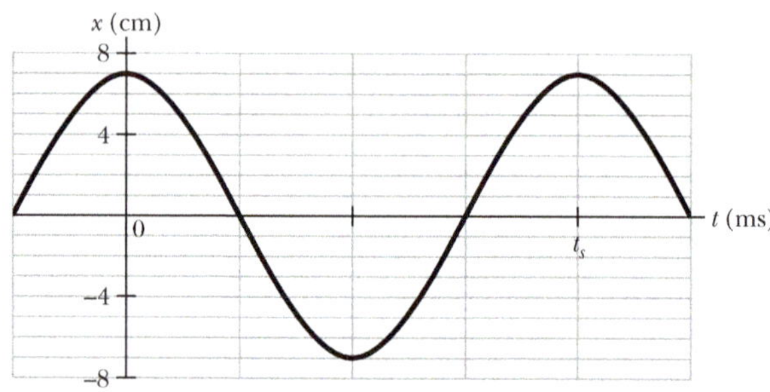

Figure 15-53 Problems 77 and 78.

78 Figure 15-53 gives the position $x(t)$ of a block oscillating in SHM on the end of a spring ($t_s = 40.0$ ms). What are (a) the speed and (b) the magnitude of the radial acceleration of a particle in the corresponding uniform circular motion?

79 Figure 15-54 shows the kinetic energy K of a simple pendulum versus its angle θ from the vertical. The vertical axis scale is set by $K_s = 10.0$ mJ. The pendulum bob has mass 0.200 kg. What is the length of the pendulum?

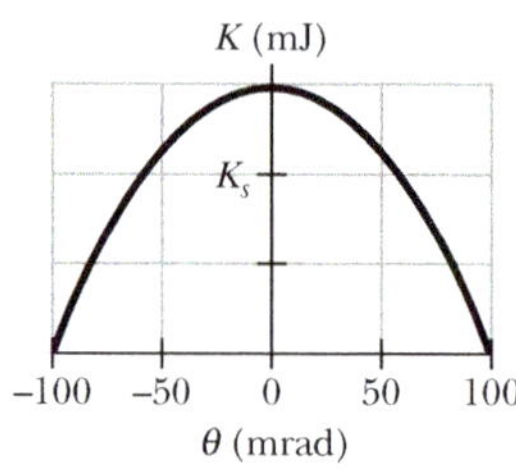

Figure 15-54 Problem 79.

80 A block is in SHM on the end of a spring, with position given by $x = x_m \cos(\omega t + \phi)$. If $\phi = \pi/5$ rad, then at $t = 0$ what percentage of the total mechanical energy is potential energy?

81 A simple harmonic oscillator consists of a 0.50 kg block attached to a spring. The block slides back and forth along a straight line on a frictionless surface with equilibrium point $x = 0$. At $t = 0$ the block is at $x = 0$ and moving in the positive x direction. A graph of the magnitude of the net force $\vec{F}$ on the block as a function of its position is shown in Fig. 15-55. The vertical scale is set by F_s = 75.0 N. What are (a) the amplitude and (b) the period of the motion, (c) the magnitude of the maximum acceleration, and (d) the maximum kinetic energy?

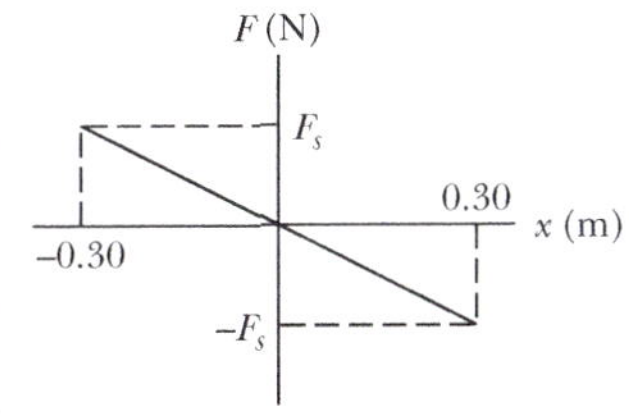

Figure 15-55 Problem 81.

82 A simple pendulum of length 20 cm and mass 5.0 g is suspended in a race car traveling with constant speed 70 m/s around a circle of radius 50 m. If the pendulum undergoes small oscillations in a radial direction about its equilibrium position, what is the frequency of oscillation?

83 The scale of a spring balance that reads from 0 to 15.0 kg is 12.0 cm long. A package suspended from the balance is found to oscillate vertically with a frequency of 2.00 Hz. (a) What is the spring constant? (b) How much does the package weigh?

84 A 0.10 kg block oscillates back and forth along a straight line on a frictionless horizontal surface. Its displacement from the origin is given by

$$x = (10\text{ cm})\cos[(10\text{ rad/s})t + \pi/2\text{ rad}].$$

(a) What is the oscillation frequency? (b) What is the maximum speed acquired by the block? (c) At what value of x does this occur? (d) What is the magnitude of the maximum acceleration of the block? (e) At what value of x does this occur? (f) What force, applied to the block by the spring, results in the given oscillation?

85 The end point of a spring oscillates with a period of 2.0 s when a block with mass m is attached to it. When this mass is increased by 2.0 kg, the period is found to be 3.0 s. Find m.

86 The tip of one prong of a tuning fork undergoes SHM of frequency 1000 Hz and amplitude 0.40 mm. For this tip, what is the magnitude of the (a) maximum acceleration, (b) maximum velocity, (c) acceleration at tip displacement 0.20 mm, and (d) velocity at tip displacement 0.20 mm?

87 A flat uniform circular disk has a mass of 3.00 kg and a radius of 70.0 cm. It is suspended in a horizontal plane by a vertical wire attached to its center. If the disk is rotated 2.50 rad about the wire, a torque of 0.0600 N·m is required to maintain that orientation. Calculate (a) the rotational inertia of the disk about the wire, (b) the torsion constant, and (c) the angular frequency of this torsion pendulum when it is set oscillating.

88 A block weighing 20 N oscillates at one end of a vertical spring for which k = 100 N/m; the other end of the spring is attached to a ceiling. At a certain instant the spring is stretched 0.30 m beyond its relaxed length (the length when no object is attached) and the block has zero velocity. (a) What is the net force on the block at this instant? What are the (b) amplitude and (c) period of the resulting simple harmonic motion? (d) What is the maximum kinetic energy of the block as it oscillates?

89 A 3.0 kg particle is in simple harmonic motion in one dimension and moves according to the equation

$$x = (5.0\text{ m})\cos[(\pi/3\text{ rad/s})t - \pi/4\text{ rad}],$$

with t in seconds. (a) At what value of x is the potential energy of the particle equal to half the total energy? (b) How long does the particle take to move to this position x from the equilibrium position?

90 A particle executes linear SHM with frequency 0.25 Hz about the point x = 0. At t = 0, it has displacement x = 0.37 cm and zero velocity. For the motion, determine the (a) period, (b) angular frequency, (c) amplitude, (d) displacement $x(t)$, (e) velocity $v(t)$, (f) maximum speed, (g) magnitude of the maximum acceleration, (h) displacement at t = 3.0 s, and (i) speed at t = 3.0 s.

91 SSM What is the frequency of a simple pendulum 2.0 m long (a) in a room, (b) in an elevator accelerating upward at a rate of 2.0 m/s^2, and (c) in free fall?

92 A grandfather clock has a pendulum that consists of a thin brass disk of radius r = 15.00 cm and mass 1.000 kg that is attached to a long thin rod of negligible mass. The pendulum swings freely about an axis perpendicular to the rod and through the end of the rod opposite the disk, as shown in Fig. 15-56. If the pendulum is to have a period of 2.000 s for small oscillations at a place where g = 9.800 m/s^2, what must be the rod length L to the nearest tenth of a millimeter?

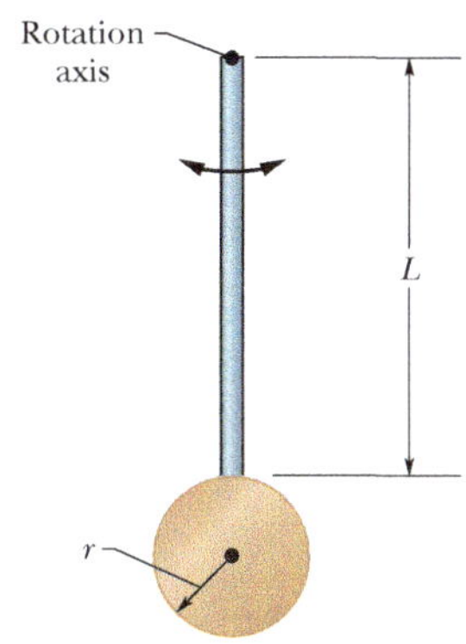

Figure 15-56 Problem 92.

93 A 4.00 kg block hangs from a spring, extending it 16.0 cm from its unstretched position. (a) What is the spring constant? (b) The block is removed, and a 0.500 kg body is hung from the same spring. If the spring is then stretched and released, what is its period of oscillation?

94 What is the phase constant for SMH with $a(t)$ given in Fig. 15-57 if the position function $x(t)$ has the form $x = x_m\cos(\omega t + \phi)$ and a_s = 4.0 m/s^2?

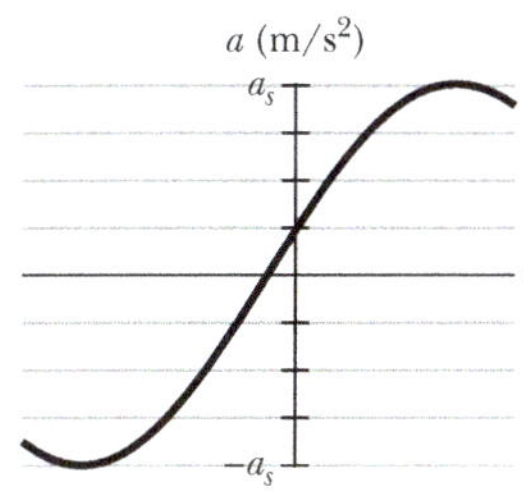

Figure 15-57 Problem 94.

95 An engineer has an odd-shaped 10 kg object and needs to find its rotational inertia about an axis through its center of mass. The object is supported on a wire stretched along the desired axis. The wire has a torsion constant κ = 0.50 N·m. If this torsion pendulum oscillates through 20 cycles in 50 s, what is the rotational inertia of the object?

96 A spider can tell when its web has captured, say, a fly because the fly's thrashing causes the web threads to oscillate. A spider can even determine the size of the fly by the frequency of the oscillations. Assume that a fly oscillates on the *capture thread* on which it is caught like a block on a spring. What is the ratio of oscillation frequency for a fly with mass m to a fly with mass 2.5m?

97 A torsion pendulum consists of a metal disk with a wire running through its center and soldered in place. The wire is mounted vertically on clamps and pulled taut. Figure 15-58a gives the magnitude τ of the torque

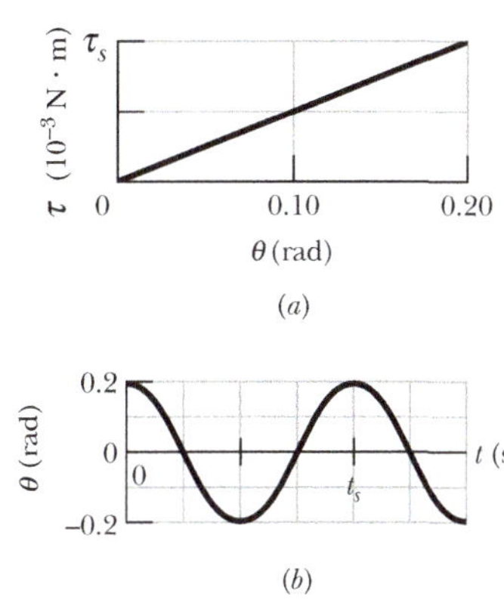

Figure 15-58 Problem 97.

needed to rotate the disk about its center (and thus twist the wire) versus the rotation angle θ. The vertical axis scale is set by $\tau_s = 4.0 \times 10^{-3}$ N·m. The disk is rotated to $\theta = 0.200$ rad and then released. Figure 15-58*b* shows the resulting oscillation in terms of angular position θ versus time t. The horizontal axis scale is set by $t_s = 0.40$ s. (a) What is the rotational inertia of the disk about its center? (b) What is the maximum angular speed $d\theta/dt$ of the disk? (*Caution:* Do not confuse the (constant) angular frequency of the SHM with the (varying) angular speed of the rotating disk, even though they usually have the same symbol ω. *Hint:* The potential energy U of a torsion pendulum is equal to $\frac{1}{2}\kappa\theta^2$, analogous to $U = \frac{1}{2}kx^2$ for a spring.)

98 When a 20 N can is hung from the bottom of a vertical spring, it causes the spring to stretch 20 cm. (a) What is the spring constant? (b) This spring is now placed horizontally on a frictionless table. One end of it is held fixed, and the other end is attached to a 5.0 N can. The can is then moved (stretching the spring) and released from rest. What is the period of the resulting oscillation?

99 For a simple pendulum, find the angular amplitude θ_m at which the restoring torque required for simple harmonic motion deviates from the actual restoring torque by 1.0%. (See "Trigonometric Expansions" in Appendix E.)

100 In Fig. 15-59, a solid cylinder attached to a horizontal spring (k = 3.00 N/m) rolls without slipping along a horizontal surface. If the system is released from rest when the spring is stretched by 0.250 m, find (a) the translational kinetic energy and (b) the rotational kinetic energy of the cylinder as it passes through the equilibrium position. (c) Show that under these conditions the cylinder's center of mass executes simple harmonic motion with period

$$T = 2\pi\sqrt{\frac{3M}{2k}},$$

where M is the cylinder mass. (*Hint:* Find the time derivative of the total mechanical energy.)

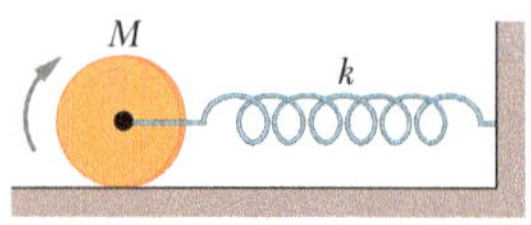

Figure 15-59 Problem 100.

101 SSM A 1.2 kg block sliding on a horizontal frictionless surface is attached to a horizontal spring with $k = 480$ N/m. Let x be the displacement of the block from the position at which the spring is unstretched. At $t = 0$ the block passes through $x = 0$ with a speed of 5.2 m/s in the positive x direction. What are the (a) frequency and (b) amplitude of the block's motion? (c) Write an expression for x as a function of time.

102 A simple harmonic oscillator consists of an 0.80 kg block attached to a spring ($k = 200$ N/m). The block slides on a horizontal frictionless surface about the equilibrium point $x = 0$ with a total mechanical energy of 4.0 J. (a) What is the amplitude of the oscillation? (b) How many oscillations does the block complete in 10 s? (c) What is the maximum kinetic energy attained by the block? (d) What is the speed of the block at $x = 0.15$ m?

103 A block sliding on a horizontal frictionless surface is attached to a horizontal spring with a spring constant of 600 N/m. The block executes SHM about its equilibrium position with a period of 0.40 s and an amplitude of 0.20 m. As the block slides through its equilibrium position, a 0.50 kg putty wad is dropped vertically onto the block. If the putty wad sticks to the block, determine (a) the new period of the motion and (b) the new amplitude of the motion.

104 A damped harmonic oscillator consists of a block (m = 2.00 kg), a spring (k = 10.0 N/m), and a damping force ($F = -bv$). Initially, it oscillates with an amplitude of 25.0 cm; because of the damping, the amplitude falls to three-fourths of this initial value at the completion of four oscillations. (a) What is the value of b? (b) How much energy has been "lost" during these four oscillations?

105 A block weighing 10.0 N is attached to the lower end of a vertical spring (k = 200.0 N/m), the other end of which is attached to a ceiling. The block oscillates vertically and has a kinetic energy of 2.00 J as it passes through the point at which the spring is unstretched. (a) What is the period of the oscillation? (b) Use the law of conservation of energy to determine the maximum distance the block moves both above and below the point at which the spring is unstretched. (These are not necessarily the same.) (c) What is the amplitude of the oscillation? (d) What is the maximum kinetic energy of the block as it oscillates?

106 A simple harmonic oscillator consists of a block attached to a spring with k = 200 N/m. The block slides on a frictionless surface, with equilibrium point $x = 0$ and amplitude 0.20 m. A graph of the block's velocity v as a function of time t is shown in Fig. 15-60. The horizontal scale is set by $t_s = 0.20$ s. What are (a) the period of the SHM, (b) the block's mass, (c) its displacement at $t = 0$, (d) its acceleration at $t = 0.10$ s, and (e) its maximum kinetic energy?

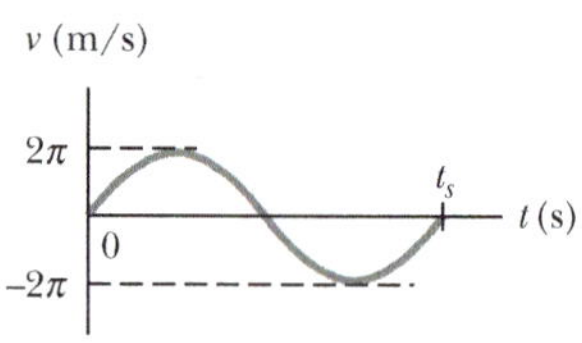

Figure 15-60 Problem 106.

107 The vibration frequencies of atoms in solids at normal temperatures are of the order of 10^{13} Hz. Imagine the atoms to be connected to one another by springs. Suppose that a single silver atom in a solid vibrates with this frequency and that all the other atoms are at rest. Compute the effective spring constant. One mole of silver (6.02×10^{23} atoms) has a mass of 108 g.

108 Figure 15-61 shows that if we hang a block on the end of a spring with spring constant k, the spring is stretched by distance h = 2.0 cm. If we pull down on the block a short distance and then release it, it oscillates vertically with a certain frequency. What length must a simple pendulum have to swing with that frequency?

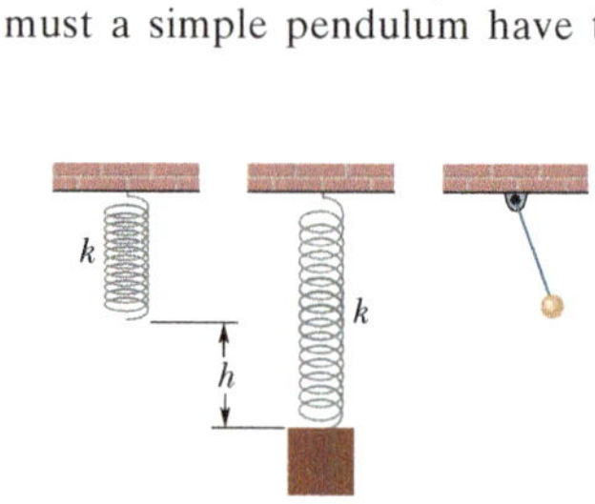

Figure 15-61 Problem 108.

109 The physical pendulum in Fig. 15-62 has two possible pivot points A and B. Point A has a fixed position but B is adjustable along the length of the pendulum as indicated by the scaling. When suspended from A, the pendulum has a period of $T = 1.80$ s. The pendulum is then suspended from B, which is moved until the pendulum again has that period. What is the distance L between A and B?

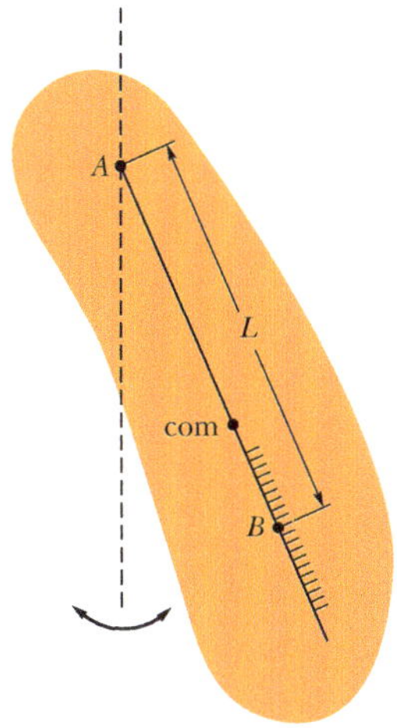

Figure 15-62 Problem 109.

110 A common device for entertaining a toddler is a *jump seat* that hangs from the horizontal portion of a doorframe via elastic cords (Fig. 15-63). Assume that only one cord is on each side in spite of the more realistic arrangement shown. When a child is placed in the seat, they both descend by a distance d_s as the cords stretch (treat them as springs). Then the seat is pulled down an extra distance d_m and released, so that the child oscillates vertically, like a block on the end of a spring. Suppose you are the safety engineer for the manufacturer of the seat. You do not want the magnitude of the child's acceleration to exceed $0.20g$ for fear of hurting the child's neck. If $d_m = 10$ cm, what value of d_s corresponds to that acceleration magnitude?

Figure 15-63 Problem 110.

111 A 2.0 kg block executes SHM while attached to a horizontal spring of spring constant 200 N/m. The maximum speed of the block as it slides on a horizontal frictionless surface is 3.0 m/s. What are (a) the amplitude of the block's motion, (b) the magnitude of its maximum acceleration, and (c) the magnitude of its minimum acceleration? (d) How long does the block take to complete 7.0 cycles of its motion?

112 In Fig. 15-64, a 2500 kg demolition ball swings from the end of a crane. The length of the swinging segment of cable is 17 m. (a) Find the period of the swinging, assuming that the system can be treated as a simple pendulum. (b) Does the period depend on the ball's mass?

Figure 15-64 Problem 112.

113 The center of oscillation of a physical pendulum has this interesting property: If an impulse (assumed horizontal and in the plane of oscillation) acts at the center of oscillation, no oscillations are felt at the point of support. Baseball players (and players of many other sports) know that unless the ball hits the bat at this point (called the "sweet spot" by athletes), the oscillations due to the impact will sting their hands. To prove this property, let the stick in Fig. 15-13a simulate a baseball bat. Suppose that a horizontal force $\vec{F}$ (due to impact with the ball) acts toward the right at P, the center of oscillation. The batter is assumed to hold the bat at O, the pivot point of the stick. (a) What acceleration does the point O undergo as a result of $\vec{F}$? (b) What angular acceleration is produced by $\vec{F}$ about the center of mass of the stick? (c) As a result of the angular acceleration in (b), what linear acceleration does point O undergo? (d) Considering the magnitudes and directions of the accelerations in (a) and (c), convince yourself that P is indeed the "sweet spot."

114 A (hypothetical) large slingshot is stretched 2.30 m to launch a 170 g projectile with speed sufficient to escape from Earth (11.2 km/s). Assume the elastic bands of the slingshot obey Hooke's law. (a) What is the spring constant of the device if all the elastic potential energy is converted to kinetic energy? (b) Assume that an average person can exert a force of 490 N. How many people are required to stretch the elastic bands?

115 What is the length of a simple pendulum whose full swing from left to right and then back again takes 3.2 s?

116 A 2.0 kg block is attached to the end of a spring with a spring constant of 350 N/m and forced to oscillate by an applied force $F = (15\text{ N})\sin(\omega_d t)$, where $\omega_d = 35$ rad/s. The damping constant is $b = 15$ kg/s. At $t = 0$, the block is at rest with the spring at its rest length. (a) Use numerical integration to plot the displacement of the block for the first 1.0 s. Use the motion near the end of the 1.0 s interval to estimate the amplitude, period, and angular frequency. Repeat the calculation for (b) $\omega_d = \sqrt{k/m}$ and (c) $\omega_d = 20$ rad/s.

CHAPTER 16

Waves—I

16-1 TRANSVERSE WAVES

Learning Objectives

After reading this module, you should be able to . . .

16.01 Identify the three main types of waves.

16.02 Distinguish between transverse waves and longitudinal waves.

16.03 Given a displacement function for a traverse wave, determine amplitude y_m, angular wave number k, angular frequency ω, phase constant ϕ, and direction of travel, and calculate the phase $kx \pm \omega t + \phi$ and the displacement at any given time and position.

16.04 Given a displacement function for a traverse wave, calculate the time between two given displacements.

16.05 Sketch a graph of a transverse wave as a function of position, identifying amplitude y_m, wavelength λ, where the slope is greatest, where it is zero, and where the string elements have positive velocity, negative velocity, and zero velocity.

16.06 Given a graph of displacement versus time for a transverse wave, determine amplitude y_m and period T.

16.07 Describe the effect on a transverse wave of changing phase constant ϕ.

16.08 Apply the relation between the wave speed v, the distance traveled by the wave, and the time required for that travel.

16.09 Apply the relationships between wave speed v, angular frequency ω, angular wave number k, wavelength λ, period T, and frequency f.

16.10 Describe the motion of a string element as a transverse wave moves through its location, and identify when its transverse speed is zero and when it is maximum.

16.11 Calculate the transverse velocity $u(t)$ of a string element as a transverse wave moves through its location.

16.12 Calculate the transverse acceleration $a(t)$ of a string element as a transverse wave moves through its location.

16.13 Given a graph of displacement, transverse velocity, or transverse acceleration, determine the phase constant ϕ.

Key Ideas

- Mechanical waves can exist only in material media and are governed by Newton's laws. Transverse mechanical waves, like those on a stretched string, are waves in which the particles of the medium oscillate perpendicular to the wave's direction of travel. Waves in which the particles of the medium oscillate parallel to the wave's direction of travel are longitudinal waves.

- A sinusoidal wave moving in the positive direction of an x axis has the mathematical form

$$y(x, t) = y_m \sin(kx - \omega t),$$

where y_m is the amplitude (magnitude of the maximum displacement) of the wave, k is the angular wave number, ω is the angular frequency, and $kx - \omega t$ is the phase. The wavelength λ is related to k by

$$k = \frac{2\pi}{\lambda}.$$

- The period T and frequency f of the wave are related to ω by

$$\frac{\omega}{2\pi} = f = \frac{1}{T}.$$

- The wave speed v (the speed of the wave along the string) is related to these other parameters by

$$v = \frac{\omega}{k} = \frac{\lambda}{T} = \lambda f.$$

- Any function of the form

$$y(x, t) = h(kx \pm \omega t)$$

can represent a traveling wave with a wave speed as given above and a wave shape given by the mathematical form of h. The plus sign denotes a wave traveling in the negative direction of the x axis, and the minus sign a wave traveling in the positive direction.

What Is Physics?

One of the primary subjects of physics is waves. To see how important waves are in the modern world, just consider the music industry. Every piece of music you hear, from some retro-punk band playing in a campus dive to the most eloquent concerto playing on the web, depends on performers producing waves and your detecting those waves. In between production and detection, the information carried by the waves might need to be transmitted (as in a live performance on the web) or recorded and then reproduced (as with CDs, DVDs, or the other devices currently being developed in engineering labs worldwide). The financial importance of controlling music waves is staggering, and the rewards to engineers who develop new control techniques can be rich.

This chapter focuses on waves traveling along a stretched string, such as on a guitar. The next chapter focuses on sound waves, such as those produced by a guitar string being played. Before we do all this, though, our first job is to classify the countless waves of the everyday world into basic types.

Types of Waves

Waves are of three main types:

1. ***Mechanical waves.*** These waves are most familiar because we encounter them almost constantly; common examples include water waves, sound waves, and seismic waves. All these waves have two central features: They are governed by Newton's laws, and they can exist only within a material medium, such as water, air, and rock.
2. ***Electromagnetic waves.*** These waves are less familiar, but you use them constantly; common examples include visible and ultraviolet light, radio and television waves, microwaves, x rays, and radar waves. These waves require no material medium to exist. Light waves from stars, for example, travel through the vacuum of space to reach us. All electromagnetic waves travel through a vacuum at the same speed $c = 299\,792\,458$ m/s.
3. ***Matter waves.*** Although these waves are commonly used in modern technology, they are probably very unfamiliar to you. These waves are associated with electrons, protons, and other fundamental particles, and even atoms and molecules. Because we commonly think of these particles as constituting matter, such waves are called matter waves.

Much of what we discuss in this chapter applies to waves of all kinds. However, for specific examples we shall refer to mechanical waves.

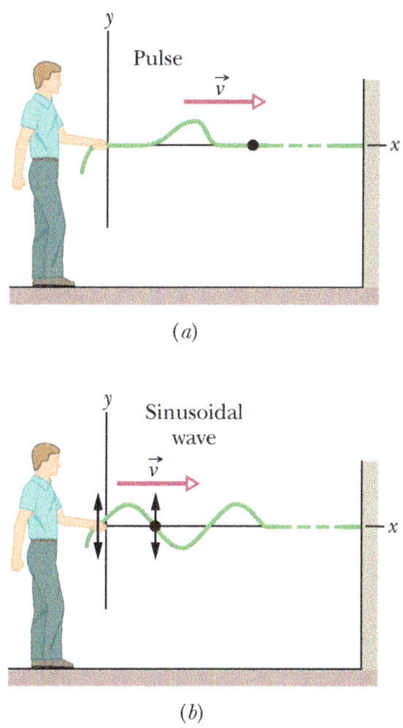

Figure 16-1 (*a*) A single pulse is sent along a stretched string. A typical string element (marked with a dot) moves up once and then down as the pulse passes. The element's motion is perpendicular to the wave's direction of travel, so the pulse is a *transverse wave*. (*b*) A sinusoidal wave is sent along the string. A typical string element moves up and down continuously as the wave passes. This too is a transverse wave.

Transverse and Longitudinal Waves

A wave sent along a stretched, taut string is the simplest mechanical wave. If you give one end of a stretched string a single up-and-down jerk, a wave in the form of a single *pulse* travels along the string. This pulse and its motion can occur because the string is under tension. When you pull your end of the string upward, it begins to pull upward on the adjacent section of the string via tension between the two sections. As the adjacent section moves upward, it begins to pull the next section upward, and so on. Meanwhile, you have pulled down on your end of the string. As each section moves upward in turn, it begins to be pulled back downward by neighboring sections that are already on the way down. The net result is that a distortion in the string's shape (a pulse, as in Fig. 16-1*a*) moves along the string at some velocity $\vec{v}$.

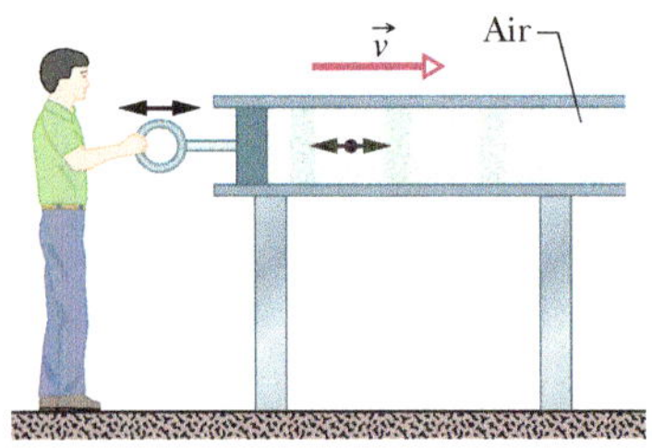

Figure 16-2 A sound wave is set up in an air-filled pipe by moving a piston back and forth. Because the oscillations of an element of the air (represented by the dot) are parallel to the direction in which the wave travels, the wave is a *longitudinal wave.*

If you move your hand up and down in continuous simple harmonic motion, a continuous wave travels along the string at velocity $\vec{v}$. Because the motion of your hand is a sinusoidal function of time, the wave has a sinusoidal shape at any given instant, as in Fig. 16-1*b*; that is, the wave has the shape of a sine curve or a cosine curve.

We consider here only an "ideal" string, in which no friction-like forces within the string cause the wave to die out as it travels along the string. In addition, we assume that the string is so long that we need not consider a wave rebounding from the far end.

One way to study the waves of Fig. 16-1 is to monitor the **wave forms** (shapes of the waves) as they move to the right. Alternatively, we could monitor the motion of an element of the string as the element oscillates up and down while a wave passes through it. We would find that the displacement of every such oscillating string element is *perpendicular* to the direction of travel of the wave, as indicated in Fig. 16-1*b*. This motion is said to be **transverse,** and the wave is said to be a **transverse wave.**

Longitudinal Waves. Figure 16-2 shows how a sound wave can be produced by a piston in a long, air-filled pipe. If you suddenly move the piston rightward and then leftward, you can send a pulse of sound along the pipe. The rightward motion of the piston moves the elements of air next to it rightward, changing the air pressure there. The increased air pressure then pushes rightward on the elements of air somewhat farther along the pipe. Moving the piston leftward reduces the air pressure next to it. As a result, first the elements nearest the piston and then farther elements move leftward. Thus, the motion of the air and the change in air pressure travel rightward along the pipe as a pulse.

If you push and pull on the piston in simple harmonic motion, as is being done in Fig. 16-2, a sinusoidal wave travels along the pipe. Because the motion of the elements of air is parallel to the direction of the wave's travel, the motion is said to be **longitudinal,** and the wave is said to be a **longitudinal wave.** In this chapter we focus on transverse waves, and string waves in particular; in Chapter 17 we focus on longitudinal waves, and sound waves in particular.

Both a transverse wave and a longitudinal wave are said to be **traveling waves** because they both travel from one point to another, as from one end of the string to the other end in Fig. 16-1 and from one end of the pipe to the other end in Fig. 16-2. Note that it is the wave that moves from end to end, not the material (string or air) through which the wave moves.

Wavelength and Frequency

To completely describe a wave on a string (and the motion of any element along its length), we need a function that gives the shape of the wave. This means that we need a relation in the form

$$y = h(x, t), \tag{16-1}$$

in which y is the transverse displacement of any string element as a function h of the time t and the position x of the element along the string. In general, a sinusoidal shape like the wave in Fig. 16-1*b* can be described with h being either a sine or cosine function; both give the same general shape for the wave. In this chapter we use the sine function.

Sinusoidal Function. Imagine a sinusoidal wave like that of Fig. 16-1*b* traveling in the positive direction of an x axis. As the wave sweeps through succeeding elements (that is, very short sections) of the string, the elements oscillate parallel to the y axis. At time t, the displacement y of the element located at position x is given by

$$y(x, t) = y_m \sin(kx - \omega t). \tag{16-2}$$

Because this equation is written in terms of position x, it can be used to find the displacements of all the elements of the string as a function of time. Thus, it can tell us the shape of the wave at any given time.

The names of the quantities in Eq. 16-2 are displayed in Fig. 16-3 and defined next. Before we discuss them, however, let us examine Fig. 16-4, which shows five "snapshots" of a sinusoidal wave traveling in the positive direction of an x axis. The movement of the wave is indicated by the rightward progress of the short arrow pointing to a high point of the wave. From snapshot to snapshot, the short arrow moves to the right with the wave shape, but the string moves *only* parallel to the y axis. To see that, let us follow the motion of the red-dyed string element at $x = 0$. In the first snapshot (Fig. 16-4a), this element is at displacement $y = 0$. In the next snapshot, it is at its extreme downward displacement because a *valley* (or extreme low point) of the wave is passing through it. It then moves back up through $y = 0$. In the fourth snapshot, it is at its extreme upward displacement because a *peak* (or extreme high point) of the wave is passing through it. In the fifth snapshot, it is again at $y = 0$, having completed one full oscillation.

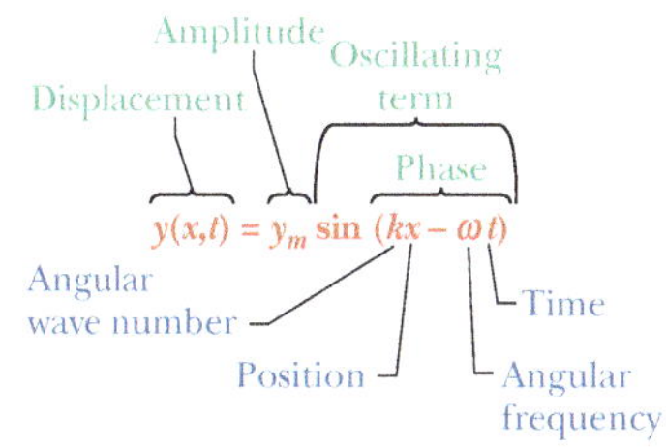

Figure 16-3 The names of the quantities in Eq. 16-2, for a transverse sinusoidal wave.

Amplitude and Phase

The **amplitude** y_m of a wave, such as that in Fig. 16-4 , is the magnitude of the maximum displacement of the elements from their equilibrium positions as the wave passes through them. (The subscript m stands for maximum.) Because y_m is a magnitude, it is always a positive quantity, even if it is measured downward instead of upward as drawn in Fig. 16-4a.

The **phase** of the wave is the *argument* $kx - \omega t$ of the sine in Eq. 16-2. As the wave sweeps through a string element at a particular position x, the phase changes linearly with time t. This means that the sine also changes, oscillating between $+1$ and -1. Its extreme positive value ($+1$) corresponds to a peak of the wave moving through the element; at that instant the value of y at position x is y_m. Its extreme negative value (-1) corresponds to a valley of the wave moving through the element; at that instant the value of y at position x is $-y_m$. Thus, the sine function and the time-dependent phase of a wave correspond to the oscillation of a string element, and the amplitude of the wave determines the extremes of the element's displacement.

Caution: When evaluating the phase, rounding off the numbers before you evaluate the sine function can throw of the calculation considerably.

Wavelength and Angular Wave Number

The **wavelength** λ of a wave is the distance (parallel to the direction of the wave's travel) between repetitions of the shape of the wave (or *wave shape*). A typical wavelength is marked in Fig. 16-4a, which is a snapshot of the wave at time $t = 0$. At that time, Eq. 16-2 gives, for the description of the wave shape,

$$y(x, 0) = y_m \sin kx. \tag{16-3}$$

By definition, the displacement y is the same at both ends of this wavelength—that is, at $x = x_1$ and $x = x_1 + \lambda$. Thus, by Eq. 16-3,

$$\begin{aligned} y_m \sin kx_1 &= y_m \sin k(x_1 + \lambda) \\ &= y_m \sin(kx_1 + k\lambda). \end{aligned} \tag{16-4}$$

A sine function begins to repeat itself when its angle (or argument) is increased by 2π rad, so in Eq. 16-4 we must have $k\lambda = 2\pi$, or

$$k = \frac{2\pi}{\lambda} \quad \text{(angular wave number)}. \tag{16-5}$$

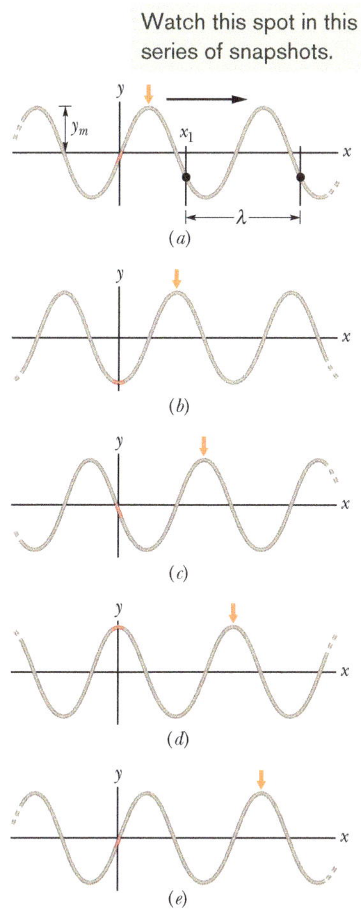

Figure 16-4 Five "snapshots" of a string wave traveling in the positive direction of an x axis. The amplitude y_m is indicated. A typical wavelength λ, measured from an arbitrary position x_1, is also indicated.

We call k the **angular wave number** of the wave; its SI unit is the radian per meter, or the inverse meter. (Note that the symbol k here does *not* represent a spring constant as previously.)

Notice that the wave in Fig. 16-4 moves to the right by $\frac{1}{4}\lambda$ from one snapshot to the next. Thus, by the fifth snapshot, it has moved to the right by 1λ.

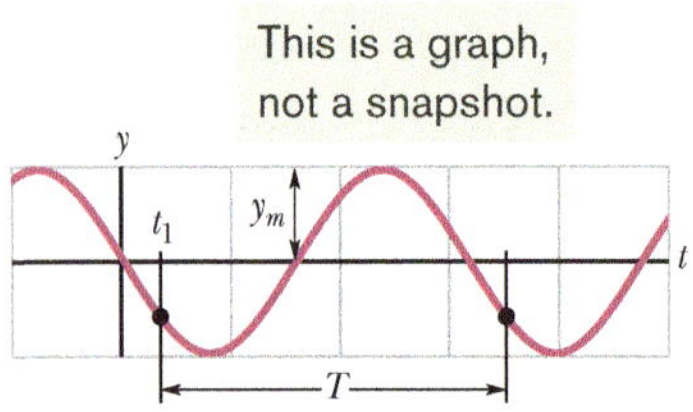

Figure 16-5 A graph of the displacement of the string element at $x = 0$ as a function of time, as the sinusoidal wave of Fig. 16-4 passes through the element. The amplitude y_m is indicated. A typical period T, measured from an arbitrary time t_1, is also indicated.

Period, Angular Frequency, and Frequency

Figure 16-5 shows a graph of the displacement y of Eq. 16-2 versus time t at a certain position along the string, taken to be $x = 0$. If you were to monitor the string, you would see that the single element of the string at that position moves up and down in simple harmonic motion given by Eq. 16-2 with $x = 0$:

$$\begin{aligned} y(0, t) &= y_m \sin(-\omega t) \\ &= -y_m \sin \omega t \qquad (x = 0). \end{aligned} \tag{16-6}$$

Here we have made use of the fact that $\sin(-\alpha) = -\sin \alpha$, where α is any angle. Figure 16-5 is a graph of this equation, with displacement plotted versus time; it *does not* show the shape of the wave. (Figure 16-4 shows the shape and is a picture of reality; Fig. 16-5 is a graph and thus an abstraction.)

We define the **period** of oscillation T of a wave to be the time any string element takes to move through one full oscillation. A typical period is marked on the graph of Fig. 16-5. Applying Eq. 16-6 to both ends of this time interval and equating the results yield

$$\begin{aligned} -y_m \sin \omega t_1 &= -y_m \sin \omega(t_1 + T) \\ &= -y_m \sin(\omega t_1 + \omega T). \end{aligned} \tag{16-7}$$

This can be true only if $\omega T = 2\pi$, or if

$$\omega = \frac{2\pi}{T} \quad \text{(angular frequency)}. \tag{16-8}$$

We call ω the **angular frequency** of the wave; its SI unit is the radian per second.

Look back at the five snapshots of a traveling wave in Fig. 16-4. The time between snapshots is $\frac{1}{4}T$. Thus, by the fifth snapshot, every string element has made one full oscillation.

The **frequency** f of a wave is defined as $1/T$ and is related to the angular frequency ω by

$$f = \frac{1}{T} = \frac{\omega}{2\pi} \quad \text{(frequency)}. \tag{16-9}$$

Like the frequency of simple harmonic motion in Chapter 15, this frequency f is a number of oscillations per unit time—here, the number made by a string element as the wave moves through it. As in Chapter 15, f is usually measured in hertz or its multiples, such as kilohertz.

Checkpoint 1

The figure is a composite of three snapshots, each of a wave traveling along a particular string. The phases for the waves are given by (a) $2x - 4t$, (b) $4x - 8t$, and (c) $8x - 16t$. Which phase corresponds to which wave in the figure?

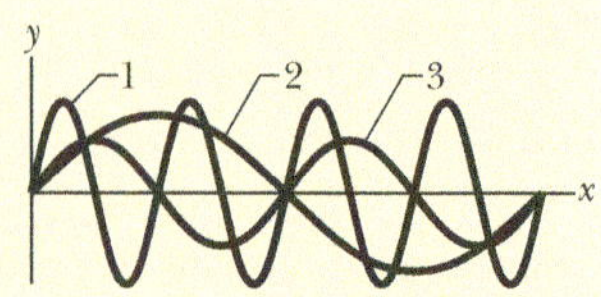

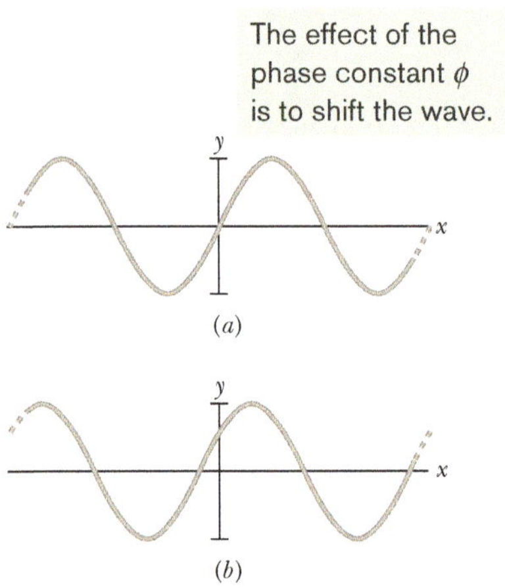

Figure 16-6 A sinusoidal traveling wave at $t = 0$ with a phase constant ϕ of (a) 0 and (b) $\pi/5$ rad.

Phase Constant

When a sinusoidal traveling wave is given by the wave function of Eq. 16-2, the wave near $x = 0$ looks like Fig. 16-6a when $t = 0$. Note that at $x = 0$, the displacement is $y = 0$ and the slope is at its maximum positive value. We can generalize Eq. 16-2 by inserting a **phase constant** ϕ in the wave function:

$$y = y_m \sin(kx - \omega t + \phi). \tag{16-10}$$

The value of ϕ can be chosen so that the function gives some other displacement and slope at $x = 0$ when $t = 0$. For example, a choice of $\phi = +\pi/5$ rad gives the displacement and slope shown in Fig. 16-6*b* when $t = 0$. The wave is still sinusoidal with the same values of y_m, k, and ω, but it is now shifted from what you see in Fig. 16-6*a* (where $\phi = 0$). Note also the direction of the shift. A positive value of ϕ shifts the curve in the negative direction of the x axis; a negative value shifts the curve in the positive direction.

The Speed of a Traveling Wave

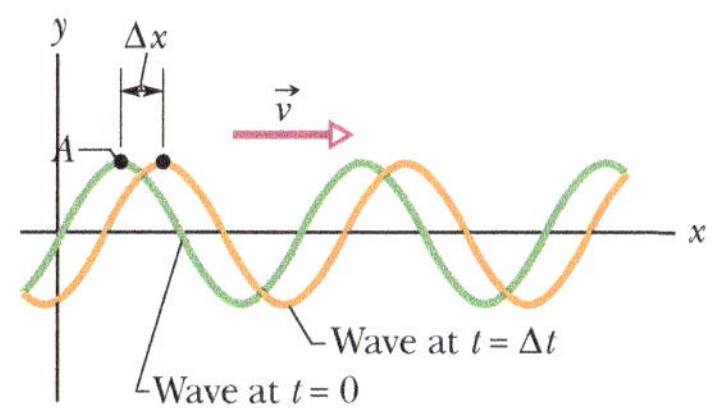

Figure 16-7 Two snapshots of the wave of Fig. 16-4, at time $t = 0$ and then at time $t = \Delta t$. As the wave moves to the right at velocity $\vec{v}$, the entire curve shifts a distance Δx during Δt. Point A "rides" with the wave form, but the string elements move only up and down.

Figure 16-7 shows two snapshots of the wave of Eq. 16-2, taken a small time interval Δt apart. The wave is traveling in the positive direction of x (to the right in Fig. 16-7), the entire wave pattern moving a distance Δx in that direction during the interval Δt. The ratio $\Delta x/\Delta t$ (or, in the differential limit, dx/dt) is the **wave speed** v. How can we find its value?

As the wave in Fig. 16-7 moves, each point of the moving wave form, such as point A marked on a peak, retains its displacement y. (Points on the string do not retain their displacement, but points on the wave *form* do.) If point A retains its displacement as it moves, the phase in Eq. 16-2 giving it that displacement must remain a constant:

$$kx - \omega t = \text{a constant}. \tag{16-11}$$

Note that although this argument is constant, both x and t are changing. In fact, as t increases, x must also, to keep the argument constant. This confirms that the wave pattern is moving in the positive direction of x.

To find the wave speed v, we take the derivative of Eq. 16-11, getting

$$k\frac{dx}{dt} - \omega = 0$$

or

$$\frac{dx}{dt} = v = \frac{\omega}{k}. \tag{16-12}$$

Using Eq. 16-5 ($k = 2\pi/\lambda$) and Eq. 16-8 ($\omega = 2\pi/T$), we can rewrite the wave speed as

$$v = \frac{\omega}{k} = \frac{\lambda}{T} = \lambda f \quad \text{(wave speed)}. \tag{16-13}$$

The equation $v = \lambda/T$ tells us that the wave speed is one wavelength per period; the wave moves a distance of one wavelength in one period of oscillation.

Equation 16-2 describes a wave moving in the positive direction of x. We can find the equation of a wave traveling in the opposite direction by replacing t in Eq. 16-2 with $-t$. This corresponds to the condition

$$kx + \omega t = \text{a constant}, \tag{16-14}$$

which (compare Eq. 16-11) requires that x *decrease* with time. Thus, a wave traveling in the negative direction of x is described by the equation

$$y(x, t) = y_m \sin(kx + \omega t). \tag{16-15}$$

If you analyze the wave of Eq. 16-15 as we have just done for the wave of Eq. 16-2, you will find for its velocity

$$\frac{dx}{dt} = -\frac{\omega}{k}. \tag{16-16}$$

The minus sign (compare Eq. 16-12) verifies that the wave is indeed moving in the negative direction of x and justifies our switching the sign of the time variable.

Consider now a wave of arbitrary shape, given by

$$y(x, t) = h(kx \pm \omega t), \tag{16-17}$$

where h represents *any* function, the sine function being one possibility. Our previous analysis shows that all waves in which the variables x and t enter into the combination $kx \pm \omega t$ are traveling waves. Furthermore, all traveling waves *must* be of the form of Eq. 16-17. Thus, $y(x, t) = \sqrt{ax + bt}$ represents a possible (though perhaps physically a little bizarre) traveling wave. The function $y(x, t) = \sin(ax^2 - bt)$, on the other hand, does *not* represent a traveling wave.

Checkpoint 2

Here are the equations of three waves:
(1) $y(x, t) = 2\sin(4x - 2t)$, (2) $y(x, t) = \sin(3x - 4t)$, (3) $y(x, t) = 2\sin(3x - 3t)$.
Rank the waves according to their (a) wave speed and (b) maximum speed perpendicular to the wave's direction of travel (the transverse speed), greatest first.

Sample Problem 16.01 Determining the quantities in an equation for a transverse wave

A transverse wave traveling along an x axis has the form given by

$$y = y_m \sin(kx \pm \omega t + \phi). \tag{16-18}$$

Figure 16-8a gives the displacements of string elements as a function of x, all at time $t = 0$. Figure 16-8b gives the displacements of the element at $x = 0$ as a function of t. Find the values of the quantities shown in Eq. 16-18, including the correct choice of sign.

KEY IDEAS

(1) Figure 16-8a is effectively a snapshot of reality (something that we can see), showing us motion spread out over the x axis. From it we can determine the wavelength λ of the wave along that axis, and then we can find the angular wave number k ($= 2\pi/\lambda$) in Eq. 16-18. (2) Figure 16-8b is an abstraction, showing us motion spread out over time. From it we can determine the period T of the string element in its SHM and thus also of the wave itself. From T we can then find angular frequency ω ($= 2\pi/T$) in Eq. 16-18. (3) The phase constant ϕ is set by the displacement of the string at $x = 0$ and $t = 0$.

Amplitude: From either Fig. 16-8a or 16-8b we see that the maximum displacement is 3.0 mm. Thus, the wave's amplitude $x_m = 3.0$ mm.

Wavelength: In Fig. 16-8a, the wavelength λ is the distance along the x axis between repetitions in the pattern. The easiest way to measure λ is to find the distance from one crossing point to the next crossing point where the string has the same slope. Visually we can roughly measure that distance with the scale on the axis. Instead, we can lay the edge of a

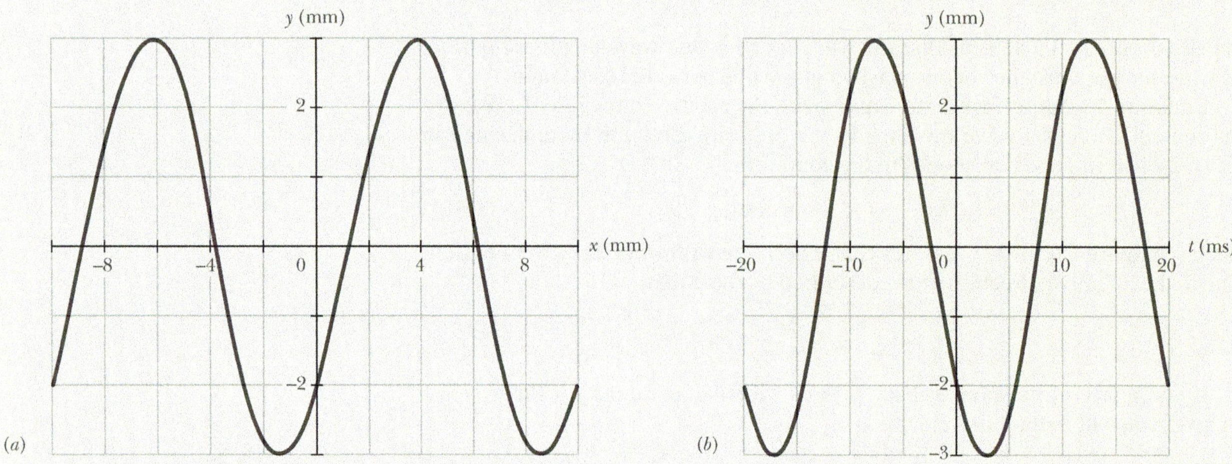

Figure 16-8 (a) A snapshot of the displacement y versus position x along a string, at time $t = 0$. (b) A graph of displacement y versus time t for the string element at $x = 0$.

paper sheet on the graph, mark those crossing points, slide the sheet to align the left-hand mark with the origin, and then read off the location of the right-hand mark. Either way we find $\lambda = 10$ mm. From Eq. 16-5, we then have

$$k = \frac{2\pi}{\lambda} = \frac{2\pi}{0.010\text{ m}} = 200\pi\text{ rad/m}.$$

Period: The period T is the time interval that a string element's SHM takes to begin repeating itself. In Fig. 16-8*b*, T is the distance along the t axis from one crossing point to the next crossing point where the plot has the same slope. Measuring the distance visually or with the aid of a sheet of paper, we find $T = 20$ ms. From Eq. 16-8, we then have

$$\omega = \frac{2\pi}{T} = \frac{2\pi}{0.020\text{ s}} = 100\pi\text{ rad/s}.$$

Direction of travel: To find the direction, we apply a bit of reasoning to the figures. In the snapshot at $t = 0$ given in Fig. 16-8*a*, note that if the wave is moving rightward, then just after the snapshot, the depth of the wave at $x = 0$ should increase (mentally slide the curve slightly rightward). If, instead, the wave is moving leftward, then just after the snapshot, the depth at $x = 0$ should decrease. Now let's check the graph in Fig. 16-8*b*. It tells us that just after $t = 0$, the depth increases. Thus, the wave is moving rightward, in the positive direction of x, and we choose the minus sign in Eq. 16-18.

Phase constant: The value of ϕ is set by the conditions at $x = 0$ at the instant $t = 0$. From either figure we see that at that location and time, $y = -2.0$ mm. Substituting these three values and also $y_m = 3.0$ mm into Eq. 16-18 gives us

$$-2.0\text{ mm} = (3.0\text{ mm})\sin(0 + 0 + \phi)$$

or

$$\phi = \sin^{-1}(-\tfrac{2}{3}) = -0.73\text{ rad}.$$

Note that this is consistent with the rule that on a plot of y versus x, a negative phase constant shifts the normal sine function rightward, which is what we see in Fig. 16-8*a*.

Equation: Now we can fill out Eq. 16-18:

$$y = (3.0\text{ mm})\sin(200\pi x - 100\pi t - 0.73\text{ rad}), \quad \text{(Answer)}$$

with x in meters and t in seconds.

Sample Problem 16.02 Transverse velocity and transverse acceleration of a string element

A wave traveling along a string is described by

$$y(x, t) = (0.00327\text{ m})\sin(72.1x - 2.72t),$$

in which the numerical constants are in SI units (72.1 rad/m and 2.72 rad/s).

(a) What is the transverse velocity u of the string element at $x = 22.5$ cm at time $t = 18.9$ s? (This velocity, which is associated with the transverse oscillation of a string element, is parallel to the y axis. Don't confuse it with v, the constant velocity at which the wave form moves along the x axis.)

KEY IDEAS

The transverse velocity u is the rate at which the displacement y of the element is changing. In general, that displacement is given by

$$y(x, t) = y_m \sin(kx - \omega t). \quad (16\text{-}19)$$

For an element at a certain location x, we find the rate of change of y by taking the derivative of Eq. 16-19 with respect to t while treating x as a constant. A derivative taken while one (or more) of the variables is treated as a constant is called a partial derivative and is represented by a symbol such as $\partial/\partial t$ rather than d/dt.

Calculations: Here we have

$$u = \frac{\partial y}{\partial t} = -\omega y_m \cos(kx - \omega t). \quad (16\text{-}20)$$

Next, substituting numerical values but suppressing the units, which are SI, we write

$$u = (-2.72)(0.00327)\cos[(72.1)(0.225) - (2.72)(18.9)]$$
$$= 0.00720\text{ m/s} = 7.20\text{ mm/s}. \quad \text{(Answer)}$$

Thus, at $t = 18.9$ s our string element is moving in the positive direction of y with a speed of 7.20 mm/s. (*Caution:* In evaluating the cosine function, we keep all the significant figures in the argument or the calculation can be off considerably. For example, round off the numbers to two significant figures and then see what you get for u.)

(b) What is the transverse acceleration a_y of our string element at $t = 18.9$ s?

KEY IDEA

The transverse acceleration a_y is the rate at which the element's transverse velocity is changing.

Calculations: From Eq. 16-20, again treating x as a constant but allowing t to vary, we find

$$a_y = \frac{\partial u}{\partial t} = -\omega^2 y_m \sin(kx - \omega t). \quad (16\text{-}21)$$

Substituting numerical values but suppressing the units, which are SI, we have

$$a_y = -(2.72)^2(0.00327)\sin[(72.1)(0.225) - (2.72)(18.9)]$$
$$= -0.0142\text{ m/s}^2 = -14.2\text{ mm/s}^2. \quad \text{(Answer)}$$

From part (a) we learn that at $t = 18.9$ s our string element is moving in the positive direction of y, and here we learn that it is slowing because its acceleration is in the opposite direction of u.

16-2 WAVE SPEED ON A STRETCHED STRING

Learning Objectives

After reading this module, you should be able to . . .

16.14 Calculate the linear density μ of a uniform string in terms of the total mass and total length.

16.15 Apply the relationship between wave speed v, tension τ, and linear density μ.

Key Ideas

- The speed of a wave on a stretched string is set by properties of the string, not properties of the wave such as frequency or amplitude.
- The speed of a wave on a string with tension τ and linear density μ is

$$v = \sqrt{\frac{\tau}{\mu}}.$$

Wave Speed on a Stretched String

The speed of a wave is related to the wave's wavelength and frequency by Eq. 16-13, but *it is set by the properties of the medium.* If a wave is to travel through a medium such as water, air, steel, or a stretched string, it must cause the particles of that medium to oscillate as it passes, which requires both mass (for kinetic energy) and elasticity (for potential energy). Thus, the mass and elasticity determine how fast the wave can travel. Here, we find that dependency in two ways.

Dimensional Analysis

In dimensional analysis we carefully examine the dimensions of all the physical quantities that enter into a given situation to determine the quantities they produce. In this case, we examine mass and elasticity to find a speed v, which has the dimension of length divided by time, or LT^{-1}.

For the mass, we use the mass of a string element, which is the mass m of the string divided by the length l of the string. We call this ratio the *linear density* μ of the string. Thus, $\mu = m/l$, its dimension being mass divided by length, ML^{-1}.

You cannot send a wave along a string unless the string is under tension, which means that it has been stretched and pulled taut by forces at its two ends. The tension τ in the string is equal to the common magnitude of those two forces. As a wave travels along the string, it displaces elements of the string by causing additional stretching, with adjacent sections of string pulling on each other because of the tension. Thus, we can associate the tension in the string with the stretching (elasticity) of the string. The tension and the stretching forces it produces have the dimension of a force—namely, MLT^{-2} (from $F = ma$).

We need to combine μ (dimension ML^{-1}) and τ (dimension MLT^{-2}) to get v (dimension LT^{-1}). A little juggling of various combinations suggests

$$v = C\sqrt{\frac{\tau}{\mu}}, \tag{16-22}$$

in which C is a dimensionless constant that cannot be determined with dimensional analysis. In our second approach to determining wave speed, you will see that Eq. 16-22 is indeed correct and that $C = 1$.

Derivation from Newton's Second Law

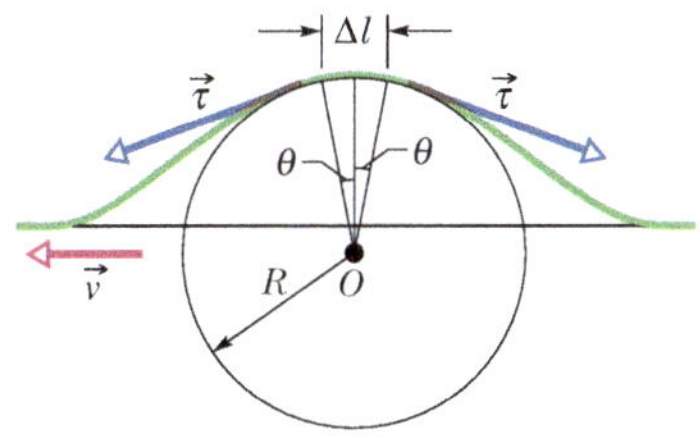

Figure 16-9 A symmetrical pulse, viewed from a reference frame in which the pulse is stationary and the string appears to move right to left with speed v. We find speed v by applying Newton's second law to a string element of length Δl, located at the top of the pulse.

Instead of the sinusoidal wave of Fig. 16-1b, let us consider a single symmetrical pulse such as that of Fig. 16-9, moving from left to right along a string with speed v. For convenience, we choose a reference frame in which the pulse remains stationary; that is, we run along with the pulse, keeping it constantly in view. In this frame, the string appears to move past us, from right to left in Fig. 16-9, with speed v.

Consider a small string element of length Δl within the pulse, an element that forms an arc of a circle of radius R and subtending an angle 2θ at the center of that circle. A force $\vec{\tau}$ with a magnitude equal to the tension in the string pulls tangentially on this element at each end. The horizontal components of these forces cancel, but the vertical components add to form a radial restoring force $\vec{F}$. In magnitude,

$$F = 2(\tau \sin\theta) \approx \tau(2\theta) = \tau\frac{\Delta l}{R} \quad \text{(force)}, \tag{16-23}$$

where we have approximated $\sin\theta$ as θ for the small angles θ in Fig. 16-9. From that figure, we have also used $2\theta = \Delta l/R$. The mass of the element is given by

$$\Delta m = \mu\,\Delta l \quad \text{(mass)}, \tag{16-24}$$

where μ is the string's linear density.

At the moment shown in Fig. 16-9, the string element Δl is moving in an arc of a circle. Thus, it has a centripetal acceleration toward the center of that circle, given by

$$a = \frac{v^2}{R} \quad \text{(acceleration)}. \tag{16-25}$$

Equations 16-23, 16-24, and 16-25 contain the elements of Newton's second law. Combining them in the form

$$\text{force} = \text{mass} \times \text{acceleration}$$

gives

$$\frac{\tau\,\Delta l}{R} = (\mu\,\Delta l)\frac{v^2}{R}.$$

Solving this equation for the speed v yields

$$v = \sqrt{\frac{\tau}{\mu}} \quad \text{(speed)}, \tag{16-26}$$

in exact agreement with Eq. 16-22 if the constant C in that equation is given the value unity. Equation 16-26 gives the speed of the pulse in Fig. 16-9 and the speed of *any* other wave on the same string under the same tension.

Equation 16-26 tells us:

> The speed of a wave along a stretched ideal string depends only on the tension and linear density of the string and not on the frequency of the wave.

The *frequency* of the wave is fixed entirely by whatever generates the wave (for example, the person in Fig. 16-1b). The *wavelength* of the wave is then fixed by Eq. 16-13 in the form $\lambda = v/f$.

Checkpoint 3

You send a traveling wave along a particular string by oscillating one end. If you increase the frequency of the oscillations, do (a) the speed of the wave and (b) the wavelength of the wave increase, decrease, or remain the same? If, instead, you increase the tension in the string, do (c) the speed of the wave and (d) the wavelength of the wave increase, decrease, or remain the same?

16-3 ENERGY AND POWER OF A WAVE TRAVELING ALONG A STRING

Learning Objective

After reading this module, you should be able to . . .

16.16 Calculate the average rate at which energy is transported by a transverse wave.

Key Idea

- The average power of, or average rate at which energy is transmitted by, a sinusoidal wave on a stretched string is given by

$$P_{avg} = \tfrac{1}{2}\mu v \omega^2 y_m^2.$$

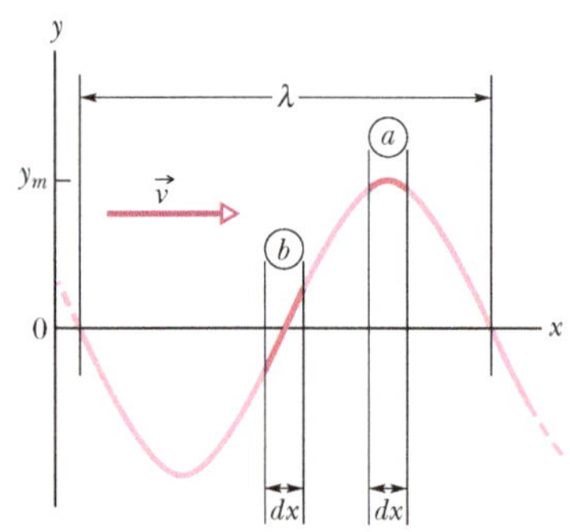

Figure 16-10 A snapshot of a traveling wave on a string at time $t = 0$. String element a is at displacement $y = y_m$, and string element b is at displacement $y = 0$. The kinetic energy of the string element at each position depends on the transverse velocity of the element. The potential energy depends on the amount by which the string element is stretched as the wave passes through it.

Energy and Power of a Wave Traveling Along a String

When we set up a wave on a stretched string, we provide energy for the motion of the string. As the wave moves away from us, it transports that energy as both kinetic energy and elastic potential energy. Let us consider each form in turn.

Kinetic Energy

A string element of mass dm, oscillating transversely in simple harmonic motion as the wave passes through it, has kinetic energy associated with its transverse velocity $\vec{u}$. When the element is rushing through its $y = 0$ position (element b in Fig. 16-10), its transverse velocity—and thus its kinetic energy—is a maximum. When the element is at its extreme position $y = y_m$ (as is element a), its transverse velocity—and thus its kinetic energy—is zero.

Elastic Potential Energy

To send a sinusoidal wave along a previously straight string, the wave must necessarily stretch the string. As a string element of length dx oscillates transversely, its length must increase and decrease in a periodic way if the string element is to fit the sinusoidal wave form. Elastic potential energy is associated with these length changes, just as for a spring.

When the string element is at its $y = y_m$ position (element a in Fig. 16-10), its length has its normal undisturbed value dx, so its elastic potential energy is zero. However, when the element is rushing through its $y = 0$ position, it has maximum stretch and thus maximum elastic potential energy.

Energy Transport

The oscillating string element thus has both its maximum kinetic energy and its maximum elastic potential energy at $y = 0$. In the snapshot of Fig. 16-10, the regions of the string at maximum displacement have no energy, and the regions at zero displacement have maximum energy. As the wave travels along the string, forces due to the tension in the string continuously do work to transfer energy from regions with energy to regions with no energy.

As in Fig. 16-1b, let's set up a wave on a string stretched along a horizontal x axis such that Eq. 16-2 applies. As we oscillate one end of the string, we continuously provide energy for the motion and stretching of the string—as the string sections oscillate perpendicularly to the x axis, they have kinetic energy and elastic potential energy. As the wave moves into sections that were previously at rest, energy is transferred into those new sections. Thus, we say that the wave *transports* the energy along the string.

The Rate of Energy Transmission

The kinetic energy dK associated with a string element of mass dm is given by

$$dK = \tfrac{1}{2}\, dm\, u^2, \tag{16-27}$$

where u is the transverse speed of the oscillating string element. To find u, we differentiate Eq. 16-2 with respect to time while holding x constant:

$$u = \frac{\partial y}{\partial t} = -\omega y_m \cos(kx - \omega t). \tag{16-28}$$

Using this relation and putting $dm = \mu\, dx$, we rewrite Eq. 16-27 as

$$dK = \tfrac{1}{2}(\mu\, dx)(-\omega y_m)^2 \cos^2(kx - \omega t). \tag{16-29}$$

Dividing Eq. 16-29 by dt gives the rate at which kinetic energy passes through a string element, and thus the rate at which kinetic energy is carried along by the wave. The dx/dt that then appears on the right of Eq. 16-29 is the wave speed v, so

$$\frac{dK}{dt} = \tfrac{1}{2}\mu v \omega^2 y_m^2 \cos^2(kx - \omega t). \tag{16-30}$$

The *average* rate at which kinetic energy is transported is

$$\left(\frac{dK}{dt}\right)_{avg} = \tfrac{1}{2}\mu v \omega^2 y_m^2 \,[\cos^2(kx - \omega t)]_{avg}$$

$$= \tfrac{1}{4}\mu v \omega^2 y_m^2. \tag{16-31}$$

Here we have taken the average over an integer number of wavelengths and have used the fact that the average value of the square of a cosine function over an integer number of periods is $\frac{1}{2}$.

Elastic potential energy is also carried along with the wave, and at the same average rate given by Eq. 16-31. Although we shall not examine the proof, you should recall that, in an oscillating system such as a pendulum or a spring–block system, the average kinetic energy and the average potential energy are equal.

The **average power,** which is the average rate at which energy of both kinds is transmitted by the wave, is then

$$P_{avg} = 2\left(\frac{dK}{dt}\right)_{avg} \tag{16-32}$$

or, from Eq. 16-31,

$$P_{avg} = \tfrac{1}{2}\mu v \omega^2 y_m^2 \quad \text{(average power).} \tag{16-33}$$

The factors μ and v in this equation depend on the material and tension of the string. The factors ω and y_m depend on the process that generates the wave. The dependence of the average power of a wave on the square of its amplitude and also on the square of its angular frequency is a general result, true for waves of all types.

Sample Problem 16.03 Average power of a transverse wave

A string has linear density $\mu = 525$ g/m and is under tension $\tau = 45$ N. We send a sinusoidal wave with frequency $f = 120$ Hz and amplitude $y_m = 8.5$ mm along the string. At what average rate does the wave transport energy?

KEY IDEA

The average rate of energy transport is the average power P_{avg} as given by Eq. 16-33.

Calculations: To use Eq. 16-33, we first must calculate angular frequency ω and wave speed v. From Eq. 16-9,

$$\omega = 2\pi f = (2\pi)(120 \text{ Hz}) = 754 \text{ rad/s}.$$

From Eq. 16-26 we have

$$v = \sqrt{\frac{\tau}{\mu}} = \sqrt{\frac{45 \text{ N}}{0.525 \text{ kg/m}}} = 9.26 \text{ m/s}.$$

Equation 16-33 then yields

$$P_{avg} = \tfrac{1}{2}\mu v \omega^2 y_m^2$$
$$= (\tfrac{1}{2})(0.525 \text{ kg/m})(9.26 \text{ m/s})(754 \text{ rad/s})^2(0.0085 \text{ m})^2$$
$$\approx 100 \text{ W}. \qquad \text{(Answer)}$$

16-4 THE WAVE EQUATION

Learning Objective

After reading this module, you should be able to . . .

16.17 For the equation giving a string-element displacement as a function of position x and time t, apply the relationship between the second derivative with respect to x and the second derivative with respect to t.

Key Idea

- The general differential equation that governs the travel of waves of all types is

$$\frac{\partial^2 y}{\partial x^2} = \frac{1}{v^2}\frac{\partial^2 y}{\partial t^2}.$$

Here the waves travel along an x axis and oscillate parallel to the y axis, and they move with speed v, in either the positive x direction or the negative x direction.

The Wave Equation

As a wave passes through any element on a stretched string, the element moves perpendicularly to the wave's direction of travel (we are dealing with a transverse wave). By applying Newton's second law to the element's motion, we can derive a general differential equation, called the *wave equation*, that governs the travel of waves of any type.

Figure 16-11*a* shows a snapshot of a string element of mass dm and length ℓ as a wave travels along a string of linear density μ that is stretched along a horizontal x axis. Let us assume that the wave amplitude is small so that the element can be tilted only slightly from the x axis as the wave passes. The force $\vec{F}_2$ on the right end of the element has a magnitude equal to tension τ in the string and is directed slightly upward. The force $\vec{F}_1$ on the left end of the element also has a magnitude equal to the tension τ but is directed slightly downward. Because of the slight curvature of the element, these two forces are not simply in opposite direction so that they cancel. Instead, they combine to produce a net force that causes the element to have an upward acceleration a_y. Newton's second law written for y components ($F_{\text{net},y} = ma_y$) gives us

$$F_{2y} - F_{1y} = dm\, a_y. \tag{16-34}$$

Let's analyze this equation in parts, first the mass dm, then the acceleration component a_y, then the individual force components F_{2y} and F_{1y}, and then finally the net force that is on the left side of Eq. 16-34.

Mass. The element's mass dm can be written in terms of the string's linear density μ and the element's length ℓ as $dm = \mu\ell$. Because the element can have only a slight tilt, $\ell \approx dx$ (Fig. 16-11*a*) and we have the approximation

$$dm = \mu\, dx. \tag{16-35}$$

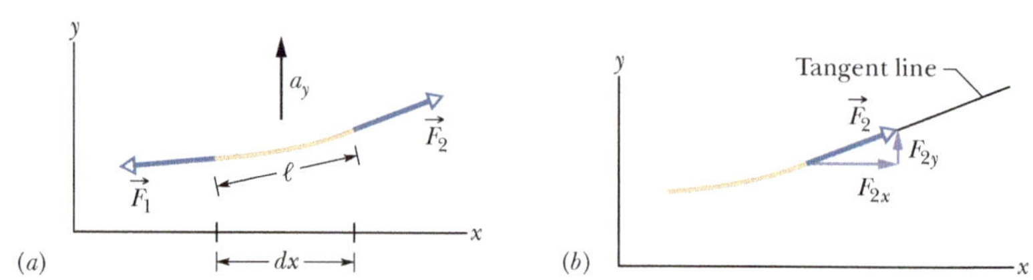

Figure 16-11 (*a*) A string element as a sinusoidal transverse wave travels on a stretched string. Forces $\vec{F}_1$ and $\vec{F}_2$ act at the left and right ends, producing acceleration $\vec{a}$ having a vertical component a_y. (*b*) The force at the element's right end is directed along a tangent to the element's right side.

Acceleration. The acceleration a_y in Eq. 16-34 is the second derivative of the displacement y with respect to time:

$$a_y = \frac{d^2y}{dt^2}. \tag{16-36}$$

Forces. Figure 16-11*b* shows that $\vec{F}_2$ is tangent to the string at the right end of the string element. Thus we can relate the components of the force to the string slope S_2 at the right end as

$$\frac{F_{2y}}{F_{2x}} = S_2. \tag{16-37}$$

We can also relate the components to the magnitude F_2 $(= \tau)$ with

$$F_2 = \sqrt{F_{2x}^2 + F_{2y}^2}$$

or

$$\tau = \sqrt{F_{2x}^2 + F_{2y}^2}. \tag{16-38}$$

However, because we assume that the element is only slightly tilted, $F_{2y} \ll F_{2x}$ and therefore we can rewrite Eq. 16-38 as

$$\tau = F_{2x}. \tag{16-39}$$

Substituting this into Eq. 16-37 and solving for F_{2y} yield

$$F_{2y} = \tau S_2. \tag{16-40}$$

Similar analysis at the left end of the string element gives us

$$F_{1y} = \tau S_1. \tag{16-41}$$

Net Force. We can now substitute Eqs. 16-35, 16-36, 16-40, and 16-41 into Eq. 16-34 to write

$$\tau S_2 - \tau S_1 = (\mu\, dx)\frac{d^2y}{dt^2},$$

or

$$\frac{S_2 - S_1}{dx} = \frac{\mu}{\tau}\frac{d^2y}{dt^2}. \tag{16-42}$$

Because the string element is short, slopes S_2 and S_1 differ by only a differential amount dS, where S is the slope at any point:

$$S = \frac{dy}{dx}. \tag{16-43}$$

First replacing $S_2 - S_1$ in Eq. 16-42 with dS and then using Eq. 16-43 to substitute dy/dx for S, we find

$$\frac{dS}{dx} = \frac{\mu}{\tau}\frac{d^2y}{dt^2},$$

$$\frac{d(dy/dx)}{dx} = \frac{\mu}{\tau}\frac{d^2y}{dt^2},$$

and

$$\frac{\partial^2 y}{\partial x^2} = \frac{\mu}{\tau}\frac{\partial^2 y}{\partial t^2}. \tag{16-44}$$

In the last step, we switched to the notation of partial derivatives because on the left we differentiate only with respect to x and on the right we differentiate only with respect to t. Finally, substituting from Eq. 16-26 ($v = \sqrt{\tau/\mu}$), we find

$$\frac{\partial^2 y}{\partial x^2} = \frac{1}{v^2}\frac{\partial^2 y}{\partial t^2} \quad \text{(wave equation)}. \tag{16-45}$$

This is the general differential equation that governs the travel of waves of all types.

16-5 INTERFERENCE OF WAVES

Learning Objectives

After reading this module, you should be able to . . .

16.18 Apply the principle of superposition to show that two overlapping waves add algebraically to give a resultant (or net) wave.

16.19 For two transverse waves with the same amplitude and wavelength and that travel together, find the displacement equation for the resultant wave and calculate the amplitude in terms of the individual wave amplitude and the phase difference.

16.20 Describe how the phase difference between two transverse waves (with the same amplitude and wavelength) can result in fully constructive interference, fully destructive interference, and intermediate interference.

16.21 With the phase difference between two interfering waves expressed in terms of wavelengths, quickly determine the type of interference the waves have.

Key Ideas

- When two or more waves traverse the same medium, the displacement of any particle of the medium is the sum of the displacements that the individual waves would give it, an effect known as the principle of superposition for waves.
- Two sinusoidal waves on the same string exhibit interference, adding or canceling according to the principle of superposition. If the two are traveling in the same direction and have the same amplitude y_m and frequency (hence the same wavelength) but differ in phase by a phase constant ϕ, the result is a single wave with this same frequency:

$$y'(x, t) = [2y_m \cos \tfrac{1}{2}\phi] \sin(kx - \omega t + \tfrac{1}{2}\phi).$$

If $\phi = 0$, the waves are exactly in phase and their interference is fully constructive; if $\phi = \pi$ rad, they are exactly out of phase and their interference is fully destructive.

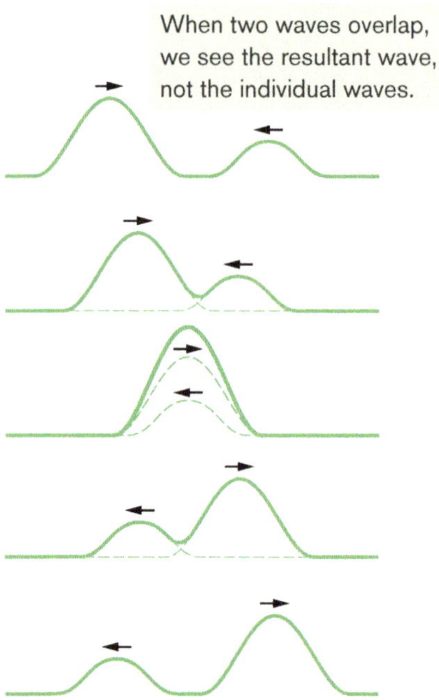

Figure 16-12 A series of snapshots that show two pulses traveling in opposite directions along a stretched string. The superposition principle applies as the pulses move through each other.

The Principle of Superposition for Waves

It often happens that two or more waves pass simultaneously through the same region. When we listen to a concert, for example, sound waves from many instruments fall simultaneously on our eardrums. The electrons in the antennas of our radio and television receivers are set in motion by the net effect of many electromagnetic waves from many different broadcasting centers. The water of a lake or harbor may be churned up by waves in the wakes of many boats.

Suppose that two waves travel simultaneously along the same stretched string. Let $y_1(x, t)$ and $y_2(x, t)$ be the displacements that the string would experience if each wave traveled alone. The displacement of the string when the waves overlap is then the algebraic sum

$$y'(x, t) = y_1(x, t) + y_2(x, t). \qquad (16\text{-}46)$$

This summation of displacements along the string means that

> Overlapping waves algebraically add to produce a **resultant wave** (or **net wave**).

This is another example of the **principle of superposition,** which says that when several effects occur simultaneously, their net effect is the sum of the individual effects. (We should be thankful that only a simple sum is needed. If two effects somehow amplified each other, the resulting nonlinear world would be very difficult to manage and understand.)

Figure 16-12 shows a sequence of snapshots of two pulses traveling in opposite directions on the same stretched string. When the pulses overlap, the resultant pulse is their sum. Moreover,

> Overlapping waves do not in any way alter the travel of each other.

Interference of Waves

Suppose we send two sinusoidal waves of the same wavelength and amplitude in the same direction along a stretched string. The superposition principle applies. What resultant wave does it predict for the string?

The resultant wave depends on the extent to which the waves are *in phase* (in step) with respect to each other—that is, how much one wave form is shifted from the other wave form. If the waves are exactly in phase (so that the peaks and valleys of one are exactly aligned with those of the other), they combine to double the displacement of either wave acting alone. If they are exactly out of phase (the peaks of one are exactly aligned with the valleys of the other), they combine to cancel everywhere, and the string remains straight. We call this phenomenon of combining waves **interference,** and the waves are said to **interfere.** (These terms refer only to the wave displacements; the travel of the waves is unaffected.)

Let one wave traveling along a stretched string be given by

$$y_1(x, t) = y_m \sin(kx - \omega t) \tag{16-47}$$

and another, shifted from the first, by

$$y_2(x, t) = y_m \sin(kx - \omega t + \phi). \tag{16-48}$$

These waves have the same angular frequency ω (and thus the same frequency f), the same angular wave number k (and thus the same wavelength λ), and the same amplitude y_m. They both travel in the positive direction of the x axis, with the same speed, given by Eq. 16-26. They differ only by a constant angle ϕ, the phase constant. These waves are said to be *out of phase* by ϕ or to have a *phase difference* of ϕ, or one wave is said to be *phase-shifted* from the other by ϕ.

From the principle of superposition (Eq. 16-46), the resultant wave is the algebraic sum of the two interfering waves and has displacement

$$\begin{aligned} y'(x, t) &= y_1(x, t) + y_2(x, t) \\ &= y_m \sin(kx - \omega t) + y_m \sin(kx - \omega t + \phi). \end{aligned} \tag{16-49}$$

In Appendix E we see that we can write the sum of the sines of two angles α and β as

$$\sin \alpha + \sin \beta = 2 \sin \tfrac{1}{2}(\alpha + \beta) \cos \tfrac{1}{2}(\alpha - \beta). \tag{16-50}$$

Applying this relation to Eq. 16-49 leads to

$$y'(x, t) = [2y_m \cos \tfrac{1}{2}\phi] \sin(kx - \omega t + \tfrac{1}{2}\phi). \tag{16-51}$$

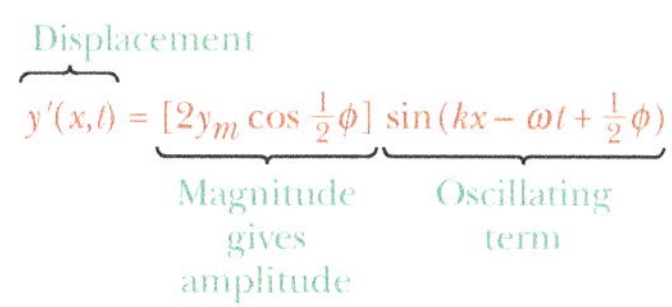

Figure 16-13 The resultant wave of Eq. 16-51, due to the interference of two sinusoidal transverse waves, is also a sinusoidal transverse wave, with an amplitude and an oscillating term.

As Fig. 16-13 shows, the resultant wave is also a sinusoidal wave traveling in the direction of increasing x. It is the only wave you would actually see on the string (you would *not* see the two interfering waves of Eqs. 16-47 and 16-48).

If two sinusoidal waves of the same amplitude and wavelength travel in the *same* direction along a stretched string, they interfere to produce a resultant sinusoidal wave traveling in that direction.

The resultant wave differs from the interfering waves in two respects: (1) its phase constant is $\frac{1}{2}\phi$, and (2) its amplitude y'_m is the magnitude of the quantity in the brackets in Eq. 16-51:

$$y'_m = |2y_m \cos \tfrac{1}{2}\phi| \quad \text{(amplitude).} \tag{16-52}$$

If $\phi = 0$ rad (or 0°), the two interfering waves are exactly in phase and Eq. 16-51 reduces to

$$y'(x, t) = 2y_m \sin(kx - \omega t) \qquad (\phi = 0). \tag{16-53}$$

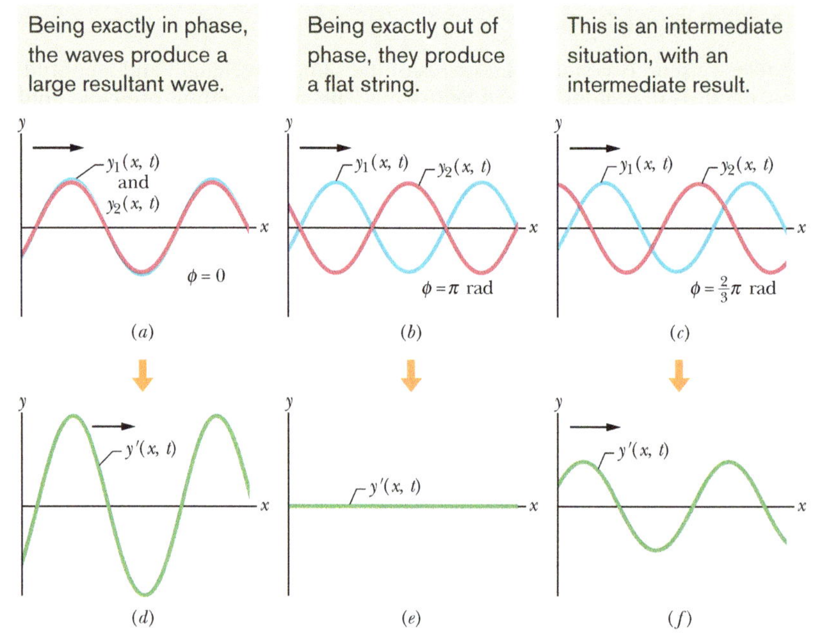

Figure 16-14 Two identical sinusoidal waves, $y_1(x, t)$ and $y_2(x, t)$, travel along a string in the positive direction of an x axis. They interfere to give a resultant wave $y'(x, t)$. The resultant wave is what is actually seen on the string. The phase difference ϕ between the two interfering waves is (a) 0 rad or 0°, (b) π rad or 180°, and (c) $\frac{2}{3}\pi$ rad or 120°. The corresponding resultant waves are shown in (d), (e), and (f).

The two waves are shown in Fig. 16-14a, and the resultant wave is plotted in Fig. 16-14d. Note from both that plot and Eq. 16-53 that the amplitude of the resultant wave is twice the amplitude of either interfering wave. That is the greatest amplitude the resultant wave can have, because the cosine term in Eqs. 16-51 and 16-52 has its greatest value (unity) when $\phi = 0$. Interference that produces the greatest possible amplitude is called *fully constructive interference.*

If $\phi = \pi$ rad (or 180°), the interfering waves are exactly out of phase as in Fig. 16-14b. Then $\cos \frac{1}{2}\phi$ becomes $\cos \pi/2 = 0$, and the amplitude of the resultant wave as given by Eq. 16-52 is zero. We then have, for all values of x and t,

$$y'(x, t) = 0 \qquad (\phi = \pi \text{ rad}). \tag{16-54}$$

The resultant wave is plotted in Fig. 16-14e. Although we sent two waves along the string, we see no motion of the string. This type of interference is called *fully destructive interference*.

Because a sinusoidal wave repeats its shape every 2π rad, a phase difference of $\phi = 2\pi$ rad (or 360°) corresponds to a shift of one wave relative to the other wave by a distance equivalent to one wavelength. Thus, phase differences can be described in terms of wavelengths as well as angles. For example, in Fig. 16-14b the waves may be said to be 0.50 wavelength out of phase. Table 16-1 shows some other examples of phase differences and the interference they produce. Note that when interference is neither fully constructive nor fully destructive, it is called *intermediate interference*. The amplitude of the resultant wave is then intermediate between 0 and $2y_m$. For example, from Table 16-1, if the interfering waves have a phase difference of 120° ($\phi = \frac{2}{3}\pi$ rad = 0.33 wavelength), then the resultant wave has an amplitude of y_m, the same as that of the interfering waves (see Figs. 16-14c and f).

Two waves with the same wavelength are in phase if their phase difference is zero or any integer number of wavelengths. Thus, the integer part of any phase difference *expressed in wavelengths* may be discarded. For example, a phase difference of 0.40 wavelength (an intermediate interference, close to fully destructive interference) is equivalent in every way to one of 2.40 wavelengths,

Table 16-1 Phase Difference and Resulting Interference Types[a]

Phase Difference, in			Amplitude of Resultant Wave	Type of Interference
Degrees	Radians	Wavelengths		
0	0	0	$2y_m$	Fully constructive
120	$\frac{2}{3}\pi$	0.33	y_m	Intermediate
180	π	0.50	0	Fully destructive
240	$\frac{4}{3}\pi$	0.67	y_m	Intermediate
360	2π	1.00	$2y_m$	Fully constructive
865	15.1	2.40	$0.60y_m$	Intermediate

[a]The phase difference is between two otherwise identical waves, with amplitude y_m, moving in the same direction.

and so the simpler of the two numbers can be used in computations. Thus, by looking at only the decimal number and comparing it to 0, 0.5, or 1.0 wavelength, you can quickly tell what type of interference two waves have.

Checkpoint 4

Here are four possible phase differences between two identical waves, expressed in wavelengths: 0.20, 0.45, 0.60, and 0.80. Rank them according to the amplitude of the resultant wave, greatest first.

Sample Problem 16.04 Interference of two waves, same direction, same amplitude

Two identical sinusoidal waves, moving in the same direction along a stretched string, interfere with each other. The amplitude y_m of each wave is 9.8 mm, and the phase difference ϕ between them is 100°.

(a) What is the amplitude y'_m of the resultant wave due to the interference, and what is the type of this interference?

KEY IDEA

These are identical sinusoidal waves traveling in the *same direction* along a string, so they interfere to produce a sinusoidal traveling wave.

Calculations: Because they are identical, the waves have the *same amplitude*. Thus, the amplitude y'_m of the resultant wave is given by Eq. 16-52:

$$y'_m = |2y_m \cos \tfrac{1}{2}\phi| = |(2)(9.8 \text{ mm}) \cos(100°/2)|$$
$$= 13 \text{ mm.} \qquad \text{(Answer)}$$

We can tell that the interference is *intermediate* in two ways. The phase difference is between 0 and 180°, and, correspondingly, the amplitude y'_m is between 0 and $2y_m$ (= 19.6 mm).

(b) What phase difference, in radians and wavelengths, will give the resultant wave an amplitude of 4.9 mm?

Calculations: Now we are given y'_m and seek ϕ. From Eq. 16-52,

$$y'_m = |2y_m \cos \tfrac{1}{2}\phi|,$$

we now have

$$4.9 \text{ mm} = (2)(9.8 \text{ mm}) \cos \tfrac{1}{2}\phi,$$

which gives us (with a calculator in the radian mode)

$$\phi = 2 \cos^{-1} \frac{4.9 \text{ mm}}{(2)(9.8 \text{ mm})}$$
$$= \pm 2.636 \text{ rad} \approx \pm 2.6 \text{ rad.} \qquad \text{(Answer)}$$

There are two solutions because we can obtain the same resultant wave by letting the first wave *lead* (travel ahead of) or *lag* (travel behind) the second wave by 2.6 rad. In wavelengths, the phase difference is

$$\frac{\phi}{2\pi \text{ rad/wavelength}} = \frac{\pm 2.636 \text{ rad}}{2\pi \text{ rad/wavelength}}$$
$$= \pm 0.42 \text{ wavelength.} \qquad \text{(Answer)}$$

WILEY PLUS Additional examples, video, and practice available at *WileyPLUS*

16-6 PHASORS

Learning Objectives

After reading this module, you should be able to . . .

16.22 Using sketches, explain how a phasor can represent the oscillations of a string element as a wave travels through its location.

16.23 Sketch a phasor diagram for two overlapping waves traveling together on a string, indicating their amplitudes and phase difference on the sketch.

16.24 By using phasors, find the resultant wave of two transverse waves traveling together along a string, calculating the amplitude and phase and writing out the displacement equation, and then displaying all three phasors in a phasor diagram that shows the amplitudes, the leading or lagging, and the relative phases.

Key Idea

- A wave $y(x, t)$ can be represented with a phasor. This is a vector that has a magnitude equal to the amplitude y_m of the wave and that rotates about an origin with an angular speed equal to the angular frequency ω of the wave. The projection of the rotating phasor on a vertical axis gives the displacement y of a point along the wave's travel.

Phasors

Adding two waves as discussed in the preceding module is strictly limited to waves with *identical* amplitudes. If we have such waves, that technique is easy enough to use, but we need a more general technique that can be applied to any waves, whether or not they have the same amplitudes. One neat way is to use phasors to represent the waves. Although this may seem bizarre at first, it is essentially a graphical technique that uses the vector addition rules of Chapter 3 instead of messy trig additions.

A **phasor** is a vector that rotates around its tail, which is pivoted at the origin of a coordinate system. The magnitude of the vector is equal to the amplitude y_m of the wave that it represents. The angular speed of the rotation is equal to the angular frequency ω of the wave. For example, the wave

$$y_1(x, t) = y_{m1} \sin(kx - \omega t) \tag{16-55}$$

is represented by the phasor shown in Figs. 16-15*a* to *d*. The magnitude of the phasor is the amplitude y_{m1} of the wave. As the phasor rotates around the origin at angular speed ω, its projection y_1 on the vertical axis varies sinusoidally, from a maximum of y_{m1} through zero to a minimum of $-y_{m1}$ and then back to y_{m1}. This variation corresponds to the sinusoidal variation in the displacement y_1 of any point along the string as the wave passes through that point. (All this is shown as an animation with voiceover in *WileyPLUS*.)

When two waves travel along the same string in the same direction, we can represent them and their resultant wave in a *phasor diagram*. The phasors in Fig. 16-15*e* represent the wave of Eq. 16-55 and a second wave given by

$$y_2(x, t) = y_{m2} \sin(kx - \omega t + \phi). \tag{16-56}$$

This second wave is phase-shifted from the first wave by phase constant ϕ. Because the phasors rotate at the same angular speed ω, the angle between the two phasors is always ϕ. If ϕ is a *positive* quantity, then the phasor for wave 2 *lags* the phasor for wave 1 as they rotate, as drawn in Fig. 16-15*e*. If ϕ is a negative quantity, then the phasor for wave 2 *leads* the phasor for wave 1.

Because waves y_1 and y_2 have the same angular wave number k and angular frequency ω, we know from Eqs. 16-51 and 16-52 that their resultant is of the form

$$y'(x, t) = y'_m \sin(kx - \omega t + \beta), \tag{16-57}$$

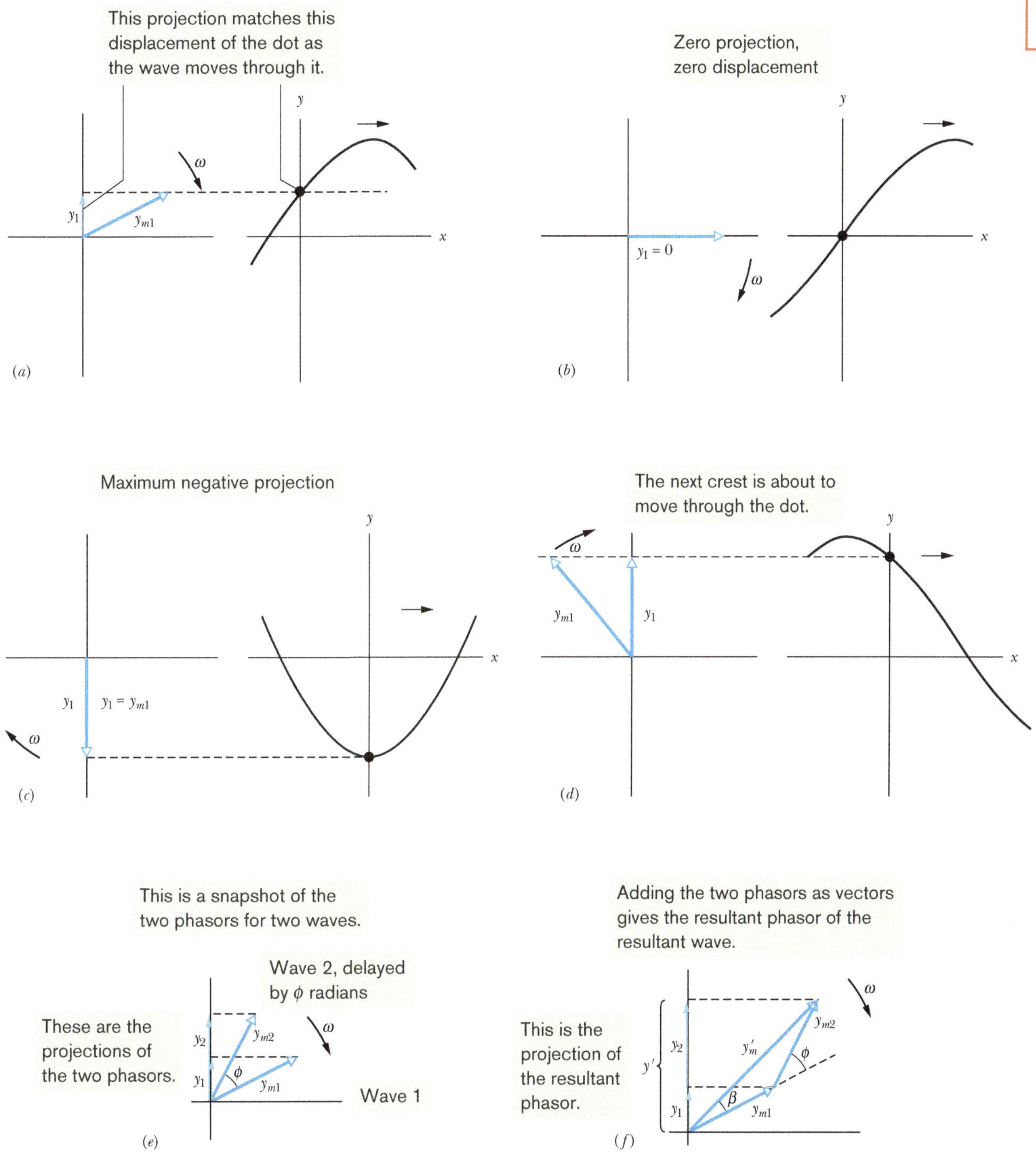

Figure 16-15 (*a*)–(*d*) A phasor of magnitude y_{m1} rotating about an origin at angular speed ω represents a sinusoidal wave. The phasor's projection y_1 on the vertical axis represents the displacement of a point through which the wave passes. (*e*) A second phasor, also of angular speed ω but of magnitude y_{m2} and rotating at a constant angle ϕ from the first phasor, represents a second wave, with a phase constant ϕ. (*f*) The resultant wave is represented by the vector sum y'_m of the two phasors.

where y'_m is the amplitude of the resultant wave and β is its phase constant. To find the values of y'_m and β, we would have to sum the two combining waves, as we did to obtain Eq. 16-51. To do this on a phasor diagram, we vectorially add the two phasors at any instant during their rotation, as in Fig. 16-15*f* where phasor y_{m2} has been shifted to the head of phasor y_{m1}. The magnitude of the vector sum equals the amplitude y'_m in Eq. 16-57. The angle between the vector sum and the phasor for y_1 equals the phase constant β in Eq. 16-57.

Note that, in contrast to the method of Module 16-5:

We can use phasors to combine waves *even if their amplitudes are different.*

Sample Problem 16.05 Interference of two waves, same direction, phasors, any amplitudes

Two sinusoidal waves $y_1(x, t)$ and $y_2(x, t)$ have the same wavelength and travel together in the same direction along a string. Their amplitudes are $y_{m1} = 4.0$ mm and $y_{m2} = 3.0$ mm, and their phase constants are 0 and $\pi/3$ rad, respectively. What are the amplitude y'_m and phase constant β of the resultant wave? Write the resultant wave in the form of Eq. 16-57.

KEY IDEAS

(1) The two waves have a number of properties in common: Because they travel along the same string, they must have the same speed v, as set by the tension and linear density of the string according to Eq. 16-26. With the same wavelength λ, they have the same angular wave number k $(= 2\pi/\lambda)$. Also, because they have the same wave number k and speed v, they must have the same angular frequency ω $(= kv)$.

(2) The waves (call them waves 1 and 2) can be represented by phasors rotating at the same angular speed ω about an origin. Because the phase constant for wave 2 is *greater* than that for wave 1 by $\pi/3$, phasor 2 must *lag* phasor 1 by $\pi/3$ rad in their clockwise rotation, as shown in Fig. 16-16*a*. The resultant wave due to the interference of waves 1 and 2 can then be represented by a phasor that is the vector sum of phasors 1 and 2.

Calculations: To simplify the vector summation, we drew phasors 1 and 2 in Fig. 16-16*a* at the instant when phasor 1 lies along the horizontal axis. We then drew lagging phasor 2 at positive angle $\pi/3$ rad. In Fig. 16-16*b* we shifted phasor 2 so its tail is at the head of phasor 1. Then we can draw the phasor y'_m of the resultant wave from the tail of phasor 1 to the head of phasor 2. The phase constant β is the angle phasor y'_m makes with phasor 1.

To find values for y'_m and β, we can sum phasors 1 and 2 as vectors on a vector-capable calculator. However, here we shall sum them by components. (They are called horizontal and vertical components, because the symbols x and y are already used for the waves themselves.) For the horizontal components we have

$$y'_{mh} = y_{m1}\cos 0 + y_{m2}\cos \pi/3$$
$$= 4.0 \text{ mm} + (3.0 \text{ mm})\cos \pi/3 = 5.50 \text{ mm}.$$

For the vertical components we have

$$y'_{mv} = y_{m1}\sin 0 + y_{m2}\sin \pi/3$$
$$= 0 + (3.0 \text{ mm})\sin \pi/3 = 2.60 \text{ mm}.$$

Thus, the resultant wave has an amplitude of

$$y'_m = \sqrt{(5.50 \text{ mm})^2 + (2.60 \text{ mm})^2}$$
$$= 6.1 \text{ mm} \qquad \text{(Answer)}$$

and a phase constant of

$$\beta = \tan^{-1}\frac{2.60 \text{ mm}}{5.50 \text{ mm}} = 0.44 \text{ rad}. \qquad \text{(Answer)}$$

From Fig. 16-16*b*, phase constant β is a *positive* angle relative to phasor 1. Thus, the resultant wave *lags* wave 1 in their travel by phase constant $\beta = +0.44$ rad. From Eq. 16-57, we can write the resultant wave as

$$y'(x, t) = (6.1 \text{ mm})\sin(kx - \omega t + 0.44 \text{ rad}). \qquad \text{(Answer)}$$

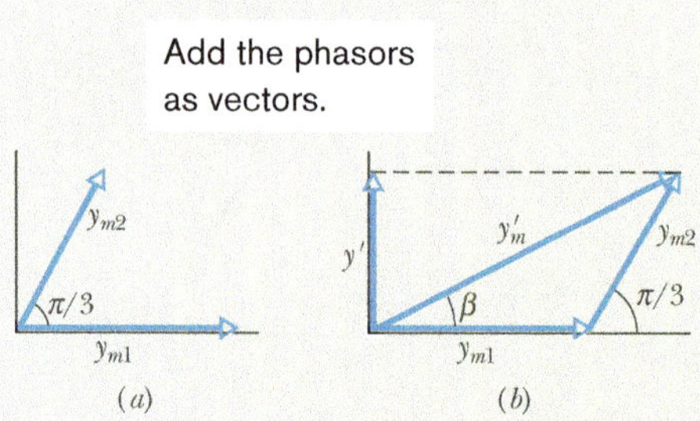

Figure 16-16 (*a*) Two phasors of magnitudes y_{m1} and y_{m2} and with phase difference $\pi/3$. (*b*) Vector addition of these phasors at any instant during their rotation gives the magnitude y'_m of the phasor for the resultant wave.

Additional examples, video, and practice available at *WileyPLUS*

16-7 STANDING WAVES AND RESONANCE

Learning Objectives

After reading this module, you should be able to . . .

16.25 For two overlapping waves (same amplitude and wavelength) that are traveling in opposite directions, sketch snapshots of the resultant wave, indicating nodes and antinodes.

16.26 For two overlapping waves (same amplitude and wavelength) that are traveling in opposite directions, find the displacement equation for the resultant wave and calculate the amplitude in terms of the individual wave amplitude.

16.27 Describe the SHM of a string element at an antinode of a standing wave.

16.28 For a string element at an antinode of a standing wave, write equations for the displacement, transverse velocity, and transverse acceleration as functions of time.

16.29 Distinguish between "hard" and "soft" reflections of string waves at a boundary.

16.30 Describe resonance on a string tied taut between two supports, and sketch the first several standing wave patterns, indicating nodes and antinodes.

16.31 In terms of string length, determine the wavelengths required for the first several harmonics on a string under tension.

16.32 For any given harmonic, apply the relationship between frequency, wave speed, and string length.

Key Ideas

- The interference of two identical sinusoidal waves moving in opposite directions produces standing waves. For a string with fixed ends, the standing wave is given by

$$y'(x, t) = [2y_m \sin kx] \cos \omega t.$$

Standing waves are characterized by fixed locations of zero displacement called nodes and fixed locations of maximum displacement called antinodes.

- Standing waves on a string can be set up by reflection of traveling waves from the ends of the string. If an end is fixed, it must be the position of a node. This limits the frequencies at which standing waves will occur on a given string. Each possible frequency is a resonant frequency, and the corresponding standing wave pattern is an oscillation mode. For a stretched string of length L with fixed ends, the resonant frequencies are

$$f = \frac{v}{\lambda} = n\frac{v}{2L}, \qquad \text{for } n = 1, 2, 3, \ldots .$$

The oscillation mode corresponding to $n = 1$ is called the *fundamental mode* or the *first harmonic*; the mode corresponding to $n = 2$ is the *second harmonic*; and so on.

Standing Waves

In Module 16-5, we discussed two sinusoidal waves of the same wavelength and amplitude traveling *in the same direction* along a stretched string. What if they travel in opposite directions? We can again find the resultant wave by applying the superposition principle.

Figure 16-17 suggests the situation graphically. It shows the two combining waves, one traveling to the left in Fig. 16-17*a*, the other to the right in Fig. 16-17*b*. Figure 16-17*c* shows their sum, obtained by applying the superposition

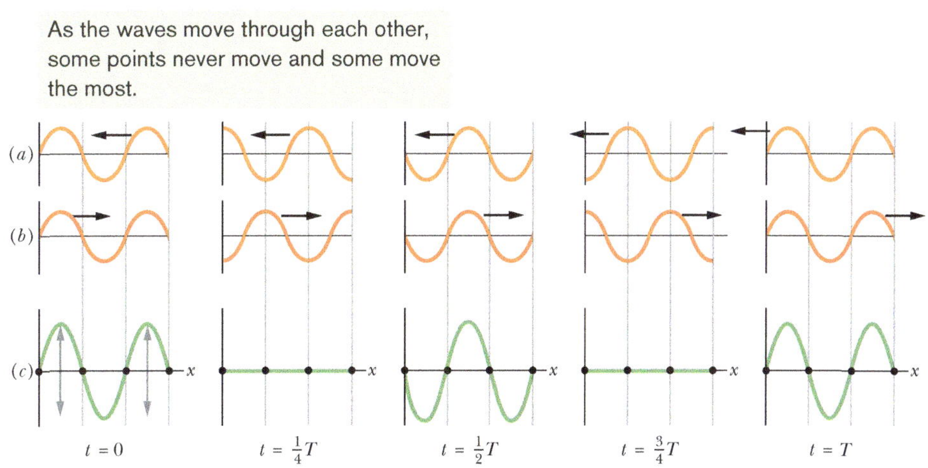

Figure 16-17 (*a*) Five snapshots of a wave traveling to the left, at the times t indicated below part (*c*) (T is the period of oscillation). (*b*) Five snapshots of a wave identical to that in (*a*) but traveling to the right, at the same times t. (*c*) Corresponding snapshots for the superposition of the two waves on the same string. At $t = 0, \frac{1}{2}T$, and T, fully constructive interference occurs because of the alignment of peaks with peaks and valleys with valleys. At $t = \frac{1}{4}T$ and $\frac{3}{4}T$, fully destructive interference occurs because of the alignment of peaks with valleys. Some points (the nodes, marked with dots) never oscillate; some points (the antinodes) oscillate the most.

principle graphically. The outstanding feature of the resultant wave is that there are places along the string, called **nodes,** where the string never moves. Four such nodes are marked by dots in Fig. 16-17*c*. Halfway between adjacent nodes are **antinodes,** where the amplitude of the resultant wave is a maximum. Wave patterns such as that of Fig. 16-17*c* are called **standing waves** because the wave patterns do not move left or right; the locations of the maxima and minima do not change.

> If two sinusoidal waves of the same amplitude and wavelength travel in *opposite* directions along a stretched string, their interference with each other produces a standing wave.

To analyze a standing wave, we represent the two waves with the equations

$$y_1(x,t) = y_m \sin(kx - \omega t) \tag{16-58}$$

and

$$y_2(x,t) = y_m \sin(kx + \omega t). \tag{16-59}$$

The principle of superposition gives, for the combined wave,

$$y'(x,t) = y_1(x,t) + y_2(x,t) = y_m \sin(kx - \omega t) + y_m \sin(kx + \omega t).$$

Applying the trigonometric relation of Eq. 16-50 leads to Fig. 16-18 and

$$y'(x,t) = [2y_m \sin kx] \cos \omega t. \tag{16-60}$$

This equation does not describe a traveling wave because it is not of the form of Eq. 16-17. Instead, it describes a standing wave.

The quantity $2y_m \sin kx$ in the brackets of Eq. 16-60 can be viewed as the amplitude of oscillation of the string element that is located at position x. However, since an amplitude is always positive and $\sin kx$ can be negative, we take the absolute value of the quantity $2y_m \sin kx$ to be the amplitude at x.

In a traveling sinusoidal wave, the amplitude of the wave is the same for all string elements. That is not true for a standing wave, in which the amplitude *varies with position*. In the standing wave of Eq. 16-60, for example, the amplitude is zero for values of kx that give $\sin kx = 0$. Those values are

$$kx = n\pi, \quad \text{for } n = 0, 1, 2, \ldots. \tag{16-61}$$

Substituting $k = 2\pi/\lambda$ in this equation and rearranging, we get

$$x = n\frac{\lambda}{2}, \quad \text{for } n = 0, 1, 2, \ldots \quad \text{(nodes)}, \tag{16-62}$$

as the positions of zero amplitude—the nodes—for the standing wave of Eq. 16-60. Note that adjacent nodes are separated by $\lambda/2$, half a wavelength.

The amplitude of the standing wave of Eq. 16-60 has a maximum value of $2y_m$, which occurs for values of kx that give $|\sin kx| = 1$. Those values are

$$kx = \tfrac{1}{2}\pi, \tfrac{3}{2}\pi, \tfrac{5}{2}\pi, \ldots$$
$$= (n + \tfrac{1}{2})\pi, \quad \text{for } n = 0, 1, 2, \ldots. \tag{16-63}$$

Substituting $k = 2\pi/\lambda$ in Eq. 16-63 and rearranging, we get

$$x = \left(n + \frac{1}{2}\right)\frac{\lambda}{2}, \quad \text{for } n = 0, 1, 2, \ldots \quad \text{(antinodes)}, \tag{16-64}$$

as the positions of maximum amplitude—the antinodes—of the standing wave of Eq. 16-60. Antinodes are separated by $\lambda/2$ and are halfway between nodes.

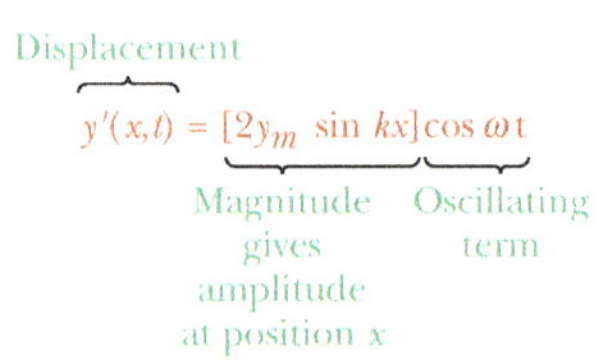

Figure 16-18 The resultant wave of Eq. 16-60 is a standing wave and is due to the interference of two sinusoidal waves of the same amplitude and wavelength that travel in opposite directions.

Reflections at a Boundary

We can set up a standing wave in a stretched string by allowing a traveling wave to be reflected from the far end of the string so that the wave travels back through itself. The incident (original) wave and the reflected wave can then be described by Eqs. 16-58 and 16-59, respectively, and they can combine to form a pattern of standing waves.

In Fig. 16-19, we use a single pulse to show how such reflections take place. In Fig. 16-19*a*, the string is fixed at its left end. When the pulse arrives at that end, it exerts an upward force on the support (the wall). By Newton's third law, the support exerts an opposite force of equal magnitude on the string. This second force generates a pulse at the support, which travels back along the string in the direction opposite that of the incident pulse. In a "hard" reflection of this kind, there must be a node at the support because the string is fixed there. The reflected and incident pulses must have opposite signs, so as to cancel each other at that point.

In Fig. 16-19*b*, the left end of the string is fastened to a light ring that is free to slide without friction along a rod. When the incident pulse arrives, the ring moves up the rod. As the ring moves, it pulls on the string, stretching the string and producing a reflected pulse with the same sign and amplitude as the incident pulse. Thus, in such a "soft" reflection, the incident and reflected pulses reinforce each other, creating an antinode at the end of the string; the maximum displacement of the ring is twice the amplitude of either of these two pulses.

There are two ways a pulse can reflect from the end of a string.

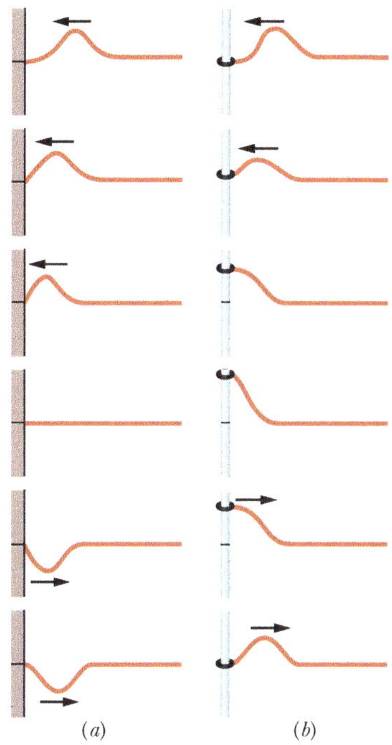

Figuer 16-19 (*a*) A pulse incident from the right is reflected at the left end of the string, which is tied to a wall. Note that the reflected pulse is inverted from the incident pulse. (*b*) Here the left end of the string is tied to a ring that can slide without friction up and down the rod. Now the pulse is not inverted by the reflection.

Checkpoint 5

Two waves with the same amplitude and wavelength interfere in three different situations to produce resultant waves with the following equations:

(1) $y'(x, t) = 4\sin(5x - 4t)$

(2) $y'(x, t) = 4\sin(5x)\cos(4t)$

(3) $y'(x, t) = 4\sin(5x + 4t)$

In which situation are the two combining waves traveling (a) toward positive x, (b) toward negative x, and (c) in opposite directions?

Standing Waves and Resonance

Consider a string, such as a guitar string, that is stretched between two clamps. Suppose we send a continuous sinusoidal wave of a certain frequency along the string, say, toward the right. When the wave reaches the right end, it reflects and begins to travel back to the left. That left-going wave then overlaps the wave that is still traveling to the right. When the left-going wave reaches the left end, it reflects again and the newly reflected wave begins to travel to the right, overlapping the left-going and right-going waves. In short, we very soon have many overlapping traveling waves, which interfere with one another.

For certain frequencies, the interference produces a standing wave pattern (or **oscillation mode**) with nodes and large antinodes like those in Fig. 16-20. Such a standing wave is said to be produced at **resonance,** and the string is said to *resonate* at these certain frequencies, called **resonant frequencies.** If the string

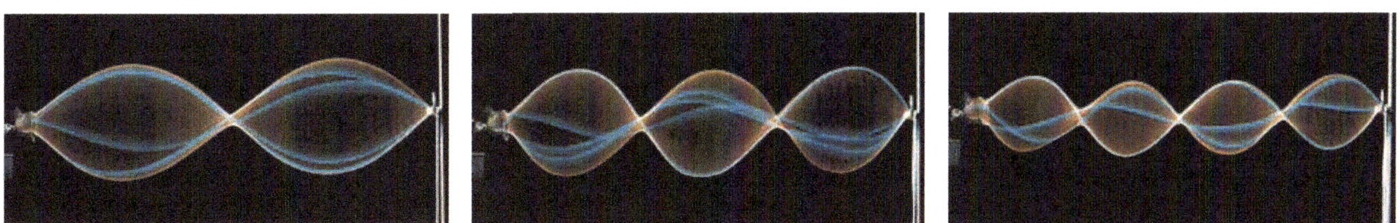

Richard Megna/Fundamental Photographs

Figure 16-20 Stroboscopic photographs reveal (imperfect) standing wave patterns on a string being made to oscillate by an oscillator at the left end. The patterns occur at certain frequencies of oscillation.

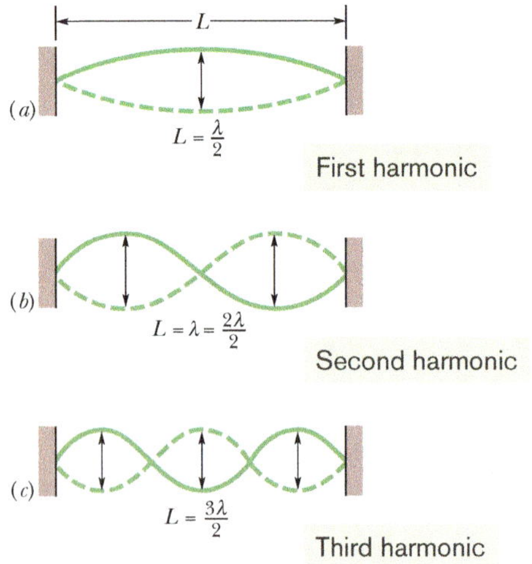

Figure 16-21 A string, stretched between two clamps, is made to oscillate in standing wave patterns. (*a*) The simplest possible pattern consists of one *loop,* which refers to the composite shape formed by the string in its extreme displacements (the solid and dashed lines). (*b*) The next simplest pattern has two loops. (*c*) The next has three loops.

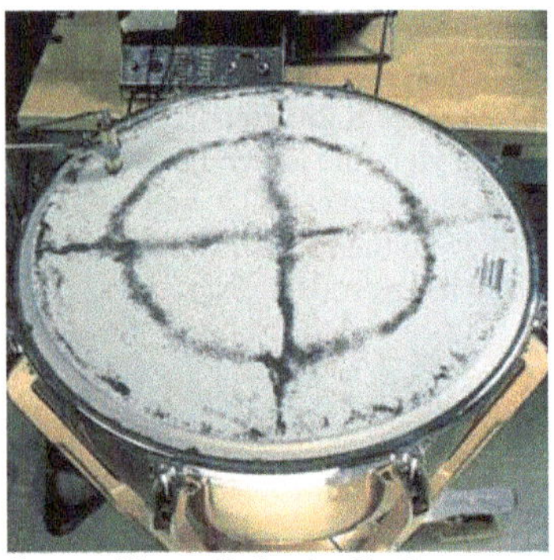

Courtesy Thomas D. Rossing, Northern Illinois University

Figure 16-22 One of many possible standing wave patterns for a kettledrum head, made visible by dark powder sprinkled on the drumhead. As the head is set into oscillation at a single frequency by a mechanical oscillator at the upper left of the photograph, the powder collects at the nodes, which are circles and straight lines in this two-dimensional example.

is oscillated at some frequency other than a resonant frequency, a standing wave is not set up. Then the interference of the right-going and left-going traveling waves results in only small, temporary (perhaps even imperceptible) oscillations of the string.

Let a string be stretched between two clamps separated by a fixed distance L. To find expressions for the resonant frequencies of the string, we note that a node must exist at each of its ends, because each end is fixed and cannot oscillate. The simplest pattern that meets this key requirement is that in Fig. 16-21*a*, which shows the string at both its extreme displacements (one solid and one dashed, together forming a single "loop"). There is only one antinode, which is at the center of the string. Note that half a wavelength spans the length L, which we take to be the string's length. Thus, for this pattern, $\lambda/2 = L$. This condition tells us that if the left-going and right-going traveling waves are to set up this pattern by their interference, they must have the wavelength $\lambda = 2L$.

A second simple pattern meeting the requirement of nodes at the fixed ends is shown in Fig. 16-21*b*. This pattern has three nodes and two antinodes and is said to be a two-loop pattern. For the left-going and right-going waves to set it up, they must have a wavelength $\lambda = L$. A third pattern is shown in Fig. 16-21*c*. It has four nodes, three antinodes, and three loops, and the wavelength is $\lambda = \frac{2}{3}L$. We could continue this progression by drawing increasingly more complicated patterns. In each step of the progression, the pattern would have one more node and one more antinode than the preceding step, and an additional $\lambda/2$ would be fitted into the distance L.

Thus, a standing wave can be set up on a string of length L by a wave with a wavelength equal to one of the values

$$\lambda = \frac{2L}{n}, \quad \text{for } n = 1, 2, 3, \ldots. \qquad (16\text{-}65)$$

The resonant frequencies that correspond to these wavelengths follow from Eq. 16-13:

$$f = \frac{v}{\lambda} = n\frac{v}{2L}, \quad \text{for } n = 1, 2, 3, \ldots. \qquad (16\text{-}66)$$

Here v is the speed of traveling waves on the string.

Equation 16-66 tells us that the resonant frequencies are integer multiples of the lowest resonant frequency, $f = v/2L$, which corresponds to $n = 1$. The oscillation mode with that lowest frequency is called the *fundamental mode* or the *first harmonic.* The *second harmonic* is the oscillation mode with $n = 2$, the *third harmonic* is that with $n = 3$, and so on. The frequencies associated with these modes are often labeled f_1, f_2, f_3, and so on. The collection of all possible oscillation modes is called the **harmonic series,** and n is called the **harmonic number** of the nth harmonic.

For a given string under a given tension, each resonant frequency corresponds to a particular oscillation pattern. Thus, if the frequency is in the audible range, you can hear the shape of the string. Resonance can also occur in two dimensions (such as on the surface of the kettledrum in Fig. 16-22) and in three dimensions (such as in the wind-induced swaying and twisting of a tall building).

Checkpoint 6

In the following series of resonant frequencies, one frequency (lower than 400 Hz) is missing: 150, 225, 300, 375 Hz. (a) What is the missing frequency? (b) What is the frequency of the seventh harmonic?

Sample Problem 16.06 Resonance of transverse waves, standing waves, harmonics

Figure 16-23 shows resonant oscillation of a string of mass $m = 2.500$ g and length $L = 0.800$ m and that is under tension $\tau = 325.0$ N. What is the wavelength λ of the transverse waves producing the standing wave pattern, and what is the harmonic number n? What is the frequency f of the transverse waves and of the oscillations of the moving string elements? What is the maximum magnitude of the transverse velocity u_m of the element oscillating at coordinate $x = 0.180$ m? At what point during the element's oscillation is the transverse velocity maximum?

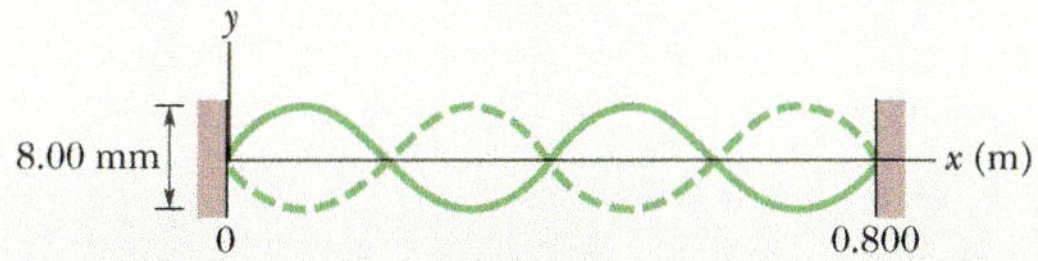

Figure 16-23 Resonant oscillation of a string under tension.

KEY IDEAS

(1) The traverse waves that produce a standing wave pattern must have a wavelength such that an integer number n of half-wavelengths fit into the length L of the string. (2) The frequency of those waves and of the oscillations of the string elements is given by Eq. 16-66 ($f = nv/2L$). (3) The displacement of a string element as a function of position x and time t is given by Eq. 16-60:

$$y'(x,t) = [2y_m \sin kx] \cos \omega t. \tag{16-67}$$

Wavelength and harmonic number: In Fig. 16-23, the solid line, which is effectively a snapshot (or freeze-frame) of the oscillations, reveals that 2 full wavelengths fit into the length $L = 0.800$ m of the string. Thus, we have

$$2\lambda = L,$$

or

$$\lambda = \frac{L}{2}. \tag{16-68}$$

$$= \frac{0.800 \text{ m}}{2} = 0.400 \text{ m}. \quad \text{(Answer)}$$

By counting the number of loops (or half-wavelengths) in Fig. 16-23, we see that the harmonic number is

$$n = 4. \quad \text{(Answer)}$$

We also find $n = 4$ by comparing Eqs. 16-68 and 16-65 ($\lambda = 2L/n$). Thus, the string is oscillating in its fourth harmonic.

Frequency: We can get the frequency f of the transverse waves from Eq. 16-13 ($v = \lambda f$) if we first find the speed v of the waves. That speed is given by Eq. 16-26, but we must substitute m/L for the unknown linear density μ. We obtain

$$v = \sqrt{\frac{\tau}{\mu}} = \sqrt{\frac{\tau}{m/L}} = \sqrt{\frac{\tau L}{m}}$$

$$= \sqrt{\frac{(325 \text{ N})(0.800 \text{ m})}{2.50 \times 10^{-3} \text{ kg}}} = 322.49 \text{ m/s}.$$

After rearranging Eq. 16-13, we write

$$f = \frac{v}{\lambda} = \frac{322.49 \text{ m/s}}{0.400 \text{ m}}$$

$$= 806.2 \text{ Hz} \approx 806 \text{ Hz}. \quad \text{(Answer)}$$

Note that we get the same answer by substituting into Eq. 16-66:

$$f = n\frac{v}{2L} = 4\frac{322.49 \text{ m/s}}{2(0.800 \text{ m})}$$

$$= 806 \text{ Hz}. \quad \text{(Answer)}$$

Now note that this 806 Hz is not only the frequency of the waves producing the fourth harmonic but also it is said to *be* the fourth harmonic, as in the statement, "The fourth harmonic of this oscillating string is 806 Hz." It is also the frequency of the string elements as they oscillate vertically in the figure in simple harmonic motion, just as a block on a vertical spring would oscillate in simple harmonic motion. Finally, it is also the frequency of the sound you would hear as the oscillating string periodically pushes against the air.

Transverse velocity: The displacement y' of the string element located at coordinate x is given by Eq. 16-67 as a function of time t. The term $\cos \omega t$ contains the dependence on time and thus provides the "motion" of the standing wave. The term $2y_m \sin kx$ sets the extent of the motion—that is, the amplitude. The greatest amplitude occurs at an antinode, where $\sin kx$ is $+1$ or -1 and thus the greatest amplitude is $2y_m$. From Fig. 16-23, we see that $2y_m = 4.00$ mm, which tells us that $y_m = 2.00$ mm.

We want the transverse velocity—the velocity of a string element parallel to the y axis. To find it, we take the time derivative of Eq. 16-67:

$$u(x,t) = \frac{\partial y'}{\partial t} = \frac{\partial}{\partial t}[(2y_m \sin kx) \cos \omega t]$$

$$= [-2y_m \omega \sin kx] \sin \omega t. \tag{16-69}$$

Here the term $\sin \omega t$ provides the variation with time and the term $-2y_m\omega \sin kx$ provides the extent of that variation. We want the absolute magnitude of that extent:

$$u_m = |-2y_m \omega \sin kx|.$$

To evaluate this for the element at $x = 0.180$ m, we first note that $y_m = 2.00$ mm, $k = 2\pi/\lambda = 2\pi/(0.400 \text{ m})$, and $\omega = 2\pi f = 2\pi(806.2 \text{ Hz})$. Then the maximum speed of the element at $x = 0.180$ m is

$$u_m = \left| -2(2.00 \times 10^{-3}\text{ m})(2\pi)(806.2\text{ Hz}) \times \sin\left(\frac{2\pi}{0.400\text{ m}}(0.180\text{ m})\right) \right|$$
$$= 6.26\text{ m/s.} \quad \text{(Answer)}$$

To determine when the string element has this maximum speed, we could investigate Eq. 16-69. However, a little thought can save a lot of work. The element is undergoing SHM and must come to a momentary stop at its extreme upward position and extreme downward position. It has the greatest speed as it zips through the midpoint of its oscillation, just as a block does in a block–spring oscillator.

Review & Summary

Transverse and Longitudinal Waves Mechanical waves can exist only in material media and are governed by Newton's laws. **Transverse** mechanical waves, like those on a stretched string, are waves in which the particles of the medium oscillate perpendicular to the wave's direction of travel. Waves in which the particles of the medium oscillate parallel to the wave's direction of travel are **longitudinal** waves.

Sinusoidal Waves A sinusoidal wave moving in the positive direction of an x axis has the mathematical form

$$y(x, t) = y_m \sin(kx - \omega t), \tag{16-2}$$

where y_m is the **amplitude** of the wave, k is the **angular wave number,** ω is the **angular frequency,** and $kx - \omega t$ is the **phase.** The **wavelength** λ is related to k by

$$k = \frac{2\pi}{\lambda}. \tag{16-5}$$

The **period** T and **frequency** f of the wave are related to ω by

$$\frac{\omega}{2\pi} = f = \frac{1}{T}. \tag{16-9}$$

Finally, the **wave speed** v is related to these other parameters by

$$v = \frac{\omega}{k} = \frac{\lambda}{T} = \lambda f. \tag{16-13}$$

Equation of a Traveling Wave Any function of the form

$$y(x, t) = h(kx \pm \omega t) \tag{16-17}$$

can represent a **traveling wave** with a wave speed given by Eq. 16-13 and a wave shape given by the mathematical form of h. The plus sign denotes a wave traveling in the negative direction of the x axis, and the minus sign a wave traveling in the positive direction.

Wave Speed on Stretched String The speed of a wave on a stretched string is set by properties of the string. The speed on a string with tension τ and linear density μ is

$$v = \sqrt{\frac{\tau}{\mu}}. \tag{16-26}$$

Power The **average power** of, or average rate at which energy is transmitted by, a sinusoidal wave on a stretched string is given by

$$P_{\text{avg}} = \tfrac{1}{2}\mu v \omega^2 y_m^2. \tag{16-33}$$

Superposition of Waves When two or more waves traverse the same medium, the displacement of any particle of the medium is the sum of the displacements that the individual waves would give it.

Interference of Waves Two sinusoidal waves on the same string exhibit **interference,** adding or canceling according to the principle of superposition. If the two are traveling in the same direction and have the same amplitude y_m and frequency (hence the same wavelength) but differ in phase by a **phase constant** ϕ, the result is a single wave with this same frequency:

$$y'(x, t) = [2y_m \cos \tfrac{1}{2}\phi] \sin(kx - \omega t + \tfrac{1}{2}\phi). \tag{16-51}$$

If $\phi = 0$, the waves are exactly in phase and their interference is fully constructive; if $\phi = \pi$ rad, they are exactly out of phase and their interference is fully destructive.

Phasors A wave $y(x, t)$ can be represented with a **phasor.** This is a vector that has a magnitude equal to the amplitude y_m of the wave and that rotates about an origin with an angular speed equal to the angular frequency ω of the wave. The projection of the rotating phasor on a vertical axis gives the displacement y of a point along the wave's travel.

Standing Waves The interference of two identical sinusoidal waves moving in opposite directions produces **standing waves.** For a string with fixed ends, the standing wave is given by

$$y'(x, t) = [2y_m \sin kx] \cos \omega t. \tag{16-60}$$

Standing waves are characterized by fixed locations of zero displacement called **nodes** and fixed locations of maximum displacement called **antinodes.**

Resonance Standing waves on a string can be set up by reflection of traveling waves from the ends of the string. If an end is fixed, it must be the position of a node. This limits the frequencies at which standing waves will occur on a given string. Each possible frequency is a **resonant frequency,** and the corresponding standing wave pattern is an **oscillation mode.** For a stretched string of length L with fixed ends, the resonant frequencies are

$$f = \frac{v}{\lambda} = n\frac{v}{2L}, \quad \text{for } n = 1, 2, 3, \ldots. \tag{16-66}$$

The oscillation mode corresponding to $n = 1$ is called the *fundamental mode* or the *first harmonic*; the mode corresponding to $n = 2$ is the *second harmonic*; and so on.

Questions

1 The following four waves are sent along strings with the same linear densities (x is in meters and t is in seconds). Rank the waves according to (a) their wave speed and (b) the tension in the strings along which they travel, greatest first:

(1) $y_1 = (3\text{ mm})\sin(x - 3t)$, (3) $y_3 = (1\text{ mm})\sin(4x - t)$,
(2) $y_2 = (6\text{ mm})\sin(2x - t)$, (4) $y_4 = (2\text{ mm})\sin(x - 2t)$.

2 In Fig. 16-24, wave 1 consists of a rectangular peak of height 4 units and width d, and a rectangular valley of depth 2 units and width d. The wave travels rightward along an x axis. Choices 2, 3, and 4 are similar waves, with the same heights, depths, and widths, that will travel leftward along that axis and through wave 1. Right-going wave 1 and one of the left-going waves will interfere as they pass through each other. With which left-going wave will the interference give, for an instant, (a) the deepest valley, (b) a flat line, and (c) a flat peak $2d$ wide?

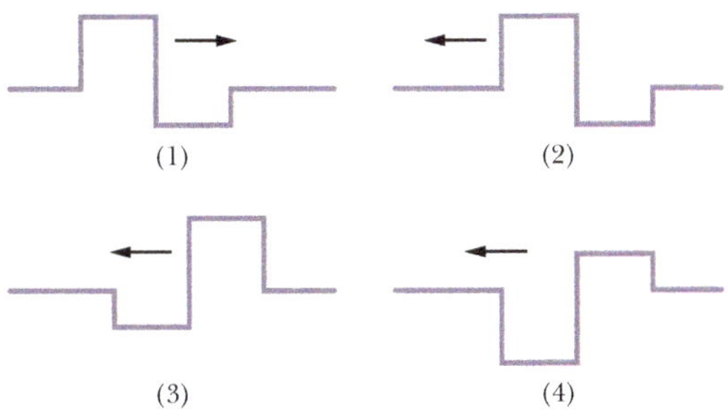

Figure 16-24 Question 2.

3 Figure 16-25a gives a snapshot of a wave traveling in the direction of positive x along a string under tension. Four string elements are indicated by the lettered points. For each of those elements, determine whether, at the instant of the snapshot, the element is moving upward or downward or is momentarily at rest. (*Hint:* Imagine the wave as it moves through the four string elements, as if you were watching a video of the wave as it traveled rightward.)

Figure 16-25b gives the displacement of a string element located at, say, $x = 0$ as a function of time. At the lettered times, is the element moving upward or downward or is it momentarily at rest?

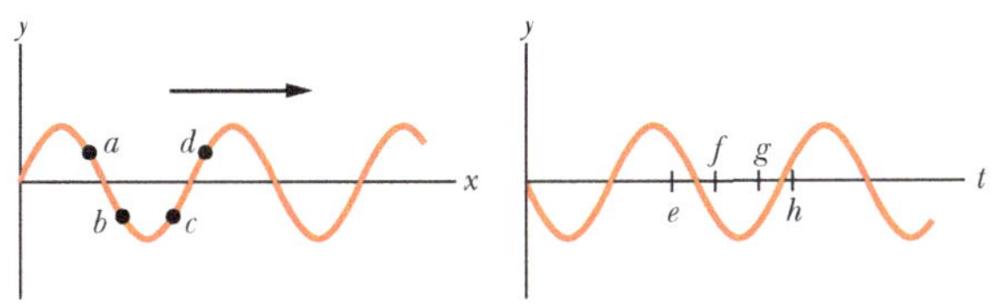

Figure 16-25 Question 3.

4 Figure 16-26 shows three waves that are *separately* sent along a string that is stretched under a certain tension along an x axis. Rank the waves according to their (a) wavelengths, (b) speeds, and (c) angular frequencies, greatest first.

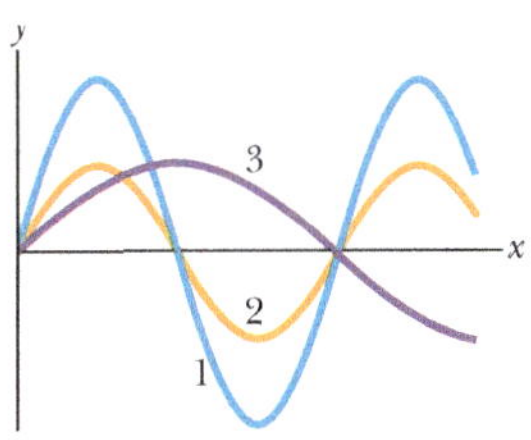

Figure 16-26 Question 4.

5 If you start with two sinusoidal waves of the same amplitude traveling in phase on a string and then somehow phase-shift one of them by 5.4 wavelengths, what type of interference will occur on the string?

6 The amplitudes and phase differences for four pairs of waves of equal wavelengths are (a) 2 mm, 6 mm, and π rad; (b) 3 mm, 5 mm, and π rad; (c) 7 mm, 9 mm, and π rad; (d) 2 mm, 2 mm, and 0 rad. Each pair travels in the same direction along the same string. Without written calculation, rank the four pairs according to the amplitude of their resultant wave, greatest first. (*Hint:* Construct phasor diagrams.)

7 A sinusoidal wave is sent along a cord under tension, transporting energy at the average rate of $P_{avg,1}$. Two waves, identical to that first one, are then to be sent along the cord with a phase difference ϕ of either 0, 0.2 wavelength, or 0.5 wavelength. (a) With only mental calculation, rank those choices of ϕ according to the average rate at which the waves will transport energy, greatest first. (b) For the first choice of ϕ, what is the average rate in terms of $P_{avg,1}$?

8 (a) If a standing wave on a string is given by

$$y'(t) = (3\text{ mm})\sin(5x)\cos(4t),$$

is there a node or an antinode of the oscillations of the string at $x = 0$? (b) If the standing wave is given by

$$y'(t) = (3\text{ mm})\sin(5x + \pi/2)\cos(4t),$$

is there a node or an antinode at $x = 0$?

9 Strings A and B have identical lengths and linear densities, but string B is under greater tension than string A. Figure 16-27 shows four situations, (a) through (d), in which standing wave patterns exist on the two strings. In which situations is there the possibility that strings A and B are oscillating at the same resonant frequency?

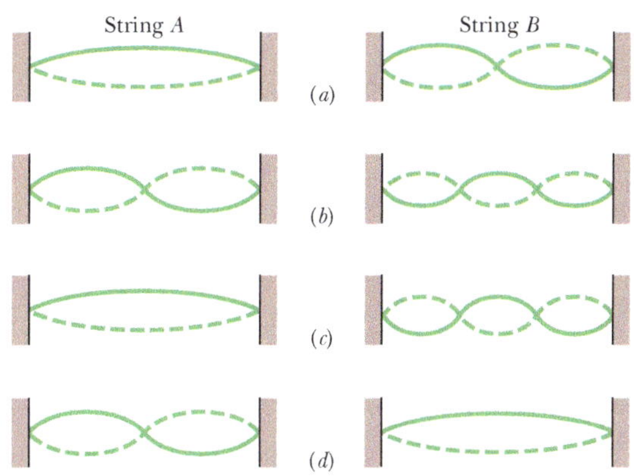

Figure 16-27 Question 9.

10 If you set up the seventh harmonic on a string, (a) how many nodes are present, and (b) is there a node, antinode, or some intermediate state at the midpoint? If you next set up the sixth harmonic, (c) is its resonant wavelength longer or shorter than that for the seventh harmonic, and (d) is the resonant frequency higher or lower?

11 Figure 16-28 shows phasor diagrams for three situations in which two waves travel along the same string. All six waves have the same amplitude. Rank the situations according to the amplitude of the net wave on the string, greatest first.

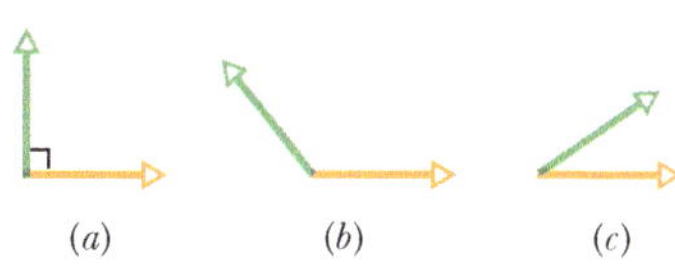

Figure 16-28 Question 11.

Problems

GO Tutoring problem available (at instructor's discretion) in *WileyPLUS* and WebAssign
SSM Worked-out solution available in Student Solutions Manual
• – ••• Number of dots indicates level of problem difficulty
Additional information available in *The Flying Circus of Physics* and at flyingcircusofphysics.com
WWW Worked-out solution is at
ILW Interactive solution is at
http://www.wiley.com/college/halliday

Module 16-1 Transverse Waves

•1 If a wave $y(x, t) = (6.0 \text{ mm}) \sin(kx + (600 \text{ rad/s})t + \phi)$ travels along a string, how much time does any given point on the string take to move between displacements $y = +2.0$ mm and $y = -2.0$ mm?

•2 *A human wave.* During sporting events within large, densely packed stadiums, spectators will send a wave (or pulse) around the stadium (Fig. 16-29). As the wave reaches a group of spectators, they stand with a cheer and then sit. At any instant, the width w of the wave is the distance from the leading edge (people are just about to stand) to the trailing edge (people have just sat down). Suppose a human wave travels a distance of 853 seats around a stadium in 39 s, with spectators requiring about 1.8 s to respond to the wave's passage by standing and then sitting. What are (a) the wave speed v (in seats per second) and (b) width w (in number of seats)?

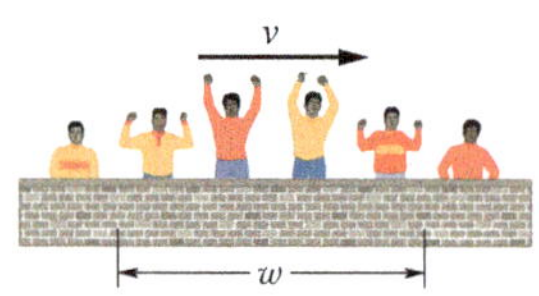

Figure 16-29 Problem 2.

•3 A wave has an angular frequency of 110 rad/s and a wavelength of 1.80 m. Calculate (a) the angular wave number and (b) the speed of the wave.

•4 A sand scorpion can detect the motion of a nearby beetle (its prey) by the waves the motion sends along the sand surface (Fig. 16-30). The waves are of two types: transverse waves traveling at $v_t = 50$ m/s and longitudinal waves traveling at $v_l = 150$ m/s. If a sudden motion sends out such waves, a scorpion can tell the distance of the beetle from the difference Δt in the arrival times of the waves at its leg nearest the beetle. If $\Delta t = 4.0$ ms, what is the beetle's distance?

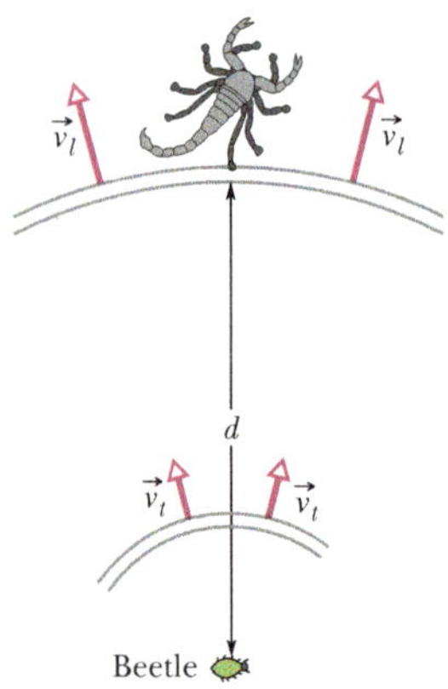

Figure 16-30 Problem 4.

•5 A sinusoidal wave travels along a string. The time for a particular point to move from maximum displacement to zero is 0.170 s. What are the (a) period and (b) frequency? (c) The wavelength is 1.40 m; what is the wave speed?

••6 GO A sinusoidal wave travels along a string under tension. Figure 16-31 gives the slopes along the string at time $t = 0$. The scale of the x axis is set by $x_s = 0.80$ m. What is the amplitude of the wave?

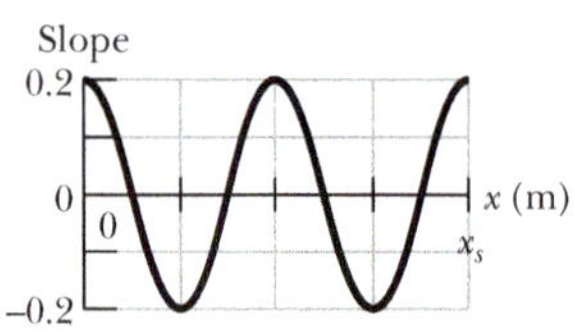

Figure 16-31 Problem 6.

••7 A transverse sinusoidal wave is moving along a string in the positive direction of an x axis with a speed of 80 m/s. At $t = 0$, the string particle at $x = 0$ has a transverse displacement of 4.0 cm from its equilibrium position and is not moving. The maximum transverse speed of the string particle at $x = 0$ is 16 m/s. (a) What is the frequency of the wave? (b) What is the wavelength of the wave? If $y(x, t) = y_m \sin(kx \pm \omega t + \phi)$ is the form of the wave equation, what are (c) y_m, (d) k, (e) ω, (f) ϕ, and (g) the correct choice of sign in front of ω?

••8 GO Figure 16-32 shows the transverse velocity u versus time t of the point on a string at $x = 0$, as a wave passes through it. The scale on the vertical axis is set by $u_s = 4.0$ m/s. The wave has the generic form $y(x, t) = y_m \sin(kx - \omega t + \phi)$. What then is ϕ? (*Caution:* A calculator does not always give the proper inverse trig function, so check your answer by substituting it and an assumed value of ω into $y(x, t)$ and then plotting the function.)

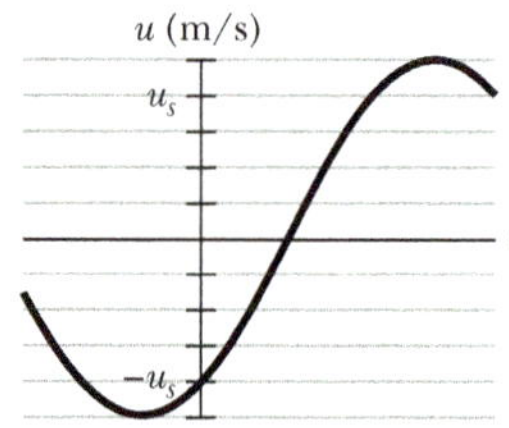

Figure 16-32 Problem 8.

••9 A sinusoidal wave moving along a string is shown twice in Fig. 16-33, as crest A travels in the positive direction of an x axis by distance $d = 6.0$ cm in 4.0 ms. The tick marks along the axis are separated by 10 cm; height $H = 6.00$ mm. The equation for the wave is in the form $y(x, t) = y_m \sin(kx \pm \omega t)$, so what are (a) y_m, (b) k, (c) ω, and (d) the correct choice of sign in front of ω?

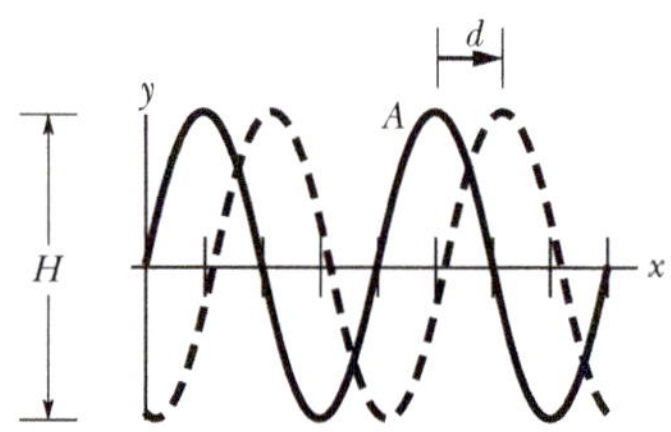

Figure 16-33 Problem 9.

••10 The equation of a transverse wave traveling along a very long string is $y = 6.0 \sin(0.020\pi x + 4.0\pi t)$, where x and y are expressed in centimeters and t is in seconds. Determine (a) the amplitude, (b) the wavelength, (c) the frequency, (d) the speed, (e) the direction of propagation of the wave, and (f) the maximum transverse speed of a particle in the string. (g) What is the transverse displacement at $x = 3.5$ cm when $t = 0.26$ s?

••11 GO A sinusoidal transverse wave of wavelength 20 cm travels along a string in the positive direction of an x axis. The displacement y of the string particle at $x = 0$ is given in Fig. 16-34 as a function of time t. The scale of the vertical axis is set by $y_s = 4.0$ cm. The wave equation is to be in the form $y(x, t) = y_m \sin(kx \pm \omega t + \phi)$. (a) At $t = 0$, is a plot of y versus x in the shape of a positive sine function or a negative sine function? What are (b) y_m, (c) k, (d) ω, (e) ϕ, (f) the sign in front of ω, and (g) the speed of the wave? (h) What is the transverse velocity of the particle at $x = 0$ when $t = 5.0$ s?

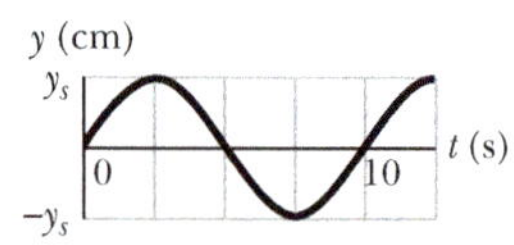

Figure 16-34 Problem 11.

••12 GO The function $y(x, t) = (15.0 \text{ cm}) \cos(\pi x - 15\pi t)$, with x in meters and t in seconds, describes a wave on a taut string. What is the transverse speed for a point on the string at an instant when that point has the displacement $y = +12.0$ cm?

••13 ILW A sinusoidal wave of frequency 500 Hz has a speed of 350 m/s. (a) How far apart are two points that differ in phase by $\pi/3$ rad? (b) What is the phase difference between two displacements at a certain point at times 1.00 ms apart?

Module 16-2 Wave Speed on a Stretched String

•14 The equation of a transverse wave on a string is

$$y = (2.0 \text{ mm}) \sin[(20 \text{ m}^{-1})x - (600 \text{ s}^{-1})t].$$

The tension in the string is 15 N. (a) What is the wave speed? (b) Find the linear density of this string in grams per meter.

•15 SSM WWW A stretched string has a mass per unit length of 5.00 g/cm and a tension of 10.0 N. A sinusoidal wave on this string has an amplitude of 0.12 mm and a frequency of 100 Hz and is traveling in the negative direction of an x axis. If the wave equation is of the form $y(x, t) = y_m \sin(kx \pm \omega t)$, what are (a) y_m, (b) k, (c) ω, and (d) the correct choice of sign in front of ω?

•16 The speed of a transverse wave on a string is 170 m/s when the string tension is 120 N. To what value must the tension be changed to raise the wave speed to 180 m/s?

•17 The linear density of a string is 1.6×10^{-4} kg/m. A transverse wave on the string is described by the equation

$$y = (0.021 \text{ m}) \sin[(2.0 \text{ m}^{-1})x + (30 \text{ s}^{-1})t].$$

What are (a) the wave speed and (b) the tension in the string?

•18 The heaviest and lightest strings on a certain violin have linear densities of 3.0 and 0.29 g/m. What is the ratio of the diameter of the heaviest string to that of the lightest string, assuming that the strings are of the same material?

•19 SSM What is the speed of a transverse wave in a rope of length 2.00 m and mass 60.0 g under a tension of 500 N?

•20 The tension in a wire clamped at both ends is doubled without appreciably changing the wire's length between the clamps. What is the ratio of the new to the old wave speed for transverse waves traveling along this wire?

••21 ILW A 100 g wire is held under a tension of 250 N with one end at $x = 0$ and the other at $x = 10.0$ m. At time $t = 0$, pulse 1 is sent along the wire from the end at $x = 10.0$ m. At time $t = 30.0$ ms, pulse 2 is sent along the wire from the end at $x = 0$. At what position x do the pulses begin to meet?

••22 A sinusoidal wave is traveling on a string with speed 40 cm/s. The displacement of the particles of the string at $x = 10$ cm varies with time according to $y = (5.0 \text{ cm}) \sin[1.0 - (4.0 \text{ s}^{-1})t]$. The linear density of the string is 4.0 g/cm. What are (a) the frequency and (b) the wavelength of the wave? If the wave equation is of the form $y(x, t) = y_m \sin(kx \pm \omega t)$, what are (c) y_m, (d) k, (e) ω, and (f) the correct choice of sign in front of ω? (g) What is the tension in the string?

••23 SSM ILW A sinusoidal transverse wave is traveling along a string in the negative direction of an x axis. Figure 16-35 shows a plot of the displacement as a function of position at time $t = 0$; the scale of the y axis is set by $y_s = 4.0$ cm. The string tension is 3.6 N, and its linear density is 25 g/m. Find the (a) amplitude, (b) wavelength, (c) wave speed, and (d) period of the wave. (e) Find the maximum transverse speed of a particle in the string. If the wave is of the form $y(x, t) = y_m \sin(kx \pm \omega t + \phi)$, what are (f) k, (g) ω, (h) ϕ, and (i) the correct choice of sign in front of ω?

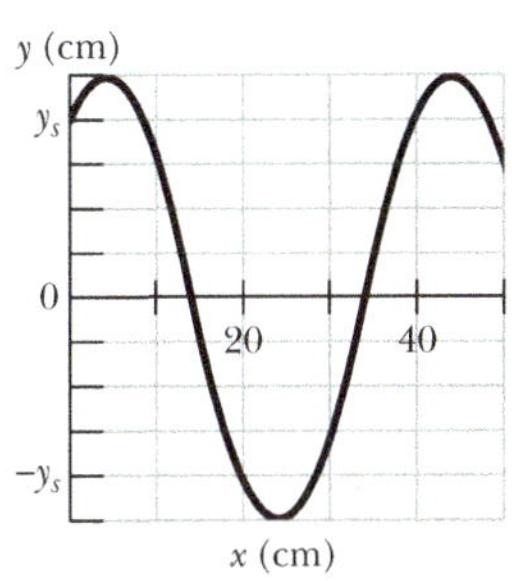

Figure 16-35 Problem 23.

•••24 In Fig. 16-36*a*, string 1 has a linear density of 3.00 g/m, and string 2 has a linear density of 5.00 g/m. They are under tension due to the hanging block of mass $M = 500$ g. Calculate the wave speed on (a) string 1 and (b) string 2. (*Hint:* When a string loops halfway around a pulley, it pulls on the pulley with a net force that is twice the tension in the string.) Next the block is divided into two blocks (with $M_1 + M_2 = M$) and the apparatus is rearranged as shown in Fig. 16-36*b*. Find (c) M_1 and (d) M_2 such that the wave speeds in the two strings are equal.

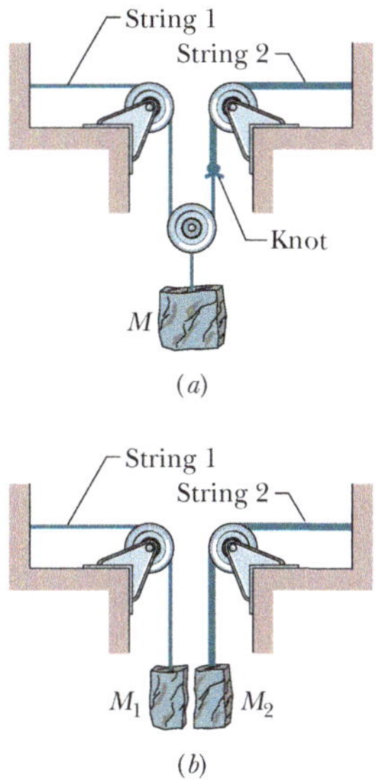

Figure 16-36 Problem 24.

•••25 A uniform rope of mass m and length L hangs from a ceiling. (a) Show that the speed of a transverse wave on the rope is a function of y, the distance from the lower end, and is given by $v = \sqrt{gy}$. (b) Show that the time a transverse wave takes to travel the length of the rope is given by $t = 2\sqrt{L/g}$.

Module 16-3 Energy and Power of a Wave Traveling Along a String

•26 A string along which waves can travel is 2.70 m long and has a mass of 260 g. The tension in the string is 36.0 N. What must be the frequency of traveling waves of amplitude 7.70 mm for the average power to be 85.0 W?

••27 GO A sinusoidal wave is sent along a string with a linear density of 2.0 g/m. As it travels, the kinetic energies of the mass elements along the string vary. Figure 16-37*a* gives the rate dK/dt at which kinetic energy passes through the string elements at a particular instant, plotted as a function of distance x along the string. Figure 16-37*b* is similar except that it gives the rate at which kinetic energy passes through a particular mass element (at a particular location), plotted as a function of time t. For both figures, the scale on the vertical (rate) axis is set by $R_s = 10$ W. What is the amplitude of the wave?

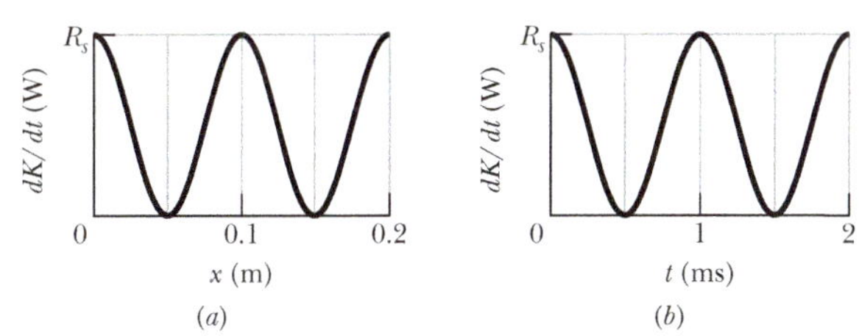

Figure 16-37 Problem 27.

Module 16-4 The Wave Equation

•**28** Use the wave equation to find the speed of a wave given by

$$y(x, t) = (3.00 \text{ mm}) \sin[(4.00 \text{ m}^{-1})x - (7.00 \text{ s}^{-1})t].$$

••**29** Use the wave equation to find the speed of a wave given by

$$y(x, t) = (2.00 \text{ mm})[(20 \text{ m}^{-1})x - (4.0 \text{ s}^{-1})t]^{0.5}.$$

•••**30** Use the wave equation to find the speed of a wave given in terms of the general function $h(x, t)$:

$$y(x, t) = (4.00 \text{ mm})\, h[(30 \text{ m}^{-1})x + (6.0 \text{ s}^{-1})t].$$

Module 16-5 Interference of Waves

•**31** SSM Two identical traveling waves, moving in the same direction, are out of phase by $\pi/2$ rad. What is the amplitude of the resultant wave in terms of the common amplitude y_m of the two combining waves?

•**32** What phase difference between two identical traveling waves, moving in the same direction along a stretched string, results in the combined wave having an amplitude 1.50 times that of the common amplitude of the two combining waves? Express your answer in (a) degrees, (b) radians, and (c) wavelengths.

••**33** GO Two sinusoidal waves with the same amplitude of 9.00 mm and the same wavelength travel together along a string that is stretched along an x axis. Their resultant wave is shown twice in Fig. 16-38, as valley A travels in the negative direction of the x axis by distance d = 56.0 cm in 8.0 ms. The tick marks along the axis are separated by 10 cm, and height H is 8.0 mm. Let the equation for one wave be of the form $y(x, t) = y_m \sin(kx \pm \omega t + \phi_1)$, where $\phi_1 = 0$ and you must choose the correct sign in front of ω. For the equation for the other wave, what are (a) y_m, (b) k, (c) ω, (d) ϕ_2, and (e) the sign in front of ω?

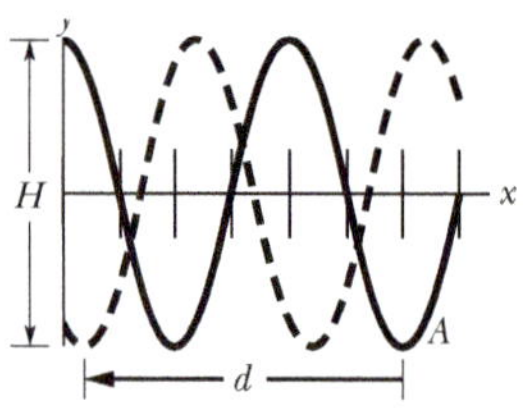

Figure 16-38 Problem 33.

•••**34** GO A sinusoidal wave of angular frequency 1200 rad/s and amplitude 3.00 mm is sent along a cord with linear density 2.00 g/m and tension 1200 N. (a) What is the average rate at which energy is transported by the wave to the opposite end of the cord? (b) If, simultaneously, an identical wave travels along an adjacent, identical cord, what is the total average rate at which energy is transported to the opposite ends of the two cords by the waves? If, instead, those two waves are sent along the *same* cord simultaneously, what is the total average rate at which they transport energy when their phase difference is (c) 0, (d) 0.4π rad, and (e) π rad?

Module 16-6 Phasors

•**35** SSM Two sinusoidal waves of the same frequency travel in the same direction along a string. If y_{m1} = 3.0 cm, y_{m2} = 4.0 cm, $\phi_1 = 0$, and $\phi_2 = \pi/2$ rad, what is the amplitude of the resultant wave?

••**36** Four waves are to be sent along the same string, in the same direction:

$$y_1(x, t) = (4.00 \text{ mm}) \sin(2\pi x - 400\pi t)$$
$$y_2(x, t) = (4.00 \text{ mm}) \sin(2\pi x - 400\pi t + 0.7\pi)$$
$$y_3(x, t) = (4.00 \text{ mm}) \sin(2\pi x - 400\pi t + \pi)$$
$$y_4(x, t) = (4.00 \text{ mm}) \sin(2\pi x - 400\pi t + 1.7\pi).$$

What is the amplitude of the resultant wave?

••**37** GO These two waves travel along the same string:

$$y_1(x, t) = (4.60 \text{ mm}) \sin(2\pi x - 400\pi t)$$
$$y_2(x, t) = (5.60 \text{ mm}) \sin(2\pi x - 400\pi t + 0.80\pi \text{ rad}).$$

What are (a) the amplitude and (b) the phase angle (relative to wave 1) of the resultant wave? (c) If a third wave of amplitude 5.00 mm is also to be sent along the string in the same direction as the first two waves, what should be its phase angle in order to maximize the amplitude of the new resultant wave?

••**38** Two sinusoidal waves of the same frequency are to be sent in the same direction along a taut string. One wave has an amplitude of 5.0 mm, the other 8.0 mm. (a) What phase difference ϕ_1 between the two waves results in the smallest amplitude of the resultant wave? (b) What is that smallest amplitude? (c) What phase difference ϕ_2 results in the largest amplitude of the resultant wave? (d) What is that largest amplitude? (e) What is the resultant amplitude if the phase angle is $(\phi_1 - \phi_2)/2$?

••**39** Two sinusoidal waves of the same period, with amplitudes of 5.0 and 7.0 mm, travel in the same direction along a stretched string; they produce a resultant wave with an amplitude of 9.0 mm. The phase constant of the 5.0 mm wave is 0. What is the phase constant of the 7.0 mm wave?

Module 16-7 Standing Waves and Resonance

•**40** Two sinusoidal waves with identical wavelengths and amplitudes travel in opposite directions along a string with a speed of 10 cm/s. If the time interval between instants when the string is flat is 0.50 s, what is the wavelength of the waves?

•**41** SSM A string fixed at both ends is 8.40 m long and has a mass of 0.120 kg. It is subjected to a tension of 96.0 N and set oscillating. (a) What is the speed of the waves on the string? (b) What is the longest possible wavelength for a standing wave? (c) Give the frequency of that wave.

•**42** A string under tension τ_i oscillates in the third harmonic at frequency f_3, and the waves on the string have wavelength λ_3. If the tension is increased to $\tau_f = 4\tau_i$ and the string is again made to oscillate in the third harmonic, what then are (a) the frequency of oscillation in terms of f_3 and (b) the wavelength of the waves in terms of λ_3?

•**43** SSM WWW What are (a) the lowest frequency, (b) the second lowest frequency, and (c) the third lowest frequency for standing waves on a wire that is 10.0 m long, has a mass of 100 g, and is stretched under a tension of 250 N?

•**44** A 125 cm length of string has mass 2.00 g and tension 7.00 N. (a) What is the wave speed for this string? (b) What is the lowest resonant frequency of this string?

•**45** SSM ILW A string that is stretched between fixed supports separated by 75.0 cm has resonant frequencies of 420 and 315 Hz, with no intermediate resonant frequencies. What are (a) the lowest resonant frequency and (b) the wave speed?

•**46** String A is stretched between two clamps separated by distance L. String B, with the same linear density and under the same tension as string A, is stretched between two clamps separated by distance $4L$. Consider the first eight harmonics of string B. For which of these eight harmonics of B (if any) does the frequency match the frequency of (a) A's first harmonic, (b) A's second harmonic, and (c) A's third harmonic?

•**47** One of the harmonic frequencies for a particular string under tension is 325 Hz. The next higher harmonic frequency is 390 Hz.

What harmonic frequency is next higher after the harmonic frequency 195 Hz?

•48 If a transmission line in a cold climate collects ice, the increased diameter tends to cause vortex formation in a passing wind. The air pressure variations in the vortexes tend to cause the line to oscillate (*gallop*), especially if the frequency of the variations matches a resonant frequency of the line. In long lines, the resonant frequencies are so close that almost any wind speed can set up a resonant mode vigorous enough to pull down support towers or cause the line to *short out* with an adjacent line. If a transmission line has a length of 347 m, a linear density of 3.35 kg/m, and a tension of 65.2 MN, what are (a) the frequency of the fundamental mode and (b) the frequency difference between successive modes?

•49 ILW A nylon guitar string has a linear density of 7.20 g/m and is under a tension of 150 N. The fixed supports are distance $D = 90.0$ cm apart. The string is oscillating in the standing wave pattern shown in Fig. 16-39. Calculate the (a) speed, (b) wavelength, and (c) frequency of the traveling waves whose superposition gives this standing wave.

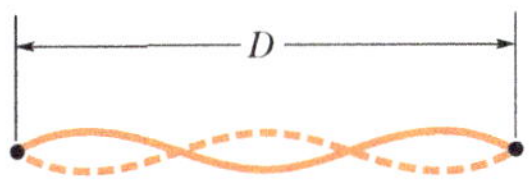

Figure 16-39 Problem 49.

••50 For a particular transverse standing wave on a long string, one of the antinodes is at $x = 0$ and an adjacent node is at $x = 0.10$ m. The displacement $y(t)$ of the string particle at $x = 0$ is shown in Fig. 16-40, where the scale of the y axis is set by $y_s = 4.0$ cm. When $t = 0.50$ s, what is the displacement of the string particle at (a) $x = 0.20$ m and (b) $x = 0.30$ m? What is the transverse velocity of the string particle at $x = 0.20$ m at (c) $t = 0.50$ s and (d) $t = 1.0$ s? (e) Sketch the standing wave at $t = 0.50$ s for the range $x = 0$ to $x = 0.40$ m.

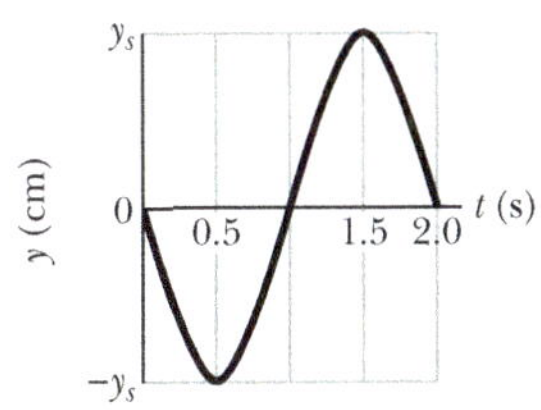

Figure 16-40 Problem 50.

••51 SSM WWW Two waves are generated on a string of length 3.0 m to produce a three-loop standing wave with an amplitude of 1.0 cm. The wave speed is 100 m/s. Let the equation for one of the waves be of the form $y(x, t) = y_m \sin(kx + \omega t)$. In the equation for the other wave, what are (a) y_m, (b) k, (c) ω, and (d) the sign in front of ω?

••52 A rope, under a tension of 200 N and fixed at both ends, oscillates in a second-harmonic standing wave pattern. The displacement of the rope is given by

$$y = (0.10 \text{ m})(\sin \pi x/2) \sin 12\pi t,$$

where $x = 0$ at one end of the rope, x is in meters, and t is in seconds. What are (a) the length of the rope, (b) the speed of the waves on the rope, and (c) the mass of the rope? (d) If the rope oscillates in a third-harmonic standing wave pattern, what will be the period of oscillation?

••53 A string oscillates according to the equation

$$y' = (0.50 \text{ cm}) \sin\left[\left(\frac{\pi}{3} \text{ cm}^{-1}\right)x\right] \cos[(40\pi \text{ s}^{-1})t].$$

What are the (a) amplitude and (b) speed of the two waves (identical except for direction of travel) whose superposition gives this oscillation? (c) What is the distance between nodes? (d) What is the transverse speed of a particle of the string at the position $x = 1.5$ cm when $t = \frac{9}{8}$ s?

••54 GO Two sinusoidal waves with the same amplitude and wavelength travel through each other along a string that is stretched along an x axis. Their resultant wave is shown twice in Fig. 16-41, as the antinode A travels from an extreme upward displacement to an extreme downward displacement in 6.0 ms. The tick marks along the axis are separated by 10 cm; height H is 1.80 cm. Let the equation for one of the two waves be of the form $y(x, t) = y_m \sin(kx + \omega t)$. In the equation for the other wave, what are (a) y_m, (b) k, (c) ω, and (d) the sign in front of ω?

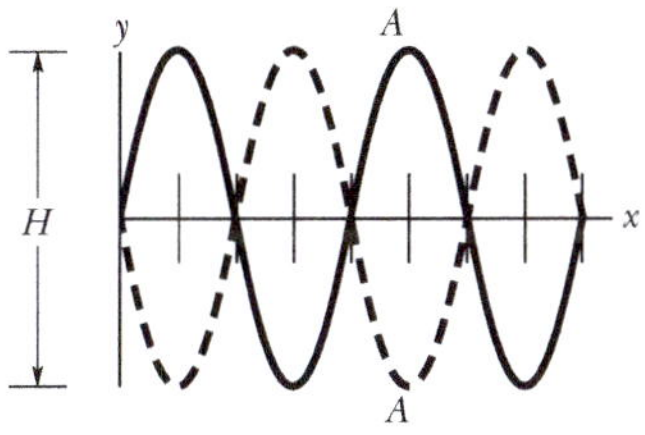

Figure 16-41 Problem 54.

••55 GO The following two waves are sent in opposite directions on a horizontal string so as to create a standing wave in a vertical plane:

$$y_1(x, t) = (6.00 \text{ mm}) \sin(4.00\pi x - 400\pi t)$$

$$y_2(x, t) = (6.00 \text{ mm}) \sin(4.00\pi x + 400\pi t),$$

with x in meters and t in seconds. An antinode is located at point A. In the time interval that point takes to move from maximum upward displacement to maximum downward displacement, how far does each wave move along the string?

••56 A standing wave pattern on a string is described by

$$y(x, t) = 0.040 (\sin 5\pi x)(\cos 40\pi t),$$

where x and y are in meters and t is in seconds. For $x \geq 0$, what is the location of the node with the (a) smallest, (b) second smallest, and (c) third smallest value of x? (d) What is the period of the oscillatory motion of any (nonnode) point? What are the (e) speed and (f) amplitude of the two traveling waves that interfere to produce this wave? For $t \geq 0$, what are the (g) first, (h) second, and (i) third time that all points on the string have zero transverse velocity?

••57 A generator at one end of a very long string creates a wave given by

$$y = (6.0 \text{ cm}) \cos \frac{\pi}{2} [(2.00 \text{ m}^{-1})x + (8.00 \text{ s}^{-1})t],$$

and a generator at the other end creates the wave

$$y = (6.0 \text{ cm}) \cos \frac{\pi}{2} [(2.00 \text{ m}^{-1})x - (8.00 \text{ s}^{-1})t].$$

Calculate the (a) frequency, (b) wavelength, and (c) speed of each wave. For $x \geq 0$, what is the location of the node having the (d) smallest, (e) second smallest, and (f) third smallest value of x? For $x \geq 0$, what is the location of the antinode having the (g) smallest, (h) second smallest, and (i) third smallest value of x?

••58 GO In Fig. 16-42, a string, tied to a sinusoidal oscillator at P and running over a support at Q, is stretched by a block of mass m. Separation $L = 1.20$ m, linear density $\mu = 1.6$ g/m, and the oscillator

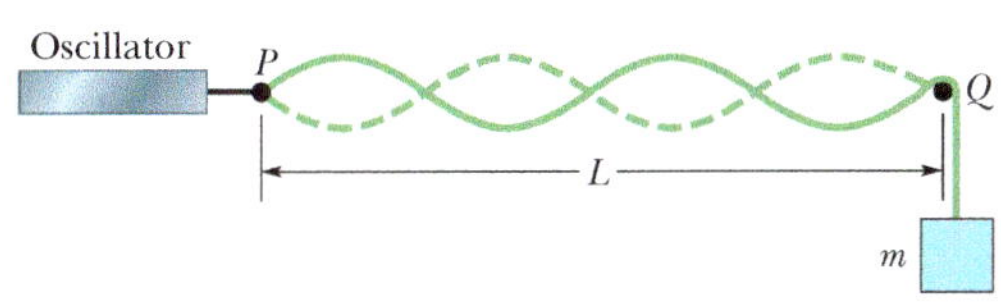

Figure 16-42 Problems 58 and 60.

frequency $f = 120$ Hz. The amplitude of the motion at P is small enough for that point to be considered a node. A node also exists at Q. (a) What mass m allows the oscillator to set up the fourth harmonic on the string? (b) What standing wave mode, if any, can be set up if $m = 1.00$ kg?

•••59 GO In Fig. 16-43, an aluminum wire, of length $L_1 = 60.0$ cm, cross-sectional area 1.00×10^{-2} cm^2, and density 2.60 g/cm^3, is joined to a steel wire, of density 7.80 g/cm^3 and the same cross-sectional area. The compound wire, loaded with a block of mass $m = 10.0$ kg, is arranged so that the distance L_2 from the joint to the supporting pulley is 86.6 cm. Transverse waves are set up on the wire by an external source of variable frequency; a node is located at the pulley. (a) Find the lowest frequency that generates a standing wave having the joint as one of the nodes. (b) How many nodes are observed at this frequency?

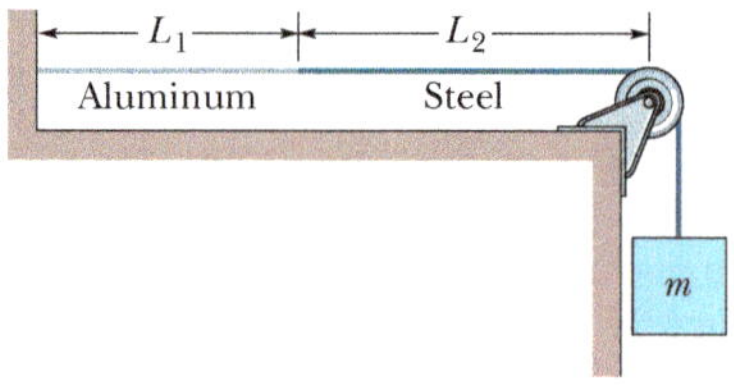

Figure 16-43 Problem 59.

•••60 GO In Fig. 16-42, a string, tied to a sinusoidal oscillator at P and running over a support at Q, is stretched by a block of mass m. The separation L between P and Q is 1.20 m, and the frequency f of the oscillator is fixed at 120 Hz. The amplitude of the motion at P is small enough for that point to be considered a node. A node also exists at Q. A standing wave appears when the mass of the hanging block is 286.1 g or 447.0 g, but not for any intermediate mass. What is the linear density of the string?

Additional Problems

61 GO In an experiment on standing waves, a string 90 cm long is attached to the prong of an electrically driven tuning fork that oscillates perpendicular to the length of the string at a frequency of 60 Hz. The mass of the string is 0.044 kg. What tension must the string be under (weights are attached to the other end) if it is to oscillate in four loops?

62 A sinusoidal transverse wave traveling in the positive direction of an x axis has an amplitude of 2.0 cm, a wavelength of 10 cm, and a frequency of 400 Hz. If the wave equation is of the form $y(x, t) = y_m \sin(kx \pm \omega t)$, what are (a) y_m, (b) k, (c) ω, and (d) the correct choice of sign in front of ω? What are (e) the maximum transverse speed of a point on the cord and (f) the speed of the wave?

63 A wave has a speed of 240 m/s and a wavelength of 3.2 m. What are the (a) frequency and (b) period of the wave?

64 The equation of a transverse wave traveling along a string is

$$y = 0.15 \sin(0.79x - 13t),$$

in which x and y are in meters and t is in seconds. (a) What is the displacement y at $x = 2.3$ m, $t = 0.16$ s? A second wave is to be added to the first wave to produce standing waves on the string. If the second wave is of the form $y(x, t) = y_m \sin(kx \pm \omega t)$, what are (b) y_m, (c) k, (d) ω, and (e) the correct choice of sign in front of ω for this second wave? (f) What is the displacement of the resultant standing wave at $x = 2.3$ m, $t = 0.16$ s?

65 The equation of a transverse wave traveling along a string is

$$y = (2.0 \text{ mm}) \sin[(20 \text{ m}^{-1})x - (600 \text{ s}^{-1})t].$$

Find the (a) amplitude, (b) frequency, (c) velocity (including sign), and (d) wavelength of the wave. (e) Find the maximum transverse speed of a particle in the string.

66 Figure 16-44 shows the displacement y versus time t of the point on a string at $x = 0$, as a wave passes through that point. The scale of the y axis is set by $y_s = 6.0$ mm. The wave is given by $y(x, t) = y_m \sin(kx - \omega t + \phi)$. What is ϕ? (*Caution:* A calculator does not always give the proper inverse trig function, so check your answer by substituting it and an assumed value of ω into $y(x, t)$ and then plotting the function.)

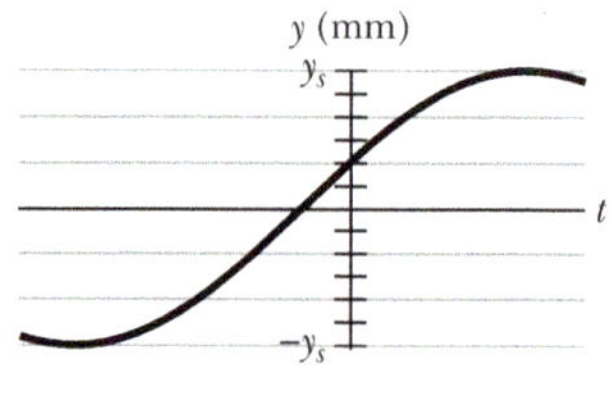

Figure 16-44 Problem 66.

67 Two sinusoidal waves, identical except for phase, travel in the same direction along a string, producing the net wave $y'(x, t) = (3.0 \text{ mm}) \sin(20x - 4.0t + 0.820 \text{ rad})$, with x in meters and t in seconds. What are (a) the wavelength λ of the two waves, (b) the phase difference between them, and (c) their amplitude y_m?

68 A single pulse, given by $h(x - 5.0t)$, is shown in Fig. 16-45 for $t = 0$. The scale of the vertical axis is set by $h_s = 2$. Here x is in centimeters and t is in seconds. What are the (a) speed and (b) direction of travel of the pulse? (c) Plot $h(x - 5t)$ as a function of x for $t = 2$ s. (d) Plot $h(x - 5t)$ as a function of t for $x = 10$ cm.

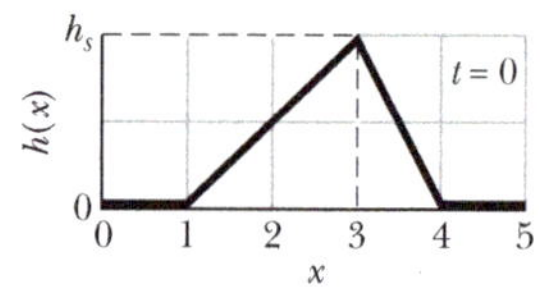

Figure 16-45 Problem 68.

69 SSM Three sinusoidal waves of the same frequency travel along a string in the positive direction of an x axis. Their amplitudes are y_1, $y_1/2$, and $y_1/3$, and their phase constants are 0, $\pi/2$, and π, respectively. What are the (a) amplitude and (b) phase constant of the resultant wave? (c) Plot the wave form of the resultant wave at $t = 0$, and discuss its behavior as t increases.

70 GO Figure 16-46 shows transverse acceleration a_y versus time t of the point on a string at $x = 0$, as a wave in the form of $y(x, t) = y_m \sin(kx - \omega t + \phi)$ passes through that point. The scale of the vertical axis is set by $a_s = 400$ m/s^2. What is ϕ? (*Caution:* A calculator does not always give the proper inverse trig function, so check your answer by substituting it and an assumed value of ω into $y(x, t)$ and then plotting the function.)

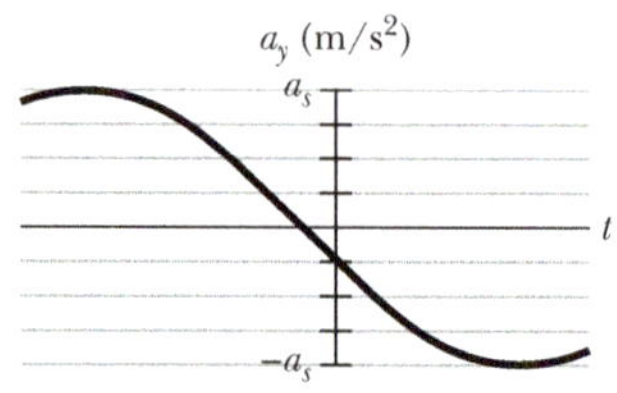

Figure 16-46 Problem 70.

71 A transverse sinusoidal wave is generated at one end of a long, horizontal string by a bar that moves up and down through a distance of 1.00 cm. The motion is continuous and is repeated regularly 120 times per second. The string has linear density 120 g/m and is kept under a tension of 90.0 N. Find the maximum value of (a) the transverse speed u and (b) the transverse component of the tension τ.

(c) Show that the two maximum values calculated above occur at the same phase values for the wave. What is the transverse displacement y of the string at these phases? (d) What is the maximum rate of energy transfer along the string? (e) What is the transverse displacement y when this maximum transfer occurs? (f) What is the minimum rate of energy transfer along the string? (g) What is the transverse displacement y when this minimum transfer occurs?

72 Two sinusoidal 120 Hz waves, of the same frequency and amplitude, are to be sent in the positive direction of an x axis that is directed along a cord under tension. The waves can be sent in phase, or they can be phase-shifted. Figure 16-47 shows the amplitude y' of the resulting wave versus the distance of the shift (how far one wave is shifted from the other wave). The scale of the vertical axis is set by $y'_s = 6.0$ mm. If the equations for the two waves are of the form $y(x, t) = y_m \sin(kx \pm \omega t)$, what are (a) y_m, (b) k, (c) ω, and (d) the correct choice of sign in front of ω?

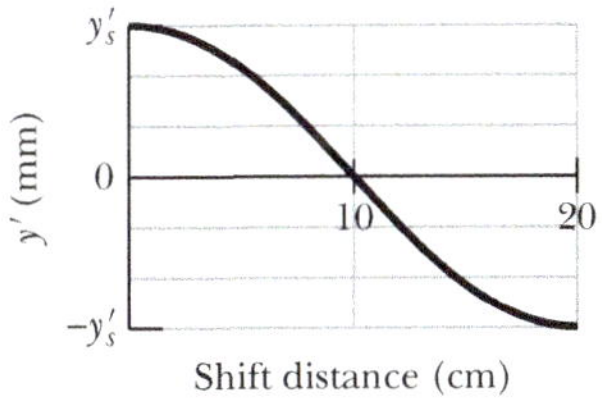

Figure 16-47 Problem 72.

73 At time $t = 0$ and at position $x = 0$ m along a string, a traveling sinusoidal wave with an angular frequency of 440 rad/s has displacement $y = +4.5$ mm and transverse velocity $u = -0.75$ m/s. If the wave has the general form $y(x, t) = y_m \sin(kx - \omega t + \phi)$, what is phase constant ϕ?

74 Energy is transmitted at rate P_1 by a wave of frequency f_1 on a string under tension τ_1. What is the new energy transmission rate P_2 in terms of P_1 (a) if the tension is increased to $\tau_2 = 4\tau_1$ and (b) if, instead, the frequency is decreased to $f_2 = f_1/2$?

75 (a) What is the fastest transverse wave that can be sent along a steel wire? For safety reasons, the maximum tensile stress to which steel wires should be subjected is 7.00×10^8 N/m^2. The density of steel is 7800 kg/m^3. (b) Does your answer depend on the diameter of the wire?

76 A standing wave results from the sum of two transverse traveling waves given by

$$y_1 = 0.050 \cos(\pi x - 4\pi t)$$

and

$$y_2 = 0.050 \cos(\pi x + 4\pi t),$$

where x, y_1, and y_2 are in meters and t is in seconds. (a) What is the smallest positive value of x that corresponds to a node? Beginning at $t = 0$, what is the value of the (b) first, (c) second, and (d) third time the particle at $x = 0$ has zero velocity?

77 SSM The type of rubber band used inside some baseballs and golf balls obeys Hooke's law over a wide range of elongation of the band. A segment of this material has an unstretched length ℓ and a mass m. When a force F is applied, the band stretches an additional length $\Delta\ell$. (a) What is the speed (in terms of m, $\Delta\ell$, and the spring constant k) of transverse waves on this stretched rubber band? (b) Using your answer to (a), show that the time required for a transverse pulse to travel the length of the rubber band is proportional to $1/\sqrt{\Delta\ell}$ if $\Delta\ell \ll \ell$ and is constant if $\Delta\ell \gg \ell$.

78 The speed of electromagnetic waves (which include visible light, radio, and x rays) in vacuum is 3.0×10^8 m/s. (a) Wavelengths of visible light waves range from about 400 nm in the violet to about 700 nm in the red. What is the range of frequencies of these waves? (b) The range of frequencies for shortwave radio (for example, FM radio and VHF television) is 1.5 to 300 MHz. What is the corresponding wavelength range? (c) X-ray wavelengths range from about 5.0 nm to about 1.0×10^{-2} nm. What is the frequency range for x rays?

79 SSM A 1.50 m wire has a mass of 8.70 g and is under a tension of 120 N. The wire is held rigidly at both ends and set into oscillation. (a) What is the speed of waves on the wire? What is the wavelength of the waves that produce (b) one-loop and (c) two-loop standing waves? What is the frequency of the waves that produce (d) one-loop and (e) two-loop standing waves?

80 When played in a certain manner, the lowest resonant frequency of a certain violin string is concert A (440 Hz). What is the frequency of the (a) second and (b) third harmonic of the string?

81 A sinusoidal transverse wave traveling in the negative direction of an x axis has an amplitude of 1.00 cm, a frequency of 550 Hz, and a speed of 330 m/s. If the wave equation is of the form $y(x, t) = y_m \sin(kx \pm \omega t)$, what are (a) y_m, (b) ω, (c) k, and (d) the correct choice of sign in front of ω?

82 Two sinusoidal waves of the same wavelength travel in the same direction along a stretched string. For wave 1, $y_m = 3.0$ mm and $\phi = 0$; for wave 2, $y_m = 5.0$ mm and $\phi = 70°$. What are the (a) amplitude and (b) phase constant of the resultant wave?

83 SSM A sinusoidal transverse wave of amplitude y_m and wavelength λ travels on a stretched cord. (a) Find the ratio of the maximum particle speed (the speed with which a single particle in the cord moves transverse to the wave) to the wave speed. (b) Does this ratio depend on the material of which the cord is made?

84 Oscillation of a 600 Hz tuning fork sets up standing waves in a string clamped at both ends. The wave speed for the string is 400 m/s. The standing wave has four loops and an amplitude of 2.0 mm. (a) What is the length of the string? (b) Write an equation for the displacement of the string as a function of position and time.

85 A 120 cm length of string is stretched between fixed supports. What are the (a) longest, (b) second longest, and (c) third longest wavelength for waves traveling on the string if standing waves are to be set up? (d) Sketch those standing waves.

86 (a) Write an equation describing a sinusoidal transverse wave traveling on a cord in the positive direction of a y axis with an angular wave number of 60 cm^{-1}, a period of 0.20 s, and an amplitude of 3.0 mm. Take the transverse direction to be the z direction. (b) What is the maximum transverse speed of a point on the cord?

87 A wave on a string is described by

$$y(x, t) = 15.0 \sin(\pi x/8 - 4\pi t),$$

where x and y are in centimeters and t is in seconds. (a) What is the transverse speed for a point on the string at $x = 6.00$ cm when $t = 0.250$ s? (b) What is the maximum transverse speed of any point on the string? (c) What is the magnitude of the transverse acceleration for a point on the string at $x = 6.00$ cm when $t = 0.250$ s? (d) What is the magnitude of the maximum transverse acceleration for any point on the string?

88 *Body armor*. When a high-speed projectile such as a bullet or bomb fragment strikes modern body armor, the fabric of the armor stops the projectile and prevents penetration by quickly spreading the projectile's energy over a large area. This spreading is done by longitudinal and transverse pulses that move *radially* from the impact point, where the projectile pushes a cone-shaped dent into the fabric. The longitudinal pulse, racing along the fibers of the fabric at speed v_l ahead of the denting, causes the fibers to thin and stretch, with material flowing radially inward into the dent. One such radial fiber is shown in Fig. 16-48*a*. Part of the projectile's energy goes into this motion and stretching. The transverse pulse, moving at a slower speed v_t, is due to the denting. As the projectile increases the dent's depth, the dent increases in radius, causing the material in the fibers to move in the same direction as the projectile (perpendicular to the transverse pulse's direction of travel). The rest of the projectile's energy goes into this motion. All the energy that does not eventually go into permanently deforming the fibers ends up as thermal energy.

Figure 16-48*b* is a graph of speed v versus time t for a bullet of mass 10.2 g fired from a .38 Special revolver directly into body armor. The scales of the vertical and horizontal axes are set by v_s = 300 m/s and t_s = 40.0 μs. Take v_l = 2000 m/s, and assume that the half-angle θ of the conical dent is 60°. At the end of the collision, what are the radii of (a) the thinned region and (b) the dent (assuming that the person wearing the armor remains stationary)?

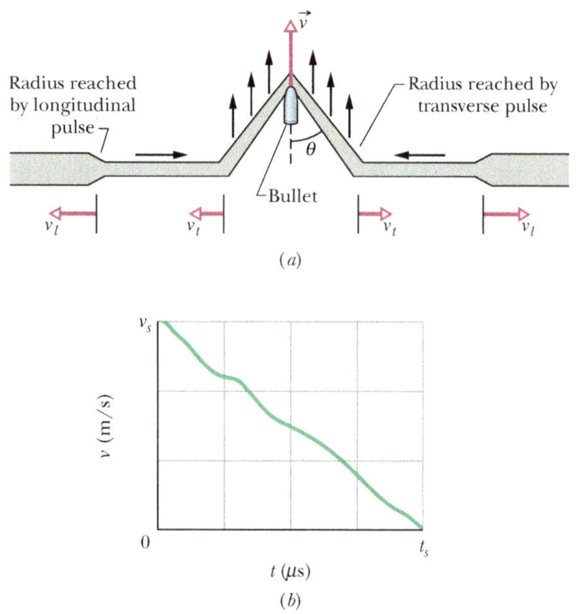

Figure 16-48 Problem 88.

89 Two waves are described by

$$y_1 = 0.30 \sin[\pi(5x - 200t)]$$

and
$$y_2 = 0.30 \sin[\pi(5x - 200t) + \pi/3],$$

where y_1, y_2, and x are in meters and t is in seconds. When these two waves are combined, a traveling wave is produced. What are the (a) amplitude, (b) wave speed, and (c) wavelength of that traveling wave?

90 A certain transverse sinusoidal wave of wavelength 20 cm is moving in the positive direction of an x axis. The transverse velocity of the particle at $x = 0$ as a function of time is shown in Fig. 16-49, where the scale of the vertical axis is set by u_s = 5.0 cm/s. What are the (a) wave speed, (b) amplitude, and (c) frequency? (d) Sketch the wave between $x = 0$ and $x = 20$ cm at $t = 2.0$ s.

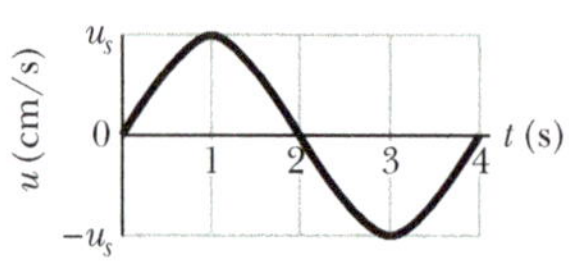

Figure 16-49 Problem 90.

91 SSM In a demonstration, a 1.2 kg horizontal rope is fixed in place at its two ends ($x = 0$ and $x = 2.0$ m) and made to oscillate up and down in the fundamental mode, at frequency 5.0 Hz. At $t = 0$, the point at $x = 1.0$ m has zero displacement and is moving upward in the positive direction of a y axis with a transverse velocity of 5.0 m/s. What are (a) the amplitude of the motion of that point and (b) the tension in the rope? (c) Write the standing wave equation for the fundamental mode.

92 Two waves,

$$y_1 = (2.50 \text{ mm}) \sin[(25.1 \text{ rad/m})x - (440 \text{ rad/s})t]$$

and
$$y_2 = (1.50 \text{ mm}) \sin[(25.1 \text{ rad/m})x + (440 \text{ rad/s})t],$$

travel along a stretched string. (a) Plot the resultant wave as a function of t for $x = 0$, $\lambda/8$, $\lambda/4$, $3\lambda/8$, and $\lambda/2$, where λ is the wavelength. The graphs should extend from $t = 0$ to a little over one period. (b) The resultant wave is the superposition of a standing wave and a traveling wave. In which direction does the traveling wave move? (c) How can you change the original waves so the resultant wave is the superposition of standing and traveling waves with the same amplitudes as before but with the traveling wave moving in the opposite direction? Next, use your graphs to find the place at which the oscillation amplitude is (d) maximum and (e) minimum. (f) How is the maximum amplitude related to the amplitudes of the original two waves? (g) How is the minimum amplitude related to the amplitudes of the original two waves?

93 A traveling wave on a string is described by

$$y = 2.0 \sin\left[2\pi\left(\frac{t}{0.40} + \frac{x}{80}\right)\right],$$

where x and y are in centimeters and t is in seconds. (a) For $t = 0$, plot y as a function of x for $0 \leq x \leq 160$ cm. (b) Repeat (a) for $t = 0.05$ s and $t = 0.10$ s. From your graphs, determine (c) the wave speed and (d) the direction in which the wave is traveling.

94 In Fig. 16-50, a circular loop of string is set spinning about the center point in a place with negligible gravity. The radius is 4.00 cm and the tangential speed of a string segment is 5.00 cm/s. The string is plucked. At what speed do transverse waves move along the string? (*Hint:* Apply Newton's second law to a small, but finite, section of the string.)

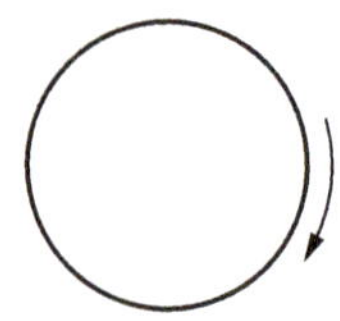

Figure 16-50 Problem 94.

95 A continuous traveling wave with amplitude A is incident on a boundary. The continuous reflection, with a smaller amplitude B, travels back through the incoming wave. The resulting interference pattern is displayed in Fig. 16-51. The standing wave ratio is defined to be

$$\text{SWR} = \frac{A + B}{A - B}.$$

The reflection coefficient R is the ratio of the power of the reflected wave to the power of the incoming wave and is thus proportional to the ratio $(B/A)^2$. What is the SWR for (a) total reflection and (b) no reflection? (c) For SWR = 1.50, what is R expressed as a percentage?

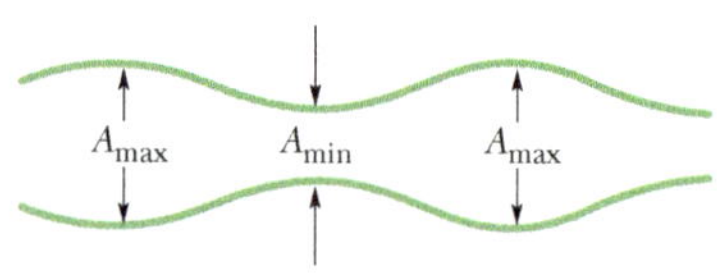

Figure 16-51 Problem 95.

96 Consider a loop in the standing wave created by two waves (amplitude 5.00 mm and frequency 120 Hz) traveling in opposite directions along a string with length 2.25 m and mass 125 g and under tension 40 N. At what rate does energy enter the loop from (a) each side and (b) both sides? (c) What is the maximum kinetic energy of the string in the loop during its oscillation?

C H A P T E R 1 7

Waves—II

17-1 SPEED OF SOUND

Learning Objectives

After reading this module, you should be able to . . .

17.01 Distinguish between a longitudinal wave and a transverse wave.

17.02 Explain wavefronts and rays.

17.03 Apply the relationship between the speed of sound through a material, the material's bulk modulus, and the material's density.

17.04 Apply the relationship between the speed of sound, the distance traveled by a sound wave, and the time required to travel that distance.

Key Idea

- Sound waves are longitudinal mechanical waves that can travel through solids, liquids, or gases. The speed v of a sound wave in a medium having bulk modulus B and density ρ is

$$v = \sqrt{\frac{B}{\rho}} \quad \text{(speed of sound).}$$

In air at 20°C, the speed of sound is 343 m/s.

What Is Physics?

The physics of sound waves is the basis of countless studies in the research journals of many fields. Here are just a few examples. Some physiologists are concerned with how speech is produced, how speech impairment might be corrected, how hearing loss can be alleviated, and even how snoring is produced. Some acoustic engineers are concerned with improving the acoustics of cathedrals and concert halls, with reducing noise near freeways and road construction, and with reproducing music by speaker systems. Some aviation engineers are concerned with the shock waves produced by supersonic aircraft and the aircraft noise produced in communities near an airport. Some medical researchers are concerned with how noises produced by the heart and lungs can signal a medical problem in a patient. Some paleontologists are concerned with how a dinosaur's fossil might reveal the dinosaur's vocalizations. Some military engineers are concerned with how the sounds of sniper fire might allow a soldier to pinpoint the sniper's location, and, on the gentler side, some biologists are concerned with how a cat purrs.

To begin our discussion of the physics of sound, we must first answer the question "What *are* sound waves?"

Sound Waves

As we saw in Chapter 16, mechanical waves are waves that require a material medium to exist. There are two types of mechanical waves: *Transverse waves* involve oscillations perpendicular to the direction in which the wave travels; *longitudinal waves* involve oscillations parallel to the direction of wave travel.

In this book, a **sound wave** is defined roughly as any longitudinal wave. Seismic prospecting teams use such waves to probe Earth's crust for oil. Ships

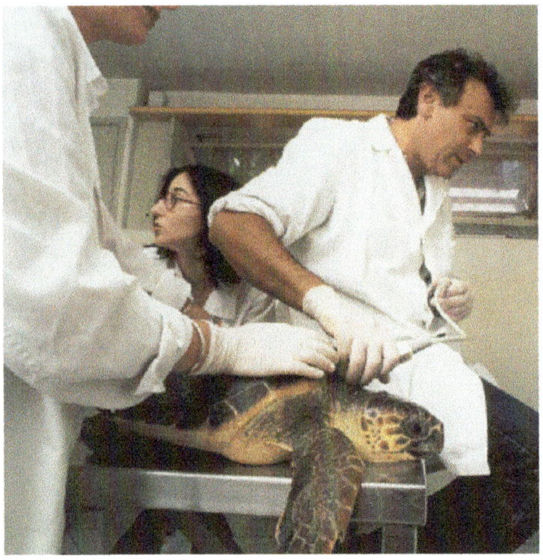

Mauro Fermariello/SPL/Photo Researchers, Inc.

Figure 17-1 A loggerhead turtle is being checked with ultrasound (which has a frequency above your hearing range); an image of its interior is being produced on a monitor off to the right.

carry sound-ranging gear (sonar) to detect underwater obstacles. Submarines use sound waves to stalk other submarines, largely by listening for the characteristic noises produced by the propulsion system. Figure 17-1 suggests how sound waves can be used to explore the soft tissues of an animal or human body. In this chapter we shall focus on sound waves that travel through the air and that are audible to people.

Figure 17-2 illustrates several ideas that we shall use in our discussions. Point S represents a tiny sound source, called a *point source,* that emits sound waves in all directions. The *wavefronts* and *rays* indicate the direction of travel and the spread of the sound waves. **Wavefronts** are surfaces over which the oscillations due to the sound wave have the same value; such surfaces are represented by whole or partial circles in a two-dimensional drawing for a point source. **Rays** are directed lines perpendicular to the wavefronts that indicate the direction of travel of the wavefronts. The short double arrows superimposed on the rays of Fig. 17-2 indicate that the longitudinal oscillations of the air are parallel to the rays.

Near a point source like that of Fig. 17-2, the wavefronts are spherical and spread out in three dimensions, and there the waves are said to be *spherical.* As the wavefronts move outward and their radii become larger, their curvature decreases. Far from the source, we approximate the wavefronts as planes (or lines on two-dimensional drawings), and the waves are said to be *planar.*

The Speed of Sound

The speed of any mechanical wave, transverse or longitudinal, depends on both an inertial property of the medium (to store kinetic energy) and an elastic property of the medium (to store potential energy). Thus, we can generalize Eq. 16-26, which gives the speed of a transverse wave along a stretched string, by writing

$$v = \sqrt{\frac{\tau}{\mu}} = \sqrt{\frac{\text{elastic property}}{\text{inertial property}}}, \tag{17-1}$$

where (for transverse waves) τ is the tension in the string and μ is the string's linear density. If the medium is air and the wave is longitudinal, we can guess that the inertial property, corresponding to μ, is the volume density ρ of air. What shall we put for the elastic property?

In a stretched string, potential energy is associated with the periodic stretching of the string elements as the wave passes through them. As a sound wave passes through air, potential energy is associated with periodic compressions and expansions of small volume elements of the air. The property that determines the extent to which an element of a medium changes in volume when the pressure (force per unit area) on it changes is the **bulk modulus** B, defined (from Eq. 12-25) as

$$B = -\frac{\Delta p}{\Delta V/V} \quad \text{(definition of bulk modulus).} \tag{17-2}$$

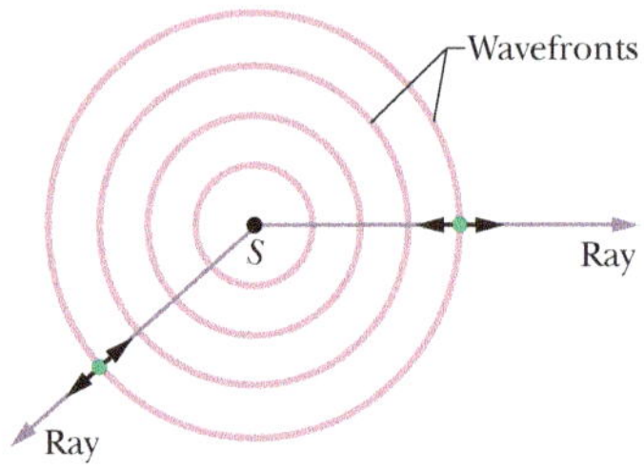

Figure 17-2 A sound wave travels from a point source S through a three-dimensional medium. The wavefronts form spheres centered on S; the rays are radial to S. The short, double-headed arrows indicate that elements of the medium oscillate parallel to the rays.

Here $\Delta V/V$ is the fractional change in volume produced by a change in pressure Δp. As explained in Module 14-1, the SI unit for pressure is the newton per square meter, which is given a special name, the *pascal* (Pa). From Eq. 17-2 we see that the unit for B is also the pascal. The signs of Δp and ΔV are always opposite: When we increase the pressure on an element (Δp is positive), its volume decreases (ΔV is negative). We include a minus sign in Eq. 17-2 so that B is always a positive quantity. Now substituting B for τ and ρ for μ in Eq. 17-1 yields

$$v = \sqrt{\frac{B}{\rho}} \quad \text{(speed of sound)} \tag{17-3}$$

as the speed of sound in a medium with bulk modulus B and density ρ. Table 17-1 lists the speed of sound in various media.

The density of water is almost 1000 times greater than the density of air. If this were the only relevant factor, we would expect from Eq. 17-3 that the speed of sound in water would be considerably less than the speed of sound in air. However, Table 17-1 shows us that the reverse is true. We conclude (again from Eq. 17-3) that the bulk modulus of water must be more than 1000 times greater than that of air. This is indeed the case. Water is much more incompressible than air, which (see Eq. 17-2) is another way of saying that its bulk modulus is much greater.

Table 17-1 The Speed of Sound[a]

Medium	Speed (m/s)
Gases	
Air (0°C)	331
Air (20°C)	343
Helium	965
Hydrogen	1284
Liquids	
Water (0°C)	1402
Water (20°C)	1482
Seawater[b]	1522
Solids	
Aluminum	6420
Steel	5941
Granite	6000

[a]At 0°C and 1 atm pressure, except where noted.
[b]At 20°C and 3.5% salinity.

Formal Derivation of Eq. 17-3

We now derive Eq. 17-3 by direct application of Newton's laws. Let a single pulse in which air is compressed travel (from right to left) with speed v through the air in a long tube, like that in Fig. 16-2. Let us run along with the pulse at that speed, so that the pulse appears to stand still in our reference frame. Figure 17-3*a* shows the situation as it is viewed from that frame. The pulse is standing still, and air is moving at speed v through it from left to right.

Let the pressure of the undisturbed air be p and the pressure inside the pulse be $p + \Delta p$, where Δp is positive due to the compression. Consider an element of air of thickness Δx and face area A, moving toward the pulse at speed v. As this element enters the pulse, the leading face of the element encounters a region of higher pressure, which slows the element to speed $v + \Delta v$, in which Δv is negative. This slowing is complete when the rear face of the element reaches the pulse, which requires time interval

$$\Delta t = \frac{\Delta x}{v}. \tag{17-4}$$

Let us apply Newton's second law to the element. During Δt, the average force on the element's trailing face is pA toward the right, and the average force on the leading face is $(p + \Delta p)A$ toward the left (Fig. 17-3*b*). Therefore, the average net force on the element during Δt is

$$\begin{aligned} F &= pA - (p + \Delta p)A \\ &= -\Delta p\, A \quad \text{(net force)}. \end{aligned} \tag{17-5}$$

The minus sign indicates that the net force on the air element is directed to the left in Fig. 17-3*b*. The volume of the element is $A\,\Delta x$, so with the aid of Eq. 17-4, we can write its mass as

$$\Delta m = \rho\,\Delta V = \rho A\,\Delta x = \rho A v\,\Delta t \quad \text{(mass)}. \tag{17-6}$$

The average acceleration of the element during Δt is

$$a = \frac{\Delta v}{\Delta t} \quad \text{(acceleration)}. \tag{17-7}$$

Figure 17-3 A compression pulse is sent from right to left down a long air-filled tube. The reference frame of the figure is chosen so that the pulse is at rest and the air moves from left to right. (*a*) An element of air of width Δx moves toward the pulse with speed v. (*b*) The leading face of the element enters the pulse. The forces acting on the leading and trailing faces (due to air pressure) are shown.

Thus, from Newton's second law ($F = ma$), we have, from Eqs. 17-5, 17-6, and 17-7,

$$-\Delta p\, A = (\rho A v\, \Delta t)\frac{\Delta v}{\Delta t}, \qquad (17\text{-}8)$$

which we can write as

$$\rho v^2 = -\frac{\Delta p}{\Delta v/v}. \qquad (17\text{-}9)$$

The air that occupies a volume V ($= Av\,\Delta t$) outside the pulse is compressed by an amount ΔV ($= A\,\Delta v\,\Delta t$) as it enters the pulse. Thus,

$$\frac{\Delta V}{V} = \frac{A\,\Delta v\,\Delta t}{Av\,\Delta t} = \frac{\Delta v}{v}. \qquad (17\text{-}10)$$

Substituting Eq. 17-10 and then Eq. 17-2 into Eq. 17-9 leads to

$$\rho v^2 = -\frac{\Delta p}{\Delta v/v} = -\frac{\Delta p}{\Delta V/V} = B. \qquad (17\text{-}11)$$

Solving for v yields Eq. 17-3 for the speed of the air toward the right in Fig. 17-3, and thus for the actual speed of the pulse toward the left.

17-2 TRAVELING SOUND WAVES

Learning Objectives

After reading this module, you should be able to . . .

17.05 For any particular time and position, calculate the displacement $s(x, t)$ of an element of air as a sound wave travels through its location.

17.06 Given a displacement function $s(x, t)$ for a sound wave, calculate the time between two given displacements.

17.07 Apply the relationships between wave speed v, angular frequency ω, angular wave number k, wavelength λ, period T, and frequency f.

17.08 Sketch a graph of the displacement $s(x)$ of an element of air as a function of position, and identify the amplitude s_m and wavelength λ.

17.09 For any particular time and position, calculate the pressure variation Δp (variation from atmospheric pressure) of an element of air as a sound wave travels through its location.

17.10 Sketch a graph of the pressure variation $\Delta p(x)$ of an element as a function of position, and identify the amplitude Δp_m and wavelength λ.

17.11 Apply the relationship between pressure-variation amplitude Δp_m and displacement amplitude s_m.

17.12 Given a graph of position s versus time for a sound wave, determine the amplitude s_m and the period T.

17.13 Given a graph of pressure variation Δp versus time for a sound wave, determine the amplitude Δp_m and the period T.

Key Ideas

- A sound wave causes a longitudinal displacement s of a mass element in a medium as given by

$$s = s_m \cos(kx - \omega t),$$

where s_m is the displacement amplitude (maximum displacement) from equilibrium, $k = 2\pi/\lambda$, and $\omega = 2\pi f$, λ and f being the wavelength and frequency, respectively, of the sound wave.

- The sound wave also causes a pressure change Δp of the medium from the equilibrium pressure:

$$\Delta p = \Delta p_m \sin(kx - \omega t),$$

where the pressure amplitude is

$$\Delta p_m = (v\rho\omega)s_m.$$

Traveling Sound Waves

Here we examine the displacements and pressure variations associated with a sinusoidal sound wave traveling through air. Figure 17-4*a* displays such a wave traveling rightward through a long air-filled tube. Recall from Chapter 16 that we can produce such a wave by sinusoidally moving a piston at the left end of

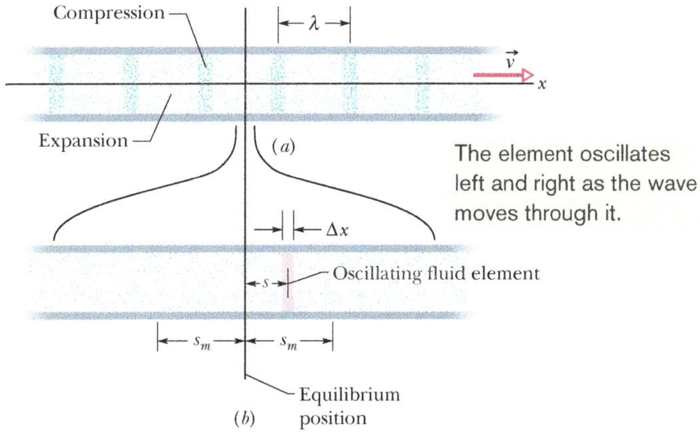

Figure 17-4 (*a*) A sound wave, traveling through a long air-filled tube with speed v, consists of a moving, periodic pattern of expansions and compressions of the air. The wave is shown at an arbitrary instant. (*b*) A horizontally expanded view of a short piece of the tube. As the wave passes, an air element of thickness Δx oscillates left and right in simple harmonic motion about its equilibrium position. At the instant shown in (*b*), the element happens to be displaced a distance s to the right of its equilibrium position. Its maximum displacement, either right or left, is s_m.

the tube (as in Fig. 16-2). The piston's rightward motion moves the element of air next to the piston face and compresses that air; the piston's leftward motion allows the element of air to move back to the left and the pressure to decrease. As each element of air pushes on the next element in turn, the right–left motion of the air and the change in its pressure travel along the tube as a sound wave.

Consider the thin element of air of thickness Δx shown in Fig. 17-4*b*. As the wave travels through this portion of the tube, the element of air oscillates left and right in simple harmonic motion about its equilibrium position. Thus, the oscillations of each air element due to the traveling sound wave are like those of a string element due to a transverse wave, except that the air element oscillates *longitudinally* rather than *transversely*. Because string elements oscillate parallel to the y axis, we write their displacements in the form $y(x, t)$. Similarly, because air elements oscillate parallel to the x axis, we could write their displacements in the confusing form $x(x, t)$, but we shall use $s(x, t)$ instead.

Displacement. To show that the displacements $s(x, t)$ are sinusoidal functions of x and t, we can use either a sine function or a cosine function. In this chapter we use a cosine function, writing

$$s(x, t) = s_m \cos(kx - \omega t). \tag{17-12}$$

Figure 17-5*a* labels the various parts of this equation. In it, s_m is the **displacement amplitude**—that is, the maximum displacement of the air element to either side of its equilibrium position (see Fig. 17-4*b*). The angular wave number k, angular frequency ω, frequency f, wavelength λ, speed v, and period T for a sound (longitudinal) wave are defined and interrelated exactly as for a transverse wave, except that λ is now the distance (again along the direction of travel) in which the pattern of compression and expansion due to the wave begins to repeat itself (see Fig. 17-4*a*). (We assume s_m is much less than λ.)

Pressure. As the wave moves, the air pressure at any position x in Fig. 17-4*a* varies sinusoidally, as we prove next. To describe this variation we write

$$\Delta p(x, t) = \Delta p_m \sin(kx - \omega t). \tag{17-13}$$

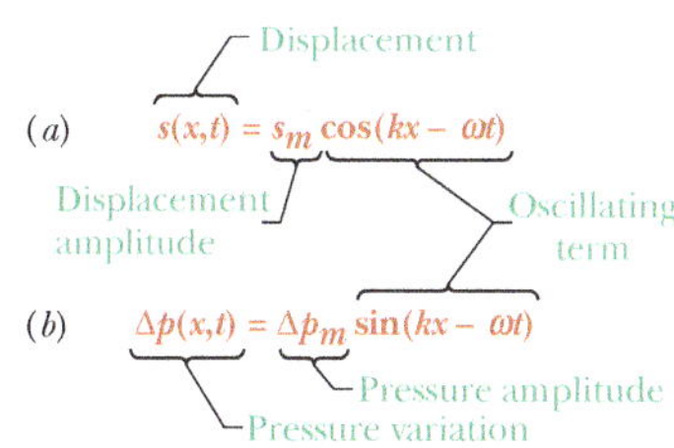

Figure 17-5 (*a*) The displacement function and (*b*) the pressure-variation function of a traveling sound wave consist of an amplitude and an oscillating term.

Figure 17-5*b* labels the various parts of this equation. A negative value of Δp in Eq. 17-13 corresponds to an expansion of the air, and a positive value to a compression. Here Δp_m is the **pressure amplitude,** which is the maximum increase or decrease in pressure due to the wave; Δp_m is normally very much less than the pressure p present when there is no wave. As we shall prove, the pressure ampli-

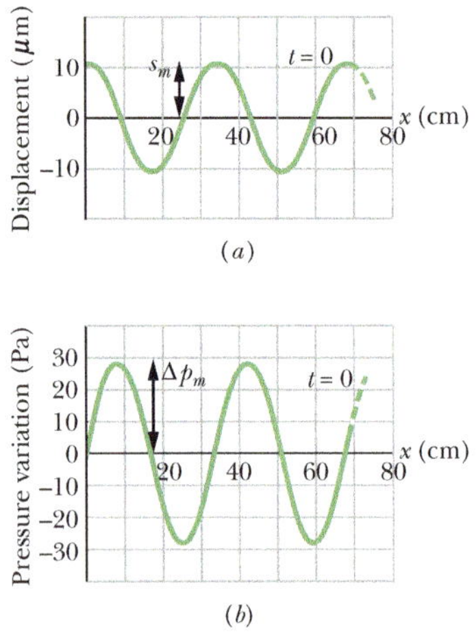

Figure 17-6 (*a*) A plot of the displacement function (Eq. 17-12) for $t = 0$. (*b*) A similar plot of the pressure-variation function (Eq. 17-13). Both plots are for a 1000 Hz sound wave whose pressure amplitude is at the threshold of pain.

tude Δp_m is related to the displacement amplitude s_m in Eq. 17-12 by

$$\Delta p_m = (v\rho\omega)s_m. \tag{17-14}$$

Figure 17-6 shows plots of Eqs. 17-12 and 17-13 at $t = 0$; with time, the two curves would move rightward along the horizontal axes. Note that the displacement and pressure variation are $\pi/2$ rad (or 90°) out of phase. Thus, for example, the pressure variation Δp at any point along the wave is zero when the displacement there is a maximum.

Checkpoint 1

When the oscillating air element in Fig. 17-4*b* is moving rightward through the point of zero displacement, is the pressure in the element at its equilibrium value, just beginning to increase, or just beginning to decrease?

Derivation of Eqs. 17-13 and 17-14

Figure 17-4*b* shows an oscillating element of air of cross-sectional area A and thickness Δx, with its center displaced from its equilibrium position by distance s. From Eq. 17-2 we can write, for the pressure variation in the displaced element,

$$\Delta p = -B\frac{\Delta V}{V}. \tag{17-15}$$

The quantity V in Eq. 17-15 is the volume of the element, given by

$$V = A\,\Delta x. \tag{17-16}$$

The quantity ΔV in Eq. 17-15 is the change in volume that occurs when the element is displaced. This volume change comes about because the displacements of the two faces of the element are not quite the same, differing by some amount Δs. Thus, we can write the change in volume as

$$\Delta V = A\,\Delta s. \tag{17-17}$$

Substituting Eqs. 17-16 and 17-17 into Eq. 17-15 and passing to the differential limit yield

$$\Delta p = -B\frac{\Delta s}{\Delta x} = -B\frac{\partial s}{\partial x}. \tag{17-18}$$

The symbols ∂ indicate that the derivative in Eq. 17-18 is a *partial derivative*, which tells us how s changes with x when the time t is fixed. From Eq. 17-12 we then have, treating t as a constant,

$$\frac{\partial s}{\partial x} = \frac{\partial}{\partial x}[s_m\cos(kx - \omega t)] = -ks_m\sin(kx - \omega t).$$

Substituting this quantity for the partial derivative in Eq. 17-18 yields

$$\Delta p = Bks_m\sin(kx - \omega t).$$

This tells us that the pressure varies as a sinusoidal function of time and that the amplitude of the variation is equal to the terms in front of the sine function. Setting $\Delta p_m = Bks_m$, this yields Eq. 17-13, which we set out to prove.

Using Eq. 17-3, we can now write

$$\Delta p_m = (Bk)s_m = (v^2\rho k)s_m.$$

Equation 17-14, which we also wanted to prove, follows at once if we substitute ω/v for k from Eq. 16-12.

Sample Problem 17.01 Pressure amplitude, displacement amplitude

The maximum pressure amplitude Δp_m that the human ear can tolerate in loud sounds is about 28 Pa (which is very much less than the normal air pressure of about 10^5 Pa). What is the displacement amplitude s_m for such a sound in air of density $\rho = 1.21\ \text{kg/m}^3$, at a frequency of 1000 Hz and a speed of 343 m/s?

KEY IDEA

The displacement amplitude s_m of a sound wave is related to the pressure amplitude Δp_m of the wave according to Eq. 17-14.

Calculations: Solving that equation for s_m yields

$$s_m = \frac{\Delta p_m}{v\rho\omega} = \frac{\Delta p_m}{v\rho(2\pi f)}.$$

Substituting known data then gives us

$$s_m = \frac{28\ \text{Pa}}{(343\ \text{m/s})(1.21\ \text{kg/m}^3)(2\pi)(1000\ \text{Hz})}$$
$$= 1.1 \times 10^{-5}\ \text{m} = 11\ \mu\text{m}. \qquad \text{(Answer)}$$

That is only about one-seventh the thickness of a book page. Obviously, the displacement amplitude of even the loudest sound that the ear can tolerate is very small. Temporary exposure to such loud sound produces temporary hearing loss, probably due to a decrease in blood supply to the inner ear. Prolonged exposure produces permanent damage.

The pressure amplitude Δp_m for the *faintest* detectable sound at 1000 Hz is 2.8×10^{-5} Pa. Proceeding as above leads to $s_m = 1.1 \times 10^{-11}$ m or 11 pm, which is about one-tenth the radius of a typical atom. The ear is indeed a sensitive detector of sound waves.

WILEY PLUS Additional examples, video, and practice available at *WileyPLUS*

17-3 INTERFERENCE

Learning Objectives

After reading this module, you should be able to . . .

17.14 If two waves with the same wavelength begin in phase but reach a common point by traveling along different paths, calculate their phase difference ϕ at that point by relating the path length difference ΔL to the wavelength λ.

17.15 Given the phase difference between two sound waves with the same amplitude, wavelength, and travel direction, determine the type of interference between the waves (fully destructive interference, fully constructive interference, or indeterminate interference).

17.16 Convert a phase difference between radians, degrees, and number of wavelengths.

Key Ideas

- The interference of two sound waves with identical wavelengths passing through a common point depends on their phase difference ϕ there. If the sound waves were emitted in phase and are traveling in approximately the same direction, ϕ is given by

$$\phi = \frac{\Delta L}{\lambda} 2\pi,$$

where ΔL is their path length difference.

- Fully constructive interference occurs when ϕ is an integer multiple of 2π,

$$\phi = m(2\pi), \qquad \text{for } m = 0, 1, 2, \ldots,$$

and, equivalently, when ΔL is related to wavelength λ by

$$\frac{\Delta L}{\lambda} = 0, 1, 2, \ldots.$$

- Fully destructive interference occurs when ϕ is an odd multiple of π,

$$\phi = (2m + 1)\pi, \qquad \text{for } m = 0, 1, 2, \ldots,$$

and

$$\frac{\Delta L}{\lambda} = 0.5, 1.5, 2.5, \ldots.$$

Interference

Like transverse waves, sound waves can undergo interference. In fact, we can write equations for the interference as we did in Module 16-5 for transverse waves. Suppose two sound waves with the same amplitude and wavelength are traveling in the positive direction of an x axis with a phase difference of ϕ. We can express the waves in the form of Eqs. 16-47 and 16-48 but, to be consistent with Eq. 17-12, we use cosine functions instead of sine functions:

$$s_1(x, t) = s_m \cos(kx - \omega t)$$

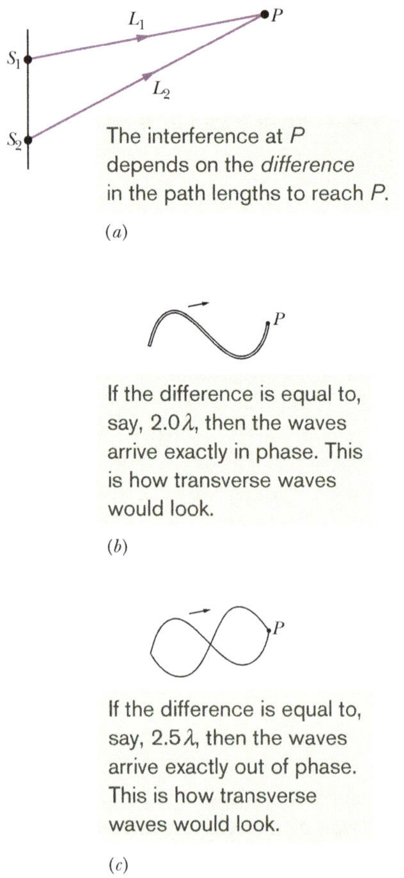

Figure 17-7 (*a*) Two point sources S_1 and S_2 emit spherical sound waves in phase. The rays indicate that the waves pass through a common point P. The waves (represented with *transverse* waves) arrive at P (*b*) exactly in phase and (*c*) exactly out of phase.

and

$$s_2(x, t) = s_m \cos(kx - \omega t + \phi).$$

These waves overlap and interfere. From Eq. 16-51, we can write the resultant wave as

$$s' = [2s_m \cos\tfrac{1}{2}\phi] \cos(kx - \omega t + \tfrac{1}{2}\phi).$$

As we saw with transverse waves, the resultant wave is itself a traveling wave. Its amplitude is the magnitude

$$s'_m = |2s_m \cos\tfrac{1}{2}\phi|. \tag{17-19}$$

As with transverse waves, the value of ϕ determines what type of interference the individual waves undergo.

One way to control ϕ is to send the waves along paths with different lengths. Figure 17-7*a* shows how we can set up such a situation: Two point sources S_1 and S_2 emit sound waves that are in phase and of identical wavelength λ. Thus, the *sources* themselves are said to be in phase; that is, as the waves emerge from the sources, their displacements are always identical. We are interested in the waves that then travel through point P in Fig. 17-7*a*. We assume that the distance to P is much greater than the distance between the sources so that we can approximate the waves as traveling in the same direction at P.

If the waves traveled along paths with identical lengths to reach point P, they would be in phase there. As with transverse waves, this means that they would undergo fully constructive interference there. However, in Fig. 17-7*a*, path L_2 traveled by the wave from S_2 is longer than path L_1 traveled by the wave from S_1. The difference in path lengths means that the waves may not be in phase at point P. In other words, their phase difference ϕ at P depends on their **path length difference** $\Delta L = |L_2 - L_1|$.

To relate phase difference ϕ to path length difference ΔL, we recall (from Module 16-1) that a phase difference of 2π rad corresponds to one wavelength. Thus, we can write the proportion

$$\frac{\phi}{2\pi} = \frac{\Delta L}{\lambda}, \tag{17-20}$$

from which

$$\phi = \frac{\Delta L}{\lambda} 2\pi. \tag{17-21}$$

Fully constructive interference occurs when ϕ is zero, 2π, or any integer multiple of 2π. We can write this condition as

$$\phi = m(2\pi), \quad \text{for } m = 0, 1, 2, \ldots \quad \text{(fully constructive interference).} \tag{17-22}$$

From Eq. 17-21, this occurs when the ratio $\Delta L/\lambda$ is

$$\frac{\Delta L}{\lambda} = 0, 1, 2, \ldots \quad \text{(fully constructive interference).} \tag{17-23}$$

For example, if the path length difference $\Delta L = |L_2 - L_1|$ in Fig. 17-7*a* is equal to 2λ, then $\Delta L/\lambda = 2$ and the waves undergo fully constructive interference at point P (Fig. 17-7*b*). The interference is fully constructive because the wave from S_2 is phase-shifted relative to the wave from S_1 by 2λ, putting the two waves *exactly in phase* at P.

Fully destructive interference occurs when ϕ is an odd multiple of π:

$$\phi = (2m + 1)\pi, \quad \text{for } m = 0, 1, 2, \ldots \quad \text{(fully destructive interference).} \tag{17-24}$$

From Eq. 17-21, this occurs when the ratio $\Delta L/\lambda$ is

$$\frac{\Delta L}{\lambda} = 0.5, 1.5, 2.5, \ldots \quad \text{(fully destructive interference).} \qquad (17\text{-}25)$$

For example, if the path length difference $\Delta L = |L_2 - L_1|$ in Fig. 17-7*a* is equal to 2.5λ, then $\Delta L/\lambda = 2.5$ and the waves undergo fully destructive interference at point P (Fig. 17-7*c*). The interference is fully destructive because the wave from S_2 is phase-shifted relative to the wave from S_1 by 2.5 wavelengths, which puts the two waves *exactly out of phase* at P.

Of course, two waves could produce intermediate interference as, say, when $\Delta L/\lambda = 1.2$. This would be closer to fully constructive interference ($\Delta L/\lambda = 1.0$) than to fully destructive interference ($\Delta L/\lambda = 1.5$).

Sample Problem 17.02 Interference points along a big circle

In Fig. 17-8*a*, two point sources S_1 and S_2, which are in phase and separated by distance $D = 1.5\lambda$, emit identical sound waves of wavelength λ.

(a) What is the path length difference of the waves from S_1 and S_2 at point P_1, which lies on the perpendicular bisector of distance D, at a distance greater than D from the sources (Fig. 17-8*b*)? (That is, what is the difference in the distance from source S_1 to point P_1 and the distance from source S_2 to P_1?) What type of interference occurs at P_1?

Reasoning: Because the waves travel identical distances to reach P_1, their path length difference is

$$\Delta L = 0. \qquad \text{(Answer)}$$

From Eq. 17-23, this means that the waves undergo fully constructive interference at P_1 because they start in phase at the sources and reach P_1 in phase.

(b) What are the path length difference and type of interference at point P_2 in Fig. 17-8*c*?

Reasoning: The wave from S_1 travels the extra distance D ($= 1.5\lambda$) to reach P_2. Thus, the path length difference is

$$\Delta L = 1.5\lambda. \qquad \text{(Answer)}$$

From Eq. 17-25, this means that the waves are exactly out of phase at P_2 and undergo fully destructive interference there.

(c) Figure 17-8*d* shows a circle with a radius much greater than D, centered on the midpoint between sources S_1 and S_2. What is the number of points N around this circle at which the interference is fully constructive? (That is, at how many points do the waves arrive exactly in phase?)

Reasoning: Starting at point a, let's move clockwise along the circle to point d. As we move, path length difference ΔL increases and so the type of interference changes. From (a), we know that is $\Delta L = 0\lambda$ at point a. From (b), we know that $\Delta L = 1.5\lambda$ at point d. Thus, there must be

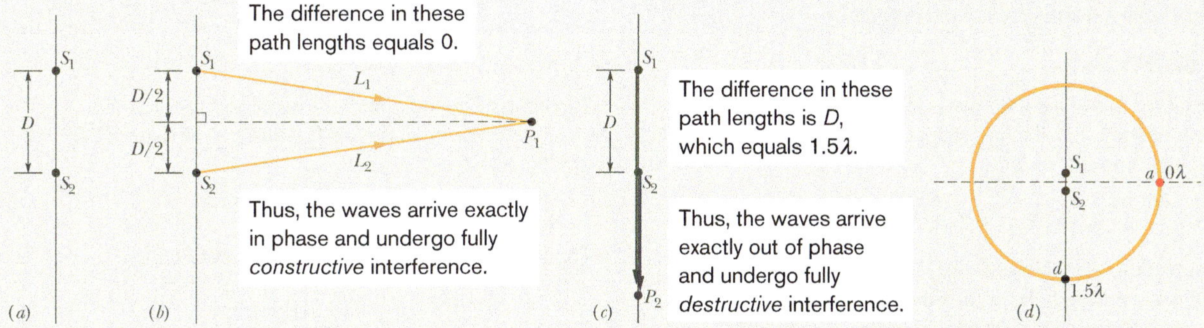

Figure 17-8 (*a*) Two point sources S_1 and S_2, separated by distance D, emit spherical sound waves in phase. (*b*) The waves travel equal distances to reach point P_1. (*c*) Point P_2 is on the line extending through S_1 and S_2. (*d*) We move around a large circle. (*Figure continues*)

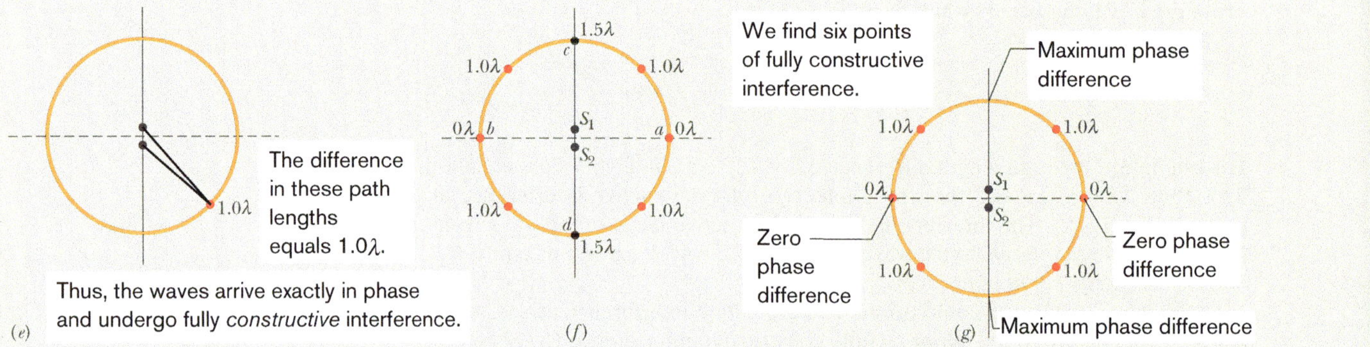

Figure 17-8 (*continued*) (*e*) Another point of fully constructive interference. (*f*) Using symmetry to determine other points. (*g*) The six points of fully constructive interference.

one point between a and d at which $\Delta L = \lambda$ (Fig. 17-8e). From Eq. 17-23, fully constructive interference occurs at that point. Also, there can be no other point along the way from point a to point d at which fully constructive interference occurs, because there is no other integer than 1 between 0 at point a and 1.5 at point d.

We can now use symmetry to locate other points of fully constructive or destructive interference (Fig. 17-8f). Symmetry about line cd gives us point b, at which $\Delta L = 0\lambda$. Also, there are three more points at which $\Delta L = \lambda$. In all (Fig. 17-8g) we have

$$N = 6. \qquad \text{(Answer)}$$

WILEY PLUS Additional examples, video, and practice available at *WileyPLUS*

17-4 INTENSITY AND SOUND LEVEL

Learning Objectives

After reading this module, you should be able to . . .

17.17 Calculate the sound intensity I at a surface as the ratio of the power P to the surface area A.

17.18 Apply the relationship between the sound intensity I and the displacement amplitude s_m of the sound wave.

17.19 Identify an isotropic point source of sound.

17.20 For an isotropic point source, apply the relationship involving the emitting power P_s, the distance r to a detector, and the sound intensity I at the detector.

17.21 Apply the relationship between the sound level β, the sound intensity I, and the standard reference intensity I_0.

17.22 Evaluate a logarithm function (log) and an antilogarithm function ($\log^{-1}$).

17.23 Relate the change in a sound level to the change in sound intensity.

Key Ideas

- The intensity I of a sound wave at a surface is the average rate per unit area at which energy is transferred by the wave through or onto the surface:

$$I = \frac{P}{A},$$

where P is the time rate of energy transfer (power) of the sound wave and A is the area of the surface intercepting the sound. The intensity I is related to the displacement amplitude s_m of the sound wave by

$$I = \tfrac{1}{2}\rho v \omega^2 s_m^2.$$

- The intensity at a distance r from a point source that emits sound waves of power P_s equally in all directions (isotropically) is

$$I = \frac{P_s}{4\pi r^2}.$$

- The sound level β in decibels (dB) is defined as

$$\beta = (10\text{ dB}) \log \frac{I}{I_0},$$

where I_0 ($= 10^{-12}$ W/m^2) is a reference intensity level to which all intensities are compared. For every factor-of-10 increase in intensity, 10 dB is added to the sound level.

Intensity and Sound Level

If you have ever tried to sleep while someone played loud music nearby, you are well aware that there is more to sound than frequency, wavelength, and speed. There is also intensity. The **intensity** I of a sound wave at a surface is the average rate per unit area at which energy is transferred by the wave through or onto the surface. We can write this as

$$I = \frac{P}{A}, \tag{17-26}$$

where P is the time rate of energy transfer (the power) of the sound wave and A is the area of the surface intercepting the sound. As we shall derive shortly, the intensity I is related to the displacement amplitude s_m of the sound wave by

$$I = \tfrac{1}{2}\rho v \omega^2 s_m^2. \tag{17-27}$$

Intensity can be measured on a detector. *Loudness* is a perception, something that you sense. The two can differ because your perception depends on factors such as the sensitivity of your hearing mechanism to various frequencies.

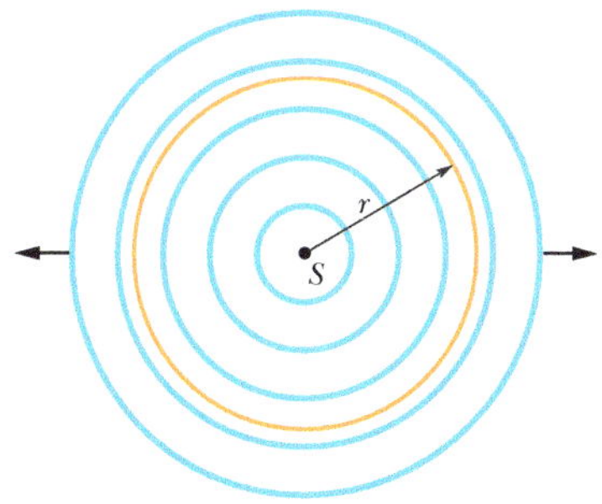

Figure 17-9 A point source S emits sound waves uniformly in all directions. The waves pass through an imaginary sphere of radius r that is centered on S.

Variation of Intensity with Distance

How intensity varies with distance from a real sound source is often complex. Some real sources (like loudspeakers) may transmit sound only in particular directions, and the environment usually produces echoes (reflected sound waves) that overlap the direct sound waves. In some situations, however, we can ignore echoes and assume that the sound source is a point source that emits the sound *isotropically*—that is, with equal intensity in all directions. The wavefronts spreading from such an isotropic point source S at a particular instant are shown in Fig. 17-9.

Let us assume that the mechanical energy of the sound waves is conserved as they spread from this source. Let us also center an imaginary sphere of radius r on the source, as shown in Fig. 17-9. All the energy emitted by the source must pass through the surface of the sphere. Thus, the time rate at which energy is transferred through the surface by the sound waves must equal the time rate at which energy is emitted by the source (that is, the power P_s of the source). From Eq. 17-26, the intensity I at the sphere must then be

$$I = \frac{P_s}{4\pi r^2}, \tag{17-28}$$

where $4\pi r^2$ is the area of the sphere. Equation 17-28 tells us that the intensity of sound from an isotropic point source decreases with the square of the distance r from the source.

Checkpoint 2

The figure indicates three small patches 1, 2, and 3 that lie on the surfaces of two imaginary spheres; the spheres are centered on an isotropic point source S of sound. The rates at which energy is transmitted through the three patches by the sound waves are equal. Rank the patches according to (a) the intensity of the sound on them and (b) their area, greatest first.

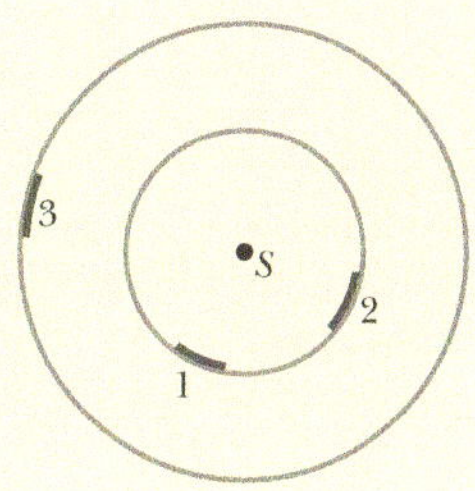

© Ben Rose

Sound can cause the wall of a drinking glass to oscillate. If the sound produces a standing wave of oscillations and if the intensity of the sound is large enough, the glass will shatter.

The Decibel Scale

The displacement amplitude at the human ear ranges from about 10^{-5} m for the loudest tolerable sound to about 10^{-11} m for the faintest detectable sound, a ratio of 10^6. From Eq. 17-27 we see that the intensity of a sound varies as the *square* of its amplitude, so the ratio of intensities at these two limits of the human auditory system is 10^{12}. Humans can hear over an enormous range of intensities.

We deal with such an enormous range of values by using logarithms. Consider the relation

$$y = \log x,$$

in which x and y are variables. It is a property of this equation that if we *multiply* x by 10, then y increases by 1. To see this, we write

$$y' = \log(10x) = \log 10 + \log x = 1 + y.$$

Similarly, if we multiply x by 10^{12}, y increases by only 12.

Thus, instead of speaking of the intensity I of a sound wave, it is much more convenient to speak of its **sound level** β, defined as

$$\beta = (10 \text{ dB}) \log \frac{I}{I_0}. \tag{17-29}$$

Here dB is the abbreviation for **decibel,** the unit of sound level, a name that was chosen to recognize the work of Alexander Graham Bell. I_0 in Eq. 17-29 is a standard reference intensity ($= 10^{-12}$ W/m^2), chosen because it is near the lower limit of the human range of hearing. For $I = I_0$, Eq. 17-29 gives $\beta = 10 \log 1 = 0$, so our standard reference level corresponds to zero decibels. Then β increases by 10 dB every time the sound intensity increases by an order of magnitude (a factor of 10). Thus, $\beta = 40$ corresponds to an intensity that is 10^4 times the standard reference level. Table 17-2 lists the sound levels for a variety of environments.

Table 17-2 Some Sound Levels (dB)

Hearing threshold	0
Rustle of leaves	10
Conversation	60
Rock concert	110
Pain threshold	120
Jet engine	130

Derivation of Eq. 17-27

Consider, in Fig. 17-4*a*, a thin slice of air of thickness dx, area A, and mass dm, oscillating back and forth as the sound wave of Eq. 17-12 passes through it. The kinetic energy dK of the slice of air is

$$dK = \tfrac{1}{2} dm\, v_s^2. \tag{17-30}$$

Here v_s is not the speed of the wave but the speed of the oscillating element of air, obtained from Eq. 17-12 as

$$v_s = \frac{\partial s}{\partial t} = -\omega s_m \sin(kx - \omega t).$$

Using this relation and putting $dm = \rho A\, dx$ allow us to rewrite Eq. 17-30 as

$$dK = \tfrac{1}{2}(\rho A\, dx)(-\omega s_m)^2 \sin^2(kx - \omega t). \tag{17-31}$$

Dividing Eq. 17-31 by dt gives the rate at which kinetic energy moves along with the wave. As we saw in Chapter 16 for transverse waves, dx/dt is the wave speed v, so we have

$$\frac{dK}{dt} = \tfrac{1}{2}\rho A v \omega^2 s_m^2 \sin^2(kx - \omega t). \tag{17-32}$$

The *average* rate at which kinetic energy is transported is

$$\left(\frac{dK}{dt}\right)_{avg} = \tfrac{1}{2}\rho A v \omega^2 s_m^2 [\sin^2(kx - \omega t)]_{avg}$$
$$= \tfrac{1}{4}\rho A v \omega^2 s_m^2. \quad (17\text{-}33)$$

To obtain this equation, we have used the fact that the average value of the square of a sine (or a cosine) function over one full oscillation is $\frac{1}{2}$.

We assume that *potential* energy is carried along with the wave at this same average rate. The wave intensity I, which is the average rate per unit area at which energy of both kinds is transmitted by the wave, is then, from Eq. 17-33,

$$I = \frac{2(dK/dt)_{avg}}{A} = \tfrac{1}{2}\rho v \omega^2 s_m^2,$$

which is Eq. 17-27, the equation we set out to derive.

Sample Problem 17.03 Intensity change with distance, cylindrical sound wave

An electric spark jumps along a straight line of length $L = 10$ m, emitting a pulse of sound that travels radially outward from the spark. (The spark is said to be a *line source* of sound.) The power of this acoustic emission is $P_s = 1.6 \times 10^4$ W.

(a) What is the intensity I of the sound when it reaches a distance $r = 12$ m from the spark?

KEY IDEAS

(1) Let us center an imaginary cylinder of radius $r = 12$ m and length $L = 10$ m (open at both ends) on the spark, as shown in Fig. 17-10. Then the intensity I at the cylindrical surface is the ratio P/A, where P is the time rate at which sound energy passes through the surface and A is the surface area. (2) We assume that the principle of conservation of energy applies to the sound energy. This means that the rate P at which energy is transferred through the cylinder must equal the rate P_s at which energy is emitted by the source.

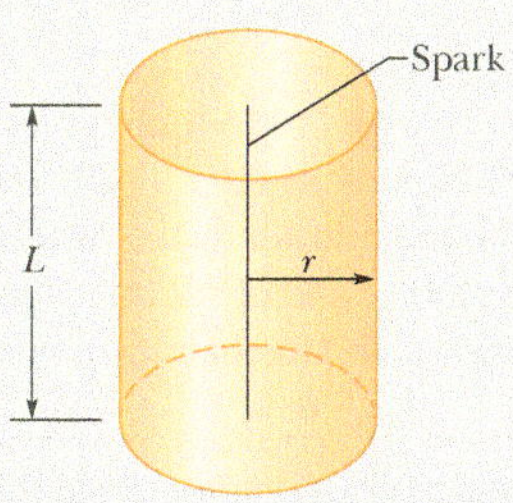

Figure 17-10 A spark along a straight line of length L emits sound waves radially outward. The waves pass through an imaginary cylinder of radius r and length L that is centered on the spark.

Calculations: Putting these ideas together and noting that the area of the cylindrical surface is $A = 2\pi r L$, we have

$$I = \frac{P}{A} = \frac{P_s}{2\pi r L}. \quad (17\text{-}34)$$

This tells us that the intensity of the sound from a line source decreases with distance r (and not with the square of distance r as for a point source). Substituting the given data, we find

$$I = \frac{1.6 \times 10^4 \text{ W}}{2\pi(12 \text{ m})(10 \text{ m})}$$
$$= 21.2 \text{ W/m}^2 \approx 21 \text{ W/m}^2. \quad \text{(Answer)}$$

(b) At what time rate P_d is sound energy intercepted by an acoustic detector of area $A_d = 2.0$ cm^2, aimed at the spark and located a distance $r = 12$ m from the spark?

Calculations: We know that the intensity of sound at the detector is the ratio of the energy transfer rate P_d there to the detector's area A_d:

$$I = \frac{P_d}{A_d}. \quad (17\text{-}35)$$

We can imagine that the detector lies on the cylindrical surface of (a). Then the sound intensity at the detector is the intensity I (= 21.2 W/m^2) at the cylindrical surface. Solving Eq. 17-35 for P_d gives us

$$P_d = (21.2 \text{ W/m}^2)(2.0 \times 10^{-4} \text{ m}^2) = 4.2 \text{ mW}. \quad \text{(Answer)}$$

WILEY PLUS Additional examples, video, and practice available at *WileyPLUS*

Sample Problem 17.04 Decibels, sound level, change in intensity

Many veteran rockers suffer from acute hearing damage because of the high sound levels they endured for years. Many rockers now wear special earplugs to protect their hearing during performances (Fig. 17-11). If an earplug decreases the sound level of the sound waves by 20 dB, what is the ratio of the final intensity I_f of the waves to their initial intensity I_i?

Figure 17-11 Lars Ulrich of Metallica is an advocate for the organization HEAR (Hearing Education and Awareness for Rockers), which warns about the damage high sound levels can have on hearing.

KEY IDEA

For both the final and initial waves, the sound level β is related to the intensity by the definition of sound level in Eq. 17-29.

Calculations: For the final waves we have

$$\beta_f = (10\text{ dB})\log\frac{I_f}{I_0},$$

and for the initial waves we have

$$\beta_i = (10\text{ dB})\log\frac{I_i}{I_0}.$$

The difference in the sound levels is

$$\beta_f - \beta_i = (10\text{ dB})\left(\log\frac{I_f}{I_0} - \log\frac{I_i}{I_0}\right). \qquad (17\text{-}36)$$

Using the identity

$$\log\frac{a}{b} - \log\frac{c}{d} = \log\frac{ad}{bc},$$

we can rewrite Eq. 17-36 as

$$\beta_f - \beta_i = (10\text{ dB})\log\frac{I_f}{I_i}. \qquad (17\text{-}37)$$

Rearranging and then substituting the given decrease in sound level as $\beta_f - \beta_i = -20$ dB, we find

$$\log\frac{I_f}{I_i} = \frac{\beta_f - \beta_i}{10\text{ dB}} = \frac{-20\text{ dB}}{10\text{ dB}} = -2.0.$$

We next take the antilog of the far left and far right sides of this equation. (Although the antilog $10^{-2.0}$ can be evaluated mentally, you could use a calculator by keying in 10^-2.0 or using the 10^x key.) We find

$$\frac{I_f}{I_i} = \log^{-1}(-2.0) = 0.010. \qquad \text{(Answer)}$$

Thus, the earplug reduces the intensity of the sound waves to 0.010 of their initial intensity (two orders of magnitude).

17-5 SOURCES OF MUSICAL SOUND

Learning Objectives

After reading this module, you should be able to . . .

17.24 Using standing wave patterns for string waves, sketch the standing wave patterns for the first several acoustical harmonics of a pipe with only one open end and with two open ends.

17.25 For a standing wave of sound, relate the distance between nodes and the wavelength.

17.26 Identify which type of pipe has even harmonics.

17.27 For any given harmonic and for a pipe with only one open end or with two open ends, apply the relationships between the pipe length L, the speed of sound v, the wavelength λ, the harmonic frequency f, and the harmonic number n.

Key Ideas

- Standing sound wave patterns can be set up in pipes (that is, resonance can be set up) if sound of the proper wavelength is introduced in the pipe.
- A pipe open at both ends will resonate at frequencies

$$f = \frac{v}{\lambda} = \frac{nv}{2L}, \qquad n = 1, 2, 3, \ldots,$$

where v is the speed of sound in the air in the pipe.

- For a pipe closed at one end and open at the other, the resonant frequencies are

$$f = \frac{v}{\lambda} = \frac{nv}{4L}, \qquad n = 1, 3, 5, \ldots.$$

Sources of Musical Sound

Musical sounds can be set up by oscillating strings (guitar, piano, violin), membranes (kettledrum, snare drum), air columns (flute, oboe, pipe organ, and the didgeridoo of Fig. 17-12), wooden blocks or steel bars (marimba, xylophone), and many other oscillating bodies. Most common instruments involve more than a single oscillating part.

Recall from Chapter 16 that standing waves can be set up on a stretched string that is fixed at both ends. They arise because waves traveling along the string are reflected back onto the string at each end. If the wavelength of the waves is suitably matched to the length of the string, the superposition of waves traveling in opposite directions produces a standing wave pattern (or oscillation mode). The wavelength required of the waves for such a match is one that corresponds to a *resonant frequency* of the string. The advantage of setting up standing waves is that the string then oscillates with a large, sustained amplitude, pushing back and forth against the surrounding air and thus generating a noticeable sound wave with the same frequency as the oscillations of the string. This production of sound is of obvious importance to, say, a guitarist.

Sound Waves. We can set up standing waves of sound in an air-filled pipe in a similar way. As sound waves travel through the air in the pipe, they are reflected at each end and travel back through the pipe. (The reflection occurs even if an end is open, but the reflection is not as complete as when the end is closed.) If the wavelength of the sound waves is suitably matched to the length of the pipe, the superposition of waves traveling in opposite directions through the pipe sets up a standing wave pattern. The wavelength required of the sound waves for such a match is one that corresponds to a resonant frequency of the pipe. The advantage of such a standing wave is that the air in the pipe oscillates with a large, sustained amplitude, emitting at any open end a sound wave that has the same frequency as the oscillations in the pipe. This emission of sound is of obvious importance to, say, an organist.

Many other aspects of standing sound wave patterns are similar to those of string waves: The closed end of a pipe is like the fixed end of a string in that there must be a node (zero displacement) there, and the open end of a pipe is like the end of a string attached to a freely moving ring, as in Fig. 16-19*b*, in that there must be an antinode there. (Actually, the antinode for the open end of a pipe is located slightly beyond the end, but we shall not dwell on that detail.)

Two Open Ends. The simplest standing wave pattern that can be set up in a pipe with two open ends is shown in Fig. 17-13*a*. There is an antinode across each

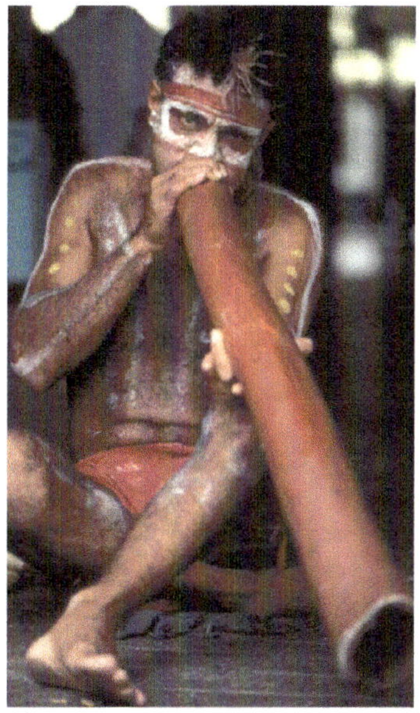

Alamy

Figure 17-12 The air column within a didgeridoo ("a pipe") oscillates when the instrument is played.

Antinodes (maximum oscillation) occur at the open ends.

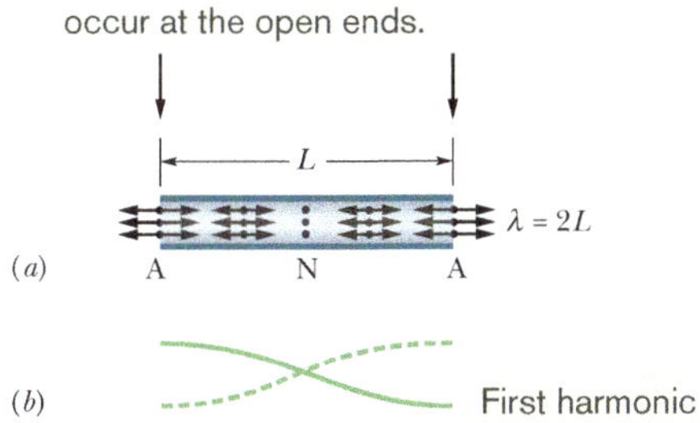

Figure 17-13 (*a*) The simplest standing wave pattern of displacement for (longitudinal) sound waves in a pipe with both ends open has an antinode (A) across each end and a node (N) across the middle. (The longitudinal displacements represented by the double arrows are greatly exaggerated.) (*b*) The corresponding standing wave pattern for (transverse) string waves.

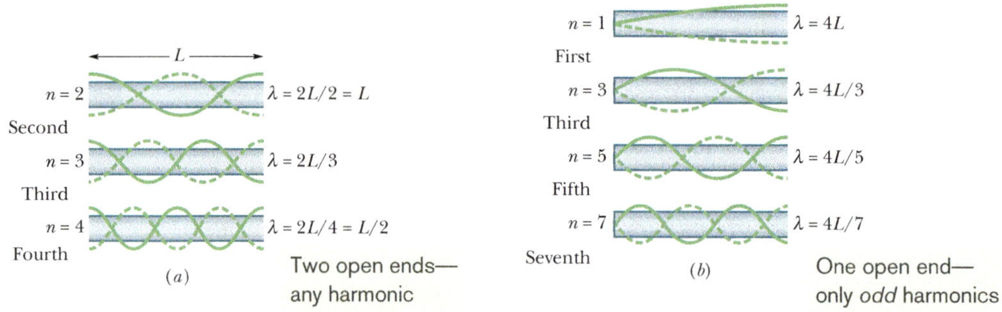

Figure 17-14 Standing wave patterns for string waves superimposed on pipes to represent standing sound wave patterns in the pipes. (*a*) With *both* ends of the pipe open, any harmonic can be set up in the pipe. (*b*) With only *one* end open, only odd harmonics can be set up.

open end, as required. There is also a node across the middle of the pipe. An easier way of representing this standing longitudinal sound wave is shown in Fig. 17-13*b*—by drawing it as a standing transverse string wave.

The standing wave pattern of Fig. 17-13*a* is called the *fundamental mode* or *first harmonic*. For it to be set up, the sound waves in a pipe of length L must have a wavelength given by $L = \lambda/2$, so that $\lambda = 2L$. Several more standing sound wave patterns for a pipe with two open ends are shown in Fig. 17-14*a* using string wave representations. The *second harmonic* requires sound waves of wavelength $\lambda = L$, the *third harmonic* requires wavelength $\lambda = 2L/3$, and so on.

More generally, the resonant frequencies for a pipe of length L with two open ends correspond to the wavelengths

$$\lambda = \frac{2L}{n}, \qquad \text{for } n = 1, 2, 3, \ldots, \tag{17-38}$$

where n is called the *harmonic number*. Letting v be the speed of sound, we write the resonant frequencies for a pipe with two open ends as

$$f = \frac{v}{\lambda} = \frac{nv}{2L}, \qquad \text{for } n = 1, 2, 3, \ldots \quad \text{(pipe, two open ends).} \tag{17-39}$$

One Open End. Figure 17-14*b* shows (using string wave representations) some of the standing sound wave patterns that can be set up in a pipe with only one open end. As required, across the open end there is an antinode and across the closed end there is a node. The simplest pattern requires sound waves having a wavelength given by $L = \lambda/4$, so that $\lambda = 4L$. The next simplest pattern requires a wavelength given by $L = 3\lambda/4$, so that $\lambda = 4L/3$, and so on.

More generally, the resonant frequencies for a pipe of length L with only one open end correspond to the wavelengths

$$\lambda = \frac{4L}{n}, \qquad \text{for } n = 1, 3, 5, \ldots, \tag{17-40}$$

in which the harmonic number n *must be an odd number*. The resonant frequencies are then given by

$$f = \frac{v}{\lambda} = \frac{nv}{4L}, \qquad \text{for } n = 1, 3, 5, \ldots \quad \text{(pipe, one open end).} \tag{17-41}$$

Note again that only odd harmonics can exist in a pipe with one open end. For example, the second harmonic, with $n = 2$, cannot be set up in such a pipe. Note also that for such a pipe the adjective in a phrase such as "the third harmonic" still refers to the harmonic number n (and not to, say, the third possible harmonic). Finally note that Eqs. 17-38 and 17-39 for two open ends contain the

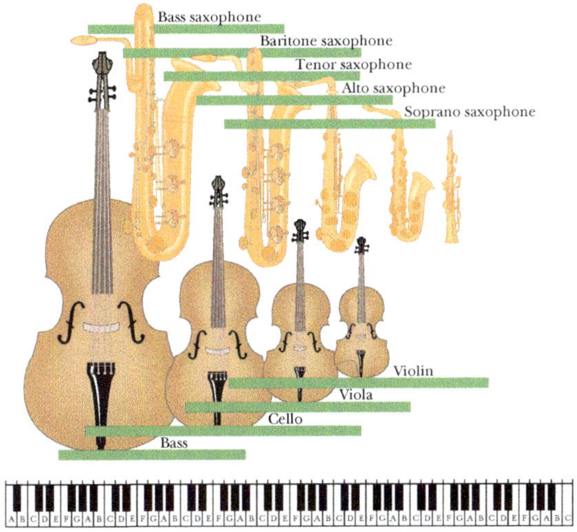

Figure 17-15 The saxophone and violin families, showing the relations between instrument length and frequency range. The frequency range of each instrument is indicated by a horizontal bar along a frequency scale suggested by the keyboard at the bottom; the frequency increases toward the right.

number 2 and any integer value of n, but Eqs. 17-40 and 17-41 for one open end contain the number 4 and only odd values of n.

Length. The length of a musical instrument reflects the range of frequencies over which the instrument is designed to function, and smaller length implies higher frequencies, as we can tell from Eq. 16-66 for string instruments and Eqs. 17-39 and 17-41 for instruments with air columns. Figure 17-15, for example, shows the saxophone and violin families, with their frequency ranges suggested by the piano keyboard. Note that, for every instrument, there is overlap with its higher- and lower-frequency neighbors.

Net Wave. In any oscillating system that gives rise to a musical sound, whether it is a violin string or the air in an organ pipe, the fundamental and one or more of the higher harmonics are usually generated simultaneously. Thus, you hear them together—that is, superimposed as a net wave. When different instruments are played at the same note, they produce the same fundamental frequency but different intensities for the higher harmonics. For example, the fourth harmonic of middle C might be relatively loud on one instrument and relatively quiet or even missing on another. Thus, because different instruments produce different net waves, they sound different to you even when they are played at the same note. That would be the case for the two net waves shown in Fig. 17-16, which were produced at the same note by different instruments. If you heard only the fundamentals, the music would not be musical.

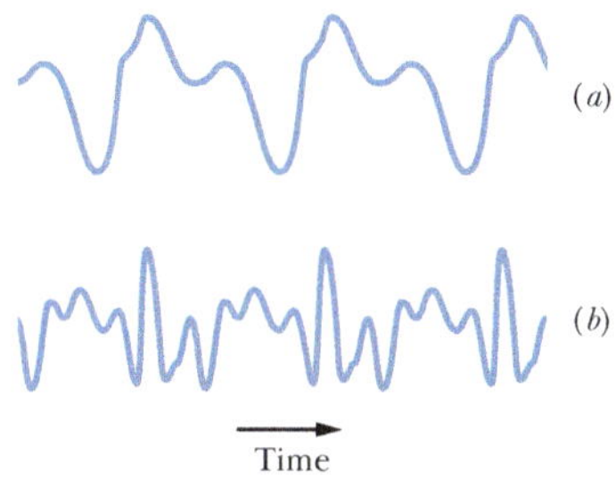

Figure 17-16 The wave forms produced by (*a*) a flute and (*b*) an oboe when played at the same note, with the same first harmonic frequency.

Checkpoint 3

Pipe A, with length L, and pipe B, with length $2L$, both have two open ends. Which harmonic of pipe B has the same frequency as the fundamental of pipe A?

Sample Problem 17.05 Resonance between pipes of different lengths

Pipe A is open at both ends and has length $L_A = 0.343$ m. We want to place it near three other pipes in which standing waves have been set up, so that the sound can set up a standing wave in pipe A. Those other three pipes are each closed at one end and have lengths $L_B = 0.500L_A$, $L_C = 0.250L_A$, and $L_D = 2.00L_A$. For each of these three pipes, which of their harmonics can excite a harmonic in pipe A?

KEY IDEAS

(1) The sound from one pipe can set up a standing wave in another pipe only if the harmonic frequencies match. (2) Equation 17-39 gives the harmonic frequencies in a pipe with two open ends (a symmetric pipe) as $f = nv/2L$, for $n = 1, 2, 3, \dots$, that is, for any positive integer. (3) Equation

17-41 gives the harmonic frequencies in a pipe with only one open end (an asymmetric pipe) as $f = nv/4L$, for $n = 1, 3, 5, \ldots$, that is, for only odd positive integers.

Pipe A: Let's first find the resonant frequencies of symmetric pipe A (with two open ends) by evaluating Eq. 17-39:

$$f_A = \frac{n_A v}{2L_A} = \frac{n_A(343 \text{ m/s})}{2(0.343 \text{ m})}$$

$$= n_A(500 \text{ Hz}) = n_A(0.50 \text{ kHz}), \quad \text{for } n_A = 1, 2, 3, \ldots.$$

The first six harmonic frequencies are shown in the top plot in Fig. 17-17.

Pipe B: Next let's find the resonant frequencies of asymmetric pipe B (with only one open end) by evaluating Eq. 17-41, being careful to use only odd integers for the harmonic numbers:

$$f_B = \frac{n_B v}{4L_B} = \frac{n_B v}{4(0.500L_A)} = \frac{n_B(343 \text{ m/s})}{2(0.343 \text{ m})}$$

$$= n_B(500 \text{ Hz}) = n_B(0.500 \text{ kHz}), \quad \text{for } n_B = 1, 3, 5, \ldots.$$

Comparing our two results, we see that we get a match for each choice of n_B:

$$f_A = f_B \quad \text{for } n_A = n_B \quad \text{with } n_B = 1, 3, 5, \ldots. \quad \text{(Answer)}$$

For example, as shown in Fig. 17-17, if we set up the fifth harmonic in pipe B and bring the pipe close to pipe A, the fifth harmonic will then be set up in pipe A. However, no harmonic in B can set up an even harmonic in A.

Pipe C: Let's continue with pipe C (with only one end) by writing Eq. 17-41 as

$$f_C = \frac{n_C v}{4L_C} = \frac{n_C v}{4(0.250L_A)} = \frac{n_C(343 \text{ m/s})}{0.343 \text{ m/s}}$$

$$= n_C(1000 \text{ Hz}) = n_C(1.00 \text{ kHz}), \quad \text{for } n_C = 1, 3, 5, \ldots.$$

From this we see that C can excite some of the harmonics of A but only those with harmonic numbers n_A that are twice an odd integer:

$$f_A = f_C \quad \text{for } n_A = 2n_C, \quad \text{with } n_C = 1, 3, 5, \ldots. \quad \text{(Answer)}$$

Pipe D: Finally, let's check D with our same procedure:

$$f_D = \frac{n_D v}{4L_D} = \frac{n_D v}{4(2L_A)} = \frac{n_D(343 \text{ m/s})}{8(0.343 \text{ m/s})}$$

$$= n_D(125 \text{ Hz}) = n_D(0.125 \text{ kHz}), \quad \text{for } n_D = 1, 3, 5, \ldots.$$

As shown in Fig. 17-17, none of these frequencies match a harmonic frequency of A. (Can you see that we would get a match if $n_D = 4n_A$? But that is impossible because $4n_A$ cannot yield an odd integer, as required of n_D.) Thus D cannot set up a standing wave in A.

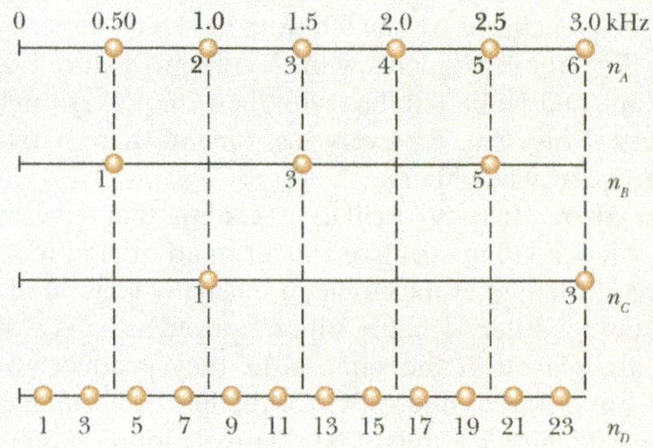

Figure 17-17 Harmonic frequencies of four pipes.

WILEY PLUS Additional examples, video, and practice available at *WileyPLUS*

17-6 BEATS

Learning Objectives

After reading this module, you should be able to . . .

17.28 Explain how beats are produced.

17.29 Add the displacement equations for two sound waves of the same amplitude and slightly different angular frequencies to find the displacement equation of the resultant wave and identify the time-varying amplitude.

17.30 Apply the relationship between the beat frequency and the frequencies of two sound waves that have the same amplitude when the frequencies (or, equivalently, the angular frequencies) differ by a small amount.

Key Idea

- Beats arise when two waves having slightly different frequencies, f_1 and f_2, are detected together. The beat frequency is

$$f_{\text{beat}} = f_1 - f_2.$$

Beats

If we listen, a few minutes apart, to two sounds whose frequencies are, say, 552 and 564 Hz, most of us cannot tell one from the other because the frequencies are so close to each other. However, if the sounds reach our ears simultaneously, what we hear is a sound whose frequency is 558 Hz, the *average* of the two combining frequencies. We also hear a striking variation in the intensity of this sound—it increases and decreases in slow, wavering **beats** that repeat at a frequency of 12 Hz, the *difference* between the two combining frequencies. Figure 17-18 shows this beat phenomenon.

Let the time-dependent variations of the displacements due to two sound waves of equal amplitude s_m be

$$s_1 = s_m \cos \omega_1 t \qquad \text{and} \qquad s_2 = s_m \cos \omega_2 t, \tag{17-42}$$

where $\omega_1 > \omega_2$. From the superposition principle, the resultant displacement is the sum of the individual displacements:

$$s = s_1 + s_2 = s_m(\cos \omega_1 t + \cos \omega_2 t).$$

Using the trigonometric identity (see Appendix E)

$$\cos \alpha + \cos \beta = 2 \cos[\tfrac{1}{2}(\alpha - \beta)] \cos[\tfrac{1}{2}(\alpha + \beta)]$$

allows us to write the resultant displacement as

$$s = 2s_m \cos[\tfrac{1}{2}(\omega_1 - \omega_2)t] \cos[\tfrac{1}{2}(\omega_1 + \omega_2)t]. \tag{17-43}$$

If we write

$$\omega' = \tfrac{1}{2}(\omega_1 - \omega_2) \qquad \text{and} \qquad \omega = \tfrac{1}{2}(\omega_1 + \omega_2), \tag{17-44}$$

we can then write Eq. 17-43 as

$$s(t) = [2s_m \cos \omega' t] \cos \omega t. \tag{17-45}$$

We now assume that the angular frequencies ω_1 and ω_2 of the combining waves are almost equal, which means that $\omega \gg \omega'$ in Eq. 17-44. We can then regard Eq. 17-45 as a cosine function whose angular frequency is ω and whose amplitude (which is not constant but varies with angular frequency ω') is the absolute value of the quantity in the brackets.

A maximum amplitude will occur whenever $\cos \omega' t$ in Eq. 17-45 has the value $+1$ or -1, which happens twice in each repetition of the cosine function. Because $\cos \omega' t$ has angular frequency ω', the angular frequency ω_{beat} at which beats occur is $\omega_{\text{beat}} = 2\omega'$. Then, with the aid of Eq. 17-44, we can write the beat angular frequency as

$$\omega_{\text{beat}} = 2\omega' = (2)(\tfrac{1}{2})(\omega_1 - \omega_2) = \omega_1 - \omega_2.$$

Because $\omega = 2\pi f$, we can recast this as

$$f_{\text{beat}} = f_1 - f_2 \quad \text{(beat frequency)}. \tag{17-46}$$

Musicians use the beat phenomenon in tuning instruments. If an instrument is sounded against a standard frequency (for example, the note called "concert A" played on an orchestra's first oboe) and tuned until the beat disappears, the instrument is in tune with that standard. In musical Vienna, concert A (440 Hz) is available as a convenient telephone service for the city's many musicians.

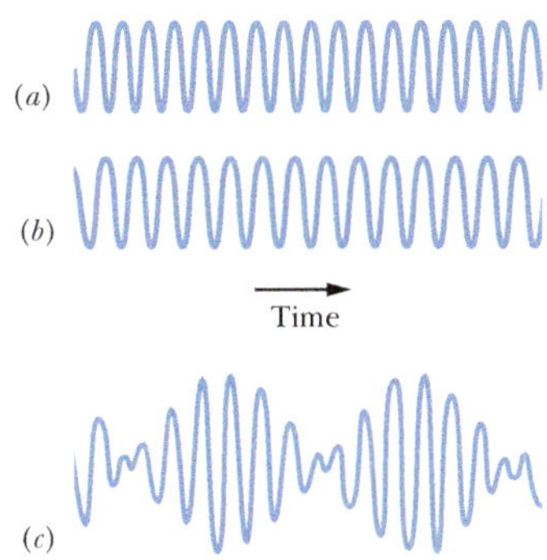

Figure 17-18 *(a, b)* The pressure variations Δp of two sound waves as they would be detected separately. The frequencies of the waves are nearly equal. *(c)* The resultant pressure variation if the two waves are detected simultaneously.

Sample Problem 17.06 Beat frequencies and penguins finding one another

When an emperor penguin returns from a search for food, how can it find its mate among the thousands of penguins huddled together for warmth in the harsh Antarctic weather? It is not by sight, because penguins all look alike, even to a penguin.

The answer lies in the way penguins vocalize. Most birds vocalize by using only one side of their two-sided vocal organ, called the *syrinx*. Emperor penguins, however, vocalize by using both sides simultaneously. Each side sets up acoustic standing waves in the bird's throat and mouth, much like in a pipe with two open ends. Suppose that the frequency of the first harmonic produced by side A is f_{A1} = 432 Hz and the frequency of the first harmonic produced by side B is f_{B1} = 371 Hz. What is the beat frequency between those two first-harmonic frequencies and between the two second-harmonic frequencies?

KEY IDEA

The beat frequency between two frequencies is their difference, as given by Eq. 17-46 ($f_{beat} = f_1 - f_2$).

Calculations: For the two first-harmonic frequencies f_{A1} and f_{B1}, the beat frequency is

$$f_{beat,1} = f_{A1} - f_{B1} = 432\text{ Hz} - 371\text{ Hz} = 61\text{ Hz.} \quad \text{(Answer)}$$

Because the standing waves in the penguin are effectively in a pipe with two open ends, the resonant frequencies are given by Eq. 17-39 ($f = nv/2L$), in which L is the (unknown) length of the effective pipe. The first-harmonic frequency is $f_1 = v/2L$, and the second-harmonic frequency is $f_2 = 2v/2L$. Comparing these two frequencies, we see that, in general,

$$f_2 = 2f_1.$$

For the penguin, the second harmonic of side A has frequency $f_{A2} = 2f_{A1}$ and the second harmonic of side B has frequency $f_{B2} = 2f_{B1}$. Using Eq. 17-46 with frequencies f_{A2} and f_{B2}, we find that the corresponding beat frequency associated with the second harmonics is

$$f_{beat,2} = f_{A2} - f_{B2} = 2f_{A1} - 2f_{B1} = 2(432\text{ Hz}) - 2(371\text{ Hz}) = 122\text{ Hz.} \quad \text{(Answer)}$$

Experiments indicate that penguins can perceive such large beat frequencies. (Humans cannot hear a beat frequency any higher than about 12 Hz — we perceive the two separate frequencies.) Thus, a penguin's cry can be rich with different harmonics and different beat frequencies, allowing the voice to be recognized even among the voices of thousands of other, closely huddled penguins.

WILEY PLUS Additional examples, video, and practice available at *WileyPLUS*

17-7 THE DOPPLER EFFECT

Learning Objectives

After reading this module, you should be able to . . .

17.31 Identify that the Doppler effect is the shift in the detected frequency from the frequency emitted by a sound source due to the relative motion between the source and the detector.

17.32 Identify that in calculating the Doppler shift in sound, the speeds are measured relative to the medium (such as air or water), which may be moving.

17.33 Calculate the shift in sound frequency for (a) a source moving either directly toward or away from a stationary detector, (b) a detector moving either directly toward or away from a stationary source, and (c) both source and detector moving either directly toward each other or directly away from each other.

17.34 Identify that for relative motion between a sound source and a sound detector, motion *toward* tends to shift the frequency up and motion *away* tends to shift it down.

Key Ideas

- The Doppler effect is a change in the observed frequency of a wave when the source or the detector moves relative to the transmitting medium (such as air). For sound the observed frequency f' is given in terms of the source frequency f by

$$f' = f\frac{v \pm v_D}{v \pm v_S} \quad \text{(general Doppler effect),}$$

where v_D is the speed of the detector relative to the medium, v_S is that of the source, and v is the speed of sound in the medium.

- The signs are chosen such that f' tends to be *greater* for relative motion toward (one of the objects moves toward the other) and *less* for motion away.

The Doppler Effect

A police car is parked by the side of the highway, sounding its 1000 Hz siren. If you are also parked by the highway, you will hear that same frequency. However, if there is relative motion between you and the police car, either toward or away from each other, you will hear a different frequency. For example, if you are driving *toward* the police car at 120 km/h (about 75 mi/h), you will hear a *higher* frequency (1096 Hz, an *increase* of 96 Hz). If you are driving *away from* the police car at that same speed, you will hear a *lower* frequency (904 Hz, a *decrease* of 96 Hz).

These motion-related frequency changes are examples of the **Doppler effect.** The effect was proposed (although not fully worked out) in 1842 by Austrian physicist Johann Christian Doppler. It was tested experimentally in 1845 by Buys Ballot in Holland, "using a locomotive drawing an open car with several trumpeters."

The Doppler effect holds not only for sound waves but also for electromagnetic waves, including microwaves, radio waves, and visible light. Here, however, we shall consider only sound waves, and we shall take as a reference frame the body of air through which these waves travel. This means that we shall measure the speeds of a source S of sound waves and a detector D of those waves *relative to that body of air.* (Unless otherwise stated, the body of air is stationary relative to the ground, so the speeds can also be measured relative to the ground.) We shall assume that S and D move either directly toward or directly away from each other, at speeds less than the speed of sound.

General Equation. If either the detector or the source is moving, or both are moving, the emitted frequency f and the detected frequency f' are related by

$$f' = f\frac{v \pm v_D}{v \pm v_S} \quad \text{(general Doppler effect)}, \tag{17-47}$$

where v is the speed of sound through the air, v_D is the detector's speed relative to the air, and v_S is the source's speed relative to the air. The choice of plus or minus signs is set by this rule:

When the motion of detector or source is toward the other, the sign on its speed must give an upward shift in frequency. When the motion of detector or source is away from the other, the sign on its speed must give a downward shift in frequency.

In short, *toward* means *shift up,* and *away* means *shift down.*

Here are some examples of the rule. If the detector moves toward the source, use the plus sign in the numerator of Eq. 17-47 to get a shift up in the frequency. If it moves away, use the minus sign in the numerator to get a shift down. If it is stationary, substitute 0 for v_D. If the source moves toward the detector, use the minus sign in the denominator of Eq. 17-47 to get a shift up in the frequency. If it moves away, use the plus sign in the denominator to get a shift down. If the source is stationary, substitute 0 for v_S.

Next, we derive equations for the Doppler effect for the following two specific situations and then derive Eq. 17-47 for the general situation.

1. When the detector moves relative to the air and the source is stationary relative to the air, the motion changes the frequency at which the detector intercepts wavefronts and thus changes the detected frequency of the sound wave.
2. When the source moves relative to the air and the detector is stationary relative to the air, the motion changes the wavelength of the sound wave and thus changes the detected frequency (recall that frequency is related to wavelength).

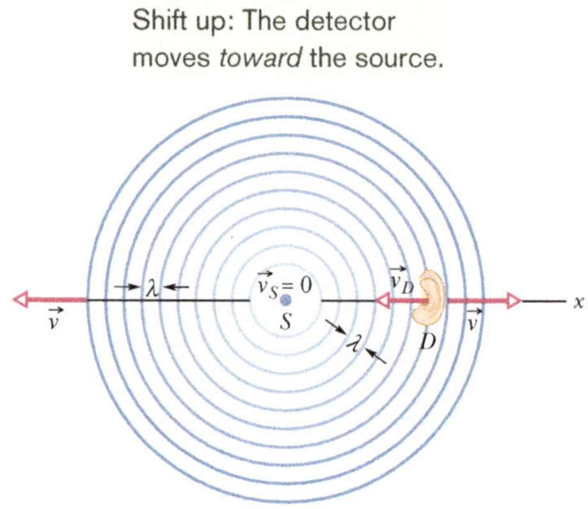

Figure 17-19 A stationary source of sound S emits spherical wavefronts, shown one wavelength apart, that expand outward at speed v. A sound detector D, represented by an ear, moves with velocity $\vec{v}_D$ toward the source. The detector senses a higher frequency because of its motion.

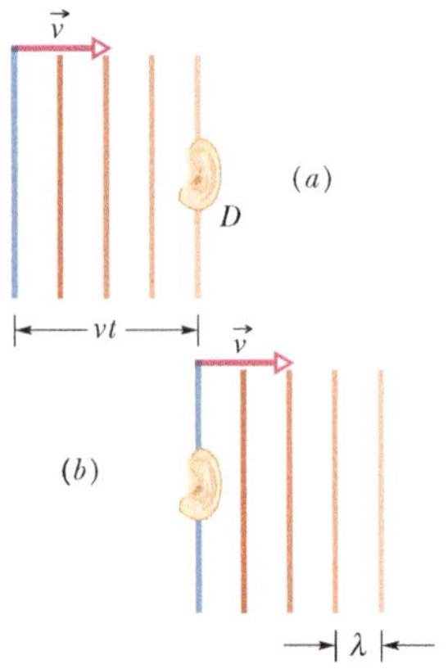

Figure 17-20 The wavefronts of Fig. 17-19, assumed planar, (a) reach and (b) pass a stationary detector D; they move a distance vt to the right in time t.

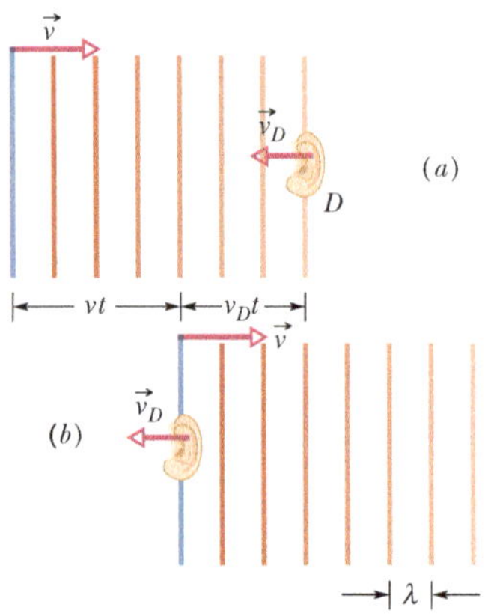

Figure 17-21 Wavefronts traveling to the right (a) reach and (b) pass detector D, which moves in the opposite direction. In time t, the wavefronts move a distance vt to the right and D moves a distance v_Dt to the left.

Detector Moving, Source Stationary

In Fig. 17-19, a detector D (represented by an ear) is moving at speed v_D toward a stationary source S that emits spherical wavefronts, of wavelength λ and frequency f, moving at the speed v of sound in air. The wavefronts are drawn one wavelength apart. The frequency detected by detector D is the rate at which D intercepts wavefronts (or individual wavelengths). If D were stationary, that rate would be f, but since D is moving into the wavefronts, the rate of interception is greater, and thus the detected frequency f' is greater than f.

Let us for the moment consider the situation in which D is stationary (Fig. 17-20). In time t, the wavefronts move to the right a distance vt. The number of wavelengths in that distance vt is the number of wavelengths intercepted by D in time t, and that number is vt/λ. The rate at which D intercepts wavelengths, which is the frequency f detected by D, is

$$f = \frac{vt/\lambda}{t} = \frac{v}{\lambda}. \tag{17-48}$$

In this situation, with D stationary, there is no Doppler effect—the frequency detected by D is the frequency emitted by S.

Now let us again consider the situation in which D moves in the direction opposite the wavefront velocity (Fig. 17-21). In time t, the wavefronts move to the right a distance vt as previously, but now D moves to the left a distance v_Dt. Thus, in this time t, the distance moved by the wavefronts relative to D is $vt + v_Dt$. The number of wavelengths in this relative distance $vt + v_Dt$ is the number of wavelengths intercepted by D in time t and is $(vt + v_Dt)/\lambda$. The *rate* at which D intercepts wavelengths in this situation is the frequency f', given by

$$f' = \frac{(vt + v_Dt)/\lambda}{t} = \frac{v + v_D}{\lambda}. \tag{17-49}$$

From Eq. 17-48, we have $\lambda = v/f$. Then Eq. 17-49 becomes

$$f' = \frac{v + v_D}{v/f} = f\frac{v + v_D}{v}. \tag{17-50}$$

Note that in Eq. 17-50, $f' > f$ unless $v_D = 0$ (the detector is stationary).

Similarly, we can find the frequency detected by D if D moves away from the source. In this situation, the wavefronts move a distance $vt - v_Dt$ relative to D in time t, and f' is given by

$$f' = f\frac{v - v_D}{v}. \tag{17-51}$$

In Eq. 17-51, $f' < f$ unless $v_D = 0$. We can summarize Eqs. 17-50 and 17-51 with

$$f' = f\frac{v \pm v_D}{v} \quad \text{(detector moving, source stationary).} \tag{17-52}$$

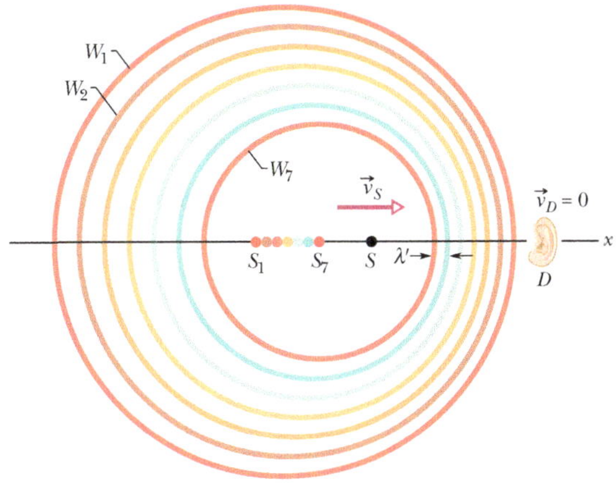

Figure 17-22 A detector D is stationary, and a source S is moving toward it at speed v_S. Wavefront W_1 was emitted when the source was at S_1, wavefront W_7 when it was at S_7. At the moment depicted, the source is at S. The detector senses a higher frequency because the moving source, chasing its own wavefronts, emits a reduced wavelength λ' in the direction of its motion.

Source Moving, Detector Stationary

Let detector D be stationary with respect to the body of air, and let source S move toward D at speed v_S (Fig. 17-22). The motion of S changes the wavelength of the sound waves it emits and thus the frequency detected by D.

To see this change, let T $(= 1/f)$ be the time between the emission of any pair of successive wavefronts W_1 and W_2. During T, wavefront W_1 moves a distance vT and the source moves a distance v_ST. At the end of T, wavefront W_2 is emitted. In the direction in which S moves, the distance between W_1 and W_2, which is the wavelength λ' of the waves moving in that direction, is $vT - v_ST$. If D detects those waves, it detects frequency f' given by

$$f' = \frac{v}{\lambda'} = \frac{v}{vT - v_ST} = \frac{v}{v/f - v_S/f}$$

$$= f\frac{v}{v - v_S}. \tag{17-53}$$

Note that f' must be greater than f unless $v_S = 0$.

In the direction opposite that taken by S, the wavelength λ' of the waves is again the distance between successive waves but now that distance is $vT + v_ST$. If D detects those waves, it detects frequency f' given by

$$f' = f\frac{v}{v + v_S}. \tag{17-54}$$

Now f' must be less than f unless $v_S = 0$.

We can summarize Eqs. 17-53 and 17-54 with

$$f' = f\frac{v}{v \pm v_S} \quad \text{(source moving, detector stationary).} \tag{17-55}$$

General Doppler Effect Equation

We can now derive the general Doppler effect equation by replacing f in Eq. 17-55 (the source frequency) with f' of Eq. 17-52 (the frequency associated with motion of the detector). That simple replacement gives us Eq. 17-47 for the general Doppler effect. That general equation holds not only when both detector and source are moving but also in the two specific situations we just discussed. For the situation in which the detector is moving and the source is stationary, substitution of $v_S = 0$ into Eq. 17-47 gives us Eq. 17-52, which we previously found. For the situation in which the source is moving and the detector is stationary, substitution of $v_D = 0$ into Eq. 17-47 gives us Eq. 17-55, which we previously found. Thus, Eq. 17-47 is the equation to remember.

Checkpoint 4

The figure indicates the directions of motion of a sound source and a detector for six situations in stationary air. For each situation, is the detected frequency greater than or less than the emitted frequency, or can't we tell without more information about the actual speeds?

	Source	Detector		Source	Detector
(a)	→	• 0 speed	(d)	←	←
(b)	←	• 0 speed	(e)	→	←
(c)	→	→	(f)	←	→

Sample Problem 17.07 Double Doppler shift in the echoes used by bats

Bats navigate and search out prey by emitting, and then detecting reflections of, ultrasonic waves, which are sound waves with frequencies greater than can be heard by a human. Suppose a bat emits ultrasound at frequency $f_{be} = 82.52$ kHz while flying with velocity $\vec{v}_b = (9.00 \text{ m/s})\hat{\text{i}}$ as it chases a moth that flies with velocity $\vec{v}_m = (8.00 \text{ m/s})\hat{\text{i}}$. What frequency f_{md} does the moth detect? What frequency f_{bd} does the bat detect in the returning echo from the moth?

KEY IDEAS

The frequency is shifted by the relative motion of the bat and moth. Because they move along a single axis, the shifted frequency is given by Eq. 17-47. Motion *toward* tends to shift the frequency *up*, and motion *away* tends to shift it *down*.

Detection by moth: The general Doppler equation is

$$f' = f\frac{v \pm v_D}{v \pm v_S}. \tag{17-56}$$

Here, the detected frequency f' that we want to find is the frequency f_{md} detected by the moth. On the right side, the emitted frequency f is the bat's emission frequency $f_{be} = 82.52$ kHz, the speed of sound is $v = 343$ m/s, the speed v_D of the detector is the moth's speed $v_m = 8.00$ m/s, and the speed v_S of the source is the bat's speed $v_b = 9.00$ m/s.

The decisions about the plus and minus signs can be tricky. Think in terms of *toward* and *away*. We have the speed of the moth (the detector) in the numerator of Eq. 17-56. The moth moves *away* from the bat, which tends to lower the detected frequency. Because the speed is in the numerator, we choose the minus sign to meet that tendency (the numerator becomes smaller). These reasoning steps are shown in Table 17-3.

We have the speed of the bat in the denominator of Eq. 17-56. The bat moves *toward* the moth, which tends to increase the detected frequency. Because the speed is in the denominator, we choose the minus sign to meet that tendency (the denominator becomes smaller).

With these substitutions and decisions, we have

$$\begin{aligned} f_{md} &= f_{be}\frac{v - v_m}{v - v_b} \\ &= (82.52 \text{ kHz})\frac{343 \text{ m/s} - 8.00 \text{ m/s}}{343 \text{ m/s} - 9.00 \text{ m/s}} \\ &= 82.767 \text{ kHz} \approx 82.8 \text{ kHz}. \qquad \text{(Answer)} \end{aligned}$$

Detection of echo by bat: In the echo back to the bat, the moth acts as a source of sound, emitting at the frequency f_{md} we just calculated. So now the moth is the source (moving *away*) and the bat is the detector (moving *toward*). The reasoning steps are shown in Table 17-3. To find the frequency f_{bd} detected by the bat, we write Eq. 17-56 as

$$\begin{aligned} f_{bd} &= f_{md}\frac{v + v_b}{v + v_m} \\ &= (82.767 \text{ kHz})\frac{343 \text{ m/s} + 9.00 \text{ m/s}}{343 \text{ m/s} + 8.00 \text{ m/s}} \\ &= 83.00 \text{ kHz} \approx 83.0 \text{ kHz}. \qquad \text{(Answer)} \end{aligned}$$

Some moths evade bats by "jamming" the detection system with ultrasonic clicks.

Table 17-3

Bat to Moth		Echo Back to Bat	
Detector	Source	Detector	Source
moth	bat	bat	moth
speed $v_D = v_m$	speed $v_S = v_b$	speed $v_D = v_b$	speed $v_S = v_m$
away	toward	toward	away
shift down	shift up	shift up	shift down
numerator	denominator	numerator	denominator
minus	minus	plus	plus

WileyPLUS Additional examples, video, and practice available at *WileyPLUS*

17-8 SUPERSONIC SPEEDS, SHOCK WAVES

Learning Objectives

After reading this module, you should be able to . . .

17.35 Sketch the bunching of wavefronts for a sound source traveling at the speed of sound or faster.

17.36 Calculate the Mach number for a sound source exceeding the speed of sound.

17.37 For a sound source exceeding the speed of sound, apply the relationship between the Mach cone angle, the speed of sound, and the speed of the source.

Key Idea

- If the speed of a source relative to the medium exceeds the speed of sound in the medium, the Doppler equation no longer applies. In such a case, shock waves result. The half-angle θ of the Mach cone is given by

$$\sin\theta = \frac{v}{v_S} \quad \text{(Mach cone angle).}$$

Supersonic Speeds, Shock Waves

If a source is moving toward a stationary detector at a speed v_S equal to the speed of sound v, Eqs. 17-47 and 17-55 predict that the detected frequency f' will be infinitely great. This means that the source is moving so fast that it keeps pace with its own spherical wavefronts (Fig. 17-23*a*). What happens when $v_S > v$? For such *supersonic* speeds, Eqs. 17-47 and 17-55 no longer apply. Figure 17-23*b* depicts the spherical wavefronts that originated at various positions of the source. The radius of any wavefront is vt, where t is the time that has elapsed since the source emitted that wavefront. Note that all the wavefronts bunch along a V-shaped envelope in this two-dimensional drawing. The wavefronts actually extend in three dimensions, and the bunching actually forms a cone called the *Mach cone*. A *shock wave* exists along the surface of this cone, because the bunching of wavefronts causes an abrupt rise and fall of air pressure as the surface passes through any point. From Fig. 17-23*b*, we see that the half-angle θ of the cone (the *Mach cone angle*) is given by

$$\sin\theta = \frac{vt}{v_S t} = \frac{v}{v_S} \quad \text{(Mach cone angle).} \tag{17-57}$$

The ratio v_S/v is the *Mach number*. If a plane flies at Mach 2.3, its speed is 2.3 times the speed of sound in the air through which the plane is flying. The shock wave generated by a supersonic aircraft (Fig. 17-24)

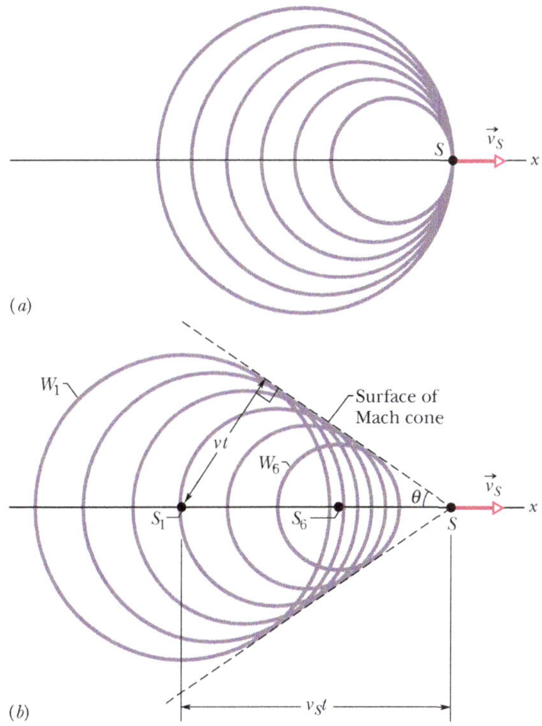

Figure 17-23 (*a*) A source of sound S moves at speed v_S equal to the speed of sound and thus as fast as the wavefronts it generates. (*b*) A source S moves at speed v_S faster than the speed of sound and thus faster than the wavefronts. When the source was at position S_1 it generated wavefront W_1, and at position S_6 it generated W_6. All the spherical wavefronts expand at the speed of sound v and bunch along the surface of a cone called the Mach cone, forming a shock wave. The surface of the cone has half-angle θ and is tangent to all the wavefronts.

U.S. Navy photo by Ensign John Gay

Figure 17-24 Shock waves produced by the wings of a Navy FA 18 jet. The shock waves are visible because the sudden decrease in air pressure in them caused water molecules in the air to condense, forming a fog.

or projectile produces a burst of sound, called a *sonic boom,* in which the air pressure first suddenly increases and then suddenly decreases below normal before returning to normal. Part of the sound that is heard when a rifle is fired is the sonic boom produced by the bullet. When a long bull whip is snapped, its tip is moving faster than sound and produces a small sonic boom—the *crack* of the whip.

Review & Summary

Sound Waves Sound waves are longitudinal mechanical waves that can travel through solids, liquids, or gases. The speed v of a sound wave in a medium having **bulk modulus** B and density ρ is

$$v = \sqrt{\frac{B}{\rho}} \quad \text{(speed of sound).} \tag{17-3}$$

In air at 20°C, the speed of sound is 343 m/s.

A sound wave causes a longitudinal displacement s of a mass element in a medium as given by

$$s = s_m \cos(kx - \omega t), \tag{17-12}$$

where s_m is the **displacement amplitude** (maximum displacement) from equilibrium, $k = 2\pi/\lambda$, and $\omega = 2\pi f$, λ and f being the wavelength and frequency of the sound wave. The wave also causes a pressure change Δp from the equilibrium pressure:

$$\Delta p = \Delta p_m \sin(kx - \omega t), \tag{17-13}$$

where the **pressure amplitude** is

$$\Delta p_m = (v\rho\omega)s_m. \tag{17-14}$$

Interference The interference of two sound waves with identical wavelengths passing through a common point depends on their phase difference ϕ there. If the sound waves were emitted in phase and are traveling in approximately the same direction, ϕ is given by

$$\phi = \frac{\Delta L}{\lambda} 2\pi, \tag{17-21}$$

where ΔL is their **path length difference** (the difference in the distances traveled by the waves to reach the common point). Fully constructive interference occurs when ϕ is an integer multiple of 2π,

$$\phi = m(2\pi), \quad \text{for } m = 0, 1, 2, \ldots, \tag{17-22}$$

and, equivalently, when ΔL is related to wavelength λ by

$$\frac{\Delta L}{\lambda} = 0, 1, 2, \ldots. \tag{17-23}$$

Fully destructive interference occurs when ϕ is an odd multiple of π,

$$\phi = (2m + 1)\pi, \quad \text{for } m = 0, 1, 2, \ldots, \tag{17-24}$$

and, equivalently, when ΔL is related to λ by

$$\frac{\Delta L}{\lambda} = 0.5, 1.5, 2.5, \ldots. \tag{17-25}$$

Sound Intensity The **intensity** I of a sound wave at a surface is the average rate per unit area at which energy is transferred by the wave through or onto the surface:

$$I = \frac{P}{A}, \tag{17-26}$$

where P is the time rate of energy transfer (power) of the sound wave and A is the area of the surface intercepting the sound. The intensity I is related to the displacement amplitude s_m of the sound wave by

$$I = \tfrac{1}{2}\rho v \omega^2 s_m^2. \tag{17-27}$$

The intensity at a distance r from a point source that emits sound waves of power P_s is

$$I = \frac{P_s}{4\pi r^2}. \tag{17-28}$$

Sound Level in Decibels The *sound level* β in *decibels* (dB) is defined as

$$\beta = (10 \text{ dB}) \log \frac{I}{I_0}, \tag{17-29}$$

where I_0 ($= 10^{-12}$ W/m^2) is a reference intensity level to which all intensities are compared. For every factor-of-10 increase in intensity, 10 dB is added to the sound level.

Standing Wave Patterns in Pipes Standing sound wave patterns can be set up in pipes. A pipe open at both ends will resonate at frequencies

$$f = \frac{v}{\lambda} = \frac{nv}{2L}, \quad n = 1, 2, 3, \ldots, \tag{17-39}$$

where v is the speed of sound in the air in the pipe. For a pipe closed at one end and open at the other, the resonant frequencies are

$$f = \frac{v}{\lambda} = \frac{nv}{4L}, \quad n = 1, 3, 5, \ldots. \tag{17-41}$$

Beats *Beats* arise when two waves having slightly different frequencies, f_1 and f_2, are detected together. The beat frequency is

$$f_{\text{beat}} = f_1 - f_2. \tag{17-46}$$

The Doppler Effect The *Doppler effect* is a change in the observed frequency of a wave when the source or the detector moves relative to the transmitting medium (such as air). For sound the observed frequency f' is given in terms of the source frequency f by

$$f' = f \frac{v \pm v_D}{v \pm v_S} \quad \text{(general Doppler effect),} \tag{17-47}$$

where v_D is the speed of the detector relative to the medium, v_S is that of the source, and v is the speed of sound in the medium. The signs are chosen such that f' tends to be *greater* for motion toward and *less* for motion away.

Shock Wave If the speed of a source relative to the medium exceeds the speed of sound in the medium, the Doppler equation no longer applies. In such a case, shock waves result. The half-angle θ of the Mach cone is given by

$$\sin\theta = \frac{v}{v_S} \quad \text{(Mach cone angle).} \tag{17-57}$$

Questions

1 In a first experiment, a sinusoidal sound wave is sent through a long tube of air, transporting energy at the average rate of $P_{avg,1}$. In a second experiment, two other sound waves, identical to the first one, are to be sent simultaneously through the tube with a phase difference ϕ of either 0, 0.2 wavelength, or 0.5 wavelength between the waves. (a) With only mental calculation, rank those choices of ϕ according to the average rate at which the waves will transport energy, greatest first. (b) For the first choice of ϕ, what is the average rate in terms of $P_{avg,1}$?

2 In Fig. 17-25, two point sources S_1 and S_2, which are in phase, emit identical sound waves of wavelength 2.0 m. In terms of wavelengths, what is the phase difference between the waves arriving at point P if (a) $L_1 = 38$ m and $L_2 = 34$ m, and (b) $L_1 = 39$ m and $L_2 = 36$ m? (c) Assuming that the source separation is much smaller than L_1 and L_2, what type of interference occurs at P in situations (a) and (b)?

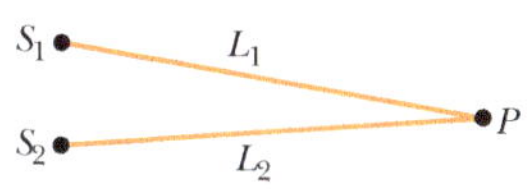

Figure 17-25 Question 2.

3 In Fig. 17-26, three long tubes (A, B, and C) are filled with different gases under different pressures. The ratio of the bulk modulus to the density is indicated for each gas in terms of a basic value B_0/ρ_0. Each tube has a piston at its left end that can send a sound pulse through the tube (as in Fig. 16-2). The three pulses are sent simultaneously. Rank the tubes according to the time of arrival of the pulses at the open right ends of the tubes, earliest first.

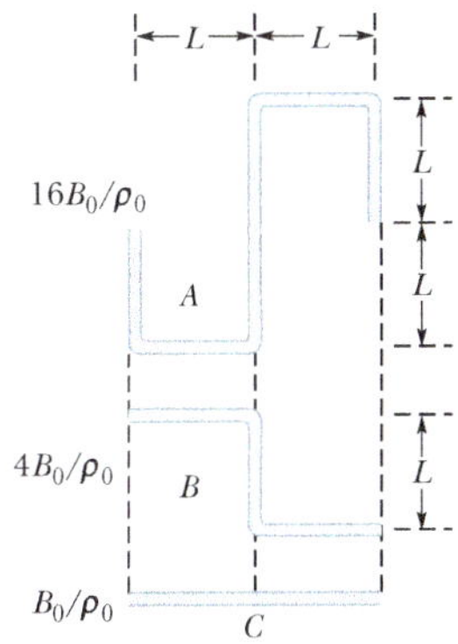

Figure 17-26 Question 3.

4 The sixth harmonic is set up in a pipe. (a) How many open ends does the pipe have (it has at least one)? (b) Is there a node, antinode, or some intermediate state at the midpoint?

5 In Fig. 17-27, pipe A is made to oscillate in its third harmonic by a small internal sound source. Sound emitted at the right end happens to resonate four nearby pipes, each with only one open end (they are *not* drawn to scale). Pipe B oscillates in its lowest harmonic, pipe C in its second lowest harmonic, pipe D in its third lowest harmonic, and pipe E in its fourth lowest harmonic. Without computation, rank all five pipes according to their length, greatest first. (*Hint:* Draw the standing waves to scale and then draw the pipes to scale.)

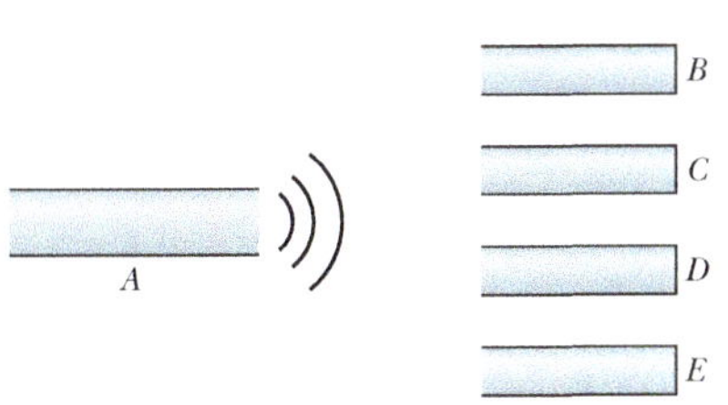

Figure 17-27 Question 5.

6 Pipe A has length L and one open end. Pipe B has length $2L$ and two open ends. Which harmonics of pipe B have a frequency that matches a resonant frequency of pipe A?

7 Figure 17-28 shows a moving sound source S that emits at a certain frequency, and four stationary sound detectors. Rank the detectors according to the frequency of the sound they detect from the source, greatest first.

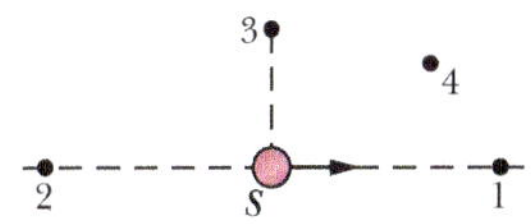

Figure 17-28 Question 7.

8 A friend rides, in turn, the rims of three fast merry-go-rounds while holding a sound source that emits isotropically at a certain frequency. You stand far from each merry-go-round. The frequency you hear for each of your friend's three rides varies as the merry-go-round rotates. The variations in frequency for the three rides are given by the three curves in Fig. 17-29. Rank the curves according to (a) the linear speed v of the sound source, (b) the angular speeds ω of the merry-go-rounds, and (c) the radii r of the merry-go-rounds, greatest first.

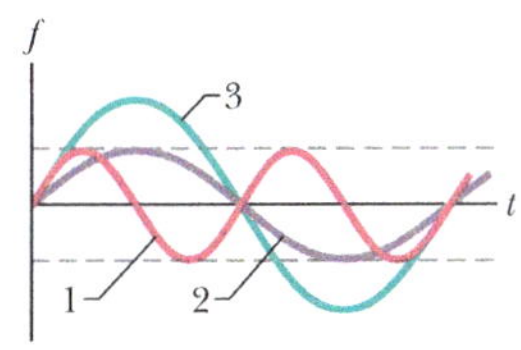

Figure 17-29 Question 8.

9 For a particular tube, here are four of the six harmonic frequencies below 1000 Hz: 300, 600, 750, and 900 Hz. What two frequencies are missing from the list?

10 Figure 17-30 shows a stretched string of length L and pipes a, b, c, and d of lengths L, $2L$, $L/2$, and $L/2$, respectively. The string's tension is adjusted until the speed of waves on the string equals the speed of sound waves in the air. The fundamental mode of oscillation is then set up on the string. In which pipe will the sound produced by the string cause resonance, and what oscillation mode will that sound set up?

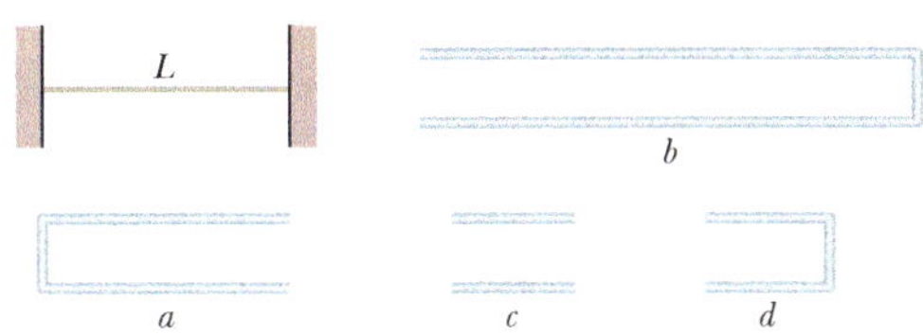

Figure 17-30 Question 10.

11 You are given four tuning forks. The fork with the lowest frequency oscillates at 500 Hz. By striking two tuning forks at a time, you can produce the following beat frequencies, 1, 2, 3, 5, 7, and 8 Hz. What are the possible frequencies of the other three forks? (There are two sets of answers.)

Problems

GO Tutoring problem available (at instructor's discretion) in *WileyPLUS* and WebAssign
SSM Worked-out solution available in Student Solutions Manual
• – ••• Number of dots indicates level of problem difficulty
Additional information available in *The Flying Circus of Physics* and at flyingcircusofphysics.com
WWW Worked-out solution is at
ILW Interactive solution is at http://www.wiley.com/college/halliday

Where needed in the problems, use

speed of sound in air = 343 m/s

and density of air = $1.21\ \text{kg/m}^3$

unless otherwise specified.

Module 17-1 Speed of Sound

•1 Two spectators at a soccer game see, and a moment later hear, the ball being kicked on the playing field. The time delay for spectator A is 0.23 s, and for spectator B it is 0.12 s. Sight lines from the two spectators to the player kicking the ball meet at an angle of 90°. How far are (a) spectator A and (b) spectator B from the player? (c) How far are the spectators from each other?

•2 What is the bulk modulus of oxygen if 32.0 g of oxygen occupies 22.4 L and the speed of sound in the oxygen is 317 m/s?

•3 When the door of the Chapel of the Mausoleum in Hamilton, Scotland, is slammed shut, the last echo heard by someone standing just inside the door reportedly comes 15 s later. (a) If that echo were due to a single reflection off a wall opposite the door, how far from the door is the wall? (b) If, instead, the wall is 25.7 m away, how many reflections (back and forth) occur?

•4 A column of soldiers, marching at 120 paces per minute, keep in step with the beat of a drummer at the head of the column. The soldiers in the rear end of the column are striding forward with the left foot when the drummer is advancing with the right foot. What is the approximate length of the column?

••5 SSM ILW Earthquakes generate sound waves inside Earth. Unlike a gas, Earth can experience both transverse (S) and longitudinal (P) sound waves. Typically, the speed of S waves is about 4.5 km/s, and that of P waves 8.0 km/s. A seismograph records P and S waves from an earthquake. The first P waves arrive 3.0 min before the first S waves. If the waves travel in a straight line, how far away did the earthquake occur?

••6 A man strikes one end of a thin rod with a hammer. The speed of sound in the rod is 15 times the speed of sound in air. A woman, at the other end with her ear close to the rod, hears the sound of the blow twice with a 0.12 s interval between; one sound comes through the rod and the other comes through the air alongside the rod. If the speed of sound in air is 343 m/s, what is the length of the rod?

••7 SSM WWW A stone is dropped into a well. The splash is heard 3.00 s later. What is the depth of the well?

••8 GO *Hot chocolate effect.* Tap a metal spoon inside a mug of water and note the frequency f_i you hear. Then add a spoonful of powder (say, chocolate mix or instant coffee) and tap again as you stir the powder. The frequency you hear has a lower value f_s because the tiny air bubbles released by the powder change the water's bulk modulus. As the bubbles reach the water surface and disappear, the frequency gradually shifts back to its initial value. During the effect, the bubbles don't appreciably change the water's density or volume or the sound's wavelength. Rather, they change the value of dV/dp—that is, the differential change in volume due to the differential change in the pressure caused by the sound wave in the water. If $f_s/f_i = 0.333$, what is the ratio $(dV/dp)_s/(dV/dp)_i$?

Module 17-2 Traveling Sound Waves

•9 If the form of a sound wave traveling through air is

$$s(x, t) = (6.0\ \text{nm}) \cos(kx + (3000\ \text{rad/s})t + \phi),$$

how much time does any given air molecule along the path take to move between displacements $s = +2.0$ nm and $s = -2.0$ nm?

•10 *Underwater illusion.* One clue used by your brain to determine the direction of a source of sound is the time delay Δt between the arrival of the sound at the ear closer to the source and the arrival at the farther ear. Assume that the source is distant so that a wavefront from it is approximately planar when it reaches you, and let D represent the separation between your ears. (a) If the source is located at angle θ in front of you (Fig. 17-31), what is Δt in terms of D and the speed of sound v in air? (b) If you are submerged in water and the sound source is directly to your right, what is Δt in terms of D and the speed of sound v_w in water? (c) Based on the time-delay clue, your brain interprets the submerged sound to arrive at an angle θ from the forward direction. Evaluate θ for fresh water at 20°C.

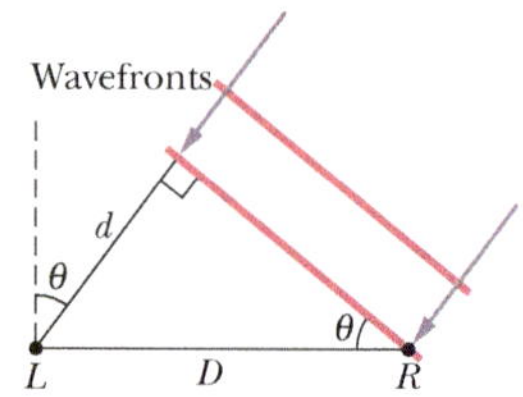

Figure 17-31 Problem 10.

•11 SSM Diagnostic ultrasound of frequency 4.50 MHz is used to examine tumors in soft tissue. (a) What is the wavelength in air of such a sound wave? (b) If the speed of sound in tissue is 1500 m/s, what is the wavelength of this wave in tissue?

•12 The pressure in a traveling sound wave is given by the equation

$$\Delta p = (1.50\ \text{Pa}) \sin \pi[(0.900\ \text{m}^{-1})x - (315\ \text{s}^{-1})t].$$

Find the (a) pressure amplitude, (b) frequency, (c) wavelength, and (d) speed of the wave.

••13 A sound wave of the form $s = s_m \cos(kx - \omega t + \phi)$ travels at 343 m/s through air in a long horizontal tube. At one instant, air molecule A at $x = 2.000$ m is at its maximum positive displacement of 6.00 nm and air molecule B at $x = 2.070$ m is at a positive displacement of 2.00 nm. All the molecules between A and B are at intermediate displacements. What is the frequency of the wave?

••14 Figure 17-32 shows the output from a pressure monitor mounted at a point along the

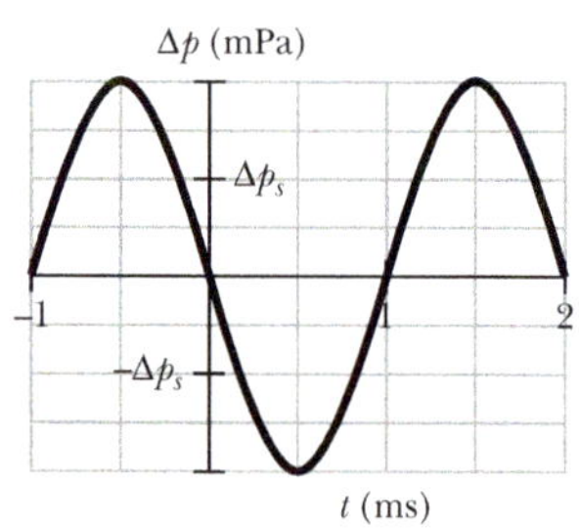

Figure 17-32 Problem 14.

path taken by a sound wave of a single frequency traveling at 343 m/s through air with a uniform density of 1.21 kg/m^3. The vertical axis scale is set by Δp_s = 4.0 mPa. If the *displacement* function of the wave is $s(x, t) = s_m \cos(kx - \omega t)$, what are (a) s_m, (b) k, and (c) ω? The air is then cooled so that its density is 1.35 kg/m^3 and the speed of a sound wave through it is 320 m/s. The sound source again emits the sound wave at the same frequency and same pressure amplitude. What now are (d) s_m, (e) k, and (f) ω?

••15 GO A handclap on stage in an amphitheater sends out sound waves that scatter from terraces of width w = 0.75 m (Fig. 17-33). The sound returns to the stage as a periodic series of pulses, one from each terrace; the parade of pulses sounds like a played note. (a) Assuming that all the rays in Fig. 17-33 are horizontal, find the frequency at which the pulses return (that is, the frequency of the perceived note). (b) If the width w of the terraces were smaller, would the frequency be higher or lower?

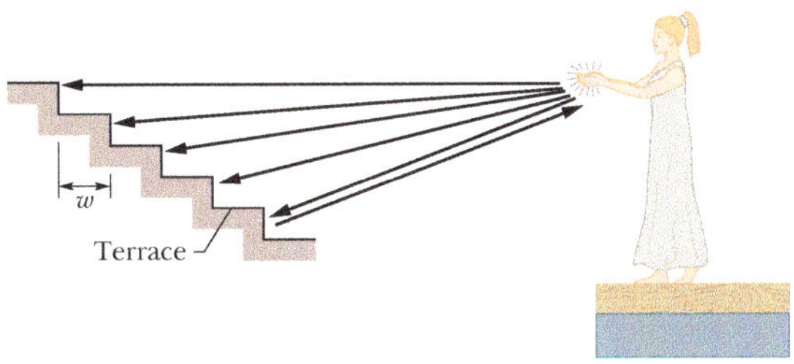

Figure 17-33 Problem 15.

Module 17-3 Interference

•16 Two sound waves, from two different sources with the same frequency, 540 Hz, travel in the same direction at 330 m/s. The sources are in phase. What is the phase difference of the waves at a point that is 4.40 m from one source and 4.00 m from the other?

••17 ILW Two loud speakers are located 3.35 m apart on an outdoor stage. A listener is 18.3 m from one and 19.5 m from the other. During the sound check, a signal generator drives the two speakers in phase with the same amplitude and frequency. The transmitted frequency is swept through the audible range (20 Hz to 20 kHz). (a) What is the lowest frequency $f_{min,1}$ that gives minimum signal (destructive interference) at the listener's location? By what number must $f_{min,1}$ be multiplied to get (b) the second lowest frequency $f_{min,2}$ that gives minimum signal and (c) the third lowest frequency $f_{min,3}$ that gives minimum signal? (d) What is the lowest frequency $f_{max,1}$ that gives maximum signal (constructive interference) at the listener's location? By what number must $f_{max,1}$ be multiplied to get (e) the second lowest frequency $f_{max,2}$ that gives maximum signal and (f) the third lowest frequency $f_{max,3}$ that gives maximum signal?

••18 GO In Fig. 17-34, sound waves A and B, both of wavelength λ, are initially in phase and traveling rightward, as indicated by the two rays. Wave A is reflected from four surfaces but ends up traveling in its original direction. Wave B ends in that direction after reflecting from two surfaces. Let distance L in the figure be expressed as a multiple q of λ: $L = q\lambda$. What are the (a) smallest and (b) second smallest values of q that put A and B exactly out of phase with each other after the reflections?

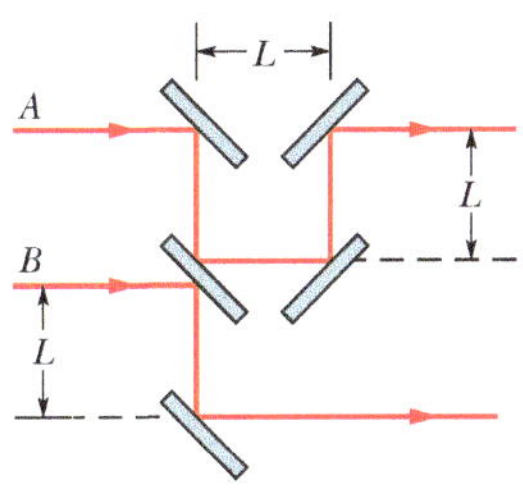

Figure 17-34 Problem 18.

••19 GO Figure 17-35 shows two isotropic point sources of sound, S_1 and S_2. The sources emit waves in phase at wavelength 0.50 m; they are separated by D = 1.75 m. If we move a sound detector along a large circle centered at the midpoint between the sources, at how many points do waves arrive at the detector (a) exactly in phase and (b) exactly out of phase?

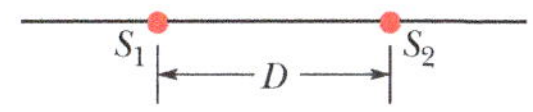

Figure 17-35 Problems 19 and 105.

••20 Figure 17-36 shows four isotropic point sources of sound that are uniformly spaced on an x axis. The sources emit sound at the same wavelength λ and same amplitude s_m, and they emit in phase. A point P is shown on the x axis. Assume that as the sound waves travel to P, the decrease in their amplitude is negligible. What multiple of s_m is the amplitude of the net wave at P if distance d in the figure is (a) $\lambda/4$, (b) $\lambda/2$, and (c) λ?

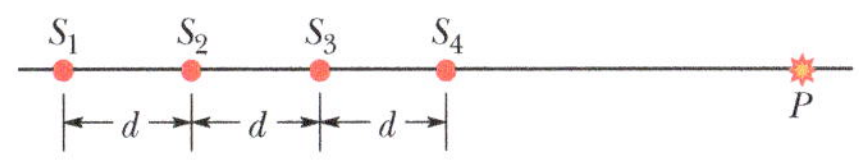

Figure 17-36 Problem 20.

••21 SSM In Fig. 17-37, two speakers separated by distance d_1 = 2.00 m are in phase. Assume the amplitudes of the sound waves from the speakers are approximately the same at the listener's ear at distance d_2 = 3.75 m directly in front of one speaker. Consider the full audible range for normal hearing, 20 Hz to 20 kHz. (a) What is the lowest frequency $f_{min,1}$ that gives minimum signal (destructive interference) at the listener's ear? By what number must $f_{min,1}$ be multiplied to get (b) the second lowest frequency $f_{min,2}$ that gives minimum signal and (c) the third lowest frequency $f_{min,3}$ that gives minimum signal? (d) What is the lowest frequency $f_{max,1}$ that gives maximum signal (constructive interference) at the listener's ear? By what number must $f_{max,1}$ be multiplied to get (e) the second lowest frequency $f_{max,2}$ that gives maximum signal and (f) the third lowest frequency $f_{max,3}$ that gives maximum signal?

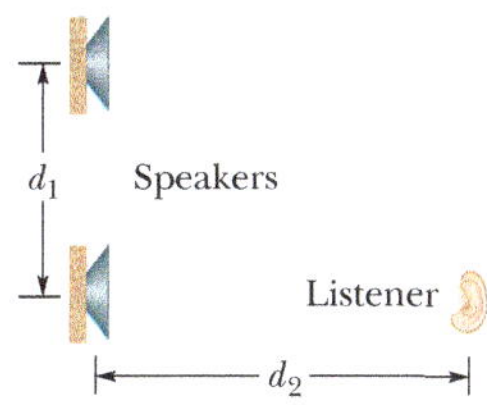

Figure 17-37 Problem 21.

••22 In Fig. 17-38, sound with a 40.0 cm wavelength travels rightward from a source and through a tube that consists of a straight portion and a half-circle. Part of the sound wave travels through the half-circle and then rejoins the rest of the wave, which goes directly through the straight portion. This rejoining results in interference. What is the smallest radius r that results in an intensity minimum at the detector?

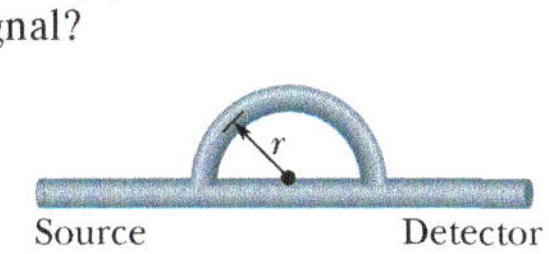

Figure 17-38 Problem 22.

•••23 GO Figure 17-39 shows two point sources S_1 and S_2 that emit sound of wavelength λ = 2.00 m. The emissions are isotropic and in phase, and the separation between

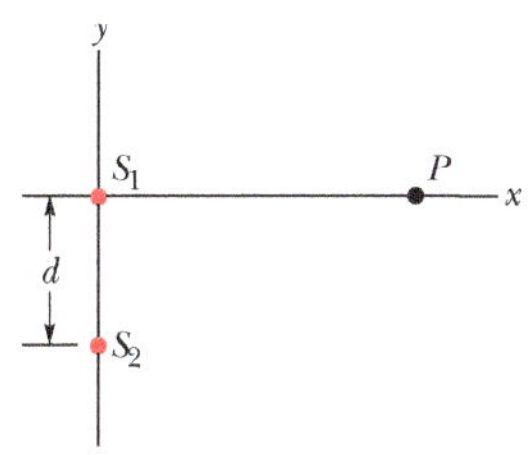

Figure 17-39 Problem 23.

the sources is $d = 16.0$ m. At any point P on the x axis, the wave from S_1 and the wave from S_2 interfere. When P is very far away ($x \approx \infty$), what are (a) the phase difference between the arriving waves from S_1 and S_2 and (b) the type of interference they produce? Now move point P along the x axis toward S_1. (c) Does the phase difference between the waves increase or decrease? At what distance x do the waves have a phase difference of (d) 0.50λ, (e) 1.00λ, and (f) 1.50λ?

Module 17-4 Intensity and Sound Level

•24 Suppose that the sound level of a conversation is initially at an angry 70 dB and then drops to a soothing 50 dB. Assuming that the frequency of the sound is 500 Hz, determine the (a) initial and (b) final sound intensities and the (c) initial and (d) final sound wave amplitudes.

•25 A sound wave of frequency 300 Hz has an intensity of 1.00 μW/m^2. What is the amplitude of the air oscillations caused by this wave?

•26 A 1.0 W point source emits sound waves isotropically. Assuming that the energy of the waves is conserved, find the intensity (a) 1.0 m from the source and (b) 2.5 m from the source.

•27 SSM WWW A certain sound source is increased in sound level by 30.0 dB. By what multiple is (a) its intensity increased and (b) its pressure amplitude increased?

•28 Two sounds differ in sound level by 1.00 dB. What is the ratio of the greater intensity to the smaller intensity?

•29 SSM A point source emits sound waves isotropically. The intensity of the waves 2.50 m from the source is 1.91×10^{-4} W/m^2. Assuming that the energy of the waves is conserved, find the power of the source.

•30 The source of a sound wave has a power of 1.00 μW. If it is a point source, (a) what is the intensity 3.00 m away and (b) what is the sound level in decibels at that distance?

•31 When you "crack" a knuckle, you suddenly widen the knuckle cavity, allowing more volume for the synovial fluid inside it and causing a gas bubble suddenly to appear in the fluid. The sudden production of the bubble, called "cavitation," produces a sound pulse—the cracking sound. Assume that the sound is transmitted uniformly in all directions and that it fully passes from the knuckle interior to the outside. If the pulse has a sound level of 62 dB at your ear, estimate the rate at which energy is produced by the cavitation.

•32 Approximately a third of people with normal hearing have ears that continuously emit a low-intensity sound outward through the ear canal. A person with such *spontaneous otoacoustic emission* is rarely aware of the sound, except perhaps in a noise-free environment, but occasionally the emission is loud enough to be heard by someone else nearby. In one observation, the sound wave had a frequency of 1665 Hz and a pressure amplitude of 1.13×10^{-3} Pa. What were (a) the displacement amplitude and (b) the intensity of the wave emitted by the ear?

•33 Male *Rana catesbeiana* bullfrogs are known for their loud mating call. The call is emitted not by the frog's mouth but by its eardrums, which lie on the surface of the head. And, surprisingly, the sound has nothing to do with the frog's inflated throat. If the emitted sound has a frequency of 260 Hz and a sound level of 85 dB (near the eardrum), what is the amplitude of the eardrum's oscillation? The air density is 1.21 kg/m^3.

••34 Two atmospheric sound sources A and B emit isotropically at constant power. The sound levels β of their emissions are plotted in Fig. 17-40 versus the radial distance r from the sources. The vertical axis scale is set by $\beta_1 = 85.0$ dB and $\beta_2 = 65.0$ dB. What are (a) the ratio of the larger power to the smaller power and (b) the sound level difference at $r = 10$ m?

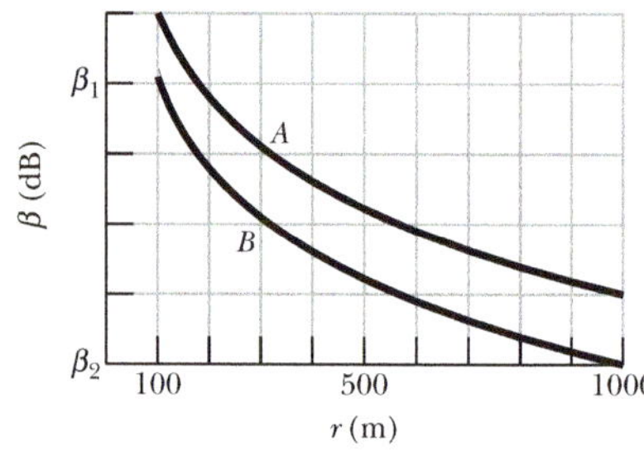

Figure 17-40 Problem 34.

••35 A point source emits 30.0 W of sound isotropically. A small microphone intercepts the sound in an area of 0.750 cm^2, 200 m from the source. Calculate (a) the sound intensity there and (b) the power intercepted by the microphone.

••36 *Party hearing.* As the number of people at a party increases, you must raise your voice for a listener to hear you against the *background noise* of the other partygoers. However, once you reach the level of yelling, the only way you can be heard is if you move closer to your listener, into the listener's "personal space." Model the situation by replacing you with an isotropic point source of fixed power P and replacing your listener with a point that absorbs part of your sound waves. These points are initially separated by $r_i = 1.20$ m. If the background noise increases by $\Delta\beta = 5$ dB, the sound level at your listener must also increase. What separation r_f is then required?

•••37 A sound source sends a sinusoidal sound wave of angular frequency 3000 rad/s and amplitude 12.0 nm through a tube of air. The internal radius of the tube is 2.00 cm. (a) What is the average rate at which energy (the sum of the kinetic and potential energies) is transported to the opposite end of the tube? (b) If, simultaneously, an identical wave travels along an adjacent, identical tube, what is the total average rate at which energy is transported to the opposite ends of the two tubes by the waves? If, instead, those two waves are sent along the *same* tube simultaneously, what is the total average rate at which they transport energy when their phase difference is (c) 0, (d) 0.40π rad, and (e) π rad?

Module 17-5 Sources of Musical Sound

•38 The water level in a vertical glass tube 1.00 m long can be adjusted to any position in the tube. A tuning fork vibrating at 686 Hz is held just over the open top end of the tube, to set up a standing wave of sound in the air-filled top portion of the tube. (That air-filled top portion acts as a tube with one end closed and the other end open.) (a) For how many different positions of the water level will sound from the fork set up resonance in the tube's air-filled portion? What are the (b) least and (c) second least water heights in the tube for resonance to occur?

•39 SSM ILW (a) Find the speed of waves on a violin string of mass 800 mg and length 22.0 cm if the fundamental frequency is 920 Hz. (b) What is the tension in the string? For the fundamental, what is the wavelength of (c) the waves on the string and (d) the sound waves emitted by the string?

•40 Organ pipe A, with both ends open, has a fundamental frequency of 300 Hz. The third harmonic of organ pipe B, with one end open, has the same frequency as the second harmonic of pipe A. How long are (a) pipe A and (b) pipe B?

•41 A violin string 15.0 cm long and fixed at both ends oscillates in its $n = 1$ mode. The speed of waves on the string is 250 m/s, and the speed of sound in air is 348 m/s. What are the (a) frequency and (b) wavelength of the emitted sound wave?

•42 A sound wave in a fluid medium is reflected at a barrier so that a standing wave is formed. The distance between nodes is 3.8 cm, and the speed of propagation is 1500 m/s. Find the frequency of the sound wave.

•43 SSM In Fig. 17-41, S is a small loudspeaker driven by an audio oscillator with a frequency that is varied from 1000 Hz to 2000 Hz, and D is a cylindrical pipe with two open ends and a length of 45.7 cm. The speed of sound in the air-filled pipe is 344 m/s. (a) At how many frequencies does the sound from the loudspeaker set up resonance in the pipe? What are the (b) lowest and (c) second lowest frequencies at which resonance occurs?

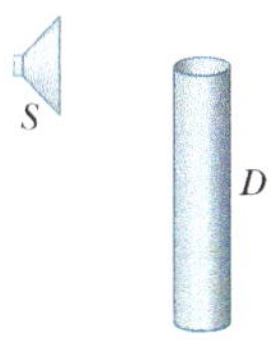

Figure 17-41 Problem 43.

•44 The crest of a *Parasaurolophus* dinosaur skull is shaped somewhat like a trombone and contains a nasal passage in the form of a long, bent tube open at both ends. The dinosaur may have used the passage to produce sound by setting up the fundamental mode in it. (a) If the nasal passage in a certain *Parasaurolophus* fossil is 2.0 m long, what frequency would have been produced? (b) If that dinosaur could be recreated (as in *Jurassic Park*), would a person with a hearing range of 60 Hz to 20 kHz be able to hear that fundamental mode and, if so, would the sound be high or low frequency? Fossil skulls that contain shorter nasal passages are thought to be those of the female *Parasaurolophus.* (c) Would that make the female's fundamental frequency higher or lower than the male's?

•45 In pipe A, the ratio of a particular harmonic frequency to the next lower harmonic frequency is 1.2. In pipe B, the ratio of a particular harmonic frequency to the next lower harmonic frequency is 1.4. How many open ends are in (a) pipe A and (b) pipe B?

••46 GO Pipe A, which is 1.20 m long and open at both ends, oscillates at its third lowest harmonic frequency. It is filled with air for which the speed of sound is 343 m/s. Pipe B, which is closed at one end, oscillates at its second lowest harmonic frequency. This frequency of B happens to match the frequency of A. An x axis extends along the interior of B, with $x = 0$ at the closed end. (a) How many nodes are along that axis? What are the (b) smallest and (c) second smallest value of x locating those nodes? (d) What is the fundamental frequency of B?

••47 A well with vertical sides and water at the bottom resonates at 7.00 Hz and at no lower frequency. The air-filled portion of the well acts as a tube with one closed end (at the bottom) and one open end (at the top). The air in the well has a density of 1.10 kg/m^3 and a bulk modulus of 1.33×10^5 Pa. How far down in the well is the water surface?

••48 One of the harmonic frequencies of tube A with two open ends is 325 Hz. The next-highest harmonic frequency is 390 Hz. (a) What harmonic frequency is next highest after the harmonic frequency 195 Hz? (b) What is the number of this next-highest harmonic? One of the harmonic frequencies of tube B with only one open end is 1080 Hz. The next-highest harmonic frequency is 1320 Hz. (c) What harmonic frequency is next highest after the harmonic frequency 600 Hz? (d) What is the number of this next-highest harmonic?

••49 SSM A violin string 30.0 cm long with linear density 0.650 g/m is placed near a loudspeaker that is fed by an audio oscillator of variable frequency. It is found that the string is set into oscillation only at the frequencies 880 and 1320 Hz as the frequency of the oscillator is varied over the range 500–1500 Hz. What is the tension in the string?

••50 GO A tube 1.20 m long is closed at one end. A stretched wire is placed near the open end. The wire is 0.330 m long and has a mass of 9.60 g. It is fixed at both ends and oscillates in its fundamental mode. By resonance, it sets the air column in the tube into oscillation at that column's fundamental frequency. Find (a) that frequency and (b) the tension in the wire.

Module 17-6 Beats

•51 The A string of a violin is a little too tightly stretched. Beats at 4.00 per second are heard when the string is sounded together with a tuning fork that is oscillating accurately at concert A (440 Hz). What is the period of the violin string oscillation?

•52 A tuning fork of unknown frequency makes 3.00 beats per second with a standard fork of frequency 384 Hz. The beat frequency decreases when a small piece of wax is put on a prong of the first fork. What is the frequency of this fork?

••53 SSM Two identical piano wires have a fundamental frequency of 600 Hz when kept under the same tension. What fractional increase in the tension of one wire will lead to the occurrence of 6.0 beats/s when both wires oscillate simultaneously?

••54 You have five tuning forks that oscillate at close but different resonant frequencies. What are the (a) maximum and (b) minimum number of different beat frequencies you can produce by sounding the forks two at a time, depending on how the resonant frequencies differ?

Module 17-7 The Doppler Effect

•55 ILW A whistle of frequency 540 Hz moves in a circle of radius 60.0 cm at an angular speed of 15.0 rad/s. What are the (a) lowest and (b) highest frequencies heard by a listener a long distance away, at rest with respect to the center of the circle?

•56 An ambulance with a siren emitting a whine at 1600 Hz overtakes and passes a cyclist pedaling a bike at 2.44 m/s. After being passed, the cyclist hears a frequency of 1590 Hz. How fast is the ambulance moving?

•57 A state trooper chases a speeder along a straight road; both vehicles move at 160 km/h. The siren on the trooper's vehicle produces sound at a frequency of 500 Hz. What is the Doppler shift in the frequency heard by the speeder?

••58 A sound source A and a reflecting surface B move directly toward each other. Relative to the air, the speed of source A is 29.9 m/s, the speed of surface B is 65.8 m/s, and the speed of sound is 329 m/s. The source emits waves at frequency 1200 Hz as measured in the source frame. In the reflector frame, what are the (a) frequency and (b) wavelength of the arriving sound waves? In the source frame, what are the (c) frequency and (d) wavelength of the sound waves reflected back to the source?

••59 GO In Fig. 17-42, a French submarine and a U.S. submarine move toward each other during maneuvers in motionless water in the North Atlantic. The French sub moves at speed v_F = 50.00 km/h, and the U.S. sub at v_{US} = 70.00 km/h. The French sub sends out a sonar signal (sound wave in water) at 1.000×10^3 Hz. Sonar waves travel at 5470 km/h. (a) What is the signal's frequency as detected by the U.S. sub? (b) What frequency is detected by the French sub in the signal reflected back to it by the U.S. sub?

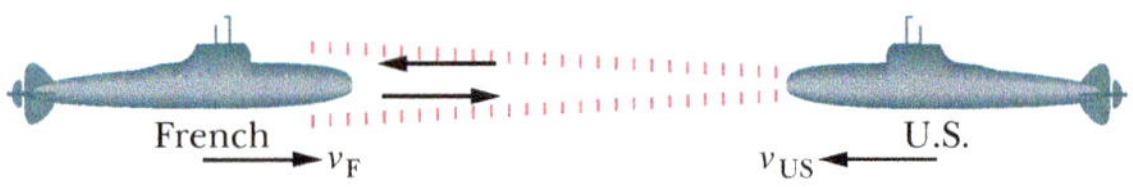

Figure 17-42 Problem 59.

••60 A stationary motion detector sends sound waves of frequency 0.150 MHz toward a truck approaching at a speed of 45.0 m/s. What is the frequency of the waves reflected back to the detector?

••61 GO A bat is flitting about in a cave, navigating via ultrasonic bleeps. Assume that the sound emission frequency of the bat is 39 000 Hz. During one fast swoop directly toward a flat wall surface, the bat is moving at 0.025 times the speed of sound in air. What frequency does the bat hear reflected off the wall?

••62 Figure 17-43 shows four tubes with lengths 1.0 m or 2.0 m, with one or two open ends as drawn. The third harmonic is set up in each tube, and some of the sound that escapes from them is detected by detector D, which moves directly away from the tubes. In terms of the speed of sound v, what speed must the detector have such that the detected frequency of the sound from (a) tube 1, (b) tube 2, (c) tube 3, and (d) tube 4 is equal to the tube's fundamental frequency?

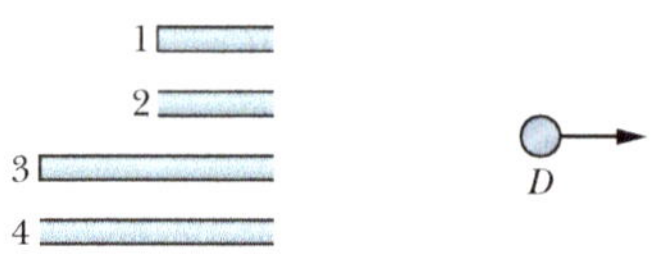

Figure 17-43 Problem 62.

••63 ILW An acoustic burglar alarm consists of a source emitting waves of frequency 28.0 kHz. What is the beat frequency between the source waves and the waves reflected from an intruder walking at an average speed of 0.950 m/s directly away from the alarm?

••64 A stationary detector measures the frequency of a sound source that first moves at constant velocity directly toward the detector and then (after passing the detector) directly away from it. The emitted frequency is f. During the approach the detected frequency is f'_{app} and during the recession it is f'_{rec}. If $(f'_{app} - f'_{rec})/f$ = 0.500, what is the ratio v_s/v of the speed of the source to the speed of sound?

•••65 GO A 2000 Hz siren and a civil defense official are both at rest with respect to the ground. What frequency does the official hear if the wind is blowing at 12 m/s (a) from source to official and (b) from official to source?

•••66 GO Two trains are traveling toward each other at 30.5 m/s relative to the ground. One train is blowing a whistle at 500 Hz. (a) What frequency is heard on the other train in still air? (b) What frequency is heard on the other train if the wind is blowing at 30.5 m/s toward the whistle and away from the listener? (c) What frequency is heard if the wind direction is reversed?

•••67 SSM WWW A girl is sitting near the open window of a train that is moving at a velocity of 10.00 m/s to the east. The girl's uncle stands near the tracks and watches the train move away. The locomotive whistle emits sound at frequency 500.0 Hz. The air is still. (a) What frequency does the uncle hear? (b) What frequency does the girl hear? A wind begins to blow from the east at 10.00 m/s. (c) What frequency does the uncle now hear? (d) What frequency does the girl now hear?

Module 17-8 Supersonic Speeds, Shock Waves

•68 The shock wave off the cockpit of the FA 18 in Fig. 17-24 has an angle of about 60°. The airplane was traveling at about 1350 km/h when the photograph was taken. Approximately what was the speed of sound at the airplane's altitude?

••69 SSM A jet plane passes over you at a height of 5000 m and a speed of Mach 1.5. (a) Find the Mach cone angle (the sound speed is 331 m/s). (b) How long after the jet passes directly overhead does the shock wave reach you?

••70 A plane flies at 1.25 times the speed of sound. Its sonic boom reaches a man on the ground 1.00 min after the plane passes directly overhead. What is the altitude of the plane? Assume the speed of sound to be 330 m/s.

Additional Problems

71 At a distance of 10 km, a 100 Hz horn, assumed to be an isotropic point source, is barely audible. At what distance would it begin to cause pain?

72 A bullet is fired with a speed of 685 m/s. Find the angle made by the shock cone with the line of motion of the bullet.

73 A sperm whale (Fig. 17-44a) vocalizes by producing a series of clicks. Actually, the whale makes only a single sound near the front of its head to start the series. Part of that sound then emerges from the head into the water to become the first click of the series. The rest of the sound travels backward through the spermaceti sac (a body of fat), reflects from the frontal sac (an air layer), and then travels forward through the spermaceti sac. When it reaches the distal sac (another air layer) at the front of the head, some of the sound escapes into the water to form the second click, and the rest is sent back through the spermaceti sac (and ends up forming later clicks).

Figure 17-44b shows a strip-chart recording of a series of clicks. A unit time interval of 1.0 ms is indicated on the chart. Assuming that the speed of sound in the spermaceti sac is 1372 m/s, find the length of the spermaceti sac. From such a calculation, marine scientists estimate the length of a whale from its click series.

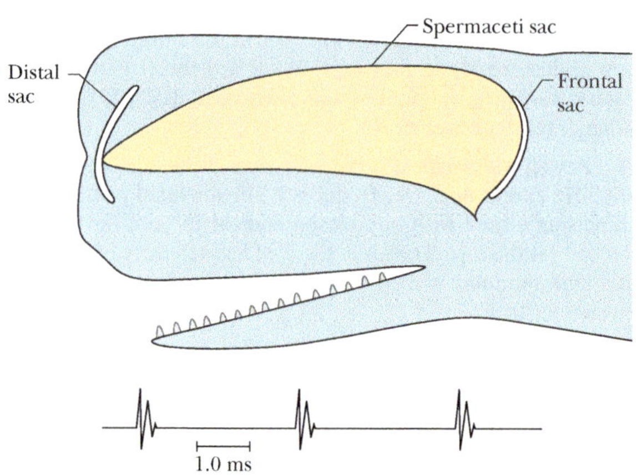

Figure 17-44 Problem 73.

74 The average density of Earth's crust 10 km beneath the continents is 2.7 g/cm^3. The speed of longitudinal seismic waves at that depth, found by timing their arrival from distant earthquakes, is 5.4 km/s. Find the bulk modulus of Earth's crust at that depth. For comparison, the bulk modulus of steel is about 16×10^{10} Pa.

75 A certain loudspeaker system emits sound isotropically with a frequency of 2000 Hz and an intensity of 0.960 mW/m^2 at a distance of 6.10 m. Assume that there are no reflections. (a) What is the intensity at 30.0 m? At 6.10 m, what are (b) the displacement amplitude and (c) the pressure amplitude?

76 Find the ratios (greater to smaller) of the (a) intensities, (b) pressure amplitudes, and (c) particle displacement amplitudes for two sounds whose sound levels differ by 37 dB.

77 In Fig. 17-45, sound waves A and B, both of wavelength λ, are initially in phase and traveling rightward, as indicated by the two rays. Wave A is reflected from four surfaces but ends up traveling in its original direction. What multiple of wavelength λ is the smallest value of distance L in the figure that puts A and B exactly out of phase with each other after the reflections?

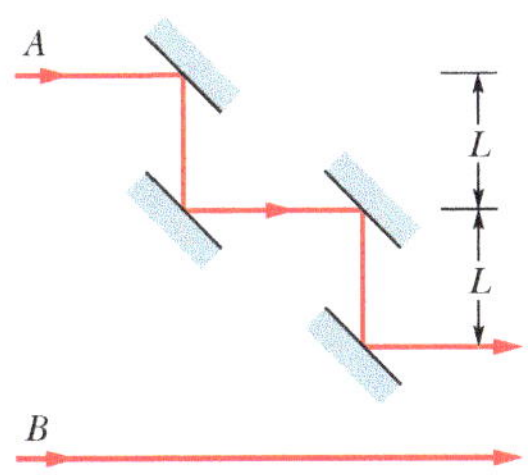

Figure 17-45 Problem 77.

78 A trumpet player on a moving railroad flatcar moves toward a second trumpet player standing alongside the track while both play a 440 Hz note. The sound waves heard by a stationary observer between the two players have a beat frequency of 4.0 beats/s. What is the flatcar's speed?

79 GO In Fig. 17-46, sound of wavelength 0.850 m is emitted isotropically by point source S. Sound ray 1 extends directly to detector D, at distance $L = 10.0$ m. Sound ray 2 extends to D via a reflection (effectively, a "bouncing") of the sound at a flat surface. That reflection occurs on a perpendicular bisector to the SD line, at distance d from the line. Assume that the reflection shifts the sound wave by 0.500λ. For what least value of d (other than zero) do the direct sound and the reflected sound arrive at D (a) exactly out of phase and (b) exactly in phase?

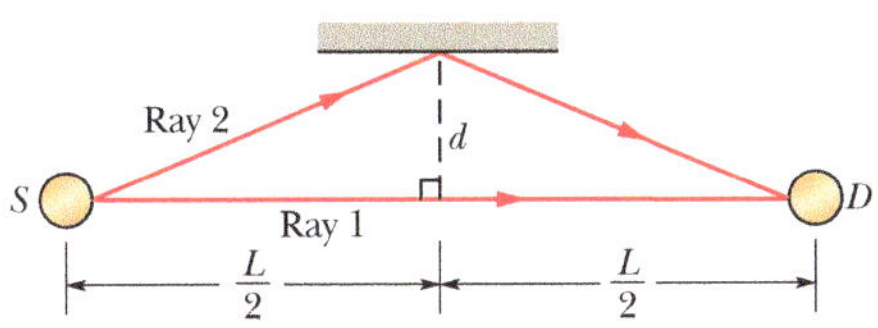

Figure 17-46 Problem 79.

80 GO A detector initially moves at constant velocity directly toward a stationary sound source and then (after passing it) directly from it. The emitted frequency is f. During the approach the detected frequency is f'_{app} and during the recession it is f'_{rec}. If the frequencies are related by $(f'_{app} - f'_{rec})/f = 0.500$, what is the ratio v_D/v of the speed of the detector to the speed of sound?

81 SSM (a) If two sound waves, one in air and one in (fresh) water, are equal in intensity and angular frequency, what is the ratio of the pressure amplitude of the wave in water to that of the wave in air? Assume the water and the air are at 20°C. (See Table 14-1.) (b) If the pressure amplitudes are equal instead, what is the ratio of the intensities of the waves?

82 A continuous sinusoidal longitudinal wave is sent along a very long coiled spring from an attached oscillating source. The wave travels in the negative direction of an x axis; the source frequency is 25 Hz; at any instant the distance between successive points of maximum expansion in the spring is 24 cm; the maximum longitudinal displacement of a spring particle is 0.30 cm; and the particle at $x = 0$ has zero displacement at time $t = 0$. If the wave is written in the form $s(x, t) = s_m \cos(kx \pm \omega t)$, what are (a) s_m, (b) k, (c) ω, (d) the wave speed, and (e) the correct choice of sign in front of ω?

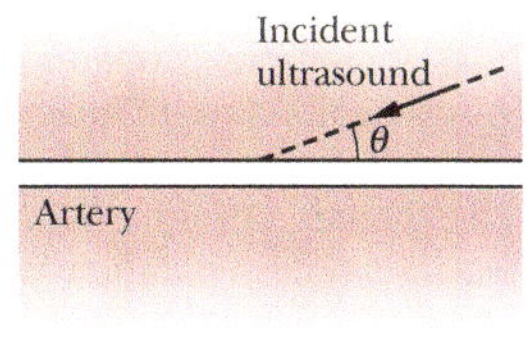

Figure 17-47 Problem 83.

83 SSM Ultrasound, which consists of sound waves with frequencies above the human audible range, can be used to produce an image of the interior of a human body. Moreover, ultrasound can be used to measure the speed of the blood in the body; it does so by comparing the frequency of the ultrasound sent into the body with the frequency of the ultrasound reflected back to the body's surface by the blood. As the blood pulses, this detected frequency varies.

Suppose that an ultrasound image of the arm of a patient shows an artery that is angled at $\theta = 20°$ to the ultrasound's line of travel (Fig. 17-47). Suppose also that the frequency of the ultrasound reflected by the blood in the artery is increased by a maximum of 5495 Hz from the original ultrasound frequency of 5.000 000 MHz. (a) In Fig. 17-47, is the direction of the blood flow rightward or leftward? (b) The speed of sound in the human arm is 1540 m/s. What is the maximum speed of the blood? (*Hint:* The Doppler effect is caused by the component of the blood's velocity along the ultrasound's direction of travel.) (c) If angle θ were greater, would the reflected frequency be greater or less?

84 The speed of sound in a certain metal is v_m. One end of a long pipe of that metal of length L is struck a hard blow. A listener at the other end hears two sounds, one from the wave that travels along the pipe's metal wall and the other from the wave that travels through the air inside the pipe. (a) If v is the speed of sound in air, what is the time interval Δt between the arrivals of the two sounds at the listener's ear? (b) If $\Delta t = 1.00$ s and the metal is steel, what is the length L?

85 An avalanche of sand along some rare desert sand dunes can produce a booming that is loud enough to be heard 10 km away. The booming apparently results from a periodic oscillation of the sliding layer of sand—the layer's thickness expands and contracts. If the emitted frequency is 90 Hz, what are (a) the period of the thickness oscillation and (b) the wavelength of the sound?

86 A sound source moves along an x axis, between detectors A and B. The wavelength of the sound detected at A is 0.500 that of the sound detected at B. What is the ratio v_s/v of the speed of the source to the speed of sound?

87 SSM A siren emitting a sound of frequency 1000 Hz moves away from you toward the face of a cliff at a speed of 10 m/s. Take the speed of sound in air as 330 m/s. (a) What is the frequency of the sound you hear coming directly from the siren? (b) What is the frequency of the sound you hear reflected off the cliff? (c) What is the beat frequency between the two sounds? Is it perceptible (less than 20 Hz)?

88 At a certain point, two waves produce pressure variations given by $\Delta p_1 = \Delta p_m \sin \omega t$ and $\Delta p_2 = \Delta p_m \sin(\omega t - \phi)$. At this point, what is the ratio $\Delta p_r/\Delta p_m$, where Δp_r is the pressure amplitude of the resultant wave, if ϕ is (a) 0, (b) $\pi/2$, (c) $\pi/3$, and (d) $\pi/4$?

89 Two sound waves with an amplitude of 12 nm and a wavelength of 35 cm travel in the same direction through a long tube, with a phase difference of $\pi/3$ rad. What are the (a) amplitude and (b) wavelength of the net sound wave produced by their interference? If, instead, the sound waves travel through the tube in opposite directions, what are the (c) amplitude and (d) wavelength of the net wave?

90 A sinusoidal sound wave moves at 343 m/s through air in the positive direction of an x axis. At one instant during the oscillations, air molecule A is at its maximum displacement in the negative direction of the axis while air molecule B is at its equilibrium position. The separation between those molecules is 15.0 cm, and the molecules between A and B have intermediate displacements in the negative direction of the axis. (a) What is the frequency of the sound wave?

In a similar arrangement but for a different sinusoidal sound wave, at one instant air molecule C is at its maximum displacement in the positive direction while molecule D is at its maximum displacement in the negative direction. The separation between the molecules is again 15.0 cm, and the molecules between C and D have intermediate displacements. (b) What is the frequency of the sound wave?

91 Two identical tuning forks can oscillate at 440 Hz. A person is located somewhere on the line between them. Calculate the beat frequency as measured by this individual if (a) she is standing still and the tuning forks move in the same direction along the line at 3.00 m/s, and (b) the tuning forks are stationary and the listener moves along the line at 3.00 m/s.

92 You can estimate your distance from a lightning stroke by counting the seconds between the flash you see and the thunder you later hear. By what integer should you divide the number of seconds to get the distance in kilometers?

93 SSM Figure 17-48 shows an air-filled, acoustic interferometer, used to demonstrate the interference of sound waves. Sound source S is an oscillating diaphragm; D is a sound detector, such as the ear or a microphone. Path SBD can be varied in length, but path SAD is fixed. At D, the sound wave coming along path SBD interferes with that coming along path SAD. In one demonstration, the sound intensity at D has a minimum value of 100 units at one position of the movable arm and continuously climbs to a maximum value of 900 units when that arm is shifted by 1.65 cm. Find (a) the frequency of the sound emitted by the source and (b) the ratio of the amplitude at D of the SAD wave to that of the SBD wave. (c) How can it happen that these waves have different amplitudes, considering that they originate at the same source?

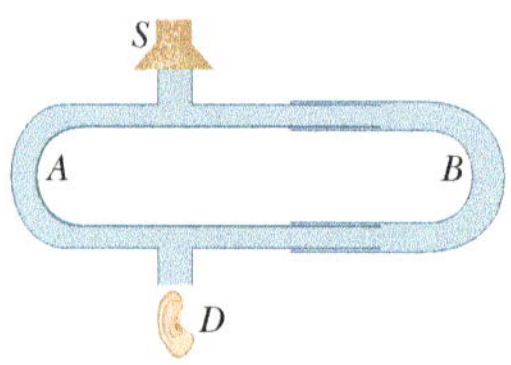

Figure 17-48 Problem 93.

94 On July 10, 1996, a granite block broke away from a wall in Yosemite Valley and, as it began to slide down the wall, was launched into projectile motion. Seismic waves produced by its impact with the ground triggered seismographs as far away as 200 km. Later measurements indicated that the block had a mass between 7.3×10^7 kg and 1.7×10^8 kg and that it landed 500 m vertically below the launch point and 30 m horizontally from it. (The launch angle is not known.) (a) Estimate the block's kinetic energy just before it landed.

Consider two types of seismic waves that spread from the impact point—a hemispherical *body wave* traveled through the ground in an expanding hemisphere and a cylindrical *surface wave* traveled along the ground in an expanding shallow vertical cylinder (Fig. 17-49). Assume that the impact lasted 0.50 s, the vertical cylinder had a depth d of 5.0 m, and each wave type received 20% of the energy the block had just before impact. Neglecting any mechanical energy loss the waves experienced as they traveled, determine the intensities of (b) the body wave and (c) the surface wave when they reached a seismograph 200 km away. (d) On the basis of these results, which wave is more easily detected on a distant seismograph?

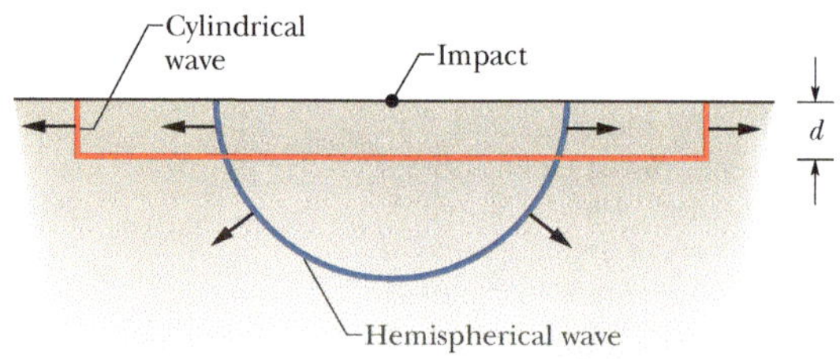

Figure 17-49 Problem 94.

95 SSM The sound intensity is 0.0080 W/m^2 at a distance of 10 m from an isotropic point source of sound. (a) What is the power of the source? (b) What is the sound intensity 5.0 m from the source? (c) What is the sound level 10 m from the source?

96 Four sound waves are to be sent through the same tube of air, in the same direction:

$$s_1(x, t) = (9.00 \text{ nm}) \cos(2\pi x - 700\pi t)$$
$$s_2(x, t) = (9.00 \text{ nm}) \cos(2\pi x - 700\pi t + 0.7\pi)$$
$$s_3(x, t) = (9.00 \text{ nm}) \cos(2\pi x - 700\pi t + \pi)$$
$$s_4(x, t) = (9.00 \text{ nm}) \cos(2\pi x - 700\pi t + 1.7\pi).$$

What is the amplitude of the resultant wave? (*Hint:* Use a phasor diagram to simplify the problem.)

97 Straight line AB connects two point sources that are 5.00 m apart, emit 300 Hz sound waves of the same amplitude, and emit exactly out of phase. (a) What is the shortest distance between the midpoint of AB and a point on AB where the interfering waves cause maximum oscillation of the air molecules? What are the (b) second and (c) third shortest distances?

98 A point source that is stationary on an x axis emits a sinusoidal sound wave at a frequency of 686 Hz and speed 343 m/s. The wave travels radially outward from the source, causing air molecules to oscillate radially inward and outward. Let us define a wavefront as a line that connects points where the air molecules have the maximum, radially outward displacement. At any given instant, the wavefronts are concentric circles that are centered on the source. (a) Along x, what is the adjacent wavefront separation? Next, the source moves along x at a speed of 110 m/s. Along x, what are the wavefront separations (b) in front of and (c) behind the source?

99 You are standing at a distance D from an isotropic point source of sound. You walk 50.0 m toward the source and observe that the intensity of the sound has doubled. Calculate the distance D.

100 Pipe A has only one open end; pipe B is four times as long and has two open ends. Of the lowest 10 harmonic numbers n_B of pipe B, what are the (a) smallest, (b) second smallest, and (c) third smallest values at which a harmonic frequency of B matches one of the harmonic frequencies of A?

101 A pipe 0.60 m long and closed at one end is filled with an unknown gas. The third lowest harmonic frequency for the pipe is 750 Hz. (a) What is the speed of sound in the unknown gas? (b) What is the fundamental frequency for this pipe when it is filled with the unknown gas?

102 A sound wave travels out uniformly in all directions from a point source. (a) Justify the following expression for the displacement s of the transmitting medium at any distance r from the source:

$$s = \frac{b}{r} \sin k(r - vt),$$

where b is a constant. Consider the speed, direction of propagation, periodicity, and intensity of the wave. (b) What is the dimension of the constant b?

103 A police car is chasing a speeding Porsche 911. Assume that the Porsche's maximum speed is 80.0 m/s and the police car's is 54.0 m/s. At the moment both cars reach their maximum speed, what frequency will the Porsche driver hear if the frequency of the police car's siren is 440 Hz? Take the speed of sound in air to be 340 m/s.

104 Suppose a spherical loudspeaker emits sound isotropically at 10 W into a room with completely absorbent walls, floor, and ceiling (an *anechoic chamber*). (a) What is the intensity of the sound at distance $d = 3.0$ m from the center of the source? (b) What is the ratio of the wave amplitude at $d = 4.0$ m to that at $d = 3.0$ m?

105 In Fig. 17-35, S_1 and S_2 are two isotropic point sources of sound. They emit waves in phase at wavelength 0.50 m; they are separated by $D = 1.60$ m. If we move a sound detector along a large circle centered at the midpoint between the sources, at how many points do waves arrive at the detector (a) exactly in phase and (b) exactly out of phase?

106 Figure 17-50 shows a transmitter and receiver of waves contained in a single instrument. It is used to measure the speed u of a target object (idealized as a flat plate) that is moving directly toward the unit, by analyzing the waves reflected from the target. What is u if the emitted frequency is 18.0 kHz and the detected frequency (of the returning waves) is 22.2 kHz?

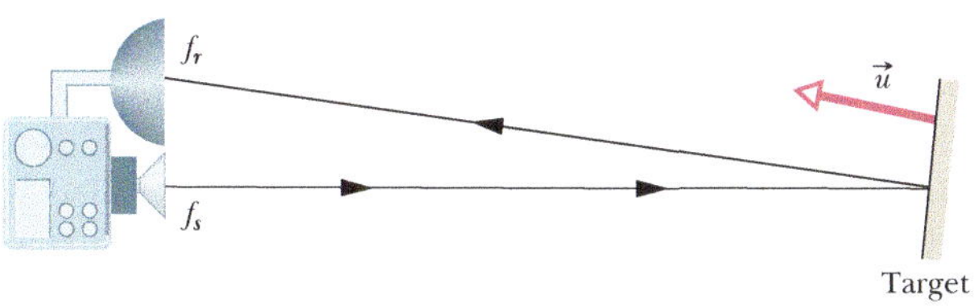

Figure 17-50 Problem 106.

107 *Kundt's method for measuring the speed of sound.* In Fig. 17-51, a rod R is clamped at its center; a disk D at its end projects into a glass tube that has cork filings spread over its interior. A

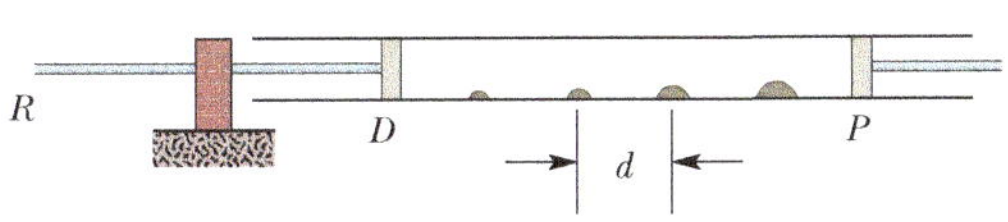

Figure 17-51 Problem 107.

plunger P is provided at the other end of the tube, and the tube is filled with a gas. The rod is made to oscillate longitudinally at frequency f to produce sound waves inside the gas, and the location of the plunger is adjusted until a standing sound wave pattern is set up inside the tube. Once the standing wave is set up, the motion of the gas molecules causes the cork filings to collect in a pattern of ridges at the displacement nodes. If $f = 4.46 \times 10^3$ Hz and the separation between ridges is 9.20 cm, what is the speed of sound in the gas?

108 A source S and a detector D of radio waves are a distance d apart on level ground (Fig. 17-52). Radio waves of wavelength λ reach D either along a straight path or by reflecting (bouncing) from a certain layer in the atmosphere. When the layer is at height H, the two waves reaching D are exactly in phase. If the layer gradually rises, the phase difference between the two waves gradually shifts, until they are exactly out of phase when the layer is at height $H + h$. Express λ in terms of d, h, and H.

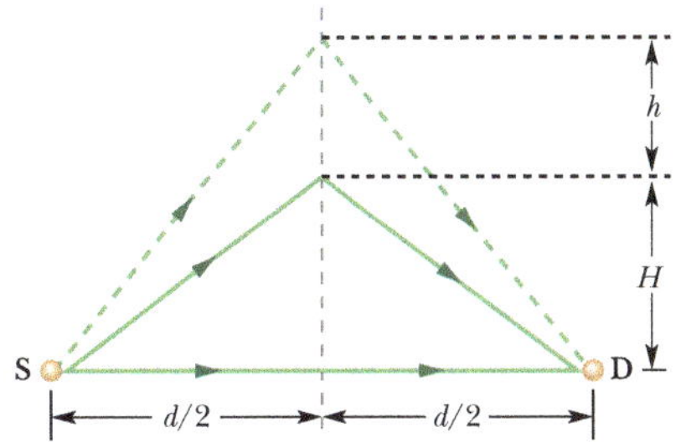

Figure 17-52 Problem 108.

109 In Fig. 17-53, a point source S of sound waves lies near a reflecting wall AB. A sound detector D intercepts sound ray R_1 traveling directly from S. It also intercepts sound ray R_2 that reflects from the wall such that the angle of incidence θ_i is equal to the angle of reflection θ_r. Assume that the reflection of sound by the wall causes a phase shift of 0.500λ. If the distances are $d_1 = 2.50$ m, $d_2 = 20.0$ m, and $d_3 = 12.5$ m, what are the (a) lowest and (b) second lowest frequency at which R_1 and R_2 are in phase at D?

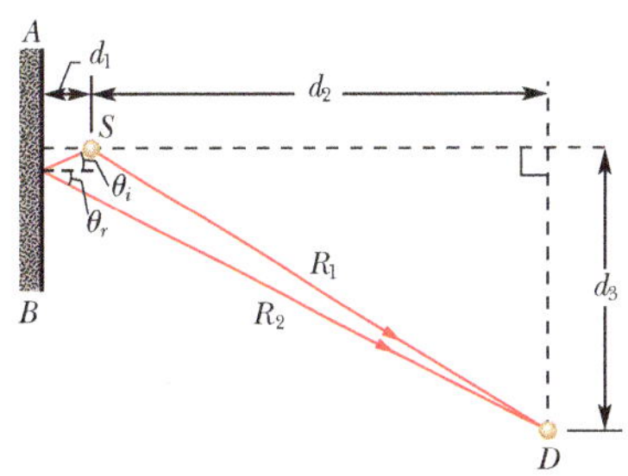

Figure 17-53 Problem 109.

110 A person on a railroad car blows a trumpet note at 440 Hz. The car is moving toward a wall at 20.0 m/s. Find the sound frequency (a) at the wall and (b) reflected back to the trumpeter.

111 A listener at rest (with respect to the air and the ground) hears a signal of frequency f_1 from a source moving toward him with a velocity of 15 m/s, due east. If the listener then moves toward the approaching source with a velocity of 25 m/s, due west, he hears a frequency f_2 that differs from f_1 by 37 Hz. What is the frequency of the source? (Take the speed of sound in air to be 340 m/s.)

CHAPTER 18

Temperature, Heat, and the First Law of Thermodynamics

18-1 TEMPERATURE

Learning Objectives

After reading this module, you should be able to . . .

18.01 Identify the lowest temperature as 0 on the Kelvin scale (absolute zero).

18.02 Explain the zeroth law of thermodynamics.

18.03 Explain the conditions for the triple-point temperature.

18.04 Explain the conditions for measuring a temperature with a constant-volume gas thermometer.

18.05 For a constant-volume gas thermometer, relate the pressure and temperature of the gas in some given state to the pressure and temperature at the triple point.

Key Ideas

- Temperature is an SI base quantity related to our sense of hot and cold. It is measured with a thermometer, which contains a working substance with a measurable property, such as length or pressure, that changes in a regular way as the substance becomes hotter or colder.
- When a thermometer and some other object are placed in contact with each other, they eventually reach thermal equilibrium. The reading of the thermometer is then taken to be the temperature of the other object. The process provides consistent and useful temperature measurements because of the zeroth law of thermodynamics: If bodies A and B are each in thermal equilibrium with a third body C (the thermometer), then A and B are in thermal equilibrium with each other.
- In the SI system, temperature is measured on the Kelvin scale, which is based on the triple point of water (273.16 K). Other temperatures are then defined by use of a constant-volume gas thermometer, in which a sample of gas is maintained at constant volume so its pressure is proportional to its temperature. We define the temperature T as measured with a gas thermometer to be

$$T = (273.16\ \text{K})\left(\lim_{\text{gas}\to 0} \frac{p}{p_3}\right).$$

Here T is in kelvins, and p_3 and p are the pressures of the gas at 273.16 K and the measured temperature, respectively.

What Is Physics?

One of the principal branches of physics and engineering is **thermodynamics,** which is the study and application of the *thermal energy* (often called the *internal energy*) of systems. One of the central concepts of thermodynamics is temperature. Since childhood, you have been developing a working knowledge of thermal energy and temperature. For example, you know to be cautious with hot foods and hot stoves and to store perishable foods in cool or cold compartments. You also know how to control the temperature inside home and car, and how to protect yourself from wind chill and heat stroke.

Examples of how thermodynamics figures into everyday engineering and science are countless. Automobile engineers are concerned with the heating of a car engine, such as during a NASCAR race. Food engineers are concerned both with the proper heating of foods, such as pizzas being microwaved, and with the proper cooling of foods, such as TV dinners being quickly frozen at a processing plant. Geologists are concerned with the transfer of thermal energy in an El Niño event and in the gradual warming of ice expanses in the Arctic and Antarctic.

Agricultural engineers are concerned with the weather conditions that determine whether the agriculture of a country thrives or vanishes. Medical engineers are concerned with how a patient's temperature might distinguish between a benign viral infection and a cancerous growth.

The starting point in our discussion of thermodynamics is the concept of temperature and how it is measured.

Temperature

Temperature is one of the seven SI base quantities. Physicists measure temperature on the **Kelvin scale,** which is marked in units called *kelvins.* Although the temperature of a body apparently has no upper limit, it does have a lower limit; this limiting low temperature is taken as the zero of the Kelvin temperature scale. Room temperature is about 290 kelvins, or 290 K as we write it, above this *absolute zero.* Figure 18-1 shows a wide range of temperatures.

When the universe began 13.7 billion years ago, its temperature was about 10^{39} K. As the universe expanded it cooled, and it has now reached an average temperature of about 3 K. We on Earth are a little warmer than that because we happen to live near a star. Without our Sun, we too would be at 3 K (or, rather, we could not exist).

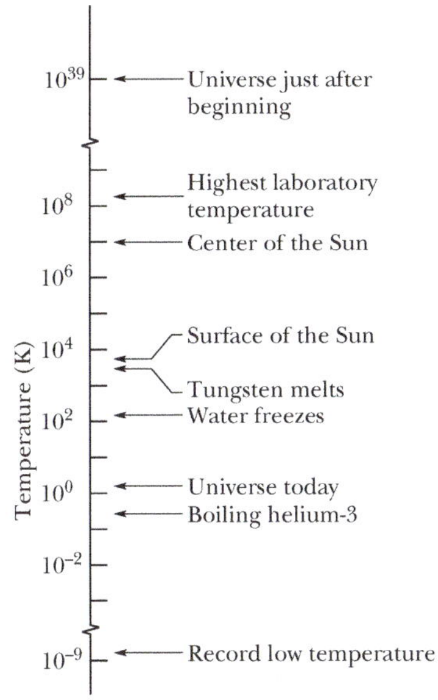

Figure 18-1 Some temperatures on the Kelvin scale. Temperature $T = 0$ corresponds to $10^{-\infty}$ and cannot be plotted on this logarithmic scale.

The Zeroth Law of Thermodynamics

The properties of many bodies change as we alter their temperature, perhaps by moving them from a refrigerator to a warm oven. To give a few examples: As their temperature increases, the volume of a liquid increases, a metal rod grows a little longer, and the electrical resistance of a wire increases, as does the pressure exerted by a confined gas. We can use any one of these properties as the basis of an instrument that will help us pin down the concept of temperature.

Figure 18-2 shows such an instrument. Any resourceful engineer could design and construct it, using any one of the properties listed above. The instrument is fitted with a digital readout display and has the following properties: If you heat it (say, with a Bunsen burner), the displayed number starts to increase; if you then put it into a refrigerator, the displayed number starts to decrease. The instrument is not calibrated in any way, and the numbers have (as yet) no physical meaning. The device is a *thermoscope* but not (as yet) a *thermometer.*

Suppose that, as in Fig. 18-3*a*, we put the thermoscope (which we shall call body T) into intimate contact with another body (body A). The entire system is confined within a thick-walled insulating box. The numbers displayed by the thermoscope roll by until, eventually, they come to rest (let us say the reading is "137.04") and no further change takes place. In fact, we suppose that every measurable property of body T and of body A has assumed a stable, unchanging value. Then we say that the two bodies are in *thermal equilibrium* with each other. Even though the displayed readings for body T have not been calibrated, we conclude that bodies T and A must be at the same (unknown) temperature.

Suppose that we next put body T into intimate contact with body B (Fig. 18-3*b*) and find that the two bodies come to thermal equilibrium *at the same reading of the thermoscope.* Then bodies T and B must be at the same (still unknown) temperature. If we now put bodies A and B into intimate contact (Fig. 18-3*c*), are they immediately in thermal equilibrium with each other? Experimentally, we find that they are.

The experimental fact shown in Fig. 18-3 is summed up in the **zeroth law of thermodynamics:**

> If bodies A and B are each in thermal equilibrium with a third body T, then A and B are in thermal equilibrium with each other.

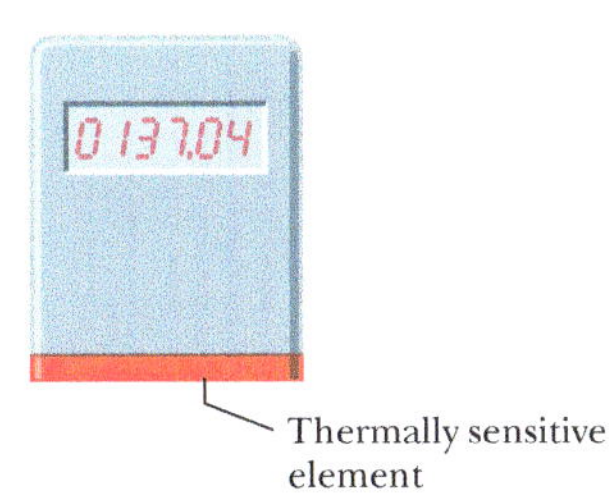

Figure 18-2 A thermoscope. The numbers increase when the device is heated and decrease when it is cooled. The thermally sensitive element could be—among many possibilities—a coil of wire whose electrical resistance is measured and displayed.

In less formal language, the message of the zeroth law is: "Every body has a property called **temperature.** When two bodies are in thermal equilibrium, their temperatures are equal. And vice versa." We can now make our thermoscope

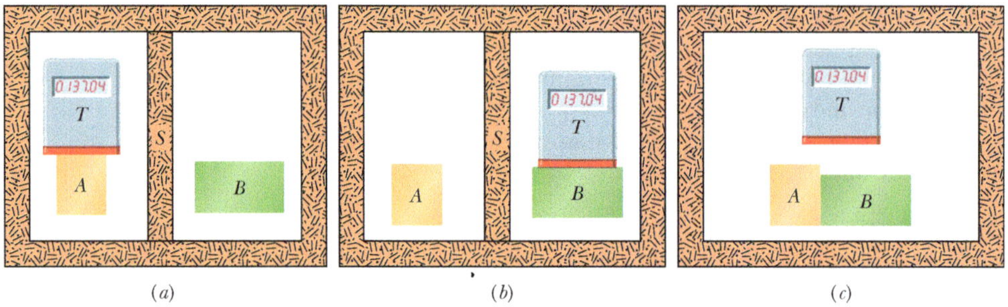

Figure 18-3 (*a*) Body *T* (a thermoscope) and body *A* are in thermal equilibrium. (Body *S* is a thermally insulating screen.) (*b*) Body *T* and body *B* are also in thermal equilibrium, at the same reading of the thermoscope. (*c*) If (*a*) and (*b*) are true, the zeroth law of thermodynamics states that body *A* and body *B* are also in thermal equilibrium.

(the third body *T*) into a thermometer, confident that its readings will have physical meaning. All we have to do is calibrate it.

We use the zeroth law constantly in the laboratory. If we want to know whether the liquids in two beakers are at the same temperature, we measure the temperature of each with a thermometer. We do not need to bring the two liquids into intimate contact and observe whether they are or are not in thermal equilibrium.

The zeroth law, which has been called a logical afterthought, came to light only in the 1930s, long after the first and second laws of thermodynamics had been discovered and numbered. Because the concept of temperature is fundamental to those two laws, the law that establishes temperature as a valid concept should have the lowest number—hence the zero.

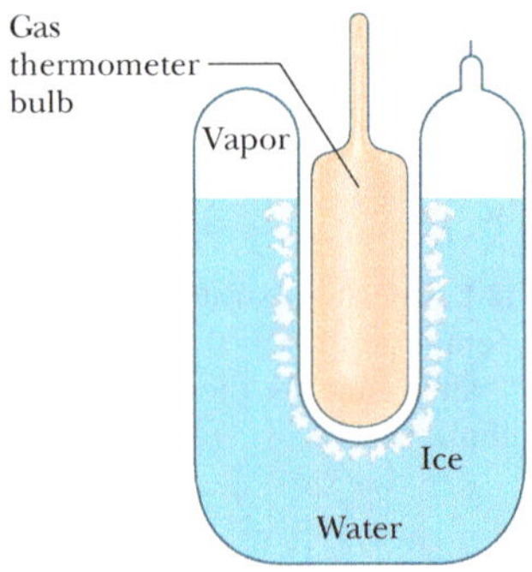

Figure 18-4 A triple-point cell, in which solid ice, liquid water, and water vapor coexist in thermal equilibrium. By international agreement, the temperature of this mixture has been defined to be 273.16 K. The bulb of a constant-volume gas thermometer is shown inserted into the well of the cell.

Measuring Temperature

Here we first define and measure temperatures on the Kelvin scale. Then we calibrate a thermoscope so as to make it a thermometer.

The Triple Point of Water

To set up a temperature scale, we pick some reproducible thermal phenomenon and, quite arbitrarily, assign a certain Kelvin temperature to its environment; that is, we select a *standard fixed point* and give it a standard fixed-point *temperature*. We could, for example, select the freezing point or the boiling point of water but, for technical reasons, we select instead the **triple point of water.**

Liquid water, solid ice, and water vapor (gaseous water) can coexist, in thermal equilibrium, at only one set of values of pressure and temperature. Figure 18-4 shows a triple-point cell, in which this so-called triple point of water can be achieved in the laboratory. By international agreement, the triple point of water has been assigned a value of 273.16 K as the standard fixed-point temperature for the calibration of thermometers; that is,

$$T_3 = 273.16\ \text{K} \quad \text{(triple-point temperature)}, \tag{18-1}$$

in which the subscript 3 means "triple point." This agreement also sets the size of the kelvin as 1/273.16 of the difference between the triple-point temperature of water and absolute zero.

Note that we do not use a degree mark in reporting Kelvin temperatures. It is 300 K (not 300°K), and it is read "300 kelvins" (not "300 degrees Kelvin"). The usual SI prefixes apply. Thus, 0.0035 K is 3.5 mK. No distinction in nomenclature is made between Kelvin temperatures and temperature differences, so we can write, "the boiling point of sulfur is 717.8 K" and "the temperature of this water bath was raised by 8.5 K."

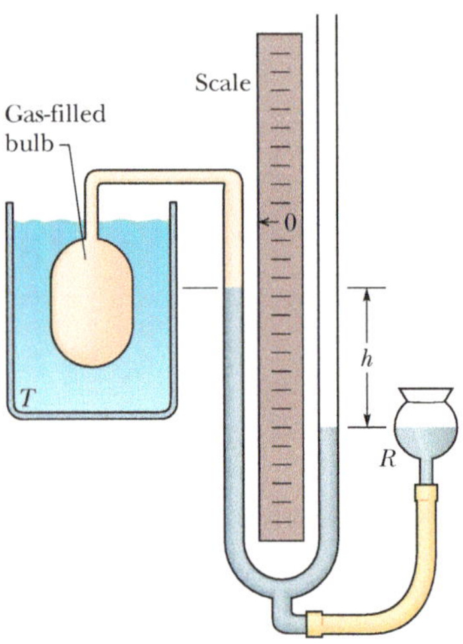

Figure 18-5 A constant-volume gas thermometer, its bulb immersed in a liquid whose temperature *T* is to be measured.

The Constant-Volume Gas Thermometer

The standard thermometer, against which all other thermometers are calibrated, is based on the pressure of a gas in a fixed volume. Figure 18-5 shows such a **constant-volume gas thermometer;** it consists of a gas-filled bulb connected by a tube to a mercury manometer. By raising and lowering reservoir *R*, the mercury

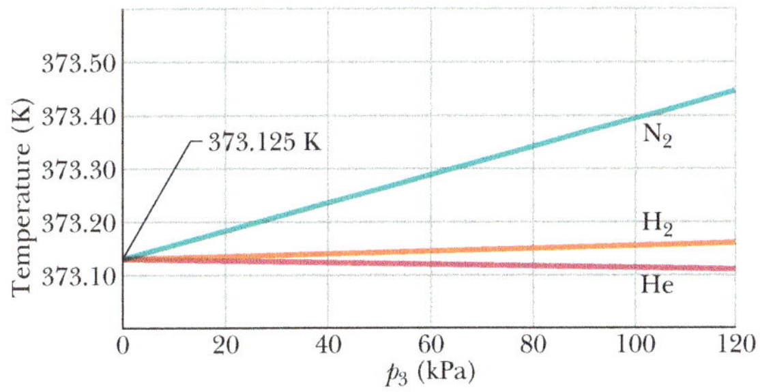

Figure 18-6 Temperatures measured by a constant-volume gas thermometer, with its bulb immersed in boiling water. For temperature calculations using Eq. 18-5, pressure p_3 was measured at the triple point of water. Three different gases in the thermometer bulb gave generally different results at different gas pressures, but as the amount of gas was decreased (decreasing p_3), all three curves converged to 373.125 K.

level in the left arm of the U-tube can always be brought to the zero of the scale to keep the gas volume constant (variations in the gas volume can affect temperature measurements).

The temperature of any body in thermal contact with the bulb (such as the liquid surrounding the bulb in Fig. 18-5) is then defined to be

$$T = Cp, \tag{18-2}$$

in which p is the pressure exerted by the gas and C is a constant. From Eq. 14-10, the pressure p is

$$p = p_0 - \rho g h, \tag{18-3}$$

in which p_0 is the atmospheric pressure, ρ is the density of the mercury in the manometer, and h is the measured difference between the mercury levels in the two arms of the tube.* (The minus sign is used in Eq. 18-3 because pressure p is measured *above* the level at which the pressure is p_0.)

If we next put the bulb in a triple-point cell (Fig. 18-4), the temperature now being measured is

$$T_3 = Cp_3, \tag{18-4}$$

in which p_3 is the gas pressure now. Eliminating C between Eqs. 18-2 and 18-4 gives us the temperature as

$$T = T_3\left(\frac{p}{p_3}\right) = (273.16\ \text{K})\left(\frac{p}{p_3}\right) \quad \text{(provisional)}. \tag{18-5}$$

We still have a problem with this thermometer. If we use it to measure, say, the boiling point of water, we find that different gases in the bulb give slightly different results. However, as we use smaller and smaller amounts of gas to fill the bulb, the readings converge nicely to a single temperature, no matter what gas we use. Figure 18-6 shows this convergence for three gases.

Thus the recipe for measuring a temperature with a gas thermometer is

$$T = (273.16\ \text{K})\left(\lim_{\text{gas}\to 0} \frac{p}{p_3}\right). \tag{18-6}$$

The recipe instructs us to measure an unknown temperature T as follows: Fill the thermometer bulb with an arbitrary amount of *any* gas (for example, nitrogen) and measure p_3 (using a triple-point cell) and p, the gas pressure at the temperature being measured. (Keep the gas volume the same.) Calculate the ratio p/p_3. Then repeat both measurements with a smaller amount of gas in the bulb, and again calculate this ratio. Continue this way, using smaller and smaller amounts of gas, until you can extrapolate to the ratio p/p_3 that you would find if there were approximately no gas in the bulb. Calculate the temperature T by substituting that extrapolated ratio into Eq. 18-6. (The temperature is called the *ideal gas temperature*.)

---

*For pressure units, we shall use units introduced in Module 14-1. The SI unit for pressure is the newton per square meter, which is called the pascal (Pa). The pascal is related to other common pressure units by

$$1\ \text{atm} = 1.01 \times 10^5\ \text{Pa} = 760\ \text{torr} = 14.7\ \text{lb/in.}^2.$$

18-2 THE CELSIUS AND FAHRENHEIT SCALES

Learning Objectives

After reading this module, you should be able to . . .

18.06 Convert a temperature between any two (linear) temperature scales, including the Celsius, Fahrenheit, and Kelvin scales.

18.07 Identify that a change of one degree is the same on the Celsius and Kelvin scales.

Key Idea

- The Celsius temperature scale is defined by

$$T_C = T - 273.15°,$$

with T in kelvins. The Fahrenheit temperature scale is defined by

$$T_F = \tfrac{9}{5}T_C + 32°.$$

The Celsius and Fahrenheit Scales

So far, we have discussed only the Kelvin scale, used in basic scientific work. In nearly all countries of the world, the Celsius scale (formerly called the centigrade scale) is the scale of choice for popular and commercial use and much scientific use. Celsius temperatures are measured in degrees, and the Celsius degree has the same size as the kelvin. However, the zero of the Celsius scale is shifted to a more convenient value than absolute zero. If T_C represents a Celsius temperature and T a Kelvin temperature, then

$$T_C = T - 273.15°. \tag{18-7}$$

In expressing temperatures on the Celsius scale, the degree symbol is commonly used. Thus, we write 20.00°C for a Celsius reading but 293.15 K for a Kelvin reading.

The Fahrenheit scale, used in the United States, employs a smaller degree than the Celsius scale and a different zero of temperature. You can easily verify both these differences by examining an ordinary room thermometer on which both scales are marked. The relation between the Celsius and Fahrenheit scales is

$$T_F = \tfrac{9}{5}T_C + 32°, \tag{18-8}$$

where T_F is Fahrenheit temperature. Converting between these two scales can be done easily by remembering a few corresponding points, such as the freezing and boiling points of water (Table 18-1). Figure 18-7 compares the Kelvin, Celsius, and Fahrenheit scales.

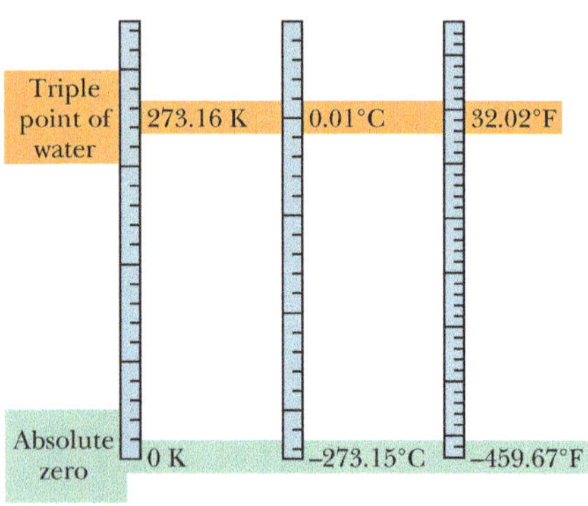

Figure 18-7 The Kelvin, Celsius, and Fahrenheit temperature scales compared.

Table 18-1 Some Corresponding Temperatures

Temperature	°C	°F
Boiling point of water[a]	100	212
Normal body temperature	37.0	98.6
Accepted comfort level	20	68
Freezing point of water[a]	0	32
Zero of Fahrenheit scale	≈ −18	0
Scales coincide	−40	−40

[a]Strictly, the boiling point of water on the Celsius scale is 99.975°C, and the freezing point is 0.00°C. Thus, there is slightly less than 100 C° between those two points.

We use the letters C and F to distinguish measurements and degrees on the two scales. Thus,

$$0°C = 32°F$$

means that 0° on the Celsius scale measures the same temperature as 32° on the Fahrenheit scale, whereas

$$5\ C° = 9\ F°$$

means that a temperature difference of 5 Celsius degrees (note the degree symbol appears *after* C) is equivalent to a temperature difference of 9 Fahrenheit degrees.

Checkpoint 1

The figure here shows three linear temperature scales with the freezing and boiling points of water indicated. (a) Rank the degrees on these scales by size, greatest first. (b) Rank the following temperatures, highest first: 50°X, 50°W, and 50°Y.

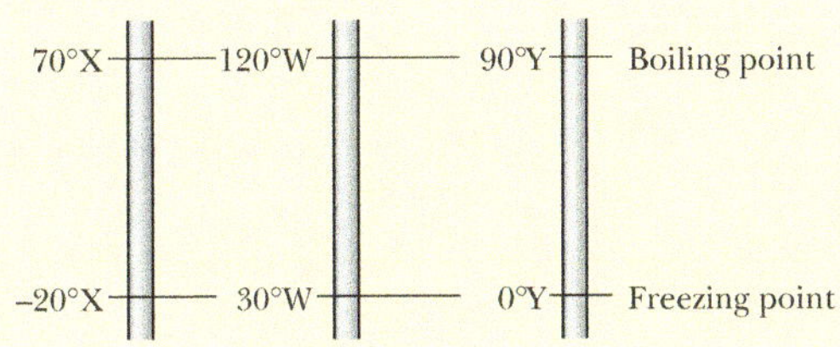

Sample Problem 18.01 Conversion between two temperature scales

Suppose you come across old scientific notes that describe a temperature scale called Z on which the boiling point of water is 65.0°Z and the freezing point is −14.0°Z. To what temperature on the Fahrenheit scale would a temperature of $T = -98.0$°Z correspond? Assume that the Z scale is linear; that is, the size of a Z degree is the same everywhere on the Z scale.

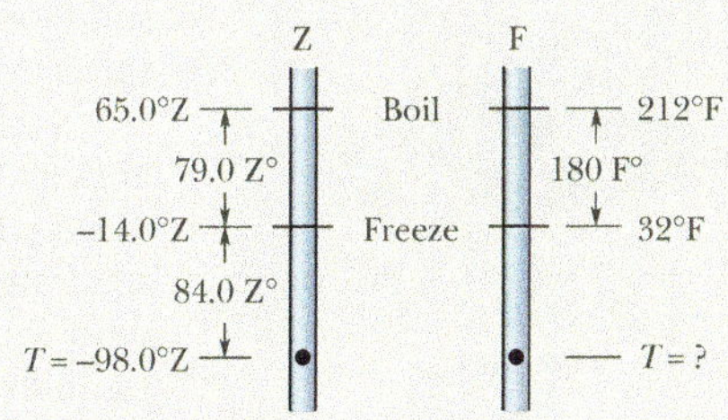

Figure 18-8 An unknown temperature scale compared with the Fahrenheit temperature scale.

KEY IDEA

A conversion factor between two (linear) temperature scales can be calculated by using two known (benchmark) temperatures, such as the boiling and freezing points of water. The number of degrees between the known temperatures on one scale is equivalent to the number of degrees between them on the other scale.

Calculations: We begin by relating the given temperature T to *either* known temperature on the Z scale. Since $T = -98.0$°Z is closer to the freezing point (−14.0°Z) than to the boiling point (65.0°Z), we use the freezing point. Then we note that the T we seek is *below this point* by −14.0°Z − (−98.0°Z) = 84.0 Z° (Fig. 18-8). (Read this difference as "84.0 Z degrees.")

Next, we set up a conversion factor between the Z and Fahrenheit scales to convert this difference. To do so, we use *both* known temperatures on the Z scale and the corresponding temperatures on the Fahrenheit scale. On the Z scale, the difference between the boiling and freezing points is 65.0°Z − (−14.0°Z) = 79.0 Z°. On the Fahrenheit scale, it is 212°F − 32.0°F = 180 F°. Thus, a temperature difference of 79.0 Z° is equivalent to a temperature difference of 180 F° (Fig. 18-8), and we can use the ratio (180 F°)/(79.0 Z°) as our conversion factor.

Now, since T is below the freezing point by 84.0 Z°, it must also be below the freezing point by

$$(84.0\ Z°)\frac{180\ F°}{79.0\ Z°} = 191\ F°.$$

Because the freezing point is at 32.0°F, this means that

$$T = 32.0°F - 191\ F° = -159°F. \qquad \text{(Answer)}$$

Additional examples, video, and practice available at *WileyPLUS*

18-3 THERMAL EXPANSION

Learning Objectives

After reading this module, you should be able to . . .

18.08 For one-dimensional thermal expansion, apply the relationship between the temperature change ΔT, the length change ΔL, the initial length L, and the coefficient of linear expansion α.

18.09 For two-dimensional thermal expansion, use one-dimensional thermal expansion to find the change in area.

18.10 For three-dimensional thermal expansion, apply the relationship between the temperature change ΔT, the volume change ΔV, the initial volume V, and the coefficient of volume expansion β.

Key Ideas

- All objects change size with changes in temperature. For a temperature change ΔT, a change ΔL in any linear dimension L is given by

$$\Delta L = L\alpha\,\Delta T,$$

in which α is the coefficient of linear expansion.

- The change ΔV in the volume V of a solid or liquid is

$$\Delta V = V\beta\,\Delta T.$$

Here $\beta = 3\alpha$ is the material's coefficient of volume expansion.

Thermal Expansion

Hugh Thomas/BWP Media/Getty Images, Inc.

Figure 18-9 When a Concorde flew faster than the speed of sound, thermal expansion due to the rubbing by passing air increased the aircraft's length by about 12.5 cm. (The temperature increased to about 128°C at the aircraft nose and about 90°C at the tail, and cabin windows were noticeably warm to the touch.)

You can often loosen a tight metal jar lid by holding it under a stream of hot water. Both the metal of the lid and the glass of the jar expand as the hot water adds energy to their atoms. (With the added energy, the atoms can move a bit farther from one another than usual, against the spring-like interatomic forces that hold every solid together.) However, because the atoms in the metal move farther apart than those in the glass, the lid expands more than the jar and thus is loosened.

Such **thermal expansion** of materials with an increase in temperature must be anticipated in many common situations. When a bridge is subject to large seasonal changes in temperature, for example, sections of the bridge are separated by *expansion slots* so that the sections have room to expand on hot days without the bridge buckling. When a dental cavity is filled, the filling material must have the same thermal expansion properties as the surrounding tooth; otherwise, consuming cold ice cream and then hot coffee would be very painful. When the Concorde aircraft (Fig. 18-9) was built, the design had to allow for the thermal expansion of the fuselage during supersonic flight because of frictional heating by the passing air.

The thermal expansion properties of some materials can be put to common use. Thermometers and thermostats may be based on the differences in expansion between the components of a *bimetal strip* (Fig. 18-10). Also, the familiar liquid-in-glass thermometers are based on the fact that liquids such as mercury and alcohol expand to a different (greater) extent than their glass containers.

Linear Expansion

If the temperature of a metal rod of length L is raised by an amount ΔT, its length is found to increase by an amount

$$\Delta L = L\alpha\,\Delta T, \tag{18-9}$$

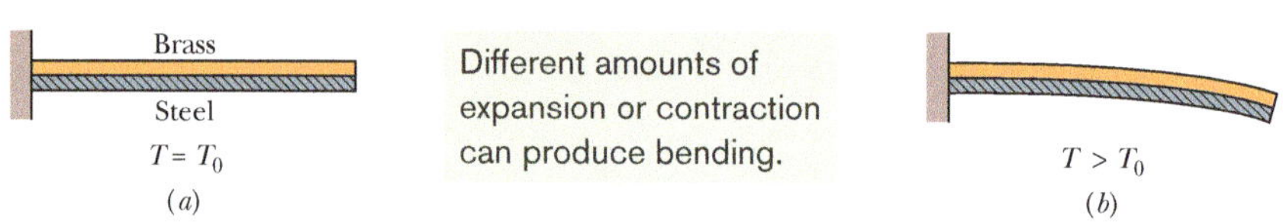

Figure 18-10 (*a*) A bimetal strip, consisting of a strip of brass and a strip of steel welded together, at temperature T_0. (*b*) The strip bends as shown at temperatures above this reference temperature. Below the reference temperature the strip bends the other way. Many thermostats operate on this principle, making and breaking an electrical contact as the temperature rises and falls.

in which α is a constant called the **coefficient of linear expansion.** The coefficient α has the unit "per degree" or "per kelvin" and depends on the material. Although α varies somewhat with temperature, for most practical purposes it can be taken as constant for a particular material. Table 18-2 shows some coefficients of linear expansion. Note that the unit C° there could be replaced with the unit K.

The thermal expansion of a solid is like photographic enlargement except it is in three dimensions. Figure 18-11*b* shows the (exaggerated) thermal expansion of a steel ruler. Equation 18-9 applies to every linear dimension of the ruler, including its edge, thickness, diagonals, and the diameters of the circle etched on it and the circular hole cut in it. If the disk cut from that hole originally fits snugly in the hole, it will continue to fit snugly if it undergoes the same temperature increase as the ruler.

Table 18-2 Some Coefficients of Linear Expansion[a]

Substance	α (10^{-6}/C°)
Ice (at 0°C)	51
Lead	29
Aluminum	23
Brass	19
Copper	17
Concrete	12
Steel	11
Glass (ordinary)	9
Glass (Pyrex)	3.2
Diamond	1.2
Invar[b]	0.7
Fused quartz	0.5

[a]Room temperature values except for the listing for ice.

[b]This alloy was designed to have a low coefficient of expansion. The word is a shortened form of "invariable."

Volume Expansion

If all dimensions of a solid expand with temperature, the volume of that solid must also expand. For liquids, volume expansion is the only meaningful expansion parameter. If the temperature of a solid or liquid whose volume is V is increased by an amount ΔT, the increase in volume is found to be

$$\Delta V = V\beta\,\Delta T, \qquad (18\text{-}10)$$

where β is the **coefficient of volume expansion** of the solid or liquid. The coefficients of volume expansion and linear expansion for a solid are related by

$$\beta = 3\alpha. \qquad (18\text{-}11)$$

The most common liquid, water, does not behave like other liquids. Above about 4°C, water expands as the temperature rises, as we would expect. Between 0 and about 4°C, however, water *contracts* with increasing temperature. Thus, at about 4°C, the density of water passes through a maximum. At all other temperatures, the density of water is less than this maximum value.

This behavior of water is the reason lakes freeze from the top down rather than from the bottom up. As water on the surface is cooled from, say, 10°C toward the freezing point, it becomes denser ("heavier") than lower water and sinks to the bottom. Below 4°C, however, further cooling makes the water then on the surface *less* dense ("lighter") than the lower water, so it stays on the surface until it freezes. Thus the surface freezes while the lower water is still liquid. If lakes froze from the bottom up, the ice so formed would tend not to melt completely during the summer, because it would be insulated by the water above. After a few years, many bodies of open water in the temperate zones of Earth would be frozen solid all year round—and aquatic life could not exist.

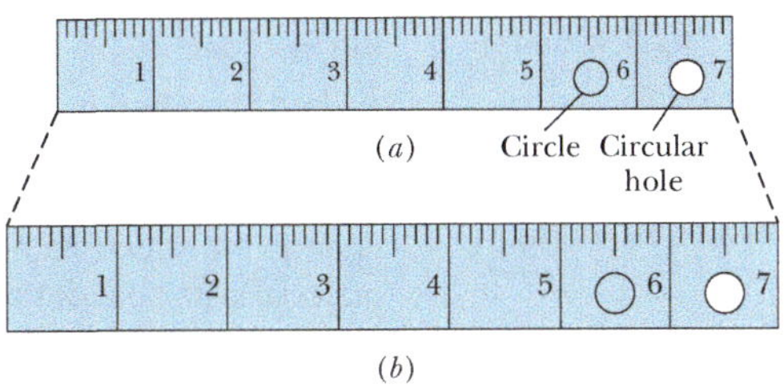

Figure 18-11 The same steel ruler at two different temperatures. When it expands, the scale, the numbers, the thickness, and the diameters of the circle and circular hole are all increased by the same factor. (The expansion has been exaggerated for clarity.)

Checkpoint 2

The figure here shows four rectangular metal plates, with sides of L, $2L$, or $3L$. They are all made of the same material, and their temperature is to be increased by the same amount. Rank the plates according to the expected increase in (a) their vertical heights and (b) their areas, greatest first.

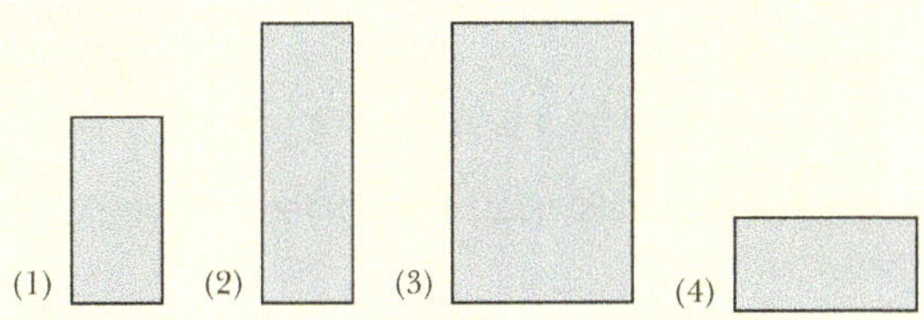

Sample Problem 18.02 Thermal expansion of a volume

On a hot day in Las Vegas, an oil trucker loaded 37 000 L of diesel fuel. He encountered cold weather on the way to Payson, Utah, where the temperature was 23.0 K lower than in Las Vegas, and where he delivered his entire load. How many liters did he deliver? The coefficient of volume expansion for diesel fuel is $9.50 \times 10^{-4}/\mathrm{C}°$, and the coefficient of linear expansion for his steel truck tank is $11 \times 10^{-6}/\mathrm{C}°$.

KEY IDEA

The volume of the diesel fuel depends directly on the temperature. Thus, because the temperature decreased, the volume of the fuel did also, as given by Eq. 18-10 ($\Delta V = V\beta\,\Delta T$).

Calculations: We find

$$\Delta V = (37\,000\ \mathrm{L})(9.50 \times 10^{-4}/\mathrm{C}°)(-23.0\ \mathrm{K}) = -808\ \mathrm{L}.$$

Thus, the amount delivered was

$$V_{\mathrm{del}} = V + \Delta V = 37\,000\ \mathrm{L} - 808\ \mathrm{L} = 36\,190\ \mathrm{L}. \quad \text{(Answer)}$$

Note that the thermal expansion of the steel tank has nothing to do with the problem. Question: Who paid for the "missing" diesel fuel?

WILEY PLUS Additional examples, video, and practice available at *WileyPLUS*

18-4 ABSORPTION OF HEAT

Learning Objectives

After reading this module, you should be able to . . .

18.11 Identify that *thermal energy* is associated with the random motions of the microscopic bodies in an object.

18.12 Identify that *heat* Q is the amount of transferred energy (either to or from an object's thermal energy) due to a temperature difference between the object and its environment.

18.13 Convert energy units between various measurement systems.

18.14 Convert between mechanical or electrical energy and thermal energy.

18.15 For a temperature change ΔT of a substance, relate the change to the heat transfer Q and the substance's heat capacity C.

18.16 For a temperature change ΔT of a substance, relate the change to the heat transfer Q and the substance's specific heat c and mass m.

18.17 Identify the three phases of matter.

18.18 For a phase change of a substance, relate the heat transfer Q, the heat of transformation L, and the amount of mass m transformed.

18.19 Identify that if a heat transfer Q takes a substance across a phase-change temperature, the transfer must be calculated in steps: (a) a temperature change to reach the phase-change temperature, (b) the phase change, and then (c) any temperature change that moves the substance away from the phase-change temperature.

Key Ideas

- Heat Q is energy that is transferred between a system and its environment because of a temperature difference between them. It can be measured in joules (J), calories (cal), kilocalories (Cal or kcal), or British thermal units (Btu), with

$$1\ \mathrm{cal} = 3.968 \times 10^{-3}\ \mathrm{Btu} = 4.1868\ \mathrm{J}.$$

- If heat Q is absorbed by an object, the object's temperature change $T_f - T_i$ is related to Q by

$$Q = C(T_f - T_i),$$

in which C is the heat capacity of the object. If the object has mass m, then

$$Q = cm(T_f - T_i),$$

where c is the specific heat of the material making up the object.

- The molar specific heat of a material is the heat capacity per mole, which means per 6.02×10^{23} elementary units of the material.

- Heat absorbed by a material may change the material's physical state—for example, from solid to liquid or from liquid to gas. The amount of energy required per unit mass to change the state (but not the temperature) of a particular material is its heat of transformation L. Thus,

$$Q = Lm.$$

- The heat of vaporization L_V is the amount of energy per unit mass that must be added to vaporize a liquid or that must be removed to condense a gas.

- The heat of fusion L_F is the amount of energy per unit mass that must be added to melt a solid or that must be removed to freeze a liquid.

Temperature and Heat

If you take a can of cola from the refrigerator and leave it on the kitchen table, its temperature will rise—rapidly at first but then more slowly—until the temperature of the cola equals that of the room (the two are then in thermal equilibrium). In the same way, the temperature of a cup of hot coffee, left sitting on the table, will fall until it also reaches room temperature.

In generalizing this situation, we describe the cola or the coffee as a *system* (with temperature T_S) and the relevant parts of the kitchen as the *environment* (with temperature T_E) of that system. Our observation is that if T_S is not equal to T_E, then T_S will change (T_E can also change some) until the two temperatures are equal and thus thermal equilibrium is reached.

Such a change in temperature is due to a change in the thermal energy of the system because of a transfer of energy between the system and the system's environment. (Recall that *thermal energy* is an internal energy that consists of the kinetic and potential energies associated with the random motions of the atoms, molecules, and other microscopic bodies within an object.) The transferred energy is called **heat** and is symbolized Q. Heat is *positive* when energy is transferred to a system's thermal energy from its environment (we say that heat is absorbed by the system). Heat is *negative* when energy is transferred from a system's thermal energy to its environment (we say that heat is released or lost by the system).

This transfer of energy is shown in Fig. 18-12. In the situation of Fig. 18-12*a*, in which $T_S > T_E$, energy is transferred from the system to the environment, so Q is negative. In Fig. 18-12*b*, in which $T_S = T_E$, there is no such transfer, Q is zero, and heat is neither released nor absorbed. In Fig. 18-12*c*, in which $T_S < T_E$, the transfer is to the system from the environment; so Q is positive.

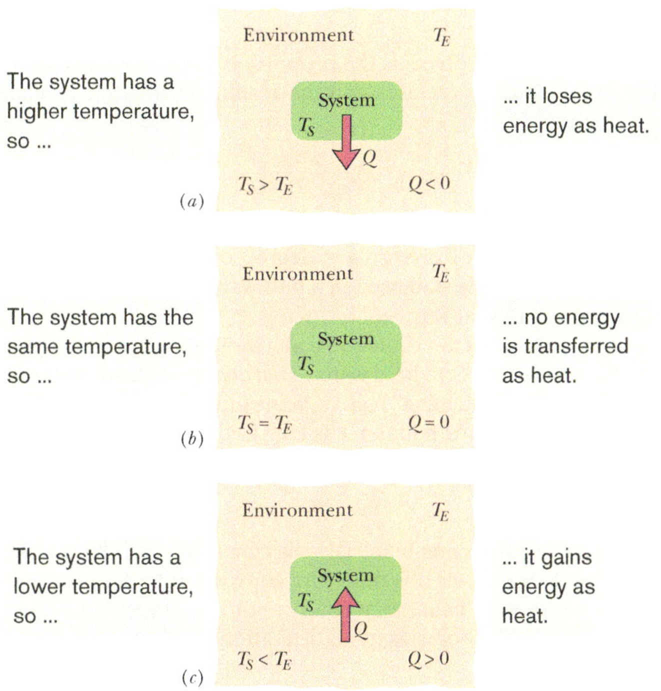

Figure 18-12 If the temperature of a system exceeds that of its environment as in (*a*), heat Q is lost by the system to the environment until thermal equilibrium (*b*) is established. (*c*) If the temperature of the system is below that of the environment, heat is absorbed by the system until thermal equilibrium is established.

We are led then to this definition of heat:

Heat is the energy transferred between a system and its environment because of a temperature difference that exists between them.

Language. Recall that energy can also be transferred between a system and its environment as *work W* via a force acting on a system. Heat and work, unlike temperature, pressure, and volume, are not intrinsic properties of a system. They have meaning only as they describe the transfer of energy into or out of a system. Similarly, the phrase "a $600 transfer" has meaning if it describes the transfer to or from an account, not what is in the account, because the account holds money, not a transfer.

Units. Before scientists realized that heat is transferred energy, heat was measured in terms of its ability to raise the temperature of water. Thus, the **calorie** (cal) was defined as the amount of heat that would raise the temperature of 1 g of water from 14.5°C to 15.5°C. In the British system, the corresponding unit of heat was the **British thermal unit** (Btu), defined as the amount of heat that would raise the temperature of 1 lb of water from 63°F to 64°F.

In 1948, the scientific community decided that since heat (like work) is transferred energy, the SI unit for heat should be the one we use for energy—namely, the **joule.** The calorie is now defined to be 4.1868 J (exactly), with no reference to the heating of water. (The "calorie" used in nutrition, sometimes called the Calorie (Cal), is really a kilocalorie.) The relations among the various heat units are

$$1 \text{ cal} = 3.968 \times 10^{-3} \text{ Btu} = 4.1868 \text{ J}. \tag{18-12}$$

The Absorption of Heat by Solids and Liquids

Heat Capacity

The **heat capacity** C of an object is the proportionality constant between the heat Q that the object absorbs or loses and the resulting temperature change ΔT of the object; that is,

$$Q = C\,\Delta T = C(T_f - T_i), \tag{18-13}$$

in which T_i and T_f are the initial and final temperatures of the object. Heat capacity C has the unit of energy per degree or energy per kelvin. The heat capacity C of, say, a marble slab used in a bun warmer might be 179 cal/C°, which we can also write as 179 cal/K or as 749 J/K.

The word "capacity" in this context is really misleading in that it suggests analogy with the capacity of a bucket to hold water. *That analogy is false,* and you should not think of the object as "containing" heat or being limited in its ability to absorb heat. Heat transfer can proceed without limit as long as the necessary temperature difference is maintained. The object may, of course, melt or vaporize during the process.

Specific Heat

Two objects made of the same material—say, marble—will have heat capacities proportional to their masses. It is therefore convenient to define a "heat capacity per unit mass" or **specific heat** c that refers not to an object but to a unit mass of the material of which the object is made. Equation 18-13 then becomes

$$Q = cm\,\Delta T = cm(T_f - T_i). \tag{18-14}$$

Through experiment we would find that although the heat capacity of a particular marble slab might be 179 cal/C° (or 749 J/K), the specific heat of marble itself (in that slab or in any other marble object) is 0.21 cal/g·C° (or 880 J/kg·K).

From the way the calorie and the British thermal unit were initially defined, the specific heat of water is

$$c = 1\ \text{cal/g}\cdot\text{C}^\circ = 1\ \text{Btu/lb}\cdot\text{F}^\circ = 4186.8\ \text{J/kg}\cdot\text{K}. \qquad (18\text{-}15)$$

Table 18-3 shows the specific heats of some substances at room temperature. Note that the value for water is relatively high. The specific heat of any substance actually depends somewhat on temperature, but the values in Table 18-3 apply reasonably well in a range of temperatures near room temperature.

Table 18-3 Some Specific Heats and Molar Specific Heats at Room Temperature

Substance	Specific Heat $\frac{\text{cal}}{\text{g}\cdot\text{K}}$	Specific Heat $\frac{\text{J}}{\text{kg}\cdot\text{K}}$	Molar Specific Heat $\frac{\text{J}}{\text{mol}\cdot\text{K}}$
Elemental Solids			
Lead	0.0305	128	26.5
Tungsten	0.0321	134	24.8
Silver	0.0564	236	25.5
Copper	0.0923	386	24.5
Aluminum	0.215	900	24.4
Other Solids			
Brass	0.092	380	
Granite	0.19	790	
Glass	0.20	840	
Ice (−10°C)	0.530	2220	
Liquids			
Mercury	0.033	140	
Ethyl alcohol	0.58	2430	
Seawater	0.93	3900	
Water	1.00	4187	

Checkpoint 3

A certain amount of heat Q will warm 1 g of material A by 3 C° and 1 g of material B by 4 C°. Which material has the greater specific heat?

Molar Specific Heat

In many instances the most convenient unit for specifying the amount of a substance is the mole (mol), where

$$1\ \text{mol} = 6.02 \times 10^{23}\ \text{elementary units}$$

of *any* substance. Thus 1 mol of aluminum means 6.02×10^{23} atoms (the atom is the elementary unit), and 1 mol of aluminum oxide means 6.02×10^{23} molecules (the molecule is the elementary unit of the compound).

When quantities are expressed in moles, specific heats must also involve moles (rather than a mass unit); they are then called **molar specific heats.** Table 18-3 shows the values for some elemental solids (each consisting of a single element) at room temperature.

An Important Point

In determining and then using the specific heat of any substance, we need to know the conditions under which energy is transferred as heat. For solids and liquids, we usually assume that the sample is under constant pressure (usually atmospheric) during the transfer. It is also conceivable that the sample is held at constant volume while the heat is absorbed. This means that thermal expansion of the sample is prevented by applying external pressure. For solids and liquids, this is very hard to arrange experimentally, but the effect can be calculated, and it turns out that the specific heats under constant pressure and constant volume for any solid or liquid differ usually by no more than a few percent. Gases, as you will see, have quite different values for their specific heats under constant-pressure conditions and under constant-volume conditions.

Heats of Transformation

When energy is absorbed as heat by a solid or liquid, the temperature of the sample does not necessarily rise. Instead, the sample may change from one *phase,* or *state,* to another. Matter can exist in three common states: In the *solid state,* the molecules of a sample are locked into a fairly rigid structure by their mutual attraction. In the *liquid state,* the molecules have more energy and move about more. They may form brief clusters, but the sample does not have a rigid structure and can flow or settle into a container. In the *gas,* or *vapor, state,* the molecules have even more energy, are free of one another, and can fill up the full volume of a container.

Melting. To *melt* a solid means to change it from the solid state to the liquid state. The process requires energy because the molecules of the solid must be freed from their rigid structure. Melting an ice cube to form liquid water is a common example. To *freeze* a liquid to form a solid is the reverse of melting and requires that energy be removed from the liquid, so that the molecules can settle into a rigid structure.

Table 18-4 Some Heats of Transformation

	Melting		Boiling	
Substance	Melting Point (K)	Heat of Fusion L_F (kJ/kg)	Boiling Point (K)	Heat of Vaporization L_V (kJ/kg)
Hydrogen	14.0	58.0	20.3	455
Oxygen	54.8	13.9	90.2	213
Mercury	234	11.4	630	296
Water	273	333	373	2256
Lead	601	23.2	2017	858
Silver	1235	105	2323	2336
Copper	1356	207	2868	4730

Vaporizing. To *vaporize* a liquid means to change it from the liquid state to the vapor (gas) state. This process, like melting, requires energy because the molecules must be freed from their clusters. Boiling liquid water to transfer it to water vapor (or steam—a gas of individual water molecules) is a common example. *Condensing* a gas to form a liquid is the reverse of vaporizing; it requires that energy be removed from the gas, so that the molecules can cluster instead of flying away from one another.

The amount of energy per unit mass that must be transferred as heat when a sample completely undergoes a phase change is called the **heat of transformation** L. Thus, when a sample of mass m completely undergoes a phase change, the total energy transferred is

$$Q = Lm. \tag{18-16}$$

When the phase change is from liquid to gas (then the sample must absorb heat) or from gas to liquid (then the sample must release heat), the heat of transformation is called the **heat of vaporization** L_V. For water at its normal boiling or condensation temperature,

$$L_V = 539 \text{ cal/g} = 40.7 \text{ kJ/mol} = 2256 \text{ kJ/kg}. \tag{18-17}$$

When the phase change is from solid to liquid (then the sample must absorb heat) or from liquid to solid (then the sample must release heat), the heat of transformation is called the **heat of fusion** L_F. For water at its normal freezing or melting temperature,

$$L_F = 79.5 \text{ cal/g} = 6.01 \text{ kJ/mol} = 333 \text{ kJ/kg}. \tag{18-18}$$

Table 18-4 shows the heats of transformation for some substances.

Sample Problem 18.03 Hot slug in water, coming to equilibrium

A copper slug whose mass m_c is 75 g is heated in a laboratory oven to a temperature T of 312°C. The slug is then dropped into a glass beaker containing a mass $m_w = 220$ g of water. The heat capacity C_b of the beaker is 45 cal/K. The initial temperature T_i of the water and the beaker is 12°C. Assuming that the slug, beaker, and water are an isolated system and the water does not vaporize, find the final temperature T_f of the system at thermal equilibrium.

KEY IDEAS

(1) Because the system is isolated, the system's total energy cannot change and only internal transfers of thermal energy can occur. (2) Because nothing in the system undergoes a phase change, the thermal energy transfers can only change the temperatures.

Calculations: To relate the transfers to the temperature changes, we can use Eqs. 18-13 and 18-14 to write

$$\text{for the water: } Q_w = c_w m_w (T_f - T_i); \tag{18-19}$$

$$\text{for the beaker: } Q_b = C_b (T_f - T_i); \tag{18-20}$$

$$\text{for the copper: } Q_c = c_c m_c (T_f - T). \tag{18-21}$$

Because the total energy of the system cannot change, the sum of these three energy transfers is zero:

$$Q_w + Q_b + Q_c = 0. \tag{18-22}$$

Substituting Eqs. 18-19 through 18-21 into Eq. 18-22 yields

$$c_w m_w(T_f - T_i) + C_b(T_f - T_i) + c_c m_c(T_f - T) = 0. \qquad (18\text{-}23)$$

Temperatures are contained in Eq. 18-23 only as differences. Thus, because the differences on the Celsius and Kelvin scales are identical, we can use either of those scales in this equation. Solving it for T_f, we obtain

$$T_f = \frac{c_c m_c T + C_b T_i + c_w m_w T_i}{c_w m_w + C_b + c_c m_c}.$$

Using Celsius temperatures and taking values for c_c and c_w from Table 18-3, we find the numerator to be

$$(0.0923 \text{ cal/g}\cdot\text{K})(75 \text{ g})(312°\text{C}) + (45 \text{ cal/K})(12°\text{C}) + (1.00 \text{ cal/g}\cdot\text{K})(220 \text{ g})(12°\text{C}) = 5339.8 \text{ cal},$$

and the denominator to be

$$(1.00 \text{ cal/g}\cdot\text{K})(220 \text{ g}) + 45 \text{ cal/K} + (0.0923 \text{ cal/g}\cdot\text{K})(75 \text{ g}) = 271.9 \text{ cal/C°}.$$

We then have

$$T_f = \frac{5339.8 \text{ cal}}{271.9 \text{ cal/C°}} = 19.6°\text{C} \approx 20°\text{C}. \qquad \text{(Answer)}$$

From the given data you can show that

$$Q_w \approx 1670 \text{ cal}, \qquad Q_b \approx 342 \text{ cal}, \qquad Q_c \approx -2020 \text{ cal}.$$

Apart from rounding errors, the algebraic sum of these three heat transfers is indeed zero, as required by the conservation of energy (Eq. 18-22).

Sample Problem 18.04 Heat to change temperature and state

(a) How much heat must be absorbed by ice of mass $m =$ 720 g at $-10°$C to take it to the liquid state at 15°C?

KEY IDEAS

The heating process is accomplished in three steps: (1) The ice cannot melt at a temperature below the freezing point—so initially, any energy transferred to the ice as heat can only increase the temperature of the ice, until 0°C is reached. (2) The temperature then cannot increase until all the ice melts—so any energy transferred to the ice as heat now can only change ice to liquid water, until all the ice melts. (3) Now the energy transferred to the liquid water as heat can only increase the temperature of the liquid water.

Warming the ice: The heat Q_1 needed to take the ice from the initial $T_i = -10°$C to the final $T_f = 0°$C (so that the ice can then melt) is given by Eq. 18-14 ($Q = cm\,\Delta T$). Using the specific heat of ice c_{ice} in Table 18-3 gives us

$$\begin{aligned} Q_1 &= c_{ice} m(T_f - T_i) \\ &= (2220 \text{ J/kg}\cdot\text{K})(0.720 \text{ kg})[0°\text{C} - (-10°\text{C})] \\ &= 15\,984 \text{ J} \approx 15.98 \text{ kJ}. \end{aligned}$$

Melting the ice: The heat Q_2 needed to melt all the ice is given by Eq. 18-16 ($Q = Lm$). Here L is the heat of fusion L_F, with the value given in Eq. 18-18 and Table 18-4. We find

$$Q_2 = L_F m = (333 \text{ kJ/kg})(0.720 \text{ kg}) \approx 239.8 \text{ kJ}.$$

Warming the liquid: The heat Q_3 needed to increase the temperature of the water from the initial value $T_i = 0°$C to the final value $T_f = 15°$C is given by Eq. 18-14 (with the specific heat of liquid water c_{liq}):

$$\begin{aligned} Q_3 &= c_{liq} m(T_f - T_i) \\ &= (4186.8 \text{ J/kg}\cdot\text{K})(0.720 \text{ kg})(15°\text{C} - 0°\text{C}) \\ &= 45\,217 \text{ J} \approx 45.22 \text{ kJ}. \end{aligned}$$

Total: The total required heat Q_{tot} is the sum of the amounts required in the three steps:

$$\begin{aligned} Q_{tot} &= Q_1 + Q_2 + Q_3 \\ &= 15.98 \text{ kJ} + 239.8 \text{ kJ} + 45.22 \text{ kJ} \\ &\approx 300 \text{ kJ}. \qquad \text{(Answer)} \end{aligned}$$

Note that most of the energy goes into melting the ice rather than raising the temperature.

(b) If we supply the ice with a total energy of only 210 kJ (as heat), what are the final state and temperature of the water?

KEY IDEA

From step 1, we know that 15.98 kJ is needed to raise the temperature of the ice to the melting point. The remaining heat Q_{rem} is then 210 kJ − 15.98 kJ, or about 194 kJ. From step 2, we can see that this amount of heat is insufficient to melt all the ice. Because the melting of the ice is incomplete, we must end up with a mixture of ice and liquid; the temperature of the mixture must be the freezing point, 0°C.

Calculations: We can find the mass m of ice that is melted by the available energy Q_{rem} by using Eq. 18-16 with L_F:

$$m = \frac{Q_{rem}}{L_F} = \frac{194 \text{ kJ}}{333 \text{ kJ/kg}} = 0.583 \text{ kg} \approx 580 \text{ g}.$$

Thus, the mass of the ice that remains is 720 g − 580 g, or 140 g, and we have

$$580 \text{ g water} \quad \text{and} \quad 140 \text{ g ice}, \quad \text{at } 0°\text{C}. \qquad \text{(Answer)}$$

18-5 THE FIRST LAW OF THERMODYNAMICS

Learning Objectives

After reading this module, you should be able to . . .

18.20 If an enclosed gas expands or contracts, calculate the work W done by the gas by integrating the gas pressure with respect to the volume of the enclosure.

18.21 Identify the algebraic sign of work W associated with expansion and contraction of a gas.

18.22 Given a p-V graph of pressure versus volume for a process, identify the starting point (the initial state) and the final point (the final state) and calculate the work by using graphical integration.

18.23 On a p-V graph of pressure versus volume for a gas, identify the algebraic sign of the work associated with a right-going process and a left-going process.

18.24 Apply the first law of thermodynamics to relate the change in the internal energy ΔE_{int} of a gas, the energy Q transferred as heat to or from the gas, and the work W done on or by the gas.

18.25 Identify the algebraic sign of a heat transfer Q that is associated with a transfer to a gas and a transfer from the gas.

18.26 Identify that the internal energy ΔE_{int} of a gas tends to increase if the heat transfer is *to* the gas, and it tends to decrease if the gas does work on its environment.

18.27 Identify that in an adiabatic process with a gas, there is no heat transfer Q with the environment.

18.28 Identify that in a constant-volume process with a gas, there is no work W done by the gas.

18.29 Identify that in a cyclical process with a gas, there is no net change in the internal energy ΔE_{int}.

18.30 Identify that in a free expansion with a gas, the heat transfer Q, work done W, and change in internal energy ΔE_{int} are each zero.

Key Ideas

- A gas may exchange energy with its surroundings through work. The amount of work W done *by* a gas as it expands or contracts from an initial volume V_i to a final volume V_f is given by

$$W = \int dW = \int_{V_i}^{V_f} p\,dV.$$

The integration is necessary because the pressure p may vary during the volume change.

- The principle of conservation of energy for a thermodynamic process is expressed in the first law of thermodynamics, which may assume either of the forms

$$\Delta E_{\text{int}} = E_{\text{int},f} - E_{\text{int},i} = Q - W \quad \text{(first law)}$$

or

$$dE_{\text{int}} = dQ - dW \quad \text{(first law)}.$$

E_{int} represents the internal energy of the material, which depends only on the material's state (temperature, pressure, and volume). Q represents the energy exchanged as heat between the system and its surroundings; Q is positive if the system absorbs heat and negative if the system loses heat. W is the work done *by* the system; W is positive if the system expands against an external force from the surroundings and negative if the system contracts because of an external force.

- Q and W are path dependent; ΔE_{int} is path independent.

- The first law of thermodynamics finds application in several special cases:

adiabatic processes:	$Q = 0, \quad \Delta E_{\text{int}} = -W$
constant-volume processes:	$W = 0, \quad \Delta E_{\text{int}} = Q$
cyclical processes:	$\Delta E_{\text{int}} = 0, \quad Q = W$
free expansions:	$Q = W = \Delta E_{\text{int}} = 0$

A Closer Look at Heat and Work

Here we look in some detail at how energy can be transferred as heat and work between a system and its environment. Let us take as our system a gas confined to a cylinder with a movable piston, as in Fig. 18-13. The upward force on the piston due to the pressure of the confined gas is equal to the weight of lead shot loaded onto the top of the piston. The walls of the cylinder are made of insulating material that does not allow any transfer of energy as heat. The bottom of the cylinder rests on a reservoir for thermal energy, a *thermal reservoir* (perhaps a hot plate) whose temperature T you can control by turning a knob.

The system (the gas) starts from an *initial state i*, described by a pressure p_i, a volume V_i, and a temperature T_i. You want to change the system to a *final state f*, described by a pressure p_f, a volume V_f, and a temperature T_f. The procedure by which you change the system from its initial state to its final state is called a *thermodynamic process.* During such a process, energy may be transferred into the system from the thermal reservoir (positive heat) or vice versa (negative heat). Also, work can be done by the system to raise the loaded piston (positive work) or lower it (negative work). We assume that all such changes occur slowly, with the result that the system is always in (approximate) thermal equilibrium (every part is always in thermal equilibrium).

Suppose that you remove a few lead shot from the piston of Fig. 18-13, allowing the gas to push the piston and remaining shot upward through a differential displacement $d\vec{s}$ with an upward force $\vec{F}$. Since the displacement is tiny, we can assume that $\vec{F}$ is constant during the displacement. Then $\vec{F}$ has a magnitude that is equal to pA, where p is the pressure of the gas and A is the face area of the piston. The differential work dW done by the gas during the displacement is

$$\begin{aligned} dW &= \vec{F}\cdot d\vec{s} = (pA)(ds) = p(A\,ds) \\ &= p\,dV, \end{aligned} \tag{18-24}$$

in which dV is the differential change in the volume of the gas due to the movement of the piston. When you have removed enough shot to allow the gas to change its volume from V_i to V_f, the total work done by the gas is

$$W = \int dW = \int_{V_i}^{V_f} p\,dV. \tag{18-25}$$

During the volume change, the pressure and temperature may also change. To evaluate Eq. 18-25 directly, we would need to know how pressure varies with volume for the actual process by which the system changes from state i to state f.

One Path. There are actually many ways to take the gas from state i to state f. One way is shown in Fig. 18-14a, which is a plot of the pressure of the gas versus its volume and which is called a p-V diagram. In Fig. 18-14a, the curve indicates that the

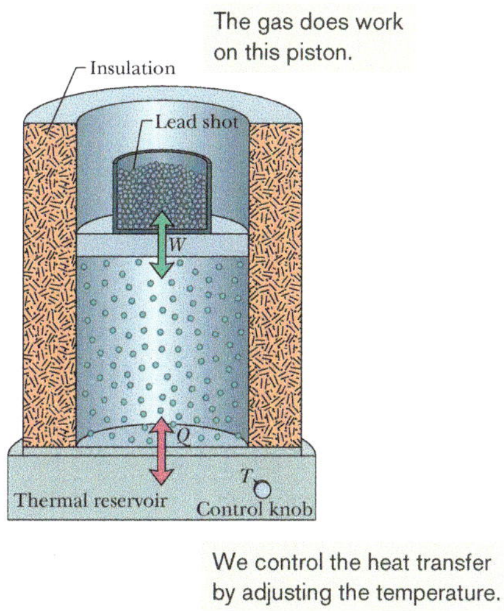

Figure 18-13 A gas is confined to a cylinder with a movable piston. Heat Q can be added to or withdrawn from the gas by regulating the temperature T of the adjustable thermal reservoir. Work W can be done by the gas by raising or lowering the piston.

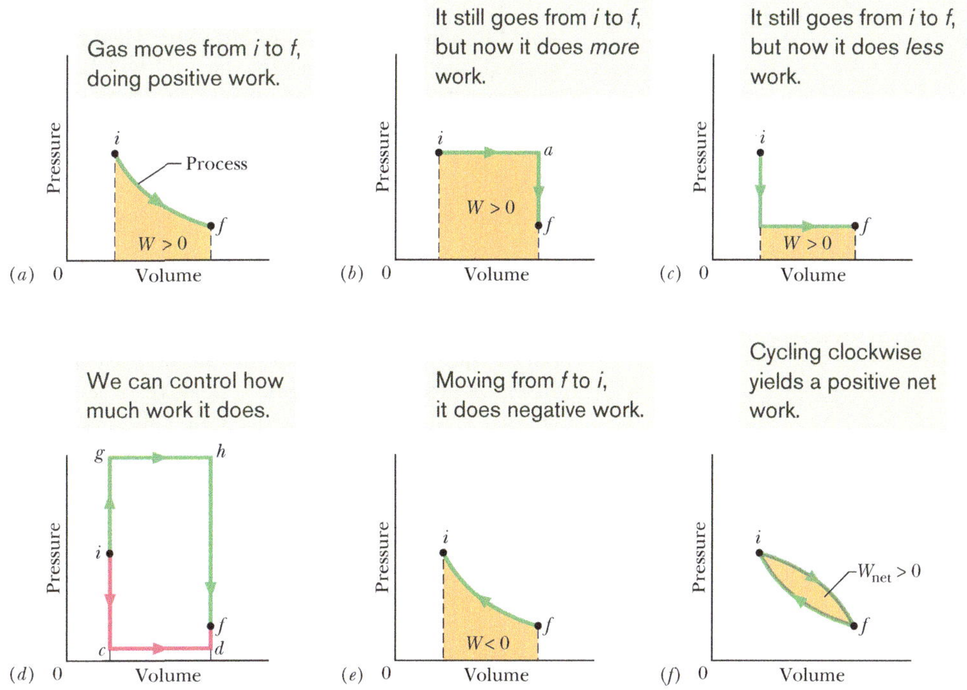

Figure 18-14 (a) The shaded area represents the work W done by a system as it goes from an initial state i to a final state f. Work W is positive because the system's volume increases. (b) W is still positive, but now greater. (c) W is still positive, but now smaller. (d) W can be even smaller (path $icdf$) or larger (path $ighf$). (e) Here the system goes from state f to state i as the gas is compressed to less volume by an external force. The work W done *by* the system is now negative. (f) The net work W_{net} done by the system during a complete cycle is represented by the shaded area.

pressure decreases as the volume increases. The integral in Eq. 18-25 (and thus the work W done by the gas) is represented by the shaded area under the curve between points i and f. Regardless of what exactly we do to take the gas along the curve, that work is positive, due to the fact that the gas increases its volume by forcing the piston upward.

Another Path. Another way to get from state i to state f is shown in Fig. 18-14*b*. There the change takes place in two steps—the first from state i to state a, and the second from state a to state f.

Step *ia* of this process is carried out at constant pressure, which means that you leave undisturbed the lead shot that ride on top of the piston in Fig. 18-13. You cause the volume to increase (from V_i to V_f) by slowly turning up the temperature control knob, raising the temperature of the gas to some higher value T_a. (Increasing the temperature increases the force from the gas on the piston, moving it upward.) During this step, positive work is done by the expanding gas (to lift the loaded piston) and heat is absorbed by the system from the thermal reservoir (in response to the arbitrarily small temperature differences that you create as you turn up the temperature). This heat is positive because it is added to the system.

Step *af* of the process of Fig. 18-14*b* is carried out at constant volume, so you must wedge the piston, preventing it from moving. Then as you use the control knob to decrease the temperature, you find that the pressure drops from p_a to its final value p_f. During this step, heat is lost by the system to the thermal reservoir.

For the overall process *iaf*, the work W, which is positive and is carried out only during step *ia*, is represented by the shaded area under the curve. Energy is transferred as heat during both steps *ia* and *af*, with a net energy transfer Q.

Reversed Steps. Figure 18-14*c* shows a process in which the previous two steps are carried out in reverse order. The work W in this case is smaller than for Fig. 18-14*b*, as is the net heat absorbed. Figure 18-14*d* suggests that you can make the work done by the gas as small as you want (by following a path like *icdf*) or as large as you want (by following a path like *ighf*).

To sum up: A system can be taken from a given initial state to a given final state by an infinite number of processes. Heat may or may not be involved, and in general, the work W and the heat Q will have different values for different processes. We say that heat and work are *path-dependent* quantities.

Negative Work. Figure 18-14*e* shows an example in which negative work is done by a system as some external force compresses the system, reducing its volume. The absolute value of the work done is still equal to the area beneath the curve, but because the gas is *compressed*, the work done by the gas is negative.

Cycle. Figure 18-14*f* shows a *thermodynamic cycle* in which the system is taken from some initial state i to some other state f and then back to i. The net work done by the system during the cycle is the sum of the *positive* work done during the expansion and the *negative* work done during the compression. In Fig. 18-14*f*, the net work is positive because the area under the expansion curve (i to f) is greater than the area under the compression curve (f to i).

Checkpoint 4

The p-V diagram here shows six curved paths (connected by vertical paths) that can be followed by a gas. Which two of the curved paths should be part of a closed cycle (those curved paths plus connecting vertical paths) if the net work done by the gas during the cycle is to be at its maximum positive value?

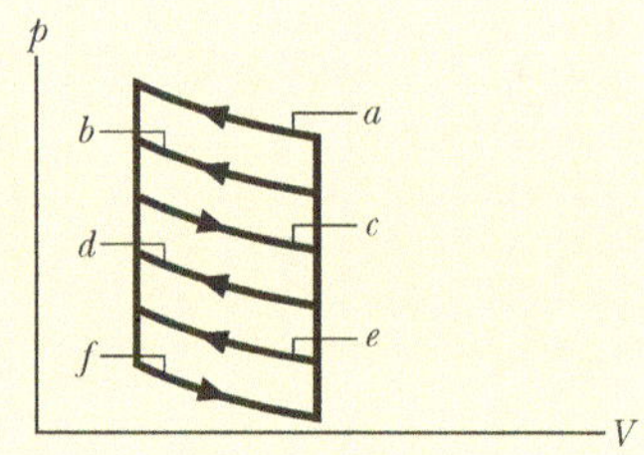

The First Law of Thermodynamics

You have just seen that when a system changes from a given initial state to a given final state, both the work W and the heat Q depend on the nature of the process. Experimentally, however, we find a surprising thing. *The quantity* $Q - W$ *is the same for all processes.* It depends only on the initial and final states and does not depend at all on how the system gets from one to the other. All other combinations of Q and W, including Q alone, W alone, $Q + W$, and $Q - 2W$, are *path dependent;* only the quantity $Q - W$ is not.

The quantity $Q - W$ must represent a change in some intrinsic property of the system. We call this property the *internal energy* E_{int} and we write

$$\Delta E_{\text{int}} = E_{\text{int},f} - E_{\text{int},i} = Q - W \quad \text{(first law).} \tag{18-26}$$

Equation 18-26 is the **first law of thermodynamics.** If the thermodynamic system undergoes only a differential change, we can write the first law as*

$$dE_{\text{int}} = dQ - dW \quad \text{(first law).} \tag{18-27}$$

The internal energy E_{int} of a system tends to increase if energy is added as heat Q and tends to decrease if energy is lost as work W done by the system.

In Chapter 8, we discussed the principle of energy conservation as it applies to isolated systems—that is, to systems in which no energy enters or leaves the system. The first law of thermodynamics is an extension of that principle to systems that are *not* isolated. In such cases, energy may be transferred into or out of the system as either work W or heat Q. In our statement of the first law of thermodynamics above, we assume that there are no changes in the kinetic energy or the potential energy of the system as a whole; that is, $\Delta K = \Delta U = 0$.

Rules. Before this chapter, the term *work* and the symbol W always meant the work done *on* a system. However, starting with Eq. 18-24 and continuing through the next two chapters about thermodynamics, we focus on the work done *by* a system, such as the gas in Fig. 18-13.

The work done *on* a system is always the negative of the work done *by* the system, so if we rewrite Eq. 18-26 in terms of the work W_{on} done *on* the system, we have $\Delta E_{\text{int}} = Q + W_{\text{on}}$. This tells us the following: The internal energy of a system tends to increase if heat is absorbed by the system or if positive work is done *on* the system. Conversely, the internal energy tends to decrease if heat is lost by the system or if negative work is done *on* the system.

Checkpoint 5

The figure here shows four paths on a p-V diagram along which a gas can be taken from state i to state f. Rank the paths according to (a) the change ΔE_{int} in the internal energy of the gas, (b) the work W done by the gas, and (c) the magnitude of the energy transferred as heat Q between the gas and its environment, greatest first.

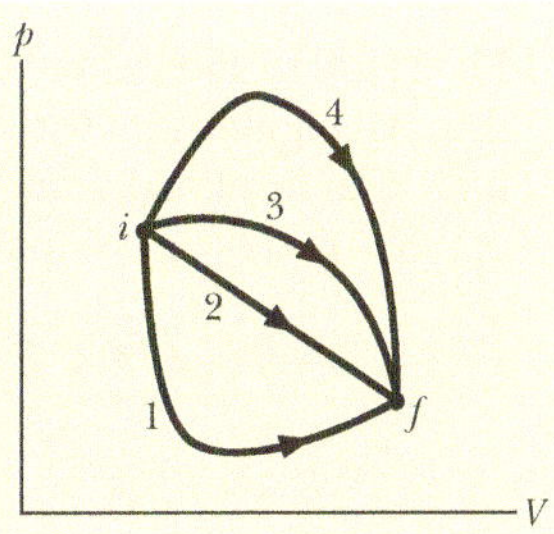

*Here dQ and dW, unlike dE_{int}, are not true differentials; that is, there are no such functions as $Q(p, V)$ and $W(p, V)$ that depend only on the state of the system. The quantities dQ and dW are called *inexact differentials* and are usually represented by the symbols $đQ$ and $đW$. For our purposes, we can treat them simply as infinitesimally small energy transfers.

We slowly remove lead shot, allowing an expansion without any heat transfer.

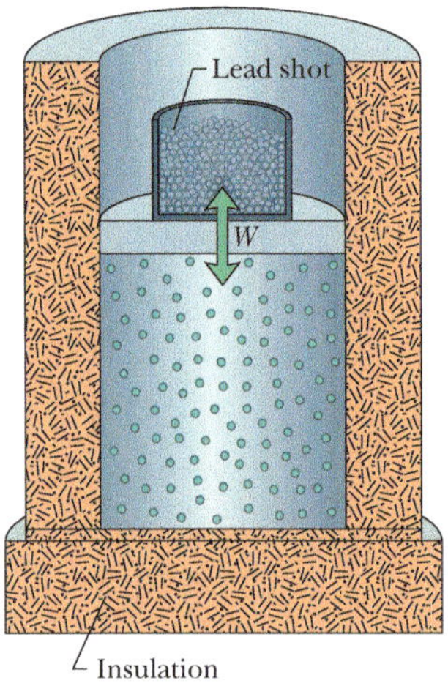

Figure 18-15 An adiabatic expansion can be carried out by slowly removing lead shot from the top of the piston. Adding lead shot reverses the process at any stage.

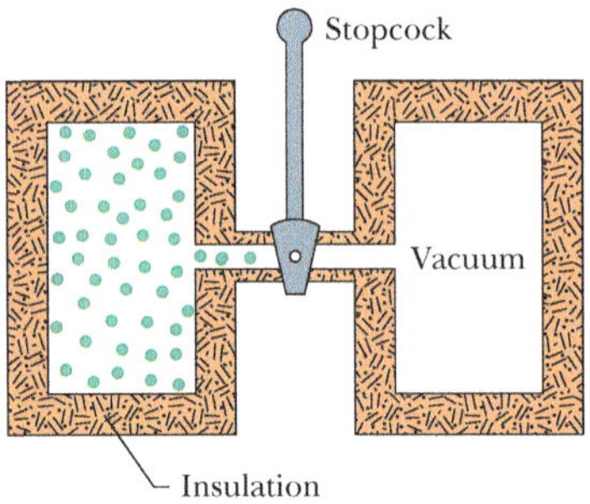

Figure 18-16 The initial stage of a free-expansion process. After the stopcock is opened, the gas fills both chambers and eventually reaches an equilibrium state.

Some Special Cases of the First Law of Thermodynamics

Here are four thermodynamic processes as summarized in Table 18-5.

1. ***Adiabatic processes.*** An adiabatic process is one that occurs so rapidly or occurs in a system that is so well insulated that *no transfer of energy as heat* occurs between the system and its environment. Putting $Q = 0$ in the first law (Eq. 18-26) yields

$$\Delta E_{\text{int}} = -W \quad \text{(adiabatic process).} \tag{18-28}$$

This tells us that if work is done *by* the system (that is, if W is positive), the internal energy of the system decreases by the amount of work. Conversely, if work is done *on* the system (that is, if W is negative), the internal energy of the system increases by that amount.

Figure 18-15 shows an idealized adiabatic process. Heat cannot enter or leave the system because of the insulation. Thus, the only way energy can be transferred between the system and its environment is by work. If we remove shot from the piston and allow the gas to expand, the work done by the system (the gas) is positive and the internal energy of the gas decreases. If, instead, we add shot and compress the gas, the work done by the system is negative and the internal energy of the gas increases.

2. ***Constant-volume processes.*** If the volume of a system (such as a gas) is held constant, that system can do no work. Putting $W = 0$ in the first law (Eq. 18-26) yields

$$\Delta E_{\text{int}} = Q \quad \text{(constant-volume process).} \tag{18-29}$$

Thus, if heat is absorbed by a system (that is, if Q is positive), the internal energy of the system increases. Conversely, if heat is lost during the process (that is, if Q is negative), the internal energy of the system must decrease.

3. ***Cyclical processes.*** There are processes in which, after certain interchanges of heat and work, the system is restored to its initial state. In that case, no intrinsic property of the system—including its internal energy—can possibly change. Putting $\Delta E_{\text{int}} = 0$ in the first law (Eq. 18-26) yields

$$Q = W \quad \text{(cyclical process).} \tag{18-30}$$

Thus, the net work done during the process must exactly equal the net amount of energy transferred as heat; the store of internal energy of the system remains unchanged. Cyclical processes form a closed loop on a p-V plot, as shown in Fig. 18-14*f*. We discuss such processes in detail in Chapter 20.

4. ***Free expansions.*** These are adiabatic processes in which no transfer of heat occurs between the system and its environment and no work is done on or by the system. Thus, $Q = W = 0$, and the first law requires that

$$\Delta E_{\text{int}} = 0 \quad \text{(free expansion).} \tag{18-31}$$

Figure 18-16 shows how such an expansion can be carried out. A gas, which is in thermal equilibrium within itself, is initially confined by a closed stopcock to one half of an insulated double chamber; the other half is evacuated. The stopcock is opened, and the gas expands freely to fill both halves of the chamber. No heat is

Table 18-5 The First Law of Thermodynamics: Four Special Cases

The Law: $\Delta E_{\text{int}} = Q - W$ (Eq. 18-26)

Process	Restriction	Consequence
Adiabatic	$Q = 0$	$\Delta E_{\text{int}} = -W$
Constant volume	$W = 0$	$\Delta E_{\text{int}} = Q$
Closed cycle	$\Delta E_{\text{int}} = 0$	$Q = W$
Free expansion	$Q = W = 0$	$\Delta E_{\text{int}} = 0$

transferred to or from the gas because of the insulation. No work is done by the gas because it rushes into a vacuum and thus does not meet any pressure.

A free expansion differs from all other processes we have considered because it cannot be done slowly and in a controlled way. As a result, at any given instant during the sudden expansion, the gas is not in thermal equilibrium and its pressure is not uniform. Thus, although we can plot the initial and final states on a p-V diagram, we cannot plot the expansion itself.

Checkpoint 6

For one complete cycle as shown in the p-V diagram here, are (a) ΔE_{int} for the gas and (b) the net energy transferred as heat Q positive, negative, or zero?

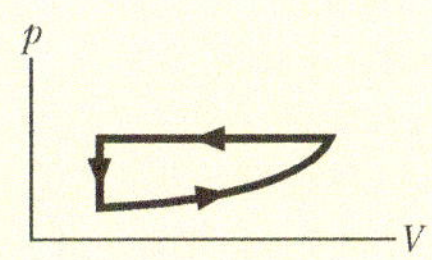

Sample Problem 18.05 First law of thermodynamics: work, heat, internal energy change

Let 1.00 kg of liquid water at 100°C be converted to steam at 100°C by boiling at standard atmospheric pressure (which is 1.00 atm or 1.01×10^5 Pa) in the arrangement of Fig. 18-17. The volume of that water changes from an initial value of 1.00×10^{-3} m^3 as a liquid to 1.671 m^3 as steam.

(a) How much work is done by the system during this process?

KEY IDEAS

(1) The system must do positive work because the volume increases. (2) We calculate the work W done by integrating the pressure with respect to the volume (Eq. 18-25).

Calculation: Because here the pressure is constant at 1.01×10^5 Pa, we can take p outside the integral. Thus,

$$W = \int_{V_i}^{V_f} p\, dV = p \int_{V_i}^{V_f} dV = p(V_f - V_i)$$

$$= (1.01 \times 10^5 \text{ Pa})(1.671 \text{ m}^3 - 1.00 \times 10^{-3} \text{ m}^3)$$

$$= 1.69 \times 10^5 \text{ J} = 169 \text{ kJ}. \quad \text{(Answer)}$$

(b) How much energy is transferred as heat during the process?

KEY IDEA

Because the heat causes only a phase change and not a change in temperature, it is given fully by Eq. 18-16 ($Q = Lm$).

Calculation: Because the change is from liquid to gaseous phase, L is the heat of vaporization L_V, with the value given in Eq. 18-17 and Table 18-4. We find

$$Q = L_V m = (2256 \text{ kJ/kg})(1.00 \text{ kg})$$

$$= 2256 \text{ kJ} \approx 2260 \text{ kJ}. \quad \text{(Answer)}$$

(c) What is the change in the system's internal energy during the process?

KEY IDEA

The change in the system's internal energy is related to the heat (here, this is energy transferred into the system) and the work (here, this is energy transferred out of the system) by the first law of thermodynamics (Eq. 18-26).

Calculation: We write the first law as

$$\Delta E_{int} = Q - W = 2256 \text{ kJ} - 169 \text{ kJ}$$

$$\approx 2090 \text{ kJ} = 2.09 \text{ MJ}. \quad \text{(Answer)}$$

This quantity is positive, indicating that the internal energy of the system has increased during the boiling process. The added energy goes into separating the H_2O molecules, which strongly attract one another in the liquid state. We see that, when water is boiled, about 7.5% (= 169 kJ/2260 kJ) of the heat goes into the work of pushing back the atmosphere. The rest of the heat goes into the internal energy of the system.

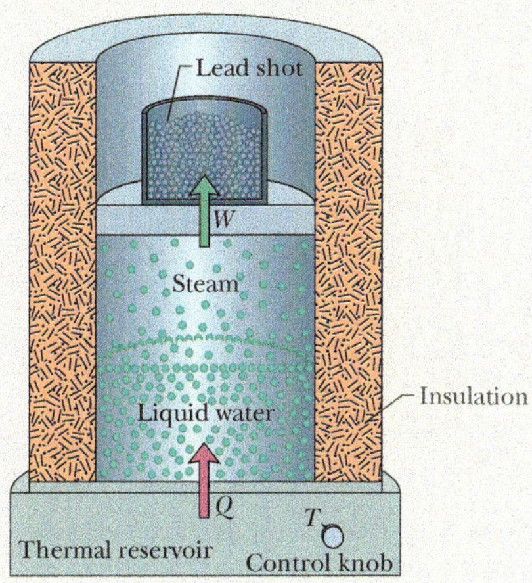

Figure 18-17 Water boiling at constant pressure. Energy is transferred from the thermal reservoir as heat until the liquid water has changed completely into steam. Work is done by the expanding gas as it lifts the loaded piston.

18-6 HEAT TRANSFER MECHANISMS

Learning Objectives

After reading this module, you should be able to . . .

18.31 For thermal conduction through a layer, apply the relationship between the energy-transfer rate P_{cond} and the layer's area A, thermal conductivity k, thickness L, and temperature difference ΔT (between its two sides).

18.32 For a composite slab (two or more layers) that has reached the steady state in which temperatures are no longer changing, identify that (by the conservation of energy) the rates of thermal conduction P_{cond} through the layers must be equal.

18.33 For thermal conduction through a layer, apply the relationship between thermal resistance R, thickness L, and thermal conductivity k.

18.34 Identify that thermal energy can be transferred by convection, in which a warmer fluid (gas or liquid) tends to rise in a cooler fluid.

18.35 In the *emission* of thermal radiation by an object, apply the relationship between the energy-transfer rate P_{rad} and the object's surface area A, emissivity ε, and *surface* temperature T (in kelvins).

18.36 In the *absorption* of thermal radiation by an object, apply the relationship between the energy-transfer rate P_{abs} and the object's surface area A and emissivity ε, and the *environmental* temperature T (in kelvins).

18.37 Calculate the net energy-transfer rate P_{net} of an object emitting radiation to its environment and absorbing radiation from that environment.

Key Ideas

- The rate P_{cond} at which energy is conducted through a slab for which one face is maintained at the higher temperature T_H and the other face is maintained at the lower temperature T_C is

$$P_{cond} = \frac{Q}{t} = kA\,\frac{T_H - T_C}{L}.$$

Here each face of the slab has area A, the length of the slab (the distance between the faces) is L, and k is the thermal conductivity of the material.

- Convection occurs when temperature differences cause an energy transfer by motion within a fluid.

- Radiation is an energy transfer via the emission of electromagnetic energy. The rate P_{rad} at which an object emits energy via thermal radiation is

$$P_{rad} = \sigma\varepsilon AT^4,$$

where σ ($= 5.6704 \times 10^{-8}\ \mathrm{W/m^2\cdot K^4}$) is the Stefan–Boltzmann constant, ε is the emissivity of the object's surface, A is its surface area, and T is its surface temperature (in kelvins). The rate P_{abs} at which an object absorbs energy via thermal radiation from its environment, which is at the uniform temperature T_{env} (in kelvins), is

$$P_{abs} = \sigma\varepsilon AT^4_{env}.$$

Heat Transfer Mechanisms

We have discussed the transfer of energy as heat between a system and its environment, but we have not yet described how that transfer takes place. There are three transfer mechanisms: conduction, convection, and radiation. Let's next examine these mechanisms in turn.

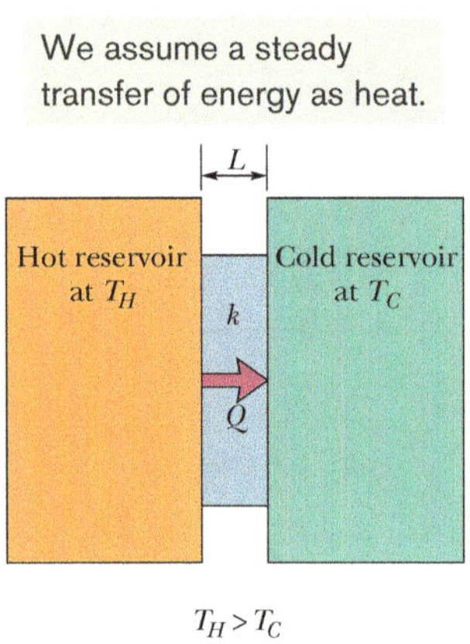

Figure 18-18 Thermal conduction. Energy is transferred as heat from a reservoir at temperature T_H to a cooler reservoir at temperature T_C through a conducting slab of thickness L and thermal conductivity k.

Conduction

If you leave the end of a metal poker in a fire for enough time, its handle will get hot. Energy is transferred from the fire to the handle by (thermal) **conduction** along the length of the poker. The vibration amplitudes of the atoms and electrons of the metal at the fire end of the poker become relatively large because of the high temperature of their environment. These increased vibrational amplitudes, and thus the associated energy, are passed along the poker, from atom to atom, during collisions between adjacent atoms. In this way, a region of rising temperature extends itself along the poker to the handle.

Consider a slab of face area A and thickness L, whose faces are maintained at temperatures T_H and T_C by a hot reservoir and a cold reservoir, as in Fig. 18-18. Let Q be the energy that is transferred as heat through the slab, from its hot face to its cold face, in time t. Experiment shows that the *conduction rate* P_{cond} (the amount of energy transferred per unit time) is

$$P_{\text{cond}} = \frac{Q}{t} = kA\,\frac{T_H - T_C}{L}, \tag{18-32}$$

in which k, called the *thermal conductivity,* is a constant that depends on the material of which the slab is made. A material that readily transfers energy by conduction is a *good thermal conductor* and has a high value of k. Table 18-6 gives the thermal conductivities of some common metals, gases, and building materials.

Table 18-6 Some Thermal Conductivities

Substance	k (W/m·K)
Metals	
Stainless steel	14
Lead	35
Iron	67
Brass	109
Aluminum	235
Copper	401
Silver	428
Gases	
Air (dry)	0.026
Helium	0.15
Hydrogen	0.18
Building Materials	
Polyurethane foam	0.024
Rock wool	0.043
Fiberglass	0.048
White pine	0.11
Window glass	1.0

Thermal Resistance to Conduction (R-Value)

If you are interested in insulating your house or in keeping cola cans cold on a picnic, you are more concerned with poor heat conductors than with good ones. For this reason, the concept of *thermal resistance* R has been introduced into engineering practice. The R-value of a slab of thickness L is defined as

$$R = \frac{L}{k}. \tag{18-33}$$

The lower the thermal conductivity of the material of which a slab is made, the higher the R-value of the slab; so something that has a high R-value is a *poor thermal conductor* and thus a *good thermal insulator.*

Note that R is a property attributed to a slab of a specified thickness, not to a material. The commonly used unit for R (which, in the United States at least, is almost never stated) is the square foot–Fahrenheit degree–hour per British thermal unit (ft^2·F°·h/Btu). (Now you know why the unit is rarely stated.)

Conduction Through a Composite Slab

Figure 18-19 shows a composite slab, consisting of two materials having different thicknesses L_1 and L_2 and different thermal conductivities k_1 and k_2. The temperatures of the outer surfaces of the slab are T_H and T_C. Each face of the slab has area A. Let us derive an expression for the conduction rate through the slab under the assumption that the transfer is a *steady-state* process; that is, the temperatures everywhere in the slab and the rate of energy transfer do not change with time.

In the steady state, the conduction rates through the two materials must be equal. This is the same as saying that the energy transferred through one material in a certain time must be equal to that transferred through the other material in the same time. If this were not true, temperatures in the slab would be changing and we would not have a steady-state situation. Letting T_X be the temperature of the interface between the two materials, we can now use Eq. 18-32 to write

$$P_{\text{cond}} = \frac{k_2 A(T_H - T_X)}{L_2} = \frac{k_1 A(T_X - T_C)}{L_1}. \tag{18-34}$$

Solving Eq. 18-34 for T_X yields, after a little algebra,

$$T_X = \frac{k_1 L_2 T_C + k_2 L_1 T_H}{k_1 L_2 + k_2 L_1}. \tag{18-35}$$

Substituting this expression for T_X into either equality of Eq. 18-34 yields

$$P_{\text{cond}} = \frac{A(T_H - T_C)}{L_1/k_1 + L_2/k_2}. \tag{18-36}$$

We can extend Eq. 18-36 to apply to any number n of materials making up a slab:

$$P_{\text{cond}} = \frac{A(T_H - T_C)}{\Sigma\,(L/k)}. \tag{18-37}$$

The summation sign in the denominator tells us to add the values of L/k for all the materials.

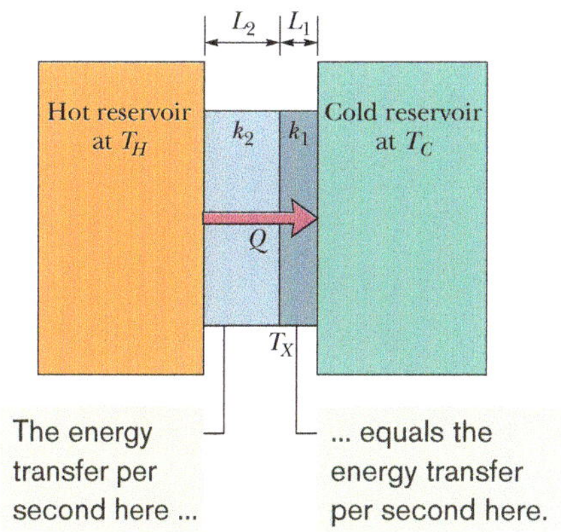

Figure 18-19 Heat is transferred at a steady rate through a composite slab made up of two different materials with different thicknesses and different thermal conductivities. The steady-state temperature at the interface of the two materials is T_X.

Checkpoint 7

The figure shows the face and interface temperatures of a composite slab consisting of four materials, of identical thicknesses, through which the heat transfer is steady. Rank the materials according to their thermal conductivities, greatest first.

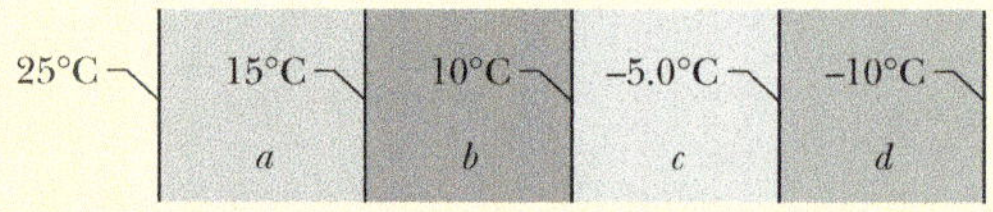

Convection

When you look at the flame of a candle or a match, you are watching thermal energy being transported upward by **convection.** Such energy transfer occurs when a fluid, such as air or water, comes in contact with an object whose temperature is higher than that of the fluid. The temperature of the part of the fluid that is in contact with the hot object increases, and (in most cases) that fluid expands and thus becomes less dense. Because this expanded fluid is now lighter than the surrounding cooler fluid, buoyant forces cause it to rise. Some of the surrounding cooler fluid then flows so as to take the place of the rising warmer fluid, and the process can then continue.

Convection is part of many natural processes. Atmospheric convection plays a fundamental role in determining global climate patterns and daily weather variations. Glider pilots and birds alike seek rising thermals (convection currents of warm air) that keep them aloft. Huge energy transfers take place within the oceans by the same process. Finally, energy is transported to the surface of the Sun from the nuclear furnace at its core by enormous cells of convection, in which hot gas rises to the surface along the cell core and cooler gas around the core descends below the surface.

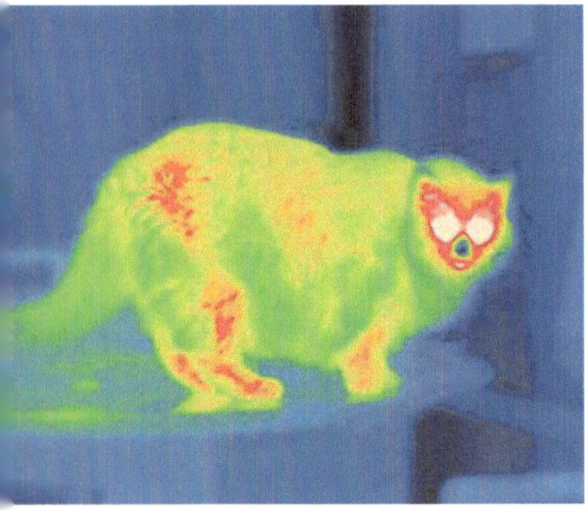

Edward Kinsman/Photo Researchers, Inc.

Figure 18-20 A false-color thermogram reveals the rate at which energy is radiated by a cat. The rate is color-coded, with white and red indicating the greatest radiation rate. The nose is cool.

Radiation

The third method by which an object and its environment can exchange energy as heat is via electromagnetic waves (visible light is one kind of electromagnetic wave). Energy transferred in this way is often called **thermal radiation** to distinguish it from electromagnetic *signals* (as in, say, television broadcasts) and from nuclear radiation (energy and particles emitted by nuclei). (To "radiate" generally means to emit.) When you stand in front of a big fire, you are warmed by absorbing thermal radiation from the fire; that is, your thermal energy increases as the fire's thermal energy decreases. No medium is required for heat transfer via radiation—the radiation can travel through vacuum from, say, the Sun to you.

The rate P_{rad} at which an object emits energy via electromagnetic radiation depends on the object's surface area A and the temperature T of that area in kelvins and is given by

$$P_{\text{rad}} = \sigma \varepsilon A T^4. \tag{18-38}$$

Here $\sigma = 5.6704 \times 10^{-8}\ \text{W/m}^2 \cdot \text{K}^4$ is called the *Stefan–Boltzmann constant* after Josef Stefan (who discovered Eq. 18-38 experimentally in 1879) and Ludwig Boltzmann (who derived it theoretically soon after). The symbol ε represents the *emissivity* of the object's surface, which has a value between 0 and 1, depending on the composition of the surface. A surface with the maximum emissivity of 1.0 is said to be a *blackbody radiator,* but such a surface is an ideal limit and does not occur in nature. Note again that the temperature in Eq. 18-38 must be in kelvins so that a temperature of absolute zero corresponds to no radiation. Note also that every object whose temperature is above 0 K—including you—emits thermal radiation. (See Fig. 18-20.)

The rate P_{abs} at which an object absorbs energy via thermal radiation from its environment, which we take to be at uniform temperature T_{env} (in kelvins), is

$$P_{abs} = \sigma\varepsilon A T_{env}^4. \quad (18\text{-}39)$$

The emissivity ε in Eq. 18-39 is the same as that in Eq. 18-38. An idealized blackbody radiator, with $\varepsilon = 1$, will absorb all the radiated energy it intercepts (rather than sending a portion back away from itself through reflection or scattering).

Because an object both emits and absorbs thermal radiation, its net rate P_{net} of energy exchange due to thermal radiation is

$$P_{net} = P_{abs} - P_{rad} = \sigma\varepsilon A(T_{env}^4 - T^4). \quad (18\text{-}40)$$

P_{net} is positive if net energy is being absorbed via radiation and negative if it is being lost via radiation.

Thermal radiation is involved in the numerous medical cases of a *dead* rattlesnake striking a hand reaching toward it. Pits between each eye and nostril of a rattlesnake (Fig. 18-21) serve as sensors of thermal radiation. When, say, a mouse moves close to a rattlesnake's head, the thermal radiation from the mouse triggers these sensors, causing a reflex action in which the snake strikes the mouse with its fangs and injects its venom. The thermal radiation from a reaching hand can cause the same reflex action even if the snake has been dead for as long as 30 min because the snake's nervous system continues to function. As one snake expert advised, if you must remove a recently killed rattlesnake, use a long stick rather than your hand.

Figure 18-21 A rattlesnake's face has thermal radiation detectors, allowing the snake to strike at an animal even in complete darkness.

Sample Problem 18.06 Thermal conduction through a layered wall

Figure 18-22 shows the cross section of a wall made of white pine of thickness L_a and brick of thickness L_d $(= 2.0L_a)$, sandwiching two layers of unknown material with identical thicknesses and thermal conductivities. The thermal conductivity of the pine is k_a and that of the brick is k_d $(= 5.0k_a)$. The face area A of the wall is unknown. Thermal conduction through the wall has reached the steady state; the only known interface temperatures are $T_1 = 25°C$, $T_2 = 20°C$, and $T_5 = -10°C$. What is interface temperature T_4?

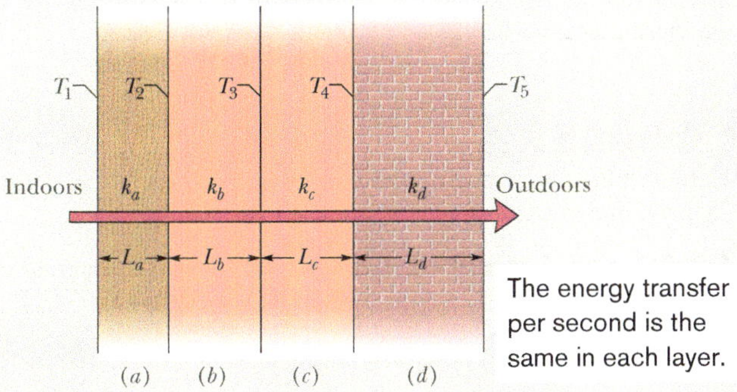

Figure 18-22 Steady-state heat transfer through a wall.

KEY IDEAS

(1) Temperature T_4 helps determine the rate P_d at which energy is conducted through the brick, as given by Eq. 18-32. However, we lack enough data to solve Eq. 18-32 for T_4. (2) Because the conduction is steady, the conduction rate P_d through the brick must equal the conduction rate P_a through the pine. That gets us going.

Calculations: From Eq. 18-32 and Fig. 18-22, we can write

$$P_a = k_a A \frac{T_1 - T_2}{L_a} \quad \text{and} \quad P_d = k_d A \frac{T_4 - T_5}{L_d}.$$

Setting $P_a = P_d$ and solving for T_4 yield

$$T_4 = \frac{k_a L_d}{k_d L_a}(T_1 - T_2) + T_5.$$

Letting $L_d = 2.0L_a$ and $k_d = 5.0k_a$, and inserting the known temperatures, we find

$$T_4 = \frac{k_a(2.0L_a)}{(5.0k_a)L_a}(25°C - 20°C) + (-10°C)$$

$$= -8.0°C. \quad \text{(Answer)}$$

WileyPLUS Additional examples, video, and practice available at *WileyPLUS*

Sample Problem 18.07 Thermal radiation by a skunk cabbage can melt surrounding snow

Unlike most other plants, a skunk cabbage can regulate its internal temperature (set at $T = 22°C$) by altering the rate at which it produces energy. If it becomes covered with snow, it can increase that production so that its thermal radiation melts the snow in order to re-expose the plant to sunlight. Let's model a skunk cabbage with a cylinder of height $h = 5.0$ cm and radius $R = 1.5$ cm and assume it is surrounded by a snow wall at temperature $T_{env} = -3.0°C$ (Fig. 18-23). If the emissivity ε is 0.80, what is the net rate of energy exchange via thermal radiation between the plant's curved side and the snow?

Figure 18-23 Model of skunk cabbage that has melted snow to uncover itself.

KEY IDEAS

(1) In a steady-state situation, a surface with area A, emissivity ε, and temperature T loses energy to thermal radiation at the rate given by Eq. 18-38 ($P_{rad} = \sigma\varepsilon AT^4$). (2) Simultaneously, it gains energy by thermal radiation from its environment at temperature T_{env} at the rate given by Eq. 18-39 ($P_{abs} = \sigma\varepsilon AT_{env}^4$).

Calculations: To find the net rate of energy exchange, we subtract Eq. 18-38 from Eq. 18-39 to write

$$P_{net} = P_{abs} - P_{rad} = \sigma\varepsilon A(T_{env}^4 - T^4). \qquad (18\text{-}41)$$

We need the area of the curved surface of the cylinder, which is $A = h(2\pi R)$. We also need the temperatures in kelvins: $T_{env} = 273\text{ K} - 3\text{ K} = 270\text{ K}$ and $T = 273\text{ K} + 22\text{ K} = 295\text{ K}$. Substituting in Eq. 18-41 for A and then substituting known values in SI units (which are not displayed here), we find

$$P_{net} = (5.67 \times 10^{-8})(0.80)(0.050)(2\pi)(0.015)(270^4 - 295^4) = -0.48\text{ W}. \qquad \text{(Answer)}$$

Thus, the plant has a net loss of energy via thermal radiation of 0.48 W. The plant's energy production rate is comparable to that of a hummingbird in flight.

WILEY PLUS Additional examples, video, and practice available at *WileyPLUS*

Review & Summary

Temperature; Thermometers Temperature is an SI base quantity related to our sense of hot and cold. It is measured with a thermometer, which contains a working substance with a measurable property, such as length or pressure, that changes in a regular way as the substance becomes hotter or colder.

Zeroth Law of Thermodynamics When a thermometer and some other object are placed in contact with each other, they eventually reach thermal equilibrium. The reading of the thermometer is then taken to be the temperature of the other object. The process provides consistent and useful temperature measurements because of the **zeroth law of thermodynamics:** If bodies A and B are each in thermal equilibrium with a third body C (the thermometer), then A and B are in thermal equilibrium with each other.

The Kelvin Temperature Scale In the SI system, temperature is measured on the **Kelvin scale,** which is based on the *triple point* of water (273.16 K). Other temperatures are then defined by use of a *constant-volume gas thermometer,* in which a sample of gas is maintained at constant volume so its pressure is proportional to its temperature. We define the *temperature* T as measured with a gas thermometer to be

$$T = (273.16\text{ K})\left(\lim_{\text{gas}\to 0} \frac{p}{p_3}\right). \qquad (18\text{-}6)$$

Here T is in kelvins, and p_3 and p are the pressures of the gas at 273.16 K and the measured temperature, respectively.

Celsius and Fahrenheit Scales The Celsius temperature scale is defined by

$$T_C = T - 273.15°, \qquad (18\text{-}7)$$

with T in kelvins. The Fahrenheit temperature scale is defined by

$$T_F = \tfrac{9}{5}T_C + 32°. \qquad (18\text{-}8)$$

Thermal Expansion All objects change size with changes in temperature. For a temperature change ΔT, a change ΔL in any linear dimension L is given by

$$\Delta L = L\alpha\,\Delta T, \tag{18-9}$$

in which α is the **coefficient of linear expansion.** The change ΔV in the volume V of a solid or liquid is

$$\Delta V = V\beta\,\Delta T. \tag{18-10}$$

Here $\beta = 3\alpha$ is the material's **coefficient of volume expansion.**

Heat Heat Q is energy that is transferred between a system and its environment because of a temperature difference between them. It can be measured in **joules** (J), **calories** (cal), **kilocalories** (Cal or kcal), or **British thermal units** (Btu), with

$$1 \text{ cal} = 3.968 \times 10^{-3} \text{ Btu} = 4.1868 \text{ J}. \tag{18-12}$$

Heat Capacity and Specific Heat If heat Q is absorbed by an object, the object's temperature change $T_f - T_i$ is related to Q by

$$Q = C(T_f - T_i), \tag{18-13}$$

in which C is the **heat capacity** of the object. If the object has mass m, then

$$Q = cm(T_f - T_i), \tag{18-14}$$

where c is the **specific heat** of the material making up the object. The **molar specific heat** of a material is the heat capacity per mole, which means per 6.02×10^{23} elementary units of the material.

Heat of Transformation Matter can exist in three common states: solid, liquid, and vapor. Heat absorbed by a material may change the material's physical state—for example, from solid to liquid or from liquid to gas. The amount of energy required per unit mass to change the state (but not the temperature) of a particular material is its **heat of transformation** L. Thus,

$$Q = Lm. \tag{18-16}$$

The **heat of vaporization** L_V is the amount of energy per unit mass that must be added to vaporize a liquid or that must be removed to condense a gas. The **heat of fusion** L_F is the amount of energy per unit mass that must be added to melt a solid or that must be removed to freeze a liquid.

Work Associated with Volume Change A gas may exchange energy with its surroundings through work. The amount of work W done *by* a gas as it expands or contracts from an initial volume V_i to a final volume V_f is given by

$$W = \int dW = \int_{V_i}^{V_f} p\,dV. \tag{18-25}$$

The integration is necessary because the pressure p may vary during the volume change.

First Law of Thermodynamics The principle of conservation of energy for a thermodynamic process is expressed in the **first law of thermodynamics,** which may assume either of the forms

$$\Delta E_{\text{int}} = E_{\text{int},f} - E_{\text{int},i} = Q - W \quad \text{(first law)} \tag{18-26}$$

or

$$dE_{\text{int}} = dQ - dW \quad \text{(first law)}. \tag{18-27}$$

E_{int} represents the internal energy of the material, which depends only on the material's state (temperature, pressure, and volume). Q represents the energy exchanged as heat between the system and its surroundings; Q is positive if the system absorbs heat and negative if the system loses heat. W is the work done *by* the system; W is positive if the system expands against an external force from the surroundings and negative if the system contracts because of an external force. *Q and W are path dependent; ΔE_{int} is path independent.*

Applications of the First Law The first law of thermodynamics finds application in several special cases:

adiabatic processes: $Q = 0, \quad \Delta E_{\text{int}} = -W$

constant-volume processes: $W = 0, \quad \Delta E_{\text{int}} = Q$

cyclical processes: $\Delta E_{\text{int}} = 0, \quad Q = W$

free expansions: $Q = W = \Delta E_{\text{int}} = 0$

Conduction, Convection, and Radiation The rate P_{cond} at which energy is *conducted* through a slab for which one face is maintained at the higher temperature T_H and the other face is maintained at the lower temperature T_C is

$$P_{\text{cond}} = \frac{Q}{t} = kA\,\frac{T_H - T_C}{L} \tag{18-32}$$

Here each face of the slab has area A, the length of the slab (the distance between the faces) is L, and k is the thermal conductivity of the material.

Convection occurs when temperature differences cause an energy transfer by motion within a fluid.

Radiation is an energy transfer via the emission of electromagnetic energy. The rate P_{rad} at which an object emits energy via thermal radiation is

$$P_{\text{rad}} = \sigma\varepsilon A T^4, \tag{18-38}$$

where σ ($= 5.6704 \times 10^{-8}$ W/m^2·K^4) is the Stefan–Boltzmann constant, ε is the emissivity of the object's surface, A is its surface area, and T is its surface temperature (in kelvins). The rate P_{abs} at which an object absorbs energy via thermal radiation from its environment, which is at the uniform temperature T_{env} (in kelvins), is

$$P_{\text{abs}} = \sigma\varepsilon A T_{\text{env}}^4. \tag{18-39}$$

Questions

1 The initial length L, change in temperature ΔT, and change in length ΔL of four rods are given in the following table. Rank the rods according to their coefficients of thermal expansion, greatest first.

Rod	L (m)	ΔT (C°)	ΔL (m)
a	2	10	4×10^{-4}
b	1	20	4×10^{-4}
c	2	10	8×10^{-4}
d	4	5	4×10^{-4}

2 Figure 18-24 shows three linear temperature scales, with the freezing and boiling points of water indicated. Rank the three scales according to the size of one degree on them, greatest first.

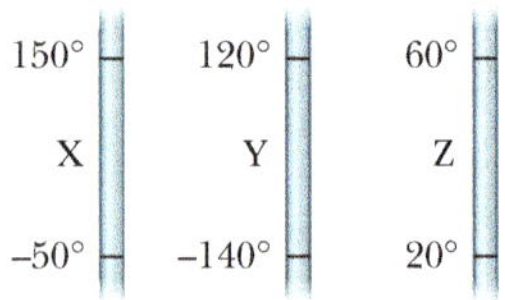

Figure 18-24 Question 2.

3 Materials A, B, and C are solids that are at their melting temperatures. Material A requires 200 J to melt 4 kg, material B requires 300 J to melt 5 kg, and material C requires 300 J to melt 6 kg. Rank the materials according to their heats of fusion, greatest first.

4 A sample A of liquid water and a sample B of ice, of identical mass, are placed in a thermally insulated container and allowed to come to thermal equilibrium. Figure 18-25*a* is a sketch of the temperature T of the samples versus time t. (a) Is the equilibrium temperature above, below, or at the freezing point of water? (b) In reaching equilibrium, does the liquid partly freeze, fully freeze, or undergo no freezing? (c) Does the ice partly melt, fully melt, or undergo no melting?

5 Question 4 continued: Graphs b through f of Fig. 18-25 are additional sketches of T versus t, of which one or more are impossible to produce. (a) Which is impossible and why? (b) In the possible ones, is the equilibrium temperature above, below, or at the freezing point of water? (c) As the possible situations reach equilibrium, does the liquid partly freeze, fully freeze, or undergo no freezing? Does the ice partly melt, fully melt, or undergo no melting?

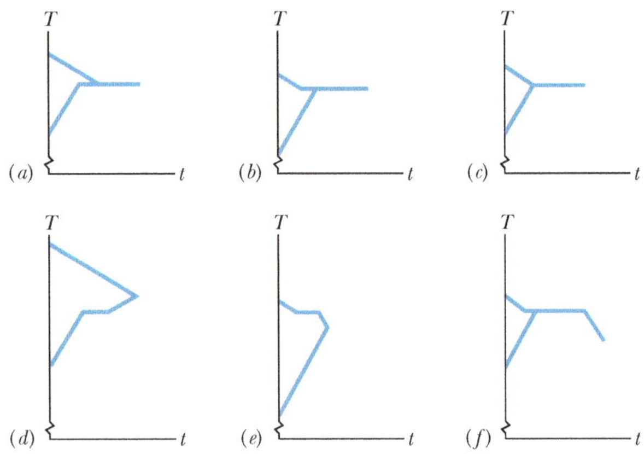

Figure 18-25 Questions 4 and 5.

6 Figure 18-26 shows three different arrangements of materials 1, 2, and 3 to form a wall. The thermal conductivities are $k_1 > k_2 > k_3$. The left side of the wall is 20 C° higher than the right side. Rank the arrangements according to (a) the (steady state) rate of energy conduction through the wall and (b) the temperature difference across material 1, greatest first.

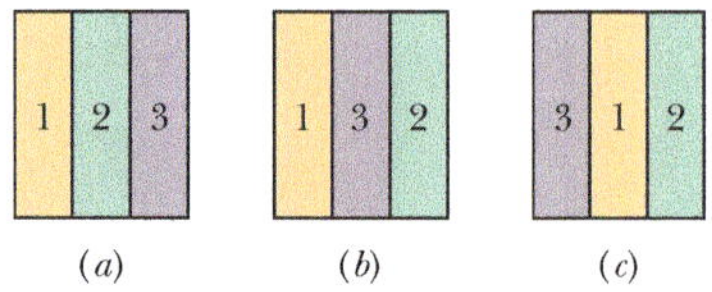

Figure 18-26 Question 6.

7 Figure 18-27 shows two closed cycles on p-V diagrams for a gas. The three parts of cycle 1 are of the same length and shape as those of cycle 2. For each cycle, should the cycle be traversed clockwise or counterclockwise if (a) the net work W done by the gas is to be positive and (b) the net energy transferred by the gas as heat Q is to be positive?

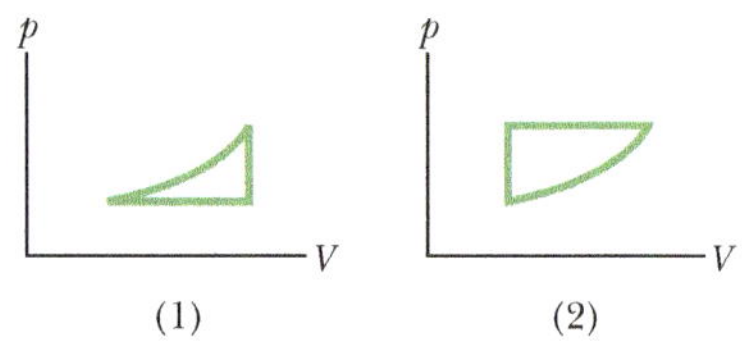

Figure 18-27 Questions 7 and 8.

8 For which cycle in Fig. 18-27, traversed clockwise, is (a) W greater and (b) Q greater?

9 Three different materials of identical mass are placed one at a time in a special freezer that can extract energy from a material at a certain constant rate. During the cooling process, each material begins in the liquid state and ends in the solid state; Fig. 18-28 shows the temperature T versus time t. (a) For material 1, is the specific heat for the liquid state greater than or less than that for the solid state? Rank the materials according to (b) freezing-point temperature, (c) specific heat in the liquid state, (d) specific heat in the solid state, and (e) heat of fusion, all greatest first.

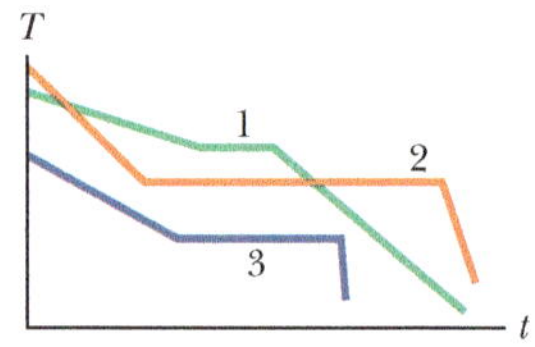

Figure 18-28 Question 9.

10 A solid cube of edge length r, a solid sphere of radius r, and a solid hemisphere of radius r, all made of the same material, are maintained at temperature 300 K in an environment at temperature 350 K. Rank the objects according to the net rate at which thermal radiation is exchanged with the environment, greatest first.

11 A hot object is dropped into a thermally insulated container of water, and the object and water are then allowed to come to thermal equilibrium. The experiment is repeated twice, with different hot objects. All three objects have the same mass and initial temperature, and the mass and initial temperature of the water are the same in the three experiments. For each of the experiments, Fig. 18-29 gives graphs of the temperatures T of the object and the water versus time t. Rank the graphs according to the specific heats of the objects, greatest first.

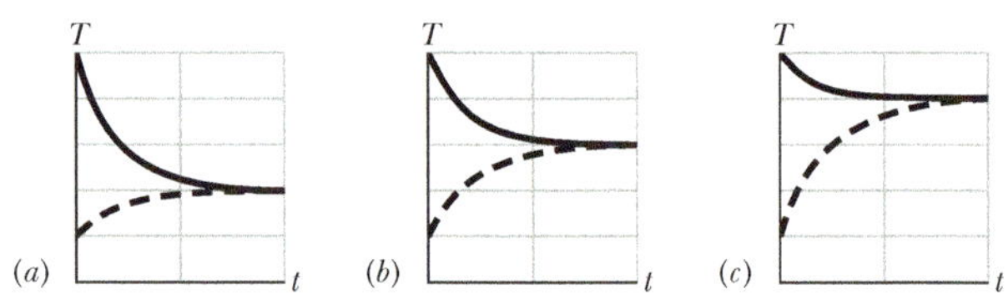

Figure 18-29 Question 11.

Problems

GO Tutoring problem available (at instructor's discretion) in *WileyPLUS* and WebAssign

SSM Worked-out solution available in Student Solutions Manual

• – ••• Number of dots indicates level of problem difficulty

Additional information available in *The Flying Circus of Physics* and at flyingcircusofphysics.com

WWW Worked-out solution is at / ILW Interactive solution is at http://www.wiley.com/college/halliday

Module 18-1 Temperature

•1 Suppose the temperature of a gas is 373.15 K when it is at the boiling point of water. What then is the limiting value of the ratio of the pressure of the gas at that boiling point to its pressure at the triple point of water? (Assume the volume of the gas is the same at both temperatures.)

•2 Two constant-volume gas thermometers are assembled, one with nitrogen and the other with hydrogen. Both contain enough gas so that $p_3 = 80$ kPa. (a) What is the difference between the pressures in the two thermometers if both bulbs are in boiling water? (*Hint:* See Fig. 18-6.) (b) Which gas is at higher pressure?

•3 A gas thermometer is constructed of two gas-containing bulbs, each in a water bath, as shown in Fig. 18-30. The pressure difference between the two bulbs is measured by a mercury manometer as shown. Appropriate reservoirs, not shown in the diagram, maintain constant gas volume in the two bulbs. There is no difference in pressure when both baths are at the triple point of water. The pressure difference is 120 torr when one bath is at the triple point and the other is at the boiling point of water. It is 90.0 torr when one bath is at the triple point and the other is at an unknown temperature to be measured. What is the unknown temperature?

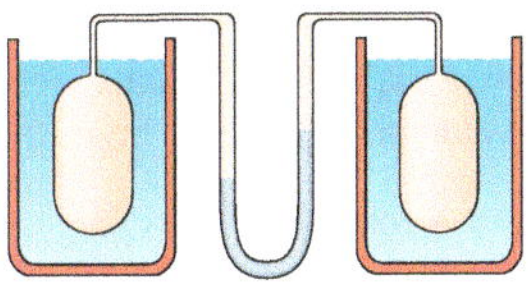

Figure 18-30 Problem 3.

Module 18-2 The Celsius and Fahrenheit Scales

•4 (a) In 1964, the temperature in the Siberian village of Oymyakon reached −71°C. What temperature is this on the Fahrenheit scale? (b) The highest officially recorded temperature in the continental United States was 134°F in Death Valley, California. What is this temperature on the Celsius scale?

•5 At what temperature is the Fahrenheit scale reading equal to (a) twice that of the Celsius scale and (b) half that of the Celsius scale?

••6 On a linear X temperature scale, water freezes at −125.0°X and boils at 375.0°X. On a linear Y temperature scale, water freezes at −70.00°Y and boils at −30.00°Y. A temperature of 50.00°Y corresponds to what temperature on the X scale?

••7 ILW Suppose that on a linear temperature scale X, water boils at −53.5°X and freezes at −170°X. What is a temperature of 340 K on the X scale? (Approximate water's boiling point as 373 K.)

Module 18-3 Thermal Expansion

•8 At 20°C, a brass cube has edge length 30 cm. What is the increase in the surface area when it is heated from 20°C to 75°C?

•9 ILW A circular hole in an aluminum plate is 2.725 cm in diameter at 0.000°C. What is its diameter when the temperature of the plate is raised to 100.0°C?

•10 An aluminum flagpole is 33 m high. By how much does its length increase as the temperature increases by 15 C°?

•11 What is the volume of a lead ball at 30.00°C if the ball's volume at 60.00°C is 50.00 cm^3?

•12 An aluminum-alloy rod has a length of 10.000 cm at 20.000°C and a length of 10.015 cm at the boiling point of water. (a) What is the length of the rod at the freezing point of water? (b) What is the temperature if the length of the rod is 10.009 cm?

•13 SSM Find the change in volume of an aluminum sphere with an initial radius of 10 cm when the sphere is heated from 0.0°C to 100°C.

••14 When the temperature of a copper coin is raised by 100 C°, its diameter increases by 0.18%. To two significant figures, give the percent increase in (a) the area of a face, (b) the thickness, (c) the volume, and (d) the mass of the coin. (e) Calculate the coefficient of linear expansion of the coin.

••15 ILW A steel rod is 3.000 cm in diameter at 25.00°C. A brass ring has an interior diameter of 2.992 cm at 25.00°C. At what common temperature will the ring just slide onto the rod?

••16 When the temperature of a metal cylinder is raised from 0.0°C to 100°C, its length increases by 0.23%. (a) Find the percent change in density. (b) What is the metal? Use Table 18-2.

••17 SSM WWW An aluminum cup of 100 cm^3 capacity is completely filled with glycerin at 22°C. How much glycerin, if any, will spill out of the cup if the temperature of both the cup and the glycerin is increased to 28°C? (The coefficient of volume expansion of glycerin is 5.1×10^{-4}/C°.)

••18 At 20°C, a rod is exactly 20.05 cm long on a steel ruler. Both are placed in an oven at 270°C, where the rod now measures 20.11 cm on the same ruler. What is the coefficient of linear expansion for the material of which the rod is made?

••19 GO A vertical glass tube of length $L = 1.280\,000$ m is half filled with a liquid at 20.000 000°C. How much will the height of the liquid column change when the tube and liquid are heated to 30.000 000°C? Use coefficients $\alpha_{glass} = 1.000\,000 \times 10^{-5}$/K and $\beta_{liquid} = 4.000\,000 \times 10^{-5}$/K.

••20 GO In a certain experiment, a small radioactive source must move at selected, extremely slow speeds. This motion is accomplished by fastening the source to one end of an aluminum rod and heating the central section of the rod in a controlled way. If the effective heated section of the rod in Fig. 18-31 has length $d = 2.00$ cm, at what constant rate must the temperature of the rod be changed if the source is to move at a constant speed of 100 nm/s?

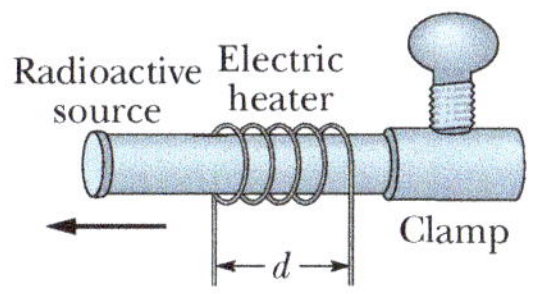

Figure 18-31 Problem 20.

•••21 SSM ILW As a result of a temperature rise of 32 C°, a bar with a crack at its center buckles upward (Fig. 18-32). The fixed distance L_0 is 3.77 m and the coefficient of linear expansion of the bar is 25×10^{-6}/C°. Find the rise x of the center.

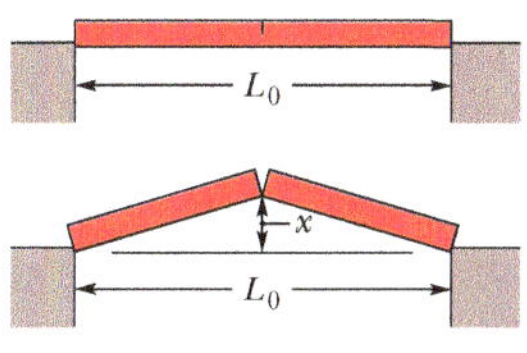

Figure 18-32 Problem 21.

Module 18-4 Absorption of Heat

•22 One way to keep the contents of a garage from becoming too cold on a night when a severe subfreezing temperature is forecast is to put a tub of water in the garage. If the mass of the water is 125 kg and its initial temperature is 20°C, (a) how much energy must the water transfer to its surroundings in order to freeze completely and (b) what is the lowest possible temperature of the water and its surroundings until that happens?

•23 SSM A small electric immersion heater is used to heat 100 g of water for a cup of instant coffee. The heater is labeled "200 watts" (it converts electrical energy to thermal energy at this rate). Calculate the time required to bring all this water from 23.0°C to 100°C, ignoring any heat losses.

•24 A certain substance has a mass per mole of 50.0 g/mol. When 314 J is added as heat to a 30.0 g sample, the sample's temperature rises from 25.0°C to 45.0°C. What are the (a) specific heat and (b) molar specific heat of this substance? (c) How many moles are in the sample?

•25 A certain diet doctor encourages people to diet by drinking ice water. His theory is that the body must burn off enough fat to raise the temperature of the water from 0.00°C to the body temperature of 37.0°C. How many liters of ice water would have to be consumed to burn off 454 g (about 1 lb) of fat, assuming that burning this much fat requires 3500 Cal be transferred to the ice water? Why is it not advisable to follow this diet? (One liter = 10^3 cm^3. The density of water is 1.00 g/cm^3.)

•26 What mass of butter, which has a usable energy content of 6.0 Cal/g (= 6000 cal/g), would be equivalent to the change in gravitational potential energy of a 73.0 kg man who ascends from sea level to the top of Mt. Everest, at elevation 8.84 km? Assume that the average g for the ascent is 9.80 m/s^2.

•27 SSM Calculate the minimum amount of energy, in joules, required to completely melt 130 g of silver initially at 15.0°C.

•28 How much water remains unfrozen after 50.2 kJ is transferred as heat from 260 g of liquid water initially at its freezing point?

••29 In a solar water heater, energy from the Sun is gathered by water that circulates through tubes in a rooftop collector. The solar radiation enters the collector through a transparent cover and warms the water in the tubes; this water is pumped into a holding tank. Assume that the efficiency of the overall system is 20% (that is, 80% of the incident solar energy is lost from the system). What collector area is necessary to raise the temperature of 200 L of water in the tank from 20°C to 40°C in 1.0 h when the intensity of incident sunlight is 700 W/m^2?

••30 A 0.400 kg sample is placed in a cooling apparatus that removes energy as heat at a constant rate. Figure 18-33 gives the temperature T of the sample versus time t; the horizontal scale is set by t_s = 80.0 min. The sample freezes during the energy removal. The specific heat of the sample in its initial liquid phase is 3000 J/kg·K. What are (a) the sample's heat of fusion and (b) its specific heat in the frozen phase?

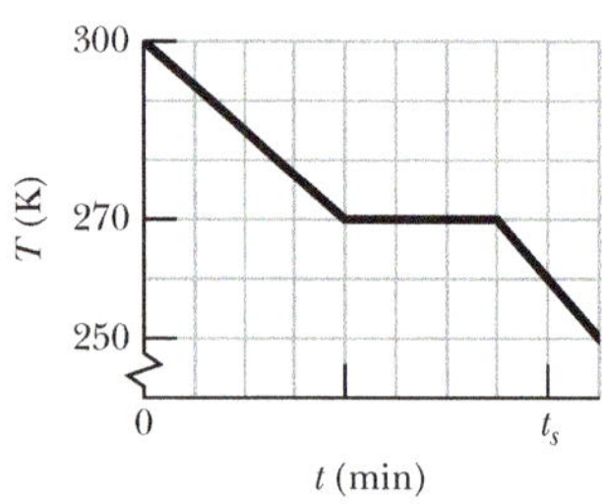

Figure 18-33 Problem 30.

••31 ILW What mass of steam at 100°C must be mixed with 150 g of ice at its melting point, in a thermally insulated container, to produce liquid water at 50°C?

••32 The specific heat of a substance varies with temperature according to the function $c = 0.20 + 0.14T + 0.023T^2$, with T in °C and c in cal/g·K. Find the energy required to raise the temperature of 2.0 g of this substance from 5.0°C to 15°C.

••33 *Nonmetric version:* (a) How long does a 2.0×10^5 Btu/h water heater take to raise the temperature of 40 gal of water from 70°F to 100°F? *Metric version:* (b) How long does a 59 kW water heater take to raise the temperature of 150 L of water from 21°C to 38°C?

••34 GO Samples A and B are at different initial temperatures when they are placed in a thermally insulated container and allowed to come to thermal equilibrium. Figure 18-34a gives their temperatures T versus time t. Sample A has a mass of 5.0 kg; sample B has a mass of 1.5 kg. Figure 18-34b is a general plot for the material of sample B. It shows the temperature change ΔT that the material undergoes when energy is transferred to it as heat Q. The change ΔT is plotted versus the energy Q per unit mass of the material, and the scale of the vertical axis is set by ΔT_s = 4.0 C°. What is the specific heat of sample A?

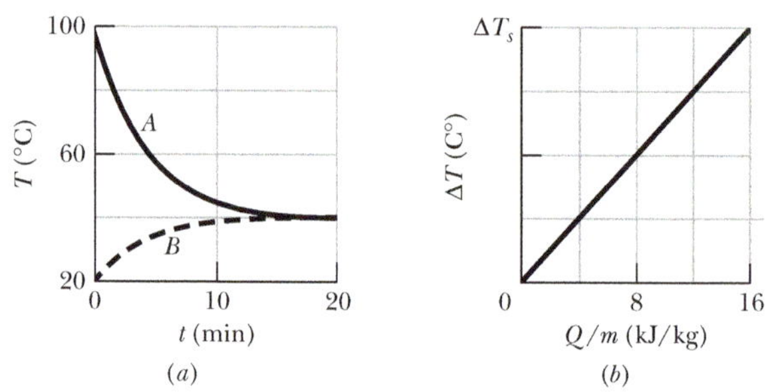

Figure 18-34 Problem 34.

••35 An insulated Thermos contains 130 cm^3 of hot coffee at 80.0°C. You put in a 12.0 g ice cube at its melting point to cool the coffee. By how many degrees has your coffee cooled once the ice has melted and equilibrium is reached? Treat the coffee as though it were pure water and neglect energy exchanges with the environment.

••36 A 150 g copper bowl contains 220 g of water, both at 20.0°C. A very hot 300 g copper cylinder is dropped into the water, causing the water to boil, with 5.00 g being converted to steam. The final temperature of the system is 100°C. Neglect energy transfers with the environment. (a) How much energy (in calories) is transferred to the water as heat? (b) How much to the bowl? (c) What is the original temperature of the cylinder?

••37 A person makes a quantity of iced tea by mixing 500 g of hot tea (essentially water) with an equal mass of ice at its melting point. Assume the mixture has negligible energy exchanges with its environment. If the tea's initial temperature is T_i = 90°C, when thermal equilibrium is reached what are (a) the mixture's temperature T_f and (b) the remaining mass m_f of ice? If T_i = 70°C, when thermal equilibrium is reached what are (c) T_f and (d) m_f?

••38 A 0.530 kg sample of liquid water and a sample of ice are placed in a thermally insulated container. The container also contains a device that transfers energy as heat from the liquid water to the ice at a constant rate P, until thermal equilibrium is reached. The temperatures T of the liquid water and the ice are given in Fig. 18-35 as functions of time t; the horizontal scale is set by $t_s = 80.0$ min. (a) What is rate P? (b) What is the initial mass of the ice in the container? (c) When thermal equilibrium is reached, what is the mass of the ice produced in this process?

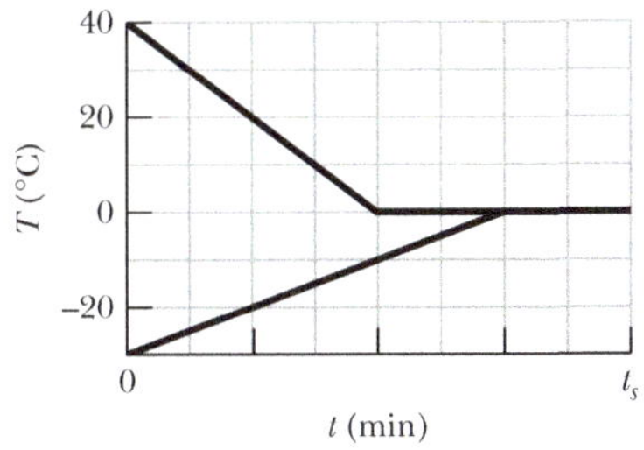

Figure 18-35 Problem 38.

••39 GO Ethyl alcohol has a boiling point of 78.0°C, a freezing point of −114°C, a heat of vaporization of 879 kJ/kg, a heat of fusion of 109 kJ/kg, and a specific heat of 2.43 kJ/kg·K. How much energy must be removed from 0.510 kg of ethyl alcohol that is initially a gas at 78.0°C so that it becomes a solid at −114°C?

••40 GO Calculate the specific heat of a metal from the following data. A container made of the metal has a mass of 3.6 kg and contains 14 kg of water. A 1.8 kg piece of the metal initially at a temperature of 180°C is dropped into the water. The container and water initially have a temperature of 16.0°C, and the final temperature of the entire (insulated) system is 18.0°C.

•••41 SSM WWW (a) Two 50 g ice cubes are dropped into 200 g of water in a thermally insulated container. If the water is initially at 25°C, and the ice comes directly from a freezer at −15°C, what is the final temperature at thermal equilibrium? (b) What is the final temperature if only one ice cube is used?

•••42 GO A 20.0 g copper ring at 0.000°C has an inner diameter of $D = 2.54000$ cm. An aluminum sphere at 100.0°C has a diameter of $d = 2.545\,08$ cm. The sphere is put on top of the ring (Fig. 18-36), and the two are allowed to come to thermal equilibrium, with no heat lost to the surroundings. The sphere just passes through the ring at the equilibrium temperature. What is the mass of the sphere?

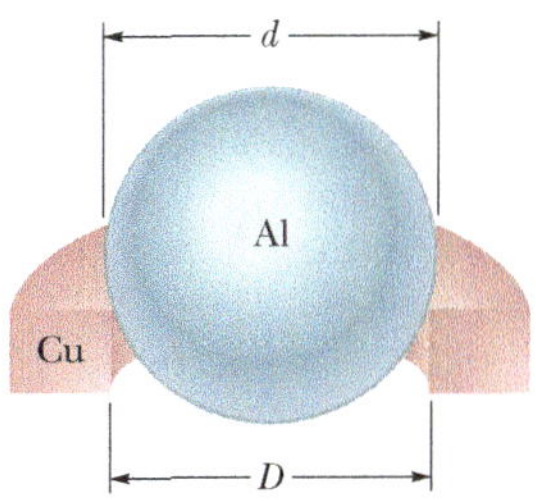

Figure 18-36 Problem 42.

Module 18-5 The First Law of Thermodynamics

•43 In Fig. 18-37, a gas sample expands from V_0 to $4.0V_0$ while its pressure decreases from p_0 to $p_0/4.0$. If $V_0 = 1.0$ m^3 and $p_0 = 40$ Pa, how much work is done by the gas if its pressure changes with volume via (a) path A, (b) path B, and (c) path C?

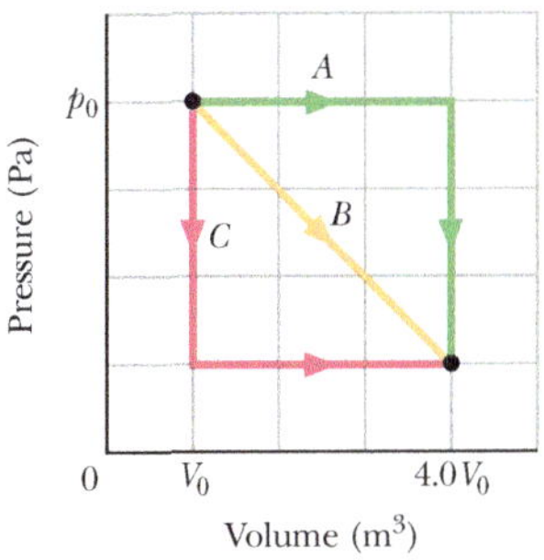

Figure 18-37 Problem 43.

•44 GO A thermodynamic system is taken from state A to state B to state C, and then back to A, as shown in the p-V diagram of Fig. 18-38a. The vertical scale is set by $p_s = 40$ Pa, and the horizontal scale is set by $V_s = 4.0$ m^3. (a)–(g) Complete the table in Fig. 18-38b by inserting a plus sign, a minus sign, or a zero in each indicated cell. (h) What is the net work done by the system as it moves once through the cycle $ABCA$?

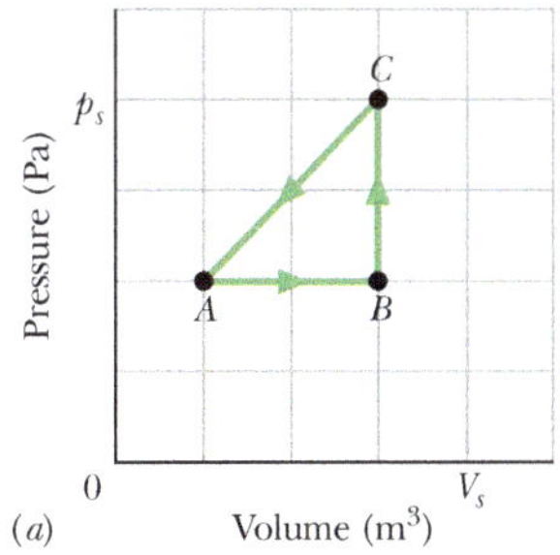

(b)

	Q	W	ΔE_{int}
$A \longrightarrow B$	(a)	(b)	+
$B \longrightarrow C$	+	(c)	(d)
$C \longrightarrow A$	(e)	(f)	(g)

Figure 18-38 Problem 44.

•45 SSM ILW A gas within a closed chamber undergoes the cycle shown in the p-V diagram of Fig. 18-39. The horizontal scale is set by $V_s = 4.0$ m^3. Calculate the net energy added to the system as heat during one complete cycle.

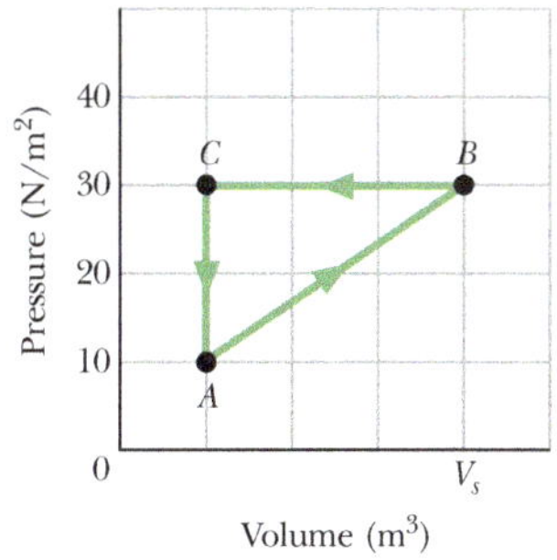

Figure 18-39 Problem 45.

•46 Suppose 200 J of work is done on a system and 70.0 cal is extracted from the system as heat. In the sense of the first law of thermodynamics, what are the values (including algebraic signs) of (a) W, (b) Q, and (c) ΔE_{int}?

••47 SSM WWW When a system is taken from state i to state f along path iaf in Fig. 18-40, $Q = 50$ cal and $W = 20$ cal. Along path ibf, $Q = 36$ cal. (a) What is W along path ibf? (b) If $W = -13$ cal for the return path fi, what is Q for this path? (c) If $E_{int,i} = 10$ cal, what is $E_{int,f}$? If $E_{int,b} = 22$ cal, what is Q for (d) path ib and (e) path bf?

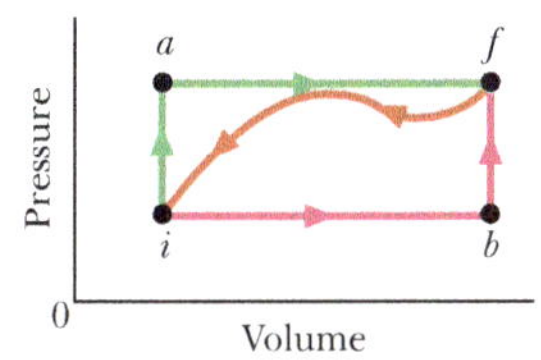

Figure 18-40 Problem 47.

••48 GO As a gas is held within a closed chamber, it passes through the cycle shown in Fig. 18-41. Determine the energy transferred by the system as heat during constant-pressure process CA if the energy added as heat Q_{AB} during constant-volume process AB is 20.0 J, no energy is transferred as heat during adiabatic process BC, and the net work done during the cycle is 15.0 J.

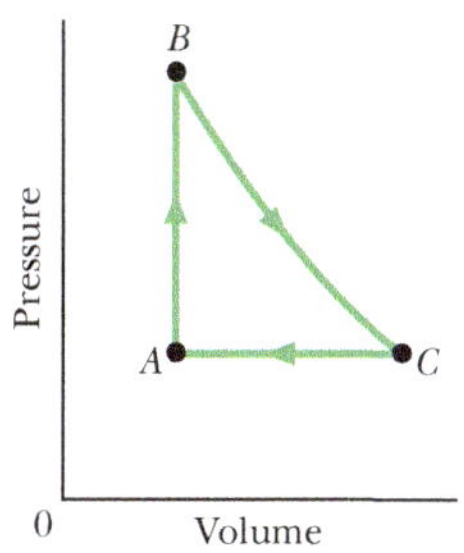

Figure 18-41 Problem 48.

••49 GO Figure 18-42 represents a closed cycle for a gas (the figure is not drawn to scale). The change in the internal energy of the gas as it moves from a to c along the path abc is -200 J. As it moves from c to d, 180 J must be transferred to it as heat. An additional transfer of 80 J to it as heat is needed as it moves from d to a. How much work is done on the gas as it moves from c to d?

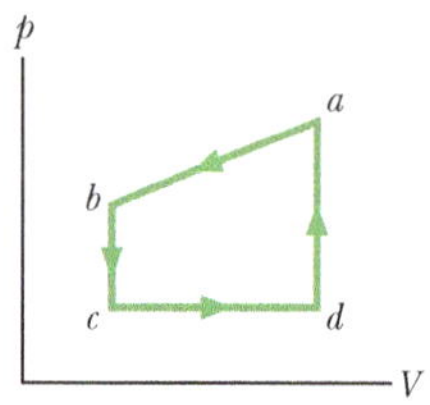

Figure 18-42 Problem 49.

••50 GO A lab sample of gas is taken through cycle $abca$ shown in the p-V diagram of Fig. 18-43. The net work done is $+1.2$ J. Along path ab, the change in the internal energy is $+3.0$ J and the magnitude of the work done is 5.0 J. Along path ca, the energy transferred to the gas as heat is $+2.5$ J. How much energy is transferred as heat along (a) path ab and (b) path bc?

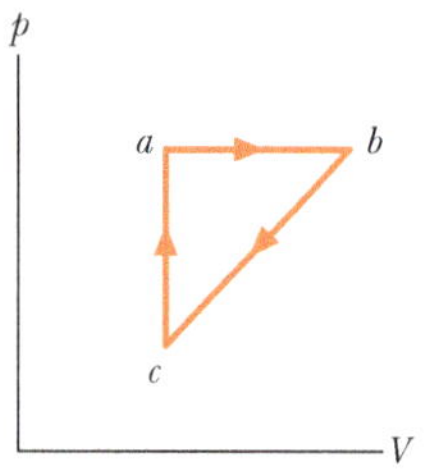

Figure 18-43 Problem 50.

Module 18-6 Heat Transfer Mechanisms

•51 A sphere of radius 0.500 m, temperature 27.0°C, and emissivity 0.850 is located in an environment of temperature 77.0°C. At what rate does the sphere (a) emit and (b) absorb thermal radiation? (c) What is the sphere's net rate of energy exchange?

•52 The ceiling of a single-family dwelling in a cold climate should have an R-value of 30. To give such insulation, how thick would a layer of (a) polyurethane foam and (b) silver have to be?

•53 SSM Consider the slab shown in Fig. 18-18. Suppose that L = 25.0 cm, A = 90.0 cm^2, and the material is copper. If T_H = 125°C, T_C = 10.0°C, and a steady state is reached, find the conduction rate through the slab.

•54 If you were to walk briefly in space without a spacesuit while far from the Sun (as an astronaut does in the movie *2001, A Space Odyssey*), you would feel the cold of space—while you radiated energy, you would absorb almost none from your environment. (a) At what rate would you lose energy? (b) How much energy would you lose in 30 s? Assume that your emissivity is 0.90, and estimate other data needed in the calculations.

•55 ILW A cylindrical copper rod of length 1.2 m and cross-sectional area 4.8 cm^2 is insulated along its side. The ends are held at a temperature difference of 100 C° by having one end in a water–ice mixture and the other in a mixture of boiling water and steam. At what rate (a) is energy conducted by the rod and (b) does the ice melt?

••56 The giant hornet *Vespa mandarinia japonica* preys on Japanese bees. However, if one of the hornets attempts to invade a beehive, several hundred of the bees quickly form a compact ball around the hornet to stop it. They don't sting, bite, crush, or suffocate it. Rather they overheat it by quickly raising their body temperatures from the normal 35°C to 47°C or 48°C, which is lethal to the hornet but not to the bees (Fig. 18-44). Assume the following: 500 bees form a ball of radius R = 2.0 cm for a time t = 20 min, the primary loss of energy by the ball is by thermal radiation, the ball's surface has emissivity ε = 0.80, and the ball has a uniform temperature. On average, how much additional energy must each bee produce during the 20 min to maintain 47°C?

Figure 18-44 Problem 56.

© Dr. Masato Ono, Tamagawa University

••57 (a) What is the rate of energy loss in watts per square meter through a glass window 3.0 mm thick if the outside temperature is −20°F and the inside temperature is +72°F? (b) A storm window having the same thickness of glass is installed parallel to the first window, with an air gap of 7.5 cm between the two windows. What now is the rate of energy loss if conduction is the only important energy-loss mechanism?

••58 A solid cylinder of radius r_1 = 2.5 cm, length h_1 = 5.0 cm, emissivity 0.85, and temperature 30°C is suspended in an environment of temperature 50°C. (a) What is the cylinder's net thermal radiation transfer rate P_1? (b) If the cylinder is stretched until its radius is r_2 = 0.50 cm, its net thermal radiation transfer rate becomes P_2. What is the ratio P_2/P_1?

••59 In Fig. 18-45a, two identical rectangular rods of metal are welded end to end, with a temperature of T_1 = 0°C on the left side and a temperature of T_2 = 100°C on the right side. In 2.0 min, 10 J is conducted at a constant rate from the right side to the left side. How much time would be required to conduct 10 J if the rods were welded side to side as in Fig. 18-45b?

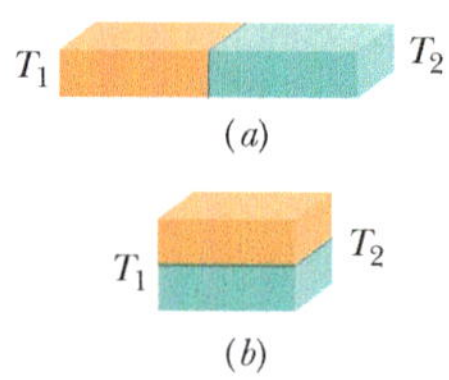

Figure 18-45 Problem 59.

••60 GO Figure 18-46 shows the cross section of a wall made of three layers. The layer thicknesses are L_1, L_2 = 0.700L_1, and L_3 = 0.350L_1. The thermal conductivities are k_1, k_2 = 0.900k_1, and k_3 = 0.800k_1. The temperatures at the left side and right side of the wall are T_H = 30.0°C and T_C = −15.0°C, respectively. Thermal conduction is steady. (a) What is the temperature difference ΔT_2 across layer 2 (between the left and right sides of the layer)? If k_2 were, instead, equal to 1.1k_1, (b) would the rate at which energy is conducted through the wall be greater than, less than, or the same as previously, and (c) what would be the value of ΔT_2?

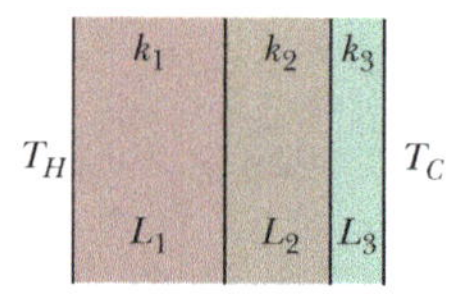

Figure 18-46 Problem 60.

••61 SSM A 5.0 cm slab has formed on an outdoor tank of water (Fig. 18-47). The air is at −10°C. Find the rate of ice formation (centimeters per hour). The ice has thermal conductivity 0.0040 cal/s·cm·C° and density 0.92 g/cm^3. Assume there is no energy transfer through the walls or bottom.

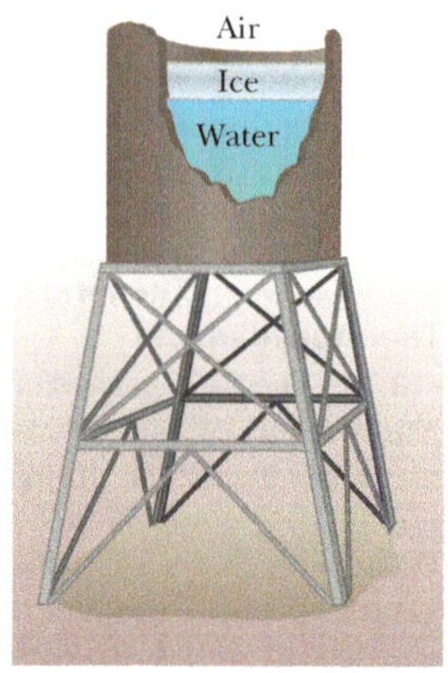

Figure 18-47 Problem 61.

••62 *Leidenfrost effect.* A water drop will last about 1 s on a hot skillet with a temperature between 100°C and about 200°C. However, if the skillet is much hotter, the drop can last several minutes, an effect named after an early investigator. The longer lifetime is due to the support of a thin layer of air and water vapor that separates the drop from the metal (by distance L in Fig. 18-48). Let L = 0.100 mm, and assume that the drop is flat with height h = 1.50 mm and bottom face area $A = 4.00 \times 10^{-6}$ m^2. Also assume that the skillet has a constant temperature T_s = 300°C and the drop has a temperature of 100°C. Water has density ρ = 1000 kg/m^3, and the supporting layer has thermal conductivity k = 0.026 W/m·K. (a) At what rate is energy conducted from the skillet to the drop through the drop's bottom surface? (b) If conduction is the primary way energy moves from the skillet to the drop, how long will the drop last?

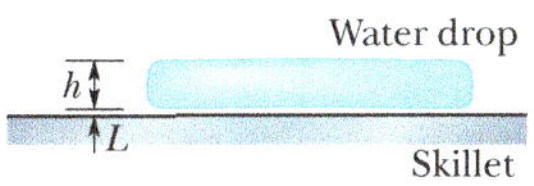

Figure 18-48 Problem 62.

••63 Figure 18-49 shows (in cross section) a wall consisting of four layers, with thermal conductivities k_1 = 0.060 W/m·K, k_3 = 0.040 W/m·K, and k_4 = 0.12 W/m·K (k_2 is not known). The layer thicknesses are L_1 = 1.5 cm, L_3 = 2.8 cm, and L_4 = 3.5 cm (L_2 is not known). The known temperatures are T_1 = 30°C, T_{12} = 25°C, and T_4 = −10°C. Energy transfer through the wall is steady. What is interface temperature T_{34}?

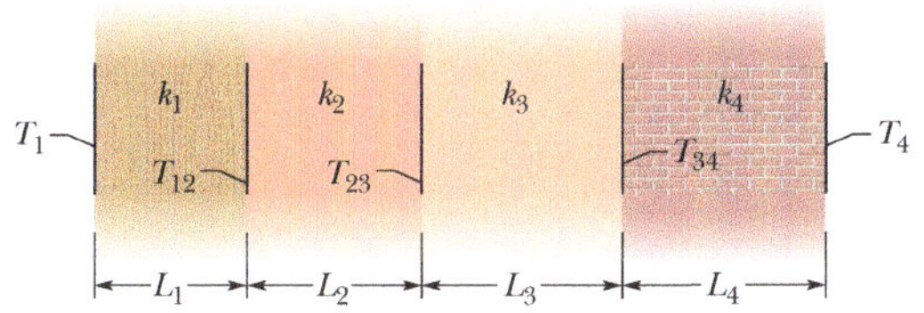

Figure 18-49 Problem 63.

••64 *Penguin huddling.* To withstand the harsh weather of the Antarctic, emperor penguins huddle in groups (Fig. 18-50). Assume that a penguin is a circular cylinder with a top surface area a = 0.34 m^2 and height h = 1.1 m. Let P_r be the rate at which an individual penguin radiates energy to the environment (through the top and the sides); thus NP_r is the rate at which N identical, well-separated penguins radiate. If the penguins huddle closely to form a *huddled cylinder* with top surface area Na and height h, the cylinder radiates at the rate P_h. If N = 1000, (a) what is the value of the fraction P_h/NP_r and (b) by what percentage does huddling reduce the total radiation loss?

Alain Torterotot/Peter Arnold/Photolibrary

Figure 18-50 Problem 64.

••65 Ice has formed on a shallow pond, and a steady state has been reached, with the air above the ice at −5.0°C and the bottom of the pond at 4.0°C. If the total depth of *ice* + *water* is 1.4 m, how thick is the ice? (Assume that the thermal conductivities of ice and water are 0.40 and 0.12 cal/m·C°·s, respectively.)

•••66 *Evaporative cooling of beverages.* A cold beverage can be kept cold even on a warm day if it is slipped into a porous ceramic container that has been soaked in water. Assume that energy lost to evaporation matches the net energy gained via the radiation exchange through the top and side surfaces. The container and beverage have temperature T = 15°C, the environment has temperature T_{env} = 32°C, and the container is a cylinder with radius r = 2.2 cm and height 10 cm. Approximate the emissivity as ε = 1, and neglect other energy exchanges. At what rate dm/dt is the container losing water mass?

Additional Problems

67 In the extrusion of cold chocolate from a tube, work is done on the chocolate by the pressure applied by a ram forcing the chocolate through the tube. The work per unit mass of extruded chocolate is equal to p/ρ, where p is the difference between the applied pressure and the pressure where the chocolate emerges from the tube, and ρ is the density of the chocolate. Rather than increasing the temperature of the chocolate, this work melts cocoa fats in the chocolate. These fats have a heat of fusion of 150 kJ/kg. Assume that all of the work goes into that melting and that these fats make up 30% of the chocolate's mass. What percentage of the fats melt during the extrusion if p = 5.5 MPa and ρ = 1200 kg/m^3?

68 Icebergs in the North Atlantic present hazards to shipping, causing the lengths of shipping routes to be increased by about 30% during the iceberg season. Attempts to destroy icebergs include planting explosives, bombing, torpedoing, shelling, ramming, and coating with black soot. Suppose that direct melting of the iceberg, by placing heat sources in the ice, is tried. How much energy as heat is required to melt 10% of an iceberg that has a mass of 200 000 metric tons? (Use 1 metric ton = 1000 kg.)

69 Figure 18-51 displays a closed cycle for a gas. The change in internal energy along path ca is −160 J. The energy transferred to the gas as heat is 200 J along path ab, and 40 J along path bc. How much work is done by the gas along (a) path abc and (b) path ab?

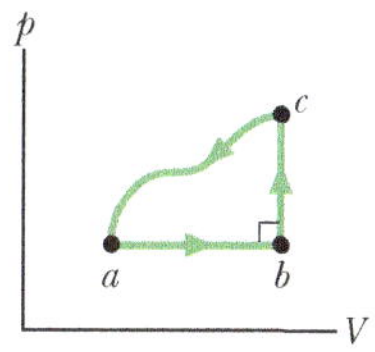

Figure 18-51 Problem 69.

70 In a certain solar house, energy from the Sun is stored in barrels filled with water. In a particular winter stretch of five cloudy days, 1.00×10^6 kcal is needed to maintain the inside of the house at 22.0°C. Assuming that the water in the barrels is at 50.0°C and that the water has a density of 1.00×10^3 kg/m^3, what volume of water is required?

71 A 0.300 kg sample is placed in a cooling apparatus that removes energy as heat at a constant rate of 2.81 W. Figure 18-52 gives the temperature T of the sam-

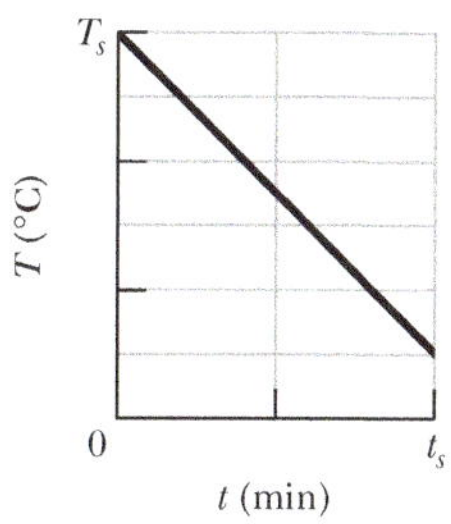

Figure 18-52 Problem 71.

ple versus time t. The temperature scale is set by $T_s = 30°C$ and the time scale is set by $t_s = 20$ min. What is the specific heat of the sample?

72 The average rate at which energy is conducted outward through the ground surface in North America is 54.0 mW/m^2, and the average thermal conductivity of the near-surface rocks is 2.50 W/m·K. Assuming a surface temperature of 10.0°C, find the temperature at a depth of 35.0 km (near the base of the crust). Ignore the heat generated by the presence of radioactive elements.

73 What is the volume increase of an aluminum cube 5.00 cm on an edge when heated from 10.0°C to 60.0°C?

74 In a series of experiments, block B is to be placed in a thermally insulated container with block A, which has the same mass as block B. In each experiment, block B is initially at a certain temperature T_B, but temperature T_A of block A is changed from experiment to experiment. Let T_f represent the final temperature of the two blocks when they reach thermal equilibrium in any of the experiments. Figure 18-53 gives temperature T_f versus the initial temperature T_A for a range of possible values of T_A, from $T_{A1} = 0$ K to $T_{A2} = 500$ K. The vertical axis scale is set by $T_{fs} = 400$ K. What are (a) temperature T_B and (b) the ratio c_B/c_A of the specific heats of the blocks?

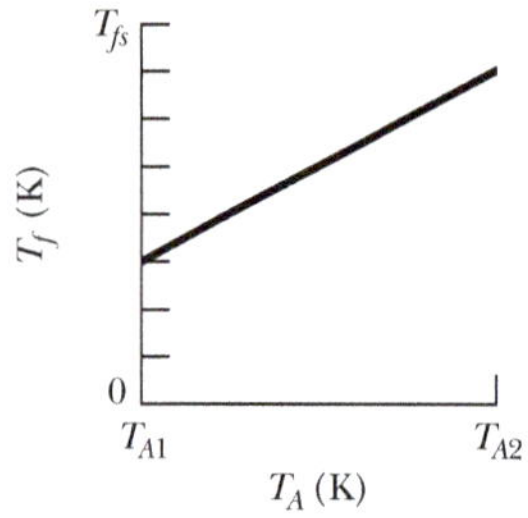

Figure 18-53 Problem 74.

75 Figure 18-54 displays a closed cycle for a gas. From c to b, 40 J is transferred from the gas as heat. From b to a, 130 J is transferred from the gas as heat, and the magnitude of the work done by the gas is 80 J. From a to c, 400 J is transferred to the gas as heat. What is the work done by the gas from a to c? (*Hint:* You need to supply the plus and minus signs for the given data.)

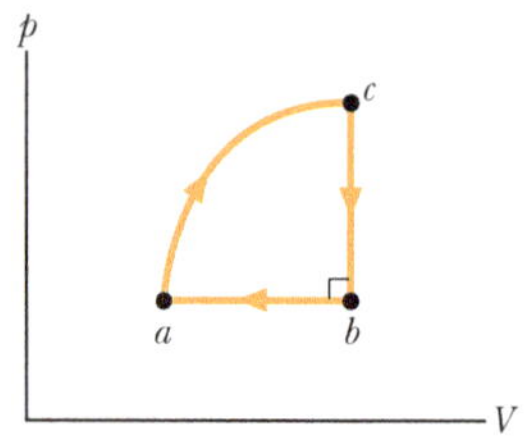

Figure 18-54 Problem 75.

76 Three equal-length straight rods, of aluminum, Invar, and steel, all at 20.0°C, form an equilateral triangle with hinge pins at the vertices. At what temperature will the angle opposite the Invar rod be 59.95°? See Appendix E for needed trigonometric formulas and Table 18-2 for needed data.

77 SSM The temperature of a 0.700 kg cube of ice is decreased to −150°C. Then energy is gradually transferred to the cube as heat while it is otherwise thermally isolated from its environment. The total transfer is 0.6993 MJ. Assume the value of c_{ice} given in Table 18-3 is valid for temperatures from −150°C to 0°C. What is the final temperature of the water?

78 GO *Icicles.* Liquid water coats an active (growing) icicle and extends up a short, narrow tube along the central axis (Fig. 18-55). Because the water–ice interface must have a temperature of 0°C, the water in the tube cannot lose energy through the sides of the icicle or down through the tip because there is no temperature change in those directions. It can lose energy and freeze only by sending energy up (through distance L) to the top of the icicle, where the temperature T_r can be below 0°C. Take $L = 0.12$ m and $T_r = -5°C$. Assume that the central tube and the upward conduction path both have cross-sectional area A. In terms of A, what rate is (a) energy conducted upward and (b) mass converted from liquid to ice at the top of the central tube? (c) At what rate does the top of the tube move downward because of water freezing there? The thermal conductivity of ice is 0.400 W/m·K, and the density of liquid water is 1000 kg/m^3.

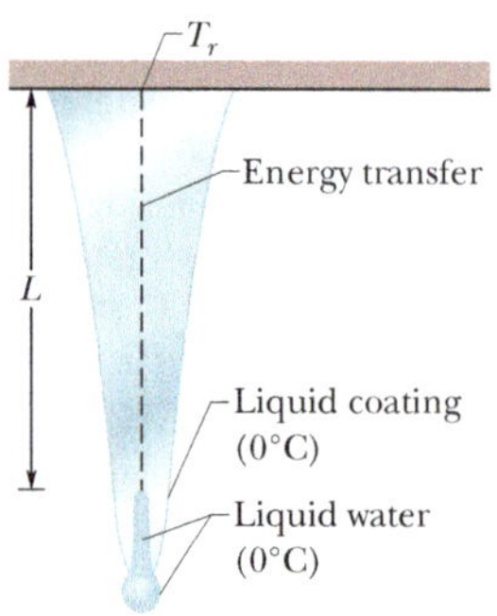

Figure 18-55 Problem 78.

79 SSM A sample of gas expands from an initial pressure and volume of 10 Pa and 1.0 m^3 to a final volume of 2.0 m^3. During the expansion, the pressure and volume are related by the equation $p = aV^2$, where $a = 10$ N/m^8. Determine the work done by the gas during this expansion.

80 Figure 18-56a shows a cylinder containing gas and closed by a movable piston. The cylinder is kept submerged in an ice–water mixture. The piston is *quickly* pushed down from position 1 to position 2 and then held at position 2 until the gas is again at the temperature of the ice–water mixture; it then is *slowly* raised back to position 1. Figure 18-56b is a p-V diagram for the process. If 100 g of ice is melted during the cycle, how much work has been done *on* the gas?

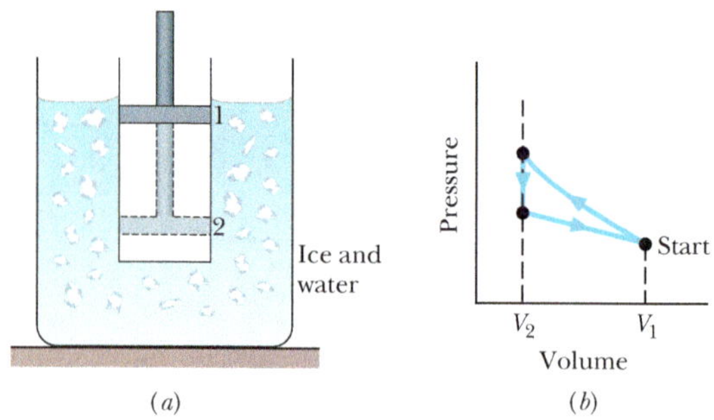

Figure 18-56 Problem 80.

81 SSM A sample of gas undergoes a transition from an initial state a to a final state b by three different paths (processes), as shown in the p-V diagram in Fig. 18-57, where $V_b = 5.00V_i$. The energy transferred to the gas as heat in process 1 is $10p_iV_i$. In terms of p_iV_i, what are (a) the energy transferred to the gas as heat in process 2 and (b) the change in internal energy that the gas undergoes in process 3?

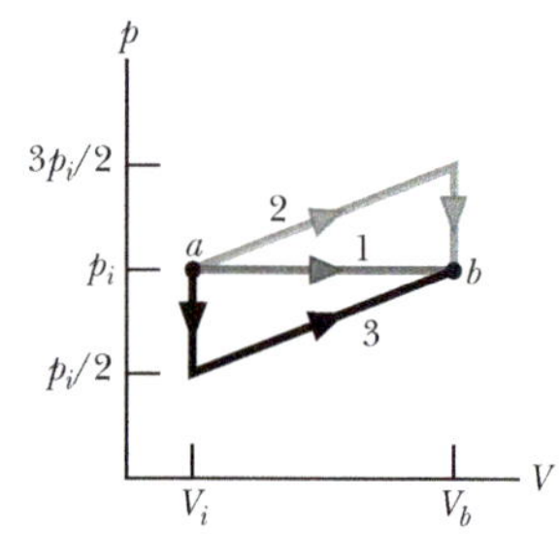

Figure 18-57 Problem 81.

82 A copper rod, an aluminum rod, and a brass rod, each of 6.00 m length and 1.00 cm diameter, are placed end to end with the aluminum rod between the other two. The free end of the copper rod is maintained at water's boiling point, and the free end of the brass rod is maintained at water's freezing point. What is the steady-state temperature of (a) the copper–aluminum junction and (b) the aluminum–brass junction?

83 SSM The temperature of a Pyrex disk is changed from 10.0°C to 60.0°C. Its initial radius is 8.00 cm; its initial thickness is 0.500 cm. Take these data as being exact. What is the change in the volume of the disk? (See Table 18-2.)

84 (a) Calculate the rate at which body heat is conducted through the clothing of a skier in a steady-state process, given the following data: the body surface area is 1.8 m^2, and the clothing is 1.0 cm thick; the skin surface temperature is 33°C and the outer surface of the clothing is at 1.0°C; the thermal conductivity of the clothing is 0.040 W/m · K. (b) If, after a fall, the skier's clothes became soaked with water of thermal conductivity 0.60 W/m · K, by how much is the rate of conduction multiplied?

85 SSM A 2.50 kg lump of aluminum is heated to 92.0°C and then dropped into 8.00 kg of water at 5.00°C. Assuming that the lump–water system is thermally isolated, what is the system's equilibrium temperature?

86 A glass window pane is exactly 20 cm by 30 cm at 10°C. By how much has its area increased when its temperature is 40°C, assuming that it can expand freely?

87 A recruit can join the semi-secret "300 F" club at the Amundsen–Scott South Pole Station only when the outside temperature is below −70°C. On such a day, the recruit first basks in a hot sauna and then runs outside wearing only shoes. (This is, of course, extremely dangerous, but the rite is effectively a protest against the constant danger of the cold.)

Assume that upon stepping out of the sauna, the recruit's skin temperature is 102°F and the walls, ceiling, and floor of the sauna room have a temperature of 30°C. Estimate the recruit's surface area, and take the skin emissivity to be 0.80. (a) What is the approximate net rate P_{net} at which the recruit loses energy via thermal radiation exchanges with the room? Next, assume that when outdoors, half the recruit's surface area exchanges thermal radiation with the sky at a temperature of −25°C and the other half exchanges thermal radiation with the snow and ground at a temperature of −80°C. What is the approximate net rate at which the recruit loses energy via thermal radiation exchanges with (b) the sky and (c) the snow and ground?

88 A steel rod at 25.0°C is bolted at both ends and then cooled. At what temperature will it rupture? Use Table 12-1.

89 An athlete needs to lose weight and decides to do it by "pumping iron." (a) How many times must an 80.0 kg weight be lifted a distance of 1.00 m in order to burn off 1.00 lb of fat, assuming that that much fat is equivalent to 3500 Cal? (b) If the weight is lifted once every 2.00 s, how long does the task take?

90 Soon after Earth was formed, heat released by the decay of radioactive elements raised the average internal temperature from 300 to 3000 K, at about which value it remains today. Assuming an average coefficient of volume expansion of 3.0×10^{-5} K^{-1}, by how much has the radius of Earth increased since the planet was formed?

91 It is possible to melt ice by rubbing one block of it against another. How much work, in joules, would you have to do to get 1.00 g of ice to melt?

92 A rectangular plate of glass initially has the dimensions 0.200 m by 0.300 m. The coefficient of linear expansion for the glass is 9.00×10^{-6}/K. What is the change in the plate's area if its temperature is increased by 20.0 K?

93 Suppose that you intercept 5.0×10^{-3} of the energy radiated by a hot sphere that has a radius of 0.020 m, an emissivity of 0.80, and a surface temperature of 500 K. How much energy do you intercept in 2.0 min?

94 A thermometer of mass 0.0550 kg and of specific heat 0.837 kJ/kg · K reads 15.0°C. It is then completely immersed in 0.300 kg of water, and it comes to the same final temperature as the water. If the thermometer then reads 44.4°C, what was the temperature of the water before insertion of the thermometer?

95 A sample of gas expands from $V_1 = 1.0$ m^3 and $p_1 = 40$ Pa to $V_2 = 4.0$ m^3 and $p_2 = 10$ Pa along path B in the p-V diagram in Fig. 18-58. It is then compressed back to V_1 along either path A or path C. Compute the net work done by the gas for the complete cycle along (a) path BA and (b) path BC.

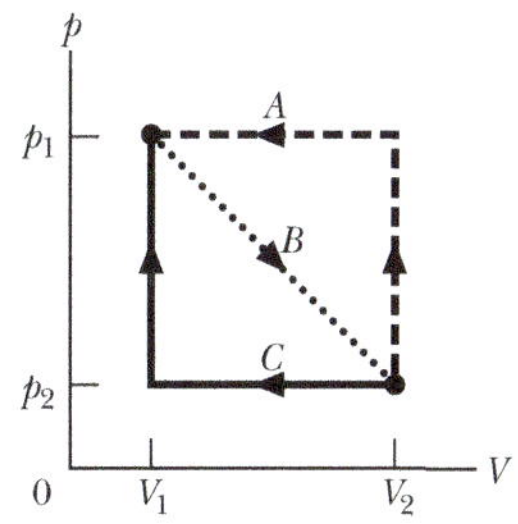

Figure 18-58 Problem 95.

96 Figure 18-59 shows a composite bar of length $L = L_1 + L_2$ and consisting of two materials. One material has length L_1 and coefficient of linear expansion α_1; the other has length L_2 and coefficient of linear expansion α_2. (a) What is the coefficient of linear expansion α for the composite bar? For a particular composite bar, L is 52.4 cm, material 1 is steel, and material 2 is brass. If $\alpha = 1.3 \times 10^{-5}$/C°, what are the lengths (b) L_1 and (c) L_2?

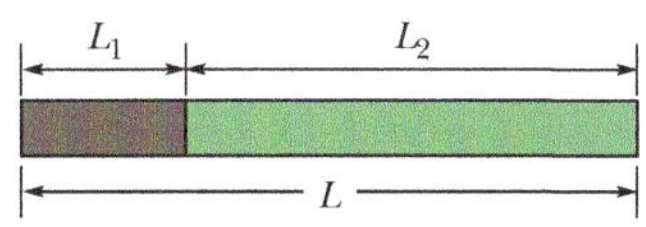

Figure 18-59 Problem 96.

97 On finding your stove out of order, you decide to boil the water for a cup of tea by shaking it in a thermos flask. Suppose that you use tap water at 19°C, the water falls 32 cm each shake, and you make 27 shakes each minute. Neglecting any loss of thermal energy by the flask, how long (in minutes) must you shake the flask until the water reaches 100°C?

98 The p-V diagram in Fig. 18-60 shows two paths along which a sample of gas can be taken from state a to state b, where $V_b = 3.0V_1$. Path 1 requires that energy equal to $5.0p_1V_1$ be transferred to the gas as heat. Path 2 requires that energy equal to $5.5p_1V_1$ be transferred to the gas as heat. What is the ratio p_2/p_1?

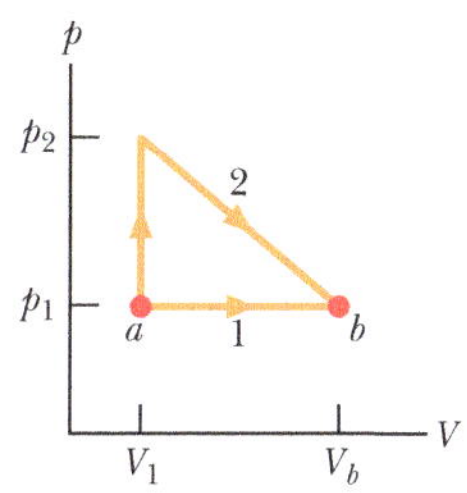

Figure 18-60 Problem 98.

99 A cube of edge length 6.0×10^{-6} m, emissivity 0.75, and temperature −100°C floats in an environment at −150°C. What is the cube's net thermal radiation transfer rate?

100 A *flow calorimeter* is a device used to measure the specific heat of a liquid. Energy is added as heat at a known rate to a stream of the liquid as it passes through the calorimeter at a known rate. Measurement of the resulting temperature difference between the inflow and the outflow points of the liquid stream enables us to compute the specific heat of the liquid. Suppose a liquid of density 0.85 g/cm^3 flows through a calorimeter at the rate of 8.0 cm^3/s. When energy is added at the rate of 250 W by means of an electric heating coil, a temperature difference of 15 C° is established in steady-state conditions between the inflow and the outflow points. What is the specific heat of the liquid?

101 An object of mass 6.00 kg falls through a height of 50.0 m and, by means of a mechanical linkage, rotates a paddle wheel that stirs 0.600 kg of water. Assume that the initial gravitational potential energy of the object is fully transferred to thermal energy of the water, which is initially at 15.0°C. What is the temperature rise of the water?

102 The Pyrex glass mirror in a telescope has a diameter of 170 in. The temperature ranges from $-16°C$ to $32°C$ on the location of the telescope. What is the maximum change in the diameter of the mirror, assuming that the glass can freely expand and contract?

103 The area A of a rectangular plate is $ab = 1.4\ m^2$. Its coefficient of linear expansion is $\alpha = 32 \times 10^{-6}/C°$. After a temperature rise $\Delta T = 89°C$, side a is longer by Δa and side b is longer by Δb (Fig. 18-61). Neglecting the small quantity $(\Delta a\, \Delta b)/ab$, find ΔA.

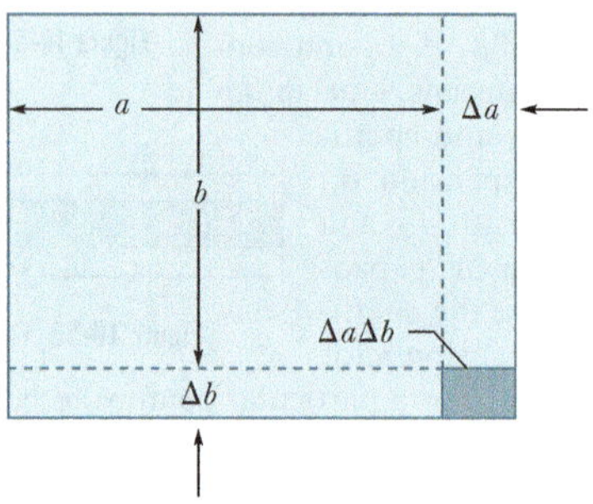

Figure 18-61 Problem 103.

104 Consider the liquid in a barometer whose coefficient of volume expansion is $6.6 \times 10^{-4}/C°$. Find the relative change in the liquid's height if the temperature changes by 12 C° while the pressure remains constant. Neglect the expansion of the glass tube.

105 A pendulum clock with a pendulum made of brass is designed to keep accurate time at 23°C. Assume it is a simple pendulum consisting of a bob at one end of a brass rod of negligible mass that is pivoted about the other end. If the clock operates at 0.0°C, (a) does it run too fast or too slow, and (b) what is the magnitude of its error in seconds per hour?

106 A room is lighted by four 100 W incandescent lightbulbs. (The power of 100 W is the rate at which a bulb converts electrical energy to heat and the energy of visible light.) Assuming that 73% of the energy is converted to heat, how much heat does the room receive in 6.9 h?

107 An energetic athlete can use up all the energy from a diet of 4000 Cal/day. If he were to use up this energy at a steady rate, what is the ratio of the rate of energy use compared to that of a 100 W bulb? (The power of 100 W is the rate at which the bulb converts electrical energy to heat and the energy of visible light.)

108 A 1700 kg Buick moving at 83 km/h brakes to a stop, at uniform deceleration and without skidding, over a distance of 93 m. At what average rate is mechanical energy transferred to thermal energy in the brake system?

CHAPTER 19

The Kinetic Theory of Gases

19-1 AVOGADRO'S NUMBER

Learning Objectives

After reading this module, you should be able to . . .

19.01 Identify Avogadro's number N_A.

19.02 Apply the relationship between the number of moles n, the number of molecules N, and Avogadro's number N_A.

19.03 Apply the relationships between the mass m of a sample, the molar mass M of the molecules in the sample, the number of moles n in the sample, and Avogadro's number N_A.

Key Ideas

- The kinetic theory of gases relates the macroscopic properties of gases (for example, pressure and temperature) to the microscopic properties of gas molecules (for example, speed and kinetic energy).
- One mole of a substance contains N_A (Avogadro's number) elementary units (usually atoms or molecules), where N_A is found experimentally to be

$$N_A = 6.02 \times 10^{23}\ \text{mol}^{-1} \quad \text{(Avogadro's number)}.$$

One molar mass M of any substance is the mass of one mole of the substance.

- A mole is related to the mass m of the individual molecules of the substance by

$$M = mN_A.$$

- The number of moles n contained in a sample of mass M_{sam}, consisting of N molecules, is related to the molar mass M of the molecules and to Avogadro's number N_A as given by

$$n = \frac{N}{N_A} = \frac{M_{sam}}{M} = \frac{M_{sam}}{mN_A}.$$

What Is Physics?

One of the main subjects in thermodynamics is the physics of gases. A gas consists of atoms (either individually or bound together as molecules) that fill their container's volume and exert pressure on the container's walls. We can usually assign a temperature to such a contained gas. These three variables associated with a gas—volume, pressure, and temperature—are all a consequence of the motion of the atoms. The volume is a result of the freedom the atoms have to spread throughout the container, the pressure is a result of the collisions of the atoms with the container's walls, and the temperature has to do with the kinetic energy of the atoms. The **kinetic theory of gases,** the focus of this chapter, relates the motion of the atoms to the volume, pressure, and temperature of the gas.

Applications of the kinetic theory of gases are countless. Automobile engineers are concerned with the combustion of vaporized fuel (a gas) in the automobile engines. Food engineers are concerned with the production rate of the fermentation gas that causes bread to rise as it bakes. Beverage engineers are concerned with how gas can produce the head in a glass of beer or shoot a cork from a champagne bottle. Medical engineers and physiologists are concerned with calculating how long a scuba diver must pause during ascent to eliminate nitrogen gas from the bloodstream (to avoid the *bends*). Environmental scientists are concerned with how heat exchanges between the oceans and the atmosphere can affect weather conditions.

The first step in our discussion of the kinetic theory of gases deals with measuring the amount of a gas present in a sample, for which we use Avogadro's number.

Avogadro's Number

When our thinking is slanted toward atoms and molecules, it makes sense to measure the sizes of our samples in moles. If we do so, we can be certain that we are comparing samples that contain the same number of atoms or molecules. The *mole* is one of the seven SI base units and is defined as follows:

> One mole is the number of atoms in a 12 g sample of carbon-12.

The obvious question now is: "How many atoms or molecules are there in a mole?" The answer is determined experimentally and, as you saw in Chapter 18, is

$$N_A = 6.02 \times 10^{23}\ \text{mol}^{-1} \quad \text{(Avogadro's number)}, \tag{19-1}$$

where mol^{-1} represents the inverse mole or "per mole," and mol is the abbreviation for mole. The number N_A is called **Avogadro's number** after Italian scientist Amedeo Avogadro (1776–1856), who suggested that all gases occupying the same volume under the same conditions of temperature and pressure contain the same number of atoms or molecules.

The number of moles n contained in a sample of any substance is equal to the ratio of the number of molecules N in the sample to the number of molecules N_A in 1 mol:

$$n = \frac{N}{N_A}. \tag{19-2}$$

(*Caution:* The three symbols in this equation can easily be confused with one another, so you should sort them with their meanings now, before you end in "N-confusion.") We can find the number of moles n in a sample from the mass M_{sam} of the sample and either the *molar mass* M (the mass of 1 mol) or the molecular mass m (the mass of one molecule):

$$n = \frac{M_{sam}}{M} = \frac{M_{sam}}{mN_A}. \tag{19-3}$$

In Eq. 19-3, we used the fact that the mass M of 1 mol is the product of the mass m of one molecule and the number of molecules N_A in 1 mol:

$$M = mN_A. \tag{19-4}$$

19-2 IDEAL GASES

Learning Objectives

After reading this module, you should be able to . . .

19.04 Identify why an ideal gas is said to be ideal.

19.05 Apply either of the two forms of the ideal gas law, written in terms of the number of moles n or the number of molecules N.

19.06 Relate the ideal gas constant R and the Boltzmann constant k.

19.07 Identify that the temperature in the ideal gas law must be in kelvins.

19.08 Sketch p-V diagrams for a constant-temperature expansion of a gas and a constant-temperature contraction.

19.09 Identify the term isotherm.

19.10 Calculate the work done by a gas, including the algebraic sign, for an expansion and a contraction along an isotherm.

19.11 For an isothermal process, identify that the change in internal energy ΔE is zero and that the energy Q transferred as heat is equal to the work W done.

19.12 On a p-V diagram, sketch a constant-volume process and identify the amount of work done in terms of area on the diagram.

19.13 On a p-V diagram, sketch a constant-pressure process and determine the work done in terms of area on the diagram.

Key Ideas

- An ideal gas is one for which the pressure p, volume V, and temperature T are related by

$$pV = nRT \quad \text{(ideal gas law)}.$$

Here n is the number of moles of the gas present and R is a constant (8.31 J/mol · K) called the gas constant.

- The ideal gas law can also be written as

$$pV = NkT,$$

where the Boltzmann constant k is

$$k = \frac{R}{N_A} = 1.38 \times 10^{-23}\ \text{J/K}.$$

- The work done *by* an ideal gas during an isothermal (constant-temperature) change from volume V_i to volume V_f is

$$W = nRT \ln \frac{V_f}{V_i} \quad \text{(ideal gas, isothermal process)}.$$

Ideal Gases

Our goal in this chapter is to explain the macroscopic properties of a gas—such as its pressure and its temperature—in terms of the behavior of the molecules that make it up. However, there is an immediate problem: which gas? Should it be hydrogen, oxygen, or methane, or perhaps uranium hexafluoride? They are all different. Experimenters have found, though, that if we confine 1 mol samples of various gases in boxes of identical volume and hold the gases at the same temperature, then their measured pressures are almost the same, and at lower densities the differences tend to disappear. Further experiments show that, at low enough densities, all real gases tend to obey the relation

$$pV = nRT \quad \text{(ideal gas law)}, \tag{19-5}$$

in which p is the absolute (not gauge) pressure, n is the number of moles of gas present, and T is the temperature in kelvins. The symbol R is a constant called the **gas constant** that has the same value for all gases—namely,

$$R = 8.31\ \text{J/mol} \cdot \text{K}. \tag{19-6}$$

Equation 19-5 is called the **ideal gas law.** Provided the gas density is low, this law holds for any single gas or for any mixture of different gases. (For a mixture, n is the total number of moles in the mixture.)

We can rewrite Eq. 19-5 in an alternative form, in terms of a constant called the **Boltzmann constant** k, which is defined as

$$k = \frac{R}{N_A} = \frac{8.31\ \text{J/mol} \cdot \text{K}}{6.02 \times 10^{23}\ \text{mol}^{-1}} = 1.38 \times 10^{-23}\ \text{J/K}. \tag{19-7}$$

This allows us to write $R = kN_A$. Then, with Eq. 19-2 ($n = N/N_A$), we see that

$$nR = Nk. \tag{19-8}$$

Substituting this into Eq. 19-5 gives a second expression for the ideal gas law:

$$pV = NkT \quad \text{(ideal gas law)}. \tag{19-9}$$

(*Caution*: Note the difference between the two expressions for the ideal gas law—Eq. 19-5 involves the number of moles n, and Eq. 19-9 involves the number of molecules N.)

You may well ask, "What is an *ideal gas*, and what is so 'ideal' about it?" The answer lies in the simplicity of the law (Eqs. 19-5 and 19-9) that governs its macroscopic properties. Using this law—as you will see—we can deduce many properties of the ideal gas in a simple way. Although there is no such thing in nature as a truly ideal gas, *all real* gases approach the ideal state at low enough densities—that is, under conditions in which their molecules are far enough apart that they do not interact with one another. Thus, the ideal gas concept allows us to gain useful insights into the limiting behavior of real gases.

(a)

(b)

Courtesy www.doctorslime.com

Figure 19-1 (*a*) Before and (*b*) after images of a large steel tank crushed by atmospheric pressure after internal steam cooled and condensed.

Figure 19-1 gives a dramatic example of the ideal gas law. A stainless-steel tank with a volume of 18 m^3 was filled with steam at a temperature of 110°C through a valve at one end. The steam supply was then turned off and the valve closed, so that the steam was trapped inside the tank (Fig. 19-1*a*). Water from a fire hose was then poured onto the tank to rapidly cool it. Within less than a minute, the enormously sturdy tank was crushed (Fig. 19-1*b*), as if some giant invisible creature from a grade B science fiction movie had stepped on it during a rampage.

Actually, it was the atmosphere that crushed the tank. As the tank was cooled by the water stream, the steam cooled and much of it condensed, which means that the number N of gas molecules and the temperature T of the gas inside the tank both decreased. Thus, the right side of Eq. 19-9 decreased, and because volume V was constant, the gas pressure p on the left side also decreased. The gas pressure decreased so much that the external atmospheric pressure was able to crush the tank's steel wall. Figure 19-1 was staged, but this type of crushing sometimes occurs in industrial accidents (photos and videos can be found on the web).

Work Done by an Ideal Gas at Constant Temperature

Suppose we put an ideal gas in a piston–cylinder arrangement like those in Chapter 18. Suppose also that we allow the gas to expand from an initial volume V_i to a final volume V_f while we keep the temperature T of the gas constant. Such a process, at *constant temperature,* is called an **isothermal expansion** (and the reverse is called an **isothermal compression**).

On a *p-V* diagram, an *isotherm* is a curve that connects points that have the same temperature. Thus, it is a graph of pressure versus volume for a gas whose temperature T is held constant. For n moles of an ideal gas, it is a graph of the equation

$$p = nRT\frac{1}{V} = (\text{a constant})\frac{1}{V}. \tag{19-10}$$

Figure 19-2 shows three isotherms, each corresponding to a different (constant) value of T. (Note that the values of T for the isotherms increase upward to the right.) Superimposed on the middle isotherm is the path followed by a gas during an isothermal expansion from state i to state f at a constant temperature of 310 K.

To find the work done by an ideal gas during an isothermal expansion, we start with Eq. 18-25,

$$W = \int_{V_i}^{V_f} p\,dV. \tag{19-11}$$

This is a general expression for the work done during any change in volume of any gas. For an ideal gas, we can use Eq. 19-5 ($pV = nRT$) to substitute for p, obtaining

$$W = \int_{V_i}^{V_f} \frac{nRT}{V}\,dV. \tag{19-12}$$

Because we are considering an isothermal expansion, T is constant, so we can move it in front of the integral sign to write

$$W = nRT\int_{V_i}^{V_f} \frac{dV}{V} = nRT\Big[\ln V\Big]_{V_i}^{V_f}. \tag{19-13}$$

By evaluating the expression in brackets at the limits and then using the relationship $\ln a - \ln b = \ln(a/b)$, we find that

$$W = nRT\ln\frac{V_f}{V_i} \quad \text{(ideal gas, isothermal process).} \tag{19-14}$$

Recall that the symbol ln specifies a *natural* logarithm, which has base *e*.

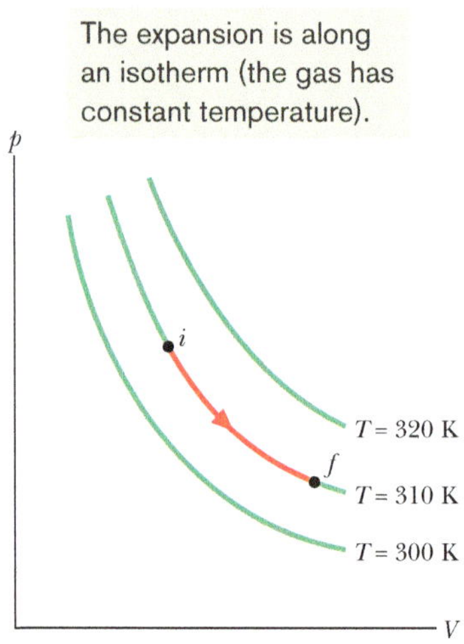

Figure 19-2 Three isotherms on a *p-V* diagram. The path shown along the middle isotherm represents an isothermal expansion of a gas from an initial state i to a final state f. The path from f to i along the isotherm would represent the reverse process—that is, an isothermal compression.

For an expansion, V_f is greater than V_i, so the ratio V_f/V_i in Eq. 19-14 is greater than unity. The natural logarithm of a quantity greater than unity is positive, and so the work W done by an ideal gas during an isothermal expansion is positive, as we expect. For a compression, V_f is less than V_i, so the ratio of volumes in Eq. 19-14 is less than unity. The natural logarithm in that equation—hence the work W—is negative, again as we expect.

Work Done at Constant Volume and at Constant Pressure

Equation 19-14 does not give the work W done by an ideal gas during *every* thermodynamic process. Instead, it gives the work only for a process in which the temperature is held constant. If the temperature varies, then the symbol T in Eq. 19-12 cannot be moved in front of the integral symbol as in Eq. 19-13, and thus we do not end up with Eq. 19-14.

However, we can always go back to Eq. 19-11 to find the work W done by an ideal gas (or any other gas) during any process, such as a constant-volume process and a constant-pressure process. If the volume of the gas is constant, then Eq. 19-11 yields

$$W = 0 \quad \text{(constant-volume process).} \tag{19-15}$$

If, instead, the volume changes while the pressure p of the gas is held constant, then Eq. 19-11 becomes

$$W = p(V_f - V_i) = p\,\Delta V \quad \text{(constant-pressure process).} \tag{19-16}$$

Checkpoint 1

An ideal gas has an initial pressure of 3 pressure units and an initial volume of 4 volume units. The table gives the final pressure and volume of the gas (in those same units) in five processes. Which processes start and end on the same isotherm?

	a	*b*	*c*	*d*	*e*
p	12	6	5	4	1
V	1	2	7	3	12

Sample Problem 19.01 Ideal gas and changes of temperature, volume, and pressure

A cylinder contains 12 L of oxygen at 20°C and 15 atm. The temperature is raised to 35°C, and the volume is reduced to 8.5 L. What is the final pressure of the gas in atmospheres? Assume that the gas is ideal.

KEY IDEA

Because the gas is ideal, we can use the ideal gas law to relate its parameters, both in the initial state i and in the final state f.

Calculations: From Eq. 19-5 we can write

$$p_iV_i = nRT_i \quad \text{and} \quad p_fV_f = nRT_f.$$

Dividing the second equation by the first equation and solving for p_f yields

$$p_f = \frac{p_iT_fV_i}{T_iV_f}. \tag{19-17}$$

Note here that if we converted the given initial and final volumes from liters to the proper units of cubic meters, the multiplying conversion factors would cancel out of Eq. 19-17. The same would be true for conversion factors that convert the pressures from atmospheres to the proper pascals. However, to convert the given temperatures to kelvins requires the addition of an amount that would not cancel and thus must be included. Hence, we must write

$$T_i = (273 + 20)\text{ K} = 293\text{ K}$$

and

$$T_f = (273 + 35)\text{ K} = 308\text{ K}.$$

Inserting the given data into Eq. 19-17 then yields

$$p_f = \frac{(15\text{ atm})(308\text{ K})(12\text{ L})}{(293\text{ K})(8.5\text{ L})} = 22\text{ atm}. \quad \text{(Answer)}$$

WileyPLUS Additional examples, video, and practice available at *WileyPLUS*

Sample Problem 19.02 Work by an ideal gas

One mole of oxygen (assume it to be an ideal gas) expands at a constant temperature T of 310 K from an initial volume V_i of 12 L to a final volume V_f of 19 L. How much work is done by the gas during the expansion?

KEY IDEA

Generally we find the work by integrating the gas pressure with respect to the gas volume, using Eq. 19-11. However, because the gas here is ideal and the expansion is isothermal, that integration leads to Eq. 19-14.

Calculation: Therefore, we can write

$$W = nRT \ln \frac{V_f}{V_i}$$

$$= (1 \text{ mol})(8.31 \text{ J/mol}\cdot\text{K})(310 \text{ K}) \ln \frac{19 \text{ L}}{12 \text{ L}}$$

$$= 1180 \text{ J}. \qquad \text{(Answer)}$$

The expansion is graphed in the p-V diagram of Fig. 19-3. The work done by the gas during the expansion is represented by the area beneath the curve if.

You can show that if the expansion is now reversed, with the gas undergoing an isothermal compression from 19 L to 12 L, the work done by the gas will be −1180 J. Thus, an external force would have to do 1180 J of work on the gas to compress it.

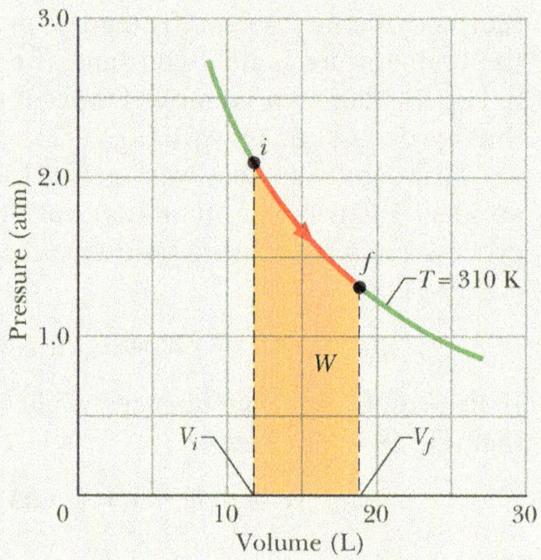

Figure 19-3 The shaded area represents the work done by 1 mol of oxygen in expanding from V_i to V_f at a temperature T of 310 K.

19-3 PRESSURE, TEMPERATURE, AND RMS SPEED

Learning Objectives

After reading this module, you should be able to . . .

19.14 Identify that the pressure on the interior walls of a gas container is due to the molecular collisions with the walls.

19.15 Relate the pressure on a container wall to the momentum of the gas molecules and the time intervals between their collisions with the wall.

19.16 For the molecules of an ideal gas, relate the root-mean-square speed v_{rms} and the average speed v_{avg}.

19.17 Relate the pressure of an ideal gas to the rms speed v_{rms} of the molecules.

19.18 For an ideal gas, apply the relationship between the gas temperature T and the rms speed v_{rms} and molar mass M of the molecules.

Key Ideas

- In terms of the speed of the gas molecules, the pressure exerted by n moles of an ideal gas is

$$p = \frac{nMv_{rms}^2}{3V},$$

where $v_{rms} = \sqrt{(v^2)_{avg}}$ is the root-mean-square speed of the molecules, M is the molar mass, and V is the volume.

- The rms speed can be written in terms of the temperature as

$$v_{rms} = \sqrt{\frac{3RT}{M}}.$$

Pressure, Temperature, and RMS Speed

Here is our first kinetic theory problem. Let n moles of an ideal gas be confined in a cubical box of volume V, as in Fig. 19-4. The walls of the box are held at temperature T. What is the connection between the pressure p exerted by the gas on the walls and the speeds of the molecules?

The molecules of gas in the box are moving in all directions and with various speeds, bumping into one another and bouncing from the walls of the box like balls in a racquetball court. We ignore (for the time being) collisions of the molecules with one another and consider only elastic collisions with the walls.

Figure 19-4 shows a typical gas molecule, of mass m and velocity $\vec{v}$, that is about to collide with the shaded wall. Because we assume that any collision of a molecule with a wall is elastic, when this molecule collides with the shaded wall, the only component of its velocity that is changed is the x component, and that component is reversed. This means that the only change in the particle's momentum is along the x axis, and that change is

$$\Delta p_x = (-mv_x) - (mv_x) = -2mv_x.$$

Hence, the momentum Δp_x delivered to the wall by the molecule during the collision is $+2mv_x$. (Because in this book the symbol p represents both momentum and pressure, we must be careful to note that here p represents momentum and is a vector quantity.)

The molecule of Fig. 19-4 will hit the shaded wall repeatedly. The time Δt between collisions is the time the molecule takes to travel to the opposite wall and back again (a distance $2L$) at speed v_x. Thus, Δt is equal to $2L/v_x$. (Note that this result holds even if the molecule bounces off any of the other walls along the way, because those walls are parallel to x and so cannot change v_x.) Therefore, the average rate at which momentum is delivered to the shaded wall by this single molecule is

$$\frac{\Delta p_x}{\Delta t} = \frac{2mv_x}{2L/v_x} = \frac{mv_x^2}{L}.$$

From Newton's second law ($\vec{F} = d\vec{p}/dt$), the rate at which momentum is delivered to the wall is the force acting on that wall. To find the total force, we must add up the contributions of all the molecules that strike the wall, allowing for the possibility that they all have different speeds. Dividing the magnitude of the total force F_x by the area of the wall ($= L^2$) then gives the pressure p on that wall, where now and in the rest of this discussion, p represents pressure. Thus, using the expression for $\Delta p_x/\Delta t$, we can write this pressure as

$$p = \frac{F_x}{L^2} = \frac{mv_{x1}^2/L + mv_{x2}^2/L + \cdots + mv_{xN}^2/L}{L^2}$$

$$= \left(\frac{m}{L^3}\right)(v_{x1}^2 + v_{x2}^2 + \cdots + v_{xN}^2), \tag{19-18}$$

where N is the number of molecules in the box.

Since $N = nN_A$, there are nN_A terms in the second set of parentheses of Eq. 19-18. We can replace that quantity by $nN_A(v_x^2)_{avg}$, where $(v_x^2)_{avg}$ is the average value of the square of the x components of all the molecular speeds. Equation 19-18 then becomes

$$p = \frac{nmN_A}{L^3}(v_x^2)_{avg}.$$

However, mN_A is the molar mass M of the gas (that is, the mass of 1 mol of the gas). Also, L^3 is the volume of the box, so

$$p = \frac{nM(v_x^2)_{avg}}{V}. \tag{19-19}$$

For any molecule, $v^2 = v_x^2 + v_y^2 + v_z^2$. Because there are many molecules and because they are all moving in random directions, the average values of the squares of their velocity components are equal, so that $v_x^2 = \frac{1}{3}v^2$. Thus, Eq. 19-19 becomes

$$p = \frac{nM(v^2)_{avg}}{3V}. \tag{19-20}$$

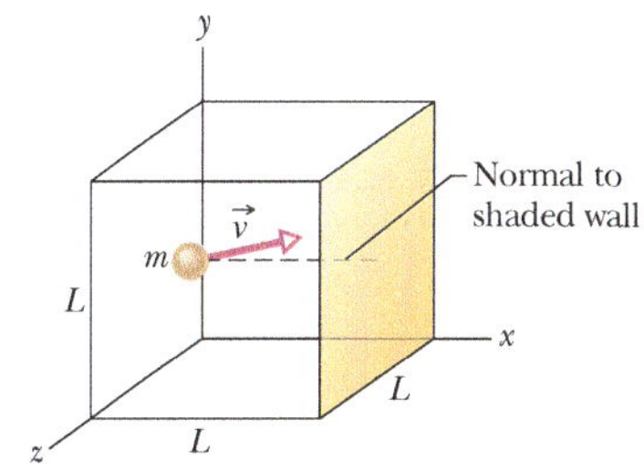

Figure 19-4 A cubical box of edge length L, containing n moles of an ideal gas. A molecule of mass m and velocity $\vec{v}$ is about to collide with the shaded wall of area L^2. A normal to that wall is shown.

Table 19-1 Some RMS Speeds at Room Temperature (T = 300 K)[a]

Gas	Molar Mass (10^{-3} kg/mol)	v_{rms} (m/s)
Hydrogen (H_2)	2.02	1920
Helium (He)	4.0	1370
Water vapor (H_2O)	18.0	645
Nitrogen (N_2)	28.0	517
Oxygen (O_2)	32.0	483
Carbon dioxide (CO_2)	44.0	412
Sulfur dioxide (SO_2)	64.1	342

[a]For convenience, we often set room temperature equal to 300 K even though (at 27°C or 81°F) that represents a fairly warm room.

The square root of $(v^2)_{avg}$ is a kind of average speed, called the **root-mean-square speed** of the molecules and symbolized by v_{rms}. Its name describes it rather well: You *square* each speed, you find the *mean* (that is, the average) of all these squared speeds, and then you take the square *root* of that mean. With $\sqrt{(v^2)_{avg}} = v_{rms}$, we can then write Eq. 19-20 as

$$p = \frac{nMv_{rms}^2}{3V}. \tag{19-21}$$

This tells us how the pressure of the gas (a purely macroscopic quantity) depends on the speed of the molecules (a purely microscopic quantity).

We can turn Eq. 19-21 around and use it to calculate v_{rms}. Combining Eq. 19-21 with the ideal gas law ($pV = nRT$) leads to

$$v_{rms} = \sqrt{\frac{3RT}{M}}. \tag{19-22}$$

Table 19-1 shows some rms speeds calculated from Eq. 19-22. The speeds are surprisingly high. For hydrogen molecules at room temperature (300 K), the rms speed is 1920 m/s, or 4300 mi/h—faster than a speeding bullet! On the surface of the Sun, where the temperature is 2×10^6 K, the rms speed of hydrogen molecules would be 82 times greater than at room temperature were it not for the fact that at such high speeds, the molecules cannot survive collisions among themselves. Remember too that the rms speed is only a kind of average speed; many molecules move much faster than this, and some much slower.

The speed of sound in a gas is closely related to the rms speed of the molecules of that gas. In a sound wave, the disturbance is passed on from molecule to molecule by means of collisions. The wave cannot move any faster than the "average" speed of the molecules. In fact, the speed of sound must be somewhat less than this "average" molecular speed because not all molecules are moving in exactly the same direction as the wave. As examples, at room temperature, the rms speeds of hydrogen and nitrogen molecules are 1920 m/s and 517 m/s, respectively. The speeds of sound in these two gases at this temperature are 1350 m/s and 350 m/s, respectively.

A question often arises: If molecules move so fast, why does it take as long as a minute or so before you can smell perfume when someone opens a bottle across a room? The answer is that, as we shall discuss in Module 19-5, each perfume molecule may have a high speed but it moves away from the bottle only very slowly because its repeated collisions with other molecules prevent it from moving directly across the room to you.

Sample Problem 19.03 Average and rms values

Here are five numbers: 5, 11, 32, 67, and 89.

(a) What is the average value n_{avg} of these numbers?

Calculation: We find this from

$$n_{avg} = \frac{5 + 11 + 32 + 67 + 89}{5} = 40.8. \quad \text{(Answer)}$$

(b) What is the rms value n_{rms} of these numbers?

Calculation: We find this from

$$n_{rms} = \sqrt{\frac{5^2 + 11^2 + 32^2 + 67^2 + 89^2}{5}} = 52.1. \quad \text{(Answer)}$$

The rms value is greater than the average value because the larger numbers—being squared—are relatively more important in forming the rms value.

Additional examples, video, and practice available at *WileyPLUS*

19-4 TRANSLATIONAL KINETIC ENERGY

Learning Objectives

After reading this module, you should be able to . . .

19.19 For an ideal gas, relate the average kinetic energy of the molecules to their rms speed.

19.20 Apply the relationship between the average kinetic energy and the temperature of the gas.

19.21 Identify that a measurement of a gas temperature is effectively a measurement of the average kinetic energy of the gas molecules.

Key Ideas

- The average translational kinetic energy per molecule in an ideal gas is

$$K_{avg} = \tfrac{1}{2}mv_{rms}^2.$$

- The average translational kinetic energy is related to the temperature of the gas:

$$K_{avg} = \tfrac{3}{2}kT.$$

Translational Kinetic Energy

We again consider a single molecule of an ideal gas as it moves around in the box of Fig. 19-4, but we now assume that its speed changes when it collides with other molecules. Its translational kinetic energy at any instant is $\frac{1}{2}mv^2$. Its *average* translational kinetic energy over the time that we watch it is

$$K_{avg} = (\tfrac{1}{2}mv^2)_{avg} = \tfrac{1}{2}m(v^2)_{avg} = \tfrac{1}{2}mv_{rms}^2, \tag{19-23}$$

in which we make the assumption that the average speed of the molecule during our observation is the same as the average speed of all the molecules at any given time. (Provided the total energy of the gas is not changing and provided we observe our molecule for long enough, this assumption is appropriate.) Substituting for v_{rms} from Eq. 19-22 leads to

$$K_{avg} = (\tfrac{1}{2}m)\frac{3RT}{M}.$$

However, M/m, the molar mass divided by the mass of a molecule, is simply Avogadro's number. Thus,

$$K_{avg} = \frac{3RT}{2N_A}.$$

Using Eq. 19-7 ($k = R/N_A$), we can then write

$$K_{avg} = \tfrac{3}{2}kT. \tag{19-24}$$

This equation tells us something unexpected:

At a given temperature T, all ideal gas molecules—no matter what their mass—have the same average translational kinetic energy—namely, $\frac{3}{2}kT$. When we measure the temperature of a gas, we are also measuring the average translational kinetic energy of its molecules.

Checkpoint 2

A gas mixture consists of molecules of types 1, 2, and 3, with molecular masses $m_1 > m_2 > m_3$. Rank the three types according to (a) average kinetic energy and (b) rms speed, greatest first.

19-5 MEAN FREE PATH

Learning Objectives

After reading this module, you should be able to . . .

19.22 Identify what is meant by mean free path.

19.23 Apply the relationship between the mean free path, the diameter of the molecules, and the number of molecules per unit volume.

Key Idea

- The mean free path λ of a gas molecule is its average path length between collisions and is given by

$$\lambda = \frac{1}{\sqrt{2}\pi d^2\, N/V},$$

where N/V is the number of molecules per unit volume and d is the molecular diameter.

Mean Free Path

Figure 19-5 A molecule traveling through a gas, colliding with other gas molecules in its path. Although the other molecules are shown as stationary, they are also moving in a similar fashion.

We continue to examine the motion of molecules in an ideal gas. Figure 19-5 shows the path of a typical molecule as it moves through the gas, changing both speed and direction abruptly as it collides elastically with other molecules. Between collisions, the molecule moves in a straight line at constant speed. Although the figure shows the other molecules as stationary, they are (of course) also moving.

One useful parameter to describe this random motion is the **mean free path** λ of the molecules. As its name implies, λ is the average distance traversed by a molecule between collisions. We expect λ to vary inversely with N/V, the number of molecules per unit volume (or density of molecules). The larger N/V is, the more collisions there should be and the smaller the mean free path. We also expect λ to vary inversely with the size of the molecules—with their diameter d, say. (If the molecules were points, as we have assumed them to be, they would never collide and the mean free path would be infinite.) Thus, the larger the molecules are, the smaller the mean free path. We can even predict that λ should vary (inversely) as the *square* of the molecular diameter because the cross section of a molecule—not its diameter—determines its effective target area.

The expression for the mean free path does, in fact, turn out to be

$$\lambda = \frac{1}{\sqrt{2}\pi d^2\, N/V} \quad \text{(mean free path).} \tag{19-25}$$

To justify Eq. 19-25, we focus attention on a single molecule and assume—as Fig. 19-5 suggests—that our molecule is traveling with a constant speed v and that all the other molecules are at rest. Later, we shall relax this assumption.

We assume further that the molecules are spheres of diameter d. A collision will then take place if the centers of two molecules come within a distance d of each other, as in Fig. 19-6a. Another, more helpful way to look at the situation is

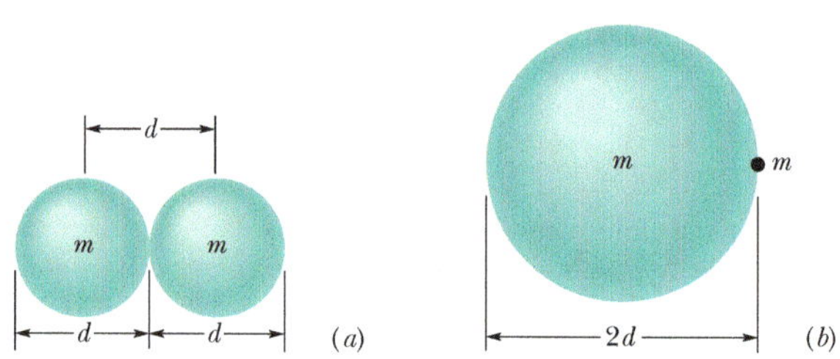

Figure 19-6 (a) A collision occurs when the centers of two molecules come within a distance d of each other, d being the molecular diameter. (b) An equivalent but more convenient representation is to think of the moving molecule as having a *radius d* and all other molecules as being points. The condition for a collision is unchanged.

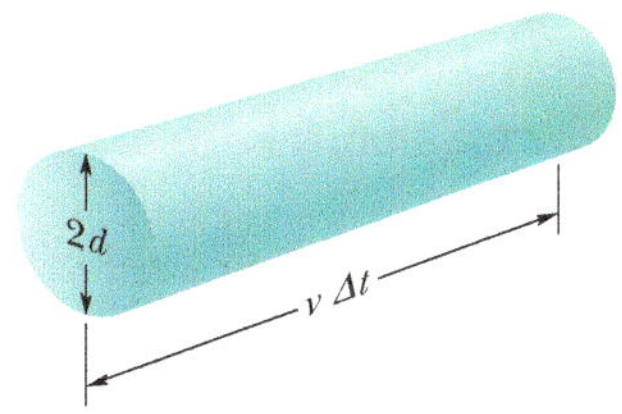

Figure 19-7 In time Δt the moving molecule effectively sweeps out a cylinder of length $v\,\Delta t$ and radius d.

to consider our single molecule to have a *radius* of d and all the other molecules to be *points,* as in Fig. 19-6*b*. This does not change our criterion for a collision.

As our single molecule zigzags through the gas, it sweeps out a short cylinder of cross-sectional area πd^2 between successive collisions. If we watch this molecule for a time interval Δt, it moves a distance $v\,\Delta t$, where v is its assumed speed. Thus, if we align all the short cylinders swept out in interval Δt, we form a composite cylinder (Fig. 19-7) of length $v\,\Delta t$ and volume $(\pi d^2)(v\,\Delta t)$. The number of collisions that occur in time Δt is then equal to the number of (point) molecules that lie within this cylinder.

Since N/V is the number of molecules per unit volume, the number of molecules in the cylinder is N/V times the volume of the cylinder, or $(N/V)(\pi d^2 v\,\Delta t)$. This is also the number of collisions in time Δt. The mean free path is the length of the path (and of the cylinder) divided by this number:

$$\lambda = \frac{\text{length of path during } \Delta t}{\text{number of collisions in } \Delta t} \approx \frac{v\,\Delta t}{\pi d^2 v\,\Delta t\,N/V}$$

$$= \frac{1}{\pi d^2\,N/V}. \tag{19-26}$$

This equation is only approximate because it is based on the assumption that all the molecules except one are at rest. In fact, *all* the molecules are moving; when this is taken properly into account, Eq. 19-25 results. Note that it differs from the (approximate) Eq. 19-26 only by a factor of $1/\sqrt{2}$.

The approximation in Eq. 19-26 involves the two v symbols we canceled. The v in the numerator is v_{avg}, the mean speed of the molecules *relative to the container.* The v in the denominator is v_{rel}, the mean speed of our single molecule *relative to the other molecules,* which are moving. It is this latter average speed that determines the number of collisions. A detailed calculation, taking into account the actual speed distribution of the molecules, gives $v_{rel} = \sqrt{2}\,v_{avg}$ and thus the factor $\sqrt{2}$.

The mean free path of air molecules at sea level is about 0.1 μm. At an altitude of 100 km, the density of air has dropped to such an extent that the mean free path rises to about 16 cm. At 300 km, the mean free path is about 20 km. A problem faced by those who would study the physics and chemistry of the upper atmosphere in the laboratory is the unavailability of containers large enough to hold gas samples (of Freon, carbon dioxide, and ozone) that simulate upper atmospheric conditions.

Checkpoint 3

One mole of gas A, with molecular diameter $2d_0$ and average molecular speed v_0, is placed inside a certain container. One mole of gas B, with molecular diameter d_0 and average molecular speed $2v_0$ (the molecules of B are smaller but faster), is placed in an identical container. Which gas has the greater average collision rate within its container?

Sample Problem 19.04 Mean free path, average speed, collision frequency

(a) What is the mean free path λ for oxygen molecules at temperature $T = 300$ K and pressure $p = 1.0$ atm? Assume that the molecular diameter is $d = 290$ pm and the gas is ideal.

KEY IDEA

Each oxygen molecule moves among other *moving* oxygen molecules in a zigzag path due to the resulting collisions. Thus, we use Eq. 19-25 for the mean free path.

Calculation: We first need the number of molecules per unit volume, N/V. Because we assume the gas is ideal, we can use the ideal gas law of Eq. 19-9 ($pV = NkT$) to write $N/V = p/kT$. Substituting this into Eq. 19-25, we find

$$\lambda = \frac{1}{\sqrt{2}\pi d^2 N/V} = \frac{kT}{\sqrt{2}\pi d^2 p}$$

$$= \frac{(1.38 \times 10^{-23}\text{ J/K})(300\text{ K})}{\sqrt{2}\pi(2.9 \times 10^{-10}\text{ m})^2(1.01 \times 10^5\text{ Pa})}$$

$$= 1.1 \times 10^{-7}\text{ m.} \qquad \text{(Answer)}$$

This is about 380 molecular diameters.

(b) Assume the average speed of the oxygen molecules is $v = 450$ m/s. What is the average time t between successive collisions for any given molecule? At what rate does the molecule collide; that is, what is the frequency f of its collisions?

KEY IDEAS

(1) Between collisions, the molecule travels, on average, the mean free path λ at speed v. (2) The average rate or frequency at which the collisions occur is the inverse of the time t between collisions.

Calculations: From the first key idea, the average time between collisions is

$$t = \frac{\text{distance}}{\text{speed}} = \frac{\lambda}{v} = \frac{1.1 \times 10^{-7}\text{ m}}{450\text{ m/s}}$$

$$= 2.44 \times 10^{-10}\text{ s} \approx 0.24\text{ ns.} \qquad \text{(Answer)}$$

This tells us that, on average, any given oxygen molecule has less than a nanosecond between collisions.

From the second key idea, the collision frequency is

$$f = \frac{1}{t} = \frac{1}{2.44 \times 10^{-10}\text{ s}} = 4.1 \times 10^9\text{ s}^{-1}. \qquad \text{(Answer)}$$

This tells us that, on average, any given oxygen molecule makes about 4 billion collisions per second.

WILEY PLUS Additional examples, video, and practice available at *WileyPLUS*

19-6 THE DISTRIBUTION OF MOLECULAR SPEEDS

Learning Objectives

After reading this module, you should be able to . . .

19.24 Explain how Maxwell's speed distribution law is used to find the fraction of molecules with speeds in a certain speed range.

19.25 Sketch a graph of Maxwell's speed distribution, showing the probability distribution versus speed and indicating the relative positions of the average speed v_{avg}, the most probable speed v_P, and the rms speed v_{rms}.

19.26 Explain how Maxwell's speed distribution is used to find the average speed, the rms speed, and the most probable speed.

19.27 For a given temperature T and molar mass M, calculate the average speed v_{avg}, the most probable speed v_P, and the rms speed v_{rms}.

Key Ideas

- The Maxwell speed distribution $P(v)$ is a function such that $P(v)\,dv$ gives the fraction of molecules with speeds in the interval dv at speed v:

$$P(v) = 4\pi\left(\frac{M}{2\pi RT}\right)^{3/2} v^2 e^{-Mv^2/2RT}.$$

- Three measures of the distribution of speeds among the molecules of a gas are

$$v_{avg} = \sqrt{\frac{8RT}{\pi M}} \quad \text{(average speed)},$$

$$v_P = \sqrt{\frac{2RT}{M}} \quad \text{(most probable speed)},$$

and

$$v_{rms} = \sqrt{\frac{3RT}{M}} \quad \text{(rms speed)}.$$

The Distribution of Molecular Speeds

The root-mean-square speed v_{rms} gives us a general idea of molecular speeds in a gas at a given temperature. We often want to know more. For example, what fraction of the molecules have speeds greater than the rms value? What fraction have speeds greater than twice the rms value? To answer such questions, we need to know how the possible values of speed are distributed among the molecules. Figure 19-8*a* shows this distribution for oxygen molecules at room temperature ($T = 300$ K); Fig. 19-8*b* compares it with the distribution at $T = 80$ K.

In 1852, Scottish physicist James Clerk Maxwell first solved the problem of finding the speed distribution of gas molecules. His result, known as **Maxwell's speed distribution law,** is

$$P(v) = 4\pi\left(\frac{M}{2\pi RT}\right)^{3/2} v^2 e^{-Mv^2/2RT}. \tag{19-27}$$

Here M is the molar mass of the gas, R is the gas constant, T is the gas temperature, and v is the molecular speed. It is this equation that is plotted in Fig. 19-8*a*, *b*. The quantity $P(v)$ in Eq. 19-27 and Fig. 19-8 is a *probability distribution function:* For any speed v, the product $P(v)\,dv$ (a dimensionless quantity) is the fraction of molecules with speeds in the interval dv centered on speed v.

As Fig. 19-8*a* shows, this fraction is equal to the area of a strip with height $P(v)$ and width dv. The total area under the distribution curve corresponds to the fraction of the molecules whose speeds lie between zero and infinity. All molecules fall into this category, so the value of this total area is unity; that is,

$$\int_0^\infty P(v)\,dv = 1. \tag{19-28}$$

The fraction (frac) of molecules with speeds in an interval of, say, v_1 to v_2 is then

$$\text{frac} = \int_{v_1}^{v_2} P(v)\,dv. \tag{19-29}$$

Average, RMS, and Most Probable Speeds

In principle, we can find the **average speed** v_{avg} of the molecules in a gas with the following procedure: We *weight* each value of v in the distribution; that is, we multiply it

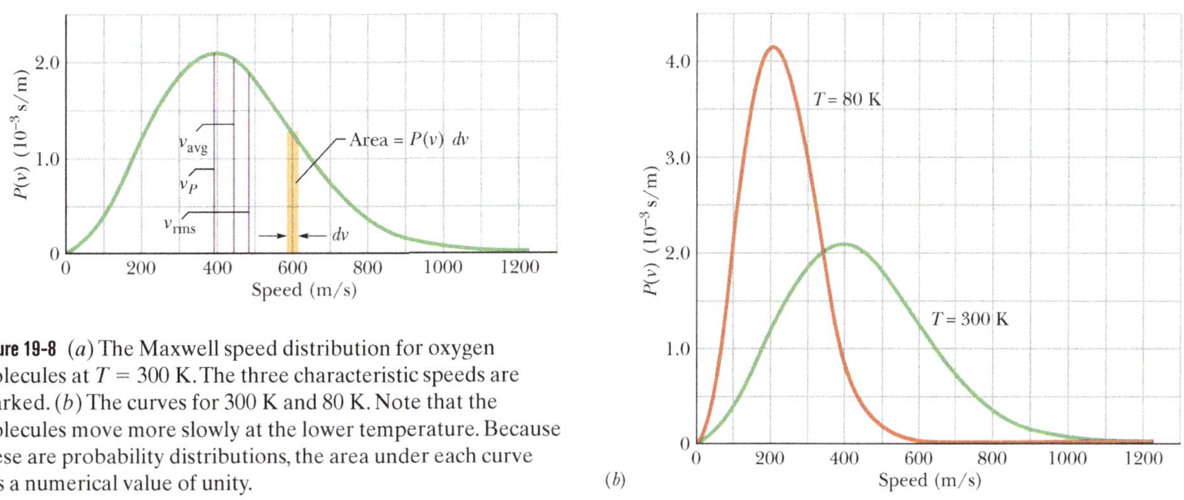

Figure 19-8 (*a*) The Maxwell speed distribution for oxygen molecules at $T = 300$ K. The three characteristic speeds are marked. (*b*) The curves for 300 K and 80 K. Note that the molecules move more slowly at the lower temperature. Because these are probability distributions, the area under each curve has a numerical value of unity.

by the fraction $P(v)\,dv$ of molecules with speeds in a differential interval dv centered on v. Then we add up all these values of $v\,P(v)\,dv$. The result is v_{avg}. In practice, we do all this by evaluating

$$v_{avg} = \int_0^\infty v\,P(v)\,dv. \tag{19-30}$$

Substituting for $P(v)$ from Eq. 19-27 and using generic integral 20 from the list of integrals in Appendix E, we find

$$v_{avg} = \sqrt{\frac{8RT}{\pi M}} \quad \text{(average speed)}. \tag{19-31}$$

Similarly, we can find the average of the square of the speeds $(v^2)_{avg}$ with

$$(v^2)_{avg} = \int_0^\infty v^2\,P(v)\,dv. \tag{19-32}$$

Substituting for $P(v)$ from Eq. 19-27 and using generic integral 16 from the list of integrals in Appendix E, we find

$$(v^2)_{avg} = \frac{3RT}{M}. \tag{19-33}$$

The square root of $(v^2)_{avg}$ is the root-mean-square speed v_{rms}. Thus,

$$v_{rms} = \sqrt{\frac{3RT}{M}} \quad \text{(rms speed)}, \tag{19-34}$$

which agrees with Eq. 19-22.

The **most probable speed** v_P is the speed at which $P(v)$ is maximum (see Fig. 19-8*a*). To calculate v_P, we set $dP/dv = 0$ (the slope of the curve in Fig. 19-8*a* is zero at the maximum of the curve) and then solve for v. Doing so, we find

$$v_P = \sqrt{\frac{2RT}{M}} \quad \text{(most probable speed)}. \tag{19-35}$$

A molecule is more likely to have speed v_P than any other speed, but some molecules will have speeds that are many times v_P. These molecules lie in the *high-speed tail* of a distribution curve like that in Fig. 19-8*a*. Such higher speed molecules make possible both rain and sunshine (without which we could not exist):

Rain The speed distribution of water molecules in, say, a pond at summertime temperatures can be represented by a curve similar to that of Fig. 19-8*a*. Most of the molecules lack the energy to escape from the surface. However, a few of the molecules in the high-speed tail of the curve can do so. It is these water molecules that evaporate, making clouds and rain possible.

As the fast water molecules leave the surface, carrying energy with them, the temperature of the remaining water is maintained by heat transfer from the surroundings. Other fast molecules—produced in particularly favorable collisions—quickly take the place of those that have left, and the speed distribution is maintained.

Sunshine Let the distribution function of Eq. 19-27 now refer to protons in the core of the Sun. The Sun's energy is supplied by a nuclear fusion process that starts with the merging of two protons. However, protons repel each other because of their electrical charges, and protons of average speed do not have enough kinetic energy to overcome the repulsion and get close enough to merge. Very fast protons with speeds in the high-speed tail of the distribution curve can do so, however, and for that reason the Sun can shine.

Sample Problem 19.05 Speed distribution in a gas

In oxygen (molar mass $M = 0.0320$ kg/mol) at room temperature (300 K), what fraction of the molecules have speeds in the interval 599 to 601 m/s?

KEY IDEAS

1. The speeds of the molecules are distributed over a wide range of values, with the distribution $P(v)$ of Eq. 19-27.
2. The fraction of molecules with speeds in a differential interval dv is $P(v)\,dv$.
3. For a larger interval, the fraction is found by integrating $P(v)$ over the interval.
4. However, the interval $\Delta v = 2$ m/s here is small compared to the speed $v = 600$ m/s on which it is centered.

Calculations: Because Δv is small, we can avoid the integration by approximating the fraction as

$$\text{frac} = P(v)\,\Delta v = 4\pi\left(\frac{M}{2\pi RT}\right)^{3/2} v^2 e^{-Mv^2/2RT}\,\Delta v.$$

The total area under the plot of $P(v)$ in Fig. 19-8*a* is the total fraction of molecules (unity), and the area of the thin gold strip (not to scale) is the fraction we seek. Let's evaluate frac in parts:

$$\text{frac} = (4\pi)(A)(v^2)(e^B)(\Delta v), \tag{19-36}$$

where

$$A = \left(\frac{M}{2\pi RT}\right)^{3/2} = \left(\frac{0.0320\text{ kg/mol}}{(2\pi)(8.31\text{ J/mol}\cdot\text{K})(300\text{ K})}\right)^{3/2}$$
$$= 2.92\times10^{-9}\text{ s}^3/\text{m}^3$$

and

$$B = -\frac{Mv^2}{2RT} = -\frac{(0.0320\text{ kg/mol})(600\text{ m/s})^2}{(2)(8.31\text{ J/mol}\cdot\text{K})(300\text{ K})}$$
$$= -2.31.$$

Substituting A and B into Eq. 19-36 yields

$$\text{frac} = (4\pi)(A)(v^2)(e^B)(\Delta v)$$
$$= (4\pi)(2.92\times10^{-9}\text{ s}^3/\text{m}^3)(600\text{ m/s})^2(e^{-2.31})(2\text{ m/s})$$
$$= 2.62\times10^{-3} = 0.262\%. \qquad \text{(Answer)}$$

Sample Problem 19.06 Average speed, rms speed, most probable speed

The molar mass M of oxygen is 0.0320 kg/mol.

(a) What is the average speed v_{avg} of oxygen gas molecules at $T = 300$ K?

KEY IDEA

To find the average speed, we must weight speed v with the distribution function $P(v)$ of Eq. 19-27 and then integrate the resulting expression over the range of possible speeds (from zero to the limit of an infinite speed).

Calculation: We end up with Eq. 19-31, which gives us

$$v_{avg} = \sqrt{\frac{8RT}{\pi M}}$$
$$= \sqrt{\frac{8(8.31\text{ J/mol}\cdot\text{K})(300\text{ K})}{\pi(0.0320\text{ kg/mol})}}$$
$$= 445\text{ m/s}. \qquad \text{(Answer)}$$

This result is plotted in Fig. 19-8*a*.

(b) What is the root-mean-square speed v_{rms} at 300 K?

KEY IDEA

To find v_{rms}, we must first find $(v^2)_{avg}$ by weighting v^2 with the distribution function $P(v)$ of Eq. 19-27 and then integrating the expression over the range of possible speeds. Then we must take the square root of the result.

Calculation: We end up with Eq. 19-34, which gives us

$$v_{rms} = \sqrt{\frac{3RT}{M}}$$
$$= \sqrt{\frac{3(8.31\text{ J/mol}\cdot\text{K})(300\text{ K})}{0.0320\text{ kg/mol}}}$$
$$= 483\text{ m/s}. \qquad \text{(Answer)}$$

This result, plotted in Fig. 19-8*a*, is greater than v_{avg} because the greater speed values influence the calculation more when we integrate the v^2 values than when we integrate the v values.

(c) What is the most probable speed v_P at 300 K?

KEY IDEA

Speed v_P corresponds to the maximum of the distribution function $P(v)$, which we obtain by setting the derivative $dP/dv = 0$ and solving the result for v.

Calculation: We end up with Eq. 19-35, which gives us

$$v_P = \sqrt{\frac{2RT}{M}}$$
$$= \sqrt{\frac{2(8.31\text{ J/mol}\cdot\text{K})(300\text{ K})}{0.0320\text{ kg/mol}}}$$
$$= 395\text{ m/s}. \qquad \text{(Answer)}$$

This result is also plotted in Fig. 19-8*a*.

19-7 THE MOLAR SPECIFIC HEATS OF AN IDEAL GAS

Learning Objectives

After reading this module, you should be able to . . .

19.28 Identify that the internal energy of an ideal monatomic gas is the sum of the translational kinetic energies of its atoms.

19.29 Apply the relationship between the internal energy E_{int} of a monatomic ideal gas, the number of moles n, and the gas temperature T.

19.30 Distinguish between monatomic, diatomic, and polyatomic ideal gases.

19.31 For monatomic, diatomic, and polyatomic ideal gases, evaluate the molar specific heats for a constant-volume process and a constant-pressure process.

19.32 Calculate a molar specific heat at constant pressure C_p by adding R to the molar specific heat at constant volume C_V, and explain why (physically) C_p is greater.

19.33 Identify that the energy transferred to an ideal gas as heat in a constant-volume process goes entirely into the internal energy (the random translational motion) but that in a constant-pressure process energy also goes into the work done to expand the gas.

19.34 Identify that for a given change in temperature, the change in the internal energy of an ideal gas is the same for *any* process and is most easily calculated by assuming a constant-volume process.

19.35 For an ideal gas, apply the relationship between heat Q, number of moles n, and temperature change ΔT, using the appropriate molar specific heat.

19.36 Between two isotherms on a p-V diagram, sketch a constant-volume process and a constant-pressure process, and for each identify the work done in terms of area on the graph.

19.37 Calculate the work done by an ideal gas for a constant-pressure process.

19.38 Identify that work is zero for constant volume.

Key Ideas

- The molar specific heat C_V of a gas at constant volume is defined as

$$C_V = \frac{Q}{n\,\Delta T} = \frac{\Delta E_{int}}{n\,\Delta T},$$

in which Q is the energy transferred as heat to or from a sample of n moles of the gas, ΔT is the resulting temperature change of the gas, and ΔE_{int} is the resulting change in the internal energy of the gas.

- For an ideal monatomic gas,

$$C_V = \tfrac{3}{2}R = 12.5\ \text{J/mol}\cdot\text{K}.$$

- The molar specific heat C_p of a gas at constant pressure is defined to be

$$C_p = \frac{Q}{n\,\Delta T},$$

in which Q, n, and ΔT are defined as above. C_p is also given by

$$C_p = C_V + R.$$

- For n moles of an ideal gas,

$$E_{int} = nC_VT \quad \text{(ideal gas)}.$$

- If n moles of a confined ideal gas undergo a temperature change ΔT due to *any* process, the change in the internal energy of the gas is

$$\Delta E_{int} = nC_V\,\Delta T \quad \text{(ideal gas, any process)}.$$

The Molar Specific Heats of an Ideal Gas

In this module, we want to derive from molecular considerations an expression for the internal energy E_{int} of an ideal gas. In other words, we want an expression for the energy associated with the random motions of the atoms or molecules in the gas. We shall then use that expression to derive the molar specific heats of an ideal gas.

Internal Energy E_{int}

Let us first assume that our ideal gas is a *monatomic gas* (individual atoms rather than molecules), such as helium, neon, or argon. Let us also assume that the internal energy E_{int} is the sum of the translational kinetic energies of the atoms. (Quantum theory disallows rotational kinetic energy for individual atoms.)

The average translational kinetic energy of a single atom depends only on the gas temperature and is given by Eq. 19-24 as $K_{avg} = \frac{3}{2}kT$. A sample of n moles of such a gas contains nN_A atoms. The internal energy E_{int} of the sample is then

$$E_{int} = (nN_A)K_{avg} = (nN_A)(\tfrac{3}{2}kT). \tag{19-37}$$

Using Eq. 19-7 ($k = R/N_A$), we can rewrite this as

$$E_{int} = \tfrac{3}{2}nRT \quad \text{(monatomic ideal gas).} \tag{19-38}$$

The internal energy E_{int} of an ideal gas is a function of the gas temperature *only*; it does not depend on any other variable.

With Eq. 19-38 in hand, we are now able to derive an expression for the molar specific heat of an ideal gas. Actually, we shall derive two expressions. One is for the case in which the volume of the gas remains constant as energy is transferred to or from it as heat. The other is for the case in which the pressure of the gas remains constant as energy is transferred to or from it as heat. The symbols for these two molar specific heats are C_V and C_p, respectively. (By convention, the capital letter C is used in both cases, even though C_V and C_p represent types of specific heat and not heat capacities.)

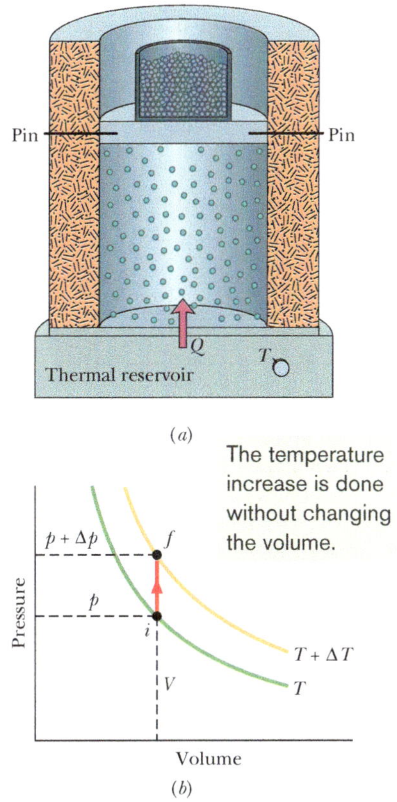

Figure 19-9 (*a*) The temperature of an ideal gas is raised from T to $T + \Delta T$ in a constant-volume process. Heat is added, but no work is done. (*b*) The process on a p-V diagram.

Molar Specific Heat at Constant Volume

Figure 19-9*a* shows n moles of an ideal gas at pressure p and temperature T, confined to a cylinder of fixed volume V. This *initial state* i of the gas is marked on the p-V diagram of Fig. 19-9*b*. Suppose now that you add a small amount of energy to the gas as heat Q by slowly turning up the temperature of the thermal reservoir. The gas temperature rises a small amount to $T + \Delta T$, and its pressure rises to $p + \Delta p$, bringing the gas to *final state* f. In such experiments, we would find that the heat Q is related to the temperature change ΔT by

$$Q = nC_V\,\Delta T \quad \text{(constant volume),} \tag{19-39}$$

where C_V is a constant called the **molar specific heat at constant volume.** Substituting this expression for Q into the first law of thermodynamics as given by Eq. 18-26 ($\Delta E_{int} = Q - W$) yields

$$\Delta E_{int} = nC_V\,\Delta T - W. \tag{19-40}$$

With the volume held constant, the gas cannot expand and thus cannot do any work. Therefore, $W = 0$, and Eq. 19-40 gives us

$$C_V = \frac{\Delta E_{int}}{n\,\Delta T}. \tag{19-41}$$

From Eq. 19-38, the change in internal energy must be

$$\Delta E_{int} = \tfrac{3}{2}nR\,\Delta T. \tag{19-42}$$

Substituting this result into Eq. 19-41 yields

$$C_V = \tfrac{3}{2}R = 12.5\ \text{J/mol}\cdot\text{K} \quad \text{(monatomic gas).} \tag{19-43}$$

As Table 19-2 shows, this prediction of the kinetic theory (for ideal gases) agrees very well with experiment for real monatomic gases, the case that we have assumed. The (predicted and) experimental values of C_V for *diatomic gases* (which have molecules with two atoms) and *polyatomic gases* (which have molecules with more than two atoms) are greater than those for monatomic gases for reasons that will be suggested in Module 19-8. Here we make the preliminary assumption that the C_V values for diatomic and polyatomic gases are greater than for monatomic gases because the more complex molecules can rotate and thus have rotational kinetic energy. So, when Q is transferred to a diatomic or polyatomic gas, only part of it goes into the translational kinetic energy, increasing the

Table 19-2 Molar Specific Heats at Constant Volume

Molecule	Example		C_V (J/mol·K)
Monatomic	Ideal		$\tfrac{3}{2}R = 12.5$
	Real	He	12.5
		Ar	12.6
Diatomic	Ideal		$\tfrac{5}{2}R = 20.8$
	Real	N_2	20.7
		O_2	20.8
Polyatomic	Ideal		$3R = 24.9$
	Real	NH_4	29.0
		CO_2	29.7

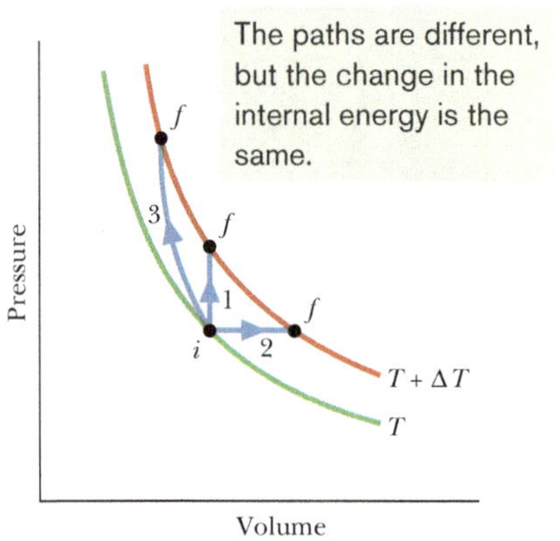

Figure 19-10 Three paths representing three different processes that take an ideal gas from an initial state i at temperature T to some final state f at temperature $T + \Delta T$. The change ΔE_{int} in the internal energy of the gas is the same for these three processes and for any others that result in the same change of temperature.

temperature. (For now we neglect the possibility of also putting energy into oscillations of the molecules.)

We can now generalize Eq. 19-38 for the internal energy of any ideal gas by substituting C_V for $\frac{3}{2}R$; we get

$$E_{\text{int}} = nC_VT \quad \text{(any ideal gas)}. \tag{19-44}$$

This equation applies not only to an ideal monatomic gas but also to diatomic and polyatomic ideal gases, provided the appropriate value of C_V is used. Just as with Eq. 19-38, we see that the internal energy of a gas depends on the temperature of the gas but not on its pressure or density.

When a confined ideal gas undergoes temperature change ΔT, then from either Eq. 19-41 or Eq. 19-44 the resulting change in its internal energy is

$$\Delta E_{\text{int}} = nC_V\,\Delta T \quad \text{(ideal gas, any process)}. \tag{19-45}$$

This equation tells us:

> A change in the internal energy E_{int} of a confined ideal gas depends on only the change in the temperature, *not* on what type of process produces the change.

As examples, consider the three paths between the two isotherms in the p-V diagram of Fig. 19-10. Path 1 represents a constant-volume process. Path 2 represents a constant-pressure process (we examine it next). Path 3 represents a process in which no heat is exchanged with the system's environment (we discuss this in Module 19-9). Although the values of heat Q and work W associated with these three paths differ, as do p_f and V_f, the values of ΔE_{int} associated with the three paths are identical and are all given by Eq. 19-45, because they all involve the same temperature change ΔT. Therefore, no matter what path is actually taken between T and $T + \Delta T$, we can *always* use path 1 and Eq. 19-45 to compute ΔE_{int} easily.

Molar Specific Heat at Constant Pressure

We now assume that the temperature of our ideal gas is increased by the same small amount ΔT as previously but now the necessary energy (heat Q) is added with the gas under constant pressure. An experiment for doing this is shown in Fig. 19-11a; the p-V diagram for the process is plotted in Fig. 19-11b. From such experiments we find that the heat Q is related to the temperature change ΔT by

$$Q = nC_p\,\Delta T \quad \text{(constant pressure)}, \tag{19-46}$$

where C_p is a constant called the **molar specific heat at constant pressure.** This C_p is *greater* than the molar specific heat at constant volume C_V, because energy must now be supplied not only to raise the temperature of the gas but also for the gas to do work—that is, to lift the weighted piston of Fig. 19-11a.

To relate molar specific heats C_p and C_V, we start with the first law of thermodynamics (Eq. 18-26):

$$\Delta E_{\text{int}} = Q - W. \tag{19-47}$$

We next replace each term in Eq. 19-47. For ΔE_{int}, we substitute from Eq. 19-45. For Q, we substitute from Eq. 19-46. To replace W, we first note that since the pressure remains constant, Eq. 19-16 tells us that $W = p\,\Delta V$. Then we note that, using the ideal gas equation ($pV = nRT$), we can write

$$W = p\,\Delta V = nR\,\Delta T. \tag{19-48}$$

Making these substitutions in Eq. 19-47 and then dividing through by $n\,\Delta T$, we find

$$C_V = C_p - R$$

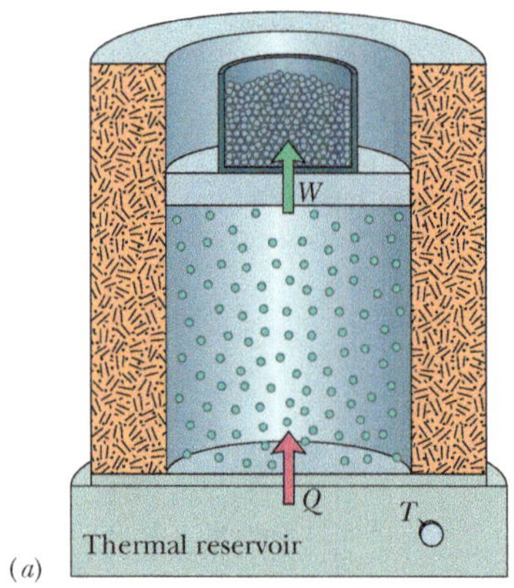

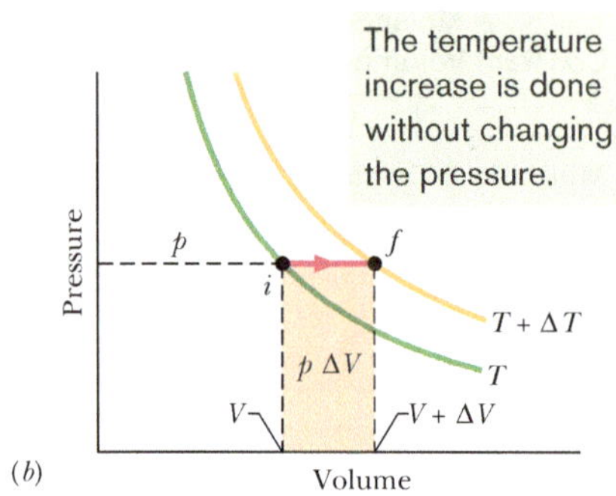

Figure 19-11 (a) The temperature of an ideal gas is raised from T to $T + \Delta T$ in a constant-pressure process. Heat is added and work is done in lifting the loaded piston. (b) The process on a p-V diagram. The work $p\,\Delta V$ is given by the shaded area.

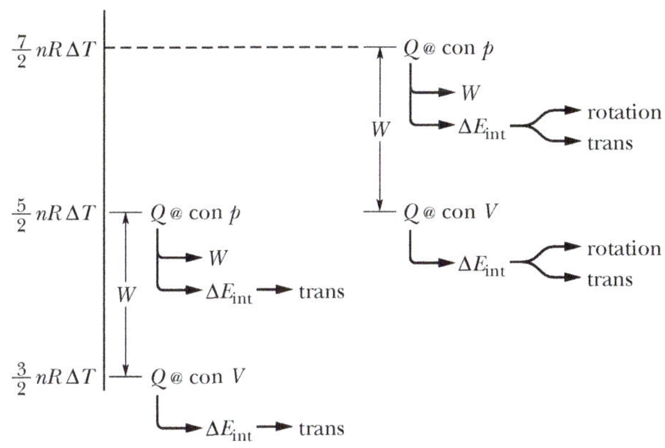

Figure 19-12 The relative values of Q for a monatomic gas (left side) and a diatomic gas undergoing a constant-volume process (labeled "con V") and a constant-pressure process (labeled "con p"). The transfer of the energy into work W and internal energy (ΔE_{int}) is noted.

and then

$$C_p = C_V + R. \tag{19-49}$$

This prediction of kinetic theory agrees well with experiment, not only for monatomic gases but also for gases in general, as long as their density is low enough so that we may treat them as ideal.

The left side of Fig. 19-12 shows the relative values of Q for a monatomic gas undergoing either a constant-volume process ($Q = \frac{3}{2}nR\,\Delta T$) or a constant-pressure process ($Q = \frac{5}{2}nR\,\Delta T$). Note that for the latter, the value of Q is higher by the amount W, the work done by the gas in the expansion. Note also that for the constant-volume process, the energy added as Q goes entirely into the change in internal energy ΔE_{int} and for the constant-pressure process, the energy added as Q goes into both ΔE_{int} and the work W.

Checkpoint 4

The figure here shows five paths traversed by a gas on a p-V diagram. Rank the paths according to the change in internal energy of the gas, greatest first.

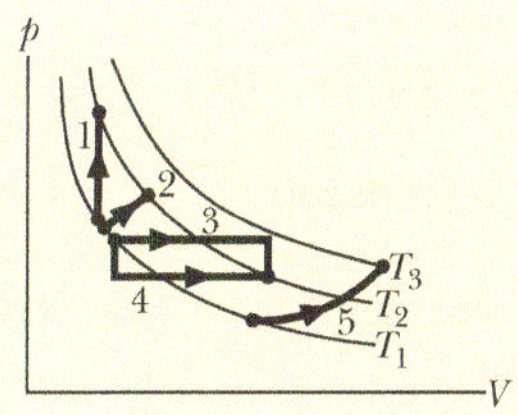

Sample Problem 19.07 Monatomic gas, heat, internal energy, and work

A bubble of 5.00 mol of helium is submerged at a certain depth in liquid water when the water (and thus the helium) undergoes a temperature increase ΔT of 20.0 C° at constant pressure. As a result, the bubble expands. The helium is monatomic and ideal.

(a) How much energy is added to the helium as heat during the increase and expansion?

KEY IDEA

Heat Q is related to the temperature change ΔT by a molar specific heat of the gas.

Calculations: Because the pressure p is held constant during the addition of energy, we use the molar specific heat at constant pressure C_p and Eq. 19-46,

$$Q = nC_p\,\Delta T, \tag{19-50}$$

to find Q. To evaluate C_p we go to Eq. 19-49, which tells us that for any ideal gas, $C_p = C_V + R$. Then from Eq. 19-43, we know that for any *monatomic* gas (like the helium here), $C_V = \frac{3}{2}R$. Thus, Eq. 19-50 gives us

$$\begin{aligned} Q &= n(C_V + R)\,\Delta T = n(\tfrac{3}{2}R + R)\,\Delta T = n(\tfrac{5}{2}R)\,\Delta T \\ &= (5.00\text{ mol})(2.5)(8.31\text{ J/mol}\cdot\text{K})(20.0\text{ C°}) \\ &= 2077.5\text{ J} \approx 2080\text{ J}. \qquad \text{(Answer)} \end{aligned}$$

(b) What is the change ΔE_{int} in the internal energy of the helium during the temperature increase?

KEY IDEA

Because the bubble expands, this is not a constant-volume process. However, the helium is nonetheless confined (to the bubble). Thus, the change ΔE_{int} is the same as *would occur* in a constant-volume process with the same temperature change ΔT.

Calculation: We can now easily find the constant-volume change ΔE_{int} with Eq. 19-45:

$$\begin{aligned}\Delta E_{int} &= nC_V\,\Delta T = n(\tfrac{3}{2}R)\,\Delta T\\ &= (5.00\text{ mol})(1.5)(8.31\text{ J/mol}\cdot\text{K})(20.0\text{ C}°)\\ &= 1246.5\text{ J} \approx 1250\text{ J}. \qquad \text{(Answer)}\end{aligned}$$

(c) How much work W is done by the helium as it expands against the pressure of the surrounding water during the temperature increase?

KEY IDEAS

The work done by *any* gas expanding against the pressure from its environment is given by Eq. 19-11, which tells us to integrate $p\,dV$. When the pressure is constant (as here), we can simplify that to $W = p\,\Delta V$. When the gas is *ideal* (as here), we can use the ideal gas law (Eq. 19-5) to write $p\,\Delta V = nR\,\Delta T$.

Calculation: We end up with

$$\begin{aligned}W &= nR\,\Delta T\\ &= (5.00\text{ mol})(8.31\text{ J/mol}\cdot\text{K})(20.0\text{ C}°)\\ &= 831\text{ J}. \qquad \text{(Answer)}\end{aligned}$$

Another way: Because we happen to know Q and ΔE_{int}, we can work this problem another way: We can account for the energy changes of the gas with the first law of thermodynamics, writing

$$\begin{aligned}W &= Q - \Delta E_{int} = 2077.5\text{ J} - 1246.5\text{ J}\\ &= 831\text{ J}. \qquad \text{(Answer)}\end{aligned}$$

The transfers: Let's follow the energy. Of the 2077.5 J transferred to the helium as heat Q, 831 J goes into the work W required for the expansion and 1246.5 J goes into the internal energy E_{int}, which, for a monatomic gas, is entirely the kinetic energy of the atoms in their translational motion. These several results are suggested on the left side of Fig. 19-12.

WILEY PLUS Additional examples, video, and practice available at *WileyPLUS*

19-8 DEGREES OF FREEDOM AND MOLAR SPECIFIC HEATS

Learning Objectives

After reading this module, you should be able to . . .

19.39 Identify that a degree of freedom is associated with each way a gas can store energy (translation, rotation, and oscillation).

19.40 Identify that an energy of $\frac{1}{2}kT$ per molecule is associated with each degree of freedom.

19.41 Identify that a monatomic gas can have an internal energy consisting of only translational motion.

19.42 Identify that at low temperatures a diatomic gas has energy in only translational motion, at higher temperatures it also has energy in molecular rotation, and at even higher temperatures it can also have energy in molecular oscillations.

19.43 Calculate the molar specific heat for monatomic and diatomic ideal gases in a constant-volume process and a constant-pressure process.

Key Ideas

- We find C_V by using the equipartition of energy theorem, which states that every degree of freedom of a molecule (that is, every independent way it can store energy) has associated with it—on average—an energy $\frac{1}{2}kT$ per molecule ($= \frac{1}{2}RT$ per mole).
- If f is the number of degrees of freedom, then

$E_{int} = (f/2)nRT$ and

$$C_V = \left(\frac{f}{2}\right)R = 4.16f\text{ J/mol}\cdot\text{K}.$$

- For monatomic gases $f = 3$ (three translational degrees); for diatomic gases $f = 5$ (three translational and two rotational degrees).

Degrees of Freedom and Molar Specific Heats

As Table 19-2 shows, the prediction that $C_V = \frac{3}{2}R$ agrees with experiment for monatomic gases but fails for diatomic and polyatomic gases. Let us try to explain the discrepancy by considering the possibility that molecules with more than one atom can store internal energy in forms other than translational kinetic energy.

Figure 19-13 shows common models of helium (a *monatomic* molecule, containing a single atom), oxygen (a *diatomic* molecule, containing two atoms), and

Table 19-3 **Degrees of Freedom for Various Molecules**

		Degrees of Freedom			Predicted Molar Specific Heats	
Molecule	Example	Translational	Rotational	Total (f)	C_V (Eq. 19-51)	$C_p = C_V + R$
Monatomic	He	3	0	3	$\frac{3}{2}R$	$\frac{5}{2}R$
Diatomic	O_2	3	2	5	$\frac{5}{2}R$	$\frac{7}{2}R$
Polyatomic	CH_4	3	3	6	$3R$	$4R$

methane (a *polyatomic* molecule). From such models, we would assume that all three types of molecules can have translational motions (say, moving left–right and up–down) and rotational motions (spinning about an axis like a top). In addition, we would assume that the diatomic and polyatomic molecules can have oscillatory motions, with the atoms oscillating slightly toward and away from one another, as if attached to opposite ends of a spring.

To keep account of the various ways in which energy can be stored in a gas, James Clerk Maxwell introduced the theorem of the **equipartition of energy:**

> Every kind of molecule has a certain number f of *degrees of freedom,* which are independent ways in which the molecule can store energy. Each such degree of freedom has associated with it—on average—an energy of $\frac{1}{2}kT$ per molecule (or $\frac{1}{2}RT$ per mole).

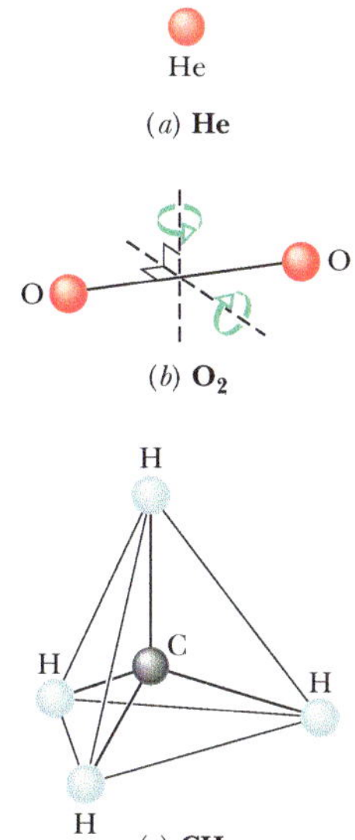

Figure 19-13 Models of molecules as used in kinetic theory: (*a*) helium, a typical monatomic molecule; (*b*) oxygen, a typical diatomic molecule; and (*c*) methane, a typical polyatomic molecule. The spheres represent atoms, and the lines between them represent bonds. Two rotation axes are shown for the oxygen molecule.

Let us apply the theorem to the translational and rotational motions of the molecules in Fig. 19-13. (We discuss oscillatory motion below.) For the translational motion, superimpose an xyz coordinate system on any gas. The molecules will, in general, have velocity components along all three axes. Thus, gas molecules of all types have three degrees of translational freedom (three ways to move in translation) and, on average, an associated energy of $3(\frac{1}{2}kT)$ per molecule.

For the rotational motion, imagine the origin of our xyz coordinate system at the center of each molecule in Fig. 19-13. In a gas, each molecule should be able to rotate with an angular velocity component along each of the three axes, so each gas should have three degrees of rotational freedom and, on average, an additional energy of $3(\frac{1}{2}kT)$ per molecule. *However,* experiment shows this is true only for the polyatomic molecules. According to *quantum theory,* the physics dealing with the allowed motions and energies of molecules and atoms, a monatomic gas molecule does not rotate and so has no rotational energy (a single atom cannot rotate like a top). A diatomic molecule can rotate like a top only about axes perpendicular to the line connecting the atoms (the axes are shown in Fig. 19-13*b*) and not about that line itself. Therefore, a diatomic molecule can have only two degrees of rotational freedom and a rotational energy of only $2(\frac{1}{2}kT)$ per molecule.

To extend our analysis of molar specific heats (C_p and C in Module 19-7) to ideal diatomic and polyatomic gases, it is necessary to retrace the derivations of that analysis in detail. First, we replace Eq. 19-38 ($E_{int} = \frac{3}{2}nRT$) with $E_{int} = (f/2)nRT$, where f is the number of degrees of freedom listed in Table 19-3. Doing so leads to the prediction

$$C_V = \left(\frac{f}{2}\right)R = 4.16f \text{ J/mol}\cdot\text{K}, \tag{19-51}$$

which agrees—as it must—with Eq. 19-43 for monatomic gases ($f = 3$). As Table 19-2 shows, this prediction also agrees with experiment for diatomic gases ($f = 5$), but it is too low for polyatomic gases ($f = 6$ for molecules comparable to CH_4).

Sample Problem 19.08 Diatomic gas, heat, temperature, internal energy

We transfer 1000 J as heat Q to a diatomic gas, allowing the gas to expand with the pressure held constant. The gas molecules each rotate around an internal axis but do not oscillate. How much of the 1000 J goes into the increase of the gas's internal energy? Of that amount, how much goes into ΔK_{tran} (the kinetic energy of the translational motion of the molecules) and ΔK_{rot} (the kinetic energy of their rotational motion)?

KEY IDEAS

1. The transfer of energy as heat Q to a gas under constant pressure is related to the resulting temperature increase ΔT via Eq. 19-46 ($Q = nC_p\,\Delta T$).
2. Because the gas is diatomic with molecules undergoing rotation but not oscillation, the molar specific heat is, from Fig. 19-12 and Table 19-3, $C_p = \frac{7}{2}R$.
3. The increase ΔE_{int} in the internal energy is the same as would occur with a constant-volume process resulting in the same ΔT. Thus, from Eq. 19-45, $\Delta E_{int} = nC_V\,\Delta T$. From Fig. 19-12 and Table 19-3, we see that $C_V = \frac{5}{2}R$.
4. For the same n and ΔT, ΔE_{int} is greater for a diatomic gas than for a monatomic gas because additional energy is required for rotation.

Increase in E_{int}: Let's first get the temperature change ΔT due to the transfer of energy as heat. From Eq. 19-46, substituting $\frac{7}{2}R$ for C_p, we have

$$\Delta T = \frac{Q}{\frac{7}{2}nR}. \tag{19-52}$$

We next find ΔE_{int} from Eq. 19-45, substituting the molar specific heat $C_V (= \frac{5}{2}R)$ for a constant-volume process and using the same ΔT. Because we are dealing with a diatomic gas, let's call this change $\Delta E_{int,dia}$. Equation 19-45 gives us

$$\Delta E_{int,dia} = nC_V\,\Delta T = n\tfrac{5}{2}R\left(\frac{Q}{\frac{7}{2}nR}\right) = \tfrac{5}{7}Q$$

$$= 0.71428Q = 714.3\ \text{J}. \qquad \text{(Answer)}$$

In words, about 71% of the energy transferred to the gas goes into the internal energy. The rest goes into the work required to increase the volume of the gas, as the gas pushes the walls of its container outward.

Increases in K: If we were to increase the temperature of a *monatomic* gas (with the same value of n) by the amount given in Eq. 19-52, the internal energy would change by a smaller amount, call it $\Delta E_{int,mon}$, because rotational motion is not involved. To calculate that smaller amount, we still use Eq. 19-45 but now we substitute the value of C_V for a monatomic gas—namely, $C_V = \frac{3}{2}R$. So,

$$\Delta E_{int,mon} = n\tfrac{3}{2}R\,\Delta T.$$

Substituting for ΔT from Eq. 19-52 leads us to

$$\Delta E_{int,mon} = n\tfrac{3}{2}R\left(\frac{Q}{n\frac{7}{2}R}\right) = \tfrac{3}{7}Q$$

$$= 0.42857Q = 428.6\ \text{J}.$$

For the monatomic gas, all this energy would go into the kinetic energy of the translational motion of the atoms. The important point here is that for a diatomic gas with the same values of n and ΔT, the same amount of energy goes into the kinetic energy of the translational motion of the molecules. The rest of $\Delta E_{int,dia}$ (that is, the additional 285.7 J) goes into the rotational motion of the molecules. Thus, for the diatomic gas,

$$\Delta K_{trans} = 428.6\ \text{J} \quad \text{and} \quad \Delta K_{rot} = 285.7\ \text{J}. \qquad \text{(Answer)}$$

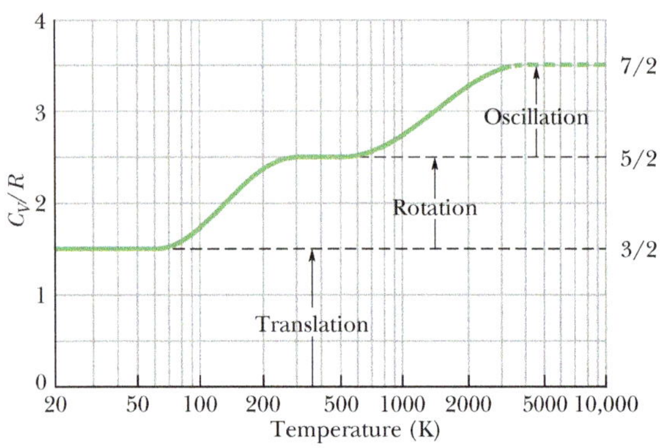

Figure 19-14 C_V/R versus temperature for (diatomic) hydrogen gas. Because rotational and oscillatory motions begin at certain energies, only translation is possible at very low temperatures. As the temperature increases, rotational motion can begin. At still higher temperatures, oscillatory motion can begin.

A Hint of Quantum Theory

We can improve the agreement of kinetic theory with experiment by including the oscillations of the atoms in a gas of diatomic or polyatomic molecules. For example, the two atoms in the O_2 molecule of Fig. 19-13*b* can oscillate toward and away from each other, with the interconnecting bond acting like a spring. However, experiment shows that such oscillations occur only at relatively high temperatures of the gas—the motion is "turned on" only when the gas molecules have relatively large energies. Rotational motion is also subject to such "turning on," but at a lower temperature.

Figure 19-14 is of help in seeing this turning on of rotational motion and oscillatory motion. The ratio C_V/R for diatomic hydrogen gas (H_2) is plotted there against temperature, with the temperature scale logarithmic to cover several orders of magnitude. Below about 80 K, we find that $C_V/R = 1.5$. This result implies that only the three translational degrees of freedom of hydrogen are involved in the specific heat.

As the temperature increases, the value of C_V/R gradually increases to 2.5, implying that two additional degrees of freedom have become involved. Quantum theory shows that these two degrees of freedom are associated with the rotational motion of the hydrogen molecules and that this motion requires a certain minimum amount of energy. At very low temperatures (below 80 K), the molecules do not have enough energy to rotate. As the temperature increases from 80 K, first a few molecules and then more and more of them obtain enough energy to rotate, and the value of C_V/R increases, until all of the molecules are rotating and $C_V/R = 2.5$.

Similarly, quantum theory shows that oscillatory motion of the molecules requires a certain (higher) minimum amount of energy. This minimum amount is not met until the molecules reach a temperature of about 1000 K, as shown in Fig. 19-14. As the temperature increases beyond 1000 K, more and more molecules have enough energy to oscillate and the value of C_V/R increases, until all of the molecules are oscillating and $C_V/R = 3.5$. (In Fig. 19-14, the plotted curve stops at 3200 K because there the atoms of a hydrogen molecule oscillate so much that they overwhelm their bond, and the molecule then *dissociates* into two separate atoms.)

The turning on of the rotation and vibration of the diatomic and polyatomic molecules is due to the fact that the energies of these motions are quantized, that is, restricted to certain values. There is a lowest allowed value for each type of motion. Unless the thermal agitation of the surrounding molecules provides those lowest amounts, a molecule simply cannot rotate or vibrate.

19-9 THE ADIABATIC EXPANSION OF AN IDEAL GAS

Learning Objectives

After reading this module, you should be able to . . .

19.44 On a p-V diagram, sketch an adiabatic expansion (or contraction) and identify that there is no heat exchange Q with the environment.

19.45 Identify that in an adiabatic expansion, the gas does work on the environment, decreasing the gas's internal energy, and that in an adiabatic contraction, work is done on the gas, increasing the internal energy.

19.46 In an adiabatic expansion or contraction, relate the initial pressure and volume to the final pressure and volume.

19.47 In an adiabatic expansion or contraction, relate the initial temperature and volume to the final temperature and volume.

19.48 Calculate the work done in an adiabatic process by integrating the pressure with respect to volume.

19.49 Identify that a free expansion of a gas into a vacuum is adiabatic but no work is done and thus, by the first law of thermodynamics, the internal energy and temperature of the gas do not change.

Key Ideas

- When an ideal gas undergoes a slow adiabatic volume change (a change for which $Q = 0$),

$$pV^\gamma = \text{a constant} \quad \text{(adiabatic process)},$$

in which $\gamma (= C_p/C_V)$ is the ratio of molar specific heats for the gas.

- For a free expansion, pV = a constant.

The Adiabatic Expansion of an Ideal Gas

We saw in Module 17-2 that sound waves are propagated through air and other gases as a series of compressions and expansions; these variations in the transmission medium take place so rapidly that there is no time for energy to be transferred from one part of the medium to another as heat. As we saw in Module 18-5, a process for which $Q = 0$ is an *adiabatic process.* We can ensure that $Q = 0$ either by carrying out the process very quickly (as in sound waves) or by doing it (at any rate) in a well-insulated container.

Figure 19-15*a* shows our usual insulated cylinder, now containing an ideal gas and resting on an insulating stand. By removing mass from the piston, we can allow the gas to expand adiabatically. As the volume increases, both the pressure and the temperature drop. We shall prove next that the relation between the pressure and the volume during such an adiabatic process is

$$pV^{\gamma} = \text{a constant} \quad \text{(adiabatic process)}, \tag{19-53}$$

in which $\gamma = C_p/C_V$, the ratio of the molar specific heats for the gas. On a *p-V* diagram such as that in Fig. 19-15*b*, the process occurs along a line (called an *adiabat*) that has the equation $p = (\text{a constant})/V^{\gamma}$. Since the gas goes from an initial state *i* to a final state *f*, we can rewrite Eq. 19-53 as

$$p_i V_i^{\gamma} = p_f V_f^{\gamma} \quad \text{(adiabatic process)}. \tag{19-54}$$

To write an equation for an adiabatic process in terms of *T* and *V*, we use the ideal gas equation ($pV = nRT$) to eliminate *p* from Eq. 19-53, finding

$$\left(\frac{nRT}{V}\right)V^{\gamma} = \text{a constant}.$$

Because *n* and *R* are constants, we can rewrite this in the alternative form

$$TV^{\gamma-1} = \text{a constant} \quad \text{(adiabatic process)}, \tag{19-55}$$

in which the constant is different from that in Eq. 19-53. When the gas goes from an initial state *i* to a final state *f*, we can rewrite Eq. 19-55 as

$$T_i V_i^{\gamma-1} = T_f V_f^{\gamma-1} \quad \text{(adiabatic process)}. \tag{19-56}$$

Understanding adiabatic processes allows you to understand why popping the cork on a cold bottle of champagne or the tab on a cold can of soda causes a slight fog to form at the opening of the container. At the top of any unopened carbonated drink sits a gas of carbon dioxide and water vapor. Because the pressure of that gas is much greater than atmospheric pressure, the gas expands out into the atmosphere when the container is opened. Thus, the gas volume increases, but that means the gas must do work pushing against the atmosphere. Because the expansion is rapid, it is adiabatic, and the only source of energy for the work is the internal energy of the gas. Because the internal energy decreases,

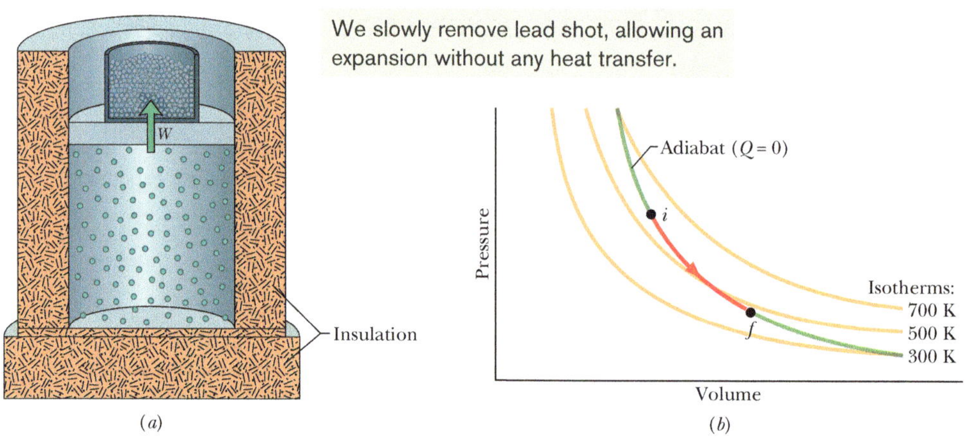

Figure 19-15 (*a*) The volume of an ideal gas is increased by removing mass from the piston. The process is adiabatic ($Q = 0$). (*b*) The process proceeds from *i* to *f* along an adiabat on a *p-V* diagram.

the temperature of the gas also decreases and so does the number of water molecules that can remain as a vapor. So, lots of the water molecules condense into tiny drops of fog.

Proof of Eq. 19-53

Suppose that you remove some shot from the piston of Fig. 19-15*a*, allowing the ideal gas to push the piston and the remaining shot upward and thus to increase the volume by a differential amount dV. Since the volume change is tiny, we may assume that the pressure p of the gas on the piston is constant during the change. This assumption allows us to say that the work dW done by the gas during the volume increase is equal to $p\,dV$. From Eq. 18-27, the first law of thermodynamics can then be written as

$$dE_{\text{int}} = Q - p\,dV. \tag{19-57}$$

Since the gas is thermally insulated (and thus the expansion is adiabatic), we substitute 0 for Q. Then we use Eq. 19-45 to substitute $nC_V\,dT$ for dE_{int}. With these substitutions, and after some rearranging, we have

$$n\,dT = -\left(\frac{p}{C_V}\right)dV. \tag{19-58}$$

Now from the ideal gas law ($pV = nRT$) we have

$$p\,dV + V\,dp = nR\,dT. \tag{19-59}$$

Replacing R with its equal, $C_p - C_V$, in Eq. 19-59 yields

$$n\,dT = \frac{p\,dV + V\,dp}{C_p - C_V}. \tag{19-60}$$

Equating Eqs. 19-58 and 19-60 and rearranging then give

$$\frac{dp}{p} + \left(\frac{C_p}{C_V}\right)\frac{dV}{V} = 0.$$

Replacing the ratio of the molar specific heats with γ and integrating (see integral 5 in Appendix E) yield

$$\ln p + \gamma \ln V = \text{a constant}.$$

Rewriting the left side as $\ln pV^{\gamma}$ and then taking the antilog of both sides, we find

$$pV^{\gamma} = \text{a constant}. \tag{19-61}$$

Free Expansions

Recall from Module 18-5 that a free expansion of a gas is an adiabatic process with *no* work or change in internal energy. Thus, a free expansion differs from the adiabatic process described by Eqs. 19-53 through 19-61, in which work is done and the internal energy changes. Those equations then do *not* apply to a free expansion, even though such an expansion is adiabatic.

Also recall that in a free expansion, a gas is in equilibrium only at its initial and final points; thus, we can plot only those points, but not the expansion itself, on a p-V diagram. In addition, because $\Delta E_{\text{int}} = 0$, the temperature of the final state must be that of the initial state. Thus, the initial and final points on a p-V diagram must be on the same isotherm, and instead of Eq. 19-56 we have

$$T_i = T_f \quad \text{(free expansion)}. \tag{19-62}$$

If we next assume that the gas is ideal (so that $pV = nRT$), then because there is no change in temperature, there can be no change in the product pV. Thus, instead of Eq. 19-53 a free expansion involves the relation

$$p_iV_i = p_fV_f \quad \text{(free expansion).} \tag{19-63}$$

Sample Problem 19.09 Work done by a gas in an adiabatic expansion

Initially an ideal diatomic gas has pressure $p_i = 2.00 \times 10^5$ Pa and volume $V_i = 4.00 \times 10^{-6}$ m^3. How much work W does it do, and what is the change ΔE_{int} in its internal energy if it expands adiabatically to volume $V_f = 8.00 \times 10^{-6}$ m^3? Throughout the process, the molecules have rotation but not oscillation.

KEY IDEAS

(1) In an adiabatic expansion, no heat is exchanged between the gas and its environment, and the energy for the work done by the gas comes from the internal energy. (2) The final pressure and volume are related to the initial pressure and volume by Eq. 19-54 ($p_iV_i^\gamma = p_fV_f^\gamma$). (3) The work done by a gas in any process can be calculated by integrating the pressure with respect to the volume (the work is due to the gas pushing the walls of its container outward).

Calculations: We want to calculate the work by filling out this integration,

$$W = \int_{V_i}^{V_f} p\, dV, \tag{19-64}$$

but we first need an expression for the pressure as a function of volume (so that we integrate the expression with respect to volume). So, let's rewrite Eq. 19-54 with indefinite symbols (dropping the subscripts f) as

$$p = \frac{1}{V^\gamma} p_iV_i^\gamma = V^{-\gamma}p_iV_i^\gamma. \tag{19-65}$$

The initial quantities are given constants but the pressure p is a function of the variable volume V. Substituting this expression into Eq. 19-64 and integrating lead us to

$$W = \int_{V_i}^{V_f} p\, dV = \int_{V_i}^{V_f} V^{-\gamma}p_iV_i^\gamma\, dV$$

$$= p_iV_i^\gamma \int_{V_i}^{V_f} V^{-\gamma}\, dV = \frac{1}{-\gamma+1} p_iV_i^\gamma \left[V^{-\gamma+1}\right]_{V_i}^{V_f}$$

$$= \frac{1}{-\gamma+1} p_iV_i^\gamma \left[V_f^{-\gamma+1} - V_i^{-\gamma+1}\right]. \tag{19-66}$$

Before we substitute in given data, we must determine the ratio γ of molar specific heats for a gas of diatomic molecules with rotation but no oscillation. From Table 19-3 we find

$$\gamma = \frac{C_p}{C_V} = \frac{\frac{7}{2}R}{\frac{5}{2}R} = 1.4. \tag{19-67}$$

We can now write the work done by the gas as the following (with volume in cubic meters and pressure in pascals):

$$W = \frac{1}{-1.4+1}(2.00 \times 10^5)(4.00 \times 10^{-6})^{1.4}$$

$$\times \left[(8.00 \times 10^{-6})^{-1.4+1} - (4.00 \times 10^{-6})^{-1.4+1}\right]$$

$$= 0.48 \text{ J}. \qquad \text{(Answer)}$$

The first law of thermodynamics (Eq. 18-26) tells us that $\Delta E_{int} = Q - W$. Because $Q = 0$ in the adiabatic expansion, we see that

$$\Delta E_{int} = -0.48 \text{ J}. \qquad \text{(Answer)}$$

With this decrease in internal energy, the gas temperature must also decrease because of the expansion.

Sample Problem 19.10 Adiabatic expansion, free expansion

Initially, 1 mol of oxygen (assumed to be an ideal gas) has temperature 310 K and volume 12 L. We will allow it to expand to volume 19 L.

(a) What would be the final temperature if the gas expands adiabatically? Oxygen (O_2) is diatomic and here has rotation but not oscillation.

KEY IDEAS

1. When a gas expands against the pressure of its environment, it must do work.
2. When the process is adiabatic (no energy is transferred as heat), then the energy required for the work can come only from the internal energy of the gas.
3. Because the internal energy decreases, the temperature T must also decrease.

Calculations: We can relate the initial and final temperatures and volumes with Eq. 19-56:

$$T_iV_i^{\gamma-1} = T_fV_f^{\gamma-1}. \tag{19-68}$$

Because the molecules are diatomic and have rotation but not oscillation, we can take the molar specific heats from

Table 19-3. Thus,

$$\gamma = \frac{C_p}{C_V} = \frac{\frac{7}{2}R}{\frac{5}{2}R} = 1.40.$$

Solving Eq. 19-68 for T_f and inserting known data then yield

$$T_f = \frac{T_i V_i^{\gamma-1}}{V_f^{\gamma-1}} = \frac{(310\text{ K})(12\text{ L})^{1.40-1}}{(19\text{ L})^{1.40-1}}$$

$$= (310\text{ K})(\tfrac{12}{19})^{0.40} = 258\text{ K}. \qquad \text{(Answer)}$$

(b) What would be the final temperature and pressure if, instead, the gas expands freely to the new volume, from an initial pressure of 2.0 Pa?

KEY IDEA

The temperature does not change in a free expansion because there is nothing to change the kinetic energy of the molecules.

Calculation: Thus, the temperature is

$$T_f = T_i = 310\text{ K}. \qquad \text{(Answer)}$$

We find the new pressure using Eq. 19-63, which gives us

$$p_f = p_i \frac{V_i}{V_f} = (2.0\text{ Pa})\frac{12\text{ L}}{19\text{ L}} = 1.3\text{ Pa}. \qquad \text{(Answer)}$$

Problem-Solving Tactics A Graphical Summary of Four Gas Processes

In this chapter we have discussed four special processes that an ideal gas can undergo. An example of each (for a monatomic ideal gas) is shown in Fig. 19-16, and some associated characteristics are given in Table 19-4, including two process names (isobaric and isochoric) that we have not used but that you might see in other courses.

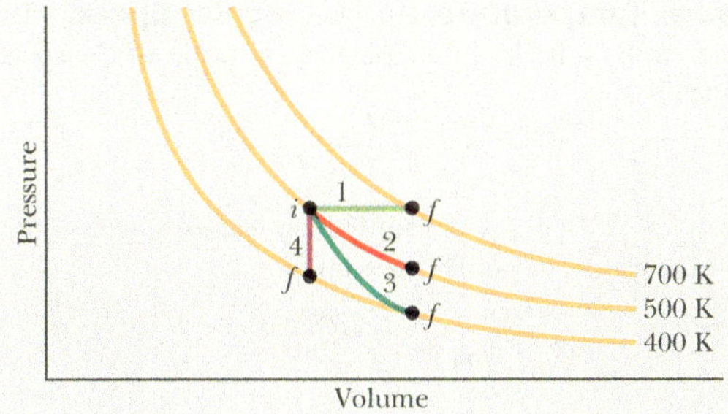

Figure 19-16 A p-V diagram representing four special processes for an ideal monatomic gas.

Checkpoint 5

Rank paths 1, 2, and 3 in Fig. 19-16 according to the energy transfer to the gas as heat, greatest first.

Table 19-4 Four Special Processes

Path in Fig. 19-16	Constant Quantity	Process Type	Some Special Results ($\Delta E_{int} = Q - W$ and $\Delta E_{int} = nC_V\,\Delta T$ for all paths)
1	p	Isobaric	$Q = nC_p\,\Delta T; W = p\,\Delta V$
2	T	Isothermal	$Q = W = nRT\ln(V_f/V_i);\ \Delta E_{int} = 0$
3	pV^{γ}, $TV^{\gamma-1}$	Adiabatic	$Q = 0;\quad W = -\Delta E_{int}$
4	V	Isochoric	$Q = \Delta E_{int} = nC_V\,\Delta T;\quad W = 0$

Review & Summary

Kinetic Theory of Gases The *kinetic theory of gases* relates the *macroscopic* properties of gases (for example, pressure and temperature) to the *microscopic* properties of gas molecules (for example, speed and kinetic energy).

Avogadro's Number One mole of a substance contains N_A (*Avogadro's number*) elementary units (usually atoms or molecules), where N_A is found experimentally to be

$$N_A = 6.02 \times 10^{23}\text{ mol}^{-1} \quad \text{(Avogadro's number)}. \qquad (19\text{-}1)$$

One molar mass M of any substance is the mass of one mole of the substance. It is related to the mass m of the individual molecules of the substance by

$$M = mN_A. \qquad (19\text{-}4)$$

The number of moles n contained in a sample of mass M_{sam}, consisting of N molecules, is given by

$$n = \frac{N}{N_A} = \frac{M_{sam}}{M} = \frac{M_{sam}}{mN_A}. \qquad (19\text{-}2, 19\text{-}3)$$

Ideal Gas An *ideal gas* is one for which the pressure p, volume V, and temperature T are related by

$$pV = nRT \quad \text{(ideal gas law)}. \tag{19-5}$$

Here n is the number of moles of the gas present and R is a constant (8.31 J/mol·K) called the **gas constant.** The ideal gas law can also be written as

$$pV = NkT, \tag{19-9}$$

where the **Boltzmann constant** k is

$$k = \frac{R}{N_A} = 1.38 \times 10^{-23}\ \text{J/K}. \tag{19-7}$$

Work in an Isothermal Volume Change The work done *by* an ideal gas during an **isothermal** (constant-temperature) change from volume V_i to volume V_f is

$$W = nRT \ln \frac{V_f}{V_i} \quad \text{(ideal gas, isothermal process)}. \tag{19-14}$$

Pressure, Temperature, and Molecular Speed The pressure exerted by n moles of an ideal gas, in terms of the speed of its molecules, is

$$p = \frac{nMv_{\text{rms}}^2}{3V}, \tag{19-21}$$

where $v_{\text{rms}} = \sqrt{(v^2)_{\text{avg}}}$ is the **root-mean-square speed** of the molecules of the gas. With Eq. 19-5 this gives

$$v_{\text{rms}} = \sqrt{\frac{3RT}{M}}. \tag{19-22}$$

Temperature and Kinetic Energy The average translational kinetic energy K_{avg} per molecule of an ideal gas is

$$K_{\text{avg}} = \tfrac{3}{2}kT. \tag{19-24}$$

Mean Free Path The *mean free path* λ of a gas molecule is its average path length between collisions and is given by

$$\lambda = \frac{1}{\sqrt{2}\pi d^2\, N/V}, \tag{19-25}$$

where N/V is the number of molecules per unit volume and d is the molecular diameter.

Maxwell Speed Distribution The *Maxwell speed distribution* $P(v)$ is a function such that $P(v)\,dv$ gives the *fraction* of molecules with speeds in the interval dv at speed v:

$$P(v) = 4\pi\left(\frac{M}{2\pi RT}\right)^{3/2} v^2 e^{-Mv^2/2RT}. \tag{19-27}$$

Three measures of the distribution of speeds among the molecules of a gas are

$$v_{\text{avg}} = \sqrt{\frac{8RT}{\pi M}} \quad \text{(average speed)}, \tag{19-31}$$

$$v_P = \sqrt{\frac{2RT}{M}} \quad \text{(most probable speed)}, \tag{19-35}$$

and the rms speed defined above in Eq. 19-22.

Molar Specific Heats The molar specific heat C_V of a gas at constant volume is defined as

$$C_V = \frac{Q}{n\,\Delta T} = \frac{\Delta E_{\text{int}}}{n\,\Delta T}, \tag{19-39, 19-41}$$

in which Q is the energy transferred as heat to or from a sample of n moles of the gas, ΔT is the resulting temperature change of the gas, and ΔE_{int} is the resulting change in the internal energy of the gas. For an ideal monatomic gas,

$$C_V = \tfrac{3}{2}R = 12.5\ \text{J/mol·K}. \tag{19-43}$$

The molar specific heat C_p of a gas at constant pressure is defined to be

$$C_p = \frac{Q}{n\,\Delta T}, \tag{19-46}$$

in which Q, n, and ΔT are defined as above. C_p is also given by

$$C_p = C_V + R. \tag{19-49}$$

For n moles of an ideal gas,

$$E_{\text{int}} = nC_VT \quad \text{(ideal gas)}. \tag{19-44}$$

If n moles of a confined ideal gas undergo a temperature change ΔT due to *any* process, the change in the internal energy of the gas is

$$\Delta E_{\text{int}} = nC_V\,\Delta T \quad \text{(ideal gas, any process)}. \tag{19-45}$$

Degrees of Freedom and C_V The *equipartition of energy* theorem states that every *degree of freedom* of a molecule has an energy $\frac{1}{2}kT$ per molecule ($= \frac{1}{2}RT$ per mole). If f is the number of degrees of freedom, then $E_{\text{int}} = (f/2)nRT$ and

$$C_V = \left(\frac{f}{2}\right)R = 4.16f\ \text{J/mol·K}. \tag{19-51}$$

For monatomic gases $f = 3$ (three translational degrees); for diatomic gases $f = 5$ (three translational and two rotational degrees).

Adiabatic Process When an ideal gas undergoes an adiabatic volume change (a change for which $Q = 0$),

$$pV^\gamma = \text{a constant} \quad \text{(adiabatic process)}, \tag{19-53}$$

in which $\gamma\,(= C_p/C_V)$ is the ratio of molar specific heats for the gas. For a free expansion, however, pV = a constant.

Questions

1 For four situations for an ideal gas, the table gives the energy transferred to or from the gas as heat Q and either the work W done by the gas or the work W_{on} done on the gas, all in joules. Rank the four situations in terms of the temperature change of the gas, most positive first.

	a	b	c	d
Q	−50	+35	−15	+20
W	−50	+35		
W_{on}			−40	+40

2 In the p-V diagram of Fig. 19-17, the gas does 5 J of work when taken along isotherm ab and 4 J when taken along adiabat bc. What is the change

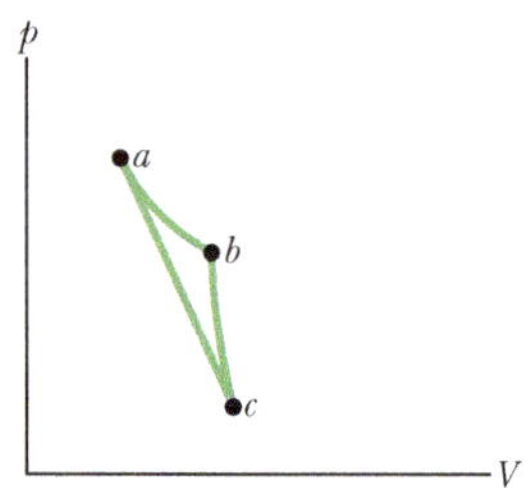

Figure 19-17 Question 2.

in the internal energy of the gas when it is taken along the straight path from a to c?

3 For a temperature increase of ΔT_1, a certain amount of an ideal gas requires 30 J when heated at constant volume and 50 J when heated at constant pressure. How much work is done by the gas in the second situation?

4 The dot in Fig. 19-18*a* represents the initial state of a gas, and the vertical line through the dot divides the p-V diagram into regions 1 and 2. For the following processes, determine whether the work W done by the gas is positive, negative, or zero: (a) the gas moves up along the vertical line, (b) it moves down along the vertical line, (c) it moves to anywhere in region 1, and (d) it moves to anywhere in region 2.

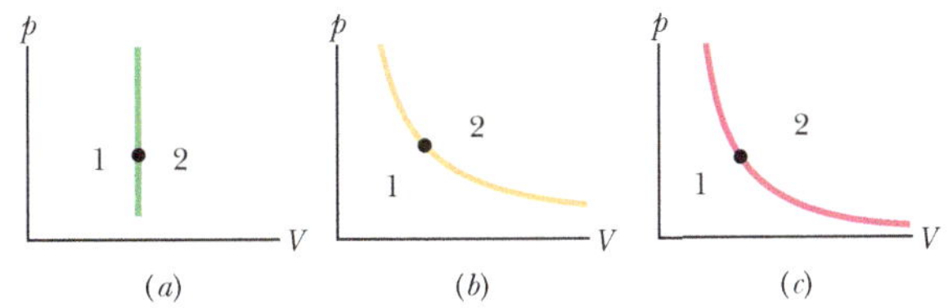

Figure 19-18 Questions 4, 6, and 8.

5 A certain amount of energy is to be transferred as heat to 1 mol of a monatomic gas (a) at constant pressure and (b) at constant volume, and to 1 mol of a diatomic gas (c) at constant pressure and (d) at constant volume. Figure 19-19 shows four paths from an initial point to four final points on a p-V diagram for the two gases. Which path goes with which process? (e) Are the molecules of the diatomic gas rotating?

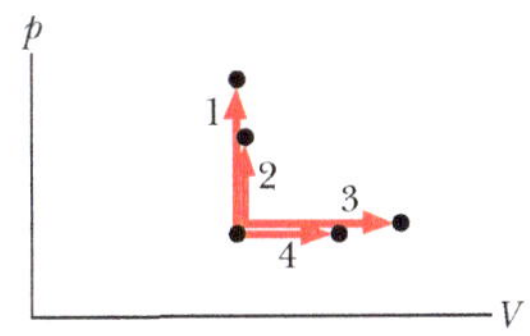

Figure 19-19 Question 5.

6 The dot in Fig. 19-18*b* represents the initial state of a gas, and the isotherm through the dot divides the p-V diagram into regions 1 and 2. For the following processes, determine whether the change ΔE_{int} in the internal energy of the gas is positive, negative, or zero: (a) the gas moves up along the isotherm, (b) it moves down along the isotherm, (c) it moves to anywhere in region 1, and (d) it moves to anywhere in region 2.

7 (a) Rank the four paths of Fig. 19-16 according to the work done by the gas, greatest first. (b) Rank paths 1, 2, and 3 according to the change in the internal energy of the gas, most positive first and most negative last.

8 The dot in Fig. 19-18*c* represents the initial state of a gas, and the adiabat through the dot divides the p-V diagram into regions 1 and 2. For the following processes, determine whether the corresponding heat Q is positive, negative, or zero: (a) the gas moves up along the adiabat, (b) it moves down along the adiabat, (c) it moves to anywhere in region 1, and (d) it moves to anywhere in region 2.

9 An ideal diatomic gas, with molecular rotation but without any molecular oscillation, loses a certain amount of energy as heat Q. Is the resulting decrease in the internal energy of the gas greater if the loss occurs in a constant-volume process or in a constant-pressure process?

10 Does the temperature of an ideal gas increase, decrease, or stay the same during (a) an isothermal expansion, (b) an expansion at constant pressure, (c) an adiabatic expansion, and (d) an increase in pressure at constant volume?

Problems

GO Tutoring problem available (at instructor's discretion) in *WileyPLUS* and WebAssign

SSM Worked-out solution available in Student Solutions Manual

• – ••• Number of dots indicates level of problem difficulty

Additional information available in *The Flying Circus of Physics* and at flyingcircusofphysics.com

WWW Worked-out solution is at / ILW Interactive solution is at http://www.wiley.com/college/halliday

Module 19-1 Avogadro's Number

•1 Find the mass in kilograms of 7.50×10^{24} atoms of arsenic, which has a molar mass of 74.9 g/mol.

•2 Gold has a molar mass of 197 g/mol. (a) How many moles of gold are in a 2.50 g sample of pure gold? (b) How many atoms are in the sample?

Module 19-2 Ideal Gases

•3 SSM Oxygen gas having a volume of 1000 cm^3 at 40.0°C and 1.01×10^5 Pa expands until its volume is 1500 cm^3 and its pressure is 1.06×10^5 Pa. Find (a) the number of moles of oxygen present and (b) the final temperature of the sample.

•4 A quantity of ideal gas at 10.0°C and 100 kPa occupies a volume of 2.50 m^3. (a) How many moles of the gas are present? (b) If the pressure is now raised to 300 kPa and the temperature is raised to 30.0°C, how much volume does the gas occupy? Assume no leaks.

•5 The best laboratory vacuum has a pressure of about 1.00×10^{-18} atm, or 1.01×10^{-13} Pa. How many gas molecules are there per cubic centimeter in such a vacuum at 293 K?

•6 *Water bottle in a hot car.* In the American Southwest, the temperature in a closed car parked in sunlight during the summer can be high enough to burn flesh. Suppose a bottle of water at a refrigerator temperature of 5.00°C is opened, then closed, and then left in a closed car with an internal temperature of 75.0°C. Neglecting the thermal expansion of the water and the bottle, find the pressure in the air pocket trapped in the bottle. (The pressure can be enough to push the bottle cap past the threads that are intended to keep the bottle closed.)

•7 Suppose 1.80 mol of an ideal gas is taken from a volume of 3.00 m^3 to a volume of 1.50 m^3 via an isothermal compression at 30°C. (a) How much energy is transferred as heat during the compression, and (b) is the transfer *to* or *from* the gas?

•8 Compute (a) the number of moles and (b) the number of molecules in 1.00 cm^3 of an ideal gas at a pressure of 100 Pa and a temperature of 220 K.

•9 An automobile tire has a volume of 1.64×10^{-2} m^3 and contains air at a gauge pressure (pressure above atmospheric pressure) of 165 kPa when the temperature is 0.00°C. What is the gauge pressure of the air in the tires when its temperature rises to 27.0°C and its volume increases to 1.67×10^{-2} m^3? Assume atmospheric pressure is 1.01×10^5 Pa.

•**10** A container encloses 2 mol of an ideal gas that has molar mass M_1 and 0.5 mol of a second ideal gas that has molar mass $M_2 = 3M_1$. What fraction of the total pressure on the container wall is attributable to the second gas? (The kinetic theory explanation of pressure leads to the experimentally discovered law of partial pressures for a mixture of gases that do not react chemically: *The total pressure exerted by the mixture is equal to the sum of the pressures that the several gases would exert separately if each were to occupy the vessel alone.* The molecule–vessel collisions of one type would not be altered by the presence of another type.)

••**11** SSM ILW WWW Air that initially occupies 0.140 m^3 at a gauge pressure of 103.0 kPa is expanded isothermally to a pressure of 101.3 kPa and then cooled at constant pressure until it reaches its initial volume. Compute the work done by the air. (Gauge pressure is the difference between the actual pressure and atmospheric pressure.)

••**12** GO *Submarine rescue.* When the U.S. submarine *Squalus* became disabled at a depth of 80 m, a cylindrical chamber was lowered from a ship to rescue the crew. The chamber had a radius of 1.00 m and a height of 4.00 m, was open at the bottom, and held two rescuers. It slid along a guide cable that a diver had attached to a hatch on the submarine. Once the chamber reached the hatch and clamped to the hull, the crew could escape into the chamber. During the descent, air was released from tanks to prevent water from flooding the chamber. Assume that the interior air pressure matched the water pressure at depth h as given by $p_0 + \rho g h$, where $p_0 = 1.000$ atm is the surface pressure and $\rho = 1024$ kg/m^3 is the density of seawater. Assume a surface temperature of 20.0°C and a submerged water temperature of −30.0°C. (a) What is the air volume in the chamber at the surface? (b) If air had not been released from the tanks, what would have been the air volume in the chamber at depth $h = 80.0$ m? (c) How many moles of air were needed to be released to maintain the original air volume in the chamber?

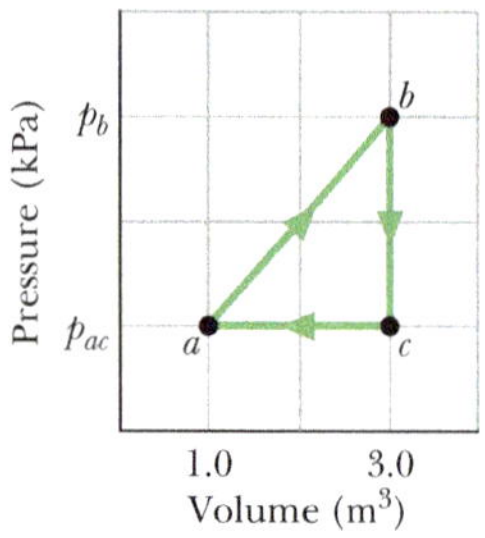

Figure 19-20 Problem 13.

••**13** GO A sample of an ideal gas is taken through the cyclic process $abca$ shown in Fig. 19-20. The scale of the vertical axis is set by $p_b = 7.5$ kPa and $p_{ac} = 2.5$ kPa. At point a, $T = 200$ K. (a) How many moles of gas are in the sample? What are (b) the temperature of the gas at point b, (c) the temperature of the gas at point c, and (d) the net energy added to the gas as heat during the cycle?

••**14** In the temperature range 310 K to 330 K, the pressure p of a certain nonideal gas is related to volume V and temperature T by

$$p = (24.9 \text{ J/K})\frac{T}{V} - (0.00662 \text{ J/K}^2)\frac{T^2}{V}.$$

How much work is done by the gas if its temperature is raised from 315 K to 325 K while the pressure is held constant?

••**15** Suppose 0.825 mol of an ideal gas undergoes an isothermal expansion as energy is added to it as heat Q. If Fig. 19-21 shows the final volume V_f versus Q, what is the gas temperature? The scale of the vertical axis is set by $V_{fs} = 0.30$ m^3, and the scale of the horizontal axis is set by $Q_s = 1200$ J.

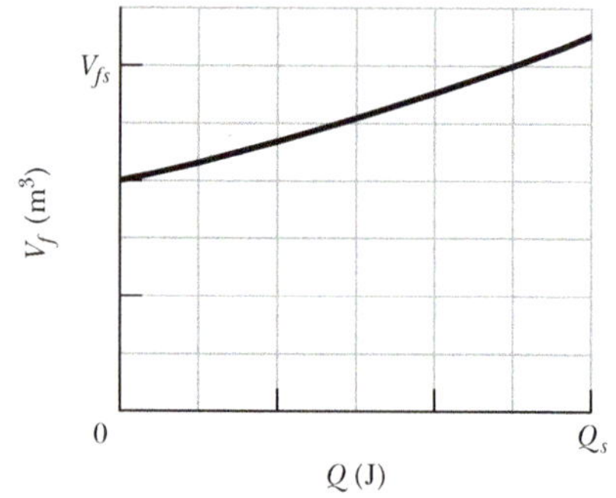

Figure 19-21 Problem 15.

•••**16** An air bubble of volume 20 cm^3 is at the bottom of a lake 40 m deep, where the temperature is 4.0°C. The bubble rises to the surface, which is at a temperature of 20°C. Take the temperature of the bubble's air to be the same as that of the surrounding water. Just as the bubble reaches the surface, what is its volume?

•••**17** GO Container A in Fig. 19-22 holds an ideal gas at a pressure of 5.0×10^5 Pa and a temperature of 300 K. It is connected by a thin tube (and a closed valve) to container B, with four times the volume of A. Container B holds the same ideal gas at a pressure of 1.0×10^5 Pa and a temperature of 400 K. The valve is opened to allow the pressures to equalize, but the temperature of each container is maintained. What then is the pressure?

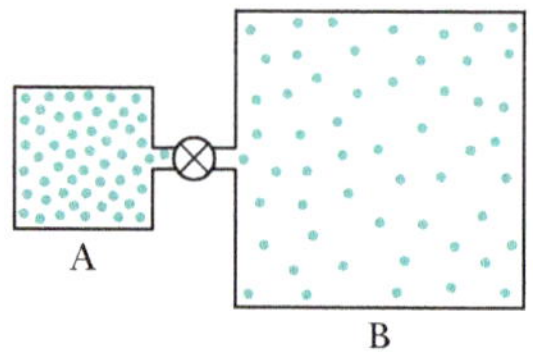

Figure 19-22 Problem 17.

Module 19-3 Pressure, Temperature, and RMS Speed

•**18** The temperature and pressure in the Sun's atmosphere are 2.00×10^6 K and 0.0300 Pa. Calculate the rms speed of free electrons (mass 9.11×10^{-31} kg) there, assuming they are an ideal gas.

•**19** (a) Compute the rms speed of a nitrogen molecule at 20.0°C. The molar mass of nitrogen molecules (N_2) is given in Table 19-1. At what temperatures will the rms speed be (b) half that value and (c) twice that value?

•**20** Calculate the rms speed of helium atoms at 1000 K. See Appendix F for the molar mass of helium atoms.

•**21** SSM The lowest possible temperature in outer space is 2.7 K. What is the rms speed of hydrogen molecules at this temperature? (The molar mass is given in Table 19-1.)

•**22** Find the rms speed of argon atoms at 313 K. See Appendix F for the molar mass of argon atoms.

••**23** A beam of hydrogen molecules (H_2) is directed toward a wall, at an angle of 55° with the normal to the wall. Each molecule in the beam has a speed of 1.0 km/s and a mass of 3.3×10^{-24} g. The beam strikes the wall over an area of 2.0 cm^2, at the rate of 10^{23} molecules per second. What is the beam's pressure on the wall?

••**24** At 273 K and 1.00×10^{-2} atm, the density of a gas is 1.24×10^{-5} g/cm^3. (a) Find v_{rms} for the gas molecules. (b) Find the molar mass of the gas and (c) identify the gas. See Table 19-1.

Module 19-4 Translational Kinetic Energy

•**25** Determine the average value of the translational kinetic energy of the molecules of an ideal gas at temperatures (a) 0.00°C and (b) 100°C. What is the translational kinetic energy per mole of an ideal gas at (c) 0.00°C and (d) 100°C?

•26 What is the average translational kinetic energy of nitrogen molecules at 1600 K?

••27 Water standing in the open at 32.0°C evaporates because of the escape of some of the surface molecules. The heat of vaporization (539 cal/g) is approximately equal to εn, where ε is the average energy of the escaping molecules and n is the number of molecules per gram. (a) Find ε. (b) What is the ratio of ε to the average kinetic energy of H_2O molecules, assuming the latter is related to temperature in the same way as it is for gases?

Module 19-5 Mean Free Path

•28 At what frequency would the wavelength of sound in air be equal to the mean free path of oxygen molecules at 1.0 atm pressure and 0.00°C? The molecular diameter is 3.0×10^{-8} cm.

•29 SSM The atmospheric density at an altitude of 2500 km is about 1 molecule/cm^3. (a) Assuming the molecular diameter of 2.0×10^{-8} cm, find the mean free path predicted by Eq. 19-25. (b) Explain whether the predicted value is meaningful.

•30 The mean free path of nitrogen molecules at 0.0°C and 1.0 atm is 0.80×10^{-5} cm. At this temperature and pressure there are 2.7×10^{19} molecules/cm^3. What is the molecular diameter?

••31 In a certain particle accelerator, protons travel around a circular path of diameter 23.0 m in an evacuated chamber, whose residual gas is at 295 K and 1.00×10^{-6} torr pressure. (a) Calculate the number of gas molecules per cubic centimeter at this pressure. (b) What is the mean free path of the gas molecules if the molecular diameter is 2.00×10^{-8} cm?

••32 At 20°C and 750 torr pressure, the mean free paths for argon gas (Ar) and nitrogen gas (N_2) are $\lambda_{Ar} = 9.9 \times 10^{-6}$ cm and $\lambda_{N_2} = 27.5 \times 10^{-6}$ cm. (a) Find the ratio of the diameter of an Ar atom to that of an N_2 molecule. What is the mean free path of argon at (b) 20°C and 150 torr, and (c) −40°C and 750 torr?

Module 19-6 The Distribution of Molecular Speeds

•33 SSM The speeds of 10 molecules are 2.0, 3.0, 4.0, . . . , 11 km/s. What are their (a) average speed and (b) rms speed?

•34 The speeds of 22 particles are as follows (N_i represents the number of particles that have speed v_i):

N_i	2	4	6	8	2
v_i (cm/s)	1.0	2.0	3.0	4.0	5.0

What are (a) v_{avg}, (b) v_{rms}, and (c) v_P?

•35 Ten particles are moving with the following speeds: four at 200 m/s, two at 500 m/s, and four at 600 m/s. Calculate their (a) average and (b) rms speeds. (c) Is $v_{rms} > v_{avg}$?

••36 The most probable speed of the molecules in a gas at temperature T_2 is equal to the rms speed of the molecules at temperature T_1. Find T_2/T_1.

••37 SSM WWW Figure 19-23 shows a hypothetical speed distribution for a sample of N gas particles (note that $P(v) = 0$ for speed $v > 2v_0$). What are the values of (a) av_0, (b) v_{avg}/v_0, and (c) v_{rms}/v_0? (d) What fraction of the particles has a speed between $1.5v_0$ and $2.0v_0$?

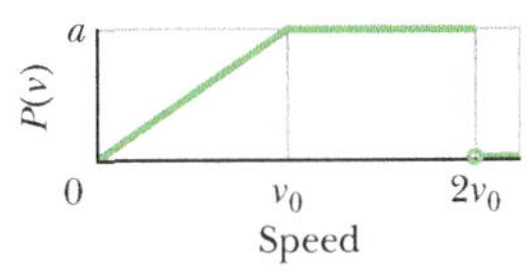

Figure 19-23 Problem 37.

••38 Figure 19-24 gives the probability distribution for nitrogen gas. The scale of the horizontal axis is set by $v_s = 1200$ m/s. What are the (a) gas temperature and (b) rms speed of the molecules?

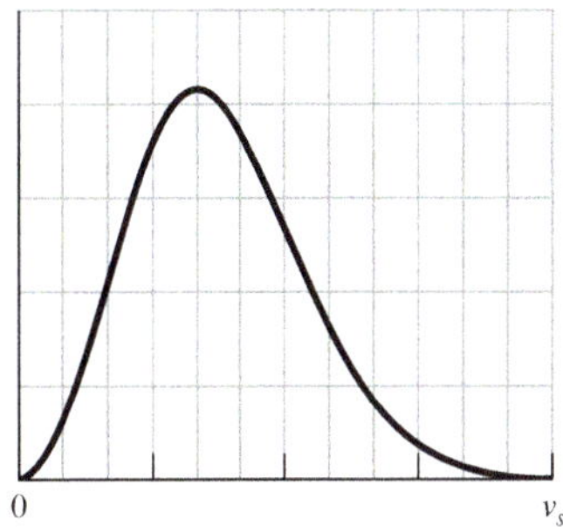

Figure 19-24 Problem 38.

••39 At what temperature does the rms speed of (a) H_2 (molecular hydrogen) and (b) O_2 (molecular oxygen) equal the escape speed from Earth (Table 13-2)? At what temperature does the rms speed of (c) H_2 and (d) O_2 equal the escape speed from the Moon (where the gravitational acceleration at the surface has magnitude $0.16g$)? Considering the answers to parts (a) and (b), should there be much (e) hydrogen and (f) oxygen high in Earth's upper atmosphere, where the temperature is about 1000 K?

••40 Two containers are at the same temperature. The first contains gas with pressure p_1, molecular mass m_1, and rms speed v_{rms1}. The second contains gas with pressure $2.0p_1$, molecular mass m_2, and average speed $v_{avg2} = 2.0v_{rms1}$. Find the mass ratio m_1/m_2.

••41 A hydrogen molecule (diameter 1.0×10^{-8} cm), traveling at the rms speed, escapes from a 4000 K furnace into a chamber containing *cold* argon atoms (diameter 3.0×10^{-8} cm) at a density of 4.0×10^{19} atoms/cm^3. (a) What is the speed of the hydrogen molecule? (b) If it collides with an argon atom, what is the closest their centers can be, considering each as spherical? (c) What is the initial number of collisions per second experienced by the hydrogen molecule? (*Hint*: Assume that the argon atoms are stationary. Then the mean free path of the hydrogen molecule is given by Eq. 19-26 and not Eq. 19-25.)

Module 19-7 The Molar Specific Heats of an Ideal Gas

•42 What is the internal energy of 1.0 mol of an ideal monatomic gas at 273 K?

••43 GO The temperature of 3.00 mol of an ideal diatomic gas is increased by 40.0 C° without the pressure of the gas changing. The molecules in the gas rotate but do not oscillate. (a) How much energy is transferred to the gas as heat? (b) What is the change in the internal energy of the gas? (c) How much work is done by the gas? (d) By how much does the rotational kinetic energy of the gas increase?

••44 GO One mole of an ideal diatomic gas goes from a to c along the diagonal path in Fig. 19-25. The scale of the vertical axis is set by $p_{ab} = 5.0$ kPa and $p_c = 2.0$ kPa, and the scale of the horizontal axis is set by $V_{bc} = 4.0$ m^3 and $V_a = 2.0$ m^3. During the transition, (a) what is the change in internal energy of the gas, and (b) how

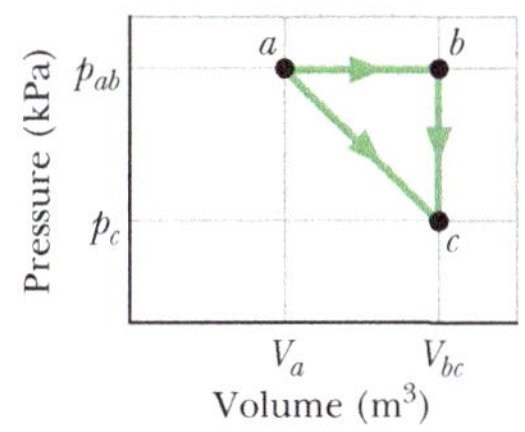

Figure 19-25 Problem 44.

much energy is added to the gas as heat? (c) How much heat is required if the gas goes from a to c along the indirect path abc?

••45 ILW The mass of a gas molecule can be computed from its specific heat at constant volume c_V. (Note that this is not C_V.) Take $c_V = 0.075$ cal/g·C° for argon and calculate (a) the mass of an argon atom and (b) the molar mass of argon.

••46 Under constant pressure, the temperature of 2.00 mol of an ideal monatomic gas is raised 15.0 K. What are (a) the work W done by the gas, (b) the energy transferred as heat Q, (c) the change ΔE_{int} in the internal energy of the gas, and (d) the change ΔK in the average kinetic energy per atom?

••47 The temperature of 2.00 mol of an ideal monatomic gas is raised 15.0 K at constant volume. What are (a) the work W done by the gas, (b) the energy transferred as heat Q, (c) the change ΔE_{int} in the internal energy of the gas, and (d) the change ΔK in the average kinetic energy per atom?

••48 GO When 20.9 J was added as heat to a particular ideal gas, the volume of the gas changed from 50.0 cm^3 to 100 cm^3 while the pressure remained at 1.00 atm. (a) By how much did the internal energy of the gas change? If the quantity of gas present was 2.00×10^{-3} mol, find (b) C_p and (c) C_V.

••49 SSM A container holds a mixture of three nonreacting gases: 2.40 mol of gas 1 with $C_{V1} = 12.0$ J/mol·K, 1.50 mol of gas 2 with $C_{V2} = 12.8$ J/mol·K, and 3.20 mol of gas 3 with $C_{V3} = 20.0$ J/mol·K. What is C_V of the mixture?

Module 19-8 Degrees of Freedom and Molar Specific Heats

•50 We give 70 J as heat to a diatomic gas, which then expands at constant pressure. The gas molecules rotate but do not oscillate. By how much does the internal energy of the gas increase?

•51 ILW When 1.0 mol of oxygen (O_2) gas is heated at constant pressure starting at 0°C, how much energy must be added to the gas as heat to double its volume? (The molecules rotate but do not oscillate.)

••52 GO Suppose 12.0 g of oxygen (O_2) gas is heated at constant atmospheric pressure from 25.0°C to 125°C. (a) How many moles of oxygen are present? (See Table 19-1 for the molar mass.) (b) How much energy is transferred to the oxygen as heat? (The molecules rotate but do not oscillate.) (c) What fraction of the heat is used to raise the internal energy of the oxygen?

••53 SSM WWW Suppose 4.00 mol of an ideal diatomic gas, with molecular rotation but not oscillation, experienced a temperature increase of 60.0 K under constant-pressure conditions. What are (a) the energy transferred as heat Q, (b) the change ΔE_{int} in internal energy of the gas, (c) the work W done by the gas, and (d) the change ΔK in the total translational kinetic energy of the gas?

Module 19-9 The Adiabatic Expansion of an Ideal Gas

•54 We know that for an adiabatic process $pV^\gamma =$ a constant. Evaluate "a constant" for an adiabatic process involving exactly 2.0 mol of an ideal gas passing through the state having exactly $p = 1.0$ atm and $T = 300$ K. Assume a diatomic gas whose molecules rotate but do not oscillate.

•55 A certain gas occupies a volume of 4.3 L at a pressure of 1.2 atm and a temperature of 310 K. It is compressed adiabatically to a volume of 0.76 L. Determine (a) the final pressure and (b) the final temperature, assuming the gas to be an ideal gas for which $\gamma = 1.4$.

•56 Suppose 1.00 L of a gas with $\gamma = 1.30$, initially at 273 K and 1.00 atm, is suddenly compressed adiabatically to half its initial volume. Find its final (a) pressure and (b) temperature. (c) If the gas is then cooled to 273 K at constant pressure, what is its final volume?

••57 The volume of an ideal gas is adiabatically reduced from 200 L to 74.3 L. The initial pressure and temperature are 1.00 atm and 300 K. The final pressure is 4.00 atm. (a) Is the gas monatomic, diatomic, or polyatomic? (b) What is the final temperature? (c) How many moles are in the gas?

••58 GO *Opening champagne.* In a bottle of champagne, the pocket of gas (primarily carbon dioxide) between the liquid and the cork is at pressure of $p_i = 5.00$ atm. When the cork is pulled from the bottle, the gas undergoes an adiabatic expansion until its pressure matches the ambient air pressure of 1.00 atm. Assume that the ratio of the molar specific heats is $\gamma = \frac{4}{3}$. If the gas has initial temperature $T_i = 5.00$°C, what is its temperature at the end of the adiabatic expansion?

••59 GO Figure 19-26 shows two paths that may be taken by a gas from an initial point i to a final point f. Path 1 consists of an isothermal expansion (work is 50 J in magnitude), an adiabatic expansion (work is 40 J in magnitude), an isothermal compression (work is 30 J in magnitude), and then an adiabatic compression (work is 25 J in magnitude). What is the change in the internal energy of the gas if the gas goes from point i to point f along path 2?

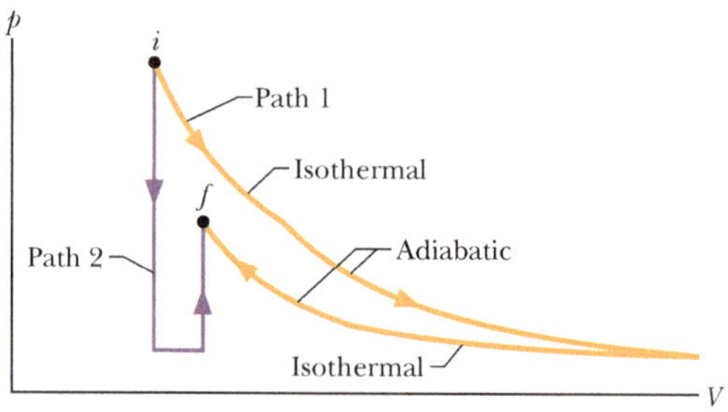

Figure 19-26 Problem 59.

••60 GO *Adiabatic wind.* The normal airflow over the Rocky Mountains is west to east. The air loses much of its moisture content and is chilled as it climbs the western side of the mountains. When it descends on the eastern side, the increase in pressure toward lower altitudes causes the temperature to increase. The flow, then called a chinook wind, can rapidly raise the air temperature at the base of the mountains. Assume that the air pressure p depends on altitude y according to $p = p_0 \exp(-ay)$, where $p_0 = 1.00$ atm and $a = 1.16 \times 10^{-4}$ m^{-1}. Also assume that the ratio of the molar specific heats is $\gamma = \frac{4}{3}$. A parcel of air with an initial temperature of -5.00°C descends adiabatically from $y_1 = 4267$ m to $y = 1567$ m. What is its temperature at the end of the descent?

••61 GO A gas is to be expanded from initial state i to final state f along either path 1 or path 2 on a p-V diagram. Path 1 consists of three steps: an isothermal expansion (work is 40 J in magnitude), an adiabatic expansion (work is 20 J in magnitude), and another isothermal expansion (work is 30 J in magnitude). Path 2 consists of two steps: a pressure reduction at constant volume and an expansion at constant pressure. What is the change in the internal energy of the gas along path 2?

•••62 GO An ideal diatomic gas, with rotation but no oscillation, undergoes an adiabatic compression. Its initial pressure and volume are

1.20 atm and 0.200 m^3. Its final pressure is 2.40 atm. How much work is done by the gas?

•••63 Figure 19-27 shows a cycle undergone by 1.00 mol of an ideal monatomic gas. The temperatures are $T_1 = 300$ K, $T_2 = 600$ K, and $T_3 = 455$ K. For $1 \rightarrow 2$, what are (a) heat Q, (b) the change in internal energy ΔE_{int}, and (c) the work done W? For $2 \rightarrow 3$, what are (d) Q, (e) ΔE_{int}, and (f) W? For $3 \rightarrow 1$, what are (g) Q, (h) ΔE_{int}, and (i) W? For the full cycle, what are (j) Q, (k) ΔE_{int}, and (l) W? The initial pressure at point 1 is 1.00 atm (= 1.013 × 10^5 Pa). What are the (m) volume and (n) pressure at point 2 and the (o) volume and (p) pressure at point 3?

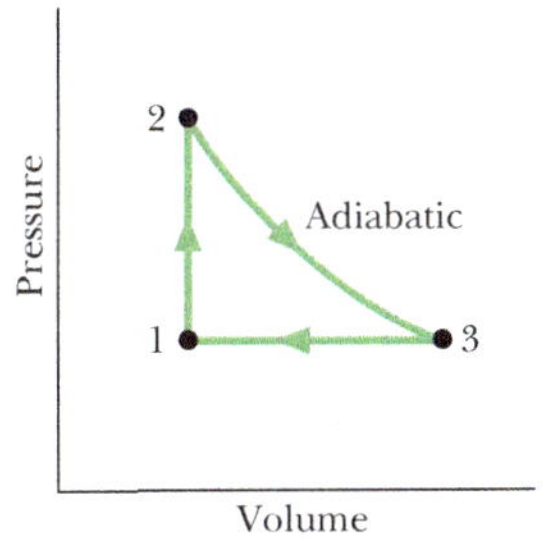

Figure 19-27 Problem 63.

Additional Problems

64 Calculate the work done by an external agent during an isothermal compression of 1.00 mol of oxygen from a volume of 22.4 L at 0°C and 1.00 atm to a volume of 16.8 L.

65 An ideal gas undergoes an adiabatic compression from $p = 1.0$ atm, $V = 1.0 \times 10^6$ L, $T = 0.0$°C to $p = 1.0 \times 10^5$ atm, $V = 1.0 \times 10^3$ L. (a) Is the gas monatomic, diatomic, or polyatomic? (b) What is its final temperature? (c) How many moles of gas are present? What is the total translational kinetic energy per mole (d) before and (e) after the compression? (f) What is the ratio of the squares of the rms speeds before and after the compression?

66 An ideal gas consists of 1.50 mol of diatomic molecules that rotate but do not oscillate. The molecular diameter is 250 pm. The gas is expanded at a constant pressure of 1.50 × 10^5 Pa, with a transfer of 200 J as heat. What is the change in the mean free path of the molecules?

67 An ideal monatomic gas initially has a temperature of 330 K and a pressure of 6.00 atm. It is to expand from volume 500 cm^3 to volume 1500 cm^3. If the expansion is isothermal, what are (a) the final pressure and (b) the work done by the gas? If, instead, the expansion is adiabatic, what are (c) the final pressure and (d) the work done by the gas?

68 In an interstellar gas cloud at 50.0 K, the pressure is 1.00 × 10^{-8} Pa. Assuming that the molecular diameters of the gases in the cloud are all 20.0 nm, what is their mean free path?

69 SSM The envelope and basket of a hot-air balloon have a combined weight of 2.45 kN, and the envelope has a capacity (volume) of 2.18 × 10^3 m^3. When it is fully inflated, what should be the temperature of the enclosed air to give the balloon a *lifting capacity* (force) of 2.67 kN (in addition to the balloon's weight)? Assume that the surrounding air, at 20.0°C, has a weight per unit volume of 11.9 N/m^3 and a molecular mass of 0.028 kg/mol, and is at a pressure of 1.0 atm.

70 An ideal gas, at initial temperature T_1 and initial volume 2.0 m^3, is expanded adiabatically to a volume of 4.0 m^3, then expanded isothermally to a volume of 10 m^3, and then compressed adiabatically back to T_1. What is its final volume?

71 SSM The temperature of 2.00 mol of an ideal monatomic gas is raised 15.0 K in an adiabatic process. What are (a) the work W done by the gas, (b) the energy transferred as heat Q, (c) the change ΔE_{int} in internal energy of the gas, and (d) the change ΔK in the average kinetic energy per atom?

72 At what temperature do atoms of helium gas have the same rms speed as molecules of hydrogen gas at 20.0°C? (The molar masses are given in Table 19-1.)

73 SSM At what frequency do molecules (diameter 290 pm) collide in (an ideal) oxygen gas (O_2) at temperature 400 K and pressure 2.00 atm?

74 (a) What is the number of molecules per cubic meter in air at 20°C and at a pressure of 1.0 atm (= 1.01 × 10^5 Pa)? (b) What is the mass of 1.0 m^3 of this air? Assume that 75% of the molecules are nitrogen (N_2) and 25% are oxygen (O_2).

75 The temperature of 3.00 mol of a gas with $C_V = 6.00$ cal/mol·K is to be raised 50.0 K. If the process is at *constant volume*, what are (a) the energy transferred as heat Q, (b) the work W done by the gas, (c) the change ΔE_{int} in internal energy of the gas, and (d) the change ΔK in the total translational kinetic energy? If the process is at *constant pressure*, what are (e) Q, (f) W, (g) ΔE_{int}, and (h) ΔK? If the process is *adiabatic*, what are (i) Q, (j) W, (k) ΔE_{int}, and (l) ΔK?

76 During a compression at a constant pressure of 250 Pa, the volume of an ideal gas decreases from 0.80 m^3 to 0.20 m^3. The initial temperature is 360 K, and the gas loses 210 J as heat. What are (a) the change in the internal energy of the gas and (b) the final temperature of the gas?

77 SSM Figure 19-28 shows a hypothetical speed distribution for particles of a certain gas: $P(v) = Cv^2$ for $0 < v \le v_0$ and $P(v) = 0$ for $v > v_0$. Find (a) an expression for C in terms of v_0, (b) the average speed of the particles, and (c) their rms speed.

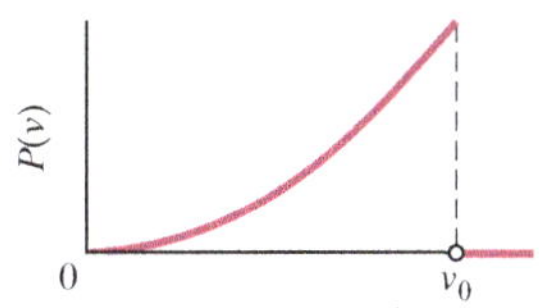

Figure 19-28 Problem 77.

78 (a) An ideal gas initially at pressure p_0 undergoes a free expansion until its volume is 3.00 times its initial volume. What then is the ratio of its pressure to p_0? (b) The gas is next slowly and adiabatically compressed back to its original volume. The pressure after compression is $(3.00)^{1/3}p_0$. Is the gas monatomic, diatomic, or polyatomic? (c) What is the ratio of the average kinetic energy per molecule in this final state to that in the initial state?

79 SSM An ideal gas undergoes isothermal compression from an initial volume of 4.00 m^3 to a final volume of 3.00 m^3. There is 3.50 mol of the gas, and its temperature is 10.0°C. (a) How much work is done by the gas? (b) How much energy is transferred as heat between the gas and its environment?

80 Oxygen (O_2) gas at 273 K and 1.0 atm is confined to a cubical container 10 cm on a side. Calculate $\Delta U_g/K_{avg}$, where ΔU_g is the change in the gravitational potential energy of an oxygen molecule falling the height of the box and K_{avg} is the molecule's average translational kinetic energy.

81 An ideal gas is taken through a complete cycle in three steps: adiabatic expansion with work equal to 125 J, isothermal contraction at 325 K, and increase in pressure at constant volume. (a) Draw a p-V diagram for the three steps. (b) How much energy is transferred as heat in step 3, and (c) is it transferred *to* or *from* the gas?

82 (a) What is the volume occupied by 1.00 mol of an ideal gas at standard conditions—that is, 1.00 atm (= 1.01 × 10^5 Pa) and 273 K? (b) Show that the number of molecules per cubic centimeter (the *Loschmidt number*) at standard conditions is 2.69 × 10^9.

83 SSM A sample of ideal gas expands from an initial pressure

and volume of 32 atm and 1.0 L to a final volume of 4.0 L. The initial temperature is 300 K. If the gas is monatomic and the expansion isothermal, what are the (a) final pressure p_f, (b) final temperature T_f, and (c) work W done by the gas? If the gas is monatomic and the expansion adiabatic, what are (d) p_f, (e) T_f, and (f) W? If the gas is diatomic and the expansion adiabatic, what are (g) p_f, (h) T_f, and (i) W?

84 An ideal gas with 3.00 mol is initially in state 1 with pressure $p_1 = 20.0$ atm and volume $V_1 = 1500\ \text{cm}^3$. First it is taken to state 2 with pressure $p_2 = 1.50p_1$ and volume $V_2 = 2.00V_1$. Then it is taken to state 3 with pressure $p_3 = 2.00p_1$ and volume $V_3 = 0.500V_1$. What is the temperature of the gas in (a) state 1 and (b) state 2? (c) What is the net change in internal energy from state 1 to state 3?

85 A steel tank contains 300 g of ammonia gas (NH_3) at a pressure of 1.35×10^6 Pa and a temperature of 77°C. (a) What is the volume of the tank in liters? (b) Later the temperature is 22°C and the pressure is 8.7×10^5 Pa. How many grams of gas have leaked out of the tank?

86 In an industrial process the volume of 25.0 mol of a monatomic ideal gas is reduced at a uniform rate from $0.616\ \text{m}^3$ to $0.308\ \text{m}^3$ in 2.00 h while its temperature is increased at a uniform rate from 27.0°C to 450°C. Throughout the process, the gas passes through thermodynamic equilibrium states. What are (a) the cumulative work done on the gas, (b) the cumulative energy absorbed by the gas as heat, and (c) the molar specific heat for the process? (*Hint:* To evaluate the integral for the work, you might use

$$\int \frac{a + bx}{A + Bx}\,dx = \frac{bx}{B} + \frac{aB - bA}{B^2}\ln(A + Bx),$$

an indefinite integral.) Suppose the process is replaced with a two-step process that reaches the same final state. In step 1, the gas volume is reduced at constant temperature, and in step 2 the temperature is increased at constant volume. For this process, what are (d) the cumulative work done on the gas, (e) the cumulative energy absorbed by the gas as heat, and (f) the molar specific heat for the process?

87 Figure 19-29 shows a cycle consisting of five paths: AB is isothermal at 300 K, BC is adiabatic with work = 5.0 J, CD is at a constant pressure of 5 atm, DE is isothermal, and EA is adiabatic with a change in internal energy of 8.0 J. What is the change in internal energy of the gas along path CD?

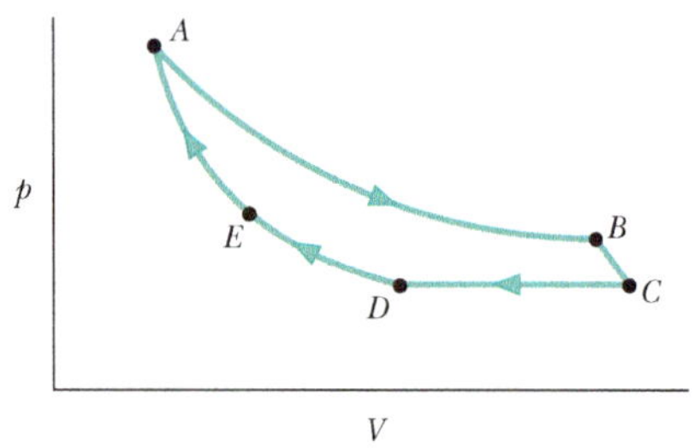

Figure 19-29 Problem 87.

88 An ideal gas initially at 300 K is compressed at a constant pressure of 25 N/m² from a volume of 3.0 m³ to a volume of 1.8 m³. In the process, 75 J is lost by the gas as heat. What are (a) the change in internal energy of the gas and (b) the final temperature of the gas?

89 A pipe of length $L = 25.0$ m that is open at one end contains air at atmospheric pressure. It is thrust vertically into a freshwater lake until the water rises halfway up in the pipe (Fig. 19-30). What is the depth h of the lower end of the pipe? Assume that the temperature is the same everywhere and does not change.

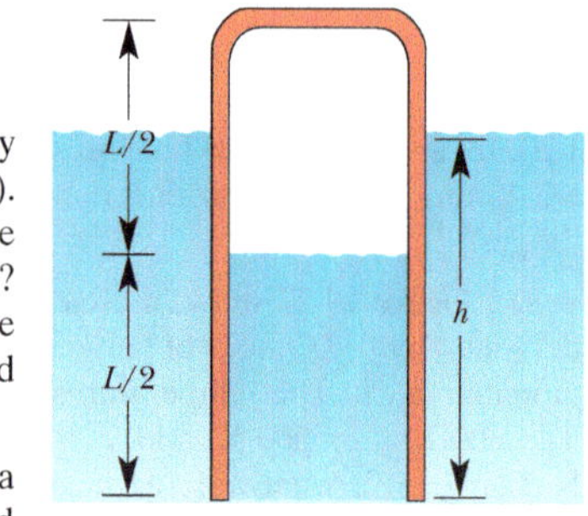

Figure 19-30 Problem 89.

90 In a motorcycle engine, a piston is forced down toward the crankshaft when the fuel in the top of the piston's cylinder undergoes combustion. The mixture of gaseous combustion products then expands adiabatically as the piston descends. Find the average power in (a) watts and (b) horsepower that is involved in this expansion when the engine is running at 4000 rpm, assuming that the gauge pressure immediately after combustion is 15 atm, the initial volume is 50 cm³, and the volume of the mixture at the bottom of the stroke is 250 cm³. Assume that the gases are diatomic and that the time involved in the expansion is one-half that of the total cycle.

91 For adiabatic processes in an ideal gas, show that (a) the bulk modulus is given by

$$B = -V\frac{dp}{dV} = \gamma p,$$

where $\gamma = C_p/C_V$. (See Eq. 17-2.) (b) Then show that the speed of sound in the gas is

$$v_s = \sqrt{\frac{\gamma p}{\rho}} = \sqrt{\frac{\gamma RT}{M}},$$

where ρ is the density, T is the temperature, and M is the molar mass. (See Eq. 17-3.)

92 Air at 0.000°C and 1.00 atm pressure has a density of 1.29×10^{-3} g/cm³, and the speed of sound is 331 m/s at that temperature. Compute the ratio γ of the molar specific heats of air. (*Hint:* See Problem 91.)

93 The speed of sound in different gases at a certain temperature T depends on the molar mass of the gases. Show that

$$\frac{v_1}{v_2} = \sqrt{\frac{M_2}{M_1}},$$

where v_1 is the speed of sound in a gas of molar mass M_1 and v_2 is the speed of sound in a gas of molar mass M_2. (*Hint:* See Problem 91.)

94 From the knowledge that C_V, the molar specific heat at constant volume, for a gas in a container is $5.0R$, calculate the ratio of the speed of sound in that gas to the rms speed of the molecules, for gas temperature T. (*Hint:* See Problem 91.)

95 The molar mass of iodine is 127 g/mol. When sound at frequency 1000 Hz is introduced to a tube of iodine gas at 400 K, an internal acoustic standing wave is set up with nodes separated by 9.57 cm. What is γ for the gas? (*Hint:* See Problem 91.)

96 For air near 0°C, by how much does the speed of sound increase for each increase in air temperature by 1 C°? (*Hint:* See Problem 91.)

97 Two containers are at the same temperature. The gas in the first container is at pressure p_1 and has molecules with mass m_1 and root-mean-square speed v_{rms1}. The gas in the second is at pressure $2p_1$ and has molecules with mass m_2 and average speed $v_{\text{avg2}} = 2v_{\text{rms1}}$. Find the ratio m_1/m_2 of the masses of their molecules.

CHAPTER 20

Entropy and the Second Law of Thermodynamics

20-1 ENTROPY

Learning Objectives

After reading this module, you should be able to . . .

20.01 Identify the second law of thermodynamics: If a process occurs in a closed system, the entropy of the system increases for irreversible processes and remains constant for reversible processes; it never decreases.

20.02 Identify that entropy is a state function (the value for a particular state of the system does not depend on how that state is reached).

20.03 Calculate the change in entropy for a process by integrating the inverse of the temperature (in kelvins) with respect to the heat Q transferred during the process.

20.04 For a phase change with a constant temperature process, apply the relationship between the entropy change ΔS, the total transferred heat Q, and the temperature T (in kelvins).

20.05 For a temperature change ΔT that is small relative to the temperature T, apply the relationship between the entropy change ΔS, the transferred heat Q, and the average temperature T_{avg} (in kelvins).

20.06 For an ideal gas, apply the relationship between the entropy change ΔS and the initial and final values of the pressure and volume.

20.07 Identify that if a process is an irreversible one, the integration for the entropy change must be done for a reversible process that takes the system between the same initial and final states as the irreversible process.

20.08 For stretched rubber, relate the elastic force to the rate at which the rubber's entropy changes with the change in the stretching distance.

Key Ideas

- An irreversible process is one that cannot be reversed by means of small changes in the environment. The direction in which an irreversible process proceeds is set by the change in entropy ΔS of the system undergoing the process. Entropy S is a state property (or state function) of the system; that is, it depends only on the state of the system and not on the way in which the system reached that state. The entropy postulate states (in part): If an irreversible process occurs in a closed system, the entropy of the system always increases.

- The entropy change ΔS for an irreversible process that takes a system from an initial state i to a final state f is exactly equal to the entropy change ΔS for any reversible process that takes the system between those same two states. We can compute the latter (but not the former) with

$$\Delta S = S_f - S_i = \int_i^f \frac{dQ}{T}.$$

Here Q is the energy transferred as heat to or from the system during the process, and T is the temperature of the system in kelvins during the process.

- For a reversible isothermal process, the expression for an entropy change reduces to

$$\Delta S = S_f - S_i = \frac{Q}{T}.$$

- When the temperature change ΔT of a system is small relative to the temperature (in kelvins) before and after the process, the entropy change can be approximated as

$$\Delta S = S_f - S_i \approx \frac{Q}{T_{avg}},$$

where T_{avg} is the system's average temperature during the process.

- When an ideal gas changes reversibly from an initial state with temperature T_i and volume V_i to a final state with temperature T_f and volume V_f, the change ΔS in the entropy of the gas is

$$\Delta S = S_f - S_i = nR \ln \frac{V_f}{V_i} + nC_V \ln \frac{T_f}{T_i}.$$

- The second law of thermodynamics, which is an extension of the entropy postulate, states: If a process occurs in a closed system, the entropy of the system increases for irreversible processes and remains constant for reversible processes. It never decreases. In equation form,

$$\Delta S \geq 0.$$

What Is Physics?

Time has direction, the direction in which we age. We are accustomed to many one-way processes—that is, processes that can occur only in a certain sequence (the right way) and never in the reverse sequence (the wrong way). An egg is dropped onto a floor, a pizza is baked, a car is driven into a lamppost, large waves erode a sandy beach—these one-way processes are **irreversible,** meaning that they cannot be reversed by means of only small changes in their environment.

One goal of physics is to understand why time has direction and why one-way processes are irreversible. Although this physics might seem disconnected from the practical issues of everyday life, it is in fact at the heart of any engine, such as a car engine, because it determines how well an engine can run.

The key to understanding why one-way processes cannot be reversed involves a quantity known as *entropy*.

Irreversible Processes and Entropy

The one-way character of irreversible processes is so pervasive that we take it for granted. If these processes were to occur *spontaneously* (on their own) in the wrong way, we would be astonished. Yet *none* of these wrong-way events would violate the law of conservation of energy.

For example, if you were to wrap your hands around a cup of hot coffee, you would be astonished if your hands got cooler and the cup got warmer. That is obviously the wrong way for the energy transfer, but the total energy of the closed system (*hands* + *cup of coffee*) would be the same as the total energy if the process had run in the right way. For another example, if you popped a helium balloon, you would be astonished if, later, all the helium molecules were to gather together in the original shape of the balloon. That is obviously the wrong way for molecules to spread, but the total energy of the closed system (*molecules* + *room*) would be the same as for the right way.

Thus, changes in energy within a closed system do not set the direction of irreversible processes. Rather, that direction is set by another property that we shall discuss in this chapter—the *change in entropy* ΔS of the system. The change in entropy of a system is defined later in this module, but we can here state its central property, often called the *entropy postulate:*

> If an irreversible process occurs in a *closed* system, the entropy S of the system always increases; it never decreases.

Entropy differs from energy in that entropy does *not* obey a conservation law. The *energy* of a closed system is conserved; it always remains constant. For irreversible processes, the *entropy* of a closed system always increases. Because of this property, the change in entropy is sometimes called "the arrow of time." For example, we associate the explosion of a popcorn kernel with the forward direction of time and with an increase in entropy. The backward direction of time (a videotape run backwards) would correspond to the exploded popcorn re-forming the original kernel. Because this backward process would result in an entropy decrease, it never happens.

There are two equivalent ways to define the change in entropy of a system: (1) in terms of the system's temperature and the energy the system gains or loses as heat, and (2) by counting the ways in which the atoms or molecules that make up the system can be arranged. We use the first approach in this module and the second in Module 20-4.

Change in Entropy

Let's approach this definition of *change in entropy* by looking again at a process that we described in Modules 18-5 and 19-9: the free expansion of an ideal gas. Figure 20-1*a* shows the gas in its initial equilibrium state *i*, confined by a closed stopcock to the left half of a thermally insulated container. If we open the stopcock, the gas rushes to fill the entire container, eventually reaching the final equilibrium state *f* shown in Fig. 20-1*b*. This is an irreversible process; all the molecules of the gas will never return to the left half of the container.

The *p*-*V* plot of the process, in Fig. 20-2, shows the pressure and volume of the gas in its initial state *i* and final state *f*. Pressure and volume are *state properties*, properties that depend only on the state of the gas and not on how it reached that state. Other state properties are temperature and energy. We now assume that the gas has still another state property—its entropy. Furthermore, we define the **change in entropy** $S_f - S_i$ of a system during a process that takes the system from an initial state *i* to a final state *f* as

$$\Delta S = S_f - S_i = \int_i^f \frac{dQ}{T} \quad \text{(change in entropy defined).} \tag{20-1}$$

Here Q is the energy transferred as heat to or from the system during the process, and T is the temperature of the system in kelvins. Thus, an entropy change depends not only on the energy transferred as heat but also on the temperature at which the transfer takes place. Because T is always positive, the sign of ΔS is the same as that of Q. We see from Eq. 20-1 that the SI unit for entropy and entropy change is the joule per kelvin.

There is a problem, however, in applying Eq. 20-1 to the free expansion of Fig. 20-1. As the gas rushes to fill the entire container, the pressure, temperature, and volume of the gas fluctuate unpredictably. In other words, they do not have a sequence of well-defined equilibrium values during the intermediate stages of the change from initial state *i* to final state *f*. Thus, we cannot trace a pressure–volume path for the free expansion on the *p*-*V* plot of Fig. 20-2, and we cannot find a relation between Q and T that allows us to integrate as Eq. 20-1 requires.

However, if entropy is truly a state property, the difference in entropy between states *i* and *f* must depend *only on those states* and not at all on the way the system went from one state to the other. Suppose, then, that we replace the irreversible free expansion of Fig. 20-1 with a *reversible* process that connects states *i* and *f*. With a reversible process we can trace a pressure–volume path on a *p*-*V* plot, and we can find a relation between Q and T that allows us to use Eq. 20-1 to obtain the entropy change.

We saw in Module 19-9 that the temperature of an ideal gas does not change during a free expansion: $T_i = T_f = T$. Thus, points *i* and *f* in Fig. 20-2 must be on the same isotherm. A convenient replacement process is then a reversible isothermal expansion from state *i* to state *f*, which actually proceeds *along* that isotherm. Furthermore, because T is constant throughout a reversible isothermal expansion, the integral of Eq. 20-1 is greatly simplified.

Figure 20-3 shows how to produce such a reversible isothermal expansion. We confine the gas to an insulated cylinder that rests on a thermal reservoir maintained at the temperature T. We begin by placing just enough lead shot on the movable piston so that the pressure and volume of the gas are those of the initial state *i* of Fig. 20-1*a*. We then remove shot slowly (piece by piece) until the pressure and volume of the gas are those of the final state *f* of Fig. 20-1*b*. The temperature of the gas does not change because the gas remains in thermal contact with the reservoir throughout the process.

The reversible isothermal expansion of Fig. 20-3 is physically quite different from the irreversible free expansion of Fig. 20-1. However, *both processes have the same initial state and the same final state and thus must have the same change in*

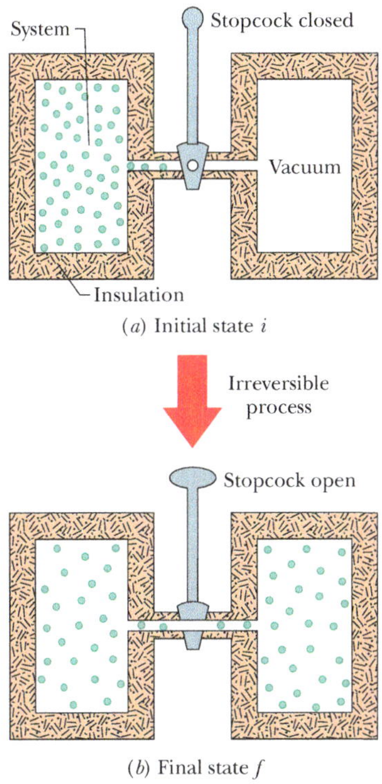

Figure 20-1 The free expansion of an ideal gas. (*a*) The gas is confined to the left half of an insulated container by a closed stopcock. (*b*) When the stopcock is opened, the gas rushes to fill the entire container. This process is irreversible; that is, it does not occur in reverse, with the gas spontaneously collecting itself in the left half of the container.

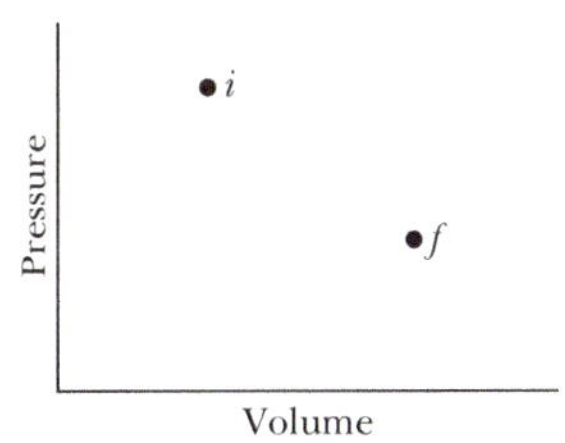

Figure 20-2 A *p*-*V* diagram showing the initial state *i* and the final state *f* of the free expansion of Fig. 20-1. The intermediate states of the gas cannot be shown because they are not equilibrium states.

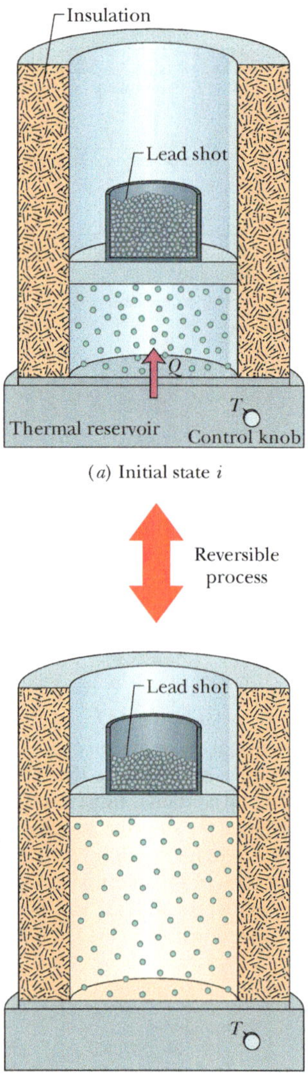

Figure 20-3 The isothermal expansion of an ideal gas, done in a reversible way. The gas has the same initial state i and same final state f as in the irreversible process of Figs. 20-1 and 20-2.

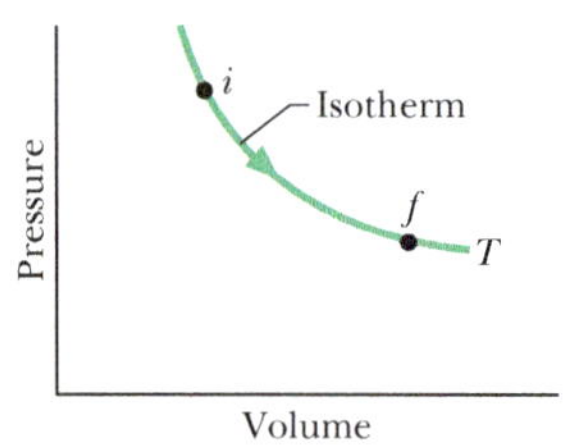

Figure 20-4 A p-V diagram for the reversible isothermal expansion of Fig. 20-3. The intermediate states, which are now equilibrium states, are shown.

entropy. Because we removed the lead shot slowly, the intermediate states of the gas are equilibrium states, so we can plot them on a p-V diagram (Fig. 20-4).

To apply Eq. 20-1 to the isothermal expansion, we take the constant temperature T outside the integral, obtaining

$$\Delta S = S_f - S_i = \frac{1}{T}\int_i^f dQ.$$

Because $\int dQ = Q$, where Q is the total energy transferred as heat during the process, we have

$$\Delta S = S_f - S_i = \frac{Q}{T} \quad \text{(change in entropy, isothermal process).} \tag{20-2}$$

To keep the temperature T of the gas constant during the isothermal expansion of Fig. 20-3, heat Q must have been energy transferred *from* the reservoir *to* the gas. Thus, Q is positive and the entropy of the gas *increases* during the isothermal process and during the free expansion of Fig. 20-1.

To summarize:

> To find the entropy change for an irreversible process, replace that process with any reversible process that connects the same initial and final states. Calculate the entropy change for this reversible process with Eq. 20-1.

When the temperature change ΔT of a system is small relative to the temperature (in kelvins) before and after the process, the entropy change can be approximated as

$$\Delta S = S_f - S_i \approx \frac{Q}{T_{avg}}, \tag{20-3}$$

where T_{avg} is the average temperature of the system in kelvins during the process.

Checkpoint 1

Water is heated on a stove. Rank the entropy changes of the water as its temperature rises (a) from 20°C to 30°C, (b) from 30°C to 35°C, and (c) from 80°C to 85°C, greatest first.

Entropy as a State Function

We have assumed that entropy, like pressure, energy, and temperature, is a property of the state of a system and is independent of how that state is reached. That entropy is indeed a *state function* (as state properties are usually called) can be deduced only by experiment. However, we can prove it is a state function for the special and important case in which an ideal gas is taken through a reversible process.

To make the process reversible, it is done slowly in a series of small steps, with the gas in an equilibrium state at the end of each step. For each small step, the energy transferred as heat to or from the gas is dQ, the work done by the gas is dW, and the change in internal energy is dE_{int}. These are related by the first law of thermodynamics in differential form (Eq. 18-27):

$$dE_{int} = dQ - dW.$$

Because the steps are reversible, with the gas in equilibrium states, we can use Eq. 18-24 to replace dW with $p\,dV$ and Eq. 19-45 to replace dE_{int} with $nC_V\,dT$. Solving for dQ then leads to

$$dQ = p\,dV + nC_V\,dT.$$

Using the ideal gas law, we replace p in this equation with nRT/V. Then we divide each term in the resulting equation by T, obtaining

$$\frac{dQ}{T} = nR\frac{dV}{V} + nC_V\frac{dT}{T}.$$

Now let us integrate each term of this equation between an arbitrary initial state i and an arbitrary final state f to get

$$\int_i^f \frac{dQ}{T} = \int_i^f nR\frac{dV}{V} + \int_i^f nC_V\frac{dT}{T}.$$

The quantity on the left is the entropy change ΔS $(= S_f - S_i)$ defined by Eq. 20-1. Substituting this and integrating the quantities on the right yield

$$\Delta S = S_f - S_i = nR\ln\frac{V_f}{V_i} + nC_V\ln\frac{T_f}{T_i}. \tag{20-4}$$

Note that we did not have to specify a particular reversible process when we integrated. Therefore, the integration must hold for all reversible processes that take the gas from state i to state f. Thus, the change in entropy ΔS between the initial and final states of an ideal gas depends only on properties of the initial state (V_i and T_i) and properties of the final state (V_f and T_f); ΔS does not depend on how the gas changes between the two states.

Checkpoint 2

An ideal gas has temperature T_1 at the initial state i shown in the p-V diagram here. The gas has a higher temperature T_2 at final states a and b, which it can reach along the paths shown. Is the entropy change along the path to state a larger than, smaller than, or the same as that along the path to state b?

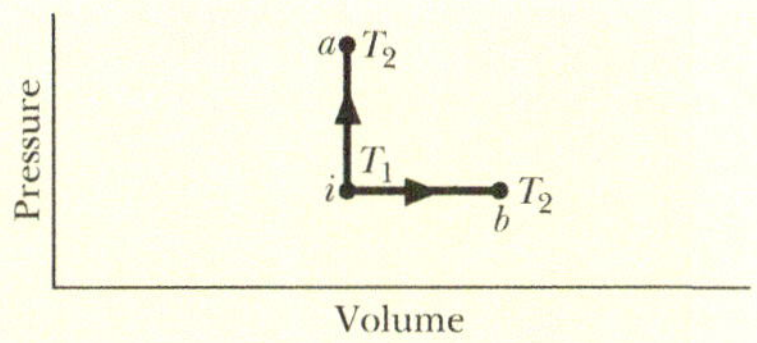

Sample Problem 20.01 Entropy change of two blocks coming to thermal equilibrium

Figure 20-5a shows two identical copper blocks of mass m = 1.5 kg: block L at temperature T_{iL} = 60°C and block R at temperature T_{iR} = 20°C. The blocks are in a thermally insulated box and are separated by an insulating shutter. When we lift the shutter, the blocks eventually come to the equilibrium temperature T_f = 40°C (Fig. 20-5b). What is the net entropy change of the two-block system during this irreversible process? The specific heat of copper is 386 J/kg·K.

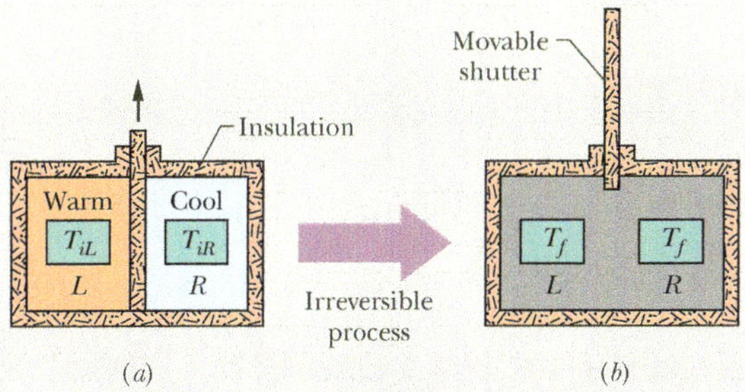

Figure 20-5 (a) In the initial state, two copper blocks L and R, identical except for their temperatures, are in an insulating box and are separated by an insulating shutter. (b) When the shutter is removed, the blocks exchange energy as heat and come to a final state, both with the same temperature T_f.

KEY IDEA

To calculate the entropy change, we must find a reversible process that takes the system from the initial state of Fig. 20-5a to the final state of Fig. 20-5b. We can calculate the net entropy change ΔS_{rev} of the reversible process using Eq. 20-1, and then the entropy change for the irreversible process is equal to ΔS_{rev}.

Calculations: For the reversible process, we need a thermal reservoir whose temperature can be changed slowly (say, by turning a knob). We then take the blocks through the following two steps, illustrated in Fig. 20-6.

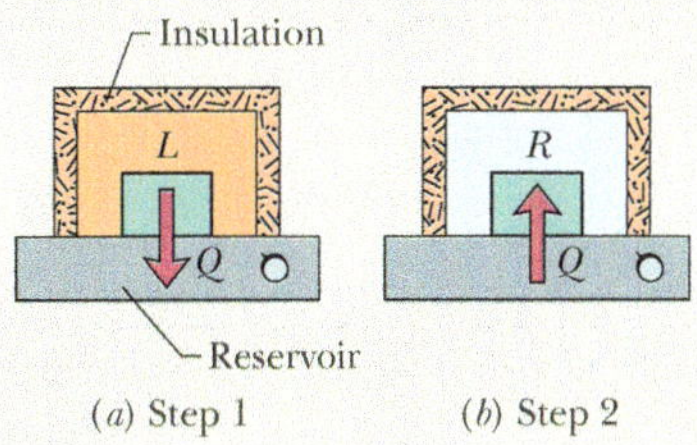

Figure 20-6 The blocks of Fig. 20-5 can proceed from their initial state to their final state in a reversible way if we use a reservoir with a controllable temperature (a) to extract heat reversibly from block L and (b) to add heat reversibly to block R.

Step 1: With the reservoir's temperature set at 60°C, put block L on the reservoir. (Since block and reservoir are at the same temperature, they are already in thermal equilibrium.) Then slowly lower the temperature of the reservoir and the block to 40°C. As the block's temperature changes by each increment dT during this process, energy dQ is transferred as heat *from* the block to the reservoir. Using Eq. 18-14, we can write this transferred energy as $dQ = mc\,dT$, where c is the specific heat of copper. According to Eq. 20-1, the entropy change ΔS_L of block L during the full temperature change from initial temperature T_{iL} (= 60°C = 333 K) to final temperature T_f (= 40°C = 313 K) is

$$\Delta S_L = \int_i^f \frac{dQ}{T} = \int_{T_{iL}}^{T_f} \frac{mc\,dT}{T} = mc \int_{T_{iL}}^{T_f} \frac{dT}{T}$$
$$= mc \ln \frac{T_f}{T_{iL}}.$$

Inserting the given data yields

$$\Delta S_L = (1.5\text{ kg})(386\text{ J/kg}\cdot\text{K}) \ln \frac{313\text{ K}}{333\text{ K}}$$
$$= -35.86\text{ J/K}.$$

Step 2: With the reservoir's temperature now set at 20°C, put block R on the reservoir. Then slowly raise the temperature of the reservoir and the block to 40°C. With the same reasoning used to find ΔS_L, you can show that the entropy change ΔS_R of block R during this process is

$$\Delta S_R = (1.5\text{ kg})(386\text{ J/kg}\cdot\text{K}) \ln \frac{313\text{ K}}{293\text{ K}}$$
$$= +38.23\text{ J/K}.$$

The net entropy change ΔS_{rev} of the two-block system undergoing this two-step reversible process is then

$$\Delta S_{rev} = \Delta S_L + \Delta S_R$$
$$= -35.86\text{ J/K} + 38.23\text{ J/K} = 2.4\text{ J/K}.$$

Thus, the net entropy change ΔS_{irrev} for the two-block system undergoing the actual irreversible process is

$$\Delta S_{irrev} = \Delta S_{rev} = 2.4\text{ J/K}. \quad \text{(Answer)}$$

This result is positive, in accordance with the entropy postulate.

Sample Problem 20.02 Entropy change of a free expansion of a gas

Suppose 1.0 mol of nitrogen gas is confined to the left side of the container of Fig. 20-1*a*. You open the stopcock, and the volume of the gas doubles. What is the entropy change of the gas for this irreversible process? Treat the gas as ideal.

KEY IDEAS

(1) We can determine the entropy change for the irreversible process by calculating it for a reversible process that provides the same change in volume. (2) The temperature of the gas does not change in the free expansion. Thus, the reversible process should be an isothermal expansion—namely, the one of Figs. 20-3 and 20-4.

Calculations: From Table 19-4, the energy Q added as heat to the gas as it expands isothermally at temperature T from an initial volume V_i to a final volume V_f is

$$Q = nRT \ln \frac{V_f}{V_i},$$

in which n is the number of moles of gas present. From Eq. 20-2 the entropy change for this reversible process in which the temperature is held constant is

$$\Delta S_{rev} = \frac{Q}{T} = \frac{nRT \ln(V_f/V_i)}{T} = nR \ln \frac{V_f}{V_i}.$$

Substituting $n = 1.00$ mol and $V_f/V_i = 2$, we find

$$\Delta S_{rev} = nR \ln \frac{V_f}{V_i} = (1.00\text{ mol})(8.31\text{ J/mol}\cdot\text{K})(\ln 2)$$
$$= +5.76\text{ J/K}.$$

Thus, the entropy change for the free expansion (and for all other processes that connect the initial and final states shown in Fig. 20-2) is

$$\Delta S_{irrev} = \Delta S_{rev} = +5.76\text{ J/K}. \quad \text{(Answer)}$$

Because ΔS is positive, the entropy increases, in accordance with the entropy postulate.

WILEY PLUS Additional examples, video, and practice available at *WileyPLUS*

The Second Law of Thermodynamics

Here is a puzzle. In the process of going from (a) to (b) in Fig. 20-3, the entropy change of the gas (our system) is positive. However, because the process is reversible, we can also go from (b) to (a) by, say, gradually adding lead shot to the piston, to restore the initial gas volume. To maintain a constant temperature, we need to remove energy as heat, but that means Q is negative and thus the entropy change is also. Doesn't this entropy decrease violate the entropy postulate: entropy always increases? No, because the postulate holds only for irreversible processes in closed systems. Here, the process is *not* irreverible and the system is *not* closed (because of the energy transferred to and from the reservoir as heat).

However, if we include the reservoir, along with the gas, as part of the system, then we do have a closed system. Let's check the change in entropy of the enlarged system *gas + reservoir* for the process that takes it from (b) to (a) in Fig. 20-3. During this reversible process, energy is transferred as heat from the gas to the reservoir—that is, from one part of the enlarged system to another. Let $|Q|$ represent the absolute value (or magnitude) of this heat. With Eq. 20-2, we can then calculate separately the entropy changes for the gas (which loses $|Q|$) and the reservoir (which gains $|Q|$). We get

$$\Delta S_{\text{gas}} = -\frac{|Q|}{T}$$

and

$$\Delta S_{\text{res}} = +\frac{|Q|}{T}.$$

The entropy change of the closed system is the sum of these two quantities: 0.

With this result, we can modify the entropy postulate to include both reversible and irreversible processes:

> If a process occurs in a *closed* system, the entropy of the system increases for irreversible processes and remains constant for reversible processes. It never decreases.

Although entropy may decrease in part of a closed system, there will always be an equal or larger entropy increase in another part of the system, so that the entropy of the system as a whole never decreases. This fact is one form of the **second law of thermodynamics** and can be written as

$$\Delta S \geq 0 \quad \text{(second law of thermodynamics)}, \tag{20-5}$$

where the greater-than sign applies to irreversible processes and the equals sign to reversible processes. Equation 20-5 applies only to closed systems.

In the real world almost all processes are irreversible to some extent because of friction, turbulence, and other factors, so the entropy of real closed systems undergoing real processes always increases. Processes in which the system's entropy remains constant are always idealizations.

Force Due to Entropy

To understand why rubber resists being stretched, let's write the first law of thermodynamics

$$dE = dQ - dW$$

for a rubber band undergoing a small increase in length dx as we stretch it between our hands. The force from the rubber band has magnitude F, is directed inward, and does work $dW = -F\,dx$ during length increase dx. From Eq. 20-2 ($\Delta S = Q/T$), small changes in Q and S at constant temperature are related by $dS = dQ/T$, or $dQ = T\,dS$. So, now we can rewrite the first law as

$$dE = T\,dS + F\,dx. \tag{20-6}$$

To good approximation, the change dE in the internal energy of rubber is 0 if the total stretch of the rubber band is not very much. Substituting 0 for dE in Eq. 20-6 leads us to an expression for the force from the rubber band:

$$F = -T\frac{dS}{dx}. \tag{20-7}$$

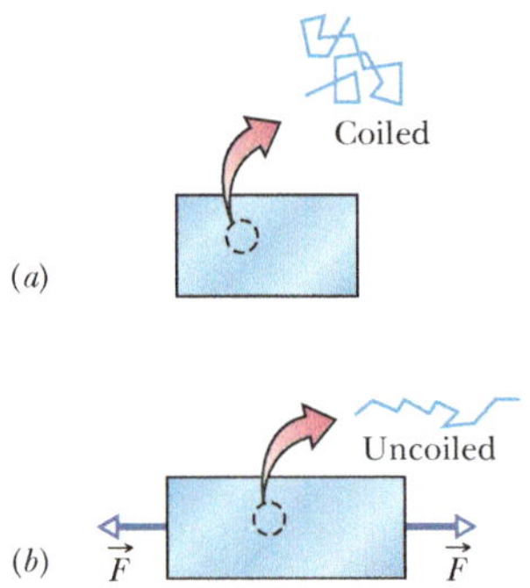

Figure 20-7 A section of a rubber band (*a*) unstretched and (*b*) stretched, and a polymer within it (*a*) coiled and (*b*) uncoiled.

This tells us that F is proportional to the rate dS/dx at which the rubber band's entropy changes during a small change dx in the rubber band's length. Thus, you can *feel* the effect of entropy on your hands as you stretch a rubber band.

To make sense of the relation between force and entropy, let's consider a simple model of the rubber material. Rubber consists of cross-linked polymer chains (long molecules with cross links) that resemble three-dimensional zig-zags (Fig. 20-7). When the rubber band is at its rest length, the polymers are coiled up in a spaghetti-like arrangement. Because of the large disorder of the molecules, this rest state has a high value of entropy. When we stretch a rubber band, we uncoil many of those polymers, aligning them in the direction of stretch. Because the alignment decreases the disorder, the entropy of the stretched rubber band is less. That is, the change dS/dx in Eq. 20-7 is a negative quantity because the entropy decreases with stretching. Thus, the force on our hands from the rubber band is due to the tendency of the polymers to return to their former disordered state and higher value of entropy.

20-2 ENTROPY IN THE REAL WORLD: ENGINES

Learning Objectives

After reading this module, you should be able to . . .

20.09 Identify that a heat engine is a device that extracts energy from its environment in the form of heat and does useful work and that in an *ideal* heat engine, all processes are reversible, with no wasteful energy transfers.

20.10 Sketch a p-V diagram for the cycle of a Carnot engine, indicating the direction of cycling, the nature of the processes involved, the work done during each process (including algebraic sign), the net work done in the cycle, and the heat transferred during each process (including algebraic sign).

20.11 Sketch a Carnot cycle on a temperature–entropy diagram, indicating the heat transfers.

20.12 Determine the net entropy change around a Carnot cycle.

20.13 Calculate the efficiency ε_C of a Carnot engine in terms of the heat transfers and also in terms of the temperatures of the reservoirs.

20.14 Identify that there are no perfect engines in which the energy transferred as heat Q from a high temperature reservoir goes entirely into the work W done by the engine.

20.15 Sketch a p-V diagram for the cycle of a Stirling engine, indicating the direction of cycling, the nature of the processes involved, the work done during each process (including algebraic sign), the net work done in the cycle, and the heat transfers during each process.

Key Ideas

- An engine is a device that, operating in a cycle, extracts energy as heat $|Q_H|$ from a high-temperature reservoir and does a certain amount of work $|W|$. The efficiency ε of any engine is defined as

$$\varepsilon = \frac{\text{energy we get}}{\text{energy we pay for}} = \frac{|W|}{|Q_H|}.$$

- In an ideal engine, all processes are reversible and no wasteful energy transfers occur due to, say, friction and turbulence.
- A Carnot engine is an ideal engine that follows the cycle of Fig. 20-9. Its efficiency is

$$\varepsilon_C = 1 - \frac{|Q_L|}{|Q_H|} = 1 - \frac{T_L}{T_H},$$

in which T_H and T_L are the temperatures of the high- and low-temperature reservoirs, respectively. Real engines always have an efficiency lower than that of a Carnot engine. Ideal engines that are not Carnot engines also have efficiencies lower than that of a Carnot engine.

- A perfect engine is an imaginary engine in which energy extracted as heat from the high-temperature reservoir is converted completely to work. Such an engine would violate the second law of thermodynamics, which can be restated as follows: No series of processes is possible whose sole result is the absorption of energy as heat from a thermal reservoir and the complete conversion of this energy to work.

Entropy in the Real World: Engines

A **heat engine,** or more simply, an **engine,** is a device that extracts energy from its environment in the form of heat and does useful work. At the heart of every engine is a *working substance.* In a steam engine, the working substance is water, in both its vapor and its liquid form. In an automobile engine the working substance is a gasoline–air mixture. If an engine is to do work on a sustained basis, the working substance must operate in a *cycle;* that is, the working substance must pass through a closed series of thermodynamic processes, called *strokes,* returning again and again to each state in its cycle. Let us see what the laws of thermodynamics can tell us about the operation of engines.

A Carnot Engine

We have seen that we can learn much about real gases by analyzing an ideal gas, which obeys the simple law $pV = nRT$. Although an ideal gas does not exist, any real gas approaches ideal behavior if its density is low enough. Similarly, we can study real engines by analyzing the behavior of an **ideal engine.**

In an ideal engine, all processes are reversible and no wasteful energy transfers occur due to, say, friction and turbulence.

We shall focus on a particular ideal engine called a **Carnot engine** after the French scientist and engineer N. L. Sadi Carnot (pronounced "car-no"), who first proposed the engine's concept in 1824. This ideal engine turns out to be the best (in principle) at using energy as heat to do useful work. Surprisingly, Carnot was able to analyze the performance of this engine before the first law of thermodynamics and the concept of entropy had been discovered.

Figure 20-8 shows schematically the operation of a Carnot engine. During each cycle of the engine, the working substance absorbs energy $|Q_H|$ as heat from a thermal reservoir at constant temperature T_H and discharges energy $|Q_L|$ as heat to a second thermal reservoir at a constant lower temperature T_L.

Figure 20-9 shows a p-V plot of the *Carnot cycle*—the cycle followed by the working substance. As indicated by the arrows, the cycle is traversed in the clockwise direction. Imagine the working substance to be a gas, confined to an insulating cylinder with a weighted, movable piston. The cylinder may be placed at will on either of the two thermal reservoirs, as in Fig. 20-6, or on an insulating slab. Figure 20-9*a* shows that, if we place the cylinder in contact with the high-temperature reservoir at temperature T_H, heat $|Q_H|$ is transferred *to* the working substance *from* this reservoir as the gas undergoes an isothermal *expansion* from volume V_a to volume V_b. Similarly, with the working substance in contact with the low-temperature reservoir at temperature T_L, heat $|Q_L|$ is transferred *from*

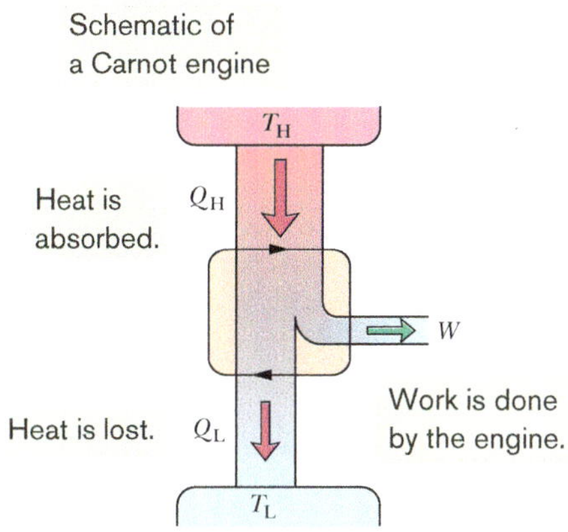

Figure 20-8 The elements of a Carnot engine. The two black arrowheads on the central loop suggest the working substance operating in a cycle, as if on a p-V plot. Energy $|Q_H|$ is transferred as heat from the high-temperature reservoir at temperature T_H to the working substance. Energy $|Q_L|$ is transferred as heat from the working substance to the low-temperature reservoir at temperature T_L. Work W is done by the engine (actually by the working substance) on something in the environment.

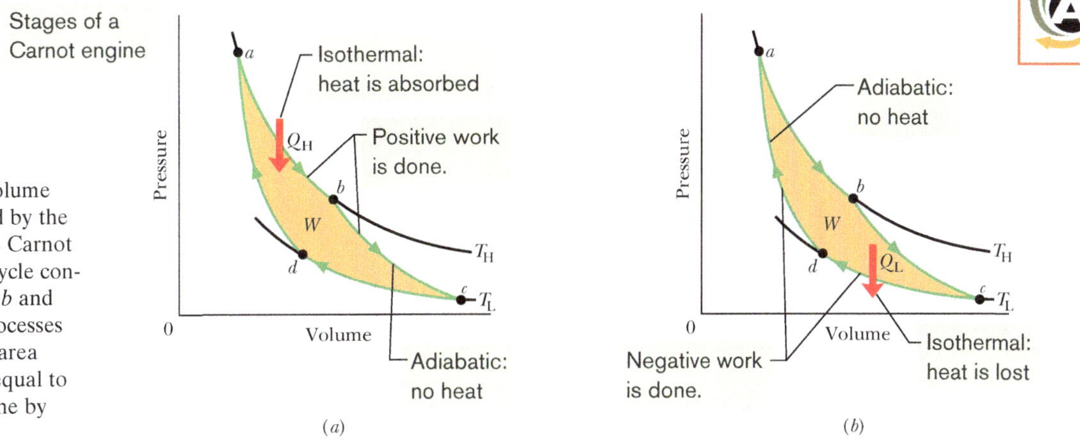

Figure 20-9 A pressure–volume plot of the cycle followed by the working substance of the Carnot engine in Fig. 20-8. The cycle consists of two isothermal (ab and cd) and two adiabatic processes (bc and da). The shaded area enclosed by the cycle is equal to the work W per cycle done by the Carnot engine.

the working substance *to* the low-temperature reservoir as the gas undergoes an isothermal *compression* from volume V_c to volume V_d (Fig. 20-9*b*).

In the engine of Fig. 20-8, we assume that heat transfers to or from the working substance can take place *only* during the isothermal processes *ab* and *cd* of Fig. 20-9. Therefore, processes *bc* and *da* in that figure, which connect the two isotherms at temperatures T_H and T_L, must be (reversible) adiabatic processes; that is, they must be processes in which no energy is transferred as heat. To ensure this, during processes *bc* and *da* the cylinder is placed on an insulating slab as the volume of the working substance is changed.

During the processes *ab* and *bc* of Fig. 20-9*a*, the working substance is expanding and thus doing positive work as it raises the weighted piston. This work is represented in Fig. 20-9*a* by the area under curve *abc*. During the processes *cd* and *da* (Fig. 20-9*b*), the working substance is being compressed, which means that it is doing negative work on its environment or, equivalently, that its environment is doing work on it as the loaded piston descends. This work is represented by the area under curve *cda*. The *net work per cycle,* which is represented by W in both Figs. 20-8 and 20-9, is the difference between these two areas and is a positive quantity equal to the area enclosed by cycle *abcda* in Fig. 20-9. This work W is performed on some outside object, such as a load to be lifted.

Equation 20-1 ($\Delta S = \int dQ/T$) tells us that any energy transfer as heat must involve a change in entropy. To see this for a Carnot engine, we can plot the Carnot cycle on a temperature–entropy (T-S) diagram as in Fig. 20-10. The lettered points *a*, *b*, *c*, and *d* there correspond to the lettered points in the p-V diagram in Fig. 20-9. The two horizontal lines in Fig. 20-10 correspond to the two isothermal processes of the cycle. Process *ab* is the isothermal expansion of the cycle. As the working substance (reversibly) absorbs energy $|Q_H|$ as heat at constant temperature T_H during the expansion, its entropy increases. Similarly, during the isothermal compression *cd*, the working substance (reversibly) loses energy $|Q_L|$ as heat at constant temperature T_L, and its entropy decreases.

The two vertical lines in Fig. 20-10 correspond to the two adiabatic processes of the Carnot cycle. Because no energy is transferred as heat during the two processes, the entropy of the working substance is constant during them.

The Work To calculate the net work done by a Carnot engine during a cycle, let us apply Eq. 18-26, the first law of thermodynamics ($\Delta E_{int} = Q - W$), to the working substance. That substance must return again and again to any arbitrarily selected state in the cycle. Thus, if X represents any state property of the working substance, such as pressure, temperature, volume, internal energy, or entropy, we must have $\Delta X = 0$ for every cycle. It follows that $\Delta E_{int} = 0$ for a complete cycle of the working substance. Recalling that Q in Eq. 18-26 is the *net* heat transfer per cycle and W is the *net* work, we can write the first law of thermodynamics for the Carnot cycle as

$$W = |Q_H| - |Q_L|. \qquad (20\text{-}8)$$

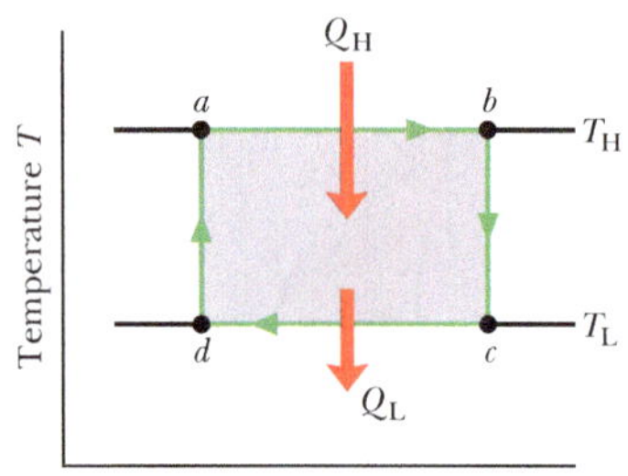

Figure 20-10 The Carnot cycle of Fig. 20-9 plotted on a temperature–entropy diagram. During processes *ab* and *cd* the temperature remains constant. During processes *bc* and *da* the entropy remains constant.

Entropy Changes In a Carnot engine, there are *two* (and only two) reversible energy transfers as heat, and thus two changes in the entropy of the working substance—one at temperature T_H and one at T_L. The net entropy change per cycle is then

$$\Delta S = \Delta S_H + \Delta S_L = \frac{|Q_H|}{T_H} - \frac{|Q_L|}{T_L}. \qquad (20\text{-}9)$$

Here ΔS_H is positive because energy $|Q_H|$ is *added to* the working substance as heat (an increase in entropy) and ΔS_L is negative because energy $|Q_L|$ is *removed from* the working substance as heat (a decrease in entropy). Because entropy is a state function, we must have $\Delta S = 0$ for a complete cycle. Putting $\Delta S = 0$ in Eq. 20-9 requires that

$$\frac{|Q_H|}{T_H} = \frac{|Q_L|}{T_L}. \qquad (20\text{-}10)$$

Note that, because $T_H > T_L$, we must have $|Q_H| > |Q_L|$; that is, more energy is extracted as heat from the high-temperature reservoir than is delivered to the low-temperature reservoir.

We shall now derive an expression for the efficiency of a Carnot engine.

Efficiency of a Carnot Engine

The purpose of any engine is to transform as much of the extracted energy Q_H into work as possible. We measure its success in doing so by its **thermal efficiency** ε, defined as the work the engine does per cycle ("energy we get") divided by the energy it absorbs as heat per cycle ("energy we pay for"):

$$\varepsilon = \frac{\text{energy we get}}{\text{energy we pay for}} = \frac{|W|}{|Q_H|} \quad \text{(efficiency, any engine).} \tag{20-11}$$

For any ideal engine we substitute for W from Eq. 20-8 to write Eq. 20-11 as

$$\varepsilon = \frac{|Q_H| - |Q_L|}{Q_H} = 1 - \frac{|Q_L|}{|Q_H|}. \tag{20-12}$$

Using Eq. 20-10 for a Carnot engine, we can write this as

$$\varepsilon_C = 1 - \frac{T_L}{T_H} \quad \text{(efficiency, Carnot engine),} \tag{20-13}$$

where the temperatures T_L and T_H are in kelvins. Because $T_L < T_H$, the Carnot engine necessarily has a thermal efficiency less than unity—that is, less than 100%. This is indicated in Fig. 20-8, which shows that only part of the energy extracted as heat from the high-temperature reservoir is available to do work, and the rest is delivered to the low-temperature reservoir. We shall show in Module 20-3 that no real engine can have a thermal efficiency greater than that calculated from Eq. 20-13.

Inventors continually try to improve engine efficiency by reducing the energy $|Q_L|$ that is "thrown away" during each cycle. The inventor's dream is to produce the *perfect engine,* diagrammed in Fig. 20-11, in which $|Q_L|$ is reduced to zero and $|Q_H|$ is converted completely into work. Such an engine on an ocean liner, for example, could extract energy as heat from the water and use it to drive the propellers, with no fuel cost. An automobile fitted with such an engine could extract energy as heat from the surrounding air and use it to drive the car, again with no fuel cost. Alas, a perfect engine is only a dream: Inspection of Eq. 20-13 shows that we can achieve 100% engine efficiency (that is, $\varepsilon = 1$) only if $T_L = 0$ or $T_H \to \infty$, impossible requirements. Instead, experience gives the following alternative version of the second law of thermodynamics, which says in short, *there are no perfect engines:*

> No series of processes is possible whose sole result is the transfer of energy as heat from a thermal reservoir and the complete conversion of this energy to work.

To summarize: The thermal efficiency given by Eq. 20-13 applies only to Carnot engines. Real engines, in which the processes that form the engine cycle are not reversible, have lower efficiencies. If your car were powered by a Carnot engine, it would have an efficiency of about 55% according to Eq. 20-13; its actual efficiency is probably about 25%. A nuclear power plant (Fig. 20-12), taken in its entirety, is an engine. It extracts energy as heat from a reactor core, does work by means of a turbine, and discharges energy as heat to a nearby river. If the power plant operated as a Carnot engine, its efficiency would be about 40%; its actual efficiency is about 30%. In designing engines of any type, there is simply no way to beat the efficiency limitation imposed by Eq. 20-13.

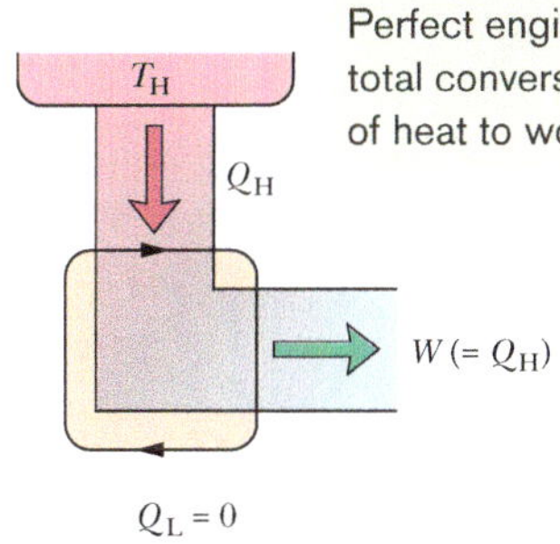

Figure 20-11 The elements of a perfect engine—that is, one that converts heat Q_H from a high-temperature reservoir directly to work W with 100% efficiency.

© Richard Ustinich

Figure 20-12 The North Anna nuclear power plant near Charlottesville, Virginia, which generates electric energy at the rate of 900 MW. At the same time, by design, it discards energy into the nearby river at the rate of 2100 MW. This plant and all others like it throw away more energy than they deliver in useful form. They are real counterparts of the ideal engine of Fig. 20-8.

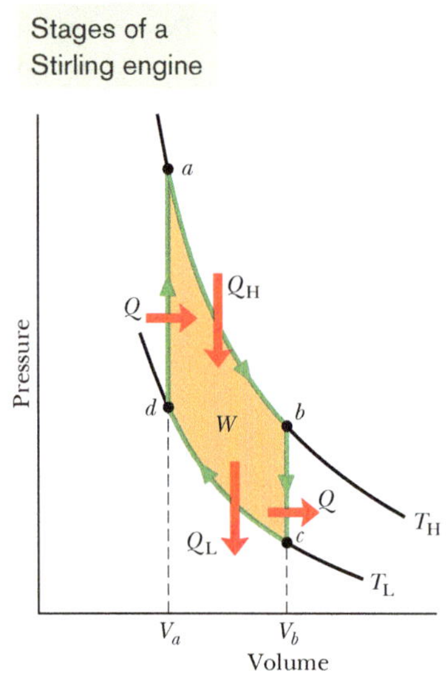

Figure 20-13 A p-V plot for the working substance of an ideal Stirling engine, with the working substance assumed for convenience to be an ideal gas.

Stirling Engine

Equation 20-13 applies not to all ideal engines but only to those that can be represented as in Fig. 20-9—that is, to Carnot engines. For example, Fig. 20-13 shows the operating cycle of an ideal **Stirling engine.** Comparison with the Carnot cycle of Fig. 20-9 shows that each engine has isothermal heat transfers at temperatures T_H and T_L. However, the two isotherms of the Stirling engine cycle are connected, not by adiabatic processes as for the Carnot engine but by constant-volume processes. To increase the temperature of a gas at constant volume reversibly from T_L to T_H (process *da* of Fig. 20-13) requires a transfer of energy as heat to the working substance from a thermal reservoir whose temperature can be varied smoothly between those limits. Also, a reverse transfer is required in process *bc*. Thus, reversible heat transfers (and corresponding entropy changes) occur in all four of the processes that form the cycle of a Stirling engine, not just two processes as in a Carnot engine. Thus, the derivation that led to Eq. 20-13 does not apply to an ideal Stirling engine. More important, the efficiency of an ideal Stirling engine is lower than that of a Carnot engine operating between the same two temperatures. Real Stirling engines have even lower efficiencies.

The Stirling engine was developed in 1816 by Robert Stirling. This engine, long neglected, is now being developed for use in automobiles and spacecraft. A Stirling engine delivering 5000 hp (3.7 MW) has been built. Because they are quiet, Stirling engines are used on some military submarines.

Checkpoint 3

Three Carnot engines operate between reservoir temperatures of (a) 400 and 500 K, (b) 600 and 800 K, and (c) 400 and 600 K. Rank the engines according to their thermal efficiencies, greatest first.

Sample Problem 20.03 Carnot engine, efficiency, power, entropy changes

Imagine a Carnot engine that operates between the temperatures $T_H = 850$ K and $T_L = 300$ K. The engine performs 1200 J of work each cycle, which takes 0.25 s.

(a) What is the efficiency of this engine?

KEY IDEA

The efficiency ε of a Carnot engine depends only on the ratio T_L/T_H of the temperatures (in kelvins) of the thermal reservoirs to which it is connected.

Calculation: Thus, from Eq. 20-13, we have

$$\varepsilon = 1 - \frac{T_L}{T_H} = 1 - \frac{300 \text{ K}}{850 \text{ K}} = 0.647 \approx 65\%. \quad \text{(Answer)}$$

(b) What is the average power of this engine?

KEY IDEA

The average power P of an engine is the ratio of the work W it does per cycle to the time t that each cycle takes.

Calculation: For this Carnot engine, we find

$$P = \frac{W}{t} = \frac{1200 \text{ J}}{0.25 \text{ s}} = 4800 \text{ W} = 4.8 \text{ kW}. \quad \text{(Answer)}$$

(c) How much energy $|Q_H|$ is extracted as heat from the high-temperature reservoir every cycle?

KEY IDEA

The efficiency ε is the ratio of the work W that is done per cycle to the energy $|Q_H|$ that is extracted as heat from the high-temperature reservoir per cycle ($\varepsilon = W/|Q_H|$).

Calculation: Here we have

$$|Q_H| = \frac{W}{\varepsilon} = \frac{1200 \text{ J}}{0.647} = 1855 \text{ J}. \quad \text{(Answer)}$$

(d) How much energy $|Q_L|$ is delivered as heat to the low-temperature reservoir every cycle?

KEY IDEA

For a Carnot engine, the work W done per cycle is equal to the difference in the energy transfers as heat: $|Q_H| - |Q_L|$, as in Eq. 20-8.

Calculation: Thus, we have

$$|Q_L| = |Q_H| - W$$
$$= 1855 \text{ J} - 1200 \text{ J} = 655 \text{ J}. \quad \text{(Answer)}$$

(e) By how much does the entropy of the working substance change as a result of the energy transferred to it from the high-temperature reservoir? From it to the low-temperature reservoir?

KEY IDEA

The entropy change ΔS during a transfer of energy as heat Q at constant temperature T is given by Eq. 20-2 ($\Delta S = Q/T$).

Calculations: Thus, for the *positive* transfer of energy Q_H from the high-temperature reservoir at T_H, the change in the entropy of the working substance is

$$\Delta S_H = \frac{Q_H}{T_H} = \frac{1855\text{ J}}{850\text{ K}} = +2.18\text{ J/K}. \quad \text{(Answer)}$$

Similarly, for the *negative* transfer of energy Q_L to the low-temperature reservoir at T_L, we have

$$\Delta S_L = \frac{Q_L}{T_L} = \frac{-655\text{ J}}{300\text{ K}} = -2.18\text{ J/K}. \quad \text{(Answer)}$$

Note that the net entropy change of the working substance for one cycle is zero, as we discussed in deriving Eq. 20-10.

Sample Problem 20.04 Impossibly efficient engine

An inventor claims to have constructed an engine that has an efficiency of 75% when operated between the boiling and freezing points of water. Is this possible?

KEY IDEA

The efficiency of a real engine must be less than the efficiency of a Carnot engine operating between the same two temperatures.

Calculation: From Eq. 20-13, we find that the efficiency of a Carnot engine operating between the boiling and freezing points of water is

$$\varepsilon = 1 - \frac{T_L}{T_H} = 1 - \frac{(0 + 273)\text{ K}}{(100 + 273)\text{ K}} = 0.268 \approx 27\%.$$

Thus, for the given temperatures, the claimed efficiency of 75% for a real engine (with its irreversible processes and wasteful energy transfers) is impossible.

WILEY PLUS Additional examples, video, and practice available at *WileyPLUS*

20-3 REFRIGERATORS AND REAL ENGINES

Learning Objectives

After reading this module, you should be able to . . .

20.16 Identify that a refrigerator is a device that uses work to transfer energy from a low-temperature reservoir to a high-temperature reservoir, and that an ideal refrigerator is one that does this with reversible processes and no wasteful losses.

20.17 Sketch a p-V diagram for the cycle of a Carnot refrigerator, indicating the direction of cycling, the nature of the processes involved, the work done during each process (including algebraic sign), the net work done in the cycle, and the heat transferred during each process (including algebraic sign).

20.18 Apply the relationship between the coefficient of performance K and the heat exchanges with the reservoirs and the temperatures of the reservoirs.

20.19 Identify that there is no ideal refrigerator in which all of the energy extracted from the low-temperature reservoir is transferred to the high-temperature reservoir.

20.20 Identify that the efficiency of a real engine is less than that of the ideal Carnot engine.

Key Ideas

- A refrigerator is a device that, operating in a cycle, has work W done on it as it extracts energy $|Q_L|$ as heat from a low-temperature reservoir. The coefficient of performance K of a refrigerator is defined as

$$K = \frac{\text{what we want}}{\text{what we pay for}} = \frac{|Q_L|}{|W|}.$$

- A Carnot refrigerator is a Carnot engine operating in reverse. Its coefficient of performance is

$$K_C = \frac{|Q_L|}{|Q_H| - |Q_L|} = \frac{T_L}{T_H - T_L}.$$

- A perfect refrigerator is an entirely imaginary refrigerator in which energy extracted as heat from the low-temperature reservoir is somehow converted completely to heat discharged to the high-temperature reservoir without any need for work.

- A perfect refrigerator would violate the second law of thermodynamics, which can be restated as follows: No series of processes is possible whose sole result is the transfer of energy as heat from a reservoir at a given temperature to a reservoir at a higher temperature (without work being involved).

Entropy in the Real World: Refrigerators

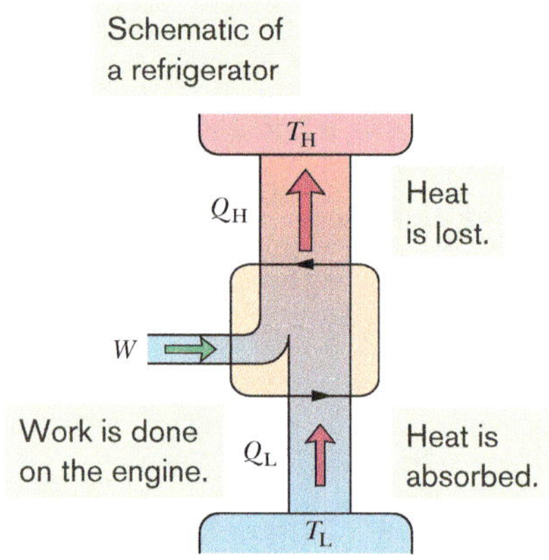

Figure 20-14 The elements of a Carnot refrigerator. The two black arrowheads on the central loop suggest the working substance operating in a cycle, as if on a p-V plot. Energy is transferred as heat Q_L to the working substance from the low-temperature reservoir. Energy is transferred as heat Q_H to the high-temperature reservoir from the working substance. Work W is done on the refrigerator (on the working substance) by something in the environment.

A **refrigerator** is a device that uses work in order to transfer energy from a low-temperature reservoir to a high-temperature reservoir as the device continuously repeats a set series of thermodynamic processes. In a household refrigerator, for example, work is done by an electrical compressor to transfer energy from the food storage compartment (a low-temperature reservoir) to the room (a high-temperature reservoir).

Air conditioners and heat pumps are also refrigerators. For an air conditioner, the low-temperature reservoir is the room that is to be cooled and the high-temperature reservoir is the warmer outdoors. A heat pump is an air conditioner that can be operated in reverse to heat a room; the room is the high-temperature reservoir, and heat is transferred to it from the cooler outdoors.

Let us consider an *ideal refrigerator:*

> In an ideal refrigerator, all processes are reversible and no wasteful energy transfers occur as a result of, say, friction and turbulence.

Figure 20-14 shows the basic elements of an ideal refrigerator. Note that its operation is the reverse of how the Carnot engine of Fig. 20-8 operates. In other words, all the energy transfers, as either heat or work, are reversed from those of a Carnot engine. We can call such an ideal refrigerator a **Carnot refrigerator.**

The designer of a refrigerator would like to extract as much energy $|Q_L|$ as possible from the low-temperature reservoir (what we want) for the least amount of work $|W|$ (what we pay for). A measure of the efficiency of a refrigerator, then, is

$$K = \frac{\text{what we want}}{\text{what we pay for}} = \frac{|Q_L|}{|W|} \quad \text{(coefficient of performance, any refrigerator),} \tag{20-14}$$

where K is called the *coefficient of performance.* For any ideal refrigerator, the first law of thermodynamics gives $|W| = |Q_H| - |Q_L|$, where $|Q_H|$ is the magnitude of the energy transferred as heat to the high-temperature reservoir. Equation 20-14 then becomes

$$K = \frac{|Q_L|}{|Q_H| - |Q_L|}. \tag{20-15}$$

Because a Carnot refrigerator is a Carnot engine operating in reverse, we can combine Eq. 20-10 with Eq. 20-15; after some algebra we find

$$K_C = \frac{T_L}{T_H - T_L} \quad \text{(coefficient of performance, Carnot refrigerator).} \tag{20-16}$$

For typical room air conditioners, $K \approx 2.5$. For household refrigerators, $K \approx 5$. Perversely, the value of K is higher the closer the temperatures of the two reservoirs are to each other. That is why heat pumps are more effective in temperate climates than in very cold climates.

It would be nice to own a refrigerator that did not require some input of work—that is, one that would run without being plugged in. Figure 20-15 represents another "inventor's dream," a *perfect refrigerator* that transfers energy as heat Q from a cold reservoir to a warm reservoir without the need for work. Because the unit operates in cycles, the entropy of the working substance does not change during a complete cycle. The entropies of the two reservoirs, however, do change: The entropy change for the cold reservoir is $-|Q|/T_L$, and that for the warm reservoir is $+|Q|/T_H$. Thus, the net entropy change for the entire system is

$$\Delta S = -\frac{|Q|}{T_L} + \frac{|Q|}{T_H}.$$

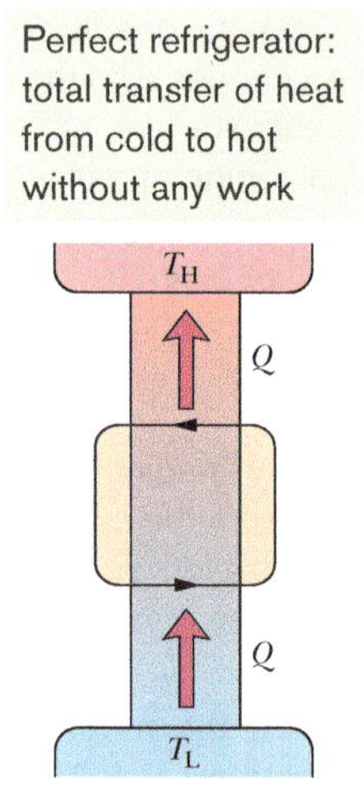

Figure 20-15 The elements of a perfect refrigerator—that is, one that transfers energy from a low-temperature reservoir to a high-temperature reservoir without any input of work.

Because $T_H > T_L$, the right side of this equation is negative and thus the net change in entropy per cycle for the closed system *refrigerator + reservoirs* is also negative. Because such a decrease in entropy violates the second law of thermodynamics (Eq. 20-5), a perfect refrigerator does not exist. (If you want your refrigerator to operate, you must plug it in.)

Here, then, is another way to state the second law of thermodynamics:

> No series of processes is possible whose sole result is the transfer of energy as heat from a reservoir at a given temperature to a reservoir at a higher temperature.

In short, *there are no perfect refrigerators.*

Checkpoint 4

You wish to increase the coefficient of performance of an ideal refrigerator. You can do so by (a) running the cold chamber at a slightly higher temperature, (b) running the cold chamber at a slightly lower temperature, (c) moving the unit to a slightly warmer room, or (d) moving it to a slightly cooler room. The magnitudes of the temperature changes are to be the same in all four cases. List the changes according to the resulting coefficients of performance, greatest first.

The Efficiencies of Real Engines

Let ε_C be the efficiency of a Carnot engine operating between two given temperatures. Here we prove that no real engine operating between those temperatures can have an efficiency greater than ε_C. If it could, the engine would violate the second law of thermodynamics.

Let us assume that an inventor, working in her garage, has constructed an engine X, which she claims has an efficiency ε_X that is greater than ε_C:

$$\varepsilon_X > \varepsilon_C \quad \text{(a claim).} \tag{20-17}$$

Let us couple engine X to a Carnot refrigerator, as in Fig. 20-16*a*. We adjust the strokes of the Carnot refrigerator so that the work it requires per cycle is just equal to that provided by engine X. Thus, no (external) work is performed on or by the combination *engine + refrigerator* of Fig. 20-16*a*, which we take as our system.

If Eq. 20-17 is true, from the definition of efficiency (Eq. 20-11), we must have

$$\frac{|W|}{|Q'_H|} > \frac{|W|}{|Q_H|},$$

where the prime refers to engine X and the right side of the inequality is the efficiency of the Carnot refrigerator when it operates as an engine. This inequality requires that

$$|Q_H| > |Q'_H|. \tag{20-18}$$

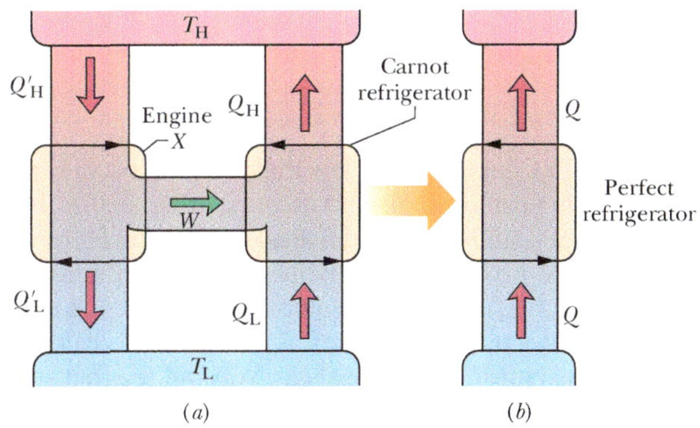

Figure 20-16 (*a*) Engine X drives a Carnot refrigerator. (*b*) If, as claimed, engine X is more efficient than a Carnot engine, then the combination shown in (*a*) is equivalent to the perfect refrigerator shown here. This violates the second law of thermodynamics, so we conclude that engine X *cannot* be more efficient than a Carnot engine.

Because the work done by engine X is equal to the work done on the Carnot refrigerator, we have, from the first law of thermodynamics as given by Eq. 20-8,

$$|Q_H| - |Q_L| = |Q'_H| - |Q'_L|,$$

which we can write as

$$|Q_H| - |Q'_H| = |Q_L| - |Q'_L| = Q. \quad (20\text{-}19)$$

Because of Eq. 20-18, the quantity Q in Eq. 20-19 must be positive.

Comparison of Eq. 20-19 with Fig. 20-16 shows that the net effect of engine X and the Carnot refrigerator working in combination is to transfer energy Q as heat from a low-temperature reservoir to a high-temperature reservoir without the requirement of work. Thus, the combination acts like the perfect refrigerator of Fig. 20-15, whose existence is a violation of the second law of thermodynamics.

Something must be wrong with one or more of our assumptions, and it can only be Eq. 20-17. We conclude that *no real engine can have an efficiency greater than that of a Carnot engine when both engines work between the same two temperatures.* At most, the real engine can have an efficiency equal to that of a Carnot engine. In that case, the real engine *is* a Carnot engine.

20-4 A STATISTICAL VIEW OF ENTROPY

Learning Objectives

After reading this module, you should be able to . . .

20.21 Explain what is meant by the configurations of a system of molecules.

20.22 Calculate the multiplicity of a given configuration.

20.23 Identify that all microstates are equally probable but the configurations with more microstates are more probable than the other configurations.

20.24 Apply Boltzmann's entropy equation to calculate the entropy associated with a multiplicity.

Key Ideas

- The entropy of a system can be defined in terms of the possible distributions of its molecules. For identical molecules, each possible distribution of molecules is called a microstate of the system. All equivalent microstates are grouped into a configuration of the system. The number of microstates in a configuration is the multiplicity W of the configuration.
- For a system of N molecules that may be distributed between the two halves of a box, the multiplicity is given by

$$W = \frac{N!}{n_1!\, n_2!},$$

in which n_1 is the number of molecules in one half of the box and n_2 is the number in the other half. A basic assumption of statistical mechanics is that all the microstates are equally probable. Thus, configurations with a large multiplicity occur most often. When N is very large (say, $N = 10^{22}$ molecules or more), the molecules are nearly always in the configuration in which $n_1 = n_2$.

- The multiplicity W of a configuration of a system and the entropy S of the system in that configuration are related by Boltzmann's entropy equation:

$$S = k \ln W,$$

where $k = 1.38 \times 10^{-23}$ J/K is the Boltzmann constant.

- When N is very large (the usual case), we can approximate $\ln N!$ with Stirling's approximation:

$$\ln N! \approx N(\ln N) - N.$$

A Statistical View of Entropy

In Chapter 19 we saw that the macroscopic properties of gases can be explained in terms of their microscopic, or molecular, behavior. Such explanations are part of a study called **statistical mechanics.** Here we shall focus our attention on a single problem, one involving the distribution of gas molecules between the two halves of an insulated box. This problem is reasonably simple to analyze, and it allows us to use statistical mechanics to calculate the entropy change for the free expansion of an ideal gas. You will see that statistical mechanics leads to the same entropy change as we would find using thermodynamics.

Figure 20-17 shows a box that contains six identical (and thus indistinguishable) molecules of a gas. At any instant, a given molecule will be in either the left or the right half of the box; because the two halves have equal volumes, the molecule has the same likelihood, or probability, of being in either half.

Table 20-1 shows the seven possible *configurations* of the six molecules, each configuration labeled with a Roman numeral. For example, in configuration I, all six molecules are in the left half of the box ($n_1 = 6$) and none are in the right half ($n_2 = 0$). We see that, in general, a given configuration can be achieved in a number of different ways. We call these different arrangements of the molecules *microstates*. Let us see how to calculate the number of microstates that correspond to a given configuration.

Suppose we have N molecules, distributed with n_1 molecules in one half of the box and n_2 in the other. (Thus $n_1 + n_2 = N$.) Let us imagine that we distribute the molecules "by hand," one at a time. If $N = 6$, we can select the first molecule in six independent ways; that is, we can pick any one of the six molecules. We can pick the second molecule in five ways, by picking any one of the remaining five molecules; and so on. The total number of ways in which we can select all six molecules is the product of these independent ways, or $6 \times 5 \times 4 \times 3 \times 2 \times 1 = 720$. In mathematical shorthand we write this product as $6! = 720$, where 6! is pronounced "six factorial." Your hand calculator can probably calculate factorials. For later use you will need to know that $0! = 1$. (Check this on your calculator.)

However, because the molecules are indistinguishable, these 720 arrangements are not all different. In the case that $n_1 = 4$ and $n_2 = 2$ (which is configuration III in Table 20-1), for example, the order in which you put four molecules in one half of the box does not matter, because after you have put all four in, there is no way that you can tell the order in which you did so. The number of ways in which you can order the four molecules is $4! = 24$. Similarly, the number of ways in which you can order two molecules for the other half of the box is simply $2! = 2$. To get the number of *different* arrangements that lead to the (4, 2) split of configuration III, we must divide 720 by 24 and also by 2. We call the resulting quantity, which is the number of microstates that correspond to a given configuration, the *multiplicity* W of that configuration. Thus, for configuration III,

$$W_{\text{III}} = \frac{6!}{4!\,2!} = \frac{720}{24 \times 2} = 15.$$

Thus, Table 20-1 tells us there are 15 independent microstates that correspond to configuration III. Note that, as the table also tells us, the total number of microstates for six molecules distributed over the seven configurations is 64.

Extrapolating from six molecules to the general case of N molecules, we have

$$W = \frac{N!}{n_1!\,n_2!} \quad \text{(multiplicity of configuration)}. \tag{20-20}$$

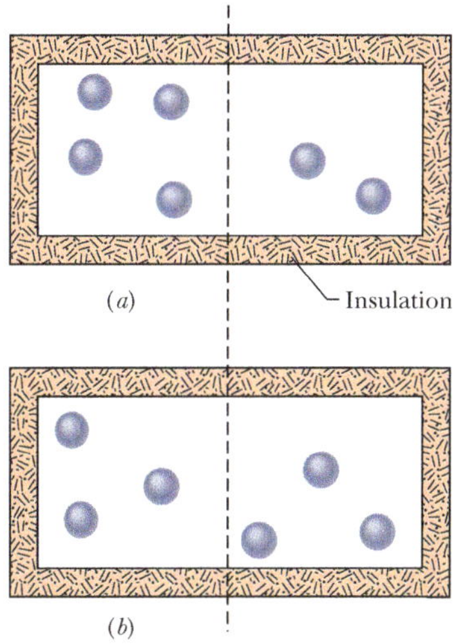

Figure 20-17 An insulated box contains six gas molecules. Each molecule has the same probability of being in the left half of the box as in the right half. The arrangement in (*a*) corresponds to configuration III in Table 20-1, and that in (*b*) corresponds to configuration IV.

Table 20-1 Six Molecules in a Box

Configuration Label	n_1	n_2	Multiplicity W (number of microstates)	Calculation of W (Eq. 20-20)	Entropy 10^{-23} J/K (Eq. 20-21)
I	6	0	1	6!/(6! 0!) = 1	0
II	5	1	6	6!/(5! 1!) = 6	2.47
III	4	2	15	6!/(4! 2!) = 15	3.74
IV	3	3	20	6!/(3! 3!) = 20	4.13
V	2	4	15	6!/(2! 4!) = 15	3.74
VI	1	5	6	6!/(1! 5!) = 6	2.47
VII	0	6	1	6!/(0! 6!) = 1	0
			Total = 64		

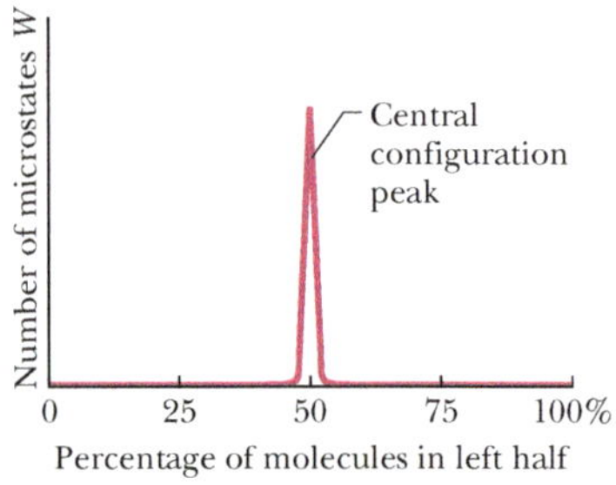

Figure 20-18 For a *large* number of molecules in a box, a plot of the number of microstates that require various percentages of the molecules to be in the left half of the box. Nearly all the microstates correspond to an approximately equal sharing of the molecules between the two halves of the box; those microstates form the *central configuration peak* on the plot. For $N \approx 10^{22}$, the central configuration peak is much too narrow to be drawn on this plot.

You should verify the multiplicities for all the configurations in Table 20-1.

The basic assumption of statistical mechanics is that

All microstates are equally probable.

In other words, if we were to take a great many snapshots of the six molecules as they jostle around in the box of Fig. 20-17 and then count the number of times each microstate occurred, we would find that all 64 microstates would occur equally often. Thus the system will spend, on average, the same amount of time in each of the 64 microstates.

Because all microstates are equally probable but different configurations have different numbers of microstates, the configurations are *not* all equally probable. In Table 20-1 configuration IV, with 20 microstates, is the *most probable configuration,* with a probability of $20/64 = 0.313$. This result means that the system is in configuration IV 31.3% of the time. Configurations I and VII, in which all the molecules are in one half of the box, are the least probable, each with a probability of $1/64 = 0.016$ or 1.6%. It is not surprising that the most probable configuration is the one in which the molecules are evenly divided between the two halves of the box, because that is what we expect at thermal equilibrium. However, it *is* surprising that there is *any* probability, however small, of finding all six molecules clustered in half of the box, with the other half empty.

For large values of N there are extremely large numbers of microstates, but nearly all the microstates belong to the configuration in which the molecules are divided equally between the two halves of the box, as Fig. 20-18 indicates. Even though the measured temperature and pressure of the gas remain constant, the gas is churning away endlessly as its molecules "visit" all probable microstates with equal probability. However, because so few microstates lie outside the very narrow central configuration peak of Fig. 20-18, we might as well assume that the gas molecules are always divided equally between the two halves of the box. As we shall see, this is the configuration with the greatest entropy.

Sample Problem 20.05 Microstates and multiplicity

Suppose that there are 100 indistinguishable molecules in the box of Fig. 20-17. How many microstates are associated with the configuration $n_1 = 50$ and $n_2 = 50$, and with the configuration $n_1 = 100$ and $n_2 = 0$? Interpret the results in terms of the relative probabilities of the two configurations.

KEY IDEA

The multiplicity W of a configuration of indistinguishable molecules in a closed box is the number of independent microstates with that configuration, as given by Eq. 20-20.

Calculations: Thus, for the (n_1, n_2) configuration (50, 50),

$$W = \frac{N!}{n_1!\,n_2!} = \frac{100!}{50!\,50!} = \frac{9.33 \times 10^{157}}{(3.04 \times 10^{64})(3.04 \times 10^{64})} = 1.01 \times 10^{29}. \quad \text{(Answer)}$$

Similarly, for the configuration (100, 0), we have

$$W = \frac{N!}{n_1!\,n_2!} = \frac{100!}{100!\,0!} = \frac{1}{0!} = \frac{1}{1} = 1. \quad \text{(Answer)}$$

The meaning: Thus, a 50–50 distribution is more likely than a 100–0 distribution by the enormous factor of about 1×10^{29}. If you could count, at one per nanosecond, the number of microstates that correspond to the 50–50 distribution, it would take you about 3×10^{12} years, which is about 200 times longer than the age of the universe. Keep in mind that the 100 molecules used in this sample problem is a very small number. Imagine what these calculated probabilities would be like for a mole of molecules, say about $N = 10^{24}$. Thus, you need never worry about suddenly finding all the air molecules clustering in one corner of your room, with you gasping for air in another corner. So, you can breathe easy because of the physics of entropy.

WILEY PLUS Additional examples, video, and practice available at *WileyPLUS*

Probability and Entropy

In 1877, Austrian physicist Ludwig Boltzmann (the Boltzmann of Boltzmann's constant k) derived a relationship between the entropy S of a configuration of a gas and the multiplicity W of that configuration. That relationship is

$$S = k \ln W \quad \text{(Boltzmann's entropy equation).} \qquad (20\text{-}21)$$

This famous formula is engraved on Boltzmann's tombstone.

It is natural that S and W should be related by a logarithmic function. The total entropy of two systems is the *sum* of their separate entropies. The probability of occurrence of two independent systems is the *product* of their separate probabilities. Because $\ln ab = \ln a + \ln b$, the logarithm seems the logical way to connect these quantities.

Table 20-1 displays the entropies of the configurations of the six-molecule system of Fig. 20-17, computed using Eq. 20-21. Configuration IV, which has the greatest multiplicity, also has the greatest entropy.

When you use Eq. 20-20 to calculate W, your calculator may signal "OVERFLOW" if you try to find the factorial of a number greater than a few hundred. Instead, you can use **Stirling's approximation** for $\ln N!$:

$$\ln N! \approx N(\ln N) - N \quad \text{(Stirling's approximation).} \qquad (20\text{-}22)$$

The Stirling of this approximation was an English mathematician and not the Robert Stirling of engine fame.

Checkpoint 5

A box contains 1 mol of a gas. Consider two configurations: (a) each half of the box contains half the molecules and (b) each third of the box contains one-third of the molecules. Which configuration has more microstates?

Sample Problem 20.06 Entropy change of free expansion using microstates

In Sample Problem 20.01, we showed that when n moles of an ideal gas doubles its volume in a free expansion, the entropy increase from the initial state i to the final state f is $S_f - S_i = nR \ln 2$. Derive this increase in entropy by using statistical mechanics.

KEY IDEA

We can relate the entropy S of any given configuration of the molecules in the gas to the multiplicity W of microstates for that configuration, using Eq. 20-21 ($S = k \ln W$).

Calculations: We are interested in two configurations: the final configuration f (with the molecules occupying the full volume of their container in Fig. 20-1b) and the initial configuration i (with the molecules occupying the left half of the container). Because the molecules are in a closed container, we can calculate the multiplicity W of their microstates with Eq. 20-20. Here we have N molecules in the n moles of the gas. Initially, with the molecules all in the left half of the container, their (n_1, n_2) configuration is $(N, 0)$. Then, Eq. 20-20 gives their multiplicity as

$$W_i = \frac{N!}{N!\,0!} = 1.$$

Finally, with the molecules spread through the full volume, their (n_1, n_2) configuration is $(N/2, N/2)$. Then, Eq. 20-20 gives their multiplicity as

$$W_f = \frac{N!}{(N/2)!\,(N/2)!}.$$

From Eq. 20-21, the initial and final entropies are

$$S_i = k \ln W_i = k \ln 1 = 0$$

and

$$S_f = k \ln W_f = k \ln(N!) - 2k \ln[(N/2)!]. \qquad (20\text{-}23)$$

In writing Eq. 20-23, we have used the relation

$$\ln \frac{a}{b^2} = \ln a - 2 \ln b.$$

Now, applying Eq. 20-22 to evaluate Eq. 20-23, we find that

$$\begin{aligned} S_f &= k \ln(N!) - 2k \ln[(N/2)!] \\ &= k[N(\ln N) - N] - 2k[(N/2)\ln(N/2) - (N/2)] \\ &= k[N(\ln N) - N - N\ln(N/2) + N] \\ &= k[N(\ln N) - N(\ln N - \ln 2)] = Nk \ln 2. \end{aligned} \quad (20\text{-}24)$$

From Eq. 19-8 we can substitute nR for Nk, where R is the universal gas constant. Equation 20-24 then becomes

$$S_f = nR \ln 2.$$

The change in entropy from the initial state to the final is thus

$$\begin{aligned} S_f - S_i &= nR \ln 2 - 0 \\ &= nR \ln 2, \end{aligned} \quad \text{(Answer)}$$

which is what we set out to show. In the first sample problem of this chapter we calculated this entropy increase for a free expansion with thermodynamics by finding an equivalent reversible process and calculating the entropy change for *that* process in terms of temperature and heat transfer. In this sample problem, we calculate the same increase in entropy with statistical mechanics using the fact that the system consists of molecules. In short, the two, very different approaches give the same answer.

WILEY PLUS Additional examples, video, and practice available at *WileyPLUS*

Review & Summary

One-Way Processes An **irreversible process** is one that cannot be reversed by means of small changes in the environment. The direction in which an irreversible process proceeds is set by the *change in entropy* ΔS of the system undergoing the process. Entropy S is a *state property* (or *state function*) of the system; that is, it depends only on the state of the system and not on the way in which the system reached that state. The *entropy postulate* states (in part): *If an irreversible process occurs in a closed system, the entropy of the system always increases.*

Calculating Entropy Change The **entropy change** ΔS for an irreversible process that takes a system from an initial state i to a final state f is exactly equal to the entropy change ΔS for *any reversible process* that takes the system between those same two states. We can compute the latter (but not the former) with

$$\Delta S = S_f - S_i = \int_i^f \frac{dQ}{T}. \quad (20\text{-}1)$$

Here Q is the energy transferred as heat to or from the system during the process, and T is the temperature of the system in kelvins during the process.

For a reversible isothermal process, Eq. 20-1 reduces to

$$\Delta S = S_f - S_i = \frac{Q}{T}. \quad (20\text{-}2)$$

When the temperature change ΔT of a system is small relative to the temperature (in kelvins) before and after the process, the entropy change can be approximated as

$$\Delta S = S_f - S_i \approx \frac{Q}{T_{avg}}, \quad (20\text{-}3)$$

where T_{avg} is the system's average temperature during the process.

When an ideal gas changes reversibly from an initial state with temperature T_i and volume V_i to a final state with temperature T_f and volume V_f, the change ΔS in the entropy of the gas is

$$\Delta S = S_f - S_i = nR \ln \frac{V_f}{V_i} + nC_V \ln \frac{T_f}{T_i}. \quad (20\text{-}4)$$

The Second Law of Thermodynamics This law, which is an extension of the entropy postulate, states: *If a process occurs in a closed system, the entropy of the system increases for irreversible processes and remains constant for reversible processes. It never decreases.* In equation form,

$$\Delta S \geq 0. \quad (20\text{-}5)$$

Engines An **engine** is a device that, operating in a cycle, extracts energy as heat $|Q_H|$ from a high-temperature reservoir and does a certain amount of work $|W|$. The *efficiency* ε of any engine is defined as

$$\varepsilon = \frac{\text{energy we get}}{\text{energy we pay for}} = \frac{|W|}{|Q_H|}. \quad (20\text{-}11)$$

In an **ideal engine,** all processes are reversible and no wasteful energy transfers occur due to, say, friction and turbulence. A **Carnot engine** is an ideal engine that follows the cycle of Fig. 20-9. Its efficiency is

$$\varepsilon_C = 1 - \frac{|Q_L|}{|Q_H|} = 1 - \frac{T_L}{T_H}, \quad (20\text{-}12, 20\text{-}13)$$

in which T_H and T_L are the temperatures of the high- and low-temperature reservoirs, respectively. Real engines always have an efficiency lower than that given by Eq. 20-13. Ideal engines that are not Carnot engines also have lower efficiencies.

A *perfect engine* is an imaginary engine in which energy extracted as heat from the high-temperature reservoir is converted completely to work. Such an engine would violate the second law of thermodynamics, which can be restated as follows: No series of processes is possible whose sole result is the absorption of energy as heat from a thermal reservoir and the complete conversion of this energy to work.

Refrigerators A refrigerator is a device that, operating in a cycle, has work W done on it as it extracts energy $|Q_L|$ as heat from a low-temperature reservoir. The coefficient of performance K of a refrigerator is defined as

$$K = \frac{\text{what we want}}{\text{what we pay for}} = \frac{|Q_L|}{|W|}. \quad (20\text{-}14)$$

A **Carnot refrigerator** is a Carnot engine operating in reverse.

For a Carnot refrigerator, Eq. 20-14 becomes

$$K_C = \frac{|Q_L|}{|Q_H| - |Q_L|} = \frac{T_L}{T_H - T_L}. \quad (20\text{-}15, 20\text{-}16)$$

A *perfect refrigerator* is an imaginary refrigerator in which energy extracted as heat from the low-temperature reservoir is converted completely to heat discharged to the high-temperature reservoir, without any need for work. Such a refrigerator would violate the second law of thermodynamics, which can be restated as follows: No series of processes is possible whose sole result is the transfer of energy as heat from a reservoir at a given temperature to a reservoir at a higher temperature.

Entropy from a Statistical View The entropy of a system can be defined in terms of the possible distributions of its molecules. For identical molecules, each possible distribution of molecules is called a **microstate** of the system. All equivalent microstates are grouped into a **configuration** of the system. The number of microstates in a configuration is the **multiplicity** W of the configuration.

For a system of N molecules that may be distributed between the two halves of a box, the multiplicity is given by

$$W = \frac{N!}{n_1!\, n_2!}, \quad (20\text{-}20)$$

in which n_1 is the number of molecules in one half of the box and n_2 is the number in the other half. A basic assumption of **statistical mechanics** is that all the microstates are equally probable. Thus, configurations with a large multiplicity occur most often.

The multiplicity W of a configuration of a system and the entropy S of the system in that configuration are related by Boltzmann's entropy equation:

$$S = k \ln W, \quad (20\text{-}21)$$

where $k = 1.38 \times 10^{-23}$ J/K is the Boltzmann constant.

Questions

1 Point i in Fig. 20-19 represents the initial state of an ideal gas at temperature T. Taking algebraic signs into account, rank the entropy changes that the gas undergoes as it moves, successively and reversibly, from point i to points a, b, c, and d, greatest first.

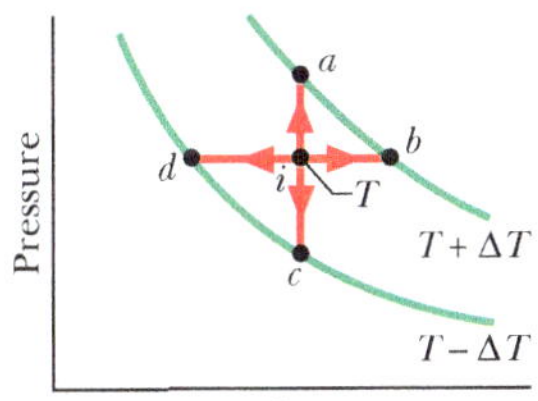

Figure 20-19 Question 1.

2 In four experiments, blocks A and B, starting at different initial temperatures, were brought together in an insulating box and allowed to reach a common final temperature. The entropy changes for the blocks in the four experiments had the following values (in joules per kelvin), but not necessarily in the order given. Determine which values for A go with which values for B.

Block	Values			
A	8	5	3	9
B	−3	−8	−5	−2

3 A gas, confined to an insulated cylinder, is compressed adiabatically to half its volume. Does the entropy of the gas increase, decrease, or remain unchanged during this process?

4 An ideal monatomic gas at initial temperature T_0 (in kelvins) expands from initial volume V_0 to volume $2V_0$ by each of the five processes indicated in the T-V diagram of Fig. 20-20. In which process is the expansion (a) isothermal, (b) isobaric (constant pressure), and (c) adiabatic? Explain your answers. (d) In which processes does the entropy of the gas decrease?

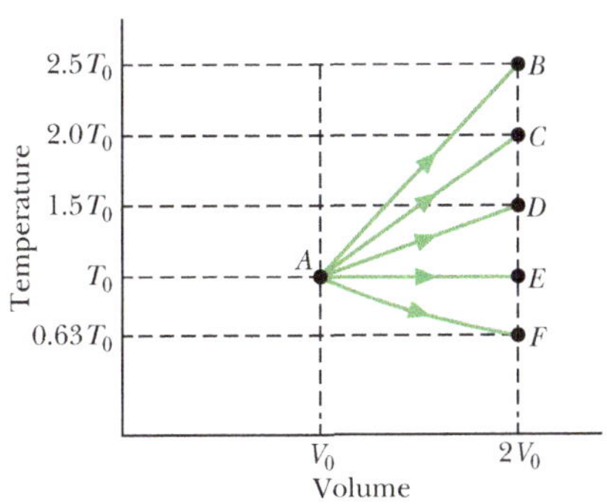

Figure 20-20 Question 4.

5 In four experiments, 2.5 mol of hydrogen gas undergoes reversible isothermal expansions, starting from the same volume but at different temperatures. The corresponding p-V plots are shown in Fig. 20-21. Rank the situations according to the change in the entropy of the gas, greatest first.

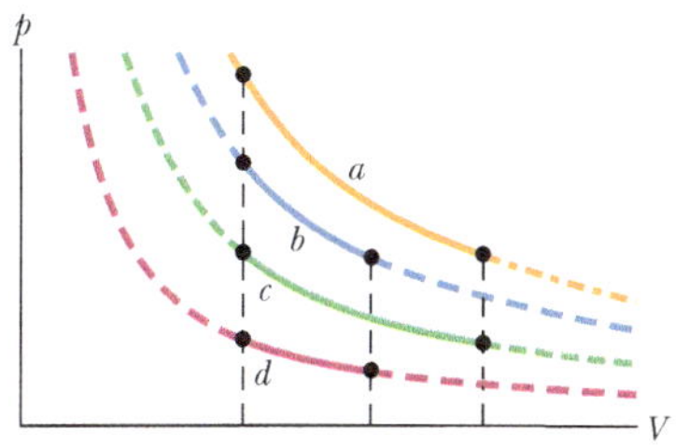

Figure 20-21 Question 5.

6 A box contains 100 atoms in a configuration that has 50 atoms in each half of the box. Suppose that you could count the different microstates associated with this configuration at the rate of 100 billion states per second, using a supercomputer. Without written calculation, guess how much computing time you would need: a day, a year, or much more than a year.

7 Does the entropy per cycle increase, decrease, or remain the same for (a) a Carnot engine, (b) a real engine, and (c) a perfect engine (which is, of course, impossible to build)?

8 Three Carnot engines operate between temperature limits of (a) 400 and 500 K, (b) 500 and 600 K, and (c) 400 and 600 K. Each engine extracts the same amount of energy per cycle from the high-temperature reservoir. Rank the magnitudes of the work done by the engines per cycle, greatest first.

9 An inventor claims to have invented four engines, each of which operates between constant-temperature reservoirs at 400 and 300 K. Data on each engine, per cycle of operation, are: engine A, $Q_H = 200$ J, $Q_L = -175$ J, and $W = 40$ J; engine B, $Q_H = 500$ J, $Q_L = -200$ J, and $W = 400$ J; engine C, $Q_H = 600$ J, $Q_L = -200$ J, and $W = 400$ J; engine D, $Q_H = 100$ J, $Q_L = -90$ J, and $W = 10$ J. Of the first and second laws of thermodynamics, which (if either) does each engine violate?

10 Does the entropy per cycle increase, decrease, or remain the same for (a) a Carnot refrigerator, (b) a real refrigerator, and (c) a perfect refrigerator (which is, of course, impossible to build)?

Problems

GO Tutoring problem available (at instructor's discretion) in *WileyPLUS* and WebAssign

SSM Worked-out solution available in Student Solutions Manual

• – ••• Number of dots indicates level of problem difficulty

Additional information available in *The Flying Circus of Physics* and at flyingcircusofphysics.com

WWW Worked-out solution is at

ILW Interactive solution is at

http://www.wiley.com/college/halliday

Module 20-1 Entropy

•1 SSM Suppose 4.00 mol of an ideal gas undergoes a reversible isothermal expansion from volume V_1 to volume $V_2 = 2.00V_1$ at temperature $T = 400$ K. Find (a) the work done by the gas and (b) the entropy change of the gas. (c) If the expansion is reversible and adiabatic instead of isothermal, what is the entropy change of the gas?

•2 An ideal gas undergoes a reversible isothermal expansion at 77.0°C, increasing its volume from 1.30 L to 3.40 L. The entropy change of the gas is 22.0 J/K. How many moles of gas are present?

•3 ILW A 2.50 mol sample of an ideal gas expands reversibly and isothermally at 360 K until its volume is doubled. What is the increase in entropy of the gas?

•4 How much energy must be transferred as heat for a reversible isothermal expansion of an ideal gas at 132°C if the entropy of the gas increases by 46.0 J/K?

•5 ILW Find (a) the energy absorbed as heat and (b) the change in entropy of a 2.00 kg block of copper whose temperature is increased reversibly from 25.0°C to 100°C. The specific heat of copper is 386 J/kg·K.

•6 (a) What is the entropy change of a 12.0 g ice cube that melts completely in a bucket of water whose temperature is just above the freezing point of water? (b) What is the entropy change of a 5.00 g spoonful of water that evaporates completely on a hot plate whose temperature is slightly above the boiling point of water?

••7 ILW A 50.0 g block of copper whose temperature is 400 K is placed in an insulating box with a 100 g block of lead whose temperature is 200 K. (a) What is the equilibrium temperature of the two-block system? (b) What is the change in the internal energy of the system between the initial state and the equilibrium state? (c) What is the change in the entropy of the system? (See Table 18-3.)

••8 At very low temperatures, the molar specific heat C_V of many solids is approximately $C_V = AT^3$, where A depends on the particular substance. For aluminum, $A = 3.15 \times 10^{-5}$ J/mol·K^4. Find the entropy change for 4.00 mol of aluminum when its temperature is raised from 5.00 K to 10.0 K.

••9 A 10 g ice cube at −10°C is placed in a lake whose temperature is 15°C. Calculate the change in entropy of the cube–lake system as the ice cube comes to thermal equilibrium with the lake. The specific heat of ice is 2220 J/kg·K. (*Hint:* Will the ice cube affect the lake temperature?)

••10 A 364 g block is put in contact with a thermal reservoir. The block is initially at a lower temperature than the reservoir. Assume that the consequent transfer of energy as heat from the reservoir to the block is reversible. Figure 20-22 gives the change in entropy ΔS of the block until thermal equilibrium is reached. The scale of the horizontal axis is set by $T_a = 280$ K and $T_b = 380$ K. What is the specific heat of the block?

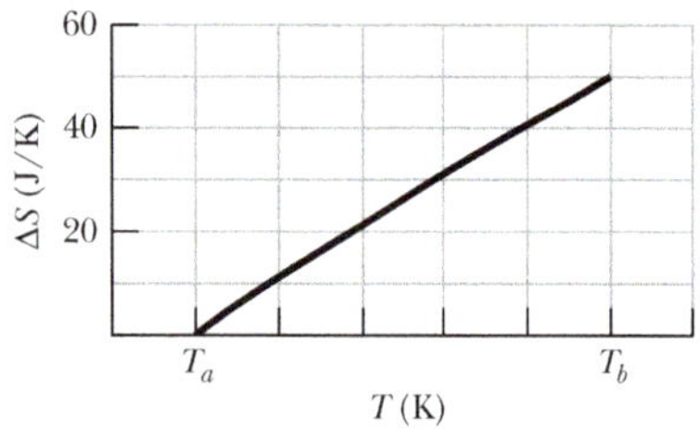

Figure 20-22 Problem 10.

••11 SSM WWW In an experiment, 200 g of aluminum (with a specific heat of 900 J/kg·K) at 100°C is mixed with 50.0 g of water at 20.0°C, with the mixture thermally isolated. (a) What is the equilibrium temperature? What are the entropy changes of (b) the aluminum, (c) the water, and (d) the aluminum–water system?

••12 A gas sample undergoes a reversible isothermal expansion. Figure 20-23 gives the change ΔS in entropy of the gas versus the final volume V_f of the gas. The scale of the vertical axis is set by $\Delta S_s = 64$ J/K. How many moles are in the sample?

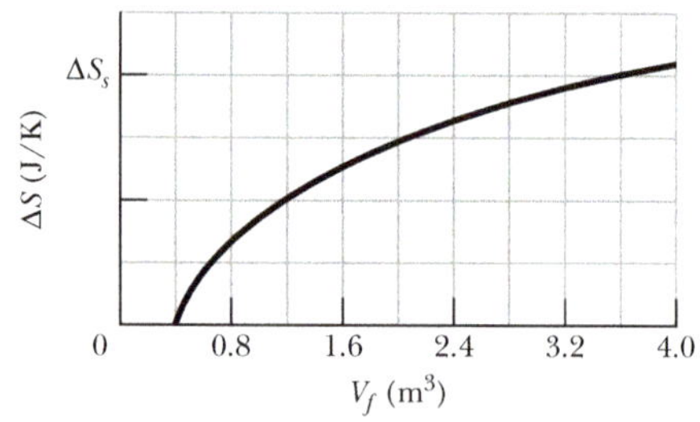

Figure 20-23 Problem 12.

••13 In the irreversible process of Fig. 20-5, let the initial temperatures of the identical blocks L and R be 305.5 and 294.5 K, respectively, and let 215 J be the energy that must be transferred between the blocks in order to reach equilibrium. For the reversible processes of Fig. 20-6, what is ΔS for (a) block L, (b) its reservoir, (c) block R, (d) its reservoir, (e) the two-block system, and (f) the system of the two blocks and the two reservoirs?

••14 (a) For 1.0 mol of a monatomic ideal gas taken through the cycle in Fig. 20-24, where $V_1 = 4.00V_0$, what is W/p_0V_0 as the gas goes from state a to state c along path abc? What is $\Delta E_{int}/p_0V_0$ in going (b) from b to c and (c) through one full cycle? What is ΔS in going (d) from b to c and (e) through one full cycle?

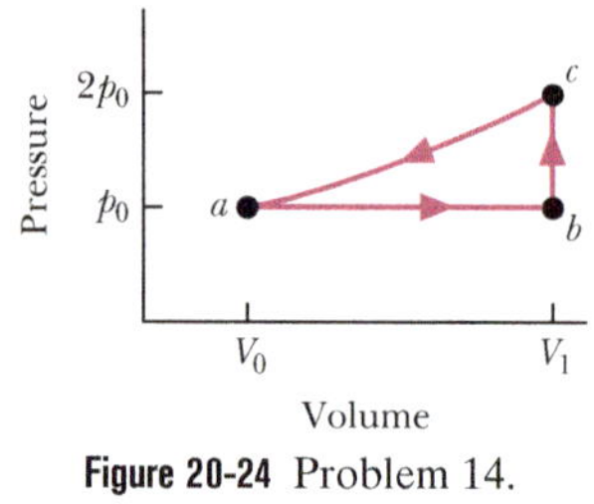

Figure 20-24 Problem 14.

••15 A mixture of 1773 g of water and 227 g of ice is in an initial equilibrium state at 0.000°C. The mixture is then, in a reversible process, brought to a second equilibrium state where the water–ice ratio, by mass, is 1.00:1.00 at 0.000°C. (a) Calculate the entropy change of the system during this process. (The heat of fusion for water is 333 kJ/kg.) (b) The system is then returned to the initial equilibrium state in an irreversible process (say, by using a Bunsen burner). Calculate the entropy change of the system during this process. (c) Are your answers consistent with the second law of thermodynamics?

••16 GO An 8.0 g ice cube at −10°C is put into a Thermos flask containing 100 cm^3 of water at 20°C. By how much has the entropy of the cube–water system changed when equilibrium is reached? The specific heat of ice is 2220 J/kg · K.

••17 In Fig. 20-25, where $V_{23} = 3.00V_1$, n moles of a diatomic ideal gas are taken through the cycle with the molecules rotating but not oscillating. What are (a) p_2/p_1, (b) p_3/p_1, and (c) T_3/T_1? For path 1 → 2, what are (d) W/nRT_1, (e) Q/nRT_1, (f) $\Delta E_{int}/nRT_1$, and (g) $\Delta S/nR$? For path 2 → 3, what are (h) W/nRT_1, (i) Q/nRT_1, (j) $\Delta E_{int}/nRT_1$, (k) $\Delta S/nR$? For path 3 → 1, what are (l) W/nRT_1, (m) Q/nRT_1, (n) $\Delta E_{int}/nRT_1$, and (o) $\Delta S/nR$?

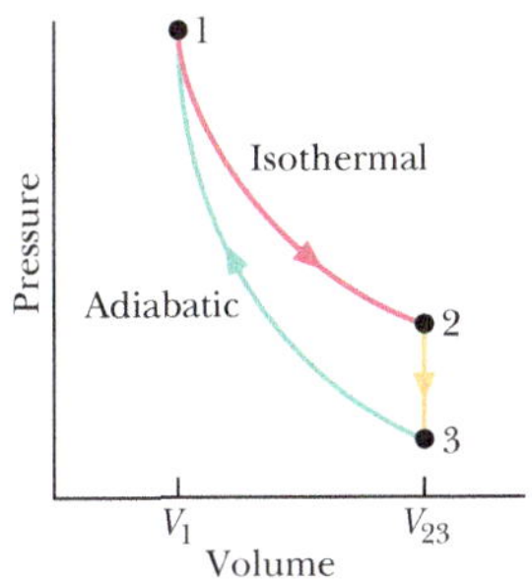

Figure 20-25 Problem 17.

••18 GO A 2.0 mol sample of an ideal monatomic gas undergoes the reversible process shown in Fig. 20-26. The scale of the vertical axis is set by $T_s = 400.0$ K and the scale of the horizontal axis is set by $S_s = 20.0$ J/K. (a) How much energy is absorbed as heat by the gas? (b) What is the change in the internal energy of the gas? (c) How much work is done by the gas?

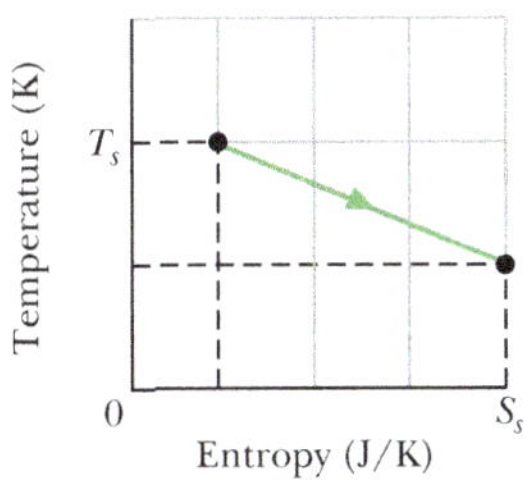

Figure 20-26 Problem 18.

•••19 Suppose 1.00 mol of a monatomic ideal gas is taken from initial pressure p_1 and volume V_1 through two steps: (1) an isothermal expansion to volume $2.00V_1$ and (2) a pressure increase to $2.00p_1$ at constant volume. What is Q/p_1V_1 for (a) step 1 and (b) step 2? What is W/p_1V_1 for (c) step 1 and (d) step 2? For the full process, what are (e) $\Delta E_{int}/p_1V_1$ and (f) ΔS? The gas is returned to its initial state and again taken to the same final state but now through these two steps: (1) an isothermal compression to pressure $2.00p_1$ and (2) a volume increase to $2.00V_1$ at constant pressure. What is Q/p_1V_1 for (g) step 1 and (h) step 2? What is W/p_1V_1 for (i) step 1 and (j) step 2? For the full process, what are (k) $\Delta E_{int}/p_1V_1$ and (l) ΔS?

•••20 Expand 1.00 mol of an monatomic gas initially at 5.00 kPa and 600 K from initial volume $V_i = 1.00$ m^3 to final volume $V_f = 2.00$ m^3. At any instant during the expansion, the pressure p and volume V of the gas are related by $p = 5.00 \exp[(V_i - V)/a]$, with p in kilopascals, V_i and V in cubic meters, and $a = 1.00$ m^3. What are the final (a) pressure and (b) temperature of the gas? (c) How much work is done by the gas during the expansion? (d) What is ΔS for the expansion? (*Hint:* Use two simple reversible processes to find ΔS.)

•••21 GO Energy can be removed from water as heat at and even below the normal freezing point (0.0°C at atmospheric pressure) without causing the water to freeze; the water is then said to be *supercooled.* Suppose a 1.00 g water drop is supercooled until its temperature is that of the surrounding air, which is at −5.00°C. The drop then suddenly and irreversibly freezes, transferring energy to the air as heat. What is the entropy change for the drop? (*Hint:* Use a three-step reversible process as if the water were taken through the normal freezing point.) The specific heat of ice is 2220 J/kg · K.

•••22 GO An insulated Thermos contains 130 g of water at 80.0°C. You put in a 12.0 g ice cube at 0°C to form a system of *ice + original water.* (a) What is the equilibrium temperature of the system? What are the entropy changes of the water that was originally the ice cube (b) as it melts and (c) as it warms to the equilibrium temperature? (d) What is the entropy change of the original water as it cools to the equilibrium temperature? (e) What is the net entropy change of the *ice + original water* system as it reaches the equilibrium temperature?

Module 20-2 Entropy in the Real World: Engines

•23 A Carnot engine whose low-temperature reservoir is at 17°C has an efficiency of 40%. By how much should the temperature of the high-temperature reservoir be increased to increase the efficiency to 50%?

•24 A Carnot engine absorbs 52 kJ as heat and exhausts 36 kJ as heat in each cycle. Calculate (a) the engine's efficiency and (b) the work done per cycle in kilojoules.

•25 A Carnot engine has an efficiency of 22.0%. It operates between constant-temperature reservoirs differing in temperature by 75.0 C°. What is the temperature of the (a) lower-temperature and (b) higher-temperature reservoir?

•26 In a hypothetical nuclear fusion reactor, the fuel is deuterium gas at a temperature of 7×10^8 K. If this gas could be used to operate a Carnot engine with $T_L = 100$°C, what would be the engine's efficiency? Take both temperatures to be exact and report your answer to seven significant figures.

•27 SSM WWW A Carnot engine operates between 235°C and 115°C, absorbing 6.30×10^4 J per cycle at the higher temperature. (a) What is the efficiency of the engine? (b) How much work per cycle is this engine capable of performing?

••28 In the first stage of a two-stage Carnot engine, energy is absorbed as heat Q_1 at temperature T_1, work W_1 is done, and energy is expelled as heat Q_2 at a lower temperature T_2. The second stage absorbs that energy as heat Q_2, does work W_2, and expels energy as heat Q_3 at a still lower temperature T_3. Prove that the efficiency of the engine is $(T_1 - T_3)/T_1$.

••29 GO Figure 20-27 shows a reversible cycle through which 1.00 mol of a monatomic ideal gas is taken. Assume that $p = 2p_0$, $V = 2V_0$, $p_0 = 1.01 \times 10^5$ Pa, and $V_0 = 0.0225$ m^3. Calculate (a) the work done during the cycle, (b) the energy added as heat during stroke *abc,* and (c) the efficiency of the cycle. (d) What is the efficiency of a Carnot engine operating between the highest and lowest temperatures that occur in the cycle? (e) Is this greater than or less than the efficiency calculated in (c)?

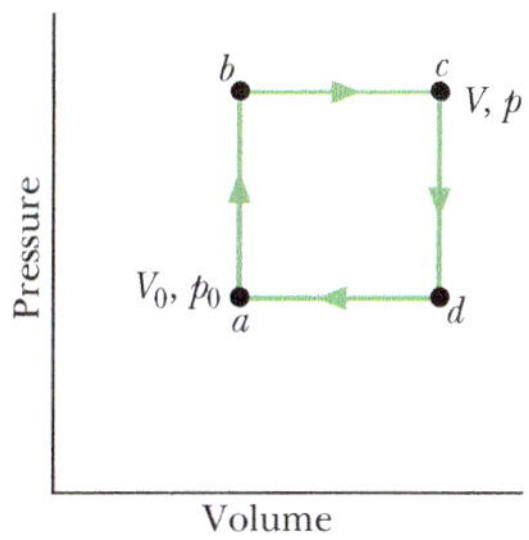

Figure 20-27 Problem 29.

••30 A 500 W Carnot engine operates between constant-temperature reservoirs at 100°C and 60.0°C. What is the rate at which energy is (a) taken in by the engine as heat and (b) exhausted by the engine as heat?

••31 The efficiency of a particular car engine is 25% when the engine does 8.2 kJ of work per cycle. Assume the process is reversible. What are (a) the energy the engine gains per cycle as heat Q_{gain} from the fuel combustion and (b) the energy the engine loses per cycle as heat Q_{lost}? If a tune-up increases the efficiency to 31%, what are (c) Q_{gain} and (d) Q_{lost} at the same work value?

••32 GO A Carnot engine is set up to produce a certain work W per cycle. In each cycle, energy in the form of heat Q_H is transferred to the working substance of the engine from the higher-temperature thermal reservoir, which is at an adjustable temperature T_H. The lower-temperature thermal reservoir is maintained at temperature $T_L = 250$ K. Figure 20-28 gives Q_H for a range of T_H. The scale of the vertical axis is set by $Q_{Hs} = 6.0$ kJ. If T_H is set at 550 K, what is Q_H?

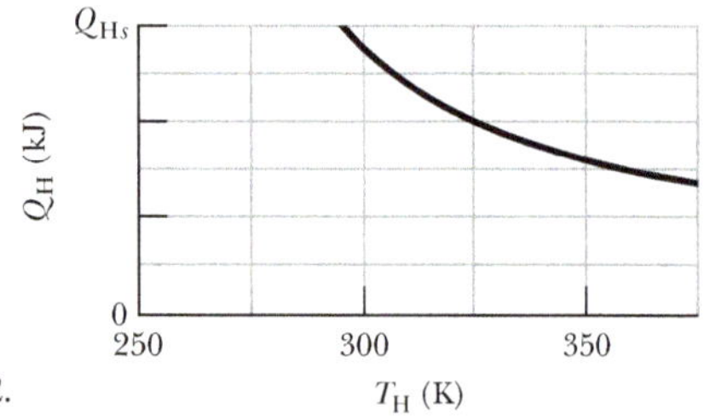

Figure 20-28 Problem 32.

••33 SSM ILW Figure 20-29 shows a reversible cycle through which 1.00 mol of a monatomic ideal gas is taken. Volume $V_c = 8.00V_b$. Process bc is an adiabatic expansion, with $p_b = 10.0$ atm and $V_b = 1.00 \times 10^{-3}$ m^3. For the cycle, find (a) the energy added to the gas as heat, (b) the energy leaving the gas as heat, (c) the net work done by the gas, and (d) the efficiency of the cycle.

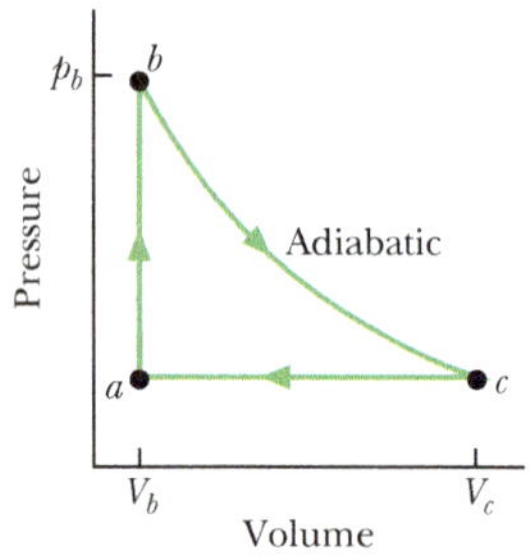

Figure 20-29 Problem 33.

••34 GO An ideal gas (1.0 mol) is the working substance in an engine that operates on the cycle shown in Fig. 20-30. Processes BC and DA are reversible and adiabatic. (a) Is the gas monatomic, diatomic, or polyatomic? (b) What is the engine efficiency?

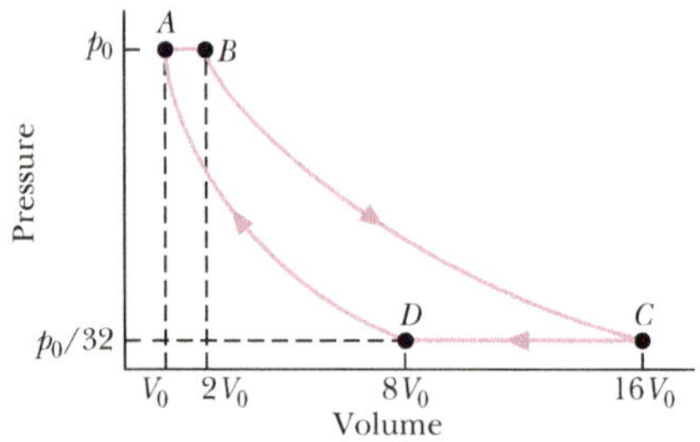

Figure 20-30 Problem 34.

•••35 The cycle in Fig. 20-31 represents the operation of a gasoline internal combustion engine. Volume $V_3 = 4.00V_1$. Assume the gasoline–air intake mixture is an ideal gas with $\gamma = 1.30$. What are the ratios (a) T_2/T_1, (b) T_3/T_1, (c) T_4/T_1, (d) p_3/p_1, and (e) p_4/p_1? (f) What is the engine efficiency?

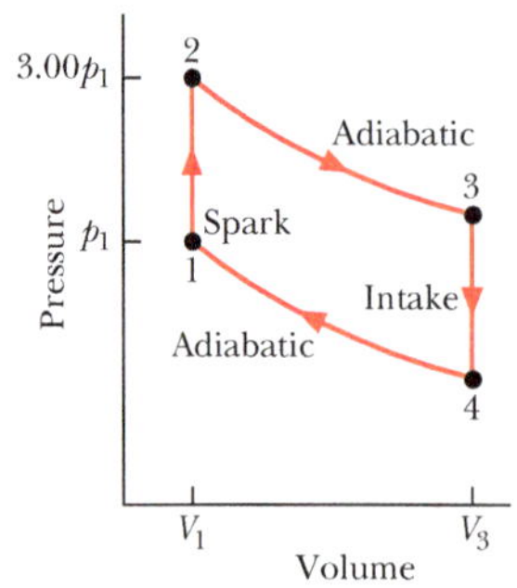

Figure 20-31 Problem 35.

Module 20-3 Refrigerators and Real Engines

•36 How much work must be done by a Carnot refrigerator to transfer 1.0 J as heat (a) from a reservoir at 7.0°C to one at 27°C, (b) from a reservoir at −73°C to one at 27°C, (c) from a reservoir at −173°C to one at 27°C, and (d) from a reservoir at −223°C to one at 27°C?

•37 SSM A heat pump is used to heat a building. The external temperature is less than the internal temperature. The pump's coefficient of performance is 3.8, and the heat pump delivers 7.54 MJ as heat to the building each hour. If the heat pump is a Carnot engine working in reverse, at what rate must work be done to run it?

•38 The electric motor of a heat pump transfers energy as heat from the outdoors, which is at −5.0°C, to a room that is at 17°C. If the heat pump were a Carnot heat pump (a Carnot engine working in reverse), how much energy would be transferred as heat to the room for each joule of electric energy consumed?

•39 SSM A Carnot air conditioner takes energy from the thermal energy of a room at 70°F and transfers it as heat to the outdoors, which is at 96°F. For each joule of electric energy required to operate the air conditioner, how many joules are removed from the room?

•40 To make ice, a freezer that is a reverse Carnot engine extracts 42 kJ as heat at −15°C during each cycle, with coefficient of performance 5.7. The room temperature is 30.3°C. How much (a) energy per cycle is delivered as heat to the room and (b) work per cycle is required to run the freezer?

••41 ILW An air conditioner operating between 93°F and 70°F is rated at 4000 Btu/h cooling capacity. Its coefficient of performance is 27% of that of a Carnot refrigerator operating between the same two temperatures. What horsepower is required of the air conditioner motor?

••42 The motor in a refrigerator has a power of 200 W. If the freezing compartment is at 270 K and the outside air is at 300 K, and assuming the efficiency of a Carnot refrigerator, what is the maximum amount of energy that can be extracted as heat from the freezing compartment in 10.0 min?

••43 GO Figure 20-32 represents a Carnot engine that works between temperatures $T_1 = 400$ K and $T_2 = 150$ K and drives a Carnot refrigerator that works between temperatures $T_3 = 325$ K and $T_4 = 225$ K. What is the ratio Q_3/Q_1?

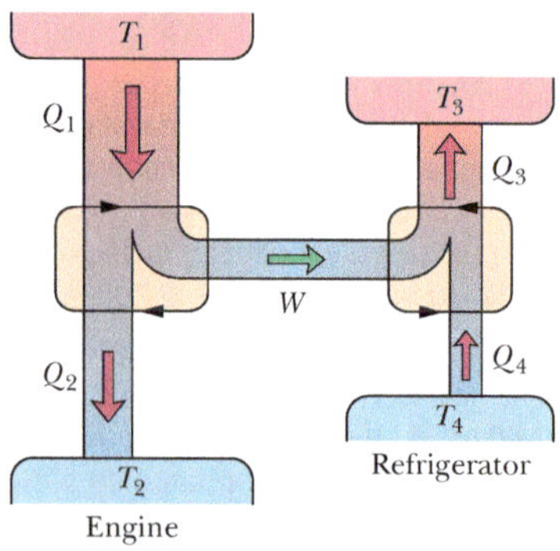

Figure 20-32 Problem 43.

••44 (a) During each cycle, a Carnot engine absorbs 750 J as heat from a high-temperature reservoir at 360 K, with the low-temperature reservoir at 280 K. How much work is done per cycle? (b) The engine is then made to work in reverse to function as a Carnot refrigerator between those same two reservoirs. During each cycle, how much work is required to remove 1200 J as heat from the low-temperature reservoir?

Module 20-4 A Statistical View of Entropy

•45 Construct a table like Table 20-1 for eight molecules.

••46 A box contains N identical gas molecules equally divided between its two halves. For $N = 50$, what are (a) the multiplicity W of the central configuration, (b) the total number of microstates, and (c) the percentage of the time the system spends in the central configuration? For $N = 100$, what are (d) W of the central configuration, (e) the total number of microstates, and (f) the percentage of the time the system spends in the central configuration? For $N = 200$, what are (g) W of the central configuration, (h) the total number of microstates, and (i) the percentage of the time the system spends in the central configuration? (j) Does the time spent in the central configuration increase or decrease with an increase in N?

•••47 SSM WWW A box contains N gas molecules. Consider the box to be divided into three equal parts. (a) By extension of Eq. 20-20, write a formula for the multiplicity of any given configuration. (b) Consider two configurations: configuration A with equal numbers of molecules in all three thirds of the box, and configuration B with equal numbers of molecules in each half of the box divided into two equal parts rather than three. What is the ratio W_A/W_B of the multiplicity of configuration A to that of configuration B? (c) Evaluate W_A/W_B for $N = 100$. (Because 100 is not evenly divisible by 3, put 34 molecules into one of the three box parts of configuration A and 33 in each of the other two parts.)

Additional Problems

48 Four particles are in the insulated box of Fig. 20-17. What are (a) the least multiplicity, (b) the greatest multiplicity, (c) the least entropy, and (d) the greatest entropy of the four-particle system?

49 A cylindrical copper rod of length 1.50 m and radius 2.00 cm is insulated to prevent heat loss through its curved surface. One end is attached to a thermal reservoir fixed at 300°C; the other is attached to a thermal reservoir fixed at 30.0°C. What is the rate at which entropy increases for the rod–reservoirs system?

50 Suppose 0.550 mol of an ideal gas is isothermally and reversibly expanded in the four situations given below. What is the change in the entropy of the gas for each situation?

Situation	(a)	(b)	(c)	(d)
Temperature (K)	250	350	400	450
Initial volume (cm^3)	0.200	0.200	0.300	0.300
Final volume (cm^3)	0.800	0.800	1.20	1.20

51 SSM As a sample of nitrogen gas (N_2) undergoes a temperature increase at constant volume, the distribution of molecular speeds increases. That is, the probability distribution function $P(v)$ for the molecules spreads to higher speed values, as suggested in Fig. 19-8*b*. One way to report the spread in $P(v)$ is to measure the difference Δv between the most probable speed v_P and the rms speed v_{rms}. When $P(v)$ spreads to higher speeds, Δv increases. Assume that the gas is ideal and the N_2 molecules rotate but do not oscillate. For 1.5 mol, an initial temperature of 250 K, and a final temperature of 500 K, what are (a) the initial difference Δv_i, (b) the final difference Δv_f, and (c) the entropy change ΔS for the gas?

52 Suppose 1.0 mol of a monatomic ideal gas initially at 10 L and 300 K is heated at constant volume to 600 K, allowed to expand isothermally to its initial pressure, and finally compressed at constant pressure to its original volume, pressure, and temperature. During the cycle, what are (a) the net energy entering the system (the gas) as heat and (b) the net work done by the gas? (c) What is the efficiency of the cycle?

53 GO Suppose that a deep shaft were drilled in Earth's crust near one of the poles, where the surface temperature is −40°C, to a depth where the temperature is 800°C. (a) What is the theoretical limit to the efficiency of an engine operating between these temperatures? (b) If all the energy released as heat into the low-temperature reservoir were used to melt ice that was initially at −40°C, at what rate could liquid water at 0°C be produced by a 100 MW power plant (treat it as an engine)? The specific heat of ice is 2220 J/kg · K; water's heat of fusion is 333 kJ/kg. (Note that the engine can operate only between 0°C and 800°C in this case. Energy exhausted at −40°C cannot warm anything above −40°C.)

54 What is the entropy change for 3.20 mol of an ideal monatomic gas undergoing a reversible increase in temperature from 380 K to 425 K at constant volume?

55 A 600 g lump of copper at 80.0°C is placed in 70.0 g of water at 10.0°C in an insulated container. (See Table 18-3 for specific heats.) (a) What is the equilibrium temperature of the copper–water system? What entropy changes do (b) the copper, (c) the water, and (d) the copper–water system undergo in reaching the equilibrium temperature?

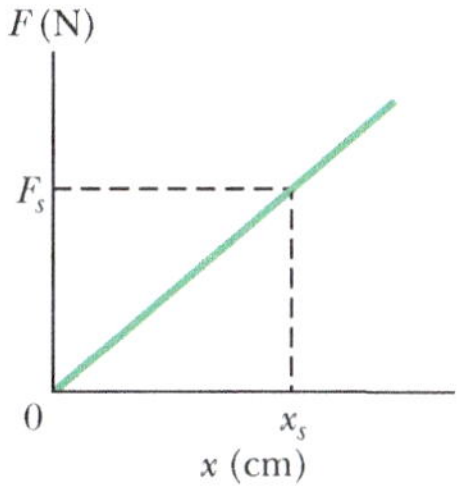

Figure 20-33 Problem 56.

56 Figure 20-33 gives the force magnitude F versus stretch distance x for a rubber band, with the scale of the F axis set by $F_s = 1.50$ N and the scale of the x axis set by $x_s = 3.50$ cm. The temperature is 2.00°C. When the rubber band is stretched by $x = 1.70$ cm, at what rate does the entropy of the rubber band change during a small additional stretch?

57 The temperature of 1.00 mol of a monatomic ideal gas is raised reversibly from 300 K to 400 K, with its volume kept constant. What is the entropy change of the gas?

58 Repeat Problem 57, with the pressure now kept constant.

59 SSM A 0.600 kg sample of water is initially ice at temperature −20°C. What is the sample's entropy change if its temperature is increased to 40°C?

60 A three-step cycle is undergone by 3.4 mol of an ideal diatomic gas: (1) the temperature of the gas is increased from 200 K to 500 K at constant volume; (2) the gas is then isothermally expanded to its original pressure; (3) the gas is then contracted at constant pressure back to its original volume. Throughout the cycle, the molecules rotate but do not oscillate. What is the efficiency of the cycle?

61 An inventor has built an engine X and claims that its efficiency ε_X is greater than the efficiency ε of an ideal engine operating between the same two temperatures. Suppose you couple engine X to an ideal refrigerator (Fig. 20-34*a*) and adjust the cycle

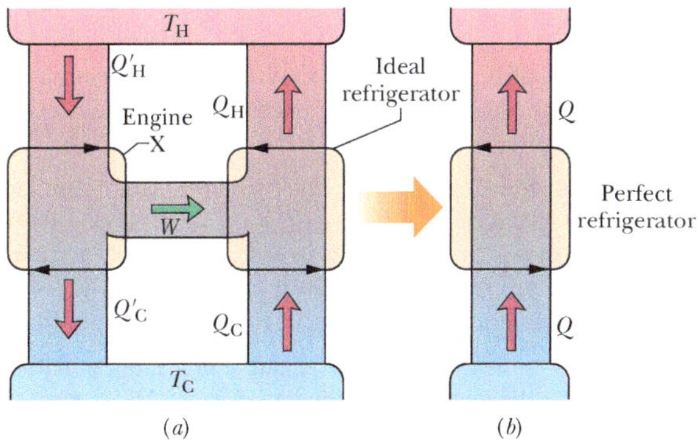

Figure 20-34 Problem 61.

of engine X so that the work per cycle it provides equals the work per cycle required by the ideal refrigerator. Treat this combination as a single unit and show that if the inventor's claim were true (if $\varepsilon_X > \varepsilon$), the combined unit would act as a perfect refrigerator (Fig. 20-34*b*), transferring energy as heat from the low-temperature reservoir to the high-temperature reservoir without the need for work.

62 Suppose 2.00 mol of a diatomic gas is taken reversibly around the cycle shown in the *T*-*S* diagram of Fig. 20-35, where $S_1 = 6.00$ J/K and $S_2 = 8.00$ J/K. The molecules do not rotate or oscillate. What is the energy transferred as heat Q for (a) path $1 \rightarrow 2$, (b) path $2 \rightarrow 3$, and (c) the full cycle? (d) What is the work W for the isothermal process? The volume V_1 in state 1 is 0.200 m^3. What is the volume in (e) state 2 and (f) state 3?

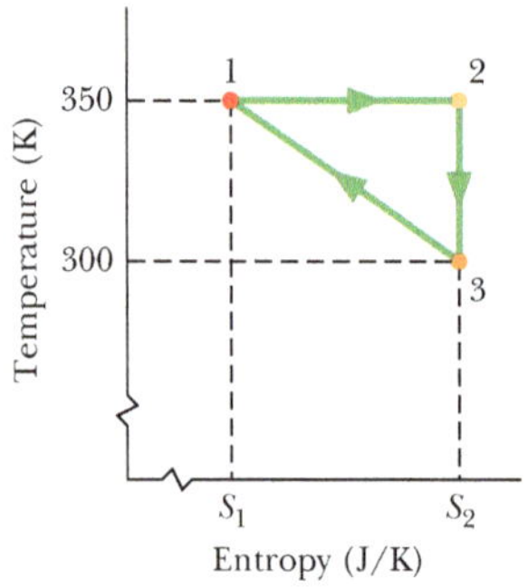

Figure 20-35 Problem 62.

What is the change ΔE_{int} for (g) path $1 \rightarrow 2$, (h) path $2 \rightarrow 3$, and (i) the full cycle? (*Hint:* (h) can be done with one or two lines of calculation using Module 19-7 or with a page of calculation using Module 19-9.) (j) What is the work W for the adiabatic process?

63 A three-step cycle is undergone reversibly by 4.00 mol of an ideal gas: (1) an adiabatic expansion that gives the gas 2.00 times its initial volume, (2) a constant-volume process, (3) an isothermal compression back to the initial state of the gas. We do not know whether the gas is monatomic or diatomic; if it is diatomic, we do not know whether the molecules are rotating or oscillating. What are the entropy changes for (a) the cycle, (b) process 1, (c) process 3, and (d) process 2?

64 (a) A Carnot engine operates between a hot reservoir at 320 K and a cold one at 260 K. If the engine absorbs 500 J as heat per cycle at the hot reservoir, how much work per cycle does it deliver? (b) If the engine working in reverse functions as a refrigerator between the same two reservoirs, how much work per cycle must be supplied to remove 1000 J as heat from the cold reservoir?

65 A 2.00 mol diatomic gas initially at 300 K undergoes this cycle: It is (1) heated at constant volume to 800 K, (2) then allowed to expand isothermally to its initial pressure, (3) then compressed at constant pressure to its initial state. Assuming the gas molecules neither rotate nor oscillate, find (a) the net energy transferred as heat to the gas, (b) the net work done by the gas, and (c) the efficiency of the cycle.

66 An ideal refrigerator does 150 J of work to remove 560 J as heat from its cold compartment. (a) What is the refrigerator's coefficient of performance? (b) How much heat per cycle is exhausted to the kitchen?

67 Suppose that 260 J is conducted from a constant-temperature reservoir at 400 K to one at (a) 100 K, (b) 200 K, (c) 300 K, and (d) 360 K. What is the net change in entropy ΔS_{net} of the reservoirs in each case? (e) As the temperature difference of the two reservoirs decreases, does ΔS_{net} increase, decrease, or remain the same?

68 An apparatus that liquefies helium is in a room maintained at 300 K. If the helium in the apparatus is at 4.0 K, what is the minimum ratio Q_{to}/Q_{from}, where Q_{to} is the energy delivered as heat to the room and Q_{from} is the energy removed as heat from the helium?

69 GO A brass rod is in thermal contact with a constant-temperature reservoir at 130°C at one end and a constant-temperature reservoir at 24.0°C at the other end. (a) Compute the total change in entropy of the rod–reservoirs system when 5030 J of energy is conducted through the rod, from one reservoir to the other. (b) Does the entropy of the rod change?

70 A 45.0 g block of tungsten at 30.0°C and a 25.0 g block of silver at −120°C are placed together in an insulated container. (See Table 18-3 for specific heats.) (a) What is the equilibrium temperature? What entropy changes do (b) the tungsten, (c) the silver, and (d) the tungsten–silver system undergo in reaching the equilibrium temperature?

71 A box contains N molecules. Consider two configurations: configuration A with an equal division of the molecules between the two halves of the box, and configuration B with 60.0% of the molecules in the left half of the box and 40.0% in the right half. For $N = 50$, what are (a) the multiplicity W_A of configuration A, (b) the multiplicity W_B of configuration B, and (c) the ratio $f_{B/A}$ of the time the system spends in configuration B to the time it spends in configuration A? For $N = 100$, what are (d) W_A, (e) W_B, and (f) $f_{B/A}$? For $N = 200$, what are (g) W_A, (h) W_B, and (i) $f_{B/A}$? (j) With increasing N, does f increase, decrease, or remain the same?

72 Calculate the efficiency of a fossil-fuel power plant that consumes 380 metric tons of coal each hour to produce useful work at the rate of 750 MW. The heat of combustion of coal (the heat due to burning it) is 28 MJ/kg.

73 SSM A Carnot refrigerator extracts 35.0 kJ as heat during each cycle, operating with a coefficient of performance of 4.60. What are (a) the energy per cycle transferred as heat to the room and (b) the work done per cycle?

74 A Carnot engine whose high-temperature reservoir is at 400 K has an efficiency of 30.0%. By how much should the temperature of the low-temperature reservoir be changed to increase the efficiency to 40.0%?

75 SSM System A of three particles and system B of five particles are in insulated boxes like that in Fig. 20-17. What is the least multiplicity W of (a) system A and (b) system B? What is the greatest multiplicity W of (c) A and (d) B? What is the greatest entropy of (e) A and (f) B?

76 Figure 20-36 shows a Carnot cycle on a *T-S* diagram, with a scale set by $S_s = 0.60$ J/K. For a full cycle, find (a) the net heat transfer and (b) the net work done by the system.

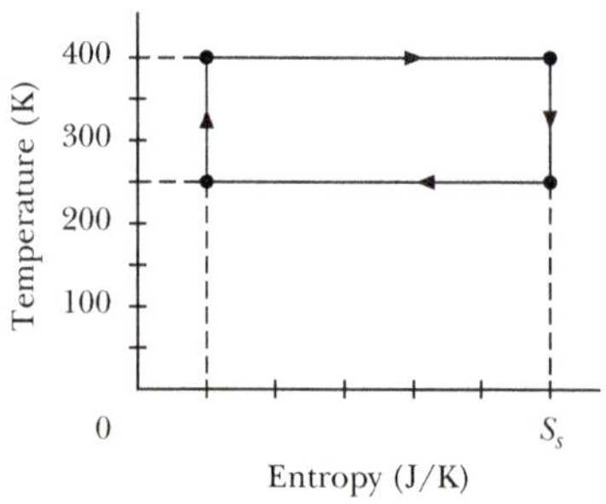

Figure 20-36 Problem 76.

77 Find the relation between the efficiency of a reversible ideal heat engine and the coefficient of performance of the reversible refrigerator obtained by running the engine backwards.

78 A Carnot engine has a power of 500 W. It operates between heat reservoirs at 100°C and 60.0°C. Calculate (a) the rate of heat input and (b) the rate of exhaust heat output.

79 In a real refrigerator, the low-temperature coils are at −13°C, and the compressed gas in the condenser is at 26°C. What is the theoretical coefficient of performance?

APPENDIX A

THE INTERNATIONAL SYSTEM OF UNITS (SI)*

Table 1 **The SI Base Units**

Quantity	Name	Symbol	Definition
length	meter	m	"... the length of the path traveled by light in vacuum in 1/299,792,458 of a second." (1983)
mass	kilogram	kg	"... this prototype [a certain platinum–iridium cylinder] shall henceforth be considered to be the unit of mass." (1889)
time	second	s	"... the duration of 9,192,631,770 periods of the radiation corresponding to the transition between the two hyperfine levels of the ground state of the cesium-133 atom." (1967)
electric current	ampere	A	"... that constant current which, if maintained in two straight parallel conductors of infinite length, of negligible circular cross section, and placed 1 meter apart in vacuum, would produce between these conductors a force equal to 2×10^{-7} newton per meter of length." (1946)
thermodynamic temperature	kelvin	K	"... the fraction 1/273.16 of the thermodynamic temperature of the triple point of water." (1967)
amount of substance	mole	mol	"... the amount of substance of a system which contains as many elementary entities as there are atoms in 0.012 kilogram of carbon-12." (1971)
luminous intensity	candela	cd	"... the luminous intensity, in a given direction, of a source that emits monochromatic radiation of frequency 540×10^{12} hertz and that has a radiant intensity in that direction of 1/683 watt per steradian." (1979)

*Adapted from "The International System of Units (SI)," National Bureau of Standards Special Publication 330, 1972 edition. The definitions above were adopted by the General Conference of Weights and Measures, an international body, on the dates shown. In this book we do not use the candela.

Table 2 Some SI Derived Units

Quantity	Name of Unit	Symbol	
area	square meter	m^2	
volume	cubic meter	m^3	
frequency	hertz	Hz	s^{-1}
mass density (density)	kilogram per cubic meter	kg/m^3	
speed, velocity	meter per second	m/s	
angular velocity	radian per second	rad/s	
acceleration	meter per second per second	m/s^2	
angular acceleration	radian per second per second	rad/s^2	
force	newton	N	$kg \cdot m/s^2$
pressure	pascal	Pa	N/m^2
work, energy, quantity of heat	joule	J	$N \cdot m$
power	watt	W	J/s
quantity of electric charge	coulomb	C	$A \cdot s$
potential difference, electromotive force	volt	V	W/A
electric field strength	volt per meter (or newton per coulomb)	V/m	N/C
electric resistance	ohm	Ω	V/A
capacitance	farad	F	$A \cdot s/V$
magnetic flux	weber	Wb	$V \cdot s$
inductance	henry	H	$V \cdot s/A$
magnetic flux density	tesla	T	Wb/m^2
magnetic field strength	ampere per meter	A/m	
entropy	joule per kelvin	J/K	
specific heat	joule per kilogram kelvin	$J/(kg \cdot K)$	
thermal conductivity	watt per meter kelvin	$W/(m \cdot K)$	
radiant intensity	watt per steradian	W/sr	

Table 3 The SI Supplementary Units

Quantity	Name of Unit	Symbol
plane angle	radian	rad
solid angle	steradian	sr

APPENDIX B

SOME FUNDAMENTAL CONSTANTS OF PHYSICS*

Constant	Symbol	Computational Value	Best (1998) Value	
			Value[a]	Uncertainty[b]
Speed of light in a vacuum	c	3.00×10^{8} m/s	2.997 924 58	exact
Elementary charge	e	1.60×10^{-19} C	1.602 176 487	0.025
Gravitational constant	G	6.67×10^{-11} m³/s²·kg	6.674 28	100
Universal gas constant	R	8.31 J/mol·K	8.314 472	1.7
Avogadro constant	N_A	6.02×10^{23} mol^{-1}	6.022 141 79	0.050
Boltzmann constant	k	1.38×10^{-23} J/K	1.380 650 4	1.7
Stefan–Boltzmann constant	σ	5.67×10^{-8} W/m²·K⁴	5.670 400	7.0
Molar volume of ideal gas at STP[d]	V_m	2.27×10^{-2} m³/mol	2.271 098 1	1.7
Permittivity constant	ϵ_0	8.85×10^{-12} F/m	8.854 187 817 62	exact
Permeability constant	μ_0	1.26×10^{-6} H/m	1.256 637 061 43	exact
Planck constant	h	6.63×10^{-34} J·s	6.626 068 96	0.050
Electron mass[c]	m_e	9.11×10^{-31} kg	9.109 382 15	0.050
		5.49×10^{-4} u	5.485 799 094 3	4.2×10^{-4}
Proton mass[c]	m_p	1.67×10^{-27} kg	1.672 621 637	0.050
		1.0073 u	1.007 276 466 77	1.0×10^{-4}
Ratio of proton mass to electron mass	m_p/m_e	1840	1836.152 672 47	4.3×10^{-4}
Electron charge-to-mass ratio	e/m_e	1.76×10^{11} C/kg	1.758 820 150	0.025
Neutron mass[c]	m_n	1.68×10^{-27} kg	1.674 927 211	0.050
		1.0087 u	1.008 664 915 97	4.3×10^{-4}
Hydrogen atom mass[c]	$m_{^1H}$	1.0078 u	1.007 825 031 6	0.0005
Deuterium atom mass[c]	$m_{^2H}$	2.0136 u	2.013 553 212 724	3.9×10^{-5}
Helium atom mass[c]	$m_{^4He}$	4.0026 u	4.002 603 2	0.067
Muon mass	m_μ	1.88×10^{-28} kg	1.883 531 30	0.056
Electron magnetic moment	μ_e	9.28×10^{-24} J/T	9.284 763 77	0.025
Proton magnetic moment	μ_p	1.41×10^{-26} J/T	1.410 606 662	0.026
Bohr magneton	μ_B	9.27×10^{-24} J/T	9.274 009 15	0.025
Nuclear magneton	μ_N	5.05×10^{-27} J/T	5.050 783 24	0.025
Bohr radius	a	5.29×10^{-11} m	5.291 772 085 9	6.8×10^{-4}
Rydberg constant	R	1.10×10^{7} m^{-1}	1.097 373 156 852 7	6.6×10^{-6}
Electron Compton wavelength	λ_C	2.43×10^{-12} m	2.426 310 217 5	0.0014

[a]Values given in this column should be given the same unit and power of 10 as the computational value.
[b]Parts per million.
[c]Masses given in u are in unified atomic mass units, where 1 u = $1.660\,538\,782 \times 10^{-27}$ kg.
[d]STP means standard temperature and pressure: 0°C and 1.0 atm (0.1 MPa).

*The values in this table were selected from the 1998 CODATA recommended values (www.physics.nist.gov).

APPENDIX C

SOME ASTRONOMICAL DATA

Some Distances from Earth

To the Moon*	3.82×10^8 m	To the center of our galaxy	2.2×10^{20} m
To the Sun*	1.50×10^{11} m	To the Andromeda Galaxy	2.1×10^{22} m
To the nearest star (Proxima Centauri)	4.04×10^{16} m	To the edge of the observable universe	$\sim 10^{26}$ m

*Mean distance.

The Sun, Earth, and the Moon

Property	Unit	Sun	Earth	Moon
Mass	kg	1.99×10^{30}	5.98×10^{24}	7.36×10^{22}
Mean radius	m	6.96×10^8	6.37×10^6	1.74×10^6
Mean density	kg/m^3	1410	5520	3340
Free-fall acceleration at the surface	m/s^2	274	9.81	1.67
Escape velocity	km/s	618	11.2	2.38
Period of rotation[a]	—	37 d at poles[b] 26 d at equator[b]	23 h 56 min	27.3 d
Radiation power[c]	W	3.90×10^{26}		

[a]Measured with respect to the distant stars.
[b]The Sun, a ball of gas, does not rotate as a rigid body.
[c]Just outside Earth's atmosphere solar energy is received, assuming normal incidence, at the rate of 1340 W/m^2.

Some Properties of the Planets

	Mercury	Venus	Earth	Mars	Jupiter	Saturn	Uranus	Neptune	Pluto[d]
Mean distance from Sun, 10^6 km	57.9	108	150	228	778	1430	2870	4500	5900
Period of revolution, y	0.241	0.615	1.00	1.88	11.9	29.5	84.0	165	248
Period of rotation,[a] d	58.7	−243[b]	0.997	1.03	0.409	0.426	−0.451[b]	0.658	6.39
Orbital speed, km/s	47.9	35.0	29.8	24.1	13.1	9.64	6.81	5.43	4.74
Inclination of axis to orbit	<28°	≈3°	23.4°	25.0°	3.08°	26.7°	97.9°	29.6°	57.5°
Inclination of orbit to Earth's orbit	7.00°	3.39°		1.85°	1.30°	2.49°	0.77°	1.77°	17.2°
Eccentricity of orbit	0.206	0.0068	0.0167	0.0934	0.0485	0.0556	0.0472	0.0086	0.250
Equatorial diameter, km	4880	12 100	12 800	6790	143 000	120 000	51 800	49 500	2300
Mass (Earth = 1)	0.0558	0.815	1.000	0.107	318	95.1	14.5	17.2	0.002
Density (water = 1)	5.60	5.20	5.52	3.95	1.31	0.704	1.21	1.67	2.03
Surface value of g,[c] m/s^2	3.78	8.60	9.78	3.72	22.9	9.05	7.77	11.0	0.5
Escape velocity,[c] km/s	4.3	10.3	11.2	5.0	59.5	35.6	21.2	23.6	1.3
Known satellites	0	0	1	2	67 + ring	62 + rings	27 + rings	13 + rings	4

[a]Measured with respect to the distant stars.
[b]Venus and Uranus rotate opposite their orbital motion.
[c]Gravitational acceleration measured at the planet's equator.
[d]Pluto is now classified as a dwarf planet.

APPENDIX D

CONVERSION FACTORS

Conversion factors may be read directly from these tables. For example, 1 degree = 2.778×10^{-3} revolutions, so $16.7° = 16.7 \times 2.778 \times 10^{-3}$ rev. The SI units are fully capitalized. Adapted in part from G. Shortley and D. Williams, *Elements of Physics,* 1971, Prentice-Hall, Englewood Cliffs, NJ.

Plane Angle

	°	′	″	RADIAN	rev
1 degree =	1	60	3600	1.745×10^{-2}	2.778×10^{-3}
1 minute =	1.667×10^{-2}	1	60	2.909×10^{-4}	4.630×10^{-5}
1 second =	2.778×10^{-4}	1.667×10^{-2}	1	4.848×10^{-6}	7.716×10^{-7}
1 RADIAN =	57.30	3438	2.063×10^{5}	1	0.1592
1 revolution =	360	2.16×10^{4}	1.296×10^{6}	6.283	1

Solid Angle

1 sphere = 4π steradians = 12.57 steradians

Length

	cm	METER	km	in.	ft	mi
1 centimeter =	1	10^{-2}	10^{-5}	0.3937	3.281×10^{-2}	6.214×10^{-6}
1 METER =	100	1	10^{-3}	39.37	3.281	6.214×10^{-4}
1 kilometer =	10^{5}	1000	1	3.937×10^{4}	3281	0.6214
1 inch =	2.540	2.540×10^{-2}	2.540×10^{-5}	1	8.333×10^{-2}	1.578×10^{-5}
1 foot =	30.48	0.3048	3.048×10^{-4}	12	1	1.894×10^{-4}
1 mile =	1.609×10^{5}	1609	1.609	6.336×10^{4}	5280	1

1 angström = 10^{-10} m
1 nautical mile = 1852 m = 1.151 miles = 6076 ft
1 fermi = 10^{-15} m
1 light-year = 9.461×10^{12} km
1 parsec = 3.084×10^{13} km
1 fathom = 6 ft
1 Bohr radius = 5.292×10^{-11} m
1 yard = 3 ft
1 rod = 16.5 ft
1 mil = 10^{-3} in.
1 nm = 10^{-9} m

Area

	METER²	cm²	ft²	in.²
1 SQUARE METER =	1	10^{4}	10.76	1550
1 square centimeter =	10^{-4}	1	1.076×10^{-3}	0.1550
1 square foot =	9.290×10^{-2}	929.0	1	144
1 square inch =	6.452×10^{-4}	6.452	6.944×10^{-3}	1

1 square mile = 2.788×10^{7} ft² = 640 acres
1 barn = 10^{-28} m²
1 acre = 43 560 ft²
1 hectare = 10^{4} m² = 2.471 acres

Volume

	METER^3	cm^3	L	ft^3	in.^{3}
1 CUBIC METER = 1	10^6	1000	35.31	6.102×10^4
1 cubic centimeter = 10^{-6}	1	1.000×10^{-3}	3.531×10^{-5}	6.102×10^{-2}
1 liter = 1.000×10^{-3}	1000	1	3.531×10^{-2}	61.02
1 cubic foot = 2.832×10^{-2}	2.832×10^4	28.32	1	1728
1 cubic inch = 1.639×10^{-5}	16.39	1.639×10^{-2}	5.787×10^{-4}	1

1 U.S. fluid gallon = 4 U.S. fluid quarts = 8 U.S. pints = 128 U.S. fluid ounces = 231 in.^{3}

1 British imperial gallon = 277.4 in.^{3} = 1.201 U.S. fluid gallons

Mass

Quantities in the colored areas are not mass units but are often used as such. For example, when we write 1 kg "=" 2.205 lb, this means that a kilogram is a *mass* that *weighs* 2.205 pounds at a location where g has the standard value of 9.80665 m/s^2.

g	KILOGRAM	slug	u	oz	lb	ton
1 gram = 1	0.001	6.852×10^{-5}	6.022×10^{23}	3.527×10^{-2}	2.205×10^{-3}	1.102×10^{-6}
1 KILOGRAM = 1000	1	6.852×10^{-2}	6.022×10^{26}	35.27	2.205	1.102×10^{-3}
1 slug = 1.459×10^4	14.59	1	8.786×10^{27}	514.8	32.17	1.609×10^{-2}
1 atomic mass unit = 1.661×10^{-24}	1.661×10^{-27}	1.138×10^{-28}	1	5.857×10^{-26}	3.662×10^{-27}	1.830×10^{-30}
1 ounce = 28.35	2.835×10^{-2}	1.943×10^{-3}	1.718×10^{25}	1	6.250×10^{-2}	3.125×10^{-5}
1 pound = 453.6	0.4536	3.108×10^{-2}	2.732×10^{26}	16	1	0.0005
1 ton = 9.072×10^5	907.2	62.16	5.463×10^{29}	3.2×10^4	2000	1

1 metric ton = 1000 kg

Density

Quantities in the colored areas are weight densities and, as such, are dimensionally different from mass densities. See the note for the mass table.

slug/ft^3	KILOGRAM/METER^3	g/cm^3	lb/ft^3	lb/in.^{3}
1 slug per foot^3 = 1	515.4	0.5154	32.17	1.862×10^{-2}
1 KILOGRAM per METER^3 = 1.940×10^{-3}	1	0.001	6.243×10^{-2}	3.613×10^{-5}
1 gram per centimeter^3 = 1.940	1000	1	62.43	3.613×10^{-2}
1 pound per foot^3 = 3.108×10^{-2}	16.02	16.02×10^{-2}	1	5.787×10^{-4}
1 pound per inch^3 = 53.71	2.768×10^4	27.68	1728	1

Time

y	d	h	min	SECOND
1 year = 1	365.25	8.766×10^3	5.259×10^5	3.156×10^7
1 day = 2.738×10^{-3}	1	24	1440	8.640×10^4
1 hour = 1.141×10^{-4}	4.167×10^{-2}	1	60	3600
1 minute = 1.901×10^{-6}	6.944×10^{-4}	1.667×10^{-2}	1	60
1 SECOND = 3.169×10^{-8}	1.157×10^{-5}	2.778×10^{-4}	1.667×10^{-2}	1

Speed

	ft/s	km/h	METER/SECOND	mi/h	cm/s
1 foot per second =	1	1.097	0.3048	0.6818	30.48
1 kilometer per hour =	0.9113	1	0.2778	0.6214	27.78
1 METER per SECOND =	3.281	3.6	1	2.237	100
1 mile per hour =	1.467	1.609	0.4470	1	44.70
1 centimeter per second =	3.281×10^{-2}	3.6×10^{-2}	0.01	2.237×10^{-2}	1

1 knot = 1 nautical mi/h = 1.688 ft/s 1 mi/min = 88.00 ft/s = 60.00 mi/h

Force

Force units in the colored areas are now little used. To clarify: 1 gram-force (= 1 gf) is the force of gravity that would act on an object whose mass is 1 gram at a location where g has the standard value of 9.80665 m/s^2.

	dyne	NEWTON	lb	pdl	gf	kgf
1 dyne =	1	10^{-5}	2.248×10^{-6}	7.233×10^{-5}	1.020×10^{-3}	1.020×10^{-6}
1 NEWTON =	10^5	1	0.2248	7.233	102.0	0.1020
1 pound =	4.448×10^5	4.448	1	32.17	453.6	0.4536
1 poundal =	1.383×10^4	0.1383	3.108×10^{-2}	1	14.10	1.410×10^2
1 gram-force =	980.7	9.807×10^{-3}	2.205×10^{-3}	7.093×10^{-2}	1	0.001
1 kilogram-force =	9.807×10^5	9.807	2.205	70.93	1000	1

1 ton = 2000 lb

Pressure

	atm	dyne/cm^2	inch of water	cm Hg	PASCAL	lb/in.^{2}	lb/ft^2
1 atmosphere =	1	1.013×10^6	406.8	76	1.013×10^5	14.70	2116
1 dyne per centimeter^2 =	9.869×10^{-7}	1	4.015×10^{-4}	7.501×10^{-5}	0.1	1.405×10^{-5}	2.089×10^{-3}
1 inch of water[a] at 4°C =	2.458×10^{-3}	2491	1	0.1868	249.1	3.613×10^{-2}	5.202
1 centimeter of mercury[a] at 0°C =	1.316×10^{-2}	1.333×10^4	5.353	1	1333	0.1934	27.85
1 PASCAL =	9.869×10^{-6}	10	4.015×10^{-3}	7.501×10^{-4}	1	1.450×10^{-4}	2.089×10^{-2}
1 pound per inch^2 =	6.805×10^{-2}	6.895×10^4	27.68	5.171	6.895×10^3	1	144
1 pound per foot^2 =	4.725×10^{-4}	478.8	0.1922	3.591×10^{-2}	47.88	6.944×10^{-3}	1

[a]Where the acceleration of gravity has the standard value of 9.80665 m/s^2.

1 bar = 10^6 dyne/cm^2 = 0.1 MPa 1 millibar = 10^3 dyne/cm^2 = 10^2 Pa 1 torr = 1 mm Hg

Energy, Work, Heat

Quantities in the colored areas are not energy units but are included for convenience. They arise from the relativistic mass–energy equivalence formula $E = mc^2$ and represent the energy released if a kilogram or unified atomic mass unit (u) is completely converted to energy (bottom two rows) or the mass that would be completely converted to one unit of energy (rightmost two columns).

	Btu	erg	ft·lb	hp·h	JOULE	cal	kW·h	eV	MeV	kg	u
1 British thermal unit =	1	1.055×10^{10}	777.9	3.929×10^{-4}	1055	252.0	2.930×10^{-4}	6.585×10^{21}	6.585×10^{15}	1.174×10^{-14}	7.070×10^{12}
1 erg =	9.481×10^{-11}	1	7.376×10^{-8}	3.725×10^{-14}	10^{-7}	2.389×10^{-8}	2.778×10^{-14}	6.242×10^{11}	6.242×10^{5}	1.113×10^{-24}	670.2
1 foot-pound =	1.285×10^{-3}	1.356×10^{7}	1	5.051×10^{-7}	1.356	0.3238	3.766×10^{-7}	8.464×10^{18}	8.464×10^{12}	1.509×10^{-17}	9.037×10^{9}
1 horsepower-hour =	2545	2.685×10^{13}	1.980×10^{6}	1	2.685×10^{6}	6.413×10^{5}	0.7457	1.676×10^{25}	1.676×10^{19}	2.988×10^{-11}	1.799×10^{16}
1 JOULE =	9.481×10^{-4}	10^{7}	0.7376	3.725×10^{-7}	1	0.2389	2.778×10^{-7}	6.242×10^{18}	6.242×10^{12}	1.113×10^{-17}	6.702×10^{9}
1 calorie =	3.968×10^{-3}	4.1868×10^{7}	3.088	1.560×10^{-6}	4.1868	1	1.163×10^{-6}	2.613×10^{19}	2.613×10^{13}	4.660×10^{-17}	2.806×10^{10}
1 kilowatt-hour =	3413	3.600×10^{13}	2.655×10^{6}	1.341	3.600×10^{6}	8.600×10^{5}	1	2.247×10^{25}	2.247×10^{19}	4.007×10^{-11}	2.413×10^{16}
1 electron-volt =	1.519×10^{-22}	1.602×10^{-12}	1.182×10^{-19}	5.967×10^{-26}	1.602×10^{-19}	3.827×10^{-20}	4.450×10^{-26}	1	10^{-6}	1.783×10^{-36}	1.074×10^{-9}
1 million electron-volts =	1.519×10^{-16}	1.602×10^{-6}	1.182×10^{-13}	5.967×10^{-20}	1.602×10^{-13}	3.827×10^{-14}	4.450×10^{-20}	10^{-6}	1	1.783×10^{-30}	1.074×10^{-3}
1 kilogram =	8.521×10^{13}	8.987×10^{23}	6.629×10^{16}	3.348×10^{10}	8.987×10^{16}	2.146×10^{16}	2.497×10^{10}	5.610×10^{35}	5.610×10^{29}	1	6.022×10^{26}
1 unified atomic mass unit =	1.415×10^{-13}	1.492×10^{-3}	1.101×10^{-10}	5.559×10^{-17}	1.492×10^{-10}	3.564×10^{-11}	4.146×10^{-17}	9.320×10^{8}	932.0	1.661×10^{-27}	1

Power

	Btu/h	ft·lb/s	hp	cal/s	kW	WATT
1 British thermal unit per hour =	1	0.2161	3.929×10^{-4}	6.998×10^{-2}	2.930×10^{-4}	0.2930
1 foot-pound per second =	4.628	1	1.818×10^{-3}	0.3239	1.356×10^{-3}	1.356
1 horsepower =	2545	550	1	178.1	0.7457	745.7
1 calorie per second =	14.29	3.088	5.615×10^{-3}	1	4.186×10^{-3}	4.186
1 kilowatt =	3413	737.6	1.341	238.9	1	1000
1 WATT =	3.413	0.7376	1.341×10^{-3}	0.2389	0.001	1

Magnetic Field

	gauss	TESLA	milligauss
1 gauss =	1	10^{-4}	1000
1 TESLA =	10^{4}	1	10^{7}
1 milligauss =	0.001	10^{-7}	1

1 tesla = 1 weber/meter^2

Magnetic Flux

	maxwell	WEBER
1 maxwell =	1	10^{-8}
1 WEBER =	10^{8}	1

APPENDIX E

MATHEMATICAL FORMULAS

Geometry

Circle of radius r: circumference $= 2\pi r$; area $= \pi r^2$.

Sphere of radius r: area $= 4\pi r^2$; volume $= \frac{4}{3}\pi r^3$.

Right circular cylinder of radius r and height h:
area $= 2\pi r^2 + 2\pi rh$; volume $= \pi r^2 h$.

Triangle of base a and altitude h: area $= \frac{1}{2}ah$.

Quadratic Formula

If $ax^2 + bx + c = 0$, then $x = \dfrac{-b \pm \sqrt{b^2 - 4ac}}{2a}$.

Trigonometric Functions of Angle θ

$$\sin\theta = \frac{y}{r} \qquad \cos\theta = \frac{x}{r}$$

$$\tan\theta = \frac{y}{x} \qquad \cot\theta = \frac{x}{y}$$

$$\sec\theta = \frac{r}{x} \qquad \csc\theta = \frac{r}{y}$$

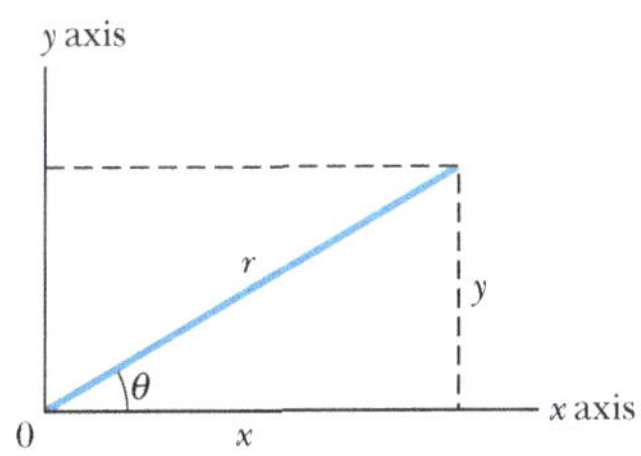

Pythagorean Theorem

In this right triangle,
$a^2 + b^2 = c^2$

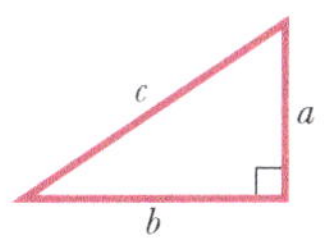

Triangles

Angles are A, B, C

Opposite sides are a, b, c

Angles $A + B + C = 180°$

$$\frac{\sin A}{a} = \frac{\sin B}{b} = \frac{\sin C}{c}$$

$c^2 = a^2 + b^2 - 2ab\cos C$

Exterior angle $D = A + C$

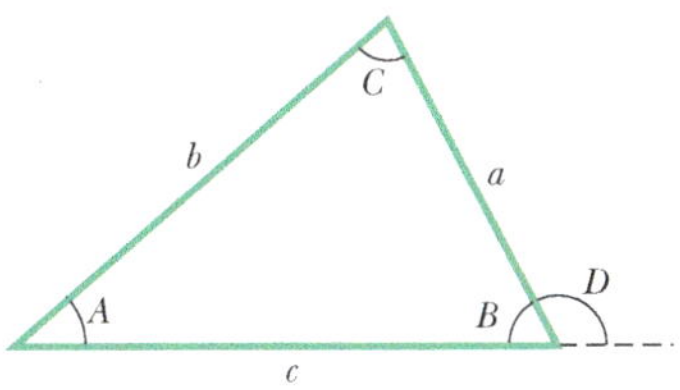

Mathematical Signs and Symbols

$=$ equals

$\approx$ equals approximately

$\sim$ is the order of magnitude of

$\neq$ is not equal to

$\equiv$ is identical to, is defined as

$>$ is greater than ($\gg$ is much greater than)

$<$ is less than ($\ll$ is much less than)

$\geq$ is greater than or equal to (or, is no less than)

$\leq$ is less than or equal to (or, is no more than)

$\pm$ plus or minus

$\propto$ is proportional to

Σ the sum of

x_{avg} the average value of x

Trigonometric Identities

$\sin(90° - \theta) = \cos\theta$

$\cos(90° - \theta) = \sin\theta$

$\sin\theta/\cos\theta = \tan\theta$

$\sin^2\theta + \cos^2\theta = 1$

$\sec^2\theta - \tan^2\theta = 1$

$\csc^2\theta - \cot^2\theta = 1$

$\sin 2\theta = 2\sin\theta\cos\theta$

$\cos 2\theta = \cos^2\theta - \sin^2\theta = 2\cos^2\theta - 1 = 1 - 2\sin^2\theta$

$\sin(\alpha \pm \beta) = \sin\alpha\cos\beta \pm \cos\alpha\sin\beta$

$\cos(\alpha \pm \beta) = \cos\alpha\cos\beta \mp \sin\alpha\sin\beta$

$$\tan(\alpha \pm \beta) = \frac{\tan\alpha \pm \tan\beta}{1 \mp \tan\alpha\tan\beta}$$

$\sin\alpha \pm \sin\beta = 2\sin\frac{1}{2}(\alpha \pm \beta)\cos\frac{1}{2}(\alpha \mp \beta)$

$\cos\alpha + \cos\beta = 2\cos\frac{1}{2}(\alpha + \beta)\cos\frac{1}{2}(\alpha - \beta)$

$\cos\alpha - \cos\beta = -2\sin\frac{1}{2}(\alpha + \beta)\sin\frac{1}{2}(\alpha - \beta)$

Binomial Theorem

$$(1 + x)^n = 1 + \frac{nx}{1!} + \frac{n(n-1)x^2}{2!} + \cdots \qquad (x^2 < 1)$$

Exponential Expansion

$$e^x = 1 + x + \frac{x^2}{2!} + \frac{x^3}{3!} + \cdots$$

Logarithmic Expansion

$$\ln(1 + x) = x - \tfrac{1}{2}x^2 + \tfrac{1}{3}x^3 - \cdots \qquad (|x| < 1)$$

Trigonometric Expansions (θ in radians)

$$\sin\theta = \theta - \frac{\theta^3}{3!} + \frac{\theta^5}{5!} - \cdots$$

$$\cos\theta = 1 - \frac{\theta^2}{2!} + \frac{\theta^4}{4!} - \cdots$$

$$\tan\theta = \theta + \frac{\theta^3}{3} + \frac{2\theta^5}{15} + \cdots$$

Cramer's Rule

Two simultaneous equations in unknowns x and y,

$$a_1x + b_1y = c_1 \quad \text{and} \quad a_2x + b_2y = c_2,$$

have the solutions

$$x = \frac{\begin{vmatrix} c_1 & b_1 \\ c_2 & b_2 \end{vmatrix}}{\begin{vmatrix} a_1 & b_1 \\ a_2 & b_2 \end{vmatrix}} = \frac{c_1b_2 - c_2b_1}{a_1b_2 - a_2b_1}$$

and

$$y = \frac{\begin{vmatrix} a_1 & c_1 \\ a_2 & c_2 \end{vmatrix}}{\begin{vmatrix} a_1 & b_1 \\ a_2 & b_2 \end{vmatrix}} = \frac{a_1c_2 - a_2c_1}{a_1b_2 - a_2b_1}.$$

Products of Vectors

Let $\hat{\text{i}}$, $\hat{\text{j}}$, and $\hat{\text{k}}$ be unit vectors in the x, y, and z directions. Then

$$\hat{\text{i}}\cdot\hat{\text{i}} = \hat{\text{j}}\cdot\hat{\text{j}} = \hat{\text{k}}\cdot\hat{\text{k}} = 1, \quad \hat{\text{i}}\cdot\hat{\text{j}} = \hat{\text{j}}\cdot\hat{\text{k}} = \hat{\text{k}}\cdot\hat{\text{i}} = 0,$$

$$\hat{\text{i}}\times\hat{\text{i}} = \hat{\text{j}}\times\hat{\text{j}} = \hat{\text{k}}\times\hat{\text{k}} = 0,$$

$$\hat{\text{i}}\times\hat{\text{j}} = \hat{\text{k}}, \quad \hat{\text{j}}\times\hat{\text{k}} = \hat{\text{i}}, \quad \hat{\text{k}}\times\hat{\text{i}} = \hat{\text{j}}$$

Any vector $\vec{a}$ with components a_x, a_y, and a_z along the x, y, and z axes can be written as

$$\vec{a} = a_x\hat{\text{i}} + a_y\hat{\text{j}} + a_z\hat{\text{k}}.$$

Let $\vec{a}$, $\vec{b}$, and $\vec{c}$ be arbitrary vectors with magnitudes a, b, and c. Then

$$\vec{a}\times(\vec{b} + \vec{c}) = (\vec{a}\times\vec{b}) + (\vec{a}\times\vec{c})$$

$$(s\vec{a})\times\vec{b} = \vec{a}\times(s\vec{b}) = s(\vec{a}\times\vec{b}) \qquad (s = \text{a scalar}).$$

Let θ be the smaller of the two angles between $\vec{a}$ and $\vec{b}$. Then

$$\vec{a}\cdot\vec{b} = \vec{b}\cdot\vec{a} = a_xb_x + a_yb_y + a_zb_z = ab\cos\theta$$

$$\vec{a}\times\vec{b} = -\vec{b}\times\vec{a} = \begin{vmatrix} \hat{\text{i}} & \hat{\text{j}} & \hat{\text{k}} \\ a_x & a_y & a_z \\ b_x & b_y & b_z \end{vmatrix}$$

$$= \hat{\text{i}}\begin{vmatrix} a_y & a_z \\ b_y & b_z \end{vmatrix} - \hat{\text{j}}\begin{vmatrix} a_x & a_z \\ b_x & b_z \end{vmatrix} + \hat{\text{k}}\begin{vmatrix} a_x & a_y \\ b_x & b_y \end{vmatrix}$$

$$= (a_yb_z - b_ya_z)\hat{\text{i}} + (a_zb_x - b_za_x)\hat{\text{j}} + (a_xb_y - b_xa_y)\hat{\text{k}}$$

$$|\vec{a}\times\vec{b}| = ab\sin\theta$$

$$\vec{a}\cdot(\vec{b}\times\vec{c}) = \vec{b}\cdot(\vec{c}\times\vec{a}) = \vec{c}\cdot(\vec{a}\times\vec{b})$$

$$\vec{a}\times(\vec{b}\times\vec{c}) = (\vec{a}\cdot\vec{c})\vec{b} - (\vec{a}\cdot\vec{b})\vec{c}$$

Derivatives and Integrals

In what follows, the letters u and v stand for any functions of x, and a and m are constants. To each of the indefinite integrals should be added an arbitrary constant of integration. The *Handbook of Chemistry and Physics* (CRC Press Inc.) gives a more extensive tabulation.

1. $\frac{dx}{dx} = 1$
2. $\frac{d}{dx}(au) = a\frac{du}{dx}$
3. $\frac{d}{dx}(u + v) = \frac{du}{dx} + \frac{dv}{dx}$
4. $\frac{d}{dx}x^m = mx^{m-1}$
5. $\frac{d}{dx}\ln x = \frac{1}{x}$
6. $\frac{d}{dx}(uv) = u\frac{dv}{dx} + v\frac{du}{dx}$
7. $\frac{d}{dx}e^x = e^x$
8. $\frac{d}{dx}\sin x = \cos x$
9. $\frac{d}{dx}\cos x = -\sin x$
10. $\frac{d}{dx}\tan x = \sec^2 x$
11. $\frac{d}{dx}\cot x = -\csc^2 x$
12. $\frac{d}{dx}\sec x = \tan x \sec x$
13. $\frac{d}{dx}\csc x = -\cot x \csc x$
14. $\frac{d}{dx}e^u = e^u\frac{du}{dx}$
15. $\frac{d}{dx}\sin u = \cos u\frac{du}{dx}$
16. $\frac{d}{dx}\cos u = -\sin u\frac{du}{dx}$

1. $\int dx = x$
2. $\int au\, dx = a\int u\, dx$
3. $\int (u + v)\, dx = \int u\, dx + \int v\, dx$
4. $\int x^m\, dx = \frac{x^{m+1}}{m+1} \quad (m \neq -1)$
5. $\int \frac{dx}{x} = \ln |x|$
6. $\int u\frac{dv}{dx}\, dx = uv - \int v\frac{du}{dx}\, dx$
7. $\int e^x\, dx = e^x$
8. $\int \sin x\, dx = -\cos x$
9. $\int \cos x\, dx = \sin x$
10. $\int \tan x\, dx = \ln |\sec x|$
11. $\int \sin^2 x\, dx = \frac{1}{2}x - \frac{1}{4}\sin 2x$
12. $\int e^{-ax}\, dx = -\frac{1}{a}e^{-ax}$
13. $\int xe^{-ax}\, dx = -\frac{1}{a^2}(ax + 1)\, e^{-ax}$
14. $\int x^2e^{-ax}\, dx = -\frac{1}{a^3}(a^2x^2 + 2ax + 2)e^{-ax}$
15. $\int_0^\infty x^ne^{-ax}\, dx = \frac{n!}{a^{n+1}}$
16. $\int_0^\infty x^{2n}e^{-ax^2}\, dx = \frac{1\cdot 3\cdot 5\cdots(2n-1)}{2^{n+1}a^n}\sqrt{\frac{\pi}{a}}$
17. $\int \frac{dx}{\sqrt{x^2 + a^2}} = \ln(x + \sqrt{x^2 + a^2})$
18. $\int \frac{x\, dx}{(x^2 + a^2)^{3/2}} = -\frac{1}{(x^2 + a^2)^{1/2}}$
19. $\int \frac{dx}{(x^2 + a^2)^{3/2}} = \frac{x}{a^2(x^2 + a^2)^{1/2}}$
20. $\int_0^\infty x^{2n+1}\, e^{-ax^2}\, dx = \frac{n!}{2a^{n+1}} \quad (a > 0)$
21. $\int \frac{x\, dx}{x + d} = x - d\ln(x + d)$

APPENDIX F

PROPERTIES OF THE ELEMENTS

All physical properties are for a pressure of 1 atm unless otherwise specified.

Element	Symbol	Atomic Number Z	Molar Mass, g/mol	Density, g/cm^3 at 20°C	Melting Point, °C	Boiling Point, °C	Specific Heat, J/(g·°C) at 25°C
Actinium	Ac	89	(227)	10.06	1323	(3473)	0.092
Aluminum	Al	13	26.9815	2.699	660	2450	0.900
Americium	Am	95	(243)	13.67	1541	—	—
Antimony	Sb	51	121.75	6.691	630.5	1380	0.205
Argon	Ar	18	39.948	1.6626×10^{-3}	−189.4	−185.8	0.523
Arsenic	As	33	74.9216	5.78	817 (28 atm)	613	0.331
Astatine	At	85	(210)	—	(302)	—	—
Barium	Ba	56	137.34	3.594	729	1640	0.205
Berkelium	Bk	97	(247)	14.79	—	—	—
Beryllium	Be	4	9.0122	1.848	1287	2770	1.83
Bismuth	Bi	83	208.980	9.747	271.37	1560	0.122
Bohrium	Bh	107	262.12	—	—	—	—
Boron	B	5	10.811	2.34	2030	—	1.11
Bromine	Br	35	79.909	3.12 (liquid)	−7.2	58	0.293
Cadmium	Cd	48	112.40	8.65	321.03	765	0.226
Calcium	Ca	20	40.08	1.55	838	1440	0.624
Californium	Cf	98	(251)	—	—	—	—
Carbon	C	6	12.01115	2.26	3727	4830	0.691
Cerium	Ce	58	140.12	6.768	804	3470	0.188
Cesium	Cs	55	132.905	1.873	28.40	690	0.243
Chlorine	Cl	17	35.453	3.214×10^{-3} (0°C)	−101	−34.7	0.486
Chromium	Cr	24	51.996	7.19	1857	2665	0.448
Cobalt	Co	27	58.9332	8.85	1495	2900	0.423
Copernicium	Cn	112	(285)	—	—	—	—
Copper	Cu	29	63.54	8.96	1083.40	2595	0.385
Curium	Cm	96	(247)	13.3	—	—	—
Darmstadtium	Ds	110	(271)	—	—	—	—
Dubnium	Db	105	262.114	—	—	—	—
Dysprosium	Dy	66	162.50	8.55	1409	2330	0.172
Einsteinium	Es	99	(254)	—	—	—	—
Erbium	Er	68	167.26	9.15	1522	2630	0.167
Europium	Eu	63	151.96	5.243	817	1490	0.163
Fermium	Fm	100	(237)	—	—	—	—
Flerovium*	Fl	114	(289)	—	—	—	—
Fluorine	F	9	18.9984	1.696×10^{-3} (0°C)	−219.6	−188.2	0.753
Francium	Fr	87	(223)	—	(27)	—	—
Gadolinium	Gd	64	157.25	7.90	1312	2730	0.234
Gallium	Ga	31	69.72	5.907	29.75	2237	0.377
Germanium	Ge	32	72.59	5.323	937.25	2830	0.322
Gold	Au	79	196.967	19.32	1064.43	2970	0.131

Element	Symbol	Atomic Number Z	Molar Mass, g/mol	Density, g/cm^3 at 20°C	Melting Point, °C	Boiling Point, °C	Specific Heat, J/(g·°C) at 25°C
Hafnium	Hf	72	178.49	13.31	2227	5400	0.144
Hassium	Hs	108	(265)	—	—	—	—
Helium	He	2	4.0026	0.1664×10^{-3}	−269.7	−268.9	5.23
Holmium	Ho	67	164.930	8.79	1470	2330	0.165
Hydrogen	H	1	1.00797	0.08375×10^{-3}	−259.19	−252.7	14.4
Indium	In	49	114.82	7.31	156.634	2000	0.233
Iodine	I	53	126.9044	4.93	113.7	183	0.218
Iridium	Ir	77	192.2	22.5	2447	(5300)	0.130
Iron	Fe	26	55.847	7.874	1536.5	3000	0.447
Krypton	Kr	36	83.80	3.488×10^{-3}	−157.37	−152	0.247
Lanthanum	La	57	138.91	6.189	920	3470	0.195
Lawrencium	Lr	103	(257)	—	—	—	—
Lead	Pb	82	207.19	11.35	327.45	1725	0.129
Lithium	Li	3	6.939	0.534	180.55	1300	3.58
Livermorium*	Lv	116	(293)	—	—	—	—
Lutetium	Lu	71	174.97	9.849	1663	1930	0.155
Magnesium	Mg	12	24.312	1.738	650	1107	1.03
Manganese	Mn	25	54.9380	7.44	1244	2150	0.481
Meitnerium	Mt	109	(266)	—	—	—	—
Mendelevium	Md	101	(256)	—	—	—	—
Mercury	Hg	80	200.59	13.55	−38.87	357	0.138
Molybdenum	Mo	42	95.94	10.22	2617	5560	0.251
Neodymium	Nd	60	144.24	7.007	1016	3180	0.188
Neon	Ne	10	20.183	0.8387×10^{-3}	−248.597	−246.0	1.03
Neptunium	Np	93	(237)	20.25	637	—	1.26
Nickel	Ni	28	58.71	8.902	1453	2730	0.444
Niobium	Nb	41	92.906	8.57	2468	4927	0.264
Nitrogen	N	7	14.0067	1.1649×10^{-3}	−210	−195.8	1.03
Nobelium	No	102	(255)	—	—	—	—
Osmium	Os	76	190.2	22.59	3027	5500	0.130
Oxygen	O	8	15.9994	1.3318×10^{-3}	−218.80	−183.0	0.913
Palladium	Pd	46	106.4	12.02	1552	3980	0.243
Phosphorus	P	15	30.9738	1.83	44.25	280	0.741
Platinum	Pt	78	195.09	21.45	1769	4530	0.134
Plutonium	Pu	94	(244)	19.8	640	3235	0.130
Polonium	Po	84	(210)	9.32	254	—	—
Potassium	K	19	39.102	0.862	63.20	760	0.758
Praseodymium	Pr	59	140.907	6.773	931	3020	0.197
Promethium	Pm	61	(145)	7.22	(1027)	—	—
Protactinium	Pa	91	(231)	15.37 (estimated)	(1230)	—	—
Radium	Ra	88	(226)	5.0	700	—	—
Radon	Rn	86	(222)	9.96×10^{-3} (0°C)	(−71)	−61.8	0.092
Rhenium	Re	75	186.2	21.02	3180	5900	0.134
Rhodium	Rh	45	102.905	12.41	1963	4500	0.243
Roentgenium	Rg	111	(280)	—	—	—	—
Rubidium	Rb	37	85.47	1.532	39.49	688	0.364
Ruthenium	Ru	44	101.107	12.37	2250	4900	0.239
Rutherfordium	Rf	104	261.11	—	—	—	—

Element	Symbol	Atomic Number Z	Molar Mass, g/mol	Density, g/cm^3 at 20°C	Melting Point, °C	Boiling Point, °C	Specific Heat, J/(g·°C) at 25°C
Samarium	Sm	62	150.35	7.52	1072	1630	0.197
Scandium	Sc	21	44.956	2.99	1539	2730	0.569
Seaborgium	Sg	106	263.118	—	—	—	—
Selenium	Se	34	78.96	4.79	221	685	0.318
Silicon	Si	14	28.086	2.33	1412	2680	0.712
Silver	Ag	47	107.870	10.49	960.8	2210	0.234
Sodium	Na	11	22.9898	0.9712	97.85	892	1.23
Strontium	Sr	38	87.62	2.54	768	1380	0.737
Sulfur	S	16	32.064	2.07	119.0	444.6	0.707
Tantalum	Ta	73	180.948	16.6	3014	5425	0.138
Technetium	Tc	43	(99)	11.46	2200	—	0.209
Tellurium	Te	52	127.60	6.24	449.5	990	0.201
Terbium	Tb	65	158.924	8.229	1357	2530	0.180
Thallium	Tl	81	204.37	11.85	304	1457	0.130
Thorium	Th	90	(232)	11.72	1755	(3850)	0.117
Thulium	Tm	69	168.934	9.32	1545	1720	0.159
Tin	Sn	50	118.69	7.2984	231.868	2270	0.226
Titanium	Ti	22	47.90	4.54	1670	3260	0.523
Tungsten	W	74	183.85	19.3	3380	5930	0.134
Unnamed	Uut	113	(284)	—	—	—	—
Unnamed	Uup	115	(288)	—	—	—	—
Unnamed	Uus	117	—	—	—	—	—
Unnamed	Uuo	118	(294)	—	—	—	—
Uranium	U	92	(238)	18.95	1132	3818	0.117
Vanadium	V	23	50.942	6.11	1902	3400	0.490
Xenon	Xe	54	131.30	5.495×10^{-3}	−111.79	−108	0.159
Ytterbium	Yb	70	173.04	6.965	824	1530	0.155
Yttrium	Y	39	88.905	4.469	1526	3030	0.297
Zinc	Zn	30	65.37	7.133	419.58	906	0.389
Zirconium	Zr	40	91.22	6.506	1852	3580	0.276

The values in parentheses in the column of molar masses are the mass numbers of the longest-lived isotopes of those elements that are radioactive. Melting points and boiling points in parentheses are uncertain.

The data for gases are valid only when these are in their usual molecular state, such as H_2, He, O_2, Ne, etc. The specific heats of the gases are the values at constant pressure.

Source: Adapted from J. Emsley, *The Elements,* 3rd ed., 1998, Clarendon Press, Oxford. See also www.webelements.com for the latest values and newest elements.

*The names and symbols for elements 114 (Flerovium, Fl) and 116 (Livermorium, Lv) have been suggested but are not official.

APPENDIX G

PERIODIC TABLE OF THE ELEMENTS

Metals

Metalloids

Nonmetals

Alkali metals IA; Noble gases 0; Transition metals; Inner transition metals

THE HORIZONTAL PERIODS

Period	IA	IIA	IIIB	IVB	VB	VIB	VIIB	VIIIB	VIIIB	VIIIB	IB	IIB	IIIA	IVA	VA	VIA	VIIA	0
1	1 H																	2 He
2	3 Li	4 Be											5 B	6 C	7 N	8 O	9 F	10 Ne
3	11 Na	12 Mg											13 Al	14 Si	15 P	16 S	17 Cl	18 Ar
4	19 K	20 Ca	21 Sc	22 Ti	23 V	24 Cr	25 Mn	26 Fe	27 Co	28 Ni	29 Cu	30 Zn	31 Ga	32 Ge	33 As	34 Se	35 Br	36 Kr
5	37 Rb	38 Sr	39 Y	40 Zr	41 Nb	42 Mo	43 Tc	44 Ru	45 Rh	46 Pd	47 Ag	48 Cd	49 In	50 Sn	51 Sb	52 Te	53 I	54 Xe
6	55 Cs	56 Ba	57-71 *	72 Hf	73 Ta	74 W	75 Re	76 Os	77 Ir	78 Pt	79 Au	80 Hg	81 Tl	82 Pb	83 Bi	84 Po	85 At	86 Rn
7	87 Fr	88 Ra	89-103 †	104 Rf	105 Db	106 Sg	107 Bh	108 Hs	109 Mt	110 Ds	111 Rg	112 Cn	113	114 Fl	115	116 Lv	117	118

Inner transition metals

Lanthanide series *	57 La	58 Ce	59 Pr	60 Nd	61 Pm	62 Sm	63 Eu	64 Gd	65 Tb	66 Dy	67 Ho	68 Er	69 Tm	70 Yb	71 Lu
Actinide series †	89 Ac	90 Th	91 Pa	92 U	93 Np	94 Pu	95 Am	96 Cm	97 Bk	98 Cf	99 Es	100 Fm	101 Md	102 No	103 Lr

Evidence for the discovery of elements 113 through 118 has been reported. See www.webelements.com for the latest information and newest elements. The names and symbols for elements 114 and 116 have been suggested but are not official.

ANSWERS

To Checkpoints and Odd-Numbered Questions and Problems

Chapter 1

P **1.** (a) 4.00×10^4 km; (b) 5.10×10^8 km^2; (c) 1.08×10^{12} km^3 **3.** (a) 10^9 μm; (b) 10^{-4}; (c) 9.1×10^5 μm **5.** (a) 160 rods; (b) 40 chains **7.** 1.1×10^3 acre-feet **9.** 1.9×10^{22} cm^3 **11.** (a) 1.43; (b) 0.864 **13.** (a) 495 s; (b) 141 s; (c) 198 s; (d) −245 s **15.** 1.21×10^{12} μs **17.** C, D, A, B, E; the important criterion is the consistency of the daily variation, not its magnitude **19.** 5.2×10^6 m **21.** 9.0×10^{49} atoms **23.** (a) 1×10^3 kg; (b) 158 kg/s **25.** 1.9×10^5 kg **27.** (a) 1.18×10^{-29} m^3; (b) 0.282 nm **29.** 1.75×10^3 kg **31.** 1.43 kg/min **33.** (a) 293 U.S. bushels; (b) 3.81×10^3 U.S. bushels **35.** (a) 22 pecks; (b) 5.5 Imperial bushels; (c) 200 L **37.** 8×10^2 km **39.** (a) 18.8 gallons; (b) 22.5 gallons **41.** 0.3 cord **43.** 3.8 mg/s **45.** (a) yes; (b) 8.6 universe seconds **47.** 0.12 AU/min **49.** (a) 3.88; (b) 7.65; (c) 156 ken^3; (d) 1.19×10^3 m^3 **51.** (a) 3.9 m, 4.8 m; (b) 3.9×10^3 mm, 4.8×10^3 mm; (c) 2.2 m^3, 4.2 m^3 **53.** (a) 4.9×10^{-6} pc; (b) 1.6×10^{-5} ly **55.** (a) 3 nebuchadnezzars, 1 methuselah; (b) 0.37 standard bottle; (c) 0.26 L **57.** 10.7 habaneros **59.** 700 to 1500 oysters

Chapter 2

CP **1.** b and c **2.** (check the derivative dx/dt) (a) 1 and 4; (b) 2 and 3 **3.** (a) plus; (b) minus; (c) minus; (d) plus **4.** 1 and 4 ($a = d^2x/dt^2$ must be constant) **5.** (a) plus (upward displacement on y axis); (b) minus (downward displacement on y axis); (c) $a = -g = -9.8$ m/s^2

Q **1.** (a) negative; (b) positive; (c) yes; (d) positive; (e) constant **3.** (a) all tie; (b) 4, tie of 1 and 2, then 3 **5.** (a) positive direction; (b) negative direction; (c) 3 and 5; (d) 2 and 6 tie, then 3 and 5 tie, then 1 and 4 tie (zero) **7.** (a) D; (b) E **9.** (a) 3, 2, 1; (b) 1, 2, 3; (c) all tie; (d) 1, 2, 3 **11.** 1 and 2 tie, then 3

P **1.** 13 m **3.** (a) +40 km/h; (b) 40 km/h **5.** (a) 0; (b) −2 m; (c) 0; (d) 12 m; (e) +12 m; (f) +7 m/s **7.** 60 km **9.** 1.4 m **11.** 128 km/h **13.** (a) 73 km/h; (b) 68 km/h; (c) 70 km/h; (d) 0 **15.** (a) −6 m/s; (b) $-x$ direction; (c) 6 m/s; (d) decreasing; (e) 2 s; (f) no **17.** (a) 28.5 cm/s; (b) 18.0 cm/s; (c) 40.5 cm/s; (d) 28.1 cm/s; (e) 30.3 cm/s **19.** −20 m/s^2 **21.** (a) 1.10 m/s; (b) 6.11 mm/s^2; (c) 1.47 m/s; (d) 6.11 mm/s^2 **23.** 1.62×10^{15} m/s^2 **25.** (a) 30 s; (b) 300 m **27.** (a) +1.6 m/s; (b) +18 m/s **29.** (a) 10.6 m; (b) 41.5 s **31.** (a) 3.1×10^6 s; (b) 4.6×10^{13} m **33.** (a) 3.56 m/s^2; (b) 8.43 m/s **35.** 0.90 m/s^2 **37.** (a) 4.0 m/s^2; (b) $+x$ **39.** (a) −2.5 m/s^2; (b) 1; (d) 0; (e) 2 **41.** 40 m **43.** (a) 0.994 m/s^2 **45.** (a) 31 m/s; (b) 6.4 s **47.** (a) 29.4 m; (b) 2.45 s **49.** (a) 5.4 s; (b) 41 m/s **51.** (a) 20 m; (b) 59 m **53.** 4.0 m/s **55.** (a) 857 m/s^2; (b) up **57.** (a) 1.26×10^3 m/s^2; (b) up **59.** (a) 89 cm; (b) 22 cm **61.** 20.4 m **63.** 2.34 m **65.** (a) 2.25 m/s; (b) 3.90 m/s **67.** 0.56 m/s **69.** 100 m **71.** (a) 2.00 s; (b) 12 cm; (c) −9.00 cm/s^2; (d) right; (e) left; (f) 3.46 s **73.** (a) 82 m; (b) 19 m/s **75.** (a) 0.74 s; (b) 6.2 m/s^2 **77.** (a) 3.1 m/s^2; (b) 45 m; (c) 13 s **79.** 17 m/s **81.** +47 m/s **83.** (a) 1.23 cm; (b) 4 times; (c) 9 times; (d) 16 times; (e) 25 times **85.** 25 km/h **87.** 1.2 h **89.** $4H$ **91.** (a) 3.2 s; (b) 1.3 s **93.** (a) 8.85 m/s; (b) 1.00 m **95.** (a) 2.0 m/s^2; (b) 12 m/s; (c) 45 m **97.** (a) 48.5 m/s; (b) 4.95 s; (c) 34.3 m/s; (d) 3.50 s **99.** 22.0 m/s **101.** (a) $v = (v_0^2 + 2gh)^{0.5}$; (b) $t = [(v_0^2 + 2gh)^{0.5} - v_0]/g$; (c) same as (a); (d) $t = [(v_0^2 + 2gh)^{0.5} + v_0]/g$, greater **103.** 414 ms **105.** 90 m **107.** 0.556 s **109.** (a) 0.28 m/s^2; (b) 0.28 m/s^2 **111.** (a) 10.2 s; (b) 10.0 m **113.** (a) 5.44 s; (b) 53.3 m/s; (c) 5.80 m **115.** 2.3 cm/min **117.** 0.15 m/s **119.** (a) 1.0 cm/s; (b) 1.6 cm/s, 1.1 cm/s, 0; (c) −0.79 cm/s^2; (d) 0, −0.87 cm/s^2, −1.2 cm/s^2

Chapter 3

CP **1.** (a) 7 m ($\vec{a}$ and $\vec{b}$ are in same direction); (b) 1 m ($\vec{a}$ and $\vec{b}$ are in opposite directions) **2.** c, d, f (components must be head to tail; $\vec{a}$ must extend from tail of one component to head of the other) **3.** (a) +, +; (b) +, −; (c) +, + (draw vector from tail of $\vec{d}_1$ to head of $\vec{d}_2$) **4.** (a) 90°; (b) 0° (vectors are parallel—same direction); (c) 180° (vectors are antiparallel—opposite directions) **5.** (a) 0° or 180°; (b) 90°

Q **1.** yes, when the vectors are in same direction **3.** Either the sequence $\vec{d}_2, \vec{d}_1$ or the sequence $\vec{d}_2, \vec{d}_2, \vec{d}_3$ **5.** all but (e) **7.** (a) yes; (b) yes; (c) no **9.** (a) $+x$ for (1), $+z$ for (2), $+z$ for (3); (b) $-x$ for (1), $-z$ for (2), $-z$ for (3) **11.** $\vec{s}, \vec{p}, \vec{r}$ or $\vec{p}, \vec{s}, \vec{r}$ **13.** Correct: c, d, f, h. Incorrect: a (cannot dot a vector with a scalar), b (cannot cross a vector with a scalar), e, g, i, j (cannot add a scalar and a vector).

P **1.** (a) −2.5 m; (b) −6.9 m **3.** (a) 47.2 m; (b) 122° **5.** (a) 156 km; (b) 39.8° west of due north **7.** (a) parallel; (b) antiparallel; (c) perpendicular **9.** (a) $(3.0\text{ m})\hat{i} - (2.0\text{ m})\hat{j} + (5.0\text{ m})\hat{k}$; (b) $(5.0\text{ m})\hat{i} - (4.0\text{ m})\hat{j} - (3.0\text{ m})\hat{k}$; (c) $(-5.0\text{ m})\hat{i} + (4.0\text{ m})\hat{j} + (3.0\text{ m})\hat{k}$ **11.** (a) $(-9.0\text{ m})\hat{i} + (10\text{ m})\hat{j}$; (b) 13 m; (c) 132° **13.** 4.74 km **15.** (a) 1.59 m; (b) 12.1 m; (c) 12.2 m; (d) 82.5° **17.** (a) 38 m; (b) −37.5°; (c) 130 m; (d) 1.2°; (e) 62 m; (f) 130° **19.** 5.39 m at 21.8° left of forward **21.** (a) −70.0 cm; (b) 80.0 cm; (c) 141 cm; (d) −172° **23.** 3.2 **25.** 2.6 km **27.** (a) $8\hat{i} + 16\hat{j}$; (b) $2\hat{i} + 4\hat{j}$ **29.** (a) 7.5 cm; (b) 90°; (c) 8.6 cm; (d) 48° **31.** (a) 9.51 m; (b) 14.1 m; (c) 13.4 m; (d) 10.5 m **33.** (a) 12; (b) $+z$; (c) 12; (d) $-z$; (e) 12; (f) $+z$ **35.** (a) −18.8 units; (b) 26.9 units, $+z$ direction **37.** (a) −21; (b) −9; (c) $5\hat{i} - 11\hat{j} - 9\hat{k}$ **39.** 70.5° **41.** 22° **43.** (a) 3.00 m; (b) 0; (c) 3.46 m; (d) 2.00 m; (e) −5.00 m; (f) 8.66 m; (g) −6.67; (h) 4.33 **45.** (a) −83.4; (b) $(1.14 \times 10^3)\hat{k}$; (c) 1.14×10^3, θ not defined, $\phi = 0°$; (d) 90.0°; (e) $-5.14\hat{i} + 6.13\hat{j} + 3.00\hat{k}$; (f) 8.54, $\theta = 130°$, $\phi = 69.4°$ **47.** (a) 140°; (b) 90.0°; (c) 99.1° **49.** (a) 103 km; (b) 60.9° north of due west **51.** (a) 27.8 m; (b) 13.4 m **53.** (a) 30; (b) 52 **55.** (a) −2.83 m; (b) −2.83 m; (c) 5.00 m; (d) 0; (e) 3.00 m; (f) 5.20 m; (g) 5.17 m; (h) 2.37 m; (i) 5.69 m; (j) 25° north of due east; (k) 5.69 m; (l) 25° south of due west **57.** 4.1 **59.** (a) $(9.19\text{ m})\hat{i}' + (7.71\text{ m})\hat{j}'$; (b) $(14.0\text{ m})\hat{i}' + (3.41\text{ m})\hat{j}'$ **61.** (a) $11\hat{i} + 5.0\hat{j} - 7.0\hat{k}$; (b) 120°; (c) −4.9; (d) 7.3 **63.** (a) 3.0 m^2; (b) 52 m^3; (c) $(11\text{ m}^2)\hat{i} + (9.0\text{ m}^2)\hat{j} + (3.0\text{ m}^2)\hat{k}$ **65.** (a) $(-40\hat{i} - 20\hat{j} + 25\hat{k})$ m; (b) 45 m **67.** (a) 0; (b) 0; (c) −1; (d) west; (e) up; (f) west **69.** (a) 168 cm; (b) 32.5° **71.** (a) 15 m; (b) south; (c) 6.0 m; (d) north **73.** (a) $2\hat{k}$; (b) 26; (c) 46; (d) 5.81 **75.** (a) up; (b) 0; (c) south; (d) 1; (e) 0 **77.** (a) $(1300\text{ m})\hat{i} + (2200\text{ m})\hat{j} - (410\text{ m})\hat{k}$; (b) 2.56×10^3 m **79.** 8.4

Chapter 4

CP **1.** (draw $\vec{v}$ tangent to path, tail on path) (a) first; (b) third **2.** (take second derivative with respect to time) (1) and (3) a_x and a_y are both constant and thus $\vec{a}$ is constant; (2) and (4) a_y is constant but a_x is not, thus $\vec{a}$ is not **3.** yes **4.** (a) v_x constant; (b) v_y initially positive, decreases to zero, and then becomes progressively more negative; (c) $a_x = 0$ throughout; (d) $a_y = -g$ throughout **5.** (a) $-(4\text{ m/s})\hat{i}$; (b) $-(8\text{ m/s}^2)\hat{j}$

Q **1.** *a* and *c* tie, then *b* **3.** decreases **5.** *a*, *b*, *c* **7.** (a) 0; (b) 350 km/h; (c) 350 km/h; (d) same (nothing changed about the vertical motion) **9.** (a) all tie; (b) all tie; (c) 3, 2, 1; (d) 3, 2, 1 **11.** 2, then 1 and 4 tie, then 3 **13.** (a) yes; (b) no; (c) yes **15.** (a) decreases; (b) increases **17.** maximum height

P **1.** (a) 6.2 m **3.** $(-2.0\text{ m})\hat{i} + (6.0\text{ m})\hat{j} - (10\text{ m})\hat{k}$ **5.** (a) 7.59 km/h; (b) 22.5° east of due north **7.** $(-0.70\text{ m/s})\hat{i} + (1.4\text{ m/s})\hat{j} - (0.40\text{ m/s})\hat{k}$ **9.** (a) 0.83 cm/s; (b) 0°; (c) 0.11 m/s; (d) −63° **11.** (a) $(6.00\text{ m})\hat{i} - (106\text{ m})\hat{j}$; (b) $(19.0\text{ m/s})\hat{i} - (224\text{ m/s})\hat{j}$; (c) $(24.0\text{ m/s}^2)\hat{i} - (336\text{ m/s}^2)\hat{j}$; (d) −85.2° **13.** (a) $(8\text{ m/s}^2)t\hat{j} + (1\text{ m/s})\hat{k}$; (b) $(8\text{ m/s}^2)\hat{j}$ **15.** (a) $(-1.50\text{ m/s})\hat{j}$; (b) $(4.50\text{ m})\hat{i} - (2.25\text{ m})\hat{j}$ **17.** $(32\text{ m/s})\hat{i}$ **19.** (a) $(72.0\text{ m})\hat{i} + (90.7\text{ m})\hat{j}$; (b) 49.5° **21.** (a) 18 cm; (b) 1.9 m **23.** (a) 3.03 s; (b) 758 m; (c) 29.7 m/s **25.** 43.1 m/s (155 km/h) **27.** (a) 10.0 s; (b) 897 m **29.** 78.5° **31.** 3.35 m **33.** (a) 202 m/s; (b) 806 m; (c) 161 m/s; (d) −171 m/s **35.** 4.84 cm **37.** (a) 1.60 m; (b) 6.86 m; (c) 2.86 m **39.** (a) 32.3 m; (b) 21.9 m/s; (c) 40.4°; (d) below **41.** 55.5° **43.** (a) 11 m; (b) 23 m; (c) 17 m/s; (d) 63° **45.** (a) ramp; (b) 5.82 m; (c) 31.0° **47.** (a) yes; (b) 2.56 m **49.** (a) 31°; (b) 63° **51.** (a) 2.3°; (b) 1.1 m; (c) 18° **53.** (a) 75.0 m; (b) 31.9 m/s; (c) 66.9°; (d) 25.5 m **55.** the third **57.** (a) 7.32 m; (b) west; (c) north **59.** (a) 12 s; (b) 4.1 m/s^2; (c) down; (d) 4.1 m/s^2; (e) up **61.** (a) 1.3×10^5 m/s; (b) 7.9×10^5 m/s^2; (c) increase **63.** 2.92 m **65.** $(3.00\text{ m/s}^2)\hat{i} + (6.00\text{ m/s}^2)\hat{j}$ **67.** 160 m/s^2 **69.** (a) 13 m/s^2; (b) eastward; (c) 13 m/s^2; (d) eastward **71.** 1.67 **73.** (a) $(80\text{ km/h})\hat{i} - (60\text{ km/h})\hat{j}$; (b) 0°; (c) answers do not change **75.** 32 m/s **77.** 60° **79.** (a) 38 knots; (b) 1.5° east of due north; (c) 4.2 h; (d) 1.5° west of due south **81.** (a) $(-32\text{ km/h})\hat{i} - (46\text{ km/h})\hat{j}$; (b) $[(2.5\text{ km}) - (32\text{ km/h})t]\hat{i} + [(4.0\text{ km}) - (46\text{ km/h})t]\hat{j}$; (c) 0.084 h; (d) 2×10^2 m **83.** (a) −30°; (b) 69 min; (c) 80 min; (d) 80 min; (e) 0°; (f) 60 min **85.** (a) 2.7 km; (b) 76° clockwise **87.** (a) 44 m; (b) 13 m; (c) 8.9 m **89.** (a) 45 m; (b) 22 m/s **91.** (a) 2.6×10^2 m/s; (b) 45 s; (c) increase **93.** (a) 63 km; (b) 18° south of due east; (c) 0.70 km/h; (d) 18° south of due east; (e) 1.6 km/h; (f) 1.2 km/h; (g) 33° north of due east **95.** (a) 1.5; (b) (36 m, 54 m) **97.** (a) 62 ms; (b) 4.8×10^2 m/s **99.** 2.64 m **101.** (a) 2.5 m; (b) 0.82 m; (c) 9.8 m/s^2; (d) 9.8 m/s^2 **103.** (a) 6.79 km/h; (b) 6.96° **105.** (a) 16 m/s; (b) 23°; (c) above; (d) 27 m/s; (e) 57°; (f) below **107.** (a) 4.2 m, 45°; (b) 5.5 m, 68°; (c) 6.0 m, 90°; (d) 4.2 m, 135°; (e) 0.85 m/s, 135°; (f) 0.94 m/s, 90°; (g) 0.94 m/s, 180°; (h) 0.30 m/s^2, 180°; (i) 0.30 m/s^2, 270° **109.** (a) 5.4×10^{-13} m; (b) decrease **111.** (a) 0.034 m/s^2; (b) 84 min **113.** (a) 8.43 m; (b) −129° **115.** (a) 2.00 ns; (b) 2.00 mm; (c) 1.00×10^7 m/s; (d) 2.00×10^6 m/s **117.** (a) 24 m/s; (b) 65° **119.** 93° from the car's direction of motion **121.** (a) 4.6×10^{12} m; (b) 2.4×10^5 s **123.** (a) 6.29°; (b) 83.7° **125.** (a) 3×10^1 m **127.** (a) $(6.0\hat{i} + 4.2\hat{j})$ m/s; (b) $(18\hat{i} + 6.3\hat{j})$ m **129.** (a) 38 ft/s; (b) 32 ft/s; (c) 9.3 ft **131.** (a) 11 m; (b) 45 m/s **133.** (a) 5.8 m/s; (b) 17 m; (c) 67° **135.** (a) 32.4 m; (b) −37.7 m **137.** 88.6 km/h

Chapter 5

CP **1.** *c*, *d*, and *e* ($\vec{F}_1$ and $\vec{F}_2$ must be head to tail, $\vec{F}_{net}$ must be from tail of one of them to head of the other) **2.** (a) and (b) 2 N, leftward (acceleration is zero in each situation) **3.** (a) equal; (b) greater (acceleration is upward, thus net force on body must be upward) **4.** (a) equal; (b) greater; (c) less **5.** (a) increase; (b) yes; (c) same; (d) yes

Q **1.** (a) 2, 3, 4; (b) 1, 3, 4; (c) 1, $+y$; 2, $+x$; 3, fourth quadrant; 4, third quadrant **3.** increase **5.** (a) 2 and 4; (b) 2 and 4 **7.** (a) M; (b) M; (c) M; (d) $2M$; (e) $3M$ **9.** (a) 20 kg; (b) 18 kg; (c) 10 kg; (d) all tie; (e) 3, 2, 1 **11.** (a) increases from initial value mg; (b) decreases from mg to zero (after which the block moves up away from the floor)

P **1.** 2.9 m/s^2 **3.** (a) 1.88 N; (b) 0.684 N; (c) $(1.88\text{ N})\hat{i} + (0.684\text{ N})\hat{j}$ **5.** (a) $(0.86\text{ m/s}^2)\hat{i} - (0.16\text{ m/s}^2)\hat{j}$; (b) 0.88 m/s^2; (c) −11° **7.** (a) $(-32.0\text{ N})\hat{i} - (20.8\text{ N})\hat{j}$; (b) 38.2 N; (c) −147° **9.** (a) 8.37 N; (b) −133°; (c) −125° **11.** 9.0 m/s^2 **13.** (a) 4.0 kg; (b) 1.0 kg; (c) 4.0 kg; (d) 1.0 kg **15.** (a) 108 N; (b) 108 N; (c) 108 N **17.** (a) 42 N; (b) 72 N; (c) 4.9 m/s^2 **19.** 1.2×10^5 N **21.** (a) 11.7 N; (b) −59.0° **23.** (a) $(285\text{ N})\hat{i} + (705\text{ N})\hat{j}$; (b) $(285\text{ N})\hat{i} - (115\text{ N})\hat{j}$; (c) 307 N; (d) −22.0°; (e) 3.67 m/s^2; (f) −22.0° **25.** (a) 0.022 m/s^2; (b) 8.3×10^4 km; (c) 1.9×10^3 m/s **27.** 1.5 mm **29.** (a) 494 N; (b) up; (c) 494 N; (d) down **31.** (a) 1.18 m; (b) 0.674 s; (c) 3.50 m/s **33.** 1.8×10^4 N **35.** (a) 46.7°; (b) 28.0° **37.** (a) 0.62 m/s^2; (b) 0.13 m/s^2; (c) 2.6 m **39.** (a) 2.2×10^{-3} N; (b) 3.7×10^{-3} N **41.** (a) 1.4 m/s^2; (b) 4.1 m/s **43.** (a) 1.23 N; (b) 2.46 N; (c) 3.69 N; (d) 4.92 N; (e) 6.15 N; (f) 0.250 N **45.** (a) 31.3 kN; (b) 24.3 kN **47.** 6.4×10^3 N **49.** (a) 2.18 m/s^2; (b) 116 N; (c) 21.0 m/s^2 **51.** (a) 3.6 m/s^2; (b) 17 N **53.** (a) 0.970 m/s^2; (b) 11.6 N; (c) 34.9 N **55.** (a) 1.1 N **57.** (a) 0.735 m/s^2; (b) down; (c) 20.8 N **59.** (a) 4.9 m/s^2; (b) 2.0 m/s^2; (c) up; (d) 120 N **61.** $2Ma/(a + g)$ **63.** (a) 8.0 m/s; (b) $+x$ **65.** (a) 0.653 m/s^3; (b) 0.896 m/s^3; (c) 6.50 s **67.** 81.7 N **69.** 2.4 N **71.** 16 N **73.** (a) 2.6 N; (b) 17° **75.** (a) 0; (b) 0.83 m/s^2; (c) 0 **77.** (a) 0.74 m/s^2; (b) 7.3 m/s^2 **79.** (a) 11 N; (b) 2.2 kg; (c) 0; (d) 2.2 kg **81.** 195 N **83.** (a) 4.6 m/s^2; (b) 2.6 m/s^2 **85.** (a) rope breaks; (b) 1.6 m/s^2 **87.** (a) 65 N; (b) 49 N **89.** (a) 4.6×10^3 N; (b) 5.8×10^3 N **91.** (a) 1.8×10^2 N; (b) 6.4×10^2 N **93.** (a) 44 N; (b) 78 N; (c) 54 N; (d) 152 N **95.** (a) 4 kg; (b) 6.5 m/s^2; (c) 13 N **97.** (a) $(1.0\hat{i} - 2.0\hat{j})$ N; (b) 2.2 N; (c) −63°; (d) 2.2 m/s^2; (e) −63°

Chapter 6

CP **1.** (a) zero (because there is no attempt at sliding); (b) 5 N; (c) no; (d) yes; (e) 8 N **2.** ($\vec{a}$ is directed toward center of circular path) (a) $\vec{a}$ downward, $\vec{F}_N$ upward; (b) $\vec{a}$ and $\vec{F}_N$ upward; (c) same; (d) greater at lowest point

Q **1.** (a) decrease; (b) decrease; (c) increase; (d) increase; (e) increase **3.** (a) same; (b) increases; (c) increases; (d) no **5.** (a) upward; (b) horizontal, toward you; (c) no change; (d) increases; (e) increases **7.** At first, $\vec{f}_s$ is directed up the ramp and its magnitude increases from $mg \sin\theta$ until it reaches $f_{s,\max}$. Thereafter the force is kinetic friction directed up the ramp, with magnitude f_k (a constant value smaller than $f_{s,\max}$). **9.** 4, 3, then 1, 2, and 5 tie **11.** (a) all tie; (b) all tie; (c) 2, 3, 1 **13.** (a) increases; (b) increases; (c) decreases; (d) decreases; (e) decreases

P **1.** 36 m **3.** (a) 2.0×10^2 N; (b) 1.2×10^2 N **5.** (a) 6.0 N; (b) 3.6 N; (c) 3.1 N **7.** (a) 1.9×10^2 N; (b) 0.56 m/s^2 **9.** (a) 11 N; (b) 0.14 m/s^2 **11.** (a) 3.0×10^2 N; (b) 1.3 m/s^2 **13.** (a) 1.3×10^2 N; (b) no; (c) 1.1×10^2 N; (d) 46 N; (e) 17 N **15.** 2° **17.** (a) $(17\text{ N})\hat{i}$; (b) $(20\text{ N})\hat{i}$; (c) $(15\text{ N})\hat{i}$ **19.** (a) no; (b) $(-12\text{ N})\hat{i} + (5.0\text{ N})\hat{j}$ **21.** (a) 19°; (b) 3.3 kN **23.** 0.37 **25.** 1.0×10^2 N **27.** (a) 0; (b) $(-3.9\text{ m/s}^2)\hat{i}$; (c) $(-1.0\text{ m/s}^2)\hat{i}$ **29.** (a) 66 N; (b) 2.3 m/s^2 **31.** (a) 3.5 m/s^2; (b) 0.21 N **33.** 9.9 s **35.** 4.9×10^2 N **37.** (a) 3.2×10^2 km/h; (b) 6.5×10^2 km/h; (c) no **39.** 2.3 **41.** 0.60 **43.** 21 m **45.** (a) light; (b) 778 N; (c) 223 N; (d) 1.11 kN **47.** (a) 10 s; (b) 4.9×10^2 N; (c) 1.1×10^3 N **49.** 1.37×10^3 N **51.** 2.2 km **53.** 12° **55.** 2.6×10^3 N **57.** 1.81 m/s **59.** (a) 8.74 N; (b) 37.9 N; (c) 6.45 m/s; (d) radially inward **61.** (a) 27 N; (b) 3.0 m/s^2 **63.** (b) 240 N; (c) 0.60 **65.** (a) 69 km/h; (b) 139 km/h; (c) yes **67.** $g(\sin\theta - 2^{0.5}\mu_k \cos\theta)$ **69.** 3.4 m/s^2 **71.** (a) 35.3 N; (b) 39.7 N; (c) 320 N **73.** (a) 7.5 m/s^2; (b) down; (c) 9.5 m/s^2; (d) down **75.** (a) 3.0×10^5 N; (b) 1.2° **77.** 147 m/s **79.** (a) 13 N; (b) 1.6 m/s^2 **81.** (a) 275 N; (b) 877 N **83.** (a) 84.2 N; (b) 52.8 N; (c) 1.87 m/s^2 **85.** 3.4% **87.** (a) 3.21×10^3 N; (b) yes **89.** (a) 222 N; (b) 334 N; (c) 311 N; (d) 311 N; (e) c, d **91.** (a) $v_0^2/(4g \sin\theta)$; (b) no **93.** (a) 0.34; (b) 0.24 **95.** (a) $\mu_k mg/(\sin\theta - \mu_k \cos\theta)$; (b) $\theta_0 = \tan^{-1}\mu_s$ **97.** 0.18 **99.** (a) 56 N; (b) 59 N; (c) 1.1×10^3 N **101.** 0.76 **103.** (a) bottom of circle; (b) 9.5 m/s **105.** 0.56

Chapter 7

CP 1. (a) decrease; (b) same; (c) negative, zero **2.** (a) positive; (b) negative; (c) zero **3.** zero

Q 1. all tie **3.** (a) positive; (b) negative; (c) negative **5.** *b* (positive work), *a* (zero work), *c* (negative work), *d* (more negative work) **7.** all tie **9.** (a) *A*; (b) *B* **11.** 2, 3, 1

P 1. (a) 2.9×10^7 m/s; (b) 2.1×10^{-13} J **3.** (a) 5×10^{14} J; (b) 0.1 megaton TNT; (c) 8 bombs **5.** (a) 2.4 m/s; (b) 4.8 m/s **7.** 0.96 J **9.** 20 J **11.** (a) 62.3°; (b) 118° **13.** (a) 1.7×10^2 N; (b) 3.4×10^2 m; (c) -5.8×10^4 J; (d) 3.4×10^2 N; (e) 1.7×10^2 m; (f) -5.8×10^4 J **15.** (a) 1.50 J; (b) increases **17.** (a) 12 kJ; (b) −11 kJ; (c) 1.1 kJ; (d) 5.4 m/s **19.** 25 J **21.** (a) $-3Mgd/4$; (b) Mgd; (c) $Mgd/4$; (d) $(gd/2)^{0.5}$ **23.** 4.41 J **25.** (a) 25.9 kJ; (b) 2.45 N **27.** (a) 7.2 J; (b) 7.2 J; (c) 0; (d) −25 J **29.** (a) 0.90 J; (b) 2.1 J; (c) 0 **31.** (a) 6.6 m/s; (b) 4.7 m **33.** (a) 0.12 m; (b) 0.36 J; (c) −0.36 J; (d) 0.060 m; (e) 0.090 J **35.** (a) 0; (b) 0 **37.** (a) 42 J; (b) 30 J; (c) 12 J; (d) 6.5 m/s, $+x$ axis; (e) 5.5 m/s, $+x$ axis; (f) 3.5 m/s, $+x$ axis **39.** 4.00 N/m **41.** 5.3×10^2 J **43.** (a) 0.83 J; (b) 2.5 J; (c) 4.2 J; (d) 5.0 W **45.** 4.9×10^2 W **47.** (a) 1.0×10^2 J; (b) 8.4 W **49.** 7.4×10^2 W **51.** (a) 32.0 J; (b) 8.00 W; (c) 78.2° **53.** (a) 1.20 J; (b) 1.10 m/s **55.** (a) 1.8×10^5 ft·lb; (b) 0.55 hp **57.** (a) 797 N; (b) 0; (c) −1.55 kJ; (d) 0; (e) 1.55 kJ; (f) *F* varies during displacement **59.** (a) 11 J; (b) −21 J **61.** −6 J **63.** (a) 314 J; (b) −155 J; (c) 0; (d) 158 J **65.** (a) 98 N; (b) 4.0 cm; (c) 3.9 J; (d) −3.9 J **67.** (a) 23 mm; (b) 45 N **69.** 165 kW **71.** −37 J **73.** (a) 13 J; (b) 13 J **75.** 235 kW **77.** (a) 6 J; (b) 6.0 J **79.** (a) 0.6 J; (b) 0; (c) −0.6 J **81.** (a) 3.35 m/s; (b) 22.5 J; (c) 0; (d) 0; (e) 0.212 m **83.** (a) -5.20×10^{-2} J; (b) −0.160 J **85.** 6.63 m/s

Chapter 8

CP 1. no (consider round trip on the small loop) **2.** 3, 1, 2 (see Eq. 8-6) **3.** (a) all tie; (b) all tie **4.** (a) *CD*, *AB*, *BC* (0) (check slope magnitudes); (b) positive direction of *x* **5.** all tie

Q 1. (a) 3, 2, 1; (b) 1, 2, 3 **3.** (a) 12 J; (b) −2 J **5.** (a) increasing; (b) decreasing; (c) decreasing; (d) constant in *AB* and *BC*, decreasing in *CD* **7.** +30 J **9.** 2, 1, 3 **11.** −40 J

P 1. 89 N/cm **3.** (a) 167 J; (b) −167 J; (c) 196 J; (d) 29 J; (e) 167 J; (f) −167 J; (g) 296 J; (h) 129 J **5.** (a) 4.31 mJ; (b) −4.31 mJ; (c) 4.31 mJ; (d) −4.31 mJ; (e) all increase **7.** (a) 13.1 J; (b) −13.1 J; (c) 13.1 J; (d) all increase **9.** (a) 17.0 m/s; (b) 26.5 m/s; (c) 33.4 m/s; (d) 56.7 m; (e) all the same **11.** (a) 2.08 m/s; (b) 2.08 m/s; (c) increase **13.** (a) 0.98 J; (b) −0.98 J; (c) 3.1 N/cm **15.** (a) 2.6×10^2 m; (b) same; (c) decrease **17.** (a) 2.5 N; (b) 0.31 N; (c) 30 cm **19.** (a) 784 N/m; (b) 62.7 J; (c) 62.7 J; (d) 80.0 cm **21.** (a) 8.35 m/s; (b) 4.33 m/s; (c) 7.45 m/s; (d) both decrease **23.** (a) 4.85 m/s; (b) 2.42 m/s **25.** -3.2×10^2 J **27.** (a) no; (b) 9.3×10^2 N **29.** (a) 35 cm; (b) 1.7 m/s **31.** (a) 39.2 J; (b) 39.2 J; (c) 4.00 m **33.** (a) 2.40 m/s; (b) 4.19 m/s **35.** (a) 39.6 cm; (b) 3.64 cm **37.** −18 mJ **39.** (a) 2.1 m/s; (b) 10 N; (c) $+x$ direction; (d) 5.7 m; (e) 30 N; (f) $-x$ direction **41.** (a) −3.7 J; (c) 1.3 m; (d) 9.1 m; (e) 2.2 J; (f) 4.0 m; (g) $(4 - x)e^{-x/4}$; (h) 4.0 m **43.** (a) 5.6 J; (b) 3.5 J **45.** (a) 30.1 J; (b) 30.1 J; (c) 0.225 **47.** 0.53 J **49.** (a) −2.9 kJ; (b) 3.9×10^2 J; (c) 2.1×10^2 N **51.** (a) 1.5 MJ; (b) 0.51 MJ; (c) 1.0 MJ; (d) 63 m/s **53.** (a) 67 J; (b) 67 J; (c) 46 cm **55.** (a) −0.90 J; (b) 0.46 J; (c) 1.0 m/s **57.** 1.2 m **59.** (a) 19.4 m; (b) 19.0 m/s **61.** (a) 1.5×10^{-2} N; (b) $(3.8 \times 10^2)g$ **63.** (a) 7.4 m/s; (b) 90 cm; (c) 2.8 m; (d) 15 m **65.** 20 cm **67.** (a) 7.0 J; (b) 22 J **69.** 3.7 J **71.** 4.33 m/s **73.** 25 J **75.** (a) 4.9 m/s; (b) 4.5 N; (c) 71°; (d) same **77.** (a) 4.8 N; (b) $+x$ direction; (c) 1.5 m; (d) 13.5 m; (e) 3.5 m/s **79.** (a) 24 kJ; (b) 4.7×10^2 N **81.** (a) 5.00 J; (b) 9.00 J; (c) 11.0 J; (d) 3.00 J; (e) 12.0 J; (f) 2.00 J; (g) 13.0 J; (h) 1.00 J; (i) 13.0 J; (j) 1.00 J; (l) 11.0 J; (m) 10.8 m; (n) It returns to $x = 0$ and stops. **83.** (a) 6.0 kJ; (b) 6.0×10^2 W; (c) 3.0×10^2 W; (d) 9.0×10^2 W **85.** 880 MW **87.** (a) $v_0 = (2gL)^{0.5}$; (b) $5mg$; (c) $-mgL$; (d) $-2mgL$ **89.** (a) 109 J; (b) 60.3 J; (c) 68.2 J; (d) 41.0 J **91.** (a) 2.7 J; (b) 1.8 J; (c) 0.39 m **93.** (a) 10 m; (b) 49 N; (c) 4.1 m; (d) 1.2×10^2 N **95.** (a) 5.5 m/s; (b) 5.4 m; (c) same **97.** 80 mJ **99.** 24 W **101.** −12 J **103.** (a) 8.8 m/s; (b) 2.6 kJ; (c) 1.6 kW **105.** (a) 7.4×10^2 J; (b) 2.4×10^2 J **107.** 15 J **109.** (a) 2.35×10^3 J; (b) 352 J **111.** 738 m **113.** (a) −3.8 kJ; (b) 31 kN **115.** (a) 300 J; (b) 93.8 J; (c) 6.38 m **117.** (a) 5.6 J; (b) 12 J; (c) 13 J **119.** (a) 1.2 J; (b) 11 m/s; (c) no; (d) no **121.** (a) 2.1×10^6 kg; (b) $(100 + 1.5t)^{0.5}$ m/s; (c) $(1.5 \times 10^6)/(100 + 1.5t)^{0.5}$ N; (d) 6.7 km **123.** 54% **125.** (a) 2.7×10^9 J; (b) 2.7×10^9 W; (c) \$$2.4 \times 10^8$ **127.** 5.4 kJ **129.** 3.1×10^{11} W **131.** because your force on the cabbage (as you lower it) does work **135.** (a) 8.6 kJ; (b) 8.6×10^2 W; (c) 4.3×10^2 W; (d) 1.3 kW

Chapter 9

CP 1. (a) origin; (b) fourth quadrant; (c) on *y* axis below origin; (d) origin; (e) third quadrant; (f) origin **2.** (a) – (c) at the center of mass, still at the origin (their forces are internal to the system and cannot move the center of mass) **3.** (Consider slopes and Eq. 9-23.) (a) 1, 3, and then 2 and 4 tie (zero force); (b) 3 **4.** (a) unchanged; (b) unchanged (see Eq. 9-32); (c) decrease (Eq. 9-35) **5.** (a) zero; (b) positive (initial p_y down *y*; final p_y up *y*); (c) positive direction of *y* **6.** (No net external force; $\vec{P}$ conserved.) (a) 0; (b) no; (c) $-x$ **7.** (a) 10 kg·m/s; (b) 14 kg·m/s; (c) 6 kg·m/s **8.** (a) 4 kg·m/s; (b) 8 kg·m/s; (c) 3 J **9.** (a) 2 kg·m/s (conserve momentum along *x*); (b) 3 kg·m/s (conserve momentum along *y*)

Q 1. (a) 2 N, rightward; (b) 2 N, rightward; (c) greater than 2 N, rightward **3.** b, c, a **5.** (a) *x* yes, *y* no; (b) *x* yes, *y* no; (c) *x* no, *y* yes **7.** (a) *c*, kinetic energy cannot be negative; *d*, total kinetic energy cannot increase; (b) *a*; (c) *b* **9.** (a) one was stationary; (b) 2; (c) 5; (d) equal (pool player's result) **11.** (a) *C*; (b) *B*; (c) 3

P 1. (a) −1.50 m; (b) −1.43 m **3.** (a) −6.5 cm; (b) 8.3 cm; (c) 1.4 cm **5.** (a) −0.45 cm; (b) −2.0 cm **7.** (a) 0; (b) 3.13×10^{-11} m **9.** (a) 28 cm; (b) 2.3 m/s **11.** $(-4.0\text{ m})\hat{i} + (4.0\text{ m})\hat{j}$ **13.** 53 m **15.** (a) $(2.35\hat{i} - 1.57\hat{j})$ m/s²; (b) $(2.35\hat{i} - 1.57\hat{j})t$ m/s, with *t* in seconds; (d) straight, at downward angle 34° **17.** 4.2 m **19.** (a) 7.5×10^4 J; (b) 3.8×10^4 kg·m/s; (c) 39° south of due east **21.** (a) 5.0 kg·m/s; (b) 10 kg·m/s **23.** 1.0×10^3 to 1.2×10^3 kg·m/s **25.** (a) 42 N·s; (b) 2.1 kN **27.** (a) 67 m/s; (b) $-x$; (c) 1.2 kN; (d) $-x$ **29.** 5 N **31.** (a) 2.39×10^3 N·s; (b) 4.78×10^5 N; (c) 1.76×10^3 N·s; (d) 3.52×10^5 N **33.** (a) 5.86 kg·m/s; (b) 59.8°; (c) 2.93 kN; (d) 59.8° **35.** 9.9×10^2 N **37.** (a) 9.0 kg·m/s; (b) 3.0 kN; (c) 4.5 kN; (d) 20 m/s **39.** 3.0 mm/s **41.** (a) $-(0.15\text{ m/s})\hat{i}$; (b) 0.18 m **43.** 55 cm **45.** (a) $(1.00\hat{i} - 0.167\hat{j})$ km/s; (b) 3.23 MJ **47.** (a) 14 m/s; (b) 45° **49.** 3.1×10^2 m/s **51.** (a) 721 m/s; (b) 937 m/s **53.** (a) 33%; (b) 23%; (c) decreases **55.** (a) +2.0 m/s; (b) −1.3 J; (c) +40 J; (d) system got energy from some source, such as a small explosion **57.** (a) 4.4 m/s; (b) 0.80 **59.** 25 cm **61.** (a) 99 g; (b) 1.9 m/s; (c) 0.93 m/s **63.** (a) 3.00 m/s; (b) 6.00 m/s **65.** (a) 1.2 kg; (b) 2.5 m/s **67.** −28 cm **69.** (a) 0.21 kg; (b) 7.2 m **71.** (a) 4.15×10^5 m/s; (b) 4.84×10^5 m/s **73.** 120° **75.** (a) 433 m/s; (b) 250 m/s **77.** (a) 46 N; (b) none **79.** (a) 1.57×10^6 N; (b) 1.35×10^5 kg; (c) 2.08 km/s **81.** (a) 7290 m/s; (b) 8200 m/s; (c) 1.271×10^{10} J; (d) 1.275×10^{10} J **83.** (a) 1.92 m; (b) 0.640 m **85.** (a) 1.78 m/s; (b) less; (c) less; (d) greater **87.** (a) 3.7 m/s; (b) 1.3 N·s; (c) 1.8×10^2 N **89.** (a) $(7.4 \times 10^3\text{ N·s})\hat{i} - (7.4 \times 10^3\text{ N·s})\hat{j}$; (b) $(-7.4 \times 10^3\text{ N·s})\hat{i}$; (c) 2.3×10^3 N; (d) 2.1×10^4 N; (e) −45° **91.** +4.4 m/s **93.** 1.18×10^4 kg **95.** (a) 1.9 m/s; (b) −30°; (c) elastic **97.** (a) 6.9 m/s; (b) 30°; (c) 6.9 m/s; (d) −30°; (e) 2.0 m/s; (f) −180° **99.** (a) 25 mm; (b) 26 mm; (c) down; (d) 1.6×10^{-2} m/s² **101.** 29 J **103.** 2.2 kg **105.** 5.0 kg **107.** (a) 50 kg/s; (b) 1.6×10^2 kg/s **109.** (a) 4.6×10^3 km; (b) 73% **111.** 190 m/s

113. 28.8 N **115.** (a) 0.745 mm; (b) 153°; (c) 1.67 mJ **117.** (a) $(2.67\text{ m/s})\hat{i} + (-3.00\text{ m/s})\hat{j}$; (b) 4.01 m/s; (c) 48.4° **119.** (a) −0.50 m; (b) −1.8 cm; (c) 0.50 m **121.** 0.22% **123.** 36.5 km/s **125.** (a) $(-1.00 \times 10^{-19}\hat{i} + 0.67 \times 10^{-19}\hat{j})$ kg·m/s; (b) 1.19×10^{-12} J **127.** 2.2×10^{-3}

Chapter 10

CP **1.** b and c **2.** (a) and (d) ($\alpha = d^2\theta/dt^2$ must be a constant) **3.** (a) yes; (b) no; (c) yes; (d) yes **4.** all tie **5.** 1, 2, 4, 3 (see Eq. 10-36) **6.** (see Eq. 10-40) 1 and 3 tie, 4, then 2 and 5 tie (zero) **7.** (a) downward in the figure ($\tau_{net} = 0$); (b) less (consider moment arms)

Q **1.** (a) *c*, *a*, then *b* and *d* tie; (b) *b*, then *a* and *c* tie, then *d* **3.** all tie **5.** (a) decrease; (b) clockwise; (c) counterclockwise **7.** larger **9.** *c*, *a*, *b* **11.** less

P **1.** 14 rev **3.** (a) 4.0 rad/s; (b) 11.9 rad/s **5.** 11 rad/s **7.** (a) 4.0 m/s; (b) no **9.** (a) 3.00 s; (b) 18.9 rad **11.** (a) 30 s; (b) 1.8×10^3 rad **13.** (a) 3.4×10^2 s; (b) -4.5×10^{-3} rad/s²; (c) 98 s **15.** 8.0 s **17.** (a) 44 rad; (b) 5.5 s; (c) 32 s; (d) −2.1 s; (e) 40 s **19.** (a) 2.50×10^{-3} rad/s; (b) 20.2 m/s²; (c) 0 **21.** 6.9×10^{-13} rad/s **23.** (a) 20.9 rad/s; (b) 12.5 m/s; (c) 800 rev/min²; (d) 600 rev **25.** (a) 7.3×10^{-5} rad/s; (b) 3.5×10^2 m/s; (c) 7.3×10^{-5} rad/s; (d) 4.6×10^2 m/s **27.** (a) 73 cm/s²; (b) 0.075; (c) 0.11 **29.** (a) 3.8×10^3 rad/s; (b) 1.9×10^2 m/s **31.** (a) 40 s; (b) 2.0 rad/s² **33.** 12.3 kg·m² **35.** (a) 1.1 kJ; (b) 9.7 kJ **37.** 0.097 kg·m² **39.** (a) 49 MJ; (b) 1.0×10^2 min **41.** (a) 0.023 kg·m²; (b) 1.1 mJ **43.** 4.7×10^{-4} kg·m² **45.** −3.85 N·m **47.** 4.6 N·m **49.** (a) 28.2 rad/s²; (b) 338 N·m **51.** (a) 6.00 cm/s²; (b) 4.87 N; (c) 4.54 N; (d) 1.20 rad/s²; (e) 0.0138 kg·m² **53.** 0.140 N **55.** 2.51×10^{-4} kg·m² **57.** (a) 4.2×10^2 rad/s²; (b) 5.0×10^2 rad/s **59.** 396 N·m **61.** (a) −19.8 kJ; (b) 1.32 kW **63.** 5.42 m/s **65.** (a) 5.32 m/s²; (b) 8.43 m/s²; (c) 41.8° **67.** 9.82 rad/s **69.** 6.16×10^{-5} kg·m² **71.** (a) 31.4 rad/s²; (b) 0.754 m/s²; (c) 56.1 N; (d) 55.1 N **73.** (a) 4.81×10^5 N; (b) 1.12×10^4 N·m; (c) 1.25×10^6 J **75.** (a) 2.3 rad/s²; (b) 1.4 rad/s² **77.** (a) −67 rev/min²; (b) 8.3 rev **81.** 3.1 rad/s **83.** (a) 1.57 m/s²; (b) 4.55 N; (c) 4.94 N **85.** 30 rev **87.** 0.054 kg·m² **89.** 1.4×10^2 N·m **91.** (a) 10 J; (b) 0.27 m **93.** 4.6 rad/s² **95.** 2.6 J **97.** (a) 5.92×10^4 m/s²; (b) 4.39×10^4 s⁻² **99.** (a) 0.791 kg·m²; (b) 1.79×10^{-2} N·m **101.** (a) 1.5×10^2 cm/s; (b) 15 rad/s; (c) 15 rad/s; (d) 75 cm/s; (e) 3.0 rad/s **103.** (a) 7.0 kg·m²; (b) 7.2 m/s; (c) 71° **105.** (a) 0.32 rad/s; (b) 1.0×10^2 km/h **107.** (a) 1.4×10^2 rad; (b) 14 s

Chapter 11

CP **1.** (a) same; (b) less **2.** less (consider the transfer of energy from rotational kinetic energy to gravitational potential energy) **3.** (draw the vectors, use right-hand rule) (a) $\pm z$; (b) $+y$; (c) $-x$ **4.** (see Eq. 11-21) (a) 1 and 3 tie; then 2 and 4 tie, then 5 (zero); (b) 2 and 3 **5.** (see Eqs. 11-23 and 11-16) (a) 3, 1; then 2 and 4 tie (zero); (b) 3 **6.** (a) all tie (same τ, same t, thus same ΔL); (b) sphere, disk, hoop (reverse order of I) **7.** (a) decreases; (b) same ($\tau_{net} = 0$, so L is conserved); (c) increases

Q **1.** *a*, then *b* and *c* tie, then *e*, *d* (zero) **3.** (a) spins in place; (b) rolls toward you; (c) rolls away from you **5.** (a) 1, 2, 3 (zero); (b) 1 and 2 tie, then 3; (c) 1 and 3 tie, then 2 **7.** (a) same; (b) increase; (c) decrease; (d) same, decrease, increase **9.** *D*, *B*, then *A* and *C* tie **11.** (a) same; (b) same

P **1.** (a) 0; (b) $(22\text{ m/s})\hat{i}$; (c) $(-22\text{ m/s})\hat{i}$; (d) 0; (e) 1.5×10^3 m/s²; (f) 1.5×10^3 m/s²; (g) $(22\text{ m/s})\hat{i}$; (h) $(44\text{ m/s})\hat{i}$; (i) 0; (j) 0; (k) 1.5×10^3 m/s²; (l) 1.5×10^3 m/s² **3.** −3.15 J **5.** 0.020 **7.** (a) 63 rad/s; (b) 4.0 m **9.** 4.8 m **11.** (a) $(-4.0\text{ N})\hat{i}$; (b) 0.60 kg·m² **13.** 0.50 **15.** (a) $-(0.11\text{ m})\omega$; (b) −2.1 m/s²; (c) −47 rad/s²; (d) 1.2 s; (e) 8.6 m; (f) 6.1 m/s **17.** (a) 13 cm/s²; (b) 4.4 s; (c) 55 cm/s; (d) 18 mJ; (e) 1.4 J; (f) 27 rev/s **19.** $(-2.0\text{ N·m})\hat{i}$ **21.** (a) $(6.0\text{ N·m})\hat{j} + (8.0\text{ N·m})\hat{k}$; (b) $(-22\text{ N·m})\hat{i}$ **23.** (a) $(-1.5\text{ N·m})\hat{i} - (4.0\text{ N·m})\hat{j} - (1.0\text{ N·m})\hat{k}$; (b) $(-1.5\text{ N·m})\hat{i} - (4.0\text{ N·m})\hat{j} - (1.0\text{ N·m})\hat{k}$ **25.** (a) $(50\text{ N·m})\hat{k}$; (b) 90° **27.** (a) 0; (b) $(8.0\text{ N·m})\hat{i} + (8.0\text{ N·m})\hat{k}$ **29.** (a) 9.8 kg·m²/s; (b) $+z$ direction **31.** (a) 0; (b) −22.6 kg·m²/s; (c) −7.84 N·m; (d) −7.84 N·m **33.** (a) $(-1.7 \times 10^2\text{ kg·m}^2\text{/s})\hat{k}$; (b) $(+56\text{ N·m})\hat{k}$; (c) $(+56\text{ kg·m}^2\text{/s}^2)\hat{k}$ **35.** (a) $48t\hat{k}$ N·m; (b) increasing **37.** (a) 4.6×10^{-3} kg·m²; (b) 1.1×10^{-3} kg·m²/s; (c) 3.9×10^{-3} kg·m²/s **39.** (a) 1.47 N·m; (b) 20.4 rad; (c) −29.9 J; (d) 19.9 W **41.** (a) 1.6 kg·m²; (b) 4.0 kg·m²/s **43.** (a) 1.5 m; (b) 0.93 rad/s; (c) 98 J; (d) 8.4 rad/s; (e) 8.8×10^2 J; (f) internal energy of the skaters **45.** (a) 3.6 rev/s; (b) 3.0; (c) forces on the bricks from the man transferred energy from the man's internal energy to kinetic energy **47.** 0.17 rad/s **49.** (a) 750 rev/min; (b) 450 rev/min; (c) clockwise **51.** (a) 267 rev/min; (b) 0.667 **53.** 1.3×10^3 m/s **55.** 3.4 rad/s **57.** (a) 18 rad/s; (b) 0.92 **59.** 11.0 m/s **61.** 1.5 rad/s **63.** 0.070 rad/s **65.** (a) 0.148 rad/s; (b) 0.0123; (c) 181° **67.** (a) 0.180 m; (b) clockwise **69.** 0.041 rad/s **71.** (a) 1.6 m/s²; (b) 16 rad/s²; (c) $(4.0\text{ N})\hat{i}$ **73.** (a) 0; (b) 0; (c) $-30t^3\hat{k}$ kg·m²/s; (d) $-90t^2\hat{k}$ N·m; (e) $30t^3\hat{k}$ kg·m²/s; (f) $90t^2\hat{k}$ N·m **75.** (a) 149 kg·m²; (b) 158 kg·m²/s; (c) 0.744 rad/s **77.** (a) 6.65×10^{-5} kg·m²/s; (b) no; (c) 0; (d) yes **79.** (a) 0.333; (b) 0.111 **81.** (a) 58.8 J; (b) 39.2 J **83.** (a) 61.7 J; (b) 3.43 m; (c) no **85.** (a) $mvR/(I + MR^2)$; (b) $mvR^2/(I + MR^2)$

Chapter 12

CP **1.** *c*, *e*, *f* **2.** (a) no; (b) at site of $\vec{F}_1$, perpendicular to plane of figure; (c) 45 N **3.** *d*

Q **1.** (a) 1 and 3 tie, then 2; (b) all tie; (c) 1 and 3 tie, then 2 (zero) **3.** *a* and *c* (forces and torques balance) **5.** (a) 12 kg; (b) 3 kg; (c) 1 kg **7.** (a) at *C* (to eliminate forces there from a torque equation); (b) plus; (c) minus; (d) equal **9.** increase **11.** *A* and *B*, then *C*

P **1.** (a) 1.00 m; (b) 2.00 m; (c) 0.987 m; (d) 1.97 m **3.** (a) 9.4 N; (b) 4.4 N **5.** 7.92 kN **7.** (a) 2.8×10^2 N; (b) 8.8×10^2 N; (c) 71° **9.** 74.4 g **11.** (a) 1.2 kN; (b) down; (c) 1.7 kN; (d) up; (e) left; (f) right **13.** (a) 2.7 kN; (b) up; (c) 3.6 kN; (d) down **15.** (a) 5.0 N; (b) 30 N; (c) 1.3 m **17.** (a) 0.64 m; (b) increased **19.** 8.7 N **21.** (a) 6.63 kN; (b) 5.74 kN; (c) 5.96 kN **23.** (a) 192 N; (b) 96.1 N; (c) 55.5 N **25.** 13.6 N **27.** (a) 1.9 kN; (b) up; (c) 2.1 kN; (d) down **29.** (a) $(-80\text{ N})\hat{i} + (1.3 \times 10^2\text{ N})\hat{j}$; (b) $(80\text{ N})\hat{i} + (1.3 \times 10^2\text{ N})\hat{j}$ **31.** 2.20 m **33.** (a) 60.0°; (b) 300 N **35.** (a) 445 N; (b) 0.50; (c) 315 N **37.** 0.34 **39.** (a) 207 N; (b) 539 N; (c) 315 N **41.** (a) slides; (b) 31°; (c) tips; (d) 34° **43.** (a) 6.5×10^6 N/m²; (b) 1.1×10^{-5} m **45.** (a) 0.80; (b) 0.20; (c) 0.25 **47.** (a) 1.4×10^9 N; (b) 75 **49.** (a) 866 N; (b) 143 N; (c) 0.165 **51.** (a) 1.2×10^2 N; (b) 68 N **53.** (a) 1.8×10^7 N; (b) 1.4×10^7 N; (c) 16 **55.** 0.29 **57.** 76 N **59.** (a) 8.01 kN; (b) 3.65 kN; (c) 5.66 kN **61.** 71.7 N **63.** (a) $L/2$; (b) $L/4$; (c) $L/6$; (d) $L/8$; (e) $25L/24$ **65.** (a) 88 N; (b) $(30\hat{i} + 97\hat{j})$ N **67.** 2.4×10^9 N/m² **69.** 60° **71.** (a) $\mu < 0.57$; (b) $\mu > 0.57$ **73.** (a) $(35\hat{i} + 200\hat{j})$ N; (b) $(-45\hat{i} + 200\hat{j})$ N; (c) 1.9×10^2 N **75.** (a) *BC*, *CD*, *DA*; (b) 535 N; (c) 757 N **77.** (a) 1.38 kN; (b) 180 N **79.** (a) $a_1 = L/2$, $a_2 = 5L/8$, $h = 9L/8$; (b) $b_1 = 2L/3$, $b_2 = L/2$, $h = 7L/6$ **81.** $L/4$ **83.** (a) 106 N; (b) 64.0° **85.** 1.8×10^2 N **87.** (a) −24.4 N; (b) 1.60 N; (c) −3.75°

Chapter 13

CP **1.** all tie **2.** (a) 1, tie of 2 and 4, then 3; (b) line *d* **3.** (a) increase; (b) negative **4.** (a) 2; (b) 1 **5.** (a) path 1 (decreased *E* (more negative) gives decreased *a*); (b) less (decreased *a* gives decreased *T*)

Q **1.** $3GM^2/d^2$, leftward **3.** Gm^2/r^2, upward **5.** *b* and *c* tie, then *a* (zero) **7.** 1, tie of 2 and 4, then 3 **9.** (a) positive *y*; (b) yes, rotates counterclockwise until it points toward particle B **11.** b, d, and f all tie, then e, c, a

P **1.** $\frac{1}{2}$ **3.** 19 m **5.** 0.8 m **7.** $-5.00d$ **9.** 2.60×10^5 km **11.** (a) $M = m$; (b) 0 **13.** 8.31×10^{-9} N **15.** (a) $-1.88d$; (b) $-3.90d$; (c) $0.489d$ **17.** (a) 17 N; (b) 2.4 **19.** 2.6×10^6 m **21.** 5×10^{24} kg **23.** (a) 7.6 m/s^2; (b) 4.2 m/s^2 **25.** (a) $(3.0 \times 10^{-7}$ N/kg$)m$; (b) $(3.3 \times 10^{-7}$ N/kg$)m$; (c) $(6.7 \times 10^{-7}$ N/kg $\cdot$ m$)mr$ **27.** (a) 9.83 m/s^2; (b) 9.84 m/s^2; (c) 9.79 m/s^2 **29.** 5.0×10^9 J **31.** (a) 0.74; (b) 3.8 m/s^2; (c) 5.0 km/s **33.** (a) 0.0451; (b) 28.5 **35.** -4.82×10^{-13} J **37.** (a) 0.50 pJ; (b) -0.50 pJ **39.** (a) 1.7 km/s; (b) 2.5×10^5 m; (c) 1.4 km/s **41.** (a) 82 km/s; (b) 1.8×10^4 km/s **43.** (a) 7.82 km/s; (b) 87.5 min **45.** 6.5×10^{23} kg **47.** 5×10^{10} stars **49.** (a) 1.9×10^{13} m; (b) $6.4R_P$ **51.** (a) 6.64×10^3 km; (b) 0.0136 **53.** 5.8×10^6 m **57.** 0.71 y **59.** $(GM/L)^{0.5}$ **61.** (a) 3.19×10^3 km; (b) lifting **63.** (a) 2.8 y; (b) 1.0×10^{-4} **65.** (a) $r^{1.5}$; (b) r^{-1}; (c) $r^{0.5}$; (d) $r^{-0.5}$ **67.** (a) 7.5 km/s; (b) 97 min; (c) 4.1×10^2 km; (d) 7.7 km/s; (e) 93 min; (f) 3.2×10^{-3} N; (g) no; (h) yes **69.** 1.1 s **71.** (a) $GMmx(x^2 + R^2)^{-3/2}$; (b) $[2GM(R^{-1} - (R^2 + x^2)^{-1/2})]^{1/2}$ **73.** (a) 1.0×10^3 kg; (b) 1.5 km/s **75.** 3.2×10^{-7} N **77.** $037\hat{j}$ μN **79.** $2\pi r^{1.5}G^{-0.5}(M + m/4)^{-0.5}$ **81.** (a) 2.2×10^{-7} rad/s; (b) 89 km/s **83.** (a) 2.15×10^4 s; (b) 12.3 km/s; (c) 12.0 km/s; (d) 2.17×10^{11} J; (e) -4.53×10^{11} J; (f) -2.35×10^{11} J; (g) 4.04×10^7 m; (h) 1.22×10^3 s; (i) elliptical **85.** 2.5×10^4 km **87.** (a) 1.4×10^6 m/s; (b) 3×10^6 m/s^2 **89.** (a) 0; (b) 1.8×10^{32} J; (c) 1.8×10^{32} J; (d) 0.99 km/s **91.** (a) Gm^2/R_i; (b) $Gm^2/2R_i$; (c) $(Gm/R_i)^{0.5}$; (d) $2(Gm/R_i)^{0.5}$; (e) Gm^2/R_i; (f) $(2Gm/R_i)^{0.5}$; (g) The center-of-mass frame is an inertial frame, and in it the principle of conservation of energy may be written as in Chapter 8; the reference frame attached to body A is noninertial, and the principle cannot be written as in Chapter 8. Answer (d) is correct. **93.** 2.4×10^4 m/s **95.** $-0.044\hat{j}$ μN **97.** $GM_Em/12R_E$ **99.** 1.51×10^{-12} N **101.** 3.4×10^5 km

Chapter 14

CP **1.** all tie **2.** (a) all tie (the gravitational force on the penguin is the same); (b) $0.95\rho_0$, ρ_0, $1.1\rho_0$ **3.** 13 cm^3/s, outward **4.** (a) all tie; (b) 1, then 2 and 3 tie, 4 (wider means slower); (c) 4, 3, 2, 1 (wider and lower mean more pressure)

Q **1.** (a) moves downward; (b) moves downward **3.** (a) downward; (b) downward; (c) same **5.** b, then a and d tie (zero), then c **7.** (a) 1 and 4; (b) 2; (c) 3 **9.** B, C, A

P **1.** 0.074 **3.** 1.1×10^5 Pa **5.** 2.9×10^4 N **7.** (b) 26 kN **9.** (a) 1.0×10^3 torr; (b) 1.7×10^3 torr **11.** (a) 94 torr; (b) 4.1×10^2 torr; (c) 3.1×10^2 torr **13.** 1.08×10^3 atm **15.** -2.6×10^4 Pa **17.** 7.2×10^5 N **19.** 4.69×10^5 N **21.** 0.635 J **23.** 44 km **25.** 739.26 torr **27.** (a) 7.9 km; (b) 16 km **29.** 8.50 kg **31.** (a) 6.7×10^2 kg/m^3; (b) 7.4×10^2 kg/m^3 **33.** (a) 2.04×10^{-2} m^3; (b) 1.57 kN **35.** five **37.** 57.3 cm **39.** (a) 1.2 kg; (b) 1.3×10^3 kg/m^3 **41.** (a) 0.10; (b) 0.083 **43.** (a) 637.8 cm^3; (b) 5.102 m^3; (c) 5.102×10^3 kg **45.** 0.126 m^3 **47.** (a) 1.80 m^3; (b) 4.75 m^3 **49.** (a) 3.0 m/s; (b) 2.8 m/s **51.** 8.1 m/s **53.** 66 W **55.** 1.4×10^5 J **57.** (a) 1.6×10^{-3} m^3/s; (b) 0.90 m **59.** (a) 2.5 m/s; (b) 2.6×10^5 Pa **61.** (a) 3.9 m/s; (b) 88 kPa **63.** 1.1×10^2 m/s **65.** (b) 2.0×10^{-2} m^3/s **67.** (a) 74 N; (b) 1.5×10^2 m^3 **69.** (a) 0.0776 m^3/s; (b) 69.8 kg/s **71.** (a) 35 cm; (b) 30 cm; (c) 20 cm **73.** 1.5 g/cm^3 **75.** 5.11×10^{-7} kg **77.** 44.2 g **79.** 6.0×10^2 kg/m^3 **81.** 45.3 cm^3 **83.** (a) 3.2 m/s; (b) 9.2×10^4 Pa; (c) 10.3 m **85.** 1.07×10^3 g **87.** 26.3 m^2 **89.** (a) 5.66×10^9 N; (b) 25.4 atm

Chapter 15

CP **1.** (sketch x versus t) (a) $-x_m$; (b) $+x_m$; (c) 0 **2.** c (a must have the form of Eq. 15-8) **3.** a (F must have the form of Eq. 15-10) **4.** (a) 5 J; (b) 2 J; (c) 5 J **5.** all tie (in Eq. 15-29, m is included in I) **6.** 1, 2, 3 (the ratio m/b matters; k does not)

Q **1.** a and b **3.** (a) 2; (b) positive; (c) between 0 and $+x_m$ **5.** (a) between D and E; (b) between $3\pi/2$ rad and 2π rad **7.** (a) all tie; (b) 3, then 1 and 2 tie; (c) 1, 2, 3 (zero); (d) 1, 2, 3 (zero); (e) 1, 3, 2 **9.** b (infinite period, does not oscillate), c, a **11.** (a) greater; (b) same; (c) same; (d) greater; (e) greater

P **1.** (a) 0.50 s; (b) 2.0 Hz; (c) 18 cm **3.** 37.8 m/s^2 **5.** (a) 1.0 mm; (b) 0.75 m/s; (c) 5.7×10^2 m/s^2 **7.** (a) 498 Hz; (b) greater **9.** (a) 3.0 m; (b) -49 m/s; (c) -2.7×10^2 m/s^2; (d) 20 rad; (e) 1.5 Hz; (f) 0.67 s **11.** 39.6 Hz **13.** (a) 0.500 s; (b) 2.00 Hz; (c) 12.6 rad/s; (d) 79.0 N/m; (e) 4.40 m/s; (f) 27.6 N **15.** (a) $0.18A$; (b) same direction **17.** (a) 5.58 Hz; (b) 0.325 kg; (c) 0.400 m **19.** (a) 25 cm; (b) 2.2 Hz **21.** 54 Hz **23.** 3.1 cm **25.** (a) 0.525 m; (b) 0.686 s **27.** (a) 0.75; (b) 0.25; (c) $2^{-0.5}x_m$ **29.** 37 mJ **31.** (a) 2.25 Hz; (b) 125 J; (c) 250 J; (d) 86.6 cm **33.** (a) 1.1 m/s; (b) 3.3 cm **35.** (a) 3.1 ms; (b) 4.0 m/s; (c) 0.080 J; (d) 80 N; (e) 40 N **37.** (a) 2.2 Hz; (b) 56 cm/s; (c) 0.10 kg; (d) 20.0 cm **39.** (a) 39.5 rad/s; (b) 34.2 rad/s; (c) 124 rad/s^2 **41.** (a) 0.205 kg$\cdot$m^2; (b) 47.7 cm; (c) 1.50 s **43.** (a) 1.64 s; (b) equal **45.** 8.77 s **47.** 0.366 s **49.** (a) 0.845 rad; (b) 0.0602 rad **51.** (a) 0.53 m; (b) 2.1 s **53.** 0.0653 s **55.** (a) 2.26 s; (b) increases; (c) same **57.** 6.0% **59.** (a) 14.3 s; (b) 5.27 **61.** (a) $F_m/b\omega$; (b) F_m/b **63.** 5.0 cm **65.** (a) 2.8×10^3 rad/s; (b) 2.1 m/s; (c) 5.7 km/s^2 **67.** (a) 1.1 Hz; (b) 5.0 cm **69.** 7.2 m/s **71.** (a) 7.90 N/m; (b) 1.19 cm; (c) 2.00 Hz **73.** (a) 1.3×10^2 N/m; (b) 0.62 s; (c) 1.6 Hz; (d) 5.0 cm; (e) 0.51 m/s **75.** (a) 16.6 cm; (b) 1.23% **77.** (a) 1.2 J; (b) 50 **79.** 1.53 m **81.** (a) 0.30 m; (b) 0.28 s; (c) 1.5×10^2 m/s^2; (d) 11 J **83.** (a) 1.23 kN/m; (b) 76.0 N **85.** 1.6 kg **87.** (a) 0.735 kg $\cdot$ m^2; (b) 0.0240 N $\cdot$ m; (c) 0.181 rad/s **89.** (a) 3.5 m; (b) 0.75 s **91.** (a) 0.35 Hz; (b) 0.39 Hz; (c) 0 (no oscillation) **93.** (a) 245 N/m; (b) 0.284 s **95.** 0.079 kg $\cdot$ m^2 **97.** (a) 8.11×10^{-5} kg $\cdot$ m^2; (b) 3.14 rad/s **99.** 14.0° **101.** (a) 3.2 Hz; (b) 0.26 m; (c) $x = (0.26\text{ m})\cos(20t - \pi/2)$, with t in seconds **103.** (a) 0.44 s; (b) 0.18 m **105.** (a) 0.45 s; (b) 0.10 m above and 0.20 m below; (c) 0.15 m; (d) 2.3 J **107.** 7×10^2 N/m **109.** 0.804 m **111.** (a) 0.30 m; (b) 30 m/s^2; (c) 0; (d) 4.4 s **113.** (a) F/m; (b) $2F/mL$; (c) 0 **115.** 2.54 m

Chapter 16

CP **1.** a, 2; b, 3; c, 1 (compare with the phase in Eq. 16-2, then see Eq. 16-5) **2.** (a) 2, 3, 1 (see Eq. 16-12); (b) 3, then 1 and 2 tie (find amplitude of dy/dt) **3.** (a) same (independent of f); (b) decrease ($\lambda = v/f$); (c) increase; (d) increase **4.** 0.20 and 0.80 tie, then 0.60, 0.45 **5.** (a) 1; (b) 3; (c) 2 **6.** (a) 75 Hz; (b) 525 Hz

Q **1.** (a) 1, 4, 2, 3; (b) 1, 4, 2, 3 **3.** a, upward; b, upward; c, downward; d, downward; e, downward; f, downward; g, upward; h, upward **5.** intermediate (closer to fully destructive) **7.** (a) 0, 0.2 wavelength, 0.5 wavelength (zero); (b) $4P_{avg,1}$ **9.** d **11.** c, a, b

P **1.** 1.1 ms **3.** (a) 3.49 m^{-1}; (b) 31.5 m/s **5.** (a) 0.680 s; (b) 1.47 Hz; (c) 2.06 m/s **7.** (a) 64 Hz; (b) 1.3 m; (c) 4.0 cm; (d) 5.0 m^{-1}; (e) 4.0×10^2 s^{-1}; (f) $\pi/2$ rad; (g) minus **9.** (a) 3.0 mm; (b) 16 m^{-1}; (c) 2.4×10^2 s^{-1}; (d) minus **11.** (a) negative; (b) 4.0 cm; (c) 0.31 cm^{-1}; (d) 0.63 s^{-1}; (e) π rad; (f) minus; (g) 2.0 cm/s; (h) -2.5 cm/s **13.** (a) 11.7 cm; (b) π rad **15.** (a) 0.12 mm; (b) 141 m^{-1}; (c) 628 s^{-1}; (d) plus **17.** (a) 15 m/s; (b) 0.036 N **19.** 129 m/s **21.** 2.63 m **23.** (a) 5.0 cm; (b) 40 cm; (c) 12 m/s; (d) 0.033 s; (e) 9.4 m/s; (f) 16 m^{-1}; (g) 1.9×10^2 s^{-1}; (h) 0.93 rad; (i) plus **27.** 3.2 mm **29.** 0.20 m/s **31.** $1.41y_m$ **33.** (a) 9.0 mm; (b) 16 m^{-1}; (c) 1.1×10^3 s^{-1}; (d) 2.7 rad; (e) plus **35.** 5.0 cm **37.** (a) 3.29 mm; (b) 1.55 rad; (c) 1.55 rad **39.** 84° **41.** (a) 82.0 m/s; (b) 16.8 m; (c) 4.88 Hz **43.** (a) 7.91 Hz; (b) 15.8 Hz; (c) 23.7 Hz **45.** (a) 105 Hz; (b) 158 m/s **47.** 260 Hz **49.** (a) 144 m/s; (b) 60.0 cm; (c) 241 Hz **51.** (a) 0.50 cm;

(b) 3.1 m^{-1}; (c) $3.1 \times 10^2\ s^{-1}$; (d) minus **53.** (a) 0.25 cm; (b) 1.2×10^2 cm/s; (c) 3.0 cm; (d) 0 **55.** 0.25 m **57.** (a) 2.00 Hz; (b) 2.00 m; (c) 4.00 m/s; (d) 50.0 cm; (e) 150 cm; (f) 250 cm; (g) 0; (h) 100 cm; (i) 200 cm **59.** (a) 324 Hz; (b) eight **61.** 36 N **63.** (a) 75 Hz; (b) 13 ms **65.** (a) 2.0 mm; (b) 95 Hz; (c) +30 m/s; (d) 31 cm; (e) 1.2 m/s **67.** (a) 0.31 m; (b) 1.64 rad; (c) 2.2 mm **69.** (a) $0.83y_1$; (b) 37° **71.** (a) 3.77 m/s; (b) 12.3 N; (c) 0; (d) 46.4 W; (e) 0; (f) 0; (g) ±0.50 cm **73.** 1.2 rad **75.** (a) 300 m/s; (b) no **77.** (a) $[k\,\Delta\ell(\ell + \Delta\ell)/m]^{0.5}$ **79.** (a) 144 m/s; (b) 3.00 m; (c) 1.50 m; (d) 48.0 Hz; (e) 96.0 Hz **81.** (a) 1.00 cm; (b) $3.46 \times 10^3\ s^{-1}$; (c) 10.5 m^{-1}; (d) plus **83.** (a) $2\pi y_m/\lambda$; (b) no **85.** (a) 240 cm; (b) 120 cm; (c) 80 cm **87.** (a) 1.33 m/s; (b) 1.88 m/s; (c) 16.7 m/s^2; (d) 23.7 m/s^2 **89.** (a) 0.52 m; (b) 40 m/s; (c) 0.40 m **91.** (a) 0.16 m; (b) 2.4×10^2 N; (c) $y(x, t) = (0.16\ m)\sin[(1.57\ m^{-1})x]\sin[(31.4\ s^{-1})t]$ **93.** (c) 2.0 m/s; (d) $-x$ **95.** (a) ∞; (b) 1.0; (c) 4.0%

Chapter 17

CP **1.** beginning to decrease (example: mentally move the curves of Fig. 17-6 rightward past the point at $x = 42$ cm) **2.** (a) 1 and 2 tie, then 3 (see Eq. 17-28); (b) 3, then 1 and 2 tie (see Eq. 17-26) **3.** second (see Eqs. 17-39 and 17-41) **4.** *a*, greater; *b*, less; *c*, can't tell; *d*, can't tell; *e*, greater; *f*, less

Q **1.** (a) 0, 0.2 wavelength, 0.5 wavelength (zero); (b) $4P_{avg,1}$ **3.** *C*, then *A* and *B* tie **5.** *E*, *A*, *D*, *C*, *B* **7.** 1, 4, 3, 2 **9.** 150 Hz and 450 Hz **11.** 505, 507, 508 Hz or 501, 503, 508 Hz

P **1.** (a) 79 m; (b) 41 m; (c) 89 m **3.** (a) 2.6 km; (b) 2.0×10^2 **5.** 1.9×10^3 km **7.** 40.7 m **9.** 0.23 ms **11.** (a) 76.2 μm; (b) 0.333 mm **13.** 960 Hz **15.** (a) 2.3×10^2 Hz; (b) higher **17.** (a) 143 Hz; (b) 3; (c) 5; (d) 286 Hz; (e) 2; (f) 3 **19.** (a) 14; (b) 14 **21.** (a) 343 Hz; (b) 3; (c) 5; (d) 686 Hz; (e) 2; (f) 3 **23.** (a) 0; (b) fully constructive; (c) increase; (d) 128 m; (e) 63.0 m; (f) 41.2 m **25.** 36.8 nm **27.** (a) 1.0×10^3; (b) 32 **29.** 15.0 mW **31.** 2 μW **33.** 0.76 μm **35.** (a) $5.97 \times 10^{-5}\ W/m^2$; (b) 4.48 nW **37.** (a) 0.34 nW; (b) 0.68 nW; (c) 1.4 nW; (d) 0.88 nW; (e) 0 **39.** (a) 405 m/s; (b) 596 N; (c) 44.0 cm; (d) 37.3 cm **41.** (a) 833 Hz; (b) 0.418 m **43.** (a) 3; (b) 1129 Hz; (c) 1506 Hz **45.** (a) 2; (b) 1 **47.** 12.4 m **49.** 45.3 N **51.** 2.25 ms **53.** 0.020 **55.** (a) 526 Hz; (b) 555 Hz **57.** 0 **59.** (a) 1.022 kHz; (b) 1.045 kHz **61.** 41 kHz **63.** 155 Hz **65.** (a) 2.0 kHz; (b) 2.0 kHz **67.** (a) 485.8 Hz; (b) 500.0 Hz; (c) 486.2 Hz; (d) 500.0 Hz **69.** (a) 42°; (b) 11 s **71.** 1 cm **73.** 2.1 m **75.** (a) 39.7 $\mu W/m^2$; (b) 171 nm; (c) 0.893 Pa **77.** 0.25 **79.** (a) 2.10 m; (b) 1.47 m **81.** (a) 59.7; (b) 2.81×10^{-4} **83.** (a) rightward; (b) 0.90 m/s; (c) less **85.** (a) 11 ms; (b) 3.8 m **87.** (a) 9.7×10^2 Hz; (b) 1.0 kHz; (c) 60 Hz, no **89.** (a) 21 nm; (b) 35 cm; (c) 24 nm; (d) 35 cm **91.** (a) 7.70 Hz; (b) 7.70 Hz **93.** (a) 5.2 kHz; (b) 2 **95.** (a) 10 W; (b) 0.032 W/m^2; (c) 99 dB **97.** (a) 0; (b) 0.572 m; (c) 1.14 m **99.** 171 m **101.** (a) 3.6×10^2 m/s; (b) 150 Hz **103.** 400 Hz **105.** (a) 14; (b) 12 **107.** 821 m/s **109.** (a) 39.3 Hz; (b) 118 Hz **111.** 4.8×10^2 Hz

Chapter 18

CP **1.** (a) all tie; (b) 50°X, 50°Y, 50°W **2.** (a) 2 and 3 tie, then 1, then 4; (b) 3, 2, then 1 and 4 tie (from Eqs. 18-9 and 18-10, assume that change in area is proportional to initial area) **3.** *A* (see Eq. 18-14) **4.** *c* and *e* (maximize area enclosed by a clockwise cycle) **5.** (a) all tie (ΔE_{int} depends on *i* and *f*, not on path); (b) 4, 3, 2, 1 (compare areas under curves); (c) 4, 3, 2, 1 (see Eq. 18-26) **6.** (a) zero (closed cycle); (b) negative (W_{net} is negative; see Eq. 18-26) **7.** *b* and *d* tie, then *a*, *c* (P_{cond} identical; see Eq. 18-32)

Q **1.** *c*, then the rest tie **3.** *B*, then *A* and *C* tie **5.** (a) *f*, because ice temperature will not rise to freezing point and then drop; (b) *b* and *c* at freezing point, *d* above, *e* below; (c) in *b* liquid partly freezes and no ice melts; in *c* no liquid freezes and no ice melts; in *d* no liquid freezes and ice fully melts; in *e* liquid fully freezes and no ice melts **7.** (a) both clockwise; (b) both clockwise **9.** (a) greater; (b) 1, 2, 3; (c) 1, 3, 2; (d) 1, 2, 3; (e) 2, 3, 1 **11.** *c*, *b*, *a*

P **1.** 1.366 **3.** 348 K **5.** (a) 320°F; (b) −12.3°F **7.** −92.1°X **9.** 2.731 cm **11.** 49.87 cm^3 **13.** 29 cm^3 **15.** 360°C **17.** 0.26 cm^3 **19.** 0.13 mm **21.** 7.5 cm **23.** 160 s **25.** 94.6 L **27.** 42.7 kJ **29.** 33 m^2 **31.** 33 g **33.** 3.0 min **35.** 13.5 C° **37.** (a) 5.3°C; (b) 0; (c) 0°C; (d) 60 g **39.** 742 kJ **41.** (a) 0°C; (b) 2.5°C **43.** (a) 1.2×10^2 J; (b) 75 J; (c) 30 J **45.** −30 J **47.** (a) 6.0 cal; (b) −43 cal; (c) 40 cal; (d) 18 cal; (e) 18 cal **49.** 60 J **51.** (a) 1.23 kW; (b) 2.28 kW; (c) 1.05 kW **53.** 1.66 kJ/s **55.** (a) 16 J/s; (b) 0.048 g/s **57.** (a) $1.7 \times 10^4\ W/m^2$; (b) 18 W/m^2 **59.** 0.50 min **61.** 0.40 cm/h **63.** −4.2°C **65.** 1.1 m **67.** 10% **69.** (a) 80 J; (b) 80 J **71.** 4.5×10^2 J/kg·K **73.** 0.432 cm^3 **75.** 3.1×10^2 J **77.** 79.5°C **79.** 23 J **81.** (a) $11p_1V_1$; (b) $6p_1V_1$ **83.** $4.83 \times 10^{-2}\ cm^3$ **85.** 10.5°C **87.** (a) 90 W; (b) 2.3×10^2 W; (c) 3.3×10^2 W **89.** (a) 1.87×10^4; (b) 10.4 h **91.** 333 J **93.** 8.6 J **95.** (a) −45 J; (b) +45 J **97.** 4.0×10^3 min **99.** −6.1 nW **101.** 1.17 C° **103.** $8.0 \times 10^{-3}\ m^2$ **105.** (a) too fast; (b) 0.79 s/h **107.** 1.9

Chapter 19

CP **1.** all but *c* **2.** (a) all tie; (b) 3, 2, 1 **3.** gas *A* **4.** 5 (greatest change in *T*), then tie of 1, 2, 3, and 4 **5.** 1, 2, 3 ($Q_3 = 0$, Q_2 goes into work W_2, but Q_1 goes into greater work W_1 and increases gas temperature)

Q **1.** *d*, then *a* and *b* tie, then *c* **3.** 20 J **5.** (a) 3; (b) 1; (c) 4; (d) 2; (e) yes **7.** (a) 1, 2, 3, 4; (b) 1, 2, 3 **9.** constant-volume process

P **1.** 0.933 kg **3.** (a) 0.0388 mol; (b) 220°C **5.** 25 molecules/cm^3 **7.** (a) 3.14×10^3 J; (b) from **9.** 186 kPa **11.** 5.60 kJ **13.** (a) 1.5 mol; (b) 1.8×10^3 K; (c) 6.0×10^2 K; (d) 5.0 kJ **15.** 360 K **17.** 2.0×10^5 Pa **19.** (a) 511 m/s; (b) −200°C; (c) 899°C **21.** 1.8×10^2 m/s **23.** 1.9 kPa **25.** (a) 5.65×10^{-21} J; (b) 7.72×10^{-21} J; (c) 3.40 kJ; (d) 4.65 kJ **27.** (a) 6.76×10^{-20} J; (b) 10.7 **29.** (a) 6×10^9 km **31.** (a) 3.27×10^{10} molecules/cm^3; (b) 172 m **33.** (a) 6.5 km/s; (b) 7.1 km/s **35.** (a) 420 m/s; (b) 458 m/s; (c) yes **37.** (a) 0.67; (b) 1.2; (c) 1.3; (d) 0.33 **39.** (a) 1.0×10^4 K; (b) 1.6×10^5 K; (c) 4.4×10^2 K; (d) 7.0×10^3 K; (e) no; (f) yes **41.** (a) 7.0 km/s; (b) 2.0×10^{-8} cm; (c) 3.5×10^{10} collisions/s **43.** (a) 3.49 kJ; (b) 2.49 kJ; (c) 997 J; (d) 1.00 kJ **45.** (a) 6.6×10^{-26} kg; (b) 40 g/mol **47.** (a) 0; (b) +374 J; (c) +374 J; (d) $+3.11 \times 10^{-22}$ J **49.** 15.8 J/mol·K **51.** 8.0 kJ **53.** (a) 6.98 kJ; (b) 4.99 kJ; (c) 1.99 kJ; (d) 2.99 kJ **55.** (a) 14 atm; (b) 6.2×10^2 K **57.** (a) diatomic; (b) 446 K; (c) 8.10 mol **59.** −15 J **61.** −20 J **63.** (a) 3.74 kJ; (b) 3.74 kJ; (c) 0; (d) 0; (e) −1.81 kJ; (f) 1.81 kJ; (g) −3.22 kJ; (h) −1.93 kJ; (i) −1.29 kJ; (j) 520 J; (k) 0; (l) 520 J; (m) 0.0246 m^3; (n) 2.00 atm; (o) 0.0373 m^3; (p) 1.00 atm **65.** (a) monatomic; (b) 2.7×10^4 K; (c) 4.5×10^4 mol; (d) 3.4 kJ; (e) 3.4×10^2 kJ; (f) 0.010 **67.** (a) 2.00 atm; (b) 333 J; (c) 0.961 atm; (d) 236 J **69.** 349 K **71.** (a) −374 J; (b) 0; (c) +374 J; (d) $+3.11 \times 10^{-22}$ J **73.** $7.03 \times 10^9\ s^{-1}$ **75.** (a) 900 cal; (b) 0; (c) 900 cal; (d) 450 cal; (e) 1200 cal; (f) 300 cal; (g) 900 cal; (h) 450 cal; (i) 0; (j) −900 cal; (k) 900 cal; (l) 450 cal **77.** (a) $3/v_0^3$; (b) $0.750v_0$; (c) $0.775v_0$ **79.** (a) −2.37 kJ; (b) 2.37 kJ **81.** (b) 125 J; (c) to **83.** (a) 8.0 atm; (b) 300 K; (c) 4.4 kJ; (d) 3.2 atm; (e) 120 K; (f) 2.9 kJ; (g) 4.6 atm; (h) 170 K; (i) 3.4 kJ **85.** (a) 38 L; (b) 71 g **87.** −3.0 J **89.** 22.8 m **95.** 1.40 **97.** 4.71

Chapter 20

CP **1.** a, b, c **2.** smaller (*Q* is smaller) **3.** c, b, a **4.** a, d, c, b **5.** b

Q **1.** *b*, *a*, *c*, *d* **3.** unchanged **5.** *a* and *c* tie, then *b* and *d* tie **7.** (a) same; (b) increase; (c) decrease **9.** A, first; B, first and second; C, second; D, neither

P **1.** (a) 9.22 kJ; (b) 23.1 J/K; (c) 0 **3.** 14.4 J/K **5.** (a) 5.79×10^4 J; (b) 173 J/K **7.** (a) 320 K; (b) 0; (c) +1.72 J/K **9.** +0.76 J/K **11.** (a) 57.0°C; (b) −22.1 J/K; (c) +24.9 J/K; (d) +2.8 J/K **13.** (a) −710 mJ/K; (b) +710 mJ/K; (c) +723 mJ/K; (d) −723 mJ/K; (e) +13 mJ/K; (f) 0 **15.** (a) −943 J/K; (b) +943 J/K; (c) yes **17.** (a) 0.333; (b) 0.215; (c) 0.644; (d) 1.10; (e) 1.10; (f) 0; (g) 1.10; (h) 0; (i) −0.889; (j) −0.889; (k) −1.10; (l) −0.889; (m) 0; (n) 0.889; (o) 0 **19.** (a) 0.693; (b) 4.50; (c) 0.693; (d) 0; (e) 4.50; (f) 23.0 J/K; (g) −0.693; (h) 7.50; (i) −0.693; (j) 3.00; (k) 4.50; (l) 23.0 J/K **21.** −1.18 J/K **23.** 97 K **25.** (a) 266 K; (b) 341 K **27.** (a) 23.6%; (b) 1.49×10^4 J **29.** (a) 2.27 kJ; (b) 14.8 kJ; (c) 15.4%; (d) 75.0%; (e) greater **31.** (a) 33 kJ; (b) 25 kJ; (c) 26 kJ; (d) 18 kJ **33.** (a) 1.47 kJ; (b) 554 J; (c) 918 J; (d) 62.4% **35.** (a) 3.00; (b) 1.98; (c) 0.660; (d) 0.495; (e) 0.165; (f) 34.0% **37.** 440 W **39.** 20 J **41.** 0.25 hp **43.** 2.03 **47.** (a) $W = N!/(n_1!\, n_2!\, n_3!)$; (b) $[(N/2)!\,(N/2)!]/[(N/3)!\,(N/3)!\,(N/3)!]$; (c) 4.2×10^{16} **49.** 0.141 J/K·s **51.** (a) 87 m/s; (b) 1.2×10^2 m/s; (c) 22 J/K **53.** (a) 78%; (b) 82 kg/s **55.** (a) 40.9°C; (b) −27.1 J/K; (c) 30.3 J/K; (d) 3.18 J/K **57.** +3.59 J/K **59.** 1.18×10^3 J/K **63.** (a) 0; (b) 0; (c) −23.0 J/K; (d) 23.0 J/K **65.** (a) 25.5 kJ; (b) 4.73 kJ; (c) 18.5% **67.** (a) 1.95 J/K; (b) 0.650 J/K; (c) 0.217 J/K; (d) 0.072 J/K; (e) decrease **69.** (a) 4.45 J/K; (b) no **71.** (a) 1.26×10^{14}; (b) 4.71×10^{13}; (c) 0.37; (d) 1.01×10^{29}; (e) 1.37×10^{28}; (f) 0.14; (g) 9.05×10^{58}; (h) 1.64×10^{57}; (i) 0.018; (j) decrease **73.** (a) 42.6 kJ; (b) 7.61 kJ **75.** (a) 1; (b) 1; (c) 3; (d) 10; (e) 1.5×10^{-23} J/K; (f) 3.2×10^{-23} J/K **77.** $e = (1 + K)^{-1}$ **79.** 6.7

Chapter 21

CP **1.** *C* and *D* attract; *B* and *D* attract **2.** (a) leftward; (b) leftward; (c) leftward **3.** (a) *a*, *c*, *b*; (b) less than **4.** −15*e* (net charge of −30*e* is equally shared)
Q **1.** 3, 1, 2, 4 (zero) **3.** *a* and *b* **5.** $2kq^2/r^2$, up the page **7.** *b* and *c* tie, then *a* (zero) **9.** (a) same; (b) less than; (c) cancel; (d) add; (e) adding components; (f) positive direction of *y*; (g) negative direction of *y*; (h) positive direction of *x*; (i) negative direction of *x* **11.** (a) +4*e*; (b) −2*e* upward; (c) −3*e* upward; (d) −12*e* upward
P **1.** 0.500 **3.** 1.39 m **5.** 2.81 N **7.** −4.00 **9.** (a) −1.00 μC; (b) 3.00 μC **11.** (a) 0.17 N; (b) −0.046 N **13.** (a) −14 cm; (b) 0 **15.** (a) 35 N; (b) −10°; (c) −8.4 cm; (d) +2.7 cm **17.** (a) 1.60 N; (b) 2.77 N **19.** (a) 3.00 cm; (b) 0; (c) −0.444 **21.** 3.8×10^{-8} C **23.** (a) 0; (b) 12 cm; (c) 0; (d) 4.9×10^{-26} N **25.** 6.3×10^{11} **27.** (a) 3.2×10^{-19} C; (b) 2 **29.** (a) −6.05 cm; (b) 6.05 cm **31.** 122 mA **33.** 1.3×10^7 C **35.** (a) 0; (b) 1.9×10^{-9} N **37.** (a) ^{9}B; (b) ^{13}N; (c) ^{12}C **39.** 1.31×10^{-22} N **41.** (a) 5.7×10^{13} C; (b) cancels out; (c) 6.0×10^5 kg **43.** (b) 3.1 cm **45.** 0.19 MC **47.** −45 μC **49.** 3.8 N **51.** (a) 2.00×10^{10} electrons; (b) 1.33×10^{10} electrons **53.** (a) 8.99×10^9 N; (b) 8.99 kN **55.** (a) 0.5; (b) 0.15; (c) 0.85 **57.** 1.7×10^8 N **59.** -1.32×10^{13} C **61.** (a) (0.829 N)$\hat{i}$; (b) (−0.621 N)$\hat{j}$ **63.** 2.2×10^{-6} kg **65.** 4.68×10^{-19} N **67.** (a) 2.72*L*; (b) 0 **69.** (a) 5.1×10^2 N; (b) 7.7×10^{28} m/s^2 **71.** (a) 0; (b) 3.43×10^9 m/s^2 **73.** (a) 2.19×10^6 m/s; (b) 1.09×10^6 m/s; (c) decrease **75.** 4.16×10^{42}

Chapter 22

CP **1.** (a) rightward; (b) leftward; (c) leftward; (d) rightward (p and e have same charge magnitude, and p is farther) **2.** (a) toward positive *y*; (b) toward positive *x*; (c) toward negative *y* **3.** (a) leftward; (b) leftward; (c) decrease **4.** (a) all tie; (b) 1 and 3 tie, then 2 and 4 tie
Q **1.** *a*, *b*, *c* **3.** (a) yes; (b) toward; (c) no (the field vectors are not along the same line); (d) cancel; (e) add; (f) adding components; (g) toward negative *y* **5.** (a) to their left; (b) no **7.** (a) 4, 3, 1, 2; (b) 3, then 1 and 4 tie, then 2 **9.** *a*, *b*, *c* **11.** *e*, *b*, then *a* and *c* tie, then *d* (zero) **13.** *a*, *b*, *c*
P **3.** (a) 3.07×10^{21} N/C; (b) outward **5.** 56 pC **7.** (1.02×10^5 N/C)$\hat{j}$ **9.** (a) 1.38×10^{-10} N/C; (b) 180° **11.** −30 cm **13.** (a) 3.60×10^{-6} N/C; (b) 2.55×10^{-6} N/C; (c) 3.60×10^{-4} N/C; (d) 7.09×10^{-7} N/C; (e) As the proton nears the disk, the forces on it from electrons e_s more nearly cancel. **15.** (a) 160 N/C; (b) 45° **17.** (a) −90°; (b) +2.0 μC; (c) −1.6 μC **19.** (a) $qd/4\pi\varepsilon_0 r^3$; (b) −90° **23.** 0.506 **25.** (a) 1.62×10^6 N/C; (b) −45° **27.** (a) 23.8 N/C; (b) −90° **29.** 1.57 **31.** (a) -5.19×10^{-14} C/m; (b) 1.57×10^{-3} N/C; (c) −180°; (d) 1.52×10^{-8} N/C; (e) 1.52×10^{-8} N/C **35.** 0.346 m **37.** 28% **39.** −5*e* **41.** (a) 1.5×10^3 N/C; (b) 2.4×10^{-16} N; (c) up; (d) 1.6×10^{-26} N; (e) 1.5×10^{10} **43.** 3.51×10^{15} m/s^2 **45.** 6.6×10^{-15} N **47.** (a) 1.92×10^{12} m/s^2; (b) 1.96×10^5 m/s **49.** (a) 0.245 N; (b) −11.3°; (c) 108 m; (d) −21.6 m **51.** 2.6×10^{-10} N; (b) 3.1×10^{-8} N; (c) moves to stigma **53.** 27 μm **55.** (a) 2.7×10^6 m/s; (b) 1.0 kN/C **57.** (a) 9.30×10^{-15} C·m; (b) 2.05×10^{-11} J **59.** 1.22×10^{-23} J **61.** $(1/2\pi)(pE/I)^{0.5}$ **63.** (a) 8.87×10^{-15} N; (b) 120 **65.** 217° **67.** 61 N/C **69.** (a) 47 N/C; (b) 27 N/C **71.** 38 N/C **73.** (a) −1.0 cm; (b) 0; (c) 10 pC **75.** +1.00 μC **77.** (a) 6.0 mm; (b) 180° **79.** 9:30 **81.** (a) −0.029 C; (b) repulsive forces would explode the sphere **83.** (a) -1.49×10^{-26} J; (b) $(-1.98 \times 10^{-26}$ N·m)$\hat{k}$; (c) 3.47×10^{-26} J **85.** (a) top row: 4, 8, 12; middle row: 5, 10, 14; bottom row: 7, 11, 16; (b) 1.63×10^{-19} C **87.** (a) (−1.80 N/C)$\hat{i}$; (b) (43.2 N/C)$\hat{i}$; (c) (−6.29 N/C)$\hat{i}$

Chapter 23

CP **1.** (a) +*EA*; (b) −*EA*; (c) 0; (d) 0 **2.** (a) 2; (b) 3; (c) 1 **3.** (a) equal; (b) equal; (c) equal **4.** 3 and 4 tie, then 2, 1
Q **1.** (a) 8 N·m^2/C; (b) 0 **3.** all tie **5.** all tie **7.** *a*, *c*, then *b* and *d* tie (zero) **9.** (a) 2, 1, 3; (b) all tie (+4*q*) **11.** (a) impossible; (b) $-3q_0$; (c) impossible
P **1.** −0.015 N·m^2/C **3.** (a) 0; (b) −3.92 N·m^2/C; (c) 0; (d) 0 **5.** 3.01 nN·m^2/C **7.** 2.0×10^5 N·m^2/C **9.** (a) 8.23 N·m^2/C; (b) 72.9 pC; (c) 8.23 N·m^2/C; (d) 72.9 pC **11.** −1.70 nC **13.** 3.54 μC **15.** (a) 0; (b) 0.0417 **17.** (a) 37 μC; (b) 4.1×10^6 N·m^2/C **19.** (a) 4.5×10^{-7} C/m^2; (b) 5.1×10^4 N/C **21.** (a) -3.0×10^{-6} C; (b) $+1.3 \times 10^{-5}$ C **23.** (a) 0.32 μC; (b) 0.14 μC **25.** 5.0 μC/m **27.** 3.8×10^{-8} C/m^2 **29.** (a) 0.214 N/C; (b) inward; (c) 0.855 N/C; (d) outward; (e) -3.40×10^{-12} C; (f) -3.40×10^{-12} C **31.** (a) 2.3×10^6 N/C; (b) outward; (c) 4.5×10^5 N/C; (d) inward **33.** (a) 0; (b) 0; (c) $(-7.91 \times 10^{-11}$ N/C)$\hat{i}$ **35.** −1.5 **37.** (a) 5.3×10^7 N/C; (b) 60 N/C **39.** 5.0 nC/m^2 **41.** 0.44 mm **43.** (a) 0; (b) 1.31 μN/C; (c) 3.08 μN/C; (d) 3.08 μN/C **45.** (a) 2.50×10^4 N/C; (b) 1.35×10^4 N/C **47.** −7.5 nC **49.** (a) 0; (b) 56.2 mN/C; (c) 112 mN/C; (d) 49.9 mN/C; (e) 0; (f) 0; (g) −5.00 fC; (h) 0 **51.** 1.79×10^{-11} C/m^2 **53.** (a) 7.78 fC; (b) 0; (c) 5.58 mN/C; (d) 22.3 mN/C **55.** $6K\varepsilon_0 r^3$ **57.** (a) 0; (b) 2.88×10^4 N/C; (c) 200 N/C **59.** (a) 5.4 N/C; (b) 6.8 N/C **61.** (a) 0; (b) $q_a/4\pi\varepsilon_0 r^2$; (c) $(q_a + q_b)/4\pi\varepsilon_0 r^2$ **63.** −1.04 nC **65.** (a) 0.125; (b) 0.500 **67.** (a) +2.0 nC; (b) −1.2 nC; (c) +1.2 nC; (d) +0.80 nC **69.** (5.65×10^4 N/C)$\hat{j}$ **71.** (a) -2.53×10^{-2} N·m^2/C; (b) $+2.53 \times 10^{-2}$ N·m^2/C **75.** 3.6 nC **77.** (a) +4.0 μC; (b) −4.0 μC **79.** (a) 693 kg/s; (b) 693 kg/s; (c) 347 kg/s; (d) 347 kg/s; (e) 575 kg/s **81.** (a) 0.25*R*; (b) 2.0*R*

Chapter 24

CP **1.** (a) negative; (b) positive; (c) increase; (d) higher **2.** (a) rightward; (b) 1, 2, 3, 5: positive; 4, negative; (c) 3, then 1, 2, and 5 tie, then 4 **3.** all tie **4.** *a*, *c* (zero), *b* **5.** (a) 2, then 1 and 3 tie; (b) 3; (c) accelerate leftward

Q **1.** $-4q/4\pi\varepsilon_0 d$ **3.** (a) 1 and 2; (b) none; (c) no; (d) 1 and 2, yes; 3 and 4, no **5.** (a) higher; (b) positive; (c) negative; (d) all tie **7.** (a) 0; (b) 0; (c) 0; (d) all three quantities still 0 **9.** (a) 3 and 4 tie, then 1 and 2 tie; (b) 1 and 2, increase; 3 and 4, decrease **11.** *a*, *b*, *c*
P **1.** (a) 3.0×10^5 C; (b) 3.6×10^6 J **3.** 2.8×10^5 **5.** 8.8 mm **7.** −32.0 V **9.** (a) 1.87×10^{-21} J; (b) −11.7 mV **11.** (a) −0.268 mV; (b) −0.681 mV **13.** (a) 3.3 nC; (b) 12 nC/m² **15.** (a) 0.54 mm; (b) 790 V **17.** 0.562 mV **19.** (a) 6.0 cm; (b) −12.0 cm **21.** 16.3 μV **23.** (a) 24.3 mV; (b) 0 **25.** (a) −2.30 V; (b) −1.78 V **27.** 13 kV **29.** 32.4 mV **31.** 47.1 μV **33.** 18.6 mV **35.** $(-12\text{ V/m})\hat{i} + (12\text{ V/m})\hat{j}$ **37.** 150 N/C **39.** $(-4.0 \times 10^{-16}\text{ N})\hat{i} + (1.6 \times 10^{-16}\text{ N})\hat{j}$ **41.** (a) 0.90 J; (b) 4.5 J **43.** −0.192 pJ **45.** 2.5 km/s **47.** 22 km/s **49.** 0.32 km/s **51.** (a) $+6.0 \times 10^4$ V; (b) -7.8×10^5 V; (c) 2.5 J; (d) increase; (e) same; (f) same **53.** (a) 0.225 J; (b) *A* 45.0 m/s², *B* 22.5 m/s²; (c) *A* 7.75 m/s, *B* 3.87 m/s **55.** 1.6×10^{-9} m **57.** (a) 3.0 J; (b) −8.5 m **59.** (a) proton; (b) 65.3 km/s **61.** (a) 12; (b) 2 **63.** (a) -1.8×10^2 V; (b) 2.9 kV; (c) −8.9 kV **65.** 2.5×10^{-8} C **67.** (a) 12 kN/C; (b) 1.8 kV; (c) 5.8 cm **69.** (a) 64 N/C; (b) 2.9 V; (c) 0 **71.** $p/2\pi\varepsilon_0 r^3$ **73.** (a) 3.6×10^5 V; (b) no **75.** 6.4×10^8 V **77.** 2.90 kV **79.** 7.0×10^5 m/s **81.** (a) 1.8 cm; (b) 8.4×10^5 m/s; (c) 2.1×10^{-17} N; (d) positive; (e) 1.6×10^{-17} N; (f) negative **83.** (a) $+7.19 \times 10^{-10}$ V; (b) $+2.30 \times 10^{-28}$ J; (c) $+2.43 \times 10^{-29}$ J **85.** 2.30×10^{-28} J **87.** 2.1 days **89.** 2.30×10^{-22} J **91.** 1.48×10^7 m/s **93.** −1.92 MV **95.** (a) $Q/4\pi\varepsilon_0 r$; (b) $(\rho/3\varepsilon_0)(1.5r_2^2 - 0.50r^2 - r_1^3 r^{-1})$, $\rho = Q/[(4\pi/3)(r_2^3 - r_1^3)]$; (c) $(\rho/2\varepsilon_0)(r_2^2 - r_1^2)$, with ρ as in (b); (d) yes **97.** (a) 38 s; (b) 2.7×10^2 days **101.** (a) 0.484 MeV; (b) 0 **103.** −1.7

Chapter 25

CP **1.** (a) same; (b) same **2.** (a) decreases; (b) increases; (c) decreases **3.** (a) *V*, *q*/2; (b) *V*/2; *q*
Q **1.** *a*, 2; *b*, 1; *c*, 3 **3.** (a) no; (b) yes; (c) all tie **5.** (a) same; (b) same; (c) more; (d) more **7.** *a*, series; *b*, parallel; *c*, parallel **9.** (a) increase; (b) same; (c) increase; (d) increase; (e) increase; (f) increase **11.** parallel, C_1 alone, C_2 alone, series
P **1.** (a) 3.5 pF; (b) 3.5 pF; (c) 57 V **3.** (a) 144 pF; (b)17.3 nC **5.** 0.280 pF **7.** 6.79×10^{-4} F/m² **9.** 315 mC **11.** 3.16 μF **13.** 43 pF **15.** (a) 3.00 μF; (b) 60.0 μC; (c) 10.0 V; (d) 30.0 μC; (e) 10.0 V; (f) 20.0 μC; (g) 5.00 V; (h) 20.0 μC **17.** (a) 789 μC; (b) 78.9 V **19.** (a) 4.0 μF; (b) 2.0 μF **21.** (a) 50 V; (b) 5.0×10^{-5} C; (c) 1.5×10^{-4} C **23.** (a) 4.5×10^{14}; (b) 1.5×10^{14}; (c) 3.0×10^{14}; (d) 4.5×10^{14}; (e) up; (f) up **25.** 3.6 pC **27.** (a) 9.00 μC; (b) 16.0 μC; (c) 9.00 μC; (d) 16.0 μC; (e) 8.40 μC; (f) 16.8 μC; (g) 10.8 μC; (h) 14.4 μC **29.** 72 F **31.** 0.27 J **33.** 0.11 J/m³ **35.** (a) 9.16×10^{-18} J/m³; (b) 9.16×10^{-6} J/m³; (c) 9.16×10^6 J/m³; (d) 9.16×10^{18} J/m³; (e) ∞ **37.** (a) 16.0 V; (b) 45.1 pJ; (c) 120 pJ; (d) 75.2 pJ **39.** (a) 190 V; (b) 95 mJ **41.** 81 pF/m **43.** Pyrex **45.** 66 μJ **47.** 0.63 m² **49.** 17.3 pF **51.** (a) 10 kV/m; (b) 5.0 nC; (c) 4.1 nC **53.** (a) 89 pF; (b) 0.12 nF; (c) 11 nC; (d) 11 nC; (e) 10 kV/m; (f) 2.1 kV/m; (g) 88 V; (h) −0.17 μJ **55.** (a) 0.107 nF; (b) 7.79 nC; (c) 7.45 nC **57.** 45 μC **59.** 16 μC **61.** (a) 7.20 μC; (b) 18.0 μC; (c) Battery supplies charges only to plates to which it is connected; charges on other plates are due to electron transfers between plates, in accord with new distribution of voltages across the capacitors. So the battery does not directly supply charge on capacitor 4. **63.** (a) 10 μC; (b) 20 μC **65.** 1.06 nC **67.** (a) 2.40 μF; (b) 0.480 mC; (c) 80 V; (d) 0.480 mC; (e) 120 V **69.** 4.9% **71.** (a) 0.708 pF; (b) 0.600 ; (c) 1.02×10^{-9} J; (d) sucked in **73.** 5.3 V **75.** 40 μF **77.** (a) 200 kV/m; (b) 200 kV/m; (c) 1.77 μC/m²; (d) 4.60 μC/m²; (e) −2.83 μC/m² **79.** (a) $q^2/2\varepsilon_0 A$

Chapter 26

CP **1.** 8 A, rightward **2.** (a)–(c) rightward **3.** *a* and *c* tie, then *b* **4.** device 2 **5.** (a) and (b) tie, then (d), then (c)
Q **1.** tie of *A*, *B*, and *C*, then tie of *A* + *B* and *B* + *C*, then *A* + *B* + *C* **3.** (a) top-bottom, front-back, left-right; (b) top-bottom, front-back, left-right; (c) top-bottom, front-back, left-right; (d) top-bottom, front-back, left-right **5.** *a*, *b*, and *c* all tie, then *d* **7.** (a) *B*, *A*, *C*; (b) *B*, *A*, *C* **9.** (a) *C*, *B*, *A*; (b) all tie; (c) *A*, *B*, *C*; (d) all tie **11.** (a) *a* and *c* tie, then *b* (zero); (b) *a*, *b*, *c*; (c) *a* and *b* tie, then *c*
P **1.** (a) 1.2 kC; (b) 7.5×10^{21} **3.** 6.7 μC/m² **5.** (a) 6.4 A/m²; (b) north; (c) cross-sectional area **7.** 0.38 mm **9.** 18.1 μA **11.** (a) 1.33 A; (b) 0.666 A; (c) J_a **13.** 13 min **15.** 2.4 Ω **17.** 2.0×10^6 (Ω·m)$^{-1}$ **19.** 2.0×10^{-8} Ω·m **21.** (1.8×10^3)°C **23.** 8.2×10^{-8} Ω·m **25.** 54 Ω **27.** 3.0 **29.** 3.35×10^{-7} C **31.** (a) 6.00 mA; (b) 1.59×10^{-8} V; (c) 21.2 nΩ **33.** (a) 38.3 mA; (b) 109 A/m²; (c) 1.28 cm/s; (d) 227 V/m **35.** 981 kΩ **39.** 150 s **41.** (a) 1.0 kW; (b) US$0.25 **43.** 0.135 W **45.** (a) 10.9 A; (b) 10.6 Ω; (c) 4.50 MJ **47.** (a) 5.85 m; (b) 10.4 m **49.** (a) US$4.46; (b) 144 Ω; (c) 0.833 A **51.** (a) 5.1 V; (b) 10 V; (c) 10 W; (d) 20 W **53.** (a) 28.8 Ω; (b) 2.60×10^{19} s^{-1} **55.** 660 W **57.** 28.8 kC **59.** (a) silver; (b) 51.6 nΩ **61.** (a) 2.3×10^{12}; (b) 5.0×10^3; (c) 10 MV **63.** 2.4 kW **65.** (a) 1.37; (b) 0.730 **67.** (a) −8.6%; (b) smaller **69.** 146 kJ **71.** (a) 250°C; (b) yes **73.** 3.0×10^6 J/kg **75.** 560 W **77.** 0.27 m/s **79.** (a) 10 A/cm²; (b) eastward **81.** (a) 9.4×10^{13} s^{-1}; (b) 2.40×10^2 W **83.** 113 min **85.** (a) 225 μC; (b) 60.0 μA; (c) 0.450 mW

Chapter 27

CP **1.** (a) rightward; (b) all tie; (c) *b*, then *a* and *c* tie; (d) *b*, then *a* and *c* tie **2.** (a) all tie; (b) R_1, R_2, R_3 **3.** (a) less; (b) greater; (c) equal **4.** (a) *V*/2, *i*; (b) *V*, *i*/2 **5.** (a) 1, 2, 4, 3; (b) 4, tie of 1 and 2, then 3
Q **1.** (a) equal; (b) more **3.** parallel, R_2, R_1, series **5.** (a) series; (b) parallel; (c) parallel **7.** (a) less; (b) less; (c) more **9.** (a) parallel; (b) series **11.** (a) same; (b) same; (c) less; (d) more **13.** (a) all tie; (b) 1, 3, 2
P **1.** (a) 0.50 A; (b) 1.0 W; (c) 2.0 W; (d) 6.0 W; (e) 3.0 W; (f) supplied; (g) absorbed **3.** (a) 14 V; (b) 1.0×10^2 W; (c) 6.0×10^2 W; (d) 10 V; (e) 1.0×10^2 W **5.** 11 kJ **7.** (a) 80 J; (b) 67 J; (c) 13 J **9.** (a) 12.0 eV; (b) 6.53 W **11.** (a) 50 V; (b) 48 V; (c) negative **13.** (a) 6.9 km; (b) 20 Ω **15.** 8.0 Ω **17.** (a) 0.004 Ω; (b) 1 **19.** (a) 4.00 Ω; (b) parallel **21.** 5.56 A **23.** (a) 50 mA; (b) 60 mA; (c) 9.0 V **25.** 3*d* **27.** 3.6×10^3 A **29.** (a) 0.333 A; (b) right; (c) 720 J **31.** (a) −11 V; (b) −9.0 V **33.** 48.3 V **35.** (a) 5.25 V; (b) 1.50 V; (c) 5.25 V; (d) 6.75 V **37.** 1.43 Ω **39.** (a) 0.150 Ω; (b) 240 W **41.** (a) 0.709 W; (b) 0.050 W; (c) 0.346 W; (d) 1.26 W; (e) −0.158 W **43.** 9 **45.** (a) 0.67 A; (b) down; (c) 0.33 A; (d) up; (e) 0.33 A; (f) up; (g) 3.3 V **47.** (a) 1.11 A; (b) 0.893 A; (c) 126 m **49.** (a) 0.45 A **51.** (a) 55.2 mA; (b) 4.86 V; (c) 88.0 Ω; (d) decrease **53.** −3.0% **57.** 0.208 ms **59.** 4.61 **61.** (a) 2.41 μs; (b) 161 pF **63.** (a) 1.1 mA; (b) 0.55 mA; (c) 0.55 mA; (d) 0.82 mA; (e) 0.82 mA; (f) 0; (g) 4.0×10^2 V; (h) 6.0×10^2 V **65.** 411 μA **67.** 0.72 MΩ **69.** (a) 0.955 μC/s; (b) 1.08 μW; (c) 2.74 μW; (d) 3.82 μW **71.** (a) 3.00 A; (b) 3.75 A; (c) 3.94 A **73.** (a) 1.32×10^7 A/m²; (b) 8.90 V; (c) copper; (d) 1.32×10^7 A/m²; (e) 51.1 V; (f) iron **75.** (a) 3.0 kV; (b) 10 s; (c) 11 GΩ **77.** (a) 85.0 Ω; (b) 915 Ω **81.** 4.0 V **83.** (a) 24.8 Ω; (b) 14.9 kΩ **85.** the cable **87.** −13 μC **89.** 20 Ω **91.** (a) 3.00 A; (b) down; (c) 1.60 A; (d) down; (e) supply; (f) 55.2 W; (g) supply; (h) 6.40 W **93.** (a) 1.0 V; (b) 50 mΩ **95.** 3 **99.** (a) 1.5 mA; (b) 0; (c) 1.0 mA **101.** 7.50 V

103. (a) 60.0 mA; (b) down; (c) 180 mA; (d) left; (e) 240 mA; (f) up **105.** (a) 4.0 A; (b) up; (c) 0.50 A; (d) down; (e) 64 W; (f) 16 W; (g) supplied; (h) absorbed

Chapter 28

CP **1.** *a*, $+z$; *b*, $-x$; *c*, $\vec{F}_B = 0$ **2.** (a) 2, then tie of 1 and 3 (zero); (b) 4 **3.** (a) electron; (b) clockwise **4.** $-y$ **5.** (a) all tie; (b) 1 and 4 tie, then 2 and 3 tie

Q **1.** (a) no, because $\vec{v}$ and $\vec{F}_B$ must be perpendicular; (b) yes; (c) no, because $\vec{B}$ and $\vec{F}_B$ must be perpendicular **3.** (a) $+z$ and $-z$ tie, then $+y$ and $-y$ tie, then $+x$ and $-x$ tie (zero); (b) $+y$ **5.** (a) $\vec{F}_E$; (b) $\vec{F}_B$ **7.** (a) $\vec{B}_1$; (b) $\vec{B}_1$ into page, $\vec{B}_2$ out of page; (c) less **9.** (a) positive; (b) $2 \to 1$ and $2 \to 4$ tie, then $2 \to 3$ (which is zero) **11.** (a) negative; (b) equal; (c) equal; (d) half-circle

P **1.** (a) 400 km/s; (b) 835 eV **3.** (a) $(6.2 \times 10^{-14}\,\text{N})\hat{k}$; (b) $(-6.2 \times 10^{-14}\,\text{N})\hat{k}$ **5.** -2.0 T **7.** $(-11.4\,\text{V/m})\hat{i} - (6.00\,\text{V/m})\hat{j} + (4.80\,\text{V/m})\hat{k}$ **9.** $-(0.267\,\text{mT})\hat{k}$ **11.** 0.68 MV/m **13.** 7.4 μV **15.** (a) $(-600\,\text{mV/m})\hat{k}$; (b) 1.20 V **17.** (a) 2.60×10^6 m/s; (b) 0.109 μs; (c) 0.140 MeV; (d) 70.0 kV **19.** 1.2×10^{-9} kg/C **21.** (a) 2.05×10^7 m/s; (b) 467 μT; (c) 13.1 MHz; (d) 76.3 ns **23.** 21.1 μT **25.** (a) 0.978 MHz; (b) 96.4 cm **27.** (a) 495 mT; (b) 22.7 mA; (c) 8.17 MJ **29.** 65.3 km/s **31.** 5.07 ns **33.** (a) 0.358 ns; (b) 0.166 mm; (c) 1.51 mm **35.** (a) 200 eV; (b) 20.0 keV; (c) 0.499% **37.** 2.4×10^2 m **39.** (a) 28.2 N; (b) horizontally west **41.** (a) 467 mA; (b) right **43.** (a) 0; (b) 0.138 N; (c) 0.138 N; (d) 0 **45.** $(-2.50\,\text{mN})\hat{j} + (0.750\,\text{mN})\hat{k}$ **47.** (a) 0.10 T; (b) 31° **49.** $(-4.3 \times 10^{-3}\,\text{N·m})\hat{j}$ **51.** 2.45 A **55.** (a) 2.86 A·m²; (b) 1.10 A·m² **57.** (a) 12.7 A; (b) 0.0805 N·m **59.** (a) 0.30 A·m²; (b) 0.024 N·m **61.** (a) -72.0 μJ; (b) $(96.0\hat{i} + 48.0\hat{k})$ μN·m **63.** (a) $-(9.7 \times 10^{-4}\,\text{N·m})\hat{i} - (7.2 \times 10^{-4}\,\text{N·m})\hat{j} + (8.0 \times 10^{-4}\,\text{N·m})\hat{k}$; (b) -6.0×10^{-4} J **65.** (a) 90°; (b) 1; (c) 1.28×10^{-7} N·m **67.** (a) 20 min; (b) 5.9×10^{-2} N·m **69.** 8.2 mm **71.** 127 u **73.** (a) 6.3×10^{14} m/s²; (b) 3.0 mm **75.** (a) 1.4; (b) 1.0 **77.** $(-500\,\text{V/m})\hat{j}$ **79.** (a) 0.50; (b) 0.50; (c) 14 cm; (d) 14 cm **81.** $(0.80\hat{j} - 1.1\hat{k})$ mN **83.** -40 mC **85.** (a) $(12.8\hat{i} + 6.41\hat{j}) \times 10^{-22}$ N; (b) 90°; (c) 173° **87.** (a) up the conducting path; (b) rim; (c) 47.1 V; (d) 47.1 V; (e) 2.36 kW **89.** $(mV/2ed^2)^{0.5}$ **91.** $n = JB/eE$

Chapter 29

CP **1.** *b*, *c*, *a* **2.** *d*, tie of *a* and *c*, then *b* **3.** *d*, *a*, tie of *b* and *c* (zero)

Q **1.** *c*, *a*, *b* **3.** *c*, *d*, then *a* and *b* tie (zero) **5.** *a*, *c*, *b* **7.** *c* and *d* tie, then *b*, *a* **9.** *b*, *a*, *d*, *c* (zero) **11.** (a) 1, 3, 2; (b) less

P **1.** (a) 3.3 μT; (b) yes **3.** (a) 16 A; (b) east **5.** (a) 1.0 mT; (b) out; (c) 0.80 mT; (d) out **7.** (a) 0.102 μT; (b) out **9.** (a) opposite; (b) 30 A **11.** (a) 4.3 A; (b) out **13.** 50.3 nT **15.** (a) 1.7 μT; (b) into; (c) 6.7 μT; (d) into **17.** 132 nT **19.** 5.0 μT **21.** 256 nT **23.** $(-7.75 \times 10^{-23}\,\text{N})\hat{i}$ **25.** 2.00 rad **27.** 61.3 mA **29.** $(80\,\mu\text{T})\hat{j}$ **31.** (a) 20 μT; (b) into **33.** $(22.3\,\text{pT})\hat{j}$ **35.** 88.4 pN/m **37.** $(-125\,\mu\text{N/m})\hat{i} + (41.7\,\mu\text{N/m})\hat{j}$ **39.** 800 nN/m **41.** $(3.20\,\text{mN})\hat{j}$ **43.** (a) 0; (b) 0.850 mT; (c) 1.70 mT; (d) 0.850 mT **45.** (a) -2.5 μT·m; (b) 0 **47.** (a) 0; (b) 0.10 μT; (c) 0.40 μT **49.** (a) 533 μT; (b) 400 μT **51.** 0.30 mT **53.** 0.272 A **55.** (a) 4.77 cm; (b) 35.5 μT **57.** (a) 2.4 A·m²; (b) 46 cm **59.** 0.47 A·m² **61.** (a) 79 μT; (b) 1.1×10^{-6} N·m **63.** (a) $(0.060\,\text{A·m}^2)\hat{j}$; (b) $(96\,\text{pT})\hat{j}$ **65.** 1.28 mm **69.** (a) 15 A; (b) $-z$ **71.** 7.7 mT **73.** (a) 15.3 μT **75.** (a) $(0.24\hat{i})$ nT; (b) 0; (c) $(-43\hat{k})$ pT; (d) $(0.14\hat{k})$ nT **79.** (a) 4.8 mT; (b) 0.93 mT; (c) 0 **83.** $(-0.20\,\text{mT})\hat{k}$ **87.** (a) $\mu_0 ir/2\pi c^2$; (b) $\mu_0 i/2\pi r$; (c) $\mu_0 i(a^2 - r^2)/2\pi(a^2 - b^2)r$; (d) 0

Chapter 30

CP **1.** *b*, then *d* and *e* tie, and then *a* and *c* tie (zero) **2.** *a* and *b* tie, then *c* (zero) **3.** *c* and *d* tie, then *a* and *b* tie **4.** *b*, out; *c*, out; *d*, into; *e*, into **5.** d and e **6.** (a) 2, 3, 1 (zero); (b) 2, 3, 1 **7.** *a* and *b* tie, then *c*

Q **1.** out **3.** (a) all tie (zero); (b) 2, then 1 and 3 tie (zero) **5.** *d* and *c* tie, then *b*, *a* **7.** (a) more; (b) same; (c) same; (d) same (zero) **9.** (a) all tie (zero); (b) 1 and 2 tie, then 3; (c) all tie (zero) **11.** *b*

P **1.** 0 **3.** 30 mA **5.** 0 **7.** (a) 31 mV; (b) left **9.** 0.198 mV **11.** (b) 0.796 m² **13.** 29.5 mC **15.** (a) 21.7 V; (b) counterclockwise **17.** (a) 1.26×10^{-4} T; (b) 0; (c) 1.26×10^{-4} T; (d) yes; (e) 5.04×10^{-8} V **19.** 5.50 kV **21.** (a) 40 Hz; (b) 3.2 mV **23.** (a) $\mu_0 iR^2\pi r^2/2x^3$; (b) $3\mu_0 i\pi R^2 r^2 v/2x^4$; (c) counterclockwise **25.** (a) 13 μWb/m; (b) 17%; (c) 0 **27.** (a) 80 μV; (b) clockwise **29.** (a) 48.1 mV; (b) 2.67 mA; (c) 0.129 mW **31.** 3.68 μW **33.** (a) 240 μV; (b) 0.600 mA; (c) 0.144 μW; (d) 2.87×10^{-8} N; (e) 0.144 μW **35.** (a) 0.60 V; (b) up; (c) 1.5 A; (d) clockwise; (e) 0.90 W; (f) 0.18 N; (g) 0.90 W **37.** (a) 71.5 μV/m; (b) 143 μV/m **39.** 0.15 V/m **41.** (a) 2.45 mWb; (b) 0.645 mH **43.** 1.81 μH/m **45.** (a) decreasing; (b) 0.68 mH **47.** (b) $L_{eq} = \Sigma L_j$, sum from $j = 1$ to $j = N$ **49.** 59.3 mH **51.** 46 Ω **53.** (a) 8.45 ns; (b) 7.37 mA **55.** 6.91 **57.** (a) 1.5 s **59.** (a) $i[1 - \exp(-Rt/L)]$; (b) $(L/R)\ln 2$ **61.** (a) 97.9 H; (b) 0.196 mJ **63.** 25.6 ms **65.** (a) 18.7 J; (b) 5.10 J; (c) 13.6 J **67.** (a) 34.2 J/m³; (b) 49.4 mJ **69.** 1.5×10^8 V/m **71.** (a) 1.0 J/m³; (b) 4.8×10^{-15} J/m³ **73.** (a) 1.67 mH; (b) 6.00 mWb **75.** 13 μH **77.** (b) have the turns of the two solenoids wrapped in opposite directions **79.** (a) 2.0 A; (b) 0; (c) 2.0 A; (d) 0; (e) 10 V; (f) 2.0 A/s; (g) 2.0 A; (h) 1.0 A; (i) 3.0 A; (j) 10 V; (k) 0; (l) 0 **81.** (a) 10 μT; (b) out; (c) 3.3 μT; (d) out **83.** 0.520 ms **85.** (a) $(4.4 \times 10^7\,\text{m/s}^2)\hat{i}$; (b) 0; (c) $(-4.4 \times 10^7\,\text{m/s}^2)\hat{i}$ **87.** (a) 0.40 V; (b) 20 A **89.** (a) 10 A; (b) 1.0×10^2 J **91.** (a) 0; (b) 8.0×10^2 A/s; (c) 1.8 mA; (d) 4.4×10^2 A/s; (e) 4.0 mA; (f) 0 **93.** 1.15 W **95.** (a) 20 A/s; (b) 0.75 A **97.** 12 A/s **99.** 3×10^{36} J **101.** 1.15 μWb

Chapter 31

CP **1.** (a) $T/2$; (b) T; (c) $T/2$; (d) $T/4$ **2.** (a) 5 V; (b) 150 μJ **3.** (a) remains the same; (b) remains the same **4.** (a) *C*, *B*, *A*; (b) 1, *A*; 2, *B*; 3, *S*; 4, *C*; (c) *A* **5.** (a) remains the same; (b) increases; (c) remains the same; (d) decreases **6.** (a) 1, lags; 2, leads; 3, in phase; (b) 3 ($\omega_d = \omega$ when $X_L = X_C$) **7.** (a) increase (circuit is mainly capacitive; increase C to decrease X_C to be closer to resonance for maximum P_{avg}); (b) closer **8.** (a) greater; (b) step-up

Q **1.** *b*, *a*, *c* **3.** (a) $T/4$; (b) $T/4$; (c) $T/2$; (d) $T/2$ **5.** *c*, *b*, *a* **7.** *a* inductor; *b* resistor; *c* capacitor **9.** (a) positive; (b) decreased (to decrease X_L and get closer to resonance); (c) decreased (to increase X_C and get closer to resonance) **11.** (a) rightward, increase (X_L increases, closer to resonance); (b) rightward, increase (X_C decreases, closer to resonance); (c) rightward, increase (ω_d/ω increases, closer to resonance) **13.** (a) inductor; (b) decrease

P **1.** (a) 1.17 μJ; (b) 5.58 mA **3.** (a) 6.00 μs; (b) 167 kHz; (c) 3.00 μs **5.** 45.2 mA **7.** (a) 1.25 kg; (b) 372 N/m; (c) 1.75×10^{-4} m; (d) 3.02 mm/s **9.** 7.0×10^{-4} s **11.** (a) 6.0; (b) 36 pF; (c) 0.22 mH **13.** (a) 0.180 mC; (b) 70.7 μs; (c) 66.7 W **15.** (a) 3.0 nC; (b) 1.7 mA; (c) 4.5 nJ **17.** (a) 275 Hz; (b) 365 mA **21.** (a) 356 μs; (b) 2.50 mH; (c) 3.20 mJ **23.** (a) 1.98 μJ; (b) 5.56 μC; (c) 12.6 mA; (d) $-46.9°$; (e) $+46.9°$ **25.** 8.66 mΩ **29.** (a) 95.5 mA; (b) 11.9 mA **31.** (a) 0.65 kHz; (b) 24 Ω **33.** (a) 6.73 ms; (b) 11.2 ms; (c) inductor; (d) 138 mH **35.** 89 Ω **37.** 7.61 A **39.** (a) 267 Ω; (b) $-41.5°$; (c) 135 mA **41.** (a) 206 Ω; (b) 13.7°; (c) 175 mA **43.** (a) 218 Ω; (b) 23.4°; (c) 165 mA **45.** (a) yes; (b) 1.0 kV **47.** (a) 224 rad/s; (b) 6.00 A; (c) 219 rad/s; (d) 228 rad/s; (e) 0.040 **49.** (a) 796 Hz; (b) no change; (c) decreased; (d) increased **53.** (a) 12.1 Ω; (b) 1.19 kW

55. 1.84 A **57.** (a) 117 μF; (b) 0; (c) 90.0 W; (d) 0°; (e) 1; (f) 0; (g) −90°; (h) 0 **59.** (a) 2.59 A; (b) 38.8 V; (c) 159 V; (d) 224 V; (e) 64.2 V; (f) 75.0 V; (g) 100 W; (h) 0; (i) 0 **61.** (a) 0.743; (b) lead; (c) capacitive; (d) no; (e) yes; (f) no; (g) yes; (h) 33.4 W **63.** (a) 2.4 V; (b) 3.2 mA; (c) 0.16 A **65.** (a) 1.9 V; (b) 5.9 W; (c) 19 V; (d) 5.9×10^2 W; (e) 0.19 kV; (f) 59 kW **67.** (a) 6.73 ms; (b) 2.24 ms; (c) capacitor; (d) 59.0 μF **69.** (a) −0.405 rad; (b) 2.76 A; (c) capacitive **71.** (a) 64.0 Ω; (b) 50.9 Ω; (c) capacitive **73.** (a) 2.41 μH; (b) 21.4 pJ; (c) 82.2 nC **75.** (a) 39.1 Ω; (b) 21.7 Ω; (c) capacitive **79.** (a) 0.577Q; (b) 0.152 **81.** (a) 45.0°; (b) 70.7 Ω **83.** 1.84 kHz **85.** (a) 0.689 μH; (b) 17.9 pJ; (c) 0.110 μC **87.** (a) 165 Ω; (b) 313 mH; (c) 14.9 μF **93.** (a) 36.0 V; (b) 29.9 V; (c) 11.9 V; (d) −5.85 V

Chapter 32

CP **1.** *d, b, c, a* (zero) **2.** *a, c, b, d* (zero) **3.** tie of *b*, *c*, and *d*, then *a* **4.** (a) 2; (b) 1 **5.** (a) away; (b) away; (c) less **6.** (a) toward; (b) toward; (c) less

Q **1.** 1 *a*, 2 *b*, 3 *c* and *d* **3.** a, decreasing; b, decreasing **5.** supplied **7.** (a) *a* and *b* tie, then *c*, *d*; (b) none (because plate lacks circular symmetry, $\vec{B}$ not tangent to any circular loop); (c) none **9.** (a) 1 up, 2 up, 3 down; (b) 1 down, 2 up, 3 zero **11.** (a) 1, 3, 2; (b) 2

P **1.** +3 Wb **3.** (a) 47.4 μWb; (b) inward **5.** 2.4×10^{13} V/m·s **7.** (a) 1.18×10^{-19} T; (b) 1.06×10^{-19} T **9.** (a) 5.01×10^{-22} T; (b) 4.51×10^{-22} T **11.** (a) 1.9 pT **13.** 7.5×10^5 V/s **17.** (a) 0.324 V/m; (b) 2.87×10^{-16} A; (c) 2.87×10^{-18} **19.** (a) 75.4 nT; (b) 67.9 nT **21.** (a) 27.9 nT; (b) 15.1 nT **23.** (a) 2.0 A; (b) 2.3×10^{11} V/m·s; (c) 0.50 A; (d) 0.63 μT·m **25.** (a) 0.63 μT; (b) 2.3×10^{12} V/m·s **27.** (a) 0.71 A; (b) 0; (c) 2.8 A **29.** (a) 7.60 μA; (b) 859 kV·m/s; (c) 3.39 mm; (d) 5.16 pT **31.** 55 μT **33.** (a) 0; (b) 0; (c) 0; (d) $\pm 3.2 \times 10^{-25}$ J; (e) -3.2×10^{-34} J·s; (f) 2.8×10^{-23} J/T; (g) -9.7×10^{-25} J; (h) $\pm 3.2 \times 10^{-25}$ J **35.** (a) -9.3×10^{-24} J/T; (b) 1.9×10^{-23} J/T **37.** (b) $+x$; (c) clockwise; (d) $+x$ **39.** yes **41.** 20.8 mJ/T **43.** (b) K_i/B; (c) $-z$; (d) 0.31 kA/m **47.** (a) 1.8×10^2 km; (b) 2.3×10^{-5} **49.** (a) 3.0 μT; (b) 5.6×10^{-10} eV **51.** 5.15×10^{-24} A·m² **53.** (a) 0.14 A; (b) 79 μC **55.** (a) 6.3×10^8 A; (b) yes; (c) no **57.** 0.84 kJ/T **59.** (a) $(1.2 \times 10^{-13}\ \text{T}) \exp[-t/(0.012\ \text{s})]$; (b) 5.9×10^{-15} T **63.** (a) 27.5 mm; (b) 110 mm **65.** 8.0 A **67.** (a) -8.8×10^{15} V/m·s; (b) 5.9×10^{-7} T·m **69.** (b) sign is minus; (c) no, because there is compensating positive flux through open end nearer to magnet **71.** (b) $-x$; (c) counterclockwise; (d) $-x$ **73.** (a) 7; (b) 7; (c) $3h/2\pi$; (d) $3eh/4\pi m$; (e) $3.5h/2\pi$; (f) 8 **75.** (a) 9; (b) 3.71×10^{-23} J/T; (c) $+9.27 \times 10^{-24}$ J; (d) -9.27×10^{-24} J

Chapter 33

CP **1.** (a) (Use Fig. 33-5.) On right side of rectangle, $\vec{E}$ is in negative *y* direction; on left side, $\vec{E} + d\vec{E}$ is greater and in same direction; (b) $\vec{E}$ is downward. On right side, $\vec{B}$ is in negative *z* direction; on left side, $\vec{B} + d\vec{B}$ is greater and in same direction. **2.** positive direction of *x* **3.** (a) same; (b) decrease **4.** *a, d, b, c* (zero) **5.** *a*

Q **1.** (a) positive direction of *z*; (b) *x* **3.** (a) same; (b) increase; (c) decrease **5.** (a) and (b) $A = 1, n = 4, \theta = 30°$ **7.** *a, b, c* **9.** *B* **11.** none

P **1.** 7.49 GHz **3.** (a) 515 nm; (b) 610 nm; (c) 555 nm; (d) 5.41×10^{14} Hz; (e) 1.85×10^{-15} s **5.** 5.0×10^{-21} H **7.** 1.2 MW/m² **9.** 0.10 MJ **11.** (a) 6.7 nT; (b) *y*; (c) negative direction of *y* **13.** (a) 1.03 kV/m; (b) 3.43 μT **15.** (a) 87 mV/m; (b) 0.29 nT; (c) 6.3 kW **17.** (a) 6.7 nT; (b) 5.3 mW/m²; (c) 6.7 W **19.** 1.0×10^7 Pa **21.** 5.9×10^{-8} Pa **23.** (a) 4.68×10^{11} W; (b) any chance disturbance could move sphere from directly above source—the two force vectors no longer along the same axis **27.** (a) 1.0×10^8 Hz; (b) 6.3×10^8 rad/s; (c) 2.1 m⁻¹; (d) 1.0 μT; (e) *z*; (f) 1.2×10^2 W/m²; (g) 8.0×10^{-7} N; (h) 4.0×10^{-7} Pa **29.** 1.9 mm/s **31.** (a) 0.17 μm; (b) toward the Sun **33.** 3.1% **35.** 4.4 W/m² **37.** (a) 2 sheets; (b) 5 sheets **39.** (a) 1.9 V/m; (b) 1.7×10^{-11} Pa **41.** 20° or 70° **43.** 0.67 **45.** 1.26 **47.** 1.48 **49.** 180° **51.** (a) 56.9°; (b) 35.3° **55.** 1.07 m **57.** 182 cm **59.** (a) 48.9°; (b) 29.0° **61.** (a) 26.8°; (b) yes **63.** (a) $(1 + \sin^2\theta)^{0.5}$; (b) $2^{0.5}$; (c) yes; (d) no **65.** 23.2° **67.** (a) 1.39; (b) 28.1°; (c) no **69.** 49.0° **71.** (a) 0.50 ms; (b) 8.4 min; (c) 2.4 h; (d) 5446 B.C. **73.** (a) $(16.7\ \text{nT}) \sin[(1.00 \times 10^6\ \text{m}^{-1})z + (3.00 \times 10^{14}\ \text{s}^{-1})t]$; (b) 6.28 μm; (c) 20.9 fs; (d) 33.2 mW/m²; (e) *x*; (f) infrared **75.** 1.22 **77.** (c) 137.6°; (d) 139.4°; (e) 1.7° **81.** (a) *z* axis; (b) 7.5×10^{14} Hz; (c) 1.9 kW/m² **83.** (a) white; (b) white dominated by red end; (c) no refracted light **85.** 1.5×10^{-9} m/s² **87.** (a) 3.5 μW/m²; (b) 0.78 μW; (c) 1.5×10^{-17} W/m²; (d) 1.1×10^{-7} V/m; (e) 0.25 fT **89.** (a) 55.8°; (b) 55.5° **91.** (a) 83 W/m²; (b) 1.7 MW **93.** 35° **97.** $\cos^{-1}(p/50)^{0.5}$ **99.** $8RI/3c$ **101.** 0.034 **103.** 9.43×10^{-10} T **105.** (a) $-y$; (b) *z*; (c) 1.91 kW/m²; (d) $E_z = (1.20\ \text{kV/m}) \sin[(6.67 \times 10^6\ \text{m}^{-1})y + (2.00 \times 10^{15}\ \text{s}^{-1})t]$; (e) 942 nm; (f) infrared **107.** (a) 1.60; (b) 58.0°

Chapter 34

CP **1.** 0.2*d*, 1.8*d*, 2.2*d* **2.** (a) real; (b) inverted; (c) same **3.** (a) *e*; (b) virtual, same **4.** virtual, same as object, diverging

Q **1.** (a) *a*; (b) *c* **3.** (a) *a* and *c*; (b) three times; (c) you **5.** convex **7.** (a) all but variation 2; (b) 1, 3, 4: right, inverted; 5, 6: left, same **9.** *d* (infinite), tie of *a* and *b*, then *c* **11.** (a) *x*; (b) no; (c) no; (d) the direction you are facing

P **1.** 9.10 m **3.** 1.11 **5.** 351 cm **7.** 10.5 cm **9.** (a) +24 cm; (b) +36 cm; (c) −2.0; (d) R; (e) I; (f) same **11.** (a) −20 cm; (b) −4.4 cm; (c) +0.56; (d) V; (e) NI; (f) opposite **13.** (a) +36 cm; (b) −36 cm; (c) +3.0; (d) V; (e) NI; (f) opposite **15.** (a) −16 cm; (b) −4.4 cm; (c) +0.44; (d) V; (e) NI; (f) opposite **17.** (b) plus; (c) +40 cm; (e) −20 cm; (f) +2.0; (g) V; (h) NI; (i) opposite **19.** (a) convex; (b) −20 cm; (d) +20 cm; (f) +0.50; (g) V; (h) NI; (i) opposite **21.** (a) concave; (c) +40 cm; (e) +60 cm; (f) −2.0; (g) R; (h) I; (i) same **23.** (a) convex; (b) minus; (c) −60 cm; (d) +1.2 m; (e) −24 cm; (g) V; (h) NI; (i) opposite **25.** (a) concave; (b) +8.6 cm; (c) +17 cm; (e) +12 cm; (f) minus; (g) R; (i) same **27.** (a) convex; (c) −60 cm; (d) +30 cm; (f) +0.50; (g) V; (h) NI; (i) opposite **29.** (b) −20 cm; (c) minus; (d) +5.0 cm; (e) minus; (f) +0.80; (g) V; (h) NI; (i) opposite **31.** (b) 0.56 cm/s; (c) 11 m/s; (d) 6.7 cm/s **33.** (c) −33 cm; (e) V; (f) same **35.** (d) −26 cm; (e) V; (f) same **37.** (c) +30 cm; (e) V; (f) same **39.** (a) 2.00; (b) none **41.** (a) +40 cm; (b) ∞ **43.** 5.0 mm **45.** 1.86 mm **47.** (a) 45 mm; (b) 90 mm **49.** 22 cm **51.** (a) −48 cm; (b) +4.0; (c) V; (d) NI; (e) same **53.** (a) −4.8 cm; (b) +0.60; (c) V; (d) NI; (e) same **55.** (a) −8.6 cm; (b) +0.39; (c) V; (d) NI; (e) same **57.** (a) +36 cm; (b) −0.80; (c) R; (d) I; (e) opposite **59.** (a) +55 cm; (b) −0.74; (c) R; (d) I; (e) opposite **61.** (a) −18 cm; (b) +0.76; (c) V; (d) NI; (e) same **63.** (a) −30 cm; (b) +0.86; (c) V; (d) NI; (e) same **65.** (a) −7.5 cm; (b) +0.75; (c) V; (d) NI; (e) same **67.** (a) +84 cm; (b) −1.4; (c) R; (d) I; (e) opposite **69.** (a) C; (d) −10 cm; (e) +2.0; (f) V; (g) NI; (h) same **71.** (a) D; (b) −5.3 cm; (d) −4.0 cm; (f) V; (g) NI; (h) same **73.** (a) C; (b) +3.3 cm; (d) +5.0 cm; (f) R; (g) I; (h) opposite **75.** (a) D; (b) minus; (d) −3.3 cm; (e) +0.67; (f) V; (g) NI **77.** (a) C; (b) +80 cm; (d) −20 cm; (f) V; (g) NI; (h) same **79.** (a) C; (b) plus; (d) −13 cm; (e) +1.7; (f) V; (g) NI; (h) same

81. (a) +24 cm; (b) +6.0; (c) R; (d) NI; (e) opposite **83.** (a) +3.1 cm; (b) −0.31; (c) R; (d) I; (e) opposite **85.** (a) −4.6 cm; (b) +0.69; (c) V; (d) NI; (e) same **87.** (a) −5.5 cm; (b) +0.12; (c) V; (d) NI; (e) same **89.** (a) 13.0 cm; (b) 5.23 cm; (c) −3.25; (d) 3.13; (e) −10.2 **91.** (a) 2.35 cm; (b) decrease **93.** (a) 3.5; (b) 2.5 **95.** (a) +8.6 cm; (b) +2.6; (c) R; (d) NI; (e) opposite **97.** (a) +7.5 cm; (b) −0.75; (c) R; (d) I; (e) opposite **99.** (a) +24 cm; (b) −0.58; (c) R; (d) I; (e) opposite **105.** (a) 3.00 cm; (b) 2.33 cm **107.** (a) 40 cm; (b) 20 cm; (c) −40 cm; (d) 40 cm **109.** (a) 20 cm; (b) 15 cm **111.** (a) 6.0 mm; (b) 1.6 kW/m^2; (c) 4.0 cm **113.** 100 cm **115.** 2.2 mm^2 **119.** (a) −30 cm; (b) not inverted; (c) virtual; (d) 1.0 **121.** (a) −12 cm **123.** (a) 80 cm; (b) 0 to 12 cm **127.** (a) 8.0 cm; (b) 16 cm; (c) 48 cm **129.** (a) $\alpha = 0.500$ rad: 7.799 cm; $\alpha = 0.100$ rad: 8.544 cm; $\alpha = 0.0100$ rad: 8.571 cm; mirror equation: 8.571 cm; (b) $\alpha = 0.500$ rad: −13.56 cm; $\alpha = 0.100$ rad: −12.05 cm; $\alpha = 0.0100$ rad: −12.00 cm; mirror equation: −12.00 cm **131.** 42 mm **133.** (b) P_n **135.** (a) $(0.5)(2 - n)r/(n - 1)$; (b) right **137.** 2.67 cm **139.** (a) 3.33 cm; (b) left; (c) virtual; (d) not inverted **141.** (a) $1 + (25\text{ cm})/f$; (b) $(25\text{ cm})/f$; (c) 3.5; (d) 2.5

Chapter 35

CP **1.** b (least n), c, a **2.** (a) top; (b) bright intermediate illumination (phase difference is 2.1 wavelengths) **3.** (a) 3λ, 3; (b) 2.5λ, 2.5 **4.** a and d tie (amplitude of resultant wave is $4E_0$), then b and c tie (amplitude of resultant wave is $2E_0$) **5.** (a) 1 and 4; (b) 1 and 4

Q **1.** (a) decrease; (b) decrease; (c) decrease; (d) blue **3.** (a) $2d$; (b) (odd number)$\lambda/2$; (c) $\lambda/4$ **5.** (a) intermediate closer to maximum, $m = 2$; (b) minimum, $m = 3$; (c) intermediate closer to maximum, $m = 2$; (d) maximum, $m = 1$ **7.** (a) maximum; (b) minimum; (c) alternates **9.** (a) peak; (b) valley **11.** c, d **13.** c

P **1.** (a) 155 nm; (b) 310 nm **3.** (a) 3.60 μm; (b) intermediate closer to fully constructive **5.** 4.55×10^7 m/s **7.** 1.56 **9.** (a) 1.55 μm; (b) 4.65 μm **11.** (a) 1.70; (b) 1.70; (c) 1.30; (d) all tie **13.** (a) 0.833; (b) intermediate closer to fully constructive **15.** 648 nm **17.** 16 **19.** 2.25 mm **21.** 72 μm **23.** 0 **25.** 7.88 μm **27.** 6.64 μm **29.** 2.65 **31.** $27 \sin(\omega t + 8.5°)$ **33.** $(17.1\ \mu\text{V/m}) \sin[(2.0 \times 10^{14}\text{ rad/s})t]$ **35.** 120 nm **37.** 70.0 nm **39.** (a) 0.117 μm; (b) 0.352 μm **41.** 161 nm **43.** 560 nm **45.** 478 nm **47.** 509 nm **49.** 273 nm **51.** 409 nm **53.** 338 nm **55.** (a) 552 nm; (b) 442 nm **57.** 608 nm **59.** 528 nm **61.** 455 nm **63.** 248 nm **65.** 339 nm **67.** 329 nm **69.** 1.89 μm **71.** 0.012° **73.** 140 **75.** $[(m + \frac{1}{2})\lambda R]^{0.5}$, for $m = 0, 1, 2, \ldots$ **77.** 1.00 m **79.** 588 nm **81.** 1.00030 **83.** (a) 50.0 nm; (b) 36.2 nm **85.** 0.23° **87.** (a) 1500 nm; (b) 2250 nm; (c) 0.80 **89.** $x = (D/2a)(m + 0.5)\lambda$, for $m = 0, 1, 2, \ldots$ **91.** (a) 22°; (b) refraction reduces θ **93.** 600 nm **95.** (a) 1.75 μm; (b) 4.8 mm **97.** $I_m \cos^2(2\pi x/\lambda)$ **99.** (a) 42.0 ps; (b) 42.3 ps; (c) 43.2 ps; (d) 41.8 ps; (e) 4 **101.** 33 μm **103.** (a) bright; (b) 594 nm; (c) Primary reason: the colored bands begin to overlap too much to be distinguished. Secondary reason: the two reflecting surfaces are too separated for the light reflecting from them to be coherent.

Chapter 36

CP **1.** (a) expand; (b) expand **2.** (a) second side maximum; (b) 2.5 **3.** (a) red; (b) violet **4.** diminish **5.** (a) left; (b) less

Q **1.** (a) $m = 5$ minimum; (b) (approximately) maximum between the $m = 4$ and $m = 5$ minima **3.** (a) A, B, C; (b) A, B, C **5.** (a) 1 and 3 tie, then 2 and 4 tie; (b) 1 and 2 tie, then 3 and 4 tie **7.** (a) larger; (b) red **9.** (a) decrease; (b) same; (c) remain in place **11.** (a) A; (b) left; (c) left; (d) right **13.** (a) 1 and 2 tie, then 3; (b) yes; (c) no

P **1.** (a) 2.5 mm; (b) 2.2×10^{-4} rad **3.** (a) 70 cm; (b) 1.0 mm **5.** (a) 700 nm; (b) 4; (c) 6 **7.** 60.4 μm **9.** 1.77 mm **11.** 160° **13.** (a) 0.18°; (b) 0.46 rad; (c) 0.93 **15.** (d) 52.5°; (e) 10.1°; (f) 5.06° **17.** (b) 0; (c) −0.500; (d) 4.493 rad; (e) 0.930; (f) 7.725 rad; (g) 1.96 **19.** (a) 19 cm; (b) larger **21.** (a) 1.1×10^4 km; (b) 11 km **23.** (a) 1.3×10^{-4} rad; (b) 10 km **25.** 50 m **27.** 1.6×10^3 km **29.** (a) 8.8×10^{-7} rad; (b) 8.4×10^7 km; (c) 0.025 mm **31.** (a) 0.346°; (b) 0.97° **33.** (a) 17.1 m; (b) 1.37×10^{-10} **35.** 5 **37.** 3 **39.** (a) 5.0 μm; (b) 20 μm **41.** (a) 7.43×10^{-3}; (b) between the $m = 6$ minimum (the seventh one) and the $m = 7$ maximum (the seventh side maximum); (c) between the $m = 3$ minimum (the third one) and the $m = 4$ minimum (the fourth one) **43.** (a) 9; (b) 0.255 **45.** (a) 62.1°; (b) 45.0°; (c) 32.0° **47.** 3 **49.** (a) 6.0 μm; (b) 1.5 μm; (c) 9; (d) 7; (e) 6 **51.** (a) 2.1°; (b) 21°; (c) 11 **53.** (a) 470 nm; (b) 560 nm **55.** 3.65×10^3 **57.** (a) 0.032°/nm; (b) 4.0×10^4; (c) 0.076°/nm; (d) 8.0×10^4; (e) 0.24°/nm; (f) 1.2×10^5 **59.** 0.15 nm **61.** (a) 10 μm; (b) 3.3 mm **63.** 1.09×10^3 rulings/mm **65.** (a) 0.17 nm; (b) 0.13 nm **67.** (a) 25 pm; (b) 38 pm **69.** 0.26 nm **71.** (a) 15.3°; (b) 30.6°; (c) 3.1°; (d) 37.8° **73.** (a) $0.7071a_0$; (b) $0.4472a_0$; (c) $0.3162a_0$; (d) $0.2774a_0$; (e) $0.2425a_0$ **75.** (a) 625 nm; (b) 500 nm; (c) 416 nm **77.** 3.0 mm **83.** (a) 13; (b) 6 **85.** 59.5 pm **87.** 4.9 km **89.** 1.36×10^4 **91.** 2 **93.** 4.7 cm **97.** 36 cm **99.** (a) fourth; (b) seventh **103.** (a) 2.4 μm; (b) 0.80 μm; (c) 2 **107.** 9

Chapter 37

CP **1.** (a) same (speed of light postulate); (b) no (the start and end of the flight are spatially separated); (c) no (because his measurement is not a proper time) **2.** (a) Eq. 2; (b) $+0.90c$; (c) 25 ns; (d) −7.0 m **3.** (a) right; (b) more **4.** (a) equal; (b) less

Q **1.** c **3.** b **5.** (a) C_1'; (b) C_1' **7.** (a) 4 s; (b) 3 s; (c) 5 s; (d) 4 s; (e) 10 s **9.** (a) a tie of 3, 4, and 6, then a tie of 1, 2, and 5; (b) 1, then a tie of 2 and 3, then 4, then a tie of 5 and 6; (c) 1, 2, 3, 4, 5, 6; (d) 2 and 4; (e) 1, 2, 5 **11.** (a) 3, tie of 1 and 2, then 4; (b) 4, tie of 1 and 2, then 3; (c) 1, 4, 2, 3

P **1.** 0.990 50 **3.** (a) 0.999 999 50 **5.** 0.446 ps **7.** 2.68×10^3 y **9.** (a) 87.4 m; (b) 394 ns **11.** 1.32 m **13.** (a) 26.26 y; (b) 52.26 y; (c) 3.705 y **15.** (a) 0.999 999 15; (b) 30 ly **17.** (a) 138 km; (b) −374 μs **19.** (a) 25.8 μs; (b) small flash **21.** (a) $\gamma[1.00\ \mu\text{s} - \beta(400\text{ m})/(2.998 \times 10^8\text{ m/s})]$; (d) 0.750; (e) $0 < \beta < 0.750$; (f) $0.750 < \beta < 1$; (g) no **23.** (a) 1.25; (b) 0.800 μs **25.** (a) 0.480; (b) negative; (c) big flash; (d) 4.39 μs **27.** $0.81c$ **29.** (a) 0.35; (b) 0.62 **31.** 1.2 μs **33.** (a) 1.25 y; (b) 1.60 y; (c) 4.00 y **35.** 22.9 MHz **37.** $0.13c$ **39.** (a) 550 nm; (b) yellow **41.** (a) 196.695; (b) 0.999 987 **43.** (a) 1.0 keV; (b) 1.1 MeV **45.** 110 km **47.** 1.01×10^7 km **49.** (a) 0.222 cm; (b) 701 ps; (c) 7.40 ps **51.** $2.83mc$ **53.** $\gamma(2\pi m/|q|B)$; (b) no; (c) 4.85 mm; (d) 15.9 mm; (e) 16.3 ps; (f) 0.334 ns **55.** (a) 0.707; (b) 1.41; (c) 0.414 **57.** 18 smu/y **59.** (a) 2.08 MeV; (b) −1.21 MeV **61.** (d) 0.801 **63.** (a) $vt \sin \theta$; (b) $t[1 - (v/c) \cos \theta]$; (c) $3.24c$ **67.** (b) $+0.44c$ **69.** (a) 1.93 m; (b) 6.00 m; (c) 13.6 ns; (d) 13.6 ns; (e) 0.379 m; (f) 30.5 m; (g) −101 ns; (h) no; (i) 2; (k) no; (l) both **71.** (a) 5.4×10^4 km/h; (b) 6.3×10^{-10} **73.** 189 MeV **75.** 8.7×10^{-3} ly **77.** 7 **79.** 2.46 MeV/c **81.** $0.27c$ **83.** (a) 5.71 GeV; (b) 6.65 GeV; (c) 6.58 GeV/c; (d) 3.11 MeV; (e) 3.62 MeV; (f) 3.59 MeV/c **85.** $0.95c$ **87.** (a) 256 kV; (b) $0.745c$ **89.** (a) $0.858c$; (b) $0.185c$ **91.** $0.500c$ **93.** (a) 119 MeV; (b) 64.0 MeV/c; (c) 81.3 MeV; (d) 64.0 MeV/c **95.** 4.00 u, probably a helium nucleus **97.** (a) 534; (b) 0.999 998 25; (c) 2.23 T **99.** (a) 415 nm; (b) blue **101.** (a) 88 kg; (b) no **103.** (a) 3×10^{-18}; (b) 2×10^{-12}; (c) 8.2×10^{-8}; (d) 6.4×10^{-6}; (e) 1.1×10^{-6}; (f) 3.7×10^{-5}; (g) 9.9×10^{-5}; (h) 0.10

Chapter 38

CP **1.** b, a, d, c **2.** (a) lithium, sodium, potassium, cesium; (b) all tie **3.** (a) same; (b)–(d) x rays **4.** (a) proton; (b) same; (c) proton **5.** same

Q **1.** (a) greater; (b) less **3.** potassium **5.** only e **7.** none **9.** (a) decreases by a factor of $(1/2)^{0.5}$; (b) decreases by a factor of 1/2 **11.** amplitude of reflected wave is less than that of incident wave **13.** electron, neutron, alpha particle **15.** all tie

P **1.** (a) 2.1 μm; (b) infrared **3.** 1.0×10^{45} photons/s **5.** 2.047 eV **7.** 1.1×10^{-10} W **9.** (a) 2.96×10^{20} photons/s; (b) 4.86×10^{7} m; (c) 5.89×10^{18} photons/m$^2 \cdot$s **11.** (a) infrared; (b) 1.4×10^{21} photons/s **13.** 4.7×10^{26} photons **15.** 170 nm **17.** 676 km/s **19.** 1.3 V; (b) 6.8×10^{2} km/s **21.** (a) 3.1 keV; (b) 14 keV **23.** (a) 2.00 eV; (b) 0; (c) 2.00 V; (d) 295 nm **25.** (a) 382 nm; (b) 1.82 eV **27.** (a) 2.73 pm; (b) 6.05 pm **29.** (a) 8.57×10^{18} Hz; (b) 3.55×10^{4} eV; (c) 35.4 keV/c **31.** 300% **33.** (a) -8.1×10^{-9}%; (b) -4.9×10^{-4}%; (c) -8.9%; (d) -66% **35.** (a) 2.43 pm; (b) 1.32 fm; (c) 0.511 MeV; (d) 939 MeV **37.** (a) 41.8 keV; (b) 8.2 keV **39.** 44° **41.** (a) 2.43 pm; (b) 4.11×10^{-6}; (c) -8.67×10^{-6} eV; (d) 2.43 pm; (e) 9.78×10^{-2}; (f) -4.45 keV **43.** (a) 2.9×10^{-10} m; (b) x ray; (c) 2.9×10^{-8} m; (d) ultraviolet **45.** (a) 9.35 μm; (b) 1.47×10^{-5} W; (c) 6.93×10^{14} photons/s; (d) 2.33×10^{-37} W; (e) 5.87×10^{-19} photons/s **47.** 7.75 pm **49.** (a) 1.9×10^{-21} kg $\cdot$ m/s; (b) 346 fm **51.** 4.3 μeV **53.** (a) 1.24 μm; (b) 1.22 nm; (c) 1.24 fm; (d) 1.24 fm **55.** (a) 15 keV; (b) 120 keV **57.** neutron **59.** (a) 3.96×10^{6} m/s; (b) 81.7 kV **67.** 2.1×10^{-24} kg $\cdot$ m/s **71.** (a) 1.45×10^{11} m^{-1}; (b) 7.25×10^{10} m^{-1}; (c) 0.111; (d) 5.56×10^{4} **73.** 4.81 mA **75.** (a) 9.02×10^{-6}; (b) 3.0 MeV; (c) 3.0 MeV; (d) 7.33×10^{-8}; (e) 3.0 MeV; (f) 3.0 MeV **77.** (a) -20%; (b) -10%; (c) $+15$% **79.** (a) no; (b) plane wavefronts of infinite extent, perpendicular to x axis **83.** (a) 38.8 meV; (b) 146 pm **85.** (a) 4.14×10^{-15} eV $\cdot$ s; (b) 2.31 eV **89.** (a) no; (b) 544 nm; (c) green

Chapter 39

CP **1.** b, a, c **2.** (a) all tie; (b) a, b, c **3.** a, b, c, d **4.** $E_{1,1}$ (neither n_x nor n_y can be zero) **5.** (a) 5; (b) 7

Q **1.** a, c, b **3.** (a) 18; (b) 17 **5.** equal **7.** c **9.** (a) decrease; (b) increase **11.** $n = 1, n = 2, n = 3$ **13.** (a) $n = 3$; (b) $n = 1$; (c) $n = 5$ **15.** b, c, and d

P **1.** 1.41 **3.** 0.65 eV **5.** 0.85 nm **7.** 1.9 GeV **9.** (a) 72.2 eV; (b) 13.7 nm; (c) 17.2 nm; (d) 68.7 nm; (e) 41.2 nm; (g) 68.7 nm; (h) 25.8 nm **11.** (a) 13; (b) 12 **13.** (a) 0.020; (b) 20 **15.** (a) 0.050; (b) 0.10; (c) 0.0095 **17.** 56 eV **19.** 109 eV **23.** 3.21 eV **25.** 1.4×10^{-3} **27.** (a) 8; (b) 0.75; (c) 1.00; (d) 1.25; (e) 3.75; (f) 3.00; (g) 2.25 **29.** (a) 7; (b) 1.00; (c) 2.00; (d) 3.00; (e) 9.00; (f) 8.00; (g) 6.00 **31.** 4.0 **33.** (a) 12.1 eV; (b) 6.45×10^{-27} kg $\cdot$ m/s; (c) 102 nm **35.** (a) 291 nm^{-3}; (b) 10.2 nm^{-1} **41.** (a) 0.0037; (b) 0.0054 **43.** (a) 13.6 eV; (b) -27.2 eV **45.** (a) $(r^4/8a^5)[\exp(-r/a)]\cos^2\theta$; (b) $(r^4/16a^5)[\exp(-r/a)]\sin^2\theta$ **47.** 4.3×10^{3} **49.** (a) 13.6 eV; (b) 3.40 eV **51.** 0.68 **59.** (b) $(2\pi/h)[2m(U_0 - E)]^{0.5}$ **61.** (b) meter$^{-2.5}$ **63.** (a) n; (b) $2\ell + 1$; (c) n^2 **65.** (a) $nh/\pi md^2$; (b) $n^2h^2/4\pi^2md^2$ **67.** (a) 3.9×10^{-22} eV; (b) 10^{20}; (c) 3.0×10^{-18} K **71.** (a) $e^2r/4\pi\varepsilon_0a^3$; (b) $e/(4\pi\varepsilon_0ma_0^3)^{0.5}$ **73.** 18.1, 36.2, 54.3, 66.3, 72.4 μeV

Chapter 40

CP **1.** 7 **2.** (a) decrease; (b)–(c) remain the same **3.** A, C, B

Q **1.** (a) 2; (b) 8; (c) 5; (d) 50 **3.** all true **5.** same number (10) **7.** 2, -1, 0, and 1 **9.** (a) 2; (b) 3 **11.** (a) n; (b) n and ℓ **13.** In addition to the quantized energy, a helium atom has kinetic energy; its total energy can equal 20.66 eV.

P **1.** 24.1° **3.** (a) 3.65×10^{-34} J $\cdot$ s; (b) 3.16×10^{-34} J $\cdot$ s **5.** (a) 3; (b) 3 **7.** (a) 4; (b) 5; (c) 2 **9.** (a) 3.46; (b) 3.46; (c) 3; (d) 3; (e) -3; (f) 30.0°; (g) 54.7°; (h) 150° **13.** 72 km/s^2 **15.** (a) 54.7°; (b) 125° **17.** 19 mT **19.** 5.35 cm **21.** 44 **23.** 42 **25.** (a) 51; (b) 53; (c) 56 **27.** (a) $(2, 0, 0, +\frac{1}{2}), (2, 0, 0, -\frac{1}{2})$; (b) $(2, 1, 1, +\frac{1}{2}), (2, 1, 1, -\frac{1}{2}), (2, 1, 0, +\frac{1}{2}), (2, 1, 0, -\frac{1}{2}), (2, 1, -1, +\frac{1}{2}), (2, 1, -1, -\frac{1}{2})$ **29.** g **31.** (a) $4p$; (b) 4; (c) $4p$; (d) 5; (e) $4p$; (f) 6 **33.** 12.4 kV **35.** (a) 35.4 pm; (b) 56.5 pm; (c) 49.6 pm **39.** 0.563 **41.** 80.3 pm **43.** (a) 69.5 kV; (b) 17.8 pm; (c) 21.3 pm; (d) 18.5 pm **45.** (a) 49.6 pm; (b) 99.2 pm **47.** 2.0×10^{16} s^{-1} **49.** 2×10^{7} **51.** 9.0×10^{-7} **53.** 7.3×10^{15} s^{-1} **55.** (a) 3.60 mm; (b) 5.24×10^{17} **57.** (a) 0; (b) 68 J **59.** 3.0 eV **61.** (a) 3.03×10^{5}; (b) 1.43 GHz; (d) 3.31×10^{-6} **63.** 186 **65.** (a) 2.13 meV; (b) 18 T **69.** (a) no; (b) 140 nm **71.** $n > 3$; $\ell = 3$; $m_\ell = +3, +2, +1, 0, -1, -2, -3$; $m_s = \pm\frac{1}{2}$ **73.** (a) 6.0; (b) 3.2×10^{6} y **75.** argon **79.** $(Ze/4\pi\varepsilon_0)(r^{-2} - rR^{-3})$

Chapter 41

CP **1.** larger **2.** a, b, and c

Q **1.** b, c, d (the latter due to thermal expansion) **3.** 8 **5.** below **7.** increase **9.** much less than **11.** b and d

P **3.** 8.49×10^{28} m^{-3} **5.** (b) 6.81×10^{27} m^{-3} eV$^{-3/2}$; (c) 1.52×10^{28} m^{-3} eV^{-1} **7.** (a) 0; (b) 0.0955 **9.** (a) 5.86×10^{28} m^{-3}; (b) 5.49 eV; (c) 1.39×10^{3} km/s; (d) 0.522 nm **11.** (a) 1.36×10^{28} m^{-3} eV^{-1}; (b) 1.68×10^{28} m^{-3} eV^{-1}; (c) 9.01×10^{27} m^{-3} eV^{-1}; (d) 9.56×10^{26} m^{-3} eV^{-1}; (e) 1.71×10^{18} m^{-3} eV^{-1} **13.** (a) 6.81 eV; (b) 1.77×10^{28} m^{-3} eV^{-1}; (c) 1.59×10^{28} m^{-3} eV^{-1} **15.** (a) 2.50×10^{3} K; (b) 5.30×10^{3} K **17.** 3 **19.** (a) 1.0; (b) 0.99; (c) 0.50; (d) 0.014; (e) 2.4×10^{-17}; (f) 7.0×10^{2} K **21.** (a) 0.0055; (b) 0.018 **25.** (a) 19.7 kJ; (b) 197 s **27.** (a) 1.31×10^{29} m^{-3}; (b) 9.43 eV; (c) 1.82×10^{3} km/s; (d) 0.40 nm **29.** 57.1 kJ **31.** (a) 226 nm; (b) ultraviolet **33.** (a) 1.5×10^{-6}; (b) 1.5×10^{-6} **35.** 0.22 μg **37.** (a) 4.79×10^{-10}; (b) 0.0140; (c) 0.824 **39.** 6.0×10^{5} **41.** 4.20 eV **43.** 13 μm **47.** (a) 109.5°; (b) 238 pm **49.** (b) 1.8×10^{28} m^{-3} eV^{-1} **53.** 3.49×10^{3} atm

Chapter 42

CP **1.** ^{90}As and ^{158}Nd **2.** a little more than 75 Bq (elapsed time is a little less than three half-lives) **3.** ^{206}Pb

Q **1.** (a) ^{196}Pt; (b) no **3.** yes **5.** (a) less; (b) greater **7.** ^{240}U **9.** no effect **11.** yes **13.** (a) all except ^{198}Au; (b) ^{132}Sn and ^{208}Pb **15.** d

P **1.** 1.3×10^{-13} m **3.** 46.6 fm **5.** (a) 0.390 MeV; (b) 4.61 MeV **7.** (a) 2.3×10^{17} kg/m^3; (b) 2.3×10^{17} kg/m^3; (d) 1.0×10^{25} C/m^3; (e) 8.8×10^{24} C/m^3 **9.** (a) 6; (b) 8 **11.** (a) 6.2 fm; (b) yes **13.** 13 km **17.** 1.0087 u **19.** (a) 9.303%; (b) 11.71% **21.** (b) 7.92 MeV/nucleon **25.** 5.3×10^{22} **27.** (a) 0.250; (b) 0.125 **29.** (a) 64.2 h; (b) 0.125; (c) 0.0749 **31.** (a) 7.5×10^{16} s^{-1}; (b) 4.9×10^{16} s^{-1} **33.** 1×10^{13} atoms **37.** 265 mg **39.** (a) 8.88×10^{10} s^{-1}; (b) 1.19×10^{15}; (c) 0.111 μg **41.** 1.12×10^{11} y **43.** 9.0×10^{8} Bq **45.** (a) 3.2×10^{12} Bq; (b) 86 Ci **47.** (a) 2.0×10^{20}; (b) 2.8×10^{9} s^{-1} **49.** (a) 1.2×10^{-17}; (b) 0 **51.** 4.269 MeV **53.** 1.21 MeV **55.** 0.783 MeV **57.** (b) 0.961 MeV **59.** 78.3 eV **61.** (a) 1.06×10^{19}; (b) 0.624×10^{19}; (c) 1.68×10^{19}; (d) 2.97×10^{9} y **63.** 1.7 mg **65.** 1.02 mg **67.** 2.50 mSv **69.** (a) 6.3×10^{18}; (b) 2.5×10^{11}; (c) 0.20 J; (d) 2.3 mGy; (e) 30 mSv **71.** (a) 6.6 MeV; (b) no **73.** (a) 25.4 MeV; (b) 12.8 MeV; (c) 25.0 MeV **75.** ^{7}Li **77.** 3.2×10^{4} y **79.** 730 cm^2 **81.** ^{225}Ac **83.** 30 MeV **89.** 27 **91.** (a) 11.906 83 u; (b) 236.2025 u **93.** 600 keV **95.** (a) 59.5 d; (b) 1.18 **97.** (a) 4.8×10^{-18} s^{-1}; (b) 4.6×10^{9} y

Chapter 43

CP **1.** c and d **2.** e
Q **1.** (a) 101; (b) 42 **3.** ^{239}Np **5.** ^{140}I, ^{105}Mo, ^{152}Nd, ^{123}In, ^{115}Pd **7.** increased **9.** less than **11.** still equal to 1
P **1.** (a) 16 day^{-1}; (b) 4.3×10^8 **3.** 4.8 MeV **5.** 1.3×10^3 kg **7.** 3.1×10^{10} s^{-1} **9.** (a) 2.6×10^{24}; (b) 8.2×10^{13} J; (c) 2.6×10^4 y **11.** −23.0 MeV **13.** (a) 253 MeV; (b) typical fission energy is 200 MeV **15.** (a) 84 kg; (b) 1.7×10^{25}; (c) 1.3×10^{25} **17.** (a) ^{153}Nd; (b) 110 MeV; (c) 60 MeV; (d) 1.6×10^7 m/s; (e) 8.7×10^6 m/s **21.** 557 W **23.** 0.99938 **25.** (b) 1.0; (c) 0.89; (d) 0.28; (e) 0.019; (f) 8 **27.** (a) 75 kW; (b) 5.8×10^3 kg **29.** 1.7×10^9 y **31.** 170 keV **33.** 1.41 MeV **35.** 10^{-12} m **37.** (a) 4.3×10^9 kg/s; (b) 3.1×10^{-4} **41.** 1.6×10^8 y **43.** (a) 24.9 MeV; (b) 8.65 megatons TNT **45.** (a) 1.8×10^{38} s^{-1}; (b) 8.2×10^{28} s^{-1} **47.** (a) 4.1 eV/atom; (b) 9.0 MJ/kg; (c) 1.5×10^3 y **49.** 14.4 kW **51.** $^{238}\text{U} + \text{n} \rightarrow {}^{239}\text{U} \rightarrow {}^{239}\text{Np} + \text{e} + \nu$, $^{239}\text{Np} \rightarrow {}^{239}\text{Pu} + \text{e} + \nu$ **55.** (a) 3.1×10^{31} protons/m^3; (b) 1.2×10^6 **57.** (a) 227 J; (b) 49.3 mg; (c) 22.7 kW

Chapter 44

CP **1.** (a) the muon family; (b) a particle; (c) $L_\mu = +1$ **2.** b and e **3.** c
Q **1.** b, c, d **3.** (a) 1; (b) positively charged **5.** a, b, c, d **7.** d **9.** c **11.** (a) lepton; (b) antiparticle; (c) fermion; (d) yes
P **1.** $\pi^- \rightarrow \mu^- + \bar{\nu}$ **3.** 2.4 pm **5.** 2.4×10^{-43} **7.** 769 MeV **9.** 2.7 cm/s **11.** (a) angular momentum, L_e; (b) charge, L_μ; (c) energy, L_μ **15.** (a) energy; (b) strangeness; (c) charge **17.** (a) yes; (b)−(d) no **19.** (a) 0; (b) −1; (c) 0 **21.** (a) K^+; (b) $\bar{n}$; (c) K^0 **23.** (a) 37.7 MeV; (b) 5.35 MeV; (c) 32.4 MeV **25.** (a) $\bar{u}\bar{u}\bar{d}$; (b) $\bar{u}\bar{d}\bar{d}$ **27.** $s\bar{d}$ **29.** (a) Ξ^0; (b) Σ^- **31.** 2.77×10^8 ly **33.** 668 nm **35.** 1.4×10^{10} ly **37.** (a) 2.6 K; (b) 976 nm **39.** (b) 5.7 H atoms/m^3 **41.** 4.57×10^3 **43.** (a) 121 m/s; (b) 0.00406; (c) 248 y **47.** 1.08×10^{42} J **49.** (a) $0.785c$; (b) $0.993c$; (c) C2; (d) C1; (e) 51 ns; (f) 40 ns **51.** (c) $r\alpha/c + (r\alpha/c)^2 + (r\alpha/c)^3 + \cdots$; (d) $r\alpha/c$; (e) $\alpha = H$; (f) 6.5×10^8 ly; (g) 6.9×10^8 y; (h) 6.5×10^8 y; (i) 6.9×10^8 ly; (j) 1.0×10^9 ly; (k) 1.1×10^9 y; (l) 3.9×10^8 ly **53.** (a) ssd; (b) $\bar{s}\bar{s}\bar{d}$

INDEX

Figures are noted by page numbers in *italics*, tables are indicated by t following the page number.

B

U

SOME PHYSICAL CONSTANTS*

Speed of light	c	2.998×10^8 m/s
Gravitational constant	G	6.673×10^{-11} N · m²/kg²
Avogadro constant	N_A	6.022×10^{23} mol^{-1}
Universal gas constant	R	8.314 J/mol · K
Mass–energy relation	c^2	8.988×10^{16} J/kg
		931.49 MeV/u
Permittivity constant	ε_0	8.854×10^{-12} F/m
Permeability constant	μ_0	1.257×10^{-6} H/m
Planck constant	h	6.626×10^{-34} J · s
		4.136×10^{-15} eV · s
Boltzmann constant	k	1.381×10^{-23} J/K
		8.617×10^{-5} eV/K
Elementary charge	e	1.602×10^{-19} C
Electron mass	m_e	9.109×10^{-31} kg
Proton mass	m_p	1.673×10^{-27} kg
Neutron mass	m_n	1.675×10^{-27} kg
Deuteron mass	m_d	3.344×10^{-27} kg
Bohr radius	a	5.292×10^{-11} m
Bohr magneton	μ_B	9.274×10^{-24} J/T
		5.788×10^{-5} eV/T
Rydberg constant	R	$1.097\,373 \times 10^7$ m^{-1}

*For a more complete list, showing also the best experimental values, see Appendix B.

THE GREEK ALPHABET

Alpha	A	α	Iota	I	ι	Rho	P	ρ
Beta	B	β	Kappa	K	κ	Sigma	Σ	σ
Gamma	Γ	γ	Lambda	Λ	λ	Tau	T	τ
Delta	Δ	δ	Mu	M	μ	Upsilon	Υ	υ
Epsilon	E	ϵ	Nu	N	ν	Phi	Φ	ϕ, φ
Zeta	Z	ζ	Xi	Ξ	ξ	Chi	X	χ
Eta	H	η	Omicron	O	o	Psi	Ψ	ψ
Theta	Θ	θ	Pi	Π	π	Omega	Ω	ω

SOME CONVERSION FACTORS*

Mass and Density

1 kg = 1000 g = 6.02×10^{26} u

1 slug = 14.59 kg

1 u = 1.661×10^{-27} kg

1 kg/m^3 = 10^{-3} g/cm^3

Length and Volume

1 m = 100 cm = 39.4 in. = 3.28 ft

1 mi = 1.61 km = 5280 ft

1 in. = 2.54 cm

1 nm = 10^{-9} m = 10 Å

1 pm = 10^{-12} m = 1000 fm

1 light-year = 9.461×10^{15} m

1 m^3 = 1000 L = 35.3 ft^3 = 264 gal

Time

1 d = 86 400 s

1 y = $365\frac{1}{4}$ d = 3.16×10^7 s

Angular Measure

1 rad = 57.3° = 0.159 rev

π rad = 180° = $\frac{1}{2}$ rev

Speed

1 m/s = 3.28 ft/s = 2.24 mi/h

1 km/h = 0.621 mi/h = 0.278 m/s

Force and Pressure

1 N = 10^5 dyne = 0.225 lb

1 lb = 4.45 N

1 ton = 2000 lb

1 Pa = 1 N/m^2 = 10 dyne/cm^2
= 1.45×10^{-4} lb/in.^{2}

1 atm = 1.01×10^5 Pa = 14.7 lb/in.^{2}
= 76.0 cm Hg

Energy and Power

1 J = 10^7 erg = 0.2389 cal = 0.738 ft · lb

1 kW · h = 3.6×10^6 J

1 cal = 4.1868 J

1 eV = 1.602×10^{-19} J

1 horsepower = 746 W = 550 ft · lb/s

Magnetism

1 T = 1 Wb/m^2 = 10^4 gauss

*See Appendix D for a more complete list.